Volkswagen New Beetle

Service Manual

1998, 1999, 2000, 2001, 2002, 2003 2004, 2005, 2006, 2007, 2008, 2009, 2010 including Convertible

[B] BentleyPublishers®
.com

BENTLEY PUBLISHERS® | Automotive Reference™

Bentley Publishers, a division of Robert Bentley, Inc.
1734 Massachusetts Avenue
Cambridge, MA 02138 USA
800-423-4595 / 617-547-4170

Information that makes the difference®

BentleyPublishers®.com

Technical contact information

We welcome your feedback. Please submit corrections and additions to our technical discussion forum at:

`http://www.BentleyPublishers.com`

Updates and corrections

We will evaluate submissions and post appropriate editorial changes online as updates or tech discussion. Appropriate updates and corrections will be added to the book in future printings. Check for updates and corrections for this book before beginning work on your vehicle. See the following web address for additional information:

`http://www.BentleyPublishers.com/updates/`

WARNING—Important Safety Notice

Do not use this manual for repairs unless you are familiar with basic automotive repair procedures and safe workshop practices. This manual illustrates the workshop procedures for some maintenance and service work. It is not a substitute for full and up-to-date information from the vehicle manufacturer or for proper training as an automotive technician. Note that it is not possible for us to anticipate all of the ways or conditions under which vehicles may be serviced or to provide cautions as to all of the possible hazards that may result.

We have endeavored to ensure the accuracy of the information in this manual. Please note, however, that considering the vast quantity and the complexity of the service information involved, we cannot warrant the accuracy or completeness of the information contained in this manual.

FOR THESE REASONS, NEITHER THE PUBLISHER NOR THE AUTHOR MAKES ANY WARRANTIES, EXPRESS OR IMPLIED, THAT THE INFORMATION IN THIS MANUAL IS FREE OF ERRORS OR OMISSIONS, AND WE EXPRESSLY DISCLAIM THE IMPLIED WARRANTIES OF MERCHANTABILITY AND OF FITNESS FOR A PARTICULAR PURPOSE, EVEN IF THE PUBLISHER OR AUTHOR HAVE BEEN ADVISED OF A PARTICULAR PURPOSE, AND EVEN IF A PARTICULAR PURPOSE IS INDICATED IN THE MANUAL. THE PUBLISHER AND AUTHOR ALSO DISCLAIM ALL LIABILITY FOR DIRECT, INDIRECT, INCIDENTAL OR CONSEQUENTIAL DAMAGES THAT RESULT FROM ANY USE OF THE EXAMPLES, INSTRUCTIONS OR OTHER INFORMATION IN THIS MANUAL. IN NO EVENT SHALL OUR LIABILITY WHETHER IN TORT, CONTRACT OR OTHERWISE EXCEED THE COST OF THIS MANUAL.

Before attempting any work on your Volkswagen, read **00 Warnings and Cautions** and any **WARNING** or **CAUTION** that accompanies a procedure in the manual. Review the **WARNINGS** and **CAUTIONS** each time you prepare to work on your Volkswagen.

Your common sense and good judgment are crucial to safe and successful service work. Read procedures through before starting them. Think about whether the condition of your car, your level of mechanical skill, or your level of reading comprehension might result in or contribute in some way to an occurrence which might cause you injury, damage your car, or result in an unsafe repair. If you have doubts for these or other reasons about your ability to perform safe repair work on your car, have the work done at an authorized Volkswagen dealer or other qualified shop.

Part numbers listed in this manual are for identification purposes only, not for ordering. Always check with your authorized Volkswagen dealer to verify part numbers and availability before beginning service work that may require new parts.

Special tools required to perform certain service operations are identified in the manual and are recommended for use. Use of improper tools may be detrimental to the car's safe operation as well as the safety of the person servicing the car.

The vehicle manufacturer will continue to issue service information updates and parts retrofits after the editorial closing of this manual. Some of these updates and retrofits will apply to procedures and specifications in this manual. We regret that we cannot supply updates to purchasers of this manual.

This manual is prepared, published and distributed by Bentley, Publishers, 1734 Massachusetts Avenue, Cambridge, Massachusetts 02138 USA. All information contained in this manual is based on information available to the publisher from Volkswagen at the time of editorial closing. Volkswagen has not reviewed and does not vouch for the accuracy or completeness of the technical specifications and work procedures described and given in this manual.

ISBN 978-0-8376-1640-7 **Editorial closing 03 / 2010** Job code: VB10-04

Library of Congress Cataloging-in-Publication Data for 2008 edition

Volkswagen New Beetle service manual : 1998, 1999, 2000, 2001, 2002, 2003, 2004, 2005, 2006, 2007, 2008, including convertible.

p. cm.

Includes index.

ISBN 978-0-8376-1559-2 (alk. paper)

1. Volkswagen New Beetle automobile--Maintenance and repair--Handbooks, manuals, etc.

I. Robert Bentley, inc.

TL215.V6V61455 2008

629.28'722--dc22

2008022196

Manufactured in the United States of America

202503R002

Foreword

For the Volkswagen owner with automotive mechanical skills, and for independent service professionals, this manual includes maintenance and repair procedures vehicles covered by this manual are likely to need during its service life, including many procedures and specifications that were available in an authorized Volkswagen dealer service department at the time this manual was prepared. The Volkswagen owner with no intention of working on his or her car will find that owning and referring to this manual will make it possible to be better informed and to more knowledgeably discuss repairs with a professional automotive technician.

For those intending to do maintenance and repair on their Volkswagen, it is essential that safety equipment be used and safety precautions observed when working on your vehicle. A minimum safety equipment list includes hand protection, eye protection, and a fire extinguisher. A selection of good quality hand tools is also needed. This includes a torque wrench to ensure that fasteners are tightened in accordance with specifications. In some cases, the text refers to special tools that are recommended or required to accomplish adjustments or repairs. These tools are often identified by their Volkswagen special tool number and illustrated.

Disclaimer

We have endeavored to ensure the accuracy of the information in this manual. When the vast array of data presented in the manual is taken into account, however, no claim to infallibility can be made. We therefore cannot be responsible for the result of any errors that may have crept into the text. Please also read **WARNING–Important Safety Notice** on the copyright page.

Prior to starting a repair procedure, read the procedure, **00 Warnings and Cautions** and the warnings and cautions that accompany the procedure. Reading a procedure before beginning work helps you determine in advance the need for specific skills, identify hazards, prepare for appropriate capture and handling of hazardous materials, and the need for particular tools and replacement parts such as gaskets.

Bentley Publishers encourages comments from the readers of this manual with regard to errors, and/or suggestions for improvement of our product. These communications have been and will be carefully considered in the preparation of this and other manuals. If you identify inconsistencies in the manual, you may have found an error. Please contact the publisher and we will endeavor to post applicable corrections on our website. Posted corrections (errata) should be reviewed before beginning work. Please see the following web address:

`http://www.BentleyPublishers.com/updates/`

Volkswagen continues to issue service information and parts retrofits after the editorial closing of this manual. Some of this updated information may apply to procedures and specifications in this manual. For the latest information, please see the following web address:

`http://erwin.vw.com`

Volkswagen offers extensive warranties, especially on components of the fuel delivery and emission control systems. Therefore, before deciding to repair a Volkswagen that may be covered wholly or in part by any warranties issued by Volkswagen of America, Inc., consult your authorized Volkswagen dealer. You may find that the dealer can make the repair either free or at minimum cost. Regardless of its age, or whether it is under warranty, your Volkswagen is both an easy car to service and an easy car to get serviced. So if at any time a repair is needed that you feel is too difficult to do yourself, a trained Volkswagen technician is ready to do the job for you.

Bentley Publishers

00 Warnings and Cautions

Please read these warnings and cautions before proceeding with maintenance and repair work.

WARNINGS—
See also ***CAUTIONS*** *on next page.*

- Read the important safety notice on the copyright page at the beginning of the book.
- Some repairs may be beyond your capability. If you lack the skills, tools and equipment, or a suitable workplace for any procedure described in this manual, we suggest you leave such repairs to an authorized Volkswagen dealer service department or other qualified shop.
- Thoroughly read each procedure and the **WARNINGS** and **CAUTIONS** that accompany the procedure. Also review posted corrections at `www.BentleyPublishers.com/updates/` before beginning work.
- If any procedure, tightening torque, wear limit, specification or data presented in this manual does not appear to be appropriate for a specific application, contact the publisher or the vehicle manufacturer for clarification before using the information in question.
- Volkswagen is constantly improving its cars. Sometimes these changes, both in parts and specifications, are made applicable to earlier models. Therefore, before starting any major jobs or repairs to components on which passenger safety may depend, consult your authorized Volkswagen dealer about technical bulletins that may have been issued.
- Do not reuse any fasteners that are worn or deformed in normal use. Many fasteners are designed to be used only once and become unreliable and may fail when used a second time. This includes, but is not limited to, nuts, bolts, washers, self-locking nuts or bolts, circlips and cotter pins. Replace these fasteners with new parts.
- Do not work under a lifted car unless it is solidly supported on stands designed for the purpose. Do not support a car on cinder blocks, hollow tiles or other props that may crumble under continuous load. Do not work under a car that is supported solely by a jack. Do not work under the car while the engine is running.
- If you are going to work under a car on the ground, make sure that the ground is level. Block the wheels to keep the car from rolling. Disconnect the battery negative (–) terminal (ground strap) to prevent others from starting the car while you are under it.
- Do not run the engine unless the work area is well ventilated. Carbon monoxide kills.
- Remove finger rings, bracelets and other jewelry so that they cannot cause electrical shorts, get caught in running machinery or be crushed by heavy parts.
- Tie long hair behind your head. Do not wear a necktie, a scarf, loose clothing, or a necklace when you work near machine tools or running engines. If your hair, clothing, or jewelry were to get caught in the machinery, severe injury could result.
- Do not attempt to work on your car if you do not feel well. You increase the danger of injury to yourself and others if you are tired, upset or have taken medication or any other substance that may keep you from being fully alert.
- Illuminate your work area adequately but safely. Use a portable safety light for working inside or under the car. Make sure the bulb is enclosed by a wire cage. The hot filament of an accidentally broken bulb can ignite spilled fuel, vapors or oil.
- Catch draining fuel, oil or brake fluid in suitable containers. Do not use food or beverage containers that might mislead someone into drinking from them. Store flammable fluids away from fire hazards. Wipe up spills at once but do not store the oily rags, which can ignite and burn spontaneously.
- Observe good workshop practices. Wear goggles when you operate machine tools or work with battery acid. Wear gloves or other protective clothing whenever the job requires working with harmful substances.
- Greases, lubricants and other automotive chemicals contain toxic substances, many of which are absorbed directly through the skin. Read the manufacturer's instructions and warnings carefully. Use hand and eye protection. Avoid direct skin contact.
- Disconnect the battery negative (–) terminal (ground strap) whenever you work on the fuel system or the electrical system. Do not smoke or work near heaters or other fire hazards. Keep an approved fire extinguisher handy.
- Friction components (such as brake pads or shoes or clutch discs) contain asbestos fibers or other friction materials. Do not create dust by grinding, sanding, or by cleaning with compressed air. Avoid breathing dust. Breathing any friction material dust can lead to serious diseases and may result in death.
- Batteries give off explosive hydrogen gas during charging. Keep sparks, lighted matches and open flame away from the top of the battery. If hydrogen gas escaping from the cap vents is ignited, it will ignite gas trapped in the cells and cause the battery to explode.
- Battery acid (electrolyte) can cause severe burns. Flush contact area with water and seek medical attention.
- Do not quick-charge the battery (for boost starting) for longer than one minute. Wait at least one minute before boosting the battery a second time.
- Connect and disconnect a battery charger only with the battery charger switched OFF.
- Connect and disconnect battery cables, jumper cables or a battery charger only with the ignition switched OFF. Do not disconnect the battery while the engine is running.
- Do not allow the battery charging voltage to exceed 16.5 volts. If the battery begins producing gas or boiling violently, reduce the charging rate. Boosting a sulfated battery at a high rate can cause an explosion.

continued on next page

Please read these warnings and cautions before proceeding with maintenance and repair work.

WARNINGS— (continued)

- The air-conditioning system is filled with a hazardous refrigerant. Make sure the system is serviced only by a trained technician using approved refrigerant recovery/recycling equipment, trained in related safety precautions, and familiar with regulations governing the discharging and disposal of automotive chemical refrigerants.
- Do not expose any part of the air-conditioning system to high temperatures such as open flame. Excessive heat increases system pressure and may cause the system to burst.
- Some aerosol tire inflators are highly flammable. Be extremely cautious when repairing a tire that may have been inflated using an aerosol tire inflator. Keep sparks, open flame or other sources of ignition away from the tire repair area. Inflate and deflate the tire at least four times before breaking the bead from the rim. Completely remove the tire from the rim before attempting any repair.
- Cars covered by this manual are equipped with a supplemental restraint system, that automatically deploys airbags and pyrotechnic seat belt reels in case of a frontal or side impact. These are explosive devices. Handled improperly or without adequate safeguards, they can be accidently activated and cause serious injury.
- The ignition system produces high voltages that can be fatal. Avoid contact with exposed terminals and use extreme care when working on a car with the engine running or the ignition switched ON.
- Place jack stands only at locations specified by the manufacturer. The vehicle lifting jack supplied with the vehicle is intended for tire changes only. Use a heavy duty floor jack to lift the vehicle before installing jack stands. See **03 Maintenance**.
- Aerosol cleaners and solvents may contain hazardous or deadly vapors and are highly flammable. Use only in a well ventilated area. Do not use on hot surfaces (engines, brakes, etc.).
- Due to risk of personal injury, be sure the engine is cold before beginning work on the cooling system.
- Use extreme care when draining and disposing of engine coolant. Coolant is poisonous and lethal to humans and pets. Pets are attracted to coolant because of its sweet smell and taste. Seek medical attention immediately if coolant is ingested.

CAUTIONS—
*See also **WARNINGS** on previous page.*

- If you lack the skills, tools and equipment, or a suitable workshop for any procedure described in this manual, we suggest you leave such repairs to an authorized Volkswagen dealer or other qualified shop.
- Volkswagen is constantly improving its cars and sometimes these changes, both in parts and specifications, are made applicable to earlier models. Therefore, part numbers listed in this manual are for reference only. Always check with your authorized Volkswagen dealer parts department for the latest information.
- Before starting a job, make certain that you have the necessary tools and parts on hand. Read the instructions thoroughly and do not attempt shortcuts. Use tools appropriate to the work and use replacement parts meeting Volkswagen specifications. Makeshift tools, parts and procedures do not make good repairs
- Do not use pneumatic and electric tools to tighten fasteners, especially on light alloy parts. Use a torque wrench to tighten fasteners to the tightening torque specification listed.
- Be mindful of the environment. Before you drain the crankcase, find out the proper way to dispose of the oil. Do not pour oil onto the ground, down a drain, or into a stream, pond or lake. Dispose of in accordance with Federal, State and Local laws.
- Do not expose electronic control modules to temperatures from a paint-drying booth or a heat lamp in excess of 203°F (95°C). Do not expose control modules to temperatures in excess of 185°F (85°C) for more than two hours.
- Before doing any electrical welding on cars equipped with ABS, disconnect the battery negative (–) terminal (ground strap) and the ABS control module connector.
- Make sure ignition is OFF before disconnecting the battery.
- Label battery cables before disconnecting. On some models, battery cables are not color coded.
- Disconnecting the battery may erase fault code(s) stored in control module memory. Use Volkswagen diagnostic equipment or equivalent to check for fault codes prior to disconnecting the battery cables. If the Check Engine light (malfunction indicator light or MIL) is illuminated or any other system faults are detected (indicated by an illuminated warning light), see an authorized Volkswagen dealer.
- If a normal or rapid charger is used to charge battery, disconnect the battery and remove from the vehicle in order to avoid damaging paint and upholstery.
- Do not quick-charge the battery (for boost starting) for longer than one minute. Wait at least one minute before boosting the battery a second time.
- Slow-charge a sealed or "maintenance free" battery at an amperage rate that is approximately 10% of the battery's ampere-hour (Ah) rating.

01 Vehicle Identification Number (VIN) Decoder

Some of the information in this manual applies only to cars of a particular model year or range of years. For example, 2005 refers to the 2005 model year but does not necessarily match the calendar year in which the car was manufactured or sold. To be sure of the model year of a particular car, check the Vehicle Identification Number (VIN) on the car.

The VIN is a unique sequence of 17 characters assigned to identify each individual car. When decoded, the VIN tells the country and year of manufacture; make, model and serial number; assembly plant and even some equipment specifications.

The Volkswagen VIN is placed at various positions on the vehicle, including a plate mounted on the top of the dashboard, on the driver's side where the number can be seen through the windshield. The 10th character is the model year code. The letters I, O, Q, U, Z and the number 0 are not used for model year designation. Examples: W for 1998, X for 1999, Y for 2000, 1 for 2001, 2 for 2002, etc. The table below explains some of the codes in the VIN for 1998 - 2010 Volkswagen New Beetle models covered by this manual.

Sample VIN: 3VWBB61C1WM031159
position: 1 2 3 4 5 6 7 8 9 10 11 12 - 17

VIN position	Description	Decoding information	
1	Manufacturing Country	3	Mexico
2	Manufacturer	V	Volkswagen
3	Vehicle Type	W	Passenger vehicle
4	Series		Varies with model year
5	Engine code		
6	Restraint system		
7-8	Model type	1C 1Y	New Beetle New Beetle Convertible
9	Check digit		0 - 9 or X, calculated by NHTSA
10	Model year	W Y Y 1 2 3 4 5 6 7 8 9 A	1998 1999 2000 2001 2002 2003 2004 2005 2006 2007 2008 2009 2010
11	Assembly plant	M	Puebla, Mexico
12-17	Serial number		Sequential production number for specific vehicle

03 Maintenance

GENERAL

All of the maintenance work described in this repair group is important and should be carried out at the specified time or mileage interval. The Owner's Manual, the Maintenance Record, and the Warranty Booklet originally supplied with the car contain the maintenance schedules that apply to your Volkswagen. Following these schedules will ensure safe and dependable operation. In addition, many of the maintenance procedures are necessary to maintain warranty protection.

NOTE —
Volkswagen is constantly updating their recommended maintenance procedures and requirements. The information contained here may not include updates or revisions made by Volkswagen since publication of the Owner's Manual, the Maintenance Record, and the Warranty Booklet supplied with the car. If there is any doubt about what procedures apply to a specific model or model year, or what intervals should be followed, remember that an authorized Volkswagen dealer has the latest maintenance information.

CAUTION —
Disconnecting the negative (–) battery cable may erase fault codes and basic settings in the engine management and automatic transmission control modules. Some driveability problems may be noticed until the system re-adapts to operating conditions. OBD II readiness codes, which may be required for emissions testing, may also be erased. Convenience electronics (alarm system, interior light control, power locks, mirrors, and windows) may need to be re-set using a VAG 1551/1552, VAS 5051/5052, or equivalent scan tool. See the **OBD On Board Diagnostics** *section at the back of this manual for more information. Also see Procedure following the interruption of voltage supply,* ***27 Engine-Electrical****.*

HOW TO USE THIS MANUAL

The manual is divided into 12 main sections, or partitions:

- **0** Maintenance
- **1** Engine
- **2** Engine Management, Exhaust, and Engine Electrical
- **3** Clutch, Transmission, and Final Drive
- **4** Suspension, Brakes, and Steering
- **5** Body–Assembly
- **6** Body–Components and Accessories
- **7** Body–Interior Trim
- **8** Heating and Air Conditioning
- **9** Electrical System
- **EWD** Electrical Wiring Diagrams
- **OBD** On Board Diagnostics

A master listing of the 12 partitions and the corresponding specific repair groups can be found on the inside front cover of this manual.

At the back of the book are various appendixes and a detailed index.

Warnings, Cautions and Notes

Throughout this manual are many passages with the headings **WARNING**, **CAUTION**, or **NOTE**. These very important headings have different meanings.

> ***WARNING —***
> *The text under this heading warns of unsafe practices that are very likely to cause injury, either by direct threat to the person(s) doing the work or by increased risk of accident or mechanical failure while driving. Warnings are always contained in a box.*

> ***CAUTION —***
> *A caution calls attention to important precautions to be observed during the repair work that will help prevent accidentally damaging the car or its parts. Cautions are always contained in a box.*

NOTE —
A note contains helpful information, tips that will help in doing a better job and completing it more easily.

Please read every **WARNING**, **CAUTION**, and **NOTE** at the front of the manual and as they appear in repair procedures. They are very important. Read them before you begin any maintenance or repair job.

Some **WARNING**s and **CAUTION**s are repeated wherever they apply. Read them all. Do not skip any. These messages are important, even to the owner who never intends to work on the car.

Work Safety

Although an automobile presents many hazards, common sense and good equipment can help ensure safety. Many accidents happen because of carelessness. Pay attention and stick to these few important safety rules.

WARNING —

- *Never run the engine in the work area unless it is well-ventilated. The exhaust should be vented to the outside. Carbon Monoxide (CO) is an odorless and colorless gas. Carbon Monoxide (CO) in the exhaust kills.*

- *Remove all neckties, scarfs, loose clothing, and jewelry when working near running engines or power tools. Tuck in shirts. Tie long hair and secure it under a cap. Severe injury can result from these things being caught in rotating parts.*

- *Remove rings, watches, and bracelets. Aside from the dangers of moving parts, metallic jewelry conducts electricity and may cause shorts, sparks, burns, or damage to the electrical system when accidentally contacting the battery or other electrical terminals.*

- *Disconnect the battery negative (–) (GND) cable whenever working on or near the fuel system, airbag system or anything that is electrically powered. Accidental electrical contact may damage the electrical system, cause fire, or result in serious personal injury.*

- *Never work under a lifted car unless it is solidly supported on jack stands that are intended for that purpose. Do not support a car on cinder blocks, bricks, or other objects that may shift or crumble under continuous load. Never work under a car that is supported only by a jack.*

- *The fuel system retains pressure even when the ignition is off. Loosen the fuel lines very slowly to allow the residual pressure to dissipate gradually. Cover fittings with a suitable shop cloth to avoid spraying fuel.*

- *Fuel is highly flammable. When working around fuel, do not smoke or work near fire hazards. Keep an approved fire extinguisher handy. Be aware that water and building heaters may have standing pilot lights.*

- *Illuminate the work area adequately and safely. Use a portable safety light for working inside or under the car. A fluorescent type light is best because it gives off less heat. If using a light with a normal incandescent bulb, use rough service bulbs to avoid breakage. The hot filament of an accidentally broken bulb can ignite spilled fuel or oil.*

- *Keep sparks, lighted matches, and open flame away from the top of the battery. Hydrogen gas emitted by the battery is highly flammable. Any nearby source of ignition may cause the battery to explode.*

- *Never lay tools or parts in the engine compartment or on top of the battery. They may fall and be difficult to retrieve, become caught in belts or other rotating parts, or cause electrical shorts and damage to the electrical system.*

IDENTIFICATION PLATES AND LABELS

Vehicle Identification Number (VIN)

◄ United States Federal law requires VIN information to be visible from outside the vehicle and at a standard location. A plate (**arrow**) with the complete 17 digit VIN is attached to the padded dash at the base of the windshield and is visible through a mask in the glass from the outside. See **01 Vehicle Identification Number (VIN) Decoder** in the front of this manual for detailed VIN decoding information.

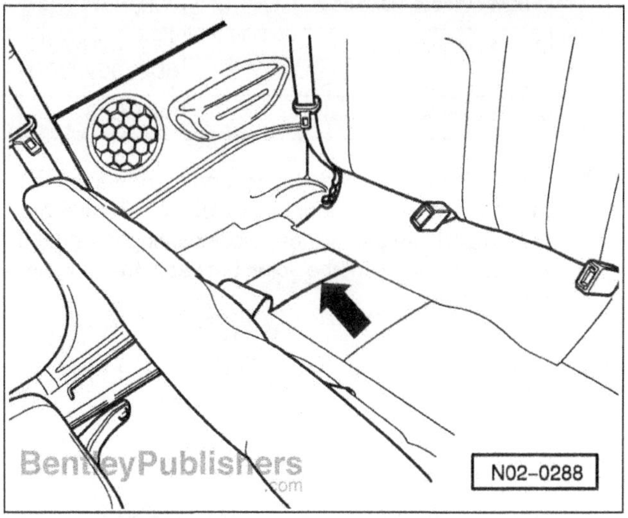

◄ The VIN (Chassis number) is also stamped into the rear floor pan under the rear seat (**arrow**). Pull up rear seat cushion and fold up. Lift carpet on right side for access.

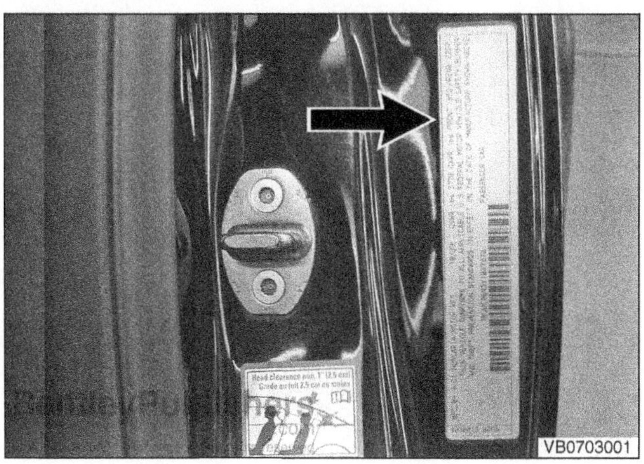

◄ Door "B" pillar certification label (**arrow**) lists full VIN, vehicle weight ratings, production date and emission compliance certification.

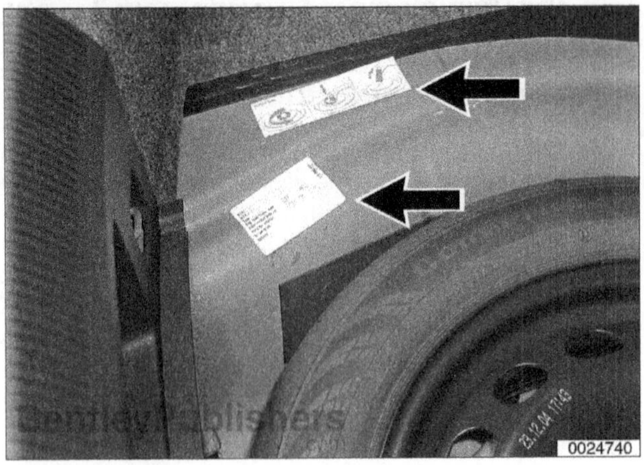

Vehicle data sticker (**lower arrow**) is located in the luggage compartment on left side of spare tire recess. The data plate contains the following vehicle information:

1. Production control number
2. Vehicle identification number
3. Model identification number
4. Model explanation/engine output
5. Engine and transmission code letters
6. Paint number/interior equipment identification number
7. Optional equipment identification numbers

Spare tire stowage information sticker (**upper arrow**) may also be attached to vehicle at this location.

Engine identification

The engines used in New Beetles covered by this manual are identified by a three letter code followed by a six digit number such as: AEG 029 452. The engine code and the engine number are located on the engine block.

Engine	Engine code
1.8L 4-cylinder turbo gasoline	APH, AWV, AWP, BKF, BNU
1.9L 4-cylinder TDI diesel	ALH, BEW
2.0L 4-cylinder gasoline	AEG, AVH, AZG, BDC, BEV, BGD
2.5L 5-cylinder gasoline	BPR, BPS

4-cylinder engine code letters and numbers are located at the rear of the cylinder block (**arrow**). This information is stamped into the cylinder block casting near the oil filter flange close to the joint between the engine and transmission.

5-cylinder engine code letters and numbers are located on back side of engine block near the joint between block and oil pan. It is directly above oil drain plug and is best viewed from underneath.

0024321

Additionally, the engine codes can often be found on a sticker on the upper portion of the engine.

- 2.0L shown, 1.8L and 1.9L are similar
- 2.5L sticker located on top of cylinder head cover

Transmission identification

The transmissions used in New Beetles covered by this manual are identified by a three letter code followed by a five digit number such as: DQY 17078. The digits indicate the transmission build date and all transmissions built on any given day will have the same number. The various transmission types and codes are listed below.

Transmission	Transmission code
5-speed manual	02J
5-speed manual	0A4
5- and 6-speed manual	02M
4-speed automatic	01M
6-speed automatic with Tiptronic	09G
Direct shift gearbox (DSG)	02E

Transmission codes

0024322A

Manual transmission codes are stamped into machined areas of case as indicated.

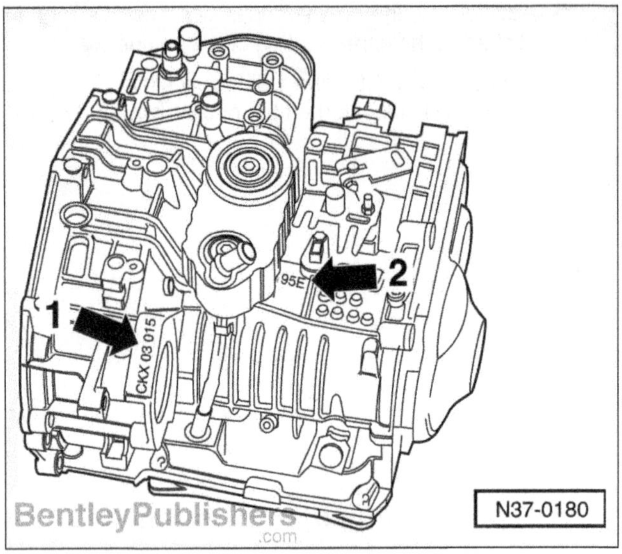

Transmission code (**arrow 1**) stamped into machined area of case for automatic transmissions. Other numbers (**arrow 2**) may be cast into the case but are not relevant to the code.

TOWING

Emergency towing

The following information is to be used by commercial tow truck operators who know how to operate their equipment safely. To prevent damage to the vehicle, read and understand all of the following information before proceeding.

CAUTION—
- *The New Beetle cannot be towed with conventional sling-type equipment or non-self-loading wheel dollies. Towing with this type of equipment will cause bumper and body panel damage.*
- *If an automatic transmission vehicle cannot be towed with wheel lift equipment in combination with self-loading wheel dollies, it must be transported with a car carrier (flatbed) to avoid damage to the transmission due to lack of lubrication.*

WARNING—
Never allow passengers to ride in a towed vehicle for any reason.

NOTE—
- *Always follow the manufacturer's recommendations for commercial towing equipment.*
- *Always be thoroughly familiar with specific towing equipment being used.*
- *Always observe all State, Provincial and Local laws regarding such items as warning signals, night illumination, speed, licensing, etc.*

Emergency towing (continued)

- Whenever possible, tow with the front wheels off the ground.
- The vehicle may be lifted at the rear and moved a short distance to a better position for front wheel lifting.
- If a vehicle with a manual transmission must be towed with the front wheels on the ground, make sure that the transmission oil has not leaked out or been drained.
- If a vehicle is to be towed on its drive wheels, the transmission and differential MUST be operable. Place the transmission in NEUTRAL.
- If any doubt exists about the condition of the transmission or differential, tow the vehicle with the drive wheels lifted, use self-loading wheel dollies or wheel lift equipment.
- Excessive towing speeds and distances may damage both manual and automatic transmissions.
- Move the vehicle only within the recommended speeds and distances.
- During any tow, the raised wheels can contact the road or other ground surfaces, so they need to rotate freely.
- The anti-theft steering column lock is not strong enough to withstand shocks transmitted from the wheels while towing. Before towing, always unlock the steering wheel with the ignition key and secure the steering wheel with a steering wheel clamp which has been specifically designed for towing service.
- To reduce the risk of serious personal injury and vehicle damage, never attempt to rock or pivot the vehicle on jack stands to allow positioning on the dolly.
- Always release the parking brake.
- When the ignition key is not available, DO NOT tow from the rear or serious damage to vehicle will result.
- The front of the vehicle must be lifted to prevent damage to the steering column anti-theft lock.
- To reduce the risk of accident and serious injury, never tow any vehicle at a speed in excess of 50 mph for any reason whenever towing with a conventional tow truck, with or without the use of a self-loading wheel dolly.
- Safe operating speeds always depend on weather, road, traffic, visibility, conditions; as well as the condition of the towed vehicle.
- A tow truck is an emergency vehicle to be used to move disabled vehicles to a suitable place of repair. It must not be used for transporting vehicles long distances.
- Towed vehicles should be raised until lifted wheels are a minimum of 6 inches from the ground and there is adequate clearance at opposite end of lifted vehicle.
- Always inspect points of attachment to the disabled vehicle. If they appear to be damaged or deteriorated, select other attachment points at a substantial structural member of the frame.

(continued on following page)

- Never allow the fuel tank to support any of the vehicle's weight during towing. Before moving the vehicle, carefully secure or remove any loose or protruding parts of damaged vehicles; such as hoods, fenders, trim, etc.
- To avoid serious vehicle damage, never lift or tow any vehicle by attaching towing chains or hooks to shock absorbers, stabilizer bars, or front strut rods.
- Never go under the vehicle while it is lifted by the towing equipment, unless the vehicle is adequately supported by safety stands.
- The safety of the operator and all others in the vicinity of the wrecker or the towed vehicle must be the first consideration at all times during a towing operation.

Vehicle hook-up, front

– Attach wheel lift equipment to wheels.

◄ Attach safety straps to wheels.

– Attach safety chains to lower control arms.

NOTE —

- *Always use a safety chain system which is completely independent of the primary lifting and towing attachments.*
- *During installation of safety chains, be careful not to damage lights, bumpers, or painted surfaces.*
- *To prevent damage to brake lines and front driveshaft boots, always position hooks and chains cautiously to prevent damage to these components.*
- *Prior to towing the vehicle, ensure that all attachment points are firmly secured.*
- *Towing clearance: 6-12 inches between tires and ground.*

Vehicle hook-up, rear

CAUTION —

- *Do not tow cars with automatic transmission from the rear.*
- *Do not tow cars with manual transmission from the rear unless transmission is in neutral and ignition is unlocked. Secure steering wheel with a steering wheel clamp which has been specifically designed for towing service.*

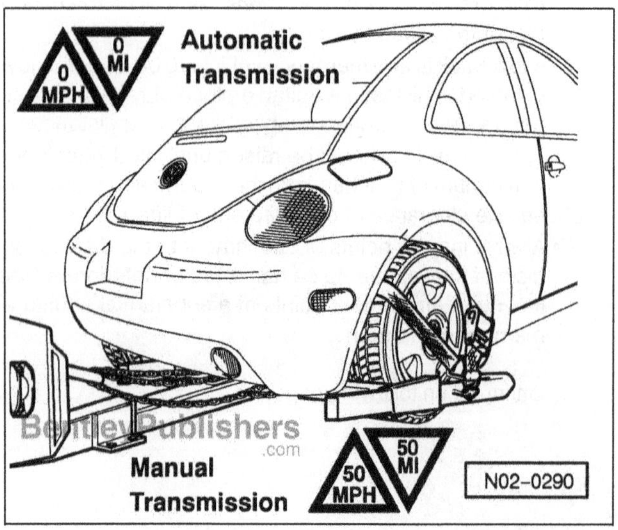

– Attach wheel lift equipment to wheels.

◄ Attach safety straps to wheels.

– Attach safety chains to axle beam.

NOTE —
Observe all notes associated with front vehicle hook-up.

Towing eyes

To be able to pull or be pulled, a towing eye must be threaded into the mount at right front of vehicle. The detachable towing eye is in the vehicle tool kit.

Pry cover off forward with flat blade screwdriver inserted at **arrow**.

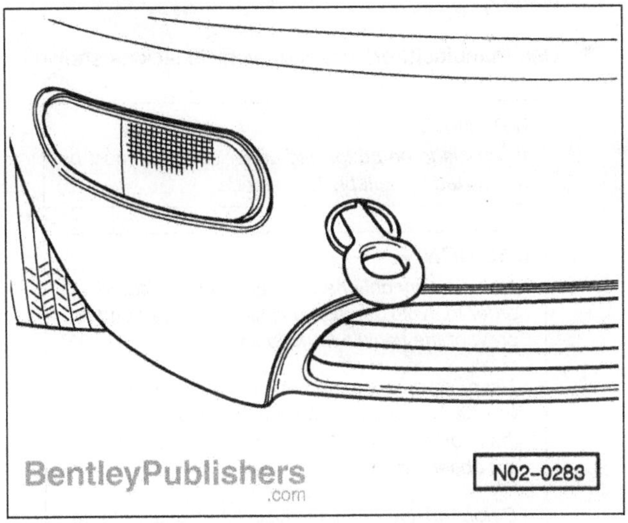

Thread towing eye fully into mount and tighten with wheel lug wrench.

- After use, unscrew towing eye and return to vehicle tool kit.
- Reinstall cover.

> ***NOTE—***
> *The rear towing eye is welded to the vehicle and is located under rear bumper on right side.*

LIFTING VEHICLE

Lifting points

New Beetles may be raised for service and repairs by floor jacks and/or hoists. For repairs that require raising the vehicle, proper jacking points must be used. There are four jack points, two on each side of the car, marked by indentations in the lower panel sheet metal just behind the front fender or just in front of rear fender. If using a floor jack or hoist, similar lift points are just inboard of the aforementioned indentations. To lift vehicle, place floor jack or hoist at appropriate lifting points as shown below. Always use a suitable rubber block between the jack and the vehicle.

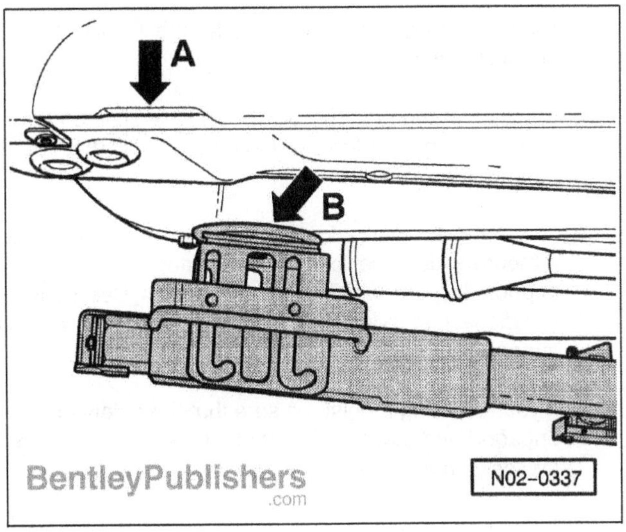

Front lifting points; when using factory supplied vehicle jack, position at indentation (**arrow A**). When using a floor jack or hoist, place pad under long member (**arrow B**) on flat surface behind curve.

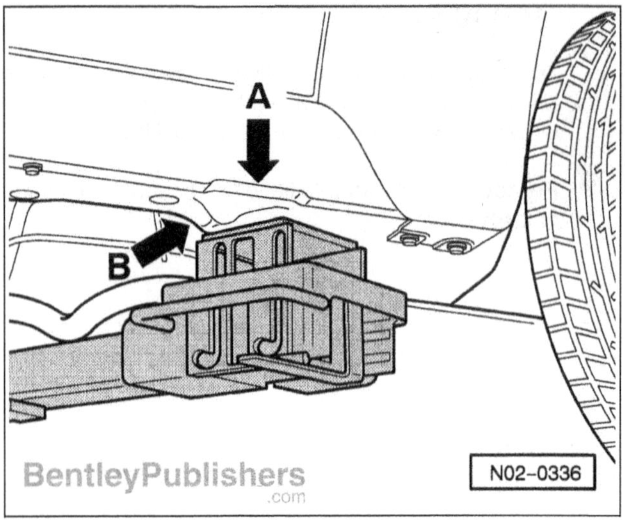

Rear lifting points; when using a factory supplied vehicle jack, position at indentation (**arrow A**). When using a floor jack or hoist, place pad under rear crossmember (**arrow B**).

> ***CAUTION —***
> *Do not lift vehicle at engine oil pan, transmission, or on front or rear axle as serious damage may result.*

- Place car on a flat level area with a surface capable of supporting jacks and jackstands, e.g. concrete.
- Block the wheels to prevent the car from rolling.
- Place the jack proper lifting points as shown previously.
- Operate the jack or hoist and raise the car slowly. If working under the car, support the weight of the car using jack stands.

Use manufacturers provided vehicle jack as shown.

> ***WARNING —***
> *If work is to be performed under vehicle it must be supported by suitable jack stands.*

> ***CAUTION —***
> - *Vehicle may only be lifted at points indicated in order to avoid damaging vehicle floor pan and to prevent the vehicle from tipping.*
> - *Never start engine and engage a gear with vehicle lifted if any drive wheel has contact with the floor! (There is danger of an accident if this is not observed).*
> - *Before driving on to a vehicle hoist, ensure there is sufficient clearance between low lying vehicle components and the hoist platform.*

Working under the vehicle

When working under any vehicle, the following points should be observed:

- Disconnect the negative (GND) battery cable so that no one else can start the car. Let others know that you will be under the vehicle.
- Place at least two jack stands under the car. A jack is a temporary lifting device and should not be used alone to support the car while you are under it. Use positively locking jack stands that are designed for the purpose of supporting a car.
- If you are using a hoist, be sure that the safety locks are engaged and that the weight of the vehicle is resting on the locks, not the hydraulic system.

– Lower the car slowly until its weight is fully supported by the jack stands or hoist. Watch to make sure that jack stands do not tip or lean as the car settles on them, and that jack stands and hoist arms are placed solidly and will not move.

> ***WARNING —***
> *Check to make sure vehicle is stable before working under car.*

– Observe all jacking precautions again when raising the car to remove the jack stands.

ENGINE COMPARTMENT MAINTENANCE

The jobs listed under this heading encompass regular maintenance and related procedures.

Sound absorber panels, removing and installing

1.8L engine cover

There are two different styles of engine covers for the 1.8L engine. One is bolted on and the other snaps into place on mounting pins.

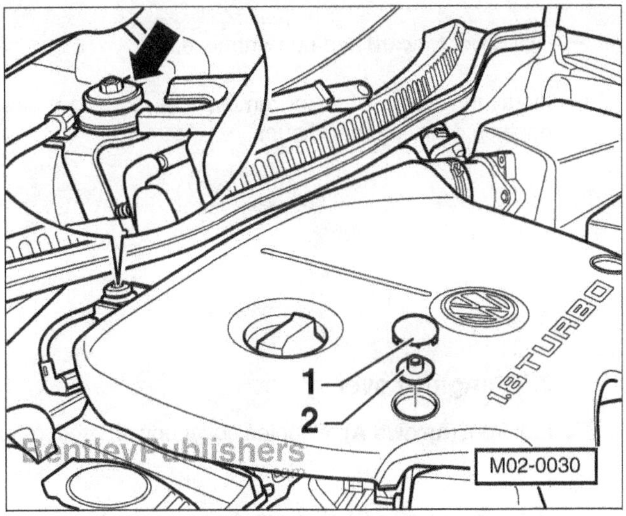

◄ Pry out plastic plugs (**1**) and remove nuts (**2**), if applicable.

– Loosen rear mounting nut (**arrow**) slightly.

– Pull cover forward and up to remove.

1.9L (code ALH) engine cover

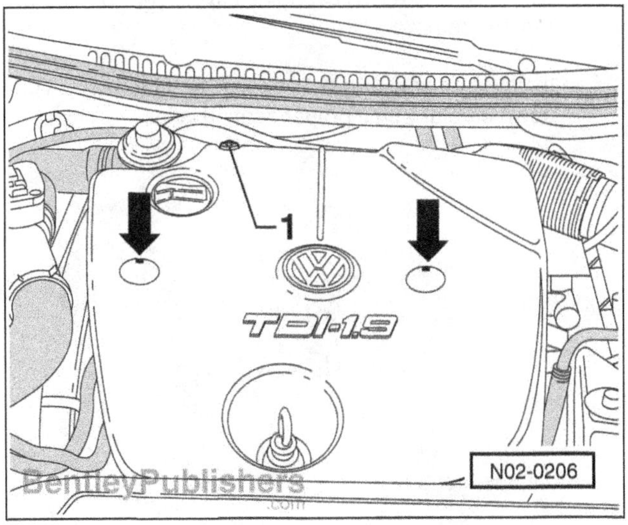

◄ Pry out plastic plugs (**arrows**) and remove nuts.

– Loosen rear mounting nut (**1**) slightly and remove dipstick.

– Pull cover forward and up to remove.

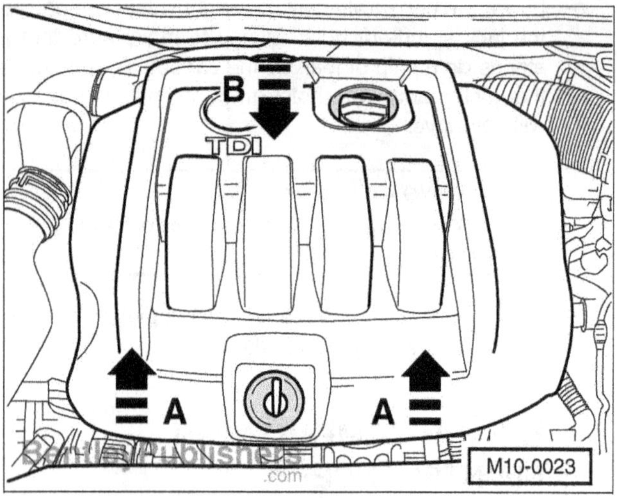

1.9L (code BEW) engine cover

< Pull engine cover up at front (**arrows A**).

- Remove by pulling cover forward (**arrow B**) off rear retaining lug.
- When installing, be sure to properly locate back of engine cover on rear retaining lug.

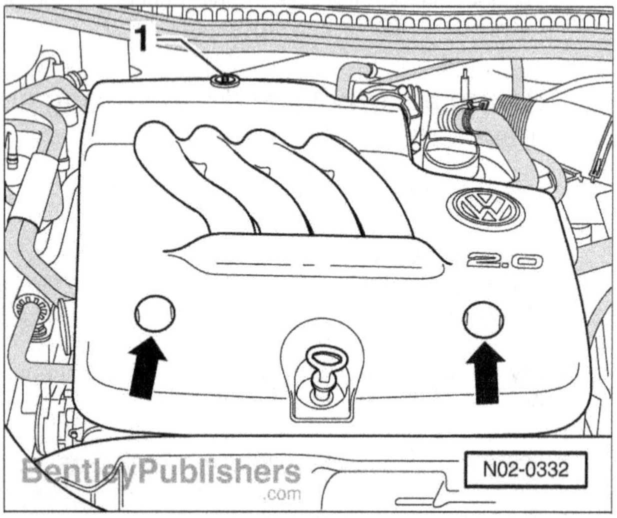

2.0L engine cover

< Pry out plastic plugs (**arrows**) and remove nuts.

- Loosen rear mounting nut (**1**) slightly and remove dipstick.
- Pull cover forward and up to remove.
- Install in reverse order of removal. Be sure to properly position cover on rear mount.

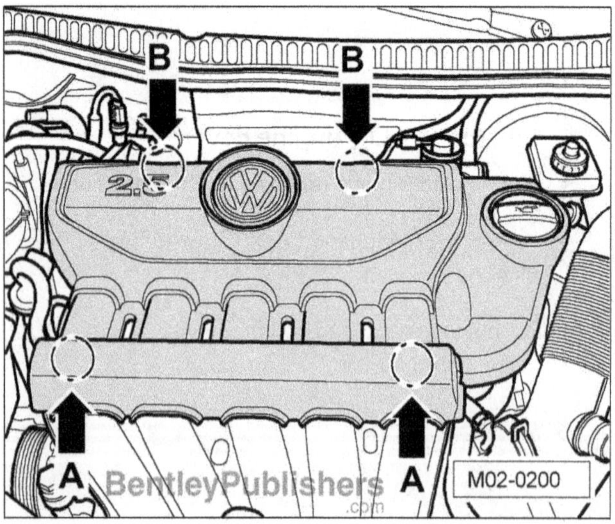

2.5L engine cover

< Pull front (**arrows A**) of engine cover upward from retainers.

- Pull rear (**arrows B**) of engine cover upward from retainers and remove.
- Install in reverse order of removal. Be sure to push panel down carefully until proper seating is felt.

Lower sound absorber panels

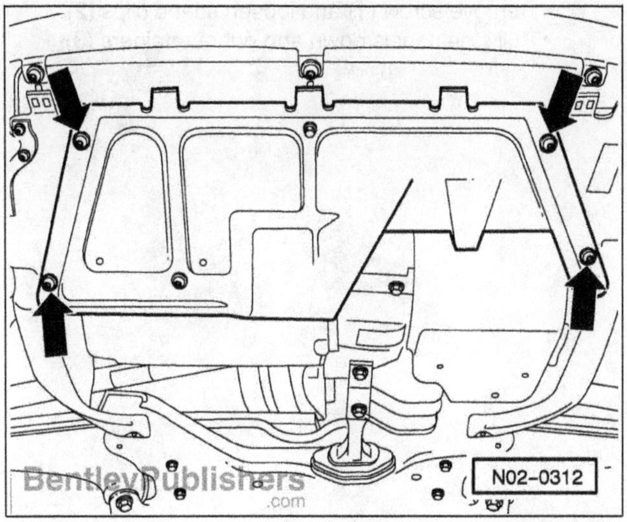

Gasoline engines, lower sound absorber panel.

- Remove 2 bolts (**arrows**) on each side.
- Pull sound absorber toward rear and remove.
- Install in reverse order of removal.

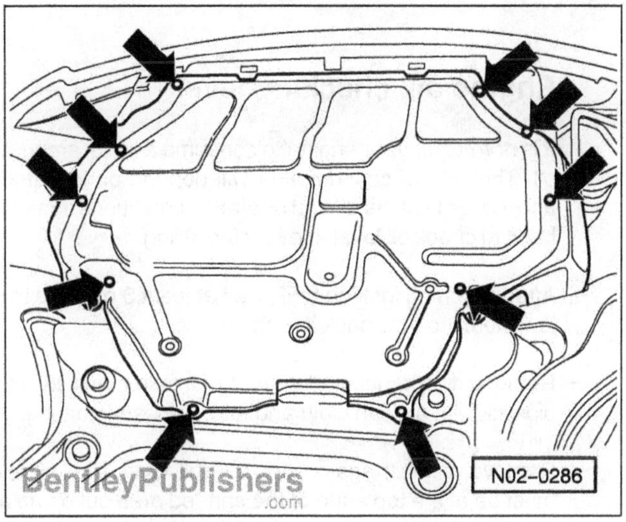

Diesel engines, lower sound absorber panel.

- Remove 4 bolts on each side.
- Remove 2 bolts in the rear.
- Pull sound absorber toward rear and remove.
- Install in reverse order of removal.

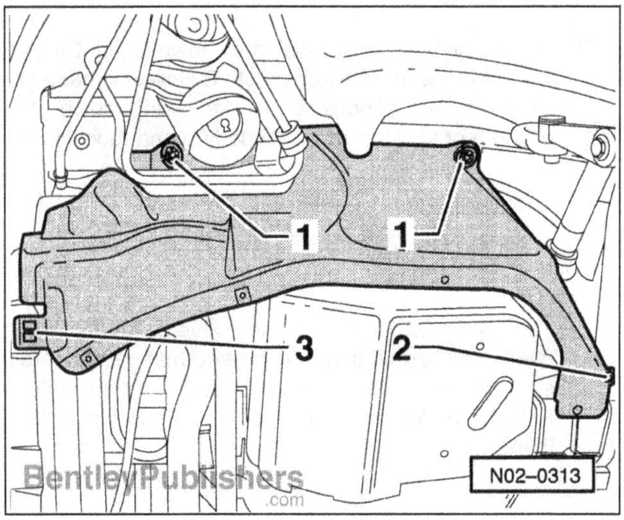

Lower driver's side sound absorber panel.

- Remove screws (**2**) and loosen speed clips (**1**).
- Pull panel down and out of retainers (**3**).

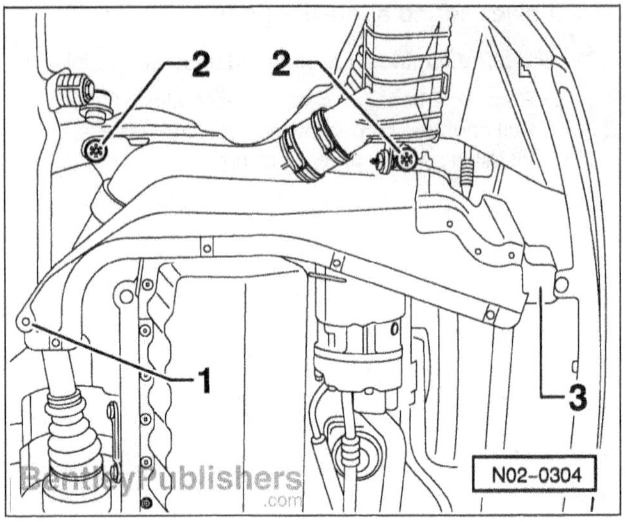

◄ Lower passenger side sound absorber panel

- Remove screw (**1**) and loosen speed clips (**2**).
- Pull side panels down and out of retainers (**3**).

Engine oil, checking level

It is normal for your engine to consume a small amount of oil. The rate of consumption will depend on the quality and viscosity of the oil and operating conditions. Make it a habit to check oil level at every fuel filling.

– After switching ignition OFF, wait at least 3 minutes to allow the oil to flow back into the oil pan.

– Remove dipstick located in front of intake manifold. Wipe dipstick with a clean cloth and fully re-insert into tube.

– Remove dipstick again and read oil level. The oil level must be at the top edge of the shaded area but on no account above the maximum level mark. If oil level is below shaded area, fill to top mark.

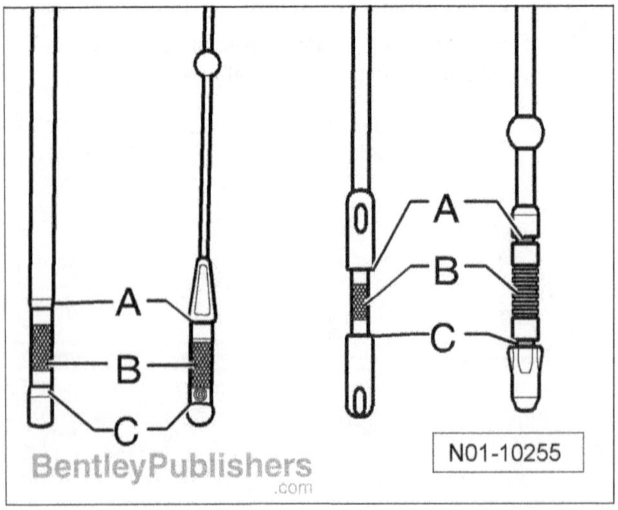

◄ Various styles of oil dipstick are shown at left. Oil level in area (**A**) must not be topped off. Oil level in shaded area (**B**) is normal and does not need to be topped up. Oil level at (**C**) or lower must be topped up and rechecked.

> ***CAUTION—***
> *Risk of damage to the catalytic converter if oil level is too high.*

Lubricating oil for the engines covered by this manual must meet certain quality and performance standards as specified by Volkswagen. These standards are sufficiently high so that generally, only synthetic oils will meet them.

Engine oil, specifications

The preferred viscosity for all engines is Society of Automotive Engineers (SAE) 5W40 (**B**), however, SAE 5W30 can be used if 5W40 is unavailable. In either situation, the container must be marked with at least one of the applicable listed specifications (**A**).

If the container of the oil in question is not marked withappropriate specification(s), the manufacturer advises that it must be assumed that the oil is not suitable for the application. Use of oils other than those specified may cause engine damage. Oil specifications and capacities differ between various engine types; consult the oil specification tables below.

Oil specifications - 1.8L 4-cylinder turbo gasoline	
Volkswagen specification	VW 502 00
Industry specification	API SL or SJ ACEA A3
Capacity (approximate)	with filter change 4.5 liters (4.8 qt.)

Oil specifications - 1.9L 4-cylinder TDI Diesel	
Volkswagen specification (1998-2003) (2004 -on)	 VW 505 00 VW 505 01
Industry specification	API CF4 or CG4 ACEA B3 or B4
Capacity (approximate)	with filter change 4.5 liters (4.8 qt.)

Oil specifications - 2.0L 4-cylinder gasoline	
Volkswagen specification	VW 502 00, 505 01 504 00/507 00
Industry specification	API SL or SJ ACEA A3
Capacity (approximate)	with filter change 4.2 liters (4.4 qt.)

Oil specifications - 2.5L 5-cylinder gasoline	
Volkswagen specification	VW 502 00, 505 01 504 00/507 00
Industry specification	API SL or SJ ACEA A3
Capacity (approximate)	with filter change 6.0 liters (6.3 qt.)

Engine oil, changing

- Warm car to normal operating temperature.
- Shut off engine and apply parking brake.
- Raise car and support on jackstands designed for the purpose.

> ***WARNING —***
> *Jack stands should be placed on hard level surface (e.g. concrete).*

- Remove lower sound absorber panel (belly pan) from beneath engine compartment. Remove drain plug from oil pan and allow oil to drain. See **17 Engine—Lubrication System** for drain plug location.

> ***WARNING —***
> *Hot oil can scald. Wear protective clothing, gloves and eye protection.*

> ***CAUTION —***
> - *Always replace oil-drain plug sealing ring anytime the drain plug is removed.*
> - *Dispose of oil properly at a facility equipped for recycling or storage.*

- Replace oil filter. See **Engine Oil filter, replacing** for applicable engine.
- Install oil drain plug with a new seal. Tighten drain plug to proper torque.

Tightening torque
- Oil drain plug to oil pan .... 30 Nm (22 ft-lb)

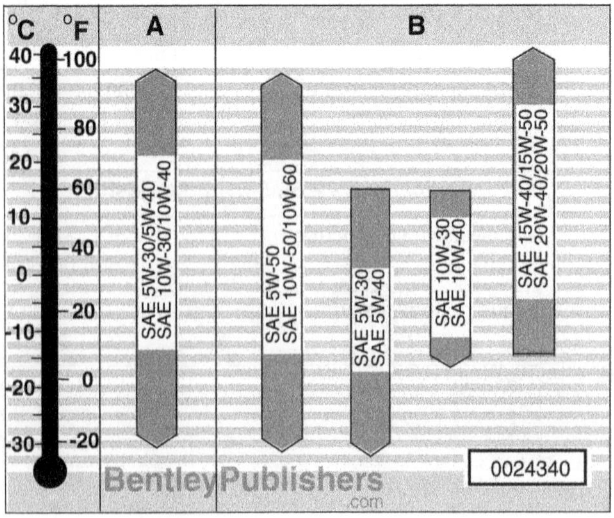

- Fill engine with proper quantity and type of oil. Oil viscosity requirement varies based on temperature. Section **A** is for energy-conserving multigrade oils. Section **B** is regular multigrade oils.

Engine oil capacity, with filter change
- 1.8L engine .... 4.5 liters (4.8 qt.)
- 1.9L engine .... 4.5 liters (4.8 qt.)
- 2.0L engine .... 4.2 liters (4.4 qt.)
- 2.5L engine .... 6.0 liters (6.3 qt.)

> ***CAUTION —***
> *The oil quantities listed above are approximate. Do not overfill engine. Overfilling can cause misfire DTC's to be stored in ECM. To prevent overfilling, add approx. 0.5L less than listed capacity, start engine and let run until engine temperature is approx. 60°C (140°F), turn OFF ignition, wait approx. 3 minutes, check dipstick and top off as necessary.*

– Run engine at idle and check for leaks.

> ***CAUTION —***
> *After changing the engine oil and the oil filter, observe the following at the first engine start:*
>
> - *The engine must only run at idling speed as long as oil pressure warning light in the instrument cluster is on. Do not rev the engine! If the engine is revved with the warning light on, the turbocharger on TDI and 1.8L turbo engines could be damaged or fail completely.*
> - *Full oil pressure is not attained until warning light has gone out. Only then can engine be revved.*

– Shut off engine and re-install sound dampening pan. Check oil level after about 3 minutes.

Engine oil filter, replacing (1.8L and 2.0L engine)

– Remove center of lower sound absorber panel as described earlier.

– Position drain pan under oil filter to catch spills.

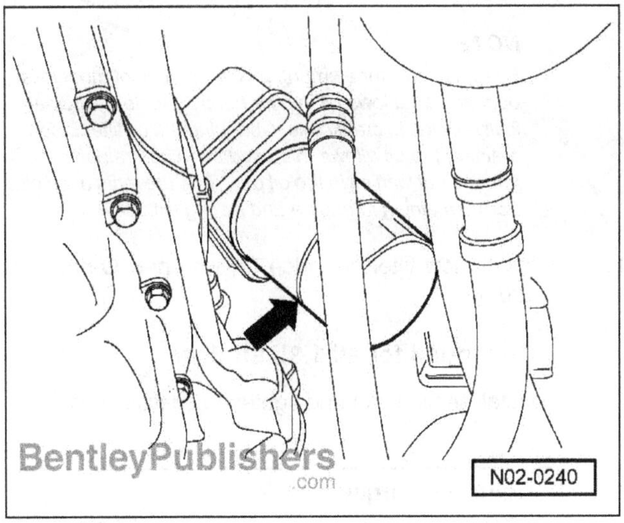

◄ Working from under engine, loosen and remove oil filter (**arrow**) with oil filter strap or suitable wrench.

– Clean sealing surface on oil filter flange.

– Lightly lubricate rubber seal with clean oil.

– Thread on new filter and tighten by hand.

> ***CAUTION —***
> *Observe all applicable regulations concerning disposal of used oil filters.*

Engine oil filter, replacing (1.9L engine)

- Remove upper sound absorber panel.
- Working from above the engine, loosen and remove oil filter sealing cap.
- Pull filter up and out.

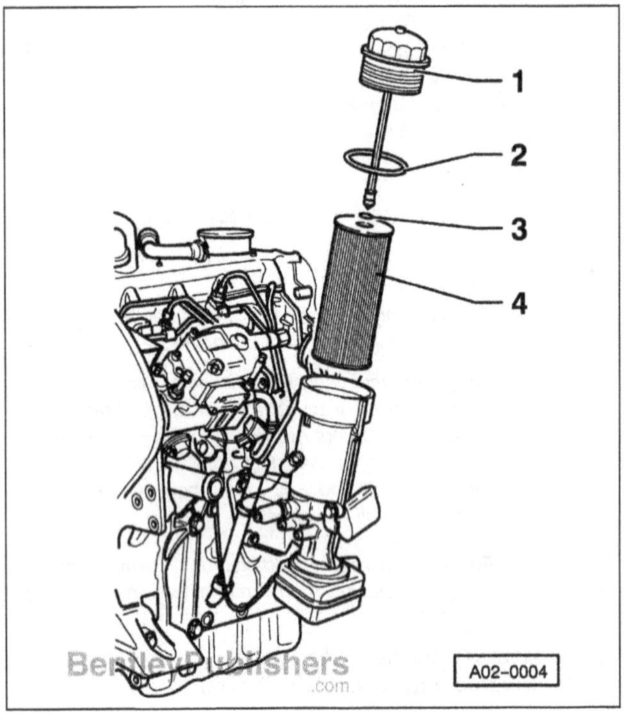

Engine code ALH (1998-2004)

◄ Remove sealing cap (**1**). Install new oil filter (**4**) with new gaskets (**2** and **3**).

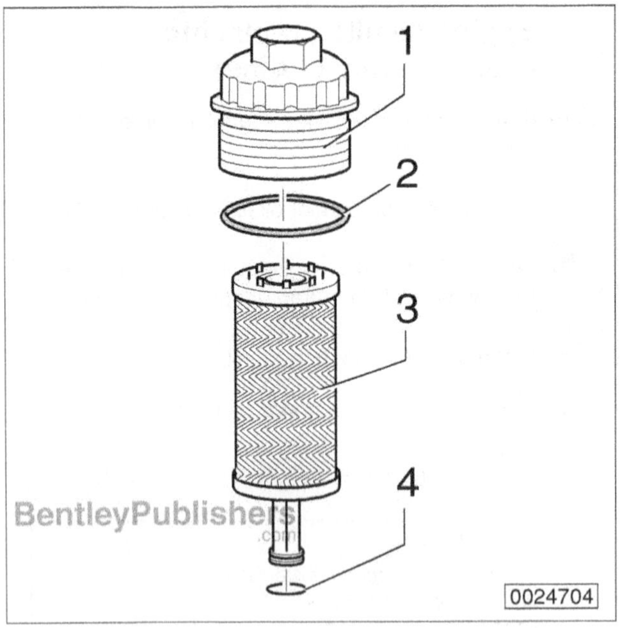

Engine code BEW (2004-on)

◄ Remove sealing cap (**1**) with a 32 mm socket and take out oil filter cartridge (**3**). Allow filter housing to drain.

NOTE —
By removing filter element, a valve in base of module is opened that allows oil in filter housing to flow automatically back into crankcase. Changing the oil filter before draining the oil allows this used oil to be drained or siphoned out with oil in the oil pan. This prevents used oil from remaining in engine and mixing with new oil.

- Install new filter cartridge (**3**) with new O-ring seals (**2** and **4**).

Continued for all 1.9L engines

- Install sealing cap and tighten to specification.

Tightening torque
• Oil filter sealing cap . . . . 25 Nm (18 ft-lb)

- Install upper sound absorber panel.

CAUTION —
Observe all applicable regulations concerning disposal of used oil filters.

Engine oil filter, replacing (2.5L engine)

Gasoline engines with valve in base of filter housing

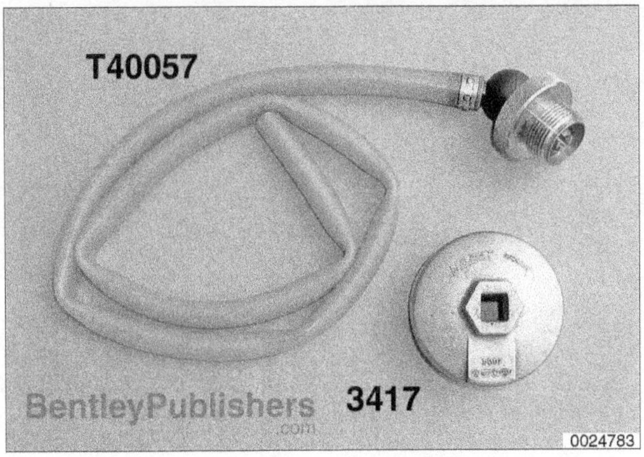

Engines with a drain valve in the base of filter housing require use of two special tools.

- Before lower filter housing (with oil filter cartridge) is removed, it must be emptied to prevent spillage onto engine using tool T40057, or equivalent.
- By screwing in oil drain adapter T40057, a drain valve is opened in bottom of oil filter housing. When oil drain adapter T40057 is unscrewed, valve automatically returns to closed position.
- Oil filter wrench 3417, or equivalent should be used to prevent damage to lower filter housing during removal and installation.

NOTE —

- *Oil filter wrench 3417 has 15 "flats".*
- *Some vehicles may require use of a 36mm socket or wrench instead of oil filter wrench 3417.*

– Warm car to normal operating temperature.

– Shut off engine and apply parking brake.

– Raise car and support on jackstands or ramps designed for the purpose or use a lift or hoist.

> ***WARNING —***
> *Jack stand and ramps should be placed on hard level surface (e.g. concrete).*

– Remove lower sound absorber panel from under engine. See **Sound absorber panels, removing and installing**.

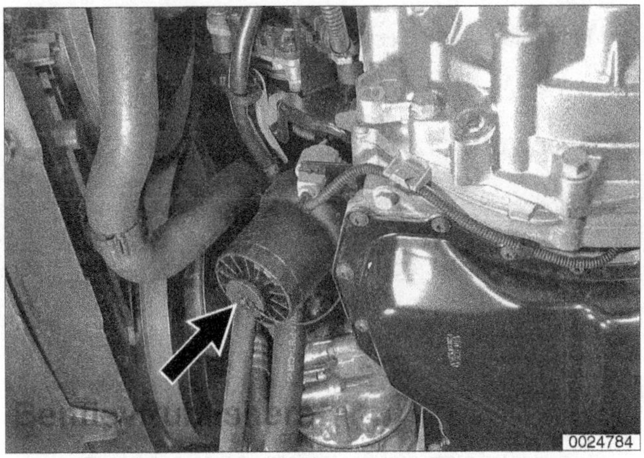

Remove threaded dust cap (**arrow**) from bottom of oil filter housing. Dust cap may be on an attached tether.

Attach oil drain adapter T40057 to bottom of oil filter housing and hold hose into an appropriate oil drain pan. Allow engine oil to drain.

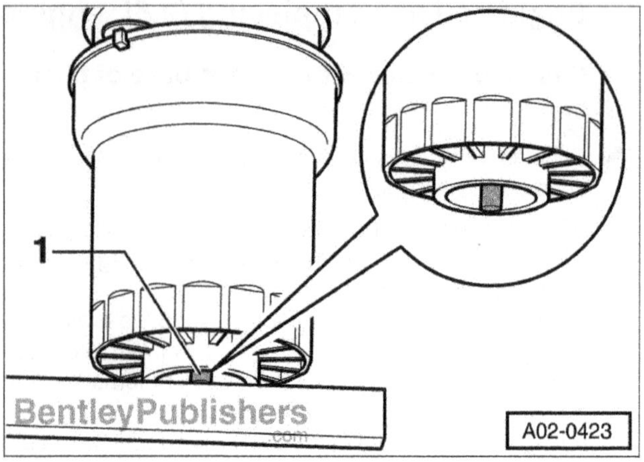

Remove oil drain adapter T40057 and inspect valve (**1**) to ensure complete closure. Valve must be flush and level.

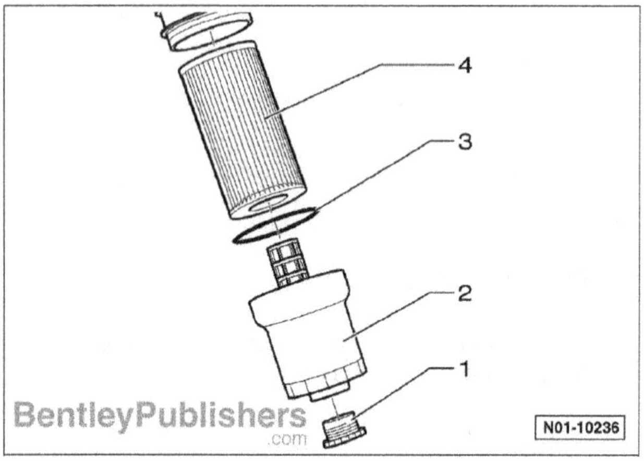

Use oil filter wrench 3417 to unscrew and remove lower oil filter housing (**2**) with oil filter cartridge (**4**).

- Clean upper and lower housings (**2**).
- Install new oil filter cartridge (**4**), O-ring (**3**) and lower oil filter housing (**2**).

Tightening torque
- Oil filter housing to engine . . . . 25 Nm (18 ft-lb)

- Hand-tighten dust cap (**1**) into lower oil filter housing (**2**).

Coolant, checking

Cooling system maintenance consists of maintaining coolant level, checking coolant freezing point, and inspecting hoses regularly.

NOTE —
Volkswagen does not require replacing the coolant as part of routine maintenance. The cooling system has been filled at the factory with a permanent coolant. For coolant system draining and refilling procedures, see **19 Engine-Cooling System.**

WARNING —
- *Hot coolant can scald. Do not work on the cooling system until it has fully cooled.*
- *Use extreme care when draining and disposing of coolant. Coolant is poisonous and lethal. Children and pets are attracted to it because of its sweet smell and taste. See a doctor or veterinarian immediately if any amount is ingested.*

CAUTION—
- *Use only Volkswagen original coolant when filling the cooling system. Use of any other coolant may be harmful to the cooling system. Do not use an anti-freeze containing phosphates.*
- *Do not use tap water in cooling system. Use distilled water only to mix coolant.*

The vehicles covered by this manual use a coolant/antifreeze that is phosphate silicate free. When supplied by Volkswagen, this coolant/antifreeze is known as G12. This coolant can be either pink or purple in color. It is permissible to mix pink and purple G12 coolant.

The advantages of G12 over earlier coolant types include improvements in corrosion protection, thermal stability, heat transfer/control, hard water tolerance and environmental protection.

CAUTION—
- *G12 (pink or purple coolant) must NEVER be mixed with G11 (green) coolant. Engine damage will result!*
- *Contamination of G12 with other types of coolants (G11) is identifiable by discoloration (i.e. brown). This mixture causes a foamy deposit to appear in the expansion tank/radiator and MUST be drained immediately.*
- *The cooling system must be completely free of this mixture before refilling with the correct type of coolant/antifreeze.*

A translucent expansion tank, or overflow reservoir, provides easy monitoring of coolant level without opening the system.

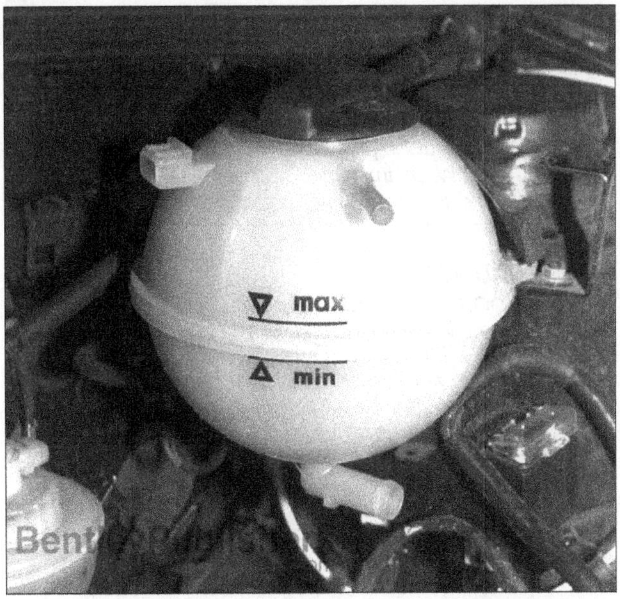

The coolant level should always be checked when engine is cold. The coolant level should be between the maximum and minimum mark on the expansion tank.

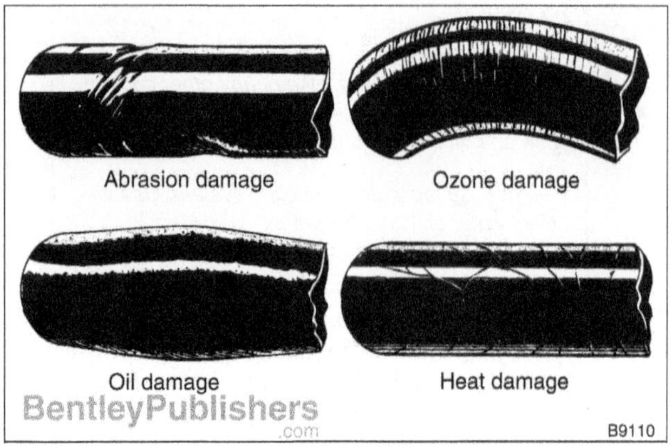

Inspect hoses by first checking that all connections are tight and dry. Coolant seepage indicates that either the hose clamp is loose, that the hose is damaged, or that the connection is dirty or corroded. Dried coolant has a chalky appearance. Check hose condition by pinching them. Hoses should be firm and springy. Replace any hose that is cracked, that has become soft and limp, or has been contaminated by oil or diesel fuel.

To check freezing point of the coolant, (and also the boiling point), use a refractometer. With the engine cold, remove cap from expansion tank and draw up some coolant with refractometer siphon. Volkswagen recommends using a 50/50 mixture of anti-freeze and water, which has a freezing point of -35°F (-38°C). If the freezing point is too high, drain a small amount of coolant from the system and add new anti-freeze until proper protection level is obtained.

NOTE —
For protection to approx. -40°C (-40°F) the percentage of anti-freeze may be increased up to 60%. The percentage of anti-freeze must not exceed 60%, as higher amounts will decrease frost protection and cooling capacity.

Brake fluid level, checking

Routine maintenance of the brake system includes maintaining an adequate level of brake fluid in the reservoir, replacing brake fluid every 2 years, checking brake pads for wear, checking hand brake function, and inspecting system for fluid leaks or other damage. See **Under Car Maintenance** for brake pad inspection and brake system inspection.

WARNING —
- *Use only new, previously unopened brake fluid conforming to US Standard FMVSS 116 DOT 4.*
- *Brake fluid is poisonous. DO NOT ingest brake fluid. Wash thoroughly with soap and water if brake fluid comes into contact with skin.*

CAUTION —
- *DO NOT let brake fluid come in contact with paint. Wash immediately with soap and water.*
- *Brake fluid absorbs moisture from the air and must be stored in an airtight container.*
- *DO NOT allow brake fluid to exceed the maximum level in the fluid reservoir.*

The brake fluid level will drop slightly as the brakes wear. Check fluid level at brake fluid reservoir, located on driver's side of engine compartment, under air cleaner connecting hose.

Brake fluid
- Type . . . . FVMSS 116 SAE DOT 4

> ***WARNING —***
> *Do not mix DOT 5 (silicone) brake fluid with DOT 4 brake fluid as severe component corrosion will result. Such corrosion could lead to brake system failure.*

Brake fluid, replacing

Brake fluid readily absorbs moisture from the atmosphere. This moisture can cause brake system corrosion and adversely affect braking performance. Therefore, the old brake should be flushed out of the system and new, fresh fluid added at least every 2 years, regardless of mileage.

> ***WARNING —***
> *See **Warnings** and **Cautions** regarding brake fluid above.*

> ***CAUTION —***
> *For vehicles with ABS/EDS, ABS/EDS/ASR, or ABS/EDS/ASRESP care must be taken to ensure that the brake fluid pressure at the VAS 5234, or equivalent, brake filler and bleeder unit must be adjusted down to 1 bar (14.8 psi).*

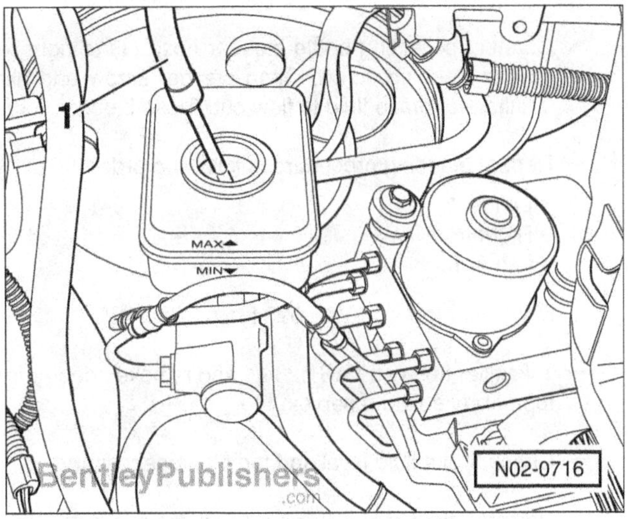

– Remove brake fluid reservoir cap.

◀ Insert suction hose from VAS 5234 (**1**) and extract as much brake fluid as possible from reservoir.

> ***NOTE —***
> *Make sure no brake fluid runs through stainer after completing the extractions sequence (the brake fluid level in the reservoir must not be lower than lower edge of strainer).*

– Install VAG 1869/2 brake pedal depressor, or use an assistant, between driver's seat and pedal and apply pressure to brake pedal.

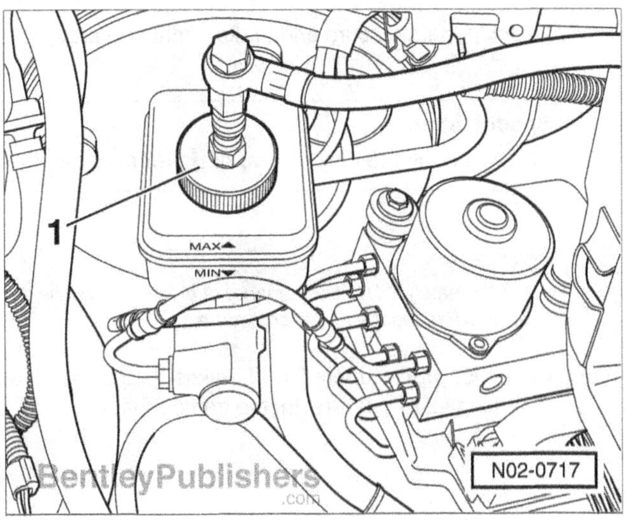

◀ Install VAS5234/1 adapter (**1**) to brake fluid reservoir. Connect filler hose from VAS 5234 to VAS 5234/1.

Vehicles with manual transmission

– Pull cover cap off clutch slave cylinder bleeder valve.

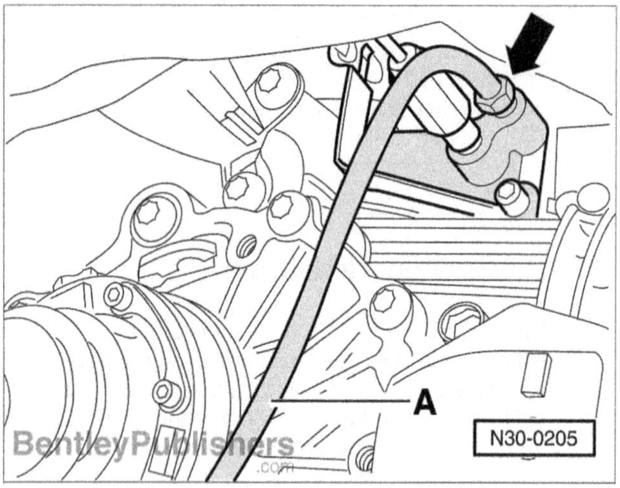

◄ Connect collecting bottle bleeder hose (**A**) to bleeder screw of clutch slave cylinder (**arrow**), open bleeder screw an allow approximately 0.1 liter to flow out. Close bleeder screw and install cover cap.

– Operate clutch pedal several times.

Continued for all vehicles

– Pull cover caps off bleeder screws at each brake caliper.

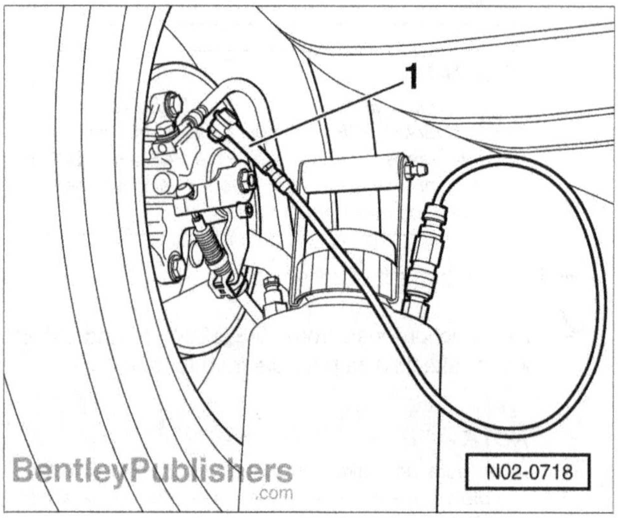

◄ Connect collecting bottle bleeder hose (**1**) to right rear caliper bleeding screw. Open bleeder screw and allow 0.2 liters of brake fluid to flow out. Close bleeder screw.

– Repeat bleeding procedure in following order:

- Left rear
- Right front
- Left front

– Move filler lever on VAS 5234 to position "B".

– Take filler hose off VAS 5234/1 and remove adapter from top of brake fluid reservoir.

– Check brake fluid level and top off as necessary.

– Reinstall fluid reservoir cover and remove brake pedal depressor.

– Check pedal pressure and brake pedal free play.

Specification
- Brake pedal free play . . . . . . max. 1/3 of pedal travel

NOTE —
Approximately 2 liters (2.1 quarts) of brake fluid will be removed during the above procedure.

– Check for proper operation of brakes during road test in a parking lot or other restricted traffic area.

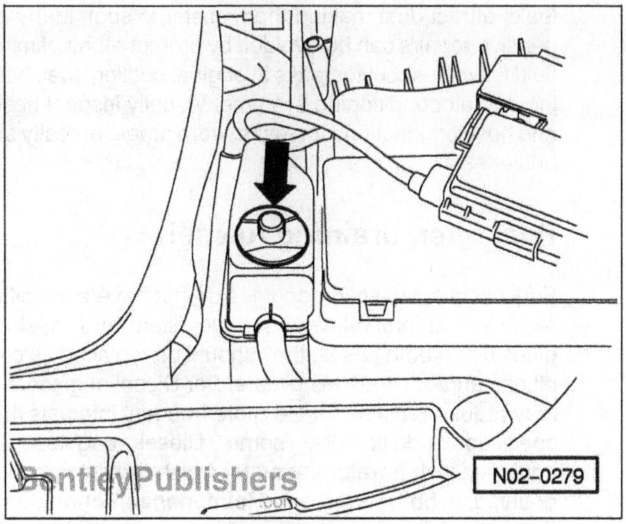

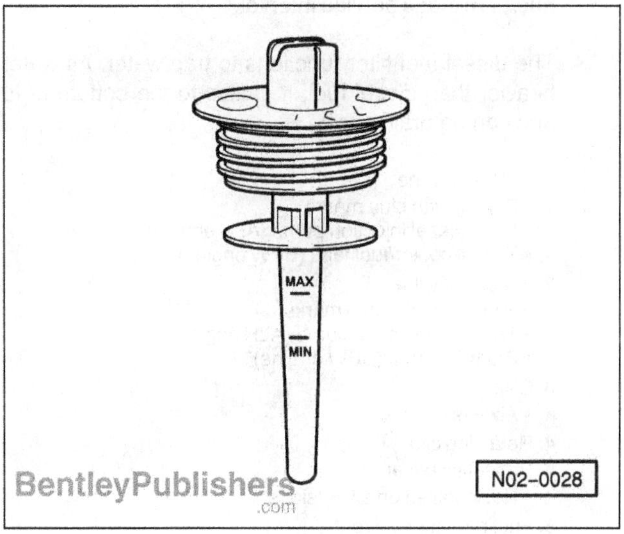

Power steering fluid level, checking

There are two methods for checking power steering fluid level.

COLD - When cold checking the power steering fluid level, the engine should not be running and the front wheels should be in the straight ahead position. Use a screwdriver to remove power steering reservoir cap (**arrow**).

Use a clean shop cloth to wipe the dipstick end and screw the cap back on hand tight only. Remove the dipstick from the fluid reservoir again and check the level on the dipstick. The level should be within ± 2 mm of the **min** mark. Adjust the fluid level if required and screw the cap back on hand tight when done.

HOT - When hot checking the power steering fluid level, start the engine and let it run until the power steering fluid temperature is above 50°C (122°F). The front wheels should be in the straight ahead position. Use a screwdriver to remove the cap. Use a clean shop cloth to wipe the dipstick end and screw the cap back on hand tight only. Remove the dipstick from the fluid reservoir again and check the level on the dipstick. The level should be between the **min** and **max** marks. Adjust the fluid level if required and screw the cap back on hand tight when done.

Power steering fluid
• Type . . . . hydraulic oil (VW part no. G 002 000)

CAUTION —
- *Use only Volkswagen hydraulic oil G 002 000 in the power steering system. Do not use ATF or other non-approved types of power steering fluid. If the wrong fluid is used, power steering components may fail.*

NOTE —
- *If fluid level is above (approx. 2 mm) the MAX mark, some fluid must be extracted.*
- *If fluid level is below (approx. 2 mm) the MIN mark, the hydraulic system must be checked for leaks. It is not sufficient to top-up with fluid.*

Fluid leaks

Check engine compartment for signs of fluid leaks. Fluid leaks attract dust making them easier to spot. Many expensive repairs can be avoided by prompt repair of minor fluid leaks. Inspect for leaks in engine, cooling, fuel, heating and air conditioning systems. Visually inspect hoses and hose connections for leaks, worn areas, porosity and brittleness.

Fuel filter, draining (diesel)

Fuel filters for gasoline engines do not have a specified replacement interval, whereas fuel filters for Diesel engines do. In both cases, they should be replaced as conditions and/or situations dictate. For Diesel engines, this may require replacement at more frequent intervals than specified. Additionally, some Diesel engines are equipped with a water separator which requires periodic draining. In both cases, see **Maintenance Schedules** at the end of this repair group for the appropriate time and/or mileage service interval.

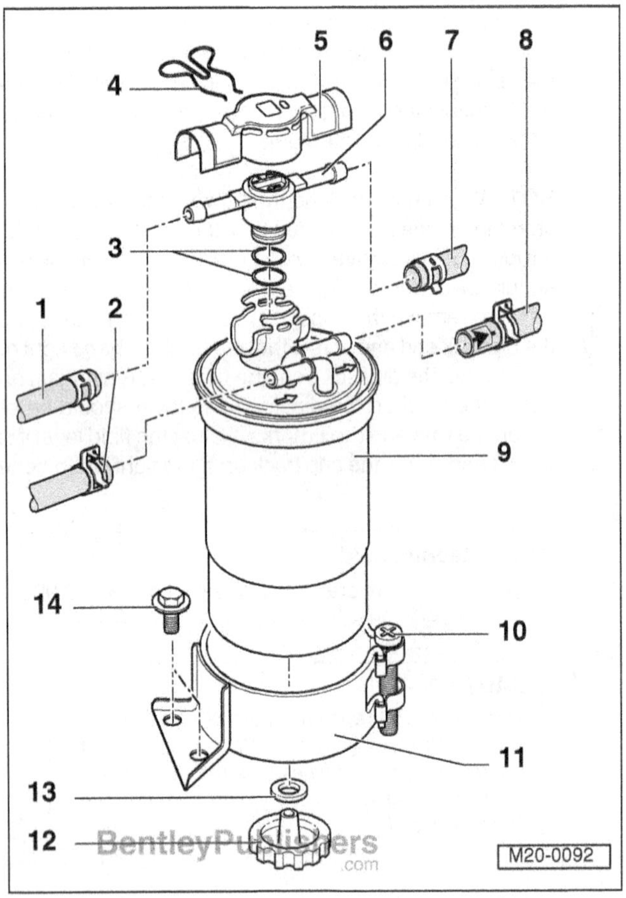

The diesel fuel filter functions to trap water. As water is heavier than diesel fuel, it settles to the bottom of filter and can be drained off.

1. Fuel return line
   - Blue or with blue marks
   - From diesel injection pump (ALH engine)
   - To fuel cooler/fuel tank (BEW engine)
2. Fuel supply line
   - White or with white marks
   - To diesel injection pump (ALH engine)
   - From fuel tank (BEW engine)
3. O-ring
   - Always replace
4. Retaining clip
5. Protective cover
   - Not installed on all versions
6. Fuel regulator valve
   - Recirculates fuel to filter at temperatures below 15°C (59°F)
   - Returns fuel to tank at temperatures above 31°C (88°F)
   - Installed position: arrow points toward fuel tank
7. Fuel return line
   - To fuel tank (ALH engine)
   - From injection pump (BEW engine)
8. Fuel supply line
   - From fuel tank (ALH engine)
   - To injection pump (BEW engine)
9. Fuel filter
   - Fill with diesel fuel before installing
   - Observe fuel flow direction marked on top
   - Do not interchange connections
10. Bolt
11. Mounting bracket
12. Water drain plug
13. Gasket
14. Bolt
    - Tighten to 10 Nm (7 ft-lb)

> ***CAUTION —***
> *Do not allow diesel fuel to contact coolant hoses or other rubber parts. Wipe diesel fuel off hoses immediately and wash with soap and water.*

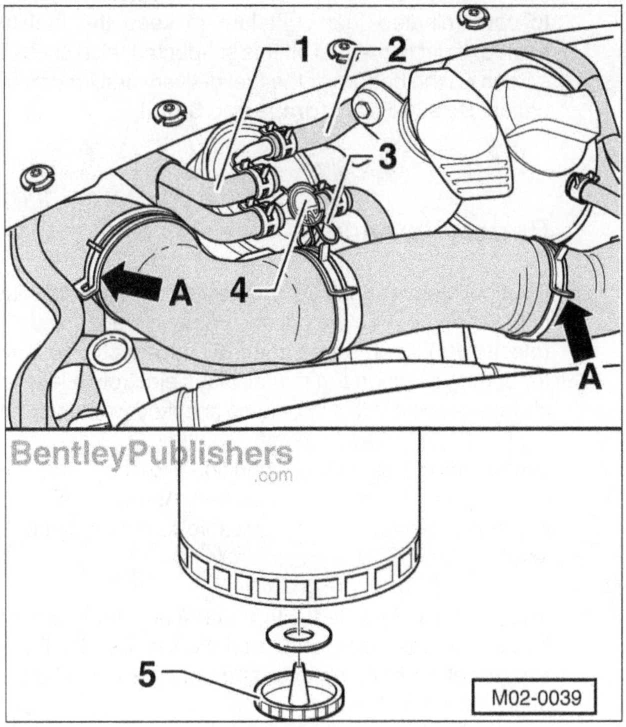

< Loosen clamps (**A**) and remove pressure tube from intercooler.

– Remove regulator valve retaining clip (**3**) and pull valve (**4**) up out of filter with fuel lines still connected.

– Loosen water drain plug (**5**) on bottom of filter and drain out approximately 0.1 liters (0.1 qt.) of fuel/water.

– Install fuel control valve (**4**) with a new O-ring.

– Attach retaining clip (**3**) and tighten drain plug (**5**).

– Install pressure tube for intercooler.

– Check fuel system for leaks (visual inspection) with the engine running. Raise engine speed several times and let idle. Fuel flow through transparent pipe must be observed.

Fuel filter, replacing (diesel)

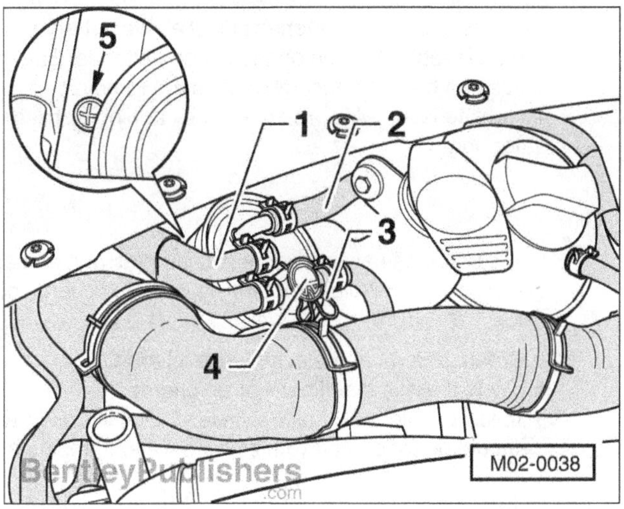

< Disconnect fuel hoses (**1**) and (**2**) from hose connections.

– Remove regulator valve retaining clip (**3**)

– Pull fuel regulator valve (**4**) up out of filter with fuel lines still connected.

– Disconnect fuel supply lines (**1**) and (**2**).

– Loosen filter bracket (**5**) and remove filter.

– Install new filter.

> ***NOTE —***
> - *The direction of fuel flow is marked by arrows on the top of the fuel filter.*
> - *It is advisable to fill new fuel filter with fresh diesel fuel to facilitate initial start-up.*

– Installation is the reverse of removal, noting the following:

- Install fuel lines properly.
- Install regulator valve with new O-ring and attach retaining clip.

– Check fuel system for leaks (visual inspection) with the engine running. Raise engine speed several times and let idle. Fuel flow through transparent pipe must be observed.

Fuel filter (gasoline engines)

Volkswagen specifies that the fuel filter used on gasoline engines is designed to last the life of the car. Replacement is only necessary if the filter becomes clogged due to contaminated fuel or failure to keep the fuel tank sealed. If dirt in the fuel filter is suspected, also check the screen on the bottom of the fuel delivery unit for contamination. **See 20 Fuel Storage and Supply.**

Battery service

Routine maintenance of the battery consists of visual checks of hold-down clamp, cable clamps, and fluid (electrolyte) level. Some batteries have a charge indicator or magic eye on top that displays electrolyte level and charge condition; this should be checked when present.

Under normal operating conditions, battery electrolyte level is generally maintenance free. At high outside temperatures, however, it is advisable to check electrolyte level.

The battery is installed with a protective cover that also has a fuse box incorporated on the top. See **27 Engine Electrical** for battery cover and battery removal and installation instructions.

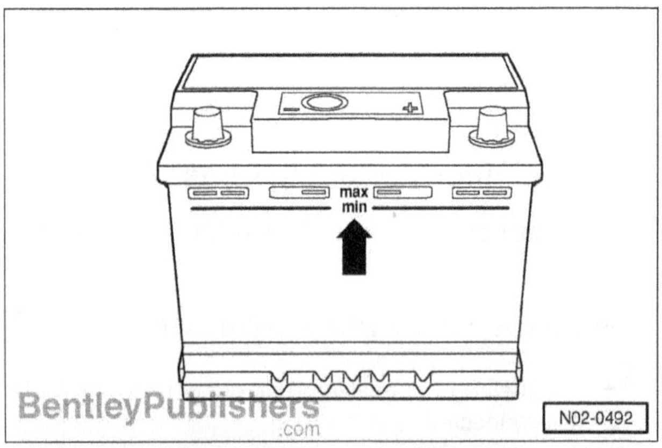

Electrolyte level should be between MIN and MAX marks on translucent case batteries or just above battery plates and their separators on opaque case batteries. Battery plates can be seen once filler caps are removed. If electrolyte level is low, replenish it by adding distilled water only.

Some batteries have a charge indicator (**arrow**) or magic eye on top that displays electrolyte level and charge condition.

- Green, charge and electrolyte level are OK.
- Black, there is insufficient or no charge.
- Yellow or colorless, electrolyte level is critically low and should be topped off with distilled water immediately.

NOTE —

- *If charge indicator on a battery over 5 years old is colorless, do not attempt to top off or recharge battery. Battery should be replaced.*
- *Air bubbles that occur normally during battery charging (even during vehicle operation) may adversely affect charge indicator reading. To obtain an accurate reading, gently tap charge indicator with a screwdriver handle or rock vehicle to displace any possible air bubbles that have formed.*

– Inspect negative (–) and positive (+) cable clamp nuts for tightness. Also check torque of battery hold-down clamp bolt.

Tightening torques

- Battery cable clamps . . . . 6 Nm (53 in-lb)
- Battery hold-down clamp . . . . 20 Nm (15 ft-lb)

When servicing battery, observe the following points and guidelines:

- On batteries with cell caps, all cell caps must be equipped with an O-ring seal.
- Electrolyte level on batteries with visible minimum and maximum markings on outside of translucent battery case is checked by visual inspection.

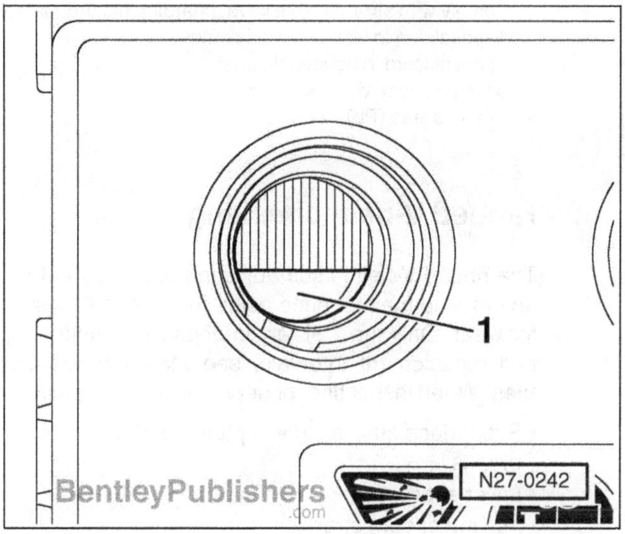

◄ Electrolyte level must be above minimum marking or just reach maximum marking.

- On batteries where exterior markings are difficult to recognize (due to installation location, battery case opaqueness or dirt), cell caps must be removed and individual cell electrolyte levels checked from inside battery.
- Electrolyte level must align with internal electrolyte level indicator (lip). This equates to external "maximum" marking on battery case.
- When electrolyte level is too low and cell plates are exposed, a loss of battery capacity (power) will result. If cell plates are not completely submerged by electrolyte, (sulfuric acid/water mixture) corrosion will occur, on plates, plate bridges and cell connectors. Optimum battery function is not possible under these conditions!
- When electrolyte level is too high, electrolyte (sulfuric acid/water mixture) may leak out and damage surrounding areas such as battery tray and engine compartment.
- Only use distilled water to top-up batteries. This prevents electrolyte impurities which cause self-discharging.
- Do not overfill battery.
- Overfilled batteries can boil over.
- Too little electrolyte reduces service life of battery.
- Excess electrolyte must be extracted using a hydrometer.
- Dispose of electrolyte (sulfuric acid) properly. Waste electrolyte must only be disposed of in appropriate waste disposal sites. Refer to local regulations pertaining to electrolyte disposal.
- Volkswagen recommends that only new generation batteries should be installed.
- For technical reasons, some batteries are equipped with sealing plugs with plastic foil. These batteries are considered maintenance free.

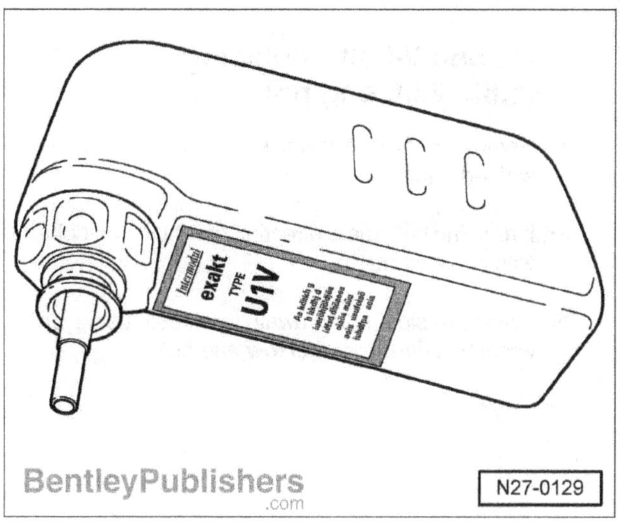

◄ A battery filling bottle such as VAS 5045 is specifically designed for the purpose and will prevent overfilling of battery.

For battery testing, service and installation procedures, see **27 Engine Electrical**.

After disconnecting and re-connecting battery, some vehicle systems may require initialization (reset of basic settings), see **27 Engine Electrical**.

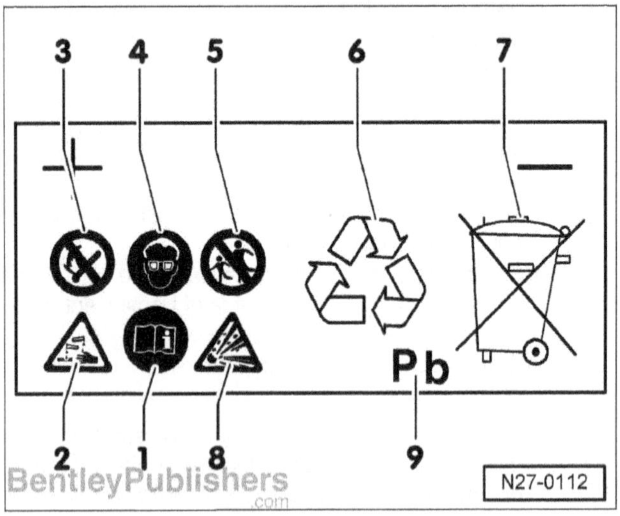

N27-0112

A warning and safety label similar to this is found on most original equipment Volkswagen batteries.

1. Follow recommendations on battery, in vehicle glovebox owners's manual, and in this manual.
2. Danger of acid burns. Follow warnings listed for working with acid.
3. Keep open flames, sparks, unenclosed light bulbs and lit smoking materials away from battery. Avoid electrical short circuits.
4. Wear eye protection.
5. Keep children away.
6. Dispose of old batteries properly. Recycle when possible. Comply with all applicable laws regarding hazardous waste disposal due to lead and acid content.
7. Do not discard in household trash.
8. Explosion hazard.
9. Contains lead (Pb).

Ribbed V-belt, checking

The ribbed V-belt is self-adjusting and does not require any routine maintenance aside from periodic inspection for wear. Only the 1.8L engine has manufacturer-specified replacement intervals, see **Maintenance Schedules**. When inspecting, observe the following points:

- Subsurface cracks, core ruptures and cross-sectional breaks
- Layer separation
- Fraying of cord strands
- Flank wear, material condition, fraying, brittleness, glassiness and cracks
- Traces of oil and grease

NOTE —
It is essential to replace the ribbed belt if it is found to be faulty.

Ribbed V-belt, replacing (1.8L, 2.0L engine)

– Remove right side lower sound absorber panel as described earlier.

– If re-using belt, mark direction of rotation on ribbed V-belt with chalk or crayon.

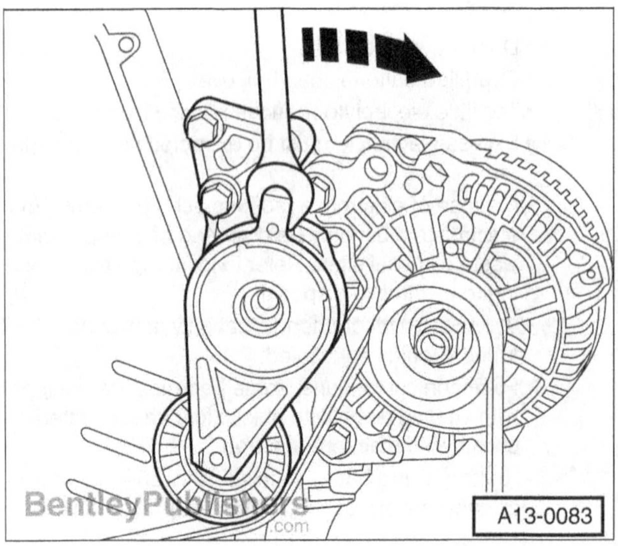

A13-0083

Swing tension roller away from belt using open end wrench in direction of **arrow** and hold.

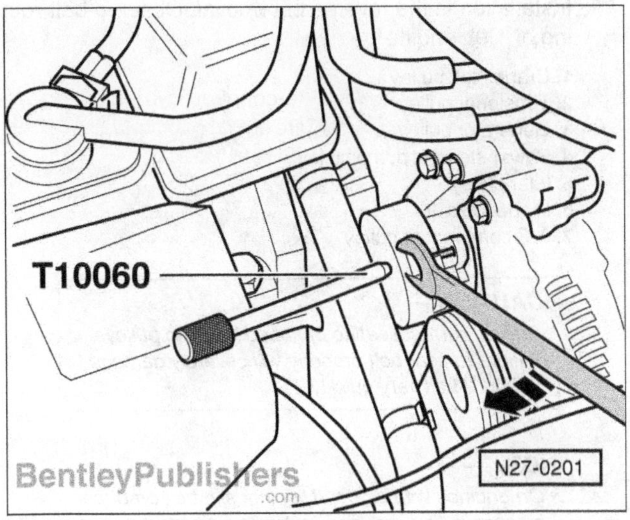

◄ Lock tensioner in released position with suitable tool.

- Special tool T10060 mandrel shown.
- Special tool 3090 connecting rod support may also be used.
- An appropriately sized drift or drill bit may also be used.

– Remove belt.

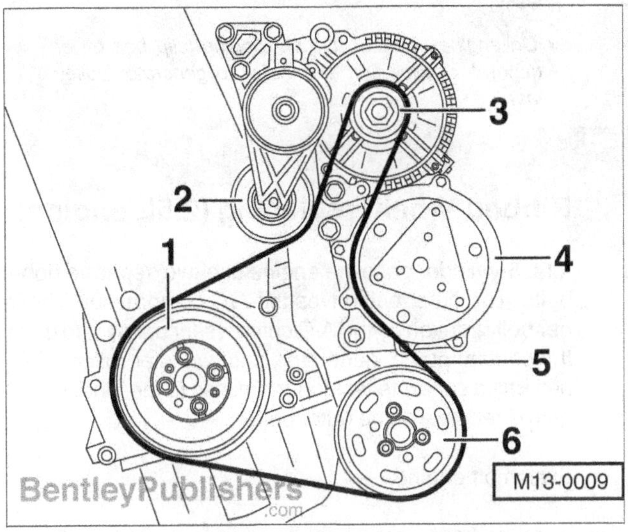

◄ Installation is the reverse of removal. Observe belt routing for 1.8L and 2.0L engines.

1. Crankshaft pulley
2. Tensioner pulley
3. Generator pulley
4. A/C compressor pulley
5. Ribbed V-belt
6. Power steering pump pulley

> ***CAUTION—***
> *Ensure correct seating of ribbed V-belt in pulleys. Failure to seat belt properly will severely damage ribbed V-belt.*

Ribbed V-belt, replacing (1.9L engine)

– Remove right side lower sound absorber panel as given earlier.

– Remove connecting pipe between intercooler and turbocharger.

– If re-using belt, mark direction of rotation on ribbed V-belt with chalk or crayon.

◄ Swing tension roller in direction of **arrow** away from belt using an open end wrench and hold while removing belt.

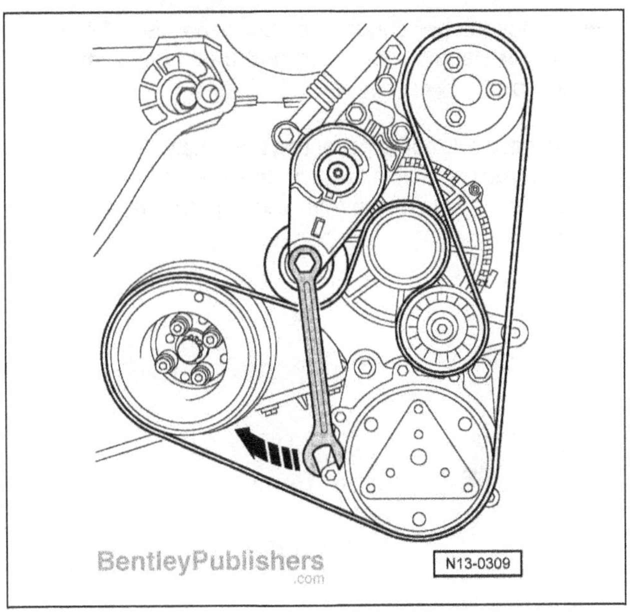

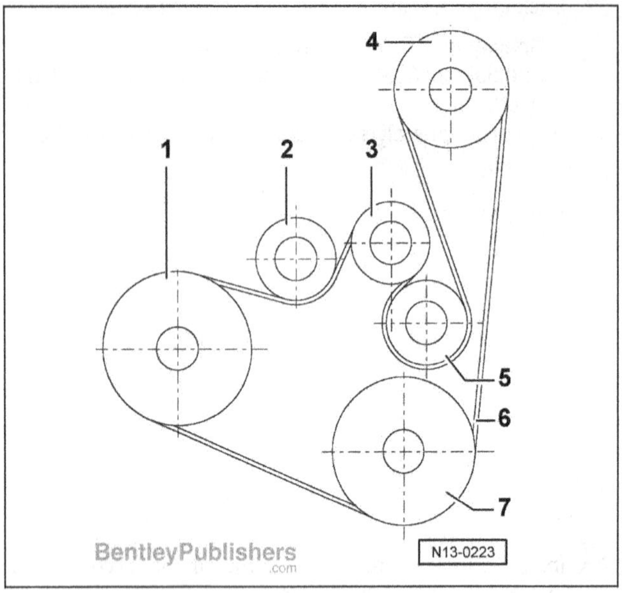

Installation is the reverse of removal. Observe belt routing of 1.9L engine.

1. Crankshaft pulley
2. Tensioner pulley
3. Generator pulley
4. Power steering pump pulley
5. Idler pulley
6. Ribbed V-belt
7. A/C compressor pulley

> ***CAUTION —***
> *Ensure correct seating of ribbed V-belt in pulleys. Failure to seat belt properly will severely damage ribbed V-belt very quickly.*

NOTE —
- *On engines with air conditioning, slip belt off of compressor pulley first and install to compressor pulley last.*
- *On engines without air conditioning, slip belt off of generator pulley first and install to generator pulley last.*

Ribbed V-belt, replacing (2.5L engine)

2.5L 5-cylinder gasoline engine use two separate ribbed belts. The outer belt drives the A/C compressor. The inner belt is driven by the A/C compressor pulley and drives the coolant pump, generator, and two idler rollers. Each belt has a separate tensioner. Access to the inner belt requires removal of the outer belt.

– Switch off engine.

– Remove front right wheel housing liner. See **66 Body–Exterior Equipment**.

– Remove upper sound absorber panel. See **Sound absorber panels, removing and installing** in this repair group.

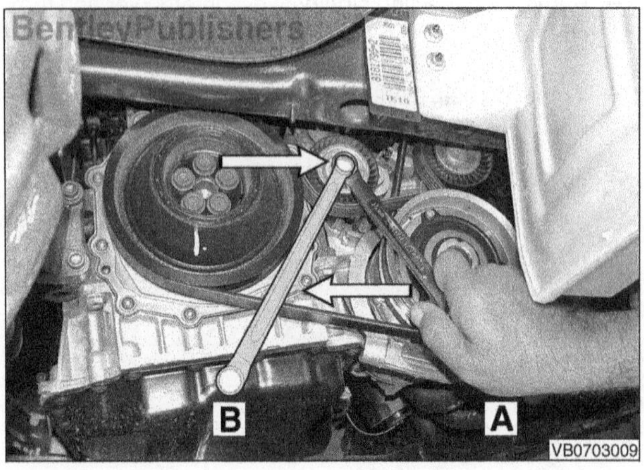

Release A/C compressor belt tension by moving tensioner (**upper arrow**) to the rear (**lower arrow**) with a 15 mm wrench. Pivot wrench from starting point (**A**) to end point (**B**).

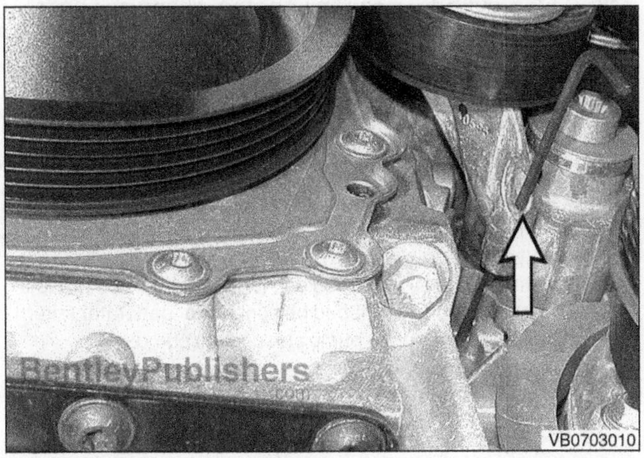

While holding wrench, align hole in tensioner (**arrow**) with hole in mount and lock with a suitable pin such as a hex key or equivalent.

NOTE—

- *Box wrench and belt shown removed for clarity.*
- *Available factory locking pin T10060A is approximately 4.5 mm (3/16 in.) in diameter and each leg is approximately 50 mm (2 in.) long.*

– Remove ribbed belt for A/C compressor.

NOTE—

If belts are to be re-used, such as for emergency spares, mark running direction prior to removal.

– Return A/C belt tensioner to original position by reversing above procedure and removing locking pin. This allows access to tensioner for coolant pump/generator and re-use of locking pin.

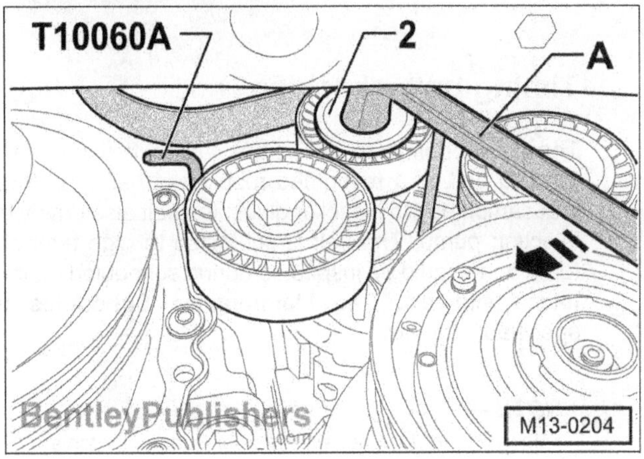

To remove ribbed belt for coolant pump/generator, partially insert locking pin (**T10060A** shown) into tensioner (**2**).

– Pivot tensioner (**2**) with a 15 mm wrench (**A**) in direction shown (**dashed arrow**). In some instances, a 15 mm hex key (Allen) wrench is needed instead of a 15 mm box wrench.

– Push locking pin (**T10060A** shown) all the way into tensioner (**2**).

– Remove ribbed belt for coolant pump/generator.

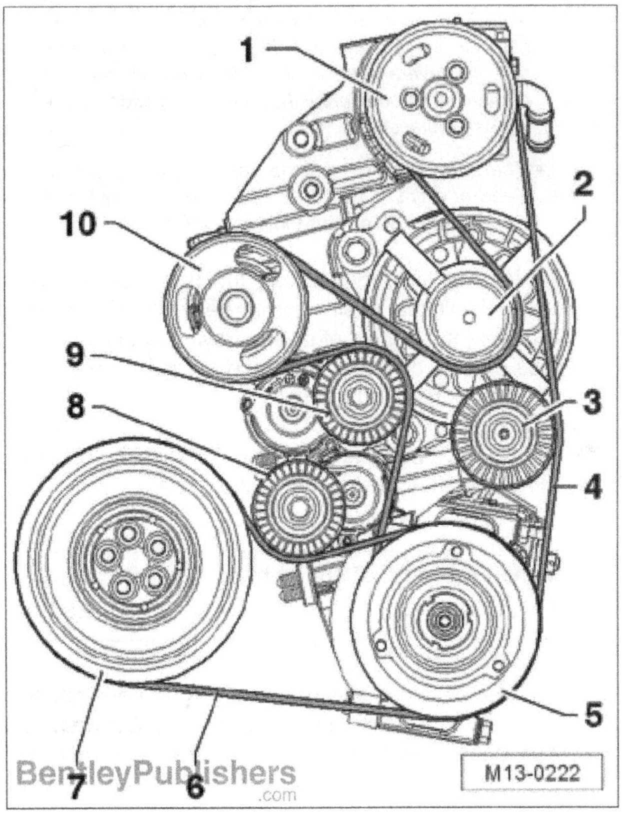

Installation is the reverse of removal noting the following points:

- When installing, note if belts are marked with running direction. Install in same direction.
- Manually turn engine at vibration damper or pulley with a suitable wrench. Inspect to be sure belts are seated completely in all pulleys.
- Be sure belts are installed with correct path.
- Install inner belt first.
- When installing inner belt, position onto coolant pump pulley (**10**), generator pulley (**2**), tensioner pulley, and power steering pump pulley (**1**) first. Install to lower idle pulley (**3**) last.
- Before installing outer ribbed belt for A/C compressor, lock tensioner with T10060A, or equivalent.

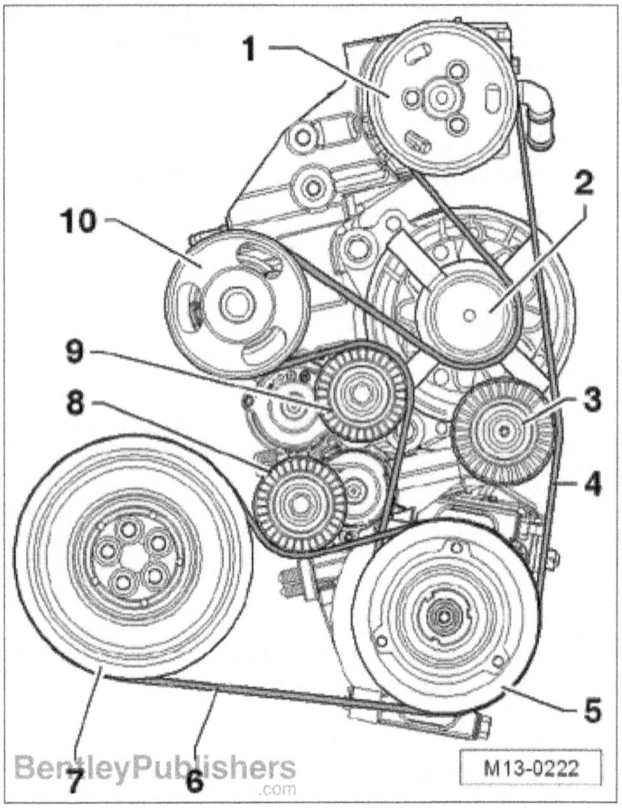

◄ **2.5L 5-cylinder gasoline engine**, ribbed belt path.

1. Power steering pump
2. Belt pulley, generator (alternator)
3. Lower idler roller
4. Ribbed belt (inner) for generator (alternator) and coolant pump
5. Belt pulley, A/C compressor
6. Ribbed belt (outer) for A/C compressor
7. Belt pulley, crankshaft
8. Tensioner roller for ribbed belt (outer) for A/C compressor
9. Tensioner roller for ribbed belt (inner) for generator (alternator), power steering pump and coolant pump
10. Belt pulley, coolant pump

Timing belt, checking

The timing belt, also called the camshaft drive belt or toothed belt, is a reinforced rubberized belt that drives that camshaft. On diesel engines, the belt also drives the injection pump. The belt is subjected to high temperatures and should be inspected during scheduled maintenance intervals. See **Maintenance Schedules** for specified intervals.

NOTE —
- *Volkswagen does not specify a replacement interval for the camshaft drive belt (also known as toothed belt). The timing belt is a wear item and should be periodically inspected and replaced. Belt failure may result in severe engine damage. Consult with an authorized VW dealer for the latest information on timing belt replacement.*
- *The 2.5L 5-cylinder engine uses a timing chain instead of a timing belt. See* **15d Cylinder Head and Valvetrain (2.5L engine)** *for more information.*

To inspect the toothed belt, the upper section of the toothed belt guard belt must first be removed. Inspect belt for tears, separation of layers, fraying of belt cords, surface cracks, or traces of oil and grease. Replace belt if any faults are found.

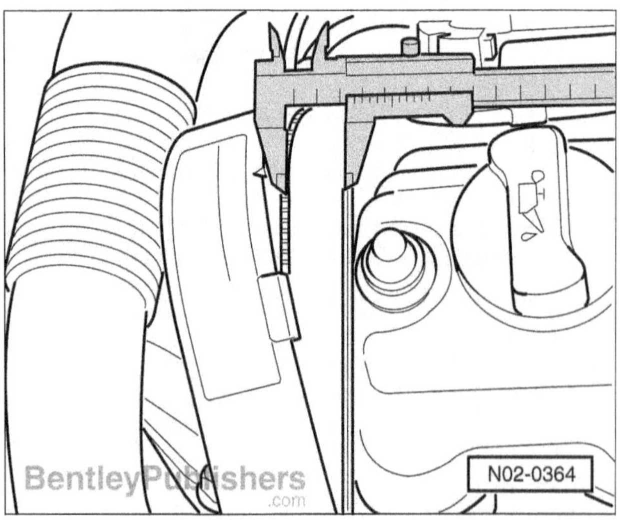

◄ The toothed belt width can also be measured to determine if replacement is necessary.

Measured width of timing belt
- Wear limit .22 mm (0.866 in.)

Toothed belts use semi-automatic tensioners that do not generally require adjustments. See **15 Cylinder Head and Valvetrain** for belt and tensioner replacement procedures.

Spark plugs, replacing (1.8L engine)

Specification

- Spark plugs (1.8L engine)
  Original equipment number 101-000-063-AA
  Manufacturers number. . . . . NGK PFR 6Q
  Gap . . . . max 0.80 mm (0.031 in)

NOTE —
Volkswagen part numbers are given for reference only. Always consult with your Volkswagen parts department or aftermarket parts specialist for the latest parts information.

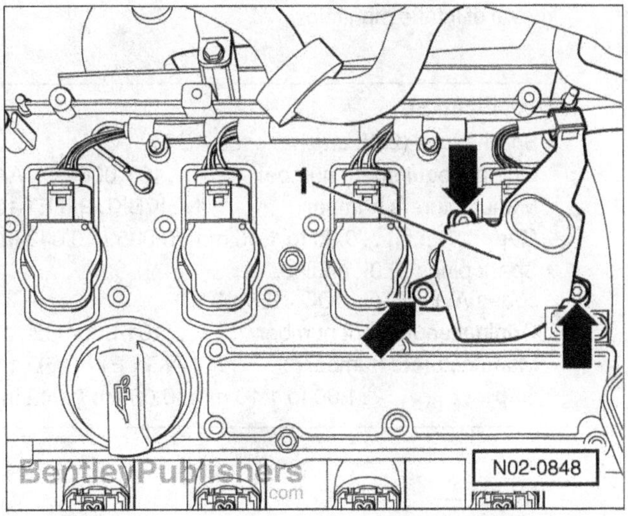

– Remove the upper sound absorber panel.

◄ On AWW, AWP, BKF or BNU engines (2001-2005) remove mounting nut for vacuum reservoir, label vacuum lines and remove reservoir. Remove three mounting nuts (**arrows**) for vacuum reservoir bracket and remove bracket (**1**).

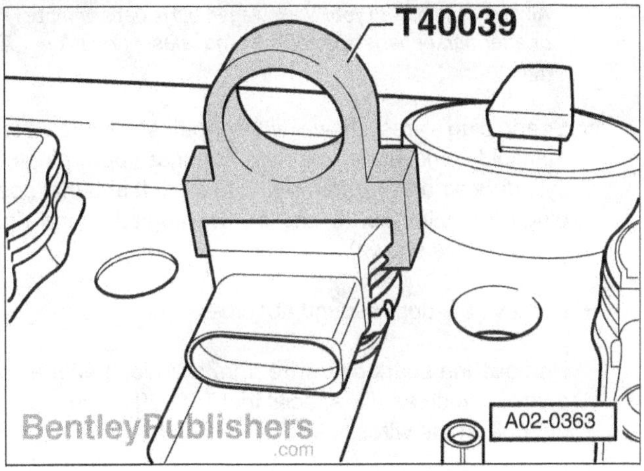

◄ Remove ignition coil.

- On APH engine (1999-2001) unbolt coil mounting bolts, remove coil and harness connector.
- On AWV, AWP, BKF or BNU engines (2001-2005) use special tool T10094 or T40039 (shown) to remove each ignition coil.

NOTE —
Two flat tipped screwdrivers may also be used to carefully pry out the ignition coils.

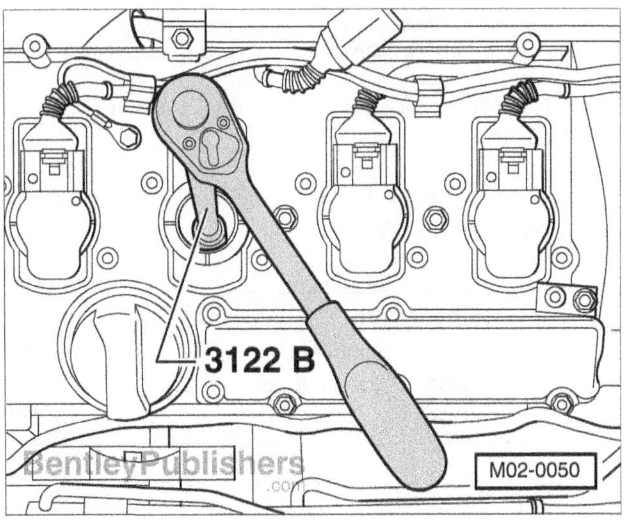

◄ Remove spark plug with spark plug socket and extension (or VW special tool 3122B).

– Check gap on new spark plug and install, re-install ignition coil and harness connector.

Tightening torque

- Spark plug to cylinder head . . . . 30 Nm (22 ft-lb)
- Ignition coil to cylinder head cover
  (APH engine) . . . . 10 Nm (7 ft-lb)

– Repeat for each remaining cylinder.

– Install upper sound absorber panel.

Spark plugs, replacing (2.0L engine)

Spark plugs are generally replaced during scheduled maintenance services. Due to the design of the 2.0L combustion chambers and shape of the cylinder head, spark plug replacement can easily be accomplished without removal of intake manifold.

Specification

- Spark plugs (2.0L engine, code AEG)
  Original equipment number 101-000-033-AA
  Manufacturers number. . . . . NGK BKUR 6 ET-10
  Gap . . . . 0.90 to 1.10 mm (0.035 to 0.043 in)
- Spark plugs (2.0L engine, codes AVH, AZG, BDC, BEV, BGD)
  Original equipment number 101-000-062-AB
  Manufacturers number. . . . . NGK PZFR 5D-11
  Gap . . . . 1.00 to 1.10 mm (0.039 to 0.043 in)

NOTE—
Volkswagen part numbers are given for reference only. Always consult with your Volkswagen parts department or aftermarket parts specialist for the latest parts information.

◄ Spark plug wires shown with upper section of intake manifold removed for clarity. Note that spark plugs for cylinders #1 and #2 are angled to the left and that spark plugs for cylinders #3 and #4 are angled to the right. (AEG engine shown)

– Remove the upper sound absorber panel.

– Remove the spark plug wire from cylinder 1 with a suitable tool such as VW special tool T10029 or T10112. Do not pull on the wires.

NOTE—
It may be necessary to unplug the injector wire and gently turn the injector in the mounts for access.

– Remove spark plug with a 5/8 in. (16 mm) spark plug socket and an extension.

– Check gap on new spark plug and install, re-install injector wire and plug wire.

Tightening torque

- Spark plug to cylinder head (AEG). . 30 Nm (22 ft-lb)
- Spark plug to cylinder head (AVH, AZG, BDC, BEV, BGD) . . . . . 25 Nm (18 ft-lb)

– Repeat for each remaining cylinder.

– Install in reverse order of removal.

Spark plugs, replacing (2.5L engine)

Specification
- Spark plugs (2.5L engine)
  Original equipment number 101-905-600
  Manufacturers number. . . . . NGK PZFR 5J-11
  Gap . . . . 1.00 to 1.10 mm (0.039 to 0.043 in)

NOTE —
Volkswagen part numbers are given for reference only. Always consult with your Volkswagen parts department or aftermarket parts specialist for the latest parts information.

– Switch off engine.

– Remove upper sound absorber panel. See **Sound absorber panels, removing and installing** in this repair group.

◄ Disconnect harness connector from secondary air injection solenoid valve (**arrow**).

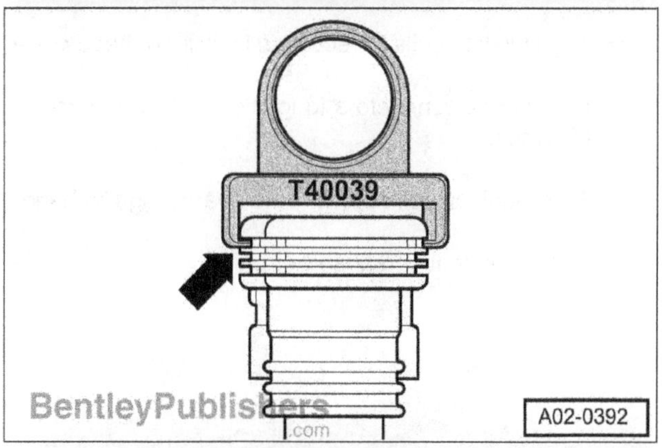

◄ Attach puller **T40039** to uppermost thick rib (**arrow**) of each ignition coil.

- *Lower ribs of coil may be damaged if they are used for removal.*
- *Harness connectors remain connected at this time.*
- *Ignition coils can carefully be pulled out using two flat tipped screwdrivers if T40039 is not available.*

– To prevent damage to wiring harness guide, do not remove wiring harness connectors until all ignition coils have been slightly pulled out of spark plug shaft in two stages.

◄ Beginning with cylinder #1, pull all ignition coils approximately 10 mm (3/8 in) out of spark plug shaft using ignition coil puller T40039 (**arrow**) or equivalent.

– Again beginning with cylinder #1, pull all ignition coils out another 10 mm (3/8 in) in the same order.

Unclip and pull off all ignition coil connectors (**A**), beginning at ignition coil #**1**.

– Remove ignition coils #**1** - **5** from spark plug shafts.

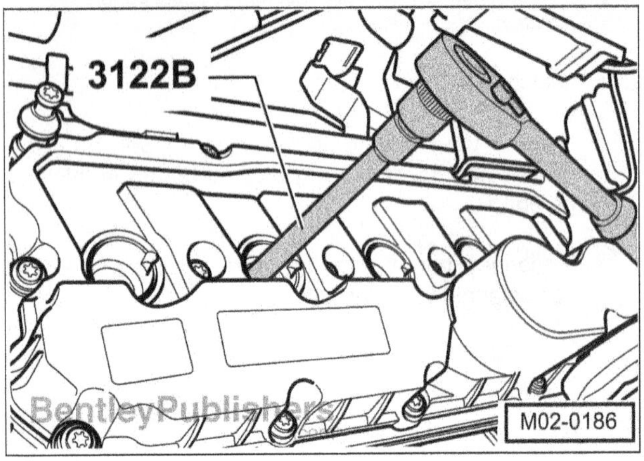

Remove and install spark plugs for each cylinder with socket **3122B** or equivalent.

NOTE—

A 16 mm or 5/8 in socket with appropriate extensions is suitable for spark plug removal and installation.

– When all spark plugs have been replaced and torqued to specification, insert all ignition coils loosely into spark plug shafts.

Tightening torque

- Spark plug to cylinder head (2.5L) . . 25 Nm (18 ft-lb)

– Align ignition coils to recesses in cylinder head cover.

– Connect all connectors to ignition coils in reverse order of removal.

– Press ignition coils evenly onto spark plugs by hand.

– Install upper sound absorber panel.

Air filter element, replacing

The air filter element should generally be replaced as specified in the **Maintenance Tables**. Under severe conditions, however, such as driving on dirt or desert roads, or in dusty areas, it is advisable to replace the element more frequently.

A snow strainer may be installed in lower section of air filter housing in area of air intake apparatus. This strainer has the function in winter months of catching snow and ice and therefore preventing blockage of the air filter element. By reaching into air intake apparatus from the front, feel upward to determine if a snow strainer is installed. Check snow strainer periodically and clean out when replacing air filter.

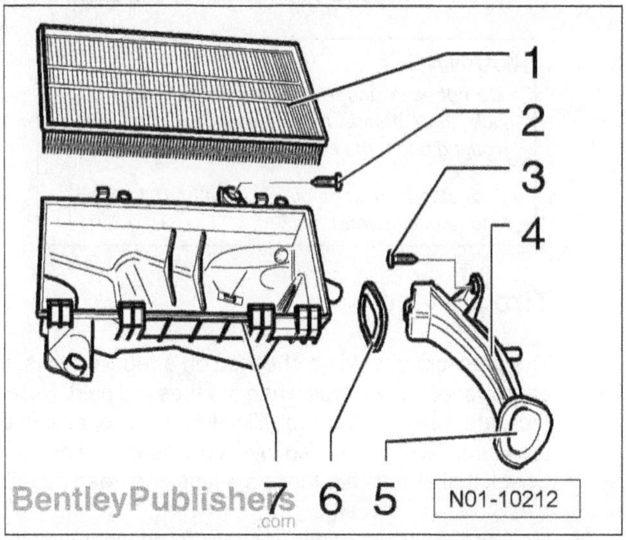

◄ Air filter assembly.

1. Air filter
2. Fastener (10 Nm)
3. Fastener (3 Nm)
4. Air intake apparatus
5. Snow strainer
6. Seal
7. Lower filter housing

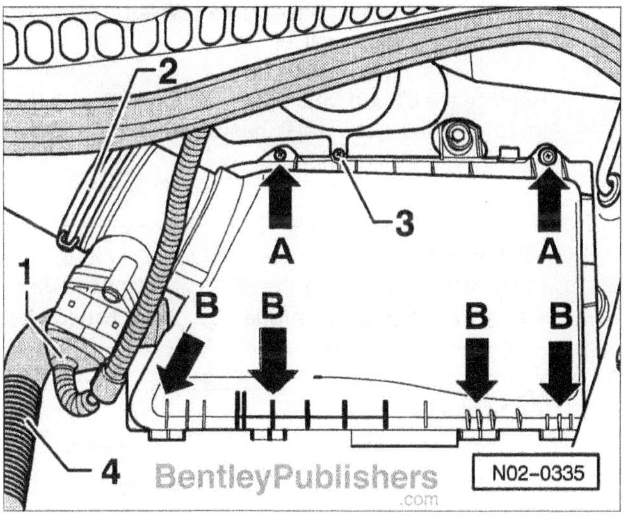

◄ Disconnect harness connector for mass air flow sensor (**1**). (2.0L AEG engine shown, others similar)

1. Mass air flow sensor harness connector
2. Intake hose clamp
3. Rear mounting screw
4. Connecting hose to secondary air pump

A. Upper/lower assembly screws
B. Retaining hinge locations

- Loosen hose clamp (**2**) on intake hose and take hose off of mass air flow sensor.
- Loosen mounting screw (**3**) at rear of air cleaner assembly, if applicable.
- Remove connecting hose (**4**) to secondary air pump, if applicable.
- Remove two screws holding upper and lower sections together (**A**).
- Lift upper filter housing and remove from lower filter housing.
- Remove old filter element.
- Clean filter housing, intake apparatus and snow screen if equipped.
- Install new filter element.
- Remainder of installation is the reverse of removal.

UNDER CAR MAINTENANCE

Under car maintenance requires that the car be raised on a lift or properly supported on jackstands. Under car maintenance should not be carried out unless proper (safe) lifting equipment is available. See **Lifting Vehicle** for more information on lifting and working under the car.

> ***WARNING —***
> - *Do not work under a car supported solely by a jack. Jack stands must always be used when working under the car.*
> - *Jack stands must be placed on a hard, level surface (e.g. concrete).*

Tire service

Tire pressure should be checked on a regular basis. It is best to check pressures when the tires are cold. Refer to the data label on the fuel filler flap for proper inflation pressures. Be sure to also check the spare tire pressure. Check that all tires are the same type and tread pattern.

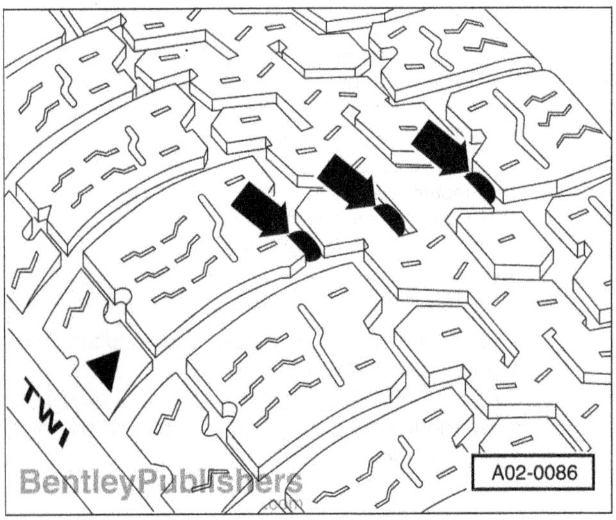

◄ Measure tire tread depth or check tread wear indicators (**arrows**) on a regular basis. Note **TWI** lettering and location **triangle** on sidewall of tire.

– Replace tire if tread wear indicators are flush with the tire tread or if the tread wear has exceeded the minimum specification listed below.

Specification
- Minimum tire tread depth . . . . 1.6 mm (2/32 in.)
- Wheel bolt tightening torque . . . . . 120 Nm (89 ft-lb)

> ***CAUTION —***
> *Wheel alignment must be checked after replacing tires to ensure maximum tire life.*

The tires should also be checked for abnormal wear patterns. The tread wear pattern on the front tires is an indication of whether the toe and camber settings need to be checked. "Feathering" on the tread indicates incorrect toe. If the tread is worn on one side, this is usually caused by incorrect camber.

To extended tire life and minimize wear, Volkswagen recommends that the tires be rotated front to rear every 10,000 miles. For more information on tires, see **44 Wheels–Tires, Wheel Alignment**.

Brake system, visual inspection

Check the brake master cylinder, vacuum brake booster (also hydraulic unit if anti-lock brake system is fitted), brake pressure regulator, and brake calipers for leaks and damage.

- Brake hoses must not be twisted.
- Brake hoses must not touch any part of the vehicle when steering is at full lock.
- Check brake hoses for porosity and deterioration, brake hoses and brake lines for chafing points.
- Check brake connections and attachments for correct seating, leaks and corrosion.
- Any faults found must be repaired.

Brake pads (front and rear), checking

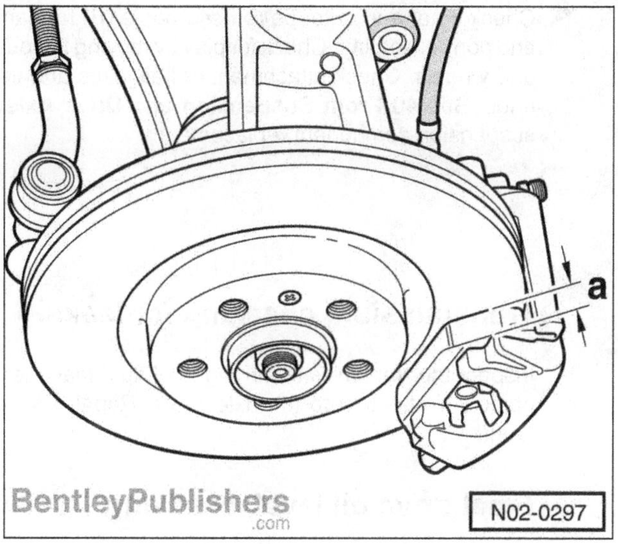

To accurately check the brake pad thickness, the wheels should be removed. With the wheel removed, measure the thickness of outer and inner pads. If pad thickness (**a**) (including backing plate) is 7 mm (0.28 in.) or less, the brake pads should be replaced.

Brake pad wear limit (front and rear)
- Minimum width
  (including backing plate) . . . . .7 mm (0.28 in.)

NOTE —
- *Brake pad wear may be minimally greater on the front passenger's side than on the driver's side. It is therefore recommended that inspection always be performed on passenger's side front wheel.*
- *When tightening wheel bolts, tighten the bolts in stages using a criss-cross pattern.*

Tightening torque
- Wheel bolt to wheel hub. . . . . 120 Nm (87 ft-lb)

Parking brake, adjusting

The parking brake acts on the rear brakes and is self-adjusting to compensate for wear. Adjustment of the parking brake is only necessary if brake components are replaced. See **46 Brakes–Mechanical Components** for parking brake adjustment and service procedures. The parking brake should begin to hold after approximately two clicks of the lever.

Drive axles, checking boots

There are inner and outer CV boots on each front axle shaft. The boots should be regularly inspected for tears, cracks, or deterioration. Once the boot is open to the weather, abrasive road debris can enter the joint and quickly destroy it. If boot replacement is carried out promptly, chances are good that the joint can be saved. See **40 Front Suspension and Drive Axles** for drive boot replacement procedures.

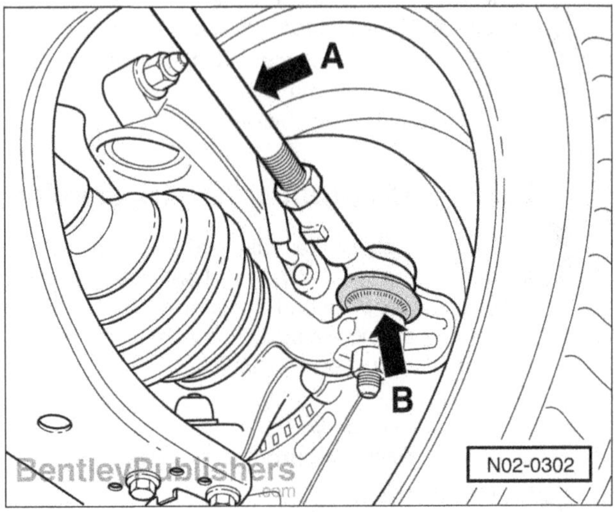

Front suspension components, checking

With vehicle raised off the ground, check ball joint dust boots for damage and correct seating. Check for play in ball joints.

< Check inner and outer tie rod end boots (**B**) for damage and correct seating. Check for play by moving tie rods (**A**) and wheels. Check attachment of ball joints and tie rod ends. See **40 Front Suspension and Drive Axles** for suspension component replacement.

Transmission, checking for leaks

Inspect the transmission for signs of fluid leakage. Pay particular attention to the axle seals. Repair any leaks found.

Final drive oil level, checking

On cars with manual transmissions, the 6-speed 09G automatic or the DSG transmission, the final drive lubricant shares a common supply with the transmission lubricant. For final drive and transmission fluid level checking, see **Transmission oil level, checking** for the appropriate transmission in this repair group.

On cars with the 01M 4-speed automatic transmission, the final drive oil level is checked by removing the speedometer drive gear from the transmission housing and using the drive gear as a dipstick. The car should be on a level surface when making the check.

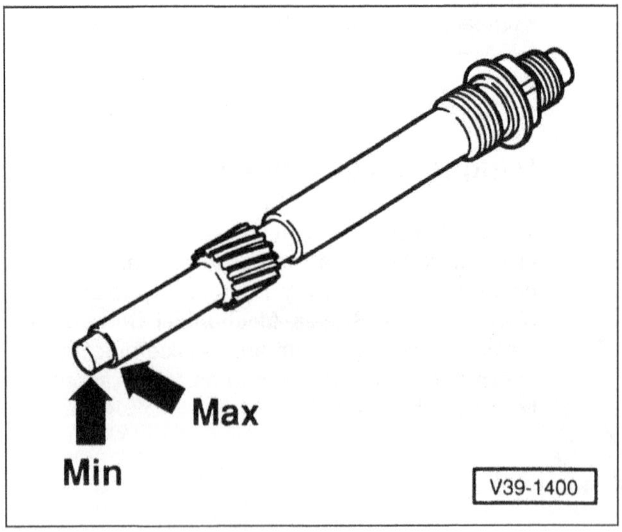

– Remove speedometer drive, wipe clean with a cloth and install again.

< Remove drive and check oil level.

- The oil level must between the MIN and MAX markings.
- The amount of oil between the MIN and MAX markings is approximately 0.1 Liter (0.1 qt).

– If oil level is too low, add oil in small amounts until proper level is reached.

Specification
- Automatic transmission (01M), final drive lubricant G 052 145 A2......... SAE 75 W90 (Synthetic oil)

NOTE —
Volkswagen part numbers are given for reference only! Always consult with your Volkswagen Parts Department or aftermarket parts specialist for the latest parts information.

– If oil level is too high, suction out excess.

– When proper level is obtained, install speedometer drive.

Transmission oil level, checking (manual transmission)

Manual transmissions codes
- 5-speed (1998-on) 02J
- 5- and 6-speed (2002-on) 02M
- 5-speed (2006-on) 0A4

02J and 02M 5- and 6-speed transmissions

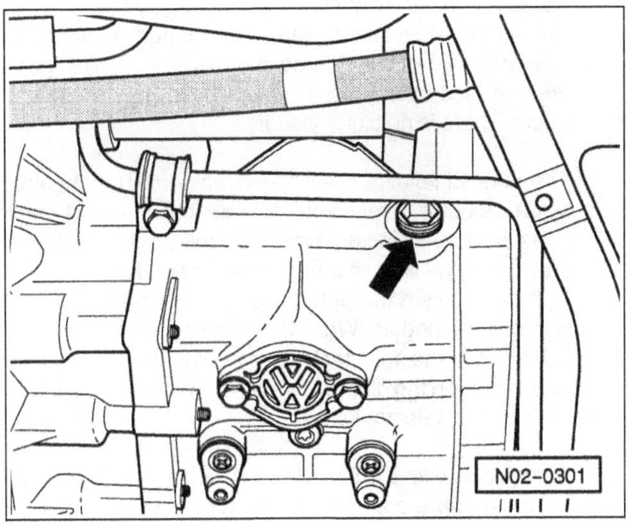

◄ Remove oil filler plug (**arrow**) with 17 mm Allen wrench.
- 02J shown, 02M similar

– Check oil level with finger.

NOTE —
Oil level must be to lower edge of filler plug hole.

– If oil level is too low, add oil through filler plug in small amounts until proper level is reached.

– Install oil filler plug.

Specification
- 02J transmission oil .... 75W90 mineral oil
  VW part no. .... G 052 726 A2
- 02M transmission oil .... 75W90 synthetic
  VW part no. .... G 052 171 A2

NOTE —
Part numbers are listed for reference only. Always check with an authorized Volkswagen parts department for the latest parts information.

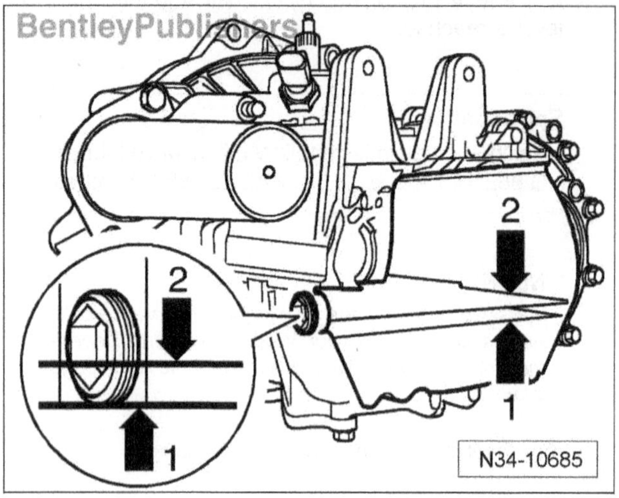

0A4 5-speed manual transmission

The gear oil level on the 0A4 transmission is above the level of the drain plug. This is due to a change in the inclination angle of the engine/transmission. Do not check gear oil by removing drain plug.

◀ Lower edge of filler hole (**arrow 1**) is located beneath new oil level (**arrow 2**).

– Gear oil level can only be checked if gear oil is drained completely and then filled up again. See **34a Manual Transmission** for gear oil replacement procedure.

Specification

- 0A4 transmission oil
  see authorized Volkswagen dealer for latest information

Transmission oil and filter, DSG (direct shift gearbox)

The 6-speed Direct Shift Automatic Gearbox (DSG) is also known as a double-clutch transmission. Engine torque is transferred to transmission by way of the dual-mass flywheel. The transmission is constructed as a 6-speed manual transmission. Alternating hydraulic actuation of the two wet multi-plate clutches permits operation similar to that of an automatic transmission. The gears are engaged automatically or manually via the Tiptronic mode. There is no clutch pedal.

The selector lever position is not transferred mechanically via selector lever cable as with the 09G automatic transmission. Selector lever positions or gear shifts are transferred by way of a separate control module for the shift mechanism through the CAN-Bus to the transmission control module. While gear shifting occurs without a cable, when the selector lever is in position "P", a parking lock acting on the differential carrier is engaged mechanically by the selector lever cable.

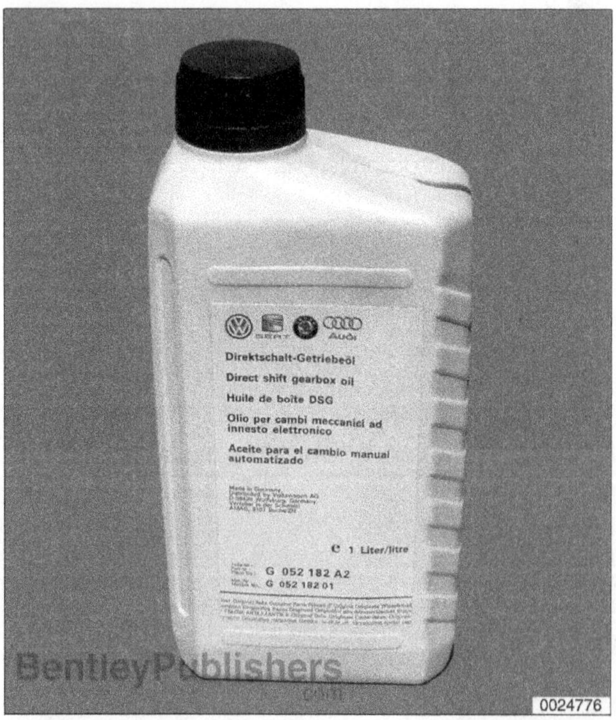

◀ On vehicles with a 6-speed DSG, gearbox section and final drive share a common fluid supply which is specifically designed for the DSG transmission. Checking oil level of gearbox also checks oil level of final drive (differential).

See repair group **34b DSG Transmission** for DSG fluid level checking, draining and filling procedures.

Transmission fluid (ATF) level, checking and adjusting (automatic transmission)

The 01M and 09G automatic transmissions do not have a dipstick. The procedure for checking ATF fluid requires measuring and maintaining a specified ATF temperature (which is neither cold nor hot, but in between) during the procedure. Special Volkswagen tools are required for this operation and it is therefore recommended that ATF level checking on an automatic transmission be left to an authorized Volkswagen dealer.

Before starting, be sure that the following checking conditions are met:

- Transmission not in limp-home mode
- ATF temperature not above approx. 30°C (86° F)
- Vehicle must be level
- Selector lever in "P"
- Center and left sound absorber panels removed

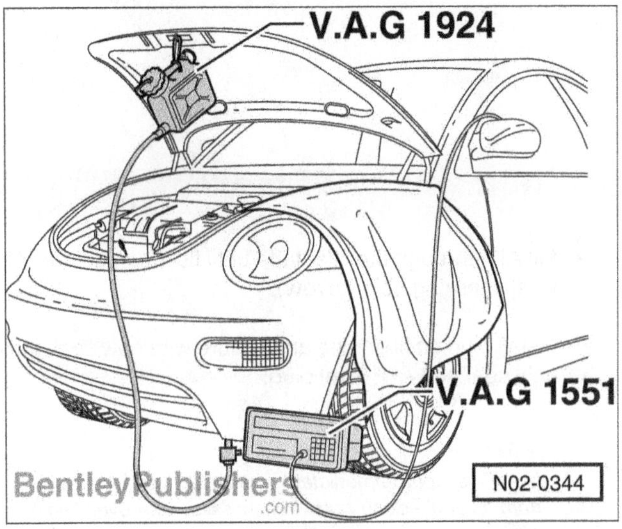

◀ Attach container of ATF filler VAG 1924 to vehicle.

– Start engine.

– Connect suitable scan tool and access Transmission Control Module. If using a Volkswagen supplied VAG 1551/1552, observe the following sequence in the scan tool display window:
  - 1 - Rapid Data
  - Address word 02 - Transmission electronics
  - Advance to Read Measuring Value Blocks - 08
  - Select display group
    01M 4-speed: display group 005
    09G 6-speed: display group 006
  - Observe display field 1 for transmission temperature (must be below 30°C at start)

– Lift vehicle.

– Remove lower sound absorber panel.

– Place container under transmission.

– Start engine and let it idle to bring ATF to test temperature.

Specification
- ATF test temperature . . . . . 35 to 45°C (95 to 113°F)

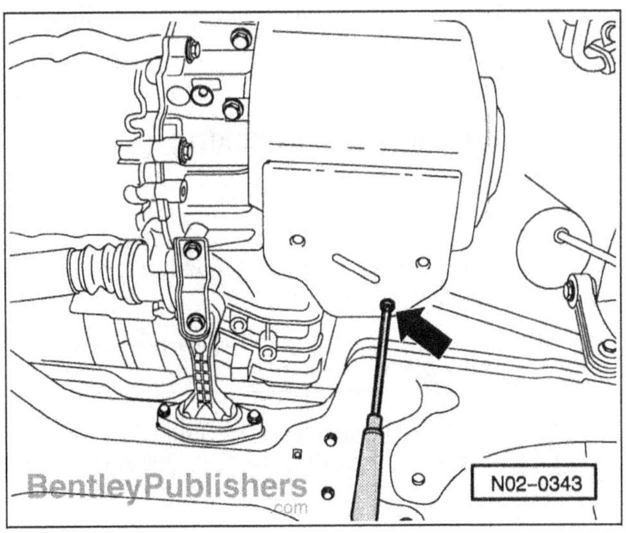

◀ Remove ATF level plug (**arrow**) from transmission oil pan.

– If ATF drips out of hole when temperature is between 35°C to 45°C, ATF level is correct and no further action is required. Install new seal on level plug and torque. This completes the ATF check.

Tightening torque
- ATF level plug with new seal. . . . . 15 Nm (11 ft-lb)

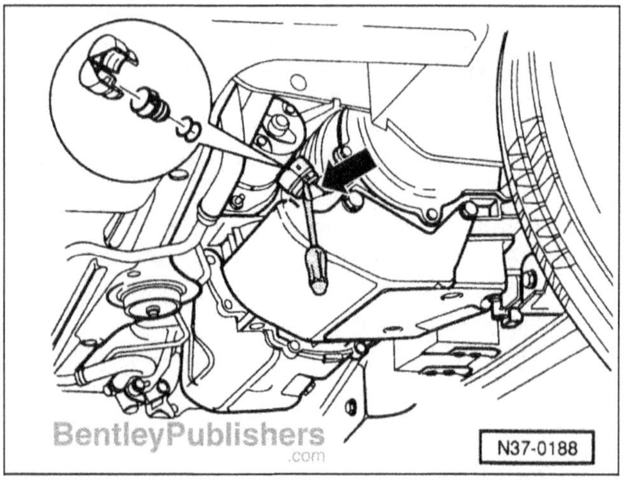

– If no ATF drips out of hole by the time the temperature reaches 45°C (113°F) then it is necessary to top off ATF.

◄ Pry off ATF filler plug and seal (**arrow**) with screwdriver. The seal will be destroyed, always replace.

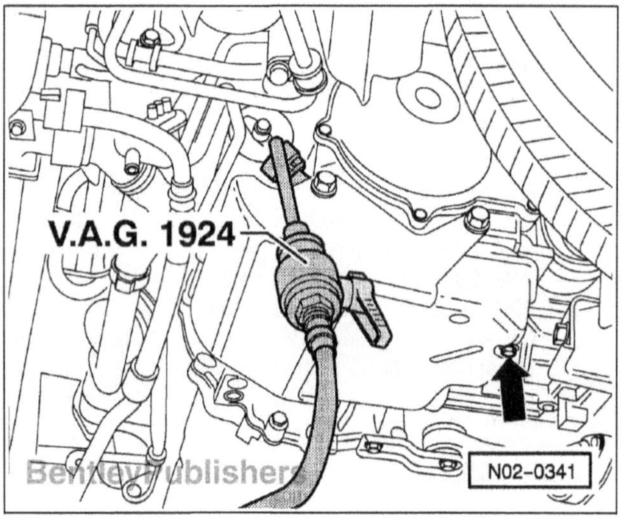

◄ Fill ATF through the filler tube until fluid comes out of the level checking hole (**arrow**).

– Install plug on filler tube and secure with new seal. Install new seal and ATF level plug.

NOTE —
If ATF ran out immediately when plug was removed from level checking hole, transmission was overfilled. Allow fluid to run out until temperature on scan tool is between 35°C and 45°C. Install new seal and ATF level plug.

– Start engine and shift through all selector lever positions with the vehicle stationary.

To change the ATF, see **37 Automatic Transmission**.

Exhaust system, checking

WARNING —
- *The exhaust system operates at extremely high temperatures. Do not touch the exhaust system while the engine is running.*
- *Allow exhaust system to cool at least one hour before touching.*

Inspect exhaust system for leaks or damage. Inspect exhaust system mounts. Replace faulty, missing or deteriorated parts.

Underbody sealant, checking

When performing the visual inspection for damage to the underbody sealant, also check the underbody, wheel housings and sill panels. Any faults found should be promptly repaired.

BODY AND INTERIOR MAINTENANCE

Body exterior

Automobile finishes are subjected to abuse from industrial fumes, corrosive road salt, acid rain, and other damaging air-born elements. Regular and correct care will contribute to maintaining and preserving the exterior of your Volkswagen.

> ***NOTE —***
> *Proper care may be a condition for upholding the new car warranty, should corrosion damage or paint defects occur.*

The best protection against environmental influences is frequent washing and waxing. How often this is required depends on the environment where the vehicle is used.

Under certain circumstances weekly washing may be necessary. Under other conditions, a monthly washing and waxing may be adequate. Even if a wax solution is used when washing your vehicle, it is advisable to protect the paint with a coat of hard wax at least twice a year. Check the paint for chips and scratches. Paint defects should be touched up soon after they occur to prevent corrosion.

After the winter, the underside of the vehicle should be thoroughly washed.

Exterior plastic and vinyl should be kept clean. Occasionally apply a colorless vinyl or leather preservative. DO NOT wax plastic or vinyl.

Exterior lights, checking

Check operation of headlights (high and low beam), marker lights, taillights, turn signal lights, brake lights, reverse lights and emergency flasher lights. Use a helper when checking lights. See **94 Lights, Accessories–Exterior**.

Interior lights, checking

Check operation of indicator and instrument cluster warning bulbs. Check operation of interior cabin illumination bulbs. See **96 Lights, Accessories-Interior**.

Airbag unit, visual inspection

All models are fitted with driver's side and passenger's side airbags. In addition, all models are fitted with side airbags built into the front seat backrest.

> ***WARNING —***
> - *The padded airbag covers on the steering wheel and instrument panel must not be covered over or have any objects affixed to them.*
> - *Do not apply any chemical treatment to airbag unit covers. Clean with a dry or water moistened cloth only.*

Inspect padded airbag unit covers on steering wheel and instrument panel and side airbag units for signs of external damage. Check with an authorized Volkswagen dealer if any faults are found.

Inspect seats for covers that may obstruct or hinder proper deployment of side airbag.

> ***WARNING —***
> *Volkswagen of America has issued a directive specifically warning against the installation of aftermarket upholstery on any vehicle equipped with side airbags. The factory-installed upholstery is designed to separate in specific places at specific rates, and in specific directions. Installation of non-factory upholstery including, but not limited to, "beads" and "sheepskins", may cause seat mounted airbags to deploy when they are not supposed to; fail to deploy when they should; or to deploy in some manner other than designed. This is a safety hazard and could result in serious injury or death to occupants of the vehicle.*

Clock and outside temperature, setting

The clock and outside temperature display are located above the rear view mirror.

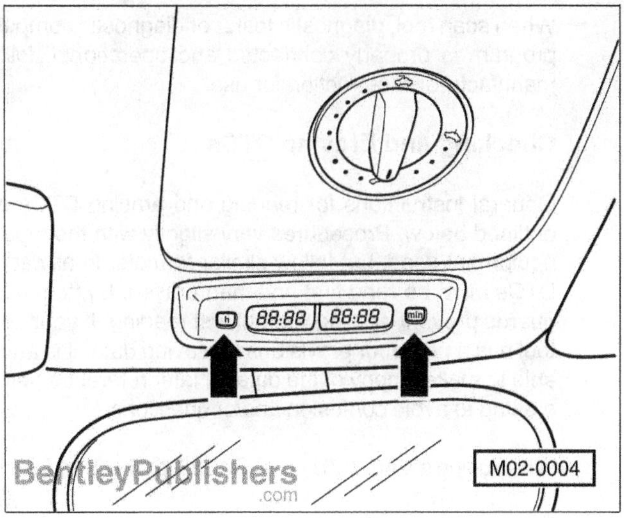

M02-0004

◀ To adjust clock display use left side button (**h**) for hours and right side button (**min**) for minutes.

– To set time to exact second, press (**min**) button until clock indicates one minute before time to be displayed. Press (**min**) button at moment when seconds indicator of accurate clock or clock to synchronize with shows a full minute.

– To change time mode (12 or 24 hour), switch engine off. Press and hold (**min**) and (**h**) buttons simultaneously for longer than 3 seconds. The display will switch to new mode.

– To change temperature display (°F or °C), switch engine on. Press and hold (**min**) and (**h**) buttons simultaneously for longer than 3 seconds. The display will switch to new mode.

Diagnostic Trouble Codes (DTCs), checking

Volkswagen's maintenance program specifies checking the DTCs using the On-board Diagnostic (OBD) system memory. The ODB system is integrated into the engine management control module and monitors emissions-related and other electrical components on the car, including the adaptive automatic transmission, the ABS system, and the airbag system. Checking the OBD memory for faults requires a diagnostic scan tool such as the VAG 1551/1552, VAS 5051/5052 or an equivalent aftermarket scan tool. More information, including a list of powertrain DTCs and control module coding can be found in the **OBD On Board Diagnostics** section of this manual.

In all cases, it is necessary to connect the scan tool or computer to the vehicle via the Data Link Connector (DLC).

Before connecting and using any scan tool, the following requirements must be met.

- Fuses #28, 29, 30, 31, 32, and 34 must be good.
- Battery voltage must be at least 11.5 Volts.
- Ground connections between engine and chassis must be good.

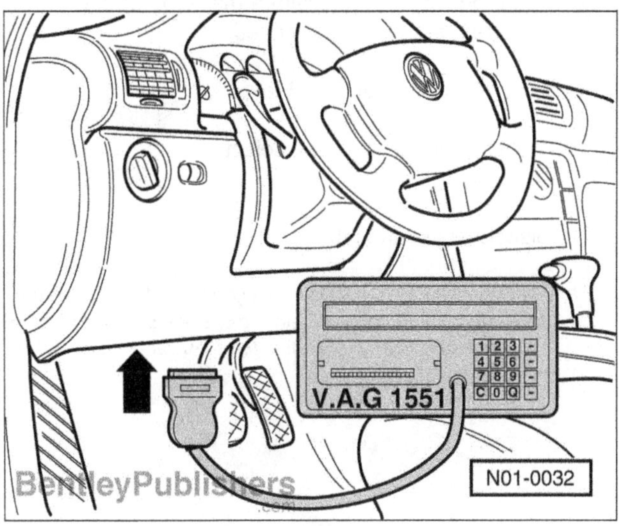

N01-0032

◀ DLC for all model years is located under the dash, to the left side of the steering column, above the driver's left leg.

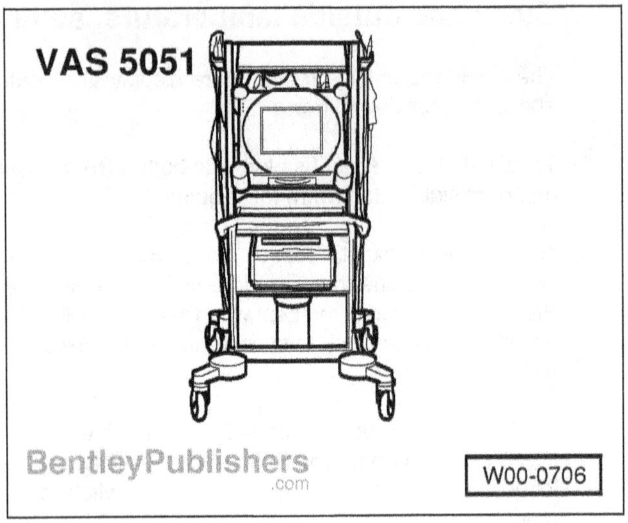

VAS 5051 (and VAS 5052) diagnostic testers replace scan tools VAG 1551 and 1552 which are now considered obsolete by Volkswagen due to their relatively limited capabilities.

– When scan tool, diagnostic tester or diagnostic computer program is properly connected and operational, follow manufacturer's instruction for use.

Checking and Erasing DTCs

General instructions for reading and erasing DTCs are outlined below. Procedures vary slightly with the type of equipment used, but follow similar formats. In all cases, DTCs must be read first and than erased. System safeguards prevent erasing without first reading. If your scan tool has a printer or provisions for saving data, it is advisable to make a copy of the data for later reference before erasing to avoid confusion and/or questions.

When using a VAG 1551 or VAG 1552, proceed as follows:

– Connect scan tool as described above.

– From the initial Menu, select:
1 - Rapid Data.

– From the Address Word Menu, select the appropriate vehicle system, see **OBD On Board Diagnostics** section (e.g. 01 - Engine Electronics).

– Advance to the Function Menu when the ECM responds with version and coding information.

– From the Function Menu, select:
02 - Check DTC memory.
Read DTCs and return and return to the Function Menu.
If no DTCs are present, scan tool will respond with a message in the display to indicate that no DTCs are recognized.

– From the Function Menu, select:
05 - Erase DTC memory.
Scan tool will respond with a message in the display to indicate that memory is erased.

– Advance to the Function Menu.

– From the Function Menu, select:
05 - End output.

– Disconnect scan tool.

After erasing DTCs, it is usually a good idea to go back into the system via Function 02, and re-check the fault memory to make sure that the codes have, in fact, have been erased.

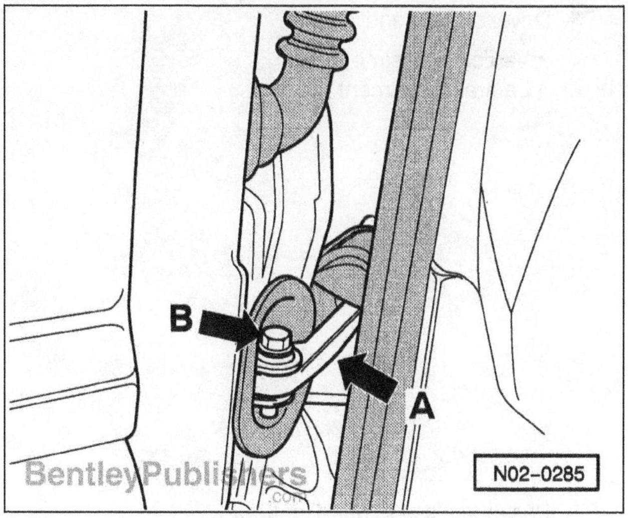

Door check straps and hinges, lubrication

The door checks straps (**arrow A**) and door hinges should be lubricated periodically. Use a lithium grease for the job. Lubricate the hinge securing bolt (**arrow B**) with lock cylinder spray. It is a good idea to clean the old grease away before applying the new lubricant.

Lubricant specification

- Door check strap . . . . VW special grease G 000 400
- Hinge securing bolt . . . . . VW lock cylinder lubricant G 000 400 01

NOTE —
Volkswagen part numbers are given for reference only! Always consult with your Volkswagen Parts Department or aftermarket parts specialist for the latest parts information.

Dust and pollen filter, replacing

All New Beetles are equipped with a ventilation air dust and pollen filter. This filter prevents most dusts and pollens from entering the passenger compartment. Volkswagen specifies replacement of this filter at regular mileage intervals. **See 80 Heating and Ventilation.**

Headlights, adjusting

Proper adjustment (aim) of the headlights requires specialized equipment which can vary from service shop to service shop. Also, adjustment parameters vary between the different State and Provincial governments.

Because of these variables, always follow the equipment manufacturers instructions along with all applicable governmental regulations when adjusting headlight aim.

Regardless of the equipment or the regulations, the following points are common and preliminary to the proper adjustment of the headlights:

- Inflate tires to correct pressures.
- Vehicle must be standing on a level surface.
- Headlight lenses and reflectors clean and undamaged with correct type and wattage bulbs.
- Vehicle must be properly loaded with approximately 75 Kg (165 lb) on driver's seat.
- Springs settled by rolling vehicle several meters (yards) or bouncing suspension.

Vehicle should be at curb weight with tool kit, jack, spare tire, etc. and 90% fuel load. If fuel tank is not at least 90% full, add weight to the trunk for compensation according to the table.

Fuel Tank Filling Compensation Table (for adjusting headlights)

Fuel level as indicated by fuel gauge	Additional weight (ballast) to add in luggage or cargo area
¼	30 Kg (66 lb)
½	20 Kg (44 lb)
¾	10 Kg (22 lb)
full	none (0)

Note:
For additional weight use containers filled with water. A container filled with 5 liters (1.3 gal) of water weighs approx. 5kg (11 lb).

Each headlight assembly is equipped with two adjustment screws.

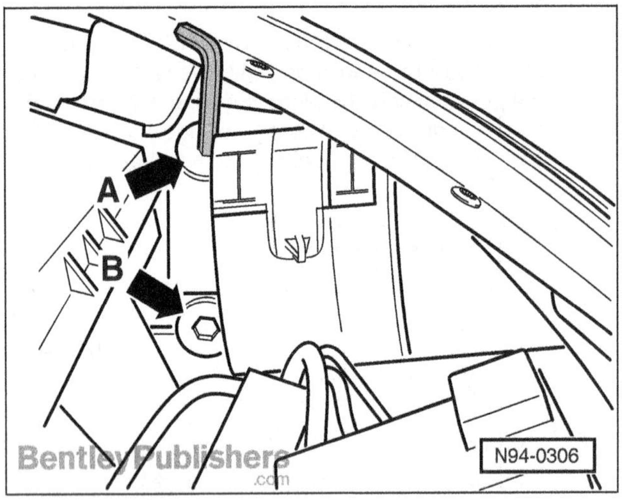

Driver's side headlight.

- Vertical adjustment (**A**)
- Lateral adjustment (**B**)

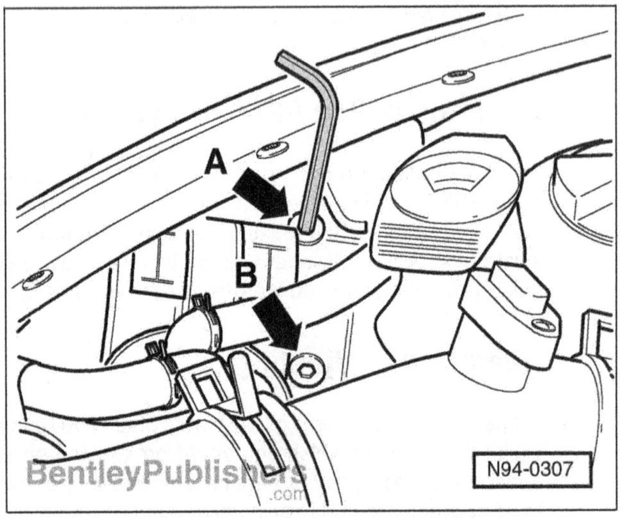

Passenger's side headlight.

- Vertical adjustment (**A**)
- Lateral adjustment (**B**)

Keys and remote controls, replacing batteries

Early production vehicles use a key with a separate remote control fob for the central locking system. Later production vehicles use a folding key with a built in remote control. In addition, some early vehicles use a key equipped with a small light.

Master key with light

Insert a coin into the slot (**2**) on the side and twist to open. Bulb and battery lift out for replacement. Observe battery polarity on reassembly. Confirm operation with button (**1**).

- Battery quantity and type: one 1.5 Volt alkaline V625U
- Bulb: one 1.5 Volt, 90 mA (0.090 A) pre-focus

NOTE —

- *Confirm battery specification above with actual battery installed in key.*
- *The master key with light is usually paired with the remote control key fob as described below.*

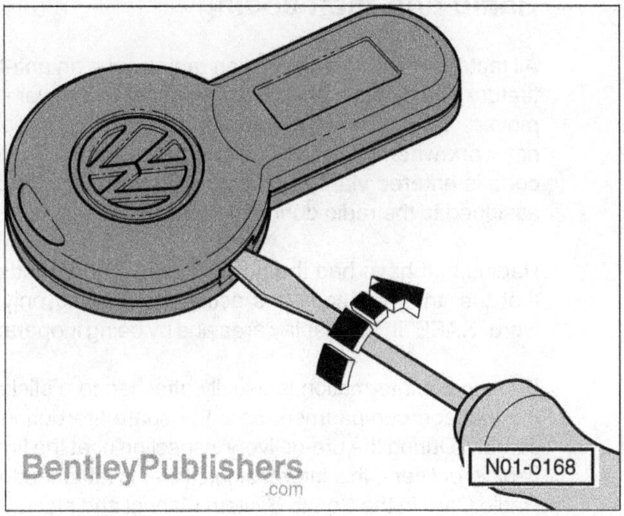

Remote control key fob

Insert small screwdriver into the slot on the side and twist (**arrow**) to open. Batteries lift out for replacement. Observe battery polarity and note spacer plate between batteries on reassembly.

- Battery quantity and type: two 3.0 Volt lithium CR2016

NOTE —
- *Confirm battery specification above with actual battery installed in key.*
- *New Beetle remote control key fobs look and operate the same as those used on some models of Golf, Jetta, Cabrio and Passat. However, they are not interchangeable due to operation on different radio frequencies.*

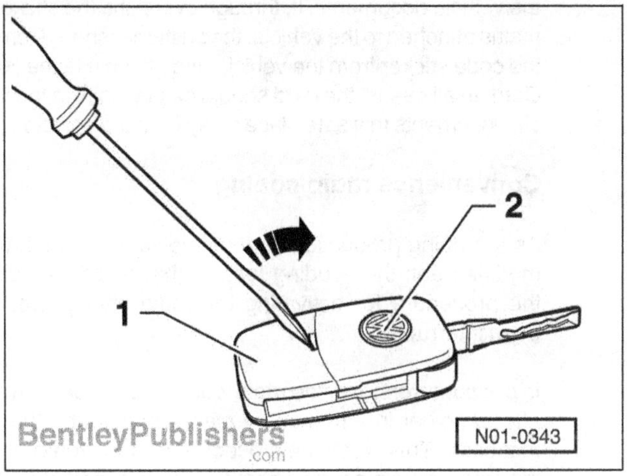

Remote control folding key

The folding key is supplied with two slightly different transmitter types. Procedures are the same for both types. Insert small screwdriver into the slot between the transmitter unit (**1**) and the main folding key body (**2**). Move screwdriver in direction of **arrow** to unclip and separate the two sections. Sections slide apart.

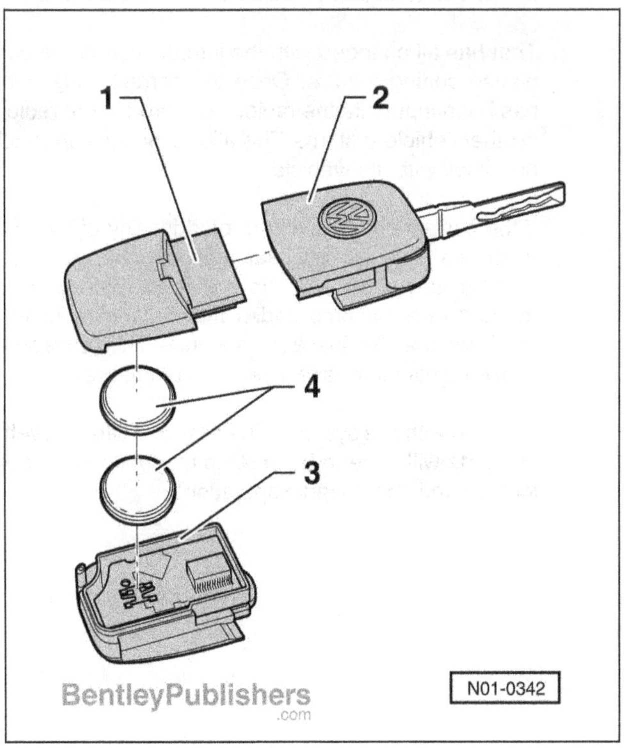

After separating the transmitter unit (**1**) and the main folding key body (**2**), remove the lower section of the transmitter (**3**) by carefully prying apart. Batteries lift out for replacement. Observe battery polarity on reassembly. Battery package (**4**) configuration depends on version.

- Battery quantity and type: two 3.0 Volt lithium CR1620 ***or***
- Battery quantity and type: one 3.0 Volt lithium CR2032

NOTE —
- *Confirm battery specification above with actual batteries installed in remote control key.*
- *Dispose of all dead batteries properly. Recycle when possible. Comply with all applicable laws regarding disposal so as not to pollute the environment.*

Radio anti-theft coding

All factory installed Volkswagen radios have an anti-theft feature that disables the radio whenever the power is removed. This feature deters theft because the radio will not work when power is reconnected unless a numeric code is entered via the radio push buttons. The code is assigned to the radio during manufacture.

Radios that have had the power interrupted will indicate that the anti-theft coding is active by showing only the word "SAFE" in the display area and by being inoperative.

Radio code information is usually attached to a sticker in the luggage compartment near the spare tire during assembly. During the pre-delivery inspection or at the time of vehicle delivery, this information sticker is attached to the Radio Card in the Sound System Manual and placed with the vehicle documents. If, through oversight, the sticker remains attached to the vehicle, the customer should remove the code sticker from the vehicle and attach it to the Radio Card, In all cases, the card should be placed with the vehicle documents in a safe place away from the vehicle.

Convenience radio coding

As a running production change, newer vehicles have a modified anti-theft coding feature that greatly simplifies the procedure for activating the radio after power has been interrupted.

In previous versions, the radio would have to be activated after a power interruption by manually entering the correct code. This is still the procedure in older vehicles and in some circumstances where the radio is replaced.

That has all changed with the introduction of the convenience coding feature. Once the correct code number has been input into the radio, it is stored in the radio and in other vehicle systems. This allows the radio to synchronize itself with "its" vehicle.

If the current supply is interrupted for any of the above-noted reasons, the radio automatically checks to see if it is still in "its" vehicle after the ignition is switched on. If it is "happy" (the systems codes match), the radio will be ready for use after just a few seconds. This procedure requires no user interface; it occurs automatically.

If a radio with this system is installed in a different vehicle, the code will have to be entered manually to cancel the lock-up and enable radio operation.

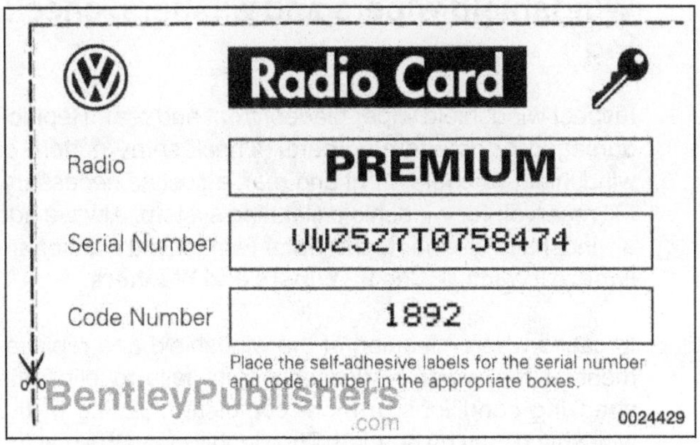

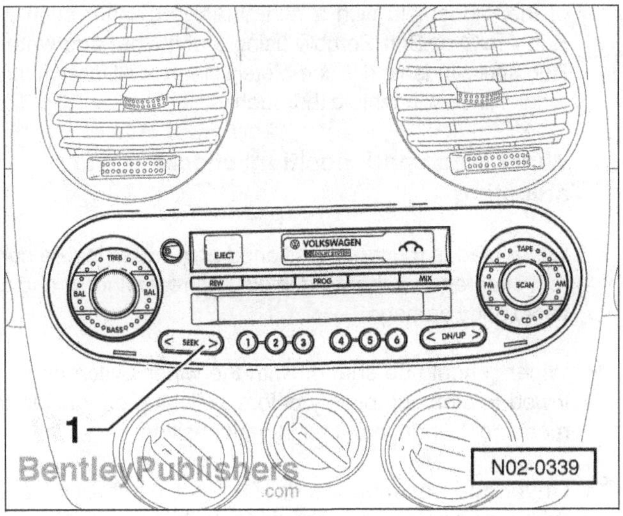

Cancelling the radio electronic lock

Obtain radio code number from radio card.

Situations that will cause the radio to be electronically locked up include:

- Disconnecting the battery ground or positive cable.
- Blown radio fuse(s).
- Removing and disconnecting for the radio for repairs to nearby systems.

– To activate fixed code, turn radio on. "SAFE" appears in display. After 3 seconds, "1000" appears in display.

Use radio station preset buttons (1 to 4) to enter code number found on radio card. Use preset button 1 to enter first digit of code, use button 2 to enter second digit, etc.

NOTE —

- *Value range for each of 4 digits is 0-9, buttons 5 and 6 are not used.*
- *The digit will increase by 1 each time you push the preset button.*
- *The number returns to 0 after 9.*

– After code is entered, press and hold right side (>) of SEEK button (**1**) longer than 2 seconds until you hear an audible signal.

– If you've entered the code correctly, "LSM" will appear in the display. The unit is now ready for use.

– If you've accidently entered a wrong code number, "SAFE" will appear in the display. Initially it will flash, then stay on. You can repeat the procedure described above again one time. The number of attempts will appear in the display. If the wrong code number is entered again, the unit will be disabled for about 1 hour. Leave the unit ON and the key in the ignition switch during this hour. After the hour has elapsed, the number of attempts will stop being displayed and you can cancel the electronic lock once again. The two attempts, one hour locked cycle will continue indefinitely.

NOTE —

If you loose your radio card, contact an authorized Volkswagen dealer.

Windshield wipers and washers, checking

Inspect windshield wiper blades, front and rear. Replace damaged or deteriorated parts. Check spray pattern of windshield washers, front and rear, adjust as necessary. Fill reservoir for windshield cleaning system. Always add a windshield cleaner to the water (windshield washer antifreeze in winter). See **92 Wipers and Washers**.

In cases where cleaning of the windshield and replacement of the wiper blades or inserts fails to eliminate streaking conditions, a more complete cleaning of the windshield may be needed. Thoroughly clean the outside of the windshield with a non-abrasive cleaner such as Bon Ami® or Soft Scrub® using a soft cloth and water. Rub until windshield is completely clean of all foreign material. Rinse windshield thoroughly.

Wiper blade park position, checking and adjusting

Wiper blades must be positioned properly to insure correct coverage on the windshield and rear window and to avoid body damage.

- Wipers should be shut off with the wiper switch and allowed to stop completely before switching off ignition to measure. Wipers must be in park position.

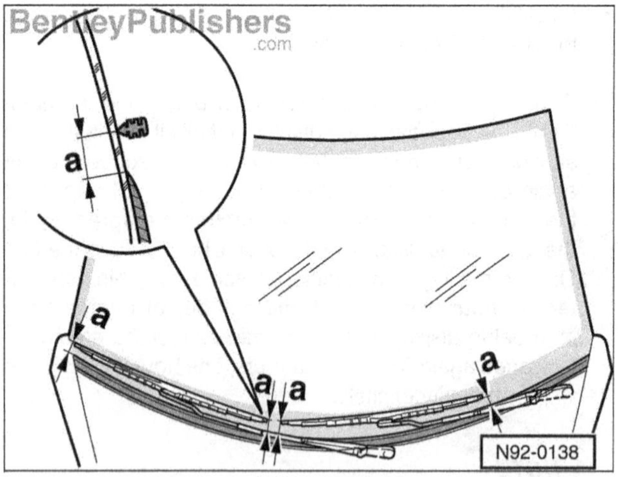

< **Driver and passenger side:** Distance **a** between blade rubber and lower edge should fall within specification.

Specification
- Dimension **a** . . . . 20 - 30 mm (0.78 - 1.18 in)

- If necessary, adjust park position by prying cover off of mounting pivot at base of wiper arm, loosening the nut, and carefully working until free.

- Reposition arm, tighten nut, and recheck position. Readjust as required. Replace plastic cover when done.

Tightening torque
- Wiper arm nut. . . . . 20 Nm (15 ft-lb)

Wiper blade contact angle, front and rear, checking and adjusting

If wipers blades become noisy or skip, and a thorough cleaning of the blades and windshield has not corrected the problem, check the contact angle of the wiper arm. Adjust as wiper arm as required.

- Shut off wipers with wiper switch and allowed to stop completely before switching off ignition to measure. Wipers must be in park position.
- Remove appropriate wiper blade and set aside.

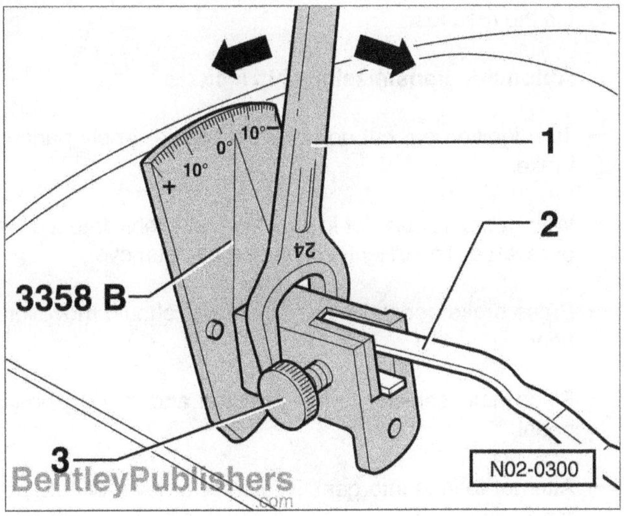

- Insert wiper arm (**2**) in adjustment tool 3358B as shown and lock with thumbscrew (**3**).
- Check contact angle and compare to table.

Wiper blade contact angle	
Driver's side	- 3 °
Passenger side	- 3 °
Tolerance	+ 2 °

- If adjustment is needed, apply 24 mm open ended wrench (**1**) to adjustment tool and twist until correct angle is reached.
- Remove wrench. Loosen and retighten thumbscrew to allow tool to centralized. Recheck angle. Repeat as needed until correct angle is reached.
- Remove adjustment tool and reinstall wiper blade.
- Check wiper operation.

QUALITY REVIEW

When you have finished working under the hood and around other areas of the vehicle, it is advisable to take a moment to quality check or review your work. This helps to insure that the operation or repair has been completed properly with all affected systems functioning within normal parameters. These may include the following:

Road test

Upon completion of the maintenance work, the vehicle should be road tested. Check the following systems during the road test:

Automatic transmission shift lock

- Turn ignition on, but do not start engine. Apply parking brake.
- With gearshift selector in park attempt to shift into a drive or reverse. The selector lever should not move.
- Press brake pedal. The selector lever should move into gear.
- Place gear selector lever in neutral and release brake pedal.
- Attempt to shift into gear. The selector lever should not move.
- Press the brake pedal. The selector lever should move into gear.
- If shift lock does not operate as described, see **37 Automatic Transmission**.

Automatic park/neutral position switch

Apply the parking brake and step on the brake pedal. Place the gear shift selector in park and attempt to start car. Repeat for all gears. The engine should only start in park or neutral. See **37 Automatic Transmission**.

Brake pedal and parking brake

Check pedal and handbrake travel (free play) and operation.

- Travel of brake pedal: max. 1/3 of overall pedal travel
- Travel of parking brake lever: 2 notches (clicks)

Manual gearshift

Check for smooth operation of the transmission and its gearshift. If it is difficult to shift the 5-speed (02J) transmission into reverse gear this may be due to a faulty back-up light switch. See **34a Manual Transmission**.

Steering

Check steering play with vehicle weight on the wheels by turning steering wheel back and forth (wheels in straight ahead position).

- Steering play: zero play with engine running

Kickdown operation

Depress accelerator pedal fully to floor. Depending on vehicle speed and engine speed, upshift is either delayed or the transmission shifts down into the next lower gear.

Air conditioner and heater

Check the full range of heater and air conditioner for proper function. Cold or warm air must flow out of vents.

Additional checks

Perform the following checks in accordance with the vehicle equipment and the road conditions available during the road test:

- Engine performance, idle speed, acceleration, cold and hot starting.
- Clutch operation, smoothness and pedal pressure.
- Brake operation, noise and function of ABS if equipped.
- Cruise control system operation.
- Radio operation, and reception.
- General vehicle handling and dynamics such as pulling to the side, cornering, vibrations in steering wheel and unusual noises.

MAINTENANCE SCHEDULES

The maintenance schedules list all of the routine maintenance specified by Volkswagen, as well as the mileage intervals at which they should be performed. In any instance where the publisher has recommended maintenance or a maintenance interval that differs from that specified by Volkswagen, such recommendations are more conservative, and still meet Volkswagen's maintenance requirements.

NOTE —

- *Aside from keeping your Volkswagen in the best possible condition, scheduled maintenance plays a role in maintaining full coverage under Volkswagen's extensive warranties. If in doubt about the terms and conditions of your car's warranty, consult an authorized Volkswagen dealer.*

- *Volkswagen continually updates maintenance schedules to suit changing conditions through the issuance of Maintenance Schedule Service Circulars. If in doubt about any of the requirements for your vehicle, consult an authorized Volkswagen dealer.*

1998 Maintenance Schedule

Miles	5	10	15	20	25	30	35	40	45	50	55	60	65	70	75	80	85	90	95	100	105
Kilometers	8	16	24	32	40	48	56	64	72	80	88	96	104	112	120	128	136	144	152	160	168
TDI																					
Engine Oil – change	•	•		•		•		•		•		•		•		•		•		•	
Engine Filter – change	•	•		•		•		•		•		•		•		•		•		•	
Water Sep. – drain	•	•		•		•		•		•		•		•		•		•		•	
Timing Belt – check condition	•	•		•		•		•		•		•		•		•		•		•	
Fuel Filter – replace				•				•				•				•				•	
Timing Belt – replace†												•									
Timing Belt Tensioner – replace***												•									
1.8T																					
Engine Oil – change	•	•	•	•	•	•	•	•	•	•	•	•	•	•	•	•	•	•	•	•	•
Engine Filter – change	•	•	•	•	•	•	•	•	•	•	•	•	•	•	•	•	•	•	•	•	•
V-Belt – replace																•					
Spark Plugs – replace								•								•					
Timing Belt – check condition								•								•					
Timing Belt – replace																					•
Timing Belt Tensioner – replace																					•
2.0																					
Engine Oil – change	•	•		•		•		•		•		•		•		•		•		•	
Engine Filter – change	•	•		•		•		•		•		•		•		•		•		•	
Spark Plugs – replace				•				•				•				•				•	
Timing Belt – check condition								•								•					
VR6																					
Engine Oil – change	•	•		•		•		•		•		•		•		•		•		•	
Engine Filter – change	•	•		•		•		•		•		•		•		•		•		•	
Spark Plugs – replace								•								•					
V6																					
Engine Oil – change	•	•	•	•	•	•	•	•	•	•	•	•	•	•	•	•	•	•	•	•	•
Engine Filter – change	•	•	•	•	•	•	•	•	•	•	•	•	•	•	•	•	•	•	•	•	•
Spark Plugs – replace								•								•					
Timing Belt – check condition								•								•					
Timing Belt – replace																					•
Timing Belt Tensioner – replace																					•
All Cars																					
W/W Fluid – check level		•		•		•		•		•		•		•		•		•		•	
Dust Pollen Filter* – replace		•		•		•		•		•		•		•		•		•		•	
Auto-Shift Lock – check		•		•		•		•		•		•		•		•		•		•	
Brake System – check		•		•		•		•		•		•		•		•		•		•	
Wheels – rotate front to rear		•		•		•		•		•		•		•		•		•		•	
Battery – check electrolyte level				•				•				•				•				•	
Manual Trans. – check for leaks				•				•				•				•				•	
Automatic Trans. – check for leaks				•				•				•				•				•	
Tires/Spare – check condition				•				•				•				•				•	
Drive Shafts – check boots				•				•				•				•				•	
OBD – check DTC memory				•				•				•				•				•	
Door Check Straps – lubricate				•				•				•				•				•	
During Road Test				•				•				•				•				•	
After Road Test				•				•				•				•				•	
Air Cleaner** – replace filter								•								•					
Cooling System – check level				•				•				•				•				•	
V-Belt – check tension/condition								•								•					
Ribbed Belt – check condition								•								•					
Front Axle – check								•								•					
Check ATF and diff. level								•								•					
Automatic Trans. AFT – replace												•									

*Where applicable **Two years or 40,000 miles ***Recommended; replace once only †Recommended

Brake Fluid – change every 2 years regardless of mileage

Air Bag System – check for function and damage after 4 years, 8 years and every 2 years thereafter

1999 Maintenance Schedule

Miles Kilometers	5 8	10 16	15 24	20 32	25 40	30 48	35 56	40 64	45 72	50 80	55 88	60 96	65 104	70 112	75 120	80 128	85 136	90 144	95 152	100 160	105 168
TDI																					
Engine Oil – change	•	•		•		•		•		•		•		•		•		•		•	
Engine Filter – change	•	•		•		•		•		•		•		•		•		•		•	
Water Sep. – drain	•	•		•		•		•		•		•		•		•		•		•	
Timing Belt – check condition	•	•		•		•		•		•		•		•		•		•		•	
Fuel Filter – replace				•				•				•				•				•	
Timing Belt – replace**												•									
Timing Belt Tensioner – replace***												•									
1.8T																					
Engine Oil – change	•	•	•	•	•	•	•	•	•	•	•	•	•	•	•	•	•	•	•	•	•
Engine Filter – change	•	•	•	•	•	•	•	•	•	•	•	•	•	•	•	•	•	•	•	•	•
V-Belt – replace																•					
Spark Plugs – replace								•								•					
Timing Belt – check condition								•								•					
Timing Belt – replace																					•
Timing Belt Tensioner – replace																					•
2.0																					
Engine Oil – change	•	•		•		•		•		•		•		•		•		•		•	
Engine Filter – change	•	•		•		•		•		•		•		•		•		•		•	
Spark Plugs – replace (all but New Beetle)				•				•				•				•				•	
Spark Plugs – replace (NB)								•								•					
Timing Belt – check condition								•								•					
VR6																					
Engine Oil – change	•	•		•		•		•		•		•		•		•		•		•	
Engine Filter – change	•	•		•		•		•		•		•		•		•		•		•	
Spark Plugs – replace								•								•					
V6																					
Engine Oil – change	•	•	•	•	•	•	•	•	•	•	•	•	•	•	•	•	•	•	•	•	•
Engine Filter – change	•	•	•	•	•	•	•	•	•	•	•	•	•	•	•	•	•	•	•	•	•
Spark Plugs – replace								•								•					
Timing Belt – check condition								•								•					
Timing Belt – replace																					•
Timing Belt Tensioner – replace																					•
All Cars																					
W/W Fluid – check level		•		•		•		•		•		•		•		•		•		•	
Auto-Shift Lock – check		•		•		•		•		•		•		•		•		•		•	
Brake System – check		•		•		•		•		•		•		•		•		•		•	
Wheels – rotate front to rear		•		•		•		•		•		•		•		•		•		•	
Battery – check electrolyte level				•				•				•				•				•	
Dust Pollen Filter – replace				•				•				•				•				•	
Manual Trans. – check for leaks				•				•				•				•				•	
Automatic Trans. – check for leaks				•				•				•				•				•	
Tires/Spare – check condition				•				•				•				•				•	
Drive Shafts – check boots				•				•				•				•				•	
OBD – check DTC memory				•				•				•				•				•	
Door Check Straps – lubricate				•				•				•				•				•	
During Road Test				•				•				•				•				•	
After Road Test				•				•				•				•				•	
Air Cleaner* – replace filter								•								•					
Cooling System – check level				•				•				•				•				•	
V-Belt – check tension/condition								•								•					
Ribbed Belt – check condition								•								•					
Front Axle – check								•								•					
Check ATF and diff. level								•								•					
Headlights – adjust				•				•				•				•				•	

*Every 40,000 miles and 4 years; except Passat and EuroVan, every 2 years **Recommended ***Recommended; replace once only

Brake Fluid – change every 2 years regardless of mileage

Air Bag System – check for function and damage after 4 years, 8 years and every 2 years thereafter

2000 Maintenance Schedule

Miles Kilometers	5 8	10 16	15 24	20 32	25 40	30 48	35 56	40 64	45 72	50 80	55 88	60 96	65 104	70 112	75 120	80 128	85 136	90 144	95 152	100 160	105 168
TDI																					
Engine Oil – change	•	•		•		•		•		•		•		•		•		•		•	
Engine Filter – change	•	•		•		•		•		•		•		•		•		•		•	
Water Sep. – drain	•	•		•		•		•		•		•		•		•		•		•	
Timing Belt – check condition	•	•		•		•		•		•		•		•		•		•		•	
Fuel Filter – replace				•				•				•				•				•	
Timing Belt – replace*								A				M				A					
1.8T																					
Engine Oil – change	•	•	•	•	•	•	•	•	•	•	•	•	•	•	•	•	•	•	•	•	•
Engine Filter – change	•	•	•	•	•	•	•	•	•	•	•	•	•	•	•	•	•	•	•	•	•
Spark Plugs – replace								•								•					
V-Belt – replace																•					
Timing Belt – check condition								•								•					
Timing Belt – replace																					•
Timing Belt Tensioner – replace																					•
2.0																					
Engine Oil – change	•	•		•		•		•		•		•		•		•		•		•	
Engine Filter – change	•	•		•		•		•		•		•		•		•		•		•	
Spark Plugs – replace (Cabrio)				•				•				•				•				•	
Spark Plugs – replace (except Cabrio)								•								•					
Timing Belt – check condition								•								•					
VR6																					
Engine Oil – change	•	•		•		•		•		•		•		•		•		•		•	
Engine Filter – change	•	•		•		•		•		•		•		•		•		•		•	
Spark Plugs – replace								•								•					
V6																					
Engine Oil – change	•	•	•	•	•	•	•	•	•	•	•	•	•	•	•	•	•	•	•	•	•
Engine Filter – change	•	•	•	•	•	•	•	•	•	•	•	•	•	•	•	•	•	•	•	•	•
Spark Plugs – replace								•								•					
Timing Belt – check condition								•								•					
Timing Belt – replace																					•
Timing Belt Tensioner – replace																					•
All Cars																					
W/W Fluid – check level		•		•		•		•		•		•		•		•		•		•	
Auto-Shift Lock – check		•		•		•		•		•		•		•		•		•		•	
Brake System – check		•		•		•		•		•		•		•		•		•		•	
Wheels – rotate front to rear		•		•		•		•		•		•		•		•		•		•	
Air Bag System **		•		•		•		•		•		•		•		•		•		•	
Battery – check electrolyte level				•				•				•				•				•	
Dust Pollen Filter – replace				•				•				•				•				•	
Cooling System – check level				•				•				•				•				•	
Manual Trans. – check for leaks				•				•				•				•				•	
Automatic Trans. – check for leaks				•				•				•				•				•	
Tires/Spare – check condition				•				•				•				•				•	
Drive Shafts – check boots				•				•				•				•				•	
OBD – check DTC memory				•				•				•				•				•	
Door Hinge				•				•				•				•				•	
Headlights – adjust				•				•				•				•				•	
During Road Test				•				•				•				•				•	
After Road Test				•				•				•				•				•	
Air Cleaner*** – replace filter								•								•					
V-Belt – check tension/condition								•								•					
Ribbed Belt – check condition								•								•					
Brake Fluid****								•								•					
Check ATF and diff. level								•								•					
Front Axle (tie rod, etc.)								•								•					

*Recommended **Every 12 months regardless of mileage ***Every 2 years: Cabrio, Passat, EuroVan. Every 4 years: Golf, Jetta, New Beetle
****Replace every 2 years regardless of mileage

2001 Maintenance Schedule

Miles Kilometers	5 8	10 16	15 24	20 32	25 40	30 48	35 56	40 64	45 72	50 80	55 88	60 96	65 104	70 112	75 120	80 128	85 136	90 144	95 152	100 160	105 168
TDI																					
Engine Oil – change	•	•		•		•		•		•		•		•		•		•		•	
Engine Filter – change	•	•		•		•		•		•		•		•		•		•		•	
Water Sep. – drain	•	•		•		•		•		•		•		•		•		•		•	
Timing Belt – check condition	•	•		•		•		•		•		•		•		•		•		•	
Fuel Filter – replace				•				•				•				•				•	
Timing Belt – replace								A				M				A					
1.8T																					
Engine Oil – change	•	•	•	•	•	•	•	•	•	•	•	•	•	•	•	•	•	•	•	•	•
Engine Filter – change	•	•	•	•	•	•	•	•	•	•	•	•	•	•	•	•	•	•	•	•	•
Spark Plugs – replace								•								•					
V-Belt – replace																•					
Timing Belt – check condition								•								•					
Timing Belt – replace																					•
Timing Belt Tensioner – replace																					•
2.0																					
Engine Oil – change	•	•		•		•		•		•		•		•		•		•		•	
Engine Filter – change	•	•		•		•		•		•		•		•		•		•		•	
Spark Plugs – replace (Cabrio)				•				•				•				•				•	
Spark Plugs – replace (except Cabrio)								•								•					
Timing Belt – check condition								•								•					
VR6																					
Engine Oil – change	•	•		•		•		•		•		•		•		•		•		•	
Engine Filter – change	•	•		•		•		•		•		•		•		•		•		•	
Spark Plugs – replace								•								•					
V6																					
Engine Oil – change	•	•	•	•	•	•	•	•	•	•	•	•	•	•	•	•	•	•	•	•	•
Engine Filter – change	•	•	•	•	•	•	•	•	•	•	•	•	•	•	•	•	•	•	•	•	•
Spark Plugs – replace								•								•					
Timing Belt – check condition								•								•					
Timing Belt Tensioner – replace																					•
Timing Belt – replace																					•
All Cars																					
W/W Fluid – check level		•		•		•		•		•		•		•		•		•		•	
Auto-Shift Lock – check operation		•		•		•		•		•		•		•		•		•		•	
Brake System – check		•		•		•		•		•		•		•		•		•		•	
Wheels – rotate from front to rear		•		•		•		•		•		•		•		•		•		•	
Air Bag System – check function/ damage*		•		•		•		•		•		•		•		•		•		•	
Battery – check electrolyte level		•		•		•		•		•		•		•		•		•		•	
Dust Pollen Filter – replace				•				•				•				•				•	
Cooling System – check level				•				•				•				•				•	
Manual Transmission – for leaks				•				•				•				•				•	
Automatic Transmission – for leaks				•				•				•				•				•	
Tire/Spare Tire – check condition				•				•				•				•				•	
Drive Shafts – check boots				•				•				•				•				•	
OBD – check DTC memory				•				•				•				•				•	
Door Hinge				•				•				•				•				•	
Headlights – check and adjust if necessary				•				•				•				•				•	
Road Test – during				•				•				•				•				•	
Road Test – after				•				•				•				•				•	
Air Cleaner – replace filter**								•								•					
V-Belt – check tension/condition								•								•					
Ribbed Belt – check condition								•								•					
Brake Fluid***								•								•					
Check ATF and diff. level								•								•					
Front Axle (tie rods, etc.)								•								•					

*Check the air bag system every 12 months, regardless of mileage. Check the air bag system at 4 years, 8 years, and then every 2 years (10, 12, 14, etc.)
**Replace the air filter every 2 years or 40,000 miles, whichever occurs first: Cabrio, Passat, EuroVan. Replace the air filter every 4 years or 40,000 miles, whichever occurs first: Golf, Jetta, New Beetle
***Replace the brake fluid every 2 years, regardless of mileage

2002 Maintenance Schedule (1.8L)

Kilometers	8	16	24	32	40	48	56	64	72	80	88	96	104	112	120	128	136	144	152	160	168
Miles	5	10	15	20	25	30	35	40	45	50	55	60	65	70	75	80	85	90	95	100	105
Engine Oil – change	•	•	•	•	•	•	•	•	•	•	•	•	•	•	•	•	•	•	•	•	•
Engine Filter – change	•	•	•	•	•	•	•	•	•	•	•	•	•	•	•	•	•	•	•	•	•
Timing Belt – check								•								•					
Timing Belt – replace																					•
Timing Belt Tensioner – replace																					•
V-belt/Ribbed Belt – check								•								•					
V-belt – replace																•					
Spark Plugs – replace								•								•					
Air Cleaner – replace filter								•								•					
W/W Fluid – check level		•		•		•		•		•		•		•		•		•		•	
Auto-shift Lock – check operation		•		•		•		•		•		•		•		•		•		•	
Brake System – check damage/ leaks, pad thickness, fluid level		•		•		•		•		•		•		•		•		•		•	
Wheels – rotate front to rear		•		•		•		•		•		•		•		•		•		•	
Airbag System – check function/ damage*		•		•		•		•		•		•		•		•		•		•	
Battery – check electrolyte level		•		•		•		•		•		•		•		•		•		•	
Engine – check for leaks				•				•				•				•				•	
Pollen Filter – replace				•				•				•				•				•	
Cooling System – check level				•				•				•				•				•	
Transmission – check for leaks**				•				•				•				•				•	
Tires/Spare – check condition				•				•				•				•				•	
Driveshaft – check boots				•				•				•				•				•	
OBD – check DTC memory				•				•				•				•				•	
Door Hinge – lube				•				•				•				•				•	
Headlights – check and adjust if necessary				•				•				•				•				•	
Exhaust System – check for damage and leaks				•				•				•				•				•	
Sunroof Frame – clean and lubricate				•				•				•				•				•	
Road Test				•				•				•				•				•	
Power Steering – check fluid level								•								•					
Brake Fluid***								•								•					
Check ATF and Differential for leaks and level; add if necessary								•								•					
Front Axle – check dust seals, ball joints, tie rods								•								•					

*Air Bag System – visual check every 12 months regardless of mileage **Manual Transmission – check level; add if necessary

***Brake Fluid – change every 2 years regardless of mileage

2002 Maintenance Schedule (1.9L)

Kilometers	8	16	24	32	40	48	56	64	72	80	88	96	104	112	120	128	136	144	152	160	168
Miles	5	10	15	20	25	30	35	40	45	50	55	60	65	70	75	80	85	90	95	100	105
Engine Oil – change	•	•		•		•		•		•		•		•		•		•		•	
Engine Filter – change	•	•		•		•		•		•		•		•		•		•		•	
Timing Belt – check		•		•		•		•		•		•		•		•		•		•	
Timing Belt – replace																•					
Timing Belt Idler – replace																•					
V-belt/Ribbed Belt – check								•								•					
Air Cleaner – replace filter								•								•					
Water Sep – drain (TDI)	•	•		•		•		•		•		•		•		•		•		•	
Fuel Filter – replace (TDI)				•				•				•				•				•	
W/W Fluid – check level		•		•		•		•		•		•		•		•		•		•	
Auto-shift Lock – check operation		•		•		•		•		•		•		•		•		•		•	
Brake System – check damage/ leaks, pad thickness, fluid level		•		•		•		•		•		•		•		•		•		•	
Wheels – rotate front to rear		•		•		•		•		•		•		•		•		•		•	
Airbag System – check function/ damage*		•		•		•		•		•		•		•		•		•		•	
Battery – check electrolyte level		•		•		•		•		•		•		•		•		•		•	
Engine – check for leaks				•				•				•				•				•	
Pollen Filter – replace				•				•				•				•				•	
Cooling System – check level				•				•				•				•				•	
Transmission – check for leaks**				•				•				•				•				•	
Tires/Spare – check condition				•				•				•				•				•	
Driveshaft – check boots				•				•				•				•				•	
OBD – check DTC memory				•				•				•				•				•	
Door Hinge – lube				•				•				•				•				•	
Headlights – check and adjust if necessary				•				•				•				•				•	
Exhaust System – check for damage and leaks				•				•				•				•				•	
Sunroof Frame – clean and lubricate				•				•				•				•				•	
Road Test				•				•				•				•				•	
Power Steering – check fluid level								•								•					
Brake Fluid***								•								•					
Check ATF and Differential for leaks and level; add if necessary								•								•					
Front Axle – check dust seals, ball joints, tie rods								•								•					

*Air Bag System – visual check every 12 months regardless of mileage **Manual Transmission – check level; add if necessary
***Brake Fluid – change every 2 years regardless of mileage

2002 Maintenance Schedule (2.0L)

Kilometers	8	16	24	32	40	48	56	64	72	80	88	96	104	112	120	128	136	144	152	160	168
Miles	5	10	15	20	25	30	35	40	45	50	55	60	65	70	75	80	85	90	95	100	105
Engine Oil – change	•	•		•		•		•		•		•		•		•		•		•	
Engine Filter – change	•	•		•		•		•		•		•		•		•		•		•	
Timing Belt – check								•								•					
V-belt/Ribbed Belt – check								•								•					
Spark Plugs – replace								•								•					
Air Cleaner – replace filter								•								•					
W/W Fluid – check level		•		•		•		•		•		•		•		•		•		•	
Auto-shift Lock – check operation		•		•		•		•		•		•		•		•		•		•	
Brake System – check damage/ leaks, pad thickness, fluid level		•		•		•		•		•		•		•		•		•		•	
Wheels – rotate front to rear		•		•		•		•		•		•		•		•		•		•	
Airbag System – check function/ damage*		•		•		•		•		•		•		•		•		•		•	
Battery – check electrolyte level		•		•		•		•		•		•		•		•		•		•	
Engine – check for leaks				•				•				•				•				•	
Pollen Filter – replace				•				•				•				•				•	
Cooling System – check level				•				•				•				•				•	
Transmission – check for leaks**				•				•				•				•				•	
Tires/Spare – check condition				•				•				•				•				•	
Driveshaft – check boots				•				•				•				•				•	
OBD – check DTC memory				•				•				•				•				•	
Door Hinge – lube				•				•				•				•				•	
Headlights – check and adjust if necessary				•				•				•				•				•	
Exhaust System – check for damage and leaks				•				•				•				•				•	
Sunroof Frame – clean and lubricate				•				•				•				•				•	
Road Test				•				•				•				•				•	
Power Steering – check fluid level								•								•					
Brake Fluid***								•								•					
Check ATF and Differential for leaks and level; add if necessary								•								•					
Front Axle – check dust seals, ball joints, tie rods								•								•					

*Air Bag System – visual check every 12 months regardless of mileage **Manual Transmission – check level; add if necessary

***Brake Fluid – change every 2 years regardless of mileage

2003 Maintenance Schedule (1.8L)

Kilometers	8	16	24	32	40	48	56	64	72	80	88	96	104	112	120	128	136	144	152	160	168
Miles	5	10	15	20	25	30	35	40	45	50	55	60	65	70	75	80	85	90	95	100	105
Engine Oil – change	•	•	•	•	•	•	•	•	•	•	•	•	•	•	•	•	•	•	•	•	•
Engine Filter – change	•	•	•	•	•	•	•	•	•	•	•	•	•	•	•	•	•	•	•	•	•
Timing Belt – check								•								•					
Timing Belt – replace																					•
Timing Belt Tensioner – replace																					•
V-belt/Ribbed Belt – check								•								•					
V-belt – replace																•					
Spark Plugs – replace								•								•					
Air Cleaner – replace filter								•								•					
W/W Fluid – check level		•		•		•		•		•		•		•		•		•		•	
Auto-Shift Lock – check operation		•		•		•		•		•		•		•		•		•		•	
Brake System – check damage/ leaks, pad thickness, fluid level		•		•		•		•		•		•		•		•		•		•	
Wheels – rotate front to rear		•		•		•		•		•		•		•		•		•		•	
Airbag System – check function/ damage*		•		•		•		•		•		•		•		•		•		•	
Battery – check electrolyte level		•		•		•		•		•		•		•		•		•		•	
Engine – check for leaks				•				•				•				•				•	
Pollen Filter – replace				•				•				•				•				•	
Cooling System – check level				•				•				•				•				•	
Transmission – check for leaks**				•				•				•				•				•	
Tires/Spare – check condition				•				•				•				•				•	
Driveshaft – check boots				•				•				•				•				•	
OBD – check DTC memory				•				•				•				•				•	
Door Hinge – lube				•				•				•				•				•	
Headlights – check and adjust if necessary				•				•				•				•				•	
Exhaust System – check for damage and leaks				•				•				•				•				•	
Sunroof Frame – clean and lubricate				•				•				•				•				•	
Road Test				•				•				•				•				•	
Power Steering – check fluid level								•								•					
Brake Fluid***								•								•					
Check ATF and Differential for leaks and level; add if necessary								•								•					
Front Axle – check dust seals, ball joints, tie rods								•								•					

*Airbag System – visual check every 12 months regardless of mileage **Manual Transmission – check level; add if necessary
***Brake Fluid – change every 2 years regardless of mileage
Soft top – check function and roll-over protection with cloth top open every 2 years or 20,000 miles (32,000 km), whichever comes first

2003 Maintenance Schedule (1.9L)

Kilometers / Miles	8 / 5	16 / 10	24 / 15	32 / 20	40 / 25	48 / 30	56 / 35	64 / 40	72 / 45	80 / 50	88 / 55	96 / 60	104 / 65	112 / 70	120 / 75	128 / 80	136 / 85	144 / 90	152 / 95	160 / 100	168 / 105
Engine Oil – change	•	•		•		•		•		•		•		•		•		•		•	
Engine Filter – change	•	•		•		•		•		•		•		•		•		•		•	
Timing Belt – check		•		•		•		•		•		•		•		•		•		•	
Timing Belt – replace																				•	
V-belt/Ribbed Belt – check								•								•					
Air Cleaner – replace filter								•								•					
Water Sep – drain (TDI)	•	•		•		•		•		•		•		•		•		•		•	
Fuel Filter – replace (TDI)				•				•				•				•				•	
W/W Fluid – check level		•		•		•		•		•		•		•		•		•		•	
Auto-Shift Lock – check operation		•		•		•		•		•		•		•		•		•		•	
Brake System – check damage/ leaks, pad thickness, fluid level		•		•		•		•		•		•		•		•		•		•	
Wheels – rotate front to rear		•		•		•		•		•		•		•		•		•		•	
Airbag System – check function/ damage*		•		•		•		•		•		•		•		•		•		•	
Battery – check electrolyte level		•		•		•		•		•		•		•		•		•		•	
Engine – check for leaks				•				•				•				•				•	
Pollen Filter – replace				•				•				•				•				•	
Cooling System – check level				•				•				•				•				•	
Transmission – check for leaks**				•				•				•				•				•	
Tires/Spare – check condition				•				•				•				•				•	
Driveshaft – check boots				•				•				•				•				•	
OBD – check DTC memory				•				•				•				•				•	
Door Hinge – lube				•				•				•				•				•	
Headlights – check and adjust if necessary				•				•				•				•				•	
Exhaust System – check for damage and leaks				•				•				•				•				•	
Sunroof Frame – clean and lubricate				•				•				•				•				•	
Road Test				•				•				•				•				•	
Power Steering – check fluid level								•								•					
Brake Fluid***								•								•					
Check ATF and Differential for leaks and level; add if necessary								•								•					
Front Axle – check dust seals, ball joints, tie rods								•								•					

*Airbag System – visual check every 12 months regardless of mileage
**Manual Transmission – check level; add if necessary
***Brake Fluid – change every 2 years regardless of mileage

2003 Maintenance Schedule (2.0L)

Kilometers / Miles	8 / 5	16 / 10	24 / 15	32 / 20	40 / 25	48 / 30	56 / 35	64 / 40	72 / 45	80 / 50	88 / 55	96 / 60	104 / 65	112 / 70	120 / 75	128 / 80	136 / 85	144 / 90	152 / 95	160 / 100	168 / 105
Engine Oil – change	•	•		•		•		•		•		•		•		•		•		•	
Engine Filter – change	•	•		•		•		•		•		•		•		•		•		•	
Timing Belt – check								•								•					
V-belt/Ribbed Belt – check								•								•					
Spark Plugs – replace								•								•					
Air Cleaner – replace filter								•								•					
W/W Fluid – check level		•		•		•		•		•		•		•		•		•		•	
Auto-Shift Lock – check operation		•		•		•		•		•		•		•		•		•		•	
Brake System – check damage/ leaks, pad thickness, fluid level		•		•		•		•		•		•		•		•		•		•	
Wheels – rotate front to rear		•		•		•		•		•		•		•		•		•		•	
Airbag System – check function/ damage*		•		•		•		•		•		•		•		•		•		•	
Battery – check electrolyte level		•		•		•		•		•		•		•		•		•		•	
Engine – check for leaks				•				•				•				•				•	
Pollen Filter – replace				•				•				•				•				•	
Cooling System – check level				•				•				•				•				•	
Transmission – check for leaks**				•				•				•				•				•	
Tires/Spare – check condition				•				•				•				•				•	
Driveshaft – check boots				•				•				•				•				•	
OBD – check DTC memory				•				•				•				•				•	
Door Hinge – lube				•				•				•				•				•	
Headlights – check and adjust if necessary				•				•				•				•				•	
Exhaust System – check for damage and leaks				•				•				•				•				•	
Sunroof Frame – clean and lubricate				•				•				•				•				•	
Road Test				•				•				•				•				•	
Power Steering – check fluid level								•								•					
Brake Fluid***								•								•					
Check ATF and Differential for leaks and level; add if necessary								•								•					
Front Axle – check dust seals, ball joints, tie rods								•								•					

*Airbag System – visual check every 12 months regardless of mileage **Manual Transmission – check level; add if necessary
***Brake Fluid – change every 2 years regardless of mileage
Soft top – check function and roll-over protection with cloth top open every 2 years or 20,000 miles (32,000 km), whichever comes first

2004 Maintenance Schedule (1.8L)

1.8 Liter Turbo Engine:

Miles (in thousands)	5	10	15	20	25	30	35	40	45	50	55	60	65	70	75	80	85	90	95	100	105
Kilometers (in thousands)	**8**	**16**	**24**	**32**	**40**	**48**	**56**	**64**	**72**	**80**	**88**	**96**	**104**	**112**	**120**	**128**	**136**	**144**	**152**	**160**	**168**
Engine: change engine oil and oil filter	•	•	•	•	•	•	•	•	•	•	•	•	•	•	•	•	•	•	•	•	•
Engine: check for leaks				•				•				•				•				•	
Battery: check				•				•				•				•				•	
Door check straps: lubricate				•				•				•				•				•	
Dust & pollen filter: replace (where applicable)				•				•				•				•				•	
Cooling system: check coolant level; add if necessary				•				•				•				•				•	
Windshield washer fluid: check level; add if necessary		•		•		•		•		•		•		•		•		•		•	
Automatic shift lock: check operation including park/neutral position switch		•		•		•		•		•		•		•		•		•		•	
Automatic transmission: check for leaks				•				•				•				•				•	
Manual transmission: check oil level; add if necessary; check for leaks				•				•				•				•				•	
Wheels: rotate front to rear		•		•		•		•		•		•		•		•		•		•	
Tires/spare wheel: check condition and pressure				•				•				•				•				•	
Brake system: check for damage/leaks, brake pad thickness, brake fluid level		•		•		•		•		•		•		•		•		•		•	
Drive shafts: check boots				•				•				•				•				•	
On-Board Diagnostic system: check fault memory; erase if necessary				•				•				•				•				•	
Sunroof frame: clean and lubricate				•				•				•				•				•	

Miles (in thousands)	5	10	15	20	25	30	35	40	45	50	55	60	65	70	75	80	85	90	95	100	105
Kilometers (in thousands)	**8**	**16**	**24**	**32**	**40**	**48**	**56**	**64**	**72**	**80**	**88**	**96**	**104**	**112**	**120**	**128**	**136**	**144**	**152**	**160**	**168**
Front axle: check dust seals on ball joints and tie rod ends; check tie rods				•				•				•				•				•	
Headlights: check & adjust if necessary				•				•				•				•				•	
Rear spoiler: lubricate (***New Beetle only***)		•		•		•		•		•		•		•		•		•		•	
Timing belt: check condition												•				•					
Timing belt and tensioning roller: Replace (***Passat, Golf, Jetta, New Beetle 1.8 T engine only***)																					•
V-belt/ribbed belt: check tension and condition; adjust if necessary								•								•					
Spark plugs: replace								•								•					
Air cleaner: replace filter								•								•					
Automatic transmission: check ATF								•								•					
Exhaust system: check for damage & leaks				•				•				•				•				•	
Power steering: check fluid level								•								•					
Automatic transmission final drive: check oil level and for leaks; add if necessary								•				•				•					
Check operation of convertible top and roll-over protection with convertible top open using VAS 5051				•				•				•				•				•	
Road Test: check kickdown, braking, steering, heating and ventilation, air conditioning, power accessories and electrical systems				•				•				•				•				•	
Every 12 months, regardless of mileage (kilometers): Perform visual check on Airbag system.																					
Every 2 years, regardless of mileage (kilometers): Change brake fluid.																					
Every 2 yrs or 20,000 miles (32,000 kilometers), whichever comes first: Check function of convertible top and check roll-over protection with convertible top open.																					

2004 Maintenance Schedule (1.9L)

1.9 Liter TDI Engine:

Miles (in thousands)	5	10	15	20	25	30	35	40	45	50	55	60	65	70	75	80	85	90	95	100	105
Kilometers (in thousands)	**8**	**16**	**24**	**32**	**40**	**48**	**56**	**64**	**72**	**80**	**88**	**96**	**104**	**112**	**120**	**128**	**136**	**144**	**152**	**160**	**168**
Engine: change engine oil and oil filter	●	●		●		●		●		●		●		●		●		●		●	
Engine: check for leaks				●				●				●				●				●	
Battery: check				●				●				●				●				●	
Door check straps: lubricate				●				●				●				●				●	
Windshield washer fluid: check level; add if necessary		●		●		●		●		●		●		●		●		●		●	
Dust & pollen filter: replace (where applicable)				●				●				●				●				●	
Automatic shift lock: check operation including park/neutral position switch		●		●		●		●		●		●		●		●		●		●	
Automatic transmission: check for leaks				●				●				●				●				●	
Manual transmission: check oil level; add if necessary; check for leaks				●				●				●				●				●	
Tires/spare wheel: check condition and pressure				●				●				●				●				●	
Wheels: rotate front to rear		●		●		●		●		●		●		●		●		●		●	
Drive shafts: check boots				●				●				●				●				●	
On-Board Diagnostic system: check fault memory; erase if necessary				●				●				●				●				●	
Brake system: check for damage/leaks, brake pad thickness, brake fluid level		●		●		●		●		●		●		●		●		●		●	
Rear spoiler: lubricate (***New Beetle only***)		●		●		●		●		●		●		●		●		●		●	

Miles (in thousands)	5	10	15	20	25	30	35	40	45	50	55	60	65	70	75	80	85	90	95	100	105
Kilometers (in thousands)	**8**	**16**	**24**	**32**	**40**	**48**	**56**	**64**	**72**	**80**	**88**	**96**	**104**	**112**	**120**	**128**	**136**	**144**	**152**	**160**	**168**
Cooling system: check coolant level; add if necessary				●				●				●				●				●	
Automatic transmission: check ATF								●								●					
Direct shift transmission (DSG): change fluid								●								●					
Front axle: check dust seals on ball joints and tie rod ends; check tie rods				●				●				●				●				●	
Air cleaner: replace filter								●								●					
Headlights: check & adjust if necessary				●				●				●				●				●	
Exhaust system: check for damage & leaks				●				●				●				●				●	
Sunroof frame: clean and lubricate				●				●				●				●				●	
Power steering: check fluid level								●								●					
Water separator: drain water	●	●		●		●		●		●		●		●		●		●		●	
Fuel filter: replace				●				●				●				●				●	
Timing belt: check																●					
Timing belt: replace																				●	
V-belt/ribbed belt: check; adjust if necessary								●								●					
Road Test: check kickdown, braking, steering, heating and ventilation, air conditioning, power accessories and electrical systems				●				●				●				●				●	
Every 12 months, regardless of mileage (kilometers): Perform visual check on Airbag system.																					
Every 2 years, regardless of mileage (kilometers): Change brake fluid.																					

2004 Maintenance Schedule (2.0L)

2.0 Liter Engine:

Miles (in thousands)	5	10	15	20	25	30	35	40	45	50	55	60	65	70	75	80	85	90	95	100	105
Kilometers (in thousands)	8	16	24	32	40	48	56	64	72	80	88	96	104	112	120	128	136	144	152	160	168
Engine: change engine oil and oil filter	•	•		•		•		•		•		•		•		•		•		•	
Engine: check for leaks				•				•				•				•				•	
Exhaust system: check for damage & leaks				•				•				•				•				•	
Battery: check				•				•				•				•				•	
Door check straps: lubricate				•				•				•				•				•	
Dust & pollen filter: replace (where applicable)				•				•				•				•				•	
Cooling system: check coolant level; add if necessary				•				•				•				•				•	
Windshield washer fluid: check level; add if necessary		•		•		•		•		•		•		•		•		•		•	
Automatic shift lock: check operation including park/neutral position switch		•		•		•		•		•		•		•		•		•		•	
Automatic transmission: check for leaks				•				•				•				•				•	
Manual transmission: check oil level; add if necessary; check for leaks				•				•				•				•				•	
Wheels: rotate front to rear		•		•		•		•		•		•		•		•		•		•	
Tires/spare wheel: check condition and pressure				•				•				•				•				•	
Brake system: check for damage/leaks, brake pad thickness, brake fluid level		•		•		•		•		•		•		•		•		•		•	
Drive shafts: check boots				•				•				•				•				•	
On-Board Diagnostic system: check fault memory; erase if necessary				•				•				•				•				•	

Miles (in thousands)	5	10	15	20	25	30	35	40	45	50	55	60	65	70	75	80	85	90	95	100	105
Kilometers (in thousands)	8	16	24	32	40	48	56	64	72	80	88	96	104	112	120	128	136	144	152	160	168
Timing belt: check																•					
Sunroof frame: clean and lubricate				•				•				•				•				•	
V-belt/ribbed belt: check tension and condition; adjust if necessary								•								•					
Spark plugs: replace								•								•					
Air cleaner: replace filter								•								•					
Rear spoiler: lubricate (***New Beetle only***)		•		•		•		•		•		•		•		•		•		•	
Automatic transmission: check ATF								•								•					
Automatic transmission final drive oil: check level and for leaks; add if necessary								•				•				•					
Headlights: check & adjust if necessary				•				•				•				•				•	
Power steering: check fluid level								•								•					
Front axle: check dust seals on ball joints and tie rod ends; check tie rods				•				•				•				•				•	
Check operation of convertible top and roll-over protection with convertible top open using VAS 5051				•				•				•				•				•	
Road Test: check kick down, braking, steering, heating and ventilation, air conditioning, power accessories and electrical systems				•				•				•				•				•	
Every 12 months, regardless of mileage (kilometers): Perform visual check on Airbag system.																					
Every 2 years, regardless of mileage (kilometers): Change brake fluid.																					
Every 2 yrs or 20,000 miles (32,000 kilometers), whichever comes first: Check function of convertible top and check roll-over protection with convertible top open.																					

2005 Maintenance Schedule (1.8L)

1.8 Liter Turbo Engine:

Miles (in thousands)	5	10	15	20	25	30	35	40	45	50	55	60	65	70	75	80	85	90	95	100
Kilometers (in thousands)	**8**	**16**	**24**	**32**	**40**	**48**	**56**	**64**	**72**	**80**	**88**	**96**	**104**	**112**	**120**	**128**	**136**	**144**	**152**	**160**
Engine: change engine oil and oil filter	•	•	•	•	•	•	•	•	•	•	•	•	•	•	•	•	•	•	•	•
Engine: check for leaks				•				•				•				•				•
Service interval display: Reset *(Where applicable)*	•	•		•		•		•		•		•		•		•		•		•
Battery: check				•				•				•				•				•
Door check straps: lubricate				•				•				•				•				•
Dust & pollen filter: replace *(Where applicable)*				•				•				•				•				•
Cooling system: check coolant level; add if necessary				•				•				•				•				•
Wiper blades: check								•								•				
Windshield washer fluid: check level; add if necessary		•		•		•		•		•		•		•		•		•		•
Automatic shift lock: check operation including park/neutral position switch		•		•		•		•		•		•		•		•		•		•
Automatic transmission: check for leaks				•				•				•				•				•
Manual transmission: check oil level; add if necessary; check for leaks				•				•				•				•				•
Wheels: rotate front to rear		•		•		•		•		•		•		•		•		•		•
Tires: check tread depth and adjust pressure if necessary		•		•		•		•		•		•		•		•		•		•
Tires/spare wheel: check condition and pressure				•				•				•				•				•
Brake system: check for damage/leaks, brake pad thickness, brake fluid level		•		•		•		•		•		•		•		•		•		•
Drive shafts: check boots				•				•				•				•				•
On-Board Diagnostic system: check fault memory; erase if necessary				•				•				•				•				•
Sunroof frame: clean and lubricate				•				•				•				•				•

Miles (in thousands)	5	10	15	20	25	30	35	40	45	50	55	60	65	70	75	80	85	90	95	100
Kilometers (in thousands)	**8**	**16**	**24**	**32**	**40**	**48**	**56**	**64**	**72**	**80**	**88**	**96**	**104**	**112**	**120**	**128**	**136**	**144**	**152**	**160**
Front axle: check dust seals on ball joints and tie rod ends; check tie rods				•				•				•				•				•
Headlights: check & adjust if necessary				•				•				•				•				•
Rear spoiler: lubricate (***New Beetle only***)		•		•		•		•		•		•		•		•		•		•
Timing belt: check condition												•				•				•
V-belt/ribbed belt: check tension and condition; adjust if necessary								•								•				
Spark plugs: replace								•								•				
Air cleaner: replace filter								•								•				
Tire pressure monitoring system sensors: check battery charge, replace wheel electronics and aluminum valve if necessary (where applicable)		•		•		•		•		•		•		•		•		•		•
Automatic transmission: check ATF								•								•				
Exhaust system: check for damage & leaks				•				•				•				•				•
Power steering: check fluid level								•								•				
Automatic transmission final drive: check oil level and for leaks; add if necessary								•				•				•				•
Check operation of convertible top and roll-over protection with convertible top open using VAS 5051				•				•				•				•				•
Road Test: check kickdown, braking, steering, heating and ventilation, air conditioning, power accessories and electrical systems				•				•				•				•				•
Every 12 months, regardless of mileage (kilometers): Perform visual check on Airbag system.																				
Every 2 years, regardless of mileage (kilometers): Change brake fluid.																				
Every 2 yrs or 20,000 miles (32,000 kilometers), whichever comes first: Check function of convertible top and check roll-over protection with convertible top open.																				

2005 Maintenance Schedule (1.9L)

1.9 Liter TDI Engine:

Miles (in thousands)	5	10	15	20	25	30	35	40	45	50	55	60	65	70	75	80	85	90	95	100
Kilometers (in thousands)	8	16	24	32	40	48	56	64	72	80	88	96	104	112	120	128	136	144	152	160
Engine: change engine oil and oil filter	•	•		•		•		•		•		•		•		•		•		•
Engine: check for leaks				•				•				•				•				•
Service interval display: Reset *(Where applicable)*	•	•		•		•		•		•		•		•		•		•		•
Battery: check				•				•				•				•				•
Door check straps: lubricate				•				•				•				•				•
Wipers: check								•								•				
Windshield washer fluid: check level; add if necessary		•		•		•		•		•		•		•		•		•		•
Dust & pollen filter: replace (*Where applicable)*				•				•				•				•				•
Automatic shift lock: check operation including park/neutral position switch		•		•		•		•		•		•		•		•		•		•
Automatic transmission: check for leaks				•				•				•				•				•
Manual transmission: check oil level; add if necessary; check for leaks				•				•				•				•				•
Tire pressure monitoring system sensors: check battery charge, replace wheel electronics and aluminum valve if necessary (*Where applicable)*		•		•		•		•		•		•		•		•		•		•
Tires: check tread depth and adjust pressure if necessary		•		•		•		•		•		•		•		•		•		•
Tires/spare wheel: check condition and pressure				•				•				•				•				•
Wheels: rotate front to rear		•		•		•		•		•		•		•		•		•		•
Drive shafts: check boots				•				•				•				•				•
On-Board Diagnostic system: check fault memory; erase if necessary				•				•				•				•				•
Brake system: check for damage/leaks, brake pad thickness, brake fluid level		•		•		•		•		•		•		•		•		•		•
Rear spoiler: lubricate (***New Beetle only***)		•		•		•		•		•		•		•		•		•		•

Miles (in thousands)	5	10	15	20	25	30	35	40	45	50	55	60	65	70	75	80	85	90	95	100
Kilometers (in thousands)	8	16	24	32	40	48	56	64	72	80	88	96	104	112	120	128	136	144	152	160
Cooling system: check coolant level; add if necessary				•				•				•				•				•
Automatic transmission: check ATF								•								•				
Direct shift transmission (DSG): change fluid								•								•				
Front axle: check dust seals on ball joints and tie rod ends; check tie rods				•				•				•				•				•
Air cleaner: replace filter								•								•				
Headlights: check & adjust if necessary				•				•				•				•				•
Exhaust system: check for damage & leaks				•				•				•				•				•
Sunroof frame: clean and lubricate				•				•				•				•				•
Power steering: check fluid level								•								•				
Water separator: drain	•	•		•		•		•		•		•		•		•		•		•
Fuel filter: replace				•				•				•				•				•
Timing belt: replace																•				
V-belt/ribbed belt: check; adjust if necessary								•								•				
Road Test: check kickdown, braking, steering, heating and ventilation, air conditioning, power accessories and electrical systems				•				•				•				•				•
Every 12 months, regardless of mileage (kilometers): Perform visual check on Airbag system.																				
Every 2 years, regardless of mileage (kilometers): Change brake fluid.																				

2005 Maintenance Schedule (2.0L)

2.0 Liter Engine:

Miles (in thousands)	**5**	**10**	**15**	**20**	**25**	**30**	**35**	**40**	**45**	**50**	**55**	**60**	**65**	**70**	**75**	**80**	**85**	**90**	**95**	**100**
Kilometers (in thousands)	**8**	**16**	**24**	**32**	**40**	**48**	**56**	**64**	**72**	**80**	**88**	**96**	**104**	**112**	**120**	**128**	**136**	**144**	**152**	**160**
Engine: change engine oil and oil filter	•	•		•		•		•		•		•		•		•		•		•
Engine: check for leaks				•				•				•				•				•
Service interval display: Reset *(Where applicable)*	•	•		•		•		•		•		•		•		•		•		•
Exhaust system: check for damage & leaks				•				•				•				•				•
Battery: check				•				•				•				•				•
Door check straps: lubricate				•				•				•				•				•
Dust & pollen filter: replace (*Where applicable)*				•				•				•				•				•
Cooling system: check coolant level; add if necessary				•				•				•				•				•
Windshield washer fluid: check level; add if necessary		•		•		•		•		•		•		•		•		•		•
Automatic shift lock: check operation including park/neutral position switch		•		•		•		•		•		•		•		•		•		•
Automatic transmission: check for leaks				•				•				•				•				•
Manual transmission: check oil level; add if necessary; check for leaks				•				•				•				•				•
Wiper blades: check								•								•				
Wheels: rotate front to rear		•		•		•		•		•		•		•		•		•		•
Tires/spare wheel: check condition and pressure				•				•				•				•				•
Brake system: check for damage/leaks, brake pad thickness, brake fluid level		•		•		•		•		•		•		•		•		•		•
Drive shafts: check boots				•				•				•				•				•
On-Board Diagnostic system: check fault memory; erase if necessary				•				•				•				•				•

Miles (in thousands)	**5**	**10**	**15**	**20**	**25**	**30**	**35**	**40**	**45**	**50**	**55**	**60**	**65**	**70**	**75**	**80**	**85**	**90**	**95**	**100**
Kilometers (in thousands)	**8**	**16**	**24**	**32**	**40**	**48**	**56**	**64**	**72**	**80**	**88**	**96**	**104**	**112**	**120**	**128**	**136**	**144**	**152**	**160**
Timing belt: check												•				•				•
Sunroof frame: clean and lubricate				•				•				•				•				•
V-belt/ribbed belt: check tension and condition; adjust if necessary								•								•				
Tire pressure monitoring system sensors: check battery charge, replace wheel electronics and aluminum valve if necessary (*Where applicable)*		•		•		•		•		•		•		•		•		•		•
Tires: check tread depth and adjust pressures if necessary		•		•		•		•		•		•		•		•		•		•
Spark plugs: replace								•								•				
Air cleaner: replace filter								•								•				
Rear spoiler: lubricate (***New Beetle only***)		•		•		•		•		•		•		•		•		•		•
Automatic transmission: check ATF								•								•				
Automatic transmission final drive oil: check level and for leaks; add if necessary								•				•				•				•
Headlights: check & adjust if necessary				•				•				•				•				•
Power steering: check fluid level								•								•				
Front axle: check dust seals on ball joints and tie rod ends; check tie rods				•				•				•				•				•
Check operation of convertible top and roll-over protection with convertible top open using VAS 5051				•				•				•				•				•
Road Test: check kick down, braking, steering, heating and ventilation, air conditioning, power accessories and electrical systems				•				•				•				•				•
Every 12 months, regardless of mileage (kilometers): Perform visual check on Airbag system.																				
Every 2 years, regardless of mileage (kilometers): Change brake fluid.																				
Every 2 yrs or 20,000 miles (32,000 kilometers), whichever comes first: Check function of convertible top and check roll-over protection with convertible top open.																				

2006 Maintenance Schedule (1.9L)

1.9 Liter TDI Engine:

Miles (in thousands)	5	10	15	20	25	30	35	40	45	50	55	60	65	70	75	80	85	90	95	100
Kilometers (in thousands)	8	16	24	32	40	48	56	64	72	80	88	96	104	112	120	128	136	144	152	160
Engine: Change engine oil and oil filter	●	●		●		●		●		●		●		●		●		●		●
Engine: Check for leaks				●				●				●				●				●
Service interval display: Reset *(Where applicable)*	●	●		●		●		●		●		●		●		●		●		●
Battery: Check				●				●				●				●				●
Door check straps: Lubricate				●				●				●				●				●
Wipers: Check								●								●				
Windshield washer fluid: Check level; add if necessary		●		●		●		●		●		●		●		●		●		●
Dust & pollen filter: Replace *(Where applicable)*				●				●				●				●				●
Automatic shift lock: Check operation including park/neutral position switch		●		●		●		●		●		●		●		●		●		●
Automatic transmission: Check for leaks				●				●				●				●				●
Manual transmission: Check oil level; add if necessary; check for leaks				●				●				●				●				●
Tire pressure monitoring system sensors: Check battery charge, replace wheel electronics and aluminum valve if necessary *(Where applicable)*		●		●		●		●		●		●		●		●		●		●
Tires: Check tread depth and adjust pressure if necessary		●		●		●		●		●		●		●		●		●		●
Tires/spare wheel: Check condition and pressure				●				●				●				●				●
Wheels: Rotate front to rear		●		●		●		●		●		●		●		●		●		●
Drive shafts: Check boots				●				●				●				●				●
On-Board Diagnostic system: Check fault memory; erase if necessary				●				●				●				●				●
Brake system: Check for damage/leaks, brake pad thickness, brake fluid level		●		●		●		●		●		●		●		●		●		●
Rear spoiler: Lubricate (***New Beetle only***)		●		●		●		●		●		●		●		●		●		●

Miles (in thousands)	5	10	15	20	25	30	35	40	45	50	55	60	65	70	75	80	85	90	95	100
Kilometers (in thousands)	8	16	24	32	40	48	56	64	72	80	88	96	104	112	120	128	136	144	152	160
Cooling system: Check coolant level; add if necessary				●				●				●				●				●
Automatic transmission: Check ATF								●								●				
Direct Shift Transmission (DSG): Change fluid and filter								●								●				
Front axle: Check dust seals on ball joints and tie rod ends; check tie rods				●				●				●				●				●
Air cleaner: Replace filter								●								●				
Headlights: Check & adjust if necessary				●				●				●				●				●
Exhaust system: Check for damage & leaks				●				●				●				●				●
Sunroof frame: Clean and lubricate				●				●				●				●				●
Power steering: Check fluid level								●								●				
Diesel particle filter: Check *(Where applicable)*																				●
Water separator: Drain water	●	●		●		●		●		●		●		●		●		●		●
Fuel filter: Replace				●				●				●				●				●
Timing belt: Replace																●				
V-belt/ribbed belt: Check; adjust if necessary								●								●				
Road Test: Check kickdown, braking, steering, heating and ventilation, air conditioning, power accessories and electrical systems				●				●				●				●				●

Every 12 months, regardless of mileage (kilometers): Perform visual check on Airbag system.

Every 2 years, regardless of mileage (kilometers): Change brake fluid.

Every 2 years, for vehicles driven less than 40,000 miles (60,000 Km) in 2 years: Replace dust and pollen filter.

Every 4 years, regardless of mileage (kilometers): Replace tire filler bottle (observe expiration date).

Every 6 years, for vehicles driven less than 60,000 miles (90,000 Km) in 6 years: Replace air filter element and clean housing.

Every 6 years, regardless of mileage (kilometers): Replace tire pressure sensors (if applicable)

2006 Maintenance Schedule (2.5L)

2.5 Liter Engine:

Miles (in thousands)	5	10	15	20	25	30	35	40	45	50	55	60	65	70	75	80	85	90	95	100
Kilometers (in thousands)	8	16	24	32	40	48	56	64	72	80	88	96	104	112	120	128	136	144	152	160
Engine: Change engine oil and oil filter	•	•		•		•		•		•		•		•		•		•		•
Engine: Check for leaks				•				•				•				•				•
Service interval display: Reset *(Where applicable)*	•	•		•		•		•		•		•		•		•		•		•
Exhaust system: Check for damage & leaks				•				•				•				•				•
Battery: Check				•				•				•				•				•
Door check straps: Lubricate				•				•				•				•				•
Dust & pollen filter: Replace (*Where applicable*)				•				•				•				•				•
Cooling system: Check coolant level; add if necessary				•				•				•				•				•
Wiper blades: Check								•								•				
Windshield washer fluid: Check level; add if necessary		•		•		•		•		•		•		•		•		•		•
Tires/spare wheel: Check condition and pressure				•				•				•				•				•
Tires: Check tread depth and adjust pressures if necessary		•		•		•		•		•		•		•		•		•		•
Brake system: Check for damage/leaks, brake pad thickness, brake fluid level		•		•		•		•		•		•		•		•		•		•
Wheels: Rotate front to rear		•		•		•		•		•		•		•		•		•		•
Spark plugs: Replace								•								•				
Sunroof frame: Clean and lubricate				•				•				•				•				•
Air cleaner: Replace filter								•								•				
Drive shafts: Check boots				•				•				•				•				•
On-Board Diagnostic system: Check fault memory; erase if necessary				•				•				•				•				•

Miles (in thousands)	5	10	15	20	25	30	35	40	45	50	55	60	65	70	75	80	85	90	95	100
Kilometers (in thousands)	8	16	24	32	40	48	56	64	72	80	88	96	104	112	120	128	136	144	152	160
Tire pressure monitoring systems sensors: check battery charge, replace wheel electronics and aluminum valve if necessary (*Where applicable*)		•		•		•		•		•		•		•		•		•		•
V-belt/ribbed belt: check tension and condition; adjust if necessary								•								•				
Automatic shift lock: check operation including park/neutral position switch		•		•		•		•		•		•		•		•		•		•
Automatic transmission: check for leaks				•				•				•				•				•
Manual transmission: check oil level; add if necessary; check for leaks				•				•				•				•				•
Direct Shift Transmission (DSG): change transmission fluid and replace filter element *(Where applicable)*								•								•				
Automatic transmission: check ATF								•								•				
Automatic transmission final drive oil: check level and for leaks; add if necessary								•								•				
Headlights: check & adjust if necessary				•				•				•				•				•
Underbody: check for damage								•								•				
Power steering: check fluid level								•								•				
Front axle: check dust seals on ball joints and tie rod ends; check tie rods				•				•				•				•				•
Road Test: check kick down, braking, steering, heating and ventilation, air conditioning, power accessories and electrical systems				•				•				•				•				•

Every 12 months, regardless of mileage (kilometers): Perform visual check on Airbag system.
Every 2 years, regardless of mileage (kilometers): Change brake fluid.
Every 2 years, for vehicles driven less than 40,000 mile (60,000 km) in 2 years : Replace dust and pollen filter.
Every 2 yrs or 20,000 miles (32,000 kilometers), whichever comes first: Check function of convertible top and check roll-over protection with convertible top open.
Every 4 years, regardless of mileage (kilometers): Replace tire filler bottle (observe expiration date).
Every 6 years, regardless of mileage (kilometers): Change tire pressure sensors (if available).

2007 - 2008 Maintenance Schedule (2.5L)

2.5L — **Miles** *(in thousands)* / **Kilometers** *(in thousands)*	5 8	10 16	20 32	30 48	40 64	50 80	60 96	70 112	80 128	90 144	100 160	110 176	120 180
Engine Oil – change oil and replace filter	•	•	•	•	•	•	•	•	•	•	•	•	•
Brake System – check brake pad thickness	•	•	•	•	•	•	•	•	•	•	•	•	•
Service Interval Display – reset *(where applicable)*	•	•	•	•	•	•	•	•	•	•	•	•	•
Service Interval Display – adapt *(for 10,000 mile /16,000 Km service interval only) (where applicable)*		•											
Windshield washer/headlight washer – check level; add if necessary		•	•	•	•	•	•	•	•	•	•	•	•
Brake System – check for damage and leaks, brake fluid level		•	•	•	•	•	•	•	•	•	•	•	•
Wheels – rotate		•	•	•	•	•	•	•	•	•	•	•	•
Automatic Shift lock - check operation including park/neutral position switch		•	•	•	•	•	•	•	•	•	•	•	•
Rear spoiler – lubricate		•	•	•	•	•	•	•	•	•	•	•	•
Tires – check tread depth and pressure; adjust if necessary		•	•	•	•	•	•	•	•	•	•	•	•
Engine – check for leaks			•		•		•		•		•		•
Engine OBD – check memory, purge if necessary			•		•		•		•		•		•
Automatic Transmission – check for leaks			•		•		•		•		•		•
Battery – check			•		•		•		•		•		•
Door Arresters – lubricate			•		•		•		•		•		•
Sunroof Frame – lubricate *(where applicable)*			•		•		•		•		•		•
Cooling System – check coolant level; add if necessary			•		•		•		•		•		•
Exhaust System – check for damage and leaks			•		•		•		•		•		•
Tires/Spare wheel – check condition and pressure		•	•	•	•	•	•	•	•	•	•	•	•
Front Axle – check dust seals on ball joints and tie rods; check tie rods			•		•		•		•		•		•
Headlights – check and adjust if necessary			•		•		•		•		•		•
Drive Shafts – check boots			•		•		•		•		•		•
Manual Transmission – check for leaks and level; add if necessary			•		•		•		•		•		•
Road Test – check kick down, braking, steering heating and ventilation, air conditioning, power accessories and electrical systems			•		•		•		•		•		•
Dust and Pollen filter – replace *(where applicable)*			•		•		•		•		•		•
Air Cleaner – replace filter and clean housing					•		•		•		•		•
Wiper Blades – check					•				•				•
Interior and exterior lights - check function					•				•				•
Underbody – check for damage			•		•		•		•		•		•
Ribbed Belt – check condition and tension; adjust if necessary					•				•				•
Automatic transmission – check ATF			•		•		•		•		•		•
Automatic transmission final drive oil – check level and for leaks; add if necessary					•				•				•
Spark Plugs – replace					•				•				•
Power steering – check fluid level *(where applicable)*					•				•				•
Direct Shift Transmission (DSG) – change fluid and replace filter *(where applicable)*					•				•				•
Convertible top - check function and rollover protection					•				•				•

Every 12 months, regardless of mileage (kilometers): Perform visual check of airbag system.
Every 2 years, regardless of mileage (kilometers): Change brake fluid and check bolting device for convertible top.
Every 2 years (for vehicles driven less than 40,000 miles (60,000 Km) in 2 years): Replace dust and pollen filter.
Every 4 years, regardless of mileage (kilometers): Tire filler bottle in tire mobility set is replaced *(Observe expiration date).*
Every 4 years, regardless of mileage (kilometers): Change air filter element and clean housing.
Every 4 years (for vehicles driven less than 40,000 miles (60,000 Km in 4 years): Change spark plugs.
Every 6 years, regardless of mileage (kilometers): Change tire pressure sensors *(If available).*

2009 - 2010 Maintenance Schedule (2.5L)

2.5L — Miles (in thousands) / Kilometers (in thousands)	10 / 16	20 / 32	30 / 48	40 / 64	50 / 80	60 / 96	70 / 112	80 / 128	90 / 144	100 / 160	110 / 176	120 / 180
Engine Oil – change oil and replace filter	•	•	•	•	•	•	•	•	•	•	•	•
Brake System – check brake pad thickness	•	•	•	•	•	•	•	•	•	•	•	•
Service Interval Display – reset *(where applicable)*	•	•	•	•	•	•	•	•	•	•	•	•
Headlight washer and rear window washer - check functionality	•	•	•		•	•	•		•	•	•	
Service label - Enter date of next service and apply to driver's side door pillar	•	•	•		•	•	•		•	•	•	
Windshield washer/headlight washer – check level; add if necessary, check functionality	•	•	•	•	•	•	•	•	•	•	•	•
Brake System – check for damage and leaks, brake fluid level		•	•	•	•	•	•	•	•	•	•	•
Wheels – rotate	•	•	•	•	•	•	•	•	•	•	•	•
Tires – check tread depth and pressure; adjust if necessary		•		•		•		•		•		•
Engine – check for leaks				•				•				•
Battery – check		•		•		•		•		•		•
Door Arresters – lubricate				•				•				•
Sunroof Frame – lubricate *(where applicable)*				•				•				•
Cooling System – check coolant level; add if necessary				•				•				•
Exhaust System – check for damage and leaks				•				•				•
Tires/Spare wheel – check condition and pressure				•				•				•
Front Axle – check dust seals on ball joints and tie rods; check tie rods				•				•				•
Headlights – check and adjust if necessary				•				•				•
Drive Shafts – check boots				•				•				•
Road Test – check kick down, braking, steering heating and ventilation, air conditioning, power accessories and electrical systems				•				•				•
Body - inspect for corrosion				•				•				•
Wiper Blades – check		•		•		•		•		•		•
Interior and exterior lights - check function				•				•				•
Underbody – check for damage				•				•				•
Ribbed Belt – check condition and tension; adjust if necessary				•				•				•
Automatic transmission – change fluid and filter				•				•				•
Spark Plugs – replace				•				•				•
Power steering – check fluid level *(where applicable)*				•				•				•
Direct Shift Transmission (DSG) – change fluid and replace filter *(where applicable)*				•				•				•
Convertible top - check function and rollover protection				•				•				•

Every 12 months, regardless of mileage (kilometers): Perform visual check of airbag system.
Every 2 years, regardless of mileage (kilometers): Change brake fluid and check bolting device for convertible top.
Every 2 years (for vehicles driven less than 40,000 miles (60,000 Km) in 2 years): Replace dust and pollen filter.
Every 4 years, regardless of mileage (kilometers): Tire filler bottle in tire mobility set is replaced *(Observe expiration date)*.
Every 4 years, regardless of mileage (kilometers): Change air filter element and clean housing.
Every 4 years (for vehicles driven less than 40,000 miles (60,000 Km in 4 years): Change spark plugs.
Every 6 years, regardless of mileage (kilometers): Change tire pressure sensors *(If available)*.

1 Engine–General

GENERAL

This general information group gives engine application information and general technical data for the engines used in the Volkswagen New Beetle. Engines are the same for both the United States and Canada.

Much of the engine repair information in **1 Engine** is organized according to engine code. It is therefore important to know the code of the engine installed in your New Beetle. For engine code location see **03 Maintenance**.

Engine	Engine code
1.8L 4-cylinder turbo gasoline	APH, AWV, AWP, BKF, BNU
1.9L 4-cylinder TDI diesel	ALH, BEW
2.0L 4-cylinder gasoline	AEG, AVH, AZG, BDC, BEV, BGD
2.5L 5-cylinder gasoline	BPR, BPS

1.8L ENGINE

0024341

The 1.8L engine is a further design evolution of the 4-cylinder gasoline engines used on previous models of the Volkswagen Passat. This new engine looks similar to the older versions, however, there have been significant design changes to this high performance 4-cylinder engine for the New Beetle.

Technical Data - 1.8L Engine

- Type . . . . 4-cylinder inline, 5 valves/cylinder
- Displacement . . . 1.8L, 1781 cc (108.7 cubic inches)
- Bore . . . . 81.0 mm
- Stroke. . . . . 86.4 mm
- Compression Ratio. 9.5:1
- Horsepower
  APH, AWV . . . . 110 kW (150 HP) @5800 RPM
  AWP. . . . . 132 kW (180 HP) @5800 RPM
- Torque
  APH, AWV . . . . 210 Nm (155 ft-lb) @1750 RPM
  AWP. . . . . 235 Nm (175 ft-lb) @1950 RPM
- Fuel requirement
  APH, AWP . . . . . . .90 AKI Plus/Mid-grade Unleaded
  AWV. . . . . 93 AKI Super Unleaded
- Engine Management . . . . Motronic ME7.5.1, OBD II

1.8L engine design features:

- Cast iron cylinder block, light alloy crossflow cylinder head and aluminum oil pan
- Crankshaft with 5 main bearings
- Dual overhead camshafts. Exhaust camshaft driven by toothed belt with tensioning dampener. Intake camshaft driven by self-adjusting chain from the exhaust camshaft
- Lightweight valvetrain, 5 valves per cylinder.
- Transverse mounting with “Pendulum” type engine mounts
- Chain-drive internal gear oil pump driven from the front of the crankshaft
- Coolant pump built into the cylinder block and driven by the toothed camshaft belt
- Thermostat integral with the cylinder block
- Turbocharger with intercooler
- Distributorless ignition with separate coils for each cylinder
- External engine accessories driven by a single poly-rib belt with tensioner
- Bosch Motronic ME7.5.1 Multi-point Sequential Fuel Injection (MFI) system with electronic accelerator system

1.9L ENGINE

The ALH engine is a design evolution of the 4-cylinder diesel engines used on previous models of the Volkswagen Golf and Jetta. Just as in the gas versions, this new engine looks similar to the older versions, but again there has been significant development. The acronym TDI stands for Turbo Direct Injection.

1.9L TDI engine design features:

- Cast iron cylinder block, light alloy cylinder head and aluminum oil pan and cylinder head cover
- Piston oil spray nozzles
- Crankshaft with 5 main bearings
- Single overhead camshaft driven by a toothed belt with semi-automatic belt tensioner and two guide pulleys
- Lightweight valvetrain, 2 valves per cylinder
- Transverse mounting with "Pendulum" type engine mounts
- Chain-drive internal gear oil pump driven from the front of the crankshaft
- Camshaft-driven vacuum pump
- Coolant pump built into the cylinder block and driven by the toothed camshaft belt
- Thermostat integral with the cylinder block
- Cartridge-type oil filter
- External engine accessories driven by a single poly-rib belt with tensioner
- Turbo Direct Injection (TDI) Diesel Fuel Injection

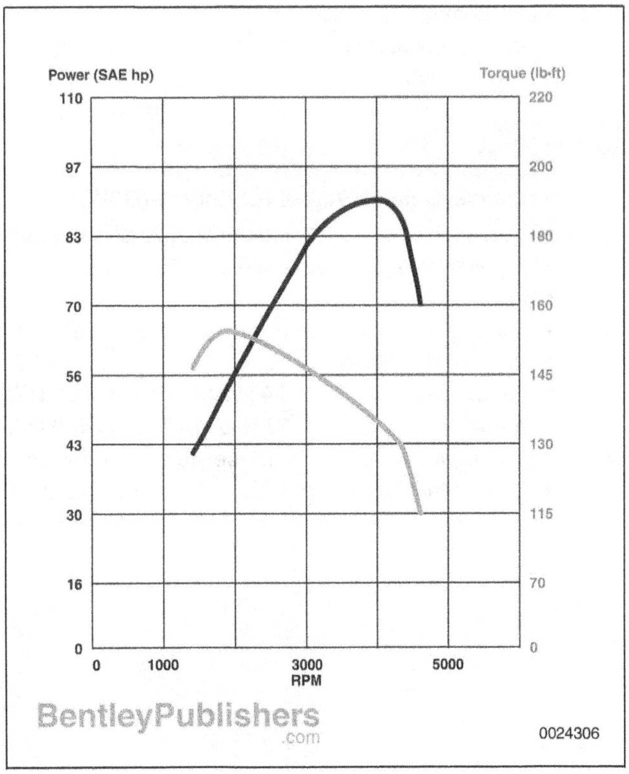

Technical Data - 1.9L TDI Engine (ALH)

- Type . . . . 4-cylinder inline, 2 valves/cylinder
- Displacement . . . 1.9L, 1896 cc (115.7 cubic inches)
- Bore . . . . 79.9 mm
- Stroke. . . . . 95.5 mm
- Compression Ratio. 19.5:1
- Horsepower . . . . 66 kW (90 HP) @4000 RPM
- Torque . . . . 149 ft-lb @1900 RPM
- Fuel requirement . . . . . . Diesel 45 Cetane minimum
- Engine Management . . . Turbo Direct Injection (TDI) Diesel OBD II

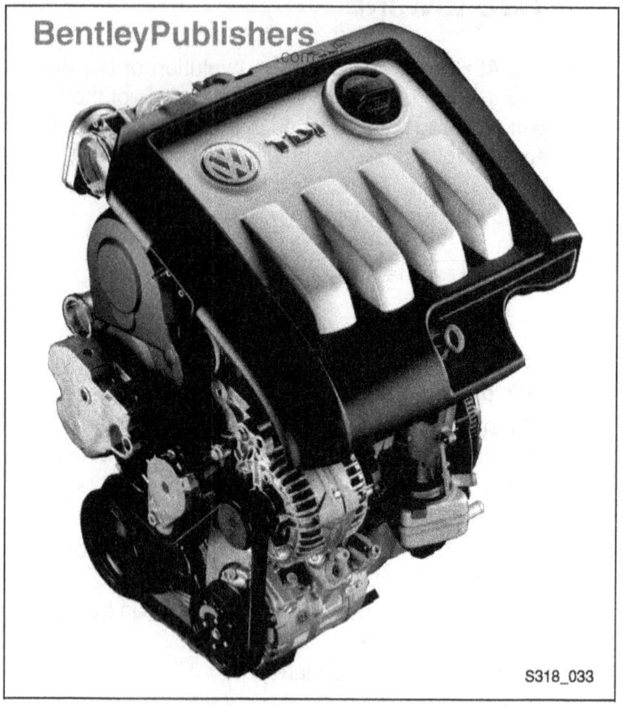

S318_033

1.9L TDI PD (engine code: BEW)

Beginning with the 2004 model year, a new TDI diesel engine was introduced. Known as the Pump Düse (PD), this engine uses pump/injectors to deliver fuel to each cylinder. The pump/injector is a pressure-generating pump that also has a solenoid valve control unit. Each cylinder has its own pump/injector located in the cylinder head.

The TDI-PD engine also has a mechanical fuel pump mounted on the end of the cylinder head to supply fuel from the tank to the pump injectors. This engine has no intermediate shaft or distributor injection pump as in the previous (engine code: ALH) TDI diesel engine.

1.9L TDI PD engine design features:

- Trapezoidal pistons and connecting rods to accommodate higher combustion pressures
- Wider timing belt with non-uniform tooth gap clearance (spacing)
- Separate mechanical fuel pump
- Expanded fuel cooling system with separate pump
- Solenoid fuel injectors with integral high pressure pumps
- Injector pressures of up to 27,846 psi (1,920 bar)
- Camshaft with four extra lobes for injector pump activation
- Camshaft position sensor and 7-pin sensor wheel
- Low combustion noise
- Low fuel consumption
- Clean emissions

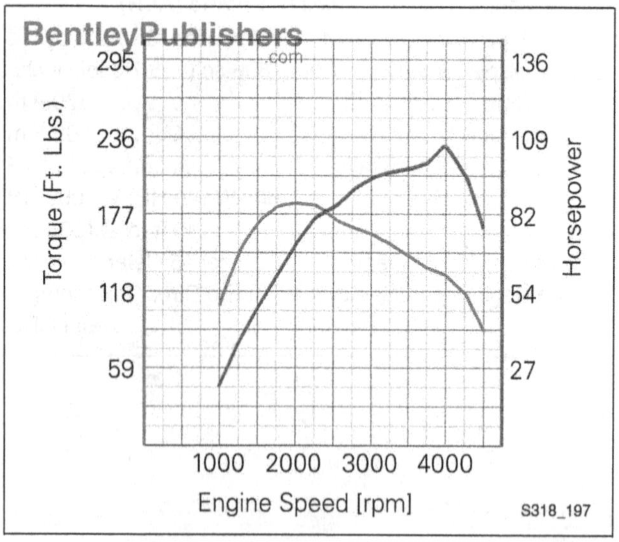

S318_197

Technical Data - 1.9L TDI PD Engine (BEW)
- Type . . . . 4-cylinder inline, 2 valves/cylinder
- Displacement . . . 1.9L, 1896 cc (115.7 cubic inches)
- Bore . . . . 79.5 mm
- Stroke. . . . . 95.5 mm
- Compression Ratio. 17.5:1
- Horsepower . . . . 74 kW (99 HP) @4000 RPM
- Torque . . . . 177 ft-lb @1800 to 2400 RPM
- Fuel requirement . . . . . . Diesel 49 Cetane minimum
- Engine Management . . . . Bosch EDC 16

2.0L ENGINE

The 2.0L engine is a design evolution of the 4-cylinder gasoline engines used on previous models of the Volkswagen Golf and Jetta. And although this new engine looks similar to the older versions, there has been significant development of the basic 4-cylinder engine for the New Beetle.

2.0L engine design features

- Cast iron cylinder block, light alloy crossflow cylinder head and aluminum oil pan
- Crankshaft with 5 main bearings
- Single overhead camshaft driven by a toothed belt with semi-automatic belt tensioner
- Lightweight valvetrain, 2 valves per cylinder
- Transverse mounting with "Pendulum" type engine mounts
- Chain-drive internal gear oil pump driven from the front of the crankshaft
- Coolant pump built into the cylinder block and driven by the toothed camshaft belt
- Thermostat integral with the cylinder block
- Two-piece aluminum intake manifold
- Tubular stainless steel exhaust manifold
- Distributorless ignition
- External engine accessories driven by a single poly-rib belt with tensioner

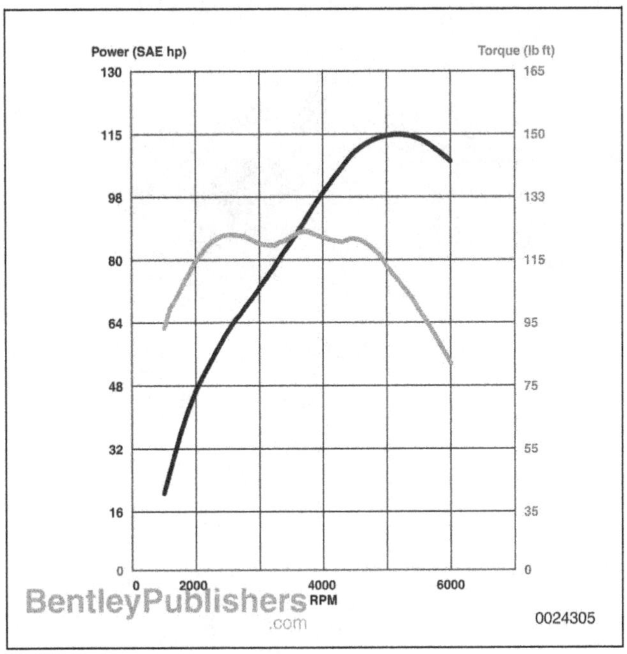

Technical Data - 2.0L Engine

- Type . . . . 4-cylinder inline, 2 valves/cylinder
- Displacement . . . 2.0L, 1984 cc (121.1 cubic inches)
- Bore . . . . 82.5 mm
- Stroke. . . . . 92.8 mm
- Compression Ratio. 10.0:1
- Horsepower . . . . 85 kW (115 HP) @5200 RPM
- Torque . . . . 122 ft-lb @2200 RPM
- Fuel requirement . . . . 87 AKI Regular Unleaded
- Engine Management
  AEG . . . . Motronic M5.9.2, OBD II
  AVH, AZG, BEV, BDC, BGD . . . . . . Motronic ME 7.5

2.5L ENGINE

The 2.5L engine was an all new design first introduced in the 2005 Jetta models. This engine is used in 2006 and later New Beetle models.

2.5L engine design features:

- Transverse 5-cylinder design
- Dual overhead camshafts
- Continuously variable intake cam
- Timing chains drive camshafts, vacuum pump and oil pump

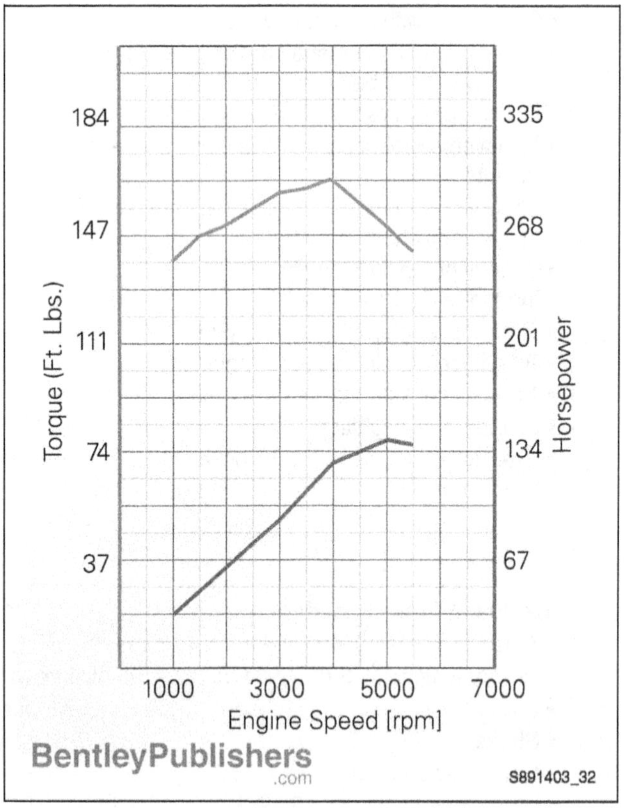

Technical Data - 2.5L Engine

- Type . . . . .5-cylinder inline, 4valves/cylinder
- Displacement . . . . . 2.5L, 2480cc (151 cubic inches)
- Bore . . . . .82.5 mm
- Stroke. . . . . .92.8 mm
- Compression Ratio. 9.5:1
- Horsepower . . . . 110 kW (150 HP) @5000 RPM
- Torque . . . . 170 ft-lb @3750 RPM
- Fuel requirement . . . . 87 AKI Regular Unleaded
- Engine Management . . . . Motronic ME 7.1.1

10 Engine–Removing and Installing

GENERAL

The engine and transmission are removed as a unit from below and separated from each other once removed from the vehicle.

The operations needed to remove the engine and transmission assembly are generally grouped into two main areas: jobs under the hood and jobs under the vehicle. The steps below follow this general sequence. Most of the operations are simple and straight forward. References to additional repair groups are provided in bold type where additional information may be helpful to complete a step.

Engine	Engine code
1.8L 4-cylinder turbo gasoline	APH, AWV, AWP, BKF, BNU
1.9L 4-cylinder TDI diesel	ALH, BEW
2.0L 4-cylinder gasoline	AEG, AVH, AZG, BDC, BEV, BGD
2.5L 5-cylinder gasoline	BPR, BPS

> ***CAUTION—***
> *Disconnecting the negative (–) battery cable will erase fault codes and basic settings in engine management and automatic transmission control unit memories. Some driveability problems may be noticed until the system re-adapts to operating conditions. OBD II readiness codes, which may be required for emissions testing, may also be erased. In some instances proper diagnosis will require use of a diagnostic scan tool. See* **OBD On Board Diagnostics** *for more information.*

NOTE—

- *It will be necessary to cut many wire tie wraps when removing the engine. The tie wraps are installed to prevent the wiring harnesses from chaffing or contacting engine parts. Be sure to make note of all tie wraps removed and install new ones during engine installation.*
- *Most of the hardware used in the vehicles covered by this repair manual is specially coated to prevent corrosion. This coating is known as "dacromet" or "delta tone" and is identified by a green tinted finish. Always use replacement hardware with the same specification.*

ENGINE, REMOVING AND INSTALLING (4-CYLINDER ENGINES)

Engine/transmission assembly, removing (4-cylinder)

– Ensure ignition is switched off. Disconnect battery negative (–) terminal and then positive (+) terminal.

NOTE—

Be sure to have the anti-theft radio code on hand before disconnecting the battery.

– Remove engine cover. See **03 Maintenance**.

– Remove power steering reservoir from battery support, but do not loosen the hoses. Temporarily tie reservoir to hood lock carrier.

– Remove the fuse holder on the top of the battery, the battery and the mounting bracket, see **27 Engine Electrical**.

– Remove air cleaner assembly and connecting hoses and ducts.

1.8L engine

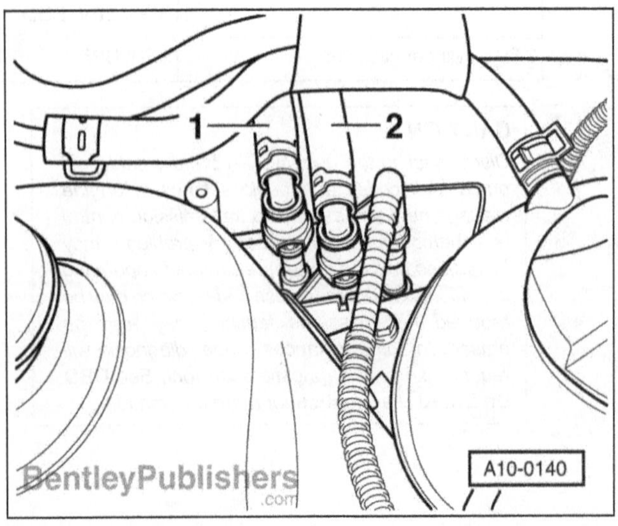

◀ Disconnect and label fuel supply line (**1**) and return line (**2**).

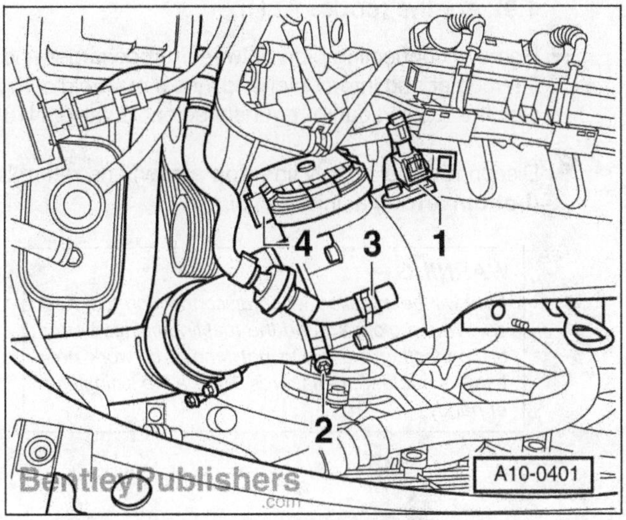

Loosen clamp (**2**) and remove connecting pipe from intercooler and intake manifold. Disconnect harness connectors for intake air temperature sensor (**1**) and throttle valve control unit (**4**). Label and remove vacuum line (**3**) at intake manifold.

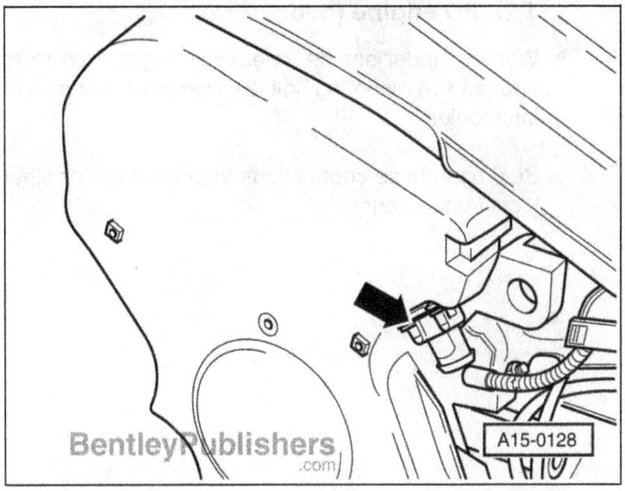

Remove harness connector (**arrow**) at hall sender.

- Unbolt vacuum canister from cylinder head cover and pull from bracket.
- Disconnect fuel injector harness connectors and unclip fuel rail from fuel distributor.
- Disconnect harness connectors and ground wires (if applicable) from all four ignition coils.

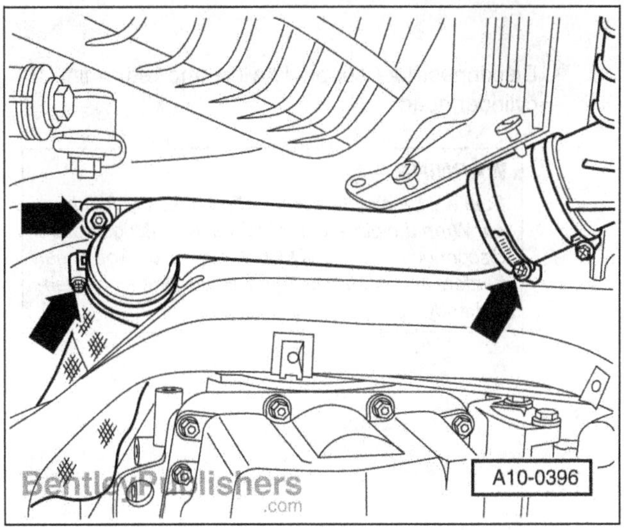

Remove air duct pipe to intercooler, located on lower longitudinal body member. Loosen hose clamps (**arrows**) and remove mounting screw (**top left arrow**).

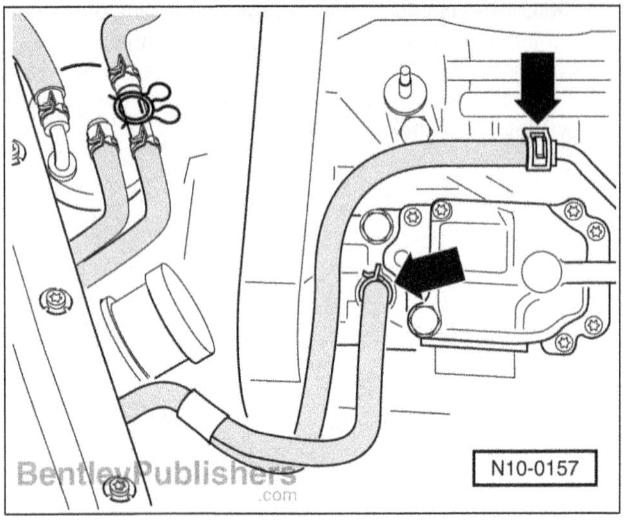

1.9L engine (code: ALH)

- Remove connecting pipe between turbocharger and intercooler and intercooler and intake manifold complete with the EGR vacuum regulator solenoid valve (N18).

◀ Disconnect fuel supply line (**top arrow**) and return line (**bottom arrow**) at injection pump.

> ***WARNING —***
> *Fuel will be expelled when disconnecting fuel hoses. Wrap a cloth around the fuel line fittings before disconnecting them. Do not smoke or work near heaters or other fire hazards. Have a fire extinguisher handy.*

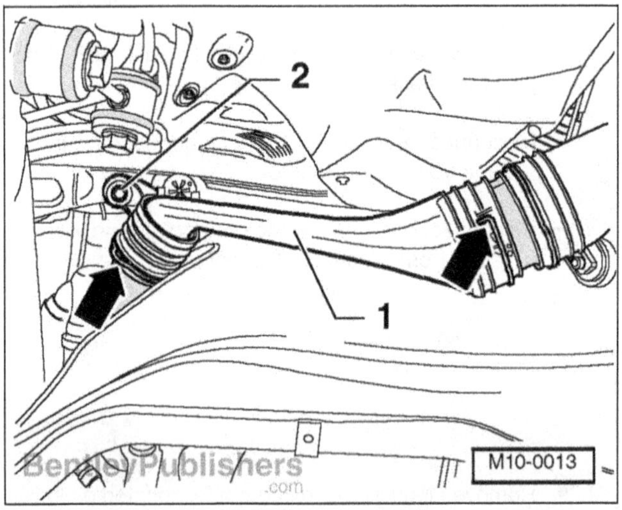

1.9L PD engine (code: BEW)

◀ Working under engine, release spring clamps (**arrows**) and remove mounting bolt (**2**). Remove lower pipe (**1**) for intercooler.

- Seal open hose connections to prevent dirt or other objects from entering.

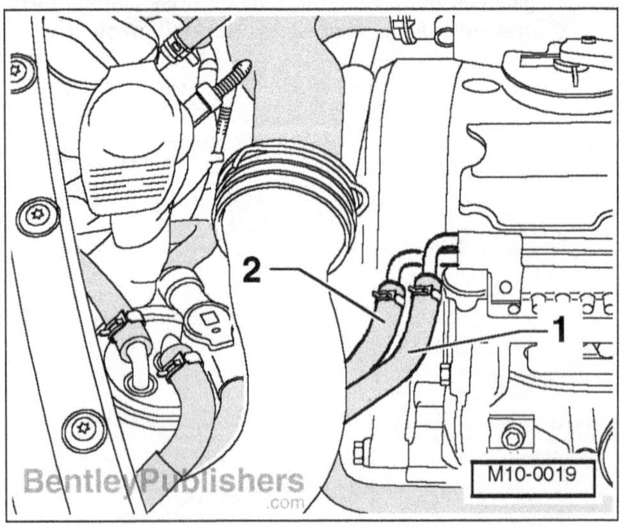

◀ Disconnect fuel supply line (**1**) and return line (**2**) from cylinder head.

> ***WARNING —***
> *Fuel will be expelled when disconnecting fuel hoses. Wrap a cloth around the fuel line fittings before disconnecting them. Do not smoke or work near heaters or other fire hazards. Have a fire extinguisher handy.*

2.0L engine (code: AEG, AVH, AZG, BDC)

Disconnect and label fuel supply line (**1**) and return line (**2**) at fuel rail, making sure to seal hoses and fittings. Disconnect accelerator cable.

> ***WARNING —***
> *Fuel will be expelled when disconnecting fuel hoses. Wrap a cloth around the fuel line fittings before disconnecting them. Do not smoke or work near heaters or other fire hazards. Have a fire extinguisher handy.*

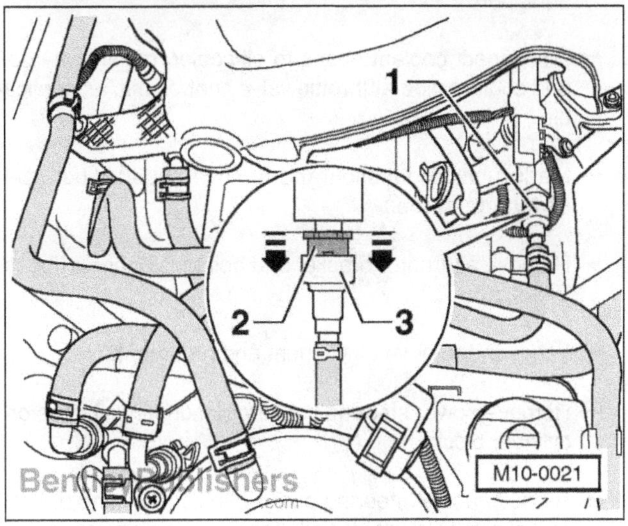

2.0L engine (code: BEV, BGD)

Disconnect fuel supply line (**1**) by pulling locking ring (**2**) toward fuel union (**3**).

> ***WARNING —***
> *Fuel will be expelled when disconnecting fuel hoses. Wrap a cloth around the fuel line fittings before disconnecting them. Do not smoke or work near heaters or other fire hazards. Have a fire extinguisher handy.*

Continued for all vehicles

– Seal all fuel lines so that fuel system is not contaminated by dirt, leaking fluids, etc. Seal off any open ports in manifold, as applicable.

– Disconnect shift linkage.
  - Manual transmission: Disconnect cables from selector mechanism and unbolt support bracket, see **34 Manual Transmission**.
  - DSG transmission: Disconnect selector cable at transmission, see **34b DSG Transmission**.
  - Automatic transmission: Disconnect selector cable at selector lever, see **37 Automatic Transmission**.

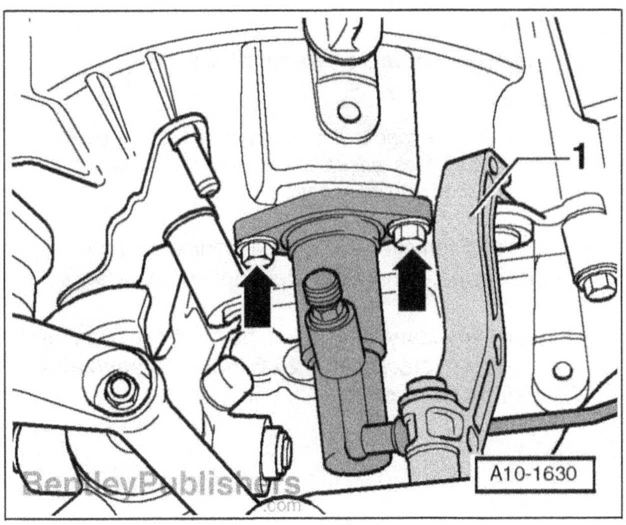

Vehicles with manual transmission

Remove bracket (**1**) if applicable, and unbolt hydraulic clutch slave cylinder (**arrows**).

– Secure slave cylinder to the side with a tie wrap. Do not disconnect the hydraulic line.

> ***CAUTION —***
> *Do not depress the clutch pedal with the slave cylinder removed, damage will result.*

– Disconnect harness connector for back-up light switch from transmission.

Continued for all vehicles

- Disconnect harness connector for speedometer vehicle speed sensor at rear of transmission, as applicable.
- Remove lower sound absorber panels from under engine. See **03 Maintenance**.
- Drain coolant, see **19 Cooling System**.

> ***WARNING —***
> *Hot coolant can scald. Drain the coolant only with engine cold.*

- Remove auxiliary cooling fan on right side of vehicle, if equipped.
- Disconnect coolant hoses to oil cooler, radiator, heater, and coolant pipe at throttle valve control unit, as applicable.
- Mark running direction and remove ribbed V-belt, see **03 Maintenance**.
- Remove alternator bracket and belt tensioner for ribbed V-belt.
- Label and remove all vacuum and breather hoses.
- Remove power steering lines from mounting clamps on cylinder block.
- Remove power steering pump with its mounting bracket and carefully lay aside with hoses remaining connected, see **48 Steering**.
- Remove retaining clamp(s) from A/C refrigerant lines. Remove A/C compressor with refrigerant lines still attached, see **87 Air Conditioning**.

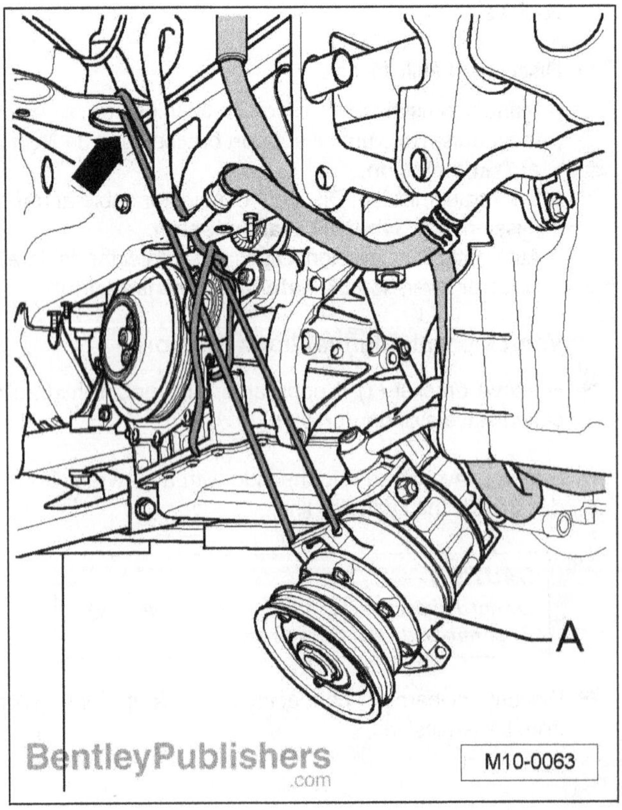

◄ Attach A/C compressor to lower section of body with a suitable support (**arrow**) so that refrigerant lines are not stressed.

- 5-cylinder shown, 4-cylinder similar

NOTE —
To prevent damage to the condenser and to the refrigerant lines/hoses, ensure that the lines and hoses are not stretched, kinked or bent.

- Label and remove all electrical connections on engine, transmission, generator (alternator) and starter.
- Unbolt secondary air injection pump, if equipped, and remove from bracket, see **26 Exhaust System and Emission Controls**.

– Raise car and support with jack stands or lift. See **03 Maintenance** for proper lifting procedure.

> ***WARNING —***
> *Observe all warnings and cautions associated with lifting vehicle in* **03 Maintenance**.

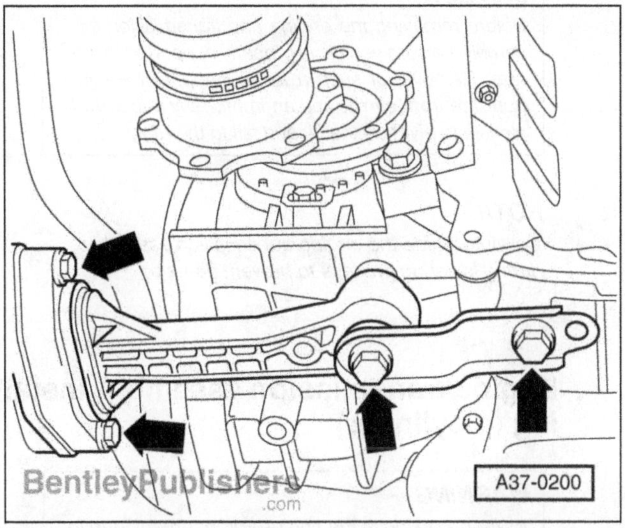

◄ Remove mounting bolts (**arrows**) and pendulum support.

– Remove right side inner CV joint protective cover from engine if equipped.

– Remove axle nuts from right side outer CV joint and unbolt both inner CV or triple-rotor joints from the transmission. Remove the right side drive axle, see **40 Front Suspension and Drive Axles.** Tie wrap left side axle up and out of way, taking care not to damage the coating on the axle.

> ***CAUTION —***
> - *Do not allow drive axle to hang by inner CV joint or triple rotor joint to avoid damaging bearing surfaces.*
> - *Do not flex inner CV or triple rotor joint more than 25° from its original angle.*

– Unbolt front exhaust pipe with catalytic converter from exhaust manifold or turbocharger. Loosen clamp at connection to center muffler and remove front pipe, see **26 Exhaust System/Emission Controls.**

– Remove oil pan guard, if equipped. See **17 Engine–Lubrication System.**

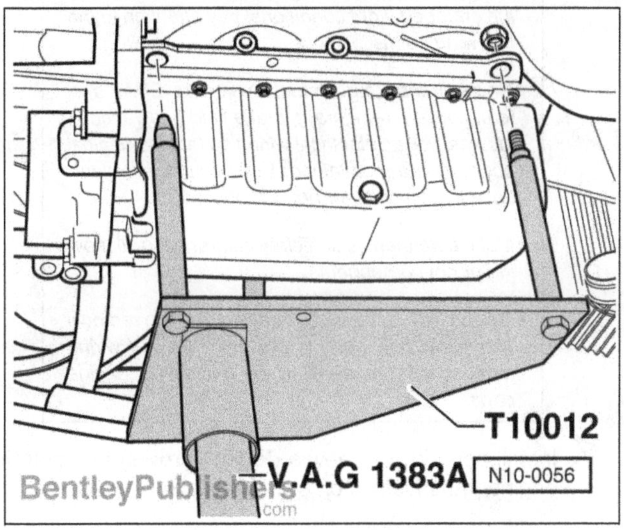

◄ Install a suitable engine supporting device from below, using supporting points.

– Remove supporting bracket for coolant hose under engine block.

– Lift supporting device slightly, so that weight of engine and transmission is on supporting device.

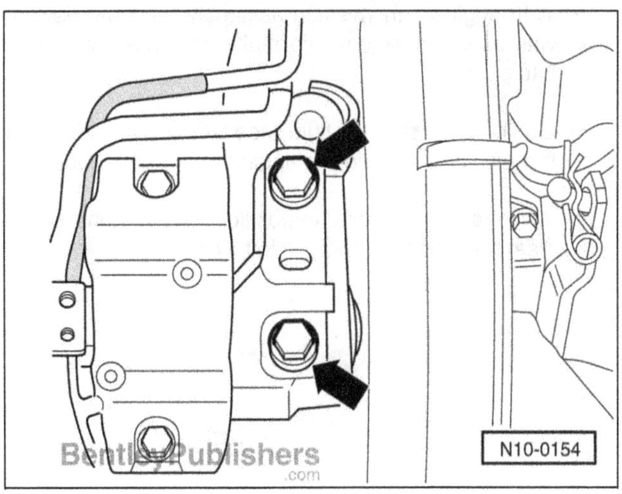

◄ Remove engine mount bolts (**arrows**).

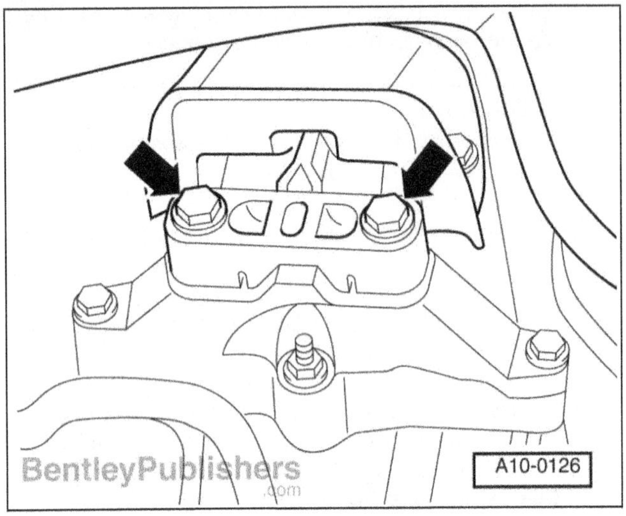

◄ Remove transmission mount bolts (**arrows**).

– Carefully lower engine and transmission assembly out of vehicle using caution to guide assembly past power steering and refrigerant lines.

> ***CAUTION —***
> *Before removing the engine and transmission assembly, be sure vehicle is properly supported in the rear. Removal of such a large amount of weight from the front can cause an improperly supported vehicle to pivot on the lift and fall to the rear.*

NOTE —
Carefully guide the engine/transmission assembly during lowering process to prevent damage.

Engine/transmission assembly, installing (4-cylinder)

> ***WARNING —***
> - *When working in the engine compartment, due to the cramped, tight conditions, pay attention to the following points*
> - *Route all lines (e.g. for fuel, hydraulics, EVAP system, coolant, refrigerant, brake fluid and vacuum, etc.) as well as electrical wiring so that the original positions are maintained. Ensure sufficient clearance to all moving or hot components.*
> - *Make sure there is sufficient clearance to all moving or hot components.*
> - *Be sure to replace all tie wraps cut during removal and install new ones to prevent lines and wiring harnesses from chaffing or contacting engine parts.*

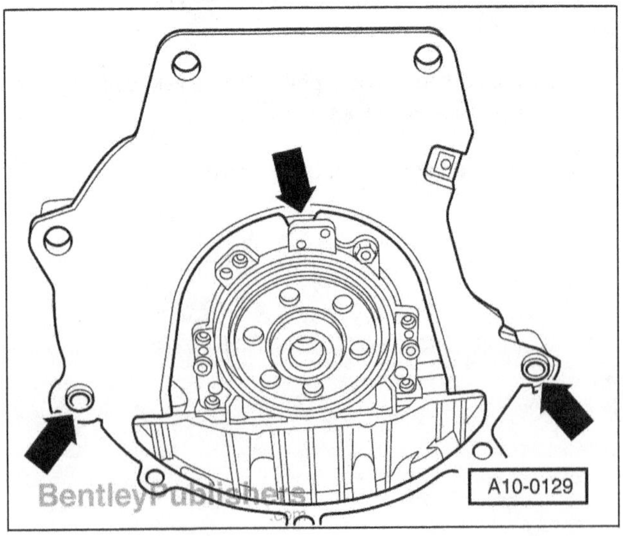

◄ Install new dowel sleeves (**lower arrows**) for centering engine/transmission on cylinder block.

– With engine/transmission assembly on same device as was used for removal, carefully raise assembly up and into position.

– Guide power steering fluid lines around transmission and the A/C lines around the engine to avoid damage.

– While raising engine into position, ensure sufficient clearance for drive axle on left side.

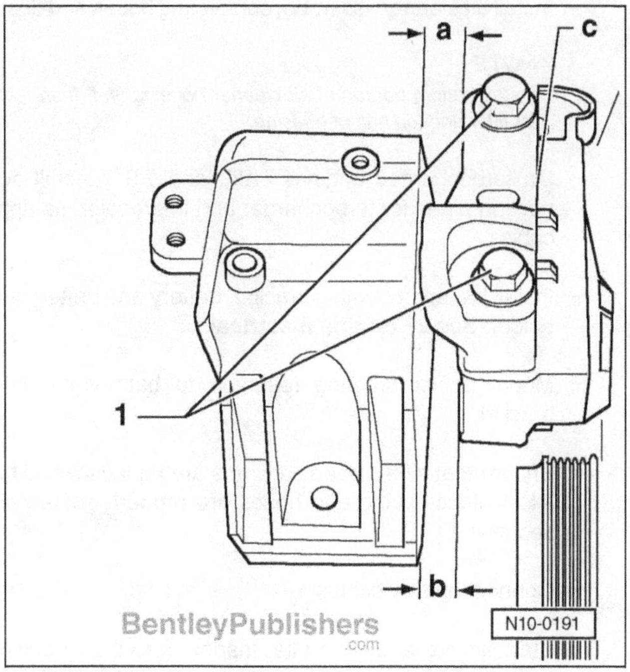

When engine/transmission assembly is in position, adjust so that mounts are aligned as specified.

Engine mount alignment

- Dimension **a** . . . . 14 mm (0.55 in.)
- Dimension **b** . . . . 10 mm (0.40 in.)

NOTE —

- *Do not stress the engine and transmission mounts during installation. Make sure they are aligned parallel to each other.*
- *Bolt heads (**1**) must be flush with edge (**c**).*

– Install engine, transmission and pendulum mounts with new bolts where indicated and torque to specification, see **Engine and transmission mounts (4-cylinder)**.

– Install front exhaust pipe with catalytic converter, see **26 Exhaust System/Emission Controls**.

– Attach left side drive shaft to transmission and install right side drive axle and protective cover, see **40 Front Suspension and Drive Axles**.

– Install oil pan guard, as applicable.

– Install A/C compressor and secure A/C lines, see **87 Air Conditioning**.

– Install power steering pump and lines, see **48 Steering**.

– Connect shift linkage.

– Install secondary air injection pump, and auxiliary fan where equipped.

– Note running direction and install ribbed V-belt, see **03 Maintenance**.

– Reinstall fuel supply and return lines.

- On AEG engines, attach accelerator cable and adjust for full throttle. Have an assistant depress throttle pedal to floor while adjusting throttle cable at manifold to insure that throttle plate is wide open.

– Install vacuum and breather hoses.

– Install heater and coolant hoses.

– Refill cooling system with appropriate ratio of distilled water and coolant and check for leaks, see **19 Engine – Cooling System**.

– Install all remaining electrical connectors and check for proper routing of wires.

– Install air cleaner assembly, connecting hoses and ducts.

NOTE —
Install bottom portion of air cleaner housing first, then air filter and top half of housing.

– On turbocharged engines (1.8L and 1.9L) install connecting pipes for turbocharger and intercooler, as applicable.

– Install battery mounting bracket, battery and battery fuse holder. See **27 Engine Electrical**.

– Mount power steering reservoir to battery mounting bracket.

– Ensure that all removed tie wraps are replaced, and that all electrical wiring and hoses are properly routed and secured.

– Connect battery cables.

– Start engine and let it idle. Inspect for oil and coolant leaks. Check for smooth operation of shifter and clutch as required. Road test as required.

– Shut off engine and install lower sound absorber panel (belly pan) and upper engine cover.

Tightening torques

• Engine to transmission bolts	
M10 bolts	
1.8L	45 Nm (33 ft-lb)
2.0L	60 Nm (44 ft-lb)
M12 bolts	
1.9L	80 Nm (59 ft-lb)
• Drive axle to transmission flange	40 Nm (30 ft-lb)
• Transmission to oil pan connecting bolts (automatic transmission only)	45 Nm (33 ft-lb)
• Triple rotor/CV joint to transmission	40 Nm (30 ft-lb)
• Torque converter to drive plate	60 Nm (44 ft-lb)
• Bolts & nuts, not specifically listed	
M6	10 Nm (7 ft-lb)
M7	15 Nm (10 ft-lb)
M8	25 Nm (18 ft-lb)
M10	40 Nm (30 ft-lb)
M12	60 Nm (44 ft-lb)

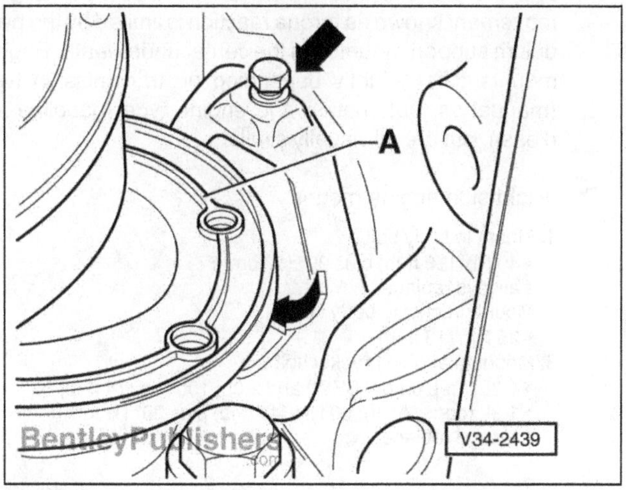

Engine/transmission, separating (4-cylinder)

- Support engine with device used to remove engine/transmission and rest transmission on work bench.
- Remove small cover plate (**A**) from behind right side axle flange if equipped.
- Remove engine oil pan bolts from transmission.
- Remove nuts from torque converter from engine side of the transmission, if equipped.
- Remove bolts holding engine and transmission together.
- Slide transmission off of engine.

NOTE —
On automatic transmissions, the torque converter should come off with transmission. Secure the torque converter to the transmission to prevent damage.

CAUTION —
- *On cars with automatic transmissions, be sure the drive plate separates cleanly from the torque converter without pulling the torque converter off of its support. Once the engine and transmission are separated, install a suitable bar across the open bell bellhousing to keep the torque converter from falling out.*
- *On cars with manual transmissions, be sure that the weight of the transmission or the engine is never supported on the transmission mainshaft. Clutch or transmission damage could result.*
- *When reattaching the transmission to the engine, check that the guide sleeves are installed into the engine block. Replace them if they are missing.*
- *Clutch or drive plate damage may result if guide sleeves are not installed.*

- Attaching engine to transmission is the reverse of separating. Observe the following points:
  - Lightly lubricate manual transmission mainshaft with molybdenum disulfide (MoS2) grease.
  - Inspect clutch components, see **30 Clutch**.
  - Install torque converter to stator support on automatic transmission before attaching.

Engine and transmission mounts (4-cylinder)

The engine and transmission assembly is suspended on its rotational axis by the engine mount on the right side and the transmission mount on the left side. Engine movement known as torque reaction is limited by the pendulum support mounted in the center underneath. Engine mounts differ slightly depending on transmission type (manual vs. automatic); and engine type (gasoline vs. diesel), but are all visually similar.

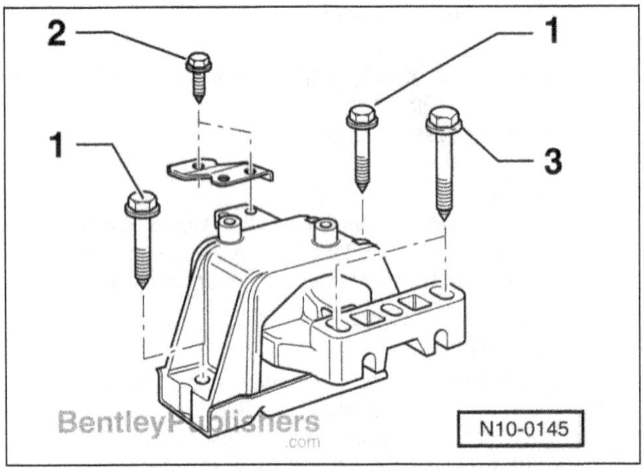

◄ Right side engine mount

1. Mount to body bolt
   - 40 Nm (30 ft-lb) plus 90° (¼ turn)
   - always replace
2. Mount bracket to body bolt
   - 25 Nm (18 ft-lb)
3. Mount to engine bracket bolt
   - 1.8L, 1.9L (code: BEW) and 2.0L: 100 Nm (74 ft-lb)
   - 1.9L (code: ALH): 60 Nm (44 ft-lb) plus 90° (¼ turn)
   - always replace

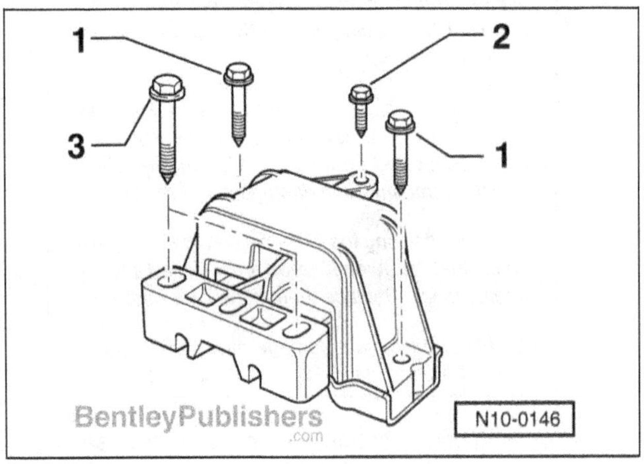

◄ Transmission mount, left side

1. Mount to body bolt
   - 40 Nm (30 ft-lb) plus 90° (¼ turn)
   - always replace
2. Mount bracket to body bolt
   - 25 Nm (18 ft-lb)
3. Mount to transmission bracket bolt
   - 1.8L, 1.9L (code: BEW) and 2.0L: 100 Nm (74 ft-lb)
   - 1.9L (code: ALH): 60 Nm (44 ft-lb) plus 90° (¼ turn)
   - always replace

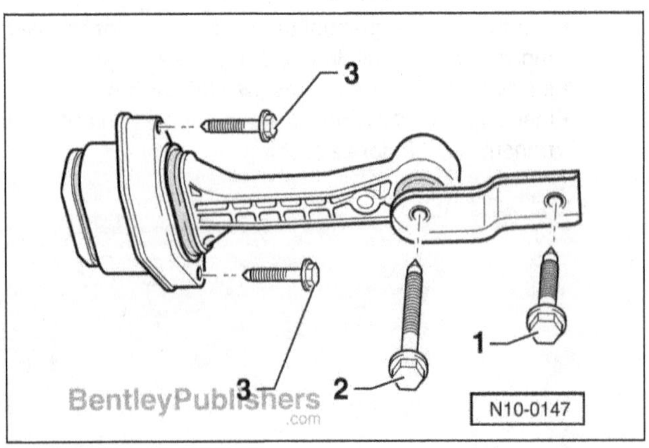

◄ Pendulum support, lower center

1. Pendulum support to transmission bolt
   - 40 Nm (30 ft-lb) plus 90° (¼ turn)
   - always replace
2. Pendulum support to transmission bracket bolt
   - 40 Nm (30 ft-lb) plus 90° (¼ turn)
   - always replace
3. Pendulum support to subframe bolt
   - 20 Nm (15 ft-lb) plus 90° (¼ turn)
   - always replace

ENGINE, REMOVING AND INSTALLING (5-CYLINDER ENGINE)

Engine/transmission assembly, removing (5-cylinder)

NOTE —
Removal of the 5-cylinder engine requires removal of the front subframe and a number of special tools. Be sure to read through entire procedure before starting job.

– Ensure ignition is switched off. Disconnect battery negative (–) terminal and then positive (+) terminal.

NOTE —
Be sure to have the anti-theft radio code on hand before disconnecting the battery.

– Remove engine cover. See **03 Maintenance**.

– Remove battery and the mounting bracket, see **27 Engine Electrical**.

– Remove air cleaner assembly and connecting ducts.

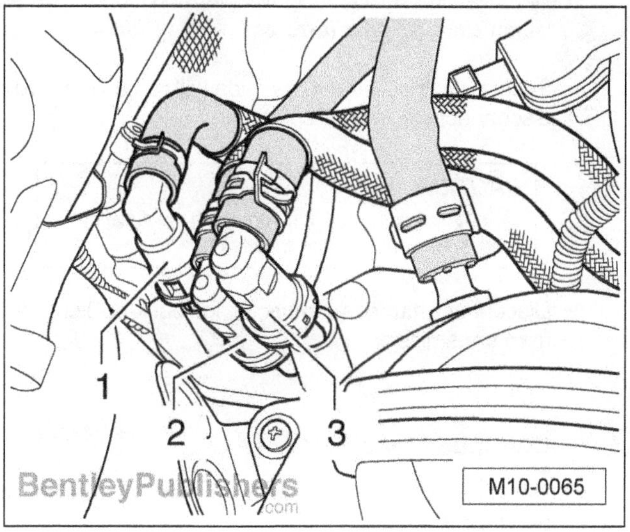

◄ Disconnect fuel supply and return lines as follows:

- Through model year 2009: Disconnect fuel supply line (**1**).
- All vehicles: Disconnect fuel vent line (**2**).
- All vehicles: Disconnect fuel return line (**3**).

> ***WARNING —***
> *Fuel will be expelled when disconnecting fuel hoses. Wrap a cloth around the fuel line fittings before disconnecting them. Do not smoke or work near heaters or other fire hazards. Have a fire extinguisher handy.*

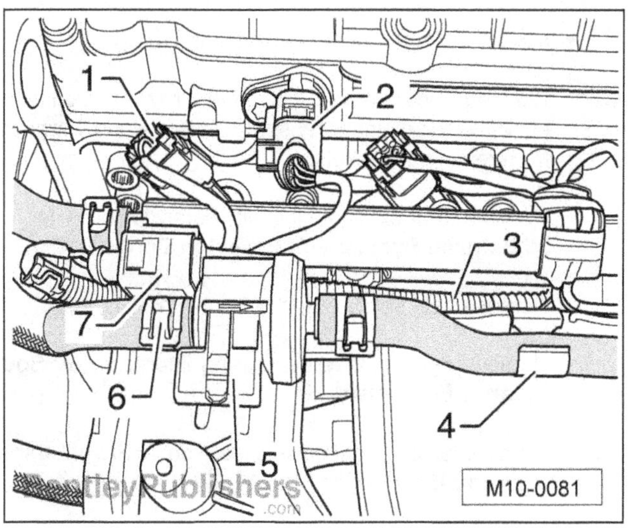

◄ Disconnect fuel injector harness connectors (**1**).

– Disconnect harness connectors (**2** and **7**).

– Remove wiring harness (**3**) from retaining strap.

– Open clamp (**6**) and disconnect hose.

– Pull all clamps (**4**) and retaining ring (**5**) from locking mechanism.

– Seal all fuel lines so that fuel system is not contaminated by dirt, leaking fluids, etc. Seal off any open ports in manifold, as applicable.

– Disconnect harness connectors from the following:
  - Ignition coils
  - Camshaft adjustment valve 1 (N205)
  - Intake manifold
  - Throttle valve control module (J338)
  - Secondary Air Injection (AIR) solenoid valve (N112)
  - Knock sensors
  - Engine Coolant Temperature (ECT) sensor (G62)

– Remove Ground (GND) wire at timing chain guard cover and expose wiring harness.

– Disconnect shift linkage.
  - Manual transmission: Disconnect cables from selector mechanism and unbolt support bracket, see **34 Manual Transmission**.
  - DSG transmission: Disconnect selector cable at transmission, see **34b DSG Transmission**.
  - Automatic transmission: Disconnect selector cable at selector lever, see **37 Automatic Transmission**.

Vehicles with manual transmission

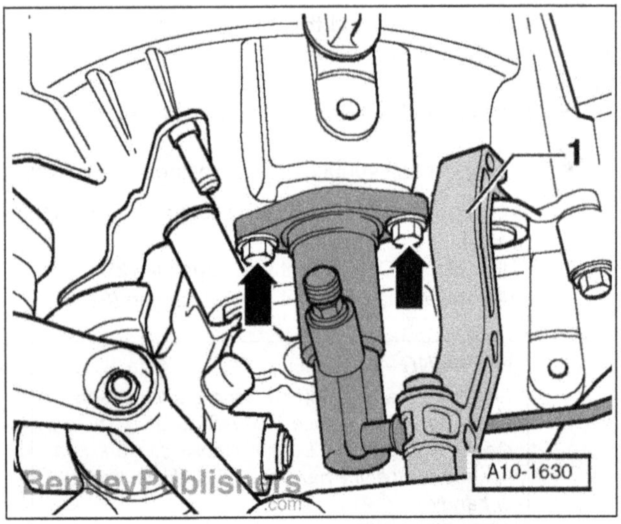

◄ Remove bracket (**1**) if applicable, and unbolt hydraulic clutch slave cylinder (**arrows**).

– Secure slave cylinder to the side with a tie wrap. Do not disconnect the hydraulic line.

> ***CAUTION —***
> *Do not depress the clutch pedal with the slave cylinder removed, damage will result.*

– Disconnect harness connector for back-up light switch from transmission.

Continued for all vehicles

– Remove wiring harness from bracket on transmission and from starter.

– Remove vacuum hose from vacuum pump.

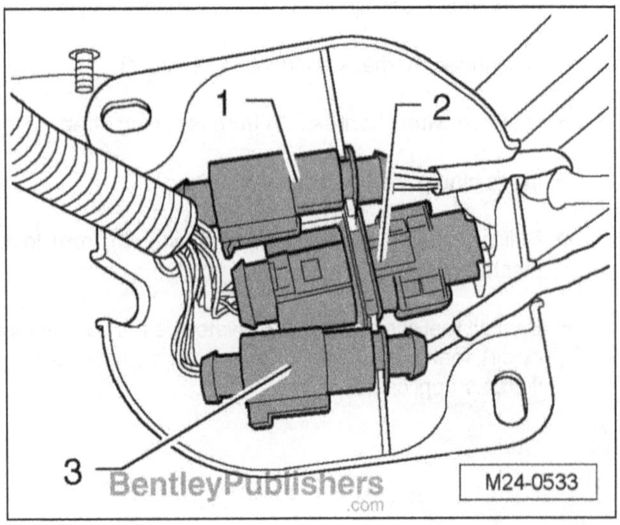

◄ Working under vehicle, disconnect harness connectors (**1, 2, 3**) for oxygen sensors. Unclip wiring harnesses from retaining brackets.

– Remove front exhaust pipe with catalytic converter. See **26 Exhaust System and Emission Controls**.

– Remove heat shield at top of front exhaust pipe.

– Remove right front wheelhousing liner. See **66 Body–Exterior Equipment**.

– Remove left and right drive axles. See **40 Front Suspension and Drive Axles**.

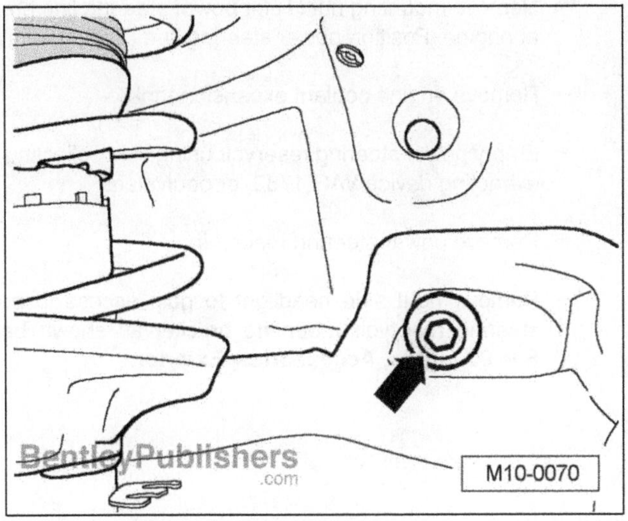

< Remove rear mounting bolt (**arrow**) from engine bracket through wheel housing.

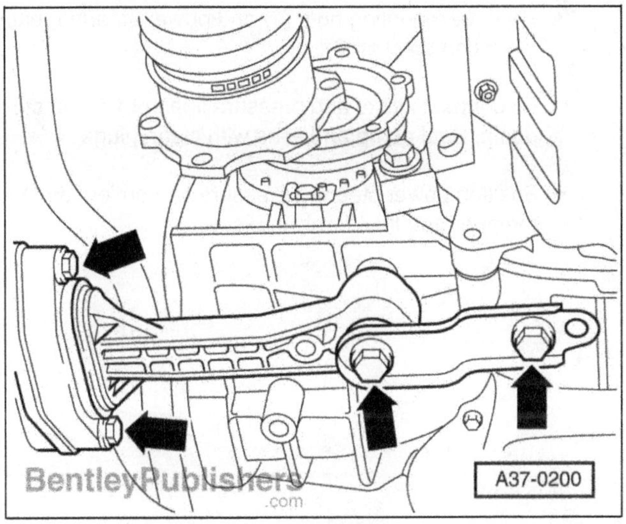

< Remove mounting bolts (**arrows**) and pendulum support.

– Remove subframe. See **40 Front Suspension and Drive Axles.**

– Drain engine coolant. See **19 Engine–Cooling System.**

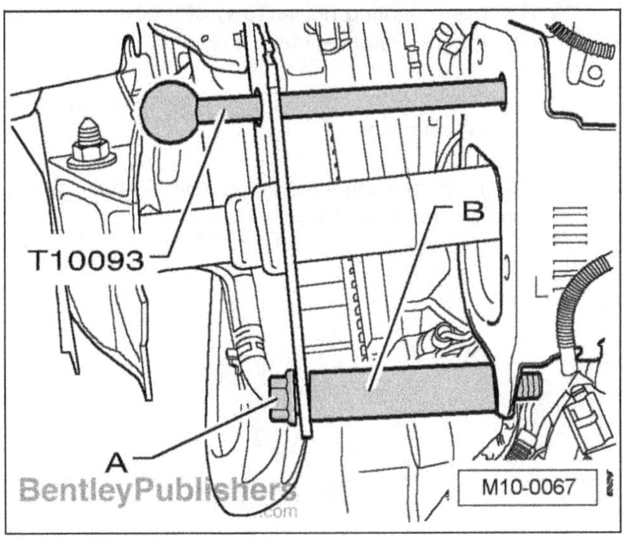

< Move front lock carrier to service position. See **50 Body–Front**. Use **T10093**, or equivalent, to slide lock carrier forward.

– Support lock carrier using the following:
  - Bolt (**A**), M8 x 100mm
  - Sleeve (**B**) 80mm long.

– Disconnect connectors from generator, engine speed sensor and oil pressure switch. Position wiring harnesses out of way.

– Disconnect both connectors at front of transmission.

– Disconnect power steering line and remove bracket for wiring harness mounted on transmission.

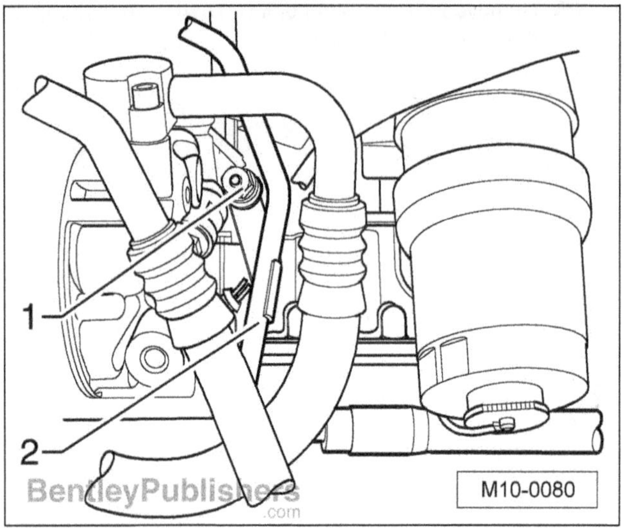

◀ Remove mounting nut (**1**) for power steering line bracket at engine. Position power steering line (**2**) out of way.

- Remove engine coolant expansion tank.
- Empty power steering reservoir using fluid collecting and extracting device VAG 1782, or equivalent.
- Remove power steering reservoir.
- Remove right side headlight to gain access to power steering reservoir mounting bracket as shown below. See **92 Lights, Accessories–Exterior**.

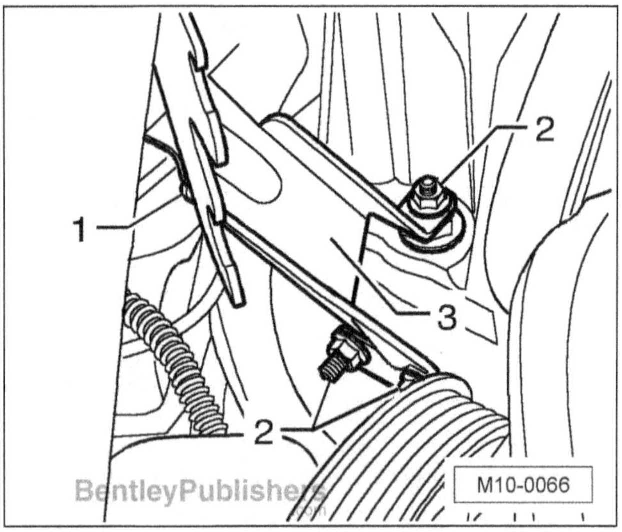

◀ Remove mounting nuts (**2**) and power steering reservoir mounting bracket (**3**).

- Disconnect intake and pressure line from power steering pump. Seal pump and lines with clean plugs.
- Position power steering pressure line on left longitudinal member and tie it firmly in place.

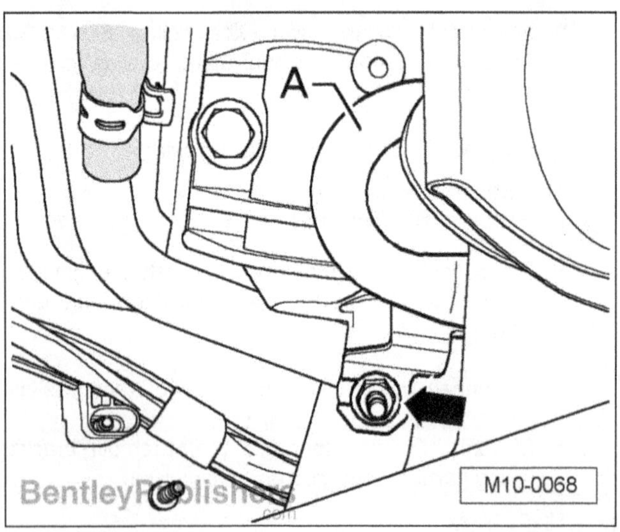

◀ Remove mounting nut (**arrow**) of intake line (**A**) at engine mount. Remove intake line.

- Disconnect both coolant hoses from heater core (quick-release couplings).
- Slightly loosen upper mounting bolts at engine and transmission (less than 1 turn).

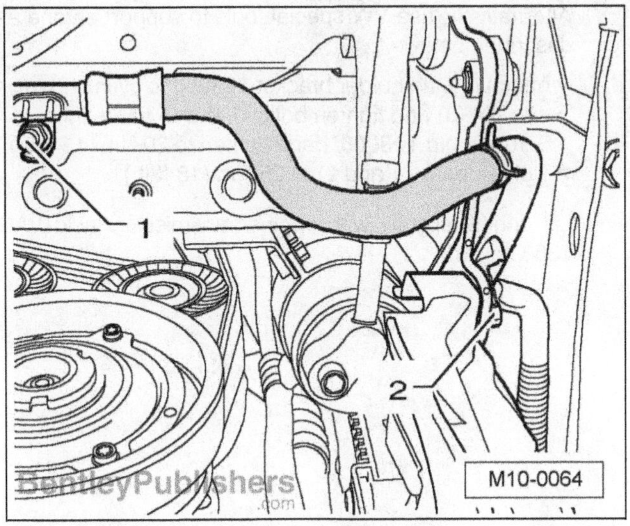

– Disconnect A/C refrigerant line from right longitudinal member at top.

> ***CAUTION —***
> *Do not disconnect A/C lines from A/C components. Doing this will cause the A/C refrigerant to discharge.*

< Remove mounting nuts (**1** and **2**). Disconnect refrigerant line from right longitudinal member at bottom.

– Remove ribbed belt. See **03 Maintenance**.

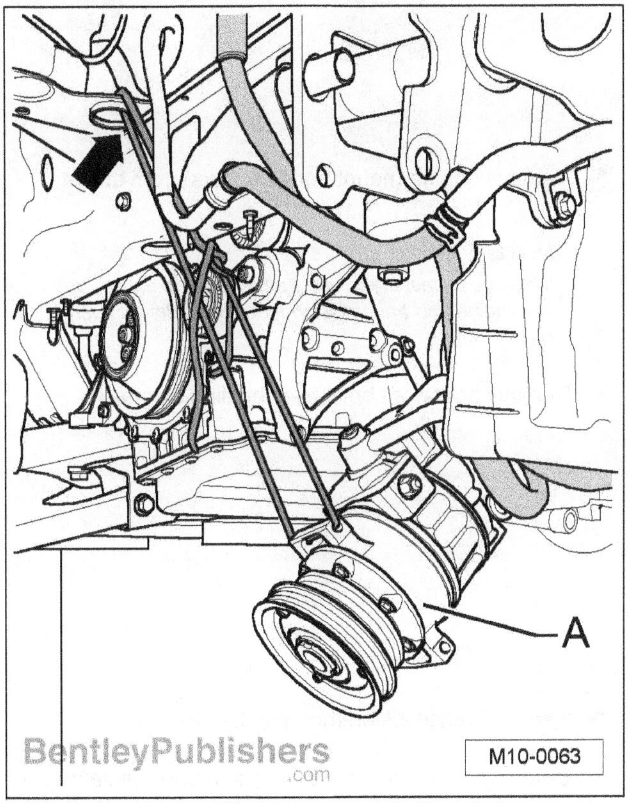

– Remove A/C compressor. See **87 Air Conditioning**.

< Attach A/C compressor to lower section of body with a suitable support (**arrow**) so that refrigerant lines are not stressed.

NOTE —
To prevent damage to the condenser and to the refrigerant lines/hoses, ensure that the lines and hoses are not stretched, kinked or bent.

– Disconnect/clamp off all other connections, coolant, vacuum and intake hoses from engine and transmission.

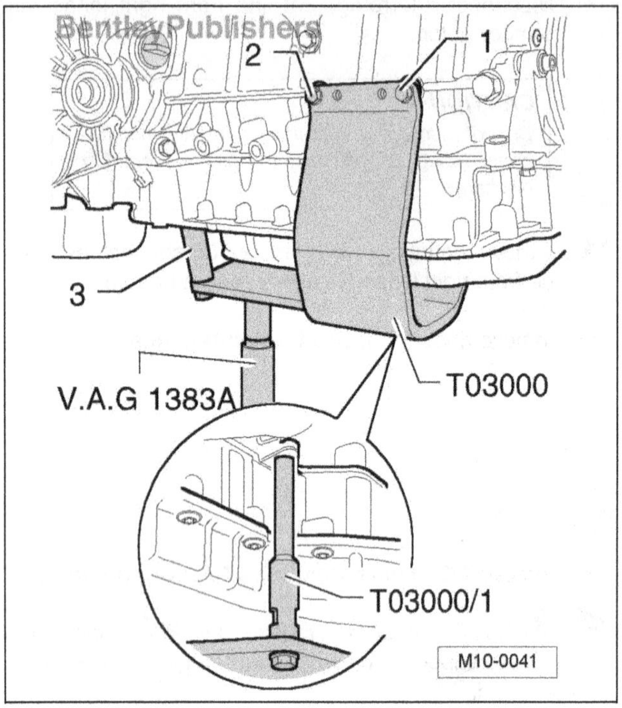

Install a suitable engine supporting device from below. Alternatively, use VW special tools to support engine as described below:

- Attach engine holder bracket **T03000** to cylinder block with pin (**3**) and tighten bolts (**1** and **2**) hand-tight.
- Screw in pin **T03000/1** and tighten to 20 Nm (15 ft-lb).
- Tighten bolts (**1** and **2**) to 25 Nm (18 ft-lb).

– Lift engine slightly with engine/transmission jack (VAG 1383 shown).

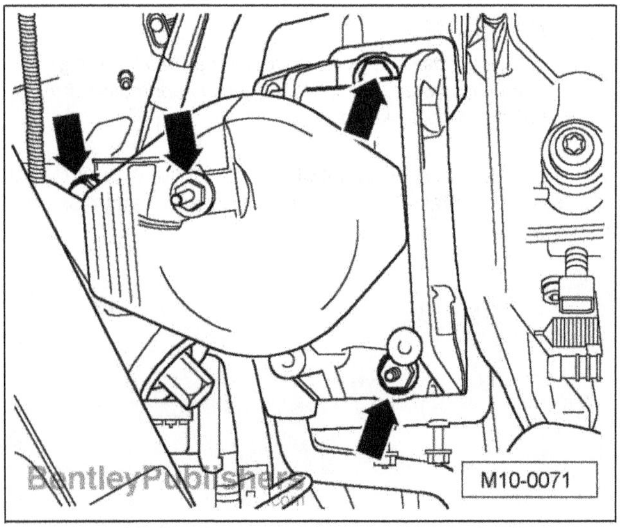

Remove engine mount bolts (**arrows**) from above.

NOTE —

It may be necessary to use a step ladder to access upper engine and transmission mounting bolts with vehicle on a lift.

– Remove both front bolts from holding fixture.

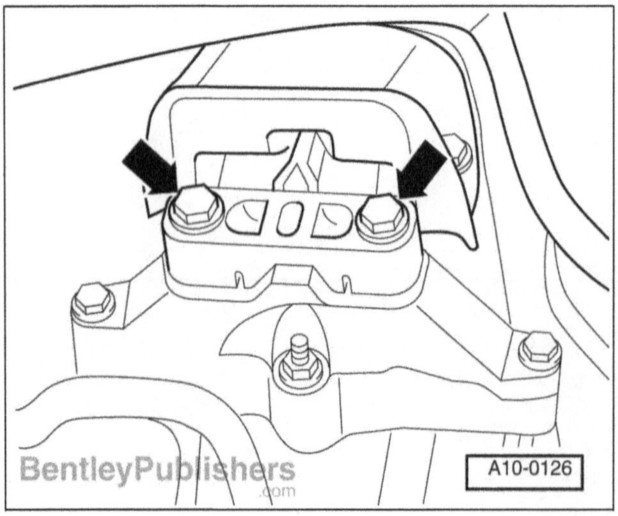

Remove transmission mount bolts (**arrows**).

– Carefully lower engine and transmission assembly forward and down out of vehicle. Use caution to guide assembly past power steering gear and A/C lines.

CAUTION —

Before removing the engine and transmission assembly, be sure vehicle is properly supported in the rear. Removal of such a large amount of weight from the front can cause an improperly supported vehicle to pivot on the lift and fall to the rear.

Engine/transmission assembly, installing (5-cylinder)

CAUTION —

- *When working in the engine compartment, due to the cramped, tight conditions, pay attention to the following points*
- *Route all lines (e.g. for fuel, hydraulics, EVAP system, coolant, refrigerant, brake fluid and vacuum, etc.) as well as electrical wiring so that the original positions are maintained. Ensure sufficient clearance to all moving or hot components.*
- *Be sure to replace all tie wraps cut during removal and install new ones to prevent lines and wiring harnesses from chaffing or contacting engine parts.*

– With engine/transmission assembly on same device as was used for removal, carefully raise assembly up and into position.

– Guide power steering fluid lines around transmission and the A/C lines around the engine to avoid damage.

– While raising engine into position, ensure sufficient clearance for drive axle on left side.

– Install engine, transmission and pendulum mounts with new bolts where indicated and torque to specification, see **Engine and Transmission Mounts (5-cylinder)**.

– Install remaining components in reverse order of removal.

Tightening torques

- Engine to transmission bolts
  - M12 bolts . . . . 80 Nm (59 ft-lb)
  - M10 bolts . . . . 40 Nm (30 ft-lb)
- Drive axle to transmission flange . . . 40 Nm (30 ft-lb)
- Transmission to oil pan connecting bolts (automatic transmission only) . . . . . 45 Nm (33 ft-lb)
- Triple rotor/CV joint to transmission . 40 Nm (30 ft-lb)
- Torque converter to drive plate. . . . . 60 Nm (44 ft-lb)
- Bolts & nuts, not specifically listed
  - M6 . . . . 10 Nm (7 ft-lb)
  - M7 . . . . 15 Nm (10 ft-lb)
  - M8 . . . . 25 Nm (18 ft-lb)
  - M10 . . . . 40 Nm (30 ft-lb)
  - M12 . . . . 60 Nm (44 ft-lb)

Engine and transmission, mounts (5-cylinder)

The engine and transmission assembly is suspended on its rotational axis by the engine mount on the right side and the transmission mount on the left side. Engine movement known as torque reaction is limited by the pendulum support mounted in the center underneath. Engine mounts may differ slightly depending on transmission type (manual vs. automatic).

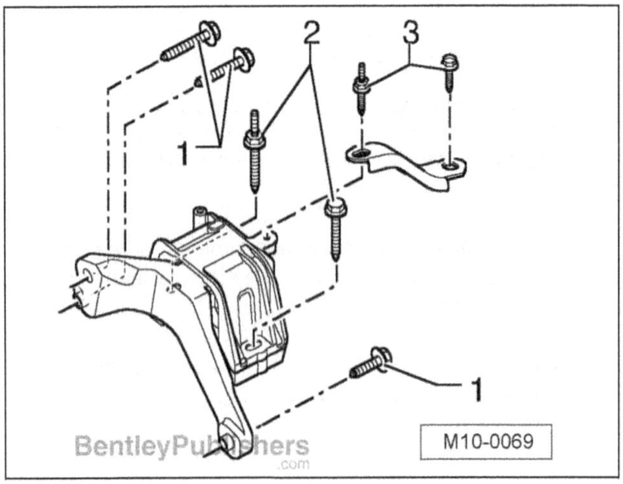

◀ Right side engine mount

1. Mount to engine bolt
   - 40 Nm (30 ft-lb) plus 90° (¼ turn)
   - always replace
2. Mount bracket to body bolt
   - 40 Nm (30 ft-lb) plus 90° (¼ turn)
   - always replace
3. Support to mount/body bolt
   - 20 Nm (15 ft-lb) plus 90° (¼ turn)
   - always replace

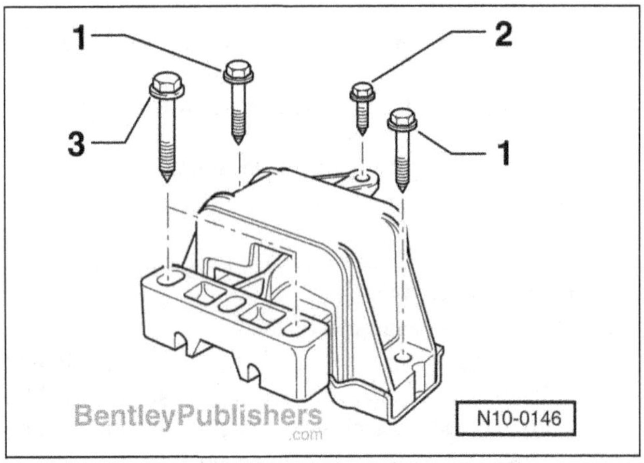

◀ Transmission mount, left side

1. Mount to body bolt
   - 40 Nm (30 ft-lb) plus 90° (¼ turn)
   - always replace
2. Mount bracket to body bolt
   - 25 Nm (18 ft-lb)
3. Mount to transmission bracket bolt
   - 60 Nm (44 ft-lb) plus 45° (⅛ turn)
   - always replace

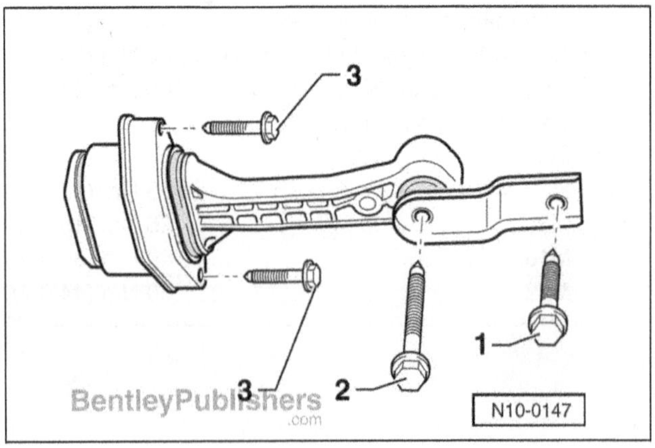

◀ Pendulum support, lower center

1. Pendulum support to transmission bolt
   - 40 Nm (30 ft-lb) plus 90° (¼ turn)
   - always replace
2. Pendulum support to transmission bracket bolt
   - 40 Nm (30 ft-lb) plus 90° (¼ turn)
   - always replace
3. Pendulum support to subframe bolt
   - 20 Nm (15 ft-lb) plus 90° (¼ turn)
   - always replace

13 Crankshaft and Cylinder Block

GENERAL

This repair group provides the special reconditioning information necessary to repair the Volkswagen short block. The information contained here is intended to be used as a reconditioning guide for the professional or experienced automotive technician. Many of the operations and specifications listed require precision measuring equipment and special tools.

Engine	Engine code
1.8L 4-cylinder turbo gasoline	APH, AWV, AWP, BKF, BNU
1.9L 4-cylinder TDI diesel	ALH, BEW
2.0L 4-cylinder gasoline	AEG, AVH, AZG, BDC, BEV, BGD
2.5L 5-cylinder gasoline	BPR, BPS

CYLINDER BLOCK OIL SEALS (4-CYLINDER)

The front crankshaft oil seal can be replaced with the engine installed. Replacement of the rear crankshaft oil seal requires that the engine be separated from the transmission.

NOTE —
In some instances, individual seals may not be available separately. See an authorized Volkswagen parts dealer, or an aftermarket parts specialist, for the latest in parts information.

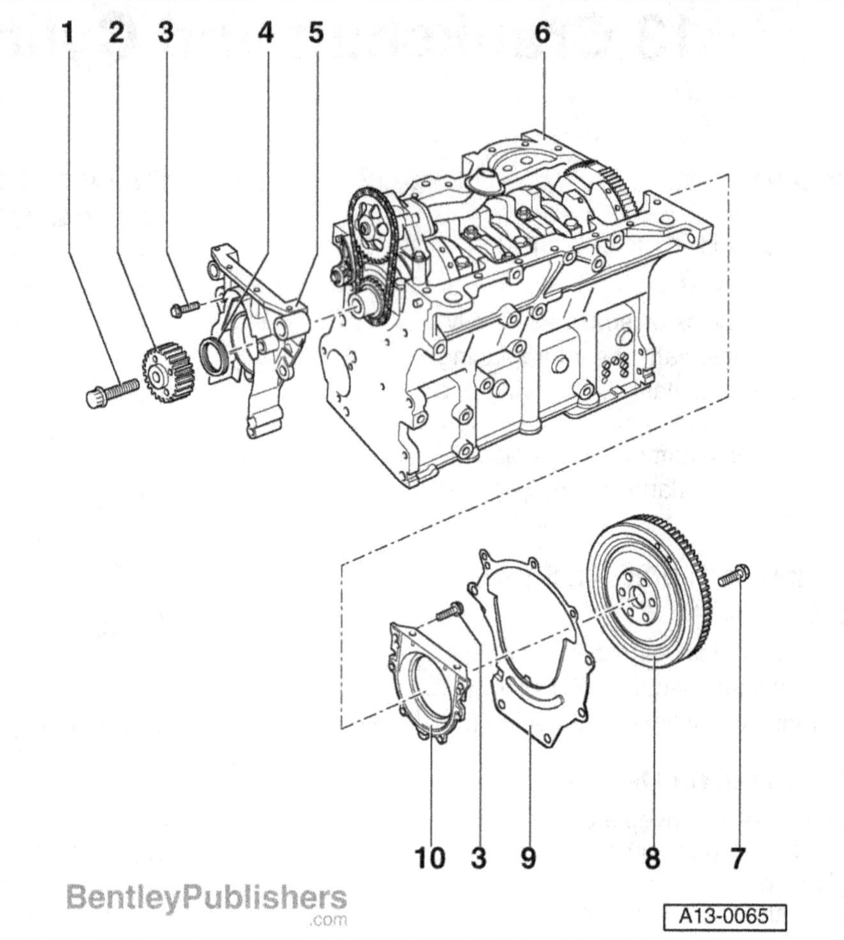

Cylinder block oil seals, assembly (4-cylinder)

1. **Bolt**
   - Tighten to:
     1.8L / 2.0L: 90 Nm (66 ft-lb)
     + ¼ turn (90°)
     1.9L: 120 Nm (88 ft-lb)
   - Always replace
2. **Crankshaft toothed belt sprocket**
3. **Bolt**
   - Tighten to 15 Nm (11 ft-lb)
4. **Front crankshaft oil seal**
5. **Front oil seal flange**
6. **Engine block**
   - Shown upside down
7. **Bolt**
   - Tighten to 60 Nm (44 ft-lb)
     + ¼ turn (90°)
   - Always replace
8. **Flywheel/driveplate**
   - Flywheel: remove and install with 3067 counterholder
9. **Intermediate plate**
   - Must be located on dowel sleeves
   - Do not damage or bend when assembling
10. **Rear oil seal flange**
    - May need to replace as complete unit (w/integral oil seal)
    - Lightly oil lip on seal
    - Install new sealing flange with guide sleeve
    - Guide sleeve protects seal during installation

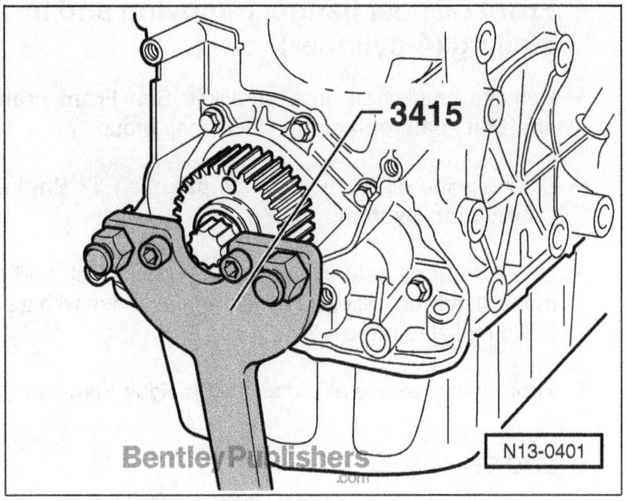

Front crankshaft oil seal, replacing

– Remove ribbed V-belt, see **03 Maintenance**.

– Remove timing belt, see **15a Cylinder Head and Valvetrain (1.8L Engine)**, **15c Cylinder Head and Valvetrain (2.0L Engine)** or **23 Fuel Injection (Diesel)**.

< Hold crankshaft stationary with a suitable counterholder and loosen crankshaft sprocket (hub) center bolt. Remove bolt and sprocket.

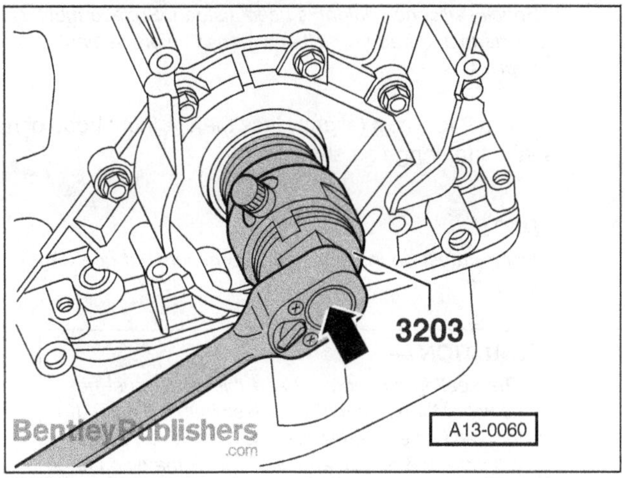

< Reinstall removed bolt into crankshaft to prevent damage to end of crankshaft when using seal extractor. Remove oil seal from flange using an appropriate seal extractor and wrench (**arrow**) or by carefully prying it out.

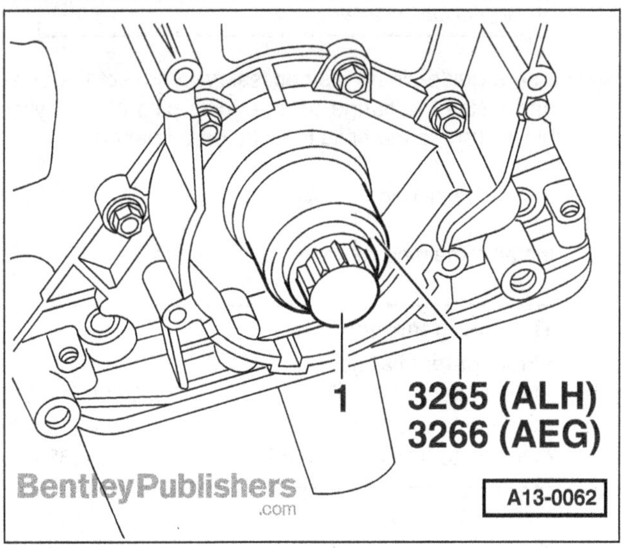

< Install new seal, lubricated with clean engine oil, with closed side facing out. Use a guide sleeve (**1**) to protect seal from sharp edges of crankshaft as necessary. Carefully press seal into place until it is fully seated.

– Align locating key on sprocket with cutout on end of crankshaft and install sprocket onto the crankshaft with a new bolt. Hold the crankshaft with the counterholder and torque to specification.

– Install remaining removed components.

> ***CAUTION —***
> *Always replace the crankshaft sprocket bolt. It is a stretch bolt designed to be used only once.*

Tightening torques

- 1.8L / 2.0L Crankshaft sprocket bolt (stretch bolt - always replace)
  - stage I . . . . 90 Nm (66 ft-lb)
  - stage II . . . . additional ¼ turn (90°)
- 1.9L Crankshaft sprocket bolt (stretch bolt - always replace)
  - stage I . . . . 120 Nm (88 ft-lb)
  - stage II . . . . additional ¼ turn (90°)

Front oil seal flange, removing and installing (4-cylinder)

– Remove crankshaft drive sprocket. See **Front crankshaft oil seal, replacing** in this repair group.

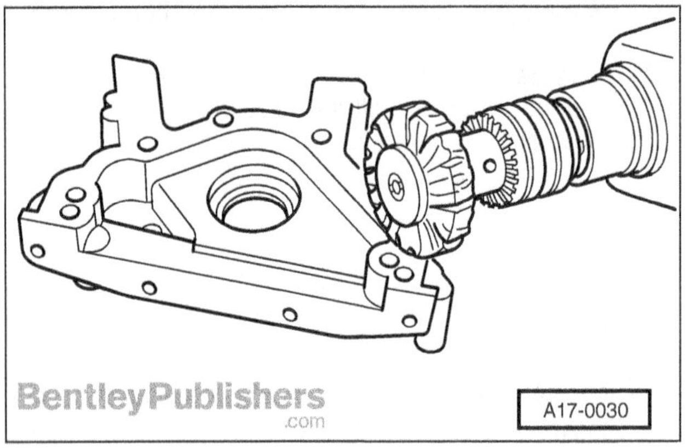

– Drain engine oil and remove oil pan. See **17 Engine–Lubrication System**.

– Unbolt front oil seal flange from cylinder block and remove. It may be necessary to lightly tap flange with a soft faced mallet to remove it.

< Thoroughly remove all old sealant residue from flange.

NOTE—
The front oil seal flange does not use a paper gasket. Special silicone sealant is used instead. Surface must be clean and free from oil and grease before sealant is applied.

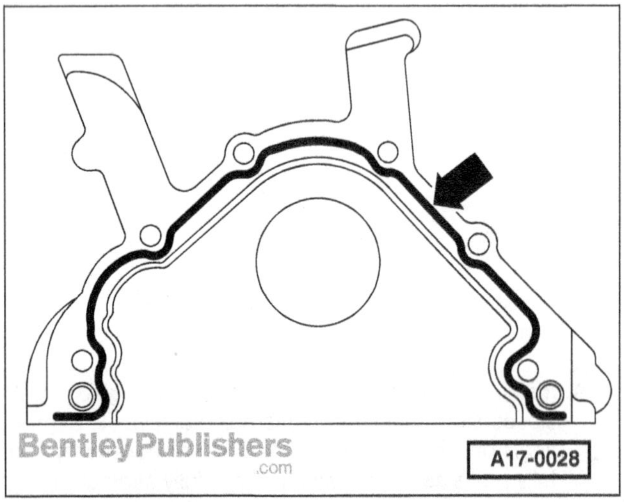

< Apply a 2 to 3 mm (slightly less than ⅛ inch) bead of new sealant to flange as shown.

NOTE—
Flange must be installed within 5 minutes of applying sealant.

> ***CAUTION—***
> *The sealing compound bead thickness must not be wider than 3 mm (slightly less than 1/8 inch). If this width is exceeded, excess sealing compound will enter the oil pan and could block the oil pump pick-up tube strainer.*

– Use a guide sleeve as necessary to protect the oil seal and install the flange on the guide pins of the cylinder block. Torque the bolts in a staggered pattern.

– Install the oil pan.

– Install remaining removed components.

Tightening torques
- Front or rear flange to cylinder block (M7) . . . . 15 Nm (11 ft-lb)
- Oil pan to cylinder block (M7) . . . . . . 15 Nm (11 ft-lb)
- Oil pan to transmission (M10) . . . . . 45 Nm (33 ft-lb)

Rear crankshaft oil seal, replacing

NOTE—
The rear crankshaft oil seal may not be available separately from the rear crankshaft oil seal flange. See an authorized Volkswagen parts dealer, or an aftermarket parts specialist for the latest in parts information.

– If engine is still in car, remove transmission as described in **34 Manual Transmission** or **37 Automatic Transmission**, as applicable.

– Separate engine from transmission as described in **10 Engine–Removing and Installing**.

– Remove flywheel or driveplate as described later.

– Remove seal from flange by carefully prying it out.

– Install new seal, lubricated with clean engine oil, with closed side facing out using a suitable seal installation tool.

– Install remaining removed components.

Rear oil seal flange, removing and installing

– If engine is still in car, remove transmission as described in **34 Manual Transmission** or **37 Automatic Transmission**, as applicable.

– If engine and transmission are out of car, separate engine from transmission as described in **10 Engine–Removing and Installing**.

– Remove flywheel or drive plate as described later.

– Drain engine oil and remove oil pan and guard, if applicable. See **17 Engine–Lubrication System**.

– Unbolt rear oil seal flange from cylinder block and remove. It may be necessary to lightly tap the flange with a soft faced mallet to remove it.

– The rear oil seal flange does not use a paper gasket. Special silicone sealant is used instead.

– Apply a 2 to 3 mm (slightly less than ⅛ inch) bead of new sealant to the flange.

NOTE —
Flange must be installed within 5 minutes of applying sealant.

CAUTION —
The sealing compound bead thickness must not be wider than 3 mm (slightly less than ⅛ inch). If this width is exceeded, excess sealing compound will enter the oil pan and could block the oil pump pick-up tube strainer.

– Use a guide sleeve as necessary to protect the oil seal and install the flange on the guide pins of the cylinder block. Torque the bolts in a staggered pattern.

– Install the oil pan and guard as applicable. See **17 Engine–Lubrication System**.

– Install remaining removed components.

CYLINDER BLOCK OIL SEALS (5-CYLINDER)

The front crankshaft oil seal can be replaced with the engine installed. Replacement of the rear crankshaft oil seal requires that the engine be separated from the transmission.

NOTE—
In some instances, individual seals may not be available separately. See an authorized Volkswagen parts dealer, or an aftermarket parts specialist, for the latest in parts information.

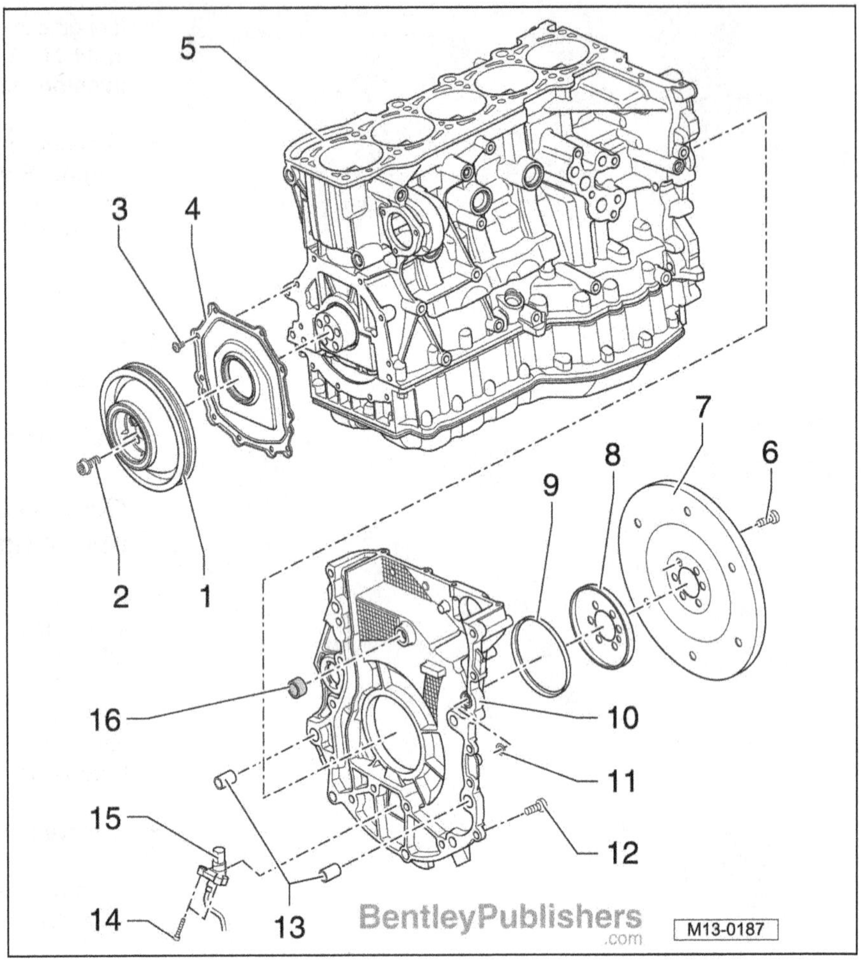

Cylinder block oil seals, assembly

1. **Belt pulley/vibration damper**
2. **Bolt**
   - Tighten to: 50 Nm (37 ft-lb) + ¼ turn (90°)
3. **Bolt**
   - Tighten to 10 Nm (7 ft-lb)
4. **Sealing flange, belt pulley side**
   - With integrated seal
5. **Cylinder block**
6. **Bolt**
   - Tighten to 60 Nm (44 ft-lb) + ¼ turn (90°)
   - Always replace
7. **Driveplate/flywheel**
   - Secure crankshaft using T40069 locking pin before removing
8. **Sensor wheel**
   - For engine speed (RPM) sensor (G28)
   - WIth locating fixture
9. **Seal, transmission side**
10. **Timing chain housing cover**
11. **O-ring**
    - Always replace
12. **Bolt**
    - Tighten to 25 Nm (18 ft-lb)
13. **Dowel sleeves**
14. **Bolt**
    - Tighten to 5 Nm (44 in-lb)
15. **Engine speed (RPM) sensor (G28)**
16. **Oil seal**
    - Always replace

Crankshaft, securing (5-cylinder)

– Remove lower sound insulation panels from underneath engine. See **03 Maintenance**.

– Remove right front wheel housing liner front section. See **66 Body–Exterior Equipment**.

– Remove ribbed belt for A/C compressor. See **03 Maintenance**.

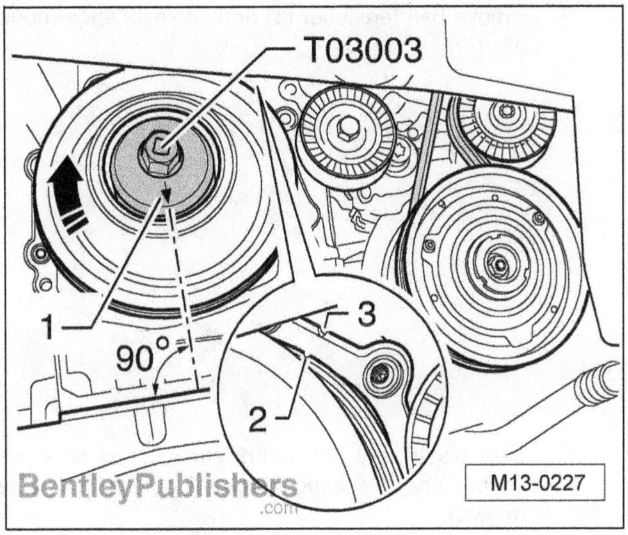

< Attach crankshaft adapter **T03003** to bolts of belt pulley.

- T03003 can only be installed in one position.

– Turn crankshaft in direction of engine rotation (**arrow**) until arrow **1** points downward, relative to engine axis.

- The crankshaft is now positioned for TDC (Top Dead Center) for cylinder #5.

NOTE —
*The TDC mark (**2**) must align with belt pulley mark (**3**).*

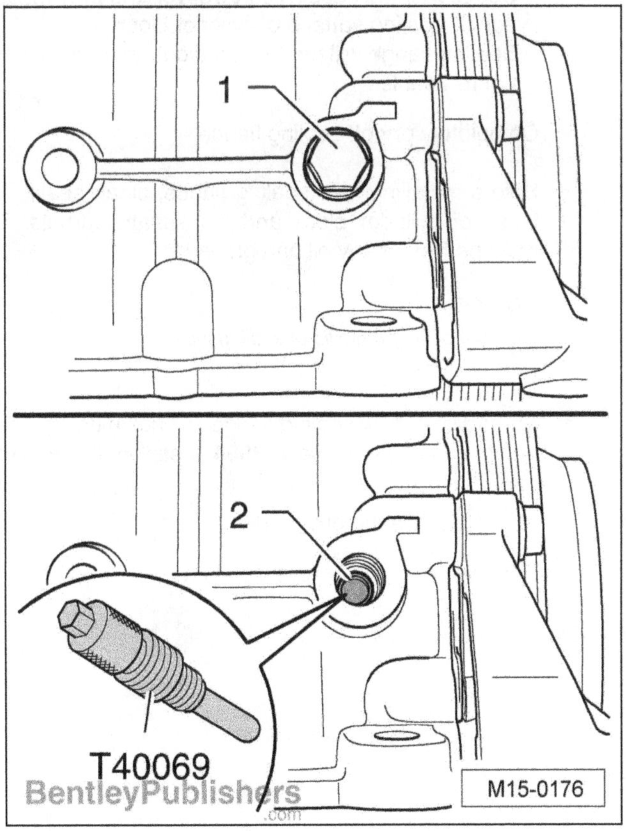

< Remove bolt (**1**) at rear of cylinder block.

– Use a flashlight and/or mirror to check if hole (**2**) in crankshaft is lined up with threaded hole in cylinder block. Rotate crankshaft slightly if need be to line up holes.

– Screw locking pin **T40069** completely into threaded hole and tighten to 10 Nm (7 ft-lb).

– Ensure that crankshaft can not be rotated.

– When repair are finished, replace bolt (**1**) in cylinder block.

Tightening torques

- **T40069** to cylinder block .... 10 Nm (7 ft-lb)
- Bolt (**1**) to cylinder block.......... 30 Nm (22 ft-lb)

Front oil seal flange, replacing (5-cylinder)

– Remove ribbed belt for A/C compressor. See **03 Maintenance**.

– Secure crankshaft, see **Crankshaft, securing (5-cylinder)** in this repair group.

– Remove crankshaft belt pulley.

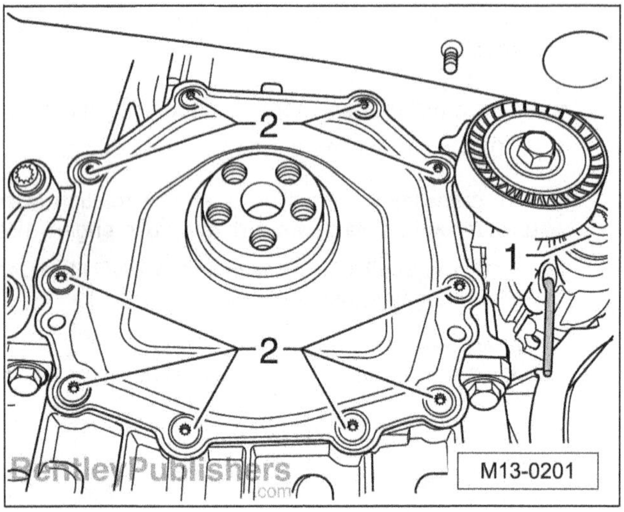

◀ Remove belt tensioner (**1**) and oil seal flange mounting bolts (**2**).

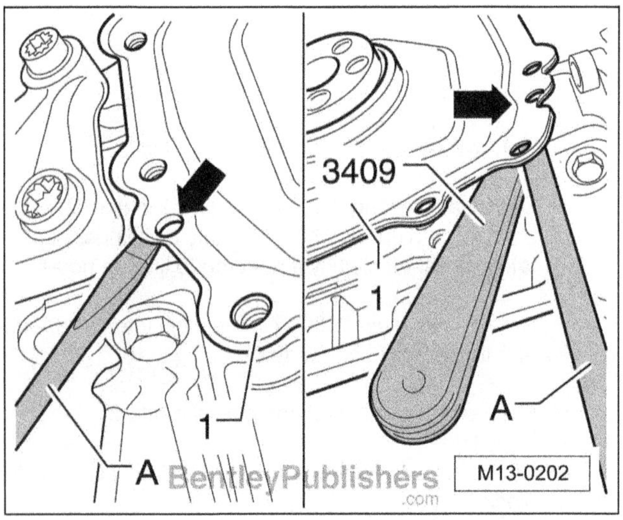

◀ Using plastic pry tool (**3409** shown) and screwdriver, carefully pry off sealing flange (**1**) at alignment sleeves (**arrows**).

- Use pry tool to support screwdriver and prevent damage to sealing surface of cylinder block.
- Sealing flange will be damaged during removal and is not re-usable.

– Completely pry off sealing flange.

– Before installing new sealing flange, clean sealing surfaces of cylinder block and crankshaft journals. They must be free of any oil and grease.

NOTE —
Do not oil or grease lip of sealing flange.

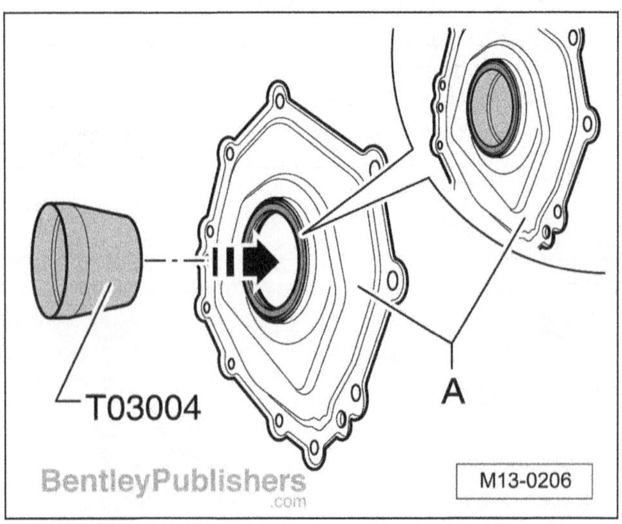

◀ Spread lip of new sealing flange as shown using tapered end of assembly sleeve **T03004**. Install in direction of **arrow**.

- Surface **A** is the outer side.

– Remove assembly sleeve T03004.

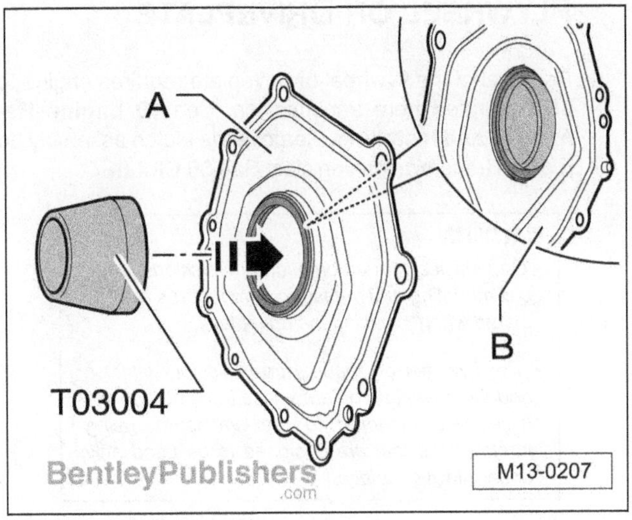

◀ Turn assembly sleeve around so tapered end is facing outward. Install in direction of **arrow**.

- Assembly sleeve must protrude approx. 3 mm on inner side (**B**) of sealing flange.

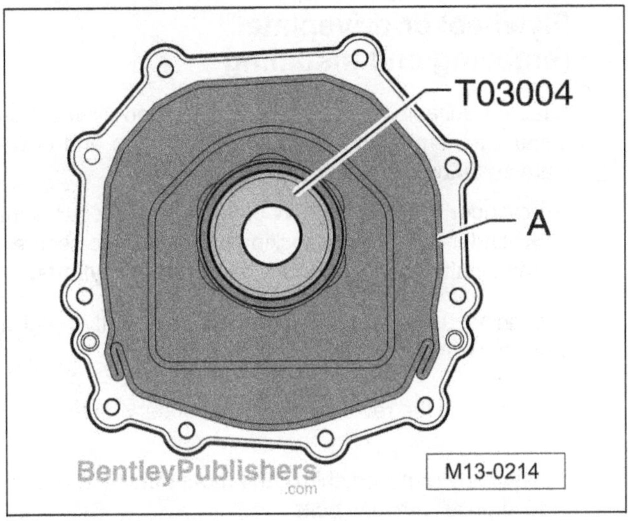

◀ Apply a 2 - 3 mm bead of sealant to the groove (A) on the inner side of the sealing flange.

- Height of sealant above sealing flange surface should be approx. 1 mm.

– Install sealing flange within 5 minutes of sealant application.

– Position assembly sleeve T03004 on crankshaft journal and press uniformly toward cylinder block.

– Remove assembly sleeve.

– Tighten sealing flange mounting bolts using a diagonal pattern.

– Remove crankshaft locking pin T40069 from cylinder block. Re-install bolt in cylinder block.

– Remaining assembly is reverse of removal.

Tightening torques

- Sealing flange to cylinder block ..... 10 Nm (7 ft-lb)
- Ribbed belt pulley to crankshaft .... 70 Nm (52 ft-lb)
- Ribbed belt tensioner to mounting bracket .... 35 Nm (26 ft-lb)
- Bolt to cylinder block .... 30 Nm (22 ft-lb)

FLYWHEEL OR DRIVEPLATE

Removal of the flywheel or driveplate requires engine to be separated from transmission. See **10 Engine–Removing and Installing**. Remove the clutch assembly on manual transmission vehicles. See **30 Clutch**.

> ***CAUTION—***
> - *On vehicles with automatic transmissions, special mounting and measuring procedures are required when installing the driveplate.*
> - *The flywheel on manual transmission vehicles and the driveplate on automatic transmission vehicles are mounted to the crankshaft using stretch bolts that are designed to be used only once. Always replace.*

Flywheel or driveplate, removing and installing

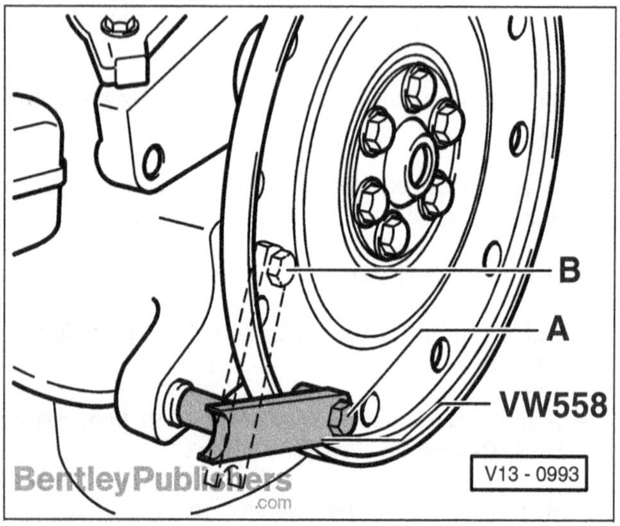

◄ Attach a suitable holder to the engine and flywheel on manual transmission vehicles or the engine and driveplate on automatic transmission vehicles.

- 4-cylinder vehicles use VW 558 (shown) or equivalent
- 5-cylinder vehicles use locking pin T40069 to secure crankshaft. See **Crankshaft, securing (5-cylinder)**.

– Loosen securing bolts diagonally and remove flywheel or drive plate and any shims.

– Installation is the reverse of removal observing the following points:

- Install any removed shims and mount the flywheel or drive plate with new bolts.
- Check driveplate clearance (as applicable) as described later.
- Attach the holder and torque diagonally to specifications.

Tightening torques
- Flywheel or driveplate to crankshaft (stretch bolt, always replace)
  - stage I . . . . 30 Nm (22 ft-lb)
  - stage II . . . . 60 Nm (44 ft-lb)
  - stage III . . . . additional ¼ turn (90°)

Driveplate clearance, adjusting (cars with automatic transmission)

Component replacement or other circumstances may necessitate adjustment of the driveplate clearance. Incorrect clearance may result in premature starter wear.

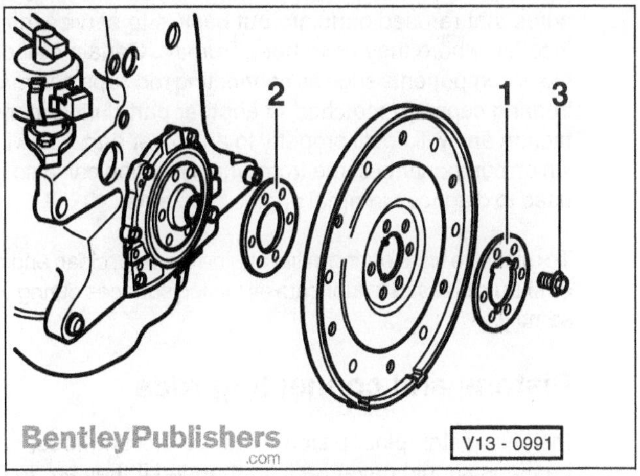

◀ Install drive plate with new bolts and backing plate (**1**), but without shim (**2**).

Tightening Torque
- Driveplate checking torque. . . . . 30 Nm (22 ft-lb)

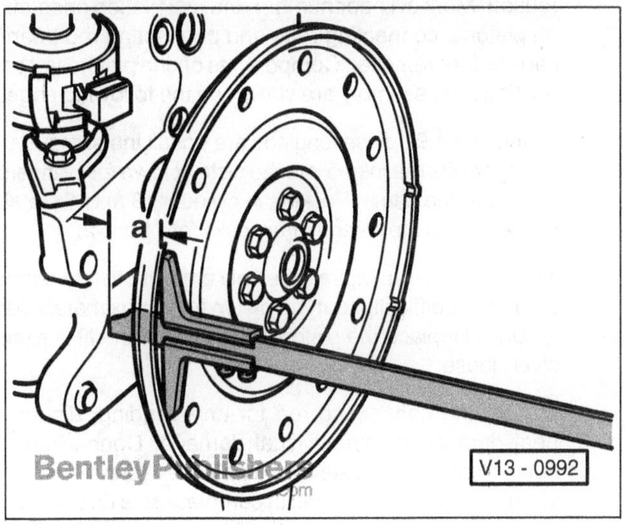

◀ Measure distance (**a**) from machined surface on cylinder block to outer edge of driveplate with a suitable measuring tool at three points. Measuring tool must fit through hole in driveplate.

– Average the three readings to get **dimension a**.

Specification
- Cylinder block to outer edge of driveplate (dimension **a**)
  4-cylinder . . . . 19.5 - 21.1 mm (0.77 - 0.83 in.)
  5-cylinder . . . . 18.8 - 20.4 mm (0.74 - 0.80 in.)

– If specification is not obtained, remove driveplate, add shim(s) and recheck.

– When specification is obtained, attach holding fixture and torque to final specification.

Tightening torques
- Driveplate to crankshaft (stretch bolt, always replace)
  stage I . . . . 30 Nm (22 ft-lb)
  stage II . . . . 60 Nm (44 ft-lb)
  stage III . . . . additional ¼ turn (90°)

CYLINDER BLOCK INTERNAL COMPONENTS

During engine block disassembly, be sure to mark the position and orientation of all parts as they are removed. This includes connecting rods, rod bearings and caps, piston pins, pistons, main bearings and caps. This ensures that re-used parts are put back in to service in the location where they have been "run-in". Certain cylinder block components such as connecting rod caps and main bearing caps are matched to another part during manufacture and will not fit properly to any other part. Knowing which components came from which location can also be used to diagnose internal engine problems.

To minimize wear during initial engine start-up, clean engine oil should be used to lubricate all friction surfaces during assembly.

Pistons and connecting rods

Pistons, piston pins, piston rings, connecting rods, and bearings should never be interchanged if they are to be reused. Mark cylinder number and installation orientation on pistons, connecting rods and connecting rod bearing caps before removal. Components of one piston and connecting rod assembly are shown on the following page.

Pistons for 1.9L diesel engines are not all the same. Valve relief pockets machined into the piston crown for cylinders 1 and 2 are the same. Pockets in cylinders 3 and 4 are also the same, but are different from cylinders 1 and 2.

The piston pin should require only a slight push to remove or install. If difficult, heat the piston to approximately 60°C (160°F). Replace the piston and the pin if the fit is excessively loose.

Inspect the connecting rod for any bending, distortion, heat damage or other visual damage. Connecting rod specifications are listed in **Table a**. Connecting rods should always be replaced in complete sets due to weight and dimensional considerations.

NOTE —

- *When checking radial clearance, reuse the old bolt or nut, and lubricate the contact surface of the nut or bolt before tightening. Tighten the nut or bolt only to the 30 Nm (22 ft-lb) specification and not the additional ¼ turn.*
- *If connecting rod radial clearance is excessive, the crankshaft connecting rod journals should be checked. If crankshaft journal diameters are within specifications, recheck radial clearance using new bearing shells.*

Piston and connecting rod, assembly

NOTE—
2.0L shown, others engines are similar.

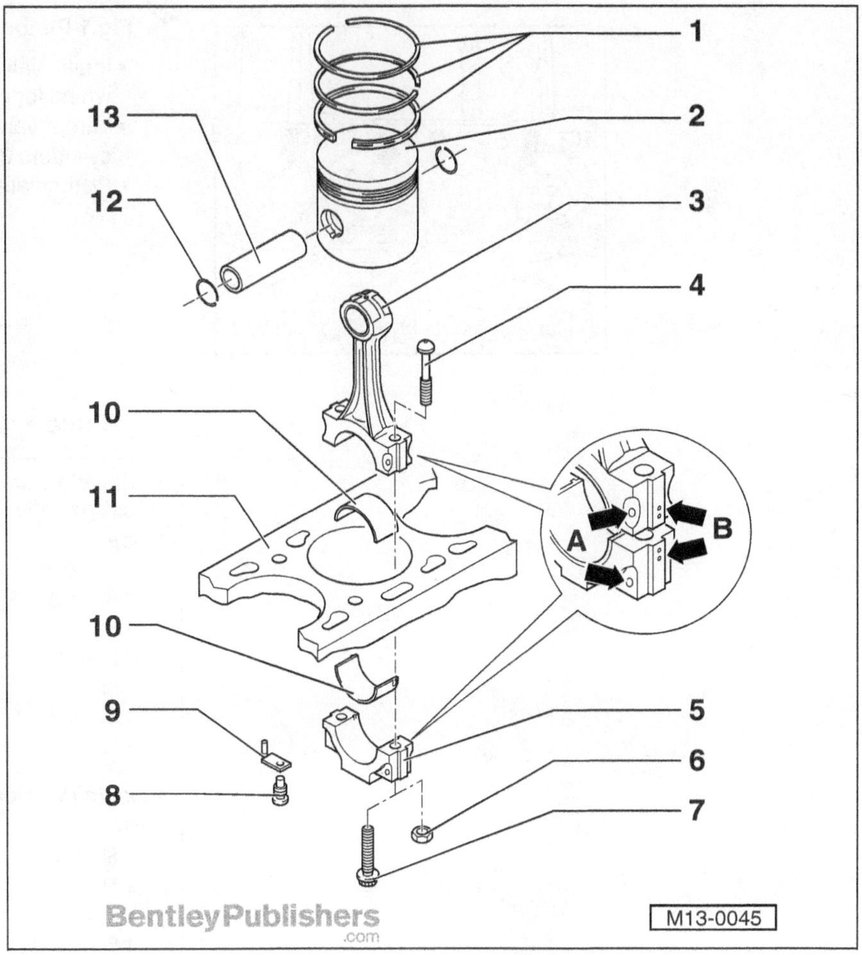

1. **Piston ring**
   - Offset gaps by 120°
   - Remove and install using piston ring pliers
   - "TOP" faces piston crown
   - 2 or 3 part oil scraper ring
2. **Piston**
   - Mark installation position and cylinder number
   - Arrow on piston crown points to belt pulley end
   - Install using piston ring clamp
   - ALH engines see **Fig.1**
3. **Connecting rod**
   - Only replace as a set
   - Mark cylinder number (**B**)
   - Installation position: mark (**A**) faces toward belt pulley end
4. **Connecting rod bolt**
   - For engine code AEG
   - Always replace
5. **Connecting rod bearing cap**
   - Note installation position
   - Mark cylinder number (**B**)
   - Installation position: mark (**A**) faces toward pulley end
6. **Connecting rod nut**
   - Tighten to 30 Nm (22 ft-lb) + ¼ turn (90°)
   - For engine codes AEG, APH
   - Always replace
   - Oil threads and contact surfaces
   - To measure radial clearance tighten to 30 Nm (22 ft-lb), but no further
7. **Connecting rod bolt**
   - Tighten to 30 Nm (22 ft-lb) + ¼ turn (90°)
   - Not for engine codes AEG, APH
   - Always replace
   - Oil threads and contact surfaces
   - To measure radial clearance tighten to 30 Nm (22 ft-lb), but no further
8. **Pressure relief valve**
   - As applicable, not on AEG engine
   - Tighten to 27 Nm (20 ft-lb)
   - Opening pressure:
     4-cylinder:2.5 to 3.2 bar (36 to 46 psi)
     5-cylinder: 1.3 to 1.6 bar (19 to 23 psi)
9. **Oil spray jet**
   - As applicable, not on AEG engine
   - For piston cooling
10. **Bearing shell**
    - Note installation position
    - Do not interchange used bearing shells
    - Ensure retaining lugs fit tightly in recesses
    - Do not rotate crankshaft when checking radial clearance
    - With oil hole for piston bolt lubrication
11. **Engine block**
12. **Circlip**
13. **Piston pin**
    - If difficult to remove, heat piston to 60°C (140°F)
    - Remove and install with VW 222a

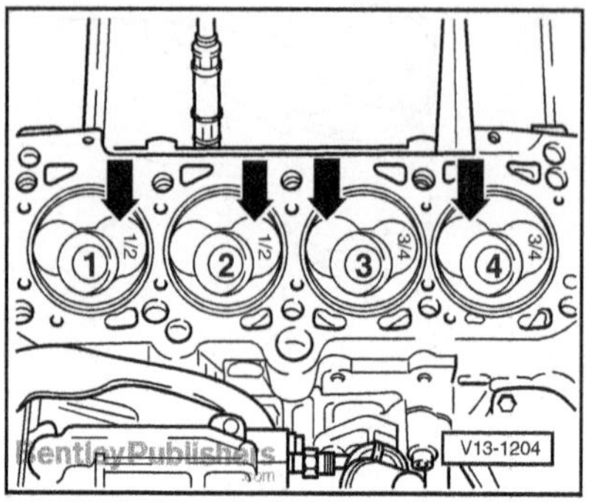

Fig.1 Pistons (1.9L diesel engine)

- Intake valve relief pocket is larger and points to flywheel for cylinders 1 and 2.
- Larger valve relief pocket points to belt pulley side for cylinders 3 and 4.
- Pistons are factory marked for proper location.

Table a. Connecting Rod Specifications

Radial clearance (Plastigage®)	
new	
1.8L (APH)	0.01–0.05 mm (0.0004–0.0020 in.)
1.9L, 2.0L	0.01–0.06 mm (0.0004–0.0024 in.)
2.5L	0.02–0.06 mm (0.0008–0.0024 in.)
wear limit	
1.9L	0.08 mm (0.0031 in.)
1.8L, 2.0L	0.12 mm (0.0047 in.)
2.5L	0.09 mm (0.0035 in.)
Axial (side) clearance	
new	
1.8L, 1.9L, 2.0L	0.05-0.31 mm (0.0020-0.0122 in.)
2.5L	0.10-0.35 mm (0.0039-0.0138 in.)
wear limit	
1.8Lm, 1.9L, 2.0L, 2.5L	0.40 mm (0.0157 in.)
Checking torque	30 Nm (22 ft-lb)
Assembly torque	30 Nm (22 ft-lb) plus ¼ turn (90°)

Piston rings

Piston ring end gaps are checked with the piston rings inserted evenly approximately 15 mm (5/8 in.) from the bottom of the cylinder. This is because wear in this area of the cylinder is negligible. **Table b** lists piston ring end gap specifications.

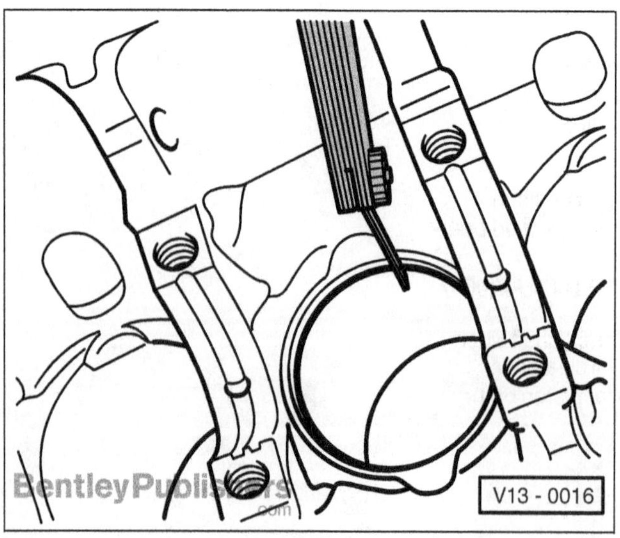

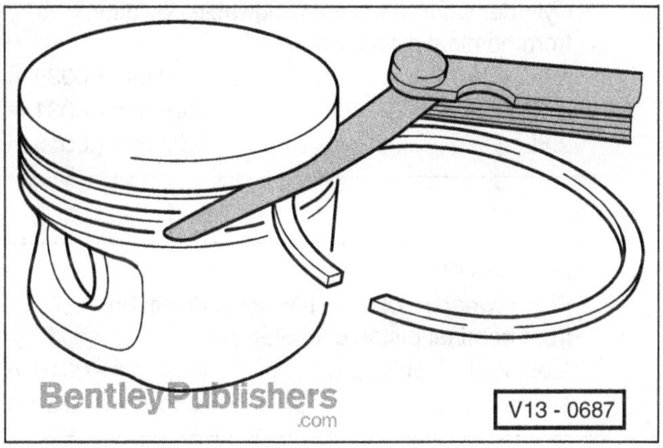

Table b. Piston Ring End Gaps

	New	Wear limit
Top (compression)		
1.8L	0.15–0.40 mm (.0059–.0157 in.)	0.8 mm (.0315 in.)
1.9L	0.20–0.40 mm (.0079–.0157 in.)	1.0 mm (.0394 in.)
2.0L, 2.5L	0.20–0.40 mm (.0079–.0157 in.)	0.8 mm (.0315 in.)
Middle (compression)		
1.8L	0.15–0.40 mm (.0059–.0157 in.)	0.8 mm (.0315 in.)
1.9L	0.20–0.40 mm (.0079–.0157 in.)	1.0 mm (.0394 in.)
2.0L, 2.5L	0.20–0.40 mm (.0079–.0157 in.)	0.8 mm (.0315 in.)
Bottom (oil scraper)		
1.8L, 1.9L	0.25–0.50 mm (.0098–.0197 in.)	1.0 mm (.0394 in.)
2.0L, 2.5L	0.25–0.50 mm (.0098–.0197 in.)	0.8 mm (.0315 in.)

Piston ring side clearance (ring to groove clearance) is checked using feeler gauges. Measure each ring in its original groove. Piston ring side clearance specifications are listed in **Table c**.

NOTE —
Piston ring groove should be thoroughly cleaned before checking ring side clearance.

Table c. Piston Ring Side Clearances

	New	Wear limit
Top (compression)		
1.8L	0.02–0.07 mm (.0007–.0027)	0.12 mm (.0047 in.)
1.9L	0.06–0.09 mm (.0024–.0035 in.)	0.25 mm (.0098 in.)
2.0L, 2.5L	0.06–0.09 mm (.0024–.0035 in.)	0.20 mm (.0079 in.)
Middle (compression)		
1.8L	0.02–0.07 mm (.0007–.0027)	0.12 mm (.0047 in.)
1.9L	0.05–0.08 mm (.0020–.0031 in.)	0.25 mm (.0098 in.)
2.0L, 2.5L	0.06–0.09 mm (.0024–.0035 in.)	0.20 mm (.0079 in.)
Bottom (oil scraper)		
1.8L	0.02–0.06 mm (.0007–.0024)	0.12 mm (.0047 in.)
1.9L, 2.0L, 2.5L	0.03–0.06 mm (.0012–.0024 in.)	0.15 mm (.0059 in.)

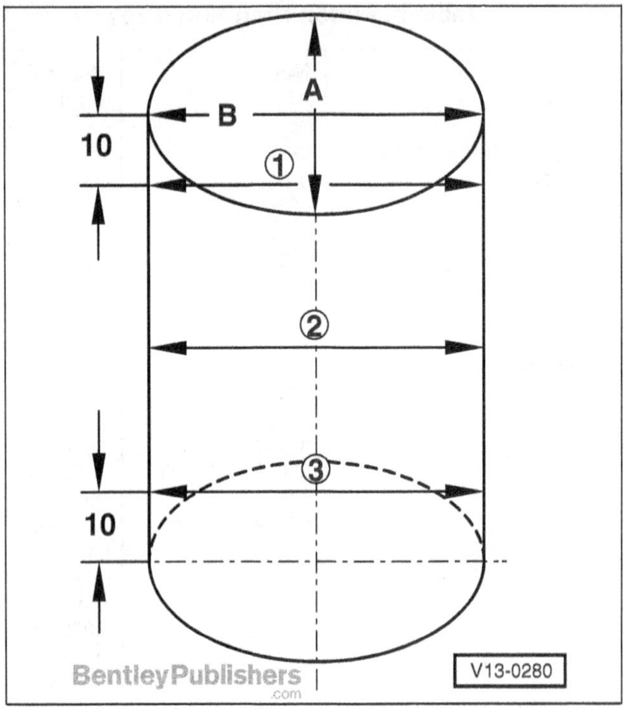

Cylinder block

Measure cylinder bores at three places; approximately the top, the middle and the bottom of the piston travel. Make measurements parallel to the crankshaft and at right angles (90°).

The top and bottom measurements should be made approximately 10 mm (3/8 in.) from the ends of the cylinder. Measure pistons from the bottom of the piston skirt and at right angles (90°) to the piston pin. Nominal piston and cylinder bore diameter specifications are given in **Table d**. Nominal piston diameters are also marked on the piston crowns.

> ***CAUTION—***
> *Mounting the bare cylinder block to an engine stand can distort its shape and cause inaccurate cylinder bore measurement. Always check cylinder bores with the block resting unstressed and without mounting brackets on a flat surface.*

Specification

- Cylinder wear–maximum allowable deviation from nominal cylinder bore
  1.8L, 1.9L . . . . 0.10 mm (.0039 in.)
  2.0L . . . . 0.08 mm (.0031 in.)
  2.5L . . . . 0.04 mm (.0016 in.)

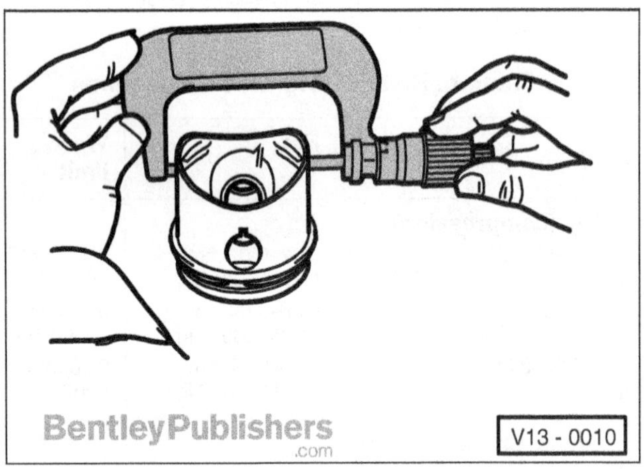

Specification

- Piston wear–maximum allowable deviation from nominal piston diameter
  1.8L, 2.0L, 2.5L . . . . 0.04 mm (.0016 in.)
  1.9L . . . . no factory specification

Table d. Piston and Cylinder Diameters

Engine	Piston diameter in mm (in.)	Cylinder bores in mm (in.)
1.8L		
standard	80.965* (3.1875)	81.01 (3.189)
1st oversize	81.465* (3.2073)	81.51 (3.209)
1.9L		
standard	79.47 (3.1287)	79.51 (3.1303)
1st oversize	79.72 (3.1386)	79.76 (3.1401)
2nd oversize	79.97 (3.1484)	80.01 (3.1500)
2.0L, 2.5L		
standard	82.465* (3.2466)	82.51 (3.2484)
1st oversize	82.965* (3.2663)	83.01 (3.2681)

* Dimension without graphite coating (thickness 0.02 mm (.0008 in.). The graphite coating wears away.

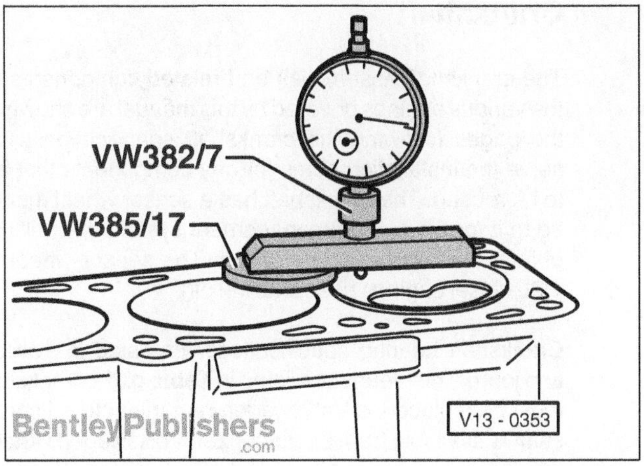

Piston projection at TDC (1.9L diesel)

Diesel engine pistons protrude slightly out of the block at TDC due to design considerations. The amount of protrusion varies slightly between engines as a result of production tolerance. In order to achieve proper compression ratios, one of three different thickness head gaskets are used to adjust compression height. Piston projection must be measured and adjusted any time components that affect it are replaced. These components include: crankshaft, piston, connecting rod and block. Measure piston height on all 4 cylinders.

Select a suitable head gasket based on the highest reading obtained from the available head gaskets listed in **Table e**. The different gaskets are identified by the number of holes punched in an identification tab on the front of the gasket.

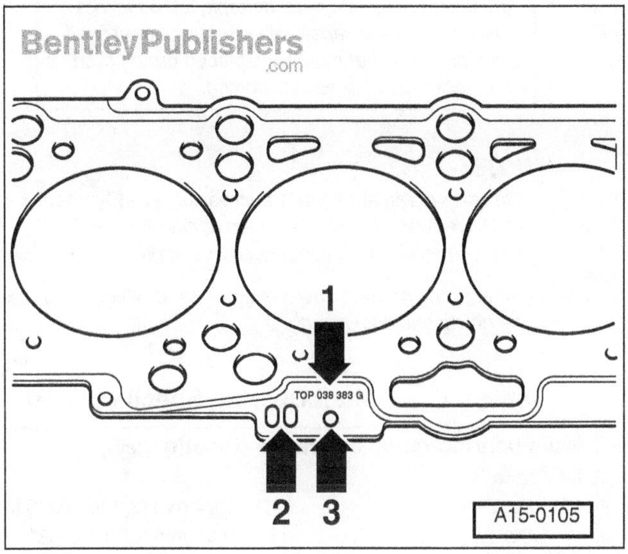

1.9L cylinder head gasket identification markings:

- **Arrow 1** = part number
- **Arrow 2** = production information (disregard)
- **Arrow 3** = identification holes

NOTE —
Cylinder measurements that differ substantially from each other could be an indication of internal engine damage.

Table e. 1.9L Diesel cylinder head gasket selection

Measured piston projection at TDC	Gasket identification	Thickness
0.91 – 1.00 mm (.0358 – .0394 in.)	1 hole	1.45 mm (.0571 in.)
1.01 – 1.10 mm (.0398 – .0433 in.)	2 holes	1.53 mm (.0602 in.)
1.11 – 1.20 mm (.0437 – .0472 in.)	3 holes	1.61 mm (.0634 in.)

Crankshaft

The crankshaft assemblies and related components for the various engines covered by this manual are shown on the pages following the crankshaft specifications. Observe the installation notes for any components that are to be reused. The crankshaft has a sensor wheel mounted to it for the engine management system that will can only be installed in one position. The sensor wheel on 1.9L diesel engines uses a dowel pin.

Crankshaft bearing specifications are listed in **Table f** and journal diameters are listed in **Table g**. If a crankshaft must be replaced, a Volkswagen remanufactured crankshaft is available from an authorized Volkswagen Dealer.

> ***CAUTION —***
> *Many of the fasteners used in the cylinder block are stretch bolts that must be replaced once loosened. Review the repair information to identify all bolts and nuts that must be replaced during cylinder block repairs or reconditioning.*

NOTE —
- *On cars with an automatic transmission, see* **Flywheel or Driveplate** *given earlier when reinstalling the driveplate. Special installation procedures apply.*
- *Attach engine to suitable engine stand when disassembling and assembling.*

Table f. Crankshaft Bearing Specifications

Main bearing radial clearance (Plastigage®)	
new parts	
1.8L, 2.0L	0.01 - 0.04 mm (.0004 -.0016 in.)
1.9L	0.03 - 0.08 mm (.0012 -.0031 in.)
2.5L	0.023 - 0.043 mm (.0009 -.0017 in.)
wear limit	
1.8L, 2.0L	0.15 mm (.0059 in.)
1.9L	0.17 mm (.0067 in.)
2.5L	0.07 mm (.0028 in.)
Crankshaft axial play (side clearance)	
new parts	
1.8L, 2.0L	0.07 - 0.23 mm (.0028 -.0091 in.)
1.9L	0.07 - 0.17 mm (.0028 -.0067 in.)
2.5L	0.07 - 0.21 mm (.0028 -.0083 in.)
wear limit	
1.8L, 2.0L, 2.5L	0.30 mm (.0118 in.)
1.9L	0.37 mm (.0146 in.)

Table g. Crankshaft Journal Diameters

Journal diameters (nominal)	mm (in.)
Basic dimension	
1.8L / 1.9L engine	
main	54.00 (2.1260) $^{-0.022\ (0.00087)}_{-0.042\ (0.00165)}$
connecting rod	47.80 (1.8819) $^{-0.022\ (0.00087)}_{-0.042\ (0.00165)}$
2.0L engine	
main	54.00 (2.1260) $^{-0.017\ (0.00067)}_{-0.037\ (0.00146)}$
connecting rod	47.80 (1.8819) $^{-0.022\ (0.00087)}_{-0.042\ (0.00165)}$
2.5L engine	
main	58.00 (2.2803) $^{-0.022\ (0.00087)}_{-0.042\ (0.00165)}$
connecting rod	47.80 (1.8819) $^{-0.022\ (0.00087)}_{-0.042\ (0.00165)}$
1st undersize (0.25 mm)	
1.8L / 1.9L engine	
main	53.75 (2.1161) $^{-0.022\ (0.00087)}_{-0.042\ (0.00165)}$
connecting rod	47.55 (1.8720) $^{-0.022\ (0.00087)}_{-0.042\ (0.00165)}$
2.0L engine	
main	53.75 (2.1161) $^{-0.017\ (0.00067)}_{-0.037\ (0.00146)}$
connecting rod	47.55 (1.8720) $^{-0.022\ (0.00087)}_{-0.042\ (0.00165)}$
2.5L engine	
main	57.75 (2.2736) $^{-0.022\ (0.00087)}_{-0.042\ (0.00165)}$
connecting rod	47.55 (1.8720) $^{-0.022\ (0.00087)}_{-0.042\ (0.00165)}$
2nd undersize (0.50 mm)	
1.8L / 1.9L engine	
main	53.50 (2.1063) $^{-0.022\ (0.00087)}_{-0.042\ (0.00165)}$
connecting rod	47.30 (1.8622) $^{-0.022\ (0.00087)}_{-0.042\ (0.00165)}$
2.0L engine	
main	53.50 (2.1063) $^{-0.017\ (0.00067)}_{-0.037\ (0.00146)}$
connecting rod	47.30 (1.8622) $^{-0.022\ (0.00087)}_{-0.042\ (0.00165)}$
2.5L engine	
main	57.50 (2.2638) $^{-0.022\ (0.00087)}_{-0.042\ (0.00165)}$
connecting rod	47.30 (1.8622) $^{-0.022\ (0.00087)}_{-0.042\ (0.00165)}$
3rd undersize (0.75 mm)	
1.8L / 1.9L engine	
main	53.25 (2.0965) $^{-0.022\ (0.00087)}_{-0.042\ (0.00165)}$
connecting rod	47.05 (1.8524) $^{-0.022\ (0.00087)}_{-0.042\ (0.00165)}$
2.0L engines	
main	53.25 (2.0965) $^{-0.017\ (0.00067)}_{-0.037\ (0.00146)}$
connecting rod	47.05 (1.8524) $^{-0.022\ (0.00087)}_{-0.042\ (0.00165)}$
2.5L engines	
main	57.25 (2.2539) $^{-0.022\ (0.00087)}_{-0.042\ (0.00165)}$
connecting rod	47.05 (1.8524) $^{-0.022\ (0.00087)}_{-0.042\ (0.00165)}$

Crankshaft assembly (1.8L, 2.0L engine)

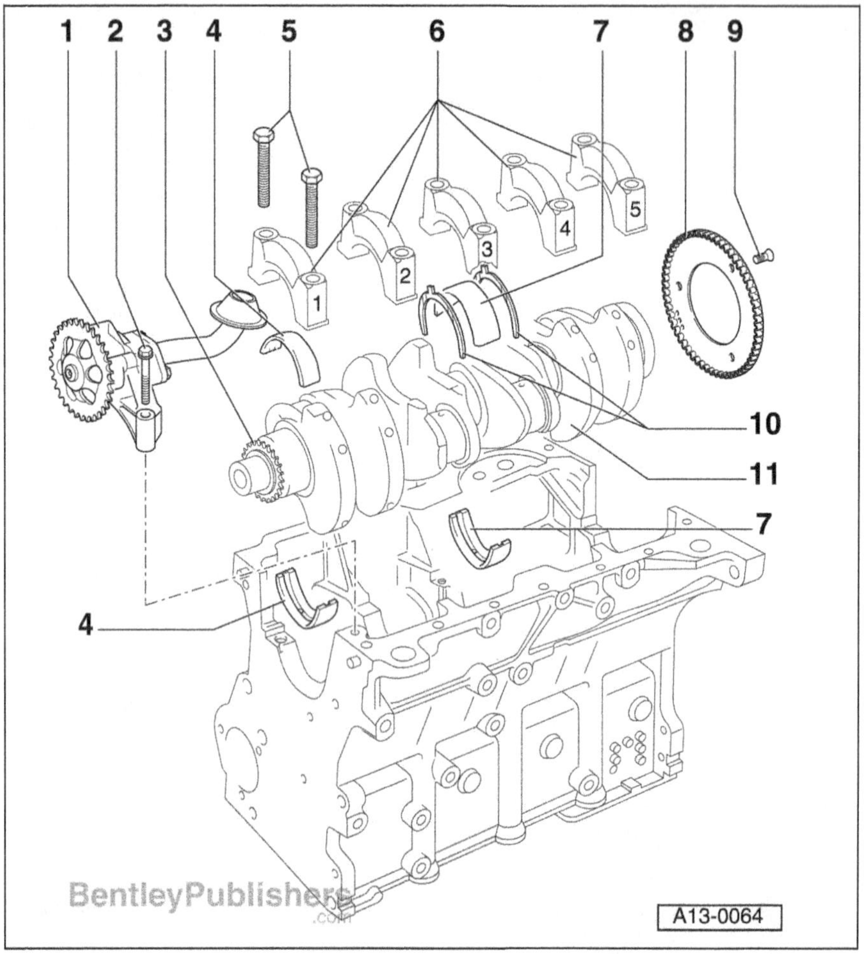

1. **Oil pump**
2. **Bolt**
   - Tighten to 15 Nm (11 ft-lb)
3. **Chain sprocket**
   - For oil pump drive
4. **Bearing caps 1, 2, 4 and 5**
   - For bearing cap without oil groove, **See Fig.1**
   - For cylinder block with oil groove
   - Do not interchange used bearing shells (mark)
5. **Bolts**
   - Tighten to 65 Nm (48 ft-lb) + ¼ turn (90°)
   - Always replace
   - Threaded along complete length
   - Tighten to 65 Nm (48 ft-lb) to measure radial clearance
6. **Bearing cap**
   - Bearing cap 1: belt pulley end
   - Bearing cap 3 with recesses for thrust washers
   - Retaining tabs for bearing shells and bearing caps must align
7. **Bearing shell 3**
   - For bearing cap without oil groove
   - For cylinder block with oil groove
   - Do not interchange used bearing shells (mark)
8. **Sensor wheel**
   - For engine speed sensor (G28)
   - Can only be installed in one position, holes are offset
9. **Bolt**
   - Tighten to 10 Nm (7 ft-lb) + ¼ turn (90°)
   - Always replace
10. **Thrust washer**
    - For bearing 3 bearing cap
    - Note installation position
11. **Crankshaft**
    - Check radial clearance with Plastigage
    - Do not rotate crankshaft when checking radial clearance

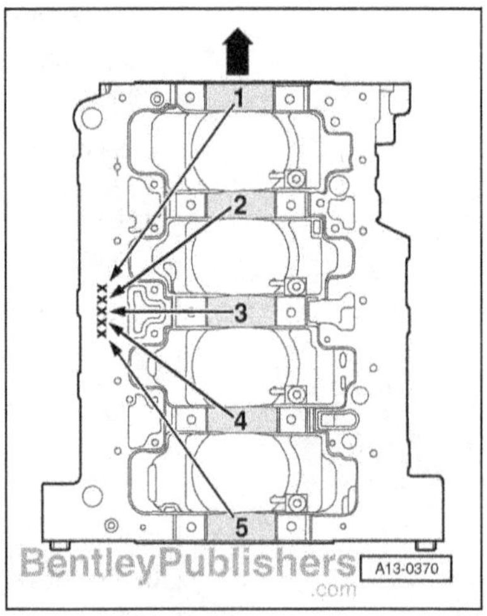

Fig. 1 Vehicles from 05/2000

- The upper crankshaft bearings are installed with the proper thickness bearing shells from the factory. Colored dots are used to identify the bearing shell thickness.

NOTE —
The arrow shows drive direction.

- Letters in the lower sealing surface of the cylinder block identify which bearing thickness must be installed at which position.

NOTE —
Lower bearing shells are always supplied as "yellow".

Main bearing color markings

Letter on block	Bearing color
S	black
R	red
G	yellow

Crankshaft assembly (1.9L engine)

1. **Bearing shells 1, 2, 4 and 5**
   - For bearing caps without oil groove
   - For engine block with oil groove
   - Do not interchange used bearing shells (mark)
2. **Bolt**
   - Tighten to 65 Nm (48 ft-lb) + ¼ turn (90°)
   - Always replace
   - Threaded along complete length
   - Tighten to 65 Nm (48 ft-lb) to measure radial clearance
3. **Bearing cap**
   - Bearing cap 1: belt pulley end
   - Bearing cap 3 with recesses for thrust washers
   - Retaining tabs for bearing shells and bearing caps must align
4. **Bearing shell 3**
   - Bearing cap without oil groove
   - For engine block with oil groove
   - Do not interchange used bearing shells (mark)
5. **Thrust washer**
   - For bearing cap 3
   - Note installation position
6. **Sensor wheel**
   - For engine speed sensor
7. **Bolt**
   - Tighten to 10 Nm (7 ft-lb) + ¼ turn (90°)
   - Always replace
8. **Dowel pin**
   - For installation position, **see Fig. 1**
9. **Crankshaft**
   - Check radial clearance with Plastigage
   - Do not rotate crankshaft when checking radial clearance
10. **Thrust washer**
    - For engine block, bearing 3
    - Note installation position

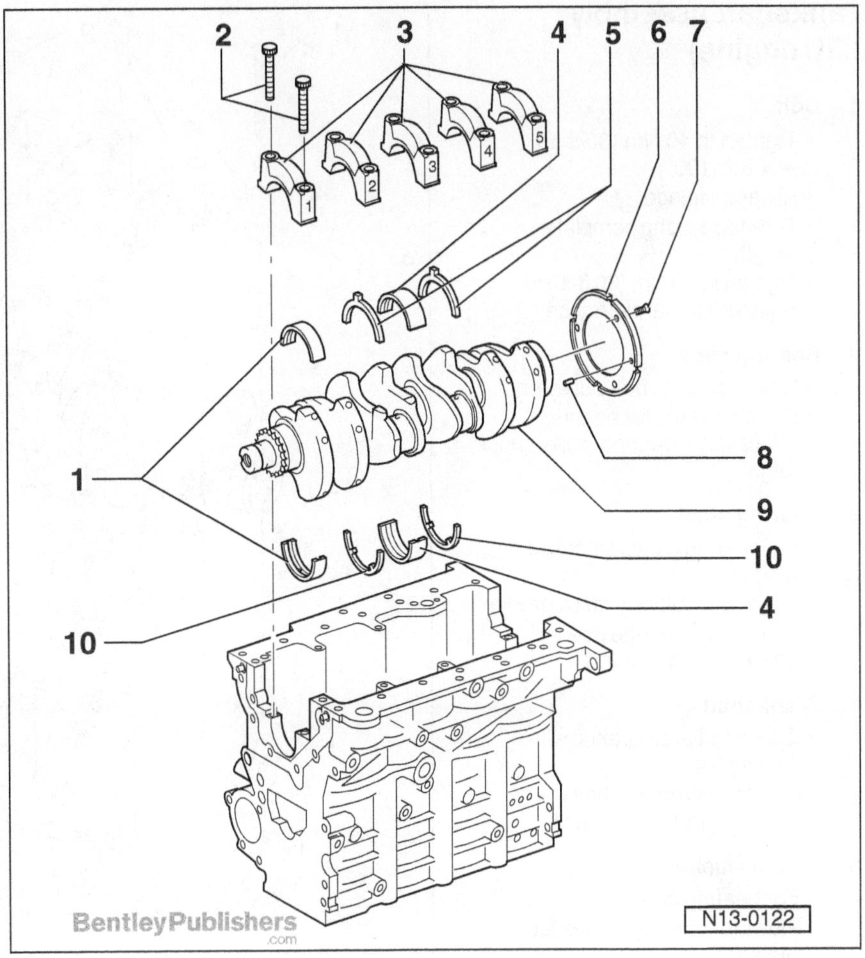

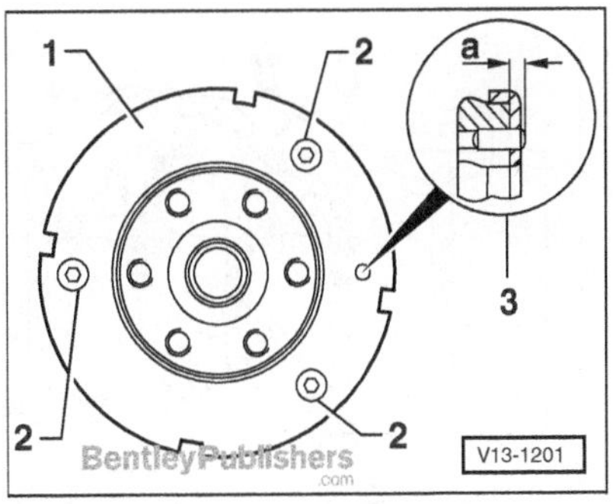

Fig.1 Dowel Pin and Sensor Wheel

- ALH engine only.
- Sensor wheel (**1**) mounted on crankshaft with screws (**2**).
- A single dowel pin (**3**) extends 2.5 – 3.0 mm (.098 –.118 in.) from the flange end, dimension (**a**).

Crankshaft assembly (2.5L engine)

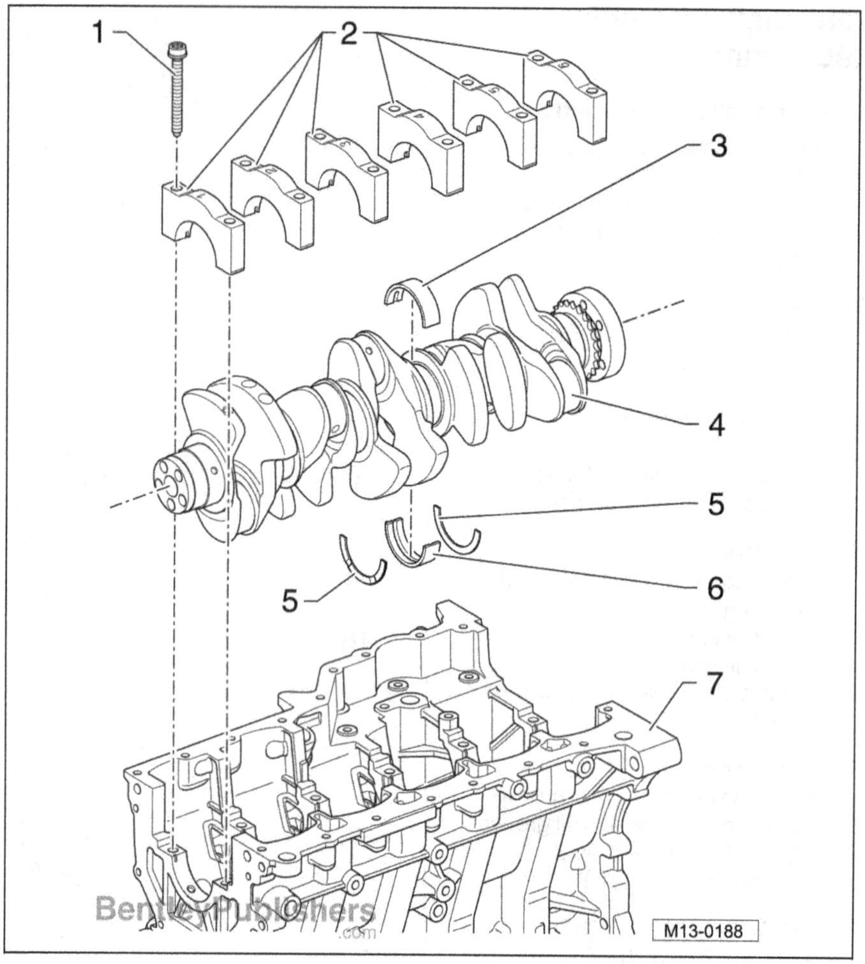

1. **Bolt**
   - Tighten to 40 Nm (30 ft-lb) + ¼ turn (90°)
   - Always replace
   - Threaded along complete length
   - Tighten to 40 Nm (30 ft-lb) to measure radial clearance
2. **Bearing caps**
   - Bearing cap 1: belt pulley end
   - Retaining tabs for bearing shells and bearing caps must align
3. **Bearing shell**
   - For bearing caps without oil groove
   - For engine block with oil groove
   - Do not interchange used bearing shells (mark)
4. **Crankshaft**
   - Check radial clearance with Plastigage
   - Do not rotate crankshaft when checking radial clearance
5. **Thrust washer**
   - For bearing cap 3
   - Side lubricating grooves face outward
6. **Bearing shell for cylinder block**
   - With lubricating groove
   - Do not interchange used bearing shells (mark)
7. **Cylinder block**

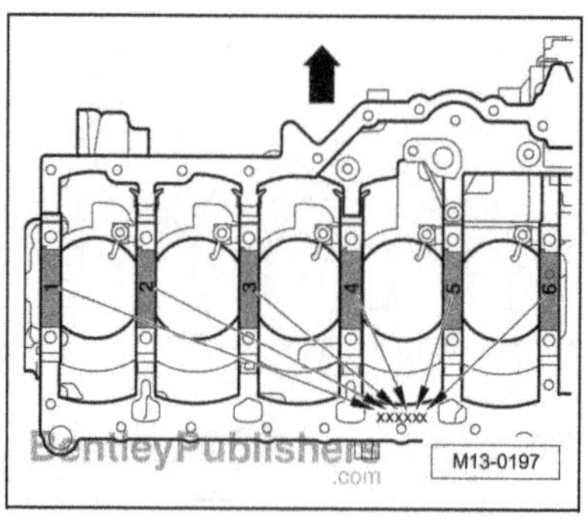

Fig. 1 Identification of top crankshaft bearing shells

- The upper crankshaft bearings are installed with the proper thickness bearing shells from the factory. Colored dots are used to identify the bearing shell thickness.

NOTE —

The arrow shows drive direction.

- Letters in the lower sealing surface of the cylinder block identify which bearing thickness must be installed at which position.

NOTE —

- *If color markings can no longer be seen, use the blue shell.*
- *Lower bearing shells are always supplied as "yellow".*

Main bearing color markings

Letter on block	Bearing color
W	white
B	blue
G	yellow

Timing chains overview (2.5L engine)

1. **Chain sprocket for exhaust cam**
2. **Chain tensioner for upper chain**
   - With guide insert
3. **Upper timing chain**
4. **Tensioning rail**
   - For timing chain tensioner
   - Attached to cylinder block
5. **Chain tensioner for lower chain**
   - With tensioning rail
6. **Chain sprocket for crankshaft**
   - Part of crankshaft
7. **Guide rail for lower chain**
   - Fastened to upper part of oil pan
8. **Chain sprocket for oil pump**
9. **Guide rail for lower chain**
10. **Lower timing chain**
11. **Double chain transfer sprocket**
12. **Guide rail for upper chain**
13. **Intake camshaft adjuster chain sprocket**

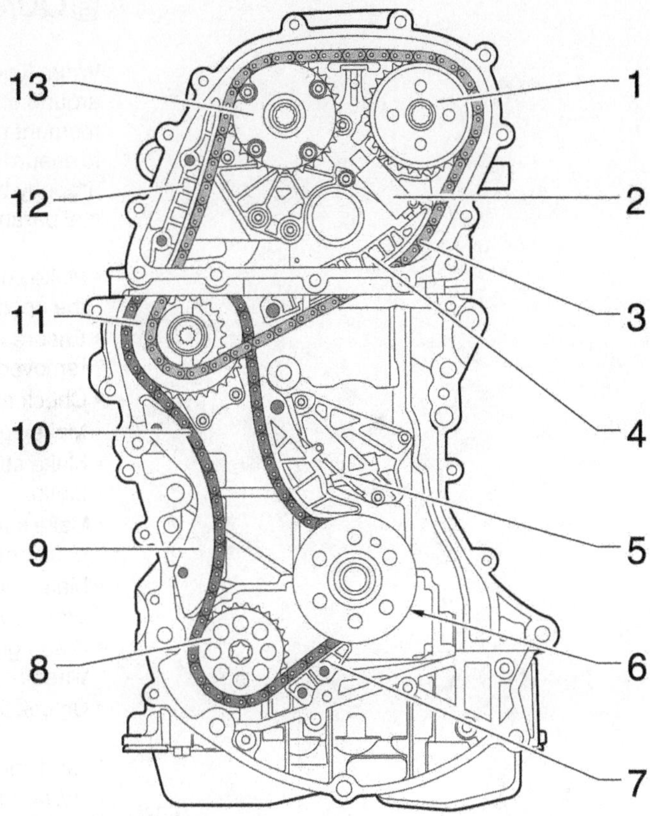

QUALITY REVIEW

When you have finished working under the hood and around other areas of the vehicle it is advisable to take a moment to quality check or review your work. This helps to ensure that the operation or repair has been completed properly with all affected systems functioning within normal parameters. This may include the following:

- Make sure that the radiator fan cycles properly and that the coolant level and concentration are correct.
- Ensure that all cable ties and hose clamps that were removed as part of the repair are replaced.
- Check and adjust all other applicable fluid levels.
- Make sure that there are no fluid leaks.
- Make sure that there are no air, vacuum or exhaust leaks.
- Make sure that all components involved in the repair are positioned correctly and function properly.
- Male sure all tools, shop cloths, fender covers, and protective tape are removed.
- Clean grease from painted surfaces and steering wheel.
- Unlock the anti-theft radio and reset the clock.

In addition to the above noted points, the ECM and TCM may need to be checked using a diagnostic scan tool. See **OBD On Board Diagnostics** section for more information.

15a Cylinder Head and Valvetrain (1.8L engine)

GENERAL

This section covers cylinder head and valvetrain service and repair work for the 1.8L 4-cylinder, 5 valve per cylinder, turbo-charged gasoline engine. For information on short block engine rebuilding and internal engine specifications, see **13 Crankshaft and Cylinder Block**.

Most of the operations described in this repair group require special equipment and experience. If you lack the skills, tools, or a suitable workplace for servicing or repairing the cylinder head, we suggest that you leave these repairs to an authorized Volkswagen dealer or other qualified shop.

Engine Codes
- APH, AWP, AWV, BKF, BNU .... 1.8L 4-cylinder turbo gasoline

DIAGNOSTIC TESTING

The tests that follow can be used to help isolate engine problems, to better understand a problem before starting expensive and extensive repairs, or to just periodically check engine condition.

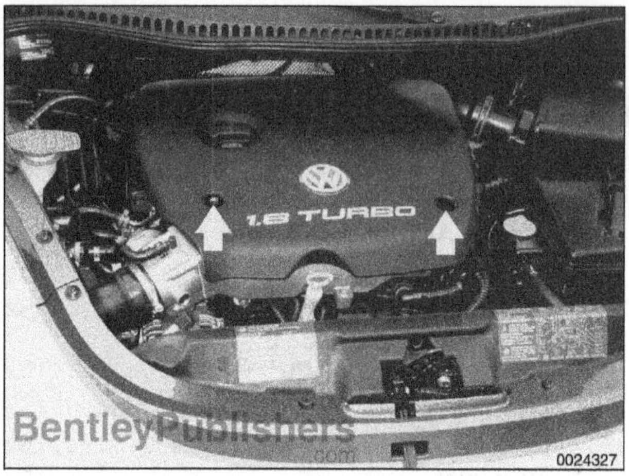

Compression test

A compression test will tell a lot about the overall condition of the engine without the need for taking it apart. Testing is relatively simple and straightforward.

< Remove two plastic caps and retaining nuts on upper engine cover. Lift cover up slightly and pull forward.

- Warm-up engine until it is a minimum of 30°C (86°F). Switch off ignition.
- Disconnect harness connectors from ignition coils and remove coils. See **03 Maintenance, Spark plugs, replacing (1.8L engine)**.
- Use compressed air to clear area around spark plugs. Remove all spark plugs and lay aside in proper order.
- Remove fuse 32 for the voltage supply to the injectors.
- Fit the compression tester into the spark plug hole.
- Have a helper:
  - depress clutch pedal fully
  - put the transmission in neutral or park
  - depress the accelerator pedal to the floor
  - crank the engine over with the starter motor

NOTE —
Cranking the engine with the ignition system disabled and components disconnected will cause Diagnostic trouble codes (DTCs) to be stored in engine management system memory.

- Engine should be cranked a minimum of 4 to 5 revolutions.
- Record readings, release pressure in the gauge and repeat procedure for each cylinder.
- Compare readings to specification.

Specification
- Compression pressures
  - new . . . . 10 to 13 bar (145 - 189 psi)
  - wear limit . . . . 7.5 bar (109 psi)
  - maximum difference between cylinders . . . . 3 bar (44 psi)

- When all cylinders have been checked, reinstall spark plugs and ignition coils; connect harness connectors for ignition coils and install the upper sound absorber panel.

Tightening torques
- Spark plug to cylinder head . . . . 30 Nm (22 ft-lb)
- Ignition coil to cylinder head cover (APH engine only) . . . . 10 Nm (7 ft-lb)

Cylinder leakdown test

The most conclusive diagnosis of low compression symptoms requires a cylinder leak-down test. Using a special tester and a supply of compressed air, each cylinder is pressurized. The rate at which the air leaks out of the cylinder, as well as the sound and location of the escaping air can more accurately pinpoint the magnitude and source of the leakage. Any engine compression diagnosis that will require major disassembly should first be confirmed by a cylinder leak-down test. Because this test requires special equipment and experience, it may be desirable to have it performed by a Volkswagen dealer or other qualified repair shop.

CYLINDER HEAD SERVICE

Many cylinder head repairs can be accomplished without removing the cylinder head from the engine. The cylinder head cover (valve cover) gasket, the camshaft, the camshaft oil seal, the valve guide oil seals, the valve springs, and the camshaft followers are all accessible with the cylinder head installed. This heading describes those repairs that can be done with the cylinder head installed.

NOTE —
Volkswagen identifies electrical components by a letter and/or a number in the electrical schematics. See **EWD Electrical Wiring Diagrams**. *These electrical identifiers are listed in parenthesis as an aid to electrical troubleshooting.*

Internal cylinder head assembly

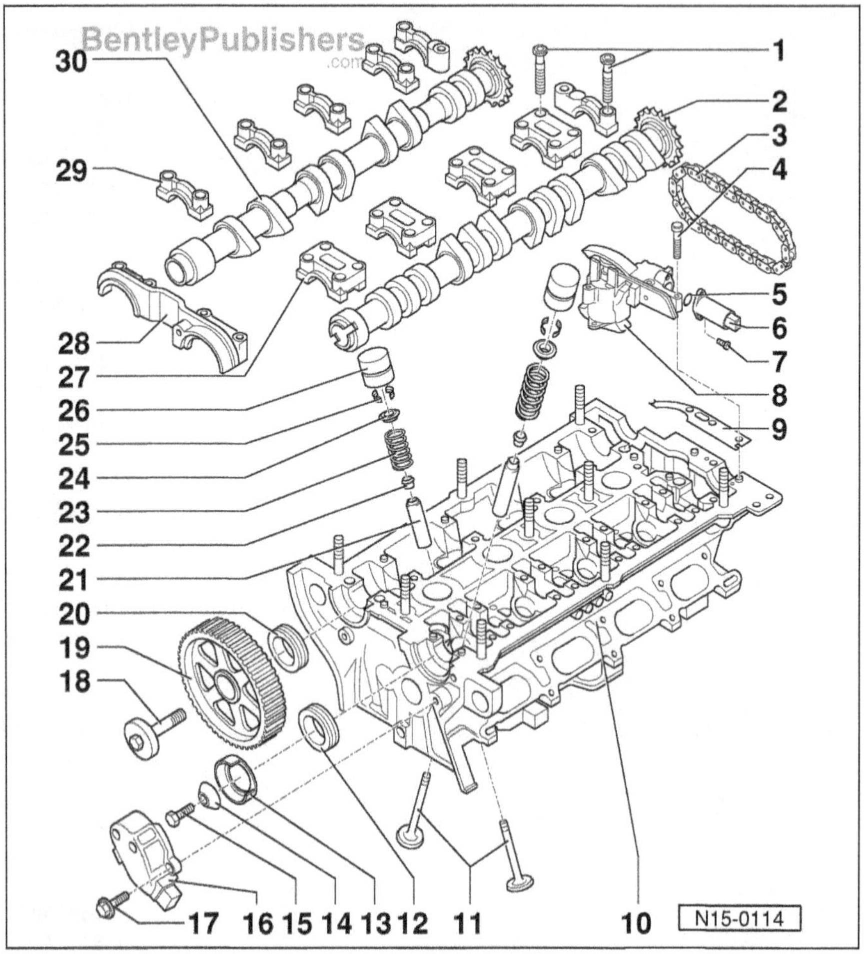

1. **Bolt**
   - 10 Nm (7 ft-lb)
2. **Intake camshaft**
   - Chain sprocket is not replaceable
   - Radial clearance checked with plastigage wear limit: 0.1 mm (0.004 in.)
   - Run out: maximum 0.035 mm (0.0014 in.)
3. **Drive chain**
   - Clean chain and sprocket and mark with ink or paint mark before removing
4. **Bolt**
   - 10 Nm (7 ft-lb)
5. **O-Ring**
   - Always replace
   - Not on APH engine
6. **Camshaft adjustment valve**
   - Not on APH engine
7. **Bolt**
   - 3 Nm (27 in-lb)
8. **Camshaft adjuster**
   - Not on APH engine
   - APH engine has chain tensioner
   - Secure with 3366 before removing
9. **Gasket**
   - Rubber/metal seal
   - Always replace
10. **Cylinder head**
11. **Valves**
    - Sodium filled (exhaust)
    - Do not rework! Only lapping is permitted
12. **Oil seal**
    - Do not oil sealing lip of PTFE seal
13. **Hood**
    - For camshaft position sensor
14. **Washer**
    - Conical
15. **Bolt**
    - 25 Nm (18 ft-lb)
16. **Camshaft position sensor (G40)**
17. **Bolt**
    - 10 Nm (7 ft-lb)
18. **Bolt**
    - 65 Nm (48 ft-lb)
    - Use counterhold 3036 to loosen/tighten
19. **Camshaft sprocket**
    - Note installed position: TDC mark for cylinder #1 visible on outside of sprocket
20. **Oil seal**
    - Do not oil sealing lip of PTFE seal
21. **Valve guide**
22. **Valve stem seal**
23. **Valve spring**
24. **Upper valve spring plate**
25. **Valve keepers**
26. **Hydraulic lifter**
    - Do not interchange
    - Equipped with hydraulic clearance compensation
    - Store with cam contact surface facing downwards
    - Oil contact surfaces
27. **Intake camshaft bearing cap**
    - Note installation position and sequence
28. **Double bearing cap**
    - Apply sealant to bearing cap/cylinder head mating surface before installing
    - Apply sealant to outer edges before installing
    - Ensure proper seating on dowel pins
29. **Exhaust camshaft bearing cap**
    - Note installation position and sequence
30. **Exhaust camshaft**
    - Chain sprocket is not replaceable
    - Radial clearance checked with plastigage wear limit: 0.1 mm (0.004 in.)
    - Run out: maximum 0.035 mm (0.0014 in.)

External cylinder head assembly

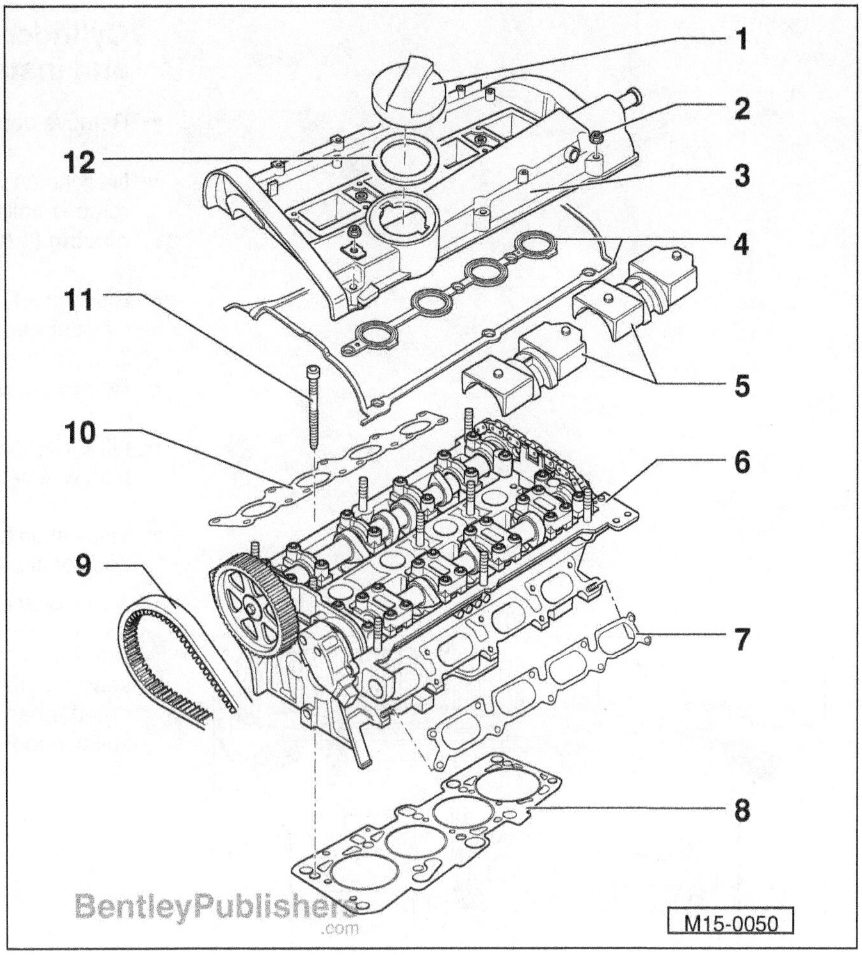

1. **Cap**
   - Replace gasket if damaged
2. **Bolt**
   - Tighten to 10 Nm (7 ft-lb)
3. **Valve cover**
4. **Gasket, valve cover**
   - Replace if damaged
   - Apply sealant to corner areas at double bearing cap and chain tensioner of before installing
5. **Oil deflector**
6. **Cylinder head**
   - Replace engine coolant if removing head
7. **Gasket, intake manifold**
   - Always replace
8. **Gasket, cylinder head**
   - Always replace
   - Metal construction
   - Installation position: part number must be readable from intake side
9. **Timing belt**
   - Mark direction of belt before removing
10. **Gasket, exhaust manifold**
    - Always replace
11. **Cylinder head bolt**
    - Always replace
    - Loosen and tighten in sequence only
12. **Gasket**
    - APH engine has connecting piece

Cylinder head (valve) cover, removing and installing

- Remove upper sound absorber panel.
- Disconnect harness connectors from ignition coils and remove coils. See **03 Maintenance, Spark plugs, replacing (1.8L engine)**.
- Disconnect crankcase breather valve hoses from cylinder head cover.
- Remove nuts from cylinder head cover.
- Lift off cylinder head cover with outer gasket and spark plug well gasket.
- Installation is reverse of removal noting the following additional points:
  - Use new gaskets for installation.

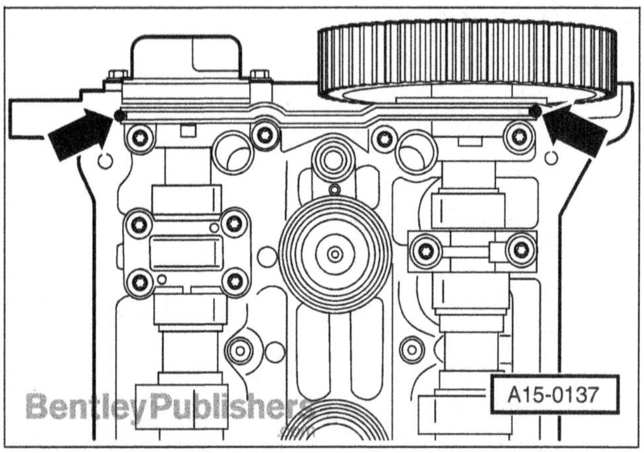

Carefully apply a thin coat of sealant to both edges of sealing surfaces between double bearing cap and cylinder head using a small screwdriver or other suitable tool.

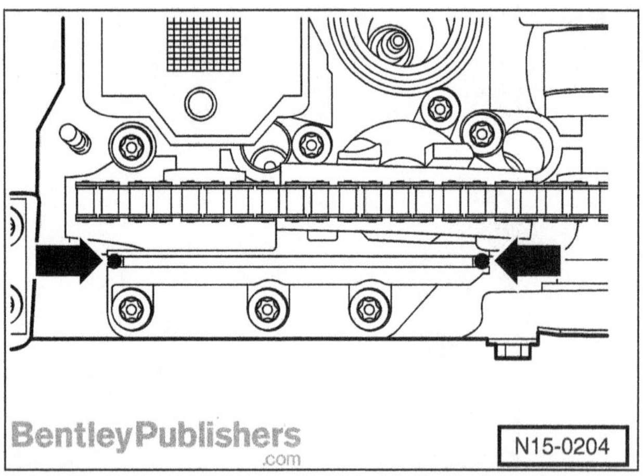

Carefully apply a thin coat of sealant to both edges of sealing surfaces between camshaft adjuster and cylinder head using a small screwdriver or other suitable tool.

Tightening torques

- Cylinder head cover to cylinder head 10 Nm (7 ft-lb)
- Ignition coil to cylinder head cover (APH engine only) . . . . 10 Nm (7 ft-lb)
- Ground wire to cylinder head cover . . 10 Nm (7 ft-lb)

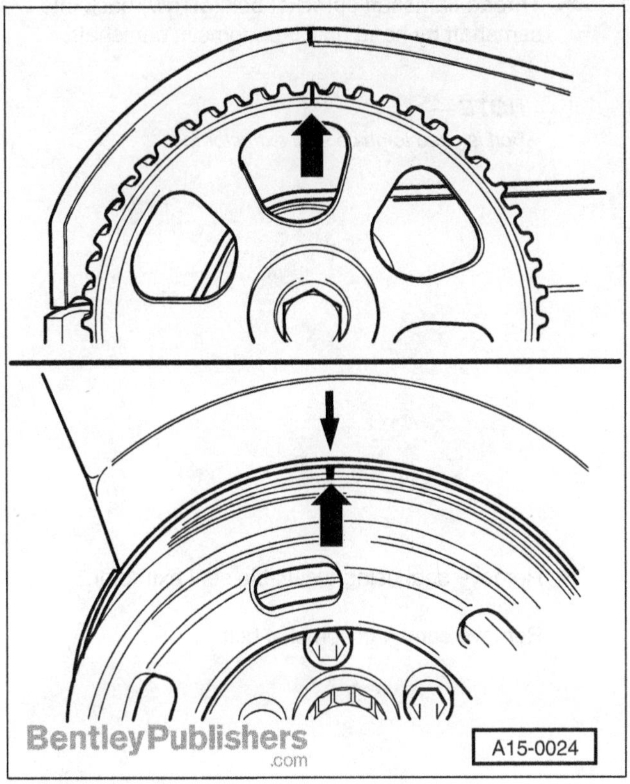

Camshaft oil seal, exhaust cam, replacing

– Remove upper sound absorber panel and unclip and remove timing belt guard, upper section.

◄ Set engine to TDC for cylinder #1 by turning crankshaft until timing mark on camshaft sprocket aligns with mark on cylinder head (valve) cover (**upper arrow**).

– Remove cylinder head (valve) cover as given earlier.

– Release tensioning roller and slide timing belt off of camshaft sprocket.

– Turn crankshaft back off of TDC slightly.

NOTE —

Turning the crankshaft away from TDC slightly minimizes the chances of contact between pistons and valves upon re-assembly.

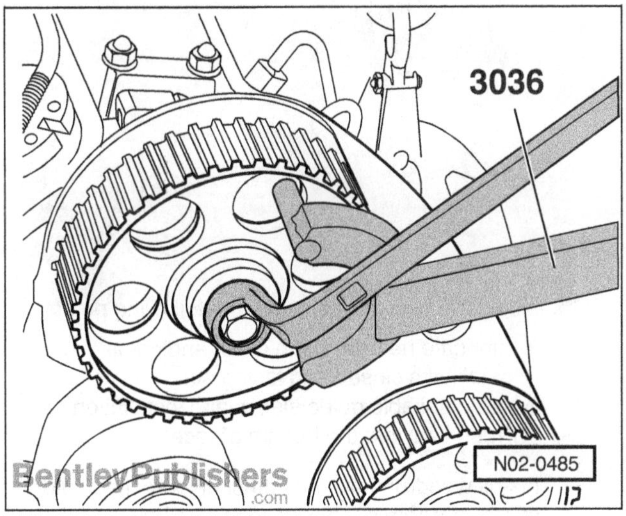

◄ Using a suitable spanner, remove camshaft sprocket bolt.

– Remove camshaft drive sprocket and woodruff key.

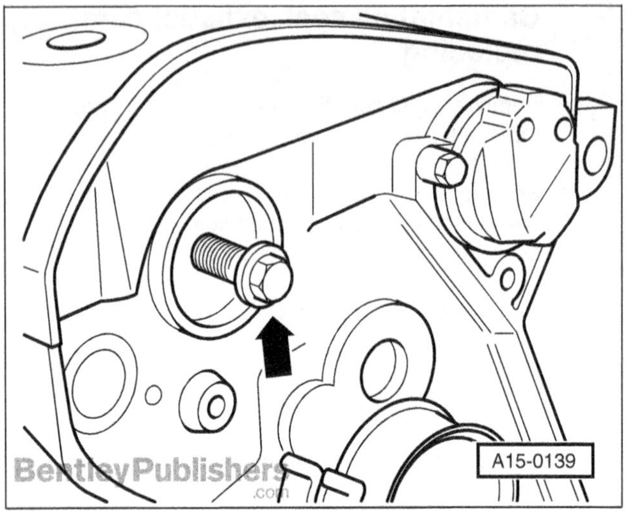

Thread camshaft sprocket bolt (**arrow**) back into camshaft by hand until it bottoms in camshaft.

NOTE —
Bolt is used to guide seal extractor tool.

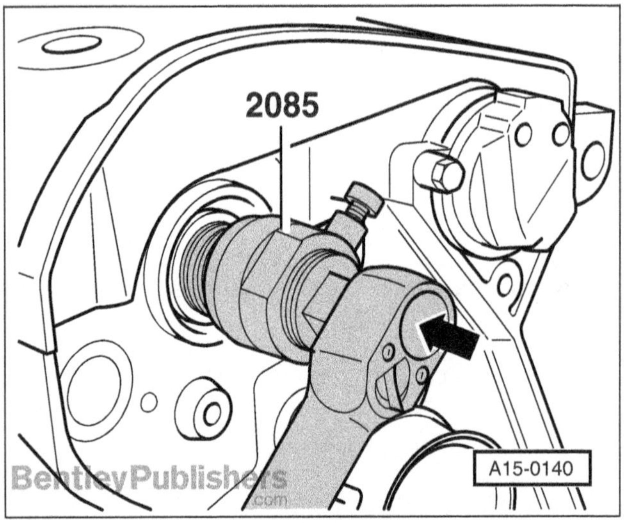

Remove seal using a suitable seal extractor.

– Remove camshaft sprocket bolt.

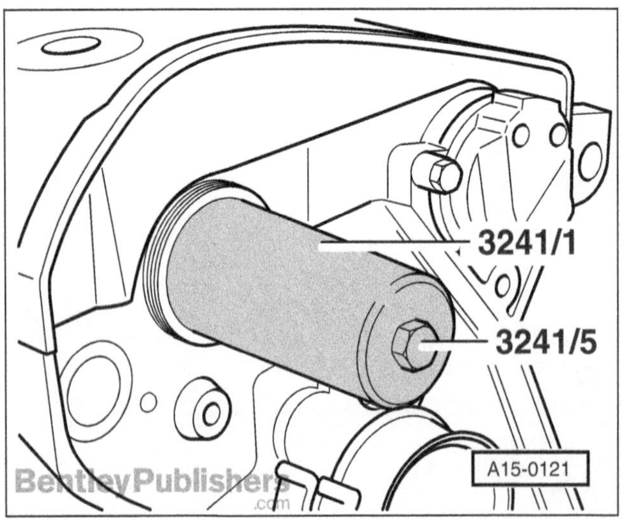

Install the new oil seal. Note the following points:

- Lubricate new seal with clean engine oil.
- Install with closed end facing out.
- Use a suitable guide sleeve and installation tool.
- Press seal into the bottom of recess.

– Install woodruff key and camshaft sprocket, hold sprocket with a spanner and torque bolt.

Tightening torque
- Camshaft sprocket to camshaft . . . . 65 Nm (48 ft-lb)

– Turn crankshaft back to TDC.

– Install timing belt and adjust tension, see **Timing Belt Assembly**, given later.

– Install cylinder head (valve) cover, breather valve hoses and ignition coils.

– Install upper timing belt guard, engine cover and any other removed components.

– Be sure to quality check your work.

Camshaft oil seal, intake cam, replacing

The camshaft position sensor is located behind the timing belt guard, upper section and is driven by the intake camshaft. A seal on the intake camshaft prevents oil leaks into the camshaft position sensor.

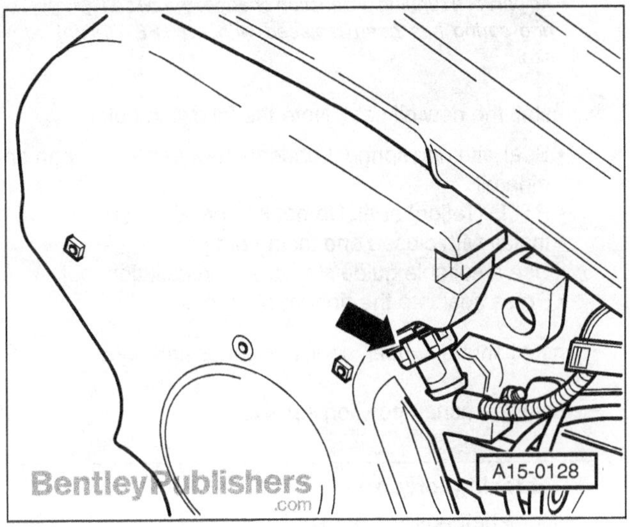

◄ Remove harness connector (**arrow**) from camshaft position sensor.

– Remove timing belt guard, upper section.

– Remove camshaft position sensor.

– Remove bolt, conical washer and shutter wheel from intake camshaft.

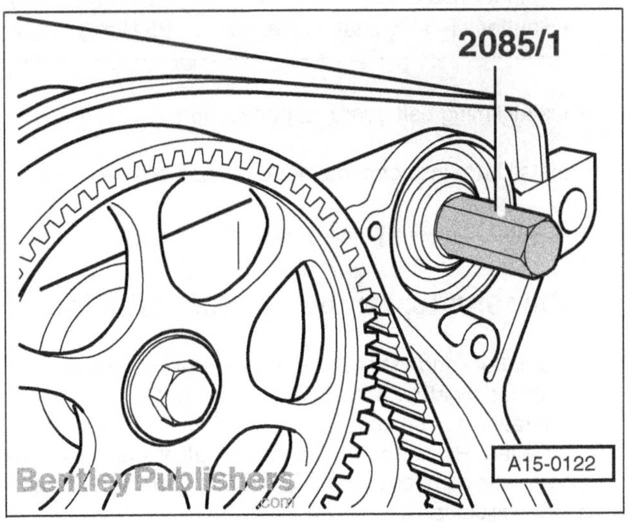

◄ Install guide tool into camshaft.

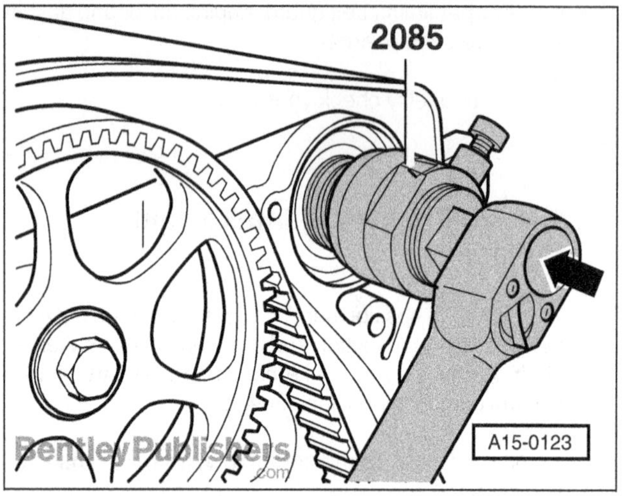

◀ Remove seal using a suitable seal extractor.

– Remove guide tool from camshaft.

NOTE —
As part of a running production change, the oil seal with ring spring has been replaced with a PTFE (Teflon) seal.

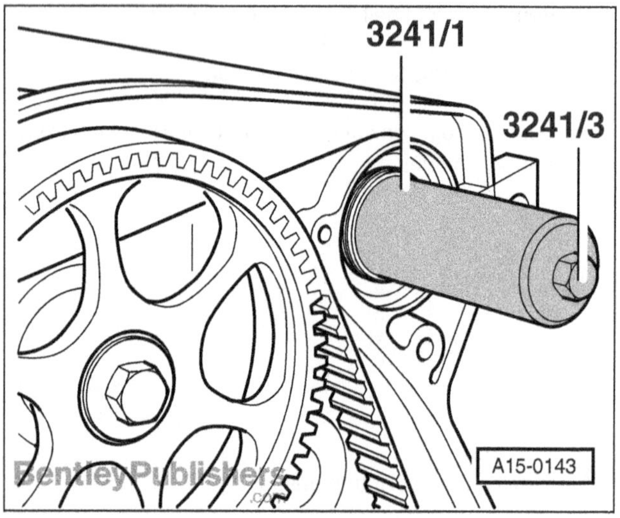

◀ Install the new oil seal. Note the following points:

- Seal with ring spring: Lubricate new seal with clean engine oil.
- PTFE (Teflon) seal: Do not lubricate.
- Install with closed end facing out.
- Use a suitable guide sleeve and installation tool.
- Press seal into the bottom of recess.

– Install shutter wheel, conical washer and bolt.

– Install camshaft position sensor.

Tightening torque
- Camshaft position sensor to cylinder head . . . . 10 Nm (7 ft-lb)
- Shutter wheel to intake camshaft . . . 25 Nm (18 ft-lb)

– Install timing belt guard, upper section.

– Install harness connector for camshaft position sensor.

– Be sure to quality check your work.

Camshafts, removing and installing

The intake and exhaust camshafts are removed and installed together with the drive chain and tensioner. This operation can be accomplished with the cylinder head installed. Removing the camshafts allows access to the hydraulic cam followers, the valve springs and the valve stem oil seals.

– Remove upper sound absorber panel and unclip and remove timing belt guard, upper section.

– Set engine to TDC for cylinder #1 by turning crankshaft until timing mark on camshaft sprocket aligns with mark on cylinder head (valve) cover as shown earlier.

– Remove cylinder head (valve) cover as given earlier.

- Release tensioning roller and slide timing belt off of camshaft sprocket.

- Turn crankshaft back off of TDC slightly.

NOTE —

Turning the crankshaft away from TDC slightly minimizes the chances of contact between pistons and valves upon re-assembly.

- Using a suitable spanner, remove the camshaft sprocket bolt as shown earlier.

- Remove camshaft drive sprocket.

- Remove camshaft position sensor, conical washer and shutter wheel.

- Lift off plastic oil deflector to expose camshafts.

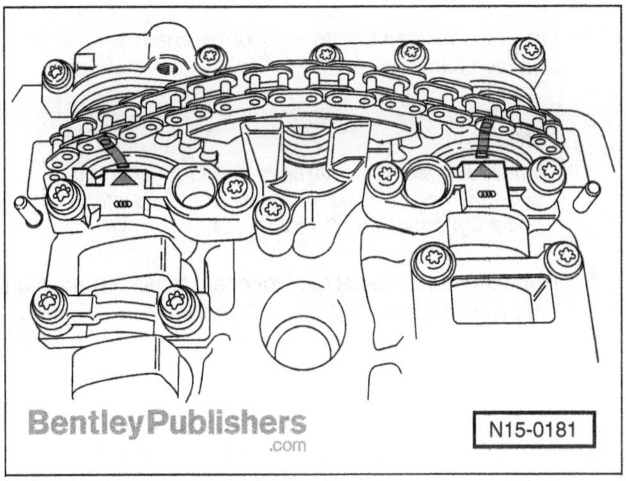

Clean drive chain and camshaft sprocket in areas of arrows on bearing caps with a suitable degreaser. Mark installed position of drive chain and camshaft chain sprocket with a permanent marker pen.

NOTE —

- *Do not mark chain with a punch mark, notch or similar.*
- *The distance between both arrows or the permanent marker marks equals 16 rollers on the drive chain.*

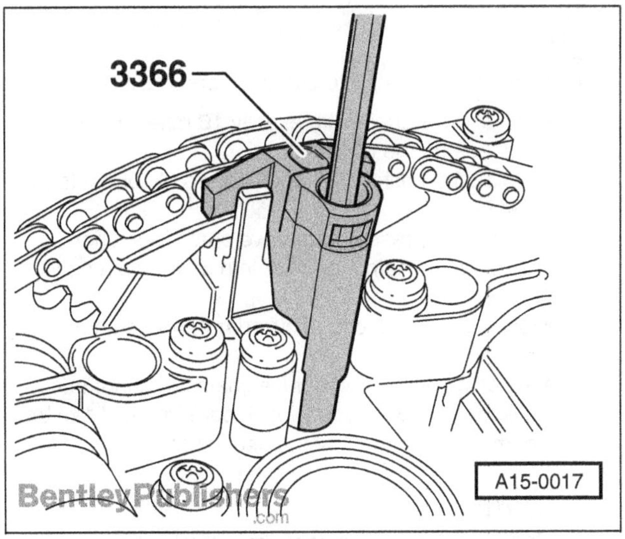

Compress drive chain tensioner.

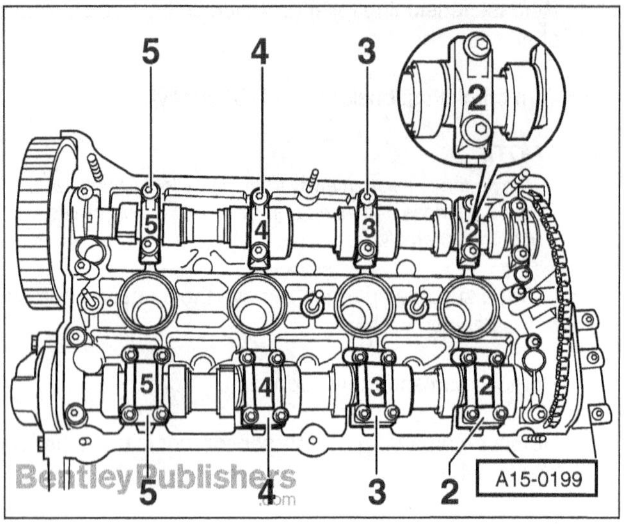

◀ Remove both camshafts together with the tensioner and the chain. Use the following sequence:

- Remove intake camshaft bearing caps 3 and 5.
- Remove exhaust camshaft bearing caps 3 and 5.
- Remove double bearing cap at sprocket end.
- Remove intake and exhaust camshaft bearing caps at chain sprocket end.
- Remove chain tensioner securing bolts.
- Remove intake camshaft bearing caps 2 and 4 alternately and evenly.
- Remove exhaust camshaft bearing caps 2 and 4 alternately and evenly.
- Remove drive chain tensioner.
- Remove drive chain tensioner compressor.

NOTE —
Mark each bearing cap before removing. Each cap is machined to each journal and must be reinstalled to that exact location to prevent damage.

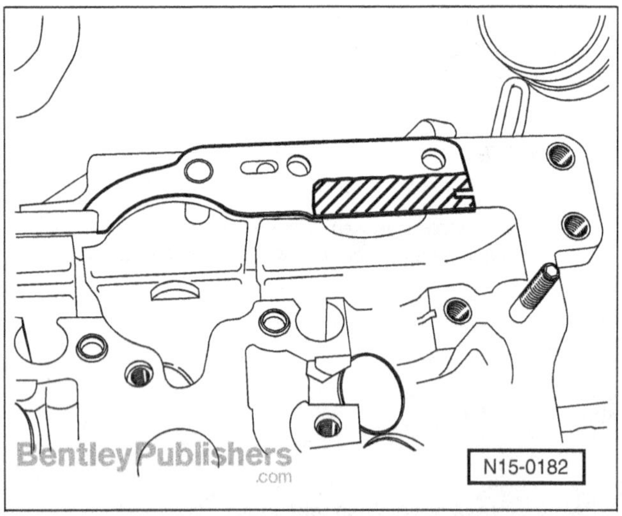

– Lift out both camshafts with drive chain and tensioner and lay aside.

NOTE —
Once removed, the camshafts should be checked for wear and other visible damage as described later under **Cylinder Head Components**.

◀ Replace rubber/metal seal for chain tensioner and lightly coat with sealant.

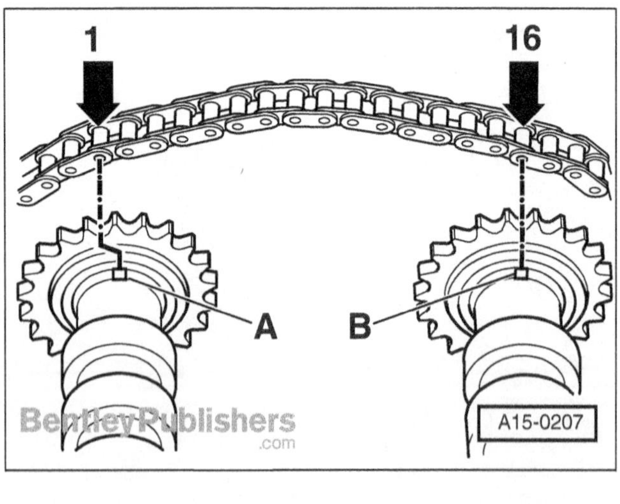

◀ Install drive chain onto camshaft sprockets.

- Install drive chain with exactly 16 rollers between the notches on the camshafts.
- Exhaust camshaft (**A**) notch is slightly offset to the inside.
- Install compressor onto drive chain tensioner.
- Slide drive chain tensioner into drive chain between camshafts.
- Oil bearing surfaces of cylinder head and both camshafts.
- Install camshafts with drive chain tensioner into cylinder head. Lobes for cylinder number #1 of both camshafts and notches must point up.

NOTE —
When re-using an original drive chain, align marks made prior to removal. In all instances, however, there must be exactly 16 rollers between notches on camshafts.

– Install drive chain tensioner over dowel sleeves. Torque to specification.

– Install bearing caps 2 and 4 for exhaust and intake camshafts over dowel sleeves. Tighten nuts alternately and evenly so that camshafts are drawn down fully and evenly into bearing saddles. Torque to specification.

Tightening torques
- Camshaft bearing cap to cylinder head . . . . 10 Nm (7 ft-lb)
- Drive chain tensioner to cylinder head . . . . 10 Nm (7 ft-lb)

> ***CAUTION —***
> - *Be sure to install bearing caps correctly. The caps are bored off-center and only fit properly one way. Caps are matched to the cylinder head by the machining process and are not available separately. Broken caps will require the replacement of the cylinder head.*
> - *When installed properly, cap markings can be read from intake side of cylinder head.*

– Install bearing caps over dowel sleeves at chain sprocket ends of both camshafts. Torque to specification.

– Remove drive chain tensioner compressor.

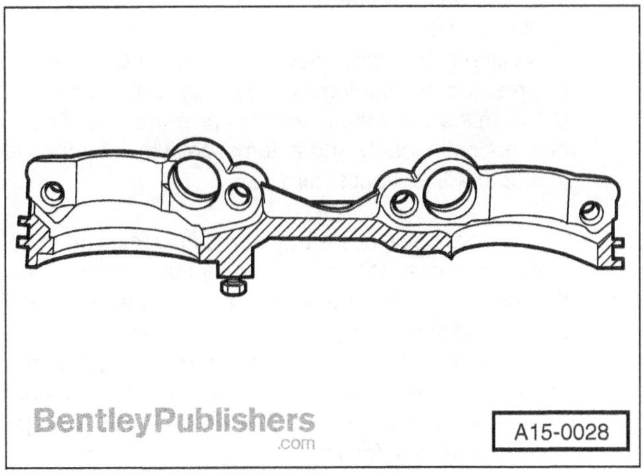

< Lightly coat double bearing cap with sealant. Install over dowel sleeves. Torque to specification.

– Install remaining bearing caps. Torque to specification.

Tightening torque
- Camshaft bearing cap to cylinder head. . . . . 10 Nm (7 ft-lb)

– Observe installed position of camshaft drive sprocket.

– Install camshaft drive sprocket using a suitable spanner as shown earlier.

NOTE —
Ensure that camshaft lobes for cylinder #1 are still pointing up.

Tightening torque
- Camshaft sprocket to camshaft . . . . 65 Nm (48 ft-lb)

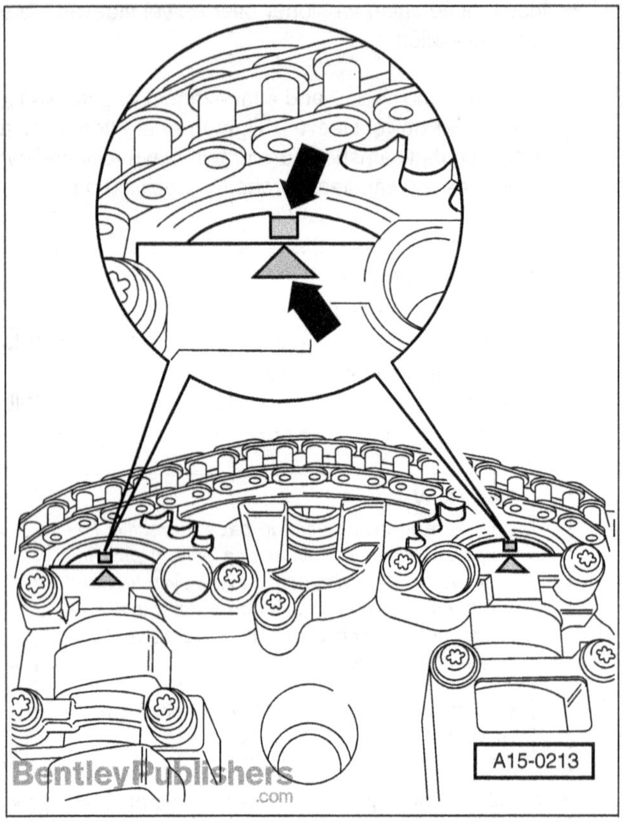

◀ Re-check position of camshafts. Make sure arrows on bearing caps align with notches on camshafts. There must be at least 16 chain rollers between notches.

– Install cylinder head (valve) cover.

– Install timing belt and adjust tension, see **Timing Belt Assembly** given later.

– Install timing belt guard, upper sound absorber panel and any other removed components.

– Be sure to quality check your work.

Hydraulic cam followers, checking

The 1.8 liter engine installed in the New Beetle is equipped with hydraulic cam followers, also known as valve lifters. The cam followers are pumped up by engine oil pressure, expanding as necessary to fill the gap between the valve stem and the camshaft lobe. This occurs continuously and automatically to keep the valve in proper adjustment at all times.

Some valve noise at start-up and during the warm-up period is normal at times due to hydraulic cam followers that have bled down while the engine was not running. Before checking noisy cam followers, check that the engine oil is new and fresh, of the correct weight, and that the level is correct. Allow 2 minutes with a warm engine running at a fast idle for the lubrication system to properly pump up the cam followers.

Cam followers should only be checked when the engine is warm. Run the engine until the radiator fan has switched on at least one time. Increase the engine speed to 2,500 rpm and hold it there for approximately 2 minutes. If the hydraulic cam followers are still noisy, shut the engine off and proceed as follows while the engine is still warm.

– Remove cylinder head (valve) cover.

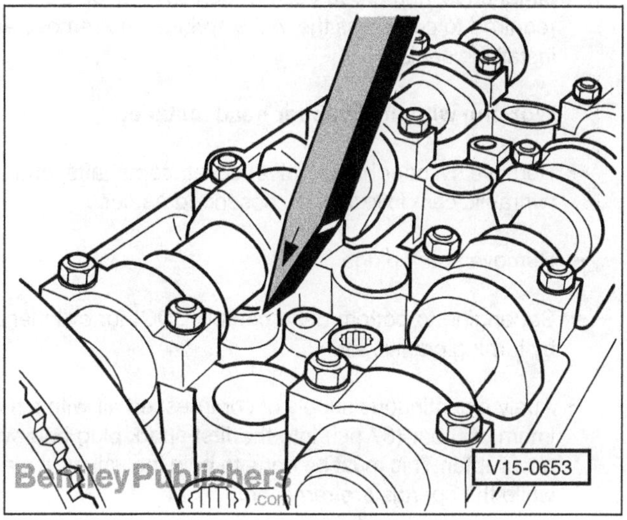

– Turn engine by hand in the running direction until all camshaft lobes of cylinder #1 are pointing approximately up.

NOTE —
Hydraulic cam follower clearance can be checked on any cam follower (lifter) that is not being depressed by the camshaft lobe.

< Check clearance between cam follower and cam lobe on cylinder #1:

- Insert a 0.2 mm (0.008 in.) feeler gauge between the camshaft lobe and the follower. If clearance exceeds 0.2 mm (0.008 in.) replace the follower.
- If clearance is less, or if no clearance is present, lightly depress cam follower with a wooden or plastic wedge and insert a feeler gauge between camshaft lobe and the follower. If clearance exceeds 0.2 mm (0.008 in.), the lifter is faulty and should be replaced.

Specification
- Hydraulic cam follower maximum clearance (1.8L engine) . . . . 0.2 mm (0.008 in.)

– Turn engine by hand and repeat procedure until all cam followers have been checked.

– Replace a faulty cam follower by removing camshaft as previously described and pulling the follower from the cylinder head. Faulty hydraulic cam followers can be replaced individually but only as a complete assembly.

CAUTION —
After installing new cam followers, the engine should not be started for at least 30 minutes. New cam followers are usually at full extended height and must be allowed to bleed down to their proper height once installed. Failure to do so may cause the valves to strike the pistons resulting in serious damage.

NOTE —
Store removed hydraulic cam followers in order on a clean surface with the camshaft contact surface facing down to minimize bleed down. Cover with a clean lint-free shop cloth.

Valve stem oil seals, replacing

The sign of faulty valve stem seals is excessive oil consumption and oil smoke from the exhaust. This is usually most noticeable during periods of high manifold vacuum such as deceleration. If compression and leak-down testing confirm the integrity of the piston rings and the turbocharger shaft seals are good, but the engine consumes oil, it is possible that faulty valve stem seals are present. It should also be noted that worn valve stem seals could be due to worn valve guides and that the worn valve guides are the major cause of the oil consumption. See **Cylinder Head Components** for more information on checking valve guides.

Replacing the valve stem oil seals requires removal of the camshaft, cam followers, and the valve springs. This can be done with the cylinder head installed or removed. In either case, numerous Volkswagen special tools are required to compress the valve springs and remove and install the seals.

Working with the cylinder head installed:

– Remove cylinder head (valve) cover, camshafts, and hydraulic cam followers as described earlier.

– Remove spark plugs.

– Set engine to bottom dead center (BDC) for cylinder #1 by turning crankshaft.

– Apply a continuous supply of compressed air with a minimum of 6 bar (87 psi) into the first spark plug hole with an adapter. This must be done to hold the valves in place while the springs are removed.

> ***CAUTION —***
> *Compressed air supply must be able to maintain at least 6 bar (87 psi) during this repair. If air supply is interrupted while valve spring is removed, valve will fall into the cylinder and may require cylinder head removal to retrieve.*

Working with the cylinder head removed:

– Secure cylinder head to the workbench. Use care to avoid damage to the head gasket surfaces.

– Remove camshaft and hydraulic cam followers as described earlier. Continue with steps below.

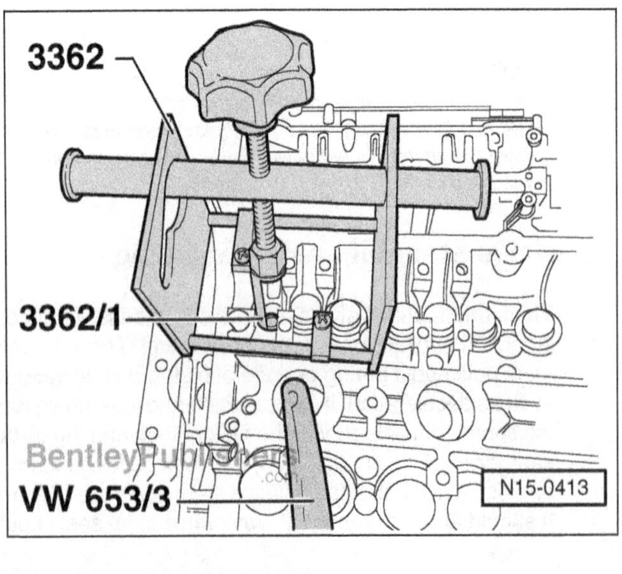

Working with cylinder head removed or installed:

◄ Install appropriate valve spring compressor tools. Compress the spring for the first cylinder and remove the spring retainer, both keepers, and the valve spring. If spring will not compress, lightly tap the tool to release the stuck keepers. Use a small magnet to retrieve the keepers.

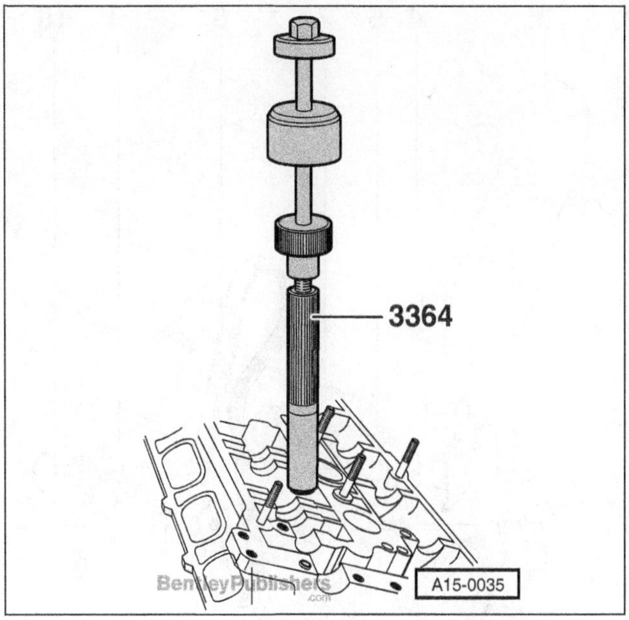

Remove valve stem seals with special slide hammer tool.

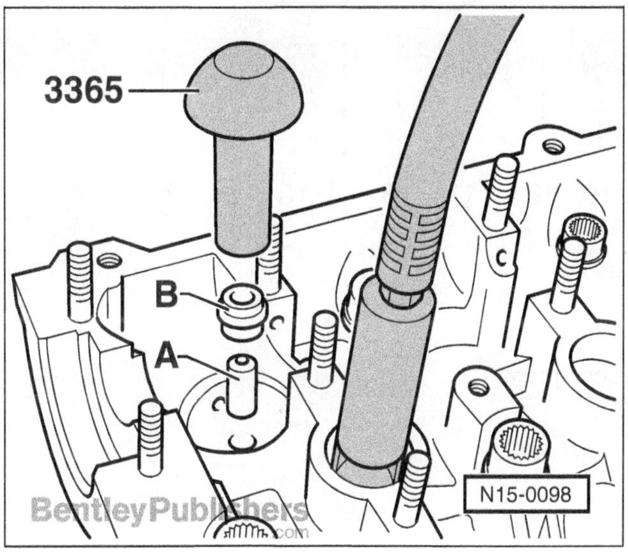

Begin installation of new seal by temporarily fitting a protective plastic fitting sleeve (**A**) over the valve stem. This sleeve is usually included with the new seal set and will prevent damage to the seal (**B**) due to the sharp edges of the keeper grooves. Lubricate new seal with clean engine oil and fit it to the installation tool.

- Push the tool (with seal) down over the valve stem until the seal is fully seated on the guide. Remove the tool and the protective fitting sleeve.
- Reinstall the valve spring, retainer and keepers.
- Repeat for the second valve on the cylinder.
- When all valve seals have been replaced on the first cylinder, transfer the compressed air adapter (if working with cylinder head installed) to the next cylinder and repeat the process until all valve seals have been replaced.
- Remaining installation is the reverse of removal.

> ***CAUTION—***
> *To prevent cylinder head or camshaft damage, be sure to follow the camshaft installation procedure exactly when tightening the camshaft bearing caps. See* **Camshaft, removing and installing**.

Tightening torque
- Spark plug to cylinder head . . . . 30 Nm (22 ft-lb)

- Be sure to quality check your work.

TIMING BELT ASSEMBLY

See **03 Maintenance** for checking the timing belt and replacement intervals, based on mileage. Although no time interval is specified, the publisher recommends replacement every 4 years. This will help prevent damage to the engine due to belt stretch and the long term effects of heat. The timing belt is also known as a toothed belt or camshaft drive belt.

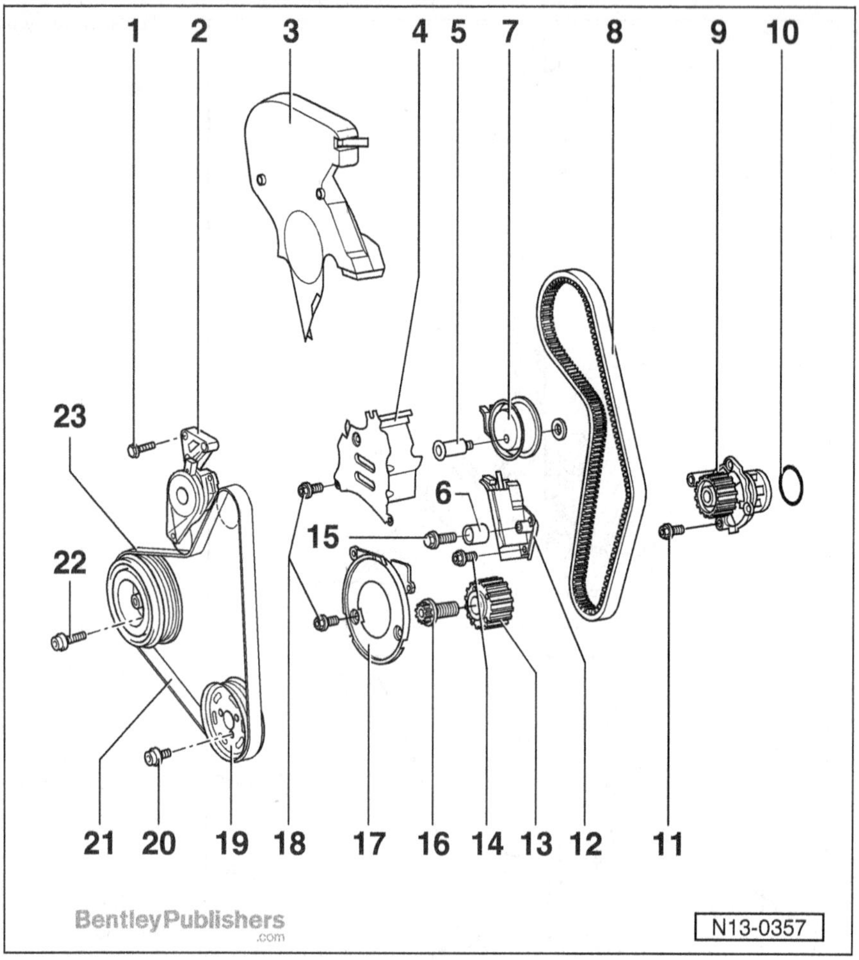

1. **Bolt**
   - Tighten to 25 Nm (18 ft-lb)
2. **Tensioner, ribbed V-belt**
   - Loosen ribbed belt by turning with open-end wrench
3. **Timing belt guard, upper**
4. **Timing belt guard, center**
5. **Bolt, fitted**
   - Tighten to 27 Nm (20 ft-lb)
6. **Idler roller**
7. **Tensioner roller**
8. **Timing belt**
   - Mark engine direction of rotation before removing
   - Check for wear
   - Do not kink
9. **Coolant pump**
10. **O-ring**
    - Always replace
11. **Bolt**
    - Tighten to 15 Nm (11 ft-lb)
12. **Tensioning damper**
13. **Crankshaft sprocket**
14. **Bolt**
    - Tighten to 15 Nm (11 ft-lb)
15. **Bolt**
    - Tighten to 20 Nm (15 ft-lb)
16. **Bolt**
    - Tighten to 90 Nm (66 ft-lb) + ¼ turn (90°)
    - Always replace
    - Threads and shoulder must be free of grease
17. **Timing belt guard, lower**
18. **Bolt**
    - Tighten to 10 Nm (7 ft-lb)
19. **Pulley**
    - For power steering pump
20. **Bolt**
    - Tighten to 25 Nm (18 ft-lb)
21. **Ribbed V-belt**
    - Mark direction of rotation before removing
22. **Bolt**
    - Tighten to 25 Nm (18 ft-lb)
23. **Belt pulley/vibration damper**
    - Can only be installed in one position, holes are offset
    - Note position when installing timing belt

Fig.1 Timing belt tensioner assembly

- Damper (**1**)
- Damper piston (**2**)
- Damper arm (**3**)
- Roller (**4**)

Timing belt, removing

– Remove the upper sound absorber panel.

– Remove the lower sound absorber panel (belly pan).

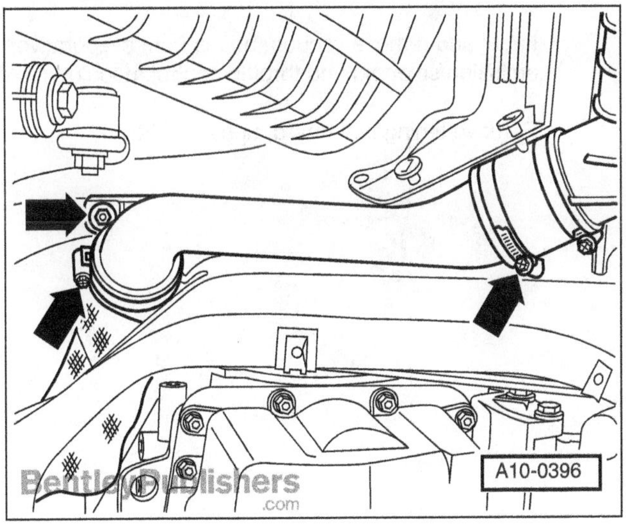

– Remove connecting pipe between intercooler and bottom right side of body longmember.

– Remove the ribbed V-belt and tensioner. See **03 Maintenance**.

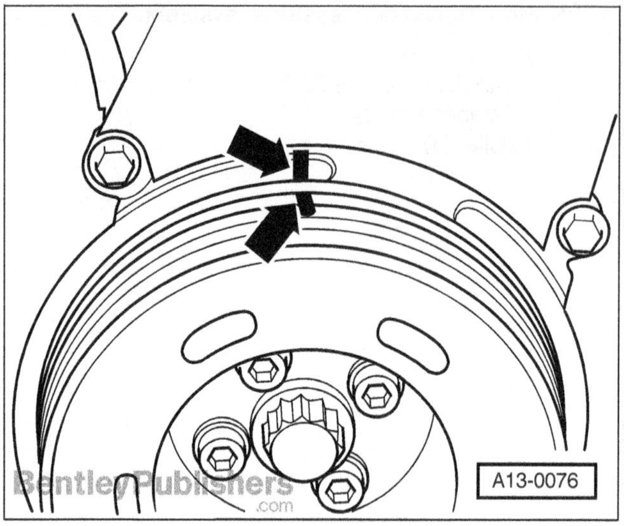

< Set engine to TDC for cylinder #1 by turning crankshaft until timing mark on camshaft sprocket aligns with mark on cylinder head (valve) cover.

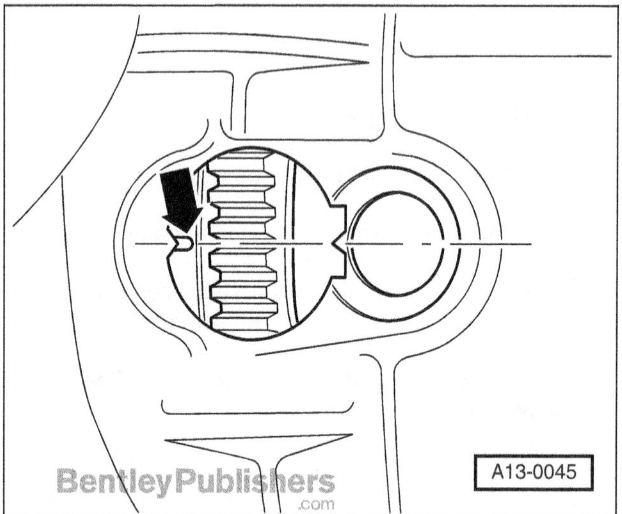

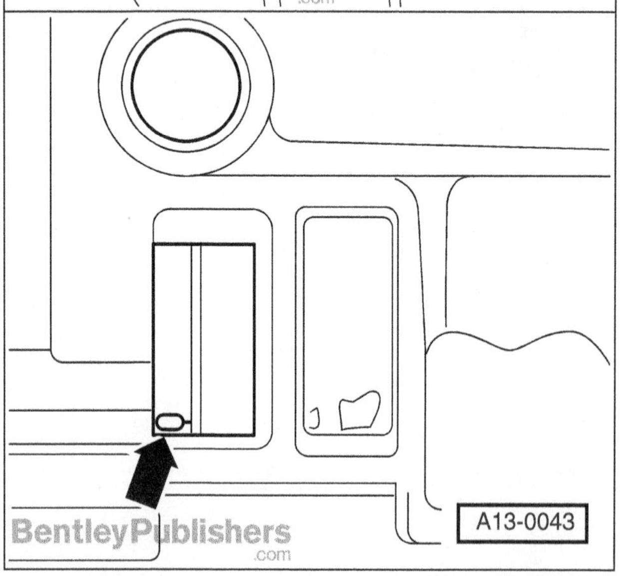

< For manual transmission vehicles, make sure mark on flywheel (**upper arrow**) shown aligned with mark on manual transmission case. For automatic transmission equipped vehicles make sure mark on torque converter (**lower arrow**) shown aligned with edge of automatic transmission case.

– Remove coolant expansion without disconnecting hoses and move to the side.

– Label and remove vacuum hoses from evaporative emission solenoid and throttle control module.

– Remove timing belt guard, upper section.

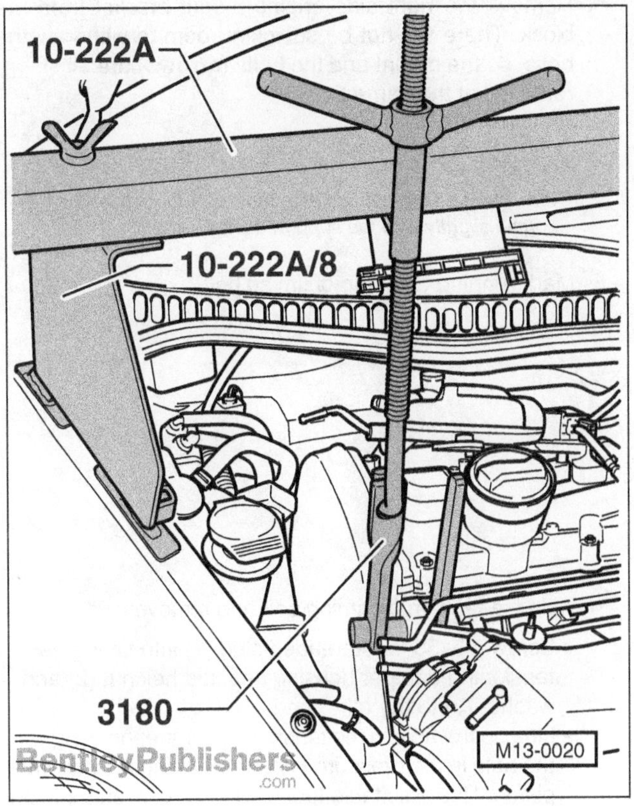

◄ Support engine with a suitable fixture that is designed to support the weight of the engine and transmission without damaging the body.

– Lift the supporting device slightly, so that the weight of the engine and transmission is on the supporting device.

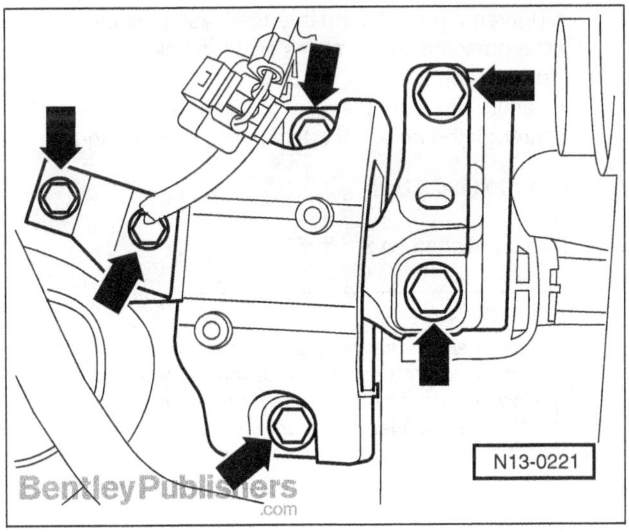

◄ Remove bolts (**arrows**) for right side engine mount and remove from body and carrier on engine.

– Remove the 4 bolts holding the belt pulley/vibration damper to the timing belt sprocket and remove the belt pulley.

– Remove the timing belt guard, center and lower sections.

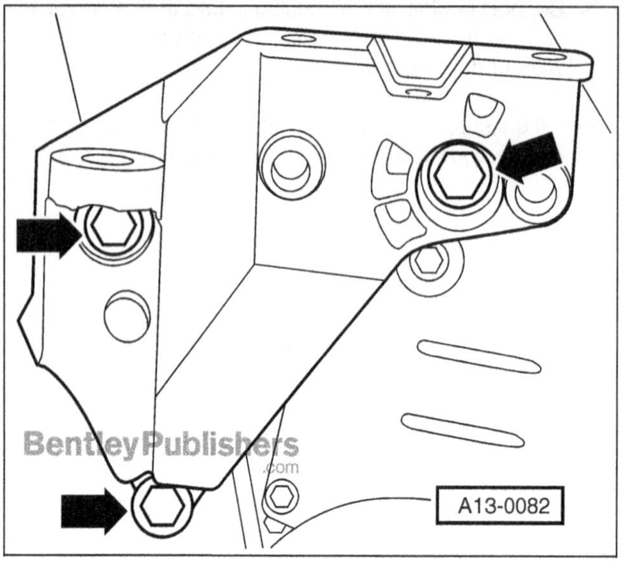

Remove the right side engine mount bracket from the block. There will not be sufficient room to withdraw the bolts, so the mount and the bolts (**arrows**) are all removed at the same time.

NOTE —
When loosening the right side engine mount, raise the engine slightly with the support device.

– Mark running direction of timing belt.

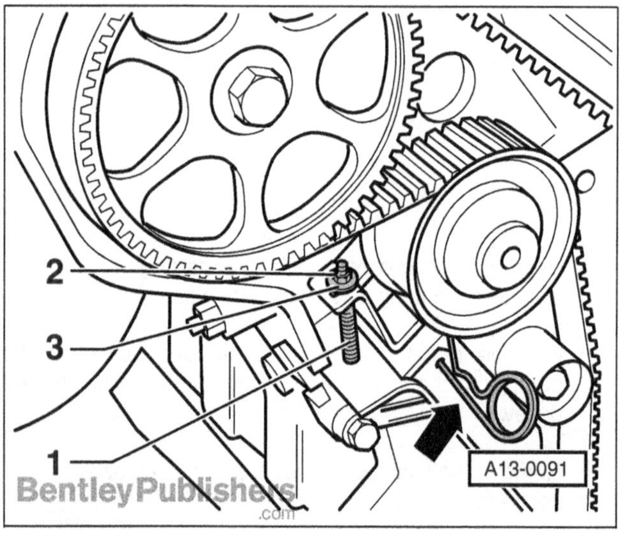

Release tension on timing belt and remove belt.

- Screw a 5x55 mm threaded stud (**1**) into timing belt tensioning damper housing and fit a hex nut (**2**) and washer (**3**) to the other end
- Line up the hole in the tensioning damper housing with the hole in the pressure piston by turning the pressure piston. Use a suitable tool to turn the pressure piston before tightening the nut on the threaded stud
- Tighten the nut on the threaded stud until the hole in the pressure piston aligns with the hole in the tension damper housing
- Secure pressure piston in position with a pin inserted through the hole in the tensioning damper housing

– Remove timing belt.

– Turn crankshaft back off of TDC slightly.

CAUTION —
Due to engine design, care must be used when turning the camshaft with the timing belt removed. If valves are allowed to open with the piston at TDC, serious internal damage will result.

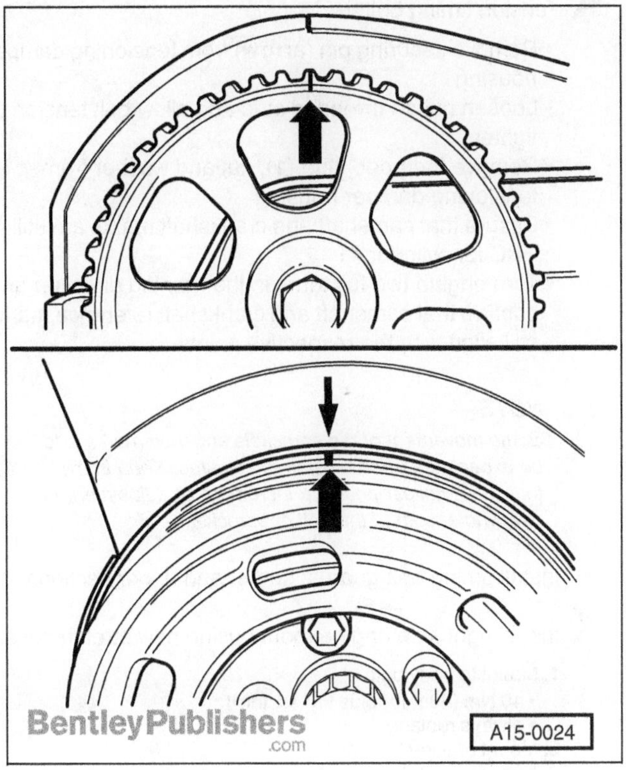

Timing belt, installing

> **CAUTION—**
> *Belt tension must not be adjusted on a hot engine. Allow to cool sufficiently before proceeding. Engine temperature must be no hotter than warm to the touch.*

– Ensure that crankshaft is NOT at TDC.

< Align the mark on the camshaft with the mark on the timing belt guard. This brings the camshaft to TDC for cylinder #1.

– Install timing belt onto the crankshaft drive sprocket.

NOTE—
If re-using an old belt, be sure to note running direction marks placed on it before removal.

– Install bolts into the right side engine mount bracket and install to the cylinder block.

Tightening torques
- Engine mount bracket to cylinder block . . . . 45 Nm (33 ft-lb)

– Install timing belt guard, lower section.

– Install the belt pulley/vibration damper to the timing belt drive sprocket.

Tightening torques
- Vibration dampener/belt pulley to timing belt drive sprocket (M8) . . . . 25 Nm (18 ft-lb)

– Turn the engine to top dead center (TDC) for cylinder number #1 by aligning mark on belt pulley with the mark on the timing belt guard.

– Ensure that camshaft is still at TDC for cylinder number 1.

– Install the timing belt onto the tensioner, coolant pump and camshaft drive sprockets and check for correct placement of belt on all sprockets.

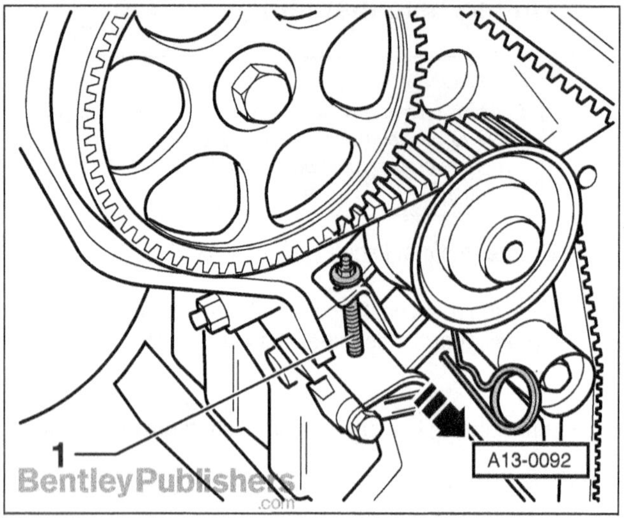

◀ Tension timing belt.

- Remove securing pin (**arrow**) from tensioning damper housing
- Loosen nut on threaded stud and allow belt tension to tighten
- Remove threaded stud (**1**), nut and washer from tensioning damper housing
- Ensure that camshaft and crankshaft marks are still at TDC for cylinder #1
- Turn engine two full turns in the running direction and confirm that camshaft and crankshaft reference marks still align with the respective points

NOTE —

Some movement of the sprockets and their marks is to be expected as belt tension is adjusted. Keep in mind that the smallest possible increment of adjustment is one whole tooth of the belt or sprocket.

– Install timing belt guards, center and upper sections.

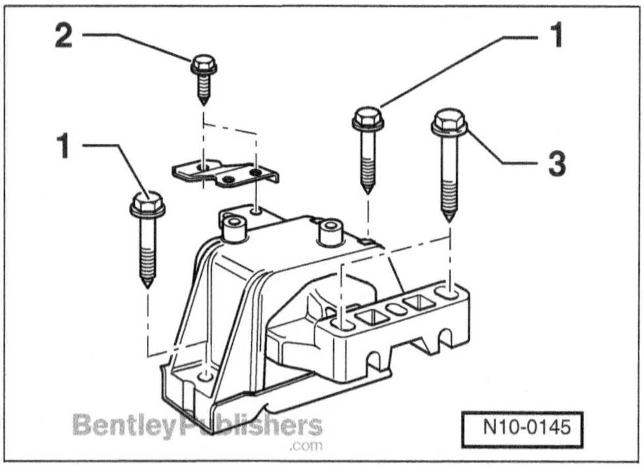

◀ Install right side engine mount using new stretch bolts.

1. Mount to body bolt
   - 40 Nm (30 ft-lb) plus 90° (¼ turn)
   - always replace
2. Mount bracket to body bolt
   - 25 Nm (18 ft-lb)
3. Mount to engine bracket bolt
   - 100 Nm (74 ft-lb)

– Remove engine support fixture.

– Install ribbed belt and tensioner.

Tightening torque

- Ribbed belt tensioner to bracket (M8) . . . . 25 Nm (18 ft-lb)

– Install connecting pipe between intercooler and bottom right side of body longmember.

– Install the lower sound absorber panels (belly pan) and the upper sound absorber panel.

– Install any remaining parts and quality check your work.

CYLINDER HEAD, REMOVING AND INSTALLING

The cylinder head can be removed and installed with the engine in the vehicle. Note that the cylinder head bolts are stretch type fasteners and should never be reused. In addition, whenever the cylinder head or the cylinder head gasket is replaced, the coolant must also be replaced, see **19 Engine–Cooling System**.

Cylinder head, removing

> ***WARNING —***
> *Do not start work on a hot engine. Allow to cool sufficiently before proceeding. Engine temperature must be no hotter than warm to the touch. Cylinder head warpage can result due to uneven cooling rates.*

NOTE —

- *Disconnecting the battery cables will erase fault codes and basic settings in the engine management and automatic transmission control unit memories. Some driveability problems may be noticeable until the system re-adapts to operating conditions. OBD II readiness codes, which may be required for emissions testing, may also be erased. See* **24 Fuel Injection–Gasoline** *for additional information. In most instances proper diagnosis will require the use of a diagnostic scan tool. See* **OBD On Board Diagnostics** *for more information.*
- *Cylinder head will be removed with the exhaust manifold and the intake manifold still attached.*

– With ignition switched off, disconnect the battery negative terminal from the battery.

NOTE —
Be sure to have the anti-theft radio code on hand before disconnecting the battery.

– Remove upper sound absorber panel.

– Remove the lower sound absorber panels (belly pan).

– Drain engine coolant, see **19 Engine–Cooling System**.

– Remove secondary air pump and mounting bracket.

– Disconnect fuel supply and fuel return lines.

> ***WARNING —***
> *Fuel will be expelled when disconnecting fuel hoses. Wrap a cloth around the fuel line fittings before disconnecting them. Do not smoke or work near heaters or other fire hazards. Have a fire extinguisher handy.*

– Disconnect fuel injector harness connectors.

– Label and disconnect vacuum connections at cylinder head.

– Seal off disconnected fuel and vacuum lines to prevent contamination.

– Unbolt the coolant connection flange from the end of the cylinder head and carefully move aside. The coolant hoses stay connected.

– Disconnect ignition coil harness connectors and ground wires.

– Disconnect all other electrical connectors as necessary and move aside.

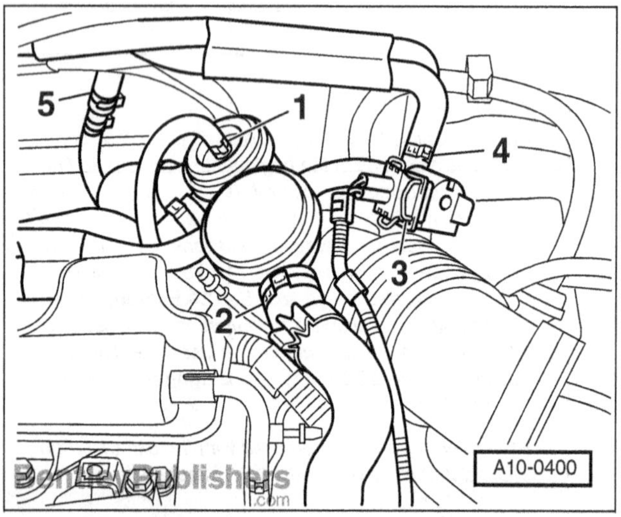

APH engine

◀ Pull off pressure control valve hose (**2**) from crankcase breather.

– Disconnect harness connector and hose (**4**) from wastegate bypass regulator valve (**3**).

– Separate hose (**5**) from solenoid valve to turbocharger on bulkhead.

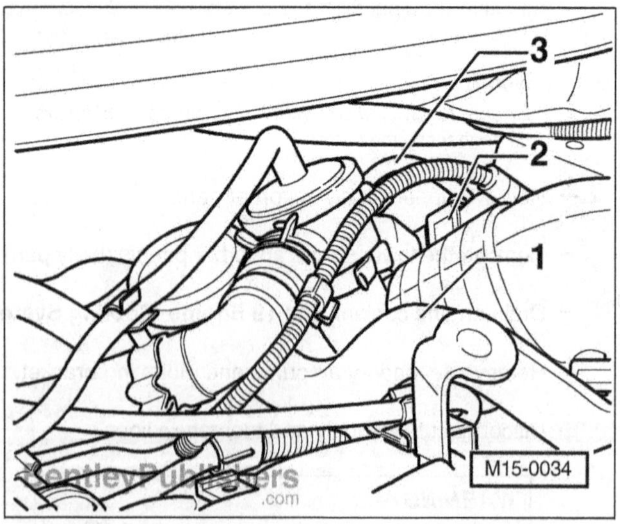

AWP, AWV, BKF, BNU engines

◀ Pull off pressure control valve hose from crankcase breather (**1**).

– Disconnect harness connector and hose (**3**) from wastegate bypass regulator valve (**2**).

– Separate hose between EVAP canister and turbocharger on bulkhead.

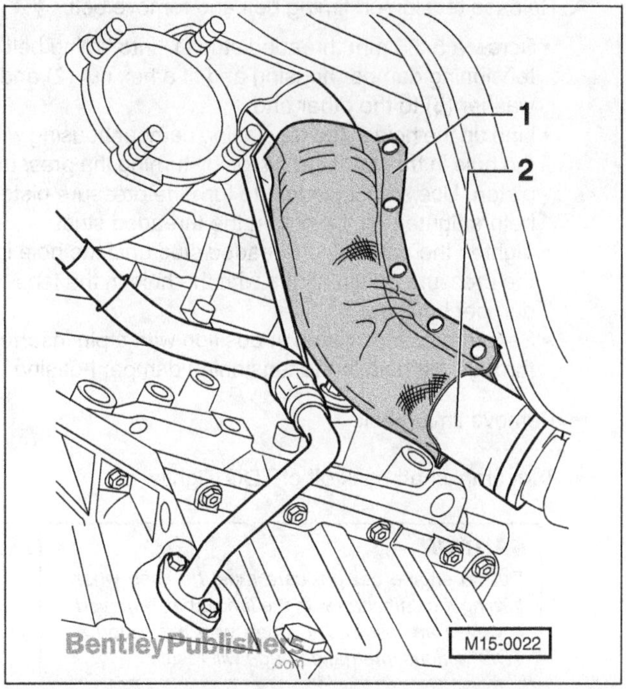

Continued for all engines

– Remove retaining clip at turbocharger inlet and remove connecting pipe.

– Observe position of heat insulation on lower turbocharger outlet pipe and remove.

◄ Remove lower turbocharger outlet pipe.

– Loosen securing bolts for turbocharger outlet pipe retaining bracket.

– Observe position of heat insulation on upper turbocharger outlet pipe and remove.

– Loosen upper turbocharger outlet pipe clamp.

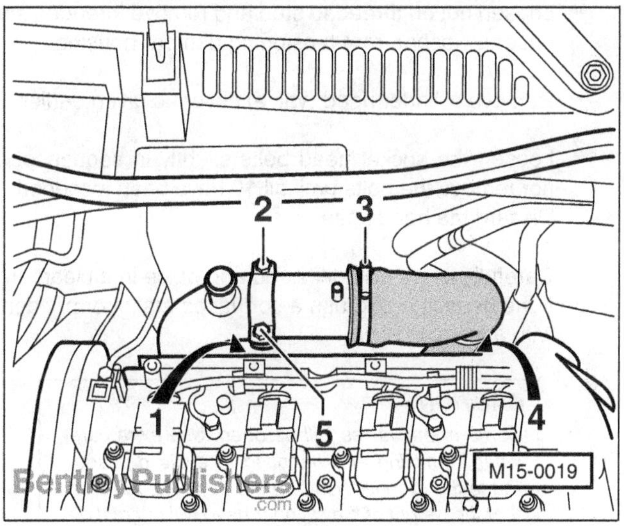

◄ Remove 2 securing bolts (**1** & **4**) for rear cylinder head heat shield and loosen retaining bracket securing bolts (**2** & **5**).

– Remove upper turbocharger outlet pipe, and rear cylinder head heat shield.

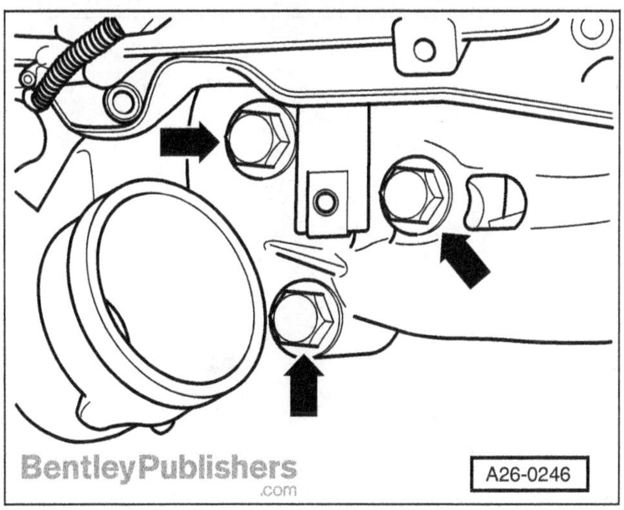

◄ Remove turbocharger to exhaust manifold securing bolts.

– Remove the ribbed V-belt and tensioner.

– Remove timing belt guard, upper section. Set engine to TDC for cylinder #1 by turning crankshaft until timing mark on camshaft sprocket aligns with mark on cylinder head (valve) cover as shown earlier.

– Mark running direction of timing belt.

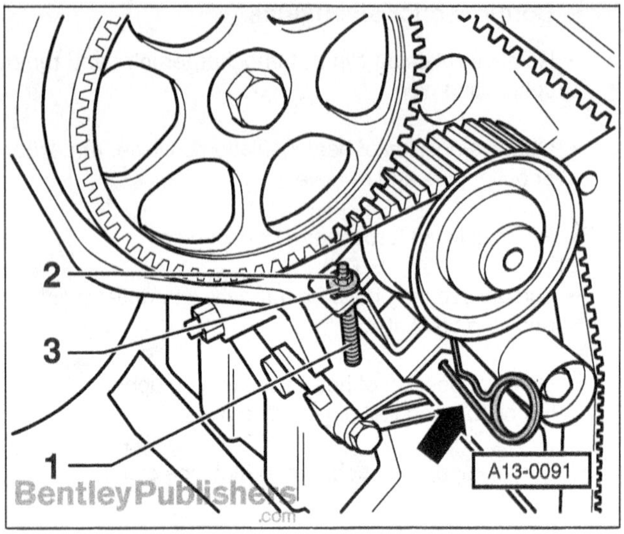

◀ Release tension on timing belt and remove belt.

- Screw a 5x55 mm threaded stud (**1**) into timing belt tensioning damper housing and fit a hex nut (**2**) and washer (**3**) to the other end.
- Line up the hole in the tensioning damper housing with the hole in the pressure piston by turning the pressure piston. Use a suitable tool to turn the pressure piston before tightening the nut on the threaded stud.
- Tighten the nut on the threaded stud until the hole in the pressure piston aligns with the hole in the tension damper housing.
- Secure pressure piston in position with a pin inserted through the hole in the tensioning damper housing.

– Remove timing belt.

– Turn crankshaft back off of TDC slightly.

> ***CAUTION—***
> *Due to engine design, care must be used when turning the camshaft with the timing belt removed. If valves are allowed to open with the piston at TDC, serious internal damage will result.*

– Loosen nut on threaded stud and remove threaded stud, nut and washer from tensioning damper housing.

– Remove cylinder head (valve) cover as given earlier.

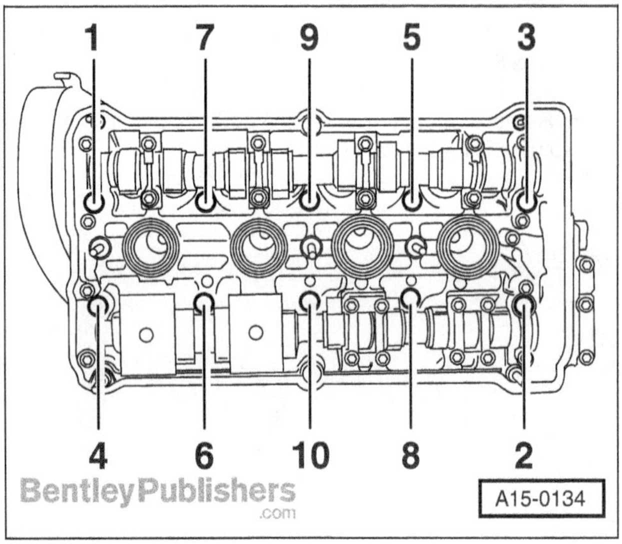

◀ Loosen the socket head bolts slightly in sequence. Do not remove the bolts until all 10 have been loosened. Discard the head bolts.

– Carefully lift off cylinder head and place in a clean area. If the head is stuck, use a soft-faced mallet or pry gently with a wooden stick.

> ***CAUTION—***
> *Some of the valves will be open. Use extra care when removing and handling to avoid damage. Place the cylinder head on the bench or table so that the weight of the cylinder head will not rest on the valves.*

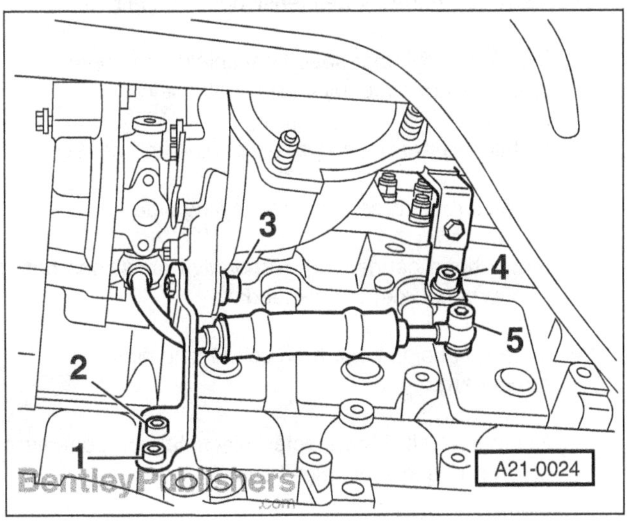

Cylinder head, installing

Before proceeding with the installation of the cylinder head, whether an original or a new unit, observe the following important points:

- Check the cylinder head and block for distortion and warpage, see **Cylinder Head Components**, given later in this section.
- Carefully clean the cylinder block and cylinder head sealing surfaces being sure to avoid scratching them during the cleaning process. Do not use metal scrapers or wire brushes.
- When using abrasive paper do not use any grades coarser than 100 grit (such as 80 grit). Lower numbers are coarser.
- If cylinder head will not be reinstalled immediately, take precautions to prevent rust from forming on the cylinder block walls and gasket sealing surfaces.
- When cleaning the old gasket material off of the cylinder block, take precautions to keep the old gasket material and the abrasive particles and dirt out of the cooling and oiling system passages. Place a clean shop cloth over the cylinders to prevent contamination from getting between the cylinder wall and the piston.
- Do not take the new cylinder head gasket out of the packaging until ready to use. Handle the new gasket with extreme care as any damage will lead to leaks.
- There MUST NOT BE any oil or coolant in the head bolt holes in the cylinder block. Any fluids in the holes creates the danger of hydrolock while torquing which could lead to structural damage of the cylinder block. Use thread chasers to remove foreign material as required. Be sure that all 10 bolt holes are clean and dry.
- When all traces of the old gasket material have been removed from the cylinder block and head, carefully remove all traces of metal particles, abrasives and lint. All gasket sealing surfaces and bolt holes must be clean and dry to ensure a proper seal.
- Always use new cylinder head bolts and a new cylinder head gasket.
- Do not use any gasket sealer on the new cylinder head gasket.
- If installing a Volkswagen supplied replacement cylinder head, be sure to inspect for and remove any plastic packaging materials used to protect the head and the open valves.

– Position the crankshaft so that pistons are NOT at TDC.

< Loosen turbocharger securing bracket bolts (**1** & **2**) slightly to facilitate cylinder head installation.

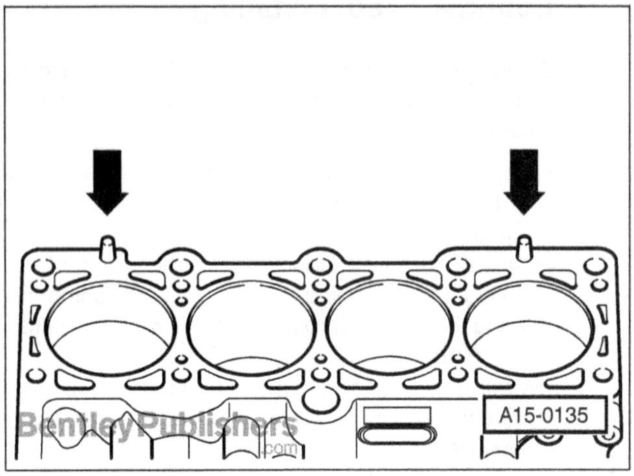

◀ Install alignment pins into bolt holes of cylinder block. Use of this tool allows for proper alignment of the head on the block, preventing slippage and damage to the head gasket.

– Install the new cylinder head gasket onto the block with the numbers and letters facing up. Be sure that none of the wiring or any vacuum hoses are caught between the cylinder head and the block.

– Carefully install the cylinder head onto the block over the alignment pins and screw in 8 of the new head bolts. Tighten the bolts hand-tight only.

– Remove the alignment pins from the bolt holes with the tool and install the remaining 2 new head bolts. Tighten these bolts hand-tight.

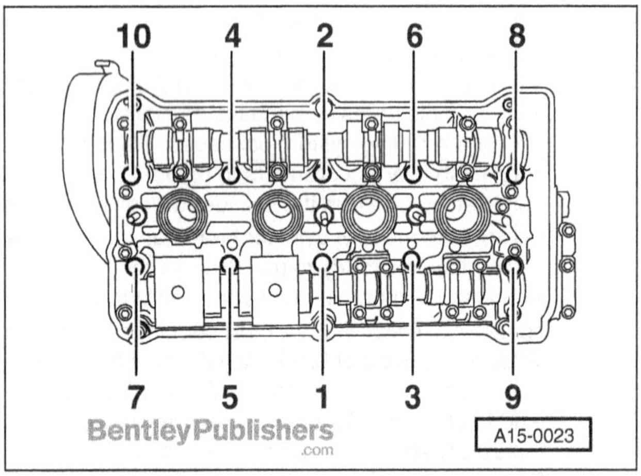

◀ Tighten the 10 cylinder head bolts in three stages following the tightening order.

Tightening torque

- Cylinder head to cylinder block, (stretch bolts – always replace)
  - stage I .... 40 Nm (30 ft-lb)
  - stage II................... additional ¼ turn (90°)
  - stage III .... additional ¼ turn (90°)

– Install new gasket and bolts between turbocharger and exhaust manifold.

– Lubricate the camshafts and followers and install the cylinder head (valve) cover.

– Install the timing belt, adjust the tension and the belt timing.

– Install the timing belt guard, upper section.

– Install the ribbed V-belt tensioner and the belt.

– Install rear cylinder head heat shield and upper turbocharger outlet pipe and heat shield.

– Install lower turbocharger outlet pipe and heat shield.

– Install turbocharger inlet connecting pipe.

– Install hose connecting evaporative emissions canister and turbocharger at bulkhead.

– Install intake air hose from connecting pipe to turbocharger.

– Reconnect all disconnected electrical components and connections including ignition coils.

– Reconnect all coolant flanges and lines.

– Reconnect all vacuum hoses connections.

– Reconnect fuel supply and return lines.

– Install secondary air pump and mounting bracket.

– Refill cooling system with fresh coolant in the appropriate ratio. See **19 Engine–Cooling System.**

> ***CAUTION—***
> - *Use only Volkswagen original coolant when filling the cooling system. Use of any other coolant may be harmful to the cooling system. Do not use coolant containing phosphates.*
> - *Do not use tap water in cooling system. Use distilled water only to mix anti-freeze.*

– Install the lower sound absorber panels (belly pan).

– Install the upper sound absorber panel.

– Reconnect the battery only after all parts/electrical connections have been reinstalled and reconnected.

– Be sure to quality check your work.

Tightening torques

- Cylinder head cover to cylinder head. 10 Nm (7 ft-lb)
- Ignition coil to cylinder head cover (APH engine only) . . . . 10 Nm (7 ft-lb)
- Ground wire to cylinder head cover . . 10 Nm (7 ft-lb)
- Heat shield to cylinder head. . . . . . . 20 Nm (15 ft-lb)
- Ribbed V-belt tensioner to bracket . . 25 Nm (18 ft-lb)
- Turbocharger to exhaust manifold (special nuts - always replace). . . . . 30 Nm (22 ft-lb)
- Coolant outlet flange to cylinder head . . . . 10 Nm (7 ft-lb)
- Secondary air pump bracket to cylinder block. . . . . 25 Nm (18 ft-lb)
- Secondary air pump to support bracket nut . . . . 10 Nm (7 ft-lb)
- Securing bolt for turbocharger connecting pipes . . . . 10 Nm (7 ft-lb)

CYLINDER HEAD COMPONENTS

This section provides the specifications and special information necessary to repair the cylinder head that has been removed from the Volkswagen 1.8L engine. Special service tools and machine shop services are required for most cylinder head repair.

NOTE —
The information given under this heading assumes that the cylinder head is removed. For cylinder head removal procedures, see **Cylinder head, removing** *given earlier.*

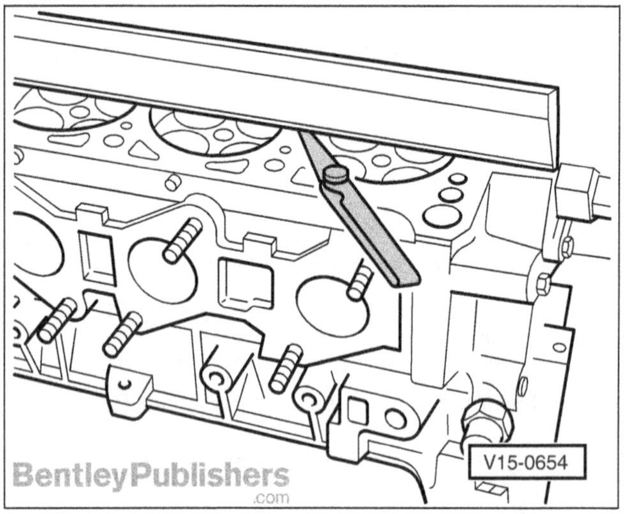

Cylinder head and camshaft

◀ Check the cylinder head for warpage and distortion with an accurate straight edge and a feeler gauge.

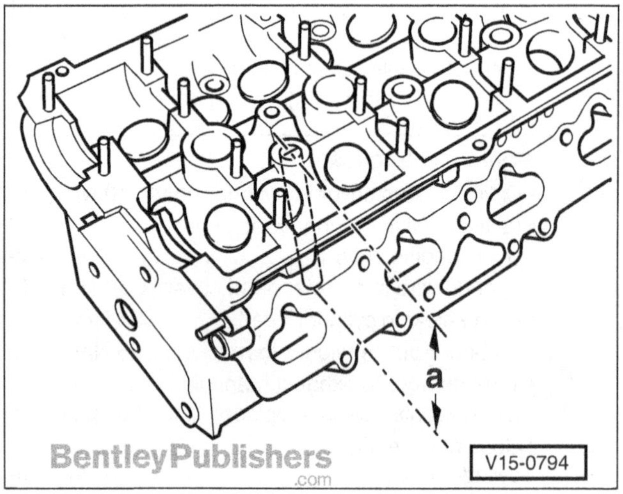

◀ The cylinder head can be resurfaced as long as the distance from the cylinder head (valve) cover seating surface to the head gasket surface is never less than specified. Measure this distance through a cylinder head bolt hole. Machining too much material off of the cylinder head surface will change the compression ratio and will affect engine emissions.

Specification
- Maximum cylinder head distortion/warpage . . . . 0.1 mm (0.004 in.)
- Minimum cylinder dimension (**a**) (valve cover gasket surface to the head gasket surface) . . . . . . 139.2 mm (5.48 in.)

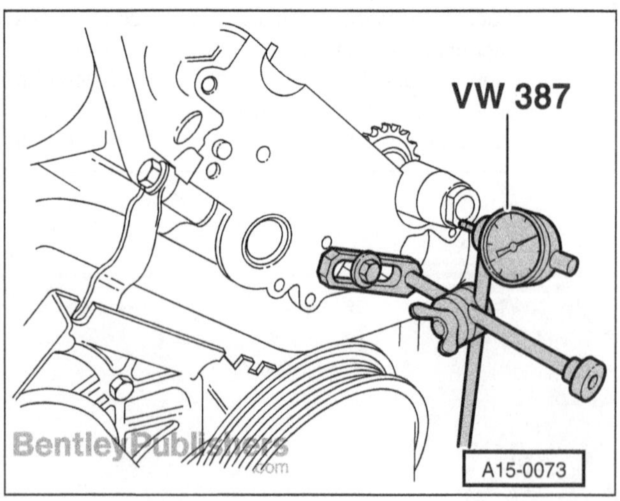

The in and out movement of the camshaft is known as axial play and is measured with the cam followers and drive chain removed and only the first and last bearing caps installed.

◀ A dial gauge is set up on the sprocket end of the camshaft and the camshaft clearance is checked by moving the camshaft as far as it can go in each direction.

Specification
- Camshaft axial clearance, maximum . . . . 0.20 mm (0.008 in.)

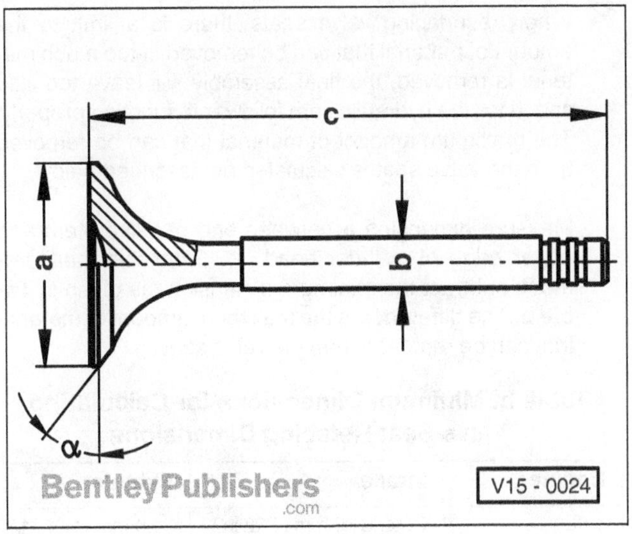

Valves

Valves should not be re-worked on a machine. Only lapping by hand with valve compound is permitted. Valve dimensions listed in **Table a** apply to the valve shown.

Table a. Valve Dimensions

Engine	1.8L
Valve head diameter (a) intake exhaust	 26.9 mm (1.059 in.) 29.9 mm (1.177 in.)
Valve stem diameter (b) intake exhaust	 5.963 mm (0.235 in.) 5.943 mm (0.234 in.)
Valve length (c) intake exhaust	 104.84-105.34 mm (4.128-4.147 in.) 103.64-104.14 mm (4.080-4.100 in.)
Valve face angle (α) intake exhaust	 45° 45°

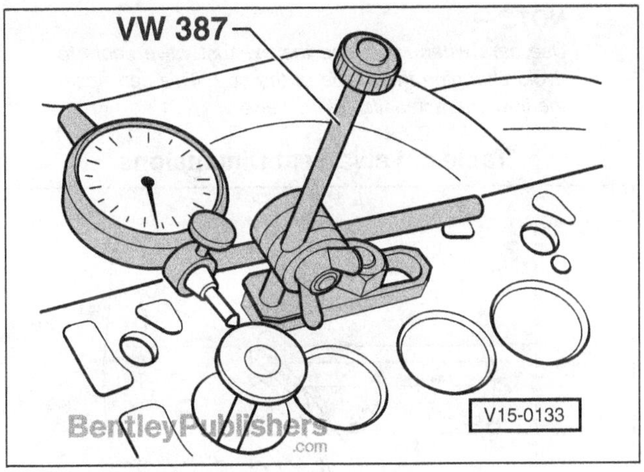

Valve guides

Special tools and a press are required to replace the valve guides. Check valve guide wear with a new valve. Always use an intake valve in an intake guide and an exhaust valve in an exhaust guide. Inspect the valve seats to ensure that the cylinder head can be reconditioned before installing new valve guides. Press out worn original valve guides with Volkswagen special driver tool 3360 from the camshaft side of the head. Replacement valve guides have a shoulder on the camshaft side to limit the installed depth and must be pressed out from the combustion chamber side.

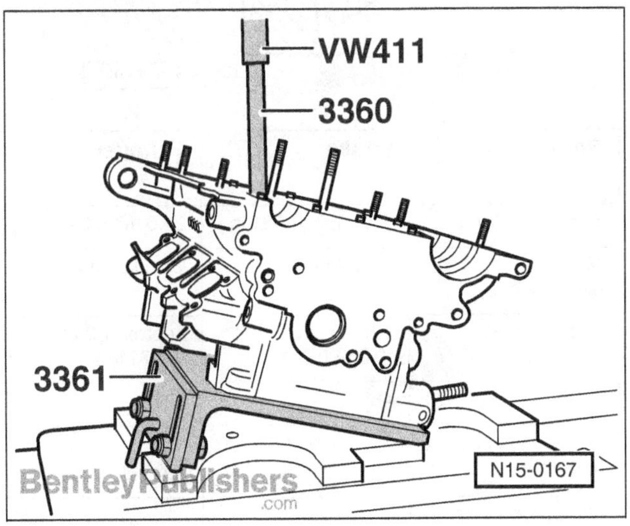

When installing new valve guides, lubricate with clean engine oil and press in from the camshaft side down to the shoulder. The cylinder head should be cold for this operation. Once the shoulder of the valve guide contacts the cylinder head, do not allow the pressure on the guide to exceed 2000 psi (1 ton) otherwise the cylinder head or the guide will be damaged. When the guide has been replaced, ream it to the proper size and continue with the reworking of the valve seats.

Specification

- Valve guide (wear limits-maximum play) with new valve
  intake valve guide .... 0.8 mm (0.031 in.)
  exhaust valve guide .... 0.8 mm (0.031 in.)

NOTE—

Due to the close tolerances found in the valve guides, it is recommended that Volkswagen special tool 3360 be used for removal and installation and special tool 3363 be used to ream the valve guide to size.

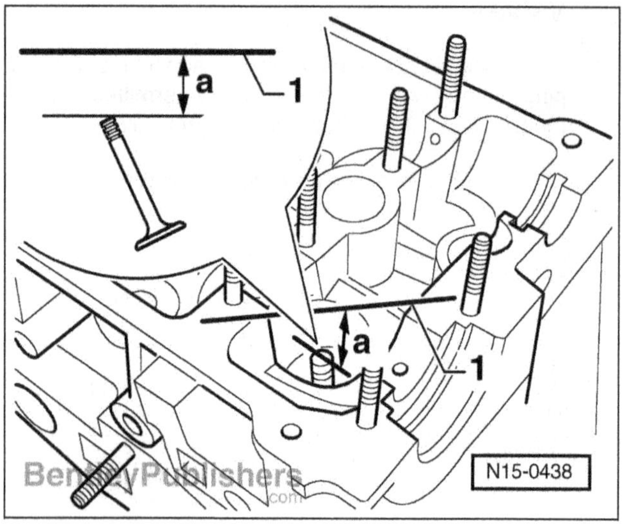

Valve seats

When resurfacing valve seats, there is a limit to the amount of material that can be removed. If too much material is removed, the final assembly will leave too little space for the hydraulic cam follower to function properly. The maximum amount of material that can be removed from the valve seat is calculated as described below.

Measure dimension **a** between end of valve stem and upper edge of cylinder head (**1**). From this measurement, subtract the minimum dimension, as given in **Table b**. The difference is the maximum amount of material that can be removed from the valve seat.

Table b. Minimum Dimensions for Calculating Valve Seat Refacing Dimensions

Engine	Intake	Exhaust
1.8L	Outer 31.0 mm (1.22 in.) Center 32.3 mm (1.27 in.)	31.9 mm (1.25 in.)

NOTE —
Use care when reworking the exhaust valve seats to avoid changing the shape of the port. This can upset the flow characteristics of the valve.

Table c. Valve Seat Dimensions

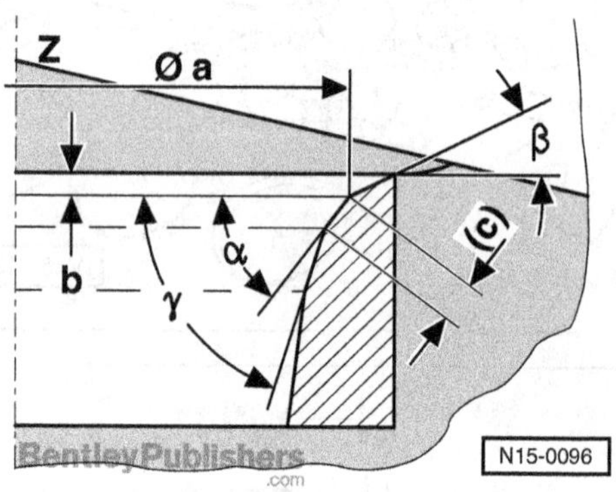

Engine: 1.8L	Intake	Exhaust
Seat diameter (**a**)	26.2 mm (1.031 in.)	29.0 mm (1.142 in.)
Maximum refacing dimension (**b**)	Calculated	Calculated
Seat width (**c**)	1.5-1.8 mm (0.059-0.071 in.)	approx. 1.8 mm (0.071 in.)
Valve seat angle (α)	45°	45°
Correction angle, upper (β)	30°	30°
Correction angle, lower (γ)	60°	60°

NOTE —
Z shows cylinder head surface reference.

15b Cylinder Head and Valvetrain (1.9L engine)

GENERAL

This section covers most cylinder head and valvetrain service and repair work for the1.9L 4-cylinder diesel engine. For information on short block engine rebuilding and internal engine specifications, see **13 Crankshaft/Cylinder Block**.

Most of the operations described in this repair group require special equipment and experience. If you lack the skills, tools, or a suitable workplace for servicing or repairing the cylinder head, we suggest that you leave these repairs to an authorized Volkswagen dealer or other qualified shop.

Engine Codes
- ALH . . . . . 1.9L 4-cylinder TDI (1998-2004)
- BEW . . . . . 1.9L 4-cylinder TDI PD (2004-2006)

DIAGNOSTIC TESTING

The tests that follow can be used to help isolate engine problems, to better understand a problem before starting expensive and extensive repairs, or to just periodically check engine condition.

Compression test

A compression test will tell a lot about the overall condition of the engine without the need for taking it apart. Testing is relatively simple and straightforward.

- Warm-up the engine until it is a minimum of 30°C (86°F).
- Remove upper sound absorber panel from engine. See **03 Maintenance**.

ALH engine

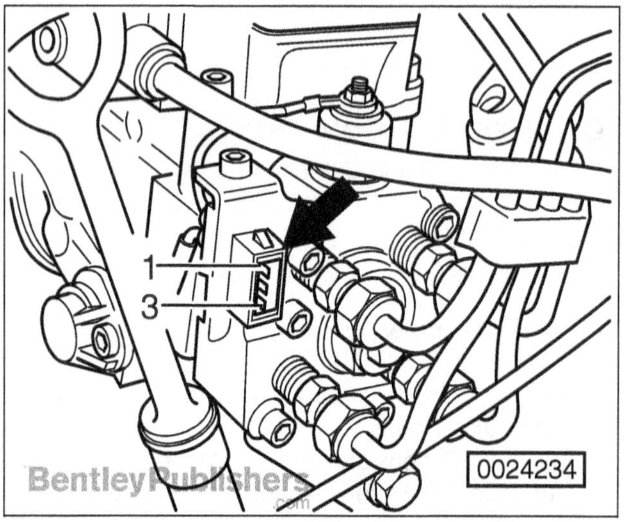

Working at diesel injection pump, disconnect harness connectors for fuel cut-off valve (**arrow**).

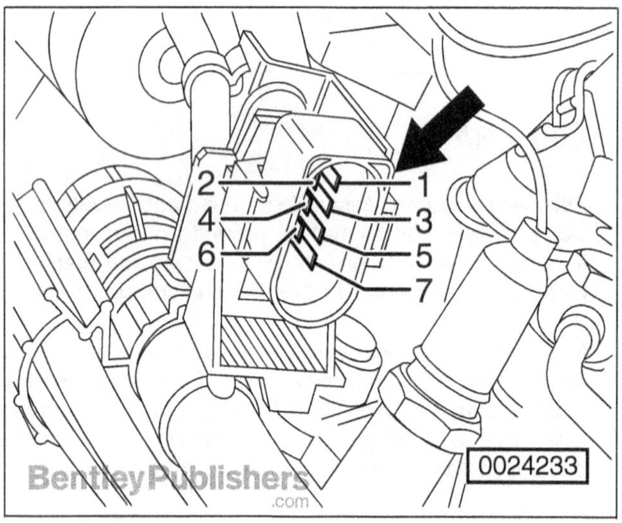

Disconnect quantity adjuster (**arrow**) from diesel injection pump.

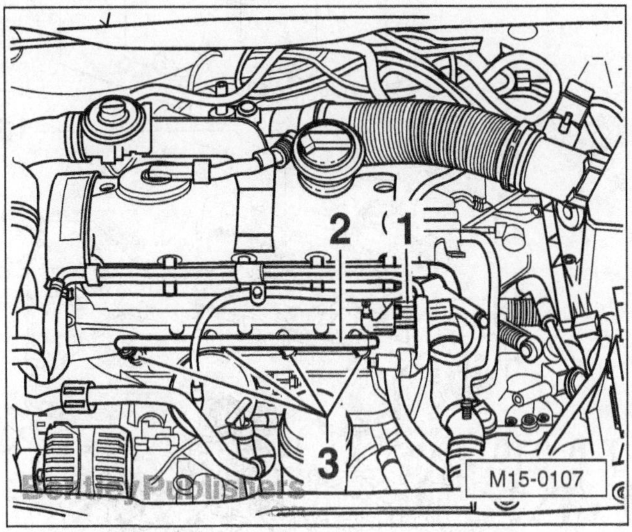

BEW engine

< Disconnect harness connector (**1**) for pump/injector units.

– Pull off connector strip (**2**) from glow plugs.

Continued for all vehicles

– Disconnect harness connector from each glow plug by gently pulling up on it. Label connectors, if necessary, and lay aside.

– Use compressed air to clear area around the glow plugs.

– Using a 10 mm deep socket and swivel or Volkswagen special tool 3220, remove all glow plugs and lay aside in proper order.

– Install compression tester into glow plug hole.

– Have a helper:
- depress clutch pedal fully
- put the transmission in neutral or park
- crank the engine over with the starter motor

NOTE —
Cranking the engine over with diesel injection system components disconnected may Diagnostic Trouble Codes (DTCs) to be stored in engine management system memory.

– Engine should be cranked a minimum of 4 to 5 revolutions.

– Record readings, release pressure in gauge and repeat procedure for each cylinder.

– Compare readings to specification.

Specification
- Compression pressures (1.9L engine)
  new . . . . 25 to 31 bar (363 - 450 psi)
  wear limit . . . . 19 bar (276 psi)
  maximum difference
  between cylinders . . . . 5 bar (73 psi)

– When all cylinders have been checked, reinstall glow plugs and harness connectors. Install upper engine cover and any other components that were removed.

Tightening torque
- Glow plug to cylinder head. . . . . 15 Nm (11 ft-lb)

CYLINDER HEAD SERVICE (ALH ENGINE)

Many cylinder head repairs can be accomplished without removing the cylinder head from the engine. The cylinder head cover (valve cover) gasket, the camshaft, the camshaft oil seal, the valve guide oil seals, the valve springs, and the camshaft followers are all accessible with the cylinder head installed.

External cylinder head assembly

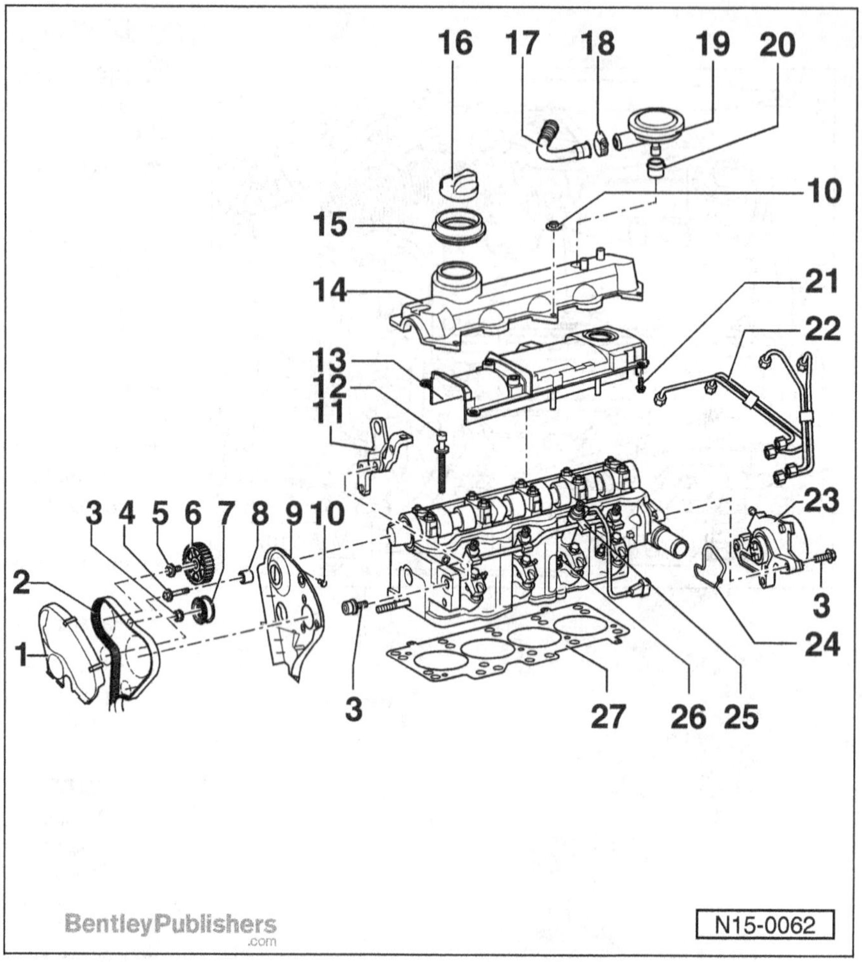

1. **Timing belt guard, upper section**
2. **Timing belt**
   - Mark direction of engine rotation before removing
   - Check for wear
   - Do not kink
3. **Nut**
   - Tighten to 20 Nm (15 ft-lb)
4. **Bolt**
   - Tighten to 20 Nm (15 ft-lb)
5. **Bolt**
   - Tighten to 45 Nm (33 ft-lb)
6. **Camshaft sprocket**
   - Drive off of camshaft taper using hammer and drift through timing belt guard openings
7. **Tensioning roller (semi-automatic, timing belt)**
8. **Idler roller**
9. **Timing belt guard, rear**
10. **Nut**
    - Tighten to 10 Nm (7 ft-lb)
11. **Lifting eye**
12. **Cylinder head bolt**
    - Always replace
    - Note sequence when loosening and tightening, see **Cylinder head, removing and installing**
13. **Oil deflector**
14. **Valve cover**
15. **Seal**
    - Replace if damaged
16. **Cap**
    - Replace seal if damaged
17. **Breather pipe**
18. **Retaining clip**
19. **Pressure regulating valve**
    - For positive crankcase ventilation valve
20. **Gasket**
    - Replace if damaged
21. **Bolt**
    - Tighten to 5 Nm (44 in-lb)
22. **Fuel injector lines**
    - Tighten to 25 Nm (18 ft-lb)
    - Remove using 3035 offset tubing wrench
    - Always remove fuel line cluster as an assembly
    - Do not bend, kink or alter shape
23. **Vacuum pump**
    - For brake booster
24. **Gasket**
    - Always replace
25. **Fuel injector**
    - Removing and installing, see **23 Diesel Fuel Injection**
26. **Glow plug**
    - Tighten to 15 Nm (11 ft-lb)
    - Checking, see **28 Ignition/Glow Plug System**
27. **Cylinder head gasket**
    - Always replace
    - If replaced; also replace entire engine coolant

Cylinder head (valve) cover, removing and installing

NOTE —
The valve cover gasket is vulcanized into the rigid aluminum cover and is not available as a separate part.

- Remove upper engine cover. See **03 Maintenance**.
- Remove connecting pipe between the intercooler and the intake hose.
- Disconnect breather hose from breather valve.
- Loosen cylinder head cover securing bolts using a diagonal patter. Remove bolts.
- Lift off cover with attached oil separator and gasket.
- When installing, tighten cover securing bolts using a diagonal pattern.

Tightening torque
- Cylinder head cover to head . . . . . . 10 Nm (89 in-lb)

Camshaft oil seal, replacing

- Remove upper engine cover. See **03 Maintenance**.
- Remove connecting pipe between the intercooler and the intake hose.
- Remove upper timing belt guard.
- Remove cylinder head (valve) cover.
- Remove brake booster vacuum pump from end of cylinder head.

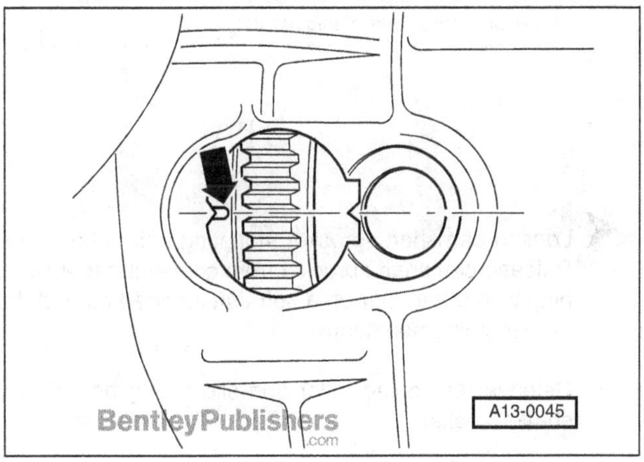

Manual transmission equipped vehicles: Set engine to TDC for cylinder #1 by turning crankshaft until timing mark on flywheel (**arrow**) aligns with mark on transmission case.

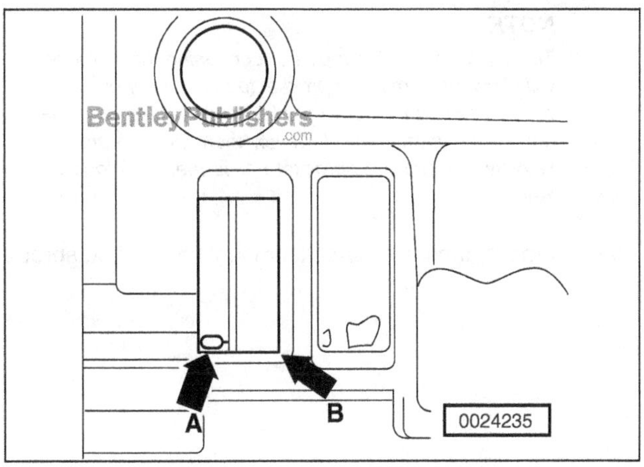

Automatic transmission equipped vehicles: Set engine to TDC for cylinder #1 by turning the crankshaft until the timing mark on the torque converter (**A**) aligns with mark on the transmission case (**B**).

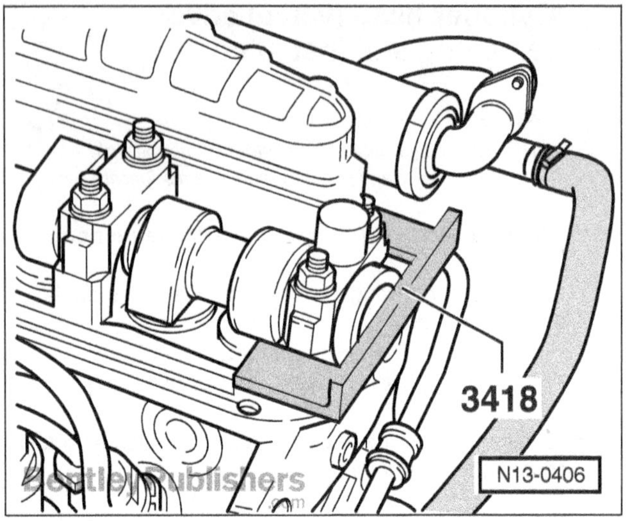

◀ Lock camshaft at TDC with setting bar (special tool 3418 shown).

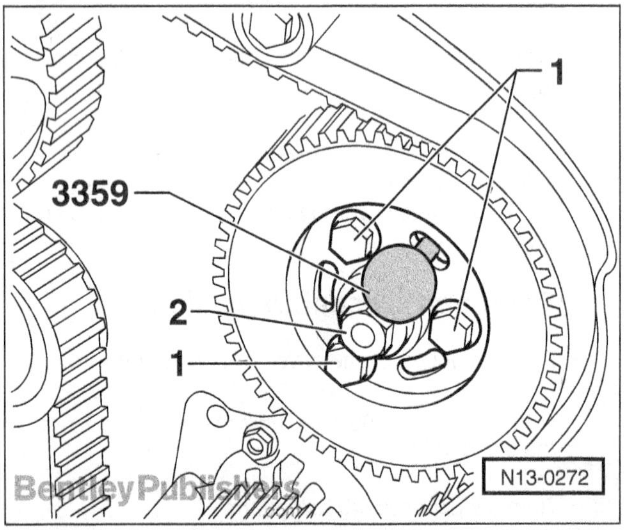

◀ Lock injection pump sprocket with an appropriate locking tool (**3359** shown).

> ***CAUTION—***
> *Do not loosen or remove the hub from the Diesel injection pump. The hub is pressed onto the tapered end of the camshaft and NOT keyed to the pump drive shaft! If removed, it will no longer be correctly indexed to the pump. Correct indexing requires special equipment and training available only at specialized Bosch service centers and is beyond the scope of this manual.*

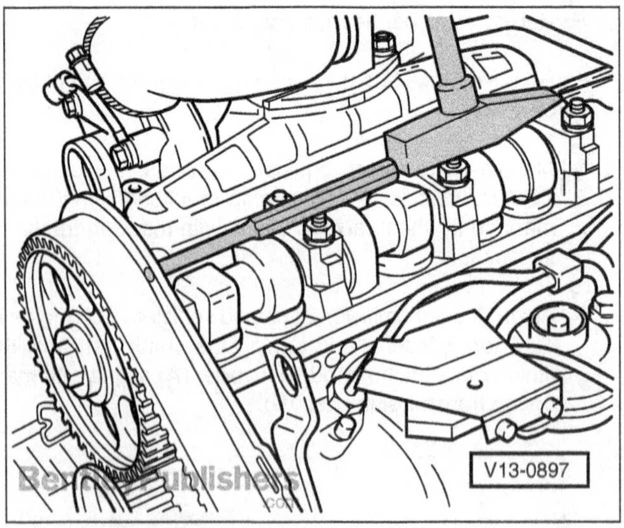

◀ Loosen camshaft sprocket mounting bolt about ½ turn. Release camshaft sprocket from camshaft taper by tapping with a hammer on a soft drift inserted through hole in rear timing belt guard.

– Release tensioning roller and slip timing belt off camshaft sprocket.

> ***NOTE—***
> *The camshaft end is tapered. Loosening the sprocket will allow the camshaft sprocket to rotate independently of the camshaft. There may be a keyway cut into the end of the camshaft. However, there is no matching keyway cut into the sprocket nor is there a Woodruff key.*

– Remove camshaft sprocket mounting bolt and sprocket.

– Loosen and remove camshaft bearing cap #1.

NOTE —
Bearing cap #1 is closest to belt end of camshaft.

– Slide seal through rear timing belt guard and off end of camshaft.

– Install new seal noting the following points:
- Lubricate new seal with clean engine oil.
- Install with closed end facing out.
- Use a suitable installation tool.
- Press seal into bottom of recess.

– Install bearing cap #1.

Tightening torque

• Camshaft bearing cap to cylinder head	20 Nm (15 ft-lb)
• Camshaft sprocket to camshaft	45 Nm (33 ft-lb)

– Install camshaft drive belt sprocket.

– Install timing belt and adjust tension, see **Timing Belt** found later in this repair group.

– Install brake booster vacuum pump, cylinder head (valve) cover, and upper timing belt guard.

– Install connecting pipe between intercooler and intake hose.

– Check that all components are installed, properly secured, and install upper engine cover.

– Be sure to quality check your work, see **Quality Review** at end of this repair group.

Cylinder head, removing and installing (ALH engine)

The cylinder head can be removed and installed with the engine in the vehicle. Note that the cylinder head bolts are stretch type fasteners and should never be reused. In addition, whenever the cylinder head or the cylinder head gasket is replaced, the coolant must also be replaced, see **19 Engine–Cooling System**.

The cylinder head will be removed with the exhaust manifold, turbo-charger and the intake manifold still attached. The plenum close-out panel at the base of the windshield will be removed to allow easier access to the cylinder head and provide space needed for an engine support.

> **WARNING —**
> *Do not start work on a hot engine. Allow to cool sufficiently before proceeding. Engine temperature must be no hotter than warm to the touch. Cylinder head warpage can result due to uneven cooling rates.*

– With ignition switched off, disconnect the battery negative terminal from the battery.

> **NOTE —**
> - *Disconnecting the negative (–) battery cable will erase fault codes and basic settings in engine management and automatic transmission control unit memories. Some driveability problems may be noticed until the system re-adapts to operating conditions. OBD II readiness codes, which may be required for emissions testing, may also be erased. In some instances proper diagnosis will require use of a diagnostic scan tool such as the Volkswagen supplied VAG 1551/1552 or 5051/5052. Use and operation of these tools is outside the scope of this repair manual. See* **OBD On Board Diagnostics** *for more information.*
> - *Be sure to have the anti-theft radio code on hand before disconnecting the battery.*

– Position crankshaft so that pistons are NOT at TDC.

– Remove upper engine cover.

– Remove air cleaner assembly and connecting pipe.

– Remove connecting pipe and hose between the intercooler and intake manifold/EGR solenoid.

– Remove connecting pipe and hoses between turbo-charger and intercooler.

– Remove lower sound absorber panel (belly pan).

– Drain engine coolant, see **19 Engine–Cooling System**.

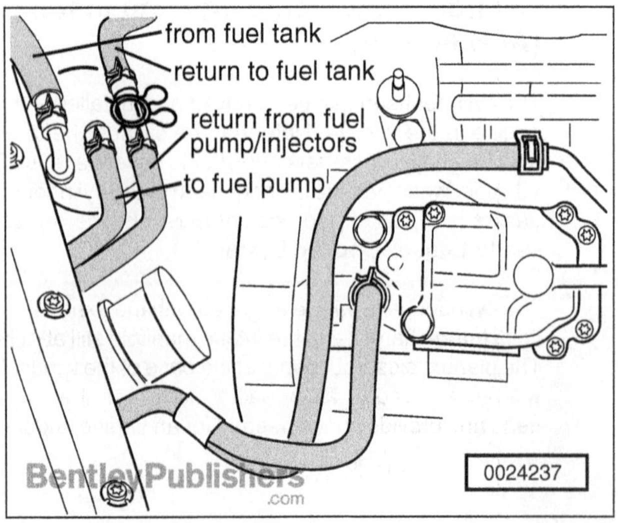

◄ Disconnect and label fuel supply and return lines at fuel filter.

> **WARNING —**
> *Fuel will be expelled when disconnecting fuel hoses. Wrap a cloth around the fuel line fittings before disconnecting them. Do not smoke or work near heaters or other fire hazards. Have a fire extinguisher handy.*

– Seal off disconnected fuel and vacuum lines to prevent contamination.

– Remove fuel filter and bracket.

– Unbolt front exhaust pipe from the turbo-charger, see **26 Exhaust System/Emission Controls**.

– Remove ribbed V-belt and tensioner. See **03 Maintenance**.

– Remove both wiper arms and plenum close-out panel at base of windshield, see **92 Wipers and Washers**.

– Remove the coolant flange from end of cylinder head and disconnect coolant hoses from EGR cooler.

– Disconnect breather and vacuum hoses from cylinder head.

– Disconnect all electrical connectors from cylinder head including glow plug bus connector.

– Remove coolant reservoir from bracket with hoses still attached and place to side.

– Remove upper timing belt guard.

– Remove cylinder head (valve) cover.

– Remove brake booster vacuum pump.

– Disconnect oil return line at turbo-charger, see **21 Turbocharger and Intercooler**.

– Disconnect turbo-charger oil supply line and support bracket.

– Remove metal fuel injector lines.

- Remove all four lines together as a set.
- Use an offset flare nut wrench such as Volkswagen special tool 3035 to loosen the fittings of the injector lines.
- Take care not to bend the lines or change the shape.

– Remove the small injector fuel return hoses.

– Install a suitable engine support fixture tool onto body with both hooks moved to belt side of engine.

– Attach center hook to lifting eye on cylinder head and tighten sufficiently to relieve tension from engine mount.

NOTE —

The lifting eye on the cylinder head will only be used temporarily to support engine while engine mount bracket is removed. A lifting eye adapter must be attached to the block after engine mount bracket is removed. The engine support must then be transferred to the adapter because the lifting eye on the cylinder head will remain attached when the head is removed.

– Remove right side engine mount and support bracket on cylinder block and lift engine slightly with engine fixture support hook.

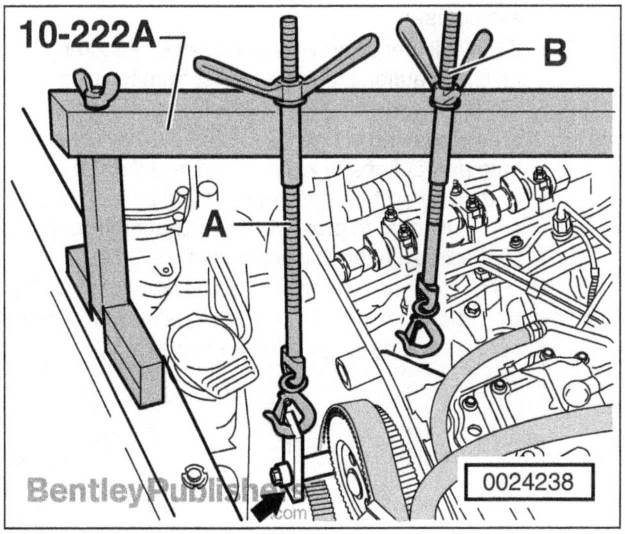

◄ Install a lifting eye adapter into engine block and attach remaining hook from engine support fixture to adapter.

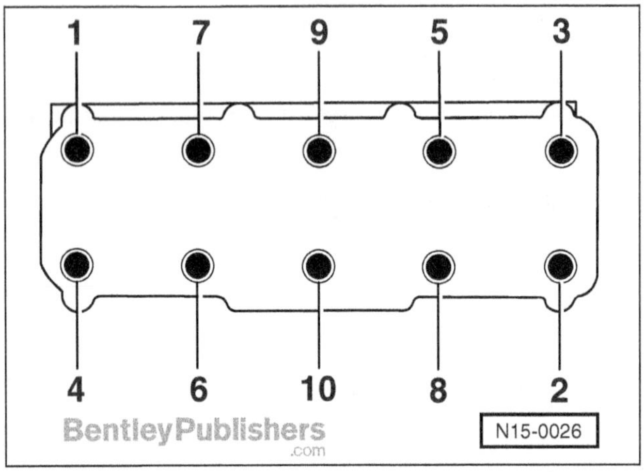

– Lift engine slightly using the engine support hook attached to adapter and remove hook from lifting eye on cylinder head.

– Loosen camshaft sprocket as shown earlier and remove timing belt from camshaft sprocket.

– Position engine to TDC for cylinder #1 and remove timing belt tensioning roller and camshaft sprocket.

– Remove upper bolt from rear timing belt guard.

< Loosen head bolts slightly in sequence but do not remove until all 10 bolts have been loosened. Discard head bolts.

– Carefully lift off cylinder head and place in a clean area. If head is stuck, use a soft-faced mallet or pry gently with a wooden stick.

> ***CAUTION—***
> *Some of the valves will be open. Use extra care when removing and handling to avoid damage. Place the cylinder head on the bench or table so that the weight of the cylinder head will not rest on the valves.*

Cylinder head, installing (ALH engine)

Before proceeding with the installation of the cylinder head, whether an original or a new unit, observe the following important points:

- Check cylinder head and block for distortion and warpage, see **Cylinder Head Components**, given later in this section.
- Carefully clean cylinder block and cylinder head sealing surfaces being sure to avoid scratching them. Do not use metal scrapers or wire brushes.
- When using abrasive paper do not use any grades coarser than 100 grit (such as 80 grit). Lower numbers are coarser.
- If the cylinder head will not be reinstalled immediately, take precautions to prevent rust from forming on cylinder block walls and gasket sealing surfaces.
- When cleaning old gasket material from cylinder block, take precautions to keep debris and dirt out of cooling and oiling system passages. Place a clean shop cloth over cylinders to prevent contamination from getting between cylinder wall and piston.
- Do not take new cylinder head gasket out of packaging until ready to use. Handle new gasket with extreme care as any damage will lead to leaks.

- There MUST NOT BE any oil or coolant in the head bolt holes in cylinder block. Any fluids in the holes creates the danger of hydrolock while torquing which could lead to structural damage of cylinder block. Use thread chasers to remove foreign material as required. Be sure that all 10 bolt holes are clean and dry.
- When all traces of the old gasket material have been removed from the cylinder block and head, carefully remove all traces of metal particles, abrasives and lint. All gasket sealing surfaces and bolt holes must be clean and dry to ensure a proper seal.
- Always use a new cylinder head gasket.
- Do not use any gasket sealer on new cylinder head gasket.
- If installing a Volkswagen supplied replacement cylinder head, be sure to inspect for and remove any plastic packaging materials used to protect the head and the open valves.

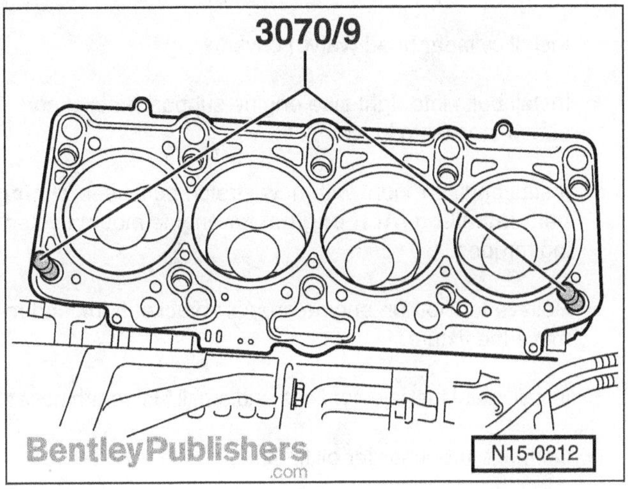

– Position crankshaft to TDC for cylinder #1 and then turn backwards until all pistons are equally spaced below TDC.

< Install two alignment pins (3070/9) into two bolt holes of cylinder block. This keeps the head from sliding off of the block and prevents damage to head gasket.

– Install new cylinder head gasket onto block with numbers and letters facing up. Be sure that none of the wiring or any vacuum hoses are caught between cylinder head and block.

– Carefully install cylinder head onto block over alignment pins and screw in 8 new head bolts. Tighten bolts hand-tight only.

– Remove alignment pins from bolt holes using special tool supplied with pins and install remaining two new head bolts. Tighten these bolts hand-tight.

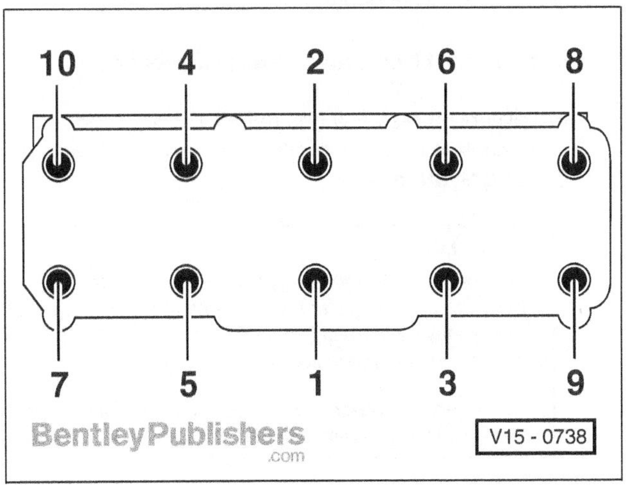

< Tighten the 10 cylinder head bolts in four stages following the tightening order.

Tightening torque

- Cylinder head to cylinder block (stretch bolt - always replace)
  - stage I .... 40 Nm (30 ft-lb)
  - stage II........................ 60 Nm (44 ft-lb)
  - stage III .... additional ¼ turn (90°)
  - stage IV .... additional ¼ turn (90°)

– Ensure that camshaft setting bar remains in place.

– Attach the available support hook from engine support fixture tool to lifting eye on cylinder head. Slightly lift support hook to relieve tension on other support hook and remove it along with adapter.

– Install upper bolt for rear timing belt guard where it is attached to cylinder head.

– Position crankshaft at TDC for cylinder #1 by turning crankshaft pulley clockwise.

– Install timing belt, sprocket and tensioner, see **Timing belt, installing (ALH engine)**, given later.

– Install brake booster vacuum pump.

– Install cylinder head (valve) cover.

– Install bolts into right side engine support bracket and attach bracket to block.

– Install engine mount with new stretch bolts, see **Timing belt, installing (ALH engine)** for engine mount tightening torques.

– Relieve tension on engine support bracket hook and remove the fixture.

– Install metal fuel injector lines and small fuel return hoses.

– Connect turbocharger oil lines.

– Attach front exhaust pipe to turbocharger with new nuts.

– Install lower sound absorber panel (belly pan).

– Install upper timing belt guard.

– Install ribbed V-belt and tensioner.

– Install plenum close-out panel and wiper arms.

– Install coolant flange and hoses to cylinder head.

– Install coolant reservoir and refill the cooling system with fresh coolant of the appropriate ratio. See **19 Engine–Cooling System**.

> ***CAUTION —***
> - *Use only Volkswagen original anti-freeze when filling the cooling system. Use of any other anti-freeze may be harmful to the cooling system. Do not use an anti-freeze containing phosphates.*
> - *Do not use tap water in cooling system. Use distilled water only to mix anti-freeze.*

– Install fuel filter, bracket and fuel hoses.

– Install breather and vacuum hoses.

– Install all electrical connectors including glow plug bus.

– Install connecting pipes between intercooler, turbocharger and intake manifold.

– Install air cleaner assembly and connecting pipe.

– Install upper sound absorber panel.

– Reconnect the battery only after all parts have been reinstalled and reconnected.

– Start engine and re-check diesel injection pump timing, see **23 Fuel Injection–Diesel**.

– Be sure to quality check your work.

Tightening torques

- Cylinder head cover to cylinder head (M6) . . . . 10 Nm (89 in-lb)
- Cylinder head to cylinder block, engine code ALH (stretch bolt - always replace)
  - stage I . . . . 40 Nm (30 ft-lb)
  - stage II . . . . 60 Nm (44 ft-lb)
  - stage III . . . . additional ¼ turn (90°)
  - stage IV . . . . additional ¼ turn (90°)
- Front exhaust pipe to exhaust manifold (special nuts - always replace)(M8) . 25 Nm (18 ft-lb)
- Coolant outlet flange to cylinder head (M6) . . . . 10 Nm (7 ft-lb)
- Rear timing belt guard to cylinder head (M6) . . . . 10 Nm (7 ft-lb)
- Ribbed belt tensioner to bracket (M8) . . . . 25 Nm (18 ft-lb)
- Timing belt tensioner (M8) . . . . 20 Nm (15 ft-lb)
- Fuel injector line flare nut . . . . 25 Nm (18 ft-lb)
- Vacuum pump to cylinder head . . . . 20 Nm (15 ft-lb)
- Camshaft sprocket to camshaft . . . . 45 Nm (33 ft-lb)
- Right side engine mount bracket to cylinder block (M10) . . . . 45 Nm (33 ft-lb)

Internal cylinder head assembly

1. **Bearing caps**
2. **Nut**
   - Tighten to 20 Nm (15 ft-lb)
3. **Camshaft**
   - Checking radial clearance with plastigage
     wear limit: 0.11 mm (0.0043 in.)
   - Run-out: max. 0.01 mm (0.0004 in.)
4. **Hydraulic lifter**
   - Do not interchange
   - Store with cam contact surface facing downwards
   - Before installing, check camshaft axial clearance, see **Cylinder Head Components**
   - Oil contact surfaces
5. **Keepers**
6. **Upper valve spring plate**
7. **Valve spring**
8. **Valve stem seal**
9. **Valve guide**
10. **Oil seal**
    - To remove and install, remove bearing cap
11. **Cylinder head**
12. **Valves**

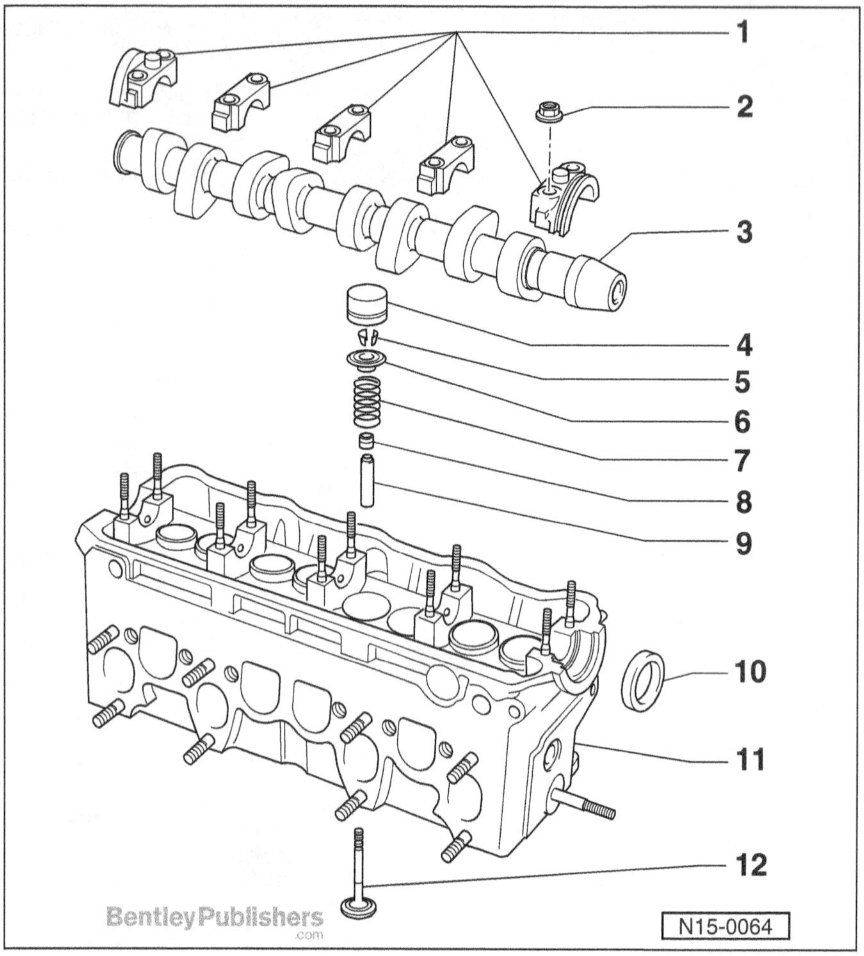

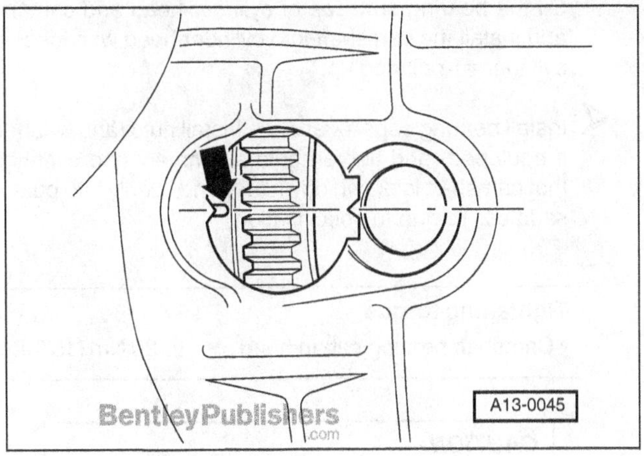

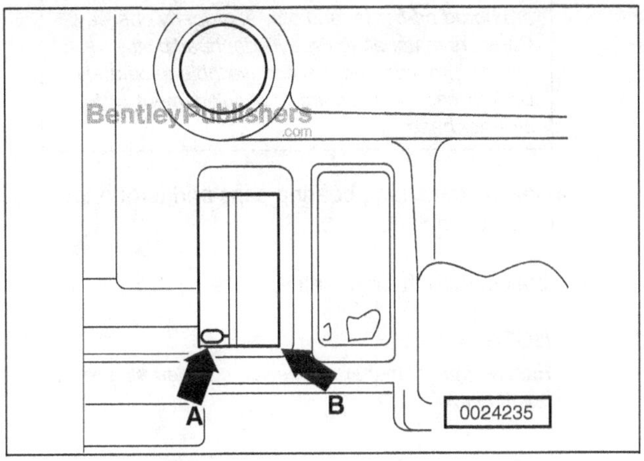

Camshaft, removing and installing

The camshaft can be removed and installed with the cylinder head installed. Removing the camshaft allows access to the hydraulic cam followers, the valve springs and the valve stem oil seals.

- Remove upper engine cover.
- Remove connecting pipe between intercooler and intake hose.
- Remove upper timing belt guard.
- Remove cylinder head (valve) cover.
- Remove brake booster vacuum pump from end of cylinder head.

< Manual transmission equipped vehicles: Set engine to TDC for cylinder #1 by turning crankshaft until timing mark on flywheel (**arrow**) aligns with mark on transmission case.

< Automatic transmission equipped vehicles: Set engine to TDC for cylinder #1 by turning the crankshaft until the timing mark on the torque converter (**A**) aligns with mark on the transmission case (**B**).

- Lock camshaft with setting bar as shown earlier.
- Loosen camshaft sprocket mounting bolt about ½ turn. Release camshaft sprocket from camshaft taper by tapping with a hammer on a soft drift inserted through the hole in rear of timing belt guard.
- Release tensioning roller and slide timing belt off camshaft sprocket.
- Remove camshaft sprocket mounting bolt and sprocket.

NOTE —
Mark each bearing cap before removing. Each cap is machined to each journal and must be reinstalled to that exact location to prevent damage.

- Loosen and remove bearing caps from positions #5, #1 and #3 in that order.

NOTE —
Position #1 is closest to the belt end of the camshaft.

- Loosen the nuts on bearing cap #2 slightly and then loosen the nuts on bearing cap #4 slightly. Loosen nuts alternately and evenly a little at a time on each of the two bearing caps. Remove nuts (and washers, if equipped), and remove bearing caps.

– Lift out camshaft and lay aside.

NOTE —
Once removed, camshaft should be checked for wear and other visible damage as described later under **Cylinder Head Components**.

– Oil the bearing surfaces of cylinder head and camshaft and install the camshaft into cylinder head with lobes for cylinder #1 pointing up.

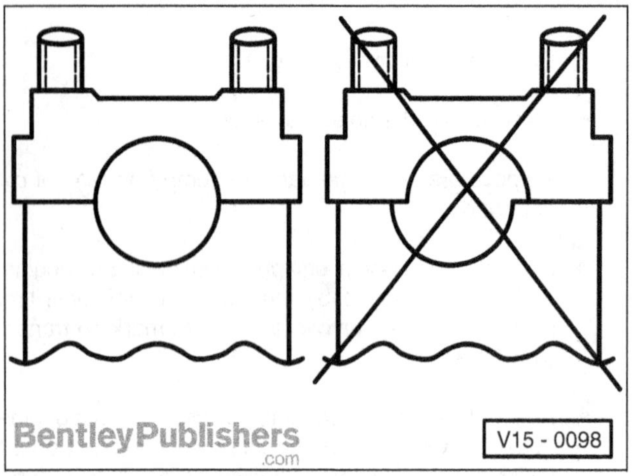

< Install bearing caps #2 and #4. Install nuts (and washers, if equipped) and tighten nuts alternately and evenly so that camshaft is drawn down fully and evenly into bearing saddles. Torque to specification.

Tightening torque
- Camshaft bearing cap to head . . . . . . 20 Nm (15 ft-lb)

CAUTION —
Be sure to install bearing caps correctly. The caps are bored off-center and only fit properly one way. Caps are matched to the cylinder head by the machining process and are not available separately. Broken caps will require the replacement of the cylinder head.

– Install all remaining bearing caps and torque to specification.

– Install camshaft drive belt sprocket.

NOTE —
Ensure that camshaft lobes for cylinder #1 are still pointing up.

Tightening torque
- Camshaft sprocket to camshaft . . . . 45 Nm (33 ft-lb)

– Install timing belt and adjust tension, see **Timing belt assembly** given earlier.

– Install brake booster vacuum pump, cylinder head (valve) cover, and upper timing belt guard.

– Install connecting pipe between intercooler and intake hose.

– Install upper engine cover.

– Be sure to quality check your work, see **Quality Review** at the end of this repair group.

Hydraulic cam followers, checking

The 1.9 liter ALH engine is equipped with hydraulic cam followers which are also known as lifters.The cam followers are pumped up by engine oil pressure, expanding as necessary to fill the gap between the valve stem and the camshaft lobe. This occurs continuously and automatically to keep the valve in proper adjustment at all times.

Some valve noise at start-up and during the warm-up period is normal at times due to hydraulic cam followers that have bled down while the engine was not running. Before checking noisy cam followers, check that the engine oil is new and fresh, of the correct weight, and that the level is correct. Allow 2 minutes with a warm engine running at a fast idle for the lubrication system to properly pump up the cam followers.

Cam followers should only be checked when the engine is fully up to operating temperature. Run the engine until the radiator fan has switched on at least one time. Increase the engine speed to 2,500 rpm and hold it there for approximately 2 minutes. If the hydraulic cam followers are still noisy, shut the engine off and proceed as follows while the engine is still warm.

– Remove cylinder head (valve) cover. See **Cylinder head (valve) cover, removing and installing**, given earlier.

– Turn the engine by hand until both camshaft lobes of cylinder #1 are pointing approximately up.

NOTE —
Hydraulic cam follower clearance can be checked on any cam follower (lifter) that is not being depressed by the camshaft lobe.

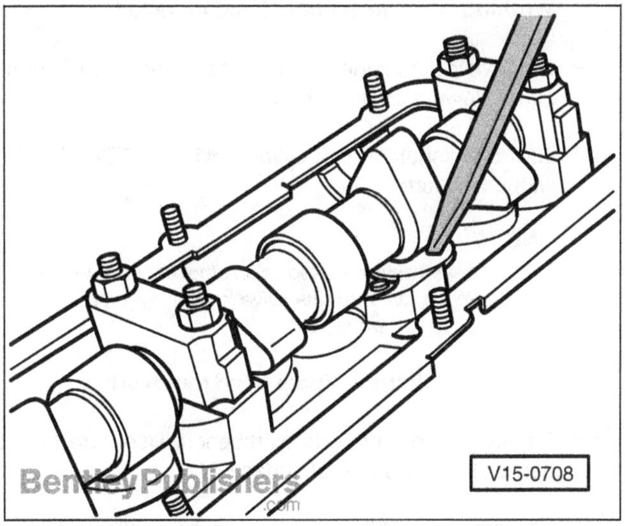

◄ Check clearance between cam follower and cam lobe on cylinder #1 by lightly depressing cam follower with a wooden or plastic wedge and inserting a feeler gauge into gap. If clearance exceeds specified limit, lifter is faulty and should be replaced.

Specification
- Hydraulic cam follower maximum clearance (ALH engine) . . . . 0.1 mm (0.004 in.)

– Turn engine by hand and repeat procedure until all cam followers have been checked.

– Replace a faulty cam follower by removing camshaft as previously described and removing follower from the cylinder head. Faulty hydraulic cam followers can be replaced individually but only as a complete assembly.

CAUTION—
After installing new cam followers, the engine should not be started for at least 30 minutes. New cam followers are usually at full extended height and must be allowed to bleed down to their proper height once installed. Failure to do so may cause valves to strike the pistons resulting in serious damage.

NOTE—
Store removed hydraulic cam followers in order on a clean surface with the camshaft contact surface facing down to minimize bleed down. Cover with a clean lint-free shop cloth.

Valve stem oil seals, replacing

The sign of faulty valve stem seals is excessive oil consumption and oil smoke from the exhaust. This is usually most noticeable during periods of high manifold vacuum such as deceleration. If compression and leak-down testing confirm the integrity of the piston rings, but the engine consumes oil, it is likely that faulty valve stem seals are present. It should also be noted that worn valve stem seals could be due to worn valve guides and that the worn valve guides are the major cause of the oil consumption. See **Cylinder Head Components** for more information on checking valve guides.

Replacing the valve stem oil seals requires removal of the camshaft, cam followers, and the valve springs. This can be done with the cylinder head installed or removed. In either case, several Volkswagen special tools are required to compress the valve springs and remove and install the seals.

Working with the cylinder head installed:

- Remove cylinder head cover, camshaft, and hydraulic cam followers as described earlier.

- Turn crankshaft so that cylinder #1 is at TDC. Continue with procedure.

NOTE—
Compressed air is not used as on other engines. When the spring is removed, the valve is properly supported by the piston crown at TDC.

Working with the cylinder head removed:

- Secure cylinder head to workbench. Use care to avoid damage to head gasket surfaces.

- Remove camshaft and hydraulic cam followers as described earlier.

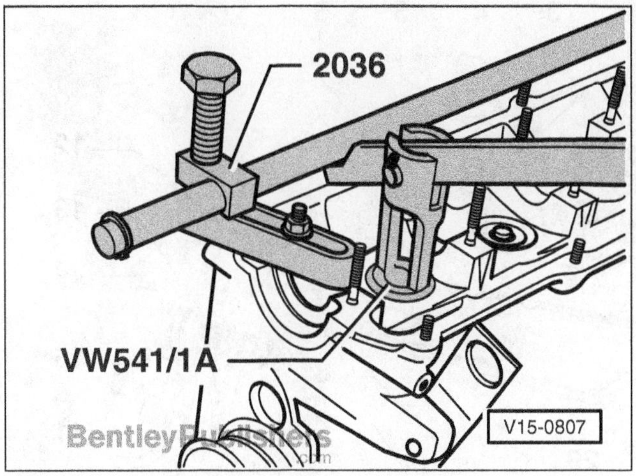

Working with cylinder head removed or installed:

◀ Install appropriate valve spring compressor tools. Compress spring for cylinder #1 and remove spring retainer, both keepers, and valve spring. If spring will not compress, lightly tap the tool to release the stuck keepers. Use a small magnet to retrieve keepers.

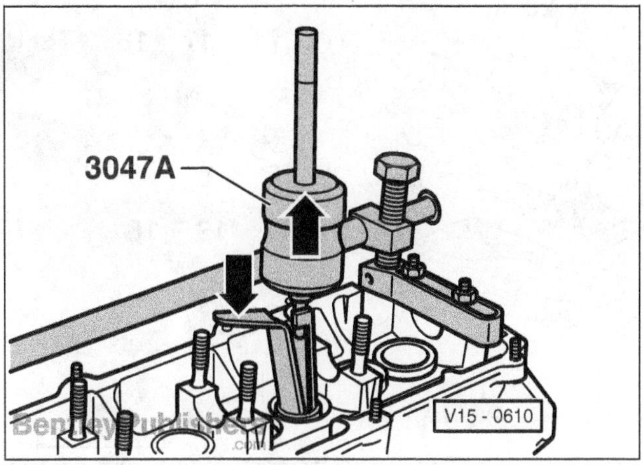

◀ Remove valve stem seals with the special slide hammer tool.

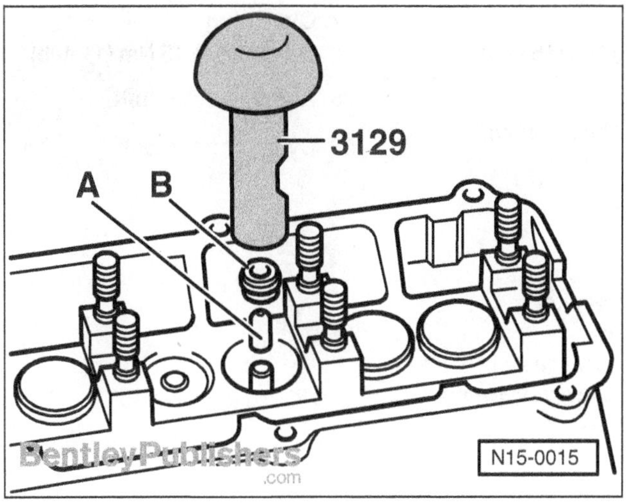

◀ Begin installation of new seal by temporarily fitting a protective plastic fitting sleeve (**A**) over valve stem. This sleeve is usually included with new seal set and will prevent seal damage to the seal (**B**) caused by sharp edges of keeper grooves. Lubricate new seal with clean engine oil and fit it to installation tool.

– Push installation tool (with seal) down over the valve stem until seal is fully seated on the guide. Remove tool and protective fitting sleeve.

– Reinstall valve spring, retainer and keepers.

– Repeat for the second valve on the cylinder.

– When both valve seals have been replaced cylinder #1, turn crankshaft to TDC for the next cylinder (if working with cylinder head installed) and repeat the process until all valve seals have been replaced.

– Remaining installation is reverse of removal.

> ***CAUTION —***
> *To prevent cylinder head or camshaft damage, be sure to follow the camshaft installation procedure when tightening the camshaft bearing caps. See* **Camshaft, removing and installing.**

CYLINDER HEAD SERVICE (BEW ENGINE)

Many cylinder head repairs can be accomplished without removing the cylinder head from the engine. The cylinder head cover (valve cover) gasket, the camshaft, the camshaft oil seal, the valve guide oil seals, the valve springs, and the pump/injector units are all accessible with the cylinder head installed.

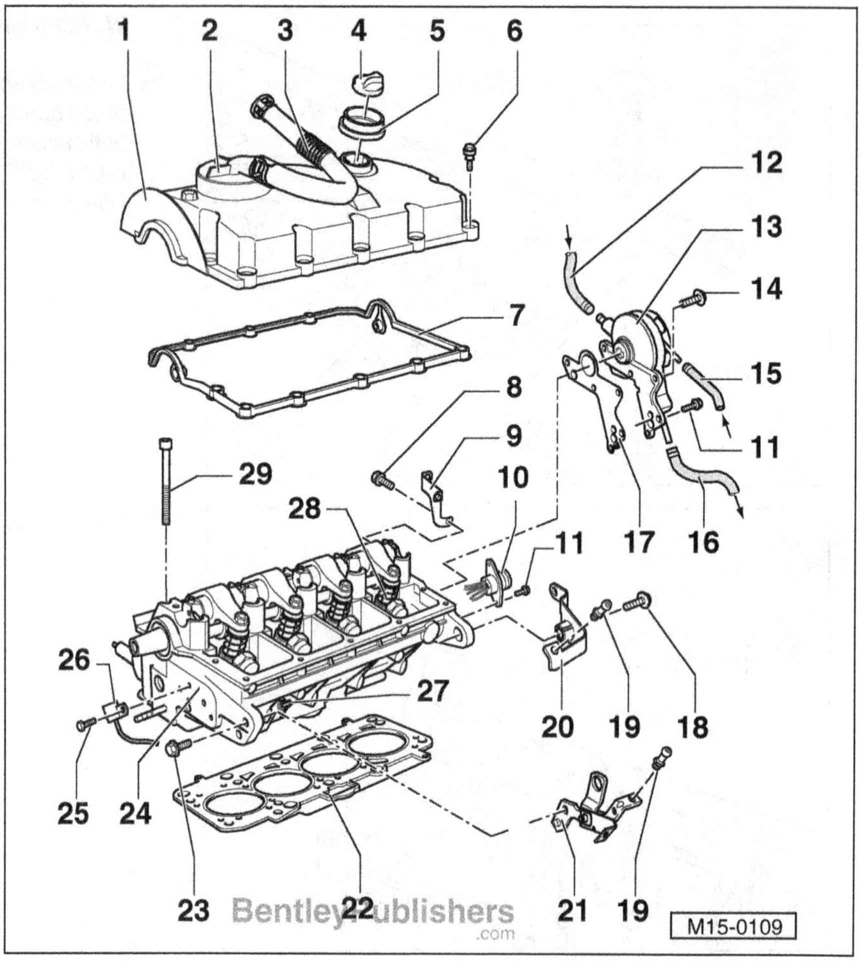

External cylinder head assembly

1. **Cylinder head (valve) cover**
2. **Pressure regulating valve**
   - For crankcase breather
3. **Vent hose**
4. **Filler cap**
5. **Seal**
   - Replace if damaged
6. **Bolt**
   - Tighten to 10 Nm (7 ft-lb)
7. **Valve cover gasket**
   - Replace if damaged or leaking
   - Seal with AMV 17400401
   - Inert nubs in cylinder head holes
8. **Bolt**
   - Tighten to 20 Nm (15 ft-lb)
9. **Lifting eye**
10. **Central connector for pump/ injector units**
11. **Bolt**
    - Tighten to 10 Nm (7 ft-lb)
12. **From brake servo**
13. **Tandem pump**
    - For fuel and vacuum supply
14. **Bolt**
    - Tighten to 25 Nm (18 ft-lb)
15. **Fuel supply hose**
    - Has white markings
16. **Fuel return hose**
    - Has blue markings
17. **Gasket**
    - Always replace
18. **Bolt**
    - Tighten to 20 Nm (15 ft-lb)
19. **Ball head bolt**
    - Tighten to 8 Nm (71 in-lb)
20. **Bracket**
    - For engine cover
21. **Lifting eye**
22. **Cylinder head gasket**
    - Always replace
    - Note markings on gasket
    - If replaced, always replace engine coolant
23. **Bolt**
    - Tighten to 20 Nm (15 ft-lb)
24. **Cylinder head**
25. **Bolt**
    - Tighten to 10 Nm (7 ft-lb)
26. **Camshaft position sensor (G40)**
    - To remove, unclip sealing grommet from rear timing belt guard
27. **Glow plug**
    - Tighten to 15 Nm (11 ft-lb)
28. **Pump/injector unit**
29. **Cylinder head bolt**

Cylinder head (valve) cover, removing and installing

NOTE —
The valve cover gasket is vulcanized into the rigid aluminum cover and is not available as a separate part.

– Remove upper engine cover. See **03 Maintenance**.

– Remove EGR vacuum regulator solenoid valve (N18).

– Disconnect harness connector for intake flap motor (V157) and position wiring out of way.

◄ Open retaining clips (**arrows**) and remove upper timing belt cover.

– Disconnect crankcase breather hose from top of cylinder head cover.

– Loosen cylinder head cover bolts in a diagonal pattern. Remove bolts and cover.

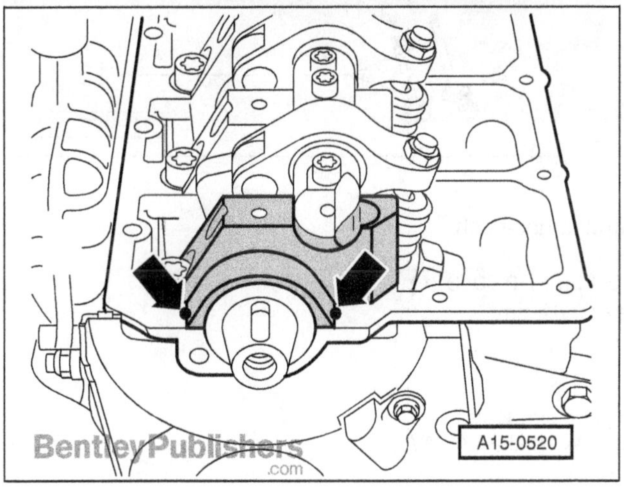

◄ Before installing cylinder head cover, coat both corners (**arrows**) of front and rear camshaft bearing cap to cylinder head sealing surfaces with a drop of sealant (approx. 5 mm (¼ inch) diameter). Use VW sealant (part no. AMV 174 004 01).

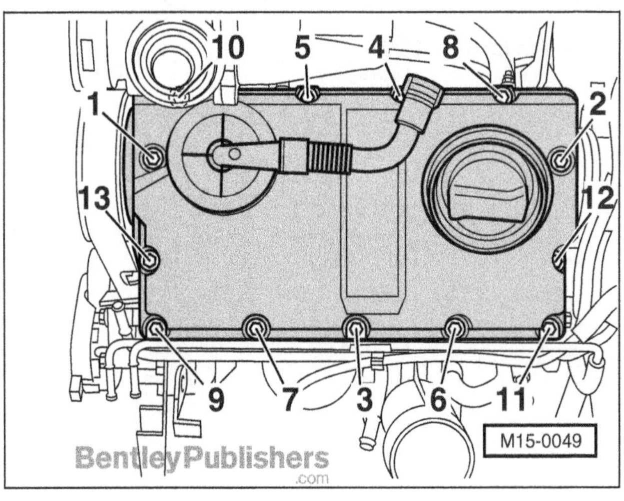

◄ Tighten cylinder head bolts in sequence shown to specified torque.

Tightening torque
- Cylinder head cover to cylinder head. 10 Nm (7 ft-lb)

– Install remaining components in reverse order of removal.

Internal cylinder head assembly

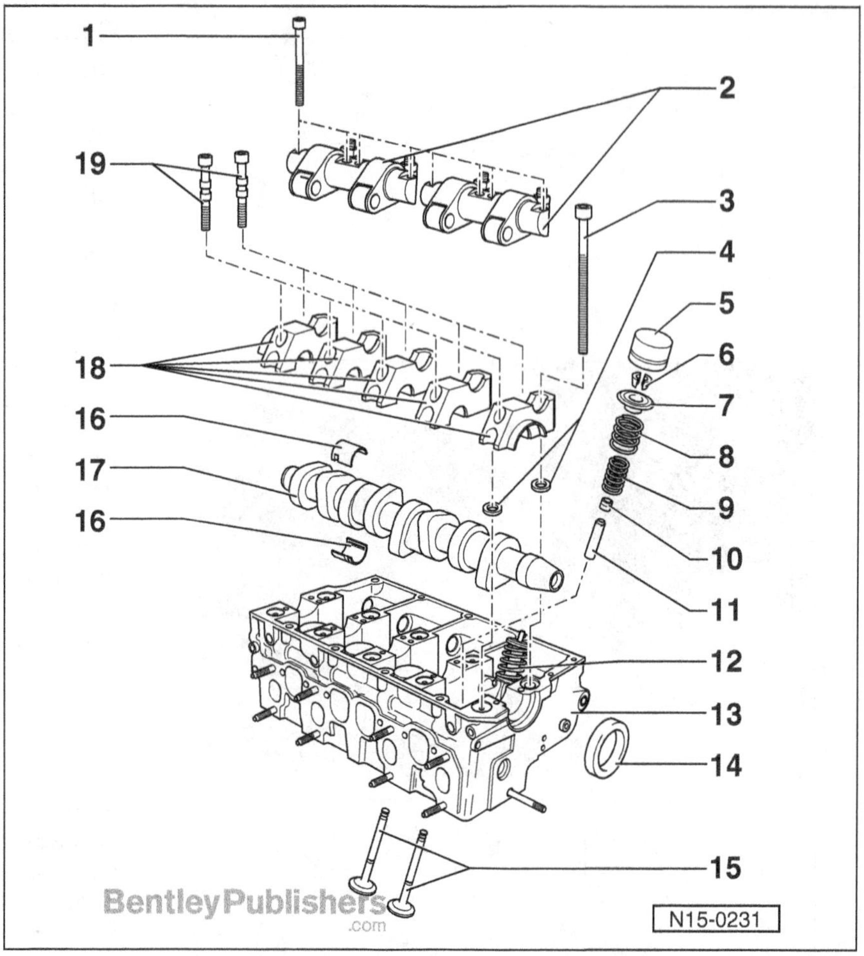

1. **Bolt**
   - Tighten to 20 Nm (15 ft-lb) + ¼ turn (90°)
   - Always replace
2. **Rocker shaft**
   - Do not interchange
3. **Cylinder head bolt**
4. **Washer**
   - Located under bearing caps
5. **Hydraulic lifter**
   - Do not interchange
   - Store with camshaft contact surface down
   - Check camshaft axial clearance before installing
   - Oil contact surfaces
   - Remove bearing shells before removing
6. **Valve keepers**
7. **Valve spring plate**
8. **Exterior valve spring**
9. **Interior valve spring**
10. **Valve stem seal**
11. **Valve guide**
    - If wear limit is exceeded, replace cylinder head
12. **Pump/injector unit**
13. **Cylinder head**
14. **Oil seal**
    - Do not grease or oil sealing lip
    - Remove any oil residue with clean cloth before installing
    - Seal groove on camshaft taper when installing with sticky tape (i.e. Selotape)
15. **Valves**
    - Replace valve when scoring can be seen at shaft and contact surfaces
    - Do not rework, only lapping-in is allowed
16. **Bearing shell**
    - Do not interchange used bearing shells
    - Mark location before removing
    - Ensure lugs seat correctly in bearing cap and cylinder head
17. **Camshaft**
    - Check radial clearance with Plastigage
    - Wear limit: 0.11 mm (0.0043 in.)
    - Run-out max.: 0.01 mm (0.0004 in.)
18. **Bearing cap**
    - Seal bearing cap surfaces #1 and #5 with VW sealant (AMV 174 004 01)
19. **Bolts**
    - Tighten to 8 Nm (71 in-lb) plus ¼ turn (90°)

TIMING BELT

The maintenance tables in **03 Maintenance** describe inspection and replacement intervals for the timing belt, also known as the camshaft drive belt or toothed belt. Regular inspection and replacement will help prevent damage to the engine due to belt stretch and the long term effects of heat.

Timing belt assembly (ALH engine)

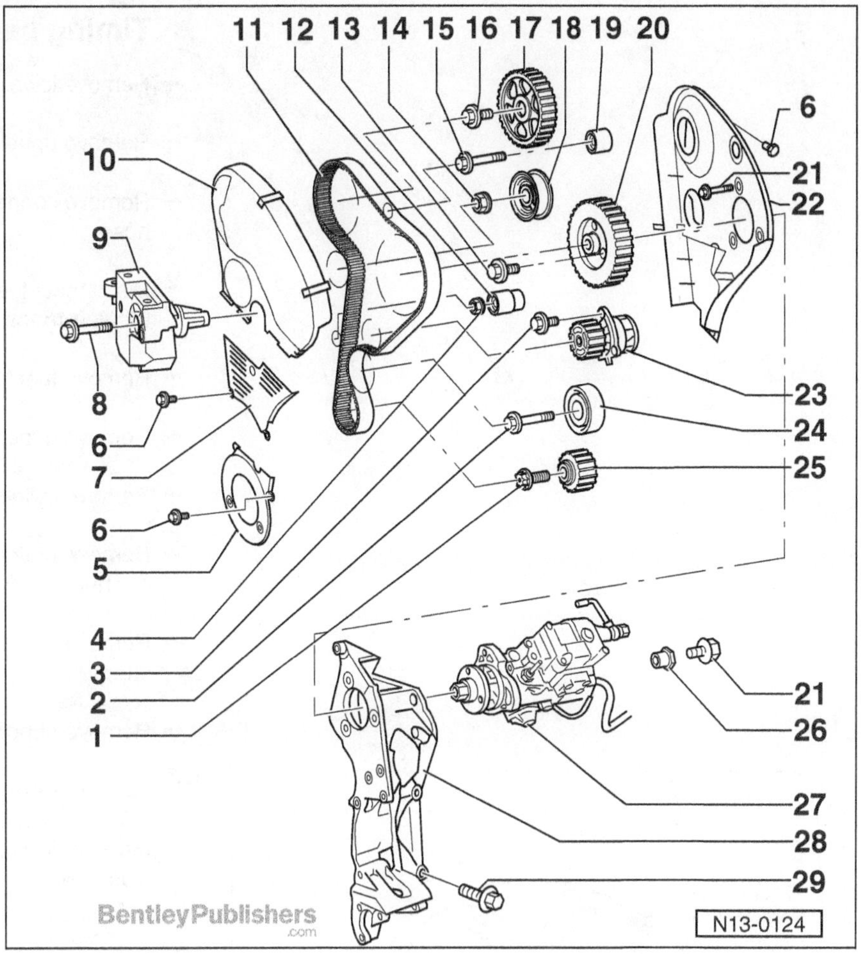

1. **Bolt**
   - Always replace
   - Tighten to 120 Nm (88 ft-lb) + ¼ turn (90°)
   - To loosen and tighten, counter-hold with 3099
   - Threads and shoulder must be free of oil and grease
   - The additional quarter turn can be obtained in several stages
2. **Bolt**
   - Always replace
   - Tighten to 40 Nm (30 ft-lb) + ¼ turn (90°)
3. **Bolt**
   - Tighten to 15 Nm (11 ft-lb)
4. **Bolt**
   - Tighten to 22 Nm (16 ft-lb)
5. **Timing belt guard, lower section**
6. **Bolt**
   - Tighten to 10 Nm (7 ft-lb)
7. **Timing belt guard, center section**
8. **Bolt**
   - Tighten to 45 Nm (33 ft-lb)
9. **Right engine bracket**
10. **Timing belt guard, upper section**
11. **Timing belt**
    - Mark direction of rotation before removing
    - Check for wear
    - Do not kink
12. **Idler wheel**
13. **Bolt**
    - Always replace
    - Tighten to 20 Nm (15 ft-lb) + ¼ turn (90°)
14. **Bolt**
    - Tighten to 20 Nm (15 ft-lb)
15. **Bolt**
    - Tighten to 20 Nm (15 ft-lb)
    - To loosen and tighten bolt, counter-hold using 3036
16. **Bolt**
    - Tighten to 45 Nm (33 ft-lb)
    - To loosen and tighten bolt, counter-hold using 3036
17. **Camshaft sprocket**
    - Drive camshaft off of taper using hammer and drift through rear timing belt guard openings
18. **Tensioning roller**
    - Semi-automatic timing belt tensioning roller
19. **Idler wheel**
20. **Injection pump sprocket**
    - Two piece construction
21. **Bolt**
    - Tighten to 30 Nm (22 ft-lb)
22. **Timing belt guard, rear**
23. **Coolant pump**
24. **Idler wheel**
    - First remove, to remove coolant pump
25. **Crankshaft timing belt sprocket**
26. **Bushing**
27. **Diesel injection pump**
28. **Assembly bracket**
    - For diesel injection pump, generator and power steering pump
    - For vehicles with A/C
29. **Bolt**
    - Tighten to 45 Nm (33 ft-lb)

Timing belt, removing (ALH engine)

- Remove lower sound absorber panel (belly pan).
- Remove upper engine cover.
- Remove connecting pipe between intercooler and intake hose.
- Disconnect fuel supply and return lines at fuel filter. See **03 Maintenance**.
- Remove fuel filter from mounting bracket.
- Remove upper timing belt guard.
- Remove cylinder head (valve) cover.
- Remove brake booster vacuum pump from end of cylinder head.
- Remove connecting pipe between intercooler and turbocharger.
- Remove ribbed V-belt and tensioner.

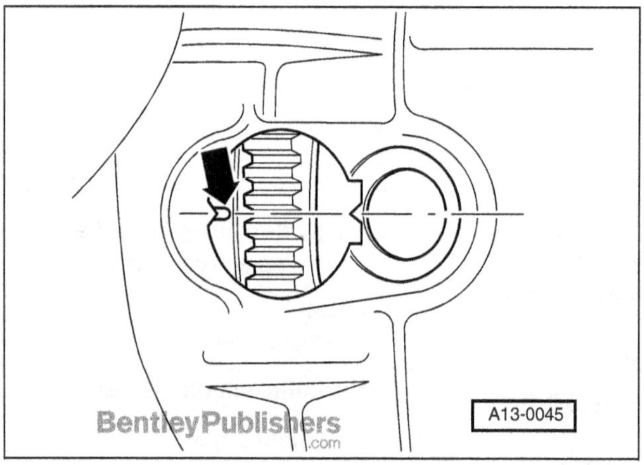

Manual transmission equipped vehicles: Set engine to TDC for cylinder #1 by turning crankshaft until timing mark on flywheel (**arrow**) aligns with mark on transmission case.

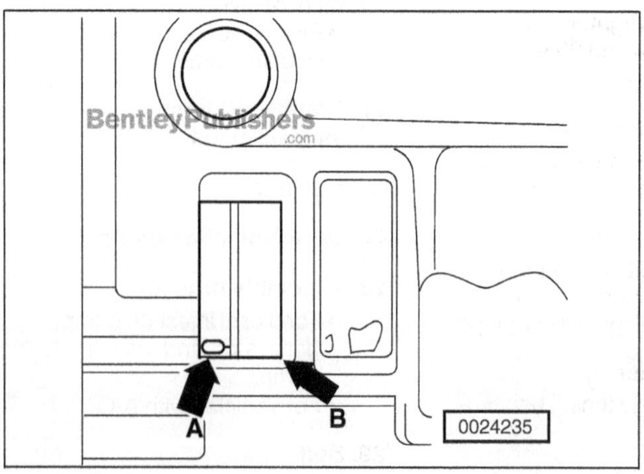

Automatic transmission equipped vehicles: Set engine to TDC for cylinder #1 by turning the crankshaft until the timing mark on the torque converter (**A**) aligns with mark on the transmission case (**B**).

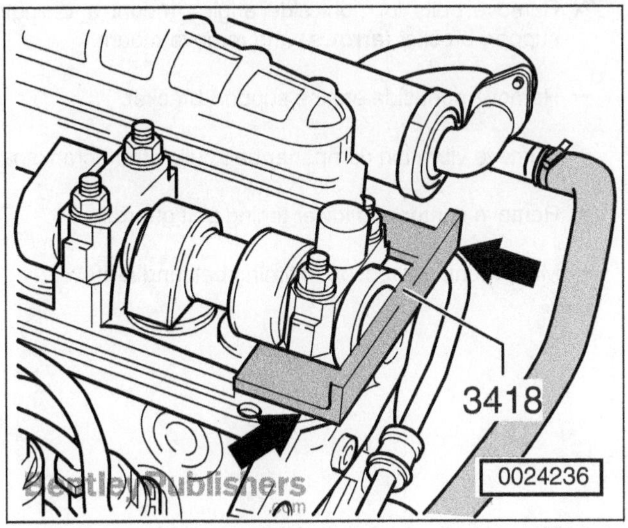

Lock camshaft with the setting bar. Use feeler gauges of equal thickness under each side (**arrows**) of setting bar (**3418**) to shim it up against camshaft. This will insure that camshaft is fully locked at TDC.

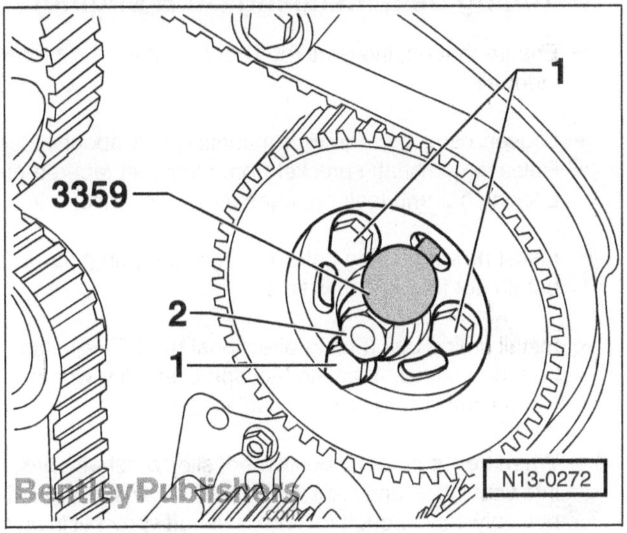

Lock injection pump sprocket with an appropriate locking tool and loosen injection pump sprocket mounting bolts.

> ***CAUTION—***
> *Do not loosen or remove the hub from the Diesel injection pump. The hub is pressed onto the tapered end of the camshaft and NOT keyed to the pump drive shaft! If removed, it will no longer be correctly indexed to the pump. Correct indexing requires special equipment and training available only at specialized Bosch service centers and is beyond the scope of this manual.*

– Loosen tensioner for timing belt.

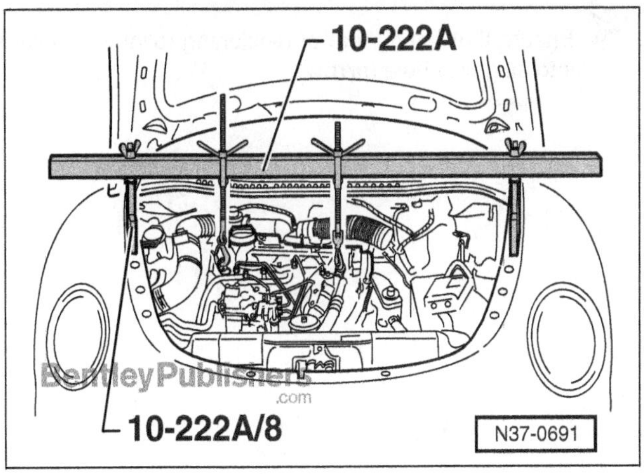

Support engine with a suitable fixture designed to support weight of the engine and transmission without damaging vehicle body.

– Lift supporting device slightly, so that weight of engine and transmission is supported.

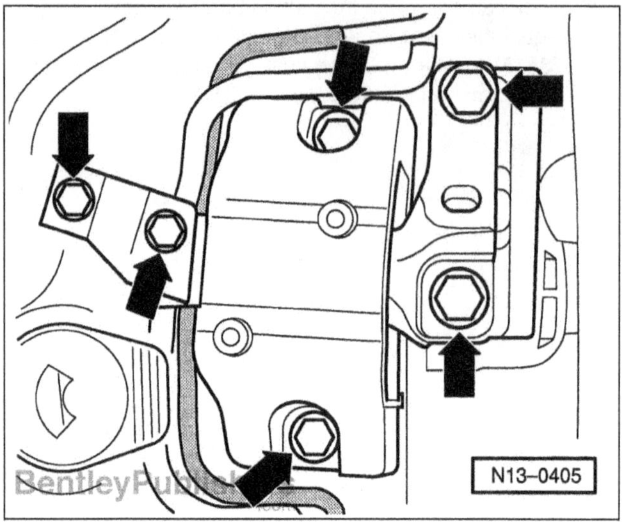

< Remove bolts for right side engine mount and engine support bracket (**arrows**) and remove mount.

- Remove right side engine support bracket.
- Remove vibration dampener/belt pulley from crankshaft.
- Remove center and lower timing belt guards.
- Mark running direction of timing belt and remove.

Timing belt, installing (ALH engine)

- Ensure that engine is at Top Dead Center (TDC) for cylinder #1.
- Loosen camshaft sprocket mounting bolt about ½ turn. Release camshaft sprocket from camshaft tapered end. Locking bar and feeler gauges remain in place.
- Install new mounting bolts into injection pump sprocket, but do not torque at this time.
- Install timing belt (noting directional mark if re-using old belt) on crankshaft timing belt sprocket, idler wheels, injection pump sprocket and coolant pump.
- When belt is properly positioned, slip camshaft sprocket into timing belt and slide onto camshaft. Install mounting bolt only tight enough to hold camshaft sprocket in place.

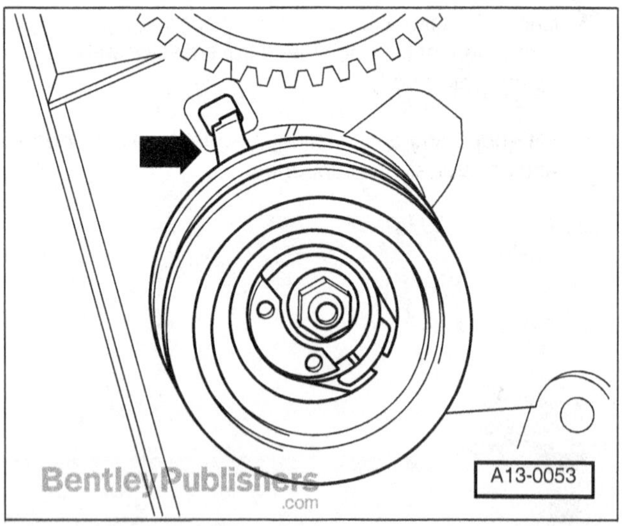

< Ensure that tab on rear of tensioning roller is positioned into oil galley hole (**arrow**).

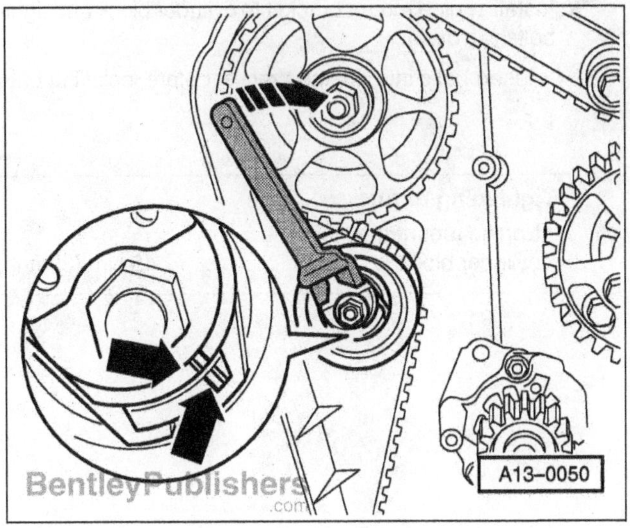

◄ Tension timing belt using a 2-pin wrench. Belt is tensioned properly when notch and raised mark are aligned (**inset arrows**).

– Tighten lock nut on tensioner.

Tightening torque
- Timing belt tensioner (M8) . . . . 20 Nm (15 ft-lb)

– Ensure that crankshaft is still at TDC.

– Tighten camshaft sprocket mounting bolt and remove setting bar and feeler gauges.

Tightening torques
- Camshaft sprocket to camshaft . . . . . 45 Nm (33 ft-lb)

NOTE —
- *Tighten the new injection pump sprocket mounting bolts to stage I only and remove the locking pin.*
- *The injection pump sprocket mounting bolts should only be used one time, since by design they have a reduced shank and are stretch bolts.*

– Turn crankshaft two rotations in the running direction and recheck timing belt tension.

– Dynamically check diesel injection pump timing, see **23 Fuel Injection (Diesel)**. If engine is removed, diesel pump injection timing must be checked and adjusted after engine has been installed.

– Return engine to TDC for cylinder #1. Install locking pin tool 3359 and tighten injection pump sprocket mounting bolts to final torque (stage II).

Tightening torque
- Injection pump sprocket - always replace
  stage I . . . . 20 Nm (15 ft-lb)
- stage II . . . . additional ¼ turn (90°)

– Install center and lower timing belt guards.

– Install vibration dampener/ribbed belt pulley onto crankshaft.

Tightening torque
- Vibration dampener/ribbed belt pulley to crankshaft
  sprocket (M8) . . . . 25 Nm (18 ft-lb)

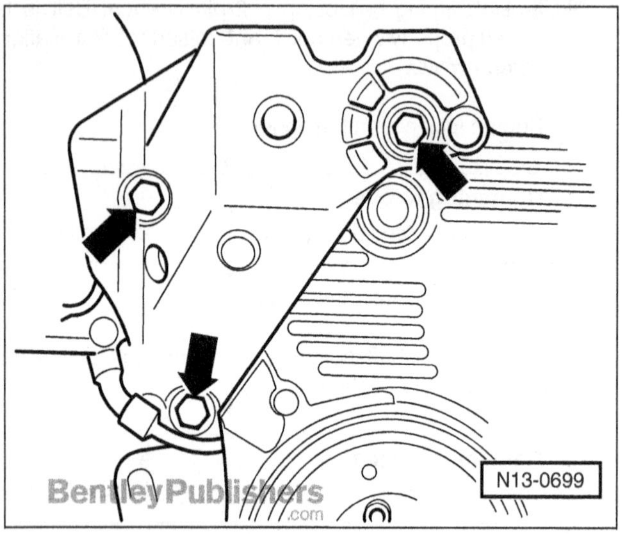

Install engine mount bracket to cylinder block and tighten bolts (**arrows**).

- Insert bolts into engine bracket before installing bracket.

Tightening torque

- Engine mount bracket to cylinder block . . . . 45 Nm (33 ft-lb)

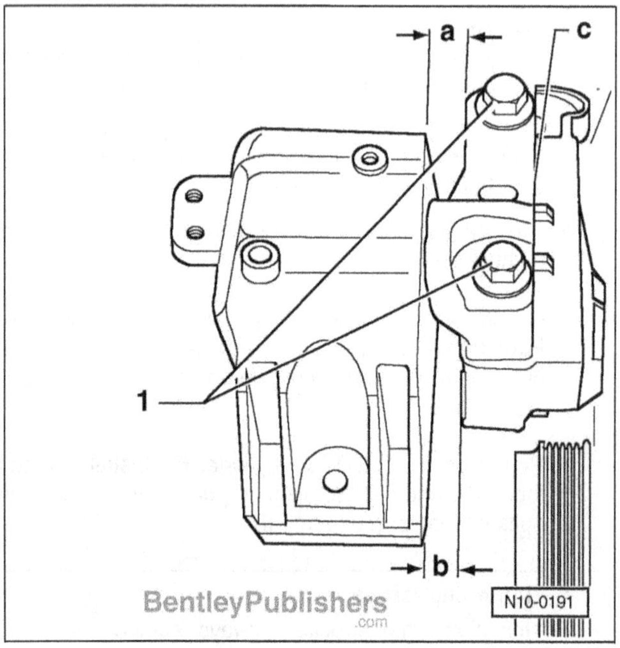

When engine/transmission assembly is in position, adjust so that mounts are aligned as specified.

Engine mount alignment

- Dimension **a** .14 mm (0.55 in.)
- Dimension **b** .10 mm (0.40 in.)

NOTE —

- *Do not stress engine and transmission mounts during installation. Make sure they are aligned parallel to each other.*
- *Bolt heads (**1**) must be flush with edge (**c**).*

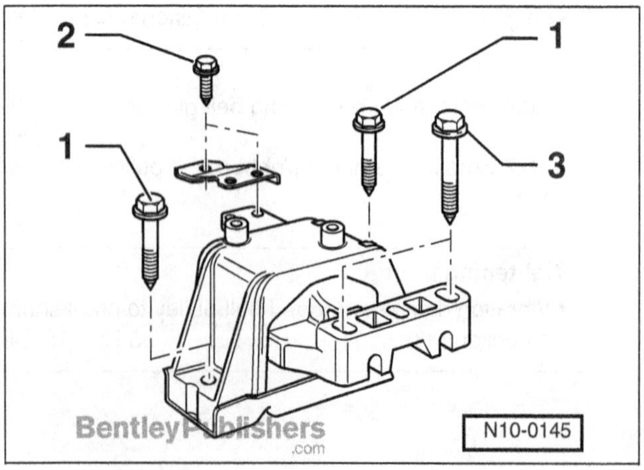

Install engine mount and tighten bolts.

1. Mount to body bolt
- 40 Nm (30 ft-lb) plus 90° (¼ turn)
- always replace

2. Mount bracket to body bolt
- 25 Nm (18 ft-lb)

3. Mount to engine bracket bolt
- 60 Nm (44 ft-lb) plus 90° (¼ turn)
- always replace

– Remove engine support device.

– Install ribbed V-belt and tensioner.

Tightening torque

- Ribbed V-belt tensioner (M8) . . . . . . 25 Nm (18 ft-lb)

– Install brake booster vacuum pump.

– Install cylinder head (valve) cover.

– Install connecting pipe between intercooler and turbocharger and between intercooler and intake hose.

– Install upper timing belt guard.

– Install fuel filter back into mounting bracket and connect fuel lines.

– Start engine and re-check diesel injection pump timing, see **23 Fuel Injection–Diesel**.

– Install lower sound absorber panel (belly pan).

– Install upper engine cover.

– Be sure to quality check your work, see **Quality Review** at the end of this repair group.

Timing belt, removing (BEW engine)

The timing belt, also known as the toothed belt or camshaft drive belt, should be inspected and replaced periodically. Inspection and replacement intervals are covered in **03 Maintenance**. Belt replacement is covered here. Regular belt inspection and replacement will prevent damage to the engine due to belt stretch or breakage.

NOTE—
Removal and installation of the timing belt requires the use of a number of special tools. Read the procedure through and have these special tools on hand before starting work.

– Remove upper sound absorber panel from engine. See **03 Maintenance**.

– Remove ribbed belt and tensioner. See **03 Maintenance**.

NOTE—
Mark running direction of ribbed belt if reusing.

– Unbolt coolant expansion tank and position out of way. Leave coolant hoses connected.

– Label fuel lines, disconnect and remove fuel filter assembly. See **03 Maintenance**.

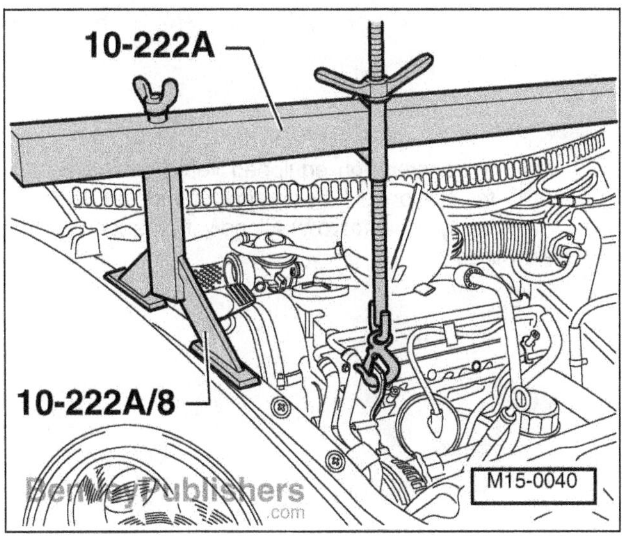

◄ Support engine using appropriate equipment. Lift engine slightly to relieve weight on engine mounts.

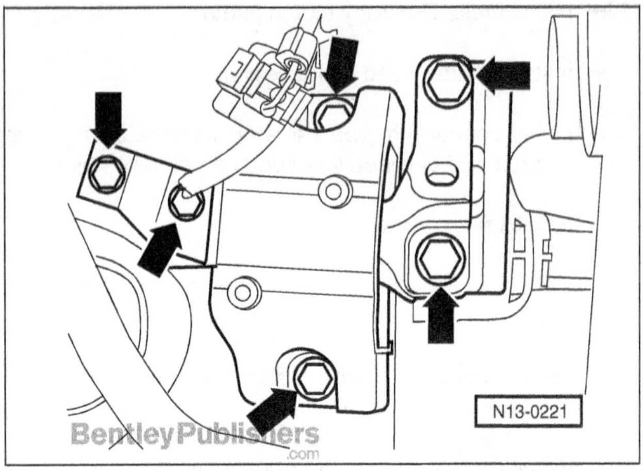

◄ Remove mounting bolts (**arrows**) from engine mount. Remove mount.

> ***WARNING —***
> *Remove engine mount only if engine is properly supported by engine support bar.*

- Remove upper timing belt guard.
- Lift vehicle using hoist. See **03 Maintenance**. if using floor jack, support vehicle with proper jackstands.
- Remove right front wheel and wheel housing liner.

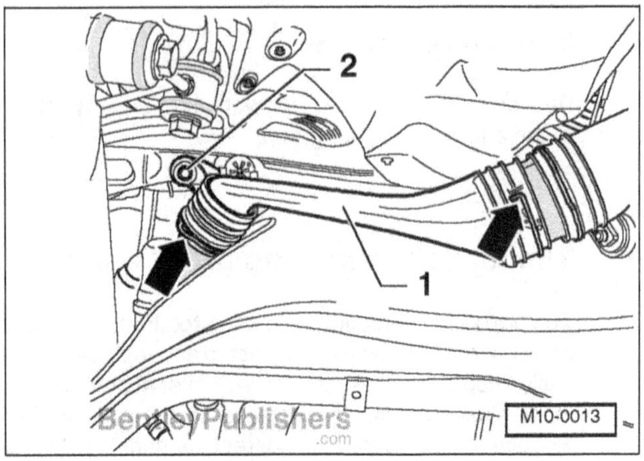

◄ Working under engine, release spring clamps (**arrows**) and remove mounting bolt (**2**). Remove lower duct (**1**) for intercooler.

- Seal open hose connections to prevent dirt or other objects from entering.
- Remove lower center and right sound insulation panels from under engine. See **03 Maintenance**.
- Raise engine slightly with supporting bar.

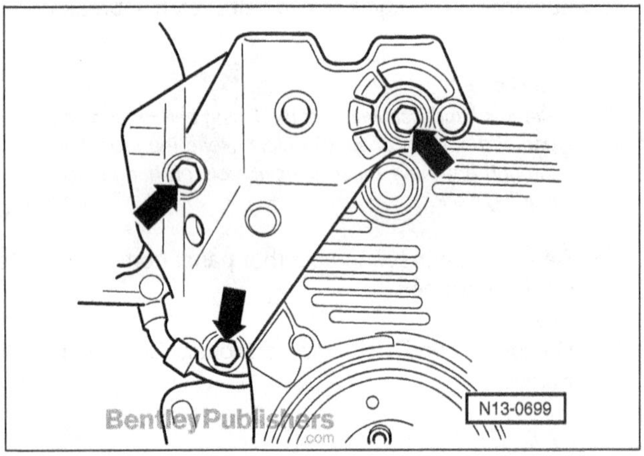

◄ Remove bolts (**arrows**) from engine mount on cylinder block and remove mount.

- Remove vibration damper/belt pulley.
- Remove lower and middle timing belt guards.

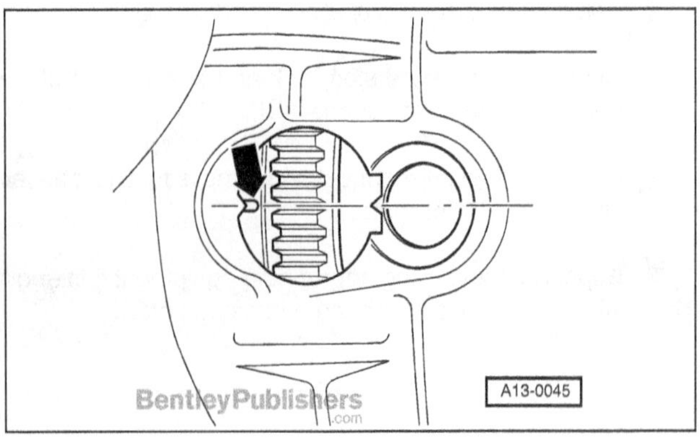

◄ Manual transmission equipped vehicles: Set engine to TDC for cylinder #1 by turning crankshaft until timing mark on flywheel (**arrow**) aligns with mark on transmission case.

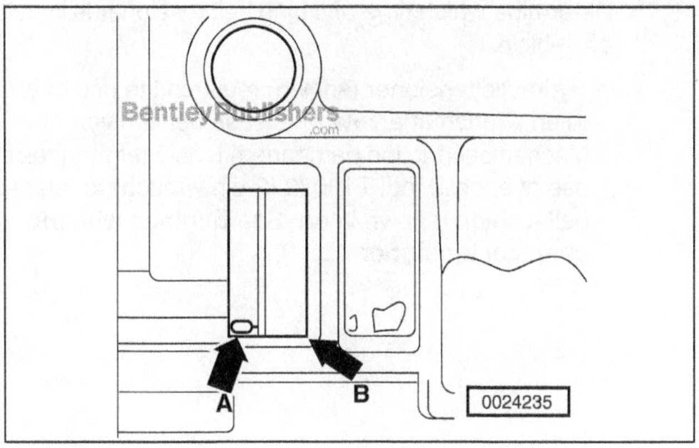

◀ Automatic transmission equipped vehicles: Set engine to TDC for cylinder #1 by turning the crankshaft until the timing mark on the torque converter (**A**) aligns with mark on the transmission case (**B**).

> ***NOTE —***
> *The engine may only be turned over using the crankshaft.*

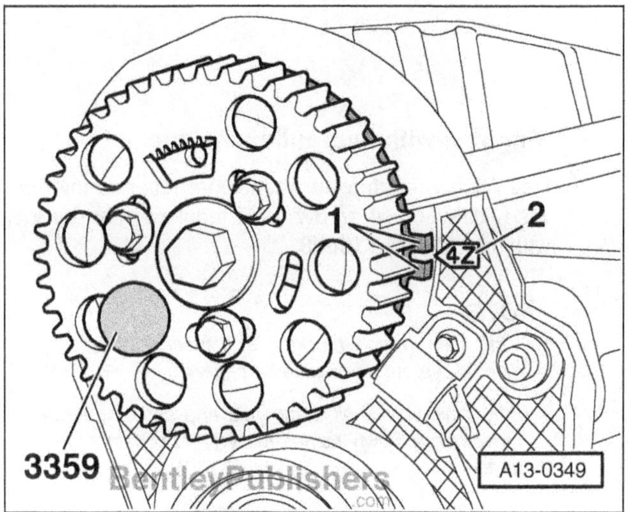

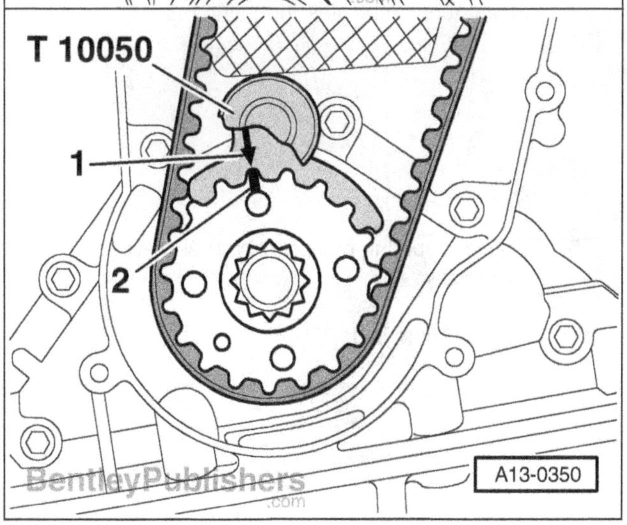

◀ Turn crankshaft further in direction of engine rotation until the mark on crankshaft sprocket is near top. (see lower illustration).

– Align camshaft sprocket sender wheel lugs (**1**) with "4Z" arrow (**2**) on the rear timing belt guard.

> ***NOTE —***
> *In the upper illustration the camshaft sprocket is shown without the timing belt on for clarity. Timing belt must remain on during this step.*

– Lock camshaft sprocket using pin 3359.

– Lock crankshaft sprocket using crankshaft lock T10050.
  - Mark on crankshaft sprocket (**2**) and on crankshaft stop (**1**) must align with one another.
  - The pin of crankshaft stop must then engage in hole in sealing flange behind crankshaft sprocket.

– Mark direction of rotation of timing belt before removing.

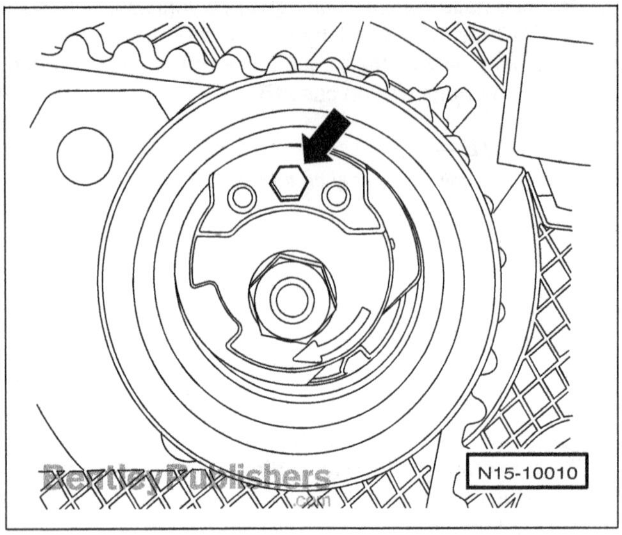

◄ Determine which type of timing belt tensioner is installed on vehicle.

- Hydraulic tensioner (shown) requires the use of an Allen wrench at **arrow** to release belt tension.
- Mechanical (friction dampened) tensioner requires the use of special tool T10020 (2-pin wrench) to release belt tension (shown later). See **Engines with mechanical tensioner**.

Engines with hydraulic tensioner

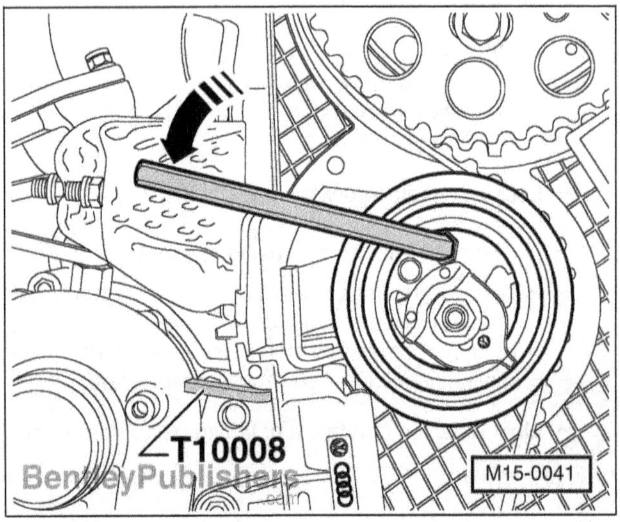

◄ Use Allen wrench to turn tensioner roller counter-clockwise (direction of **arrow**) until tensioner can be locked with special tool **T10008**.

NOTE —

- *If the Allen wrench (hex key) is not inserted sufficiently it could slip out and cause damage.*
- *The tensioner is oil dampened and should only be moved back slowly using an even and constant pressure.*

– Loosen tensioner roller mounting nut.

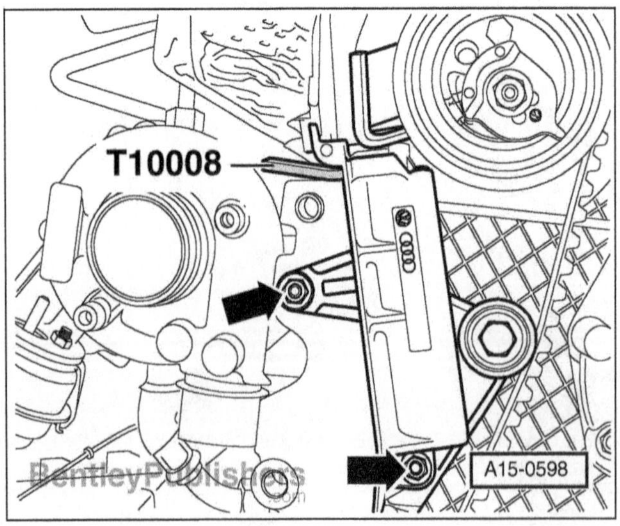

◄ Remove mounting bolts (**arrows**) and hydraulic tensioner assembly.

– Remove timing belt.

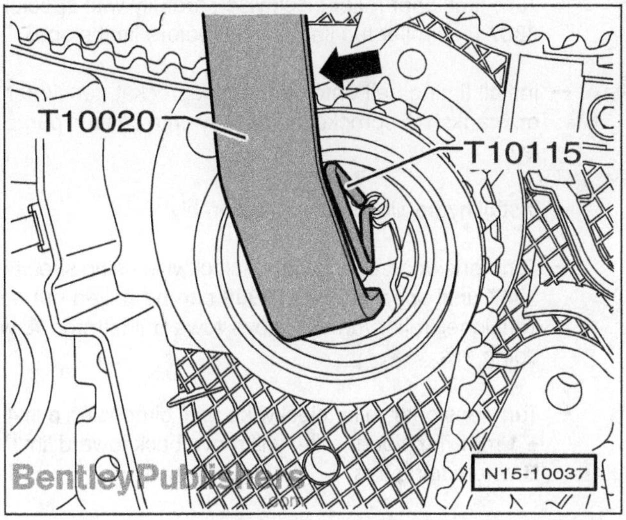

Engines with mechanical tensioner

– Loosen belt tensioner roller mounting nut.

◄ Turn tensioner roller counter-clockwise (direction of **arrow**) with 2-pin wrench (special tool T10020 or T10004) until roller can be locked in place with pin T10115, or equivalent.

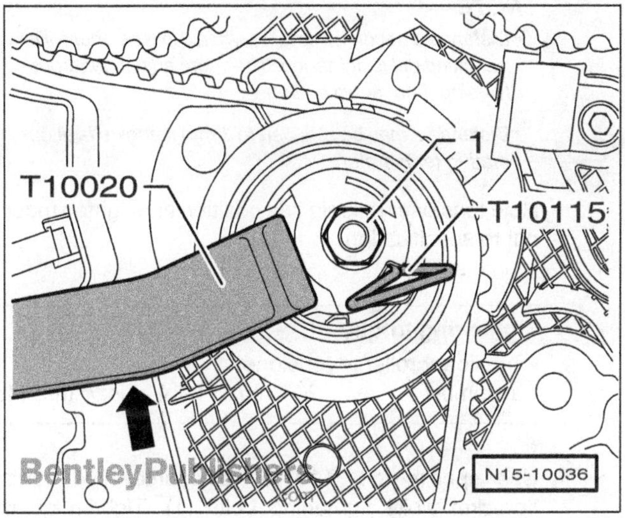

◄ Turn tensioner roller with 2-pin wrench clockwise (direction of **arrow**) to stop and tighten roller mounting nut (**1**) by hand.

– Remove timing belt from coolant pump and then from remaining pulleys.

Timing belt, installing (BEW engine)

Before beginning make sure of the following:

- Engine is cold
- Camshaft sprocket is locked with pin 3359
- Crankshaft sprocket is locked with T10050

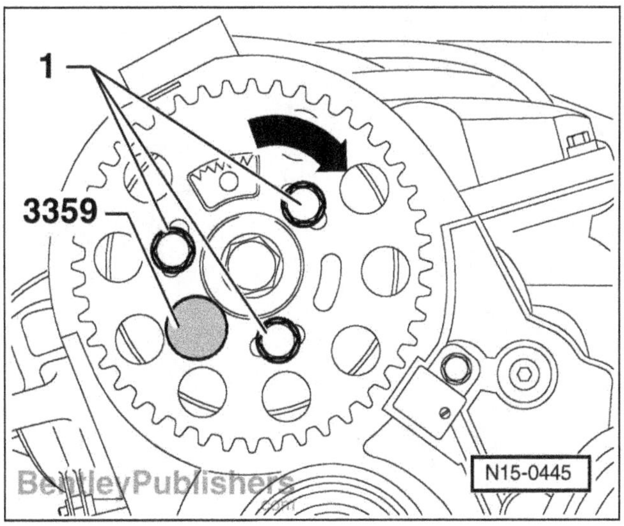

Engines with hydraulic tensioner

◄ Loosen camshaft sprocket mounting bolts (**1**) until sprocket can be moved slightly in direction of **arrow**. Move sprocket to limit stop.

NOTE —
Locking pin 3359 must remain in position.

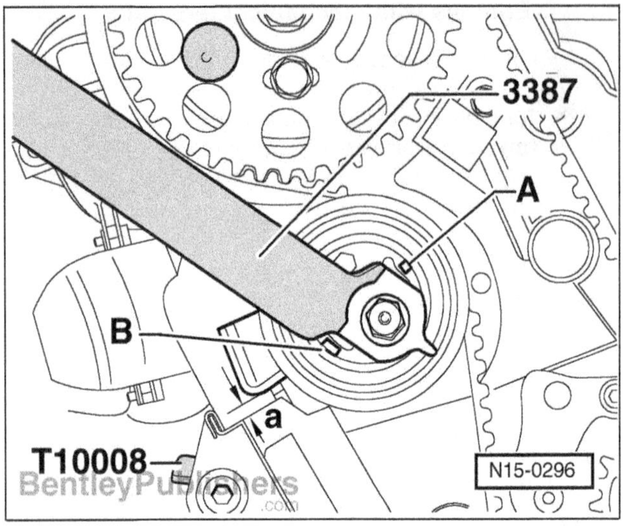

◀ Turn tensioner roller clockwise carefully with special tool **3387** until roller lug lies directly before limit stop **B**.

– Install timing belt onto camshaft sprocket, tensioner roller, crankshaft sprocket and finally onto coolant pump pulley.

– Install hydraulic tensioner assembly.

– Turn tensioner roller counter-clockwise using special tool **3387** until special tool **T10008** can be pulled out easily. Tensioner roller lug will move toward limit stop **A** when doing this.

– Turn tensioner roller clockwise until dimension **a** is 4 mm ± 1 mm. Tensioner roller will move back toward limit stop **B** when doing this.

NOTE —

- *Distance **a** becomes smaller when the tensioner roller mounting nut is tightened. Therefore adjust generously using 4mm specification.*
- *Distance **a** may be reduced to 1 mm if engine is at operating temperature.*

– Hold tensioner roller in this position and tighten mounting nut to specification.

Tightening torque

- Tensioner roller to tensioner assembly . . . . 20 Nm (15 ft-lb) plus ⅛ turn (45°)

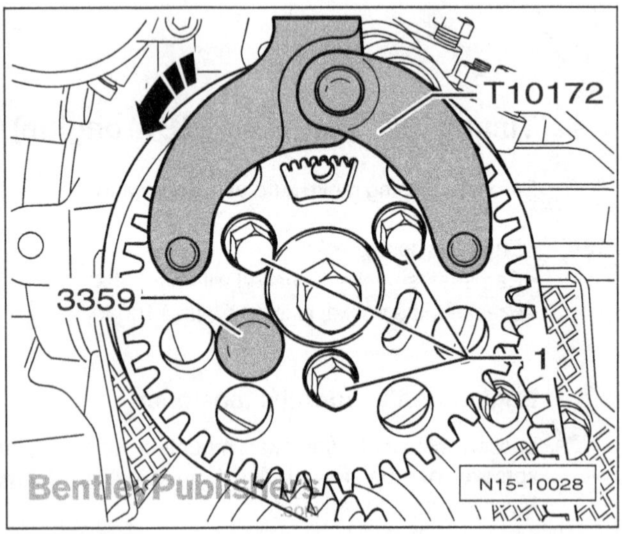

◀ Use special tool T10172 to counter-hold camshaft sprocket while tightening bolts (**1**). Tighten camshaft sprocket mounting bolts to specification.

Tightening torque

- Camshaft sprocket to camshaft . . . . 25 Nm (18 ft-lb)

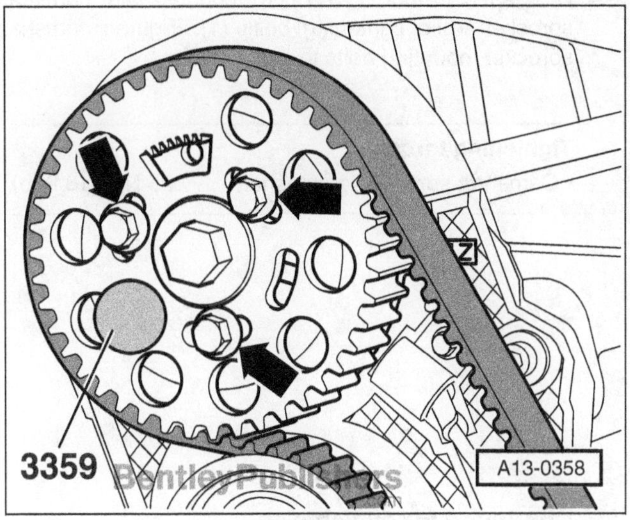

Engines with mechanical tensioner

< Rotate camshaft sprocket so that mounting bolts are centered in slots (**arrows**).

– Install timing belt onto crankshaft sprocket, tensioner roller, camshaft sprocket, idler wheel and finally onto coolant pump pulley.

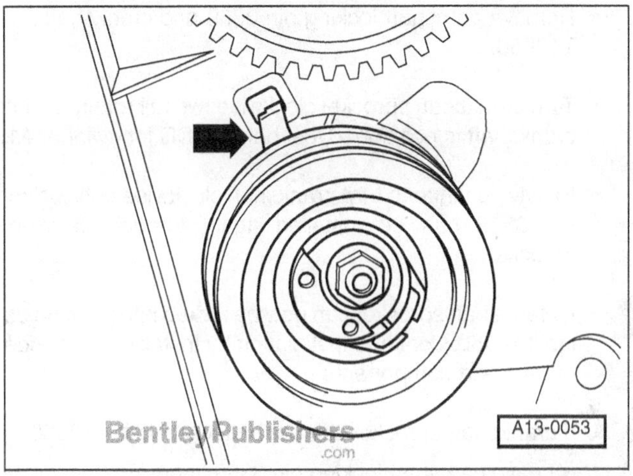

< Make sure tensioner roller locating tab (**arrow**) is properly positioned in rear timing belt guard.

– Loosen tensioner roller mounting nut and remove locking pin T10115.

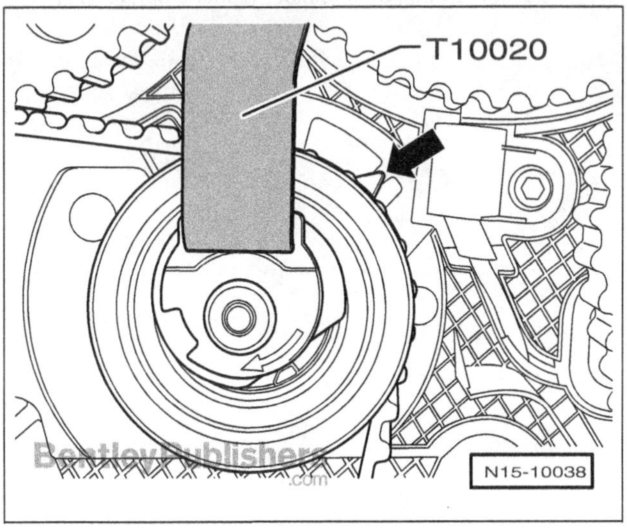

< Use 2-pin wrench (VW special tool T10020) to turn tensioner roller clockwise until indicator is centrally located in gap (**arrow**) of base plate. Make sure mounting nut does not move with tensioner.

– Hold tensioner roller in this position and tighten mounting nut to specification.

Tightening torque
- Tensioner roller to tensioner assembly .... 20 Nm (15 ft-lb) plus 1/8 turn (45°)

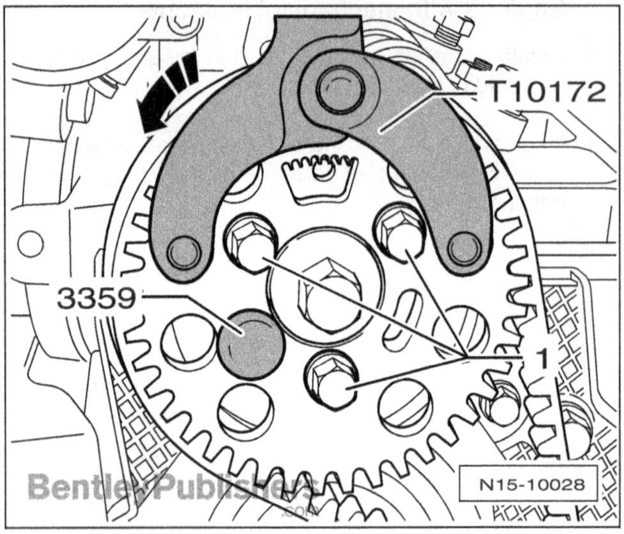

- Use special tool T10172 to counter-hold camshaft sprocket while tightening bolts (**1**). Tighten camshaft sprocket mounting bolts to specification.

Tightening torque

- Camshaft sprocket to camshaft . . . . 25 Nm (18 ft-lb)

Continued for all vehicles

- Remove camshaft locking pin 3359 and crankshaft lock T10050.
- Turn crankshaft sprocket clockwise two full rotations until crankshaft is positioned just before TDC for cylinder #1.
- Slowly continue to turn crankshaft clockwise until locking pin 3359 can be installed again through camshaft sprocket.
- If crankshaft sprocket can now be locked into position using T10050, continue installation by installing remainder of removed components.

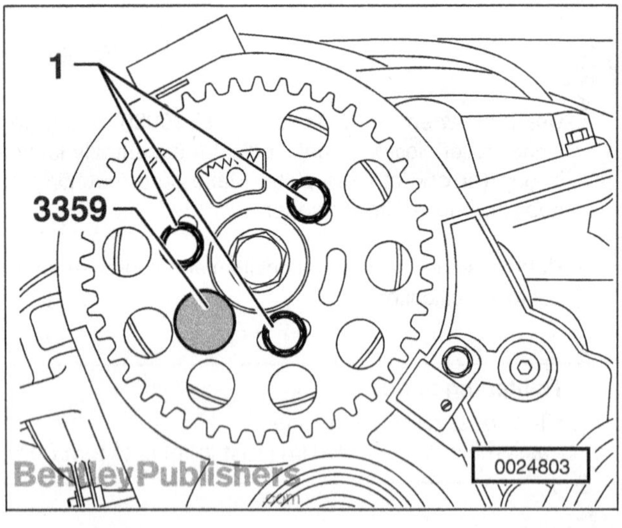

- If crankshaft sprocket can not be locked with T10050:
  - Leave camshaft locking pin 3359 in position.
  - Loosen camshaft sprocket mounting bolts (**1**).

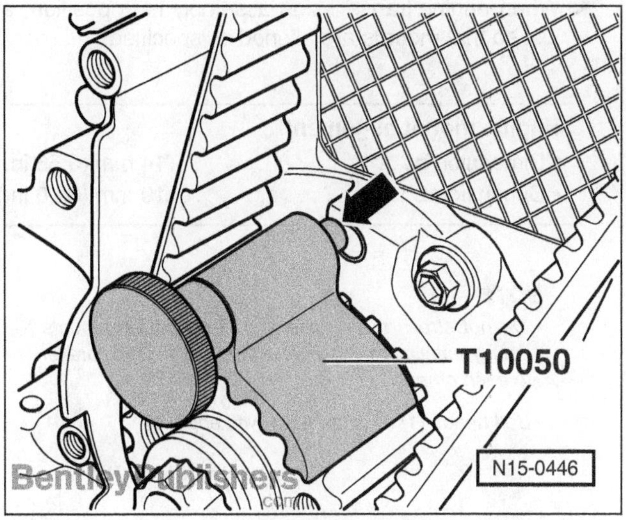

– Install crankshaft sprocket lock T10050 without pushing locking pin into sealing flange hole.

◄ Turn crankshaft counter-clockwise slightly until crankshaft locking pin is positioned just before sealing flange hole (**arrow**).

– Slowly turn crankshaft sprocket clockwise until locking pin engages in sealing flange hole.

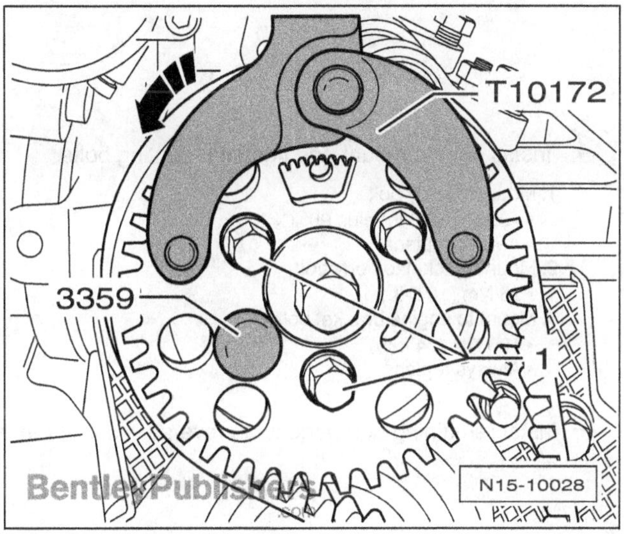

◄ Use special tool T10172 to counter-hold camshaft sprocket while tightening bolts (**1**). Tighten camshaft sprocket mounting bolts to specification.

Tightening torque
- Camshaft sprocket to camshaft . . . . 25 Nm (18 ft-lb)

– Remove camshaft locking pin 3359 and crankshaft lock T10050.

– Turn crankshaft sprocket clockwise two full rotations and check that crankshaft is positioned at TDC for cylinder #1.
  - If not OK, repeat steps above until crankshaft sprocket is positioned at TDC for cylinder #1.
  - If OK, continue installing removed component as described below.

– Install middle and lower timing belt guard.

– Install V-belt pulley/vibration damper.

Tightening torque
- V-belt pulley/vibration damper to crankshaft sprocket . . . . 10 Nm (7 ft-lb) plus ¼ turn (90°)

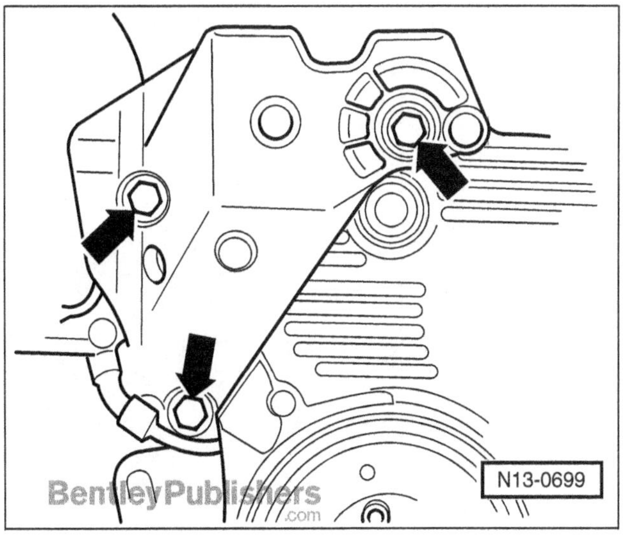

◄ Install engine mount bracket to cylinder block and tighten bolts (**arrows**).
  - Insert bolts into engine bracket before installing the bracket.

Tightening torque
- Engine mount bracket to cylinder block . . . . 45 Nm (33 ft-lb)

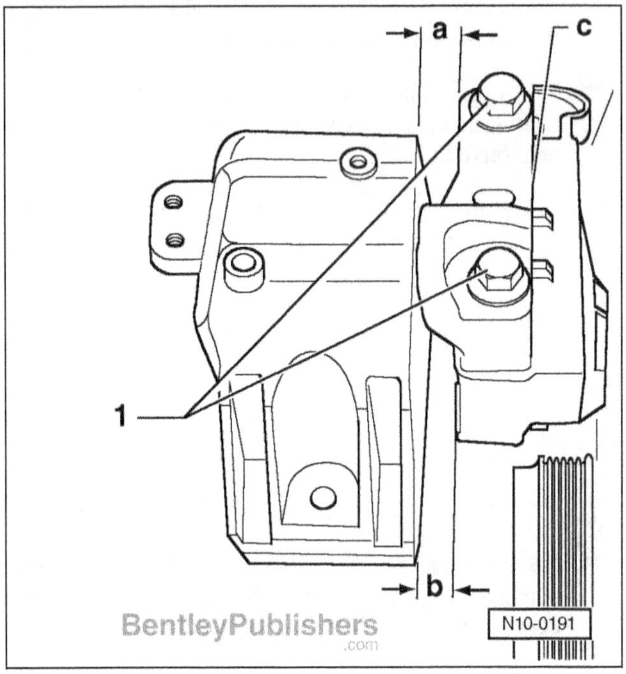

◀ When engine/transmission assembly is in position, adjust so that mounts are aligned as specified.

Engine mount alignment

- Dimension **a** . . . . 14 mm (0.55 in.)
- Dimension **b** . . . . 10 mm (0.40 in.)

NOTE —

- *Do not stress the engine and transmission mounts during installation. Make sure they are aligned parallel to each other.*
- *Bolt heads (**1**) must be flush with edge (**c**).*

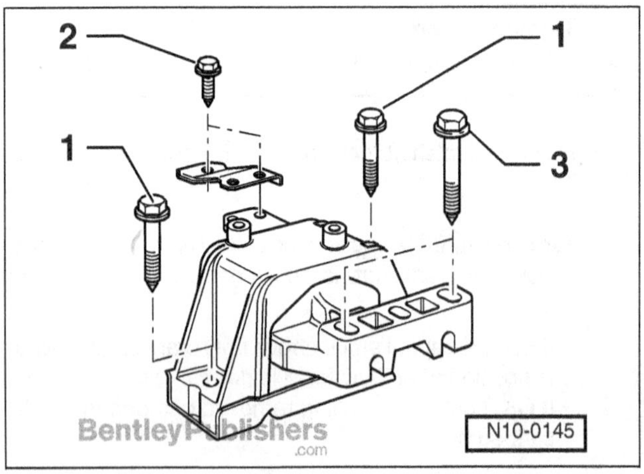

◀ Install engine mount and tighten mounting bolts.

1. Mount to body bolt
   - 40 Nm (30 ft-lb) plus 90° (¼ turn)
   - Always replace
2. Mount bracket to body bolt
   - 25 Nm (18 ft-lb)
3. Mount to engine bracket bolt
   - 100 Nm (74 ft-lb)
   - Always replace

– Instal remaining components in reverse order of removal.

CYLINDER HEAD COMPONENTS

This section provides the specifications and special information necessary to repair the 1.9L cylinder head that has been removed from the engine. Special service tools and machine shop services are required for most cylinder head repair.

NOTE —
- *The information given under this heading assumes that the cylinder head is removed. For cylinder head removal procedures, see* **Cylinder head, removing** *given earlier.*

Cylinder Head and Camshaft

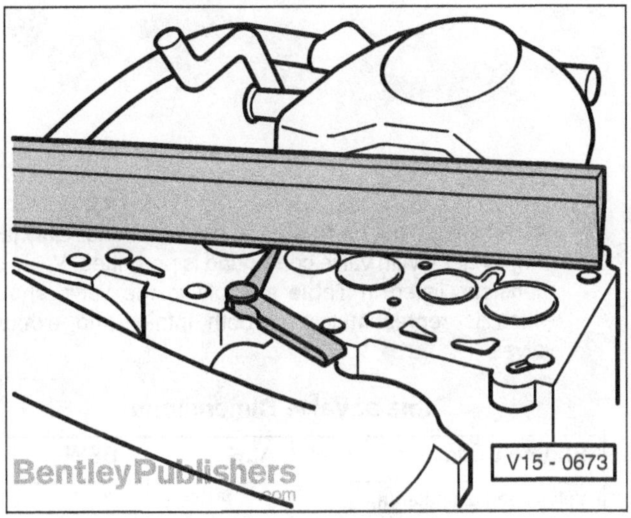

Check the cylinder head for warpage and distortion with an accurate straight edge and a feeler gauge. Resurfacing of the cylinder head is specifically not recommended by Volkswagen. Excessively warped or distorted cylinder heads must be replaced.

Specification
- Maximum cylinder head distortion/warpage (1.9L) .... 0.1 mm (0.004 in.)

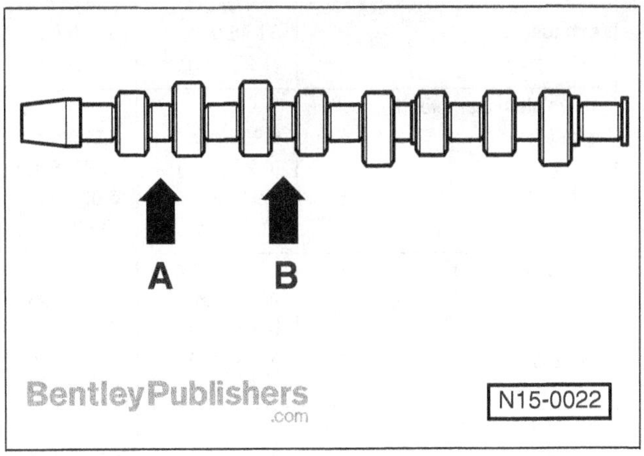

The camshaft for the ALH engine is easily identified by markings cast into it during manufacture.

Specification
- Valve timing at 1 mm valve lift, (ALH engine)
  - Intake opens ATDC .... 16°
  - Intake closes ABDC........................ 25°
  - Exhaust opens BBDC .... 28°
  - Exhaust closes BTDC .... 19°

The in and out movement of the camshaft is known as axial play and is measured with the cam followers removed and only the first and last bearing caps installed.

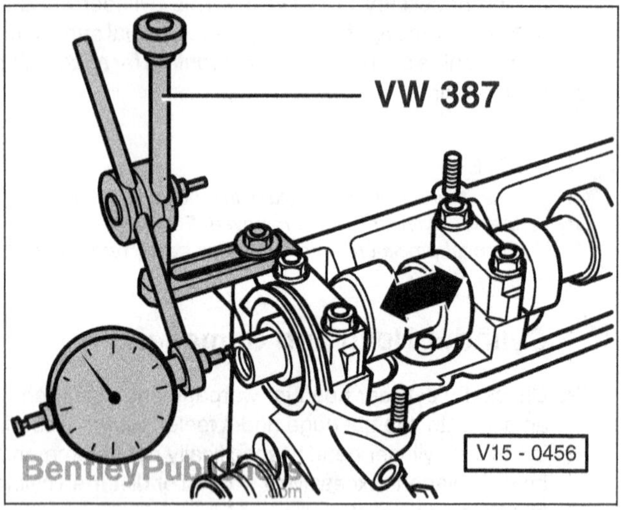

A dial gauge is set up on the sprocket end of the camshaft and the camshaft clearance is checked by moving the camshaft as far as it can go in each direction.

Specification

- Camshaft axial clearance (1.9L) 0.15 mm (0.006 in.)

Valves

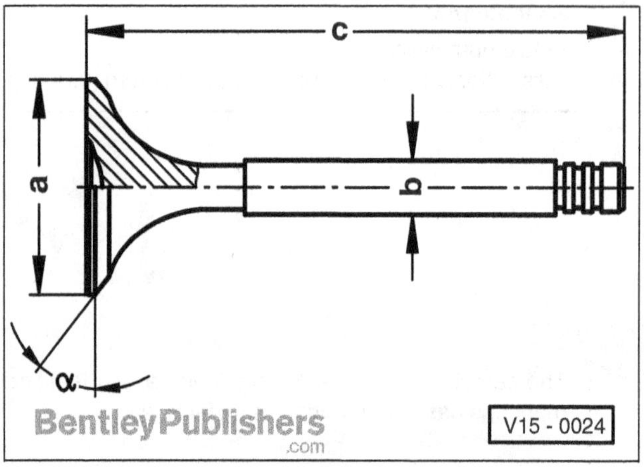

Valves should not be re-worked on a machine. Only lapping by hand with valve compound is permitted. Valve dimensions listed in **Table a** apply to the valve shown which is representative of both intake and exhaust valves.

Table a. Valve Dimensions

Engine Code	ALH	BEW
Valve head diameter (**a**)		
intake	35.95 mm (1.415 in.)	35.95 mm (1.415 in.)
exhaust	31.45 mm (1.238 in.)	31.45 mm (1.238 in.)
Valve stem diameter (**b**)		
intake	6.963 mm (0.2741 in.)	6.980 mm (0.2748 in.)
exhaust	6.943 mm (0.2733 in.)	6.956 mm (0.2739 in.)
Valve length (**c**)		
intake	96.85 mm (3.813 in.)	89.95 mm (3.541 in.)
exhaust	96.85 mm (3.813 in.)	89.95 mm (3.541 in.)
Valve face angle (α)		
intake	45°	45°
exhaust	45°	45°

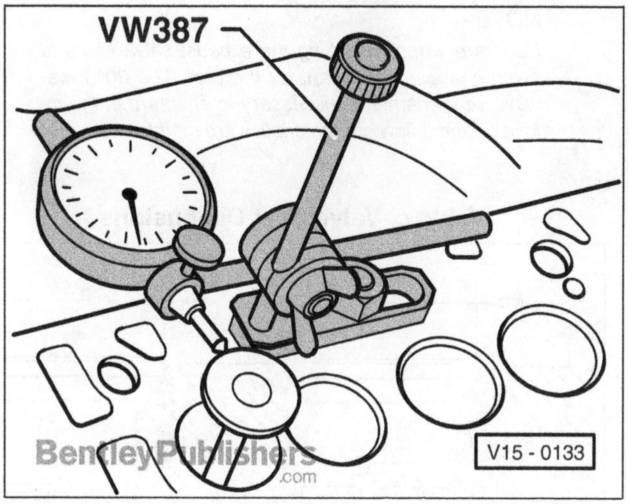

Valve guides

Special tools and a press are required to replace the valve guides. Check valve guide wear with a new valve. Always use an intake valve in an intake guide and an exhaust valve in an exhaust guide. Inspect valve seats to ensure that cylinder head can be reconditioned before installing new valve guides. Press out worn original valve guides with Volkswagen special tool 3121 from camshaft side of head. Replacement valve guides have a shoulder on the camshaft side to limit installed depth and must be pressed out from the combustion chamber side.

When installing new valve guides, lubricate with clean engine oil and press in from camshaft side down to shoulder. The cylinder head must be cold for this operation. Once the shoulder of valve guide contacts the cylinder head, do not allow pressure on the guide to exceed 2000 psi (1 ton) otherwise cylinder head or guide will be damaged. When guide has been replaced, ream it to the proper size and continue with reworking of valve seats.

Specification

- Valve guide (1.9L) (wear limits-maximum play) with new valve
  intake and exhaust valve guide . . 1.3 mm (0.051 in.)

NOTE —
Due to the close tolerances found in valve guides, it is recommended that Volkswagen special tool 3121 be used for removal and installation and special tool 3120 be used to ream valve guide to size. Be sure to use appropriate cutting fluid during reaming operation.

Valve seats

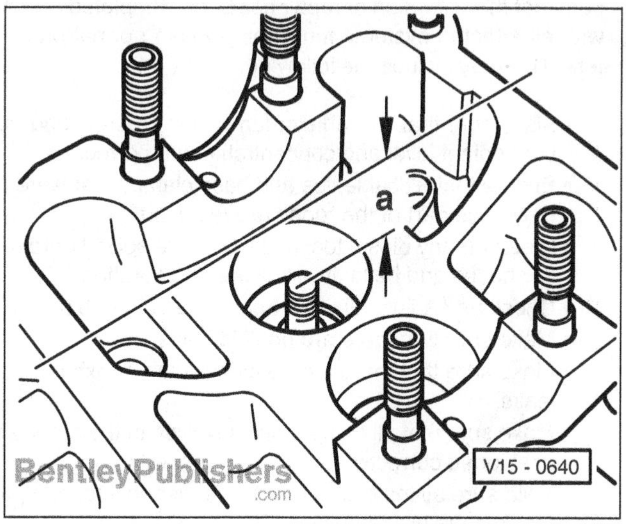

When resurfacing valve seats, there is a limited amount of material that can be removed to bring the seat back into specification. If too much material is removed, the final assembly will leave too little space for the hydraulic cam follower to function properly. The maximum refacing dimension, that is, the maximum amount of material that can be removed from the valve seat, is calculated from the measurement shown.

Measure dimension **a**, and subtract the minimum dimension, as given in **Table b**. The difference is the maximum amount of material that can be removed from valve seat.

Table b. Minimum Dimensions for Calculating Valve Seat Refacing Dimensions

Engine Code	Intake	Exhaust
ALH	35.8 mm (1.409 in.)	36.1 mm (1.421 in.)

NOTE—
Use care when reworking the exhaust valve seats to avoid changing the shape of the port. The 30° lower valve seat chamfer is necessary to ensure that the intake channel flow characteristics are maintained.

Table c. Valve Seat Dimensions

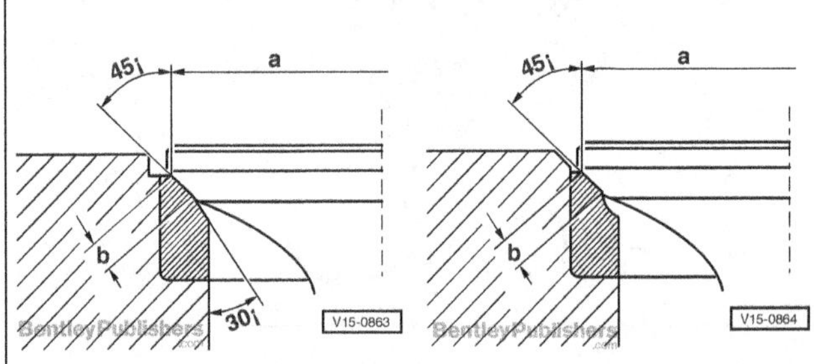

Engine: 1.9L	Intake	Exhaust
Seat diameter (**a**)	35.7 mm (1.406 in.)	31.4 mm (1.236 in.)
Seat width (**b**)	1.6 mm (0.063 in.)	2.7 mm (0.106 in.)
Valve seat angle	45°	45°
Correction angle, lower	30°	N/A

NOTE—
*Cylinder head surface reference line is (**Z**).*

QUALITY REVIEW

When you have finished working under the hood and around other areas of the vehicle it is advisable to take a moment to quality check or review your work. This helps to ensure that the operation or repair has been completed properly with all affected systems functioning within normal parameters. This may include the following:

- Make sure that the radiator fan cycles properly and that the coolant level and concentration are correct.
- Ensure that all cable ties and hose clamps that were removed as part of the repair are replaced.
- Clean off any diesel fuel that may have spilled onto rubber hoses and belts. It will cause deterioration.
- Check and adjust all other applicable fluid levels.
- Make sure that there are no fluid leaks.
- Make sure that there are no air, vacuum or exhaust leaks.
- Make sure that all components involved in the repair are positioned correctly and function properly.
- Male sure all tools, shop cloths, fender covers, and protective tape are removed.
- Clean grease from painted surfaces and steering wheel.
- Unlock the anti-theft radio and reset the clock.

In addition to the above noted points, the ECM and TCM may need to be checked using a diagnostic scan tool.

15c Cylinder Head and Valvetrain (2.0L engine)

TABLES

GENERAL

This section covers cylinder head and valvetrain service and repair work for the 2.0L 4-cylinder, 2 valve per cylinder, gasoline engine. For information on short block engine rebuilding and internal engine specifications, see **13 Crankshaft/Cylinder Block**.

Most of the operations described in this repair group require special equipment and experience. If you lack the skills, tools, or a suitable workplace for servicing or repairing the cylinder head, we suggest that you leave these repairs to an authorized Volkswagen dealer or other qualified shop.

Engine Codes

- AEG, AVH, AZG, BDC, BEV, BGD . . . . 2.0L 4-cylinder gasoline

DIAGNOSTIC TESTING

The tests that follow can be used to help isolate engine problems, to better understand a problem before starting expensive and extensive repairs, or to just periodically check engine condition.

Compression test

A compression test will tell a lot about the overall condition of the engine without the need for taking it apart. Testing is relatively simple and straightforward.

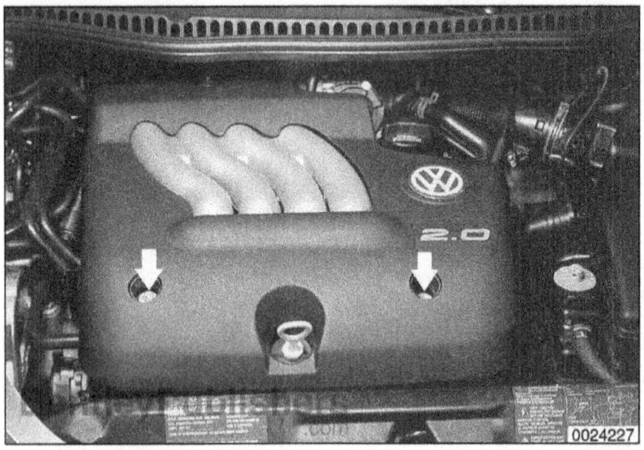

◀ Remove two plastic caps and retaining nuts (**arrows**) on engine cover. Loosen the rear retaining nut slightly. Remove dipstick and lift cover up slightly and pull forward. (AEG engine shown, other engines are similar)

– Replace dipstick and warm-up the engine until it is a minimum of 30°C (86°F).

– Disconnect spark plug wires from spark plugs by gently pulling on the boot, do not pull on the wire. Label all wires to avoid confusion during reinstallation.

– Use compressed air to clear the area around the spark plugs. Remove all spark plugs and lay aside in proper order.

◀ Disable ignition system by unplugging harness connector (**arrow**) from ignition coil power output stage (AEG engine shown, secondary air pump removed for clarity).

> ***CAUTION —***
> *Failure to disable the ignition system during testing can result in damage to the Motronic ECM or the ignition power output stage due to the high voltage developed in the ignition coil.*

– Remove fuse 32 for fuel injector voltage supply.

– Fit compression tester into spark plug hole.

– Have a helper:

- depress clutch pedal fully
- put transmission in neutral or park
- depress accelerator pedal to floor
- crank engine over with starter motor

NOTE —
Cranking the engine with the ignition system disabled and components disconnected may cause Diagnostic trouble codes (DTCs) to be stored in engine management system memory.

– Engine should be cranked a minimum of 4 to 5 revolutions.

– Record readings, release pressure in the gauge and repeat procedure for each cylinder.

– Compare readings to specification.

Specification
- Compression pressures (2.0L engine)
  new .... 10 to 13 bar (145 - 189 psi)
  wear limit .... 7.5 bar (109 psi)
- maximum difference
  between cylinders .... 3 bar (44 psi)

– When all cylinders have been checked, reinstall spark plugs and wires, connect harness connector for ignition coil power output stage and install the upper engine cover.

Tightening torque
- Spark plug to cylinder head .... 30 Nm (22 ft-lb)

Cylinder leakdown test

The most conclusive diagnosis of low compression symptoms requires a cylinder leak-down test. Using a special tester and a supply of compressed air, each cylinder is pressurized. The rate at which the air leaks out of the cylinder, as well as the sound and location of the escaping air can more accurately pinpoint the magnitude and source of the leakage. Any engine compression diagnosis that will require major disassembly should first be confirmed by a cylinder leak-down test. Because this test requires special equipment and experience, it may be desirable to have it performed by a Volkswagen dealer or other qualified repair shop.

CYLINDER HEAD SERVICE

Many cylinder head repairs can be accomplished without removing the cylinder head from the engine. The cylinder head cover (valve cover) gasket, the camshaft, the camshaft oil seal, the valve guide oil seals, the valve springs, and the camshaft followers are all accessible with the cylinder head installed. This heading describes those repairs that can be done with the cylinder head installed.

Internal cylinder head assembly

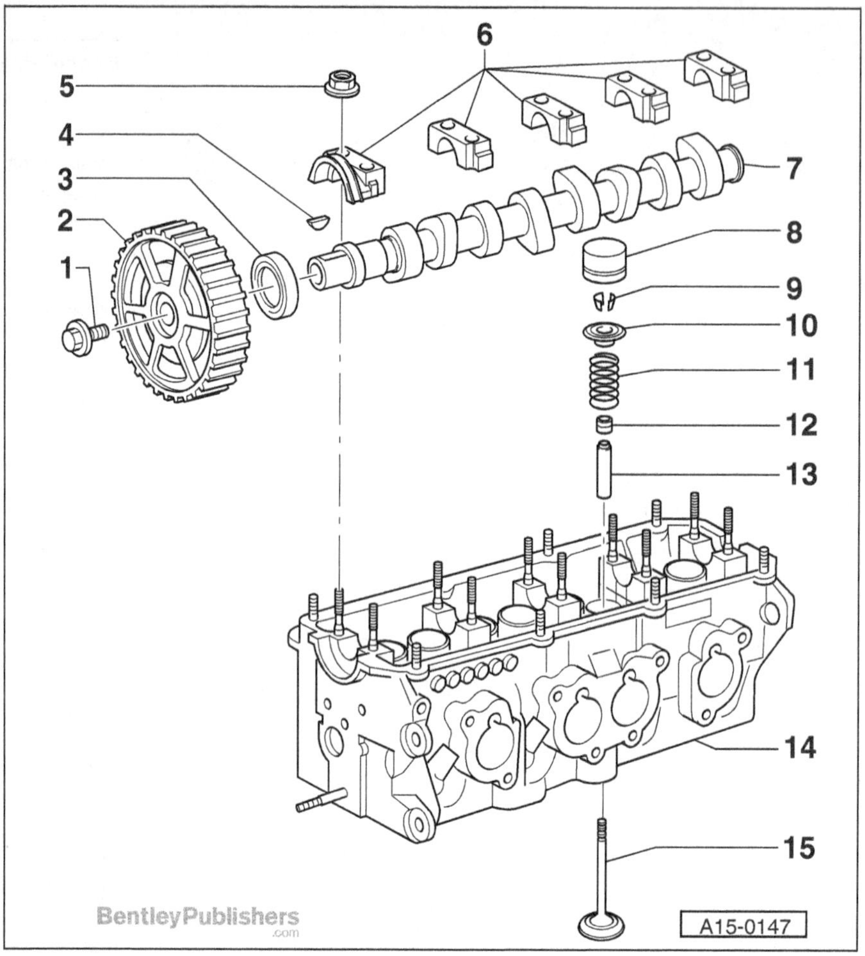

1. **Camshaft sprocket bolt**
   - Tighten to 100 Nm (74 ft-lb)
   - Use 3415 counterholder to loosen and tighten
2. **Camshaft sprocket**
3. **Oil seal**
4. **Woodruff key**
   - Ensure tight fit
5. **Bearing cap nut**
   - Tighten to 20 Nm (15 ft-lb)
6. **Bearing cap**
   - Lightly coat bearing cap 1 cylinder head mating surface with sealant
7. **Camshaft**
   - Radial clearance, checking with plastigage
     Wear limit: 0.1 mm (.004 in.)
   - Run-out: max. 0.05 mm (.002 in.)
8. **Hydraulic lifter**
   - Do not interchange
   - Equipped with hydraulic clearance compensation
   - Store with cam contact surface facing downwards
   - Oil contact surfaces
9. **Keepers**
10. **Valve spring retainer, upper**
11. **Valve spring**
12. **Valve stem seal**
13. **Valve guide**
    - Service version with collar
14. **Cylinder head**
15. **Intake and exhaust valves**
    - Do not rework! Only lapping is permitted

External cylinder head assembly

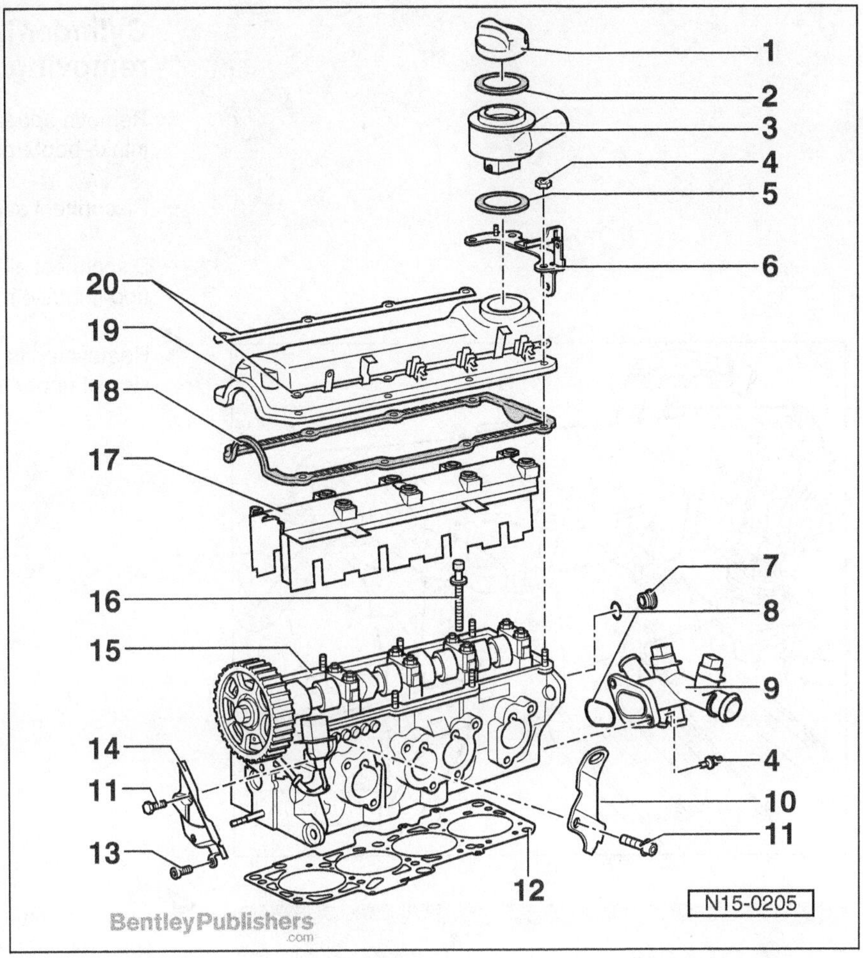

1. **Cap**
2. **Gasket**
3. **Vent housing**
   - Turn clockwise to remove
4. **Nut**
   - Tighten to 10 Nm (7 ft-lb)
5. **Gasket**
   - Replace if damaged
6. **Bracket**
7. **Sealing plug**
   - Tighten to 15 Nm (11 ft-lb)
   - Always replace
8. **Oil seal**
   - Always replace
9. **Coolant connection flange**
10. **Lifting eye**
11. **Bolt**
    - Tighten to 20 Nm (15 ft-lb)
12. **Cylinder head gasket**
    - Always replace
    - Replace engine coolant if replacing gasket
13. **Bolt**
    - Tighten to 15 Nm (11 ft-lb)
14. **Timing belt guard, rear**
15. **Cylinder head**
    - Replace engine coolant if removing head
16. **Cylinder head bolt**
    - Always replace
17. **Oil deflector**
18. **Valve cover gasket**
    - Replace if damaged
    - Before installing gasket coat bearing cap 1 cylinder head mating surface with sealant
19. **Valve cover**
20. **Reinforcing strip**

Cylinder head (valve) cover, removing and installing

– Remove upper sound absorber panel and disconnect intake boot from throttle housing.

– Disconnect accelerator cable from throttle body.

– Disconnect all hoses and electrical connectors from upper intake manifold.

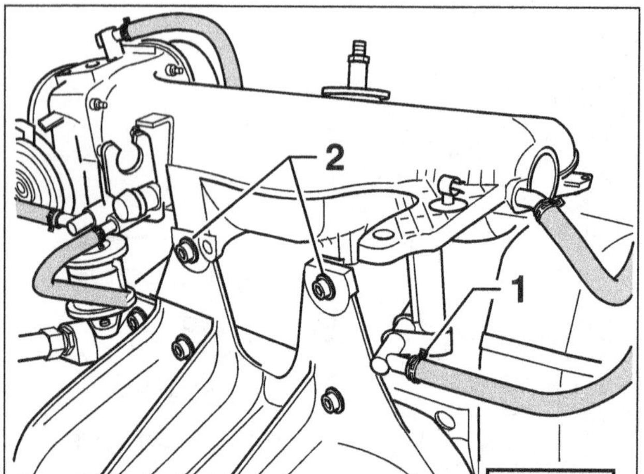

N15-0207

< Remove bolts (**2**) from warm air deflector plate on back side of upper intake manifold.

0024228

< Remove upper intake manifold bolts (**arrows**). Carefully separate upper manifold from lower manifold. Cover open intake runner openings in lower manifold with clean shop rags.

– Disconnect crankcase breather valve from cylinder head cover.

– Remove nuts from cylinder head cover.

– Unclip and remove upper camshaft drive belt cover. Remove small protection cover from back of camshaft drive sprocket.

– Remove both reinforcing strips from cylinder head cover.

– Lift off cylinder head cover and gasket.

– Installation is reverse of removal noting the following additional points:

- Be sure to remove shop rags from lower intake manifold runner openings.
- Use new gaskets where appropriate.
- Apply a small amount of sealer to the area where the cylinder head cover gasket and the number one camshaft bearing cap meet.

Tightening Torques

- Cylinder head cover to cylinder head (M6) . . . . 10 Nm (89 in-lb)
- Upper intake manifold to lower manifold (M8) . . . . 20 Nm (15 ft-lb)
- Warm air deflector to upper intake manifold (M8) . . . . 25 Nm (18 ft-lb)

Camshaft oil seal, replacing

– Remove upper engine cover and unclip and remove timing belt guard, upper section.

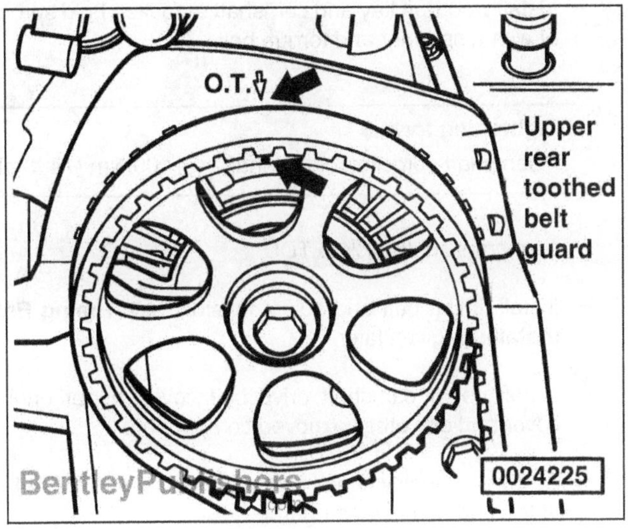

< Set engine to TDC for cylinder 1 by turning crankshaft until timing mark on camshaft sprocket aligns with mark on timing belt upper rear guard (**arrows**).

– Release tensioning roller and slide timing belt off of camshaft sprocket.

– Turn crankshaft back off of TDC slightly.

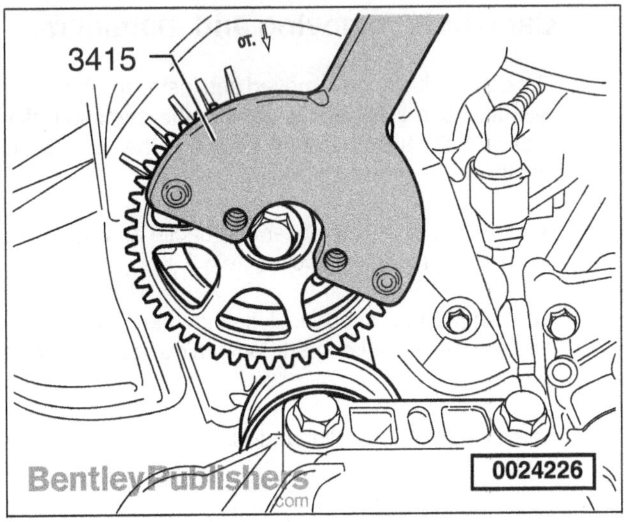

< Using a suitable spanner wrench, remove camshaft sprocket bolt.

– Remove camshaft drive sprocket and woodruff key.

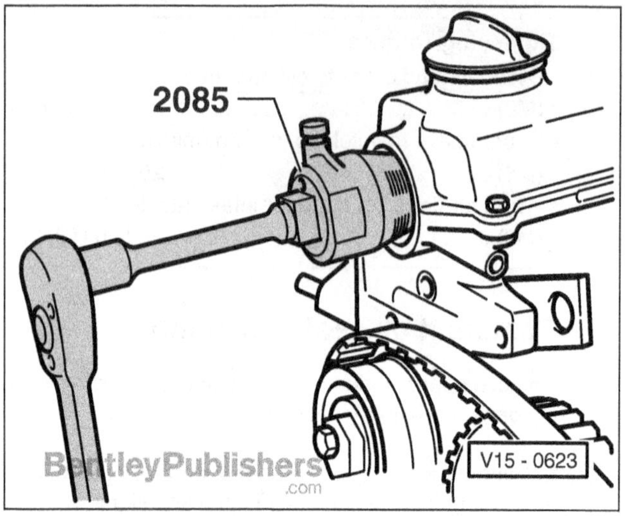

Remove seal using a suitable seal extractor.

NOTE —
As part of a running production change, the oil seal with ring spring has been replaced with a PTFE (Teflon) seal.

– Install new oil seal noting the following points:
  - Seal with ring spring: Lubricate new seal with clean engine oil.
  - PTFE (Teflon) seal: Do not lubricate.
  - Install with closed end facing out.
  - Use a suitable guide sleeve and installation tool.
  - Press seal into the bottom of recess.

– Install woodruff key and camshaft sprocket, hold sprocket with a spanner and torque bolt.

Tightening torque
- Camshaft sprocket to camshaft . . . 100 Nm (74 ft-lb)

– Turn crankshaft back to TDC.

– Install timing belt and adjust tension, see **Timing Belt, installing**, given later.

– Install upper camshaft drive belt cover, upper engine cover and any other removed components.

– Be sure to quality check your work, see **Quality Review** at the end of this repair group.

Camshaft, removing and installing

The camshaft can be removed and installed with the cylinder head installed. Removing the camshaft allows access to the hydraulic cam followers, the valve springs and the valve stem oil seals.

– Remove upper engine cover and unclip and remove upper camshaft drive belt cover.

– Turn engine by hand until timing mark on camshaft sprocket aligns with mark on timing belt upper rear guard.

– Release tensioning roller and slide timing belt off of camshaft sprocket.

– Turn crankshaft back off of TDC slightly.

NOTE —
Turning the crankshaft away from TDC slightly minimizes the chances of contact between pistons and valves upon re-assembly.

– Using a suitable spanner, remove the camshaft sprocket bolt.

– Remove camshaft drive sprocket and woodruff key.

– Remove cylinder head (valve) cover and gasket as described previously.

– Lift off plastic oil deflector to expose camshaft.

– Loosen and remove the bearing caps from positions 5, 1 and 3 in that order. Position 1 is closest to the belt end of the camshaft.

> ***NOTE —***
> *Mark each bearing cap before removing. Each cap is machined to each journal and must be reinstalled to that exact location to prevent damage.*

– Loosen nuts on bearing cap 2 slightly and then loosen nuts on bearing cap 4 slightly. Loosen nuts alternately and evenly a little at a time on each of the two bearing caps. Remove nuts (and washers, if equipped), and remove bearing caps.

– Lift out camshaft and lay aside.

> ***NOTE —***
> - *Once removed, the camshaft should be checked for wear and other visible damage as described later under* **Cylinder Head Components**.
> - *Bearing caps are bored off-center. If necessary, test install caps without camshaft to confirm installation direction.*

– Oil bearing surfaces of cylinder head and camshaft and install camshaft into cylinder head with lobes for cylinder number 1 pointing up.

– Install bearing caps 2 and 4. Install nuts (and washers, if equipped) and tighten them alternatively and evenly so that the camshaft is drawn down fully and evenly into bearing saddles. Torque to specification.

Tightening torque
- Camshaft bearing cap to cylinder head. . . . . 20 Nm (15 ft-lb)

> ***CAUTION —***
> *Be sure to install bearing caps correctly. The caps are bored off-center and only fit properly one way. Caps are matched to the cylinder head by the machining process and are not available separately. Broken caps will require the replacement of the cylinder head.*

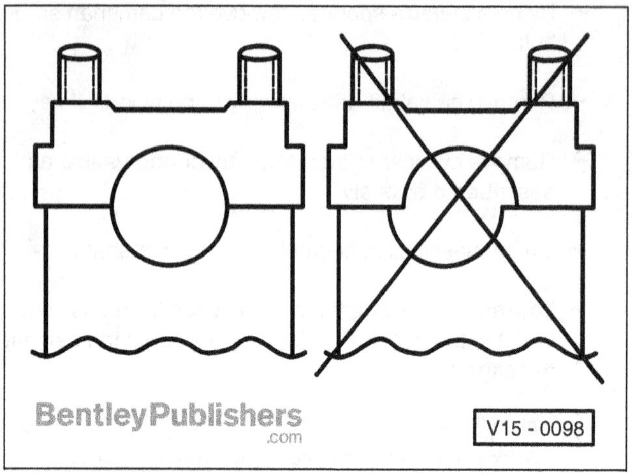

< Install bearing cap number 1 with a small amount of sealer on cylinder head mating surface. Install all remaining bearing caps and torque to specification.

– Install the plastic oil deflector.

– Install the camshaft drive sprocket and woodruff key using a suitable spanner wrench as shown earlier.

NOTE —
Ensure that camshaft lobes for cylinder #1 are still pointing up.

Tightening torque
- Camshaft sprocket to camshaft . . . 100 Nm (74 ft-lb)

– Install cylinder head (valve) cover and rear upper timing belt guard.

– Install toothed camshaft drive belt and adjust tension, see **Timing Belt, installing** given later.

– Install upper timing belt cover, upper engine cover and any other removed components.

– Be sure to quality check your work, see **Quality Review** at the end of this repair group.

Hydraulic cam followers, checking

The 2.0 liter engine installed in the New Beetle is equipped with hydraulic cam followers which are also known as lifters.The cam followers are pumped up by engine oil pressure, expanding as necessary to fill the gap between the valve stem and the camshaft lobe. This occurs continuously and automatically to keep the valve in proper adjustment at all times.

Some valve noise at start-up and during the warm-up period is normal at times due to hydraulic cam followers that have bled down while the engine was not running. Before checking noisy cam followers, check that the engine oil is new and fresh, of the correct weight, and that the level is correct. Allow 2 minutes with a warm engine running at a fast idle for the lubrication system to properly pump up the cam followers.

Cam followers should only be checked when the engine is warm. Run the engine until the radiator fan has switched on at least one time. Increase the engine speed to 2,500 rpm and hold it there for approximately 2 minutes. If the hydraulic cam followers are still noisy, shut the engine off and proceed as follows while the engine is still warm.

– Remove cylinder head (valve) cover. See **Cylinder head (valve) cover, removing and installing**, given earlier.

– Turn engine by hand until both camshaft lobes of cylinder #1 are pointing approximately up.

NOTE —
Hydraulic cam follower clearance can be checked on any cam follower (lifter) that is not being depressed by the camshaft lobe.

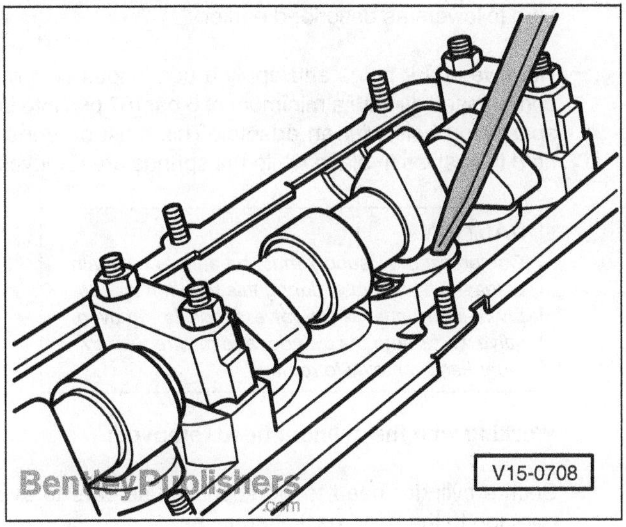

< Check clearance between cam follower and cam lobe on cylinder #1 by lightly depressing cam follower with a wooden or plastic wedge and inserting a feeler gauge into the gap. If clearance exceeds specified limit, the lifter is faulty and should be replaced.

Specification

- Hydraulic cam follower maximum clearance (2.0L engine) . . . . 0.2 mm (0.008 in.)

– Turn engine by hand and repeat procedure until all cam followers have been checked.

– Replace a faulty cam follower by removing camshaft as previously described and pulling the follower from the cylinder head. Faulty hydraulic cam followers can be replaced individually but only as a complete assembly.

CAUTION —
After installing new cam followers, the engine should not be started for at least 30 minutes. New cam followers are usually at full extended height and must be allowed to bleed down to their proper height once installed. Failure to do so may cause the valves to strike the pistons resulting in serious damage.

NOTE —
Store removed hydraulic cam followers in order on a clean surface with the camshaft contact surface facing down to minimize bleed down. Cover with a clean lint-free shop cloth.

Valve stem oil seals, replacing

The sign of faulty valve stem seals is excessive oil consumption and oil smoke from the exhaust. This is usually most noticeable during periods of high manifold vacuum such as deceleration. If compression and leak-down testing confirm the integrity of the piston rings, but the engine consumes oil, it is likely that faulty valve stem seals are present. It should also be noted that worn valve stem seals could be due to worn valve guides and that the worn valve guides are the major cause of the oil consumption. See **Cylinder Head Components** for more information on checking valve guides.

Replacing the valve stem oil seals requires removal of the camshaft, cam followers, and the valve springs. This can be done with the cylinder head installed or removed. In either case, several Volkswagen special tools are required to compress the valve springs and remove and install the seals.

Working with the cylinder head installed:

- Remove cylinder head cover, camshaft, and hydraulic cam followers as described earlier.

- Remove spark plugs and apply a continuous supply of compressed air with a minimum of 6 bar (87 psi) into first spark plug hole with an adapter. This must be done to hold the valves in place while the springs are removed.

> ***CAUTION —***
> *Compressed air supply must be able to maintain at least 6 bar (87 psi) during this repair. If air supply is interrupted while valve spring is removed, valve will fall into the cylinder and may require cylinder head removal to retrieve.*

Working with the cylinder head removed:

- Secure cylinder head to workbench. Use care to avoid damage to the head gasket surfaces.

- Remove camshaft and hydraulic cam followers as described earlier.

Working with cylinder head removed or installed:

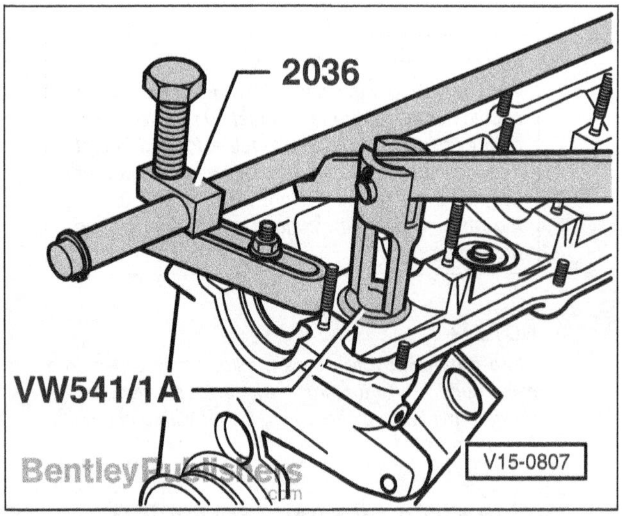

◀ Install appropriate valve spring compressor tools. Compress the spring for the first cylinder and remove the spring retainer, both keepers, and the valve spring. If spring will not compress, lightly tap tool to release stuck keepers. Use a small magnet to retrieve the keepers.

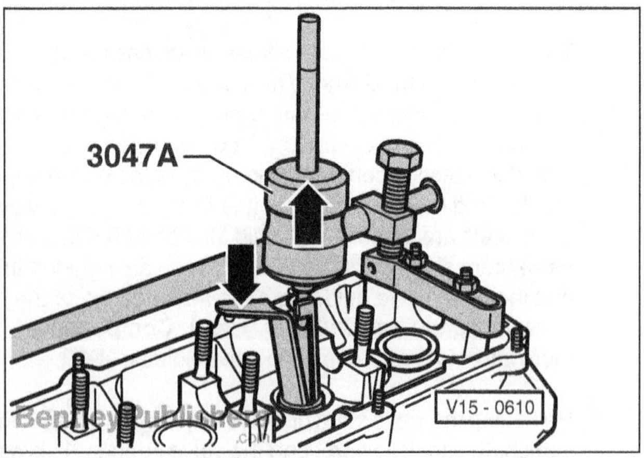

◀ Remove valve stem seals with special slide hammer tool.

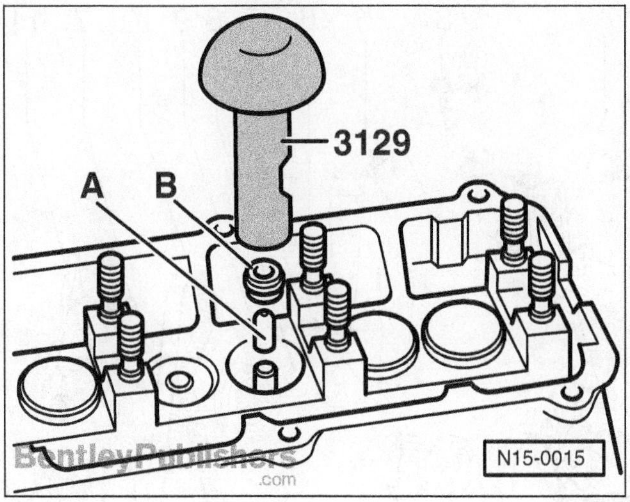

◄ Begin installation of new seal by temporarily fitting a protective plastic fitting sleeve (**A**) over valve stem. This sleeve is usually included with the new seal (**B**) set and will prevent damage to seal due to the sharp edges of the keeper grooves. Lubricate new seal with clean engine oil and fit it to the installation tool.

– Push installation tool (with seal) down over valve stem until seal is fully seated on guide. Remove tool and protective fitting sleeve.

– Reinstall valve spring, retainer and keepers.

– Repeat for second valve on cylinder.

– When both valve seals have been replaced on first cylinder, transfer compressed air adapter (if working with cylinder head installed) to next cylinder and repeat process until all valve seals have been replaced.

– Remaining installation is reverse of removal.

> ***CAUTION —***
> *To prevent cylinder head or camshaft damage, be sure to follow the camshaft installation procedure when tightening the camshaft bearing caps. See* **Camshaft, removing and installing**.

Tightening torque
- Spark plug to cylinder head . . . . 30 Nm (22 ft-lb)

– Be sure to quality check your work, see **Quality Review** at the end of this repair group.

TIMING BELT

The maintenance tables in **03 Maintenance** describe inspection and replacement intervals for the timing belt, also known as the camshaft drive belt or toothed belt. Regular inspection and replacement will help prevent damage to the engine due to belt stretch and the long term effects of heat.

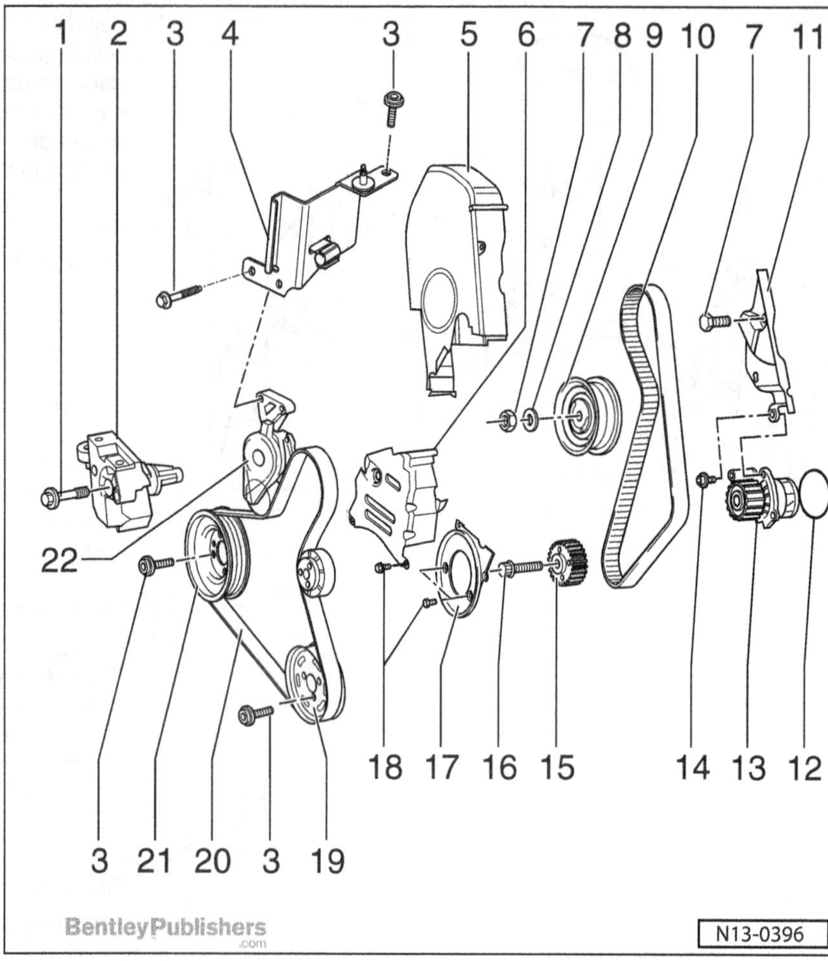

Timing belt, assembly

1. **Bolt**
   - Tighten to 45 Nm (33 ft-lb)
2. **Engine support**
3. **Bolt**
   - Tighten to 25 Nm (18 ft-lb)
4. **Bracket**
5. **Timing belt guard, upper section**
6. **Timing belt guard, center section**
7. **Nut**
   - Tighten to 20 Nm (15 ft-lb)
8. **Washer**
9. **Tensioner**
   - Semi-automatic timing belt tensioning roller
10. **Timing belt**
    - Mark engine direction of rotation before removing
    - Check for wear
    - Do not kink
11. **Timing belt guard, rear**
12. **O-ring**
    - Always replace
13. **Coolant pump**
14. **Bolt**
    - Tighten to 15 Nm (11 ft-lb)
15. **Timing belt sprocket (crankshaft)**
16. **Bolt**
    - Tighten to 90 Nm (66 ft-lb) + ¼ turn (90°)
    - Always replace
    - Threads and shoulder must be free of grease
17. **Timing belt guard, lower section**
18. **Bolt**
    - Tighten to 10 Nm (7 ft-lb)
19. **Pulley**
    - For power steering pump
20. **Ribbed belt**
    - Mark direction of rotation before removing
21. **Belt pulley/vibration damper**
    - Can only be installed in one position, holes are offset
    - Note position when installing timing belt
22. **Ribbed belt, tensioning device**
    - Loosen ribbed belt by turning with open-ended wrench

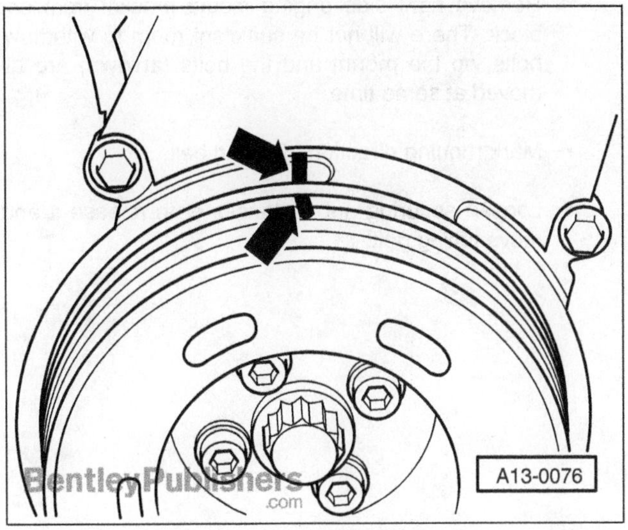

Timing belt, removing

- Remove lower sound absorber panel (belly pan).
- Remove upper engine cover.
- Remove ribbed V-belt and tensioner.
- ◀ Turn engine to top dead center, TDC, for cylinder #1. Mark on pulley (**lower arrow**) aligns with mark on timing belt guard, center section (**upper arrow**).
- Remove upper timing belt cover.

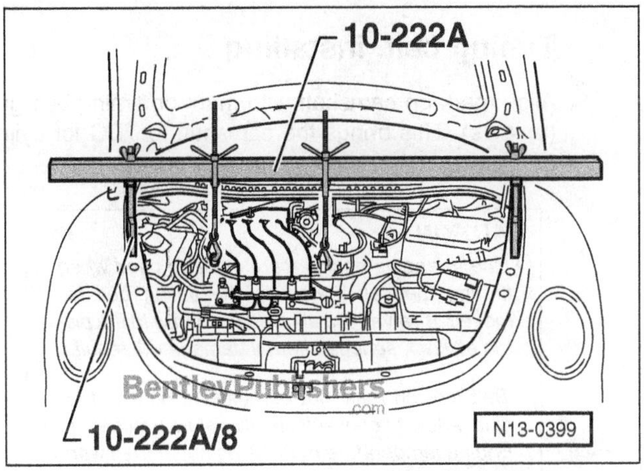

- ◀ Support the engine with a suitable fixture that is designed to support the weight of the engine and transmission without damaging the body.
- Lift supporting device slightly, so that weight of engine and transmission is on supporting device.

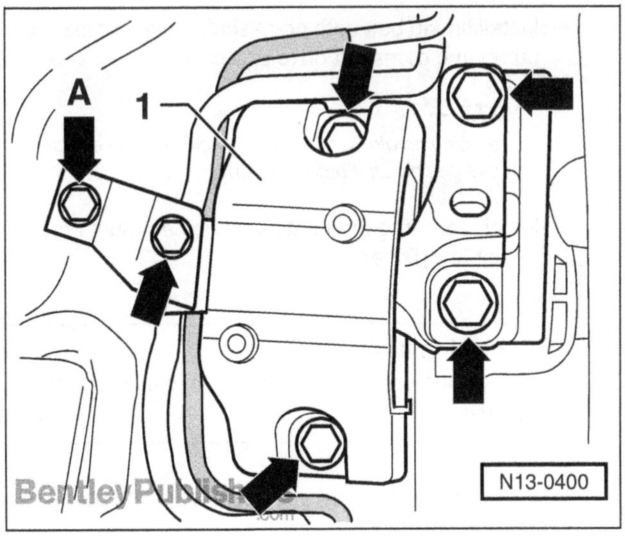

- ◀ Remove 5 bolts (**arrows**) for right side engine mount from body and carrier on engine, and loosen body bolt (**arrow A**).
- Remove 4 bolts holding belt pulley to timing belt sprocket and remove belt pulley.
- Remove center and lower sections of timing belt guard.

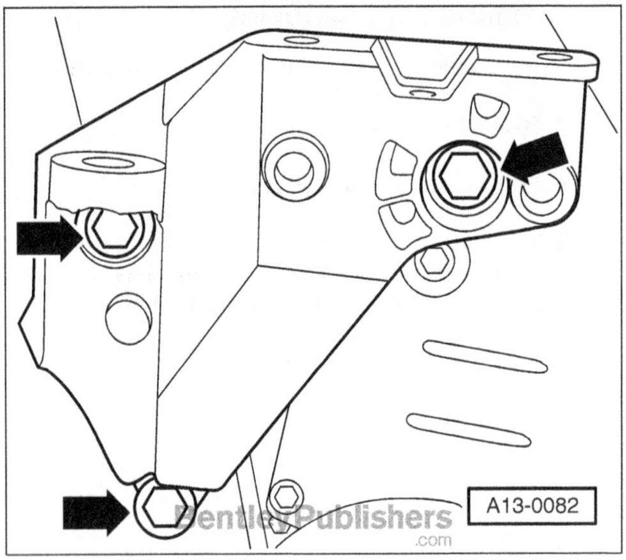

◀ Remove right side engine mount bracket from engine block. There will not be sufficient room to withdraw the bolts, so the mount and the bolts (**arrows**) are all removed at same time.

- Mark running direction of timing belt.

- Loosen securing nut on tensioner to release it and remove timing belt.

Timing belt, installing

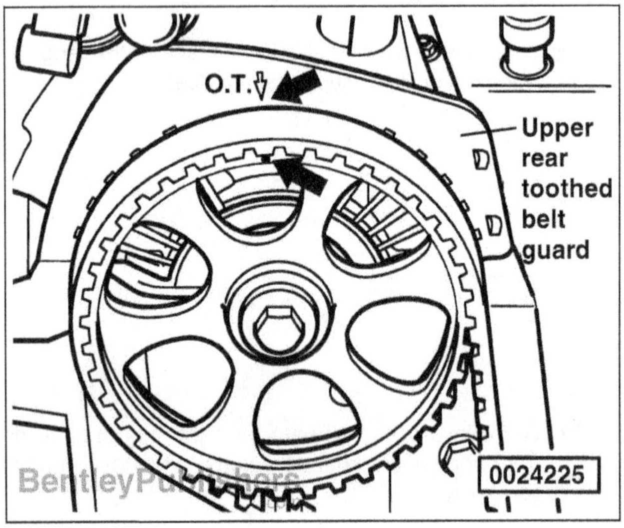

◀ Align mark on camshaft with mark on timing belt guard (**arrows**). This brings the camshaft to TDC for cylinder #1.

> ***CAUTION —***
> - *Due to engine design, care must be used when turning the camshaft with the timing belt removed. If valves are allowed to open with the piston at TDC, serious internal damage will result.*
> - *Belt tension must not be adjusted on a hot engine. Allow to cool sufficiently before proceeding. Engine temperature must be no hotter than warm to the touch.*

- Install timing belt onto crankshaft drive sprocket, coolant pump and camshaft drive sprocket.

NOTE —
If re-using an old belt, be sure to note running direction marks placed on it before removal.

- Install bolts into the right side engine mount bracket and mount to cylinder block.

Tightening torques
- Right side engine mount bracket to cylinder block (M10) . . . . 45 Nm (33 ft-lb)

- Install center and lower sections of timing belt guard.

- Install vibration dampener/ribbed belt pulley to timing belt drive sprocket.

Tightening torques
- Vibration dampener/ribbed belt pulley to timing belt drive sprocket (M8) . . . . . 25 Nm (18 ft-lb)

– Ensure that crankshaft is at TDC for cylinder 1 by aligning mark on belt pulley with mark on timing belt guard, center section.

– Ensure that camshaft is still at TDC for cylinder #1.

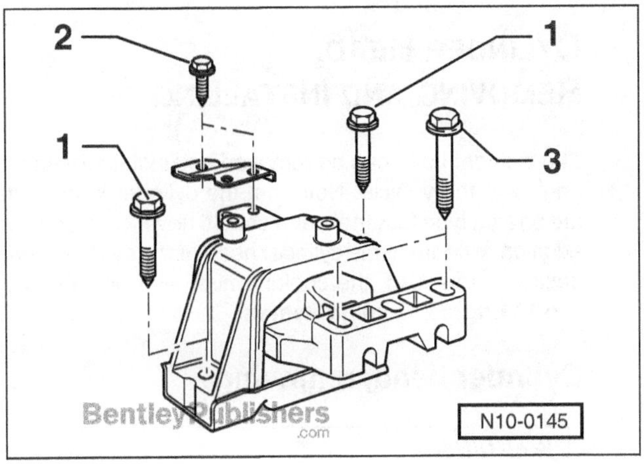

◀ Install right side engine mount using new stretch bolts.

1. Mount to body bolt
   - 40 Nm (30 ft-lb) plus 90° (¼ turn)
   - Always replace
2. Mount bracket to body bolt
   - 25 Nm (18 ft-lb)
   - Always replace
3. Mount to engine bracket bolt
   - 100 Nm (74 ft-lb)
   - Always replace

– Remove engine support fixture.

– Install timing belt onto tensioner and check for correct placement of belt on all sprockets. Securing nut should be just tight enough to allow movement of center section.

– Ensure that camshaft and crankshaft are still at TDC.

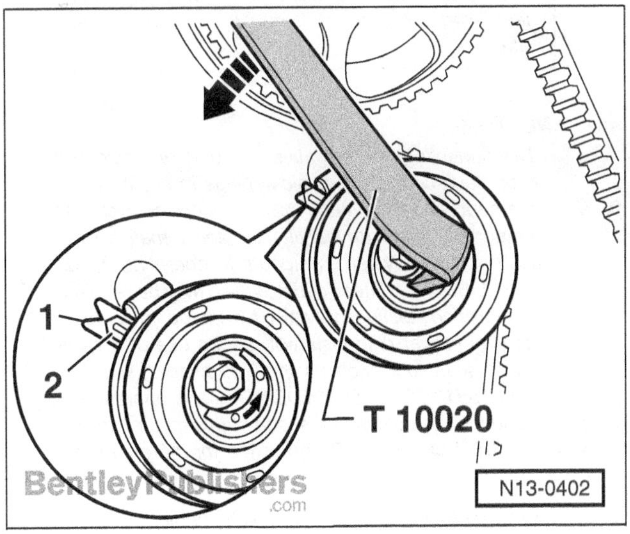

◀ Tension timing belt by turning center section eccentric of tensioner with a 2 pin spanner wrench until notch (**1**) and indictor pointer (**2**) line up. When marks line up, tighten securing nut.

Tightening torques
- Timing belt tensioner (M8) . . . . 20 Nm (15 ft-lb)

– Ensure that marks on camshaft sprocket and dampener/pulley still align with their respective TDC marks. If they do not align, repeat until they do.

NOTE —
Some movement of the sprockets and their marks is to be expected as belt tension is adjusted. Keep in mind that the smallest possible increment of adjustment is one whole tooth of the belt or sprocket.

– With all marks in proper alignment, rotate the crankshaft two revolutions in the running direction and re-check belt tension marks on the tensioner.

– Install timing belt guard, upper section.

– Install ribbed belt tensioner and belt.

Tightening torque
- Ribbed belt tensioner to bracket (M8) . . . . 25 Nm (18 ft-lb)

– Install lower sound absorber panel (belly pan) and upper engine cover.

– Install any remaining parts and quality check your work, see **Quality Review** at end of this repair group.

CYLINDER HEAD, REMOVING AND INSTALLING

The cylinder head can be removed and installed with the engine in the vehicle. Note that the cylinder head bolts are stretch type fasteners and should never be reused. In addition, whenever the cylinder head or the cylinder head gasket is replaced, the coolant must also be replaced, see **19 Engine–Cooling System**.

Cylinder head, removing

WARNING —
Do not start work on a hot engine. Allow to cool sufficiently before proceeding. Engine temperature must be no hotter than warm to the touch. Cylinder head warpage can result due to uneven cooling rates.

NOTE —
- *Disconnecting the negative (–) battery cable will erase fault codes and basic settings in engine management and automatic transmission control unit memories. Some driveability problems may be noticed until the system re-adapts to operating conditions. OBD II readiness codes, which may be required for emissions testing, may also be erased. In some instances proper diagnosis will require use of a diagnostic scan tool such as the Volkswagen supplied VAG 1551/1552 or 5051/5052. Use and operation of these tools is outside the scope of this repair manual. See* **OBD On Board Diagnostics** *for more information.*
- *Cylinder head will be removed with exhaust manifold and lower section of intake manifold still attached.*

– With ignition switched off, disconnect battery negative (–) terminal.

NOTE —
Be sure to have the anti-theft radio code on hand before disconnecting the battery.

– Position crankshaft so that pistons are NOT at TDC.

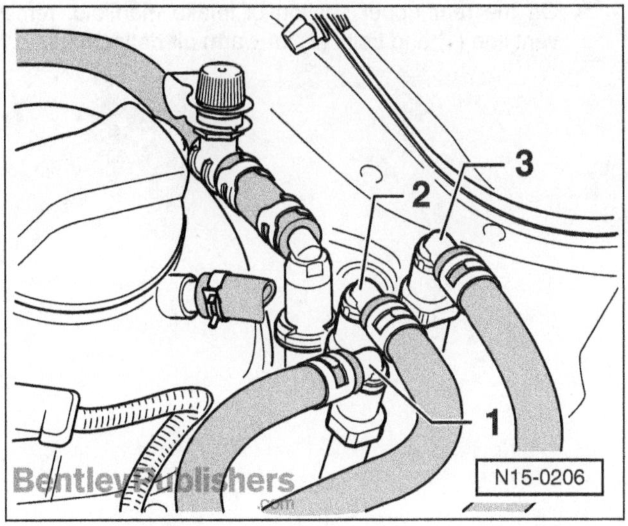

– Remove upper engine cover.

– Remove lower sound absorber panel (belly pan).

– Drain engine coolant, see **19 Engine–Cooling System**.

◄ Disconnect fuel supply line (**3**), fuel return line (**2**) and vacuum connection (**1**) for leak detection pump (LDP).

> ***WARNING —***
> *Fuel will be expelled when disconnecting fuel hoses. Wrap a cloth around the fuel line fittings before disconnecting them. Do not smoke or work near heaters or other fire hazards. Have a fire extinguisher handy.*

– Seal off disconnected fuel and vacuum lines to prevent contamination.

– Remove air cleaner assembly and connecting hose to throttle housing.

– AEG engine: Disconnect accelerator cable from throttle control module linkage and cable retainers.

– Unplug wiring harness connector from throttle control module.

– Remove the 2 coolant hoses from throttle control module that connect to the reservoir and cylinder head.

– Disconnect vacuum hose connection for power brake booster on back of intake manifold and any remaining connections on the intake manifold.

◄ Loosen and remove bolts (**arrows**) holding upper and lower sections of intake manifold together.

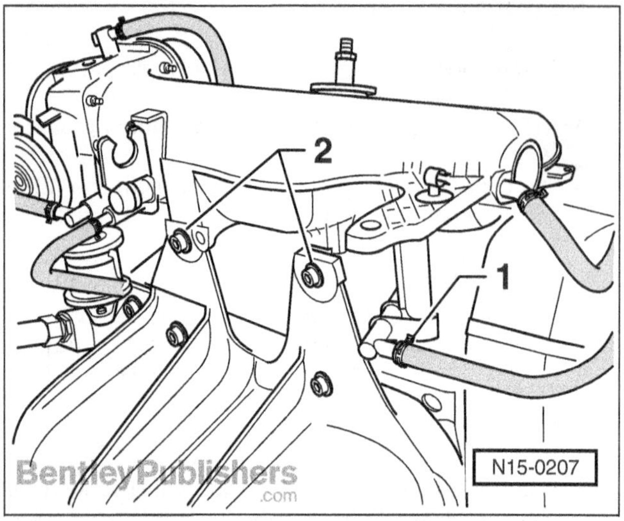

On the rear upper section of intake manifold, remove vent line (**1**) and bolts (**2**) for warm air deflector plate.

Lift off upper section of intake manifold and lay aside. Block off exposed open ports (**arrow**) on lower section with wide tape or other suitable material.

Remove mounting bolts (**arrows**) for coolant connection flange from end of cylinder head and carefully move aside. The coolant hoses stay connected.

< Disconnect intake (**A**) and pressure hose (**B**) from secondary air injection pump. Unbolt nuts (**arrows**) holding pump to bracket and disconnect harness connector (**C**). Remove pump.

– Disconnect the following electrical components/connections:
  - Fuel injectors (and unclip the wiring from the holders)
  - Camshaft position sensor
  - Spark plug wires

– Remove ribbed V-belt from the tensioner and remove tensioner.

– Remove upper timing belt cover.

– Release timing belt tensioning roller and take timing belt off camshaft sprocket.

– Remove upper bolt from rear timing belt guard where it is attached to cylinder head.

– Unbolt front exhaust pipe from exhaust manifold, see **26 Exhaust System and Emission Controls**.

– Remove cylinder head (valve) cover as described earlier.

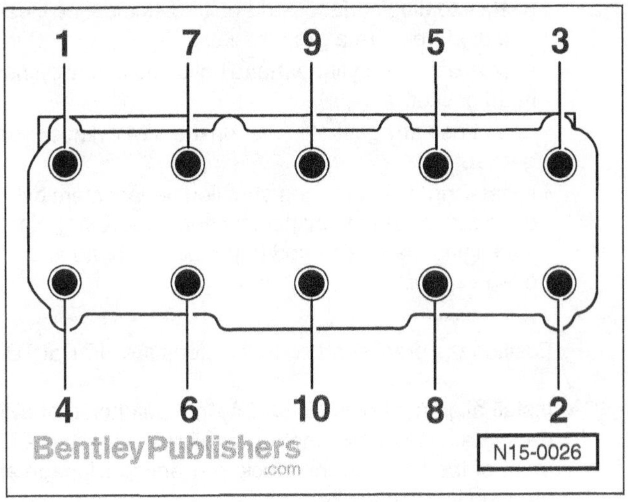

< Loosen socket head bolts slightly in sequence. Do not remove bolts until all 10 have been loosened. Discard head bolts.

– Carefully lift off cylinder head and place in a clean area. If head is stuck, use a soft-faced mallet or pry gently with a wooden stick.

> ***CAUTION —***
> *Some of the valves will be open. Use extra care when removing and handling to avoid damage. Place the cylinder head on the bench or table so that the weight of the cylinder head will not rest on the valves.*

Cylinder head, installing

Before proceeding with the installation of the cylinder head, whether an original or a new unit, observe the following important points:

- Check the cylinder head and block for distortion and warpage, see **Cylinder Head Components**, given later in this section.
- Carefully clean the cylinder block and cylinder head sealing surfaces being sure to avoid scratching them during the cleaning process. Do not use metal scrapers or wire brushes.

- When using abrasive paper do not use any grades coarser than 100 grit (such as 80 grit). Lower numbers are coarser.
- If cylinder head will not be reinstalled immediately, take precautions to prevent rust from forming on the cylinder block walls and gasket sealing surfaces.
- When cleaning the old gasket material off of the cylinder block, take precautions to keep the old gasket material and the abrasive particles and dirt out of the cooling and oiling system passages. Place a clean shop cloth over the cylinders to prevent contamination from getting between the cylinder wall and the piston.
- Do not take the new cylinder head gasket out of the packaging until ready to use. Handle the new gasket with extreme care as any damage will lead to leaks.
- There MUST NOT BE any oil or coolant in the head bolt holes in the cylinder block. Any fluids in the holes creates the danger of hydrolock while torquing which could lead to structural damage of the cylinder block. Use thread chasers to remove foreign material as required. Be sure that all 10 bolt holes are clean and dry.
- When all traces of the old gasket material have been removed from the cylinder block and head, carefully remove all traces of metal particles, abrasives and lint. All gasket sealing surfaces and bolt holes must be clean and dry to ensure a proper seal.
- Always use new cylinder head bolts and a new cylinder head gasket.
- Do not use any gasket sealer on the new cylinder head gasket.
- If installing a Volkswagen supplied replacement cylinder head, be sure to inspect for and remove any plastic packaging materials used to protect the head and the open valves.

– Position the crankshaft so that pistons are NOT at TDC.

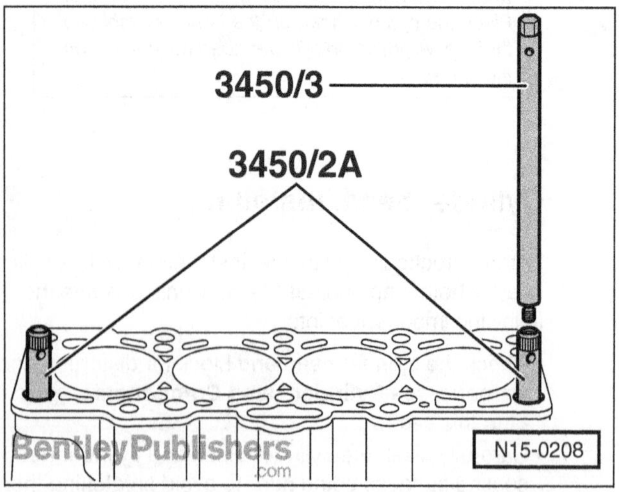

◄ Install alignment pins (**3450/2A**) into bolt holes of cylinder block. Use of alignment pins allows for proper alignment of the head on the block. preventing slippage and damage to head gasket.

– Install new cylinder head gasket onto block with numbers and letters facing up. Be sure that no wiring or vacuum hoses are caught between cylinder head and block.

– Carefully install cylinder head onto block over alignment pins and screw in 8 new head bolts. Tighten bolts hand-tight only.

– Remove alignment pins from the bolt holes with the supplied tool (**3450/3**) and install the remaining 2 new head bolts. Tighten these bolts hand-tight.

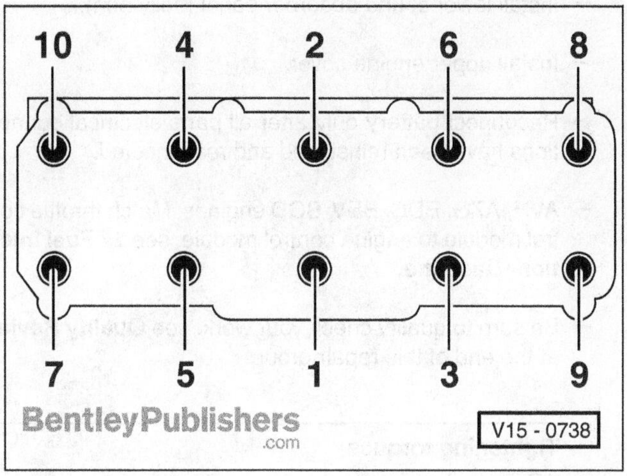

◄ Tighten the 10 cylinder head bolts in three stages following the tightening order.

Tightening torque

- Cylinder head to cylinder block (stretch bolt – always replace)
  - stage I . . . . 40 Nm (30 ft-lb)
  - stage II . . . . additional ¼ turn (90°)
  - stage III . . . . additional ¼ turn (90°)

– Lubricate camshaft and followers and install cylinder head (valve) cover.

– Install timing belt, adjust tension and belt timing.

– Install upper bolt for rear timing belt guard where it is attached to cylinder head.

– Install upper timing belt cover.

– Install ribbed V-belt tensioner and belt.

– Remove whatever material was used to seal exposed lower section of intake manifold and install upper section of intake manifold. Re-attach vacuum fitting and warm air deflector plate.

– Reconnect all coolant flanges and lines.

– Reconnect all vacuum hoses connections.

– Reconnect fuel supply and return hoses and vacuum hose to leak detection pump.

– AEG engine: Reconnect accelerator cable and check for smooth operation and full throttle.

– Reconnect all disconnected electrical components and connections including spark plug wires.

– Refill cooling system with fresh coolant in the appropriate ratio. See **19 Engine–Cooling System.**

CAUTION —

- *Use only Volkswagen original anti-freeze when filling the cooling system. Use of any other anti-freeze may be harmful to the cooling system. Do not use an anti-freeze containing phosphates.*
- *Do not use tap water in cooling system. Use distilled water only to mix anti-freeze.*

– Install air cleaner assembly and connecting hoses.

– Attach front exhaust pipe to exhaust manifold with new nuts. See **26 Exhaust System and Emission Controls**.

- Install lower sound absorber panel (belly pan).
- Install upper engine cover.
- Reconnect battery only after all parts/electrical connections have been reinstalled and reconnected.
- AVH, AZG, BDC, BEV, BGD engines: Match throttle control module to engine control module, see **24 Fuel Injection–Gasoline**.
- Be sure to quality check your work, see **Quality Review** at the end of this repair group.

Tightening torques

- Cylinder head cover to cylinder head (M6) . . . . 10 Nm (89 in-lb)
- Upper intake manifold bolts to lower manifold (M8) . . . . 20 Nm (15 ft-lb)
- Warm air deflector to upper intake manifold (M8) . . . . 25 Nm (18 ft-lb)
- Cylinder head to cylinder block (stretch bolt - always replace)
  - stage I . . . . 40 Nm (30 ft-lb)
  - stage II . . . . additional ¼ turn (90°)
  - stage III . . . . additional ¼ turn (90°)
- Front exhaust pipe to exhaust manifold (special nuts - always replace)(M10) . . . . 40 Nm (30 ft-lb)
- Coolant outlet flange to cylinder head (M6) . . . . 10 Nm (89 in-lb)
- Rear timing belt guard to cylinder head (M8) . . . . 20 Nm (15 ft-lb)
- Ribbed belt tensioner to bracket (M8) . . . . 25 Nm (18 ft-lb)
- Timing belt tensioner (M8) . . . . 20 Nm (15 ft-lb)
- Secondary air pump to support bracket nut (M6) . . . . 10 Nm (89 in-lb)

CYLINDER HEAD COMPONENTS

This section provides the specifications and special information necessary to repair the cylinder head that has been removed from the Volkswagen 2.0L engine. Special service tools and machine shop services are required for most cylinder head repair.

NOTE —
The information given under this heading assumes that the cylinder head is removed. For cylinder head removal procedures, see **Cylinder head, removing** *given earlier.*

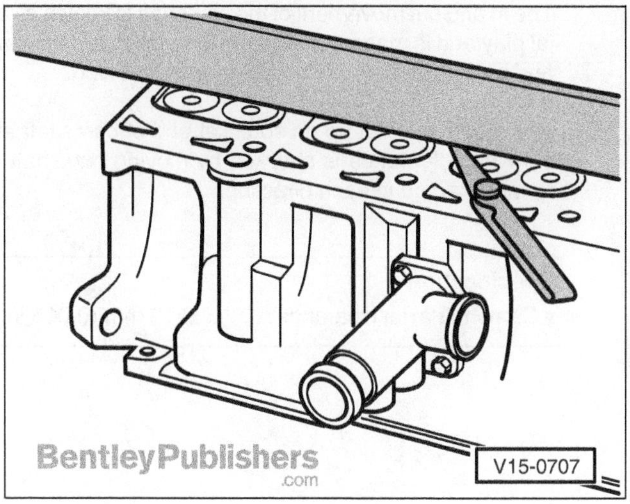

Cylinder Head and Camshaft

Check the cylinder head for warpage and distortion with an accurate straight edge and a feeler gauge. The cylinder head can be resurfaced provided that the distance from the valve cover gasket surface to the head gasket surface is never less than specified. Resurfacing the cylinder head will require that the valves be set deeper into the seat by the same amount as was removed by the surfacing operation. This can be done by reworking the valve seats. Machining too much material off of this surface will change the compression ratio and will affect engine emissions.

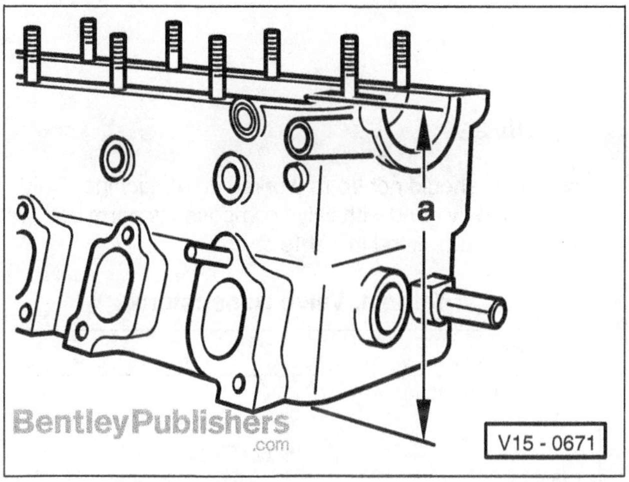

Specification

- Maximum cylinder head distortion/warpage . . . . 0.1 mm (0.004 in.)
- Minimum cylinder dimension (**A**) (valve cover gasket surface to the head gasket surface) . . . . . . 132.6 mm (5.22 in.)

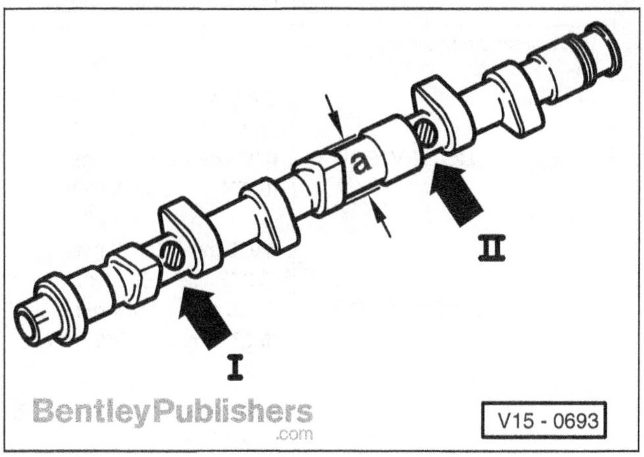

The camshaft for the 2.0L engine is easily identified by markings cast into it during manufacture.

Specification

- Valve timing at 1 mm valve lift

Intake opens ATDC
AEG . . . . 4.6°
AVH, AZG . . . . 6.2°

Intake closes ABDC
AEG . . . . 40.85°
AVH, AZG . . . . 42.45°

Exhaust opens BBDC
AEG . . . . 37.4°
AVH, AZG . . . . 35.8°

Exhaust closes BTDC
AEG . . . . 1.15°
AVH, AZG . . . . 0.45°

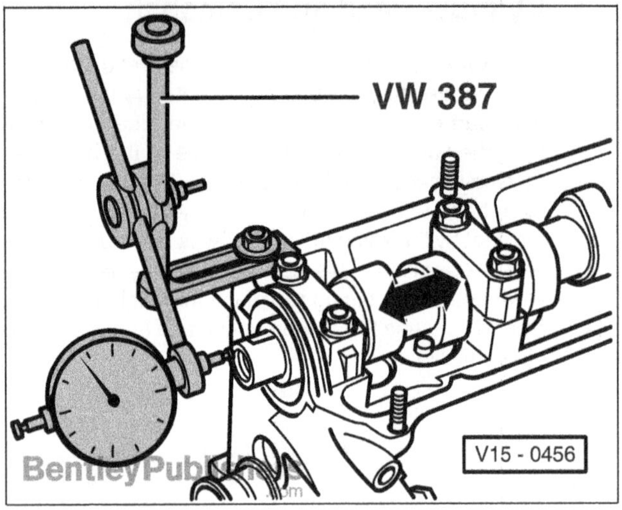

The in and out movement of the camshaft is known as axial play and is measured with the cam followers removed and only the first and last bearing caps installed.

A dial gauge is set up on sprocket end of camshaft and camshaft clearance is checked by moving camshaft as far as it can go in each direction.

Specification

- Camshaft axial clearance...... 0.15 mm (0.006 in.)

Valves

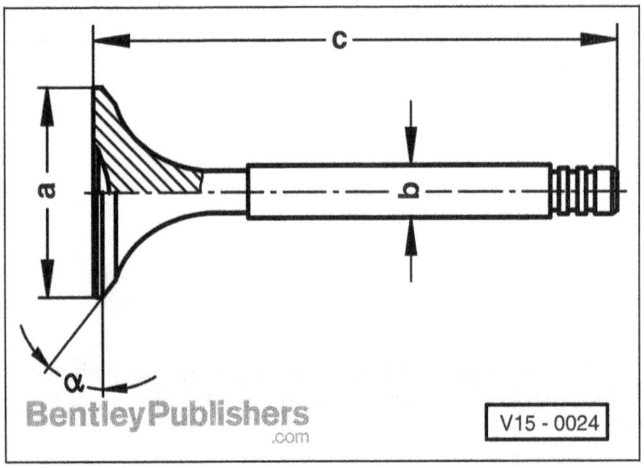

Valves should not be re-worked on a machine. Only lapping by hand with valve compound is permitted. Valve dimensions listed in **Table a**.

Table a. Valve Dimensions

Engine	**2.0L**
Valve head diameter (**a**)	
intake	39.5 mm ± 0.15 mm (1.555 in. ± 0.006 in)
exhaust	32.9 mm ± 0.15 mm (1.295 in. ± 0.0008 in.)
Valve stem diameter (**b**)	
intake	
AEG	6.920 mm ± 0.020 mm (0.272 in. ± 0.0008 in.)
AVH, AZG, BDC, BEV, BGD	6.980 mm ± 0.0072 mm (0.2748 in. ± 0.00002 in.)
exhaust	
AEG	6.920 mm ± 0.020 mm (0.272 in. ± 0.0008 in.)
AVH, AZG, BDC, BEV, BGD	6.965 mm ± 0.007 mm (0.2742 in. ± 0.00002 in.)
Valve length (**c**)	
intake	91.85 mm (3.616 in.)
exhaust	91.15 mm (3.588 in.)
Valve face angle (α)	
intake	45°
exhaust	45°

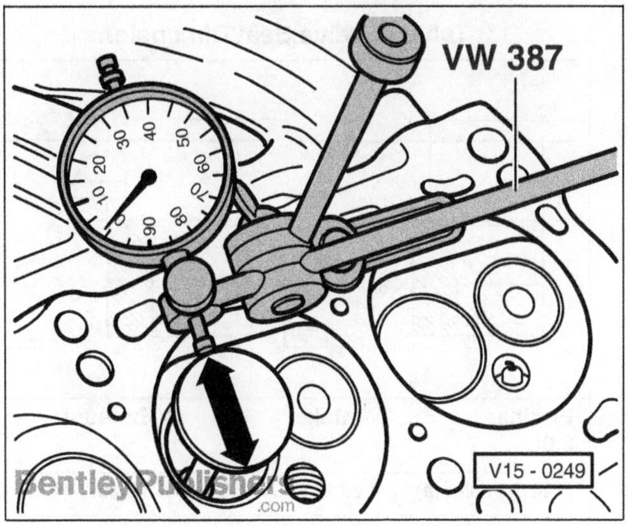

Valve guides

Special tools and a press are required to replace the valve guides. Check valve guide wear with a new valve. Always use an intake valve in an intake guide and an exhaust valve in an exhaust guide. Inspect the valve seats to ensure that the cylinder head can be reconditioned before installing new valve guides. Press out worn original valve guides with Volkswagen special driver tool 3121 from the camshaft side of the head. Replacement valve guides have a shoulder on the camshaft side to limit the installed depth and must be pressed out from the combustion chamber side.

When installing new valve guides, lubricate with clean engine oil and press in from the camshaft side down to the shoulder. The cylinder head should be cold for this operation. Once the shoulder of the valve guide contacts the cylinder head, do not allow the pressure on the guide to exceed 2000 psi (1 ton) otherwise the cylinder head or the guide will be damaged. When the guide has been replaced, ream it to the proper size and continue with the reworking of the valve seats.

Specification

- Valve guide (wear limits-maximum play) with new valve
  intake valve guide . . . . 1.0 mm (0.039 in.)
  exhaust valve guide . . . . 1.3 mm (0.051 in.)

NOTE —

Due to the close tolerances found in the valve guides, it is recommended that Volkswagen special tool 3121 be used for removal and installation and special tool 3120 be used to ream the valve guide to size.

Valve seats

When resurfacing valve seats, there is a limit to the amount of material that can be removed to bring the seat back into specification. If too much material is removed, the final assembly will leave too little space for the hydraulic cam follower to function properly. The maximum refacing dimension, that is, the maximum amount of material that can be removed from the valve seat, is calculated from the measurement shown.

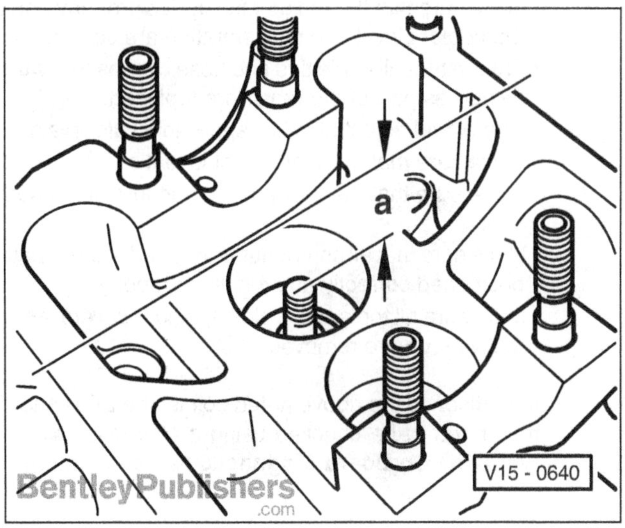

Measure dimension (**a**), and subtract the minimum dimension, as given in **Table b**. The difference is the maximum amount of material that can be removed from the valve seat.

Table b. Minimum Dimensions for Calculating Valve Seat Refacing Dimensions

Engine	Intake	Exhaust
2.0L	33.8 mm (1.331 in.)	34.1 mm (1.343 in.)

Table c. Valve Seat Dimensions

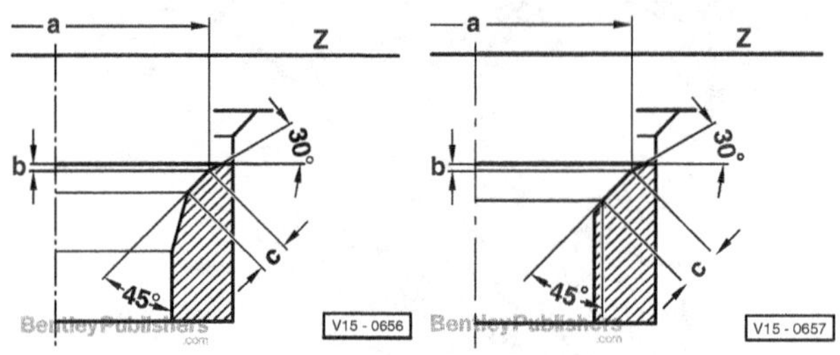

Engine: 2.0L	Intake	Exhaust
Seat diameter (**a**)	39.2 mm (1.543 in.)	32.4 mm (1.276 in.)
Maximum refacing dimension (**b**)	Calculated	Calculated
Seat width (**c**)	approx. 2.0 mm (0.079 in.)	approx. 2.4 mm (0.094 in.)
Valve seat angle	45°	45°
Correction angle, upper	30°	30°

NOTE —
Z shows cylinder head surface reference.

QUALITY REVIEW

When you have finished working under the hood and around other areas of the vehicle it is advisable to take a moment to quality check or review your work. This helps to ensure that the operation or repair has been completed properly with all affected systems functioning within normal parameters. This may include the following:

- Make sure that the radiator fan cycles properly and that the coolant level and concentration are correct.
- Ensure that all cable ties and hose clamps that were removed as part of the repair are replaced.
- Check and adjust all other applicable fluid levels.
- Make sure that there are no fluid leaks.
- Make sure that there are no air, vacuum or exhaust leaks.
- Make sure that all components involved in the repair are positioned correctly and function properly.
- Male sure all tools, shop cloths, fender covers, and protective tape are removed.

In addition to the above noted points, the ECM and TCM may need to be checked using a diagnostic scan tool. See **OBD On Board Diagnostic** section for more information.

15d Cylinder Head and Valvetrain (2.5L engine)

GENERAL

This section covers cylinder head and valvetrain assembly for the 2.5 liter gasoline engine. Most of the operations described in this repair group require special equipment and experience. If you lack the skills, tools, or a suitable workplace for servicing or repairing the cylinder head, then leave these repairs to an authorized Volkswagen dealer or other qualified shop.

Engine Codes
- BPR, BPS . . . . 2.5L 5-cylinder gasoline

WARNING —
- *When using an exchanged cylinder head with camshaft installed, lubricate the contact surfaces between the lifters and cam lobes before installing cylinder head cover.*
- *Remove the plastic protectors installed to protect the open valves immediately before installing the cylinder head.*
- *After replacing cylinder head, replace all coolant.*
- *Cylinder heads with cracks between the valve seats can continue to be used without reducing service life, as long as the cracks have a width of max. 0.5 mm (0.02 in.).*
- *Do not mill the valve seats in the cylinder head; only lapping of the valves is permissible.*
- *After installing the camshafts, do not start the engine for approx. 30 minutes. This allows the hydraulic lifters to seat themselves. Otherwise the valves will crash into the pistons.*
- *After working on the valvetrain and lifters, slowly rotate the crankshaft by hand at least 2 full revolutions before starting engine to be sure that valves do not strike the pistons.*
- *Always replace gaskets and seals.*

DIAGNOSTIC TESTING

The tests that follow can be used to help isolate engine problems, to better understand a problem before starting expensive and extensive repairs, or to just periodically check engine condition.

Compression Test

A compression test will tell a lot about the overall condition of the engine without the need for taking it apart. Testing is relatively simple and straightforward.

– Warm-up engine until oil temperature is a minimum of 30°C (86°F).

– Remove upper engine cover. See **03 Maintenance**.

– Remove fuse for voltage supply to Motronic Engine Control Module (ECM) from fuse box. Fuse designations vary by model year. See **EWD Electrical Wiring Diagrams** for specific fuse designations.

> ***NOTE —***
> *Removing fuses for Motronic ECM will prevent fuel injectors from injecting fuel into engine during compression test.*

– Remove ignition coil and spark plug for cylinder #1, see **03 Maintenance**.

◀ Install compression tester into spark plug hole for cylinder #1.

– Have a helper:

- Depress clutch pedal fully (manual transmission).
- Put transmission in neutral or park.
- Crank engine over with starter motor.

> ***NOTE —***
> *Cranking engine over with fuel injection system components disconnected may Diagnostic Trouble Codes (DTCs) to be stored in engine management system memory.*

– Crank engine a minimum of 4 to 5 revolutions.

– Record readings, release pressure in gauge and repeat procedure for each cylinder.

– Compare readings to specification.

Specification
- Compression pressures (2.5L engine)
  new . . . . 9 to 13 bar (131 - 189 psi)
  wear limit . . . . 8 bar (116 psi)
- maximum difference
  between cylinders . . . . 3 bar (44 psi)

– When all cylinders have been checked, reinstall spark plugs and wires, connect harness connector for ignition coil power output stage and install the upper engine cover.

Tightening torque
- Spark plug to cylinder head . . . . 30 Nm (22 ft-lb)

Cylinder leakdown test

The most conclusive method for diagnosing low compression symptoms is a cylinder leak-down test. Using a special tester and a supply of compressed air, each cylinder is pressurized. The rate at which air leaks out of cylinder, as well as the sound and location of escaping air can more accurately pinpoint the magnitude and source of leakage. Any engine compression diagnosis that will require major disassembly should first be confirmed by a cylinder leak-down test. Because this test requires special equipment and experience, it should be performed by a Volkswagen dealer or other qualified repair shop. Leak-down limit specifications are usually supplied by the equipment manufacturer.

CYLINDER HEAD SERVICE

The following page contains an assembly view of internal and external cylinder head components. Removal and installation of the cylinder head cover (valve cover) and timing chain guard cover are also covered in this section. The 2.5L engine does not use a timing belt, but instead uses a timing chain. Timing chain removal and installation is beyond the scope of this repair manual. For more in-depth repair information, including timing chain removal and installation, contact the publisher of this manual at www.BentleyPublishers.com for an electronic edition of our "Official Factory Repair Manual".

Cylinder head, assembly

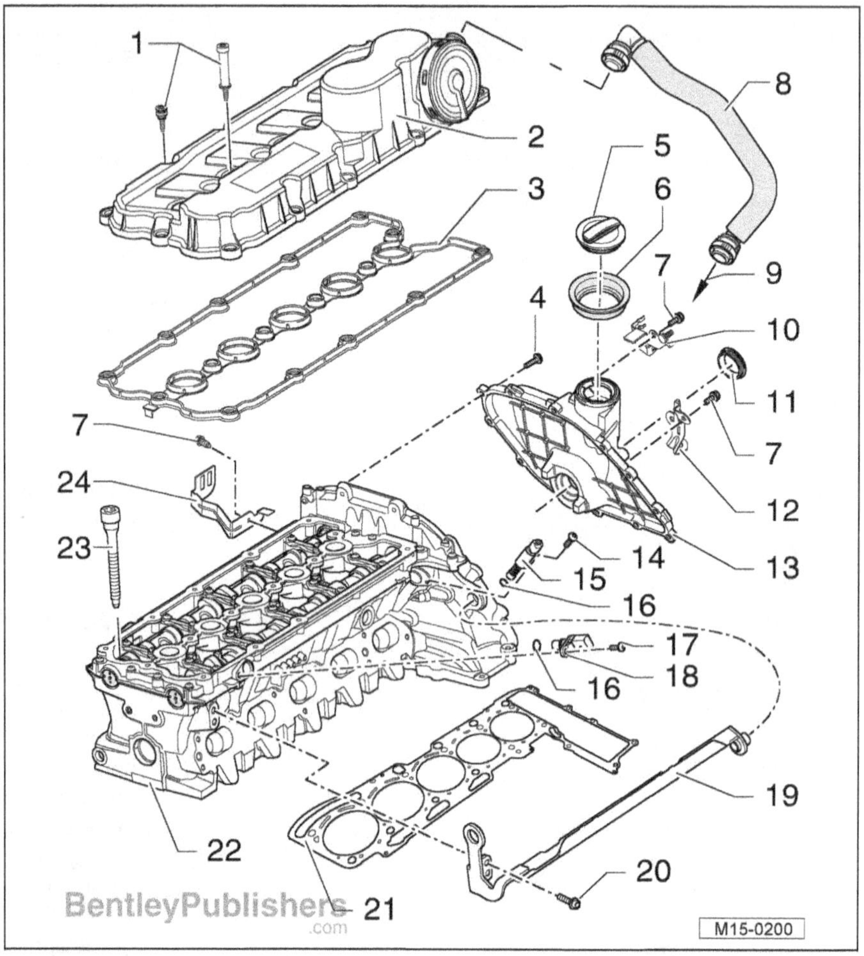

NOTE —
Volkswagen identifies electrical components by a letter and/or a number in the electrical schematics. See ***EWD Electrical Wiring Diagrams****. These electrical identifiers are listed in parentheses as an aid to electrical troubleshooting.*

1. **Bolt**
   - 10 Nm (7 ft-lb)
2. **Cylinder head cover**
   - With pressure regulator valve for crankcase ventilation
   - Observe tightening sequence when installing, see **Cylinder head cover, removing and installing**
3. **Cylinder head cover gasket**
   - Replace if damaged or leaking
4. **Bolt**
   - 10 Nm (7 ft-lb)
5. **Oil filler cap**
6. **Gasket**
   - Replace if damaged or leaking
7. **Bolt**
   - 10 Nm (7 ft-lb)
8. **Bleeder hose for crankcase ventilation**
   - Ensure seated tightly
   - Squeeze securing ring to disengage
9. **To intake hose/intake manifold**
10. **Harness bracket**
11. **Oil seal**
    - See **Chain guard cover seal, replacing** in this repair group
12. **Harness bracket**
13. **Chain guard cover**
    - See **Chain guard cover, removing and installing**
14. **Screw**
    - 2 Nm (18 in-lb)
15. **Camshaft adjustment valve (N205)**
16. **O-ring**
    - Replace if damaged
17. **Bolt**
    - 10 Nm (7 ft-lb)
18. **Camshaft position sensor (G40)**
19. **Motor transport strap**
20. **Bolt**
    - 25 Nm (18 ft-lb)
21. **Cylinder head gasket**
    - Always replace
    - After replacing, replace entire amount of coolant
22. **Cylinder head**
    - Check for distortion as shown below
    - Reworking sealing surface is not possible
23. **Cylinder head bolt**
    - Always replace
    - Tighten to 40 Nm (30 ft-lb) plus additional 180° (½ turn)
24. **Harness bracket**

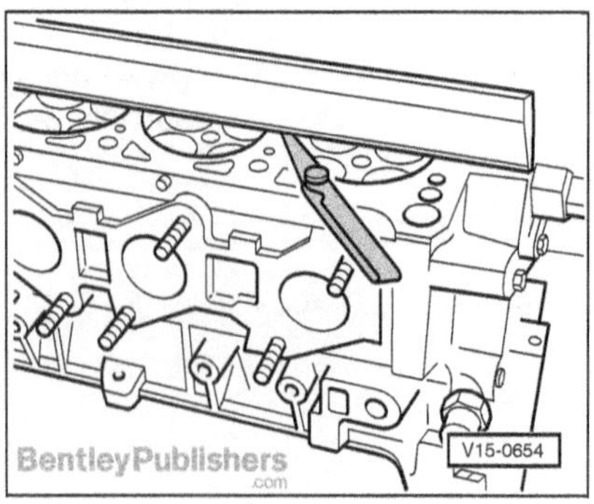

◄ Using straight edge and feeler gauge, check cylinder head for warpage at multiple points.

- Maximum permissible warpage: 0.05 mm (0.0020 in)

Cylinder head cover, removing and installing

- Remove upper engine cover. See **03 Maintenance**.

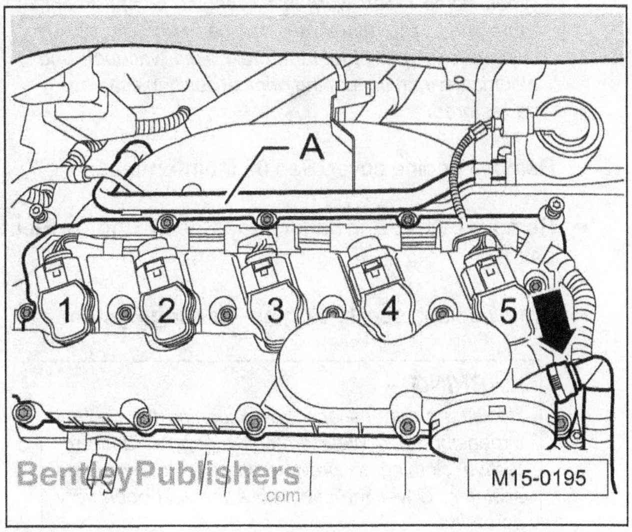

◀ Disconnect air hose for crankcase ventilation (**arrow**).

- Remove connecting pipe (**A**) for secondary air system.
- Remove ignition coils (**1** through **5**) using ignition coil puller T40039. See **03 Maintenance**.

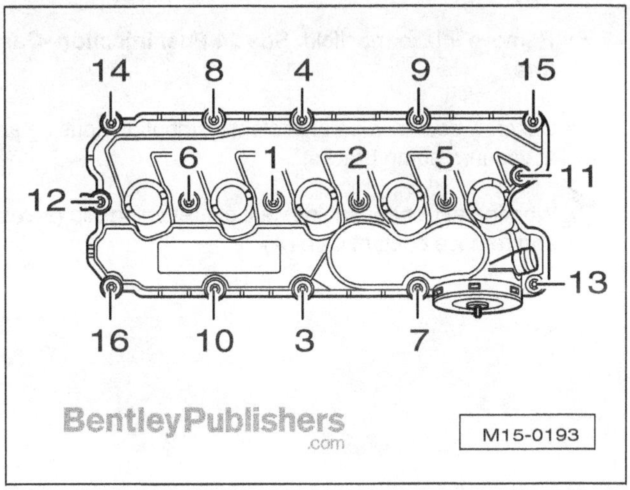

◀ Remove bolts for cylinder head cover in reverse order of tightening sequence (shown) and remove cylinder head cover.

- For removal, first loosen bolt #16, then #15, then #14 and so on.

- Clean sealing surfaces of cylinder head cover.
- Replace cylinder head cover gasket if damaged or leaking.
- Install cylinder head cover and tighten bolts in sequence shown (1 through 16).

Tightening torque
- Cylinder head cover to cylinder head. 10 Nm (7 ft-lb)

- Install remaining components in reverse order of removal.

Chain guard cover, removing and installing

NOTE —
When doing repair work in the engine compartment, route lines of all types (e.g. for fuel, hydraulic, EVAP system, coolant, refrigerant, brake fluid, vacuum) and electrical wiring so that the original positions can easily be restored.

– Remove engine cover. See **03 Maintenance**.

– Remove battery and battery tray. See **27 Engine Electrical**.

– Drain coolant, see **19 Engine–Cooling System**.

WARNING —
Hot steam may escape when opening the coolant expansion tank. Wear protective goggles and protective clothing to prevent damage to eyes and scalding. Cover the cap with a rag and open very carefully.

– Remove intake manifold. See **24 Fuel Injection–Gasoline**.

– Unscrew coolant pipe at coolant distribution housing and at vacuum pump bracket.

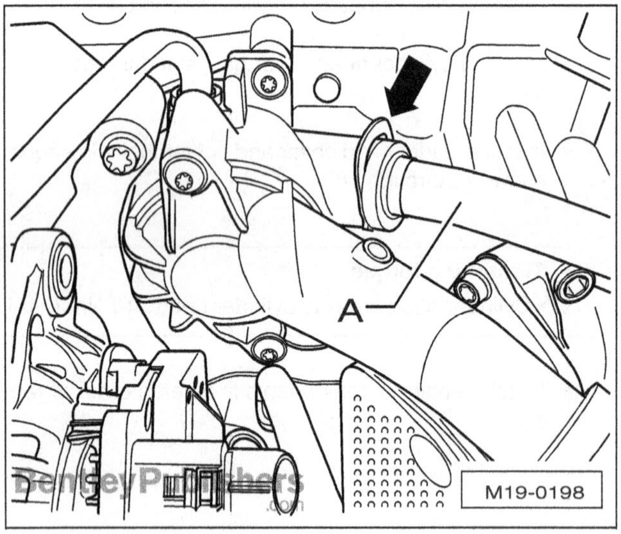

◀ Working at front of engine, pull out retaining clip (**arrow**) and remove coolant pipe (**A**).

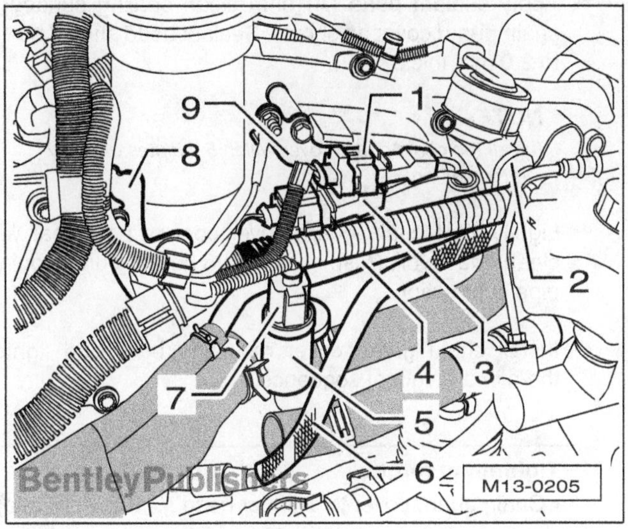

◀ Disconnect harness connectors (**1**) and (**3**).

– Pull off pressure pipe (**2**) from secondary air injection (AIR) valve.

– Remove rear coolant pipe (**4**).

– Pull off vacuum hose (**6**) at vacuum pump and harness connector (**7**).

– Unscrew brackets (**8**) and (**9**) and place wiring harness with pressure pipe aside.

– Position bracket for knock sensor wiring harness, located on secondary air injection (AIR) valve, slightly toward rear.

– Remove coolant distribution housing (**5**) and set it aside with coolant hoses connected.

– Remove all mounting bolts from chain guard cover.

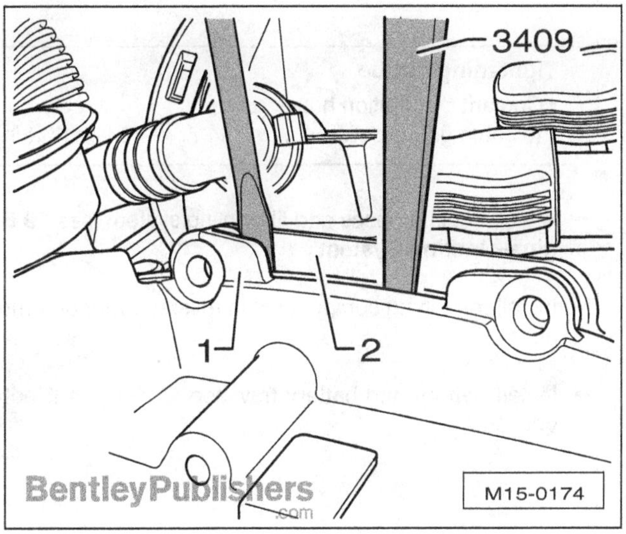

◀ Pry chain guard cover (**1**) evenly off of cylinder head (**2**) at upper and lower notches intended for this purpose.

NOTE —

Do not damage the sealing surfaces of chain guard cover. If necessary, use plastic trim removal wedge (3409) to pry guard free.

– Remove sealant residue from chain guard cover and cylinder head sealing surface. A rotating plastic brush can be used for cleaning residue.

NOTE —

Make sure that no sealant residue gets inside engine.

– Clean pipe connections at cylinder head. If necessary, remove coolant deposits with a brass wire brush or with a fine sandpaper (min. grit 100).

Chain guard cover, installing

– Replace chain guard cover coolant pipe seal. See **Chain guard cover seal, replacing** in this repair group.

– Clean sealing surfaces; they must be free of oil and grease.

– Cut sealant tube nozzle so that nozzle opening is approximately 1 mm (0.04 in.) diameter.

NOTE —

See a Volkswagen authorized parts department or aftermarket parts specialist for proper chain guard cover sealant.

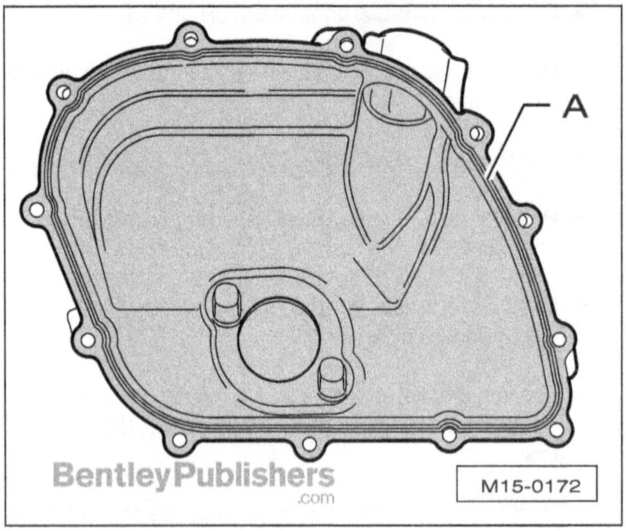

◄ Apply sealant bead (**A**) onto clean sealing surface of chain guard cover as shown. Sealant bead should be 1.5 to 2.0 mm thick.

> ***NOTE —***
> *Install the chain guard cover within 5 minutes of applying the sealant.*

– Lightly coat chain guard cover coolant pipe seal with clean engine oil. Push chain guard housing onto coolant pipe connection.

– Install chain guard cover mounting bolts and tighten them in a diagonal sequence.

Tightening torque
- Chain guard cover to cylinder head . . 10 Nm (7 ft-lb)

– Replace seal in coolant distribution housing and install housing to chain guard cover.

Tightening torque
- Coolant distribution housing to chain guard cover. . . . . 10 Nm (7 ft-lb)

– Attach coolant hoses and fill cooling system, see **19 Engine–Cooling System**.

– Install remaining components in reverse order of removal.

– Install battery and battery tray, see **27 Engine Electrical**.

Chain guard cover seal, replacing

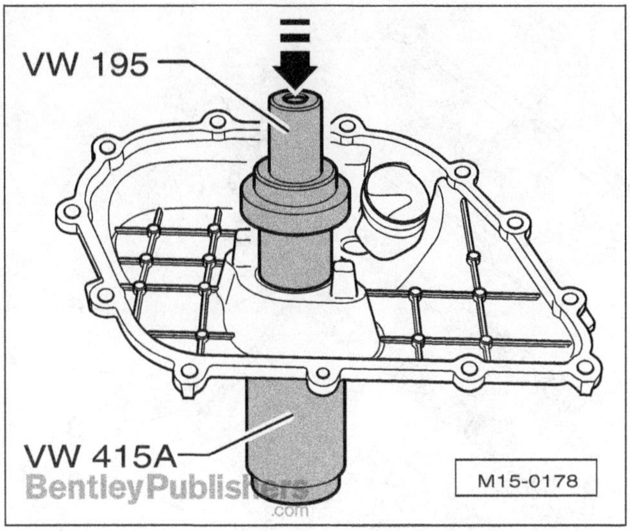

– Remove chain guard cover. See **Chain guard cover, removing and installing** in this repair group.

< Drive out old seal using bearing driver (special tool VW 195) and support (special tool VW 415A).

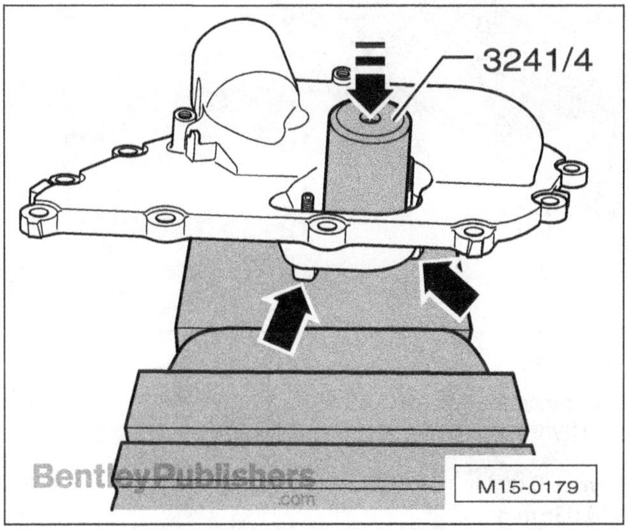

< Install new chain guard cover seal using fitting sleeve (special tool 3241/4).

– Support chain guard cover and connections (**lower arrows**) on a secure backing surface and press in new seal up to stop.

– Install chain guard cover.

Camshaft adjustment valve, removing and installing

< Disconnect harness connector (**arrow**) from camshaft adjustment valve (N205).

NOTE —
Volkswagen identifies electrical components by a letter and/or a number in the electrical schematics. See **EWD Electrical Wiring Diagrams**. *These electrical identifiers are listed in parentheses as an aid to electrical troubleshooting.*

– Remove mounting bolt and pull camshaft adjustment valve out of cylinder head.

– Install in reverse order of removal, noting the following:
- Camshaft adjustment valve and housing must be free of dirt and other contamination.
- Remove new camshaft adjustment valve from packaging just prior to installation.

VALVETRAIN ASSEMBLY

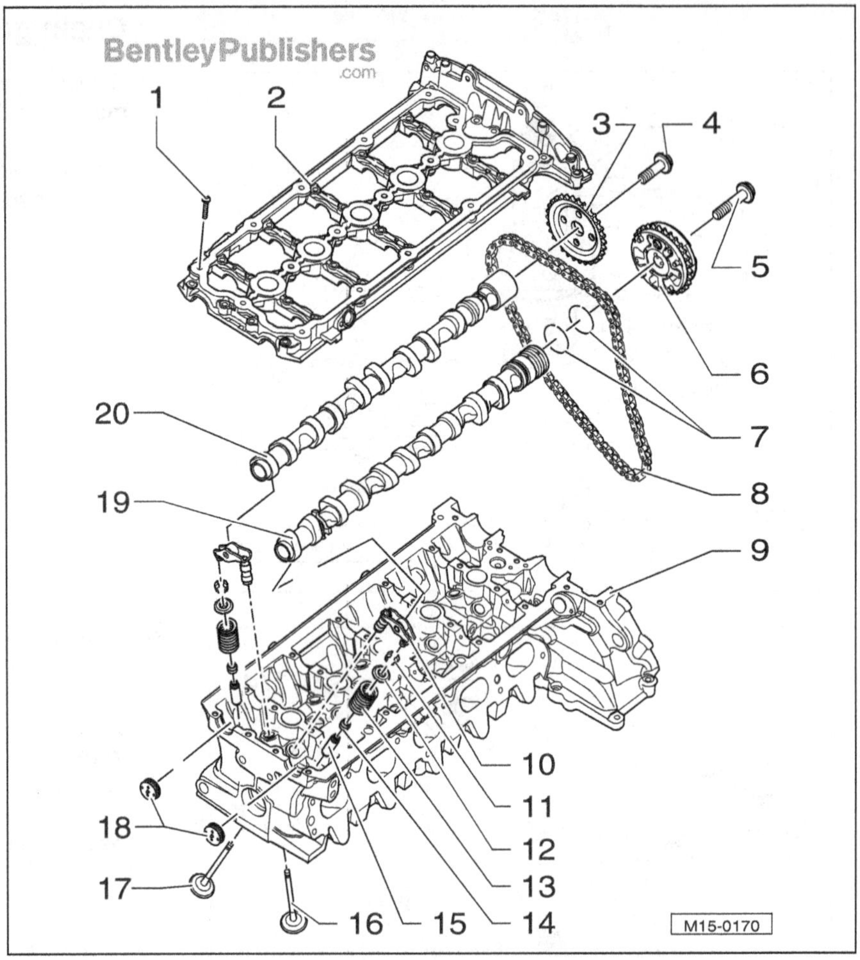

1. **Bolt**
   - Tighten to 8 Nm (71 in-lb) plus additional 90° (¼) turn
2. **Bearing bracket**
   - With integrated camshaft bearings
   - Clean sealing surface
   - Reworking is not permitted
   - Remove old sealant residue
3. **Chain sprocket**
   - For exhaust camshaft
4. **Bolt**
   - Tighten to 60 Nm (44 ft-lb) plus additional 90° (¼) turn
   - Always replace
5. **Bolt**
   - Tighten to 60 Nm (44 ft-lb) plus additional 90° (¼) turn
   - Always replace
6. **Camshaft adjuster**
   - For intake camshaft
7. **Oil seals**
   - For camshaft adjuster
8. **Timing chain**
9. **Cylinder head**
   - Do not reface valve seats, only lapping is permitted
10. **Hydraulic valve lifter**
    - Do not interchange
11. **Lubricate contact surface**
12. **Valve keepers**
13. **Upper valve spring plate**
14. **Valve spring**
15. **Valve stem seal**
16. **Valve guide**
17. **Intake valve**
    - Do not rework, only lapping is permitted
18. **Exhaust valve**
    - Do not rework, only lapping is permitted
19. **Sealing plugs**
    - Always replace
20. **Intake camshaft**
    - Check radial clearance using Plastigage (with hydraulic valve lifter removed)
    - Wear limit: 0.1 mm (0.004 in.)
    - Run-out: max. 0.035 mm (0.00138 in.)
    - Axial clearance: max. 0.17 mm (0.0067 in.)
21. **Exhaust camshaft**
    - Check radial clearance using Plastigage (with hydraulic valve lifter removed)
    - Wear limit: 0.1 mm (0.004 in.)
    - Run-out: max. 0.035 mm (0.00138 in.)
    - Axial clearance: max. 0.17 mm (0.0067 in.)

17 Engine–Lubrication System

17

GENERAL

Proper engine lubrication relies on a constant supply of oil, fed to the moving parts under pressure. Pressure is supplied by a gear-type oil pump located inside the engine oil pan. Engine oil returns to the oil pan by way of internal passages in the cylinder block and head. The returned oil collects in the oil pan where it is stored for eventual pickup by the oil pump. Additionally, engines with turbo-chargers have external oil supply and return lines.

Engine	Engine code
1.8L 4-cylinder turbo gasoline	APH, AWV, AWP, BKF, BNU
1.9L 4-cylinder TDI diesel	ALH, BEW
2.0L 4-cylinder gasoline	AEG, AVH, AZG, BDC, BEV, BGD
2.5L 5-cylinder gasoline	BPR, BPS

NOTE —
For changing the oil and filter, see **03 Maintenance**.

Oil Pressure Warning System

To prevent serious engine damage, a dynamic oil pressure warning system warns the driver of insufficient oil pressure. This system includes a single oil pressure switch, an electric control unit, a warning light and a buzzer. The electronic control unit with buzzer and warning light is integrated into the instrument cluster.

Oil pressure warning system, checking

With the ignition off, the single oil pressure switch is open (no continuity between the terminal connection and the switch body). When the ignition is switched on, but the engine is not started, the oil pressure warning light should come on for approximately 3 seconds and then go out.

When the engine is started and the oil pressure rises, the oil pressure switch should close at the specified pressure completing a circuit to ground. When the engine reaches approximately 1,500 rpm, the electronic control unit checks to see if the ground is complete and if so, the system is functioning normally. If the ground is not complete, the warning light and buzzer are activated. Several small time delays are programmed into the system to prevent accidental warnings due to normal minor fluctuations in pressure.

> ***CAUTION —***
> *If the warning indicators stay on after the engine is started, or while driving, always assume that there is insufficient oil pressure. Check oil level and test oil pressure before proceeding.*

A quick-check of the system can be accomplished as follows:

- Check instrument cluster warning light.
  - Turn key on, but do not start engine.
  - Warning light should come on for about 3 seconds and then go out.
- Turn key off. Disconnect and ground the wire at oil pressure switch.
- Turn key on, but do not start engine.
- Warning light should flash and the buzzer should beep 3 times.
- Turn key off. Remove oil pressure switch wire from ground but do not re-connect to oil pressure switch.
- Start engine and let idle.
  - Warning light and buzzer should not come on.

– Increase engine speed to approx. 1,500-2,000 rpm.
  - Warning light and buzzer should come on after a 2-3 second delay.

If the above tests give results as noted, the oil pressure warning system is functioning within normal parameters. If the above test results deviate from specification, continue by testing the oil pressure and the oil pressure switch, given later.

NOTE —
Since oil temperature can have a considerable influence on oil pressure, it may be advisable to conduct the above noted tests on both a cold and a warmed-up engine.

Oil pressure and oil pressure switch, testing

It is possible for the oil pressure warning system to be functioning properly and still give an occasional indication of a malfunction. In this case, it is advisable to check the oil pressure switch. This switch is located on the oil filter flange for both gas and diesel engines, see **Oil filter flange, assembly**, given later.

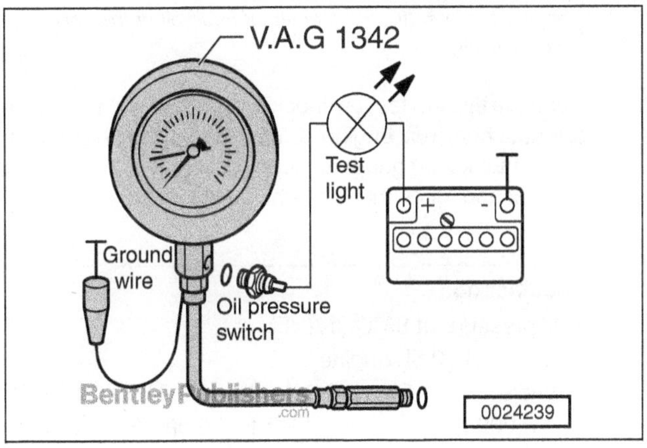

The oil pressure switch can be accurately tested by removing it from the oil filter flange and temporarily installing it into an oil pressure tester such as the Volkswagen supplied VAG 1342.

– Disconnect wire from oil pressure switch and leave it unplugged for this test.

– For 2.5L engine only, place front lock carrier into service position. See **50 Body–Front**.

– Remove oil pressure switch and thread it into tester.

– Thread oil pressure tester hose into hole previously occupied by switch.

– Connect ground wire from tester to a good ground.

– Connect a test light from battery positive to oil pressure switch terminal.

– Start engine and let it warm up.

– When radiator fan has cycled at least once, accelerate engine until test light comes on and observe pressure gauge. The light must come on at the value specified below. If light does not illuminate at specified value, switch is defective and should be replaced.

< Factory installed and original equipment replacement oil pressure switches can be identified by their color and pressure rating (**arrow**) marked on the switch body. 1.4 bar switch shown.

Specification

- Oil pressure switch closing pressure
  - 1.8L / 2.0L / 2.5L engine
    - black switch, (1.4 bar) . . . . . . 1.2-1.6 bar (18-24 psi)
  - 1.9L engine (ALH)
    - gray switch, (0.9 bar). . . . . 0.75-1.05 bar (11-15 psi)
  - 1.9L engine (BEW)
    - brown switch, (0.7 bar) . . 0.55-0.85 bar (8-12.5 psi)

NOTE —

Incorrect engine oil viscosity can influence engine oil pressures. Before performing repairs, ensure that the engine has a sufficient quantity of fresh oil of the correct viscosity.

– Continue system test by checking oil pressure. With tester still attached, run engine at specified speeds and compare readings on gauge to specification. If specification is not reached, oil pump deficiencies are indicated.

Specification

- Oil pressure at 80°C (176°F)
  - 1.8L / 2.0L /2.5L engine
    - at idle . . . . minimum 2.0 bar (29 psi)
    - at 2,000 rpm . . . . 2.7 to 4.5 bar (39 to 66 psi)
    - maximum .7.0 bar (103 psi)
  - 1.9L engine
    - at idle . . . . no factory specification available
    - at 2,000 rpm . . . . minimum 2.0 bar (29 psi)
    - maximum .7.0 bar (103 psi)

– When testing is completed, remove the test light and tester. Install the oil pressure switch and the wiring.

NOTE —

If an oil pressure switch seal is leaking, a new sealing washer should be installed. The sealing washer is held captive on the switch and must be cut off. New sealing rings are available from authorized Volkswagen parts dealers or aftermarket parts specialists.

LUBRICATION SYSTEM COMPONENTS (4-CYLINDER)

Lubrication system, assembly

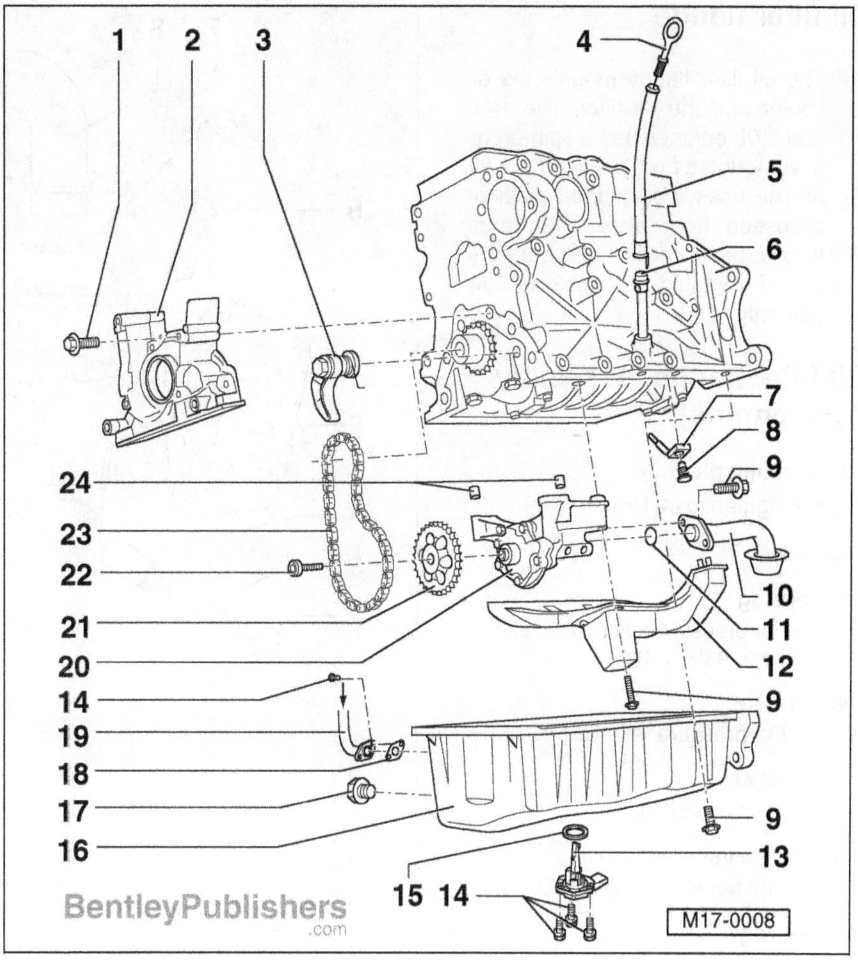

1. **Bolt**
   - Tighten to 15 Nm (11 ft-lb)
2. **Sealing flange**
   - Insert with silicone sealer D176 404 A2
3. **Chain tensioner (with tensioning rail)**
   - Tighten to 15 Nm (11 ft-lb)
   - To install, first pretension spring
4. **Dipstick**
   - Oil level must NOT be above MAX mark!
5. **Guide**
   - Pull off to extract oil
6. **Guide tube**
7. **Oil spray jet**
   - For piston cooling
   - Not on AEG engine
8. **Pressure relief valve**
   - Tighten to 27 Nm (20 ft-lb)
   - Opening pressure: 2.5-3.2 bar
   - Not on AEG engine
9. **Bolt**
   - Tighten to 15 Nm (11 ft-lb)
10. **Suction pipe**
    - Clean strainer if soiled
11. **O-ring**
    - Always replace
12. **Baffle plate**
13. **Oil level thermal sensor (G266)**
    - Not installed on all engines
14. **Bolts**
    - Tighten to 10 Nm (7 ft-lb)
15. **O-ring**
    - Always replace
    - Coat with oil before installing
16. **Oil pan**
    - Clean oiling surfaces before installing
    - Install with silicone sealant D 176 404 A2
17. **Oil drain plug**
    - Tighten to 30 Nm (22 ft-lb)
18. **Gasket**
    - 1.8L engine only
    - Always replace
19. **Oil return pipe**
    - 1.8L engine only
    - From turbocharger
20. **Oil pump**
    - With 12 bar (174 psi) pressure relief valve
    - Before installing, ensure that both centering dowels are installed
    - Replace if any running surfaces or gears are scored
21. **Chain sprocket**
22. **Bolt**
    - Tighten to 25 Nm (18 ft-lb)
23. **Chain**
    - Mark installed position before removing
24. **Centering dowels**

Oil filter flange

The oil filter flange mounts the oil cooler and the oil filter. The 1.8L and 2.0L engines use a spin-on oil filter mounted from below. The 1.9L engine uses a cartridge-type filter accessed from above. All three flanges have valves that bypass the filter if it should be unable to flow correctly.

Oil filter flange, assembly (1.8L engine)

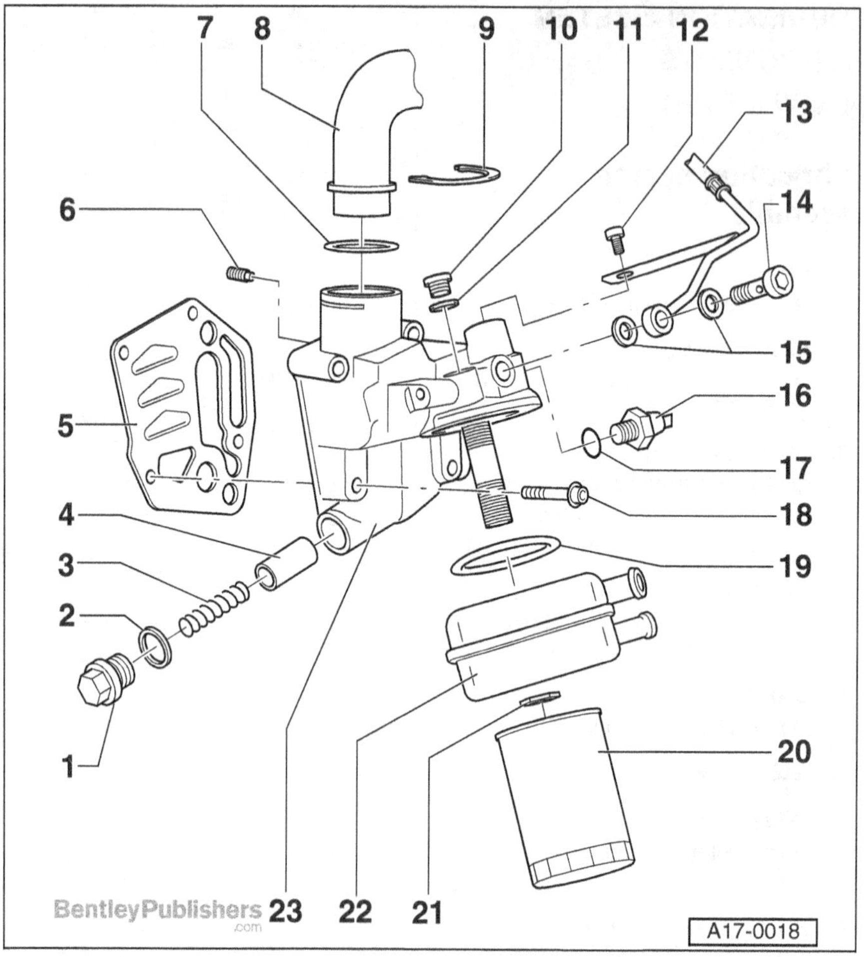

1. **Sealing plug**
   - Tighten to 40 Nm (30 ft-lb)
2. **Seal**
3. **Spring**
   - For pressure relief valve, approx. 4 bar (58 psi)
4. **Piston**
   - For pressure relief valve
5. **Gasket**
   - Always replace
6. **Check valve**
   - Tighten to 8 Nm (71 in-lb)
7. **O-ring**
   - Always replace
   - Install onto sealing cap and shoulder (item 8)
8. **Pipe**
9. **Retaining clip**
10. **Sealing plug**
    - Tighten to 15 Nm (11 ft-lb)
11. **Seal**
12. **Bolt**
    - Tighten to 20 Nm (15 ft-lb)
13. **Oil supply pipe**
    - To turbocharger
14. **Banjo bolt**
    - Tighten to 30 Nm (22 ft-lb)
15. **Gasket**
    - Always replace
16. **Oil pressure switch (1.4 bar)**
    - Tighten to 25 Nm (18 ft-lb)
17. **Seal**
18. **Nut**
    - Tighten to 15 Nm (11 ft-lb) + ¼ turn (90°) further
    - Always replace
19. **Gasket**
    - Always replace
    - Fit into groove on oil cooler
20. **Oil filter**
    - Loosen with strap wrench
    - Hand tighten
21. **Nut**
    - Tighten to 25 Nm (18 ft-lb)
22. **Oil cooler**
    - Installed position: must be turned fully clockwise (viewed from below) until touching its own bracket
23. **Oil filter bracket**
    - With pressure relief valve, approx. 4 bar (58 psi)

Oil filter flange, assembly (1.9L engine)

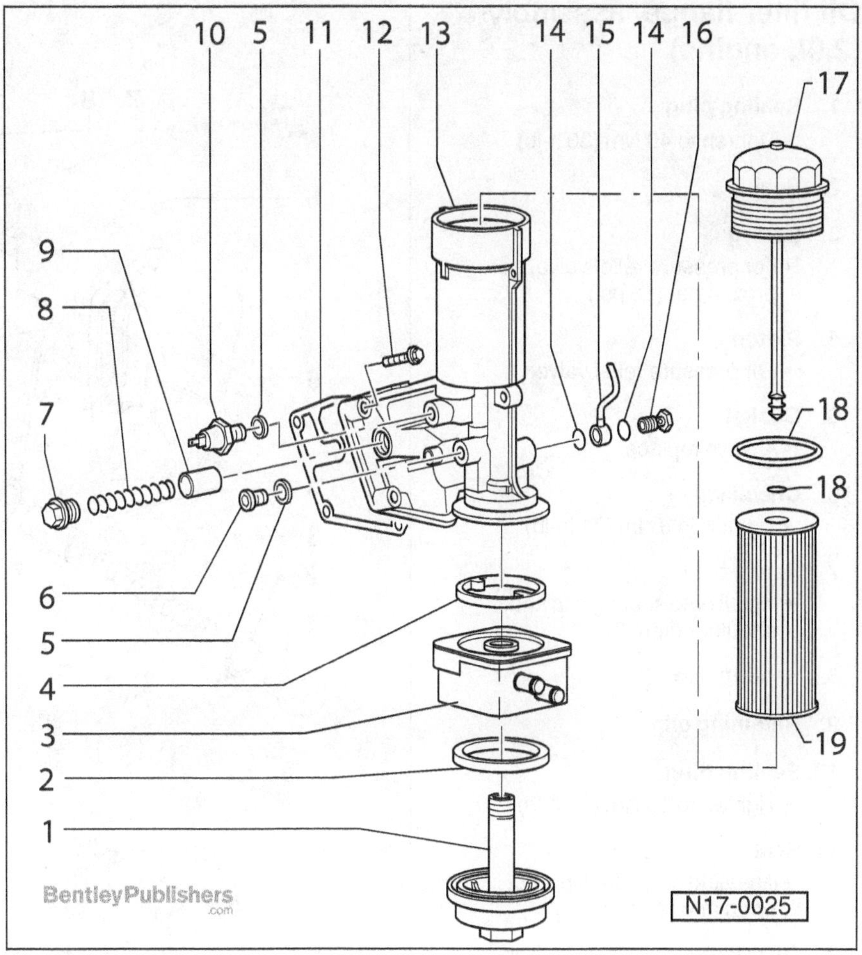

1. **Sealing plug**
   - Tighten to 25 Nm (18 ft-lb)
2. **Gasket**
   - Always replace
3. **Oil cooler**
   - Installed position: must be turned fully clockwise (viewed from below) until touching its own bracket
4. **Gasket**
   - Always replace
   - Install into oil cooler groove
5. **Seal**
   - If seal leaks, cut open and replace
6. **Plug**
   - Tighten to 10 Nm (7 ft-lb)
7. **Sealing plug**
   - Do NOT open
8. **Spring**
   - For pressure relief valve, approx. 5 bar (72.5 psi)
9. **Piston**
   - For pressure relief valve, approx. 5 bar (72.5 psi)
10. **Oil pressure switch**
    - Gray, 0.9 bar (ALH engine) Tighten to 25 Nm (18 ft-lb)
    - Brown, 0.7 bar (BEW engine) Tighten to 20 Nm (15 ft-lb)
11. **Gasket**
    - Always replace
12. **Bolt**
    - Tighten to 15 Nm (11 ft-lb) + ¼ turn (90°) further
    - Always replace
13. **Oil filter bracket**
14. **Seal**
    - Always replace
15. **Oil supply pipe**
    - To turbocharger
16. **Banjo bolt**
    - Tighten to 20 Nm (15 ft-lb)
17. **Sealing cap**
    - Tighten to 25 Nm (18 ft-lb)
    - Loosen and tighten using 3417 oil filter wrench
18. **O-ring**
    - Always replace
19. **Oil filter element**

Oil filter flange, assembly (2.0L engine)

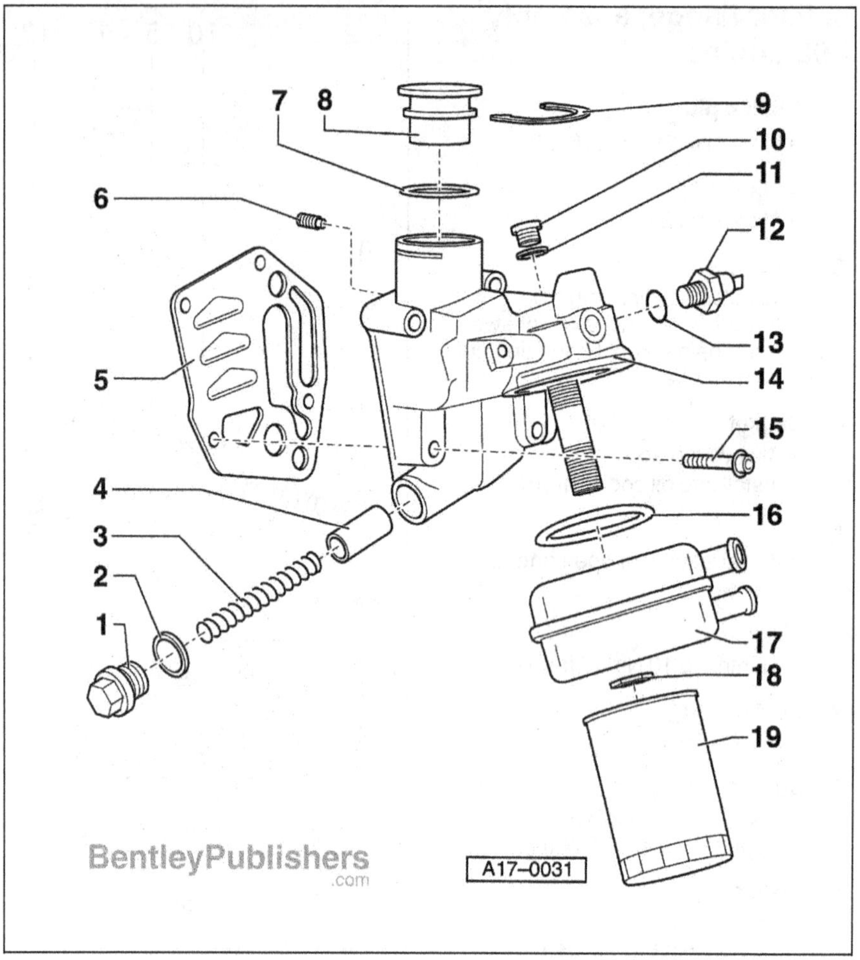

1. **Sealing plug**
   - Tighten to 40 Nm (30 ft-lb)
2. **Seal**
3. **Spring**
   - For pressure relief valve, approx. 4 bar (58 psi)
4. **Piston**
   - For pressure relief valve
5. **Gasket**
   - Always replace
6. **Check valve**
   - Tighten to 8 Nm (71 in-lb)
7. **Seal**
   - Install onto sealing cap and shoulder (item 8)
8. **Sealing cap**
9. **Retaining clip**
10. **Sealing plug**
    - Tighten to 15 Nm (11 ft-lb)
11. **Seal**
    - If sealing ring is leaking, cut open and replace
12. **Oil pressure switch (1.4 bar)**
    - Tighten to 25 Nm (18 ft-lb)
    - Black
13. **Seal**
    - If sealing ring is leaking, cut open and replace
14. **Oil filter flange**
15. **Nut**
    - Tighten to 15 Nm (11 ft-lb) + ¼ turn (90°) further
    - Always replace
16. **Gasket**
    - Always replace
    - Fit into groove on oil cooler
17. **Oil cooler**
    - Installed position: must be turned fully clockwise (viewed from below) until touching its own bracket
18. **Nut**
    - Tighten to 25 Nm (18 ft-lb)
19. **Oil filter**
    - Loosen with strap wrench
    - Hand tighten
    - Observe installation instructions on oil filter

Oil pan, removing

The engine oil pan is a cast aluminum design bolted to both the cylinder block and the transmission. This design uses silicone sealant instead of a conventional gasket at the joint to the cylinder block. Some engines also have a two-piece (upper and lower) oil pan.

– Remove lower sound absorber panel (belly pan).

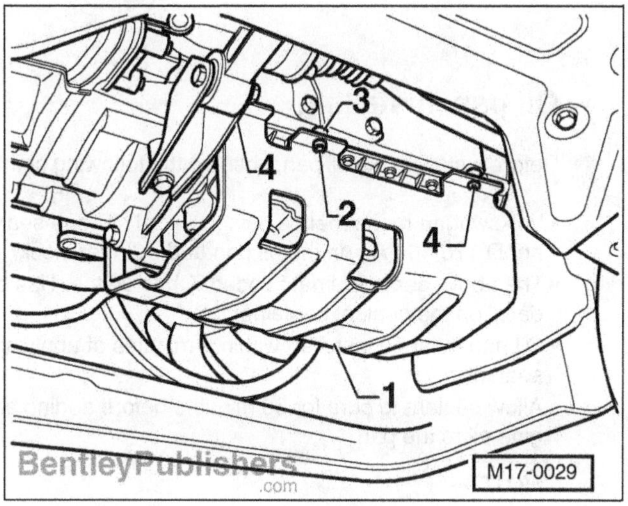

< Engines with oil pan guard. Remove bolts (**2**) and sleeves (**3**) from front and rear. Press spreader pin out in an upward direction and remove spreader rivet (**4**). Remove oil pan guard (**1**).

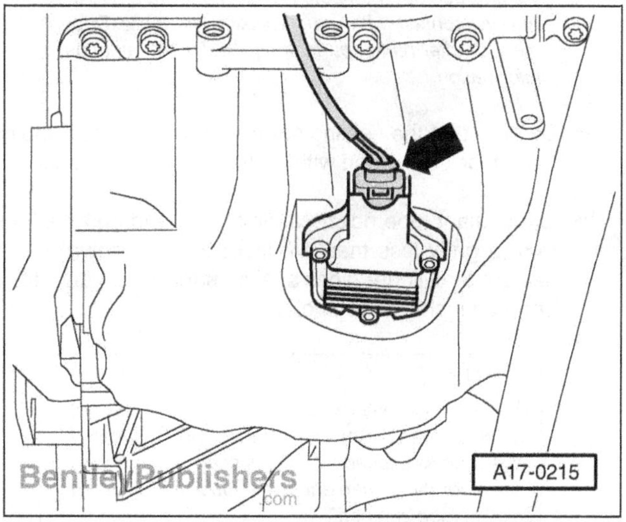

< Engines with oil level thermal sensor (G266). Disconnect harness connector (**arrow**) from sensor.

– Remove oil drain plug and drain oil into a suitable container.

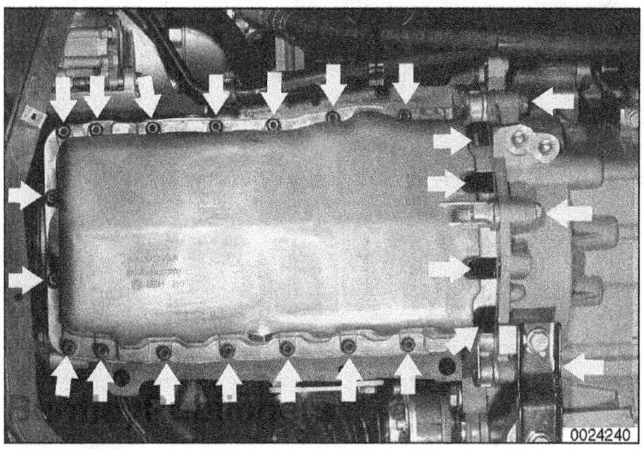

< Remove M10 bolts (3 **arrows** on right) holding the oil pan to the transmission.

– Remove the M7 bolts (**arrows**) holding the oil pan to the cylinder block.

– Remove oil pan. It may be necessary to free oil pan by tapping it lightly with a rubber hammer.

– Remove old sealant residue from cylinder block with a flat scraper.

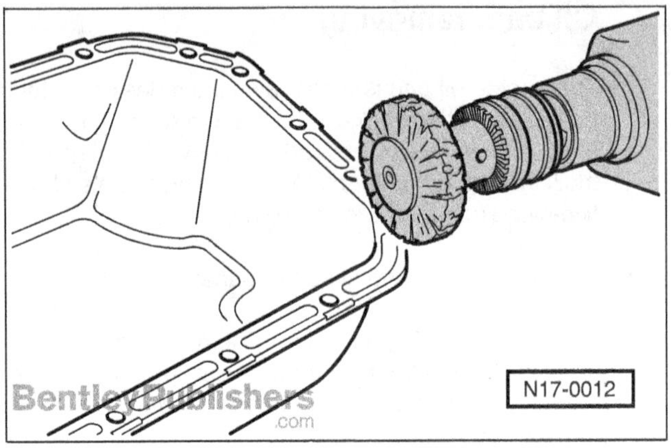

Remove old sealant residue from oil pan with a suitable rotating brush/pad.

> ***WARNING —***
> *Always wear suitable eye protection when using a rotating brush/pad.*

– Clean sealing surfaces on oil pan and cylinder block so that there are no traces of oil or grease.

Oil pan, installing

Before installing the oil pan, observe the following points:

- Volkswagen recommends using special silicone sealant D 176 404 A2 on the oil pan and cylinder block.
- The above sealant is date coded. Observe the "Use by date" on the sealant container.
- Oil pan must be installed within 5 minutes of applying sealant.
- Allow sealant to cure for 30 minutes before adding engine oil to the pan.

NOTE —
Volkswagen part numbers are given for reference only! Always consult with your Volkswagen Parts Department or aftermarket parts specialist for the latest parts information.

– Be sure that the sealing surfaces on the oil pan and cylinder block are clean with no traces of oil or grease.

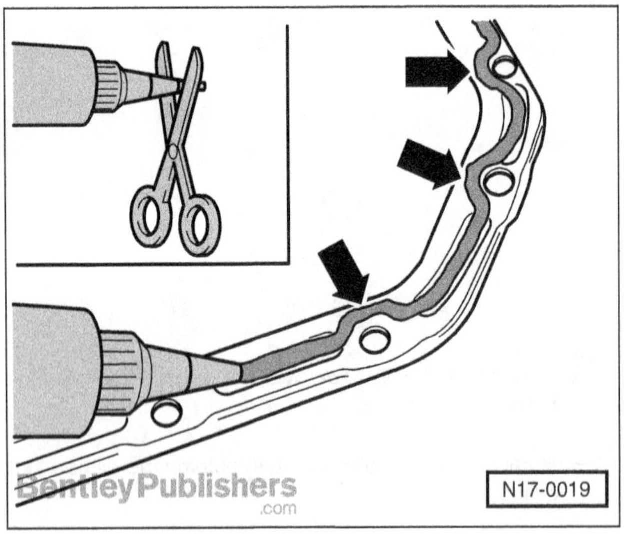

Cut sealant tube nozzle at first mark and apply a 2 to 3 mm (slightly less than 1/8 inch) bead (**arrows**) of new sealant to cylinder block sealing surface. Be sure to run bead on inside of bolt holes.

> ***CAUTION —***
> *The sealing compound bead thickness must not be wider than 3 mm (slightly less than 1/8 inch). If this width is exceeded, excess sealing compound will enter the oil pan and could block the oil pump pickup tube strainer.*

– Immediately install oil pan to cylinder block and loosely install M7 bolts using a diagonal pattern to tighten.

NOTE —
If the engine and transmission are separated for this operation, install the oil pan flush with the flywheel edge of the cylinder block.

– Install and lightly tighten M10 bolts securing oil pan to transmission.

– Diagonally tighten M7 oil pan bolts further, but not to final torque.

– Torque M10 bolts securing oil pan to transmission.

– Diagonally torque M7 bolts securing oil pan to cylinder block.

Tightening torques

- Oil pan to transmission (M10) . . . . . 45 Nm (33 ft-lb)
- Oil pan to cylinder block (M7). . . . . . 15 Nm (11 ft-lb)
- Oil drain plug (M14) . . . . 30 Nm (22 ft-lb)
- Oil pan guard to oil pan . . . . 37 Nm (27 ft-lb)

– Install oil drain plug.

– Install lower sound absorber panel (belly pan) and oil pan guard, if applicable.

– Connect harness connector for oil thermal sensor, if applicable

– Wait 30 minutes after oil pan has been installed and refill with proper viscosity and quantity of engine oil. See **03 Maintenance** for recommended oil types.

– Be sure to quality check your work, see **Quality Review** at the end of this repair group.

Oil pump

The oil pump is located inside the engine oil pan and is bolted to the cylinder block below the front of the crankshaft. The oil pump is driven by a short chain connected directly to the crankshaft. A spring loaded tensioner takes up any slack in the chain.

There is normally no need to remove and inspect the oil pump unless oil pressure is inadequate. See **Oil pressure and oil pressure switch, testing**.

There oil pump had no serviceable parts, however the pump can be disassembled and visually inspected for scoring and wear on the gears and running surfaces. If abnormal scoring is found, the pump must be replaced.

– Remove oil pan and baffle plate to access oil pump.

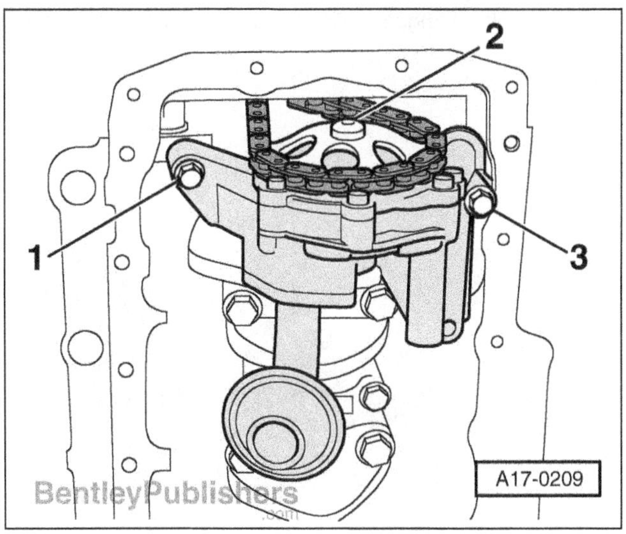

< Remove bolt (**2**) and pull chain sprocket off oil pump shaft. Remove bolts (**1** and **3**) and oil pump.

– Install in reverse order of removal.

Tightening torques

- Oil pump to cylinder block (M7) . . . . 15 Nm (11 ft-lb)
- Sprocket to oil pump (M8) . . . . 25 Nm (18 ft-lb)
- Oil pump cover to body (M6) . . . . 10 Nm (7 ft-lb)
- Suction pipe to body (M7) . . . . 15 Nm (11 ft-lb)

Oil spray nozzles

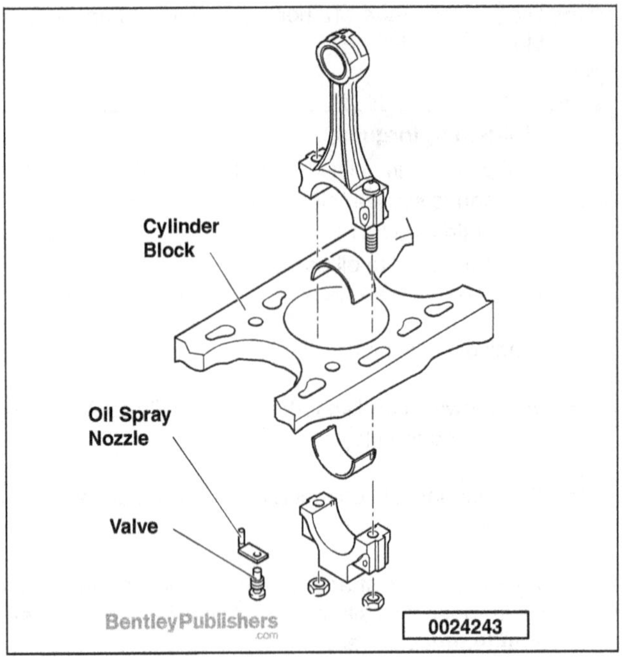

With the exception of the 2.0L AEG engine, engines are equipped with oil spray nozzles attached to the cylinder block at the base of the cylinder bore. The oil spray nozzles atomize a fine mist of engine oil and direct it against the bottom of the piston for additional cooling and lubrication. The oil spray nozzle is regulated by the valve that secures it to the cylinder block. The valve only opens to allow flow when oil pressure is greater than 2.5 to 3.2 bar (37 to 47 psi). To access the oil spray nozzles, the oil pan must first be removed.

Oil cooler

The oil cooler is an oil-to-coolant heat exchanger. Engine oil flows through one part of the cooler and gives up heat to the engine coolant flowing through the other part of the cooler. During warm-up, the process is reversed and engine coolant gives up heat to the engine oil because the engine coolant gets warm first. This speeds the warm-up process and gets the engine to operating temperature more quickly.

NOTE—
- *The oil cooler should be replaced if metal particles or metal shavings are found in the engine oil. The cooler cannot be cleaned.*
- *The cooler is a potential source of leakage between the lubrication system and the cooling system and should be considered whenever such leakage is suspected.*

Remove the cooler only if it needs to be inspected or replaced. Check to see that there is adequate space for the coolant hose connections and that the cooler is properly positioned.

Tightening torque
- Oil cooler sealing cap or nut to filter flange . . . . 25 Nm (18 ft-lb)

LUBRICATION SYSTEM COMPONENTS (5-CYLINDER)

Lubrication system, assembly

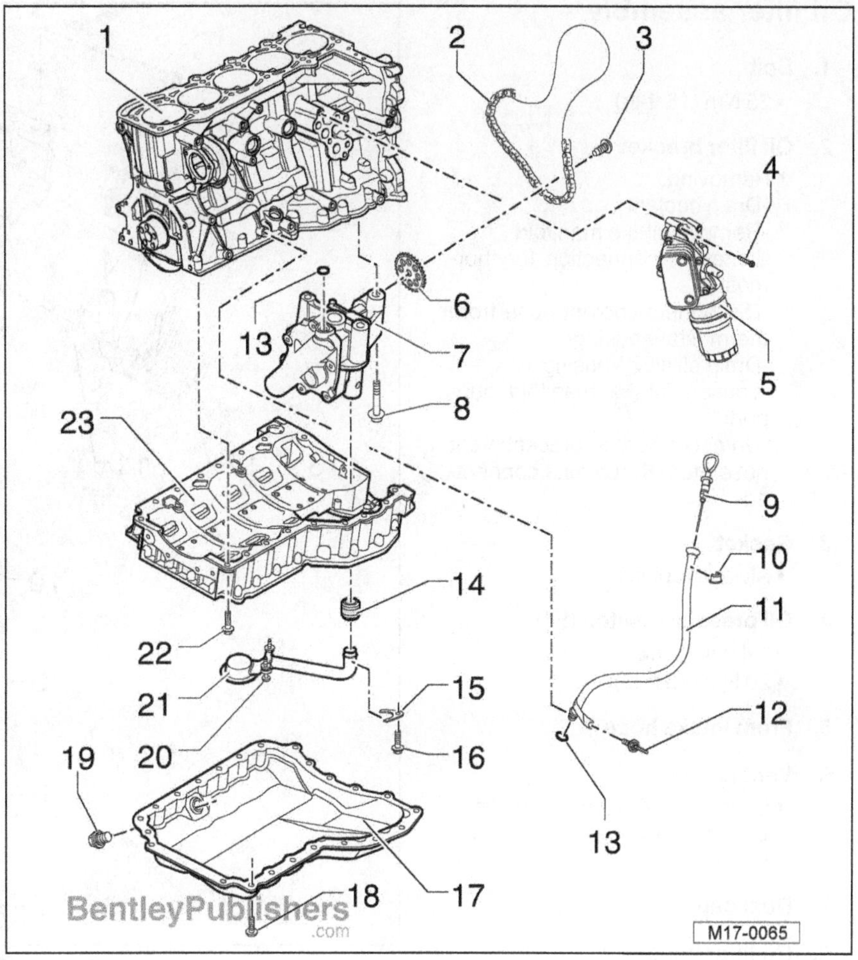

1. **Cylinder block**
2. **Oil pump drive chain**
3. **Bolt**
   - Tighten to 20 Nm (15 ft-lb) + ¼ turn (90°)
4. **Bolt**
   - Tighten to 25 Nm (18 ft-lb)
5. **Oil filter bracket**
6. **Chain sprocket**
   - Lettering points outward
   - With anti-twist mechanism
7. **Oil pump**
8. **Bolt**
   - Tighten to 25 Nm (18 ft-lb)
9. **Oil dipstick**
   - Oil level must not be above MAX mark
10. **Retaining ring**
    - Clipped to intake manifold
11. **Guide tube**
12. **Double bolt**
    - Tighten to 25 Nm (18 ft-lb)
13. **O-ring**
    - Always replace
14. **Seal**
    - Always replace
15. **Bracket**
16. **Bolt**
    - Tighten to 10 Nm (7 ft-lb)
17. **Oil pan, lower section**
18. **Bolt**
    - Tighten to 10 Nm (7 ft-lb)
19. **Oil drain plug**
    - Tighten to 30 Nm (22 ft-lb)
    - With permanent washer
    - Always replace
20. **Decoupling element**
    - Tighten bolts to 10 Nm (7 ft-lb)
21. **Oil pickup tube**
22. **Bolt**
    - Tighten to 25 Nm (18 ft-lb)
23. **Oil pan, upper part**

Oil filter assembly

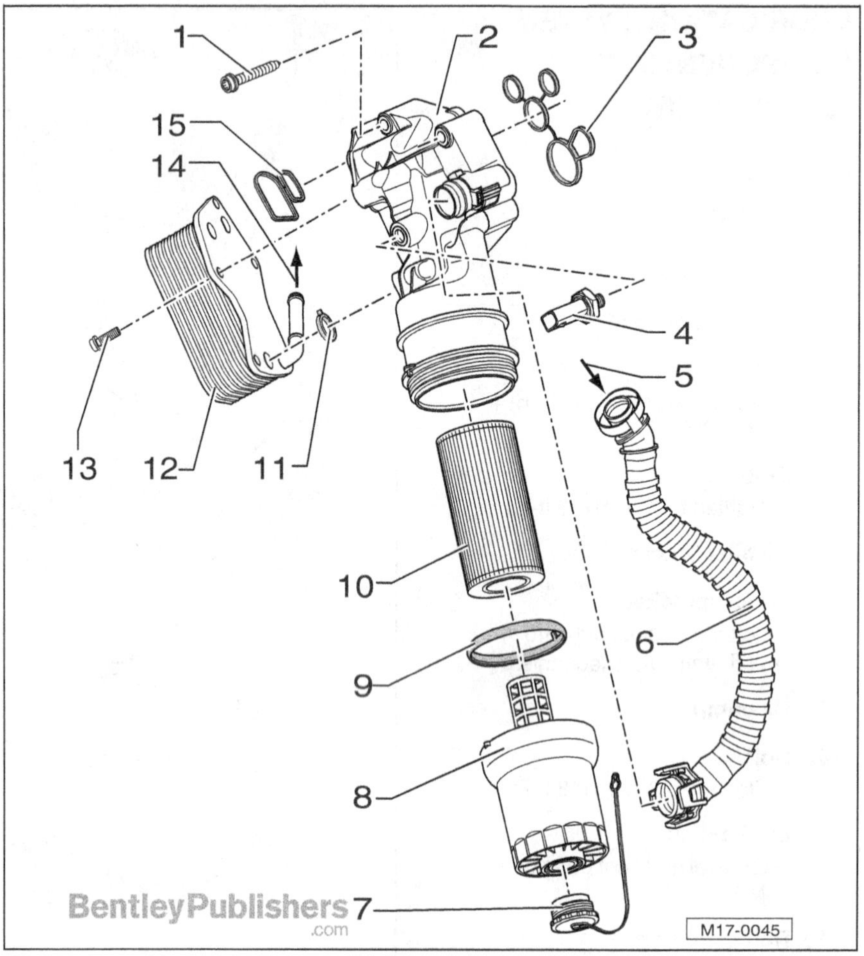

1. **Bolt**
   - 25 Nm (18 ft-lb)
2. **Oil filter bracket**
   - Removing:
     -Drain coolant
     -Remove intake manifold
     -Unfasten connection for thermostat
     -Disconnect coolant hose from thermostat housing
     -Drain oil filter housing
     -Loosen intake manifold support
     -Unfasten oil filter bracket; vent hose (item **6**) remains connected
3. **Gasket**
   - Always replace
4. **Oil pressure switch (F1)**
   - 1.4 bar, black
   - 20 Nm (15 ft-lb)
5. **From intake hose**
6. **Vent hose**
   - Because of 4-pin retainer, disconnect only after oil filter is removed
7. **Dust cap**
8. **Oil filter housing**
   - 25 Nm (18 ft-lb)
   - Use oil filter wrench 3417 for removal and installation
9. **Oil seal**
   - Always replace
   - Lubricate with oil
10. **Oil filter element**
    - Drain oil filter housing before removing
11. **Gasket**
    - Always replace
12. **Oil cooler**
    - Make sure there is sufficient clearance for surrounding components
13. **Bolt**
    - 25 Nm (18 ft-lb)
14. **To thermostat housing**
15. **Gasket**
    - Always replace

Oil pan, lower section, removing and installing

– Remove lower sound insulation and drain engine oil, see **03 Maintenance**.

– Remove oil pan lower section mounting bolts.

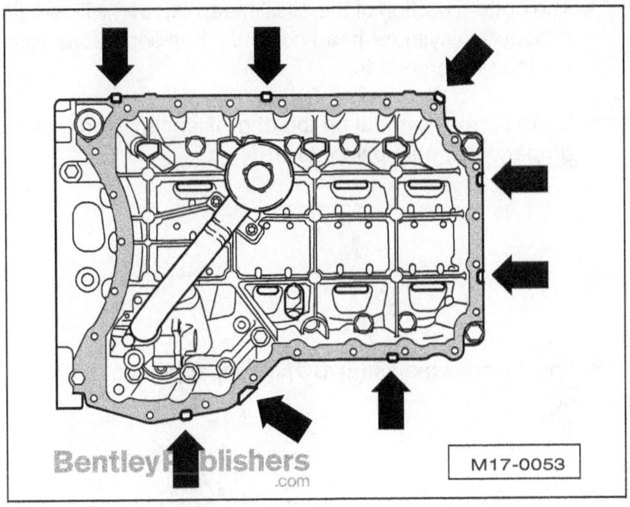

◄ Remove oil pan lower part by pressing on lift-out pockets (**arrows**) of oil pan upper part. Do not damage sealing surfaces.

– Remove any sealant from oil pan. A rotating plastic brush may be used for this. Be sure to wear safety goggles. Make sure sealing surfaces are completely clean.

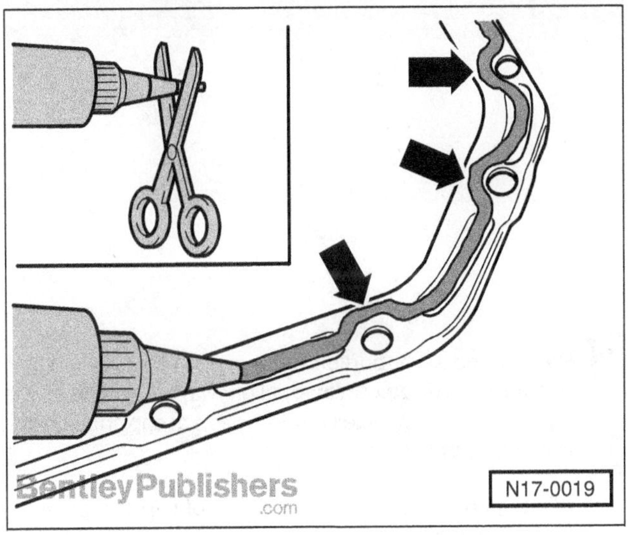

◄ Cut silicon sealant tube nozzle so that nozzle opening diameter is approx. 1 mm. Apply a 1.5 to 2 mm wide bead of silicon sealant to oil pan lower section sealing surface. Apply sealant on inner side of oil pan lower section mounting holes (**arrows**).

NOTE —
If sealant bead is too thick, the excess sealant could get in oil pan and block oil suction pipe screen.

– The oil pan lower section must be installed and tightened within 5 minutes after application of silicon sealant.

– Tighten oil pan lower section mounting bolts in a diagonal sequence to specified torque.

Tightening torque
- Oil pan lower section to upper section . . . . 10 Nm (7 ft-lb)

– After installing oil pan, allow sealant to dry for approx. 30 minutes before adding engine oil.

– Fill engine with oil and replace lower sound absorber panel.

Oil filler cap, repairing

The oil filler cap on 2.5L engines may come apart when removing engine cover. Replacement of the cap is not necessary if it can be repaired as described below.

17-A063

◀ The bottom portion of the oil filler cap (**arrow**) will remain attached to cylinder head cover if it has separated from the top portion of cap.

– Make sure oil symbol on top of oil filler cap is oriented properly (i.e. not upside down).

17-A065

◀ Locate index marks (**arrow**) on upper portion of oil filler cap.

17-A066

◀ Position the lower portion of the cap so that the molded parallel lines (**black arrows**) align with the handle on the top portion of cap. Index marks on top of cap should align with **white arrow**.

– Align upper portion to lower portion of cap and press firmly together.

– Oil filter cap should now operate properly.

QUALITY REVIEW

When you have finished working under the hood and around other areas of the vehicle, it is advisable to take a moment to quality check or review your work. This helps to insure that the operation or repair has been completed properly with all affected systems functioning within normal parameters.

19 Engine–Cooling System

TABLES

GENERAL

This section covers repairs and troubleshooting for the engine cooling system. For heater core and related heating and air conditioning components, see **80 Heating and Air Conditioning**. For information on the engine oil cooler, see **17 Engine-Lubrication**. For information on the ATF cooler used on vehicles with automatic transmissions, see **37 Automatic Transmission**.

Engine	Engine code
1.8L 4-cylinder turbo gasoline	APH, AWV, AWP, BKF, BNU
1.9L 4-cylinder TDI diesel	ALH, BEW
2.0L 4-cylinder gasoline	AEG, AVH, AZG, BDC, BEV, BGD
2.5L 5-cylinder gasoline	BPR, BPS

Warnings and cautions

The following warnings and cautions should be observed when working on the cooling system.

> ***WARNING —***
> - *Hot coolant can scald and result in serious personal injury. Do not work on the cooling system until it has fully cooled.*
> - *At normal operating temperature the cooling system is pressurized. Allow the system to cool as long as possible before opening, a minimum of an hour, then release the cap slowly to allow safe release of the pressure.*

> ***WARNING —***
> - *Releasing the cooling system pressure lowers the coolant's boiling point, and the coolant may boil suddenly. Use heavy gloves and wear eye and face protection to guard against scalding.*
> - *Use extreme care when working at or near the radiator cooling fans when the engine is hot. The fan can come on at any time, even if the ignition key is switched off.*
> - *Use extreme care when draining and disposing of engine coolant. Coolant is poisonous and lethal. Children and pets are attracted to coolant because of its sweet smell and taste. See a doctor or veterinarian immediately if any amount of coolant is ingested.*
> - *Disposal of used coolant must be done in conformance with all applicable local, state and federal laws. Do not pour into the ground or down drains or sewers. Contamination of ground water will result.*

> ***CAUTION —***
> *Avoid adding cold water to the cooling system while the engine is hot or overheated. If it is necessary to add coolant to a hot system, do so only with the engine running and the coolant pump operating.*

Coolant, pump and thermostat

Volkswagen uses only phosphate and silicate free anti-freeze/coolant. When supplied by Volkswagen, this coolant is known as G12 (red color) or G12+ (purple color). Due to its special characteristics the G12 or G12+ coolant should never be mixed with G11 (green coolant). See **03 Maintenance** for additional information and cautions.

For best overall performance, the coolant additive proportion must be at least 40%, but not more than 60% to maintain the antifreeze protection and cooling efficiency.

Distilled water should always be used due to the various chemicals and minerals usually associated with tap water. Under no circumstances should 100% coolant/antifreeze be used. This will generally result in overheating and poor heater operation due to the poor heat transfer characteristics of pure anti/freeze. For accurate testing of antifreeze/coolant protection, a refractometer such as VW tool CEN5021 should be used, see **Coolant/antifreeze protection, testing**. A 50/50 mixture is generally preferred and will provide freeze protection to -34°F (-37°C).

An impeller-type coolant pump is mounted into cylinder block. The pump is crankshaft-driven by the same toothed belt that drives the camshaft and circulates coolant through the system whenever the engine is running.

The thermostat controls coolant flow through the radiator. When the engine is cold, the thermostat restricts flow through the radiator to allow the engine to reach operating temperature quicker. As the engine heats up, the thermostat opens and coolant circulates through the whole system, including the radiator.

Radiator and cooling fan

A radiator cooling fan assembly provides auxiliary air flow through the radiator. The fan assembly is electrically operated and thermostatically controlled so that it runs only when extra air flow is required to maintain proper coolant temperature.

The radiator is a cross-flow design constructed of an aluminum core and plastic side tanks. A translucent expansion tank, or overflow reservoir, provides for the expansion of the coolant at higher temperatures, easy monitoring of the coolant level and a convenient location for adding coolant.

An electric radiator cooling fan assembly is controlled by a thermoswitch located in the side of the radiator. At higher coolant temperatures, the switch closes to start the cooling fans. Any time the coolant temperature rises to higher than normal levels, the fan will start and continue to run until coolant temperature returns to a normal range. On vehicles with air conditioning, rising refrigerant pressures will also activate the cooling fans.

> ***WARNING —***
> *The electric cooling fans can come on at any time, even if the key is out of the ignition. To avoid personal injury, use extreme caution when working at or near the cooling fan if the engine is hot. As a safety precaution, always disconnect the harness connector from the fan when working around the radiator, fans and associated components.*

Coolant temperature/level warning lights

The New Beetle is not equipped with a conventional coolant temperature gauge. In place of the gauge is a symbol on the instrument cluster (**arrow**) that changes color to provide information about cooling system operation.

The coolant temperature and level symbol will change color to indicate the following cooling system functions:

- Steady BLUE
  May come on when the engine is first started to indicate a cold engine.
- Flashing BLUE
  Comes on to indicate a malfunction within the electrical section of the cooling system.
- Steady RED
  Comes on when the ignition is first switched on as a functional check and self-test, flashes several times and goes out.
- Flashing RED
  Comes on when driving to indicate a low coolant level condition or an overheat situation. A warning buzzer may also sound. Pull vehicle to side of road as soon as possible and determine cause of warning light.
- Not Illuminated (no light)
  If self test is OK, no light indicates normal cooling system operation.

The coolant temperature and level warning lights have Onboard Diagnostic (OBD) capabilities included with the diagnostic capabilities of the instrument cluster. In order to access these capabilities, a scan tool such as the Volkswagen VAG 1551 or 1552 is required. Due to the specialized nature of this particular tool, its use and operation are outside the scope of this manual.

TROUBLESHOOTING

When investigating the cause of overheating or coolant loss, begin with a visual inspection of the system. Check the coolant level and inspect for evidence of coolant leaks.

The system becomes pressurized at normal operating temperatures. Leaks may prevent the system from becoming pressurized and allow the coolant to boil at a lower temperature. If visual evidence is inconclusive, a cooling system pressure test will determine whether the system leaks and may help to indicate the source.

Table a. Cooling system troubleshooting

Symptom	Probable cause	Corrective action
1. Engine overheats, red overheat warning light flashes	a. Low coolant level	a. Fill expansion tank to the MAX mark on a cold engine. Check cooling system for leaks with pressure tester.
	b. Incorrect coolant concentration	b. Adjust coolant concentration to a mix of 50% coolant additive and 50% distilled water.
	c. Poor air flow through radiator	c. Check for debris (leaves, bugs, etc.) buildup on front of radiator. Clean radiator exterior using compressed air.
	d. Radiator fan not switching on	d. Test thermoswitch and fan. Replace faulty parts.
	e. Faulty radiator cap	e. Pressure test radiator cap. Replace faulty cap.
	f. Faulty thermostat	f. Remove and test thermostat. Replace faulty thermostat.
	g. Coolant pump faulty	g. Remove and inspect for spun impeller or pulley hub.
	h. Radiator hose restricted (lower hose may collapse at high engine or highway speeds)	h. Check hoses for soft, spongy areas. Replace faulty hoses.
	i. Clogged radiator	i. Clean or replace faulty radiator.
	j. Internal engine mechanical fault	j. Check internal engine condition, see **15 Cylinder Head and Valvetrain**.
2. Blue low temperature warning light on, poor heater output	a. Incorrect coolant concentration	a. Adjust coolant concentration to a mix of 50% coolant additive and 50% distilled water.
	b. Faulty thermostat	b. Remove and test thermostat. Replace faulty thermostat.
	c. Radiator fan not switching off	c. Test thermoswitch. Replace faulty parts.
3. Blue low temperature warning light on, heater output normal	a. Faulty instrument cluster or coolant temperature sensor	a. Test instrument cluster and sensor. Replace parts as required.
4. Blue and red warning lights operate normally, poor heater output	a. Installed position of heater hoses reversed	a. Inspect routing of heater hoses, see **80 Heating and Ventilation.**
	b. Heater hose restriction	b. Inspect heater hoses for ply separation and clogging. Replace hoses as required.
	c. Heat core restricted or clogged	c. Clean or replace faulty heater core.
	d. Heater or A/C ventilation controls or flaps not operating correctly	d. Check operation and adjustment of controls, see **80 Heating and Ventilation**
5. Blue warning light flashes	a. Malfunction with gauge, Engine Coolant Temperature sensor (ECT), or wiring	a. Visually inspect ECT wiring in engine compartment. See **Cooling System, Components** for ECT location. b. Troubleshoot with suitable scan tool.

If the cooling system holds pressure, the most probable causes of overheating are an electrical fault with the cooling fans, a faulty thermostat, or poor coolant circulation due to restrictions in the system.

Table a lists overheating and underheating symptoms, their probable causes, and suggested corrective actions. The bold type refers to areas outside of this repair group for suggested repairs.

NOTE —

- *Coolant also circulates through the heater core in the passenger compartment. For problems associated with the heater core and the heating system, see* **80 Heating and Ventilation** *or* **87 Air Conditioning**.
- *Overheating problems may also be caused by an engine fault that leaks hot combustion gases into the cooling system. See* **13 Crankshaft and Cylinder Block** *for additional information concerning the cylinder block.*

Diagnostic checks

The following checks and tests should generally be performed prior to any major tear down or repair work. This will help insure that the full extent or nature of the problem is known beforehand and may prevent unneeded repair work.

Coolant/antifreeze concentration, testing

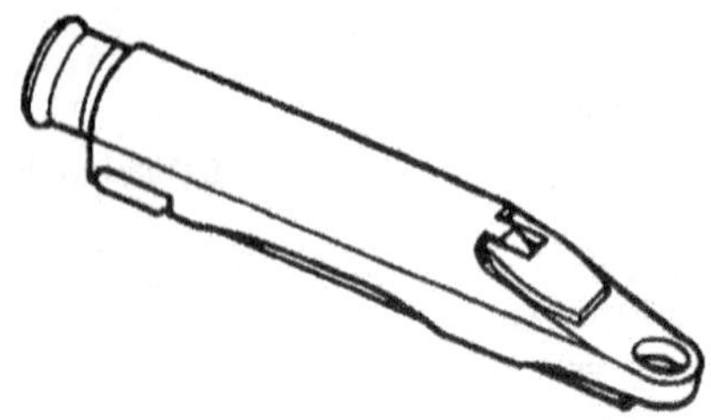

19-A093

< Using a refractometer (VW tool CEN5021), take a small amount of coolant and place it onto measuring prism.

- Close the cover and hold tester up to a light source.
- Look into the eyepiece and read the scale for ethylene glycol. The measure point is where the black and white come together. A 50/50 mixture is generally preferred and will provide freeze protection to -34°F (-37°C). See **Coolant, pump and thermostat** given earlier.
- Rinse the tester clean.
- Periodically check calibration by placing distilled water onto the prism and closing the cover. The measure point should be at 32°F (0°C) which is the freezing point for pure water. If this reading is not obtained, follow the instructions that came with tester for calibration and adjustment.

Cooling system, pressure testing

A pressure test uses a special tester in place of the cap on the expansion tank to pressurize the system and simulate normal operating conditions. If the system is unable to hold pressure, the engine will overheat more easily and fluid will be lost.

NOTE —
A pressure test also checks for internal leakage. Some of the common sources of internal leakage include a faulty cylinder head gasket, a cracked cylinder head, and a cracked cylinder block.

- With engine at normal operating temperature, pressure test the system using a cooling system pressure tester with suitable adapters.
- Observe the gauge reading. If the pressure drops, there is a leak in the system. A rapid drop may indicate a faulty cylinder head gasket if no external leaks can be found. Remove the spark plugs or glow plugs to inspect for coolant in the cylinders.

– Using the correct tester adapter, test coolant expansion tank cap opening pressure. Compare opening pressure to specification and replace the cap if opening pressure is outside of specification.

NOTE —
It is not unusual for opening pressure to be slightly above specification when pressure is first applied due to slight sticking of the internal valve. Subsequent pressure applications should open the valve within specification, otherwise, replace the cap.

Specification
- Cooling system test pressures
  testing pressure (max) . . . . 1.25 bar (18 psi)
  Cap opening pressure . . . . 1.4 - 1.6 bar (20 - 23 psi)

– Inspect gasket in expansion tank cap for cuts and damage.

Cooling fans and thermoswitch, testing

New Beetles use a pair of 2-speed electric cooling fans to pull air across the radiator (and A/C condenser where equipped). A thermo-switch and a fan control module (**B**) are used to control fan operation. The engine will overheat if the electric cooling fans are not operating properly, especially in traffic where speeds are low and there is insufficient natural air flow across the radiator. A/C operation will also be affected and damage to the system could result. In normal operation, the fans will switch on and off according to engine coolant temperature at the radiator thermo-switch or whenever the A/C is on.

NOTE —
*Radiator drain is shown at (**C**) and fog light harness connector is shown at (**A**).*

WARNING —
Use extreme care when working at or near the radiator cooling fans when the engine is hot. The fan can come on at any time, even if the ignition key is switched off.

NOTE —
If the cooling fans fail to operate normally use the following procedure to help determine where the problem lies.

– Check fuses #3 and #8 in the holder on battery cover and fuse #16 in the dash fuse box. Replace as necessary. See **EWD Electrical Wiring Diagrams** for wiring diagram circuit information.

– Disconnect harness connector from radiator fan thermo-switch on lower left side of radiator.

– Using a fused jumper wire, bridge the red wire and red/white wire. Both fans should run on the first speed.

– Remove jumper wire from the red/white wire and move it to the red/yellow wire so that the red wire and red/yellow wire are jumped. Both fans should run on the second speed.

– If the fans run with the jumpers installed, This indicates the fans are functionng properly and the radiator thermo-switch may be faulty. Switch opening and closing specifications are given below for more accurate testing of the thermo-switch. Drain coolant before replacing a faulty switch.

Tightening torque
- Thermo-switch to radiator . . . . 35 Nm (26 ft-lb)

– If the fans do not run, check for battery voltage at the red wire in the harness connector. No voltage at harness connector indicates a blown fuse or a break in wiring from the fuse. See **EWD Electrical Wiring Diagrams**.

– If the fans do not run and voltage is present at the thermo-switch harness connector, reconnect jumper to the red wire and red/white wire.

– Unplug left side fan motor and plug a suitable test light into terminals 1 and 2 of harness connector. The light should light.

– Repeat for right side fan motor. The light should light.

– If the light does not come on for steps 8 and 9 check for broken or shorted wiring between the fan control module and the thermo-switch or between the fan control module and the fans. See **EWD Electrical Wiring Diagrams**. If the wiring is in order, then the fan control module is faulty. If the test light comes on, this indicates that the radiator fan is faulty and should be replaced.

The above tests check first speed fan operation. The second speed can be checked by connecting the jumper to the red wire and red/yellow wire at the thermo-switch harness and a test light to terminals 1 and 3 of the fan harness connectors. As with the first speed, if the light does not come on for steps 8 and 9, check for broken or shorted wiring between the fan control module and the thermo-switch or between the fan control module and the fans. See **EWD Electrical Wiring Diagrams**. If the wiring is in order, than the fan control module is faulty. If the test light comes on, this indicates that the radiator fan is faulty and should be replaced.

Specification

- Radiator fan thermo-switch operating temperatures
  Stage I
  switch on. . . . . .197° - 206°F (92° - 97°C)
  switch off. . . . . .183° - 195°F (84° - 91°C)
  Stage II
  switch on. . . . . .210° - 221°F (99° - 105°C)
  switch off. . . . . .195° - 208°F (91° - 98°C)

Engine coolant temperature sensor, testing

The engine coolant temperature (ECT) sensor supplies a signal to the engine control module (ECM) for coolant temperature. Failure of the sensor can result in poor running symptoms. The sensor can be tested by reading the resistance across the specified terminals. The ECT also supplies a signal to the coolant temperature gauge in the instrument cluster.

< ECT sensor (**1**) is located on coolant flange, to left of cylinder head. Remove harness connector from sensor.

- Later style sensor shown on 2000 m.y. vehicle.

– Use digital multimeter to read resistance (Ω) across sensor terminals specified below.

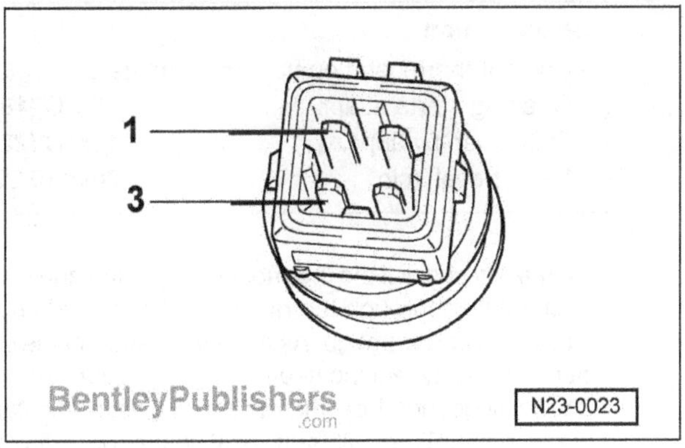

< Early style ECT sensor used on 1998-1999 vehicles.

- Measure resistance at terminals **1** and **3** of sensor.

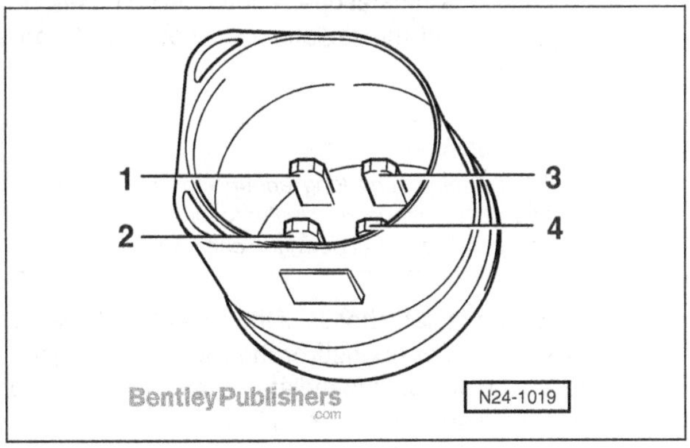

< Later style ECT sensor used on 2000-on vehicles.

- Measure resistance at terminals **3** and **4** of sensor.

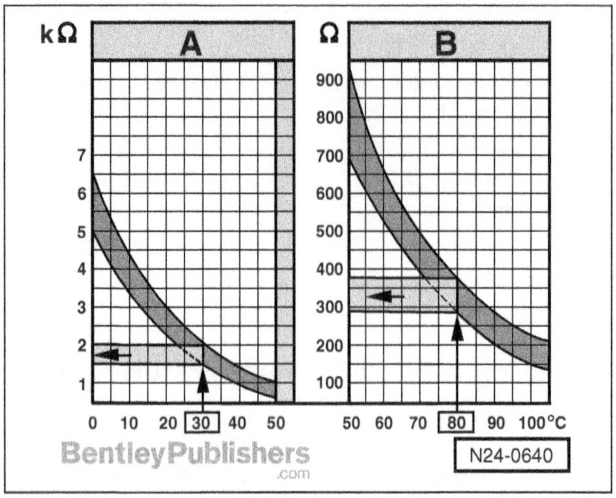

– Measure temperature of engine coolant using diagnostic scan tool or thermometer.

Compare ECT sensor resistance and coolant temperature readings to graph.

- Scale **A** shows resistance for coolant temperature range 0 - 50°C (32 - 122°F).
- Scale **B** shows resistance for coolant temperature range 50 - 100°C (122 - 212°F).
- Example:
  30°C corresponds to 1500 - 2000 Ω
  80°C corresponds to 275 - 375 Ω

Thermostat, testing

A thermostat that is stuck open will cause the engine to warm up slowly and generally run below normal operating temperature at highway speed. A thermostat that is stuck closed will restrict coolant flow to the radiator and cause overheating.

Accurate testing of the thermostat requires removal and observation while being warmed in a container of hot water. An accurate thermometer is used to note the temperatures at which it begins to open and at which it is fully open.

Specification

- Coolant thermostat operating parameters
  Opening starts at approx.. . . . . 85°C (185°F)
  Fully open by approx.. . . . . 105°C (221°F)
  Valve travel, min. . . . . 7mm (0.28 in.)

To quickly check if the thermostat is opening and if coolant is circulating through the radiator, allow a cold engine to run at idle and warm up. As the temperature of the coolant rises, carefully feel the heater and expansion tank hoses. They will get hot. Feel the radiator hoses, particularly the lower hose. They will stay relatively cool to the touch. When the thermostat opens, the radiator hoses will quickly get hot and the radiator fan should cycle on shortly thereafter.

CAUTION—
If the engine runs long enough for the warning light in the instrument cluster to indicate an overheat condition, shut down the engine immediately.

Check the radiator hoses. If they are not hot, then the thermostat has not opened or the radiator is clogged. If the hoses are hot, but the radiator (cooling) fans did not come on, troubleshoot the cooling fans as per instructions given earlier in this section.

COOLING SYSTEM, DRAINING AND FILLING

> ***CAUTION—***
> *Always use a mixture of genuine Volkswagen coolant/anti-freeze and distilled water to avoid the formation of harmful deposits in the cooling system. Use of coolant/anti-freeze with phosphate compounds or tap water can be harmful to the cooling system.*

When refilling the engine coolant, keep in mind the following points:

- Use only Volkswagen G12, G12+, or equivalent phosphate and silicate free coolant/anti-freeze identified by its red or purple color.
- Under no circumstances should G12/G12+ be mixed with any other type of coolant/anti-freeze. If the fluid in the expansion tank is brown, mixing with other types has occurred and the system must be flushed and coolant replaced.
- Under no circumstances should straight coolant/anti-freeze or straight water be used. A 50/50 mix of coolant/anti-freeze to water provides freeze protection to -34°F (-35°C) and boiling protection to 226°F (108°C). The coolant/anti-freeze percentage should always be a minimum of 40% and a maximum of 60%.
- Hoses are secured with spring clamps. If replacement is needed, always replace with spring clamps due to their ability to expand with engine heat and remain tight.
- Most hoses have alignment marks to ensure correct positioning on the appropriate fitting; always line up the marks to ensure stress free installation.

– Remove lower sound absorber panel (belly pan).

– With engine fully cold, remove cap from the coolant expansion tank.

– Position a drain pan under left front part of the radiator.

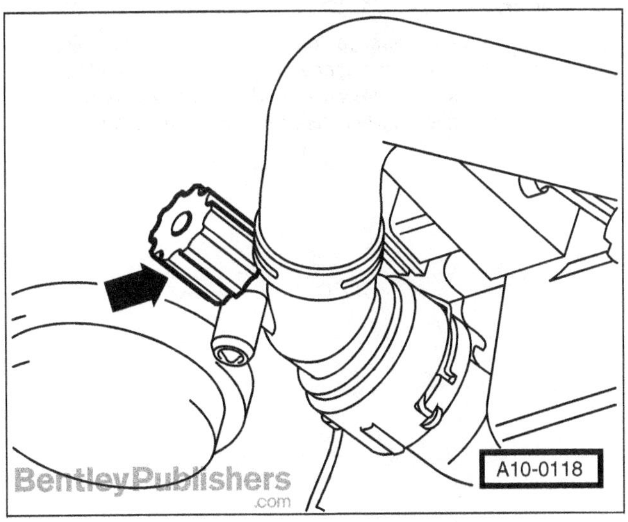

◄ Unscrew drain plug (**arrow**) at lower radiator hose fitting and allow coolant to drain.

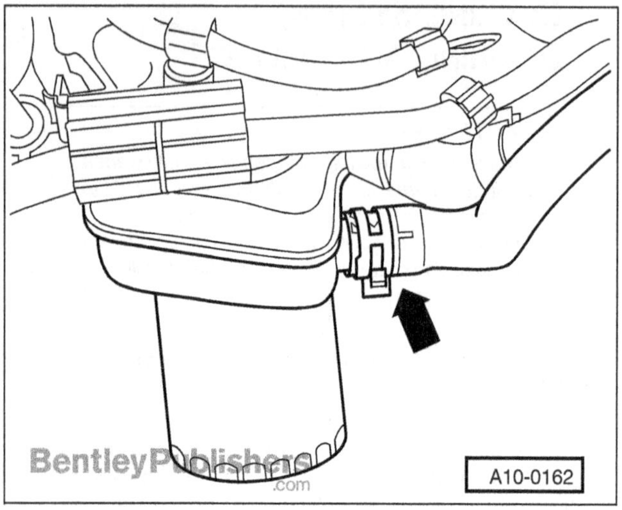

Remove lower hose (**arrow**) from oil cooler to drain cylinder block and allow coolant to drain.

> ***NOTE —***
> *Dispose of used coolant properly.*

- Close drain plug on lower radiator hose fitting.
- Reconnect the lower oil cooler hose.
- Slowly fill expansion tank with the appropriate mixture of coolant/anti-freeze while allowing the air to escape.

Specification

- Cooling system capacity

1.8L / 2.0L engine	5.3 quarts (5.0 liters)
1.9L engine	6.3 quarts (6.0 liters)
2.5L engine	9.5 quarts (9.0 liters)

> ***NOTE —***
> *If a cooling system pressure tester is available, connect it to the expansion tank and pressurize the system. This will force the coolant/anti-freeze past the thermostat and into many of the hoses that drained and will speed up the entire process.*

- Install expansion tank cap, start engine and allow to run until the radiator fans cycle at least once.

> ***NOTE —***
> *If the expansion tank empties completely during warm-up, shut off the engine, carefully remove the expansion tank cap and refill to the **max** marks on the side. Restart the engine and continue until the fans cycle.*

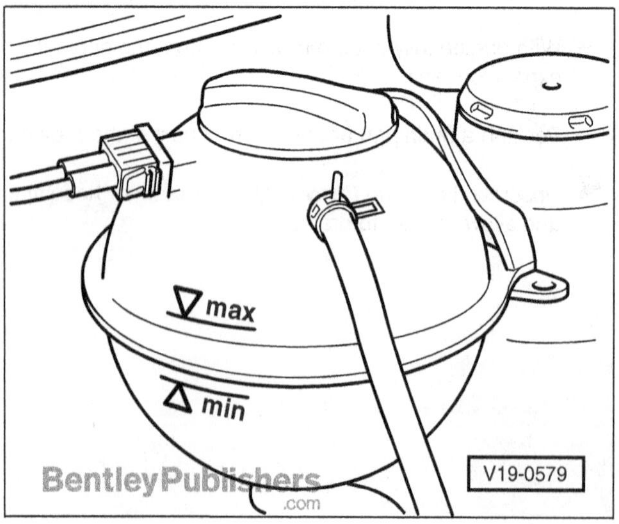

Recheck the level in the expansion tank when the engine has cooled.

> ***NOTE —***
> *The coolant/anti-freeze level should be at the **max** mark with the engine at operating temperature and between the **min** and **max** marks when cold. The final level is best checked when the cooling system is fully cold.*

COOLING SYSTEM, COMPONENTS (4-CYLINDER)

> ***NOTE —***
> *Volkswagen identifies electrical components by a letter and/or a number in the electrical schematics. See* **EWD Electrical Wiring Diagrams.** *These electrical identifiers are listed in parenthesis as an aid to electrical troubleshooting.*

Cooling system, components (1.8L)

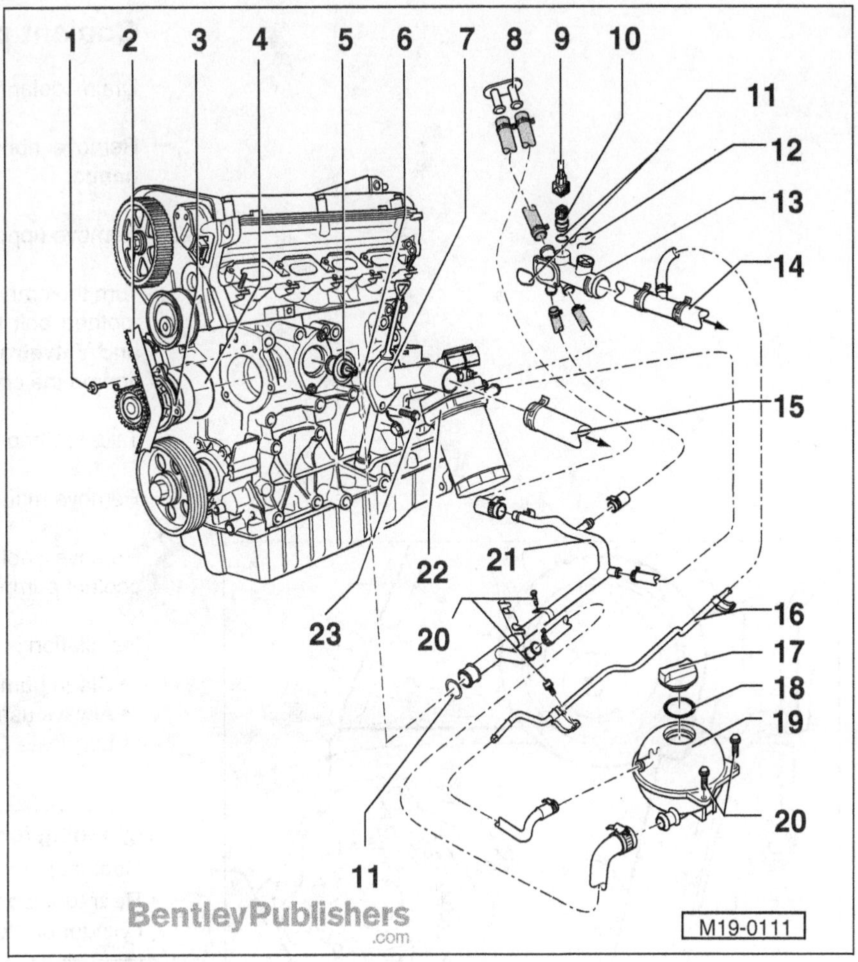

1. **Bolt**
   - Tighten to 15 Nm (11 ft-lb)
2. **Toothed belt**
3. **Coolant pump**
   - Check pulley for ease of movement
   - If damaged or leaking replace complete assembly
4. **O-ring**
   - Always replace
5. **Thermostat**
   - Checking: heat-up thermostat in water
   - Opening starts at approx. 189°F (87°C) and ends at approx. 216°F (102°C)
   - Opening lift: 7 mm minimum
6. **O-ring**
   - Always replace
7. **Connection**
8. **To heat exchanger**
9. **4-pin harness connector**
   - Square style through 1999 m.y.
   - Round style from 2000 m.y.
10. **Engine coolant temperature sensor (G62)**
    - For engine control module
    - With engine coolant temperature warning light sensor (G2)
    - If necessary, release cooling system pressure before removing
    - See **Engine coolant temperature sensor, testing** earlier in this repair group
11. **O-ring**
    - Always replace
12. **Retaining clip**
    - Ensure clip is securely seated
13. **Connection**
14. **Upper coolant hose**
15. **Lower coolant hose**
16. **Upper coolant pipe**
    - Secured to intake manifold
17. **Expansion tank cap**
    - Test pressure: 1.4-1.6 bar (20.3-23.2 psi)
18. **O-ring**
    - Replace if damaged
19. **Expansion tank**
20. **Bolt**
    - Tighten to 20 Nm (15 ft-lb)
21. **Coolant pipe**
22. **Oil cooler**
23. **Bolt**
    - Tighten to 15 Nm (11 ft-lb)

Coolant pump, replacing (1.8L)

- Drain coolant as previously described.
- Remove ribbed V-belt and tensioner, see **03 Maintenance**.
- Remove upper and center toothed belt guards.
- Turn the crankshaft to TDC for cylinder 1 and release the toothed belt tensioning roller, see **15a Cylinder Head and Valvetrain**. The toothed belt should be left in position on the crankshaft sprocket.
- Take toothed belt off coolant pump drive sprocket.
- Remove mounting bolts from rear toothed belt guard.

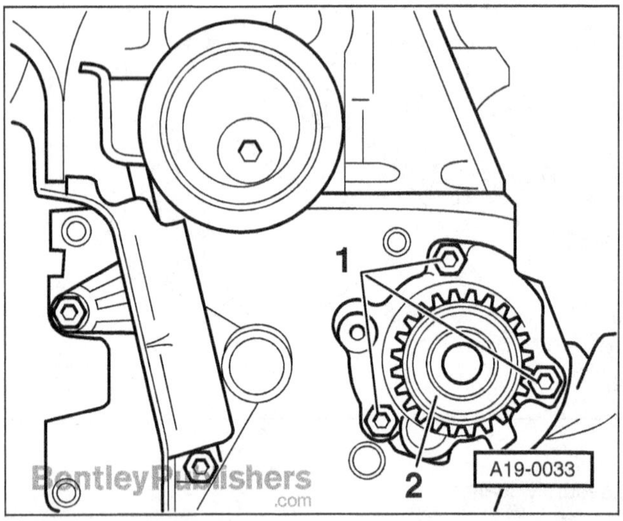

◄ Remove coolant pump mounting bolts (**1**) and remove coolant pump (**2**).

- Installation is reverse of removal, noting the following:
  - Clean pump mating surface on cylinder block.
  - Always use new O-rings and hardware as indicated.
  - Lubricate O-rings with coolant before installing.

Tightening torques

- Coolant pump to cylinder block ..... 15 Nm (11 ft-lb)
- Rear toothed belt guard to cylinder block .... 20 Nm (15 ft-lb)

- Fill cooling system as described previously.
- Start engine and check for leaks. Be sure to quality check your work.

Cooling system, components (1.9L, code: ALH)

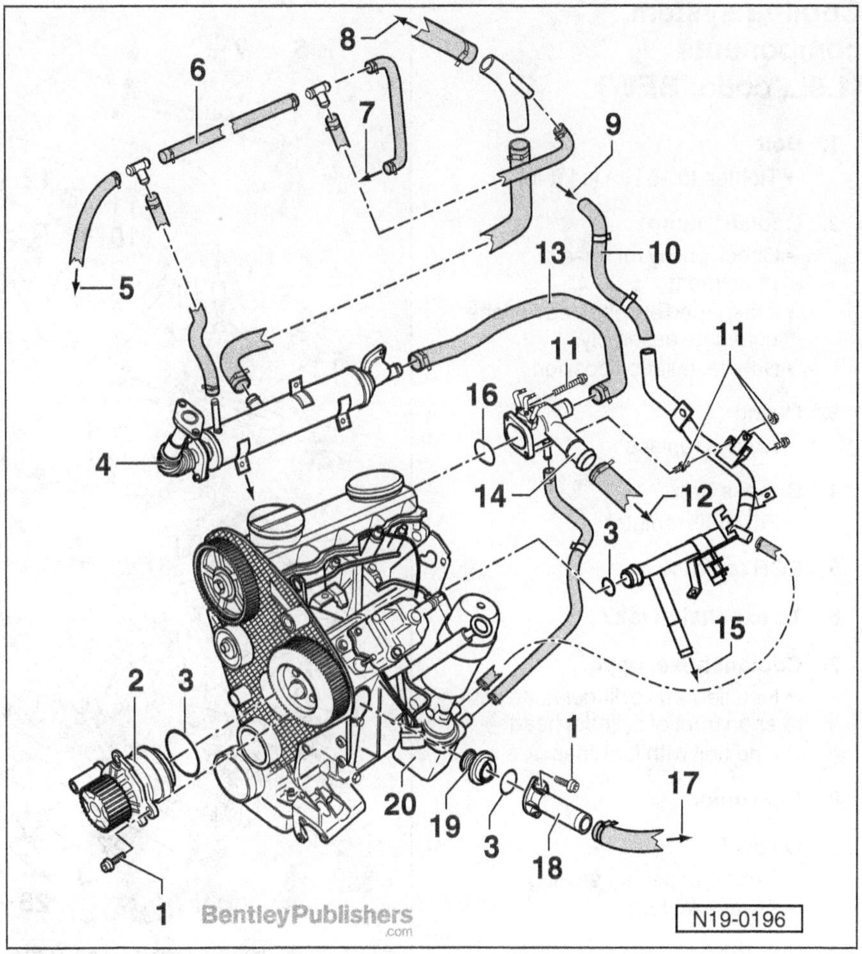

1. **Bolt**
   - Tighten to 25 Nm (18 ft-lb)
2. **Coolant pump**
   - Check pulley for ease of movement
   - If damaged and leaking replace complete assembly
   - Note installation position
3. **O-ring**
   - Always replace
4. **EGR cooler**
5. **To expansion tank**
6. **Coolant hose, upper**
7. **To cylinder head**
8. **To heater core**
9. **From heater core**
10. **Coolant pipe**
11. **Bolt**
    - Tighten to 10 Nm (7 ft-lb)
12. **To expansion tank, lower**
13. **EGR cooler supply**
14. **Flange**
    - With glow plugs for auxiliary heater
15. **To top of radiator**
    - Clipped to the firewall
16. **O-ring**
    - Ensure it is seated securely
    - Always replace
17. **To bottom of radiator**
18. **Connection**
    - For thermostat
19. **Coolant thermostat**
    - Checking: heat-up thermostat in water
    - Opening starts at approx. 185°F (85°C)
    - Ends approx. 221°F (105°C)
    - Opening lift: 7 mm minimum
20. **Oil cooler**

Cooling system, components (1.9L, code: BEW)

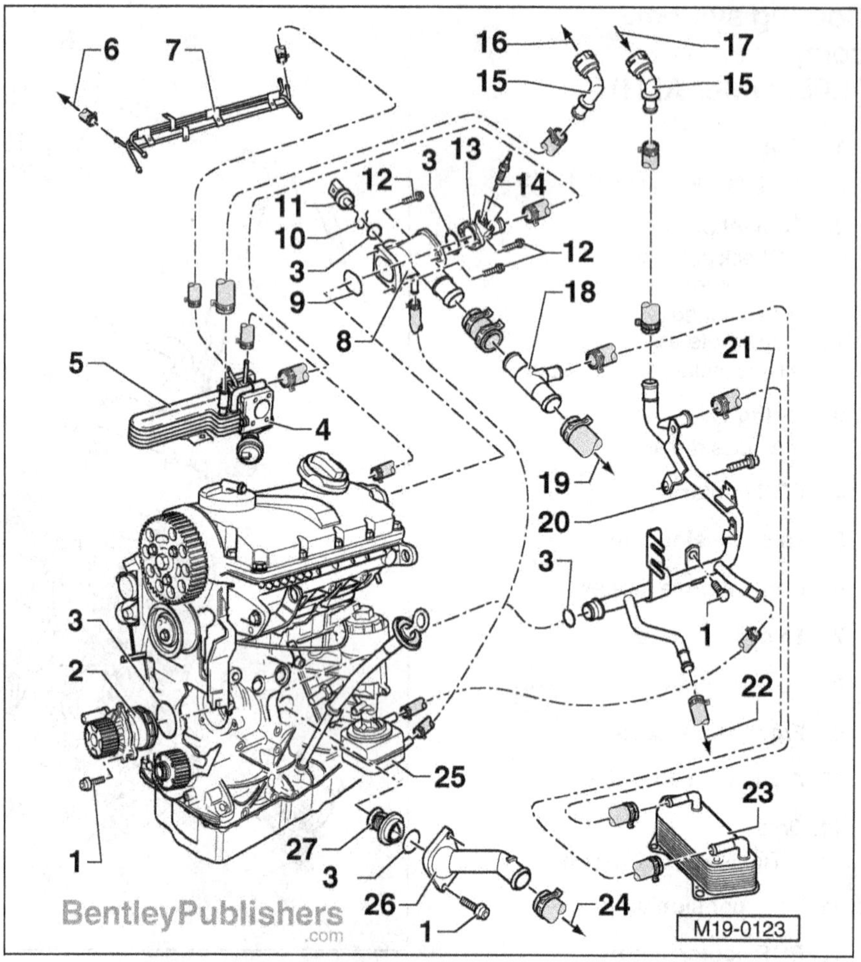

1. **Bolt**
   - Tighten to 15 Nm (11 ft-lb)
2. **Coolant pump**
   - Check pulley for ease of movement
   - If damaged and leaking replace complete assembly
   - Note installation position
3. **O-ring**
   - Always replace
4. **Bypass flap**
   - For EGR radiator
5. **EGR radiator**
6. **To expansion tank**
7. **Coolant hose, upper**
   - Installed with cylinder head cover on front of cylinder head
   - One unit with fuel lines
8. **Pipe union**
9. **Oil seal**
   - Check for secure seating
   - Always replace
10. **Securing clip**
    - Check for secure seating
11. **Engine coolant temperature (ECT) sensor (G62)**
    - Also has ECT sensor for coolant gauge (G2)
12. **Bolts**
    - Tighten to 10 Nm (7 ft-lb)
13. **Heating flange**
14. **Glow plugs**
    - Tighten to 25 Nm (18 ft-lb)
    - For auxiliary heater
15. **Quick disconnect coupler**
16. **To heater core**
17. **From heater core**
18. **Distributor**
19. **To top of radiator**
20. **Coolant pipe**
21. **Bolt**
    - Tighten to 40 Nm (30 ft-lb)
22. **To bottom of expansion tank**
23. **Transmission oil cooler**
24. **To bottom of radiator**
25. **Oil cooler**
26. **Pipe union**
27. **Coolant thermostat**
    - Checking: heat-up thermostat in water
    - Opening starts at approx. 185°F (85°C)
    - Ends approx. 221°F (105°C)
    - Opening lift: 7 mm minimum

Coolant pump, replacing (1.9L)

– Drain coolant as previously described.

– Remove ribbed V-belt and tensioner, see **03 Maintenance**.

– Disconnect fuel supply and return lines at fuel filter.

– Remove connecting pipe between intercooler and intake manifold.

– Remove fuel filter and mounting bracket.

– Remove upper and center toothed belt guards.

– Turn crankshaft to TDC for cylinder 1 and release toothed belt tensioning roller, see **15b Cylinder Head and Valvetrain**. Remove toothed belt from camshaft, injection pump and coolant pump sprockets, but leave it in position on crankshaft sprocket.

– ALH engines: Remove bolt from idler pulley and work the pulley free from cylinder block. Discard bolt.

– Remove coolant pump mounting bolts and carefully lift coolant pump from cylinder block between engine mount support and rear toothed belt guard.

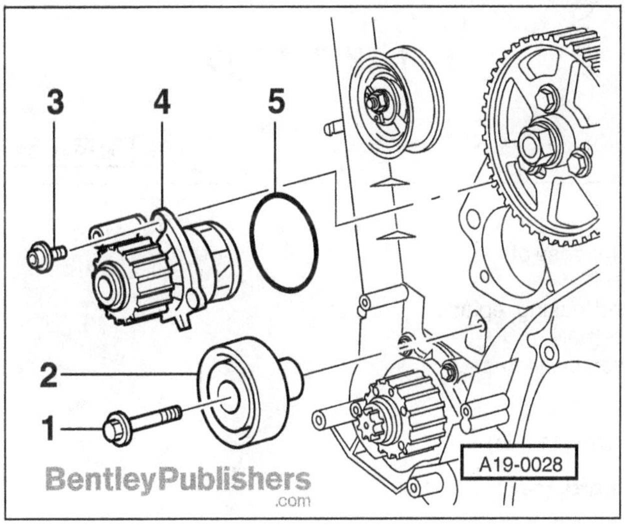

Coolant pump installation position

1. **Bolt**
   • Tighten to 40 Nm (30 ft-lb) plus ¼ turn (90°)
   • Always replace
2. **Idler pulley**
   • ALH engine only
3. **Bolt**
   • Tighten to 15 Nm (11 ft-lb)
4. **Coolant pump**
   • Installation position: plug in housing points downward
5. **O-ring seal**

– Installation is reverse of removal, noting the following:
- Clean pump mating surface on cylinder block.
- Always use new O-rings and hardware as indicated.
- Lubricate O-rings with coolant before installing.
- Fill cooling system as described previously.
- Start engine and check for leaks Be sure to quality check your work.

Tightening torques
- Coolant pump to cylinder block (M7) 15 Nm (11 ft-lb)
- Idler pulley bolt. . . . . (M10)(always replace)
  stage I . . . . 40 Nm (30 ft-lb)
  stage I . . . . additional ¼ turn (90°)
- Rear toothed belt guard to
  cylinder block (M8). . . . . 20 Nm (15 ft-lb)

Cooling system, components (2.0L)

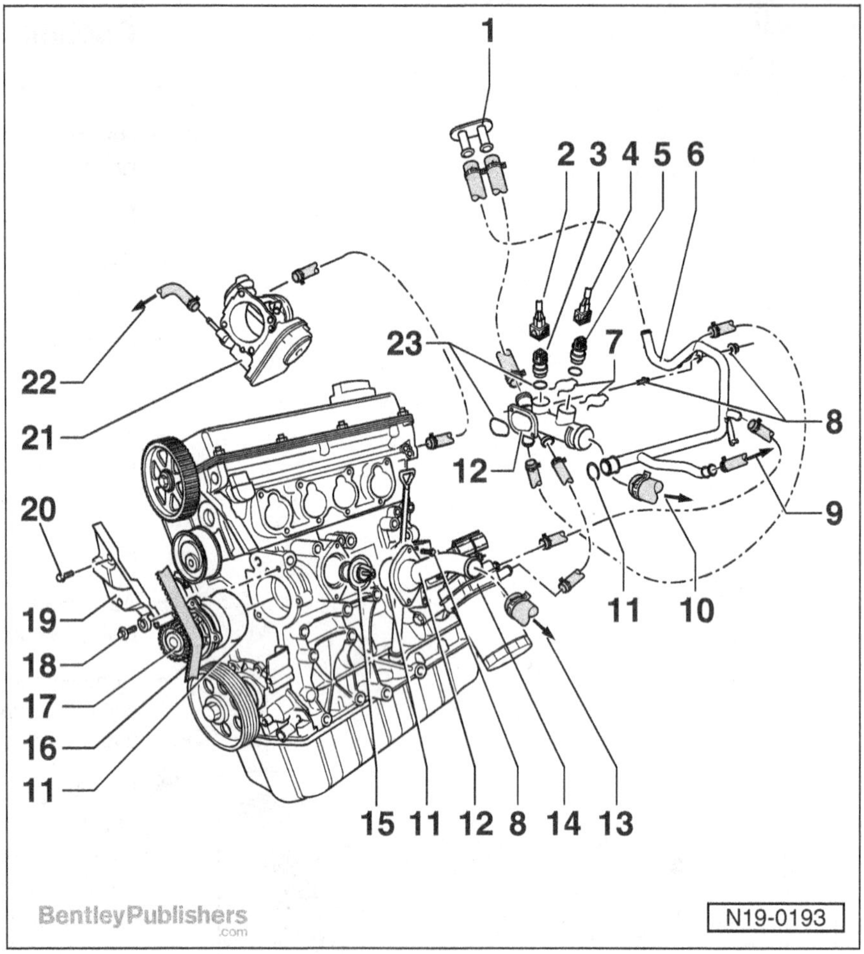

1. **Heater core**
   - For passenger compartment heat
2. **4-pin harness connector**
   - Square style through 1999 m.y.
   - Round style from 2000 m.y.
3. **Engine coolant temperature sensor (G62)**
   - For engine control module
   - With engine coolant temperature warning light sensor (G2)
   - If necessary, release cooling system pressure before removing
   - See **Engine coolant temperature sensor, testing** earlier in this repair group
4. **2-pin harness connector**
   - For AEG engine with A/C only
5. **A/C cut-out thermal switch (F163)**
   - For AEG engine with A/C only
6. **Coolant pipe**
7. **Retaining clip**
   - Ensure clip is securely seated
8. **Stepped stud and nut**
   - Tighten to 10 Nm (7 ft-lb)
9. **To expansion tank**
10. **To top of radiator**
11. **O-ring**
    - Replace
12. **Flange**
13. **To bottom of radiator**
14. **Oil cooler**
15. **Engine coolant thermostat**
    - Checking: heat-up thermostat in water
    - Opening starts at approx. 187°F (86°C)
    - Opening lift: 7 mm minimum
16. **Toothed belt**
    - Mark engine direction of rotation before removing
    - Check for wear
    - Do not kink
17. **Coolant pump**
    - Check pulley for ease of movement
    - If damaged and leaking replace complete assembly
    - Note installation position
18. **Bolt**
    - Tighten to 15 Nm (11 ft-lb)
19. **Toothed belt guard, rear**
20. **Bolt**
    - 20 Nm (15 ft-lb)
21. **Throttle control module**
    - Heated by engine coolant
22. **To expansion tank, upper**
23. **O-ring**
    - Replace if damaged

Coolant pump, replacing (2.0L)

– Drain coolant as previously described.

– Remove ribbed V-belt and tensioner, see **03 Maintenance**.

– Remove upper and center toothed belt guards.

– Turn the crankshaft to TDC for cylinder 1 and release the toothed belt tensioning roller, see **15c Cylinder Head and Valvetrain**. The toothed belt should be left in position on the crankshaft sprocket.

– Take toothed belt off coolant pump drive sprocket.

– Remove mounting bolts from rear toothed belt guard.

– Remove coolant pump mounting bolts and remove coolant pump.

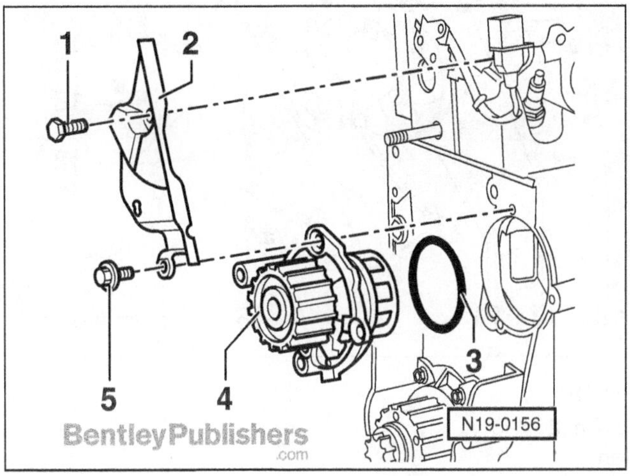

Coolant pump installation position

1. Bolt
- Tighten to 20 Nm (15 ft-lb)

2. Rear toothed belt guard

3. O-ring seal

4. Coolant pump
- Installed position: plug in housing points downward

5. Bolt
- Tighten to 15 Nm (11 ft-lb)

– Installation is reverse of removal, noting the following:
- Clean pump mating surface on cylinder block.
- Always use new O-rings and hardware as indicated.
- Lubricate O-rings with coolant before installing.

Tightening torque
- Coolant pump to cylinder block . . . . 15 Nm (11 ft-lb)
- Rear toothed belt guard to cylinder block . . . . 20 Nm (15 ft-lb)

– Fill cooling system as described previously.

– Start engine and check for leaks. Be sure to quality check your work.

Radiator and cooling fans, assembly (1.8L)

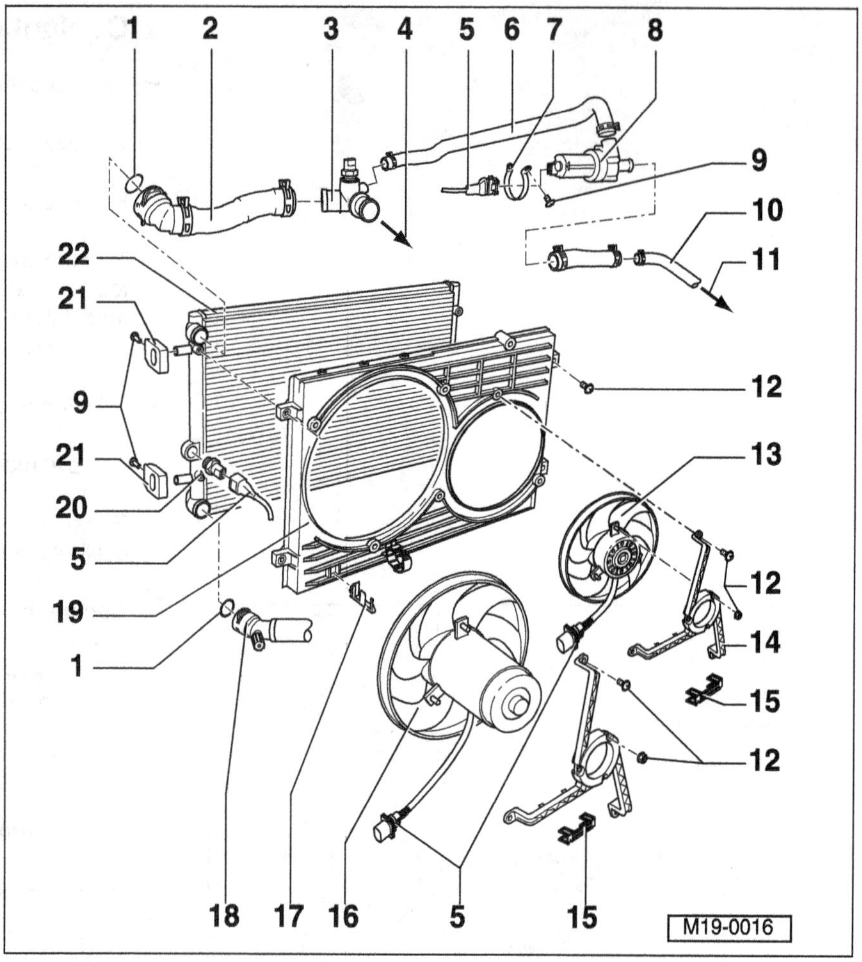

1. **O-ring**
   - Replace if damaged
2. **Coolant hose, upper**
3. **Junction piece**
   - With A/C cut-out thermal switch (F163)
4. **To engine coolant temperature (ECT) sensor (G62)**
   - See **Engine coolant temperature sensor, testing** earlier in this repair group
5. **Connector**
6. **Upper coolant hose**
   - Secured to radiator with retaining clip
7. **Bracket**
   - Secured at intake air boot
8. **After-run coolant pump**
   - Continues to run up to 10 minutes after ignition is switched off
9. **Bolt**
   - Tighten to 15 Nm (11 ft-lb)
10. **Coolant pipe**
    - Secured with retaining clamps at right long member
11. **To turbocharger**
12. **Screw**
    - Tighten to 10 Nm (7 ft-lb)
13. **Additional fan**
14. **Fan ring**
15. **Retaining clip**
16. **Radiator fan**
17. **Bracket**
    - For radiator fan connector
18. **Coolant hose, lower**
    - Attached to radiator with retaining clip
19. **Air shroud**
20. **Thermoswitch (F18)**
    - Tighten to 35 Nm (26 ft-lb)
    - For electric fan
    - Stage 1 switching temperatures:
      ON: 198 to 207°F (92 to 97°C)
      OFF: 183 to 196°F (84 to 91°C)
    - Stage 2 switching temperatures:
      ON: 210 to 221°F (99 to 105°C)
      OFF: 196 to 208°F (91 to 98°C)
21. **Bracket**
    - For radiator
22. **Radiator**
    - If replaced, flush cooling system and use new coolant

Radiator and cooling fans, assembly (1.9L / 2.0L)

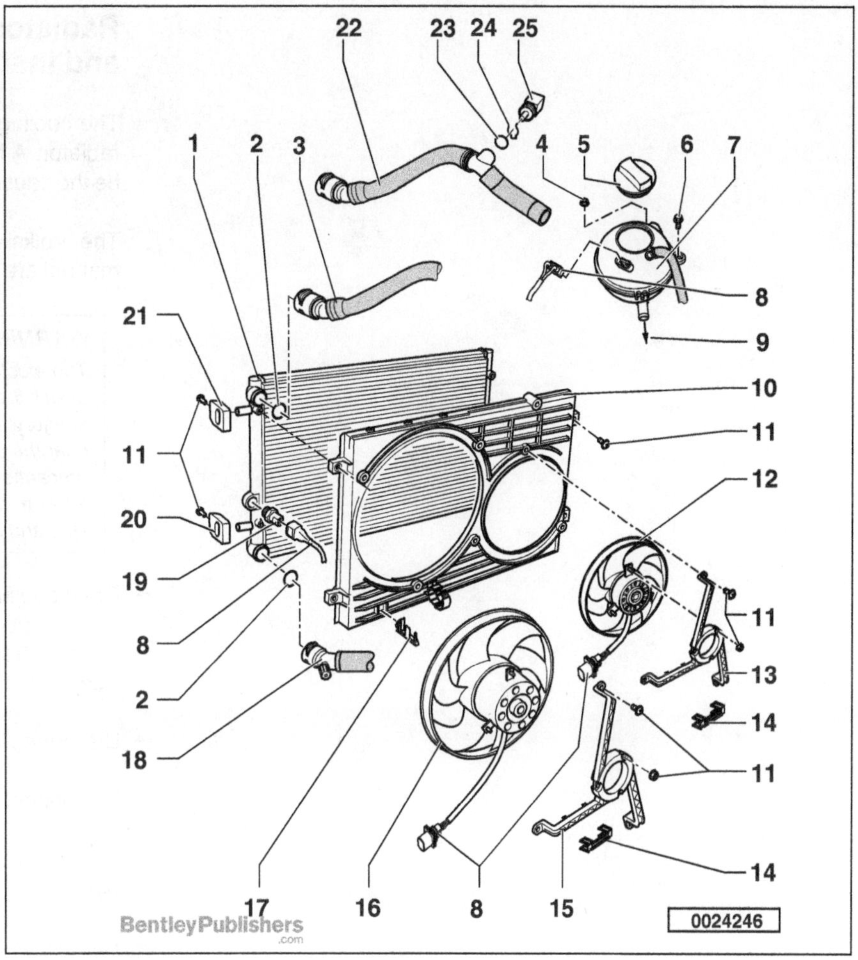

1. **Radiator**
   - If replaced, flush cooling system and use new coolant
2. **O-ring**
   - Always replace
3. **Coolant hose, upper**
   - Attached to radiator with retaining clip
   - 2.0L engine only
4. **Nut**
   - Tighten to 4 Nm (35 in-lb)
5. **Cap**
6. **Bolt**
   - Tighten to 2 Nm (18 in-lb)
7. **Expansion tank**
8. **Connector**
9. **To coolant hose**
10. **Air shroud**
11. **Bolt**
    - Tighten to 10 Nm (7 ft-lb)
12. **Additional fan**
13. **Fan ring**
14. **Retaining clip**
    - Ensure it is seated securely
15. **Fan ring**
16. **Radiator fan**
17. **Bracket**
    - For radiator fan connector
18. **Coolant hose, lower**
    - Attached to radiator with retaining clip
19. **Thermoswitch (F18)**
    - Tighten to 35 Nm (26 ft-lb)
    - For electric fan
    - Stage 1 switching temperatures:
      ON: 198 to 207°F (92 to 97°C)
      OFF: 183 to 196°F (84 to 91°C)
    - Stage 2 switching temperatures:
      ON: 210 to 221°F (99 to 105°C)
      OFF: 196 to 208°F (91 to 98°C)
20. **Bracket**
    - For radiator, lower
21. **Bracket**
    - For radiator, upper
    - Note installation position
22. **Coolant hose, upper**
    - Attached to radiator with retaining clip
    - 1.9L engine only
23. **O-ring**
    - Always replace
    - 1.9L engine only
24. **Retaining clip**
    - Ensure it is seated securely
    - 1.9L engine only
25. **Engine coolant temperature sensor (G62)**
    - 1.9L engine shown
    - 2.0L ECT located on coolant hose flange bolted to engine block
    - See **Engine coolant temperature sensor, testing** earlier in this repair group

Radiator and cooling fans, removing and installing (4-cylinder)

The cooling fans provide additional air flow through the radiator. A faulty cooling fan motor or thermoswitch may be the cause of insufficient air flow and overheating.

The Volkswagen New Beetle models covered by this manual are equipped with dual two-speed radiator fans.

> ***WARNING —***
> *The electric cooling fans can come on at any time, even if the key is out of the ignition. To avoid personal injury, use extreme caution when working at or near the cooling fan if the engine is hot. As a safety precaution, always disconnect the harness connector from the fan when working around the radiator, fans and associated components.*

- Drain coolant as previously described.
- Remove front body section and move lock carrier to service position, see **50 Body-Front**.
- Disconnect coolant hoses from radiator.
- Disconnect thermo-switch harness connector.
- Remove ribbed V-belt and tensioner, see **03 Maintenance**.
- Remove radiator fans from the ducting.
- Remove retaining clamps from A/C hoses near condenser.

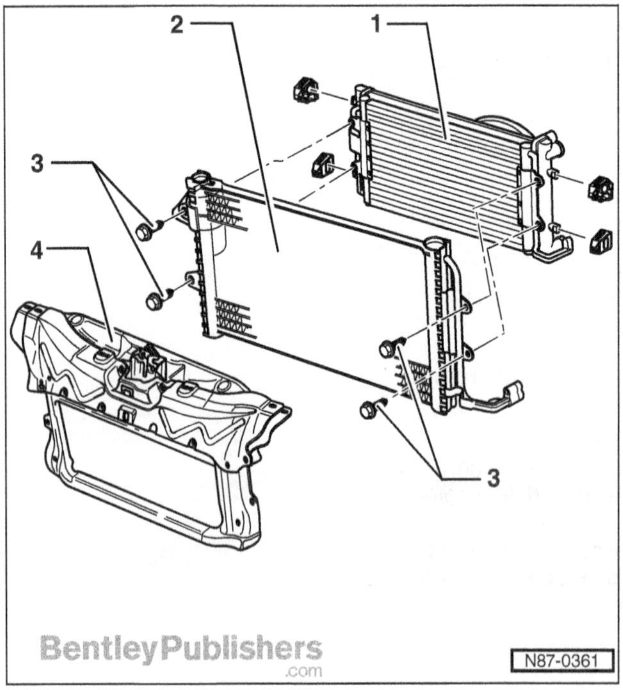

< Remove radiator mounting bolts (**3**) from both sides and slide radiator slightly to rear.

- Unbolt A/C condenser from radiator, support condenser and remove radiator.

> ***NOTE —***
> *Do not discharge the A/C refrigerant. Support the A/C condenser with wire and avoid stretching or kinking the hoses.*

- Installation is the reverse of removal.

Tightening torque
- A/C condenser to radiator .... 8 Nm 71 in-lb)

- Fill the cooling system as described previously. Start engine and check for leaks. Be sure to quality check your work.

Thermostat, replacing (4-cylinder)

A thermostat mounted in the cylinder block provides primary control of coolant temperature by regulating the flow of coolant to the radiator. The thermostat remains closed when coolant is below the specified temperature allowing the engine to reach operating temperature quickly and providing immediate heat to the passenger compartment. Thermostats for the 1.8L / 2.0L engine and the 1.9L engine operate within the same temperature ranges, but are not interchangeable.

– Drain coolant as previously given.

– Disconnect lower radiator hose from flange on cylinder block.

– Remove the outlet flange mounting bolts and remove the flange and O-ring seal.

– Remove the thermostat:

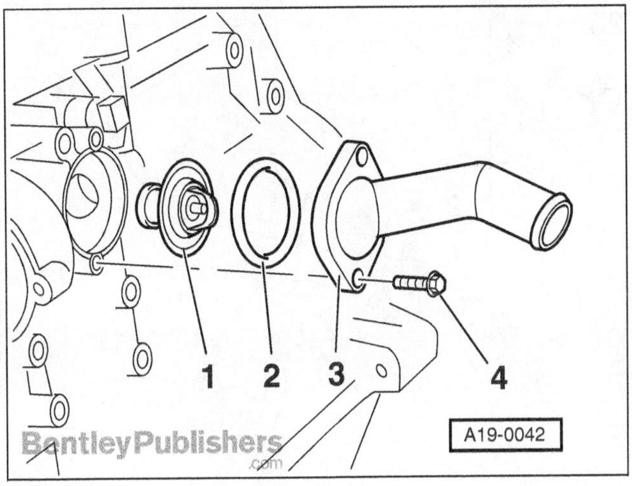

◀ 1.8L / 2.0L engines carefully pull the thermostat (**1**) out.

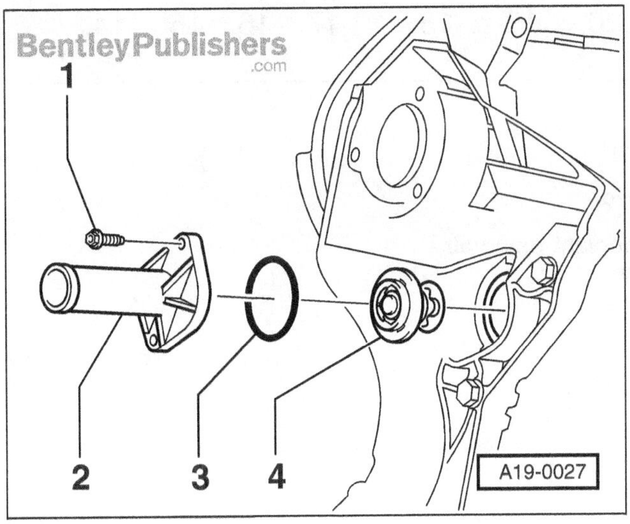

◀ 1.9L engines, rotate the thermostat (**4**) ¼ turn, (90°), counterclockwise and pull out.

– Install thermostat into cylinder block:

- 1.8L / 2.0L engines, fit thermostat into cylinder block.
- 1.9L engines, fit thermostat into cylinder block and rotate ¼ turn (90°) clockwise.

NOTE —

On 2.0L engines the support brace on the thermostat must be almost vertical when installed.

– Remaining installation is the reverse of removal noting the following:

- Clean mating surface on cylinder block before installing outlet flange.
- Always use new O-rings with coolant before installing.
- Fill cooling system as described previously.

Tightening torque

- Thermostat flange to engine block. . 15 Nm (11 ft-lb)

– Start engine and check for leaks. Be sure to quality check your work.

COOLING SYSTEM COMPONENTS (5-CYLINDER)

Cooling system components (2.5L)

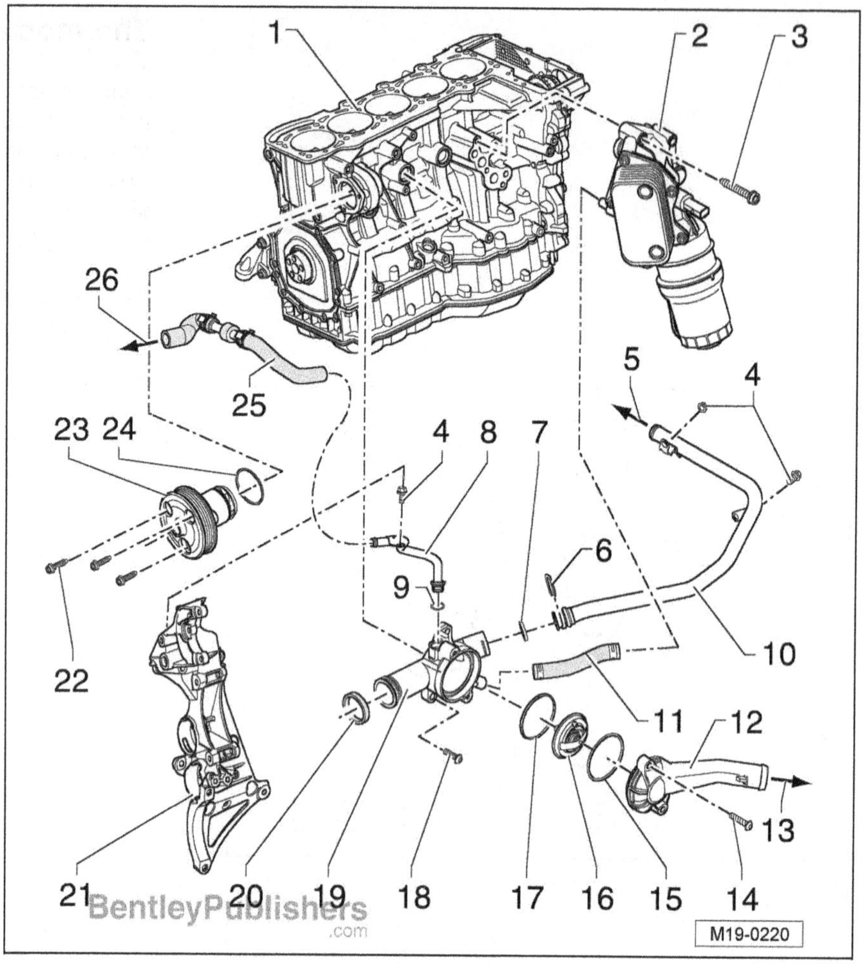

1. **Cylinder block**
2. **Oil cooler**
3. **Bolt**
   - Tighten to 25 Nm (18 ft-lb)
4. **Bolt**
   - Tighten to 10 Nm (7 ft-lb)
5. **To bottom of heat exchanger**
6. **Retaining clip**
7. **O-ring**
   - Always replace
8. **Front coolant line**
9. **O-ring**
   - Always replace
10. **Coolant line**
11. **Connecting hose**
12. **Connection piece**
13. **To bottom of radiator**
14. **Bolt**
    - Tighten to 5 Nm (44 in-lb)
15. **O-ring**
    - Always replace
16. **Coolant thermostat**
    - Checking: heat-up thermostat in water
    - Opening starts at approx. 185°F (85°C)
    - Ends approx. 216°F (102°C)
    - Opening lift: 7 mm minimum
17. **Seal**
    - Always replace
18. **Bolt**
    - Tighten to 25 Nm (18 ft-lb)
19. **Thermostat housing**
20. **Seal**
    - Always replace
21. **Accessory bracket**
22. **Bolt**
    - Tighten to 10 Nm (7 ft-lb)
23. **Coolant pump**
24. **O-ring**
    - Always replace
25. **Connecting hose**
26. **To bottom of coolant reservoir tank**

Cooling system components on upper chain guard (2.5L)

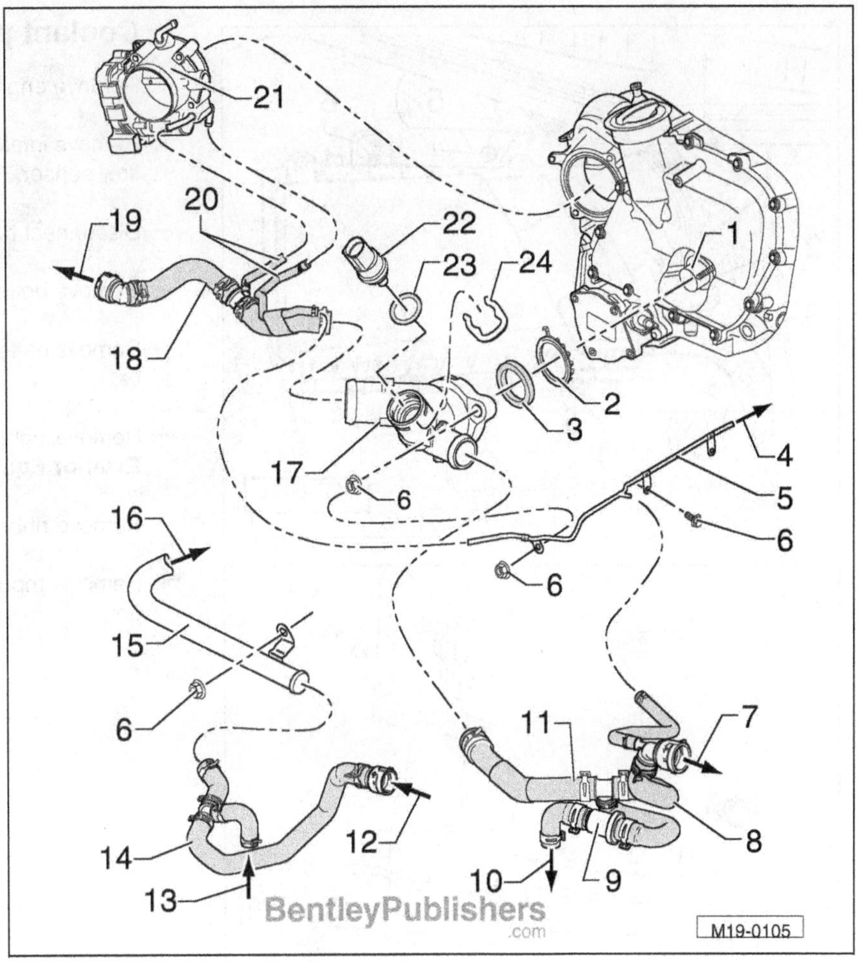

NOTE —
Volkswagen identifies electrical components by a letter and/or a number in the electrical schematics. See **EWD Electrical Wiring Diagrams.** *These electrical identifiers are listed in parenthesis as an aid to electrical troubleshooting.*

1. **Pipe connection**
2. **Circlip**
3. **Seal**
4. **To top of coolant reservoir tank**
5. **Rear coolant pipe**
6. **Bolt**
   - Tighten to 10 Nm (7 ft-lb)
7. **To top of heat exchanger**
8. **Supply hose**
9. **Bypass thermostat**
   - Only for vehicles with automatic transmission
10. **To ATF cooler**
    - Only for vehicles with automatic transmission
11. **Supply hose**
12. **From bottom of heat exchanger**
13. **From ATF cooler**
    - Only for vehicles with automatic transmission
14. **Return hose**
15. **Coolant line**
16. **To coolant regulator housing**
17. **Coolant distributor housing**
18. **Supply hose**
19. **To top of radiator**
20. **Coolant hoses**
    - For throttle valve control module (J338) heating
21. **Throttle valve control module (J338)**
22. **Engine coolant temperature (ECT) sensor (G62)**
23. **O-ring**
    - Always replace
24. **Retaining clip**

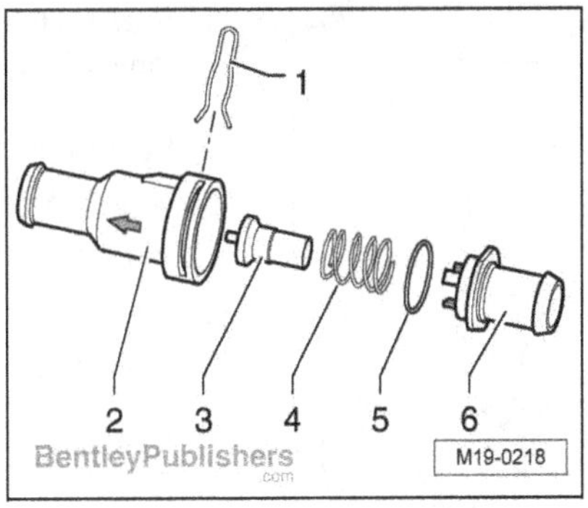

◄ Bypass thermostat assembly

1. Spring bracket
2. Lower housing
   - Note installation position: arrow points toward ATF cooler
3. Thermostat
   - Checking: heat-up thermostat in water
   - Opening starts at approx. 167°F (75°C)
   - Ends approx. 185°F (85°C)
   - Opening lift: 7 mm minimum
4. Spring
5. O-ring
6. Upper housing

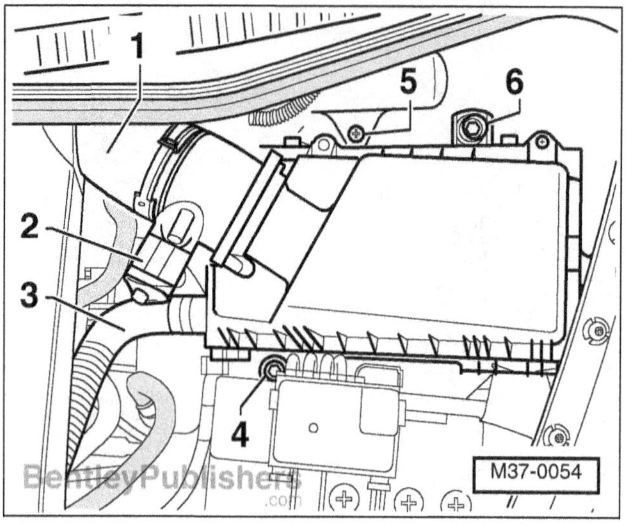

Coolant pump, replacing (2.5L)

– Remove engine cover. See **03 Maintenance**.

◀ Remove intake hose (**1**) and connector (**2**) from mass air flow sensor.

– Disconnect hose (**3**) from air filter housing.

– Remove bolts (**4**) and (**6**) and remove air filter housing.

– Remove battery and battery tray. See **27 Engine Electrical**.

– Remove right front wheel housing liner. See **66 Body–Exterior Equipment**.

– Remove ribbed belt. See **03 Maintenance**.

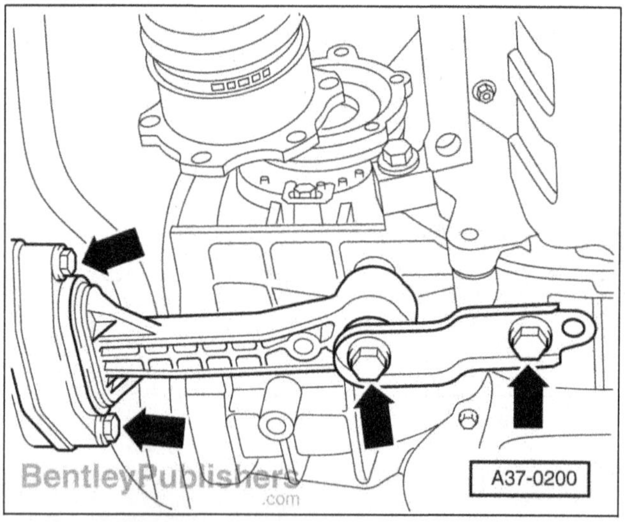

◀ Remove mounting bolts (**arrows**) and pendulum support.

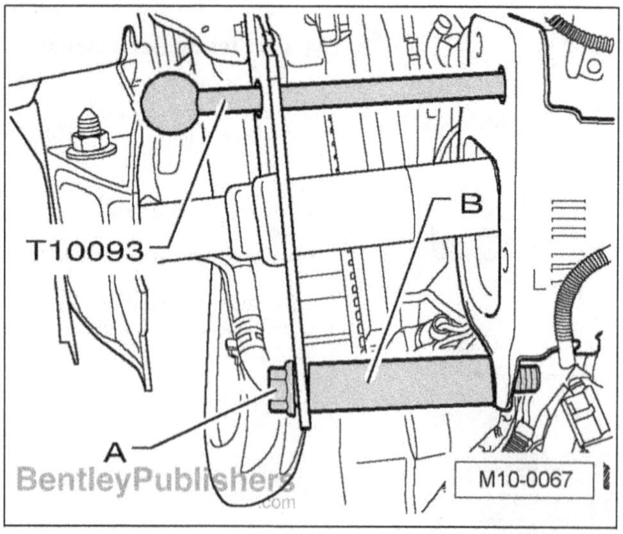

◀ Move front lock carrier to service position. See **50 Body–Front**. Use **T10093**, or equivalent, to slide lock carrier forward.

– Support lock carrier using the following:

- Bolt (**A**), M8 x 100mm
- Sleeve (**B**) 80mm long.

– Empty power steering reservoir using fluid collecting and extracting device VAG 1782, or equivalent.

– Remove power steering reservoir.

– Remove right side headlight to gain access to power steering reservoir mounting bracket as shown below. See **92 Lights, Accessories–Exterior**.

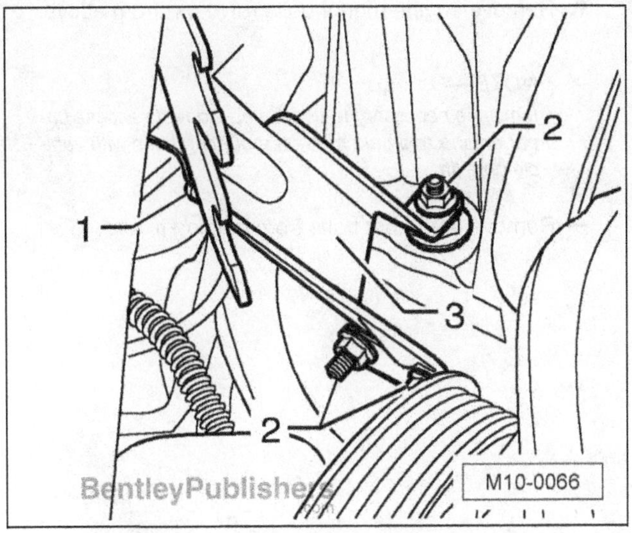

◀ Remove mounting nuts (**2**) and power steering reservoir mounting bracket (**3**).

– Disconnect intake and pressure line from power steering pump. Seal pump and lines with clean plugs.

– Position power steering pressure line on left longitudinal member and tie it firmly in place.

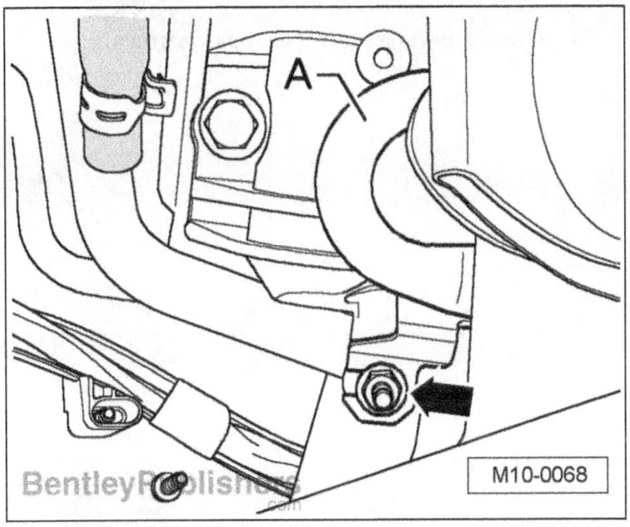

◀ Remove mounting nut (**arrow**) of intake line (**A**) at engine mount. Remove intake line.

– Slightly loosen upper mounting bolts at engine and transmission (less than 1 turn).

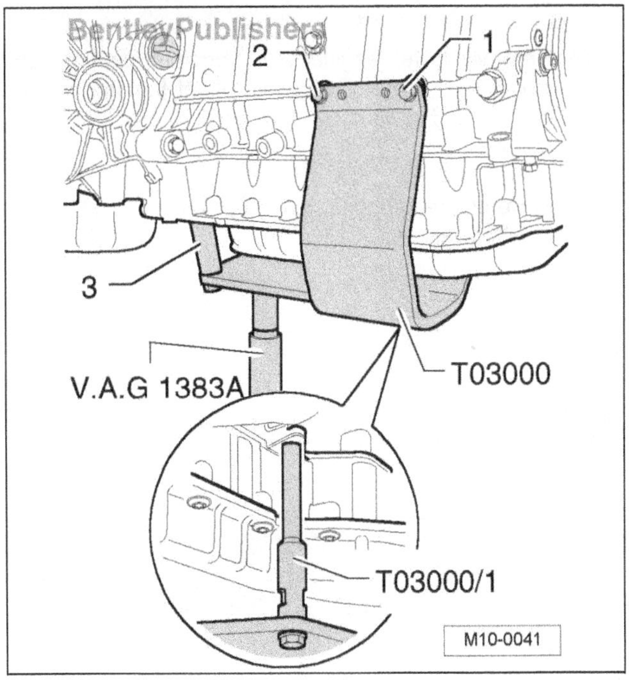

◀ Install a suitable engine supporting device from below. Alternatively, use VW special tools to support engine as described below:

- Attach engine holder bracket **T03000** to cylinder block with pin (**3**) and tighten bolts (**1 and 2**) hand-tight.
- Screw in pin **T03000/1** and tighten to 20 Nm (15 ft-lb).
- Tighten bolts (**1 and 2**) to 25 Nm (18 ft-lb).

– Lift engine slightly with engine/transmission jack (VAG 1383 shown).

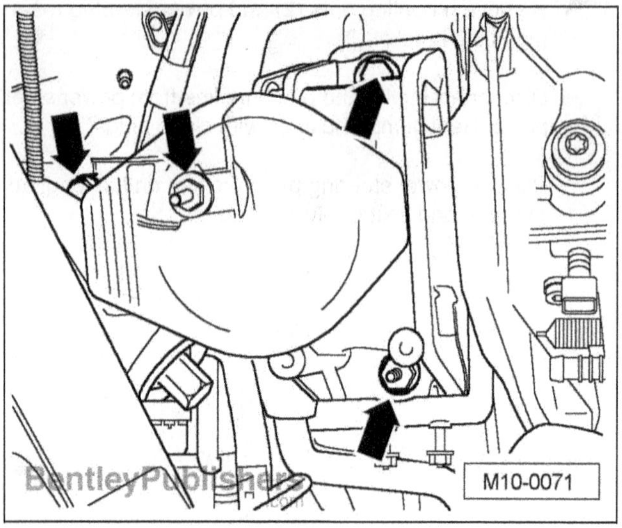

◄ Remove engine mount bolts (**arrows**) from above.

NOTE —

It may be necessary to use a step ladder to access upper engine and transmission mounting bolts with vehicle on a lift.

– Remove both front bolts from holding fixture.

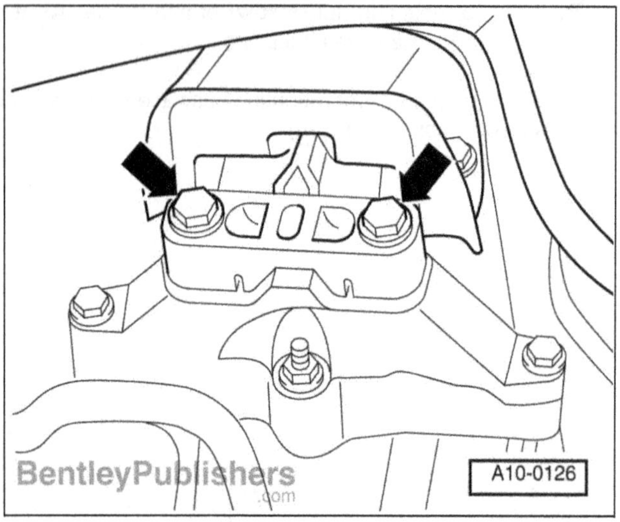

◄ Remove transmission mount bolts (**arrows**).

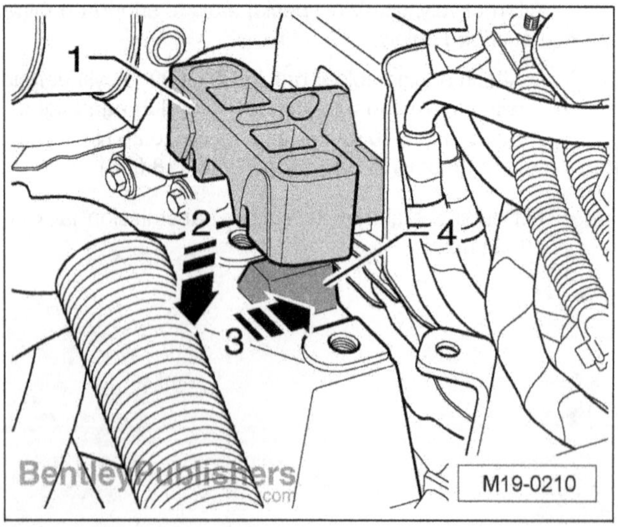

◄ Carefully lower engine with transmission only as far downward (**arrow 2**) until pin (**4**) of transmission plate can be slid toward left (**arrow 3**) out of transmission mount.

– Move engine/transmission assembly until it stops at left longitudinal member.

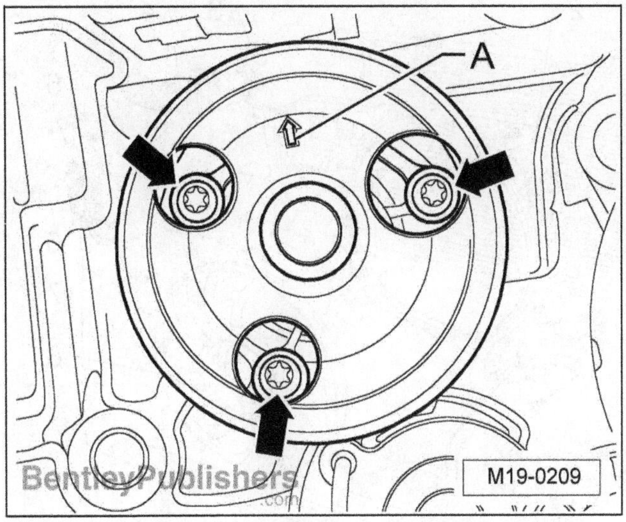

◄ Rotate belt pulley so that arrow **A** (relative to engine axis) points upward vertically.

– Remove mounting bolts (**arrows**) for coolant pump.

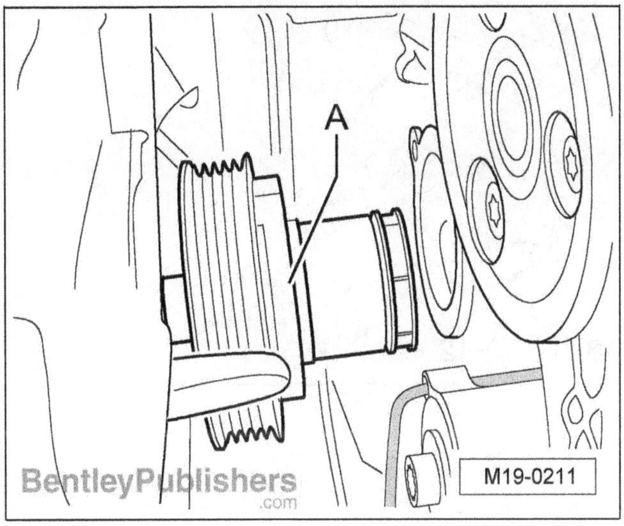

◄ Pull coolant pump (**A**) out and guide it up between longitudinal member and cylinder block to remove.

– Installation is in the reverse order of removal. Be sure to note the following:

> ***CAUTION—***
> - *When working in the engine compartment, due to the cramped, tight conditions, pay attention to the following points*
> - *Route all lines (e.g. for fuel, hydraulics, EVAP system, coolant, refrigerant, brake fluid and vacuum, etc.) as well as electrical wiring so that the original positions are maintained. Ensure sufficient clearance to all moving or hot components.*
> - *Be sure to replace all tie wraps cut during removal and install new ones to prevent lines and wiring harnesses from chaffing or contacting engine parts.*

– Replace O-ring for coolant pump.

– Note installation position of coolant pump before installing.

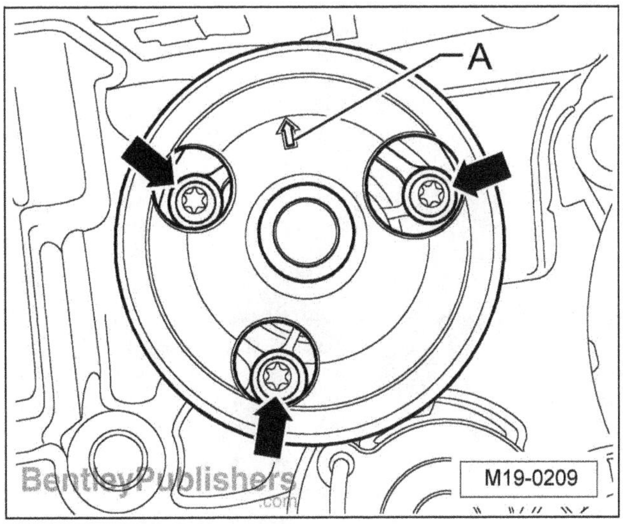

◄ Rotate belt pulley so that arrow **A** (relative to engine axis) points upward vertically. Holes in belt pulley must be lined up with mounting bolt holes (**arrows**). Tighten coolant pump mounting bolts.

Tightening torque

- Thermostat flange to engine block. . 15 Nm (11 ft-lb)

– First, firmly bolt engine mount to engine, then align engine/transmission assembly free of tension by shaking assembly before tightening engine mount bolts to final torque. See **10 Engine–Removing and Installing** for engine mount alignment and tightening torques.

Radiator and cooling fans, assembly (2.5L)

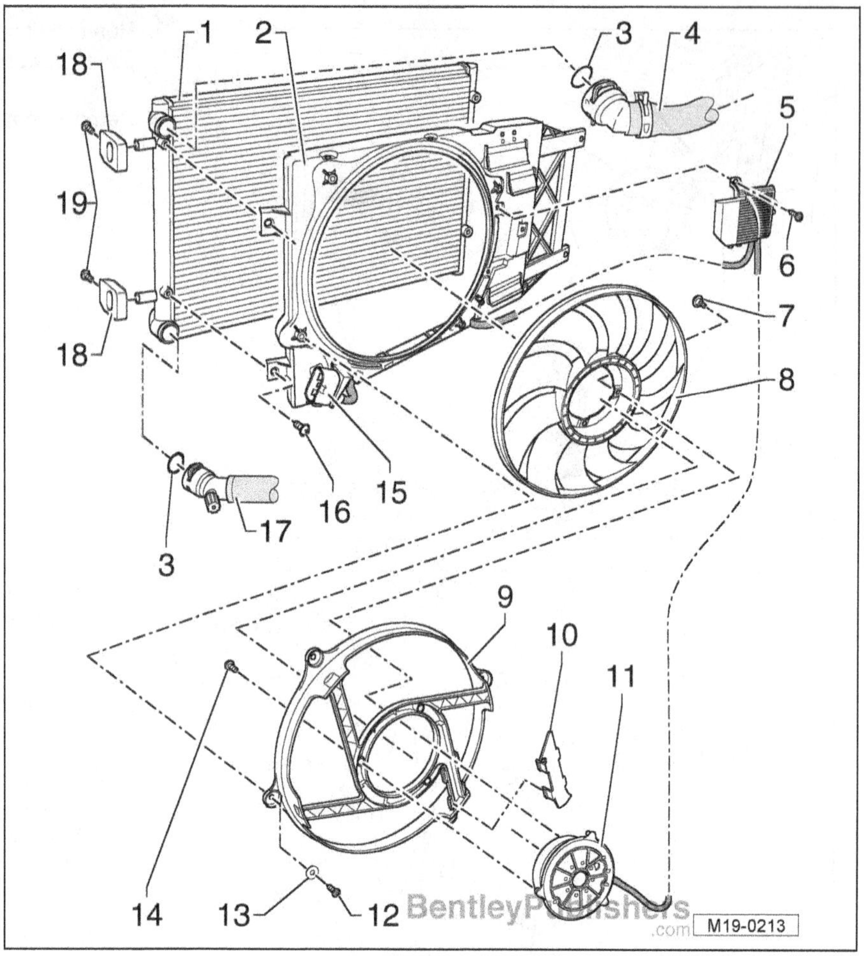

1. **Radiator**
   - Replace all coolant if removing/installing
2. **Air shroud**
3. **O-ring**
   - Replace if damaged
4. **Coolant hose, upper**
5. **Coolant fan control module (J293)**
6. **Bolt**
   - Tighten to 5 Nm (44 in-lb)
7. **Bolt**
8. **Fan**
9. **Fan mounting ring**
10. **Cover**
11. **Radiator fan motor**
12. **Bolt**
    - Tighten to 5 Nm (44 in-lb)
13. **Washer**
14. **Bolt**
15. **Connector**
16. **Bolt**
    - Tighten to 5 Nm (44 in-lb)
17. **Lower coolant hose**
18. **Bracket**
19. **Bolt**
    - Tighten to 5 Nm (44 in-lb)

Radiator, removing and installing (2.5L)

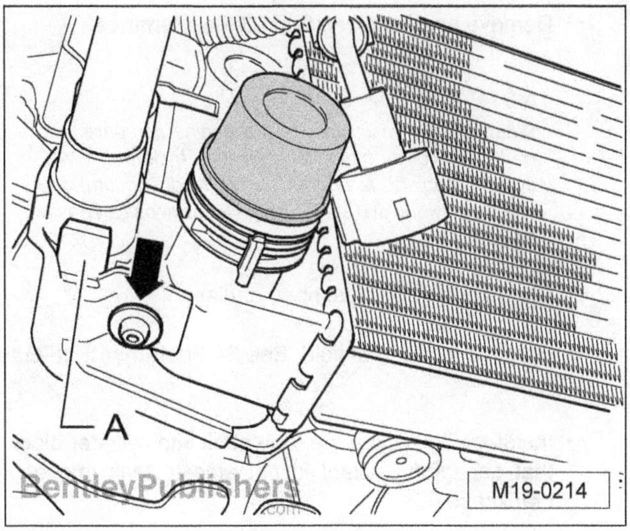

- Drain coolant. See **Cooling System, Draining and Filling** in this repair group.
- Remove coolant fan. See **Radiator and coolant fans assembly (2.5L)** in this repair group.

◄ Remove mounting bolt (**arrow**) for A/C refrigerant line (**A**).

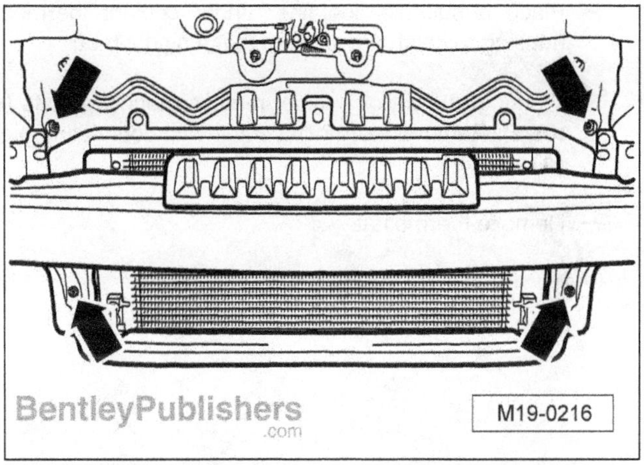

◄ Remove mounting bolts (**arrows**) for radiator.

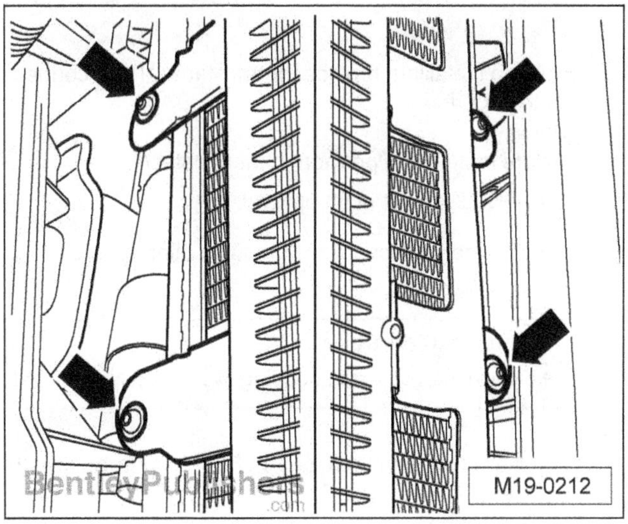

◄ Remove mounting bolts (**arrows**) for A/C condenser.

- Secure A/C condenser to lock carrier with safety wire or heavy duty zip ties.
- Remove radiator downward.
- Install in reverse order of removal.

Tightening torque

- Radiator to lock carrier . . . . 5 Nm (44 in-lb)
- Condenser to radiator . . . . 5 Nm (44 in-lb)
- Intake air elbow to radiator . . . . 5 Nm (44 in-lb)

Thermostat, removing and installing (2.5L)

– Remove engine cover. See **03 Maintenance**.

> ***NOTE —***
> *When doing repair work in the engine compartment, route lines of all types (e.g. for fuel, hydraulic, EVAP system, coolant, refrigerant, brake fluid, vacuum) and electrical wiring so that the original positions can easily be restored.*

– Drain coolant, as described earlier.

– Remove intake manifold. See **24 Fuel Injection–Gasoline**.

– Install oil dipstick guide tube back into cylinder block so that escaping coolant is not able to leak into engine crankcase.

– Place a suitable container under coolant thermostat housing to catch any coolant that may drain out.

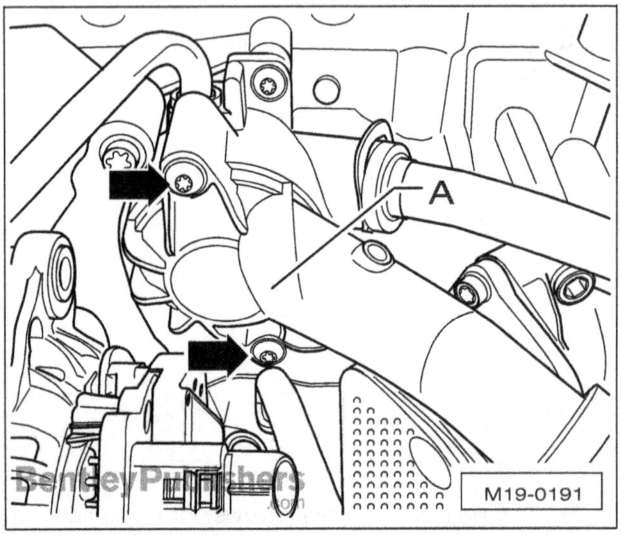

◄ Working at front of engine, remove mounting bolts (**arrows**). Remove thermostat cover with connecting piece (**A**).

– Remove thermostat.

– When installing, replace thermostat seal and connecting piece O-ring.

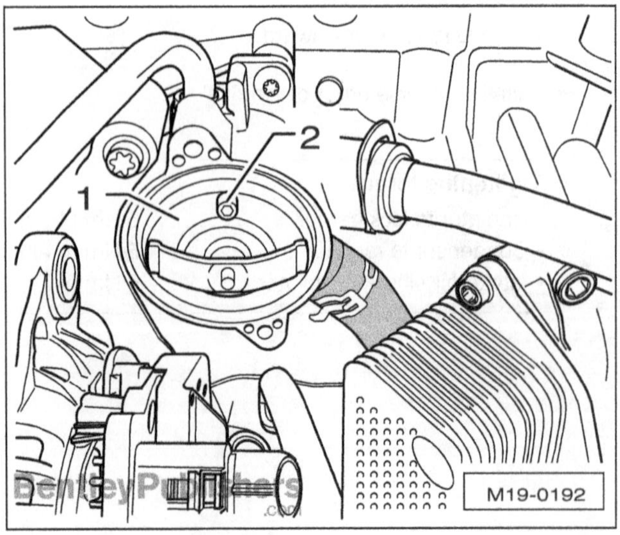

◄ Note installation position of thermostat (**1**).
- Valve (**2**) must point upward.

– Install thermostat cover with connecting piece.

Tightening torque
- Thermostat cover to housing . . . . 5 Nm (44 in-lb)

– Install intake manifold. See **24 Fuel Injection–Motronic**.

– Fill cooling system as described earlier.

– Install engine cover.

2 Fuel, Ignition, and Exhaust Systems

GENERAL

This general information group covers application information and system descriptions for the repair groups listed under 2 Fuel, Ignition and Exhaust Systems.

NOTE —
- *For general information on the battery, starter and alternator, see* **27 Engine Electrical**.
- *For emission control system application information, see* **26 Exhaust System and Emission Controls**.

FUEL SUPPLY

The plastic fuel tank is mounted beneath the rear of the vehicle. On gasoline engines, an electric fuel pump with integral level sensor is submersed in the tank. On diesel engines, only the fuel the level sensor is mounted in the tank, as the engine-mounted diesel injection pump handles fuel delivery to the engine.

This gasoline fuel pump/fuel level sensor assembly is called the fuel delivery unit. An inlet strainer provides course filtration and prevents dirt and debris from entering the fuel system. Built into the fuel delivery unit is a check valve and a relief valve. The check valve is on the outlet side of the pump and holds pressure in the system after the engine is shut off. The relief valve prevents high pressure from damaging the system. The check valve and relief valve are not replaceable.

The fuel tank is designed to prevent overfilling and allow for fuel expansion. The fuel cap contains a valve to prevent a vacuum from forming in the tank.

On all engines, fuel is supplied to the engine and excess fuel is returned to the tank through special plastic fuel lines. In addition, diesel engines with automatic transmissions use a fuel cooler on the supply line to the pump.

FUEL INJECTION

Various types of engine management systems are used on the engines covered in this manual as noted below.

Engine Management Systems

- 1.8L 4-cylinder turbo gasoline
  APH, AWP, AWV, ... Bosch Motronic ME 7.5
- 1.8L 4-cylinder turbo gasoline
  BKF, BNU, ...Bosch Motronic ME 7.1.1
- 1.9L 4-cylinder turbo diesel
  ALH, BEW . . . . TDI Diesel Direct Fuel Injection
- 2.0L 4-cylinder gasoline
  AEG . . . . Bosch Motronic 5.9.2
- 2.0L 4-cylinder gasoline
  AVH, AZG, BDC, BEV, BGD Bosch Motronic ME 7.5
- 2.5L 5-cylinder gasoline
  BPR, BPS . . . . .Bosch Motronic ME 7.1.1

Motronic 5.9.2 engine management

The AEG (2.0 liter) engine is equipped with the Bosch Motronic 5.9.2 engine management system. This version complies with the federal and state government mandated On-Board Diagnostic (OBD) II standards.

Basic fuel metering is determined by engine speed and engine load. The Engine Control Module (ECM) receives engine speed and crankshaft position information from the engine speed and reference sensor, and engine load information from the Mass Air Flow sensor (MAF). The ECM then meters fuel to the engine by sequentially triggering the fuel injectors at a rate proportional to engine speed and load. The length of time the injectors remain open determines fuel quantity.

The ECM uses the same information to determine the correct ignition firing point. A small output signal is generated and sent to the Power Output Stages built into the Ignition Coils. The Camshaft Position Sensor (CMP), identifies cylinder number 1 firing position for cylinder-selective injection and ignition knock control. The Engine Coolant Temperature sensor (ECT), supplies engine temperature information to the ECM.

The Throttle Actuator Control Module (TACM), combines 4 functions. Three are used on the input side: the Throttle Position Sensor (TPS), which signals throttle angle; the Closed Throttle Position switch (CTP), which signals fully closed throttle plate; and the Throttle Position Feedback sensor, which signals position of the electric motor used for idle stabilization and cruise control function.

Motronic 5.9.2, overview

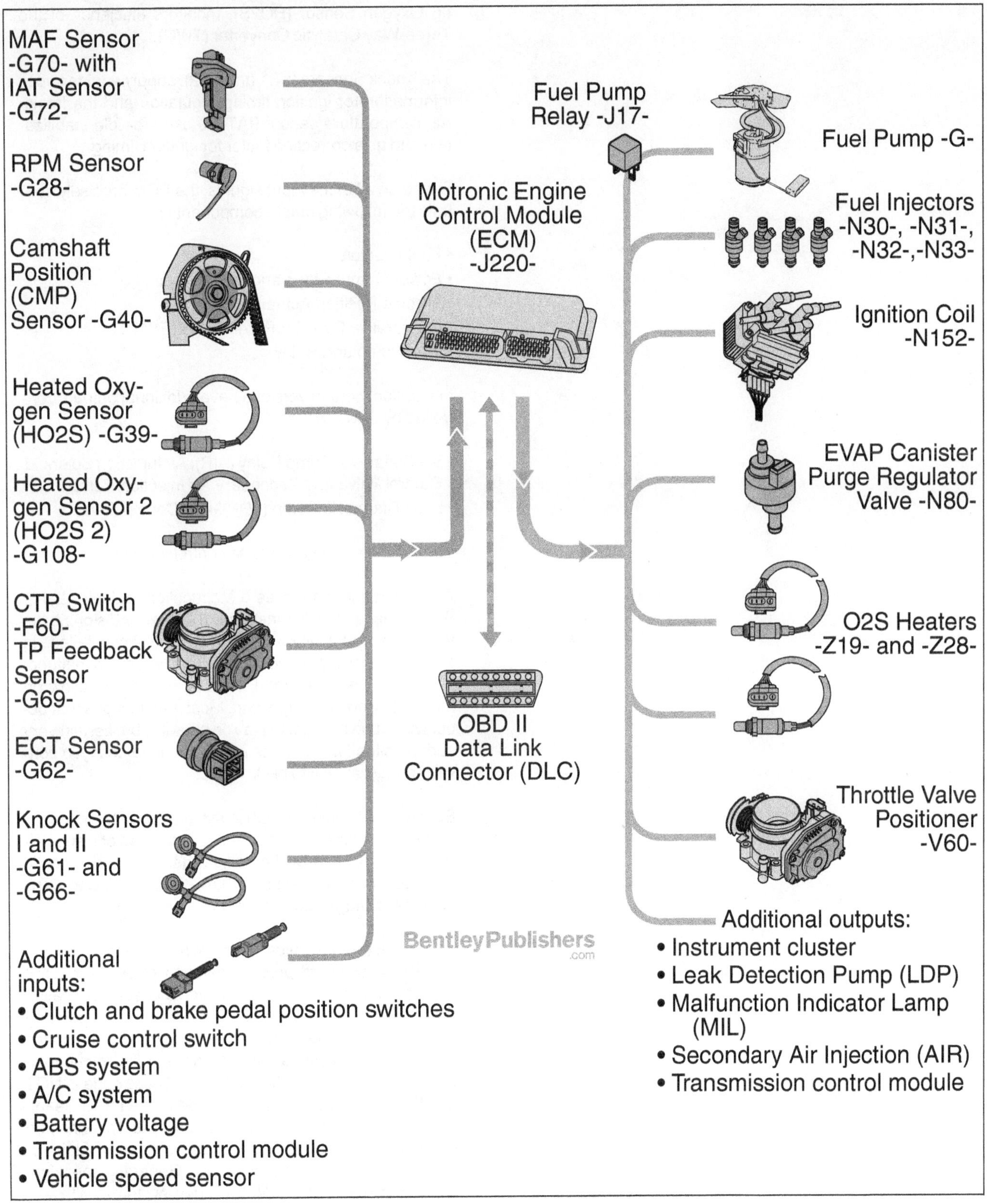

The Pre-Catalyst Heated Oxygen Sensor (HO2S), signals combustion efficiency, and the Post-Catalyst Heated Oxygen Sensor (HO2S), monitors efficiency of the Three-Way Catalytic Converter (TWC).

The Knock Sensors (KS1 and KS2), supply engine knock information for ignition timing regulation and the Intake Air Temperature sensor (IAT), is used for idle stabilization and as a correction factor for ignition timing.

Based on all of the input signals, the ECM precisely controls the following output components:

- Fuel Injectors
- Power Output Stage and Ignition Coil
- Throttle Position Actuator
- Evaporative Canister Purge Valve (EVAP)
- Fuel Pump and Relay

In addition, certain versions have additional outputs controlled by the ECM:

- Secondary Air Pump Relay (AIR); Air Injection Solenoid Control Valve and Secondary Air Injection Pump
- Leak Diagnosis Pump (LDP); and Control Relay

The ECM also controls cruise control functions.

A warning light known as a Malfunction Indicator Lamp (MIL), is located in the instrument cluster and signals the vehicle operator when certain systems have failed. In some instances, it will come on if relatively small malfunctions have been recorded, such as running very low on fuel or leaving off the gas cap. Most malfunctions that occur will cause the light to stay on steadily, but certain very serious situations such as an overheating catalytic converter may cause it to blink.

Because the Motronic 5.9.2 system is quite complex, proper diagnosis and repair requires the use of a specialized scan tool such as the Volkswagen factory diagnostic scant tool or equivalent aftermarket scan tool. See **OBD On Board Diagnostics** for more information.

In addition to diagnostics, Motronic 5.9.2 systems include the ability to record proper operation of as many as 8 monitored functions. These functional checks are called readiness codes and are set when the proper operating parameters are reached at least one time. These codes are important because some areas with emissions testing check these codes first. If the readiness code is NOT set, the vehicle will not pass the specialized test. Readiness codes may be erased if the battery is disconnected or runs low, if the ECM is disconnected, or if faults or DTCs are erased with a scan tool. Generally speaking, the readiness codes in Volkswagen systems will reset themselves after the vehicle has been started and driven under varying conditions several times. They can also be reset using the Volkswagen scan tool.

Diesel Turbo Direct Injection (TDI)

The 1.9L engine is equipped with the Diesel Turbo Direct Injection System that features engine controls that closely resemble those of a gasoline engine. The TDI system combines sophisticated computer control of fuel management and emissions with system monitoring and diagnostics. An exhaust driven turbocharger and an intercooler work with a specially designed combustion chamber and cylinder head to produce more efficient combustion and lower fuel consumption. This also results in reduced engine noise and increased power.

In addition, the accelerator pedal is directly connected to the Engine Control Module (ECM) via a potentiometer. This eliminates the need for an accelerator cable and is known as "drive by wire". The ECM also controls all glow plug functions, auxiliary coolant glow plug functions and cruise control.

Basic fuel metering is determined by engine speed and engine load. When the key is switched on and the engine is started, the fuel cut-off valve opens and fuel flows into the injection pump. The ECM receives engine RPM from the engine speed sensor and engine load information from the Throttle Position Sensor (TPS). These signals are modified by the Engine Coolant Temperature sensor (ECT), the Mass Air Flow Sensor (MAF), and the Fuel Temperature Sensor. The ECM then signals the quantity adjuster which allows the proper quantity of fuel to be metered sequentially to the mechanical injectors. The ECM also uses the same information to determine the correct moment of injection which is the ignition firing point. An output signal is generated and sent to the cold start valve at the correct time.

The ECM monitors the operation of the quantity adjuster via a signal received from the modulating piston displacement sensor and makes corrections to the fuel quantity as required.

The ECM monitors the operation of the cold start valve via a signal received from the needle lift sensor attached to injector number 3 and again makes corrections as required.

Turbocharger boost pressure is regulated by the ECM based on a signal from the Intake Air Temperature sensor (IAT), the Mass Air Flow (MAF) sensor, and an ambient pressure signal from an internal Barometric Pressure Sensor (BARO). The wastegate bypass regulator valve receives output from the ECM and controls a vacuum signal sent to the turbocharger wastegate.

In order to prevent possible damage from simultaneous application of the brake, clutch and accelerator pedals, the brake pedal switch, brake light switch, and clutch pedal switch signal their respective positions to the ECM.

Diesel Turbo Direct Injection, overview

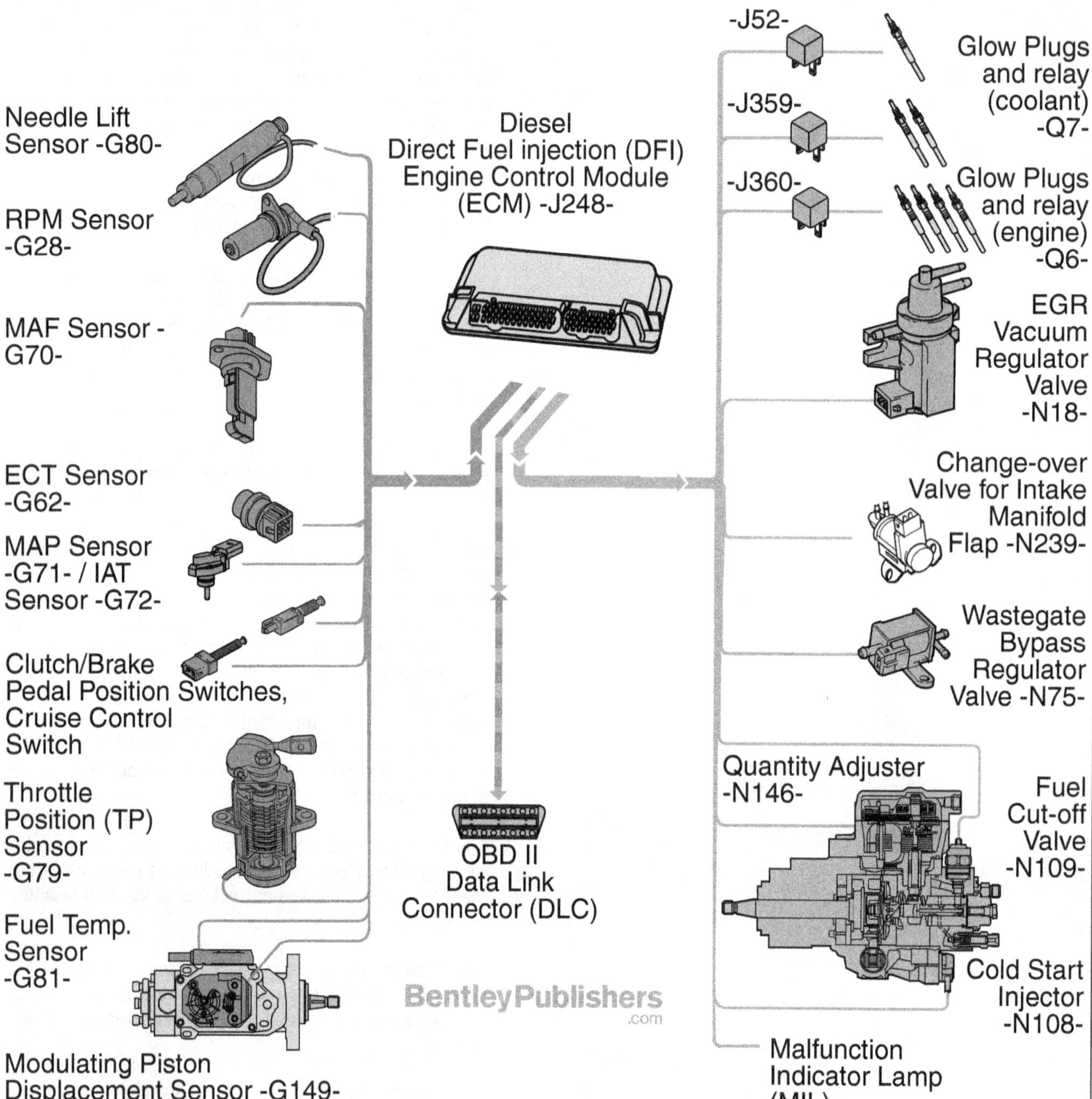

Additional inputs:
- A/C system
- Battery voltage
- Transmission control module
- Vehicle speed sensor

Additional outputs:
- Automatic transmission control module
- Instrument cluster

These signals are also used by the ECM during in cruise control mode.

Based on all of the input signals, the ECM precisely controls the following output components:

- Quantity Adjuster
- Cold Start Injector
- Fuel Cut-off Valve
- Wastegate Bypass Regulator Valve
- EGR Vacuum Regulator Valve
- Intake Manifold Change-over Valve

In addition, the ECM uses various sensors to control these additional outputs:

- Glow Plugs
- Glow Plug Relay
- Auxiliary Heater Coolant Glow Plugs
- Coolant Glow Plug Relay

When the ECM receives the signal to shut down, a signal is sent to a flap valve mounted in the intake tract. This valve, known as the intake manifold change-over valve, closes and blocks the flow of air to the still-turning engine. Blocking the flow of air during the shut-down cycle reduces the abruptness associated with stopping a diesel engine.

Tailpipe emissions are further reduced by a two-way oxidation-type catalytic converter.

TDI equipped New Beetles have two warning lights in the instrument cluster to advise of system status. The glow plug indicator light operates when the key is first turned on to indicate the need to wait for the glow plugs to preheat the combustion chambers. The Malfunction Indicator Lamp (MIL), lights if a failure of a monitored component occurs. In some serious failure modes, both warning indicators can be lit.

Owing to the complex nature, most diagnosis and repair of the TDI system requires the use of specialized scan tools such as the Volkswagen factory diagnostic scan tool or equivalent aftermarket scan tool. See **OBD On Board Diagnostics** for more information.

Motronic ME 7 engine management

Most gasoline engines covered by this manual are equipped with a torque-based Bosch Motronic ME7 engine management system. These systems fully comply with the federal and state government mandated On-Board Diagnostic (OBD) II standards.

In previous engine management systems, demands for torque and efficiency are handled by separate sub-systems. The interaction between all of the different subsystems has become increasingly complicated as new demands are placed on the engine management system. These demands can include control of transmission, traction control and vehicle dynamics in addition to enhanced links with existing subsystems.

The Bosch Motronic ME 7 engine management systems use a new internal circuit design to calculate engine torque. The Engine Control Module (ECM) receives all of the relevant signal data from the input sensors and various other sources and computes the required amount of engine torque needed based on pre-established priorities. All torque and efficiency demands are defined as mathematical variables.

The ECM is then able to obtain the calculated engine torque through precise coordination of the following output components:

- Throttle Valve Control Module (throttle angle)
- Wastegate Regulator Solenoid Valve (boost control)
- Fuel Injectors (injection time and fuel cut-off)
- Power Output Stages with Ignition Coils (ignition timing)

Previous Motronic versions modified basic fuel and ignition maps as required to match the air flow of the mechanical throttle valve. ME 7 systems integrate a mechanically decoupled accelerator pedal and throttle valve (drive-by-wire) system to allow the ECM processor complete control of engine torque generation as a response to external, internal and efficiency demands. Simply stated, the ECM controls all three main combustion requirements: fuel, spark and air, whereas previous systems only controlled fuel and spark.

Motronic ME 7 overview

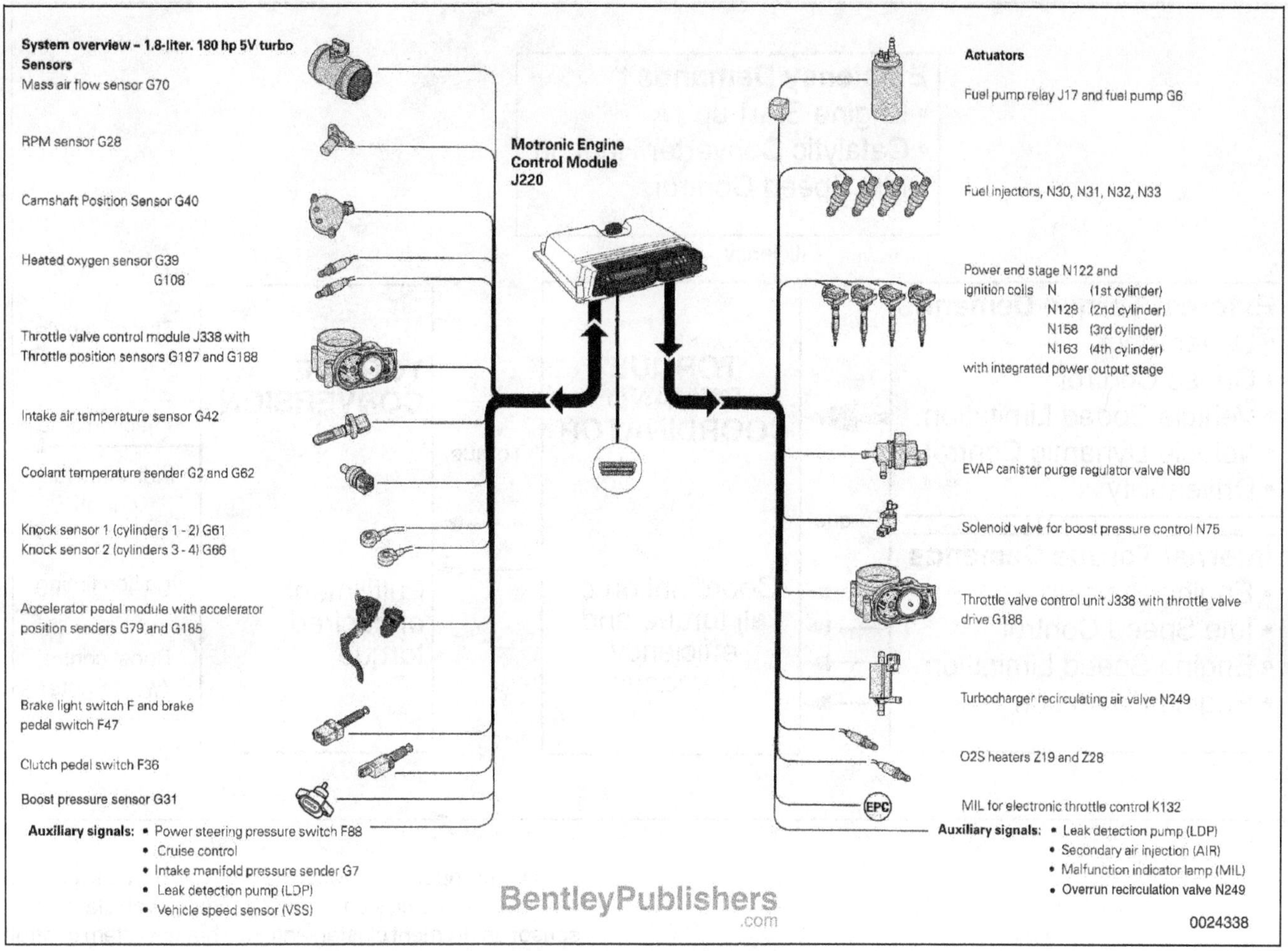

Main sensor inputs to the ME 7 ECM (J220) include:

- Mass air flow sensor (G70)
- Speed/reference sensor (G28)
- Camshaft position sensor (G40)
- Pre-Catalyst heated oxygen sensor (G39)
- Post-Catalyst heated oxygen sensor (G108)
- Throttle valve control module (J338) with position sensors (G187) and (G188)
- Intake air temperature sensor (G42)
- Coolant temperature sensors (G2) and (G62)
- Knock sensor, cylinders 1-2, (G61)
- Knock sensor, cylinders 3-4, (G66)
- Accelerator pedal position sensors (G79) and (G185)
- Boost pressure sensor (G31)
- Brake light (F) and pedal switches (F47)
- Clutch pedal switch (F36)
- Leak Detection Pump (V144)

NOTE —

Volkswagen identifies electrical components by a letter and/or a number in the electrical schematics. See **97 Wiring Diagrams, Fuses and Relays**. *These electrical identifiers are listed in parenthesis as an aid to electrical troubleshooting.*

Motronic ME 7 system structure

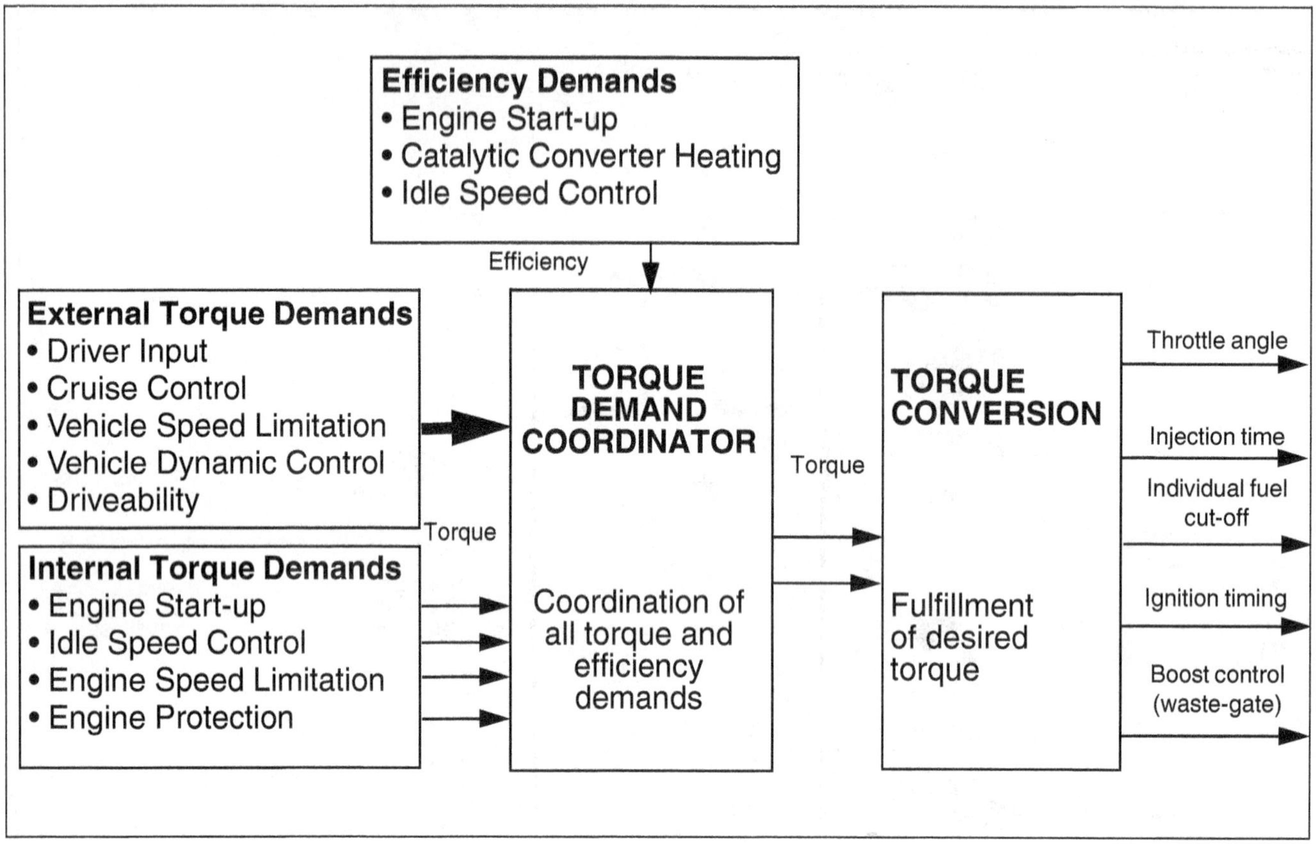

Additional inputs are received from the cruise control switches, transmission control module, vehicle speed sensor, instrument cluster, anti-lock brake system control module, fan control module, airbag control module and air conditioning system.

Based on the calculations of the torque demand coordinator in the central processor of the ECM, signals are sent to the torque conversion processor for activation of the outputs required to achieve the requested torque.

The primary outputs used to achieve the ECM-requested torque are:

- Throttle valve control module (J338) with throttle valve drive motor (G186)
- Fuel Injectors (N30, N31, N 32, N33)
- Ignition Coil Packs with Power Output Stages (N70, N127, N291, N292)
- Wastegate Regulator Solenoid Valve (N75)

Additional outputs controlled by the ECM include:

- Fuel Pump Relay (J17)
- Evaporative Canister Purge Valve (N80)
- Secondary Air Injection Pump Relay (J299)
- Secondary Air Injection Solenoid Valve (N112)
- Leak Detection Pump (V144)
- Overrun Recirculation Valve (N249)
- Oxygen Sensor Heaters

A Malfunction Indicator Lamp (MIL), is located in the instrument cluster and signals the vehicle operator when certain emissions-related systems have in some way malfunctioned. In some instances, it will come on if relatively small malfunctions have been recorded, such as running very low on fuel or leaving off the gas cap. Most malfunctions that occur will cause the light to stay on steadily, but certain very serious situations such as an overheating catalytic converter may cause it to blink. An EPC indicator light also located in the instrument cluster will signal the vehicle operator of certain malfunctions within the electronic accelerator pedal system.

Due to the complex nature of the Motronic ME 7 systems, proper diagnosis and repair requires the use of a specialized scan tool such as the Volkswagen factory diagnostic scan tool or equivalent aftermarket scan tool. See **OBD On Board Diagnostics** for more information.

In addition to diagnostics, Motronic ME 7 systems include the ability to record proper operation of as many as 8 monitored functions. These functional checks are called readiness codes and are set when the proper operating parameters are reached at least one time. These codes are important because some areas with emissions testing check these codes first. If the readiness code is NOT set, the vehicle will not pass the specialized test. Readiness codes may be erased if the battery is disconnected or runs low, if the ECM is disconnected, or if faults or DTCs are erased with a scan tool. Generally speaking, the readiness codes in Volkswagen systems will reset themselves after the vehicle has been started and driven under varying conditions several times. They can also be reset using the scan tool.

Electronic accelerator system

0024339

The ME 7 engine management systems use electronic control to operate the throttle valve rather than the usual mechanical throttle cable. Pedal resistance and operation matches cable designs so that the operator is unlikely to differentiate. In the event of a failure with either one of the accelerator pedal position sensors, the other sensor signal would be used in a limited capacity with functions such as cruise control deactivated. If both sensors malfunction, the ECM will limit engine speed to approximately 1200 rpm.

Ignition System

The ignition function is handled through the fuel injection/engine management system on all engines.

On gasoline engines, the Motronic ECM computes ignition timing based on inputs from the various sensors. Crankshaft position and speed are the main inputs to the ECM used for starting. The other sensors are used to adapt the basic timing map for varying operating conditions. Motronic engines incorporate adaptive knock control to adjust the ignition timing for individual cylinders. See **28 Ignition System–Gasoline** for more information.

On diesel engines, the ECM uses the same information that was used to compute the fuel quantity to determine the correct moment of injection. When the fuel is injected into the combustion chamber, heat generated by the high compression ratio causes the fuel to spontaneously burn. This is the ignition firing point. There is no separate ignition system as such. In spite of this high compression ratio, cold engines do not have sufficient heat to burn the fuel. For this reason, glow plugs are used to provide supplementary heat. See **23 Fuel Injection–Diesel** for more information.

Emission Controls

Gas and diesel engines use different emission controls. Most functions are integral with the engine management systems and are monitored by the On-Board Diagnostic (OBD) system. Most of these functions cannot be isolated. For additional information on those functions, see **26 Exhaust System and Emission Controls**.

20 Fuel Storage and Supply

GENERAL

This repair group covers the fuel supply and storage systems, including fuel pump testing and replacement. For general fuel system descriptions and overviews, see **2 General Information**.

NOTE —
1.9L diesel engine fuel filter replacement is covered in **03 Maintenance***.*

Engine	Engine code
1.8L 4-cylinder turbo gasoline	APH, AWV, AWP, BKF, BNU
1.9L 4-cylinder TDI diesel	ALH, BEW
2.0L 4-cylinder gasoline	AEG, AVH, AZG, BDC, BEV, BGD
2.5L 5-cylinder gasoline	BPR, BPS

Safety precautions

Please read and be familiar with the following warnings and cautions before working on the fuel pump, fuel tank or fuel lines.

WARNING —

- *Always disconnect negative (–) battery cable and cover terminal with an insulated material whenever working on any fuel related component.*
- *Gasoline and diesel fuel are dangerous to your health. Wear suitable hand and skin protection when working on fuel system. Do not breath fuel vapors. Always work in a well-ventilated area.*
- *Fuel and fuel vapors will be present during many of the operations described in this repair group. Do not smoke or create sparks. Be aware of pilot lights in gas operated equipment (heating systems, water heaters, etc.). Have an approved fire extinguisher handy.*
- *Gasoline fuel supply systems are designed to maintain pressure in the system after the engine is turned off. Fuel will be expelled under pressure as fuel lines are disconnected. This can be a fire hazard, especially if the engine is warm. Always wrap a clean shop rag around any fuel line fitting before loosening or disconnecting it.*
- *Exercise extreme caution when using spray-type cleaners on a warm engine. Observe all manufacturer recommendations.*

CAUTION —

- *Cleanliness is essential when working on any part of the fuel system. Thoroughly clean fuel line unions and hose fittings before disconnecting them. Use only clean tools.*
- *Use only spring-type clamps on fuel hoses. Do not use screw-type hose clamps, they do not expand and contract with engine heat and will leak.*
- *Keep removed components clean. Seal or cover them with plastic or paper, especially if the repair cannot be completed immediately. Seal open fuel supply and return lines to prevent contamination.*
- *When replacing parts, install only new, clean components. Seals and O-rings should always be replaced rather than reused.*
- *Poor diesel engine performance in ambient air temperatures below freezing (32°F (0°C)) may be caused by diesel fuel gelling, a result of improperly winterized/conditioned diesel fuel provided by fuel suppliers. Water contamination in diesel fuel can also cause similar conditions as a result of ice accumulation in the diesel fuel system. Use a cold weather diesel fuel conditioner as necessary. Volkswagen recommends the use of Stanadyne™ one shot performance formula all season diesel fuel conditioner. Add 8 oz. (235 ml) per 15 gallons (60 liters) of diesel fuel, directly to diesel fuel tank.*

FUEL PUMP TROUBLESHOOTING (GASOLINE ENGINES)

The fuel supply system is an integral part of the operation of the fuel injection system. Problems such as a no-start condition, hesitation, or stalling may be due to poor fuel delivery. The fuel pump itself is not directly monitored as part of the OBD II system, however, the fuel pump relay is monitored and several Diagnostic Trouble Codes (DTCs) are associated with it. In addition, the fuel injection system will try to adjust for poor fuel delivery and the resultant lean running condition may exceed the system's ability to compensate. This will store DTCs in the system memory. Suspected fuel pump problems should first be investigated with an appropriate diagnostic scan tool.

There are some preliminary tests that can be used to determine if the fuel pump or its electrical circuits are causing problems. Some of the tests described below require special test equipment such as a fuel pressure gauge and fittings.

The electrical current that operates the fuel pump is controlled by a relay and protected by a fuse. This helps to handle the high current load of the pump and also ensures that the pump will not continue to run in the event of an accident or if the engine stalls. If, for any reason, electric power to run the pump is interrupted, the engine will not run.

The fuel pump is located in the fuel tank and is part of the fuel delivery unit.

Begin troubleshooting with a simple check of the fuel pump electrical circuit. The pump should run while cranking the engine with the starter. If necessary, remove the floor cover and access plate in the luggage compartment and listen or feel to verify that the fuel pump is running. If the pump does not run, see **Fuel pump, checking electrical circuit**.

> ***CAUTION —***
> *In cold weather, water in the fuel may freeze in the pump causing the circuit to overload and the fuse to fail. Be sure to check the fuel pump fuse in the dashboard fuse/relay panel. See* **EWD Electrical Wiring Diagrams** *for fuse locations.*

If the fuel pump runs, check fuel pump delivery rate as described later in this section. The test will indicate whether further tests are necessary. This is especially important on higher mileage cars, where normal pump wear may decrease delivery volume. Also check for correct pump installation and for a clogged or restricted pump inlet screen.

The electric fuel pump operates only when the car is running or being started. Because many of the fuel pump and fuel injection tests require that the pump be operated with the engine off, the fuel pump relay can be temporarily bypassed during testing.

All tests assume that there is sufficient fuel in the tank. If there is any doubt regarding fuel quantity, add more. The fuel gauge may not indicate true fuel level. Do not run the fuel pump when the tank is empty.

The 5-cylinder, 2.5L engine (code: BPR, BPS) and the 2.0L 4-cylinder engine (code: BEV, BGD) require the fuel system to be bled any time it has been opened for testing. See **Fuel system, bleeding** in this repair group.

Fuel pump, operating for testing

The procedure below uses a temporary wiring connection to bypass the fuel pump fuse and run the pump directly from the battery by way of the fuse panel connection.

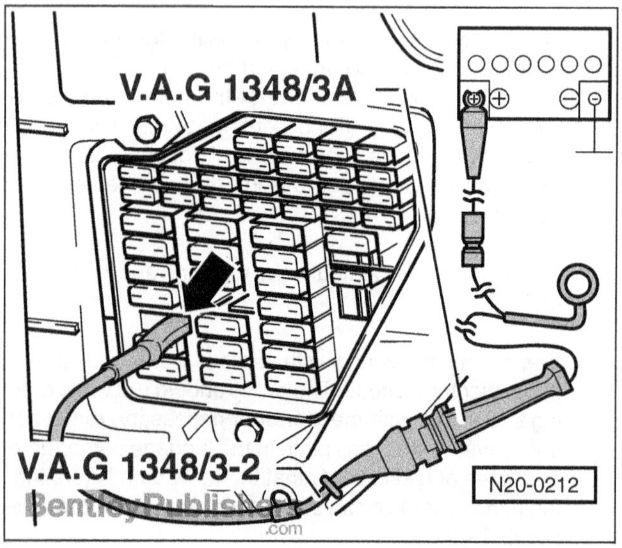

The preferred method is to use a remote switch such as Volkswagen special tool VAG 1348/3A and VAG 1348/3-2. You can also accomplish the same thing with a homemade fused jumper wire and an in-line switch.

> ***CAUTION —***
> - *A homemade jumper wire with a switch should be at least 1.5mm metric wire size (14 gauge AWG) and, for safety, include a 15-amp in-line fuse.*
> - *To avoid damaging the fuse panel sockets, the ends of the jumper wires should be flat-blade connectors that are the same size as the fuse blades.*
> - *Connect and disconnect the remote switch or jumper wire only with the switch in the off position.*

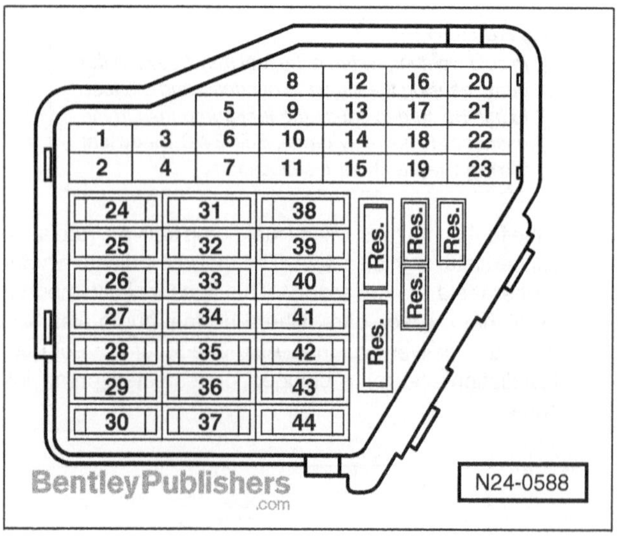

Remove fuel pump fuse from dashboard fuse panel.

- See **EWD Electrical Wiring Diagrams** to determine fuel pump fuse.
- Fuse #28 supplies fuel pump power for 1.8L and 2.0L engines covered by this manual.

– With ignition switched off, connect one lead of fused jumper wire to rear terminal of fuel pump fuse holder.

– Connect other lead of fused jumper to positive terminal of battery.

– Turn jumper wire switch on to run fuel pump.

- If the pump does not run, remove jumper lead from rear terminal of fuse holder and put it in on front terminal.
- If fuel pump still does not run, the problem is most likely in the wiring to fuel pump or the pump itself is faulty. See **Fuel pump, checking electrical circuit**.
- If fuel pump runs only with jumper connected, the relay, the engine control module (ECM), the circuit wiring or fuel pump fuse is faulty. See **EWD Electrical Wiring Diagrams**.

Fuel pump, checking electrical circuit

The fuel pump receives power from the fuel pump relay which is energized by the engine control module (ECM). The test given below checks for power at the fuel pump.

– Check that battery is fully charged and that fuel pump fuse is good.

- See **EWD Electrical Wiring Diagrams** to determine fuel pump fuse.
- Fuse #28 supplies fuel pump power for 1.8L and 2.0L engines covered by this manual.

– Fold down rear seat, pull back floor cover and remove access plate to fuel delivery unit.

– While a helper listens at the fuel pump, turn ignition key on. The pump may run briefly. If not, turn key to start position and crank engine. If fuel pump runs, the circuit is probably operating correctly. If fuel pump does not run, proceed as described below.

NOTE—
Listen carefully for fuel pump to run, in operation it is barely audible.

– Remove fuel pump fuse and run fuel pump as described in **Fuel pump, operating for testing**.

- If fuel pump runs only with jumper connected, the relay, the engine control module (ECM), the circuit wiring or fuel pump fuse is faulty. See **EWD Electrical Wiring Diagrams**.
- If fuel pump does not run, leave jumper connected and continue.

– Disconnect 4-pin harness connector from fuel delivery unit.

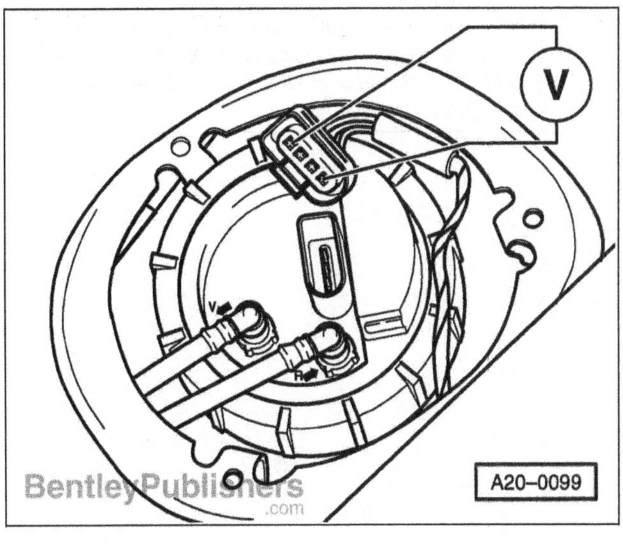

◄ Check for voltage across terminals 1 and 4 of fuel delivery unit harness connector.

- Terminal 1 is located at flat end of connector
- Terminal 4 is located at round end of connector

– If voltage is not present at connector, check wiring between dashboard fuse panel and fuel pump harness connector. See **EWD Electrical Wiring Diagrams**.

– If voltage is present, remove the pump and check for open circuits and internal grounds in the delivery unit. If no faults can be found, the fuel pump is probably faulty.

Specification

- Voltage at fuel pump . . . . 10.5 to 12.6 volts
- Fuel pump amperage (current) draw with engine at idle
  1.8L / 2.0L .8 amps maximum
  2.5L . . . . .9 amps maximum

Fuel pump, checking delivery rate

The tests given below require an accurate fuel pressure gauge with a shutoff valve and a range of 0-6 bar (approximately 0-100 psi). Volkswagen special tool VAG 1318 with appropriate adapters, or equivalent, can be used to check fuel delivery rate.

– Connect jumper wire with switch to fuel pump circuit as described in **Fuel pump, operating for testing**, given earlier.

– Remove fuel tank cap.

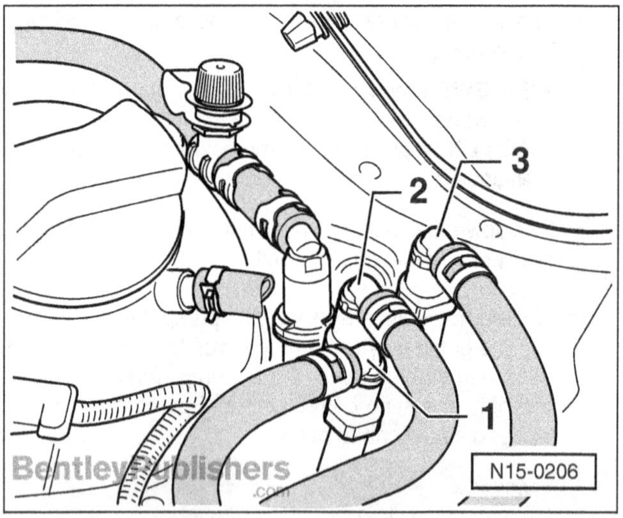

◄ 4-cylinder engines: Working in engine compartment, disconnect fuel supply line (**3**) from supply pipe and connect it to fuel pressure gauge.

> ***WARNING —***
> *Fire Hazard! Fuel will be expelled under pressure when fuel lines are disconnected. Do not smoke or work near heaters or other fire hazards. Keep a fire extinguisher handy.*

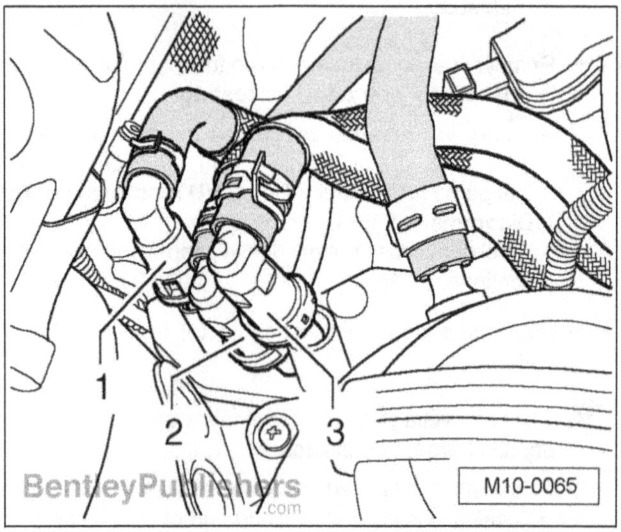

◄ 5-cylinder engines: Working in engine compartment, disconnect fuel supply line (**3**) from supply pipe and connect it to fuel pressure gauge.

> ***WARNING —***
> *Fire Hazard! Fuel will be expelled under pressure when fuel lines are disconnected. Do not smoke or work near heaters or other fire hazards. Keep a fire extinguisher handy.*

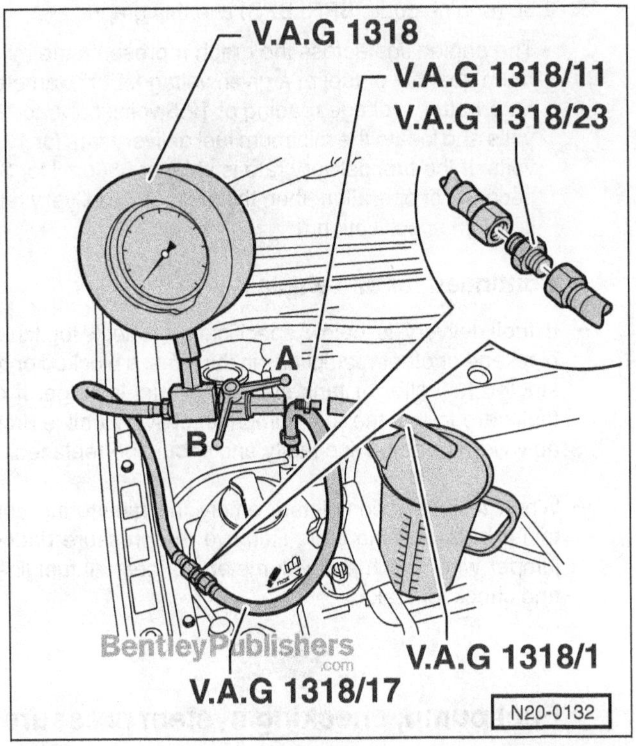

Connect fuel pressure gauge to supply pipe with appropriate adapters.

- Connect one end of a length of hose to valve side of gauge and place other end into a clean, fuel resistant measurement container with at least 1 liter (1 quart) capacity.

- Open the pressure gauge valve.

- Operate fuel pump with jumper wire switch while slowly closing valve on gauge until fuel pressure reads as specified.
  - 1.8L (APH, AWV, AWP, BKF, BNU) . . . 3 bar (44 psi)
  - 2.0L (AEG, AVH, AZG, BDC) . . . . 3 bar (44 psi)
  - 2.0L (BEV, BGD) . . . . .3.5 bar (51 psi)
  - 2.5L (BPR) .3.5 bar (51 psi)
  - 2.5L (BPS). . . . . .4.0 bar (58 psi)

- Without moving the valve, shut off fuel pump using jumper wire and empty fuel measuring container into another container.

- Connect voltmeter to battery and record voltage reading.

- Operate fuel pump with jumper wire switch for exactly 30 seconds and note quantity of fuel expelled into container.

- Subtract 2 volts from battery voltage reading and compare measured fuel quantity using appropriate graph shown below.

NOTE —

Fuel pump delivery volume is dependent on voltage available at fuel pump. For purposes of this test, the factory assumes a 2 volt drop between battery and fuel pump.

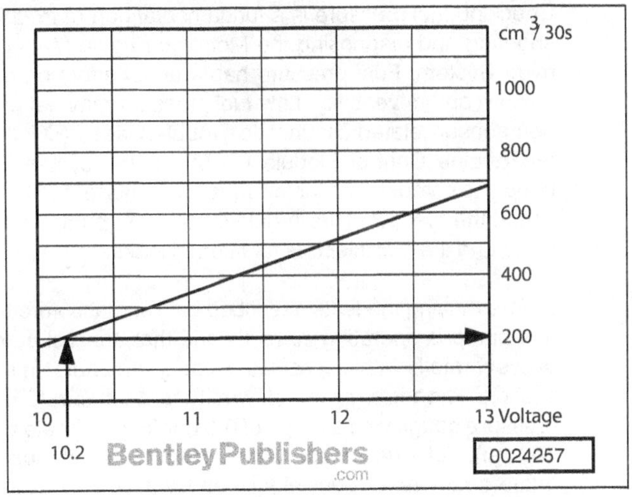

1.8L (engine code: APH, AWV, AWP, BKF, BNU) and 2.0L (engine code: AEG, AVH, AZG, BDC) use this graph.

- The angled line across the graph represents the minimum quantity of fuel at a given voltage. For example, for a battery voltage reading of 12.2 volts, subtract 2 volts and locate the minimum fuel delivery rate for 10.2 volts. If the fuel delivery rate is at least 200cm^3 for 30 seconds of operation, then the fuel pump delivery rate is within specification.

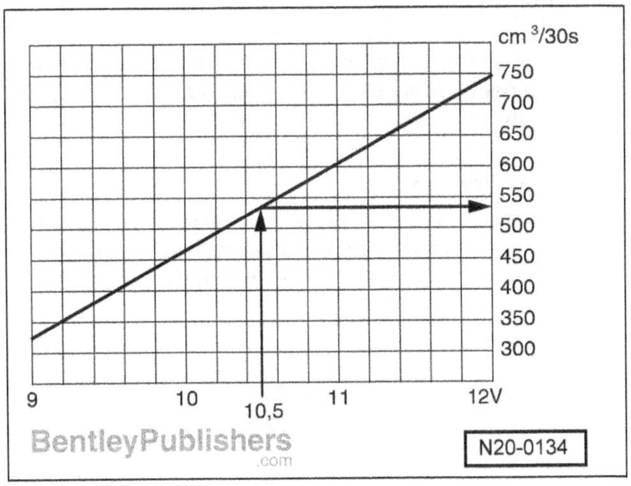

2.0L (engine code: BEV, BGD) use this graph.

- The angled line across the graph represents the minimum quantity of fuel at a given voltage. For example, for a battery voltage reading of 12.5 volts, subtract 2 volts and locate the minimum fuel delivery rate for 10.5 volts. If the fuel delivery rate is at least 540cm^3 for 30 seconds of operation, then the fuel pump delivery rate is within specification.

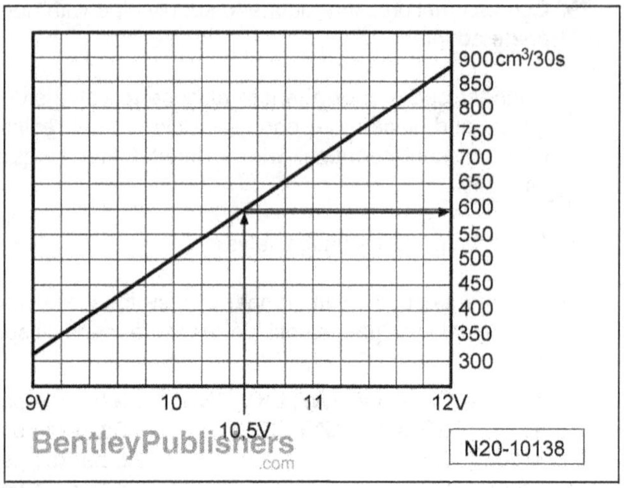

2.5L (engine code: BPR, BPS) use this graph.

- The angled line across the graph represents the minimum quantity of fuel at a given voltage. For example, for a battery voltage reading of 12.5 volts, subtract 2 volts and locate the minimum fuel delivery rate for 10.5 volts. If the fuel delivery rate is at least 580cm^3 for 30 seconds of operation, then the fuel pump delivery rate is within specification.

Continued for all vehicles

– If fuel delivery is below specification, check for kinks, blockage or other restrictions in the lines, a blocked or restricted fuel filter or tank strainer, or fuel leakage. If no faults are found, the fuel pump/fuel delivery unit is probably worn or otherwise faulty and should be replaced.

– When testing is completed, empty the measuring container back into the tank. Remove the pressure gauge, jumper wire switch and volt meter, connect all fuel lines and check for leaks.

Fuel pump, checking system pressures

Checking fuel pressure is a fundamental part of troubleshooting and diagnosing the Motronic Engine Management System. Fuel pressure has a direct effect on fuel mixture and driveability. Low fuel pressure may set oxygen sensor-related Diagnostic Trouble Codes (DTC) in the Engine Control Module (ECM) as the system attempts to compensate for low pressure. This test will also check the fuel pressure regulator, the fuel pump check valve and the fuel injectors for internal leaks.

Before making the tests described below, make sure the fuel pump is operating correctly and that the fuel pump delivery rate is within specification as given earlier in this section. The tests given below require an accurate fuel pressure gauge with a range of 0-6 bar (approximately 0-100 psi). Volkswagen special tool VAG 1318 with appropriate adapters, or equivalent, can be used to measure fuel pressure.

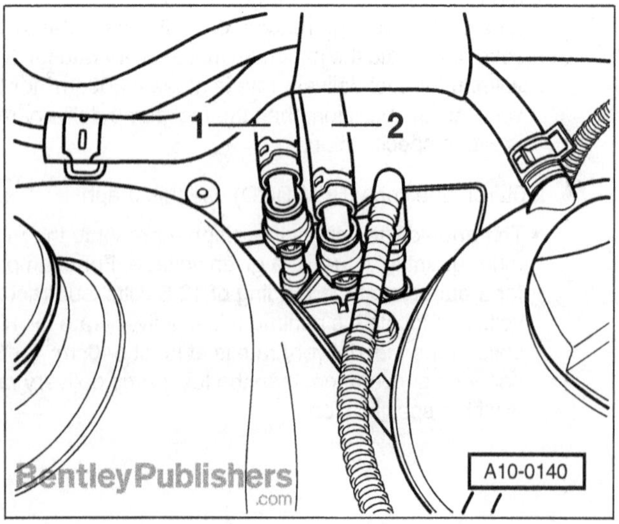

1.8L engine: Working in the engine compartment, disconnect fuel supply line (**1**) and catch leaking fuel with a cloth.

> ***WARNING —***
> *Fire Hazard! Fuel will be expelled under pressure when fuel lines are disconnected. Do not smoke or work near heaters or other fire hazards. Keep a fire extinguisher handy.*

2.0L engine: Working in engine compartment, disconnect fuel supply line (**arrow**) and catch leaking fuel with a cloth.

> ***WARNING —***
> *Fire Hazard! Fuel will be expelled under pressure when fuel lines are disconnected. Do not smoke or work near heaters or other fire hazards. Keep a fire extinguisher handy.*

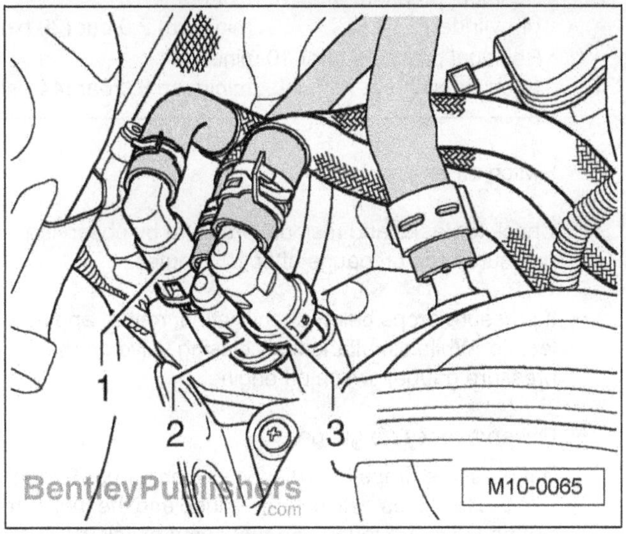

2.5L engine: Working in engine compartment, disconnect fuel supply line (**3**) and catch leaking fuel with a cloth.

> ***WARNING —***
> *Fire Hazard! Fuel will be expelled under pressure when fuel lines are disconnected. Do not smoke or work near heaters or other fire hazards. Keep a fire extinguisher handy.*

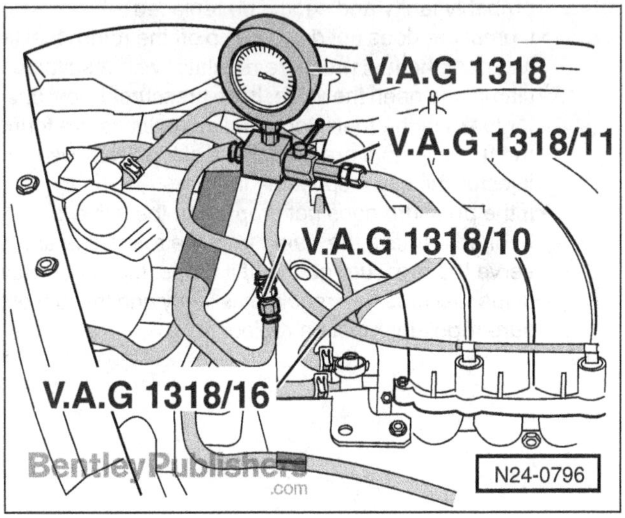

Connect fuel pressure gauge between fuel supply line and fuel rail. Make sure that if gauge is equipped with a valve, it is in the open position.

– Start engine, let it run at idle speed and make sure that there are no fuel leaks.

> ***NOTE —***
> *If engine does not start, operate fuel pump with a jumper wire as given earlier in* **Fuel pump, operating for testing**.

– Observe fuel pressure on gauge with engine at idle:
  - If system pressure is too high, check for a blocked or restricted fuel line from the fuel pressure regulator. Check the vacuum line to the fuel pressure regulator. If the fuel line has no restrictions and the vacuum line is properly connected, the fuel pressure regulator is probably faulty and should be replaced.
  - If system pressure is too low, check for leaks in the fuel supply lines. Also check for restrictions in the fuel supply lines or a clogged fuel filter. If there are no leaks or restrictions, the fuel pressure regulator is probably faulty and should be replaced.

- 4-cylinder engines: If fuel pressure is OK, disconnect and plug the small vacuum hose on the fuel pressure regulator and check that the fuel pressure increases as specified.
  - If fuel pressure does not increase, check the vacuum hose for kinks, restrictions and proper connections.

Specification
- Fuel pressure at idle (4-cylinder) . . . . approx. 2.0 bar (29 psi)
- Fuel pressure at idle (5-cylinder) . . . . approx. 4.0 bar (58 psi)
- Fuel pressure with regulator hose disconnected (4-cylinder) . . . . approx. 3.0 bar (44 psi)
- Residual pressure after 10 minutes (4-cylinder) . . . . minimum 2.0 bar (29 psi)
- Residual pressure after 10 minutes (5-cylinder) . . . . minimum 3.0 bar (44 psi)

- Switch off engine.

- Check for leaks and residual pressure by observing pressure drop on gauge after 10 minutes.

- If pressure drops below specification, restart engine and let idle. While simultaneously closing valve on the pressure gauge, switch off engine.

- Observe pressure gauge.
  - If pressure drops below specification, check for leaks in the fuel lines between the gauge and the fuel pump. If no leaks are found, the fuel pump check valve is probably faulty and should be replaced.
  - If pressure does not drop, clamp off the return fuel line just past the fuel pressure regulator with a suitable clamp and open the valve. If the pressure drops, check for leaks in the fuel injector rail. If no leaks are found, the fuel injectors are probably leaking. Remove the injector rail and inspect the injectors.
  - If the pressure does not drop when the valve is opened, remove the clamp from the return line and observe the pressure gauge. If it drops, the check valve in the fuel pressure regulator is faulty and the fuel pressure regulator must be replaced.

FUEL PUMP (FUEL DELIVERY UNIT)

The fuel delivery unit is mounted in the top of the tank and retained by a threaded retaining ring. The 2.5L 5-cylinder engine (code: BPR) has a metal fuel tank and the fuel delivery unit is located underneath a cover which is mounted to the fuel tank with 10 nuts.

NOTE —
The procedure given below also applies to removing and installing the fuel level sensor on 1.9L diesel engines.

Fuel delivery unit, removing and installing

– With the ignition switched off, disconnect the negative (–) battery cable.

NOTE —
Be sure to have the anti-theft radio code on hand before disconnecting the battery.

– Fold rear seat bottom forward, pull back floor cover and remove the access plate to the fuel delivery unit.

– Disconnect the 4-pin harness connector from the fuel delivery unit.

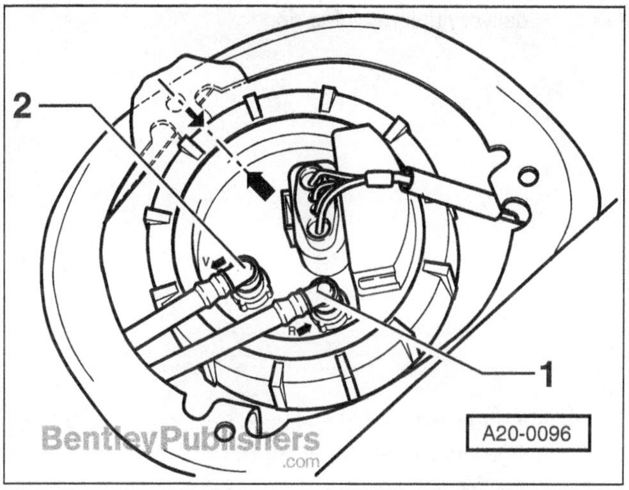

◄ Disconnect the fuel supply (**2**) and return lines (**1**) and wrap with a cloth to prevent fuel spillage.

> ***WARNING —***
> *Fuel will be discharged and dangerous fuel vapors will be present. Work in a well ventilated area. Do not disconnect wires that could cause sparks. Do not smoke or work near heaters or other fire hazards. Keep a fire extinguisher handy.*

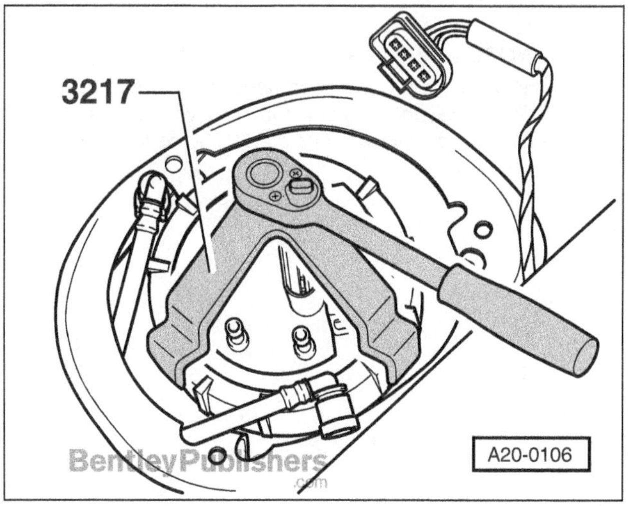

◄ Unscrew and remove the fuel pump retaining ring using appropriate special tool.

- Special tool 3217 for plastic retaining ring (vehicles with 4-cylinder engines)
- Special tool 3087 or T10202 for 2.5L (BPS) with plastic fuel tank

– Carefully lift the delivery unit from the tank. Empty fuel delivery unit into a suitable container once it is removed.

– Installation is the reverse of removal noting the following points:

- Inspect and clean the intake screen as needed.
- Do not bend the fuel gauge sending unit arm when installing.
- Inspect the fuel tank seal and replace if damaged or distorted.
- Moisten the inner lip of fuel delivery unit seal with fuel when installing.
- Check for leaks before installing access plate.

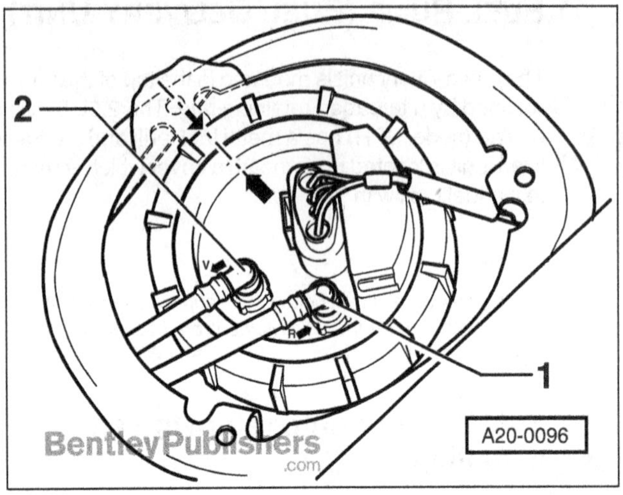

4-cylinder: Position retaining ring for fuel delivery unit with marks (**arrows**) aligned as shown.

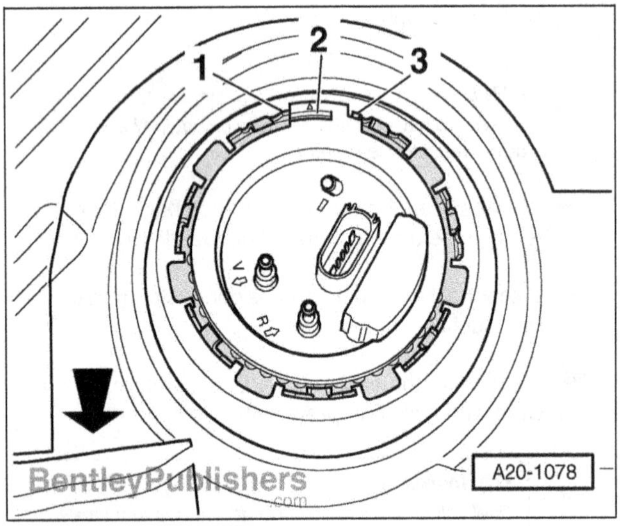

5-cylinder with plastic fuel tank: Position tab (**2**) on fuel delivery unit between tabs (**1** and **3**) on fuel tank as shown.

- **Arrow** points toward front of vehicle.
- Use special tool 3087 or T10202 to remove/install fuel delivery unit retaining ring

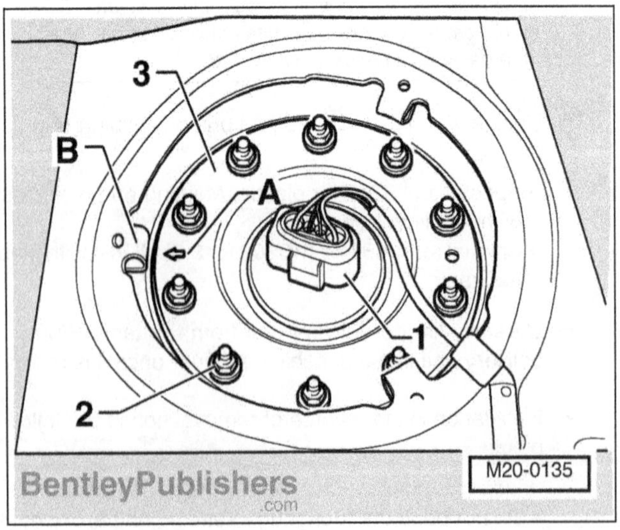

5-cylinder with metal fuel tank: Position arrow (**A**) on fuel delivery unit cover (**3**) at tab (**B**).

- Tab (**B**) is oriented toward rear of vehicle.

NOTE —

Metal fuel tanks have been introduced for vehicles with SULEV (Super Ultra Low Emission Vehicles) requirements. This began with the 2.5L engine (code: BPR) from m.y. 2006.

Fuel gauge sensor, removing and installing

– Remove fuel delivery unit as described previously.

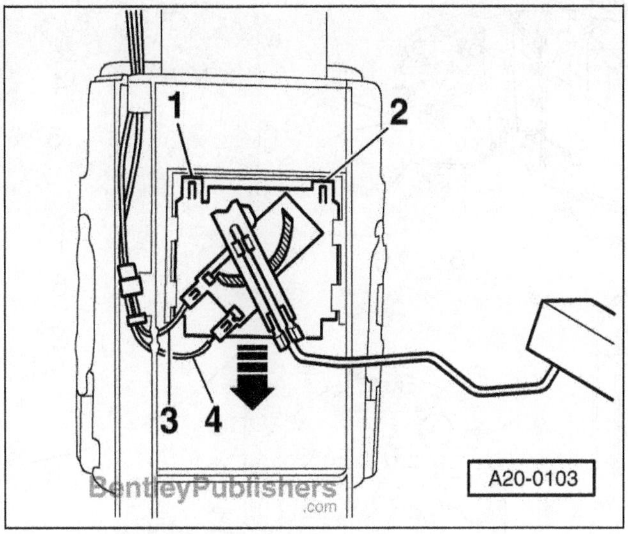

– Release and pull off terminals for wires (**3** and **4**).
  - Note wire positions for installation later.

– Carefully pry off retaining straps (**1** and **2**) using a screwdriver.

– Remove fuel gauge sensor downward (**arrow**).

– To install, insert sensor into guides on fuel delivery unit and carefully press up until engaged.

– Install wires onto sensor terminals.

– Install fuel delivery unit.

EVAPORATIVE EMISSIONS (GASOLINE ENGINES)

All New Beetles with gasoline engines are equipped with evaporative emissions system monitoring. This system is referred to as the Leak Detection System.

The ECM determines when the appropriate conditions are met and activates the Leak Detection Pump (LDP) to pump a small amount of air pressure into the fuel tank and lines. The pump runs until a pre-determined pressure or time is reached. If the system does not hold the pressure, a fault or DTC is stored in the ECM memory.

Basic troubleshooting should begin with a visual inspection of the evaporative system hoses and components with particular attention paid to the fuel cap and seal. Further testing will require use of a scan tool such as Volkswagen special tool VAG 1551/1552 or VAS 5051/5052 or equivalent diagnostic scan tool.

NOTE —

Volkswagen identifies electrical components by a letter and/or a number in the electrical schematics. See **97 Wiring Diagrams, Fuses and Relays**. *These electrical identifiers are listed in parenthesis as an aid to electrical troubleshooting.*

Leak detection system components

NOTE —

5-cylinder shown, 4-cylinder similar.

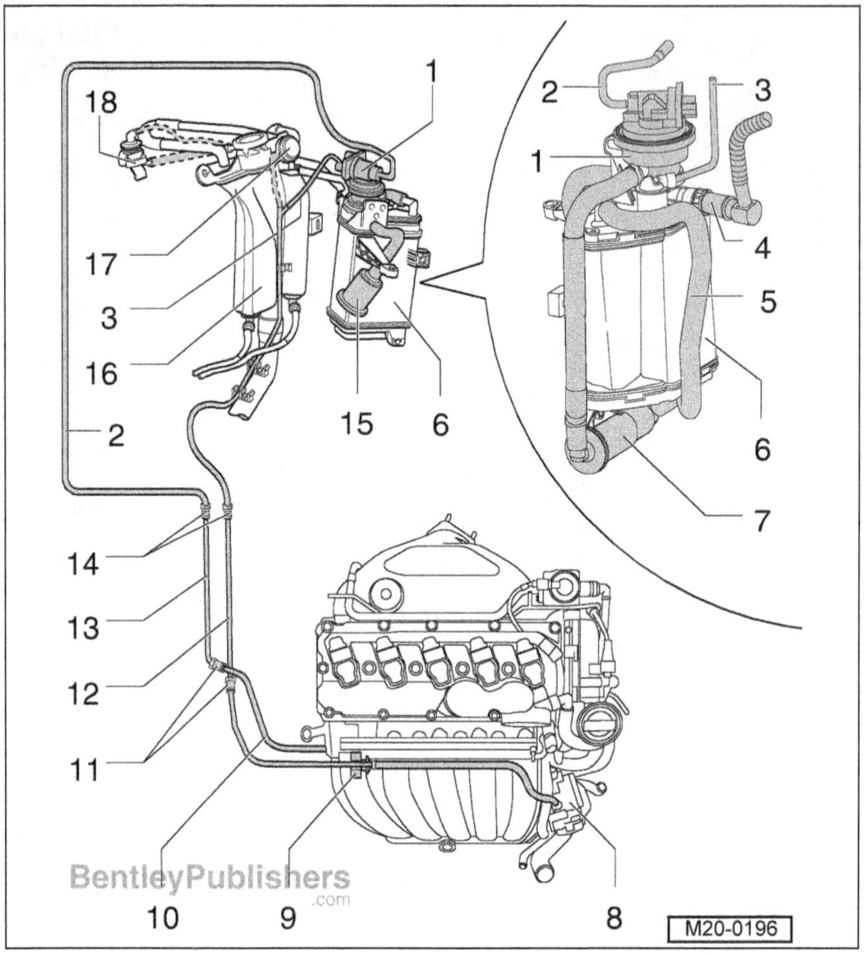

1. **Leak detection pump (V144)**
   - On EVAP canister
   - Under right rear wheel housing liner
2. **Vacuum line**
   - From leak detection pump to intake manifold
3. **Vent line**
   - To EVAP canister purge regulator valve
4. **Vent line**
5. **Connecting line**
   - From leak detection pump
6. **Evaporative emissions (EVAP) canister**
   - In rear wheel housing under wheel housing liner
7. **Auxiliary filter**
8. **Throttle valve control module (J338)**
9. **EVAP canister purge regulator valve (N80)**
10. **Vacuum line**
11. **Separating point**
12. **Vent line**
13. **Vacuum line**
14. **Separating point**
15. **Air filter housing**
    - For leak detection pump
16. **Filler tube**
17. **Pressure retention valve**
    - Only for 2.5L (BPS) engine
    - For fuel tank ventilation
18. **Gravity valve**
    - Only for 2.5L (BPS) engine

Leak detection pump and EVAP canister (4-cylinder)

1. **Leak detection pump (V144)**
   - On EVAP canister
2. **3-pin harness connector**
3. **Vacuum line**
   - To intake pipe upper section
   - Press together at front to release
4. **Vent line**
   - From fuel tank change-over valve
5. **To EVAP canister purge regulator valve (N80)**
   - In engine compartment on air cleaner
6. **Evaporative emissions (EVAP) canister**
   - In right rear wheel housing under liner
   - To remove and install: remove the taillight
7. **Bolt**
   - Tighten to 10 Nm (7 ft-lb)
8. **Bolt**
   - Tighten to 3 Nm (27 in-lb)
9. **Air cleaner**
   - For leak detection pump
10. **Connecting hose**
    - Suction side
11. **Connecting hose**
    - Pressure side

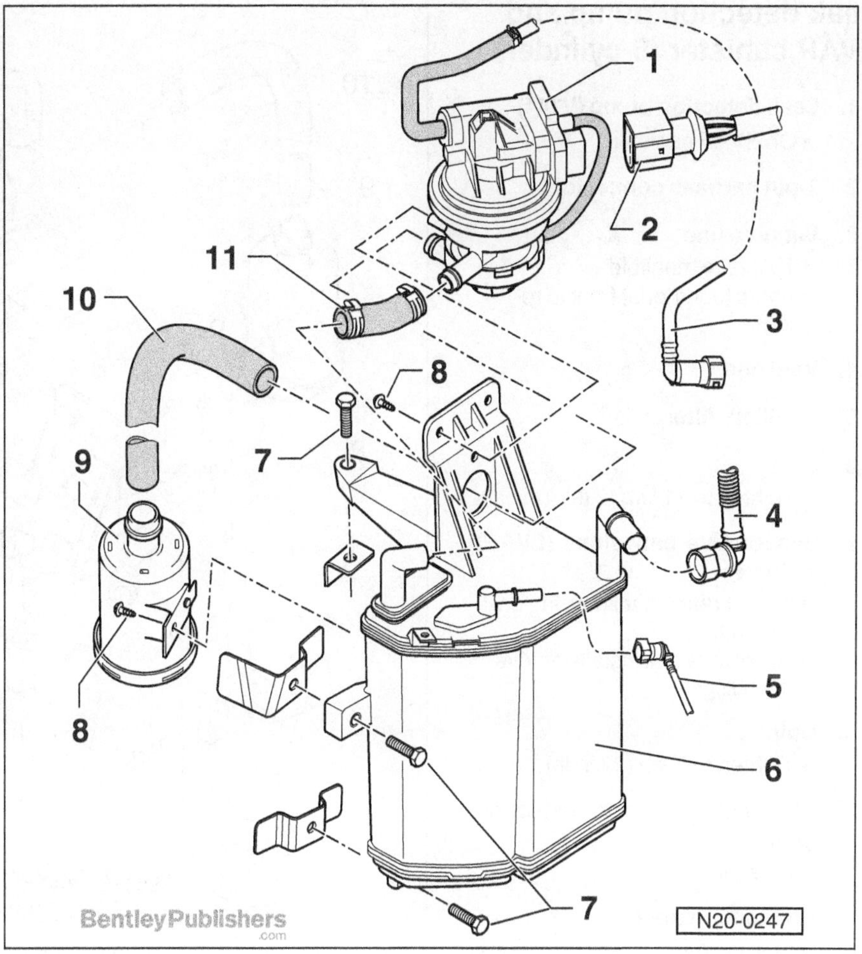

Leak detection pump and EVAP canister (5-cylinder)

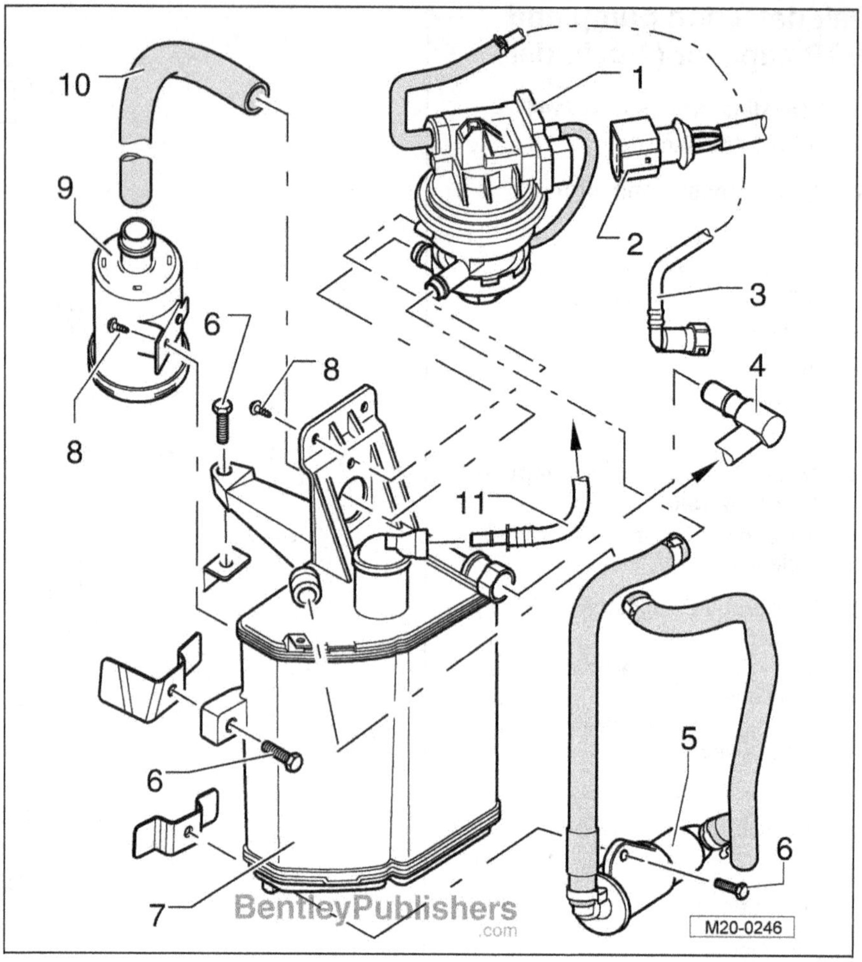

1. **Leak detection pump (V144)**
   - On EVAP canister
2. **3-pin harness connector**
3. **Vacuum line**
   - To intake manifold
   - Press together at front to re-lease
4. **Vent line**
5. **Auxiliary filter**
6. **Bolt**
   - Tighten to 11 Nm (8 ft-lb)
7. **Evaporative emissions (EVAP) canister**
   - In right rear wheel housing un-der liner
   - To remove and install: remove the taillight
8. **Bolt**
   - Tighten to 3Nm (27 in-lb)
9. **Air filter for leak detection pump**
   - Clean if soiled
10. **Connecting hose**
    - Intake side
11. **To EVAP canister purge regula-tor valve (N80)**
    - In engine compartment

FUEL TANK AND LINES

WARNING —

- *If working on a lift or hoist be sure to support front of vehicle before removing fuel tank. Removal of the fuel tank and associated components can unbalance vehicle sufficiently to cause vehicle to fall.*
- *When removing or installing fuel tank, the tank should be as empty as possible to facilitate easier handling.*

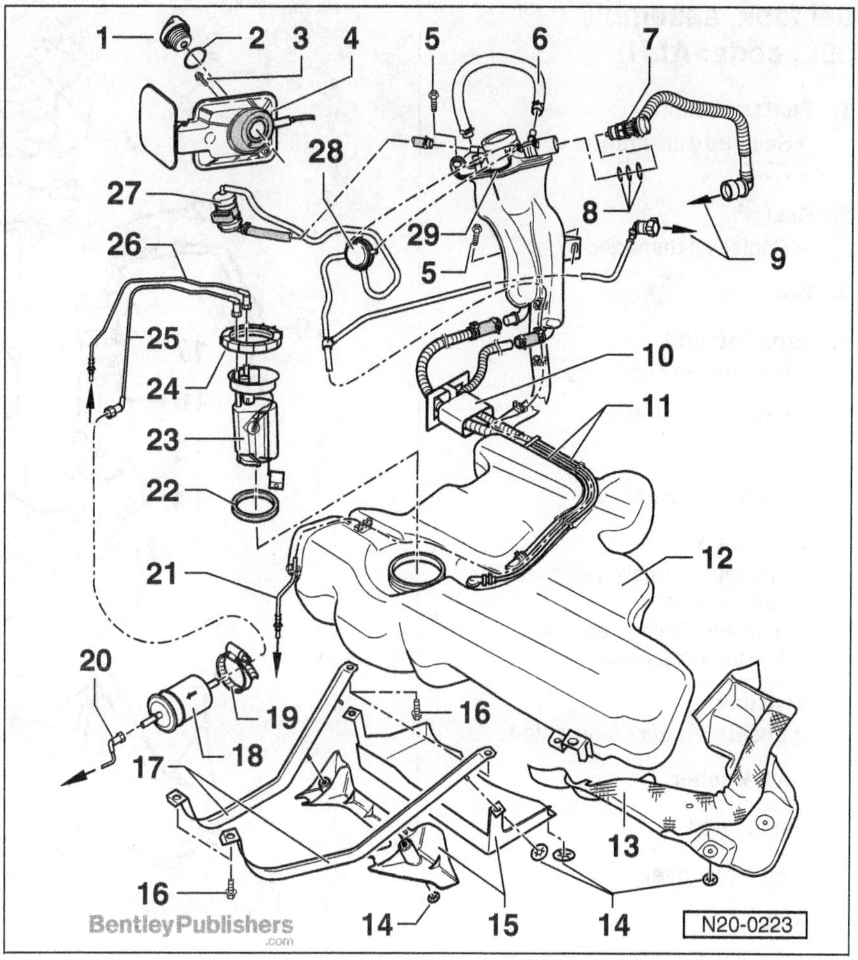

Fuel tank, assembly (1.8L, 2.0L engines)

1. **Gas cap**
2. **Seal**
   - Replace if damaged
3. **Bolt**
   - Tighten to 1 Nm (9 in-lb)
4. **Tank flap unit**
   - With rubber cup
5. **Bolt**
   - Tighten to 10 Nm (7 ft-lb)
6. **Vent pipe**
7. **Change-over valve**
   - Removing: unclip out of side support
   - Before installing, remove sealing cap (item 1)
8. **O-ring**
   - Replace if damaged
9. **EVAP canister**
10. **Wiring routing**
    - In floor pan long member
11. **Vent pipe**
    - Ensure it is securely seated
    - Secure using spring clips
12. **Fuel tank**
    - Support with VAG 1383A engine/transmission jack when removing
13. **Heat shield**
    - For fuel tank
14. **Clamping washer**
15. **Fuel tank cover**
16. **Bolt**
    - Tighten to 20 Nm (15 ft-lb) plus ¼ turn (90°)
17. **Mounting straps**
    - Note differing lengths
    - Installed position: attaching points (holes) point in direction of travel
18. **Fuel filter**
    - Installed position: arrow points in direction of flow
19. **Screw clip**
20. **Supply line**
    - Black
    - To fuel rail
    - Ensure it is securely seated
21. **Vent line**
    - White
    - Clipped onto top of fuel tank
    - With EVAP canister purge regulator valve (N80)
    - Ensure it securely seated
22. **Sealing ring**
    - Replace if damaged
    - Coat with fuel when installing
23. **Fuel delivery unit**
    - Note installed position
    - Clean strainer if soiled
24. **Retaining ring**
    - Remove and install using special tool:
      3217 for plastic ring
      T10202 for metal ring
25. **Supply line**
    - Black
    - Clipped to side of fuel tank
    - Ensure it is securely seated
26. **Return line**
    - Blue or blue markings
    - Clipped to side of fuel tank
    - Ensure it is securely seated
    - From fuel rail
27. **Gravity overflow valve**
    - Check valve for flow
      valve vertical: OPEN
      valve tilted 45°: CLOSED
28. **Pressure retention valve**
    - For fuel reservoir ventilation
29. **Ground connection**
    - Ensure it is securely seated

Fuel tank, assembly (1.9L, code: ALH)

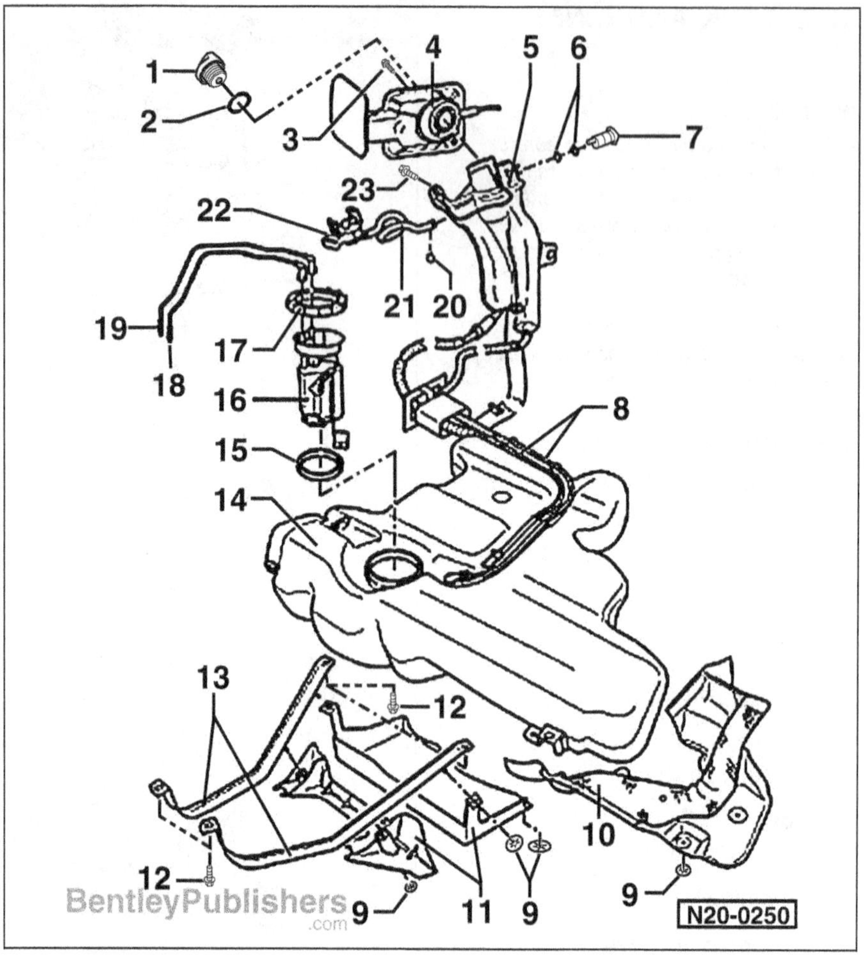

1. **Fuel tank cap**
   - Secured with strap from m.y. 2000
2. **Seal**
   - Replace if damaged
3. **Bolt**
4. **Tank flap unit**
   - With rubber cup
5. **Ground connection**
6. **O-ring**
   - Always replace
7. **Vent valve**
   - To remove: unclip valve out of side support
   - Before installing, remove fuel tank cap (item 1)
8. **Vent line**
   - Ensure it is securely seated
9. **Lock washer**
10. **Heat shield**
11. **Fuel tank cover**
12. **Bolt**
    - Tighten to 25 Nm (18 ft-lb)
13. **Mounting straps**
    - Note differing lengths
14. **Fuel tank**
    - Support with VAG 1383A engine/transmission jack when removing
15. **Seal**
    - Replace if damaged
    - Moisten with fuel when installing
16. **Fuel gauge sensor**
    - Note installed position on fuel tank
17. **Retaining ring**
    - Remove and install using special tool 3217
18. **Supply line**
    - Black
    - To fuel filter
    - Clipped onto fuel tank
    - Ensure line is securely seated during installation
    - To remove from flange, press down on tabs at connection
19. **Return line**
    - Blue or blue markings
    - From fuel filter
    - Clipped onto fuel tank
    - Ensure line is securely seated during installation
    - To remove from flange, press down on tabs at connection
20. **Clamp**
    - Always replace
21. **Hose**
22. **Gravity overflow valve**
    - To remove, unclip upwards out of support
    - Check valve for flow
      valve vertical: OPEN
      valve tilted 45°: CLOSED
23. **Bolt**
    - Tighten to 10 Nm (7 ft-lb)

Fuel tank, assembly (1.9L, code: BEW)

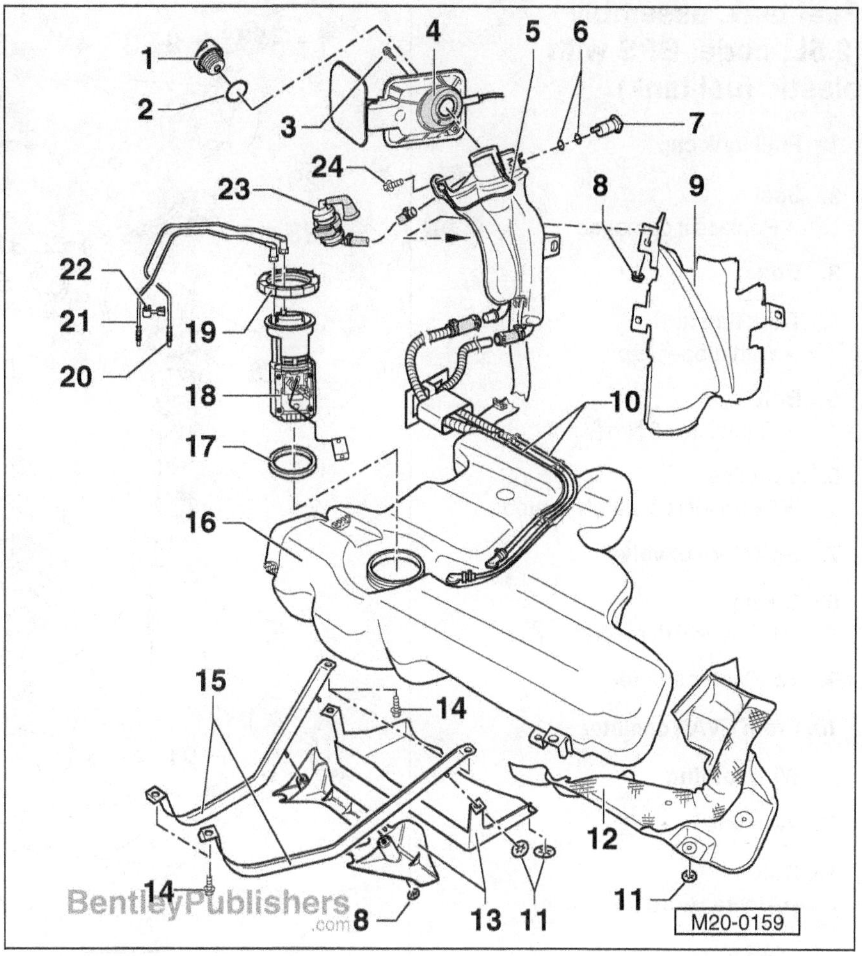

1. **Fuel tank cap**
2. **Seal**
   - Replace if damaged
3. **Bolt**
4. **Tank flap unit**
   - With rubber cup
5. **Ground connection**
6. **O-ring**
   - Always replace
7. **Vent valve**
   - To remove: unclip valve out of side support
   - Before installing, remove fuel tank cap (item 1)
8. **Nut**
   - Tighten to 6 Nm (53 in-lb)
9. **Protective cover**
   - Note installed position
   - May not be installed on all vehicles
10. **Vent line**
    - Ensure it is securely seated
11. **Lock washer**
12. **Heat shield**
13. **Fuel tank cover**
14. **Bolt**
    - Tighten to 20 Nm (15 ft-lb) plus ¼ turn (90°)
    - Always replace
15. **Mounting straps**
    - Note differing lengths
16. **Fuel tank**
    - Support with VAG 1383A engine/transmission jack when removing
17. **Seal**
    - Replace if damaged
    - Install seal "dry" into fuel tank opening
    - Moisten inner edge of seal with fuel when installing fuel delivery unit
18. **Fuel delivery unit**
    - Note installed position on fuel tank
19. **Retaining ring**
    - Remove and install using special tool:
      3217 for plastic ring
      T10202 for metal ring
20. **Supply line**
    - Black
    - To fuel filter
    - Clipped onto fuel tank
    - Ensure it is securely seated
    - To remove from flange, press down on tabs at connection
21. **Return line**
    - Blue or blue markings
    - From fuel filter
    - Clipped onto fuel tank
    - Ensure it is securely seated
    - To remove from flange, press down on tabs at connection
22. **Clamp**
23. **Gravity overflow valve**
    - To remove, unclip upwards out of support
    - Check valve for flow
      valve vertical: OPEN
      valve tilted 45°: CLOSED
24. **Bolt**
    - Tighten to 10 Nm (7 ft-lb)

Fuel tank, assembly (2.5L, code: BPS with plastic fuel tank)

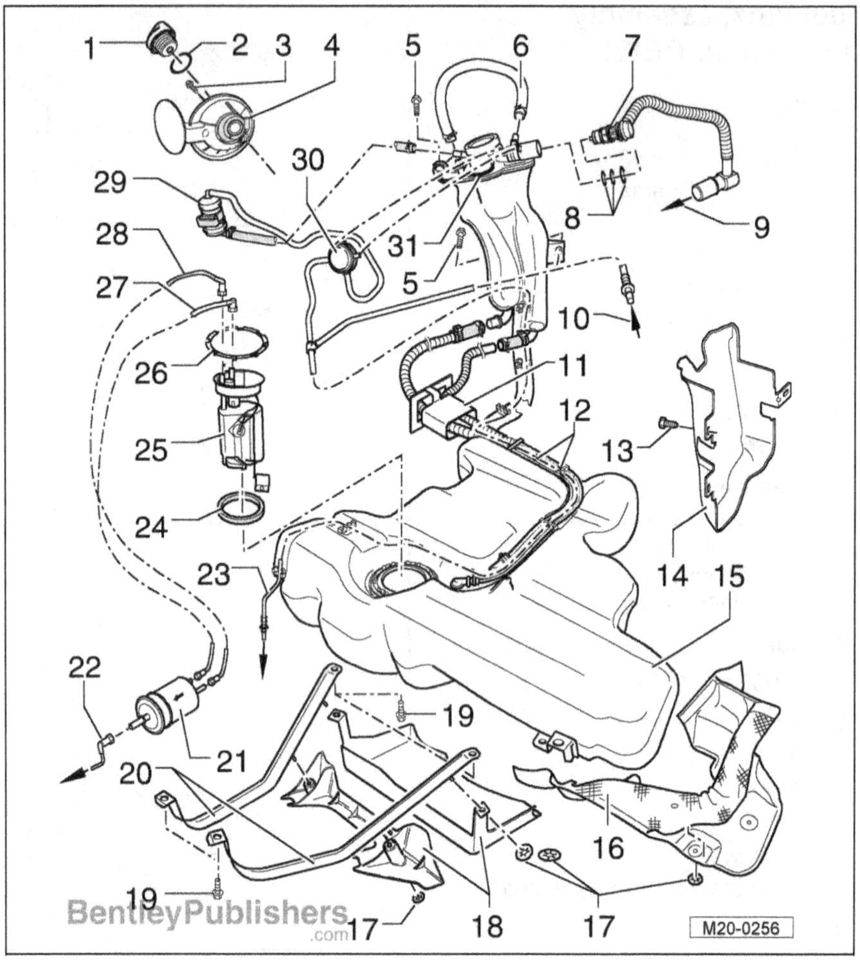

1. **Fuel tank cap**
2. **Seal**
   - Replace if damaged
3. **Bolt**
4. **Tank flap unit**
   - With rubber cup
5. **Bolt**
   - Tighten to 10 Nm (7 ft-lb)
6. **Vent line**
   - Ensure it is securely seated
7. **Switch-over valve**
8. **O-ring**
   - Replace if damaged
9. **To EVAP canister**
10. **From EVAP canister**
11. **Wire routing**
12. **Vent line**
13. **Bolt**
    - Tighten to 10 Nm (7 ft-lb)
14. **Protective plate**
15. **Fuel tank**
    - Support with VAG 1383A engine/transmission jack when removing
16. **Heat shield**
17. **Lock washer**
18. **Fuel tank cover**
19. **Bolt**
    - Tighten to 20 Nm (15 ft-lb) plus ¼ turn (90°)
    - Always replace
20. **Mounting straps**
    - Note differing lengths
21. **Fuel filter**
    - Located under fuel tank
    - Arrow points in direction of flow
    - Note position of retaining clamp
22. **Supply line**
    - Black
    - To fuel rail
    - Ensure it is securely seated
    - To disconnect from fuel filter squeeze retaining collar and pull
23. **Vent line**
    - White
    - Clipped to top of fuel tank
    - Ensure it is securely seated
    - To EVAP canister purge regulator valve
24. **Seal**
    - Replace if damaged
    - Install seal "dry" into fuel tank opening
    - Moisten inner edge of seal with fuel when installing fuel delivery unit
25. **Fuel delivery unit**
    - Note installed position on fuel tank, see **Fuel delivery unit, removing and installing**
26. **Retaining ring**
    - Remove and install using special tool T10202
27. **Supply line**
    - Black
    - To fuel filter
    - Ensure it is securely seated
28. **Return line**
    - Blue or blue markings
    - From fuel filter
29. **Gravity overflow valve**
    - To remove, unclip upwards out of support
    - Check valve for flow
      valve vertical: OPEN
      valve tilted 45°: CLOSED
30. **Pressure retention valve**
    - For fuel tank ventilation
31. **Ground connection**
    - Ensure clean tight connection

Fuel tank, assembly (2.5L, code: BPR with metal fuel tank)

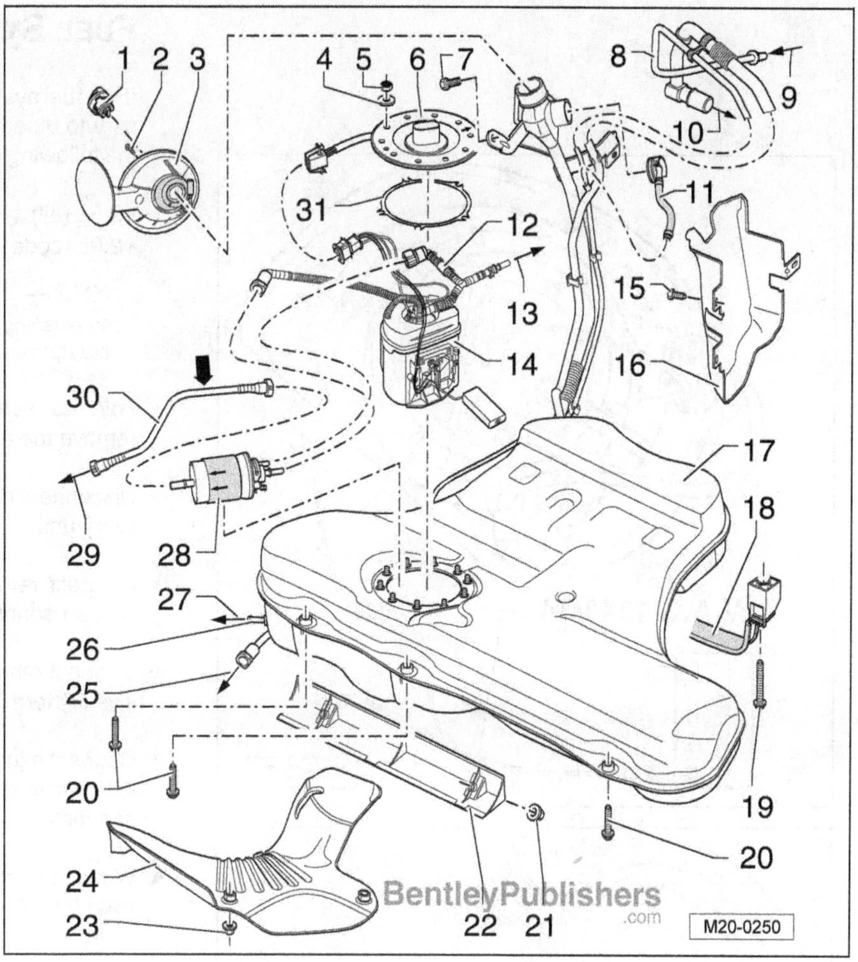

NOTE —
Metal fuel tanks have been introduced for vehicles with SULEV (Super Ultra Low Emission Vehicles) requirements. This began with the 2.5L engine (code: BPR) from m.y. 2006.

1. **Fuel tank cap**
2. **Bolt**
3. **Tank flap unit**
   - With rubber cup
4. **Washer**
5. **Nut**
   - Tighten to 7 Nm (62 in-lb)
6. **Cover**
   - Note installation position, see **Fuel delivery unit, removing and installing**
7. **Bolt**
   - Tighten to 10 Nm (7 ft-lb)
8. **Vent line**
   - White
   - Ensure it is securely seated
9. **From EVAP canister**
10. **To EVAP canister**
11. **One-way valve**
    - Forces flow in direction of filler neck
12. **Fuel pump check valve**
13. **To suction jet pump**
    - Suction jet pump is located on fuel tank and used to bleed tank
14. **Fuel delivery unit**
    - Note installed position
    - When installing, rotate unit in bayonet mount clockwise until it contacts stop
    - Clean strainer if soiled
15. **Bolt**
    - Tighten to 10 Nm (7 ft-lb)
16. **Protective plate**
17. **Fuel tank**
    - Support with VAG 1383A engine/transmission jack when removing
18. **Mounting strap**
    - With rubber profile and spacer sleeve
19. **Bolt**
    - Tighten to 20 Nm (15 ft-lb) plus ¼ turn (90°)
    - Always replace
20. **Bolt**
    - Tighten to 20 Nm (15 ft-lb) plus ¼ turn (90°)
    - Always replace
21. **Nut**
    - Tighten to 3 Nm (27 in-lb)
22. **Trim panel**
23. **Nut**
    - Tighten to 3 Nm (27 in-lb)
24. **Heat shield**
25. **To EVAP canister purge regulator valve**
26. **Connection for supply line**
27. **To fuel rail**
28. **Fuel filter**
    - Located under fuel tank
    - Arrow points in direction of flow
    - Note position of retaining clamp
    - Also has fuel pressure regulator
29. **To connection in fuel tank**
30. **Supply line**
    - Black
    - To fuel rail
    - Ensure it is securely seated
    - To disconnect from fuel filter squeeze retaining collar and pull
31. **Seal**
    - Replace if damaged
    - Note installation position: bulge rests on fuel tank

FUEL SYSTEM, BLEEDING

If the fuel system has been opened for service it is necessary to bleed the fuel system for vehicles equipped with the following engines:

- 2.5L (all)
- 2.0L (code: BEV, BGD)

NOTE —
Not bleeding the fuel system can cause damage to the catalytic converter.

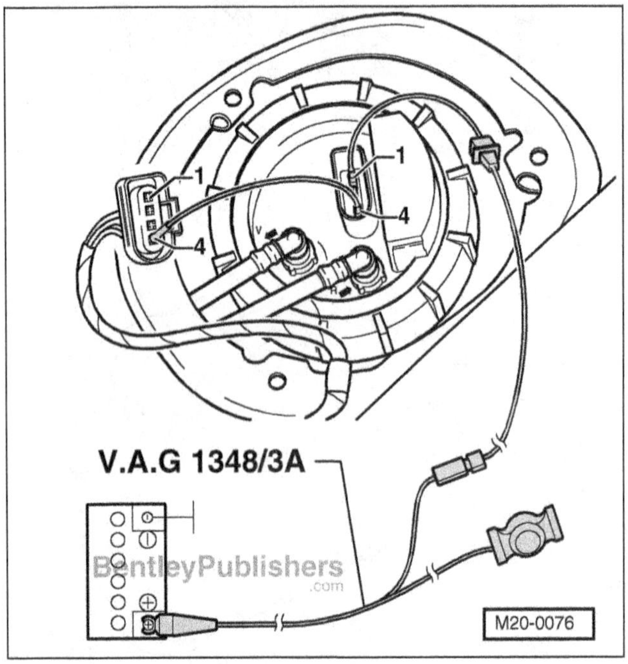

– Fold rear seat bottom forward, pull back floor cover and remove the access plate to the fuel delivery unit.

– Disconnect the 4-pin harness connector from the fuel delivery unit.

< Connect remote control unit (VAG 1348/3A or similar) with an adapter cable to terminal **1** of fuel delivery unit.

– Connect other end of remote control unit to battery positive (+) terminal.

– Connect a jumper wire (or adapter from VAG 1594 C test kit) between fuel delivery unit terminal **4** and harness connector terminal **4**.

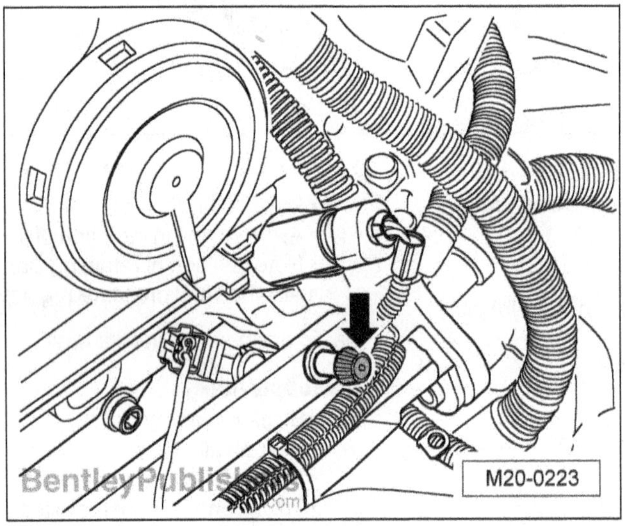

< Working in engine compartment, remove cover cap (**arrow**) from fuel system bleeder valve on fuel rail.

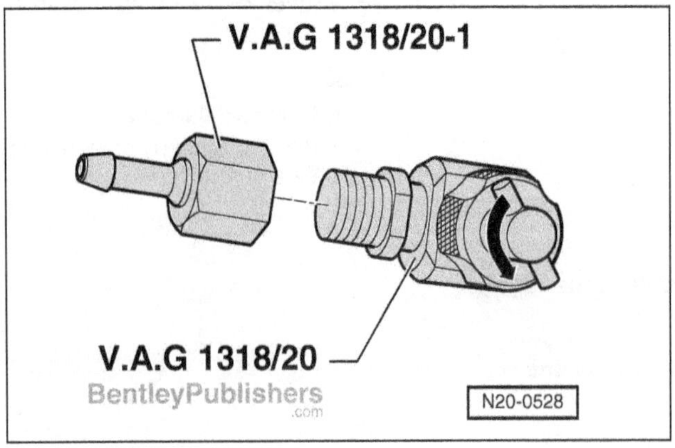

< Screw adapter VAG 1318/20-1 onto VAG 1318/20.

– Open valve (**arrow**) of VAG 1318/20.

– Install VAG 1318/20 with VAG 1318/20-1 onto fuel bleeder valve.

◀ Connect hose from fluid extractor (VAS 5226 or equivalent).

– Turn valve (T-handle of VAG 1318/20) clockwise (closed) to stop.

– Check adapter and hose connection for leaks.

– Operate fuel pump remote control (VAG 1348/3A) until fuel flows out of bleeder valve without bubbles.

– Turn valve (T-handle of VAG 1318/20) counter-clockwise until completely open again.

– Clamp off hose from suction pump VAS 5226 (i.e. with hose clamps 3094) and pull hose off from adapter (VAG 1318/20-1).

– Remove adapter (VAG 1318/20) from fuel rail bleeder valve.

– Install cap back on bleeder valve. Fuel system is now bled.

ELECTRONIC ACCELERATOR (E-GAS) SYSTEM

Most Volkswagen New Beetles covered by this manual use an electronic accelerator (E-Gas) system. The exception to this is the 2.0L (code: AEG) engine (1998-2000) that uses a traditional accelerator cable.

This repair group provides an overview of the E-Gas system. For information of the components and operation of the Bosch engine management systems, see **2 Fuel, Ignition and Exhaust Systems, 23 Fuel Injection–Diesel** or **24 Fuel Injection–Gasoline.**

In the E-Gas system, the throttle valve is not operated via a cable from the accelerator pedal. There is no mechanical connection between the accelerator pedal and the throttle valve.

The position of the accelerator pedal is communicated to the engine control module via two sensors for accelerator pedal position (changeable resistances; stored in one housing) that are connected with the accelerator pedal. The position of the accelerator pedal (driver's intention/demand) is a main input value for the Engine Control Module (ECM).

Operation of the throttle valve occurs throughout the entire RPM and load range via an electric motor (throttle valve actuator) in the throttle valve control module. The throttle valve is operated by the throttle valve actuator according to the specifications of the ECM.

With the engine at standstill and the ignition switched on, the ECM activates the throttle valve actuator exactly according to the specifications of the accelerator pedal position sensor. This means, if the pedal is depressed halfway, the throttle valve actuator opens the throttle valve to the same degree; the throttle valve is then opened halfway.

With the engine running (under load), the ECM can open and close the throttle valve independently of the accelerator pedal position sensor. In this way, the throttle valve can already be completely open, even though the accelerator pedal has only been depressed half way. This is advantageous since charging losses at the throttle valve can be avoided. Furthermore, it results in significantly better emissions and consumption values under certain load conditions.

It would be wrong to assume that "E-Gas" consists of only one or two components. E-Gas is much more of a system containing all components that contribute to determining, regulating and monitoring the throttle valve position (e.g. accelerator pedal position sensor, throttle valve control module, Electronic Power Control (EPC) warning lamp, Engine Control Module).

Malfunction light for accelerator activation

NOTE —

The malfunction light for power accelerator activation is also known as the Electronic Power Control (EPC) warning lamp.

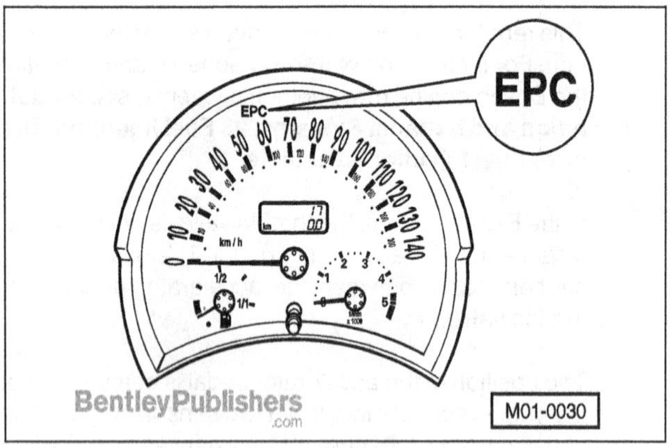

A functional check of the EPC warning lamp can be conducted by first switching on the ignition and noting if the EPC warning lamp comes on.

The EPC light should go out once the engine is started and running at idle speed if there are no Diagnostic Trouble Codes (DTCs) relating to the E-Gas system. Further diagnosis of the EPC warning light involves the use of Volkswagen special tools which is outside the scope of this manual.

E-Gas pedal components

1. **Bracket**
2. **Connector**
   - Black, 5-pin
3. **Nut**
   - Tighten to 10 Nm (7 ft-lb)
4. **Accelerator pedal position sensor (G79)**
   - Not adjustable
   - Communicates the driver's intentions to the control module
   - Remove cover in footwell before removing sensor

   NOTE —
   Volkswagen identifies electrical components by a letter and/or a number in the electrical schematics. See **EWD Electrical Wiring Diagrams**. *These electrical identifiers are listed in parenthesis as an aid to electrical troubleshooting.*
5. **Bracket**
   - For cover in footwell
   - Clipped on at accelerator pedal position sensor

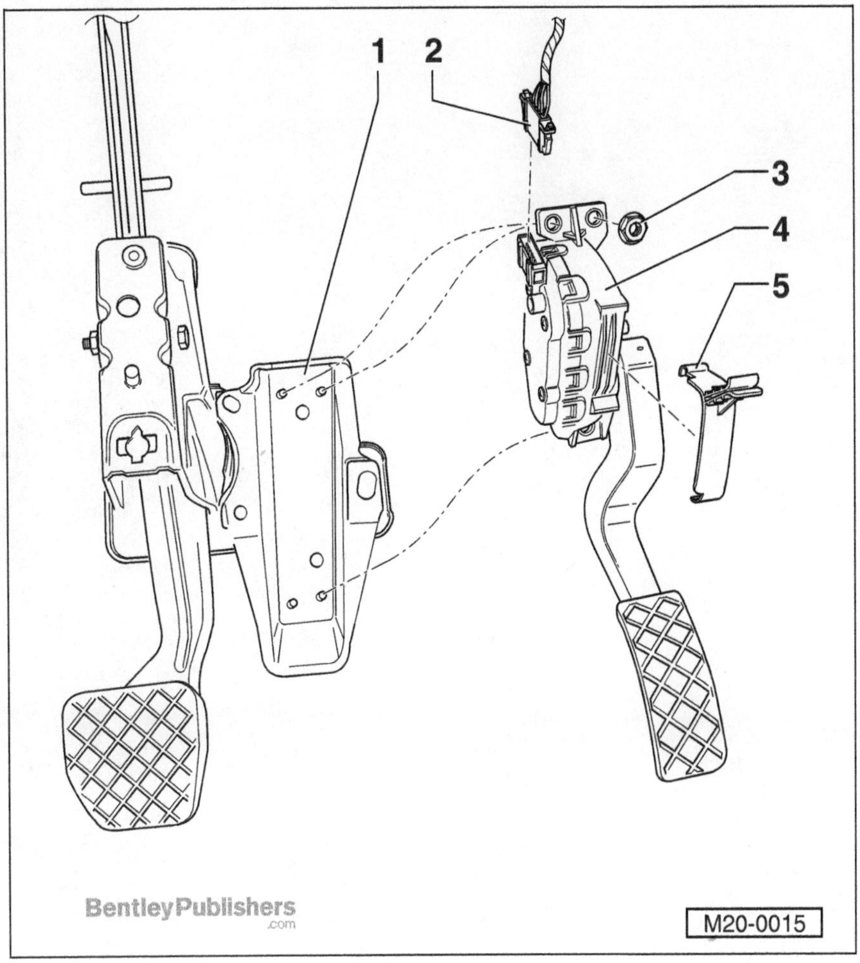

21 Turbocharger and Intercooler

21

.GENERAL

This section covers turbocharger and intercooler service and repair for the 1.8L gasoline and 1.9L diesel engine. Turbocharger boost pressure is not adjustable and separate parts for the turbocharger are not generally available from Volkswagen. The turbocharger and exhaust manifold are integral components. If the turbocharger is found to be faulty, the complete turbocharger/exhaust manifold assembly must be replaced.

Read the following cautions and warnings before servicing the turbocharger or exhaust system.

CAUTION—

- *Thoroughly clean all joints, pipe unions and connections, before disconnecting or reconnecting the turbocharger or any of its related components.*
- *Take measures to prevent dust and dirt contamination. Cover all components with dust-free paper or seal them in plastic bags. Do not use cloth material. Avoid nearby use of compressed air. Do not move the car or work in dusty conditions while the turbocharger is open or removed. Install only clean components. Do not remove components from packaging until just before installation. Avoid use of components that have been stored loose.*

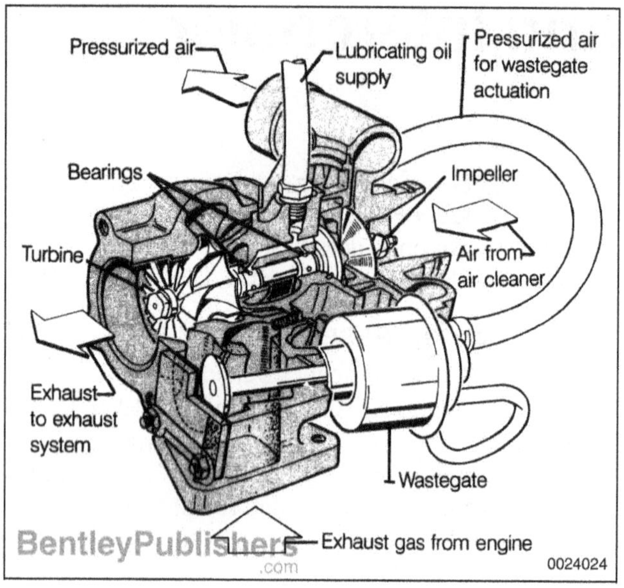

WARNING —
The turbocharger and related components operate at very high temperature. Always allow the system to cool or use proper protective clothing to prevent severe burns.

The exhaust-driven turbocharger vanes spin at very high speeds and are precisely balanced. The turbocharger is cooled and lubricated with engine oil. The boost pressure is limited by a mechanical wastegate. When the boost pressure exceeds a predetermined level, the wastegate opens to bypass some of the exhaust gases around the turbine. As the turbocharger compresses the incoming air, it also raises the air temperature. Output from the turbocharger is routed to an air-to-air radiator called an intercooler. The intercooler removes the excess heat allowing for greater boost pressure.

TURBOCHARGER DIAGNOSTICS

Due to the extremely high speeds and temperatures, proper lubrication of the turbocharger is critical. Small oil passages are drilled into the turbocharger housing and bearings. Seals are placed around the shaft at each end to prevent oil from entering the compressor and turbine housing. When the turbocharger bearings and seals become worn, the lubricating oil can slip past the seals and into the combustion chamber or exhaust system. The result is blue-gray oil smoke from the tailpipe. When the bearings become severely worn, the turbocharger compressor and turbine impeller may contact the turbocharger housing, making loud screeching or thumping noises.

The major cause of turbocharger bearing and seal failure is "coking", or baking of the oil in the turbocharger's oil passages. During operation, the temperature of the turbocharger housing can exceed the boiling point of the oil running through it. Because the oil is constantly circulating, it stays cool enough to prevent coking. If the oil pipes are clogged or if the engine is turned off before the turbocharger has had time to cool down, the oil can quickly turn to coke. When these carbon deposits form, oil flow to the turbocharger is reduced. In addition, this substance is abrasive and can wear the seals, bearing and shafts prematurely.

If engine performance is low, and low boost pressure is suspected, first check for any loose or broken hoses or fittings. Make sure the control line to the wastegate is not blocked, loose, or leaking. Make sure there are no restrictions in the intake air system, such as a dirty air filter element or damage to the intercooler. Check the exhaust system for restrictions such as a damaged catalytic converter or clogged muffler. If no visible faults can be found, check the turbo boost pressure with a diagnostic scan tool.

NOTE—
Suspected engine performance problems must first be investigated with an appropriate diagnostic scan tool. Diagnostic Trouble Codes (DTCs) stored in ECM memory must be repaired and the DTCs erased before proceeding. See **OBD On Board Diagnostics** *section for more information.*

Turbo boost pressure, checking (1.8L, code: APH, BKF, BNU)

Boost pressure can be measured with Volkswagen digital boost pressure gauge VAG1397A or equivalent.

NOTE—
Pressure measurements as specified by Volkswagen are given in "absolute' values. This system as used for automotive purposes does not recognize vacuum. The starting point for measurements is 1 atmosphere (1000 mbar). Values commonly referred to as vacuum are simply pressures less than 1 atmosphere.

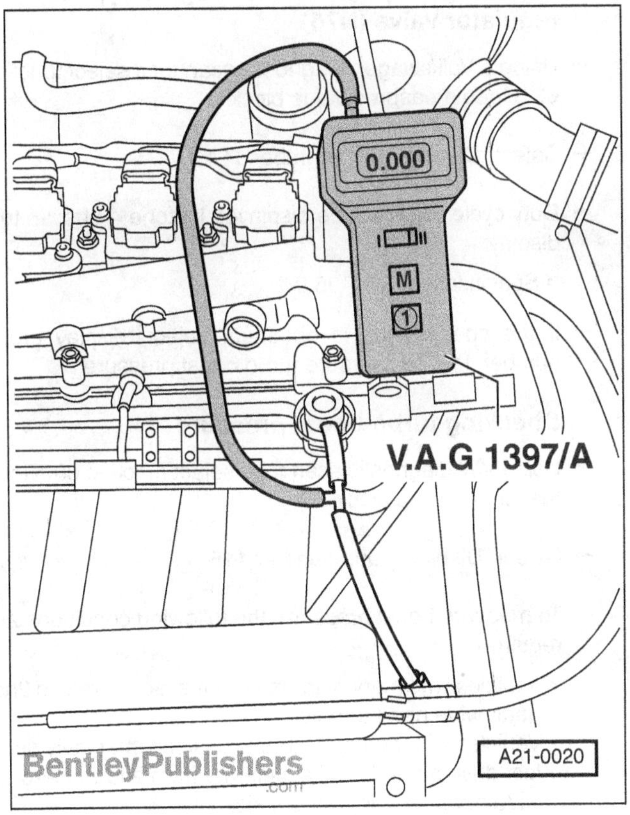

◀ Connect boost pressure gauge.

— Observe the following preliminary conditions:
- Intake and exhaust systems free of leaks and restrictions
- Engine temperature at 60°C (140°) minimum
- Engine compression OK. See **15a Cylinder Head and Valvetrain (1.8L engine)**
- No DTCs in ECM memory

— Road test the vehicle under the following conditions:
- For safety reasons, road test with a helper operating and/or observing the pressure gauge.
- Road test in an area free from vehicular and pedestrian traffic.
- Observe all applicable traffic laws.

— Accelerate at full throttle from 2000 rpm in 3rd gear (manual transmission) or 2nd gear (automatic transmission).

— Observe and record pressure gauge reading at 3000 rpm while accelerating at full throttle.

— Compare to specification.

Specification
- 1.8L, code: APH. . . . . .1300 to 1400 mbar
- 1.8L, code: BKF, BNU . . . . .1500 to 1600 mbar

— If specified values are not reached or are exceeded, additional testing using an appropriate diagnostic scan tool will be necessary and is beyond the scope of this manual.

Turbo boost pressure, checking (1.8L, code: AWP, AWV)

Boost pressure can be measured with a diagnostic scan tool. See **OBD On Board Diagnostics** section for more scan tool information.

NOTE —
Turbo boost pressure is measured under full load, while either driving or testing on a chassis dynometer.

Requirements

- Engine oil temperature 80°C (176°F)

Checking duty cycle of wastegate bypass regulator valve (N75)

— Using a Volkswagen diagnostic scan tool, select function 08 "Read measured value block".

— Select "Display group number 114".

— Duty cycle value will be displayed in zone 4 of scan tool display.

- Specification: 5.0 - 95.0%

— If the specification is obtained, select "display group number 115" to measure turbo boost pressure.

Checking turbo boost pressure

— Using VW diagnostic scan tool, select function 08 "Read measured value block".

— Select "Display group number 115".

— To measure boost pressure, the following conditions are required:

- Vehicles with manual transmissions, accelerate in 2nd gear, wide open throttle.
- Vehicles with automatic transmissions, manually select drive mode 3 and accelerate from a low speed at wide open throttle, without kickdown.

— At wide open throttle between 1800 - 2300 rpm, read boost pressure actual value in display zone 4. Compare to specified values.

Specification
- 1.8L, code: AWP .... 1700 to 2000 mbar
- 1.8L, code: AWV .... 1350 to 1750 mbar
- Maximum deviation .... 100 mbar

— If boost pressure has not fully built up, or if value is beyond specification, repeat test.

— Select function 06 "End Output" on diagnostic scan tool.

Charge pressure bypass valve, checking (1.8L)

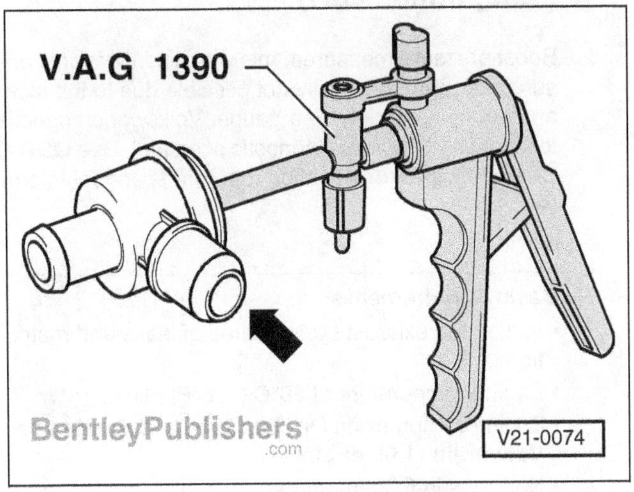

◀ Connect hand vacuum pump VAG 1390 to vacuum connection of charge pressure bypass valve.

– Operate hand vacuum pump.

– Charge pressure bypass valve (**arrow**) must open.

– Ventilate valve on hand vacuum pump.

– Charge pressure bypass valve (**arrow**) must close.

– If the valve does not open or close, replace charge pressure bypass valve.

> ***NOTE —***
> *The charge pressure bypass valve connections must be secured with spring-type clamps.*

Charge pressure control valve, checking (1.8L)

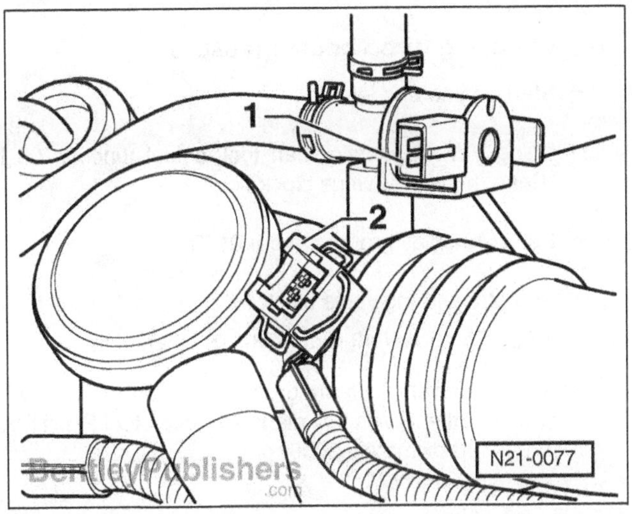

Requirements:
- Engine oil temperature 60°C (140°F) min.
- No leaks on intake and exhaust systems.

◀ Disconnect harness connector (**2**) from wastegate bypass regulator valve (**1**).

– Have a helper start engine and raise to maximum speed (just below RPM "redline") briefly with throttle pedal.

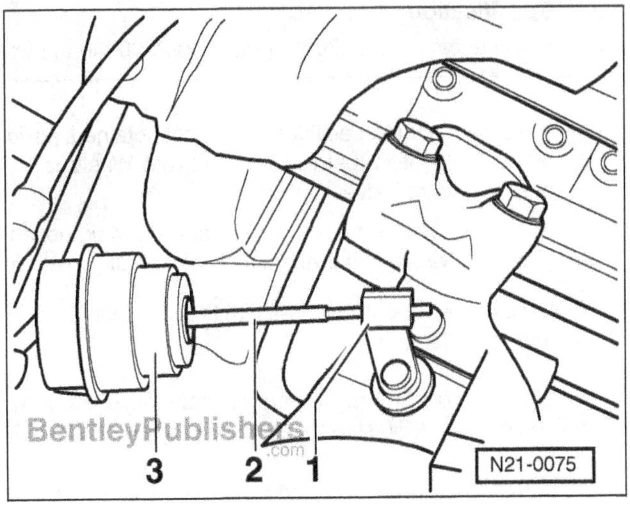

◀ Check to see if operating rod (**2**) for charge pressure control valve moves.

– If operating rod does not move, check charge pressure control valve lever (**1**) for ease of operation. If valve is rusted, replace turbocharger.

– If operating rod does not move, even though lever moves freely, replace turbocharger.

Turbo boost pressure, checking (1.9L, code: ALH)

Boost pressure measurements with a conventional pressure gauge are generally not possible due to the lack of an attachment fitting for a gauge. Volkswagen specifies testing with a factory diagnostic scan tool. See **OBD On Board Diagnostics** section for more scan tool information.

Basic Requirements

- Intake and exhaust systems free of leaks and restrictions
- Engine temperature at 80°C (176°F) minimum
- Engine compression OK. See **15b Cylinder Head and Valvetrain (1.9L engine)**
- No DTCs in ECM memory

Checking turbo boost pressure

– Start engine and let idle.

– Using VW diagnostic scan tool, select function 08 for "Read measured value block".

– Select "Display group number 011".

– Read measuring value block 11.

- Specification: 840 rpm, 938-978 mbar, 92%.

– Accelerate vehicle in 3rd gear from 1500 rpm using wide open throttle. Press VW diagnostic scan tool PRINT button at approximately 3000 rpm.

Specification

- ALH engine . . . . . 3000 rpm, 1700-2200 mbar, 92%.

– If boost pressure specification was not obtained, perform output diagnostic test mode to activate Wastegate Bypass regulator Valve.

- Rod must move back and forth at lest 3-4 times, as long as vacuum is present in the reservoir.

– If rod moves and boost pressure is not obtained, replace turbocharger unit.

– If the rod doesn't move and solenoid valve is not functioning, check Wastegate Bypass Regulator Valve (N75).

– If rod doesn't not move, but solenoid valve operates, test rod movement using a vacuum pump.

– If rod does not move easily, replace turbocharger assembly.

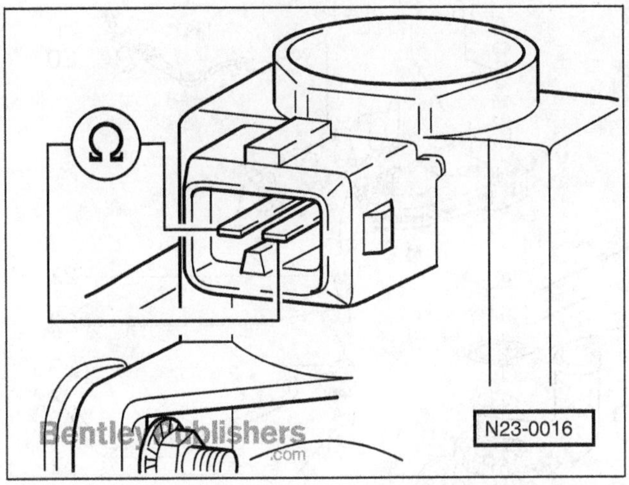

Wastegate bypass regulator valve (N75), checking (1.9L, code: ALH)

– Switch ignition off.

– Remove electrical harness connector at Wastegate Bypass Regulator Valve (N75).

< Measure resistance between terminals of valve.

• Specification: 14 to 20 Ω

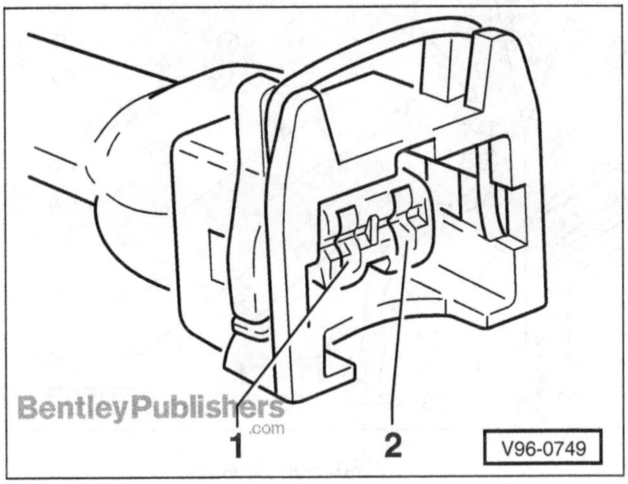

< If the specification is obtained, switch ignition on and measure supply voltage between terminal 1 (**1**) and ground.

• Specification: approximate battery voltage.

– If specification is not met, test wiring, see **EWD Electrical Wiring Diagrams.**

– If wiring tests are ok, replace engine control module.

TURBOCHARGER, ASSEMBLY (1.8L)

When replacing the turbocharger, the oil return and supply pipes/lines should be removed and thoroughly cleaned using a stiff brush and solvent. If the pipes cannot be completely cleaned, they should be replaced. The engine oil and oil filter should also be changed. If the impeller blades are damaged, inspect the intake connecting pipes and the intercooler for debris and clean or replace as needed. Be sure to check the boost pressure once installation is complete.

NOTE —

- *If the turbocharger is being replaced because of excessive exhaust smoke, there will probably be some residual oil in the exhaust upon start-up. Be sure to test drive vehicle long enough to clear left over oil from exhaust system.*
- *Volkswagen recommended high temperature, anti-seize compound for all operations: part no. G 052 112 A3.*
- *Volkswagen part numbers are given for reference only. Always consult with your Volkswagen Parts Department or aftermarket parts specialist for the latest parts information.*

Turbocharger, assembly (1.8L)

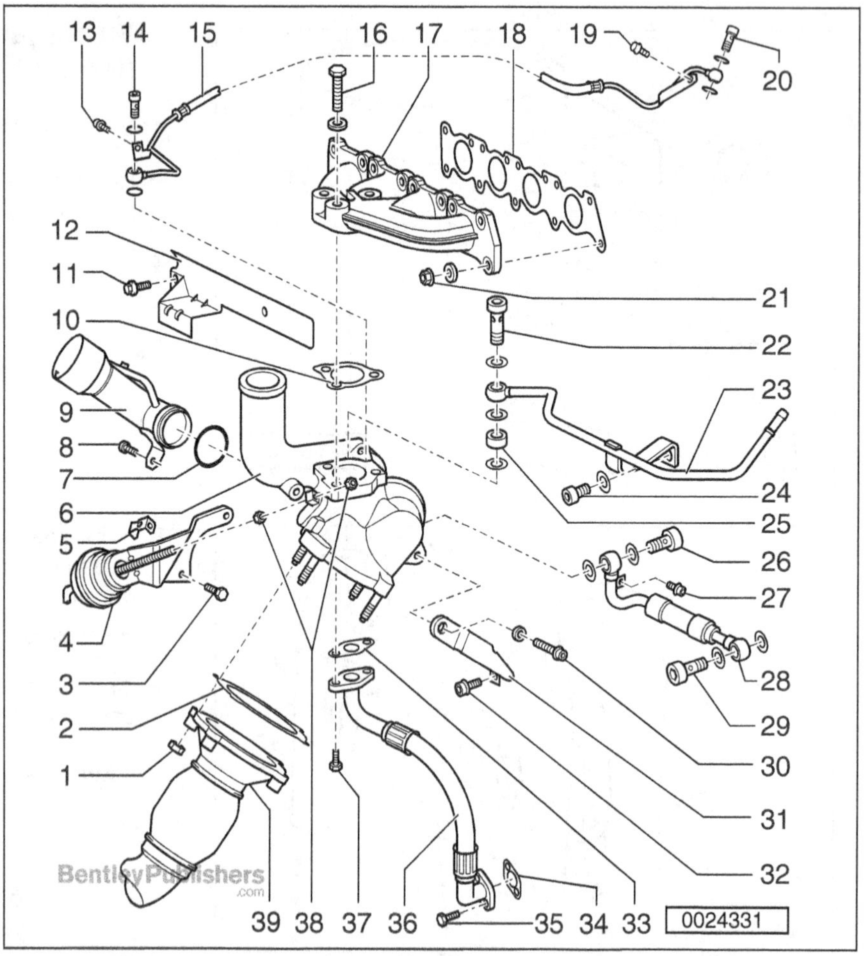

1. **Nut**
   - Tighten to 40 Nm (30 ft-lb)
   - Always replace
   - Coat threads with high temperature anti-seize
2. **Gasket**
   - Always replace
3. **Bolt**
   - Do not loosen
   - Tighten to 10 Nm (7 ft-lb)
4. **Pressure unit (servo) for wastegate**
   - Integral with turbocharger
5. **Clip**
6. **Turbocharger**
   - With integral wastegate and servo
   - Manufacturer: KKK
   - Allow engine to idle for at least one minute after installing turbocharger to ensure adequate oil supply
7. **O-ring**
   - Always replace
8. **Bolt**
   - Tighten to 10 Nm (7 ft-lb)
9. **Intake pipe**
   - In from air cleaner
10. **Gasket**
    - Always replace
    - Note installation position
11. **Bolt**
    - Tighten to 20 Nm (15 ft-lb)
12. **Heat shield**
13. **Bolt**
    - Tighten to 10 Nm (7 ft-lb)
14. **Banjo bolt**
    - Tighten to 30 Nm (22 ft-lb)
15. **Oil supply pipe**
    - From oil filter flange
16. **Bolt**
    - Tighten to 30 Nm (22 ft-lb)
    - Always replace
    - Coat threads with high temperature anti-seize
17. **Exhaust manifold**
18. **Gasket**
    - Always replace
    - Note installation position
19. **Bolt**
    - Tighten to 20 Nm (15 ft-lb)
20. **Banjo bolt**
    - Tighten to 30 Nm (22 ft-lb)
21. **Nut**
    - Tighten to 25 Nm (18 ft-lb)
    - Always replace
    - Coat threads with high temperature anti-seize
22. **Banjo bolt**
    - Tighten to 35 Nm (26 ft-lb)
23. **Coolant return pipe**
24. **Bolt**
    - Tighten to 25 Nm (18 ft-lb)
25. **Spacer sleeve**
26. **Banjo bolt**
    - Tighten to 35 Nm (26 ft-lb)
27. **Bolt**
    - Tighten to 10 Nm (7 ft-lb)
28. **Coolant supply pipe**
29. **Banjo bolt**
    - Tighten to 35 Nm (26 ft-lb)
30. **Bolt, special**
    - Tighten to 30 Nm (22 ft-lb)
31. **Support bracket**
    - Between turbocharger and cylinder block
32. **Bolt**
    - Tighten to 25 Nm (18 ft-lb)
33. **Gasket**
    - Always replace
34. **Gasket**
    - Always replace
35. **Bolt**
    - Tighten to 10 Nm (7 ft-lb)
36. **Oil return pipe**
    - To oil pan
37. **Bolt**
    - Tighten to 10 Nm (7 ft-lb)
38. **Adjusting nut**
    - Do not loosen or alter setting
    - Tighten to 10 Nm (7 ft-lb)
39. **Front exhaust pipe**

Turbocharger connecting pipes (1.8L, code: APH)

NOTE —
Volkswagen identifies electrical components by a letter and/or a number in the electrical schematics. See **EWD Electrical Wiring Diagrams**. *These electrical identifiers are listed in parenthesis as an aid to electrical troubleshooting.*

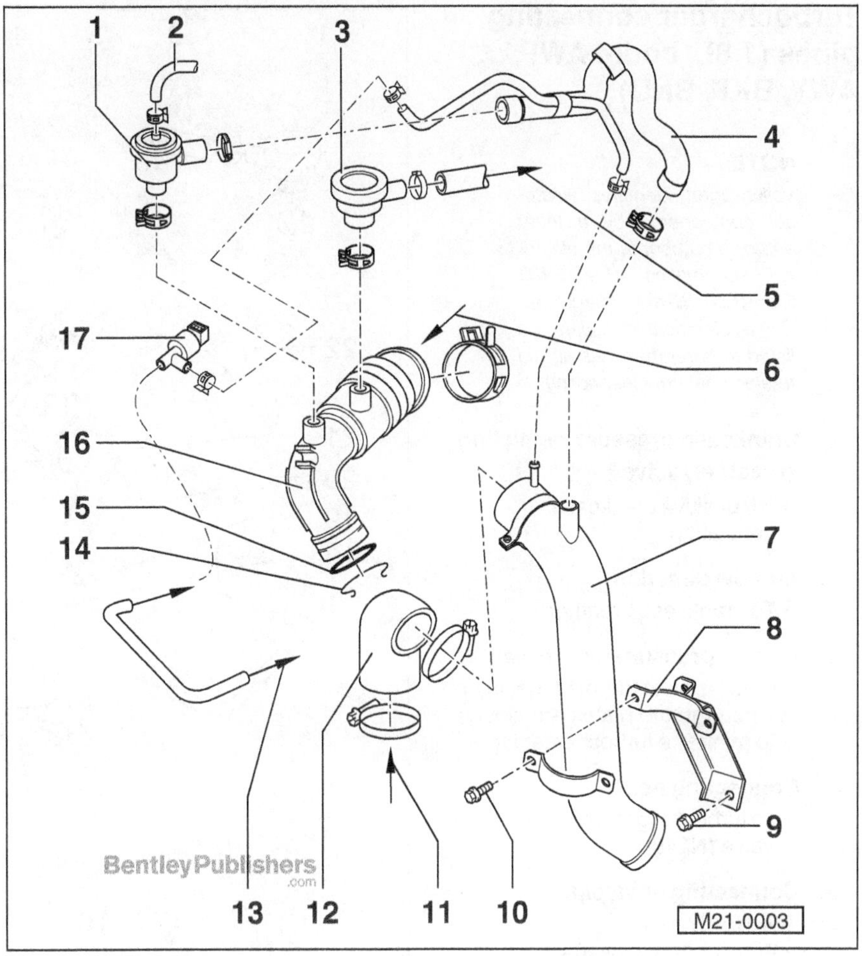

1. **Charge pressure bypass valve**
   - Returns air to turbocharger inlet when throttle plates are closed to minimize turbocharger lag
2. **Connecting hose**
   - To turbocharger recirculating valve (N249)
3. **Crankcase pressure regulating (breather) valve**
   - For positive crankcase ventilation
4. **Connecting hose/pipe**
   - From charge pressure bypass valve to upper air duct
5. **Air flow from crankcase ventilation housing**
6. **Air flow in from air cleaner**
7. **Connecting pipe**
   - Turbocharger outlet
   - Between turbocharger and intercooler
8. **Support bracket**
9. **Bolt**
   - Tighten to 25 Nm (18 ft-lb)
10. **Bolt**
    - Tighten to 10 Nm (7 ft-lb)
11. **Air flow out from turbocharger**
12. **Connecting elbow**
    - Between turbocharger outlet and connecting pipe
13. **Connecting hose**
    - From wastegate bypass regulator valve (N75) to pressure unit (servo) for wastegate
14. **Retaining ring**
15. **O-ring**
    - Always replace
16. **Intake hose**
    - From air cleaner assembly to turbocharger inlet
17. **Wastegate bypass regulator valve (N75)**
    - Controlled (pulsed) by ECM

Turbocharger connecting pipes (1.8L, code: AWP, AWV, BKF, BNU)

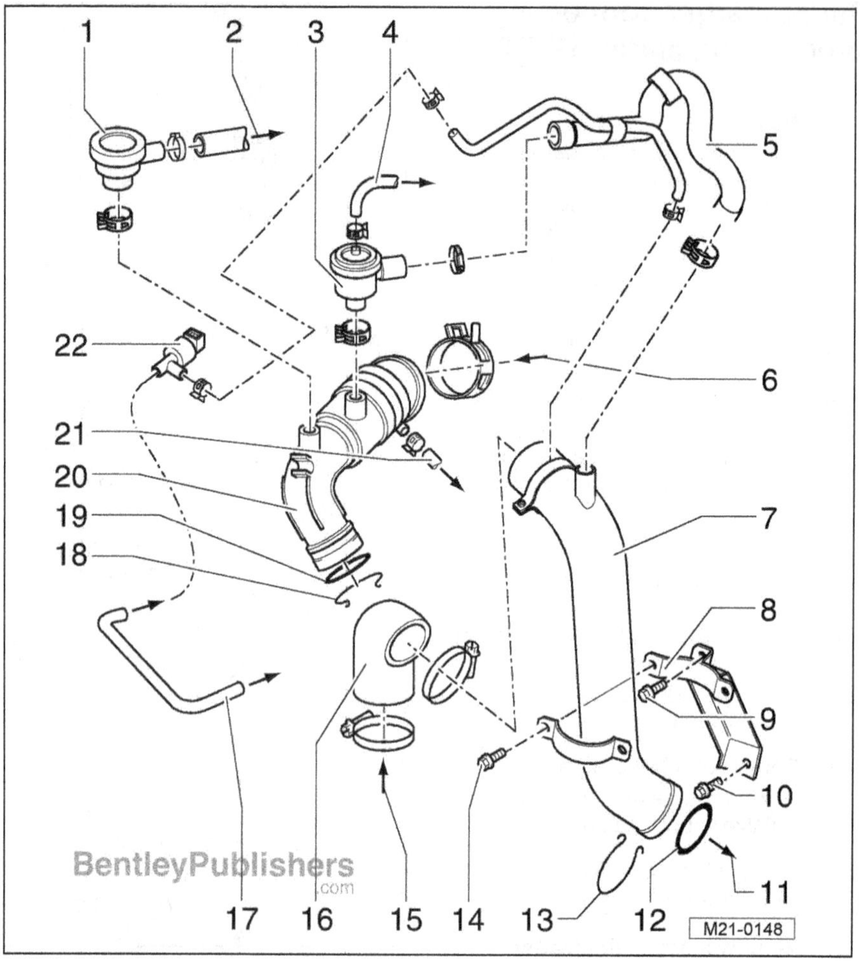

NOTE —

Volkswagen identifies electrical components by a letter and/or a number in the electrical schematics. See **EWD Electrical Wiring Diagrams**. *These electrical identifiers are listed in parenthesis as an aid to electrical troubleshooting.*

1. **Crankcase pressure regulating (breather) valve**
   - For positive crankcase ventilation
2. **Air flow direction**
   - To crankcase breather
3. **Charge pressure bypass valve**
   - Returns air to turbocharger inlet when throttle plates are closed to minimize turbocharger lag
4. **Connecting hose**
   - To turbocharger recirculating valve (N249)
5. **Connecting hose/pipe assembly**
   - From charge pressure bypass valve to turbocharger outlet connecting pipe
6. **Air flow in from air cleaner**
7. **Connecting pipe**
   - Turbocharger outlet
   - Between turbocharger and intercooler
8. **Support bracket**
9. **Bolt**
   - Tighten to 40 Nm (30 ft-lb)
10. **Bolt**
    - Tighten to 25 Nm (18 ft-lb)
11. **Air flow out to intercooler**
12. **O-ring**
    - Always replace
13. **Circlip**
    - Always replace
14. **Bolt**
    - Tighten to 10 Nm (7 ft-lb)
15. **Air flow out from turbocharger**
16. **Connecting elbow**
    - Between turbocharger outlet and connecting pipe
17. **Connecting hose**
    - From wastegate regulator valve (N75) to pressure unit (servo) for wastegate
18. **Retaining ring**
19. **O-ring**
    - Always replace
20. **Intake hose**
21. **From air cleaner assembly to turbocharger inlet**
22. **Wastegate bypass regulator valve (N75)**
    - Controlled (pulsed) by ECM

Turbocharger hose diagram (1.8L, code: APH)

NOTE —

Volkswagen identifies electrical components by a letter and/or a number in the electrical schematics. See **EWD Electrical Wiring Diagrams.** *These electrical identifiers are listed in parenthesis as an aid to electrical troubleshooting.*

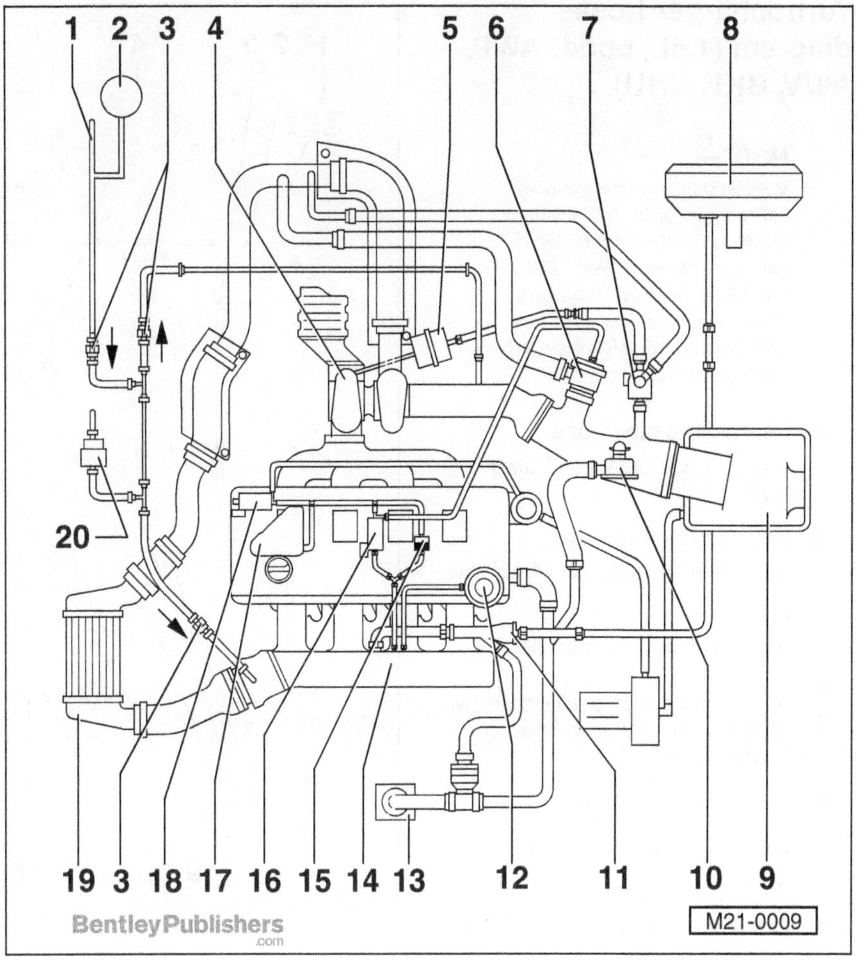

1. **Connecting pipe/hose**
   - From gravity valve connection on fuel tank
2. **Vacuum reservoir**
   - Location: in right rear wheel well
3. **Check valve**
   - For activated charcoal filter system
   - Note installation position: arrow points in direction of flow (to engine)
4. **Turbocharger assembly**
5. **Pressure unit (servo) for wastegate**
6. **Overrun recirculation valve**
7. **Wastegate bypass regulator valve (N75)**
8. **Power brake servo**
9. **Air cleaner assembly with mass air flow sensor (G70)**
10. **Crankcase pressure regulating (breather) valve**
11. **Vacuum booster (amplifier)**
12. **Fuel pressure regulator**
13. **Crankcase ventilation housing**
14. **Intake manifold and throttle valve control module (J338)**
15. **Check valve**
    - Note installation position: arrow points in direction of flow
16. **Overrun recirculation valve solenoid (N249)**
17. **Vacuum reservoir**
    - Location: on cylinder head (valve) cover
18. **Secondary air injection solenoid valve (N112)**
19. **Intercooler**
20. **EVAP canister purge valve (N80)**
    - Location: above right front wheel well lining

Turbocharger hose diagram (1.8L, code: AWP, AWV, BKF, BNU)

> ***NOTE —***
> *Volkswagen identifies electrical components by a letter and/or a number in the electrical schematics. See* **EWD Electrical Wiring Diagrams.** *These electrical identifiers are listed in parenthesis as an aid to electrical troubleshooting.*

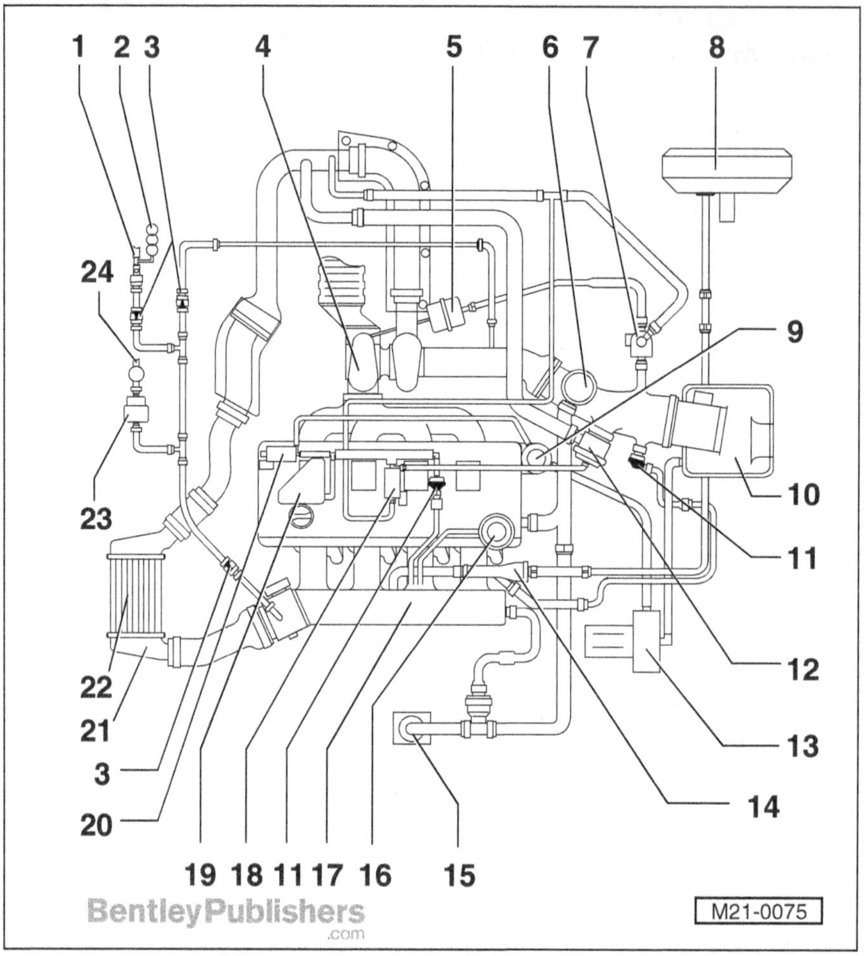

1. **Connecting pipe/hose**
   - From Leak Detection Pump (V144)
2. **Vacuum reservoir**
   - Location: in right rear wheel well
3. **Check valve**
   - For activated charcoal filter system
   - Note installation position: arrow points in direction of flow (to engine)
4. **Turbocharger assembly**
5. **Pressure unit (servo) for wastegate**
6. **Crankcase pressure regulating (breather) valve**
7. **Wastegate bypass regulator valve (N75)**
8. **Power brake servo**
9. **Combination valve**
   - For secondary air injection system
10. **Air cleaner assembly with mass air flow sensor (G70)**
11. **Check valve**
    - Note installation position: arrow points in direction of flow (to engine)
12. **Overrun recirculation valve**
13. **Secondary Air injection pump motor (V101)**
14. **Vacuum booster (amplifier)**
15. **Check valve**
    - Note installation position: arrow points in direction of flow
16. **Fuel pressure regulator**
17. **Intake manifold and throttle valvecontrol module (J338)**
18. **Turbocharger recirculating valve (N249)**
19. **Vacuum reservoir**
    - Location: on cylinder head (valve) cover
20. **Secondary air injection solenoid valve (N112)**
21. **Charge pressure sensor (G31)**
22. **Intercooler**
23. **EVAP canister purge valve (N80)**
    - Location: above right front wheel well lining
24. **Connecting hose**
    - To EVAP canister

TURBOCHARGER, ASSEMBLY (1.9L)

When replacing the turbocharger, the oil return and supply pipes/lines should be removed and thoroughly cleaned using a stiff brush and solvent. If the pipes cannot be completely cleaned, they should be replaced. The engine oil and oil filter should also be changed. If the impeller blades are damaged, inspect the intake connecting pipes and the intercooler for debris and clean or replace as needed. Be sure to check the boost pressure once installation is complete.

NOTE —
- *If the turbocharger is being replaced because of excessive exhaust smoke, there will probably be some residual oil in the exhaust upon start-up. Be sure to test drive vehicle long enough to clear left over oil from exhaust system.*
- *Volkswagen recommended high temperature, anti-seize compound for all operations: part no. G 052 112 A3.*

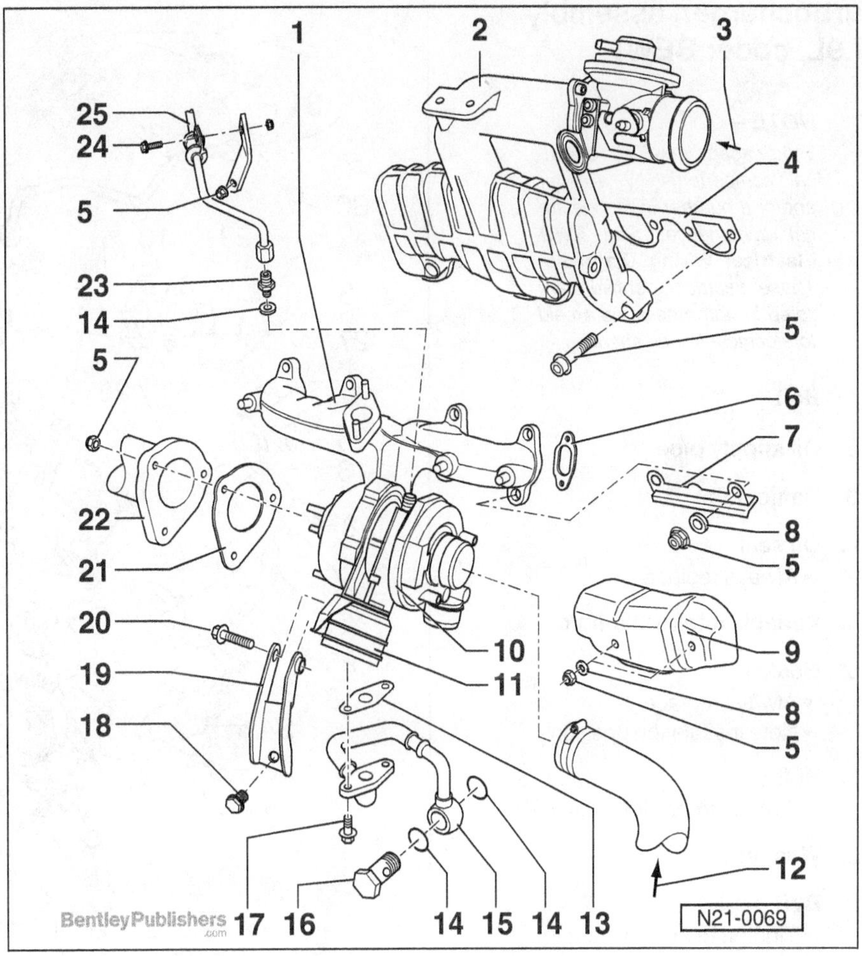

Turbocharger, assembly (1.9L, code: ALH)

1. **Exhaust manifold**
   - Integral with turbocharger
2. **Intake manifold**
   - Shown with EGR valve and manifold flap
3. **Air flow in from intercooler**
4. **Gasket**
   - Always replace
   - Coated (beaded) side faces intake manifold
5. **Bolt**
   - Tighten to 25 Nm (18 ft-lb)
6. **Gasket**
   - Always replace
   - Note installation position
7. **Bracket**
   - For heat shield
8. **Washer**
9. **Heat shield**
10. **Turbocharger**
    - Integral with exhaust manifold
    - Allow engine to idle for at least one minute after installing turbocharger to ensure adequate oil supply
11. **Pressure unit (servo) for wastegate**
12. **Air flow in from air cleaner**
13. **Gasket**
    - Always replace
14. **Oil seal**
    - Always replace
15. **Oil return pipe**
    - To cylinder block
16. **Banjo bolt**
    - Tighten to 25 Nm (18 ft-lb)
17. **Bolt**
    - Tighten to 15 Nm (11 ft-lb)
18. **Bolt**
    - Tighten to 40 Nm (30 ft-lb)
19. **Bracket**
20. **Bolt**
    - Tighten to 25 Nm (18 ft-lb)
21. **Gasket**
    - Always replace
22. **Front exhaust pipe**
23. **Fitting**
24. **Bolt**
    - Tighten to 10 Nm (7 ft-lb)
25. **Oil supply pipe**
    - From oil filter flange

Turbocharger, assembly (1.9L, code: BEW)

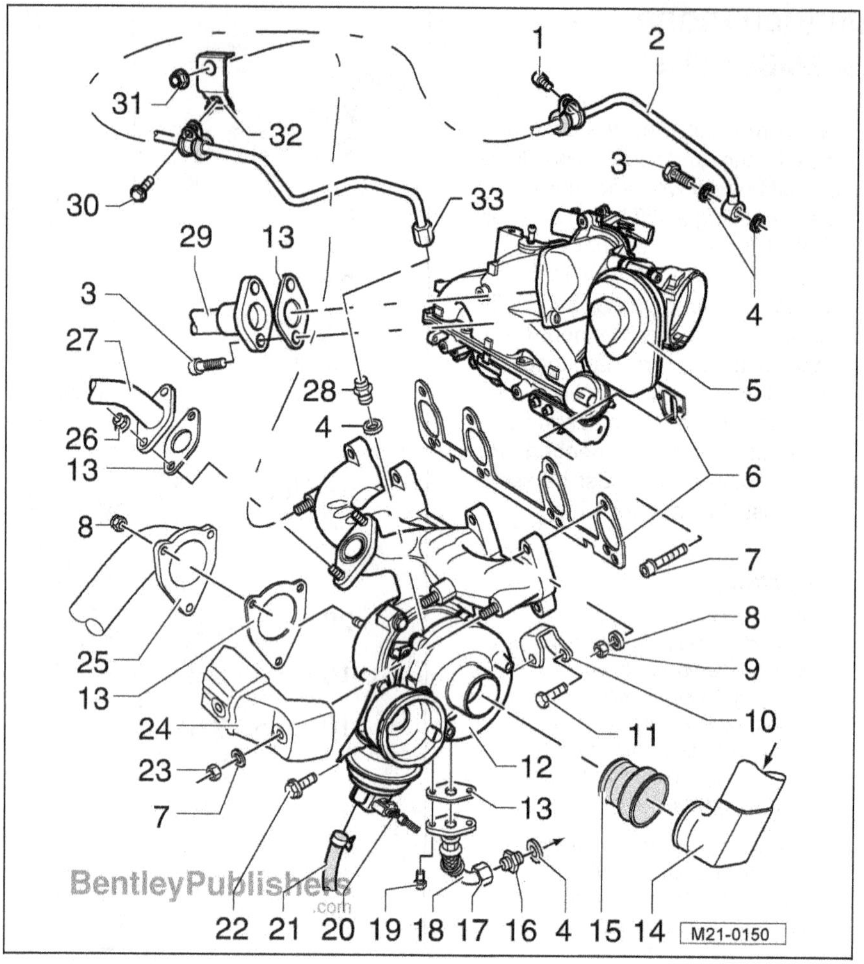

NOTE —
Volkswagen identifies electrical components by a letter and/or a number in the electrical schematics. See **EWD Electrical Wiring Diagrams**. *These electrical identifiers are listed in parenthesis as an aid to electrical troubleshooting.*

1. **Bolt**
2. **Oil supply pipe**
3. **Banjo bolt**
4. **Oil seal**
   - Always replace
5. **Variable intake manifold**
6. **Gasket**
   - Always replace
   - Note installation position
7. **Bolt**
   - Tighten to 22 Nm (16 ft-lb)
8. **Washer**
9. **Bolt**
   - Tighten to 25 Nm (18 ft-lb)
10. **Support**
11. **Bolt**
    - Tighten to 40 Nm (30 ft-lb)
    - Hand tighten all bolts first when installing
12. **Turbocharger**
    - Integral with exhaust manifold
    - Allow engine to idle for at least one minute after installing turbocharger to ensure adequate oil supply
13. **Gasket**
    - Always replace
14. **Connecting pipe**
    - From air filter to turbocharger
15. **Coupler**
16. **Pipe union**
    - Tighten to 40 Nm (30 ft-lb)
17. **Union nut**
    - Tighten to 30 Nm (22 ft-lb)
18. **Oil return pipe**
    - To cylinder block
19. **Bolt**
    - Tighten to 17 Nm (13 ft-lb)
20. **Connector**
    - For turbocharger vane position sensor (G500)
21. **Vacuum hose**
    - To wastegate bypass regulator valve (N75)
22. **Nut**
    - Tighten to 20 Nm (15 ft-lb)
23. **Nut**
    - Tighten to 20 Nm (15 ft-lb)
    - Always replace
24. **Heat shield**
25. **Front exhaust pipe**
26. **Nut**
    - Tighten to 22 Nm (16 ft-lb)
    - Oil threaded studs with G 052 112 A3
27. **Connecting pipe**
    - Hand tighten all nuts/bolts first when installing
28. **Pipe union**
    - Tighten to 30 Nm (22 ft-lb)
    - Always replace
    - Oil threads and contact surfaces with G 052 112 A3
29. **Connecting pipe**
    - Hand tighten all nuts/bolts first when installing
30. **Bolt**
31. **Nut**
32. **Bracket**
33. **Union nut**

Turbocharger vacuum hose connections (1.9L, code: ALH)

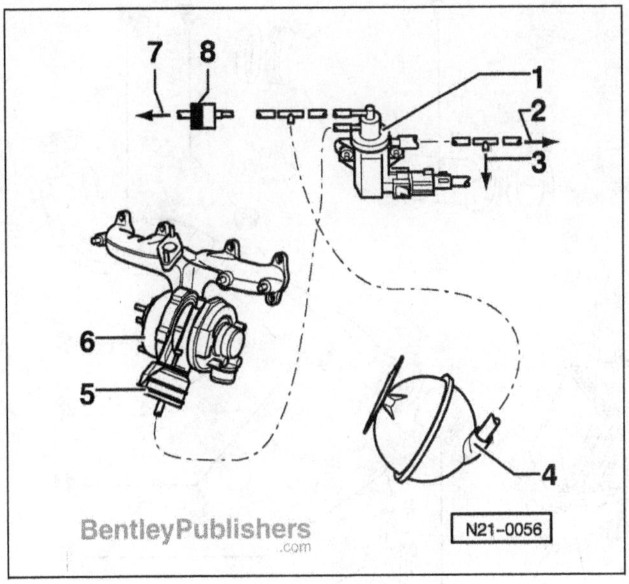

1. Wastegate regulator valve
2. Connection to EGR regulator valve
3. Connection to air filter
4. Vacuum reservoir
5. Pressure unit (servo) for wastegate
6. Exhaust manifold with turbocharger
7. Connection to change-over valve for intake manifold
8. Check valve

Turbocharger vacuum hose connections (1.9L, code: BEW)

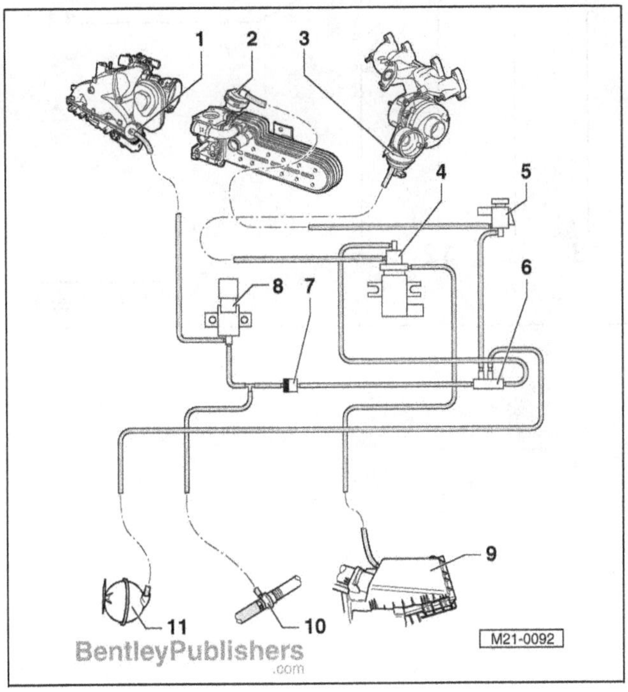

1. Vacuum unit
   - For variable intake manifold
2. Vacuum unit
   - For bypass flap for exhaust gas recirculation cooler
3. Vacuum unit
   - For charge pressure control, cannot be replaced
4. Wastegate bypass regulator valve (N75)
5. Exhaust gas recirculation (EGR) cooler switch-over valve (N345)
6. Distributor
7. Non-return valve
8. intake manifold change-over valve (N156)
9. Air cleaner
10. Vacuum connection
    - Between tandem pump and brake servo
    - With check valve
11. Vacuum reservoir

Solenoid valve identification (1.9L, code: BEW)

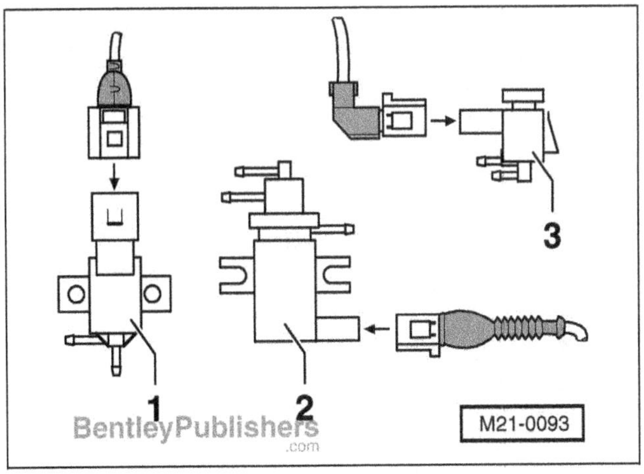

1. Intake manifold change-over valve (N156)
   - With straight protective sleeve for harness connector
2. Wastegate bypass regulator valve (N75)
   - With rubber sleeve for harness connector
3. Exhaust gas recirculation (EGR) cooler switch-over valve (N345)
   - With angled protective sleeve for harness connector

INTERCOOLER

The intercooler is used to reduce the temperature associated with compressing the air by the turbocharger. This reduction in air temperature increases the density of the air which contributes to a reduction of Nitrous Oxide (NOS) gases and an improvement in performance.

NOTE —
Volkswagen identifies electrical components by a letter and/or a number in the electrical schematics. See **EWD Electrical Wiring Diagrams**. *These electrical identifiers are listed in parenthesis as an aid to electrical troubleshooting.*

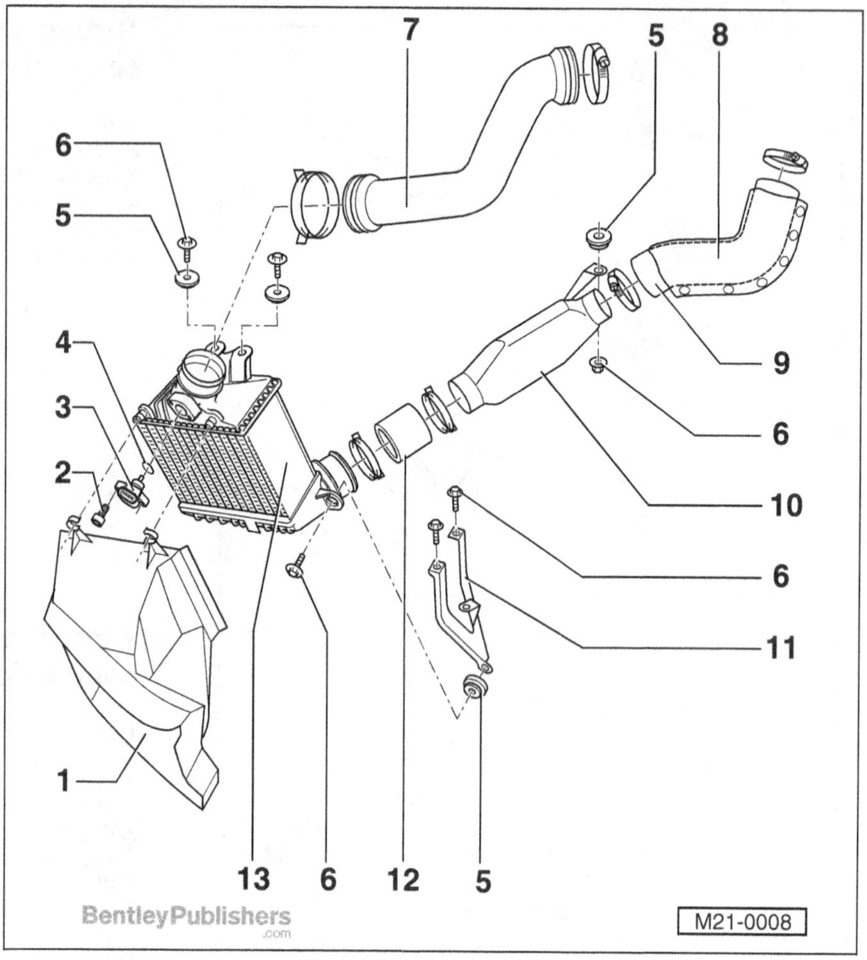

Intercooler assembly (1.8L)

1. **Air ducting**
   - Clip in at intercooler
2. **Bolt**
   - Tighten to 10 Nm (7 ft-lb)
3. **Boost pressure sensor (G31)**
4. **O-ring**
   - Always replace
5. **Mounting grommet**
   - Rubber with metal sleeve
6. **Bolt**
   - Tighten to 10 Nm (7 ft-lb)
7. **Connecting pipe**
   - Between intercooler and intake manifold
   - Lower
8. **Heat shield**
   - Must be positioned to enclose upper hose clamp
9. **Connecting hose**
10. **Connecting pipe**
    - Between intercooler and turbocharger
11. **Support bracket**
12. **Connecting hose**
13. **Intercooler**

Intercooler assembly (1.9L, code: ALH)

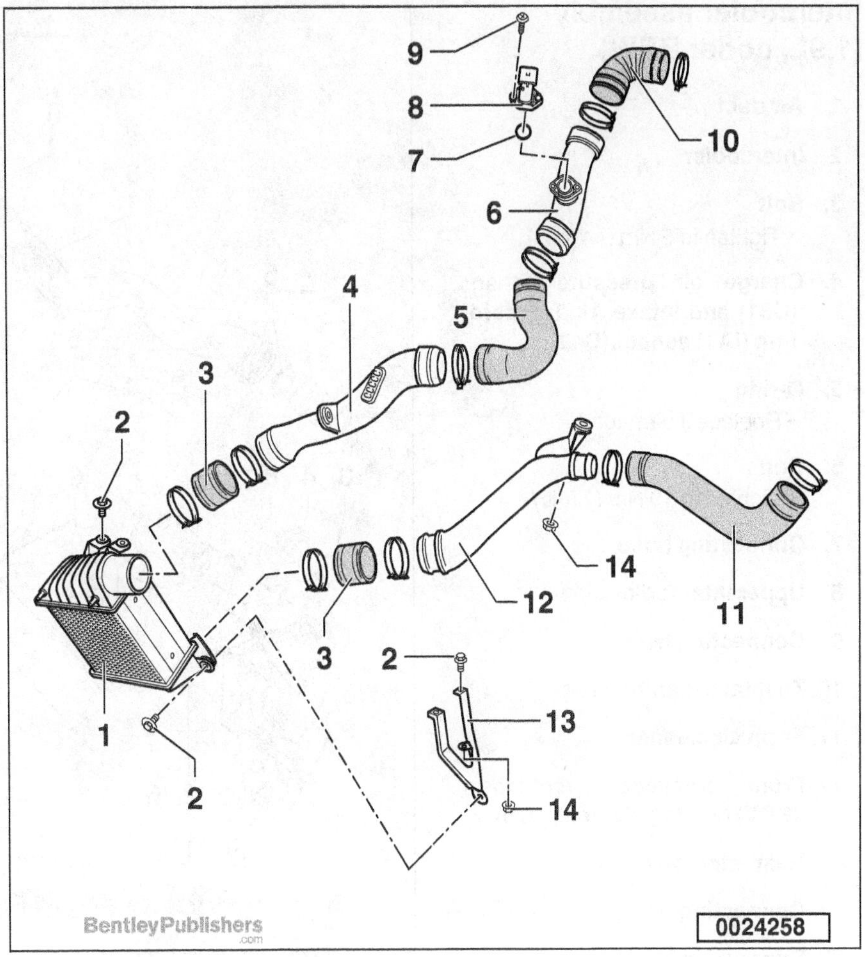

1. **Intercooler**
2. **Bolt**
   - Tighten to 10 Nm (7 ft-lb)
3. **Connecting hose**
4. **Connecting pipe**
   - Between intercooler and intake manifold
   - Lower
5. **Connecting hose**
6. **Connecting pipe**
   - Between intercooler and intake manifold
   - Upper
7. **O-ring**
   - Replace if damaged
8. **Intake Air Temperature (IAT) sensor (G72)**
9. **Bolt**
   - Tighten to 5 Nm (44 in-lb)
10. **Connecting hose**
    - From intercooler to intake manifold
11. **Connecting hose**
12. **Connecting pipe**
    - Between turbocharger and intercooler
13. **Support bracket**
14. **Nut**
    - Tighten to 10 Nm (7 ft-lb)

Intercooler assembly (1.9L, code: BEW)

1. **Air duct**
2. **Intercooler**
3. **Bolt**
   - Tighten to 5 Nm (44 in-lb)
4. **Charge air pressure sensor (G31) and Intake Air Temperature (IAT) sensor (G42)**
5. **O-ring**
   - Replace if damaged
6. **Bolt**
   - Tighten to 10 Nm (7 ft-lb)
7. **Connecting hose**
8. **Upper intercooler pipe**
9. **Connecting hose**
10. **To intake manifold flap**
11. **From air cleaner**
12. **From crankcase ventilation (PCV) heating element (N79)**
13. **Rear intercooler pipe**
14. **Connecting hose**
15. **Support bracket**
16. **Connecting hose**
17. **Turbocharger**
18. **Turbocharger lower pipe**

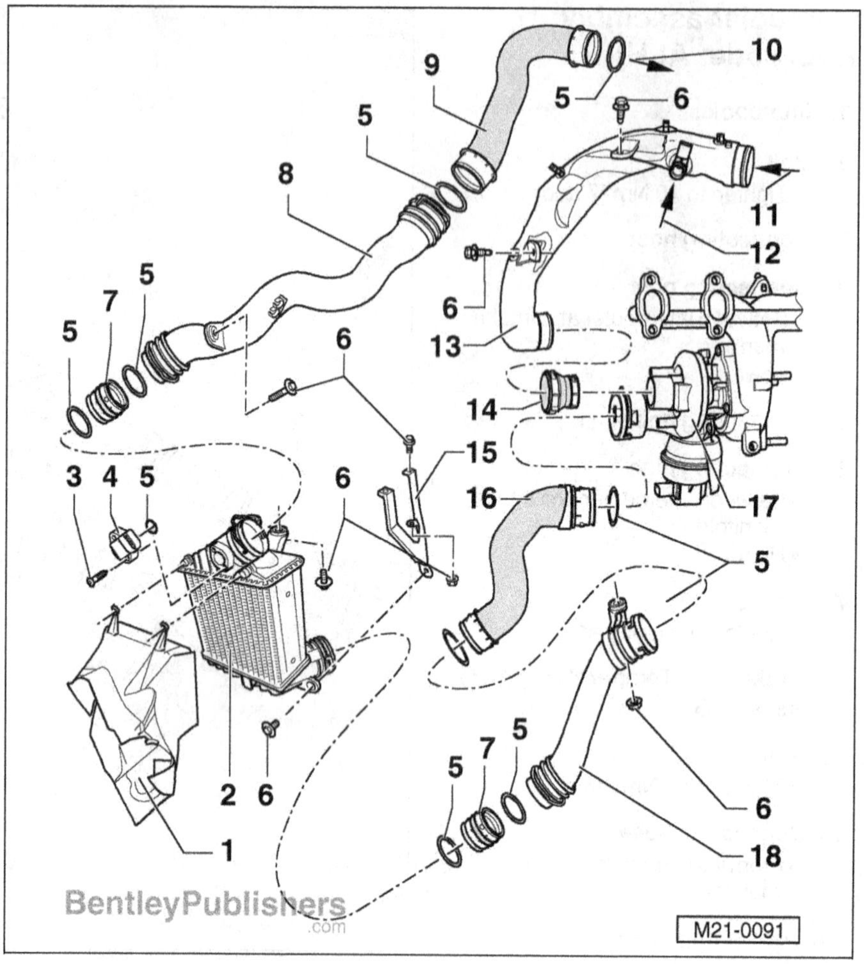

23 Fuel Injection–Diesel

TABLES

GENERAL

This repair group covers the diesel fuel injection system for the 1.9L Turbo Direct Injection (TDI) diesel engines.

Engine applications

- 1.9L TDI, code: ALH 1998-2004
- 1.9L TDI PD, code: BEW 2004-2006

CAUTION —

- *Repairs and adjustments to the injection pump (1.9L, code: ALH) require specialized equipment. Faulty pumps must be serviced by the manufacturer. Internal problems usually require replacement of the pump.*
- *Adjustments and repairs to the fuel injection system should be made carefully. Always clean fuel unions before removing lines*
- *Diesel fuel is damaging to rubber. Wipe off any fuel that spills on hoses, wiring, and rubber steering and suspension parts and wash with soap and water. If coolant hoses have been contaminated with diesel fuel, they must be replaced.*

> ***CAUTION —***
> *Disconnecting the negative (–) battery cable may erase fault codes and basic settings in the engine management and automatic transmission control modules. Some driveability problems may be noticed until the system re-adapts to operating conditions. OBD II readiness codes, which may be required for emissions testing, may also be erased. Convenience electronics (alarm system, interior light control, power locks, mirrors, and windows) may need to be re-set using a diagnostic scan tool.*

Engine control module

Diesel engine management is controlled by an Electronic Control Module (ECM). The ECM is located at the very rear of the engine compartment in the air plenum on the driver's side. Due to the shape of the body in this area, the engine control module is accessed through a trim panel on top of the dash.

Engine control module coding

The Engine Control Module must "know" what equipment is installed in the vehicle. This process is known as coding and must be performed whenever the ECM is replaced, as new ECMs are un-coded. If the ECM is moved between vehicles such as might be the case for diagnostic testing, the coding will need to be changed if equipment is different. Coding memory is, however, retained when the battery is disconnected. If a new ECM is installed without being coded, the engine will usually run poorly and the automatic transmission and anti-lock braking system (if equipped) will not function properly. The MIL will be illuminated.

This coding process is done using a fdiagnostic scan tool such as the Volkswagen VAG 1551/1552, VAS 5051/5052, or equivalent aftermarket scan tool. The appropriate code is electronically programmed into the ECM using the scan tool. See **OBD On Board Diagnostics** for more information.

TROUBLESHOOTING

Proper operation of the diesel engine requires a supply of clean fuel under pressure, an unrestricted supply of air, and properly timed fuel delivery. Fuel must be of the proper grade and properly winterized for cold-start and cold-running conditions. Dirt and water can interfere with combustion and will cause problems.

Good compression is especially necessary for the engine to run well. Engine mechanical faults leading to poor compression can cause problems that may seem to be injection related. Engine lubricating oil that leaks past worn piston rings, valve stem seals or turbocharger seals may produce exhaust smoke that also can be mistaken for fuel injection problems.

A faulty glow plug system may contribute to cold starting problems. The battery must have the correct electrical capacity and the starter must be able to turn the engine fast enough (generally at least 150 rpm) to begin compression ignition. See **27 Engine Electrical System**. Insufficient engine power can also be the result of turbocharger or intercooler malfunctions. See **21 Turbocharger and Intercooler**.

Because the diesel engine speed is controlled by the amount of fuel injected and not by the amount of air admitted past a throttle plate, the air intake must be unrestricted. A clogged air filter, faulty turbocharger, or restricted intercooler can lower the air flow and reduce power output.

The injection pump, under the control of the Engine Control Module (ECM), produces the high pressure necessary to open the injectors and spray fuel. Fuel leaks or air in the fuel lines may reduce system pressure. Since power is controlled by the amount of fuel injected, low pressure may cause sluggish performance.

For cold starting, when the compression heat necessary for combustion is dissipated quickly by a cold engine, injection timing is automatically advanced by the ECM to give the fuel more time to burn. At higher rpm, there is less time for fuel to burn, so injection timing is advanced to start the fuel burning sooner. Since the injection pump is driven by the engine and externally mounted, its precise fuel metering and timing can be degraded by such things as a loose timing belt, worn sprocket, or loose mounting bolts. Correct Cetane rating as specified by the vehicle's Owner's Manual will profoundly influence starting and overall performance as well.

Most of the in-depth troubleshooting of this electronically controlled engine management system will require the use of a diagnostic scan tool. However, there are several basic tests that should not be overlooked. **Table a** lists symptoms, their probable causes, and corrective actions.

NOTE —
Engine performance problems should first be investigated with a diagnostic scan tool. Diagnostic Trouble Codes (DTCs) stored in ECM memory must be re-set and the DTCs erased before proceeding. See ***OBD On Board Diagnostics*** *section for more information.*

Diagnostic quick checks

These tests are used to isolate and diagnose diesel fuel system problems. It is assumed that the glow plug system is functioning within normal parameters and that injection pump timing and valve timing are correct.

Table a. TDI Diesel Fuel Injection Troubleshooting

Problem	Probable cause	Corrective action
1. Engine does not start	**a.** Cranking speed too low or will not crank	**a.** Repair starting system or charge/replace battery. See **27 Engine Electrical Systems.**
	b. No fuel or wrong fuel in fuel tank	**b.** Verify sufficient fuel of correct type in tank. Add as necessary.
	c. No voltage at fuel cut-off valve on injection pump. Fuel cut-off valve loose or faulty	**c.** Valve should click each time ignition is turned on and off. If not, check for voltage at connector. See **EWD Electrical Wiring Diagrams.** Replace a faulty solenoid.
	d. No voltage at glow plug bus or glow plug(s) faulty	**d.** The glow plug circuit is activated by the Engine Control Module. See testing and repair information.
	e. Excessive air in fuel system	**e.** Check fuel supply lines from fuel tank for cracks and connections for leaks, especially at and around fuel filter and injection pump inlet.
	f. Injection pump not delivering fuel	**f.** Check for basic fuel delivery to engine. Check for cracked or damaged lines, clogged fuel filter or other fuel supply problems.
	g. Injection timing incorrect	**g.** Adjust injection timing. Requires diagnostic scan tool.
	h. Faulty injectors	**h.** Test and, if necessary, replace injectors.
	i. Engine mechanical faults	**i.** Test compression. See **15b Cylinder Head and Valvetrain (1.9L engine).**
	j. Faulty injection pump	**j.** Replace pump.
	k. Fault in engine speed sensor (G28)	**k.** Test sensor at connector.
	l. Fault in ECM or other engine management system component	**l.** Interrogate permanent fault memory. Repair or replace components as required. Requires diagnostic scan tool.
2. Glow plug warning light not working	**a.** Bulb burned out or malfunction in glow plug relay circuit	**a.** Test and repair as necessary.
3. Idle speed incorrect, rough or irregular	**a.** Accelerator pedal or linkage binding	**a.** Remove driver's side trim panels under dash and inspect pedal and linkage for free movement. Repair as required.
	b. Excessive air in fuel system	**b.** Check fuel supply line from fuel tank for cracks, kinks, or leaks.
	c. Clogged fuel filter, or fuel return line and injector pipes leaking, dirty, kinked, or damaged at connectors	**c.** Inspect and, if necessary, replace lines and hoses, replace fuel filter. See **03 Maintenance.**
	d. Faulty injectors	**d.** Test and, if necessary, replace injector.
	e. Injection timing incorrect	**e.** Adjust injection timing. Requires diagnostic scan tool.
	f. Engine mechanical faults	**f.** Test compression. See **15b Cylinder Head and Valvetrain (1.9L engine).**
	g. Faulty injection pump	**g.** Replace pump.

Table a. TDI Diesel Fuel Injection Troubleshooting

Problem	Probable cause	Corrective action
4. Smoky exhaust (black, blue, or white)	**a.** Faulty injectors	**a.** Check and, if necessary, replace injectors.
	b. Injection timing incorrect	**b.** Adjust injection timing. Requires diagnostic scan tool.
	c. Turbocharger leaking internally	**c.** Inspect and repair, or replace turbocharger.
	d. Engine mechanical faults	**d.** Test compression. See **15b Cylinder Head and Valvetrain (1.9L engine)**.
	e. Faulty injection pump	**e.** Replace pump.
	f. Fault in ECM or other engine management system component	**f.** Interrogate permanent fault memory. Repair or replace components as required. Requires diagnostic scan tool.
5. Poor power output, slow acceleration or top speed	**a.** Accelerator pedal or linkage binding	**a.** Remove driver's side trim panels under dash and inspect pedal and linkage for free movement. Repair as required.
	b. Air filter dirty or restricted	**b.** Clean or replace air filter. Inspect housing and ducting for debris. See **03 Maintenance**.
	c. Clogged fuel filter; or fuel return line and injection pipes leaking, dirty, kinked, or damaged at connections	**c.** Check for basic fuel delivery to engine. Inspect and, if necessary, replace lines and hoses, replace fuel filter. See **03 Maintenance**.
	d. Excessive air in fuel system	**d.** Check fuel supply line from fuel tank for cracks, kinks, or leaks, especially at and around fuel filter and injection pump inlet.
	e. Faulty injectors	**e.** Check and, if necessary, replace injectors.
	f. Injection timing incorrect	**f.** Adjust injection timing. Requires diagnostic scan tool.
	g. Engine mechanical faults	**g.** Test compression. See **15b Cylinder Head and Valvetrain (1.9L engine)**.
	h. Faulty injection pump	**h.** Replace pump.
	i. Fault in ECM or other engine management system component	**i.** Interrogate permanent fault memory. Repair or replace components as required. Requires diagnostic scan tool.
6. Engine runs at a high constant idle and speed does not vary	**a.** Throttle position sensor (G79), malfunctioning	**a.** Interrogate permanent fault memory. Repair or replace components as required. Requires diagnostic scan tool.
7. Excessive fuel consumption	**a.** Air filter dirty or restricted	**a.** Clean or replace air filter element. Inspect housing and ducting for debris. See **03 Maintenance**.
	b. Fuel leaks	**b.** Check and, if necessary, replace or tighten all pipes, hoses and connections.
	c. Fuel return pipe blocked or restricted	**c.** Check return line for kinks and dents; replace faulty lines. If line is clogged, blow it out with compressed air, then bleed fuel system.
	d. Faulty injectors	**d.** Test and, if necessary, repair or replace injectors.
	e. Injection timing incorrect	**e.** Adjust injection timing. Requires diagnostic scan tool.
	f. Faulty injection pump	**f.** Replace pump.
	g. Fault in ECM or other engine management system component	**g.** Interrogate permanent fault memory. Repair or replace components as required. Requires diagnostic scan tool.

Fuel shut-off valve, checking

The diesel engine is stopped by cutting off its fuel supply. This is done by means of an electrical solenoid located on the diesel injection pump and that is closed with power off. An electrical signal from the engine control module (key ON) opens the valve. If the valve is not working, the engine will not start. Conversely, a faulty valve could also allow the engine to continue running after the key is switched off.

– Turn ignition key on and off without operating starter. Listen for valve to operate.
  - The fuel cut-off valve should click open when the key is turned on and click closed when the key is turned off. Do not confuse this sound with continuous sounds coming from other electronic components on pump.

– If valve does not operate, check for voltage at harness connector with key on.
  - If voltage is not present, check for wiring faults. See **EWD Electrical Wiring Diagrams.**
  - If voltage is present, disconnect battery ground (GND) strap from battery negative (–) terminal and proceed with step 3. See Caution at the beginning of this repair group regarding battery disconnection.

> ***CAUTION —***
> *Before disconnecting battery cables, be sure to obtain radio anti-theft code.*

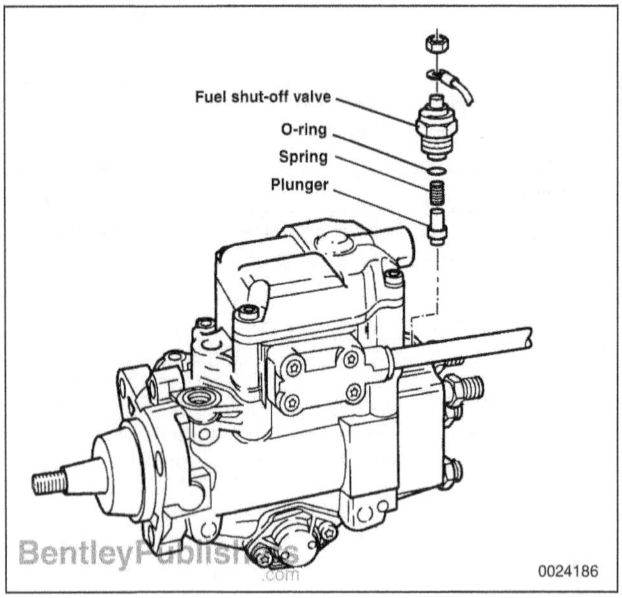

◄ Clean area around fuel shut-off valve and remove valve, spring and plunger.

> ***WARNING —***
> *Fuel will be expelled. Do not smoke or work near heaters or other fire hazards. Have a fire extinguisher handy.*

– Clean and inspect solenoid plunger and seat. Slide plunger into the valve without spring and check for free movement. If plunger does not move freely, replace valve assembly.

– Carefully check for dirt and metal particles in valve bore of injection pump. The slightest amount of debris can cause malfunctions.

– Install fuel cut-off valve using a new O-ring.

– Reconnect battery negative (–) terminal.

Tightening Torque
- Fuel shut-off valve to injection pump . . . . 40 Nm (30 ft-lb)

– Be sure to quality check your work.

Basic fuel delivery, checking

This test checks if fuel is being delivered to the pump and fuel injectors. Perform this test when experiencing hard starting or no-start problems.

– Check visually for fuel leaks on injector pump and especially around injector line unions.

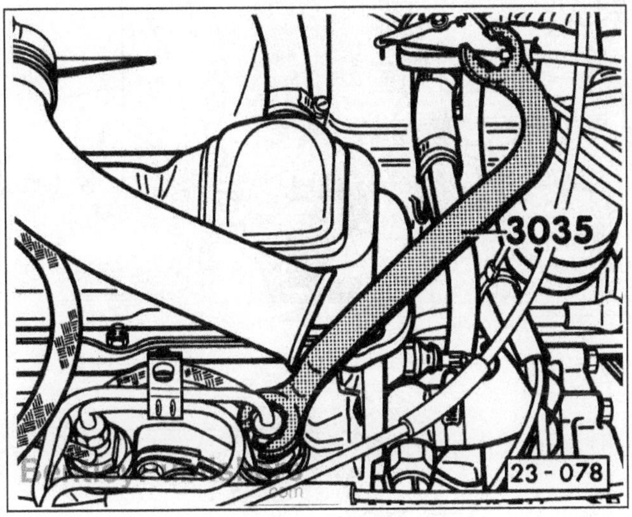

< If no leaks are found, clean the fuel union nut for cylinder No. 1 fuel injector pipe at the injector (fuel line at cylinder No. 4 shown). Slightly loosen the fuel union nut.

> ***WARNING —***
> - *Loosen nut only about one-half turn to limit amount of fuel leakage. The fuel injection system operates at very high pressure. Keep hands and eyes clear. Wear heavy gloves and eye protection. Wrap a cloth around the nut when loosening.*
> - *Fuel will be expelled. Do not smoke or work near heaters or other fire hazards. Have a fire extinguisher handy.*

– Crank engine and look for fuel to run out of loosened union nut at injector.

> ***CAUTION —***
> *Diesel fuel is damaging to rubber. Wipe off any fuel that spills on hoses, wiring, and rubber steering and suspension parts and wash with soap and water. If coolant hoses have been contaminated with diesel fuel, they must be replaced.*

– If no fuel is observed, the most likely cause is a restriction or air leak in fuel lines between fuel tank and injection pump, or possibly a clogged fuel filter. It is also possible that vehicle has run out of fuel regardless of fuel gauge readings. If vehicle has a semi-transparent plastic fuel pipe on the supply side of diesel injection pump, check for movement of fuel in the line.

NOTE —
A stream of tiny air bubbles moving in the line is normal (unlike previous versions of the diesel engine).

– If fuel is observed, but engine still fails to start, consult **Table a**, given earlier, for additional troubleshooting.

NOTE —
It is not necessary to test for basic fuel delivery on the remaining three cylinders due to the way the TDI system operates.

– When finished testing, tighten fuel union nuts and check for leaks. Clean up any spilled fuel.

Tightening torque
- Fuel injector union nut to injector or pump . . . . 25 Nm (18 ft-lb)

FUEL INJECTION COMPONENTS (1.9L, CODE: ALH)

NOTE —
Volkswagen identifies electrical components by a letter and/or a number in the electrical schematics. See **EWD Electrical Wiring Diagrams.** *These electrical identifiers are listed in parenthesis as an aid to electrical troubleshooting.*

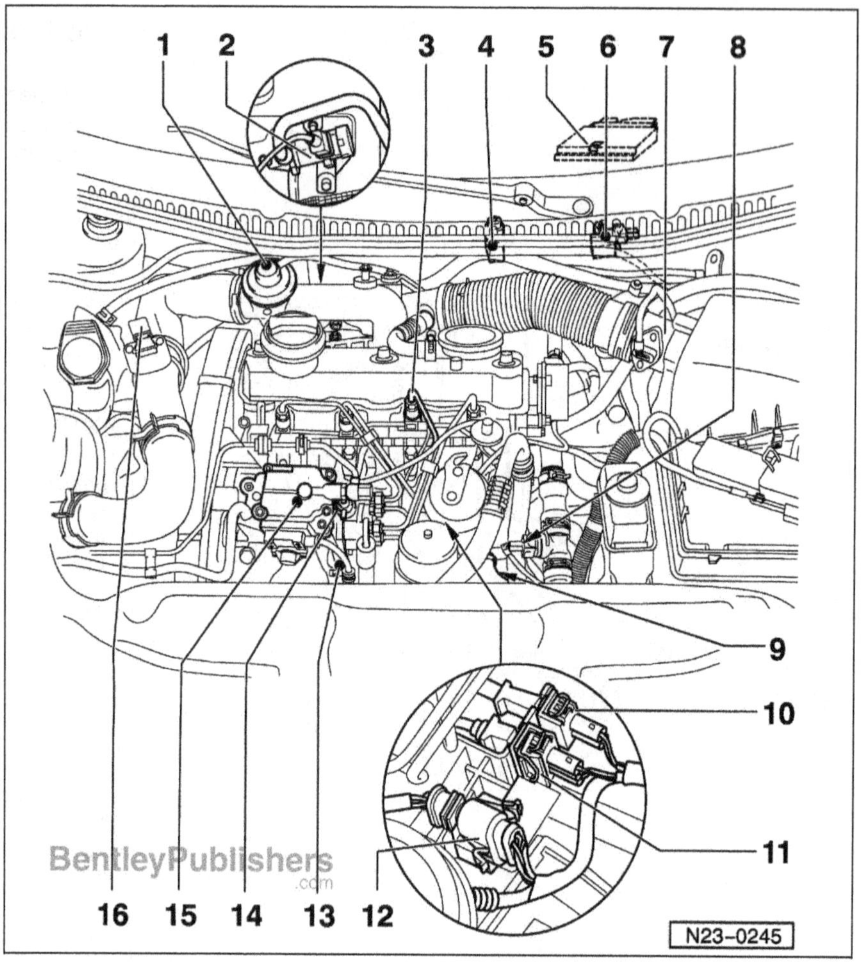

Fuel injection system, components

1. **EGR valve**
2. **Intake manifold change-over valve (N239)**
3. **Injector with needle lift sensor (G80)**
4. **EGR vacuum regulator solenoid valve (N18)**
5. **Diesel direct injection system ECM (J248)**
   - With BARO sensor (F96)
6. **Wastegate bypass regulator valve (N75)**
7. **Mass air flow sensor (G70)**
8. **Engine coolant temperature sensor (G62)**
9. **Engine speed sensor (G28)**
10. **Harness connector**
    - For needle lift sensor (G80)
11. **Harness connector**
    - For engine speed sensor (G28)
12. **Harness connector**
    - For fuel temperature sensor (G81)
    - For quantity adjuster (N146)
    - For modulating piston displacement sensor (G149)
    - For fuel shut-off valve (N109)
    - For cold start injector
13. **Cold start injector**
14. **Fuel shut-off valve (N109)**
15. **Injection pump quantity adjuster**
    - With fuel temperature sensor
    - With quantity adjuster (N146)
    - With modulating piston displacement sensor (G149)
16. **Manifold absolute pressure sensor (G71) and intake air temperature sensor (G72)**

Diesel injection components not shown:

- Brake pedal switch (F47) and brake light switch (F); located in one housing on brake pedal.
- Throttle position sensor (G79); located in footwell on throttle pedal.
  In case of malfunction, the scan tool display is Throttle Position Sensor (G69)
- Clutch pedal switch (F36); located in footwell on clutch pedal.

Fuel injectors, checking

The 1.9L TDI diesel engine uses a two-spring injector that enables the fuel to be injected in two stages for "softer" combustion. In addition, the injectors are a 5-hole design that spray fuel from 5 ports in a more lateral direction rather than in a single conical pattern. Injector opening pressures are approximately 220 to 230 bar (3190-3335 psi). All 4 injectors are functionally the same. However, injector for cylinder #3 carries the needle lift sensor (G80). This injector must always be installed in cylinder #3 to prevent erroneous DTCs or performance problems from occurring.

The signs of injector trouble usually appear as misfiring and knocking noises from one or more cylinders, engine overheating, loss of power, smoky black exhaust, and excessive blue smoke during cold starting.

> ***CAUTION—***
> *When working on the injectors, everything must be kept absolutely clean. Clean all pipe unions before disconnecting.*

A defective injector (or a weak cylinder) can often be identified by simply "turning it off". This can be done by relieving the fuel pressure so that the injector can not open and inject fuel into the cylinder. Idle speed may not substantially change, however, because the ECM will often be able to compensate for the weak cylinder.

– Clean area around fuel unions where fuel lines attach to injectors.

– Start engine and let it run at idle.

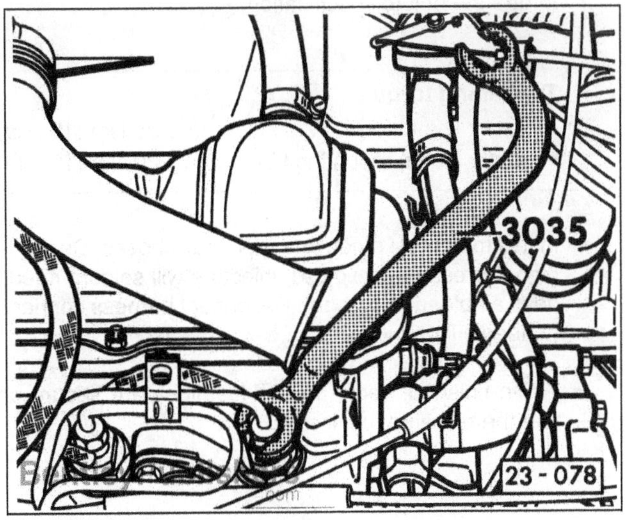

◄ Wrap a cloth around the union nut for cylinder #1 fuel injector at the injector. Slightly loosen the nut. Cylinder #1 is closest to the drive belt.

- If a particular cylinder is defective, the noise or smoke will cease when union nut is loosened.
- If the problem was roughness, the roughness will be affected very little or not at all because that cylinder was already malfunctioning. Tighten union nut.

> ***WARNING—***
> - *Loosen the nut only about one half turn to limit the amount of fuel leakage. The fuel injection system operates at very high pressures. Keep hands and eyes clear. Wear heavy duty gloves and eye protection. Wrap a cloth around the nut when loosening. Fuel will be expelled.*
> - *Do not smoke or work near heaters or other fire hazards.*
> - *Have a fire extinguisher handy.*

– Repeat procedure for other suspected cylinders.

– When finished, verify that all fuel lines are secure and that there are no fuel leaks. Thoroughly clean any spilled fuel.

> ***CAUTION —***
> *Diesel fuel is damaging to rubber. Wipe off any fuel that spills on hoses, wiring, and rubber steering and suspension parts and wash with soap and water. If coolant hoses are contaminated with diesel fuel, they must be replaced.*

Fuel injectors, replacing

– Unscrew fuel pipe union nuts at injection pump and injectors. Remove injector fuel pipes as one assembly taking care not to bend them.

– Remove fuel return hoses from between injectors.

– Unplug harness connector for the needle lift sensor (G80) on the injector for cylinder #3.

– Loosen and remove bolt, special washer and hold down clamp for each injector. Lift out injectors.

– Remove small copper sealing washer from tip of injector or from injector port in head and discard.

NOTE —
The crush-type sealing washers must always be replaced to prevent compression leaks.

– Install injector with a new copper sealing washer and torque the bolt to specification.

Tightening torque
- Injector line union. . . . . 25 Nm (18 ft-lb)
- Injector clamp mounting bolt (M8) . . 20 Nm (15 ft-lb)

– Install metal fuel pipes and fuel return hoses. Due to the higher pressures involved, injectors will seldom need to have air bleed from them. Reconnect harness connector to needle lift sensor.

– When finished, verify that all fuel lines are secure and that there are no fuel leaks.

Fuel injectors, pressure testing

– Remove injector as described previously.

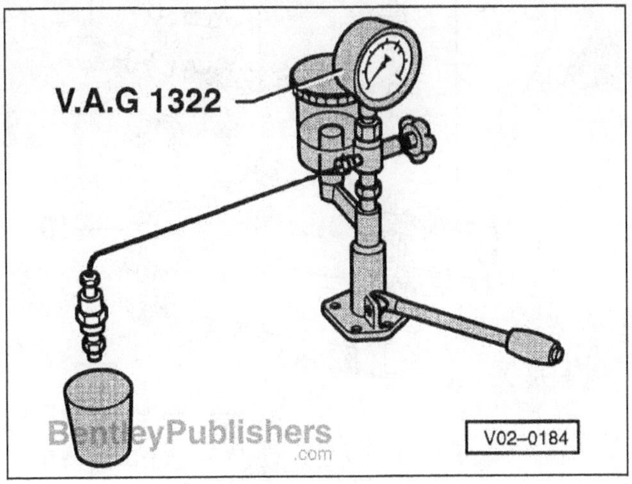

< Install fuel injector on pressure tester according to manufacturer's directions. Place a suitable container under injector to catch the spray.

> ***WARNING —***
> *When testing fuel injectors, make sure that the high pressure discharge does not contact the hands or any other bare skin. The high pressure spray can be injected directly into the skin causing severe injuries or health complications. Always wear eye protection!*

– Pump lever to increase pressure. Read opening pressure when spray begins.

Injector Opening Pressure

- New . . . . 220–230 bar (3190–3335 psi)
- Wear limit . . . . 200 bar (2900 psi)

< Spray should discharge evenly from all 5 ports.

> ***WARNING —***
> *Fire hazard! No smoking! Do not have anything in the area that can ignite the spray discharged by the injectors. This includes overhead heating units and hidden sources such as water heaters with standing pilot lights.*

– Pump lever slowly to a pressure of approximately 150 bar (2175 psi) for 10 seconds. Check for leaks from injector tip. No fuel should leak.

– If any injector fails to meet opening pressure or leak specifications, it should be replaced. No repairs are possible to the fuel injectors.

Injection pump

The diesel fuel injection pump is controlled by the ECM and performs three primary functions:

- It takes the fuel from the tank and pressurizes it to over 3000 psi.
- It precisely times the delivery of fuel to the injectors.
- It accurately meters the quantity of fuel delivered to the injectors.

The pump is driven by the same toothed drive belt that drives the camshaft at 1/2 crankshaft speed. Internal parts are lubricated and cooled by the diesel fuel. No routine maintenance is required or possible because no internal repair parts are available.

Because the Turbo Direct Injection system is fully electronic, the ECM can detect and store most faults associated with the injection pump. These faults are known as DTCs and are held in the ECM memory. Most DTCs will turn on the malfunction indicator light (MIL) and/or the glow plug light, alerting the driver to this condition. In addition, the ECM has diagnostic capabilities that allow an experienced technician with a scan tool to view certain operating parameters of the injection pump and other parts of the system. These must be investigated before a decision is made to replace an injection pump.

Injection pump assembly

> ***NOTE —***
> *Volkswagen identifies electrical components by a letter and/or a number in the electrical schematics. See* **97 Wiring Diagrams, Fuses and Relays**. *These electrical identifiers are listed as an aid to electrical troubleshooting.*

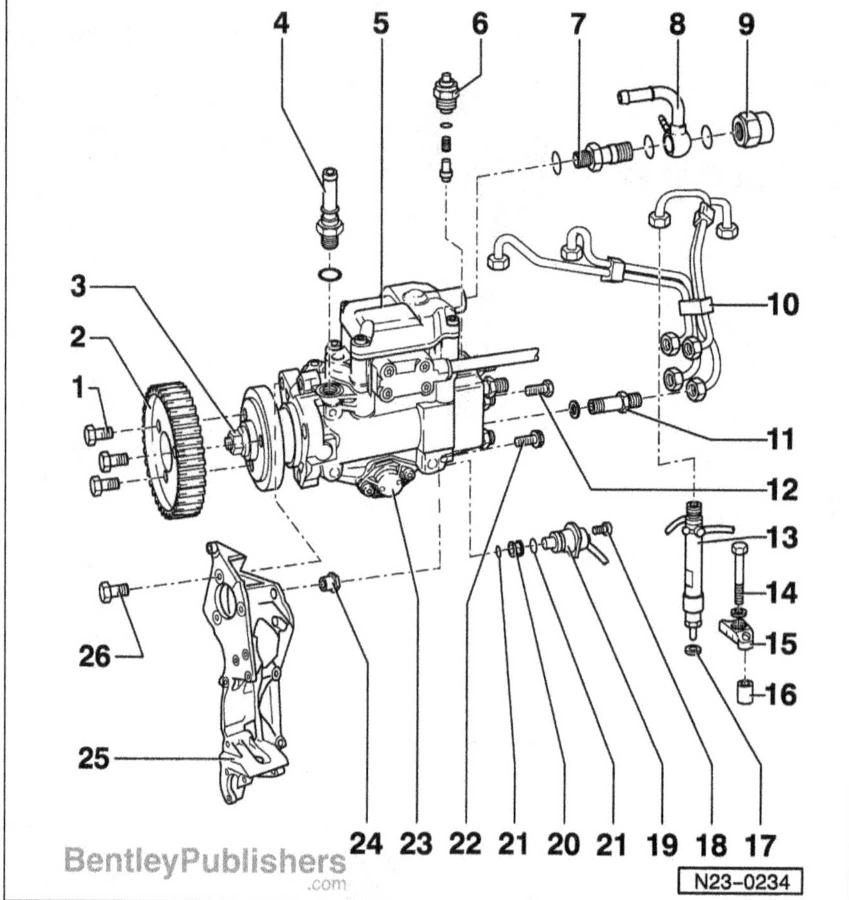

1. **Bolts**
   - Tighten to 20 Nm (15 ft-lb) plus ¼ turn (90°)
   - Always replace
2. **Injection pump sprocket**
3. **Injection pump sprocket nut**
   - Do not loosen for any reason!
4. **Union**
   - Supply line from fuel filter
   - Tighten to 25 Nm (18 ft-lb)
5. **Diesel injection pump**
   - With quantity adjuster (N146)
   - With modulating piston displacement sensor (G149)
   - With Fuel Temperature Sensor (G81)
6. **Fuel shut off valve (N109)**
   - Tighten to 40 Nm (30 ft-lb)
7. **Union**
   - Return line from diesel injection pump
8. **Return line**
   - To control valve on fuel filter
9. **Cap nut**
   - Tighten to 25 Nm (18 ft-lb)
10. **Fuel injector lines**
    - Tighten to 25 Nm (18 ft-lb)
    - Always remove as a complete assembly
    - Do not bend, kink or alter shape
11. **Union**
    - To injector lines
    - With check valve
    - Tighten to 45 Nm (33 ft-lb)
12. **Bolt**
    - Tighten to 25 Nm (18 ft-lb)
13. **Fuel injector**
    - Cylinder # 3 with needle lift sensor (G80)
14. **Bolt**
    - Tighten to 20 Nm (15 ft-lb)
15. **Injector retainer**
    - Use special washer only
16. **Injector retainer mounting**
    - Fits into machined hole in cylinder head
17. **Sealing washer**
    - Always replace
18. **Bolt**
    - Bolt
    - Tighten to 10 Nm (7 ft-lb)
19. **Cold Start Injector (N108)**
    - Controls start of injection (timing)
20. **Strainer**
21. **O-ring**
    - Always replace
22. **Bolt**
    - Tighten to 25 Nm (18 ft-lb)
23. **Cover**
24. **Rear sleeve nut**
25. **Assembly bracket**
26. **Bolt**
    - Tighten to 25 Nm (18 ft-lb)

Injection pump, removing and installing

Removal of the injection pump requires partial removal of the toothed belt driving the camshaft and the injection pump. Special Volkswagen tools, including a diagnostic scan tool, are required to remove and install the diesel injection pump and properly set it up.

- Remove upper engine cover.
- Remove connecting pipes between intercooler and intake hose. Remove intercooler and turbocharger.
- Remove fuel supply and return lines at injector pump.
- Remove injector fuel lines as an assembly using a flare nut wrench as shown earlier.

> ***WARNING —***
> *Fuel will be expelled. Do not smoke or work near heaters or other fire hazards. Have a fire extinguisher handy.*

NOTE —

Cover all open fuel fittings with a clean cloth to prevent contamination.

- Remove upper toothed belt guard.
- Remove vacuum pump from end of cylinder head.
- Remove cylinder head (valve) cover.

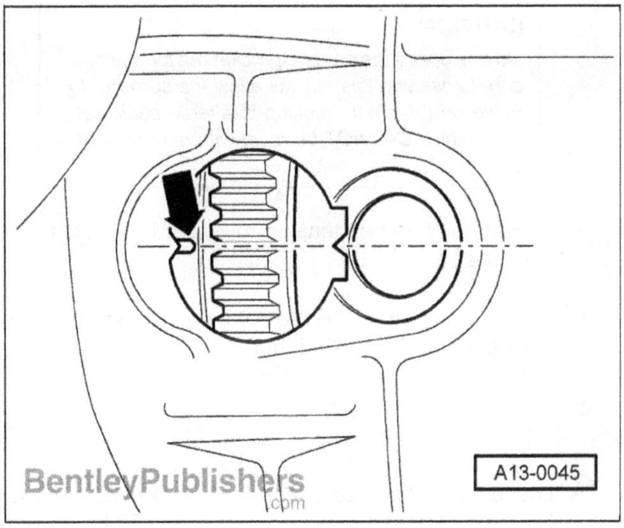

Manual transmission equipped vehicles: Set engine to top dead center (TDC), for cylinder #1 by turning crankshaft until timing mark on flywheel (**arrow**) aligns with mark on transmission case.

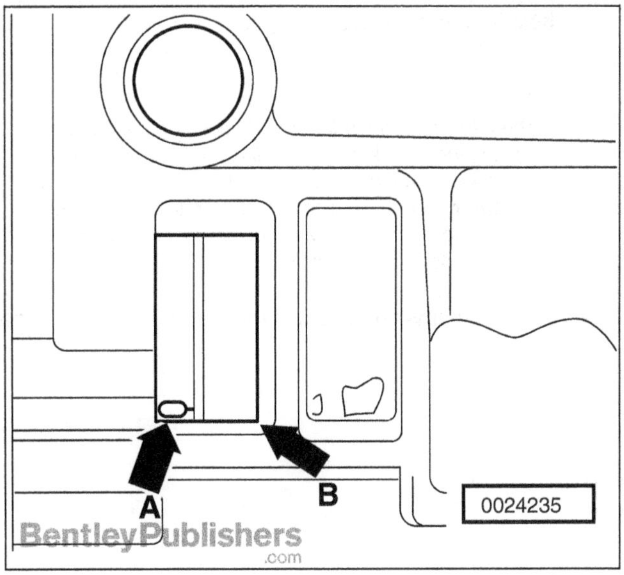

Automatic transmission equipped vehicles: Set engine to top dead center (TDC), for cylinder #1 by turning crankshaft until timing mark on torque converter (**A**) aligns with mark on transmission case (**B**).

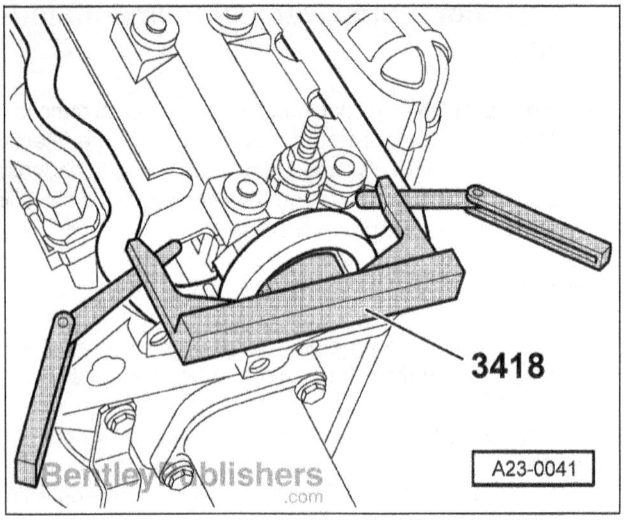

◀ Lock camshaft at TDC with a setting bar such as Volkswagen special tool 3418 and centralize using feeler gauges.

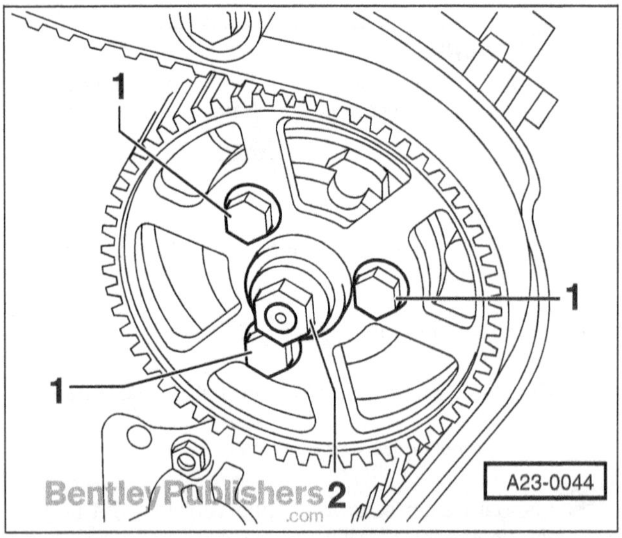

◀ Remove injection pump sprocket mounting bolts (**1**) and discard. Be sure not to loosen center sprocket nut (**2**).

> ***CAUTION—***
> *Do not loosen the center sprocket nut for any reason. Loosening this nut will allow the sprocket to move on the shaft resulting in altered basic settings which CANNOT be reset with normal workshop equipment.*

– Loosen toothed belt tension roller nut to relieve toothed belt tension.

– Slip toothed belt off injection pump sprocket and remove sprocket from pump shaft hub.

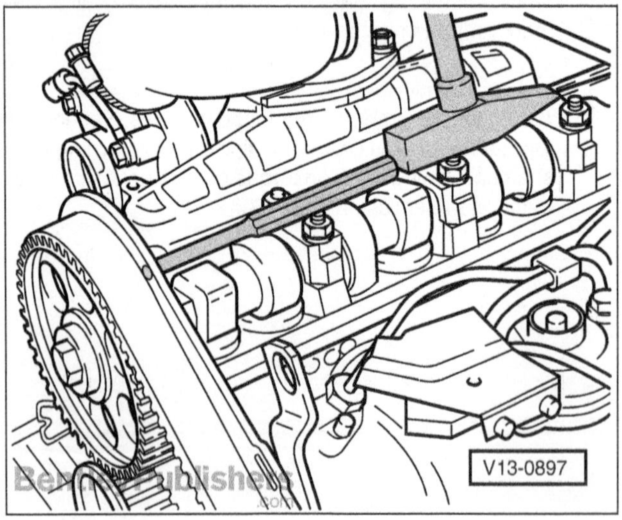

◀ Loosen camshaft sprocket mounting bolt about ½ turn. Release camshaft sprocket from camshaft taper by tapping with a hammer on a soft drift inserted through the hole in rear of toothed belt guard.

NOTE—
The camshaft end is tapered. Loosening the bolt and tapping the sprocket will allow the camshaft sprocket to rotate independently of the camshaft. There may be a keyway cut into the end of the camshaft. However, there is no matching keyway cut into the sprocket nor is there a woodruff key installed.

– Remove camshaft sprocket mounting bolt and sprocket.

Disconnect the harness connector (**arrow**) for the diesel pump electronics and unclip the connector from the retaining bracket.

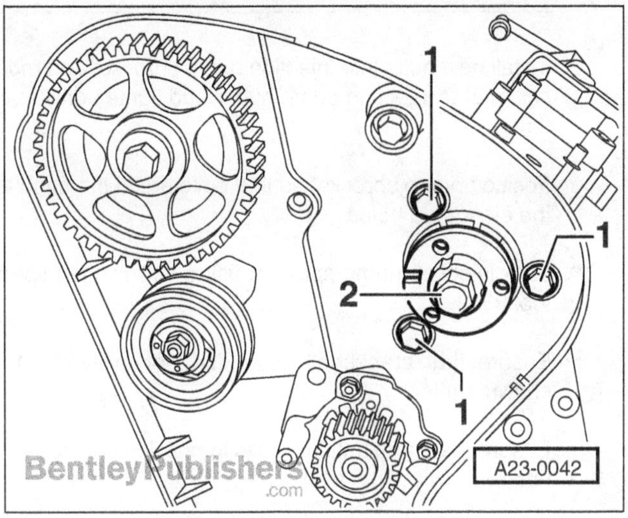

Remove front injection pump mounting bolts (**1**) from assembly bracket. Be sure not to loosen center sprocket nut (**2**).

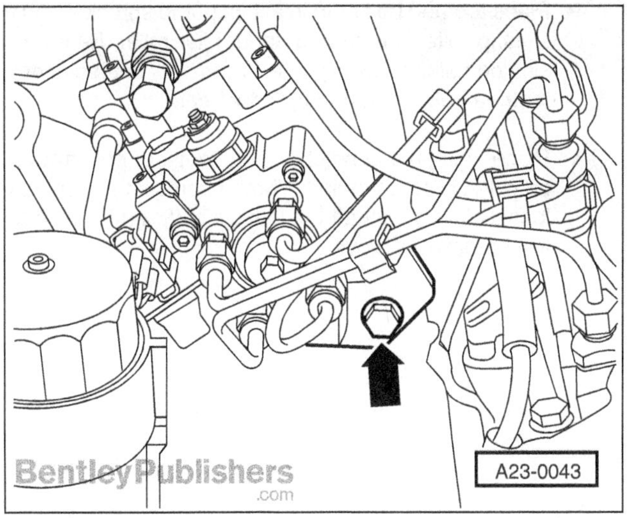

Remove rear injection pump mounting bolt (**arrow**) from support bracket.

- Carefully slide pump shaft out of assembly bracket and lift pump out.
- If pump will not be immediately reinstalled, take steps to insure that no dirt can get into any of pump openings or injector unions. Store pump so that diesel fuel in the pump will not drain out.
- Be sure to quality check your work.

Injection pump, installing

Before installing a new injection pump or if reusing the original, remove all packaging materials and plugs sealing port openings.

– Install injection pump into assembly bracket and hand tighten rear injection pump mounting bolt.

– Position pump in assembly bracket. Install and torque front injection pump mounting bolts and torque rear injection pump mounting bolt.

Tightening torque
- Injection pump bolts, front (M8) .... 25 Nm (18 ft-lb)
- Injection pump bolt, rear (M8) ..... 25 Nm (18 ft-lb)

– Install new bolts into injection pump sprocket and mount sprocket to injection pump hub. Hand tighten only at this time.

– Position pump sprocket so that bolts are in the middle of the elongated holes

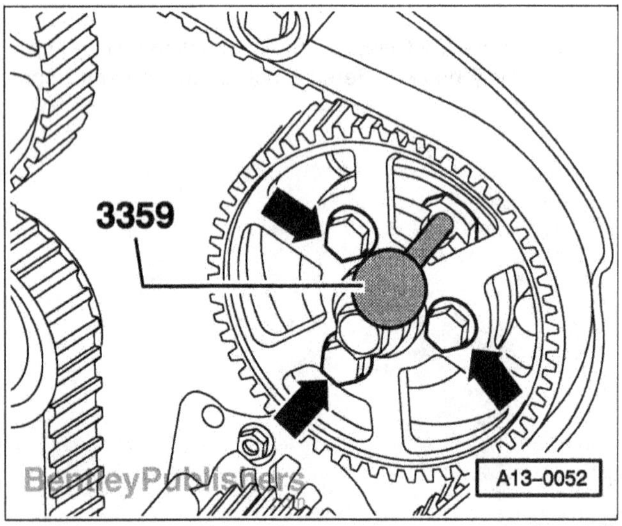

◄ Lock injection pump sprocket using Volkswagen special tool 3359.

– Ensure that crankshaft is at TDC for cylinder #1 as shown earlier.

– Ensure that camshaft is locked in position as shown earlier.

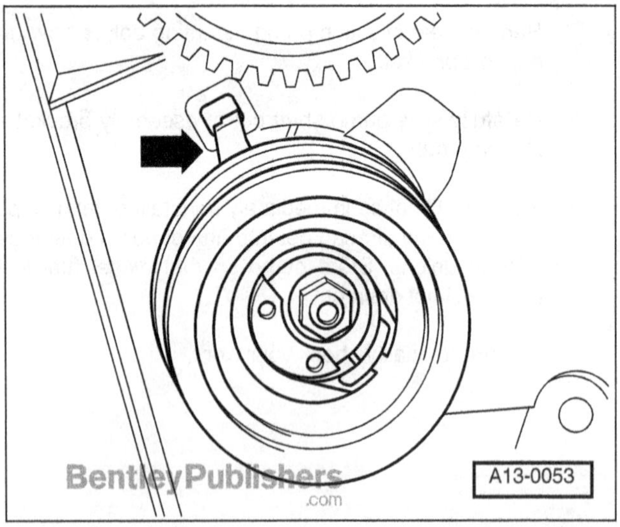

◄ Place toothed belt onto injection pump sprocket and tensioning roller and thread under idler roller. Be sure tensioning roller retaining hook is correctly seated in hole in toothed belt guard and cylinder head.

– Place camshaft sprocket into toothed belt and slide sprocket onto camshaft. Thread sprocket mounting bolt loosely into camshaft but leave it sufficiently loose to allow sprocket to move.

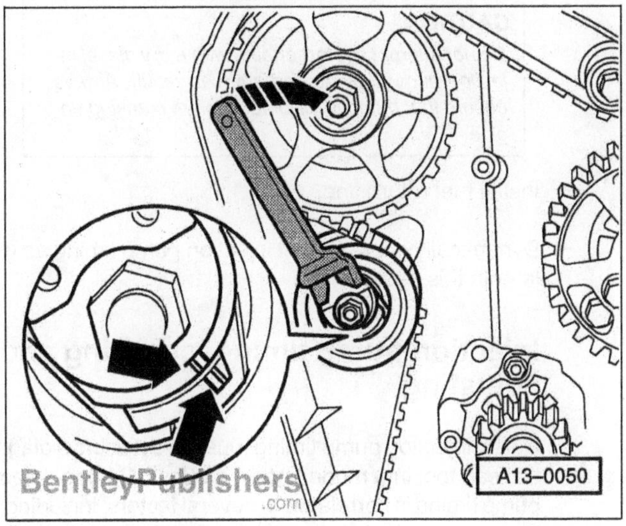

◀ Tension toothed belt using a two-pin spanner in the center holes of the eccentric belt tensioner. Belt is tensioned properly when notch and raised mark are aligned **(inset arrow)**.

– Tighten lock nut on tensioner while maintaining belt tension.

Tightening torque
- Toothed belt tensioner (M8) . . . . 20 Nm (15 ft-lb)

– Ensure that crankshaft is still at TDC.

– Tighten new injection pump sprocket mounting bolts to **stage I** specification only.

NOTE—
- *Tighten the new injection pump sprocket mounting bolts to stage I only at this time. Final torque (stage II) will be done after timing has been dynamically checked/adjusted.*
- *Injection pump sprocket mounting bolts must only be used one time, since by design, they have a reduced shank and are stretch bolts.*

Tightening torque
- Injection pump sprocket bolt - always replace
  stage I . . . . 20 Nm (15 ft-lb)
  stage II. . . . . additional ¼ turn (90°)
- Camshaft sprocket bolt. . . . . 45 Nm (33 ft-lb)

– Tighten camshaft sprocket mounting bolt and remove setting bar and feeler gauges.

– Remove locking pin from injection pump sprocket.

– Turn crankshaft two rotations in the running direction and recheck toothed belt tension.

– Connect harness connector for diesel pump electronics and slip connector into retaining bracket.

– Install cylinder head (valve) cover.

– Install vacuum pump.

– Install injector fuel line assembly.

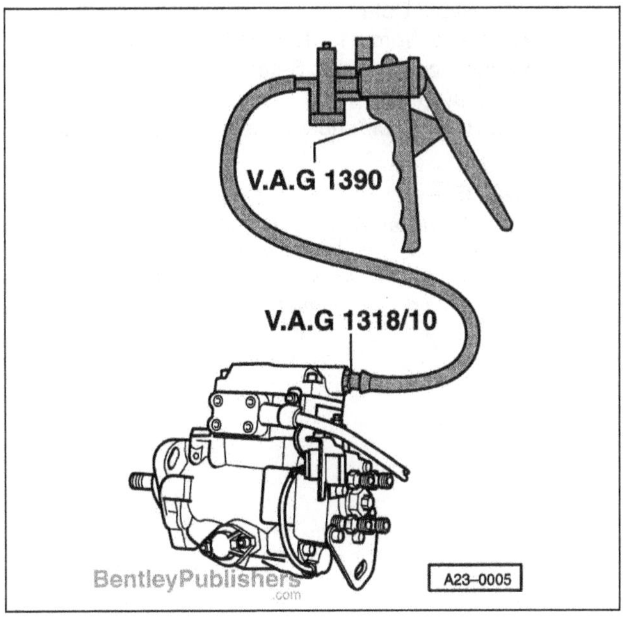

◀ Install fuel supply line and draw fuel into pump with a hand operated vacuum pump.

NOTE—
Do not draw diesel fuel into hand operated vacuum pump made of plastic!

> ***CAUTION —***
> *Do not attempt to start engine with a dry diesel injection pump, severe damage can result. Always ensure that pump is full of fuel before cranking engine.*

– Install fuel return line.

– Dynamically check diesel injection pump timing as given later in this section.

Injection pump timing, checking and adjusting

Diesel injection pump timing must be set using a diagnostic scan tool in a mode known as Basic Setting. Injection pump timing is correlated to several factors, including fuel temperature. Failure to properly set injection pump timing will result in hard starting, poor performance, poor fuel economy, excessive noise, smoking and increased emissions. The vehicle may not be drivable.

If timing adjustment is necessary, special adjustment tools and new pump sprocket mounting bolts are required.

The following instructions assume that the individual technician already has a good working knowledge of the operation and use of a diagnsotic scan tool.

– Connect scan tool to vehicle Data Link Connector (DLC) located in lower left side of the dash panel. The scan tool will be powered up when plugged into the DLC.

– Start the engine and let it idle.

System in Basic Setting	000
XX 53 XX XX XX XX XX XX 90 XX	

◄ Observe the following sequence in the scan tool display window:

- 1 – Rapid Data
- Address Word 01 – Engine Electronics
- Function 04 – Basic Settings
- Display Group – 000
- The display will show 10 fields (groups) of information.

> ***NOTE —***
> *Data is only available in the format shown in Fig. 19 through use of VW diagnostic scan tool VAG 1551/1552. Other scan tools may provide the same data, but in a different format.*

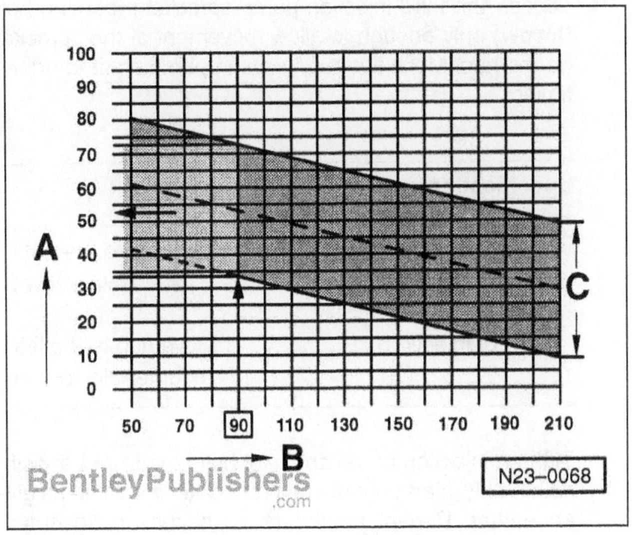

◄ Use the graph to find timing value. Observe the number in field 9 on the scan tool display. This is a fuel temperature value and is assigned **B** in the table. The timing value must fall within shaded area of the table and is assigned the value **A**. If value is within shaded area, no adjustment is needed. When value is obtained, shut off engine.

- If this procedure was done for diagnosis purposes, and timing value is within specification, remove scan tool.
- If this procedure was done as part of injection pump removal and installation, and timing value is not within specification, continue procedure steps.

Example:

- This example assumes the engine is running. Note that the fuel temperature value in the scan tool display window, field 9, is 90. This is value **B**. Look on the table along the B axis until you see 90. Follow the line from 90 up into the shaded area. The shaded area is the range of timing values. Find the center of the shaded area and look to the left for the value on the **A** axis. In this case it is 53. Adjust the pump until the value in the scan tool display field 2 reads 53.

– If timing adjustment is required, access injection pump sprocket. See **Injection pump, removing** given earlier for appropriate steps.

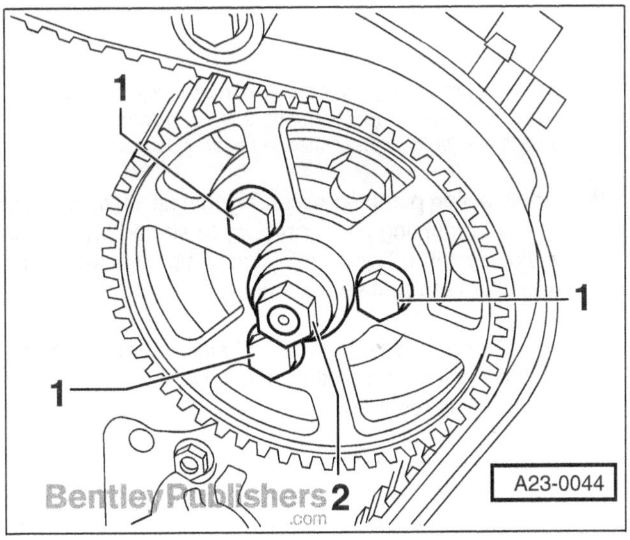

◄ Loosen two of the injection pump sprocket mounting bolts (**1**). Be sure not to loosen center sprocket nut (**2**).

NOTE —

It is permissible to counterhold the pump shaft during loosening using a 22 mm wrench on the center nut. Under no circumstances should the center nut be loosened or used as a counterholder during tightening sequences.

CAUTION —

Do not loosen the center sprocket nut for any reason. Loosening this nut will allow the sprocket to move on the shaft resulting altered basic settings which CANNOT be reset with normal workshop equipment.

Loosen the third injection pump sprocket mounting bolt (**arrow**) only enough to allow movement of the sprocket on the hub. Move the hub/injection pump shaft to adjust timing.

Specification
- Advance timing (start of injection) . . . . Move hub to the right (clockwise)
- Retard timing (start of injection) . . . . Move hub to the left (counter-clockwise)

– Tighten injection pump sprocket bolts to **stage I** specification only, start engine and recheck timing value as given earlier. Repeat procedure as necessary to obtain correct timing value.

Tightening torque
- Injection pump sprocket bolt - always replace
  stage I . . . . 20 Nm (15 ft-lb)
  stage II. . . . . additional ¼ turn (90°)

– When a satisfactory timing value is obtained, remove and replace 3 pump sprocket with new bolts (**arrows**) one at a time.

NOTE —
If bolts were already replaced as part of injection pump installation and/or replacement, it is not necessary to install new bolts as described in the above step.

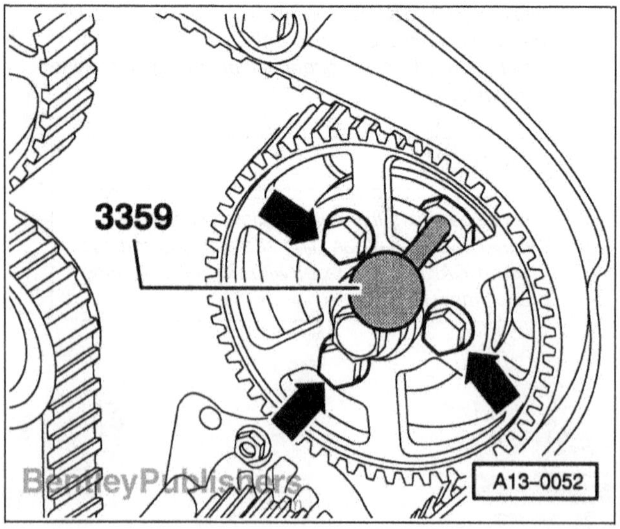

Lock injection pump sprocket. Tighten all injection pump sprocket mounting bolts (**arrows**) to stage I value if not previously done. Tighten injection pump sprocket mounting bolts to **stage II** specification.

– With scan tool still connected, check for any Diagnostic Trouble Codes (DTCs) that may have been set. Repair and erase as required.

– Disconnect scan tool.

– Install all remaining fasteners, fuel lines, electrical connectors and components that were removed.

– Verify that all fuel lines are secure and that there are no leaks.

– Install upper sound absorber panel.

– Be sure to quality check your work.

Intake manifold assembly

The intake manifold tuning valve is a flap that closes for approximately 3 seconds when the engine stops, blocking the air to the engine. This reduces the suddenness of the diesel engine shut-down. The valve reopens once the engine has stopped.

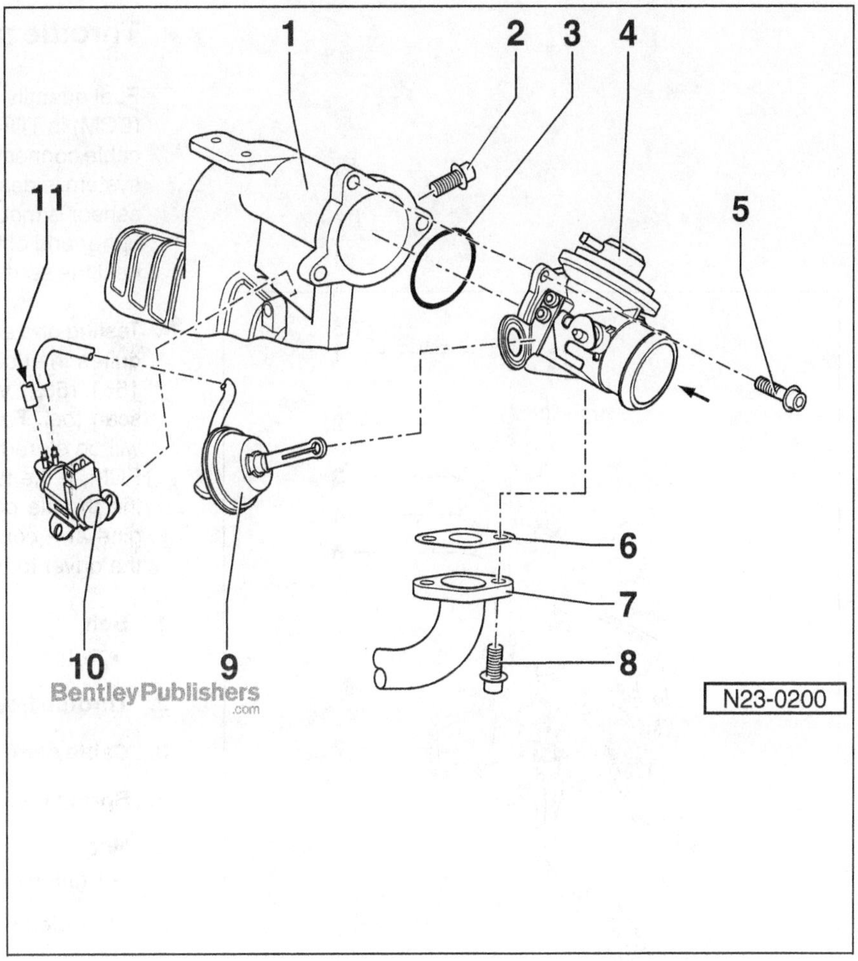

1. **Intake manifold**
2. **Bolt**
   - Tighten to 10 Nm (7 ft-lb)
3. **O-ring**
   - Always replace
4. **Connecting flange**
   - With EGR valve
   - With intake manifold change-over valve
5. **Bolt**
   - Tighten to 10 Nm (7 ft-lb)
6. **Gasket**
   - Always replace
7. **Connecting pipe**
   - From EGR cooler
8. **Bolt**
   - Tighten to 25 Nm (18 ft-lb)
9. **Vacuum servo**
10. **Intake manifold tuning vacuum control valve**
11. **Vacuum supply**
    - From vacuum source

Throttle position sensor

Fuel quantity is controlled by the Engine Control Module (ECM) in TDI equipped vehicles. There is no accelerator cable connection between the pedal and the engine. This system is described as "drive by wire". A throttle position sensor is mounted on the inside of the bulkhead near the upper end of the pedal. A very short cable and cam connect the sensor and the pedal.

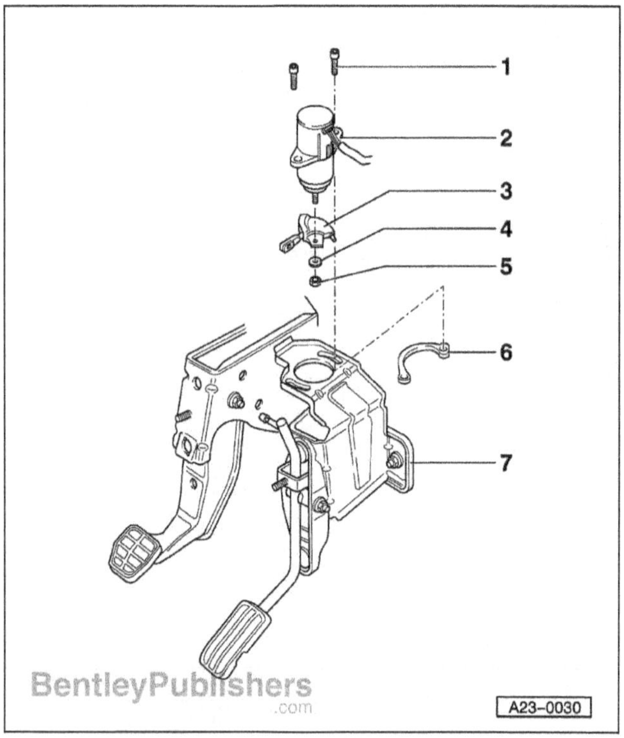

◀ Testing and adjustment of the throttle position sensor requires the use of Volkswagen diagnostic scan tool VAG 1551/1552, VAS 5051/5052, or equivalent aftermarket scan tool. Faults related to the throttle position sensor will be stored as diagnostic trouble codes (DTCs) in the ECM. If the fault is severe enough, the ECM will ignore the throttle position sensor completely and run the engine at a constant high idle speed. This should enable the driver to reach the nearest workshop.

1. **Bolt**
   - Tighten to 10 Nm (7 ft-lb)
2. **Throttle Position Sensor (G79)**
3. **Cable cam**
4. **Spring washer**
5. **Nut**
   - Tighten to 10 Nm (7 ft-lb)
6. **Threaded retainer plate**
7. **Pedal mounting bracket**
   - Bracket supports accelerator and brake pedals

GLOW PLUGS (1.9L, CODE: ALH)

The TDI diesel engine uses four engine glow plugs. Because of the TDI combustion chamber design, the glow plugs are installed in a vertical manner and are fitted with push-on electrical connectors similar to spark plug connectors. The glow plugs assist combustion during cold starts by providing additional heat for ignition. Due to the efficiency of the TDI engine design, the additional heat that the glow plugs provide is not needed until temperatures drop to approximately 48°F (9°C). This function is known as preglow. In addition to preglowing, the glow plugs will continue to operate after the engine is started. This is known as afterglowing and is used to reduce engine noise, improve idle quality and reduce hydrocarbon emissions. The afterglow function is enabled after every start, both hot and cold until the specified time has expired or the engine speed exceeds 2,500 rpm.

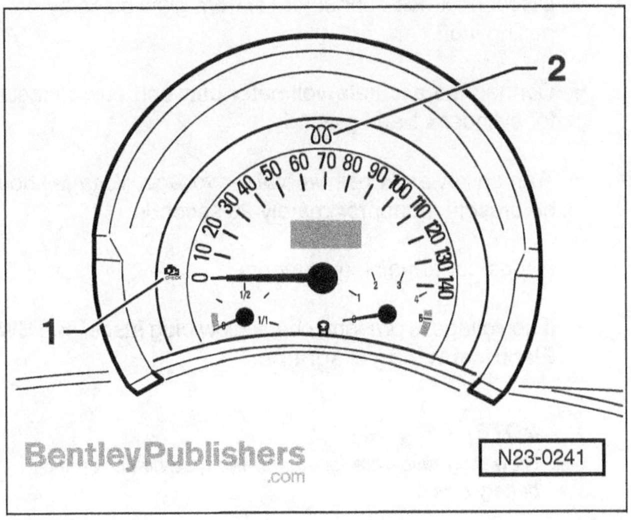

The glow plug system is controlled by the Engine Control Module (ECM) and the Engine Coolant Temperature (ECT) sensor. The period of preglow depends on the temperature of the engine coolant as determined by the ECT. A dash mounted indicator light (**2**) signals the driver that preglowing is operating. Preglowing is not linked to operation of the driver's door as in some earlier Volkswagen diesel systems.

> ***WARNING —***
> *Disconnecting engine management components for this procedure will cause diagnostic trouble codes to be stored in the engine control module. Erasing these DTCs and resetting the readiness code will require using a diagnostic scan tool.*

NOTE —
See **EWD Electrical Wiring Diagrams** *for a complete wiring diagram of the glow plug electrical circuits.*

Service precautions

Ignition system service and repair work must be carried out carefully. The ignition system contains sensitive electronic components. To guard against system damage, and for general safety, the following warnings apply to any ignition system troubleshooting, maintenance or repair work.

> ***WARNING —***
> *Glow plug systems operate at high current levels. For this reason, always insure that electrical connections that have been disturbed during testing are properly secured when done.*

Glow plug system, testing (through May 2001)

Basic checks of the glow plug system are similar to earlier diesel systems.

– Remove upper engine cover.

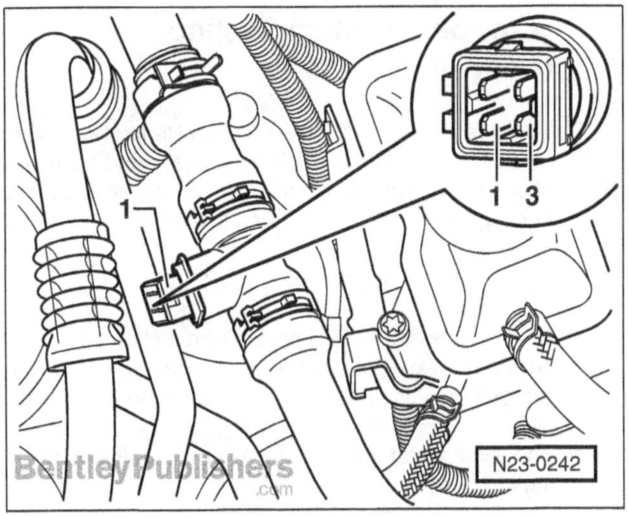

Disconnect harness connector from Engine Coolant Temperature (ECT) sensor (**1**).

NOTE —
Disconnecting the ECT harness connector simulates a cold engine when ignition key is switched on.

◄ Disconnect bus connector (**1**) from glow plugs by gently pulling it off.

– Connect an accurate voltmeter between bus connector for cylinder #1 and ground.

– Turn on key and observe battery voltage. Voltage should be present for approximately 20 seconds.

– Repeat for remaining cylinders.

– If no voltage is present, check glow plug fuses, see **EWD Electrical Wiring Diagrams.**

NOTE —
Glow plug fuses are located in the fuse holder on the battery cover.

– If voltage is present, switch off ignition.

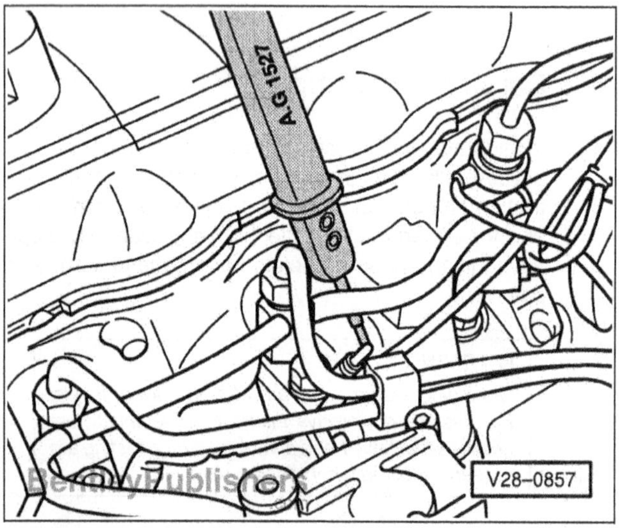

◄ Connect a suitable test light to battery positive and touch other end to disconnected terminal on glow plug. The low resistance of glow plug will cause test light to light up if the glow plug is good.

– Replace defective glow plugs using a 10mm deep socket and suitable extension or use Volkswagen special tool 3220. Torque replacement glow plugs to specification.

Tightening torque
- TDI glow plug in cylinder head . . . . . . 15 Nm (11 ft-lb)

NOTE —
Advanced testing of the glow plug system requires the use of a diagnostic scan tool.

Glow plug system, testing (from June 2001)

Vehicles from m.y. 2002 use a Glow Period Control Module (J370). It is not possible to optically and acoustically check the activation. Checking is only possible by connecting a current pick-up clamp to the glow plug wiring.

– Remove upper engine cover.

– Start engine and run at idle. Be sure that all electrical accessories (e.g. lights, rear window heating, etc.) and A/C are switched off. Engine must be also be cold.

– The system test must be done within 3 minutes of starting cold engine.

– Connect multimeter in sequence to glow plug wiring using a inductive current pick-up clamp. Compare to specification.

Specification

- Glow plug current (each) . . . . min. 6 amps

– If the specification is not met, check wiring to glow plug.

– After repairs are completed, check DTC memory using diagnostic scan tool and erase DTC memory.

Coolant glow plugs, testing and replacing

Because of its outstanding efficiency, the ALH TDI engine develops very little waste heat. In certain circumstances, there may not be sufficient heat with manual transmission vehicles to warm the vehicle interior. An auxiliary heater consisting of three coolant glow plugs is installed to provide additional coolant heat when needed. The coolant glow plugs are operated by two coolant glow plug relays which are controlled by the engine control module. The coolant glow plug relays are located under the hood.

The coolant glow plugs can operate in one of three modes depending on the amount of heat required. Coolant glow plug #1 can operate separately, or #2 and #3 together, or all three at the same time.

Complete testing of the coolant glow plug system requires the use of a diagnostic scan tool. The coolant glow plugs can, however, be checked in the same manner as the glow plugs used for starting.

– Visually inspect the fuse for the coolant glow plugs in the fuse holder on battery cover. See **97 Fuses, Relays and Component Locations**. Replace as required.

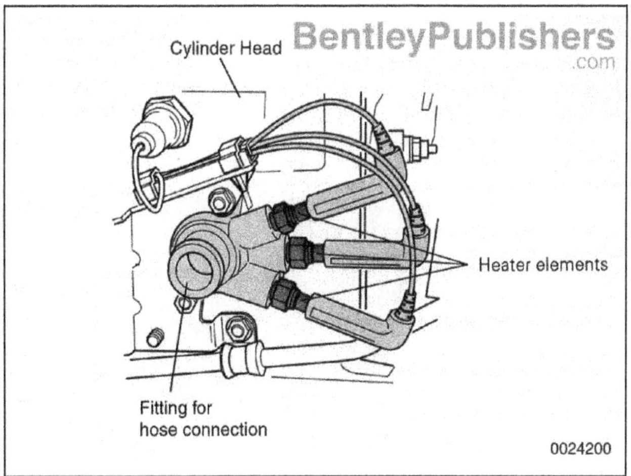

< Pull the electrical connectors off of the coolant glow plugs.

– Connect a suitable test light to battery positive and touch the other end to the end of the coolant glow plug. The test light will glow due to the low resistance of glow plug if it is good. Faulty coolant glow plugs will not light test light.

– If defective glow plugs are found, unscrew and replace as necessary. Coolant will leak out of the system when unscrewing glow plugs.

> ***WARNING —***
>
> *• Hot coolant can scald and result in serious personal injury. Do not work on the cooling system until it has fully cooled.*
>
> *• At normal operating temperature the cooling system is pressurized. Allow the system to cool as long as possible before opening—a minimum of an hour—then release the cap very slowly to allow safe release of pressure.*
>
> *• Releasing the cooling system pressure lowers the coolant's boiling point, and the coolant may boil suddenly. Use heavy gloves and wear eye and face protection to guard against scalding.*
>
> *• Use extreme care when working at or near the cooling fan when the engine is hot. The fan can come on at any time–even if the ignition key is off.*
>
> *• Use extreme care when draining and disposing of engine coolant. Coolant is poisonous and lethal. Children and pets are attracted to coolant because of its sweet smell and taste. See a doctor or veterinarian immediately if any amount of coolant is ingested.*

– Re-attach electrical connectors and replace lost coolant.

FUEL INJECTION COMPONENTS (1.9L, CODE: BEW)

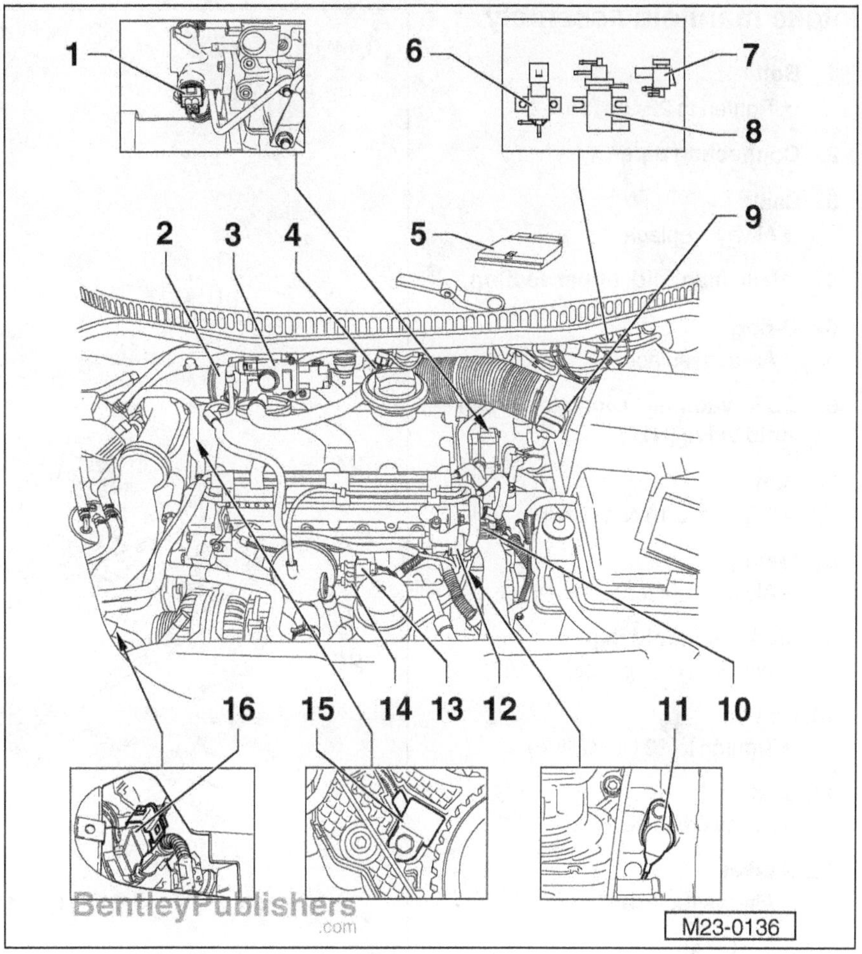

> ***NOTE —***
> *Volkswagen identifies electrical components by a letter and/or a number in the electrical schematics. See* **EWD Electrical Wiring Diagrams.** *These electrical identifiers are listed in parenthesis as an aid to electrical troubleshooting.*

Fuel injection system, components

1. **Engine coolant temperature (ECT) sensor (G62)**
   - WIth integral engine coolant temperature sensor (G2), for coolant temperature gauge
2. **Intake manifold flap**
   - With motor for intake flap (V157)
3. **EGR vacuum regulator solenoid valve (N18)**
4. **Positive Crankcase Ventilation (PCV) heating element (N79)**
   - For vehicles driven in cold climates
5. **Engine Control Module (ECM) (J248)**
   - With integral barometric pressure sensor (F96)
6. **Intake manifold tuning valve (N156)**
7. **Exhaust Gas Recirculation (EGR) cooler with switch-over valve (N345)**
8. **Wastegate bypass regulator valve (N75)**
   - Checking, see **21 Turbocharger and Intercooler**
9. **Mass Air Flow (MAF) sensor (G70)**
10. **Harness connector**
    - For pump injectors #1 to #4 (N240, N241, N242, N243)
11. **Engine speed (RPM) sensor (G28)**
12. **Fuel temperature sensor (G81)**
13. **Harness connector**
    - For camshaft position sensor
14. **Harness connector**
    - For engine speed (RPM) sensor
15. **Camshaft Position (CMP) sensor (G40)**
16. **Charge air pressure sensor (G31) and Intake Air Temperature (IAT) sensor (G42)**

Intake manifold assembly

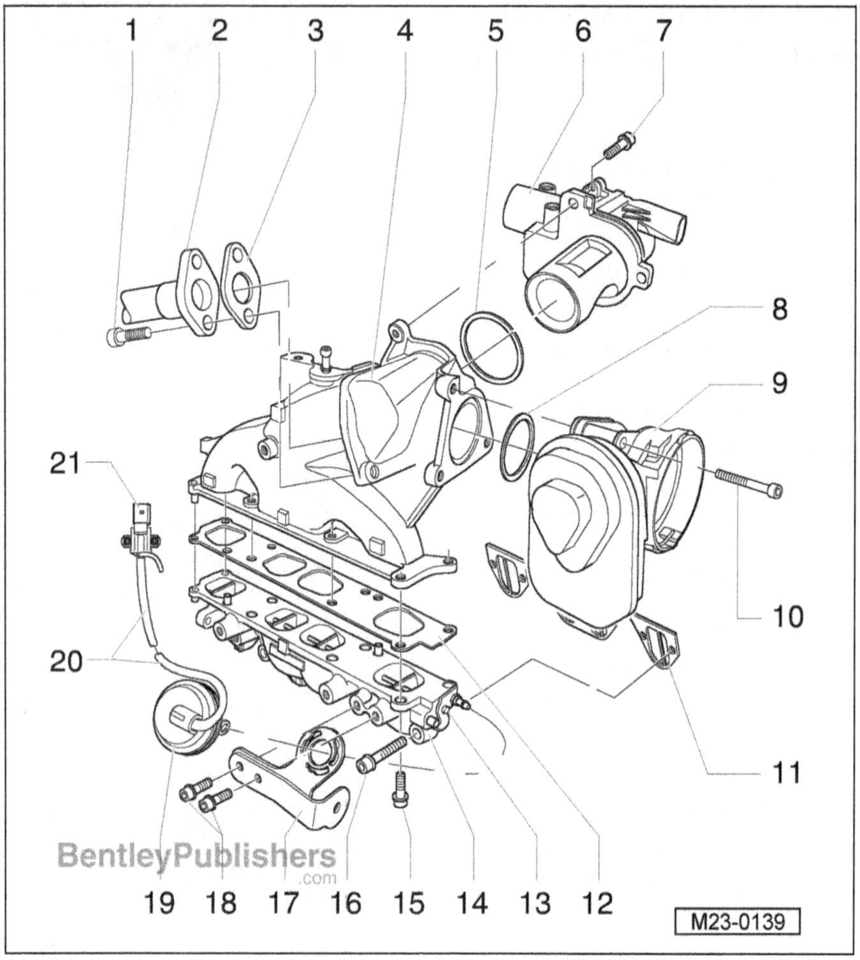

1. **Bolt**
   - Tighten to 22 Nm (16 ft-lb)
2. **Connecting pipe**
3. **Seal**
   - Always replace
4. **Intake manifold, upper section**
5. **O-ring**
   - Always replace
6. **EGR vacuum regulator solenoid valve (N18)**
7. **Bolt**
   - Tighten to 10 Nm (7 ft-lb)
8. **O-ring**
   - Always replace
9. **Intake manifold flap**
   - With intake flap motor (V157)
10. **Bolt**
    - Tighten to 10 Nm (7 ft-lb)
11. **Seal**
    - Always replace
12. **Gasket**
    - Always replace
13. **Selector shaft**
    - Do not stress axially, danger of damage
14. **Intake manifold, lower section**
    - With selector shaft
15. **Bolt**
    - Tighten to 10 Nm (7 ft-lb)
    - Tighten in sequence as shown below
16. **Bolt**
    - Tighten to 22 Nm (16 ft-lb)
17. **Bracket**
18. **Bolt**
    - Tighten to 10 Nm (7 ft-lb)
19. **Vacuum unit**
    - When pushing the vacuum unit lifter on the selector shaft conical journal, the selector shaft must be supported with an appropriate wedge on the lower part of the intake manifold
20. **Vacuum hose**
21. **Intake manifold tuning valve (N156)**

Intake manifold, upper and lower sections, tightening sequence of bolts

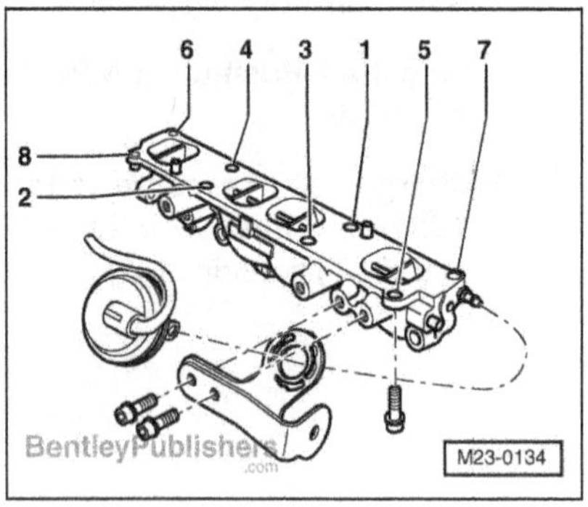

Tighten bolts in order shown (**1** through **8**)

- Tighten bolts to 10 Nm (7 ft-lb)
- Always replace gasket

Pump injector assembly

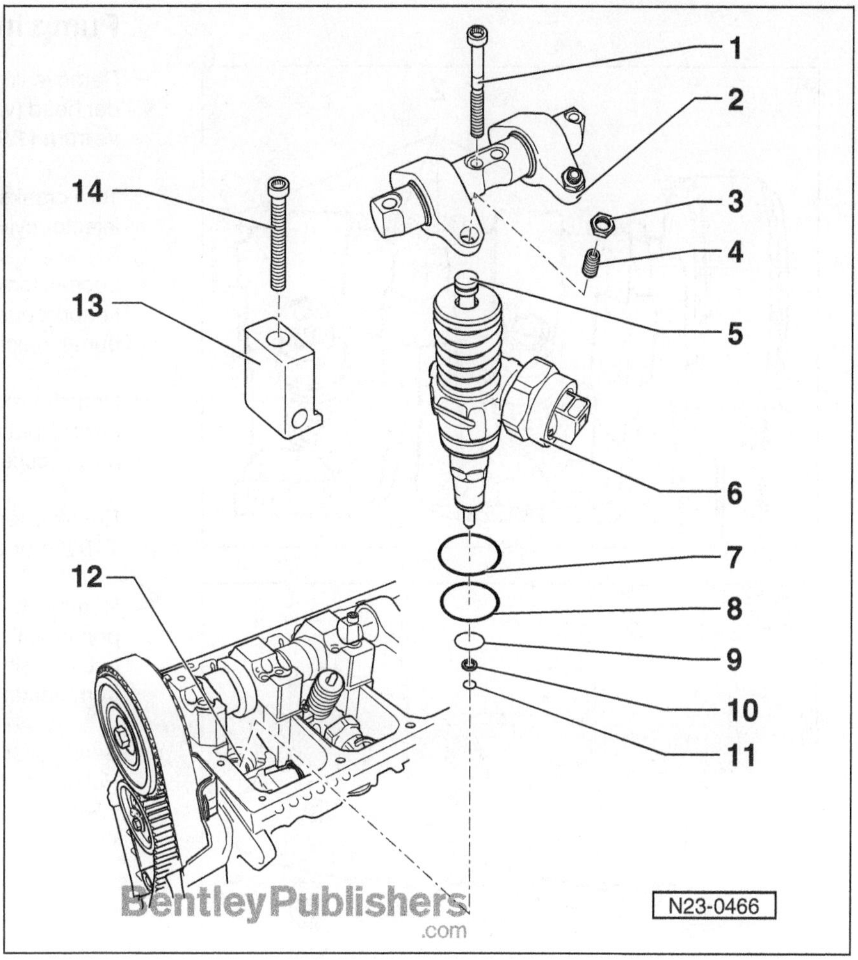

The pump injector (*Pumpe Düse)* is a pressure generating pump that also contains a solenoid valve control unit and a fuel injector. Each engine cylinder has its own pump injector. Utilization of pump injectors allows for elimination of the high pressure fuel line and distributor injection pump, as used on the earlier (code: ALH) diesel engine.

1. **Bolt**
   - Tighten to 20 Nm (15 ft-lb) plus an additional 90° (¼ turn)
   - Always replace
2. **Rocker arm**
3. **Lock nut**
   - Tighten to 30 Nm (22 ft-lb)
4. **Adjustment screw**
   - Always replace
5. **Ball pin**
   - Always replace
6. **Pump injector unit**
7. **O-ring**
   - Replace with special tool T10056/1
8. **O-ring**
   - Replace with special tool T10056/2
9. **O-ring**
   - Replace with special tool T10056/3
10. **Heat protection seal**
    - Always replace
11. **Circlip**
12. **Cylinder head**
13. **Tension block**
14. **Bolt**
    - Tighten to 12 Nm (9 ft-lb) plus an additional 270° (¾ turn)
    - Always replace

Pump injector, removing and installing

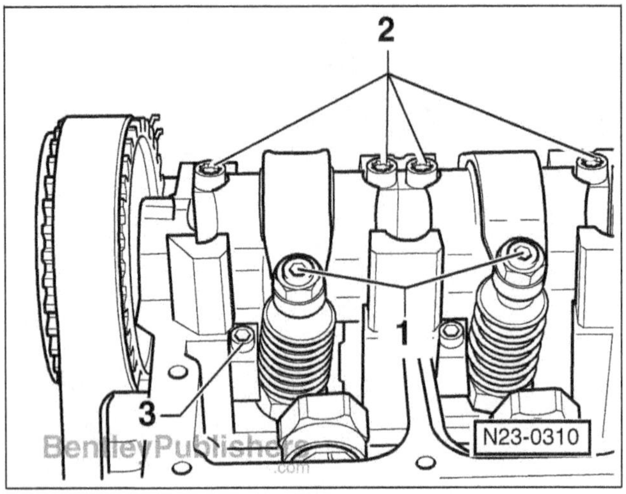

– Remove timing belt guard, upper outer section and cylinder head (valve) cover. See **15b Cylinder Head and Valvetrain (1.9L Engine)**.

– Turn crankshaft until cam lobe pair for appropriate pump injector cylinder points upward.

< Loosen lock nut of adjustment screw (**1**) and unscrew until appropriate rocker arm lies against plunger spring of pump injector.

– Remove rocker arm securing bolts (**2**) in an outward-to-inward pattern with socket 3410 or equivalent and remove rocker arm shaft.

– Remove tensioning block securing bolts (**3**) with socket T10054 or equivalent and remove tensioning block.

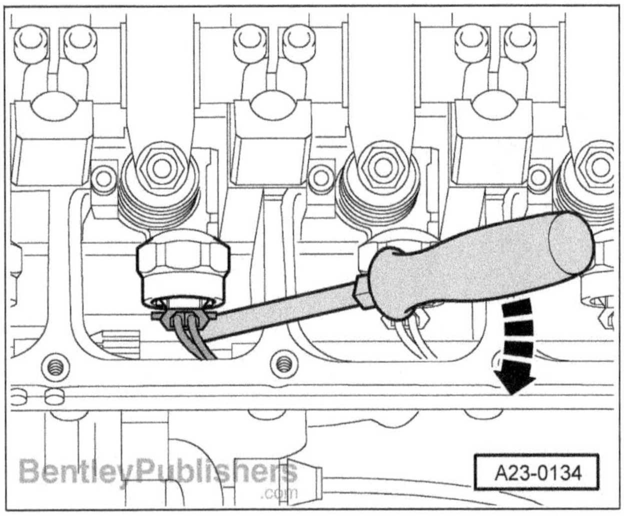

< Remove harness connector from pump injector by gently prying with screwdriver. Support opposite side of connector using light finger pressure to prevent connector from becoming wedged.

– Seat puller T10055 into slot on side of pump injector. Puller fits into slot where tensioning block retains pump injector.

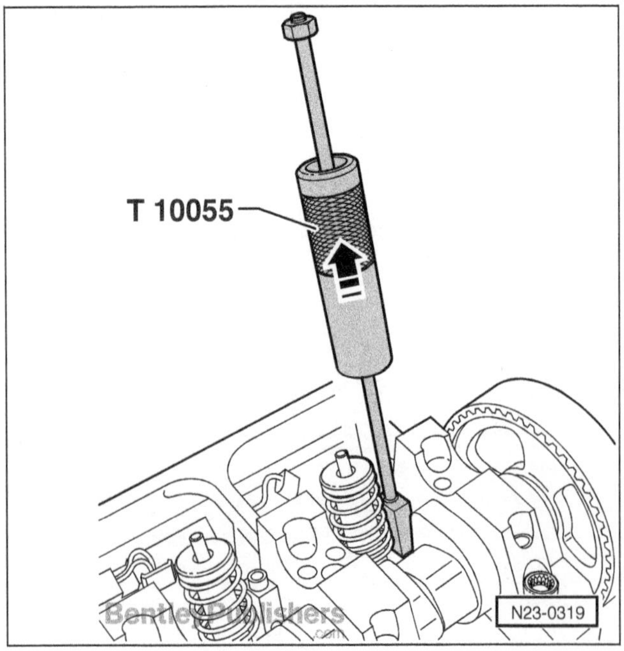

< Pull pump injector out of cylinder head seat (**arrow**) with gentle taps using special tool **T10055**, or equivalent.

Pump injector, installing

NOTE —

- *If a new pump injector is installed, also replace the appropriate adjustment screw in the rocker arm.*
- *Any time that work is performed which requires pump injector to be adjusted, clean the adjustment screw in rocker arm and pump injector ball pin and checked for signs of wear. If wear is evident, replace the ball pin and adjustment screw.*
- *New pump injectors, as supplied by Volkswagen, have O-rings and heat shield seals supplied.*

– Replace heat shield seal and O-rings if old pump injector is reused.

– Check that three O-rings, heat shield seal and circlip are seated correctly before installing pump injector. The O-rings must not be twisted.

– Oil all three O-rings and carefully install pump injector down into cylinder head.

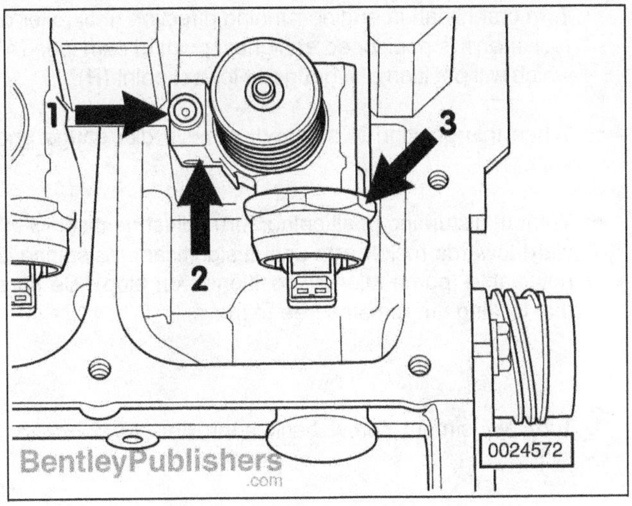

< Push pump injector (**3**) evenly and squarely into cylinder head as far as possible.

– Install tensioning block (**2**) into slot on side of pump injector.

NOTE —

If the pump injector is not at right angles (square) to the tensioning block, the securing bolt may loosen. This can damage the pump injector and / or the cylinder head.

– Preliminarily align pump injector (**3**) in position shown by screwing a new securing bolt (**1**) into tensioning block (**2**) until pump injector can still just be turned easily.

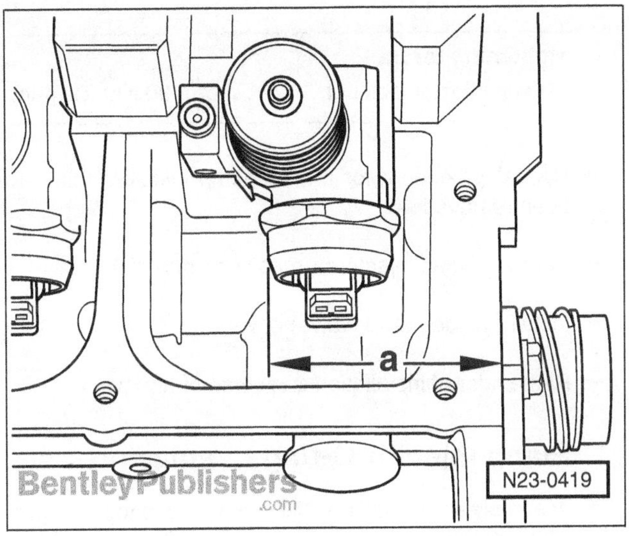

< Align pump injector at right angles with camshaft bearing caps. Check dimension **a** from outer edge of cylinder head to rounded edge of appropriate pump injector with a vernier caliper with a measuring range of at least 400 mm (16 in). Cylinder #4 is shown.

– Adjust and align pump injector position for each cylinder, as required, according to specification below.

Pump injector positions	
Cylinder	**Dimension a**
1	333.0 ± 0.8 mm (13.110 ± 0.031 in)
2	245.0 ± 0.8 mm (9.646 ± 0.031 in)
3	153.6 ± 0.8 mm (6.047 ± 0.031 in)
4	65.6 ± 0.8 mm (2.583 ± 0.031 in)

– When pump injector is properly aligned, tighten new securing bolt in tensioning block.

Tightening torques (always replace bolts)

- Tensioning block to cylinder head . . . . 12 Nm (9 ft-lb) + ¾ turn (270°)
- Rocker arm shaft to cylinder head, tighten in sequence as described below . . 20 Nm (15 ft-lb) + ¼ turn (90°)

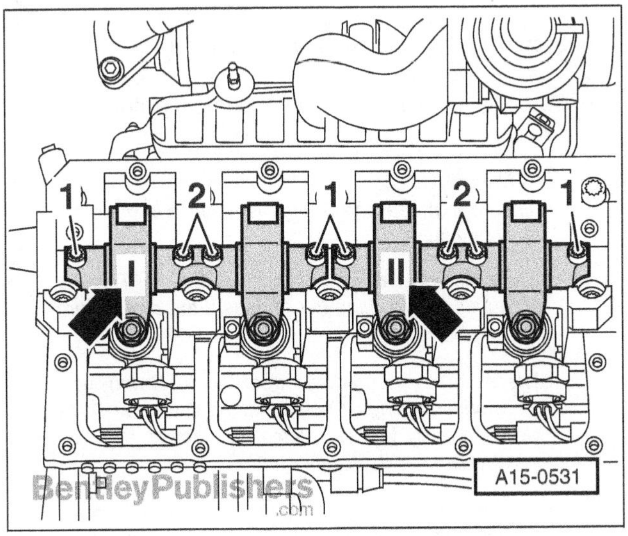

< Install rocker arm shaft (**arrows**, as required) with new securing bolts and tighten in sequence.

- Step 1: Hand tighten inner bolts (**2**), and then outer bolts (**1**).
- Step 2: Torque inner bolts (**2**) and outer bolts (**1**) to specification.

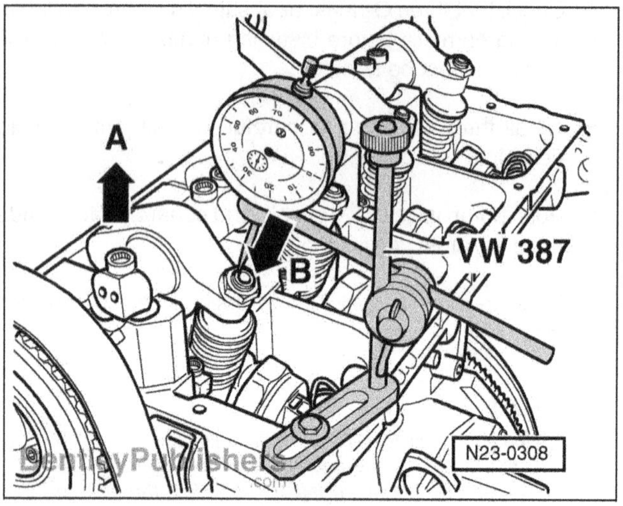

- Mount dial gauge support (VW 387) to cylinder head as shown and place dial gauge against adjustment screw of pump injector.

- Turn crankshaft in engine running direction until roller of rocker arm is positioned at highest point of cam lobe (**A**) which will position dial gauge at lowest point (**B**).

- When this position is attained, remove dial gauge and support.

- Without disturbing positioning, turn adjustment screw inward towards rocker arm until a significant resistance is noticeable (pump injector positioned on stop). Be sure that locking nut remains free to move.

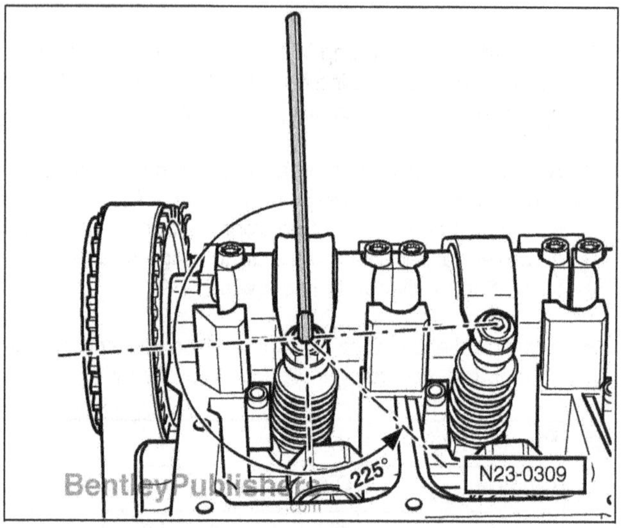

- Turn adjustment screw back approximately 225° from stop.

- Hold adjustment screw in this position and tighten lock nut.

Tightening torque
- Pump injector lock nut . . . . 30 Nm (22 ft-lb)

- Repeat procedure for any/all pump injectors that have been removed.

- Connect pump injector harness connector(s).

- Install cylinder head (valve) cover.

- Remainder of installation is reverse of removal.

Pump injector O-rings, replacing

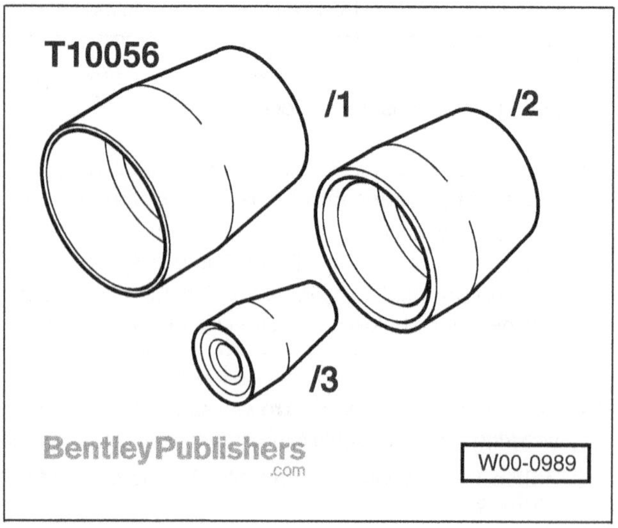

- Assembly sleeve set T10056 is required to replace pump injector O-rings.

- Remove pump injector as previously described.

- Carefully pry old O-rings off of pump injector.

NOTE —
- *Ensure that seat for O-rings is clean and does not have any burrs.*
- *Use assembly sleeves to install the O-rings. There is a danger of damaging the O-rings if the sleeves are not used.*
- *There has been a gradual introduction of O-rings with no colored markings. Ensure that the O-rings are allocated correctly to their respective grooves; ring size decreases towards nozzle.*
- *Prevent the O-rings from rolling when sliding them on. Do not twist O-rings when seating in pump injector.*

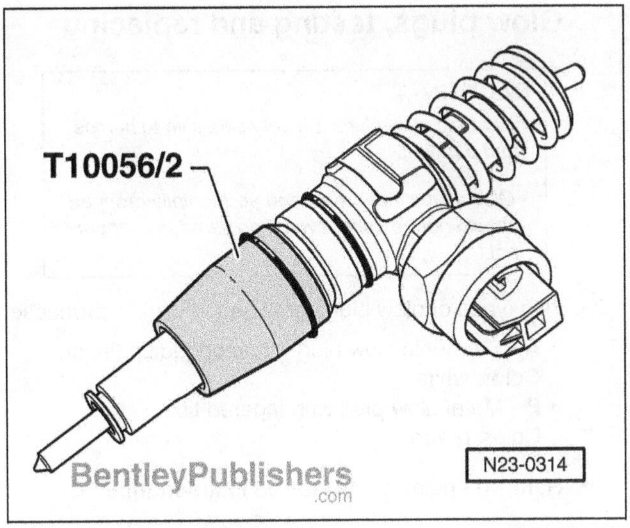

– Pull heat insulating shield and circlip off end of pump injector.

– Clean O-ring seating surfaces on pump injector very carefully.

< Slide appropriate assembly sleeve as far as possible onto pump injector. Middle O-ring shown at left.

Pump injector O-ring special tool designation	
Top O-ring (largest size)	T10056/1
Middle O-ring	T10056/2
Bottom O-ring (smallest size)	T10056/3

– Slide O-ring carefully onto assembly sleeve and into seat on pump injector. Make sure O-ring is not twisted.

– Remove assembly sleeve. Repeat process for remaining O-rings.

– Install new heat insulating shield and circlip on end of pump injector.

– Repeat procedure for remaining pump injectors as necessary.

– Install pump injector. See **Pump injector, removing and installing** in this repair group.

GLOW PLUGS (1.9L, CODE: BEW)

The 1.9L TDI-PD diesel engine uses four glow plugs. The glow plugs assist combustion during cold starts by providing additional heat for ignition. Due to the efficiency of the TDI engine design, the additional heat that glow plugs provide is not needed until temperatures drop to approximately 48°F (9°C). This function is known as preglow. In addition to preglowing, the glow plugs will continue to operate after the engine is started. This is known as afterglowing or extended glow period and is used to reduce engine noise, improve idle quality and reduce hydrocarbon emissions. The afterglow function is enabled after every start, both hot and cold, regardless of whether or not preglowing has occurred, until the specified time has expired or the engine speed exceeds 2,500 rpm.

The glow plug system is controlled by the engine control module (ECM) and the engine coolant temperature (ECT) sensor. The period of preglow is dependent on the engine coolant temperature as determined by the ECT.

The instrument cluster glow plug warning light lights up while the glow plugs are preglowing. When the warning light goes off the engine can be started. If the warning light does not light up, the engine is already at operating temperature and can be started immediately.

If glow plug warning light does not light up when engine is cold, then there is a fault in the system. If glow plug warning light flashes this indicates there is a malfunction in the system. Check system for DTCs using a diagnostic scan tool.

Glow plugs, testing and replacing

> ***CAUTION—***
> - *Ceramic glow plugs are very sensitive to bumps and bending.*
> - *Glow plugs that have been accidentally dropped should not be used, even if no damage is apparent.*

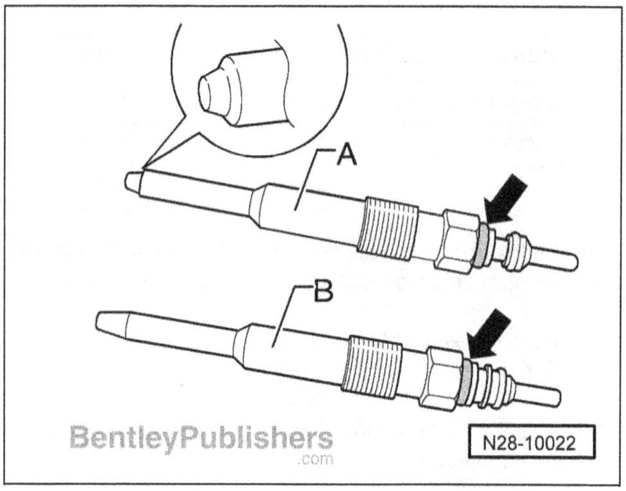

Two types of glow plugs are used in vehicle production.

- **A** - Sermonic glow plug with stepped tip (inset)
  Color: white
- **B** - Metal glow plug with tapered tip
  Color: green

– Remove engine cover, see **03 Maintenance**.

– Disconnect harness connectors from glow plugs.

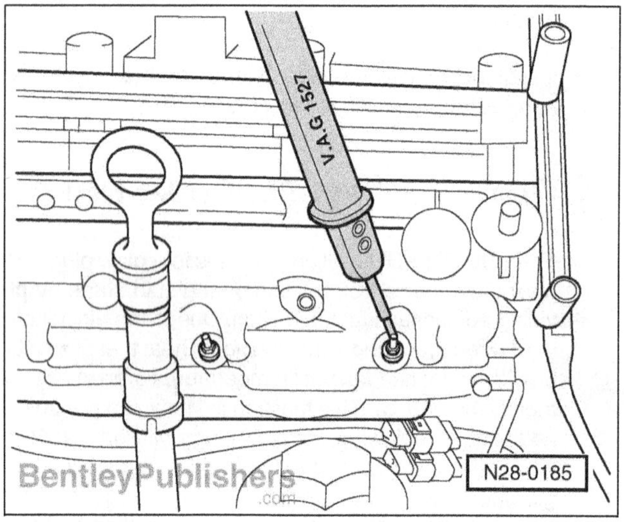

Connect power supply wire of LED voltage tester (V.A.G 1527 shown) to battery positive (+). Touch test probe of LED voltage tester to each glow plug. The low resistance of glow plug will provide a ground path for test light and cause it to light if glow plug is good.

– If LED does not light up, replace glow plug.

– Remove and install glow plugs using VW tool 3220 (hinged socket), or equivalent.

> ***NOTE—***
> *A 10 mm, 6-point deep socket, a universal joint and an appropriate extension can also be used to remove and install glow plugs.*

Tightening torque
- Glow plug to cylinder head. . . . . 15 Nm (11 ft-lb)

24 Fuel Injection–Gasoline

24

GENERAL

This repair group covers the fuel injection/engine management systems for the vehicles covered by this manual that are equipped with gasoline engines. More general information on the fuel injection systems can be found **2 Fuel, Ignition and Exhaust Systems**.

Engine Management Systems

- 1.8L 4-cylinder turbo gasoline
  code: APH, AWP, AWV, ... Bosch Motronic ME 7.5
- 1.8L 4-cylinder turbo gasoline
  code: BKF, BNU, ... Bosch Motronic ME 7.1.1
- 2.0L 4-cylinder gasoline
  code: AEG . . . . Bosch Motronic 5.9.2
- 2.0L 4-cylinder gasoline
  code: AVH, AZG,
  BDC, BEV, BGD . . . . Bosch Motronic ME 7.5
- 2.5L 5-cylinder gasoline
  code: BPR, BPS . . . . Bosch Motronic ME 7.1.1

Special testing equipment may be necessary for some of the tests and repair procedures given here.

NOTE —

- *Bosch Motronic engine management systems which is OBD II compliant. It is recommended that fault diagnosis and troubleshooting be carried out using Volkswagen factory diagnostic scan tool or equivalent aftermarket scan tool.*
- *Electronic accelerator system (E-Gas), fuel pump and fuel tank testing and repair along with EVAP information is covered in* **20 Fuel Storage and Supply**.
- *Ignition system testing and repair is covered in* **28 Ignition System–Gasoline.**
- *Related systems such as Exhaust Gas Recirculation (EGR) and catalytic converters are covered in* **26 Exhaust System and Emission Controls**.

CAUTION —

Disconnecting the negative (–) battery cable may erase fault codes and basic settings in the engine management and automatic transmission control modules. Some driveability problems may be noticed until the system re-adapts to operating conditions. OBD II readiness codes, which may be required for emissions testing, may also be erased. Convenience electronics (alarm system, interior light control, power locks, mirrors, and windows) may need to be re-set using a Volkswagen factory diagnostic scan tool or equivalent aftermarket scan tool.

The Motronic systems combine ignition, fuel injection and electronic accelerator pedal functions into one system managed by a single Engine Control Module (ECM). The system is fully adaptive and features built-in diagnostics that are capable of detecting and storing coded fault information. There are no basic adjustments or settings that can be made to the system without specialized equipment.

Safety precautions

The following warnings and cautions should be adhered to whenever working on the engine management system.

WARNING —

- *Fuel may be discharged during fuel system repairs. Do not smoke or work near heaters or other fire hazards. Have a fire extinguisher handy. Work only in a well-ventilated area.*
- *Wear suitable hand and eye protection when working with gasoline. Prolonged contact with fuel can cause illness and skin disorders.*
- *Fuel hoses in engine compartment must only be secured with spring-type clips. The use of clamp or screw-type clips will damage fuel lines and cause fuel leaks.*

> **CAUTION—**
> - *Connect and disconnect wires and test equipment only with the ignition switched off.*
> - *Before making any electrical tests that require the engine to be cranked using the starter, disable the ignition system as described in* **28 Ignition System.**
> - *Do not use sealants containing silicones. Particles of silicone drawn into the engine will not be burned in the engine and will damage the heated oxygen sensor. Use only sealants and engine chemicals marked as safe for oxygen sensors.*
> - *During testing, it is possible for the Engine Control Module to recognize and store a diagnostic trouble code (DTC) that was due solely to the testing procedure. Therefore, after completing repairs, the DTC memory must be checked and erased as necessary with a Volkswagen factory diagnostic scan tool or equivalent aftermarket scan tool.*

ON BOARD DIAGNOSTICS (OBD)

On Board Diagnostics found on the Bosch Motronic engine management systems monitor many aspects of engine operation. The current generation OBD II is integrated into all Volkswagen New Beetle gasoline engines. For a general description of the system and an overview of operation, see **2 Fuel, Ignition, and Exhaust Systems.**

Because of the large number of operating parameters monitored and the vast quantity of data available, fault diagnosis can only be properly carried out using a Volkswagen factory diagnostic scan tool or equivalent aftermarket scan tool. The engine control module or ECM also supports certain limited functions in a government-mandated generic scan tool mode. These functions are standardized for all OBD II compliant vehicles, but are not as comprehensive as those found by using the Volkswagen factory diagnostic scan tool.

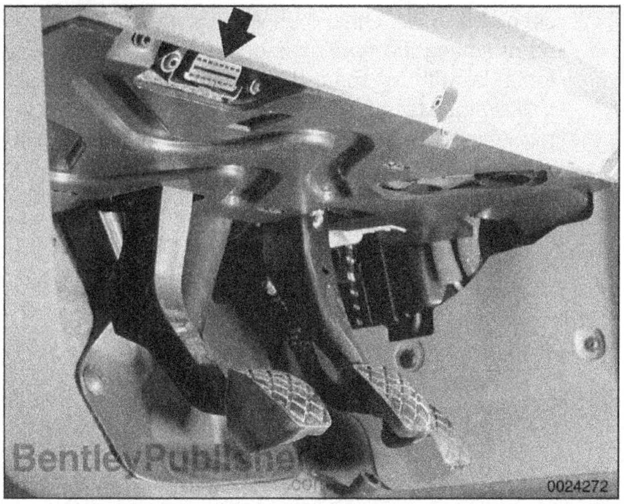

Access for the scan tool is through the Data Link Connector (DLC) (**arrow**) under the lower left side of the instrument panel.

During repairs where the battery must be disconnected, consideration must be given to the effect that this will have on the engine management system.

- Diagnostic Trouble Codes may be erased.
- Readiness codes may be erased.
- Fault counters that monitor certain functions will be reset to zero.
- Adaptive leaning values will be set to default.

Engine Control Module (ECM)

The engine control module is located at the very rear of the engine compartment in the air plenum on the left side. The ECM is accessed through a trim panel on top of the dash.

Adaptation

The Motronic engine management systems are adaptive. Idle speed, ignition timing, injection timing and quantity automatically compensate and adapt to changes in the engine due to wear and operating conditions. Minor problems such as mixture changes due to small vacuum leaks can be eliminated. As a result, idle speed, fuel mixture (CO%) and ignition timing are non-adjustable.

Engine control module coding

The ECM for the Motronic ME7.5.1 system does not require coding as in previous ME7 versions. Equipment matching for other systems is automatically accomplished when using a factory (VAS 5051/5052) diagnostic scan tool.

Readiness Codes

Federal and state legislation requires that all auto manufacturers build into their engine management systems the ability to check for proper operation of up to 8 functions. For the Motronic engine management system, the exact number depends on the equipment level of each particular vehicle. Most of the time, the readiness code will set itself after the appropriate conditions have been met. These may include a cold or a hot start, operation at a certain load or speed for a certain period of time, or operation at different temperatures. After repairs, it may be advantageous to confirm repairs rather than wait to see if they set on their own. Other than waiting, the only other method of setting the readiness codes on the Motronic systems is with a scan tool. This method requires a Volkswagen factory diagnostic scan tool or equivalent aftermarket scan tool, and extensive knowledge of its operation and capabilities. See the **OBD On Board Diagnostics** section for more information on readiness codes.

NOTE —

- *The appropriate code is electronically programmed into the ECM according to instructions provided by the scan tool manufacturer.*
- *Volkswagen identifies electrical components by a letter and/or a number in the electrical schematics. See* **EWD Electrical Wiring Diagrams**. *These electrical identifiers are listed in parenthesis as an aid to electrical troubleshooting.*

1.8L Fuel Injection System

Fuel injection components (1.8L, code: APH, AWP, AWV)

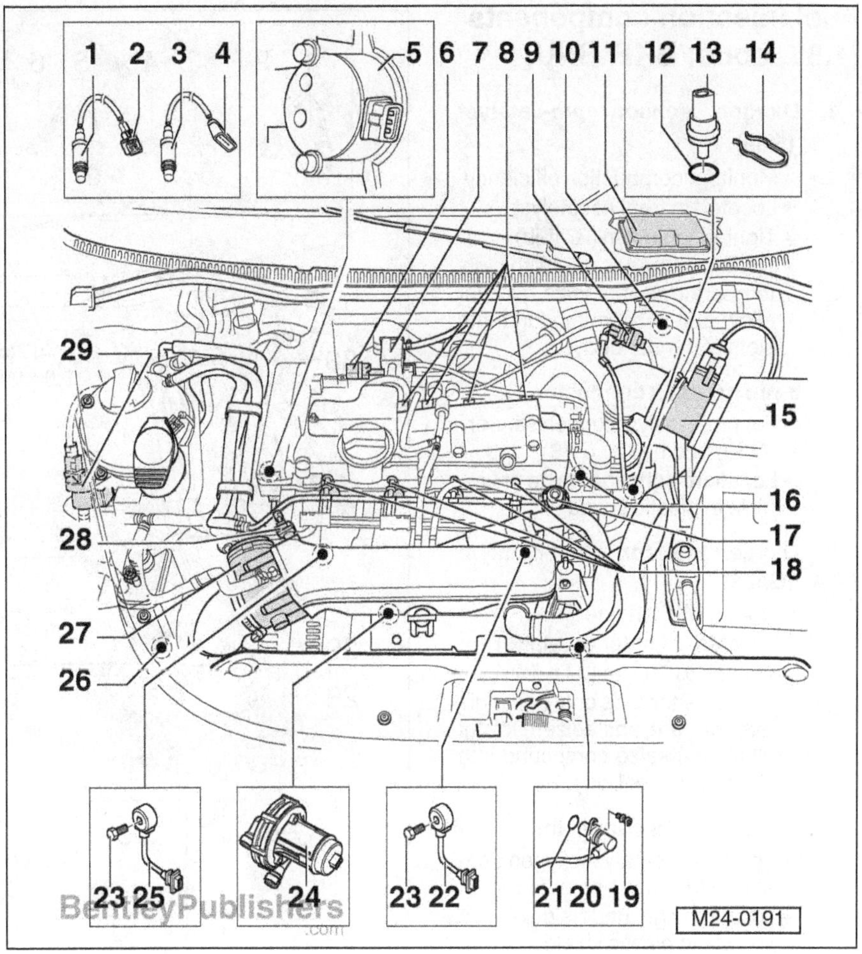

1. **Oxygen sensor, pre-catalyst (G39)**
   - Monitors combustion efficiency
   - Located in front of catalyst
   - Tighten to 50 Nm (37 ft-lb)
   - Lubricate threads only with high temperature anti-seize. Do not allow anti-seize compound into slots of sensor body.
2. **6-pin harness connector**
   - For pre-catalyst oxygen sensor (G39) and heater (Z19)
   - Location: right underside of vehicle in plastic shield
3. **Oxygen sensor, post-catalyst (G130)**
   - Located behind catalyst
   - Tighten to 50 Nm (37 ft-lb)
   - Lubricate threads only with high temperature anti-seize. Do not allow anti-seize compound into slots of sensor body.
4. **4-pin harness connector**
   - For post-catalyst oxygen sensor (G130) and heater (Z29)
   - Location: right underside of vehicle in plastic shield
5. **Camshaft position sensor (G40)**
   - Location: on front of cylinder head near intake camshaft
   - Shutter wheel (rotor) mounted on camshaft
6. **Secondary air injection solenoid valve (N112)**
7. **Overrun recirculation valve solenoid for turbocharger (N249)**
8. **Ignition coils (N70, N127, N291, N292)**
   - Separate coil for each cylinder
9. **Wastegate bypass regulator valve (N75)**
10. **Protective cover**
    - For secondary air pump relay (J229)
11. **Motronic engine control module (J220)**
    - With integrated barometric pressure sensor (F96)
12. **O-ring**
    - Replace if damaged
13. **Engine coolant temperature sensor (G62)**
    - Combined with sensor (G2) for instrument panel warning light
    - AWP/AWV shown, APH has square harness connector
    - See **19 Engine–Cooling System** for testing
14. **Retaining clip**
    - Ensure clip is securely seated
15. **Mass air flow sensor (G70)**
16. **Camshaft adjuster**
    - AWP/AWV shown, APH has timing chain tensioner
17. **Fuel pressure regulator**
    - Location: on fuel rail
18. **Fuel injectors (N30 to N33)**
19. **Bolt**
    - Tighten to 10 Nm (7ft-lb)
20. **Engine speed sensor (G28)**
    - On cylinder block near oil filter
21. **O-ring**
    - Replace if damaged
22. **Knock sensor 2 (G66)**
23. **Bolt**
    - Tighten to 20 Nm (15 ft-lb)
    - Correct torque is critical
24. **Secondary air injection pump motor (V101)**
25. **Knock sensor 1 (G61)**
26. **Charge air pressure sensor (G31)**
    - Mounted to intercooler
27. **Throttle valve control module (J338)**
    - For Electronic Pedal Control, EPC, (E-Gas system)
28. **Intake air temperature sensor (G42)**
29. **Evaporative emissions (EVAP) canister purge regulator valve (N80)**

Fuel injection components (1.8L, code: BKF, BNU)

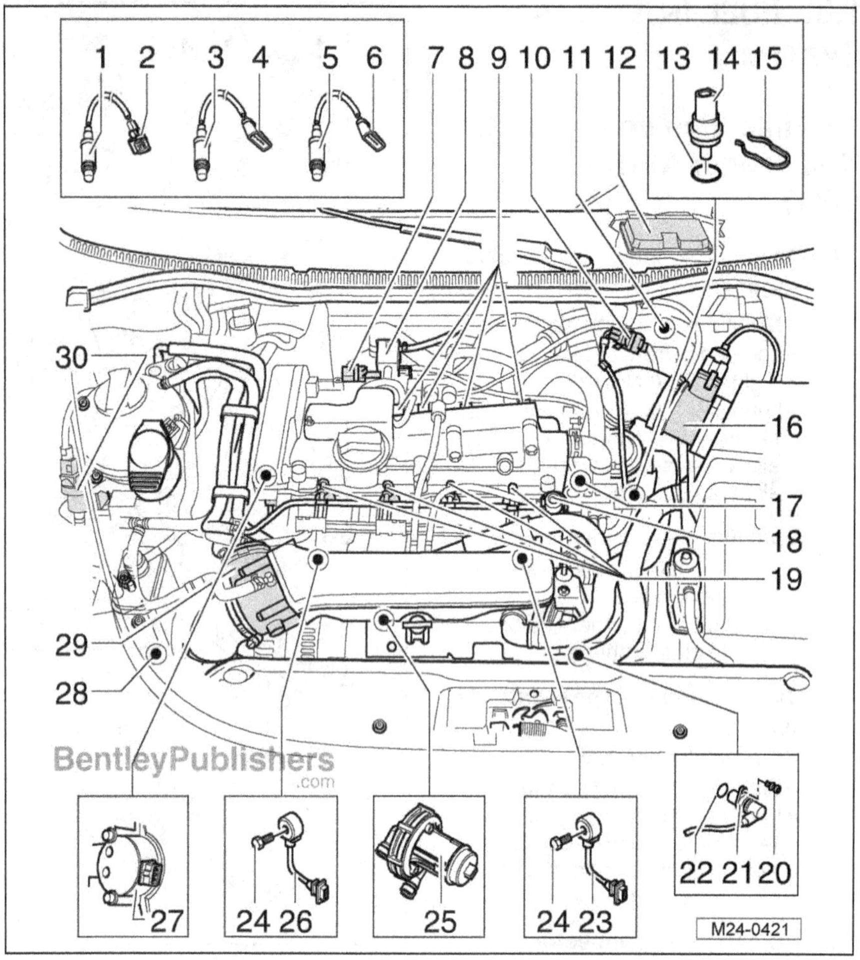

1. **Oxygen sensor, pre-catalyst (G39)**
   - Monitors combustion efficiency
   - Located in front of catalyst
   - Tighten to 50 Nm (37 ft-lb)
   - Lubricate threads only with high temperature anti-seize. Do not allow anti-seize compound into slots of sensor body.
2. **6-pin harness connector**
   - For pre-catalyst oxygen sensor (G39) and heater (Z19)
   - Location: right underside of vehicle in plastic shield
3. **Oxygen sensor, post-catalyst (G465)**
   - Not on BKF engine
   - Located in center of catalyst
   - Tighten to 50 Nm (37 ft-lb)
   - Lubricate threads only with high temperature anti-seize. Do not allow anti-seize compound into slots of sensor body.
4. **4-pin harness connector**
   - For center-catalyst oxygen sensor (G465) and heater (Z59)
   - Location: right underside of vehicle in plastic shield
5. **Oxygen sensor, post-catalyst (G285)**
   - Monitors catalyst efficiency
   - Located behind catalyst
   - Tighten to 50 Nm (37 ft-lb)
   - Lubricate threads only with high temperature anti-seize. Do not allow anti-seize compound into slots of sensor body.
6. **4-pin harness connector**
   - For post-catalyst oxygen sensor (G285) and heater (Z29)
   - Location: right underside of vehicle in plastic shield
7. **Secondary air injection solenoid valve (N112)**
8. **Overrun recirculation valve for turbocharger (N249)**
9. **Ignition coils (N70, N127, N291, N292)**
   - Separate coil for each cylinder
10. **Wastegate bypass regulator valve (N75)**
11. **Protective cover**
    - For secondary air pump relay
12. **Motronic engine control module (J220)**
    - With integrated barometric pressure sensor (F96)
13. **O-ring**
    - Replace if damaged
14. **Engine coolant temperature sensor (G62)**
    - Combined with sensor (G2) for instrument panel warning light
    - See **19 Engine–Cooling System** for testing
15. **Retaining clip**
    - Ensure clip is securely seated
16. **Mass air flow sensor (G70)**
    - With intake air temperature sensor (G42)
17. **Camshaft adjuster**
18. **Fuel pressure regulator**
    - Location: on fuel rail
19. **Fuel injectors (N30 to N33)**
20. **Bolt**
    - Tighten to 10 Nm (7ft-lb)
21. **Engine speed sensor (G28)**
    - On cylinder block near oil filter
22. **O-ring**
    - Replace if damaged
23. **Knock sensor 2 (G66)**
24. **Bolt**
    - Tighten to 20 Nm (15 ft-lb)
    - Correct torque is critical
25. **Secondary air injection pump motor (V101)**
26. **Knock sensor 1 (G61)**
27. **Camshaft position sensor (G40)**
28. **Charge air pressure sensor (G31)**
    - Mounted to intercooler
29. **Throttle valve control module (J338)**
    - For Electronic Pedal Control, EPC, (E-Gas system)
30. **EVAP canister purge regulator valve (N80)**

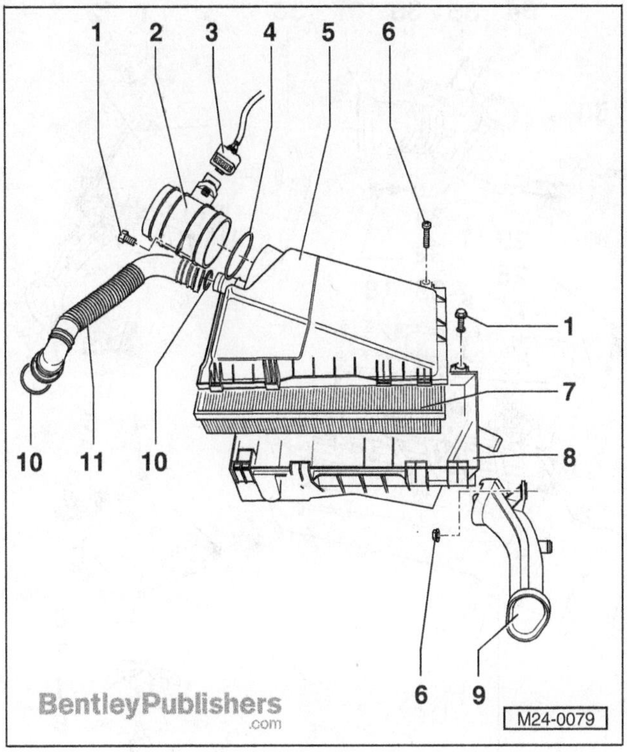

Air cleaner assembly (1.8L)

1. Bolt
   - Tighten to 10 Nm (7 ft-lb)
2. Mass air flow sensor (G70)
   - Gold plated terminals
   - BKF, BNU engines also include intake air temperature sensor (G42)
3. 5-pin connector
   - All terminals gold plated
4. Seal
   - Replace if damaged
5. Air cleaner housing, upper section
6. Bolt
   - Tighten to 6 Nm (52 in-lb)
7. Filter element
8. Air cleaner housing, lower section
9. Duct, intake air
10. O-ring
    - Replace if damaged
11. Intake hose
    - From air cleaner housing upper section to secondary air injection pump intake
    - Ensure hose is seated tightly
    - Press together at front to release

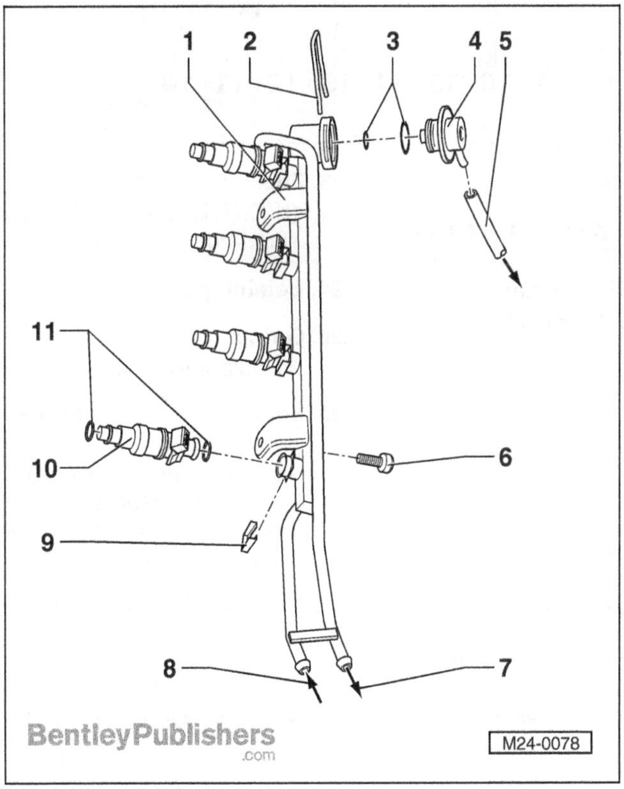

Fuel injector and fuel rail assembly (1.8L)

1. Fuel rail
2. Retaining clip
   - Ensure it is securely seated
3. O-ring
   - Replace if damaged
4. Fuel pressure regulator
   - With residual check valve
5. Vacuum hose
   - To intake manifold
6. Bolt
   - Tighten to 10 Nm (7 ft-lb)
7. Fuel return line/hose
   - Secure with spring clips only
   - Ensure coupling is securely seated
   - To fuel delivery unit in fuel tank
8. Fuel supply line/hose
   - Secure with spring clips only
   - Ensure coupling is securely seated
   - From fuel filter
9. Retaining clip
   - Ensure it is securely seated
10. Fuel injectors (N30, N31, N32, N33)
    - 2-pin harness connector
11. O-ring
    - Always replace
    - Moisten with clean engine oil or fuel before installing

Intake manifold assembly (1.8L)

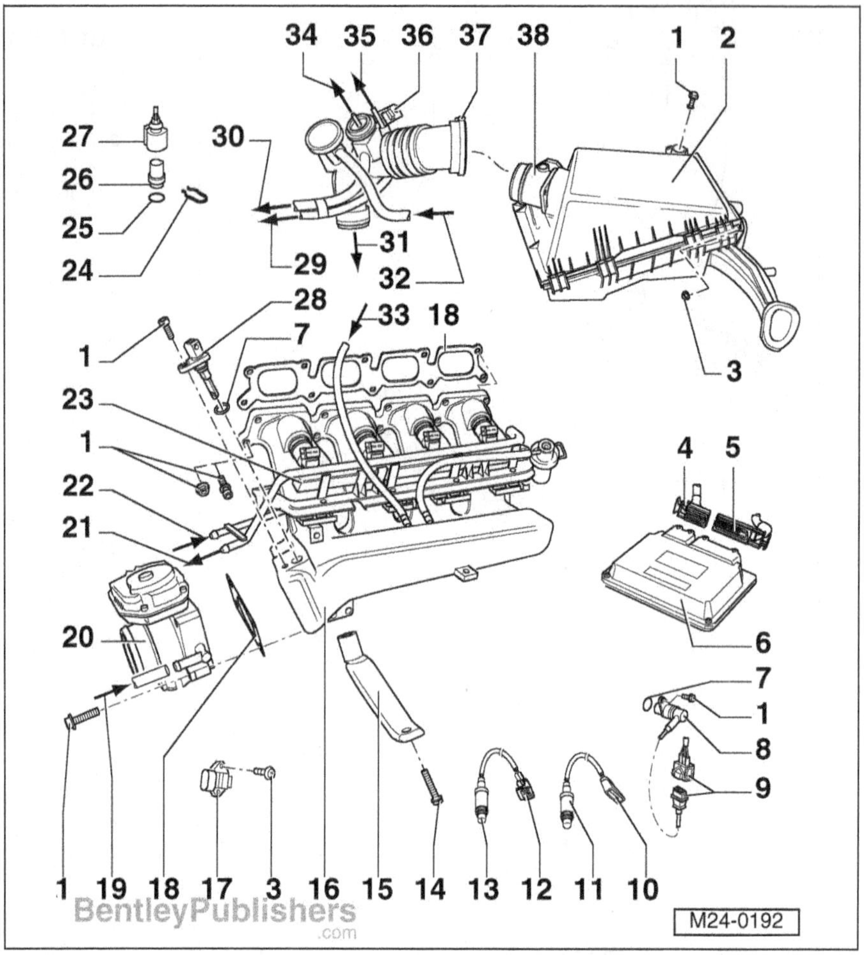

1. **Bolt**
   - Tighten to 10 Nm (7 ft-lb)
2. **Air filter assembly**
3. **Bolt**
   - Tighten to 6 Nm (52 in-lb)
4. **40-pin harness connector (terminals 82 to 121)**
   - Only disconnect or connect with ignition switched OFF
5. **81-pin connector (terminals 1 to 81)**
   - Only disconnect or connect with ignition switched OFF
6. **Motronic engine control module (J220)**
7. **O-ring**
   - Always replace
8. **Engine speed sensor (G28)**
   - On cylinder block near oil filter
   - Inductive type sensor
9. **3-pin harness connector**
   - For engine speed sensor
10. **4-pin harness connector**
    - Contacts 3 and 4 are gold plated
11. **Oxygen sensor, post-catalyst (G130)**
    - Monitors catalytic converter efficiency
    - Located in rear of catalyst
    - Tighten to 50 Nm (37 ft-lb)
    - Lubricate threads only with high temperature anti-seize. Do not allow anti-seize compound into slots of sensor body.
12. **6-pin harness connector**
    - All contacts gold plated
13. **Oxygen sensor, pre-catalyst (G39)**
    - Monitors combustion efficiency
    - With additional monitoring
    - Located in front exhaust pipe
    - Tighten to 50 Nm (37 ft-lb)
    - Lubricate threads only with high temperature anti-seize. Do not allow anti-seize compound into slots of sensor body.
14. **Bolt**
    - Tighten to 20 Nm (15 ft lb)
15. **Support bracket**
16. **Intake manifold**
17. **Charge air pressure sensor (G31)**
    - Mounted to intercooler
    - 4-pin harness connector
18. **Gasket**
    - Always replace
19. **Connecting hose**
    - From EVAP canister purge regulator valve (N80)
20. **Throttle valve control module (J338)**
    - Must be matched to Motronic ME7.5.1 engine control module (J220) with a scan tool if replaced
    - 6-pin harness connector
21. **Fuel return line/hose**
    - Secure with spring clips only
    - To fuel delivery unit in fuel tank
22. **Fuel supply line/hose**
    - Secure with spring clips only
    - From fuel filter
23. **Fuel rail, with fuel injectors (N30-N33) and fuel pressure regulator**
24. **Retaining clip**
25. **O-ring**
    - Replace if damaged
26. **Engine coolant temperature sensor (G62)**
    - Combined with sensor (G2) for instrument panel warning light
    - See **19 Engine–Cooling System** for testing
27. **4-pin harness connector**
    - Black
    - For engine coolant temperature sensor (G62/G2)
    - ECT sensor contacts 1and 3 are gold plated
28. **Intake air temperature sensor (G42)**
    - Incorporated with MAF for BKF, BNU engines
    - 2-pin harness connector

(continued on next page)

Intake manifold assembly (1.8L) continued

29. **Connecting hose**
    - From turbocharger to intercooler connecting pipe
    - To overrun recirculation valve
30. **Connecting hose**
    - From turbocharger to intercooler connecting pipe
    - To wastegate bypass regulator valve (N75)
31. **Connecting hose**
    - To turbocharger air intake
32. **Connecting hose**
    - From crankcase breather
33. **Connecting hose**
    - From non-return valve
    - To overrun recirculation valve solenoid (249)
34. **Connecting hose**
    - From overrun recirculation valve
    - To overrun recirculation valve solenoid (N249)
35. **Connecting hose**
    - To pressure unit (servo) for wastegate on turbocharger
    - From wastegate bypass regulator valve (N75)
36. **Wastegate bypass regulator valve (N75)**
37. **Intake hose**
    - Between turbocharger intake and mass air flow sensor/air cleaner assembly
    - AWP, AWV engine shown, others similar
38. **Mass air flow sensor (G70)**
    - 5-pin harness connector
    - Terminal 1 position not used
    - All contacts on sensor and harness connector are gold plated
    - With intake air temperature sensor (G42) for BKF, BNU engines

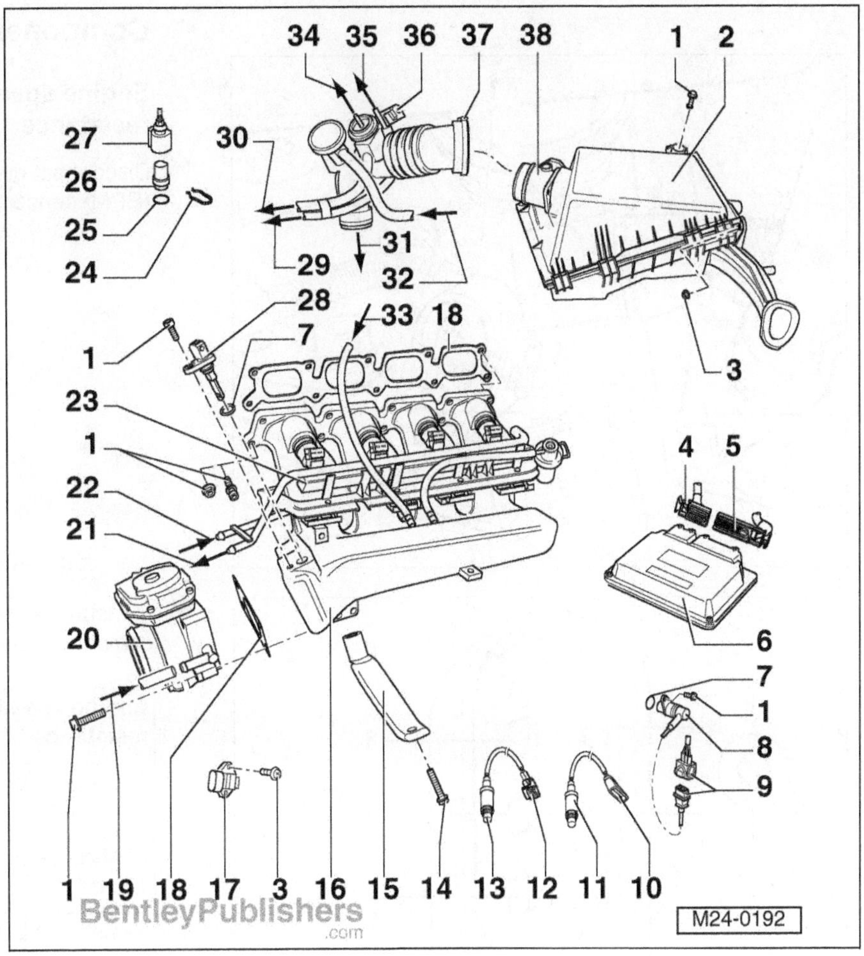

Component testing (1.8L)

Engine speed (RPM) sensor (G28), measuring resistance

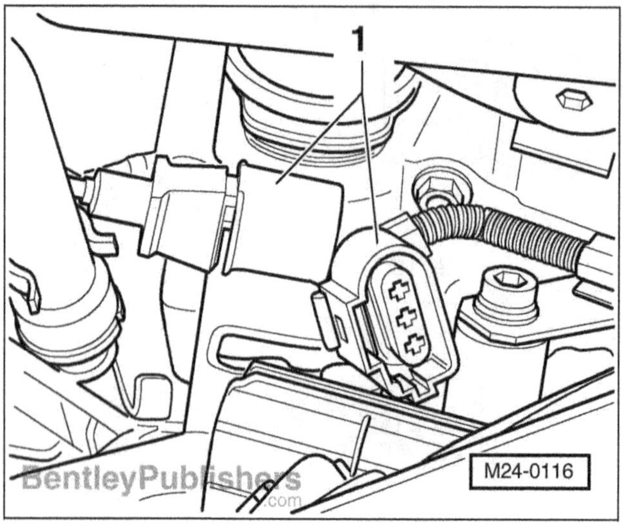

◄ Disconnect grey 3-pin connector (**1**) from engine speed (RPM) sensor.

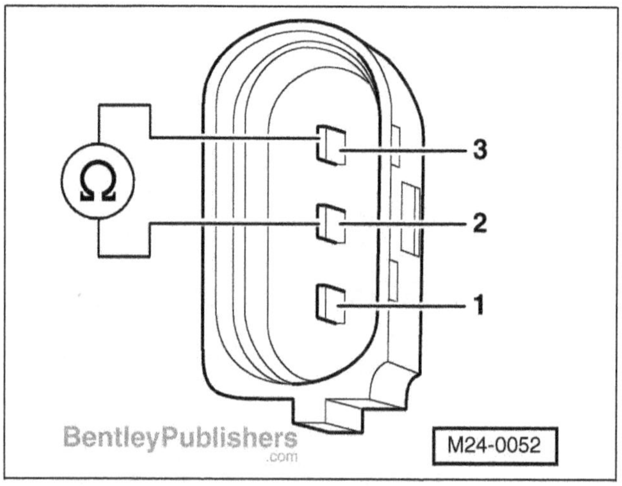

◄ Measure resistance between the various terminals, on component side of connector.

Engine speed (RPM) sensor resistance	Specification
Resistance between terminals 1 and 3	∞ Ω (no continuity)
Resistance between terminals 1 and 2	∞ Ω (no continuity)
Resistance between terminals 2 and 3	730 to 1000 Ω

Fuel injectors, checking resistance

Requirements:

- Ignition off
- Ambient temperature 68°F (20°C)

NOTE —
Fuel injector resistance will vary with temperature. At higher temperatures the resistance increases.

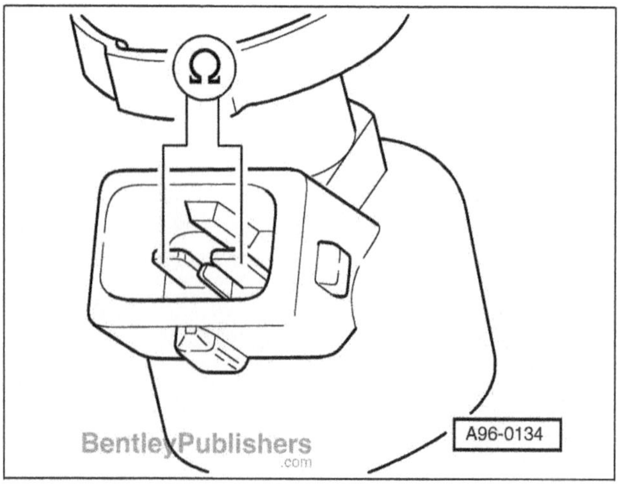

◄ Disconnect fuel injector harness connectors. Measure resistance across two terminals as shown.

Specification
- 1.8L fuel injector resistance . . . . 12.0 to 17.0 Ω

– If specified value is not obtained, replace fuel injector.

Intake air temperature (IAT) sensor (G42), measuring resistance

Engine code: APH, AWV, AWP

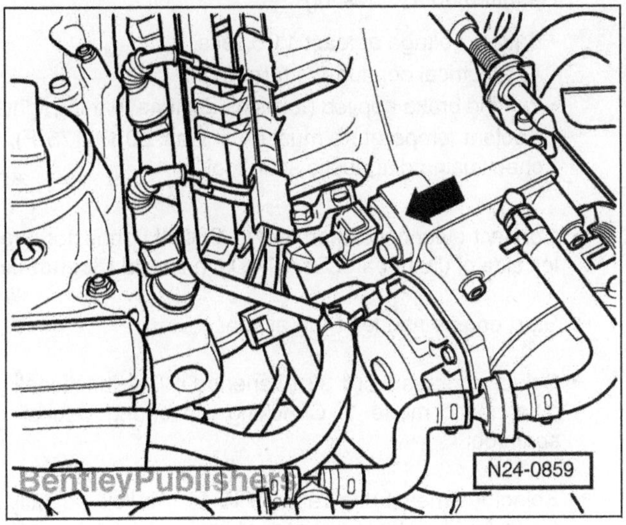

◀ Disconnect harness connector from IAT sensor (**arrow**) and measure resistance between two terminals of sensor.

– Measure temperature of intake air using diagnostic scan tool or thermometer.

Engine code: BKF, BNU

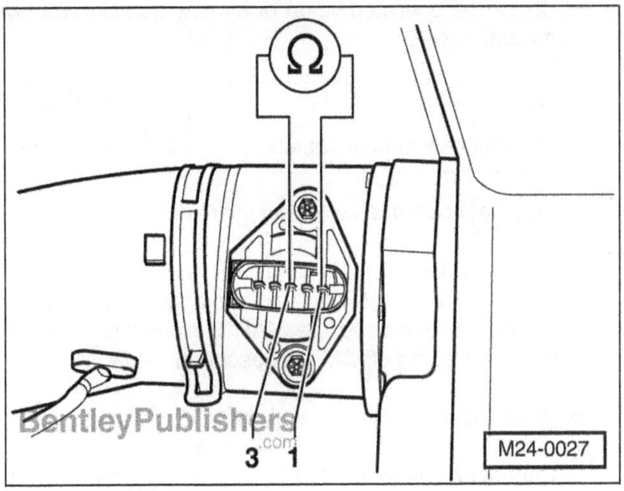

◀ Disconnect harness connector from MAF/IAT sensor near air filter housing. Measure resistance between terminals **1** and **3** of sensor.

– Measure temperature of intake air using diagnostic scan tool or thermometer.

continued for all 1.8L engines

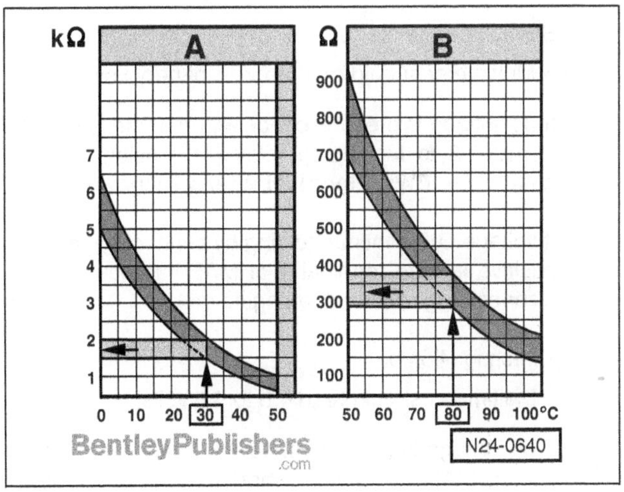

◀ Compare IAT sensor resistance and intake air temperature readings to graph.

- Scale **A** shows resistance for intake air temperature range 0 - 50°C (32 - 122°F).
- Scale **B** shows resistance for intake air temperature range 50 - 100°C (122 - 212°F).
- Example:
  30°C corresponds to 1.5 - 2.0 k Ω
  80°C corresponds to 275 - 375 Ω

Mass air flow (MAF) sensor (G70), testing

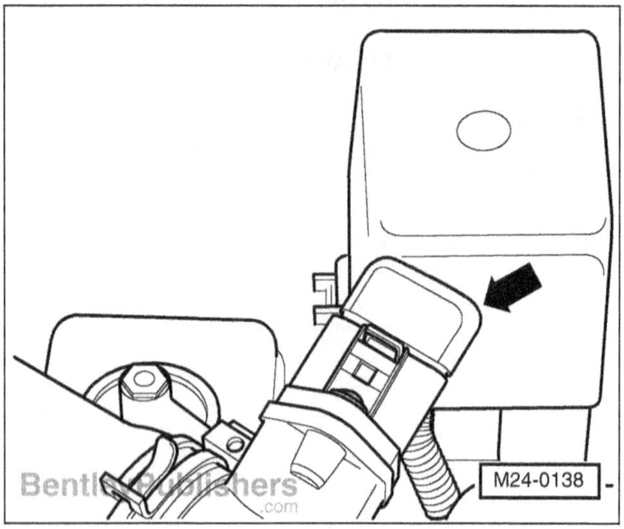

MAF sensor (**arrow**) is located near the air filter housing.

Requirements for testing:

- Battery voltage at least 11.5 volts.
- All electrical consumers turned off.
- Parking brake applied (turns off daytime running lights).
- Coolant temperature must be at least 80°C (176°F), check using diagnostic scan tool.

– Connect diagnostic scan tool to OBD II connector under left side of dashboard. See **OBD On Board Diagnostics**.

– Start engine and let it run at idle.

– Select address word 33 (Generic OBD II functionality), "Diagnostic mode 1: Checking measuring values" on scan tool.

– Select the measuring value, "PID 16: Air flow quantity at Mass Air Flow (MAF) sensor at idle".

– Check the specified value of air flow quantity displayed on scan tool.

Diagnostic test PID 16	Specification
MAF air flow quantity at idle	2.00 to 5.00 g/sec.

– End diagnosis and switch off ignition.

– If specification was not obtained, replace MAF. If specified value was obtained and a diagnostic trouble code still exists relating to the MAF, then continue checking electrical wiring of MAF as described below.

– Disconnect wiring harness from MAF.

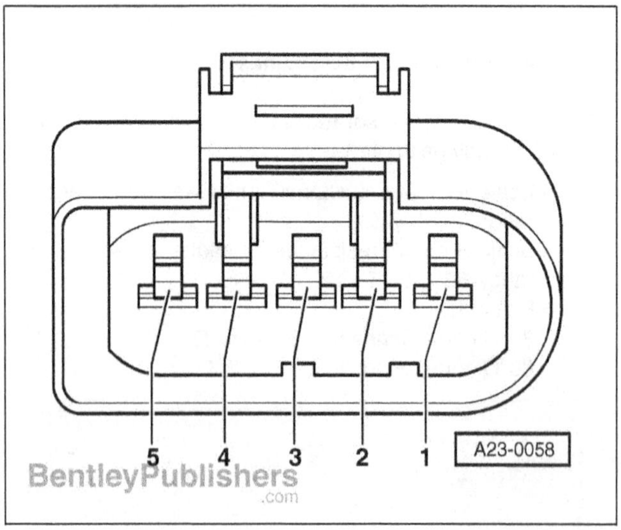

Connect multimeter positive lead to terminal **2** of MAF harness connector. Connect negative lead to engine ground (–).

– Briefly operate starter.

- Multimeter should read approx. battery voltage.

– If no voltage is present, check wiring harness from MAF terminal 2 to fuel pump relay (J17) using wiring diagrams. See **EWD Electrical Wiring Diagrams**.

– If voltage OK, continue testing by connecting multimeter positive lead to terminal **4** of MAF harness connector. Connect negative lead to engine ground (–).

– Switch ignition on.

- Multimeter should read 4.5 to 5.5 volts.

– If no voltage is present, check wiring harness from the various MAF terminals for short circuits to each other, to ground (–), or battery positive (+).

– Repair wiring or replace MAF as necessary.

2.0L FUEL INJECTION SYSTEM

NOTE —
Volkswagen identifies electrical components by a letter and/or a number in the electrical schematics. See **EWD Electrical Wiring Diagrams**. *These electrical identifiers are listed in parenthesis as an aid to electrical troubleshooting.*

Fuel injection components (2.0L, code: AEG)

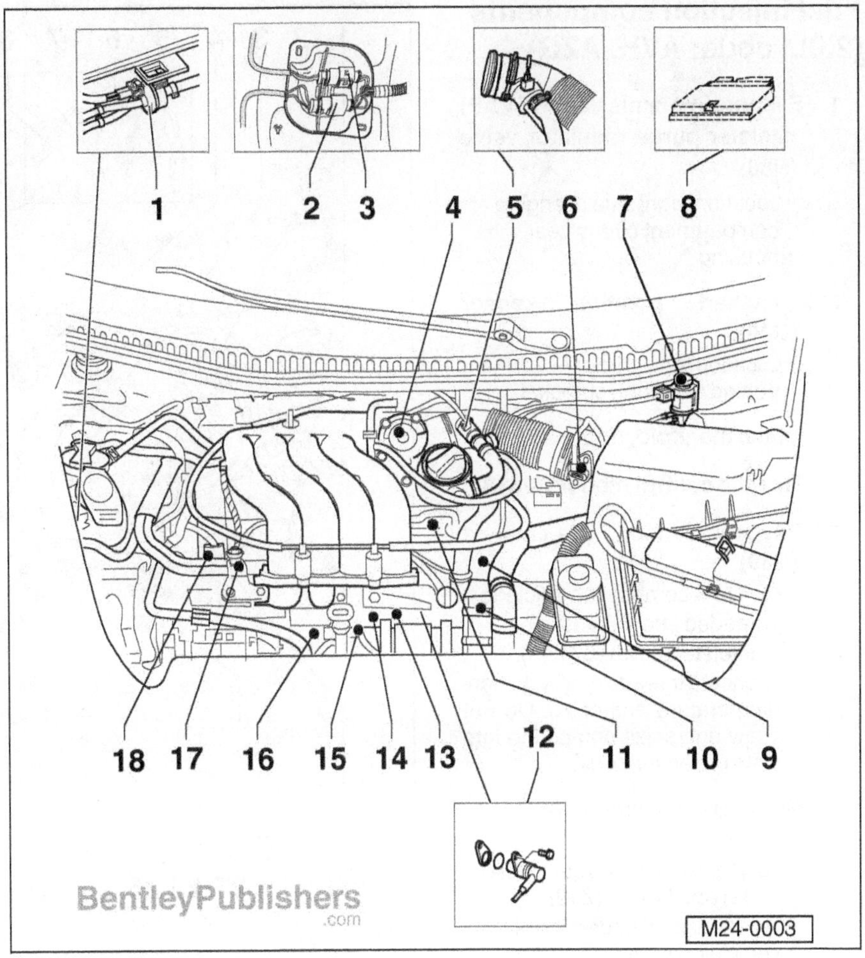

1. **Evaporative emissions (EVAP) canister purge regulator valve (N80)**
   - Location: right side of engine compartment over wheel housing
2. **4-pin harness connector**
   - Black
   - For pre-catalyst oxygen sensor (G39) and heater (Z19)
   - Location: right underside of vehicle in plastic shield
3. **4-pin harness connector**
   - Brown
   - For post-catalyst oxygen sensor (G130) and heater (Z29)
   - Location: right underside of vehicle in plastic shield
4. **Throttle control module (J338)**
5. **Positive crankcase ventilation (PCV) heating element (G42)**
   - For crankcase breather
   - Not installed on all engines
6. **Mass air flow sensor (G70), with intake air temperature sensor (G42)**
7. **Secondary air injection solenoid valve (N112)**
8. **Motronic engine control module (J220)**
9. **Engine coolant temperature sensor (G62)**
   - Combined with sensor (G2) for instrument panel warning light
   - Testing, see **19 Engine–Cooling System**
10. **Secondary air injection pump motor (V101)**
11. **Fuel injectors (N30 to N33)**
12. **Engine speed sensor (G28)**
    - Location: on cylinder block near oil filter
    - Inductive type sensor
13. **Ignition coils (N, N128)**
    - Combined with power output stage (N122)
14. **3-pin harness connector**
    - For engine speed sensor (G28)
15. **Knock sensor 2 (G66)**
    - Brown connector
16. **Knock sensor 1 (G61)**
    - Black connector
17. **Fuel pressure regulator**
    - Location: on fuel rail
18. **Camshaft position sensor (G40)**
    - Location: on cylinder head behind camshaft sprocket

Fuel injection components (2.0L, code: AVH, AZG)

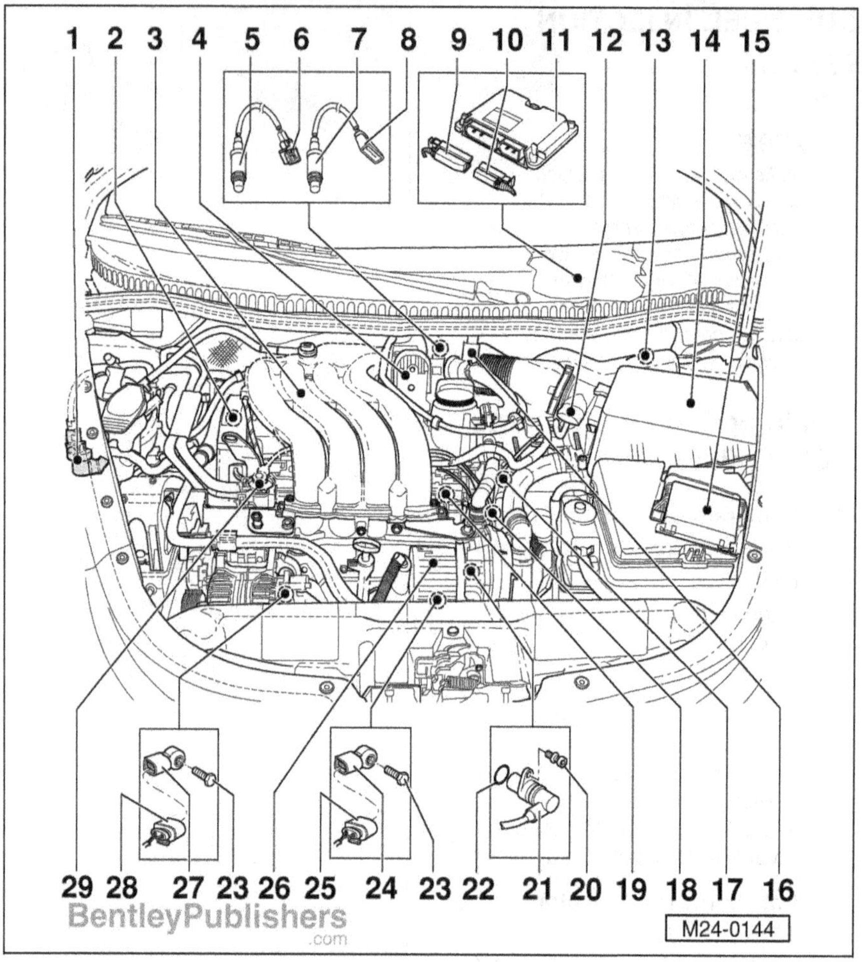

1. **Evaporative emissions (EVAP) canister purge regulator valve (N80)**
   - Location: right side of engine compartment over wheel housing
2. **Camshaft position sensor (G40)**
   - Location: on cylinder head behind camshaft sprocket
3. **Intake manifold, upper section**
4. **Throttle control module (J338)**
5. **Oxygen sensor, pre-catalyst (G39)**
   - Monitors combustion efficiency
   - Threaded into exhaust pipe
   - Tighten to 50 Nm (37 ft-lb)
   - Lubricate threads only with high temperature anti-seize. Do not allow anti-seize compound into slots of sensor body.
6. **6-pin harness connector**
   - Black
   - For pre-catalyst oxygen sensor (G39) and heater (Z19)
   - Location: right underside of vehicle in plastic shield
7. **Oxygen sensor, post-catalyst (G130)**
   - Monitors catalytic converter efficiency
   - Threaded into rear of catalytic converter
   - Tighten to 50 Nm (37 ft-lb)
   - Lubricate threads only with high temperature anti-seize. Do not allow anti-seize compound into slots of sensor body.
8. **4-pin harness connector**
   - Brown
   - For post-catalyst oxygen sensor (G130) and heater (Z29)
   - Location: right underside of vehicle in plastic shield
9. **81-pin harness connector**
   - Only disconnect and connect with ignition switched off
10. **40-pin harness connector**
    - Only disconnect and connect with ignition switched off
11. **Motronic engine control module (J220)**
12. **Mass air flow sensor (G70), with intake air temperature sensor (G42)**
13. **Protective housing for relays**
    - Contains secondary air injection pump relay (J299)
14. **Air cleaner**
15. **Protective housing for fuses**
16. **Positive crankcase ventilation heating element (N79)**
17. **Engine coolant temperature sensor (G62)**
    - Combined with sensor (G2) for instrument panel warning light
18. **Ignition coils with power output stage (N70, N127, N291, N292)**
19. **Fuel injectors (N30 to N33)**
20. **Bolt**
    - Tighten to 10 Nm (7 ft-lb)
21. **Engine speed sensor (G28)**
    - Location: on cylinder block near oil filter
22. **O-ring**
    - Replace if damaged
23. **Bolt**
    - Tighten to 20 Nm (15 ft-lb)
    - Correct torque is critical for proper operation of knock sensor
24. **Knock sensor 2 (G66)**
    - Black connector
25. **2-pin harness connector**
    - For Knock Sensor 2 (G66)
26. **Secondary air injection pump motor (V101)**
27. **Knock sensor 1 (G61)**
    - Black connector
28. **2-pin harness connector**
    - For Knock Sensor 1 (G61)
29. **Fuel pressure regulator**
    - Location: on fuel rail

Fuel injection system components (2.0L, code: BDC, BGD)

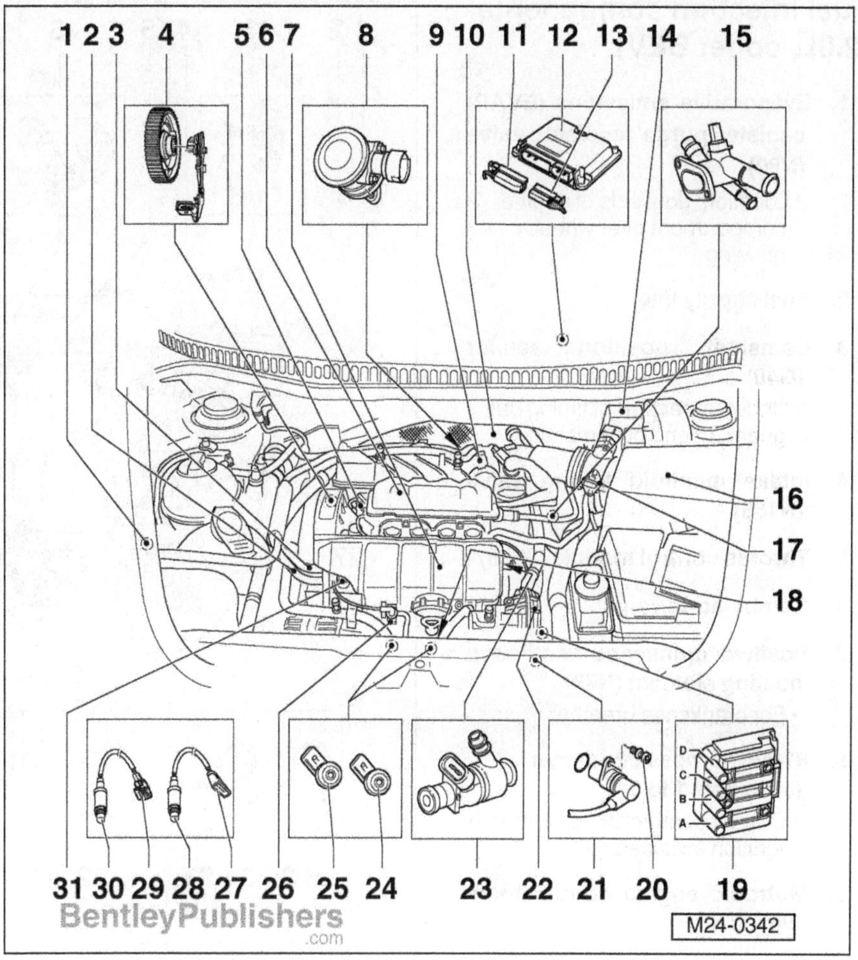

1. **Evaporative emissions (EVAP) canister purge regulator valve (N80)**
   - Location: right side of engine compartment over wheel housing
2. **Fuel return line**
3. **Fuel supply line**
4. **Camshaft position sensor (G40)**
   - Under upper part of timing belt guard, at end of camshaft
5. **Intake manifold tuning valve (N156)**
6. **Intake manifold, upper section**
7. **Intake manifold, lower section**
8. **Combination valve**
9. **Throttle control module (J338)**
10. **Positive crankcase ventilation heating element (N79)**
    - For crankcase breather
    - Not installed on all engines
11. **81-pin connector (terminals 1 to 81)**
    - Only disconnect or connect with ignition switched OFF
12. **Motronic engine control module (J220)**
13. **40-pin harness connector (terminals 82 to 121)**
    - Only disconnect or connect with ignition switched OFF
14. **Protective housing for relays**
15. **Engine coolant temperature sensor (G62)**
    - Combined with sensor (G2) for instrument panel warning light
    - Testing, see **19 Engine–Cooling System**
16. **Air cleaner**
17. **Mass air flow sensor (G70), with intake air temperature sensor (G42)**
18. **3-pin harness connector**
    - For engine speed sensor
19. **Ignition coils with power output stage (N70, N127, N291, N292)**
20. **Bolt**
    - Tighten to 10 Nm (7 ft-lb)
21. **Engine speed sensor (G28)**
22. **Secondary air injection pump motor (V101)**
23. **Fuel injectors (N30 to N33)**
24. **Knock sensor 2 (G66)**
25. **Knock sensor 1(G61)**
26. **Intake manifold tuning valve 2 (N261)**
27. **4-pin harness connector**
    - Contacts 3 and 4 are gold plated
28. **Oxygen sensor, post-catalyst (G130)**
    - Monitors catalytic converter efficiency
    - Threaded into rear of catalytic converter
    - Tighten to 50 Nm (37 ft-lb)
    - Lubricate threads only with high temperature anti-seize. Do not allow anti-seize compound into slots of sensor body.
29. **6-pin harness connector**
    - Black
    - For pre-catalyst oxygen sensor (G39) and heater (Z19)
    - Location: right underside of vehicle in plastic shield
30. **Oxygen sensor, pre-catalyst (G39)**
    - Monitors combustion efficiency
    - Threaded into exhaust pipe
    - Tighten to 50 Nm (37 ft-lb)
    - Lubricate threads only with high temperature anti-seize. Do not allow anti-seize compound into slots of sensor body.
31. **Fuel pressure regulator**
    - BDC: located on fuel rail
    - BGD: located on fuel filter

Fuel injection components (2.0L, code: BEV)

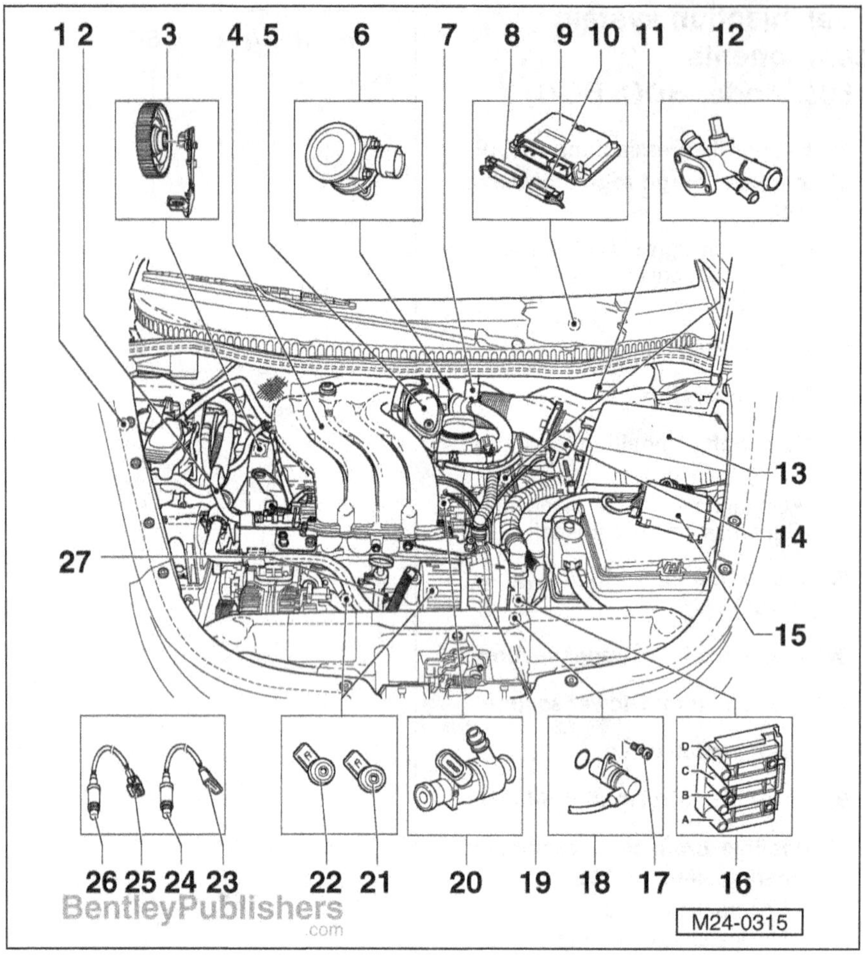

1. **Evaporative emissions (EVAP) canister purge regulator valve (N80)**
   - Location: right side of engine compartment over wheel housing
2. **Fuel supply line**
3. **Camshaft position sensor (G40)**
   - Under upper part of timing belt guard, at end of camshaft
4. **Intake manifold tuning valve (N156)**
5. **Throttle control module (J338)**
6. **Combination valve**
7. **Positive crankcase ventilation heating element (N79)**
   - For crankcase breather
8. **81-pin harness connector (terminals 1 to 81)**
   - Only disconnect or connect with ignition switched OFF
9. **Motronic engine control module (J220)**
10. **40-pin harness connector (terminals 82 to 121)**
    - Only disconnect or connect with ignition switched OFF
11. **Protective housing for relays**
12. **Engine coolant temperature sensor (G62)**
    - Combined with sensor (G2) for instrument panel warning light
    - Testing, see **19 Engine–Cooling System**
13. **Air cleaner**
14. **Mass air flow sensor (G70), with intake air temperature sensor (G42)**
15. **Protective housing for fuses**
16. **Ignition coils with power output stage (N70, N127, N291, N292)**
17. **Bolt**
    - Tighten to 10 Nm (7 ft-lb)
18. **Engine speed sensor (G28)**
19. **Secondary air injection pump motor (V101)**
20. **Fuel injectors (N30 to N33)**
21. **Knock sensor 2 (G66)**
22. **Knock sensor 1(G61)**
23. **4-pin harness connector**
    - Contacts 3 and 4 are gold plated
24. **Oxygen sensor, post-catalyst (G130)**
    - Monitors catalytic converter efficiency
    - Threaded into rear of catalytic converter
    - Tighten to 50 Nm (37 ft-lb)
    - Lubricate threads only with high temperature anti-seize. Do not allow anti-seize compound into slots of sensor body.
25. **6-pin harness connector**
    - For pre-catalyst oxygen sensor (G39) and heater (Z19)
    - Location: right underside of vehicle in plastic shield
26. **Oxygen sensor, pre-catalyst (G39)**
    - Monitors combustion efficiency
    - Threaded into exhaust pipe
    - Tighten to 50 Nm (37 ft-lb)
    - Lubricate threads only with high temperature anti-seize. Do not allow anti-seize compound into slots of sensor body.
27. **3-pin harness connector**
    - For engine speed sensor

Air cleaner assembly (2.0L)

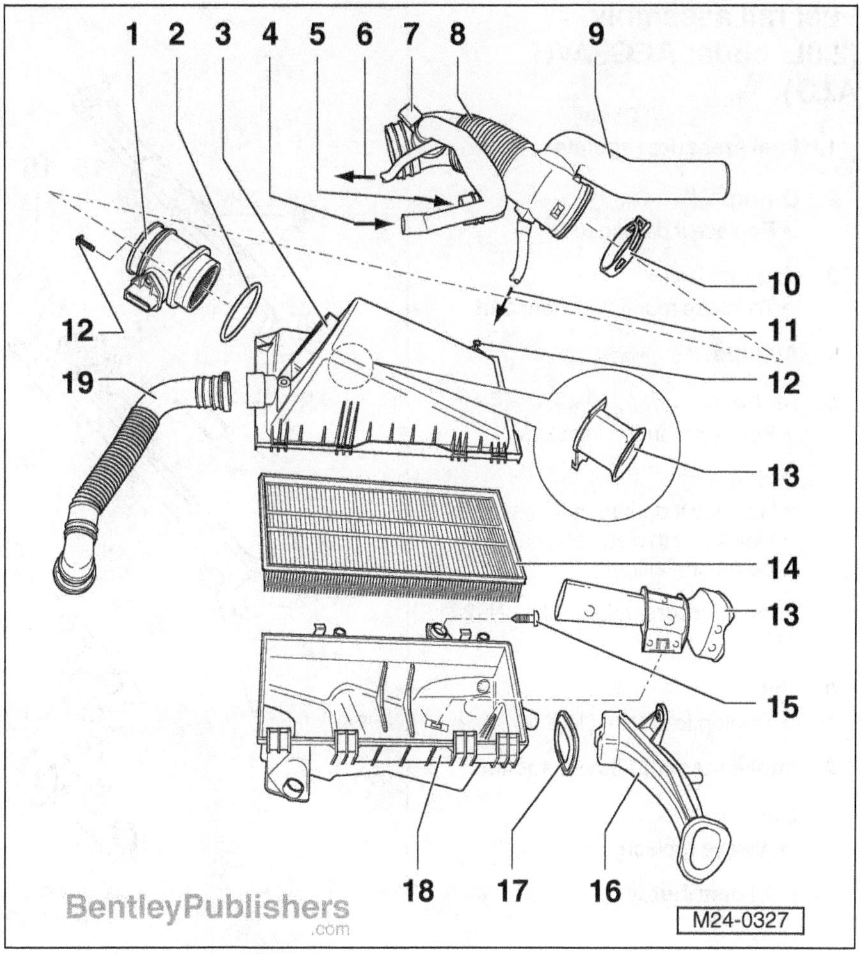

1. **Mass air flow sensor (G70, with intake air temperature sensor (G42)**
2. **Seal**
   - Replace if damaged
3. **Air cleaner housing, upper section**
4. **Intake manifold hose**
5. **From cylinder head cover**
   - For crankcase breather
6. **Vacuum hose**
   - To upper intake manifold
7. **Positive crankcase ventilation heating element (N79)**
   - For crankcase breather
   - Not installed on all engines
8. **Intake hose**
9. **Resonator**
10. **Spring-type clip**
11. **Pipe union**
    - For air shrouded injectors
12. **Bolt**
    - Tighten to 6 Nm (53 in-lb)
13. **Air duct**
14. **Filter element**
15. **Bolt**
    - Tighten to 10 Nm (7 ft-lb)
16. **Air duct**
17. **Seal**
    - Replace if damaged
18. **Air cleaner housing, lower**
19. **Intake hose**
    - For secondary air pump

Fuel rail assembly (2.0L, code: AEG, AVH, AZG)

1. **Fuel pressure regulator**
2. **O-ring**
   - Replace if damaged
3. **Vacuum hose**
   - To intake manifold, upper part
4. **Air pipe**
5. **Air hose**
   - For air shrouded injectors
6. **O-ring**
   - Replace if damaged
   - Moisten with clean engine oil before installing
7. **Fuel injectors (N30, N31, N32, N33)**
8. **Nut**
   - Tighten to 20 Nm (15 ft-lb)
9. **Intake manifold, lower section**
10. **Gasket**
    - Always replace
11. **Fuel distributor**
12. **Bolt**
    - Tighten to 10 Nm (7 ft-lb)
13. **Retaining clips**
14. **Wiring guide**
15. **Connection**
    - Return flow line
16. **Connection**
    - Supply line
17. **To intake air duct**

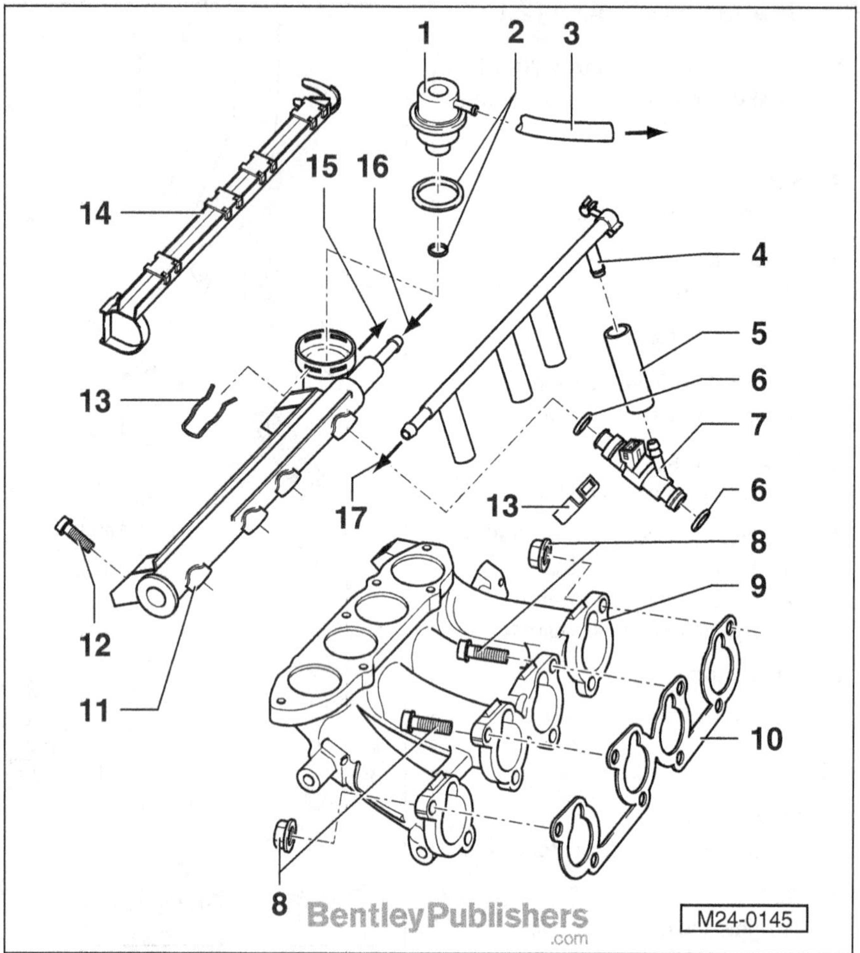

Fuel rail assembly (2.0L, code: BDC)

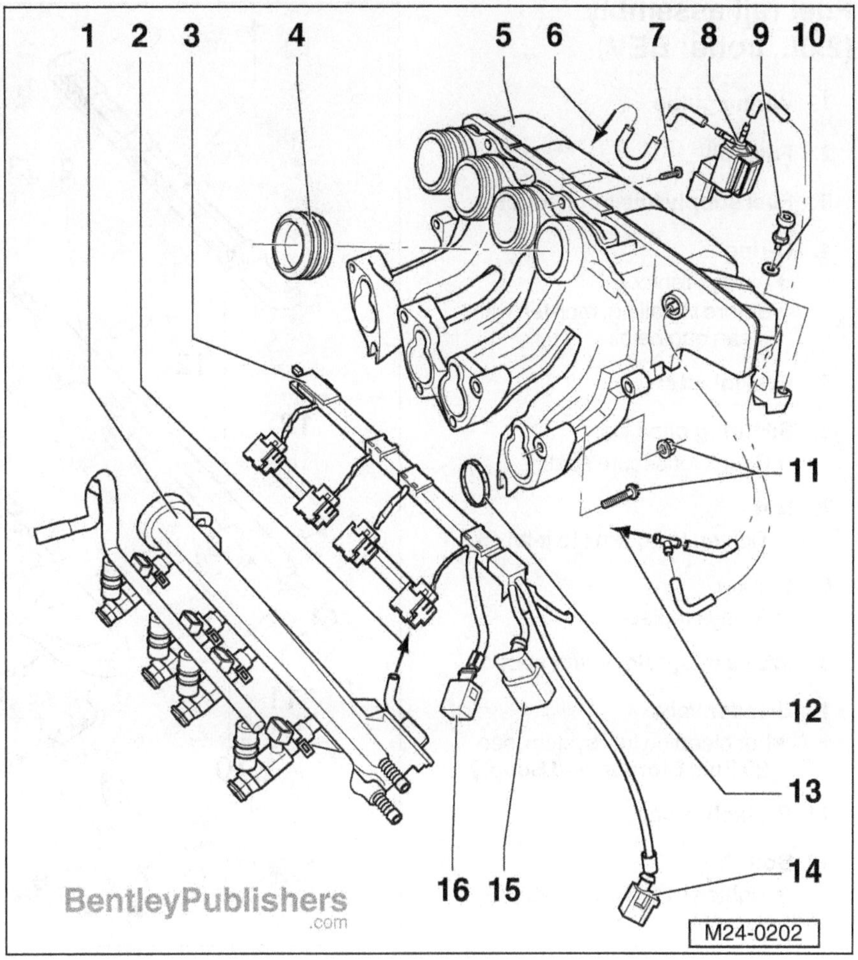

1. **Fuel rail with injectors**
2. **Vacuum hose**
   - To intake manifold, upper part
3. **Wiring guide**
4. **Seal**
   - Replace if damaged
5. **Intake manifold, lower section**
   - With intake manifold change-over valve
6. **To vacuum positioning element**
   - On front side of lower intake manifold
7. **Bolt**
   - Tighten to 3 Nm (27 in-lb)
8. **Intake manifold tuning valve 2 (N261)**
9. **Bolt**
   - Tighten to 5 Nm (44 in-lb)
10. **Washer**
11. **Nut**
    - Tighten to 25 Nm (18 ft-lb)
12. **To upper intake manifold**
13. **Seal**
    - Always replace
14. **2-pin connector**
    - For knock sensor 1 (G61)
15. **3-pin connector**
    - For camshaft position sensor (G40)
16. **2-pin connector**
    - For intake manifold tuning valve 2 (N261)

Fuel rail assembly (2.0L, code: BEV)

1. **Wiring guide**
2. **Fuel rail**
3. **Fuel supply line connection**
4. **O-ring**
   - Always replace
   - Before installing, moisten with clean engine oil
5. **Fuel injector**
6. **Securing clip**
   - Check for secure seating
7. **Nut**
   - Tighten to 20 Nm (15 ft-lb)
8. **Gasket**
   - Always replace
9. **Intake manifold, lower section**
10. **Bleeder valve**
    - For bleeding fuel system, see **20 Fuel Storage and Supply**
11. **Protective cap**
12. **Bolt**
    - Tighten to 10 Nm (7 ft-lb)

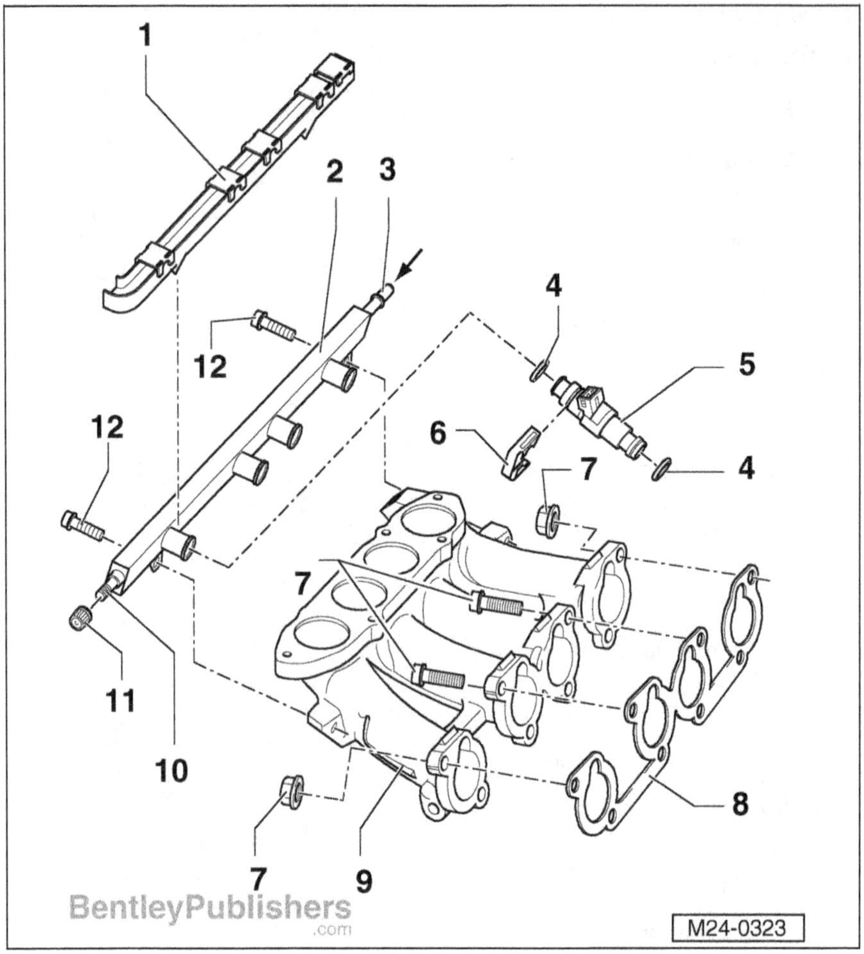

Fuel rail assembly (2.0L, code: BGD)

1. **Fuel rail**
2. **Protective cap**
3. **Bleeder valve**
   - For bleeding fuel system, see **20 Fuel Storage and Supply**
4. **Wiring guide**
5. **Seal**
   - Replace if damaged
6. **Bolt**
   - Tighten to 10 Nm (7 ft-lb)
7. **Intake manifold, lower section**
   - With intake manifold change-over valve
8. **Bolt**
   - Tighten to 3 Nm (27 in-lb)
9. **To vacuum positioning element**
   - On front side of lower intake manifold
10. **Intake manifold tuning valve 2 (N261)**
11. **Bolt**
    - Tighten to 5 Nm (44 in-lb)
12. **Washer**
13. **Nut**
    - Tighten to 25 Nm (18 ft-lb)
14. **To upper intake manifold**
15. **Seal**
    - Always replace
16. **Fuel supply line connection**
17. **O-ring**
    - Always replace
    - Before installing, moisten with clean engine oil
18. **Fuel injector**
19. **Securing clip**
    - Check for secure seating

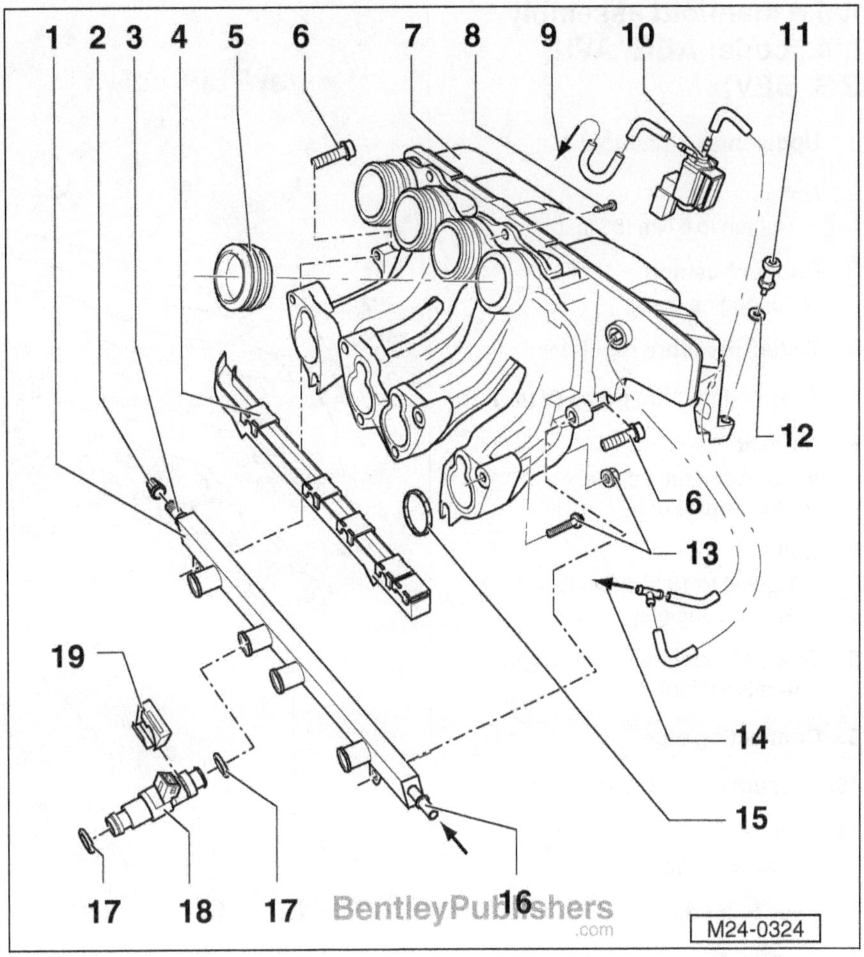

Intake manifold assembly (2.0L, code: AEG, AVH, AZG, BEV)

1. **Upper intake manifold**
2. **Nut**
   - Tighten to 6 Nm (53 in-lb)
3. **Rubber bushing**
   - For engine cover
4. **To fuel pressure regulator**
5. **To leak detection pump (V144)**
6. **Bracket**
   - For accelerator cable
   - AEG engine only
7. **Bolt**
   - Tighten to 10 Nm (7 ft-lb)
   - AEG engine only
8. **To EVAP canister purge regulator valve (N80)**
9. **Connecting pipe**
10. **Vacuum line to brake servo**
11. **Seal**
    - Always replace
12. **Hose to intake**
13. **Pipe union for coolant hoses**
14. **Throttle valve control module (J338)**
15. **Gasket**
    - Always replace
    - Note installation position

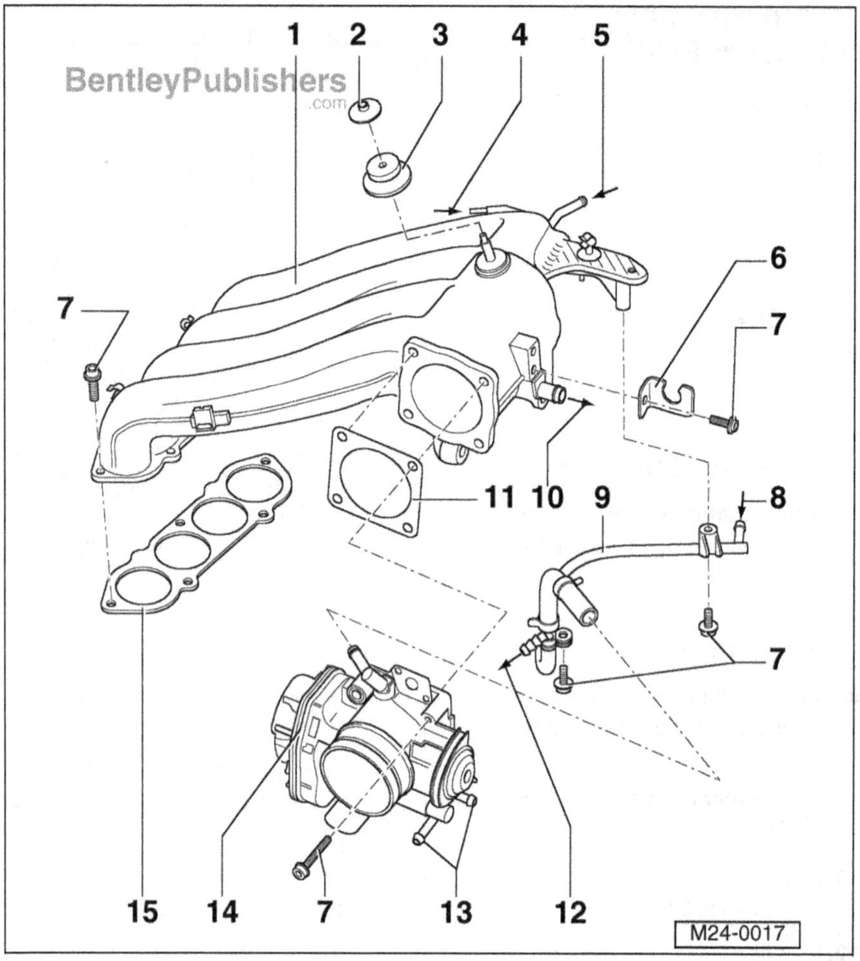

Intake manifold assembly (2.0L, code: BDC, BGD)

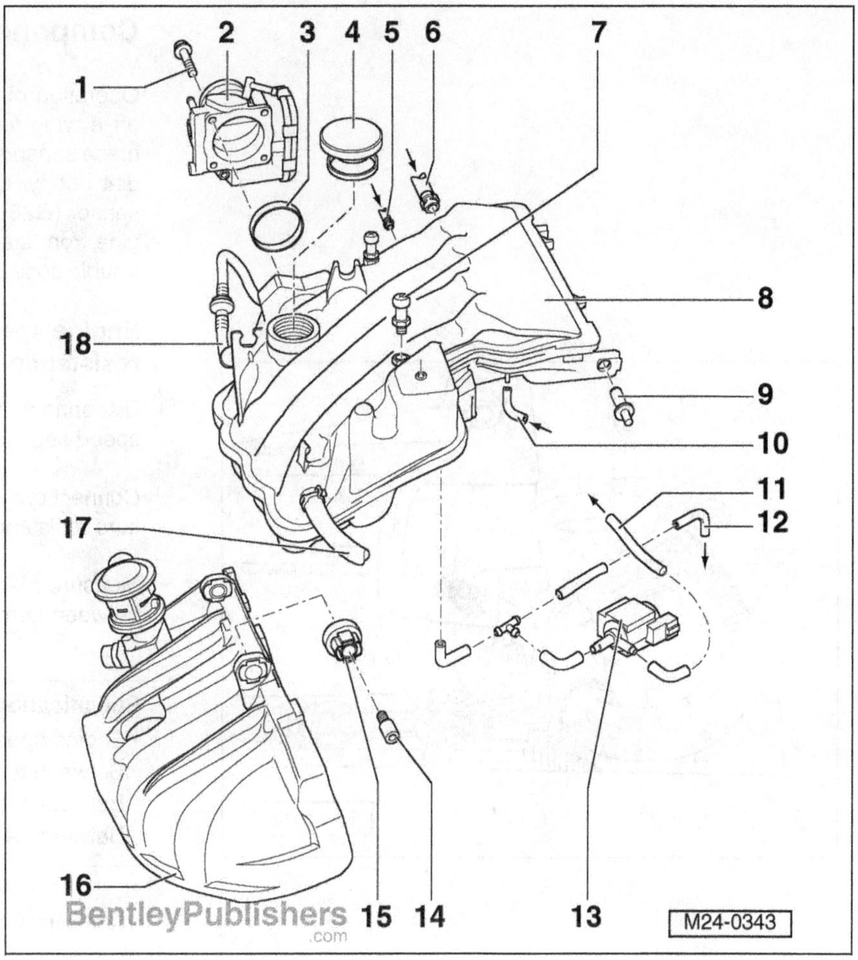

1. **Bolt**
   - Tighten to 8 Nm (71 in-lb)
2. **Throttle valve control module (J338)**
3. **Seal**
   - Replace if damaged
4. **Pressure relief valve**
   - Protects intake manifold from damage caused by excess pressure
5. **Vacuum line**
   - To positive crankcase ventilation heating element (N79)
6. **Hose**
   - To positive crankcase ventilation heating element (N79)
7. **Engine cover mounting pin**
   - Tighten to 5Nm (44 in-lb)
8. **Upper intake manifold**
   - With intake manifold change-over valve
9. **Spreader rivet**
10. **Vacuum hose**
    - BDC engine only
11. **To vacuum positioning element**
    - On underside of upper intake manifold
12. **To T-piece**
13. **Intake manifold tuning valve (N156)**
14. **Mounting stud**
    - Tighten to 8Nm (71 in-lb)
15. **Rubber bushing**
16. **Warm air collector plate**
    - With intake manifold support and combination valve
17. **To leak detection pump (V144)**
18. **To brake servo**

Component testing (2.0L)

Operation of the engine management system depends on a wide variety of input sensors. Failure of most of these sensors will allow the engine to continue to run, albeit poorly. One exception to this is the engine speed sensor (G28). Failure of this sensor will prevent the engine from starting and running and will set a diagnostic trouble code in the ECM.

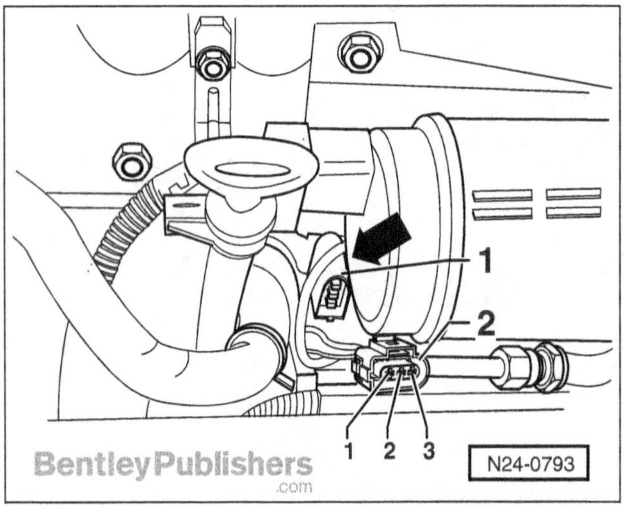

Engine speed (RPM) sensor (G28), measuring resistance

◄ Disconnect 3-pin harness connector (**1**) from engine speed sensor near the secondary air injection pump.

– Connect ohmmeter between terminals 2 and 3 and measure resistance.

– Measure resistance between terminals 1 and 3, and between terminals 1 and 2.

Specification

- Engine speed sensor resistance values
  - Between terminals 2 and 3 . . . . 480 to 1000 Ω
  - Between terminals 1 and 2 . . . . infinity Ω
  - Between terminals 1 and 3 . . . . infinity Ω

– If specification is not obtained, replace engine speed sensor.

– If specification is obtained, and engine will not start, additional testing with a suitable diagnostic scan tool will be required.

Fuel injectors, checking resistance

Requirements:

- Ignition off
- Ambient temperature 68°F (20°C)

NOTE —
Fuel injector resistance will vary with temperature. At higher temperatures the resistance increases.

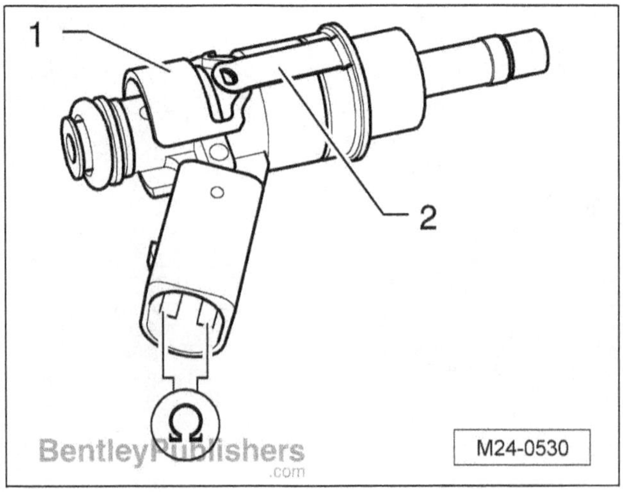

◄ Disconnect fuel injector harness connectors. Measure resistance across two terminals as shown.

Specification

- 1.8L fuel injector resistance . . . . 14.0 to 20.0 Ω

– If specified value is not obtained, replace fuel injector.

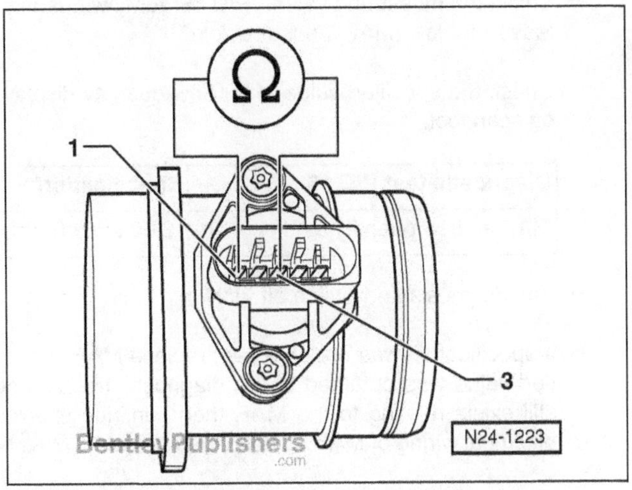

Intake air temperature (IAT) sensor (G42), measuring resistance

◄ Disconnect harness connector from MAF/IAT sensor near air filter housing. Measure resistance between terminals **1** and **3** of sensor.

– Measure temperature of intake air using diagnostic scan tool or thermometer.

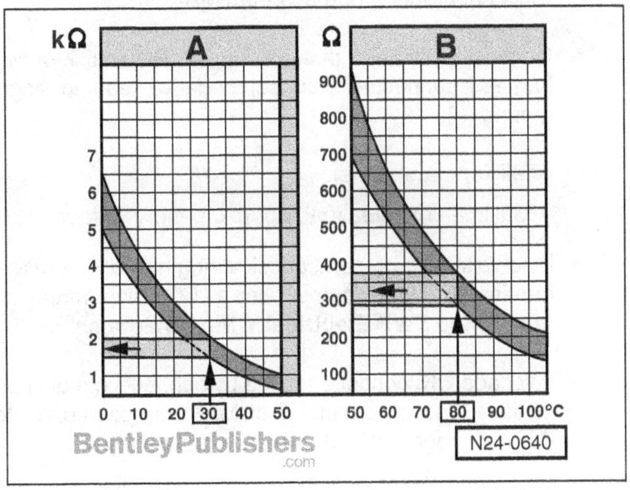

◄ Compare IAT sensor resistance and intake air temperature readings to graph.

- Scale **A** shows resistance for intake air temperature range 0 - 50°C (32 - 122°F).
- Scale **B** shows resistance for intake air temperature range 50 - 100°C (122 - 212°F).
- Example:
  30°C corresponds to 1.5 - 2.0 k Ω
  80°C corresponds to 275 - 375 Ω

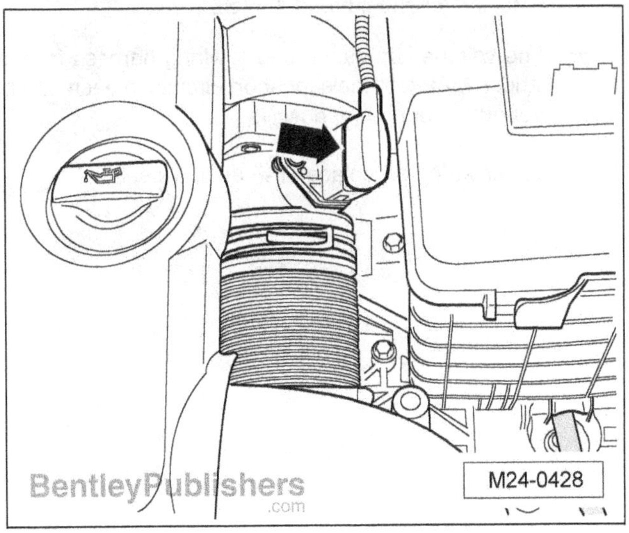

Mass air flow (MAF) sensor (G70), testing

◄ MAF sensor (**arrow**) is located near the air filter housing.

Requirements for testing:

- Battery voltage at least 11.5 volts.
- All electrical consumers turned off.
- Parking brake applied (turns off daytime running lights).
- Coolant temperature must be at least 80°C (176°F), check using diagnostic scan tool.

– Connect diagnostic scan tool to OBD II connector under left side of dashboard. See **OBD On Board Diagnostics**.

– Start engine and let it run at idle.

– Select address word 33 (Generic OBD II functionality), “Diagnostic mode 1: Checking measuring values” on scan tool.

– Select the measuring value, "PID 16: Air flow quantity at Mass Air Flow (MAF) sensor at idle".

– Check the specified value of air flow quantity displayed on scan tool.

Diagnostic test PID 16	Specification
MAF air flow quantity at idle	2.00 to 5.00 g/sec.

– End diagnosis and switch off ignition.

– If specification was not obtained, replace MAF. If specified value was obtained and a diagnostic trouble code still exists relating to the MAF, then continue checking electrical wiring of MAF as described below.

– Disconnect wiring harness from MAF.

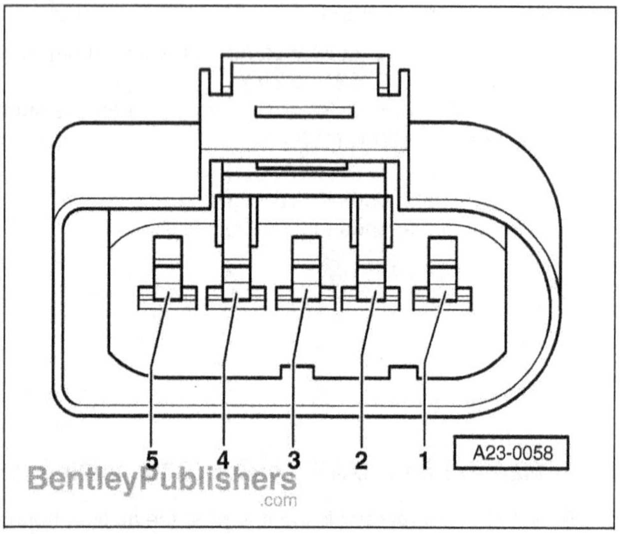

< Connect multimeter positive lead to terminal **2** of MAF harness connector. Connect negative lead to engine ground (–).

– Briefly operate starter.

- Multimeter should read approx. battery voltage.

– If no voltage is present, check wiring harness from MAF terminal 2 to fuel pump relay (J17) using wiring diagrams. See **EWD Electrical Wiring Diagrams**.

– If voltage OK, continue testing by connecting multimeter positive lead to terminal **4** of MAF harness connector. Connect negative lead to engine ground (–).

– Switch ignition on.

- Multimeter should read 4.5 to 5.5 volts.

– If no voltage is present, check wiring harness from the various MAF terminals for short circuits to each other, to ground (–), or battery positive (+).

– Repair wiring or replace MAF as necessary.

2.5L FUEL INJECTION SYSTEM

NOTE —
Volkswagen identifies electrical components by a letter and/or a number in the electrical schematics. See **EWD Electrical Wiring Diagrams.** *These electrical identifiers are listed in parenthesis as an aid to electrical troubleshooting.*

Fuel injection components (2.5L, code: BPR, BPS)

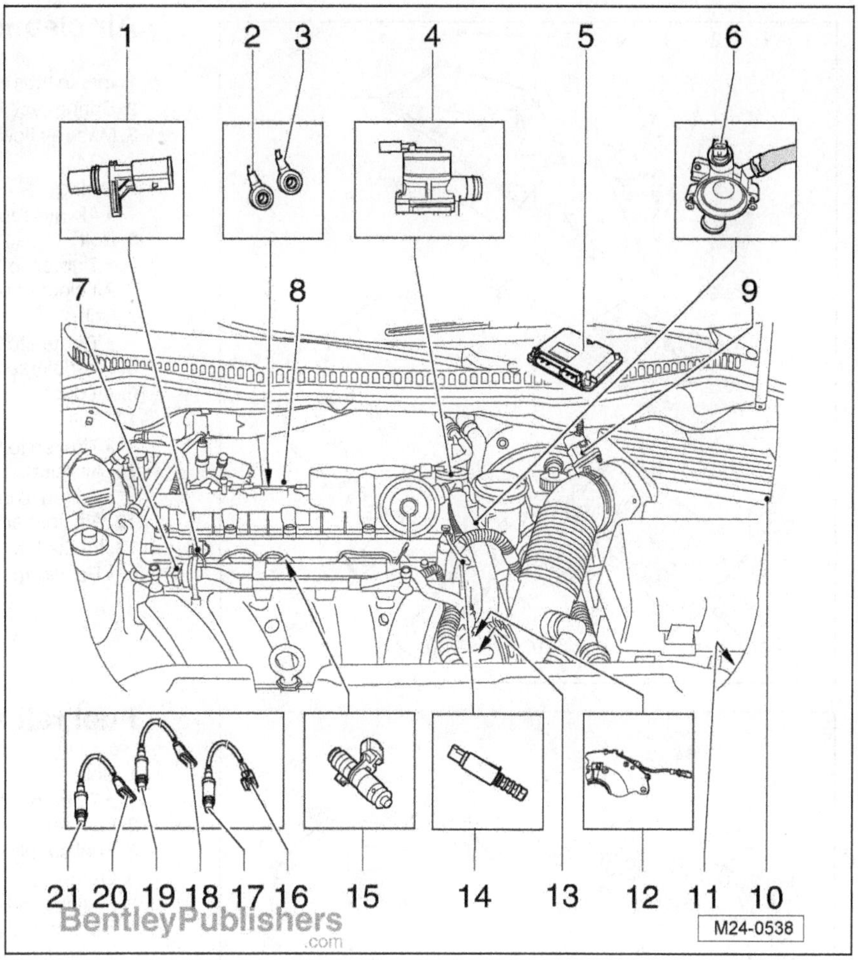

1. **Camshaft position sensor (G40)**
2. **Knock sensor 2 (G66)**
3. **Knock sensor 1 (G61)**
4. **Secondary air injection solenoid valve (N112)**
5. **Motronic engine control module (J220)**
6. **Engine coolant temperature sensor (G62)**
   - Combined with sensor (G2) for instrument panel warning light
   - Testing, see **19 Engine–Cooling System**
7. **EVAP canister purge regulator valve (N80)**
8. **Ignition coils (N, N127, N291, N292, N323)**
   - Combined with power output stage
9. **Mass air flow sensor (G70), with intake air temperature sensor (G42)**
10. **Auxiliary relay carrier above central electric panel**
    - Location of:
      Secondary air injection pump relay (J299)
      Power supply relay (J317)
11. **Secondary air injection pump motor (V101)**
12. **Engine speed sensor (G28)**
13. **Throttle control module (J338)**
14. **Camshaft adjustment valve 1 (N205)**
15. **Fuel injectors (N30, N31, N32, N33, N83)**
16. **6-pin harness connector**
    - Black
    - For pre-catalyst oxygen sensor (G39) and heater (Z19)
17. **Oxygen sensor, pre-catalyst (G39)**
    - Monitors combustion efficiency
    - Threaded into exhaust pipe
    - Tighten to 55 Nm (41 ft-lb)
    - Lubricate threads only with high temperature anti-seize. Do not allow anti-seize compound into slots of sensor body.
18. **4-pin harness connector**
    - Black
    - Only for engine code BPR
    - For center-catalyst oxygen sensor (G465) and heater (Z59)
19. **Oxygen sensor, center-catalyst (G465)**
    - Only for engine code BPR
    - Threaded into center of catalytic converter
    - Tighten to 55 Nm (41 ft-lb)
    - Lubricate threads only with high temperature anti-seize. Do not allow anti-seize compound into slots of sensor body.
20. **4-pin harness connector**
    - Brown
    - For post-catalyst oxygen sensor (G130) and heater (Z29)
21. **Oxygen sensor, post-catalyst (G130)**
    - Monitors catalytic converter efficiency
    - Threaded into rear of catalytic converter
    - Tighten to 55 Nm (41 ft-lb)
    - Lubricate threads only with high temperature anti-seize. Do not allow anti-seize compound into slots of sensor body.

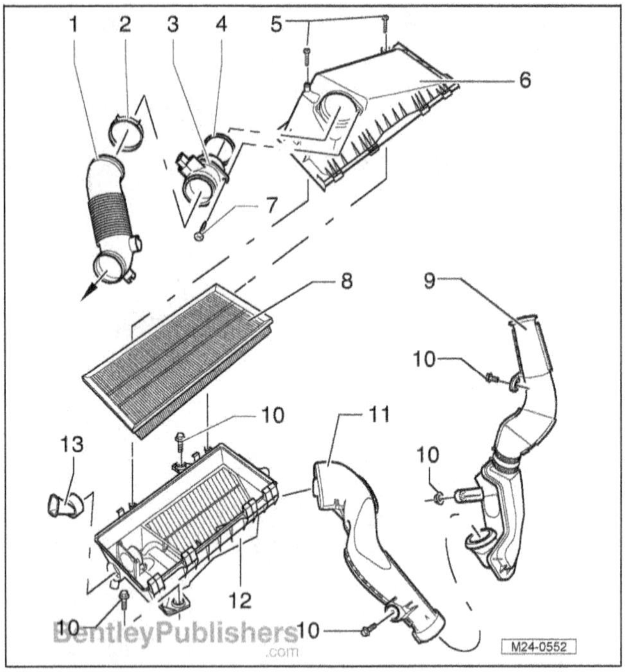

Air cleaner assembly (2.5L)

1. Intake tube
2. Spring-type clip
3. Mass air flow sensor (G70), with intake air temperature sensor (G42)
4. O-ring
   - Always replace
5. Bolt
   - Tighten to 2 Nm (18 in-lb)
6. Air cleaner housing, upper section
7. Bolt
   - Tighten to 2 Nm (18 in-lb)
8. Filter element
9. Air duct
10. Bolt
    - Tighten to 8 Nm (71 in-lb)
11. Air duct
    - Bolted to battery tray
12. Air cleaner housing, lower
13. Intake tube
    - For warm air

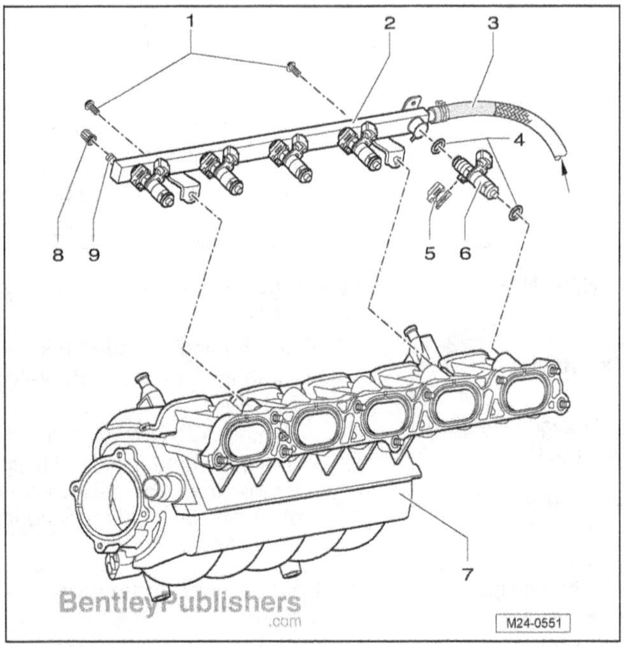

Fuel rail assembly (2.5L)

1. Bolt
   - Tighten to 3.5 Nm (31 in-lb)
2. Fuel rail
3. Fuel supply line
4. O-ring
   - Always replace
5. Retaining clip
   - Check for secure seating
6. Fuel injector
7. Intake manifold
8. Protective cap
9. Bleeder valve

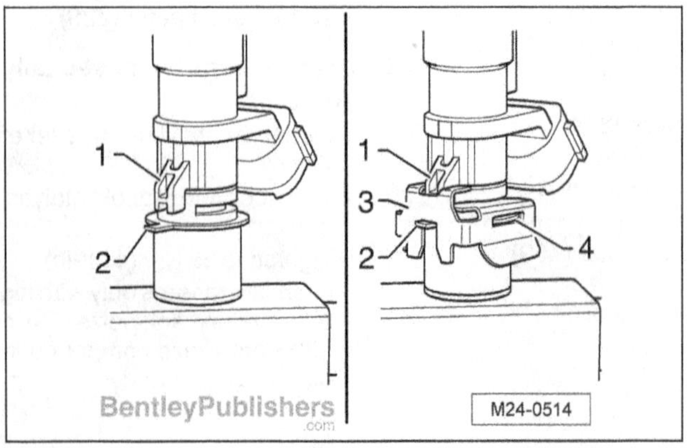

Fuel injector installation details

- Install new O-rings for fuel injectors and coat them lightly with clean engine oil.
- Press fuel injector into fuel rail with tabs (**1** and **2**) aligned.
- Push retaining clip (**3**) as shown into groove of fuel injector.
- Position shoulder (**4**) correctly in cutout of retaining clip on both sides.
- Verify that all fuel injectors are securely seated.

Intake manifold assembly (2.5L)

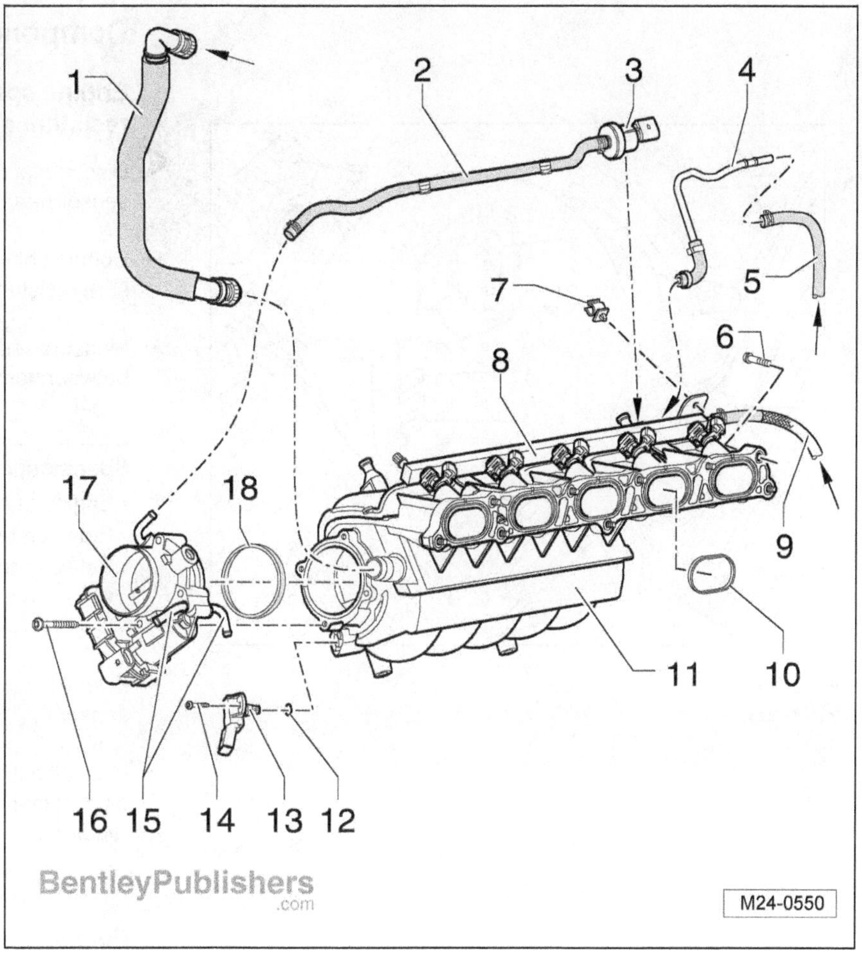

1. **Hose for crankcase ventilation**
   - From cylinder head cover
2. **Ventilation hose**
3. **EVAP canister purge regulator valve (N80)**
4. **Line**
5. **Vacuum hose**
6. **Bolt**
   - Tighten to 9 Nm (80 in-lb)
7. **Bracket**
8. **Fuel rail**
9. **Fuel supply line**
10. **Gasket**
    - Always replace
    - Note installation position: casting mark faces upward
11. **Intake manifold**
12. **O-ring**
    - Always replace
13. **Manifold absolute pressure sensor (G71)**
14. **Bolt**
    - Tighten to 3.5 Nm (31 in-lb)
15. **Coolant connections**
16. **Bolt**
    - Tighten to 6.5 Nm (58 in-lb)
17. **Throttle valve control module (J338)**
    - With throttle drive for electronic power control (G186)
    - With throttle angle sensor 1 (G187)
    - With throttle angle sensor 2 (G188)
    - During replacement, erase adaptation values and adapt ECM to J338 using VAS 5051/5052 scan tool or equivalent
18. **Seal**
    - Replace if damaged

Component testing (2.5L)

Engine speed (RPM) sensor (G28), measuring resistance

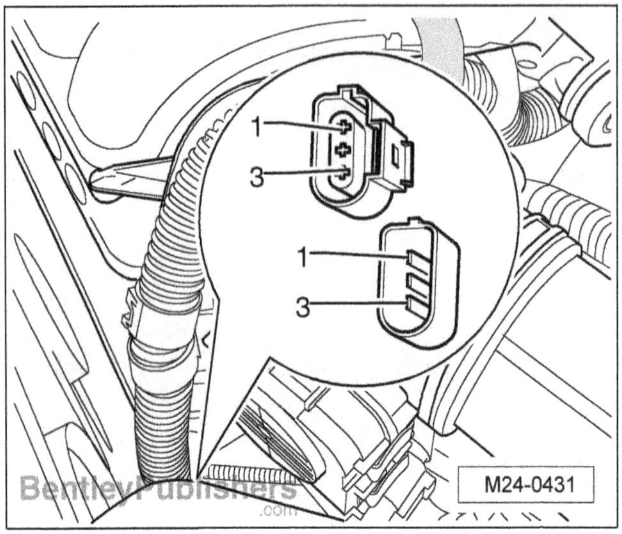

◀ Disconnect 3-pin harness connector from engine speed sensor near the throttle valve control module.

– Connect ohmmeter between terminals 2 and 3 and measure resistance.

– Measure resistance between terminals 1 and 3, and between terminals 1 and 2.

Specification

- Engine speed sensor resistance values
  Between terminals 2 and 3 .... 730 to 1000 Ω
  Between terminals 1 and 2 .... infinity Ω
  Between terminals 1 and 3 .... infinity Ω

– If specification is not obtained, replace engine speed sensor.

– If specification is obtained, and engine will not start, additional testing with a suitable diagnostic scan tool will be required.

Fuel injectors, checking resistance

Requirements:

- Ignition off
- Ambient temperature 68°F (20°C)

NOTE —
Fuel injector resistance will vary with temperature. At higher temperatures the resistance increases.

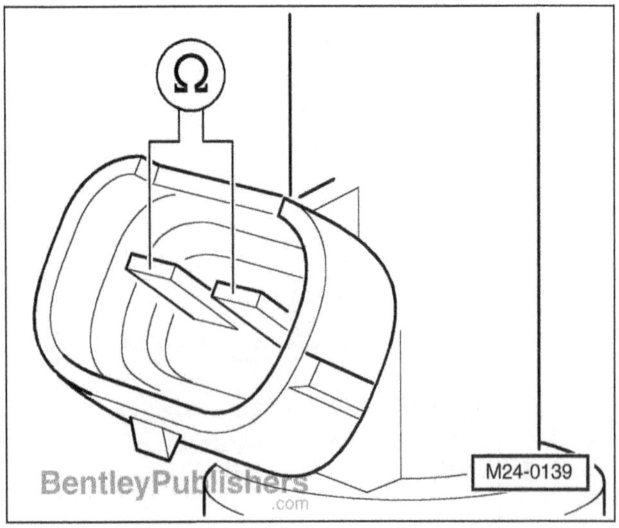

◀ Disconnect fuel injector harness connectors. Measure resistance across two terminals as shown.

Specification

- 1.8L fuel injector resistance .... 12.0 to 20.0 Ω

– If specified value is not obtained, replace fuel injector.

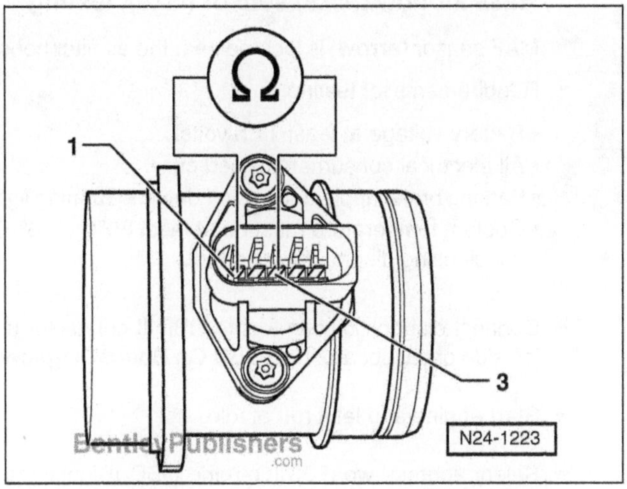

Intake air temperature (IAT) sensor (G42), measuring resistance

◄ Disconnect harness connector from MAF/IAT sensor near air filter housing. Measure resistance between terminals **1** and **3** of sensor.

– Measure temperature of intake air using diagnostic scan tool or thermometer.

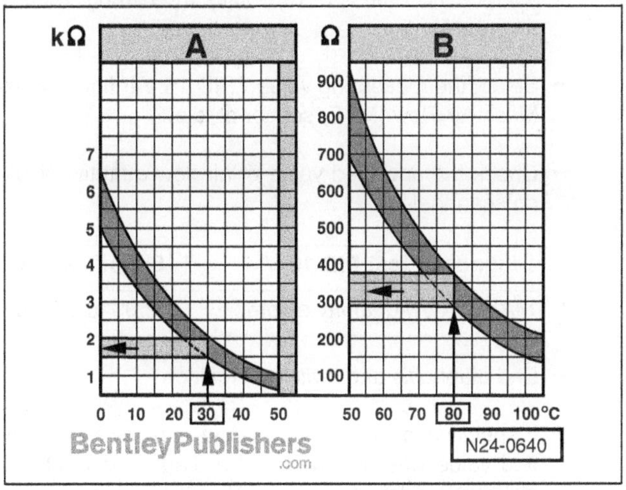

◄ Compare IAT sensor resistance and intake air temperature readings to graph.

- Scale **A** shows resistance for intake air temperature range 0 - 50°C (32 - 122°F).
- Scale **B** shows resistance for intake air temperature range 50 - 100°C (122 - 212°F).
- Example:
  30°C corresponds to 1.5 - 2.0 k Ω
  80°C corresponds to 275 - 375 Ω

Manifold absolute pressure (MAP) sensor (G71), testing

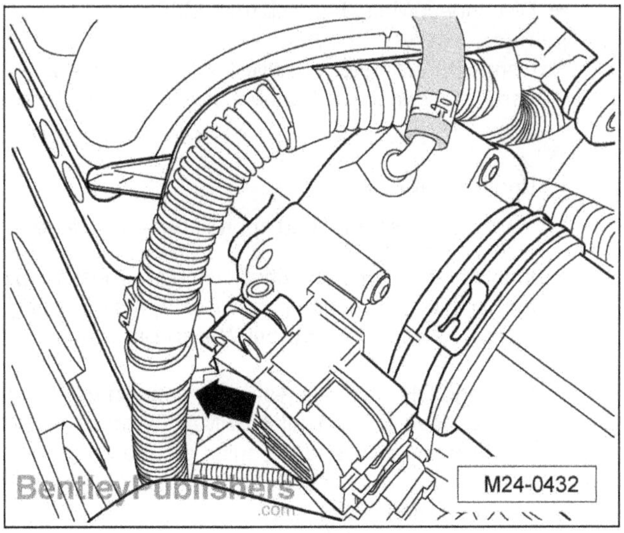

◄ MAP sensor (**arrow**) is located near the air filter housing.

Requirements for testing:

- Battery voltage at least 11.5 volts.
- All electrical consumers turned off.
- Parking brake applied (turns off daytime running lights).
- Coolant temperature must be at least 80°C (176°F), check using diagnostic scan tool.

– Connect diagnostic scan tool to OBD II connector under left side of dashboard. See **OBD On Board Diagnostics**.

– Start engine and let it run at idle.

– Check the specified value of manifold absolute pressure displayed on scan tool.

Diagnostic text	Specification
Intake manifold pressure	approx. 100 kPa

– End diagnosis and switch off ignition.

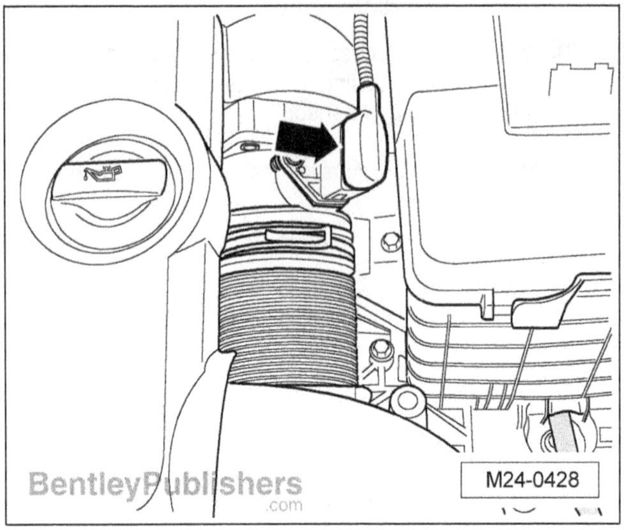

Mass air flow (MAF) sensor (G70), testing

◀ MAF sensor (**arrow**) is located near the air filter housing.

Requirements for testing:

- Battery voltage at least 11.5 volts.
- All electrical consumers turned off.
- Parking brake applied (turns off daytime running lights).
- Coolant temperature must be at least 80°C (176°F), check using diagnostic scan tool.

– Connect diagnostic scan tool to OBD II connector under left side of dashboard. See **OBD On Board Diagnostics**.

– Start engine and let it run at idle.

– Select address word 33 (Generic OBD II functionality), "Diagnostic mode 1: Checking measuring values" on scan tool.

– Select the measuring value, "PID 16: Air flow quantity at Mass Air Flow (MAF) sensor at idle".

– Check the specified value of air flow quantity displayed on scan tool.

Diagnostic test PID 16	Specification
MAF air flow quantity at idle	2.00 to 5.00 g/sec.

– End diagnosis and switch off ignition.

– If specification was not obtained, replace MAF. If specified value was obtained and a diagnostic trouble code still exists relating to the MAF, then continue checking electrical wiring of MAF as described below.

– Disconnect wiring harness from MAF.

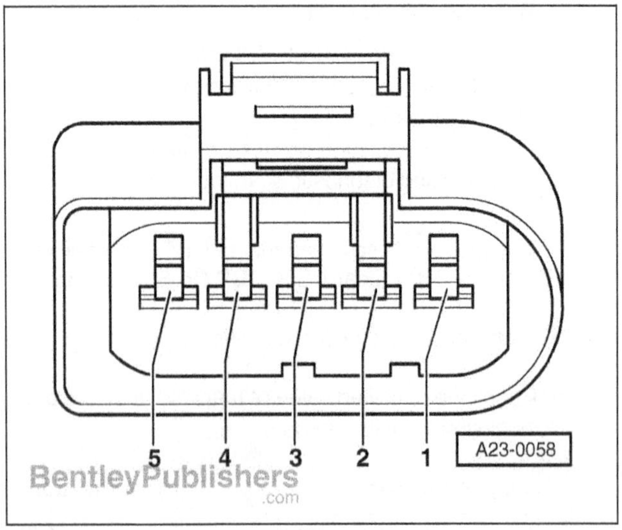

◀ Connect multimeter positive lead to terminal **2** of MAF harness connector. Connect negative lead to engine ground (–).

– Briefly operate starter.

- Multimeter should read approx. battery voltage.

– If no voltage is present, check wiring harness from MAF terminal 2 to fuel pump relay (J17) using wiring diagrams. See **EWD Electrical Wiring Diagrams**.

– If voltage OK, continue testing by connecting multimeter positive lead to terminal **4** of MAF harness connector. Connect negative lead to engine ground (–).

– Switch ignition on.

- Multimeter should read 4.5 to 5.5 volts.

– If no voltage is present, check wiring harness from the various MAF terminals for short circuits to each other, to ground (–), or battery positive (+).

– Repair wiring or replace MAF as necessary.

26 Exhaust System and Emission Controls

GENERAL

This repair group covers repair and troubleshooting for the emission control systems and removal and installation of exhaust components.

Engine	Engine code
1.8L 4-cylinder turbo gasoline	APH, AWV, AWP, BKF, BNU
1.9L 4-cylinder TDI diesel	ALH, BEW
2.0L 4-cylinder gasoline	AEG, AVH, AZG, BDC, BEV, BGD
2.5L 5-cylinder gasoline	BPR, BPS

NOTE —

- *All cars have a self-diagnostic program to detect emissions-related problems and store coded fault information in electronic memory. When diagnosing an emissions-related faults, first check for stored fault codes using a diagnostic scan tool.*
- *The cars covered by this manual have a vacuum hose routing diagram (sticker) located in the engine compartment which can be helpful when working with vacuum hoses.*

Service precautions

To help guard against personal injury or damage to car components, the following warnings and cautions apply when servicing the exhaust system.

WARNING —

- *Exhaust gases are colorless, odorless, and very toxic. Run the engine only in a well-ventilated area. Immediately repair any leaks in the exhaust system or structural damage to the car body that might allow exhaust gases to enter the passenger compartment.*
- *The exhaust system, catalytic converter and other emission control systems operate at very high temperatures. Allow these components to cool before servicing, or wear protective clothing to prevent burns. Do not use flammable chemicals near a hot catalytic converter.*
- *Old corroded exhaust system components crumble easily and often have exposed sharp edges. To avoid injury, wear eye protection and heavy gloves when working with such parts.*
- *Do not work under a lifted car unless it is solidly supported on jack stands designed for that purpose. Never work under a car that is supported solely by a jack.*

CAUTION —

Disconnecting the negative (–) battery cable may erase fault codes and basic settings in the engine management and automatic transmission control modules. Some driveability problems may be noticed until the system re-adapts to operating conditions. OBD II readiness codes, which may be required for emissions testing, may also be erased.

EXHAUST SYSTEM

For safe and proper exhaust system operation, all components must be free of holes and all connections must be airtight. Check the system immediately if it becomes noisy or an exhaust odor is detected inside the car.

NOTE —

Excessive exhaust system back pressure can cause driveability problems such as a rough idle or stalling. Back pressure problems are caused by external damage such as crushed or collapsed parts, or internal damage such as a plugged catalytic converter.

Exhaust system replacement

Use these illustrations as a guide for removal and installation. Tightening torques and other relevant installation information are given in the component lists.

A liberal application of penetrating oil to cold exhaust system nuts, bolts and slip joints will make removal easier. New fasteners, clamps and rubber mounts are always recommended when replacing exhaust components. Gaskets should be replaced whenever flange joints are disconnected. Use high-temp anti-seize compound on threaded fasteners to extend their service life and make future replacement easier. Some slight smoking and/or odor is normal as new exhaust system parts become hot for the first time.

> ***WARNING —***
> *Inspect the exhaust system heat shields on the car underbody and repair or replace any damaged parts. Heat shields protect the car occupants, undercoating, and various other components from excessive heat. Damaged or missing shields, particularly those above the catalytic converter, will increase interior temperatures and create a fire hazard.*

NOTE —

- *After performing repairs on the exhaust system, ensure that the system is not under stress and that it has sufficient clearance from the body. If necessary, loosen the double clamps and align the muffler and exhaust pipe so sufficient clearance is maintained to the body while the support rings are evenly loaded.*
- *When performing repairs, always replace self-locking nuts.*
- *Volkswagen identifies electrical components by a letter and/or a number in the electrical schematics. See* **EWD Electrical Wiring Diagrams**. *These electrical identifiers are listed in parenthesis as an aid to electrical troubleshooting.*

Front exhaust pipe and catalyst (1.8L engine)

1. **Turbocharger**
2. **Gasket**
   - Always replace
3. **Front exhaust pipe with catalyst**
4. **Oxygen sensor 1 (G39)**
   - Tighten to 50 Nm (37 ft-lb)
   - Coat threads with high-temp anti-seize compound; do not allow compound to enter slots on sensor body
5. **6-pin connector**
   - Black
   - For oxygen sensor and oxygen sensor heating
6. **4-pin connector**
   - Brown
   - For oxygen sensor and oxygen sensor heating
7. **Oxygen sensor 2 (G130)**
   - Tighten to 50 Nm (37 ft-lb)
   - Coat threads with high-temp anti-seize compound; do not allow compound to enter slots on sensor body
8. **To center muffler**
9. **Nut**
   - Tighten to 40 Nm (30 ft-lb)

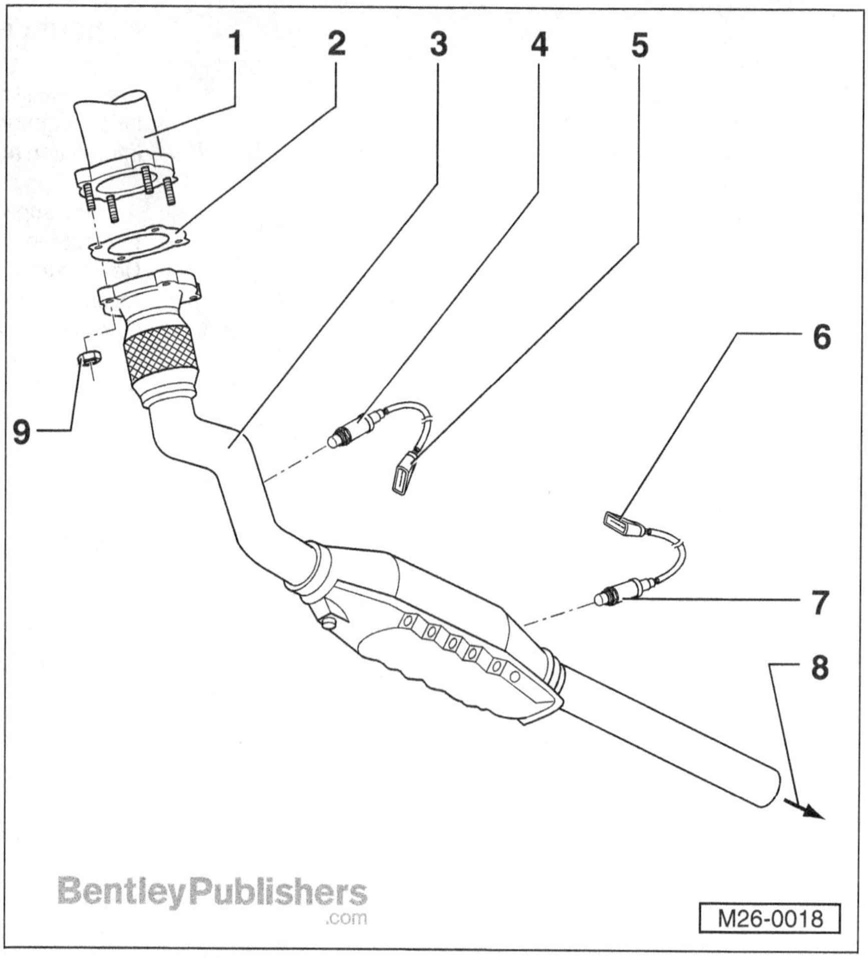

Front exhaust pipe and catalyst (1.9L ALH engine)

1. **From turbocharger**
2. **Gasket**
   - Always replace
3. **Nut**
   - Tighten to 25 Nm (18 ft-lb)
4. **Front exhaust pipe with catalytic converter**
5. **To rear muffler**
6. **Harness connector**
7. **Front support**
8. **Bolt**
   - Tighten to 25 Nm (18 ft-lb)
9. **Heated oxygen sensor (G39) before catalyst**
   - Tighten to 50 Nm (37 ft-lb)
   - Grease threads with special hot bolt anti-seize lubricant. Do not get anti-seize into slots of sensor body

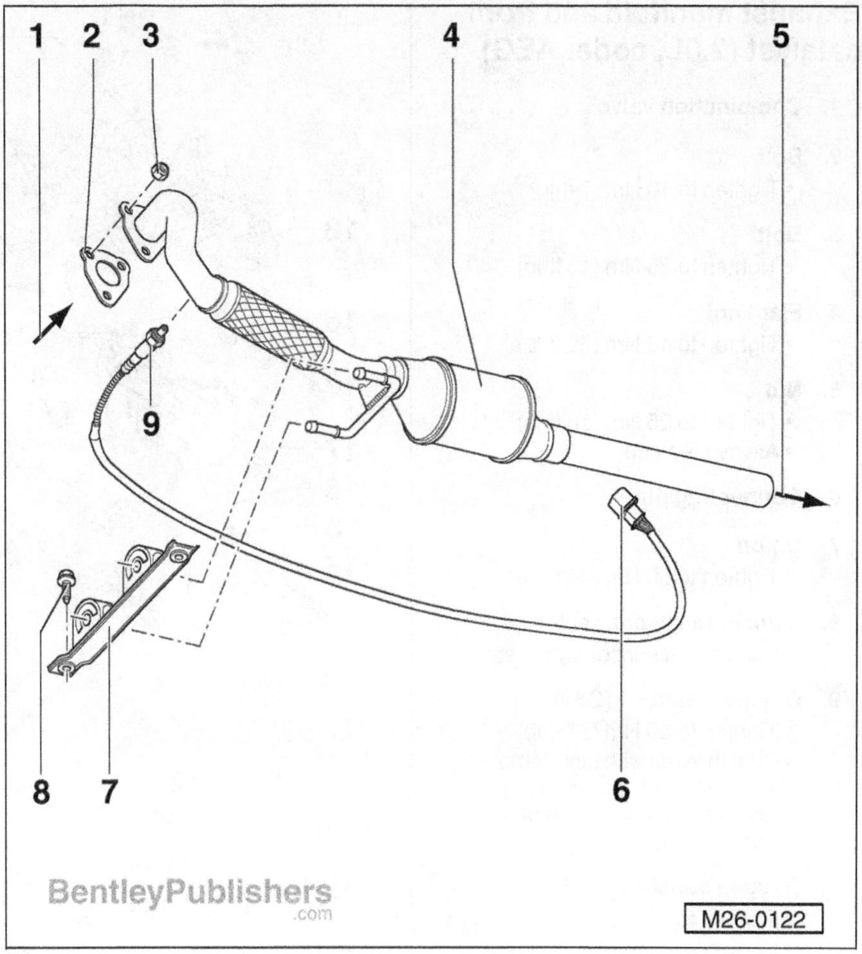

Exhaust manifold and front catalyst (2.0L, code: AEG)

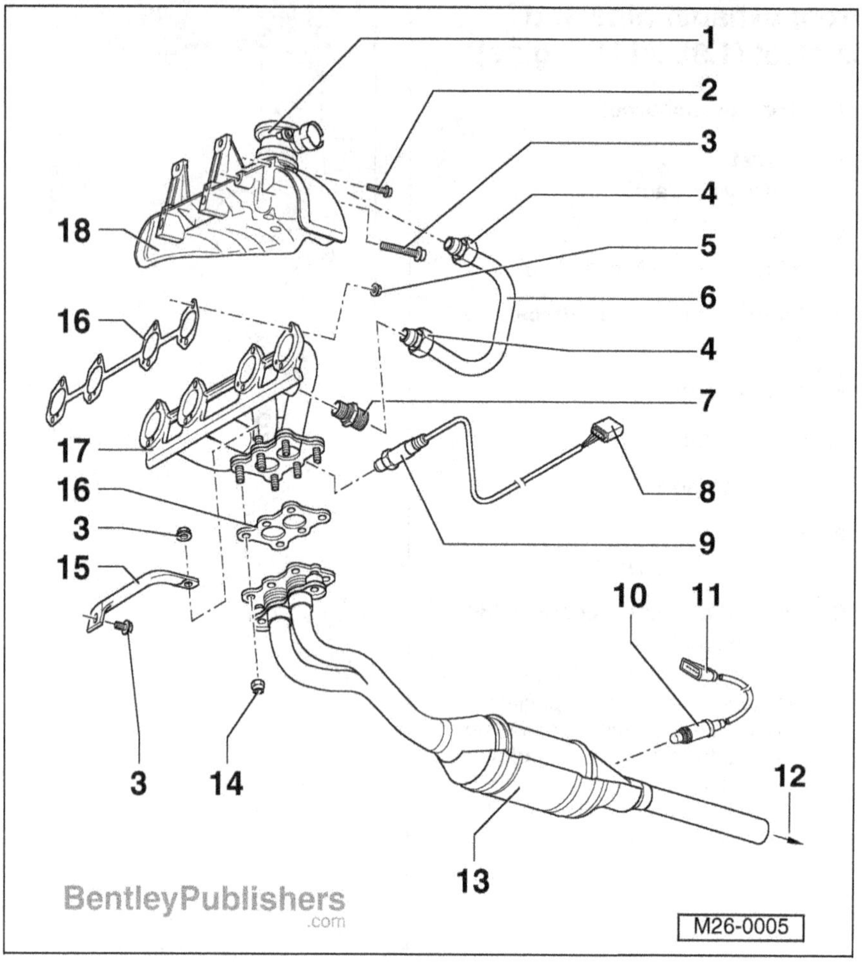

1. **Combination valve**
2. **Bolt**
   - Tighten to 10 Nm (7 ft-lb)
3. **Bolt**
   - Tighten to 25 Nm (18 ft-lb)
4. **Flare nut**
   - Tighten to 40 Nm (30 ft-lb)
5. **Nut**
   - Tighten to 25 Nm (18 ft-lb)
   - Always replace
6. **Connecting pipe**
7. **Union**
   - Tighten to 60 Nm (44 ft-lb)
8. **4-pin harness connector**
   - For oxygen sensor and heater
9. **Oxygen sensor 1 (G39)**
   - Tighten to 50 Nm (37 ft-lb)
   - Coat threads with high-temp anti-seize compound; do not allow compound to enter slots on sensor body
10. **Oxygen sensor 2 (G130)**
    - Tighten to 50 Nm (37 ft-lb)
    - Coat threads with high-temp anti-seize compound; do not allow compound to enter slots on sensor body
11. **4-pin harness connector**
    - For Oxygen sensor and Oxygen sensor heating
12. **To center muffler**
13. **Front exhaust pipe with catalyst**
14. **Nut**
    - Tighten to 40 Nm (30 ft-lb)
    - Always replace
15. **Exhaust manifold support brace**
16. **Gasket**
    - Always replace
17. **Exhaust manifold**
18. **Warm air collector plate**

Exhaust manifold and front catalyst (2.0L, code: AVH, AZG, BDC, BEV, BGD)

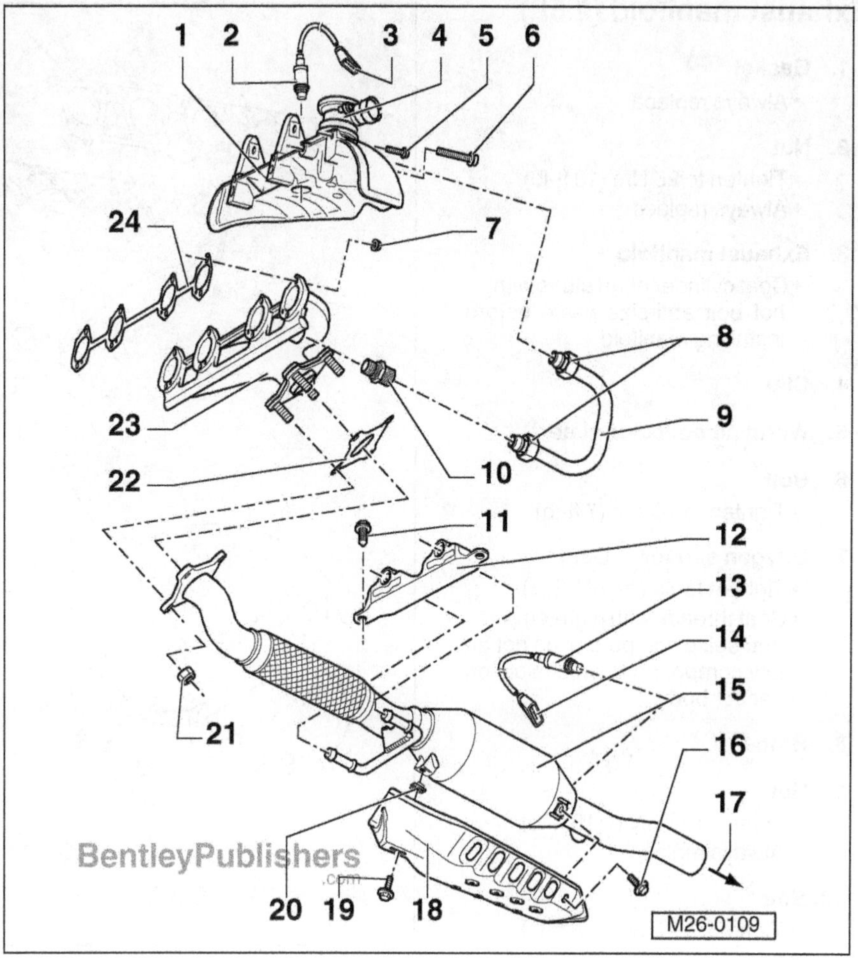

1. **Warm air collector plate**
2. **Oxygen sensor 1 (G39)**
   - Tighten to 50 Nm (37 ft-lb)
   - Coat threads with high-temp anti-seize compound; do not allow compound to enter slots on sensor body
3. **6-pin harness connector**
   - For Oxygen sensor and Oxygen sensor heating
4. **Combination valve**
5. **Bolt**
   - Tighten to 10 Nm (7 ft-lb)
6. **Bolt**
   - Tighten to 25 Nm (18 ft-lb)
7. **Nut**
   - Tighten to 25 Nm (18 ft-lb)
   - Always replace
8. **Flare nut**
   - Tighten to 40 Nm (30 ft-lb)
9. **Connecting pipe for secondary air combination valve**
10. **Union**
    - Tighten to 60 Nm (44 ft-lb)
11. **Bolt**
    - Tighten to 25 Nm (18 ft-lb)
12. **Mounting**
13. **Oxygen sensor 2 (G130)**
    - Tighten to 50 Nm (37 ft-lb)
    - Coat threads with high-temp anti-seize compound; do not allow compound to enter slots on sensor body
14. **4-pin harness connector**
    - For Oxygen sensor and Oxygen sensor heating
15. **Front exhaust pipe with catalyst**
16. **Bolt**
    - Tighten to 10 Nm (7 ft-lb)
17. **To center muffler**
18. **Heat shield**
19. **Bolt**
    - Tighten to 5 Nm (44 in-lb)
20. **Speed nut**
21. **Nut**
    - Tighten to 25 Nm (18 ft-lb)
    - Always replace
22. **Gasket**
    - Always replace
23. **Exhaust manifold**
24. **Gasket**
    - Always replace

Exhaust manifold (2.5L)

1. **Gasket**
   - Always replace
2. **Nut**
   - Tighten to 25 Nm (18 ft-lb)
   - Always replace
3. **Exhaust manifold**
   - Coat cylinder head studs with hot bolt anti-size paste before installing manifold
4. **Clip**
5. **Warm air collector plate**
6. **Bolt**
   - Tighten to 10 Nm (7 ft-lb)
7. **Oxygen sensor 1 (G39)**
   - Tighten to 55 Nm (41 ft-lb)
   - Coat threads with high-temp anti-seize compound; do not allow compound to enter slots on sensor body
8. **Brace**
9. **Nut**
   - Tighten to 25 Nm (18 ft-lb)
   - Always replace
10. **Bolt**
    - Tighten to 25 Nm (18 ft-lb)
    - ace

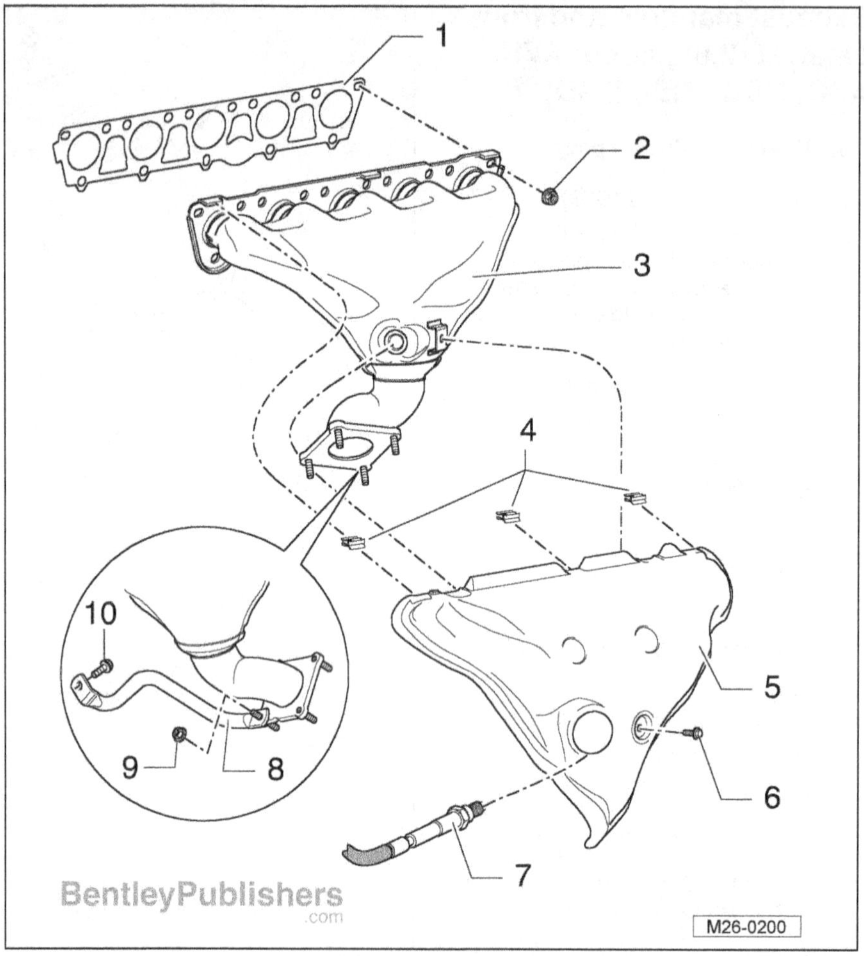

Front exhaust pipe and catalyst (2.5L)

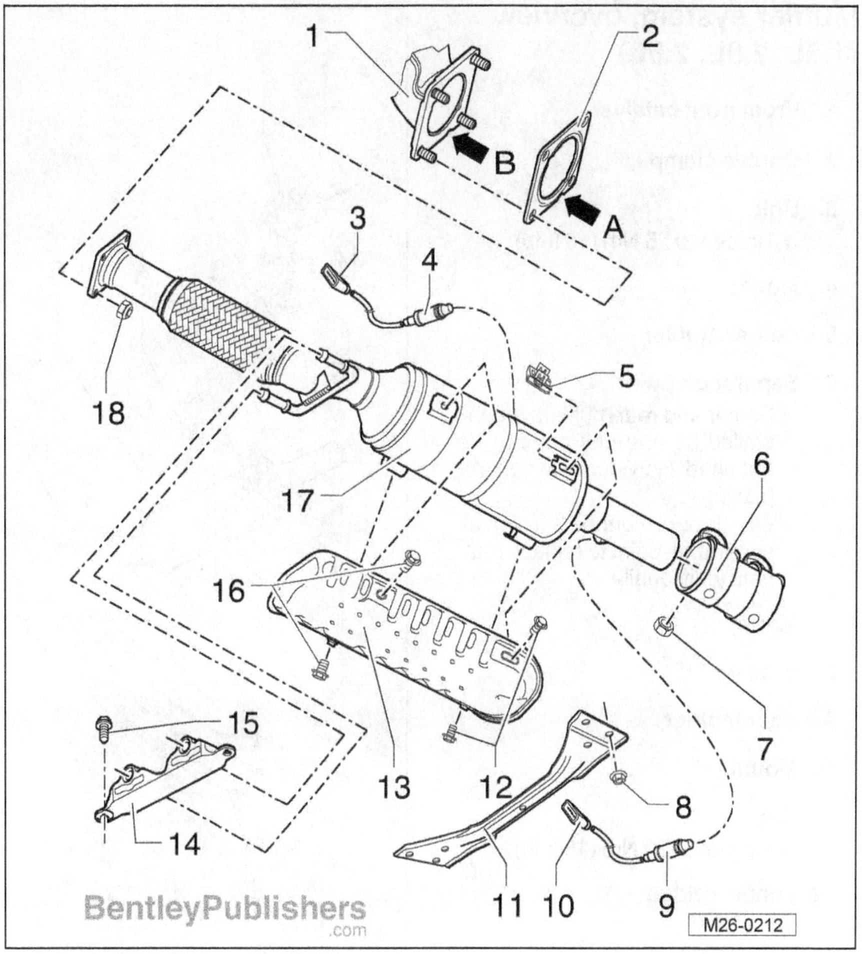

1. **Exhaust manifold**
2. **Gasket**
   - Always replace
   - Note installation position: Recess in gasket (**arrow A**) fits into recess in manifold (**arrow B**)
3. **4-pin harness connector**
   - BPR engine only
   - Black
   - For oxygen sensor and oxygen sensor heating
4. **Oxygen sensor 3 (G465)**
   - BPR engine only
   - Tighten to 55 Nm (41 ft-lb)
   - Coat threads with high-temp anti-seize compound; do not allow compound to enter slots on sensor body
5. **Spring nut**
6. **Double clamp**
   - Tighten bolts evenly
   - Bolts must not project beyond bottom edge of clamp
7. **Nut**
   - Tighten to 25 Nm (18 ft-lb)
8. **Nut**
   - Tighten to 23 Nm (17 ft-lb)
9. **Oxygen sensor 2 (G130)**
   - Tighten to 50 Nm (37 ft-lb)
   - Coat threads with high-temp anti-seize compound; do not allow compound to enter slots on sensor body
10. **4-pin harness connector**
    - Brown
    - For oxygen sensor and oxygen sensor heating
11. **Front cross member**
12. **Bolt**
    - Tighten to 5 Nm (44 in-lb)
13. **Heat shield**
    - Not installed on all vehicles
14. **Mounting**
15. **Bolt**
    - Tighten to 25 Nm (18 ft-lb)
16. **Bolt**
    - Tighten to 10 Nm (7 ft-lb)
17. **Front exhaust pipe with catalyst**
    - Protect from shock and impact damage
    - With flex joint
    - Flex joint must not bend more than 10°
    - Install exhaust system free of stress
18. **Nut**
    - Tighten to 25 Nm (18 ft-lb)
    - Always replace

Muffler system, overview (1.8L, 2.0L, 2.5L)

1. **From front catalyst**
2. **Double clamp**
3. **Bolt**
   - Tighten to 25 Nm (18 ft-lb)
4. **Mount**
5. **Center muffler**
6. **Separation point**
   - Center and rear mufflers are installed as one unit but can be obtained individually when replacing
   - Saw through connecting pipe at separation point to replace mufflers individually
7. **Mount**
8. **Bracket**
9. **Rear muffler**
10. **Mount**
11. **Nut**
    - Tighten to 20 Nm (15 ft-lb)
12. **Tunnel bridge**
13. **Nut**
    - Tighten to 40 Nm (30 ft-lb)
14. **Washer**

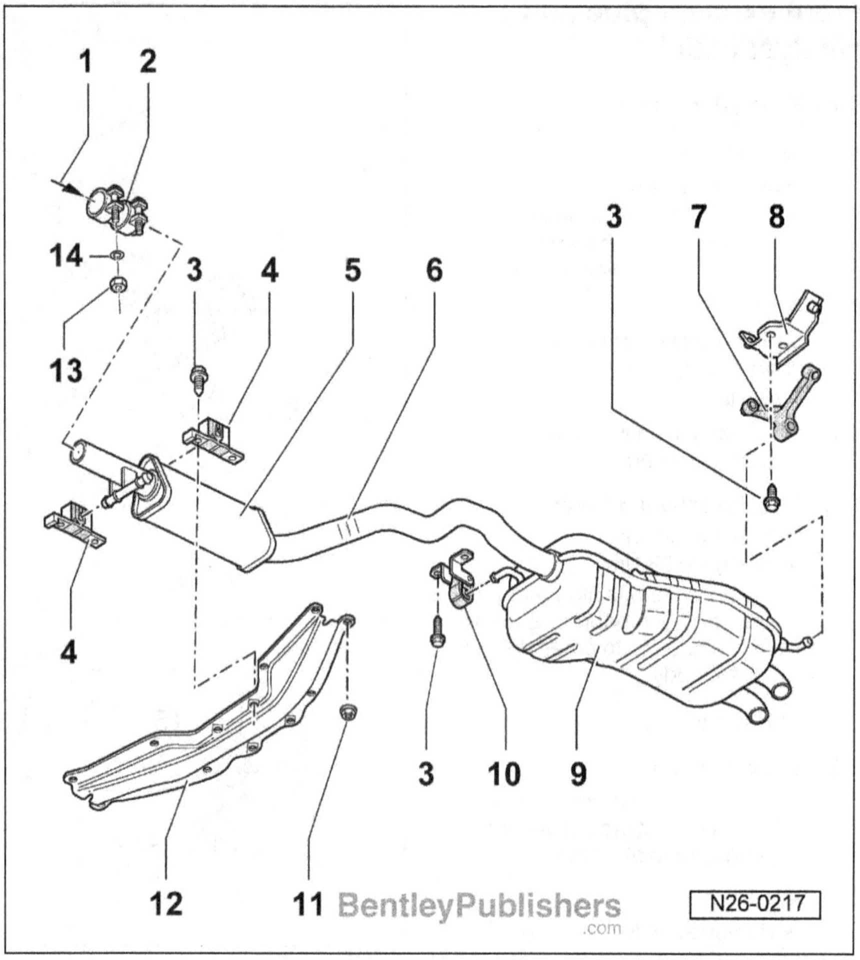

Muffler system, overview (1.9L)

1. **Nut**
   - Tighten to 40 Nm (30 ft-lb)
2. **From catalyst**
3. **Double clamp**
4. **Bolt**
   - Tighten to 25 Nm (18 ft-lb)
5. **Mount**
6. **Mount**
7. **Bracket**
8. **Mount**
9. **Rear muffler**
10. **Separation point**
    - Center and rear mufflers are installed as one unit but can be obtained individually when replacing
    - Saw through connecting pipe at separation point to replace mufflers individually
11. **Nut**
    - Tighten to 25 Nm (18 ft-lb)
12. **Tunnel bridge**

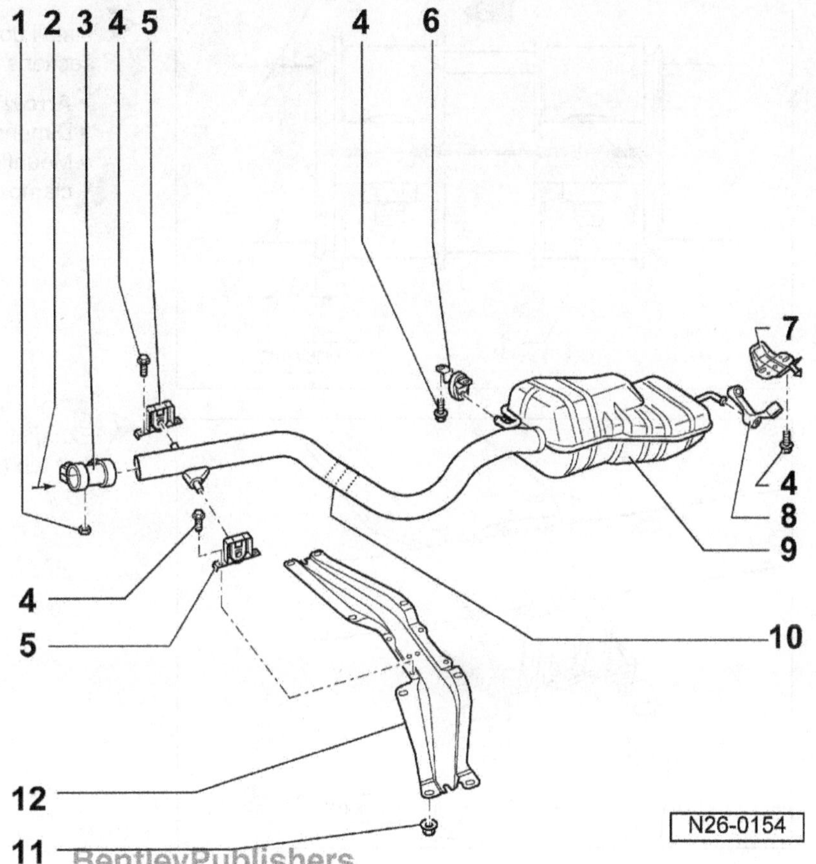

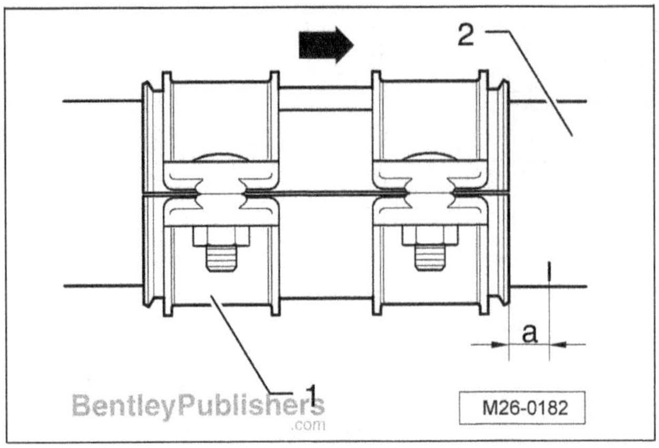

Exhaust system, installation details

Install double clamp that attaches front and rear exhaust sections as shown.

- **Arrow** points in direction of travel
- Dimension **a** = approx. 5 mm
- Mounting bolts should not extend below bottom of clamp

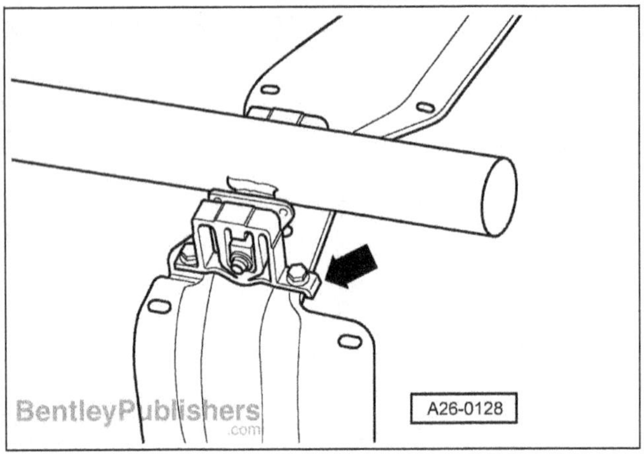

Exhaust mount located behind double clamp. Ensure that tab (**arrow**) on foot of mount points forward.

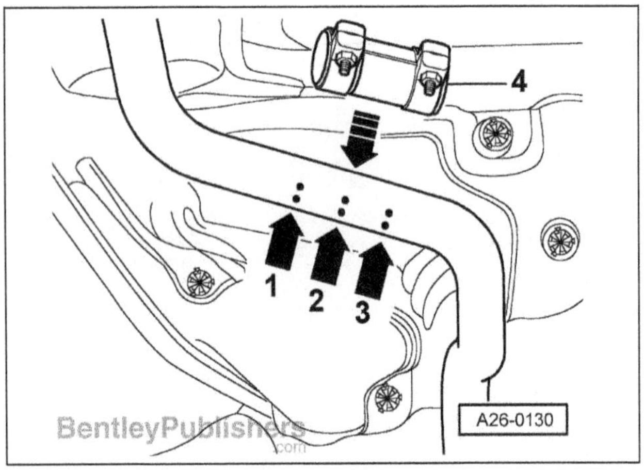

Exhaust pipe separation point. Cut pipe at right angles at **arrow 2**. Repair clamp (**4**) is then positioned between **arrow 1** and **arrow 3**.

Tightening torque
- Exhaust pipe repair clamp bolts. . . . 40 Nm (30 ft-lb)

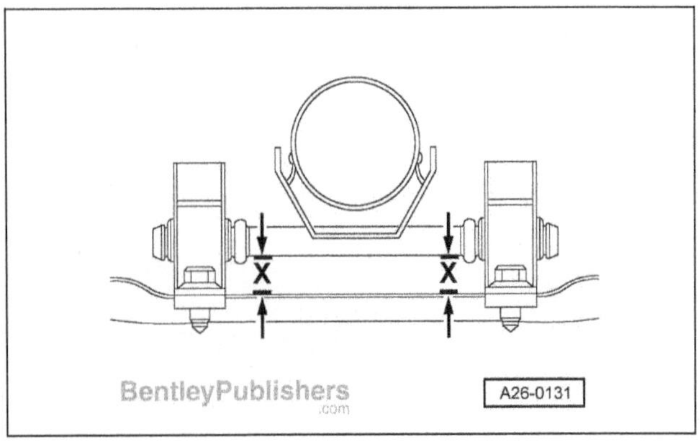

Ensure that rear muffler is aligned stress free. Mounting pins on exhaust pipe must run parallel with tunnel bridge (dimension X should be equal on left and right sides).

Exhaust manifold, removing (2.0L)

NOTE —
2.0L engine code: AEG shown. Other engine codes are similar.

– Remove engine cover.

< Remove intake air hose (**arrow**).

– Remove pressure and vacuum hoses from secondary air injection valve (combination valve). See **Secondary air injection system, components**, given later.

NOTE —
The combination valve is located behind the upper intake manifold.

– Remove connecting pipe between combination valve and exhaust manifold. See **Exhaust manifold and front catalyst (2.0L)**, given earlier.

– Remove warm air collector plate. See **Exhaust manifold and front catalyst (2.0L)**, given earlier.

< Remove retaining bolts (**arrow**) and right side inner CV joint protective cover (if installed).

– Unbolt right side inner CV joint and secure drive axle with stiff wire.

– Remove nuts securing front exhaust pipe to exhaust manifold.

– Unbolt exhaust manifold support brace from engine block.

– Remove exhaust manifold bolts (qty. 8) and manifold with support brace still attached.

– Installation is the reverse of the removal noting the following:

- Replace all gaskets and self-locking nuts/bolts.
- See **Exhaust manifold and front catalyst (2.0L)** given earlier, for tightening torques.
- Use high-temp anti-seize compound on threaded fasteners to extend their service life and make future replacement easier.

– Be sure to quality check your work, see **Quality Review** at the end of this repair group.

EMISSION CONTROLS

> *NOTE —*
> *Volkswagen identifies electrical components by a letter and/or a number in the electrical schematics. See* **EWD Electrical Wiring Diagrams**. *These electrical identifiers are listed in parenthesis as an aid to electrical troubleshooting.*

Evaporative emission (EVAP) controls, schematic diagram

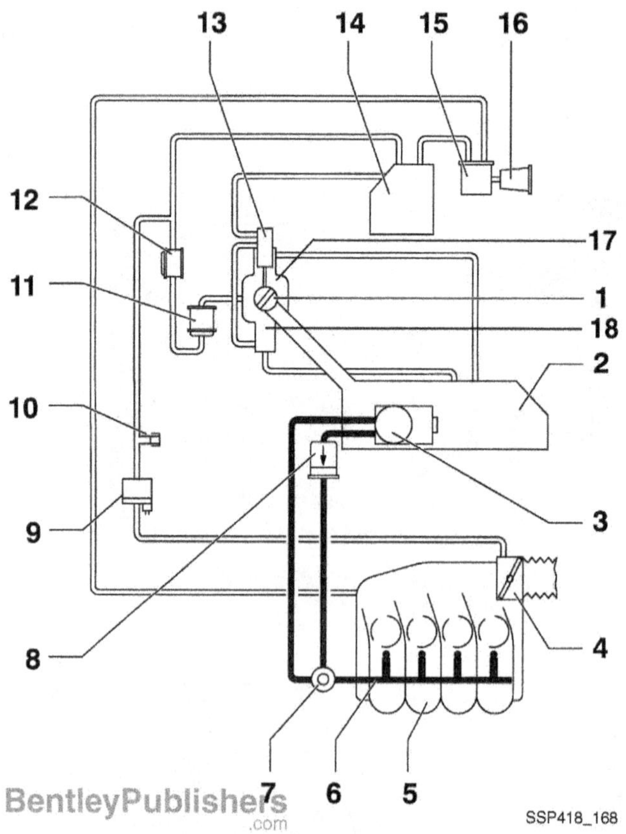

1. **Filler flap**
   - Location: at top of fuel filler neck
2. **Fuel tank**
3. **Fuel pump**
   - Location: in fuel delivery unit
4. **Throttle valve**
5. **Intake manifold**
6. **Fuel rail with injectors**
7. **Fuel pressure regulator**
8. **Fuel filter**
9. **EVAP canister purge regulator valve (N80)**
10. **Pressure test port (IM 240)**
11. **Roll-over valve**
12. **Pressure holding valve**
13. **Breather valve**
14. **EVAP canister**
15. **Leak detection pump (LDP)**
16. **LDP filter**
17. **Operating breather bottle**
18. **Filler breather bottle**

Positive crankcase ventilation (PCV)

The PCV system traps crankcase vapors and routes them back into the intake air stream to be burned.

(1.8L engine)

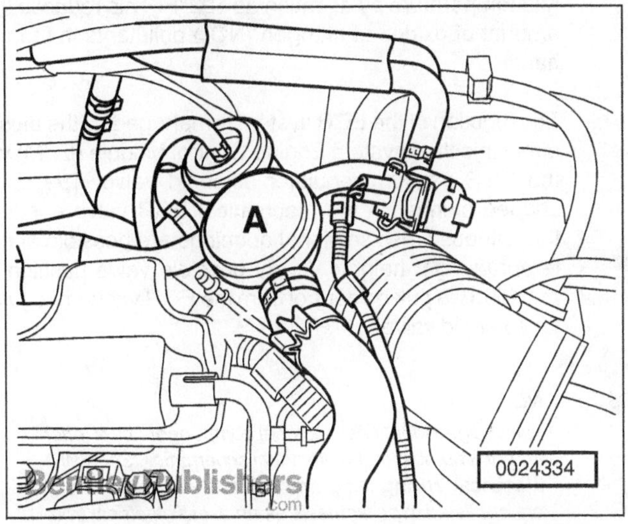

◀ The system consists of a breather hose between the crankcase and the intake air boot. This breather hose has a pressure regulating valve (**A**) at the intake air boot.

(2.0L engine)

◀ The system consists simply of a breather hose (**arrow**) between the valve cover and the intake air boot. The valve cover contains a flame trap to prevent ignition of the crankcase vapors in the event of a backfire. To prevent icing during cold weather, a heating element is integrated into the breather hose.

> ***CAUTION—***
> *Replace PCV hoses only with parts designed for PCV or fuel system service. Conventional vacuum and heater hoses deteriorate rapidly when exposed to oil vapors and combustion gasses.*

Most crankcase ventilation problems result when the hoses or valves become clogged with oily residues. Restrictions create excessive crankcase pressure that can eventually cause driveability problems. PCV system service is limited to inspecting and cleaning the breather valves, hoses, and replacing faulty parts.

Exhaust gas recirculation (EGR) (1.9L)

The EGR system installed on the 1.9L diesel engine reduces emissions by directing a small quantity of exhaust gases back into the intake manifold to dilute the air/fuel mixture effectively. An EGR cooler reduces the exhaust gas temperature by as much as 122°F. This reduces the amount of oxides of nitrogen (NO_x) pollutants in the exhaust.

The function of the EGR system is managed by the diesel direct injection system Engine Control Module (ECM) via the EGR vacuum regulator solenoid valve. The cone shaped plunger in the mechanical EGR valve ensures that various cross sectional openings are possible at different plunger heights. Every possible valve position is provided via pulsed control from the EGR vacuum regulator solenoid valve.

NOTE —
Volkswagen identifies electrical components by a letter and/or a number in the electrical schematics. See **EWD Electrical Wiring Diagrams**. *These electrical identifiers are listed in parenthesis as an aid to electrical troubleshooting.*

EGR vacuum regulator solenoid valve (N18), checking (1.9L)

NOTE —
1.9L engine code: ALH shown, other engines similar.

- Remove engine cover.
- Disconnect pipe between charge air cooler and intake manifold at intake manifold.
- Disconnect vacuum hose from EGR valve.
- Connect hand vacuum pump, such as VAG 1390 or equivalent, to EGR valve.

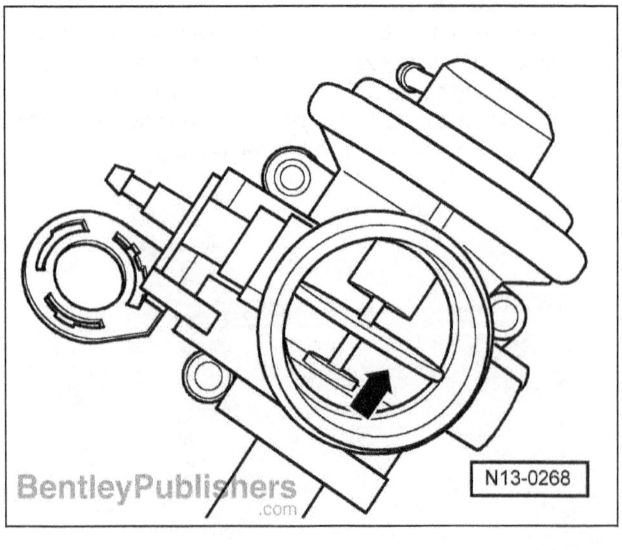

- Operate pump and observe if membrane rod moves.
- Disconnect hand vacuum pump hose from EGR valve.
- Membrane rod must move back to its original position.
- If the membrane rod did not move then the EGR valve is faulty and should be replaced. If the rod does move correctly then the problem may be with the EGR vacuum regulator solenoid valve. Check the wiring to the solenoid valve and check for a duty cycle (on-off) signal at the valve. If no faults can be found, the signal from the ECM may be missing or the valve or ECM may be faulty.

EGR system and components (1.9L, code: ALH)

1. **Intake manifold**
2. **O-ring**
   - Always replace
3. **Throttle valve**
   - With EGR valve and control flap
4. **EGR vacuum regulator solenoid valve (N18)**
   - Can only replace with intake support assembly
   - See **Vacuum hose connections (1.9L engine)**
5. **Bolt**
   - Tighten to 10 Nm (7 ft-lb)
6. **Intake air from air cleaner**
7. **Gasket**
   - Always replace
8. **Bolt**
   - Tighten to 25 Nm (18 ft-lb)
9. **Connecting pipe**
   - EGR cooler to intake support
10. **Connecting pipe**
    - EGR cooler to exhaust elbow
11. **Exhaust manifold**
12. **Nut**
    - Tighten to 25 Nm (18 ft-lb)
13. **To heat exchanger**
14. **EGR cooler**
15. **Connection**
    - From heater
16. **Connection**
    - From coolant expansion tank

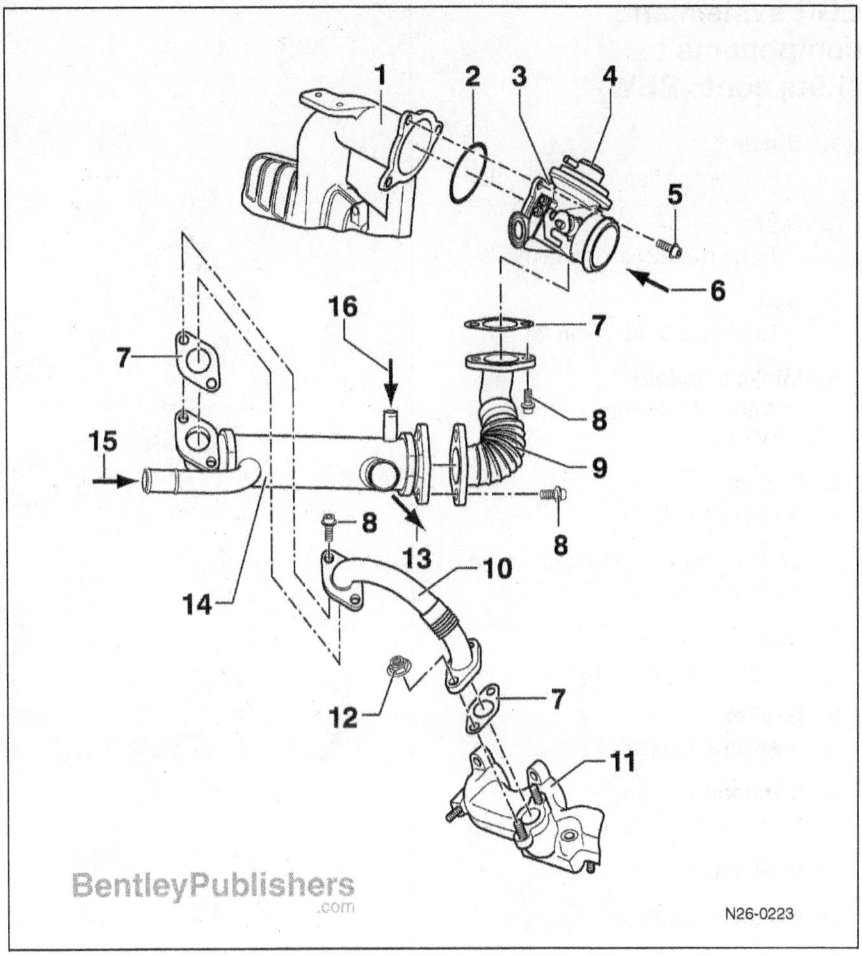

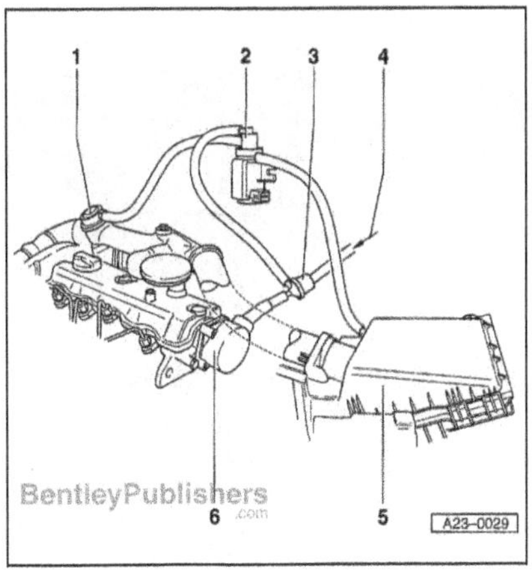

Vacuum hose connections (1.9L engine)

1. EGR valve
2. EGR vacuum regulator solenoid valve (N18)
3. Check valve
4. from Brake servo
5. Air cleaner
6. Vacuum pump

EGR system and components (1.9L, code: BEW)

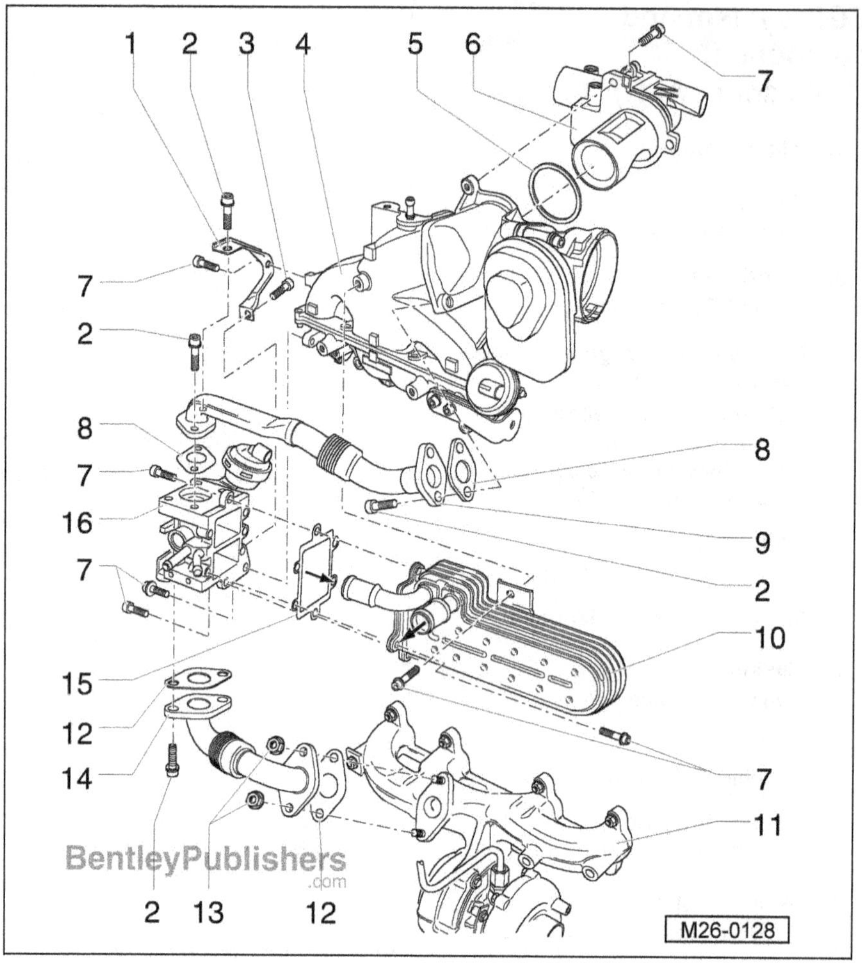

1. **Bracket**
   - For bypass flap for EGR cooler
2. **Bolt**
   - Tighten to 22 Nm (16 ft-lb)
3. **Bolt**
   - Tighten to 5 Nm (44 in-lb)
4. **Intake manifold**
   - With motor for intake flap (V157)
5. **O-ring**
   - Always replace
6. **EGR vacuum regulator solenoid valve (N18)**
7. **Bolt**
   - Tighten to 10 Nm (7 ft-lb)
8. **Gasket**
   - Always replace
9. **Connecting pipe**
   - For intake manifold bypass flap
10. **EGR cooler**
11. **Exhaust manifold**
12. **Gasket**
    - Always replace
13. **Bolt**
    - Tighten to 22 Nm (16 ft-lb)
14. **Connecting pipe**
    - For turbocharger bypass flap
15. **Gasket**
    - Always replace
16. **Turbocharger bypass flap**
    - Allows exhaust gas flow into EGR cooler when EGR switch-over valve (N345) is activated

Secondary air injection system

The gasoline engines covered by this manual have secondary air injection (AIR) systems These systems use an air pump to inject fresh air behind the exhaust valves for approximately 100 seconds during cold starting.

Also, after each subsequent engine start (up to 96°C (205°F) max. engine coolant temperature), the secondary air injection system will, after a 20 second delay, switch in for 10 seconds during idle and is then monitored by the system.

The purpose of the AIR system is to reduce exhaust emissions during engine warm-up when the Motronic engine management system is in open loop. The AIR system produces an oxygen rich exhaust gas, causes afterburning and reduces the duration of the catalyst heat-up phase. Activation initiates from the Motronic ECM via the secondary air injection pump relay to the secondary air injection solenoid valve and combination valve.

The system consists of the electric air pump, the vacuum-operated shut-off valve (combination valve), the inlet (solenoid) valve, and the related duct work.

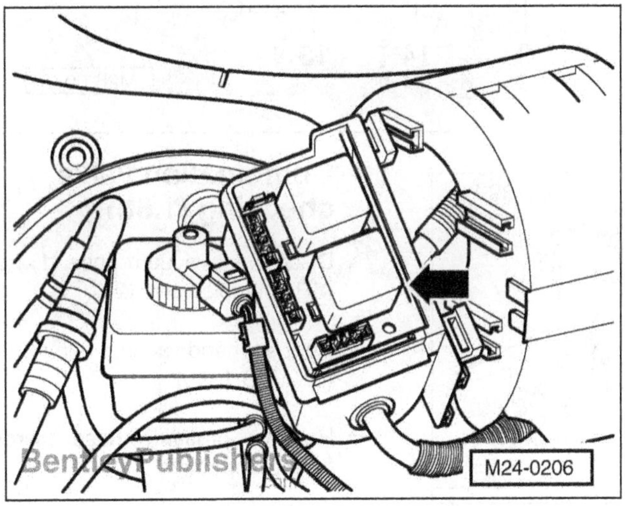

The secondary air injection pump relay (J299) (**arrow**) is located in the relay housing at rear of engine compartment.

NOTE —

- *Volkswagen identifies electrical components by a letter and/or a number in the electrical schematics. See* **EWD Electrical Wiring Diagrams**. *These electrical identifiers are listed in parenthesis as an aid to electrical troubleshooting.*
- *If the following DTCs (Diagnostic Trouble Codes) are present:*
  P0410 - Sec. Air Inj. (AIR) Sys., Malfunction
  P0411 - Sec. Air Inj. (AIR) Sys., Incorrect Flow Detected
  P1200 - Recirculating Valve for Turbocharger (N249) Mechanical malfunction
  P1423 - Bank 1 Sec. Air Inj. (AIR) Sys., throughput too small
  The cause may be cracked or broken vacuum hoses. Replace secondary air system vacuum hoses as necessary.
- *See* **OBD On Board Diagnostics** *for more information on DTCs.*

Secondary air injection system, components (1.8L engine)

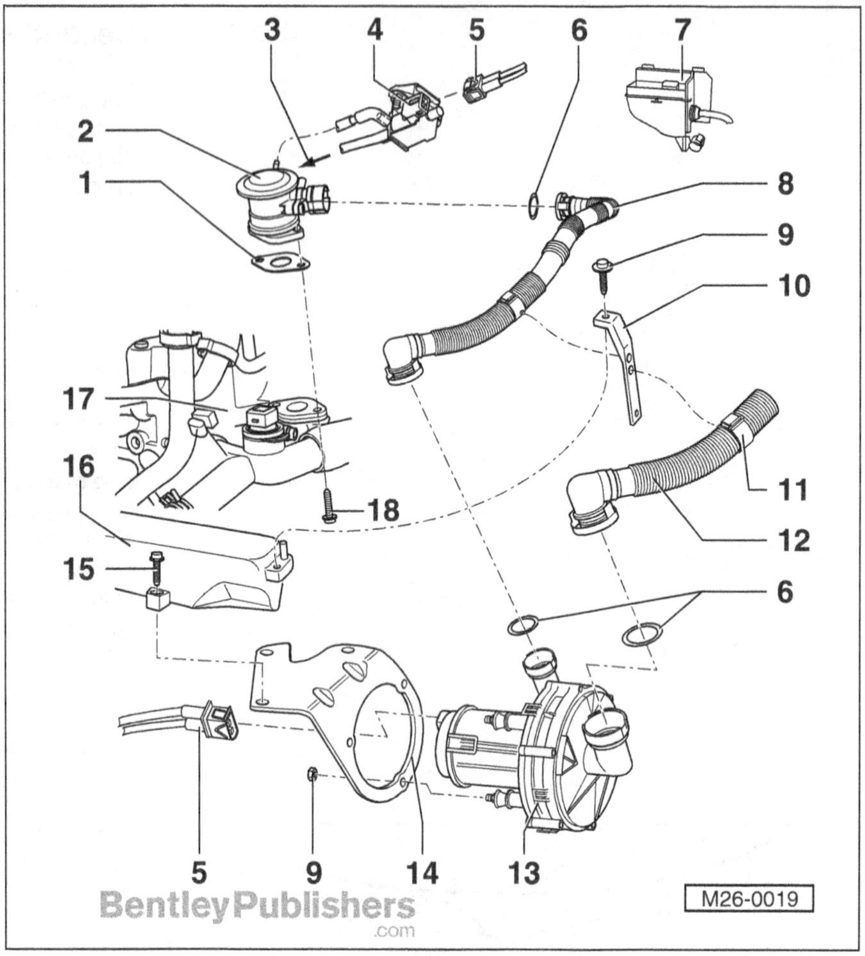

1. **Gasket**
   - Always replace
2. **Combination valve**
   - See **Combination valve, checking (1.8L)**
3. **Vacuum hose**
   - To distributor piece at vacuum reservoir on cylinder head cover
4. **Secondary air injection solenoid valve (N112)**
   - Fastened to cylinder head cover
5. **2-pin harness connector**
6. **O-ring**
   - Replace if damaged
7. **Secondary air injection pump motor relay (J299)**
   - In relay panel at rear of engine compartment
8. **Pressure hose**
   - Ensure hose is tightly sealed
   - Press together at front to release
9. **Bolt**
   - Tighten to 10 Nm (7 ft-lb)
10. **Bracket**
11. **Retaining clip**
12. **Intake hose**
    - From upper air filter housing
    - Ensure hose is tightly sealed
    - Press together at front to release
13. **Secondary air injection pump motor (V101)**
14. **Bracket**
    - For secondary air pump motor
15. **Bolt**
    - Tighten to 25 Nm (18 ft-lb)
16. **Intake pipe**
17. **Cylinder head**
18. **Bolt**
19. **Tighten to 10 Nm (7 ft-lb)**

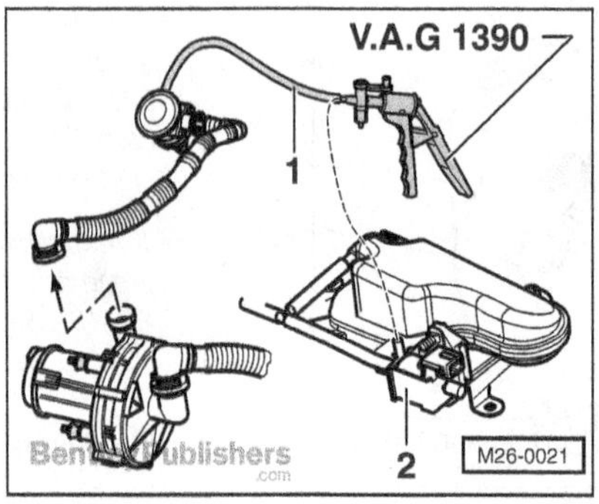

Combination valve, checking (1.8L)

◀ Disconnect vacuum hose (**1**) from AIR solenoid valve (**2**).

– Connect hand vacuum pump. (i.e. VAG 1390)

– Remove pressure hose (**arrow**) from pump motor.

– Blow into pressure hose with slight pressure. Do not use compressed air.

– Both valves must be closed.

– Operate hand vacuum pump.

– Combination valve should open, replace if doesn't open.

Secondary air injection system, components (2.0L, code: AEG, AVH, AZG, BEV)

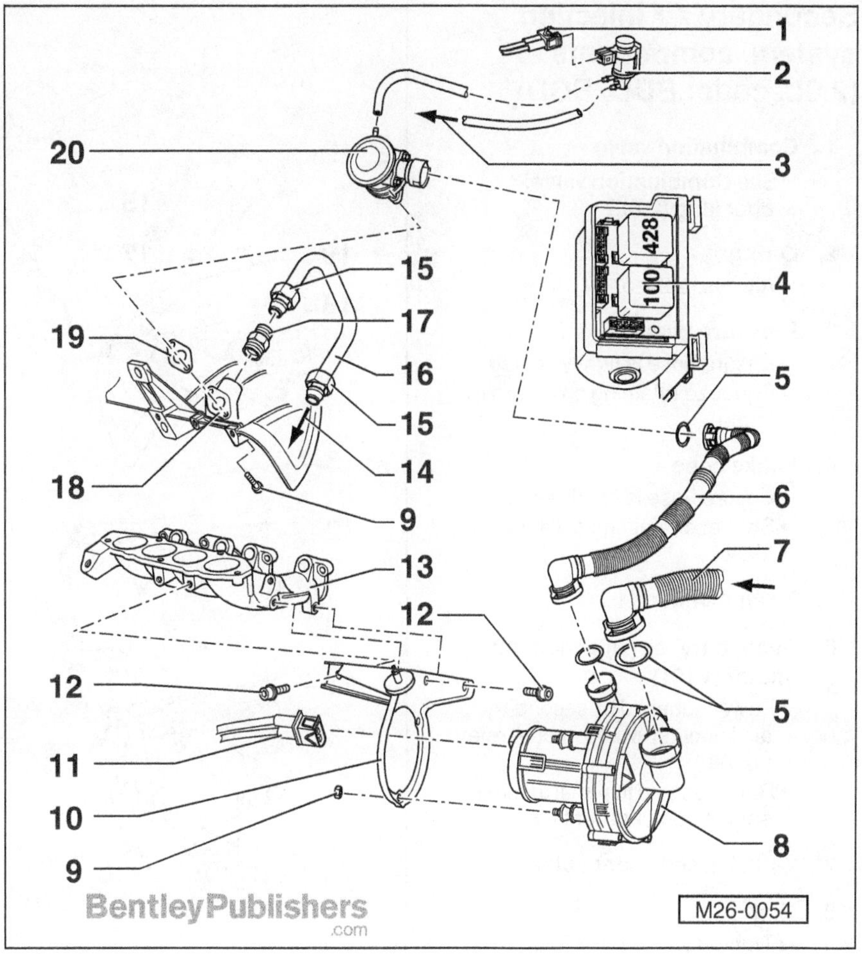

1. **2-pin harness connector**
   - Through 1999 m.y. 2.0L, code: AEG engine only
2. **Secondary air injection solenoid valve**
   - Through 1999 m.y. 2.0L, code: AEG engine only
   - Fastened to close-out panel
3. **to Brake booster**
4. **Secondary air injection pump motor relay (J299)**
   - In relay panel at rear of engine compartment
5. **O-ring**
   - Always replace
6. **Pressure hose**
   - Ensure hose is tightly sealed
   - Squeeze retaining collar to release
7. **Intake hose**
   - Ensure hose is tightly sealed
   - Squeeze retaining collar to release
8. **Secondary air injection pump motor (V101)**
9. **Bolt**
   - Tighten to 10 Nm (7 ft-lb)
10. **Bracket**
    - Attached to intake manifold
11. **2-pin harness connector**
12. **Bolt**
    - Tighten to 25 Nm (18 ft-lb)
13. **Intake manifold, lower section**
14. **To union on exhaust manifold**
15. **Flare nut**
    - Tighten to 30 Nm (22 ft-lb)
16. **Connecting pipe for combination valve**
17. **Union**
    - Tighten to 35 Nm (26 ft-lb)
18. **Warm air collector plate**
    - With mount for combination valve
19. **Gasket**
    - Always replace
20. **Combination valve**
    - See **Combination valve, checking (2.0L)**
    - No connection for secondary air inlet on 2.0L, code: AEG from 2000 m.y., and all other 2.0L engines

Secondary air injection system, components (2.0L, code: BDC, BGD)

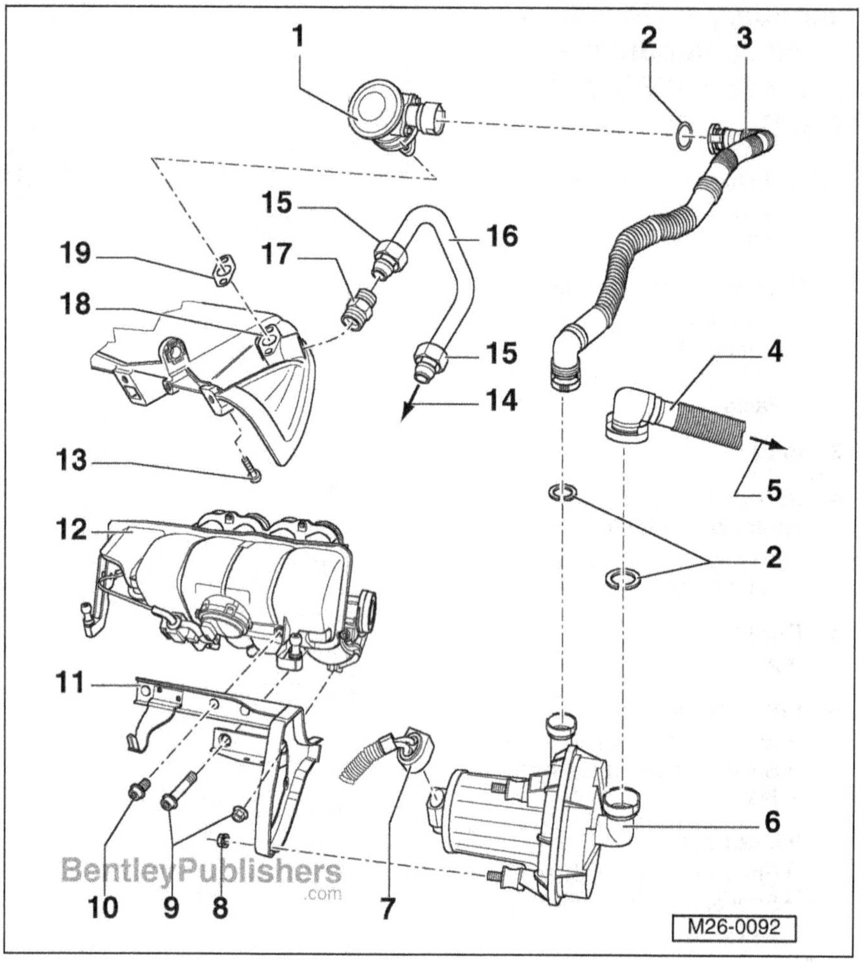

1. **Combination valve**
   - See **Combination valve, checking (2.0L)**
2. **O-ring**
   - Always replace
3. **Pressure hose**
   - Ensure hose is tightly sealed
   - Squeeze retaining collar to release
4. **Intake hose**
   - Ensure hose is tightly sealed
   - Squeeze retaining collar to release
5. **To air cleaner**
6. **Secondary air injection pump motor (V101)**
   - Power supplied by secondary air injection pump motor relay (J299)
   - Relay located in panel at rear of engine compartment
7. **2-pin harness connector**
8. **Bolt**
   - Tighten to 10 Nm (7 ft-lb)
9. **Bolt**
   - Tighten to 25 Nm (18 ft-lb)
10. **Bolt**
    - Tighten to 15 Nm (11 ft-lb)
11. **Bracket**
    - Attached to intake manifold
12. **Intake manifold, lower section**
13. **Bolt**
    - Tighten to 10 Nm (7 ft-lb)
14. **To union on exhaust manifold**
15. **Flare nut**
    - Tighten to 40 Nm (30 ft-lb)
16. **Connecting pipe for combination valve**
17. **Union**
    - Tighten to 60 Nm (44 ft-lb)
18. **Warm air collector plate**
    - With mount for combination valve
19. **Gasket**
    - Always replace

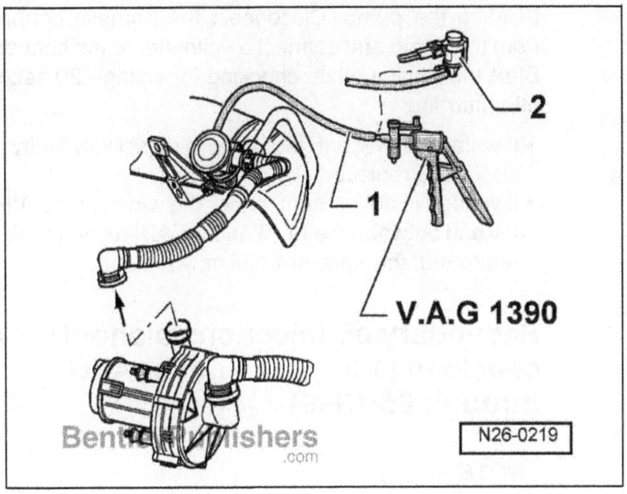

Combination valve checking (2.0L)

2.0L AEG engine through 05/1999

- Disconnect vacuum hose (**1**) from AIR solenoid valve (**2**)
- Connect hand vacuum pump. (i.e. VAG 1390)
- Remove pressure hose (**arrow**) from pump motor.
- Blow into pressure hose with slight pressure Do not use compressed air!
- Both valves must be closed.
- Operate hand vacuum pump.
- Combination valve should open, replace if doesn't open.

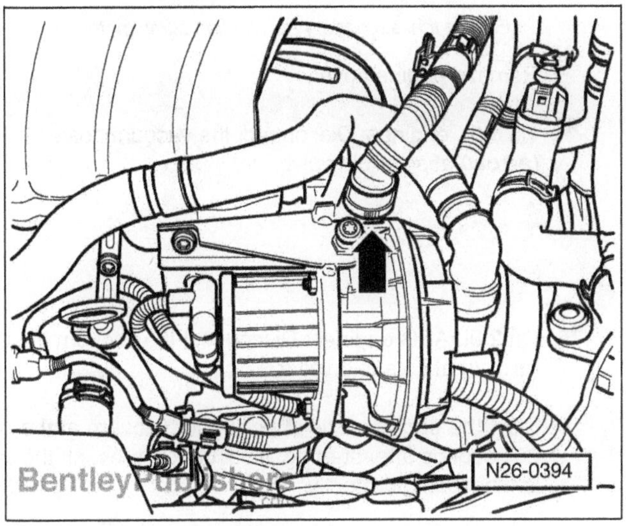

2.0L AEG engine from 06/1999, AVH, AZG, BDC, BEV, BGD

- Remove engine cover.
- Remove pressure hose (**arrow**) from secondary air injection pump motor.
- Blow into pressure hose with light pressure.
  - Both valves must remain closed.
- Blow into pressure hose with higher pressure.
  - Combination valve must open

NOTE —
Do not use compressed air to test combination valve, use mouth to blow into pressure hose.

- If valve doesn't open, replace.

Secondary air injection pump, testing

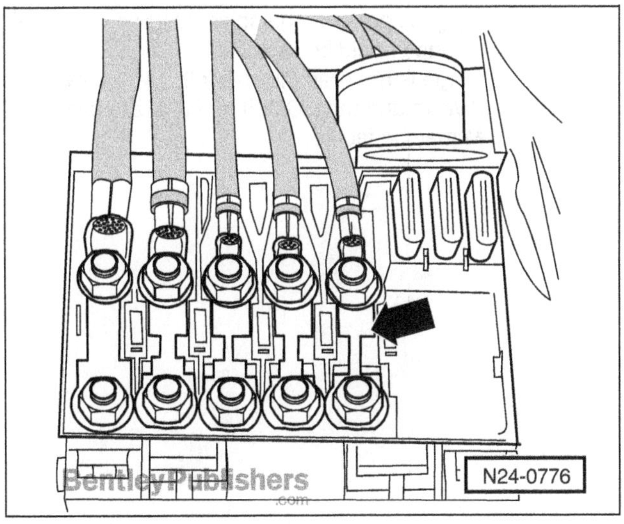

- Remove pressure hose (**arrow**) from secondary air injection pump motor.
- With the engine fully warm, start and allow the engine to idle. Approximately 20 seconds after starting, the air pump should run for about five seconds and secondary air should be felt coming from the pump outlet.
  - If the air pump does not run, check the fuse (**arrow**) on top of the battery.
  - If fuse is OK check the secondary air injection pump relay (J299). See **97 Fuses, Relays and Component Locations**.
- If no faults are found, continue testing.

– If the secondary air injection pump does not run and no wiring faults can be found, check the signal from the ECM to the pump. Disconnect the harness connector from the pump and connect a voltmeter to the connector. Start the engine while checking for voltage 20 seconds after starting.
  - If voltage is present, the pump is most likely faulty and should be replaced.
  - If voltage is not present, check the wiring at the AIR relay and between the ECM and the AIR relay. If no faults are found, the relay or ECM may be faulty.

Secondary air injection solenoid valve, checking (1.8L, 2.0L, code: AEG through 05/1999)

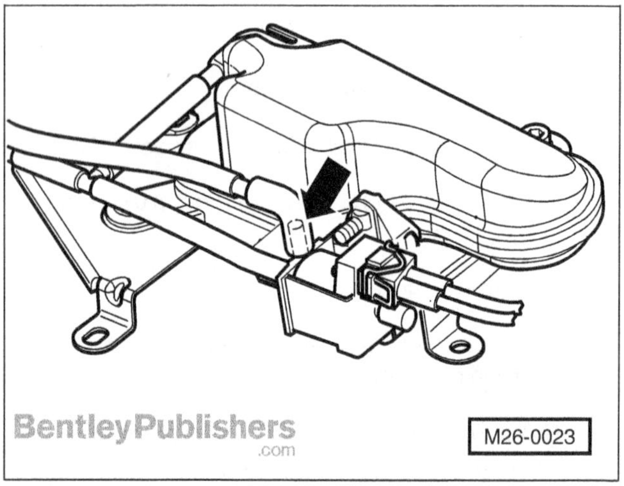

NOTE —
- *Engine oil temperature should be between 5–33°C (41–91°F).*
- *Ensure that secondary air injection pump motor is OK.*

– Remove engine cover.

◄ On 1.8L engines: Disconnect the vacuum hose (**arrow**) at solenoid valve.

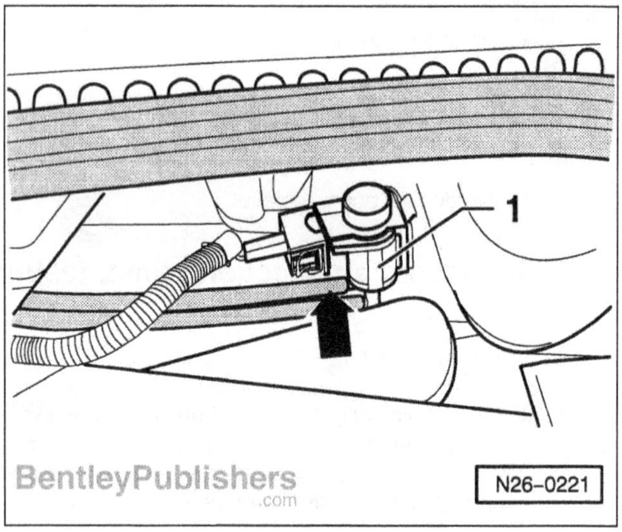

◄ On 2.0L AEG engines: Disconnect the vacuum hose (**arrow**) at solenoid valve.

– Start the engine while checking for vacuum at the hose 20 seconds after starting. If vacuum is present, the valve is functioning correctly.

– If vacuum is not present, shut off engine and disconnect the harness connector from the solenoid valve and connect a voltmeter to the harness connector. Start the engine while checking for voltage 20 seconds after starting.
  - If voltage is present, the solenoid valve is most likely faulty and should be replaced.
  - If voltage is not present, check the wire between solenoid valve and the ECM. If no faults are found, the ECM may be faulty.

Combination valve, removing and installing (1.8L, 2.0L)

NOTE —
• Always replace all seals and gaskets.

• Installation is best accomplished using a helper.

– Remove engine cover.

– Remove intake hose between air cleaner and secondary air injection pump motor.

– Remove pressure and vacuum hoses from combi-valve (1.8l APH and 2.0L AEG engines).

– Unbolt connecting pipe from combination valve.

– Working underneath the car, remove protective cover from above right inner CV joint (if applicable) as shown earlier.

– Unbolt connecting pipe from exhaust manifold.

– Unscrew combination valve from warm air collector plate using 4 mm ball-headed allen wrench.

– To install, thread front bolts through the warm air collector plate from below.

– Have helper hold fastener tight from above and position combination valve during installation of second bolt.

Tightening torque
• Combination valve to warm air collector plate. . . . . 10 Nm (7 ft-lb)

Catalytic converter

The three-way catalytic converter chemically reduces pollutants in the engine exhaust. A properly operating converter provides a 90-95 percent reduction of the three major exhaust gas pollutants (nitrogen (NO_x), carbon monoxide (CO), and unburned hydrocarbons (HC).

The catalytic converter works correctly only when the air/fuel ratio is kept within a very narrow range. The oxygen sensors provides feedback for control of the air/fuel mixture.

The catalytic converter itself does not require any routine maintenance. However, any problem that increases converter temperature beyond its normal operating range (incorrect ignition timing or air/fuel mixture, engine misfire, prolonged idling or extended high engine loads) can damage the converter. Reduced power, stalling, exhaust system rattles and excessive emissions are symptoms that may be caused by a faulty catalytic converter.

Catalytic converters on gasoline engines covered by this manual use a second heated oxygen sensor to monitor the operation of the catalyst. If the converter fails to properly oxidize the spent combustion gasses, the heated oxygen sensor will detect this malfunction and a fault code (DTC) will be stored in the ECM. Beginning in 2006, the 2.5L engine (code: BPR) uses a third oxygen sensor in the center of the catalytic converter (Super Ultra Low Emissions (SULEV) requirement).

Exhaust leaks anywhere in the system, but especially between the cylinder head and the catalytic converter, can cause running problems, false DTCs or converter failure.

> ***WARNING —***
> *Do not operate the starter for long periods if the engine fails to start. Extended cranking may allow excess fuel to enter the catalytic converter, creating a fire hazard and possibly damaging the converter.*

Accurately testing the function of the catalytic converter requires an infrared exhaust gas analyzer to check the air/fuel mixture adjustment (% CO) at the exhaust system test port upstream of the catalytic converter, and then again at the tailpipe. If the converter is working properly, the tailpipe reading will be lower than the test port reading. If the reading is the same or only slightly less, the converter is probably faulty.

QUALITY REVIEW

When you have finished working under the hood and around other areas of the vehicle, it is advisable to take a moment to quality check or review your work. This helps to insure that the operation or repair has been completed properly with all affected systems functioning within normal parameters. These may include the following:

- Ensure that the exhaust system is not under stress, and that it has sufficient clearance from the body. If necessary, loosen the double clamps and align the muffler and exhaust pipe so sufficient clearance is maintained to the body while the support rings are evenly loaded.
- Confirm that all heat shields are in place.
- Make sure that there are no air, vacuum or exhaust leaks.
- Make sure that all components involved in the repair are positioned correctly and function properly.
- Make sure all tools and shop cloths are removed.

27 Engine Electrical

TABLES

GENERAL

This repair group covers battery, starter, and generator (alternator) troubleshooting and repair.

NOTE —
- *The alternator is identified as generator by the vehicle manufacturer. Car makers must by law use uniform words and terms (nomenclature) to describe certain components and or systems. The uniform word for alternator is generator. Therefore, the component labeled "Generator (GEN)" in the wiring diagrams is the alternator.*
- *Battery service, including electrolyte filling can be found in* **03 Maintenance**.
- *Wiring diagrams for the battery, charging system and starter motor are given in* **EWD Electrical Wiring Diagrams**.

The six-cell, 12-volt lead-acid battery capacity is rated in Ampere/hours (Ah) and cold cranking amps (CCA). The Ah rating is determined by the average amount of current the battery can deliver over time without dropping below a specified voltage. The CCA rating is determined by the battery's ability to deliver starting current at 0°F (–18°C). The battery is installed in the engine compartment, behind the left headlight assembly.

The charging system consists of a belt-driven 14-volt generator and a voltage regulator. The voltage regulator, which is mounted in the generator, also serves as the generator brush holder. The charging system provides the current necessary to keep the battery charged and to operate the vehicle's electrical accessories.

Please read the following warnings and cautions before doing any work on any parts of the engine electrical system.

WARNING —

- *Wear goggles, rubber gloves, and a rubber apron when working around batteries and battery acid (electrolyte). Battery acid contains sulfuric acid and can cause skin irritation and burning. If acid is spilled on your skin or clothing, flush the area at once with large quantities of water. If electrolyte gets into your eyes, bathe them with large quantities of clean water for several minutes and call a physician.*
- *Batteries that are being charged or are fully charged give off explosive hydrogen gas. Keep sparks and open flames away. Do not smoke.*

CAUTION —

- *Disconnecting the negative (–) battery cable may erase fault codes and basic settings in the engine management and automatic transmission control modules. Some driveability problems may be noticed until the system re-adapts to operating conditions. OBD II readiness codes, which may be required for emissions testing, may also be erased.Convenience electronics (alarm system, interior light control, power locks, mirrors, and windows) may need to be re-set using a diagnostic scan tool.*
- *Do not disconnect the battery cables while the engine is running. The generator will be damaged.*
- *Never operate the generator with its output terminal (B+ or 30) disconnected and the other terminals connected. Never short, bridge, or ground any terminals of the charging system.*
- *Always disconnect the negative (–) battery cable when working at or near the generator. Battery voltage is always present at the rear of the generator, even with the ignition key off.*
- *Always disconnect the battery cables during battery charging. This will prevent damage to the generator and any solid-state components. Do not exceed 16.5 volts at the battery.*
- *Never reverse the battery terminals. Even a momentary wrong connection can damage the generator or electrical components.*
- *Replace the battery if the case is cracked or leaking. Electrolyte can damage the car. If electrolyte is spilled, clean the area with a solution of baking soda and water.*

BATTERY SERVICE

If the battery discharges when the vehicle is not driven, there may be a constant drain or current draw causing it to discharge when the ignition is off. Depending on the draw and the condition of the battery, a full discharge can happen overnight or it may take a few weeks. Although a small static drain on the battery is normal (for example to operate the clock or radio memory), a large drain such as a relay sticking on or a faulty switch will cause the battery to quickly discharge. Make a static current draw test as the first step when experiencing battery discharge.

If the current draw on the battery is not excessive and the battery still discharges, the condition of the battery should be tested. Battery testing determines the state of battery charge. The most common methods are open-circuit and load voltage testing. Batteries with filler caps can also be tested by checking the specific gravity of the electrolyte. Inexpensive specific gravity testers are available at most auto supply stores.

Some New Beetles are equipped with batteries with a central gas venting system with anti-flash protection. The anti-flash protection consists of a small round fiberglass mat with a diameter of approximately 15 mm and a thickness of 2 mm. Its purpose is to vent gasses that form in the battery. The gasses that form while charging can vent centrally through an opening on the upper side of the battery cover. The anti-flash protection is also installed to prevent ignition of the flammable gasses in the battery

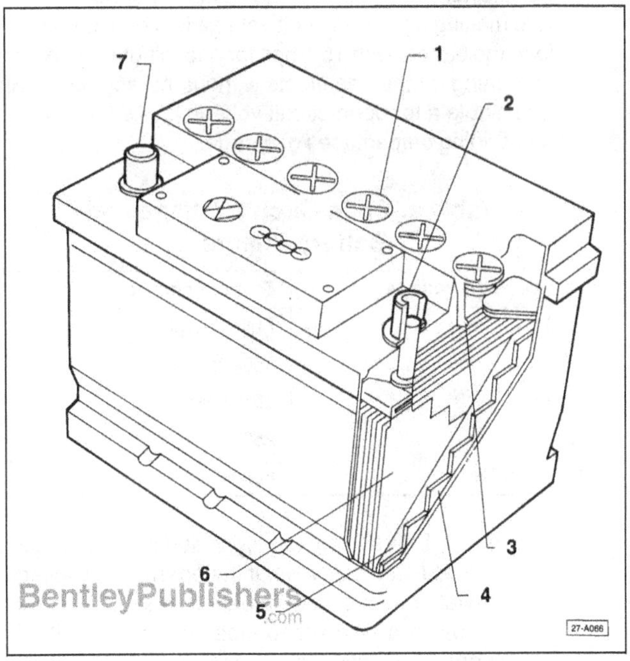

Battery

1. Filler cap
2. Positive (B+) terminal
3. Electrolyte level indicator
4. Negative plate (grey)
5. Separator (insulator)
6. Positive plate (dark brown)
7. Negative (–) terminal

CAUTION—

- *Use new generation batteries with central gas venting when replacing the battery.*
- *Always use genuine VW battery cell caps. Caps must be fitted with an O-ring seal.*
- *Before diagnosing electrical problems visually inspect the battery for clean and tight connections at all terminals.*

Static current draw, checking

– Make sure the ignition and all electrical accessories are switched off.

– Disconnect the negative (–) cable from the battery. See the Cautions at the beginning of this repair group regarding battery disconnection.

NOTE —
Be sure to have the anti-theft radio code on hand before disconnecting the battery.

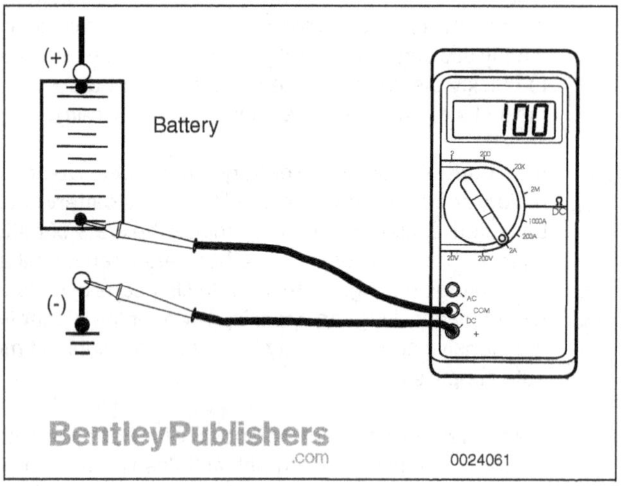

< Connect a digital ammeter between the battery negative post and the negative battery cable and measure the current draw.

A range of about 0 to 100 milliamps is normal, depending on the number of accessories that need constant power. A current of 500 milliamps (0.5 amp) or more indicates a problem. To determine the circuit or component causing the problem, remove one fuse at a time until the current drops to a normal range. Use the wiring diagrams shown in **EWD Electrical Wiring Diagrams** to locate wiring or component faults.

Open-circuit voltage test

An open-circuit voltage test checks battery voltage by connecting an accurate digital voltmeter to the battery posts after disconnecting the battery ground cable. Before making an open-circuit voltage test on a battery, first load the battery with 15 amps for one minute, for example by turning on the headlights without the engine running. See **Table a** for open-circuit voltage levels and their corresponding percentages of charge.

Table a. Open-Circuit Voltage and Battery Charge

Open-circuit voltage	State of charge
12.6 V or more	Fully charged
12.4 V	75% charged
12.2 V	50% charged
12.0 V	25% charged
11.7 V or less	Fully discharged

The battery is in satisfactory condition if the open-circuit voltage is at least 12.4 volts. If the open-circuit voltage is at this level or above, but the battery still lacks power for starting, make a load voltage test to determine the battery's service condition. If the open-circuit voltage is below 12.4 volts, recharge the battery. If the battery cannot be recharged to at least 75%, it should be replaced.

Load voltage testing

A load voltage battery test is made by connecting a specific resistive load to the battery terminals and then measuring the battery's voltage. The test requires a special tester and can generally be performed quickly and inexpensively by an authorized VW dealer or other qualified repair facility.

The battery should be fully charged and at room temperature for the most accurate results. If the equipment is available, disconnect the negative (–) battery cable. Then apply the specified load for 15 seconds and measure the battery's voltage. If the voltage is below that listed, the battery should be replaced. **Table b** lists load current and minimum voltages for original-equipment VW batteries.

> ***WARNING —***
> *Always wear protective goggles and clothing when performing a load test.*

Table b. Battery Load Current and Minimum Voltage

Battery Capacity Amp-hour (Ah)	Cold Cranking Amps (CCA)	Load Current	Minimum voltage
36 Ah	340	100	10.0
40 Ah - 49 Ah	220	200	9.2
50 Ah - 60 Ah	265 - 280	200	9.4
61 Ah - 80 Ah	300 - 380	300	9.0
81 Ah - 110 Ah	380 - 500	300	9.5

Battery charging

Discharged batteries can be recharged using a battery charger. Prolonged charging causes gassing that will evaporate the electrolyte to a level that can damage the battery.

Always read and follow the instructions provided by the battery charger's manufacturer. A slow-charging rate (10% of battery capacity) is best to prevent battery damage caused by overheating.

> ***WARNING —***
> - *The gasses given off by the battery during charging are explosive. Do not smoke. Keep open flames away from the top of the battery, and prevent electrical sparks by turning off the battery charger before connecting or disconnecting it.*
> - *Battery cell caps must not be removed while charging.*
> - *Batteries are damaged by quick charging. It should only be used as a last alternative when slow charging is not possible.*
> - *Never quick charge a totally discharged battery.*
> - *Do not allow the battery charging rate to exceed 16.5 volts.*
> - *If the battery begins gassing (boiling) violently when charging, reduce the charging rate immediately.*
> - *Ensure that the battery is charged in a well ventilated area.*
> - *Precision tools must not be kept in the same room where batteries are being charged. Tools may corrode due to chemical reaction.*

Battery, disconnecting terminals

1998-2002

– Switch ignition off.

◄ Press locking tabs (**arrows**) together to open hinged battery cover. Fold the top part of the battery blanket up.

– Disconnect battery ground (GND) strap from battery negative (–) terminal.

> ***CAUTION —***
> *Be sure to have the anti-theft radio code on hand before disconnecting the battery.*

– Open fuse cover on top of battery and remove carefully.

◄ Remove rear wiring cover by pulling upward.

– Remove mounting nut from positive (+) battery terminal lead (**arrow**). Position battery cover out of the way.

Tightening torque
- Cable terminals on top of battery . . . . 6 Nm (53 in-lb)

> ***NOTE —***
> *Incorrect tightening torque of cable terminals on top of battery could cause a no-start condition.*

– Remove positive (+) battery clamp from battery and position cable out of the way.

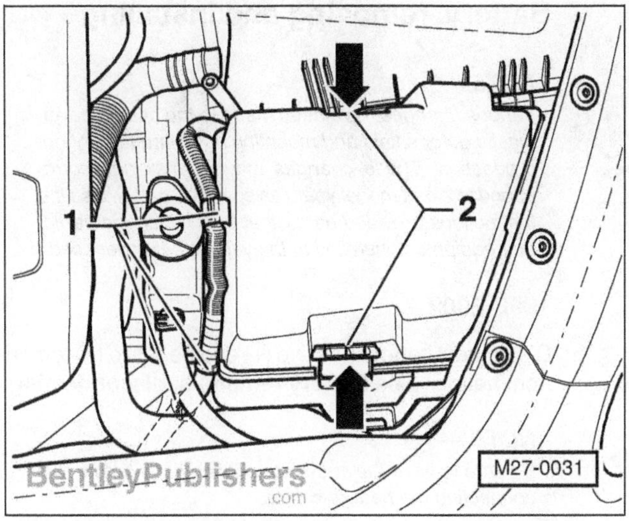

2003 - on

◄ Unclip battery positive (+) cable from retaining clips (**1**).

– Press locking tabs (**arrows**) together to remove battery cover. Fold the top part of the battery blanket up.

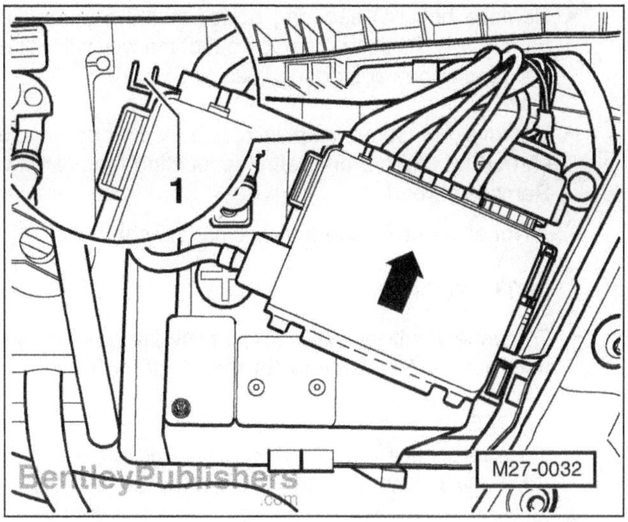

◄ Unclip fuse holder from battery box by pressing lock (**1**) with a screwdriver and pushing fuse holder toward rear (**arrow**).

> ***NOTE—***
> *Illustration shows vehicle with Tiptronic automatic transmission. Other vehicles may vary based on equipment installed.*

– Disconnect negative (–) battery terminal.

– Disconnect positive (+) battery terminal.

Continued for all vehicles

– Reconnect battery positive (+) cable first and negative (–) cable last.

Tightening torque
- Battery cable to battery . . . . 6 Nm (53 in-lb)

> ***CAUTION—***
> *Do not use excessive force to attach battery cable, damage to battery case can result.*

– Install battery cover and fuse box in reverse order of removal.

– After connecting battery cables do the following:

- Check for Diagnostic Trouble Codes (DTCs) using a diagnostic scan tool. See **OBD On Board Diagnostics** section at rear of manual.
- Completely open and close all power windows.
- Re-set radio anti-theft function. See **91 Radio**.
- Re-set clock.
- Check operation of all other electrical accessories.
- See **Procedure following interruption of voltage supply**.

Battery, removing and installing

NOTE —
Various changes have been made to the battery cover, plastic outer shield and mounting hardware throughout production. These changes may, or may not, correspond to exact model years as specified below. Use the instructions provided below that best match the particular equipment installed in the vehicle being serviced.

1998-2002

– Disconnect battery positive (+) and negative (–) terminals from battery. See **Battery terminals, disconnecting**.

NOTE —
Be sure to have the anti-theft radio on hand before disconnecting the battery cables.

◄ Remove bolt (**A**) securing power steering reservoir to battery tray. Position reservoir out of the way without disconnecting power steering lines.

– Remove nut securing plastic battery shroud (**B**), pull shroud off of front and left side locating tabs (**arrows**). Remove shroud.

• Not all models have a plastic battery shroud.

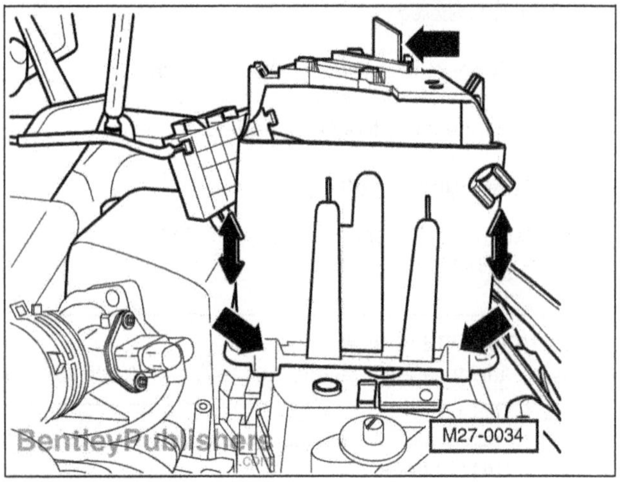

2003 - 2004

– Disconnect battery positive (+) and negative (–) terminals from battery. See **Battery terminals, disconnecting**.

NOTE —
Be sure to have the anti-theft radio on hand before disconnecting the battery cables.

◄ Remove battery shield by pressing retaining clips (**arrows**) and lifting upward (**double arrows**).

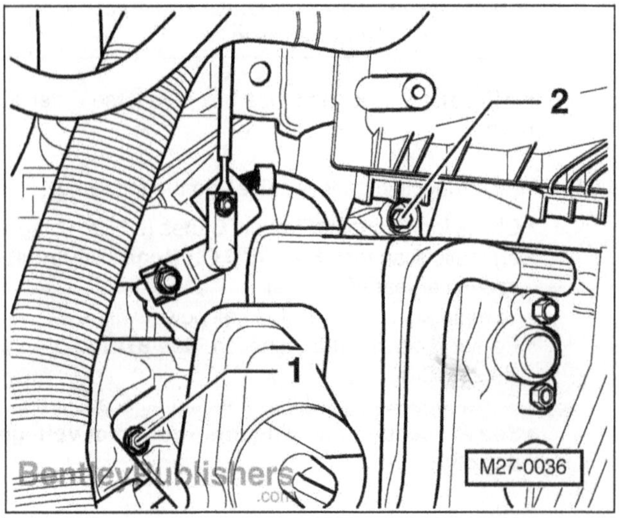

◄ Remove bolt (**1**) securing power steering reservoir to battery tray. Position reservoir out of the way without disconnecting power steering lines.

– Remove clip holder for battery positive (+) cable at power steering reservoir mounting bolt.

– Remove bolt (**2**) securing air cleaner assembly to battery tray. If necessary, remove remaining air cleaner mounting bolts and move air cleaner out of way to access battery for removal.

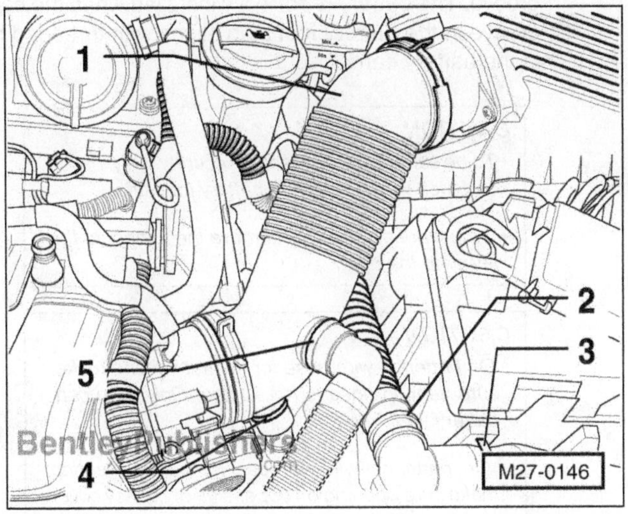

(2006 - on, 2.5L engine)

⮜ Remove connecting hoses (**2**, **4** and **5**) from intake hose (**1**). Remove intake hose.

– Unclip retainer (**3**) from battery box.

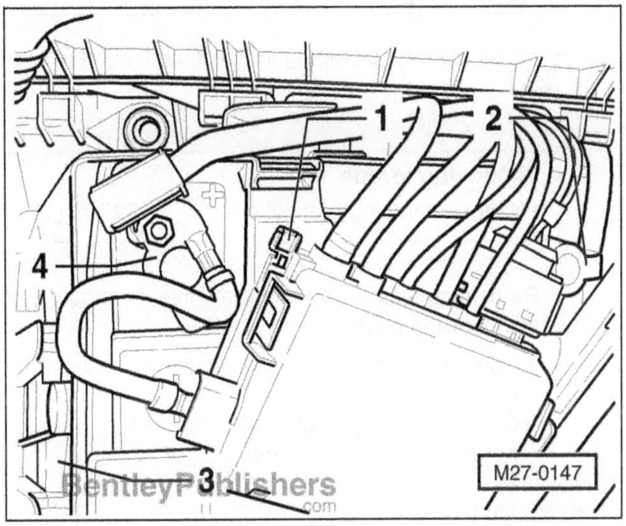

⮜ Depress retainer (**1**) and slide fuse panel cover out from mounts and set aside.

– Disconnect battery positive (+) and negative (–) terminals from battery. See **Battery terminals, disconnecting**.

NOTE —

Be sure to have the anti-theft radio on hand before disconnecting the battery cables.

– Unhook forward part of battery box (**3**) and remove upwards.

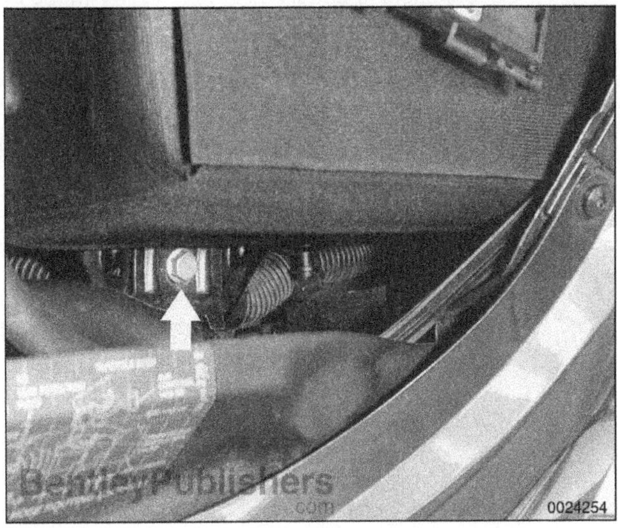

Continued for all models

⮜ Remove battery clamping bracket (**arrow**).

– Remove battery with insulating blanket (if applicable).

• Slide battery to left and then up to remove.

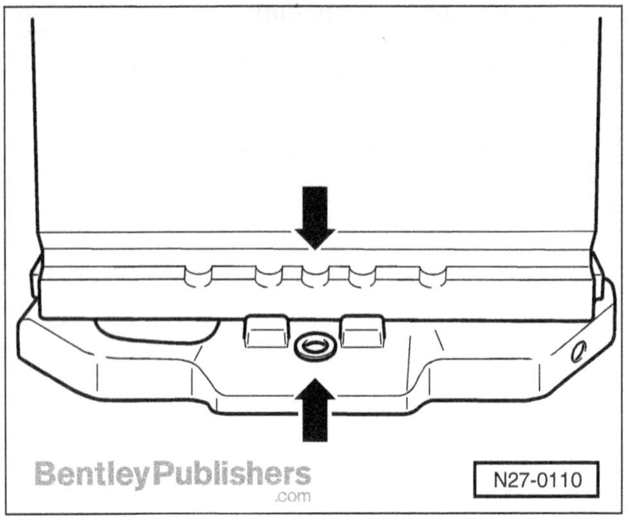

Install battery so lug of battery tray engages in battery groove. The battery is correctly installed when the center groove in the battery foot strip aligns with threaded hole in the battery carrier.

> ***CAUTION —***
> - *To be sure the battery seats securely, only install batteries with a 10.5 mm battery foot strip.*
> - *It must not be possible to move the battery to the left or to the right.*

> ***CAUTION —***
> - *On batteries with hose for central venting make sure venting hose is not kinked. Only then can battery vent freely.*
> - *On batteries without hose for central venting make sure opening on upper side of battery cover is not blocked.*

– Install battery clamping bracket and tighten securing bolt.

Tightening torque
- Battery clamping bracket bolt. . . . . . 22 Nm (16 ft-lb)

– Remaining installation is the reverse of removal, noting the following points:
  - Be sure to properly position plastic battery shroud on locating tabs, as applicable.
  - Connect battery positive (+) cable before installing negative (–) battery clamp.

Tightening torque
- Battery cable to battery . . . . 6 Nm (53 in-lb)

– After connecting battery cables do the following:
  - Check for Diagnostic Trouble Codes (DTCs) using a diagnostic scan tool. See **OBD On Board Diagnostics** section at rear of manual.
  - Completely open and close all power windows.
  - Re-set radio anti-theft function. See **91 Radio**.
  - Re-set clock.
  - Check operation of all other electrical accessories.
  - See **Procedure following interruption of voltage supply**.

> ***CAUTION —***
> *After installing, check if the battery is seated securely. If the battery is not secured properly, the following will result:*
> - *Shortened battery life due to vibration damage*
> - *Battery cell and plate damage*
> - *Electrolyte leakage due to battery case damage*
> - *Poor collision safety*

Procedure following interruption of voltage supply

NOTE —
If the voltage supply is interrupted, the Engine Control Module (ECM) loses all the "learned" and/or "basic" values. As a result, driveability problems may occur depending on vehicle mileage.

– Connect diagnostic scan tool to connector under left side of dash. See **OBD On Board Diagnostics** section at rear of manual.

– Switch ON ignition for at least 10 seconds.

– Switch OFF ignition.

– Use the diagnostic scan tool to adapt the ECM to the throttle valve control module.

– For vehicles equipped with automatic transmissions, the kick down shift point must also be adapted using the diagnostic scan tool.

– Test drive until any driveability problems are eliminated. During the road test the following operating conditions must be fulfilled:

- The engine coolant temperature must exceed 80°C (176°F).
- Once the required temperature is reached, the following operating conditions must be attained several times.
- Closed throttle position
- Part throttle position
- Wide open throttle position
- Overrun

Battery tray, removing and installing

– Remove battery as described earlier in this repair group.

4-cylinder

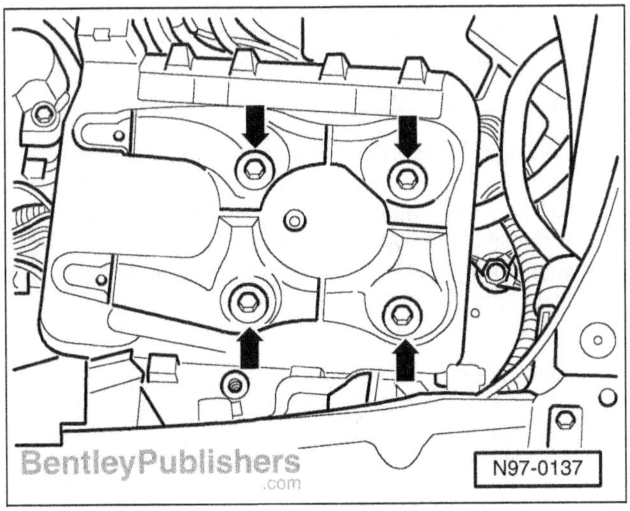

◄ Remove mounting bolts (**arrows**) and battery tray.

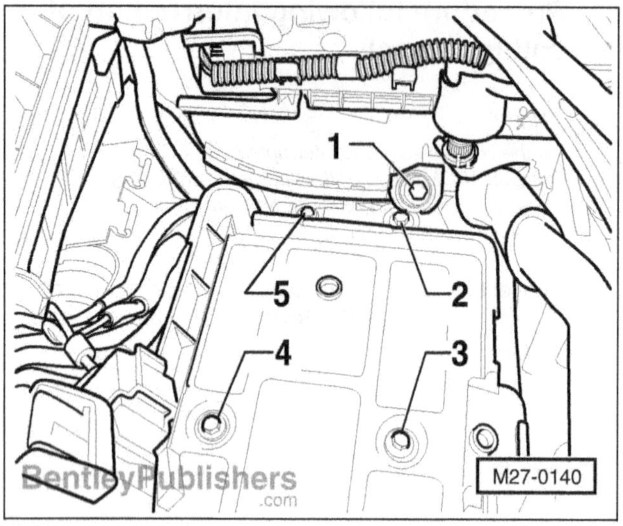

5-cylinder

- Remove mounting bolts (**1 - 5**) and battery tray.
- Install in reverse order of removal.

Tightening torque
- Battery tray to body .... 6 Nm (53 in-lb)

CHARGING SYSTEM SERVICE

Charging system trouble is indicated by an illuminated generator warning light on the instrument panel, or by an under or overcharged battery.

The generator generates electrical current by electrical induction. When the engine is running, part of the current it produces energizes its electromagnetic field. When starting, some other current must be provided to initially energize the field and begin the current generating process. This current is provided by the battery through the generator warning light in the instrument cluster. If the lamp burns out, the generator will not charge the battery properly.

CAUTION—
- *Always disconnect the negative (–) battery cable before servicing the charging system. The large red wire at the rear of the generator comes directly from the battery and is not fuse protected.*
- *Disconnecting the negative (–) battery cable may erase fault codes and basic settings in the engine management and automatic transmission control modules. Some driveability problems may be noticed until the system re-adapts to operating conditions. OBD II readiness codes, which may be required for emissions testing, may also be erased.*

As a quick-check, measure the voltage across the battery terminal with the key off and then again with the engine running. The battery voltage should be about 12.6 volts with key off and approximately 14.0 volts with the engine running. If the voltage does not increase when the engine is running, there is a fault in the charging system.

NOTE—
The regulated voltage (engine running) should be roughly between 13.8 and 14.5 volts, depending on temperature and operating conditions. If the voltage is much higher than 14.5 volts, the voltage regulator is most likely faulty.

Charging system, testing

– Inspect the ribbed V-belt (accessory belt) for cracking, glazing or wear. Replace the belt if any defects are found. Check that belt is not slipping and that belt tension is correct. See **03 Maintenance**.

– Make sure battery is fully charged and capable of holding a charge. Make sure battery terminals are clean and tight.

– Check that charge warning lamp comes on when key is on. If light comes on, check for battery voltage between ground and terminal B+. If light does not come on, repair any wiring or bulb faults before continuing test.

> ***NOTE —***
> *The charge warning light must come on when ignition is switched on or system may not charge. If any faults are found, see* **EWD Electrical Wiring Diagrams** *for electrical schematics.*

– Check for battery voltage between ground and terminal **B+** (large red wire) at back of generator. Check that wire is securely fastened.

– Turn ignition key on and check for battery voltage between terminal **D+** and ground. If voltage is not present, check wiring from warning light in instrument cluster.

– If no faults are found up to this point, have the generator tested with a load tester, such as the Sun VAT40.

> ***NOTE —***
> *If a load test is not possible, an imperfect output test can be done by running the engine at about 2000 rpm and turning on several electrical consumers such as blower fan, lights and rear defrost, wipers, etc. When loaded, the battery should still have at least 12 volts.*

Generator overview (1.8L, 2.0L)

1. **Belt tensioner**
2. **Mounting bracket**
3. **Socket head bolts**
   - M10 x 45mm
   - Tighten to 45 Nm (33 ft lb)
4. **Bolts**
   - M3 x 18mm
5. **End cap**
6. **Screws**
   - M4 x 25mm
7. **Voltage regulator**
   - To remove and install, remove items 4, 5, 6
   - See **Carbon brushes, checking**
8. **Generator**
   - Tighten nut of B+ wire to generator to 15 Nm (11 ft-lb)
9. **Ribbed belt**
   - Removing and installing, see **03 Maintenance**
10. **Bolts**
    - M8 x 85mm
    - Tighten to 25 Nm (18 ft-lb)
11. **Bolts**
    - M8 x 85mm
    - Tighten to 25 Nm (18 ft-lb)

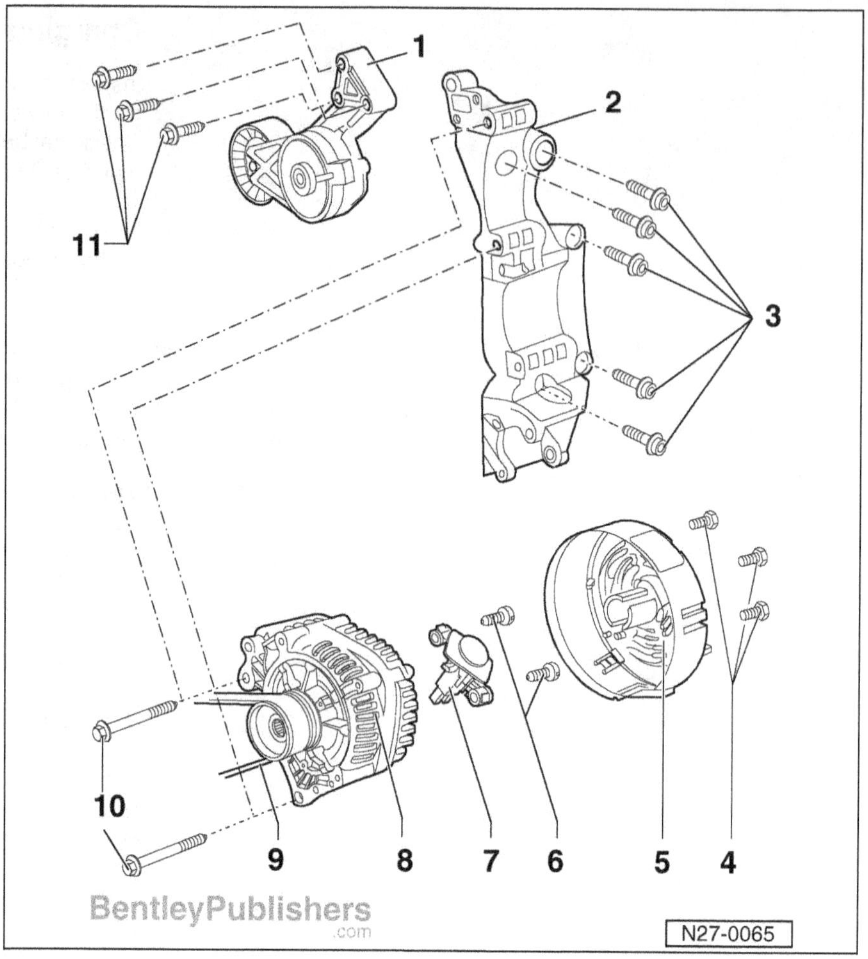

Generator overview (1.9L, code: ALH)

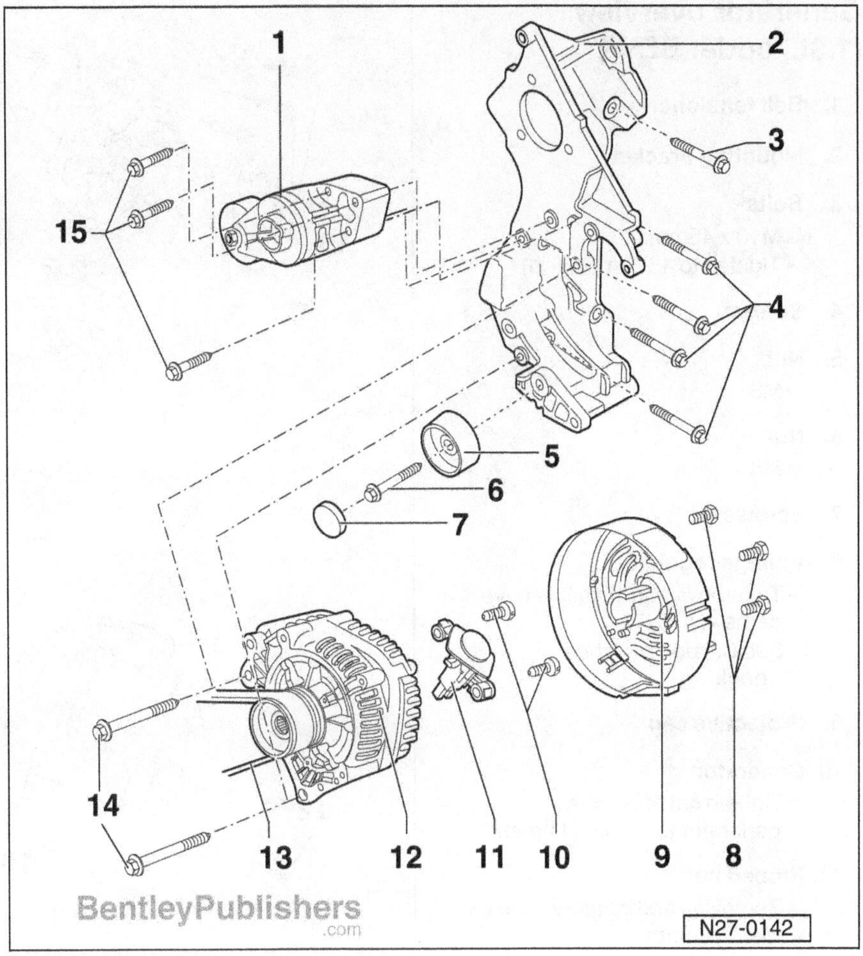

1. **Belt tensioner**
2. **Mounting bracket**
3. **Bolts**
   - M10 x 45mm
   - Tighten to 45 Nm (33 ft lb)
4. **Bolts**
   - M10 x 65mm
   - Tighten to 45 Nm (33 ft lb)
5. **Relay roller**
6. **Bolt**
   - M8 x 50mm
   - Tighten to 25 Nm (18 ft-lb)
7. **Protective cap**
8. **Bolts**
   - M3 x 18mm
9. **End cap**
10. **Screws**
    - M4 x 25mm
11. **Voltage regulator**
    - To remove and install, remove items 8, 9, 10
    - See **Carbon brushes, checking**
12. **Generator**
    - Tighten nut of B+ wire to generator to 15 Nm (11 ft-lb)
13. **Ribbed belt**
    - Removing and installing, see **03 Maintenance**
14. **Bolts**
    - M8 x 85mm
    - Tighten to 25 Nm (18 ft-lb)
15. **Bolts**
    - M8 x 45mm
    - Tighten to 25 Nm (18 ft-lb)

Generator overview (1.9L, code: BEW)

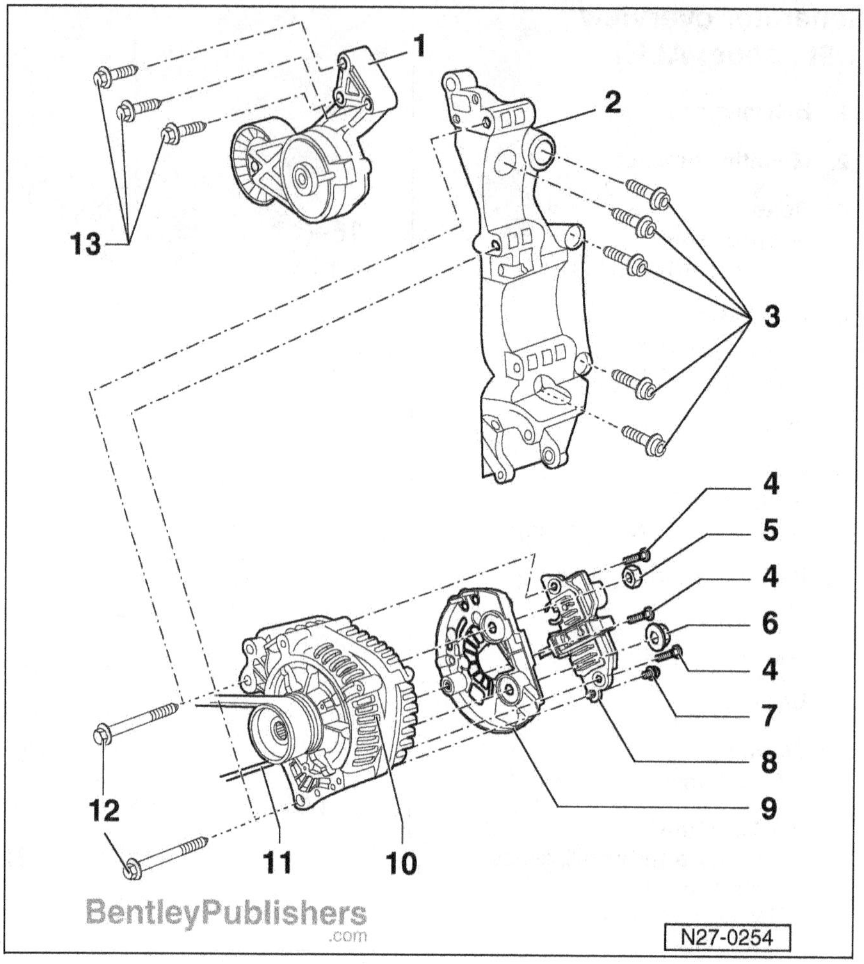

1. **Belt tensioner**
2. **Mounting bracket**
3. **Bolts**
   - M10 x 45mm
   - Tighten to 45 Nm (33 ft lb)
4. **Screws**
5. **Nut**
   - M8
6. **Nut**
   - M8
7. **Screws**
8. **Voltage regulator**
   - To remove and install, remove items 4, 5, 6, 7
   - See **Carbon brushes, checking**
9. **Protective cap**
10. **Generator**
    - Tighten nut of B+ wire to generator to 15 Nm (11 ft-lb)
11. **Ribbed belt**
    - Removing and installing, see **03 Maintenance**
12. **Bolts**
    - M8 x 85mm
    - Tighten to 25 Nm (18 ft-lb)
13. **Bolts**
    - M8 x 45mm
    - Tighten to 25 Nm (18 ft-lb)

Generator overview (2.5L)

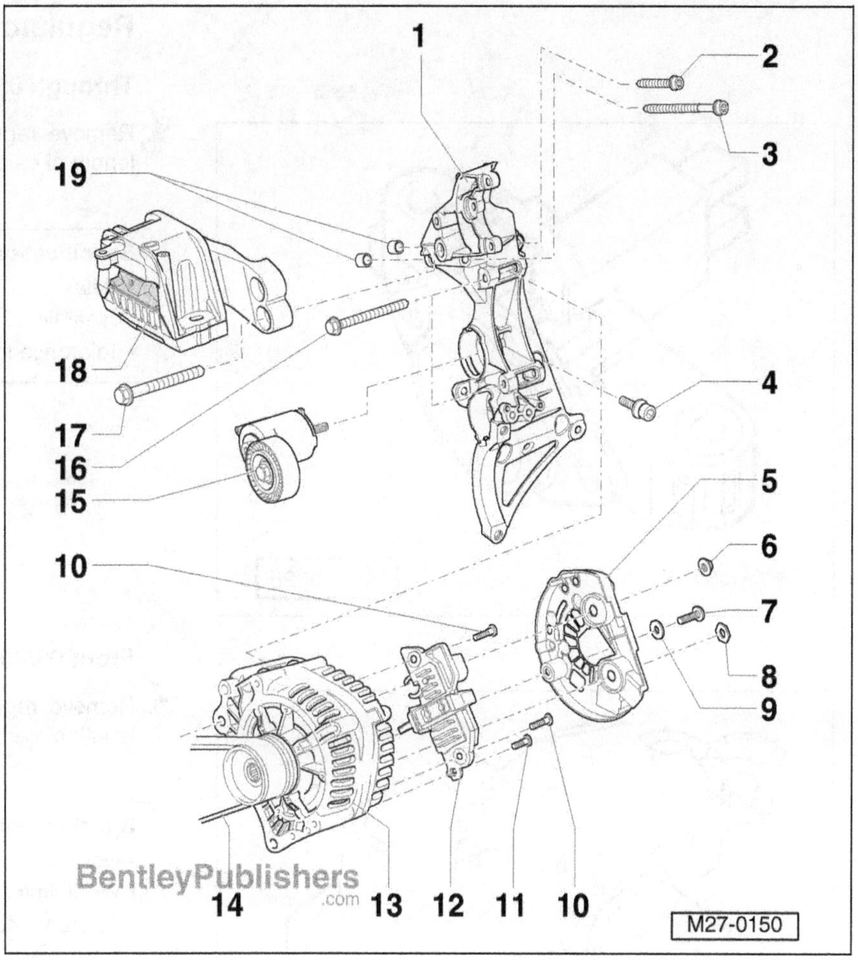

1. **Mounting bracket**
2. **Bolt**
   - Tighten to 25 Nm (18 ft-lb)
3. **Bolt**
   - Tighten to 25 Nm (18 ft-lb)
4. **Bolt**
   - Tighten to 25 Nm (18 ft-lb)
5. **Protective cap**
6. **Nut**
7. **Screw**
   - Tighten to 4.5 Nm (39 in-lb)
8. **Nut**
   - M8
9. **Washer**
   - M5
10. **Screws**
    - M4 x 19mm
    - Tighten to 2 Nm (18 in-lb)
11. **Screws**
    - M4 x 13mm
    - Tighten to 2 Nm (18 in-lb)
12. **Voltage regulator**
    - To remove and install, remove items 5 through 11
    - See **Carbon brushes, checking**
13. **Generator**
    - Tighten nut of B+ wire to generator to 15 Nm (11 ft-lb)
14. **Ribbed belt**
    - Removing and installing, see **03 Maintenance**
15. **Belt tensioner**
16. **Bolt**
    - Tighten to 25 Nm (18 ft-lb)
17. **Bolt**
    - Tighten to 40 Nm (30 ft-lb) plus ¼ turn (90°)
    - Always replace
18. **Engine mount**
19. **Bushings**
    - Quantity: 2

Regulator carbon brushes, checking

Through 05/99

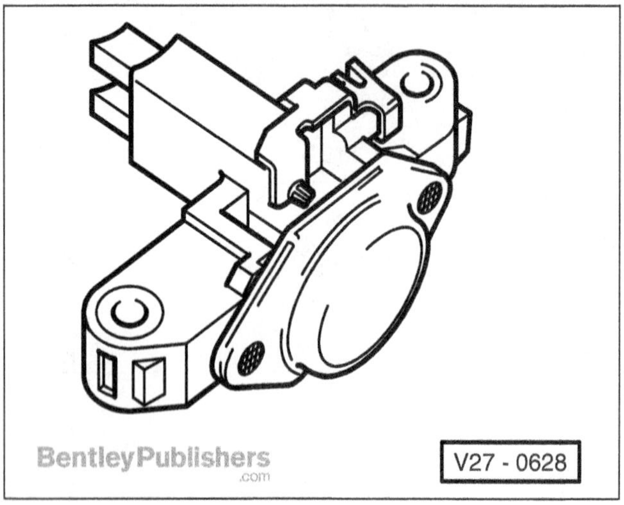

Remove regulator from back of generator and check length of carbon brushes. Replace if worn.

Specification - regulator carbon brush length
- New . . . . 12 mm (0.5 in.)
- Wear limit . . . . 5 mm (0.19 in.)
- Tolerance to one another . . . . 1 mm (0.03 in.)

From 06/99

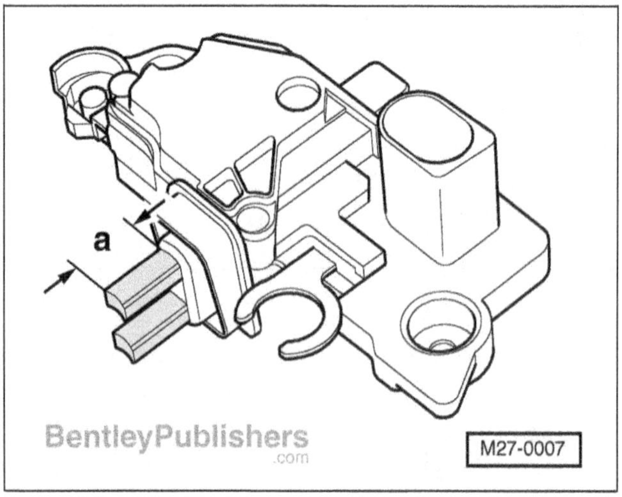

Remove regulator from back of generator and check length of carbon brushes. Replace if worn.

Specification - regulator carbon brush length
- New . . . . 12 mm (0.5 in.)
- Wear limit . . . . 5 mm (0.19 in.)
- Tolerance to one another . . . . 1 mm (0.03 in.)

Generator, removing and installing (4-cylinder)

– Disconnect the battery ground (GND) strap from battery negative (–) terminal. See **Battery terminals, disconnecting**.

NOTE —
Be sure to have the anti-theft radio on hand before disconnecting the battery cables.

– Mark the running direction on ribbed V-belt and remove belt from generator pulley. See **03 Maintenance**.

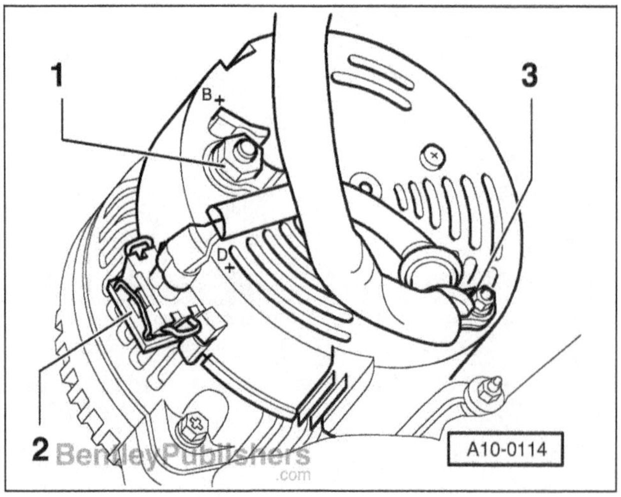

Disconnect wiring from rear of generator.
- **1** - B+ (30) connection
- **2** - Harness connector
- **3** - Harness retainer

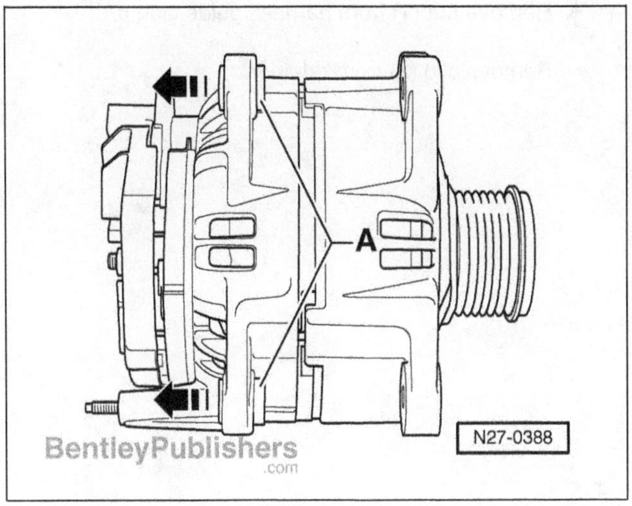

N27-0388

– Remove upper and lower generator mounting bolts and remove generator from its bracket.

< Before installing generator, drive threaded sleeves (**A**) approximately 4 mm in direction of **arrows**.

– Installation is reverse of removal. See **Generator overview** given earlier for tightening torques.

– Be sure to quality check your work, see **Quality Review** at the end of this repair group.

Generator, removing and installing (5-cylinder)

– Disconnect the battery ground (GND) strap from battery negative (–) terminal. See **Battery terminals, disconnecting**.

NOTE —

Be sure to have the anti-theft radio on hand before disconnecting the battery cables.

– Lift vehicle and support with jackstands. Remove lower sound insulation.

– Remove right front wheel and wheel housing liner.

– Mark the running direction on the outer and inner ribbed V-belts and remove belts. See **03 Maintenance**.

– Secure belt tensioners with locking pin T10060A, or equivalent.

– Place lock carrier in service position. See **50 Body–Front**.

– Remove right headlight retainer. See **94 Lights, Accessories–Exterior**.

– Remove A/C compressor. See **87 Air Conditioning**.

NOTE —

Removal of the A/C compressor allows the generator to be removed from below engine.

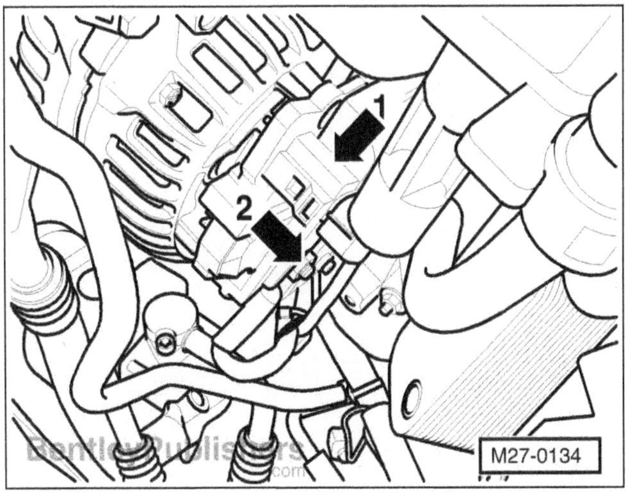

M27-0134

< Disconnect wiring from rear of generator.

- **1** - Harness connector
- **2** - B+ (30) connection

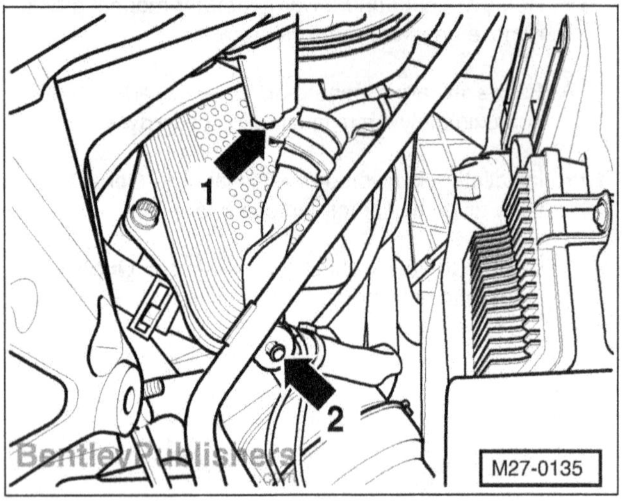

◀ Remove nut (**1**) from harness guide clamp.

– Remove clip (**2**) from retainer.

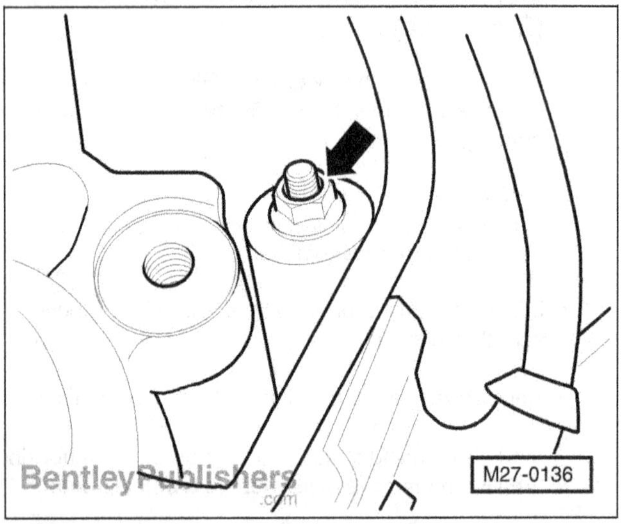

◀ Remove nut (**arrow**) from pressure line bracket.

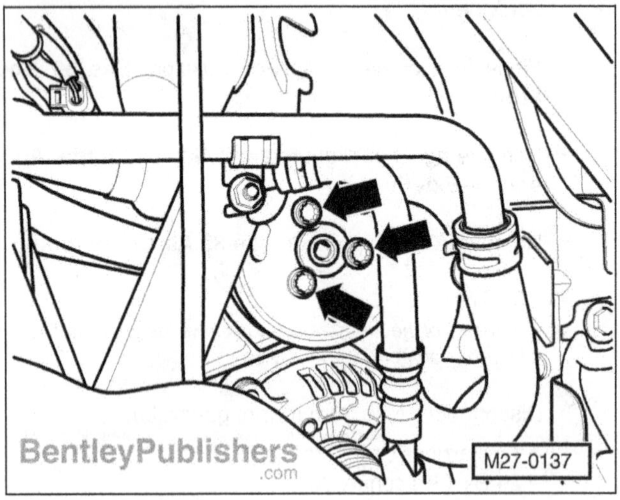

◀ Remove bolts (**arrows**) from ribbed V-belt pulley.

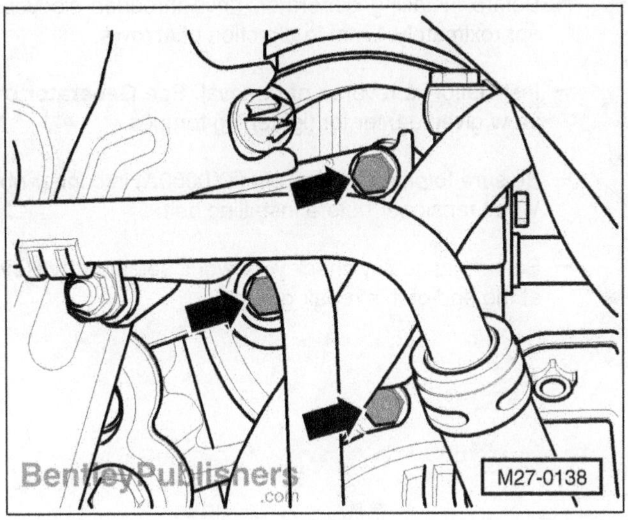

Remove mounting bolts (**arrow**) from power steering pump.

- Move power steering pump out of way and secure to vehicle with safety wire. Power steering lines remain attached.

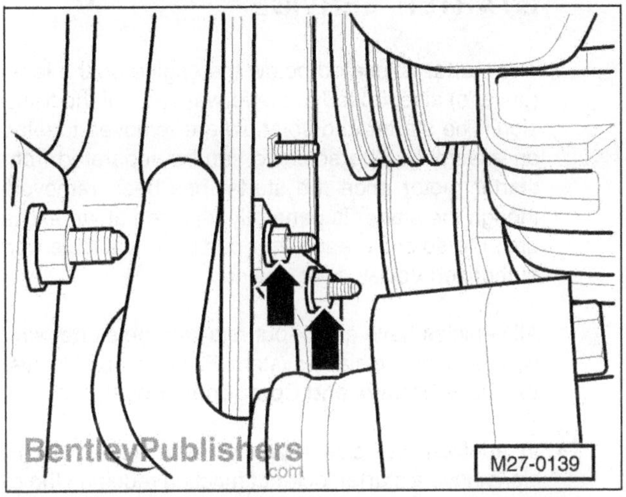

Loosen mounting nuts (**arrows**) for A/C receiver drier. Do not remove nuts.

NOTE —

Loosening receiver drier mounting nuts allows the drier to be moved slightly to allow clearance for removal of generator.

- Remove locking pin (T10060A) from inner ribbed V-belt tensioner.

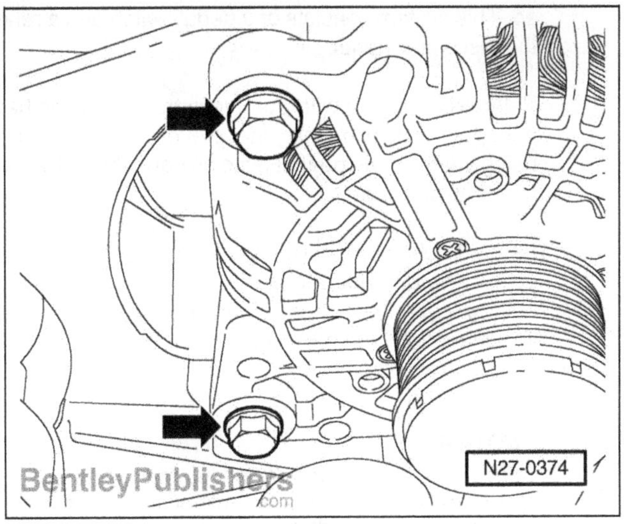

Remove mounting bolts (**arrows**) for generator. Remove generator downward.

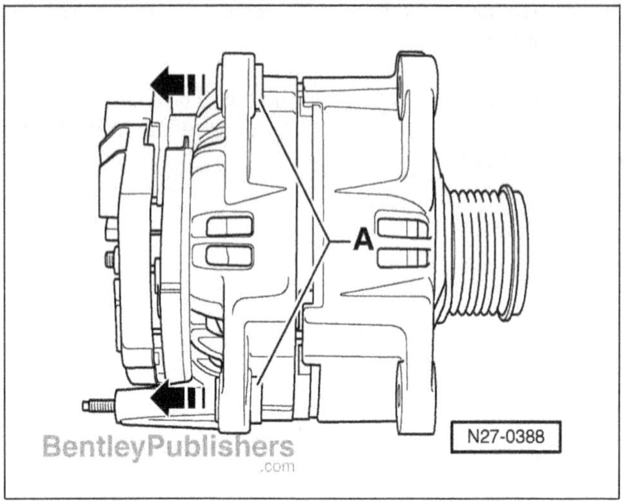

Before installing generator, drive threaded sleeves (**A**) approximately 4 mm in direction of **arrows**.

- Installation is reverse of removal. See **Generator overview** given earlier for tightening torques.
- Be sure to place locking pin (T10060A) into inner ribbed V-belt tensioner before installing belt.
- Be sure to quality check your work, see **Quality Review** at the end of this repair group.

STARTER SERVICE

The starter is located below the engine on the left-hand (driver's) side, bolted to the forward part of the transmission. The starter and solenoid are removed together as an assembly. The solenoid can be separated from the starter motor once the starter has been removed. Although the starter is generally replaced as an exchange unit, the solenoid and other parts are available from an authorized Volkswagen dealer.

All vehicles have a lock-out relay for the starter which is controlled by the alarm system. For location of relays, see **97 Fuses, Relays and Component Locations**.

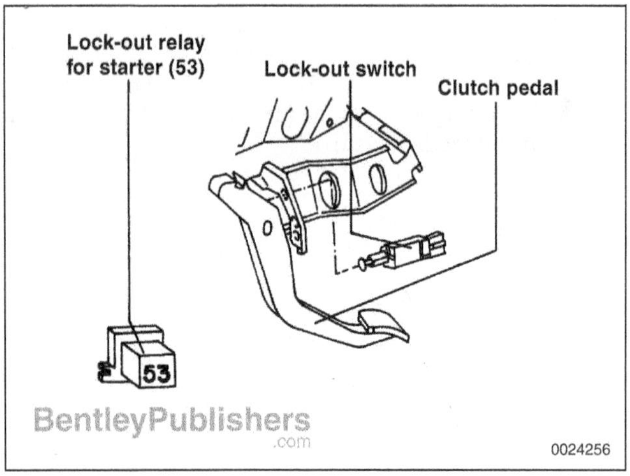

In addition, vehicles equipped with manual transmissions have a starter lock-out function requiring the clutch pedal to be depressed before the starter can be activated. This system consists of a pedal switch and a relay on the fuse-relay panel.

Vehicles equipped with automatic transmissions have a park/neutral position relay that prevents the vehicles from being started if the gear selector is not in the P (Park) or N (Neutral) position.

Before troubleshooting the starter, make sure the battery is fully charged and the battery cables and ground connections are free of corrosion and in good condition. Troubleshooting information for the starting system appears in **Table c**.

NOTE —

Starter cranking speed is affected by engine oil viscosity, especially in cold weather. Make sure the correct oil is in the engine. See **03 Maintenance.**

Table c. Starter System Troubleshooting

Symptom	Probable cause	Corrective action
1. Starter does not operate when ignition switch is turned to START	a. Ignition switch or wire leading from ignition switch to solenoid faulty (less than 8 volts to solenoid)	a. Test for voltage at terminal 50 of solenoid with ignition at START. If not at least 8 volts, test for voltage at terminal 50 of ignition switch with switch at START. Replace ignition switch or eliminate open circuit between ignition switch and solenoid.
	b. Automatic transmission Park/Neutral position (PNP) relay faulty	b. Test relay. See **97 Fuses, Relays and Component Locations** and **37 Automatic Transmission.**
	c. Starter lock-out switch or relay faulty (manual transmission only)	c. Test switch and relay. See **97 Fuses, Relays and Component Locations.**
2. Starter turns slowly or fails to turn engine	a. Dirty, loose, or corroded starter connections	a. Clean, and tighten connections. If necessary check voltage drop between the battery and the starter as described in **9 Electrical System—General.**
	b. Dirty, loose, or corroded ground strap between engine and body	b. Remove and clean or replace ground strap
	c. Starter worn or faulty	c. Repair or replace starter
3. Starter operates, but does not turn engine	a. Flywheel or driveplate teeth missing or damaged	a. Replace flywheel or driveplate. See **13 Crankshaft and Cylinder Block.**
	b. Starter drive or armature shaft faulty	b. Repair or replace starter
	c. Solenoid mechanism faulty	c. Replace starter solenoid

Starter, removing and installing

– Disconnect battery ground (GND) strap from battery negative (–) terminal. See **Battery terminals, disconnecting**.

NOTE —

Be sure to have the anti-theft radio code on hand before disconnecting the battery.

– Remove lower center and left side sound absorber panels.

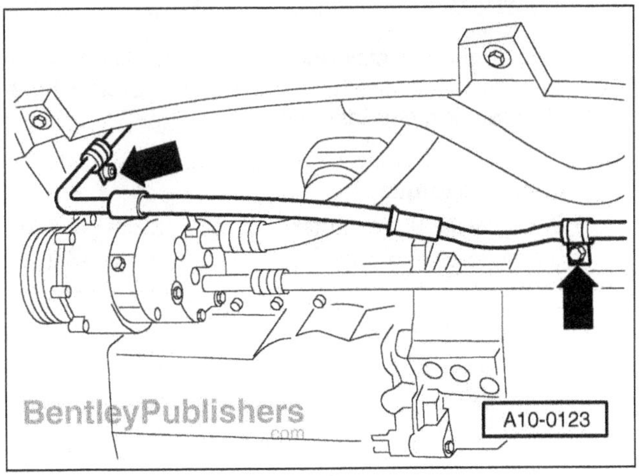

< Remove power steering pump line from retainer (**right arrow**).

– Remove battery and battery tray as described earlier in this repair group.

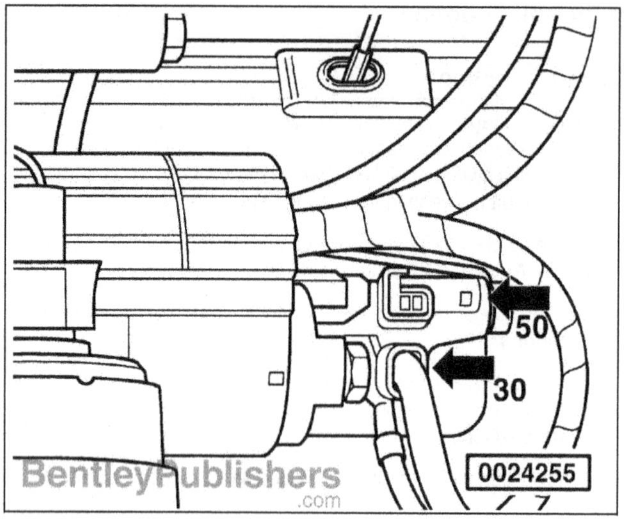

– Pry open cable retainer on top of starter (as applicable). Release retainer and slide toward rear of starter to remove.

◄ Release and disconnect starter switch harness connector (terminal 50) and battery positive (+) cable (terminal 30) (**arrows**) from starter.

Starter terminal designation	
• Switched battery positive (+) from ignition/starter switch	terminal 50
• Battery positive (+), hot at all times	terminal 30

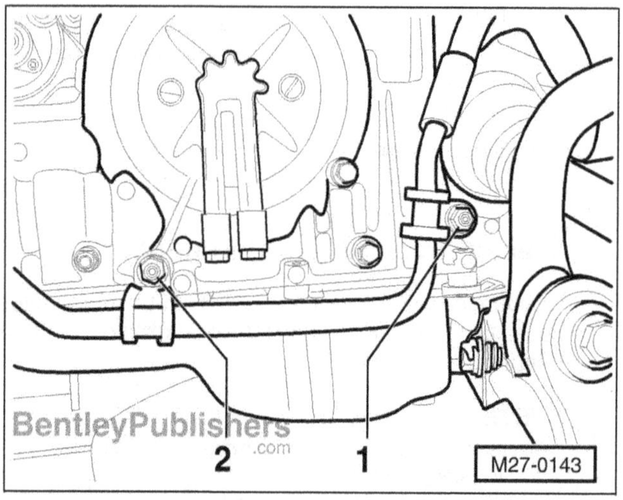

◄ Remove any other electrical harness retainers, as applicable.

- 5-cylinder engine shown. Remove nuts (**1** and **2**) from electrical harness retaining clamps.

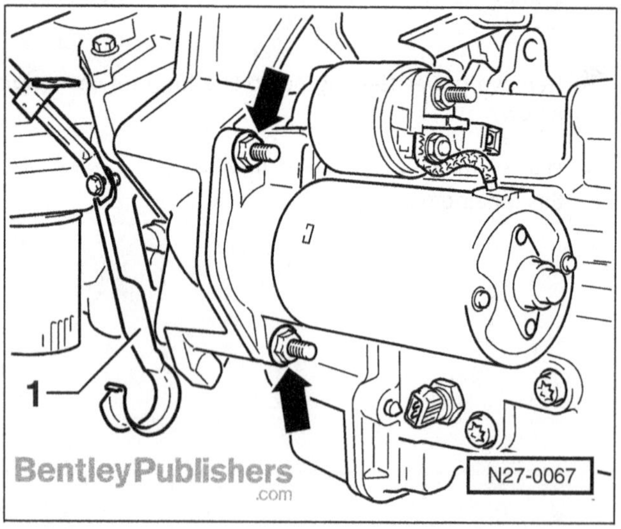

◄ Remove coolant hose bracket (**1**) and starter mounting bolts (**arrows**).

– Remove starter. Installation is the reverse of removal. Be sure to install a new starter bushing, see **Starter bushing, removing and installing**.

Tightening torque	
• Electrical cables to starter	13 Nm (10 ft-lb)
• Starter to engine block	60 Nm (44 ft-lb)

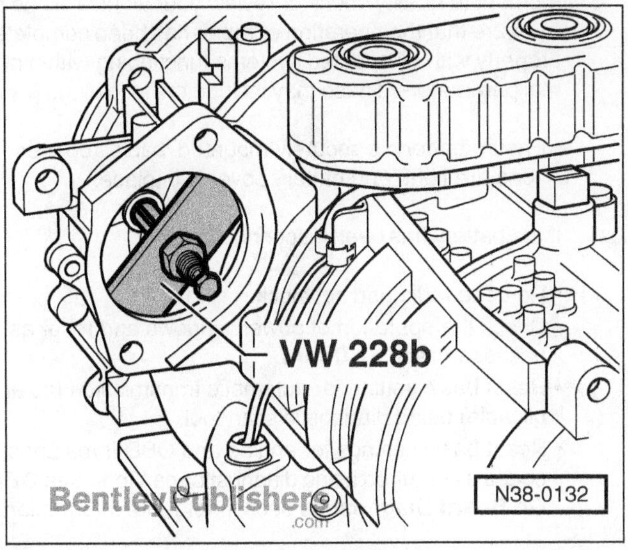

Starter bushing, removing and installing

– Remove starter as described previously.

◄ Use Volkswagen special tool VW 228b, or similar to extract the starter bushing.

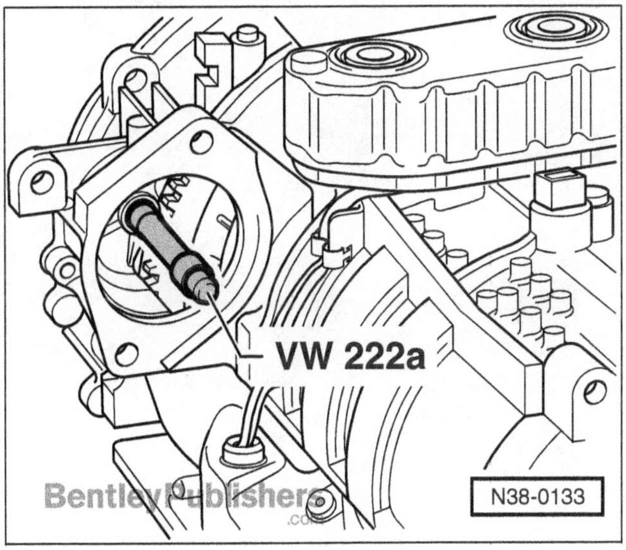

◄ Use Volkswagen special tool VW 222a or similar to install a new starter bushing.

QUALITY REVIEW

When you have finished working under the hood and around other areas of the vehicle, it is advisable to take a moment to quality check or review your work. This helps to insure that the operation or repair has been completed properly with all affected systems functioning within normal parameters. These may include the following:

- Ensure battery is securely mounted, cables connections are tight, and battery cover is in place.

If the battery has been disconnected:

- Re-code radio and set clock.
- Check the operation of power windows and reset as necessary, see **57 Doors**.
- Reset basic setting for automatic transmission (as applicable) using diagnostic scan tool.
- Reset basic settings for engine and OBD II readiness codes as required using diagnostic scan tool. See **OBD On Board Diagnostics** section for more information.

28 Ignition System–Gasoline

28

GENERAL

The ignition functions on the gasoline engines covered by this manual are handled by the Motronic engine management system. The Engine Control Module (ECM) computes ignition timing based on inputs from various sensors. Crankshaft speed from the crankshaft position sensor and engine load from the mass air flow sensor are the two main inputs used to calculate the basic ignition timing point. The other sensors, including the two knock sensors, are used by the ECM to modify and correct the ignition timing based on changing operating conditions. The 1.8L and 2.5L distributor-less ignition systems feature individual ignition coil packs with integral power output stages for each cylinder. The 2.0L engines use a single ignition coil pack with power output stage.

> ***WARNING —***
> *Disconnecting the negative (–) battery cable may erase fault codes and basic settings in the engine management and automatic transmission control modules. Some driveability problems may be noticed until the system re-adapts to operating conditions. OBD II readiness codes, which may be required for emissions testing, may also be erased. Convenience electronics (alarm system, interior light control, power locks, mirrors, and windows) will need to be re-set using a diagnostic scan tool. See **OBD On Board Diagnostics** section for more information.*

NOTE —
*Removal and installation of the spark plugs and individual ignition coils is covered in **03 Maintenance**.*

Engine	Engine code
1.8L 4-cylinder turbo gasoline	APH, AWV, AWP, BKF, BNU
2.0L 4-cylinder gasoline	AEG, AVH, AZG, BDC, BEV, BGD
2.5L 5-cylinder gasoline	BPR, BPS

Service precautions

Ignition system service and repair work must be carried out carefully. The ignition system contains sensitive electronic components. To guard against system damage, and for personal and general safety, the following warnings and cautions apply to any ignition system troubleshooting, maintenance or repair work.

WARNING —
- *Ignition systems operate in a dangerous voltage range that could prove to be fatal if exposed terminals or live parts are contacted. Use extreme caution when working on a vehicle with the ignition on or the engine running.*
- *Connect and disconnect ignition system wires, terminal connectors and test equipment leads only while the ignition is off unless specifically instructed otherwise. Do not touch or disconnect any of the high voltage wires from the coil pack or spark plugs while the engine is running or being cranked by the starter.*

Disabling ignition system

If the engine must be turned at starter speed without starting, the ignition system should be disabled to prevent the discharge of dangerously high voltage. This should be done any time repairs or maintenance are performed on the engine with the ignition key on or if the starter needs to be operated without running the engine.

– Remove upper sound absorber panel.

1.8L and 2.5L engines

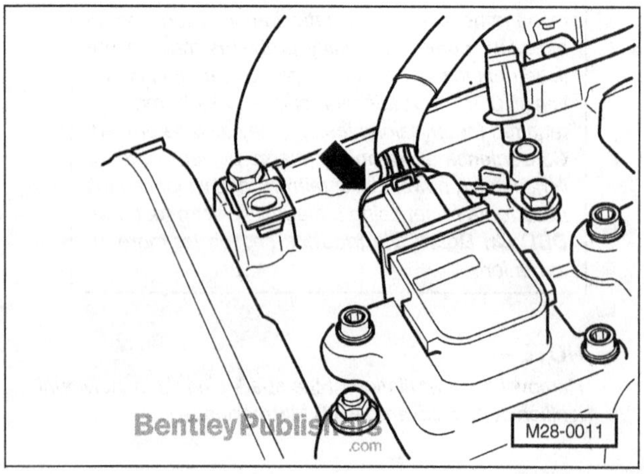

Disconnect 4-pin harness connector (**arrow**) from each ignition coil power output stage.

- 1.8L, code: APH shown, others are similar.

2.0L engine

◄ Disconnect 4-pin harness connector (**arrow**) from ignition coil power output stage.

NOTE —
Secondary air pump removed for clarity.

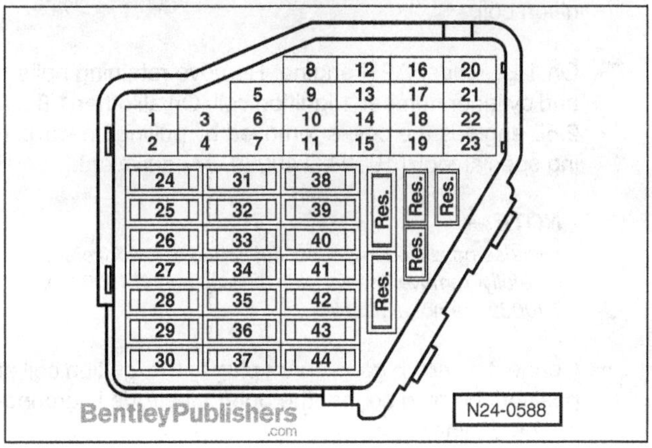

Continued for all engines

◄ Remove fuse for fuel injectors.

- 1.8L and 2.0L remove fuse #32
- 2.5L remove fuse #29

IGNITION SYSTEM TROUBLESHOOTING

Poor driveability may have a variety of causes. The fault may lie with the ignition system, the fuel system, parts of the emission control system, or a combination of the three. Because of the interrelated functions of these systems, it is often difficult to know where to begin looking for problems. For this reason, when troubleshooting always consider these systems in unison, as one major system.

A complete failure of the ignition system to produce spark at the spark plugs is self-evident. For other problems such as rough idle, misfiring, or poor starting the cause may not be so clear. The engine management system has a built in diagnostic circuit that monitors the ignition system components and detects and stores system faults. Before troubleshooting the ignition system, always check for diagnostic trouble codes (DTCs) using a suitable diagnostic scan tool. See **OBD On Board Diagnostics** section for more information.

NOTE —
- *Continued operation of the engine with a misfire will result in serious damage to the catalytic converter.*
- *For 2.0L engines, if one or more of the following DTCs is present:*
  P1355 - Ignition activation cyl. 1 Open circuit
  P1358 - Ignition activation cyl. 2 Open circuit
  P1361 - Ignition activation cyl. 3 Open circuit
  P1364 - Ignition activation cyl. 4 Open circuit
  It may be necessary to install a new ground wire from the ignition coil to ground #12 located under battery. Refer to VW technical bulletin 01 07 28, dated March 6, 2007 for more information.

Quick-check of ignition system

The first step in troubleshooting a no-start condition is to determine whether the problem is caused by the ignition system or some other system, such as a fuel delivery problem. This is done by checking that the spark plugs are firing. If no spark is present, a more detailed testing of the ignition system is necessary.

NOTE—
For accurate test results, the battery should be fully charged.

– Remove upper sound absorber panel.

1.8L and 2.5L engines

– Disconnect 4-pin harness connector from cylinder #1 ignition coil.

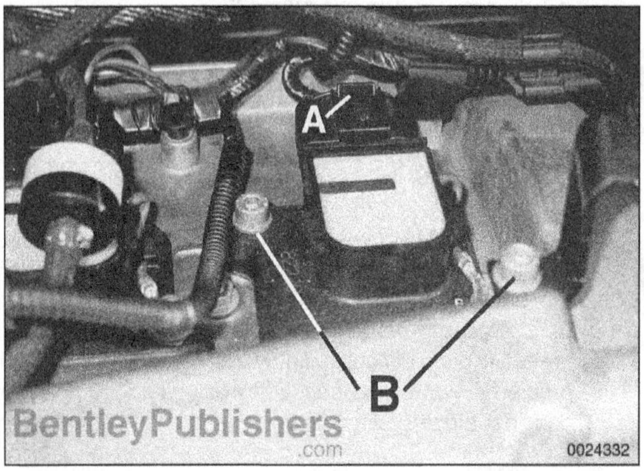

< On 1.8L, code: APH engine: Remove retaining bolts (**B**) and cylinder number 1 ignition coil. On all other 1.8L and 2.5L engines the coil is removed by pulling upwards using special tool T10094, T40039, or equivalent.

NOTE—
It is also possible to use two flat tipped screwdrivers to carefully remove the ignition coils if tool T10094 or T40039 are not available.

– Connect a known good spark plug to the ignition coil and position the plug so that the outer electrode is grounded on the engine.

– Connect the 4-pin harness connector to the ignition coil.

CAUTION—
Never let the spark plug gap (between electrode and ground) exceed 5 mm (0.20 in.) when checking for a spark, as the power output stage may be damaged. Always ground the spark plug on the engine when checking for spark.

2.0L engine

– Remove spark plug wire from spark plug. Connect plug wire to a known good spark plug.

– Position plug so that outer electrode is grounded on engine.

Continued for all engines

– While a helper actuates the starter, look and listen for a spark at the plug. A bright blue spark indicates a healthy ignition system. If there is no spark, make sure engine speed sensor plug is connected properly and the speed sensor is secured in mount on cylinder block.

– If the above inspection reveals no obvious defects, then test ignition coil output stage and ignition coil as described later in this repair group.

> ***WARNING —***
> - *Do not hold the spark plug or its connector during the test, even if using insulated pliers.*
> - *If ignition system failure is not the problem, the engine may start during this test. Be prepared to switch the ignition off immediately. Running the engine with a spark plug wire disconnected may damage the catalytic converter.*

Specification
- Ignition firing order, (4-cylinder) 1-3-4-2
- Ignition firing order, (5-cylinder) 1-2-4-5-3

Ignition coil and power output stage, testing (1.8L)

Individual ignition coils each with a power output stage are mounted above each spark plug. The power output stage takes the low power signal from the ECM and boosts it to a level usable by the ignition coils. The power output stage cannot be directly tested, but the ECM signal along with power and ground can be easily tested.

– Remove 4-pin harness connector from ignition coil as shown earlier.

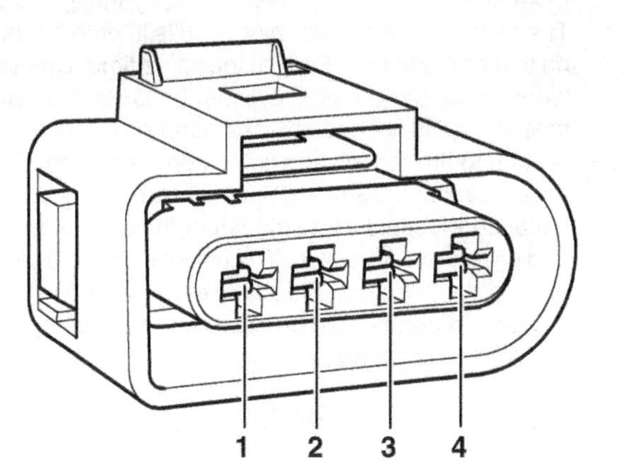

N24-1641

◄ Connect voltmeter to terminals **1** and **4**.

– Switch on ignition and read voltage.

– Repeat for terminals **1** and **2**.

– If voltage is present, proceed to next step. If no voltage is present, check for breaks or open circuits in the wiring, see **EWD Electrical Wiring Diagrams**.

Specification
- Voltage supply
  power output stage. . . . . 11.5 Volts minimum

– Switch off ignition.

– Remove fuse 32 (for fuel injectors) as shown earlier.

– Connect an LED test light only (such as VAG 1527B) to the disconnected connector at terminals **2** and **3**.

– Operate starter and check for a signal from the ECM. LED test light must flicker.

- If LED test light flickers, signal from ECM is good and ignition coil with power output stage should be replaced.
- If test light does not flicker, check for break in wiring, see **EWD Electrical Wiring Diagrams.** If no faults are found, check ECM for DTCs with a suitable scan tool.

– Repeat procedure for each ignition coil.

> ***CAUTION —***
> - *Use a high-quality digital automotive multimeter and an LED test light to make the tests. An analog (swing-needle) meter and conventional test light should not be used as they can permanently damage electronic components.*
> - *Always connect and disconnect the ECM connector and meter probes with the ignition off to avoid damage to electronic components.*

NOTE —
- *Ignition coils and power output stages are a combined unit.*
- *Ignition coil primary and secondary resistance cannot be measured.*

Ignition coil and power output stage, testing (2.0L)

Two double-ended ignition coils are mounted in a common housing together with an power output stage. The power output stage takes the low power signal from the ECM and boosts it to a level usable by the ignition coils. The signal to fire a spark plug results in both spark plugs on that particular coil firing at the same time. One spark is "wasted" by the un-used cylinder because it occurs at a time when it is not needed. The other simultaneous spark is used to fire the mixture in the normal manner. The un-used spark causes no additional wear on the spark plug because it occurs at a time when there is minimal heat and pressure on the plug. The power output stage cannot be directly tested, but the signal from the ECM along with power and ground can be easily tested. The ignition coil can also be checked.

– Remove harness connector from ignition coil power output stage as shown earlier.

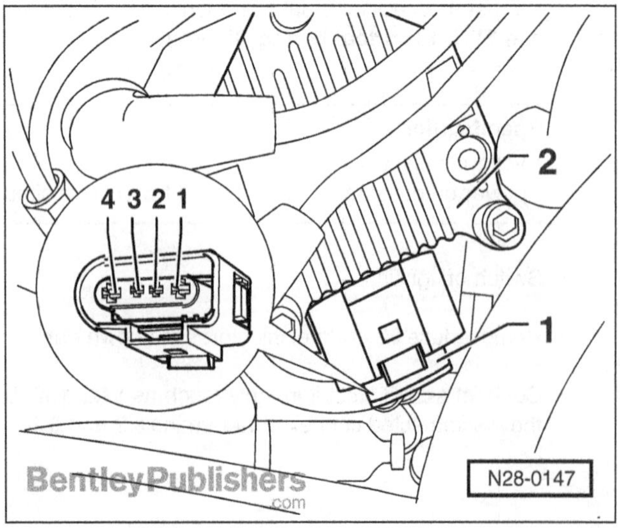

◄ On AEG engine: Connect a voltmeter to disconnected connector at terminals **2** and **4**.

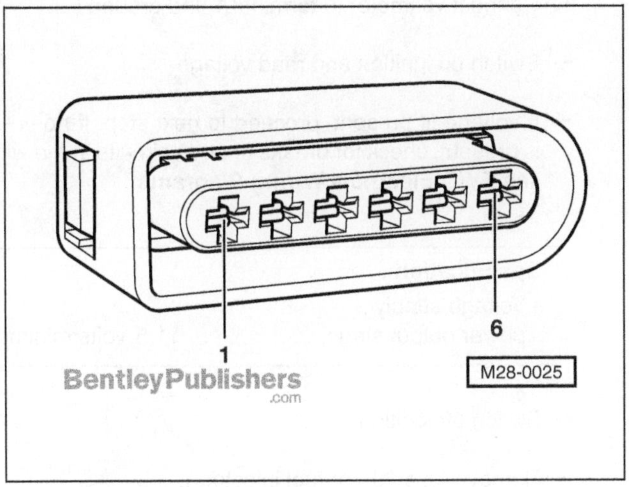

On AVH, AZG, BDC, BEV, BGD engines: Connect a voltmeter to disconnected connector at terminals **1** and **6**.

– Switch on ignition and read voltage. If voltage is present, proceed to next step. If no voltage is present, check for breaks or open circuits in the wiring, see **EWD Electrical Wiring Diagrams.**

Specification

- Voltage supply
  power output stage. . . . . 11.5 Volts minimum

– Switch off ignition.

– Remove fuse 32 (for fuel injectors) as shown earlier.

– Connect an LED test light only (such as VAG 1527B) to the disconnected connector at terminals 2 and 4 (AEG) or terminals 1 and 6 (AVH, AZG, BDC, BEV, BGD).

– Operate starter and check for a signal from the ECM. LED test light must flicker.

– Repeat steps 6 and 7 with the LED test light in terminals 3 and 4.

- If LED test light flickers, signal from ECM is good and ignition coil with power output stage should be replaced.
- If test light does not flicker, check for break in wiring, see **EWD Electrical Wiring Diagrams.** If no faults are found, check ECM for DTCs with a suitable scan tool.

CAUTION —

- *Use a high-quality digital automotive multimeter and an LED test light to make the tests. An analog (swing-needle) meter and conventional test light should not be used as they can permanently damage electronic components.*

- *Always connect and disconnect the ECM connector and meter probes with the ignition off to avoid damage to electronic components.*

Ignition coil and power output stage, testing (2.5L)

Individual ignition coils each with a power output stage are mounted above each spark plug. The power output stage takes the low power signal from the ECM and boosts it to a level usable by the ignition coils. The power output stage cannot be directly tested, but the ECM signal along with power and ground can be easily tested.

– Remove 4-pin harness connector from ignition coil as shown earlier.

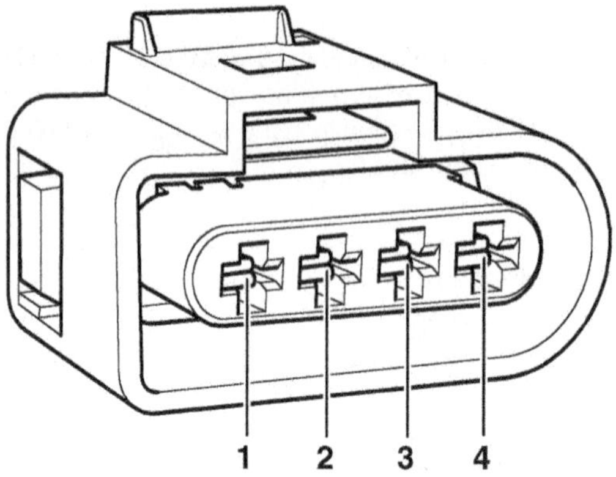

◄ Connect voltmeter to terminal **1** and ground (–).

– Switch on ignition and read voltage.

– If voltage is present, proceed to next step. If no voltage is present, check for breaks or open circuits in the wiring, see **EWD Electrical Wiring Diagrams**.

Specification

- Voltage supply
  power output stage. . . . . 11.5 Volts minimum

– Switch off ignition.

– Remove fuse 29 (for fuel injectors).

– Connect an LED test light only (such as VAG 1527B) to the disconnected connector at terminals **1** and **3**.

– Operate starter and check for a signal from the ECM. LED test light must flicker.

- If LED test light flickers, signal from ECM is good and ignition coil with power output stage should be replaced.
- If test light does not flicker, check for break in wiring, see **EWD Electrical Wiring Diagrams.** If no faults are found, check ECM for DTCs with a suitable scan tool.

– Repeat procedure for each ignition coil.

CAUTION—

- *Use a high-quality digital automotive multimeter and an LED test light to make the tests. An analog (swing-needle) meter and conventional test light should not be used as they can permanently damage electronic components.*
- *Always connect and disconnect the ECM connector and meter probes with the ignition off to avoid damage to electronic components.*

IGNITION SYSTEM COMPONENTS

The ignition system on the gasoline engines covered by this manual require little or no periodic maintenance due to the distributor-less nature of their design. Most of the components are monitored by the engine management system and a failure will cause a fault (DTC) to be stored in the engine control module. Engine misfiring can even result in a DTC being stored due to the potential for damage to the catalytic converter.

NOTE—

Volkswagen identifies electrical components by a letter and/or a number in the electrical schematics. See **EWD Electrical Wiring Diagrams**. *These electrical identifiers are listed in parenthesis as an aid to electrical troubleshooting.*

Ignition components (1.8L, code: APH)

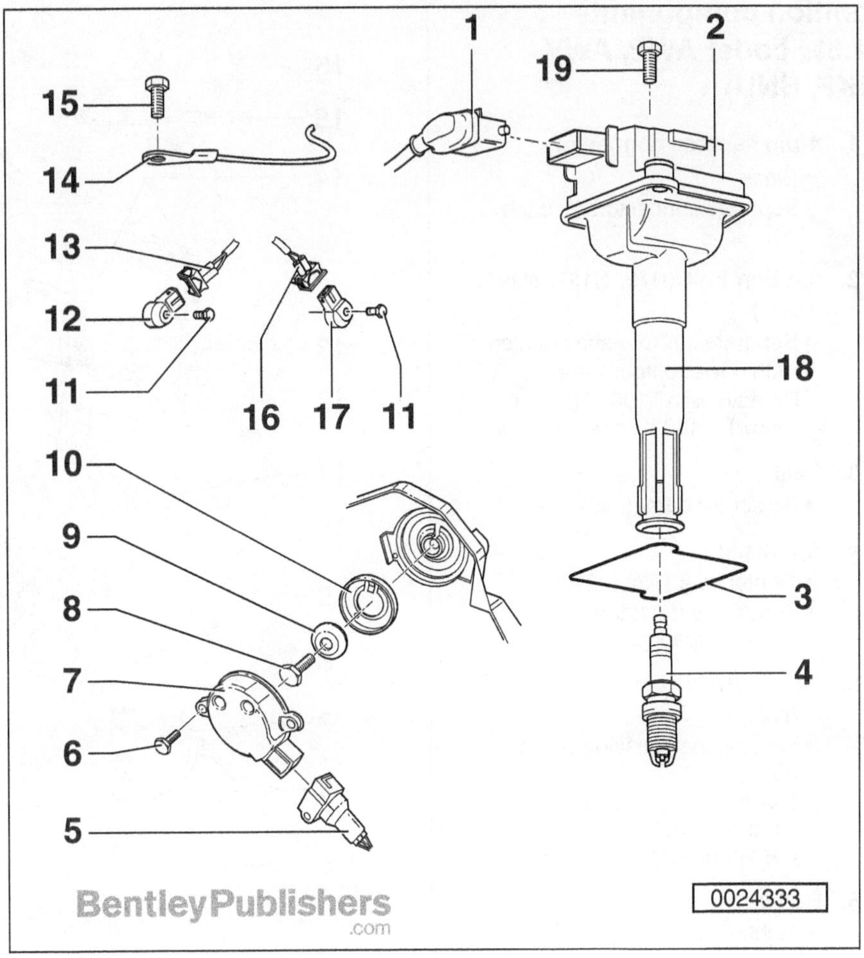

1. **4-pin harness connector**
   - Black
   - Separate connectors for each coil
2. **Ignition coil (N70, N127, N291, N292)**
   - Separate coil for each cylinder
   - With power output stage
3. **Seal**
   - Replace if damaged
4. **Spark plug**
   - Tighten to 30 Nm (22 ft-lb)
   - Remove and install with 5/8 inch (16 mm) spark plug socket
5. **3-pin harness connector**
   - Black
   - For camshaft position sensor (G40)
   - Test for voltage signal (min. 4.5 V) at terminals 1 and 3 of harness connector with ignition on
6. **Bolt**
   - Tighten to 10 Nm (7 ft lb)
7. **Camshaft position sensor (G40)**
   - Mounted on front of cylinder head near intake camshaft
   - Shutter wheel (rotor) mounted on camshaft
8. **Bolt**
   - Tighten to 25 Nm (18 ft lb)
9. **Washer**
   - Conical shape
10. **Shutter wheel**
    - For camshaft position sensor
    - Note installation position
11. **Bolt**
    - Tighten to 20 Nm (15 ft lb)
    - Correct torque is critical for proper operation of knock sensors
12. **Knock sensor 1 (G61)**
    - Sensor and connector terminals are gold-plated
13. **2-pin harness connector**
    - Black
    - Terminals gold plated
14. **Ground wire**
    - Loosen and tighten only with ignition switched off
15. **Bolt**
    - Tighten to 10 Nm (7 ft lb)
    - Loosen and tighten only with ignition switched off
16. **2-pin harness connector**
    - Black
    - Terminals gold plated
17. **Knock sensor 2 (G66)**
    - Sensor and connector terminals are gold-plated
18. **Spark plug connector**
    - RF suppressor
    - Resistance 2000Ω
19. **Mounting bolt**
    - Tighten to 10 Nm (7 ft lb)

Ignition components (1.8L, code: AWP, AWV, BKF, BNU)

1. **4-pin harness connector**
   - Black
   - Separate connectors for each coil
2. **Ignition coil (N70, N127, N291, N292)**
   - Separate coil for each cylinder
   - With power output stage
   - Remove with T10094 (shown below), T40039, or equivalent
3. **Seal**
   - Replace if damaged
4. **Spark plug**
   - Tighten to 30 Nm (22 ft-lb)
   - Remove and install with 5/8 inch (16 mm) spark plug socket
5. **3-pin harness connector**
   - Black
   - For camshaft position sensor (G40)
   - Test for voltage signal (min. 4.5 V) at terminals 1 and 3 of harness connector with ignition on
6. **Bolt**
   - Tighten to 10 Nm (7 ft lb)
7. **Camshaft position sensor (G40)**
   - Mounted on front of cylinder head near intake camshaft
   - Shutter wheel (rotor) mounted on camshaft
8. **Bolt**
   - Tighten to 25 Nm (18 ft lb)
9. **Washer**
   - Conical shape
10. **Hood**
    - For camshaft position sensor (G40)
11. **Bolt**
    - Tighten to 20 Nm (15 ft lb)
    - Correct torque is critical for proper operation of knock sensors
12. **Knock sensor 1 (G61)**
    - Sensor and connector terminals are gold-plated
13. **Knock sensor 2 (G66)**
    - Sensor and connector terminals are gold-plated

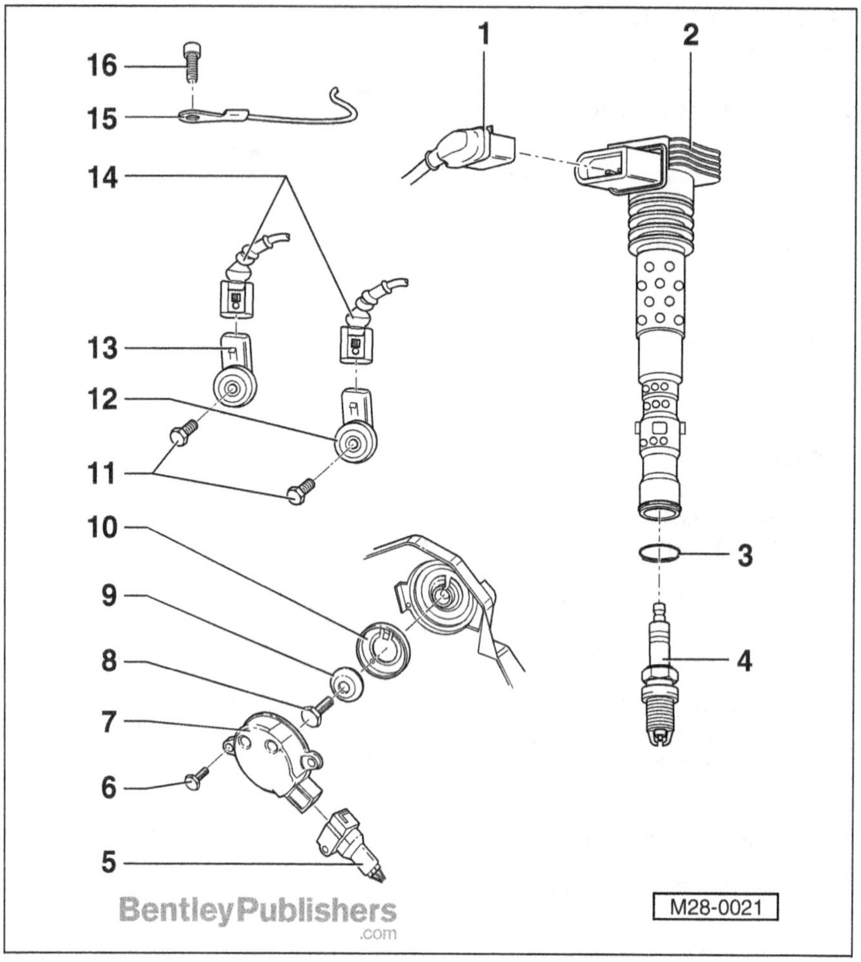

14. **2-pin harness connector**
    - Black
    - Terminals gold plated
15. **Ground wire**
    - Loosen and tighten only with ignition switched off
16. **Bolt**
    - Tighten to 10 Nm (7 ft lb)
    - Loosen and tighten only with ignition switched off

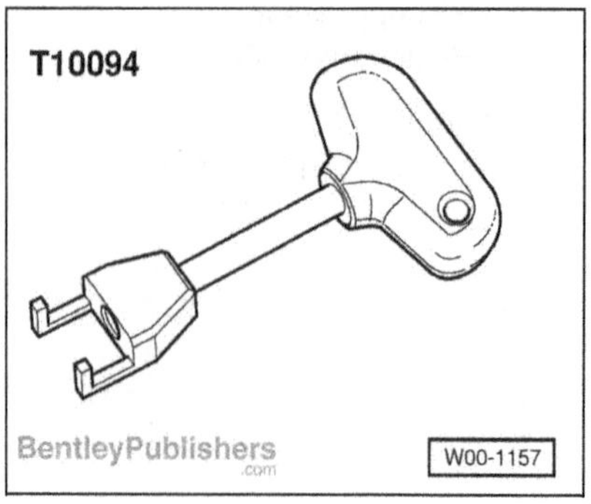

VAG T10094 Special Tool for removing ignition coils (1.8L, code: AWP, AWV, BKF, BNU)

NOTE —

Two flat tipped screwdrivers may also be used to carefully remove the coils if special tool T10094 is not available.

Ignition components (2.0L, code: AEG)

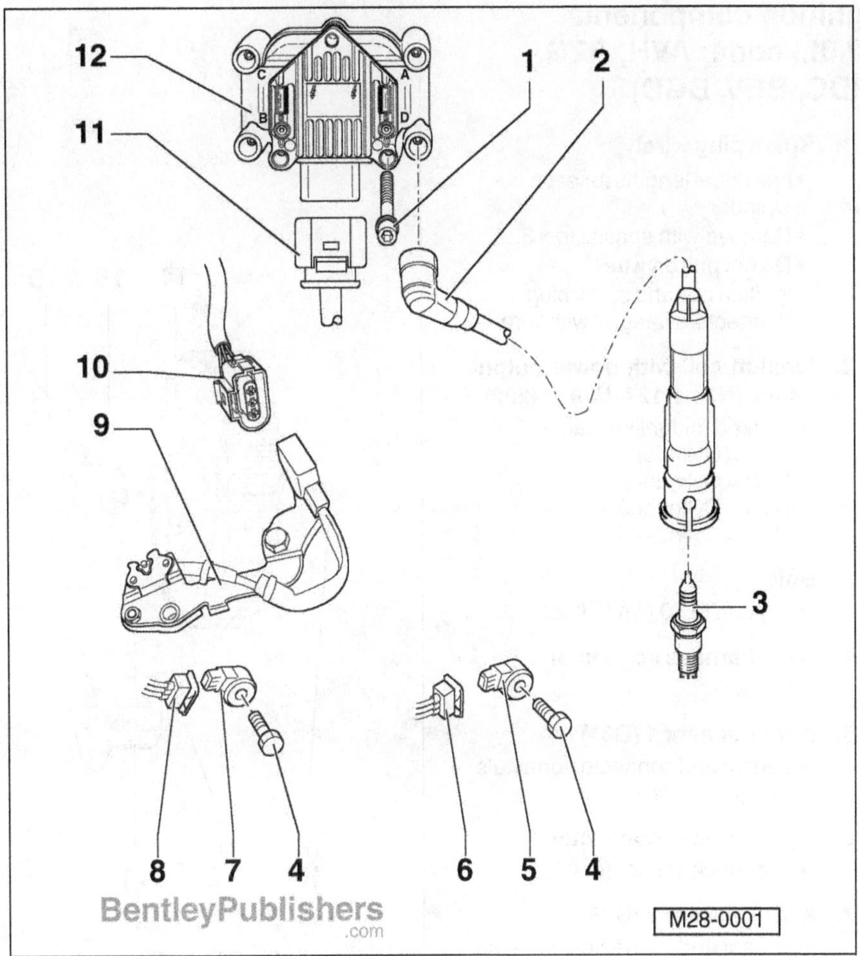

1. **Bolt**
   - Tighten to 10 Nm (7 ft lb)
2. **Spark plug wire**
   - Individual lengths for each cylinder
   - Remove with special tool 3277
   - Do not pull on wire
   - Ignition coil and spark plug connectors integral with wire
   - Resistance 4 - 8 kΩ
3. **Spark plug**
   - Tighten to 30 Nm (22 ft lb)
   - Remove and install with 5/8 inch (16 mm) spark plug socket
4. **Bolt**
   - Tighten to 20 Nm (15 ft-lb)
   - Tightening torque influences knock sensor function
5. **Knock sensor 2 (G66)**
   - Sensor and connector terminals are gold-plated
6. **3-pin harness connector**
   - Terminals gold plated
   - Brown for knock sensor 2 (G66)
7. **Knock sensor 1 (G61)**
   - Sensor and connector terminals are gold-plated
8. **3-pin harness connector**
   - Terminals gold plated
   - Black for knock sensor 1 (G61)
9. **Camshaft position sensor (G40)**
   - Mounted on cylinder head behind camshaft sprocket
   - Shutter wheel (rotor) mounted on camshaft sprocket
10. **3-pin harness connector**
    - Black
    - For camshaft position sensor (G40)
    - Test for voltage signal (min. 4.5 V) at terminals 1 and 3 of harness connector with ignition on
11. **5-pin harness connector**
    - Black
12. **Ignition coils (N, N128)**
    - With power output stage (N122)
    - Marked for ignition cables:
      A = Cylinder 1
      B = Cylinder 2
      C = Cylinder 3
      D = Cylinder 4

Ignition components (2.0L, code: AVH, AZG, BDC, BEV, BGD)

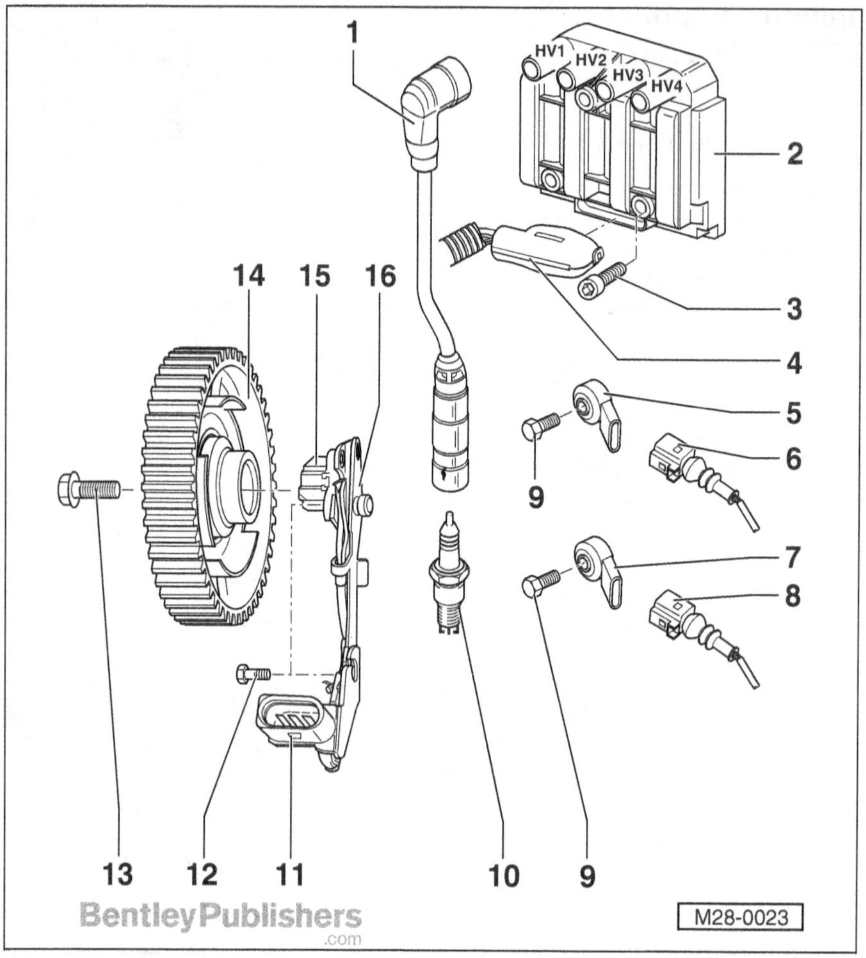

1. **Spark plug wire**
   - Individual lengths for each cylinder
   - Remove with special tool 3277
   - Do not pull on wire
   - Ignition coil and spark plug connectors integral with wire
2. **Ignition coil with power output stage (N70, N127, N291, N292)**
   - Marked for ignition cables:
     HV1 = Cylinder 1
     HV2 = Cylinder 2
     HV3 = Cylinder 3
     HV4 = Cylinder 4
3. **Bolt**
   - Tighten to 10 Nm (7 ft-lb)
4. **6-pin harness connector**
   - Black
5. **Knock sensor 1 (G61)**
   - Sensor and connector terminals are gold-plated
6. **2-pin harness connector**
   - For knock sensor (G61)
7. **Knock sensor 2 (G66)**
   - Sensor and connector terminals are gold-plated
8. **2-pin harness connector**
   - For knock sensor (G66)
9. **Bolt**
   - Tighten to 20 Nm (15 ft-lb)
   - Tightening torque influences knock sensor function
10. **Spark plug**
    - Tighten to 25 Nm (18 ft-lb)
    - Remove and install with 5/8 inch (16 mm) spark plug socket
11. **3-pin harness connector**
    - Black
    - For camshaft position sensor (G40)
    - Test for voltage signal (min. 4.5 V) at terminals 1 and 3 of harness connector with ignition on
12. **Bolt**
    - Tighten to 10 Nm (7 ft-lb)
13. **Bolt**
    - Tighten to 100 Nm (71 ft-lb)
14. **Camshaft sprocket**
    - With rotor for camshaft position sensor (G40)
15. **Camshaft position sensor (G40)**
    - Mounted on cylinder head behind camshaft sprocket
    - Shutter wheel (rotor) mounted on camshaft sprocket
16. **Bracket**
    - For camshaft position sensor (G40)

Ignition components (2.5L)

1. **Harness connector**
   - For knock sensor 2 (G66)
   - Green
   - Gold plated contacts
2. **Harness connector**
   - For knock sensor 1 (G61)
   - Gray
   - Gold plated contacts
3. **Knock sensor 2 (G66)**
   - Note installation position: cable connection points 45° downward right on outside
   - Sensor and connector terminals are gold-plated
4. **Knock sensor 1 (G61)**
   - Note installation position: cable connection points downward vertically
   - Sensor and connector terminals are gold-plated
5. **Bolt**
   - Tighten to 20 Nm (15 ft lb)
   - Correct torque is critical for proper operation of knock sensors
6. **Ignition coil (N70, N127, N291, N292, N323)**
   - Separate coil for each cylinder
   - With power output stage
   - Remove with T40039, or equivalent
7. **Spark plug**
   - Tighten to 25 Nm (16 ft-lb)
   - Remove and install with 5/8 inch (16 mm) spark plug socket
8. **Bolt**
   - Tighten to 10 Nm (7 ft lb)
9. **Camshaft position sensor (G40)**
   - Mounted on front of cylinder head near intake camshaft
   - Shutter wheel (rotor) mounted on camshaft
   - Test for voltage signal (min. 4.5 V) at terminals 1 and 3 of harness connector with ignition on
10. **O-ring**
    - Replace if damaged
    - Tighten to 10 Nm (7 ft lb)
    - Loosen and tighten only with ignition switched off

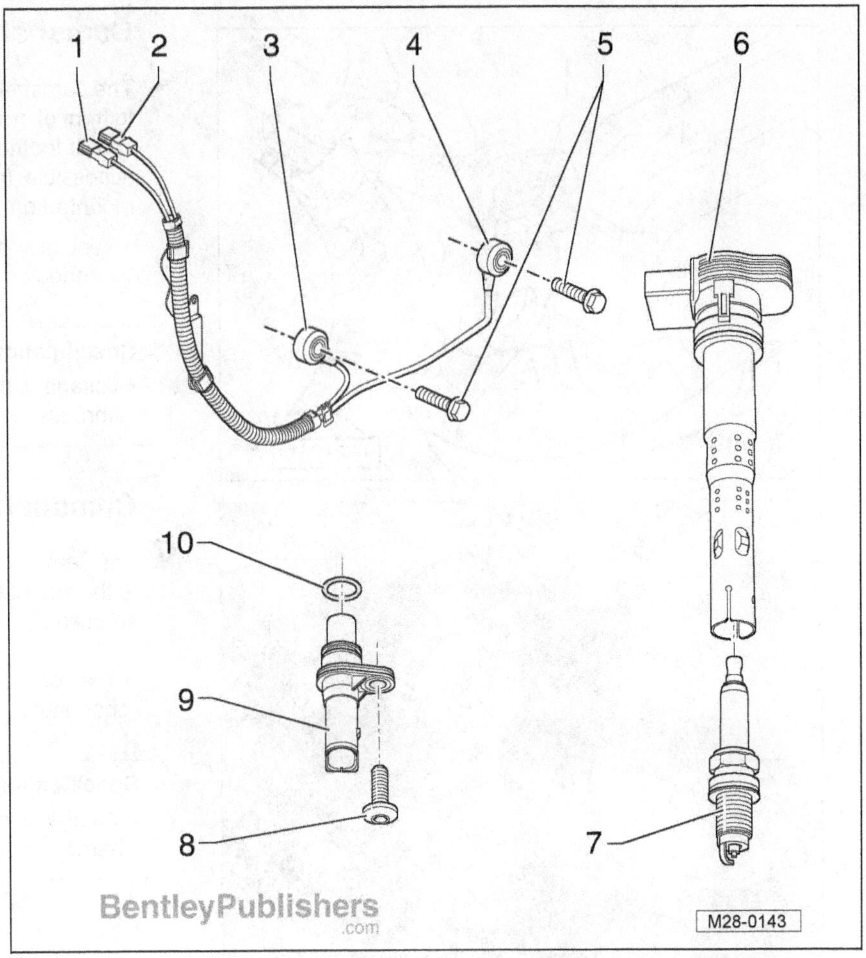

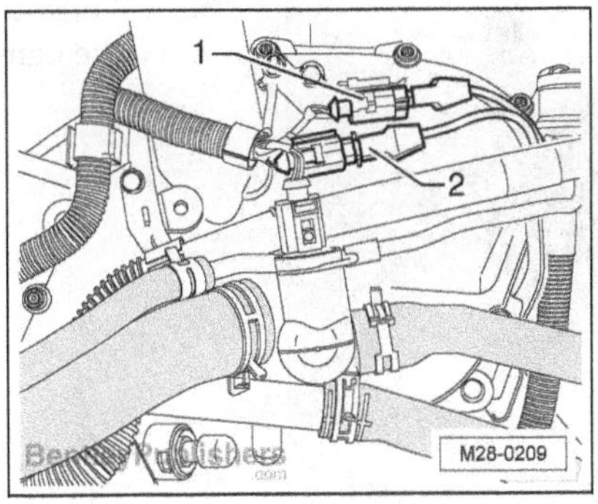

Installation position of harness connectors for knock sensors.

1. Green connector for knock sensor 1 (G61)
2. Gray connector for knock sensor 2 (G66)

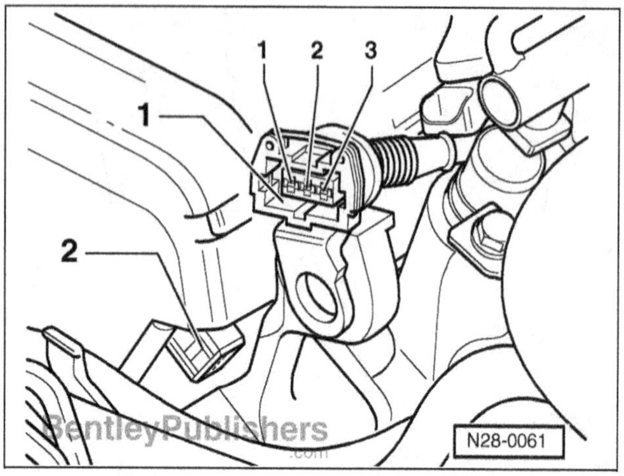

Camshaft position sensor (1.8L)

The camshaft position sensor (G40) (item **2**) is mounted to front of cylinder head on intake side behind upper section of toothed belt guard. The harness connector (**1**) is accessible from the side. A shutter wheel or rotor is mounted on the end of the intake camshaft.

- Test for voltage signal at terminals **1** and **3** of harness connector, with ignition on.

Specification

- Voltage signal at camshaft position sensor (G40) harness connector with ignition on. . . . . . .min. 4.5 V

Camshaft position sensor (2.0L)

The camshaft position sensor (G40) (**arrow**) is mounted to the end of the cylinder head behind the camshaft drive sprocket.

- Test for voltage signal at terminals 1 and 3 of harness connector, with ignition on.

Specification

- Voltage signal at camshaft position sensor (G40) harness connector with ignition on. . . . . . .min. 4.5 V

A shutter wheel (**arrow**) or rotor is permanently mounted on the camshaft drive sprocket.

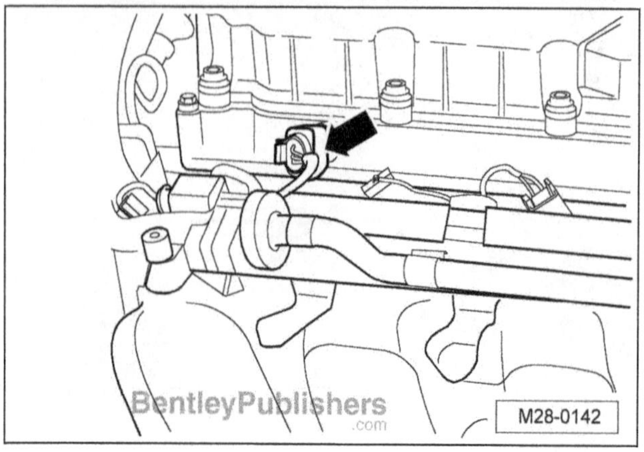

Camshaft position sensor (2.5L)

The camshaft position sensor (G40) (**arrow**) is mounted to the front of the cylinder head near the fuel rail and EVAP canister purge regulator valve (N80).

- Test for voltage signal at terminals 1 and 3 of harness connector, with ignition on.

Specification

- Voltage signal at camshaft position sensor (G40) harness connector with ignition on. . . . . . .min. 4.5 V

30 Clutch

GENERAL

Servicing of the clutch assembly requires that the transmission be removed from the engine. Special tools and equipment are required to remove the transmission and service the clutch. See **34 Manual Transmission** for transmission removal procedures. Read the procedures through to fully understand the scope and nature of the job.

Manual transmission codes

- 02J 5-speed 1.8L, 1.9L, 2.0L
- 02M 5-speed . . . . . 1.8L, 150 hp
- 02M 6-speed . . . . . 1.8L, 180 hp
- 0A4 5-speed 2.5L

WARNING —
- *The cars covered by this manual use an airbag system that automatically deploys the airbags in the event of a frontal or side impact. The airbags are inflated by an explosive device. Handled improperly or without adequate safeguards, the system can be very dangerous. Special precautions must be observed prior to any work at or near the steering wheel or steering column, including the pedal assembly. See* ***69 Seatbelts, Airbags****.*
- *To guard against personal injury or airbag system failure, only trained Volkswagen service technicians should test, disassemble or service the airbag system.*

Clutch Actuating Mechanism

Pedal cluster, assembly

1. **Bulkhead**
2. **Seal**
3. **Clutch pedal mounting bracket**
4. **Pivot bolt**
5. **Accelerator and brake pedal mounting bracket**
6. **Self-locking nut**
   - Tighten to 25 Nm (18 ft-lb)
   - Always replace
7. **Connecting plate**
8. **Self-locking nut**
   - Tighten to 25 Nm (18 ft-lb)
   - Always replace
9. **Bushing**
10. **Fulcrum pin**
11. **Clutch pedal**
12. **Retainer**
13. **Self-locking nut**
    - Tighten to 25 Nm (18 ft-lb)
    - Always replace
14. **Clutch master cylinder**
15. **Supply hose**
    - from brake fluid reservoir
16. **Over-center spring**
17. **Over-center spring mount**
18. **Clutch pedal stop**
19. **Self-locking nut**
    - Tighten to 25 Nm (18 ft-lb)
    - Always replace
20. **Clutch pedal switch**
    - Rotate 45° to remove
    - Electrical testing (4-pin):
      Check resistance at terminals 2 and 3 of switch
      Clutch pedal up: max 1.5 Ω
      Clutch down: infinite Ω
      Check voltage at connector terminal 2 and ground:
      Ignition on, min. 11.5 V
    - Electrical testing (2-pin):
      Check resistance as described above at terminals 1 and 2
      Check voltage at connector terminal 1 and ground:
      Ignition on, min. 11.5 V

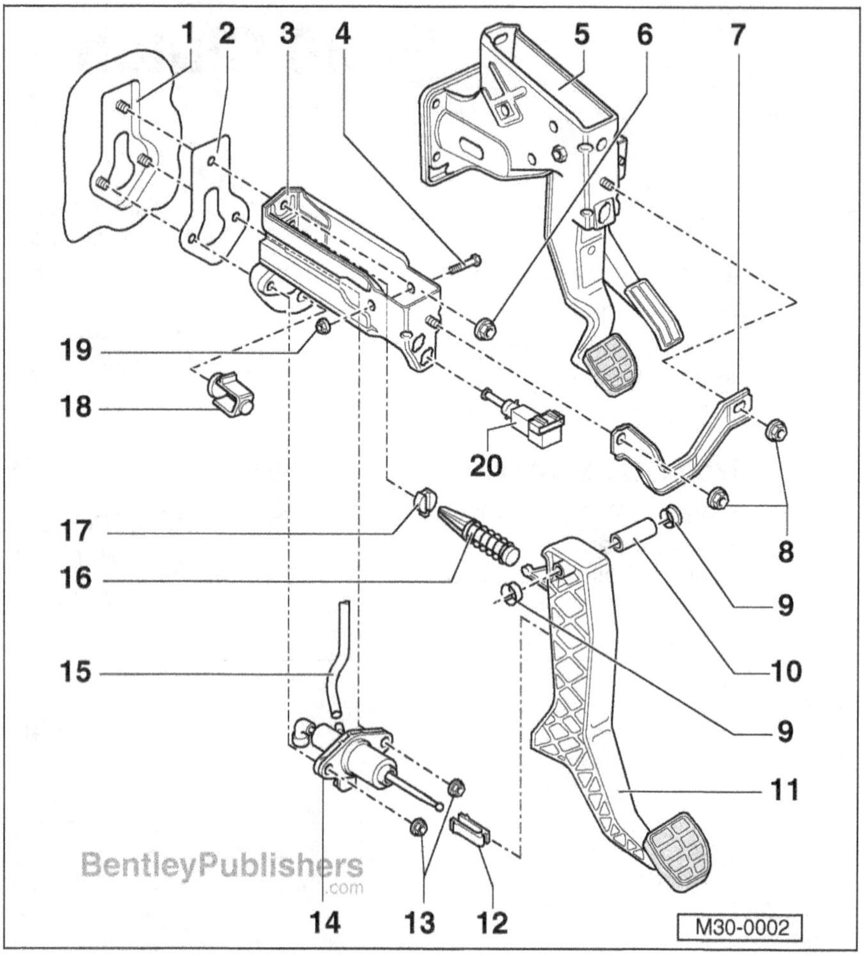

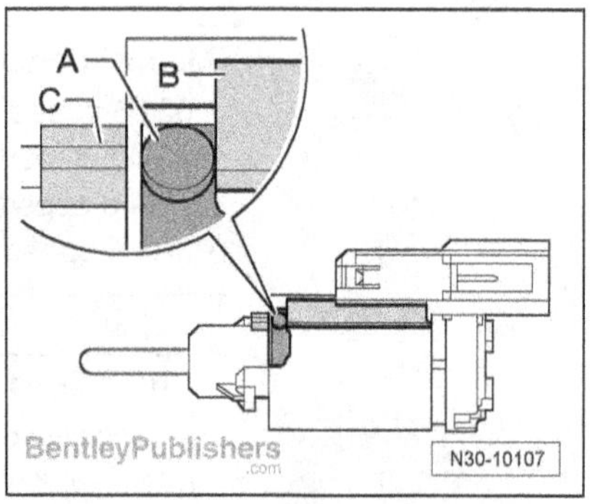

Clutch pedal switch, installing (0A4 transmission)

◄ Before installing clutch pedal switch, make sure of the following:

- Stop **A** must be located in front of section **B**.
- Stop **C** will then be located directly below take-up hook.

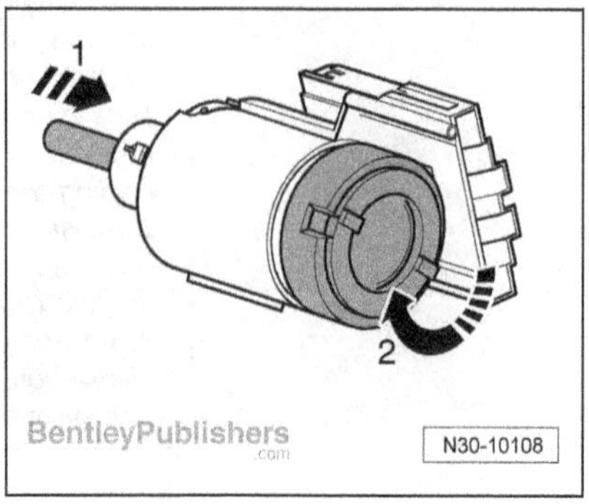

◄ Push in plunger (**1**) until it stops.

– At the same time, rotate upper section in direction of **arrow 2** until it stops.

– Pull out plunger completely.

– Install switch into clutch pedal and rotate 45° clockwise.

– Connect harness connector.

Clutch pedal over-center spring, removing and installing

The clutch pedal mechanism has an over-center spring to assist with clutch pedal operation.

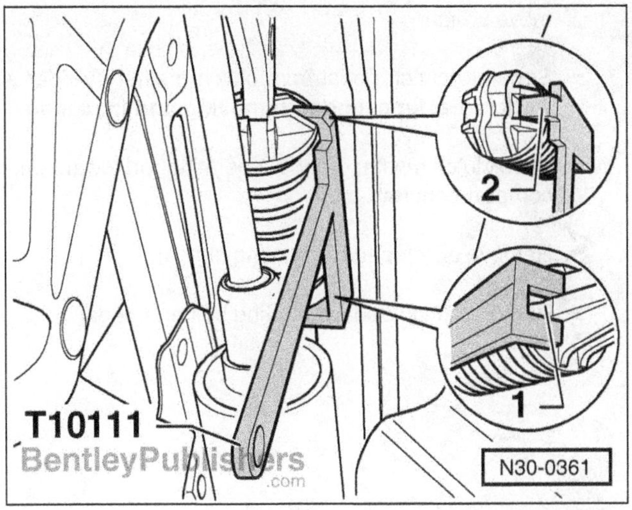

– Remove drivers side lower trim to access pedal assembly. See **70 Trim–Interior**.

◄ Install retaining clamp **T10111** onto over-center spring from right side.

- Make sure retaining clamp is placed correctly on the over-center spring.
- The retaining clamp claws must make contact with rear (**1**) of over-center spring.

– Press down on the clutch pedal lightly.

– Install retaining clamp lug (**2**) into recess at front of over-center spring.

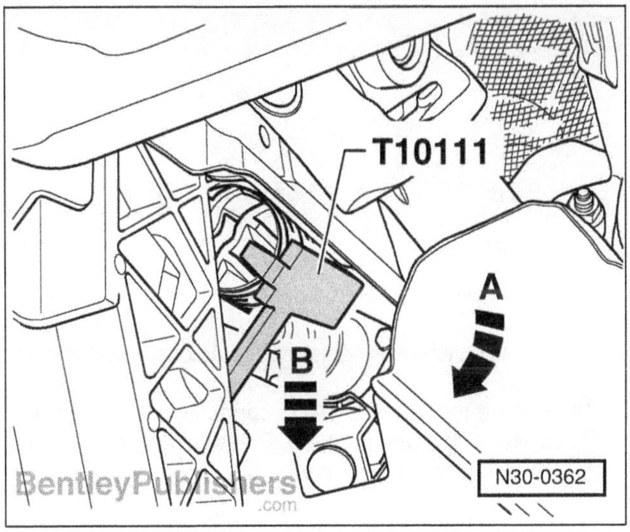

◄ Press clutch pedal down until over-center spring can be removed from mounting bracket and clutch pedal.

– Rotate over-center spring using retaining clamp (**T10111**) in direction of **arrow A** and remove downward in direction of **arrow B**.

– Release clutch pedal.

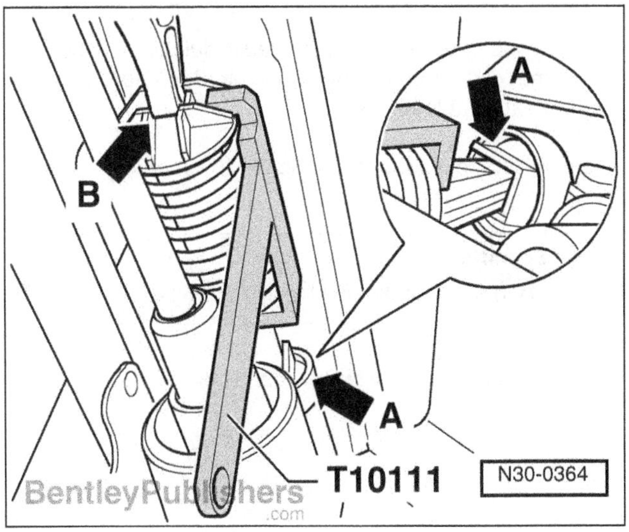

◄ To install over-center spring, pull clutch pedal toward passenger compartment and insert over-center spring (using **T10111**) into rear mount (**arrow A**).

– Press clutch pedal down until over-center spring seats on mounting lug (**arrow B**) of clutch pedal.

– Remove retaining clamp while moving clutch pedal to rest position.

– Install drivers side lower trim to access pedal assembly. See **70 Trim–Interior**.

Clutch pedal, removing and installing

– Remove clutch pedal over-center spring as described earlier in this repair group.

– Rotate clutch pedal switch 45° counter-clockwise and remove switch.

– Separate clutch pedal from clutch master cylinder, see **Clutch master cylinder, removing and installing**.

– Push clutch master cylinder operating rod toward engine compartment until stop.

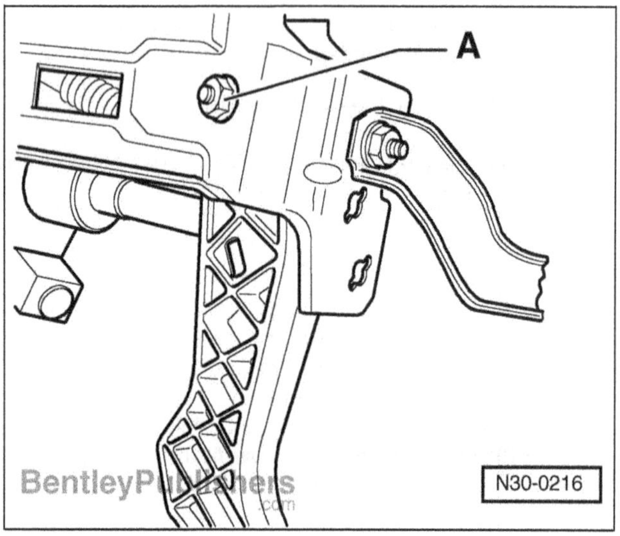

< Remove clutch pedal mounting nut (**A**).

– Remove clutch pedal mounting bolt and pedal.

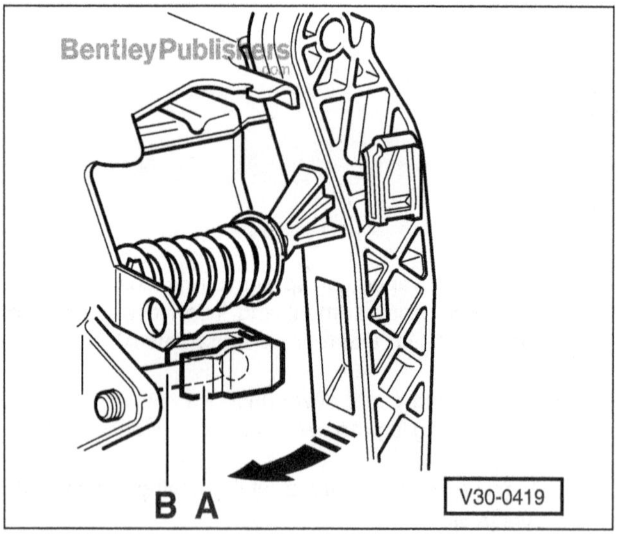

– Install clutch pedal and mounting bolt.

– Install clutch over-center spring as described earlier.

< Ensure that retainer (**A**) is correctly installed on master cylinder operating rod (**B**). Press clutch pedal down (**arrow**) to engage retainer.

– Install clutch pedal over-center spring.

– Bleed clutch system if hydraulic fluid level is below MIN marking on brake fluid reservoir. See **Clutch hydraulic system, bleeding**.

– Fill brake fluid reservoir up to MAX marking.

Specification
- Hydraulic (brake) fluid . . . . .DOT 4

– Be sure to quality check your work.

CLUTCH HYDRAULIC SYSTEM

Clutch hydraulic system, assembly

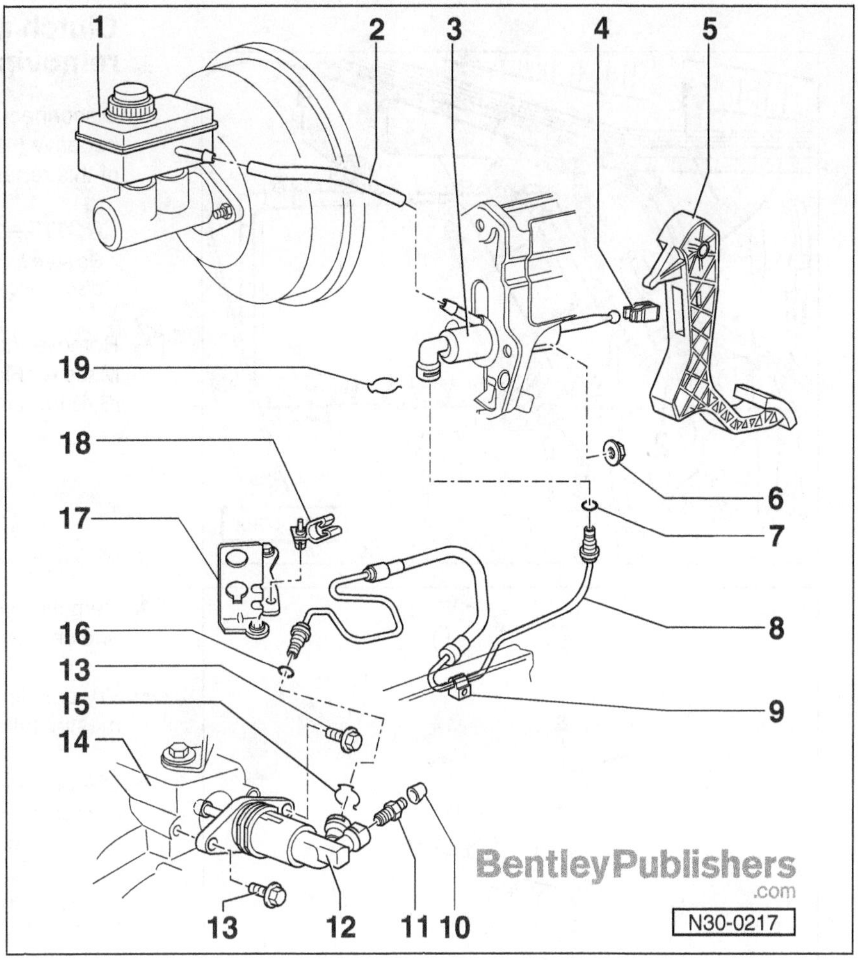

1. **Hydraulic (brake) fluid reservoir**
2. **Supply hose**
3. **Master cylinder**
4. **Retainer**
   - Replace only with master cylinder removed
   - See **Clutch master cylinder retainer, removing**
5. **Clutch pedal**
6. **Self-locking nut**
   - Tighten to 25 Nm (18 ft-lb)
   - Always replace
7. **O-ring**
   - Coat with hydraulic fluid
   - Pull into line/hose connection
8. **Hydraulic line/hose assembly**
9. **Bracket**
   - Attached to body
10. **Dust cap**
11. **Bleeder valve**
    - 02J shown, 02M/0A4 have bleeder on top slave cylinder
12. **Slave cylinder**
    - See **Clutch slave cylinder, removing and installing**
13. **Bolts**
    - 02J trans.: 25 Nm (18 ft-lb)
    - 02M/0A4 trans.: 20 Nm (15 ft-lb)
14. **Transmission**
15. **Clip**
16. **O-ring**
    - Coat with hydraulic fluid
    - Pull onto/hose connection
17. **Cable support bracket**
    - May not be installed on all versions
18. **Hose bracket**
    - Secure to cable support bracket
19. **Clip**

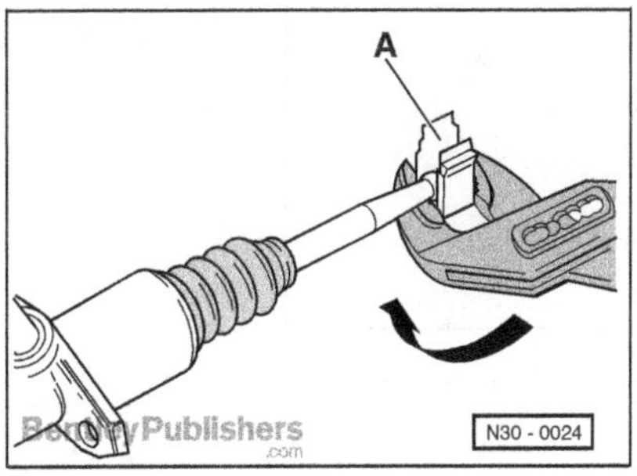

Clutch master cylinder retainer, removing

◄ Pry retainer off in direction of **arrow**.

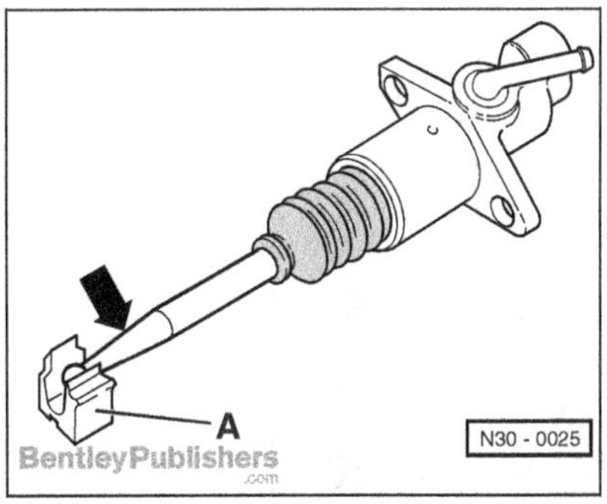

Master cylinder operating rod retainer, installing

◄ Press operating rod in direction of **arrow**.

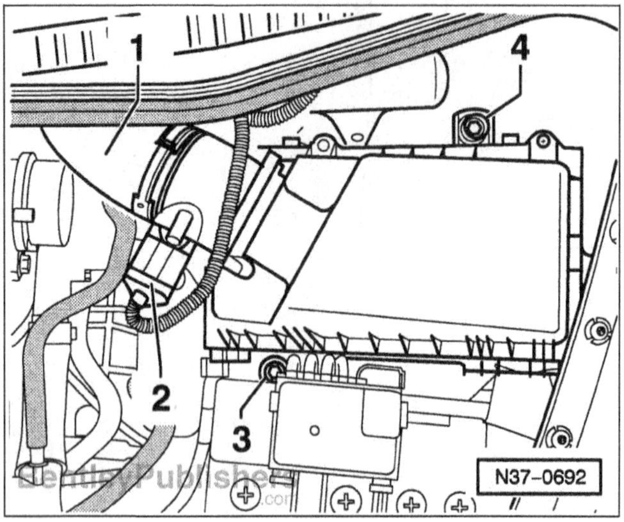

Clutch master cylinder, removing and installing

- Disconnect battery ground (GND) strap from battery negative (–) terminal. See the **Cautions** at the beginning of this repair group regarding battery disconnection.

NOTE—

Be sure to have the anti-theft radio code on hand before disconnecting the battery.

◄ Remove intake air hose (**1**) harness connector (**2**) from Mass Air Flow (MAF) sensor, and remove mounting bolts (**3**,**4**) for air cleaner and lift out air cleaner assembly.

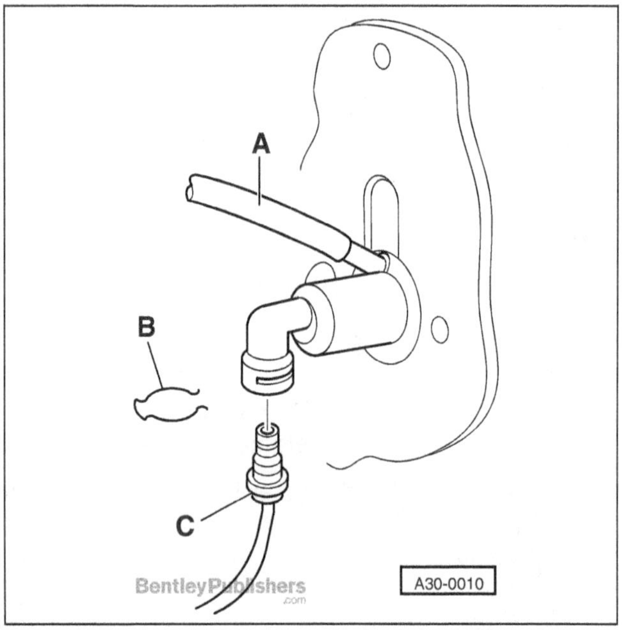

◄ Remove hydraulic fluid (brake) reservoir supply hose (**A**) and seal end to prevent dirt or moisture from entering.

- Pry out clip for brake line (**B**) and hose assembly (**C**) at master cylinder.

- Disconnect hydraulic brake line/hose assembly from rear of master cylinder and seal end to prevent dirt or moisture from entering.

- Remove drivers side lower trim to access pedal assembly. See **70 Trim–Interior**.

- Rotate clutch pedal switch 45° counter-clockwise and remove switch.

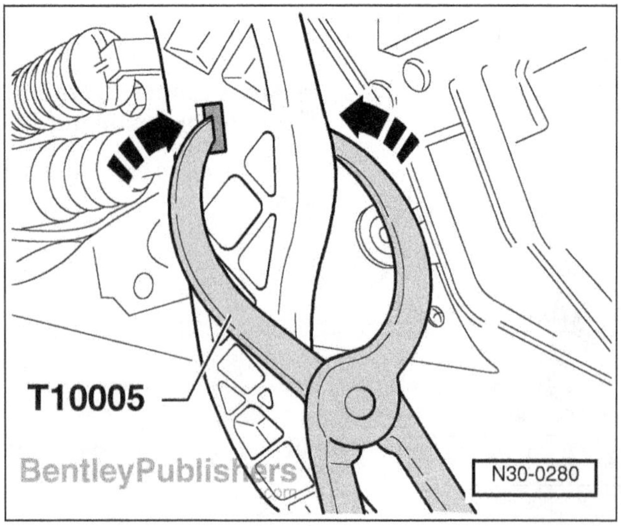

◄ Separate master cylinder operating rod from clutch pedal as follows:

- Pull clutch pedal back slightly toward passenger compartment.
- Position Volkswagen special tool **T10005** (pliers) or similar, in recess on sides of pedal (**arrows**) and squeeze to release pedal from operating rod.

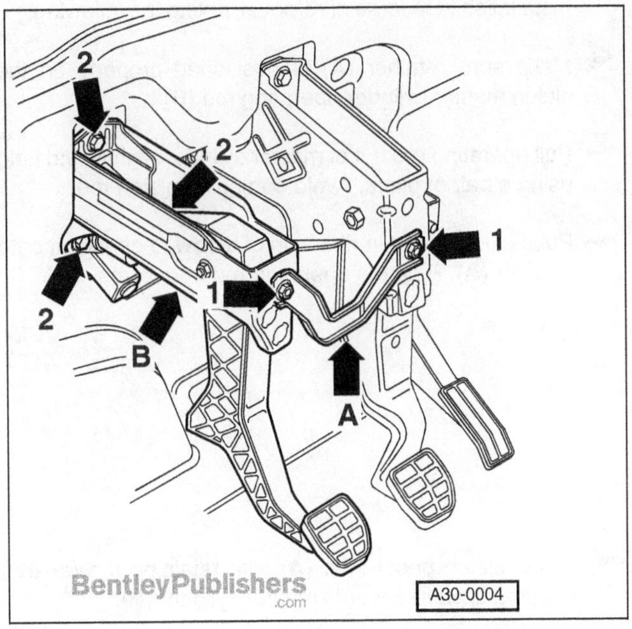

Remove mounting nuts (**1**) for clutch/brake pedal connecting plate (**A**)

– Remove mounting nuts (**2**) for clutch pedal mounting bracket (**B**). Remove mounting bracket.

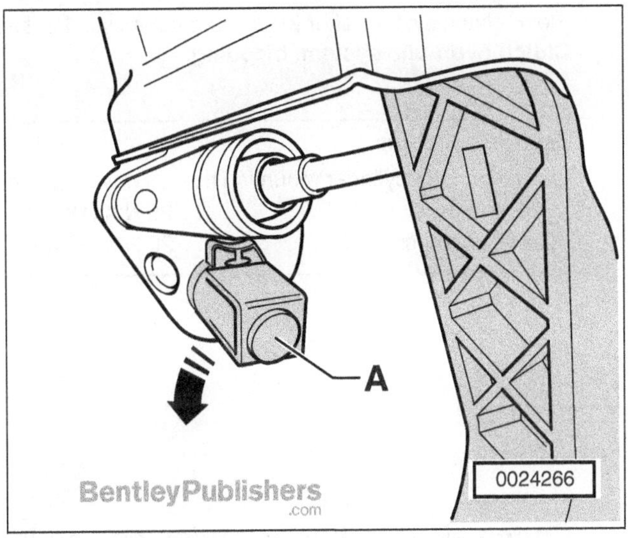

Turn clutch pedal stop (**A**) counterclockwise (**arrow**) and remove.

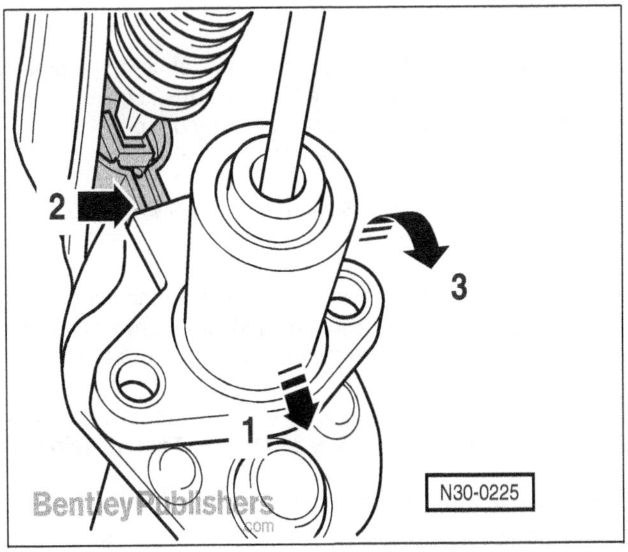

Push master cylinder down (**1**) onto stop and swing downward (**3**) out of mounting bracket (**2**).

NOTE —
*The master cylinder must not be blocked in the upper area by the over-center spring mount (**arrow 2** above).*

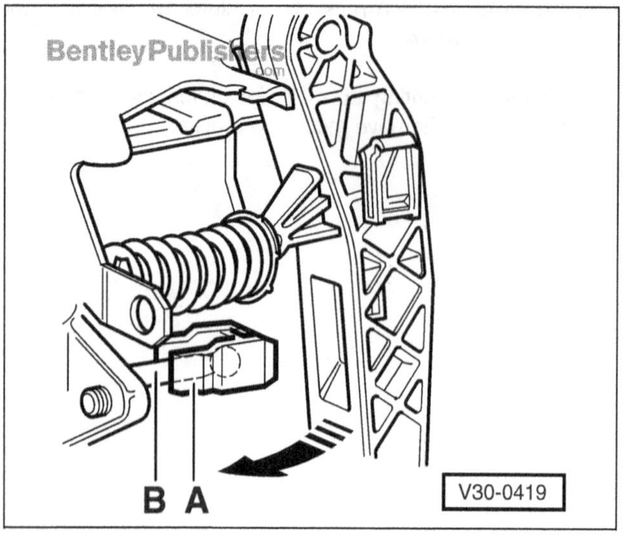

Installation is reverse of removal, noting the following:

◄ Make sure retainer (**A**) is positioned properly on the clutch master cylinder operating rod (**B**).

– Pull operating rod out of master cylinder slightly and hold using a pair of pliers. Avoid damaging piston rod.

– Push clutch pedal in direction of **arrow** to engage it onto retainer (**A**). Release operating rod from pliers.

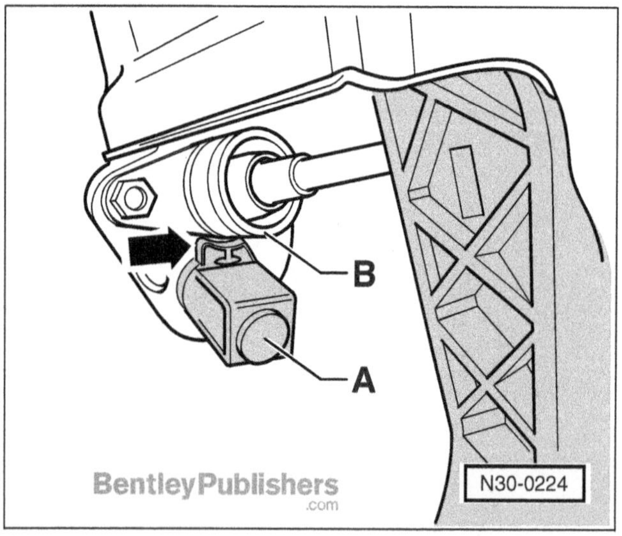

◄ Install clutch pedal stop (**A**) with retaining tab (**arrow**) pointing upward towards master cylinder (**B**).

– Bleed clutch system after installing master cylinder, see **Clutch hydraulic system, bleeding**.

Tightening torque

- Clutch master cylinder mounting nuts (always replace)
  02J transmission . . . . 25 Nm (18 ft-lb)
  02M/04A transmission . . . . 20 Nm (15 ft-lb)

Clutch slave cylinder, removing and installing

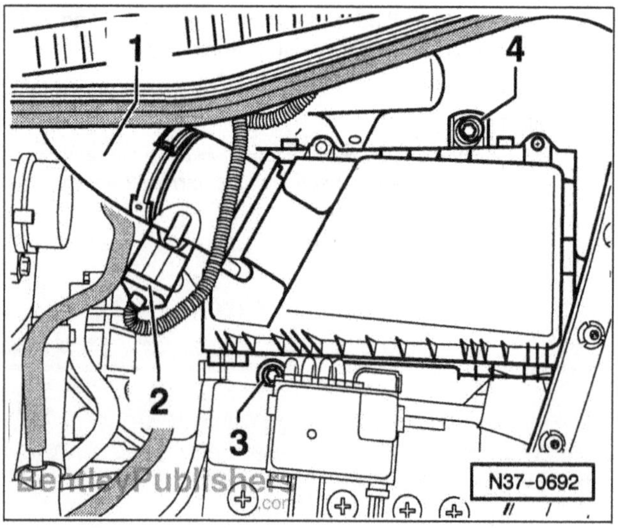

– Disconnect battery ground (GND) strap from battery negative (–) terminal. See the **Cautions** at the beginning of this repair group regarding battery disconnection.

NOTE—
Be sure to have the anti-theft radio code on hand before disconnecting the battery.

◄ Remove intake air hose (**1**) harness connector (**2**) from Mass Air Flow (MAF) sensor, and remove mounting bolts (**3**,**4**) for air cleaner and lift out air cleaner assembly.

– Disconnect shift linkage, if necessary, to gain access to clutch slave cylinder. See **34 Manual Transmission**.

– Place a large lint-free cloth under the slave cylinder to absorb any leaking fluid during removal.

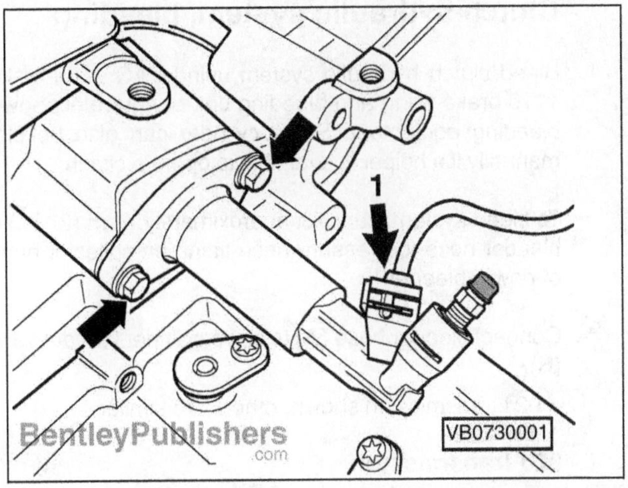

02J transmission

◀ Remove retaining clip and fluid supply line (**arrow 1**). Seal fluid line opening.

– Remove mounting bolts (**arrows**) and slave cylinder.

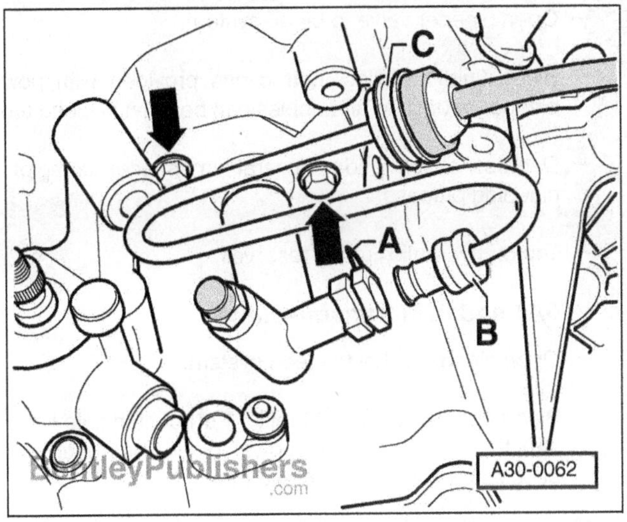

02M transmission

◀ Remove retaining clip (**A**) which secures fluid line to slave cylinder.

– Remove fluid pipe/hose assembly from housing (**C**) on transmission.

– Separate pipe/hose assembly (**B**) from slave cylinder and seal fluid line openings.

– Remove mounting bolts (**arrows**) and remove slave cylinder.

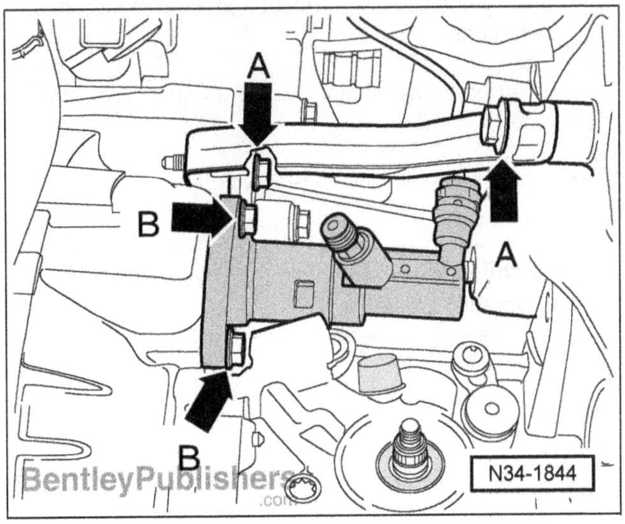

0A4 transmission

◀ Remove transmission support (**A**).

– Remove retaining clip for hydraulic fluid line.

– Pull pipe/hose assembly from slave cylinder and seal fluid line openings.

– Remove mounting bolts (**arrows B**) and remove slave cylinder.

Continued for all vehicles

> ***CAUTION—***
> *Do not press clutch pedal with slave cylinder removed.*

– Install slave cylinder in reverse order of removal.

Tightening torques
- 02J slave cylinder to trans. . . . . 25 Nm (18 ft-lb)
- 02M/0A4 slave cylinder to trans. . . . 20 Nm (15 ft-lb)

– Bleed clutch hydraulic system as described below.

Clutch hydraulic system, bleeding

Bleed clutch hydraulic system using VAG 1869 or US 1116 brake filling and bleeding unit or equivalent power bleeding equipment. Slave cylinder can also be bled manually if a helper is available to operate clutch.

– To bleed system, connect approximately 7 cm (27 in.) of bleeder hose to pressure hose fitting on collector bottle of power bleeder.

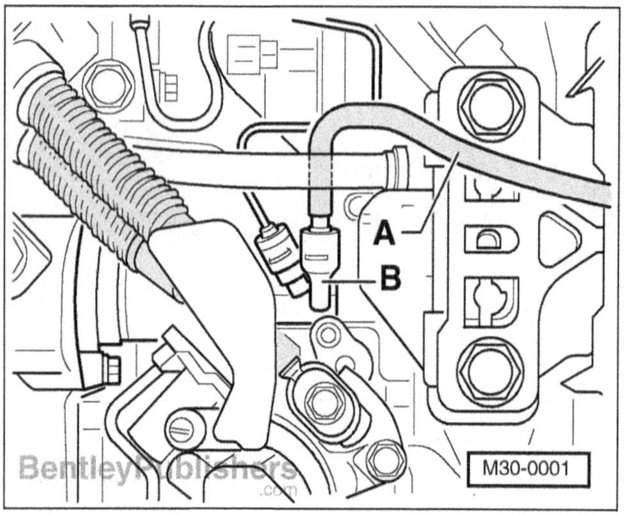

< Connect bleeder hose (**A**) to slave cylinder bleeder valve (**B**).

- 02J transmission shown, others are similar.

02J transmission

– Open bleeder valve to bleed system.

– Bleed fluid following instructions provided with power bleeder or until no air bubbles can be seen in bleed tube.

– Depress clutch pedal several times after completing bleeding process.

– Top off hydraulic fluid in reservoir.

0A4 and 02M transmissions

– Open bleeder valve to bleed system.

– Bleed approximately 100 ml (.11 qt) of fluid following instructions provided with power bleeder or until no air bubbles can be seen in bleed tube.

– Quickly operate clutch pedal from stop to stop ten to fifteen times.

– Bleed off an additional 50 ml (.05 qt) of fluid and close valve.

– Top off hydraulic fluid in reservoir.

Specification
- Hydraulic (brake) fluid .... DOT 4

CLUTCH RELEASE MECHANISM

The clutch slave cylinder can be replaced with the transmission installed in the car. Replacement of the other clutch release components require that the transmission be removed from the car. See **34 Manual Transmission** for transmission removal and installation procedures.

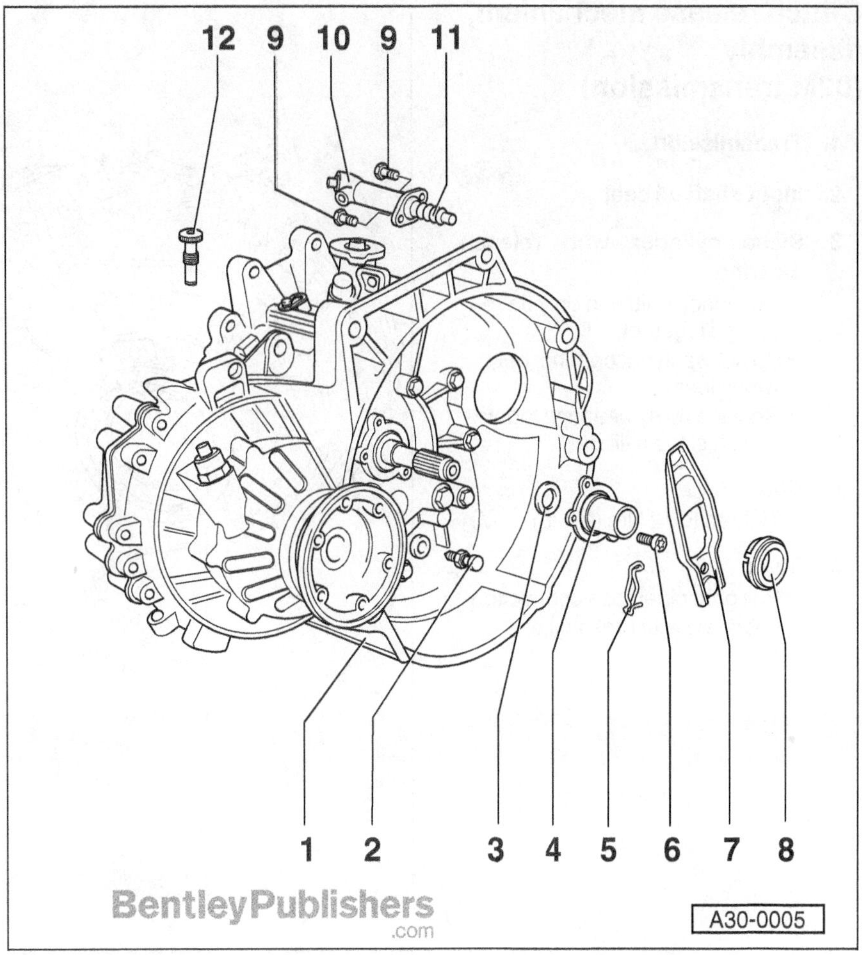

Clutch release mechanism, assembly (02J, 0A4 transmission)

1. **Transmission**
2. **Ball stud**
   - Tighten to 25 Nm (18 ft lb)
   - Lubricate with MoS_2
3. **Input shaft oil seal**
4. **Guide sleeve**
   - With vulcanized O-ring
   - If damaged, replace guide sleeve and O-ring
   - Lubricate guide sleeve in area of release bearing with MoS_2 grease
5. **Retaining spring**
   - Secure to clutch release lever
6. **Bolt**
   - Tighten to 20 Nm (15 ft-lb)
7. **Clutch release lever**
8. **Release bearing**
   - Do not wash bearing out, only wipe
   - Replace noisy bearings
   - Lubricate contact surfaces at release lever with MoS_2 grease
9. **Bolt**
   - Tighten to 20 Nm (15 ft-lb)
10. **Slave cylinder**
11. **Plunger**
    - Grease end of plunger with MoS_2 grease
12. **Assembly bolt**
    - Secures clutch release lever when installing transmission
    - Remove after trans. installation
    - An M8 x 35 bolt can be used in place of assembly bolt

Release bearing and lever, removing and installing

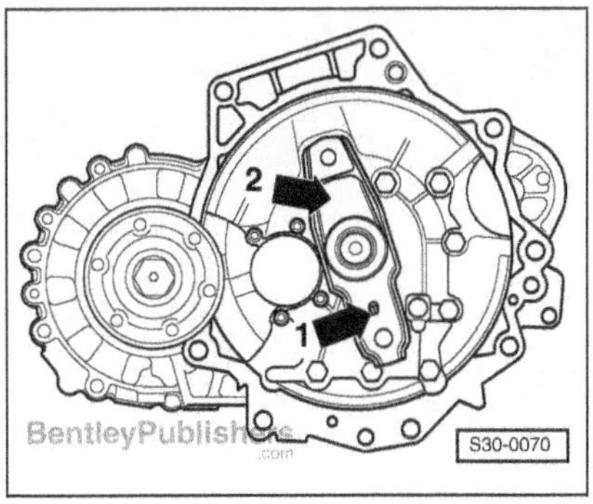

◄ Disengage spring (**arrow 1**) to remove clutch release lever (**arrow 2**).

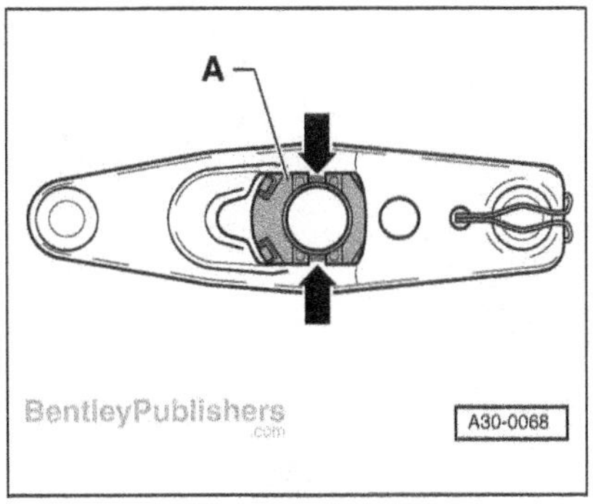

◄ Press retaining tabs (**arrows**) together or rear side of clutch release lever.

– Remove release bearing (**A**) from release lever.

– To install release bearing, press clutch release lever until retaining tabs (**arrows**) engage.

Clutch release mechanism, assembly (02M transmission)

1. **Transmission**
2. **Input shaft oil seal**
3. **Slave cylinder with release bearing**
   - One piece unit, can only be replaced together
   - Do not wash out bearing, only wipe clean
   - Replace noisy bearings together with slave cylinder
4. **Bolt**
   - Tighten to 12 Nm (9 ft-lb)
5. **Seal**
   - Pull onto pipe/hose connection
   - Lubricate with brake fluid

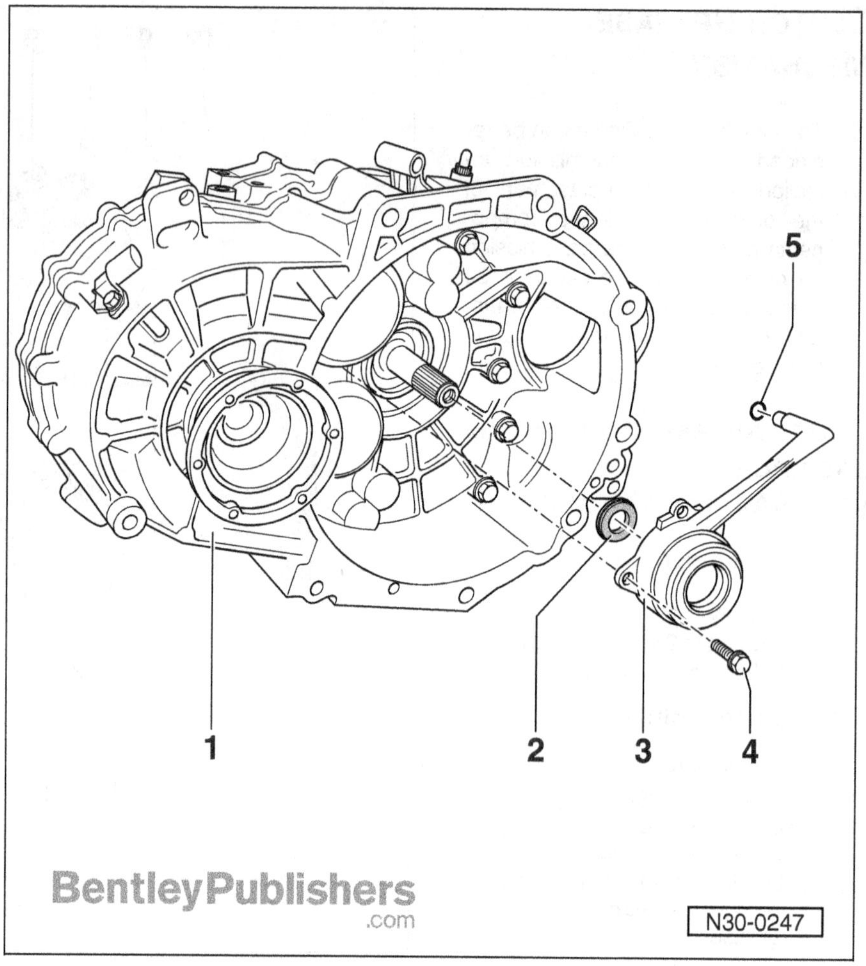

CLUTCH, SERVICING (TRANSMISSION REMOVED)

Clutch assembly

1. **Flywheel**
   - Make sure centering pins fit tightly
   - Contact surfaces for clutch disc must be free of grooves, oil and grease
   - Removing/installing, see **13 Crankshaft and Cylinder Block**
2. **Clutch disc**
   - Installation position with two-piece flywheel: shorter hub end (**arrow**) points toward pressure plate
   - Installation position with one-piece flywheel: spring cage faces pressure plate
   - Clean and lightly grease splines
   - After light greasing, move clutch plate back and forth on input shaft until hub moves freely
   - See **Clutch disc, centering pressure plate, removing/installing**
3. **Pressure plate**
   - 02J pressure plate shown, 02M/0A4 similar
   - Clean contact surface only
   - See **Diaphragm spring, checking ends**
4. **Bolt**
   - Select correct flywheel bolt for two- or one-piece flywheel
   - Loosen and tighten gradually and diagonally
   - Two-piece flywheel
     02J transmission:
     Tighten to 13 Nm (10 ft-lb)
     02M transmission:
     Tighten to 22 Nm (16 ft-lb)
   - One-piece flywheel:
     Tighten to 20 Nm (15 ft-lb)
   - 0A4 transmission:
     Sachs clutch: 20 Nm (15 ft-lb)
     Luk clutch: 13 Nm (10 ft-lb)

NOTE —

- *Replace clutch plates and pressure plates that have damaged or loose rivets.*
- *Select the correct clutch plate and pressure plate according to engine code.*

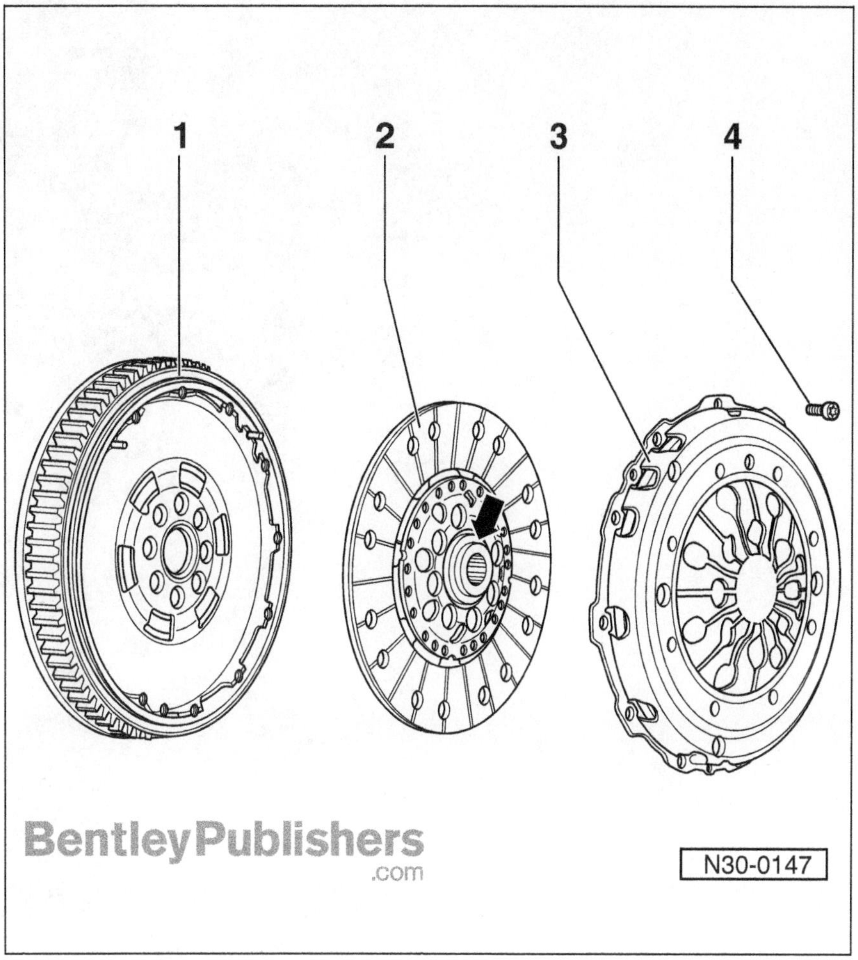

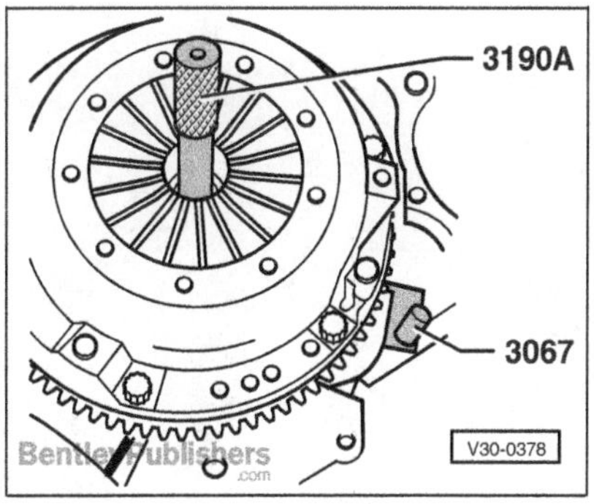

Clutch disc, centering pressure plate, removing/installing

- Use VW special tool 3190A, T10097, or equivalent, to center clutch disc
- Loosen and tighten bolts gradually and diagonally
- When removing, reverse position of flywheel locking tool (VW 3067 shown)
- Pressure plate contact surface and clutch disc must make full contact with flywheel.

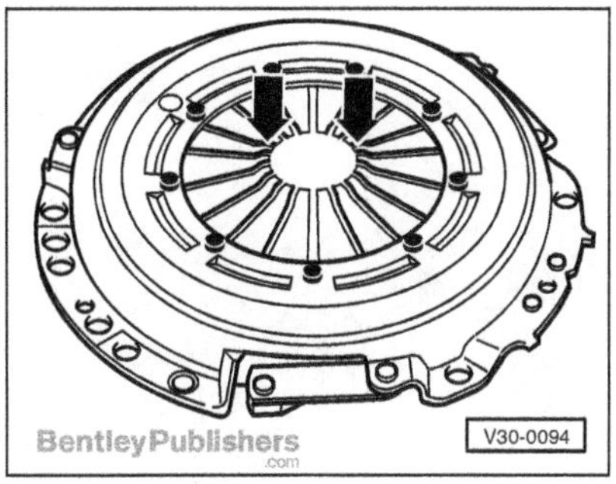

Diaphragm spring, checking ends

- Check pressure plate diaphragm spring for wear (arrows)
- Maximum wear: up to half of original diaphragm spring thickness

34a Manual Transimission

GENERAL

This repair group covers repair information for the cable-operated gear shift mechanism and the removal and installation of the various manual transmissions. This repair group does not cover transmission teardown and disassembly.

Manual transmission applications

- 02J 5-speed 1.8L, 1.9L, 2.0L
- 02M 5-speed . . . . 1.8L, 150 hp
- 02M 6-speed . . . . 1.8L, 180 hp
- 0A4 5-speed 2.5L

NOTE —
- *To check manual transmission oil, see* **03 Maintenance**.

CAUTION —
- *Before working on the transmission or gear selector lever mechanism disconnect the negative (–) battery cable.*
- *Disconnecting the negative (–) battery cable may erase fault codes and basic settings in the engine management and automatic transmission control modules. Some driveability problems may be noticed until the system re-adapts to operating conditions. OBD II readiness codes, which may be required for emissions testing, may also be erased.*

The manual transmissions covered by this manual use a hydraulic clutch release mechanism and a cable operated shift mechanism. **Table a** lists the engine applications, gear ratios, and other data for the 02J and 02M manual transmissions. **Table b** lists the engine applications, final drive ratios, and other data for the 0A4 manual transmission. The transmission code and production number can be found stamped on the transmission case near the shift linkage.

Table a. 02J/02M manual transmission specifications

Designation	**02J**					**02M**	
Code letters	EBQ, EGD, EGX	FBW	DQY, EBJ	EGR	DZQ, EBP, EGT, EKG, EZK, FBV	ERR	FML, FZQ
Engine application	1.8L, 150hp	1.8L, 150hp	1.9L, 90hp	1.9L, 90hp	2.0L, 115hp	1.8L, 180 hp	1.8L, 180 hp
Ratio: Z_2:Z_1							
Final drive	3.938	3.938	3.389	3.389	4.235	3.875 / 3.100*	3.944 / 3.087*
1st gear	3.300	3.300	3.788	3.788	3.778	3.417	3.357
2nd gear	1.944	1.944	2.118	2.118	2.118	2.105	2.087
3rd gear	1.308	1.308	1.360	1.360	1.360	1.481	1.469
4th gear	1.029	1.029	0.971	0.971	1.029	1.152	1.150
5th gear	0.837	0.837	0.756	0.756	0.837	1.167	1.194
6th gear						0.970	0.975
Reverse gear	3.060	3.060	3.600	3.600	3.600	4.107	3.990
Lubricant							
Capacity	2.0 liters (2.1 qt.)					2.3 liters (2.4 qt.)	
Specification	SAE 75W90 mineral oil VW part # G 052 726 A2					SAE 75W90 synthetic oil VW part # G 052 171 A2	
Clutch control	hydraulic						
Clutch disc diameter	225 mm	225 mm	228 mm	228 mm	215 mm	240 mm	
Ratio, overall in top gear	3.296	3.296	2.562	2.562	3.545	3.007	3.010

* 3.875 is final drive ration for 1st through 4th gears, 3.100 is final drive ratio for 5th, 6th, and reverse gears.

Table b. 0A4 manual transmission specifications

Designation	**0A4**	
Code letters	HJE	JCW
Manufacture date	07/05 to 11/05	11/05 - on
Engine application	2.5L	
Ratio: Z_2:Z_1		
Final drive	3.647	
Lubricant capacity	1.8 liters (2.0 qt.)	
Lubricant specification	see authorized Volkswagen dealer	

GEAR SELECTOR MECHANISM

Access to the gear selector mechanism may be difficult without partially removing the exhaust system, see **26 Exhaust System and Emission Controls.**

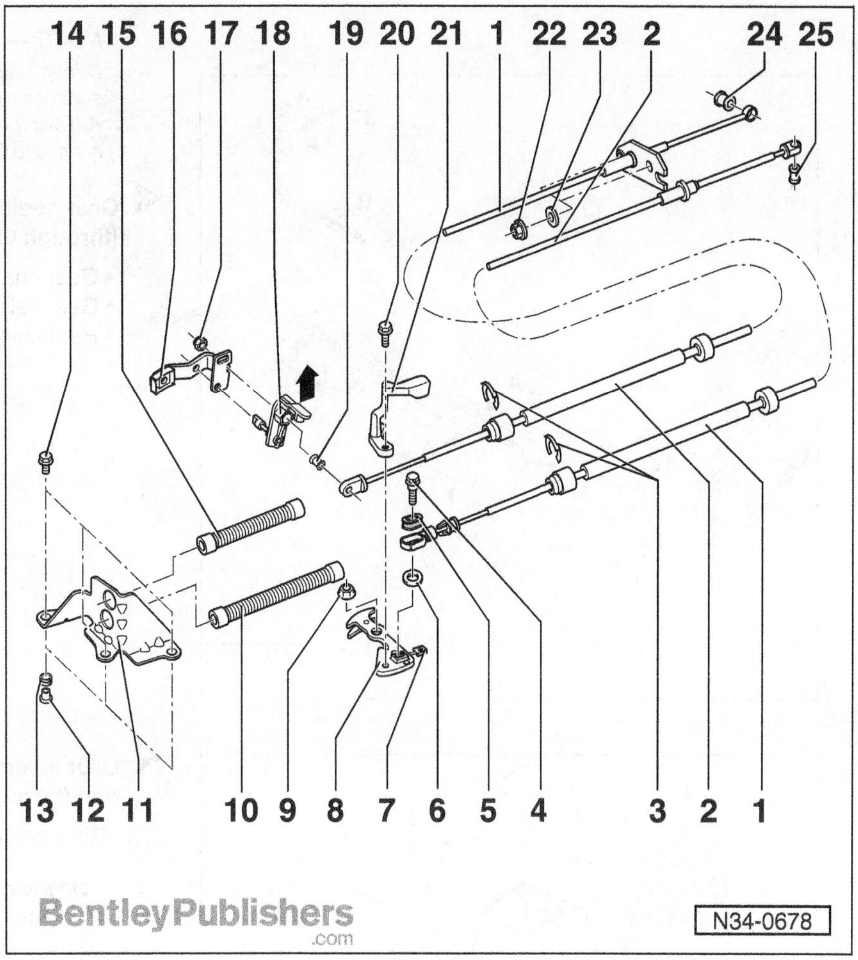

Gear selector cables, assembly (through 05/99)

1. **Gear change cable**
   - See **Gear selector mechanism, installation position**
2. **Gate selector cable**
   - Pull off actuating arm (item 18) by pulling tab in direction of **arrow**
   - Install before installing gear selector cable
   - See **Gear selector mechanism, installation position**
3. **Circlip**
   - When removing do not damage boot
4. **Bolt**
   - Tighten to 25 Nm (18 ft-lb)
   - With shoulder
   - Gear selector cable to gear change lever
5. **Rubber bushing**
   - For gear change cable to gear selector lever
   - To replace boot press out of end piece
6. **Washer**
   - Install between gear selector cable and selector lever
7. **Square nut**
   - Insert into gear selector lever
8. **Gear selector lever**
   - Only fits in one position
   - After installing adjust gear selector mechanism
   - See **Gear selector lever and relay lever, installation position**
9. **Self-locking nut**
   - Tighten to 25 Nm (18 ft lb)
   - For gear selector lever
10. **Boot**
    - Carefully pull over gear selector cable end piece
    - Use MoS_2 grease
11. **Cable support bracket**
12. **Spacer**
13. **Bushing**
    - Cable support bracket mounting to transmission
14. **Bolt**
    - Tighten to 25 Nm (18 ft-lb)
    - For cable support bracket (item 11)
15. **Boot**
    - Carefully pull over gear selector cable end piece
    - Use MoS_2 grease
16. **Relay lever**
    - Remove before removing clutch controls
    - See **Gear selector lever and relay lever, installation position**
17. **Self-locking nut**
    - Tighten to 15 Nm (11 ft-lb)
    - For relay lever to actuating arm
    - Only loosen to adjust selector mechanism
18. **Actuating arm**
    - With pivot pin to mount relay lever
19. **Rubber bushing**
    - For gate selector cable to actuating arm
    - Press in and out with a drift
20. **Bolt**
    - Tighten to 25 Nm (18 ft-lb)
    - For balance weight
21. **Balance weight**
    - Remove to adjust selector mechanism
22. **Self-locking nut**
    - Tighten to 15 Nm (11 ft-lb)
    - For selector cable to selector housing
23. **Rubber washer**
24. **Rubber bushing**
    - For gear selector cable/gear change lever
    - Press in and out with a drift
25. **Rubber bushing**
    - For gate selector cable/selector bracket
    - Press in and out with a drift

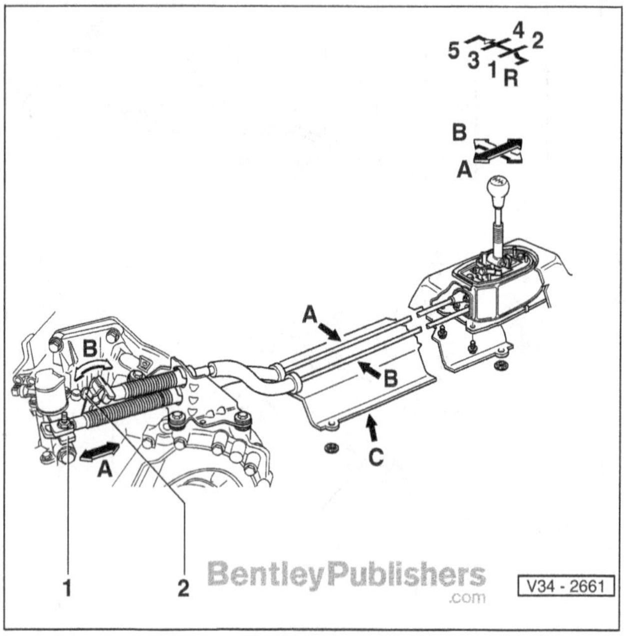

NOTE —
During reassembly, lubricate all mounting and contact surfaces with Polyuric grease (VW part no. G 052 142 A2). As the final step, adjust the gear selector mechanism, see **Gear selector mechanism, adjusting**.

Gear selector mechanism, installation position (through 05/99)

- Gear change cable (**A**) controls forward/back travel.
- Gear selector cable (**B**) controls side-to-side travel.
- Heat shield (**C**) holds cables in position.

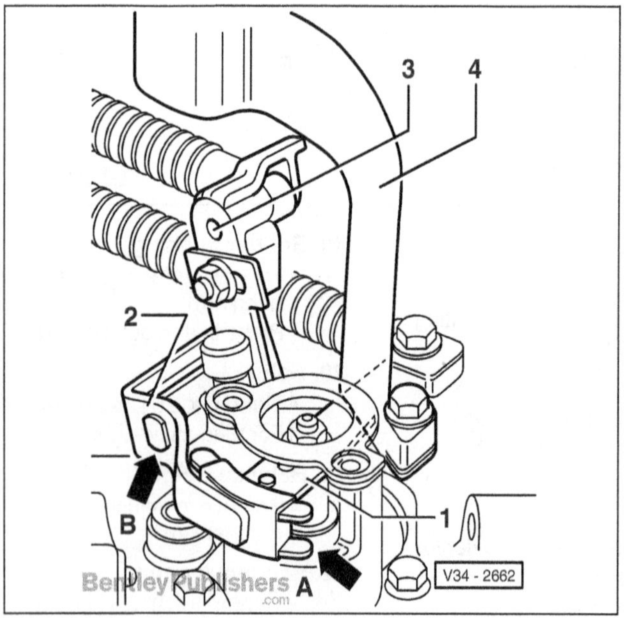

Gear selector lever and relay lever, installation position (through 05/99)

1. **Gear selector lever**
2. **Relay lever**
    - Locates in guide rail of gear shift lever (**arrow A**)
    - Lubricate with MoS_2 grease
3. **Actuating arm**
    - Relay lever only fits in one position (**arrow B**)
4. **Balance weight**
    - Bolt on to gear selector lever

Gear selector cables, assembly (from 06/99)

NOTE —
02M transmission shown, 02J (from 06/99) and 0A4 are similar.

1. **Gear selector cable**
   - Press onto gear shift guide
   - See **Installation position of transmission selector lever/relay lever**
2. **Gate selector cable**
   - On relay lever
   - See **Installation position of transmission selector lever/relay lever**
3. **Circlip**
4. **Circlip**
   - Do not damage cable when removing
5. **Shift housing**
6. **Bolt**
   - Tighten to 25 Nm (18 ft-lb)
7. **Support bracket**
8. **Rubber bushing**
   - Support bracket mounting to transmission
9. **Distance piece**
10. **Cable locking mechanism**
    - For gate selector cable to relay lever
    - Not necessary to be replaced
    - Secured with securing clip
11. **Relay lever**
12. **Bushing**
13. **Circlip**
14. **Shoe**
15. **Transmission selector lever**
    - With balance weight
    - Install so that master spline aligns with selector shaft
    - After installing adjust gear selector mechanism
16. **Hex nut**
    - Tighten to 20 Nm (15 ft-lb)
    - Self locking, always replace
17. **Cable locking mechanism**
    - For gear selector cable to gear selector lever
    - Not necessary to be replaced
    - Secured with securing clip

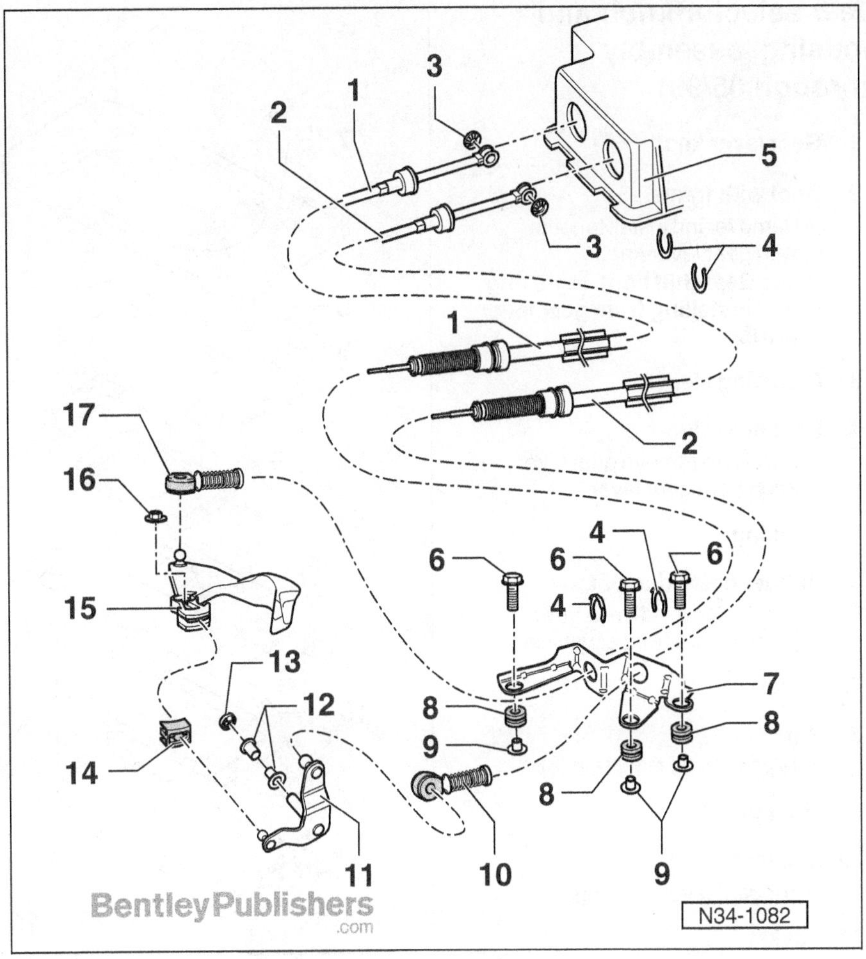

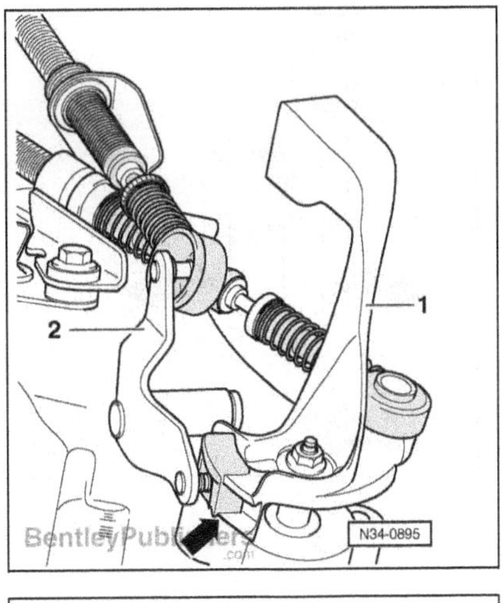

Installation position of transmission selector lever/relay lever (02J transmission, from 06/99)

1. Gear selector lever with balance weight
2. Relay lever - locates in guide rail of gear selector lever via shoe (**arrow**)

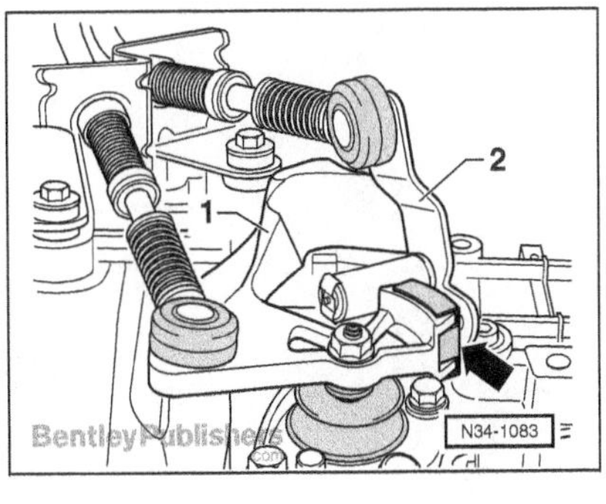

Installation position of transmission selector lever/relay lever (02M transmission)

1. Gear selector lever with balance weight
2. Relay lever - locates in guide rail of gear selector lever via shoe (**arrow**)

Gear selector knob and housing, assembly (through 05/99)

1. **Gear lever knob**
2. **Boot with frame**
   - Remove and install together with gear lever knob
   - See **Gearshift boot, removing and installing from gear lever knob**
3. **Clamping sleeve**
4. **Retaining clip**
   - Cut clip to remove gear lever knob from gear lever
5. **Bushing**
6. **Rubber noise damper**
   - Installation position: Arrows point toward front of vehicle
7. **Cover**
8. **Nut**
   - Tighten to 25 Nm (18 ft-lb)
9. **Floor panel**
10. **Retainer**
    - For gear lever housing
11. **Gasket**
    - Between gear lever housing and floor
    - Self-adhesive
    - Bond to gear lever housing
12. **Gasket**
    - Between selector cables and gear lever housing
13. **Gear lever housing**
14. **Gasket**
    - Self-adhesive
    - Bond underneath gear lever housing
15. **Cover plate**
16. **Bolt**
    - Tighten to 25 Nm (18 ft-lb)

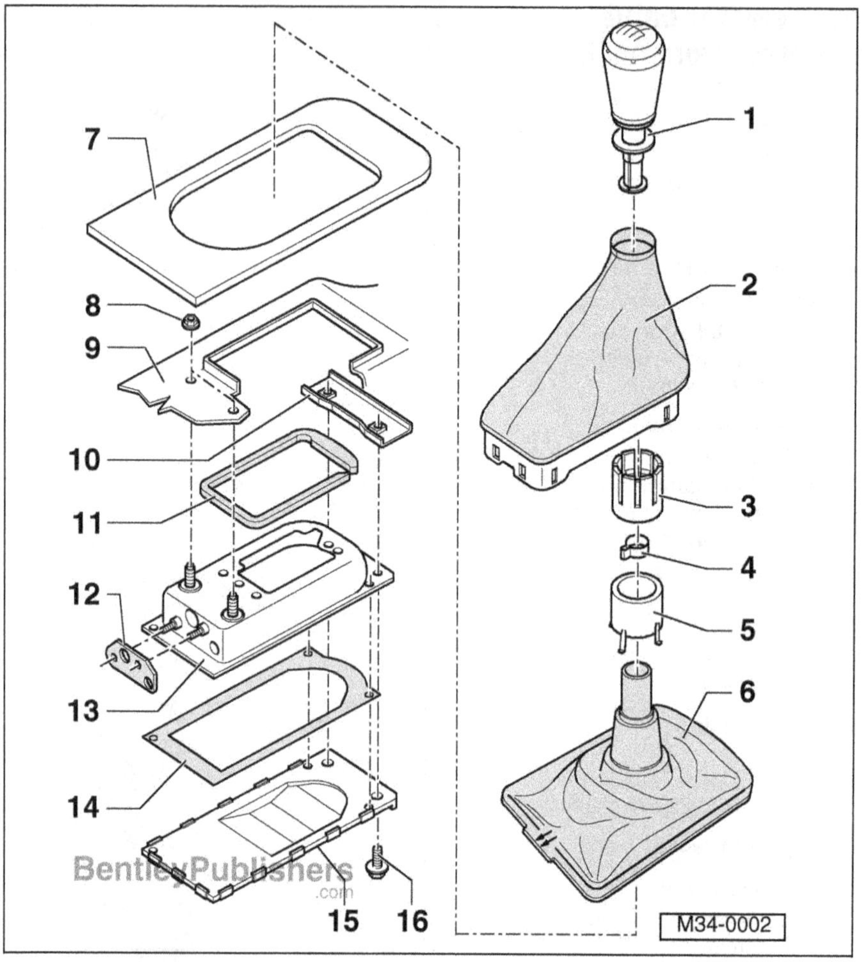

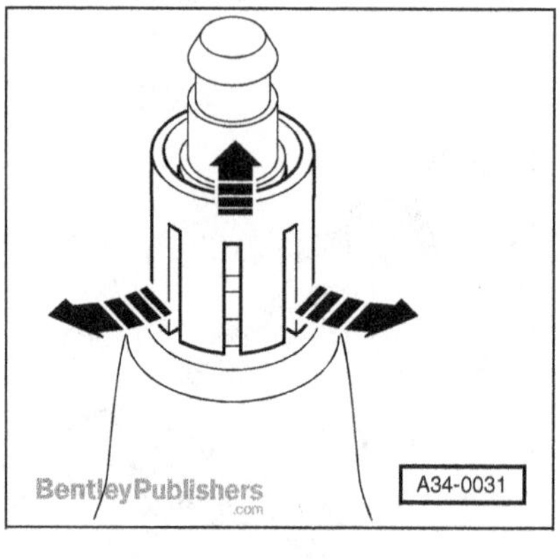

Gearshift boot, removing and installing from gear lever knob

- Cut retaining clip to remove gear lever knob from gear lever
- Turn boot inside out.
- Carefully pry clamping sleeve legs with small screwdriver in direction of lower arrows while simultaneously pulling gear lever knob out.
- When installing insert gear lever knob in boot.
- Push sleeve onto gear lever knob and engage.
- Install knob and boot together on gear lever.

Gear selector knob and cover, assembly (from 06/99)

1. **Gear lever knob**
   - Remove and install together with cover
2. **Cover**
   - Remove and install together with gear lever knob
   - See **Shift lever cover, separating and assembling to shift lever knob**
3. **Sleeve**
4. **Clamp**
   - To secure shift lever knob on shift lever
   - Secure on shift lever knob using pliers
5. **Frame**
6. **Rubber noise damper**

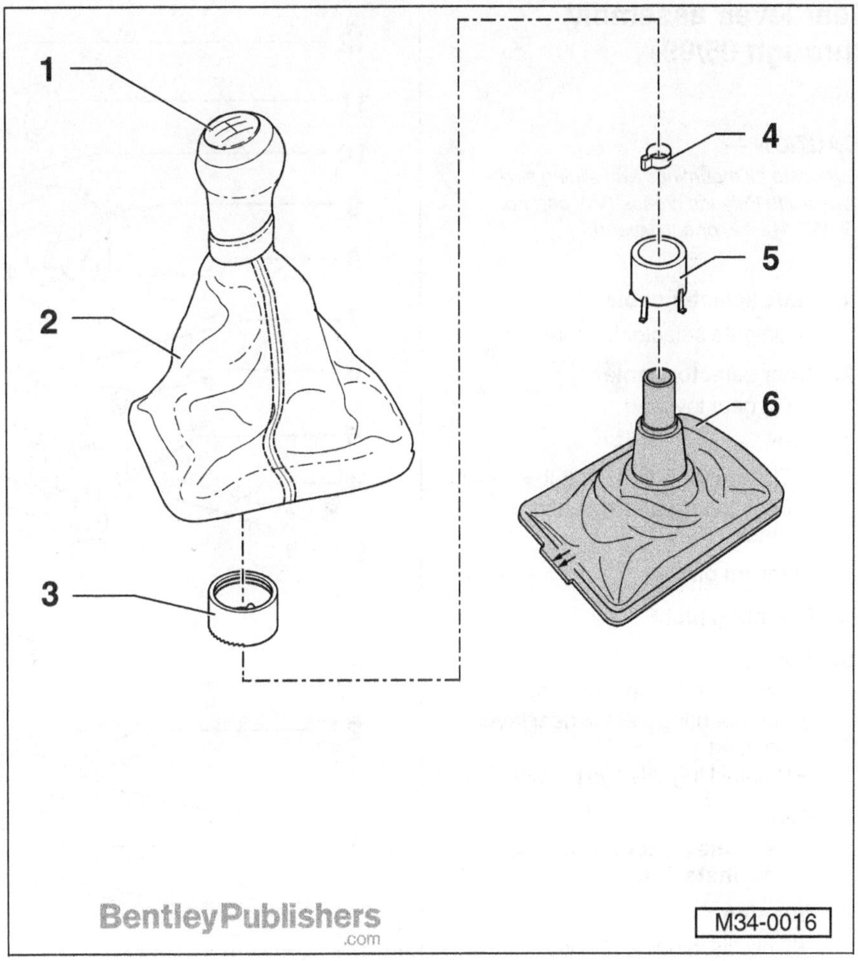

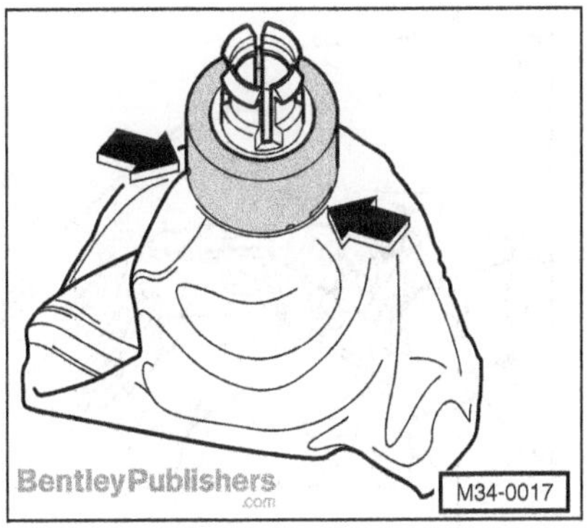

Shift lever cover, separating and assembling to shift lever knob
- Turn cover inside out
- Carefully pry clamping sleeve (**arrow**) off with a screwdriver and pull shift knob out.
- When installing insert gear lever knob in boot.
  Push sleeve onto gear lever knob and engage.
  Install knob and boot together on gear lever.

Gear lever, assembly (through 05/99)

CAUTION—
Lubricate all mountings and sliding surfaces with Polyuric grease (VW part no. G 052 142 A2, or equivalent).

1. **Gate selector cable**
   - On gate selector bracket
2. **Gear selector cable**
   - On gear lever
3. **Bolt**
   - Tighten to 15 Nm (11 ft-lb)
   - For fulcrum pin to mounting plate
4. **Fulcrum pin**
5. **Mounting plate**
6. **Nut**
   - Tighten to 20 Nm (15 ft-lb)
   - For mounting plate to gear lever housing
   - Self-locking, Always replace
7. **Cap**
   - See **Gate selector housing cap, installing**
8. **Bushing**
   - Only fits in one position
9. **Gate selector housing**
   - Install into mounting plate after installing complete gear lever
10. **Bushing**
11. **Spring**
12. **Guide bushing**
    - Position on bushing (item 8)
13. **Circlip**
    - See **Gear lever circlip, removing**
14. **Spacer sleeve**
15. **Spring**
16. **Gear lever guide**
17. **Lock plate**
    - See **Lock plate for gate selector bracket and gate selector housing, removing**
18. **Damping washer**
19. **Bushing**
    - Only fits in one position
20. **Gate selector bracket**
21. **Spring clip**
22. **Gear lever**

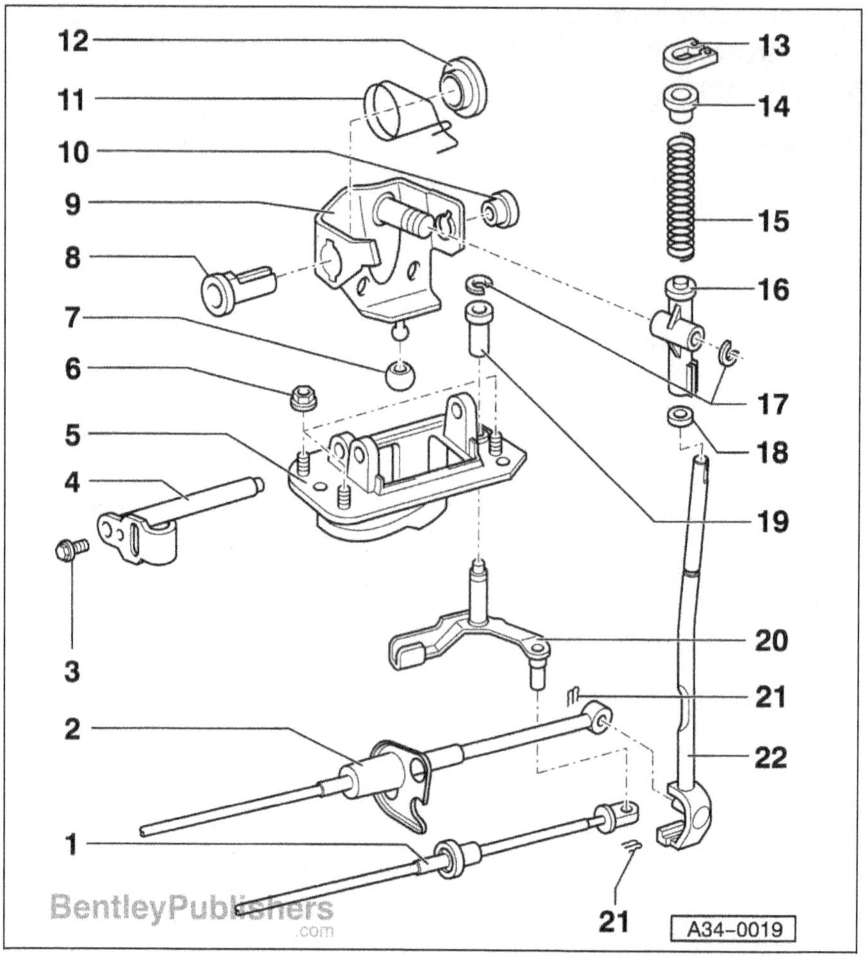

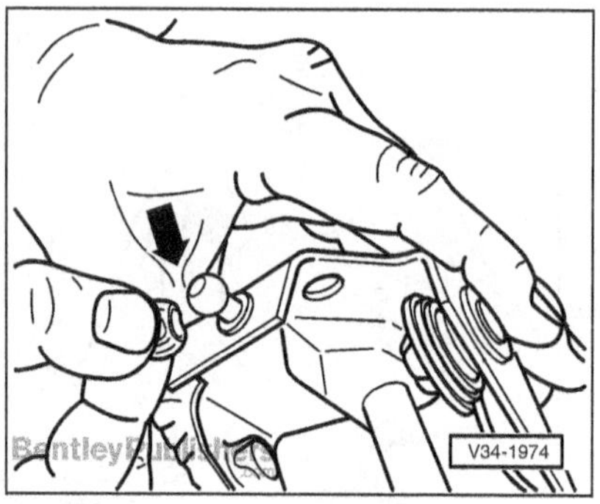

Gate selector housing cap, installing

- Press on from side (**arrow**).
- Press opening together to form a slot.

NOTE—
If necessary, heat cap with hot water to soften.

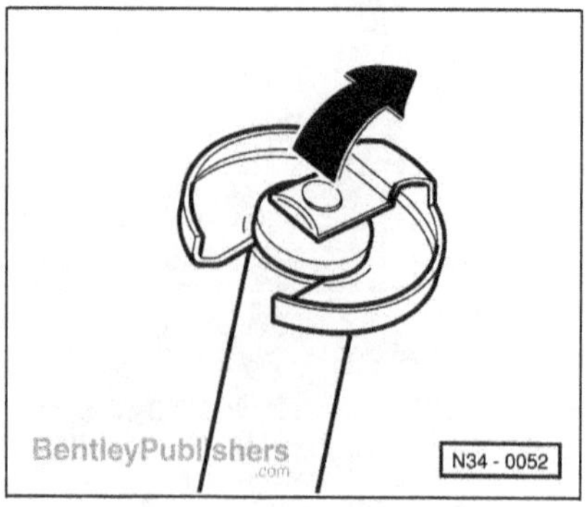

Lock plate for gate selector bracket and gate selector housing, removing

- Lift tab in direction of **arrow**.

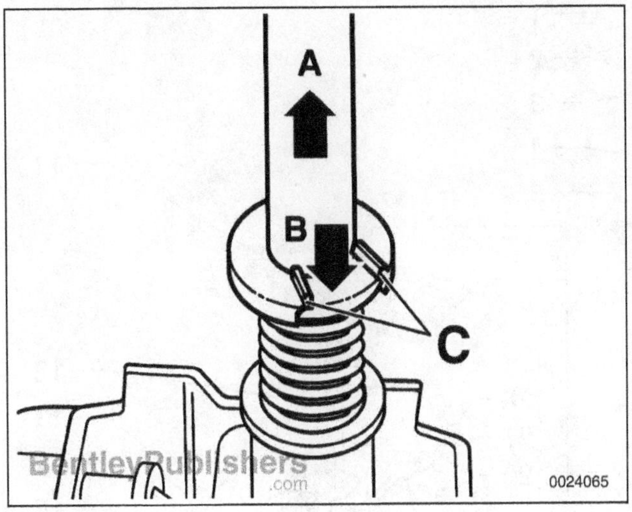

Gear lever circlip, removing

- Pull gear lever up (**arrow A**) and simultaneously use a screwdriver to push spacer bushing down (**arrow B**) to stop. Remove circlip (**C**).

NOTE —

Do not tilt spacer bushing when pushing down. Carefully release pressure at groove in shift lever for circlip

Gear lever, assembly (from 06/99)

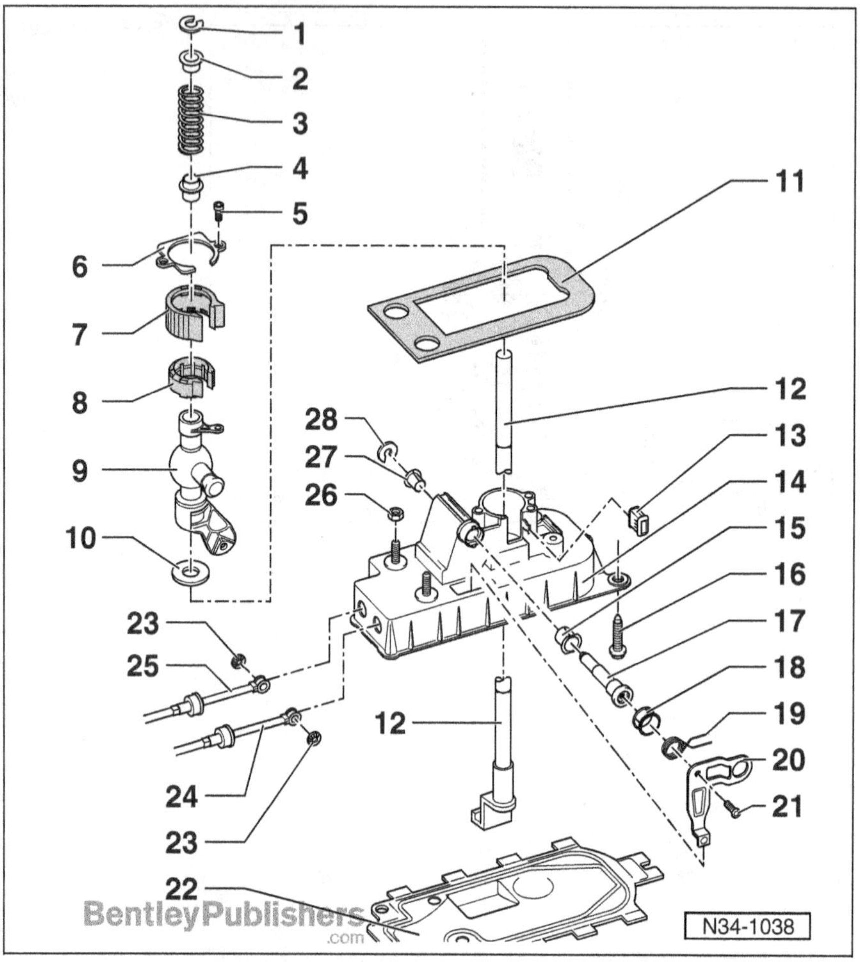

NOTE —
Lubricate all mountings and sliding surfaces with grease (VW part no. G 000 450 02, or equivalent).

1. **Circlip**
   - See **Gear lever circlip, removing**
2. **Bushing**
3. **Spring**
4. **Bushing**
5. **Torx screw**
   - Tighten to 5 Nm (44 in-lb)
6. **Cover plate**
7. **Damper**
8. **Bearing shell**
9. **Gear shift guide**
10. **Damping washer**
11. **Gasket**
12. **Transmission shift lever**
13. **Damper**
14. **Shift housing**
15. **Bushing**
16. **Bolt**
    - Tighten to 25 Nm (18 ft-lb)
17. **Fulcrum pin**
18. **Guide bushing**
19. **Spring**
    - See **Installing spring**
20. **gate selector bracket**
21. **Torx screw**
    - Tighten to 5 Nm (44 in-lb)
22. **Base plate**
    - Open tabs to remove
23. **Lock washer**
24. **Gate selector cable**
25. **Gear selector cable**
    - Press onto gear selector guide
26. **Hex nut**
    - tighten to 25 Nm (18 ft-lb)
27. **Bushing**
    - Only fits in one position
28. **Lock washer**

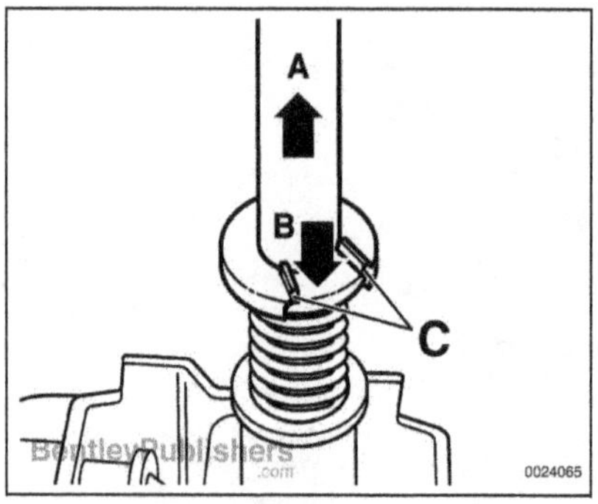

Gear lever circlip, removing

- Pull gear lever up (**arrow A**) and simultaneously use a screwdriver to push spacer bushing down (**arrow B**) to stop. Remove circlip (**C**).

NOTE —
Do not tilt spacer bushing when pushing down. Carefully release pressure at groove in gear lever for circlip.

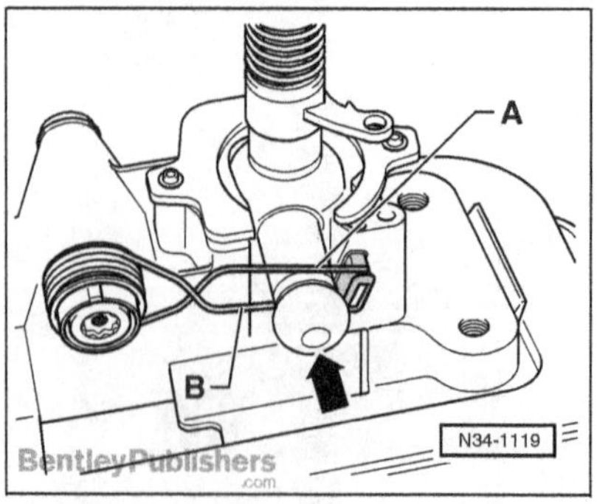

Installing spring

- Insert spring so that spring extension (**A**) lies above pin (**arrow**)
- Pull spring extension (**B**) down so that it reaches below pin (**arrow**).

Gear selector mechanism, adjusting (through 05/99)

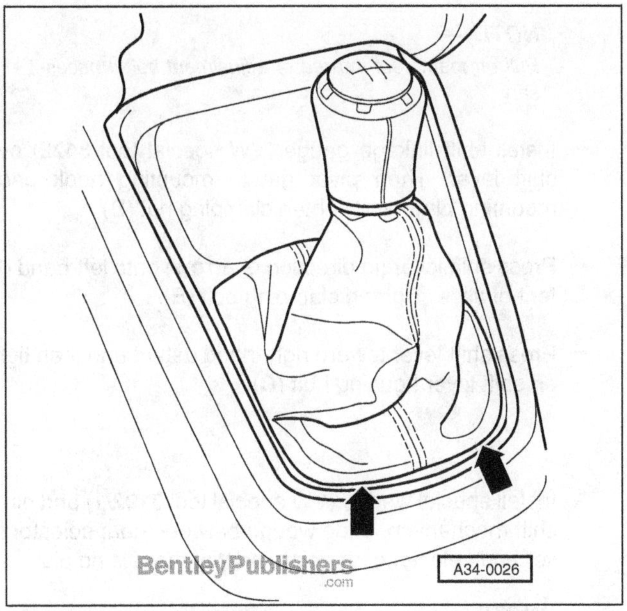

NOTE —

Special VW tools are required to accurately adjust the shift mechanism. Read the procedure through to determine what tools will be necessary.

– Place transmission in neutral.

◄ Carefully pry boot off center console (**arrows**).

– Pull boot to rear off center console.

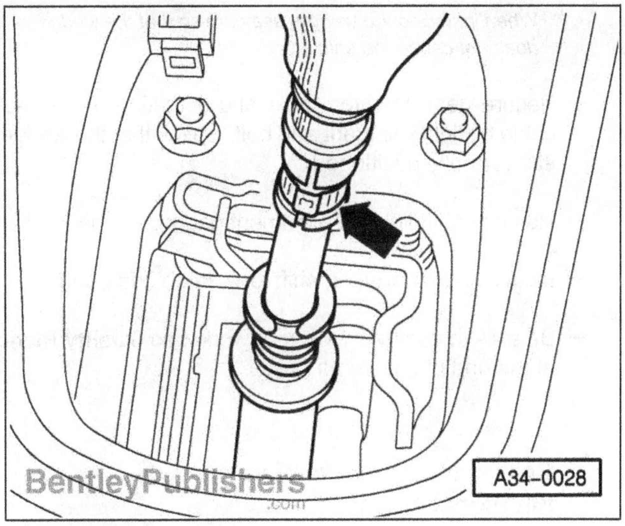

◄ Cut off retaining clip (**arrow**) under boot and pull gear lever knob off with boot.

– Remove rubber cover, if installed.

– Working at the transmission, remove balance weight from shift lever, as described earlier.

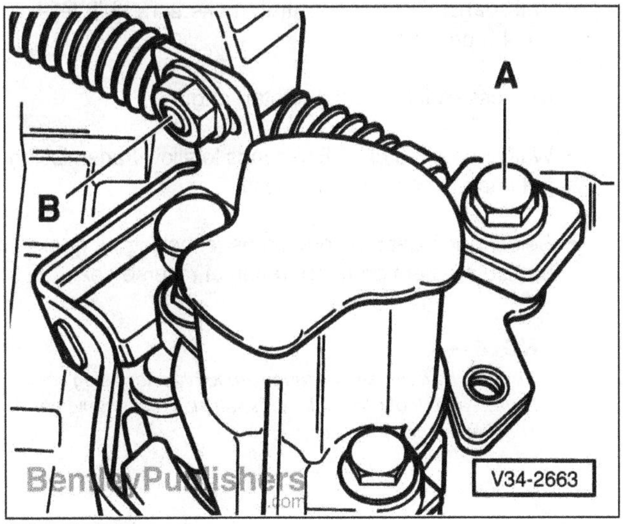

◄ Loosen bolt (**A**) and nut (**B**) for gear selector cables. Gear selector cables and actuating arm/gate selector cable should now move freely in elongated holes.

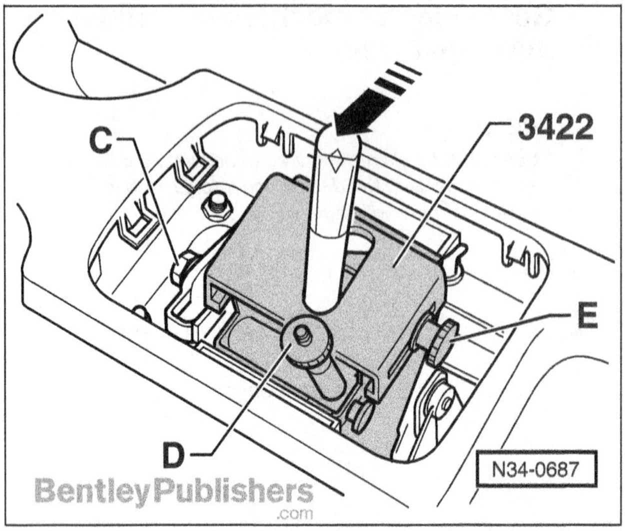

◀ Working in shift lever housing, loosen adjustment bolt (**C**).

> ***NOTE —***
> *Pull off foam seal in area of adjustment bolt if necessary.*

- Install shift linkage gauge (VW special tool **3422**) over shift lever. Then pivot gauge mounting hook under mounting plate and tighten clamping nut (**D**).
- Press shift lever (in direction of **arrow**) onto left-hand detent of slide. Tighten clamping bolt (**E**).
- Press shift lever toward right-hand detent and then tighten shift lever housing bolt (**C**).

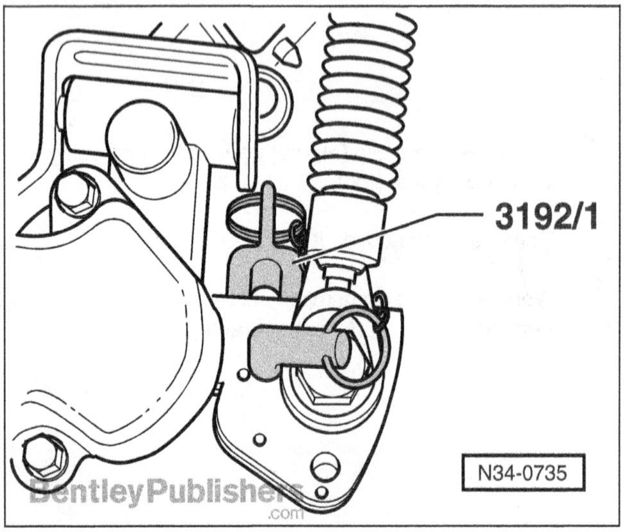

◀ Install special wedge (VW special tool **3192/1**) and pin to shift mechanism. Slide wedge between gear selector lever and selector cover plate so that there is no play.

> ***NOTE —***
> *When installing the wedge, use care so that the wedge does not cause the shift lever to lift.*

- Secure gear selector cable and actuating arm/selector cable by tightening nut and bolt. Check that the wedge is still correctly positioned.
- Remove shift linkage alignment gauge, wedge, and pin.
- Install balance weight, shift lever knob and boot.
- Be sure to quality check your work, see **Quality Review** at the end of this repair group.

Gear selector mechanism, functional check

- With transmission in neutral, make sure shift lever is in 3rd/4th gear gate
- Depress clutch pedal and start engine.
- Wait approximately 3-6 seconds to allow transmission input shaft to come to a standstill.
- Select each gear several times, checking all gears. Pay particular attention to operation of reverse gear.

> ***NOTE —***
> *Should any gear fail to engage smoothly after being selected repeatedly, the selector shaft play (lift) should be checked as described below.*

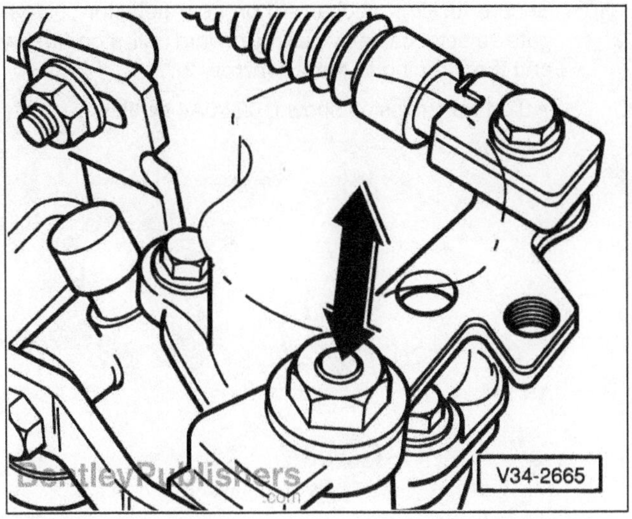

Have a helper select 1st gear and then press shift lever to left stop, then release. At same time, observe selector shaft on transmission. The selector shaft must move approximately 1 mm (0.04 in.).

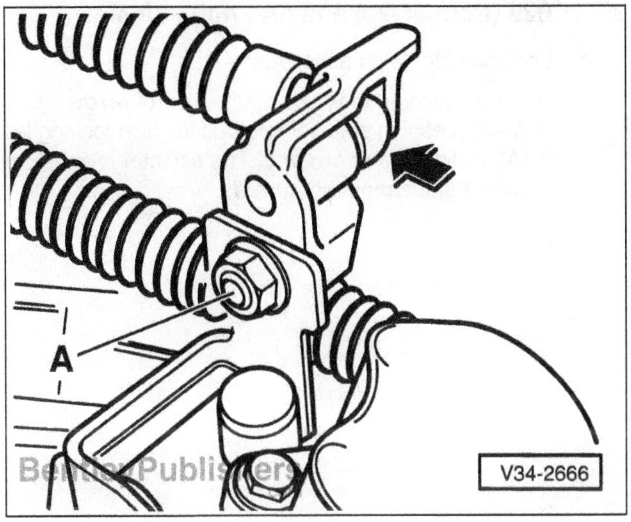

If any faults are found, disengage 1st gear, then loosen nut (**A**) at actuating arm/gate selector cable mounting. Take up gate selector cable play by lightly pressing on actuating arm (**arrow**) and retighten nut.

NOTE —
Gate selector cable has some play at mounting due to transfer elements, i.e. minimal play at mounting is normal.

Gear selector mechanism, adjusting (from 06/99)

Special tools

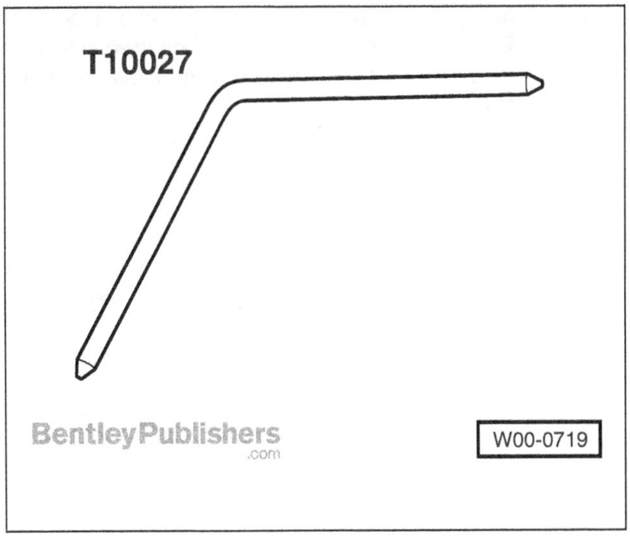

T10027 connecting pin

Requirements

The following points are essential to ensure correct selector adjustment:

- Moving parts of the selector mechanism and transmission elements must be in proper condition.
- Selector mechanism must move freely.
- The transmission, clutch and clutch mechanism must also be in proper condition.
- Transmission in neutral.

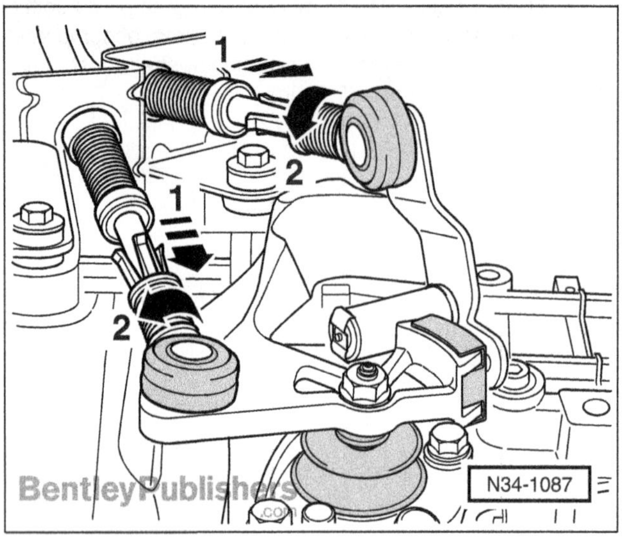

Secure locking mechanism on gear selector cable and gate selector cable by pulling forward until stop (**arrow 1**) and then turning to the left (**arrow 2**).

- 02M transmission shown, 02J/0A4 similar.

02J (from 06/99) and 0A4 transmission

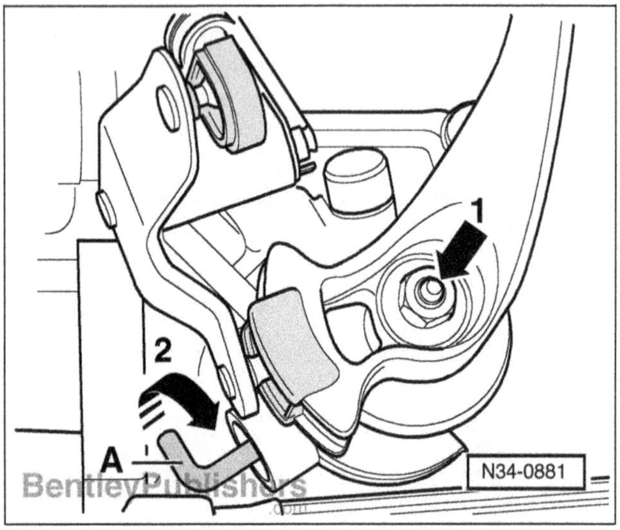

Lock selector lever as follows:

- Press selector shaft down (direction of **arrow 1**).
- While pressing selector shaft down, turn locking lever (**A**) in direction of **arrow 2**. The selector lever is now locked and cannot be moved.

02M transmission

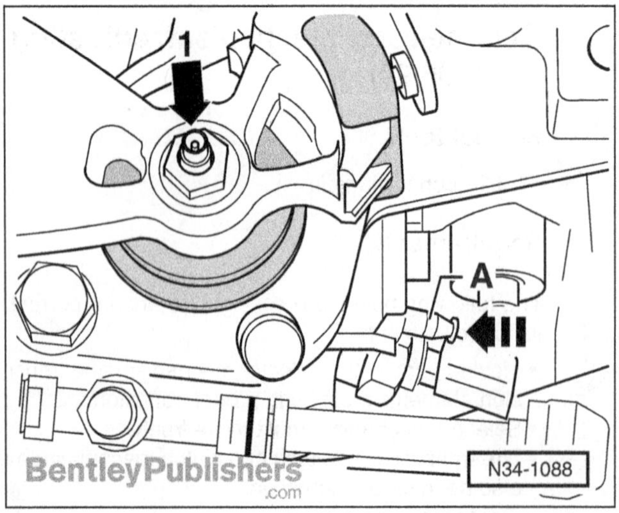

Lock selector lever as follows:

- Press selector shaft down (direction of **arrow 1**).
- While pressing selector shaft down, push locking pin (**A**) into transmission (direction of **dashed arrow**) until it engages. The selector lever is now locked and cannot be moved.

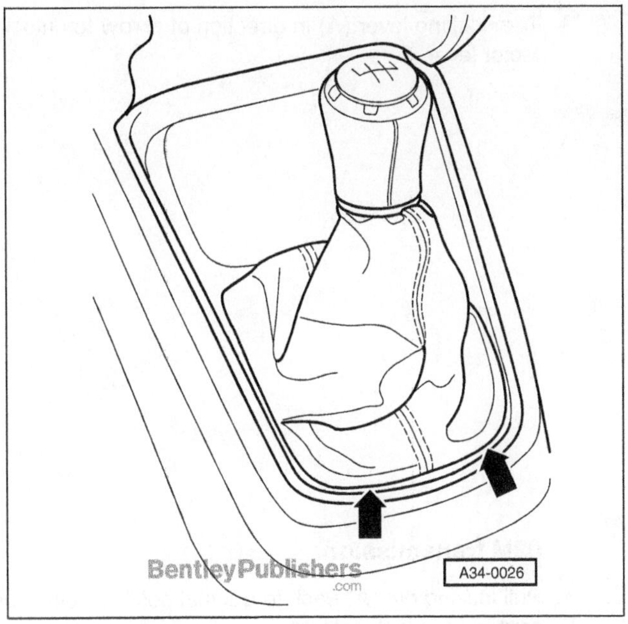

Continued for all versions

Pull shift lever cover off center console (**arrows**).

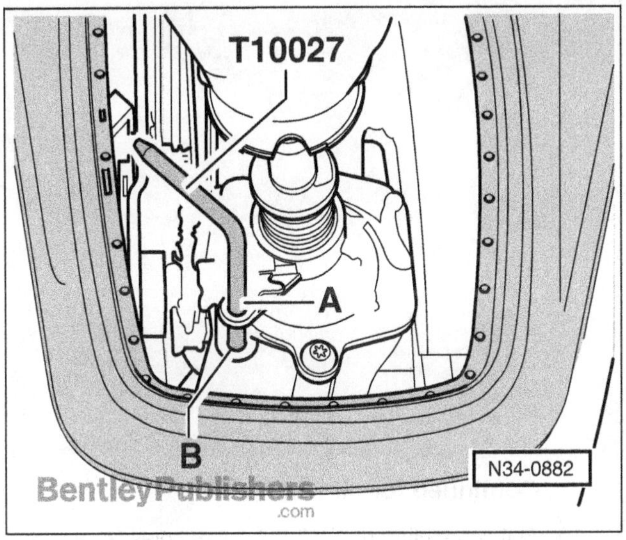

Set gear shift lever as follows:

- Guide shift lever in neutral to left into gate for 1st/2nd gear.
- Install connecting pin T10027 through hole (**A**) and into hole (**B**).

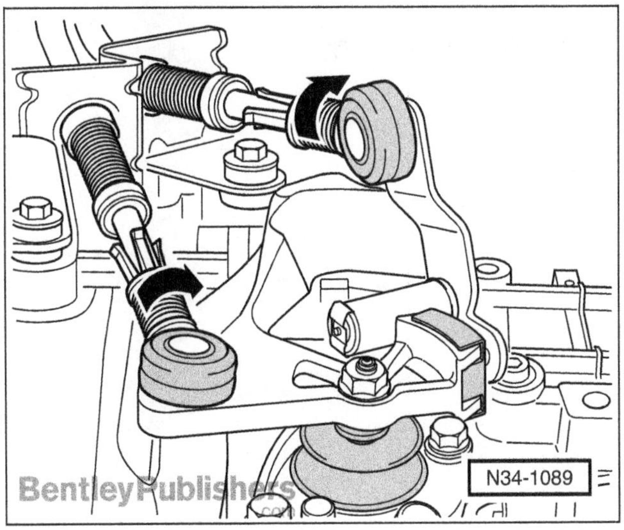

Turn locking mechanism on gear selector cable and gate selector cable to right onto stop (direction of **arrows**). The spring presses the locking mechanism into normal position.

- 02M transmission shown, 02J/0A4 similar.

02J (from 06/99) and 0A4 transmission

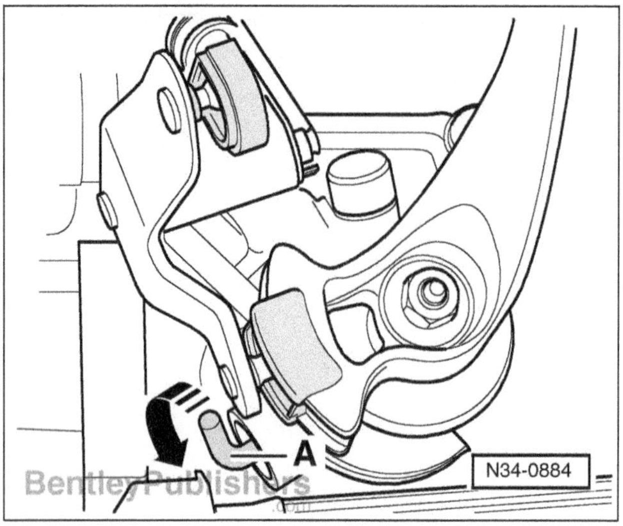

◄ Turn locking lever (**A**) in direction of **arrow** to unlock selector lever.

02M transmission

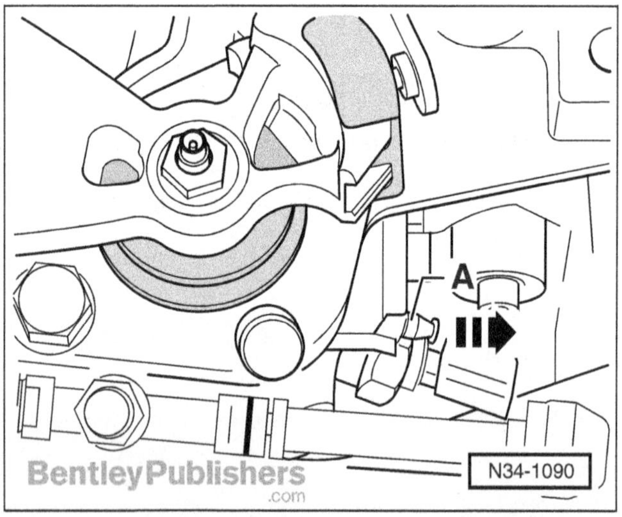

◄ Pull locking pin (**A**) back to original position (direction of **arrow**) to unlock selector lever.

Continued for all versions

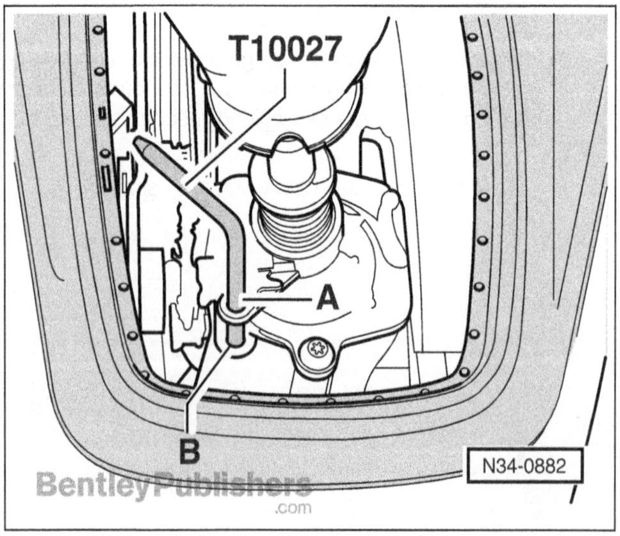

◄ Pull connecting pin T10027 out of holes (**A** and **B**).

– Clip cover into frame.

Gear selector mechanism, functional check

- With transmission in neutral, make sure shift lever is in 3rd/4th gear gate
- Depress clutch pedal and start engine.
- Wait approximately 3-6 seconds to allow transmission input shaft to come to a standstill.
- Select each gear several times, checking all gears. Pay particular attention to operation of reverse gear.

> ***NOTE —***
> *Should any gear fail to engage smoothly after being selected repeatedly, the selector shaft play (lift) should be checked as described below.*

- Have a helper select 1st gear and then press shift lever to left stop, then release. At same time, observe selector shaft on transmission. The selector shaft must move downward.

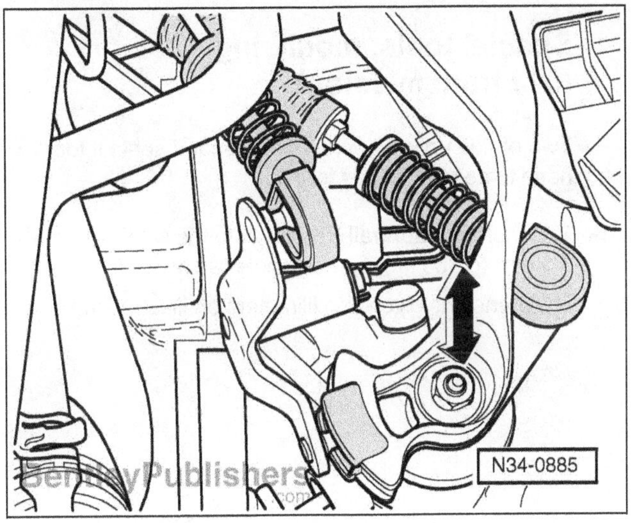

02J (from 06/99) and 0A4 transmission

When moving the gear lever, selector shaft must move approx. 1 mm (in direction of **arrow**). If not, re-adjust selector mechanism.

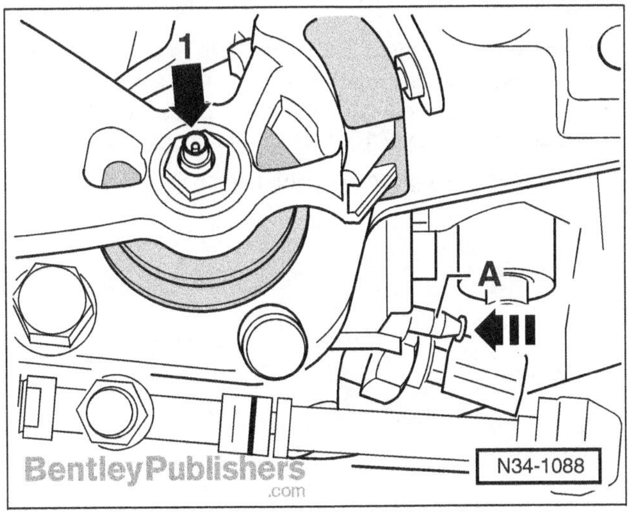

02M transmission

If shift mechanism is correctly adjusted, lock pin (**A**) can now be pushed into transmission (direction of **arrow**) and thus into selector shift gate. If not, re-adjust selector mechanism.

Transmission, Removing and Installing

This procedure describes the removal and installation of the manual transmission. Special engine lifting and jacking equipment is needed to support and reposition the engine as the transmission is removed from below.

Volkswagen recommended special tools
- VW 457/1 support rails
- 10-222A transmission support kit
- 10-222A/8 adapter
- 3282 transmission support
- 3282/8 adjustment plate
- 3300A engine support
- 3336 transmission lifting beam
- VAG 1331 torque wrench
- VAG 1332 torque wrench
- VAG 1383A engine/transmission jack
- VAS 5024 mounting pliers

Special tools, modifying (02J transmission)

Two of the Volkswagen recommended special tools will need to be modified as follows.

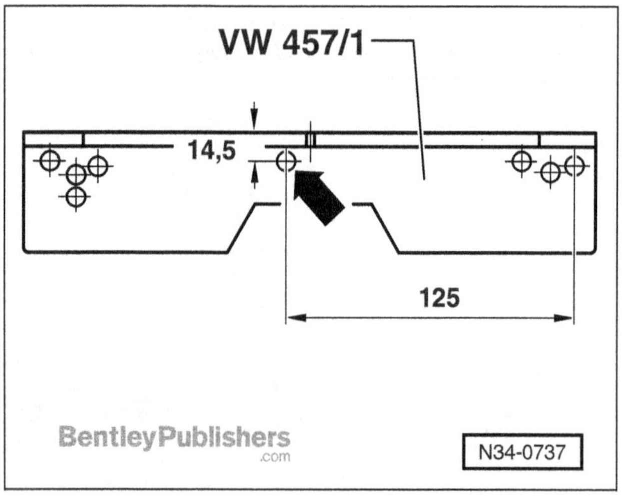

To secure support rail VW 457/1 to the subframe drill an 8.5 mm (0.335 in.) diameter hole (**arrow**).

- Dimensions given in millimeters on illustration.

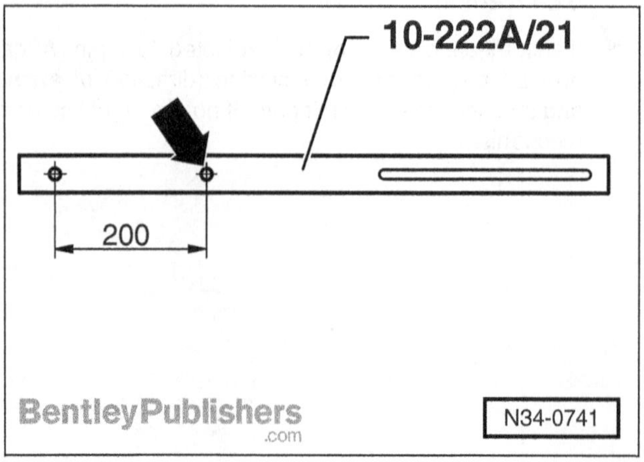

The engine support bridge 10-222A is used to support the engine/transmission assembly. Drill a 12.5 mm (0.50 in.) diameter hole (**arrow**) in the guide 10-222A/21 to mount the adapter 10-222A/8 to the guide 10-222A/21.

- Dimensions given in millimeters on illustration.

02J transmission, removing

– Remove engine cover.

– Disconnect battery ground (GND) strap from battery negative (–) terminal.

NOTE —
Be sure to have the anti-theft radio code on hand before disconnecting the battery.

– Remove power steering reservoir from battery bracket but do not loosen hoses. Tie reservoir up to upper radiator support.

– Disconnect battery positive (+) cable and remove battery and battery tray. See **27 Engine Electrical**.

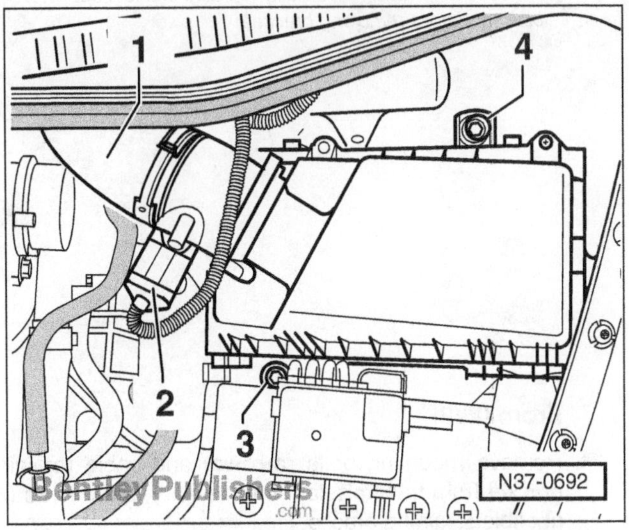

◀ Remove intake hose (**1**) and harness connector (**2**) from Mass Air Flow (MAF) sensor.

– Remove air cleaner mounting bolts (**3** and **4**) and pull out air cleaner.

Through 05/99

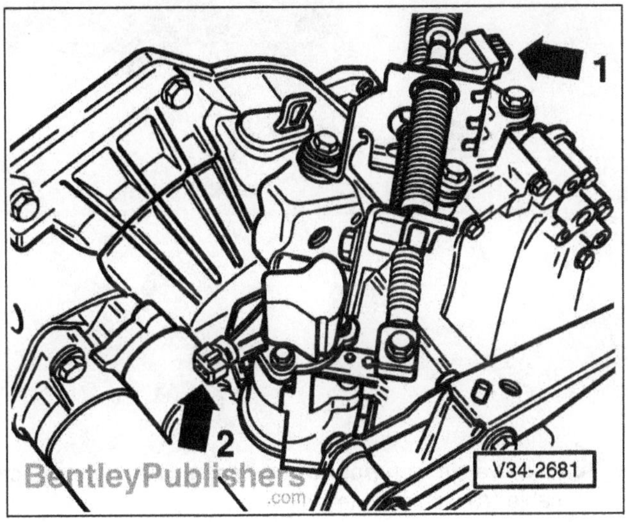

◀ Disconnect harness connector from vehicle speed sensor (**arrow 1**) and back-up light switch (**arrow 2**).

NOTE —
A defective back-up light switch may cause difficulty engaging reverse gear. Volkswagen technical bulletin number 34 06 05, dated Dec. 8, 2006, describes a procedure to test the switch mechanism ball end for damage and ease of movement. A new version of the switch is available from your authorized Volkswagen parts department.

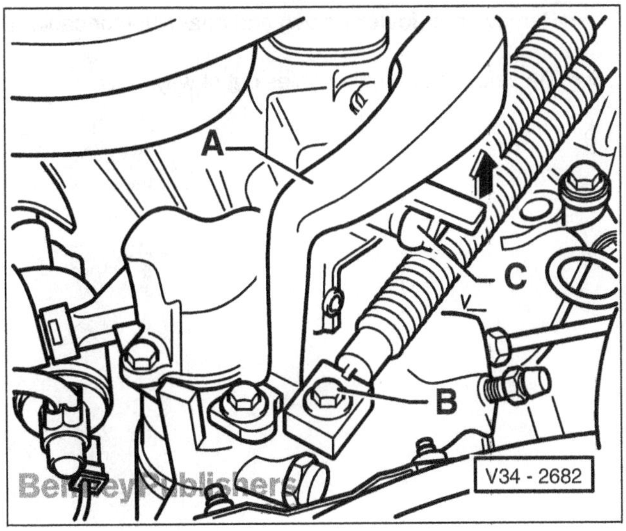

◀ Remove balance weight (**A**) and gear selector cable (**B**) from gear selector lever.

– Take off gear selector cable with washer and square nut.

– Lift (**arrow**) actuating arm/relay lever (**C**) upward and remove gate selector cable.

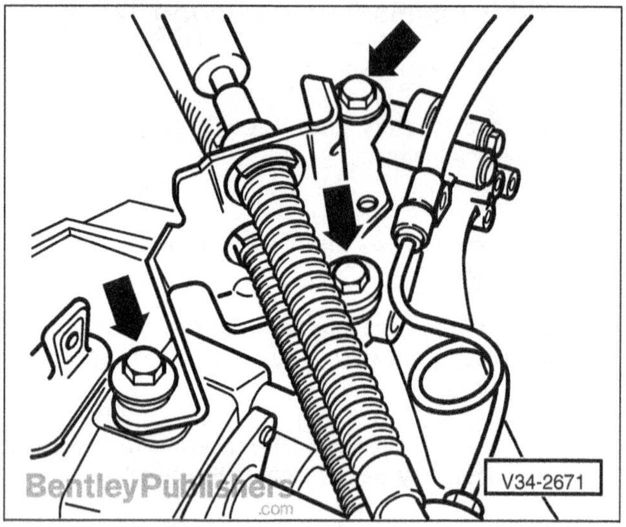

Remove mounting bolts (**arrows**) and cable retaining bracket from transmission.

NOTE —

If necessary unclip hydraulic hose at cable support bracket before removing bracket.

From 06/99

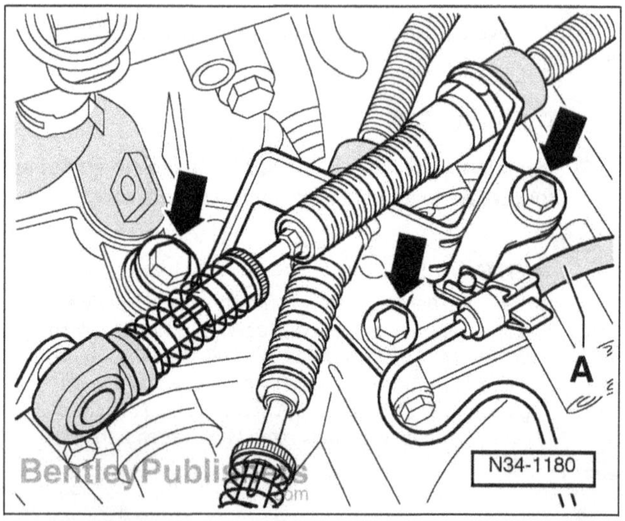

Remove mounting bolts (**arrows**) and cable retaining bracket from transmission. If necessary unclip hose (**A**) at cable retaining bracket first.

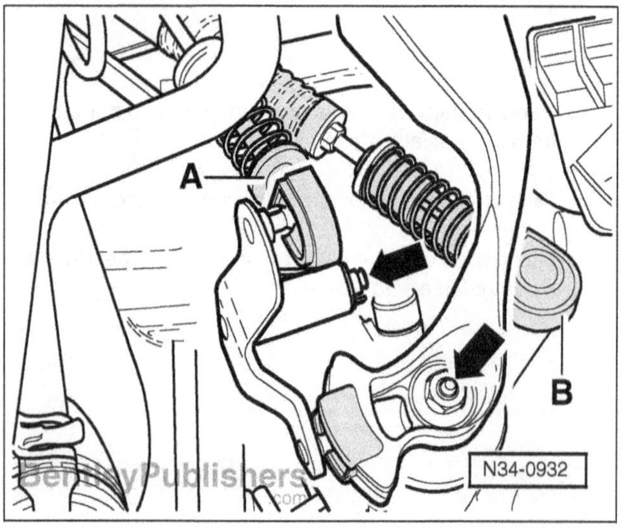

Remove circlip (**upper arrow**) and pull off gate selector cable (**A**) with relay lever.

- Remove nut (**lower arrow**) and gear selector cable (**B**).
- Tie selector and gate cables out of way.

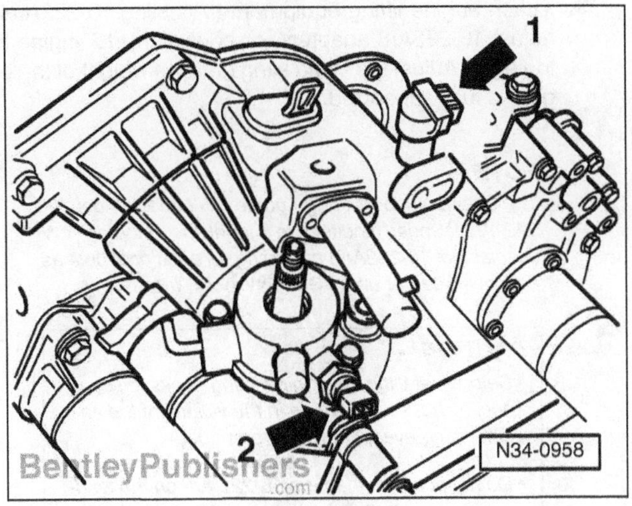

Remove harness connector from vehicle speed sensor (**arrow 1**) and back-up light switch (**arrow 2**).

> ***NOTE —***
> *A defective back-up light switch may cause difficulty engaging reverse gear. Volkswagen technical bulletin number 34 06 05, dated Dec. 8, 2006, describes a procedure to test the switch mechanism ball end for damage and ease of movement. A new version of the switch is available from your authorized Volkswagen parts department.*

Continued for all versions

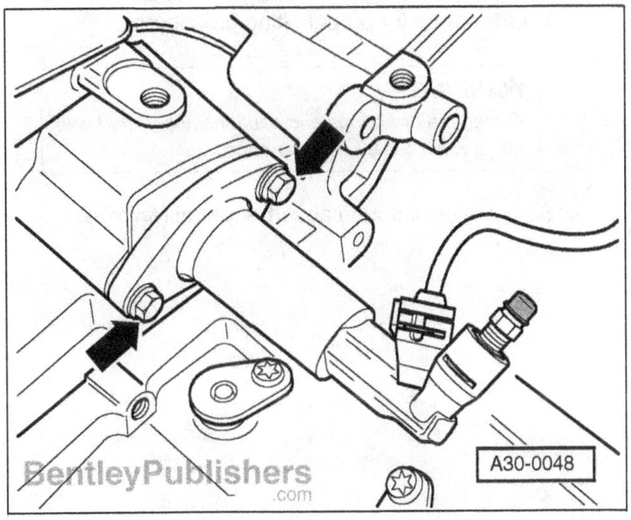

Remove slave cylinder mounting bolts (**arrows**) and secure to side with wire. Do not disconnect hydraulic line.

> ***CAUTION —***
> *Do not depress clutch pedal with slave cylinder removed.*

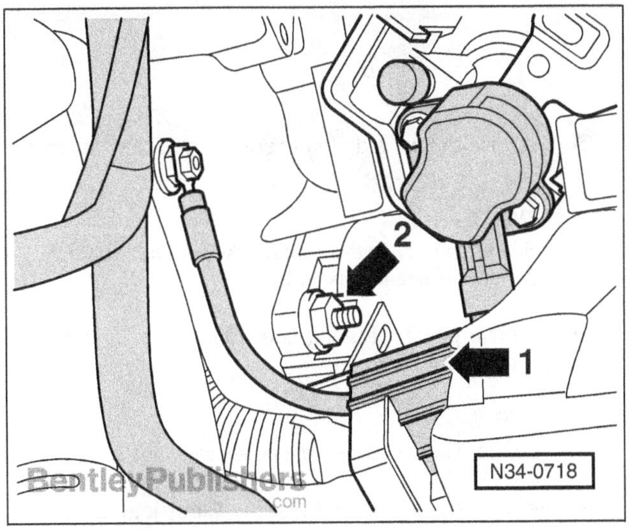

Remove cable retainer (**arrow 1**) and upper mounting bolt (**arrow 2**) on starter.

- Remove ground strap at engine-to-transmission upper securing bolt.
- Remove upper engine-to-transmission mounting bolts.
- On vehicles with 1.9L TDI (ALH engine), remove intake air duct between intake air cooler and turbocharger, see **21 Turbocharger and Intercooler**.

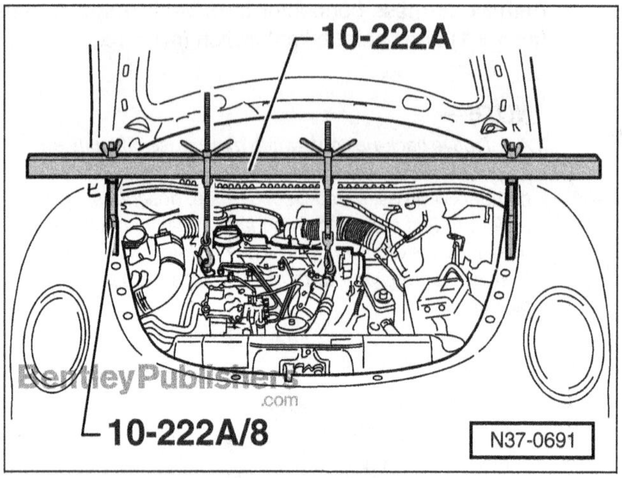

◀ Attach engine lifting equipment (VW sling 10-222A and adjust 10-222A/8 adapters, or equivalent) to engine lifting eyes. Adjust the lifting sling until the weight of the engine is fully supported.

NOTE —

The engine support bridge guide (VW special tool 10-222 A/21) must modified to mount the adapter (VW special tool 10-222A/8) by drilling an additional hole as described earlier under **Special tools, modifying**.

> ***CAUTION —***
>
> - *Before installing the engine lifting hooks, disconnect all hoses and wiring in the vicinity of the engine lifting eyes, to prevent damage.*
> - *Do not position adapters 10-222 A/8 on the fender mounting bolts.*

– Raise car and support with jack stands or lift. See **03 Maintenance** for proper lifting procedure.

> ***WARNING —***
>
> *Observe all warnings and cautions associated with lifting vehicle in* **03 Maintenance**.

◀ Remove lower sound absorber panel (**arrows**).

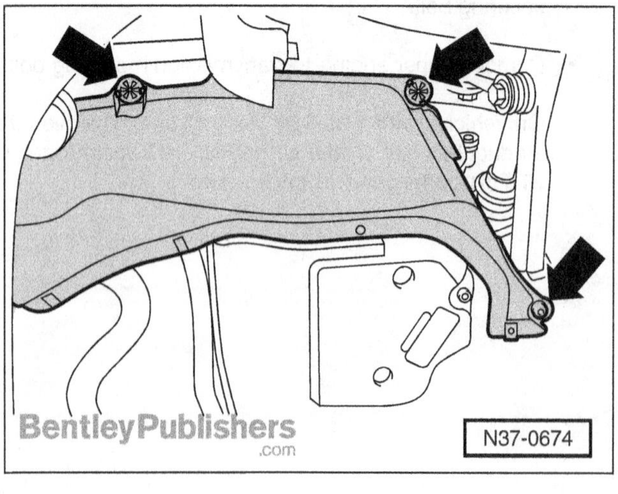

◀ Remove drivers side lower sound insulator (**arrows**).

NOTE —

Sound insulator fasteners may best be removed and installed with needle nose pliers.

– Remove power steering line from starter and transmission mounts.

NOTE —

Use care not to damage the power steering line.

– Remove starter, see **27 Engine Electrical**.

- Remove right inner constant velocity joint protective cover (**arrows**) from engine if installed.

- Turn wheels all the way to the left.

- Disconnect drive axles from transmission and tie up out of way. See **40 Front Suspension and Drive Axles**.

NOTE —
When tying up axle shafts use care not to damage the paint on the body panels or the axle boots.

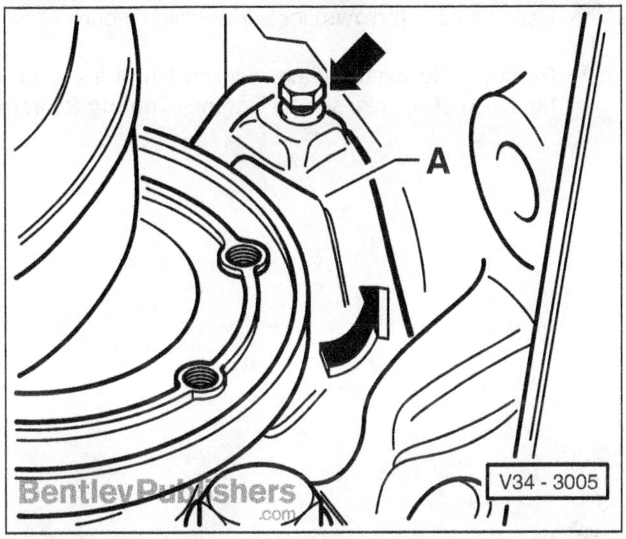

- Remove flywheel small cover plate (**A**) behind right axle flange (**arrows**), if installed.

- If installed, remove lower flywheel cover plate.

- Separate exhaust system at front exhaust pipe, see **26 Exhaust System and Emission Controls**.

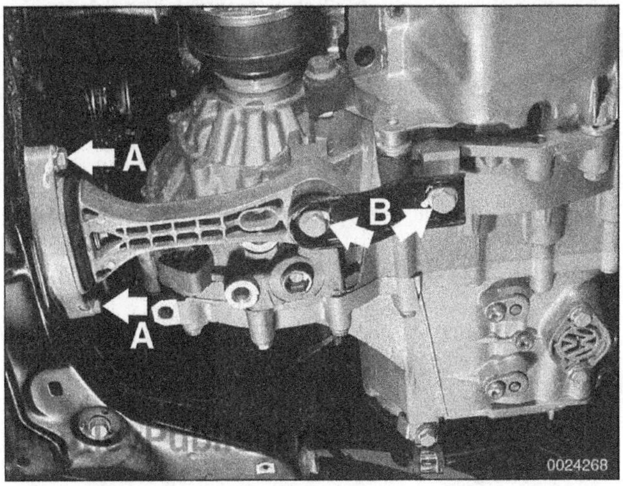

- Remove mounting bolts (**A** and **B**) and engine pendulum mount.

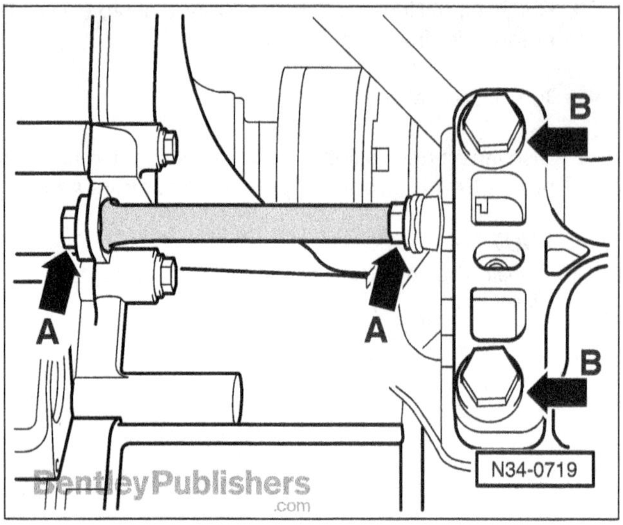

◀ Remove transmission support (**arrows A)** from transmission and bolts (**arrows B**) from left assembly mount.

- Transmission support tightening torques
  (**arrow A**) . . . . 20 Nm (15 ft-lb) plus ¼ turn (90°)
  (**arrow B**) . . . . 40 Nm (30 ft-lb) plus ¼ turn (90°)

– Tilt engine/transmission assembly by lowering it via left-side spindle of support bar (VW 10-222A, or equivalent).

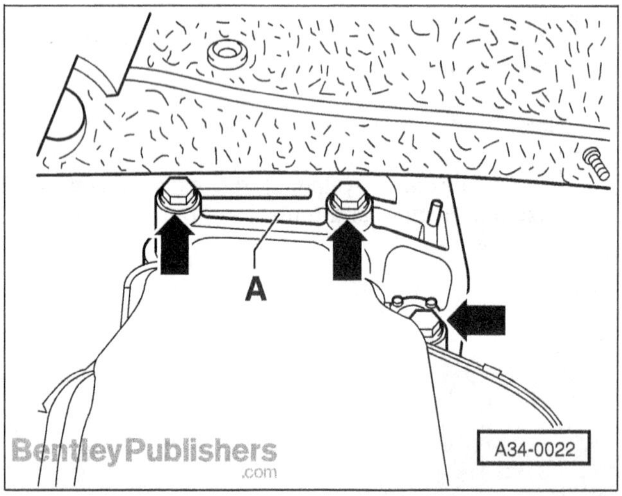

◀ Remove bolts (**arrows**) for transmission mount.

– Remove additional radiator cooling fan, if installed, on right side of vehicle, see **19 Engine–Cooling System**.

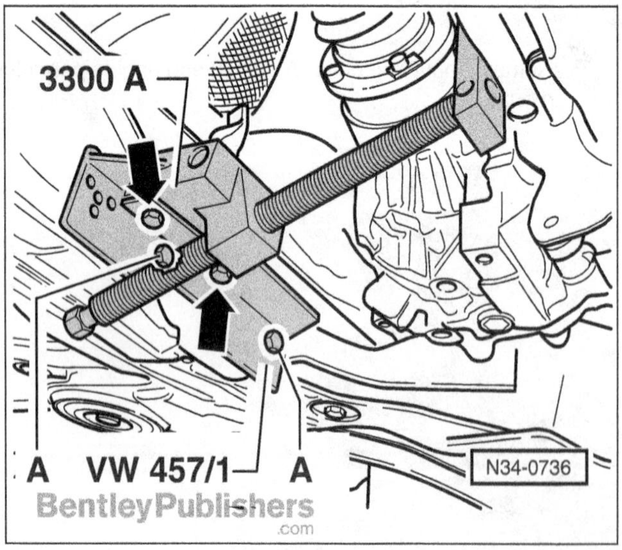

◀ Bolt support rail (VW 457/1 or equivalent) to the pendulum support securing holes (**arrows**) on subframe.

NOTE —
- *Volkswagen special tool 457/1 must be modified as described earlier under* **Special tools, modifying**.
- *Spacers totaling a thickness of 6 mm must be inserted between the subframe and the support rail VW 457/1.*
- *Volkswagen special tool T10036 can be used instead of VW 457/1.*

– Install support (VW 3300A) to support rail (VW 457/1) and secure.

– Press engine/transmission assembly forward carefully.

NOTE —
Do not damage power steering line when moving engine/transmission assembly.

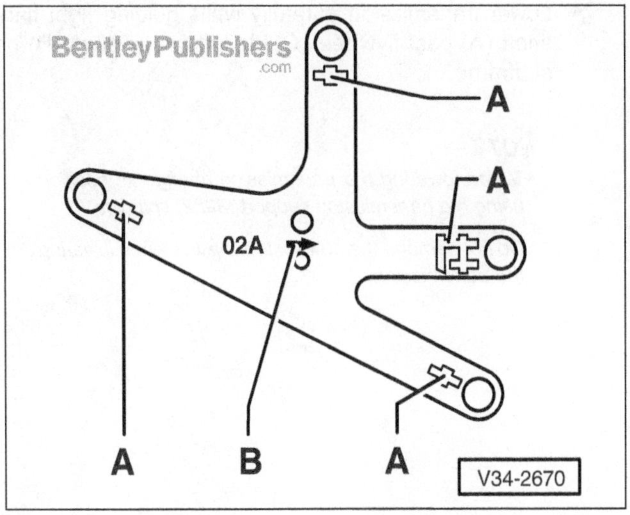

◀ Assemble transmission jack with support 3282, adjustment plate 3282/8 (same as for 02A transmission) and support elements. Arrow (**B**) should point toward front of vehicle.

- Place adjustment plate 3282/8 on transmission support 3282. (Adjustment plate only fits in one position).
- Position transmission support arms according to holes in adjustment plate.

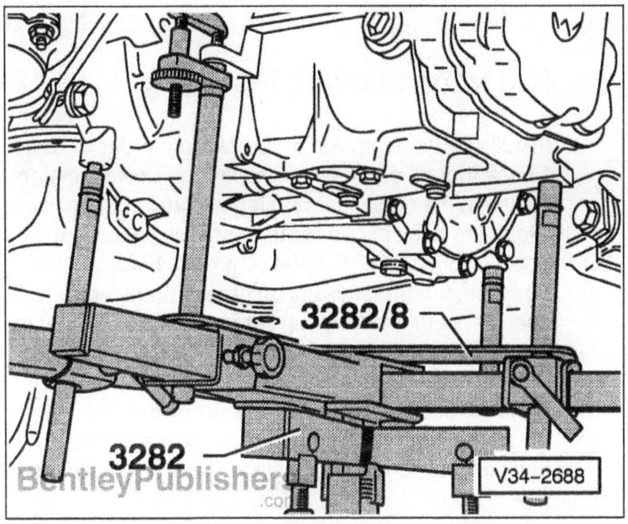

◀ Place transmission jack under vehicle.

– Align adjustment plate parallel to transmission and lock safety supports on transmission jack.

– Remove lower engine/transmission securing bolts.

> ***WARNING —***
> *Before unbolting lower engine-to-transmission mounting bolts be sure to support the weight of the vehicle at all four corners with jack stands designed for that purpose. Removal of the transmission can upset the weight balance of the vehicle and cause it to fall off the lift.*

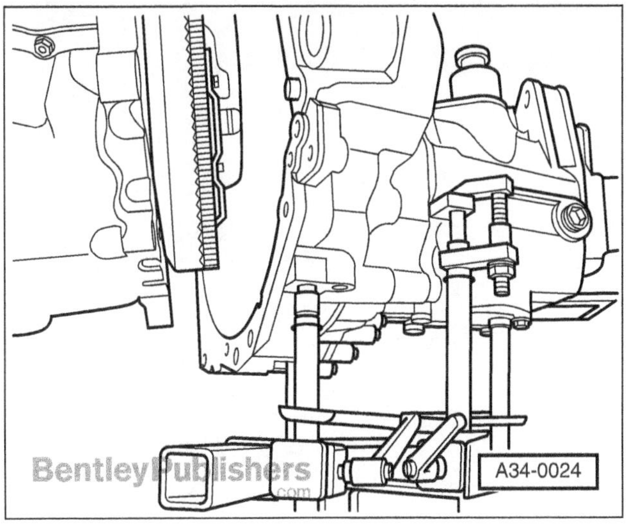

◀ Press transmission off alignment dowel sleeves and carefully swing toward subframe.

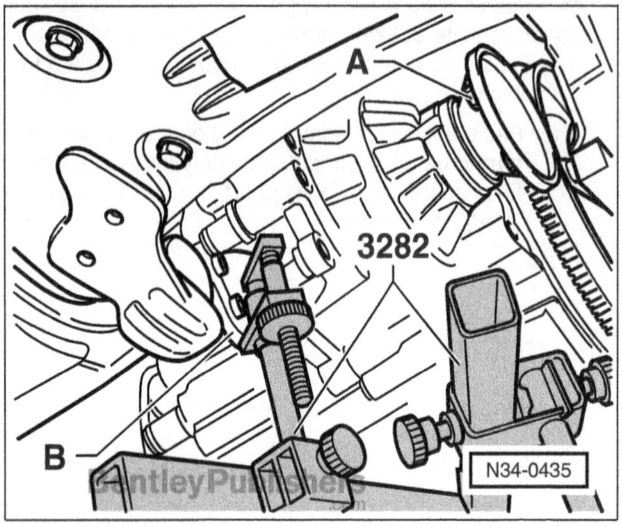

Lower transmission carefully while guiding right flange shaft (**A**) past flywheel. Guide left flange shaft (**B**) past subframe.

NOTE —

- *When lowering the transmission change its position using the transmission support (3282) spindles.*
- *Do not damage the power steering line when lowering the transmission.*

02J transmission, installing

Install the transmission in the reverse order of removal. Check that all engine-to-transmission dowel pins are installed in the engine block. Replace any that are missing. Clean the hub splines of the transmission input shaft and apply a light coat of G 000 100 grease before installing. If replacing the transmission assembly, transfer the vehicle speed sensor, the back-up light switch, and the transmission relay lever to the new transmission.

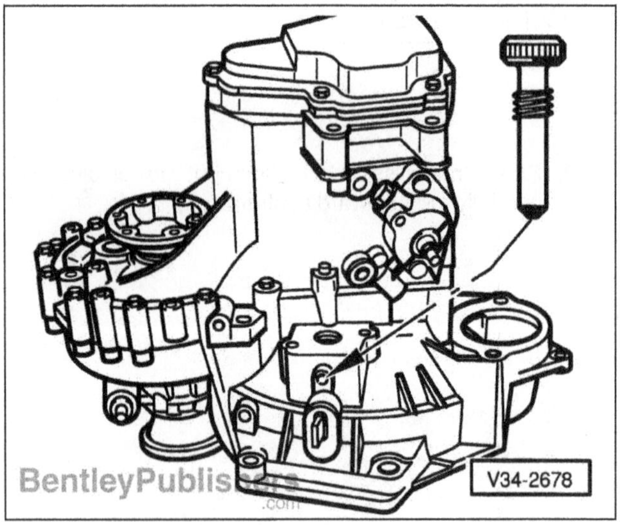

Before installing the transmission to the engine, press the clutch release lever toward the transmission case and secure the lever in position using a M8 x 35 bolt.

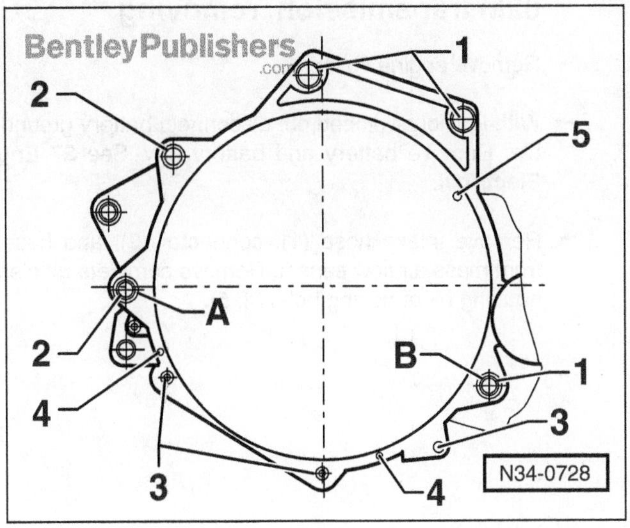

Transmission-to-engine bolt specifications and tightening torques are listed in **Table b**.

Table c. Transmission to engine fasteners

Fastener	Size	Qty.	Tightening torque
1	M12 X 55	3	80 Nm (59 ft-lb)
2*	M12 X 150	2	80 Nm (59 ft-lb)
3**	M10 X 50	3	45 Nm (33 ft-lb)
4***	M7 X 12	2	10 Nm (7 ft-lb)
5****	M7 X 12	1	10 Nm (7 ft-lb)

* Also starter to transmission
** Only on engines with an aluminum oil pan
*** Large cover plate for flywheel, only on engines with sheet steel oil pan (painted black)
**** Small cover plate for flywheel

Remaining transmission installation tightening torques are given below.

Tightening torques (02J transmission)

- Axle shaft to axle flange . . . . 40 Nm (30 ft-lb)
- Balance weight to transmission selector lever . . . . 25 Nm (18 ft-lb)
- Clutch slave cylinder mounting bolts 25 Nm (18 ft-lb)
- CV joint to drive flange on transmission . . . . 40 Nm (30 ft-lb)
- CV joint protective cover to engine. . 35 Nm (26 ft-lb)
- Engine-to-transmission bolts (M10) . 60 Nm (44 ft-lb)
- Engine-to-transmission bolts (M12) . 80 Nm (59 ft-lb)
- Selector cable support bracket to transmission. . . . . 25 Nm (18 ft-lb)
- Selector cable bolt to transmission lever . . . . 25 Nm (18 ft-lb)
- Starter to transmission. . . . . 80 Nm (59 ft-lb)

For engine and transmission mount torques and installation procedures, see **10 Engine–Removing and Installing**.

02M transmission, removing

– Remove engine cover.

– With ignition switched off, disconnect battery ground cable. Remove battery and battery tray. See **27 Engine Electrical**.

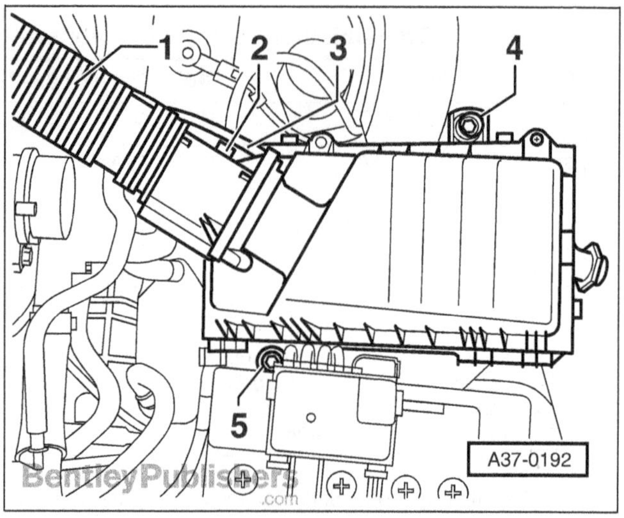

◄ Remove intake hose (**1**), connector (**2**), and hose (**3**) from mass air flow sensor. Remove complete air cleaner housing by removing bolts (**4**, **5**).

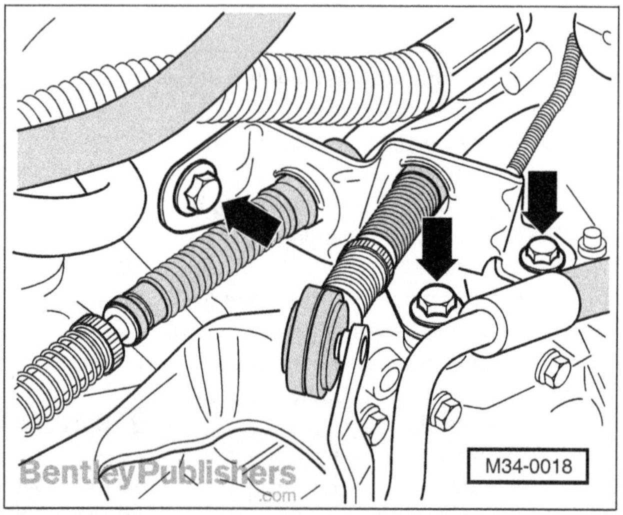

◄ Remove cable support bracket from transmission (**arrows**).

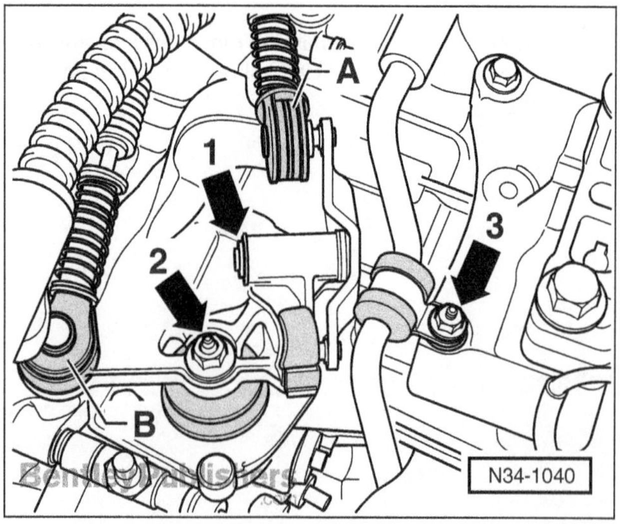

◄ Pull lock washer (**arrow 1**) from gear selector cable relay lever (**A**). Remove gear selector lever by removing nut (**arrow 2**). Remove power steering line (**arrow 3**).

NOTE —
If necessary use puller to remove selector lever.

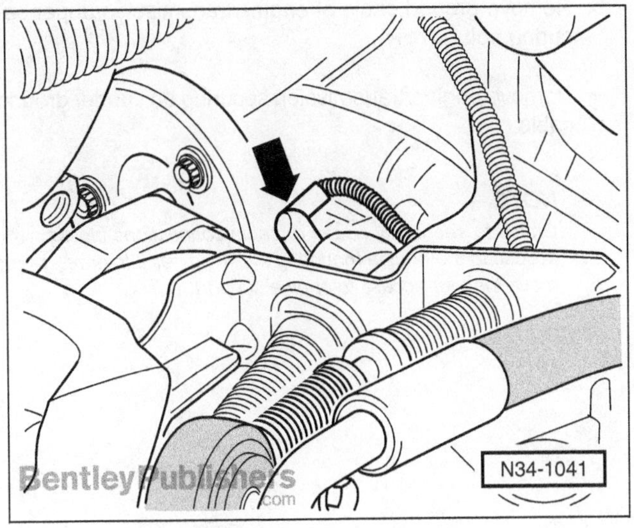

Remove electrical harness connector from vehicle speed sensor (**arrow**).

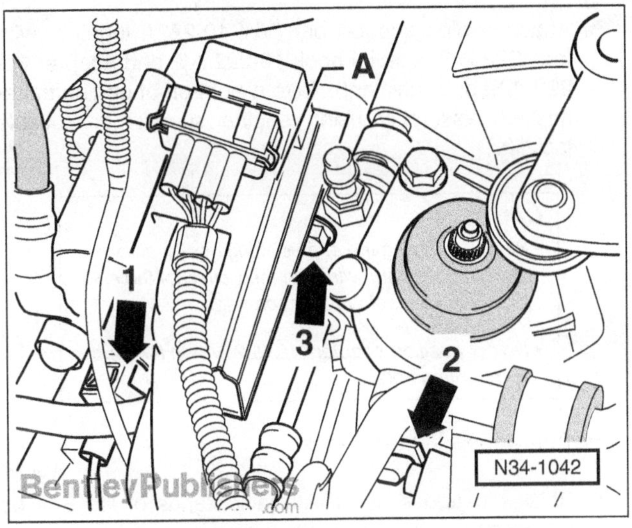

Remove wiring (**arrow 1**) from starter. Disconnect reverse light harness connector (**arrow 2**). Remove cable retainer (**arrow 3**) on starter. Remove upper securing bolt on starter.

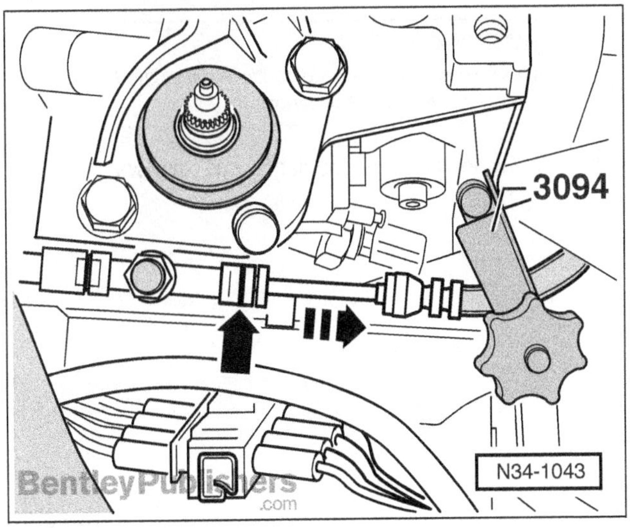

Clamp clutch slave cylinder hose with hose clamp to prevent leakage. Pull retainer (**arrow**) and remove pipe/hose in direction of **dashed arrow**.

NOTE —

Do not depress clutch pedal with hose clamped.

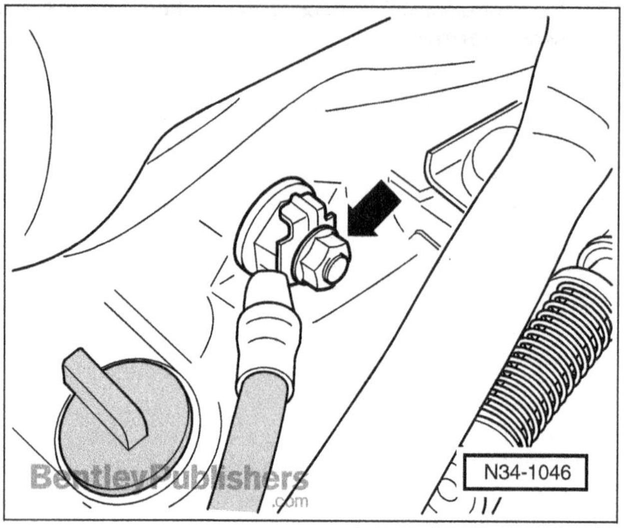

◄ Remove ground cable at engine/transmission upper securing bolt (**arrow**).

– Remove engine/transmission securing bolt under ground cable.

NOTE —

Check to see if any hoses or cable connections block access to the engine mounting eye. If so, reroute or remove then for access to engine support.

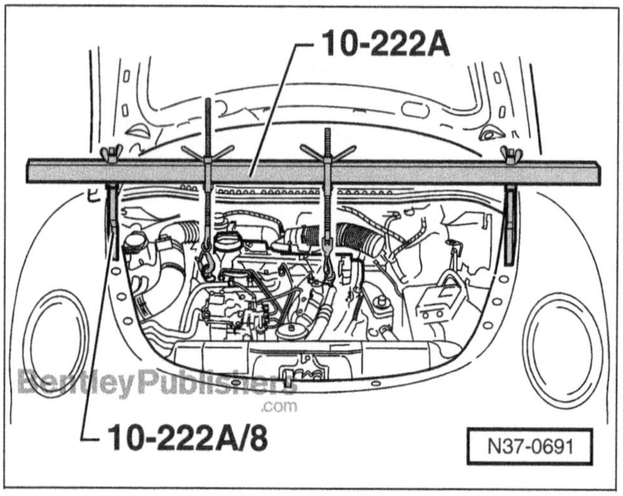

◄ Attach engine support bar (VW 10-222A together with legs 222 A/8, special hook 10-222 A/2 and adapter 10-222 A/3, or equivalent). Take weight off of engine/transmission assembly using engine support or suitable jack/lift.

CAUTION —

- *Before installing the engine lifting hooks, disconnect all hoses and wiring in the vicinity of the engine lifting eyes, to prevent damage.*
- *Do not position adapters 10-222 A/8 on the fender mounting bolts.*

– Turn wheels left to full lock.

– Raise car and support with jack stands or lift. See **03 Maintenance** for proper lifting procedure.

WARNING —

Observe all warnings and cautions associated with lifting vehicle in ***03 Maintenance****.*

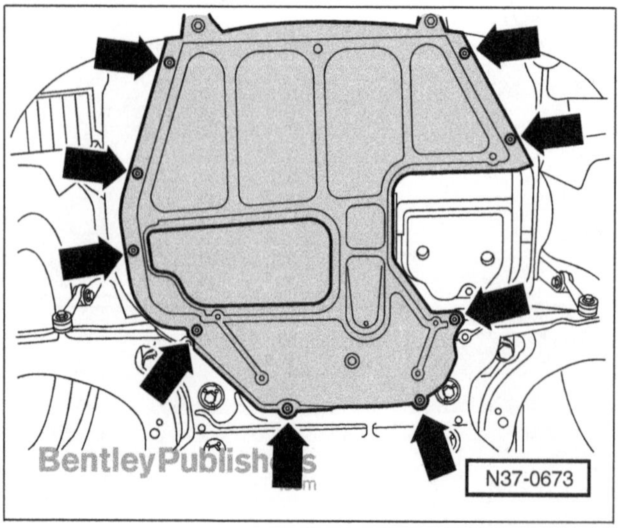

◄ Remove lower sound insulation tray (**arrows**).

◀ Remove driver's side lower sound insulation tray (**arrows**).

– Remove passenger side lower sound insulation tray.

◀ Remove right inner constant velocity joint protective cover (**arrows**) from engine if installed.

– Disconnect drive axles from transmission and tie up out of way. See **40 Front Suspension and Drive Axles**.

NOTE —
When tying up axle shafts use care not to damage the paint on the body panels or the axle boots.

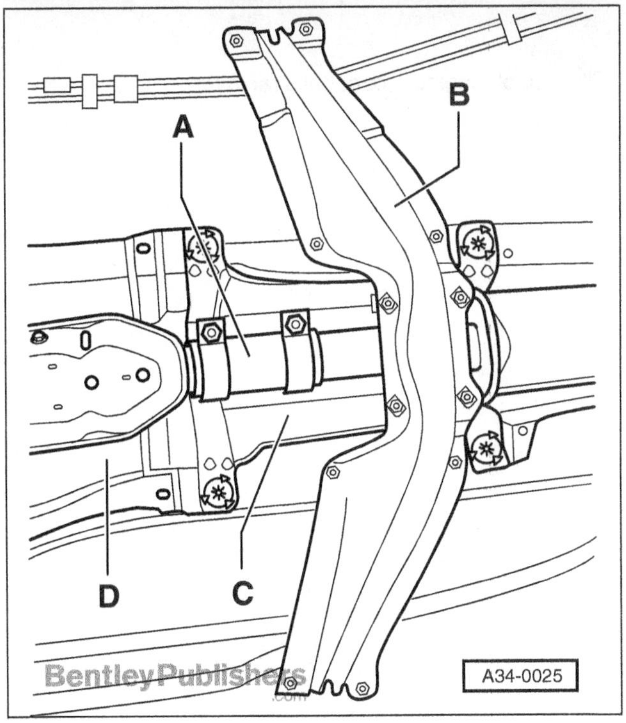

◀ Disconnect front exhaust connection (**A**). See **26 Exhaust System and Emission Controls**.

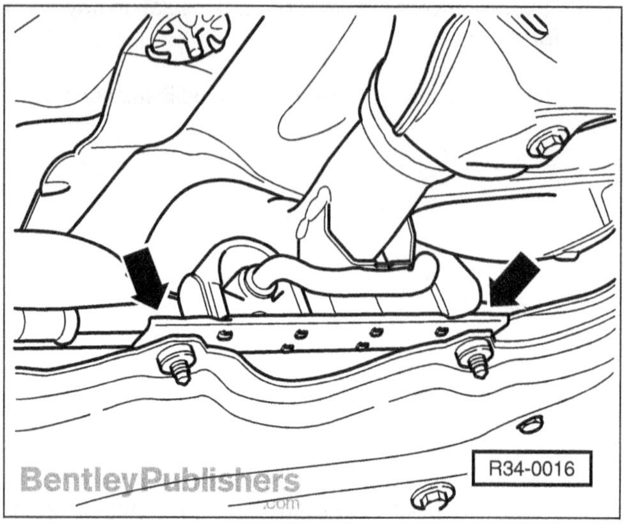

Remove bolts (**arrows**) and unhook exhaust support elements.

– Secure front exhaust pipe to body using stiff wire.

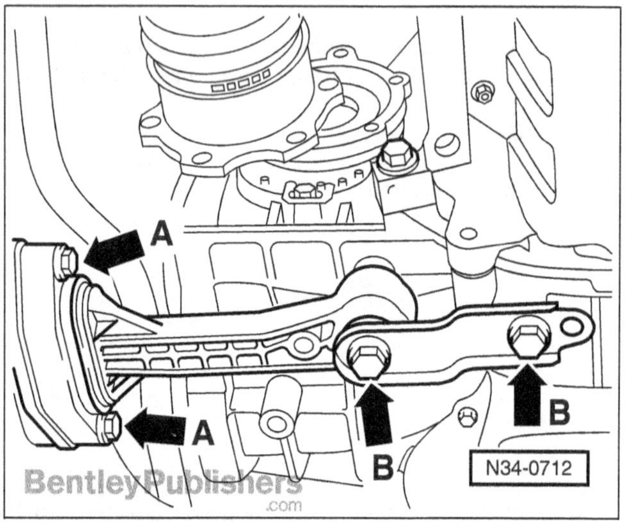

Remove mounting bolts (**arrows A** and **B**) and engine pendulum mount.

– Remove power steering line from starter and transmission.

– Remove starter. See **27 Engine Electrical**.

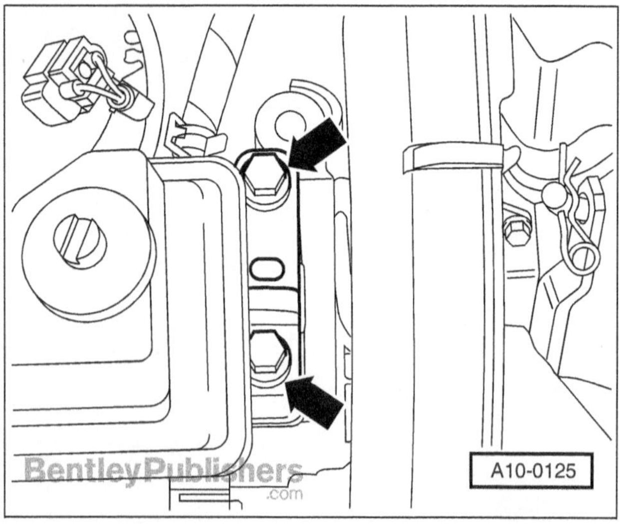

Unbolt engine mount bolts (**arrows**).

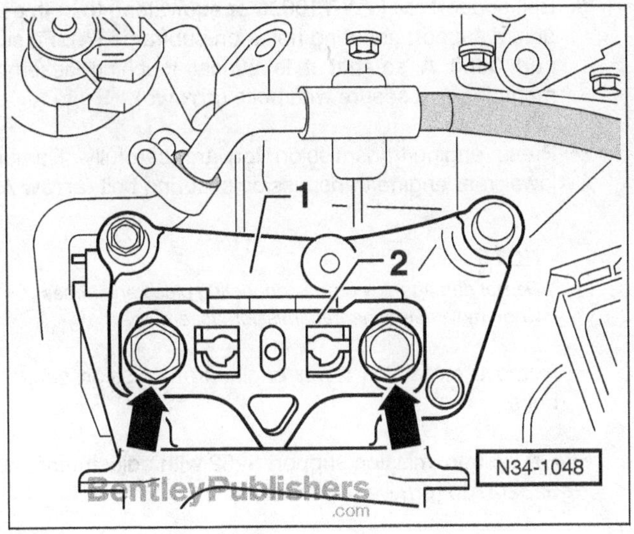

Remove mounting bolts (**arrows**) and mount bracket (**2**) from left engine/transmission mount (**1**).

– Carefully tilt engine and transmission by lowering left-hand spindle of engine support bar approximately 60 mm (2.4 in.).

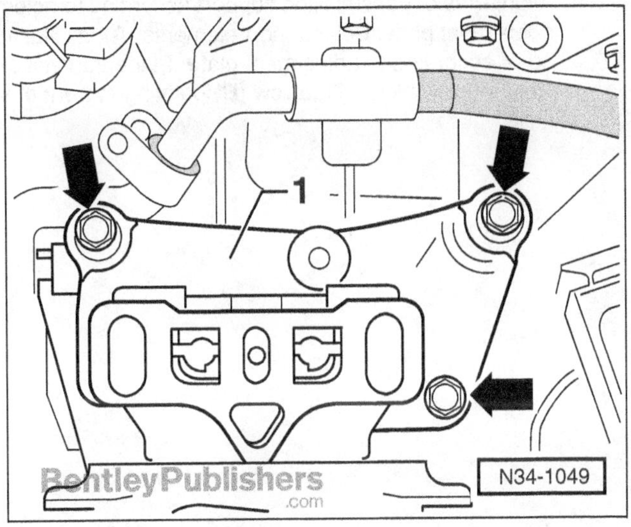

Remove mounting bolts (**arrows**) and left engine/transmission mount (**1**) from transmission.

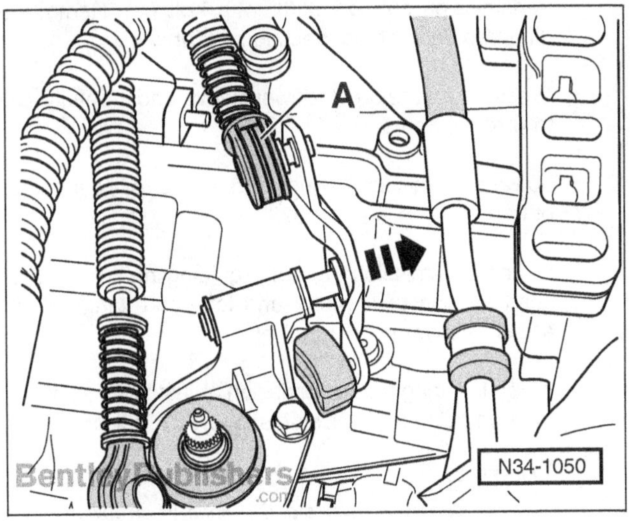

Pull off gate selector cable (**A**) in direction of **arrow**.

– Tie gate and gear selector cables up out of way.

– Tilt engine/transmission assembly further by lowering left spindle of engine support bar (10-222 A).

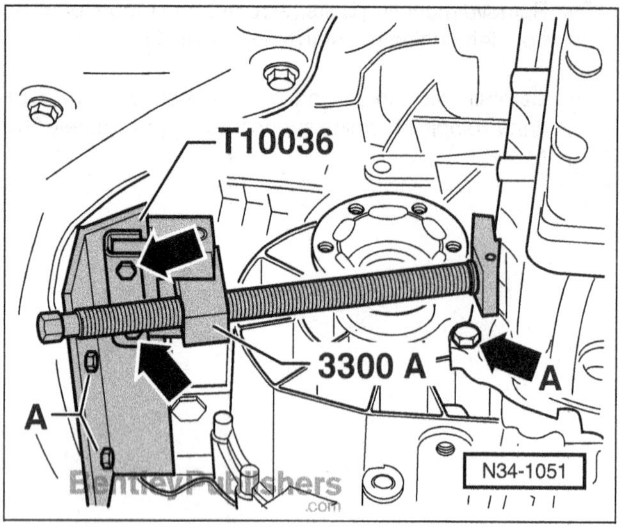

◀ Bolt support rail (VW **T10036** or equivalent) to both pendulum support securing holes on subframe (**A**). Fit support **3300 A** so that it is almost touching assembly mounting and secure with bolts (**arrows**).

– Press engine/transmission forward carefully. Remove lower rear engine/transmission securing bolt (**arrow A**).

NOTE—

Do not damage any of the connecting pipes and hoses when moving engine/transmission assembly.

– Remove remaining lower engine/transmission securing bolts.

– Set up transmission support 3282 with adjustment plate 3282/27 to remove transmission.

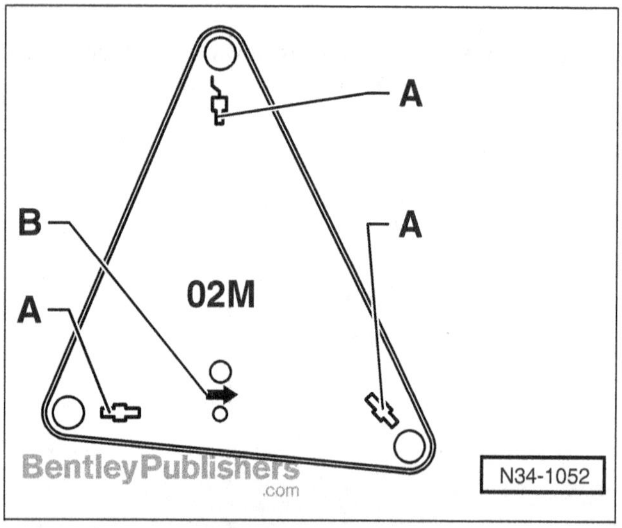

◀ Align arms of transmission support according to holes in adjustment plate. Bolt support elements (**A**), as illustrated or according to adjustment plate. Place transmission jack under vehicle with arrow (**B**) pointing to front of vehicle.

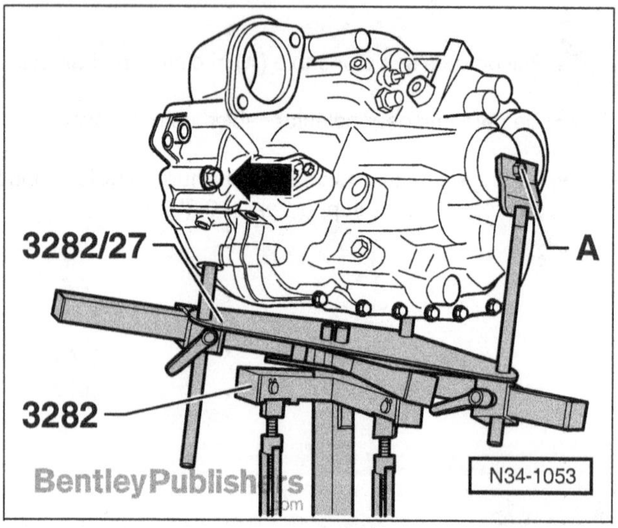

◀ Align adjustment plate parallel to transmission. Secure transmission support **3282** with bolt (**A**). Remove engine/transmission connecting bolt (**arrow**).

– Press transmission off dowel sleeve and carefully swing toward subframe.

– Lower transmission slightly and incline to lower spindles of transmission support **3282**.

– Lower transmission carefully, guiding right-hand flange shaft in area of flywheel and left-hand flange shaft in area of subframe.

– When lowering the transmission adjust position of transmission using transmission support **3282** spindles.

NOTE—

Do not damage power steering lines when lowering transmission.

02M transmission, installing

Install the transmission in the reverse order of removal. Check that all engine-to-transmission dowel pins are installed in the engine block. Replace any that are missing. Clean the hub splines of the transmission input shaft and apply a light coat of G 000 100 grease before installing. If replacing the transmission assembly, transfer the vehicle speed sensor, the back-up light switch, and the transmission relay lever to the new transmission. Be sure to check the following:

- Check transmission oil level, see **03 Maintenance**.
- Bleed clutch system.
- Adjust gear selector mechanism.

Tightening torques

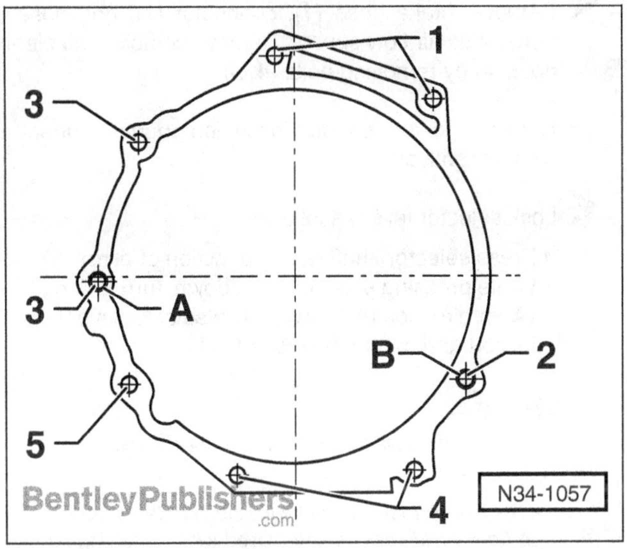

Transmission-to-engine bolt specifications and tightening torques are listed in **Table c**.

Table d. Transmission to engine fasteners

Fastener	Size	Qty.	Tightening torque
1	M12 X 55	2	80 Nm (59 ft-lb)
2	M12 X 55	1	80 Nm (59 ft-lb)
3*	M10 X 105	2	80 Nm (59 ft-lb)
4	M10 X 50	2	40 Nm (30 ft-lb)
5	M10 X 50	1	40 Nm (30 ft-lb)

* Also starter to transmission
Items **A** and **B** are dowel sleeves

Remaining 02M transmission installation tightening torques are given below.

Tightening torques (02M transmission)

Always replace all stretch bolts
(stretch bolts have plus ¼ turn (90°) specification)

- Axle shaft to axle flange . . . . 40 Nm (30 ft-lb)
- Balance weight to transmission selector lever . . . . 25 Nm (18 ft-lb)
- Clutch slave cylinder mounting bolts . 12 Nm (9 ft-lb)
- CV joint to drive flange on transmission . . . . 40 Nm (30 ft-lb)
- CV joint protective cover to engine . 35 Nm (26 ft-lb)
- Engine-to-transmission bolts (M10) . 40 Nm (30 ft-lb)
- Engine-to-transmission bolts (M12) . 80 Nm (59 ft-lb)
- Selector cable support bracket to transmission. . . . . 25 Nm (18 ft-lb)
- Selector cable bolt to transmission lever . . . . 25 Nm (18 ft-lb)
- Starter to transmission . . . . 60 Nm (44 ft-lb)

For engine and transmission mount torques and installation procedures, see **10 Engine–Removing and Installing**.

GEAR OIL, CHECKING AND REPLACING (0A4 TRANSMISSION)

The gear oil level on the 0A4 transmission is above the level of the drain plug. This is due to a change in the inclination of the angle of the engine/transmission. The gear oil level cannot be checked by removing the drain plug. The gear oil must be drained completely to check exact quantity.

– Remove engine cover.

– With ignition switched off, disconnect battery ground cable. Remove battery and battery tray. See **27 Engine Electrical**.

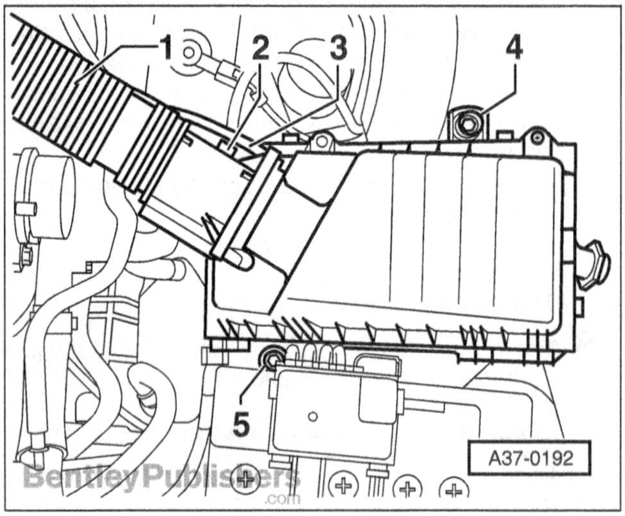

◄ Remove intake hose (**1**), connector (**2**), and hose (**3**) from mass air flow sensor. Remove complete air cleaner housing by removing bolts (**4**, **5**).

– Remove lower sound insulation panel under engine/transmission.

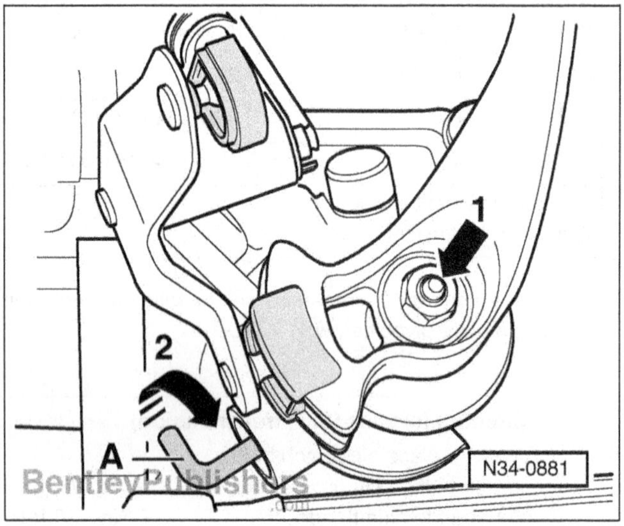

◄ Lock selector lever as follows:

- Press selector shaft down (direction of **arrow 1**).
- While pressing selector shaft down, turn locking lever (**A**) in direction of **arrow 2**. The selector lever is now locked and cannot be moved.

NOTE —

- *To drain gear oil, the selector lever journal must be removed.*
- *The selector lever must be locked in order to prevent it from accidentally being moved while removing selector lever journal.*

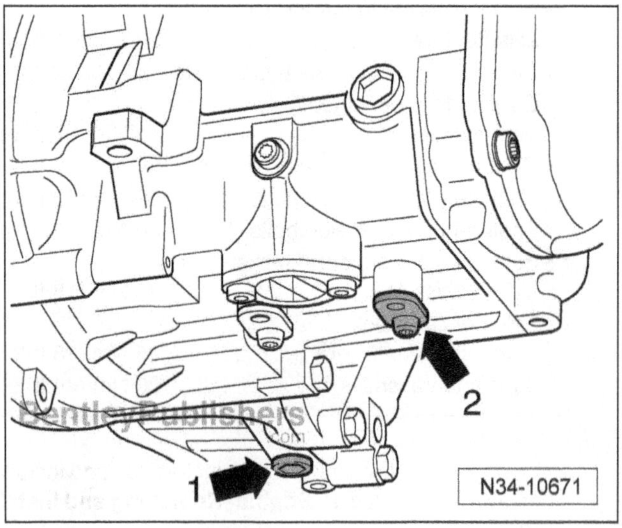

◄ Remove drain plug (**arrow 1**) and selector shaft journal (**arrow 2**). Drain gear oil into a clean container.

– Install selector shaft journal with new O-ring.

– Install oil drain plug.

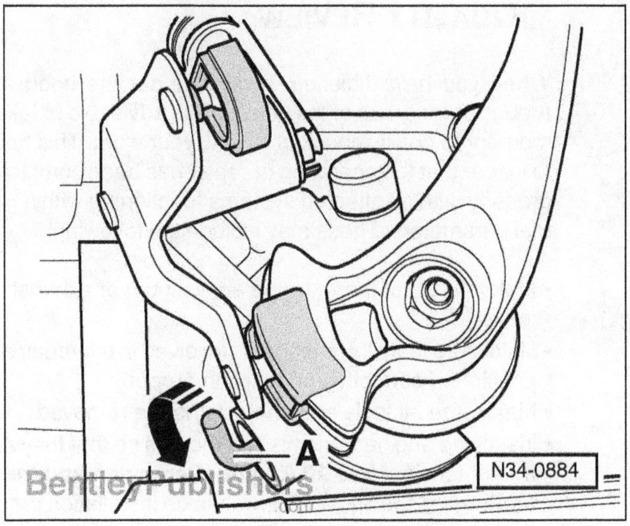

< Turn locking lever (**A**) in direction of **arrow** to unlock selector lever.

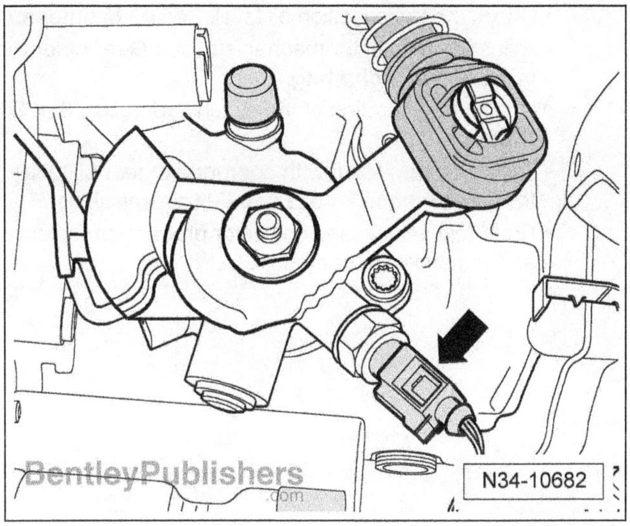

< Remove back-up light switch (**arrow**).

– Connect hose to fill transmission gear oil into back-up light switch mounting hole.

  • Hose should be approx. 600 mm (24 inches) long.

Specification

• 0A4 gear oil .... see authorized Volkswagen dealer
• Fill quantity.................... 1.8L (1.9 quarts)

– Install removed components in reverse order of removal.

QUALITY REVIEW

When you have finished working under the hood and around other areas of the vehicle, it is advisable to take a moment to quality check or review your work. This helps to insure that the operation or repair has been completed properly with all affected systems functioning within normal parameters. These may include the following:

- Make sure that there are no air, vacuum or exhaust leaks.
- Make sure that all components involved in the repair are positioned correctly and function properly.
- Make sure all tools and shop cloths are removed.
- Install the engine/transmission mounts so that they are free of tension. See **10 Engine–Removing and Installing** for specific engine/transaxle installation procedures.
- Check the transmission oil level. See **03 Maintenance**.
- Adjust gear selector mechanism, see **Gear selector mechanism, adjusting**.
- Check the operation of the clutch hydraulics. See **30 Clutch**.
- Top off transmission with appropriate gear oil, see **03 Maintenance** for 02J or 02M transmissions.
- Road test vehicle and check for proper gear selection and transmission operation.

34b DSG Transmission

GENERAL

This repair group covers fluid service and repair information for external components of the Direct Shift Gearbox (DSG). Special tools and equipment are required for the removal, installation and internal repair of the DSG. These repairs are outside the scope of this manual.

The DSG is also known as a double-clutch transmission. Engine torque is transferred to the transmission via the dual-mass flywheel. This transmission is constructed like a 6-speed manual transmission. Alternating hydraulic actuation of the two wet multi-plate clutches permits operation similar to that of an automatic transmission, i.e. the gears are engaged automatically, or manually via the Tiptronic mode. There is no clutch pedal.

Transmission code
- 6-speed direct shift gearbox 02E

Table a lists engine applications, gear ratios, and other data for the DSG transmission. The transmission code letters can be found on the vehicle data plates. See **03 Maintenance** for location of identification plates and labels.

Table a. DSG transmission specifications

Designation	**02E**						
Code letters	GHG	GPX	GYS	HLJ	HQJ	HQP	HXV
Engine application	1.9L, 90hp	1.9L, 90hp	1.9L, 90hp	1.9L, 90hp	1.9L, 90hp	1.9L, 90hp	1.9L, 90hp
Model year	2004	2004	2004	2005	2005	2006 - on	2006 - on
Ratio: Z_2:Z_1							
Final drive 1st through 4th	4.375	4.375	4.375	4.375	4.375	4.375	4.375
Final drive 5th through 6th	3.333	3.333	3.333	3.333	3.333	3.333	3.333
1st gear	3.462	3.462	3.462	3.462	3.462	3.462	3.462
2nd gear	2.050	2.050	2.050	2.050	2.050	2.050	2.050
3rd gear	1.300	1.300	1.300	1.300	1.300	1.300	1.300
4th gear	0.902	0.902	0.902	0.902	0.902	0.902	0.902
5th gear	0.914	0.914	0.914	0.914	0.914	0.914	0.914
6th gear	0.756	0.756	0.756	0.756	0.756	0.756	0.756
Reverse gear	3.989	3.989	3.989	3.989	3.989	3.989	3.989
Lubricant							
Capacity	2004 - 2005, new fill: 6.4L (6.8 quarts), refill: 4.5L (4.8 quarts)					2006 - on, new fill: 7.2L (7.4 quarts), refill 5.5L (5.8 quarts)	
Specification	DSG oil, VW part # G 052 182 A2						

Fluid Service

The DSG transmission oil requires changing as part of regular maintenance. Also replace transmission oil after repairs requiring removal of transmission oil pan.

DSG transmission oil and filter, changing

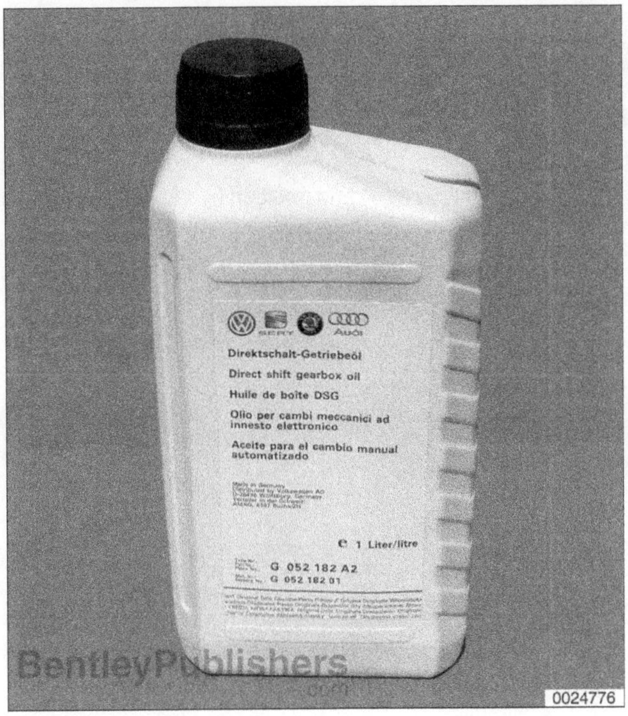

0024776

◀ The DSG transmission and final drive share a common fluid supply which is specifically designed for the DSG. Checking oil level of gearbox also checks oil level of final drive (differential).

> ***WARNING —***
> - *Do not start engine if there is little or no oil in transmission such as immediately following repairs or after severe loss or leakage of transmission oil.*
> - *If there is little or no oil in transmission, add sufficient quantity with engine cold and not running before proceeding with a diagnosis.*
> - *Use only oil specified for direct shift gearboxes (DSG). Always consult your Volkswagen Parts Department or aftermarket parts specialist for the latest information. Use of oils other than those specified will lead to malfunctions or failure of the transmission.*
> - *Observe all rules for cleanliness when working on the DSG.*
> - *Special tools are required for some operations. Review all repair procedures before starting.*
> - *Wear eye protection and acid-resistant protection gloves.*

NOTE —
- *Transmission oil level changes with transmission oil temperature.*
- *Checking oil level when temperature is too low (cold) leads to overfilling. Checking oil level check when temperature is too high (hot) will cause insufficient filling.*
- *If DSG oil level is too high or too low it will adversely affect operation of transmission.*
- *Always replace sealing ring for drain/check plug.*

Before beginning maintenance work involving the fluid in the DSG, observe the following:

- Ignition, A/C, and all electrical consumers switched off.
- Vehicle in a level, horizontal position.
- Scan tool or scan tool computer program connected.
- At the start, ATF temperature not above approximately 35°C (95°F).

- Remove upper sound absorber panel. See **03 Maintenance**.
- Remove air filter housing. See **03 Maintenance**.
- Remove battery and battery tray. See **27 Engine Electrical**.
- Raise vehicle and remove lower sound absorber panel. See **03 Maintenance**.
- Place a suitable drip tray under transmission to catch spillage.
- Lower vehicle enough to access top of transmission.

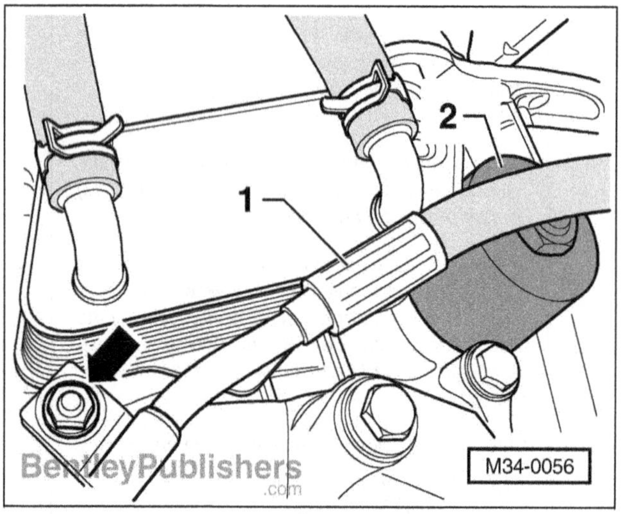

Remove nut (**arrow**) and position hydraulic line (**1**) out of way to access DSG filter housing (**2**).

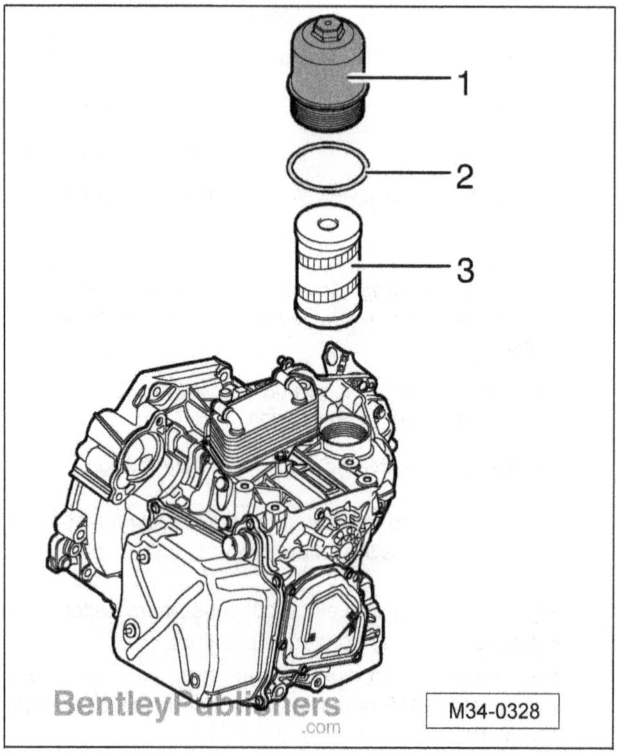

Unscrew filter housing (**1**) from transmission to the point where it is just ready to take off, but do not remove. Tilt filter housing slightly to allow air in so that fluid can flow back into transmission minimizing spillage.

NOTE —
Removing DSG oil filter before draining oil allows fluid in filter housing to drain back into transmission.

- Remove filter housing (**1**), clean and install new O-ring (**2**). Coat new O-ring with DSG transmission oil.
- Remove old filter (**3**).

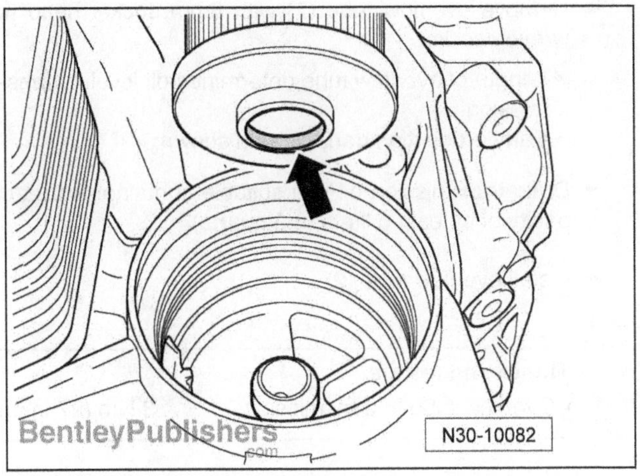

◀ Coat O-ring in suction collar (**arrow**) of a new filter with DSG transmission oil.

– Install new filter with suction collar (**arrow**) down, install and tighten filter housing.

Tightening torque
- DSG filter housing to transmission. . 20 Nm (15 ft-lb)

– Re-install battery tray and battery. See **27 Engine Electrical**.

– Raise vehicle enough to access bottom of transmission.
  - Vehicle must be level.

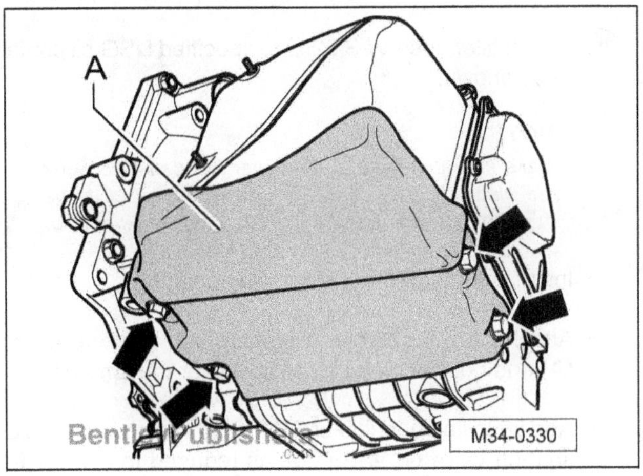

◀ Remove mounting bolts (**arrows**) and slide protection shield (**A**) from front of transmission.

– Position a drain pan of at least 6 liters (6.3 quarts) capacity under transmission.

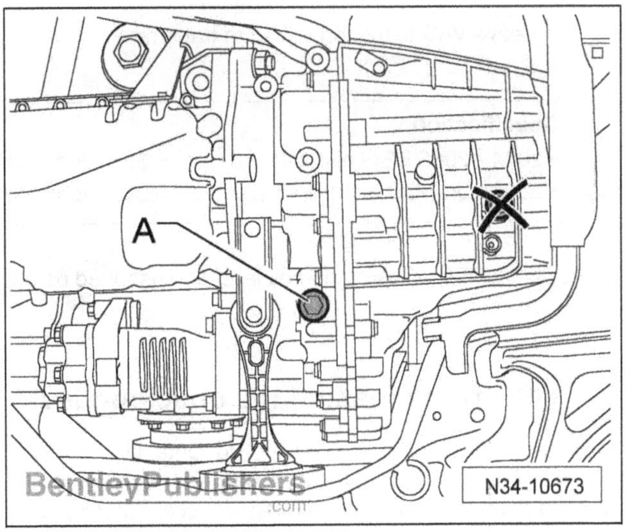

◀ Remove drain plug (**A**).
- DSG transmissions built up to production date September 20, 2004, have two plugs.
- Remove drain plug (**A**) near the pendulum mount for all applications.
- The early production drain plug (**X**) should also be removed, as applicable, for quicker, more complete draining.
- Do not interchange plugs.

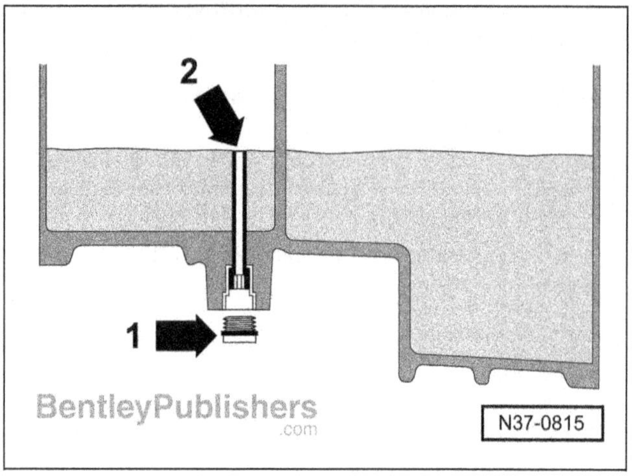

◀ Remove overflow tube (**2**) with 8mm socket head hex wrench/socket.

- Length of overflow tube determines oil level in transmission.
- Early production transmission shown.

– Drain transmission oil into suitable container with a capacity of at least 6 liters (6.3 quarts).

– Install overflow tube (**2**).

Tightening torque

- Overflow tube to transmission . . . . . . 3 Nm (27 in-lb)

– Shake DSG oil container thoroughly before opening.

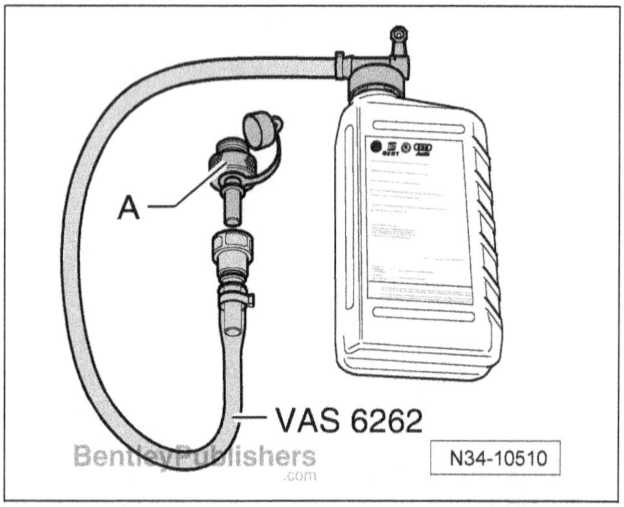

◀ Attach filler hose VAS 6262 to specified DSG oil container as shown.

NOTE —

Make sure that hose and adapter fitting of VAS 6262 are clean. Do not allow DSG transmission oil to be mixed or contaminated by any other type of oil or fluid.

– Install adapter (**A**) into drain plug opening.

– Attach VAS 6262 hose assembly with bottle to adapter (**A**) and hold as high as possible over transmission.

– Add specified quantity of DSG oil to transmission. Close shut-off valve on VAS 6262 as required to change fluid containers.

- Leave VAS 6262 connected to transmission.

Specification

- 2004 - 2005 DSG oil refill. . . . . 4.5L (4.8 quarts)
- 2006 - on DSG oil refill. . . . . 5.5L (5.8 quarts)

– Check DSG transmission oil level as described below.

DSG transmission oil, checking level

– Connect diagnostic scan tool to Data Link Connector (DLC). See **OBD On Board Diagnostics** at the back of manual for more information on scan tool operation. Start program so that DSG oil temperature can be read.

– Start engine and shift once through all selector lever positions, remaining in each shifted position for approximately 3 seconds, with vehicle at a standstill. Move shifter lever back into "P" position and leave engine running. Do not switch off engine.

> ***WARNING —***
> - *Radiator cooling fan can be activated at any time. Be sure that there is no chance of personal contact or contact with oil filler apparatus!*
> - *Use care with engine running while the vehicle is not supported by it's own wheels.*

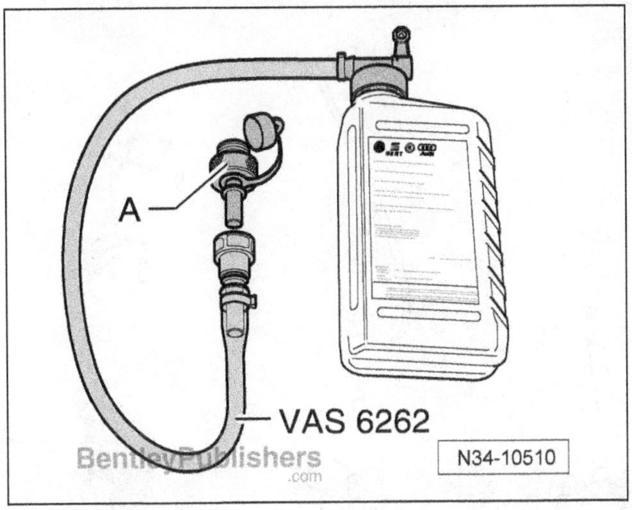

When transmission oil temperature reads between 35°C and 45°C (95°F and 113°F) on scan tool display, remove disconnect from VAS 6262 hose, leaving adapter (**A**) in place. Allow excess oil to drain out. See **DSG transmission oil and filter, changing** for installation instructions of VAS 6262.

NOTE —
- *Every 30 seconds, a small surge of oil will exit overflow tube, independent of the height of the transmission oil level. This is due to cool oil impulse to clutches. This surge of oil is not indicative of excessive transmission oil level and must **not** be considered when determining transmission oil level.*
- *Drained transmission oil must not be reused. Dispose of oil in accordance with all applicable regulations.*

- As soon as oil stops flowing out and is only dripping out (taking 30 second surge into account), remove adapter fitting (**A**) from drain plug opening.
  - If a small amount of fluid drips out when transmission fluid temperature is between 35°C and 45°C (95°F and 113°F), transmission oil level is correct. This occurs because fluid expands as it heats up. In this case, no transmission fluid needs to be added; fluid level is correct.
  - If no fluid has dripped out by the time the oil temperature reaches 45°C (113°F), level is low and needs to be topped up. Add a small amount of oil and check level again.
- Install drain plug with new seal.

Tightening torque
- Drain plug to transmission . . . . 45 Nm (33 ft-lb)

- Switch ignition off.
- Move drip tray and/or drain pan from under vehicle.
- Install removed components in reverse order of removal.
- Lower vehicle, disconnect scan tool or scan tool computer program.

Component Overviews

The next few pages contain overviews of the various components of the DSG transmission. These are intended to be a guide for locating the various components. For more detailed removal and installation procedures, please consult the Official Factory Repair Manual, electronic edition, available from the publishers at www.bentleypublishers.com.

NOTE —

Volkswagen identifies electrical components by a letter and/or a number in the electrical schematics. See **EWD Electrical Wiring Diagrams.** *These electrical identifiers are listed in parenthesis as an aid to electrical troubleshooting.*

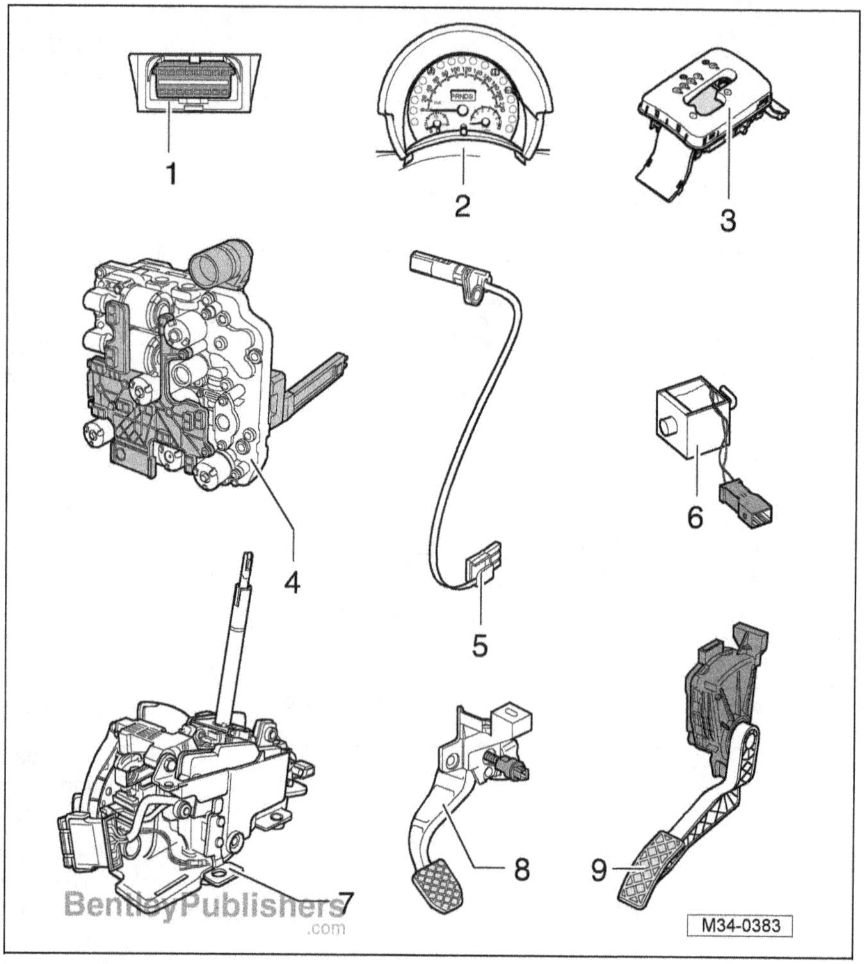

Electrical components

1. **Data link connector (DLC) (T16)**
   - Under left side of dashboard
2. **Transmission range display (Y6)**
   - All displays lit if transmission is operating in emergency mode. Check for diagnostic trouble codes.
   - Can only be replaced together with instrument cluster
3. **Gear selector with integrated components**
   - Shift pattern illumination (L101)
   - Selector lever (E313)
   - Tiptronic switch (F189)
   - Selector lever park position lock switch (F319)
   - Selector lever sensor system control module (J587)
4. **DSG mechatronic unit (J743)**
   - Bolted to front of transmission housing and covered by transmission oil pan

   Integrated components:

   - Solenoid valves (N88, N89, N90, N91, N92)
   - Pressure control valves (N215, N216, N217, N218, N233, N371)
   - Hydraulic pressure sensors (G193, G194)
   - ATF temperature sensors (G93, G510)
   - Transmission output speed sensors (G195 and G196)
   - Drive axle speed sensors (G501 and G502)
   - Gear position distance sensors (G487, G488, G489, G490)
   - Electronic control module
5. **Transmission input speed sensor (G182) and clutch oil temperature sensor (G509)**
   - Bolted to transmission housing under DSG mechatronic unit
6. **Shift lock solenoid (N110) integrated with gear selector**
7. **Selector lever (E313)**
   - With tiptronic switch (F189)
   - With selector lever sensor system control module (J587)
8. **Brake light switch (F) and brake pedal switch (F47)**
   - Data transfer via CAN-bus
9. **Throttle position sensor (G79)**
   - With kickdown switch (F8)
   - Data transfer via CAN-bus

Transmission components

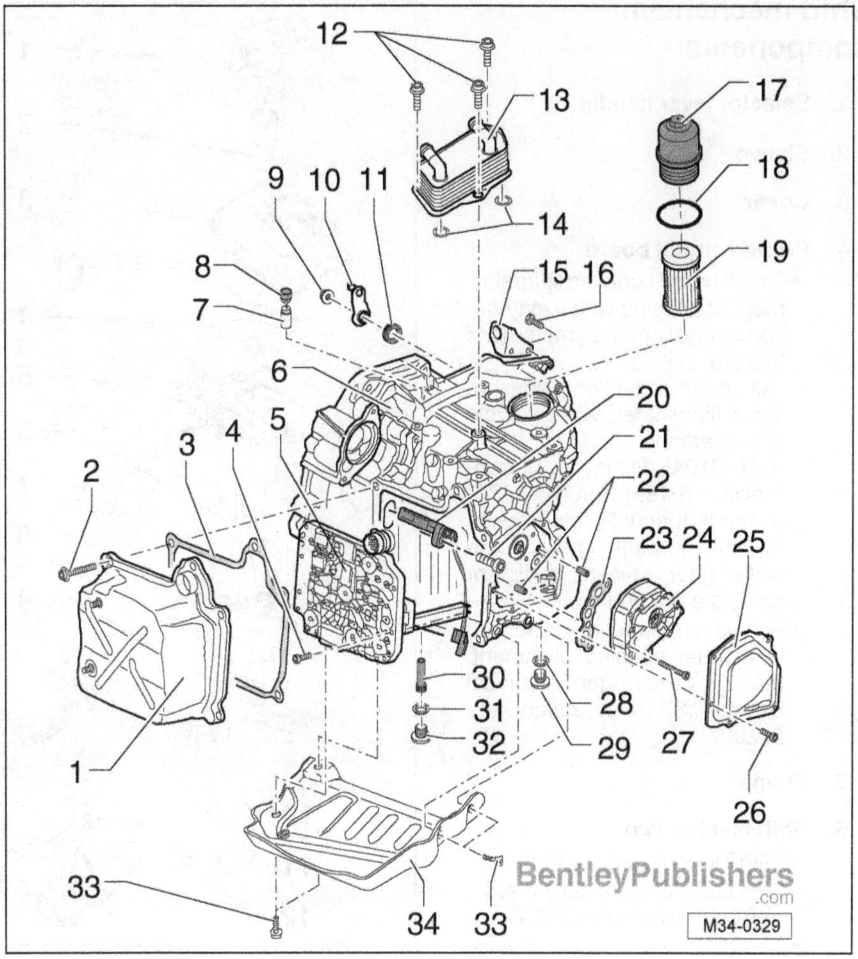

1. **Oil pan**
2. **Bolt**
   - 10 Nm (7 ft-lb)
   - Quantity: 5
   - Always replace
3. **Gasket**
   - Always replace
4. **Bolt**
   - 5 Nm (44 in-lb) plus 90° (¼ turn)
   - Quantity: 10
   - Always replace
5. **Mechatronic unit (J743)**
6. **Transmission housing**
7. **Ventilation pipe**
   - Pressed into transmission
   - Always replace after disconnecting
8. **Ventilation cap**
9. **Nut**
   - 20 Nm (15 ft-lb)
   - Always replace
10. **Lever/selector shaft**
    - Insert so that groove matches up to shift rod
11. **Sealing ring for selector lever**
    - Installation position: Lettering points outward toward lever
12. **Bolts**
    - 20 Nm (15 ft-lb) plus 90° (¼ turn)
    - Always replace
13. **Transmission oil cooler**
14. **O-rings**
    - Always replace
15. **Bracket**
    - For selector lever cable
16. **Bolts**
    - 20 Nm (15 ft-lb) plus 90° (¼ turn)
    - Always replace
    - Quantity: 2
17. **Filter housing**
    - 20 Nm (15 ft-lb)
18. **O-ring**
    - Always replace
19. **Transmission oil filter**
    - Always replace with every oil change.
20. **Transmission input speed (RPM) sensor (G182)**
    - With clutch oil temperature sensor (G509)
21. **Bolt, 10 Nm**
    - 10 Nm (7 ft-lb)
    - Always replace
22. **Alignment pin**
    - Quantity: 2
    - For oil pump alignment
23. **Gasket**
    - Always replace
24. **Transmission oil pump**
25. **Cap for transmission oil pump**
    - With vulcanized gasket
    - Always replace
26. **Bolt**
    - 8 Nm (71 in-lb)
    - Always replace
    - Quantity: 4
27. **Bolt**
    - 5 Nm (44 in-lb) plus 90° (¼ turn)
    - Quantity: 4
    - Always replace
    - Loosen and tighten in diagonal sequence
28. **Seal**
    - Always replace
29. **Drain plug**
    - 45 Nm (33 ft-lb)
    - No longer installed as of September 2004
    - The oil can be drained through inspection/drain plug
30. **Overflow tube**
    - 3 Nm (27 in-lb)
    - Plastic
31. **Seal**
    - Always replace
32. **Inspection/drain plug**
    - 45 Nm (33 ft-lb)
33. **Bolt**
    - 32 Nm (24 ft-lb)
34. **Protective plate**

Shift mechanism components

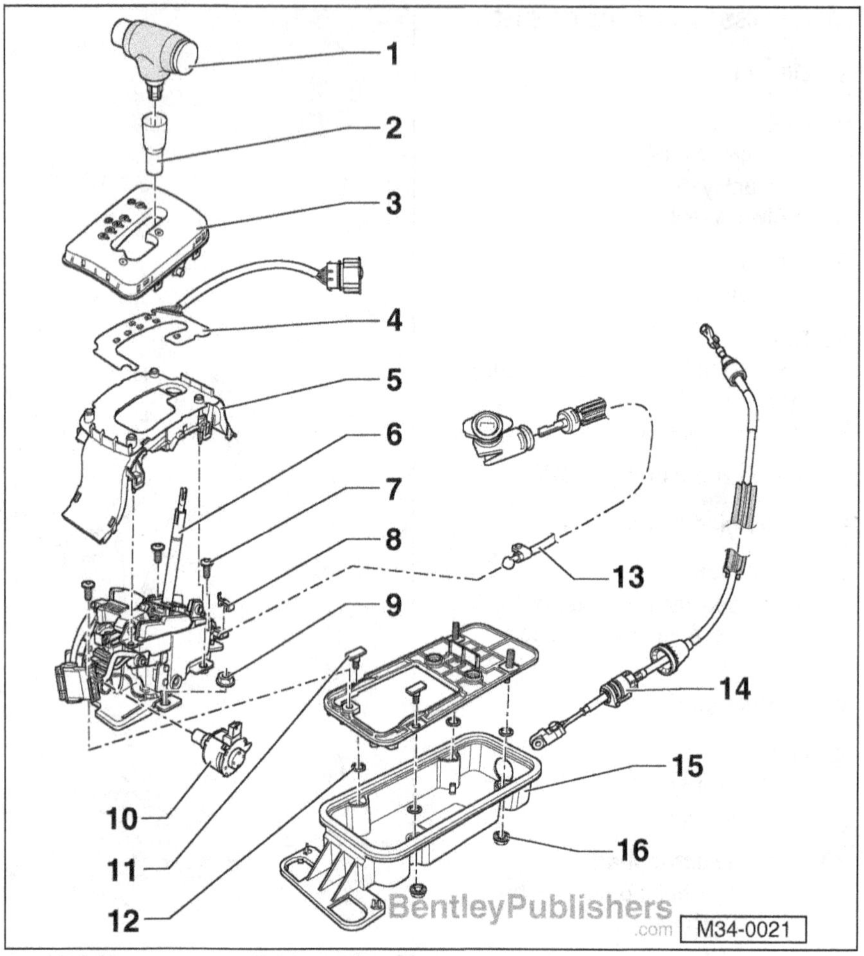

1. **Selector lever handle**
2. **Sleeve**
3. **Cover**
4. **Printed circuit board**
   - Circuit board contains tiptronic magnet housing which may become misaligned. Symptoms of this are:
     -All gear selection indicator lights illuminated while vehicle is in operation
     -DTC 01045 displayed, Tiptronic switch (F189) stored in TCM
     -Cannot manually shift transmission through gear range while gear selector indicator lights are illuminated
   - For more information on tiptronic magnet housing alignment, refer to Volkswagen Technical Bulletin #37 07 07, dated Feb. 2, 2007.
5. **Frame**
6. **Shift mechanism**
   - With Tiptronic switch (F189)
   - With selector lever sensor system control module (J587)
   - Complete assembly must be replaced if above components fail
7. **Bolt**
   - 6 Nm (53 in-lb)
   - Quantity: 3
8. **Circlip**
9. **Nut**
10. **Shift lock solenoid (N110)**
11. **Bolt**
12. **Seal**
13. **Ignition lock cable**
    - Must not be kinked
14. **Selector lever cable**
    - Do not bend or kink
    - If protective sleeve is damaged, cable must be replaced
    - Do not grease cable eyelet or ball socket
15. **Housing**
    - With glued-in seal
16. **Nut**
    - Self-locking
    - 10 Nm (7 ft-lb)
    - Quantity: 4
    - Always replace

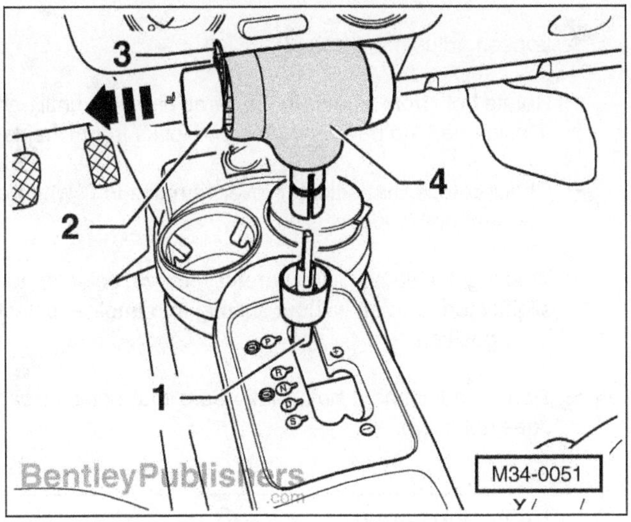

Selector lever handle, removing and installing

- Place selector lever in PARK (P) position.
- Slide sleeve (**1**) downward.
- Pull button (**2**) in direction of **arrow** until it stops.
- Secure button from springing back using a wire clip (**3**).
  - Wire clip can be made from a paper clip or similar.
- Remove selector lever handle (**4**).
- Before installing handle, make sure selector lever is still in P position.
- Press handle onto selector lever shaft up to stop.
- Remove wire bracket (**3**).
- Press in button (**2**) to secure it to shaft. Button should spring back out to stop.
- Rotate and slide sleeve (**1**) upward until tabs in sleeve engage and lock audibly into slots of handle.

Shift mechanism, checking

- Start with selector lever in PARK (P). Press release button and pull selector lever out of P approximately 5 mm (¼ in) toward rear and hold, but do not shift into R.
  - Release selector lever. Check to see that selector springs back into P by itself.
- Place selector lever in NEUTRAL (N). Press release button and pull selector lever out of N approximately 5 mm (¼ in) toward rear and hold, but do not shift into D.
  - Release selector lever. Check to see that selector springs back into N by itself.
- Place selector lever in NEUTRAL (N). Press release button and push selector lever out of N approximately 5 mm (¼ in) toward front and hold, but do not shift into R.
  - Release selector lever. Check to see that selector springs back into N by itself.
- If selector lever does not spring back into specified position, adjust selector cable. See **Selector cable, adjusting** in this repair group.
- Move selector lever to Tiptronic position ("+" and "–"). Check to see that display in instrument cluster changes from "P R N D S" to "6 5 4 3 2 1". Check that "6 5 4 3 2 1" display in instrument cluster changes to appropriate gear when selecting higher (+) or lower gears (–) in Tiptronic position.

Selector cable, adjusting

– Place selector lever in PARK (P) position.

– Remove air filter assembly. See **03 Maintenance**.

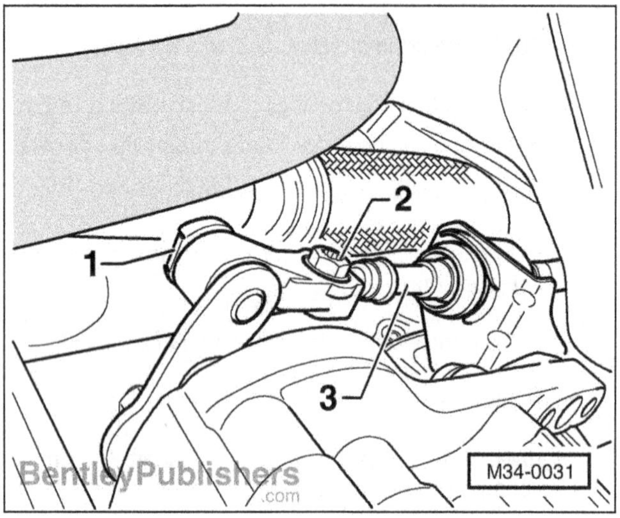

◄ Loosen adjustment bolt (**2**).

– Rotate both front wheels in the same direction until transmission parking pawl engages and locks front wheels.

– Check to see that selector lever remains in P when adjustment bolt is loose.

– Working inside vehicle, carefully move selector lever slightly fore and aft without shifting into another selector lever position.

– Tighten adjustment bolt. Make sure that selector cable does not move.

Tightening torque
- Adjustment bolt to selector lever cable .... 13 Nm (10 ft-lb)

37 Automatic Transmission

37

GENERAL

Cars covered by this manual with automatic transmissions are equipped with either the 01M 4-speed or the 09G 6-speed automatic transmission. Both transmissions are controlled electro-hydraulically and feature adaptive programming and On-Board Diagnostic (OBD) capabilities. **Table a** and **b** list the available factory specifications for the two automatic transmissions.

The transmission code and production number can be found stamped on the transmission case near the Automatic Transmission Fluid (ATF) cooler, see **ATF Cooler**, given later. The transmission code letters can also be found on the vehicle data plate.

NOTE—
ATF checking and filling procedures are covered in **03 Maintenance**.

Automatic transmission application

- 01M 4-speed . . . . 1998 - 2005
- 09G 6-speed . . . . 2003 - on

CAUTION—
- *Before working on the transmission or gear selector mechanism, disconnect the negative (–) battery cable.*
- *Disconnecting the negative (–) battery cable may erase fault codes and basic settings in the engine management and automatic transmission control modules. Some driveability problems may be noticed until the system re-adapts to operating conditions. OBD II readiness codes, which may be required for emissions testing, may also be erased. Convenience electronics (alarm system, interior light control, power locks, mirrors, and windows) will need to be re-set using a diagnostic scan tool.*

Table a. 01M Automatic transmission specifications

Designation		01M		
Code letters		DYQ, ECV, ELY, EPG, FDF	DMP, ECN, ELT, EPB, EPP, FDB	EFB, ECP, ELU, ELZ, EPC, EPH, FDC, FDG
Engine applications		2.0L	1.9L	1.8L, 2.0L
Ratio:	Final drive	4.875	4.875	4.533
	Intermediate drive	1.033	0.978	0.978
	1st gear	2.714	2.714	2.714
	2nd gear	1.551	1.441	1.441
	3rd gear	1.000	1.000	1.000
	4th gear	0.679	0.742	0.742
	Reverse	2.111	2.884	2.884
Transmission Lubricant	Initial filling	5.3 Liters (5.6 qt.)		
	Oil change	approx. 3.0 Liters (3.2 qt.)		
	Specification	VW ATF, part # G 052 162 A2		
Final drive Lubricant	Initial filling	0.75 Liters (0.8 qt.)		
	Oil change	Filled for life, no change		
	Specification	Synthetic oil, SAE 75/90W VW part # G 052 145 A2		
Drive shaft		tripodic		

Table b. 09G Automatic transmission specifications

Designation		09G		
Code letters		GHE, HGA, HTT	GGZ, HFZ, HMD	HHG, HTU
Engine application		1.8L	2.0L	2.5L
Transmission Lubricant	Initial filling	7.0 Liters (7.4 qt.)		
	Oil change	approx. 3.0 Liters (3.2 qt.)		
	Specification	VW ATF, part # G 055 025 A2		
Final drive Lubricant	Initial filling	Included with transmission fill		
	Oil change	Filled for life, no change		

BASIC SETTINGS

The automatic Transmission Control Module (TCM) and the Engine Control Module (ECM) share data concerning engine and transmission operation. One piece of data that the TCM may not always "know" is the full throttle position. This is known as a basic setting. The basic setting influences automatic transmission shifting and is set at the time that the vehicle is new. The basic setting will be lost if the battery is disconnected or runs down, or if the TCM is disconnected. It will change if the TCM or the throttle control module is replaced. In these circumstances, the basic settings must be restored to insure proper automatic transmission operation.

Basic settings, restoring

Basic settings can be restored using a suitable diagnostic tool and following the manufacturer's instructions or by using the procedure outlined below. In either case, the kickdown switch on the accelerator cable must function properly.

– Switch on the ignition but do not start the engine.

– Push the accelerator pedal all the way to the floor and hold it there for a minimum of 5 seconds.

> ***NOTE —***
> *Be sure that floor mats do not prevent the accelerator pedal from being completely depressed.*

– Release the accelerator pedal and switch off the ignition.

Basic requirements

– Check that the ATF level is correct. Check that the fluid is clean and of the correct type. See **03 Maintenance**.

> ***CAUTION —***
> *The automatic transmissions covered by this manual use special VW-only ATF. See* **Table a** *or* **b** *for specifications.*

– Make a visual check of the components. Check the wiring and harness connectors for loose, damaged or corroded wiring. Check that all related grounds are firmly connected and in good condition. Consult the wiring diagrams shown in **EWD Electrical Wiring Diagrams**.

– Check the shift mechanism for proper function. See **Shift Mechanism**.

– Review the list given below.

- Engine control module (ECM) replaced
- New, reconditioned, or used engine installed
- New or altered/adjusted throttle housing
- New or altered/adjusted throttle position sensor
- Transmission control module (TCM) replaced
- Throttle valve control module replaced

If any of the above conditions are met, the TCM must be reset to its "basic settings" using a diagnostic scan tool or the procedure described above. If the TCM basic settings are not re-established, driveability problems may be encountered.

If no faults are found up to this point, the next logical step is to have an authorized VW dealer check for faults using the special VW diagnostic scan tool, VAG 1551/1552 or 5051/5052. If the transmission problem is electrical/electronic in nature, specific DTCs will most likely be stored in memory.

AUTOMATIC TRANSMISSION FLUID (ATF), REPLACING

> ***NOTE —***
> *ATF checking and filling procedures are covered in* **03 Maintenance**.

– Remove lower sound insulation panel. Remove sound insulation from left front longitudinal member. See **03 Maintenance**.

– Place container under transmission to catch ATF.

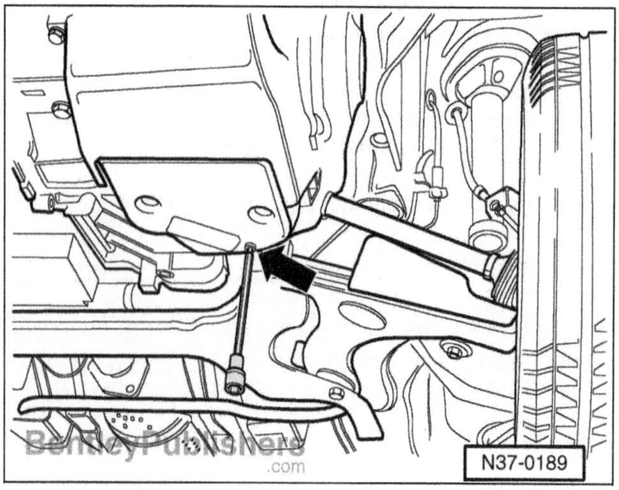

< Remove ATF level plug (**arrow**) from transmission oil pan.

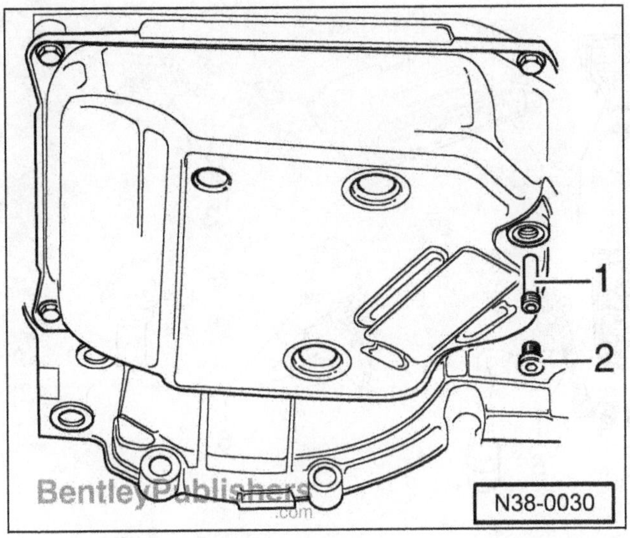

◀ Remove overflow tube (**1**) through level hole.

- Drain ATF.
- Install overflow pipe (**1**) and screw level plug (**2**) in hand-tight.

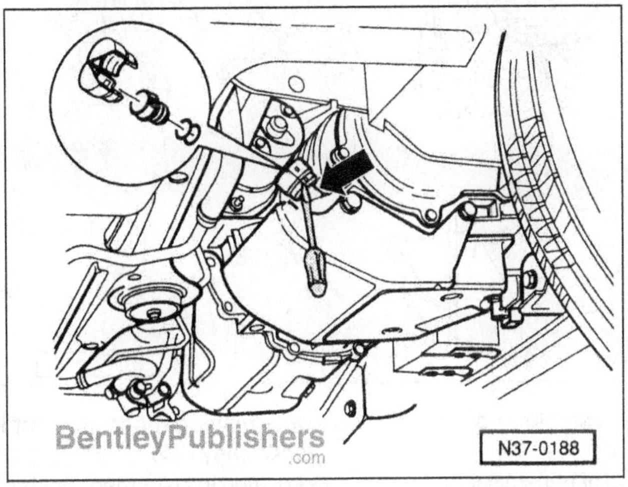

◀ Pry off ATF filler plug and seal (**arrow**) with screwdriver. The seal will be destroyed, always replace.

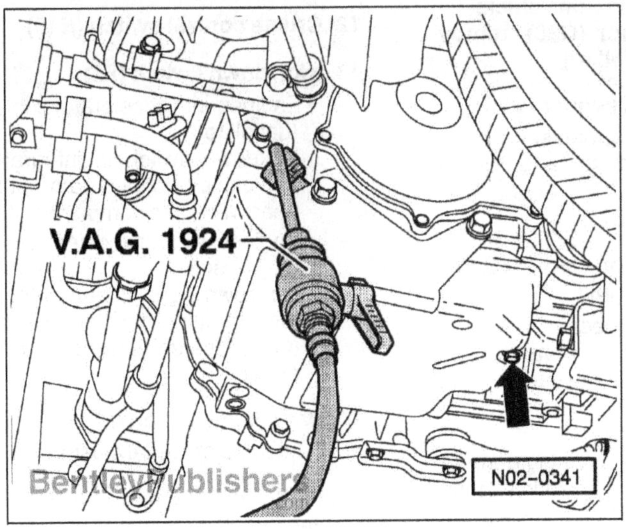

◀ Fill with 3 liters of ATF through filler pipe using **VAG 1924**, or equivalent. See **Table a** or **b** for ATF part number.

NOTE —

Volkswagen part numbers are given for reference only! Always consult with your Volkswagen Parts Department or aftermarket parts specialist for the latest parts information.

- Start engine and shift through all selector lever positions with vehicle stationary.
- Check and top off ATF level. See **03 Maintenance**.

AUTOMATIC TRANSMISSION ELECTRICAL COMPONENTS

Electrical components (01M)

NOTE —
Volkswagen identifies electrical components by a letter and/or a number in the electrical schematics. See **EWD Electrical Wiring Diagrams**. *These electrical identifiers are listed in parenthesis as an aid to electrical troubleshooting.*

- *If engine or transmission control modules are replaced, the system must be returned to basic setting.*

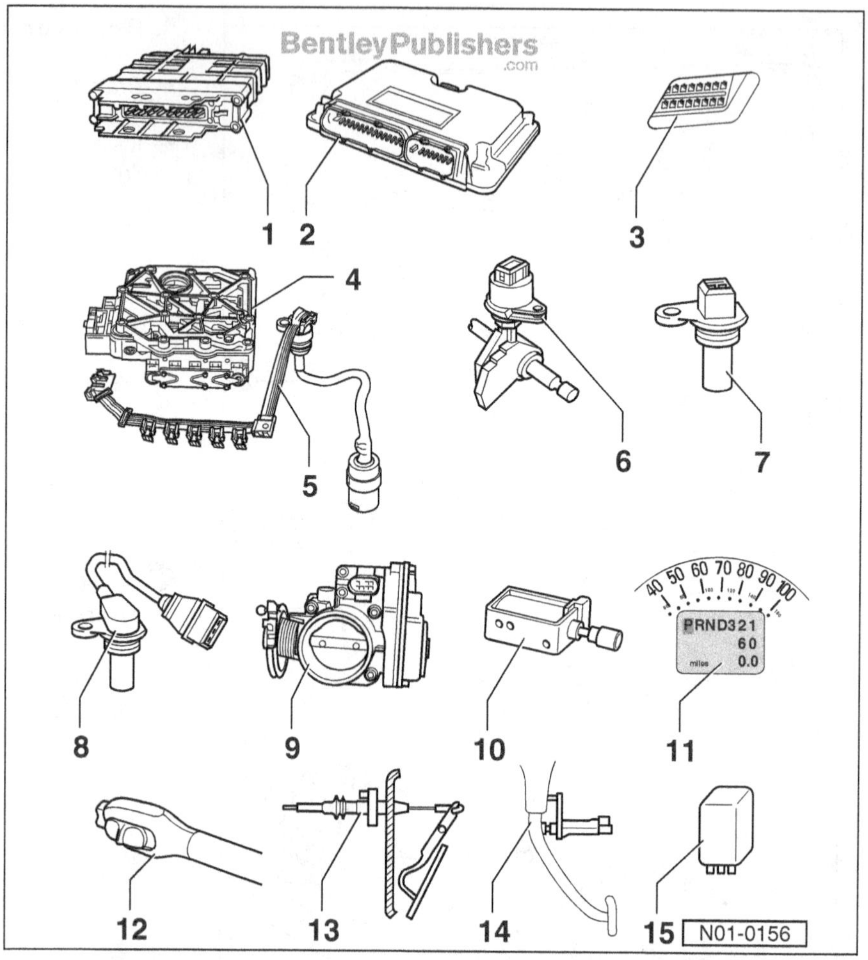

1. **Transmission control module (J217)**
   - Located in center plenum
   - Factory coded, no user changes possible
2. **Engine control module**
   - Located in plenum, driver's side
   - Removing and installing, see **23 Fuel Injection–Diesel** or **24 Fuel Injection–Motronic**
3. **Data Link Connector (DLC)**
   - Located under dashboard on driver's side
4. **Valve body**
   - Located above oil pan
   - Solenoid valves (N88, N89, N90, N91, N92, N93, N94) are attached to the valve body
   - Checked by On-Board Diagnostics (OBD)
5. **Conductor strip with integrated transmission fluid temperature sensor (G93)**
   - Located in oil pan, on valve body
   - Checked by OBD
   - Can be replaced without removing valve body or transmission
6. **Multi-function transmission range switch (F125)**
   - See **Multi-function transmission range switch (F125) replacing**
   - Checked by OBD
7. **Transmission vehicle speed sensor (G38)**
   - Located on top of transmission
   - See **Transmission vehicle speed sensor (G38), removing and installing**
8. **Vehicle speed sensor (G68)**
   - See **Vehicle speed sensor (G68), removing and installing**
   - Checked by OBD
9. **Throttle position sensor (G69)**
   - Checked by OBD
   - Gasoline engines:
     located on throttle valve housing, part of throttle valve control module (J338)
   - Diesel engines:
     throttle valve signal is generated in the throttle position sensor (G79) located near accelerator pedal
10. **Shift Lock Solenoid (N110)**
    - Location, see **Shift mechanism, assembly**
    - Checked by OBD
11. **Transmission range selector lever display (Y5)**
    - Located in instrument cluster
12. **Cruise control switch (E45)**
13. **Kick down switch (F8)**
    - Engines with accelerator cable (gasoline):
      kickdown switch is integrated into cable and located on bulkhead in engine compartment
    - Engines without accelerator cable (diesel):
      kickdown signal is generated in the throttle position sensor (G79) located near accelerator pedal
14. **Brake light switch (F)**
    - Located on pedal cluster
15. **Park/Neutral Position (PNP) Relay (J226)**
    - Located on additional relay panel under instrument panel, left side
    - Marked with number "175"

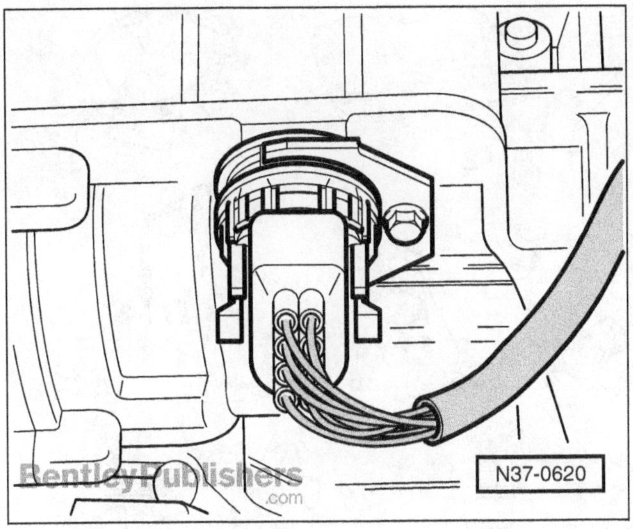

Multi-function transmission range switch (F125) replacing (01M transmission)

- Switch off ignition, remove harness connector, bolt, and retaining bracket.
- When installing, replace seal.
- Tighten retaining clamp bolt to 10 Nm (7 ft-lb).

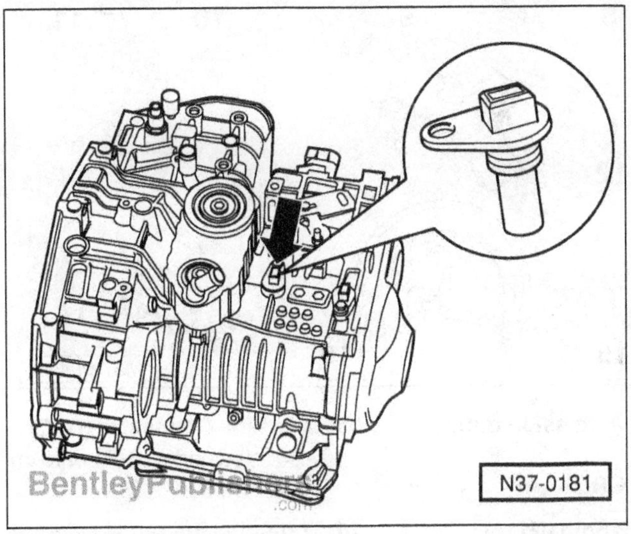

Transmission vehicle speed sensor (G38), removing and installing (01M transmission)

- Switch off ignition, remove harness connector and retaining bolt.
- When installing, replace seal.
- Tighten retaining clamp bolt to 10 Nm (7 ft-lb).

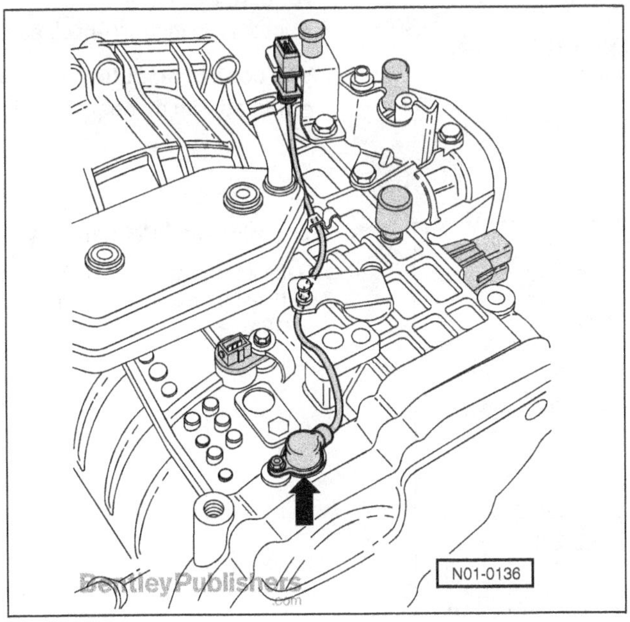

Vehicle speed sensor (G68), removing and installing (01M transmission)

- When transmission is installed the sensor is covered by left transmission mount.
- Transmission must be unbolted from mounts and lowered approximately 2.5 in. to access sensor, see **Transmission, Removing and Installing (01M)**, given later.
- Disconnect harness connector from sensor and remove sensor retaining bolt.
- When installing, replace seal.
- Tighten retaining bolt to 10 Nm (7 ft-lb).

Electrical components (09G)

NOTE —
Volkswagen identifies electrical components by a letter and/or a number in the electrical schematics. See **EWD Electrical Wiring Diagrams**. *These electrical identifiers are listed in parenthesis as an aid to electrical troubleshooting.*

- *If engine or transmission control modules are replaced, the system must be returned to basic setting.*

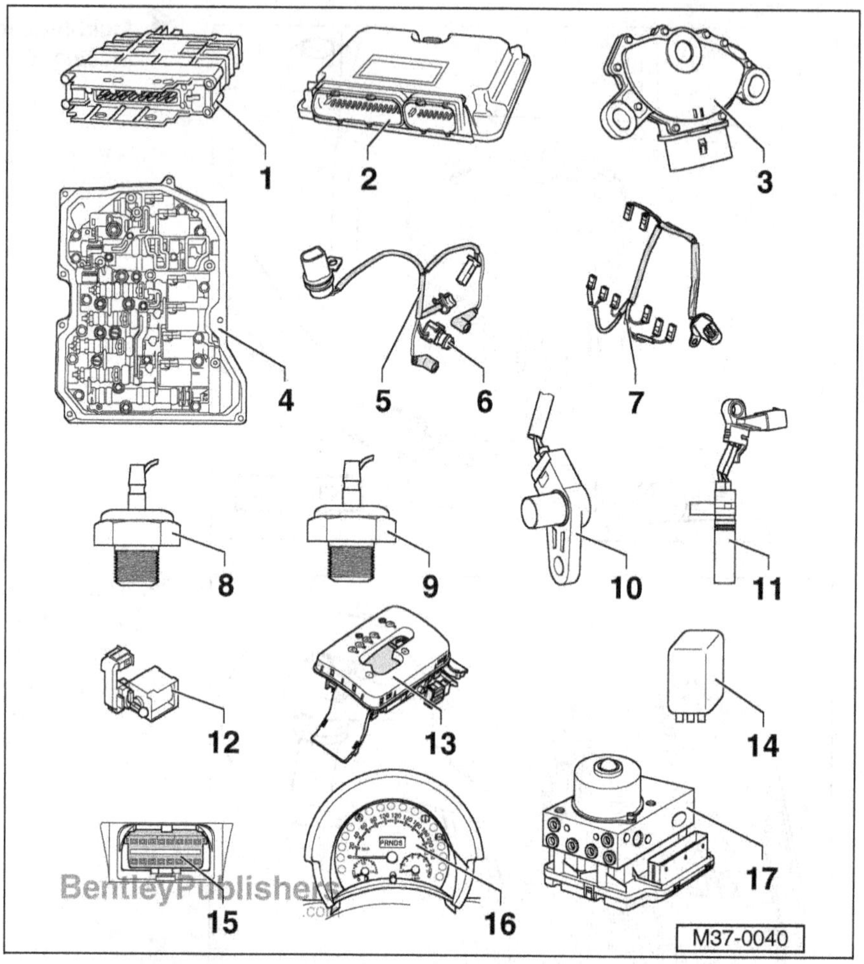

1. **Transmission control module (J217)**
   - Located in center plenum
   - Factory coded, no user changes possible
2. **Engine control module**
   - Located in plenum, driver's side
   - Removing and installing, see **23 Fuel Injection–Diesel** or **24 Fuel Injection–Motronic**
3. **Multi-function transmission range switch (F125)**
   - Located on front left side of transmission
   - Checked by OBD
4. **Valve body**
   - Located above oil pan
   - Solenoid valves (N88, N89, N90, N91, N92, N93, N94) are attached to the valve body
   - Valves are checked by OBD
5. **Wiring harness**
   - 8-pin
   - With transmission fluid sensor (G93)
6. **Transmission fluid sensor (G93)**
   - Checked by OBD
7. **Wiring harness**
   - 14-pin
8. **Sender 1 for hydraulic pressure sensor (G193)**
   - In valve body
9. **Sender 2 for hydraulic pressure sensor (G194)**
   - In valve body
10. **Sensor for transmission RPM (G182)**
    - Checked by OBD
11. **Sender for transmission output RPM (G195)**
    - Checked by OBD
12. **Shift Lock Solenoid (N110)**
    - Checked by OBD
13. **Tiptronic switch (F189)**
    - Checked by OBD
14. **Starter interlock relay (J207)**
    - Checked by OBD
15. **Data Link Connector (DLC)**
    - Located under dashboard on driver's side
16. **Transmission range selector lever display (Y5)**
    - Located in instrument cluster
17. **ABS control module**
    - Marked with number "175"

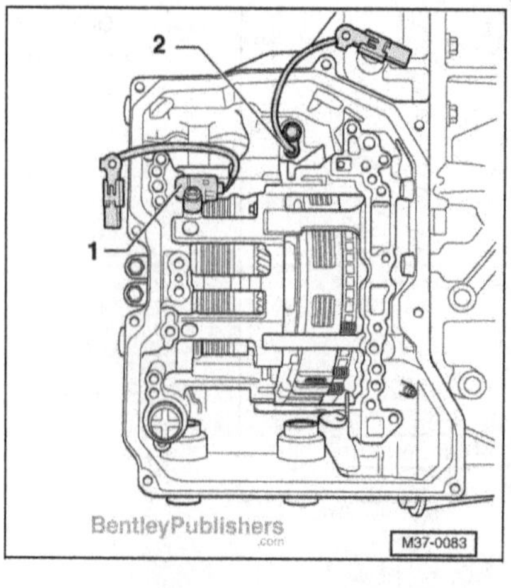

◀ **Transmission RPM sensors**

1. Sensor for transmission RPM (G182)
2. Sender for transmission output RPM (G195)

SHIFT MECHANISM

Selector lever cable, checking and adjusting

– Switch ignition off and set parking brake.

– Open hood and locate selector lever cable and selector shaft lever on transmission.

– Use a screwdriver or 10mm open end wrench to pry selector lever cable off selector shaft lever and move cable aside so that the end is free to move. See appropriate illustration at left, depending on transmission type.

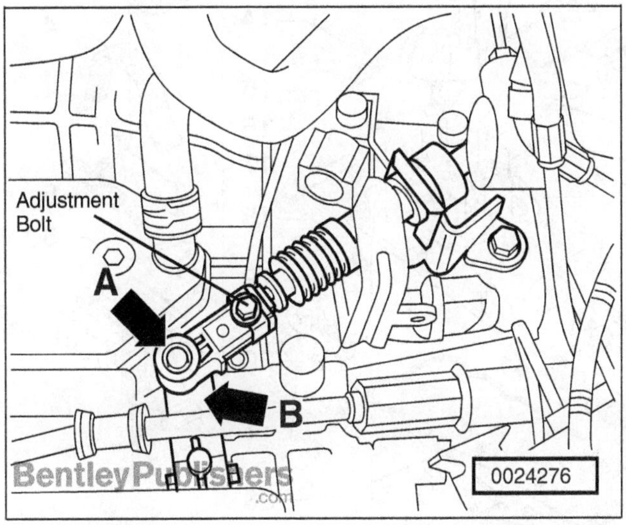

01M 4-speed transmission.

- Selector lever cable (**A**)
- Selector shaft lever (**B**)

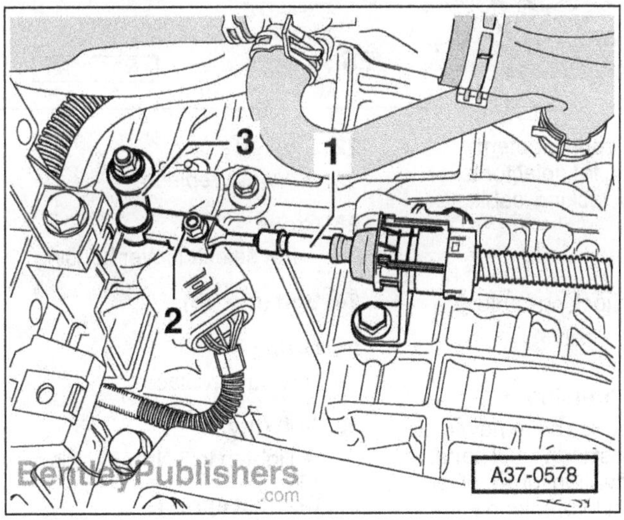

09G 6-speed transmission.

- Selector lever cable (**1**)
- Adjustment bolt (**2**)
- Selector shaft lever (**3**)

– Move selector lever from "**P**" to "**1**" and check that shift mechanism and selector lever cable move freely. If necessary replace selector lever cable or service shift mechanism.

– Check protective bellows at front of selector cable for damage. If damaged, replace selector cable with bellows.

– To adjust cable, install selector lever cable on selector shaft lever and move selector lever to "**P**".

– Loosen adjustment bolt.

– Place selector shaft lever into position "**P**" on transmission then tighten adjustment bolt. Shift through all gear positions and check for smooth operation.

NOTE —
*With selector lever in "**P**" position, the transmission locking lever must engage, locking both front wheels.*

Tightening torque
- Selector lever adjustment bolt (01M) . 5 Nm (44 in-lb)
- Selector lever adjustment bolt (09G) . 13Nm (10 ft-lb)

Shift mechanism, assembly (01M)

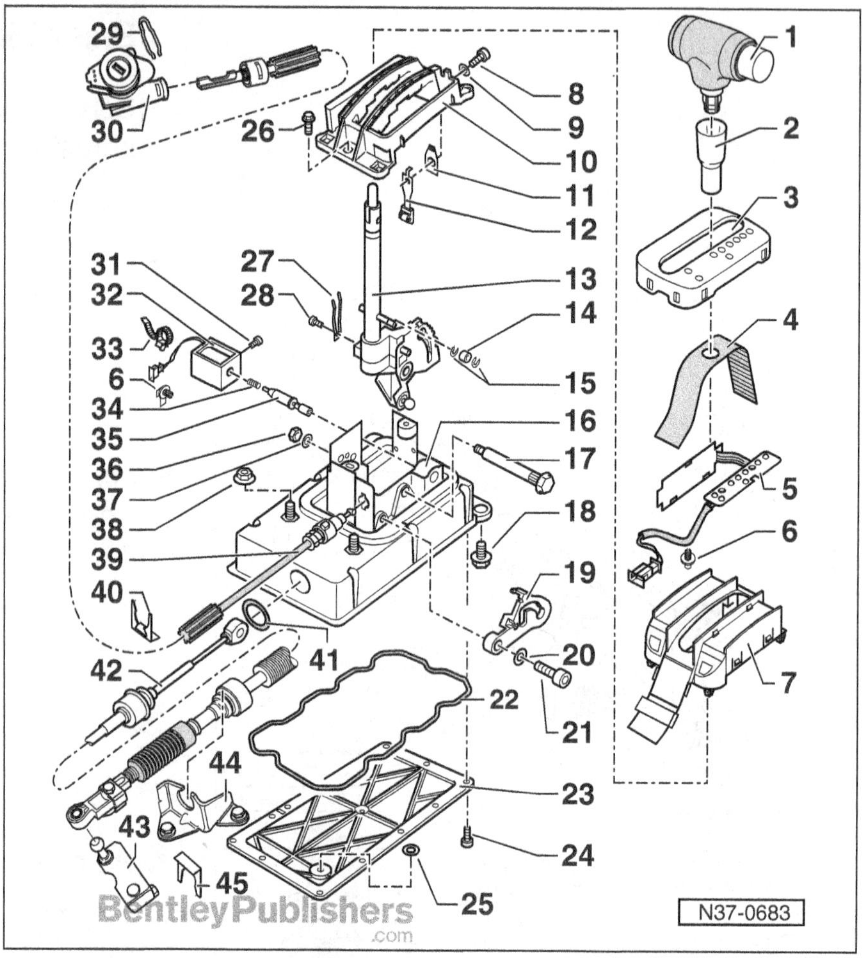

NOTE —
Lubricate mounting points and contact surfaces with grease (VW part no. G 052 142 A2).

1. **Selector lever handle**
   - Removing: push sleeve (item 2) down and while depressing knob, remove handle upward
   - Installing: while depressing knob, install handle until lock engages. Push sleeve up to lock.
2. **Sleeve**
   - Prevents selector lever handle from being pulled off
3. **Cover**
   - With selector indicator
   - Snapped into bracket (item 7)
4. **Cover strip**
   - Insert in bracket (item 7)
5. **Selector lever position display**
   - Printed circuit for selector lever position illuminated display is held in position by clips
6. **Retainer**
   - To position connector in bracket (item 7)
7. **Bracket**
   - Carefully pry off at corners
   - Push into cover strip (item 4)
   - Installation position: ribs must face down
   - Carefully position printed circuits and wiring (item 5)
   - When installing be careful not to damage contact springs (item 27) for selector lever position display
8. **Bolt**
   - Tighten to 7 Nm (62 in-lb)
   - Attaches locating spring (item 12) and plate (item 11) to bracket (item 7)
9. **Washer**
10. **Locking segment**
    - Place on centering elements of selector lever housing
11. **Plate**
12. **Locating spring with roller**
    - Engages in detent in selector lever
13. **Selector lever**
    - With pull rod, spring, detent, contact spring for selector lever position and locking cable release
14. **Roller**
    - For releasing locking cable
15. **Circlip**
16. **Selector lever housing**
    - Does not need not be removed to replace most individual parts, with the exception of selector lever and shift lock solenoid
17. **Fulcrum pin**
    - Apply grease to shank
    - Do not turn when installing selector lever
18. **Bolt**
    - Tighten to 25 Nm (18 ft-lb)
19. **Locking lever**
    - For locking cable
    - For locking selector lever in "**P**"
20. **Washer**
21. **Bolt**
22. **Gasket**
    - Always replace
23. **Cover**
    - For selector lever housing
24. **Bolt (qty. 14)**
25. **O-ring**
    - Always replace
26. **Bolt (qty. 2)**
    - Tighten to 8 Nm (71 in-lb)
27. **Contact spring**
    - For selector lever position display
28. **Bolt**
    - Tighten to 4 Nm (35 in-lb)
    - Place selector lever in "**P**" to loosen and tighten
29. **Clip**
30. **Steering lock**

(continued on next page)

Shift mechanism, assembly (01M continued)

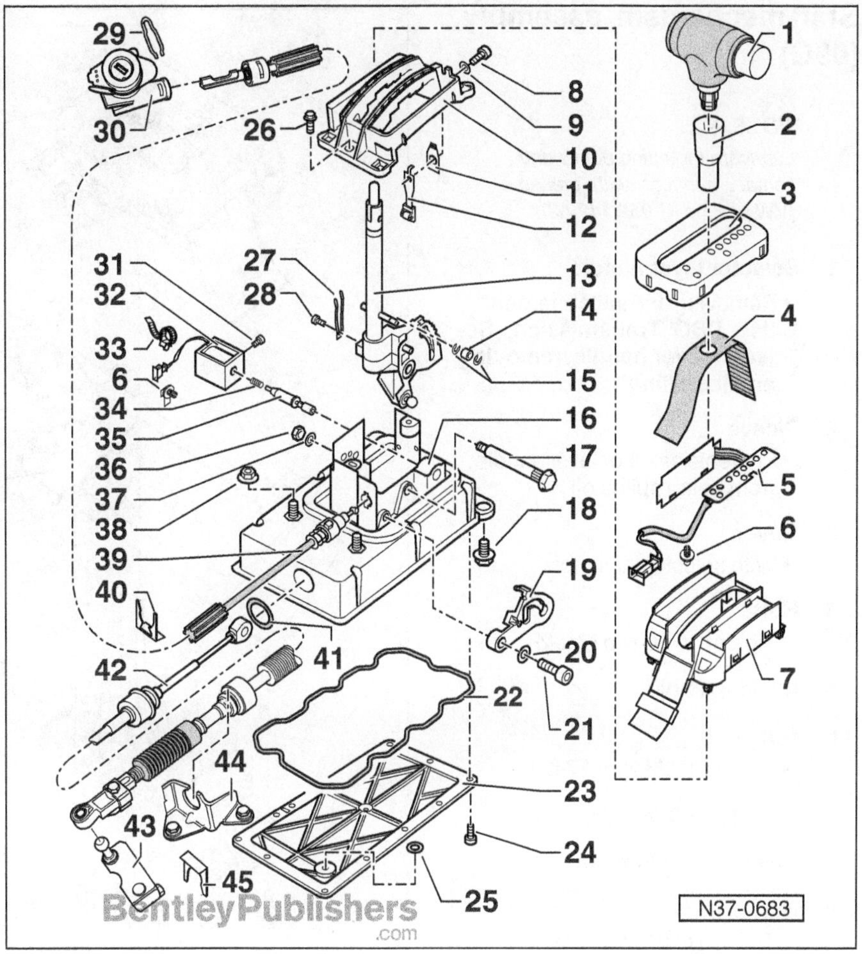

31. Bolt (qty. 2)
- Attaches solenoid to selector lever housing

32. Shift lock solenoid (N110)
- Checked via OBD
- Remove/install only with selector lever in position "**1**"
- Look out for spring and locking pin when removing
- Install together with locking pin and spring while moving selector lever slightly back and forth

33. Cable tie
- Attaches wiring/solenoid to selector lever housing
- Always replace

34. Spring

35. Locking pin

36. Nut
- Tighten to 13 Nm (10 ft-lb)
- Always replace

37. Washer

38. Nut (qty. 2)
- Tighten to 25 Nm (18 ft-lb)
- Always replace

39. Shift-lock cable
- Do not kink
- See **Automatic Shift Lock**, given later

40. Clip
- Always replace
- Mounts selector lever cable to selector lever housing
- Install angled end toward inside of selector housing

41. Gasket
- Always replace

42. Selector lever cable
- Do not bend or kink
- If outer cable sleeve is damaged, replace selector lever cable.
- Lightly grease ball socket and cable ends before installing

43. Lever
- For selector shaft

44. Support bracket
- For mounting selector lever cable on transmission

45. Clip
- Always replace

Shift mechanism, assembly (09G)

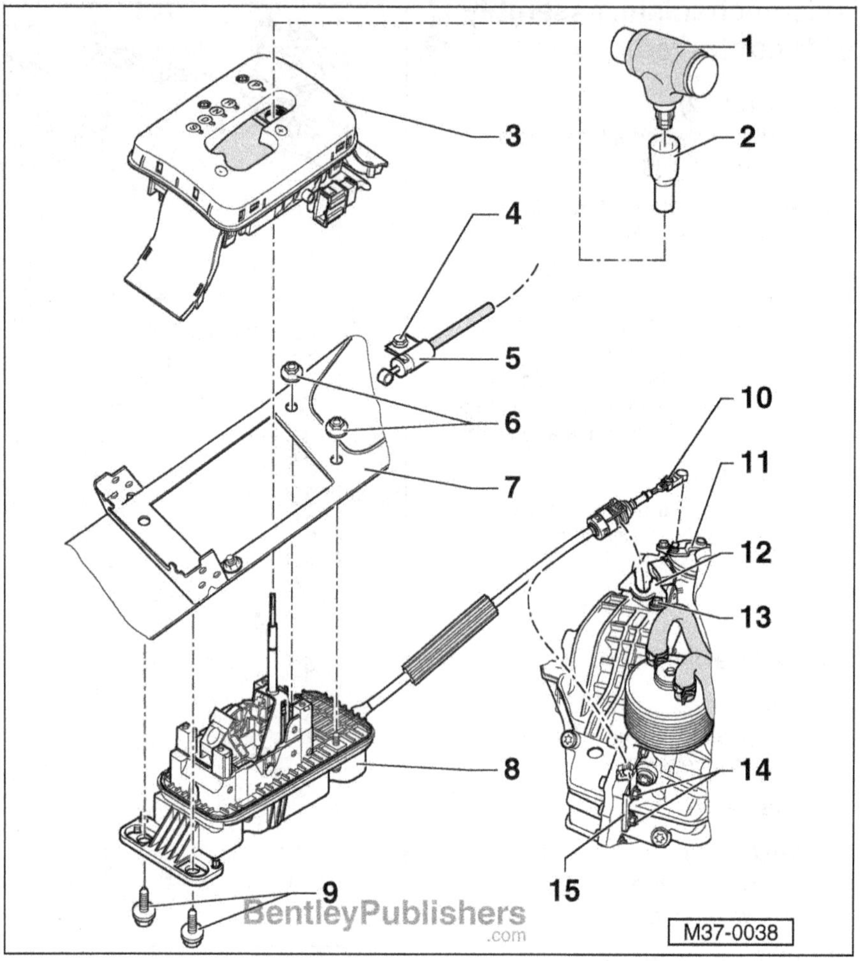

NOTE —
Lubricate mounting points and contact surfaces with grease (VW part no. G 052 142 A2).

1. **Selector lever handle**
   - Removing and installing: see **34b DSG Transmission, Selector lever handle, removing and installing**
2. **Sleeve**
   - Prevents selector lever handle from being pulled off
3. **Cover**
   - With selector indicator
4. **Bolt**
   - Tighten to 10 Nm (7 ft-lb)
5. **Locking cable**
6. **Nut**
   - Tighten to 23 Nm (17 ft-lb)
7. **Floor plate/body**
8. **Selector mechanism**
9. **Bolt**
   - Tighten to 23 Nm (17 ft-lb)
10. **Bolt**
    - Tighten to 13 Nm (10 ft-lb)
    - For selector cable adjustment
11. **Lever**
    - For selector shaft
12. **Pivot**
13. **Bolt**
    - Tighten to 23 Nm (17 ft-lb)
14. **Nut**
    - Tighten to 8 Nm (71 in-lb)
15. **Bracket**
    - With clip for selector lever cable

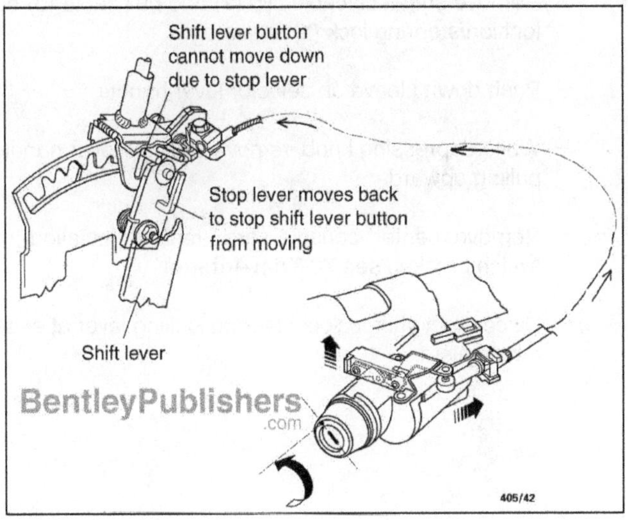

AUTOMATIC SHIFT LOCK

Automatic shift-lock is used on cars with automatic transmissions. Turning the ignition key on and off operates a bowden cable to lock and unlock the selector lever.

◄ When the key is turned to the off position (shift lever in **P**), the cable pushes the stop lever into the shift lever. This locks the lever button and also allows the ignition key to be withdrawn from the lock cylinder. The key can only be removed from the ignition with the lever in the "**P**" position. With the key out of the ignition, the selector lever cannot be shifted out of "**P**" position.

When the key is in the **ON** position, the cable pulls the stop lever away from the shift lever, allowing the shift lever to be moved out of park.

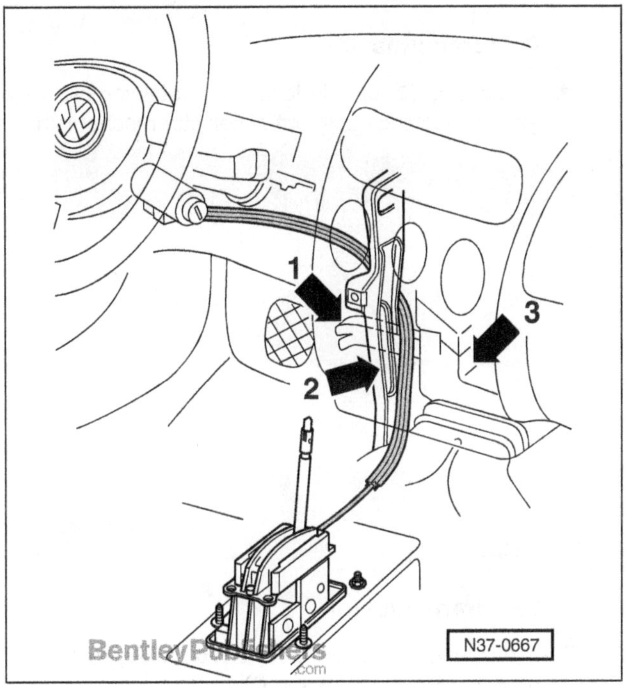

Shift-lock cable, removing and installing

◄ The shift-lock cable is routed through the dashboard and under the center console. Be sure to route the cable correctly when installing.

1. Footwell vents
2. Instrument panel support
3. Heater unit

– Disconnect battery ground (GND) strap from battery negative (–) terminal. See the **Cautions** at the beginning of this repair group regarding battery disconnection.

NOTE —
Be sure to have the anti-theft radio code on hand before disconnecting the battery.

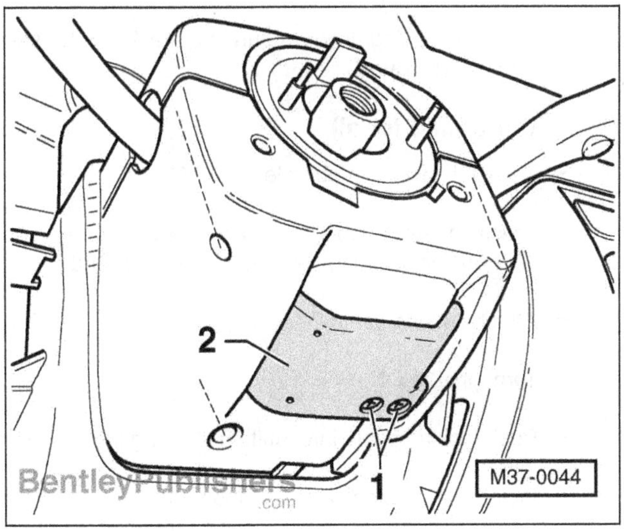

◄ Remove steering wheel, handle for steering height and reach adjustment and cover for ignition/starter lock, see **48 Steering**.

- Mounting bolts (**1**) shown for steering height and reach adjustment lever (**2**).

– Turn ignition **ON** and move selector lever to "**P**".

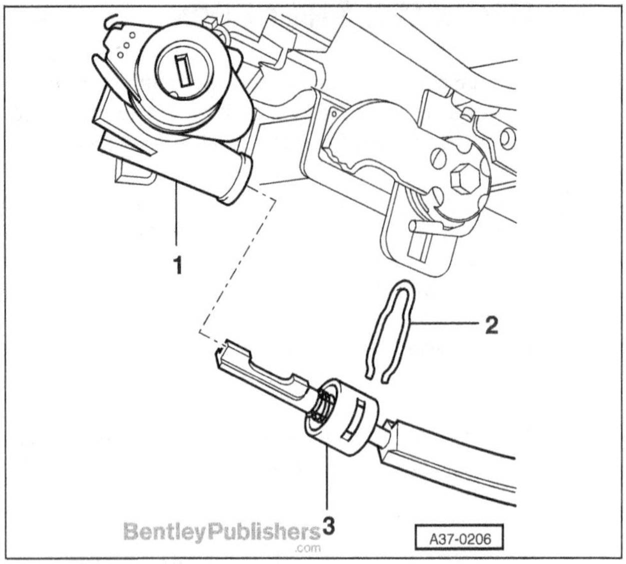

◄ Remove shift-lock cable clip (**2**) and pull cable (**3**) out of ignition/steering lock (**1**) assembly.

- Push down sleeve on selector lever handle.
- While depressing knob, remove selector lever handle by pulling upward.
- Remove center console, arm rest (if installed), and switch bracket, see **70 Trim–Interior**.
- Disconnect shift-lock cable from locking lever at selector mechanism.

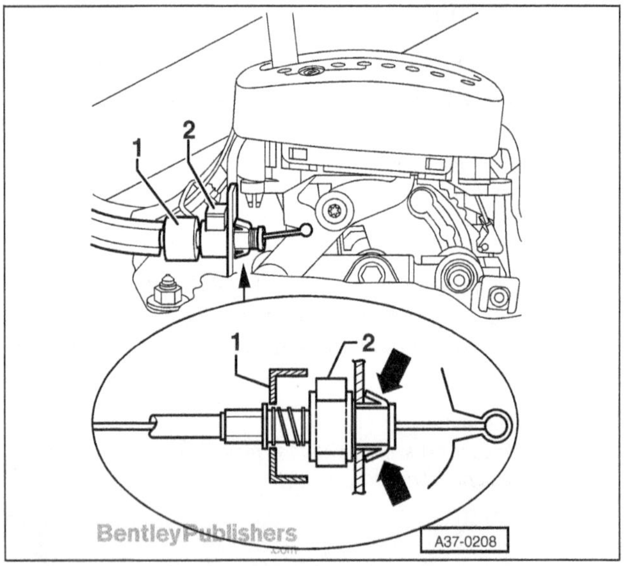

01M transmission

◄ Press tabs (**2**) of shift lock cable retaining clip together and pull locking cable out of selector mechanism.

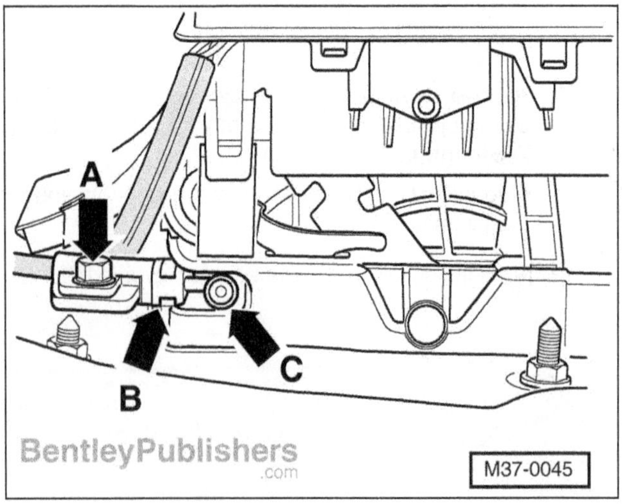

09G transmission

◄ Press locking cable slightly out of cable sleeve and engage locking lever (**arrow C**).

- Engage locking lever mounting bracket into selector housing (**arrow B**).

Continued for all

- To install, move selector lever to "**P**".
- Press shift-lock cable into support bracket on selector lever housing until tabs on retaining clip spread.
- Attach locking cable in locking lever.
- Turn ignition **ON**.
- Push locking cable into ignition/steering lock assembly.
- Install clip onto cable and test operation.

Shift-lock cable, adjusting (01M)

> ***CAUTION —***
> *Follow the adjustment procedure exactly as outlined.*

- Remove ignition key.

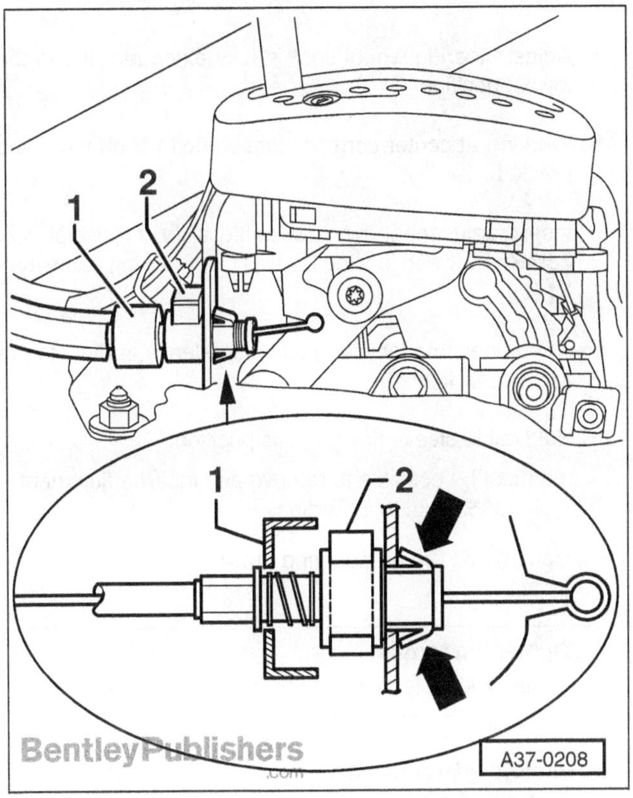

◄ Working at center console, slide cable sleeve (**1**) forward and release clip (**2**) by pushing up (**arrows**).

- Carefully pry up at four corners of shift lever bracket and remove.

NOTE —
Be careful not to damage the contact spring for shift lever position indicator.

- Slide 0.8 mm feeler gauge between locking lever and selector lever roller.

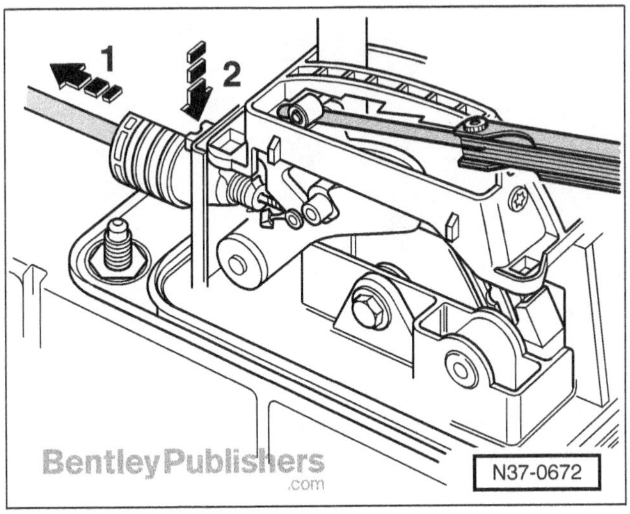

◄ Pull outer sleeve of shift-lock cable slightly forward (**1**) in direction of travel and push red clip down (**2**) until it engages.

- Push sliding sleeve over clip.
- Check for proper ignition key removal.
- Replace center console, arm rest (if applicable), and switch bracket.

Shift-lock cable, adjusting (09G)

Adjustment requirements:

- Locking cable in installation position
- Center console removed
- Selector lever in position "P"
- Ignition key removed

– Adjust steering wheel so it is fully extended and in the lowest position.

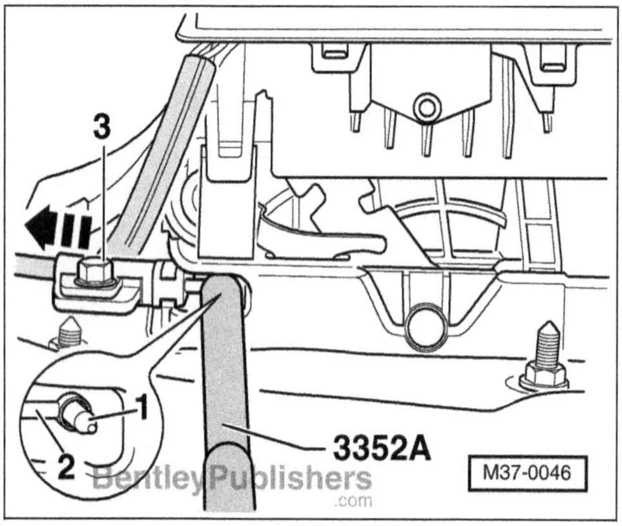

◄ Working at center console, loosen bolt (**3**) on mounting bracket.

– Insert adjustment pin for shiftlock (VW special tool **3352A**) between locking lever (**1**) and locking cable eyelet (**2**).

– To compensate for play, pull cable sleeve in direction of **arrow** until it stops.

– Hold cable sleeve firmly in this position.

- It must be possible to remove and insert adjustment pin (**3352A**) in this position.

– Tighten bolt (**3**) for mounting bracket

Tightening torque

- Shift-lock cable mounting bracket bolt 10 Nm (7 ft-lb)

– Remove adjustment pin.

– Check function of shift-lock cable.

TRANSMISSION, REMOVING AND INSTALLING (01M)

This procedure describes the removal and installation of the 4-speed automatic transmission (01M). Special engine lifting and jacking equipment is needed to support and reposition the engine as the transmission is removed from below.

Volkswagen recommended special tools
- VW 457/1 support rails
- 10-222A transmission support kit
- 10-222A/8 adapter
- 3094 hose clamps
- 3282 transmission support
- 3282/2 adjustment plate
- 3300A engine support
- 3336 transmission lifting beam
- VAG 1331 torque wrench
- VAG 1332 torque wrench
- VAG 1383A engine/transmission jack
- V/175 socket

Special tools, modifying

Two of the Volkswagen recommended special tools will need to be modified as follows.

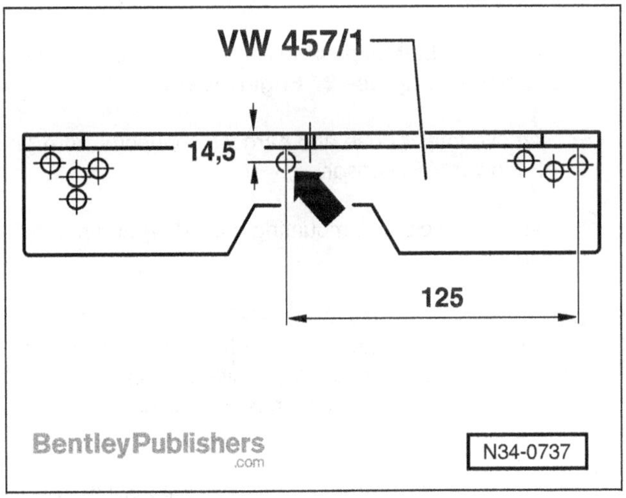

◄ To secure support rail VW 457/1 to the subframe a new hole (**arrow**) is required.

- Dimensions provided in millimeters.

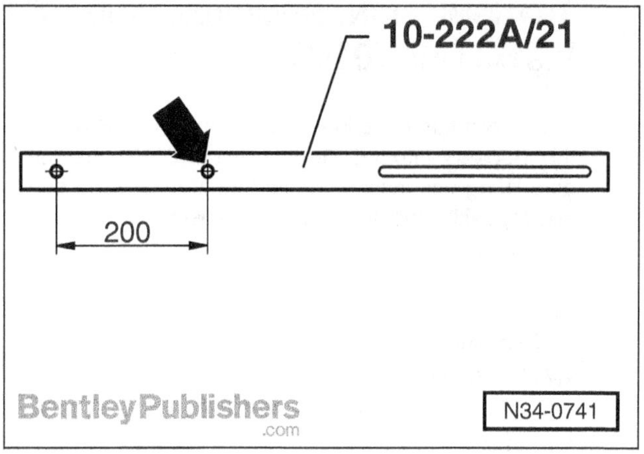

The engine support bridge 10-222A is used to support the engine/transmission assembly. Drill a 12.5 mm (0.50 in.) diameter hole (**arrow**) in the guide 10-222A/21.

- Dimensions provided in millimeters.

Transmission, removing

- Remove engine cover.
- Disconnect battery ground (GND) strap from battery negative (–) terminal. See the **Cautions** at the beginning of this repair group regarding battery disconnection.

NOTE —

Be sure to have the anti-theft radio code on hand before disconnecting the battery.

- Remove power steering reservoir from battery bracket but do not loosen hoses. Tie reservoir up to upper radiator support.
- Disconnect battery positive (+) cable and remove battery and battery tray. See **27 Engine Electrical**.
- Remove intake hose and harness connector from Mass Air Flow (MAF) sensor.

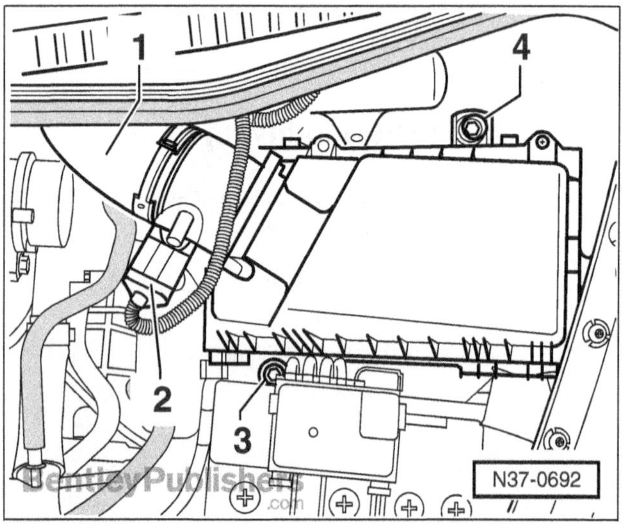

Remove air cleaner mounting bolts (**3**,**4**) and pull out air cleaner.

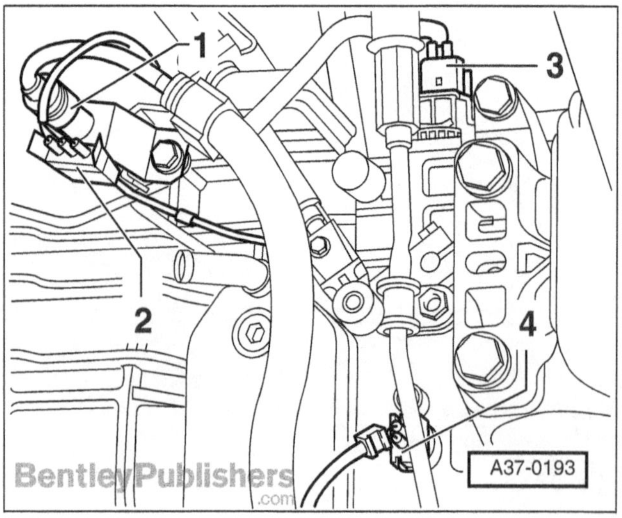

Disconnect harness connector from solenoid valves (**1**), vehicle speed sensor (**2**), multi-function transmission range sensor (**3**), and transmission vehicle speed sensor (**4**).

- Remove wiring harness from retainer on transmission and move aside.
- Remove bracket for power steering hose with retainer for wiring harness from transmission.

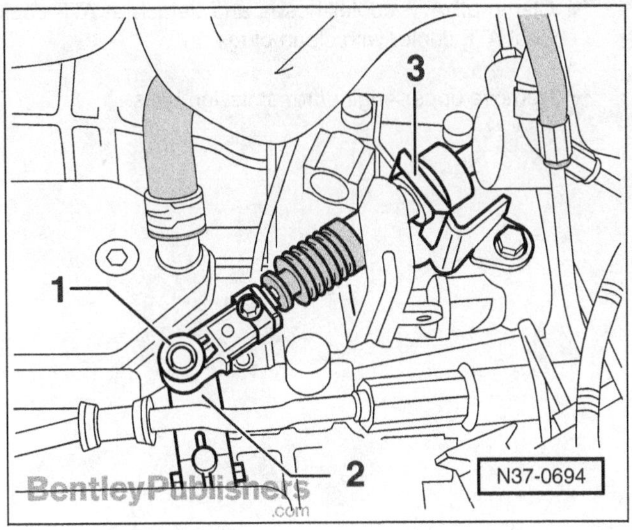

◀ Move selector lever to "**P**". Using a screwdriver, pry selector lever cable (**1**) off selector shaft lever (**2**).

– Remove clip (**3**) at selector lever cable support bracket and remove selector lever cable.

NOTE —

Do not kink selector lever cable.

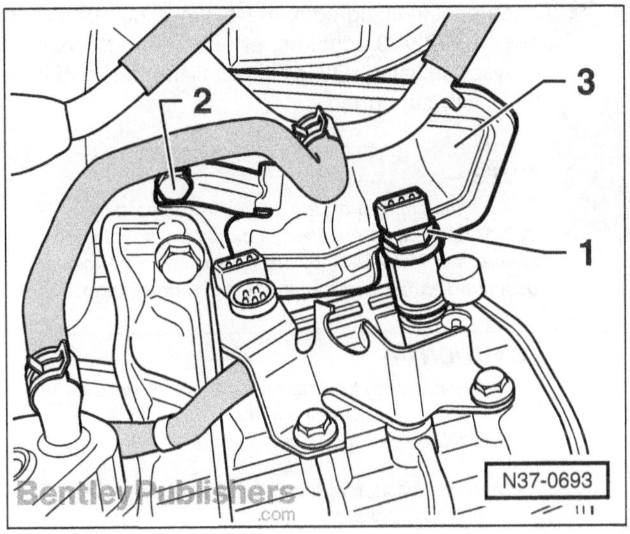

◀ On vehicles with 2.0L engine, remove vehicle speed sensor (**1**), mounting bolt (**2**) and heat shield (**3**) for exhaust manifold.

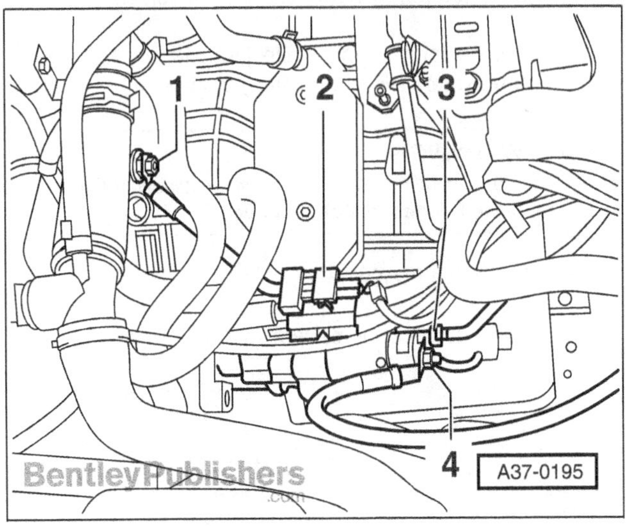

◀ For all vehicles, disconnect ground cable (**1**) from upper engine/transmission bolt, electrical connections at starter motor (**3**, **4**) and harness connector (**2**) on top of starter motor.

– Pull harness connector out of retainer on top of starter motor and remove retainer.

– Remove upper starter mounting bolt.

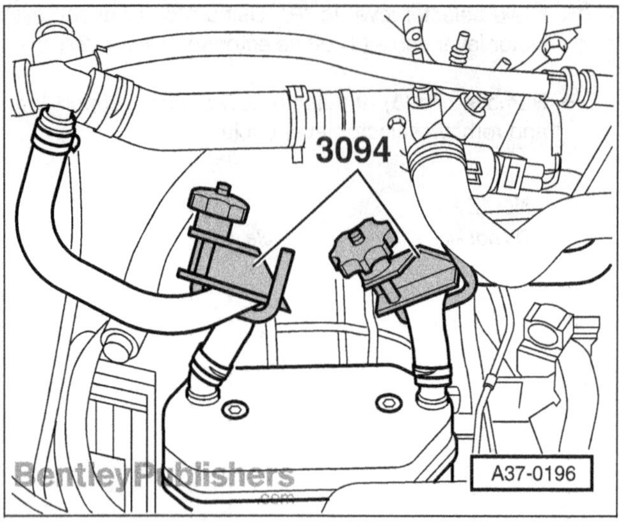

< Clamp-off ATF cooler hoses and detach at ATF cooler. Seal ATF cooler with clean plugs.

– Remove upper engine/transmission bolts.

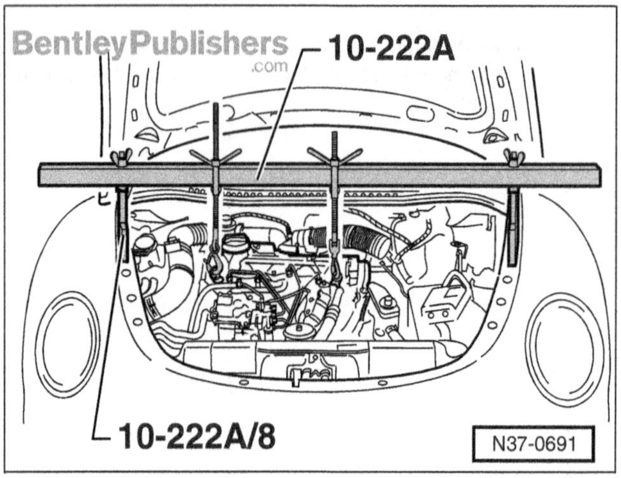

< Attach engine lifting equipment (VW sling 10-222A and adjust 10-222A/8 adapters, or equivalent) to engine lifting eyes. Adjust the lifting sling until the weight of the engine is fully supported.

NOTE —
The engine support bridge guide (VW special tool 10-222 A/21) must be modified to mount the adapter (VW special tool 10-222A/8) by drilling an additional hole as described earlier under **Special tools, modifying**.

CAUTION —
- *Before installing the engine lifting hooks, disconnect all hoses and wiring in the vicinity of the engine lifting eyes, to prevent damage.*
- *Do not position adapters 10-222 A/8 on the fender mounting bolts.*

– Loosen left front wheel bolts and lift vehicle.

– Raise car and support with jack stands or lift. See **03 Maintenance** for proper lifting procedure.

WARNING —
Observe all warnings and cautions associated with lifting vehicle in **0 Maintenance**.

– Remove left front wheel.

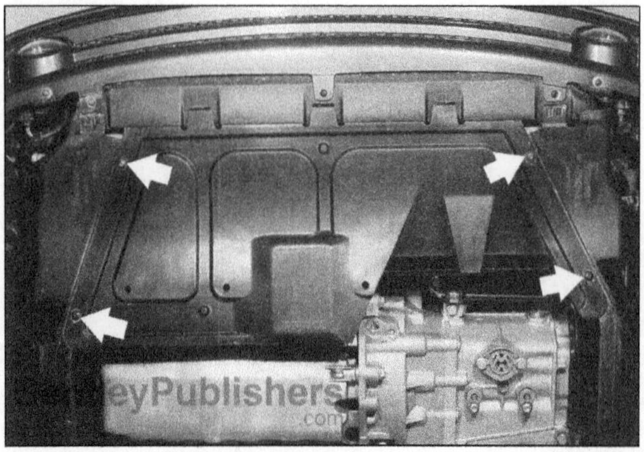

< Remove lower sound absorber panel (**arrows**).

– Remove drivers side lower sound insulator (**arrows**).

NOTE —
Sound insulator fasteners may best be removed and installed with needle nose pliers.

– On vehicles with 1.9L TDI engine, remove intake air duct between intake air cooler and turbocharger, see **21 Turbocharger and Intercooler**.

– Remove passenger side lower sound insulator.

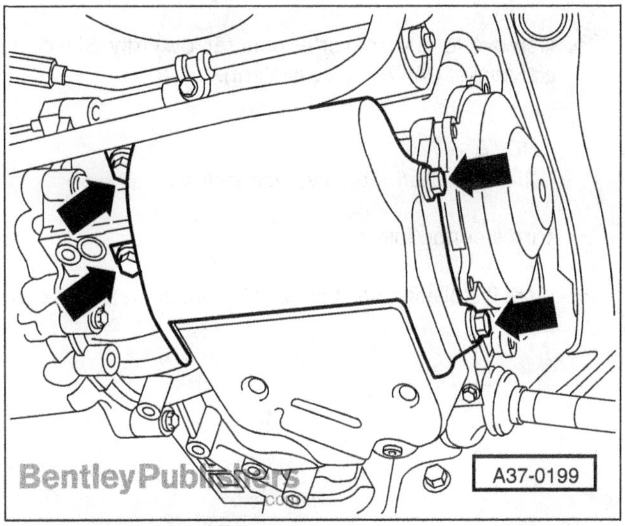

◄ Remove ATF pan protective cover (**arrows**).

– Remove power steering line from starter and transmission mounts.

NOTE —
Use care not to damage the power steering line.

– Remove starter, see **27 Engine Electrical**.

◄ Remove right inner constant velocity joint protective cover (**arrows**) from engine if installed.

– Disconnect drive shafts from transmission flanges and tie up as high as possible.

NOTE —
When tying up axle shafts use care not to damage the paint on the body panels or the axle boots.

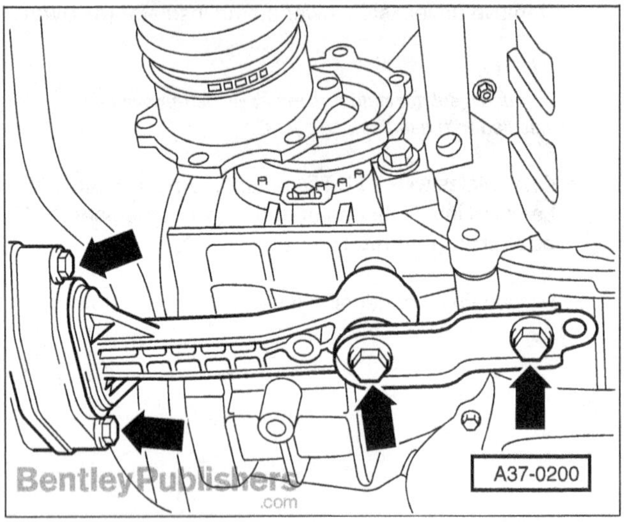

◀ Remove bolts (**arrows**) for pendulum mount.

- Remove additional radiator cooling fan, if installed, on right side of vehicle, see **19 Engine–Cooling System**.

- Remove cap for torque converter nut cover.

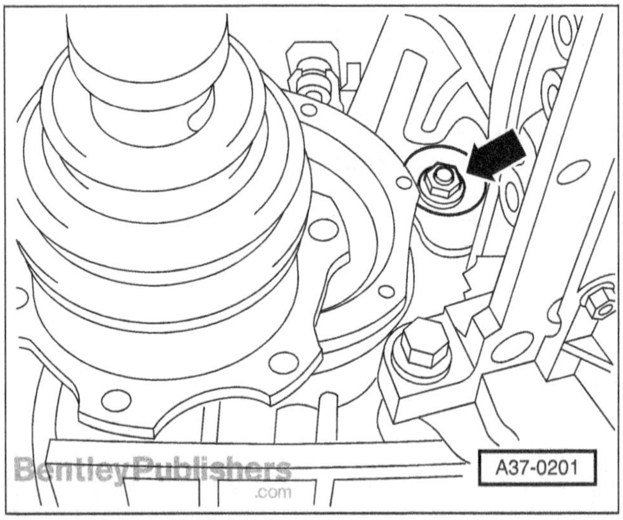

◀ Remove torque converter nuts (**arrow**) (qty. 3) with special socket (V/175 or equivalent).

NOTE —

Turn crankshaft 120° to access each nut.

- Turn steering fully to left.

- Mark installation position of axle joint boots on left control arm.

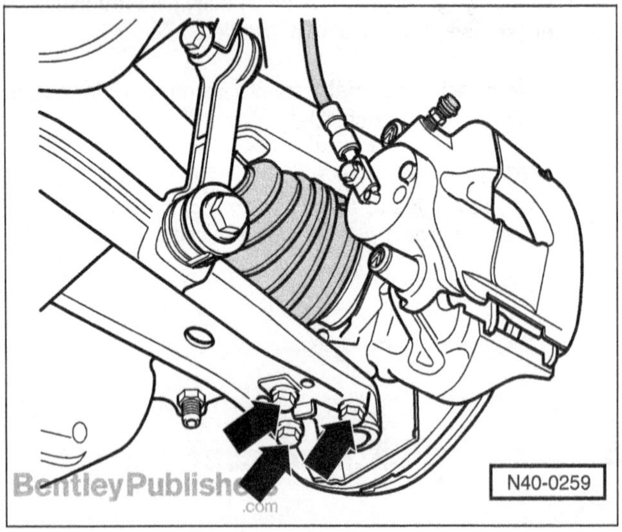

◀ Remove left control arm bolts (**arrows**).

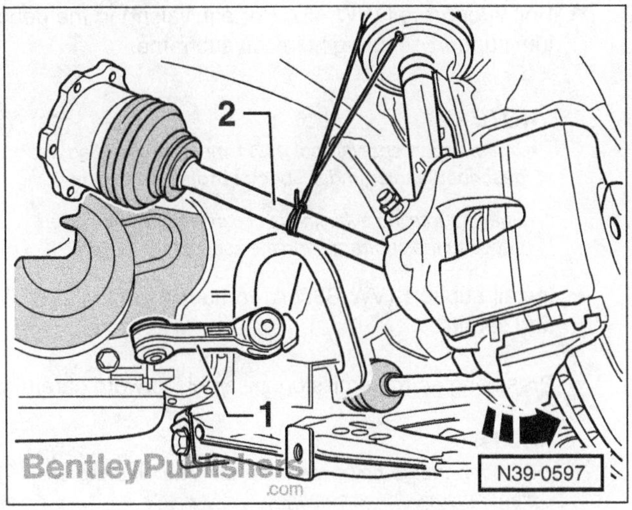

Unbolt left stabilizer coupling rod (**1**) from control arm and turn coupling rod upward. Swing wheel bearing housing outward (**arrow**) and guide left drive shaft (**2**) out between subframe and transmission. Lift drive shaft and secure to suspension strut with wire.

– Disconnect exhaust at front double clamp. Remove front exhaust support if necessary, see **26 Exhaust System/Emission Controls**.

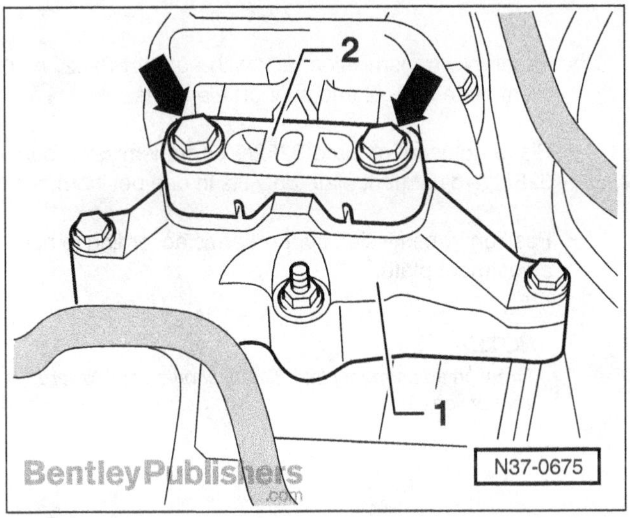

Remove mounting bolts (**arrows**) for left engine/transmission mount (**2**).

– Carefully tilt engine/transmission assembly by lowering left-side spindle of support bar (VW 10-222A, or equivalent) by approximately 60 mm (2.3 in.).

Remove left support (**1**) from transmission.

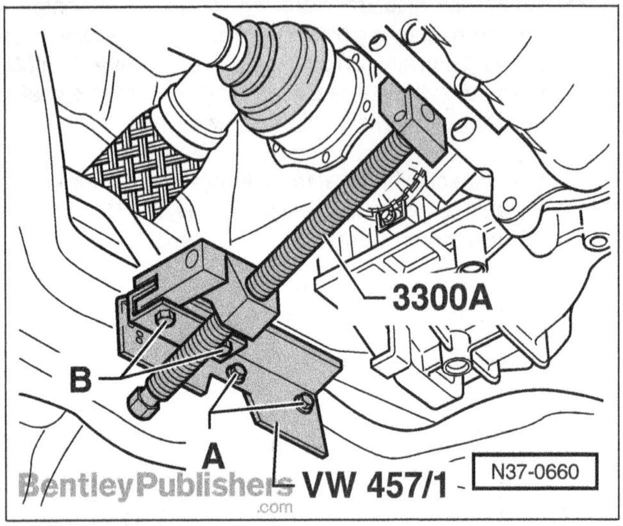

Bolt support rail (VW 457/1 or equivalent) to the pendulum support securing holes on subframe.

NOTE —

- *Volkswagen special tool 457/1 must be modified as described earlier under* **Special tools, modifying**.
- *Spacers totaling a thickness of 6 mm must be inserted between the subframe and the support rail VW 457/1.*

- Install support (VW 3300A) to support rail (VW 457/1) and secure.
- Press engine/transmission assembly forward carefully.

NOTE —

Do not damage power steering line when moving engine/transmission assembly.

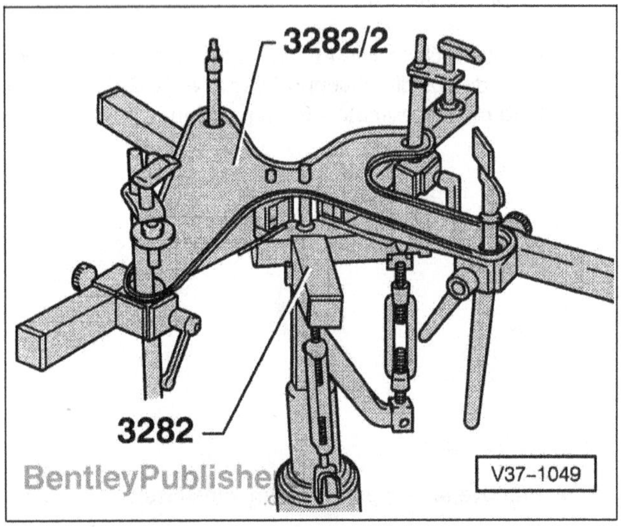

Assemble transmission jack with support 3282, adjustment plate 3282/2 and support elements.

- Place adjustment plate 3282/2 on transmission support 3282. (Adjustment plate only fits in one position).
- Position transmission support arms according to holes in adjustment plate.

NOTE —

Arrow on adjustment plate 3282/2 should point toward front of vehicle.

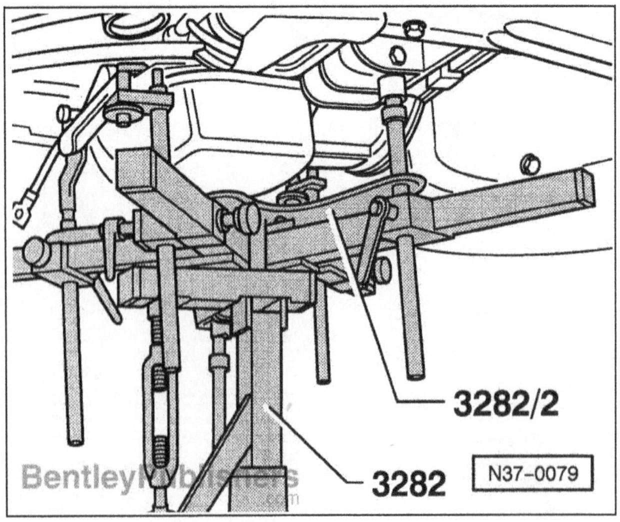

Place transmission jack under vehicle.

- Align adjustment plate parallel to transmission and lock safety supports on transmission jack.

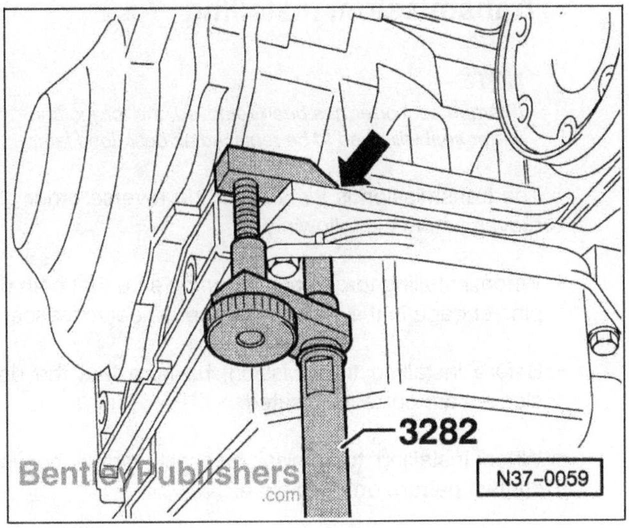

◄ Place safety support pin on oil pan and secure it to transmission housing. (**arrow**)

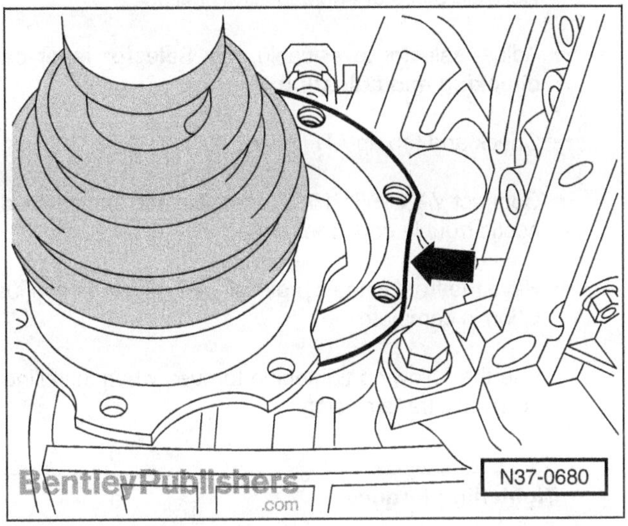

◄ Turn right transmission output flange to the right until flat section (**arrow**) is vertical.

– Remove lower engine/transmission securing bolts.

> ***WARNING —***
> *Before unbolting lower engine-to-transmission mounting bolts be sure to support the weight of the vehicle at all four corners with jack stands designed for that purpose. Removal of the transmission can upset the weight balance of the vehicle and cause it to fall off the lift.*

– Separate transmission from engine while pushing torque converter away from drive plate.

NOTE —
On automatic transmissions, the torque converter should come off with transmission. Secure torque converter to transmission to prevent damage.

– Lower transmission slightly.

– While lowering transmission, guide power steering pressure line past transmission.

– Turn transmission and carefully lower.

– While lowering transmission, guide right axle joint at engine past support 3300 A.

NOTE —
Be careful not to let multi-function switch contact engine/transmission mount.

Transmission, installing

NOTE —
If torque converter has been removed, the torque converter seal will need to be replaced as described later.

The transmission is installed in the reverse order of removal, noting the following:

– When installing torque converter, be sure that both drive pins engage in the ATF pump inner wheel recesses.

– Before installing transmission, be sure that the dowel sleeves are correctly located.

– When installing transmission, check torque converter contact pattern on drive plate.

– Replace selector cable locking circlip.

– Adjust selector lever cable, see **Selector lever cable, checking and adjusting**.

– Check and top up ATF level.

– Connect VAG 1551/1552, or equivalent, and check diagnostic trouble code memory.

– Have the front wheels professionally aligned once installation is complete.

Use the following tightening torques when installing the automatic transmission.

Tightening torques
- Drive shaft to transmission drive flange . . . . 40 Nm (30 ft-lb)
- Torque converter to drive plate. . . . . 60 Nm (44 ft-lb)
- Transmission to engine (M12) . . . . . 80 Nm (59 ft-lb)
- Transmission to oil pan (M10) . . . . . 40 Nm (30 ft-lb)
- Axle joint to control arm. . . . . . 20 Nm (15 ft-lb) plus ¼ turn (90°)
- Coupling rod to control arm . . . . 45 Nm (33 ft-lb)
- Left mount to transmission. . . . . 25 Nm (18 ft-lb)
- Wheel bolts to wheel hub. . . . . 120 Nm (89 ft-lb)
- Protective cap to drive shaft at engine . . . . 33 Nm (24 ft-lb)

For engine and transmission mount tightening torques and installation procedures, see **10 Engine–Removing and Installing**.

TORQUE CONVERTER

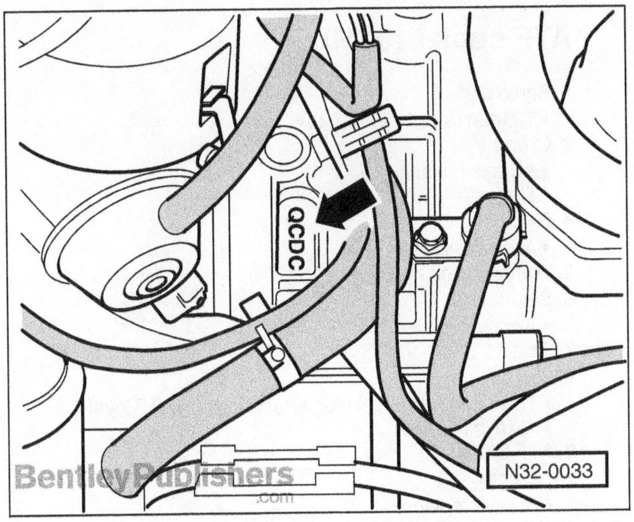

◄ When replacing the torque converter, always note the torque converter code number. The code number can be found in one of two places, either on the torque converter itself, or viewed though access hole (**arrow**) with transmission installed. The replacement torque converter must have the same code number, as it is matched to the transmission/engine application.

> ***CAUTION —***
> *When installing the torque converter, make sure that both drive pins engage in the recesses in the ATF pump inner wheel.*

If the torque converter has become contaminated by particles resulting from wear or abrasion, or in the course of a major overhaul of the transmission, always flush the torque converter with clean ATF. The torque converter can be flushed with commercially-available extraction equipment, such as EZ1 fluid evacuator, or Volkswagen special tools VAG 1358A and VAG 1358 A/1 probe.

Torque converter oil seal, replacing (transmission removed)

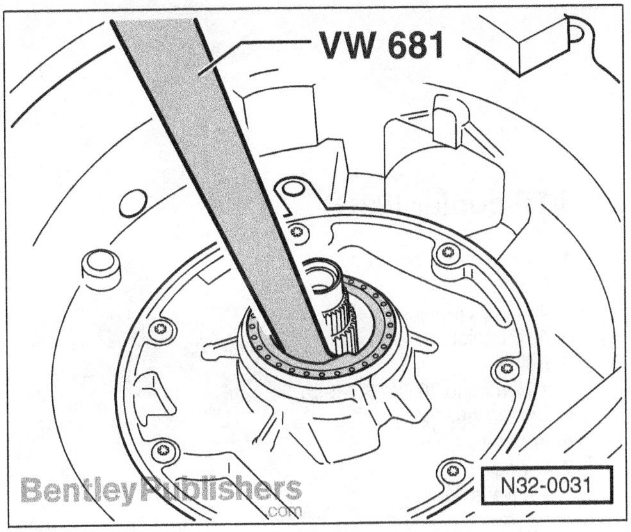

◄ Remove the torque converter seal using a seal extractor (**VW 681** shown).

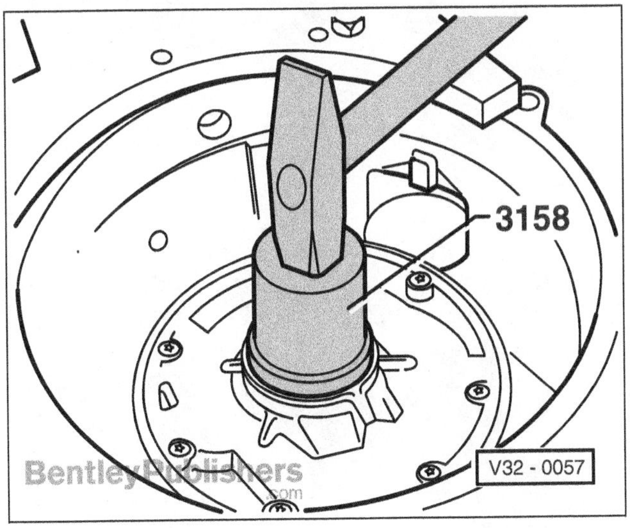

◄ Using a seal driver (**3158** shown), drive seal in until flush.

ATF COOLER

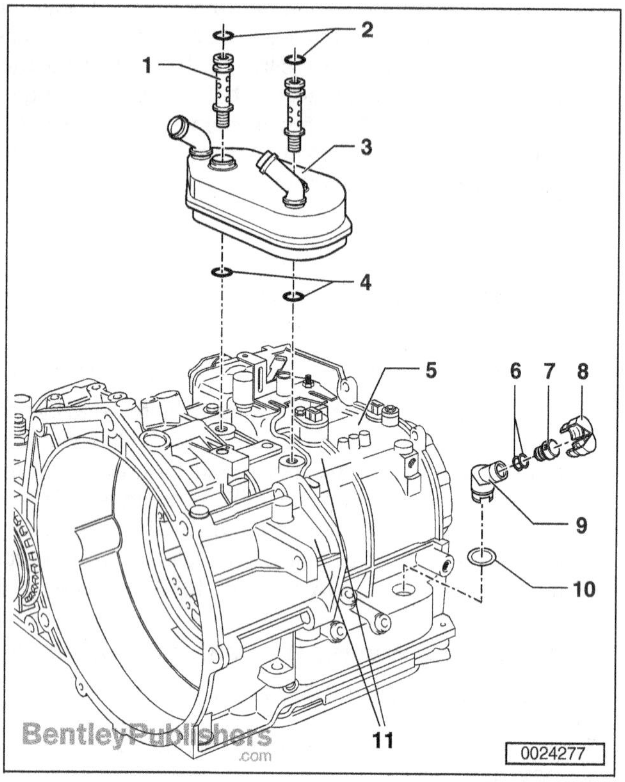

ATF cooler (01M)

1. Banjo bolt
   - Tighten to 35 Nm (26 ft-lb)
2. O-ring
   - Always replace
3. ATF cooler
4. O-ring
   - Always replace
5. Transmission housing
6. Seal
   - Always replace
7. Plug
8. Cap
   - To secure ATF level plug after checking ATF level
   - Always replace
9. ATF filler tube
10. O-ring
    - Always replace
11. Location of transmission code and production number

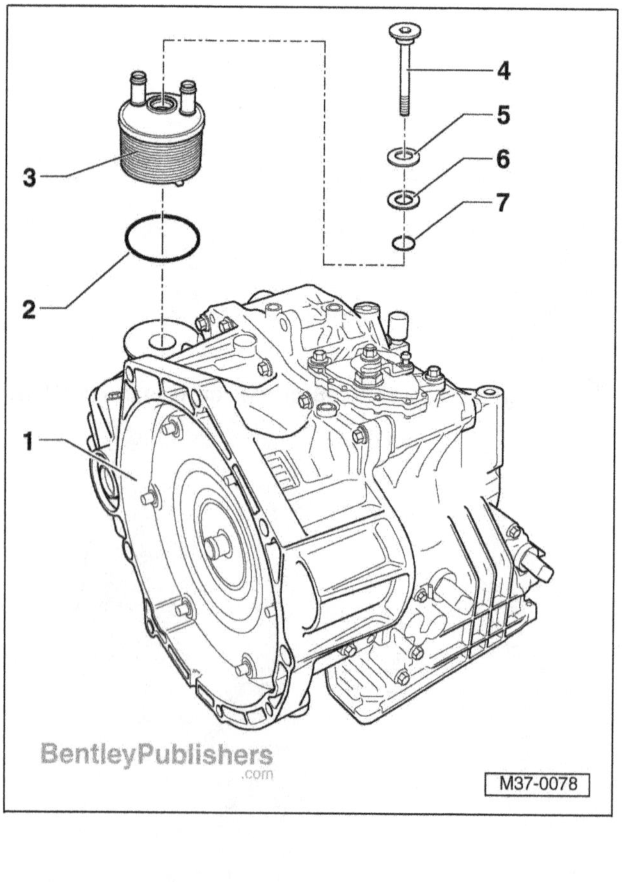

ATF cooler (09G)

1. Transmission housing
2. O-ring
   - Always replace
3. ATF cooler
4. Bolt
   - Tighten to 36 Nm (27 ft-lb)
5. Dished washer
6. Washer
7. O-ring
   - Always replace

40 Front Suspsension and Drive Axles

GENERAL

Special tools, equipment and procedures are required for most front suspension repair and component replacement. In addition, front wheel alignment is almost always disturbed when suspension components are removed or replaced.

This section covers repairs to the front suspension and related components. For rear suspension servicing and repair information, see **42 Rear Suspension**. For wheel alignment specifications, see **44 Wheels, Tires, Wheel Alignment**.

Also covered are the drive axle assemblies with Constant Velocity (CV) joints as installed in vehicles with manual transmissions, as well as drive axles with triple roller joints installed in vehicles with automatic transmissions.

WARNING —

- *Do not re-use any fasteners that are worn or deformed in normal use. Most fasteners are designed to be used only once and become unreliable and may fail when used a second time. This includes, but is not limited to, nuts, bolts, washers, circlips cotter pins, self-locking nuts and bolts. For replacements, always use new parts.*

- *Do not reinstall bolts and nuts coated with undercoating wax, as correct tightening torque cannot be ensured. Always clean the threads of removed bolts and nuts that are to be re-used with a suitable solvent before installation. Ensure that new parts are clean.*

WARNING —

- *Do not use standard nuts and bolts in place of green colored parts. The green color denotes a special anti-corrosion process known as "dacromet" or "delta-tone". Torque specifications are given based on use of these parts only.*

- *Always replaced rusted or corroded bolts, nuts and washers even if not specifically indicated.*

- *DO NOT attempt to straighten or weld suspension struts, wheel bearing housings, control arms or any other wheel locating or load bearing components of the front suspension.*

CAUTION —

If a vehicle has to be moved (rolled) after removing the drive axle, install an outer constant velocity joint into the wheel bearing/housing and tighten the nut to 50 Nm (37 ft-lb). Otherwise the wheel bearing will be damaged.

SUBFRAME AND CONTROL ARMS

The control arms and the subframe are mounted in replaceable rubber bushings that are subject to wear. The subframe supports the steering gear and the lower engine/transmission pendulum mount. Always understand and observe the warnings and cautions listed above before working on the suspension components.

Subframe/stabilizer bar/control arm assembly

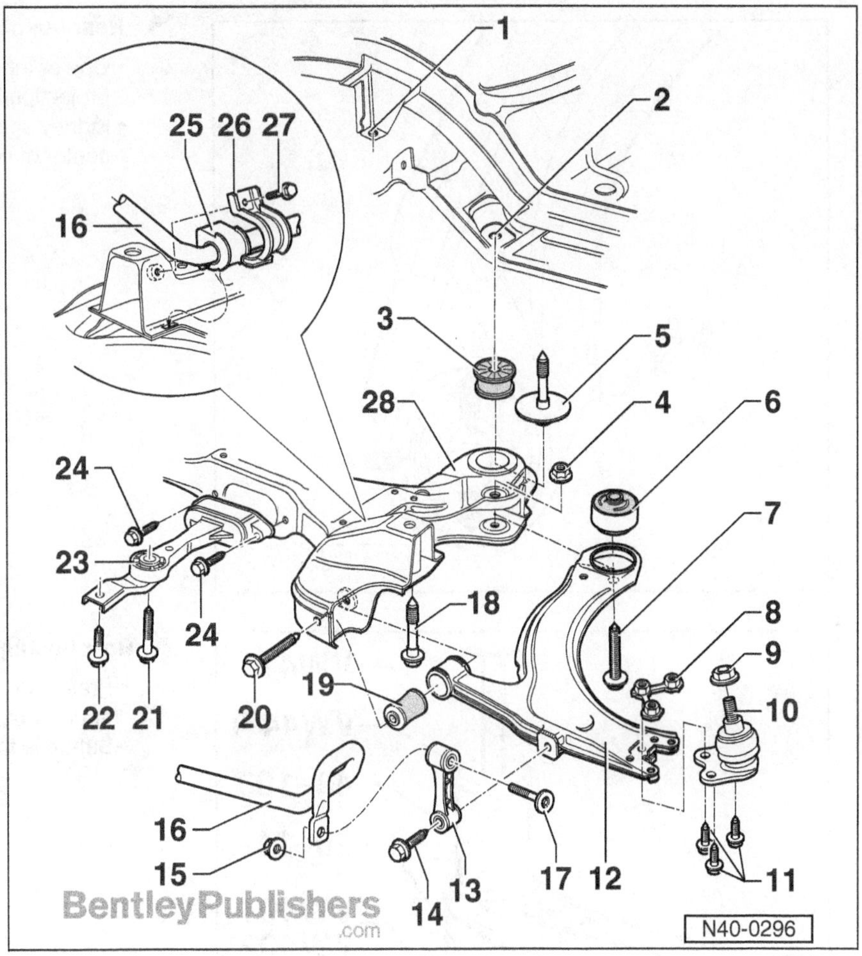

1. **Subframe retaining bracket (on body)**
2. **Welded nut (in body)**
3. **Rear bonded rubber bushing for subframe**
4. **Nut, self-locking**
   - Always replace
5. **Bolt**
   - Always replace
   - Tighten to 100 Nm (74 ft-lb) + ¼-turn (90°)
6. **Rear bonded rubber bushing for control arm**
   - See **Rear bushing for control arm, installed position**
   - See **Rear bushing for control arm, pressing out/in**
7. **Bolt**
   - Always replace
   - Tighten to 70 Nm (52 ft-lb) + ¼-turn (90°)
8. **Nut plate**
9. **Nut, self-locking**
   - Always replace
   - Tighten to 45 Nm (33 ft-lb)
10. **Ball joint**
    - See **Ball joint, checking axial and radial play**
11. **Bolts**
    - Always replace
    - Tighten to 20 Nm (15 ft-lb) + ¼-turn (90°)
12. **Control arm**
13. **Stabilizer connecting link**
14. **Bolt**
    - Tighten to 45 Nm (33 ft-lb)
15. **Nut, self-locking**
    - Always replace
    - Tighten to 30 Nm (22 ft-lb)
16. **Stabilizer bar**
    - Subframe must be lowered to remove and install
17. **Bolt**
18. **Bolt**
    - Always replace
    - Tighten to 100 Nm (74 ft-lb) + ¼-turn (90°)
19. **Front bonded rubber bushing for control arm**
    - See **Pressing out front bonded rubber bushing for control arm**
    - See **Pressing in front bonded rubber bushing for control arm**
20. **Bolt**
    - Always replace
    - Tighten to 70 Nm (52 ft-lb) + ¼-turn (90°)
21. **Bolt**
    - Always replace
    - Tighten to 40 Nm (30 ft-lb) + ¼-turn (90°)
22. **Bolt**
    - Always replace
    - Tighten to 40 Nm (30 ft-lb) + ¼-turn (90°)
23. **Pendulum support**
24. **Bolts**
    - Always replace
    - Tighten to 20 Nm (15 ft-lb) + 1/4-turn (90°)
25. **Stabilizer bar bushing**
26. **Stabilizer bar mounting bracket**
27. **Bolt**
    - Tighten to 25 Nm (18 ft-lb)
28. **Subframe**

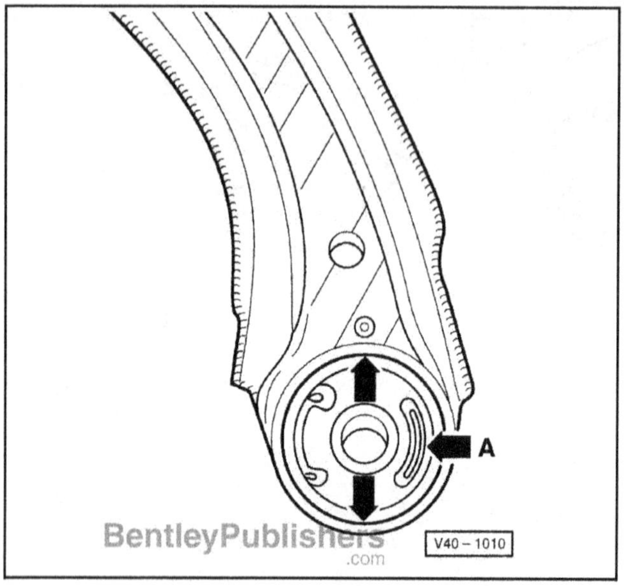

Rear bushing for control arm, installed position

- One of the two arrows on bushing must point toward projection in control arm.
- Kidney shaped opening (**arrow A**) must point toward center of vehicle.

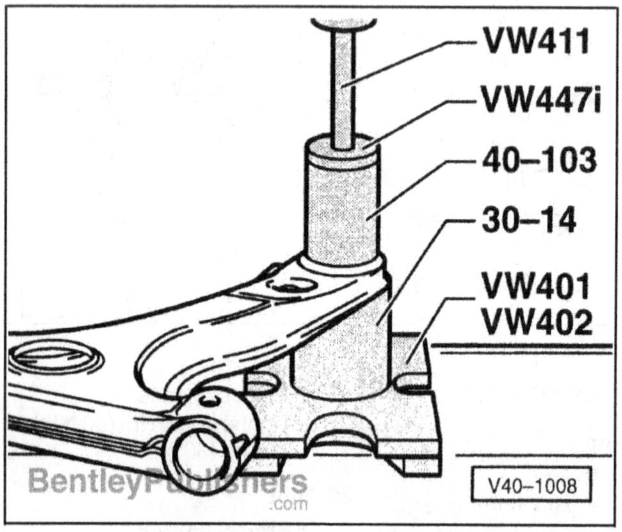

Rear bushing for control arm, pressing out/in

- Press the bushing using Volkswagen special tools or equivalent, as shown.
- Set-up is the same for removal and installation.

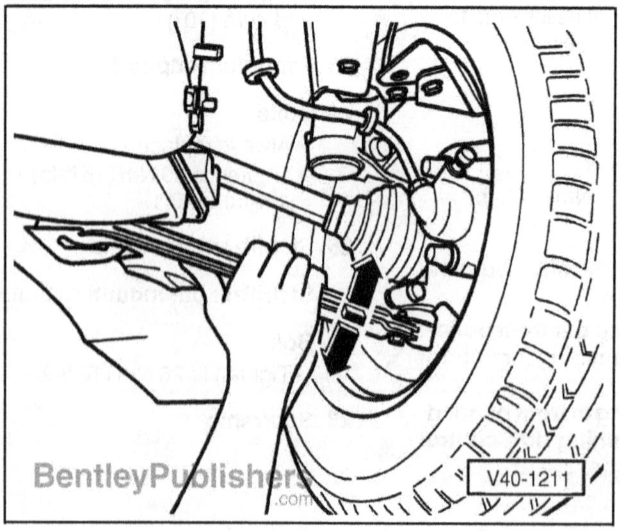

Ball joint, checking axial and radial play

- Forcibly pull ball joint down and press up again (**arrows**).
- Forcibly push lower part of wheel in and out.
- There must not be any perceptible play as a result of either check. Do not confuse with upper suspension strut mount or wheel bearing play.
- Replace ball joint if dust boot is damaged.

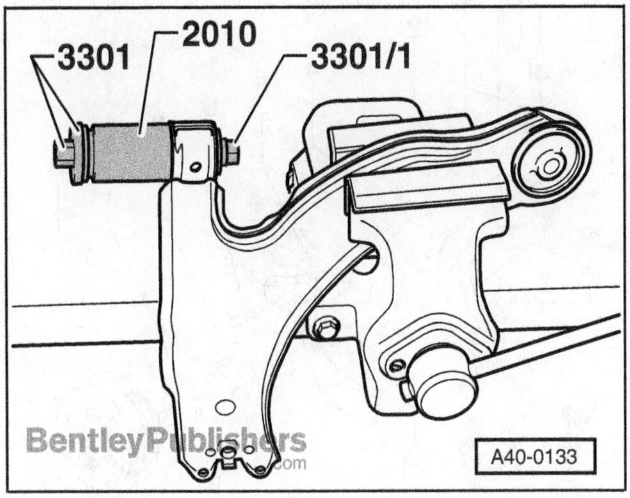

Pressing out front bonded rubber bushing for control arm

- Support control arm securely in vise as shown.

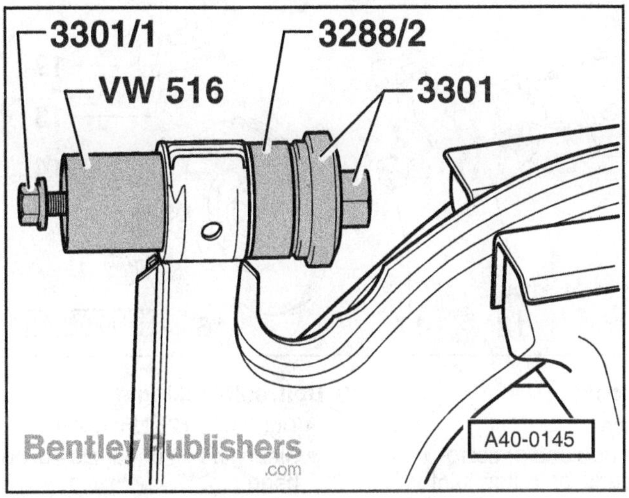

Pressing in front bonded rubber bushing for control arm

- Before installing front bushing, lubricate with acid-free lubricant (e.g. soft soap).
- Never use grease.

WHEEL BEARINGS

Removal and/or replacement of the front wheel bearings requires special tools and equipment. Note that the wheel bearing is destroyed any time it is removed. Always understand and observe the warnings and cautions listed at the beginning of this section.

Front wheel bearing housing assembly

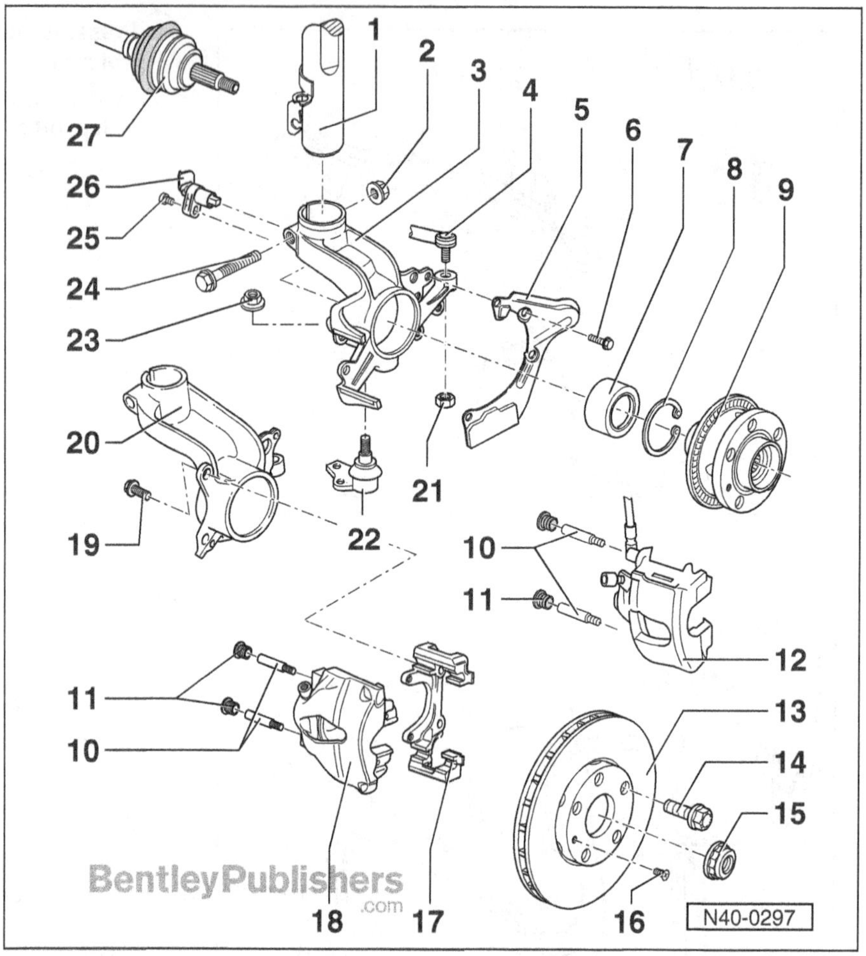

1. **Suspension strut**
2. **Nut, self-locking**
   - Always replace
   - Tighten to 50 Nm (37 ft-lb) + ¼-turn (90°)
   - Never less than 90°
   - Turning angle tolerance 90°–120°
3. **Wheel bearing housing (1.9L / 2.0L)**
4. **Tie rod end**
5. **Splash plate**
6. **Bolt**
   - Tighten to 10 Nm (7 ft-lb)
7. **Wheel bearing**
   - Always replace (bearing is destroyed when removed)
   - See **Wheel bearing, pressing out**
   - See **Wheel bearing, pressing in**
8. **Circlip**
   - Make sure clip is correctly seated
9. **Wheel hub with ABS wheel speed sensor rotor**
   - Rotor welded to wheel hub
   - See **Wheel hub, pressing out of housing**
   - See **Wheel hub, pressing into wheel bearing**
10. **Guide pins**
    - Tighten to 27.5 Nm (20 ft-lb)
11. **Protective caps**
12. **Brake caliper (1.9L / 2.0L)**
    - Do not loosen the brake hose when working on the front suspension
    - Do not allow the caliper to hang by the brake hose. The unsupported weight can stretch and damage the hose. Suspend the brake caliper using a piece of wire or similar.
13. **Ventilated brake disc (rotor)**
14. **Wheel (lug) bolts**
    - Tighten to 120 Nm (87 ft-lb)
15. **Axle nut, self-locking**
    - Always replace
    - Any paint residue and/or corrosion on threads of outer joint must be removed before nut is installed
    - See **Axle nut, final tightening - recommended procedure**
    - See **Axle nut, final tightening - alternate procedure**
16. **Screw**
    - Tighten to 4 Nm (35 in-lb)
17. **Brake carrier (1.8L / 2.5L)**
18. **Brake caliper (1.8L / 2.5L)**
    - Do not loosen the brake hose when working on the front suspension
    - Do not allow the caliper to hang by brake hose. The unsupported weight can stretch and damage hose. Suspend brake caliper using a piece of wire or similar.
19. **Bolt, self-locking**
    - Tighten to 125 Nm (92 ft-lb)
    - Clean ribs on under side of bolt head
20. **Wheel bearing housing (1.8L / 2.5L)**
21. **Nut, self-locking**
    - Always replace
    - Tighten to 50 Nm (37 ft-lb)
22. **Ball joint**
23. **Nut, self-locking**
    - Always replace
    - Tighten to 45 Nm (33 ft-lb)
24. **Bolt**
    - Always replace
25. **Bolt**
    - Tighten to 10 Nm (7 ft-lb)
26. **ABS wheel speed sensor**
27. **Drive axle**
    - See **Drive axle/constant velocity joint, pressing out**

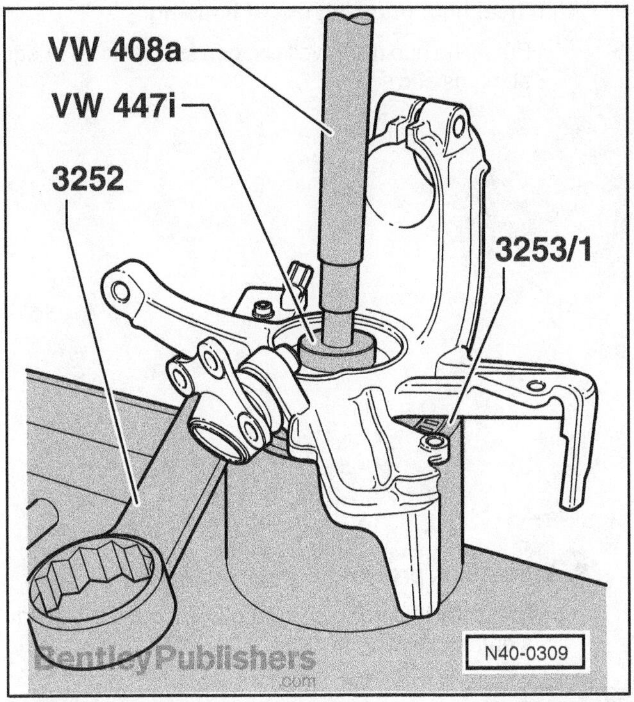

Wheel bearing, pressing out

- Press the bushing using Volkswagen special tools or equivalent, as shown.

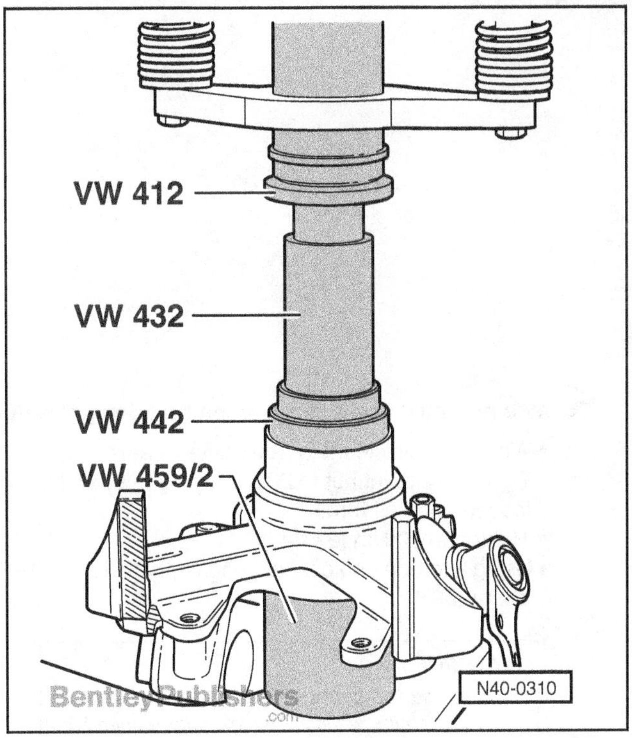

Wheel bearing, pressing in

- Install circlip in housing with open end at the bottom of the wheel bearing housing.
- Be sure that circlip is properly seated.
- Press the bushing using Volkswagen special tools or equivalent, as shown.

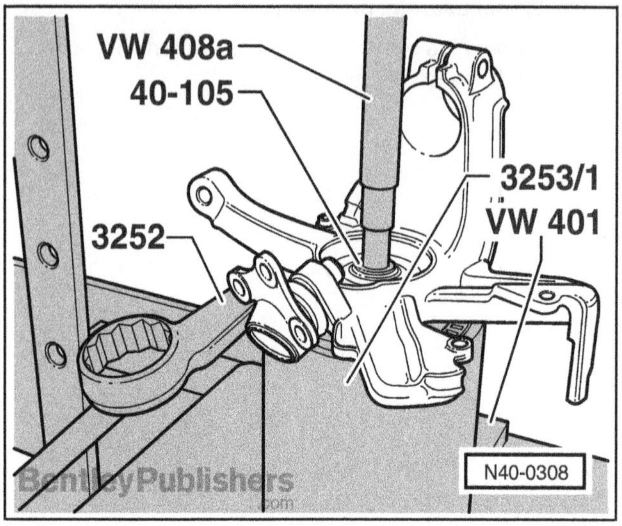

Wheel hub, pressing out of housing

- Press the hub using Volkswagen special tools or equivalent, as shown.

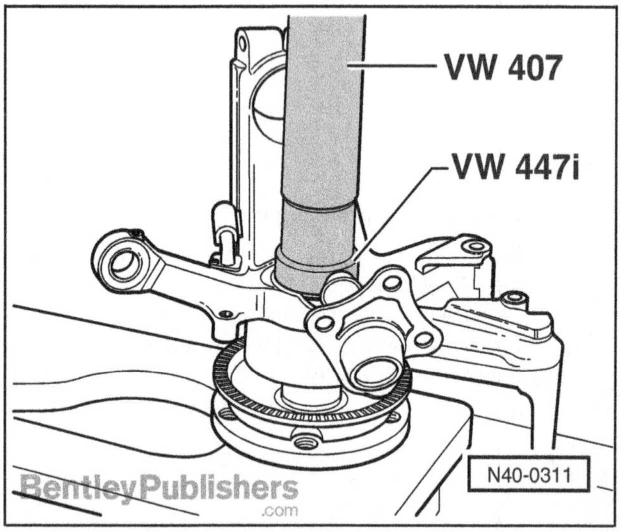

Wheel hub, pressing into wheel bearing

- Press the hub using Volkswagen special tools or equivalent, as shown.

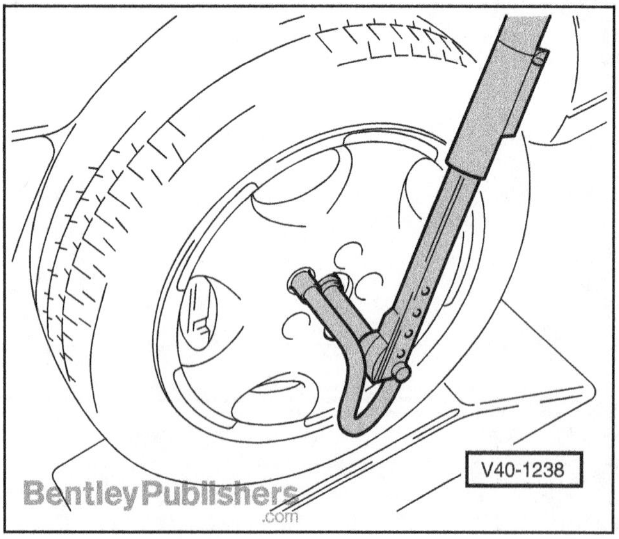

Axle nut, final tightening - recommended procedure

- Weight of vehicle MUST be on its wheels.
- Tighten 12-point nut to 200 Nm (148 ft-lb) and then loosen one half turn.
- Tighten to 50 Nm (37 ft-lb).
- Using a torque angle measuring wrench, tighten 1/6 of a turn (60°)

> ***WARNING —***
> *Loosen and tighten axle nuts only when weight of vehicle is on the wheels. The leverage required for this operation is sufficient to topple the vehicle to the ground if supported on a lift or jack stands.*

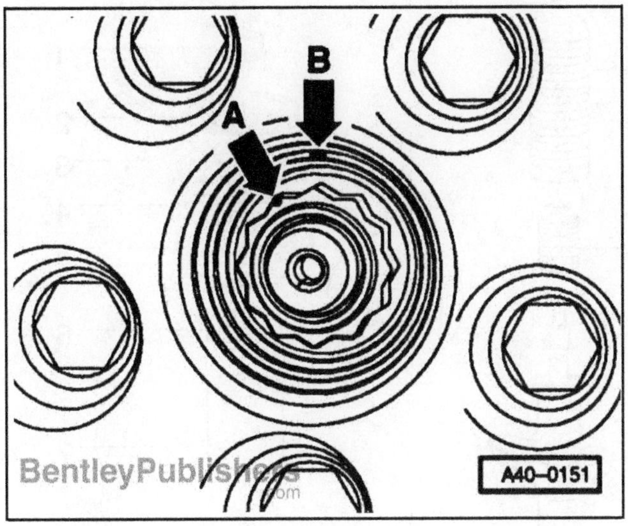

Axle nut, final tightening - alternate procedure

- Weight of vehicle MUST be on its wheels.
- Tighten 12-point nut to 300 Nm (221 ft-lb) and then loosen one full turn.
- Tighten to 50 Nm (37 ft-lb).
- Make a mark on nut point with line (arrow A).
- Make a second mark on the wheel hub (arrow B) one nut point away from the first. (The distance between each nut point on a 12-point nut is 30°)
- Tighten the axle nut until both marks line up.

> ***WARNING —***
> *Loosen and tighten axle nuts only when weight of vehicle is on the wheels. The leverage required for this operation is sufficient to topple the vehicle to the ground if supported on a lift or jack stands.*

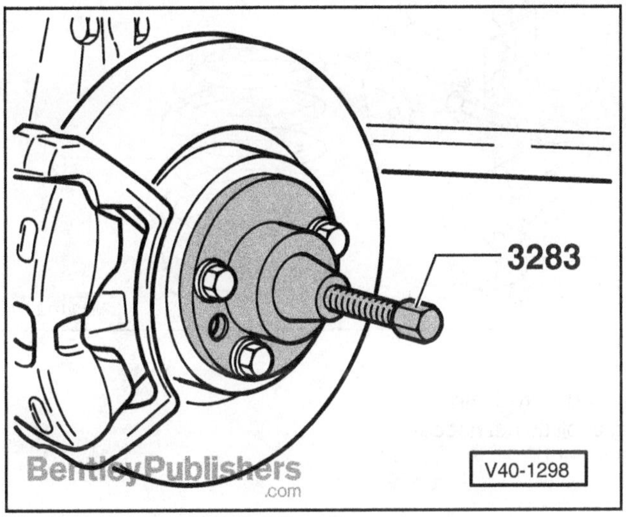

Drive axle/constant velocity joint, pressing out

- Press end of axle out using Volkswagen special tools or equivalent, as shown.
- Be sure there is sufficient clearance before pressing.

Suspension Struts

Removal and/or replacement of the front suspension struts requires special tools and equipment. Note that the strut is a sealed unit integral with the shock absorber. Always understand and observe the warnings and cautions listed at the beginning of this section.

Front suspension strut assembly

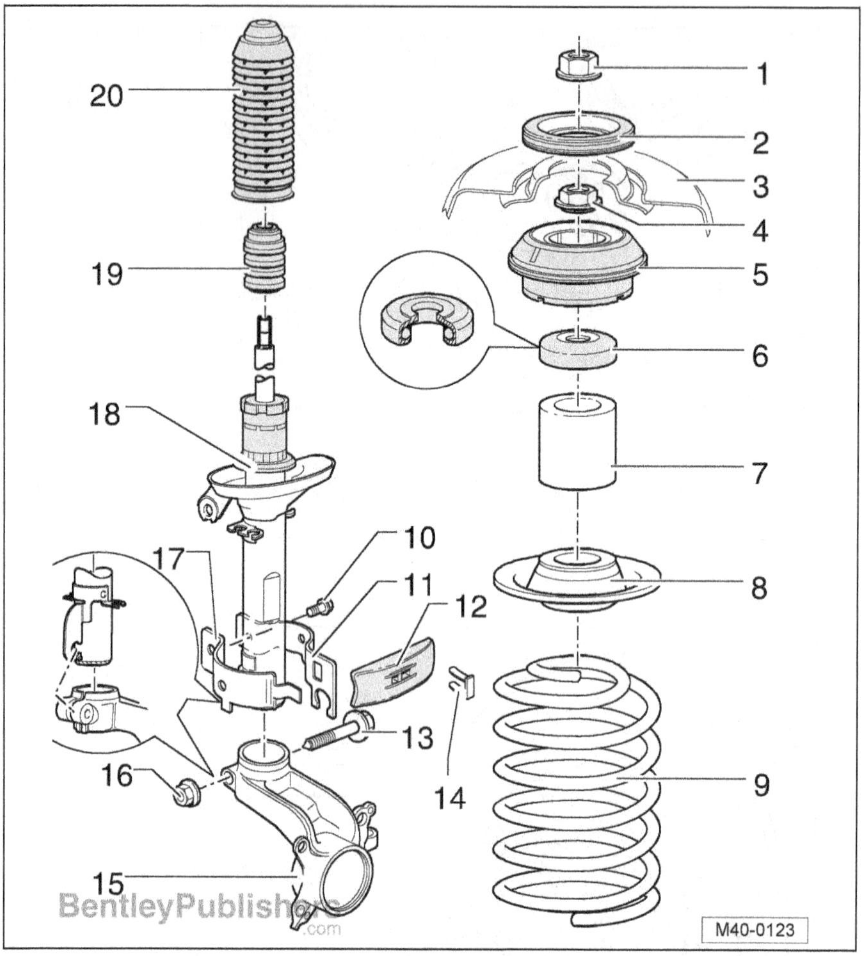

1. **Outer nut, self-locking**
   - Always replace
   - Tighten to 60 Nm (44 ft-lb)
2. **Stop plate**
3. **Suspension strut turret**
   - Welded to inner wheel housing
4. **Inner nut**
   - Tighten to 60 Nm (44 ft-lb)
5. **Suspension strut bearing**
6. **Axial groove ball bearing**
7. **Bushing**
   - Only for Mexican vehicles
8. **Spring seat, upper**
9. **Coil spring**
   - Note color-code marking for identification
   - Outer surface of spring must not be damaged
10. **Bolt**
11. **Bracket**
12. **Heat shield**
13. **Bolt**
    - Always replace
    - Bolt must face in direction of travel
14. **Clip**
15. **Wheel bearing housing**
16. **Nut, self-locking**
    - Always replace
    - Tighten to 60 Nm (44 ft-lb) + ¼ turn (90°)
    - Never less than 90°
    - Turning angle tolerance 90°–120°
17. **Clamp**
18. **Shock absorber**
    - Can be replaced individually
    - Slight traces of oil do not necessitate replacement
19. **Bump stop**
20. **Protective sleeve**

Front suspension strut, removing and installing

> ***WARNING —***
> - *Suspension strut springs are compressed and under pressure when installed on struts.*
> - *DO NOT attempt to disassemble or repair suspension without proper tools and experience. Serious injury will result from using improper tools or procedures.*
> - *Suspension strut is removed and installed as an assembly. DO NOT disassemble or attempt repair while assembly is still installed in vehicle.*
> - *Front wheel alignment may be altered during suspension strut repairs. Wheel alignment should be checked and adjusted any time repairs are made to the suspension strut area.*

– Raise vehicle and properly support.

> ***WARNING —***
> - *Be sure to use jack stands designed for the purpose.*
> - *Jack stands should be on a level hard surface.*
> - *If a vehicle lift/hoist is used, be sure vehicle is lifted according to manufacturer's instructions.*

– Remove wheel.

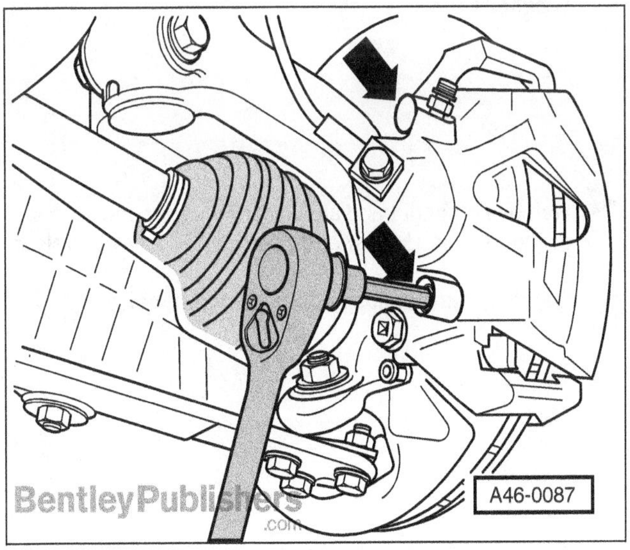

◄ Remove both guide pins (**arrows**) from brake caliper.

– Remove brake caliper and suspend using wire.

NOTE —
- *Suspend the brake caliper using a piece of wire or similar.*
- *Do not allow the caliper to hang by the brake hose. The unsupported weight can stretch and damage the hose.*

– Remove connecting link from control arm.

– Disconnect ABS wheel speed sensor wiring from suspension strut (where applicable).

– When removing right side suspension strut the following additional work is required
- Remove lower sound absorber panel (belly pan).
- Vehicles with TDI engine, remove intake air duct between charge air cooler and turbocharger.
- Remove right side noise insulation.
- Disconnect drive axle from transmission drive flange.

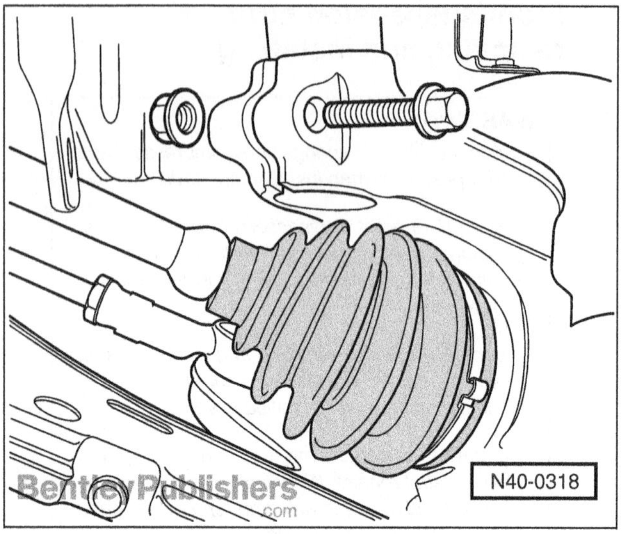

For left and right sides, remove bolt and nut from wheel bearing housing and suspension strut connection.

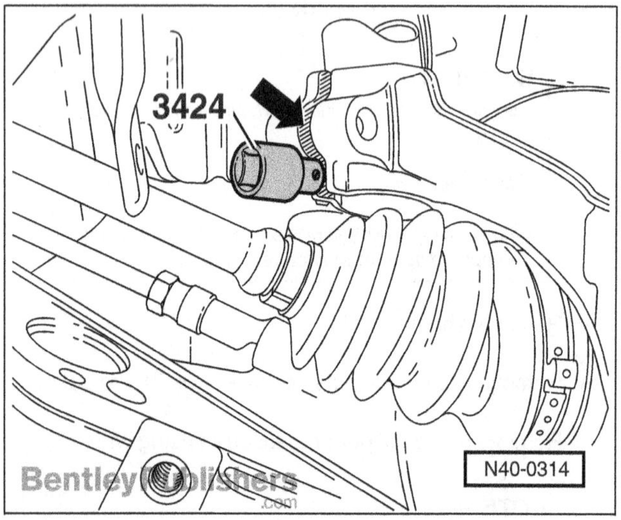

Insert Volkswagen special spreader tool **3424** into gap (**arrow**) in wheel bearing housing and turn ¼-turn in either direction to spread apart.

- Push brake disc in direction of suspension strut by hand to prevent shock absorber tube from canting in wheel bearing housing opening.
- Pull wheel bearing housing down and clear of shock absorber.
- Remove wiper arms and cowl panel, see **92 Wipers and Washers.**

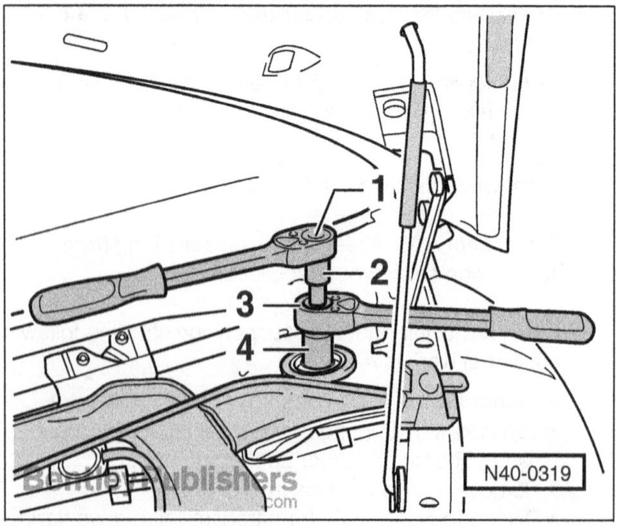

Remove outer nut and stop plate for upper shock absorber mount using Volkswagen special tools or equivalent.

- Carefully remove suspension strut from below.

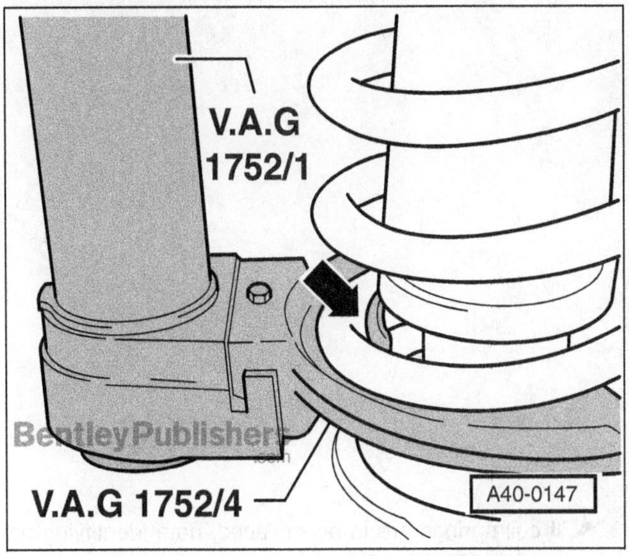

Coil spring, removing and installing

◄ Using VAG1752 spring compressor or equivalent, compress coil spring until upper spring seat is free.

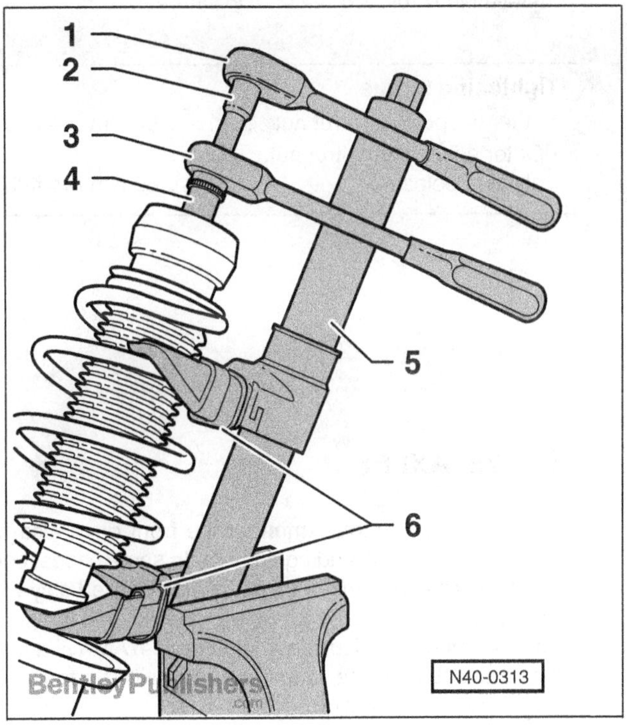

◄ With coil spring properly compressed, remove inner nut.

– Remove components of suspension strut.

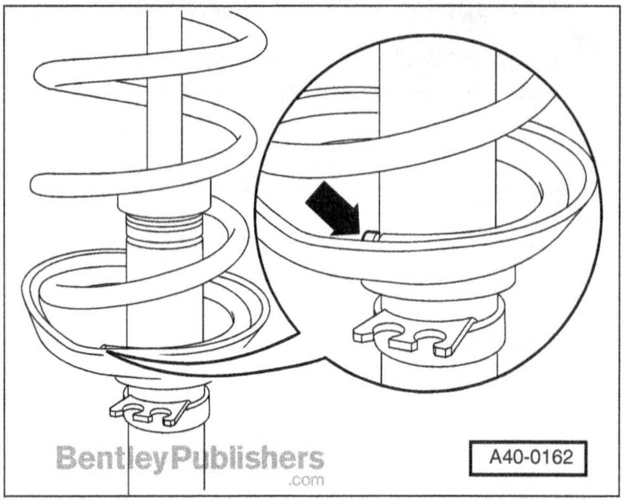

When re-installing coil spring, fit end of coil against stop (**arrow**) in lower spring seat on strut.

If coil springs are to be replaced, note identifying paint marks (**arrows**) on coils for proper matching.

– Installation is the reverse of disassembly.

Tightening torque

- Inner suspension strut nut . . . . 60 Nm (44 ft-lb)
- Outer suspension strut nut, always replace . . . . 60 Nm (44 ft-lb)

DRIVE AXLES

Removal and/or replacement of the front drive axle requires special tools and equipment. In some cases, constant velocity and triple rotor joints are supplied only with axles. Lubricate all axle joints with special high temperature molybdenum-disulfide "moly" grease. Always understand and observe the warnings and cautions listed at the beginning of this section.

Specification

- Molybdenum-disulfide constant velocity and triple rotor joint lubricant
  VW part # G 000 603 . . . . 90 G (3.2 oz.)
  VW part # G 000 633 . . . . 120 G (4.2 oz.)

Drive axles with constant velocity joints

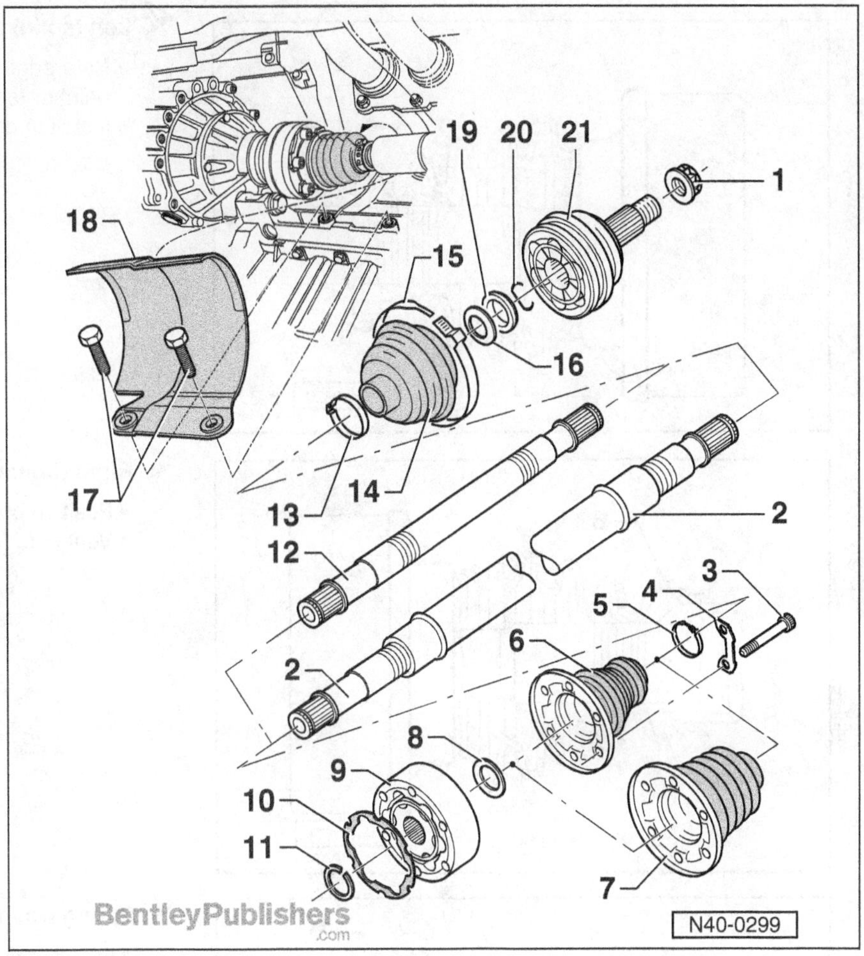

1. **Axle nut, self-locking**
   - Always replace
   - Any paint residue and/or corrosion on threads of outer joint must be removed before nut is installed
   - For tightening procedure see **Front wheel bearing housing assembly** given earlier
2. **Axle shaft, right side (tube shaft)**
3. **Bolt**
   - Tighten to 40 Nm (30 ft-lb)
4. **Lock plate**
5. **Clamp**
   - Always replace
6. **CV joint boot, inner**
   - Material: Hytrel (Polyelastomer)
   - No vent hole
   - Check for tears and chafing
   - Drive off CV joint using drift
7. **CV joint boot, inner**
   - Material: Rubber
   - With vent hole
   - Check for tears and chafing
   - Drive off CV joint using drift
   - See **Left (solid) axle CV joint boot, installed position**
   - See **Right (hollow) axle CV joint boot, installed position**
8. **Spring washer**
   - See **Spring washer on inner CV joints, installed position**
9. **Inner CV joint**
   - Only replace complete unit
   - Support hub and press off of axle
   - Install with chamfer on inner diameter of hub facing stop on axle shaft
   - Press up to stop
   - Install new circlip
10. **Gasket**
    - Always replace
    - Adhesive surface on CV joint must be free of oil and grease
    - Remove protective foil and stick gasket onto CV joint.
11. **Circlip**
12. **Axle shaft, left side (solid shaft)**
13. **Clamp**
    - Always replace
14. **CV joint boot, outer**
    - Check for tears and chafing
    - Material: Hytrel (Polyelastomer)
15. **Clamp**
    - Always replace
16. **Spring washer**
    - See **Spring washer and thrust washer on outer CV joints, installed position**
17. **Bolts**
    - Tighten to 35 Nm (26 ft-lb)
18. **Heat shield**
19. **Thrust washer**
    - See **Spring washer and thrust washer on outer CV joints, installed position**
20. **Circlip**
    - Always replace
    - Install in shaft groove
21. **Outer CV joint**
    - Only replace complete unit
    - Drive off axle with plastic/brass/aluminum hammer
    - Drive onto shaft up to stop using plastic/brass/aluminum hammer

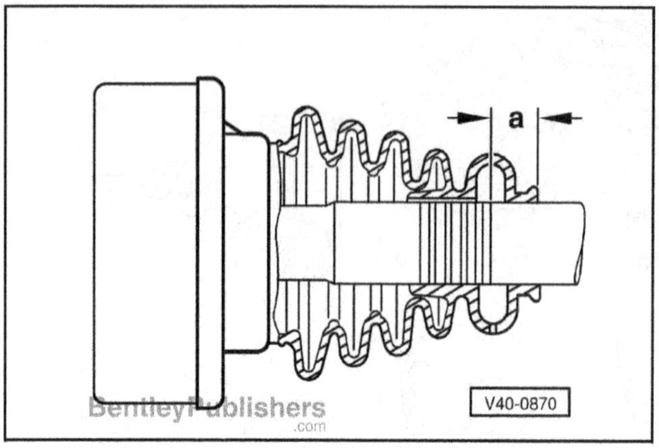

Left (solid) axle CV joint boot, installed position

- Note edge location of boot before disassembly with paint or tape mark. Do not mark by scratching axle.
- Install to dimension (a); 17 mm (0.669 in)

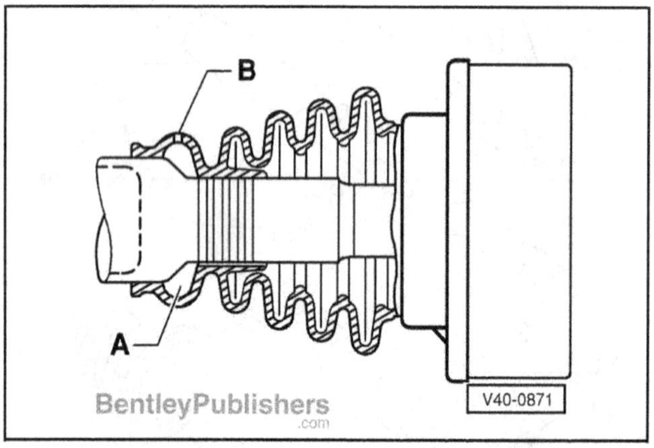

Right (hollow) axle CV joint boot, installed position

- Position boot to allow for vent chamber (**A**).
- Vent hole (**B**) must open into vent chamber (**A**).

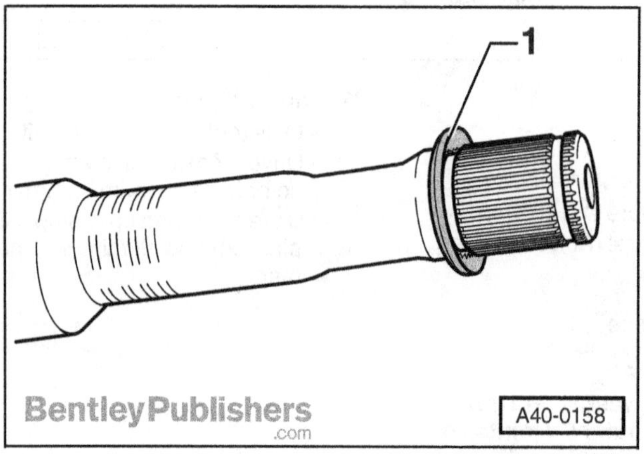

Spring washer on inner CV joints, installed position

- Install spring washer (**1**) into position as shown.

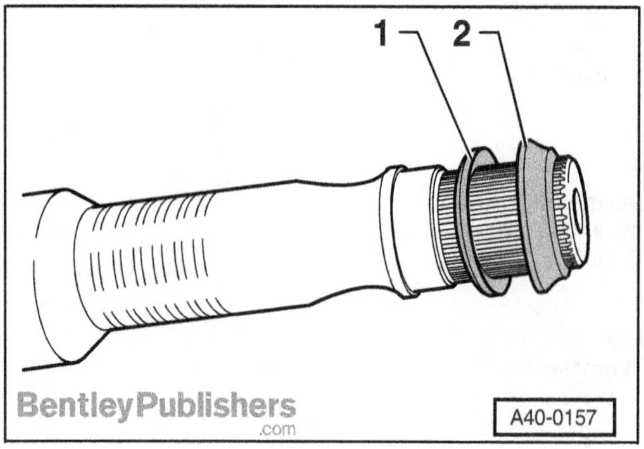

Spring washer and thrust washer on outer CV joints, installed position

- Install spring washer (**1**) and thrust washer (**2**) into position as shown.

Drive axles with triple-roller joint (AAR 2900)

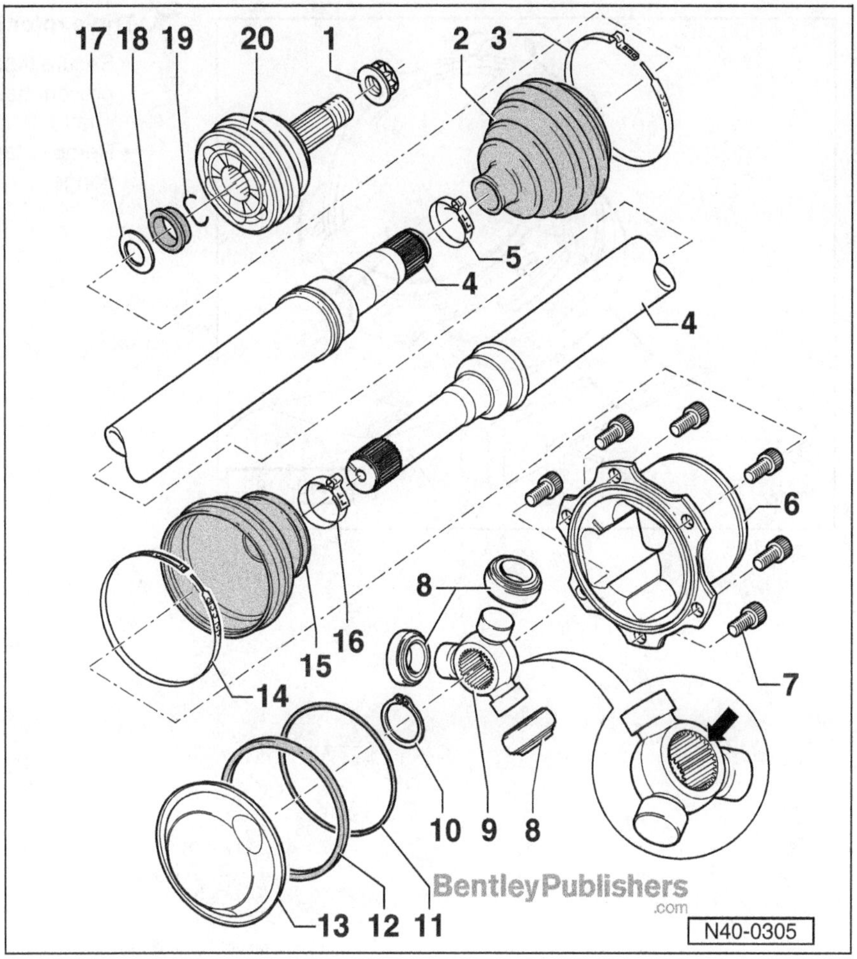

1. **Axle nut, self-locking**
   - Always replace
   - Any paint residue and/or corrosion on threads of outer joint must be removed before nut is installed
   - For tightening procedure see **Front wheel bearing housing assembly** given earlier
2. **CV joint boot, outer**
   - Check for tears and chafing
3. **Clamp**
   - Always replace
4. **Axle shaft**
5. **Clamp**
   - Always replace
6. **Triple rotor joint housing**
   - Also called "tripot" and "tripod" joints
   - Only replace complete unit
   - Use caution on re-assembly to prevent axle from inadvertently falling apart, See **Triple rotor joint, installation precautions**
7. **Bolt**
   - Tighten to 40 Nm (30 ft-lb)
8. **Rollers**
9. **Triple-rotor star (hub)**
   - Chamfer faces opposite end of drive axle
10. **Circlip**
    - Always replace
    - Install in shaft groove
11. **O-ring**
    - Not required for assembly
12. **Rectangular section seal**
    - Seal is part of repair kit; not installed in production
13. **Cover**
    - Destroyed when disassembling
    - Not required for assembling; not supplied as replacement part
14. **Clamp**
    - For triple-rotor joint
    - Only installed to left side on production vehicles
15. **Triple-rotor joint boot**
16. **Clamp**
17. **Spring washer**
    - Installation position, see **Drive axles with constant velocity joints, assembly** given earlier
18. **Thrust washer**
    - Installation position, see **Drive axles with constant velocity joints, assembly** given earlier
19. **Circlip**
    - Always replace
    - Install in shaft groove
20. **Outer constant velocity (CV) joint**
    - Only replace complete unit
    - Drive off of axle with plastic/brass/aluminum hammer
    - Drive onto shaft up to stop using plastic/brass/aluminum hammer

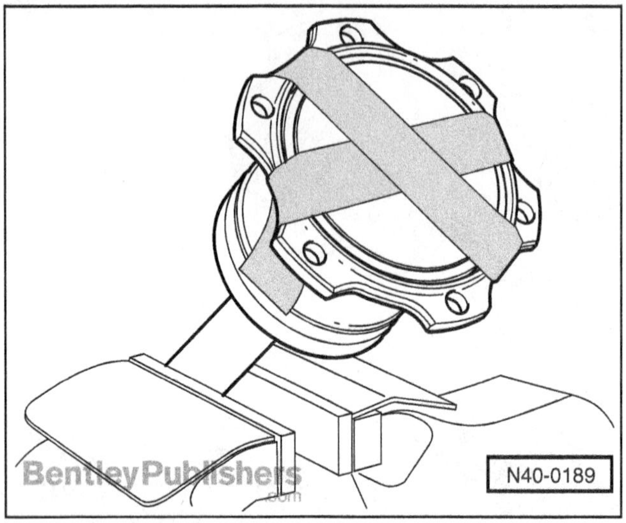

Triple rotor joint, installation precautions

- Secure triple rotor joint with adhesive tape as shown to prevent housing from accidently sliding down axle shaft.
- Remove tape just prior to bolting securely to drive flange.

Drive axles with triple-roller joint (AAR 2000)

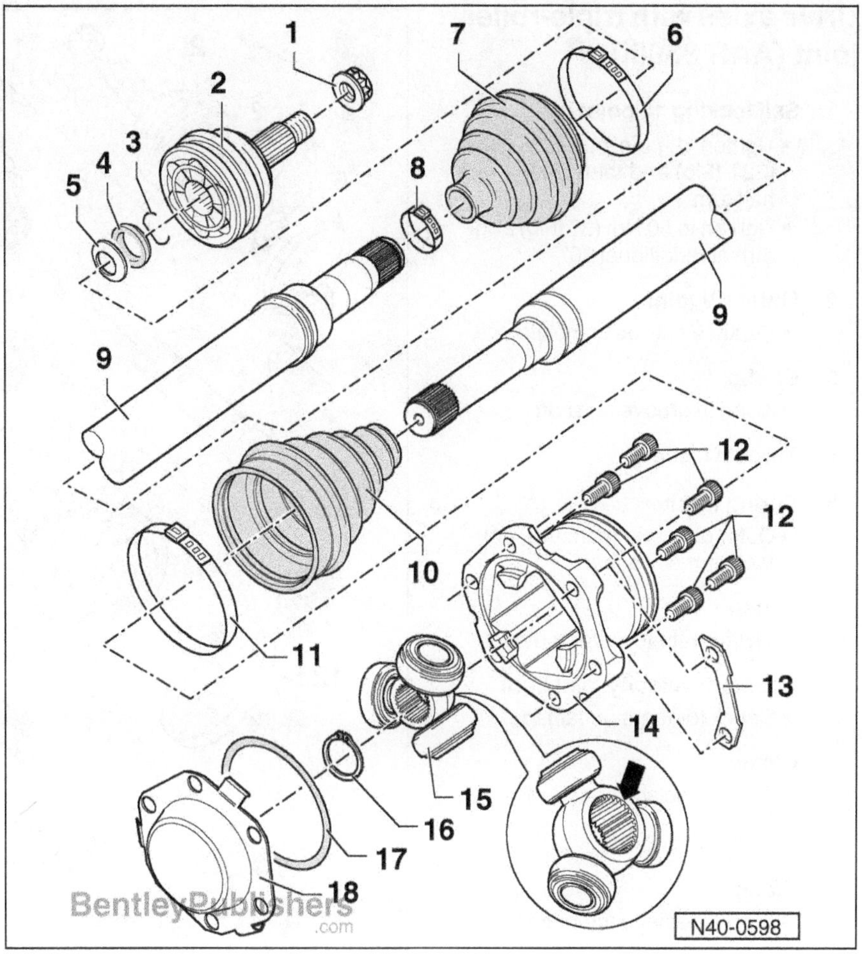

1. **Self locking 12 point nut**
   - Tighten 12-point nut to 200 Nm (221 ft-lb) and then loosen one half turn
   - Tighten to 50 Nm (37 ft-lb) and turn an additional 60°
2. **Outer CV joint**
   - Replace only as complete unit
3. **Circlip**
   - Insert in groove on shaft
4. **Thrust washer**
5. **Spring washer**
   - Outer diameter (concave) contacts thrust washer
6. **Clamp**
   - Always replace after removal
7. **Constant velocity joint boot**
   - Check for tears or damage
8. **Clamp**
   - Always replace after removal
9. **Axle shaft**
10. **Triple rotor joint boot**
    - Check for tears or damage
11. **Clamp**
    - Always replace after removal
12. **Multi point socket head bolt**
    - 100 mm joints: tighten to 40 Nm (30 ft-lb)
    - 108 mm joints: tighten to 80 Nm (59 ft-lb)
13. **Backing piece**
    - Only installed on joints with 108 mm diameter
14. **Triple rotor joint housing**
15. **Triple rotor star with rollers**
    - The chamfer (**arrow**) points toward axle shaft splines.
16. **Circlip**
    - Always replace
17. **Seal**
    - Always replace
18. **Cover**
    - Always replace

Drive axles with triple-roller joint (AAR 3300i)

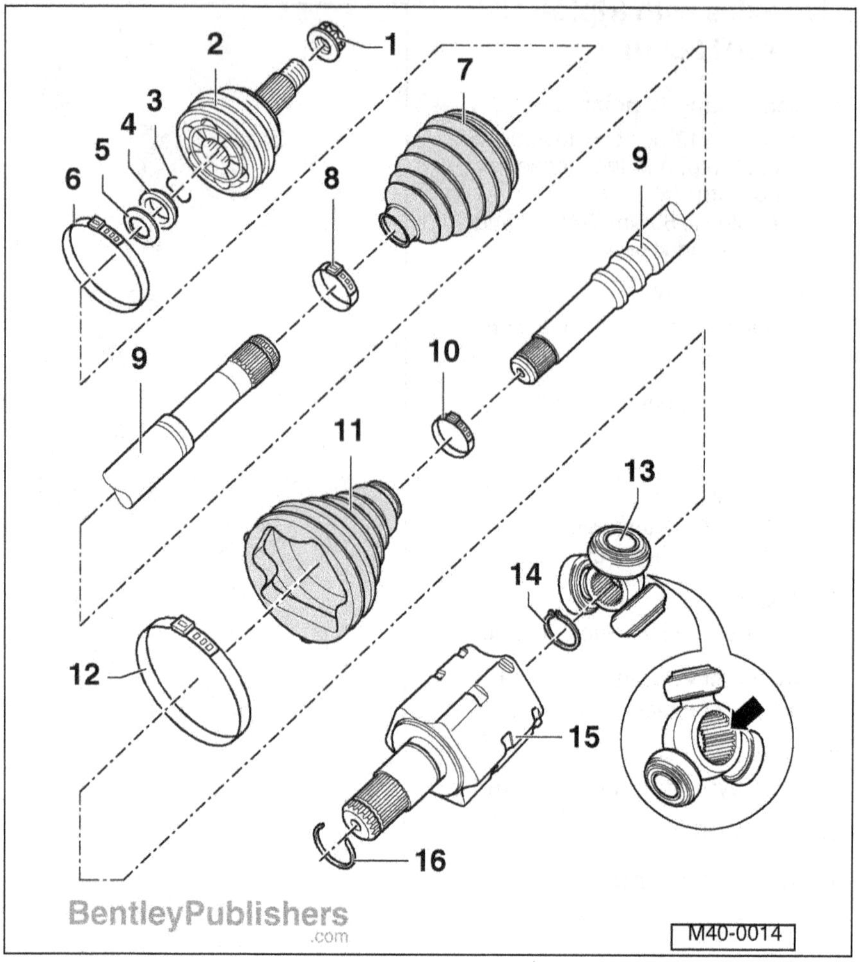

1. **Self locking 12 point nut**
   - Tighten 12-point nut to 200 Nm (221 ft-lb) and then loosen one half turn
   - Tighten to 50 Nm (37 ft-lb) and turn an additional 60°
2. **Outer CV joint**
   - Replace only as complete unit
3. **Circlip**
   - Insert in groove on shaft
4. **Thrust washer**
5. **Spring washer**
   - Outer diameter (concave) contacts thrust washer
6. **Clamp**
   - Always replace after removal
7. **Constant velocity joint boot**
   - Check for tears or damage
8. **Clamp**
   - Always replace after removal
9. **Axle shaft**
10. **Clamp**
    - Always replace after removal
11. **Triple rotor joint boot**
    - Check for tears or damage
12. **Clamp**
    - Always replace after removal
13. **Triple rotor star with rollers**
    - The chamfer (**arrow**) points toward axle shaft splines.
14. **Circlip**
    - Always replace
    - Insert in groove on shaft
15. **Triple rotor joint housing**
16. **Circlip**
    - Always replace
    - Insert in groove on shaft

42 Rear Suspension

GENERAL

Special tools, equipment and procedures are required for most rear suspension repair and component replacement. In addition, wheel alignment is almost always disturbed when suspension components are removed or replaced.

This section covers repairs to the rear suspension and related components. For front suspension servicing and repair information, see **40 Front Suspension**. For wheel alignment specifications, see **44 Wheels, Tires, Wheel Alignment**.

> ***WARNING —***
> - *Do not re-use any fasteners that are worn or deformed in normal use. Most fasteners are designed to be used only once and become unreliable and may fail when used a second time. This includes, but is not limited to, nuts, bolts, washers, circlips, cotter pins, self-locking nuts and bolts. For replacements, always use new parts.*
> - *Do not reinstall bolts and nuts coated with undercoating wax, as correct tightening torque cannot be ensured. Always clean the threads of removed bolts and nuts that are to be re-used with a suitable solvent before installation. Ensure that new parts are clean.*

WARNING —
- *Do not use standard nuts and bolts in place of green colored parts. The green color denotes a special anti-corrosion process known as "dacromet" or "delta-tone". Torque specifications are given based on use of these parts only.*
- *Always replaced rusted or corroded bolts, nuts and washers even if not specifically indicated.*
- *DO NOT attempt to straighten or weld suspension struts, wheel bearing housings, control arms or any other wheel locating or load bearing components of the front suspension.*

REAR SUSPENSION

The beam-type rear axle is a one-piece welded assembly consisting of an axle beam, trailing arms, stabilizer bar, and coil spring seat. Rear wheel stub axles are bolted to the trailing arms. Coil springs are mounted low on the beam with the upper spring seats attached to the rear body structure. Gas-pressure telescopic rear shock absorbers are bolted to the outside of the wheel wells and are easily removed independent of the springs. Rear wheel bearings are sealed into the hub assembly and are not serviceable.

Rear suspension overview

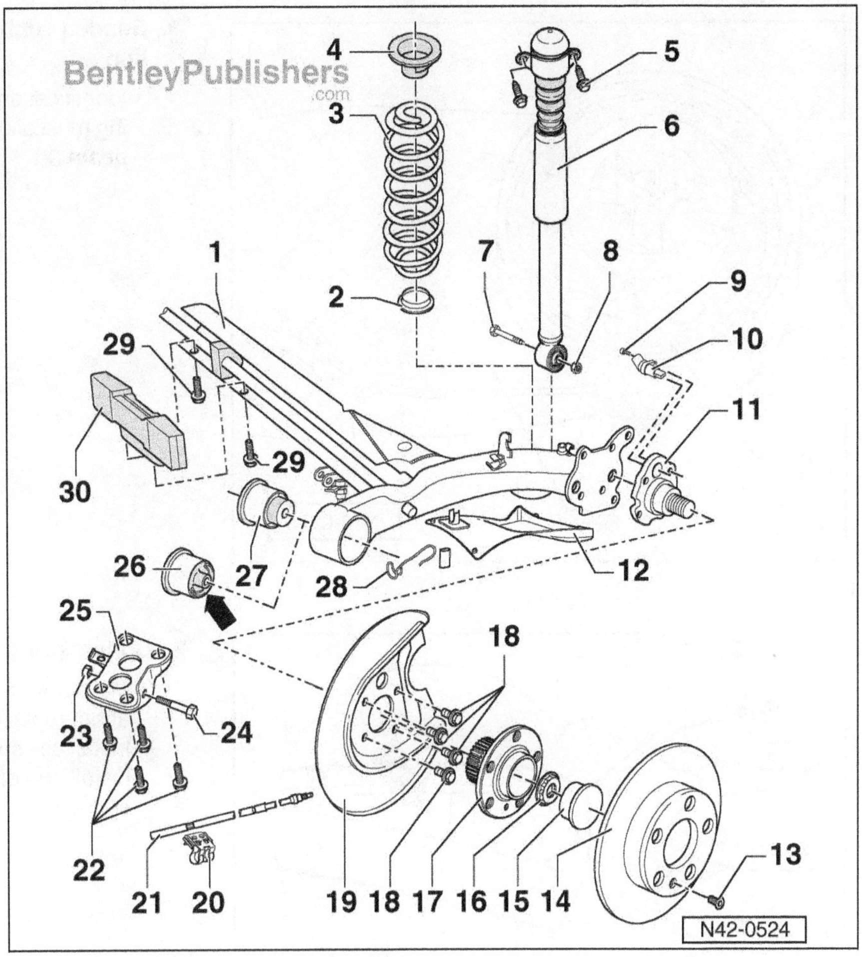

1. **Axle beam**
   - Stub axle contact surfaces and threaded holes must be free of paint and dirt
2. **Base plate**
   - Material: zinc
   - Check for damage
3. **Coil spring**
   - Note color-code marking for identification
   - Outer surface of spring must not be damaged
4. **Spring seat, upper**
   - See **Upper spring seat and coil spring, installed position**
5. **Upper shock bolt**
   - Always replace
   - Tighten to 75 Nm (55 ft-lb)
6. **Shock absorber**
   - Gas pressure
   - Can be replaced individually
7. **Lower shock bolt**
   - Always replace
   - Tighten to 55 Nm (41 ft-lb)
8. **Nut**
   - Always replace
   - Axle beam must be in middle position when tightening nut to insure proper shock bushing preload
   - Load rear of vehicle with weight of one person when tightening
9. **Bolt**
   - Tighten to 10 Nm (7 ft-lb)
10. **ABS wheel speed sensor**
11. **Stub axle**
    - Do not attempt to straighten stub axle
    - Do not attempt to re-cut threads
12. **Stone protection plate**
13. **Screw**
    - Tighten to 4 Nm (35 in-lb)
14. **Brake disc (rotor)**
15. **Dust cap**
    - Always replace
    - Proper function and long service life can only be ensured by perfect seal of new dust cap
16. **Nut, self-locking**
    - Always replace
    - Tighten to 175 Nm (129 ft-lb)
17. **Wheel bearing/hub unit with ABS wheel speed sensor rotor**
    - Wheel bearing, wheel hub and wheel speed sensor rotor are installed together in housing
    - Wheel bearing/hub unit is maintenance and adjustment free
18. **Bolts**
    - Always replace
    - Tighten to 60 Nm (44 ft-lb)
19. **Splash shield**
20. **Parking brake cable bracket**
21. **Parking brake cable**
22. **Bolts**
    - Always replace
    - Tighten to 75 Nm (55 ft-lb)
    - If threads in welded nut in longitudinal member (rear body structure) are damaged, repair using Heli-Coil
    - Do not repair more than one welded nut on each side with Heli-Coils
23. **Nut, self-locking**
    - Always replace
    - Tighten to 80 Nm (59 ft-lb)
24. **Bolt**
    - Always replace
    - Tighten to 80 Nm (59 ft-lb)
25. **Rear axle mounting bracket**
    - Check and if necessary adjust rear axle total toe after installation
    - If possible do not loosen when removing rear axle
26. **Hydraulic bonded rubber bushing**
    - Bulge (**arrow**) points toward vehicle exterior
    - If bushing leaks it must be replaced
27. **Bonded rubber bushing**
    - See **Bonded rubber axle beam bushing, installed position**
28. **Parking brake cable bracket**
29. **Bolts**
    - Always replace
30. **Vibration absorber**

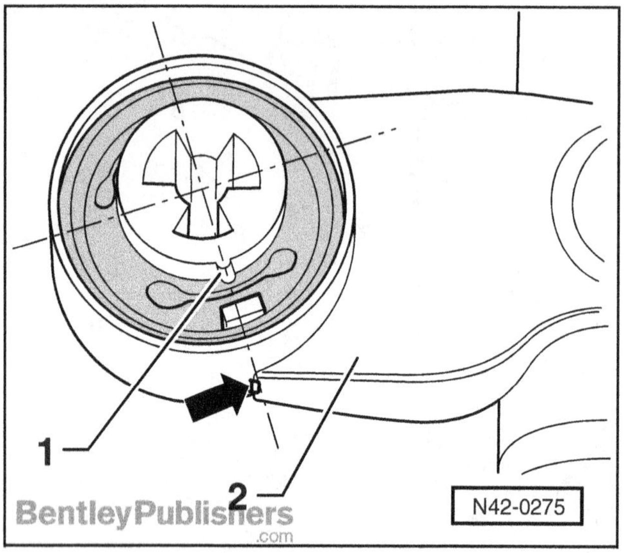

Bonded rubber axle beam bushing, installed position

- Identification mark (**1**) on face of bonded rubber bushing must align with edge of trailing arm (arrow) on axle beam (**2**).

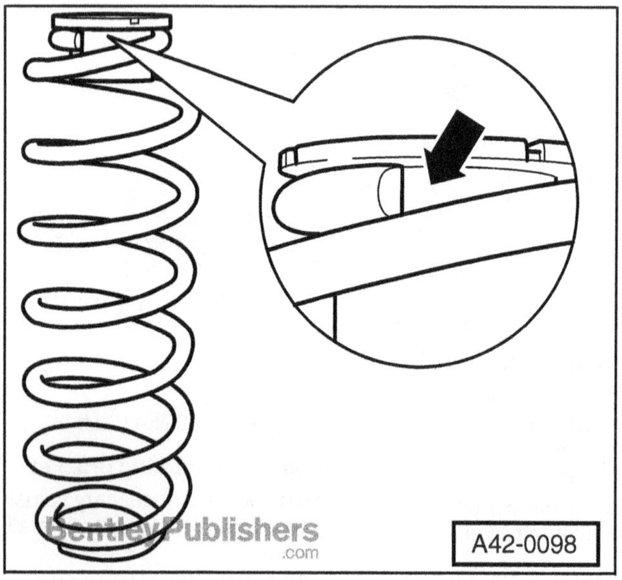

Upper spring seat and coil spring, installed position

- Make sure that zinc lower spacer bushing is not damaged; replace as required.
- Install coil spring and upper spring seat together.
- Install end of coil spring against stop (**arrow**) in upper spring seat.

REAR SHOCK ABSORBERS

Rear shock absorber assembly

1. **Gas-filled shock absorber**
   - Can be replaced individually
   - Slight traces of oil do not necessitate replacement
   - Note markings for identification
2. **Protective cap**
3. **Protective tube**
4. **Bump stop**
5. **Upper shock absorber mount**
6. **Nut, self-locking**
   - Always replace
   - Tighten to 25 Nm (18 ft-lb)
7. **Protective sleeve**

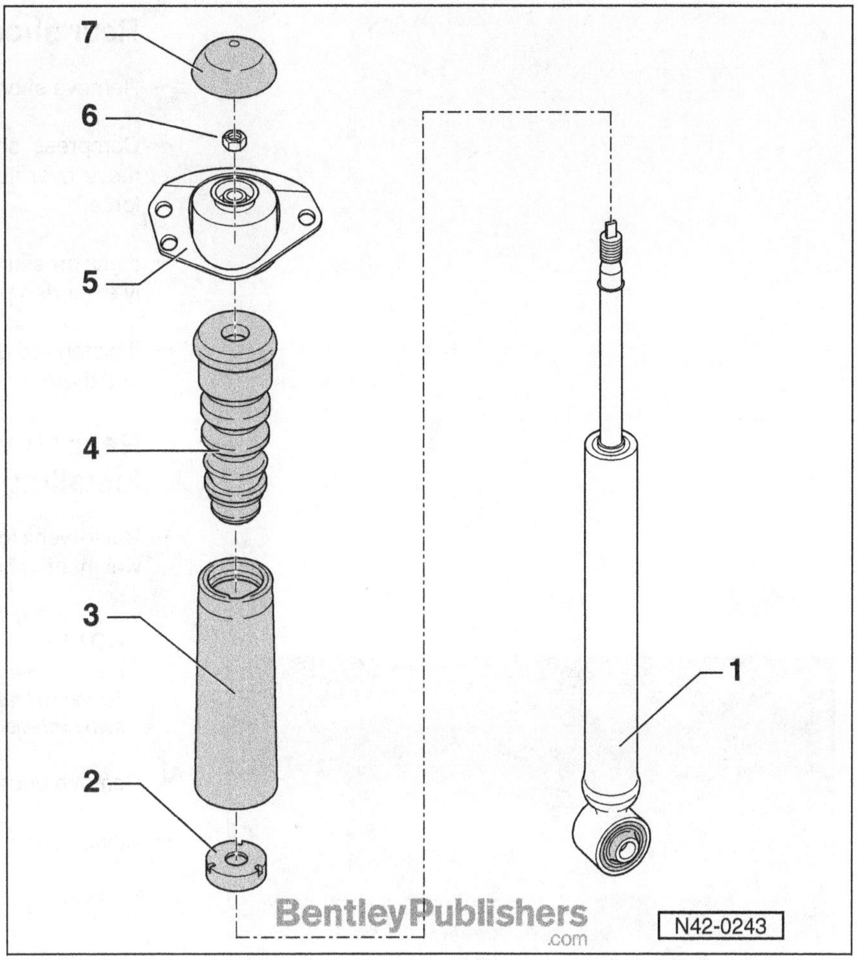

Rear shock absorber, checking

- Remove shock absorber as described below.
- Compress shock absorber by hand. Piston rod must move over its complete length smoothly and with even force.
- If gas pressure is sufficient, piston rod will return to its fully extended position.
- If piston rod does not return to its fully extended position and there is no loss of oil, then shock absorber is still OK.

Rear shock absorber, removing and installing

- Raise vehicle slightly with a lift or floor jack, but leave weight of vehicle on the wheels.

NOTE —

Rear suspension must not be allowed to extend fully. Spring will fall out when shock absorber is unbolted if some vehicle weight is not supported by the rear wheel.

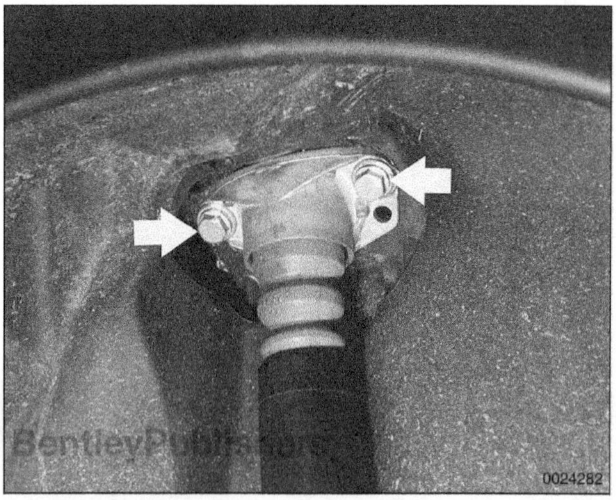

< Remove upper shock mounting bolts (**arrows**).

- Raise vehicle slightly to relieve coil spring pressure.

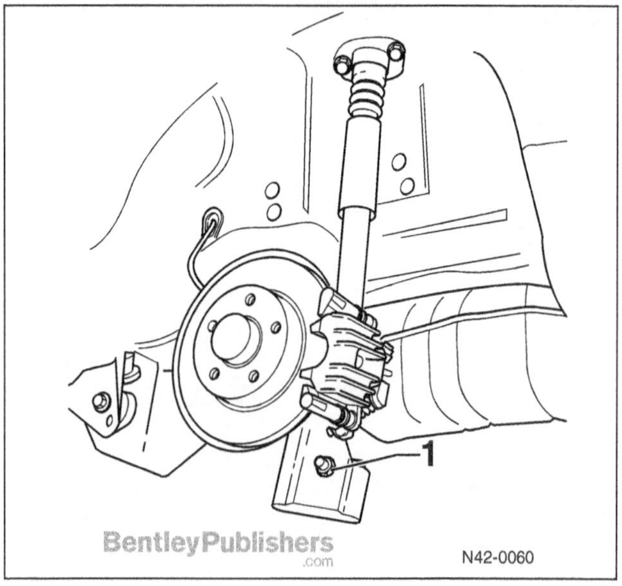

< Remove lower shock absorber bolt (**1**) on rear axle.

- Remove shock absorber.

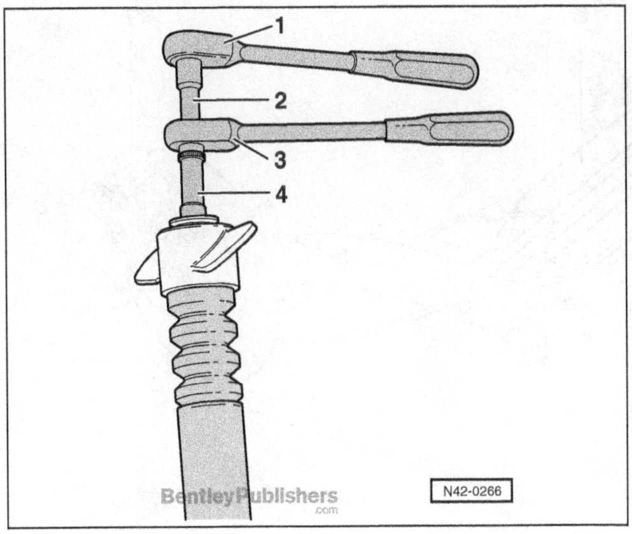

Remove upper shock mount.

– Installation is the reverse of removal.

Tightening torque
- Upper shock bolt (M10) always replace . . . . 75 Nm (55 ft-lb)
- Lower shock bolt and nut (M10) always replace . . . . 60 Nm (44 ft-lb)

Rear coil spring, removing and installing

– Disconnect harness connector for ABS wheel speed sensor (where applicable).

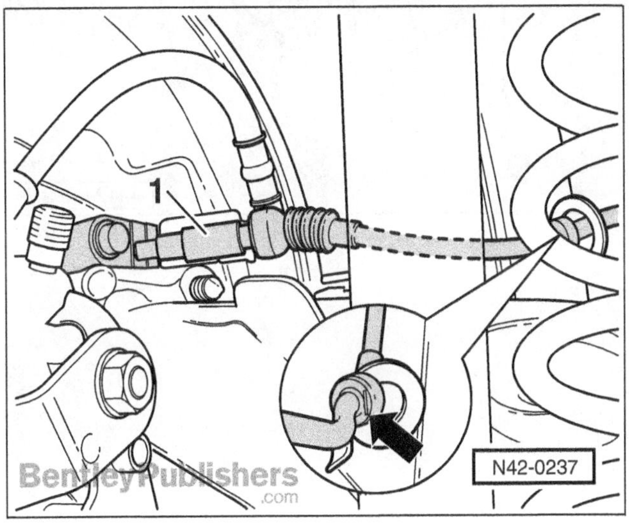

Unclip wheel speed sensor (**1**) wiring from retainer (**arrow**).

– Raise vehicle slightly with a lift or floor jack, but leave vehicle resting on its wheels.

– Remove upper shock mounting bolts as shown earlier.

– Raise vehicle further to relieve coil spring pressure.

– Remove coil spring.

– Installation is the reverse of removal, noting the following:
  - Make sure spacer bushing (zinc) is not damaged, replace if necessary.
  - Install spring together with spring seat.
  - Observe spring installation position, given earlier.
  - Press rear axle upward using transmission jack.
  - Using new bolts, bolt shock absorber to body.

> ***WARNING —***
> *If vehicle is supported on a vehicle lift or hoist, ensure that transmission jack does not push up on the suspension sufficient to topple the vehicle off.*

Tightening torque
- Upper shock bolt (M10) always replace . . . . 75 Nm (55 ft-lb)

Rear Wheel Bearings

Inner and outer rear wheel bearings along with the hub and ABS speed sensor rotor are combined into a single integrated assembly. This assembly must be replaced as a complete unit as there are no individually serviceable components.

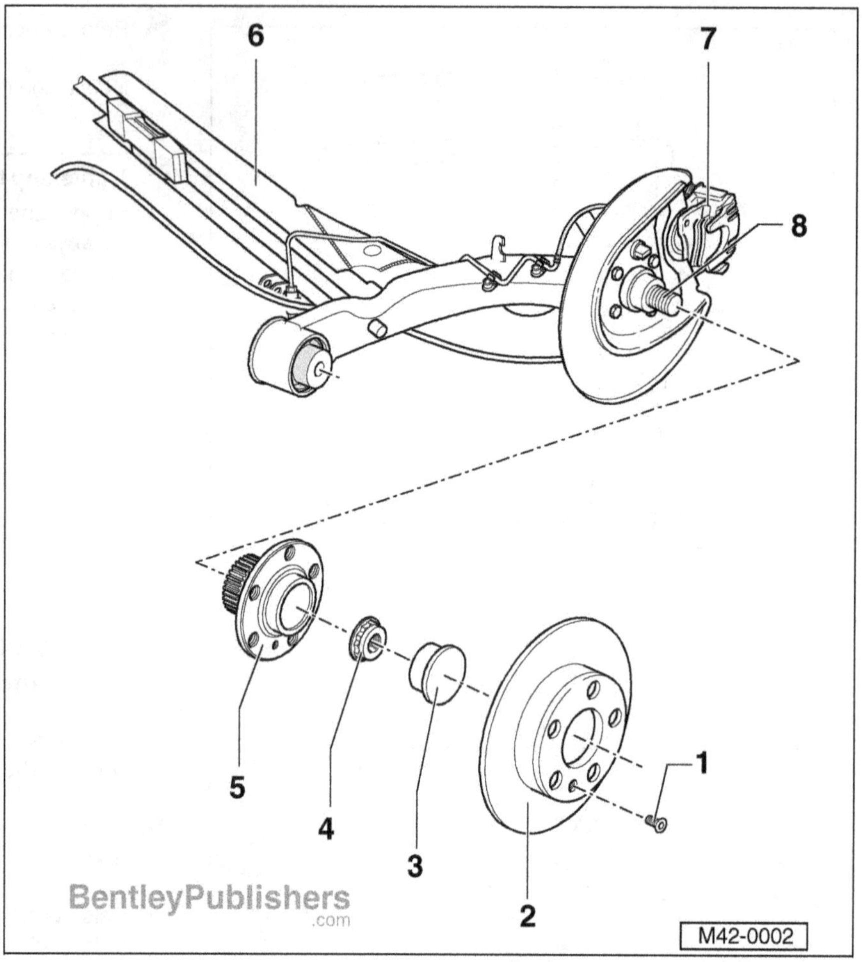

Rear wheel bearing/hub unit, assembly

1. **Screw**
   - Tighten to 4 Nm (35 in-lb)
2. **Brake disc (rotor)**
3. **Dust cap**
   - Always replace
   - Proper function and long service life can only be ensured by the perfect seal of a new dust cap
4. **Nut, self-locking**
   - Always replace
   - Tighten to 175 Nm (129 ft-lb)
5. **Wheel bearing/hub unit with ABS wheel speed sensor rotor**
   - Wheel bearing, wheel hub and wheel speed sensor rotor are installed together in housing
   - Wheel bearing/hub unit is maintenance and adjustment free
6. **Axle beam assembly**
7. **Rear brake caliper**
8. **Stub axle**
   - Do not attempt to straighten stub axle
   - Do not attempt to re-cut threads

Rear wheel bearing/hub unit, removing

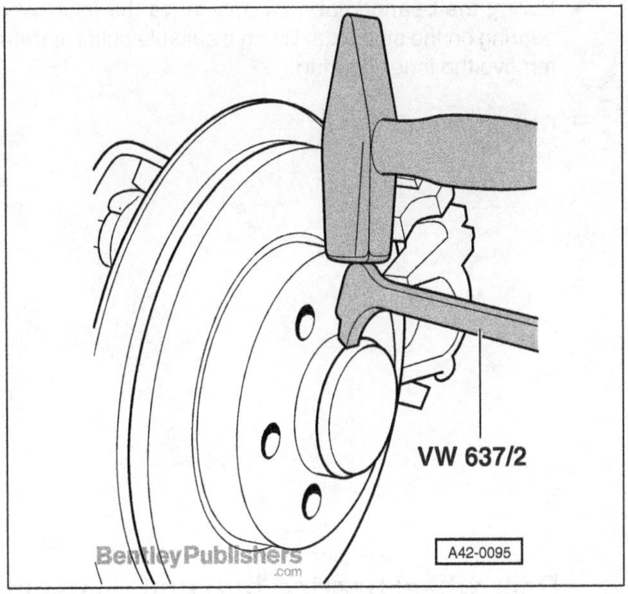

- Remove wheel assembly.

◄ Remove rear hub dust cap from seat on rotor.

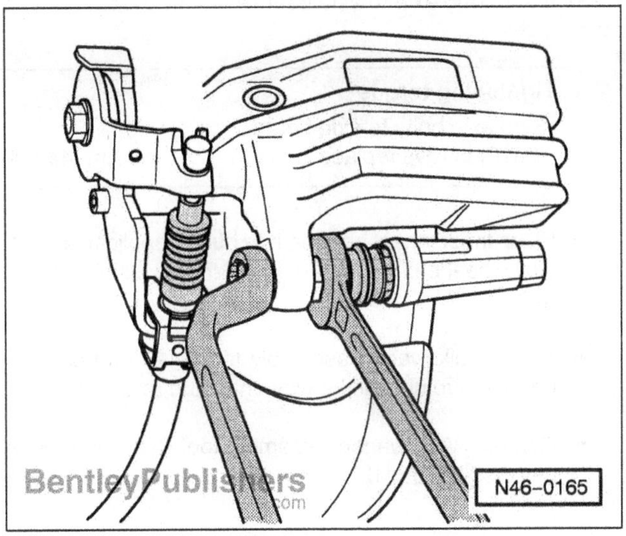

◄ Counterhold pins and remove mounting bolts from brake caliper.

> **CAUTION —**
> *Do not allow the caliper to hang by the brake hose. The unsupported weight can stretch and damage the hose.*

- Remove phillips head screw from brake disc (rotor) and remove brake disc.

- Remove self-locking nut at center of hub assembly.

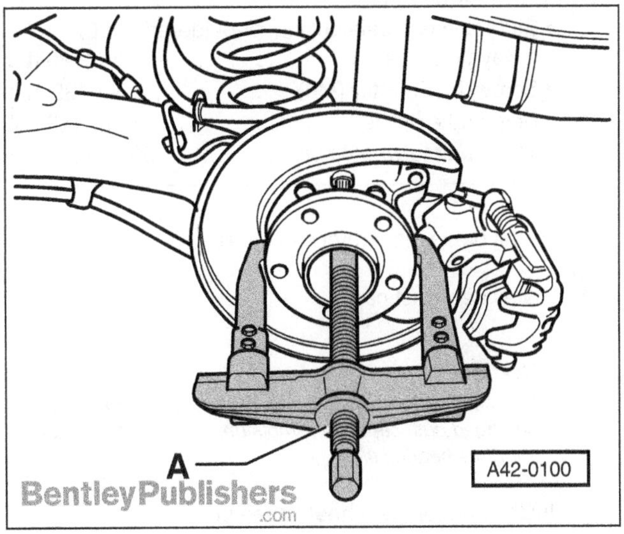

◄ Using a suitable puller, remove bearing/hub unit.

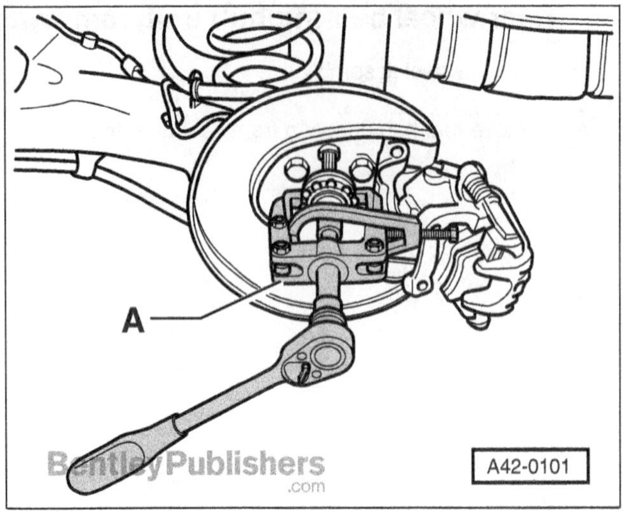

◄ Pulling the bearing/hub unit may leave the inner wheel bearing on the stub axle. Using a suitable puller, carefully remove the inner bearing.

– Remove stub axle.

Rear wheel bearing/hub unit, installing

– Install stub axle to axle beam with new bolts.

Tightening torque

- Stub axle bolts to axle beam (M10) always replace .... 60 Nm (44 ft-lb)

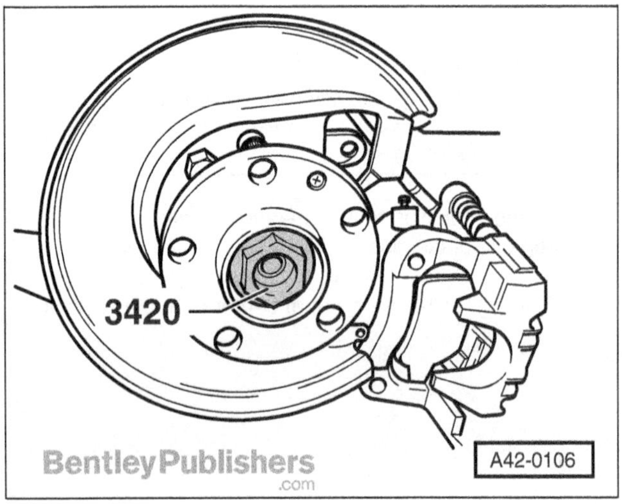

– Place inner wheel bearing into hub (if applicable) and install wheel bearing/hub unit as far as possible onto stub axle.

◄ Attach Volkswagen assembly tool 3420 and pull wheel bearing/hub unit all the way onto stub axle.

– Remove Volkswagen assembly tool 3420 and install a new self-locking 12-point nut.

Tightening torques

- Stub axle nut, self-locking, shouldered (M20) always replace .... 175 Nm (129 ft-lb)
- Screw, brake rotor (M6) .... 4 Nm (35 in-lb)
- Bolt, brake caliper .... 65 Nm (48 ft-lb)
- Wheel bolt to wheel hub .... 120 Nm (87 ft-lb)

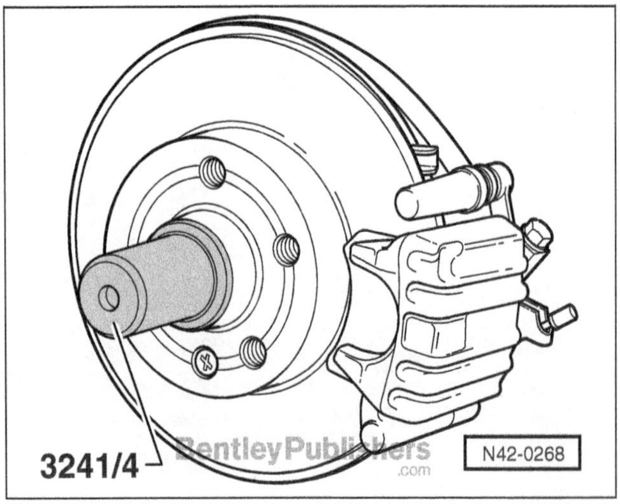

– Install brake disc (rotor) and caliper.

◄ Install new dust cap.

NOTE—

Use care when installing new dust cap as dented and damaged dust caps allow moisture contamination that shortens bearing life.

– Install and torque wheel assembly.

44 Wheels–Tires, Wheel Alignment

44

GENERAL

This repair group covers basic tire, wheel, and wheel alignment information. Also covered here are wheel alignment specifications to be used in conjunction with professional alignment tools and measuring equipment.

Wheels and tires

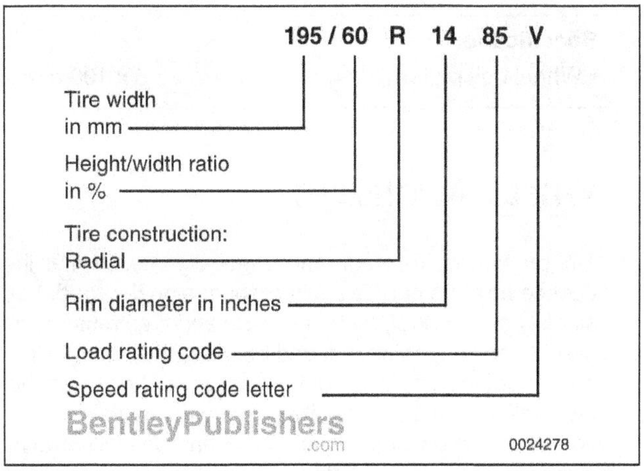

Wheels and tires approved by the manufacturer have been matched to the vehicle and contribute largely to road handling and driving characteristics. To retain the handling characteristics, it is recommended that the tires be replaced only with tires having the same specifications with regard to size, design, load carrying capacity, speed rating, tread pattern, tread depth, etc. This information can be found on the tire's sidewall. Various tire and wheel applications can be found in **Table a**.

Volkswagen recommends that the tires be rotated front to back, with the tires remaining on the same side of the vehicle. Only when tires show unusual wear should they be rotated diagonally. See **03 Maintenance** for maintenance schedules regarding tire rotation.

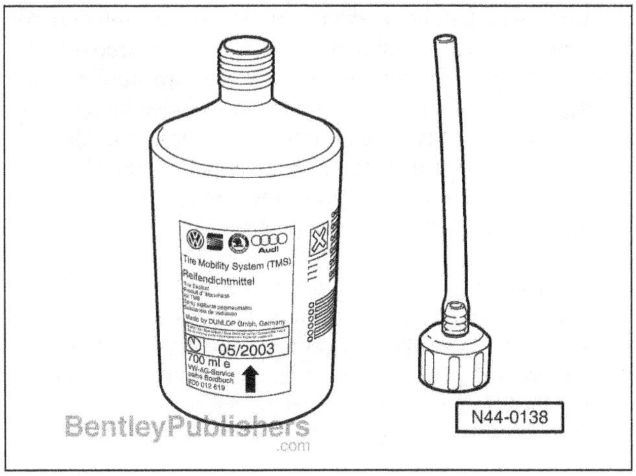

The New Beetle convertible models are equipped with a flat tire repair kit. The repair kit is located in the luggage compartment under the cargo liner.

The tire sealant in the flat tire kit has a limited shelf life. The bottle is marked with a minimum shelf life date (**arrow**). The sealant bottle should be replaced if this date has passed. The bottle should also be replaced if it has been opened and/or partially used.

Table a. Wheel and tire applications

Application	Tire size	Wheel	Offset (ET) in mm	Snow chains permissible
All season	195/65 R 15 91 T	6 J x 15	38	Yes
	195/65 R 15 91 H	6 J x 15	38	Yes
	195/65 R 15 91 V	6 J x 15	38	Yes
	205/55 R 16 91 H	6½ J x 16	42	No
	205/55 R 16 91 V	6½ J x 16	42	No
	225/45 R 17 91 W	7 J x 17 (2002-on)	38	No
	225/45 R 17 91 W	7 1/2 J x 17	38	No
Winter	195/65 R 15 91 Q/T	6 J x 15	38	Yes
	205/55 R 16 91 T/H	5½ J x 16	36	Yes
Temporary spare	T 125/70 R 18 98 M	3½ J x 18	38	No

Tightening torque
- Wheel bolt to hub (diagonally and evenly) . . . . 120 Nm (87 ft-lb)

Specification
- Wheel bolt pattern . . . . 5 x 100 mm

WHEEL ALIGNMENT

Tire pressures, tire wear, and wheel alignment will all influence how the car feels and responds on the road. For stability and control, all four wheels and tires must be in good condition, balanced, and be properly aligned. Precise wheel alignment can only be accomplished when the tires, the suspension, and the steering are in good condition. Reputable wheel alignment technicians will always inspect the front and rear suspension and the steering for worn parts before an alignment, and will recommend that any necessary repairs be made before proceeding. The important front wheel alignment angles are camber, caster, and toe. In the rear, the important angles are camber, toe and thrust line. Although rear alignment angles are not generally adjustable, they should be checked because of the effect that they have on rear tire wear and straight-line stability of the vehicle.

Camber is the angle that the wheels tilt from vertical when viewed from front or rear. Wheels which tilt out at the top have positive (+) camber. Wheels that tilt in at the top have negative (–) camber. On the vehicles covered by this manual, camber is non-adjustable and corrections must be made by moving the front subframe slightly.

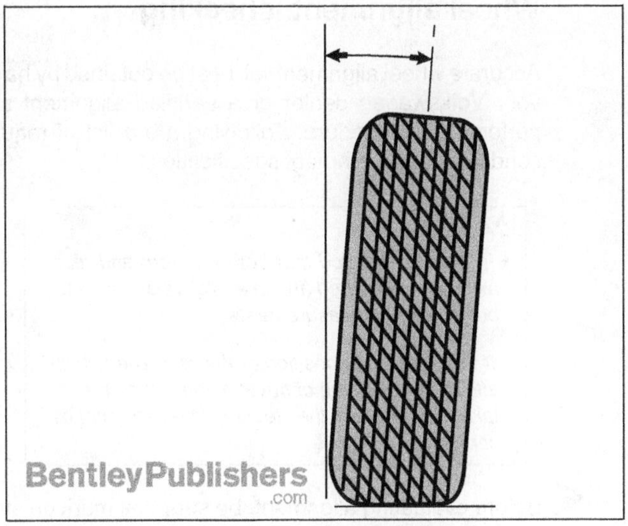

Camber influences cornering, directional stability, and tire wear. Different camber on the two front wheels may cause the car to pull to one side. Incorrectly adjusted camber will cause uneven tire wear.

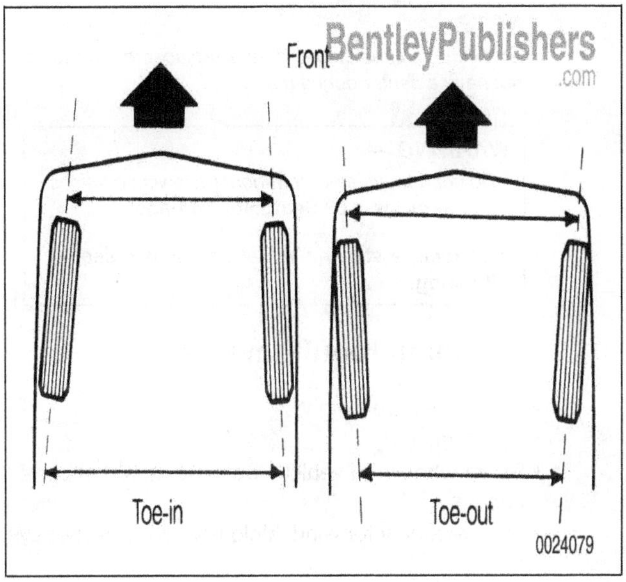

Caster is the angle at which the steering axis deviates from vertical. Most cars are designed with positive caster, which improves directional stability and tends to make the steering more self-centering. Caster angle should be checked as part of wheel alignment, but it is not adjustable on cars covered by this manual.

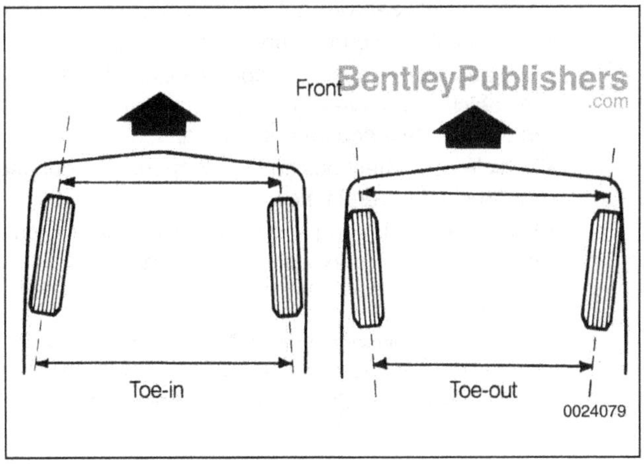

Toe is a measurement of the amount that two wheels on the same axle point toward each other (toe-in) or away from each other (toe-out). Toe affects directional stability and tire wear. Toe also affects response to steering input. Too much toe will cause tires to "scrub" and to wear unevenly and quickly. Too little toe may cause the car to be less stable and wander at highway speeds.

Wheel alignment, checking

Accurate wheel alignment will best be obtained by having your Volkswagen dealer or a certified alignment shop perform the procedure. Following are a list of required conditions and alignment specifications.

CAUTION—
- *It is recommended that both the front and rear axles are measured (thrust angle) when carrying out alignment measurements.*
- *If the installation position of the rear axle (which affects the direction of travel of the vehicle) is not taken into account the steering wheel may not be centered.*

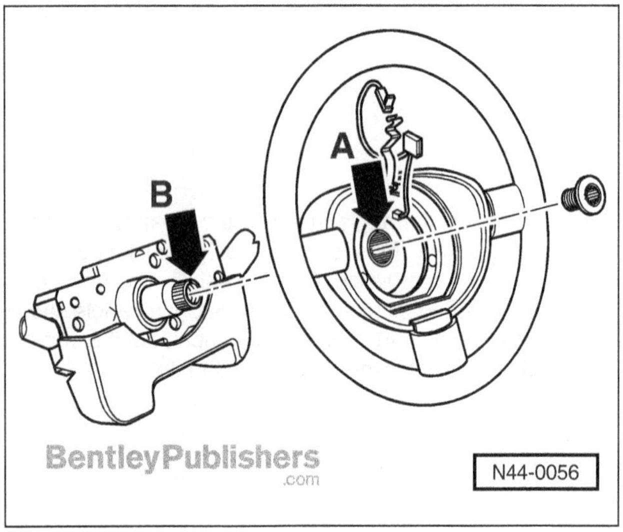

Before evaluating alignment, be sure that mark on steering wheel (**A**) and steering column (**B**) are aligned.

NOTE—
Steering columns supplied as a replacement parts do not have a center punch mark.

WARNING—
- *Do not remove air bag without reviewing cautions and warnings in* **69 Seatbelts, Airbags**.
- *To remove steering wheel and airbag, see* **48 Steering**.

Alignment requirements

- Vehicle at curb weight
- Fuel tank full
- Spare wheel and vehicle tools are stored at correct locations
- Fluid reservoir for windshield/headlight washer system full
- Equal tread depth for both tires on each axle, with difference between two not more than 2 mm (0.079 in.)
- Tires inflated to correct pressure
- Test equipment must be properly adjusted and attached to vehicle. Observe all instructions, cautions and warnings of the test equipment manual
- Vehicle accurately positioned, suspension bounced and rocked several times
- Suspension, steering and steering linkage in proper condition; without excessive play or damage

NOTE—
The vehicle alignment should not be checked before the vehicle has completed 1000-2000 km (600-1200 miles) so that the coil springs can have a chance to settle.

Wheel alignment specifications

Table b. Wheel alignment data—front suspension

Suspension (version)	**Base suspension (G18, G49, G80)**	**Sport suspension (1GD, G16, G73, G87)**
Total toe [1] (wheels not pressed)	0° ± 10'	
Camber [1] (in straight-ahead position)	–30' ± 30'	–33' ± 30'
Maximum permissible difference between left and right	max. 30'	
Toe angle difference at 20° lock from left to right	+1° 30' ± 20'	+1° 31' ± 20'
Caster (not adjustable)	+7° 40' ± 30'	+7° 50' ± 30'
Maximum permissible difference between left and right	30'	30'

[1] Camber adjustments are not possible. Corrections are possible by moving the subframe slightly. Adjustment approx. 10' to 15'.

Table c. Wheel alignment data—rear suspension

Suspension (version)	**Base suspension (UB1, UC2)**	**Sport suspension (1JC, UC4)**
Camber	–1°27' ± 10'	
Maximum permissible difference between left and right	30'	
Total toe (with specified camber)	+20' ± 10'	
Maximum permissible deviation from direction of travel (thrust angle)	20'	

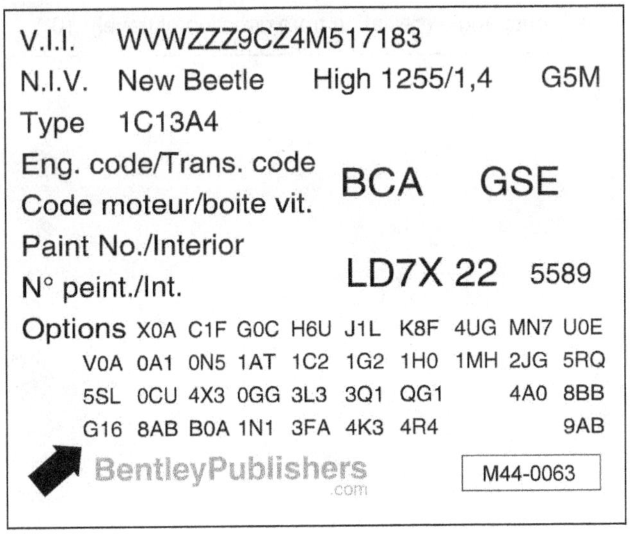

M44-0063

Suspension type

Various suspension versions are available and the data label in spare wheel recess or in the glovebox owner's manual identifies the suspension type (**arrow**).

Thrust angle, calculating

Thrust angle (deviation from direction of travel) can be calculated using one of the two following examples. The result is the deviation of the actual running direction from the longitudinal center line of the vehicle.

If alignment readings for rear toe are both positive or both negative (+/+ or -/-), subtract smaller figure from larger figure to determine total toe angle, then divide by 2.

Example:

- Left rear wheel track. . . . . + 15'
- Right rear wheel track . . . . + 5'

15' - 5' = 10'

10' ÷ 2 = 5'

- Thrust angle (deviation from direction of travel): 5'

If alignment readings for rear toe are different signs, one positive and one negative (+/-), add two figures to determine total toe angle, then divide by 2.

Example:

- Left rear wheel track. . . . . + 15'
- Right rear wheel track . . . . - 5'

15' + 5' = 20'

10' ÷ 2 = 10'

- Thrust angle (deviation from direction of travel): 10'

---

45 Anti-Lock Brakes (ABS)

GENERAL

This repair group covers the components of the anti-lock brake system (ABS). There are several different versions of the ABS system that contain various other systems:

- **EDL** - Electronic Differential Lock
- **ASR** - Automatic Slip Regulation
- **ESP** - Electronic Stability Program

ESP allows the ABS system to monitor and implement vehicle dynamics to maintain vehicle handling. Brake pedal pressure is boosted pneumatically by a vacuum brake booster in this dual-diagonal system.

NOTE —

- *The ABS system features built-in diagnostic circuitry that detects and stores Diagnostic Trouble Code (DTC) information. When the system detects a fault, a DTC is generated and stored in the ABS control module's permanent memory. These fault codes can be accessed using a diagnostic scan tool.*
- *If an ABS malfunction is stored, the ABS warning light comes on. ABS and other functions may be switched off depending on the nature of the fault. However, normal vacuum and hydraulic brake function is retained.*

> ***CAUTION —***
> *Disconnecting the negative (–) battery cable may erase fault codes and basic settings in the engine management and automatic transmission control modules. Some driveability problems may be noticed until the system re-adapts to operating conditions. OBD II readiness codes, which may be required for emissions testing, may also be erased. Convenience electronics (alarm system, interior light control, power locks, mirrors, and windows) may need to be re-set using a diagnostic scan tool.*

> ***WARNING —***
> - *ABS is a vehicle safety system; appropriate knowledge and special equipment are necessary to properly work on the system.*
> - *The ABS system must be bled after repairs requiring opening of the brake hydraulic system. See* **47 Brakes-Hydraulic System.**
> - *If you drive an ABS-equipped vehicle after a fault has been detected by the On-Board Diagnostics, keep in mind that the function of the brake system may be limited and there is a risk of accident. The brake pressure at the rear wheels is no longer controlled by the Electronic Brake Distribution (EBD) function. This can result in excessive braking at the rear, and the vehicle may skid unexpectedly under braking.*

Special precautions for ABS equipped vehicles

The ABS control module processes precise, low-level electronic signals and can be very sensitive to changes in the power supply and the environment. The following precautions apply to vehicles with ABS.

> ***WARNING —***
> - *When driving or riding in an airbag-equipped vehicle, NEVER hold the scan tool or other test equipment in your hands or lap while in motion. Objects between you and the airbag increase the risk of injury in an accident.*
> - *During a test drive in an air-bag equipped vehicle, test equipment must always be fastened to and operated from the rear seat by a helper.*
> - *Brake fluid is poisonous!*

CAUTION —

- *When working on the electrical sections of the ABS, only disconnect harness connectors with the ignition switched off.*

- *If the ABS warning light illuminates, there is a fault in the system. Determine the cause of the malfunction using the On-Board Diagnostic (OBD) program. Check the fault memory with a suitable diagnostic scan tool before proceeding.*

- *Certain faults will only be detected and stored as DTCs at speeds greater than 20 kph (13 mph).*

- *The ABS and brake system warning lights are not capable of monitoring all brake system functions; visual inspections are still required for the system.*

- *Brake fluid must not contact any painted surfaces; it will remove paint.*

- *Brake fluid absorbs moisture from the surrounding air and must be replaced every two years. Use only new brake fluid that complies with Federal Motor Vehicle Safety Standards 116, DOT 4.*

- *Do not use silicone-based brake fluids (DOT 5). This fluid is incompatible with the brake system and even the smallest trace can cause severe corrosion.*

- *Absolute cleanliness is required when working on the ABS system. Do not use any products that contain mineral oil such as oil, grease, etc.*

- *Thoroughly clean all connections and the area immediately around them before loosening. Do not use aggressive cleaning agents such as brake cleaner, gasoline or thinners.*

- *Place removed parts on a clean surface and cover them. Only use clean plastic covering or paper. Do not use rags that may contain lint.*

- *Carefully seal or cover opened components if repairs cannot be completed immediately.*

- *After separating the control module from the hydraulic unit, use the transportation protection for the valve dome.*

- *Only install replacement parts from original packaging. Only remove the parts from the package immediately before installation.*

- *Do not work with compressed air nearby whenever the hydraulic system is open. Do not move the vehicle with the hydraulic system open.*

- *Be very careful not to let any brake fluid enter any wiring harness connectors.*

- *Do not re-use fasteners that are worn or deformed from normal use.*

ABS COMPONENTS

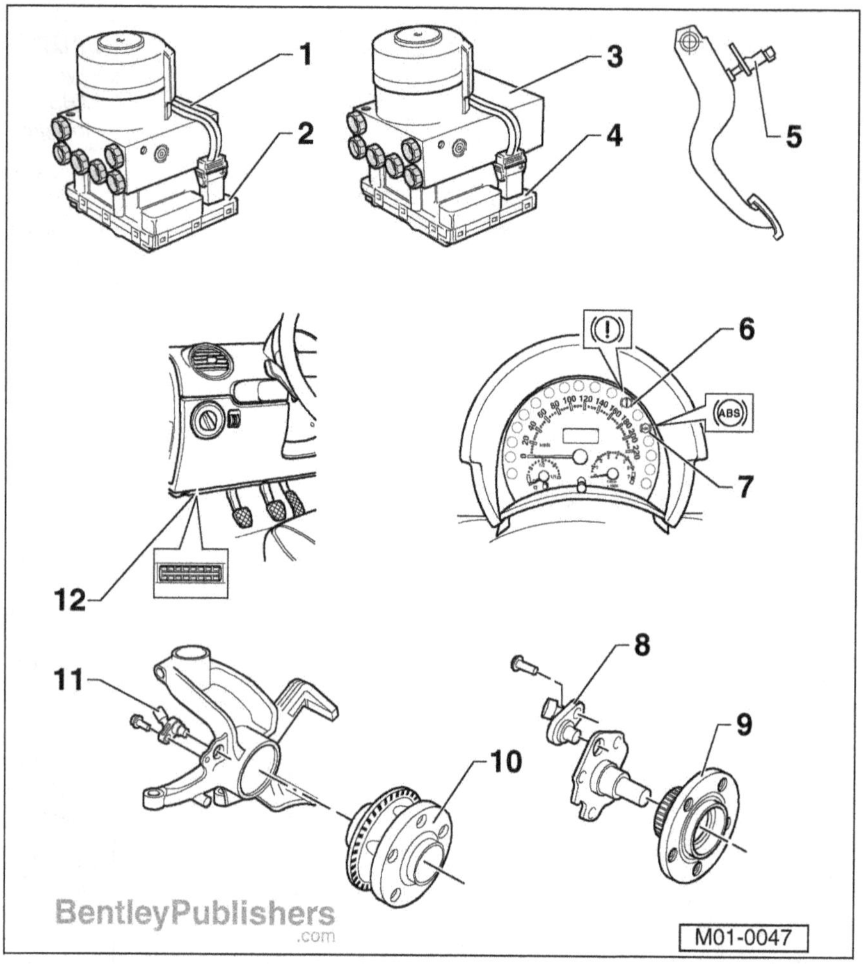

For details on braking system components, see **46 Brakes–Mechanical Components** or **47 Brakes–Hydraulic System.**

NOTE—

Volkswagen identifies electrical components by a letter and/or a number in the electrical schematics. See **EWD Electrical Wiring Diagrams**. *These electrical identifiers are listed in parenthesis as an aid to electrical troubleshooting.*

ABS electrical components (ABS, ABS/EDL, ABS/EDL/ASR, Mark 20 IE, through 09/00)

1. **ABS hydraulic unit (N55)**
   - For ABS only systems
   - With ABS hydraulic pump (V64)
   - With valve block containing inlet/outlet valves
   - If exchanging, use plugs from new unit to seal openings
2. **ABS control module (J104)**
   - Mounted to underside of hydraulic unit
   - New control modules (replacement parts) are not coded and must be coded after installation
   - Disconnect harness connector only with ignition switched off
   - Check for Diagnostic Trouble Codes (DTCs) with scan tool before disconnecting harness connector
3. **ABS hydraulic unit (N55)**
   - For ABS/EDL or ABS/EDL/ASR system
   - With ABS hydraulic pump (V64)
   - With valve block containing inlet/outlet valves
   - If exchanging, use plugs from new unit to seal openings
4. **ABS/EDL or ABS/EDL/ASR control module (J104)**
   - Mounted to underside of hydraulic unit
   - New control modules (replacement parts) are not coded and must be coded after installation
   - Disconnect harness connector only with ignition switched off
   - Check for Diagnostic Trouble Codes (DTCs) with scan tool before disconnecting harness connector
5. **Brake light switch (F)**
   - See **Brake light switch** in this repair group
6. **Brake system warming light (K118)**
   - See **Warning light functions**
7. **ABS warning light (K47)**
   - See **Warning light functions**
8. **ABS wheel speed sensor, rear**
   - Clean inner surface in wheel bearing housing before installing and lubricate with grease (VW part no. G 000 650 or equivalent)
   - Check to be sure wiring is not twisted in wheel well before connecting
   - Right rear (G44)
   - Left rear (G46)
9. **Rear wheel hub**
   - With ABS wheel speed sensor rotor welded to hub
10. **Front wheel hub**
    - With ABS wheel speed sensor rotor welded to hub
11. **ABS wheel speed sensor, front**
    - Clean inner surface in wheel bearing housing before installing and lubricate with grease (VW part no. G 000 650 or equivalent)
    - Check to be sure wiring is not twisted in wheel well before connecting
    - Right front (G45)
    - Left front (G47)
12. **Data link connector**
    - Located below left side of dashboard

ABS electrical components (ABS/EDL/ASR/ESP, Mark 20 IE, through 09/00)

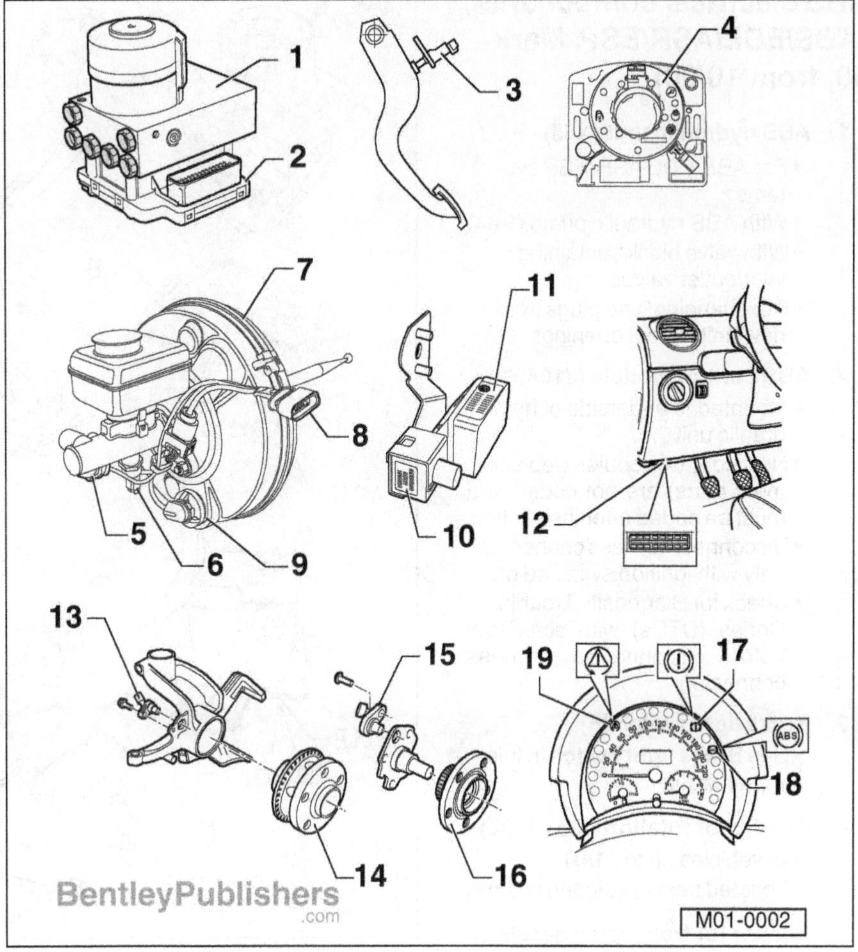

1. **ABS hydraulic unit (N55)**
   - For ABS/EDL/ASR/ESP systems
   - With ABS hydraulic pump (V64)
   - With valve block containing inlet/outlet valves
   - If exchanging, use plugs from new unit to seal openings
2. **ABS control module (J104)**
   - Mounted to underside of hydraulic unit
   - New control modules (replacement parts) are not coded and must be coded after installation
   - Disconnect harness connector only with ignition switched off
   - Check for Diagnostic Trouble Codes (DTCs) with scan tool before disconnecting harness connector
3. **Brake light switch (F)**
   - See **Brake light switch** in this repair group
4. **Steering angle sensor (G85)**
   - Located on steering column, under steering wheel and turn signal switch
5. **Brake pressure sensor 1 (G201)**
6. **Brake pressure sensor 2 (G214)**
7. **Brake booster**
   - With brake recognition switch for ESP (F83)
   - With magnetic coil for brake pressure (N247), in brake booster
8. **14-pin connector**
9. **Cable bushing**
10. **Sensor for transverse acceleration (G200)**
11. **Sender for rotation rate (G202)**
12. **Data link connector**
    - Located below left side of dashboard
13. **ABS wheel speed sensor, front**
    - Clean inner surface in wheel bearing housing before installing and lubricate with grease (VW part no. G 000 650 or equivalent)
    - Check to be sure wiring is not twisted in wheel well before connecting
    - Right front (G45)
    - Left front (G47)
14. **Front wheel hub**
    - With ABS wheel speed sensor rotor welded to hub
15. **ABS wheel speed sensor, rear**
    - Clean inner surface in wheel bearing housing before installing and lubricate with grease (VW part no. G 000 650 or equivalent)
    - Check to be sure wiring is not twisted in wheel well before connecting
    - Right rear (G44)
    - Left rear (G46)
16. **Rear wheel hub**
    - With ABS wheel speed sensor rotor welded to hub
17. **Brake system warming light (K118)**
    - See **Warning light functions**
    - USA vehicles have "BRAKE" inscription instead of symbol
18. **ABS warning light (K47)**
    - See **Warning light functions**
    - USA vehicles have "ABS" inscription instead of symbol
19. **ASR/ESP control lamp (K155)**

ABS electrical components (ABS/EDL/ASR/ESP, Mark 60, from 10/00)

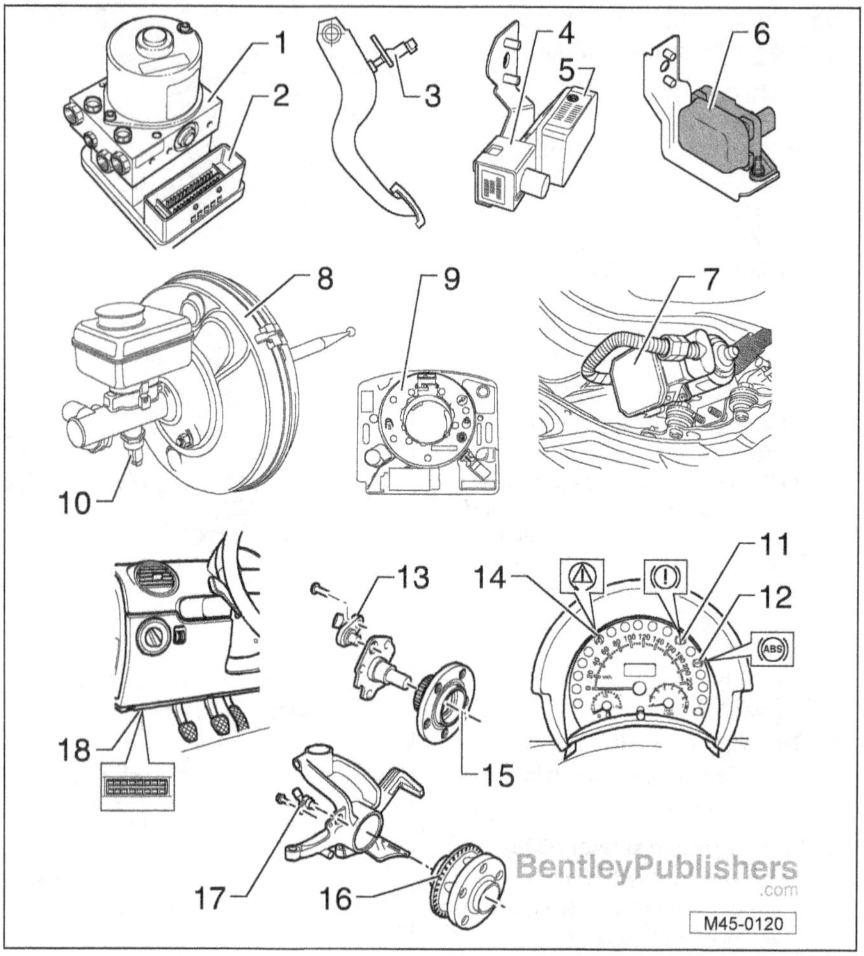

1. **ABS hydraulic unit (N55)**
   - For ABS/EDL/ASR/ESP systems
   - With ABS hydraulic pump (V64)
   - With valve block containing inlet/outlet valves
   - If exchanging, use plugs from new unit to seal openings
2. **ABS control module (J104)**
   - Mounted to underside of hydraulic unit
   - New control modules (replacement parts) are not coded and must be coded after installation
   - Disconnect harness connector only with ignition switched off
   - Check for Diagnostic Trouble Codes (DTCs) with scan tool before disconnecting harness connector
3. **Brake light switch (F)**
   - See **Brake light switch** in this repair group
4. **Sender for rotation rate (G202)**
   - In vehicles up to 11/01
   - Located next to steering column
5. **Sensor for transverse acceleration (G200)**
   - In vehicles up to 11/01
   - Located next to steering column
6. **ESP sensor unit (G419)**
   - In vehicles from 12/01
   - Located next to steering column
   - Combines G200 and G202 into one unit
7. **Brake system vacuum pump (V192)**
   - Only in vehicles with gasoline engine and automatic transmission
   - Located at left mounting
8. **Brake booster**
9. **Steering angle sensor (G85)**
   - Located on the steering column, under steering wheel and turn signal switch
10. **Brake pressure sensor 1 (G201)**
11. **Brake system warming light (K118)**
    - See **Warning light functions**
    - USA vehicles have "BRAKE" inscription instead of symbol
12. **ABS warning light (K47)**
    - See **Warning light functions**
    - USA vehicles have "ABS" inscription instead of symbol
13. **ASR/ESP control lamp (K155)**
14. **ABS wheel speed sensor, rear**
    - Clean inner surface in wheel bearing housing before installing and lubricate with grease (VW part no. G 000 650 or equivalent)
    - Check to be sure wiring is not twisted in wheel well before connecting
    - Right rear (G44)
    - Left rear (G46)
15. **Rear wheel hub**
    - With ABS wheel speed sensor rotor welded to hub
16. **Front wheel hub**
    - With ABS wheel speed sensor rotor welded to hub
17. **ABS wheel speed sensor, front**
    - Clean inner surface in wheel bearing housing before installing and lubricate with grease (VW part no. G 000 650 or equivalent)
    - Check to be sure wiring is not twisted in wheel well before connecting
    - Right front (G45)
    - Left front (G47)
18. **Data link connector**
    - Located below left side of dashboard

Brake light switch

The brake light switch is located above the brake pedal. Different style brake light switches have been used. Be sure to check with an authorized Volkswagen parts specialist for the latest parts information. Part numbers are provided below for reference only.

Application
- Early style brake light switch
  part number 191 945 515 B......... through 2001
- Later style brake light switch
  part number 1J0 945 511 D............from 2001
  part number 1J0 945 511 E.......... 2006 - 2007
  part number 1C0 945 511 A
  (green switch with white tip plunger) .....from 2007

Early style switches can be identified by their large (approx. 10 mm) plunger tip button and square plunger shaft. Later style switches can be identified by their small (approx. 4 mm) plunger tip and round plunger shaft.

Removing

– Remove trim panel above pedal assembly.

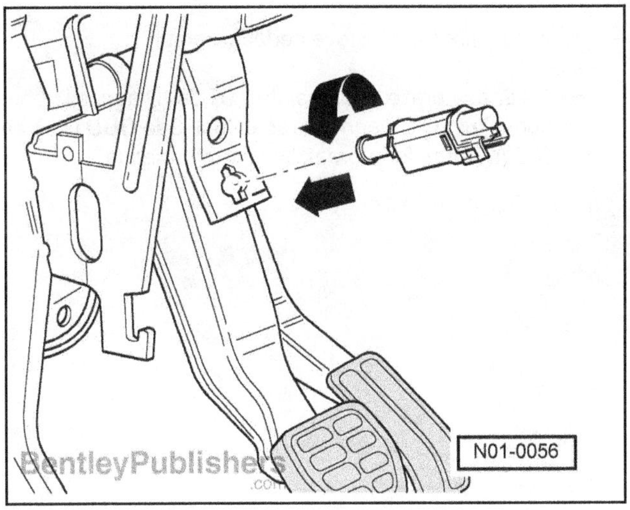

N01-0056

◄ Disconnect harness connector and rotate brake switch counterclockwise to remove.
- Early style switches (though 2001) rotate 90°
- Later style switches (from 2001) rotate 45°

Installing/adjusting (early style brake switch)

– Apply a light coat of lubricant (G 052 142 A2) to end of switch plunger.

– Pull switch plunger out fully.

– Push brake pedal down fully by hand and hold.

– Install switch in brake pedal assembly and rotate 90° clockwise to lock in place.

– Slowly release brake pedal to stop.

– Connect harness connector to switch and check function.

– Install trim panel above pedal assembly.

– Clear any brake pedal related DTCs (Diagnostic trouble codes) using a diagnostic scan tool. See OBD On Board Diagnostics for more scan tool information.

Installing/adjusting (later style brake switch)

NOTE —
- *Some later style brake switches are not re-usable, once they are removed. If switch is defective, replace with latest version.*
- *The latest style switch (green with white tip, part # 1C0 945 511 A) is reusable. Installation instructions are provided below for this switch.*

– Apply a light coat of lubricant (G 052 142 A2) to end of switch plunger.

– Leave pedal in rest position.
  - Do not depress pedal or pull up.

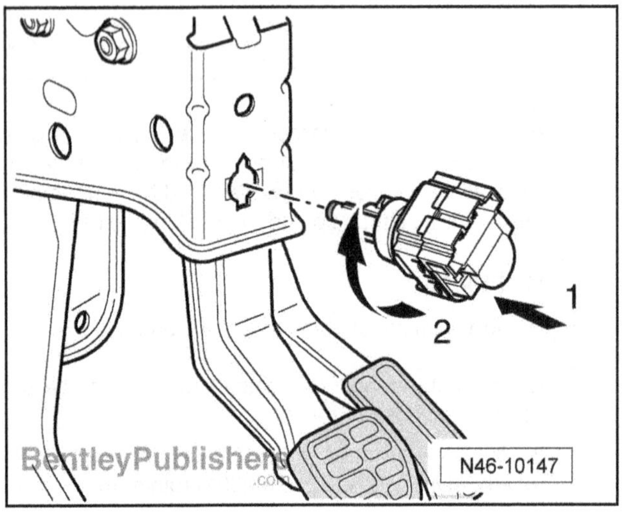

< Slowly insert switch fully into mounting hole in direction of **arrow 1**.
  - Switch only fits in one way.

– While inserting switch, press plunger against brake pedal to adjust position.

– Rotate switch 45° clockwise (**arrow 2**) to lock in position.

– Connect harness connector to switch and check function.

– Install trim panel above pedal assembly.

– Clear any brake pedal related DTCs (Diagnostic trouble codes) using a diagnostic scan tool. See **OBD On Board Diagnostics** for more scan tool information.

Warning light functions

NOTE —

Not all warning lights described below are installed in all vehicles.

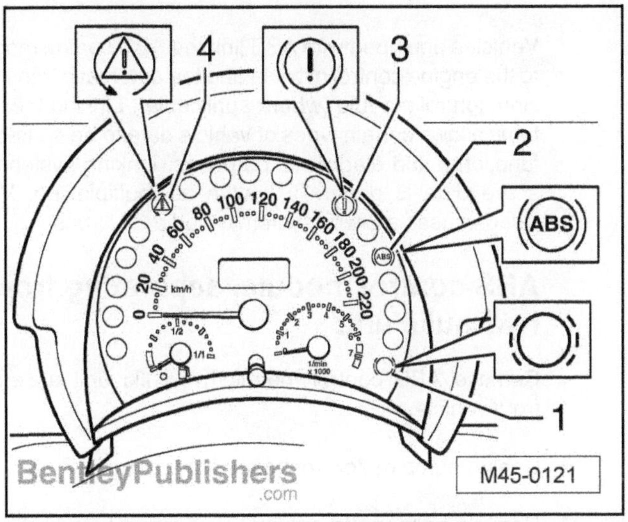

M45-0121

◀ Brake system warning light functions:

1. Brake pad wear indicator lamp (K32)
- From 10/00 production
- Should light briefly upon starting vehicle
- If light stays on, or comes on during driving, brake pads may be worn. Check brake pad thickness, see **03 Maintenance**.

2. ABS warning lamp (K47)
- Should light briefly upon starting vehicle
- If light stays on, or comes on during driving, the problem could be one or more of the following:
  - Voltage supply to ABS control module less than 10 volts
  - Malfunction with the ABS system
  - ABS wheel speed sensor malfunction
  - Connection from instrument cluster to ABS control module has been interrupted
  - Faulty instrument cluster
- The ABS system can remain switched off if there is a malfunction with the system, but the hydraulic brake system remains fully operational.

3. Brake system warning lamp (K118)
- Upon start of vehicle, if K47 goes off but K118 stays on, the cause could be:
  - Brake fluid level low
  - Malfunction has occurred in the activation of K118
- Upon start of vehicle, if both K47 and K118 stay on, the ABS system if faulty and the EBA (engine braking assist) (if equipped) has failed

CAUTION —

If ABS warning lamp (K47) and brake system warning lamp (K118) are both on after starting vehicle, it is possible that the rear wheels will lock up earlier. Drive with care in this situation.

4. ASR/ESP control lamp (K155)
- Upon start of vehicle, if K155 does not go out the cause could be:
  - Short to positive in ASR/ESP button (E256)
  - Malfunction in the activation of K155
  - Connection from instrument cluster to ABS control module has been interrupted
  - The ASR/ESP system was switched off by the ASR/ESP button (E256)
- If K155 flashes while vehicle is being driven, the system is in controlled (normal) operation
- If K47 and K118 are flashing and K155 is on continuously, the ABS control module has not been coded porperly.

ABS CONTROL MODULE

The ABS control module is attached to the bottom of the ABS hydraulic unit located in the engine compartment. Control unit replacement is only possible if the hydraulic unit has been removed.

Vehicles equipped with ABS link the ABS control module to the engine control module and the automatic transmission control module (where applicable). Linking the systems allows certain types of vehicle data to be shared for functional and diagnostic purposes. Linking systems to share data is generally known as multiplexing. Volkswagen uses a 2-wire system called a CAN-bus.

ABS control module, separating from hydraulic unit

– Remove ABS control module/hydraulic unit assembly from vehicle.

– Unplug pump motor from control module.

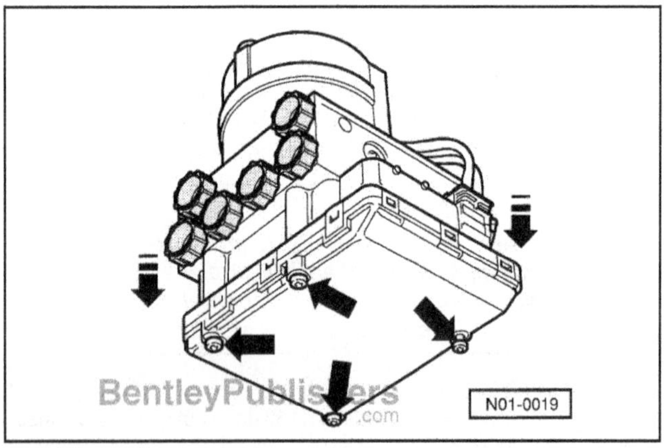

< Remove bolts (**solid arrows**) from control module.

– Pull control module and hydraulic unit straight apart (**dashed arrows**) to separate.

– When installing, always use new attaching bolts.

> ***CAUTION —***
>
> - *Only remove sealing plugs on the new hydraulic unit just before installing the corresponding brake line.*
>
> - *If the sealing plugs are removed too early, brake fluid can escape. It can then no longer be guaranteed that the unit is sufficiently filled or adequately bled.*
>
> - *Cover control module magnetic coils with a plastic covering. Do not use rags.*
>
> - *After separating the control module and hydraulic unit, use transportation protection for the valve dome.*
>
> - *When assembling the control module and hydraulic unit, be sure that the hydraulic unit valve dome is not tilted against the control module solenoid valves.*

Tightening torques

- Control module to hydraulic unit always replace (M4) . . . . 4 Nm (35 in-lb)
- Hydraulic unit bolt to bracket (M6) . . . 8 Nm (71 in-lb)
- Master cylinder nut to booster (M8) . 20 Nm (15 ft-lb)
- Brake lines to ABS unit (M10/M12) . 14 Nm (10 ft-lb)
- Mounting bracket to body (M8). . . . . 20 Nm (15 ft-lb)

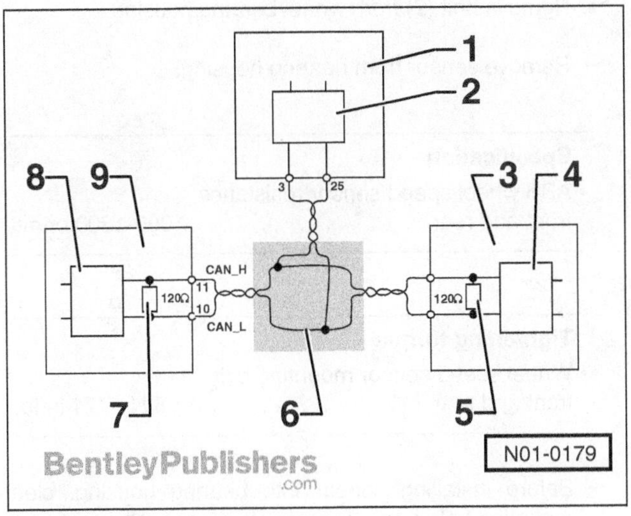

Control module systems

1. Automatic transmission control module (TCM)
2. Communications circuitry
3. Engine control module (ECM)
4. Communications circuitry
5. Terminating impedance
6. CAN-bus multiplexing
7. Terminating impedance
8. Communications circuitry
9. Anti -lock brakes control module (ABS)

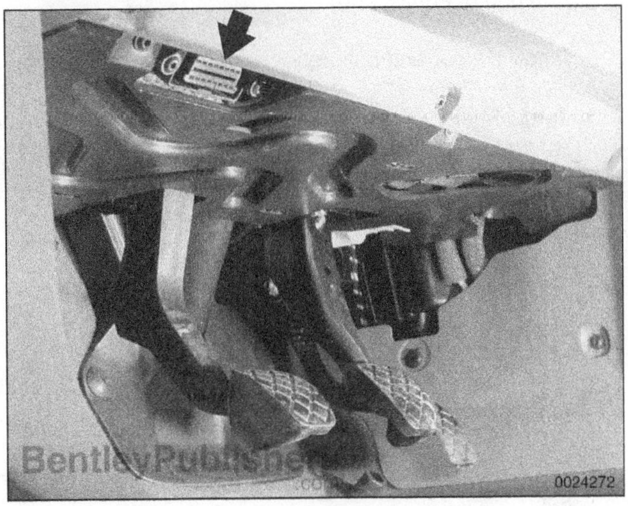

Most service and repair to the ABS system requires the use of a diagnostic scan tool such as the Volkswagen VAG 1551/1552, 5051/5052 or equivalent. Access is via the Data Link Connector (DLC) (**arrow**). This includes coding of replacement control modules. Because of the specialized nature of these repairs, servicing should be referred to an authorized Volkswagen dealer or qualified independent repair shop.

WHEEL SPEED SENSORS

Each wheel has a separate speed sensor to supply data to the ABS control module. Each sensor is also monitored by the control module and malfunctions will cause a DTC to be stored. The system checks the sensors for proper resistance when the ignition is first switched on and again for proper signal when the vehicle is in motion.

Front wheel speed sensor, removing and installing

– Raise car and support with jack stands or lift. See **03 Maintenance** for proper lifting procedure.

> ***WARNING —***
> *Observe all warnings and cautions associated with lifting vehicle in* **03 Maintenance**.

– Disconnect harness connector from wheel speed sensor.

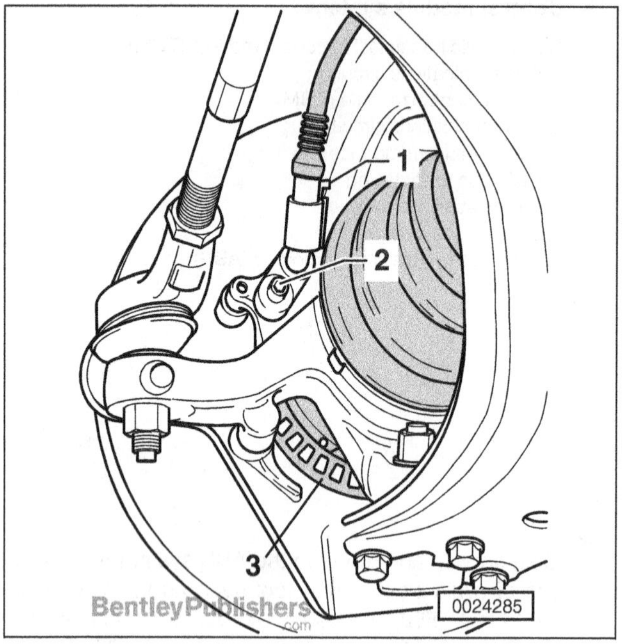

◄ Remove bolt (**2**) from wheel bearing housing.

– Remove sensor from bearing housing.

Specification
- ABS wheel speed sensor resistance, front and rear . . . . 1000-1300 ohms

Tightening torque
- Wheel speed sensor mounting bolt, front and rear, M6. . . . . 8 Nm (71 in-lb)

– Before installing sensor into bearing housing, clean mounting hole and lubricate with grease (VW part no. G 000 650 or equivalent).

– Install sensor and bolt, reconnect harness connector.

– Turn wheels fully to left and then to right. Check sensor wiring for interference and clearance.

Rear wheel speed sensor, removing and installing

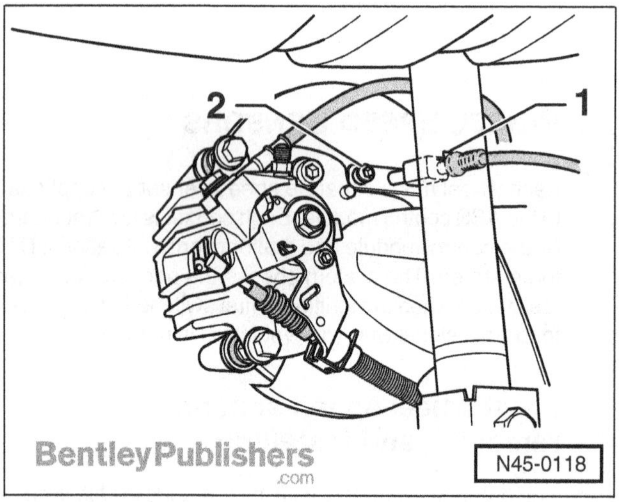

– Raise car and support with jack stands or lift. See **03 Maintenance** for proper lifting procedure.

– Disconnect harness connector from wheel speed sensor.

◄ Remove ABS wheel sensor mounting bolt (**2**) from stub axle.

– Pull ABS wheel speed sensor out of stub axle.

– Before installing sensor into bearing housing, clean mounting hole and lubricate with grease (VW part no. G 000 650 or equivalent).

– Install sensor and bolt, reconnect harness connector.

– Check sensor wiring for interference and clearance around suspension components.

ELECTRONIC STABILIZATION PROGRAM

The Electronic Stabilization Program (ESP) is an optional component of the ABS system that works to stabilize the vehicle in understeering and oversteering situations. By measuring wheel speed (via the ABS wheel speed sensors), input from body mounted sensors (accelerometers), and a steering angle sensor, the control module can communicate with other vehicle systems to reduce the danger of skidding and increasing steering control in dangerous situations.

The ESP system indicates function through a flashing warning lamp on the vehicles instrument panel. If regulating, the ESP warning lamp will flash three times per second.

Rotational rate sensor sender and transverse acceleration sensor, removing and installing

– Disconnect battery.

◄ Remove screws (**1**). Unclip top of upper foot well trim panel (**2**). Remove screws (**3**). Unclip top of side footwell trim panel (**4**).

◄ Remove screws (**arrows**) and lower trim panel.

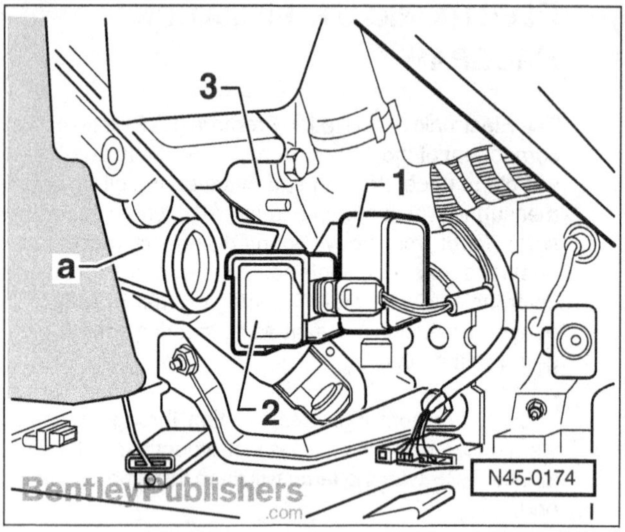

◀ Rotational rate sender (G202) (**1**) and the transverse acceleration sensor (G200) (**2**) are located on the steering column bearing bracket (**3**).

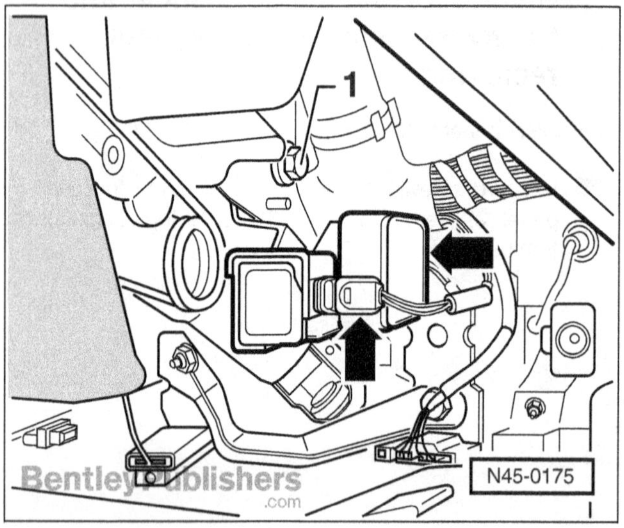

◀ Remove electrical harness connectors (**arrows**) from rotational rate sender, and transverse acceleration sensor.

Rotational rate sender (G202), removing and installing

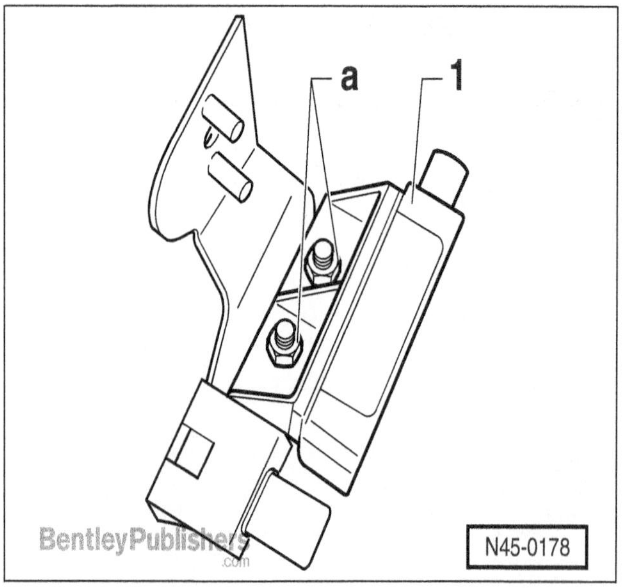

◀ Remove nuts (**A**) from rotational rate sender (**1**) and remove sender.

– To install, insert sender and align straight. Tighten retaining nuts securely.

Tighening torques

- Rotational rate sender to bracket . . . . 9 Nm (71 in-lb)
- Hex bolt to steering column mounting bracket . . . . 20 Nm (15 ft-lb)

CAUTION —
Strong vibrations can damage rotational rate sender.

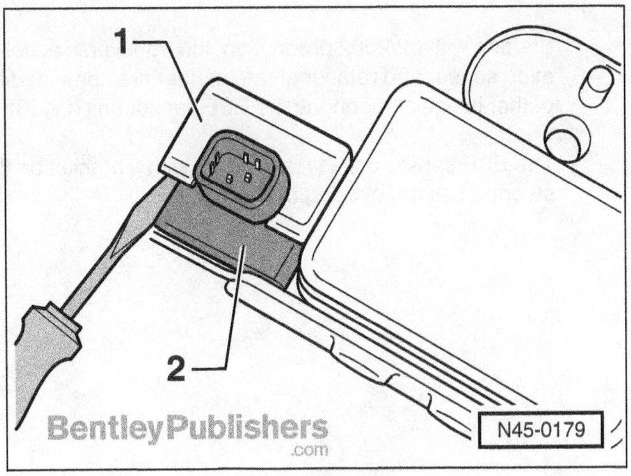

Transverse acceleration sensor, removing and installing

Using a screwdriver, lift retainer (**1**) off transverse acceleration sensor, and remove sensor (**2**).

> ***CAUTION—***
> *Strong vibrations can damage the transverse acceleration sensor.*

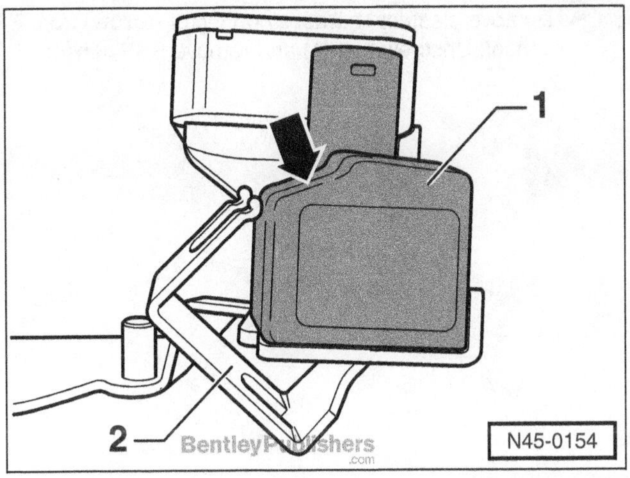

To install transverse acceleration sensor, insert sensor into bracket.

- The slanted side (**arrow**) of the acceleration sensor (**1**) must face retainer (**2**).
- Slide retainer (**2**) over the transverse acceleration sensor (**1**) until it engages.

– Installation is reverse of removal, making sure to perform basic settings of transverse acceleration sensor and rotational rate sender, using a diagnostic scan tool.

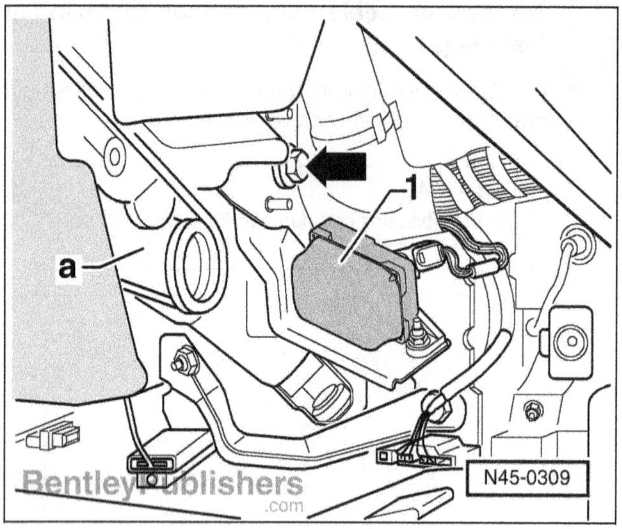

ESP sensor unit, removing and installing (from 02/2002)

Starting with 02/2002 production, the transverse acceleration sensor and rotational rate sender are combined together in one component: the ESP sensor unit (G419).

◀ The ESP sensor unit (**1**) is secured with a bracket on the steering column bearing bracket (**A**).

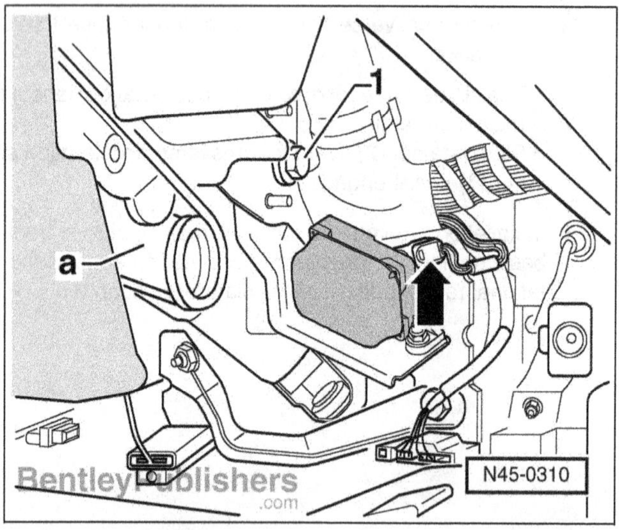

◀ Remove electrical harness connector (**arrow**) from ESP sensor. Unscrew bolt (**1**) and remove ESP sensor.

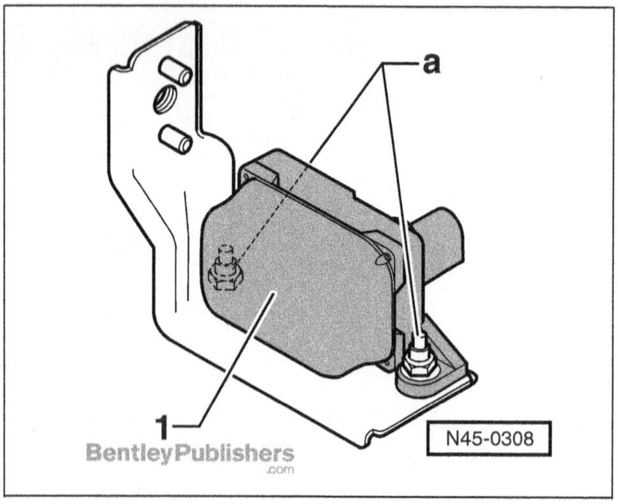

◀ Remove nuts (**A**) from ESP sensor unit and remove unit.

> ***CAUTION —***
> *Sharp shaking motions can destroy the ESP sensor unit.*

– Installation is reverse of removal, making sure to perform basic settings of the ESP sensor unit using a diagnostic scan tool.

NOTE —

When the ESP sensor unit is replaced, a zero compensation of the transverse acceleration sensor and rotational rate sender, must be performed using a diagnostic scan tool.

Tightening torques

- ESP sensor unit to bracket. . . . . 9 Nm (80 in-lb)
- Hex bolt to steering column mounting bracket . . . . 20 Nm (15 ft-lb)

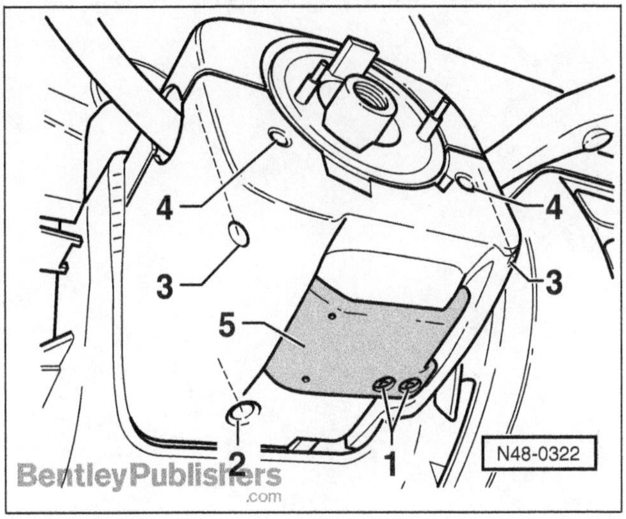

Steering angle sensor, removing and installing

Removing

- Turn wheels to straight ahead position.
- Remove air bag module and steering wheel, see **69 Seatbelts, Airbags**.
- Remove lower storage compartment, and footwell covering, see **70 Trim–Interior**.

◄ Remove bolts (**1**). Remove tilt/slide adjustment lever (**5**). Remove bolts (**2**, **3**, **4**). Take off lower steering column switch trim.

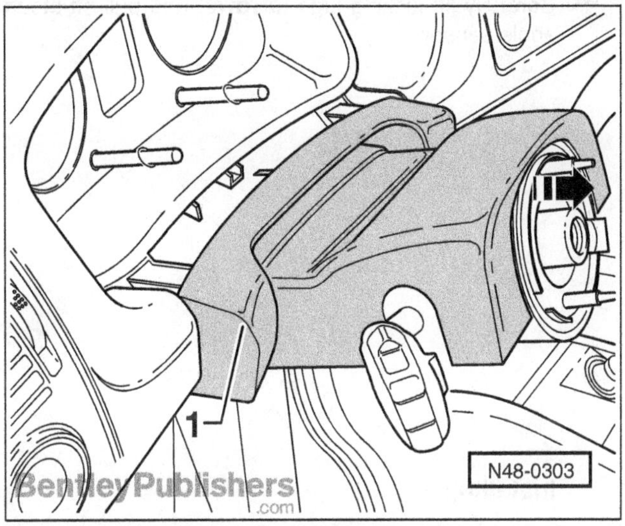

◄ Remove upper steering column switch trim (**1**).

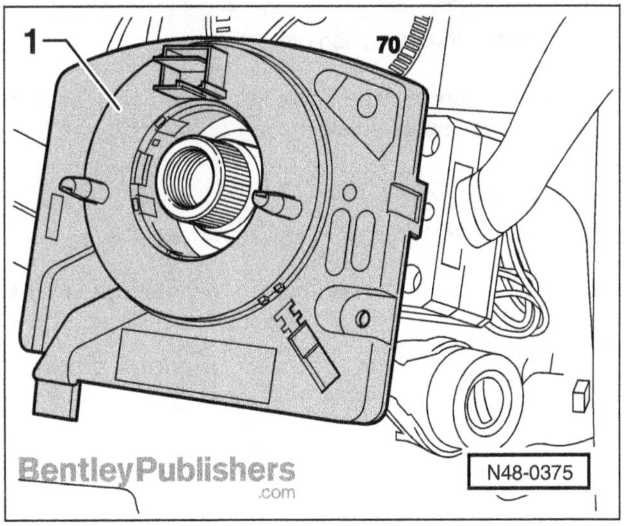

◄ The steering angle sensor (G85) is installed in a housing (**1**) together with the coil connector.

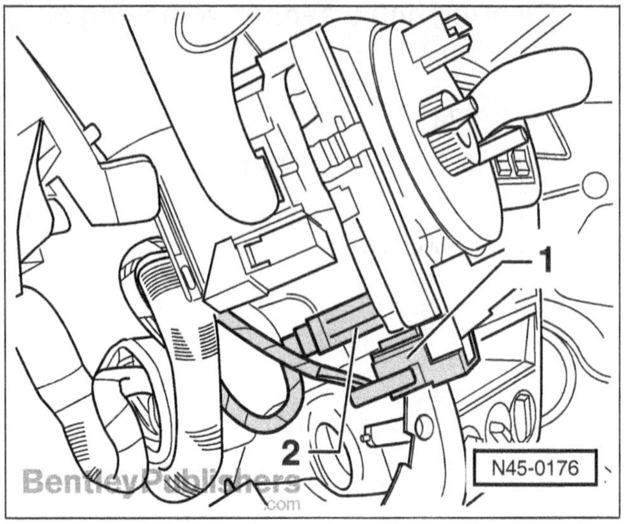

◄ Disconnect harness connectors (**1**) and (**2**).

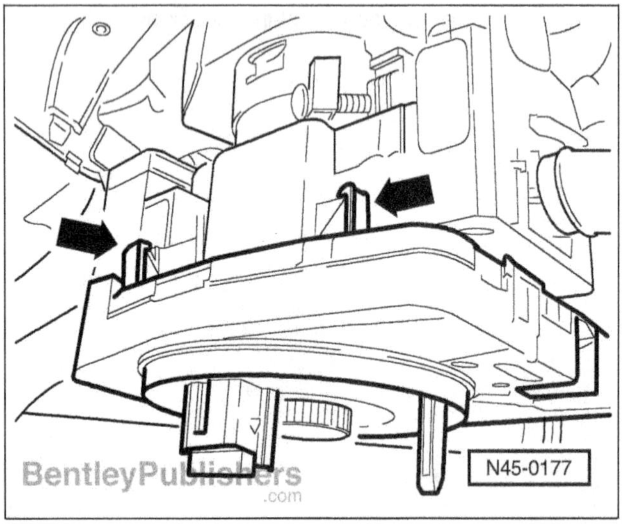

◄ Carefully lift locking lugs (**arrows**) and pull off steering angle sensor.

Installing

- Install steering angle sensor until lugs engage.
- Remove transportation protection.

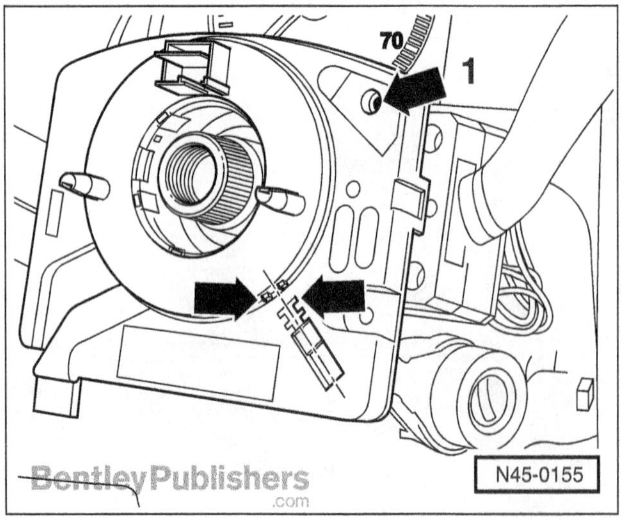

◄ Yellow mark must be visible in hole (**arrow 1**). The marks (**arrows**) must also align.

NOTE —
Make sure it does not move from centered position.

- Installation is reverse order of removal. Make sure to perform steering angle sensor basic settings using a diagnostic scan tool.

46 Brakes–Mechanical Components

GENERAL

This repair groups covers service and repair to the brake friction components; brake pads and rotors. Also included here is parking brake service.

Cars covered by this manual are equipped with disc brakes on all four wheels. Some cars with four-wheel disc brakes are also equipped with anti-lock brakes (ABS).

NOTE —

- *For brake caliper and rear brake wheel cylinder repair information, see* **47 Brakes–Hydraulic System**.
- *For information on the ABS hydraulic unit and ABS service, see* **45 Anti-Lock Brakes (ABS)**.
- *Brake fluid should be flushed from the system every two years. See* **47 Brakes–Hydraulic System**.

WARNING —

- *A properly functioning brake system is essential to safe driving. If the red brake/parking warning light or ABS warning light illuminates while driving, it is imperative that the system be given a thorough check, even if braking action still seems satisfactory. The brakes should be inspected regularly.*
- *Brake fluid is poisonous. Wear safety glasses when working with brake fluid, and wear rubber gloves to prevent brake fluid from entering the bloodstream through cuts or scratches. Do not siphon brake fluid by mouth.*
- *New brake pads and shoes require some break-in. Allow for slightly longer stopping distances for the first 100 to 150 miles of city driving, and avoid hard stops.*

46

CAUTION —
- *All brake work must be done with cleanliness, careful attention to specifications, and proper working procedures. If you lack the skills, the tools, or a clean workplace for servicing the brake system, we suggest you leave these repairs to an authorized VW dealer or other qualified shop.*
- *After replacing brake components, depress the brake pedal firmly several times to seat the brakes in their normal operating position.The pedal should be firm and at its normal height, if not, further work is required before driving vehicle.*
- *Brake fluid is very damaging to paint. Immediately wipe up any brake fluid that spills on the vehicle.*

Brake specifications

The specifications for the various front and rear brake rotors, pads and calipers are provided in the table below.

Table a. Brake specifications

Front brakes	
FS III	Pr. no.: 1ZM, 1ZP
Brake pad thickness wear limit	14 mm 7 mm
Brake disc diameter	280 mm
Brake disc thickness wear limit	22 mm 19 mm
Brake caliper piston diameter	54 mm
FN3	Pr. no.: 1ZD, 1ZE
Brake pad thickness wear limit	14 mm 7 mm
Brake disc diameter	288 mm
Brake disc thickness wear limit	25 mm 22 mm
Brake caliper piston diameter	54 mm
Rear brakes	
C38	Pr. no.: 1KK, 1KS, 1KV
Brake pad thickness wear limit	12 mm 7 mm
Brake disc diameter	232 mm
Brake disc thickness wear limit	9 mm 7 mm
Brake caliper piston diameter	38 mm

Front brake allocation
- FS III. 1.9L, 2.0L
- FN3 . . . . 1.8L, 2.5L

FRONT BRAKES

Two types of front discs are used on the cars covered by this manual. Repair procedures vary depending on caliper/rotor application. See the appropriate (FS III or FN 3) assembly view.

Rotors should be inspected for cracks, scoring, glazing and warpage. Rotors must be replaced (in pairs) if either disc is worn below the minimum thickness specification or if any of the above listed defects are found.

Front brake assembly (FS III)

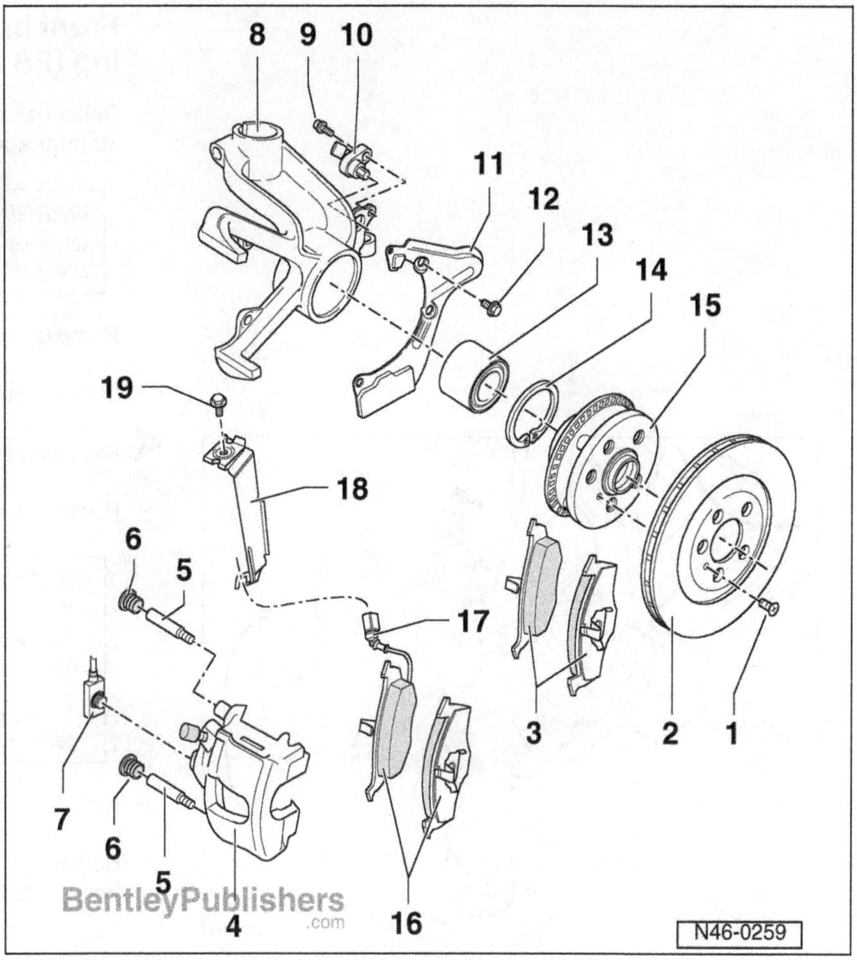

1. **Screw**
   - Tighten to 4 Nm (35 in-lb)
2. **Brake disc**
   - Diameter: 280 mm (11.02 in.)
   - Thickness: 22 mm (0.866 in.)
   - Wear limit: 19 mm (0.748 in.)
   - When worn always replace on both sides
   - Remove brake caliper before removing disc
   - Never remove brake discs from hub by using force. If necessary use penetrating fluid, otherwise brake discs can be damaged.
3. **Brake pads**
   - Through model year 2000
   - Thickness 14 mm (0.551 in.)
   - Wear limit: 7 mm (0.276 in.) including backing plate
   - Checking thickness, see **03 Maintenance**
   - Always replace all pads on one axle at same time
4. **Brake caliper**
   - Do not loosen hydraulic line when replacing pads
5. **Guide pin**
   - Tighten to 28 Nm (21 ft-lb)
6. **Protective cap**
7. **Brake hose with union and banjo bolt**
   - Tighten to 35 Nm (26 ft-lb)
8. **Wheel bearing housing**
9. **Socket-head bolt**
   - Tighten to 8 Nm (71 in-lb)
10. **ABS wheel speed sensor**
    - Before inserting sensor, clean hole inner surface and coat with grease (VW part no. G 000 650)
11. **Splash shield**
12. **Bolt**
    - Tighten to 10 Nm (7 ft-lb)
13. **Wheel bearing**
    - Replace each time after removing
    - Pressing out and in, see **40 Front Suspension and Drive Axles**
14. **Circlip**
15. **Wheel hub with rotor**
    - Removing and installing, see **40 Front Suspension and Drive Axles**
16. **Brake pads**
    - From model year 2001, with wear indicator
    - Thickness 14 mm (0.551 in.)
    - Wear limit: 7 mm (0.276 in.) including backing plate
    - Checking thickness, see **03 Maintenance**
    - Always replace all pads on one axle at same time
17. **Connector**
    - From model year 2001, with wear indicator pads
18. **Bracket**
    - From model year 2001, with wear indicator pads
19. **Bolt**
    - From model year 2001, with wear indicator pads

Front brake pads, removing and installing (FS III)

- Raise car and support with jack stands or lift. See **03 Maintenance** for proper lifting procedure.

> ***WARNING —***
> *Observe all warnings and cautions associated with lifting vehicle in* **0 Maintenance**.

- Remove front wheels.
- Remove guide pin protective caps.

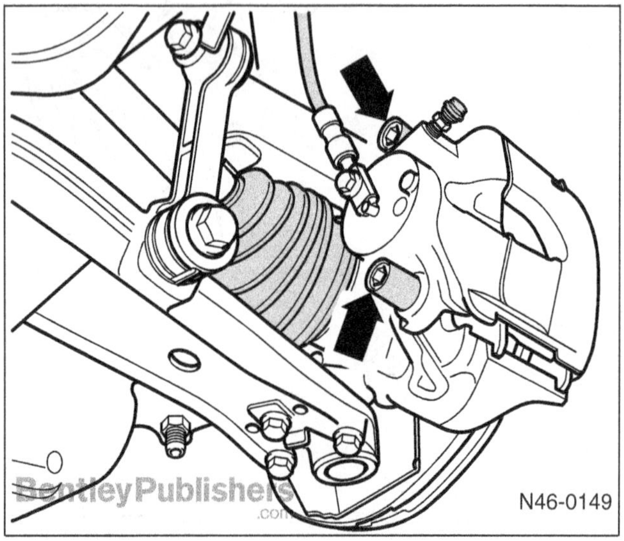

N46-0149

- Remove both guide pins (**arrows**) from brake caliper.
- Remove brake caliper housing and hang up with wire.

> ***CAUTION —***
> - *Do not allow caliper to hang by the brake hose. The unsupported weight can stretch and damage the hose.*
> - *Do not disconnect the brake hose from the caliper when removing brake pads.*

- Remove brake pads from brake caliper housing.
- Before installing new pads extract some of the brake fluid from the brake fluid reservoir.

> ***WARNING —***
> *Brake fluid is poisonous. Never siphon by mouth.*

> ***CAUTION —***
> *A full brake fluid reservoir may allow brake fluid to flow out and cause damage when pistons are pushed back.*

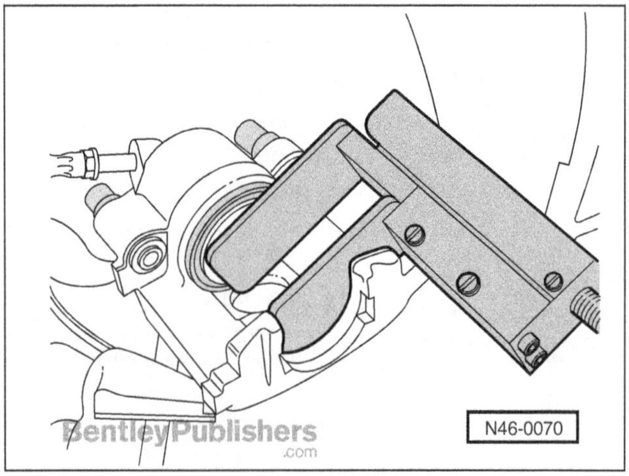

N46-0070

- Press piston into caliper housing.
- Install brake pads with retaining spring in brake caliper housing.

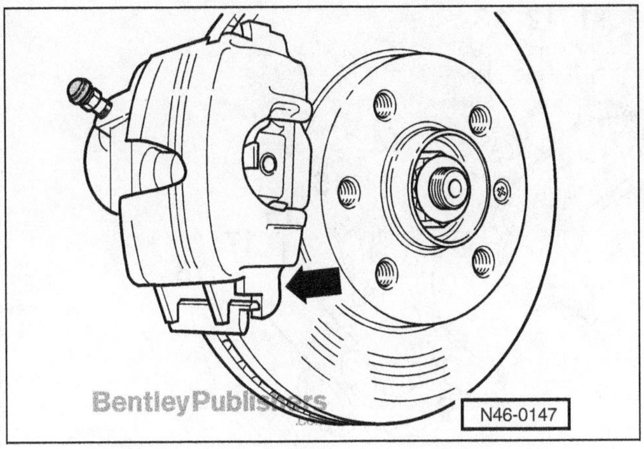

< Install brake caliper onto wheel bearing housing.

- Position lower brake caliper housing on first.
- The brake caliper housing tab (**arrow**) must be behind the wheel bearing housing guide.

– Mount brake caliper housing to brake carrier with both guide pins and install guide pin protective caps.

– Install wheels and lower vehicle.

Tightening torques

- Brake caliper to bearing housing . . . 28 Nm (21 ft-lb)
- Wheel bolt to hub . . . . 120 Nm (87 ft-lb)

– Depress brake pedal firmly several times to seat brake pads to brake disc. Check brake fluid level.

Front brake assembly (FN3)

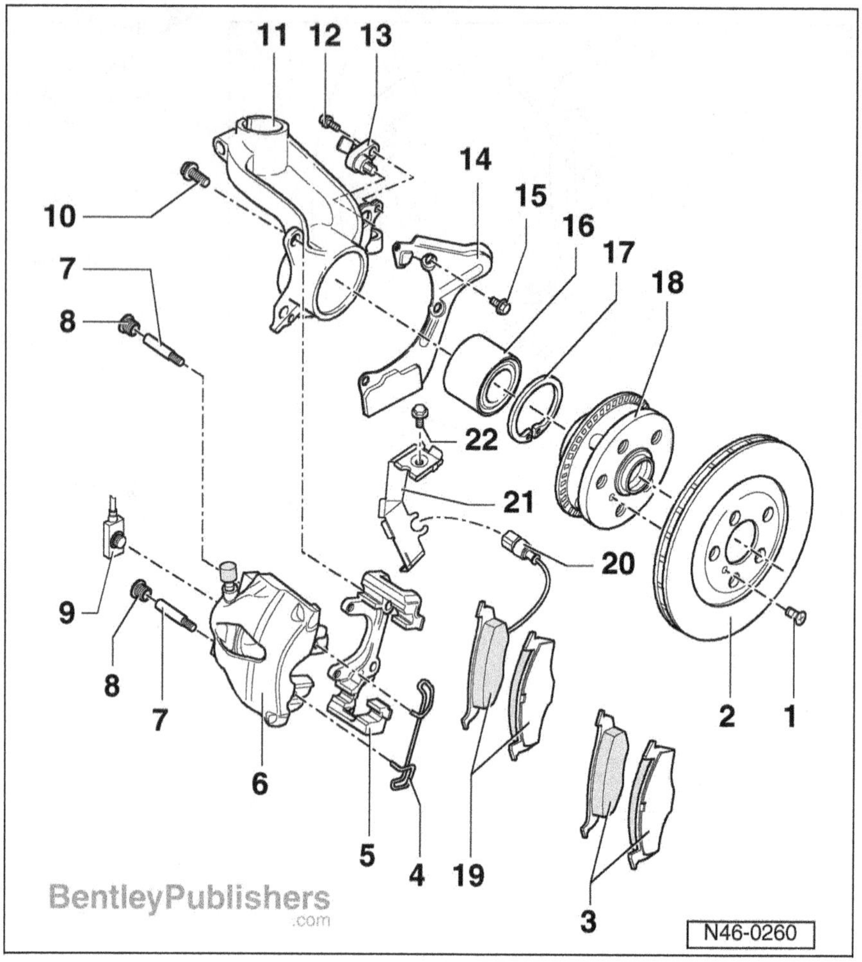

1. **Screw**
   - Tighten to 4 Nm (35 in-lb)
2. **Brake disc**
   - Diameter: 288 mm (11.34 in.)
   - Thickness: 25 mm (0.984 in.)
   - Wear limit: 22 mm (0.866 in.)
   - When worn always replace brake disc pairs on both sides
3. **Brake pads**
   - Through model year 2000
   - Thickness: 14 mm (0.551 in.)
   - Wear limit: 7 mm (0.276 in.) including backing plate
   - Checking thickness, see **03 Maintenance**
   - Always replace all pads on one axle at same time
4. **Retaining spring**
   - Insert in both brake caliper housings
5. **Brake carrier**
   - Supplied as replacement part
   - If protective caps are damaged, use repair kit
   - Use grease packet supplied to lubricate guide pins
6. **Brake caliper housing**
   - Do not disconnect brake hose when changing brake pads
7. **Guide pins**
   - Tighten to 28 Nm (21 ft-lb)
8. **Protective cap**
9. **Brake hose with union and banjo bolt**
   - Tighten to 35 Nm (26 ft-lb)
10. **Ribbed bolt**
    - Tighten to 125 Nm (92 ft-lb)
11. **Wheel bearing housing**
12. **Socket-head bolt**
    - Tighten to 8 Nm (71 in-lb)
13. **ABS wheel speed sensor**
    - Before inserting sensor, clean hole inner surface and coat with grease (VW part no. G 000 650)
14. **Splash shield**
15. **Bolt**
    - Tighten to 10 Nm (7 ft-lb)
16. **Wheel bearing**
    - Replace each time after removing
    - Removing and installing, see **40 Front Suspension and Drive Axles**
17. **Circlip**
18. **Wheel hub with rotor**
    - Removing and installing, see **40 Front Suspension and Drive Axles**
19. **Brake pads**
    - From model year 2001, with wear indicator
    - Thickness 14 mm (0.551 in.)
    - Wear limit: 7 mm (0.276 in.) including backing plate
    - Checking thickness, see **03 Maintenance**
    - Always replace all pads on one axle at same time
20. **Connector**
    - From model year 2001, with wear indicator pads
21. **Bracket**
    - From model year 2001, with wear indicator pads
22. **Bolt**
    - From model year 2001, with wear indicator pads

Front brake pads, removing and installing (FN3)

– Raise car and support with jack stands or lift. See **03 Maintenance** for proper lifting procedure.

> ***WARNING —***
> *Observe all warnings and cautions associated with lifting vehicle in* **03 Maintenance**.

– Remove front wheels.

– Remove guide pin protective caps.

– Using a screwdriver, pry brake pad retaining spring out of brake caliper housing and remove.

> ***WARNING —***
> *Wear appropriate eye protection when removing retaining spring.*

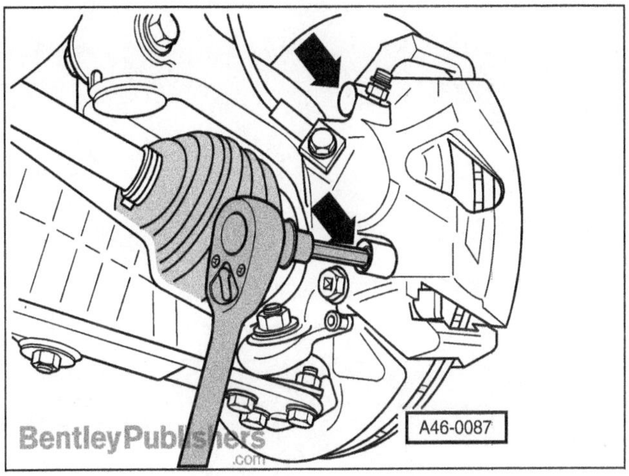

◄ Remove both guide pins (**arrows**) from brake caliper.

– Remove brake caliper housing and hang up with wire.

> ***CAUTION —***
> - *Do not allow caliper to hang by the brake hose. The unsupported weight can stretch and damage the hose.*
> - *Do not disconnect the brake hose from the caliper when removing brake pads.*

– Remove brake pads.

– Clean brake caliper housing, especially bonding surface for brake pad. Surfaces must be free of adhesive or grease residue.

– Before installing new pads extract some of the brake fluid from the brake fluid reservoir.

> ***WARNING —***
> *Brake fluid is poisonous. Never siphon brake fluid by mouth.*

> ***CAUTION —***
> *A full brake fluid reservoir may allow brake fluid to flow out and cause damage when pistons are pushed back.*

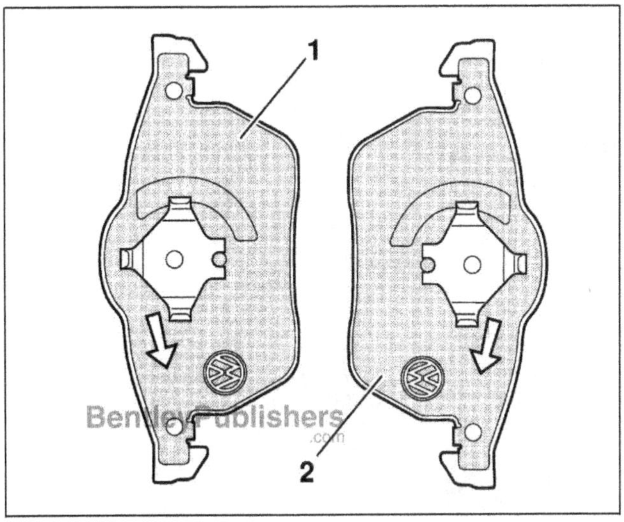

◄ Be sure to install correct brake pads into left and right caliper. **Arrow** on backing plate must point downward when installed in caliper.

– Press piston into caliper housing and install brake pads.

– Install outer brake pad in brake caliper.

– Pull protective foil off outer brake pad backing plate.

– Install brake caliper housing with both guide pins to brake carrier.

– Install guide pin protective caps.

– Insert retaining spring into brake caliper housing.

– Install wheels and lower vehicle.

Tightening torques
- Brake caliper to bearing housing . . . 28 Nm (21 ft-lb)
- Wheel bolt to hub . . . . 120 Nm (87 ft-lb)

– Depress brake pedal firmly several times to seat brake pads to brake disc. Check brake fluid level and add as necessary.

REAR BRAKES

Rear disc brake assembly

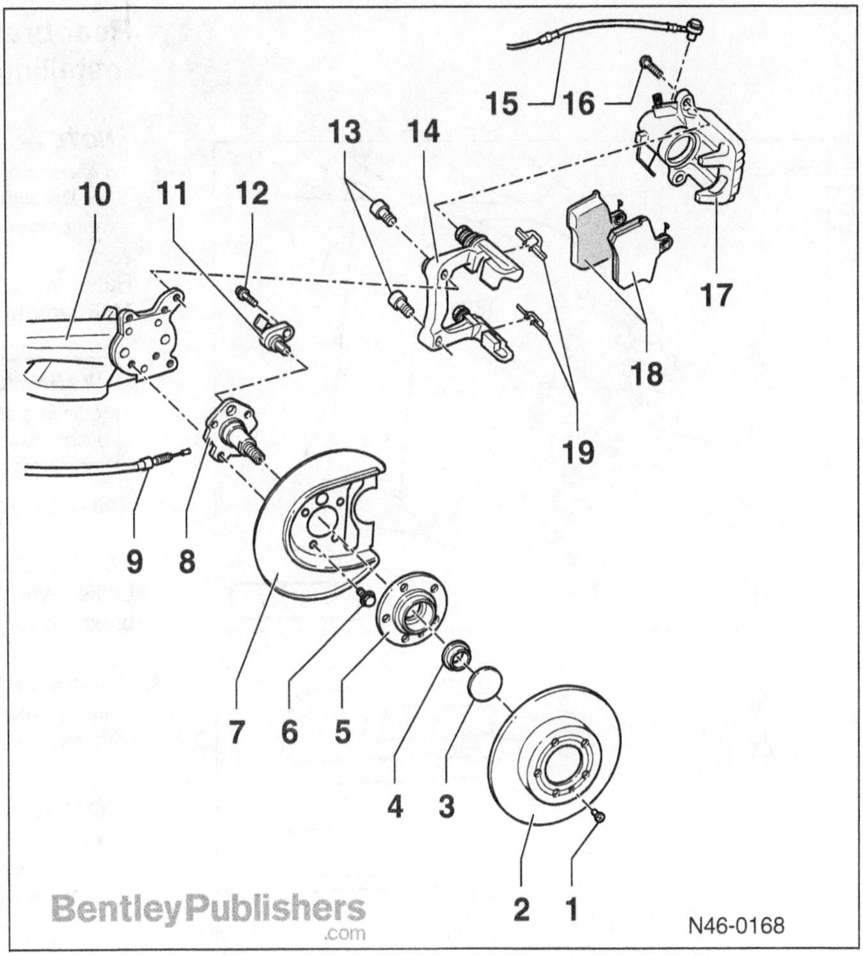

1. **Screw**
   - Tighten to 4 Nm (35 in-lb)
2. **Brake disc**
   - Diameter: 232 mm (9.13 in.)
   - Thickness: 9 mm (0.354 in.)
   - Wear limit: 7 mm (0.275 in.)
   - When worn, always replace on both sides.
3. **Cap**
   - Removing and installing, see **42 Rear Suspension**
4. **Self-locking nut**
   - Tighten to 175 Nm (129 ft-lb)
   - Always replace after removing
5. **Wheel hub with wheel bearing and rotor**
   - Always replace after removing
   - Only replace as complete unit
   - Removing and installing, see **42 Rear Suspension**
6. **Bolt**
   - Tighten to 60 Nm (44 ft-lb)
   - With dished spring washer
7. **Splash shield**
8. **Stub axle**
9. **Parking brake cable**
10. **Axle beam**
11. **ABS wheel speed sensor**
    - Before inserting sensor, clean installation hole inner surface and coat with grease (VW part no. G 000 650)
12. **Socket-head bolt**
    - Tighten to 8 Nm (71 in-lb)
13. **Socket-head bolt**
    - Tighten to 65 Nm (48 ft-lb)
14. **Brake carrier with guide pins and protective cap**
    - Supplied as replacement part assembled with sufficient grease on guide pins
    - If protective caps or guide pins are damaged use repair kit
    - Use grease packet supplied to lubricate guide pins.
15. **Brake hose with union and banjo bolt**
    - With seals
    - Tighten to 35 Nm (26 ft-lb)
    - Do not disconnect brake hose when changing brake pads
16. **Self-locking bolt**
    - Tighten to 35 Nm (26 ft-lb)
    - Always replace
17. **Brake caliper**
    - Adjust parking brake cable after maintenance or replacement.
    - Do not pull parking brake before adjusting it
18. **Brake pads**
    - Thickness: 12 mm (0.472 in.)
    - Wear limit: 7 mm (0.276 in.) including backing plate
    - Checking thickness, see **0 Maintenance**
    - Always replace all pads on one axle at same time
19. **Pad retaining springs**
    - Always replace springs when changing pads

Rear brake pads, removing and installing

NOTE —
If re-using brake pads, mark pad position before removing. Reinstall brake pads in their original position to prevent uneven braking.

– Raise car and support with jack stands or lift. See **03 Maintenance** for proper lifting procedure.

> **WARNING —**
> *Observe all warnings and cautions associated with lifting vehicle in* **03 Maintenance**.

– Remove rear wheels.

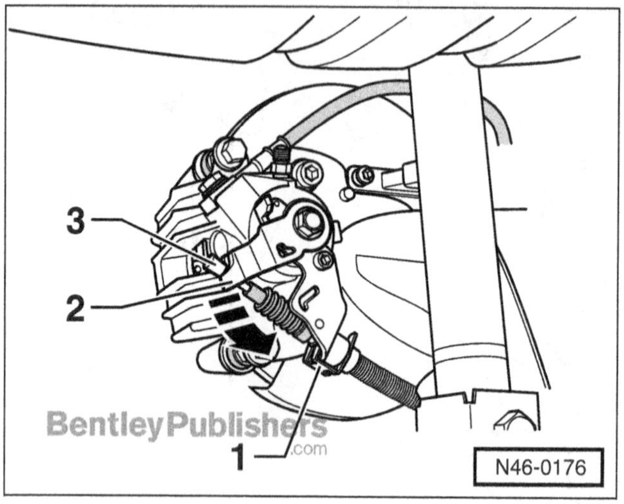

◄ Remove retaining clip (**1**) for parking brake cable. Press brake lever (**2**) downward (**arrow**) and unhook parking brake cable (**3**).

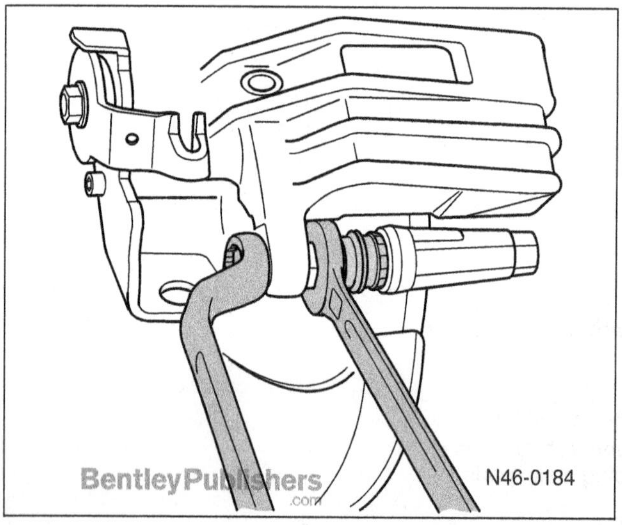

◄ Remove bolts from brake caliper housing while counterholding guide pins. Remove brake caliper and hang up with wire.

> **CAUTION —**
> - *Do not allow caliper to hang by the brake hose. The unsupported weight can stretch and damage the hose.*
> - *Do not disconnect the brake hose from the caliper when removing brake pads.*

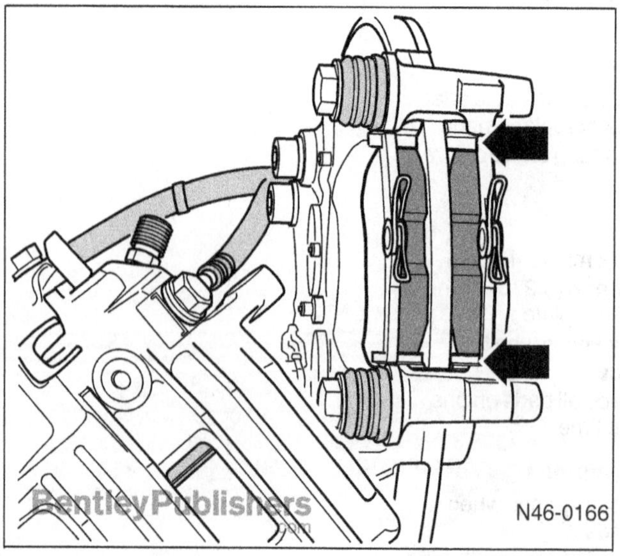

◄ Remove brake pads and pad retaining springs (**arrows**).

– Clean brake caliper housing, especially bonding surface for brake pad. Bonding surfaces must be free of adhesive and grease residues.

– Before installing new pads extract some of the brake fluid from the brake fluid reservoir.

> **WARNING —**
> *Brake fluid is poisonous. Never siphon brake fluid by mouth.*

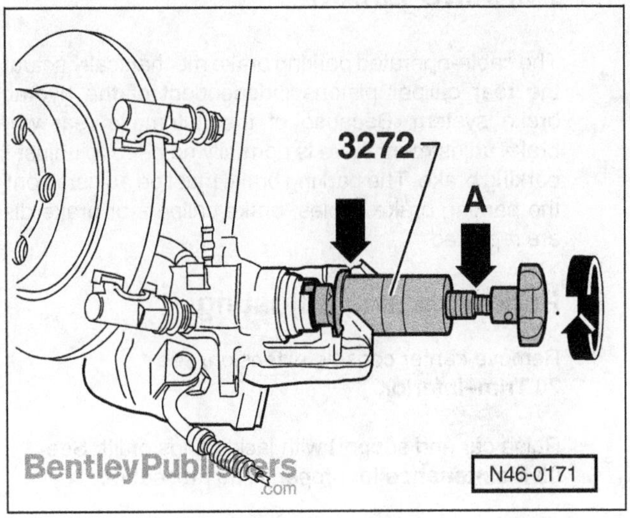

> ***CAUTION—***
> - *A full brake fluid reservoir may allow brake fluid to flow out and cause damage when pistons are pushed back.*
> - *Use a bleeder bottle or plastic bottle which is used only for brake fluid.*

< Reset caliper's automatic adjustment mechanism by turning piston clockwise while pushing in.

> ***CAUTION—***
> - *If the piston is not reset correctly, or if the brake pedal is operated with the caliper removed, the automatic adjustment mechanism will be destroyed.*
> - *Always remove some brake fluid from the reservoir before resetting and pushing the caliper piston in. When the piston is pushed in, fluid is forced up into the reservoir.*

– Pull protective foil off outer brake pad backing plate.

– Insert brake pads and brake pad retaining springs into brake carrier.

– Install brake caliper housing with new self-locking bolts.

> ***WARNING—***
> *Always replace the self-locking caliper mounting bolts with the ones in the repair kit.*

Tightening torque
- Brake caliper to brake carrier . . . . . . 35 Nm (26 ft-lb)

– Connect parking brake cable to lever on caliper and install retaining clip.

– Adjust parking brake if necessary, see **Parking brake, adjusting** given later.

> ***NOTE—***
> *Operate the foot brake first after adjusting the parking brake.*

– Install rear wheels and lower vehicle.

– Depress brake pedal firmly several times to seat brake pads to brake disc. Check brake fluid level and add as necessary.

PARKING BRAKE

The cable-operated parking brake mechanically actuates the rear caliper pistons independent of the hydraulic brake system. Because of the automatic rear wheel brake adjustment there is normally no need to adjust the parking brake. The parking brake must be adjusted only if the parking brake cables, brake calipers or brake discs are replaced.

Parking brake, adjusting

- Remove center console extension, see **70 Trim–Interior**.
- Raise car and support with jack stands or lift. See **03 Maintenance** for proper lifting procedure.

> ***WARNING —***
> *Observe all warnings and cautions associated with lifting vehicle in* **03 Maintenance**.

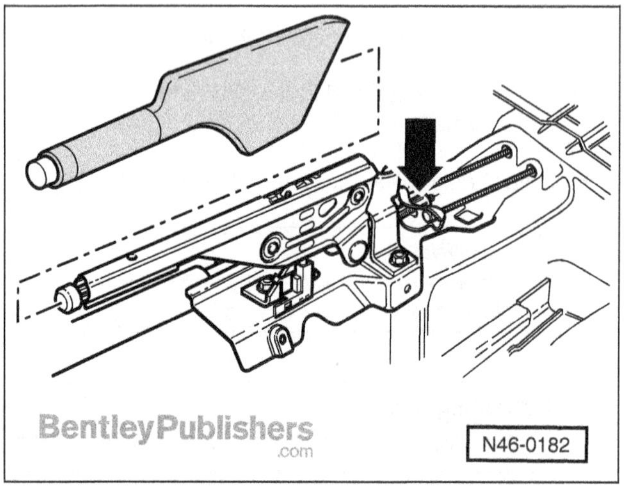

- Release parking brake.
- Firmly depress brake pedal once.
- Pull parking brake lever to fourth notch.
- ◄ Tighten parking brake cable adjusting nut (**arrow**) until both rear wheels are difficult to turn by hand.
- Release parking brake and check that both wheels turn freely. If necessary turn adjusting nut back slightly.
- Install center console extension.
- Lower vehicle.

Parking brake cables, removing and installing

- Remove center console extension, see **70 Trim–Interior**.
- Release parking brake.
- Loosen parking brake adjustment nut until parking brake cable can be unhooked from compensator.
- Raise car and support with jack stands or lift. See **03 Maintenance** for proper lifting procedure.

> ***WARNING —***
> *Observe all warnings and cautions associated with lifting vehicle in* **03 Maintenance**.

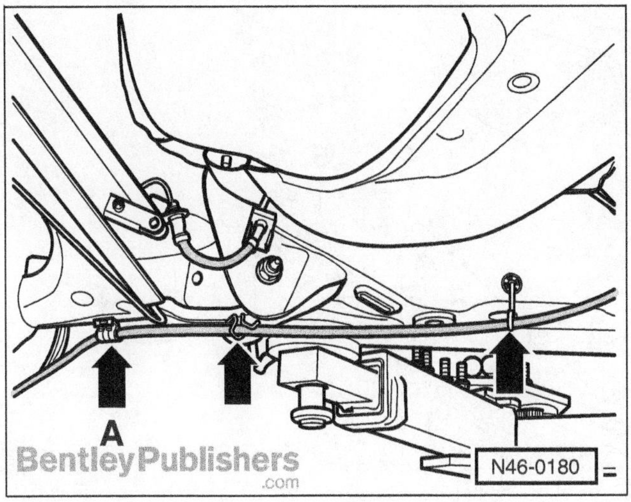

- Remove retaining clip for parking brake cable. Press brake lever downward and unhook parking brake cable as shown earlier.

◀ Unclip parking brake cable from retainers (**arrows** and **arrow A**).

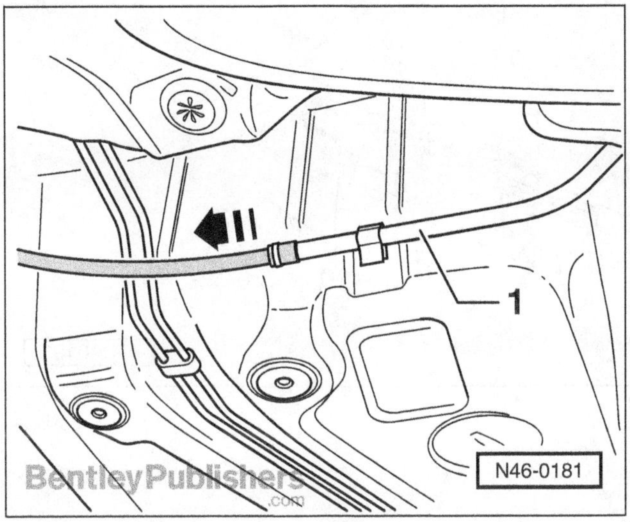

◀ Pull parking brake cable out of guide tube (**1**).

- Begin installation by sliding parking brake cable into guide tube.

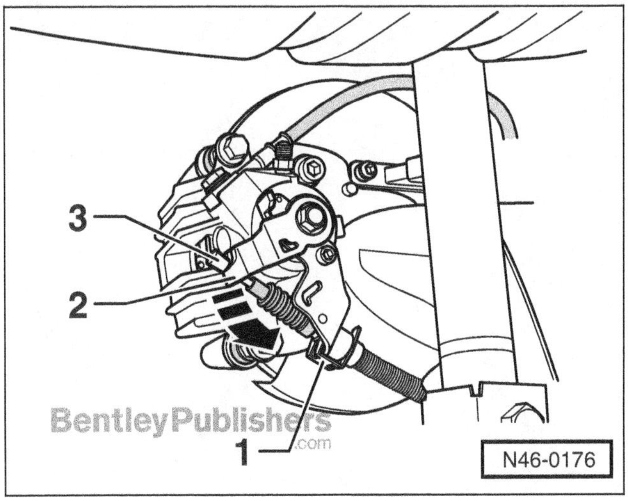

◀ Connect parking brake cable (**3**) to lever (**2**) on caliper and install retaining clip (**1**).

- Clip parking brake cable into retainer on rear axle.

> ***NOTE —***
> *Parking brake cable clamp ring must lie in middle of clip.*

- Hook parking brake cable into remaining retainers.
- Hook parking brake cable into compensator at parking brake lever.
- Install parking brake cable adjustment nut.
- Adjust parking brake, see **Parking brake, adjusting** given earlier.
- Lower vehicle.
- Install center console extension.

Parking brake lever assembly

1. **Parking brake lever**
   - Before removing remove center console
2. **Circlip**
3. **Parking brake lever trim**
   - Pull off toward front
4. **Nuts**
   - Tighten to 25 Nm (18 ft-lb)
5. **Adapter**
   - For vehicles with center armrest
6. **Pull rod**
7. **Compensator**
8. **Adjusting nut**
9. **Fulcrum pin**
   - Holds pull rod (item 6) in parking brake
10. **Parking brake cables**
11. **Parking brake warning light switch**

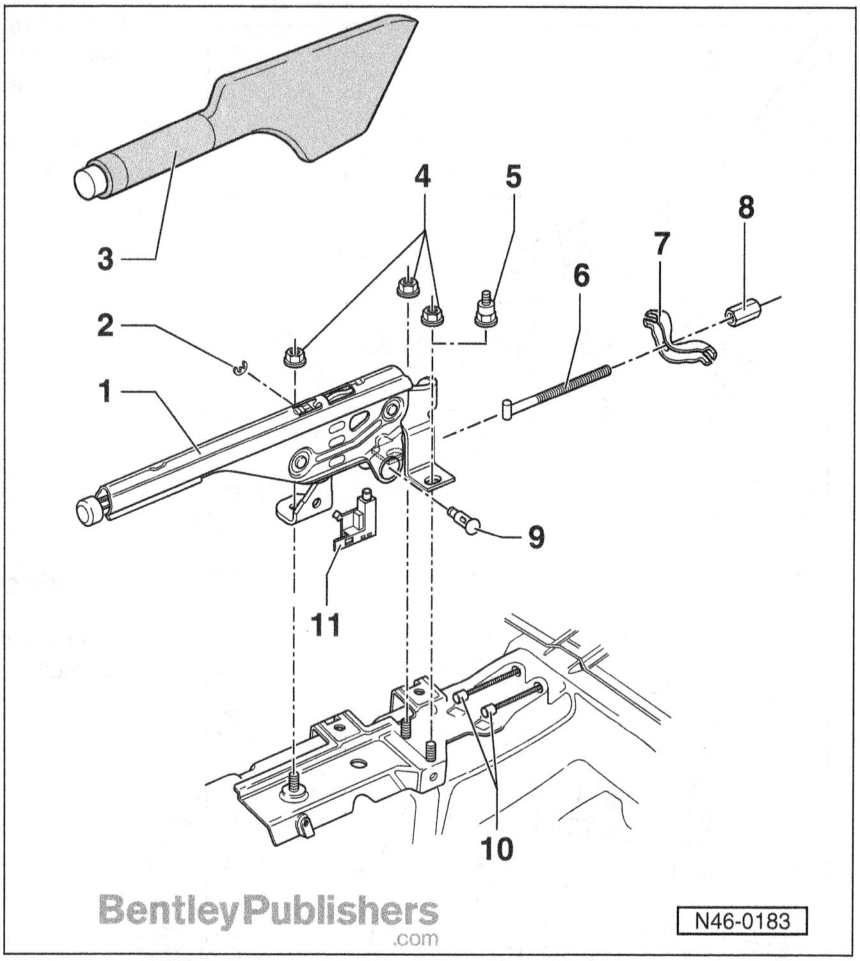

47 Brakes–Hydraulic System

GENERAL

This repair group covers service and repair to the hydraulic and vacuum-assist brake system components. Also included here is brake bleeding. Cars covered by this manual are equipped with front and rear disc brakes. Some cars are also equipped with anti-lock brakes (ABS).

> ***CAUTION —***
> *This repair group does not cover the anti-lock braking system hydraulic unit (with integral master cylinder). For information on the ABS system see* **45 Anti-Lock Brakes (ABS)**.

Brake service precautions

The following warnings and cautions should be read before servicing the brake hydraulic system:

WARNING —

- *A properly functioning brake system is essential to safe driving. If the red brake/parking warning light or the ABS warning light illuminates while driving, it is imperative that the system be given a thorough check, even if braking action still seems satisfactory. The brakes should be inspected regularly.*
- *Brake fluid absorbs moisture from the air and must be replaced every two years. Use only new, approved brake fluid that complies with MVSS 116 DOT 4. Do not use silicone-based brake fluid (DOT 5). Even the smallest traces can cause severe corrosion in the brake system.*
- *Brake fluid is poisonous. Wear safety glasses and rubber gloves when working with brake fluid. Prevent brake fluid from entering the bloodstream through cuts or scratches. Do not siphon brake fluid by mouth.*

CAUTION —

- *Always disconnect battery negative (–) cable when working at or near pedal cluster.*
- *Disconnecting the negative (–) battery cable may erase fault codes and basic settings in the engine management and automatic transmission control modules. Some driveability problems may be noticed until the system re-adapts to operating conditions. OBD II readiness codes, which may be required for emissions testing, may also be erased. Convenience electronics (alarm system, interior light control, power locks, mirrors, and windows) may need to be re-set using a diagnostic scan tool.*
- *Before disconnecting the battery be sure to obtain the anti-theft radio code.*
- *After replacing brake components, depress the brake pedal firmly several times to seat the brakes in their normal operating position. The pedal should be firm and at its normal height, if not, further work is required before driving vehicle.*
- *Brake fluid is very damaging to paint.*

FRONT BRAKE CALIPER

Two types of front calipers are used on the cars covered by this manual. Repair procedures vary depending on caliper application. See the appropriate (FS III or FN3) assembly view.

Front caliper applications
- 1.9L engine .... FS III
- 2.0L engine .... FS III
- 1.8L engine FN3
- 2.5L engine FN3

NOTE—
- *Install all parts in repair kit.*
- *New brake calipers are filled with brake fluid and are pre-bled.*
- *When assembling, apply a thin coat of grease (VW part no. G 052 150 A2) to brake cylinders, pistons and seals.*

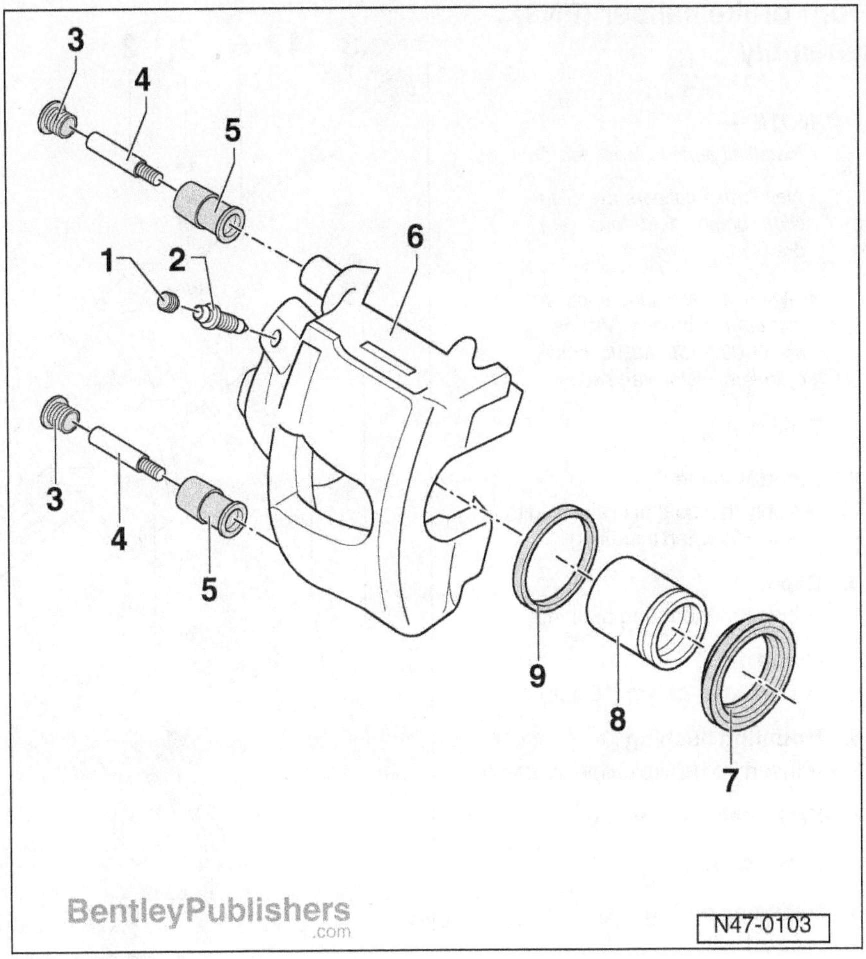

Front brake caliper (FS III) assembly

1. **Dust cap**
2. **Bleeder valve**
3. **Caps**
4. **Guide pins**
   - Tighten to 28 Nm (21 ft-lb)
5. **Mounting bushing**
6. **Brake caliper housing**
7. **Piston seal**
8. **Piston**
9. **Dust seal**
   - Do not damage when inserting piston

Front brake caliper (FN3) assembly

NOTE —
- *Install all parts in repair kit.*
- *New brake calipers are filled with brake fluid and are pre-bled.*
- *When assembling, apply a thin coat of grease (VW part no. G 052 150 A2) to brake cylinders, pistons and seals.*

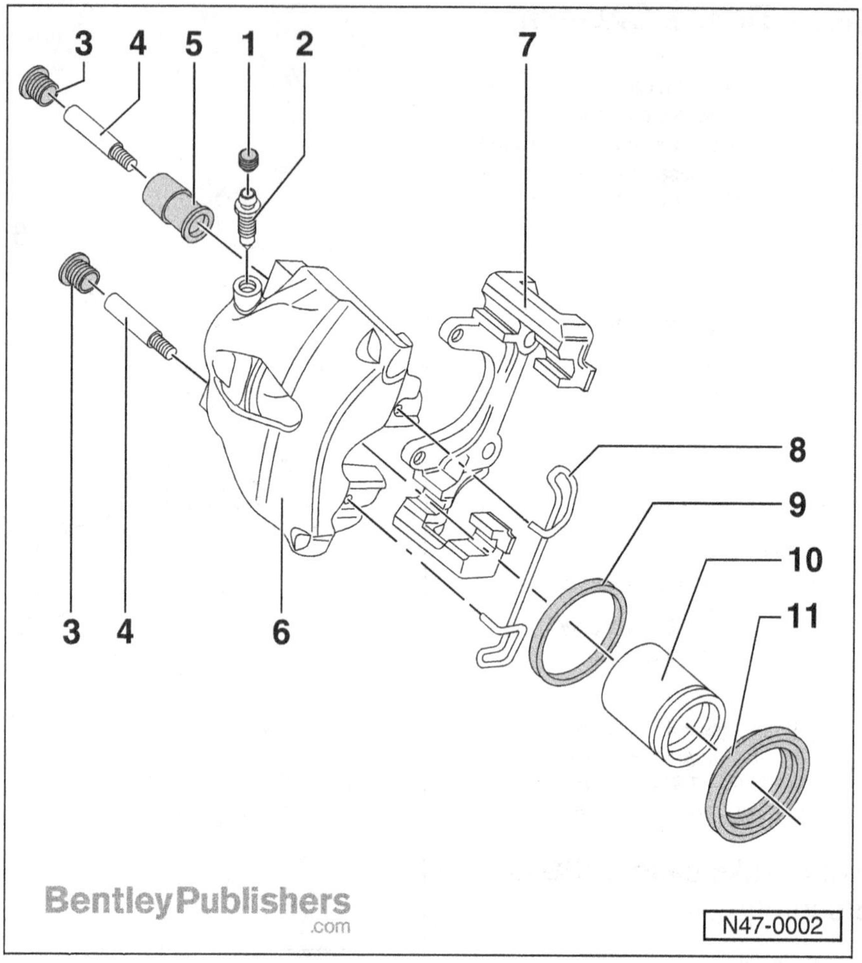

1. **Dust cap**
2. **Bleeder valve**
   - Apply thin coat of brake fluid to threads when installing
3. **Cap**
   - Insert in mounting bushing
4. **Guide pin**
   - Tighten to 25 Nm (18 ft-lb)
5. **Mounting bushing**
   - Insert into brake caliper housing
6. **Brake caliper housing**
7. **Brake carrier**
8. **Retaining spring**
   - Insert with both ends in holes in brake caliper housing
9. **Piston seal**
10. **Piston**
    - Apply thin coat of brake fluid to piston before inserting
    - Piston diameter: 54 mm (2.125 in.)
11. **Dust boot**
    - Do not damage when inserting piston

Front caliper piston, removing and installing

– Remove brake caliper and pads as described in **46 Brakes–Mechanical Components**.

< Force piston out of brake caliper housing using compressed air blown into the brake line bore (**arrow**).

> ***CAUTION —***
> *• Place a piece of wood in the recess to prevent damaging the piston.*
>
> *• Use only enough air pressure to force piston out.*

> ***WARNING —***
> *Always wear safety goggles when working with compressed air.*

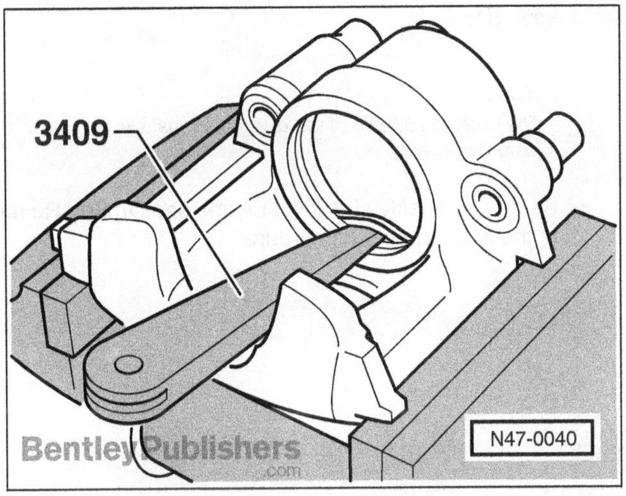

< Carefully remove piston seal using a plastic wedge (special tool **3409** shown).

> ***CAUTION —***
> *When removing, use care to ensure that the cylinder bore is not damaged.*

– Clean caliper and piston with fresh brake fluid.

– Install new piston seal in caliper.

– Lubricate piston and cylinder bore lightly using grease (VW part no. G 052 150 A2).

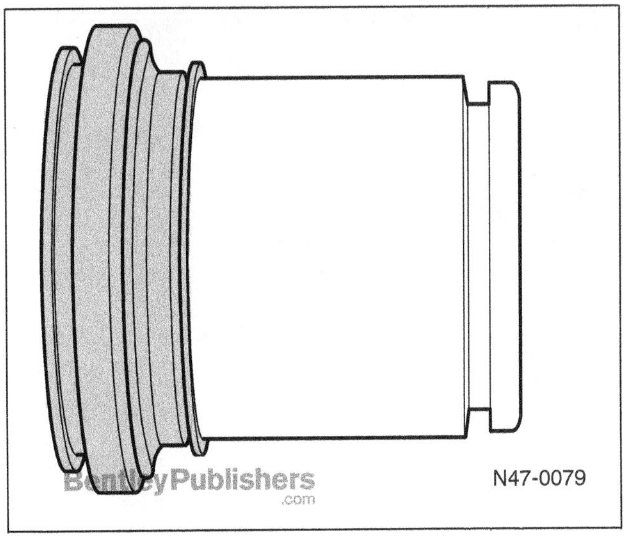

< Install dust boot with outer sealing lip on piston.

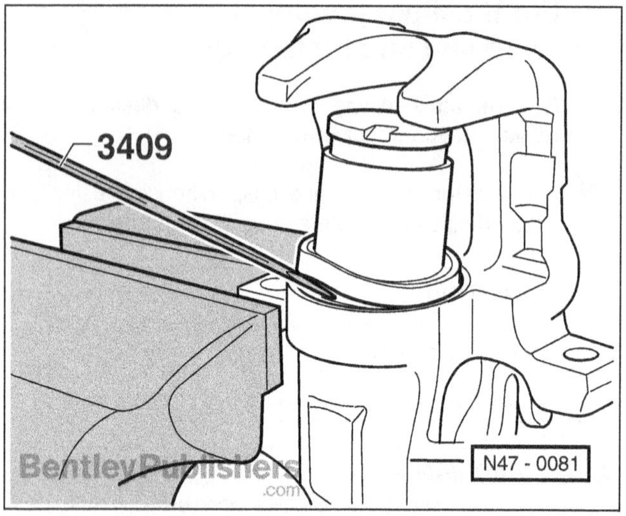

◄ Hold piston in front of caliper and insert inner sealing lip into grove in cylinder using a plastic wedge.

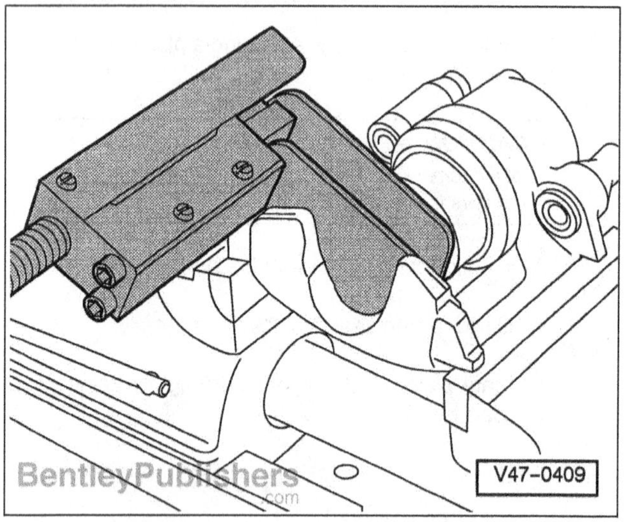

◄ Press piston into brake caliper housing using piston resetting tool.

NOTE —

The outer sealing lip of the dust boot must slip into the piston groove.

– Be sure to quality check your work, see **Quality Review** at the end of this repair group.

REAR BRAKES

Rear brake caliper assembly

NOTE —
- *Install all parts in repair kit.*
- *New brake calipers are filled with brake fluid and are pre-bled.*
- *When assembling, apply a thin coat of grease (VW part no. G052 150 A2) to brake cylinders, pistons and seals.*
- *When repairing, be sure to pre-bleed brake calipers (without brake pads) before installing, see* **Rear caliper piston, removing and installing.**

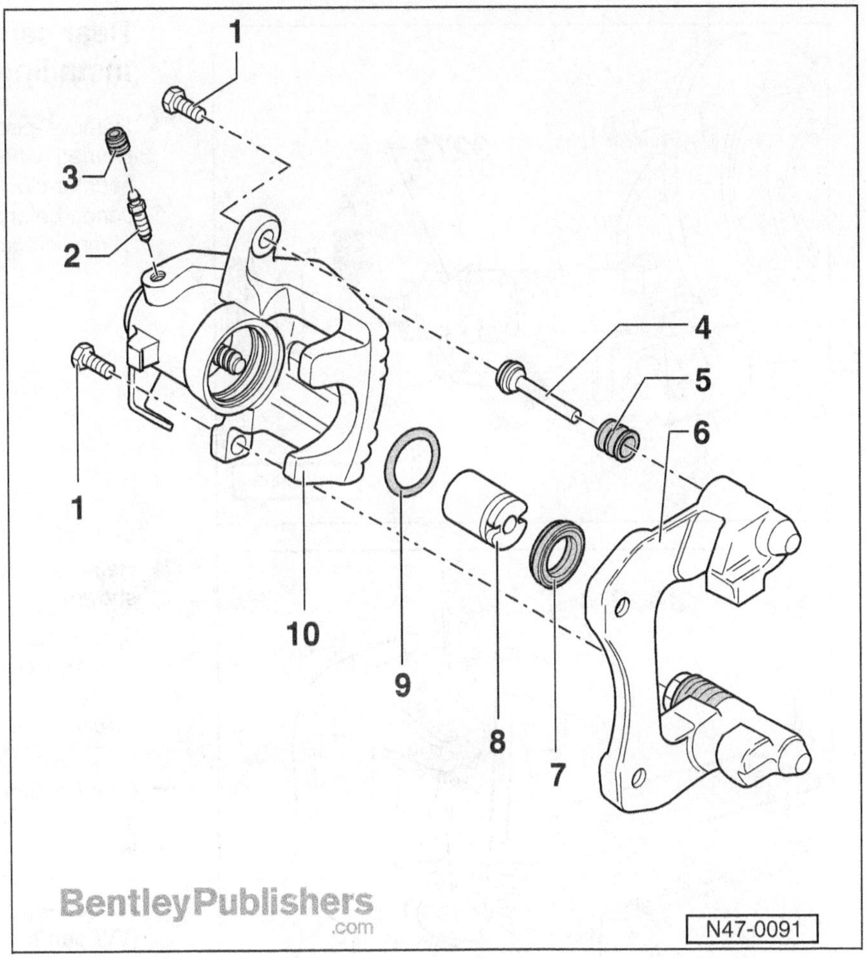

1. **Self-locking bolt**
   - Tighten to 35 Nm (26 ft-lb)
   - Always replace
   - When loosening and tightening, counter-hold on guide pin
2. **Bleeder screw**
   - Apply thin coat of grease to threads before installing
3. **Dust cap**
4. **Guide pins**
   - Apply thin coat of grease before installing protective cap
5. **Protective cap**
   - Pull onto brake carrier and guide pin
6. **Brake carrier with guide pin and protective cap**
   - Replacement part is assembled with sufficient grease on guide pins.
   - If protective caps or guide pins are damaged, install repair kit. Use grease packet supplied to lubricate guide pins.
7. **Dust boot**
   - Pull outer sealing lip onto piston
8. **Piston with automatic adjustment**
   - Apply thin coat of grease before installing
9. **Piston seal**
10. **Brake caliper housing**
    - With parking brake cable lever
    - If fluid is leaking at parking brake cable lever, replace brake caliper housing
    - After repairing, pre-bleed caliper housing, see **Rear caliper piston, removing and installing**

Rear caliper piston, removing and installing

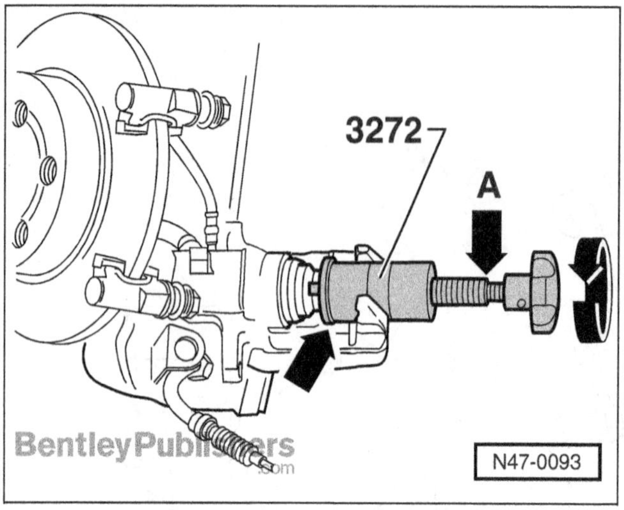

◀ Remove piston from brake caliper housing by turning knurled wheel on piston resetting and removal tool counter-clockwise. Install tool with collar (**arrow**) positioned before piston. If piston is difficult to move, use a 13mm wrench on flat spots of tool (**arrow A**)

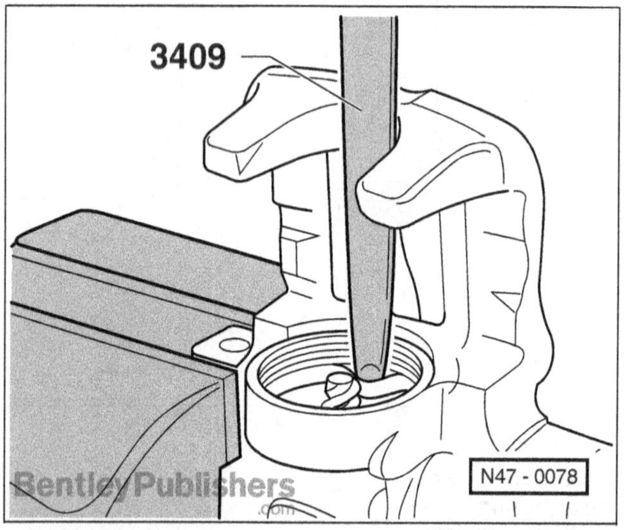

◀ Remove seal using plastic wedge (special tool **3409** shown).

> ***CAUTION —***
> *When removing, use care to ensure that the cylinder bore is not damaged.*

- Clean caliper and piston with fresh brake fluid.
- Install new piston seal in caliper.
- Lubricate piston and cylinder bore lightly using grease (VW part no. G 052 150 A2).

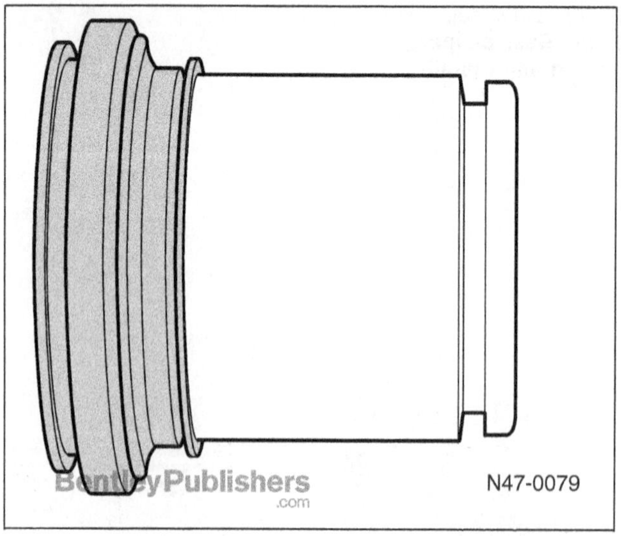

◀ Install dust boot with outer sealing lip on piston.

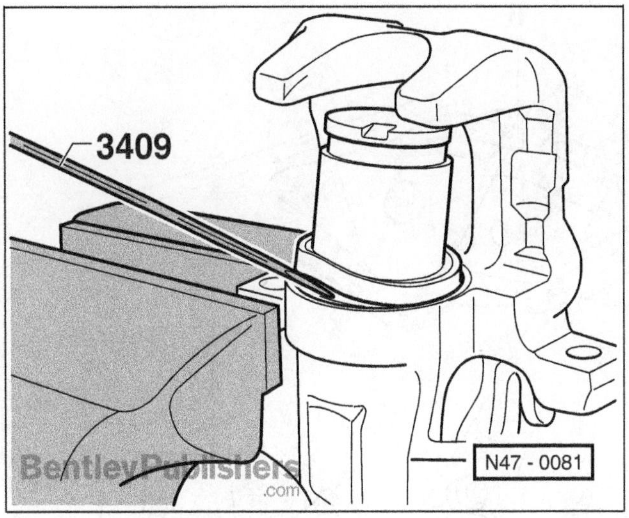

Hold piston in front of caliper and insert inner sealing lip into grove in cylinder using a plastic wedge (VAG 3409 shown).

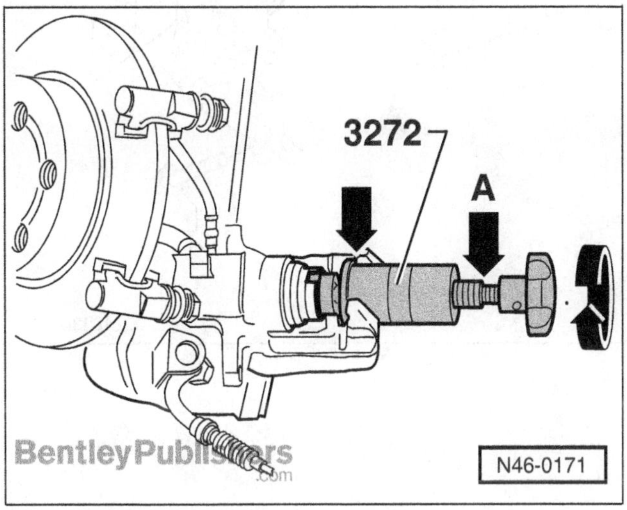

Reset caliper's automatic adjustment mechanism by turning the piston clockwise (**rotational arrow**) while pushing piston in.

- Install tool with collar (**arrow**) positioned before the piston. If piston is difficult to move, use a 13mm wrench on flat spots (**arrow A**) of tool.

> ***CAUTION—***
> - *If the piston is not reset correctly, or if the brake pedal is operated with the caliper removed, the automatic adjustment mechanism will be destroyed.*
> - *Always remove some brake fluid from the reservoir before resetting and pushing the caliper piston in. When the piston is pushed in, fluid is forced up into the reservoir.*

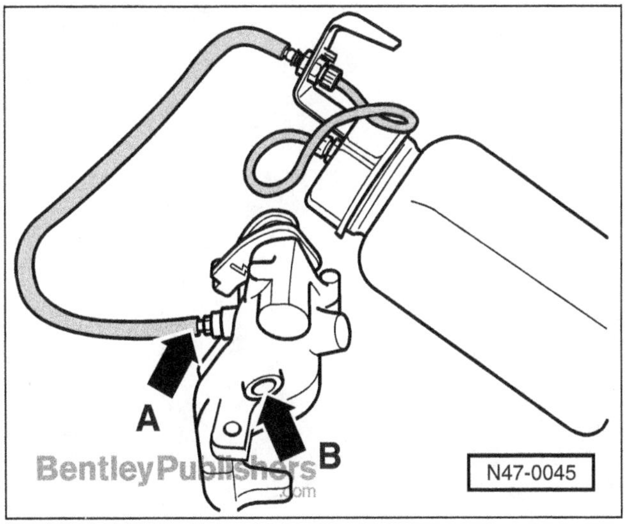

Pre-bleed caliper by opening bleeder screw (**arrow A**) and fill a standard bleeder bottle with brake fluid until bubble-free fluid flows from brake hose connection (**arrow B**).

– Close bleeder screw.

– Be sure to quality check your work, see **Quality Review** at the end of this repair group.

Brake pressure regulator assembly

1. **Brake pressure regulator**
2. **Socket-head bolt**
   - Tighten to 21 Nm (15 ft-lb)
3. **Bolt**
   - Tighten to 21 Nm (15 ft-lb)
4. **Nut**
   - Tighten to 21 Nm (15 ft-lb)
5. **Spring**
6. **Mounting bracket**
7. **Bolt**
   - Tighten to 16 Nm (12 ft-lb)

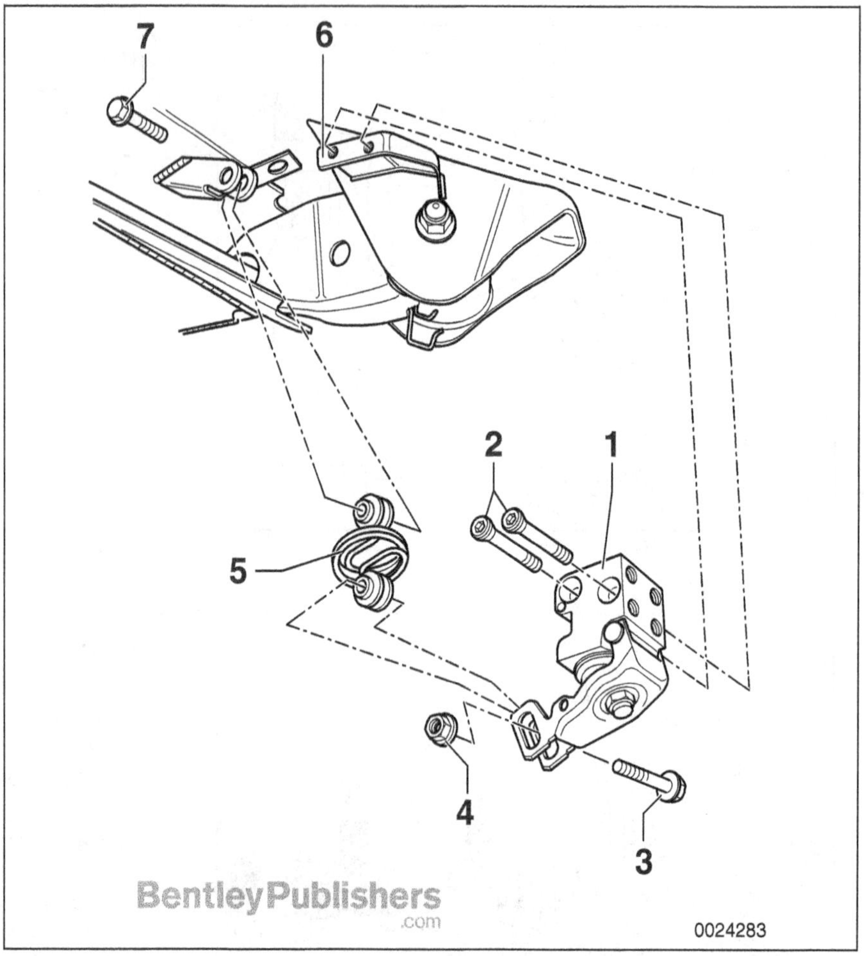

Brake pressure regulator, checking

Cars without ABS have a brake pressure regulator mounted on a bracket attached to the rear axle. This load-sensing regulator is controlled from the rear axle by a spring. When the axle changes position, due to a heavy load or during hard braking, the pressure regulator varies the pressure to the rear brakes.

> ***NOTE —***
> *The brake pressure regulator should be checked and if necessary adjusted following repairs to the rear suspension, or when there is excessive rear brake wear.*

With the car resting on all four wheels and the fuel tank full, observe the pressure regulator while a helper depresses and quickly releases the brake pedal. The regulator should move slightly when the brake is quickly released. If not, the regulator is faulty and should be replaced.

> ***NOTE —***
> *For this check to be accurate, the car should be emptied of all cargo and occupants, except driver.*

Testing and adjusting the pressure-regulating function requires measuring brake system pressure at each wheel caliper using two pressure gauges. Because of the need for this specialized equipment, we recommend having this test performed by an authorized Volkswagen dealer.

BRAKE BLEEDING

The procedure given here applies to all vehicles covered in this manual with and without ABS.

The brake system can be bled using a pressure bleeder or manually using a helper. Pressure bleeding, if the equipment is available, is the fastest. Manual bleeding requires a helper, but is easy and requires no special tools.

> ***CAUTION—***
>
> - *Brake fluid absorbs moisture from the air and must be replaced every two years, see* **03 Maintenance**.
> - *Use only new, approved brake fluid that complies with MVSS 116 DOT 4. Do not use silicone-based brake fluid (DOT 5). Even the smallest traces can cause severe corrosion in the brake system.*
> - *Brake fluid is poisonous. Wear safety glasses and rubber gloves when working with brake fluid. Prevent brake fluid from entering the bloodstream through cuts or scratches. Do not siphon brake fluid by mouth.*
> - *Brake fluid is very damaging to paint. Immediately wipe up any brake fluid that spills on the vehicle.*
> - *The brake fluid level in the reservoir must not fall below the MIN mark during bleeding.*

NOTE—

- *On non-ABS cars, the brake pressure regulator lever must be pressed toward the rear during rear brake bleeding.*
- *Depress the brake pedal several times during the bleeding operation.*

Special points for vehicles with Mark 60 ABS system

- *Vehicles with ABS/EDL, ABS/EDL/ASR, and ABS/EDL/ASR/ESP that have had the hydraulic unit/control module replaced, or have had one chamber of the unit drained of fluid, require that the hydraulic unit/control module be bled using a diagnostic scan tool via "basic settings" function.*
- *Vehicles with ABS/EDL/ASR/ESP require that zero compensation of sender 1 for brake booster (G201) be performed using a diagnostic scan tool.*

Bleeding, with pressure bleeder

Pressure bleeding using Volkswagen special tool US1116 or equivalent is the preferred method of removing air from all brake systems because it does a more thorough purge.

– Connect pressure bleeder with regulated compressed air supply to fluid reservoir according to manufacturer's instructions.

> ***CAUTION —***
> *Do not exceed filling pressure of 1 bar (14.5 psi) when filling brake fluid using pressure bleeder US 1116. The brake system will not be completely bled if excessive pressure is used.*

– Connect hose from bleeder bottle to caliper/wheel cylinder bleeder screw and bleed brakes in following sequence:

- Right rear caliper
- Left rear caliper
- Right front caliper
- Left front caliper

– Brake fluid should be allowed to flow from the bleeder valve/screw until it runs clear.

Bleeding, manually

– Connect hose from bleeder bottle to brake bleeder screw at the right rear caliper or wheel cylinder.

> ***CAUTION —***
> *The brake fluid level in the reservoir must not fall below the MIN mark during bleeding.*

– Have a helper pump the brake pedal several times and then hold pedal down.

– Open bleeder screw at the caliper or wheel cylinder and collect fluid.

– Close bleeder screw and then release brake pedal. Repeat operation until brake fluid runs clear and flows without air bubbles.

– Repeat the above procedure at the remaining wheels, using the following sequence:

- Left rear caliper
- Right front caliper
- Left front caliper

MASTER CYLINDER/ BRAKE BOOSTER

Master cylinder/brake booster, assembly

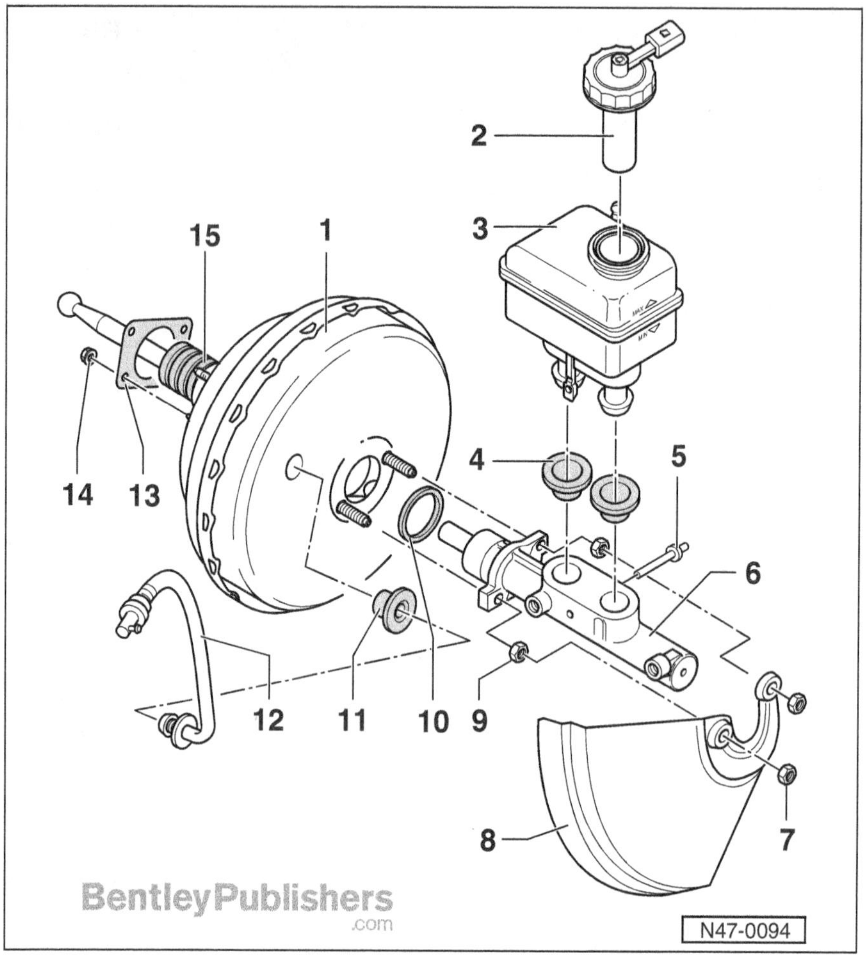

1. **Brake Booster**
   - Can be replaced separately from master cylinder
   - Gasoline engine: vacuum supplied by intake manifold
   - Diesel engine: pump supplies vacuum, See **Vacuum pump for brake booster (1.9L engine), removing and installing**
   - Functional check: With engine off, depress brake pedal firmly several times (to exhaust vacuum in unit). Depress brake pedal with average foot pressure, hold and start engine. If brake booster is working properly, pedal will give slightly under foot.
   - If booster is not functioning replace complete brake booster
   - Check valve (in vacuum hose), see **Vacuum check valve, checking**
2. **Cap**
3. **Brake fluid reservoir**
4. **Sealing plug**
   - Moisten with brake fluid and press into reservoir
5. **Retainer pin**
   - Install through brake master cylinder
6. **Brake master cylinder**
   - Cannot be repaired. If faulty, replace as complete unit.
7. **Self-locking nut**
   - Tighten to 20 Nm (15 ft-lb)
   - Always replace
8. **Heat shield**
   - Not installed on all vehicles
9. **Self-locking nut**
   - Tighten to 20 Nm (15 ft-lb)
   - Always replace
10. **Seal**
    - Always replace
11. **Sealing plug**
12. **Vacuum hose**
    - Insert into brake booster
13. **Gasket**
    - For brake booster
14. **Self-locking nut**
    - Tighten to 20 Nm (15 ft-lb)
    - Always replace
15. **Boot**
    - Must be seated correctly or may cause air noise.

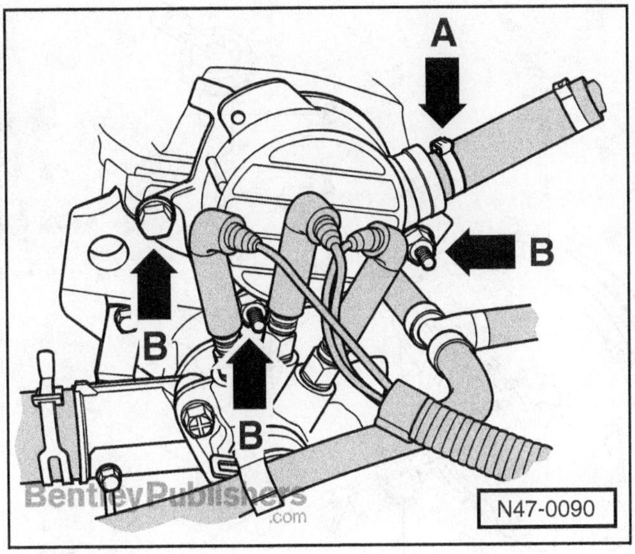

Vacuum pump for brake booster (1.9L engine), removing and installing

- Remove retainer for wiring harness from vacuum pump.
- Loosen vacuum hose clamp (**arrow A**) and remove hose from pump.
- Remove bolts (**arrow B**) on flange and remove pump.
- When installing vacuum pump be sure that follower engages correctly with camshaft.
- Install bolts (**arrows B**) on cylinder head flange and secure vacuum hose with clamp (**arrow A**).

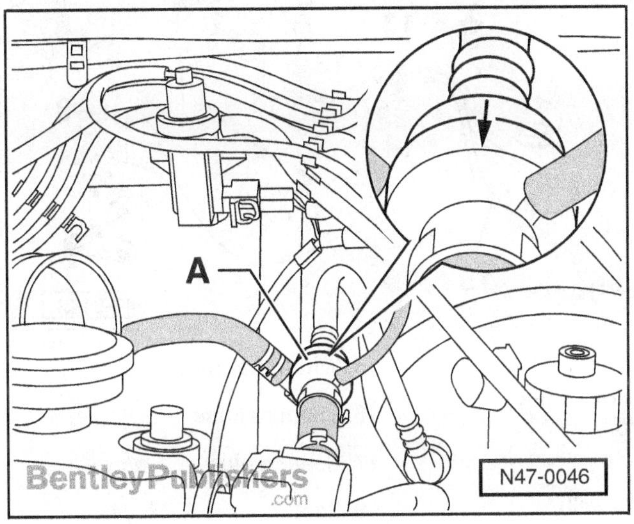

Vacuum check valve, checking

- Air must pass through in direction of arrow.
- Check valve (**A**) must stay closed (no air passes through) if supplied from opposite direction.

Master cylinder/brake booster, assembly (Mark 20 ABS/EDL/ASR/ESP, through 09/00)

NOTE —
Volkswagen identifies electrical components by a letter and/or a number in the electrical schematics. See **EWD Electrical Wiring Diagrams**. *These electrical identifiers are listed in parenthesis as an aid to electrical troubleshooting.*

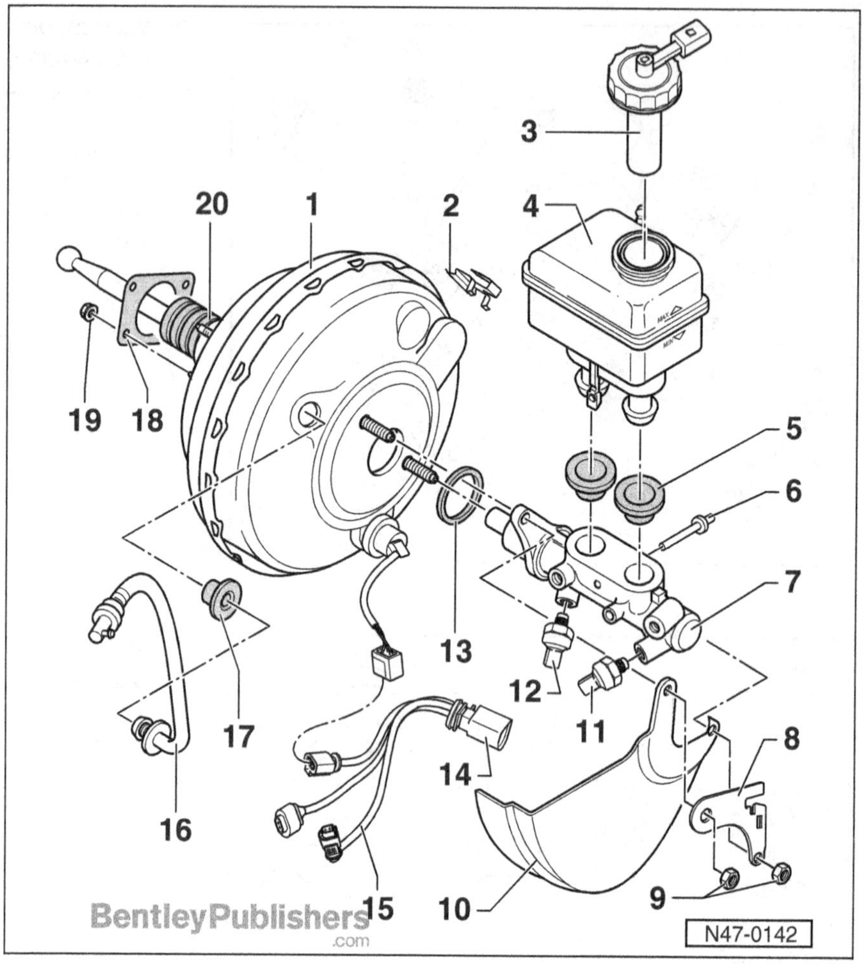

1. **Brake Booster**
   - Can be replaced separately from master cylinder
   - Gasoline engine: vacuum supplied by intake manifold
   - Diesel: pump supplies vacuum, See **Vacuum pump for brake booster (1.9L engine), removing and installing**
   - Functional check:
     With engine off, depress brake pedal firmly several times (to exhaust vacuum in unit).
     Depress brake pedal with average foot pressure, hold and start engine. If brake booster is working properly, pedal will give slightly under foot.
   - If booster is not functioning replace complete brake booster
2. **Retainer**
   - Secured to edge of brake servo
   - used to secure connector housing -14-
3. **Cap**
4. **Brake fluid reservoir**
5. **Sealing plug**
   - Moisten with brake fluid and press into brake reservoir
6. **Retainer pin**
   - Insert through brake master cylinder
7. **Brake master cylinder**
   - Cannot be repaired, replace if faulty
8. **Retainer plate**
   - Used to secure the wiring harness
9. **Self-locking nut**
   - Tighten to 20 Nm (15 ft-lb)
   - Always replace
10. **Heat shield**
    - Only for vehicles with
    - 1.8L engine
11. **Sender 1 for brake booster (G201)**
    - See **Sender 1 for brake booster and sensor 2 for brake pressure, removing and installing**
12. **Sensor 2 for brake pressure (G241)**
    - See **Sender 1 for brake booster and sensor 2 for brake pressure, removing and installing**
13. **Seal**
    - Always replace
14. **Connector housing**
15. **Wiring harness**
16. **Vacuum hose**
17. **Sealing plug**
18. **Seal**
    - For brake servo
19. **Self-locking nut**
    - Tighten to 20 Nm (15 ft-lb)
    - Always replace
20. **Boots**
    - Make sure it seats correctly

Master cylinder/brake booster, assembly (Mark 60 ABS/EDL/ASR/ESP, from 10/00)

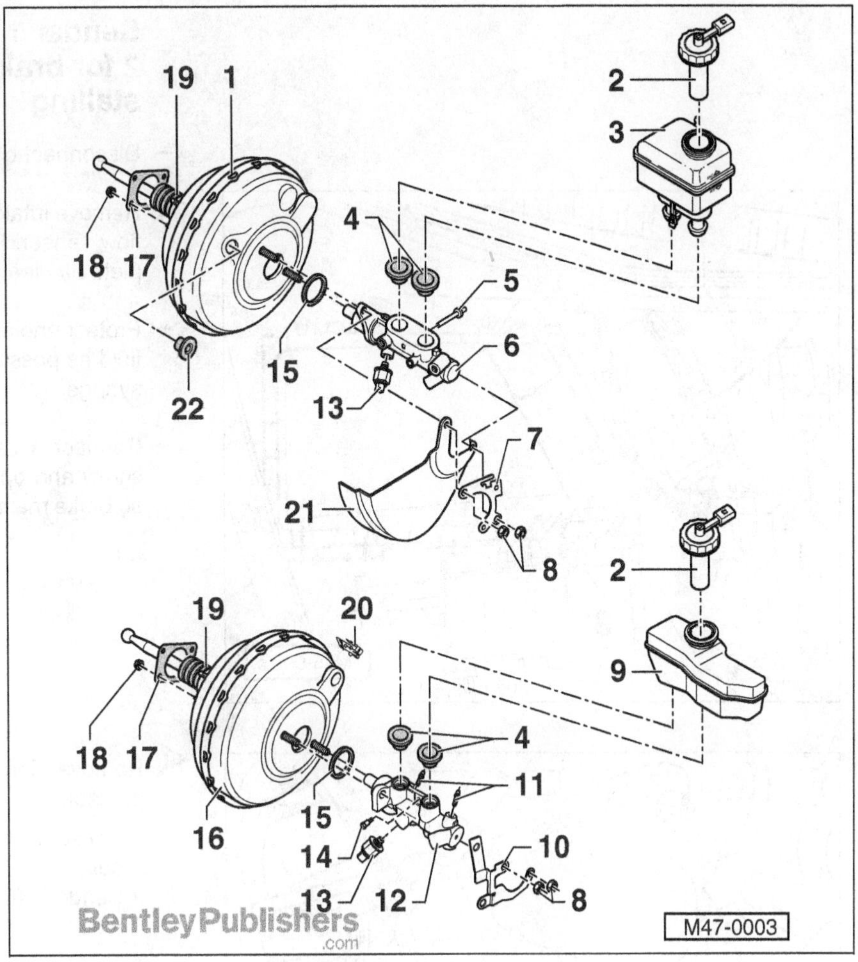

1. **Brake Booster**
   - Can be replaced separately from master cylinder
   - Gasoline engine: vacuum supplied by intake manifold
   - Diesel engine: pump supplies vacuum, See **Brake system vacuum pump -V192-, removing and installing**
   - Functional check:
     With engine off, depress brake pedal firmly several times (to exhaust vacuum in unit).
     Depress brake pedal with average foot pressure, hold and start engine. If brake booster is working properly, pedal will give slightly under foot.
   - If booster is not functioning replace complete brake booster
2. **Sealing Cap**
3. **Brake fluid reservoir, left hand drive**
4. **Sealing plug**
   - Moisten with brake fluid and press into reservoir
5. **Retainer pin**
   - Install through brake master cylinder
6. **Brake master cylinder, left hand drive**
   - Cannot be repaired. If faulty, replace as complete unit.
7. **Retaining plate**
   - Used to secure the wiring harness
8. **Self-locking nut**
   - Tighten to 20 Nm (15 ft-lb)
   - Always replace
9. **Brake fluid reservoir, right hand drive**
10. **Retaining plate**
    - Used to secure the wiring harness
11. **Bleed screw**
12. **Brake master cylinder, right hand drive**
    - Cannot be repaired. If faulty, replace as complete unit.
13. **Sender 1 for brake booster (G201)**
    - See **Sender 1 for brake booster and sensor 2 for brake pressure, removing and installing**
14. **Bolt**
    - To secure brake fluid reservoir to brake master cylinder
15. **Seal**
    - Always replace.
16. **Brake servo, right hand drive**
17. **Seal**
    - For brake servo
18. **Self-locking nut**
    - Tighten to 20 Nm (15 ft-lb)
    - Always replace
19. **Boots**
20. **Boots**
    - Make sure it seats correctly
21. **Heat shield**
22. **Sealing plug**

Sender 1 for brake booster and sensor 2 for brake pressure, removing and installing

– Disconnect battery.

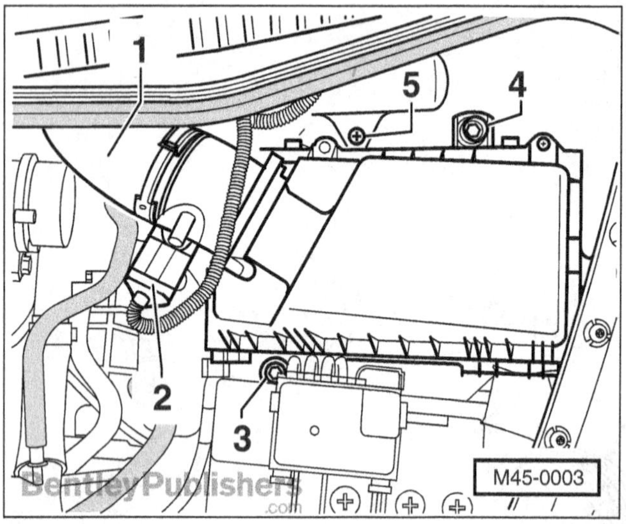

◄ Remove intake hose (**1**) and connector (**2**) from mass air flow sensor. Remove bolts (**3**, **4**, and **5**). Remove complete air cleaner assembly.

– Protect engine bay with shop rag. Draw as much brake fluid as possible from reservoir using a suction bottle or syringe.

– Connect a catch can to left front brake caliper bleed screw and open. Operate brake pedal at least 15 times so brake master cylinder empties and close bleed screw.

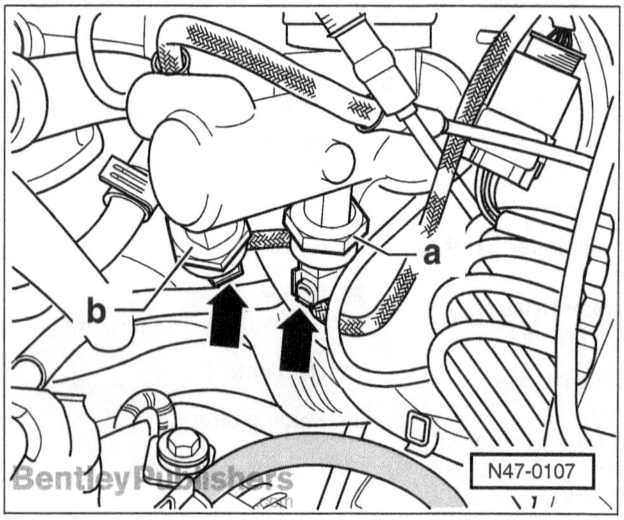

◄ Remove connector (**arrows**) from appropriate brake pressure sensor, and remove sensor.

- Sensor 2 (**a**) for brake pressure (G214) in primary circuit.
- Sender 1 (**b**) for brake booster (G201) in secondary circuit.

Tightening torques
- Sender 1 to master cylinder .... 25 Nm (18 ft-lb)
- Sensor 2 to master cylinder .... 25 Nm (18 ft-lb)

– Installation is reverse of removal. Make sure to fill and bleed brake system after service.

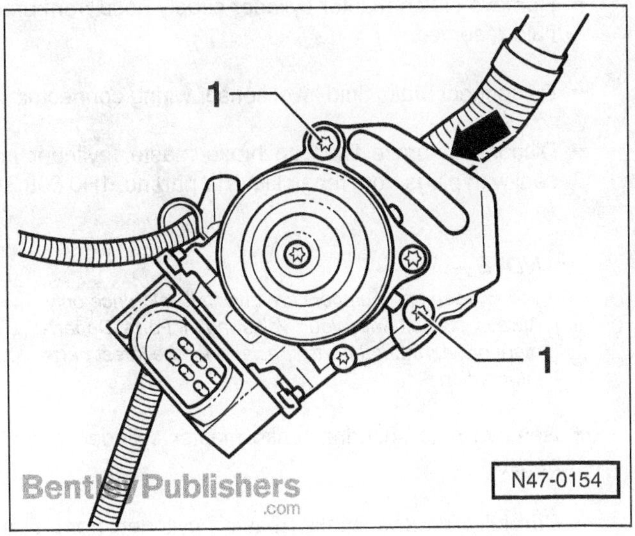

Brake system vacuum pump, removing and installing

There are no provisions for internal repairs to the brake system vacuum pump (V192). If a malfunction is detected, pump must be replaced.

◄ Remove bolt (**1**) on bracket for brake system vacuum pump. Pull hose (**arrow**) off brake system vacuum pump. Take out brake system vacuum pump.

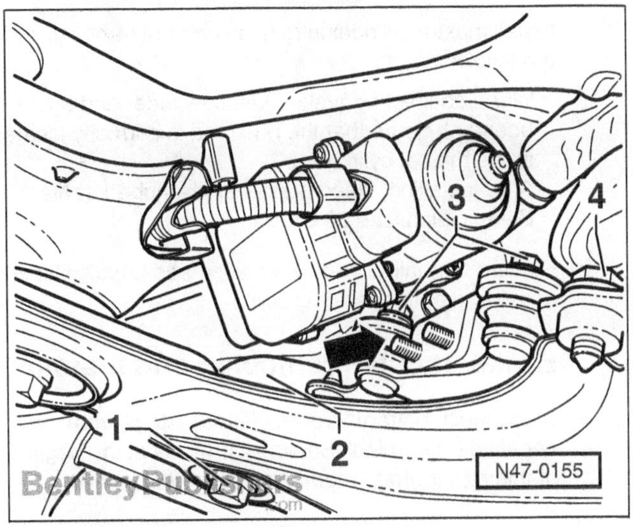

◄ Bracket for brake system vacuum pump. Remove lock nut (**2**) from bolt (**1**). Remove bolt (**4**) and bracket (**arrow**). Remove bolts (**3**) to disassemble bracket further.

Tightening torques

- Nut (**2**) to bracket .... 20 Nm (15 ft-lb)
- Bolts (**3**) to bracket .... 8 Nm (71 in-lb)
- Bolt (**4**) to bracket .... 20 Nm (15 ft-lb)
- Torx socket head bolt for brake system vacuum pump to bracket .... 8 Nm (71 in-lb)

– Installation is reverse of removal.

Brake master cylinder, removing and installing

– Spread sufficient lint-free cloths in area of plenum chamber and on engine and transmission to catch any escaping brake fluid.

> ***WARNING —***
> - *Brake fluid is poisonous.*
> - *To draw off brake fluid from the reservoir, use a bleeder bottle which is used only for brake fluid. Never siphon brake fluid by mouth.*

– Using bleeder bottle, draw off as much brake fluid as possible from brake fluid reservoir.

– Clamp clutch master cylinder supply hose with special tool (VW 3094 or equivalent).

– Remove clutch master cylinder supply hose from brake fluid reservoir.

– Disconnect brake fluid level sensor wiring connector.

– Disconnect brake lines on brake master cylinder and seal with plugs from repair kit, (VW part no. 1H0 698 311 A).

> ***NOTE —***
> *Volkswagen part numbers are given for reference only. Always consult with your Volkswagen Parts Department or aftermarket parts specialist for the latest parts information.*

– Remove nuts securing brake master cylinder to brake booster.

– Carefully remove brake master cylinder from brake booster.

– Install master cylinder in reverse order of removal, noting the following:

- When connecting brake master cylinder and brake booster be sure that the push rod is correctly located in the master cylinder.
- After installing, bleed brakes as described earlier. Bleed clutch, see **30 Clutch**.

– Be sure to quality check your work, see **Quality Review** at the end of this repair group.

Brake booster, removing and installing

– Disconnect battery ground (GND) strap from battery negative (–) terminal. See the **Cautions** at the beginning of this repair group regarding battery disconnection.

> ***NOTE —***
> *Be sure to have the anti-theft radio code on hand before disconnecting the battery.*

Vehicles with manual transmission

– Clamp clutch master cylinder supply hose with special tool (VW 3094 or equivalent).

– Remove clutch master cylinder supply hose from brake fluid reservoir.

All vehicles

– Remove ABS control module and hydraulic unit, see **45 Anti-Lock Brakes (ABS)**.

– Disconnect vacuum hose from brake booster.

Vehicles with Diesel TDI engine

– Remove wiring harness retainer from vacuum pump.

All vehicles

– Remove trim below instrument panel.

Vehicles with manual transmission

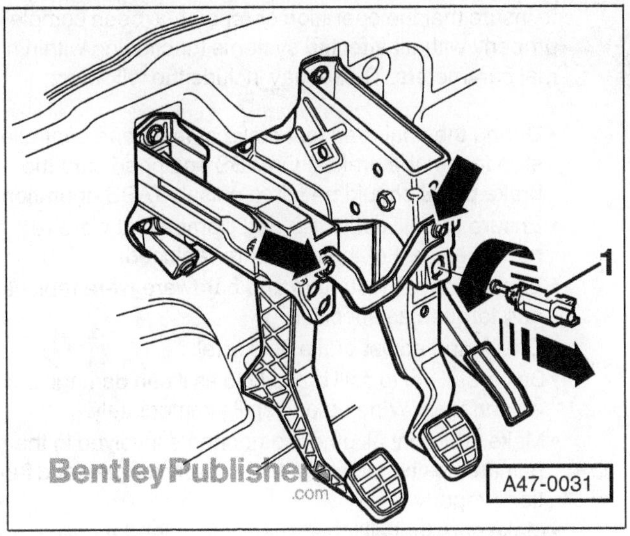

< Remove connecting plate between clutch and brake pedals.

All vehicles

– Remove brake light switch and separate brake pedal from brake booster.

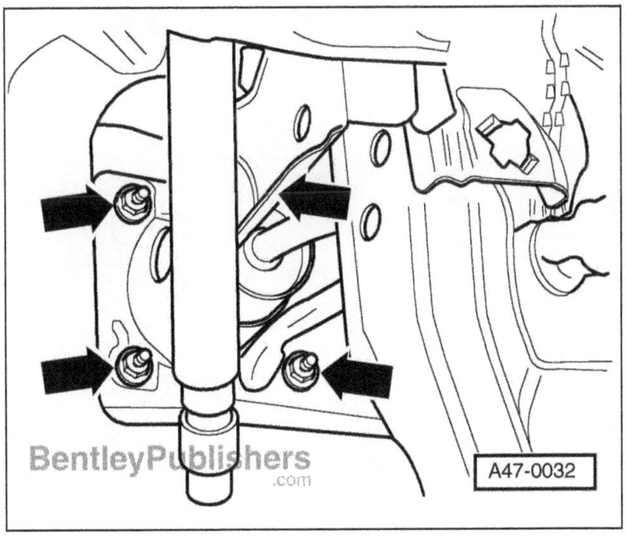

< Remove brake booster mounting nuts (**arrows**).

– Guide brake booster with brake master cylinder out forward to remove.

– Install brake master cylinder with brake booster in reverse order of removal noting the following:

- Adjust brake light switch, see **45 Anti-Lock Brakes (ABS)**.

– Be sure to quality check your work, see **Quality Review** at the end of this repair group.

QUALITY REVIEW

When you have finished working under the hood and around other areas of the vehicle, it is advisable to take a moment to quality check or review your work. This helps to insure that the operation or repair has been completed properly with all affected systems functioning within normal parameters. These may include the following:

- During the final road test make at least one controlled stop to test the brakes. On ABS equipped cars the brake pedal should be felt to pulsate (ABS operation).
- Ensure that all cable ties and clamps that were removed as part of the repair are replaced.
- Ensure that all fasteners and hardware were replaced and torqued as specified.
- Check and adjust brake fluid level.
- Be careful not to spill brake fluid as it can damage painted surfaces. Wipe up any spills immediately.
- Make sure that all other components involved in the repair are positioned correctly, properly torqued and function properly.
- Make sure that all tools, shop cloths, fender covers and protective tapes are removed before closing the hood.
- Clean grease and fingerprints from painted surfaces, steering wheel and shifter.

48 Steering

GENERAL

Power assisted rack and pinion steering is standard on New Beetles. A vane type pump driven by the ribbed V-belt provides hydraulic pressure to the steering gear. Service of the steering gear, including repair of housing seal leaks is by replacement only. The power steering gear is available only as a complete assembly.

Special tools, equipment and procedures are required for most steering repair and component replacement. In addition, front wheel alignment is almost always disturbed when tie rods or steering gear are removed or replaced.

WARNING —

- *Do not re-use any fasteners that are worn or deformed in normal use. Most fasteners are designed to be used only once and become unreliable and may fail when used a second time. This includes, but is not limited to, nuts, bolts, washers, circlips, cotter pins, self-locking nuts and bolts. For replacements, always use new parts.*
- *Do not reinstall bolts and nuts coated with undercoating wax as correct tightening torque cannot be ensured. Always clean the threads of removed bolts and nuts that are to be re-used with a suitable solvent before installation. Ensure that new parts are clean.*

> ***WARNING —***
> - *Do not use standard nuts and bolts in place of green colored parts. The green color denotes a special anti-corrosion process known as "dacromet" or "delta-tone". Torque specifications are given based on use of these parts only.*
> - *Always replaced rusted or corroded bolts, nuts and washers even if not specifically indicated.*
> - *DO NOT attempt to straighten or weld suspension struts, wheel bearing housings, control arms or any other wheel locating or load bearing components of the front suspension.*

> ***CAUTION —***
> *Disconnecting the negative (–) battery cable may erase fault codes and basic settings in the engine management and automatic transmission control modules. Some driveability problems may be noticed until the system re-adapts to operating conditions. OBD II readiness codes, which may be required for emissions testing, may also be erased. Convenience electronics (alarm system, interior light control, power locks, mirrors, and windows) may need to be re-set using a diagnostic scan tool.*

Special precautions for airbag equipped vehicles

Airbag systems can enhance passenger safety in the event of a collision. To accomplish this task, they rely on sensitive electronic circuits and fast-acting deployment devices. Special service precautions must be followed to prevent possible serious bodily injury to the repair technician and others involved in the handling and storage of airbag components and to insure proper operation in the event of a collision.

STEERING WHEEL

Vehicles covered by this manual are equipped with multiple airbags. The driver's airbag unit is installed in the center area of the steering wheel. Because removal of the steering wheel is often required for repairs to the steering column, the driver's airbag unit will be covered in this section. For further details on the airbag unit, see **69 Seatbelts, Airbags.**

WARNING —

- *Before working on any airbag, steering wheel or steering column component, always **disconnect the battery first.***
- *No waiting time is required after disconnecting the battery.*
- *Airbags are inflated by an explosive device. Handled improperly or without adequate safeguards and training, the system can be very dangerous. Special precautions must be observed prior to any work at or near the steering wheel and steering column, including the pedal assembly.*
- *The airbag is a vehicle safety system. To guard against personal injury or airbag system failure, only trained Volkswagen service technicians should test, disassemble or service the airbag system.*
- *Airbag units that have been dropped onto a hard surface must not be installed. Always replace any airbag system component that has been mechanically or physically damaged (example: dented or cracked).*
- *Make sure that no one is in the passenger compartment when connecting the battery Ground (GND) strap to the battery negative (–) terminal.*
- *A technician must be electrostatically discharged before picking up or touching any airbag unit. This is accomplished by touching a suitable metal ground such as a water or heating pipe or metal frame. If in the vehicle, this is accomplished by touching a suitable chassis ground such as a door latch or striker.*
- *Install new airbag unit as soon as the unit is removed from the packaging. Reinstall in packaging if work will not be completed immediately.*
- *Do not leave any undeployed airbag unit unattended. If work is interrupted, store airbag unit in a secure location where it cannot be disturbed.*
- *Undeployed airbag units that have been replaced must only be stored and shipped in packaging designed specifically for the purpose such as that found with the replacement unit.*
- *Undeployed airbag units that have been replaced must be properly identified as such.The storage and transportation of airbag units must be in accordance with all applicable federal, state and local rules and regulations.*
- *Airbag units that have been removed during the course of repairs must be stored with the padded side facing up and in a secure location where it cannot be disturbed.*
- *Observe all cautions, warnings and notes before starting repairs involving airbag systems.*

Steering wheel with airbag, assembly

1. **Steering wheel**
2. **Harness connector for airbag**
3. **Locking lugs**
   - Release from rear of steering wheel
4. **Airbag unit**
5. **Bolt, multi-point socket-head**
   - Tighten to 60 Nm (44 ft-lb) with locking compound
   - Can be reused up to 5 times
   - Mark with center punch after each installation
6. **Securing plate**
7. **Spiral spring with slip ring**
8. **Steering column trim**
9. **Spring clip**

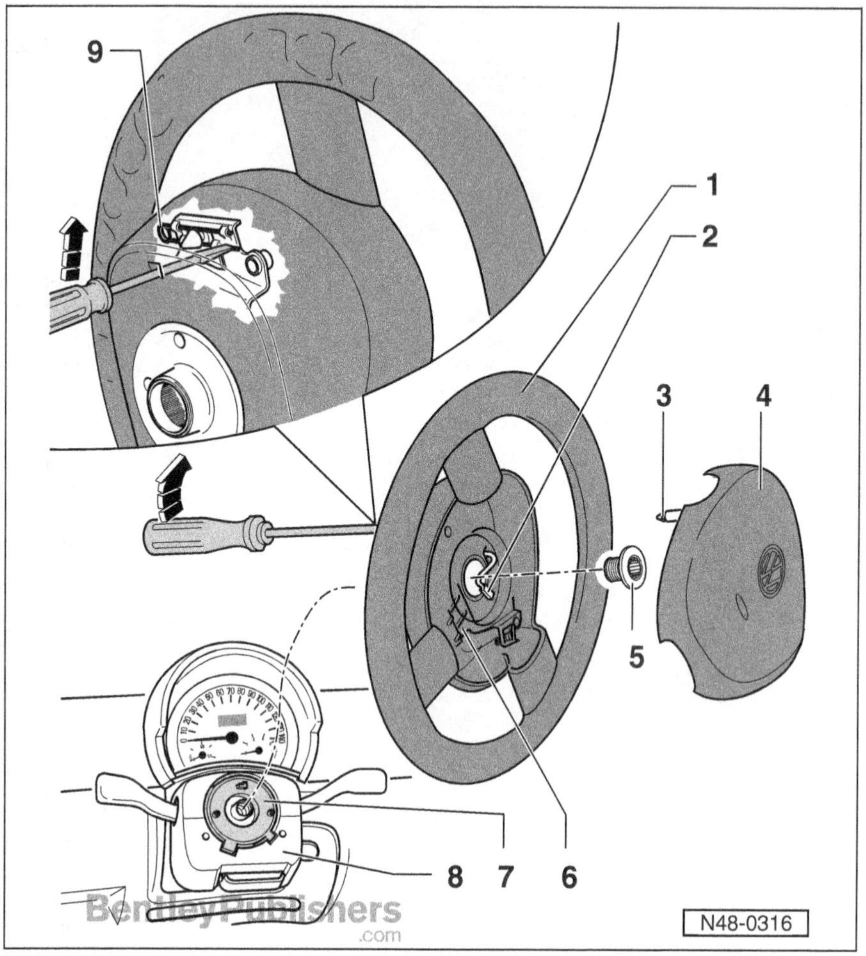

Steering wheel with airbag, removing

Removal of the steering wheel airbag unit is accomplished by inserting a tool into openings in the dashboard side of the steering wheel.

– Disconnect battery ground (GND) strap from battery negative (–) terminal. See the **Cautions** at the beginning of this repair group regarding battery disconnection.

NOTE —
Be sure to have the anti-theft radio code on hand before disconnecting the battery.

– Remove instrument cluster, see **70 Trim–Interior**.

– Release steering column adjustment.

- Turn steering wheel until a spoke is vertical. Extend steering wheel fully and move to uppermost position.

- Secure steering column adjustment.

- Using a screwdriver approximately 175 mm (7 in.) long, insert approximately 45 mm (1¾ in.) from reverse side into hole on dashboard side of steering wheel hub.

NOTE —
If a screwdriver with a very short handle is used, it is possible to eliminate step 2 and leave the instrument cluster in place.

- Press screwdriver in direction of arrow to press back spring clip and release locking lug of airbag unit.

- Turn steering wheel back ½ turn (180°) and release second locking lug on opposite side.

- Turn steering wheel to center position (wheels straight ahead).

- Disconnect harness connector from airbag unit.

- Place removed airbag unit in a secure location with padding side facing up.

Steering wheel with airbag, installing

- Install instrument cluster, if applicable.

- Reconnect harness connector for airbag unit.

- Position airbag unit on steering wheel and snap into place.

NOTE —
Airbag unit locking lugs must be heard and felt to snap into place.

- Switch ignition on.

WARNING —
- *Make sure the passenger compartment is not occupied before connecting the battery ground (GND) strap.*
- *Observe all cautions, warnings and notes when completing repairs involving airbag systems.*

- Connect battery ground (GND) strap to battery negative (–) terminal.

Steering wheel without airbag, assembly

1. **Steering wheel**
   - Installed position, see **Steering wheel and steering column, installed position**
2. **Bolt, multi-point socket-head**
   - Tighten to 60 Nm (44 ft-lb) with locking compound
   - Can be reused up to 5 times
   - Mark with center punch after each installation
3. **Bolts, Torx®**
4. **Vibration damper**
   - Installed on vehicles equipped with ALH engine and automatic transmission
5. **Bolts, Torx®**
   - Tighten to 5 Nm (44 in-lb)
6. **Airbag unit**
7. **Securing plate**

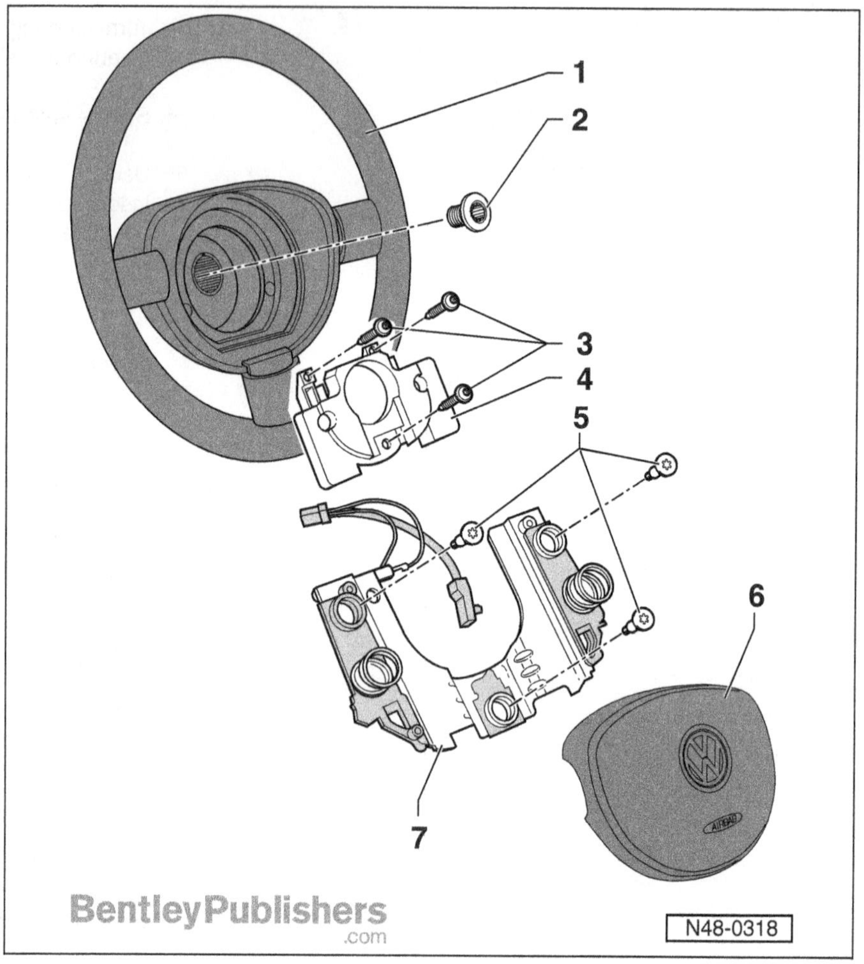

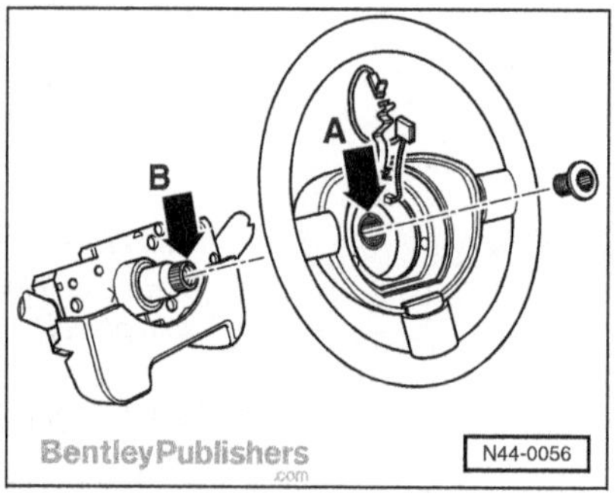

Steering wheel and steering column, installed position

- Center punch mark on steering wheel (**A**) must align with center punch mark on steering column (**B**).
- Replacement steering columns do not have center punch marks.

STEERING COLUMN

Standard equipment on the New Beetle includes a steering column that is height adjustable and telescopic. Defective steering columns must be replaced as a complete assembly as no internal repair parts are available from Volkswagen. There are two different versions of the steering column as described below.

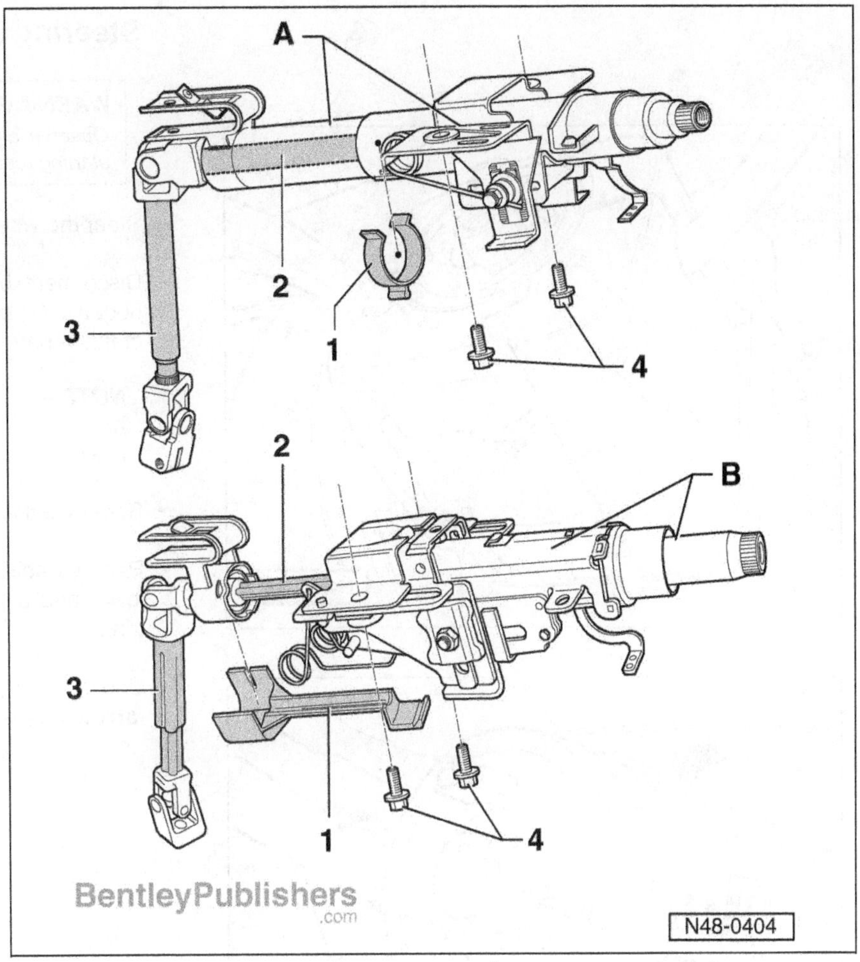

Steering column, assembly

A. Nacam steering column
- Through 01/00 production

B. Presta steering column
- From 02/00 production

NOTE —

Check steering column version before removal. If replacement is necessary, use the same type as originally installed.

1. **Transportation securing device**
   - Remove after column has been installed into vehicle
2. **Steering column**
   - See **Steering column, securing before removal**
   - See **Steering column, inspecting for damage**
3. **Steering column universal joint shaft**
4. **Bolts**
   - Tighten to 22 Nm (16 ft-lb)

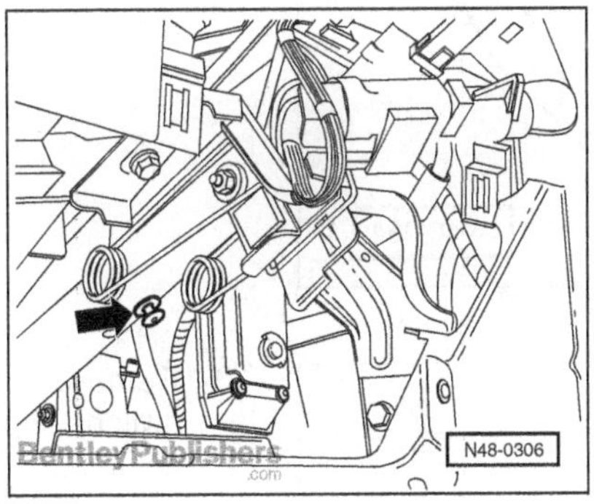

Steering column, securing before removing

- Assembly aid is required to prevent upper and lower sections of steering column from pulling apart during removal.
- Splined sections can separate and cause rattling noises unless installed in original positions.
- Push or pull steering column slightly to align holes and insert suitable clip or pin into hole (**arrow**).

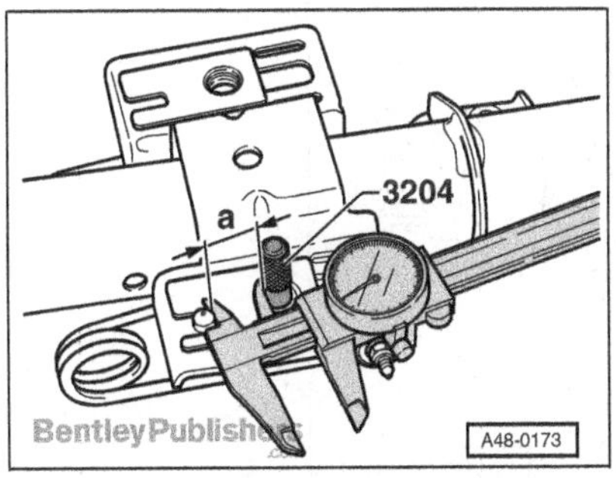

Steering column, inspecting for damage

- Visually inspect for damage.
- Disconnect steering column universal joint shaft and check for smooth operation (without binding) of steering shaft.
- Check that steering column can be adjusted for full range of height and telescopic adjustment.
- Check gap (**a**) with alignment pin inserted in hole. Maximum distance from edge to edge is 23 mm (0.906 in).

Steering lock housing, removing

> ***WARNING —***
> *Observe all cautions, warnings and notes before starting repairs involving airbag systems.*

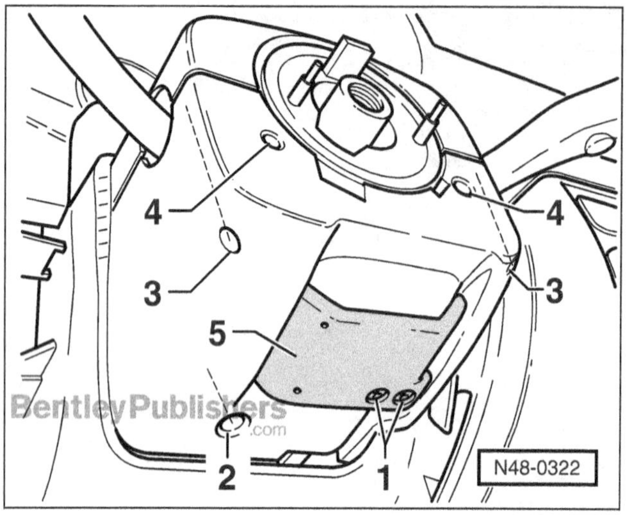

– Steer the wheels to the straight ahead position.

– Disconnect battery ground (GND) strap from battery negative (–) terminal. See the **Cautions** at the beginning of this repair group regarding battery disconnection.

> ***NOTE —***
> *Be sure to have the anti-theft radio code on hand before disconnecting the battery.*

– Remove airbag and steering wheel as described earlier.

◄ Remove adjustment handle (**5**), handle screws (**1**) and lower and upper trim retaining screws (**2**,**3**,**4**). Remove trim.

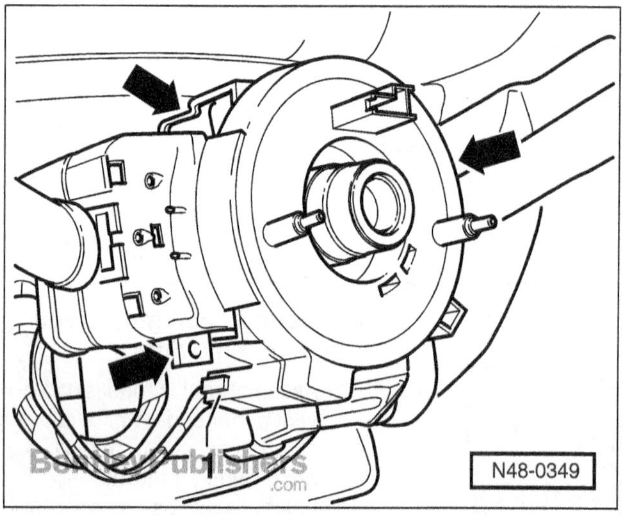

◄ Unplug harness connector (**1**) and release locking lugs (**arrows**) on spiral spring/slip ring and pull off.

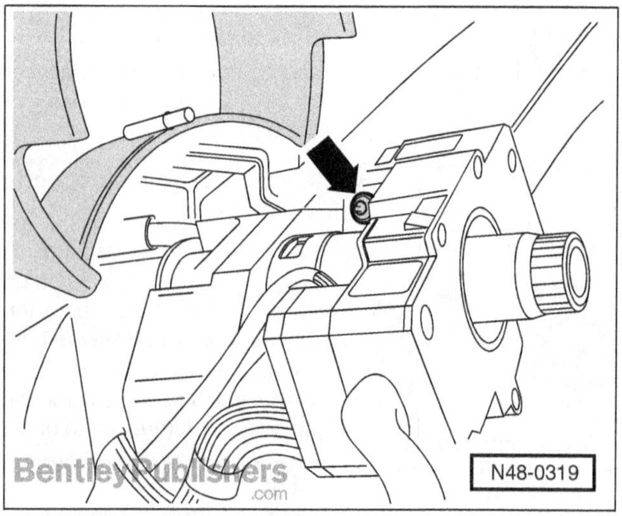

◄ Remove securing bolt (**arrow**) for steering column switch, unplug switch assembly and slide off steering column.

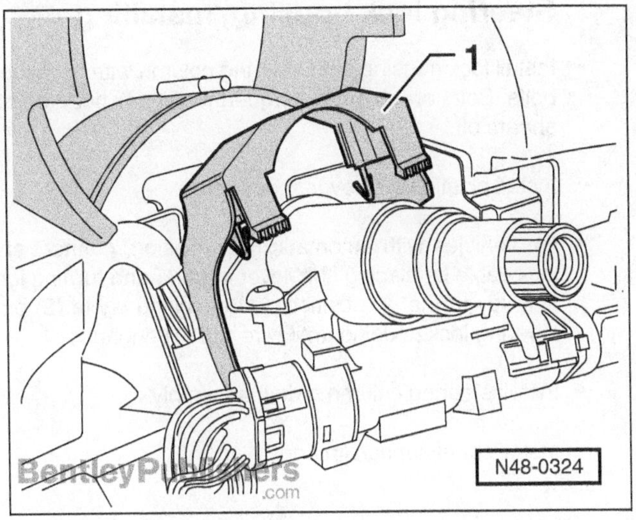

◀ Remove plastic cover (**1**) over shear bolts.

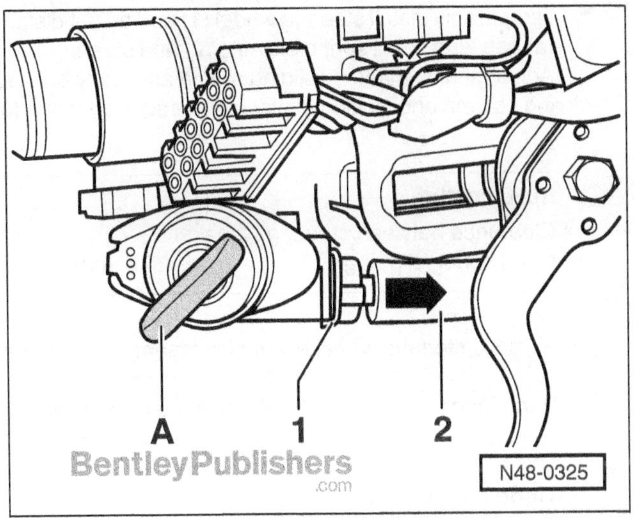

◀ On vehicles with automatic transmission, unhook shift lock cable by placing shift lever in park, and turning ignition key to the on position. Press wire clip (**1**) upward or downward (depending on installed position) while pulling out on locking cable (**arrow**).

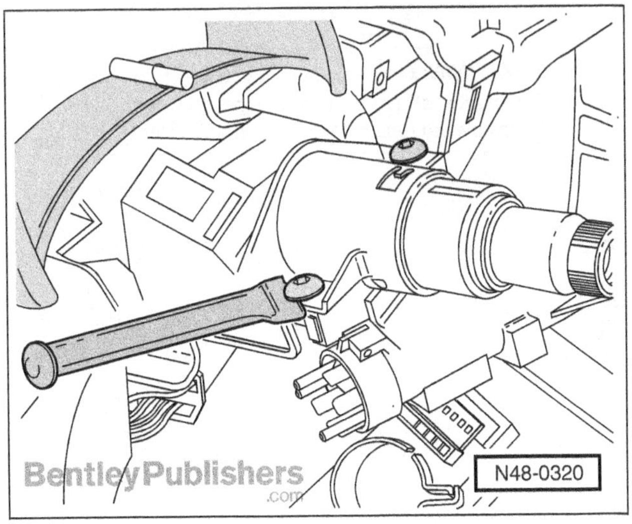

◀ Cut off shear head bolts using a suitable sharp chisel.

– Unplug harness connector on ignition switch and slide lock housing off of column.

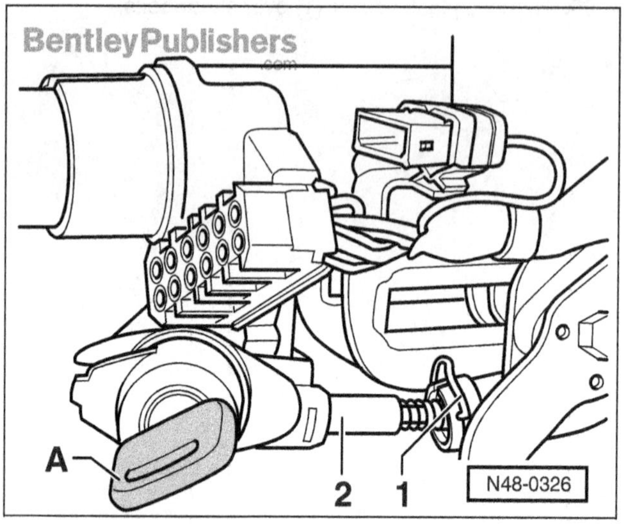

Steering lock housing, installing

– Install lock housing onto steering column with new shear bolts. Bolts are correctly torqued when hex head on bolt shears off.

– Install plastic cover over shear bolts.

◄ On vehicles with automatic transmission, connect shift lock cable by placing shift lever in park, and turning ignition key to the on position. Slide locking cable (**2**) onto steering lock housing until wire clip (**1**) engages.

– Install steering column switch assembly.

– Install spiral spring/slip ring.

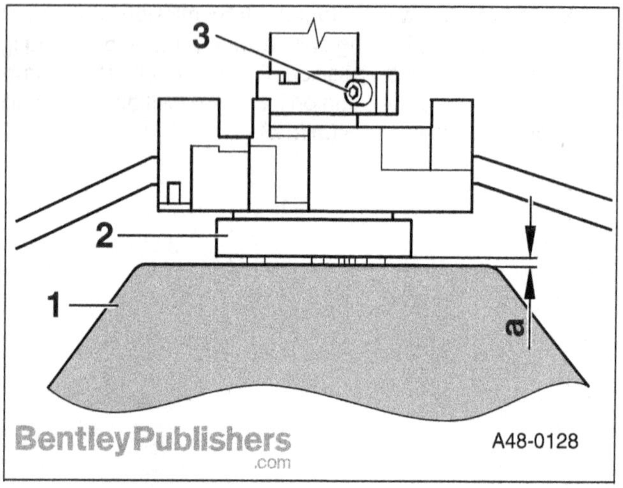

◄ Temporarily install steering wheel (**1**) and adjust distance (**a**) between spiral spring/slip ring (**2**) and steering wheel. Adjustment is made by sliding switch assembly on steering column and locking in place with securing screw (**3**).

Specification

- Clearance between steering wheel and spiral spring/slip ring. . . . . approx. 2½ mm (0.1 in)

– Remove steering wheel and install steering column trim.

– Remainder of installation is the reverse of removal.

Tightening torque

- Steering wheel bolt. . . . . 50 Nm (37 ft-lb)

WARNING —

Observe all cautions, warnings and notes when completing repairs involving airbag systems.

Ignition lock cylinder and switch, removing and installing

> ***WARNING —***
> *Observe all cautions, warnings and notes before starting repairs involving airbag systems.*

— Disconnect battery ground strap.

— Remove airbag unit and steering wheel as described earlier.

— Remove upper and lower steering column trim as described earlier.

— Insert ignition key into lock cylinder and turn to ignition **ON** position.

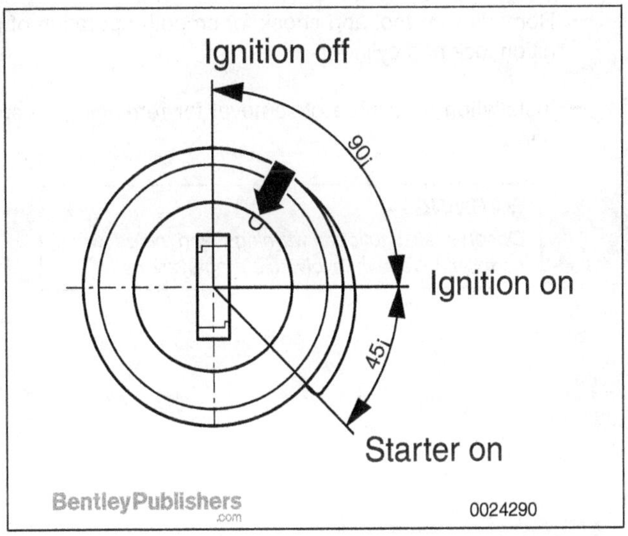

◀ Insert a wire tool approximately 1.2 mm (0.047 in) in diameter into the opening (**arrow**) in the face of the lock cylinder.

NOTE —
A large paper clip or a 3/64 inch drill bit are suitable for inserting into lock cylinder opening.

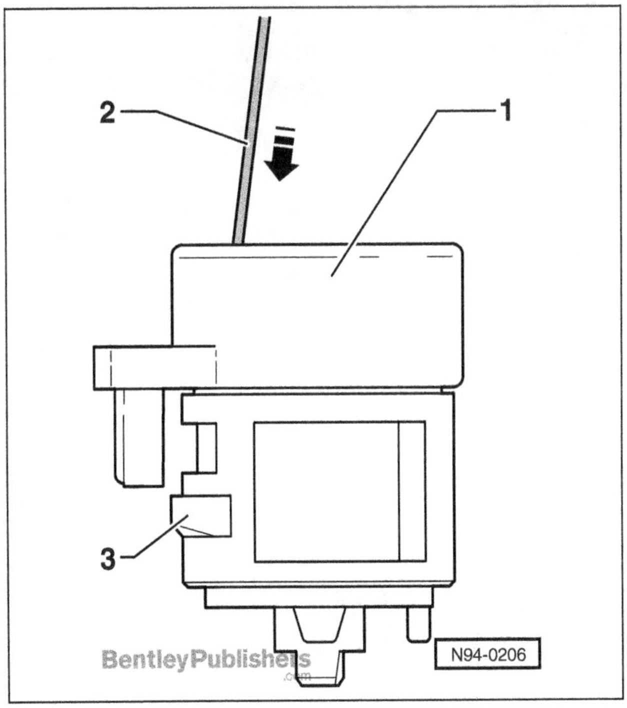

◀ Release lock cylinder by gently pulling on key while pushing in (**arrow**) on wire tool (**2**).

— Disconnect harness connector from ignition switch.

— Remove sealant (paint) from threaded holes of ignition switch securing screws.

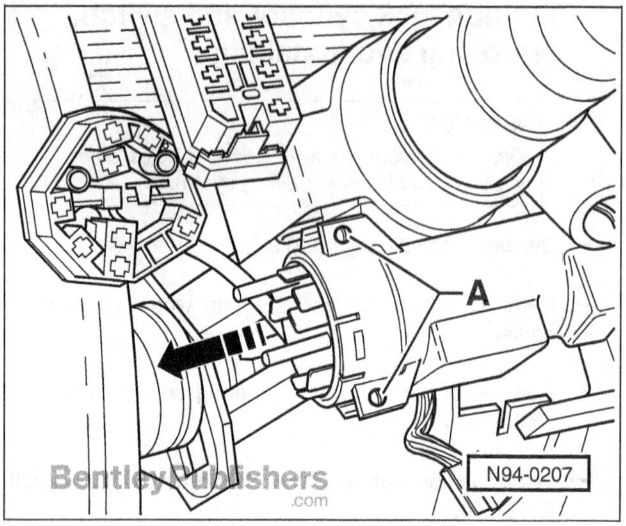

◀ Carefully remove two ignition switch (**A**) securing screws and pull ignition switch out from steering lock housing.

> ***NOTE —***
> - *When installing ignition/starter lock switch ensure that it is in the same position as the lock cylinder, e.g. "ignition on".*
> - *Be sure to use locking compound on ignition switch securing screws when installing.*

– Install ignition lock cylinder by inserting key into lock cylinder and turning key into the same position used on removal.

– Insert wire tool into opening and slide assembly fully into ignition lock housing until seated.

– Remove wire tool and check for smooth operation of ignition lock and cylinder.

– Installation is reverse of removal for remaining components.

> ***WARNING —***
> *Observe all cautions, warnings and notes when completing repairs involving airbag systems.*

Steering column crossmember, assembly

1. **Steering column crossmember**
2. **Adjustment brackets**
3. **Hex bolts**
   - Tighten to 25 Nm (18 ft-lb)
4. **Support, right**
5. **Bolts**
6. **Speed nuts**
7. **Support, left**
8. **Steering column assembly**
9. **Steering lock housing**
10. **Sealing plug**
11. **Support for bulkhead**
12. **Bolts**
    - Tighten to 8 Nm (71 in-lb)

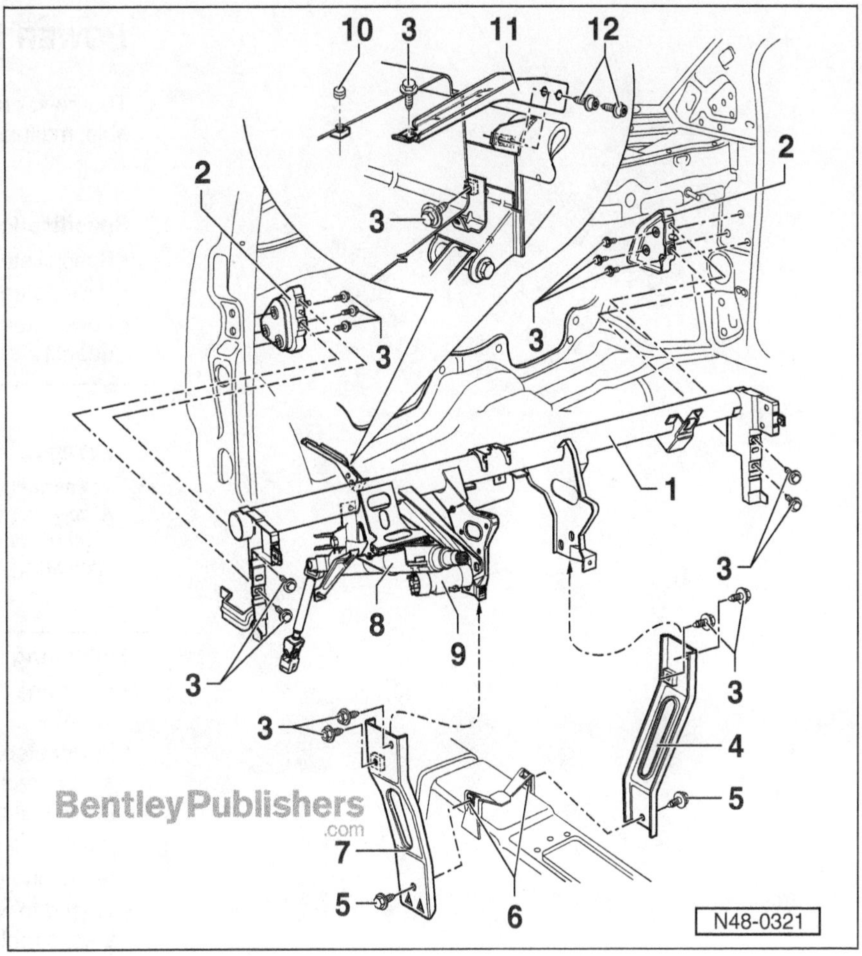

POWER STEERING SYSTEM

The power rack and pinion steering gear is not serviceable, malfunctioning units must be replaced.

Specifications

- Special steering rack grease . . . . . . AOF 063 000 04
- Power steering hydraulic oil . . . . G 002 000
- Power steering hydraulic oil quantity . . . . 0.7-0.9 L (11.8-15.2 oz.)

NOTE —
Volkswagen part numbers are given for reference only! Always consult with your Volkswagen Parts Department or aftermarket parts specialist for the latest parts information.

Tightening torques

- Subframe to body (always replace) (M14) . . . . .100 Nm (74 ft-lb) + ¼ turn (90°)
- Power steering gear to subframe always replace (M8). .20 Nm (15 ft-lb) + ¼ turn (90°)
- Banjo bolt to steering gear
  pressure line (M14) . . . . 38 Nm (28 ft-lb)
  return line (M16). . . . . 45 Nm (33 ft-lb)
- Tie rod to wheel bearing housing (M10) always replace . . . . . .20 Nm (15 ft-lb) + ¼ turn (90°)
- Heat shield to steering gear (M8). . . 22 Nm (16 ft-lb)
- Tie rod ball joint to tie rod (lock nut) (M14) . . . . 50 Nm (37 ft-lb)
- Tie rod to steering rack (M14) . . . . . 75 Nm (55 ft-lb)
- Steering column to cross member
  (M6) . . . . 10 Nm (7 ft-lb)
  (M8) . . . . 22 Nm (16 ft-lb)
- Steering column to steering gear (M8) . . . . 30 Nm (22 ft-lb)
- Universal joint to steering gear pinion always replace (M8). . . . . 30 Nm (22 ft-lb)
- Pendulum support to transmission see **10 Engine–Removing and Installing**

Power steering gear, assembly

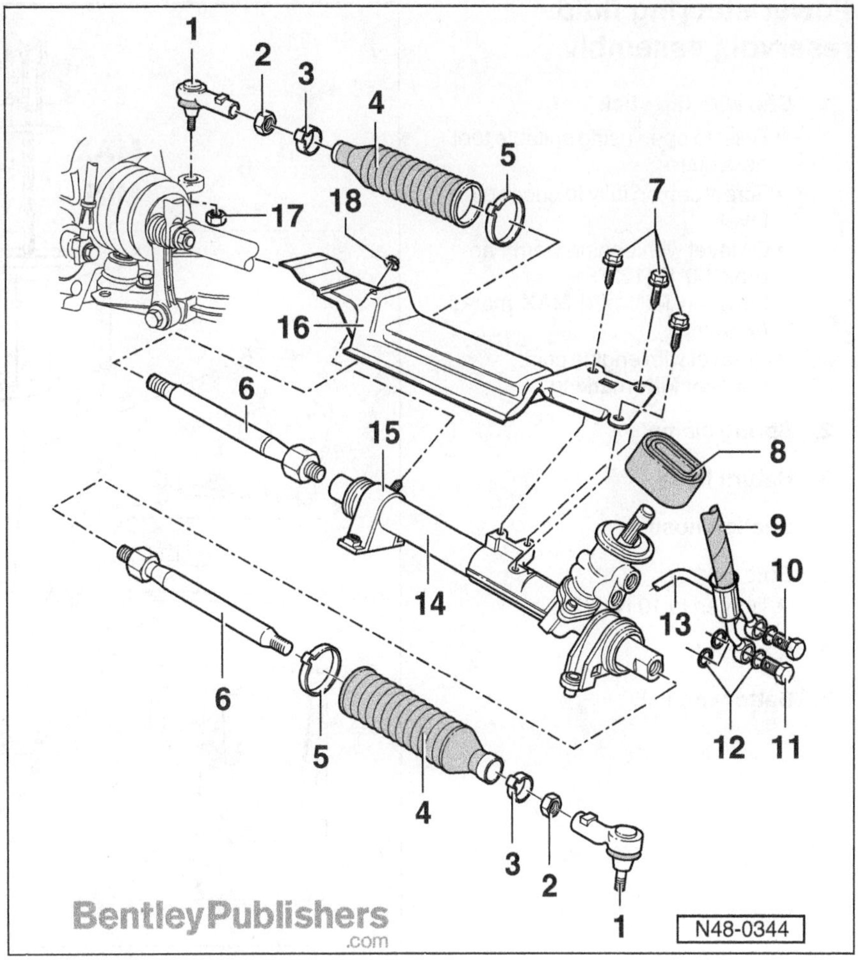

1. **Tie rod end**

   NOTE—
   Tie rod end design has been changed during the course of model production. Previous and new pieces cannot be mixed.

2. **Lock nut**
   - Tighten to 50 Nm (37 ft lb)
3. **Clamp**
   - Always replace
4. **Boot**
   - Must not be twisted while adjusting toe
   - Remove steering gear to replace
5. **Clamp**
   - Always replace
6. **Tie rod**
   - Tighten to 75 Nm (55 ft-lb)
   - Replacement parts are supplied preset
   - Check vehicle alignment if replaced

   NOTE—
   Tie rod design has been modified during model production, previous and new tie-rods cannot be mixed.

7. **Bolts**
   - Tighten to 22 Nm (16 ft-lb)
8. **Gasket**
9. **Hydraulic pressure hose**
10. **Banjo bolt**
    - Power steering return line
    - Tighten to 45 Nm (33 ft-lb)
11. **Banjo bolt**
    - Power steering pressure line
    - Tighten to 38 Nm (28 ft-lb)
12. **Seals**
    - Always replace
13. **Hydraulic return line**
14. **Power steering gear**
    - See **Power steering rack, determining center position**
15. **Mounting bracket with rubber bushing**
    - Replace if threads in welded nut are damaged
16. **Heat shield**
17. **Nut, self-locking**
    - Tighten to 20 Nm (15 ft-lb) plus ¼ turn (90°)
    - Always replace
18. **Nut**
    - Tighten to 22 Nm (16 ft-lb)

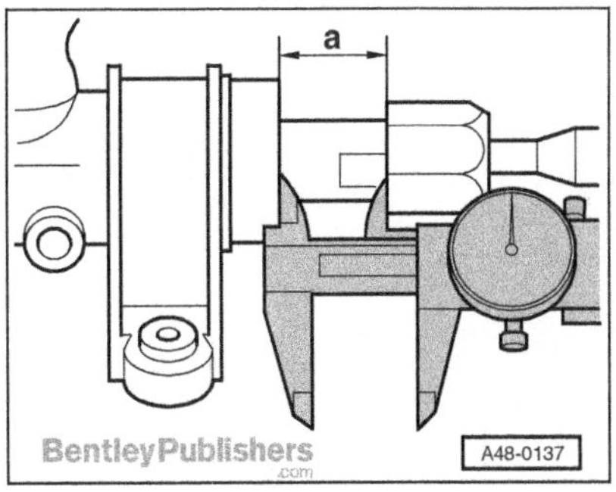

◀ **Power steering rack, determining center position**

- Center position must be identified before installation.
- Slide rack until dimension (**a**) is obtained.
- Dimension (**a**) = 30.5 mm (1.20 in).

Power steering fluid reservoir, assembly

1. **Cap with dip stick**
   - Twist to open using suitable tool as required
   - Screw cap on fully to check oil level
   - Oil level with engine warm, approx. 50°C (122°F): between **MIN** and **MAX** marks (**arrow A**)
   - Oil level with engine cold: not over **MIN** marking
2. **Spring clamp**
3. **Return hose**
4. **Suction hose**
5. **Bolt**
   - Tighten to 10 Nm (7 ft-lb)
6. **Reservoir**
7. **Battery tray**

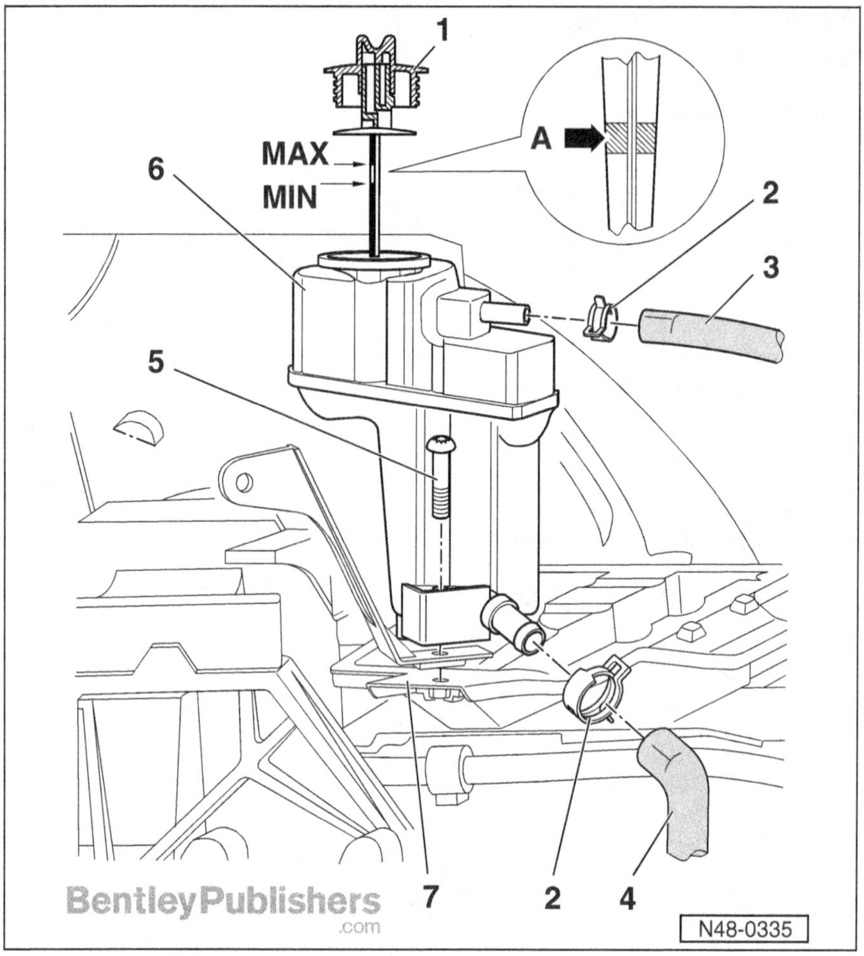

Power steering hydraulic lines and hoses (4-cylinder)

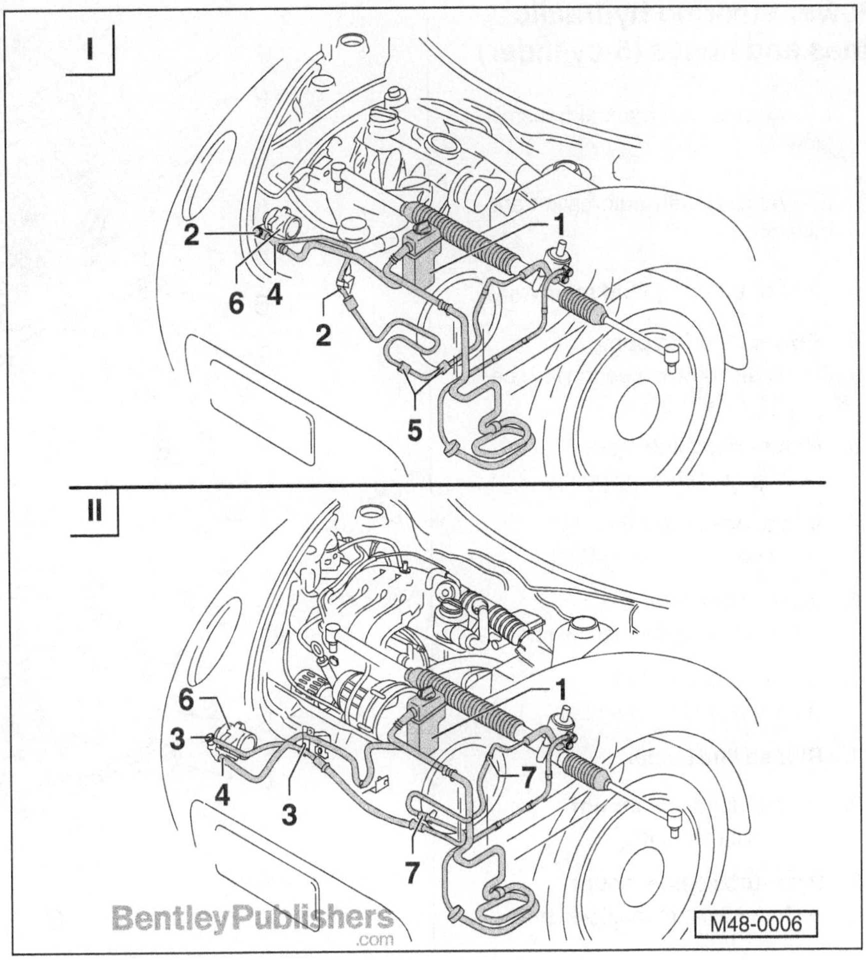

I - Vehicles with power steering pump in upper section

II - Vehicles with power steering pump in lower section

1. **Power steering fluid reservoir**
2. **Pressure hose assembly**
   - 1.9L engine with A/C
3. **Pressure hose assembly**
   - 1.8L, 1.9L without A/C, and 2.0L engine
4. **Suction hose**
5. **Flexible pressure hose**
   - 1.9L engine with A/C and manual transmission
6. **Power steering pump**
   - Fill with power steering fluid before installing
7. **Flexible pressure hose**
   - vehicles with automatic transmission

Power steering hydraulic lines and hoses (5-cylinder)

I - Vehicles with manual transmission

II - Vehicles with automatic transmission

1. **Power steering fluid reservoir**
2. **Power steering pump**
   - Fill with power steering fluid before installing
3. **Pressurized line mount**
   - On power steering pump
4. **Intake hose mount**
   - On power steering pump
5. **Intake hose mount**
   - On right longitudinal member
6. **Return hose mount**
   - On longitudinal member
7. **Ribbed line cooler**
8. **Pressurized line mount**
   - On cylinder block
9. **Pressurized line mount**
   - On 5-speed manual transmission (0A4)
10. **Pressurized line mount**
    - On automatic transmission (0A9G)

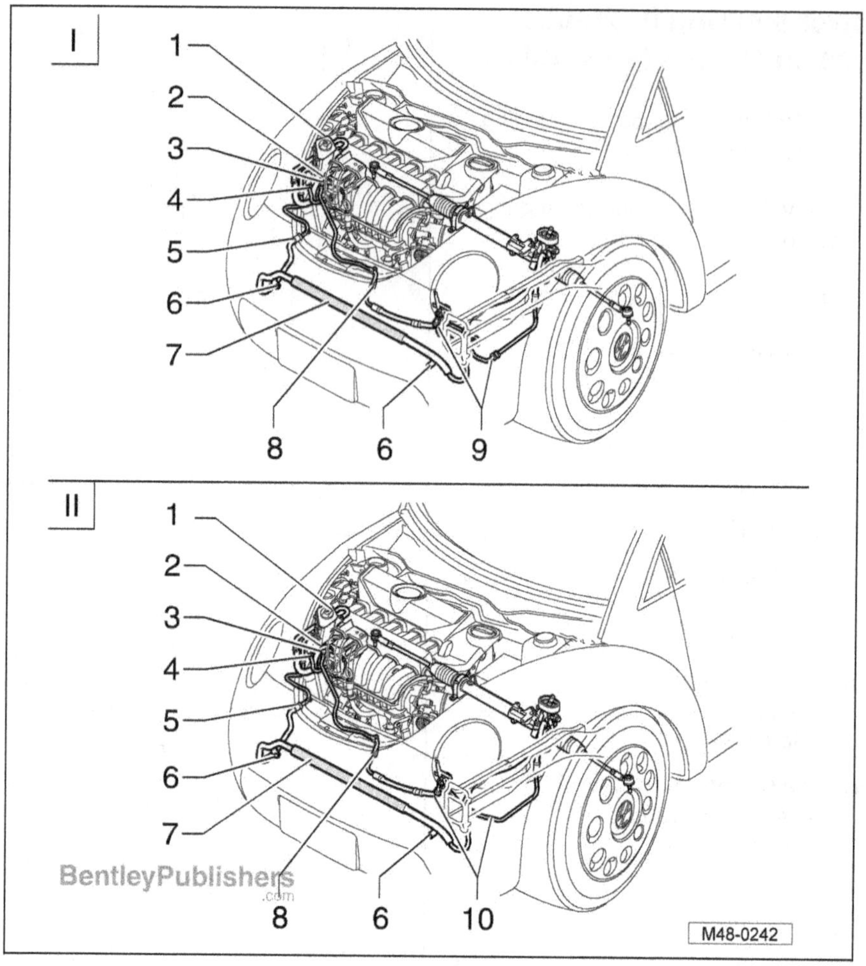

Power steering pump assembly (pump in lower mounting bracket)

NOTE —
Different power steering pump positions and belt routings are used depending on engine type and equipment level.

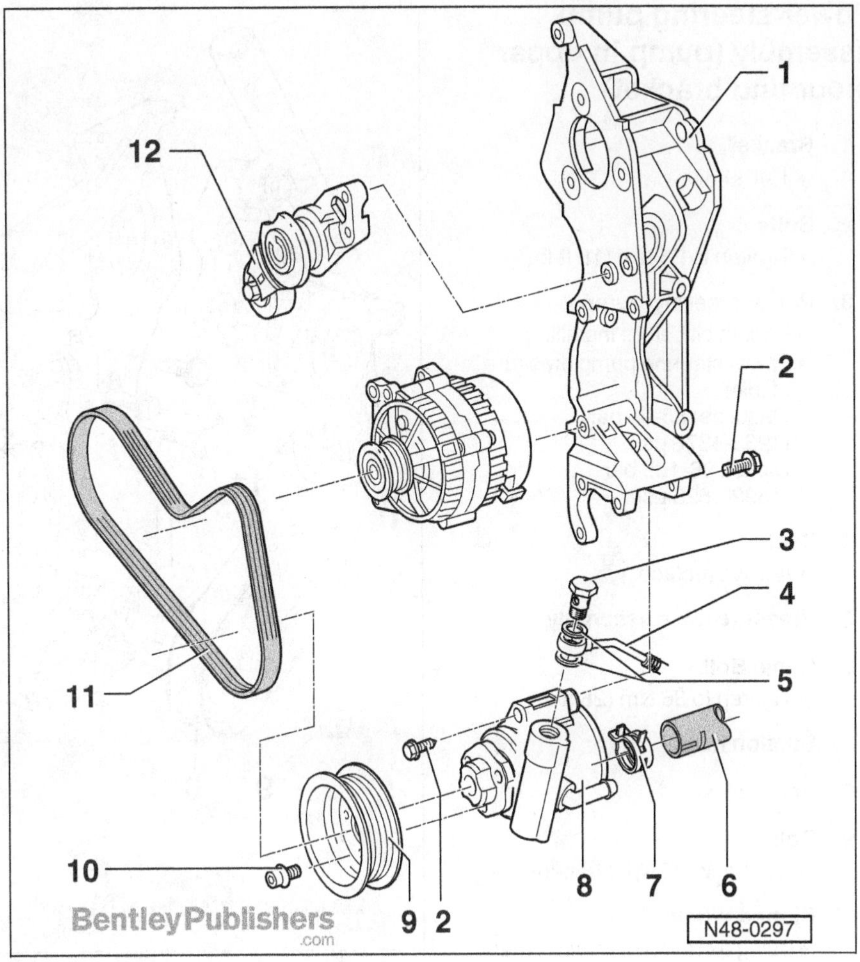

1. **Bracket**
   - 1.9L shown
2. **Bolts**
   - Tighten to 25 Nm (18 ft-lb)
3. **Banjo bolt**
   - Tighten to 38 Nm (28 ft-lb)
4. **Pressure hose assembly**
5. **Seals**
   - Always replace
6. **Suction hose**
7. **Spring clamp**
8. **Power steering pump**
   - Fill with oil before installing
   - Power steering pump pressure at idle:
     Gasoline: 85-95 bar (1233-1378 psi)
     Diesel: 96-105 bar (1392-1523 psi)
9. **Belt pulley**
10. **Bolt**
    - Tighten to 25 Nm (18 ft-lb)
11. **Ribbed V-belt**
    - Using a crayon or marker, mark the direction of travel before removing the ribbed belt
    - Reinstalling a used belt in the opposite direction could damage the belt
    - Make sure belt is correctly seated in the pulleys when installing
12. **Ribbed V-belt tensioner**

Power steering pump assembly (pump in upper mounting bracket)

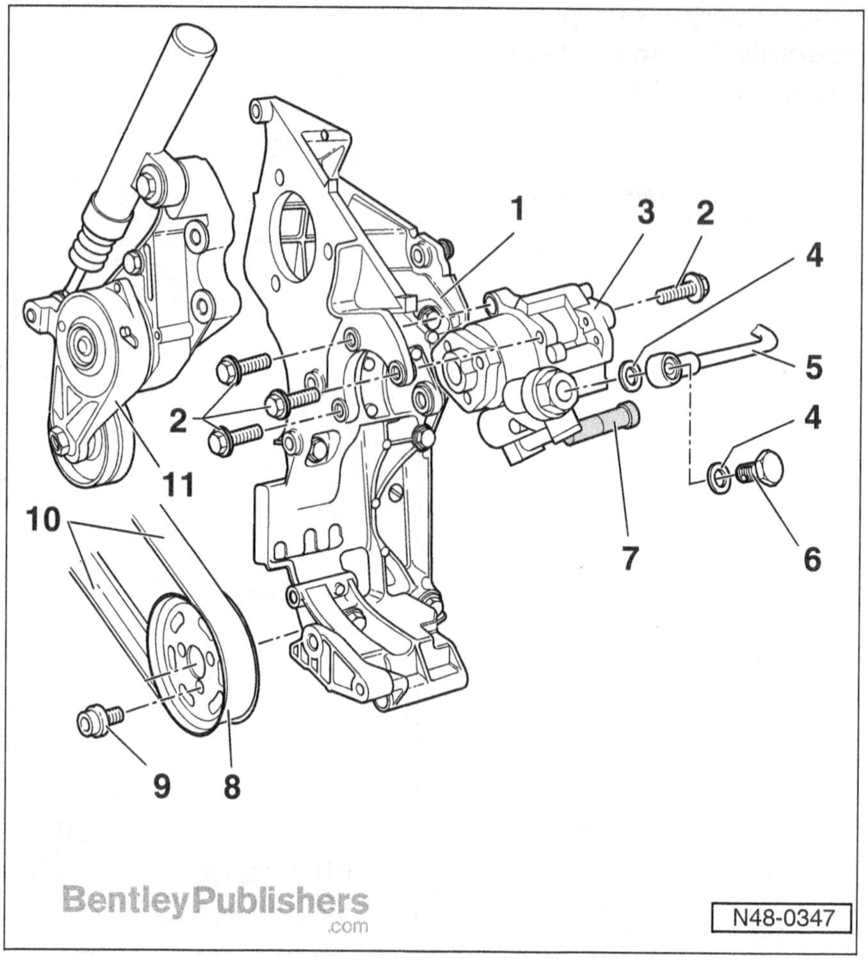

1. **Bracket**
   - 1.9L shown
2. **Bolts**
   - Tighten to 25 Nm (18 ft-lb)
3. **Power steering pump**
   - Fill with oil before installing
   - Power steering pump pressure at idle:
     Gasoline: 85-95 bar (1233-1378 psi)
     Diesel: 96-105 bar (1392-1523 psi)
4. **Seals**
   - Always replace
5. **Pressure hose assembly**
6. **Banjo Bolt**
   - Tighten to 38 Nm (28 ft-lb)
7. **Suction hose**
8. **Belt pulley**
9. **Bolt**
   - Tighten to 25 Nm (18 ft-lb)
10. **Ribbed V-belt**
    - Using a crayon or marker, mark the direction of travel before removing the ribbed belt
    - Reinstalling a used belt in the opposite direction could damage the belt
    - Make sure belt is correctly seated in the pulleys when installing
11. **Ribbed V-belt tensioner**

50 Body–Front

GENERAL

This repair group covers the front body section, front fender and radiator support removal and installation. The radiator support can be removed from the car as one unit to facilitate repairs, such as engine removal.

> ***WARNING —***
> *Disconnecting the negative (–) battery cable may erase fault codes and basic settings in the engine management and automatic transmission control modules. Some driveability problems may be noticed until the system re-adapts to operating conditions. OBD II readiness codes, which may be required for emissions testing, may also be erased. Convenience electronics (alarm system, interior light control, power locks, mirrors, and windows) may need to be re-set using a diagnostic scan tool.*

NOTE —
Volkswagen refers to the radiator support as a lock carrier.

FRONT BODY COMPONENTS

The front fenders are both part of the front clip which must be removed as a complete assembly.

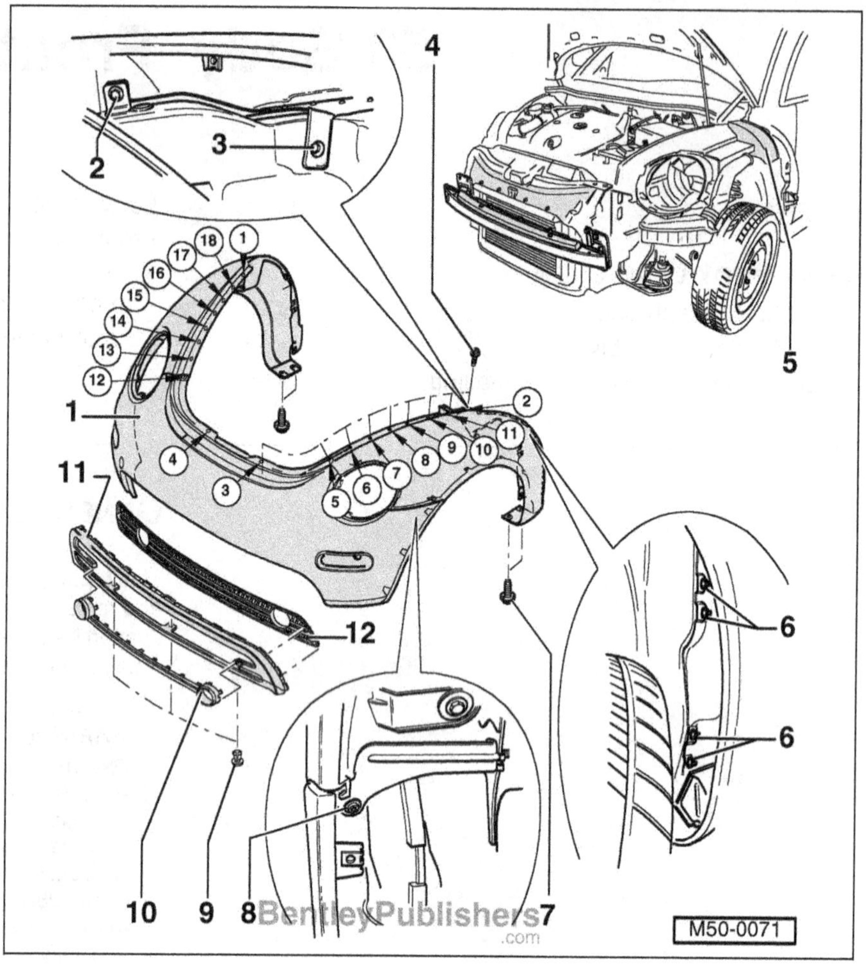

Front body assembly

1. **Front clip**
   - Use of low-tack tape on front quarter panels above fender mounting point is recommended to prevent damage

   Removing:

   - Remove headlight assembly, see **94 Lights, Accessories–Exterior**
   - Remove wheel housing liners, see **66 Body–Exterior Equipment**
   - Remove wheel housing bolts (items 2, 3, and 8)
     Remove nuts (item 6)
     Remove screws (items 4 and 7)
   - Disconnect fog light, turn signal, and temperature sensor harness connectors
   - Using a helper, pull front clip outward slightly on left and right and forward off vehicle

   Installing:

   - Before installing fender mounting screws, check condition of anti-friction foil at contact areas and replace if necessary
   - Ensure proper seating of foam cushion (item 5) on both sides of long member and make sure it engages properly on headlight bracket
   - Install upper fender mounting screws (item 4) in order shown by numbers in circles

2. **Bolt**
   - With nut
   - Tighten to 6 Nm (53 in-lb)

3. **Combi bolt**
   - Tighten to 6 Nm (53 in-lb)

4. **Screw**
   - Qty. 18
   - Tighten to 7.5 Nm (66 in-lb)
   - Install with locking fluid (VW part # D 185 400 A2)

5. **Foam cushion**
   - When installing the front clip, ensure the foam cushion is installed as shown on both sides

6. **Hex nut**
   - Tighten to 6 Nm (53 in-lb)

7. **Bolt**
   - Tighten to 7.5 Nm (66 in-lb)
   - Qty. 4
   - Install with locking fluid (VW part # D 185 400 A2)

8. **Bolt**
   - With nut
   - Tighten to 6 Nm (53 in-lb)
   - Install with locking fluid (VW part # D 185 400 A2)

9. **Spreader clip**
   - Qty. 3
   - To remove: Press spreader pin through clip

10. **Face plate**

11. **Front spoiler**

12. **Ventilation grid**

Front fender assembly

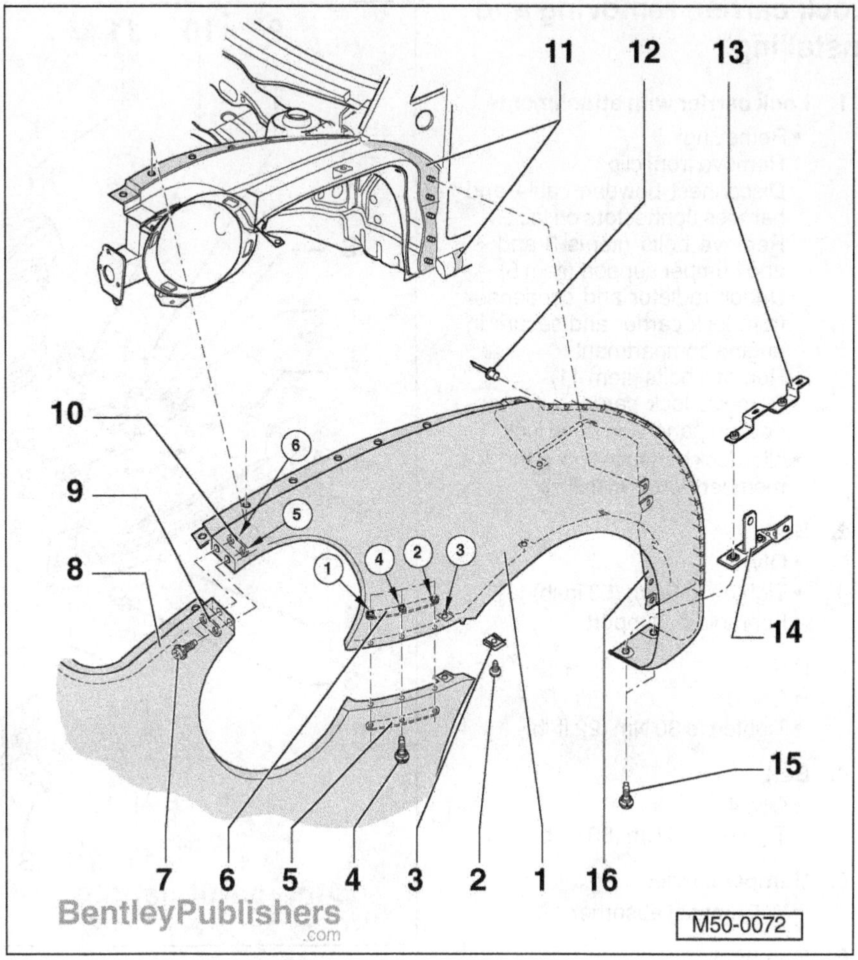

1. **Fender**
   - Remove with front clip as complete assembly
   - Before installing fender mounting screws, check condition of anti-friction foil at contact areas and replace if necessary
2. **Bolt**
   - Install with locking fluid (VW part # D 185 400 A2)
3. **Clip**
4. **Bolts**
   - Install with locking fluid (VW part # D 185 400 A2)
   - Qty. 3
   - Tighten to 6 Nm (53 in-lb)
   - Install in order shown by numbers in circles
5. **Reinforcement**
   - With holes
6. **Reinforcement**
   - With weld nuts
7. **Bolts**
   - Install with locking fluid (VW part # D 185 400 A2)
   - Qty. 2
   - Tighten to 6 Nm (53 in-lb)
8. **Front bumper cover**
9. **Reinforcement**
   - With holes
10. **Reinforcement**
    - With weld nuts
11. **Anti-friction foil**
    - Check condition, replace if necessary
12. **Pop rivet**
    - Qty. 19
13. **Bracket**
    - For convertible models
    - Bolted to outside of side member
14. **Bracket**
    - For sedan models
    - Bolted to outside of side member
15. **Bolt**
    - Qty. 2
    - Tighten to 6 Nm (53 in-lb)
16. **Close-out panel**
    - Riveted to fender

Lock carrier, removing and installing

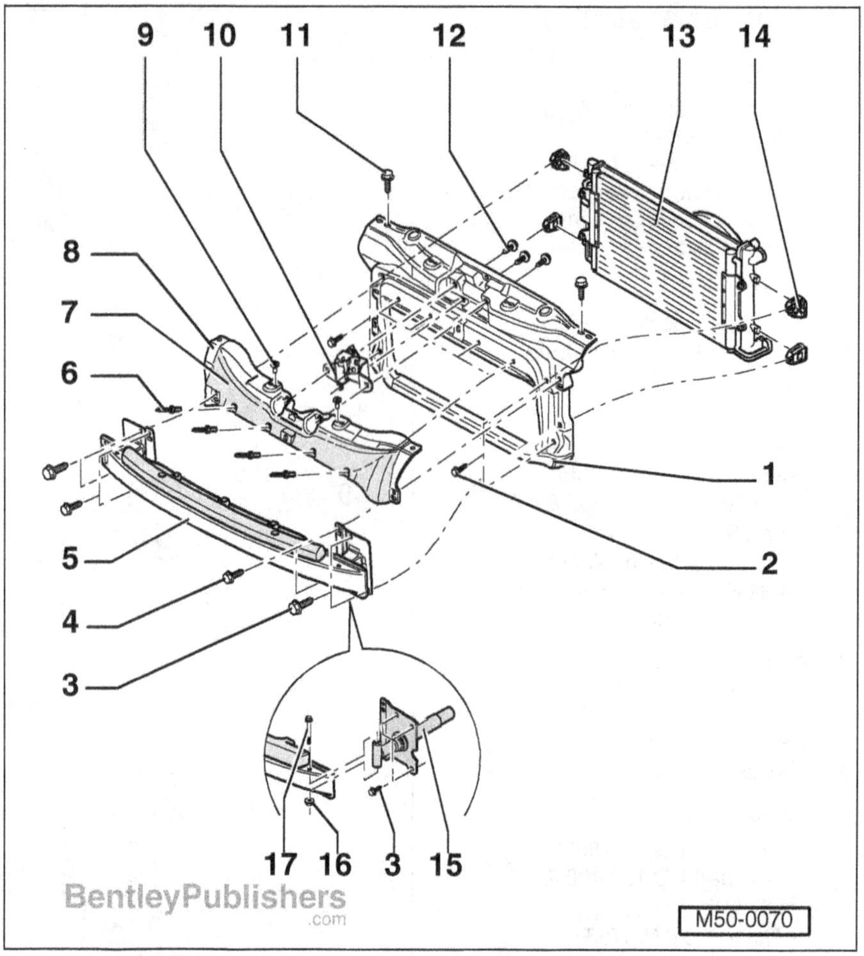

1. **Lock carrier with attachments**
   - Removing:
     Remove front clip
     Disconnect bowden cable and harness connectors on lock
     Remove bolts (items 3 and 4) and bumper support (item 5)
     Unbolt radiator and condenser from lock carrier and secure in engine compartment
     Remove bolts (item 11)
     Remove lock carrier with connecting panel and hood lock
   - Align lock carrier to floor pan members when installing
2. **Bolt**
   - Qty. 4
   - Tighten to 6 Nm (53 in-lb)
   - For radiator support
3. **Bolt**
   - Qty. 6
   - Tighten to 30 Nm (22 ft-lb)
4. **Bolt**
   - Qty. 4
   - Tighten to 9 Nm (80 in-lb)
5. **Bumper carrier**
   - With impact absorber
6. **Pop rivet**
   - Qty. 4
7. **Insulation**
8. **Connecting panel**
   - Fastened to lock carrier with blind rivets
9. **Blind rivet nut**
   - Qty. 2
10. **Hood lock**
    - Removing and installing, see **55 Hood and Lids**
11. **Bolt**
    - Tighten to 20 Nm (15 ft-lb)
12. **Bolt (self-locking)**
    - Tighten to 12 Nm (9 ft-lb)
    - When reusing: clean bolt threads with a wire brush and nut threads with thread chaser; coat bolt with locking fluid (VW part no. D 185 400 A 2)
13. **Radiator and condenser**
14. **Bracket**
15. **Impact absorber**
16. **Nut**
17. **Bolt**

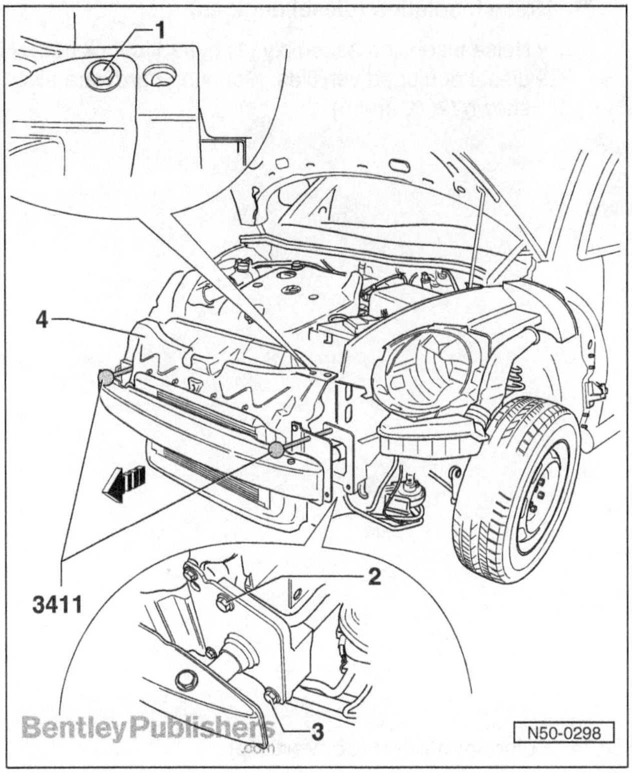

Lock carrier, service position

The lock carrier can be moved forward to allow easier access for various service procedures.

– Remove front clip as described earlier.

– Remove front bumper, see **63 Bumpers**.

– Disconnect bowden cable on hood lock.

– Remove one lock carrier mounting bolt (**2**) from each floor pan member and install VW special tool 3411 (guide rods), or equivalent. Remove remaining lock carrier mounting bolts (**1**, **3**) and pull lock carrier forward.

Front noise insulation

Gasoline and diesel equipped vehicles have different lower sound absorption panels (belly pans).

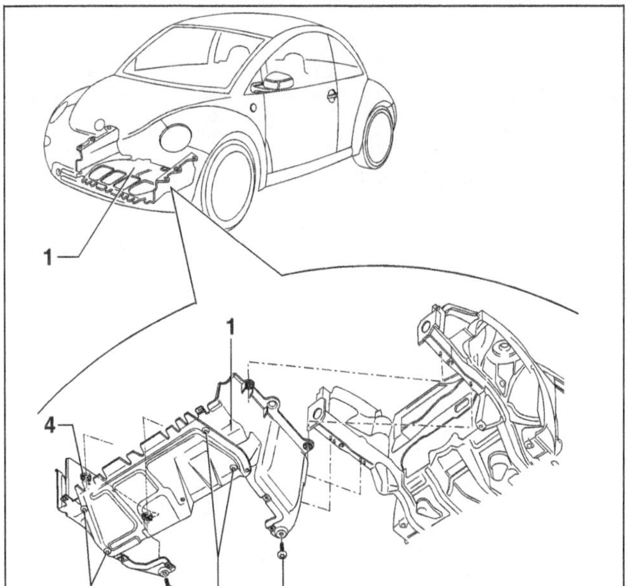

Noise insulation (gasoline engines)

- Noise insulation assembly (**1**) bolted to lock carrier in gasoline equipped vehicles. Mounting hardware also shown (**2**, **3**, and **4**).

NOTE —

*See **03 Maintenance** for more details on removal and installation of various lower front sound absorber panels.*

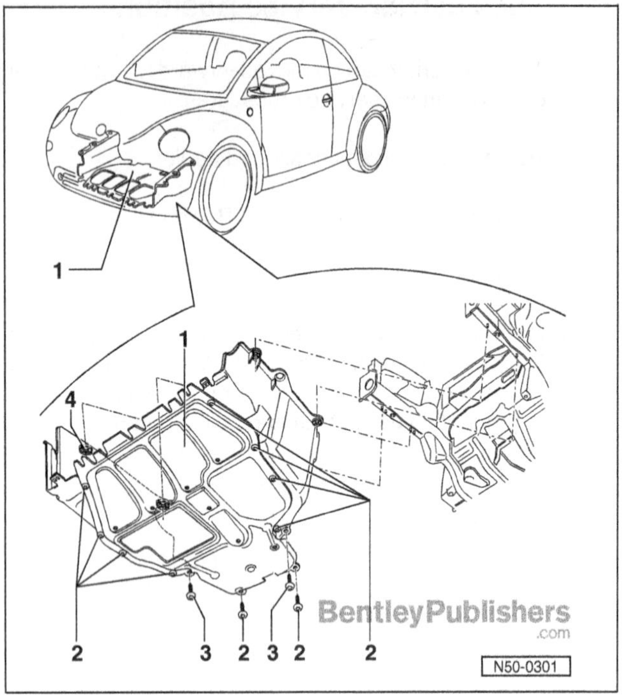

Noise insulation (diesel engines)

- Noise insulation assembly (**1**) bolted to lock carrier in diesel equipped vehicles. Mounting hardware also shown (**2**, **3**, and **4**).

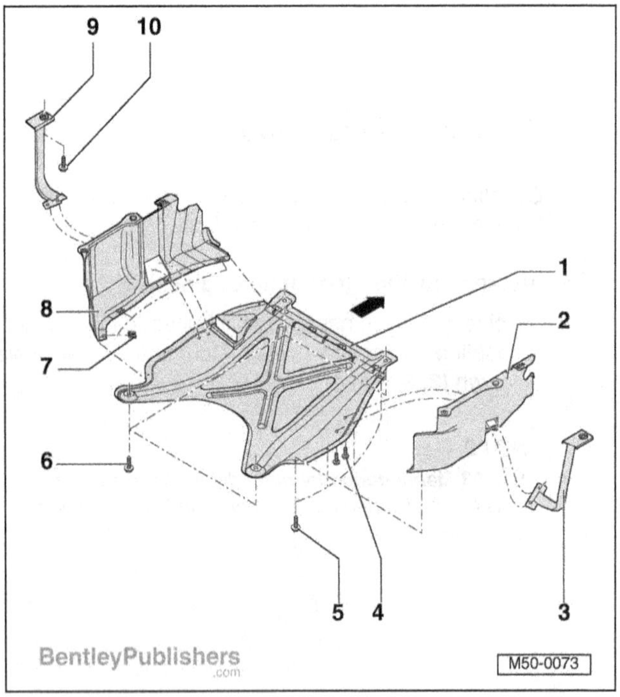

Noise insulation (convertible)

- **Arrow** points toward front of vehicle.

1. Center noise insulation panel
2. V-belt guard
3. Right support bracket
4. Bolt
   - Qty. 4
5. Bolt
   - Qty. 8
6. Bolt
   - Qty. 4
7. Speed nut
   - Qty. 8
8. Engine screening
9. Left support bracket
10. Bolt
    - Qty. 2

Plenum close-out panel, removing and installing

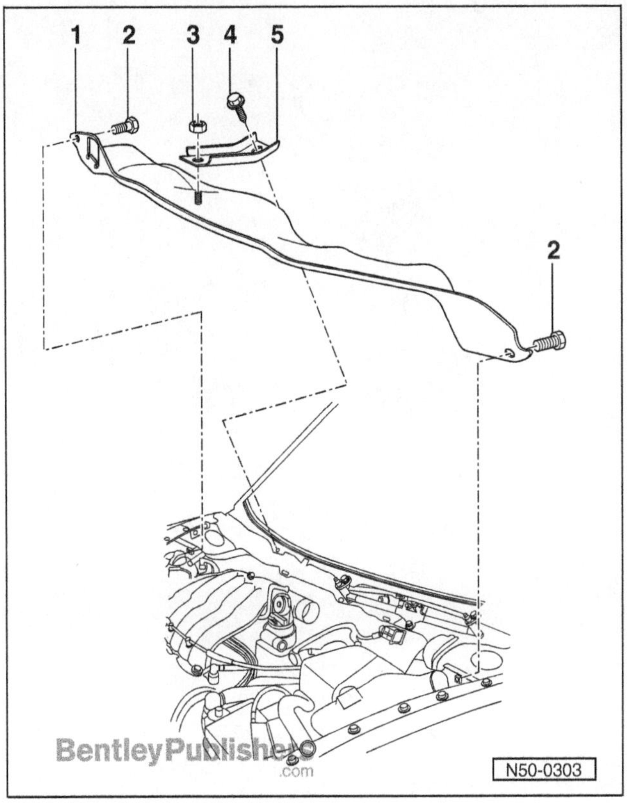

- Remove windshield wiper arms.
- Remove seal from front edge of close-out panel.
- Remove plastic plenum cover panel (outer).
- If necessary, remove engine components from front of close-out panel.

◄ Remove mounting bolt (**4**) and nut (**3**), then remove bracket (**5**).

- Remove mounting bolts on outer edges of close-out panel and remove panel.
- Installation is reverse of removal.

Counterbalance weight (convertible)

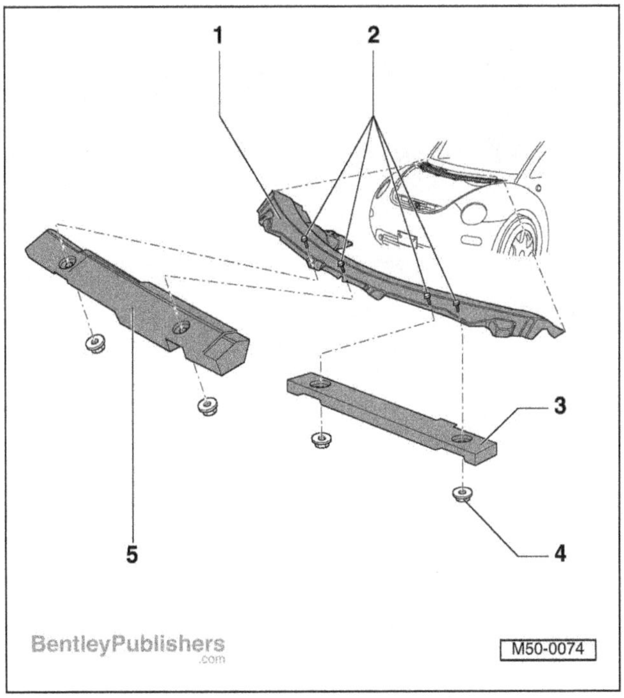

◄ Convertible models have counterbalance weights located in the plenum chamber at the rear of the engine compartment.

1. Bulkhead carrier
2. Bolt
   - Qty. 4
3. Balance weight
4. Nut
   - Qty. 4
   - Tighten to 5 Nm (44 in-lb)
5. Balance weight

53 Body–Rear

GENERAL

The rear clip is removed as a complete unit. It will be necessary to have a helper when pulling the assembly away from the car.

NOTE —

Use of low-tack tape on rear quarter panels above fender mounting point is recommended to prevent damage to paint.

REAR FENDER COMPONENTS

Rear end assembly (sedan)

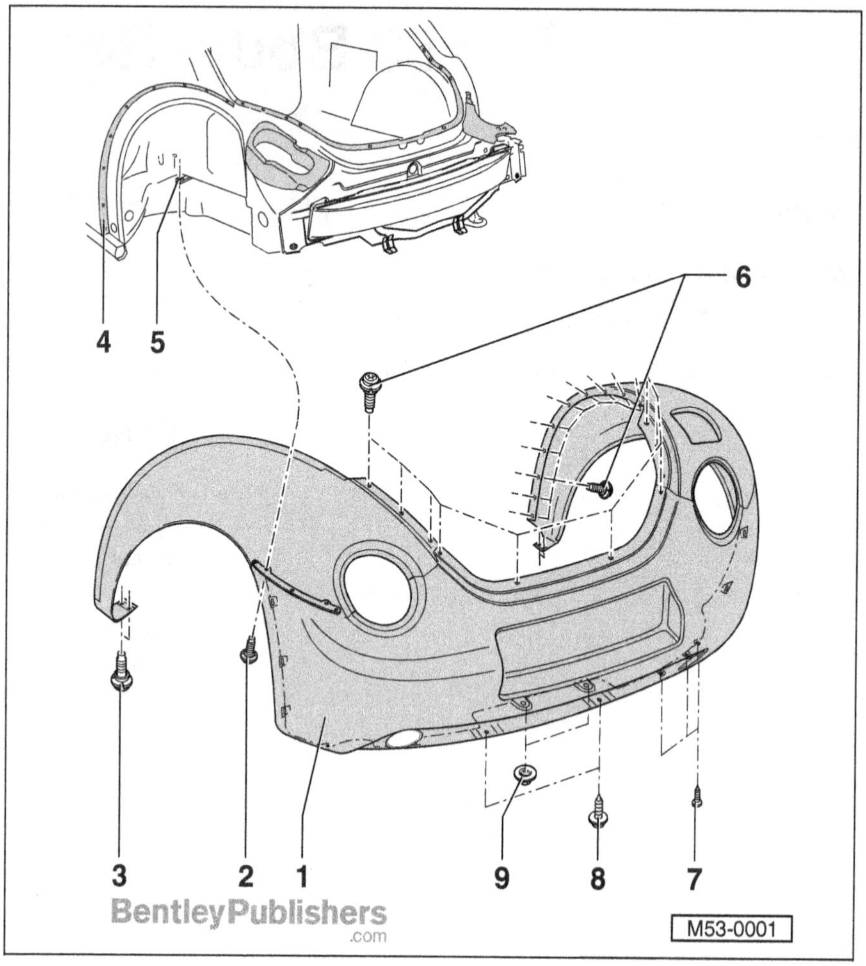

1. **Rear clip**

   Removing:

   - Remove rear wheel housing liners, see **66 Body–Exterior Equipment**.
   - Remove tail light assembly, see **94 Lights, Accessories–Exterior**.
   - Remove fuel tank filler lid, see **55 Hoods and Lids**.
   - Remove bolts (**2**, **7**, and **8**). Remove nuts (**9**).
   - Remove bolts (**3** and **6**).
   - Pull rear end on left and right hand sides outwards off threaded studs and carry away with aid of a second mechanic.

   Installing:

   - Before installing, check condition of anti-friction foil on fender contact surfaces, replace if necessary.

2. **Bolt**
   - Tighten to 6 Nm (53 in-lb)
   - Install with locking fluid (VW part # D 185 400 A2)

3. **Bolt**
   - Tighten to 7.5 Nm (66 in-lb)
   - Qty. 3
   - Install with locking fluid (VW part # D 185 400 A2)

4. **Anti-friction foil**

5. **Captive weld nuts**

6. **Bolt**
   - Tighten to 7.5 Nm (66 in-lb)
   - Qty. 28
   - Install with locking fluid (VW part # D 185 400 A2)

7. **Bolt**
   - Tighten to 1.2 Nm (10 in-lb)
   - Qty. 3

8. **Bolt**
   - Tighten to 7.5 Nm (66 in-lb)
   - Qty. 2
   - Install with locking fluid (VW part # D 185 400 A2)

9. **Spring nut**
   - Qty. 2

Rear end assembly (convertible)

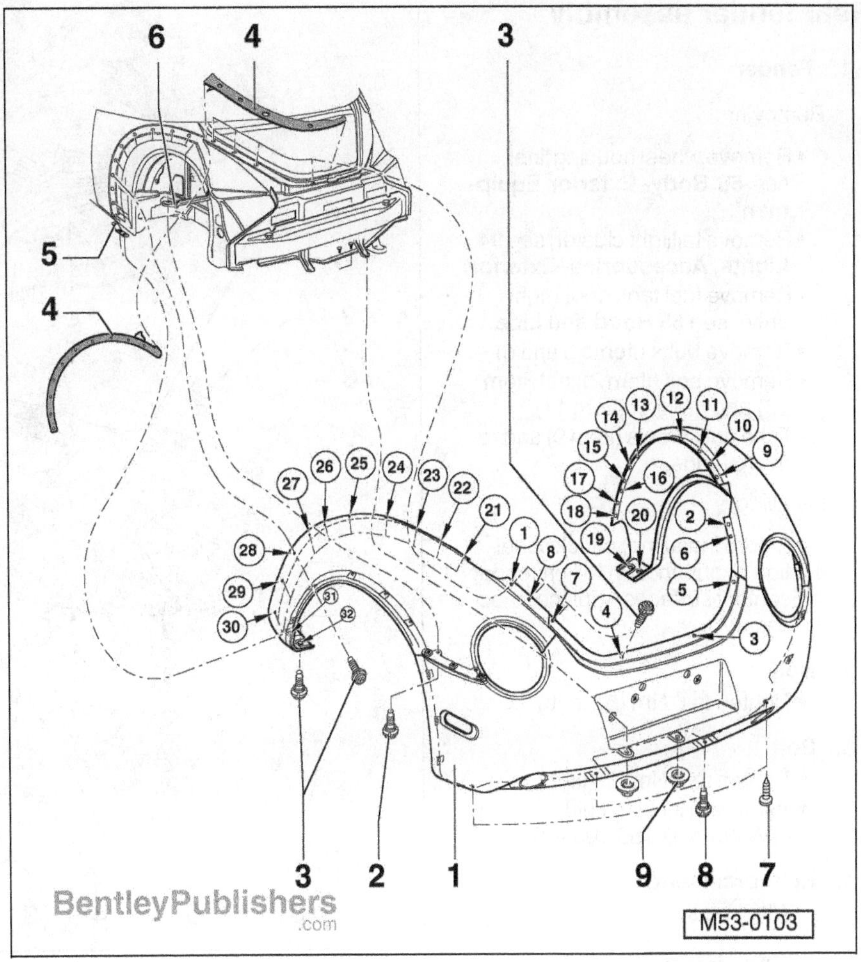

1. **Rear clip**

   Removing:

   - Remove rear wheel housing liners, see **66 Body–Exterior Equipment**.
   - Remove tail light assembly, see **94 Lights, Accessories–Exterior**.
   - Remove fuel tank filler lid, see **55 Hoods and Lids**.
   - Remove bolts (**2**, **7**, and **8**). Remove nuts (**9**).
   - Remove bolts (**3** and **6**).
   - Pull rear end on left and right hand sides outwards off threaded studs and carry away with aid of a second mechanic.

   Installing:

   - Before installing, check condition of anti-friction foil on fender contact surfaces, replace if necessary.
   - Tighten fender mounting bolts (item 3) in order shown by numbers in circles

2. **Bolt**
   - Tighten to 5 Nm (44 in-lb)
   - Install with locking fluid (VW part # D 185 400 A2)

3. **Bolt**
   - Tighten to 5 Nm (44 in-lb)
   - Qty. 32
   - Install with locking fluid (VW part # D 185 400 A2)

4. **Anti-friction foil**

5. **Bracket**

6. **Bracket**

7. **Bolt**
   - Tighten to 1.2 Nm (10 in-lb)
   - Qty. 2

8. **Bolt**
   - Tighten to 7.5 Nm (66 in-lb)
   - Qty. 2
   - Install with locking fluid (VW part # D 185 400 A2)

9. **Spring nut**
   - Qty. 2

Rear fender assembly

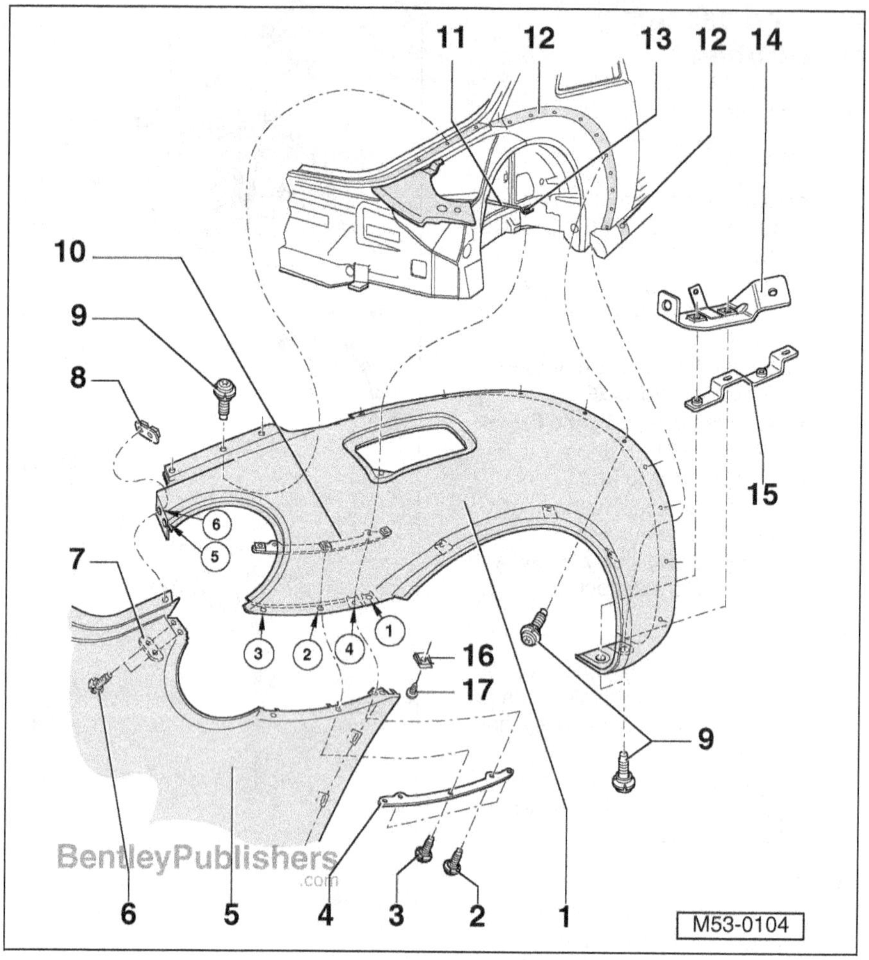

1. **Fender**

   Removing:

   - Remove wheel housing liner, see **66 Body–Exterior Equipment.**
   - Remove taillight cluster, see **94 Lights, Accessories–Exterior**
   - Remove fuel tank door (right only), see **55 Hood and Lids.**
   - Remove bolts (items 3 and 6)
   - Remove bolt (item 2), nut (item 13) and bracket (item 11).
   - Take out screws (item 9) and remove fender

   Installing:

   - Before installing, check condition of anti-friction foil on fender contact surfaces, replace if necessary.

2. **Bolt**
   - Tighten to 6 Nm (53 in-lb)
3. **Bolt**
   - Tighten to 6 Nm (53 in-lb)
   - Install with locking fluid (VW part # D 185 400 A2)
4. **Reinforcement**
   - With holes
5. **Bumper cover, rear**
   - Removing, see **63 Bumpers**
6. **Bolt**
   - Tighten to 6 Nm (53 in-lb)
7. **Reinforcement**
   - With holes
8. **Reinforcement**
   - With welded nuts
9. **Screw**
   - Tighten to 7.5 Nm (67 in-lb)
   - Qty. 14
   - Install with locking fluid (VW part # D 185 400 A2)
10. **Reinforcement**
    - With weld nuts
11. **Bracket**
    - Bolted to front of floor pan member
12. **Anti-friction foil**
13. **Nut**
14. **Bracket**
    - Sedan models
    - Bolted to outside of side member
15. **Bracket**
    - For convertible models
16. **Clip**
17. **Bolt**
    - Follow tightening order shown by numbers in circles

55 Hood and Lids

GENERAL

The front hood and rear hatch are easily removed with the aid of a helper. Where applicable, the support strut(s) should first be disconnected. Be sure to mark the lid location in reference to the hinges for proper alignment during installation.

55

FRONT HOOD

Hood assembly

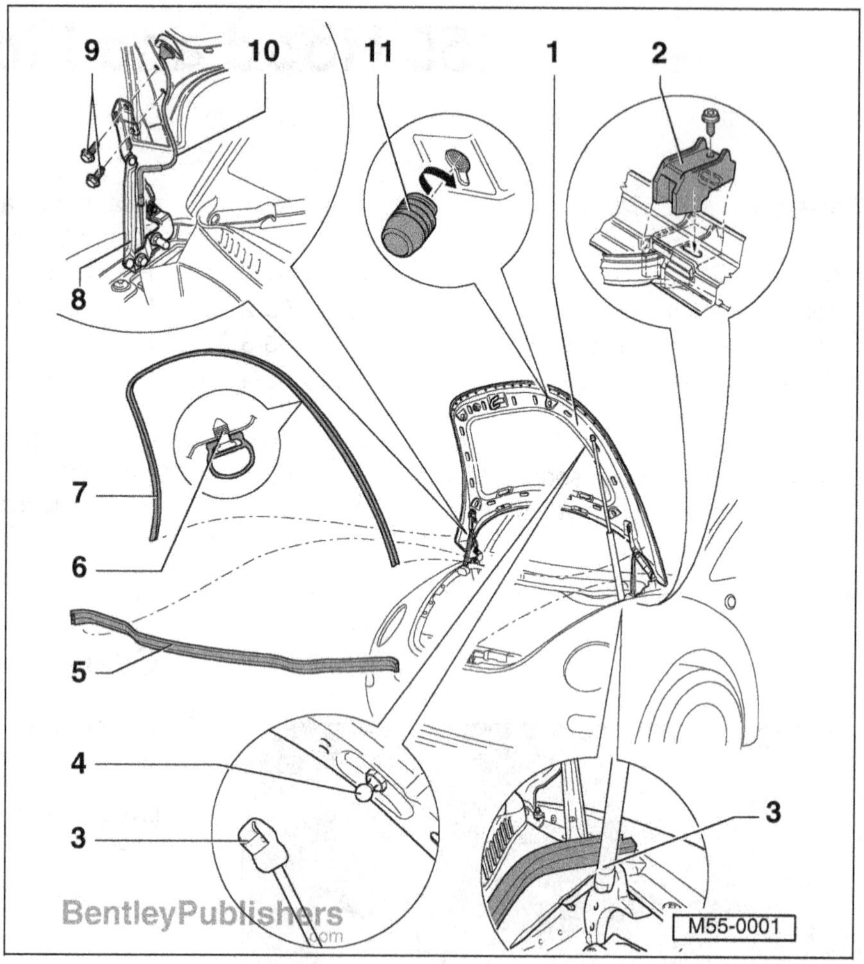

1. **Hood**
   - Removing:
     Remove windshield washer hoses and jets.
     Take out bolts (item 9) and remove hood.
2. **Cushion**
3. **Hood strut**
   - From 1999-on
4. **Ball stud**
   - Tighten to 21 Nm (15 ft-lb)
5. **Plenum seal**
   - Pushed onto flange
6. **Clip**
7. **Hood seal**
8. **Hinge**
   - Hood can be aligned between fenders by moving hinges within the slotted holes.
   - After installation or adjusting, treat hinges and bolts with corrosion protection.
9. **Bolt**
   - Tighten to 23 Nm (17 ft-lb)
10. **Hose**
    - For windshield washer spray jet
    - Removing, see **92 Wipers and Washers**
11. **Rubber stop**
    - Adjust hood height to fenders using rubber stops
    - Location may vary

Front gas strut, removing (from 1999-on)

– Support lid.

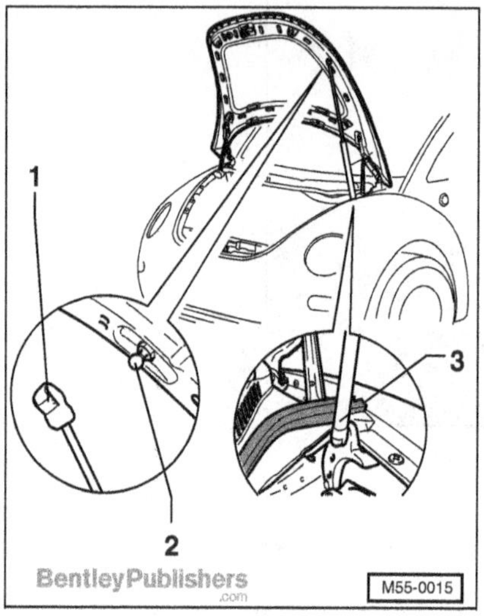

< Pry up retaining clip (**1**) with a screwdriver and pull gas strut off ball stud (**2**).

NOTE —
*If gas strut is to be reused, retaining clip (**1**) must not be taken out of ball socket completely, otherwise retaining clip will be damaged.*

1. Retaining clip
2. Ball stud
   - 21 Nm (15 ft-lb)
3. Gas strut

Hood release cable assembly

1. **Hood release (bowden) cable**
   - Removing:
     Remove release lever (item 4).
     Remove cable from clip at lock (item 10) and engine compartment (item 9).
     Remove lower A-pillar trim, see **70 Trim–Interior**.
     Tie string (approx. 39 in.) to cable at hood release and pull cable out at release lever end.
   - Use the same string to pull new cable into place.
2. **Grommet**
3. **Spreader nut**
4. **Release lever**
   - See **Release lever, removing and installing**
5. **Self-tapping screw**
6. **Mounting bracket (sedan)**
   - Cable engaged
7. **Nut**
8. **Mounting bracket (convertible)**
9. **Hood release cable bracket**
10. **Hood lock**

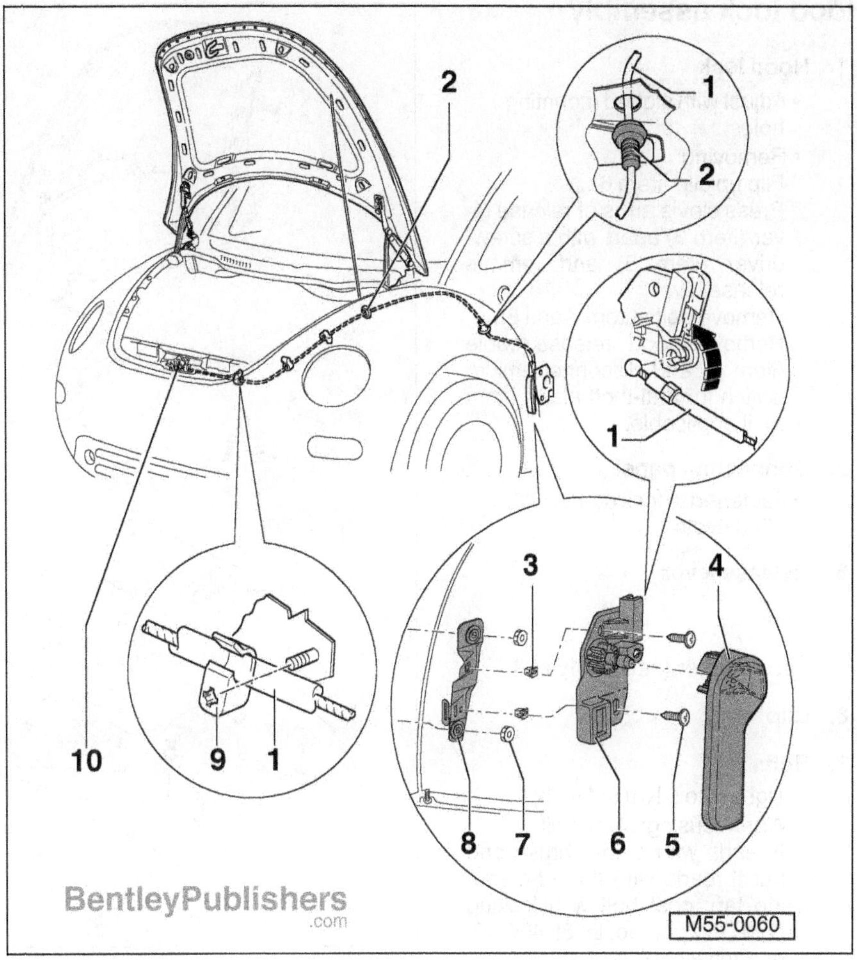

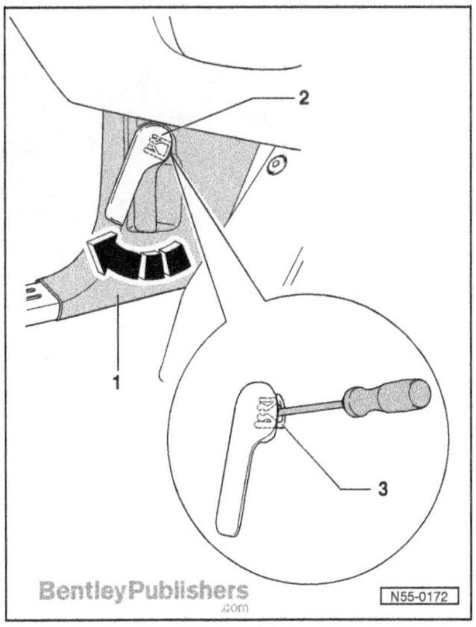

Release lever, removing and installing

- Pull back release lever (**1**) approx. 2 cm (0.8 in.).
- Insert small screwdriver into gap between release lever and clip (**3**).
- Pull release lever again and remove clip from release lever (clip falls behind trim).
- To install, push clip (**3**) completely into release lever and press release lever onto bracket.
- Check operation of release lever.

Hood lock assembly

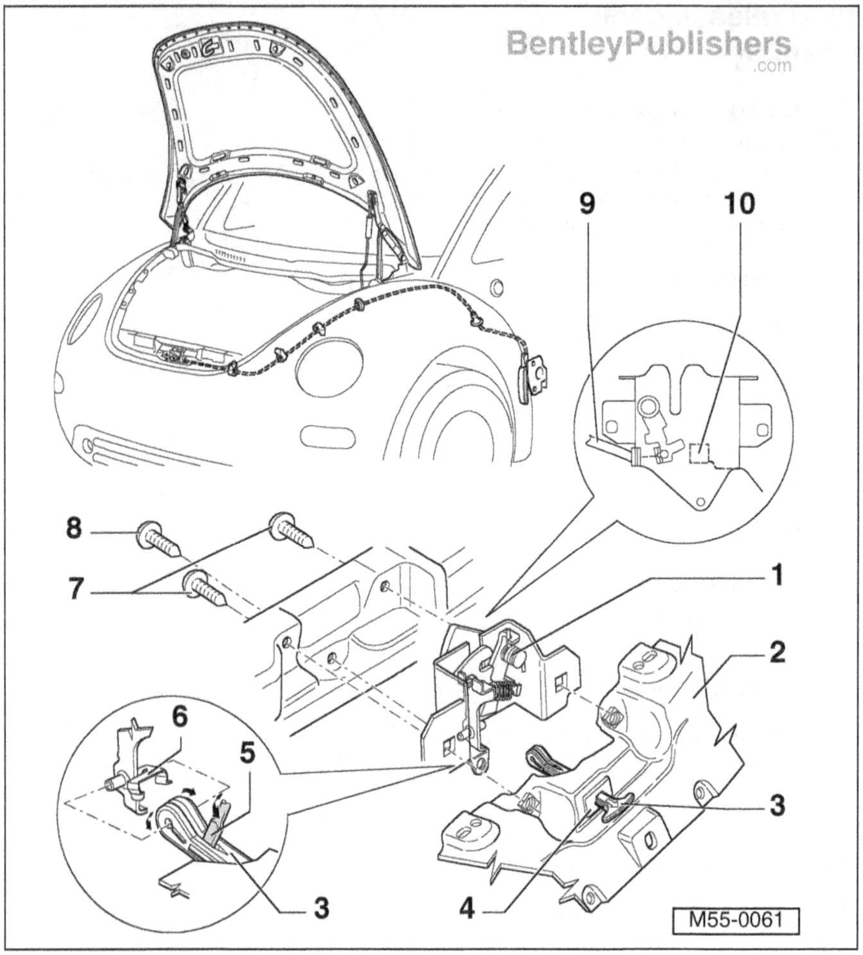

1. **Hood lock**
   - Adjust with slotted mounting holes
   - Removing:
     Flip up clip (item 6).
     Press clevis arms of release lever (item 3) apart with a screwdriver (item 5) and remove release lever.
     Remove bolts (item 7 and 8).
     Remove hood release cable (item 9) and disconnect micro switch for anti-theft alarm (item 9), if applicable.
2. **Connecting panel**
   - Fastened to lock carrier with blind rivets
3. **Release lever**
4. **Clip**
5. **Screwdriver (assembly aid)**
6. **Clip**
7. **Bolts**
   - Tighten to 5 Nm (44 in-lb)
   - When reusing: clean bolt threads with a wire brush and nut threads with thread cleaning tap; coat bolt with locking fluid (VW part no. D185 400 A2, or equivalent)
8. **Bolt**
   - Tighten to 15 Nm (11 ft lb)
   - When reusing: clean bolt threads with a wire brush and nut threads with thread cleaning tap; coat bolt with locking fluid (VW part no. D185 400 A2, or equivalent)
9. **Hood release cable**
10. **Micro switch**

REAR LIDS

Rear hatch assembly (sedan)

1. **Rear hatch**
2. **Bolt**
   - Tighten to 20 Nm (15 ft-lb)
3. **Hinge**
4. **Bolt**
   - Tighten to 8 Nm (71 in-lb)
5. **Nut**
   - Tighten to 8 Nm (71 in-lb)
6. **Gasket**
7. **Bolt**
   - Tighten to 23 Nm (17 ft-lb)
8. **Lock plate**
   - Adjust using slotted holes
9. **Bolt**
   - Tighten to 10 Nm (89 in-lb)
10. **Adjustment buffer**
11. **Ball stud**
    - Tighten to 21 Nm (15 ft-lb)
12. **Gas-filled strut**
13. **Bolt**
    - Tighten to 10 Nm (89 in-lb)
14. **Rubber buffer**
15. **Rubber buffer**

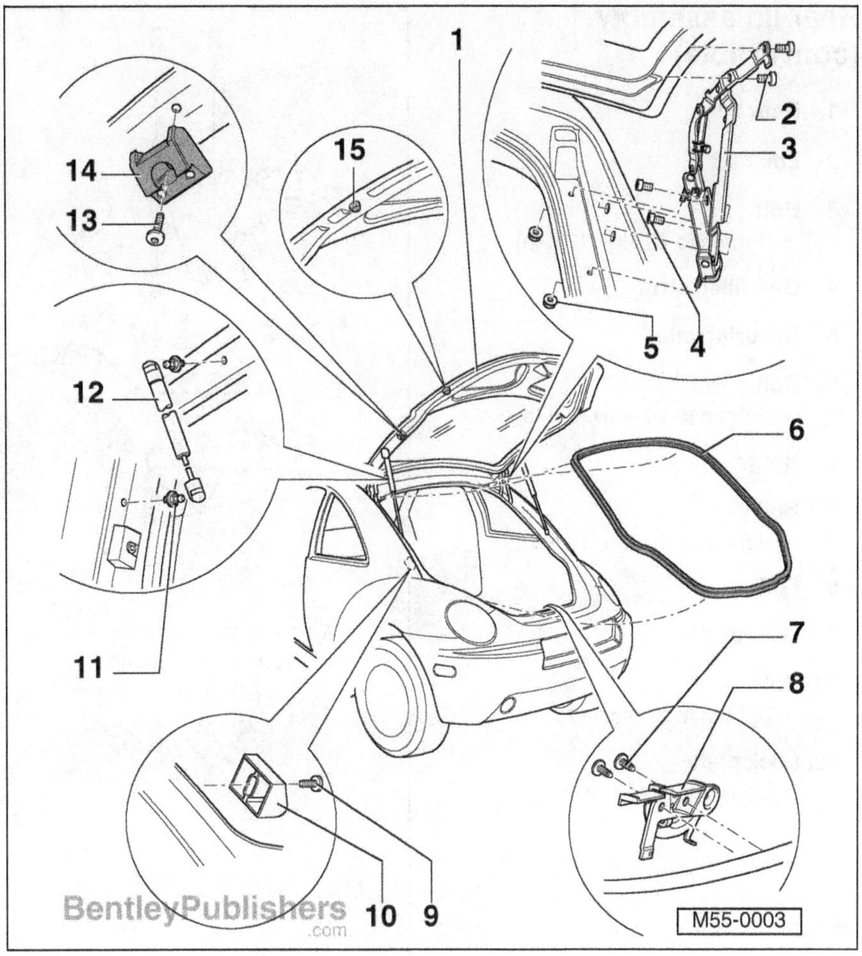

Rear lid assembly (convertible)

1. **Rear lid**
2. **Lock**
3. **Bolt**
   - Tighten to 23 Nm (17 ft-lb)
4. **Gas-filled strut**
5. **Securing clip**
6. **Bolt**
   - Tighten to 20 Nm (15 ft-lb)
7. **Hinge**
8. **Bolt**
   - Tighten to 20 Nm (15 ft-lb)
9. **Foil**
10. **Gasket**
11. **Bolt**
    - Tighten to 23 Nm (17 ft-lb)
12. **Lock plate**
    - Adjust using slotted holes

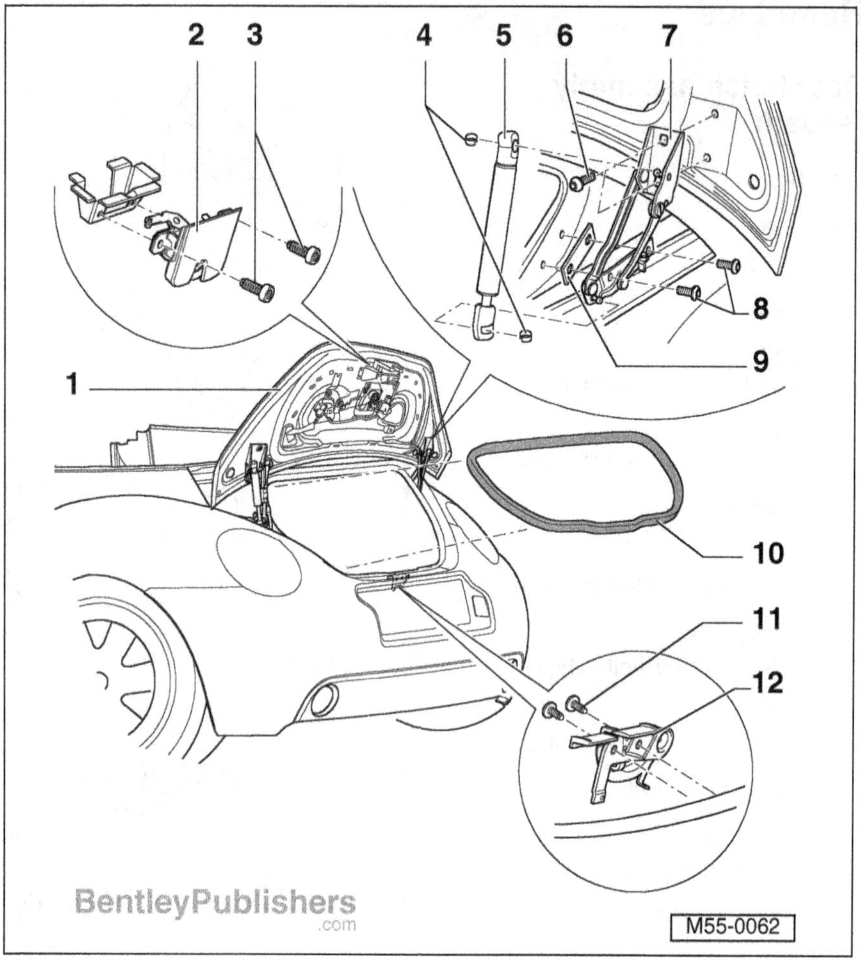

Rear gas strut, removing (convertible)

– Support lid.

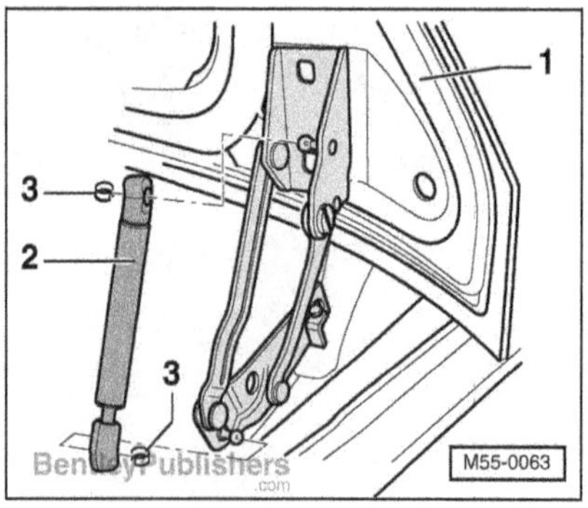

◄ Pry up retaining clip (**3**) with a screwdriver and pull gas strut (**2**) off ball stud.

> ***NOTE —***
> *If gas strut is to be reused, retaining clip (**3**) must not be taken out of ball socket completely, otherwise retaining clip will be damaged.*

1. Rear lid
2. Gas strut
3. Retaining clip

Rear hatch, adjusting (sedan)

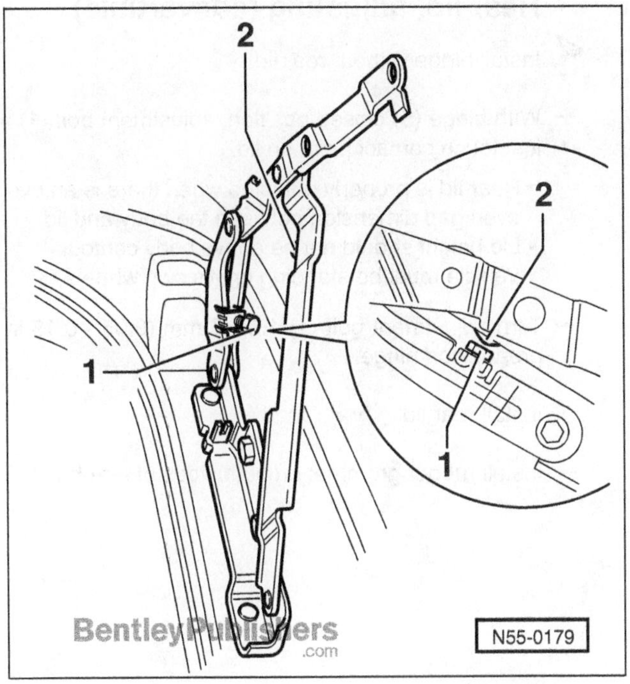

◄ Install hinge without rear hatch.

- When hinge is in closed position, set screw (**1**) must just touch contact stop (**2**) without pressure.

– Turn hinge set screws on both sides 1.5 - 3 mm (0.05 - 0.15 in.) to pre-tension hinge.

– Install rear hatch.

– Install rubber grommet on both sides of rear hatch.

◄ Install adjustment buffer (**1**) in channel loosely, to aid adjustment.

– Slide buffer up to top position.

– Close rear hatch completely and open again.

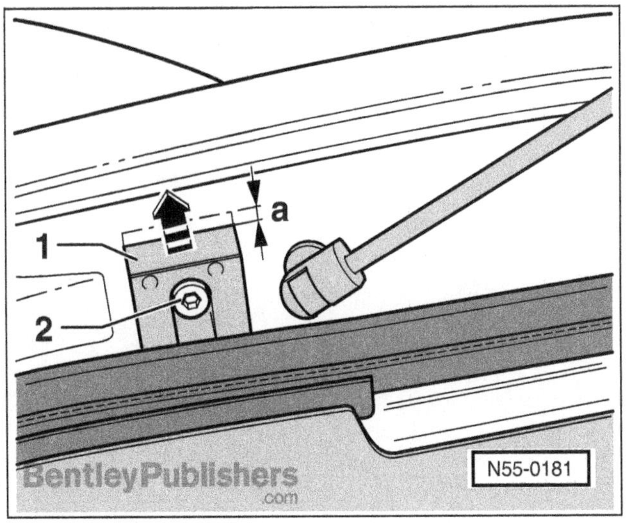

◄ Slide adjustment buffer (**1**) up 1 to 2 mm (dimension **a**).

– Check that rear hatch closes without too much force and doesn't have any up or down movement when in locked position. When unlocking, the lid will pop up slightly due to pre-tension on hinge.

Tightening torque

- Rear hatch adjustment buffer. . . . . . . 10 Nm (7 ft-lb)

Rear lid, adjusting (convertible)

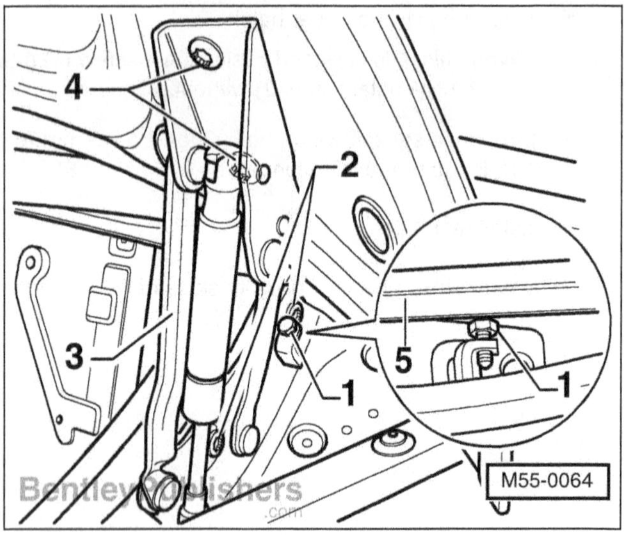

◀ Install hinge without rear lid.

– With hinge (**5**) closed position, adjustment bolt (**1**) must just touch contact point on lid.

- Rear lid is properly adjusted when there is an overall even gap dimension between the body and lid.
- Lid height should not be above body contour.
- Vehicle must be standing on it's own wheels

– Turn adjustment bolt (**1**) 1.5 - 3 mm (0.05 - 0.15 in.) to pre-tension hinge.

– Install rear lid.

– Install rubber grommet on both sides of rear hatch.

Rear hatch lock assembly (through 08/2001)

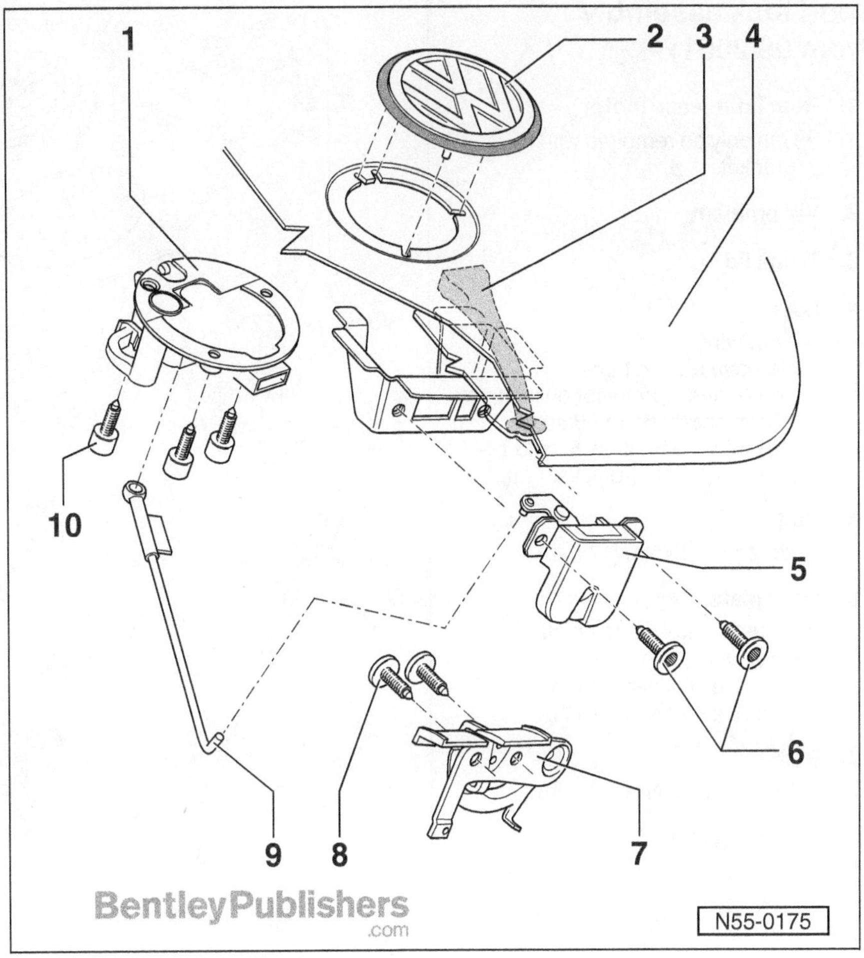

1. **Lock cylinder carrier**
   - The lock cylinder can only be removed with the carrier
2. **VW emblem**
   - Assemble with carrier
   - Remove with lock cylinder carrier
3. **Water drain channel (rubber)**
   - Push fit onto drain connection of locking cylinder carrier

CAUTION—
Ensure the rubber drain channel is routed correctly, otherwise water will drain into lock and trunk.

4. **Rear hatch**
5. **Lock**
   - Removing:
     Remove rear hatch trim, see **70 Trim–Interior.**
     Disconnect connector on lock.
     Unclip operating rod and remove lock from rear hatch.
6. **Bolt**
   - Tighten to 23 Nm (17 ft-lb)
7. **Lock plate**
   - Adjust using slotted holes
   - Removing:
     Remove lock carrier trim, see **70 Trim–Interior**.
     Remove bolts (item 8).
8. **Bolt**
   - Tighten to 23 Nm (17 ft-lb)
9. **Operating rod**
10. **Bolts**
    - Tighten to 6 Nm (53 in-lb)

Hood lock assembly (from 09/2001)

1. **Rear lid release motor**
   - Can only be removed with bracket
2. **VW emblem**
3. **Trunk lid**
4. **Lock**
   - Removing:
     Remove rear lid trim.
     Disconnect connector on lock.
     Unclip operating rod (item 8).
     Unscrew bolts (item 5) and remove lock from trunk lid.
5. **Bolt**
   - Tighten to 23 Nm (17 ft-lb)
6. **Lock plate**
   - Adjust using slotted holes
   - Removing:
     Remove lock carrier trim.
     Remove screws (item 7).
7. **Bolt**
   - Tighten to 23 Nm (17 ft-lb)
8. **Operating rod**

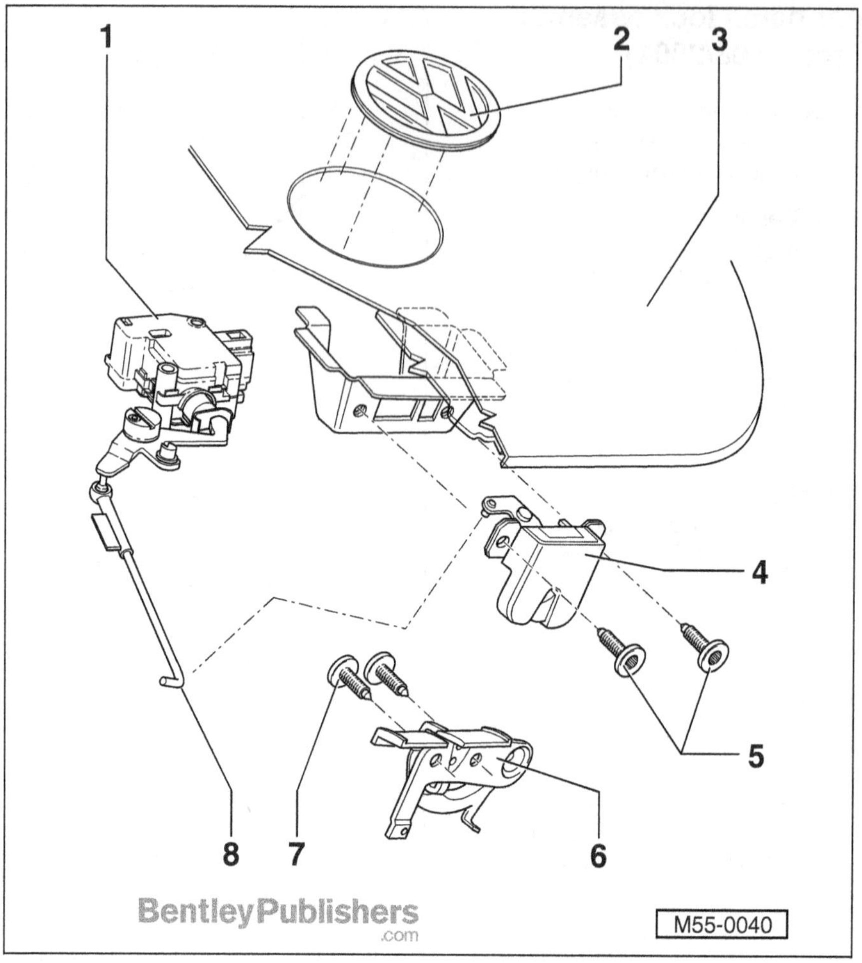

Rear lock cylinder carrier (power), removing and installing (through 08/2001)

– Remove rear hatch trim, see **70 Trim–Interior**.

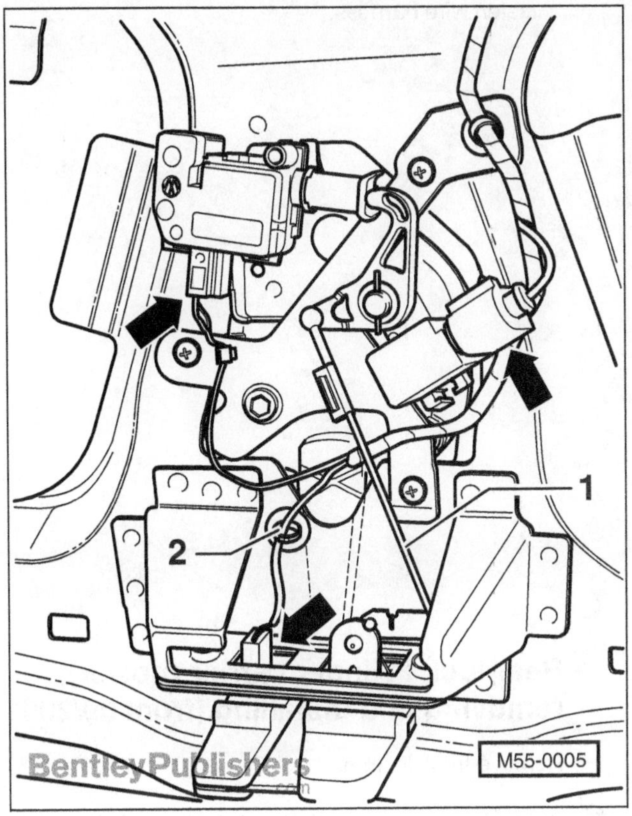

◄ Unplug harness connectors (**arrows**) and release from retaining clip (**2**). Unclip operating rod for lock carrier (**1**).

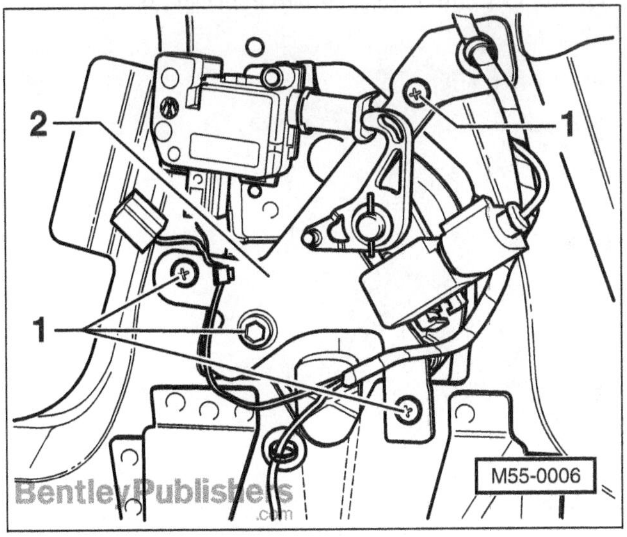

◄ Remove mounting screws (**1**) and electrical wire support panel (**2**).

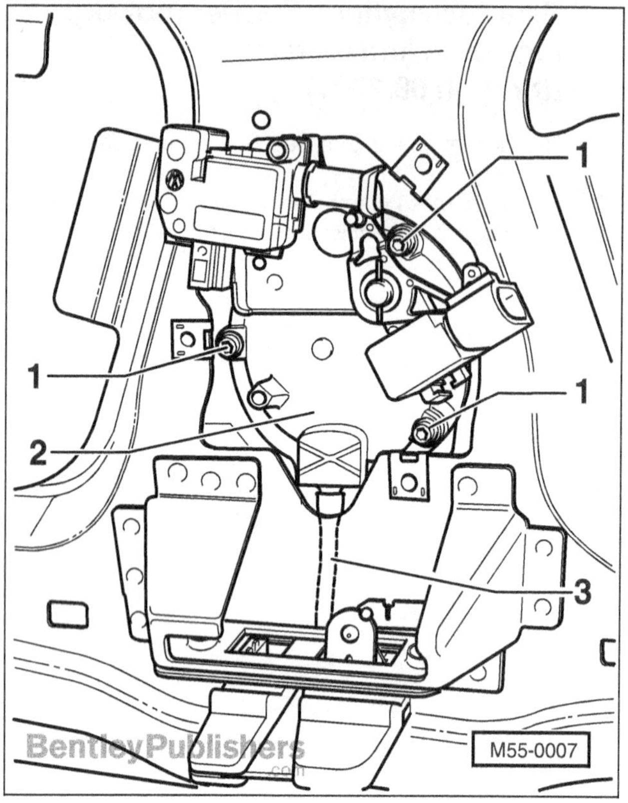

◀ Remove lock carrier mounting bolts (**1**) and disconnect water drain channel. Pull lock cylinder carrier (**2**) off rear hatch and then away from outer hand grip.

– Assembly is the reverse of removal, noting the following:
  - Replace tie wrap for water drain channel and clips to fasten wire harness.

Rear lock cylinder carrier (power), removing and installing (from 09/2001)

– Remove trunk lid trim.

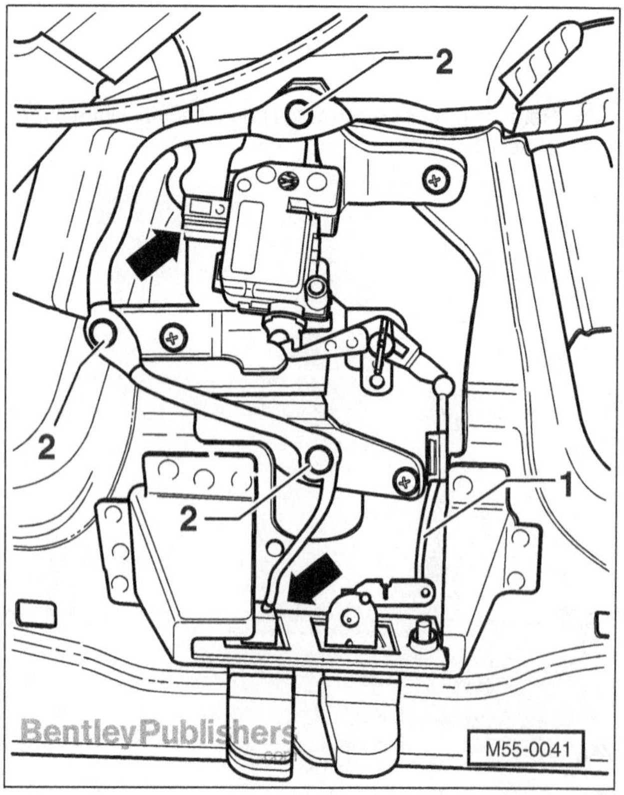

◀ Separate connectors (**arrows**). Unclip operating rod (**1**) for lock carrier. release wire from clip (**2**).

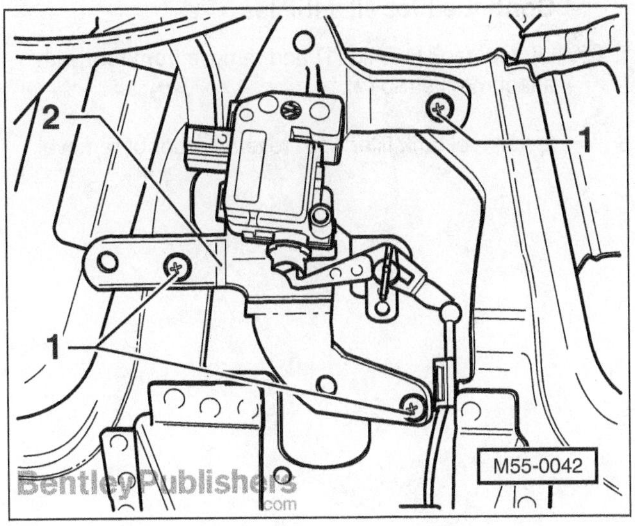

◄ Remove screws (**1**) and remove electrical wire support panel (**2**).

– Installation is reverse of removal, noting the following.
  - Replace wiring harness retaining clips.

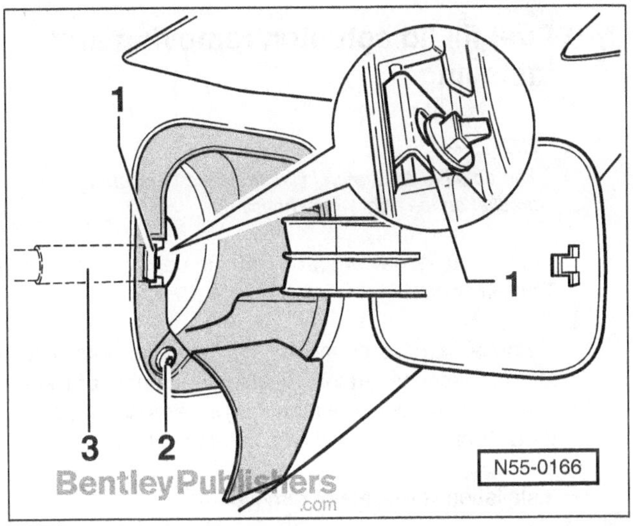

Fuel tank filler lid, removing and installing

– Remove right tail light, see **94 Lights, Accessories–Exterior**.

– Remove filler cap.

◄ Pull retaining ring (**1**) (behind the sealing lip) out of rubber cup (**3**). Remove mounting screw (**2**) in lower corner of fuel fill shroud.

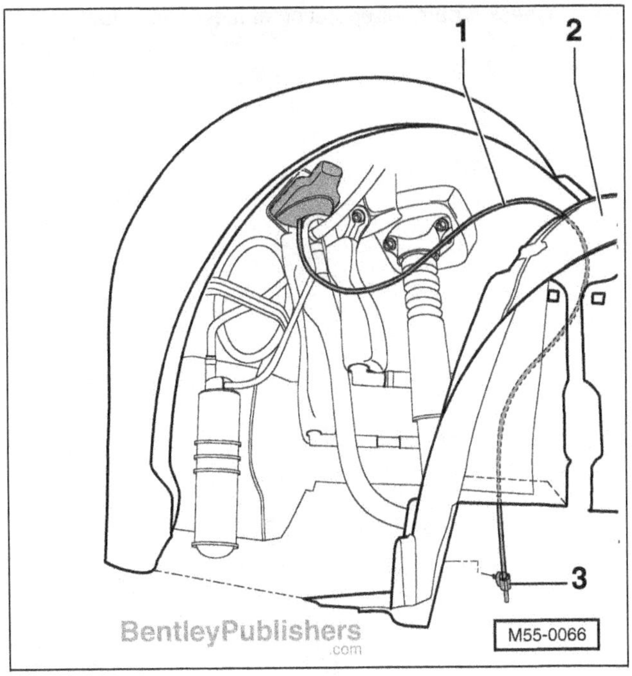

Convertible

– Remove right rear wheel housing liner, see **66 Body–Exterior Equipment**.

◄ Remove hose (**1**) from bracket (**3**) and allow to hang free.

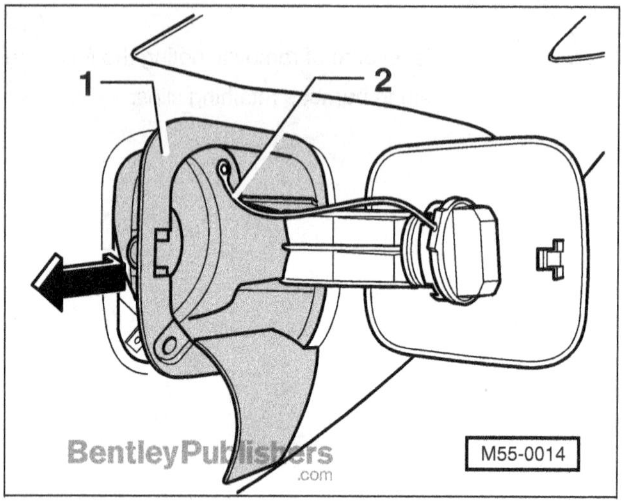

Continued for all vehicles

- Lift fuel tank filler lid (**1**) and remove from rear with attachment strap (**2**).

- Install fuel tank filler lid in reverse order of removal.

Fuel fill lid actuator, removing and installing

NOTE —

The actuator for the fuel fill lid release is located on a bracket on the right wheel housing.

- Remove luggage compartment trim on right side, see **70 Trim–Interior**.

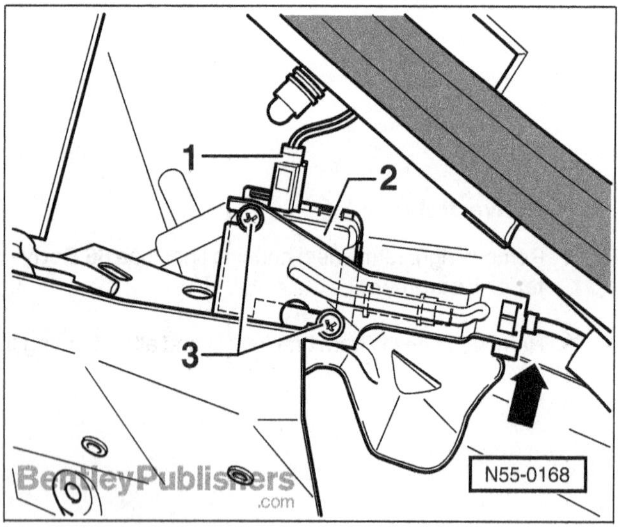

- Remove harness connector (**1**) from fuel door actuator motor. Disconnect release cable at retaining bracket (arrow). Remove actuator mounting screws (**3**), remove motor (**2**) from bracket and disconnect cable at motor.

- Installation is reverse of removal.

- Check for proper operation of fuel fill lid actuator.

57 Doors, Central Locking

GENERAL

This repair group covers removal and installation of the door assemblies, the door glass and window regulators, and service to the door lock assemblies including the central locking system. Special tools and equipment are required for some operations. For removal of interior door panels see **70 Trim–Interior**.

> ***WARNING —***
> *Disconnecting the negative (–) battery cable may erase fault codes and basic settings in the engine management and automatic transmission control modules. Some driveability problems may be noticed until the system re-adapts to operating conditions. OBD II readiness codes, which may be required for emissions testing, may also be erased. Convenience electronics (alarm system, interior light control, power locks, mirrors, and windows) may need to be re-set using a diagnostic scan tool.*

Power windows

Models equipped with power windows feature the ability to raise and lower both front windows by using the key in the front door lock. By holding the key in the locked position for more than 1 second, either window that was in the down position will be raised to the closed position in addition to locking the vehicle. This feature is known as convenience close.

57

As a ventilation aid in hot weather, both front windows can be lowered. By holding the key in the unlock position for more than 2 seconds, both front windows will be lowered at the same time in addition to unlocking the vehicle.

Power windows also feature pinch protection and one-touch up and one-touch down operation. The power windows will remain functional for approximately 10 minutes after the ignition is shut off provided that a door is not opened. The one-touch up feature is deactivated when the ignition is switched off.

Central locking

All models are equipped with central locking that is activated by insertion of the key into an outside door, by activation of a locking switch from within, or by the keyless remote.

When opening a locked vehicle from the outside with the key, there are two options. If the key is turned once, only that particular door will unlock. If the key is quickly turned twice, both doors will be unlocked. This safety feature is known as selective unlocking.

DOORS

The doors of the New Beetle are designed with a split hinge. The hinge pin is locked by a grub screw that allows the hinge to slide apart once removed. After disconnecting the appropriate electrical connectors and the door check, the complete door can be easily removed from the vehicle as required. Most of the internal components of the doors including the window regulators and window glass are assembled into a modular assembly known as a door assembly carrier.

Door assembly (sedan)

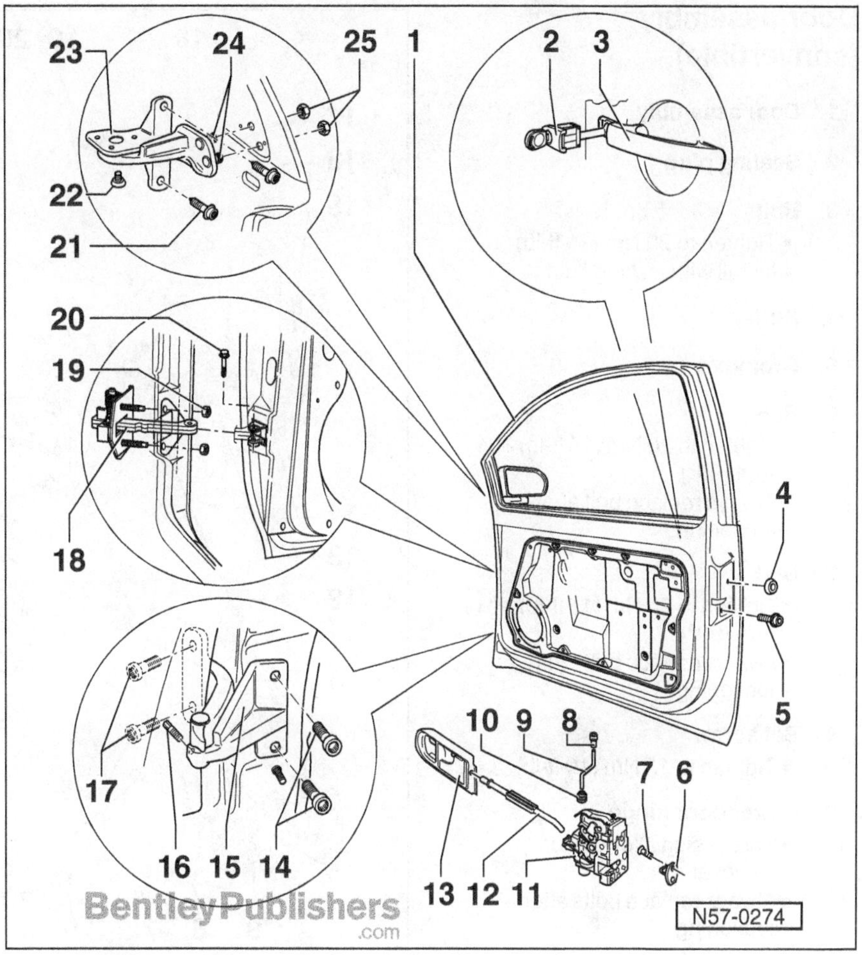

1. **Door assembly**
2. **Lock cylinder housing**
3. **Door handle**
4. **Sealing plug**
5. **Bolt**
   - Tighten to 20 Nm (15 ft-lb)
   - Install with locking fluid (VW part # D 185 400 A2)
6. **Striker plate**
7. **Bolt**
   - Tighten to 20 Nm (15 ft-lb)
8. **Locking button**
9. **Locking rod**
10. **Rubber boot**
11. **Door lock assembly**
    - Lubricate lock with silicone spray to eliminate squeaks
12. **Interior release cable**
13. **Interior release handle**
14. **Lower hinge bolt, body side**
    - Tighten to 20 Nm (15 ft-lb) + ¼ turn (90°)
    - Always replace bolt after loosening
15. **Lower door hinge**
    - Split design allows door removal
    - Always replace bolts after loosening
16. **Grub screw**
    - Secures split hinge sections
    - Concealed under plastic cap
    - Tighten to 13 Nm (10 ft-lb)
17. **Lower hinge bolt, door side**
    - Tighten to 20 Nm (15 ft-lb) + ¼ turn (90°)
    - Always replace bolts after loosening
18. **Door check assembly**
19. **Nut**
    - Tighten to 9 Nm (80 in-lb)
20. **Pin**
    - Tighten 4.5 Nm (39 in-lb)
21. **Upper hinge bolt, body side**
    - Tighten to 20 Nm (15 ft-lb) + ¼ turn (90°)
    - Always replace bolts after loosening
22. **Bolt**
    - Tighten to 27 Nm (20 ft-lb)
23. **Upper door hinge**
    - Split design allows door removal
    - Always replace bolts after loosening
24. **Upper hinge bolts, door side**
    - Always replace bolt after loosening
25. **Nuts**
    - Tighten to 20 Nm (15 ft-lb) + ¼ turn (90°)
    - Always replace bolt after loosening

Door assembly (convertible)

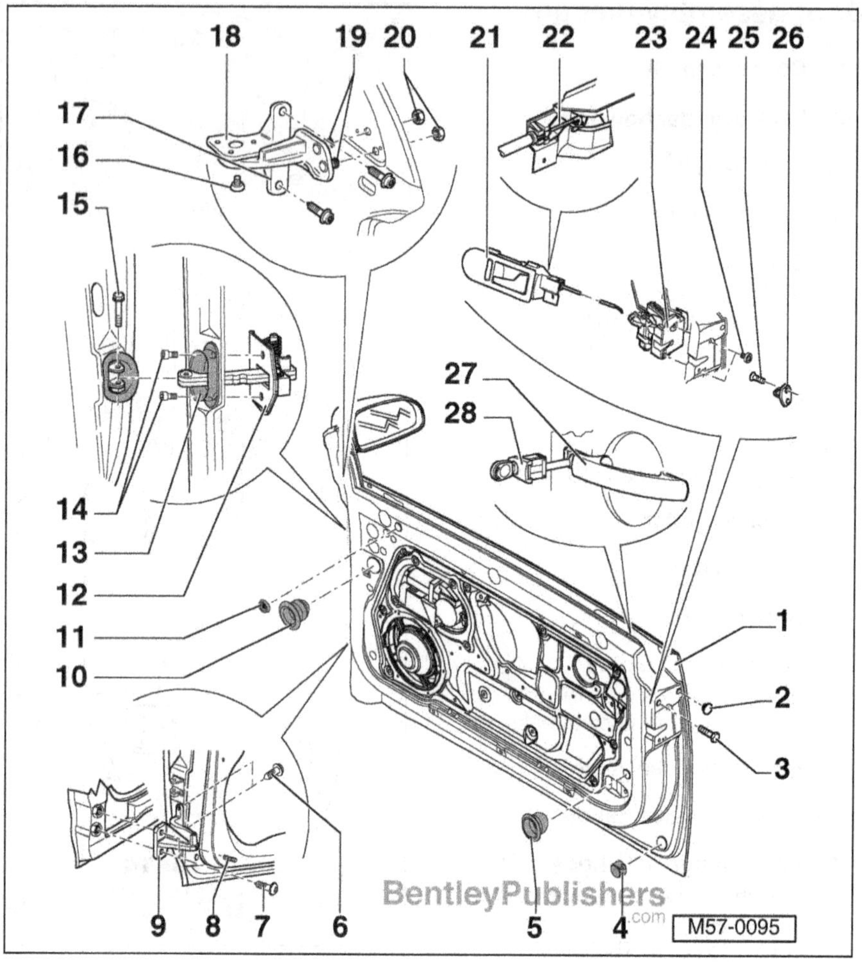

1. **Door assembly**
2. **Sealing plug**
3. **Bolt**
   - Tighten to 20 Nm (15 ft-lb)
   - Install with locking fluid
4. **Buffer**
5. **Grommet**
6. **Bolt**
   - Tighten to 20 Nm (15 ft-lb) + ¼ turn (90°)
   - Always replace bolt after loosening
7. **Bolt**
   - Tighten to 20 Nm (15 ft-lb) + ¼ turn (90°)
   - Always replace bolt after loosening
8. **Set screw**
   - Tighten to 13 Nm (10 ft-lb)
9. **Lower door hinge**
   - Split design allows door removal
   - Always replace bolts after loosening
10. **Grommet**
11. **Grommet**
12. **Door check**
13. **Cover**
14. **Bolt**
    - Tighten to 10 Nm (7 ft-lb)
15. **Pin**
    - Tighten 4.5 Nm (39 in-lb)
16. **Bolt**
    - Tighten to 20 Nm (15 ft-lb) + ¼ turn (90°)
    - Always replace bolt after loosening
    - Use special tools 3320 and 3220/2 to loosen/tighten
17. **Bolt**
    - Tighten to 27 Nm (20 ft-lb)
18. **Upper door hinge**
    - Split design allows door removal
    - Always replace bolts after loosening
19. **Upper hinge bolts**
    - Always replace bolts after loosening
20. **Nuts**
    - Tighten to 20 Nm (15 ft-lb) + ¼ turn (90°)
    - Always replace nuts after loosening
21. **Interior release handle**
22. **Hood release cable**
23. **Door lock assembly**
24. **Bolt**
    - Tighten to 20 Nm (15 ft-lb)
    - Install with locking fluid
25. **Bolt**
26. **Striker plate**
27. **Door handle**
28. **Lock cylinder housing**

Door, removing and installing

Vehicles through 05/2000

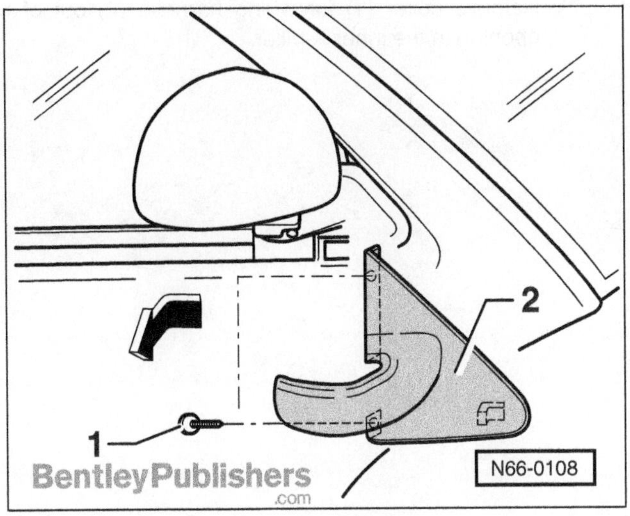

Remove bolt (**1**) from mirror hinge cover (**2**) and swing hinge cover out (**arrow**) from rear to remove.

Vehicles from 06/2000

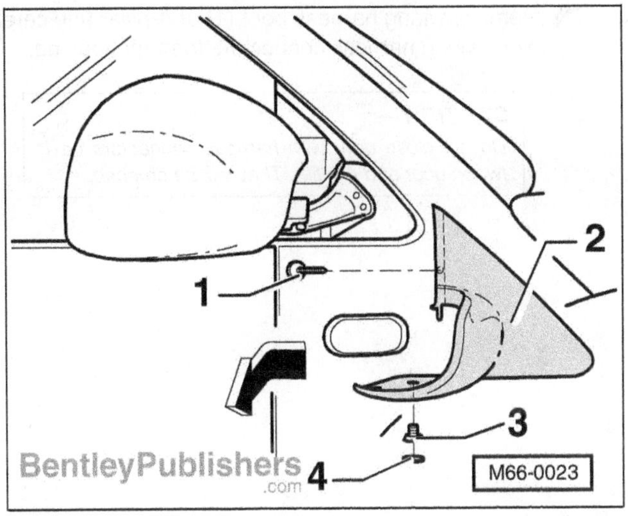

Use screwdriver to carefully remove trim (**4**)

- Remove mounting bolts (**1** and **3**) and triangular panel (**2**) toward rear (**arrow**).

Continued for all vehicles

- If removing left side door, remove hood release lever. See **55 Hoods and Lids**.
- Remove lower A-pillar trim, see **70 Body–Interior Trim.**

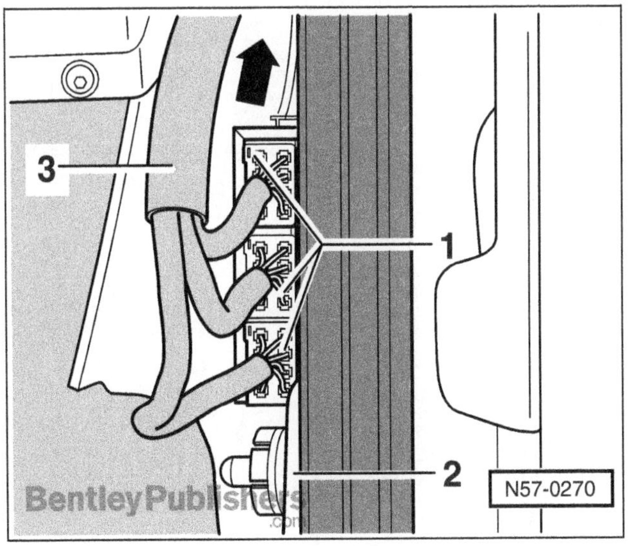

Disconnect multi-pin harness connectors (**1**) on lower A-pillar (**2**).

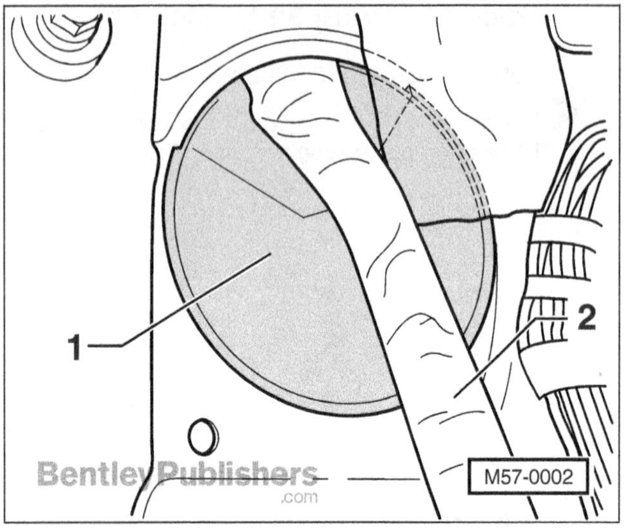

◄ Remove cover (**1**) for wiring harness (**2**) out of body opening in the inner A-pillar.

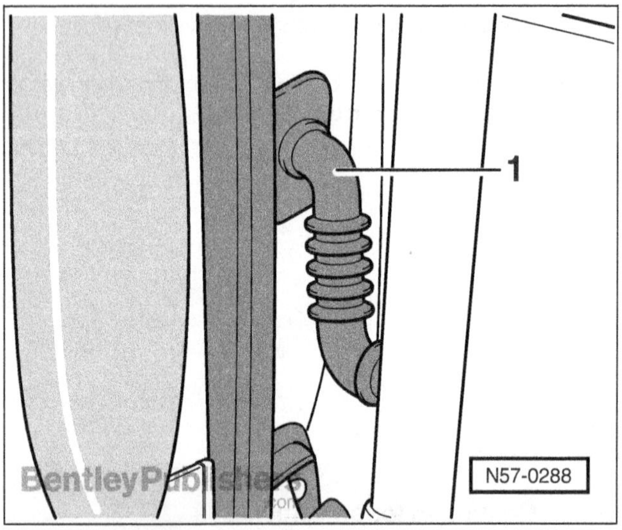

◄ Remove wiring harness boot (**1**) at A-pillar and carefully work wiring harness connectors through opening.

> ***CAUTION —***
> *Do not close door with harness connectors between door and A-pillar. They will be crushed.*

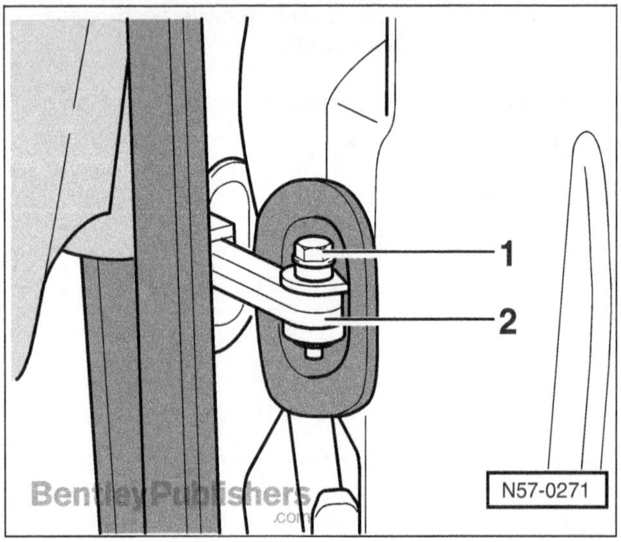

◄ Remove pin (**1**) from door check (**2**).

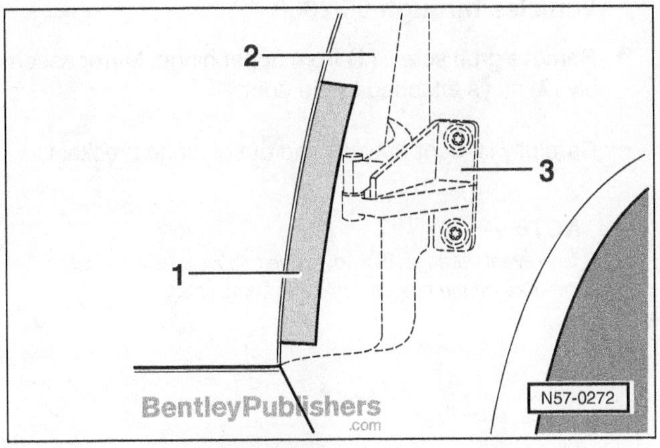

Apply textile reinforced adhesive tape (**1**) (duct tape or equivalent) to fender (**2**) to prevent paint damage at lower hinge (**3**) area during removal.

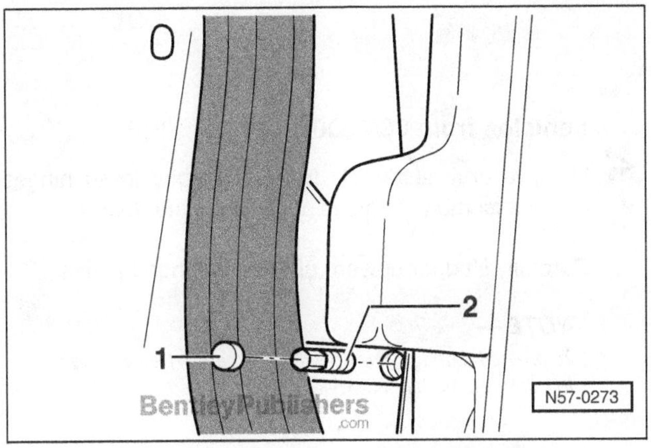

Remove plastic cap (**1**) from grub screw (**2**) at lower hinge and remove grub screw.

If access to grub screw through gap between door and A-pillar is not possible or too difficult, proceed as follows:

- Remove front wheel housing liner. See **66 Body–Exterior Equipment**.

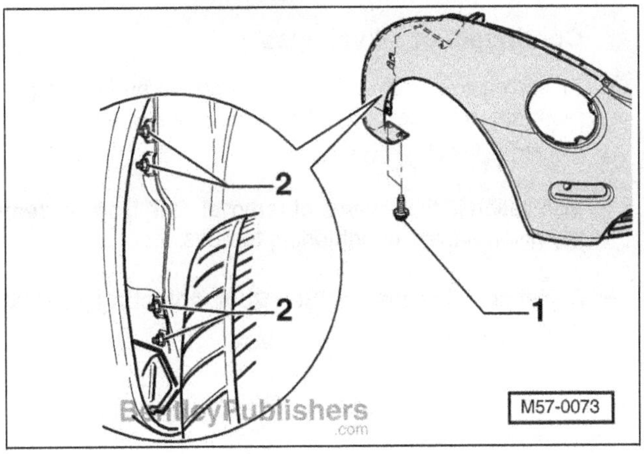

Remove fender mounting bolt (**1**) and nuts (**2**) to allow front portion of fender to be pulled away from body. Alternatively, remove front body assembly, see **50 Body–Front**.

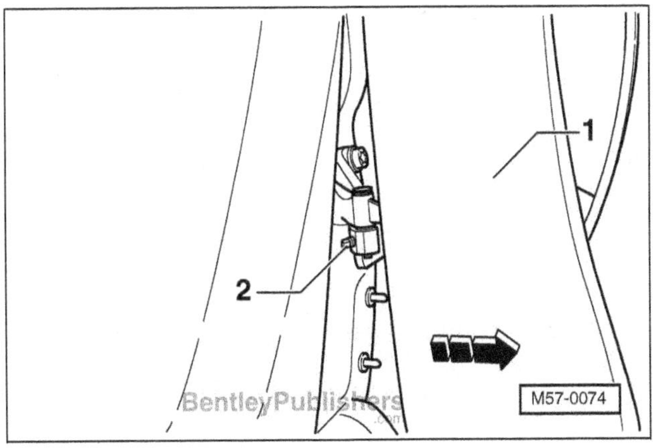

Place plastic wedge between fender (**1**) and body to allow access to grub screw (**2**).

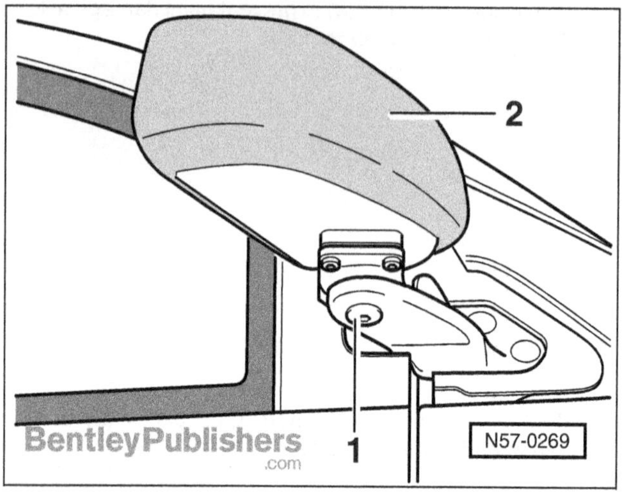

Vehicles through 05/2000

◄ Remove grub screw (**1**) from upper hinge. Mirror assembly (**2**) stays attached to the door.

– Carefully lift door upward and out of hinge brackets.

> ***NOTE —***
> *To prevent damage, it is advisable to have a helper guide the door off the hinges while it is being lifted.*

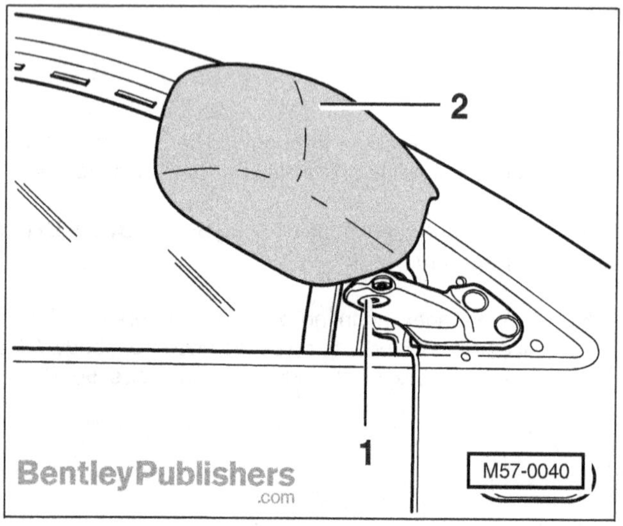

Vehicles from 06/2000

◄ Remove grub screw (1) from upper and lower hinges. Mirror assembly (**2**) stays attached to the door.

– Carefully lift door upward and out of hinge brackets.

> ***NOTE —***
> *To prevent damage, it is advisable to have a helper guide the door off the hinges while it is being lifted.*

Continued for all vehicles

– Place removed door on suitable padded surface to protect paint from damage. If stored vertically, also protect the lower edge.

– Installation is the reverse of removal. See **Door assembly** given earlier for tightening torques.

– Adjust door if necessary. See **Door, adjusting** given later in this repair group.

Door, adjusting

To correct door adjustment, loosen hinges. Attempting to correct door adjustment by measures such as raising the door upward without loosening the hinges generally have no effect and only distort the door and/or the A-pillar, making further adjustment impossible.

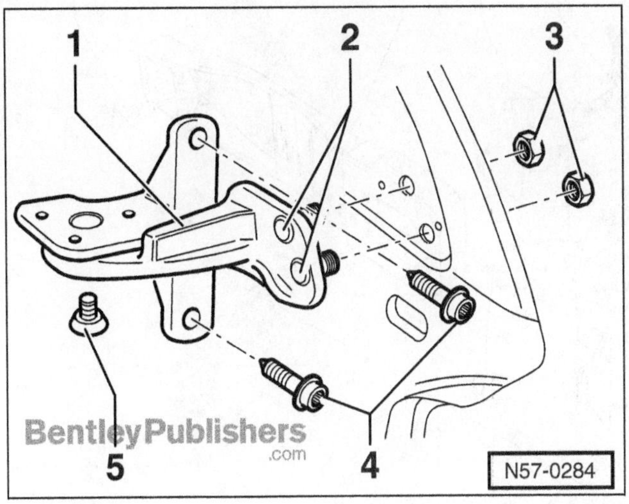

Door hinge, upper, assembly (through 05/2000)

1. Door hinge
2. Hinge stud
   - Always replace
3. Hinge nut
   - Always replace
   - Tighten to 20 Nm (15 ft-lb) + ¼ turn (90°)
4. Hinge bolt
   - Always replace
   - Tighten to 20 Nm (15 ft-lb) + ¼ turn (90°)
5. Grub screw
   - Tighten to 27 Nm (20 ft-lb)

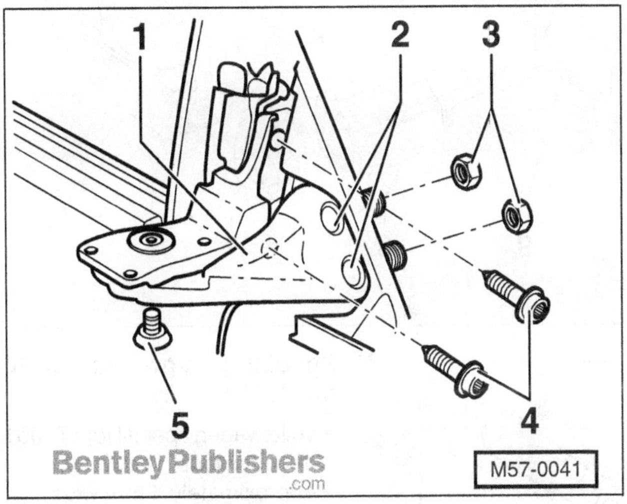

Door hinge, upper, assembly (from 06/2000)

1. Door hinge
2. Hinge stud
   - Always replace
3. Hinge nut
   - Always replace
   - Tighten to 20 Nm (15 ft-lb) + ¼ turn (90°)
4. Hinge bolt
   - Always replace
   - Tighten to 20 Nm (15 ft-lb) + ¼ turn (90°)
5. Grub screw
   - Tighten to 27 Nm (20 ft-lb)

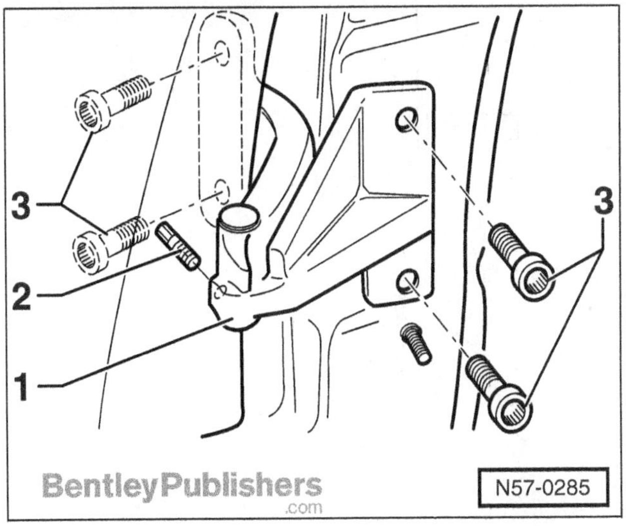

Door hinge, lower, assembly

1. Door hinge
2. Grub screw
   - Tighten to 13 Nm (10 ft-lb)
3. Hinge bolt
   - Always replace
   - Tighten to 20 Nm (15 ft-lb) + ¼ turn (90°)

Door handle and lock assembly

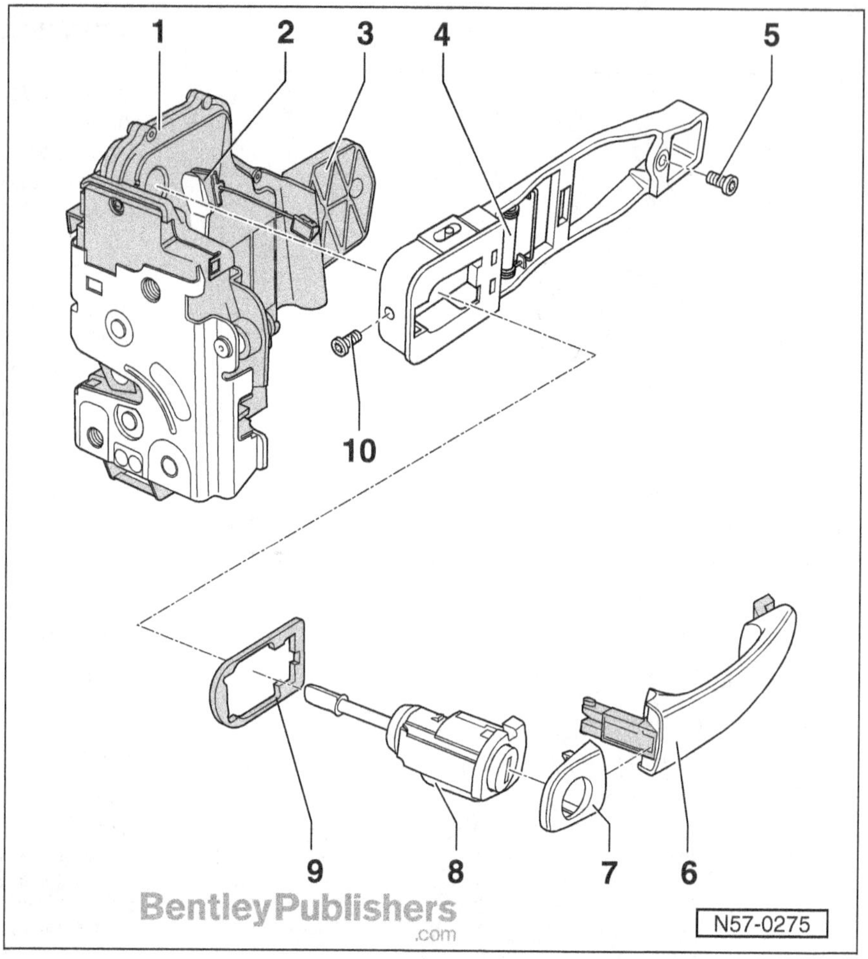

1. **Door lock**
   - Can only be removed with the door carrier assembly
2. **Cable**
   - Outside door handle release
3. **Retaining bracket**
   - Bolted and riveted to carrier assembly and door lock
4. **Support plate**
5. **Screw, polygon socket**
   - Requires special socket insert tool T10011 or equivalent, see **Special polygon driver tool for door repairs**
   - Tighten to 8.5 Nm (76 in-lb)
6. **Door handle, outside**
7. **Lock cylinder cover (trim)**
8. **Lock cylinder housing**
   - Only supplied with lock cylinder and keys.
9. **Gasket**
   - Plastic
10. **Screw, polygon socket**
    - Requires special socket insert tool T10011 or equivalent, see **Special polygon driver tool for door repairs**
    - Tighten to 4.5 Nm (40 in-lb)
    - Loosening this screw releases the lock cylinder housing, allowing it to be pulled out of support plate
    - Loosen only, do not remove

CAUTION —
Loosen polygon socket screw only. If screw is removed, the door lock must be removed to realign.

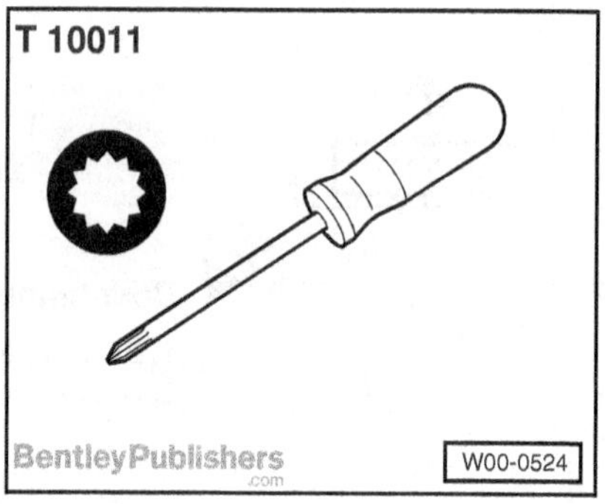

Special polygon driver tool for door repairs

- Volkswagen special tool T10011, or equivalent
- Approximately 4 mm wide

Lock cylinder housing, removing

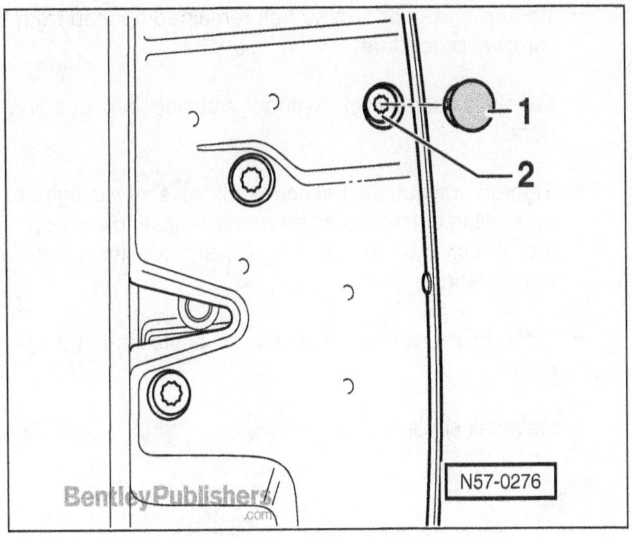

- Carefully pry out sealing plug (**1**).
- Pull door handle open and hold.

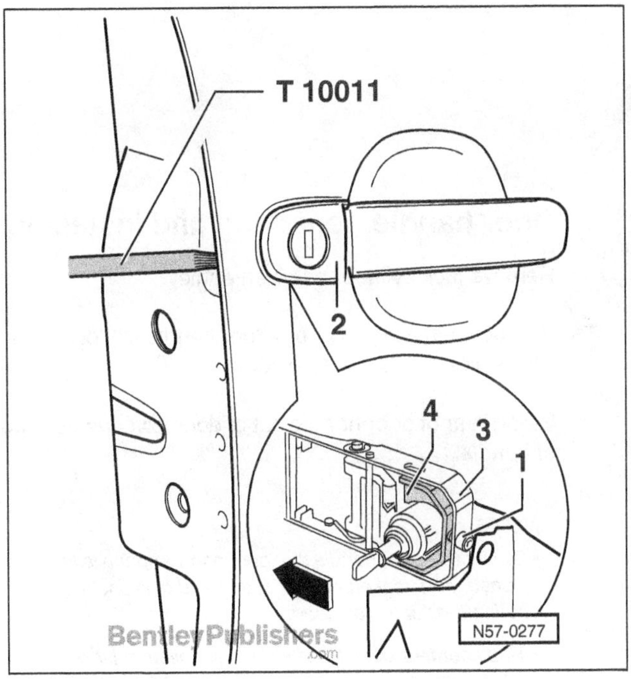

- Loosen polygon socket screw (**1**) to the stop with Volkswagen special tool T10011 or equivalent. Lock cylinder housing (**2**) is released when screw is loosened.

> ***CAUTION—***
> *Loosen polygon socket screw only. If screw is removed, the door lock must be removed to reinstall.*

NOTE—
The door handle will remain open on its own if the polygon socket screw is loosened properly.

- Pull lock cylinder housing straight out.

NOTE—
Inserting the key into lock cylinder aids in removing the lock cylinder housing.

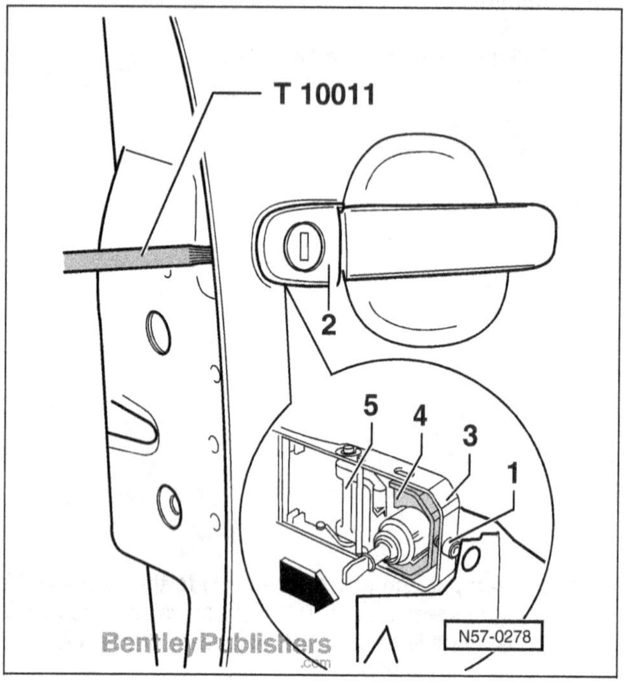

Lock cylinder housing, installing

- Ensure that door handle has remained opened from the removal procedure.
- Carefully install lock cylinder housing into opening in door.
- Tighten polygon socket screw (**3**). As screw is tightened, an audible noise should be heard indicating correct seating of lock cylinder housing (**2**) and tensioning of door handle spring.
- Operate key and door handle to ensure correct operation.
- Reinstall sealing plug.

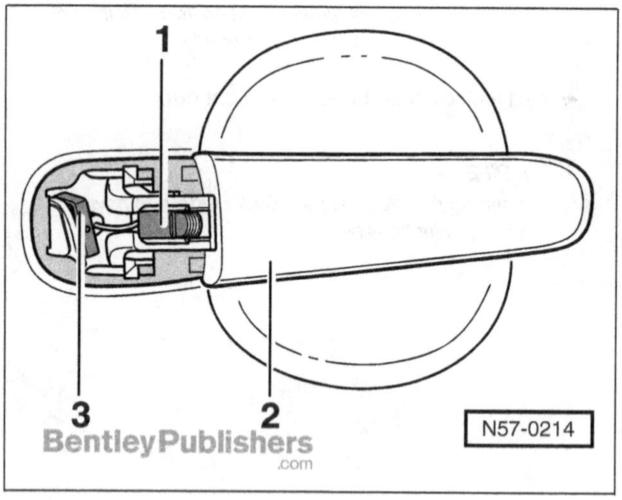

Door handle, removing and installing

- Remove lock cylinder as given earlier.
- Remove cable clip (**1**) from door handle (**2**) lock release with a pick or a small screwdriver.
- Swing rear of door handle out of door, pivoting from front of handle.

NOTE —
- *By pivoting door handle out, door handle spring is tensioned and locked. After door handle is installed, spring tension is released.*
- *Replacement door handles are supplied finished in primer only and must be painted prior to installation.*

- Slip front of door handle into door opening and swing rear of door handle into position against door.
- Snap cable clip into slot in door handle taking care not to pull cable tight.

NOTE —
When installing cable clip, do not pull lock operating lever while pushing on door handle.

- Install lock cylinder housing.

Door lock, removing

The window regulator, door glass, door speaker and door lock are secured to the door assembly carrier. The door lock can only be removed with the door assembly carrier.

NOTE —
Window glass must be operable in order to remove door assembly carrier. Any malfunction that prevents the windows (manual or power) from being lowered must be corrected before proceeding.

– Remove door trim panel. **See 70 Trim–Interior.**

– Remove lock cylinder housing as given earlier.

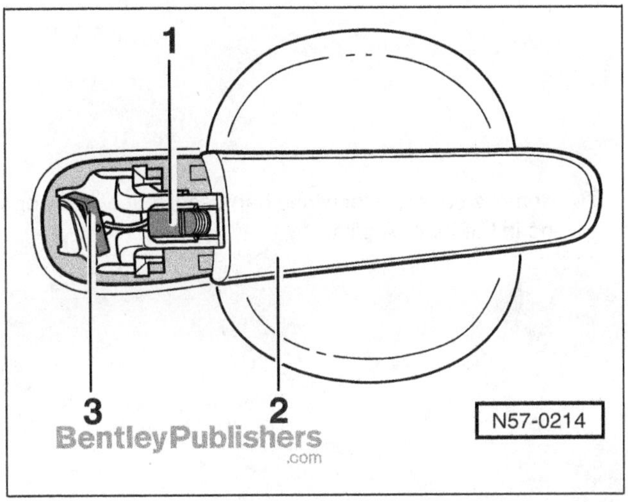

◄ Remove cable clip (**1**) from door handle.

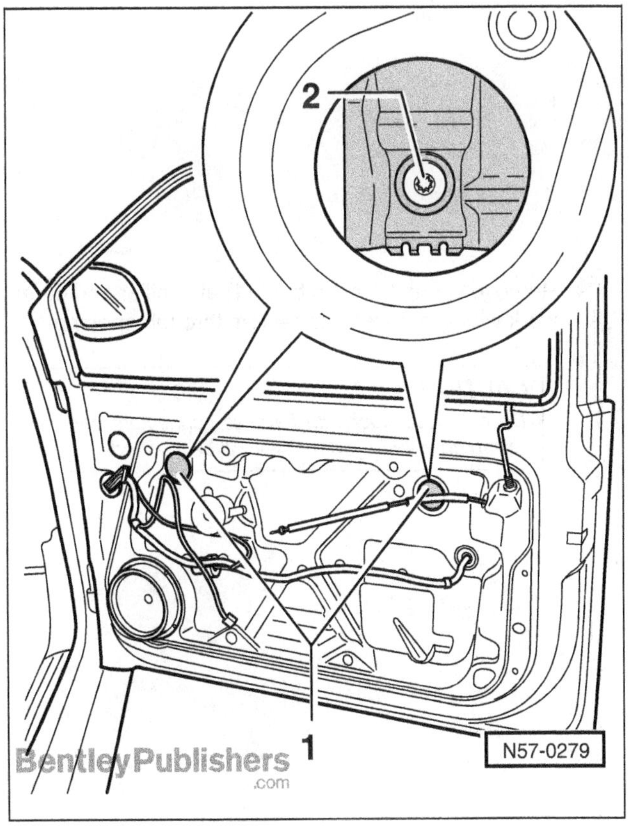

◄ Pry out sealing plugs (**1**) from door assembly carrier to gain access to window glass securing bolts (**2**).

– Lower window until window glass securing bolts are accessible.

– Loosen window glass securing bolts and carefully press glass securing clamps apart.

– Push window glass upward in the track. Secure the window glass in up position with strips of suitable adhesive tape stuck to inside glass, looped over door frame and stuck to the outside glass.

– Remove hood release lever on left side, as required. See **55 Hoods and Lids**.

– Remove lower A-pillar trim. See **70 Body–Interior Trim.**

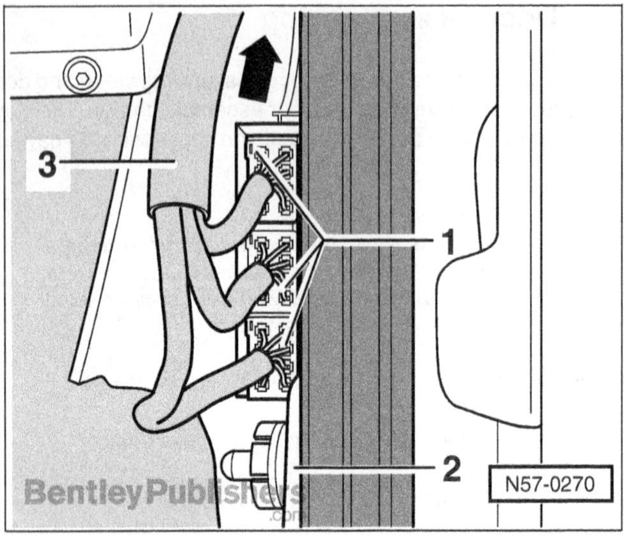

Disconnect multi-pin harness connectors (**1**) and harness cover on lower A-pillar (**2**).

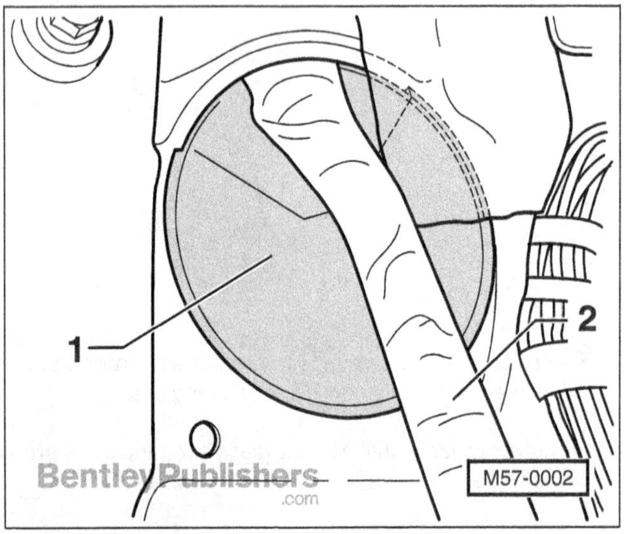

Remove cover (**1**) for wiring harnesses out of body opening in the inner A-pillar.

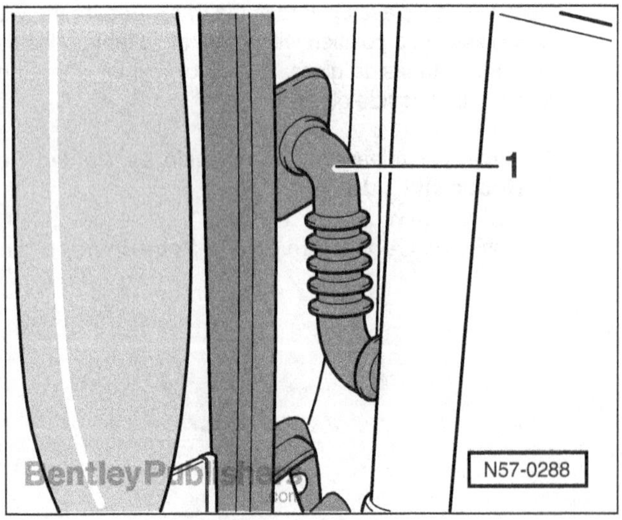

Remove wiring harness boot (**1**) at A-pillar and carefully work wiring harness connectors through opening.

> ***CAUTION—***
> *Do not close door with harness connectors between door and A-pillar. They will be crushed.*

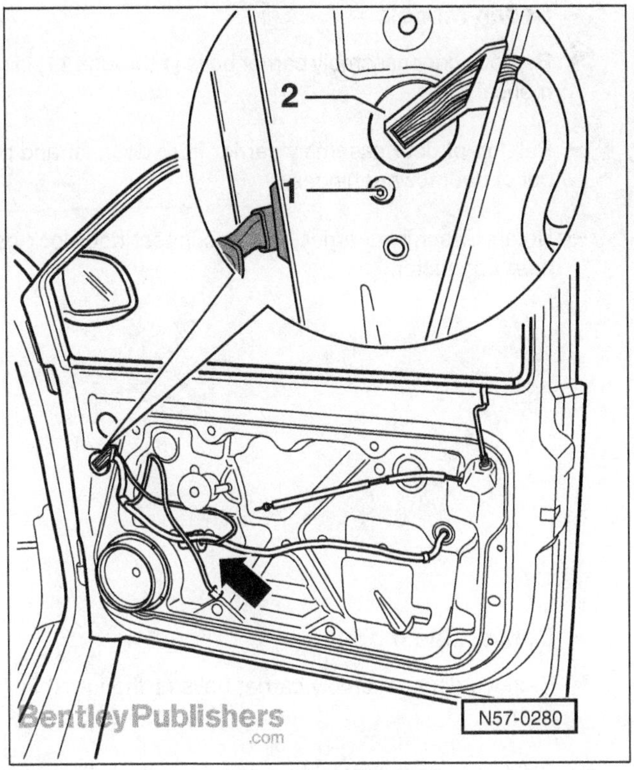

Disconnect harness connector (**arrow**) for mirror. Disconnect wiring harness guide panel (**2**). Guide wiring harness with boot through door frame.

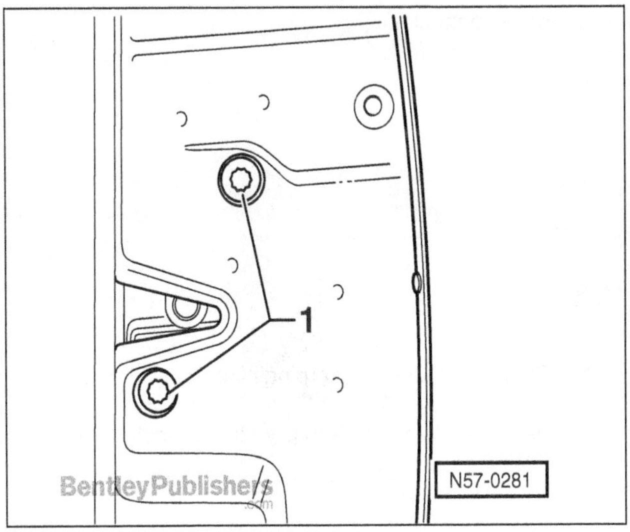

Remove door lock mounting bolts (**1**).

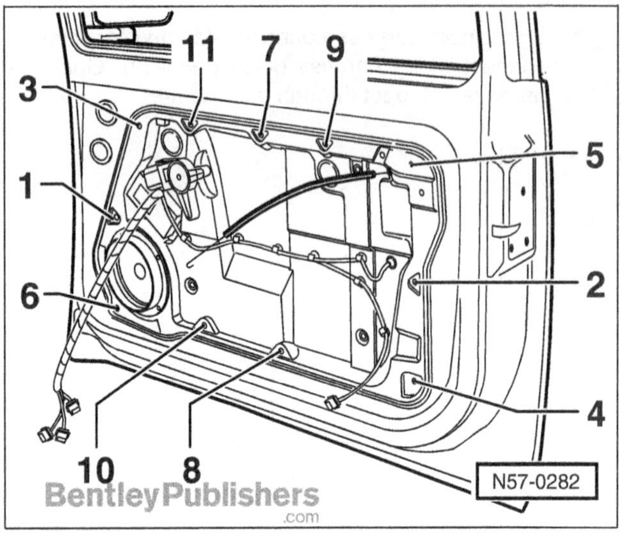

Sedan models

◀ Remove door assembly carrier bolts (**1** through **11**) in numerical order.

- Pull top of door assembly carrier from door, lift and pull out of door toward hinges.
- Rotate assembly carrier and disconnect door lock harness connector.

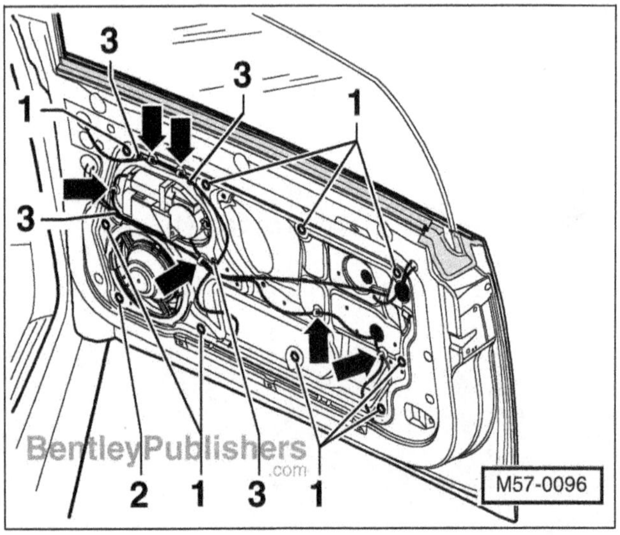

Convertible models

◀ Remove door assembly carrier bolts (**1** though **3**) in sequence.

- Pull top of door assembly carrier from door, lift and pull out of door toward hinges.
- Rotate assembly carrier and disconnect door lock harness connector.

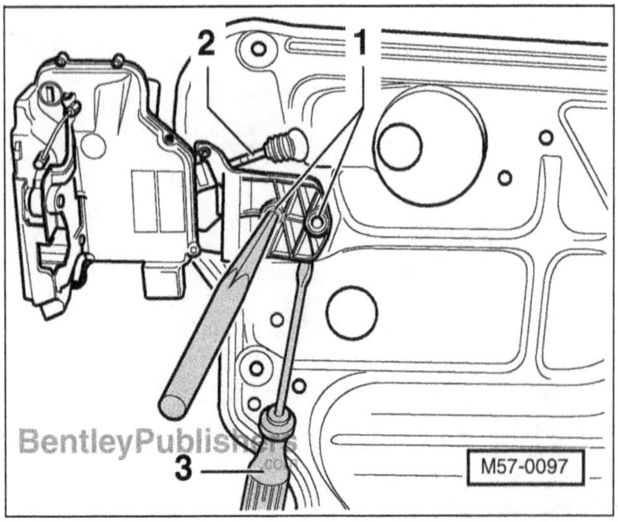

◀ Drive out door lock securing clips (**1**) with a drift.

- Use a screwdriver (**3**) to pry off door lock assembly with retaining bracket.

NOTE —

The door lock retaining bracket is a not included with new door lock assemblies. It is secured to the door lock with a bolt and a pop rivet.

- Pull door release cable (**2**) out of door enough to disconnect it from back of lock assembly.

Sedan models

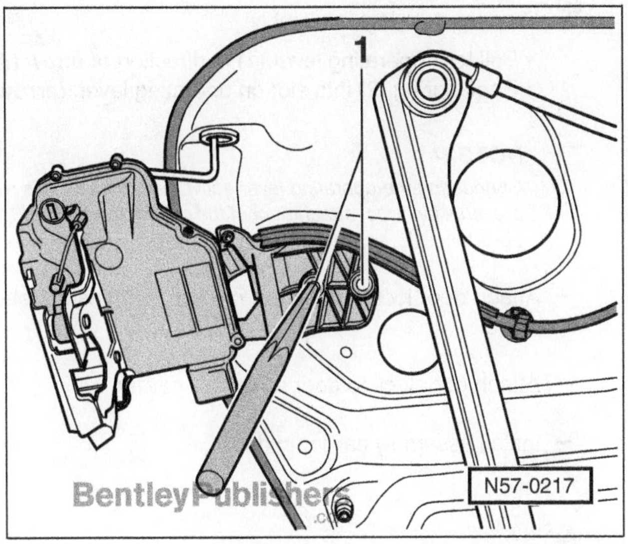

Drive out door lock securing clips (**1**) with a drift.

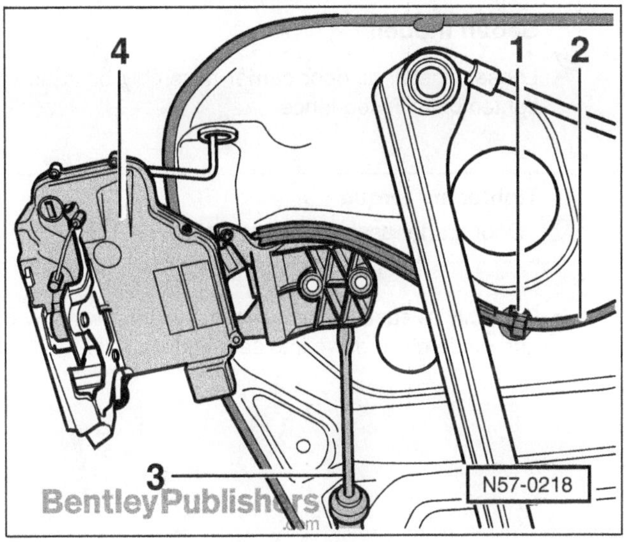

Disconnect inside release cable (**2**) from clip (**1**) on assembly carrier and carefully pry door lock (**4**) off assembly carrier using a screwdriver (**3**).

NOTE —

The door lock retaining bracket is a not included with new door lock assemblies. It is secured to the door lock with a bolt and a pop rivet.

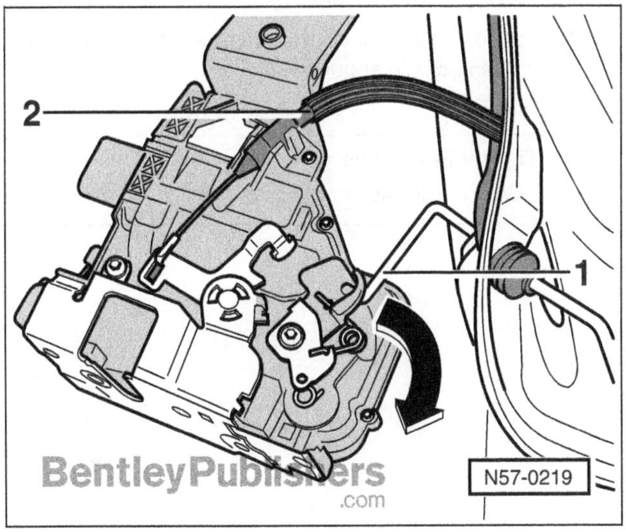

Disconnect door lock rod (**1**) and inside release cable (**2**).

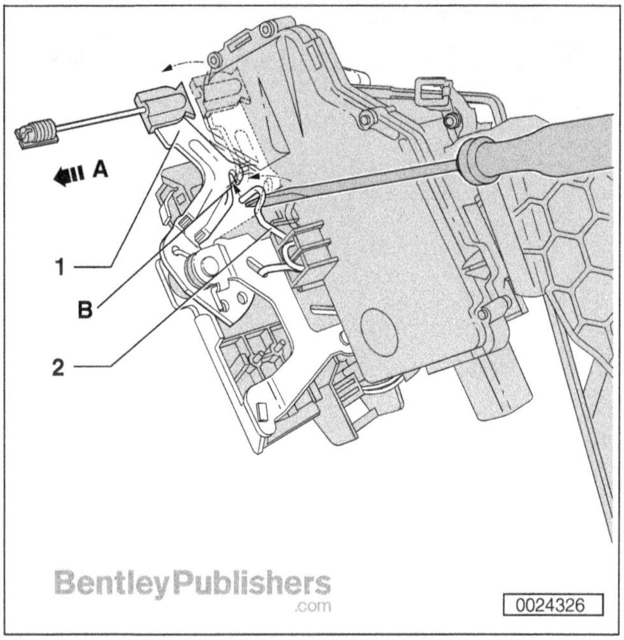

Door lock, installing

◀ Prepare door lock assembly:

- Pull lock operating lever (**1**) in direction of arrow (**A**).
- Hook spring (**2**) into slot on operating lever, (**arrow B**)

NOTE —

Connecting the operating lever activates the lock. This prevents incorrect attachment of the outside handle cable later.

– Attach door lock rod (sedan models) and inside release cable.

– Attach door lock to door assembly carrier.

– Install assembly carrier in door.

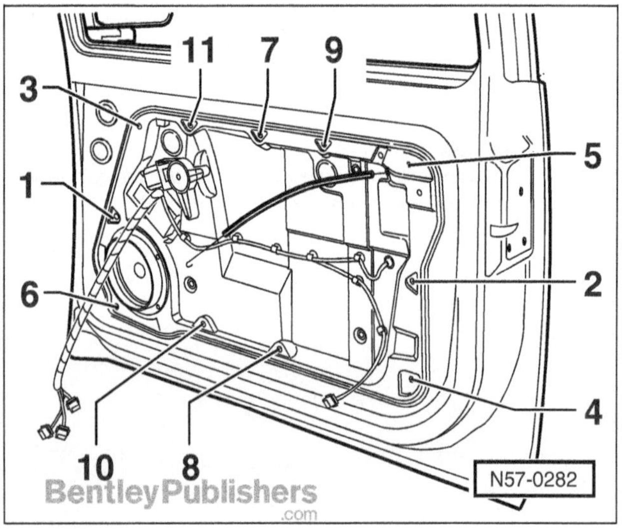

Sedan models

◀ Loosely install all door carrier assembly bolts and then tighten bolts in sequence.

Tightening torque

- Door carrier assembly bolts (M6) . . . . 8 Nm (71 ft-lb)

– Carefully release adhesive tape securing window glass and guide glass down into securing clamps.

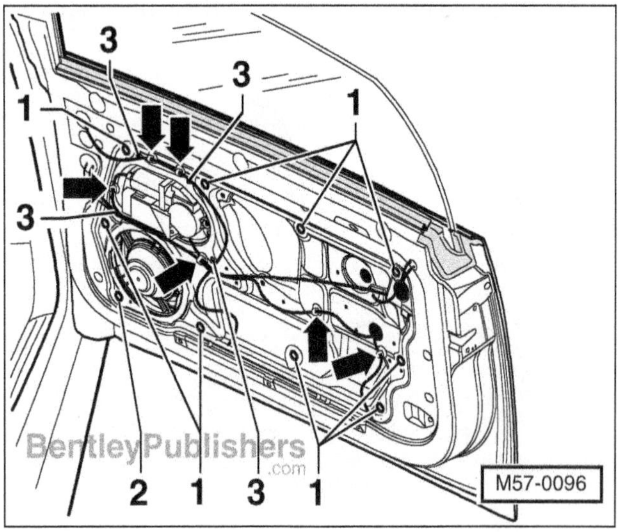

Convertible models

◀ Loosely install all door carrier assembly bolts and then tighten bolts in sequence.

Tightening torque

- Door carrier assembly bolts (M6) . . . . 8 Nm (71 ft-lb)

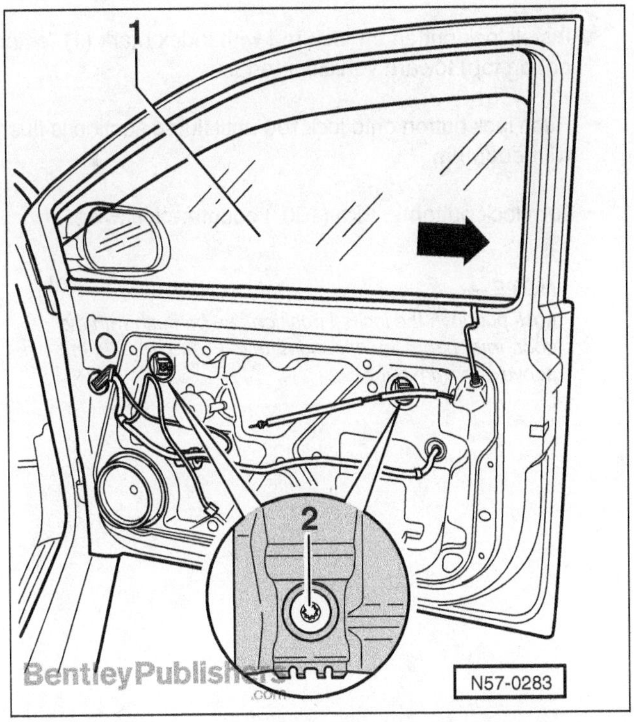

Gently press window (**1**) in direction of arrow and tighten securing clamp bolts (**2**).

- Sedan model shown, convertible model is similar.

Tightening torques

- Window glass securing bolts (sedan) . . . . 10 Nm (89 in-lb)
- Window glass securing bolts (convertible) . . . . 8 Nm (71 in-lb)

– Remainder of installation is the reverse of disassembly.

– Check operation of window before final installation of door trim panel.

– Check operation of door release handle, door lock and outer door handle.

Lock button, left door, removing and installing

> ***CAUTION—***
> *Failure to follow this procedure may result in loosening of the locking rods in the doors requiring complete disassembly to repair.*

– Remove lock button by turning ½ turn (180°) counter-clockwise.

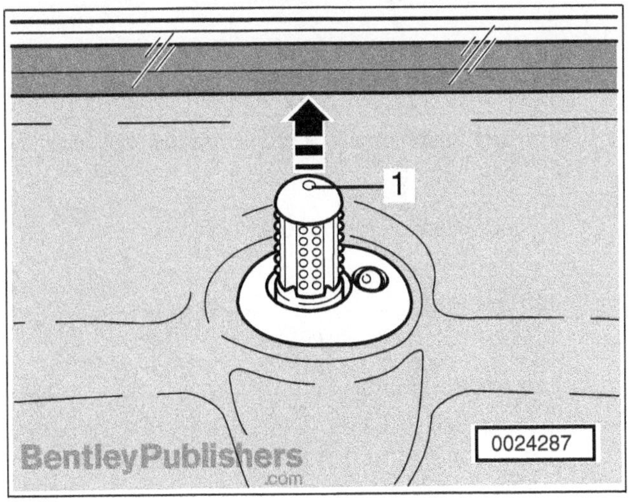

With index mark (**1**) turned to the inside, pull lock button upward and off.

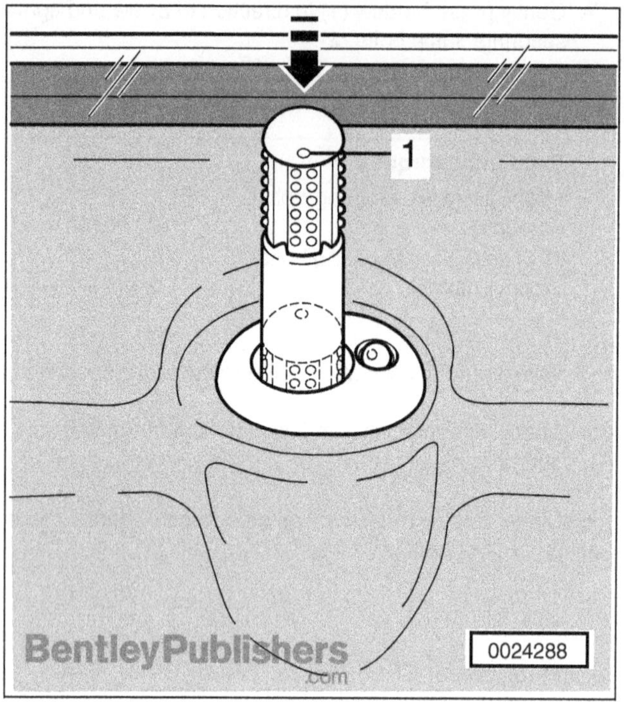

- Install lock button on lock rod with index mark (**1**) (small dot on top) toward vehicle interior.
- Push lock button onto lock rod until fluted section is flush with door trim.
- Turn lock button ½ turn (180°) counter-clockwise.

NOTE —
Lock button in the locked position can be flush with the door trim panel to approximately 2 mm (0.078 in.) above door trim panel.

Lock button, right door, removing and installing

CAUTION —
Failure to follow this procedure may result in loosening of the locking rods in the doors requiring complete disassembly to repair

- Remove lock button by turning ½ turn (180°) clockwise.
- Pull lock button upward and off.
- Install lock button on lock rod with index mark (small dot on top) toward vehicle interior.
- Push lock button onto lock rod until fluted section is flush with door trim.
- Turn lock button ½ turn (180°) clockwise.

NOTE —
Lock button in the locked position can be flush with the door trim panel to approximately 2 mm (0.078 in.) above door trim panel.

Door window assembly (sedan)

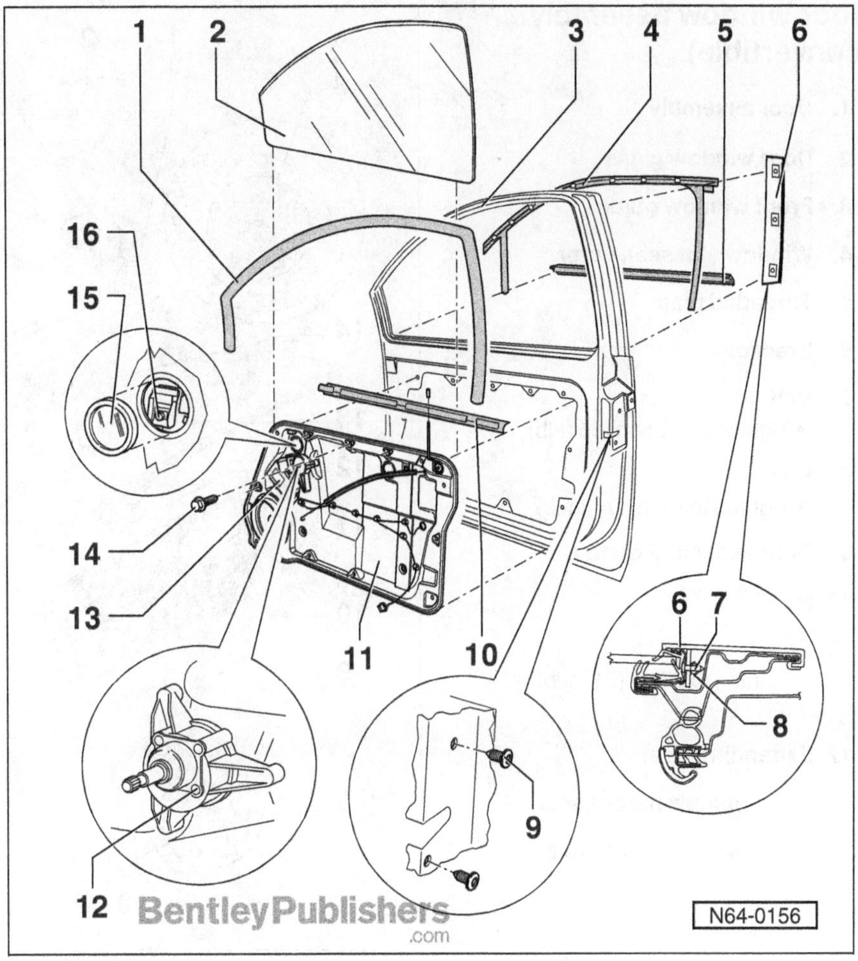

1. **Window channel, inner**
   - Pushed into door frame opening
2. **Door window glass**
3. **Door assembly**
4. **Window channel, outer**
   - Pressed into door frame opening
5. **Window slot seal, outer**
   - Pushed onto door frame
6. **Trim**
7. **Screw**
8. **Spring clip**
9. **Bolt**
   - Tighten to 20 Nm (15 ft-lb)
   - Install with locking compound
10. **Window slot seal, inner**
    - Pushed onto door frame
11. **Door assembly carrier**
    - Window regulator is part of assembly carrier
12. **Crank drive for manual window regulator**
    - Fastened to assembly carrier form the back with bolt
13. **Door control module with motor for window regulator (J386 and V15)**
    - Bolted to assembly carrier from the back
14. **Bolt**
    - Tighten to 8 Nm (71 in-lb)
15. **Rubber sealing plug**
16. **Bolt**
    - Tighten to 10 Nm (89 in-lb)

Door window assembly (convertible)

1. **Door assembly**
2. **Door window glass**
3. **Front window guide**
4. **Window slot seal, outer**
5. **Threaded plate**
6. **Bracket**
7. **Bolt**
   - Tighten to 20 Nm (15 ft-lb)
8. **Bolt**
   - Tighten to 8 Nm (71 in-lb)
9. **Door assembly carrier**
10. **Nut**
11. **Bolt**
    - Tighten to 8 Nm (71 in-lb)
    - Qty.: 9
12. **Expanding rivet**
13. **Inner window recess seal**
14. **Window regulator motor**

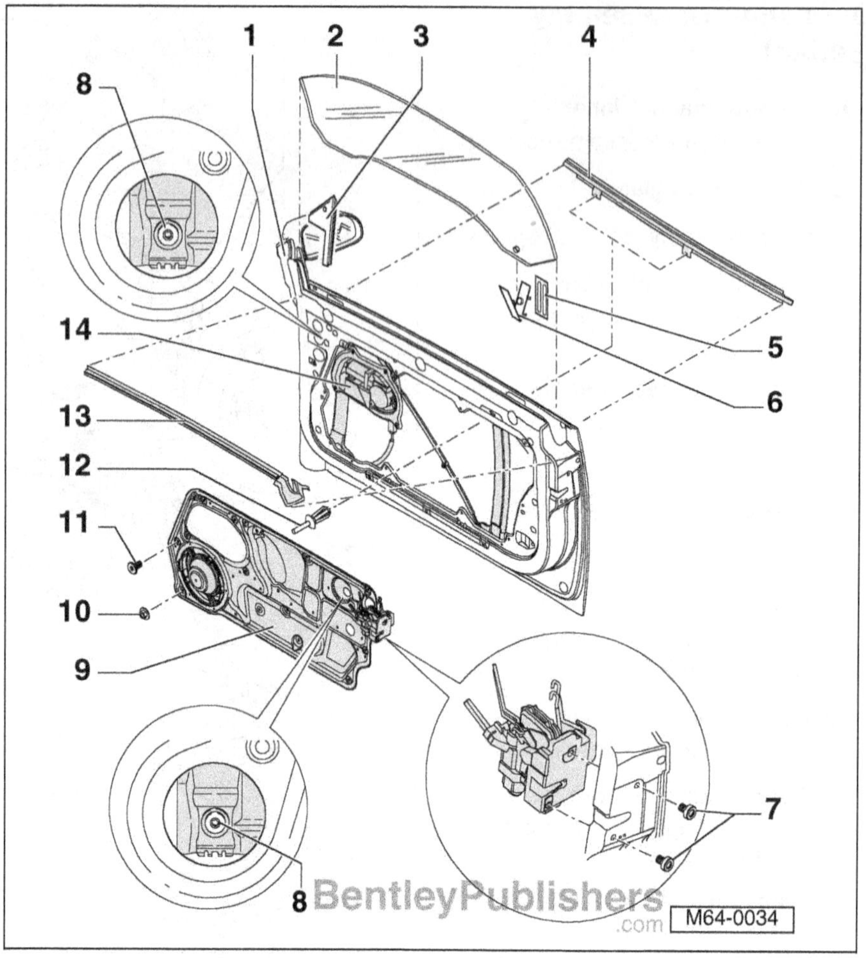

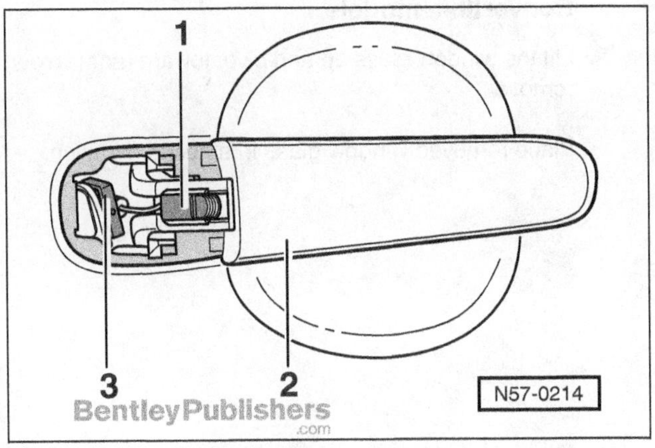

Door window glass, removing

NOTE —

Window glass must be operable in order to remove door assembly carrier. Any malfunction that prevents the windows (manual or power) from being lowered should be corrected before proceeding.

– Remove door trim panel. **See 70 Trim–Interior.**

– Remove lock cylinder housing as given earlier.

◄ Remove cable clip (**1**) from door handle.

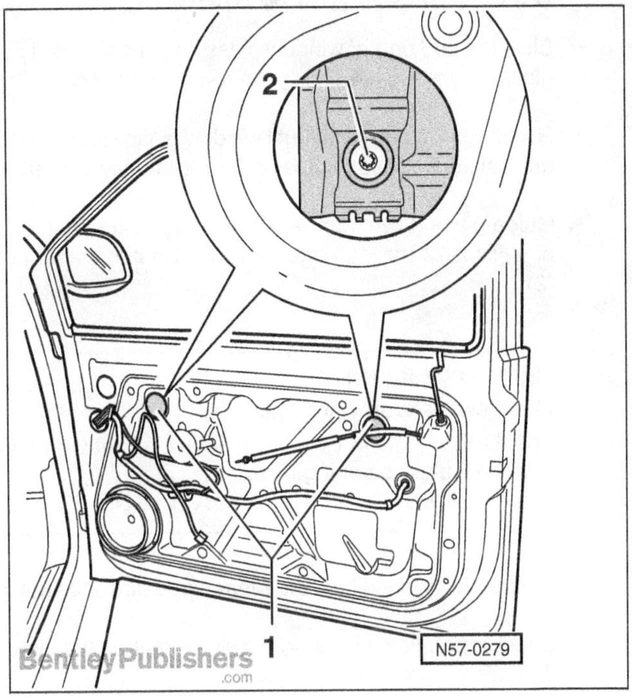

◄ Pry out sealing plugs (**1**) in door assembly carrier to gain access to window glass securing bolts (**2**).

- Sedan shown, convertible is similar.

– Raise or lower window until glass securing bolts (**2**) are accessible.

– Loosen window glass securing bolts and carefully press glass securing clamps apart.

NOTE —

- *If electric window regulator is not working, the door assembly carrier and regulator can be removed to slide window down or up to access securing bolts.*
- *If glass is being replaced due to breakage, remove door assembly carrier as given earlier and remove all broken shards of glass to prevent rattling noises in door.*

– Carefully lower the window regulator.

Sedan models

◄ Lift the window glass up and pivot forward (**arrow**) to remove.

– Place removed window glass in a secure location.

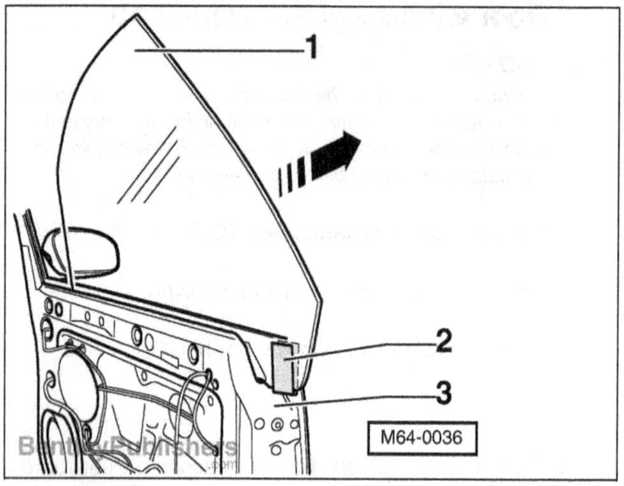

Convertible models

- Lift the window glass up and pivot toward rear (**arrow**) to remove.
- Place removed window glass in a secure location.

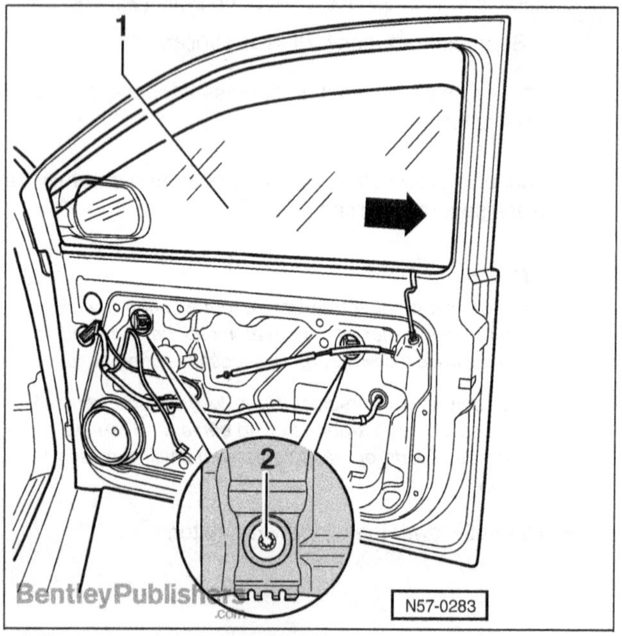

Door window glass, installing

- Slide forward part of window glass into opening and pivot glass downward, reversing the removal procedure.
- Raise window regulator until window glass securing bolts are visible in access holes of door assembly carrier.
- Guide glass down into securing clamps and adjust window glass at trailing edge (**arrow**). Tighten clamp bolts (**2**).

Tightening torques

- Window glass securing bolts (sedan) . . . . 10 Nm (89 in-lb)
- Window glass securing bolts (convertible) . . . . 8 Nm (71 in-lb)

- Remainder of installation is the reverse of disassembly.
- Check operation of window before final installation of door trim panel.

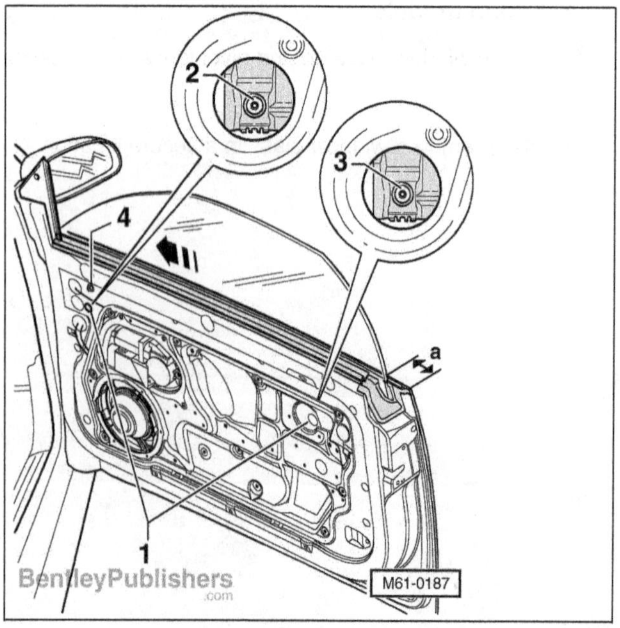

Door window glass, adjusting (convertible)

NOTE —
An incorrectly adjusted door window may cause damage to convertible top.

Rear adjustment (X-axis)

- Raise or lower window glass until securing bolts (**2** and **3**) are visible in access holes (**1**) of door assembly carrier.
- Loosen securing bolts (**2** and **3**) and nut (**4**) enough to adjust window.
- Position window so that dimension **a** = 65 mm (2.56 in.).
- Tighten securing bolts (**2** and **3**) and nut (**4**) to 8 Nm (71 in-lb).

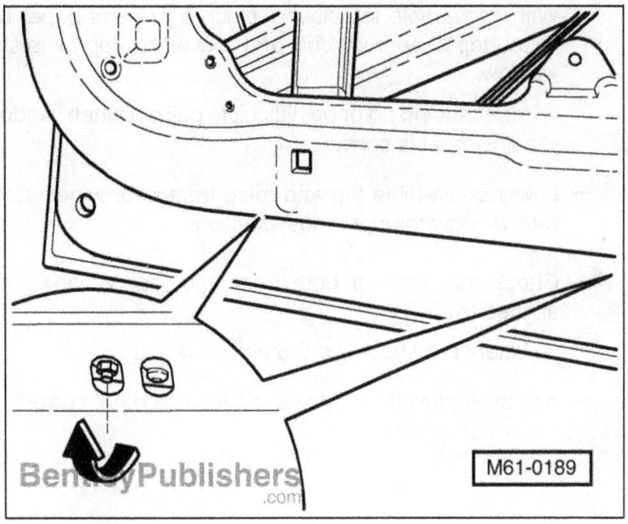

Left - right adjustment (Y-axis)

NOTE—
Adjust Y-axis if there is a problem with the door window sealing properly against convertible top when window is completely closed.

< Loosen adjustment nuts at bottom of door and slide bolts/nuts simultaneously outward (**arrow**) to adjust plus (+) Y-axis.

– Loosen adjustment nuts at bottom of door and slide bolts/nuts simultaneously inward to adjust minus (–) Y-axis.

– Tighten adjustment nuts.

– Check that window seals securely against convertible top when completely closed.

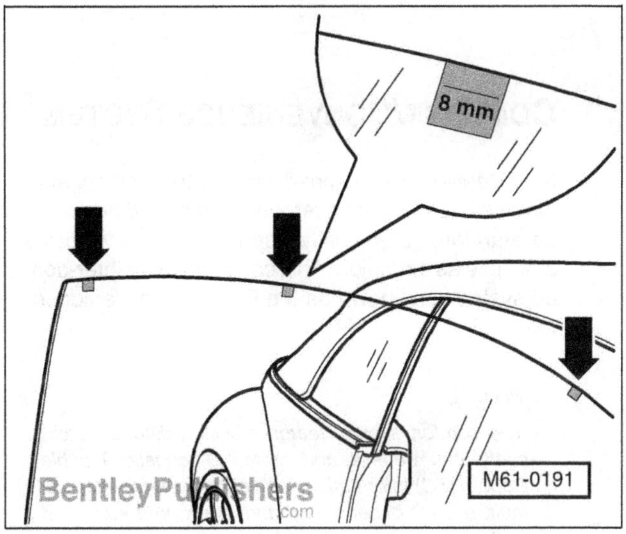

Up - down adjustment (Z-axis)

NOTE—
Adjust Z-axis if door window height is too high or too low when window is completely closed.

< Apply adhesive tape to top of window at three reference points (**arrows**).

– Make a mark on each piece of tape where the convertible top seal meets the window. The mark should be 8 to 10 mm below the top edge of the window.

– With convertible top closed, raise window completely and check that marks on tape are a consistent distance from edge of convertible top.

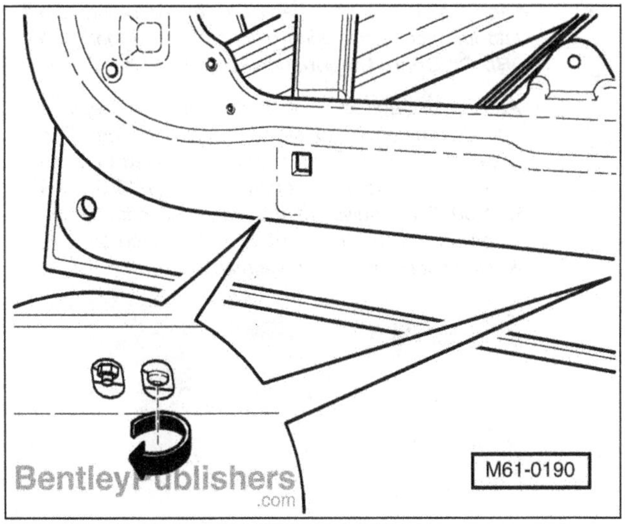

< To adjust window height, turn adjustment screws at bottom of door.

- Turn screws in clockwise direction (**arrow**) to raise window height (plus (+) Z-axis).
- Turn screws in counter-clockwise direction to lower window height (minus (–) Z-axis).
- A 360° turn = 1mm of adjustment.

Door window function test

- With convertible top closed, pinch a piece of paper between top of door window and convertible top by raising window.
  - Paper should rip or be difficult to pull out when window adjustment is correct.

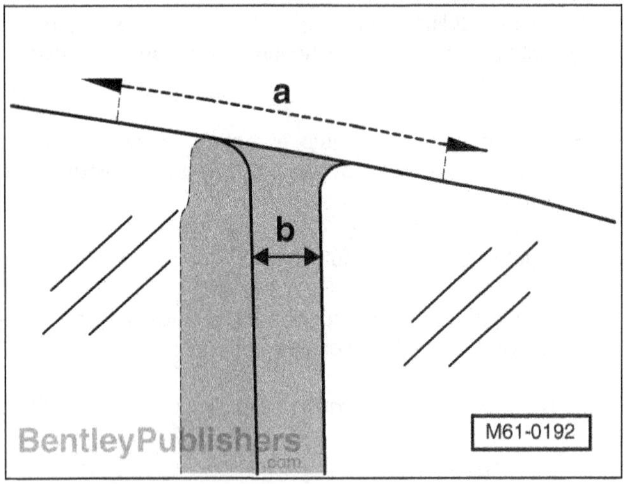

- Lower convertible top and raise front door window and rear window to upper most positions.
- Check that tops of both front and rear windows are aligned (**a**).
  - Distance **b** should be 9.5mm ± 0.5 mm.
- Adjust window if height specifications are not correct.

COMFORT/CONVENIENCE SYSTEM

New Beetles are equipped with central locking, anti-theft alarm system, keyless remote system and power mirrors as standard. Electrically operated power windows are available as an option. This combination of inter-connected systems is known as the Comfort/Convenience System.

NOTE—

- *The comfort system features built-in diagnostic circuitry that detects and stores Diagnostic Trouble Code (DTC) information. When the system detects a fault, a DTC is generated and stored in the comfort system central control module memory. This may result in failure of the system to operate properly. Proper diagnosis of fault codes can be accessed through the data link connector using diagnostic scan tool. See **OBD On Board Diagnostics** section.*
- *Additional diagnostic information is available in a data-stream format to aid in troubleshooting. This valuable information can only be accessed by scan tools such as the Volkswagen VAG 1551/1552, VAS 5051/5052, or equivalent aftermarket scan tool. Evaluation and interpretation of this information fall outside the scope of this service manual.*

Comfort/convenience system, component overview

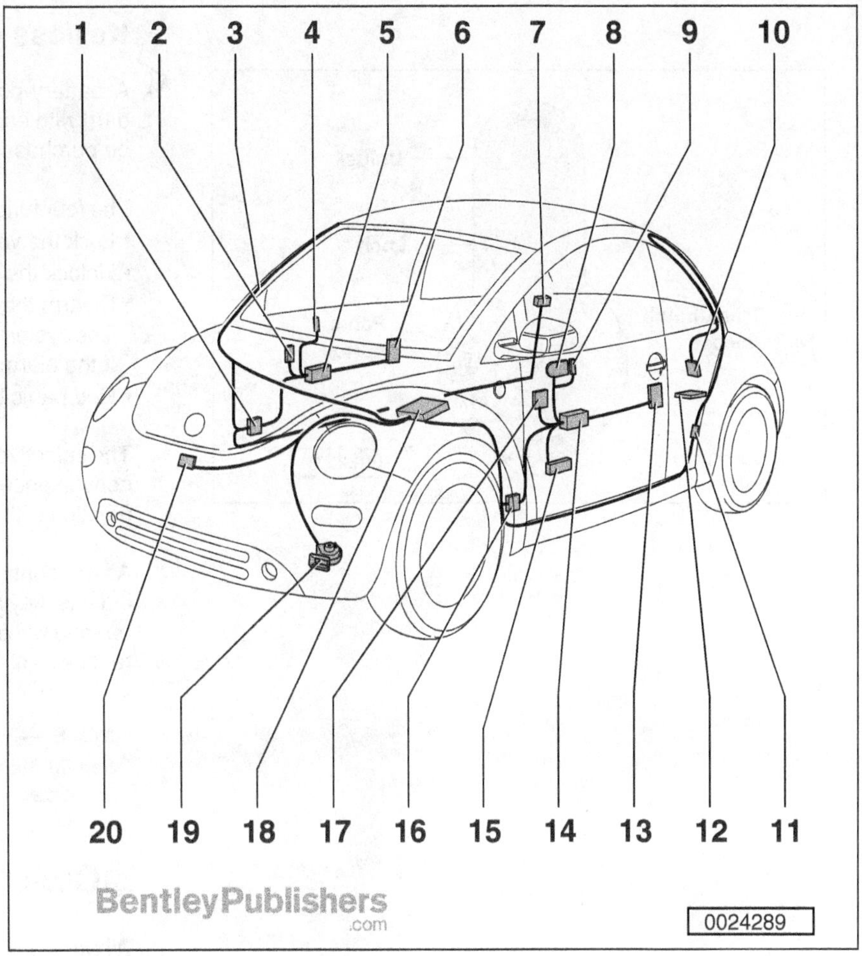

1. **Connector station (T10l)**
   - Color: black
   - Location: right side lower A-pillar
2. **Switch, right power window (E107)**
   - Location: right front door panel
3. **Antenna wire (R47)**
   - Comfort system remote receiving
   - Location: upper A-pillar
   - U.S. market cars located in left A-pillar
4. **Switch, central locking (E198)**
   - Location: right front door
5. **Door control module with motor for window regulator, right (J387 and V148)**
   - Operating functions: power windows, central locking
   - Linked to central control module via CAN-bus
   - Integrated into right side power window motor
6. **Lock unit, front door (F122)**
   - Location: inside right front door, secured to door assembly carrier
7. **Motor, fuel tank filler flap (V155)**
   - Location: inside vehicle above right rear wheel housing
8. **Switch, mirror adjustment (E43 & E48)**
   - Adjustment switch (E43)
   - Selector switch (E48)
   - Location: Left front door trim panel
9. **Switch, central locking (E150)**
   - Location: left front door
10. **Lock unit, rear hatch (V151)**
    - Location: inside rear hatch
11. **Harness connector (T5)**
    - Color: black
    - Location: inside vehicle above left rear wheel housing
12. **Rear hatch striker plate**
13. **Lock unit, left door (F220)**
    - Location: inside left front door, secured to door assembly carrier.
14. **Door control module with motor for window regulator, left (J386 and V147)**
    - Operating functions: power windows, central locking, power mirrors
    - Linked to central control module via CAN-bus
    - Location: integrated into left side power window motor
15. **Switch, rear hatch release (E188) and fuel tank filler flap release (E204)**
    - Location: in left door panel
16. **Connector station (T10k)**
    - Color: black
    - Location: left side lower A pillar
17. **Switch, left and right power windows (E40 & E81)**
    - Location: left front door panel
18. **Comfort system central control module (J393)**
    - Operating functions: keyless remotes, anti-theft alarm, interior lights, electrical system interface
    - Linked to door control modules via CAN-bus
    - Location: under instrument panel, left of steering column.
19. **Horn, anti-theft alarm (H8)**
    - Location: in right side of air plenum accessible from engine compartment.
20. **Switch, hood alarm (F120)**
    - Location: in hood lock carrier

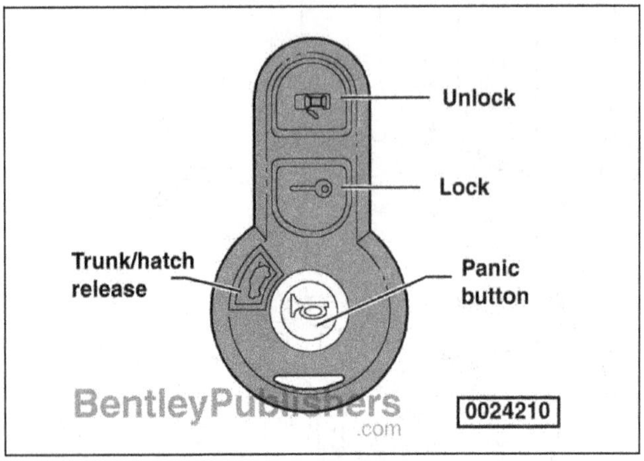

Keyless remote control

A battery-powered keyless remote transmitter is standard with each New Beetle. Up to 4 keyless remotes can be purchased and programmed into the system.

The four functions of the keyless remote transmitter are:

- Lock the vehicle and arm the alarm.
- Unlock the vehicle and disarm the alarm.
- Disarm the alarm and open the trunk or hatch. The system will rearm when the trunk or hatch is closed if the alarm system had previously been armed.
- Red panic button to trigger the alarm at any time.

The selective unlock feature of the central locking and the convenience close feature of the power windows do not function with the keyless remote system.

As a security feature, if the vehicle has been unlocked using the keyless remote transmitter and a door is not opened within approximately 30 seconds, the system will re-lock all of the doors and re-arm the alarm system.

NOTE —
*See **03 Maintenance** for keyless remote control battery replacement.*

QUALITY REVIEW

When you have finished working on the vehicle, it is advisable to take a moment to quality check or review your work. This helps to insure that the operation or repair has been completed properly with all affected systems functioning within normal parameters. These may include the following:

- Ensure that all cable ties and clamps that were removed as part of the repair are replaced.
- Ensure that all fasteners and hardware were replaced and torqued as specified.
- Make sure that all other components involved in the repair are positioned correctly and function properly.
- Make sure that all tools, shop cloths, fender covers and protective tapes are removed before closing the hood.
- Clean grease and fingerprints from painted surfaces, steering wheel and shifter.
- Unlock anti-theft radio and reset the clock.

60 Sunroof

GENERAL

The power sunroof is a two-way design that tilts open or slides back into the roof. The sunroof is a glass panel with an interior sliding shade (headliner). An electronic control unit is combined with the sunroof motor and regulates the motor operation.

Function

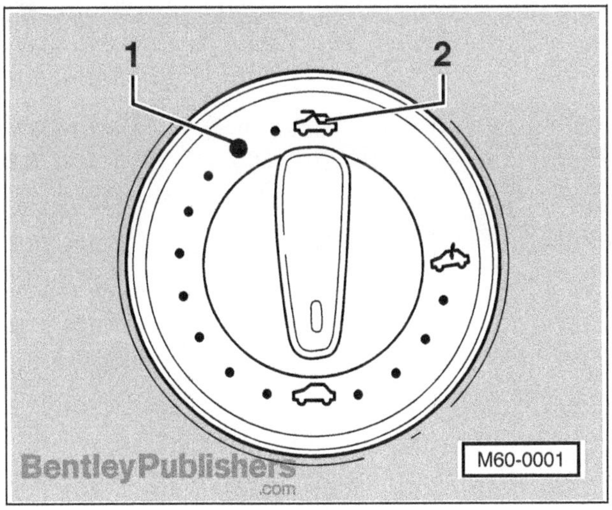

The sunroof is opened and closed by switching the ignition on and turning the rotary switch. In addition, the sunroof may be tilted or closed by pushing or pulling the switch. After switching off the ignition the sunroof can be opened or closed until either the driver's or front passenger's door is opened. Marks on the rotary switch are for preset opening positions of the glass panel. Any other intermediate opening position can be selected by turning the switch clockwise. To tilt or close the sunroof, push or pull the switch until desired position is reached.

Additionally, there is an emergency lock function in the sunroof switch. If there are lock problems, the sunroof can be locked by pressing the pre-selector actuator, which must be in the "sunroof closed" position. During emergency lock function the closing force limiter of the sunroof is switched off.

The drive for the sunroof is also protected by a running time limit to prevent overheating. This protection mechanism responds after an uninterrupted actuation of approx. 2 minutes. After a cool down phase the mechanism is then operational again.

60

SUNROOF

Sunroof with glass panel, assembly

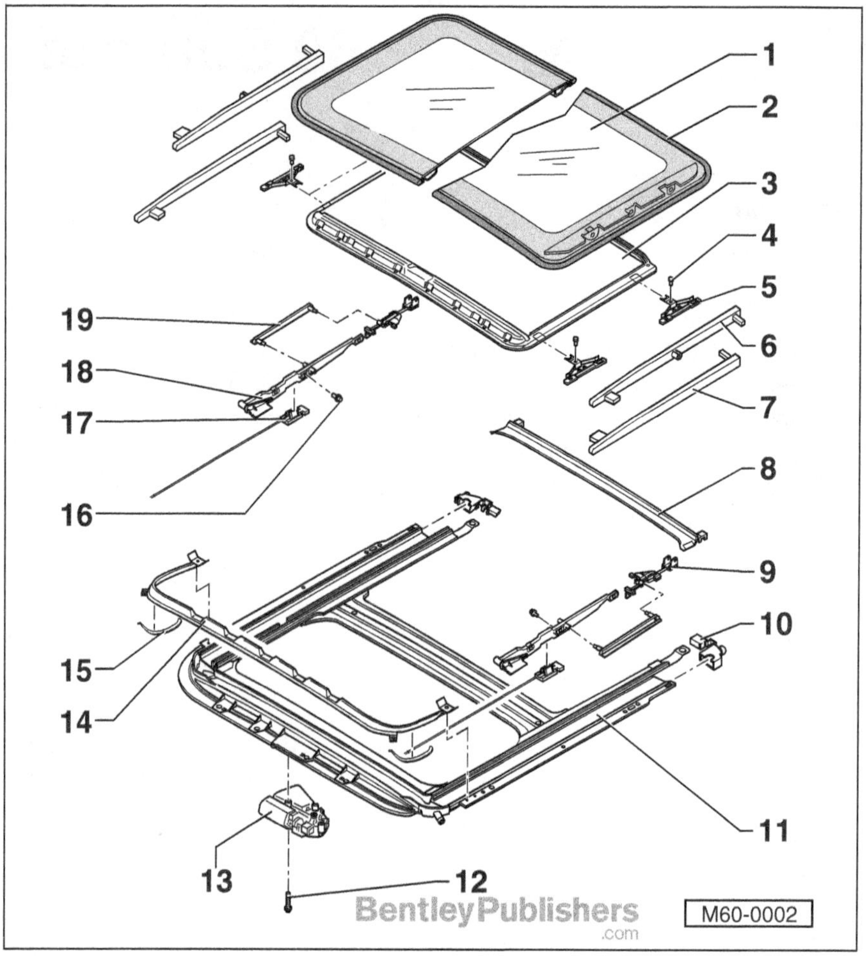

1. **Glass panel for sunroof**
   - Single pane safety glass
2. **Panel seal**
   - Seal joint located centrally at rear
3. **Sliding headliner**
4. **Countersunk screws**
   - Always replace
5. **Slide**
   - If sunroof operation is noisy or binding, check slides for damage or missing foam spacers. See VW technical bulletin number 60 07 04, dated May 18, 2007 for more information.
6. **Upper trim**
7. **Lower trim**
8. **Water channel**
   - Remove together with slotted guide rail
9. **Guide locking hook**
10. **End piece**
    - Use butyl adhesive sealing cord AKL 450 005 05 to seal.
11. **Carrier unit**
    - If required, guide channels are to be greased with G 000 450 02 only, otherwise correct operation cannot be guaranteed
12. **Bolt**
13. **Drive motor**
14. **Wind deflector**
15. **Spring**
16. **Torx screw**
    - T25
    - Tighten to 4.5 Nm (39 in-lb)
17. **Rear guide**
    - Removed together with slotted guide rail
18. **Slotted guide rail**
19. **Spacer**
    - Remove together with slotted guide rail

Glass panel, removing

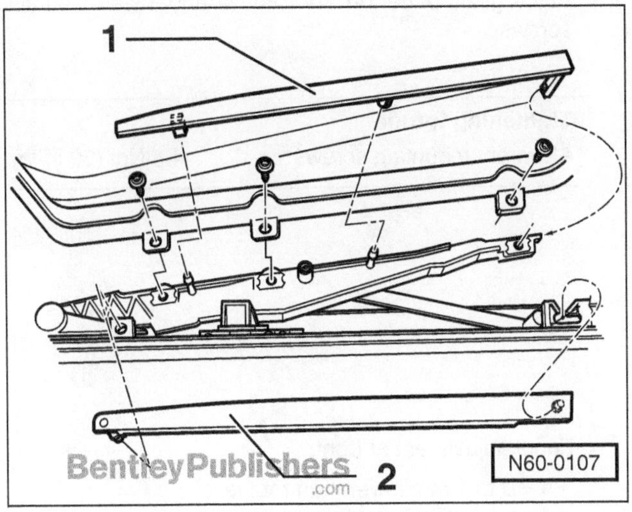

- Slide headliner to rear.
- Tilt sunroof open.
- Unclip lower trim (**2**) in forward area inwards to center of vehicle and unhook at rear.
- Unclip upper trim (**1**) at front and center and unhook at rear.

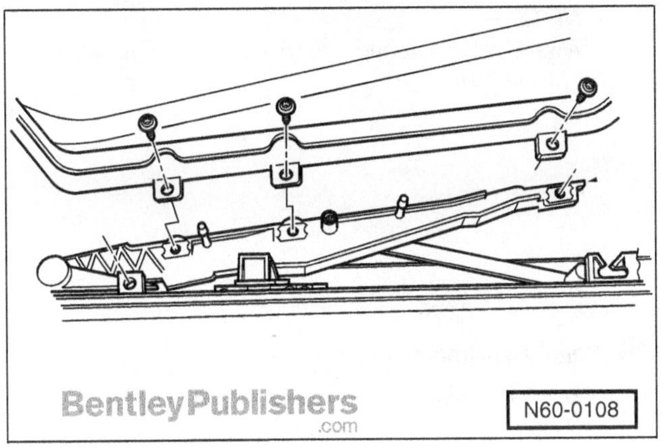

Remove mounting screws (Torx T25). Remove sunroof panel by lifting out upwards.

NOTE—
Sunroof must be not moved to "open" position when removed. The water channel will not be compressed by the glass.

Glass panel, installing and adjusting

NOTE—
Panel must be installed in "0" position (panel closed)

"0" position:

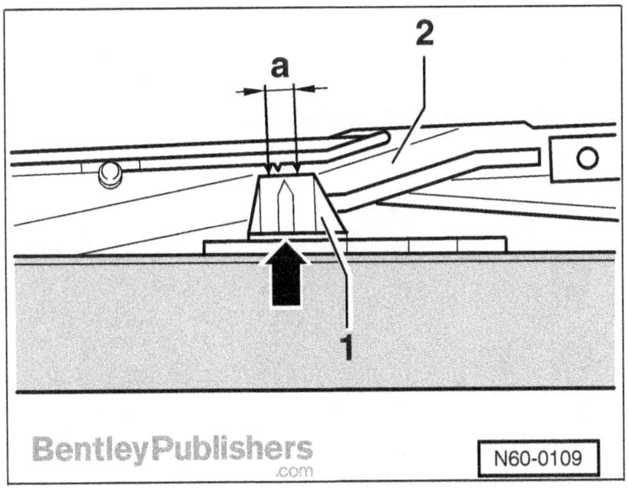

The mark (**arrow**) on the rear of the guide upper section (**1**) must lie within marks (**a**) on both sides of the guide rail gate (**2**).

NOTE—
*The slotted guide rail (**2**) must be located in the guide rails (cannot be moved by hand). If this is not the case, adjust parallel running.*

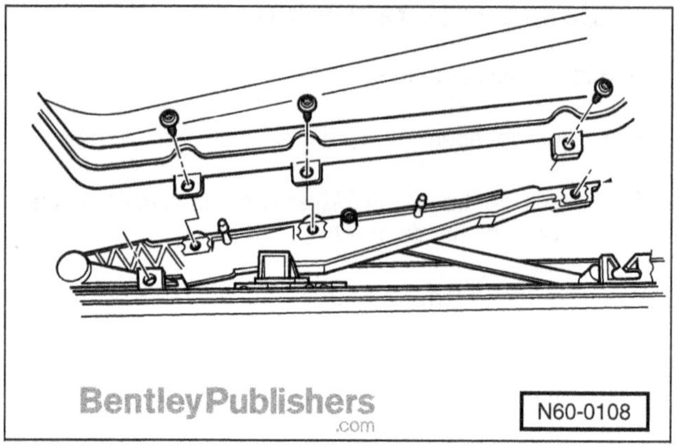

◀ Place glass panel on from above and install mounting screws.

Tightening torque

- Sunroof mounting screws. . . . . 4.5 Nm (39 in-lb)

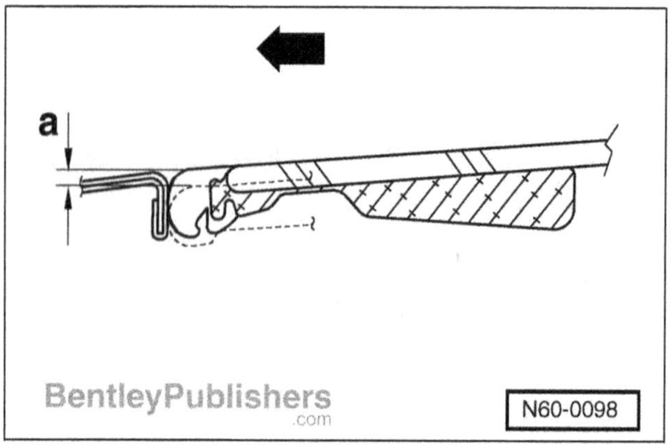

◀ Panel adjustment at front:

- **a** = 0 to 1 mm lower than roof level.
- **Arrow** = direction of travel (forwards)

NOTE —

Always adjust left and right hand glass panel sides symmetrically.

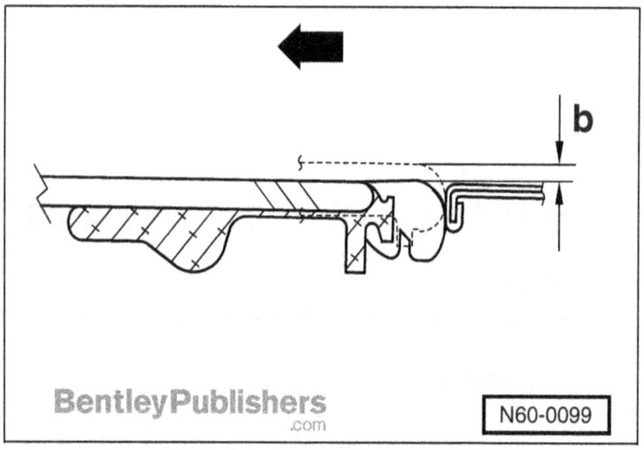

◀ Panel adjustment at rear:

- **b** = 0 to 1 mm higher than roof level.
- **Arrow** = direction of travel (forwards)

NOTE —

Always adjust left and right hand glass panel sides symmetrically.

– Tilt sunroof open.

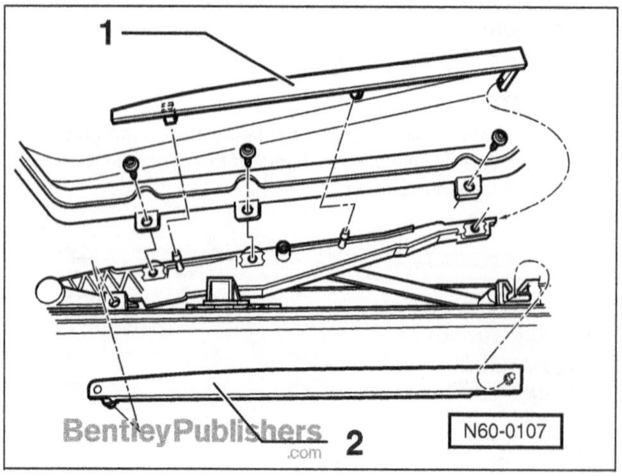

◀ Hook in upper trim (**1**) at rear and clip front and center in.

– Hook in lower trim (**2**) at rear and clip in at front.

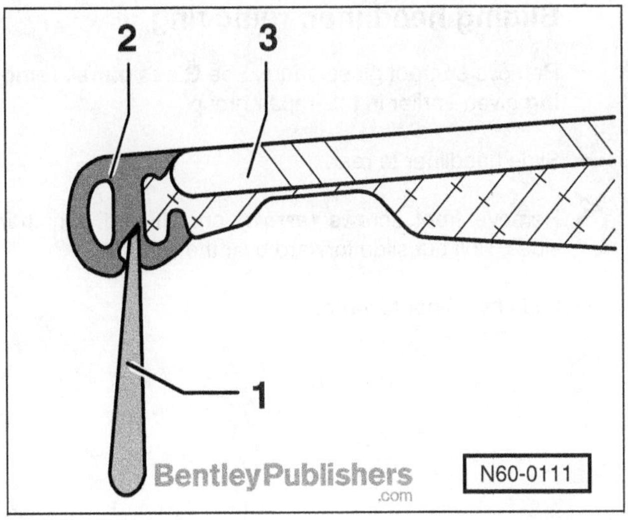

Glass panel seal, adjusting

◀ Check that the pre-tension between panel seal (**2**) and body is uniform using a 0.3 mm thick strip of cardboard (i.e. business card). Cardboard should slide through the gap with firm resistance.

– If adjustment is needed, remove glass panel (**3**) to adjust seal. See **Sunroof glass panel, removing** given earlier in this repair group.

– If seal gap is too large (insufficient pre-tension), the seal can be pushed apart with a plastic trim wedge (**1**).

– If seal gap is too small (excessive pre-tension), the seal can be closed together with a plastic trim wedge (**1**).

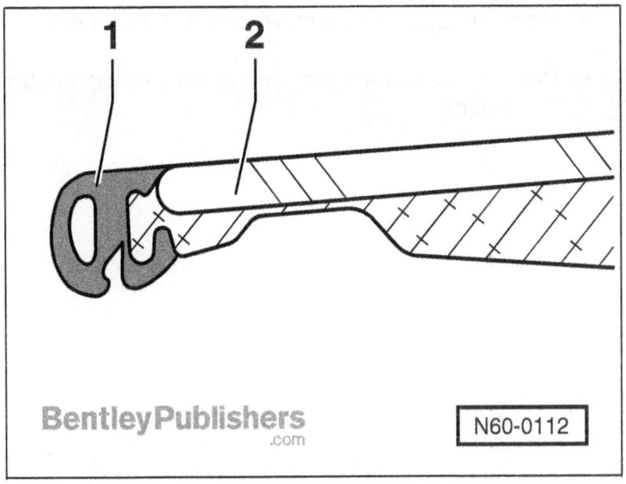

Glass panel seal, replacing

– Remove glass panel for sunroof.

◀ Pull seal (**1**) off glass panel.

– Install new seal onto glass panel (**2**) from bottom to top, starting at center rear.

– Cut panel seal to length after installing.

NOTE —
To ease fitting of seal, coat panel edge with a soapy solution.

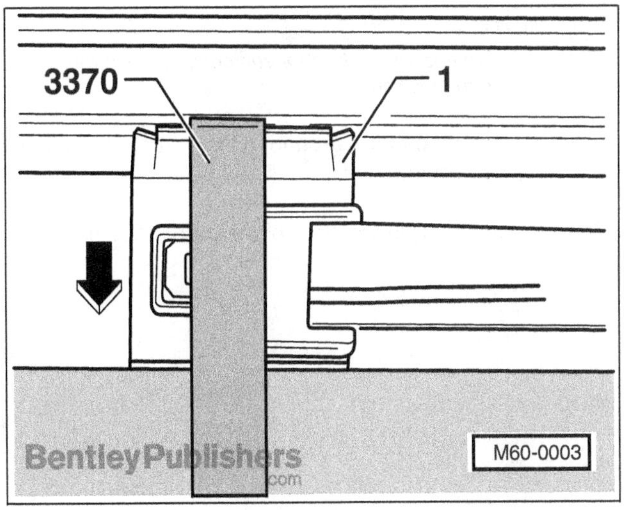

Wind deflector, removing

Sunroof must be completely open.

◀ Insert assembly hooks (special tool 3370) between edge of roof and wind deflector mount. Release wind deflector mount (**1**) from locking devices on left and right sides by pulling on assembly hook tool.

– Pry wind deflector mounting out from guide rail (left and right sides) using a screwdriver.

– Take out wind deflector.

NOTE —
When installing, ensure that wind deflector mount engages in assembly panel in the prescribed position.

Sliding headliner, removing

– Remove sunroof glass panel. See **Glass panel, removing** given earlier in this repair group.

– Slide headliner to rear.

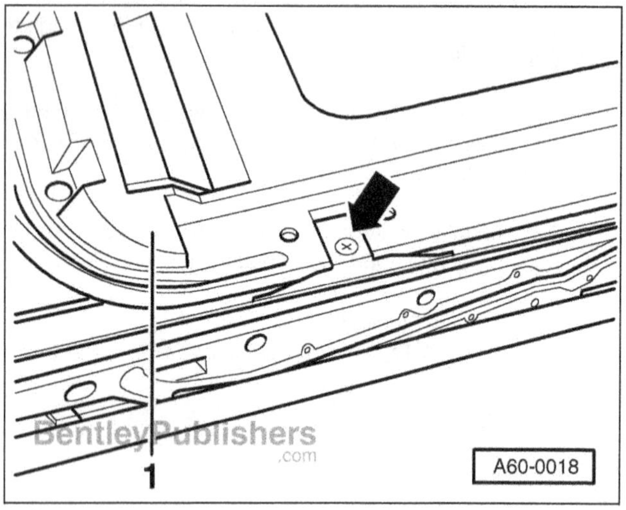

◄ Remove front screws (**arrow**) on left and right hand sides. Pull out slide forward over the stop (**1**).

– Slide headliner forward.

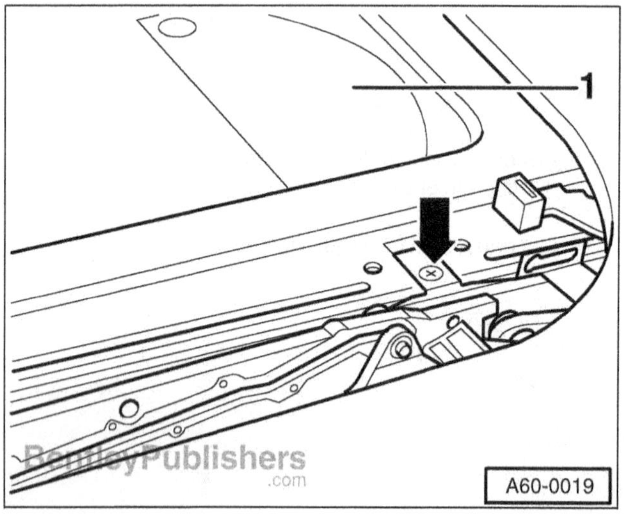

◄ Remove rear screws (**arrow**) on left and right sides.

– Pull out slide toward rear and remove sliding headliner (**1**) upward.

Drive motor, removing

NOTE —
Remove and install drive for sunroof only when roof is closed "0" position.

– Disconnect battery ground (GND) strap from battery negative (–) terminal.

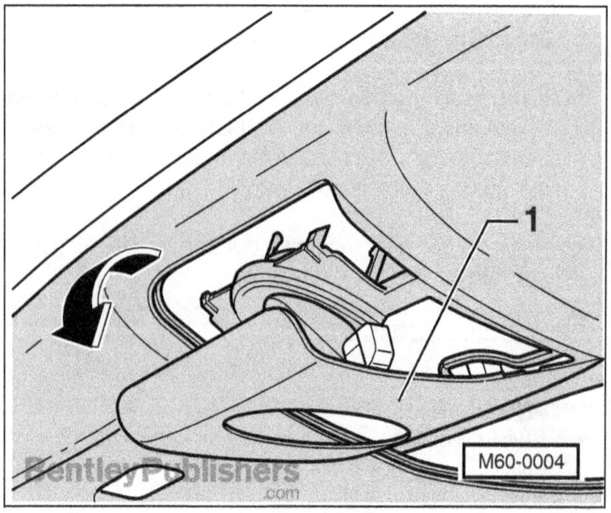

◄ Unclip trim (**1**) in direction of (**arrow**) and remove.

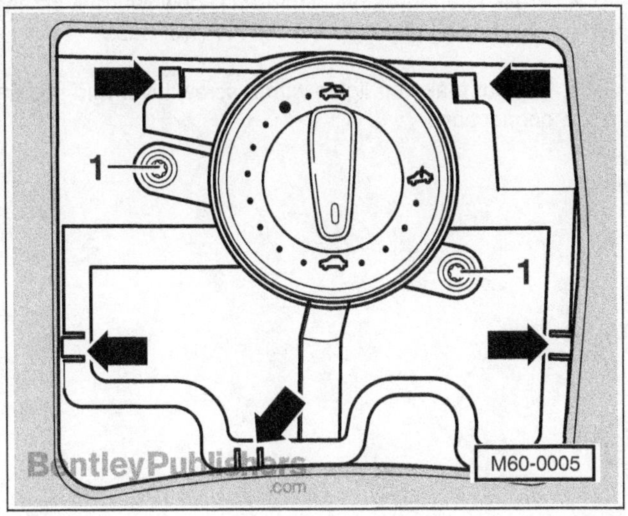

Remove screws (**1**) and carefully release the locking lugs (**arrows**) of the automatic preselector retaining frame.

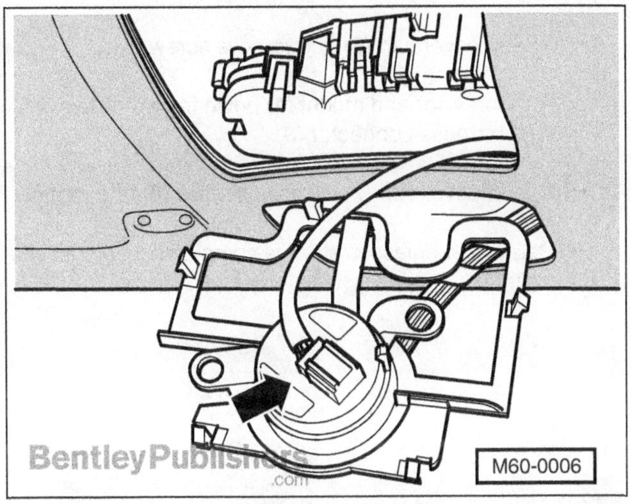

Disconnect harness connector (**arrow**).

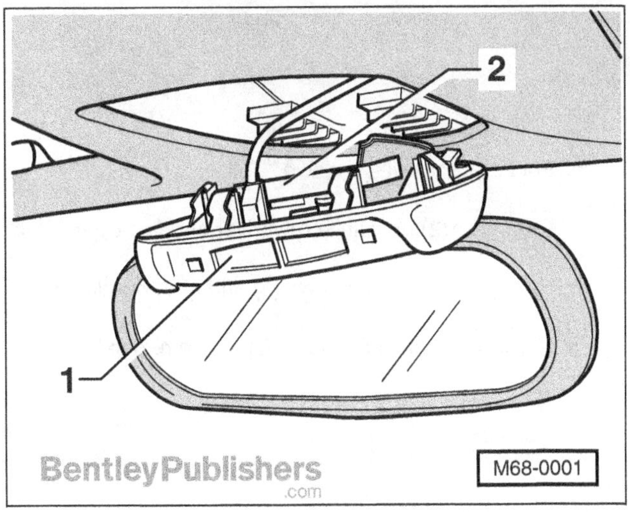

Pull rear edge of trim (with clock) (**1**) down off roof trim and unhook trim from front edge.

- Disconnect harness connector (**2**) for interior light and clock.

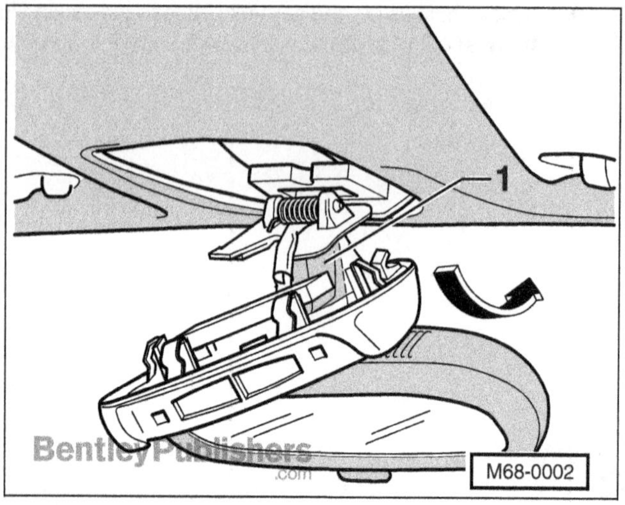

◄ Press interior rear view mirror (**1**) downward at an angle off retaining plate (spring clip in mirror base).

– Pry out make-up lights with a screwdriver and separate connections.

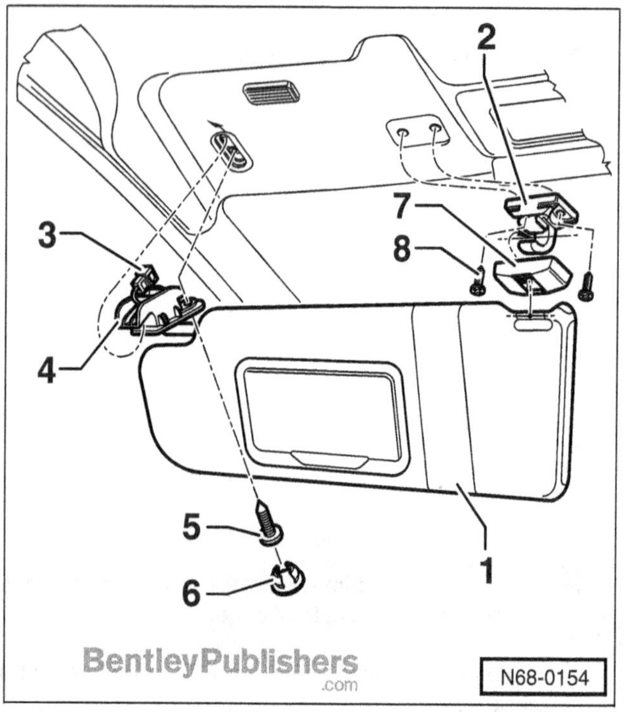

◄ Unclip sun visor (**1**) from mount (**2**).

– Pry off cover cap (**6**) and remove screw (**5**).

– Pull sun visor and mount (**4**) down from roof and disconnect harness connector (**3**).

– Pry off cover cap (**7**), remove screws (**8**) and mount (**2**).

– Remove sunglasses storage compartment on left side of roof.

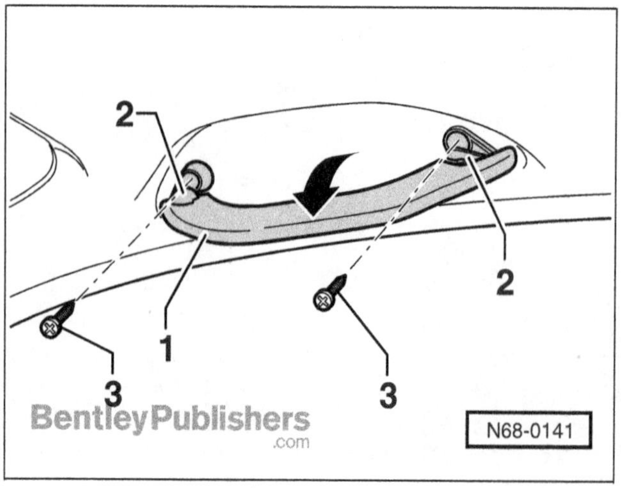

◄ Fold grab handle (**1**) down.

– Pry cover (**2**) up with screwdriver and fold open.

– Remove mounting screws (**3**) and grab handle (**1**).

– Remove left and right hand A-pillar trim. See **70 Trim–Interior**.

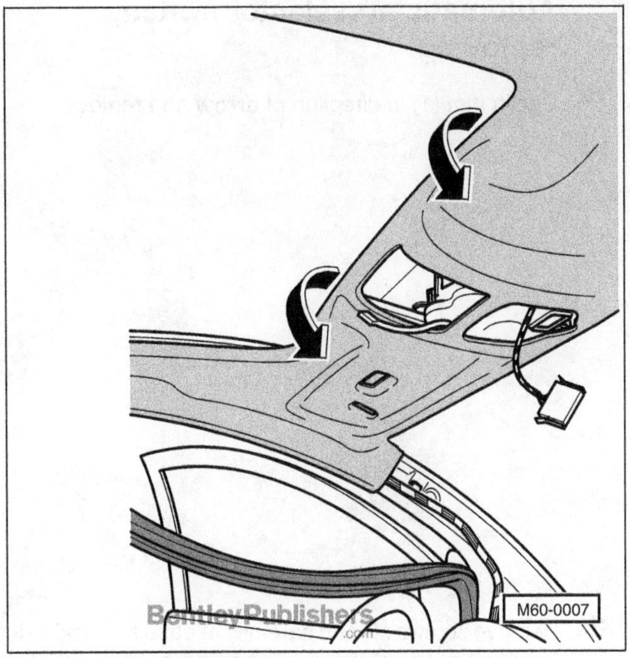

Carefully unhook molded headliner at front of roof cut-out (**arrows**) and pull down until drive motor mounting bolts can be accessed. Disconnect harness connector.

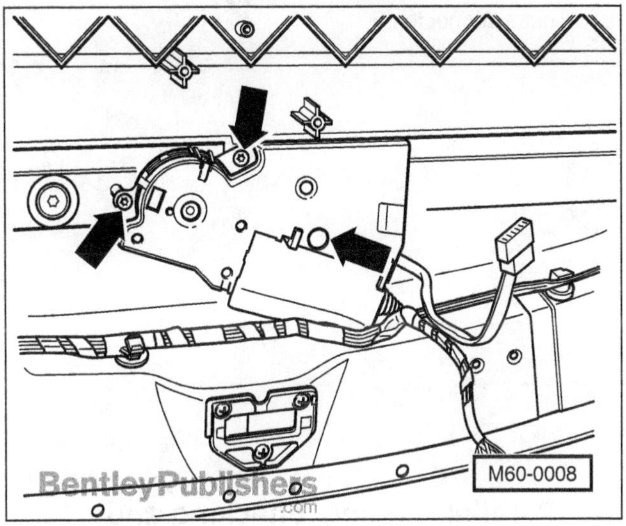

Remove mounting bolts (**arrows**) and drive motor.

Tightening torque
- Sunroof motor to body (always replace) . . . . 3.5 Nm (31 in-lb)

Drive motor, adjusting (setting "0" position)

There may be a requirement to adjust the "0" position if the drive was removed when not in "0" position or the sunroof was closed using the emergency key.

Drive removed but still connected electrically.

- Select "roof titled" with rotary switch.
- Select "roof closed" with rotary switch.
- Select "roof opened" with rotary switch.
- Select "roof closed" with rotary switch.
- Install drive in "0" position with sunroof closed.

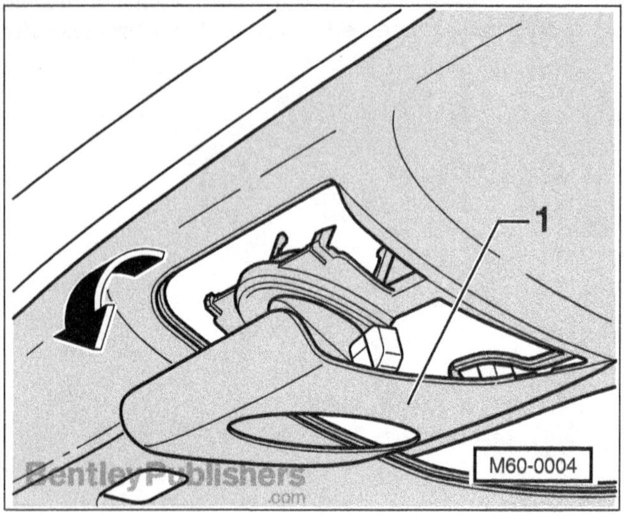

Automatic preselector switch, removing

Unclip trim (**1**) in direction of **arrow** and remove.

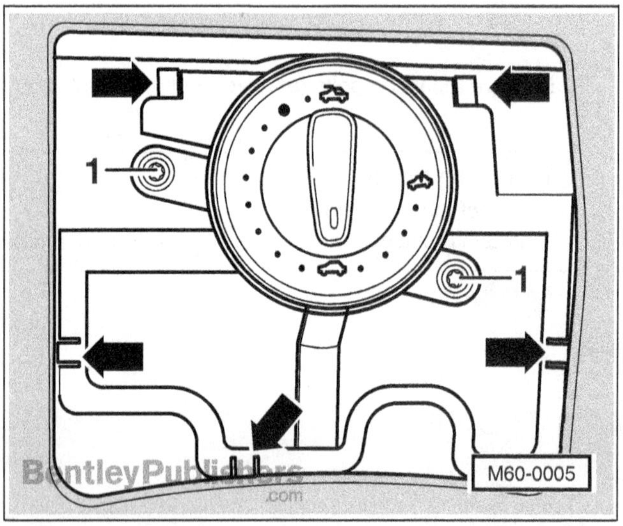

Remove screws (**1**) and carefully release the locking lugs (**arrows**) of the switch retaining frame. Disconnect harness connector.

Parallel running, checking and adjusting

– Remove glass panel for sunroof.

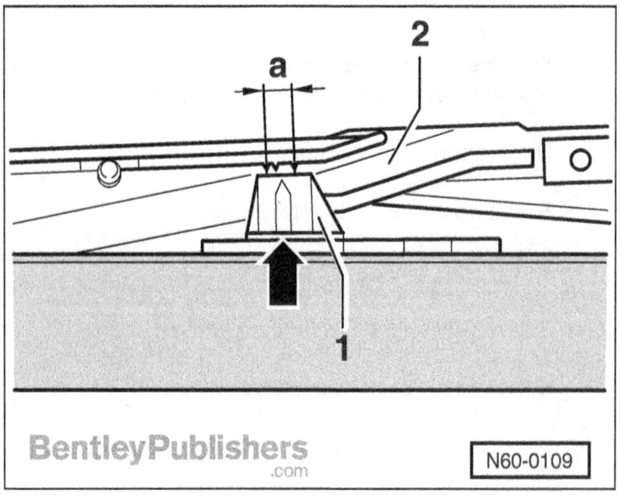

The mark (**arrow**) on the rear of the guide upper section (**1**) must lie within the marks (**a**) on both sides on the guide rail gate (**2**).

NOTE —

*The slotted guide rail (**2**) must be located in the guide rails (cannot be moved by hand).*

- If specification is not met, slide guide upper part (**1**) from front to rear center between the marks.

– Install drive in this position ("0" position). Then check "0" position.

Carrier unit, removing and installing

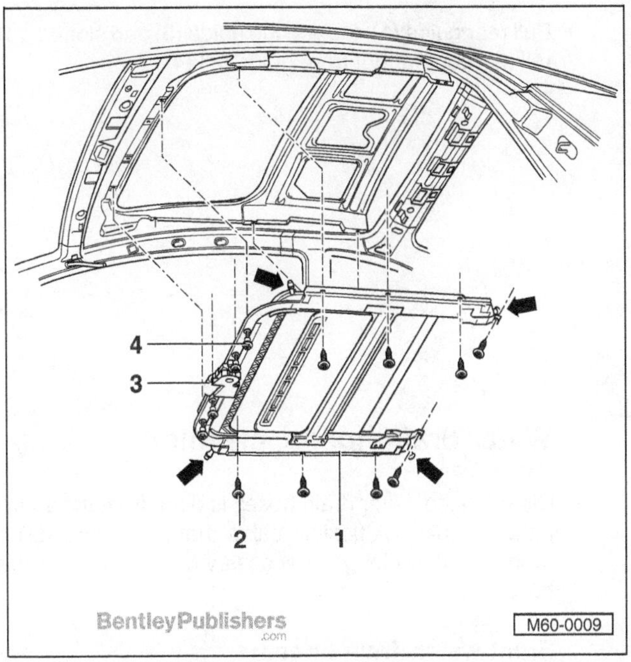

Removing

– Remove automatic preselector switch.

– Remove molded headliner.

◄ Pull water drain hoses (**arrows**) off carrier unit (**1**).

- Remove bolts (**2**) and (**4**) (with sleeves) and lift carrier unit out of vehicle with help of second mechanic.

Installing

– Insert carrier unit (**1**), align with bolt holes and install mounting screws (**2** and **4**) with sleeves.

– Check electrical consumers, electrical wiring and connections on roof and reinstate if necessary.

– Tighten mounting screws for carrier unit (**1**), starting at sunroof drive (**3**) on left/right hand sides and moving rearward.

- Tighten mounting screws to 8 Nm (71 in-lb).

– Install water drain hoses (**arrows**).

– Connect harness connector to drive motor (**3**).

Slotted guide, removing

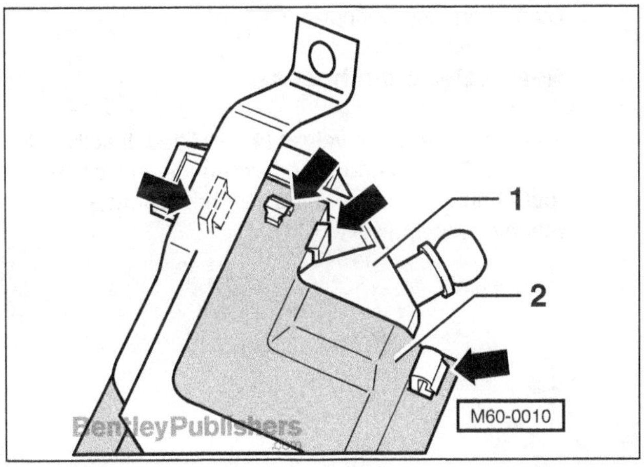

– Remove glass sunroof panel.

– Remove assembly unit.

– Remove sunroof drive motor. See **Drive motor, removing** in this repair group.

◄ Unclip end cap (**1**) at **arrows** from guide rail (**2**) and remove.

> ***NOTE —***
> *When installing, seal end cap (1) to guide rail with butyl adhesive sealing cord (part # AKL 450 005 05).*

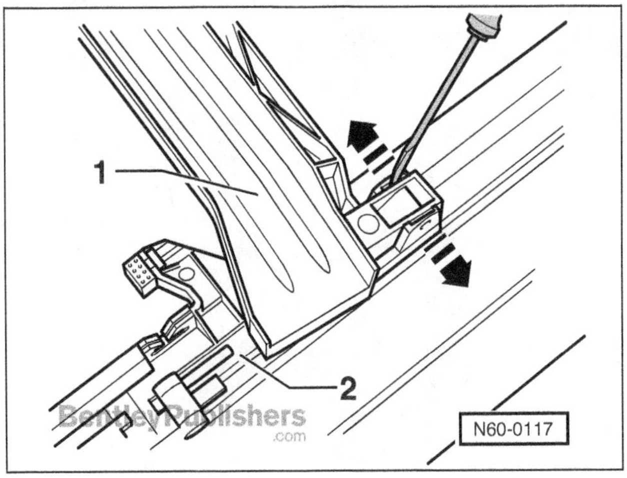

◄ Unclip water channel (**1**) from locking lug guide (**2**) on both sides carefully with a screwdriver and take off.

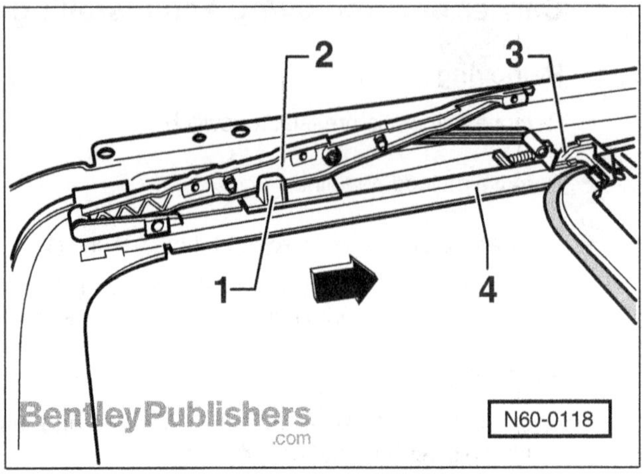

Slide rear guide (**1**) in direction of **arrow** into guide rail (**4**) until the locking lug guide (**3**) is released.

- Pull rear guide (**1**), locking lug guide (**3**) and slotted guide rail (**2**) together out from guide rail (**4**).

Water drain hoses, cleaning

Cleaning the water drain hoses is done from the sunroof panel opening. A flexible cable that is approx. 150 cm (approx. 60 in.) long is necessary for cleaning the drain hose.

Front water drain hoses:

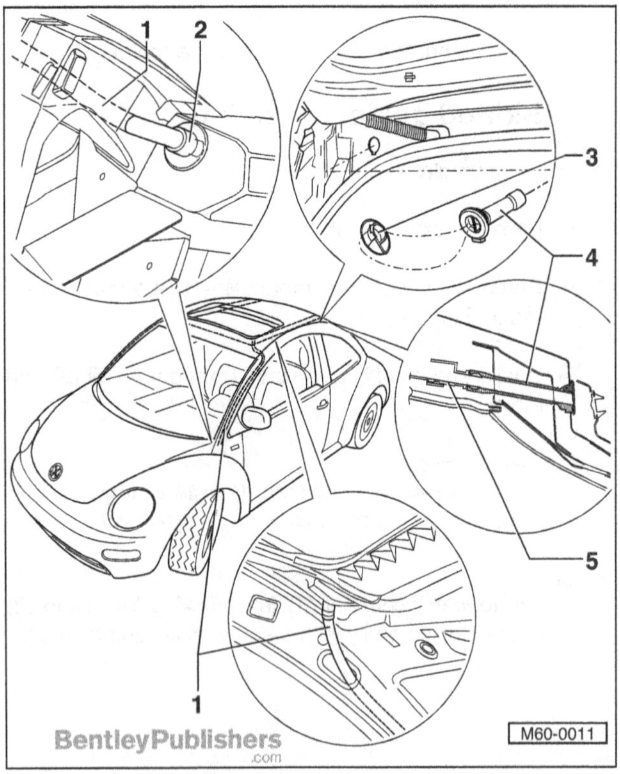

The front water drain hoses (**1**) run in the A-pillars and end with a water drain valve (**2**) in the plenum chamber. Clean from sunroof cut-out.

Rear water drain hoses

The rear water drain valves (**4**) are fitted directly onto the sunroof (**5**) water drains and exits under the tailgate in the roof channel. The cleaning can be performed from the rear by removing cover (**3**).

61 Convertible Top

GENERAL

This repair group covers the convertible top maintenance and components of the power convertible top hydraulic system. For convertible top removal and installation, as well as electrical component locations, see **Appendix B and C** at the back of this manual.

Maintenance

Under normal conditions the convertible top is windproof and waterproof. In extremely bad weather, such as torrential or continuous rain, a few drops of water could possibly get inside the vehicle. This is normal and should not be a cause of concern.

To keep the convertible top in good condition, the following points should be adhered to:

- Always clean convertible top when it gets dirty. Remove any bird droppings or other residue as soon as possible.
- If the vehicle will be outside for extended periods of time without being driven, a car cover is recommended, especially if parking under a tree.
- Too much direct sunlight on the convertible top could bleach it out.
- Only open the convertible top when it is clean. Otherwise particles of grit can scratch the rear window and damage the top fabric.
- Remove any ice and snow before opening top. Do not use a sharp object or anything that could scratch the top to remove ice and snow. A normal ice scraper can be used to clear the rear window.
- Only open top when it is completely dry. Moisture on the top can cause unsightly marks and mildew.
- Do not leave convertible top open for extended periods (more than a week or two). It is best to close the top occasionally to prevent any discoloration of the folds of the fabric and to allow creases to smooth out.
- The convertible top should not be used at temperatures below 32°F (0°C).
- Do place items on the shelf behind the rear seat. Objects could damage convertible top when it is opened.

> ***WARNING —***
> * *Never attempt to open or close convertible top when vehicle is moving.*
> * *Do not drive vehicle if convertible top is not secured properly when open, or completely latched and locked when closed.*
> * *Do not open or close convertible top if rollover protection supports (at top of rear seat) have been deployed.*
> * *Do not open or close convertible top when vehicle is on a lift or is raised with jack.*
> * *When opening and closing convertible top, make sure there is no risk of injury from the moving parts of the frame, etc. Never touch the top mechanism when operating top.*

Emergency closing

If the power convertible top cannot be closed due to a malfunction or loss of electrical power, it can be closed manually by following the procedure below.

> ***NOTE —***
> *The power convertible top should not be opened manually.*

- Switch the ignition off and engage the parking brake.
- Press the lower part of the side flap in direction of **arrow**. Hold the flap in this position.
- Unclip the red release cord (**A**) from its holder and pull upward.
- Repeat the last two steps on the other side of vehicle.
- Open the rear lid.
- Working in left side of luggage compartment, pry off round plastic cover for hydraulic pressure screw.

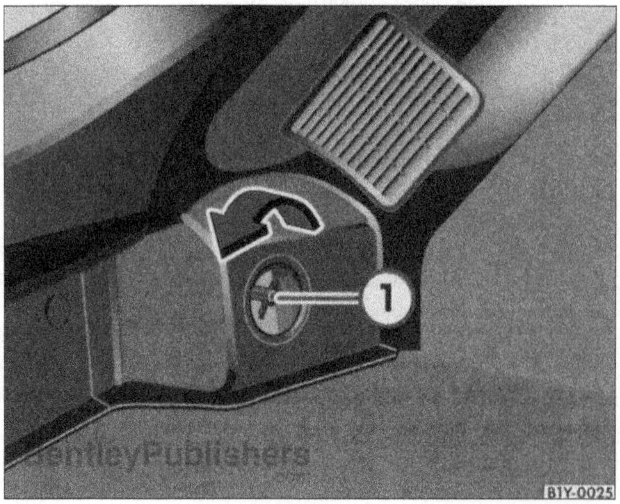

- Turn screw (**1**) on hydraulic reservoir counter-clockwise (**arrow**) to stop to release hydraulic pressure.
- Close top using both hands to lift and push it forward.
- Lock top in place.
- Turn screw (**1**) on hydraulic reservoir clockwise until it stops.
- Replace plastic cover for hydraulic pressure screw.

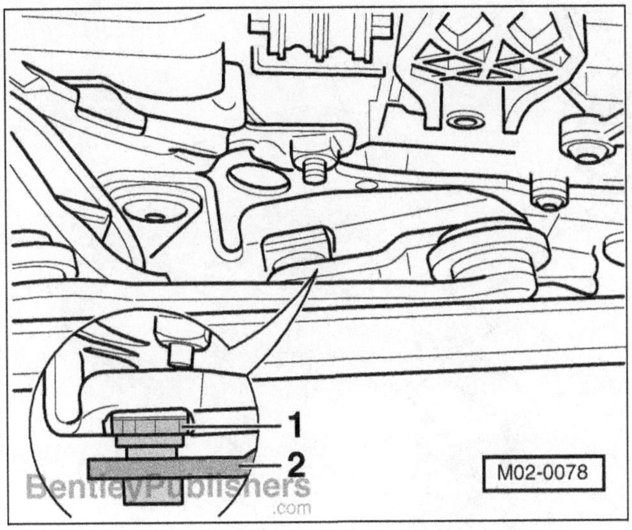

End position, checking

When the convertible top is fully opened, it is held in position by two locking devices. These locking devices are designed to hold the top in position in the event of an accident. The function of the locking devices should be checked periodically.

– Switch ignition on.

– Open convertible top completely.

– Check to make sure convertible warning light is off.

– Turn off ignition switch.

< Check that locking device (**2**) encloses pin (**1**) on both rear sides of top.

• Right rear side shown.

Power window pinch protection, resetting

If power has been interrupted to the convertible top control module (J256), including disconnection of the battery, the convertible top operation switch (E137) in the center console may be deactivated when battery is re-connected. It may be necessary to reset the power window pinch protection to restore proper function of convertible top switch.

– Switch ignition on.

– Cycle (lower and lift) all windows one time using the "all" window switch located in the driver's door panel.

– If "all" window switch is inoperative, cycle (lower and lift) windows one time using the key in the driver's door lock.

– Convertible top operation switch (E137) should now be operational.

Convertible top open/close system, resetting

If convertible top is opened and closed repeatedly in a short time span, the top operating system may cease to function due to overload protection provided by a circuit breaker. It can take up to 37 minutes for the system to reset itself.

– Reset power window pinch protection as described above.

– Allow 37 minutes for system to reset and check operation again.

CONVERTIBLE TOP OVERVIEW

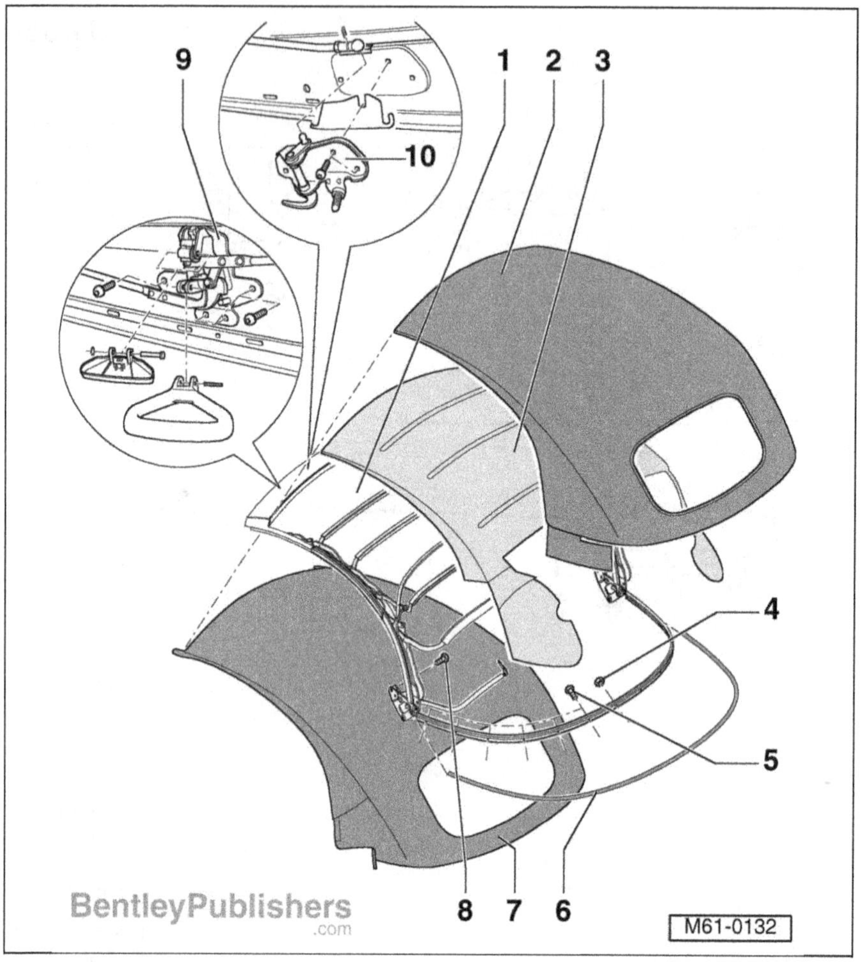

1. **Convertible top frame**
2. **Top cover**
   - Rear window is part of top cover
   - Replace entire cover if window is damaged
3. **Padding**
4. **Nut**
   - Tighten to 10 Nm (7 ft-lb)
5. **Bolt**
   - Qty: 12
   - Tighten to 10 Nm (7 ft-lb)
6. **Sealing band**
7. **Headliner**
8. **Bolt**
   - Tighten to 50 Nm 37 ft-lb
9. **Front latch**
   - If convertible top handle does not return to rest position after closing top, check mounting screw for handle and tighten if necessary. Use Loctite 242 on mounting screw threads to secure.
   - Tighten mounting screw to 24 Nm (18 ft-lb)
10. **Front retaining latch**
    - If noises or rattles are heard at convertible top latch, it may be necessary to install a rubber grommet (part # 1Y0871469B) onto top locator pin.
      See VW technical bulletin # 61 06 01, dated March 14, 2006 for more information.

HYDRAULIC SYSTEM

Convertible top electric/hydraulic components

NOTE—
Volkswagen identifies electrical components by a letter and/or a number in the electrical schematics. See **EWD Electrical Wiring Diagrams.** *These electrical identifiers are listed in parenthesis as an aid to electrical troubleshooting.*

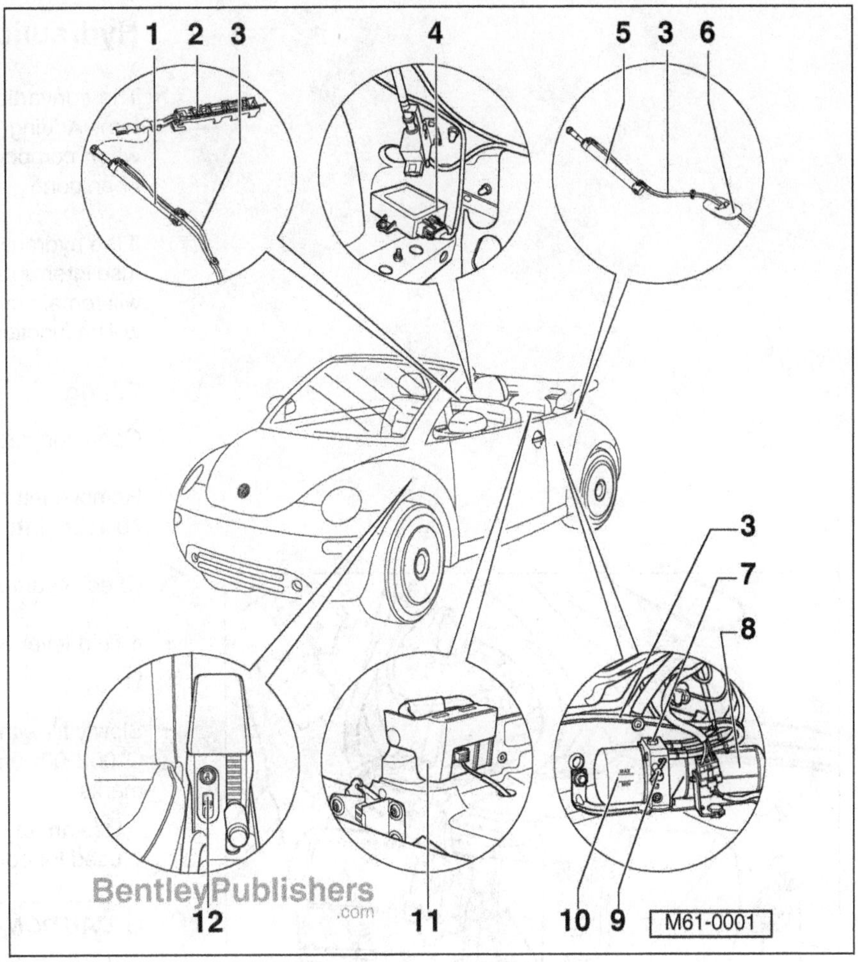

1. **Right hydraulic cylinder**
2. **Convertible top open position switch (F171)**
   - Installed only on right side
   - Unclip from hydraulic cylinder to remove
3. **Hydraulic hoses**
4. **Convertible top control module (J256)**
   - In right rear of luggage compartment
5. **Left hydraulic cylinder**
6. **Buffer plugs**
7. **Oil filler plug**
   - Tighten to 5 Nm (44 in-lb)
8. **Hydraulic pump (V118)**
9. **Pressure screw**
   - Opening screw will release pressure from hydraulic system and allow top to be closed by hand
10. **Hydraulic fluid fill indicator**
    - Fluid level should be between min and max marks with top open
11. **Contact switch**
    - Right shown (F156)
    - Left (F155)
12. **Convertible top operation switch (E137)**
    - Located in center console

Hydraulic system, bleeding and filling

The convertible top hydraulic system is a "closed" system. Adding hydraulic fluid should only be necessary when components have been replaced or repairs have been done.

If the hydraulic pump is electrically overloaded, a thermo fuse interrupts the circuit to prevent damage. The system will remain off for a maximum of 35 seconds. After that it will be functional again.

Filling

- Open convertible top completely.
- Remove left side luggage compartment trim panel. See **70 Trim–Interior**.

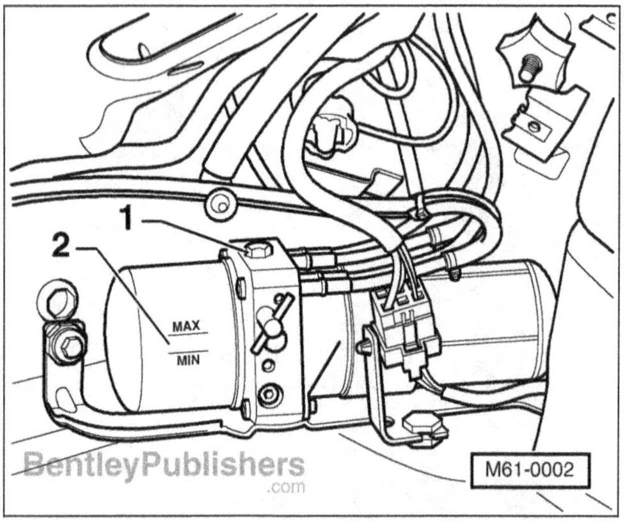

- Check hydraulic oil level (**2**).
- If fluid level is below "MIN" mark, unscrew oil filler plug (**1**).
- Slowly fill with convertible top hydraulic oil (VW part # G 002 00) only until level is between "MIN" and MAX" marks.
  - Use an appropriate filler container that will only be used for convertible top hydraulic oil.

> ***CAUTION—***
> *Do not fill hydraulic fluid above "MAX" mark. Damage can result from over filling.*

- Install oil filler plug.

Tightening torque
- Oll filler plug to hydraulic pump . . . . . 5 Nm (44 in-lb)

Bleeding

- Unscrew oil filler plug at hydraulic pump.
- Switch ignition on and off again. Leave key inserted in ignition.
- Use convertible top switch to open and close top 4 to 5 times.
  - The system will bleed itself automatically.
  - Convertible top must open and close smoothly.
- Install oil filler plug.

Tightening torque
- Oll filler plug to hydraulic pump . . . . . 5 Nm (44 in-lb)

Hydraulic pump, removing and installing

> ***CAUTION—***
> *• Hydraulic system is sensitive to any ingress of dirt.*
>
> *• As soon as system is opened, cover or seal components so that no dirt can enter system.*

– Disconnect battery ground (GND) strap from battery negative (–) terminal.

NOTE—
Be sure to have the anti-theft radio code on hand before disconnecting the battery.

– Remove left rear luggage compartment trim. See **70 Trim–Interior**.

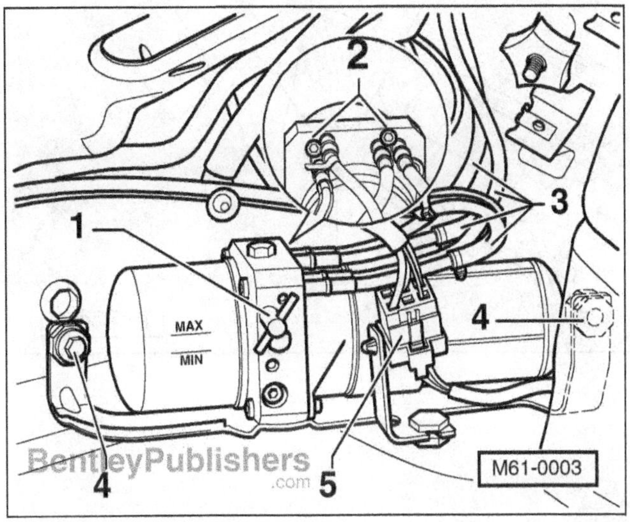

◄ Disconnect harness connector (**5**) from hydraulic pump.

– Unscrew pressure screw (**1**). Tighten screw after hydraulic pressure is released.

– Mark hydraulic lines (**3**) before disconnecting.

– Remove mounting bolts (**2**) and carefully pull hydraulic lines out of openings.

– Disconnect harness connector (**5**).

– Remove mounting nut (**4**) and pump/reservoir assembly.

– Install in reverse order of removal, noting the following:
  - Check for damage and proper fitting of O-rings when installing hydraulic lines.
  - After installing pump, bleed hydraulic system. See **Hydraulic system, filling and bleeding** in this repair group.

Hydraulic cylinder assembly

1. **Hydraulic cylinder**
2. **Hydraulic hose**
3. **Locking washer**
4. **Hydraulic hose**
5. **Retaining clip**
6. **Bracket**
7. **Pin**
8. **Bolt**
   - Tighten to 23 Nm (17 ft-lb)
9. **Convertible top open position switch (F171)**
   - Only installed on right side
10. **Screwdriver**
11. **Ball head**
12. **Cap**
13. **Securing clip**
14. **Piston rod**

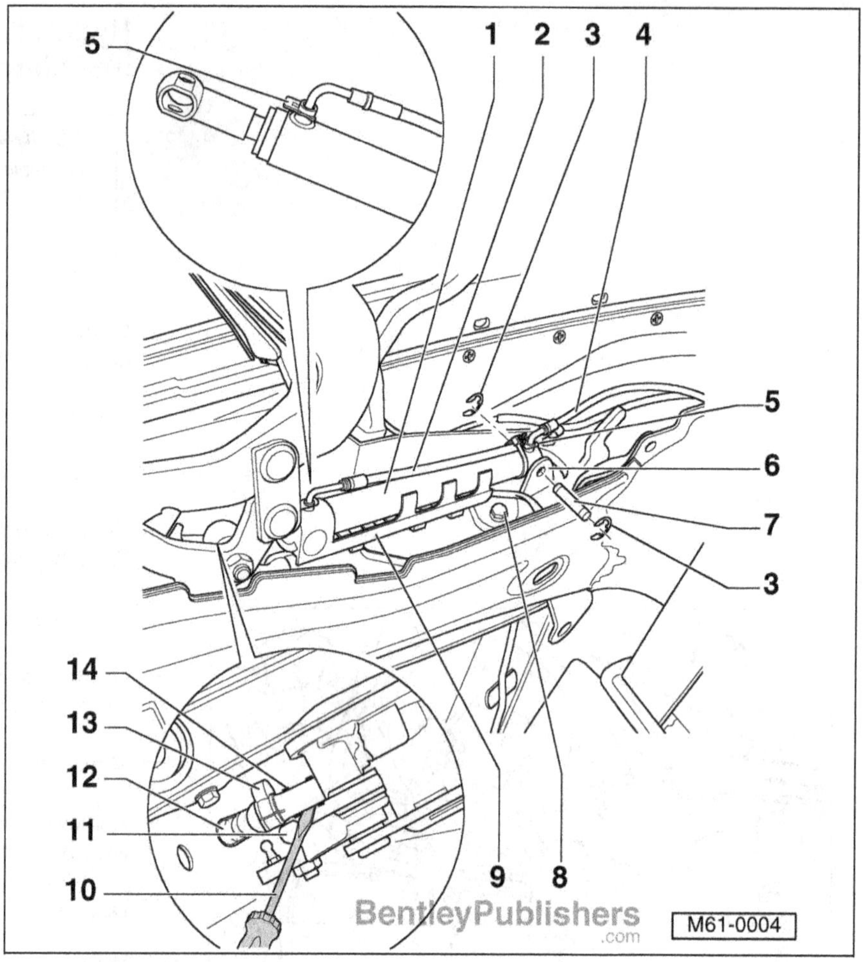

Hydraulic cylinder, removing and installing

- Open convertible top.
- Remove side trim. See **70 Trim–Interior**.
- Open cap (**12**) of convertible top open position switch (**9**).
- Pry securing clip (**13**) off piston rod (**14**).
- Pry piston rod (**14**) off ball head using screwdriver (**10**).
- Close convertible top by hand.
- Remove bolts (**8**).
- Remove retaining clip (**5**).
- Remove hydraulic hoses (**2** and **4**).
- Install in reverse order of removal.

63 Bumpers

GENERAL

The bumpers consist of a cross-member that is bolted directly to the body structure. The front bumper cover is part of the front end assembly (front clip) and must be removed with the fenders attached, see **50 Body–Front**. The rear bumper cover can be removed without the fenders as shown later, however it may be easier to remove the entire rear clip (bumper cover and fenders) as an assembly.

BUMPERS, SERVICING

Front bumper assembly

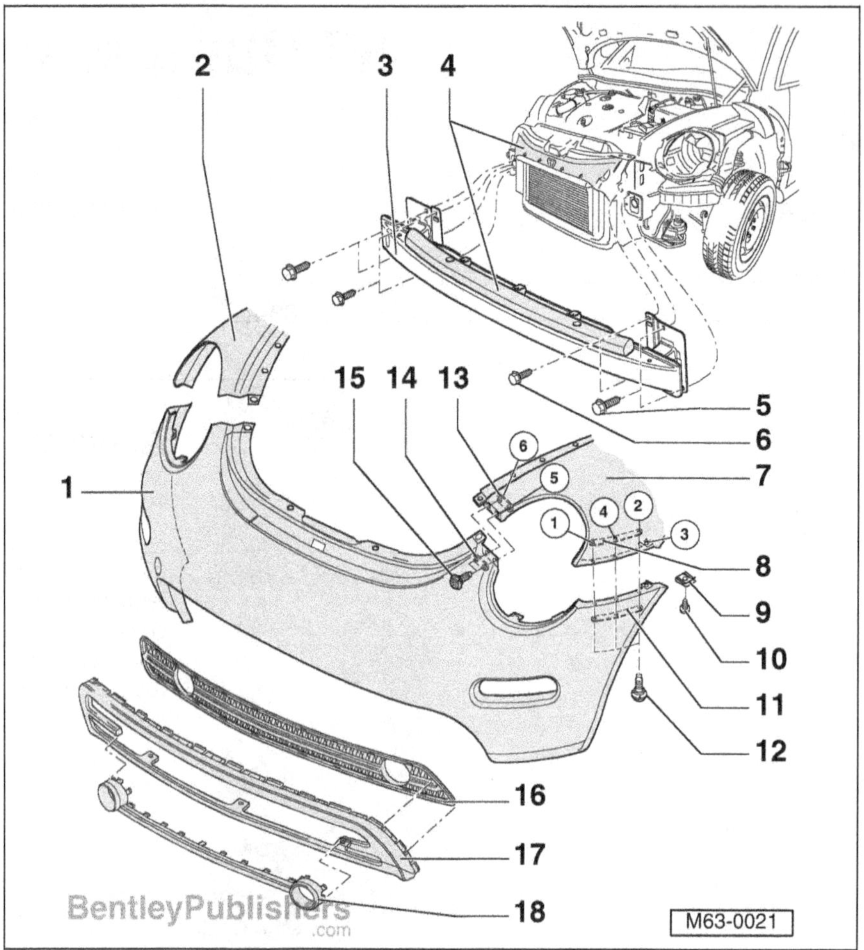

1. **Bumper cover**
   - Remove as complete assembly with fenders, see **50 Body–Front**
2. **Fender**
3. **Bumper impact beam**
4. **Insulation**
   - Bonded to bumper with butyl adhesive sealing cord AKL 450 005 05
5. **Bolt**
   - Qty. 4
   - Tighten to 30 Nm (22 ft-lb)
6. **Bolt**
   - Qty. 4
   - Tighten to 9 Nm (80 in-lb)
7. **Fender**
8. **Reinforcement**
   - Qty. 2
   - With welded captive nuts
9. **Speed nut**
10. **Bolt**
    - Install with locking fluid D 185 400 A2
    - Tighten in order shown by numbers in circles
11. **Reinforcement**
    - Qty. 2
12. **Bolt**
    - Qty. 6
    - Tighten to 6 Nm (53 in-lb)
    - Install with locking fluid D 185 400 A2
    - Tighten in order shown by numbers in circles
13. **Reinforcement**
    - Qty. 2
    - With welded captive nuts
14. **Reinforcement**
    - Qty. 2
    - With welded captive nuts
15. **Bolt**
    - Qty. 4
    - Tighten to 6 Nm (53 in-lb)
    - Install with locking fluid D 185 400 A2
    - Tighten in order shown by numbers in circles
16. **Grille**
    - Clipped into bumper cover
17. **Front spoiler**
18. **Face plate**
    - Foglight style shown

Rear bumper assembly

1. **Bumper cover**

   Removing:

   - Remove wheel housing liners, see **66 Body–Exterior Equipment**
   - Remove tail light assemblies, see **94 Lights, Accessories–Exterior**
   - Disconnect harness connector for licence plate light (below left tail light)
   - Remove mounting bolts and speed nuts on left and right
   - Pull cover off mounts

2. **Bolt**
   - Qty. 6
   - Tighten to 6 Nm (53 in-lb)
   - Tighten in order shown by numbers in circles

3. **Reinforcement**
   - Qty. 2

4. **Bolt**
   - Qty. 6
   - Install with locking fluid D 185 400 A2

5. **Speed nut**

6. **Reinforcement**
   - With welded captive nuts
   - Qty. 2

7. **Left insulation**
   - Bonded to bumper with butyl adhesive sealing cord AKL 450 005 05

8. **Right insulation**
   - Bonded to bumper with butyl adhesive sealing cord AKL 450 005 05

9. **Bumper impact beam**

10. **Bolt**
    - Qty. 4
    - Tighten to 30 Nm (22 ft-lb)

11. **Fender**

12. **Bolt**
    - Qty. 4
    - Tighten to 6 Nm (53 in-lb)
    - Install with locking fluid D 185 400 A2
    - Tighten in order shown by numbers in circles

13. **Reinforcement**
    - Qty. 2

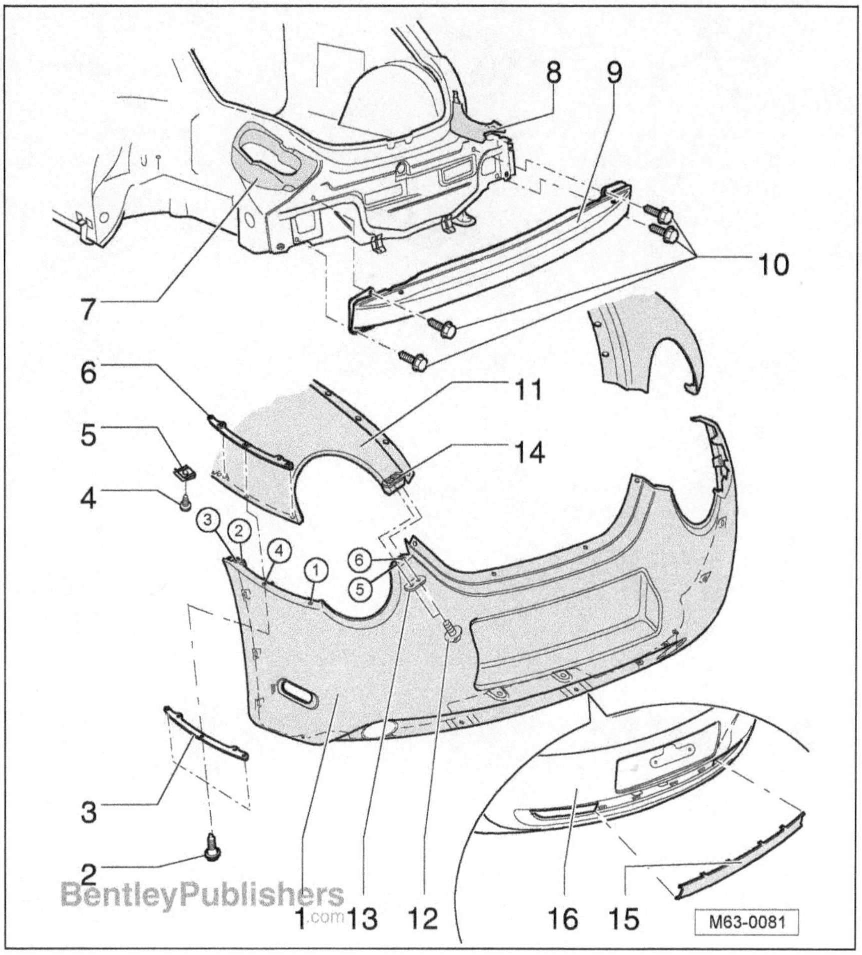

14. **Reinforcement**
    - Qty. 2

15. **Ventilation grid**

16. **Cover**

66 Body–Exterior Equipment

GENERAL

This repair group covers various exterior equipment including side-view mirrors, wheel housing liners, roof moldings, rear spoiler and convertible wind deflector.

SIDE-VIEW MIRROR

Mirror glass, removing and installing

> ***WARNING —***
> *Use protective eye wear and gloves.*

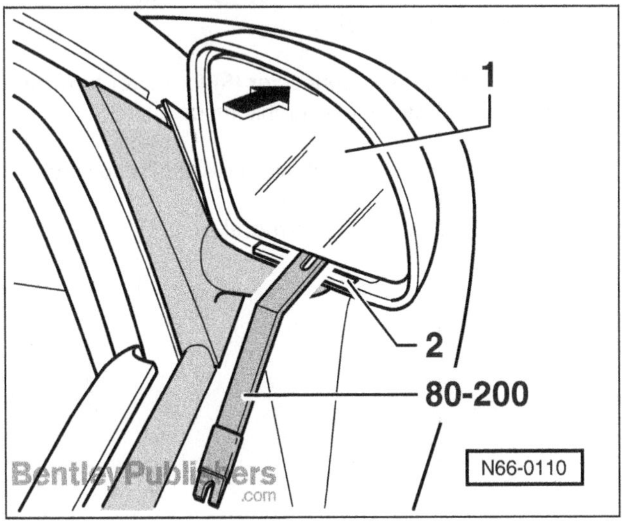

◄ Protect bottom mirror housing edge from paint damage with reinforced tape (**2**).

– Pivot top of mirror glass (**1**) into housing in direction of **arrow**.

– Use VW special tool 80-200 to carefully pry mirror from mounting tab in housing.

– Disconnect harness connector for heated mirror and remove mirror glass.

– To install, connect harness connector and press mirror glass at center onto mounting bracket in housing.

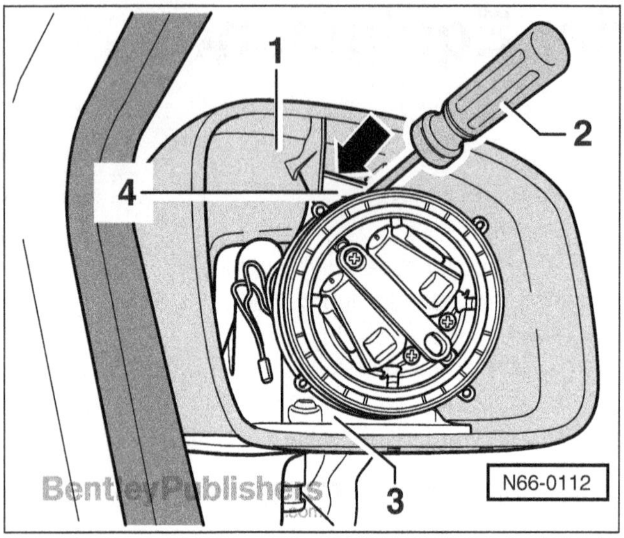

Mirror housing, removing and installing

– Remove mirror glass as described earlier.

Vehicles through 05/2000

◄ Press screwdriver (**2**) in direction of **arrow** to release housing retaining clip (**4**) on top of actuator bracket.

– Pull mirror housing (**1**) upward off mirror carrier (**3**).

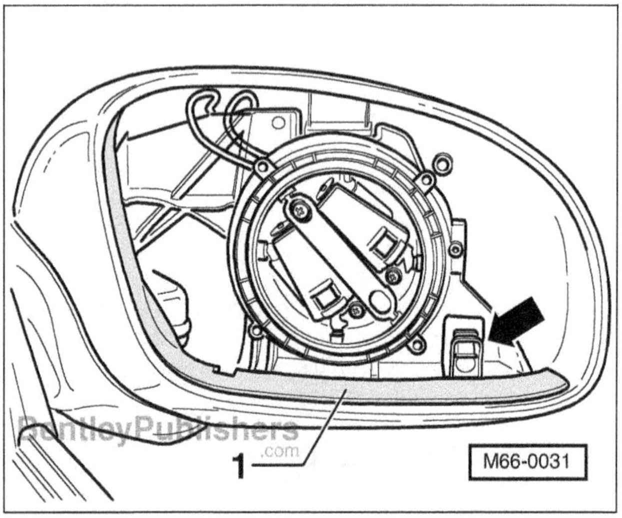

Vehicles from 06/2000

◄ Loosen locking mechanism (**arrow**) with screwdriver to unclip trim (**1**) from mirror housing.

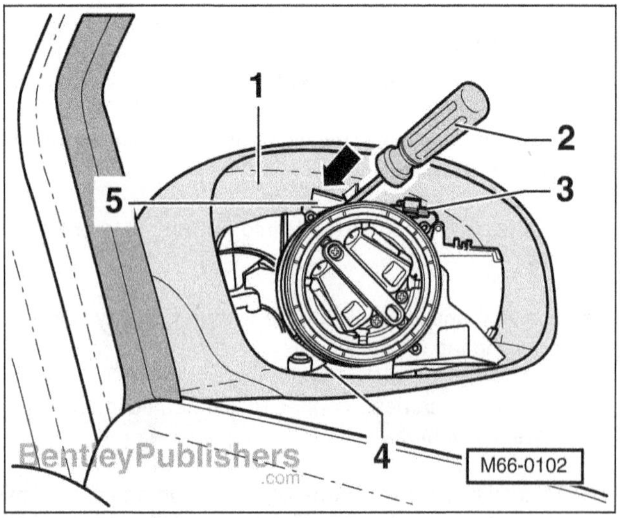

◄ Press screwdriver (**2**) in direction of **arrow** causing clip (**4**) to release the mirror housing (**1**).

– Turn the mirror housing (**1**) counterclockwise to unclip it from bracket rear retainer.

– Disconnect harness connector (**5**) for mirror turn signal lamp, as applicable (convertible models).

– Pull mirror housing (**1**) off mirror carrier (**3**) upward.

– Install in reverse order of removal.

Side-view mirror assembly, removing and installing

– Remove front door trim, see **70 Trim–Interior**.

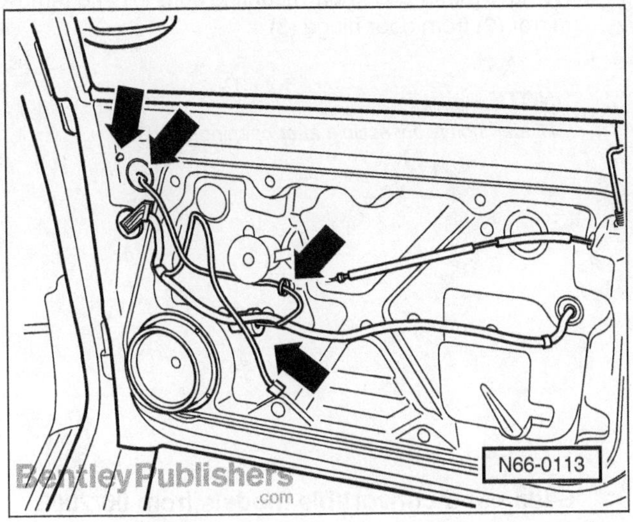

Sedan models

< Disconnect harness connectors (**arrows**) in door.

– Tape harness connectors out of way so they don't get damaged if door is closed.

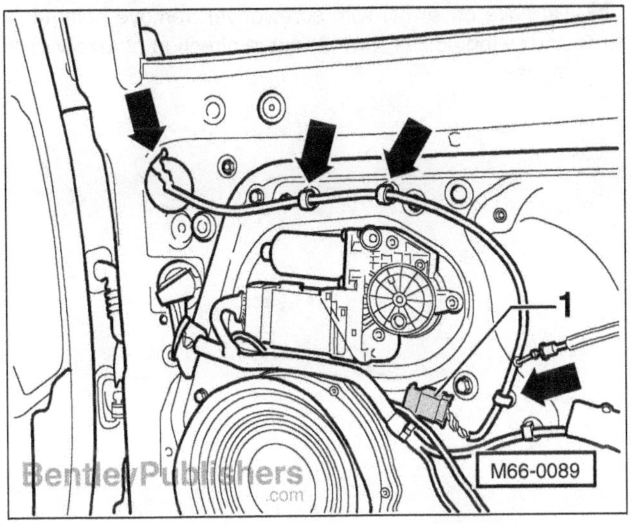

Convertible models

< Unclip mirror wire harness from retainers (**arrows**).

– Disconnect harness connector (**1**) for mirror.

– Tape harness connector out of way so it doesn't get damaged if door is closed.

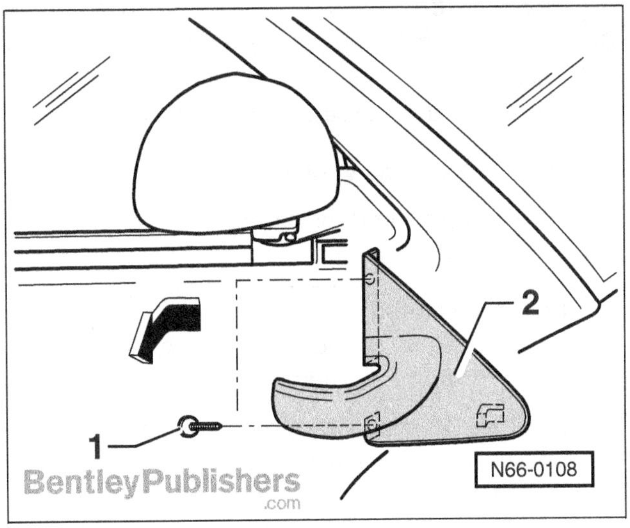

Sedan models through 05/2000

< Remove screws (**1**) and pull off hinge cover (**2**) from the rear (**arrow**).

NOTE —
Hinge cover bolts are accessible after opening the door.

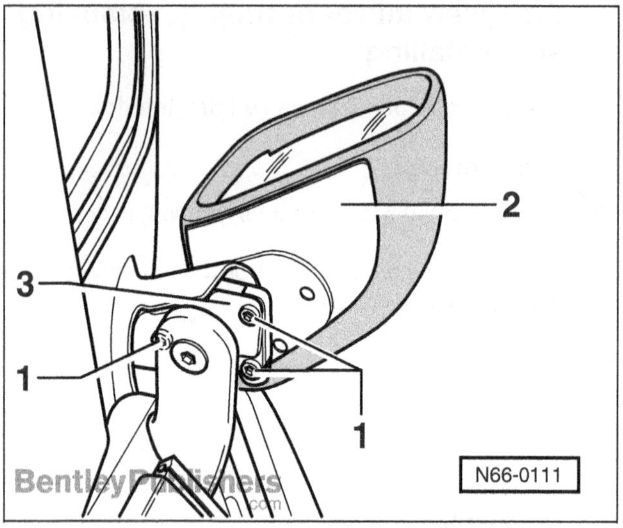

◀ Remove mirror assembly mounting bolts (**1**) and remove mirror (**2**) from door hinge (**3**).

NOTE —
Hidden bolt is accessible after opening the door.

Sedan and convertible models from 06/2000

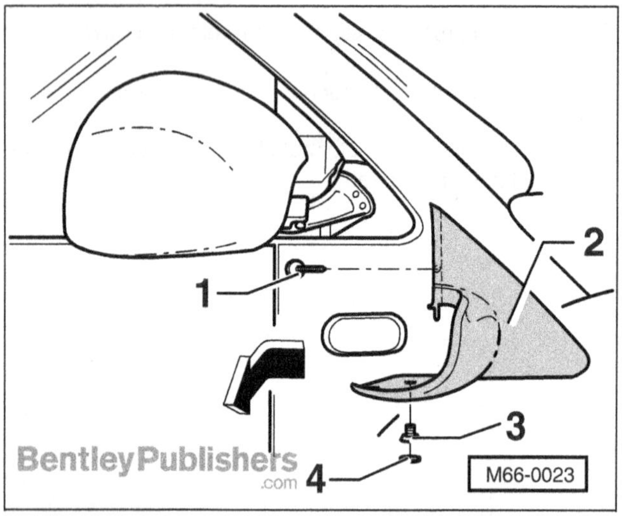

◀ Remove cover (**4**) with screwdriver. remove bolts (**1**, **3**) and swing corner trim (**2**) out in direction of **arrow** to remove.

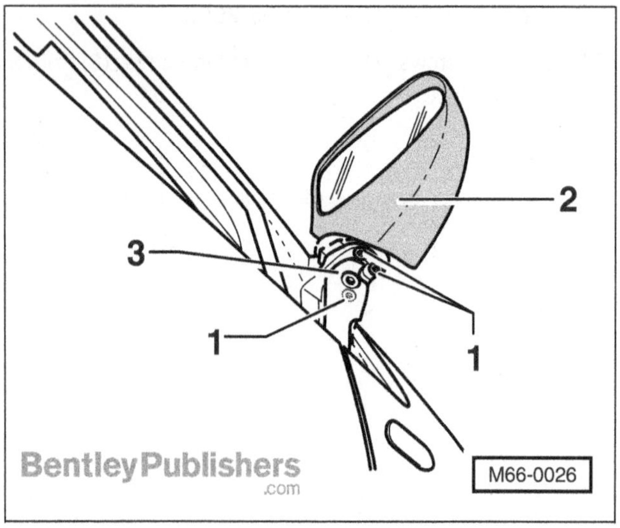

◀ Remove bolts (**1**) and remove outside mirror (**2**) from door hinge (**3**).

NOTE —
Hidden bolt is accessible after opening the door.

Continued for all vehicles

– Install in reverse order of removal.

Tightening torque
- Mirror assembly to hinge . . . . 10 Nm (89 in-lb)

– When installing mirror assembly, ensure electrical wire in window channel is routed correctly, otherwise window can catch on the wire when opening and closing window.

– Check operation of power mirror.

Side-view mirror turn signal light, removing and installing

NOTE —
The side-view mirror turn signal bulbs are long-lasting light emitting diodes (LED). If they are damaged, the entire turn signal light assembly must be replaced, as described below.

– Turn off all electrical consumers, switch off ignition and remove key.

– Remove mirror glass and side-view mirror assembly as described earlier in this repair group.

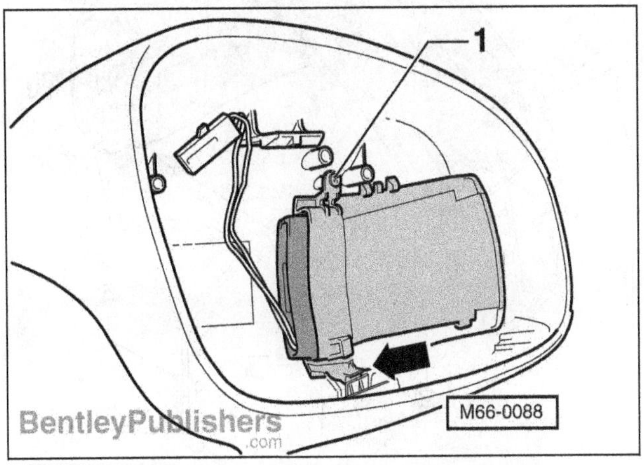

◄ Remove bolt (**1**) and unclip catch (**arrow**).

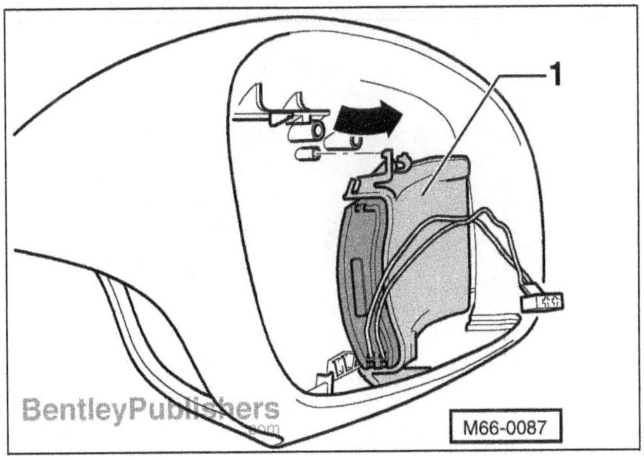

◄ Swing turn signal lamp (**1**) in direction of **arrow** and remove.

– Install in reverse order of removal.

Tightening torque
- Turn signal to mirror assembly . . . . 1 Nm (9 in-lb)

WHEEL HOUSING

Front wheel housing liner assembly

1. **Wheel housing liner**
   - Remove wheel and mounting screws (item 2) to remove liner
2. **Screw**
   - Qty.: 12
   - Tighten to 4 Nm (35 in-lb)
3. **Spreader nut**
4. **Speed nut**
   - On bumper cover

   NOTE —

   Wheel housing liners may differ slightly depending on equipment, with regard to position and number of mounting elements.

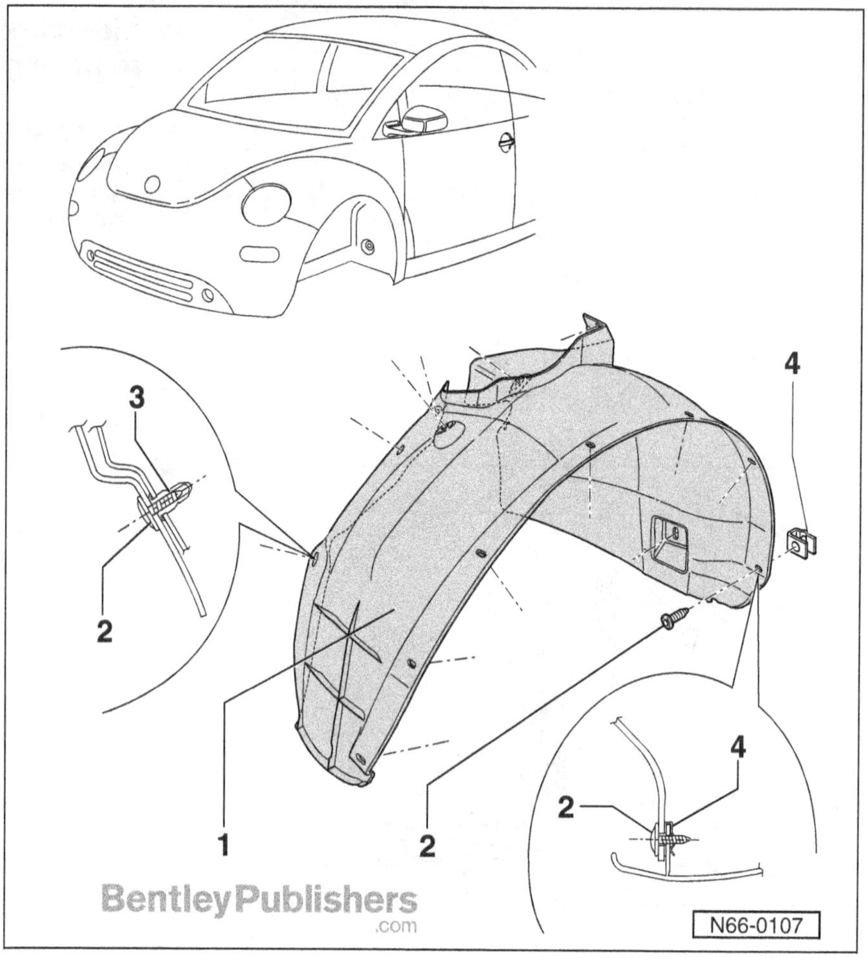

Rear wheel housing liner assembly

1. **Wheel housing liner**
   - Remove wheel and mounting screws and nuts to remove liner
2. **Nut**
3. **Screw**
   - Qty.: 14
   - Tighten to 4 Nm (35 in-lb)
4. **Spreader nut**
5. **Cover**
   - Right side only
6. **Speed nut**
7. **Speed nut**
   - On bumper cover or fender

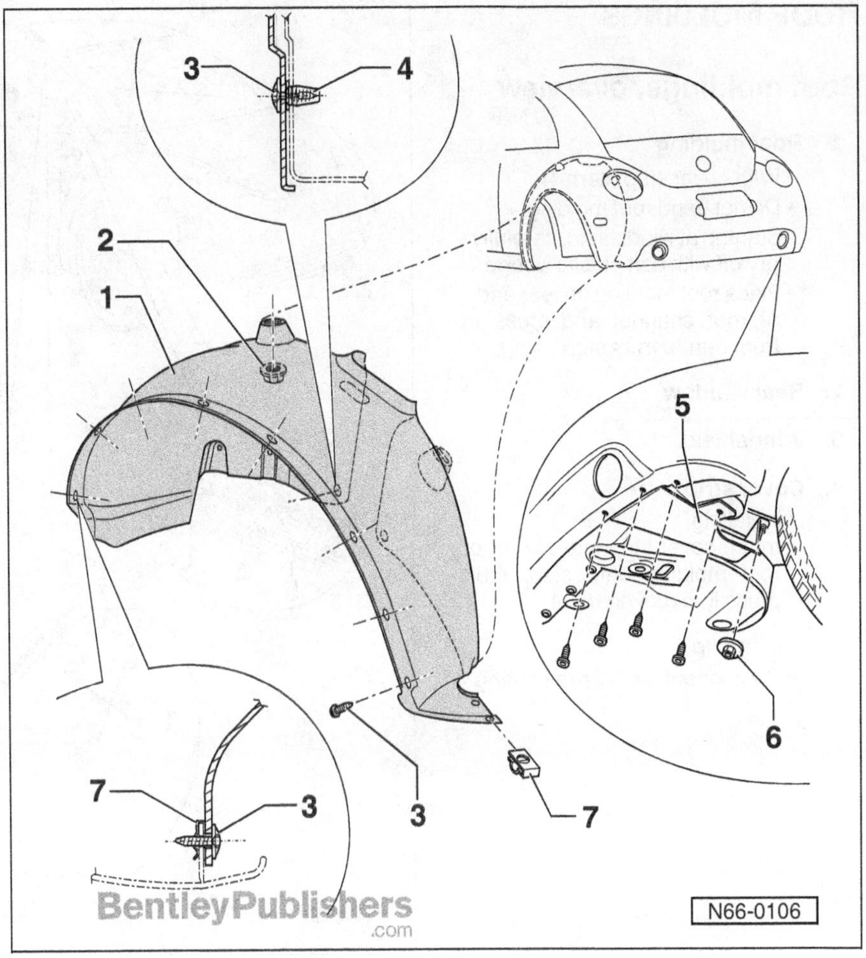

ROOF MOLDINGS

Roof moldings, overview

1. **Roof molding**
   - With cover strip (item 4).
   - Do not bend roof molding
   - Starting at windshield, carefully pry off with soft plastic wedge
   - Place roof molding on rear end of roof channel and press in from rear to front stop
2. **Rear window**
3. **Windshield**
4. **Cover strip**
   - Installing:
     Insert cover strip into pocket of roof molding onto stop, then press into roof channel
5. **Sealing lip**
   - Component part of pre-coating

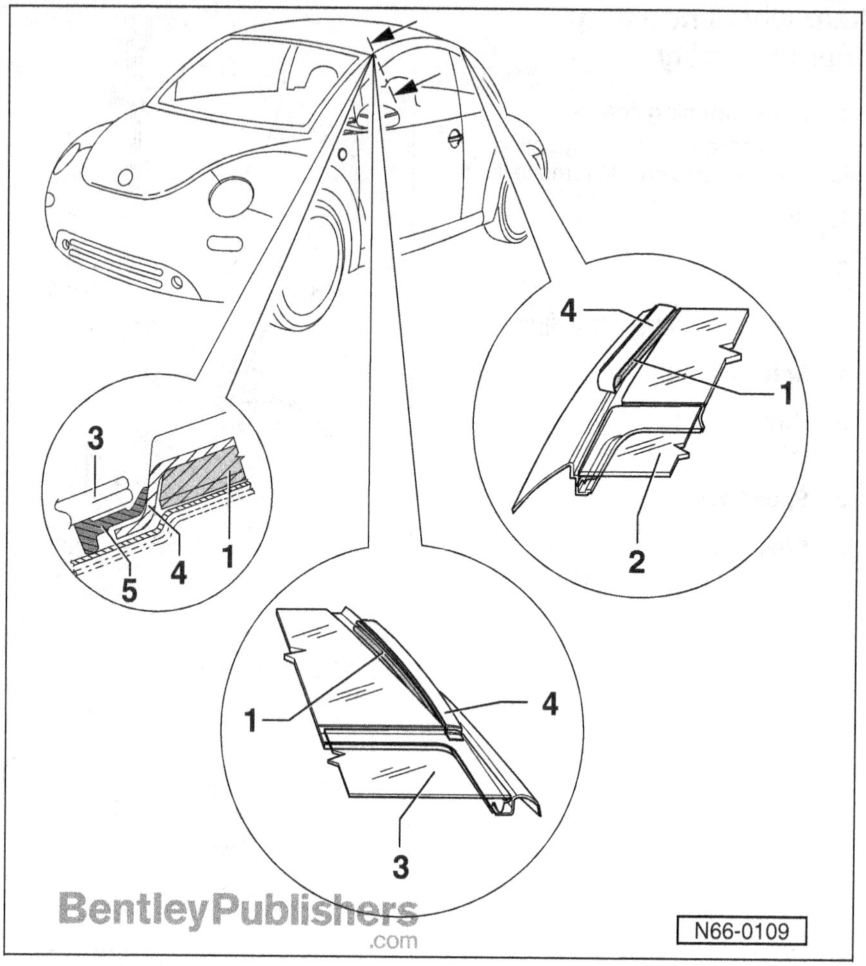

REAR SPOILER

Function

Certain models come equipped with an electronically controlled spoiler to aid in high speed vehicle stability.The spoiler raises automatically when vehicle speed reaches a predetermined speed. On vehicles produced through 05/2001 the spoiler is activated at 94 mph (150 km/h) and lowers when the vehicle is slowed to 81 mph (130 km/h). On vehicles produced after 06/2001 the spoiler is activated at 45 mph (72 km/h) and lowers at 43 mph (70 km/h).

The rear spoiler is controlled by the rear spoiler control module (J223). The control module will recognize malfunctions related to the rear spoiler system and store the information as Diagnostic Trouble Codes (DTCs) in a permanent memory. These DTCs can be read through the vehicle On Board Diagnostic system, using a diagnostic scan tool. See **OBD On Board Diagnostics** for more information.

Since spoiler function is based upon vehicle aerodynamics, the sunroof will close automatically when the spoiler is deployed based on a predetermined road speed.

Function of the rear spoiler system is indicated via the rear spoiler warning lamp (**arrow**). The rear spoiler warning lamp will also light up if the spoiler is raised manually.

Raising and lowering rear spoiler manually

The rear spoiler switch is located below the dash panel, close to the steering column and is used to manually raise and lower the rear spoiler. Manually raising and lowering the rear spoiler can only be performed when the vehicle is stationary, or when the vehicle is travelling at less then the rear spoiler activation speed.

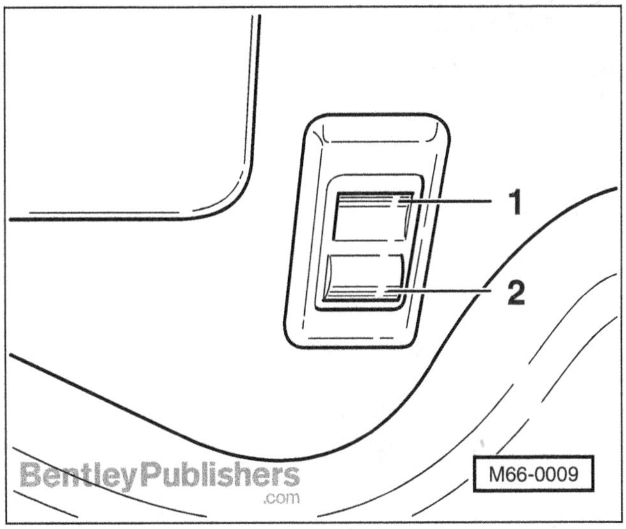

When the ignition is switched on and the switch is operated in position (**1**), the rear spoiler will raise fully in one movement. Lowering is performed by operating the switch in position (**2**) and the spoiler will lower step by step until fully closed.

WARNING —
- *Danger of being caught/trapped!*
- *Make sure there is sufficient clearance when raising and lowering the rear spoiler manually.*

When working on the spoiler it is advisable to raise the rear spoiler before beginning repairs on the unit.

NOTE —
Do not press the raised spoiler in or pull out by hand.

Rear spoiler assembly, overview

NOTE —
Volkswagen identifies electrical components by a letter and/or a number in the electrical schematics. See **EWD Electrical Wiring Diagrams**. *These electrical identifiers are listed in parenthesis as an aid to electrical troubleshooting.*

1. **Rear spoiler**
   - With gate guide, upper and lower guide tubes and cables.
   - Grease gate guides with G 000 450 02
2. **Rear spoiler control module**
3. **Lower guide tube**
4. **Rear spoiler motor (V52)**
5. **Rear spoiler switch (E127)**
   - Under dash panel near steering column
   - For raising and lowering the spoiler manually
6. **Rear spoiler adjustment warning lamp**
7. **Rubber buffer**

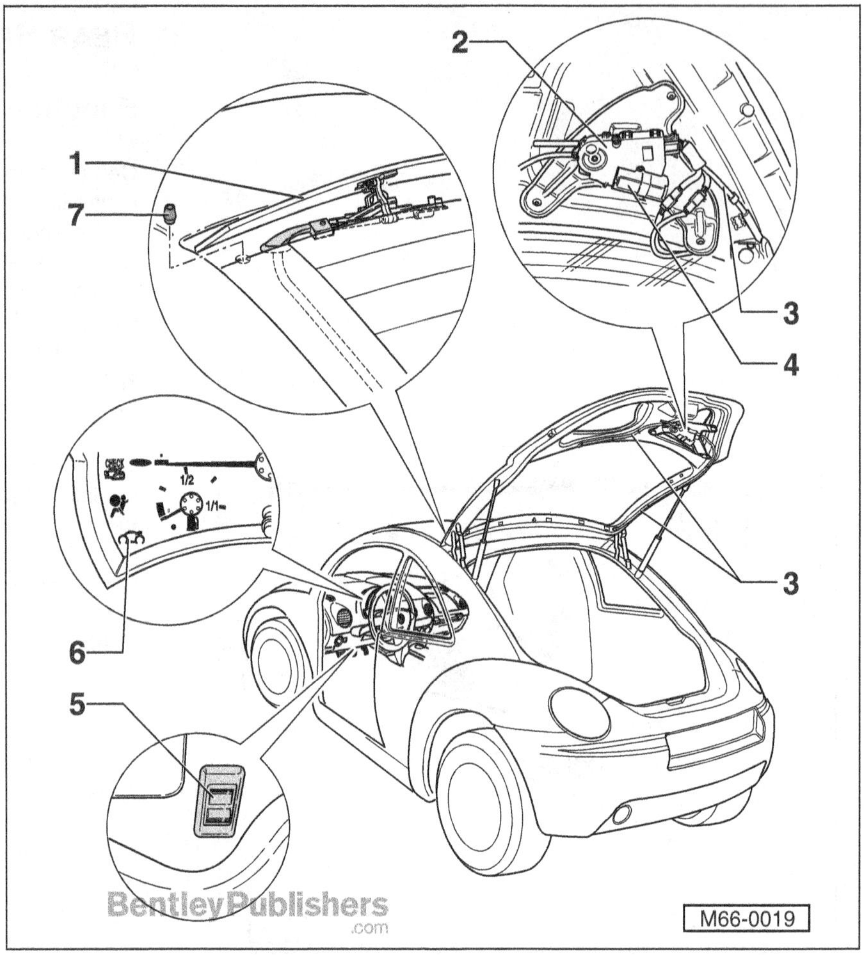

Rear spoiler rails, lubricating

Grease rear spoiler rails (**arrows**) with G 052 778 grease.

Rear spoiler assembly, removing and installing

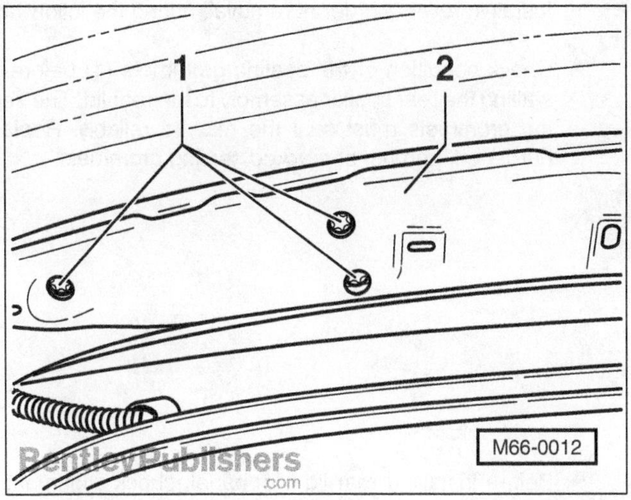

- Raise rear spoiler manually.
- Remove rear hatch lower trim. See **70 Trim–Interior**.
- Remove rear spoiler drive as described later in this repair group.

◀ Working inside top edge of rear lid (**2**), remove spoiler mounting bolts (**1**) on the left and right hand sides.

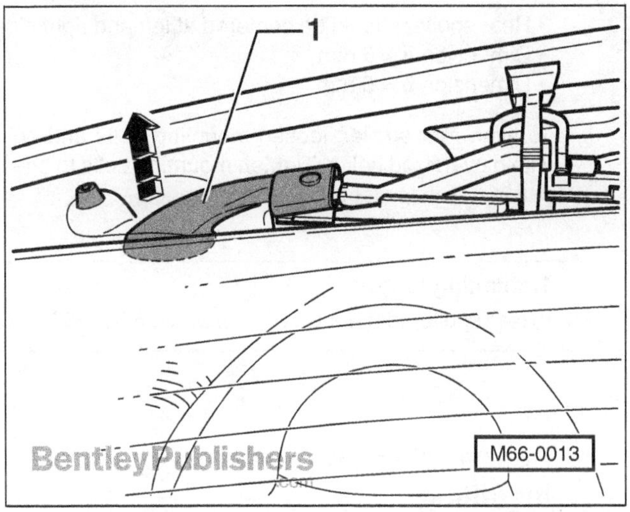

◀ Unclip the sealing grommets (**1**) on the left and right hand sides of the rear lid (**arrow**).

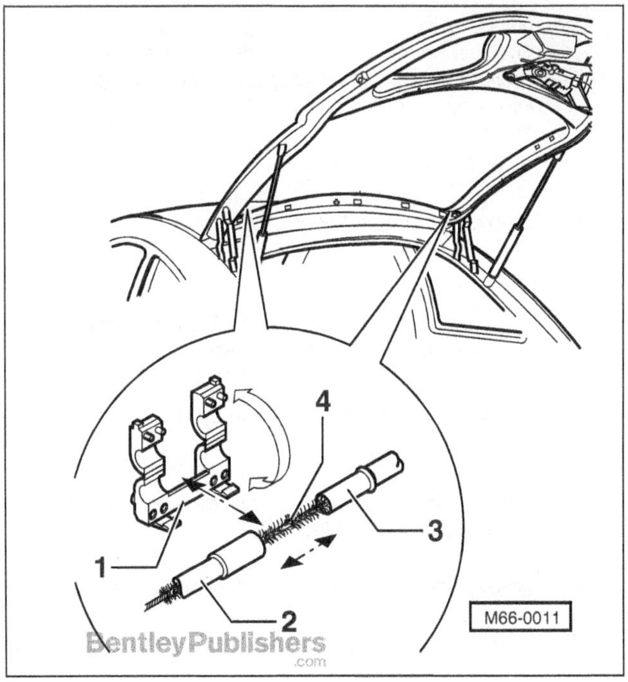

◀ Open clips (**1**) on left and right hand sides.

NOTE —
Replace damaged clips (1) when re-installing.

- Take rear spoiler assembly off vehicle. Unhook lower guide tube (**2**) and upper guide tube (**3**) and pull apart on left and right hand sides when taking the rear spoiler assembly off.
- Pull cable (**4**) out of left and right hand sides of lower guide tube (**2**) while removing the rear spoiler assembly.

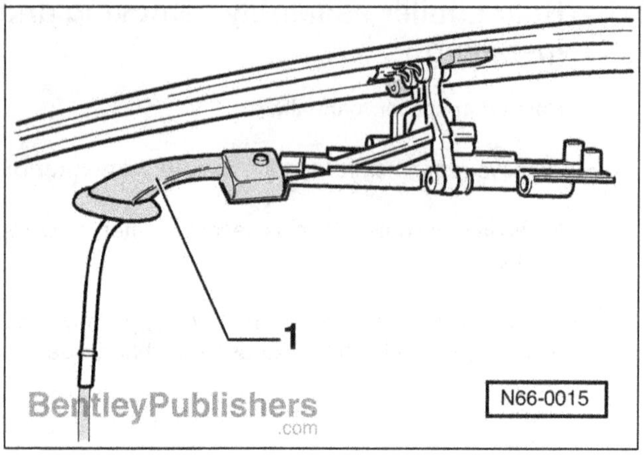

– Install in reverse order of removal, noting the following:

◀ Check condition of the sealing grommets (**1**) before installing the rear spoiler assembly to the rear lid. The sealing grommets must seal the rear lid reliably. Replace hardened, porous, or cracked sealing grommets.

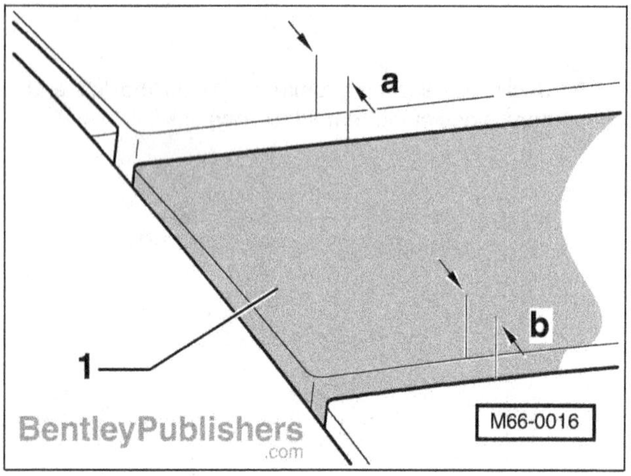

◀ Before installing rear lid trim panel, check alignment of rear spoiler (**1**) at lid and correct if necessary.

- Rear spoiler should be centered at left and right sides.
- Dimension **a** = 6 mm
- Dimension **b** = 5 mm

– To align rear spoiler, loosen mounting bolts and adjust within oversized holes. Tighten mounting bolts to specification when spoiler is correctly aligned.

Tightening torque
- Rear spoiler to mount. . . . . 6 Nm (53 in-lb)

Rear spoiler drive, removing and installing

NOTE —
The rear spoiler control module (J223) and the rear spoiler motor (V52) are integrated as one component, known as the rear spoiler drive.

– Raise rear spoiler manually.

– Remove rear hatch lower trim. See **70 Trim-Interior**.

WARNING —
Disconnect battery Ground (GND) strap before beginning work on the electrical system.

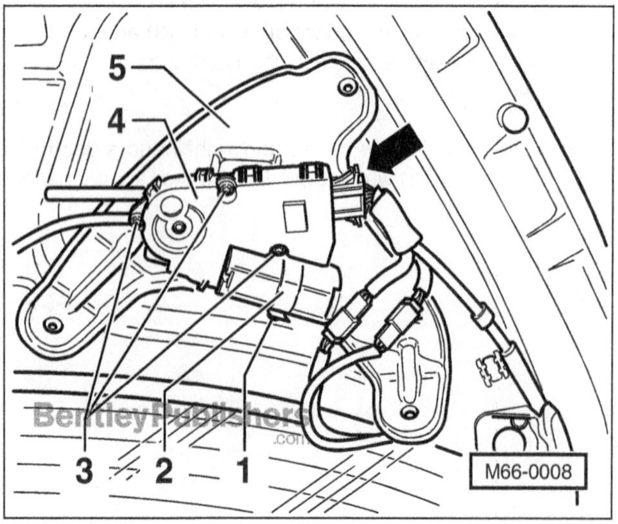

◀ Pull harness connector (**arrow**) off rear spoiler drive (**4**).

NOTE —
Wiring harness connectors can be unhooked from bracket (5) at lower guide tube.

– Remove screw (**1**) and open clip (**2**) (unhook).

– Remove screws (**3**) and take drive (**4**) off lower bracket (**5**).

– Installation is reverse of removal, noting the following:

– Check parallel running and re-adjust if necessary before installing the rear spoiler drive.

– Tighten spoiler drive mounting bolts to specification.

Tightening torque
• Spoiler drive to body . . . . 2.5 Nm (22 in-lb)

– After installing rear spoiler drive, lower rear spoiler completely using manual switch, then manually raise again fully via the switch to initiate drive limit stop positions.

Parallel running, checking and adjusting

NOTE —
• A malfunctioning rear spoiler or binding in the rear spoiler mechanism can occur if the parallel running is not correctly aligned. When operating rear spoiler manually it must raise and lower smoothly, in one continuous motion.

• Reasons for non-parallel running are wear in the rear spoiler drive cables or a worn spoiler drive.

– Raise spoiler manually.

Checking

NOTE —
Illustration shows the gate guide on the lefts hand side

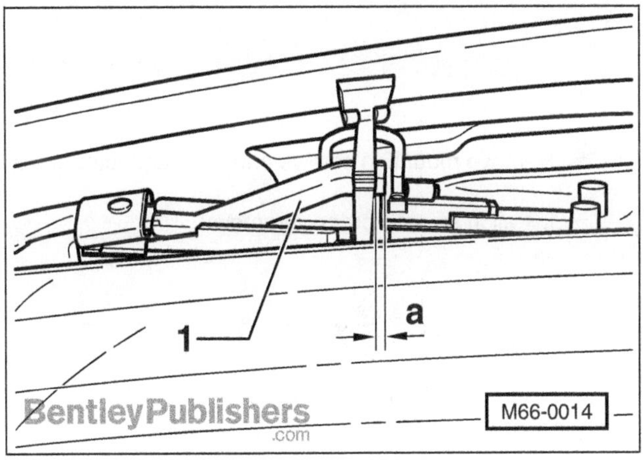

◄ The left and right hand side gate guides (**1**) must be in the same position (parallel). When the rear spoiler is raised, dimension (**a**) on both gate guides must be the same.

Adjusting

– Remove rear hatch lower trim. See **70 Trim–Interior**.

– Remove rear spoiler drive.

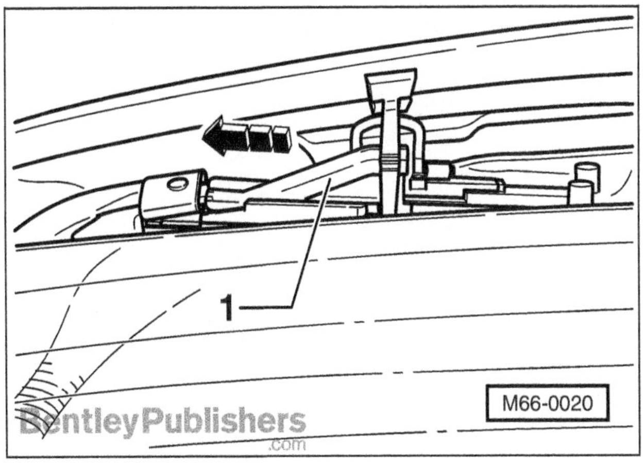

◄ Move the left and right hand gate guides (**1**) toward the outer sides of the vehicle (**arrow**) onto their respective limit stops by hand (rear spoiler raised fully).

– Install rear spoiler drive.

– After installing rear spoiler drive, lower rear spoiler completely using manual switch, then manually raise again fully via the switch to initiate drive limit stop positions.

WIND DEFLECTOR (CONVERTIBLE)

Wind deflector bearing and bracket, removing and installing

– Switch ignition off.

– If folding wind deflector is installed, remove it.

– Remove rear seat bottom and backrest. See **72 Seats**.

– Remove rear side trim. See **70 Trim–Interior**.

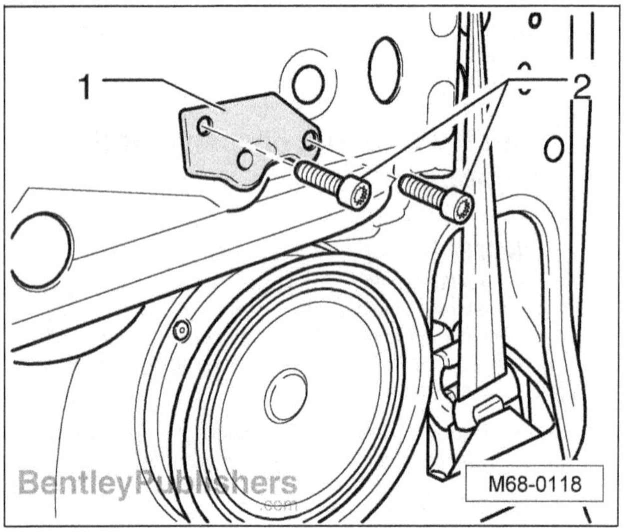

◄ Remove mounting bolts (**2**) for wind deflector bearing (**1**) and bearing.

– Install in reverse order of removal.

Tightening torque
- Wind deflector bearing to body . . . . . 10 Nm (7 ft-lb)

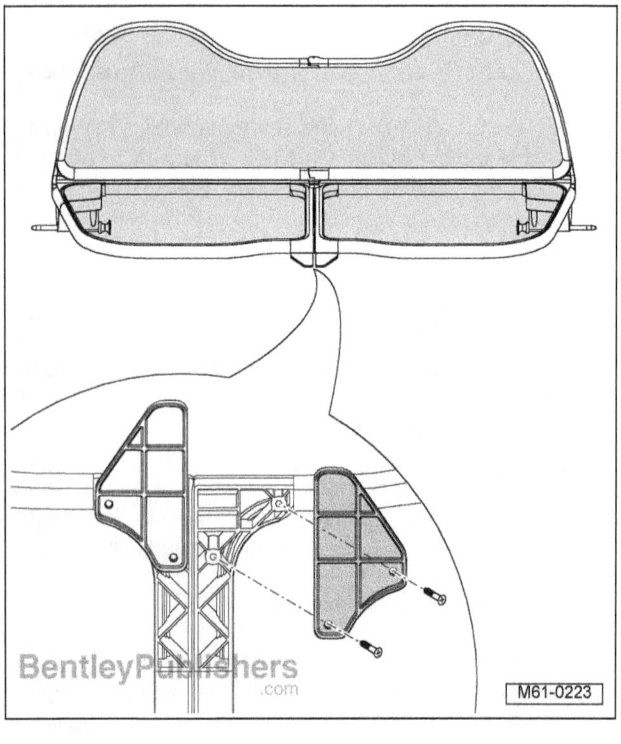

◄ Remove mounting screws and wind deflector bracket.

– Install in reverse order of removal.

69 Seatbelts, Airbags

GENERAL

This repair group covers emergency tensioning (pyrotechnic) seat belt assemblies and airbag system components. This repair group does not cover airbag system or pyrotechnic seat belt fault diagnosis or repair. Service and repair to these systems requires special test equipment, knowledge and training and should only be carried out by an authorized Volkswagen dealer. Before starting repairs involving these systems, always read and observe all warnings, cautions and notes.

> ***WARNING —***
> *Disconnecting the negative (–) battery cable may erase fault codes and basic settings in the engine management and automatic transmission control modules. Some driveability problems may be noticed until the system re-adapts to operating conditions. OBD II readiness codes, which may be required for emissions testing, may also be erased. Convenience electronics (alarm system, interior light control, power locks, mirrors, and windows) may need to be re-set using a diagnostic scan tool.*

69

0024312

SEATBELTS

Volkswagen New Beetles use an inertia reel-type lap/shoulder belt combination to restrain the front seat passengers. In addition, these belts are self-tensioning through the use of a pyrotechnic device. Deployment of the pyrotechnic tensioner is by means of either a mechanical trigger or in conjunction with the airbag system. Rear seat passengers are protected by an inertia reel-type lap/shoulder belt combination. A warning light (**arrow**) in the instrument cluster illuminates when the ignition is first switched on as a reminder to buckle up.

Special precautions for pyrotechnic seatbelt equipped vehicles

Emergency tensioning front seat belt systems can enhance passenger safety in the event of a collision provided that the seat belts are used. Special service precautions must be followed to prevent possible serious bodily injury to the repair technician and others involved in the handling and storage of pyrotechnic seatbelt components and to insure proper operation in the event of a collision.

WARNING —

- *Before working on any seatbelt components linked to airbag systems, **disconnect the battery first.***
- *Pyrotechnic seat belts are activated by an explosive device. Handled improperly or without adequate safeguards and training, the system can be very dangerous. Special precautions must be observed when working on or near the seatbelts.*
- *Seat belts are vehicle safety systems. To guard against personal injury or system failure, only trained Volkswagen service technicians should test, disassemble or service the pyrotechnic seat belt systems.*
- *Install new seatbelt unit as soon as the unit is removed from the packaging. Reinstall in packaging if work will not be completed immediately.*
- *Do not leave any pyrotechnic seatbelt unit unattended. If work is interrupted, store seatbelt unit in a secure location where it cannot be disturbed.*
- *The storage and transportation of airbag units must be in accordance with all applicable Federal, State and Local rules and regulations.*
- *Observe all cautions, warnings and notes before starting repairs involving airbag systems.*
- *Belt tensioners that have been deployed can be disposed of as normal scrap.*
- *Do not use tools with a hammer-type action to disassemble the belt tensioner.*
- *The pyrotechnic seatbelt propellant has no expiration date, and it has an unlimited, maintenance-free life.*
- *The belt tensioner unit must not be exposed to grease, cleaning solutions or similar substances.*
- *Belt tensioner units must not be exposed to temperatures above 100°C (212°F), even for short periods.*
- *Belt tensioner components may not be opened or repaired; always use new parts.*
- *Belt tensioner units which have been dropped onto a hard surface must not be installed into a vehicle.*
- *Belt tensioner units that show evidence of mechanical or physical damaged (dents, cracks) must be replaced.*
- *Pyrotechnic seatbelt units which have not been deployed should be marked and returned to the manufacturer for disposal using original seatbelt shipping container.*

Seatbelts, inspecting

Any time that a New Beetle is involved in an accident, all of the seatbelts should be checked for damage and proper operation.

> ***WARNING —***
> *After every accident the seat belt system must be inspected systematically. If damage is found when inspecting the check points, the customer must be advised regarding the necessity of replacing the seatbelts.*

Webbing (belt material), checking

- Pull belt completely out.
- Inspect webbing for dirt and soiling; wash as needed using a mild soap solution.

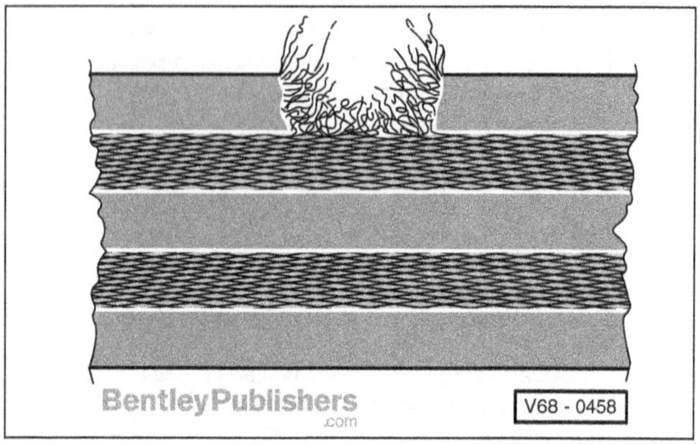

Seatbelt webbing showing example of cuts, chafing and tears. If vehicle was in an accident replace complete seatbelt assembly with buckle. If vehicle was not in an accident and damage is from other sources, it is ok to replace only the belt.

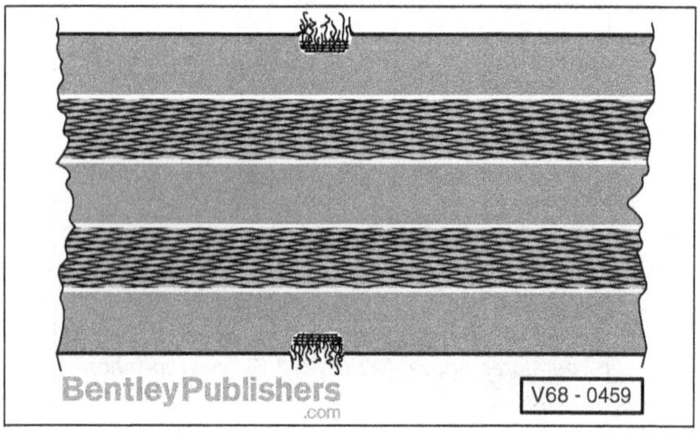

Seatbelt webbing showing example of torn edges. If vehicle was in an accident replace complete seatbelt assembly with buckle. If vehicle was not in an accident and damage is from other sources, it is permissible to replace only the belt.

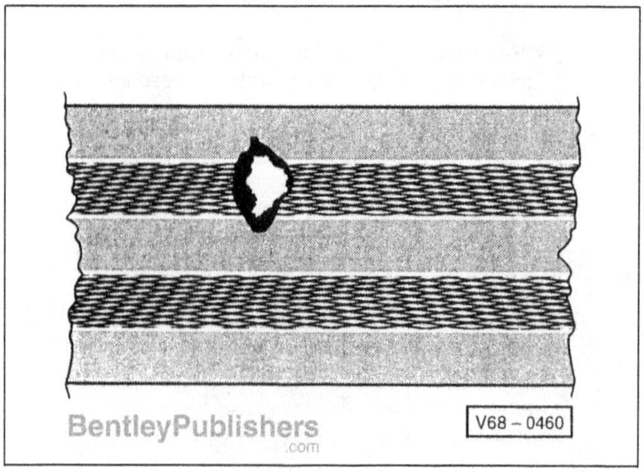

Seatbelt webbing showing example of burns from cigarettes etc. It is permissible to replace only the belt in this case.

Inertia reel, checking

- Pull belt out of reel with a firm and sudden jerk. Seatbelt should lock immediately.
- Road test from 20 kph (12 mph) by applying brakes suddenly to simulate an emergency stop. Seatbelt should lock immediately.
- If belt does not lock, check mounting position and correct as necessary.
- If belt does not lock and mounting is in order, replace complete seatbelt assembly.

> ***WARNING —***
> *For safety reasons, the road test should be done on a traffic-free stretch of road or parking lot to ensure safety to the vehicle and driver as well as other vehicles and pedestrians.*

Belt buckle latch, checking

- Inspect belt latch and buckle for cracks and breaks.
- If there is any indication of cracking or breakage, replace complete seatbelt assembly
- Inspect belt latch and buckle for proper operation by inserting latch into buckle a minimum of 5 times. Latch should engage buckle firmly and securely with an audible clicking sound and should not pull apart.
- If latch and buckle do not engage smoothly or firmly, or come apart when pulled, replace complete seatbelt assembly.

Belt buckle release, checking

- Engage latch and buckle with no tension on the belt (slack) and push release button a minimum of 5 times. Latch must spring out of belt buckle on its own.
- If latch fails to spring out of belt buckle, replace complete seatbelt assembly.

Belt guides and latch tongue, checking

- Inspect plastic guides for deformation, scoring and fractures due to accident. Do not confuse smooth wear marks which are normal due to frequent use.
- If deformation, scoring or fractures are present, replace complete seatbelt assembly.

Seatbelt mounting components and anchorage points, checking

- Check latch and bracket for deformation and/or stretching.
- Check function of height adjuster.
- Check anchorage points on B-pillar, seat and floor for distortion, damage and proper torque.
- If distortion or damage is present, repair or replace components as required. Torque fasteners to specification.

Automatic retractor mechanism, checking

- Check retractor mechanism for smooth operation when seatbelt is pulled out and smooth operation when belt is reeled in.
- Ensure that seatbelts equipped with child seat feature operate properly when seatbelt is pulled completely out.
- Ensure that seatbelt is not twisted.
- If retractor mechanism does not operate properly, replace seat belt retractor.
- If front pyrotechnic seatbelts do not retract, they have been deployed and are no longer usable. They must be replaced.

WARNING —
Observe all cautions, warnings and notes before starting repairs involving airbag systems.

CAUTION —
Before starting any body disassembly, straightening or other body repair work, the belt tensioning units must be removed.

Front seatbelt, assembly (sedan)

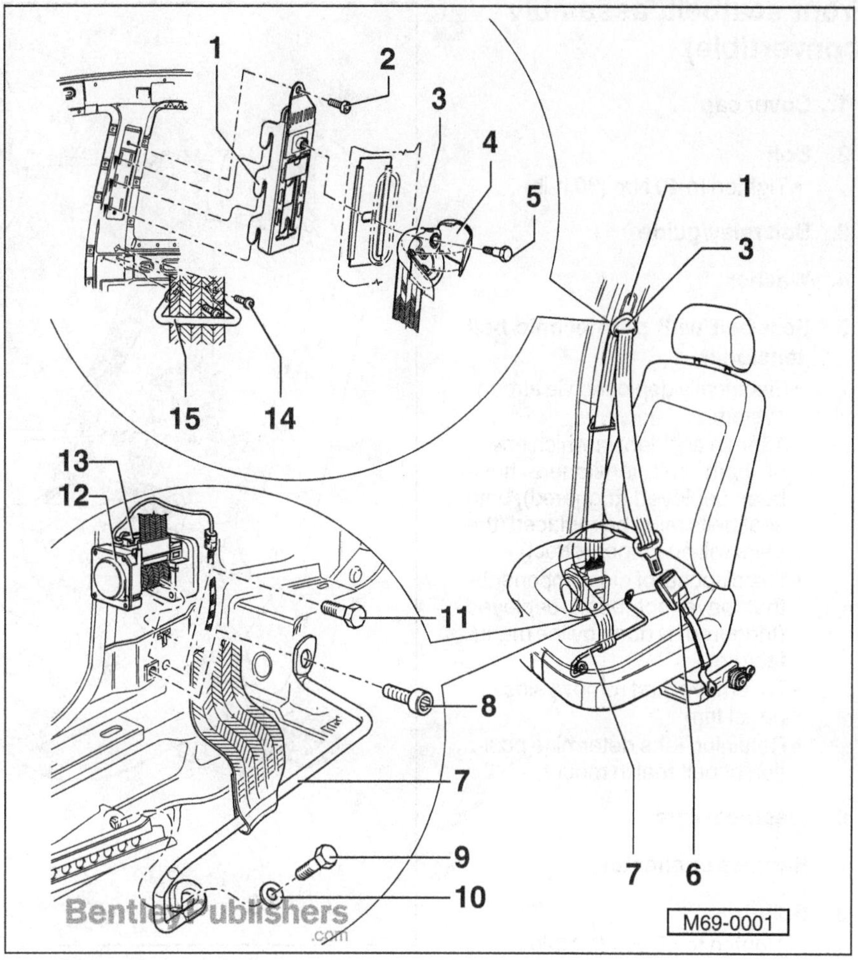

1. **Belt height adjuster**
   - To remove height adjustment mechanism, first remove trim on top of B-pillar
2. **Screw**
   - Tighten to 23 Nm (17 ft-lb)
3. **Belt relay/guide**
4. **Belt relay/guide cover**
5. **Bolt**
   - Tighten to 40 Nm (30 ft-lb)
6. **Front belt buckle (latch)**
7. **Anchor rail**
8. **Bolt**
   - Tighten to 23 Nm (17 ft-lb)
9. **Bolt**
   - Tighten to 40 Nm (30 ft-lb)
10. **Washer**
11. **Bolt**
    - Tighten to 40 Nm (30 ft-lb)
    - Loosen bolt to prevent belt tensioner from deploying during repairs (safe mode)
    - Tighten bolt to return tensioner to operational condition after repairs
12. **Seat belt with pyrotechnic belt tensioner**
    - Two different pyrotechnic seatbelt tensioners are installed depending on vehicle production date:
      Mechanically activated (m.y. 1998)
      Electrically deployed via airbag system (from m.y. 1999)
    - After an accident in which one or both belt tensioners have been deployed (triggered), both seat belts must be replaced (the belts will no longer retract)
    - The disposal of old components that have not been deployed (triggered) is done by the manufacturer
    - Retaining tabs determine position of belt reel in mount
13. **Connector**
    - For pyrotechnic belt tensioner only
14. **Screws**
    - Tighten to 1.5 Nm (13 in-lb)
15. **Belt guide**

Front seatbelt, assembly (convertible)

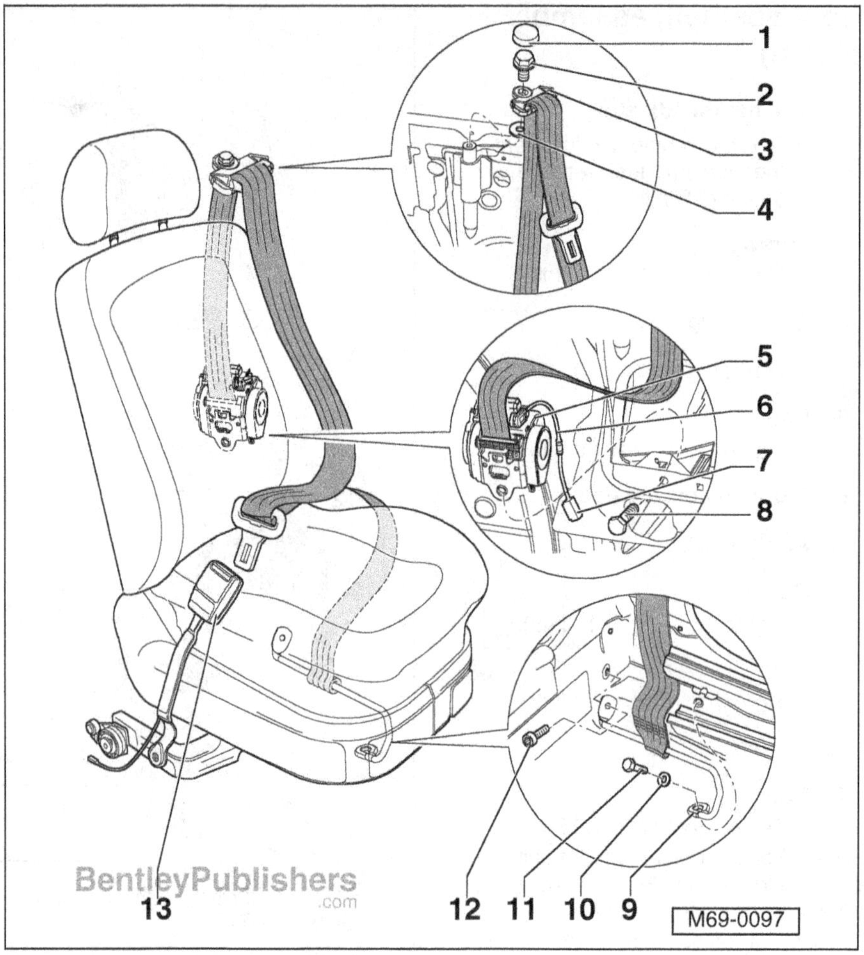

1. **Cover cap**
2. **Bolt**
   - Tighten to 40 Nm (30 ft-lb)
3. **Belt relay/guide**
4. **Washer**
5. **Seat belt with pyrotechnic belt tensioner**
   - Electrically deployed via airbag system
   - After an accident in which one or both belt tensioners have been deployed (triggered), both seat belts must be replaced (the belts will no longer retract)
   - The disposal of old components that have not been deployed (triggered) is done by the manufacturer
   - To remove, first remove side panel trim
   - Retaining tabs determine position of belt reel in mount
6. **Electrical wire**
7. **Harness connector**
8. **Bolt**
   - Tighten to 40 Nm (30 ft-lb)
9. **Anchor rail**
10. **Washer**
11. **Bolt**
    - Tighten to 40 Nm (30 ft-lb)
12. **Bolt**
    - Tighten to 23 Nm (17 ft-lb)
13. **Front belt buckle (latch)**

Rear seat belts, assembly (sedan)

1. **Belt inertia reel**
   - Retaining tabs determine position of belt reel in mount
   - To remove, first remove quarter panel trim, then remove belt inertia reel (item 1), belt relay (item 4), and lower anchorage (item 10)
2. **Special washer**
3. **Spacer sleeve**
4. **Belt relay/guide**
5. **Belt relay/guide cover**
6. **Bolt**
   - Tighten to 40 Nm (30 ft-lb)
7. **Belt buckle (latch)**
8. **Bolt**
   - Tighten to 40 Nm (30 ft-lb)
9. **Sealing washer**
10. **Lower anchorage**
11. **Bolt**
    - Tighten to 40 Nm (30 ft-lb)
12. **Bolt**
    - Tighten to 40 Nm (30 ft-lb)

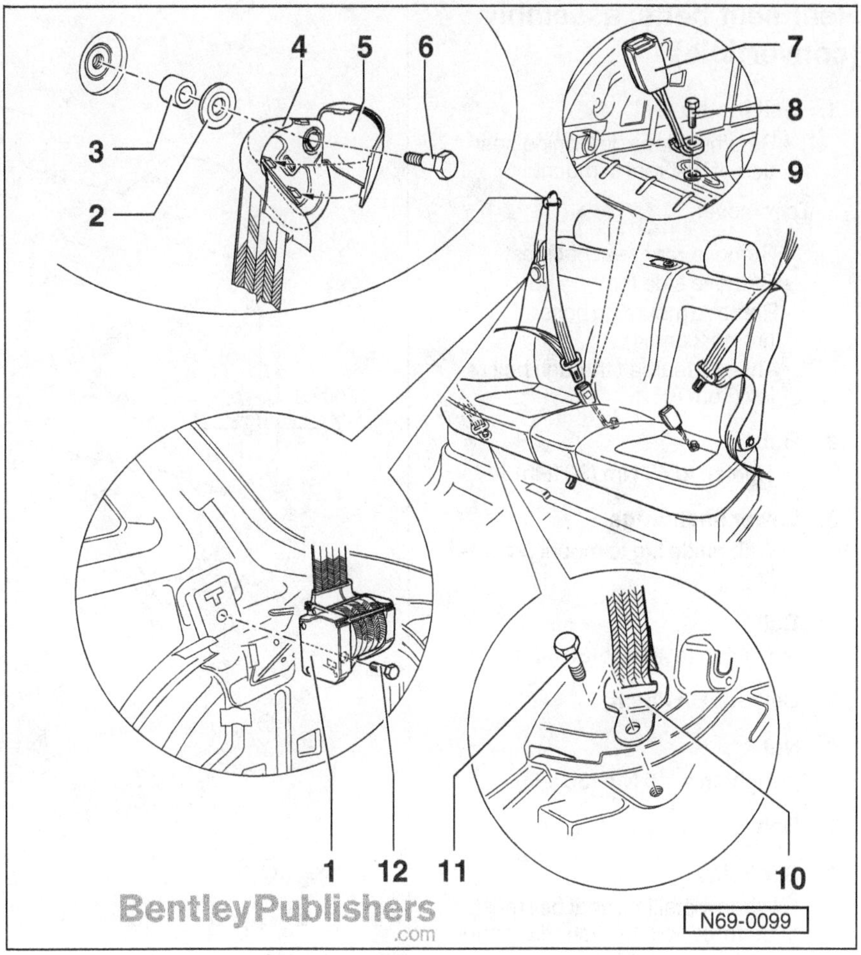

Rear seat belts, assembly (convertible)

1. **Belt inertia reel**
   - Retaining tabs determine position of belt reel in mount

   To remove:

   - Remove rear seat backrest
   - Remove side trim
   - Remove mounting bolts (items 2 and 4)
   - Thread seatbelt through belt relay (item 8)
2. **Bolt**
   - Tighten to 40 Nm (30 ft-lb)
3. **Lower anchorage**
   - With guide lug to mount properly
4. **Bolt**
   - Tighten to 40 Nm (30 ft-lb)
5. **Belt buckle (latch)**
6. **Nut**
   - Tighten to 40 Nm (30 ft-lb)
7. **Belt**
8. **Belt relay**
   - When installing, seat belt must be threaded through rear panel from bottom to top

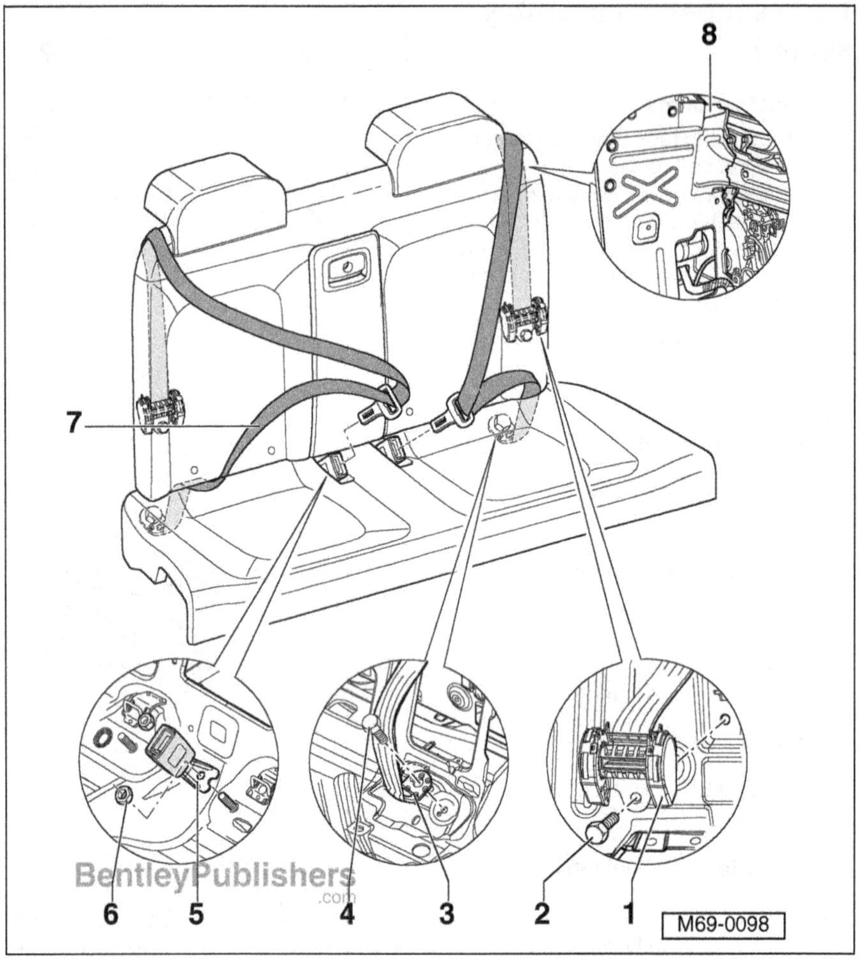

LATCH system "Lower Anchors and Tethers for CHildren"

The LATCH system for "Lower Anchors and Tethers for CHildren" consists of the combination of rear upper tethers and rear lower anchors used to secure a child safety seat.

Except for early 1998 models, all vehicles have anchor hardware installed.

Vehicles with production date 09/00 and later are provided with the LATCH guidance fixtures. See certification label on driver's side B-pillar.

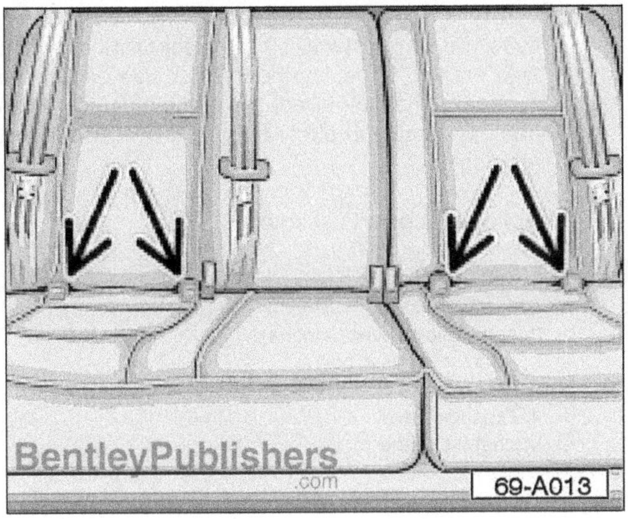

Vehicles manufactured as of 01/2001 are equipped with buttons (**arrows**) to identify the rear lower anchor points.

– **Rear upper tethers**
  - Tether anchor location points are located on the back side of the rear passenger's seat.

Rear lower anchors
  - Illustration shows location of rear lower anchors (**arrow**) for rear seat.

NOTE —
- *Rear lower anchors secure the child restraint in the seat without using the vehicle's safety belt. Anchors provide a secure and easy to use attachment and minimize the possibility of improper child restraint instalation.*
- *Rear lower anchor attachment points are located between the rear seat back and rear seat cushion for the outboard seating positions.*

Automatic rollover support system (convertible)

The automatic rollover support system consists of:

- Two supports integrated into the rear seat backrest
- Seat belts
- Reinforced windshield frame
- Electronic control unit that works together with the air-bag system

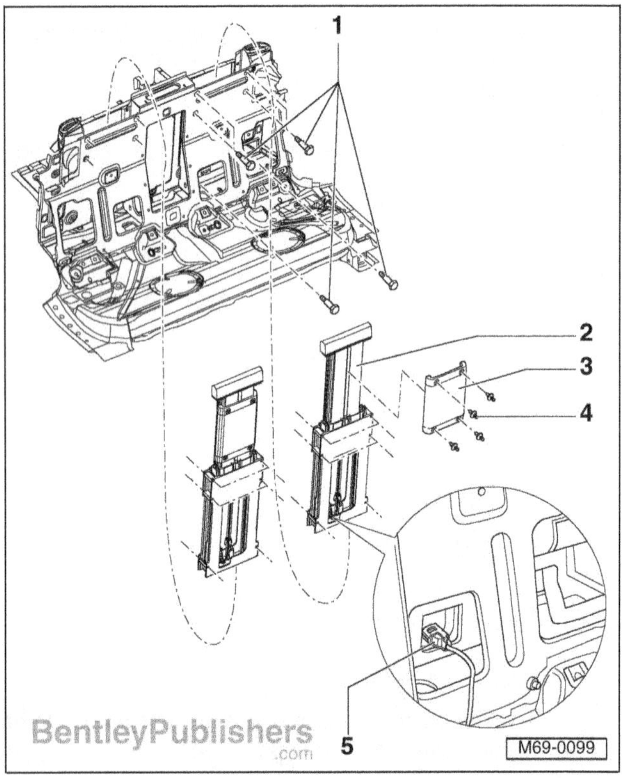

The two supports integrated into the rear seat backrest will deploy upwards if the vehicle tips sideways to a degree and rate that is more than the amount programmed into the electronic control module. The supports can also deploy in frontal, side and rear-end collisions when the deceleration registered by the electronic control unit is high enough to deploy the front or side airbags.These supports, when deployed, and the reinforced windshield frame help protect vehicle occupants in case of a rollover accident.

Rollover support assembly

1. Bolt
   - Tighten to 29 Nm (21 ft-lb)
2. Automatic rollover supports
   - Shown in raised position
3. Impact bar
4. Expanding rivet
5. Harness connector

Rollover support, lowering

If the rollover supports have been deployed, they can be retracted downward, if necessary.

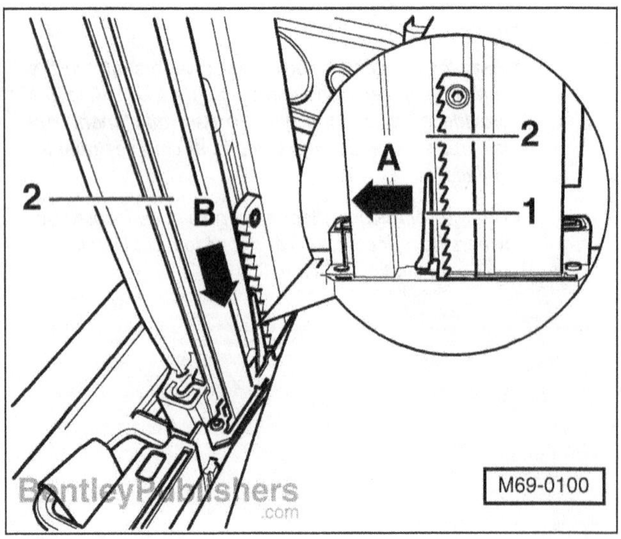

> ***CAUTION —***
> *If the rollover supports have been deployed, do not operate convertible top until lowering them.*

Release locking latch (**1**) by pushing in direction of **arrow A**.

- Push rollover support down (**arrow B**) halfway and let go of locking latch.
- Push rollover support down completely until it securely latches in place.

AIRBAGS

Diagnostics, component testing and repair of the airbag system should only be carried out by properly trained Volkswagen technicians using specialized test equipment.

0024314

When the ignition key is turned on, the airbag warning light (**arrow**) will illuminate for approximately 5 seconds. The illuminated light indicates the self-test of the air bag electronic control module, and all of the related electronic components. If all monitored systems are operating normally, the light will go out. If the warning light does not go out after approximately 5 seconds, or if it comes on at any time while driving, the system has detected a fault and will not operate. In this case, the vehicle should be inspected by an authorized Volkswagen dealer.

If the vehicle has been involved in an accident where the airbag was deployed, Volkswagen specifies that the following components must be replaced:

- All airbag units that have been deployed
- Passenger side airbag support
- Airbag control module and sensors
- Airbag spiral spring/slip ring
- Deployed front seatbelt tensioners

In addition, the following must also be replaced if necessary:

- All damaged components including, but not limited to, seat frames, covers and rear seatbelts

In the event of an accident where the airbags have not been deployed, observe the airbag system warning light. If the airbag system warning light does not indicate a fault, it is not necessary to replace any airbag components.

Special precautions for airbag equipped vehicles

Special service precautions must be followed to prevent possible serious bodily injury to the repair technician and others involved in the handling and storage of airbag components and to insure proper operation in the event of a collision.

WARNING —

- *Before working on any airbag, steering wheel or steering column component, always* ***disconnect the battery first.***
- *No waiting time is required after disconnecting the battery. Airbag systems can be worked on immediately.*
- *Airbags are inflated by an explosive device. Handled improperly or without adequate safeguards and training, the system can be very dangerous. Special precautions must be observed prior to any work at or near the steering wheel and steering column including the pedal assembly.*
- *The airbag is a vehicle safety system. To guard against personal injury or airbag system failure, only factory trained Volkswagen service technicians should test, disassemble or service the airbag system.*
- *Airbag units that have been dropped onto a hard surface must not be installed. Always replace any airbag system component that has been mechanically or physically damaged (example: dented or cracked).*
- *Make sure that no one is in the passenger compartment when connecting the battery Ground (GND) strap.*
- *A technician must be electrostatically discharged before picking up or touching any airbag unit. This is accomplished by touching a suitable metal ground such as a water or heating pipe or metal frame. If in the vehicle, this is accomplished by touching a suitable chassis ground such as a door latch or striker.*
- *Install new airbag unit as soon as the unit is removed from the packaging. Reinstall in packaging if work will not be completed immediately.*
- *Do not leave any undeployed airbag unit unattended. If work is interrupted, store airbag unit in a secure location where it cannot be disturbed.*
- *Undeployed airbag units that have been replaced must only be stored and shipped in packaging designed specifically for the purpose, such as packaging found with the replacement unit.*
- *Undeployed airbag units that have been replaced must be properly identified as such.*
- *The storage and transportation of airbag units must be in accordance with all applicable Federal, State and Local rules and regulations.*
- *Airbag units that have been removed during the course of repairs must be stored with the padded side facing up and in a secure location where it cannot be disturbed.*
- *Observe all cautions, warnings and notes before starting repairs involving airbag systems.*

Airbag system, overview

NOTE —
If replacing any airbag unit or airbag control module, consult your authorized Volkswagen dealer parts department for proper registration procedure.

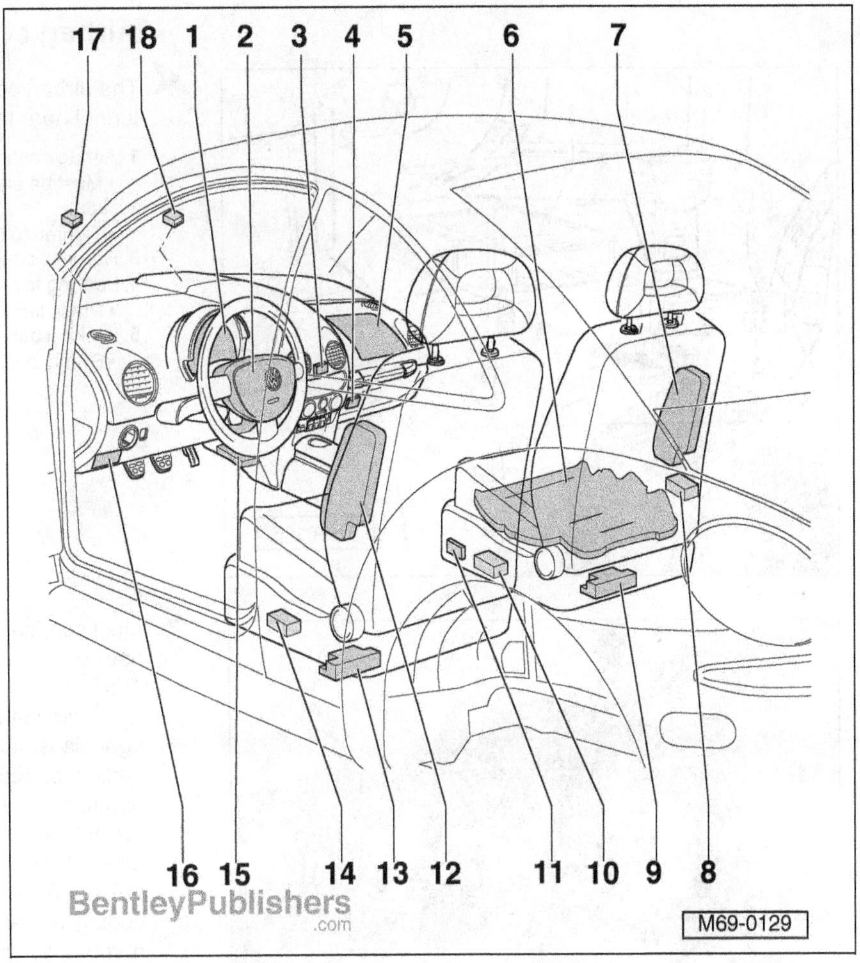

1. **Airbag Malfunction Indicator Lamp (K75)**
   - Location: in instrument cluster
2. **Airbag unit, driver's side**
   - Location: in steering wheel
3. **Front passenger airbag disabled indicator light (K145)**
   - From 05/2003
4. **Front passenger airbag OFF key switch**
   - For non-US market vehicles
5. **Airbag unit, front passenger**
   - Location: in padded dash, right side
   - Activated via seat occupant recognition from 05/2003
6. **Seat occupied recognition mat**
   - USA only
   - From 05/2003
7. **Airbag unit, side, front passenger**
   - Location: in right front seat backrest frame
8. **Impact sensor, side airbag, front passenger**
   - From 05/2003
9. **Impact sensor, side airbag, front passenger**
   - Through 04/2003
10. **Seat occupied recognition control module (J706)**
    - USA only
11. **Seat occupied recognition pressure sensor (G452)**
    - USA only
12. **Airbag unit, side, driver**
    - Location: in left front seat backrest frame
13. **Impact sensor, side airbag, front passenger**
    - From 05/2003
14. **Impact sensor, side airbag, front passenger**
    - Through 04/2003
15. **Airbag Control Module (J234)**
    - Location: on tunnel under center console
16. **Data Link Connector (DLC)**
    - Location: left lower side of instrument panel
17. **Front left crash sensor (G283)**
    - USA only, from 05/2003
    - Mounted at front of left long member
18. **Front right crash sensor (G284)**
    - USA only, from 05/2003
    - Mounted at front of right long member

Airbag control module

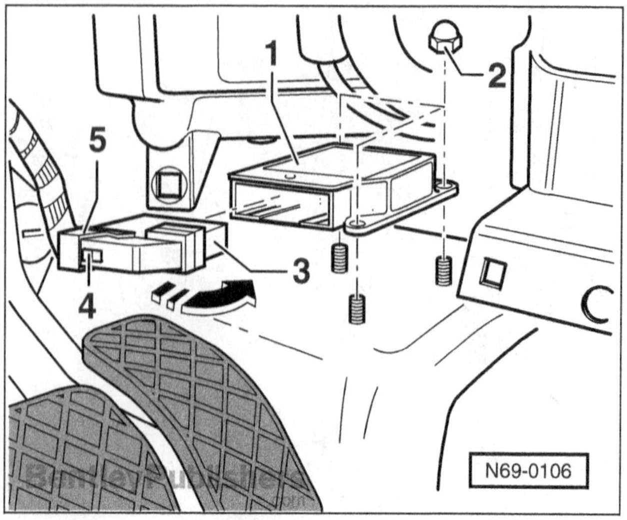

The airbag control module is mounted to the floor on the tunnel near the base of the bulkhead.

1. Airbag control module
   - Must be electronically coded before installing
2. Nuts
   - Tighten to 6 Nm (53 in-lb)
3. Harness connector
4. Locking tab
   - Press tab to release locking bar
5. Locking bar
   - Swing in direction of arrow to release harness connector

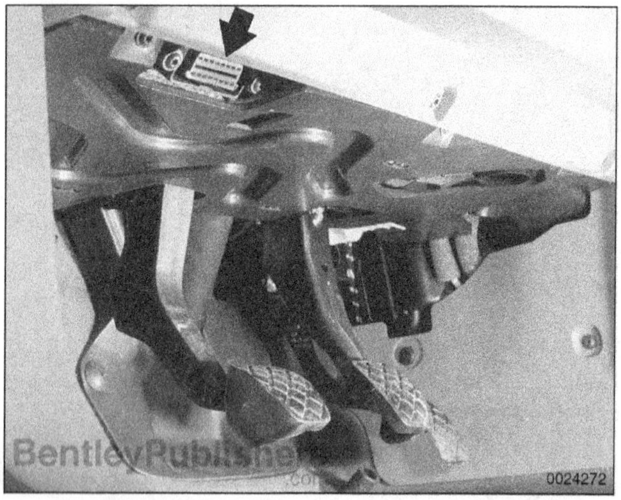

Most service and repair to the airbag system requires the use of a scan tool such as the Volkswagen VAG 1551/1552, VAS 5051/5052 or equivalent aftermarket diagnostic scan tool or computer program. Access for diagnosis is via the data link connector (**arrow**) including coding of replacement control modules. Replacement airbag control modules are shipped uncoded and will not function when installed. Because of the specialized nature of these repairs, see Volkswagen dealer or qualified independent repair shop. Proper airbag control module coding is derived from the airbag control module part number as reported by the scan tool. Control module coding is outside the scope of this manual.

WARNING —
- *Observe all cautions, warnings and notes before starting repairs involving airbag systems.*
- *Disconnect battery Ground (GND) strap before disconnecting or removing the airbag control module.*

Airbag unit in steering wheel, assembly

WARNING —
- *Observe all cautions, warnings and notes before starting repairs involving airbag systems.*
- *Observe all cautions, warnings and notes when completing repairs involving airbag systems.*

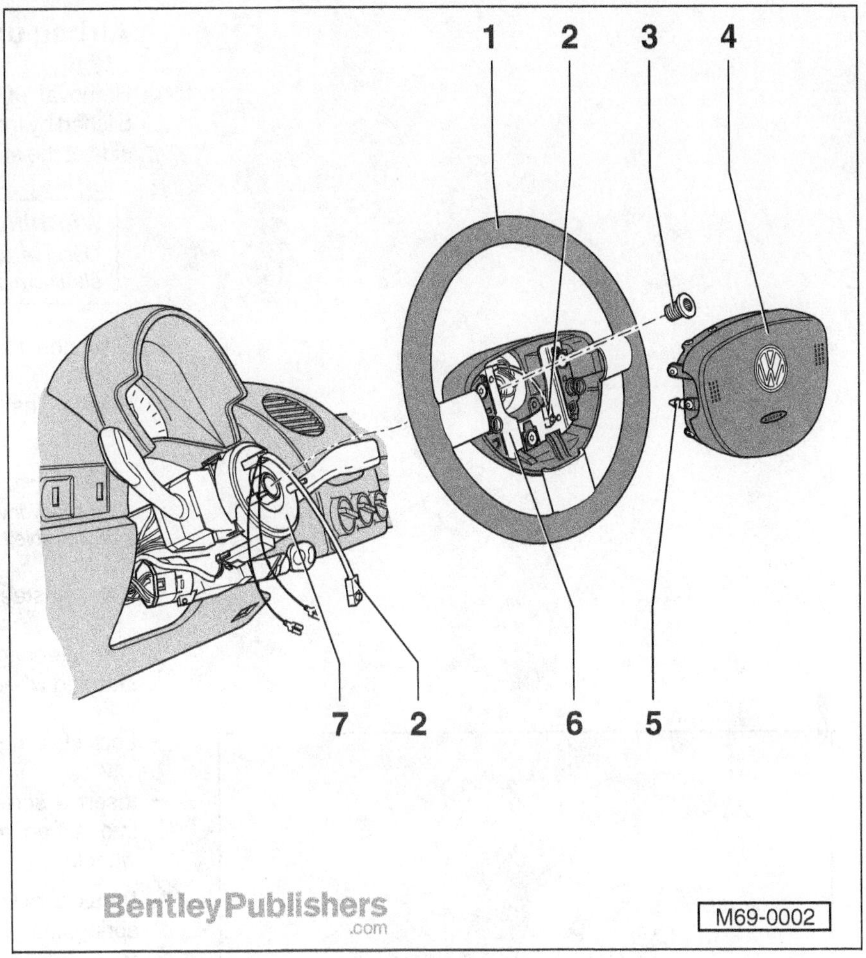

1. **Steering wheel**
2. **Harness connector for airbag**
3. **Bolt, multi-point socket-head**
   - Tighten to 60 Nm (44 ft-lb)
   - With locking compound
   - Can be used up to 5 times
   - Mark with center punch after each installation
4. **Airbag unit**
5. **Locking lugs**
   - Release from rear of steering wheel
6. **Securing plate**
7. **Spiral spring with slip ring**

Airbag unit in steering wheel, removing

Removal of the steering wheel airbag unit is accomplished by inserting a tool into openings in the dashboard side of the steering wheel.

> ***WARNING —***
> *Observe all cautions, warnings and notes before starting repairs involving airbag systems.*

– Disconnect battery ground strap.

– Remove instrument cluster, see **90 Instruments**.

NOTE —
If a screwdriver with a very short handle is used, it may be possible to leave the instrument cluster in place.

– Release steering column adjustment lever.

– Turn steering wheel until a spoke is vertical. Extend steering wheel fully and move to uppermost position.

– Lock steering column adjustment lever.

– Insert a screwdriver approximately 70 mm (3 in.) long into airbag release hole on the rear side of steering wheel.

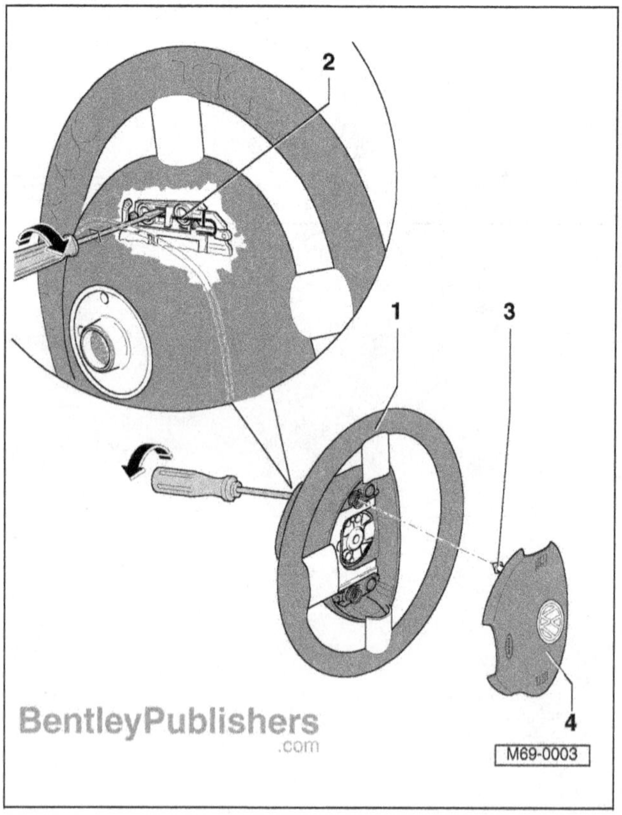

◄ Press screwdriver in direction of **arrow** to press back spring clip (**2**) and release locking lug (**3**) of airbag unit (**4**).

– Turn steering wheel back ½ turn (180°) and release second locking lug on opposite side.

– Turn steering wheel to center position (wheels straight ahead).

> ***WARNING —***
> *A technician must be electrostatically discharged before picking up or touching any airbag unit. This is accomplished by touching a suitable metal ground such as a water or heating pipe or metal frame. If in the vehicle, this is accomplished by touching a suitable chassis ground such as a door latch or striker.*

– Pull airbag away from steering wheel and disconnect harness connector from airbag unit.

– Place removed airbag unit in a secure location with padded side facing up.

Airbag unit in steering wheel, installing

– Install instrument cluster as required.

– Reconnect harness connector for airbag unit.

– Position airbag unit on steering wheel and snap into place.

NOTE —
Airbag unit locking lugs must be heard and felt to snap into place.

– Close doors and switch ignition on.

CAUTION —
Make sure the passenger compartment is not occupied before connecting the battery Ground (GND) strap.

WARNING —
Observe all cautions, warnings and notes when completing repairs involving airbag systems.

– Connect battery ground strap.

Airbag spiral spring/slip ring, removing and installing

WARNING —
Observe all cautions, warnings and notes before starting repairs involving airbag systems.

– Turn the wheels to the straight ahead position.

– Disconnect battery ground strap.

– Remove airbag unit in steering wheel as described earlier.

– Remove steering wheel.

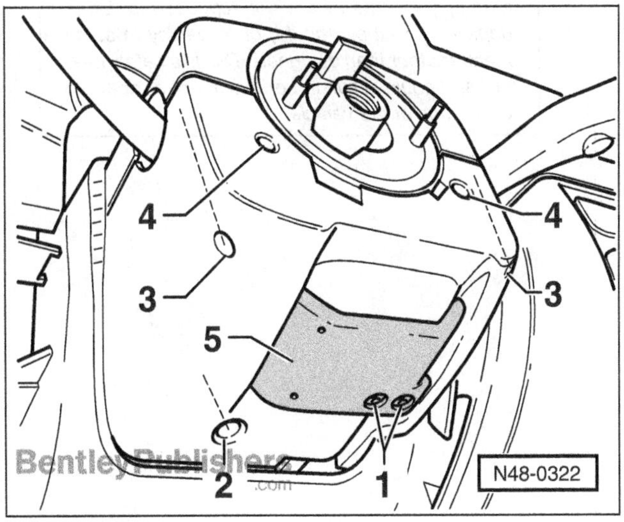

◀ Remove screws (**1**), adjustment lever (**5**), securing screws (**2**, **3**, **4**) and lower and upper steering column switch trim.

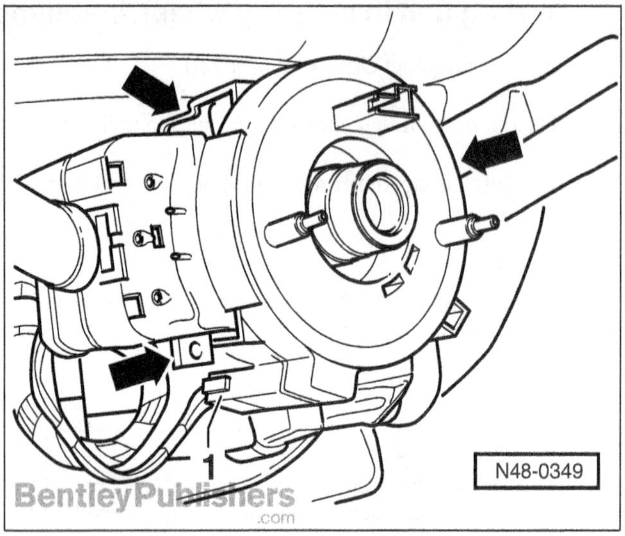

◀ Unplug harness connector (**1**) and release locking lugs (**arrows**) on spiral spring/slip ring and pull off.

– Installation is the reverse of removal.

> ***WARNING —***
> *Observe all cautions, warnings and notes when completing repairs involving airbag systems.*

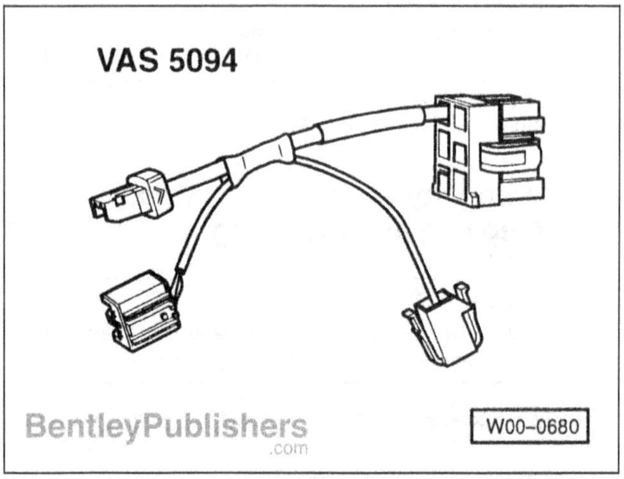

Side airbag units

◀ Both driver's and passenger's front seats are equipped with airbags mounted in the seat backrest frames. These airbags are designed to deploy under certain side impact conditions and are triggered by separate sensors mounted to the front floor crossmembers under the front seats. Before removing the front seats, a special safety harness must be installed to prevent accidental deployment of the airbag units.

> ***WARNING —***
> - *Observe all cautions, warnings and notes before starting and when completing repairs involving airbag system.*
> - *Volkswagen of America specifically warns against the installation of aftermarket upholstery on any vehicle equipped with side airbags. The factory-installed upholstery is designed to separate in specific places at specific rates, and in specific directions. Installation of non-factory upholstery including, but not limited to, "beads" and "sheepskins", may cause seat mounted airbags to deploy when they are not supposed to; fail to deploy when they should; or to deploy in some manner other than designed. This is a safety hazard and could result in serious injury or death to occupants of the vehicle.*

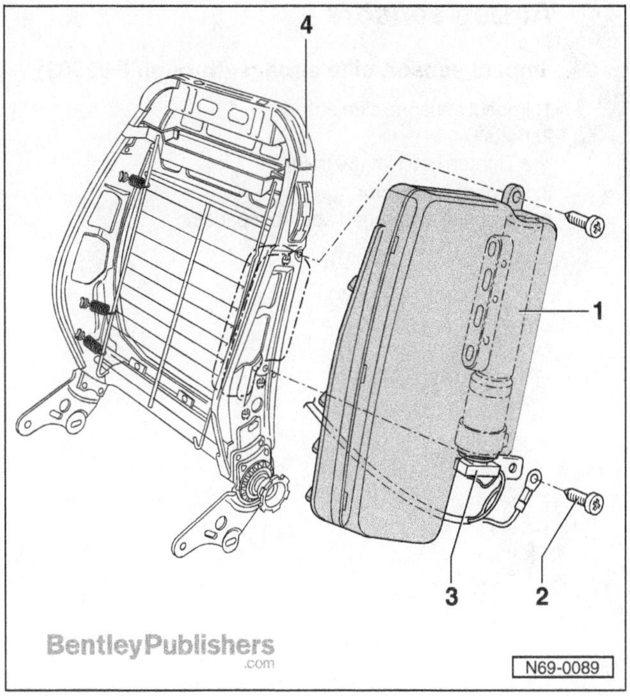

Side airbag assembly (through 04/2003)

1. Side airbag unit
   - Driver's side shown, passenger side similar
2. Bolt
   - Tighten to 7 Nm (62 in-lb)
3. Harness connector
4. Seat backrest frame

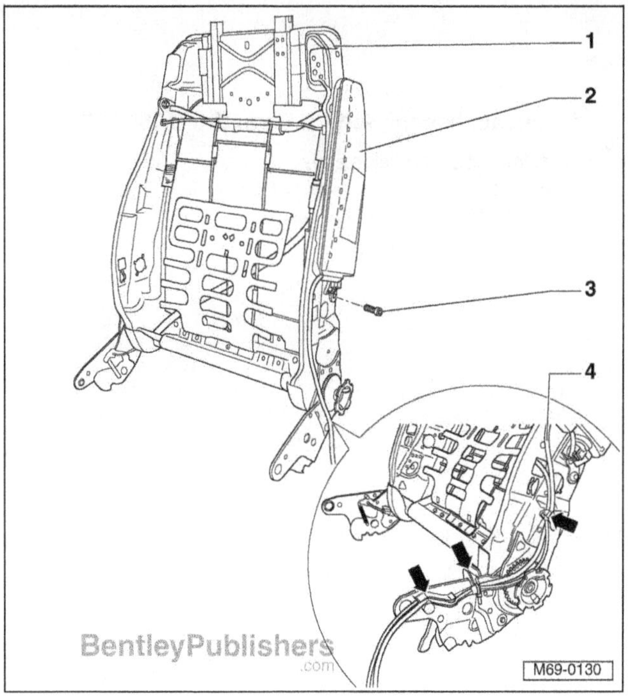

Side airbag assembly (from 05/2003)

1. Seat backrest frame
2. Side airbag unit
   - Driver's side shown, passenger side similar
3. Bolt
   - Tighten to 10 Nm (7 ft-lb)
4. Harness connector

Airbag sensors

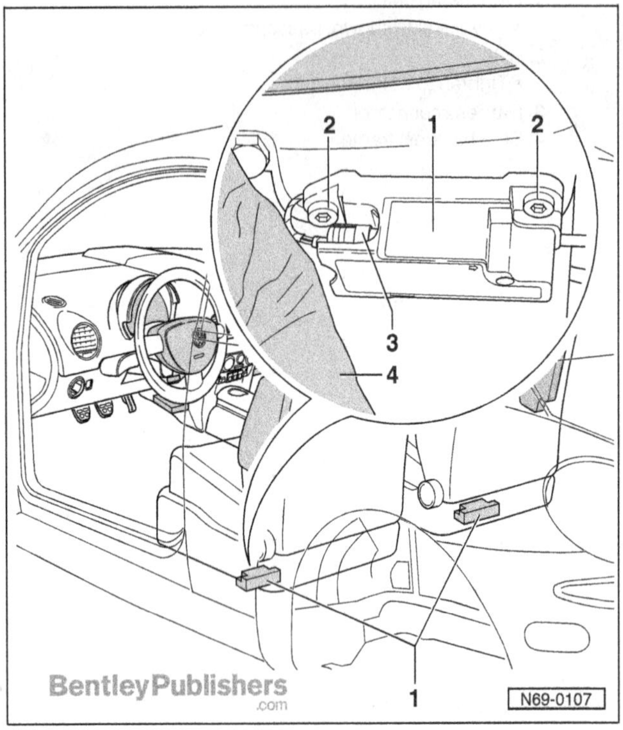

Impact sensor, side airbags (through 04/2003)

1. Impact sensors, side airbag
2. Bolts
    - Tighten to 9 Nm (80 in-lb)
3. Harness connector
4. Carpeting and sound absorber padding

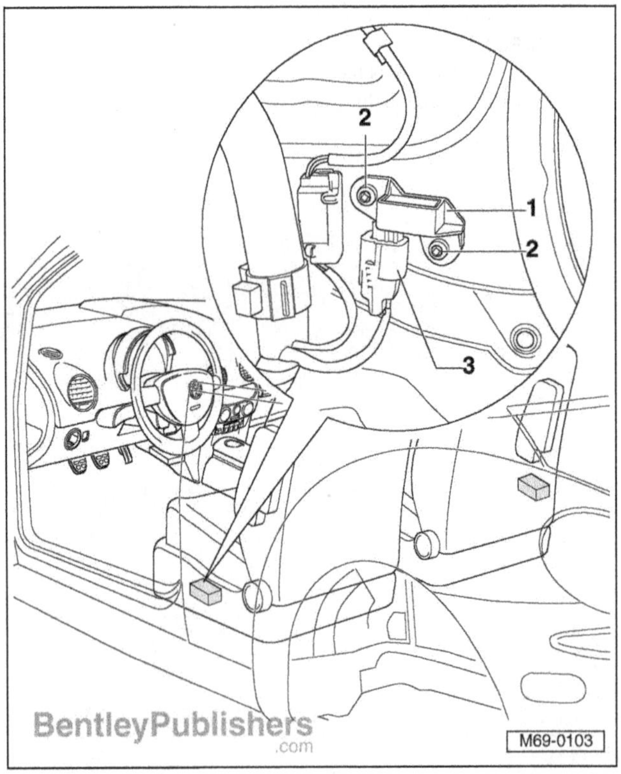

Impact sensor, side airbags (from 05/2003)

1. Impact sensor, side airbag
2. Bolts
    - Tighten to 9 Nm (80 in-lb)
3. Harness connector

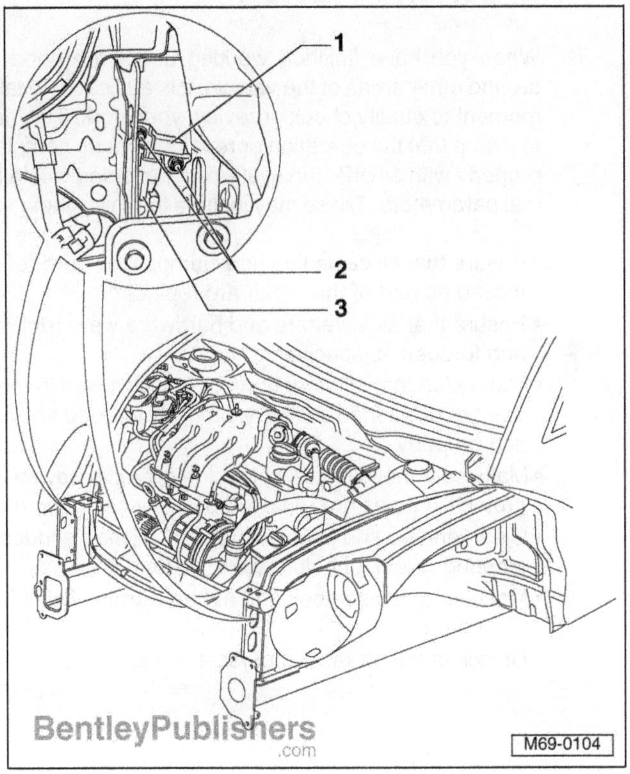

Impact sensors, front

1. Front impact sensors
   - Left front (G283)
   - Right front (G284)
2. Nuts
   - Tighten to 6 Nm (53 in-lb)
3. Harness connector

Passenger front airbag

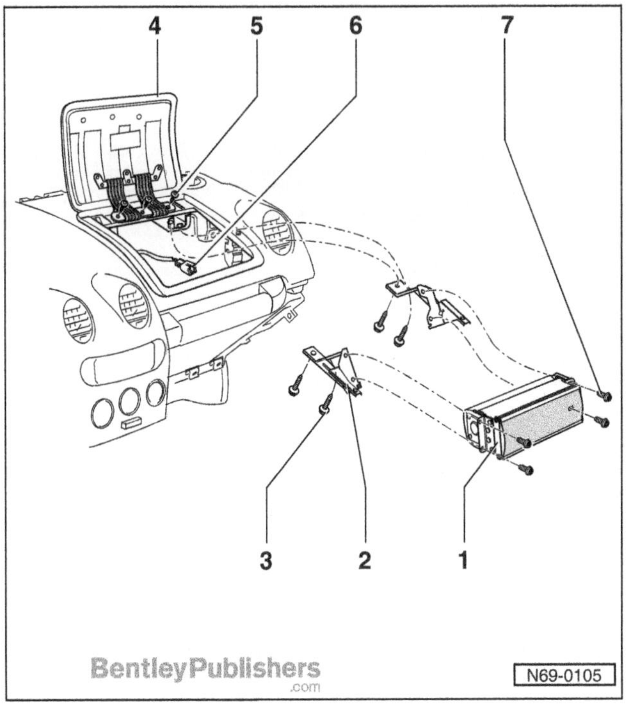

Airbag unit, front passenger, assembly

1. Airbag unit, front passenger
2. Supports
   - Always replace if airbag has been deployed
3. Bolts
   - Tighten to 12 Nm (9 ft-lb)
4. Cover panel
   - Carefully pry up from bottom edge to release
   - Swing up and out of the way
5. Screw
   - Tighten to 4 Nm (35 in-lb)
6. Airbag harness connector
7. Screws
   - Tighten to 12 Nm (9 ft-lb)

QUALITY REVIEW

When you have finished working under the hood and around other areas of the vehicle, it is advisable to take a moment to quality check or review your work. This helps to insure that the operation or repair has been completed properly with all affected systems functioning within normal parameters. These may include the following:

- Ensure that all cable ties and clamps that were removed as part of the repair are replaced.
- Ensure that all fasteners and hardware were replaced and torqued as specified.
- Make sure that all other components involved in the repair are positioned correctly, properly torqued and function properly.
- Make sure that all tools, shop cloths, fender covers and protective tapes are removed before closing the hood.
- Clean grease and fingerprints from painted surfaces, steering wheel, upholstery and shifter.
- Make sure that seatbelt and airbag warning lights function correctly.
- Unlock anti-theft radio and reset clock.

70 Trim–Interior

GENERAL

This repair group covers interior equipment and trim panels, including center console and instrument panel removal. For interior lights and accessories see **96 Lights, Accessories–Interior**.

CENTER CONSOLE

Center console, removing and installing

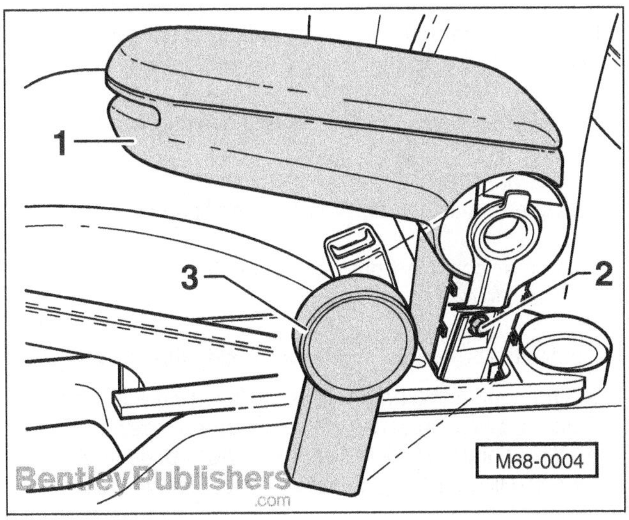

Remove armrest, if applicable:

- Unclip trim (**3**) from center console (**1**).
- Remove bolt (**2**) and take off armrest.
- Tighten bolt to 18 Nm (13 ft-lb).

70

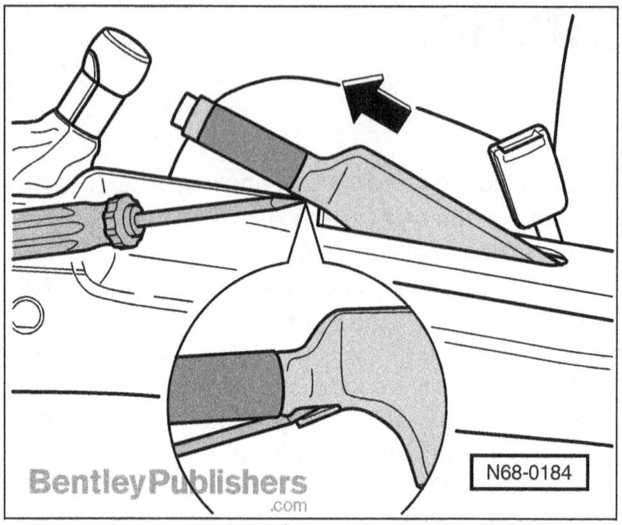

Pull up parking brake lever and use a screwdriver to disengage trim around lever. Pull trim forward (**arrow**) to remove.

– Manual transmission vehicles:
  - Carefully pry shift boot out from center console.
  - Turn shift boot inside out to expose retaining clamp.
  - Remove retaining clamp and shift knob with shift boot, see **34 Manual Transmission**.

– For automatic transmission vehicles, see **37 Automatic Transmission** for removing selector lever position display and related components.

– For a guide to center console components, refer to appropriate **Center console assembly**.

– Remove ashtrays, if applicable, and cupholders by pulling upward out of center console.

– Remove center console cover mounting hardware.

– Remove console cover and disconnect harness connector for cigarette lighter/12V accessory outlet.

> ***WARNING —***
> - *Disconnect the battery Ground strap (GND) from negative (–) battery terminal before working on the electrical system.*
> - *Disconnecting the negative (–) battery cable may erase fault codes and basic settings in the engine management and automatic transmission control modules. Some driveability problems may be noticed until the system re-adapts to operating conditions. OBD II readiness codes, which may be required for emissions testing, may also be erased. Convenience electronics (alarm system, interior light control, power locks, mirrors, and windows) may need to be re-set using a diagnostic scan tool.*

– Remove center console insert mounting screws and insert.

– Pry off cover caps on either side of console and remove screws.

– Remove remaining console mounting screws.

– Lift center console from rear and carefully pull upward over shift lever.

– Installation is reverse of removal, noting the following:
  - After connecting battery, check all vehicle equipment (radio, clock, power windows) for proper operation.
  - After connecting battery, the readiness code for the ECM will need to be re-set using a diagnostic scan tool.

Center console assembly (through 01/2001)

1. **Center console**
2. **Bolt**
3. **Screw**
   - Qty. 2
4. **Cover cap**
   - Qty. 2
5. **Screw**
   - Qty. 2
6. **Rubber grommet**
7. **Ashtray**
8. **Cupholder**
   - Qty. 2
9. **Screw**
   - Qty. 2
10. **Top cover**
11. **Center console insert**
12. **Screw**
    - Qty. 2
13. **Cigarette lighter**
14. **Cup holder**
15. **Screw**
16. **Screw**
    - Qty. 2

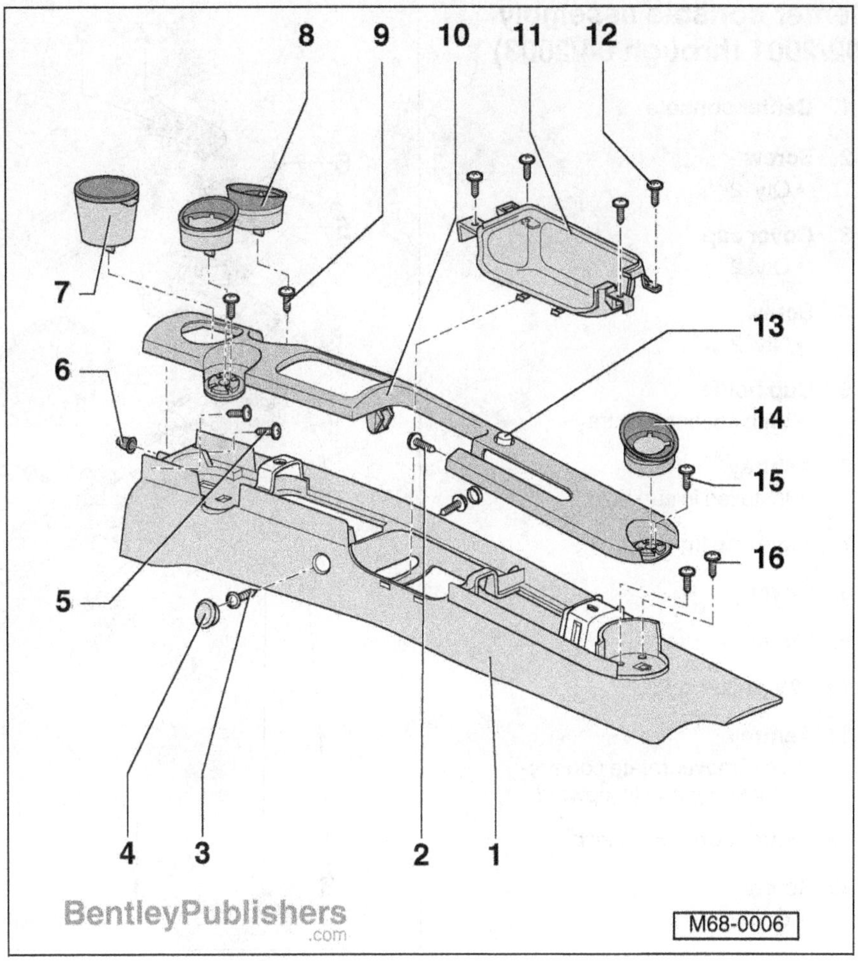

Center console assembly (02/2001 through 04/2003)

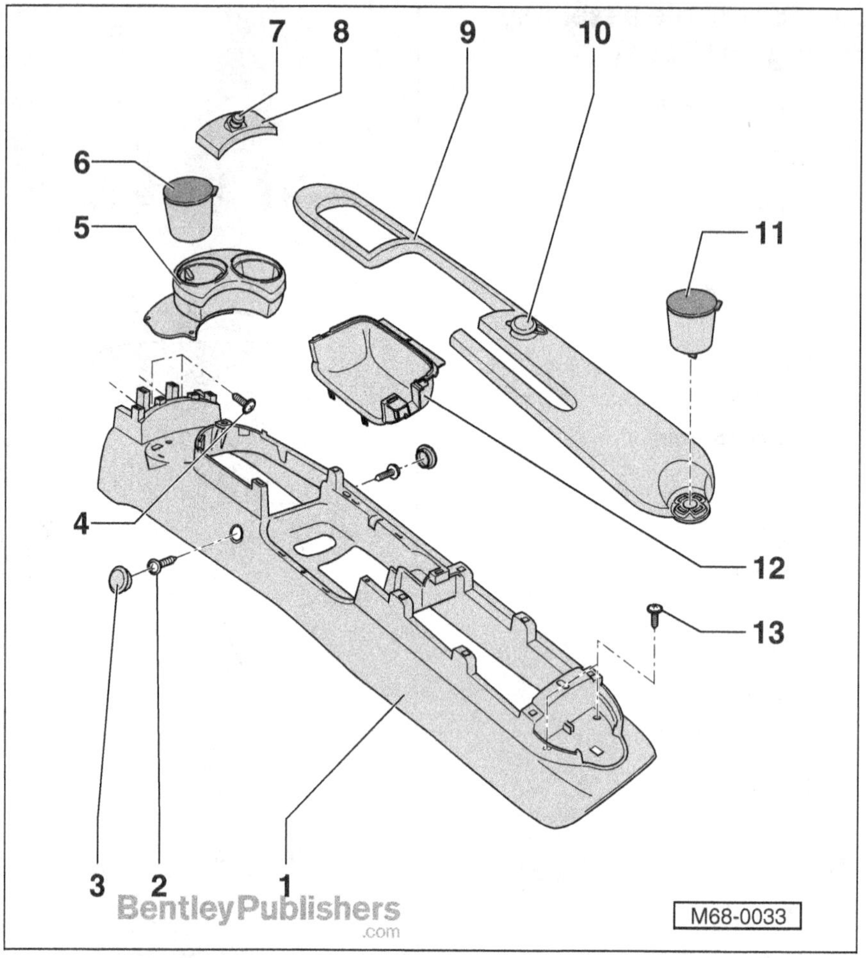

1. **Center console**
2. **Screw**
   - Qty. 2
3. **Cover cap**
   - Qty. 2
4. **Screw**
   - Qty. 2
5. **Cup holder**
   - With optional ashtray
6. **Ashtray**
   - Installed in cup holder
7. **Cigarette lighter**
8. **Cover**
9. **Cover**
10. **12V power outlet**
11. **Ashtray**
    - To remove, rotate counter-clockwise and lift upward
12. **Center console insert**
13. **Screw**
    - Qty. 2

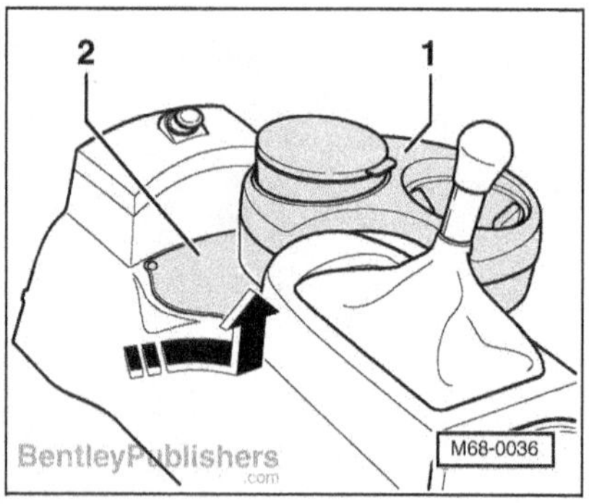

Cup holder, removing

◄ Slide cup holder (**1**) toward right (**arrow**) until stop.

– Disengage cup holder floor plate (**2**) from center console and remove cup holder with floor plate.

Center console assembly, convertible (05/2003 through 06/2005)

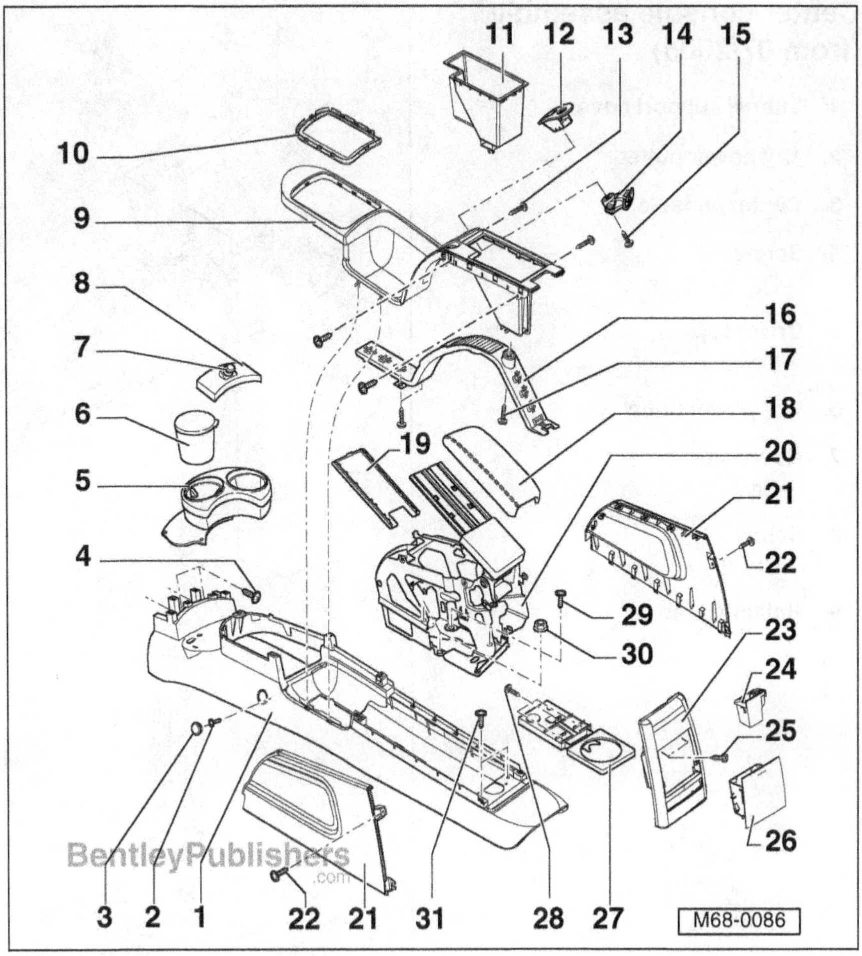

1. **Center console**
2. **Screw**
   - Qty. 2
3. **Cover cap**
   - Qty. 2
4. **Screw**
   - Qty. 2
5. **Cup holder**
   - With optional ashtray
6. **Ashtray**
   - Installed in cup holder
7. **Cigarette lighter**
8. **Cover**
9. **Cover**
10. **Retaining plate for shift lever cover**
11. **Storage compartment**
12. **Cover**
    - Not installed on all vehicles
13. **Screw**
    - Qty. 4
14. **Armrest lock**
    - Not installed on all vehicles
15. **Screw**
16. **Parking brake lever guide**
17. **Screw**
    - Qty. 3
18. **Center armrest padding**
19. **Cover**
20. **Armrest**
21. **Cover**
22. **Screw**
    - Qty. 2
23. **Cover**
    - Redesigned cover available from Volkswagen
24. **Ashtray insert**
    - Redesigned ashtray insert available from Volkswagen
25. **Screw**
    - Qty. 2
26. **Ashtray**
    - Redesigned ashtray insert available from Volkswagen
27. **Cup holder**
28. **Screw**
    - Qty. 3
29. **Screw**
    - Qty. 2
30. **Nut**
    - Qty. 2
31. **Screw**
    - Qty. 2

Center console assembly (from 07/2005)

1. **Tunnel support cover**
2. **12V power outlet**
3. **Center console**
4. **Screw**
   - Qty. 2
5. **Cover cap**
   - Qty. 2
6. **12V power outlet**
7. **Cover cap**
   - Qty. 2
8. **Screw**
   - Qty. 2
9. **Retaining tab**

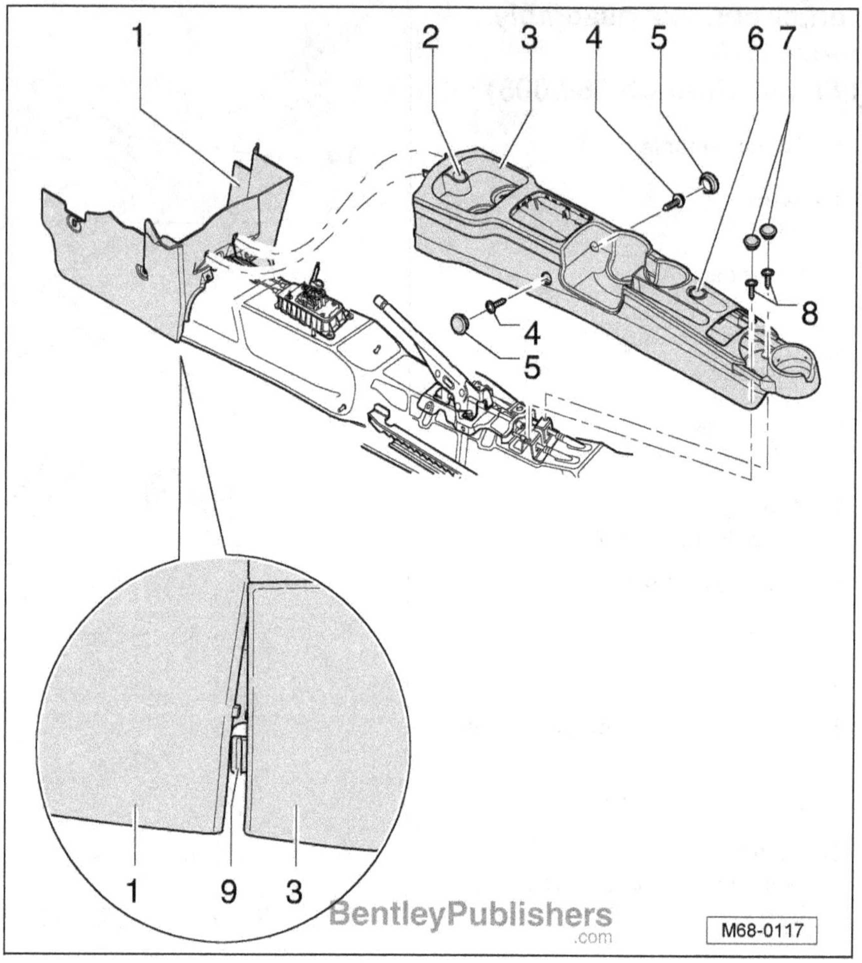

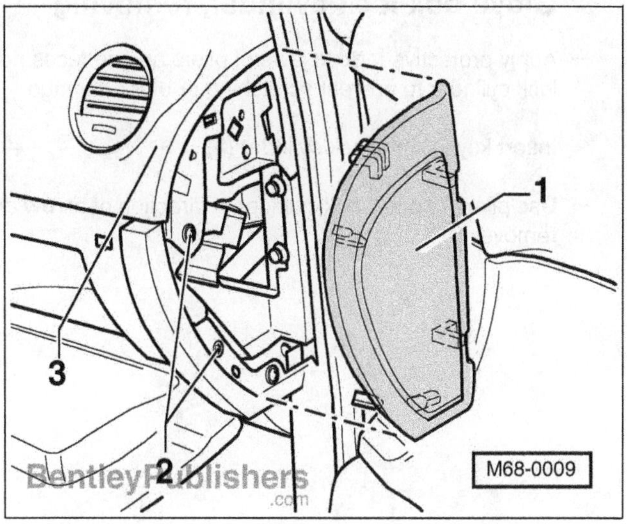

INTERIOR EQUIPMENT

Glove box, removing

- Carefully pry off right side trim panel cover (**1**) at mounting clips.

- Remove glove box mounting screws (**2**).

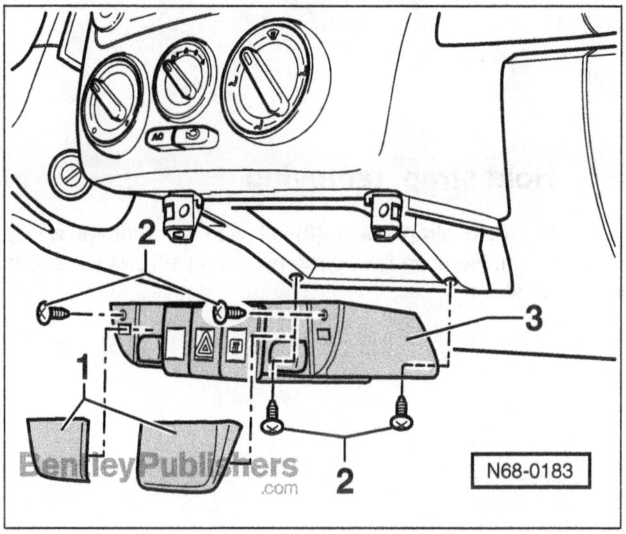

- Pry off lower switch console covers (**1**) and remove mounting screws (**2**).

- Pull switch console (**3**) down from bracket.

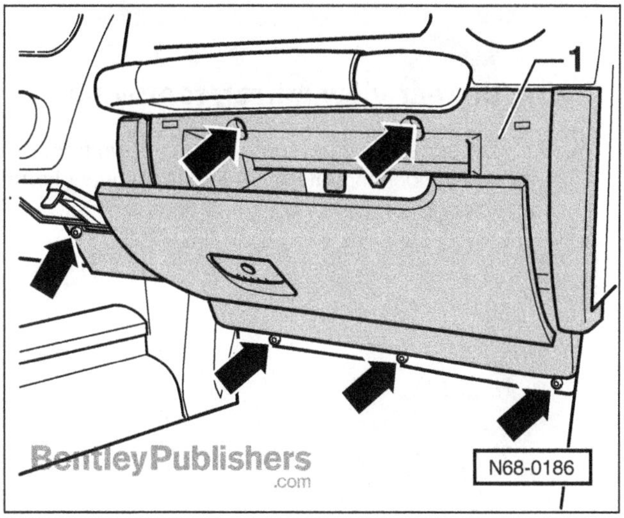

- Remove the glovebox retaining screws (**arrow**) and the glove box (**1**).

- Disconnect harness connector for glove box light.

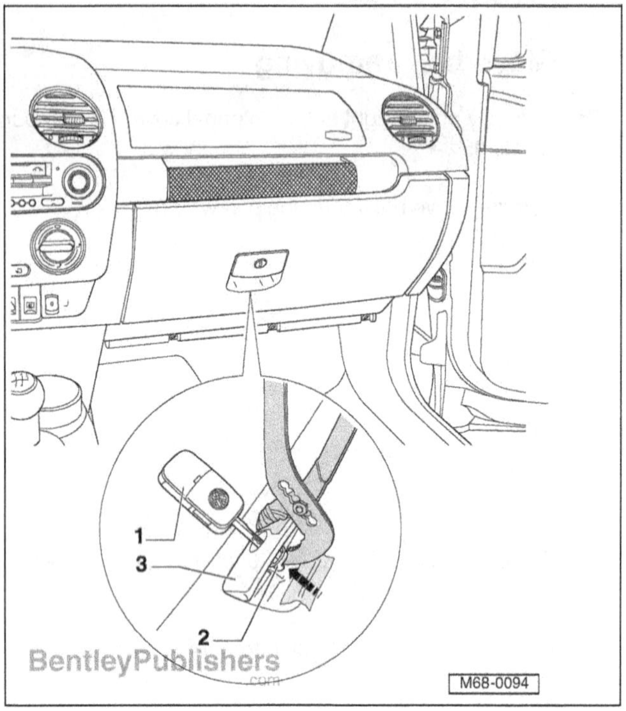

Glove box lock cylinder, removing

- Apply protective tape to jaws of pliers and surfaces near lock cylinder to prevent scratching or other damage.

< Insert key (**1**) into lock cylinder (**2**).

- Use pliers to push lock cylinder in direction of **arrow** and remove.

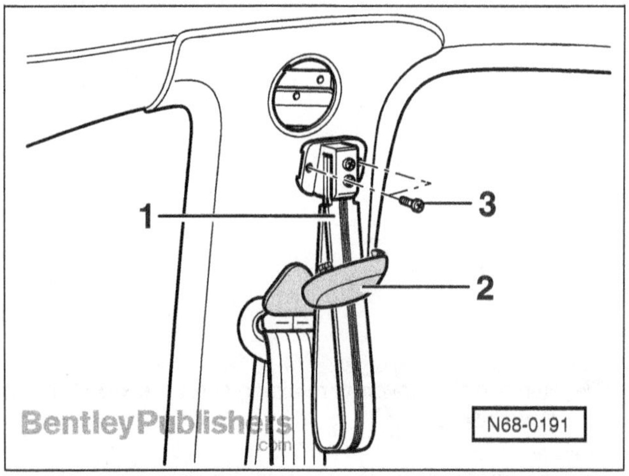

Hold strap, removing

< Pry hold strap cover (**2**) off with screwdriver and pull down. Remove hold strap mounting screws (**3**) and take off strap (**1**).

Inside rear view mirror, removing

< Pull rear edge of mirror mount trim (**1**) down from headliner and unhook trim at front edge. Disconnect harness connector (**2**) for interior light from mirror mount.

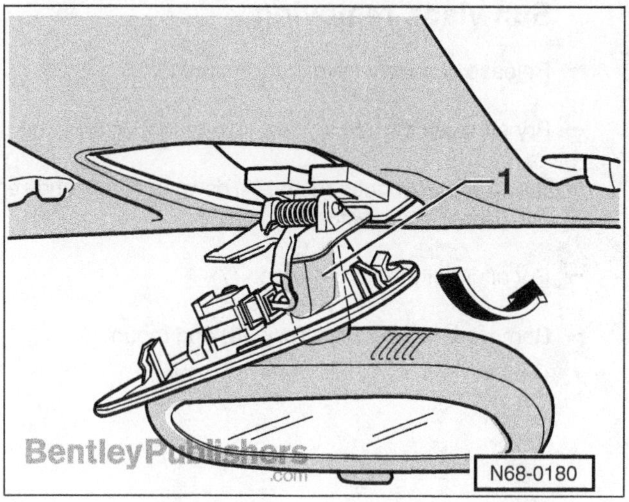

Press mirror (**1**) downward (**arrow**) at an angle off retaining plate (spring clip in mirror base).

Instrument panel handle, removing

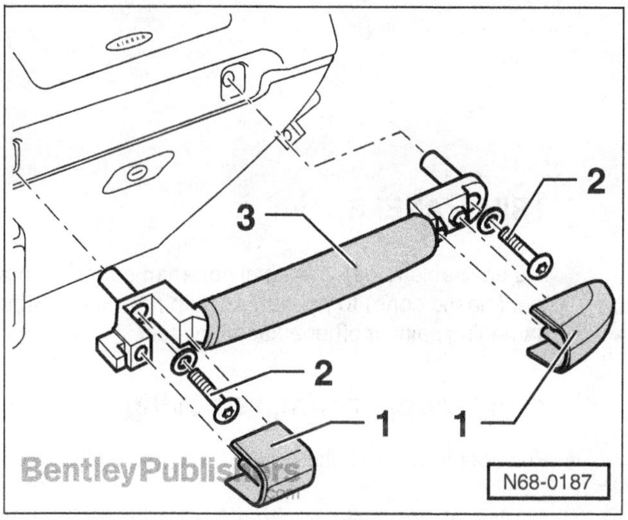

Pull off handle mount covers (**1**). Remove mounting screws (**2**) and pull handle (**3**) out of instrument panel.

Tightening torque
- Handle to instrument panel . . . . 8 Nm (71 in-lb)

Roof handle, removing

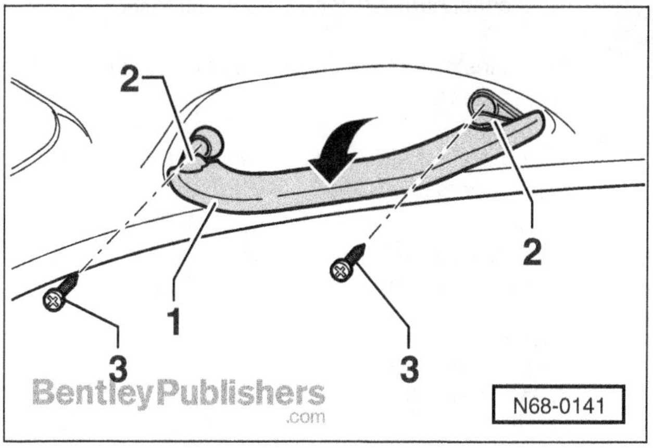

Fold handle (**1**) down. Pry cover (**2**) up with screwdriver. Remove mounting screws (**3**) and handle.

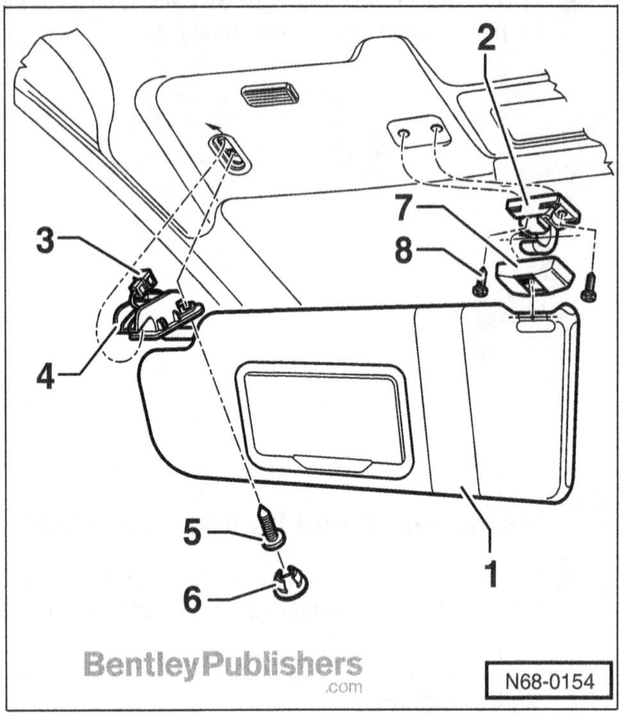

Sun visor, removing

- Release sun visor from inner mount (**2**).
- Pry off cover cap (**5**) and remove mounting screw (**6**).
- Swing sun visor outer mount (**4**) down from roof and separate harness connector (**3**).
- Pry off inner mount cover cap (**7**).
- Remove inner mount screws (**8**) and mount (**2**).

TRIM PANELS

It is advisable to use a special non-marring trim removal tool (plastic bone) to pry various trim panels out without damaging paint or other adjacent panels.

A-pillar upper trim, removing

- Pull door inner seal off A-pillar.
- Starting at top, unclip trim from A-pillar. Lift up to remove.
- Disconnect harness connector for speaker.
- Before installing trim, check clips on A-pillar and replace as necessary.

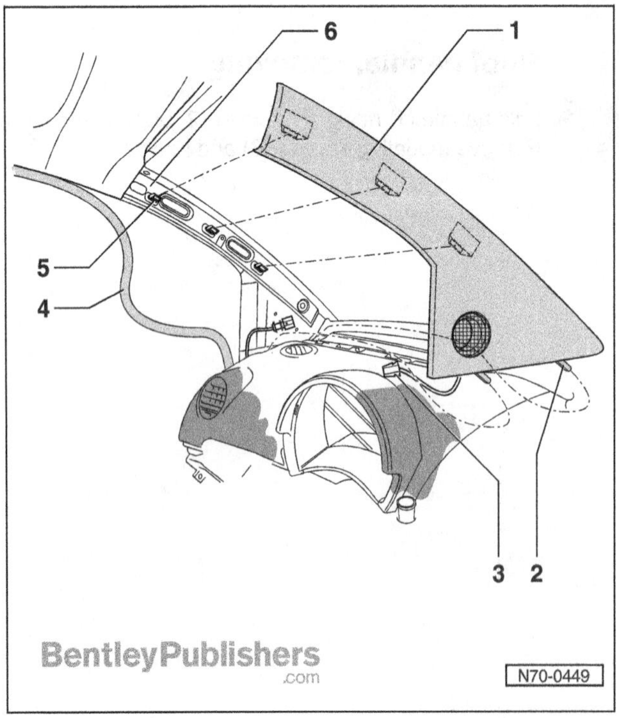

After installing trim, make sure door seal is seated correctly.

1. Trim
- Round grill (sedan through 1998)
- Triangular grill (sedan from 1999)
- No grill (convertible)

2. Retaining pin
3. Speaker connector
4. Door inner seal
5. Clip
6. A-pillar

A-pillar lower trim, removing

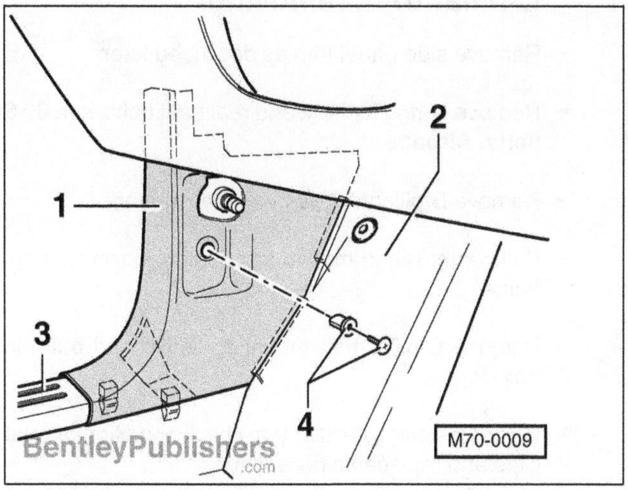

Driver's side

- Remove hood release lever as described in **55 Hoods and Lids**.
- Remove center part of clip and clip (**4**).
- Unclip top part of A-pillar lower trim (**1**) and release bottom from sill panel trim (**3**).

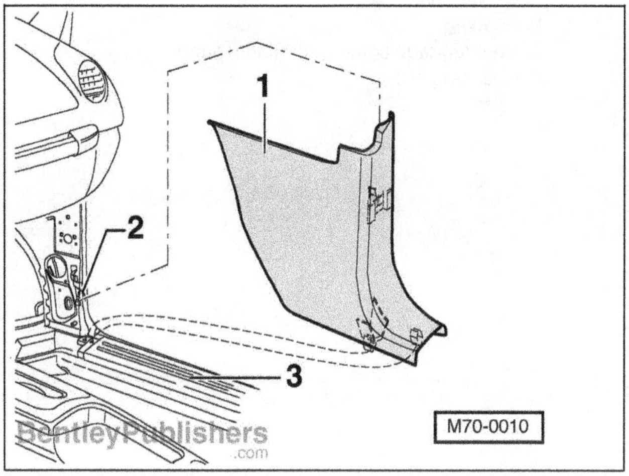

Passenger side

- Unclip top part of A-pillar lower trim (**1**) and release bottom from sill panel trim (**3**).

B-pillar trim, removing

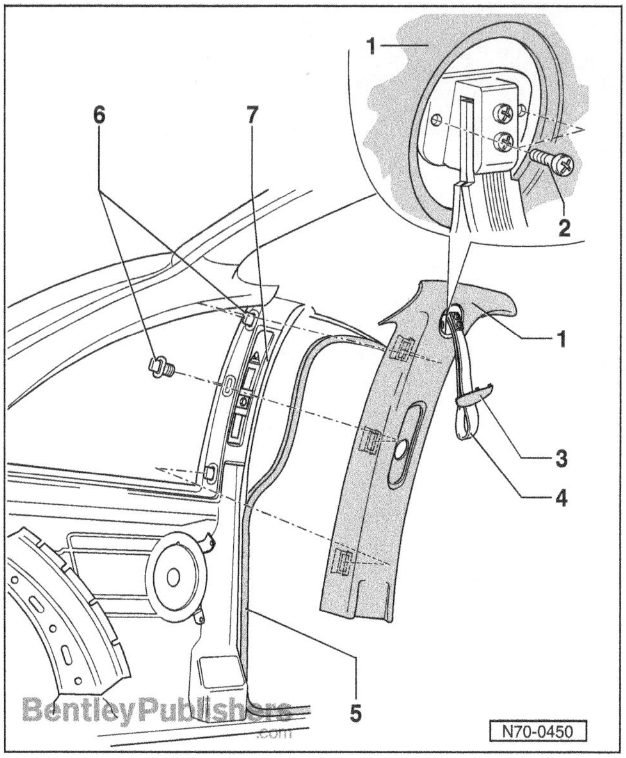

- Remove rear side panel as described later.
- Remove D-ring for front seat belt, see **69 Seatbelts, Airbags**.
- Pull inner door seal off B-pillar.
- Pry off hold strap cover and remove hold strap as described earlier.
- Pry B-pillar trim off at sides and then in center.
- Before installing B-pillar trim check condition of retaining clips and replace as necessary.

1. B-pillar trim
2. Bolt
3. Cover
4. Hold strap
   - Not installed on all vehicles
5. Door inner seal
6. Securing clips
7. B-pillar

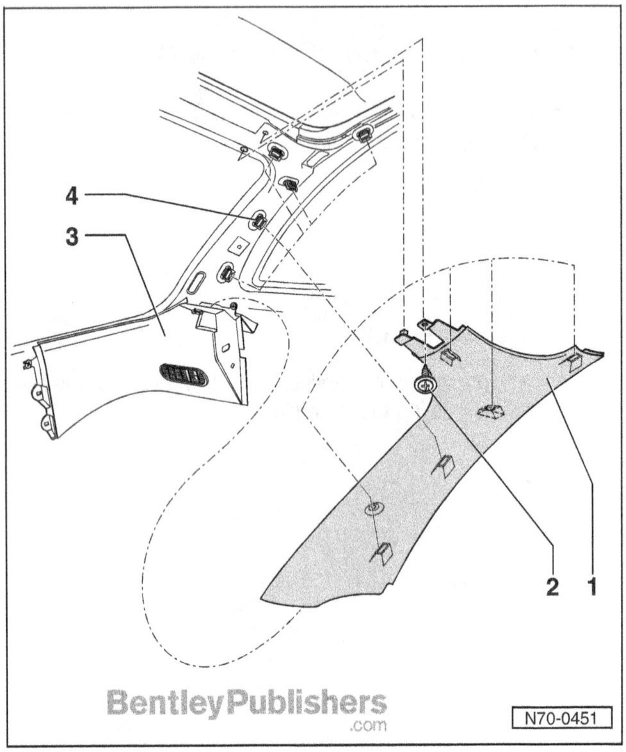

C-pillar trim, removing

- Remove side panel trim as described later.
- Remove D-ring for front and rear seat belts, see **69 Seatbelts, Airbags**.
- Remove B-pillar trim as described earlier.
- Pull center rear trim strip (clipped on) downward off roof frame.
- Remove C-pillar trim mounting screw and pull trim off from top.

Before installing C-pillar trim check condition of retaining clips and replace as necessary.

1. C-pillar trim
2. Screw
3. Rear luggage compartment trim panel
4. Retaining clip

Door panel assembly

NOTE —
Sedan shown, convertible is similar.

1. **Door panel**
   - Sedan shown, convertible similar
2. **Screw**
   - Qty. 3
3. **Screw**
   - Qty. 3
4. **Handle cover**
5. **Window crank**
   - For manual windows
6. **Window switch**
   - For power windows
7. **Window inner slot seal**
8. **Locking knob**
9. **Clip**
   - Qty. 3
10. **Screw**
    - Qty. 3

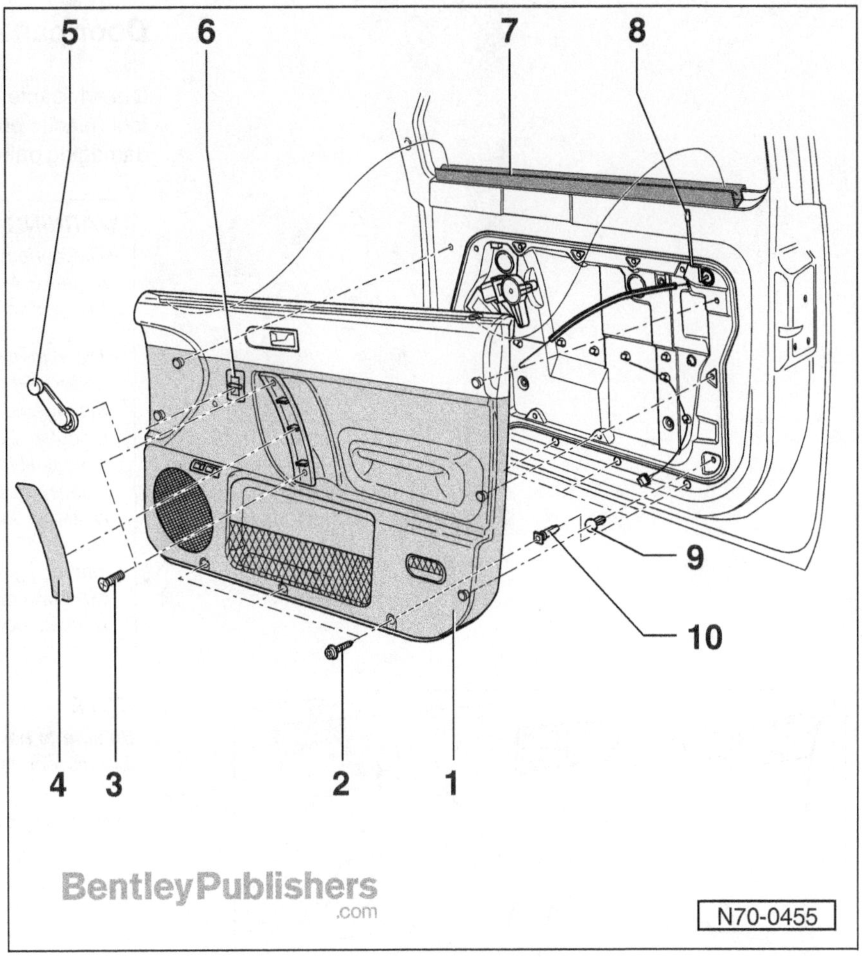

Door panel, removing and installing

It is advisable to use a special non-marring trim removal tool (plastic bone) to pry various trim panels out without damaging paint or other adjacent panels.

> ***WARNING —***
> - *Disconnect the battery ground strap (GND) from negative (–) battery terminal before working on the electrical system.*
> - *Disconnecting the negative (–) battery cable may erase fault codes and basic settings in the engine management and automatic transmission control modules. Some driveability problems may be noticed until the system re-adapts to operating conditions. OBD II readiness codes, which may be required for emissions testing, may also be erased. Convenience electronics (alarm system, interior light control, power locks, mirrors, and windows) may need to be re-set using a diagnostic scan tool.*

NOTE —
Be sure to have the anti-theft radio code on hand before disconnecting the battery.

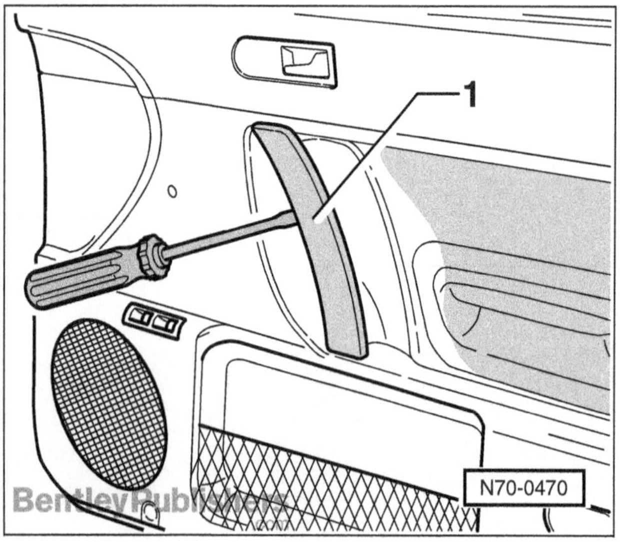

◄ Carefully pry off door handle trim.

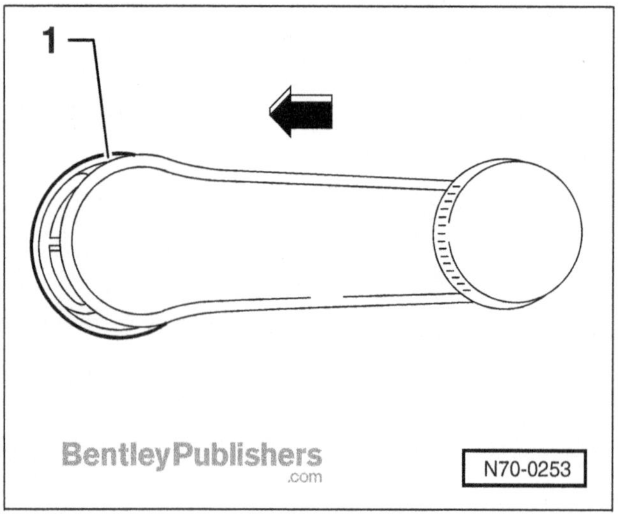

◄ On cars with manual windows, slide spacer (**1**) behind window crank handle in direction of **arrow**. Retaining clip will unlock allowing window crank to be removed.

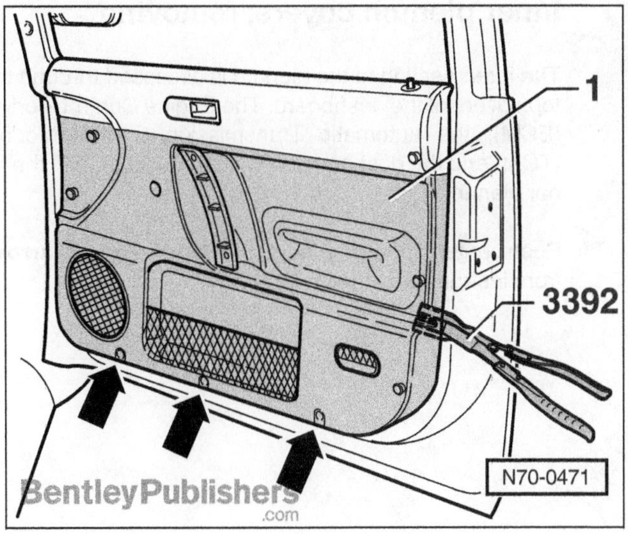

◄ Remove door panel mounting screws (**arrows**) and unclip door panel at sides. Lift door trim upward off window slot.

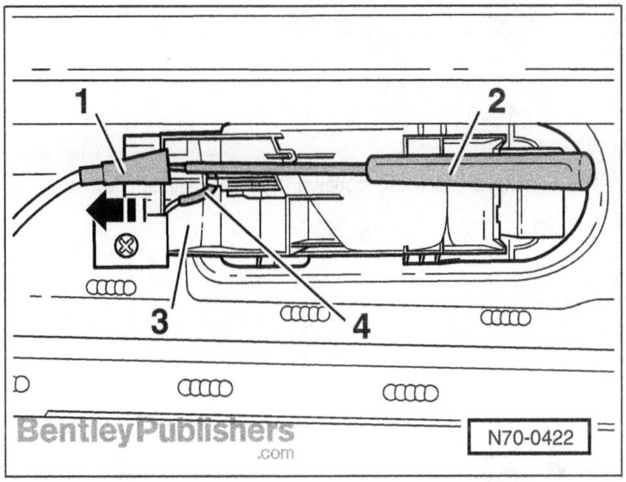

◄ Unclip cable guide (**1**) from door latch with screwdriver (**2**), pull cable out from retainer (**3**) and unhook cable hook (**4**).

– Disconnect harness connectors from door panel switches.

– Install door panel in reverse order of removal, noting the following:
  - Check door trim clips before installing and replace if necessary.
  - After connecting battery, check operation of vehicle equipment (radio, clock, power windows).

Driver's side footwell cover, removing

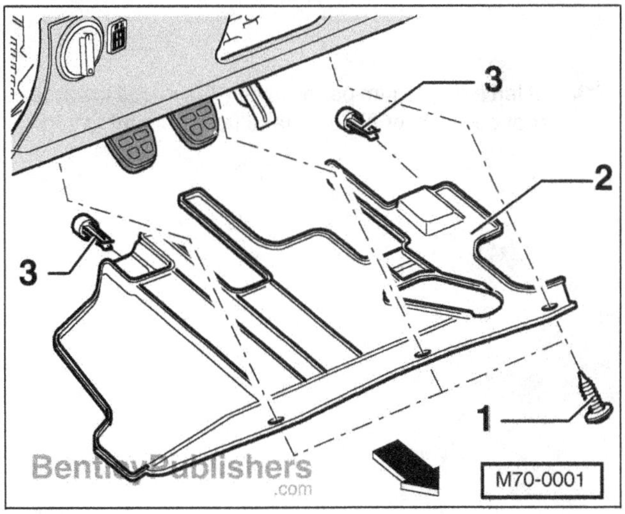

◄ Remove footwell cover mounting screws (**1**). Pull footwell cover (**2**) to rear off retainers (**3**).

> ***NOTE —***
> *The footwell cover was discontinued during the 2006 model year.*

Inner plenum covers, removing

The inner section of the plenum is accessed through the top covers of the dashboard. The Engine Control Module (ECM), the automatic Transmission Control Module (TCM) and the dust and pollen filter are located in the inner plenum.

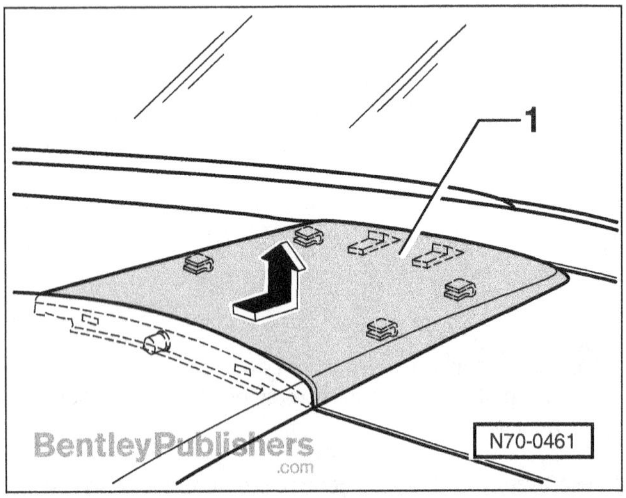

Push center plenum cover (**1**) forward and lift out (**arrow**) from instrument panel.

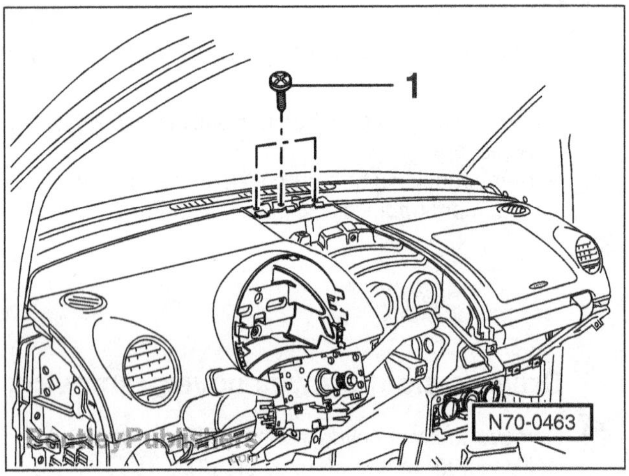

Remove retaining screws (**1**) in center of plenum.

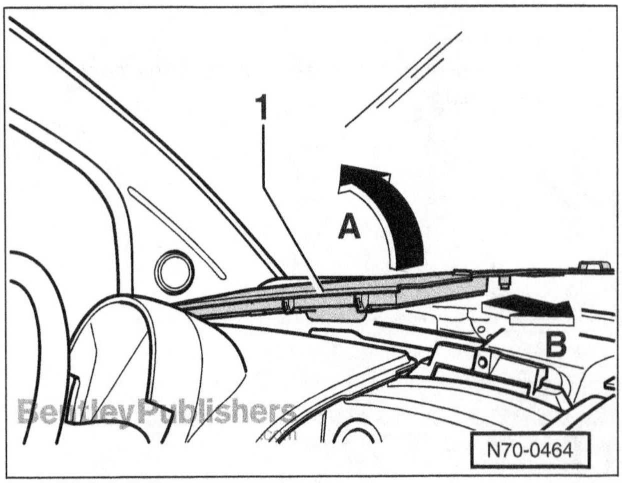

Lift left side plenum panel cover (**1**) and pull towards center of car (**arrow A**) to release from retainer in A-pillar.

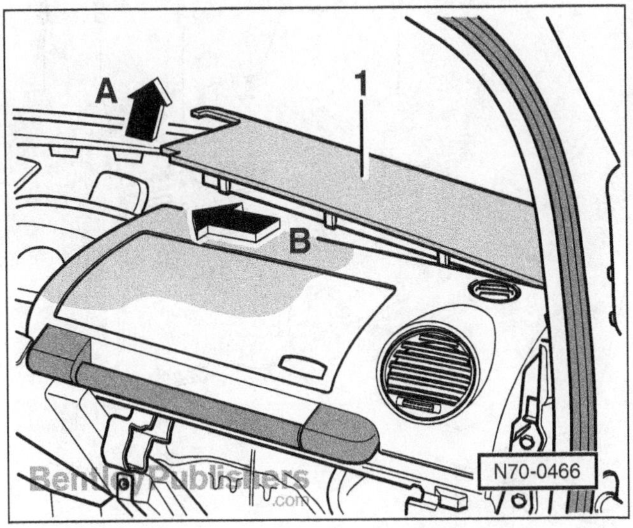

◀ Lift right side plenum panel cover (**1**) and pull towards center of car (**arrows A, B**) to release from retainer in A-pillar.

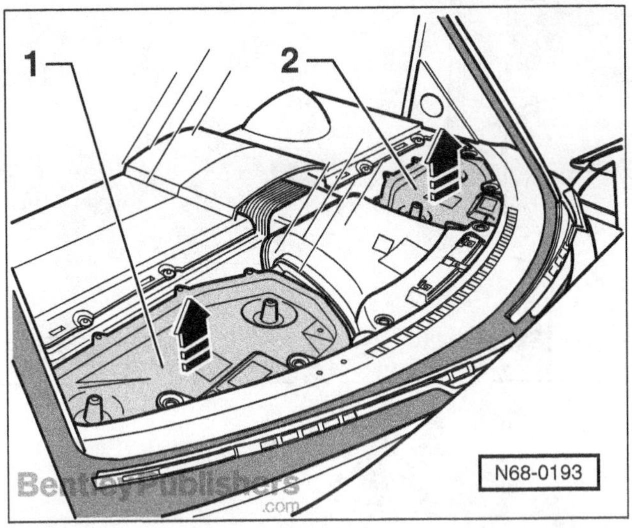

◀ Remove mounting bolts from right and/or left close-out panel and pull panels (**1** and/or **2**) upward out of plenum to remove.

Rear side panels (sedan)

1. **Rear side panel**

   Removing:

   - Remove rear seat cushion and backrest
   - Pull off inner door seal from lower B-pillar
   - Pull side panel off at clips (items **8** and **5**)
   - Lift upward at clip (item **7**)

2. **Rear seat belt with tensioner**

3. **Clip**
   - Attached to side window flange
   - May be necessary to attach to window flange for side panel installation

4. **Securing point**
   - For clips (item **3**)

5. **Clip**
   - Check for damage and replace if necessary

6. **Inner door seal**
   - Check for proper position after installing side panel

7. **Clip**

8. **Clip**

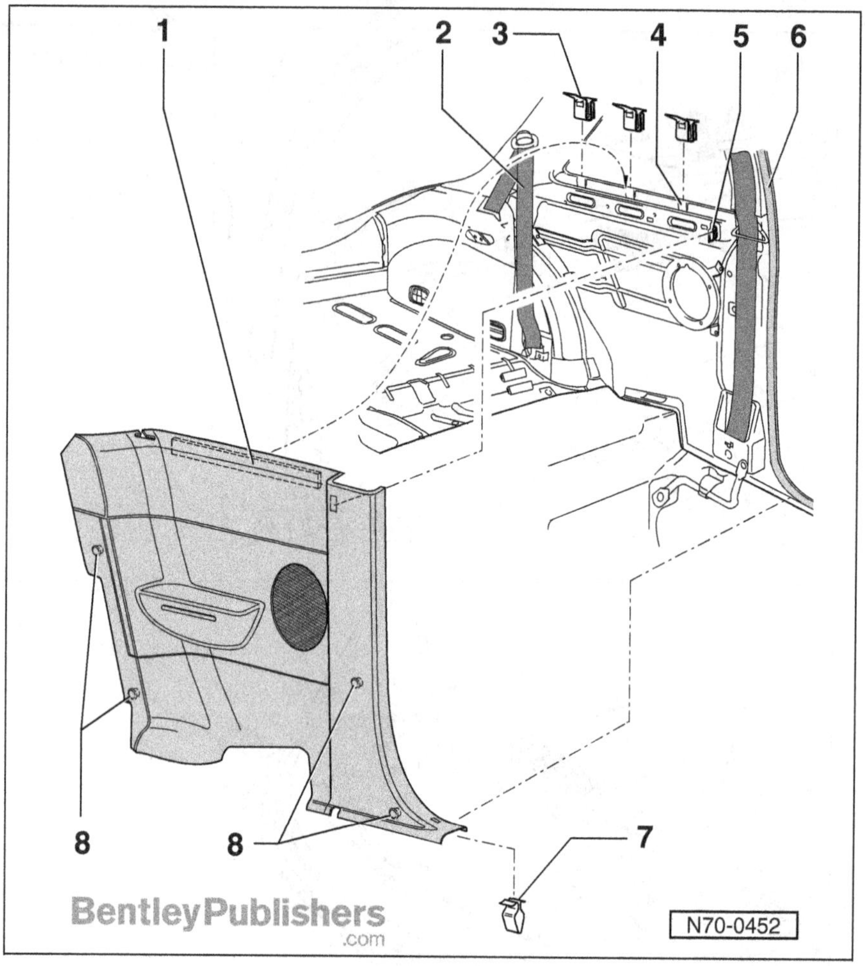

Rear side panels (convertible)

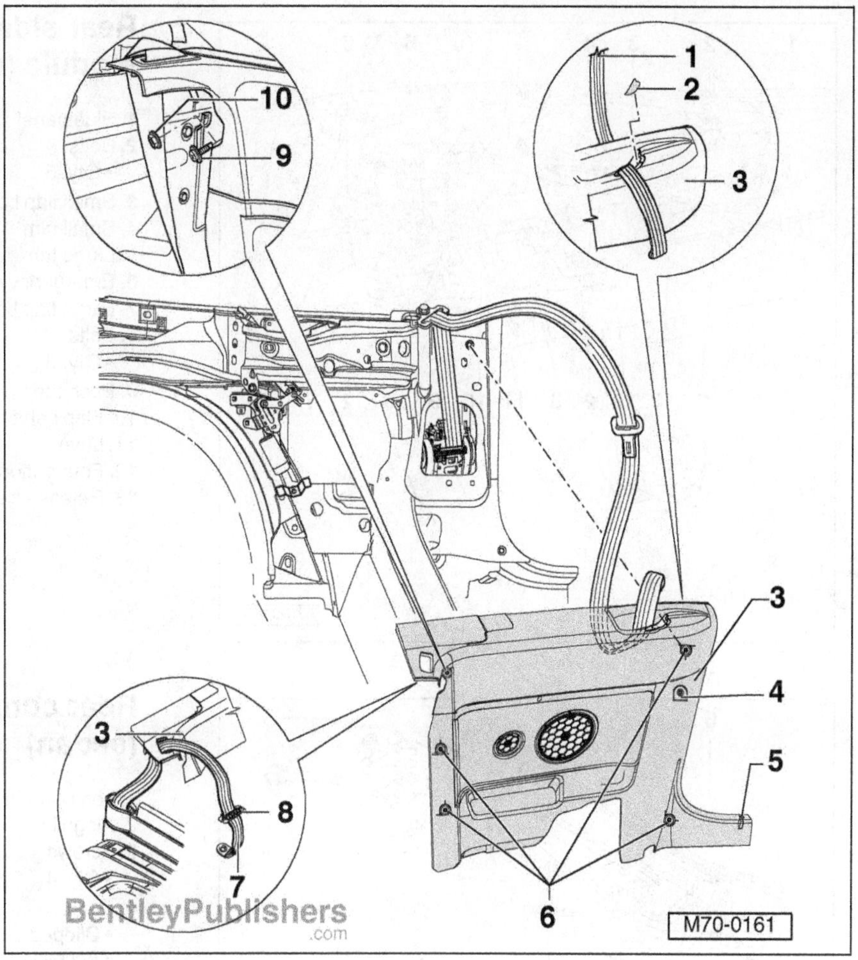

1. **Front seat belt with tensioner**
2. **Cover**
3. **Rear side panel**

   Removing:

   - Remove rear seat cushion and backrest.
   - Remove bolt (**9**) or nut (**10**) as applicable.
   - Pull off inner door seal from lower B-pillar.
   - Pull side panel off at clips (**5** and **6**) and remove upward.
   - Guide front seatbelt (**1**) and rear seatbelt (**7**) through side panel.
   - Disconnect harness connectors from convertible top servo motor and loudspeaker.
   - Unclip release cable from ball head pin, pry off spring clip and remove release cable from retaining plate. See **Cable detail** below.
   - Remove cover (**2**) and guide seatbelt (**1**) through opening.
   - Remove rear seatbelt relay. See **69 Seatbelts, Airbags**.
   - Unclip cover (**8**) and guide seatbelt (**7**) through opening.

   Installing:

   - Install in reverse order of removal.
4. **Guide**
5. **Clip**
   - Replace if damaged
6. **Clip**
   - Replace if damaged
7. **Rear seatbelt**
8. **Cover**
9. **Bolt**
   - Early production
10. **Nut**
    - Late production

Cable detail

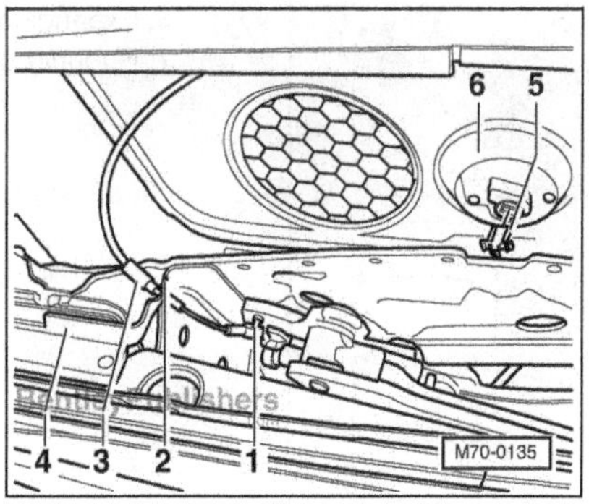

◄ Unclip release cable (**3**) from ball head pin (**1**).

– Pry off spring clip (**2**) and remove release cable (**3**) from retaining plate (**4**).

– Disconnect harness connector (**5**) for loudspeaker (**6**).

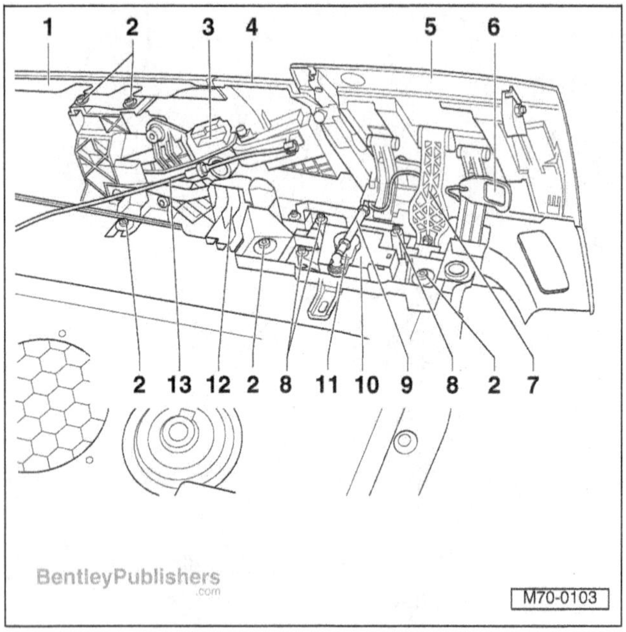

Rear side panel, flap positioning module (convertible)

1. Side panel
2. Bolts
   - Qty. 5
3. Small flap base
4. Small trim
5. Large trim
6. Emergency release
7. Large flap base
8. Bolts
   - Qty. 3
9. Push rod
10. Flap servomotor
11. Lever
12. Foundation frame for flap positioning module
13. Release cable

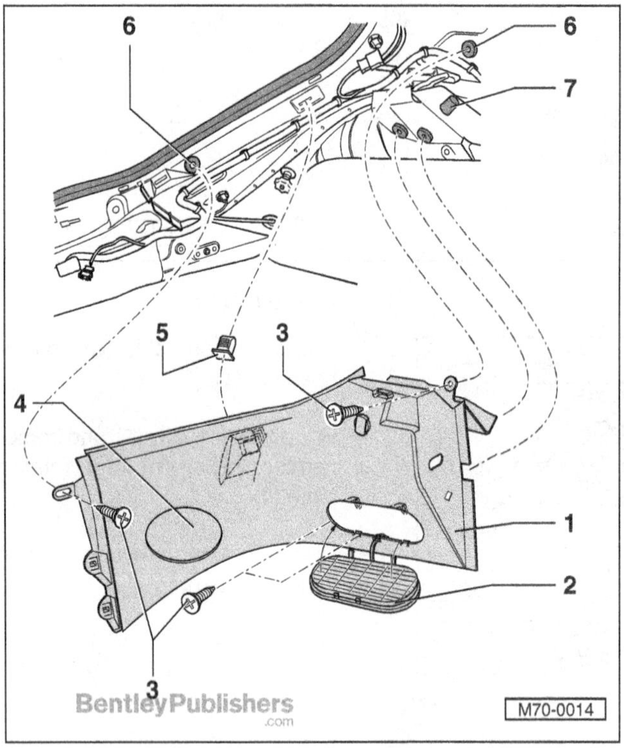

Rear compartment trim, left side (sedan)

1. Left rear trim panel
2. Air grille
3. Screws
   - Qty. 4
4. Cover
   - Clipped on
   - Access to tail light
   - Access to emergency fuel filler lid unlock
5. Clip
6. Expanding nut
7. Rubber buffer

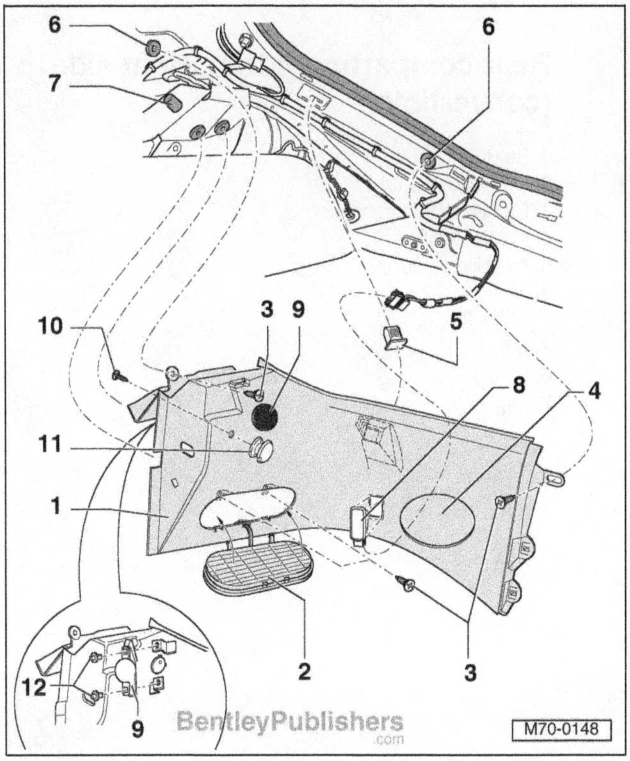

Rear compartment trim, right side (sedan)

1. Right rear trim panel
2. Air grille
3. Screws
   - Qty. 3
4. Cover
   - Clipped on
   - Access to tail light
5. Clip
6. Expanding nuts
7. Rubber buffer
8. Interior light
9. Speaker
   - Only for vehicles with park assist system
10. Bolt
    - Only for vehicles with luggage compartment net
11. Pin nut
    - Only for vehicles with luggage compartment net
12. Rivets
    - Qty. 2

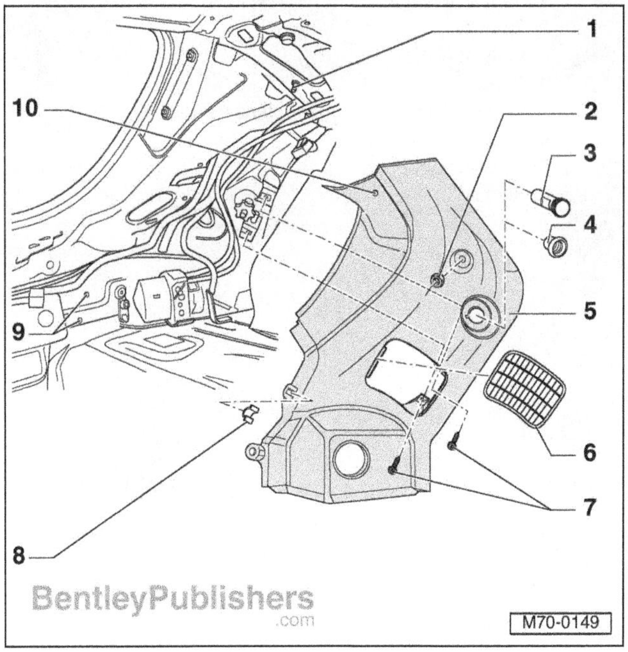

Rear compartment trim, left side (convertible)

1. Stud
2. Combination nut
3. 12V power outlet
4. Plastic plug
5. Left rear trim panel
6. Air grille
7. Screws
   - Qty. 2
8. Clip
9. Stud
10. Clip

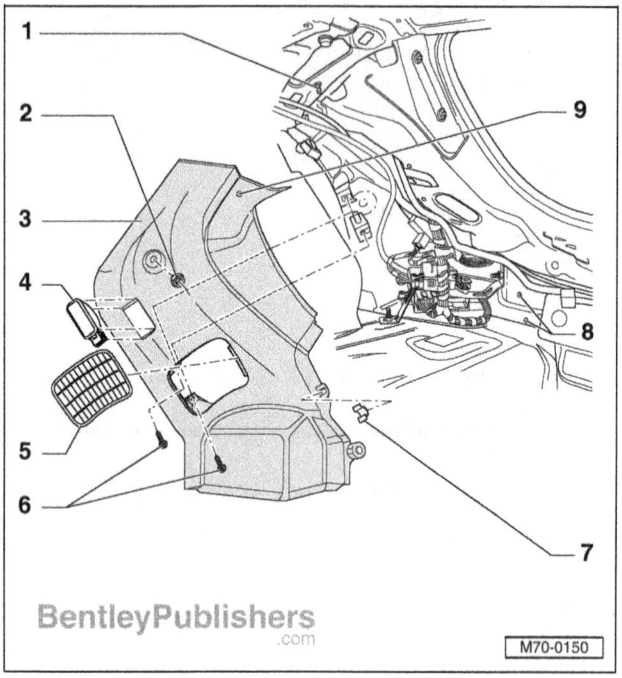

Rear compartment trim, right side (convertible)

1. Stud
2. Combination nut
3. Right rear trim panel
4. Interior light
5. Air grille
6. Screws
   - Qty. 2
7. Clip
8. Stud
9. Clip

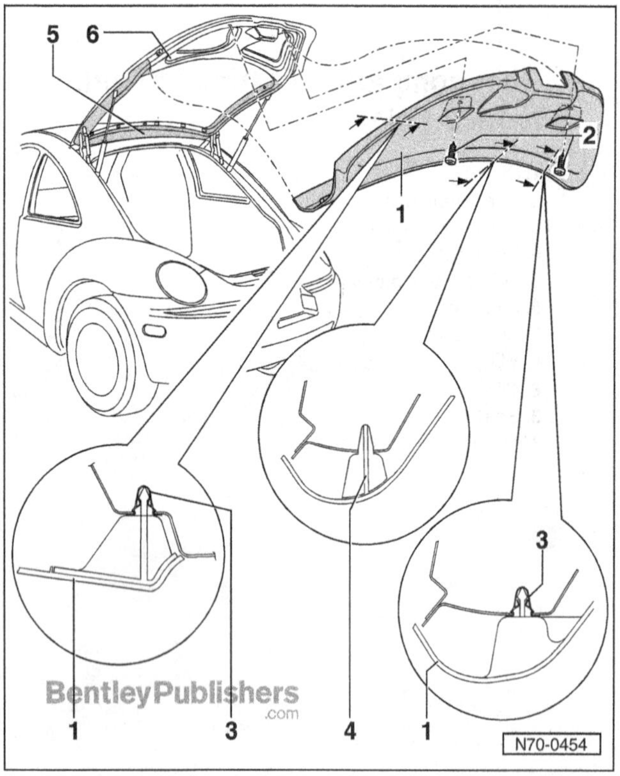

Rear lid lower trim (sedan)

1. Rear lid lower trim
2. Screws
   - Qty. 2
3. Clips
   - Check for damage and replace as necessary
4. Stud
5. Rear lid upper trim
6. Rear lid

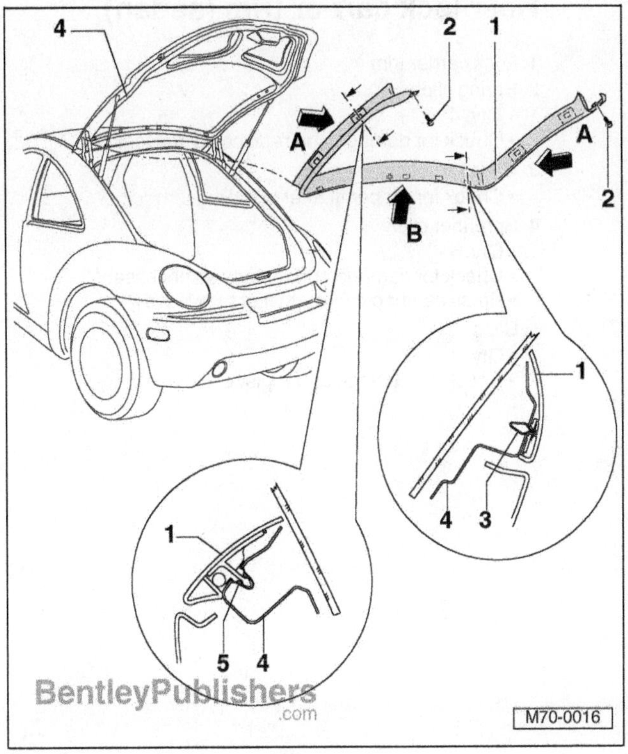

Rear lid upper trim (sedan)

1. Rear hatch upper trim
- To remove, first remove rear hatch lower trim
- Pry out plug (item **2**)
- Unclip trim at sides (**arrow A**) and then top (**arrow B**)

2. Screw
- Qty. 2

3. Clip
- Check for damage and replace as necessary

4. Rear hatch

5. Clip
- For mounting stud to trim
- Check for damage and replace as necessary

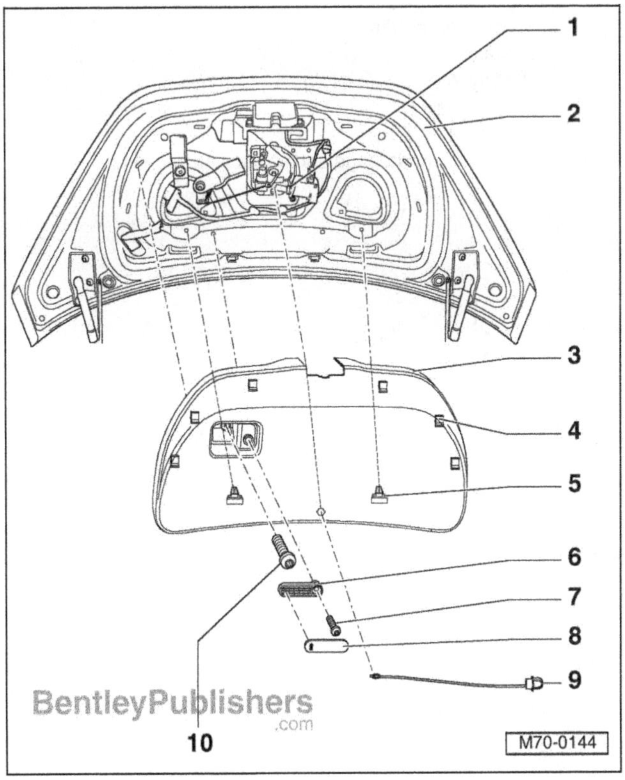

Rear lid trim (convertible)

1. Lid lock

2. Rear lid

3. Rear lid trim

4. Clips
- Qty. 6
- Check for damage and replace as necessary

5. Clips
- Qty. 2
- Installed on vehicles from 12/2004
- Check for damage and replace as necessary

6. Inside release handle

7. Bolt

8. Cover

9. Plug with plastic cable
- Installed on vehicles from 12/2003

10. Bolt

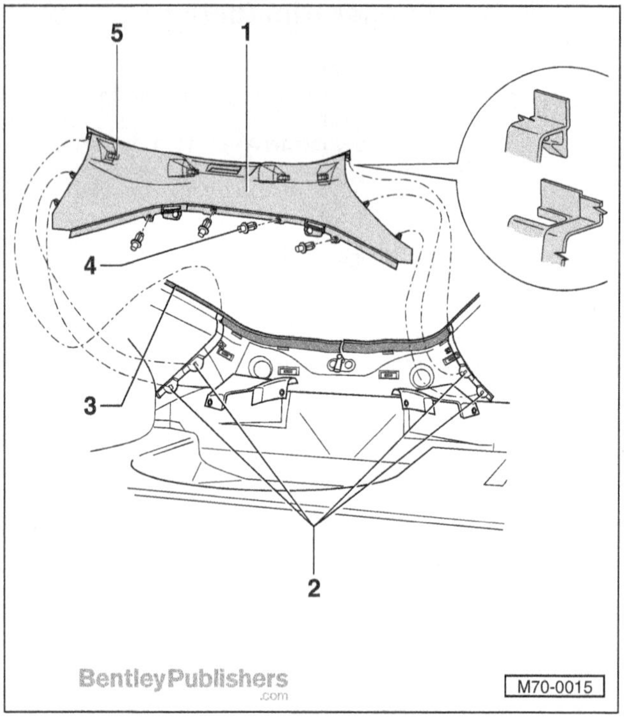

Rear lock carrier trim (sedan)

1. Lock carrier trim
2. Spring clips
   - Qty. 4
   - Check for damage and replace as necessary
3. Rear compartment seal
   - Check for proper fit after installing trim
4. Spreader clips
   - Qty. 4
   - Check for damage and replace as necessary
   - Press center pin through first to remove
5. Clips
   - Qty. 4
   - Check for damage and replace as necessary

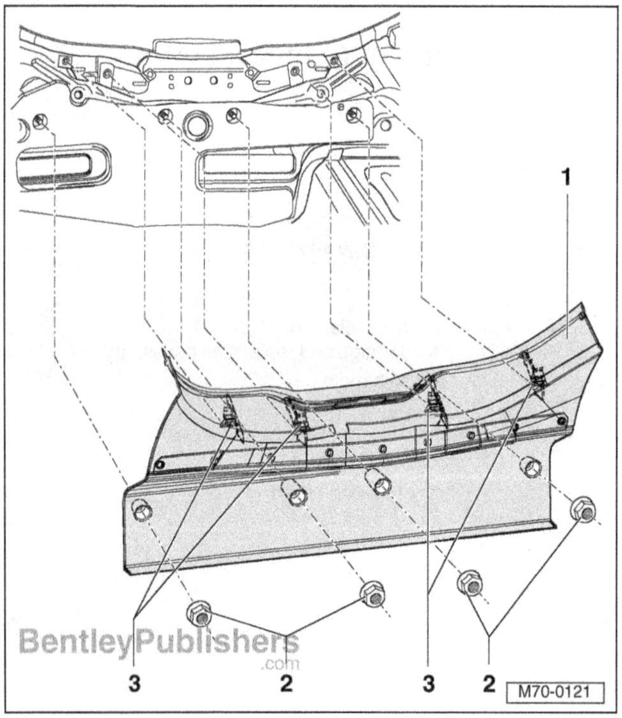

Rear lock carrier trim (convertible)

1. Lock carrier trim
2. Nuts
   - Qty. 4
3. Spring clips
   - Qty. 4
   - Check for damage and replace as necessary

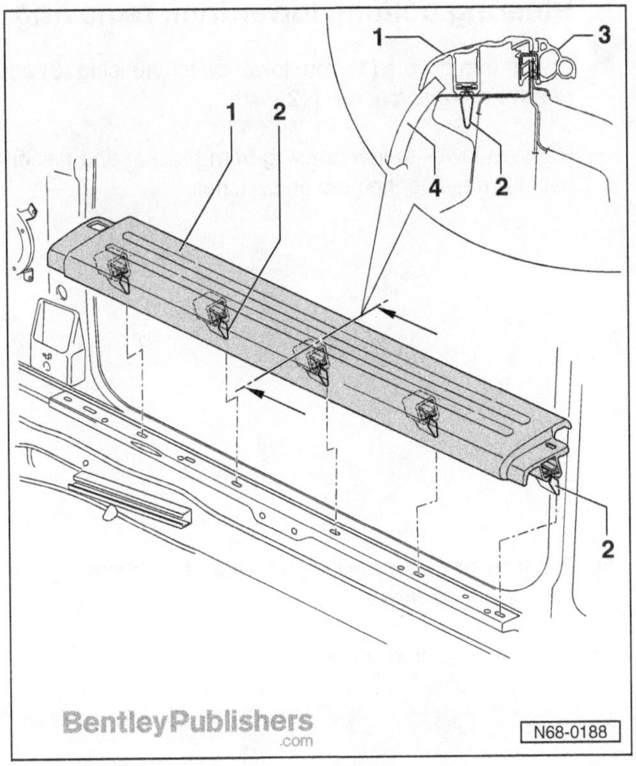

Sill panel trim (sedan)

1. Sill panel trim
   - Removing:
     Remove lower A-pillar and side panel trim first
     Unclip sill panel trim by pulling upward
2. Clip
   - Qty. 5
   - Check for damage and replace as necessary
3. Door inner seal
   - Make sure seal is seated correctly after installing sill panel trim
4. Carpet

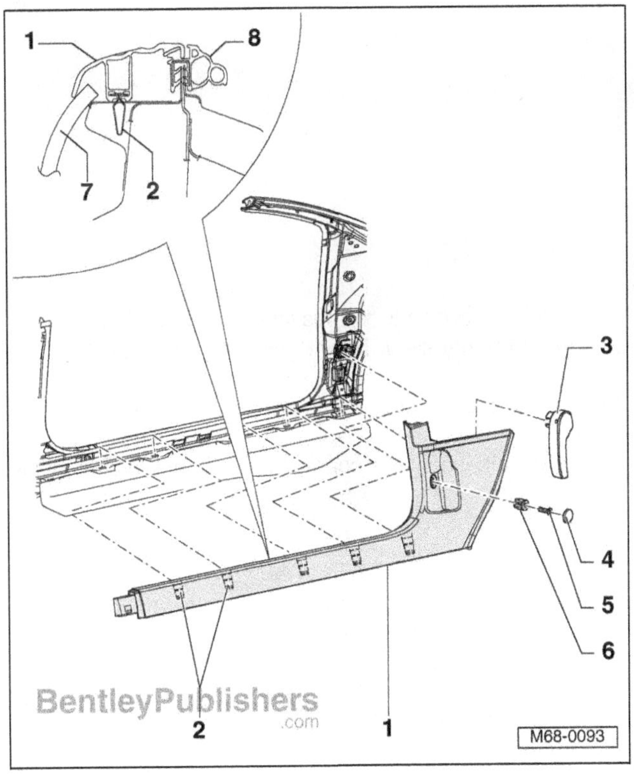

Sill panel trim (convertible)

1. Sill panel trim
   - Removing:
     Remove lower A-pillar and side panel trim first
     Unclip sill panel trim by pulling upward
2. Clips
   - Qty. 6
   - Check for damage and replace as necessary
3. Hood release lever
4. Cap
5. Bolt
6. Expanding nut
7. Carpet
8. Door inner seal
   - Make sure seal is seated correctly after installing sill panel trim

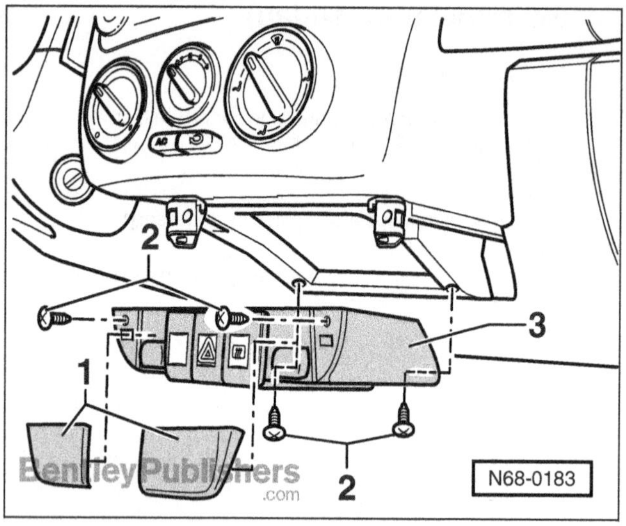

Steering column lower trim, removing

- Pry off trim covers (**1**) from lower switch housing (**3**) and remove mounting screws (**2**).

- Remove lower switch housing from bracket and disconnect harness connectors on switches.

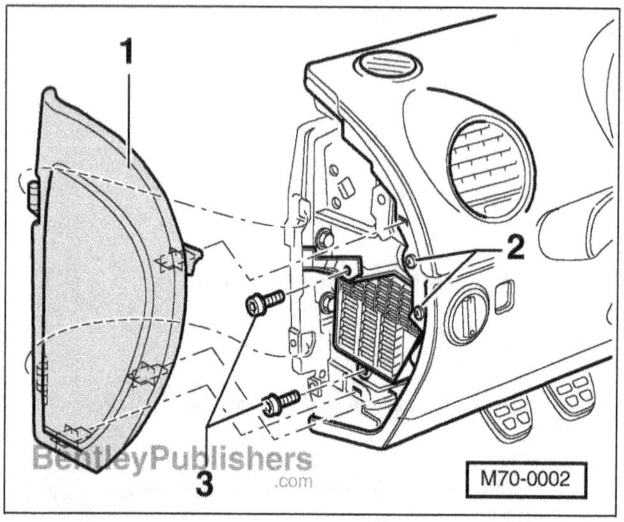

- Pull fuse box cover (**1**) on left side of instrument panel from retaining clips.

- Remove mounting screws (**2**).

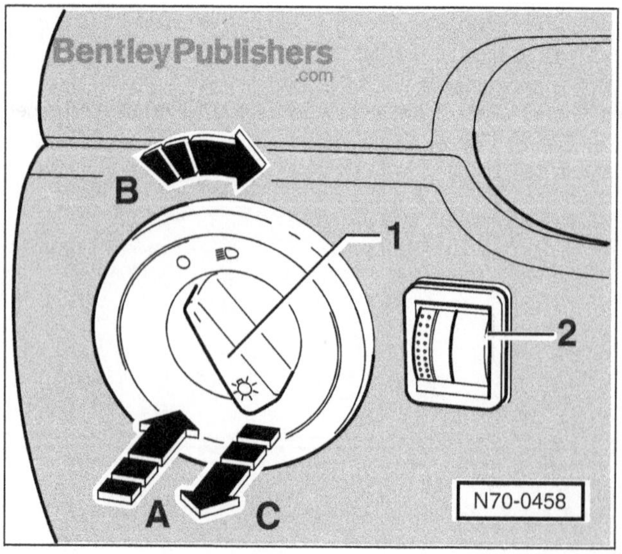

- Turn light switch to “0” position, push knob in (**arrow A**), turn to right (**arrow B**) and pull out (**arrow C**).

- Unclip and remove instrument panel dimmer switch (**2**).

- Disconnect harness connectors and set switches aside.

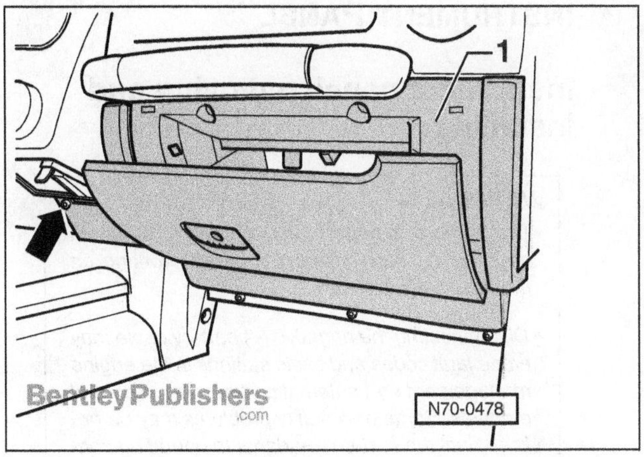

◀ Remove screw (**arrow**) under cap on left side of glove box.

– Remove screws from steering column lower trim. Unclip top of trim and remove from instrument panel.

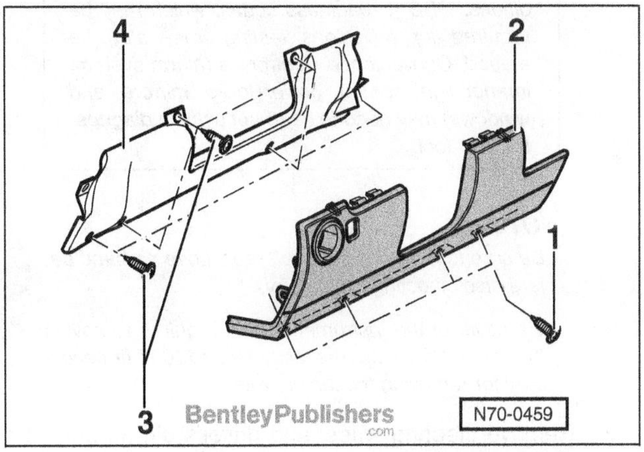

◀ Remove mounting screws (**1**) and trim panel (**2**),

– Remove mounting screws (**3**) and kneebar (**4**).

Tunnel support trim, removing

– Remove center console as shown earlier.

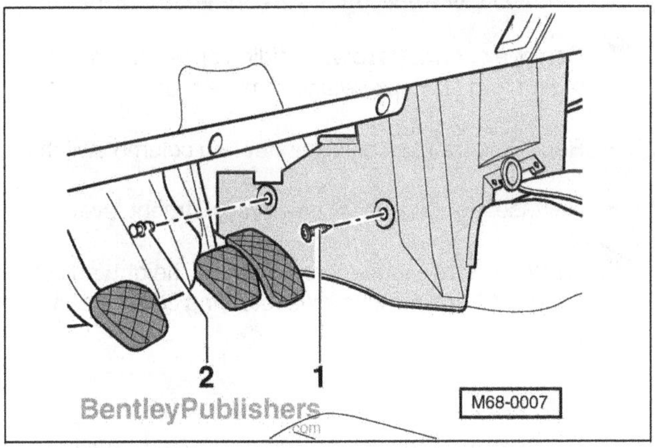

◀ Remove screw (**1**) and clip (**2**) from left side trim panel.

◀ Remove screw (**1**) on right side of trim panel (**2**) and pull panel toward rear (**arrow**) to remove.

INSTRUMENT PANEL

Instrument panel, removing and installing

> ***WARNING —***
> - *Disconnect the battery Ground strap (GND) from negative (–) battery terminal before working on the electrical system.*
> - *Disconnecting the negative (–) battery cable may erase fault codes and basic settings in the engine management and automatic transmission control modules. Some driveability problems may be noticed until the system re-adapts to operating conditions. OBD II readiness codes, which may be required for emissions testing, may also be erased. Convenience electronics (alarm system, interior light control, power locks, mirrors, and windows) may need to be re-set using a diagnostic scan tool.*

NOTE —
- *Be sure to have the anti-theft radio code on hand before disconnecting the battery.*
- *Removal of the instrument panel requires removing the radio. Volkswagen special tool T10057 is necessary for removing the factory radio.*

– Remove steering wheel and driver's airbag unit as described in **48 Steering** and **69 Seatbelts, Airbags**.

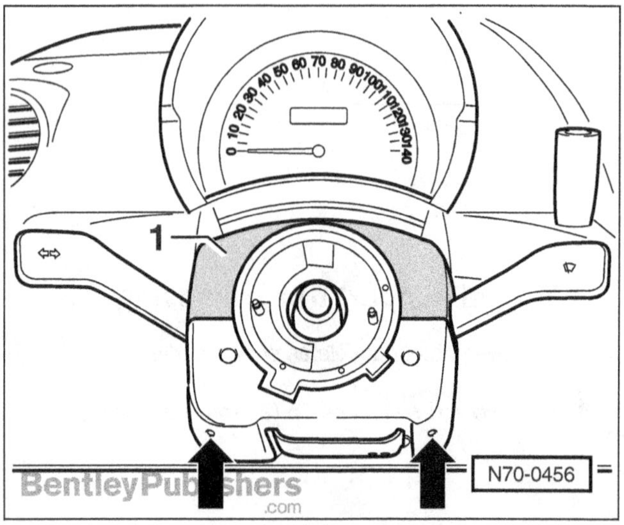

◄ Remove screws (**arrows**) holding upper steering column switch trim (**1**) in place and remove trim.

– Remove screws from lower steering column switch trim.

– Release steering wheel height adjustment lever.

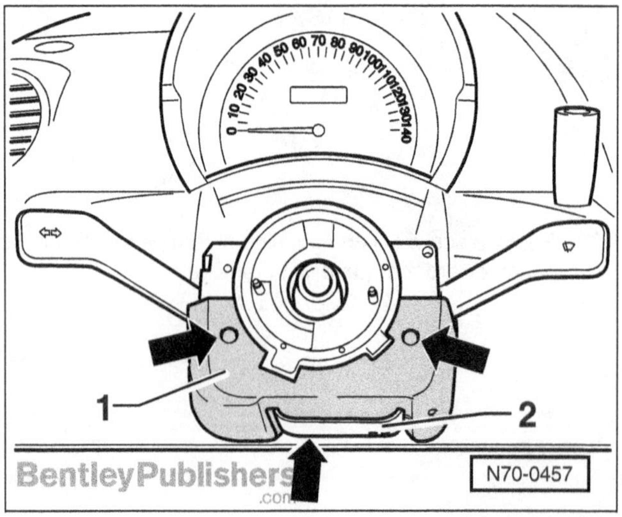

◄ Remove screws (**arrows**), release height adjustment handle (**2**) and remove lower steering column switch trim (**1**).

NOTE —
Remove the height adjustment handle (two screws) for easier removal of the lower trim.

– Remove bolt behind steering column switch assembly.

– Disconnect harness connectors from steering column switch assembly.

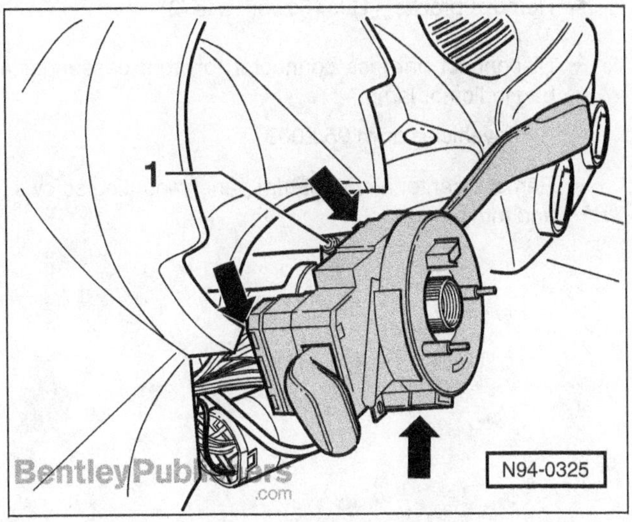

◀ Remove steering column switch assembly. Remove bolt (**1**) and disconnect harness connectors (**arrows**).

– Remove driver's side footwell cover as shown earlier.

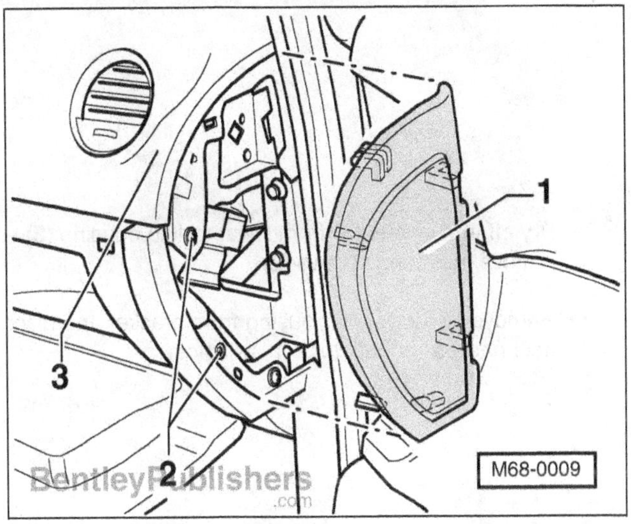

◀ Pull off right side instrument panel cover (**1**) and remove mounting screws (**2**) for glovebox trim.

– Remove lower switch housing as shown earlier.

– Remove inner plenum covers from top of instrument panel. See **Inner plenum covers, removing**, given earlier.

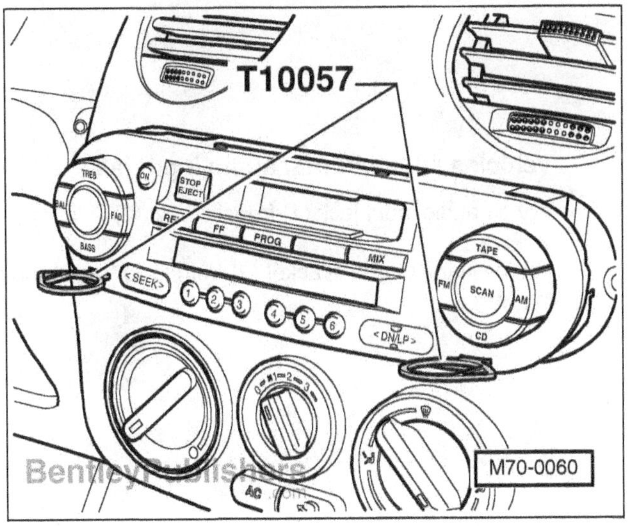

◀ Slide Volkswagen radio release special tools T10057 into slots on front of radio until they engage.

– Pull radio out of instrument panel using grip rings of release tools and disconnect harness connectors on back of radio.

NOTE —
- *Do not push radio release tools to the side or tilt them.*
- *Press the locating tabs on the side of the radio inward to remove the release tools.*
- *Do not install radio with release tools in place.*

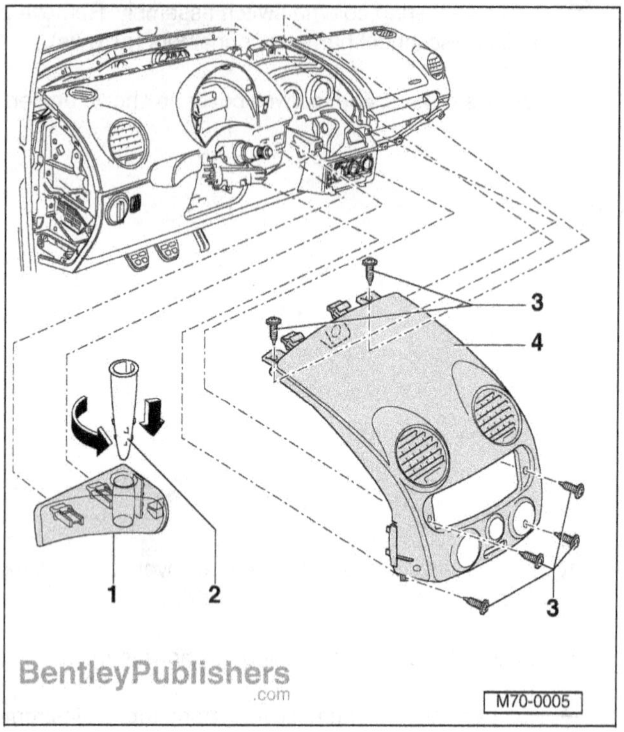

◀ Remove bracket (**1**) for flower vase (**2**).

– Disconnect harness connector for front passenger air-bag indicator lamp.

  - For vehicles from 05/2003

– Remove center air outlet trim panel mounting screws (**3**) and trim panel.

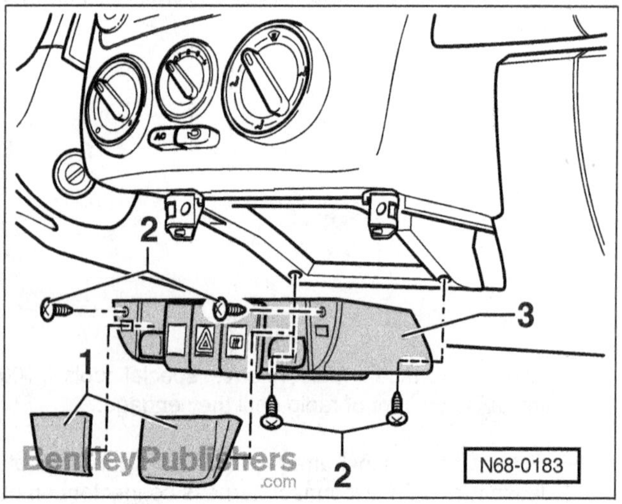

◀ Pry off trim covers (**1**) from lower switch housing (**3**) and remove mounting screws (**2**).

– Remove lower switch housing from bracket and disconnect harness connectors on switches.

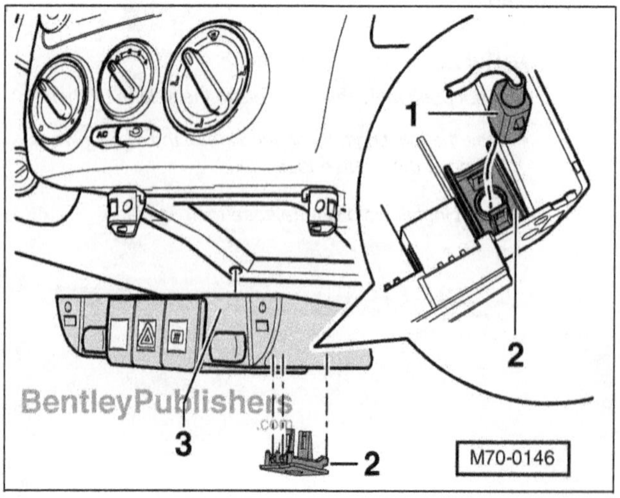

Vehicles with external audio input jack

◀ Pry off audio input jack (**1**) from bracket (**2**)

– If necessary, pry off bracket (**2**) from lower switch housing (**3**).

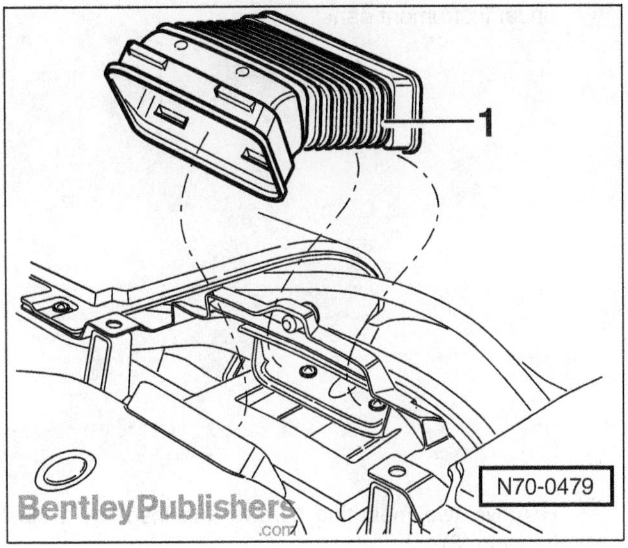

Continued for all vehicles

◀ Remove air duct (**1**) from center of instrument panel.

– Remove glovebox and disconnect harness connector for light, see **Glove box, removing**.

– Disconnect harness connector for key switch for airbag, if applicable.

– Remove fuse box cover on left side of dash and mounting screws as shown earlier.

– Remove light switch and instrument dimmer switch as shown earlier.

– Remove lower steering column trim panel and kneebar as shown earlier.

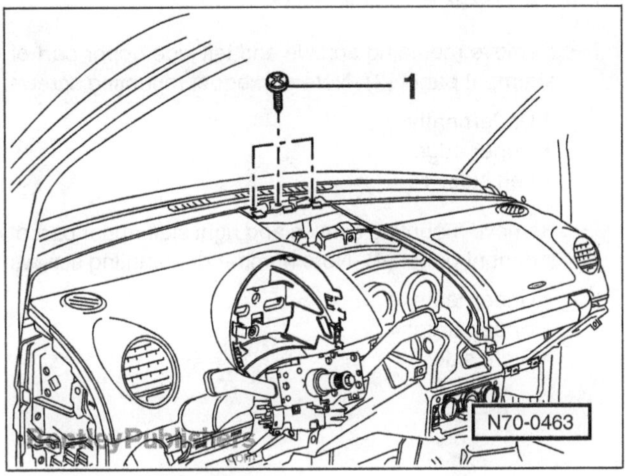

◀ Remove mounting screws (**1**) on top of instrument panel.

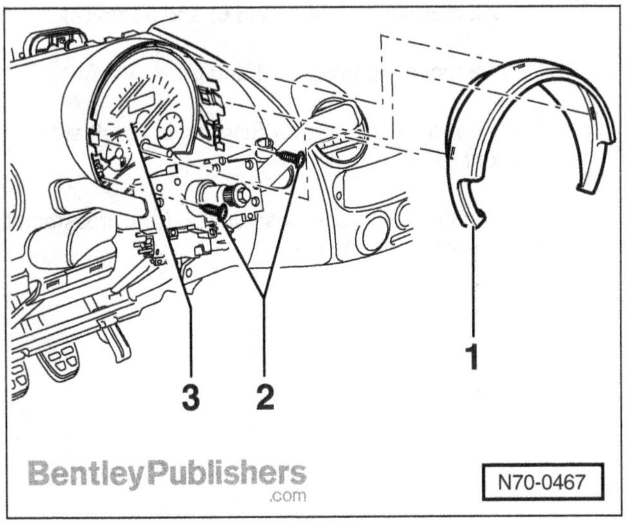

◀ Unclip trim (**1**) from instrument cluster, remove cluster mounting screws (**2**) and cluster (**3**). Disconnect multi-pin connector on back of cluster.

– Remove instrument panel handle as shown earlier.

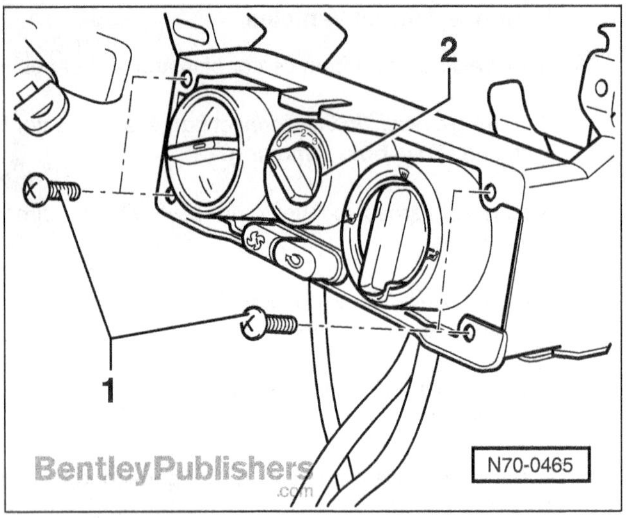

Remove mounting screws (**1**) for heating and ventilation control unit (**2**). Press control unit with attached cables under instrument panel.

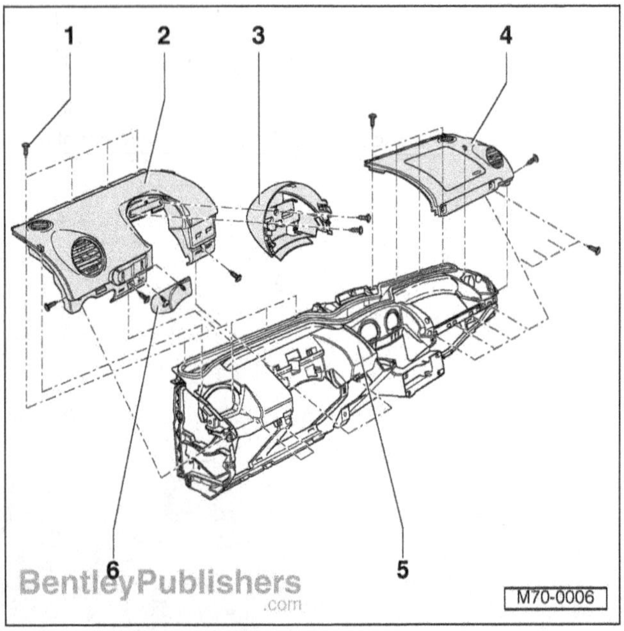

Remove two mounting screws and instrument cluster housing (**3**).

– Remove small trim panel (**6**).

– Remove mounting screws and left side upper part of instrument panel (**2**). Note number of mounting screws:

- Underneath: 2
- Upper edge: 4
- Left side: 1

– Remove mounting screws and right side upper part of instrument panel (**4**). Note number of mounting screws:

- Underneath: 4
- Upper edge: 4
- Right side: 1

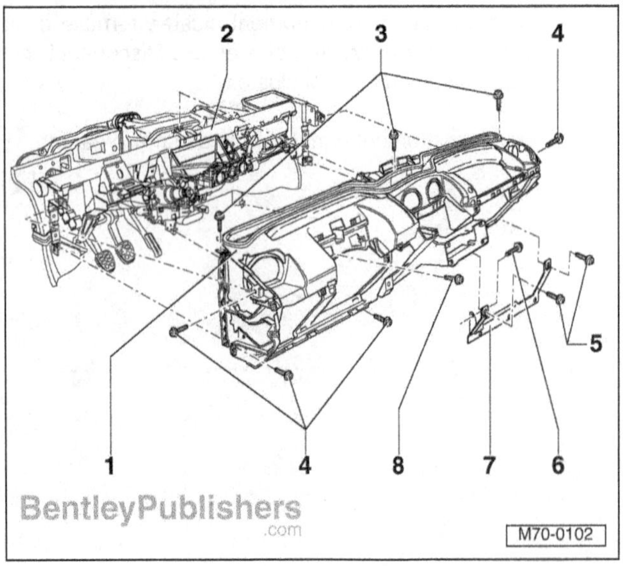

Remove mounting bolts (**5** and **6**) and frame (**7**).

– Remove mounting bolts (**3**, **4** and **8**).

– Remove mounting bolts for fuse panel at left side of instrument panel.

– Pull off instrument panel base unit (**1**) from crossmember (**2**) and remove from vehicle.

– Install instrument panel in reverse order of removal, noting the following:
  - Always replace steering wheel mounting bolt when installing steering wheel.
  - After connecting battery check operation of vehicle equipment (radio, clock, power windows).
  - After connecting battery, readiness code for the ECM will need to be re-set using a diagnostic scan tool.
  - If airbag malfunction indicator lamp (MIL) signals a malfunction after installing instrument panel, the Diagnostic Trouble Code (DTC) memory must be erased and checked again with a diagnostic scan tool.

QUALITY REVIEW

When you have finished working under the hood and around other areas of the vehicle, it is advisable to take a moment to quality check or review your work. This helps to insure that the operation or repair has been completed properly with all affected systems functioning within normal parameters. These may include the following:

- Ensure that all cable ties and clamps that were removed as part of the repair are replaced.
- Ensure that all fasteners and hardware were replaced and torqued as specified.
- Make sure that all other components involved in the repair are positioned correctly, properly torqued and function properly.
- Make sure that all tools, shop cloths, fender covers and protective tapes are removed.
- Clean grease and fingerprints from painted surfaces, steering wheel, upholstery and shifter.
- Make sure that seatbelt and airbag warning lights function correctly.
- Unlock anti-theft radio and reset the clock.

72 Seats

FRONT SEATS

Both driver and passenger front seats are equipped with airbags mounted in the seatback frame. These airbags are designed to deploy under certain side impact conditions and are triggered by separate sensors mounted to front floor crossmembers under the front seats. Before removing front seats, a safety device (VAS 5094) must be installed to prevent accidental deployment of the airbag units.

> ***WARNING —***
> - *Observe all cautions, warnings and notes before starting repairs involving airbag systems, see* **69 Seatbelts, Airbags**.
> - *Volkswagen of America specifically warns against the installation of aftermarket upholstery on any vehicle equipped with side airbags. The factory-installed upholstery is designed to separate in specific places at specific rates, and in specific directions. Installation of non-factory upholstery including, but not limited to, "beads" and "sheepskins", may cause seat mounted airbags to deploy when they are not supposed to; fail to deploy when they should; or to deploy in some manner other than designed. This is a safety hazard and could result in serious injury or death to occupants of the vehicle.*

72

Front seat, removing and installing

- Disconnect the battery ground (–) strap before working on the electrical system.

> ***CAUTION —***
> *Disconnecting the negative (–) battery cable may erase fault codes and basic settings in the engine management and automatic transmission control modules. Some driveability problems may be noticed until the system re-adapts to operating conditions. OBD II readiness codes, which may be required for emissions testing, may also be erased. Convenience electronics (alarm system, interior light control, power locks, mirrors, and windows) may need to be re-set using a diagnostic scan tool.*

NOTE —
- *The removal and installation procedures may have to be modified slightly depending on equipment variations.*
- *Obtain the radio code before disconnecting the battery.*

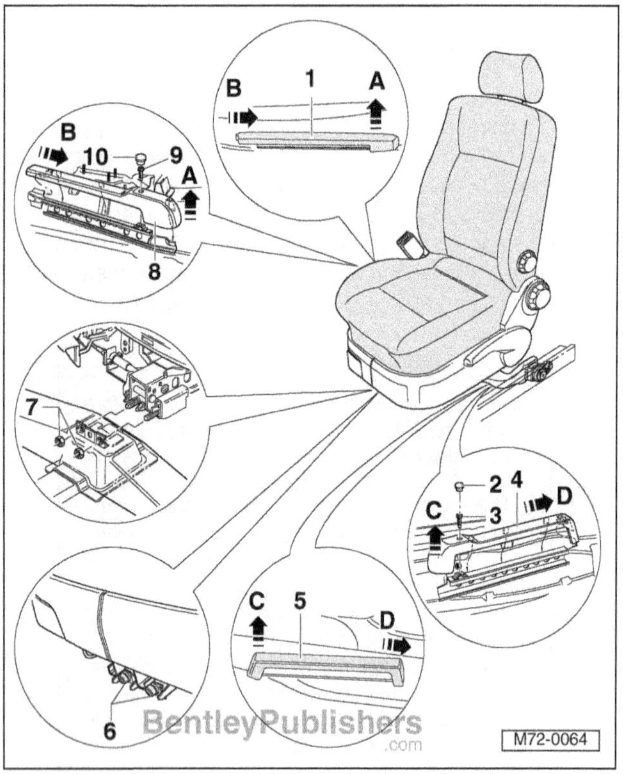

◄ Slide seat forward.

Early version

- Unclip cover strip (**1**) from inner seat rail in direction of **arrow A** and remove toward rear (**arrow B**).
- Unclip cover strip (**5**) from outer seat rail in direction of **arrow C** and remove toward rear (**arrow D**).

Late version

- Unclip cap (**10**) from inner seat rail and remove bolt (**9**).
- Unclip cover strip (**8**) at rear (**arrow A**) and remove toward rear (**arrow B**).
- Unclip cap (**2**) from outer seat rail and remove bolt (**3**).
- Unclip cover strip (**4**) at rear (**arrow C**) and remove toward front (**arrow D**).

Continued for all versions

- Slide seat to rear.
- Remove mounting bolts (**6**) or nuts (**7**).

Tightening torque
- Front seat to body . . . . 23 Nm (17 ft-lb)

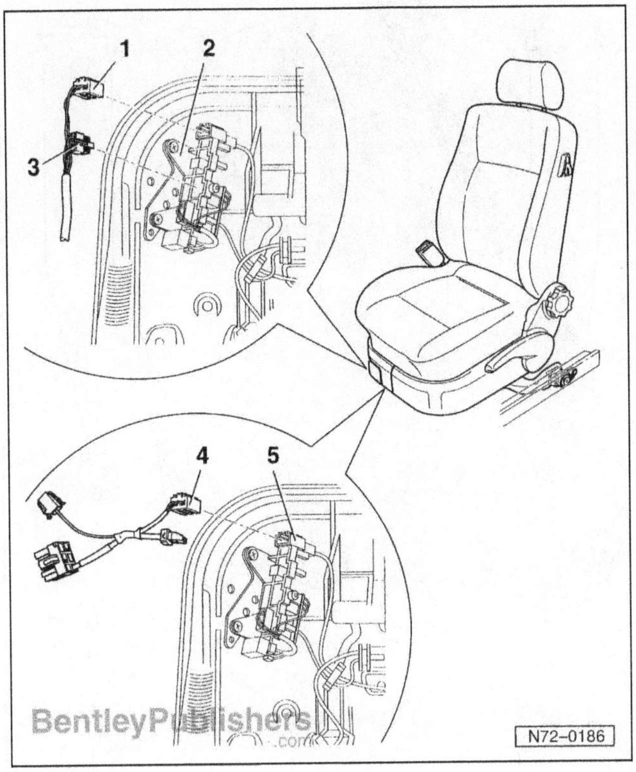

- Push seat toward rear out of guide rails.

< Separate side airbag wiring harness connector (**1**) at seat connector station (**2**).

- Separate connector (**3**) for seat heating (as applicable) and seat adjustment connector (as applicable).

- Insert connector (**4**) of special tool VAS 5094 into connector housing (**5**).

> ***WARNING —***
> *Failure to install VAS 5094 could result in deployment of the side airbag.*

- Install in reverse order of removal, noting the following:
  - After all components are installed, switch ignition on.
  - Connect battery ground strap.
  - After connecting the battery ground (–), ensure proper operation of vehicle equipment (radio, clock, power windows).

> ***WARNING —***
> *Make sure the passenger compartment is not occupied before connecting the battery ground strap.*

Front seat assembly

1. **Seat trim, right**
2. **Screw**
3. **Cover**
4. **Seat**
5. **Seatback**
   - Removal:
     Remove seat from vehicle
     Remove adjusting knob (item 7)
     Remove adjusting lever (item 10)
     Remove trim (items 1 and 11)
     Unclip and remove covers (item 3 and 6)
   - See **Front seatback, removing and installing**
6. **Cover**
7. **Adjusting knob for seatback recline**
8. **Cap**
   - Qty. 2
9. **Screw**
   - Qty. 2
10. **Adjusting lever for seat height**
11. **Seat trim, left**
12. **Screw**

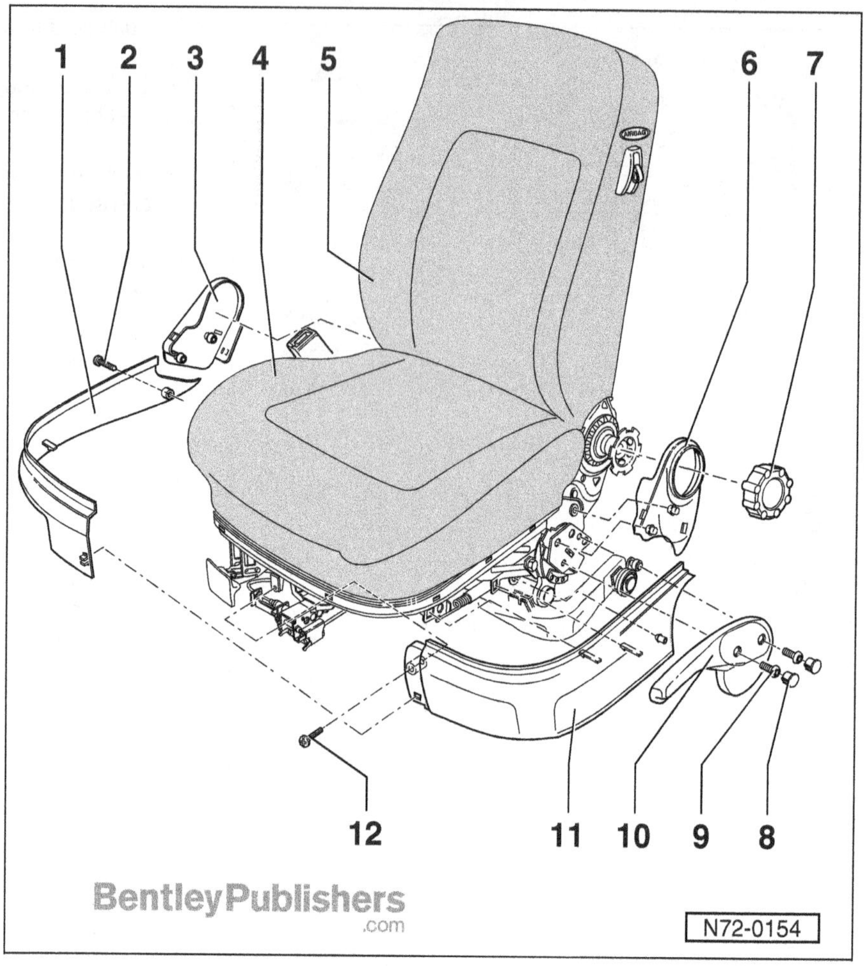

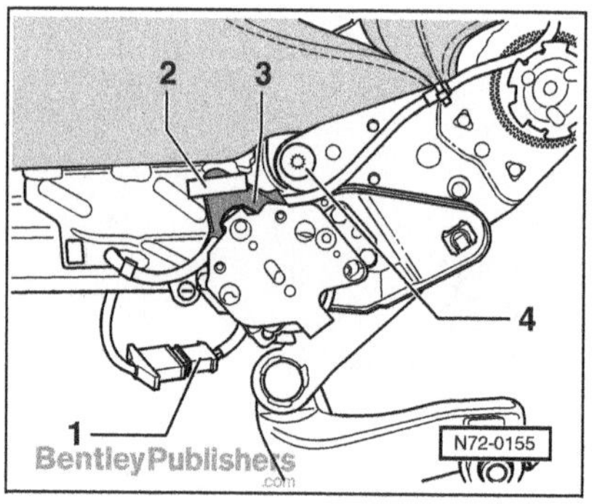

Front seatback, removing and installing

– Unclip wiring guides under seat.

– Disconnect wiring connector (**1**) for backrest heating, if equipped.

– Pry out retaining clip (**2**) and remove guide for side airbag wiring.

– Remove bolts (**4**) from both sides and take off seatback.

– When installing, bolts (**4**) must be secured with thread locking compound and tightened to 20 Nm (15 ft-lb).

Head restraint guides, removing and installing

Remove head restraint from seatback by pulling the restraint up until it stops. Then, press the release button (**arrow**) on the guide while pulling the head restraint up and out.

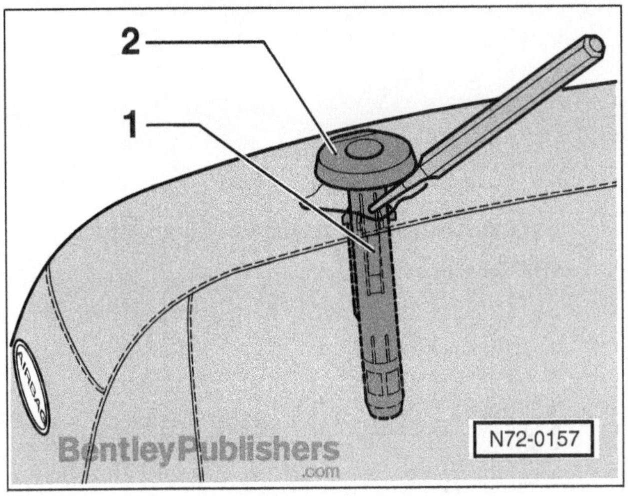

Push seatback cushion down below head restraint guide (**2**) and release retaining tab (**1**) with screwdriver or mandrel.

- Pull out backrest guide.
- Installation is reverse of removal.

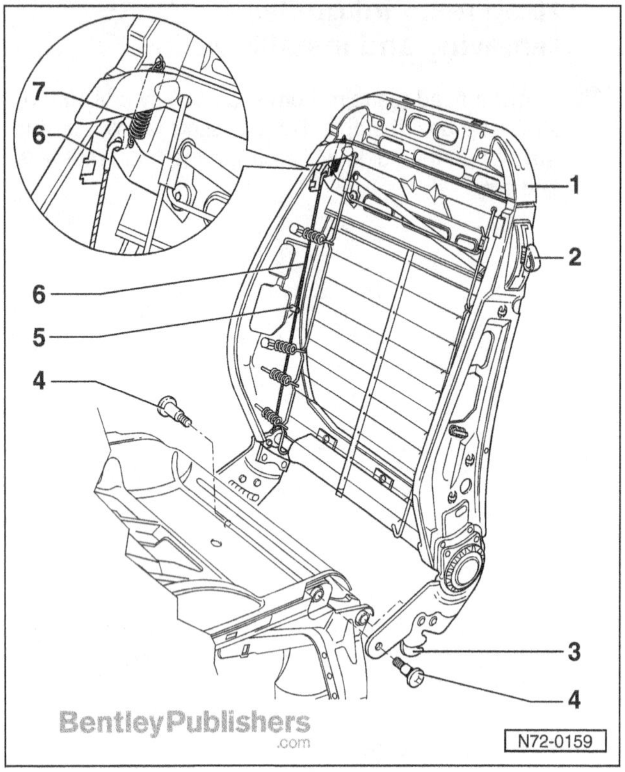

Front seatback frame assembly (through 04/2003)

1. Seatback frame
2. Release lever
3. Locking hook
4. Mounting bolts
   - Tighten to 20 Nm (15 ft-lb)
   - Install with locking compound
5. Cable guide
6. Cable
7. Spring

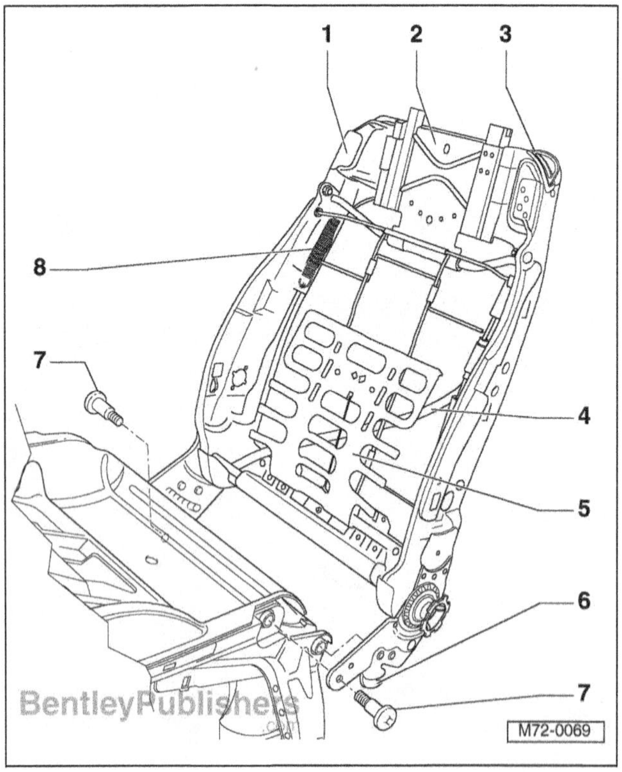

Front seatback frame assembly (from 05/2003)

1. Seatback frame
2. Active headrest guide
3. Actuating lever
4. Release cable
5. Lumbar support
6. Catches
7. Mounting bolts
   - Tighten to 20 Nm (15 ft-lb)
   - Install with locking compound
8. Spring

Front seat adjustment cable, removing and installing

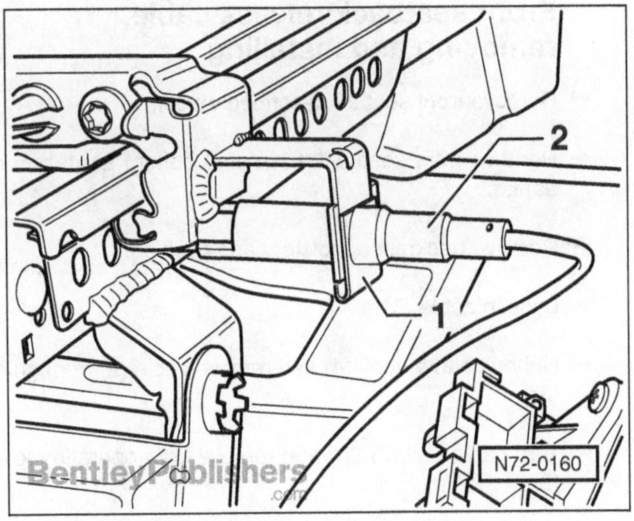

The front seat adjustment cable allows the seat to be moved forward or backward using the lever on the lower front corner of the seat.

- Remove seat as described earlier.
- Remove retaining clip (**1**) and lock pin (**2**) from bracket under seat.

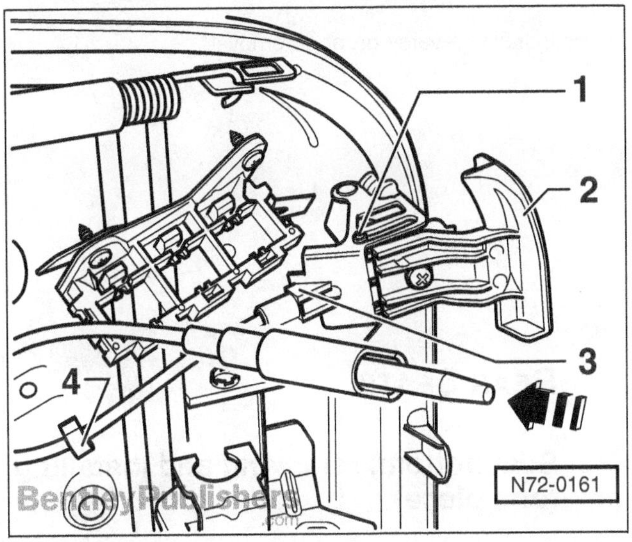

- Press lock pin (**arrow**) into cable housing and unhook cable at bolt (**1**) on operating lever (**2**).
- Press out bolt (**1**) and unclip adjustment lever mount (**3**).
- Unclip cable guide (**4**) from seat frame and remove cable.

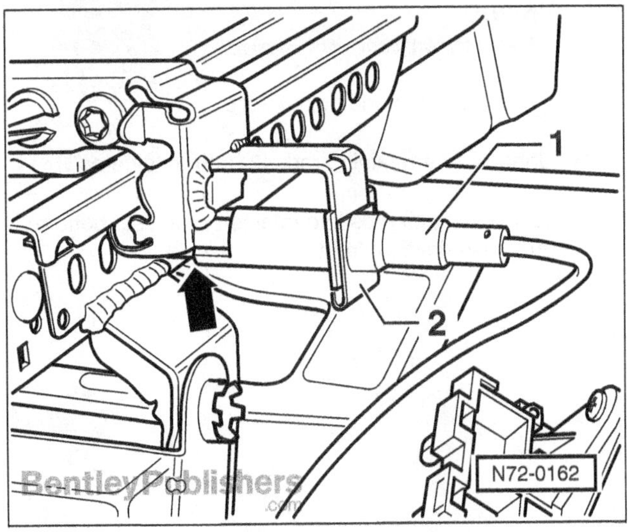

- Installation is reverse of removal, noting the following:
  - Press lock pin housing (**1**) in up to stop in bracket (**2**) and position lock pin in opening (**arrow**) below bracket.
  - Install lock pin retaining clip.

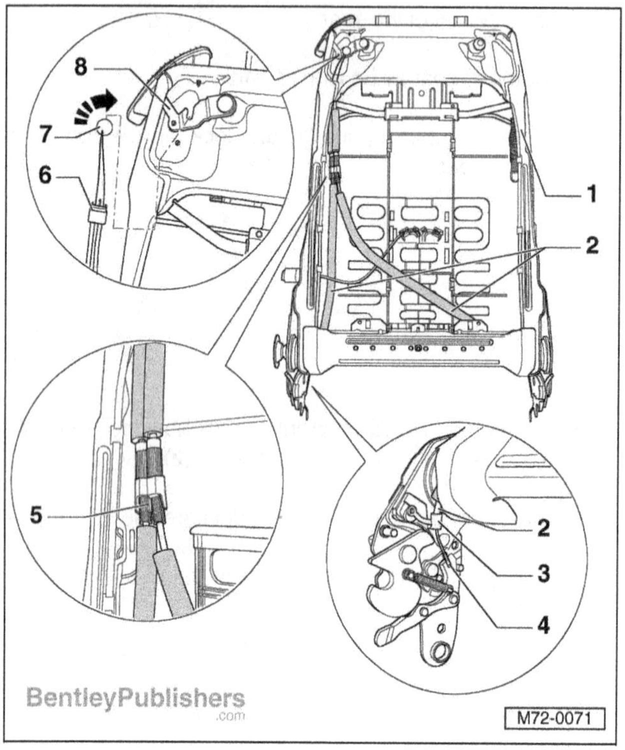

Front seatback release cable, removing and installing

- Remove front seat as described earlier.
- Remove front seatback from seat bottom as described earlier.
- Remove backrest upholstery and cushion.

◀ Loosen cable (**5**).

- Unhook cable hook (**4**) and remove cable (**2**) from guide (**3**).
- Turn cable ball (**7**) 90° and remove from operating lever (**8**).
- Unclip clip (**6**) and remove cable from backrest frame (**1**).
- Install in reverse order of removal.

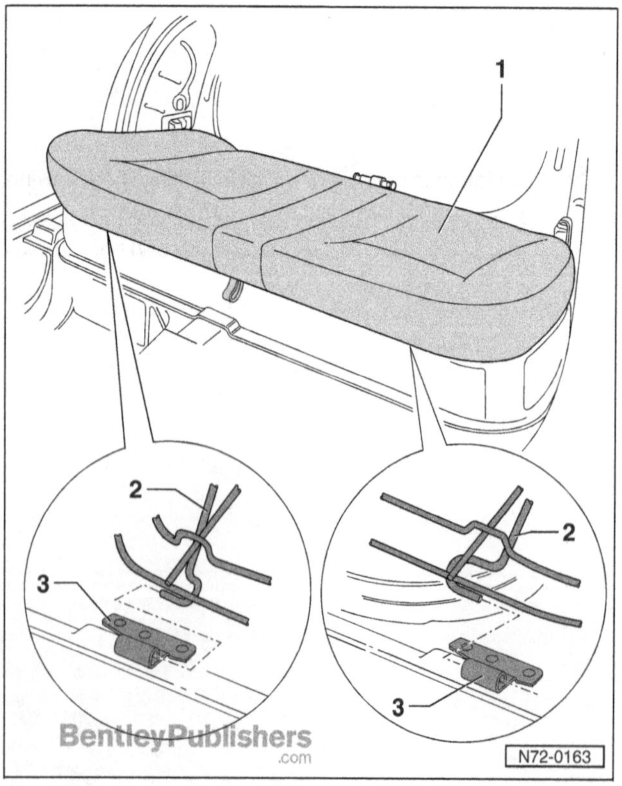

REAR SEATS

Seat bottom, removing and installing (one piece)

NOTE —
Sedan shown, convertible similar.

- Lift seat bottom and pull forward.

◀ Unhook seat retaining rods (**2**) from retainers (**3**).

- To install, hook rods into retainers, fold seat bottom back and push down at front.

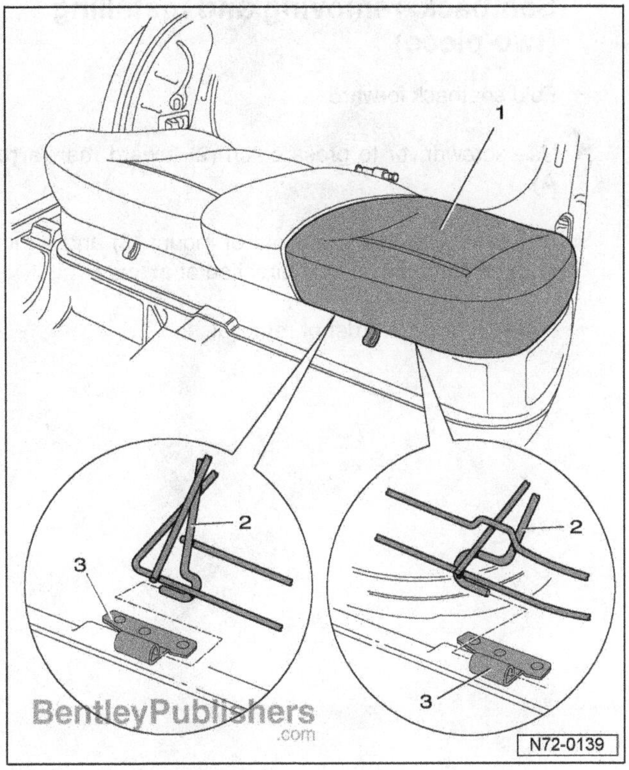

Seat bottom, removing and installing (two piece)

- Lift seat bottom and pull forward.
- ◄ Unhook seat retaining rods (**2**) from retainers (**3**).
- To install, hook rods into retainers, fold seat bottom back and push down at front.

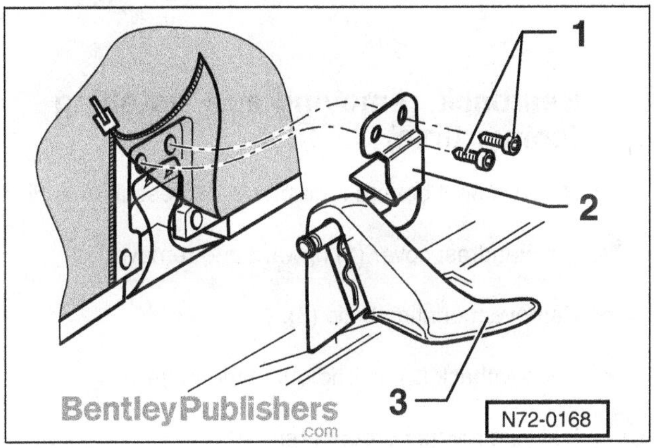

Seatback, removing and installing (one piece)

- ◄ Remove bolts (**1**) from retainer (**2**) at rear center of back-rest.
- Fold seat bottom forward.
- Fold backrest down.

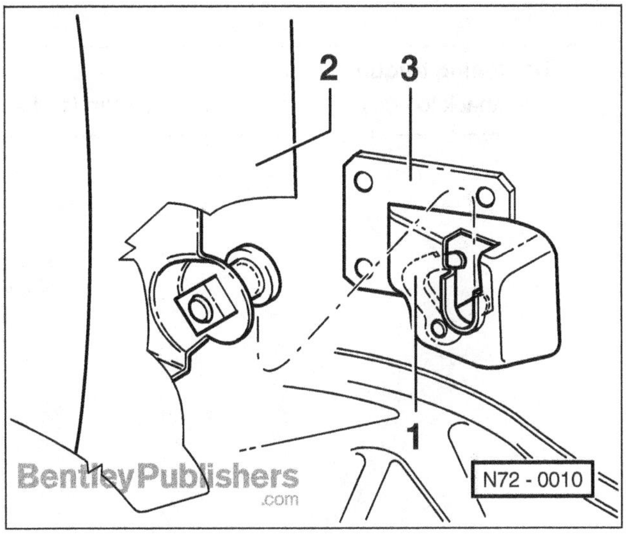

- ◄ Using a screwdriver, press back seatback retaining hooks (**1**) on right and left sides and pull seatback (**2**) up out of mount (**3**).
- Installation is reverse of removal.

Tightening torque
- Seatback mount to body . . . . 20 Nm (15 ft-lb)

Seatback, removing and installing (two piece)

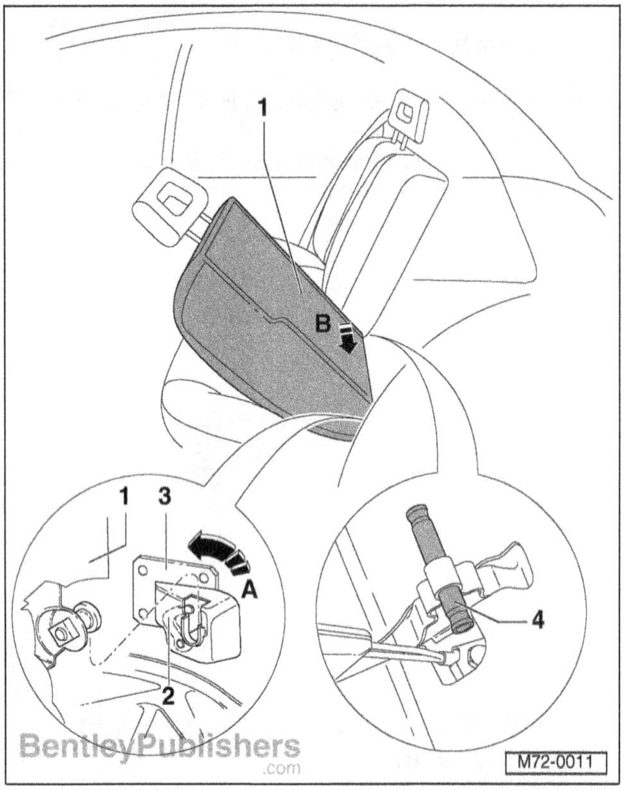

- Fold seatback forward.
- Use screwdriver to press catch (**2**) toward rear (**arrow A**).
- Pull seatback (**1**) up and out of mount (**3**) and remove from center anchor (**4**) in direction of **arrow B**.
- Install in reverse order of removal.

Seatback, removing and installing (convertible)

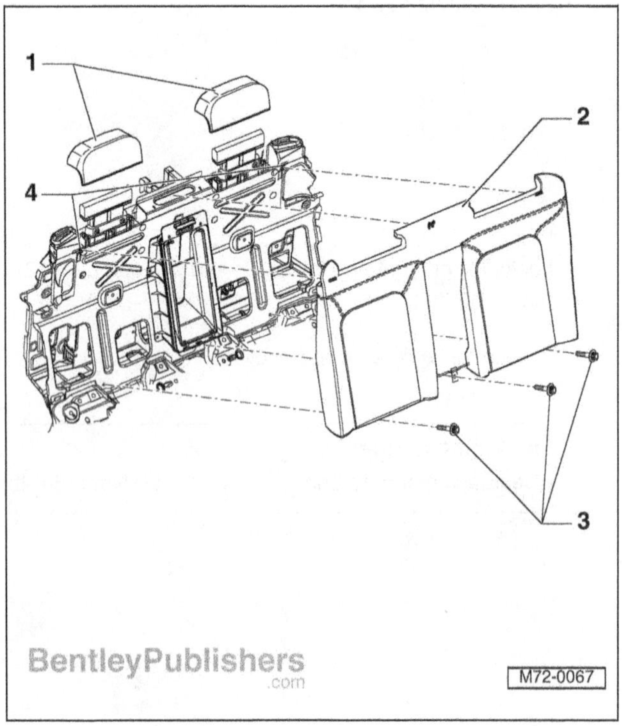

- Remove seat bottom (one piece) as described earlier.
- Pull headrest cover (**1**) upward and remove.
- Remove mounting bolts (**3**).
- Pull seatback (**2**) up and out of mount (**4**).
- Install in reverse order of removal.

Tightening torque
- Seatback to body . . . . 20 Nm (15 ft-lb)

Seatback release mechanism assembly

> *NOTE —*
> *One piece shown, two piece seatback is similar.*

1. Release rod
2. Lock hook
3. Lock pin
4. Rear panel
5. Padding
6. Cover
7. Padding wire
8. Release button
9. Relay rod
10. Clip

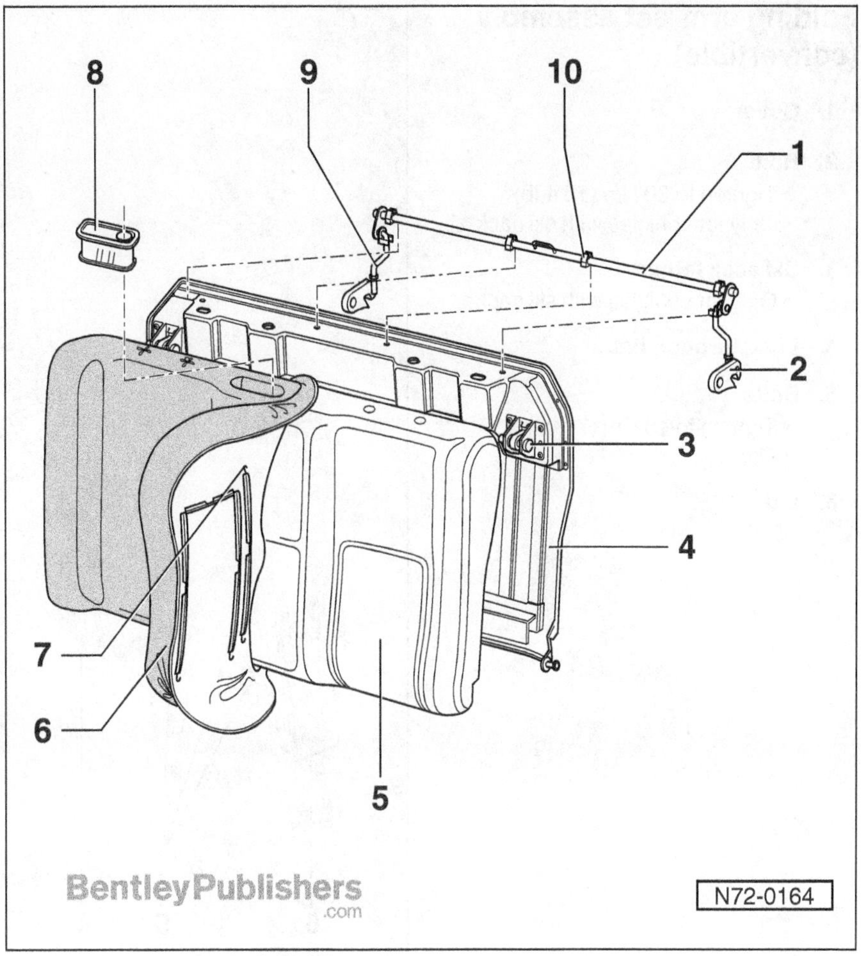

Folding armrest assembly (convertible)

1. **Cover**
2. **Bolt**
   - Tighten to 20 Nm (15 ft-lb)
   - Only for vehicles with ski sack
3. **Ski sack tether**
   - Only for vehicles with ski sack
4. **Load through hatch**
5. **Bolts**
   - Tighten to 10 Nm (7 ft-lb)
   - Qty. 4
6. **Lid**

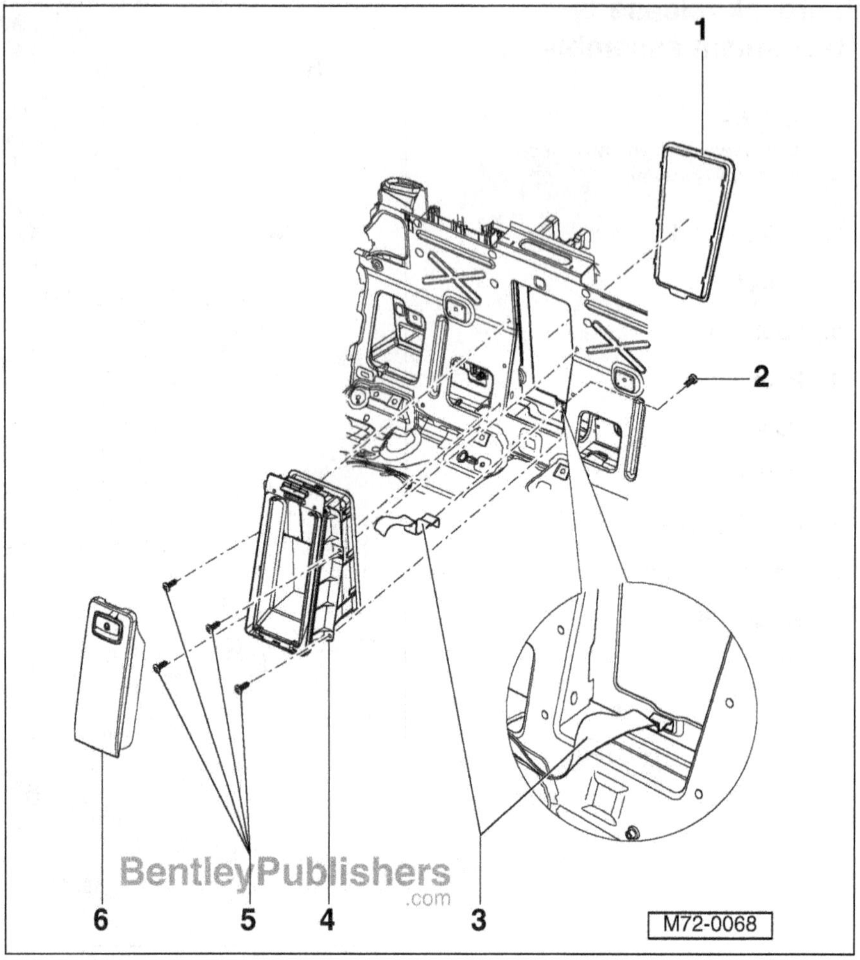

80 Heating and Ventilation

GENERAL

This section primarily applies to vehicles without air conditioning (A/C). For those vehicles with A/C, the refrigerant must first be discharged using specialized equipped before working on most of the heating and ventilation components, see **87 Air Conditioning**.

> ***WARNING —***
> *The cooling system is pressurized when the engine is warm. Wear gloves and other protection and carefully release system pressure if necessary, before performing repairs.*

> ***CAUTION —***
> *Disconnecting the negative (–) battery cable may erase fault codes and basic settings in the engine management and automatic transmission control modules. Some driveability problems may be noticed until the system re-adapts to operating conditions. OBD II readiness codes, which may be required for emissions testing, may also be erased. Convenience electronics (alarm system, interior light control, power locks, mirrors, and windows) may need to be re-set using a diagnostic scan tool.*

HEATING AND VENTILATION CONTROLS

Passenger compartment components, overview

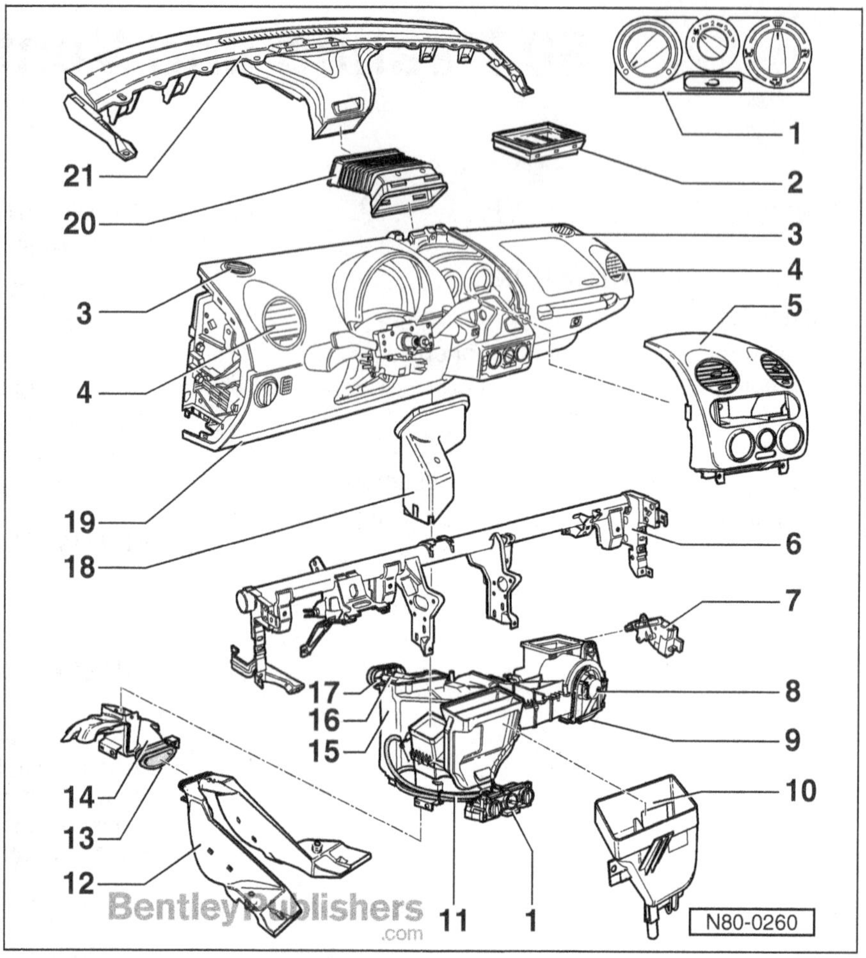

1. **Heating and ventilation controls**
   - With blower fan switch (E9)
   - With recirculation switch (E184)
2. **Dust and pollen filter**
   - Covers outside air inlet to passenger compartment
   - Filters all incoming air
3. **Side window air outlet**
   - See **Side window fixed air vents, removing and installing**
4. **Side air outlets**
   - See **Side window movable air vents, removing and installing**
5. **Center air outlet console**
6. **Instrument panel cross member**
   - See **Instrument panel crossmember, loosening and tightening**
7. **Servo motor for recirculation door (V154)**
8. **Blower fan (V2)**
9. **Blower fan series resistance pack (N24)**
   - With thermal fuse
10. **Intermediate duct**
11. **Cables**
    - Color code yellow - air distribution control knob to central flap
    - Color code green - air distribution control knob to footwell/defrost flap
    - Color code beige - temperature control knob to temperature flap
12. **Rear footwell duct**
13. **Gasket**
14. **Connecting duct**
15. **Heater assembly**
16. **Heater core**
17. **Heater core/bulkhead seal**
    - Note installed position
18. **Defroster duct**
19. **Instrument panel**
20. **Center air outlet duct**
21. **Defroster air outlet panel**

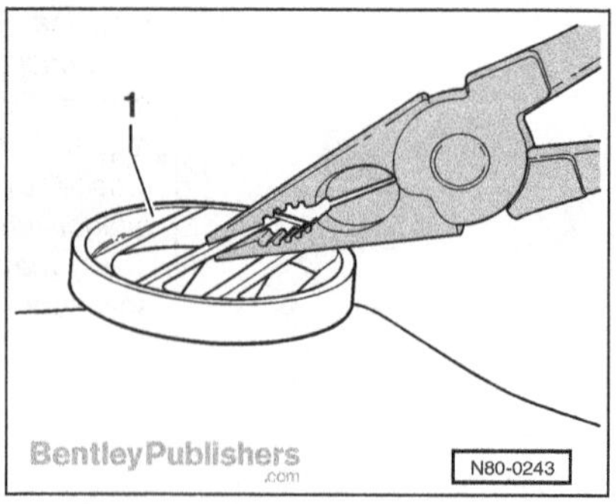

◀ Side window fixed air vents, removing and installing

- Protect vanes of vent before removing.
- Apply cloth tape to the pliers or vanes or use needle nose pliers with plastic or rubber protective jaws.
- Carefully remove air outlet (**1**) by pulling out with needle nose pliers.
- Snap into place to install.

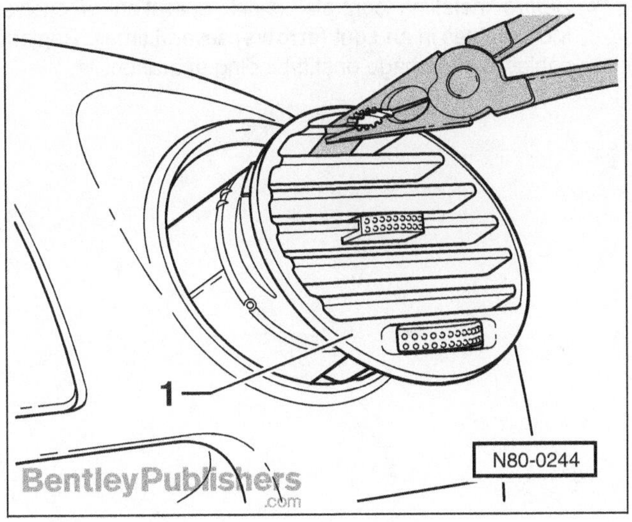

Side window movable air vents, removing and installing

- Apply cloth tape to the pliers or vanes or use needle nose pliers with plastic or rubber protective jaws.
- Carefully remove air outlet (**1**) by pulling out with needle nose pliers.
- Snap into place to install.

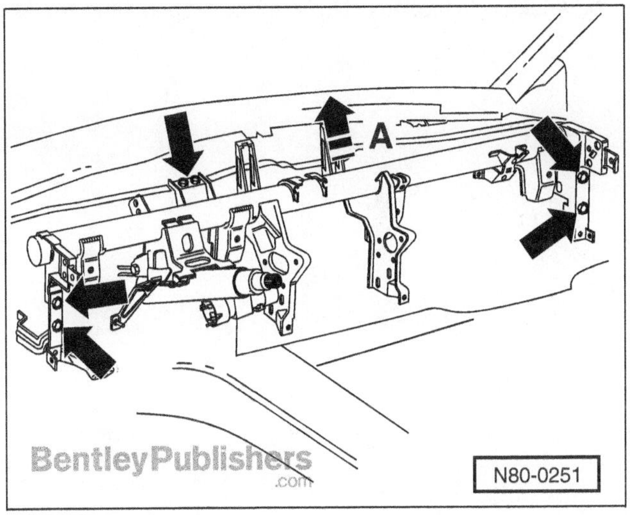

Instrument panel crossmember, loosening and tightening

- Remove securing bolts (**arrows**).
  - Tightening torque: 25 Nm (18 ft-lb).
- To access or remove heating and ventilation unit, lift cross member in direction of (**arrow A**) and support unit.
- To prevent damage to the steering column, work should be done with aid of another technician.
- If removed component(s) are not to be reinstalled immediately, re-secure cross member to bulkhead.
- Any wiring harnesses released from their retainers must be secured again at the same point, using the same method of securing.

Heating and ventilation controls, removing and installing

- Remove center air outlet console as described later.
- Remove mounting screws (**arrows**) and lift controls from opening in dash (**arrow A**).
- Unclip cables and detach from controls.
- Disconnect harness connector and remove controls.

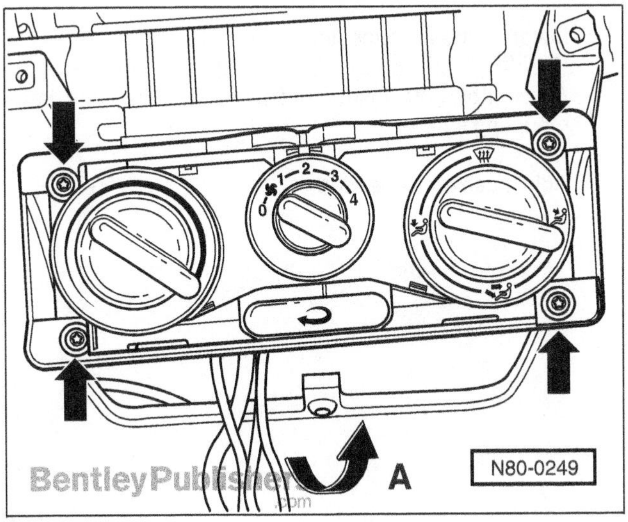

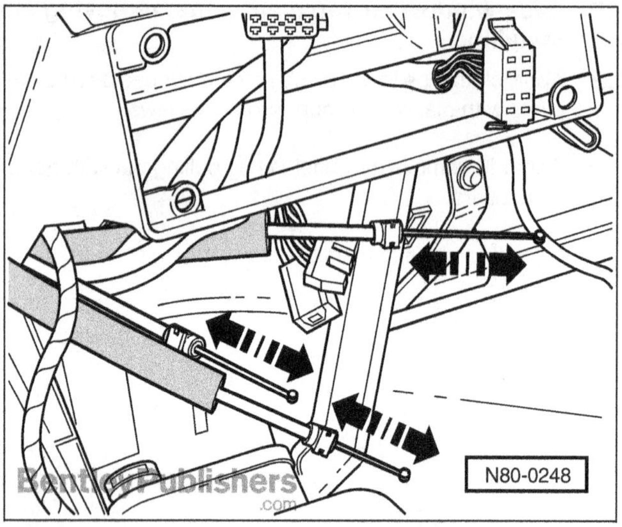

◀ Before installing controls, check operation of cables. Move cables in and out (**arrows**) several times. Replace cables with damage or stiff/binding operation.

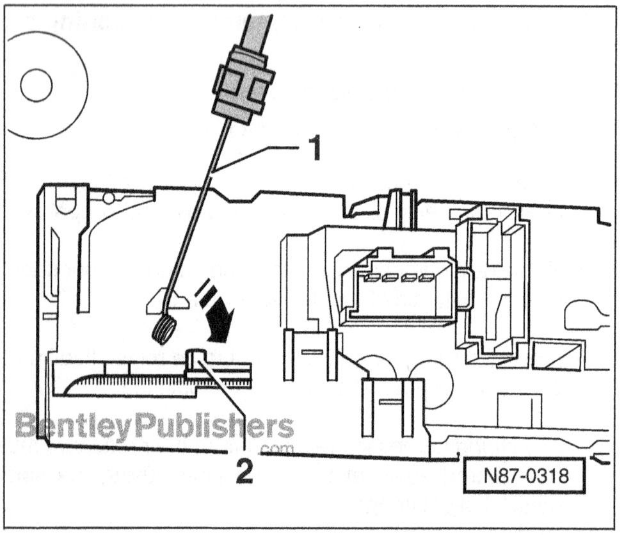

◀ Attach color coded cable (**1**) to correct actuating arm (**2**).

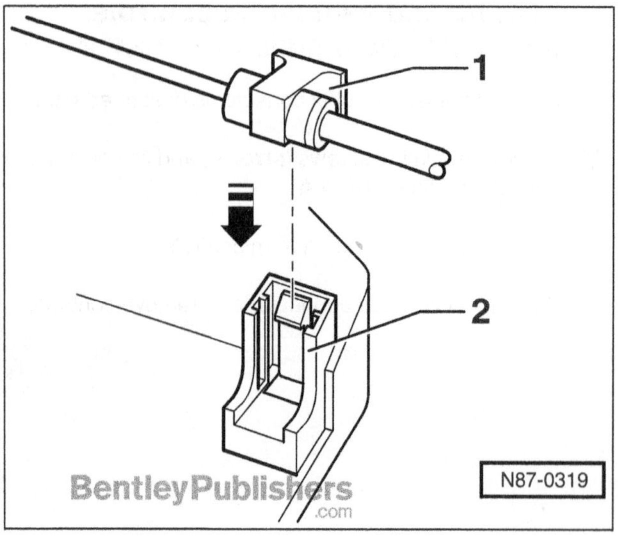

◀ Snap each cable (**1**) into appropriate retainer (**2**).

- Install harness connector.
- Fit controls into instrument panel opening and install mounting screws.
- Adjust cables as needed.

Heating and ventilation controls, assembly

1. **Heating and ventilation controls)**
   - With blower fan switch (E9)
   - With recirculation switch (E184)
2. **Illumination filter ring**
3. **Trim panel**
4. **Fresh air control lever light (L16)**
   - Clear bulb 12V, 1.2W
   - Use small (approx. 4mm i.d.) rubber hose to remove bulb. Press hose onto bulb far enough to grip it and pull.
5. **Rotary control knob**
   - For blower fan speed
   - Removing: use pliers with suitable plastic or rubber protection on jaws
6. **Rotary control knob**
   - For air distribution
   - Removing: Use pliers with suitable plastic or rubber protection on jaws
7. **Rotary control knob**
   - For interior temperature regulation
   - Removing: Use pliers with suitable plastic or rubber protection on jaws
8. **Recirculation switch (E184)**
   - Integral with controls

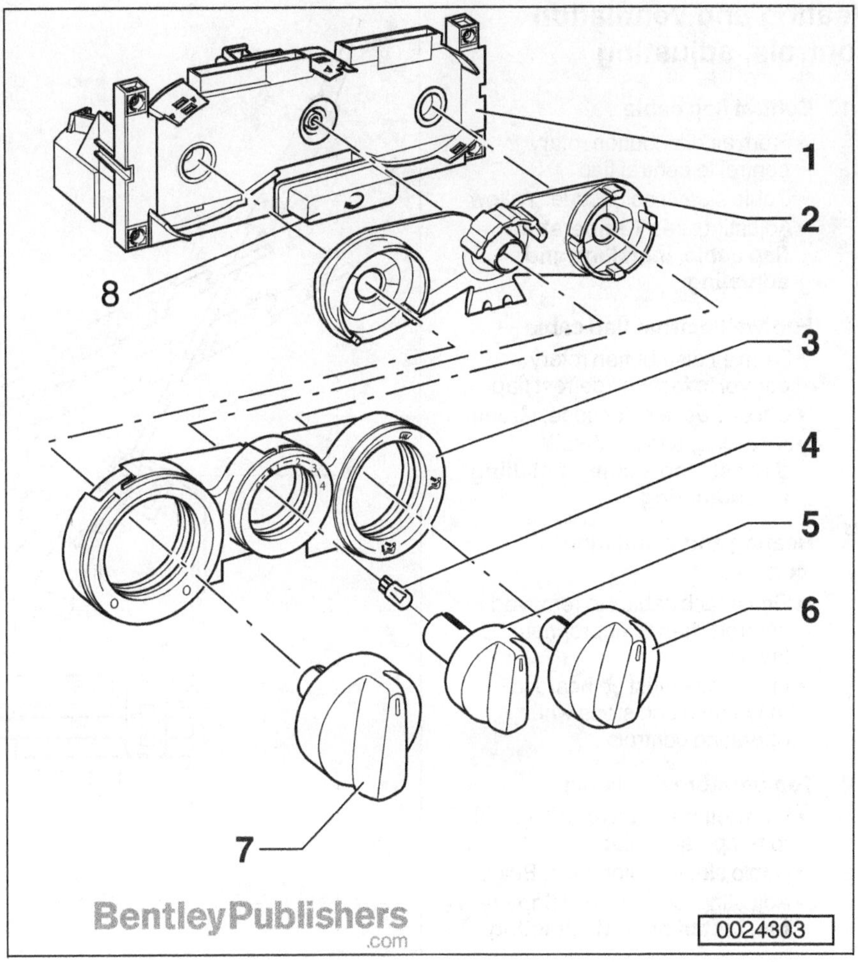

Heating and ventilation controls, adjusting

1. **Central flap cable**
   - From air distribution rotary control to central flap
   - Cable sleeve color code: Yellow
   - Adjusting, see **Temperature flap cable, installing and adjusting**
2. **Footwell/defrost flap cable**
   - From air distribution rotary control to footwell/defrost flap
   - Cable sleeve color code: Green
   - Adjusting, see **Footwell/ defrost flap cable, installing and adjusting**
3. **Heating and ventilation controls**
   - First attach cables to removed controls then to appropriate levers
   - All air flaps must be heard to move onto end stops when operating controls
4. **Temperature flap cable**
   - From temperature rotary control to temperature flap
   - Cable sleeve color code: Beige
   - Adjusting, see **Central flap cable, installing and adjusting**

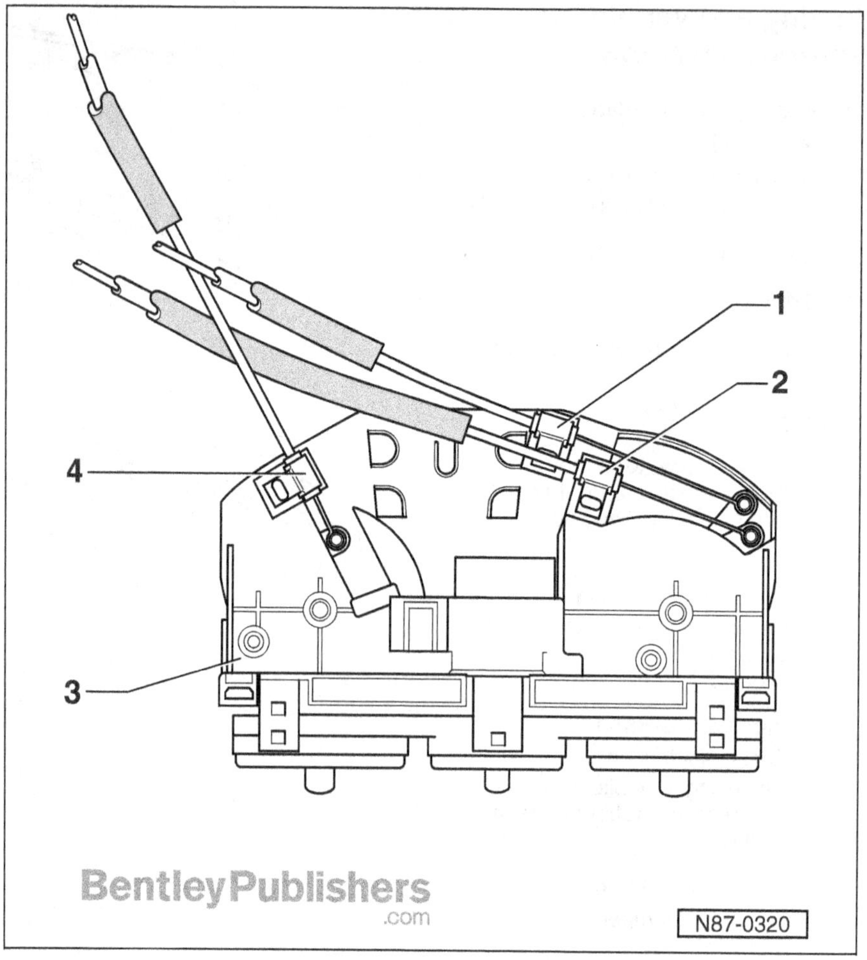

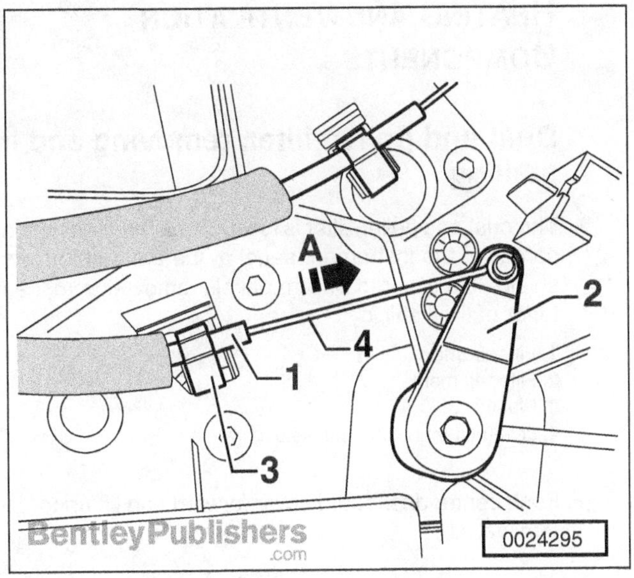

Temperature flap cable, installing and adjusting

- Cable sleeve color code: beige.
- Install/adjust with controls installed and connecting duct removed.

– Turn temperature rotary control knob fully counter-clockwise against stop.

– Attach center wire of cable (**4**) to temperature flap lever (**2**).

– Push temperature flap lever (**2**) against stop (**arrow A**) and secure outer cable (**1**) with clip (**3**).

– Turn temperature rotary control knob fully to left and right against stops.

– When turning rotary control knob both end stops must be reached.

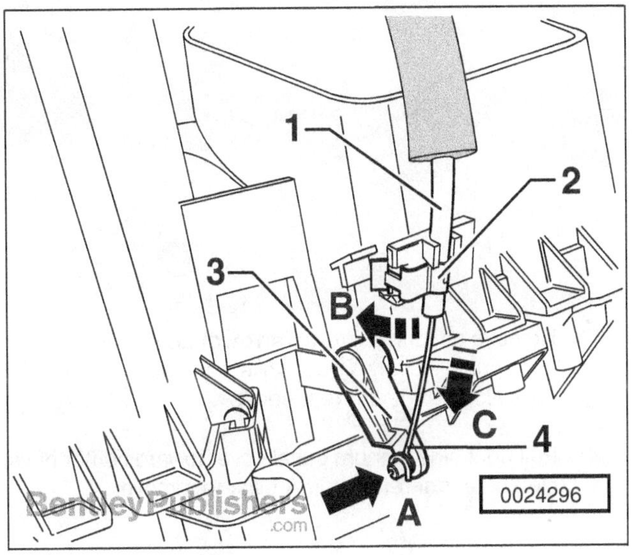

Footwell/defrost flap cable, installing and adjusting

- Cable sleeve color code: green
- Install/adjust with controls installed.

– Turn air distribution rotary control knob fully counter-clockwise against stop.

– Attach center wire of cable (**4**) to the footwell/defrost flap lever (**3**) at arrow (**A**).

– Press footwell/defrost flap lever against stop (**arrow C**) and secure outer cable (**1**) with clip (**2**) (**arrow B**).

– Turn air distribution rotary control knob fully to left and right against stops.

– When turning the rotary control knob both end stops must be reached.

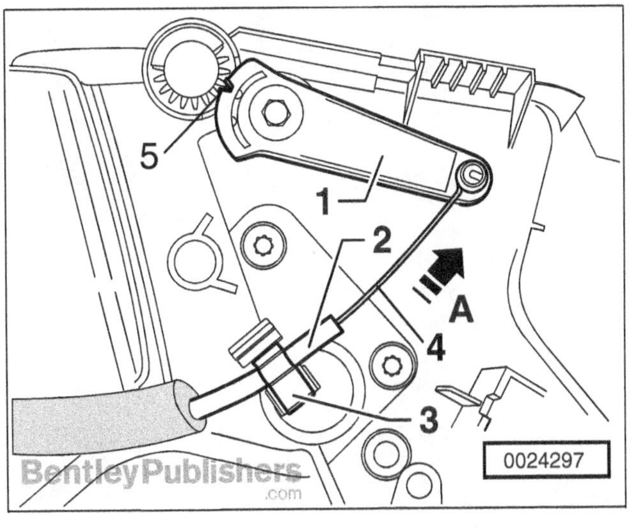

Central flap cable, installing and adjusting

- Cable sleeve color code: yellow.
- Install/adjust with controls installed, connecting duct removed and flap lever correctly indexed (**5**).

– Turn air distribution rotary control knob fully counter-clockwise against stop.

– Attach center wire of cable (**4**) to the central flap lever (**1**).

– Press central flap lever against stop (**arrow A**) and attach outer cable (**2**) with clip (**3**).

– Turn air distribution rotary control knob fully to left and right against stops.

– When turning rotary control knob both end stops must be reached.

HEATING AND VENTILATION COMPONENTS

Dust and pollen filter, removing and installing

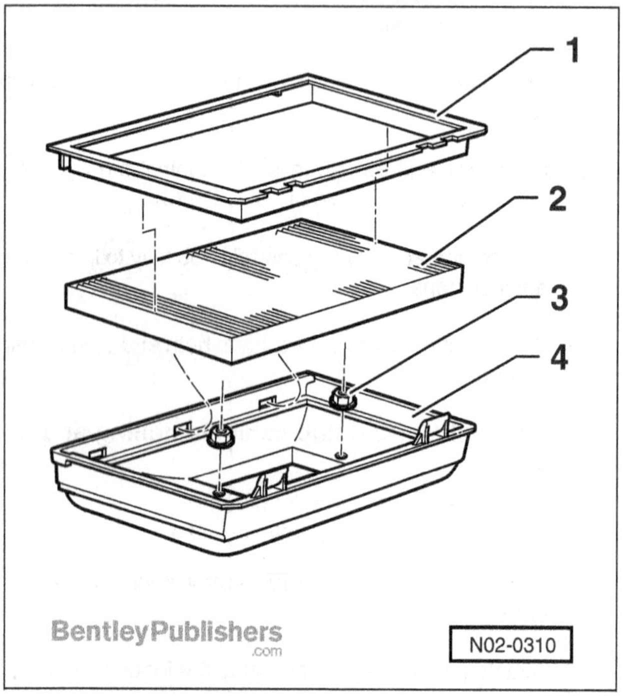

The dust and pollen filter is located near the right forward corner of the instrument panel at the base of the windshield. It is covered by trim panels removed from the interior of the vehicle.

1. Filter frame
2. Filter element
3. Plastic nut
4. Lower filter housing with seal

- Push center dashboard cover forward and lift up to remove.

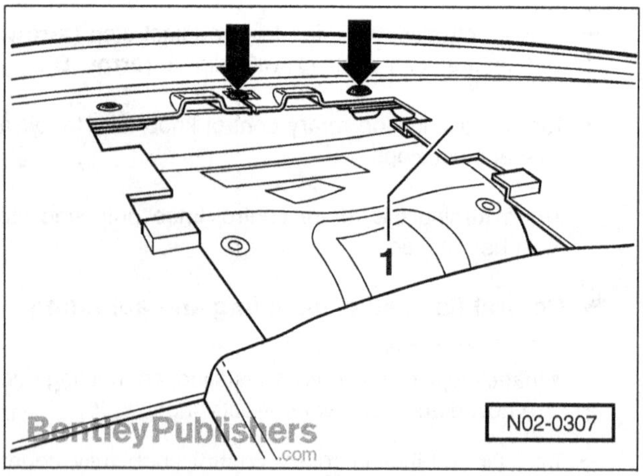

Remove securing screws (**arrows**) and lift right side plenum panel cover (**1**) up while carefully releasing from clips along front and rear edges.

- Pull right side plenum panel cover toward center of vehicle and out of retainer in A-pillar to remove.

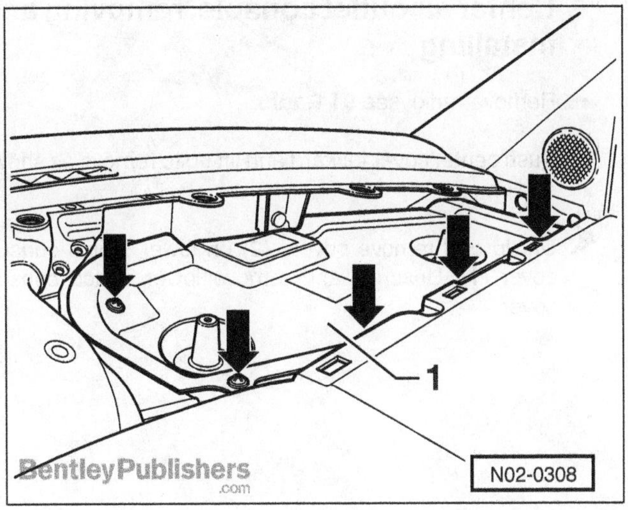

Remove screws (**arrows**) and lift right side air plenum close-out panel from air plenum.

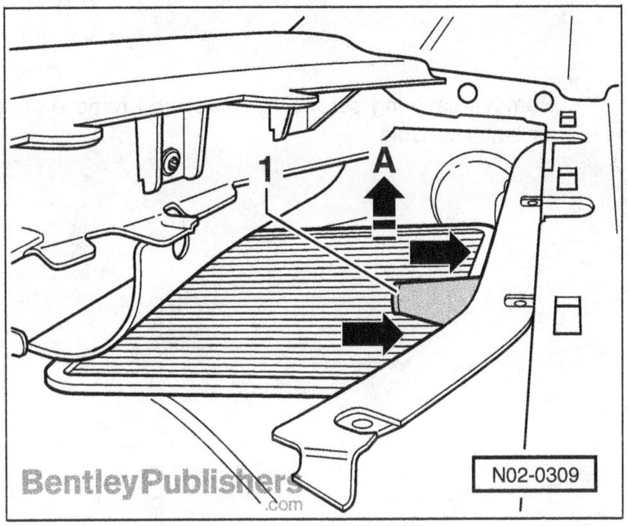

Remove rubber grommet (**1**) above filter element. Press tabs of filter housing (**arrows**) (not the filter frame) towards the rear of the vehicle and lift (**arrow A**) the filter element with the frame up and out.

- Install new filter element into frame. Make sure the ends of the frame fit into the first fold of filter element.
- Press frame with filter into housing and snap into retainers.
- Install remaining components in the reverse of removal.

Blower fan, removing and installing

- Remove glove box, see **70 Trim–Interior**

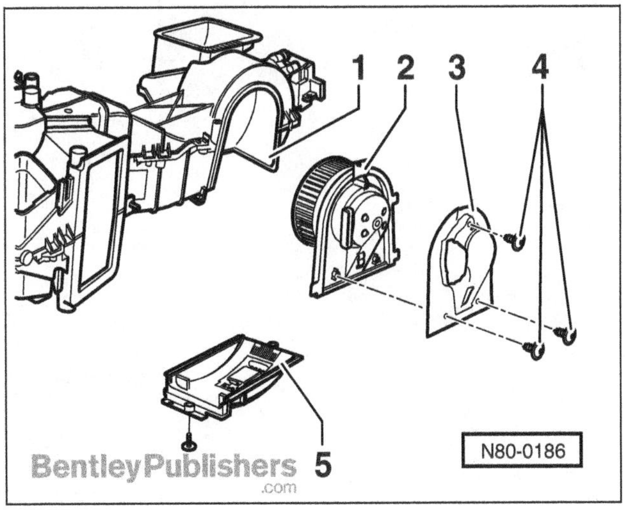

Remove screws (**4**) and cover (**3**) where applicable.

- Remove lower cover containing series resistor pack (**5**) and disconnect harness connector.
- Disconnect harness connector from blower fan (**2**).
- Pull blower fan out and downward.
- Installation is reverse of removal.

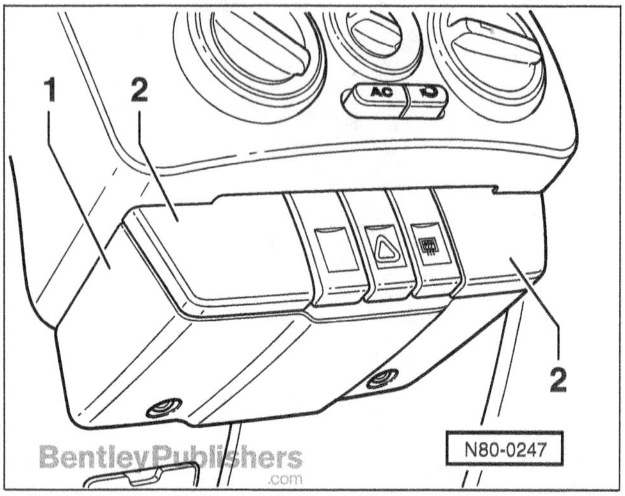

Center air outlet console, removing and installing

– Remove radio, see **91 Radio**.

– Push center cover forward and lift up to remove as shown earlier.

◄ Unclip and remove covers (**2**) on lower switch console cover (**1**). Unscrew and remove lower switch console cover.

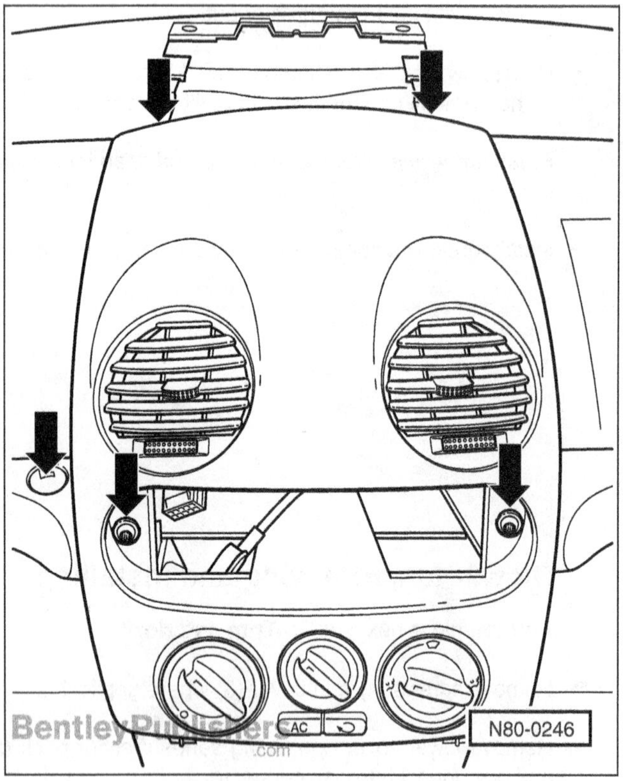

◄ Remove securing screws (**arrows**) and remove center air outlet console.

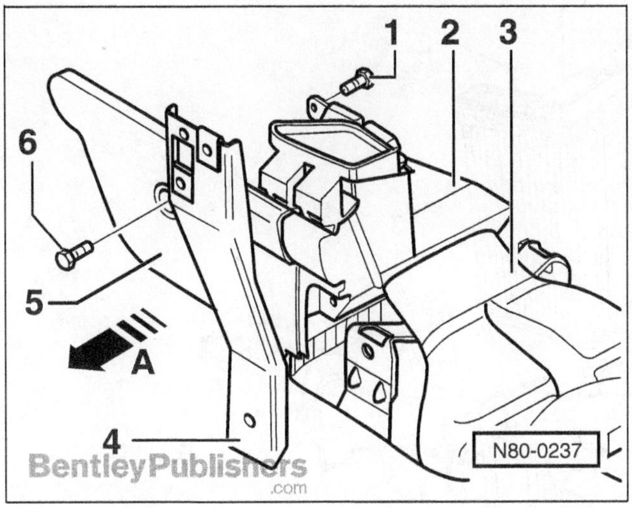

Connecting duct, assembly

1. Bolt
2. Connecting duct for footwell vents
3. Rear duct
4. Retaining bracket
5. Cover
6. Bolt

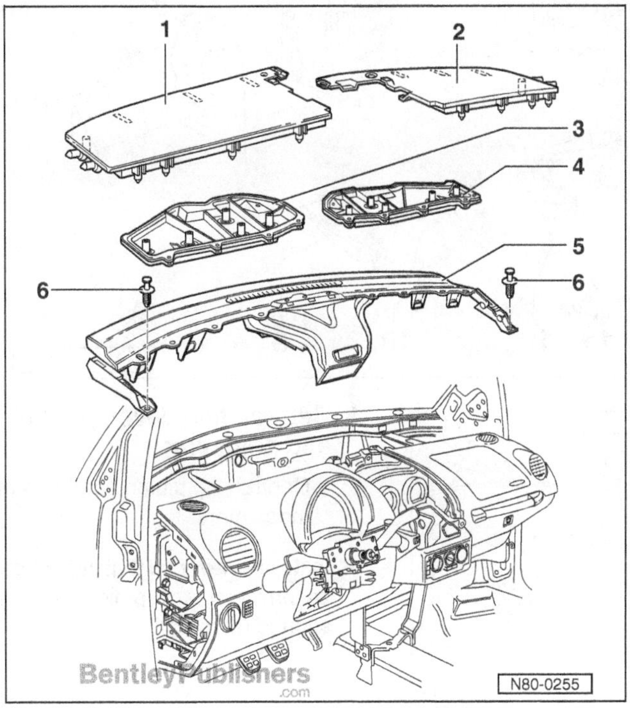

Defroster air outlet panel, assembly

1. Plenum panel cover
   - Driver side (left)
2. Plenum panel cover
   - Passenger side (right)
3. Air plenum closeout panel
   - Driver side (left)
4. Air plenum closeout panel
   - Passenger side (right)
5. Defroster air outlet panel and duct
6. Spreader rivets

Heater assembly, overview

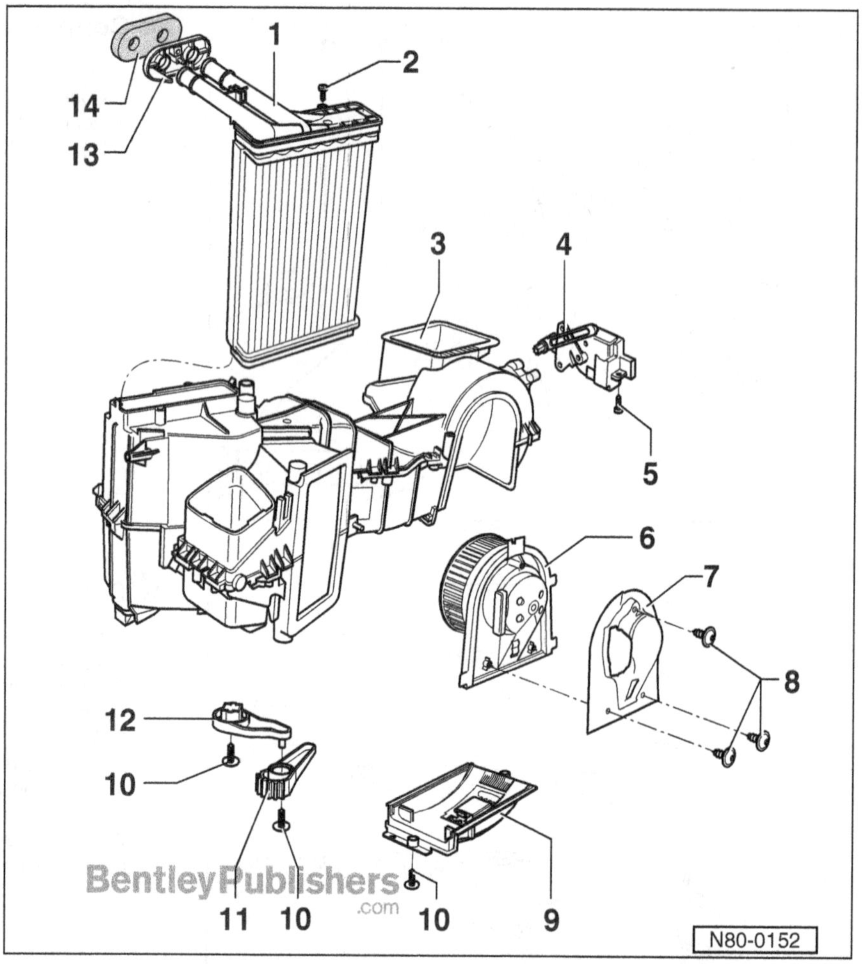

1. **Heater core**
   - Drain coolant before removal
   - Always use new coolant after replacement
   - Secured to housing with retainer clips
2. **Screw, self-tapping**
   - Use if retainer clips break
3. **Air distribution housing**
   - With air distribution flaps
4. **Servo motor for recirculation flap (V154)**
5. **Screw, self-tapping**
6. **Blower fan (V2)**
   - Updated fan available from Volkswagen with reinforced center hub. Part #1J1 819 021B is described in technical bulletin #80-01-01, dated July 18, 2001.
7. **Cover**
8. **Screw, self-tapping**
9. **Series resistor pack (N24)**
10. **Screw, self-tapping**
11. **Central flap lever**
    - Installed position, see **Central flap lever, installing and adjusting**
12. **Temperature flap lever**
13. **Base plate**
14. **Heater core bulkhead seal**
    - Installed position, see **Heater core bulkhead seal, installed position**

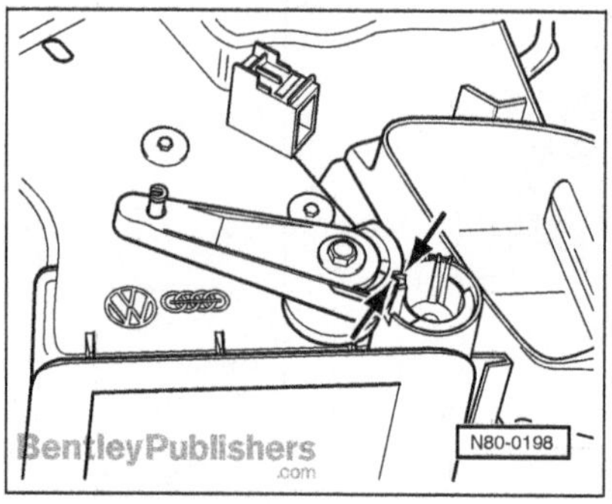

◄ Central flap lever, installing and adjusting

- Central flap pinion shaft has an index mark.

– Install the lever so that index notches on lever and pinion shaft are aligned (**arrows**).

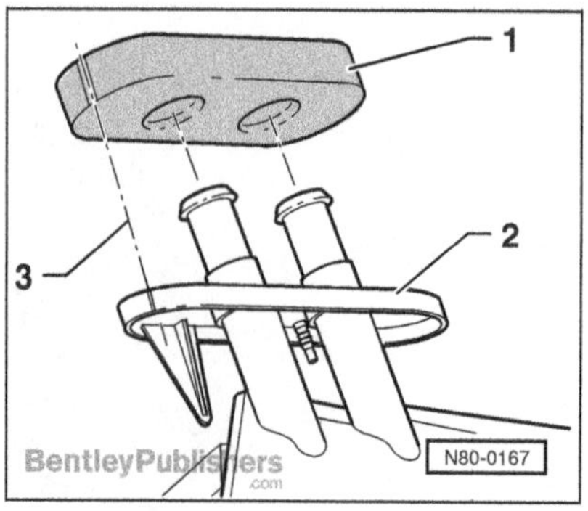

◄ Heater core bulkhead seal, installed position

- Seal between heater core and bulkhead has a positioning mark on the side.

– Install seal (**1**) so that the notches on the seal and notches on the base plate (**2**) are aligned (**3**).

Recirculation flap servo, removing and installing

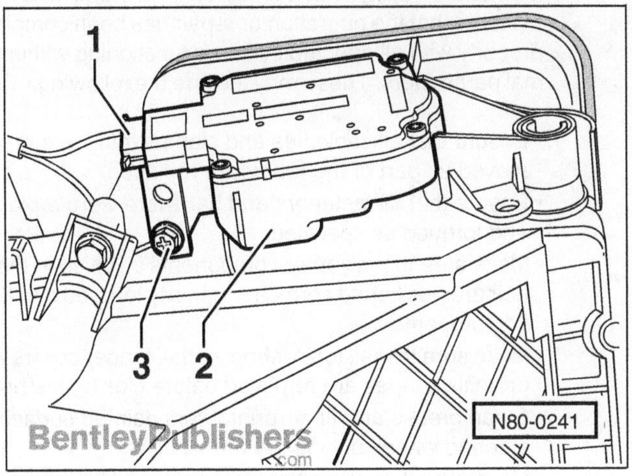

– Remove glove box, see **70 Trim–Interior**.

– Disconnect harness connector (**1**) from flap servo (**2**).
  - Remove securing screw (**3**) and swing flap servo downwards.

– Disconnect flap servo from recirculation flap lever.

– Before installing a new flap servo, attach harness connector to the new servo and switch on ignition.

NOTE—
Replacement servos are supplied with operating levers in the "recirculation" position.

– Operate switch for recirculation flap and observe lever position.

– When center position of travel is reached, quickly unplug the harness connector and switch off ignition.

– Install servo onto recirculation flap shaft.

NOTE—
If it is difficult to align and fit the servo to the recirculation flap shaft, remove the blower fan and position the flap and shaft as required.

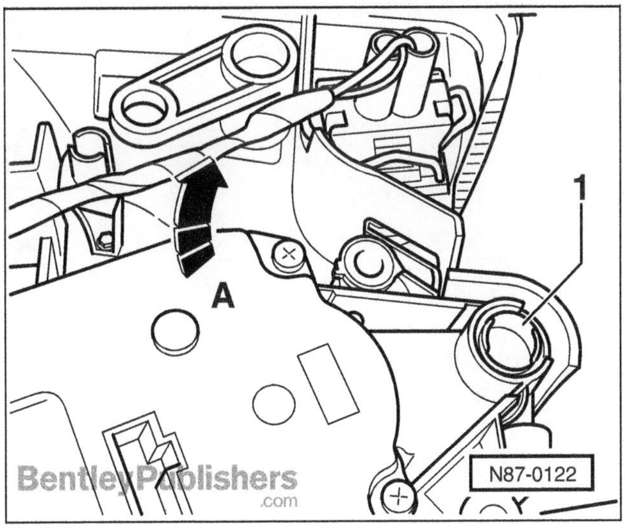

– When servo has been installed onto recirculation flap shaft (**1**), move motor into mounting position (**arrow A**) and install securing screw.

– Install harness connector.

– Install glove box and remainder of removed components.

Quality Review

When you have finished working under the hood and around other areas of the vehicle, it is advisable to take a moment to quality check or review your work. This helps to insure that the operation or repair has been completed properly with all affected systems functioning within normal parameters. These may include the following:

- Ensure that all cable ties and clamps that were removed as part of the repair are replaced.
- Ensure that all fasteners and hardware were replaced and torqued as specified.
- Make sure that all other components involved in the repair are positioned correctly, properly torqued and function properly.
- Make sure that all tools, shop cloths, fender covers and protective tapes are removed before closing the hood.
- Clean grease and fingerprints from painted surfaces, steering wheel, upholstery and shifter.
- Make sure that warm/hot air is delivered by the appropriate air vents when the heater is switched on.
- Unlock anti-theft radio and reset the clock.

87 Air Conditioning

Tables

GENERAL

This section covers the Air Conditioning (A/C) refrigerant system and related components. Service and adjustment to the ventilation control cables is covered in **80 Heating and Ventilation.**

> ***CAUTION —***
> *Disconnecting the negative (–) battery cable may erase fault codes and basic settings in the engine management and automatic transmission control modules. Some driveability problems may be noticed until the system re-adapts to operating conditions. OBD II readiness codes, which may be required for emissions testing, may also be erased. Convenience electronics (alarm system, interior light control, power locks, mirrors, and windows) may need to be re-set using a diagnostic scan tool.*

The refrigerant system of the New Beetle is designed to work only with a refrigerant known as R-134a. In this section, the refrigerant will always be referred to as R-134a, howev-

87

er, this chemical is marketed and sold under several trade designations such as: Tetrafluoroethane, CH_2F CF_3, H-KW 134a, SUVA® 134a, and ARCTON® 134a.

Other refrigerants may be available in the marketplace, however, Volkswagen specifically recommends that no other refrigerant be used. Serious damage will usually result and performance will be reduced.

Special precautions for A/C systems

New Beetle air conditioning systems are filled with refrigerant R-134a under pressure. The unique characteristics of this chemical necessitate special servicing and handling procedures which may be governed by Federal, State, Provincial and Local regulations.

WARNING —

- *Work in a well ventilated area. Refrigerant gases are heavier than air, displace oxygen and may cause suffocation in areas of poor air circulation, for example under the vehicle.*
- *Pressurized R134a refrigerant in the presence of oxygen may form a combustible mixture. Never introduce compressed air into any R-134a container (full or empty), capped off A/C component, or piece of service equipment.*
- *As of January 1, 1992 any person who services a motor vehicle air conditioner in the USA MUST, by law, be properly trained and certified and use approved refrigerant recycling equipment. Technicians must complete an EPA approved recycling course to be certified.*
- *State, Provincial and Local governments may have additional requirements regarding air conditioning servicing. Always comply with all applicable laws and regulations.*
- *The A/C system is filled with refrigerant gas which is under pressure.*
- *Avoid breathing refrigerant vapors. Exposure may irritate eyes, nose and throat.*
- *Always be careful that refrigerant does not come in contact with your skin. Always wear hand and eye protection (gloves and goggles) when working around the A/C system. If refrigerant has come in contact with your skin or eyes:*
  - *Do not rub skin or eyes.*
  - *Immediately flush with cool water for 15 minutes.*
  - *Rush to a doctor or hospital.*
  - *Do not attempt to treat yourself.*
- *Keep refrigerant containers stored below 50°C (122°F) and use care to avoid dropping.*
- *DO NOT warm refrigerant containers with an open flame. If refrigerant needs to be warmed, place bottom of tank in warm water.*

WARNING —

- *Do not expose any component of the A/C system to high temperatures above 80°C (176°F) or open flames. Excessive heat will cause system pressure increases which could burst the system.*
- *Switch on exhaust/ventilation systems when working on the refrigerant system.*
- *Keep refrigerant away from open flames. Poisonous gas will be produced if it burns. Do not smoke when refrigerant gases are present for the same reason.*
- *Electric welding near refrigerant hoses causes R-134a to decompose from ultraviolet light. Discharge system before electric welding.*
- *Always use an Underwriter's Laboratory (UL) approved refrigerant recovery/recycling/recharging unit such as Kent-Moore ACR4, or equivalent, whenever servicing an R-134a A/C system.*
- *Do not steam clean condensers or evaporators. Use only cold water or compressed air on the outside of the component.*

CAUTION —

- *Refrigerant oils used for the R-134a system and R-12 (FREON®) systems are NOT compatible. Use only the specified synthetic oil (Polyalkylene Glycol/PAG) for the R-134a refrigerant system.*
- *R-134a refrigerant system oil (PAG oil) absorbs moisture very rapidly. Moisture combines with the refrigerant to form acids which will damage the system. Use only the specified oil from a sealed container and ALWAYS reseal oil container immediately after use.*
- *Only use R-134a in New Beetle A/C systems. Never use substitute or "drop-in" replacement refrigerants. Vehicle safety will be compromised and severe damage to vehicle and servicing equipment will result.*
- *Immediately plug open connections on A/C components to prevent dirt and especially moisture contamination. Likewise, DO NOT remove new components from packaging until ready to install. Immediately tighten component connections after installation.*
- *Always use separate refrigerant recovery/recycling/recharging servicing equipment for R-12 and R-134a systems. DO NOT use one piece of equipment for both R-12 and R-134a systems. The residual traces of refrigerant will contaminate and damage the equipment.*

> **CAUTION—**
> - *R-134a and R-12 systems use different size service fittings. NEVER use adaptors to convert an R-12 fitting to R-134a size or R-134a fitting to R-12 size.*
> - *R-134a and R-12 A/C components including compressor, hoses, O-rings, evaporator, condenser, receiver-drier, etc. are NOT interchangeable. Components of the R-134a system are identified by lettering (R-134a) or by a green label (or stripe).*
> - *Discharge A/C system before removing any A/C system component.*
> - *Always replace damaged and/or leaking A/C system components. Do not attempt repair by soldering or welding.*
> - *Always reinstall caps over A/C service valves.*

A/C refrigerant circuit

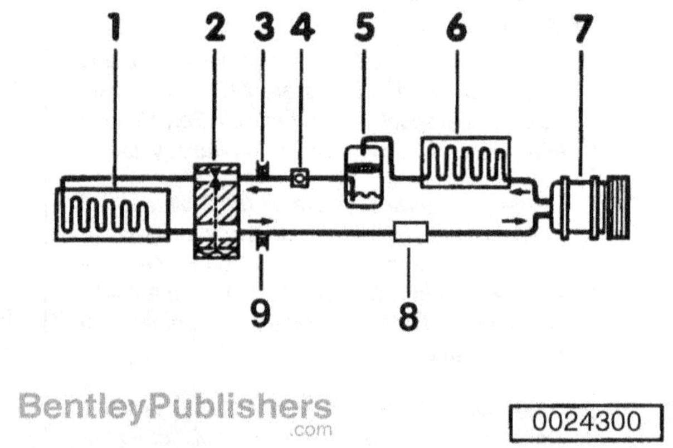

1. Evaporator
2. Expansion valve
3. High pressure service valve
4. Sight glass (if equipped)
5. Receiver drier
6. Condenser
7. Compressor
8. Damper
9. Low pressure service valve

Several components make up the air conditioning system found in the New Beetle. The variable displacement compressor compresses the R-134a refrigerant and moves it from the discharge port in a gaseous state to the condenser. The condenser is mounted in the front of the vehicle where air from the radiator fans or the motion of the vehicle passes through it. The motion of the air removes the heat from the refrigerant and a change of state from gas to liquid occurs. The liquid refrigerant flows to the receiver-drier where it is temporarily stored and a desiccant removes any moisture. The refrigerant then moves past a sight glass (if applicable) and a high pressure service port to the expansion valve unit. The expansion valve has a small internal orifice which sprays the liquid refrigerant into the evaporator. The resulting change of state from liquid to gas and the associated pressure drop causes the evaporator to become cold. Air from inside the passenger compartment is passed over the evaporator by a fan where it gives up its heat and becomes cold. It is then distributed through the vent system. The gaseous refrigerant passes back through a port on the expansion valve, past a low pressure service port and damper and finally back to the suction side of the compressor to be compressed again.

The compressor runs when needed by means of an electro-magnetic clutch which is controlled by several relays. A pressure switch in the refrigerant line also controls operation of the compressor clutch and the radiator cooling fans. A pressure relief valve is mounted to the compressor cylinder head plate. Switches for operation are mounted within easy reach of the driver.

A/C REFRIGERANT SYSTEM

The air conditioning system is comprised of components that can be grouped into several areas. This heading covers those components that are used to handle and contain the R-134a refrigerant.

A/C refrigerant system, component overview

NOTE —
Volkswagen identifies electrical components by a letter and/or a number in the electrical schematics. See **EWD Electrical Wiring Diagrams**. *These electrical identifiers are listed in parenthesis as an aid to electrical troubleshooting.*

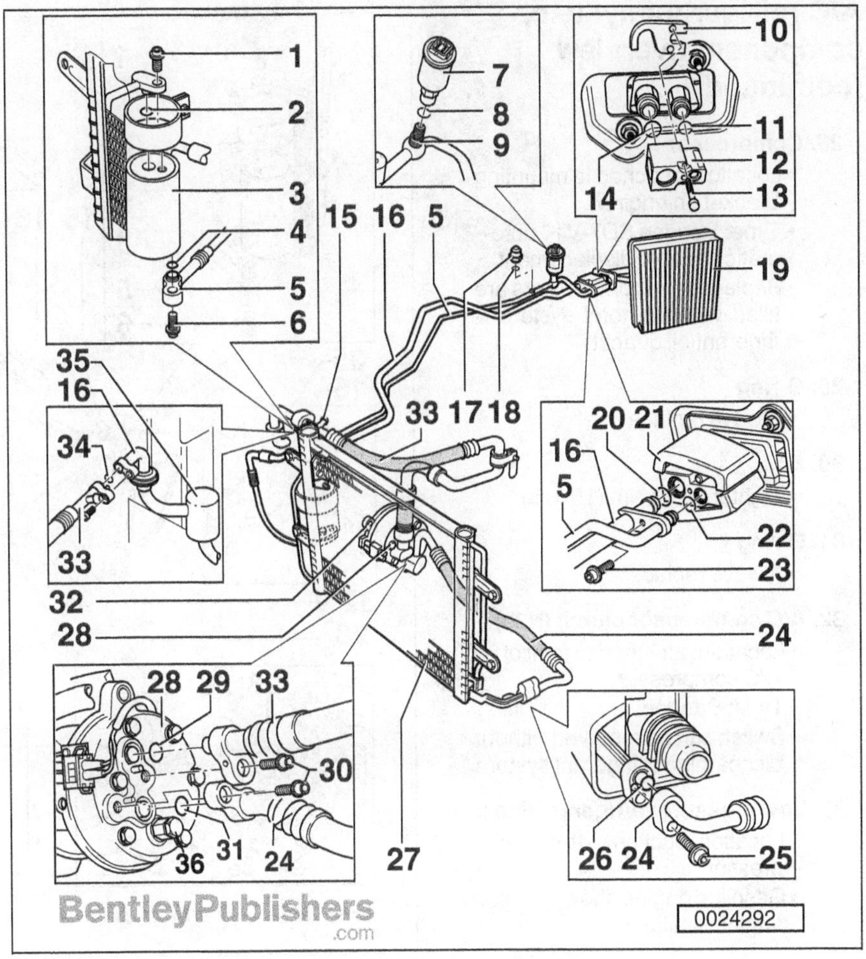

1. **Sealing plug**
   - Tighten to 15 Nm (11 ft-lb)
2. **O-ring**
   - Always replace
3. **Receiver drier**
   - Attached to right side of condenser
4. **O-ring**
   - Always replace
5. **Liquid refrigerant line**
   - Between receiver drier outlet and expansion valve inlet
6. **Bolt**
   - Tighten to 15 Nm (11 ft-lb)
7. **A/C pressure switch (F129)/ high pressure sensor (G65)**
   - Location: liquid refrigerant line near expansion valve
   - Switch can be removed without discharging refrigerant system
   - See **Component details**
8. **O-ring**
   - Always replace
9. **Threaded fitting**
10. **Reinforcement plate**
11. **O-ring**
    - Always replace
12. **Expansion valve**
    - Near bulkhead on right side of engine compartment
13. **Bolt**
    - Tighten to 8 Nm (71 in-lb)
14. **O-ring**
    - Always replace
15. **Low pressure service valve port**
    - Location: low pressure refrigerant line II near top of condenser
16. **Low pressure refrigerant line I**
    - Location: outlet side of expansion valve
17. **High pressure service valve port**
    - Location: liquid refrigerant line near expansion valve
    - Only use Kent Moore ACR4 or equivalent
18. **O-ring**
    - Always replace
19. **Evaporator**
    - Location: in heater/evaporator housing
20. **O-ring**
    - Always replace
21. **Insulation**
    - Location: wrapped around expansion valve
22. **O-ring**
    - Always replace
23. **Bolt**
    - Tighten to 8 Nm (71 in-lb)
24. **High pressure refrigerant line**
    - Location: discharge side of compressor to condenser
25. **Bolt**
    - Tighten to 15 Nm (11 ft lb)
26. **O-ring**
    - Always replace
27. **Condenser**
    - Location: in front of radiator

(continued on next page)

A/C refrigerant system, component overview (continued)

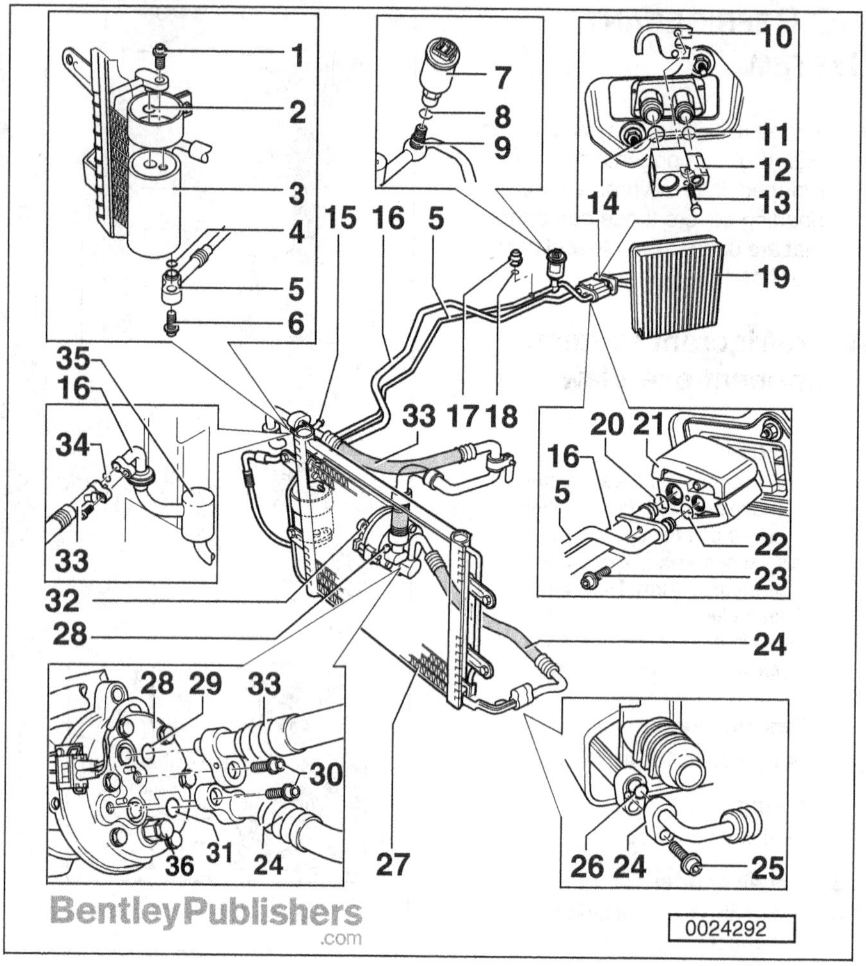

28. Compressor
- Location: attached to mounting bracket on engine
- Type: Sanden SD7-V16 automatic variable displacement
- Replacement compressors are filled with the total system refrigerant oil quantity

29. O-ring
- Always replace

30. Bolt
- Tighten to 20 Nm (15 ft-lb)

31. O-ring
- Always replace

32. A/C compressor clutch (N25)
- Location: attached to front of A/C compressor
- Type: Sanden
- Switch can be removed without discharging refrigerant system

33. Low pressure refrigerant line II
- Location: suction side of compressor
- Gasoline engine shown, diesel engine similar

34. O-ring
- Always replace

35. Damper
- Location: built into low pressure line I (item 16)

36. Pressure relief valve
- Location: on compressor cylinder head plate
- See **Component details**

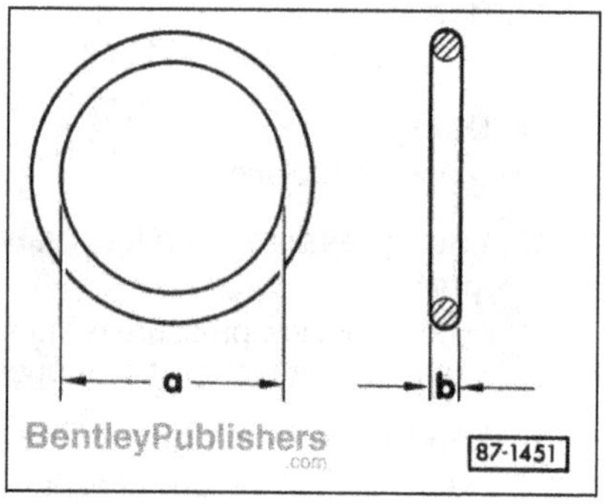

O-rings
- Note inner diameter (**a**) and thickness (**b**).
- Lubricate with PAG oil before installing.
- Always replace during repairs, never reuse.
- Only use O-rings that are compatible with R-134a.
- May be color coded green, red, violet or black.

Component details

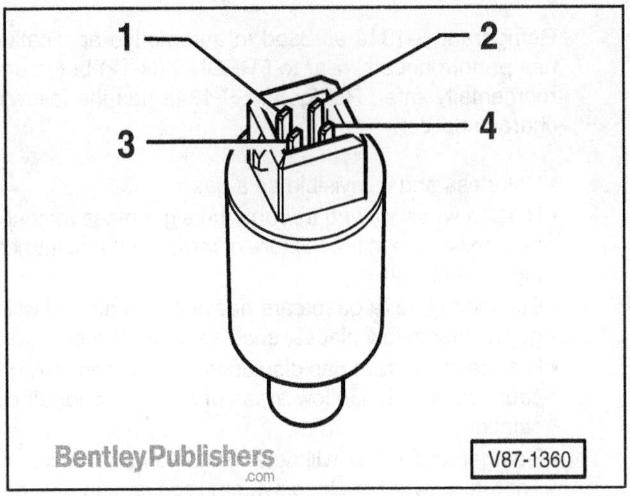

A/C pressure switch (F129)

- Not installed on all vehicles.
- Can be replaced without discharging refrigerant due to schraeder valve in threaded fitting.
- Switches A/C clutch off when refrigerant pressure exceeds 32 bar (464 psi) and resets when pressure drops to 24 bar (348 psi).
- Switches A/C clutch off when refrigerant pressure falls below 1.2 bar (17.4 psi).
- Switches radiator coolant fan to second speed when pressure reaches 16 bar (232 psi).
- Terminals 1 and 2 switch A/C clutch.
- Terminals 3 and 4 switch radiator coolant fan.

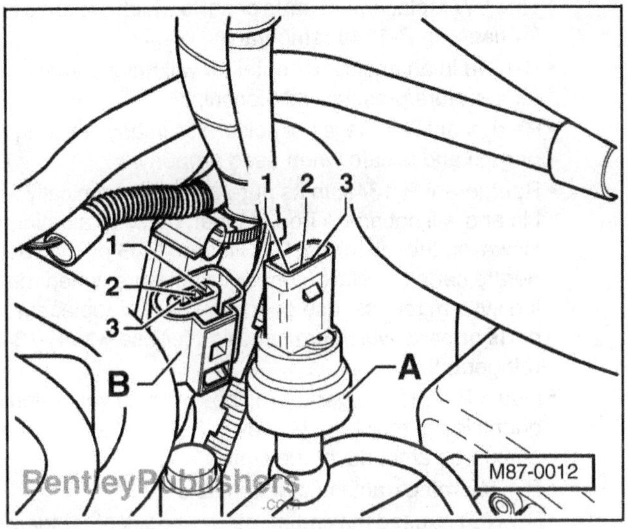

A/C high pressure sensor (G65)

- Not installed on all vehicles.
- Transmits a square wave signal to the coolant fan control module (J293) at a rate that varies according to refrigerant system pressure.
- Coolant fan control module (J293) determines the A/C refrigerant system pressure based on signal from G65, and controls A/C clutch (N25) and coolant fan operation.
- Switches radiator coolant fan to second speed when refrigerant pressure increases.
- Switches off A/C clutch when refrigerant pressure is too excessive (i.e. when engine cooling is insufficient).
- Switches off A/C clutch when refrigerant pressure is too low.
- Terminal 1 = ground
- Terminal 2 = signal output
- Terminal 3 = switched battery positive voltage (B+) from ignition switch

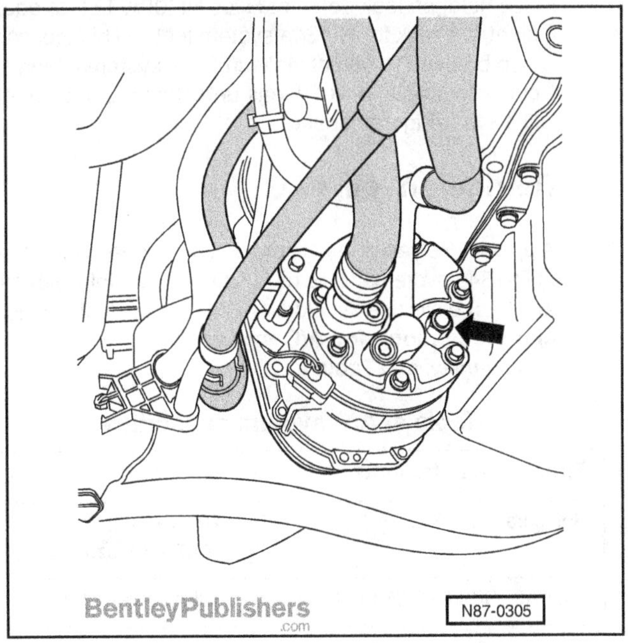

Pressure relief valve

- Protects refrigerant circuit from over-pressure.
- Opens to vent pressures over 40 bar (580 psi) and then closes to prevent total refrigerant loss.
- Adhesive plate (**arrow**) is pushed out if valve has opened.

A/C refrigerant R-134a

Refrigerant R-134a as used in automotive applications has performance similar to FREON® (R-12) but is environmentally safer. Refrigerant R-134a has the following characteristics:

- Colorless and is invisible as a gas.
- R-134a when viewed through the sight glass may appear milky due to the mixture of refrigerant and lubricating oil (PAG oil).
- Escaped R-134a gases are heavier than air and will gather first in low places, such as under the car.
- R-134a refrigerant gas displaces oxygen and may cause suffocation in low areas or areas of poor air circulation.
- Refrigerant R-134a will deteriorate some plastics. Therefore, when making system repairs, use only genuine VW replacement parts or parts which are specified for use with R-134a refrigerant.
- R-134a in an enclosed container will have a specific temperature/pressure relationship.
- Refrigerant R-134a is not poisonous in any state (liquid or gas) and is safe when used properly.
- Refrigerant R-134a, in its pure state, is chemically stable and will not attack iron, copper, brass or aluminum. However, the mixture of R-134a and PAG oil may deteriorate certain metals (copper). Therefore, when making system repairs, use only genuine VW replacement parts or parts which are specified for use with R-134a refrigerant.
- Liquid R-134a refrigerant will absorb only very minute quantities of moisture. However, R-134a vapor can absorb large amounts of moisture.
- R-134a refrigerant is not flammable.
- DO NOT exceed maximum rated capacity of refrigerant stored in containers.
- Use halogen leak detector Hitec HI400A-TEL or equivalent to check for R-134a system leaks. This tool can also be used to detect leaks in R-12 systems. Many currently available R-12 leak detectors cannot detect R-134a refrigerant leaks.

Refrigerant oil, PAG

A special Polyalkylene Glycol (PAG) synthetic oil is used in R-134a systems. This oil is NOT compatible with mineral based oils used in older R-12 systems. Refer to **Special precautions for A/C systems**. For specifications and quantity, see **Table a**.

Table a. A/C system capacities

Fluid & Specification	Quantity
Refrigerant, R-134a	700 g + 50 g (24.7 + 1.8 oz.)
Oil, PAG, SP-10 (G 052 154 A2)	135 cc (4.6 oz.)

NOTE —

- *Volkswagen part numbers are given for reference only. Always consult with your Volkswagen Parts Department or aftermarket parts specialist for the latest parts information.*
- *A label on the lock carrier in the engine compartment indicates the refrigerant type and capacity and should be consulted before starting any servicing or repairs.*

Refrigerant lines and hoses

Lines and hoses are fastened together with threaded couplings and fittings. They are retained to the bodywork or components with specially isolated hose clamps. During servicing, all couplings, fittings and related fasteners must be torqued to specification.

A/C refrigerant system, replacing components

Replacement A/C compressors, evaporators and condensers supplied by the authorized Volkswagen dealer parts departments are filled with Nitrogen. If gas (Nitrogen) does not escape when component is first opened, the component may be faulty (leaking), do not install. Additionally, replacement A/C compressors are filled with the total refrigerant oil quantity needed for the entire refrigerant system.

Always replace the receiver drier whenever the refrigerant system has been left open. Install immediately after opening to prevent moisture contamination of drier desiccant. Moreover, keep refrigerant system and other replacement components sealed for as long as possible to minimize the chance of dirt and moisture contamination. Even the slightest amount of trapped moisture can freeze within the system causing blockage and erratic performance. Always plug any open refrigerant line connections to prevent dirt and moisture contamination during the repair especially if the repair will not be completed immediately.

If the system has been discharged due to a damaged or leaking component (refrigerant hose/line, compressor, evaporator, condenser, etc.), flush the refrigerant system first with compressed air, then with nitrogen (available locally) and collect the oil that runs out. This will remove the refrigerant oil which may be saturated with moisture.

> ***CAUTION —***
> *DO NOT flush R-134a refrigerant system with R-12. R-12 is not compatible with refrigerant R-134a and PAG oil and will cause total system contamination!*

If the compressor is not replaced after flushing the system, fill the compressor with the correct type and quantity of refrigerant oil for the total system as specified in **Table a**. If a new Sanden compressor is used, do not add any additional oil as the total amount of oil required is already in the compressor. Refrigerant oil is distributed in most major parts of the refrigerant system. For approximate capacities, see **Table b**.

Table b. Refrigerant oil distribution

Component	Approximate amount of oil in component - % of total
Compressor	67.5 cc (2.3 oz) - 50%
Condenser	13.5 cc (0.46 oz) - 10%
Suction line	13.5 cc (0.46 oz) - 10%
Discharge line	--
Evaporator	27 cc (0.92 oz) - 20%
Receiver-drier	13.5 cc (0.46 oz) - 10%

WARNING —

- *Dispose of contaminated refrigerant oil following laws governing hazardous waste disposal. Do not combine any refrigerant oil with any other old oils such as engine oil or transmission fluid unless approved by the appropriate governing body.*
- *Observe all cautions, warnings and notes before starting repairs involving air conditioning systems.*

Condenser assembly, overview

1. **Radiator**
   - Drain coolant before removal
   - Always use new coolant after replacement
2. **Condenser**
   - Discharge and recover refrigerant before removal
   - Disconnect refrigerant lines at condenser
   - Seal open refrigerant lines immediately
   - Inspect for bent and obstructed fins and physical damage before installing
   - Always replace O-rings for refrigerant lines
3. **Bolts**
   - Tighten to 8 Nm (71 in-lb)
4. **Lock carrier**
   - Bring into service position for condenser removal
5. **Brackets**
   - For radiator, upper
   - Note installation position
6. **Brackets**
   - For radiator, lower
   - Note installation position

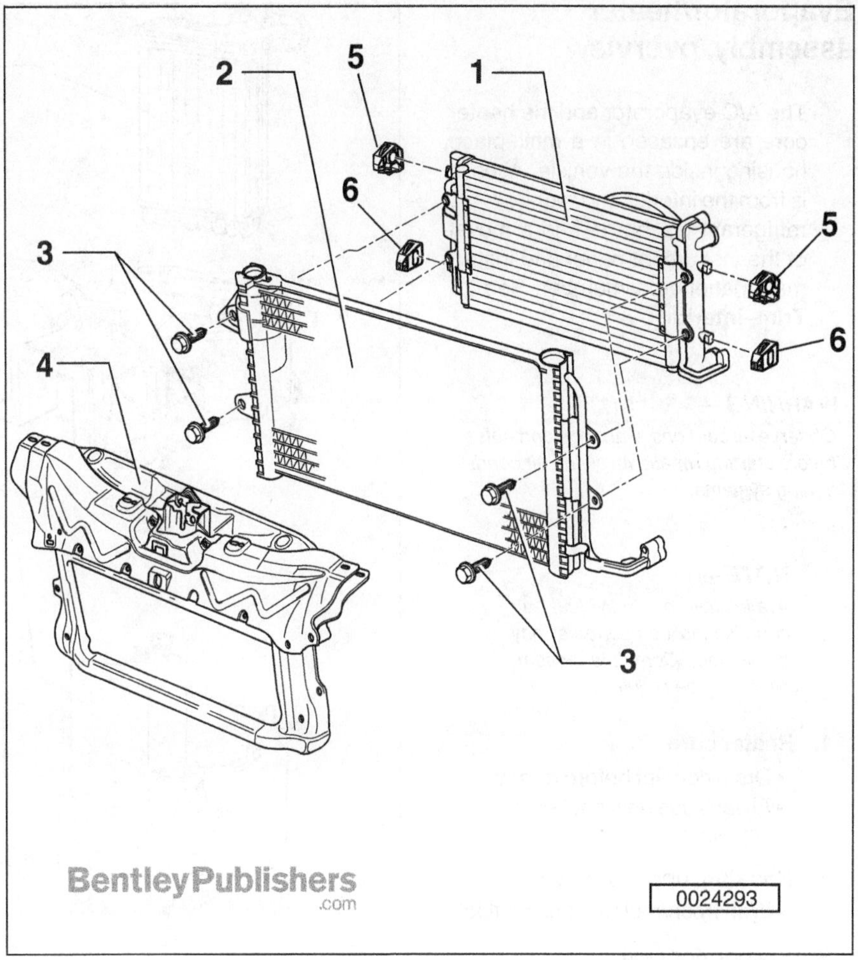

Evaporator/heater assembly, overview

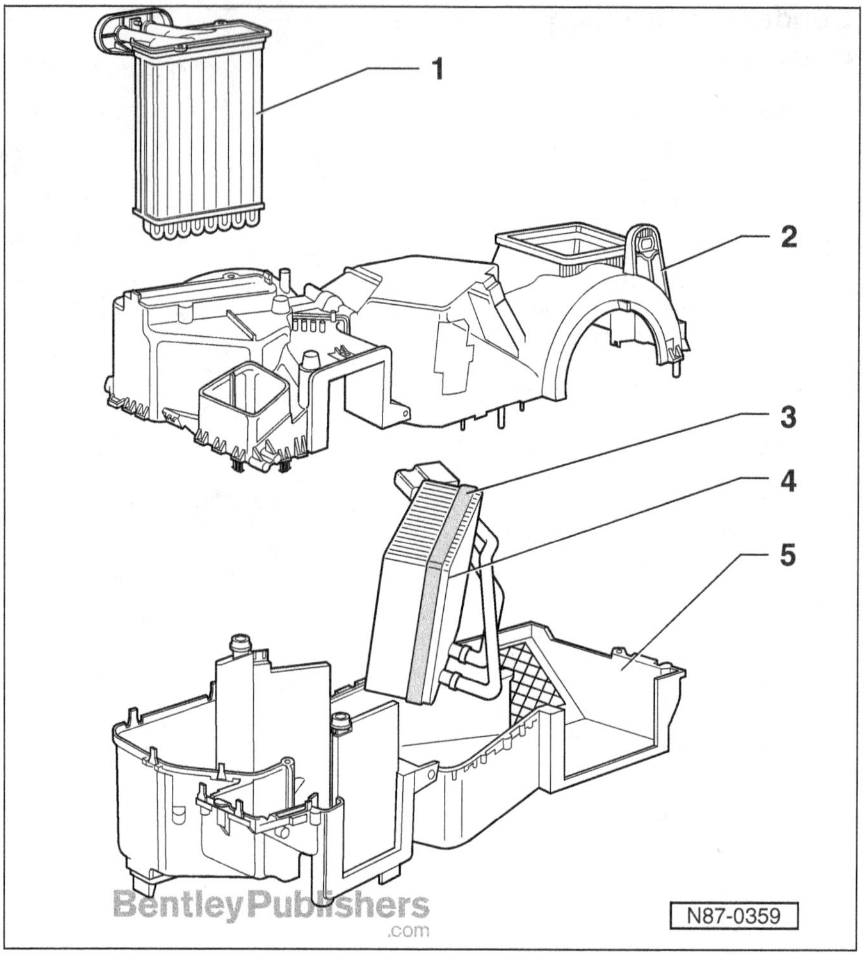

The A/C evaporator and the heater core are encased in a multi-piece housing inside the vehicle. Access is from the interior and requires A/C refrigerant discharge and removal of the instrument panel and instrument panel crossmember. See **70 Trim–Interior**.

WARNING —
Observe all cautions, warnings and notes before starting repairs involving air conditioning systems.

NOTE —
Availability of individual air conditioning components may be limited. Complete assemblies may be required.

1. **Heater core**
   - Drain coolant before removal
   - Always use new coolant after replacement
2. **Housing, upper section**
   - With fresh/recirculating air flap
3. **Evaporator seal**
   - Install seal to top and sides only, not to bottom
   - Secure with glue
   - Condensation MUST be free to drain out from the bottom of the housing under the evaporator.
4. **Evaporator**
   - Discharge and recover refrigerant before removal
   - Disconnect refrigerant lines at expansion valve
   - Seal open refrigerant lines immediately
   - Inspect for bent and obstructed fins and physical damage before installing
   - Always replace O-rings for refrigerant lines
5. **Housing, lower section**

A/C system odor, checking

An unpleasant smell or "musty" odor can be emitted from any automotive air conditioning system under certain conditions. Many times this "musty" odor is caused by residual condensatation in and around the A/C evaporator mixing with airborne pollutants. This environment is suitable for the formation and growth of bacteria and/or fungi which can cause undesirable odors. Condensation is normally formed as part of the air conditioning process and must be allowed to flow out of the system.

To minimize the growth of bacteria and/or fungi, there are several steps that can be performed:

- Water drain(s) in the bottom of the plenum chamber(s) under the hood must be open and free of any debris.
- The bulkhead between the engine compartment and the passenger compartment must be sealed along with any cable grommets passing through the bulkhead.
- The evaporator water (condensate) drain must open freely. This valve is located behind the bulkhead insulation.

If the above steps fail to correct an odor problem, professional preparation and treatment using a commercially available product such as Airsept Air Conditioning Treatment™ should be performed. Special tools, equipment and training are required to properly apply Airsept™ products. These procedures are outside the scope of this repair manual.

A/C REFRIGERANT SYSTEM SERVICING

In accordance with SAE and industry standards, New Beetle air conditioners are equipped with service fittings that accept standard R-134a charging equipment. Connecting gauges to the refrigerant system and measuring the various pressures during operation is a valuable and accurate way to troubleshoot. The pressures and associated temperatures in the A/C system will vary depending on engine speed (rpm), coolant fan speed, engine coolant temperature, A/C clutch engagement, outside temperature, humidity, etc. and can give an accurate indication of system performance.

> ***WARNING —***
> *Observe all cautions, warnings and notes before starting repairs involving air conditioning systems.*

A/C refrigerant system, testing with pressure gauges

Due to the constant temperature/pressure relationship of refrigerant R-134a, approximate high side and low side system temperature can be determined based on system pressures, see **Table c**.

Table c. R-134a temperature/pressure relationship

Temperature in °C (°F)	Pressure in bar (psi)
-30 (-22)	0.0 (0.0)
-20 (-4)	0.3 (4.4)
-10 (14)	1.0 (14.5)
0 (32)	1.9 (27.5)
10 (50)	3.1 (45.0)
20 (68)	4.7 (68.2)
30 (86)	6.7 (97.2)
40 (104)	9.1 (132.0)
50 (122)	12.2 (177.0)
60 (140)	15.8 (229.0)
70 (158)	20.2 (293.0)

Commercially available manifold gauge sets are specifically calibrated for R-134a and should be connected to the system according to the manufacturers instructions. All commercially available refrigerant recovery, recycling and recharging equipment also have gauges that are appropriate for R-134a and are suitable for system testing.

Pressure and temperature measurements should be made using the following conditions:

- Engine running at 1,500 rpm with A/C switched on.
- Blower fan control set to highest speed.
- A/C temperature control set to maximum cooling and recirculation switched on.
- All passenger compartment air outlets open fully.

For pressure and temperature specifications of the various components, see **Table d**.

Table d. Pressure and temperature specifications[a]

<table>
<tr><th>Component</th><th>Refrigerant state</th><th>Approximate pressure</th><th>Approximate temperature</th></tr>
<tr><td>Evaporator, inlet to outlet</td><td>Vapor to gas</td><td rowspan="3">1.2 bar (17.4 psi)[b]</td><td>-7°C (19°F)[c]</td></tr>
<tr><td>Low pressure service valve port</td><td>Gas</td><td rowspan="2">-1°C (30°F)</td></tr>
<tr><td>Compressor, low pressure (suction) side</td><td>Gas</td></tr>
<tr><td>Compressor, high pressure (discharge) side</td><td>Gas</td><td rowspan="6">14 bar (203 psi)</td><td>65°C (149°F)</td></tr>
<tr><td>Condenser</td><td>Gas to vapor to liquid</td><td>55°C (131°F) at outlet</td></tr>
<tr><td>Receiver drier</td><td>Liquid</td><td rowspan="3">55°C (131°F)</td></tr>
<tr><td>Sight glass (where equipped)</td><td>Liquid</td></tr>
<tr><td>High pressure service valve port</td><td>Liquid</td></tr>
<tr><td>Expansion valve</td><td>Liquid to vapor</td><td>55°C (131°F) reducing to -7°C (19°F)</td></tr>
<tr><td>Low pressure service valve port</td><td>Gas</td><td>Inlet: 14 bar (203 psi)
Outlet: 1.2 bar (17.4 psi)</td><td>Inlet: 55°C (131°F)
Outlet: -7°C (19°F)</td></tr>
</table>

a. Pressure and temperature specifications are based on: 1) Engine speed at 1,500 rpm, 2) Blower fan set to high speed, 3) A/C controls set to maximum cooling.

b. Pressure maintained in the refrigerant system by the variable displacement compressor despite variables in temperature, load and engine speeds (RPM).

c. Temperature maintained in the refrigerant system by the variable displacement compressor despite variable in temperature, load and engine speeds (RPM).

A/C refrigerant system, discharging and charging

Before starting repairs involving opening of the refrigerant system, the refrigerant must first be removed. It is not permissible to vent the refrigerant into the atmosphere for environmental, legal and economic reasons. In addition, due to the lack of a sight glass in most vehicles, a measured amount of refrigerant must be installed to insure that the system is fully charged. Equipment suitable for this purpose is available from several manufacturers. All refrigerant servicing equipment attaches to the high and low pressure service valve ports, but exact connection details and operation may differ slightly between individual equipment manufacturers. Always follow the equipment manufacturer's instructions regarding connections to the vehicle.

After discharging and recovering the refrigerant, proceed with repairs as required. Do not leave the refrigerant system open. Seal openings to keep dirt and especially moisture from entering. When repairs have been completed, the system must be evacuated for a minimum of 30 minutes. This will "pull the system down" by creating a vacuum. It will remove all traces of the old refrigerant ensuring an accurate recharge. More importantly, however, it will remove any moisture that may have entered the system.

When the system has been repaired and properly evacuated, add refrigerant oil if required and recharge according to the equipment manufacturers instructions. Consult the A/C system capacities table given earlier.

After the system has been recharged, switch on the ignition, but do not start the engine. Switch on the A/C and manually rotate the A/C compressor approximately 10 turns before starting the engine to insure adequate compressor lubrication. Start the engine with the A/C switched OFF. After idle speed has stabilized, switch A/C ON and let the engine idle with the compressor running for a minimum of two minutes before raising engine speed. This will allow the refrigerant oil to properly circulate to all parts of the system.

A/C refrigerant system, flushing

Certain circumstances may require complete removal of the refrigerant and refrigerant oil. This process is known as flushing and is accomplished with nitrogen and compressed air. Compressed nitrogen with regulators and adapters are available locally. Flushing is usually needed under the following conditions:

- Refrigerant oil is dark and viscous (thick) or shows metallic particles.
- Too much refrigerant oil is known to be in the system following compressor replacement.
- Unclear or do not know how much refrigerant oil is in the system.
- Moisture, dirt or other impurities have entered the refrigerant system, (i.e. following an accident).
- Unable to pull a constant vacuum during evacuation of a leak-free system due to excessive moisture in system.
- Refrigerant system has been open longer than the time required for normal repairs, (i.e. following an accident).
- Based on temperature and pressure measurements, system is diagnosed with moisture contamination.
- Compressor is replaced.

If flushing the system, observe the following points:

- When using compressed nitrogen always use a pressure regulator and the proper adaptor hoses and fittings (available locally).
- During flushing, use existing exhaust/ventilation systems to draw off the gas mixture escaping from the A/C system.
- Use compressed air and nitrogen (available locally) to remove moisture, impurities and old refrigerant oil from A/C refrigerant system.
- First blow out old refrigerant oil and dirt with compressed air, then dry components with nitrogen.
- DO NOT blow compressed air and nitrogen through the compressor or expansion valve. Only blow compressed air and nitrogen through disconnected, free flowing components (i.e. disconnected hose, condenser, evaporator, etc.)

- DO NOT blow compressed air and nitrogen into a capped off A/C component. Pressurized R-134a refrigerant in the presence of oxygen may form a combustible mixture.
- Always flush components in opposite direction of refrigerant flow.
- Flush evaporator through the low pressure line with the high pressure line removed.
- If any component has dark thick deposits that cannot be removed with compressed air, replace component.
- Thin light gray deposits in refrigerant lines and hoses are normal and do not impair the function of the system.
- Always replace receiver drier and restrictor after flushing.
- Dispose of contaminated refrigerant (PAG) oil following laws governing hazardous waste disposal. Do not combine PAG oil with any other old oils such as engine oil or transmission fluid.

WARNING —
- *DO NOT flush R-134a refrigerant system with R-12. R-12 is not compatible with R-134a refrigerant and PAG oil and will cause total system contamination.*
- *DO NOT blow compressed air and nitrogen through the compressor or expansion valve. Only blow compressed air and nitrogen through disconnected, free flowing components (i.e. disconnected hose, condenser, evaporator, etc.).*
- *DO NOT blow compressed air and nitrogen into a capped off A/C component. Pressurized R-134a refrigerant in the presence of oxygen may form a combustible mixture.*

A/C refrigerant system, checking for leaks

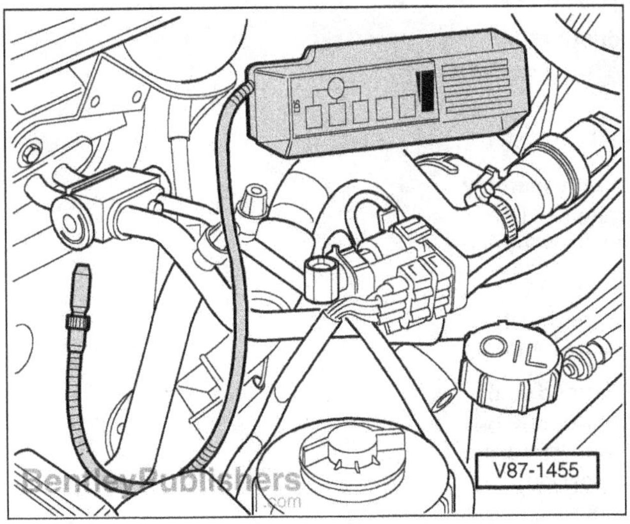

If it is determined that the refrigerant system is low on refrigerant and the source of the leak is not readily apparent, use of a halogen leak detector may help in locating the leak.

NOTE —
- *Refrigerant gas dissipates very quickly. To make the job easier, avoid drafty or windy areas when checking for leaks.*
- *If the refrigerant system is discharged (empty), it is permissible to temporarily recharge the system with approx. 100 g (3.5 oz) of refrigerant in order to check for leaks. When the source of the leak is found, this refrigerant must be recovered from the system before repairs are started.*

In some instances, very small leaks may not be readily apparent due to location or other circumstances. In these situations, commercially available refrigerant dye may be useful in locating the source of the leak.

A/C, HEATING AND VENTILATION CONTROLS

WARNING —
Observe all cautions, warnings and notes before starting repairs involving air conditioning systems.

NOTE —
- *For vehicles without air conditioning, see* **80 Heating and Ventilation**.
- *Volkswagen identifies electrical components by a letter and/or a number in the electrical schematics. See* **EWD Electrical Wiring Diagrams**. *These electrical identifiers are listed in parenthesis as an aid to electrical troubleshooting.*

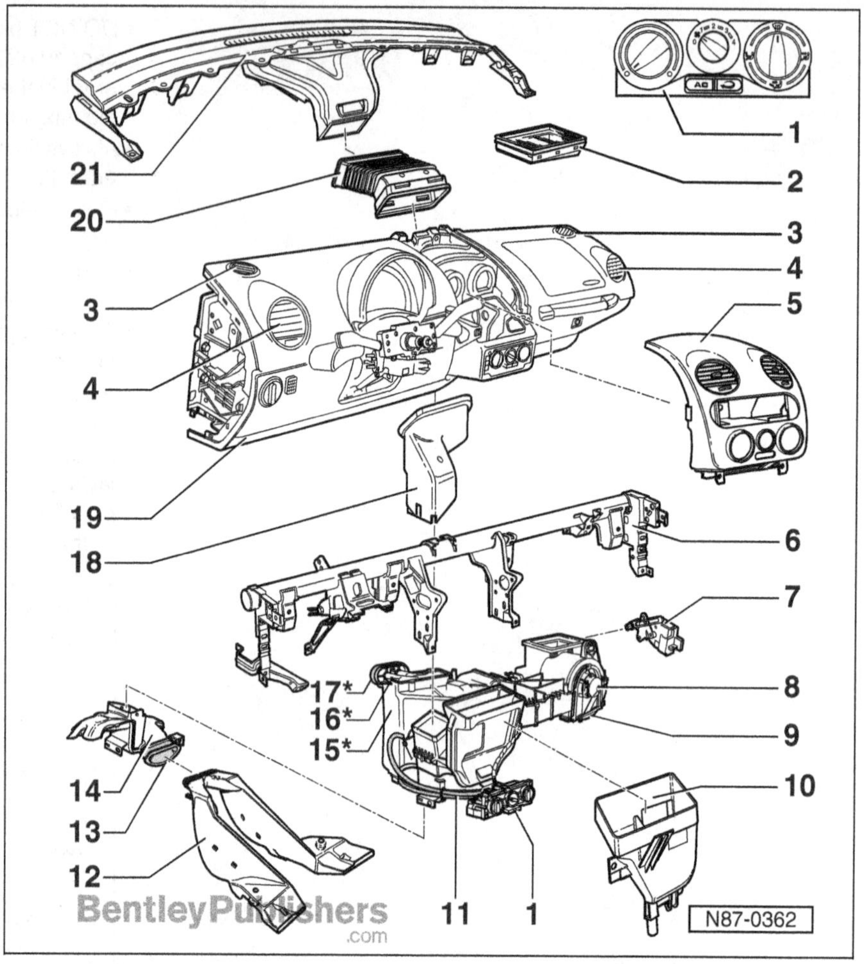

A/C passenger compartment components, overview

1. **A/C, heating and ventilation controls**
   - With A/C switch (E35)
   - With blower fan switch (E9)
   - With recirculation switch (E184)
2. **Dust and pollen filter**
   - Covers outside air inlet to passenger compartment
   - Cleans all incoming air
3. **Side window air outlet**
4. **Side air outlets**
5. **Center air outlet console**
6. **Instrument panel cross member**
7. **Servo motor for recirculation door (V154)**
8. **Blower fan (V2)**
   - Updated fan available from Volkswagen with reinforced center hub. Part #1J1 819 021B is described in technical bulletin #80-01-01, dated July 18, 2001.
9. **Blower fan series resistance pack (N24)**
   - With thermal fuse
10. **Intermediate duct**
11. **Control cables**

    Color coding:
    - Yellow - central flap
    - Green - footwell/defrost flap
    - Beige - temperature flap
12. **Rear footwell duct**
13. **Gasket**
14. **Connecting duct**
15. **Evaporator/heater assembly**
    - Refrigerant system must be discharged before removing.
16. **Heater core**
    - Refrigerant system must be discharged before removing.
17. **Heater core/bulkhead seal**
    - Refrigerant system must be discharged before removing.
    - Note installed position
18. **Defroster duct**
19. **Instrument panel**
20. **Center air outlet duct**
21. **Defroster air outlet panel**

A/C, heating and ventilation controls, removing and installing

> ***WARNING —***
> *Observe all cautions, warnings and notes before starting repairs involving air conditioning systems.*

- Remove center air outlet console, see **80 Heating and Ventilation.**

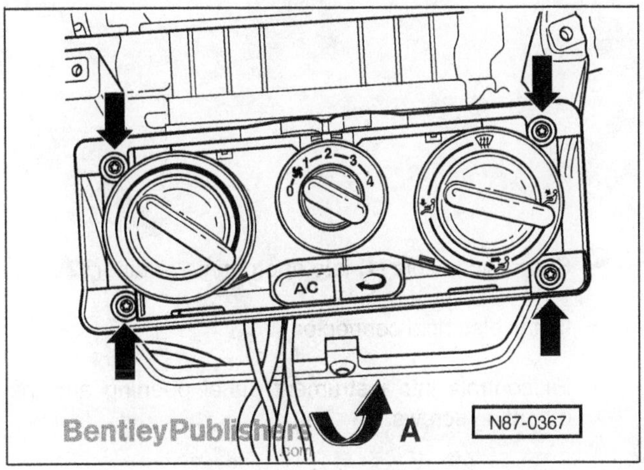

◄ Remove mounting screws (**arrows**) and lift (**arrow A**) controls from opening in dash.

- Unclip cables and detach from controls.

- Disconnect electrical connector and remove controls.

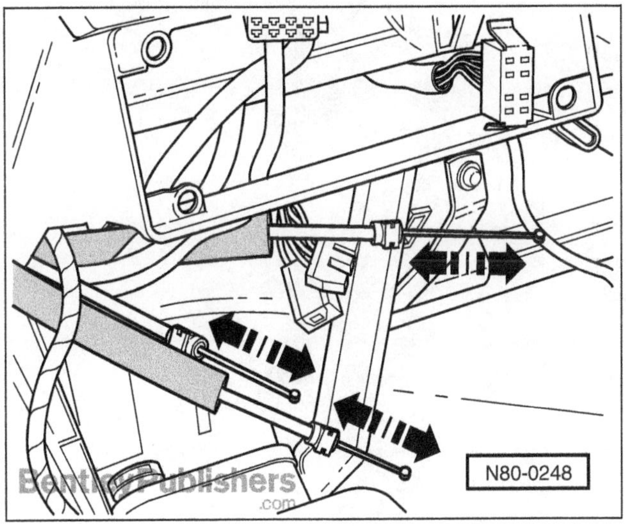

◄ Before installing controls, check operation of cables by moving in direction of **arrows**.

- Replace cables with damage or stiff/binding operation.

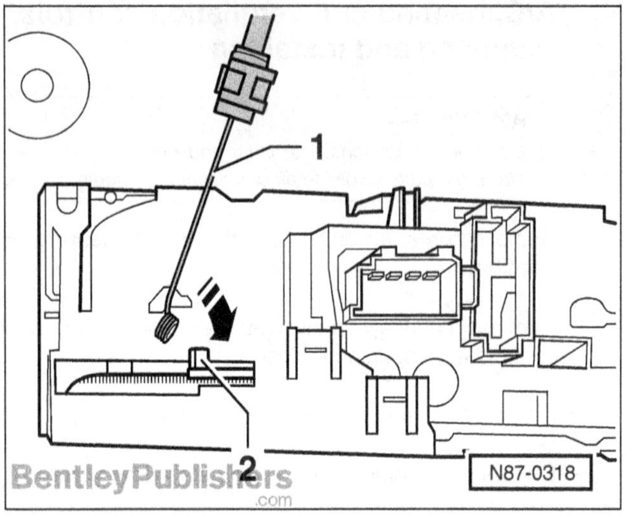

Attach color coded cable (**1**) to correct actuating arm (**2**).

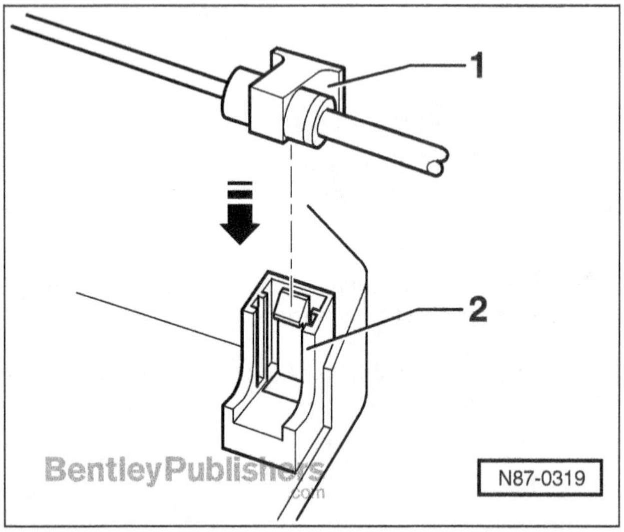

Snap each cable (**1**) into appropriate retainer (**2**).

- Install electrical connector.
- Fit controls into instrument panel opening and install mounting screws.
- Adjust cables as needed.

A/C, heating and ventilation controls, assembly

NOTE —
Volkswagen identifies electrical components by a letter and/or a number in the electrical schematics. See **EWD Electrical Wiring Diagrams**. *These electrical identifiers are listed in parenthesis as an aid to electrical troubleshooting.*

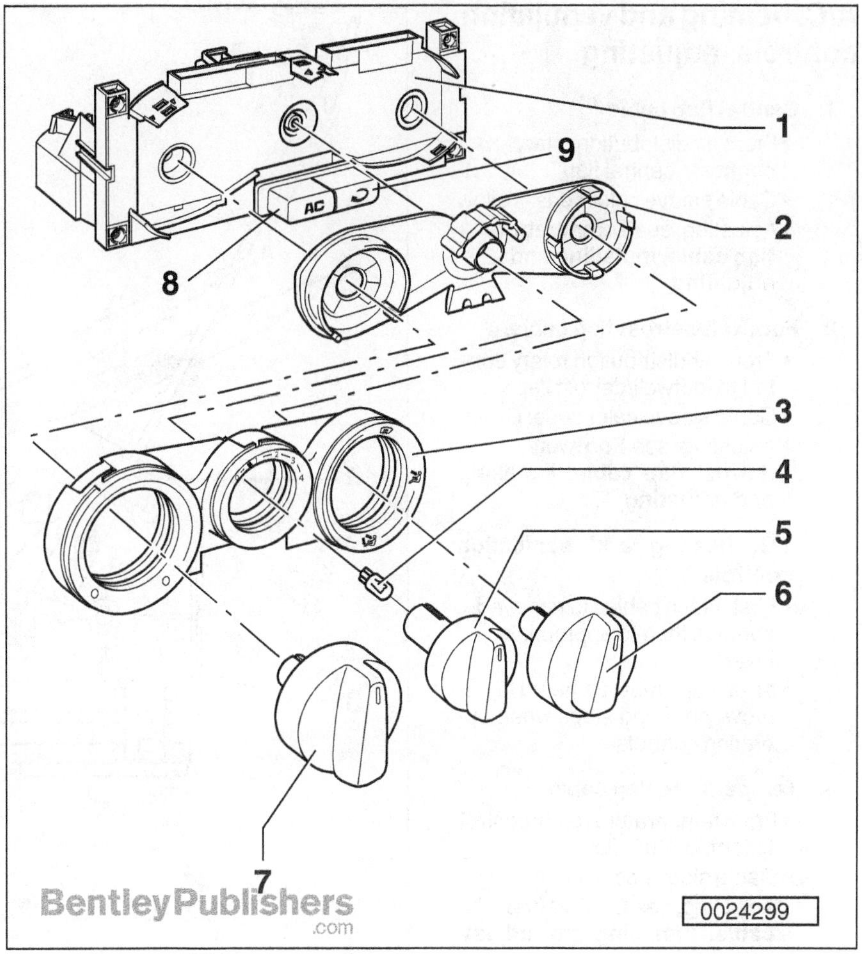

1. **A/C, heating and ventilation controls**
   - With A/C switch (E35)
   - With blower fan switch (E9)
   - With recirculation switch (E184)
2. **Illumination filter ring**
3. **Trim panel**
4. **Fresh air control lever light (L16)**
   - Clear bulb 12V, 1.2W
   - Use small (approx. 4mm i.d.) rubber hose to remove bulb. Press hose onto bulb far enough to grip it and pull.
5. **Rotary control knob**
   - For blower fan speed
   - Removing: use pliers with suitable plastic or rubber protection on jaws
6. **Rotary control knob**
   - For air distribution
   - Removing: Use pliers with suitable plastic or rubber protection on jaws
7. **Rotary control knob**
   - For interior temperature regulation
   - Removing: Use pliers with suitable plastic or rubber protection on jaws
8. **A/C switch (E35)**
   - Integral with controls
9. **Recirculation switch (E184)**
   - Integral with controls

A/C, heating and ventilation controls, adjusting

1. **Central flap cable**
   - From air distribution rotary control to central flap
   - Cable sleeve color code: Yellow
   - Adjusting, see **Temperature flap cable, installing and adjusting**
2. **Footwell/defrost flap cable**
   - From air distribution rotary control to footwell/defrost flap
   - Cable sleeve color code: Green
   - Adjusting, see **Footwell/ defrost flap cable, installing and adjusting**
3. **A/C, heating and ventilation controls**
   - First attach cables to removed controls then to appropriate levers
   - All air flaps must be heard to move onto end stops when operating controls
4. **Temperature flap cable**
   - From temperature rotary control to temperature flap
   - Cable sleeve color code: Beige
   - Adjusting, see **Central flap cable, installing and adjusting**

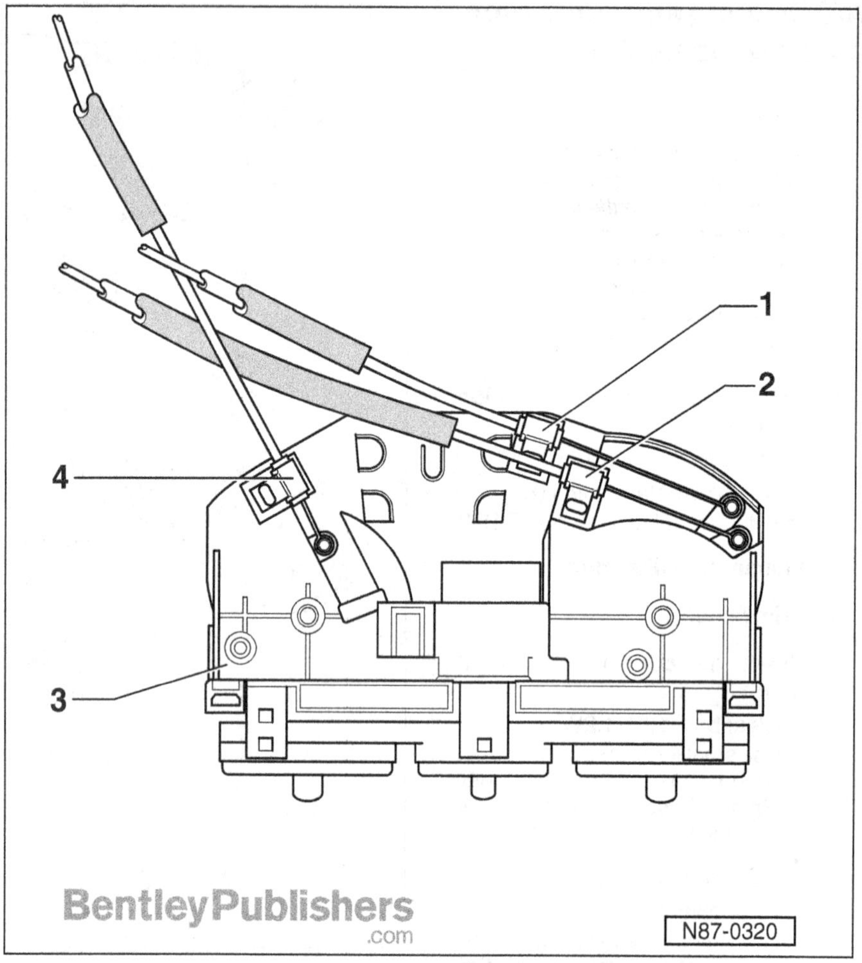

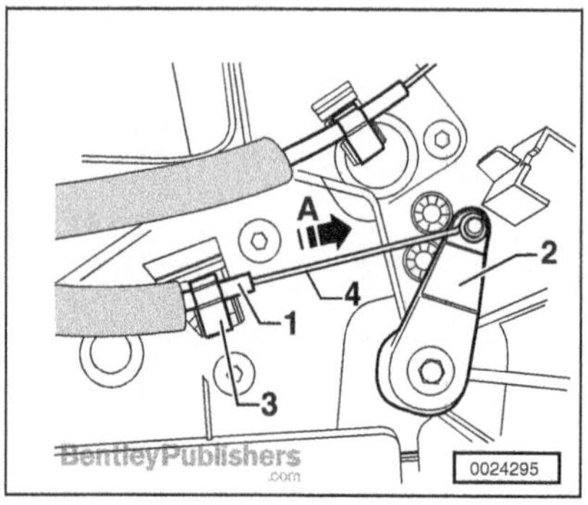

Temperature flap cable, installing and adjusting

- Cable sleeve color code: beige.
- Install/adjust with controls installed and connecting duct removed.
- Turn temperature rotary control knob fully counter-clockwise against stop.
- Attach center wire of cable (**4**) to the temperature flap lever (**2**).
- Push temperature flap lever (**2**) against stop (**arrow A**) and secure outer cable (**1**) with clip (**3**).
- Turn temperature rotary control knob fully to left and right against stops.
- When turning the rotary control knob both end stops must be reached.

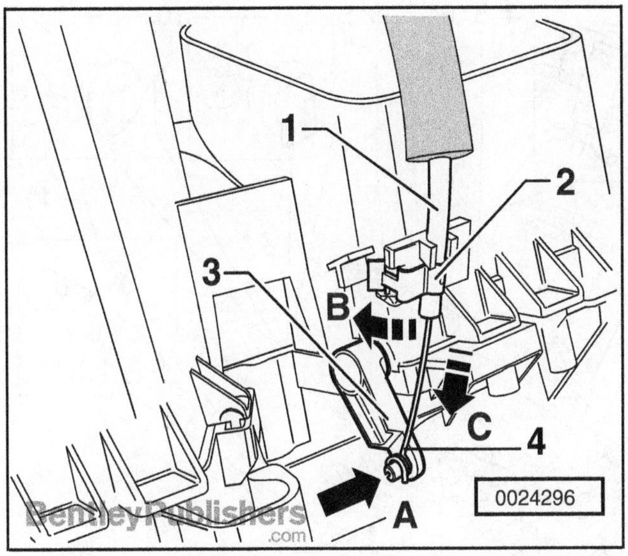

Footwell/defrost flap cable, installing and adjusting

- Cable sleeve color code: green.
- Install/adjust with controls installed.
- Turn air distribution rotary control knob fully counter-clockwise against stop.
- Attach center wire of cable (**4**) to the footwell/defrost flap lever (**3**) at arrow (**A**).
- Press footwell/defrost flap lever against stop (**arrow C**) and secure outer cable (**1**) with clip (**2**) (**arrow B**).
- Turn air distribution rotary control knob fully to left and right against stops.
- When turning the rotary control knob both end stops must be reached.

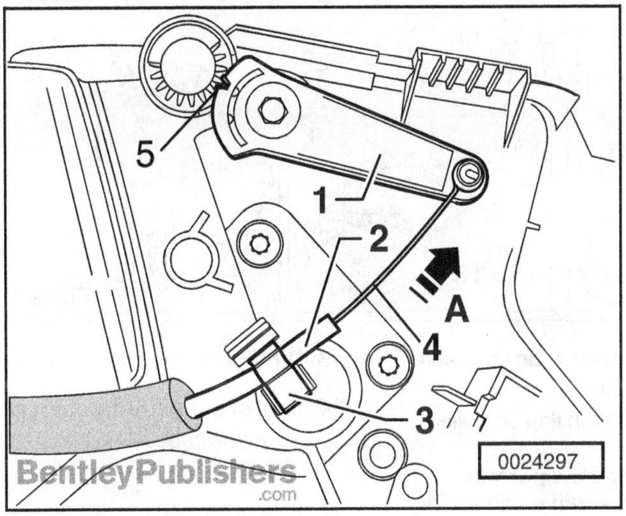

Central flap cable, installing and adjusting

- Cable sleeve color code: yellow.
- Install/adjust with controls installed, connecting duct removed and flap lever correctly indexed (**5**).
- Turn air distribution rotary control knob fully counter-clockwise against stop.
- Attach center wire of cable (**4**) to the central flap lever (**1**).
- Press central flap lever against stop (**arrow A**) and attach outer cable (**2**) with clip (**3**).
- Turn air distribution rotary control knob fully to left and right against stops.
- When turning the rotary control knob both end stops must be reached.

A/C COMPONENTS, UNDERHOOD

WARNING —
Observe all cautions, warnings and notes before starting repairs involving air conditioning systems.

NOTE —
- *Components identified with an asterisk (*) can only be removed after discharging the refrigerant.*
- *Volkswagen identifies electrical components by a letter and/or a number in the electrical schematics. See* **EWD Electrical Wiring Diagrams**. *These electrical identifiers are listed in parenthesis as an aid to electrical troubleshooting.*

A/C components, underhood overview

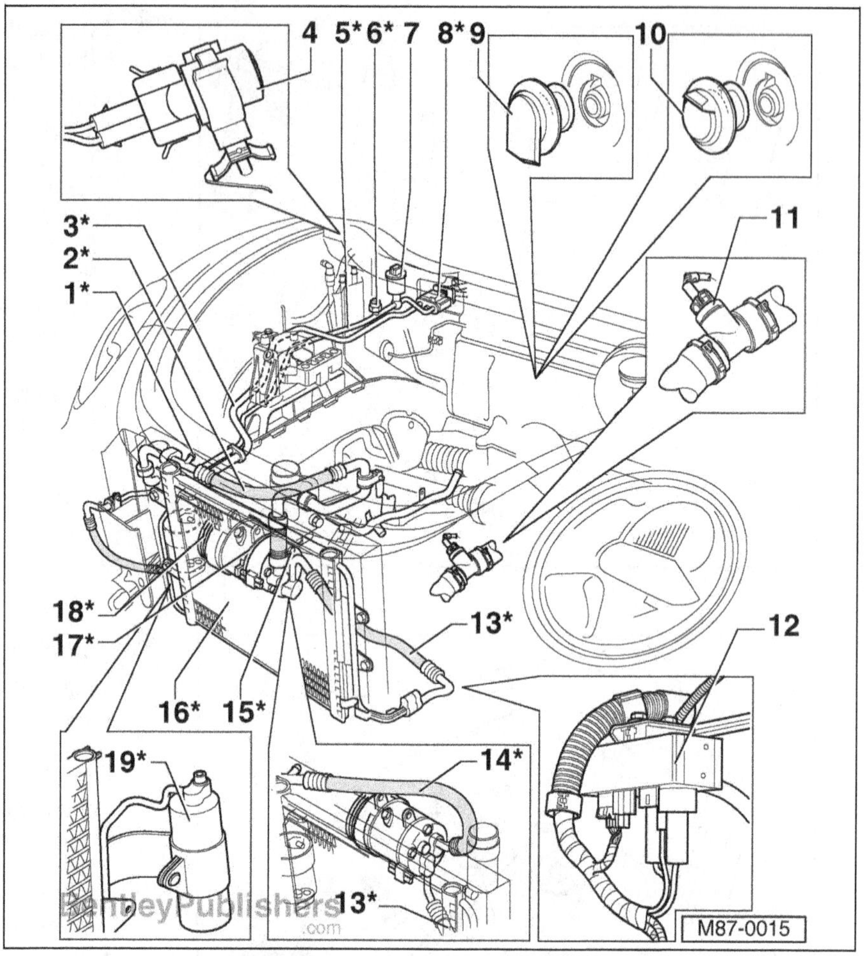

1. **Damper**
2. **Low pressure refrigerant line II**
   - 2.0L engine shown
3. **Low pressure refrigerant line I**
4. **Ambient air temperature switch (F38)**
   - Switches off A/C clutch (N25) at low ambient air temperatures to prevent damage
   - Switches clutch off at -1°C (30°F)
   - Switches back on at 7°C (45°F)
5. **Liquid refrigerant line**
6. **High pressure service valve port**
7. **A/C pressure switch (F129)**
8. **Expansion valve**
9. **Evaporator water drain valve (from 06/2000)**
   - Checking, see **Evaporator water drain valve, checking (from 06/2000)**
10. **Evaporator water drain valve (through 05/2000)**
    - Under insulation flap on bulkhead
    - Checking, see **Evaporator water drain valve, checking (through 05/2000)**
11. **A/C thermal cut-out switch (F163)**
    - Switches off A/C clutch (N25) at excessively high coolant temperature
    - Switches clutch off at 119°C (246°F)
    - Switches clutch back on at 112°C (234°F)
12. **Coolant fan control module (J293)**
    - Location: lower left side of engine compartment near radiator
    - Controls all A/C clutch functions and A/C related radiator fan functions
13. **High pressure refrigerant line**
14. **Low pressure refrigerant line II**
    - Engine code ALH version shown
15. **Pressure relief valve**
16. **Condenser**
17. **Compressor**
18. **A/C compressor clutch (N25)**
19. **Receiver drier**

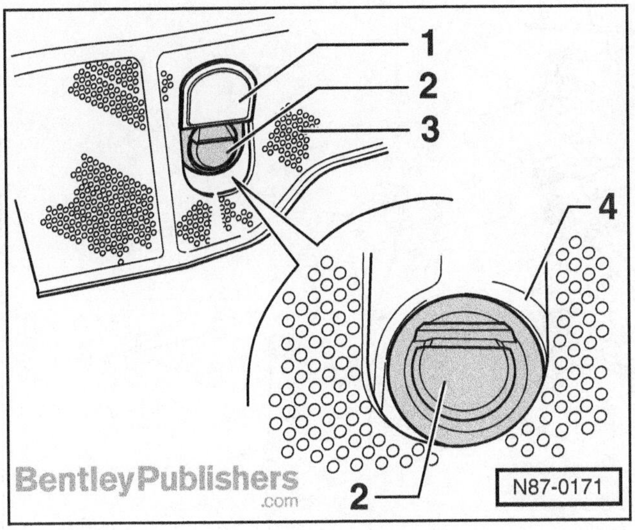

Evaporator water drain valve, checking (through 05/2000)

- Fold cover (**1**) in bulkhead insulation mat (**3**) upwards.
- Remove drain valve (**2**) from opening in bulkhead (**4**).
- The water drain valve flap and body must not be stuck together.
- Insulation mat must not be deformed or damaged in the area of water drain valve.
- The opening in the drain valve body must point downwards with the flap hinge at the top.
- When insulation cover (**1**) is closed it must be flush with insulation matting (**3**). If insulation cover (**1**) is pushed in too far, the water drain valve flap can become jammed.

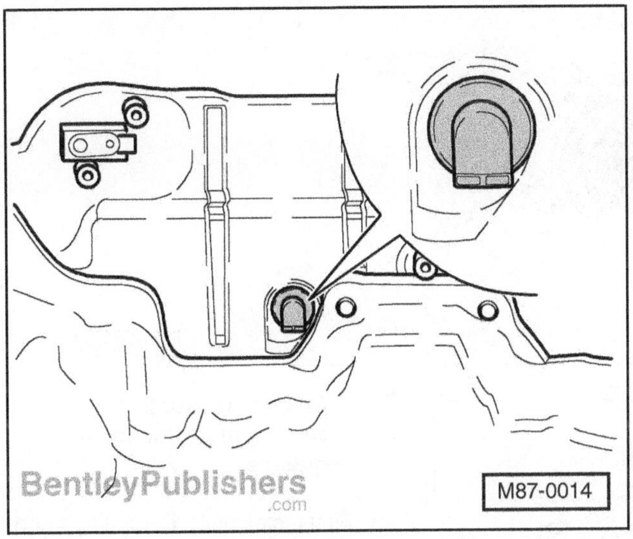

Evaporator water drain valve, checking (from 06/2000)

- Design of new evaporator drain valve is essentially maintenance free.

A/C compressor and mount (1.8L)

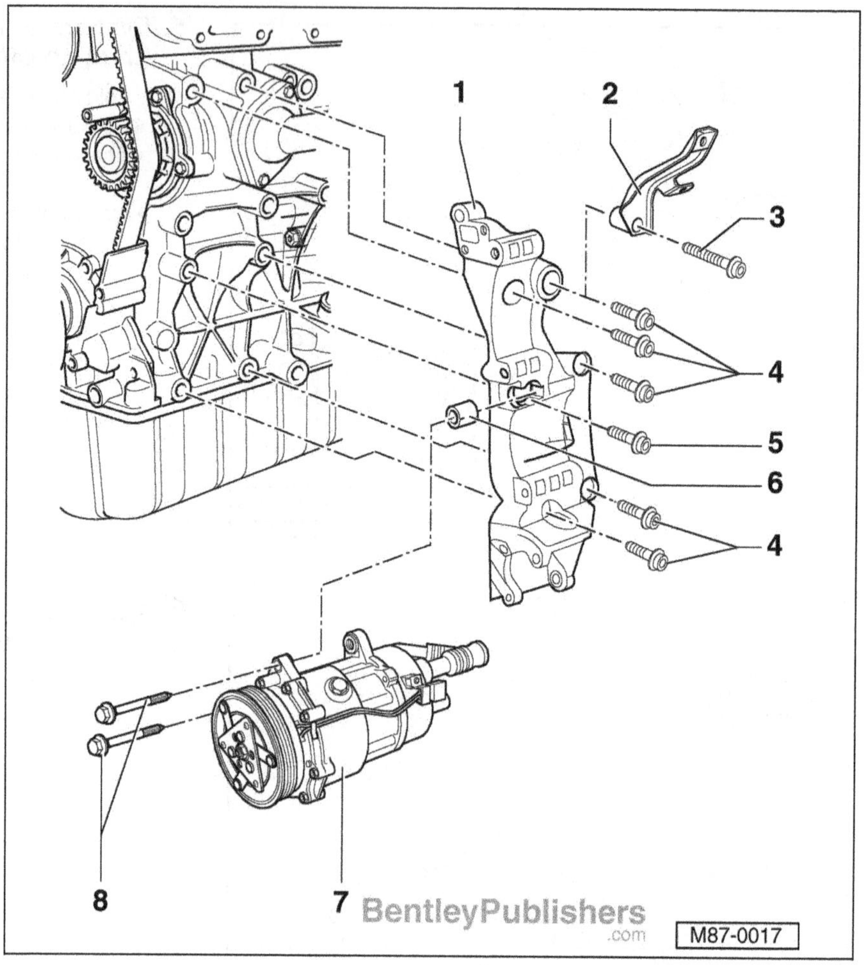

1. **Mounting bracket**
   - Compressor bracket and related components can be removed and installed without having to open the refrigerant circuit
   - Mounting for generator, A/C compressor and power steering pump
   - Remove compressor from bracket, swing away and secure in engine compartment with wire
2. **Intake manifold bracket**
3. **Bolt**
   - M10 x 70
   - Tighten to 45 Nm (33 ft-lb)
   - Note installation sequence
4. **Bolts**
   - M10 x 45
   - Tighten to 45 Nm (33 ft-lb)
   - Note installation sequence
5. **Bolt**
   - Not installed on all versions
   - M10 x 45
   - Tighten to 45 Nm (33 ft-lb)
   - Note installation sequence
6. **Bushing**
   - Not installed on all versions
7. **A/C compressor**
   - Sanden SD7-V16
   - With A/C clutch (N25)
8. **Bolts**
   - M10 x 112
   - Tighten to 45 Nm (33 ft-lb)

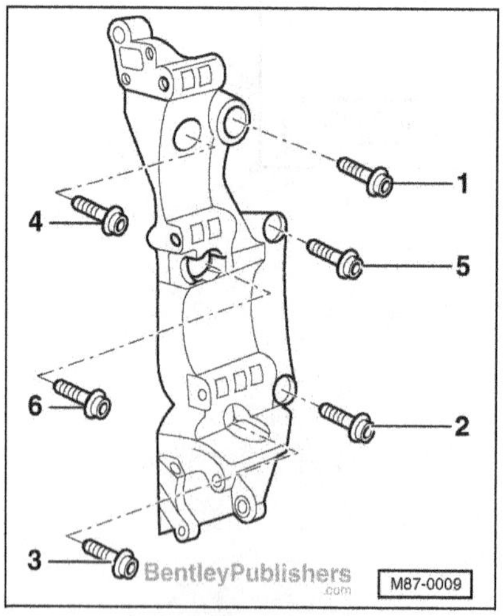

1.8L compressor bracket mounting bolts, installation sequence

– First install bolts **1** and **2** by hand.

– Install remaining bolts (**3, 4, 5** and **6** as applicable).
   - Bolt **6** is not used in all applications

– Tighten all bolts to 45 Nm (33 ft-lb).

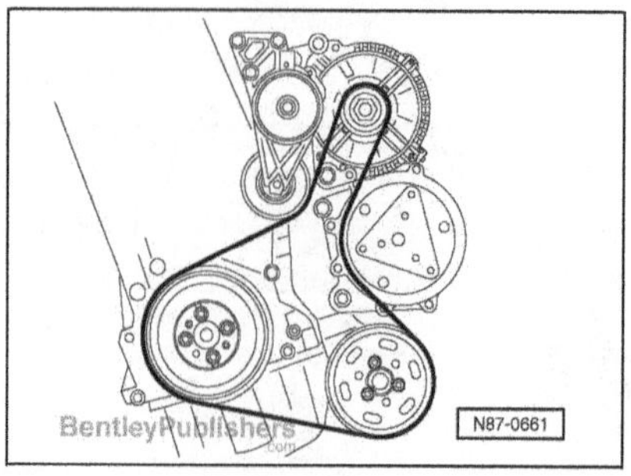

Ribbed V-belt routing

- Always mark the running direction before removing and install in the same direction.
- When installing the belt ensure it is correctly seated in the pulleys.
- See **03 Maintenance** for ribbed V-belt removal and installation.

A/C compressor and mount (1.9L, ALH engine)

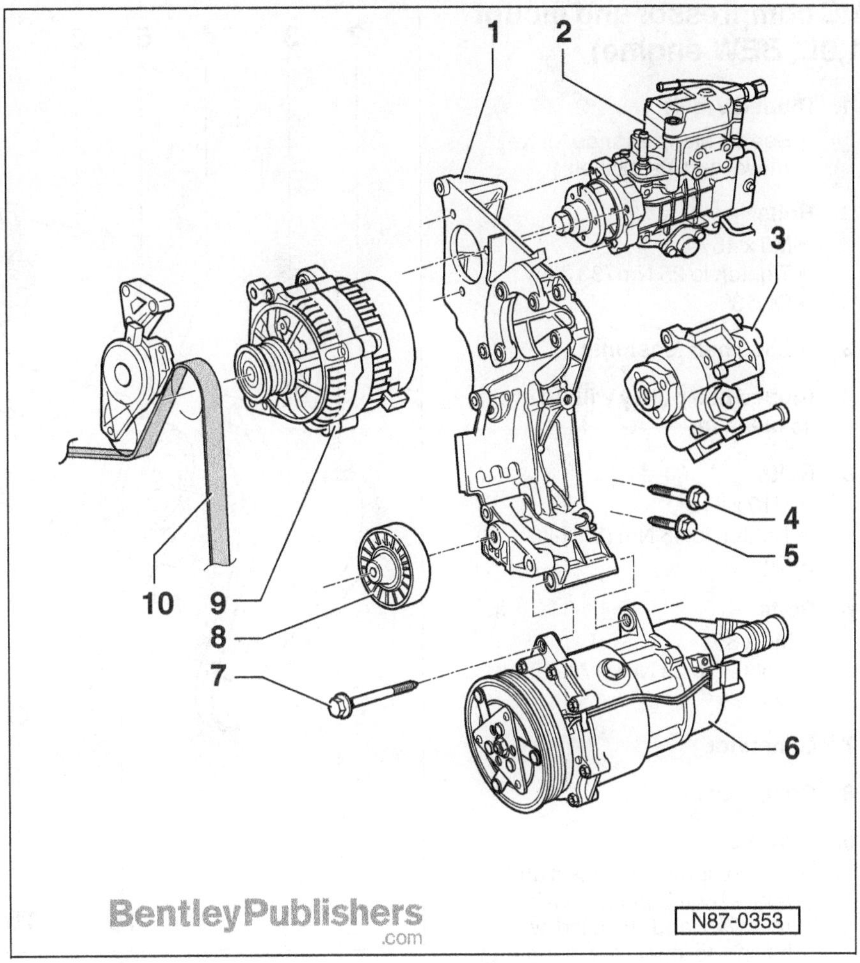

1. **Bracket**
   - The compressor bracket and related components can be removed and installed without having to open the refrigerant circuit.
   - Mounting for generator, compressor, power steering pump, diesel injection pump.
   - Remove compressor from bracket, swing away and secure in engine compartment with wire.
2. **Diesel injection pump**
3. **Power steering pump**
4. **Bolts**
   - Tighten to 45 Nm (33 ft-lb)
5. **Bolts**
   - Tighten to 45 Nm (33 ft-lb)
6. **Compressor**
7. **Bolt**
   - Tighten to 45 Nm (33 ft-lb)
8. **Idler pulley**
   - See **Idler pulley, assembly**
9. **Generator (GEN) and pulley**
10. **Ribbed V-belt**
    - See **Ribbed V-belt routing**
    - See **03 Maintenance** for removal and installation

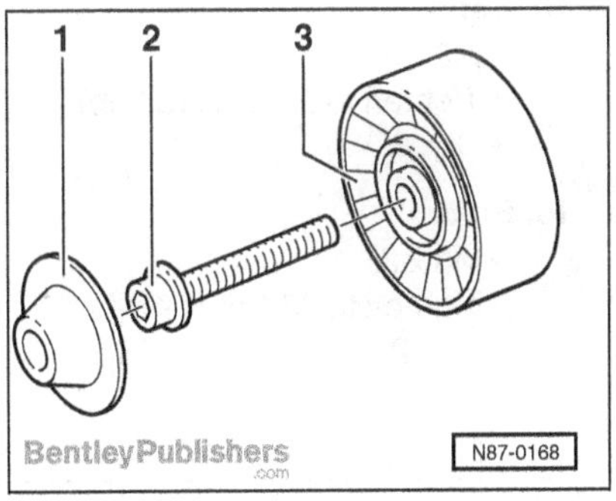

Idler pulley, assembly

1. Protective cap
2. Bolt
   - Tighten to 25 Nm (18 ft-lb)
3. Idler roller with bearings

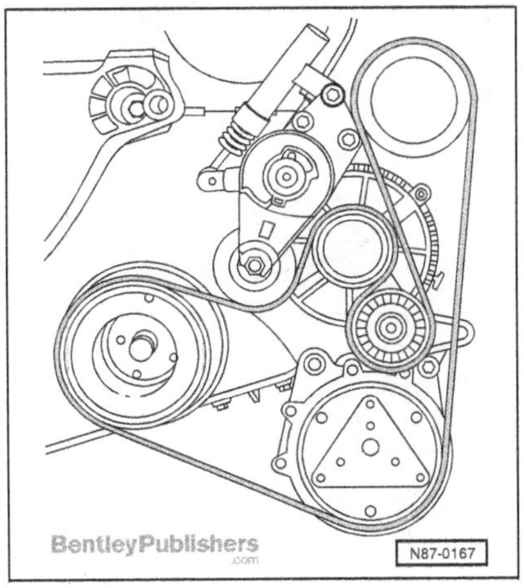

Ribbed V-belt routing

- Always mark the running direction before removing and install in the same direction.
- When installing the belt ensure it is correctly seated in the pulleys.
- See **03 Maintenance** for ribbed V-belt removal and installation.

A/C compressor and mount (1.9L, BEW engine)

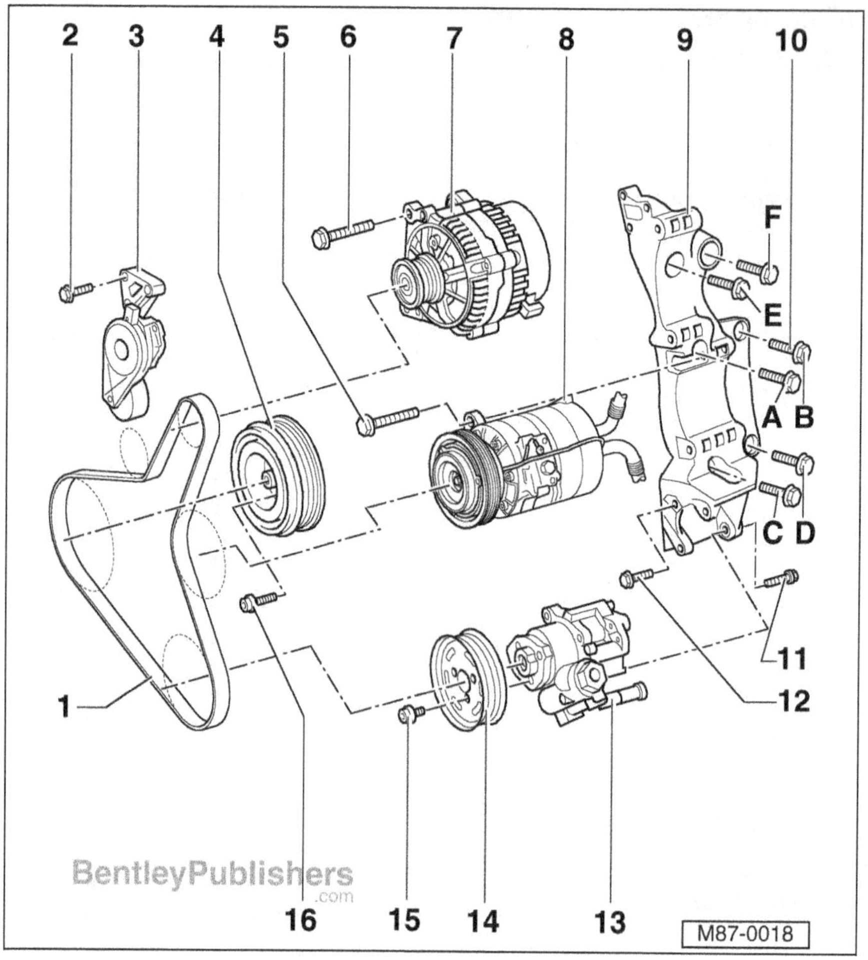

1. **Ribbed V-belt**
   - See **03 Maintenance** for removal and installation
2. **Bolts**
   - M8 x 45
   - Tighten to 25 Nm (33 ft-lb)
   - Qty: 3
3. **Ribbed belt tensioner**
4. **Ribbed belt pulley with harmonic balancer**
5. **Bolts**
   - M10 x 112
   - Tighten to 45 Nm (33 ft-lb)
   - Qty: 2
6. **Bolts**
   - M8 x 85
   - Tighten to 23 Nm (17 ft-lb)
   - Qty: 2
7. **Generator**
8. **Compressor**
9. **Bracket**
   - The compressor bracket and related components can be removed and installed without having to open the refrigerant circuit.
   - Mounting for generator, compressor, and power steering pump.
   - Remove compressor from bracket, swing away and secure in engine compartment with wire.
   - Tighten mounting bolts in sequence, starting with **A**.
10. **Bolts**
    - M10 x 45
    - Tighten to 45 Nm (33 ft-lb)
    - Qty 6
11. **Bolt**
    - M8 x 40
    - Tighten to 25 Nm (18 ft-lb)
12. **Bolt**
    - M8 x 40
    - Tighten to 25 Nm (18 ft-lb)
13. **Power steering pump**
14. **Ribbed V-belt pulley**
15. **Bolts**
    - M8 x 14
    - Tighten to 25 Nm (18 ft-lb)
    - Qty: 3
16. **Bolts**
    - M8 x 85
    - Tighten to 23 Nm (17 ft-lb)
    - Qty: 2

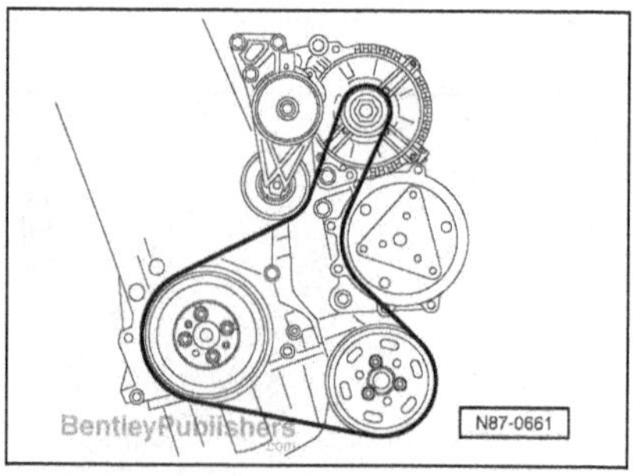

Ribbed V-belt routing

- Always mark the running direction before removing and install in the same direction.
- When installing the belt ensure it is correctly seated in the pulleys.
- See **03 Maintenance** for ribbed V-belt removal and installation.

A/C compressor and mount (2.0L)

1. **Tensioning roller, ribbed V-belt**
   - See **03 Maintenance** for removal and installation of ribbed V-belt
2. **Generator (GEN) and pulley**
3. **Bracket**
   - Compressor bracket and related components can be removed and installed without having to open the refrigerant circuit
   - Mounting for generator, A/C compressor and power steering pump
   - Remove compressor from bracket, swing away and secure in engine compartment with wire
4. **Bolts**
   - Tighten to 50 Nm (37 ft-lb)
5. **Power steering pump**
6. **A/C compressor**
7. **Bolts**
   - Tighten to 45 Nm (33 ft-lb)

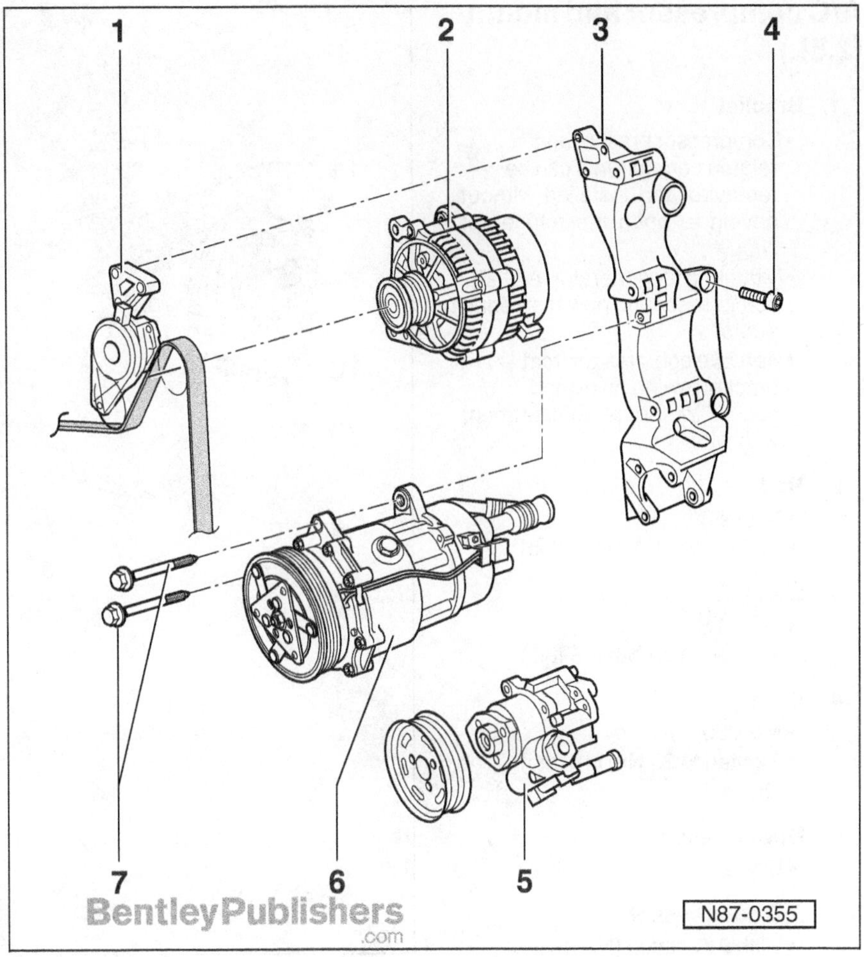

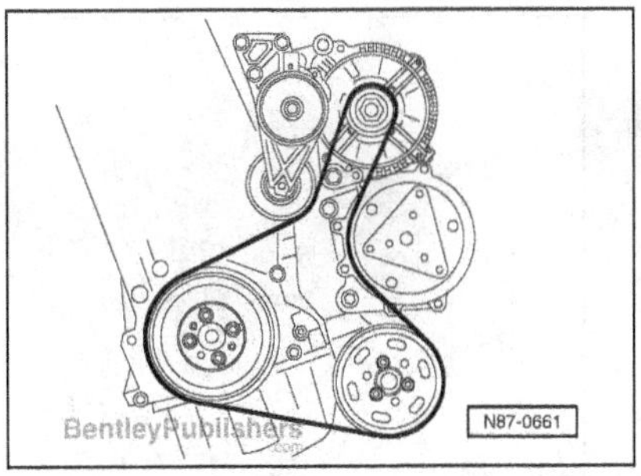

Ribbed V-belt routing

- Always mark the running direction before removing and install in the same direction.
- When installing the belt ensure it is correctly seated in the pulleys.
- See **03 Maintenance** for ribbed V-belt removal and installation.

A/C compressor and mount (2.5L)

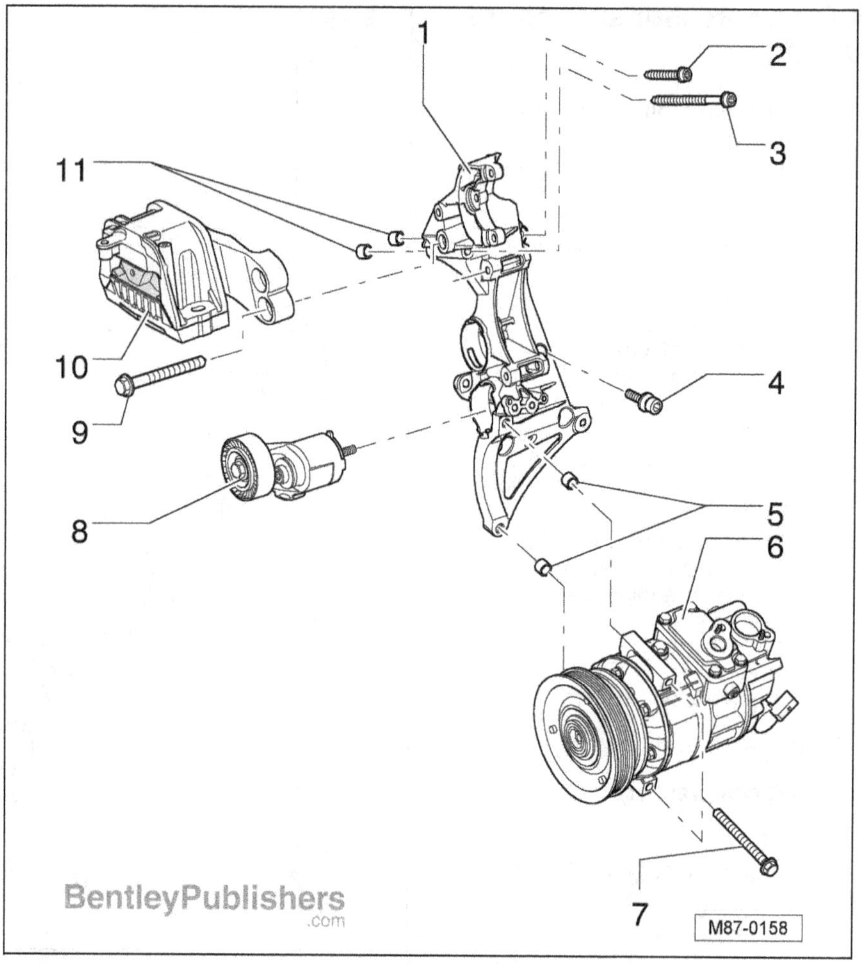

1. **Bracket**
   - Compressor bracket and related components can be removed and installed without having to open the refrigerant circuit
   - Mounting for generator, A/C compressor and power steering pump
   - Remove compressor from bracket, swing away and secure in engine compartment with wire
2. **Bolt**
   - M8 x 60
   - Tighten to 25 Nm (18 ft-lb)
3. **Bolt**
   - M8 x 110
   - Tighten to 25 Nm (18 ft-lb)
4. **Bolts**
   - M8 x 30
   - Tighten to 25 Nm (18 ft-lb)
   - Qty: 4
5. **Dowel sleeves**
   - Qty: 2
6. **A/C compressor**
   - With A/C clutch (N25)
7. **Bolts**
   - M8 x 85
   - Tighten to 25 Nm (18 ft-lb)
   - Qty: 3
8. **Tensioning roller, ribbed V-belt**
   - See **03 Maintenance** for removal and installation of ribbed V-belt
9. **Bolt**
   - For engine mount
   - Tighten to 40 Nm (30 ft-lb) plus 90° (¼ turn)
   - Always replace
   - Qty: 2
   - See **10 Engine–Removing and Installing** for more details
10. **Engine mount**
11. **Dowel sleeves**
    - Qty: 2

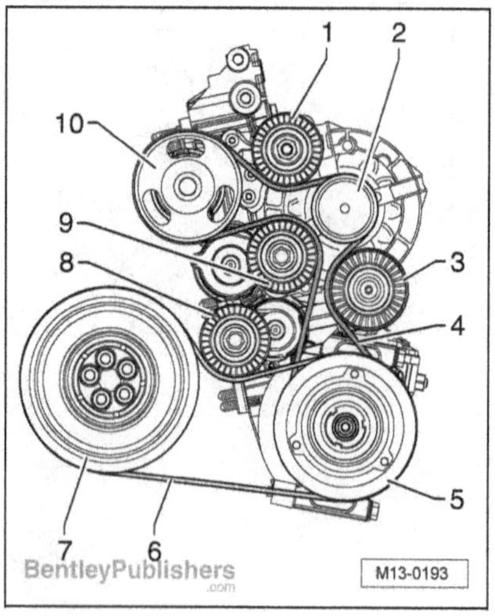

◄ **Ribbed V-belts, routing**

- Always mark the running direction before removing and install in the same direction.
- When installing the belt ensure it is correctly seated in the pulleys.
- See **03 Maintenance** for ribbed V-belt removal and installation.

1. Upper idler pulley
2. Generator pulley
3. Lower idler pulley
4. inner ribbed V-belt for generator and coolant pump
5. A/C compressor
6. Outer ribbed V-belt for A/C compressor
7. Crankshaft pulley
8. Tensioner for outer belt
9. Tensioner for inner belt
10. Coolant pump pulley

A/C compressor overview

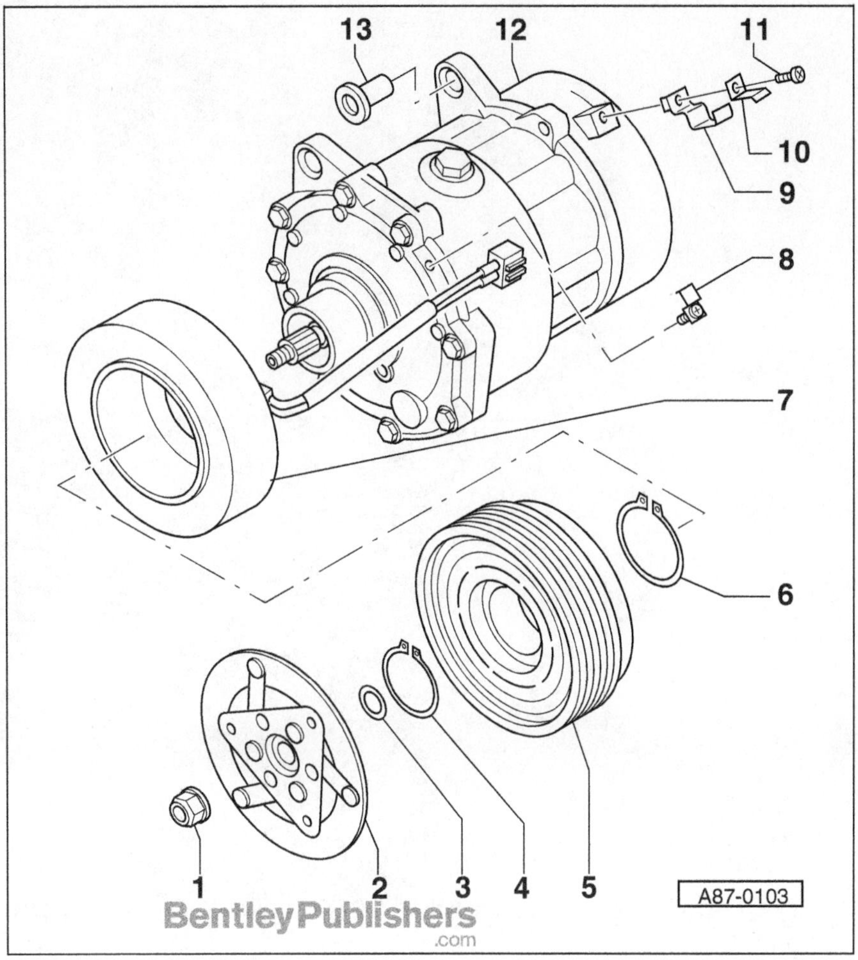

1. **Nut, self locking**
   - Always replace
   - Tighten to 15 Nm (11 ft-lb)
2. **Clutch plate**
3. **Shims**
   - Used to adjust gap between clutch plate and pulley
4. **Circlip**
   - Always replace
   - Note installation position: flat side faces compressor
   - Ensure correct seating in groove
5. **Clutch pulley**
   - Refrigerant circuit does not need to be discharged to service A/C clutch.
   - If A/C clutch is to be serviced without removing compressor, remove intercooler first (1.8L engine).
6. **Circlip**
   - Always replace
   - Note installation position: flat side faces compressor
   - Ensure correct seating in groove
7. **Clutch coil**
   - Thermo-fuse integrated into the clutch coil protects the A/C clutch in the event of overheating due to a binding compressor. The clutch coil circuit is interrupted.
   - Note installation position: locking tab on coil into recess on compressor
8. **Wiring harness securing clip**
   - With screw
9. **Terminal connector securing clip**
10. **Terminal connector securing clip**
11. **Screw**
12. **Compressor**
    - Sanden
13. **Threaded bushing**

9 Electrical System

GENERAL

This group covers a brief description of the principal parts of the electrical system. Also covered here is general electrical system troubleshooting.

Loose or poor connections

Loose, corroded or poor connections can cause numerous electrical problems and create DTCs (Diagnostic Trouble Codes). Be sure to check all electrical connections carefully, including multi-pin harness connectors at control modules and components. It is possible for an individual wire connector to slip out of place within a multi-pin connector and not be noticed unless inspected very carefully.

Voltage and polarity

Volkswagen electrical systems are 12-volt direct current (DC) negative-ground systems. A voltage regulator controls the output of the alternator to approximately 13.5 volts. All circuits are grounded by direct or indirect connection to the negative (–) terminal of the battery. A number of ground connections throughout the car connect the wiring harness to chassis ground. These circuits are completed by the battery cable or ground strap between the body and the battery negative (–) terminal.

9

Electrical system safety precautions

Please read the following warnings and cautions before doing any work on your electrical system.

WARNING —

- *Ignition systems operate in a dangerous voltage range that could prove to be fatal if exposed terminals or live parts are contacted. Use extreme caution when working on a vehicle with the ignition on or the engine running.*

- *On cars equipped with Airbags, special precautions apply to any electrical system testing or repair. The airbag unit is an explosive device and must be handled with extreme care. Before starting any work on an airbag equipped car, refer to the warnings and cautions in* **69 Seatbelts, Airbags**.

- *Before operating the starter without starting the engine (as when making a compression test), disable ignition system as described in* **28 Ignition System–Gasoline**.

CAUTION —

- *Always switch the ignition off and disconnect the negative (–) battery cable before removing any electrical components.*

- *Before disconnecting battery be sure to obtain radio anti-theft code.*

- *Connect and disconnect ignition system wires, multiple connectors, and ignition test equipment leads only while the ignition is switched off.*

- *Do not disconnect battery while the engine is running. Never reverse the battery terminal connections. Even a momentary wrong connection can damage the alternator or electrical components. If the polarity markings on the battery are not visible, confirm the polarity of battery using a voltmeter.*

- *Always remove the battery cables before quick-charging the battery. Never use a quick-charger as a booster for starting the car. Do not exceed 16.5 volts at the battery.*

- *Many solid-state modules operate on very low current and can be permanently damaged if exposed to static discharge. Always handle the modules using proper static prevention equipment and techniques.*

- *Always switch a test meter to the appropriate function and range before making test connections.*

- *Disconnect the battery before doing any electric welding on the car.*

- *Do not wash the engine while it is running, or anytime the ignition is switched on.*

- *Do not try to start the engine of a car which has been heated above 176°F (80°C), (for example, in a paint drying booth) until allowing it to cool to normal temperature.*

Electrical test equipment

Many of the electrical tests described in this manual call for measuring voltage, current or resistance using a digital multimeter (DMM). DMMs are preferred for precise measurements and for electronics work because they are generally more accurate than analog meters. The DMM is also safe for most solid state components whereas an analog meter can damage some components.

An LED test light is a safe, inexpensive tool that can be used to perform many simple electrical tests that would otherwise require a multimeter. The LED indicates when voltage is present between any two test-points in a circuit.

> ***CAUTION—***
> - *Choose test equipment carefully. Use a meter with at least 10 megohm input impedance, or an LED test light. An analog meter (swing-needle) or a test light with a normal incandescent bulb may draw enough current to damage sensitive electronic components.*
> - *An analog meter must not be used to measure resistance on solid state components such as control units or time delay relays.*
> - *Always disconnect the battery before making resistance (ohm) measurements on the circuit.*

Wiring diagrams, fuses and relays

Nearly all parts of the wiring harness connect to components of the electrical system with keyed, push-on connectors that lock into place. Notable exceptions are the heavy battery cables and the alternator wiring.

With the exception of the charging system, all electrical power is routed from the ignition switch or the battery through the fuse panel, located in the passenger compartment behind the driver side left knee bar. Fuses prevent excessive current from damaging components and wiring. Fuses are color coded to indicate their different current capacities. Most relays are electro-mechanical switches that operate on low current to switch a high-current circuit on and off.

The wiring diagrams shown in **EWD Electrical Wiring Diagrams** are organized according to model year and engine type. **97 Fuses, Relays and Component Locations** contains information and locations for fuses, relays and various electrical components.

ELECTRICAL TROUBLESHOOTING

Four things are required for current to flow in any electrical circuit: a voltage source, wires or connections to transport the voltage, a consumer or device that uses the electricity, and a connection to ground or a return to the voltage source. Most problems can be found using only a digital multimeter (volt/ohm/amp meter) to check for voltage supply, for breaks in the wiring (infinite resistance/no continuity), or for a path to ground that completes the circuit.

Electric current is logical in its flow, always moving from the voltage source toward ground. Keeping this in mind, electrical faults can be located through a process of elimination. When troubleshooting a complex circuit, separate the circuit into smaller parts. Be sure to analyze the problem. Use the wiring diagrams to determine the most likely cause of the problem. Get an understanding of how the circuit works by following the circuit from ground back to the power source.

> ***CAUTION —***
> *When making test connections at connectors and components, use care to avoid spreading or damaging the connectors or terminals. Some electrical tests may require jumper wires to bypass components. When connecting jumper wires, use blade connectors at the wire ends that match the size of the terminal being tested. The small internal contacts are easily spread apart, and this can cause intermittent or faulty connections that can lead to more problems.*

Voltage and ground, checking

Checking for the presence of voltage or ground is usually the first step in troubleshooting a problem circuit. For example, if a parking light does not work, a check for voltage at the bulb socket will quickly determine if the circuit is functioning properly or if the bulb itself is faulty. If voltage and ground are found at the socket, then the bulb is most likely faulty.

Another valuable troubleshooting technique is a voltage drop test. This is a good test to perform if current is flowing through the circuit but the circuit is not operating correctly. Sluggish wipers or dim headlights are examples of this. A voltage drop test will help to pinpoint a corroded ground strap or a faulty switch. Normally, there should be less than 1 volt drop across most wires or closed switches. A voltage drop across a connector or short cable should not exceed 0.5 volts.

A voltage drop is caused by higher than normal resistance in a circuit. This additional resistance actually decreases or stops the flow of current. Some common sources of voltage drops are faulty wires or switches, dirty or corroded connections or contacts, and loose or corroded ground wires and ground connections.

A voltage drop can be checked only when current is flowing through the circuit, such as by operating the starter motor or turning on the headlights. Making a voltage drop test requires measuring the voltage in the circuit and comparing it to what the voltage should be. Since these measurements are usually small, a digital voltmeter should be used to ensure accurate readings. If a voltage drop is suspected, turn the circuit on and measure the voltage at the circuit's load.

NOTE —

- *A voltage drop test is generally more accurate than a simple resistance check because the resistances involved are often too small to measure with most ohmmeters. For example, a resistance as small as 0.02 ohms would result in a 3 Volt drop in a typical 150 amp starter circuit (150 amps x 0.02 ohms = 3 volts).*
- *Keep in mind that voltage with the key on and voltage with the engine running are not the same. With the ignition on and the engine off (battery voltage), voltage should be approximately 12.6 volts. With the engine running (charging voltage), voltage should be approximately 14.0 volts. Measure voltage at the battery with the ignition on and then with the engine running to get exact measurements.*

Voltage, measuring

– Set voltage meter to 20V DC scale and connect negative lead to a reliable ground point on car.

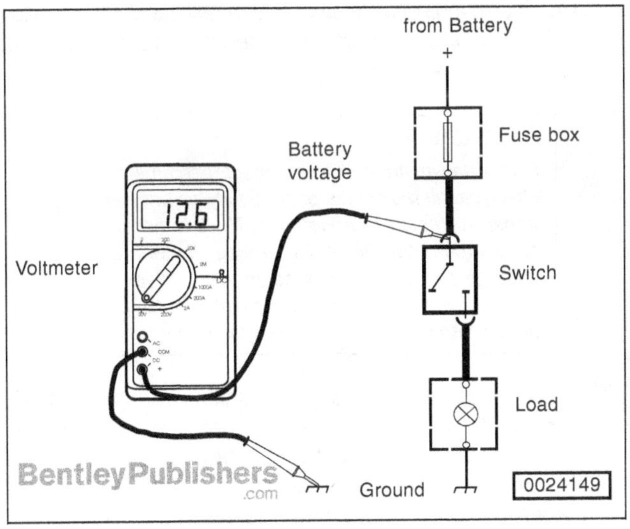

◄ Connect voltmeter positive lead to the point in the circuit you wish to measure.

– If a reading is obtained, there is voltage at that point in the circuit. The voltage reading should not deviate more than 1 volt from voltage at battery. If the voltage is less than this, there is probably a fault in the circuit, such as a corroded connector or a loose ground wire.

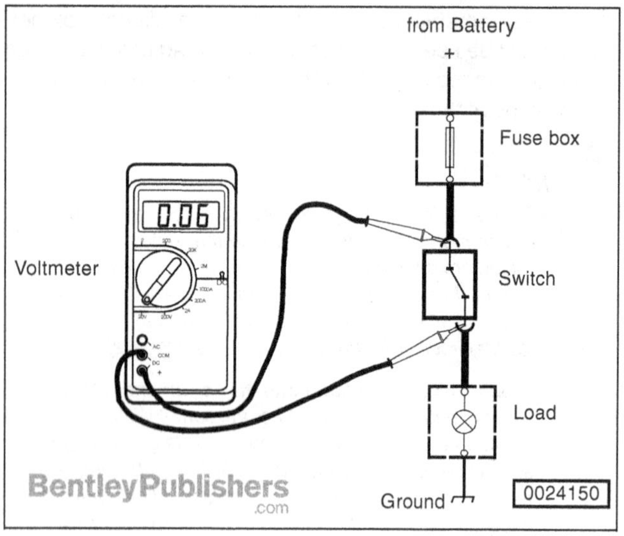

Voltage drop, testing

- Connect digital voltmeter positive lead to positive (+) connector of component to be tested.

- Connect voltmeter negative lead to the negative (–) connector of component being tested.

- With power on and circuit working, meter shows the voltage drop (difference between the two points). This value should not exceed 1 volt.

> ***NOTE —***
> *The maximum voltage drop in an automotive circuit, as recommended by the Society of Automotive Engineers (SAE), is as follows: 0 Volts for small wire connections; 0.1 Volts for high current connections; 0.2 Volts for high current cables; and 0.3 Volts for switch or solenoid contacts. On longer wires or cables, the drop may be slightly higher. In any case, a voltage drop of more than 1.0 Volt usually indicates a problem.*

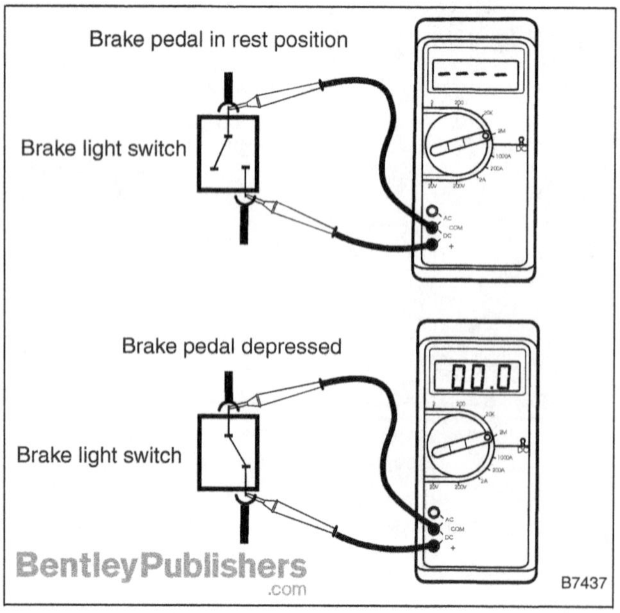

Continuity, checking

The continuity test can be used to check the basic integrity of a circuit or switch. Because most automotive circuits are designed to have little or no resistance, a circuit or part of a circuit can be easily checked for faults using an ohmmeter. An open circuit or a circuit with high resistance will not allow current to flow. A circuit with little or no resistance allows current to flow easily.

> ***CAUTION —***
> *Do not use an analog (swing-needle) ohmmeter to check circuit resistance or continuity on any electronic (solid-state) components. The internal power source used in most analog meters can damage solid state components. Use only a high quality digital ohmmeter having high input impedance when checking electronic components.*

- When checking continuity, the ignition should be off. On circuits that are powered at all times, the battery should be disconnected. Using the appropriate wiring diagram, a circuit can be easily tested for faulty connections, wires, switches, relays, and engine sensors by checking for continuity.

Short circuits, checking

A short circuit is exactly what the name implies. The circuit takes a shorter path than it was designed to take. The most common short that causes problems is a short to ground where the insulation on a positive (+) wire wears away and the metal wire is exposed. When the wire rubs against a metal part of the car or other ground source, the circuit is shorted to ground. If the exposed wire is live (positive battery voltage), the direct current flow to ground will blow a fuse or damage an unfused circuit.

> ***CAUTION—***
> - *On circuits protected with large fuses (25 amp and greater), the wires or circuit components may be damaged before the fuse blows. Always check for damage before replacing fuses of this rating.*
> - *When replacing blown fuses, use only fuses having the correct rating. Always confirm the correct fuse rating printed on the fuse panel cover.*

Short circuit test with voltmeter

- Remove blown fuse from circuit.
- Disconnect harness connector from circuit load or consumer.

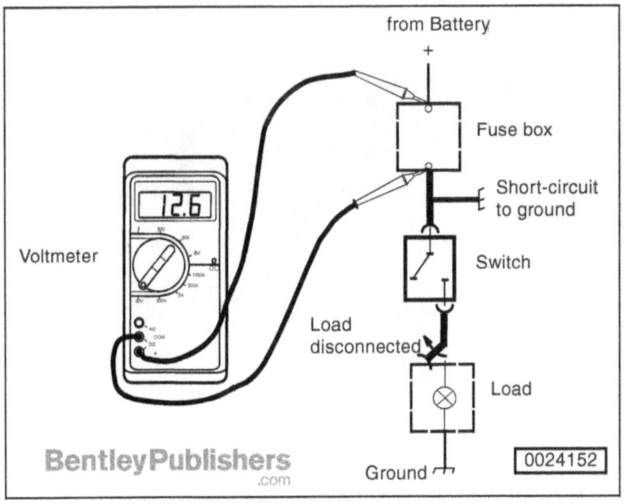

- ◄ Using a voltmeter, connect test leads across fuse terminals. Make sure power is present in circuit. If necessary, turn key on.
- If voltage is indicated at voltmeter, there is a short to ground somewhere in the circuit.
- If voltage is not indicated, work from wire harness nearest to fuse panel and move or wiggle wires while observing meter. Continue to move down harness until meter displays a reading. This is the location of the short to ground.
- Inspect wire harness at this point for any faults. If no faults are visible, carefully slice open harness cover or wire insulation for further inspection. Repair any faults found.

90 Instruments

TABLES

GENERAL

An electronic speedometer with an LCD digital odometer display is used on all models. The speedometer receives an electronic signal from a Hall sender on the transmission. Mileage is permanently held in memory and will be retained if power to the instrument cluster is interrupted.

Individual components for the instrument cluster are not available from Volkswagen and therefore disassembly of the cluster is not recommended. The illumination for the instrument cluster is provided by several blue light emitting diodes (LEDs) that are not serviceable. Remanufactured instrument cluster assemblies are available through the Volkswagen parts department.

NOTE—
Some versions of the instrument cluster may use a miniature light bulb to provide illumination for certain displays. These bulbs are replaceable separately.

CAUTION—
Disconnecting the negative (–) battery cable may erase fault codes and basic settings in the engine management and automatic transmission control modules. Some driveability problems may be noticed until the system re-adapts to operating conditions. OBD II readiness codes, which may be required for emissions testing, may also be erased. Convenience electronics (alarm system, interior light control, power locks, mirrors, and windows) may need to be re-set using a diagnostic scan tool.

90

INSTRUMENT CLUSTER

NOTE —

- *Vehicles with manual transmission have the trip odometer display.*
- *Vehicles with automatic transmission have trip odometer display and gear selector lever position display.*
- *2006 New Beetle and New Beetle convertible models may experience a no-start situation if the ambient air temperature is below 40°F (4°C). The following DTC may be present:*
  P1570 - Control Module Locked
  Volkswagen technical bulletin #90-08-03, dated Feb. 6, 2008 describes checking the version and, if necessary, replacement of the instrument cluster with the latest version to remedy this situation.

Malfunction recognition and display

The instrument cluster has On-Board Diagnostic (OBD) capability, which is an aid to troubleshooting.

If the instrument cluster control module detects a malfunction with the odometer or speedometer that is non-repairable, then "dEF" will appear in the trip recorder display. Replacement of the instrument cluster will be necessary.

If the instrument cluster control module detects a malfunction and nothing is displayed on the trip recorder display, carry out the following work before removing the instrument cluster:

– Check DTC memory with a diagnostic scan tool.

– Read service interval display values and odometer reading with a diagnostic scan tool and note the values.

If the malfunction indicates that the instrument cluster requires replacement, the recorded values can be entered in the new instrument cluster.

Instrument cluster, removing and installing

WARNING —

- *Special safety precautions apply to vehicles equipped with airbags.*
- *Refer to airbag **CAUTIONS** and **WARNINGS** in* **69 Seatbelts, Airbags**.

CAUTION—

- *Disconnect battery ground (GND) strap from battery negative (–) terminal.*
- *Be sure to have the anti-theft radio code on hand before disconnecting the battery.*
- *Disconnecting the negative (–) battery cable may erase fault codes and basic settings in the engine management and automatic transmission control modules. Some driveability problems may be noticed until the system re-adapts to operating conditions. OBD II readiness codes, which may be required for emissions testing, may also be erased. Convenience electronics (alarm system, interior light control, power locks, mirrors, and windows) may need to be re-set using a diagnostic scan tool.*
- *See* **Malfunction recognition and display** *given earlier before removing instrument cluster.*

– Release steering wheel position lock, pull steering wheel out completely and lock again in lowest position.

NOTE—

For ease of illustration the steering wheel is not shown in the following illustrations. It is not necessary to remove the steering wheel to remove the instrument cluster.

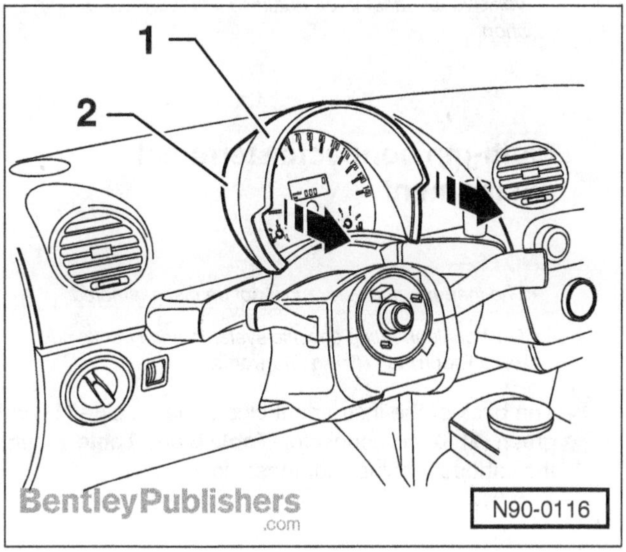

N90-0116

◄ Pull trim ring (**1**) from instrument cluster mounting clips (**2**).

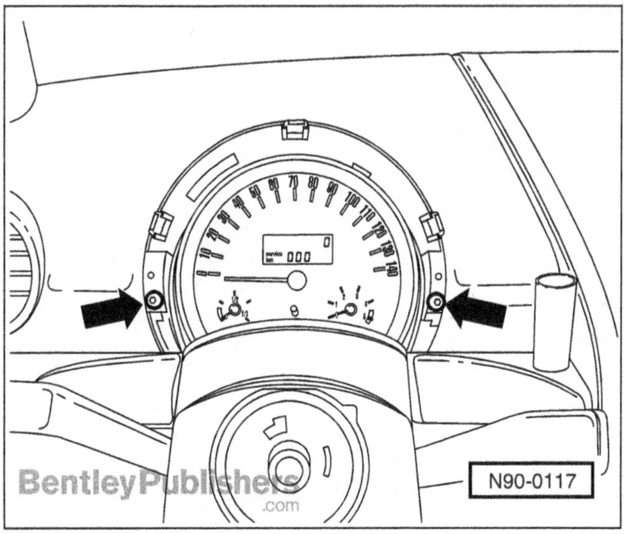

N90-0117

◄ Remove instrument cluster mounting screws (**arrows**).

– Tilt instrument cluster slightly toward passenger compartment.

– Disconnect harness connectors at rear of instrument cluster.

– Remove instrument cluster.

– Installation is reverse of removal. After completing installation check function of instrument cluster.

Through 1999 model year

– Use a diagnostic scan tool (i.e. VAG 1551/1552 or equivalent) to code the instrument cluster. It may be necessary to record the part number of the instrument cluster and provide this to an authorized Volkswagen dealer to obtain a coding number for the instrument cluster control module.

From 2000 model year

– Perform necessary instrument cluster adaptations and coding, including coding the Data Bus On Board Diagnostic Interface (J533) and immobilizer adaptation using the VAS 5051/5052 diagnostic scan tool.

NOTE—
- *Due to the large number of instrument cluster variations it always advisable to order a replacement cluster using the part number on the back of the cluster.*
- *Replacement clusters can have the odometer mileage adjusted to match the mileage of the original cluster. Because of the specialized nature of these repairs, servicing should be referred to an authorized Volkswagen dealer or qualified independent repair shop.*

Multi-pin connector terminal assignment

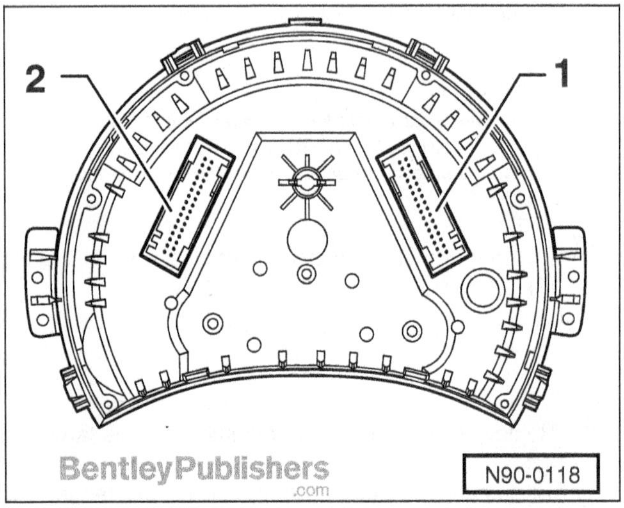

NOTE—
- *The instrument cluster must not be disassembled.*
- *For troubleshooting specific systems and circuits see* **EWD Electrical Wiring Diagrams.**

The back of the instrument cluster has a blue (**1**) and a green (**2**) 32-pin connector. **Table b** and **Table c** identify the various terminals for these connectors.

Table a. Instrument cluster 32-pin connector (blue) terminal identification (through 06/1999)

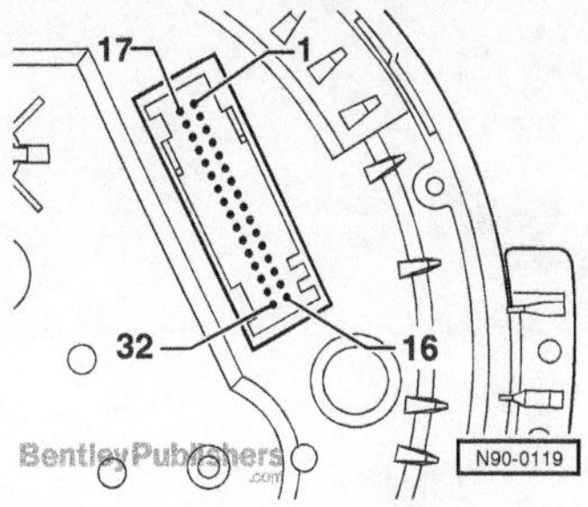

Terminal	Circuit
1	Terminal 15, positive
2	Right turn signal indicator light (K94)
3	Output signal 1 from electronic speedometer
4	Open
5	Fuel gauge (G1)
6	Airbag Malfunction Indicator Lamp (MIL)
7	Terminal 31, sender Ground (GND)
8	Engine coolant temperature gauge (G3)
9	Terminal 31, Ground (GND)
10	Oil pressure warning lamp
11	Engine RPM signal
12	Generator warning light (K2), terminal 61
13	Glow plug indicator light (K29) (Diesel only)
14	Rear fog light indicator light (K13) (where applicable)
15	Open
16	Warning light for rear hatch/trunk lid open (K116)
17	Headlight high beam indicator light (K1), terminal 56a
18	Left turn signal indicator light (K65)
19	ABS warning light (K47)
20	Instrument cluster illumination, terminal 58b
21	Driver's door open warning buzzer
22	Engine Coolant Level (ECL) sensor (G32)
23	Terminal 30, (B+)
24	Terminal 31, Ground (GND)
25	On-Board Diagnostic (K-wire)
26	Parking light right (warning buzzer for "lights on")
27	Parking light left (warning buzzer for "lights on")
28	Speedometer Vehicle Speed Sensor (VSS) (G22)
29	Brake fluid level warning lamp
30	Ignition key warning buzzer, terminal 86s
31	Seat belt warning lamp/buzzer
32	Malfunction Indicator Light (MIL)

Table b. Instrument cluster 32-pin connector (green) terminal identification (through 06/1999)

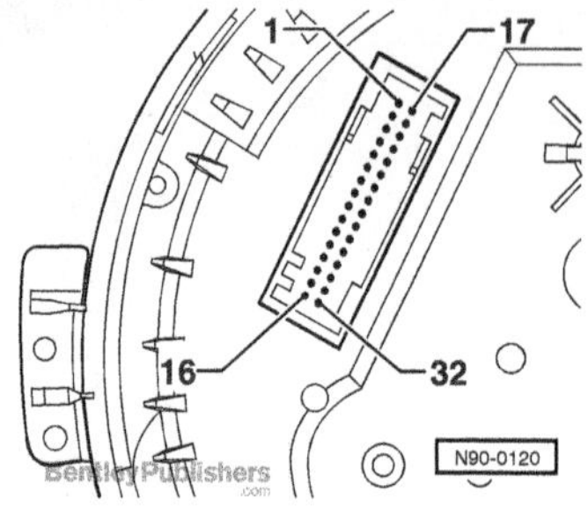

Terminal	Circuit
1	Open
2	Reading coil for immobilizer 1 (where applicable)
3	Open
4	Open
5	W-wire
6	Rear spoiler
7	Brake pad wear indicator (where applicable)
8	Open
9	Open
10	Fuel reserve warning (certain countries only)
11	Signal for vehicle stationary
12	A/C clutch cut-off - engine coolant temperature signal
13	Warning lamp for brake system and handbrake (K14)
14	ABS/EDL warning lamp
15	Electronic Power Control (EPC) warning lamp
16	Open
17	Reading coil for immobilizer 2 (where applicable)
18	Open
19	Open
20	Open
21	Open
22	Open
23	Open
24	Open
25	Open
26	Open
27	Open
28	Open
29	Open
30	Output signal 2 from electronic speedometer
31	Selector lever display (automatic transmission)
32	Open

Table c. Instrument cluster 32-pin connector (blue) terminal identification (from 07/1999)

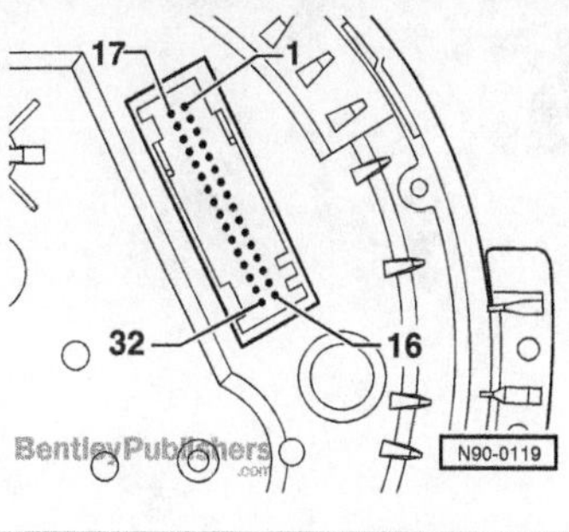

Terminal	Circuit
1	Terminal 15, B+ (ignition on)
2	Right turn signal indicator light (K94)
3	Output signal 1 from electronic speedometer
4	Open
5	Fuel gauge (G1)
6	Airbag Malfunction Indicator Lamp (MIL)
7	Terminal 31, sender Ground (GND)
8	Engine coolant temperature gauge (G3)
9	Terminal 31, Ground (GND)
10	Oil pressure switch (F1)
11	Open
12	Generator warning light (K2), terminal 61
13	Open
14	Open
15	Open
16	Warning light for rear lid unlocked (K116)
17	Headlight high beam indicator light (K1), terminal 56a
18	Left turn signal indicator light (K65)
19	Open
20	Instrument cluster illumination, terminal 58b
21	Driver's door open warning buzzer
22	Engine Coolant Level (ECL) sensor (G32)
23	Terminal 30, (B+)
24	Terminal 31, Ground (GND)
25	On-Board Diagnostic (K-wire)
26	Parking light right (warning buzzer for “lights on”)
27	Parking light left (warning buzzer for “lights on”)
28	Speedometer Vehicle Speed Sensor (VSS) (G22)
29	Brake fluid level warning lamp
30	Ignition key warning buzzer, terminal 86s
31	Seat belt warning lamp/buzzer
32	Open

Table d. Instrument cluster 32-pin connector (green) terminal identification (from 07/1999)

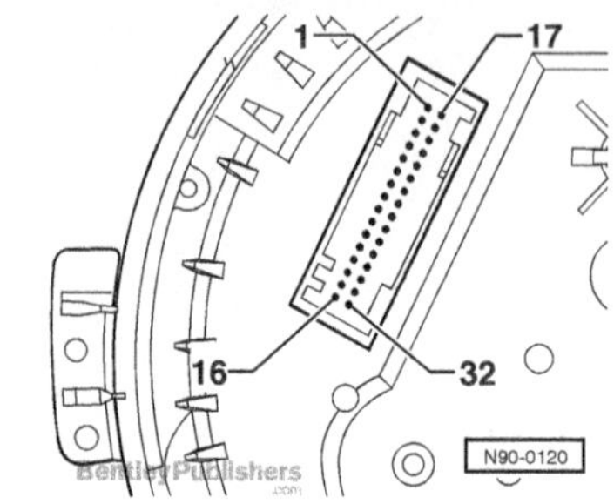

Terminal	Circuit
1	Open
2	Reading coil for immobilizer 1 (where applicable)
3	Open
4	Open
5	W-wire
6	Rear spoiler
7	Brake pad wear indicator
8	Open
9	Open
10	Fuel reserve warning (where applicable)
11	Signal for vehicle stationary
12	Open
13	Warning lamp for brake system and handbrake (K14)
14	Open
15	Open
16	Open
17	Reading coil for immobilizer 2 (where applicable)
18	Open
19	Interface - CAN data Hi-bus
20	Interface - CAN data Lo-bus
21	Open
22	Open
23	Open
24	Open
25	Open
26	Open
27	Interface - On Board Diagnostic CAN data Hi-bus
28	Interface - On Board Diagnostic CAN data Lo-bus
29	Open
30	Output signal 2 from electronic speedometer
31	Selector lever display (automatic transmission)
32	Open

91 Radio

GENERAL

This section covers the Volkswagen factory installed stereo radio sound systems. For specific circuit tracing, see **EWD Electrical Wiring Diagrams**.

> ***CAUTION —***
> *Disconnecting the negative (–) battery cable may erase fault codes and basic settings in the engine management and automatic transmission control modules. Some driveability problems may be noticed until the system re-adapts to operating conditions. OBD II readiness codes, which may be required for emissions testing, may also be erased. Convenience electronics (alarm system, interior light control, power locks, mirrors, and windows) may need to be re-set using a diagnostic scan tool.*

> ***WARNING —***
> - *Be sure to have the anti-theft radio code on hand before disconnecting the battery.*
> - *Before working on any part of the electrical system, always disconnect the battery ground strap first.*

91

> ***CAUTION —***
> - *The radio is wired to the vehicle alarm system. If the alarm system is armed, removing the radio will activate the alarm even if the proper tools are used. Do not attempt to remove the radio without disarming the alarm.*
> - *The factory installed connectors are designed for genuine Volkswagen radios. If installing a different radio, remember that the radio may not fit properly into the space provided, the electrical connections may not be compatible and different terminals may be needed.*
> - *Keep in mind that factory installed radios are electronically linked to the ignition, alarm and data link circuits.*

Radio reception

Radio stations send out electromagnetic signals from transmitting towers. When these signals move past an automobile, a small electrical impulse is induced in the antenna and sent to the radio for detection and amplification. The radio takes these small electrical impulses and converts them to a level sufficient to operate a speaker.

AM (amplitude modulation) radio signals travel in two ways; ground waves and sky waves. Ground waves travel through the air and follow the curve of the earth. Sky waves spread up into the sky until they reach the ionosphere where they are reflected back to earth. Depending on the power of the transmitting station, this reflection or "skip", enables AM radio waves to travel great distances. FM (frequency modulation) radio waves are not reflected by the atmosphere and travel in what is known as "line of sight". Therefore, FM broadcasts do not travel as far as AM broadcasts.

Radio reception will be affected by the height of the transmitting antenna, station power and conditions between radio stations and the vehicle radio. Radio signals are also affected by the weather, mountains, buildings, tunnels and other barriers. Best reception is when the vehicle antenna can "see" the station's transmitting antenna. That is, however, rarely the case. Normally, the signal picked up by the vehicle antenna has been reflected by many solid objects. This makes the transmission path longer or shorter and delays or weakens the signal.

Several impediments can prevent optimum radio reception in any vehicle and can include:

Fading

- Signal fading is typical on AM when driving through an underpass or near large buildings or objects. FM does not tend to fade as much as AM. In the same location where AM fades, FM may come in strong and clear because the shorter radio waves are reflected by metal objects such as buildings and bridges.

Flutter-fence effect

- In weak FM reception areas you may hear short pops of hissing background noise with otherwise good reception of the radio program. This "flutter" noise is like the sound burst that occurs when passing poles or posts close to the side of the road. This flutter effect may cause the stereo indicator to flicker because the signal has fallen below the minimum level to operate the stereo decoder. In even more remote areas such as in the desert or in hilly or mountain areas, it may not be possible to receive some AM or FM radio stations at all.

Multi-Path Cancellation

- Flutter and distortion is also caused by a mixing of several signals coming from different directions as a result of reflection from various objects. Mixing or cancellation effects often happen in cities even when close to the radio station transmitting towers and when transmitting towers are in close proximity to each other.

Interference

- Ignition or accessory interference, flutter, distortion and background noise can be caused by the ignition system or electrical accessories in the vehicle. In addition to internal sources, noise can also be caused by outside sources such as electrical power lines, other radio wave transmitters and other vehicles.

Interference suppression

The majority of the electrical consumers in the vehicle have radio suppression built in. All vehicles also include additional suppression measures in the following areas and components:

- Coolant fans (V7)
- Windshield wiper motor (V)
- Ground (GND) strap from left wheel housing to fender

RADIO SYSTEM

Radio system, description

The radio system consists of a radio head and cassette tape or CD player with bass loudspeakers (woofers) in the front doors and rear side panels. Domed treble loudspeakers (tweeters) are installed in each A-pillar. The system also includes an amplified roof mounted antenna.

A 6-Disc CD changer is available as optional equipment, and (if equipped) is located in left side of the cargo compartment behind the rear seat. All vehicles are pre-wired from the factory for this unit only.

Volkswagen New Beetle radios are capable of communicating with the factory diagnostic scan tool (VAG 1551/1552, VAS 5051/5052) through the Data Link Connector (DLC). Radios can store Diagnostic Trouble Codes, (DTCs), and must be coded via the scan tool in order to function properly. The correct code varies depending on the type of vehicle, the type of antenna and the CD changer, if applicable.

NOTE —
Radio malfunctions and equipment changes requiring the use of scan tool VAG 1551/1552 or VAS 5051/5052 should be handled by an authorized Volkswagen dealer or other qualified repair facility.

CAUTION —
Observe all cautions and warnings in **9 Electrical System** *before starting repairs involving the electrical system.*

NOTE —
- *Before troubleshooting or servicing, the technician must be familiar with the function and operation of the radio system. Read the radio owner's manual and review all radio, tape player and CD changer functions.*
- *Only factory approved accessory radios and radio equipment (available from Volkswagen of America, Inc.) should be installed. This ensures proper installation and minimizes risk of damaging vehicle electrical system and On-Board Diagnostic functions.*

Satellite radio was available as an option beginning in 04/2004 for sedan models and 07/2004 for convertible models. This option is known as Satellite Digital Audio Radio Service (SDRAS).

Various standard and optional radio systems were installed throughout production. The list below provides approximate application dates.

Radio systems
- Premium V . . . . 1998 to early 2004
- Premium V Monsoon . . . . 10/2000 to early 2004
- Premium 5.5. . . . . .early 2004 - on
- Premium 5.5 Monsoon. . . . . .early 2004 - on
- Premium 5.5 SDARS . . . . .sedan from 04/2004
  convertible from 07/2004
- Premium 5.5 Monsoon
  and SDARS . . . . .sedan from 04/2004
  convertible from 07/2004

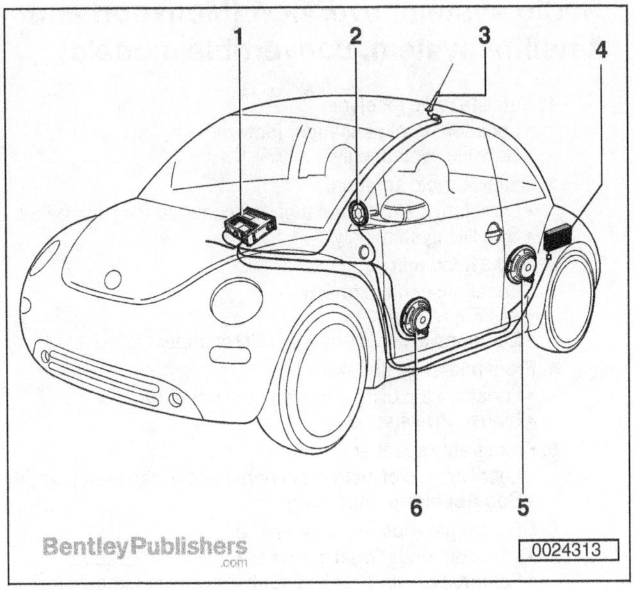

Radio system, overview (standard system)

1. Radio head unit
   - Location: center console
   - AM/FM receiver
   - Equipped to control optional CD changer
2. Front treble speaker (tweeter)
   - Location: base of A-pillar
   - See **Speakers, replacing**
3. Amplified antenna
   - Location: center rear of roof
4. CD changer (optional equipment)
   - Location:
     Through 07/2002: left rear area of cargo compartment
     From 08/2002: under front center armrest
   - Factory pre-wired for VW supplied changers only
5. Rear bass speaker (woofer)
   - Location: rear side panels
   - Diameter: 168 mm (approx. 6 5/8 in.)
   - See **Speakers, replacing**
6. Bass speaker (woofer)
   - Location: front door panel
   - Diameter: 168 mm (approx. 6-5/8 in.)
   - See **Speakers, replacing**

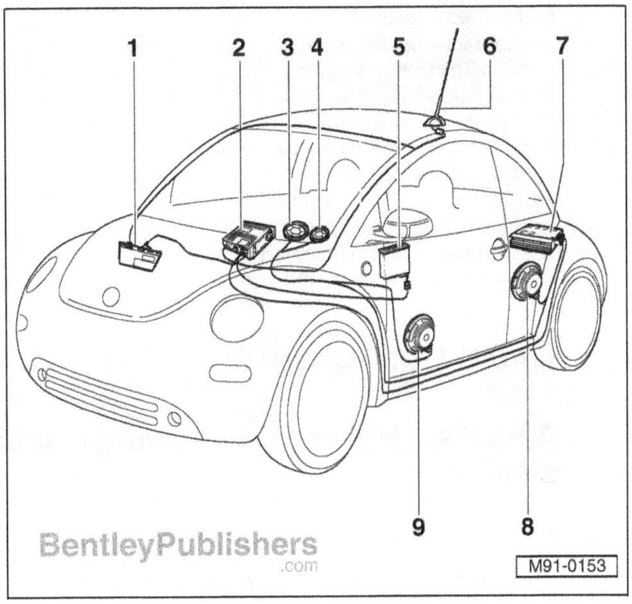

Radio system, overview (Monsoon and satellite system)

1. Satellite radio receiver
   - Location: front passenger footwell
   - Satellite system only
2. Radio head unit
   - Location: center console
   - AM/FM receiver
   - Equipped to control optional CD changer
3. Front midrange speaker
   - Location: top of instrument panel under trim panel
   - Monsoon system only
4. Front treble speaker (tweeter)
   - Location: top of instrument panel under trim panel
   - See **Speakers, replacing**
5. CD changer (optional equipment)
   - Location: under front center armrest from 08/2002
   - Factory pre-wired for VW supplied changers only
6. Amplified antenna
   - Location: center rear of roof
7. Amplifier
   - Location: behind rear lock carrier trim panel
   - Monsoon system only
8. Rear bass speaker (woofer)
   - Location: rear side panels
   - Diameter: 168 mm (approx. 6-5/8 in.)
   - See **Speakers, replacing**
9. Front bass speaker (woofer)
   - Location: front door panel
   - Diameter: 220 mm (approx. 8-21/32 in.)
   - See **Speakers, replacing**

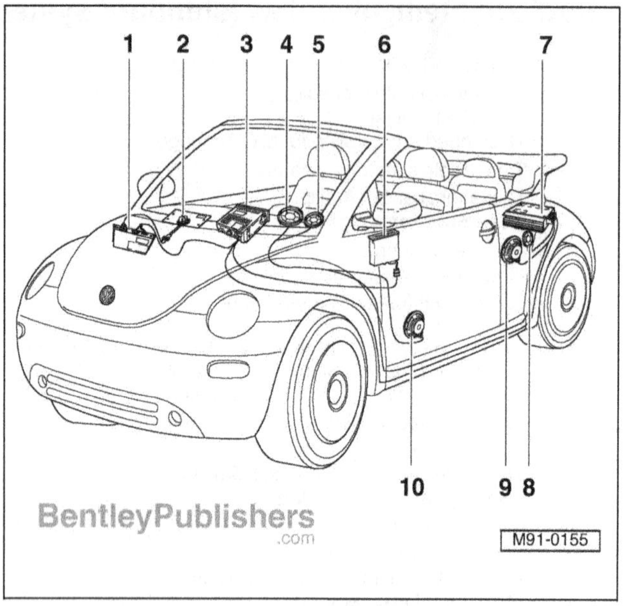

Radio system, overview (Monsoon and satellite system, convertible models)

1. Satellite radio receiver
   - Location: front passenger footwell
   - Satellite system only
2. Satellite radio antenna
   - Location: instrument panel, top center, under trim panel
   - Satellite system only
3. Radio head unit
   - Location: center console
   - AM/FM receiver
   - Equipped to control optional CD changer
4. Front midrange speaker
   - Location: top of instrument panel under trim panel
   - Monsoon system only
5. Front treble speaker (tweeter)
   - Location: top of instrument panel under trim panel
   - See **Speakers, replacing**
6. CD changer (optional equipment)
   - Location: under front center armrest from 08/2002
   - Factory pre-wired for VW supplied changers only
7. Amplifier
   - Location: behind rear lock carrier trim panel
   - Monsoon system only
8. Rear treble speaker
   - Location: rear side panels
   - See **Speakers, replacing**
9. Rear bass speaker (woofer)
   - Location: rear side panels
   - Diameter: 168 mm (approx. 6-5/8 in.)
   - See **Speakers, replacing**
10. Front bass speaker (woofer)
    - Location: front door panel
    - Diameter: 220 mm (approx. 8-21/32 in.)
    - See **Speakers, replacing**

RADIO COMPONENTS

Amplifier (Monsoon), removing and installing

Sedan models

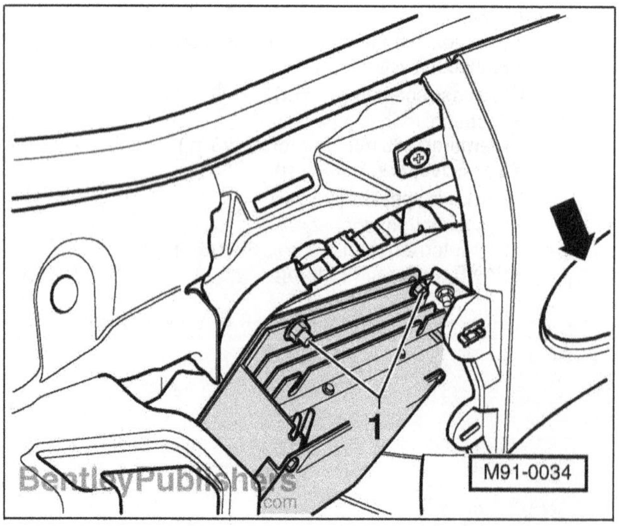

– Be sure to have radio anti-theft code available before removing radio.

– Disconnect battery ground (GND) strap from battery negative (–) terminal. See the **Caution** at the beginning of this repair group regarding battery disconnection.

– Remove rear lock carrier trim. See **70 Trim–Interior**.

◄ Remove access cover (**arrow**) from left rear cargo compartment trim panel.

– Reach through access cover and disconnect amplifier harness connectors.

– Remove mounting bolts (**1**) and slide amplifier up and out from mounting bracket.

Convertible models

– Be sure to have radio anti-theft code available before removing radio.

– Disconnect battery ground (GND) strap from battery negative (–) terminal. See the **Caution** at the beginning of this repair group regarding battery disconnection.

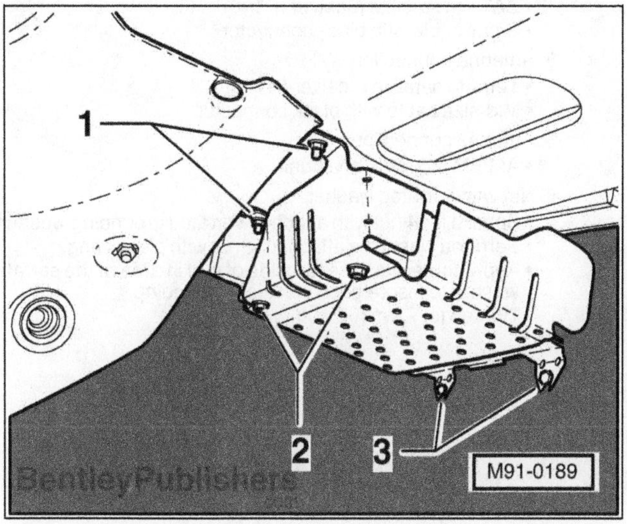

< Working in left upper part of cargo compartment, remove amplifier mounting screws (**3**) and nuts (**1**).

- To remove amplifier mounting bracket, remove mounting nuts (**2**).

– Disconnect harness connectors and remove amplifier.

Continued for all models

– Install amplifier in reverse order of removal.

Antenna (standard)

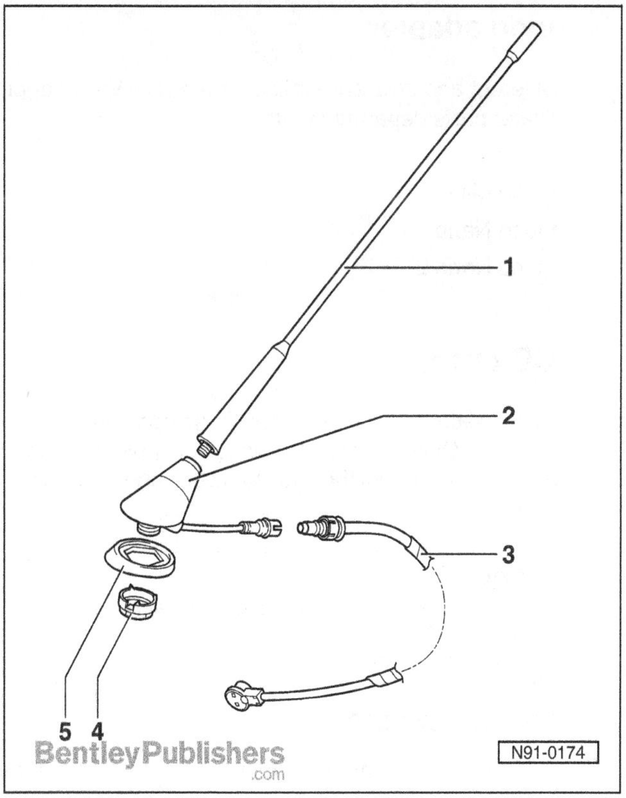

1. Antenna mast
2. Antenna base
   - With amplifier
   - Lower molded headliner at rear to access base
3. Antenna cable
   - From roof antenna to radio head unit
   - Current to operate the amplifier is supplied from the radio head unit through the center coaxial conductor. No separate 12 volt power lead is used.
4. Nut with serrated washer
   - Special M14 nut with attached serrated grounding washer
   - Serrated washer is attached to nut with plastic ring
   - Apply contact grease to inside of roof in area of the serrated washer to insure good electrical grounding
   - Tighten to: 7 Nm (62 in lb)
5. Gasket
   - Ensure correct positioning to provide water tight seal.

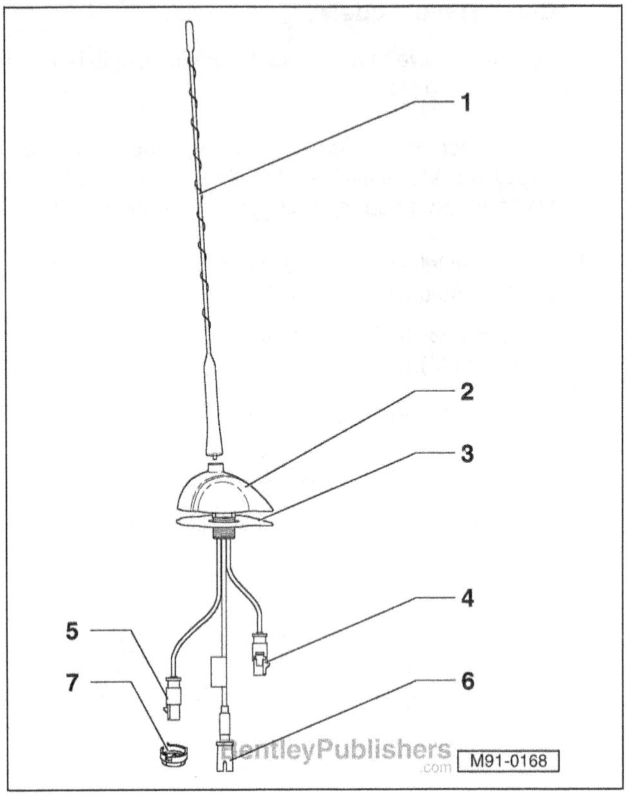

Antenna (satellite)

1. Antenna mast
2. Antenna base
   - With amplifier
   - Lower molded headliner at rear to access base
3. Foam gasket
   - Ensure correct positioning to provide water tight seal.
4. Antenna connection
   - SAT - to satellite receiver (R146)
   - Small cable with black connector
5. Antenna connection
   - Terr - to satellite receiver (R146)
   - Mid-size cable with black connector
6. Antenna connection
   - AM/FM - to radio head unit
7. Nut with serrated washer
   - Special M14 nut with attached serrated grounding washer
   - Serrated washer is attached to nut with plastic ring
   - Apply contact grease to inside of roof in area of the serrated washer to insure good electrical grounding
   - Tighten to: 7 Nm (62 in lb)

Ipod adapter

Different adapters are available from your Volkswagen dealer parts department for:

- Ipod
- Ipod Mini
- Ipod Nano
- Ipod Nano 2

CD changer

New Beetles are factory wired for an optional 6 disc CD changer. Only genuine Volkswagen supplied changers are compatible with the radio head unit and the additional wiring.

> ***CAUTION —***
> *Do not use novelty CDs or any disk that is not a full size CD in the CD changer. Damage to disk and/or changer could result.*

Through 07/2002

Installed in left rear cargo compartment just behind the rear seat.

From 08/2002

Installed under the front center armrest.

CD changer, installing (cargo compartment)

NOTE —
- *Always confirm CD changer and radio compatibility with your Volkswagen parts department or aftermarket parts specialist.*
- *Carefully review all of the following instructions before proceeding with installation.*
- *Never install CD changer in a location, position, or manner other than specified by these instructions.*

> ***CAUTION —***
> *Observe all cautions and warnings in* **9 Electrical System** *before starting repairs involving the electrical system.*

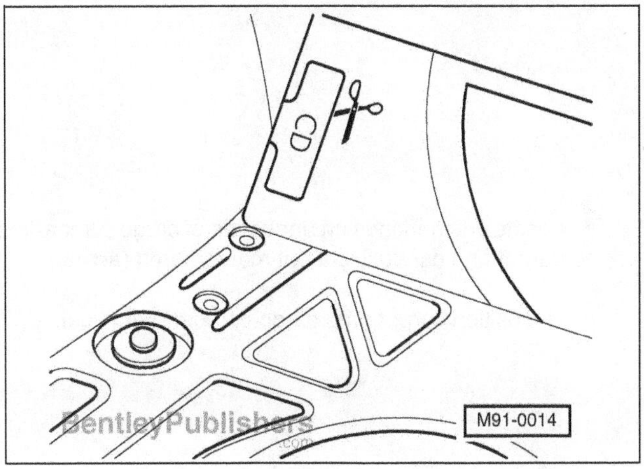

Remove all items from cargo compartment. Lift cargo compartment carpet to expose floor of compartment.

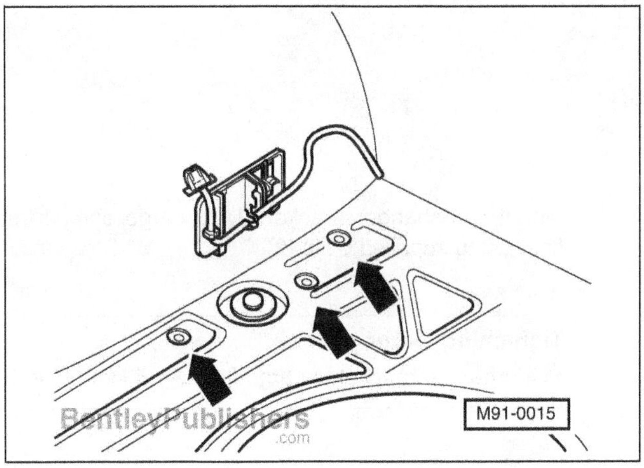

Note location of three CD changer bracket mounting holes (**arrows**) on the cargo compartment floor.

- Locate CD changer cable on left side of cargo compartment.

Early version cable

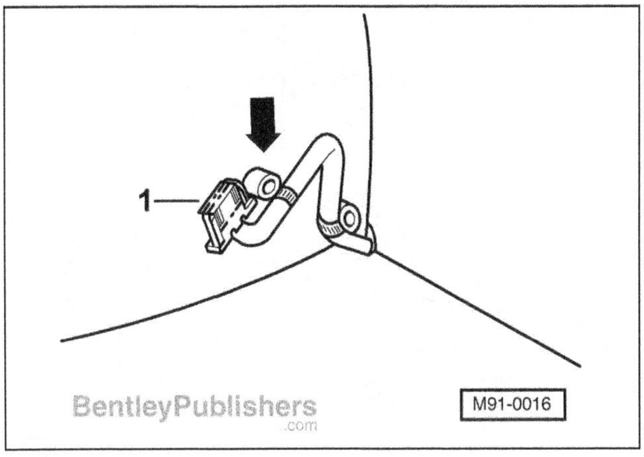

CD changer cable is secured by a clip on left side of cargo compartment. Remove clip (**arrow**) and fold back connector (**1**).

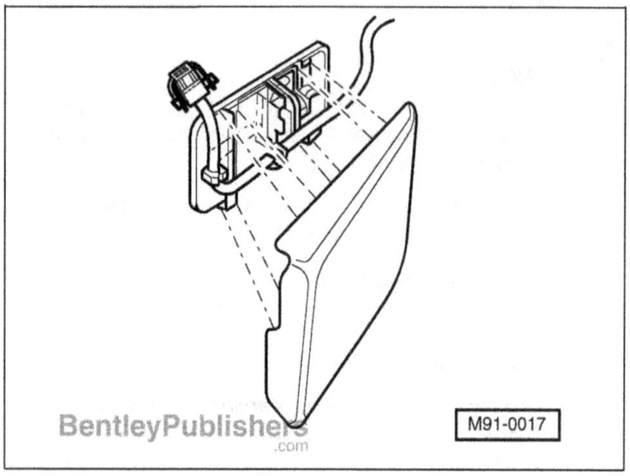

Late version cable

∢ CD changer cable is secured inside cover. Unclip cover and pull out harness connector.

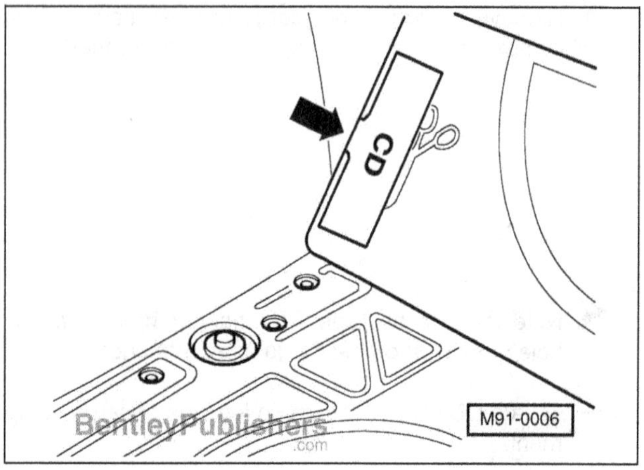

∢ Locate cut markings on underside of cargo compartment carpet and carefully cut out marked area (**arrow**).

– Reposition carpet onto cargo compartment floor.

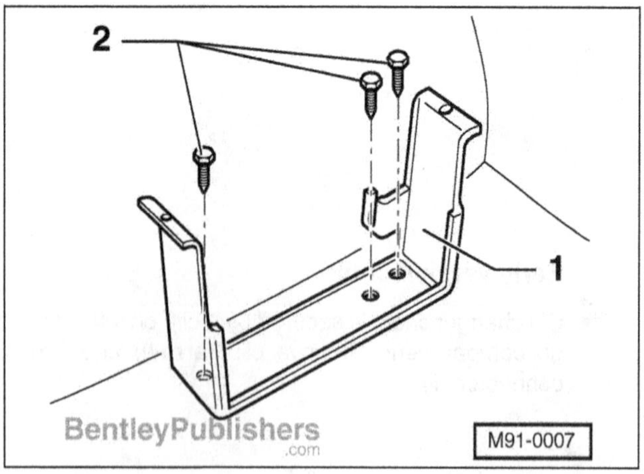

∢ Mount CD changer bracket (**1**) to cargo compartment floor using supplied bolts (**2**).

Tightening torque

- CD changer bracket to cargo floor . 4.5 Nm (40 in-lb)

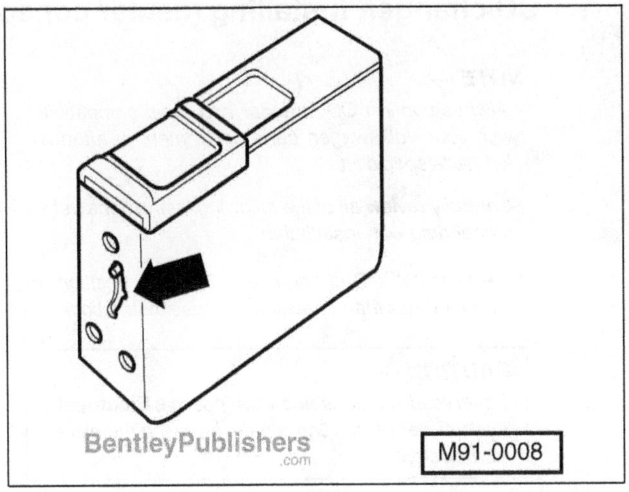

◀ Check CD changer angle of installation switches. Both switches must be in vertical (V) position for this application. Switches are located on the side (**arrow**) and bottom of the CD changer.

– If switches need to be reset, loosen securing screws, slide both switches to "V" position and tighten screws.

Tightening torque
- CD changer position switch screw . . . 2 Nm (18 in-lb)

NOTE —
- *CD playback will skip if switches are not set correctly.*
- *Always check and/or set switches before installation.*
- *Switches must move completely to their end positions.*

– Cover position switch openings with adhesive seals (if supplied).

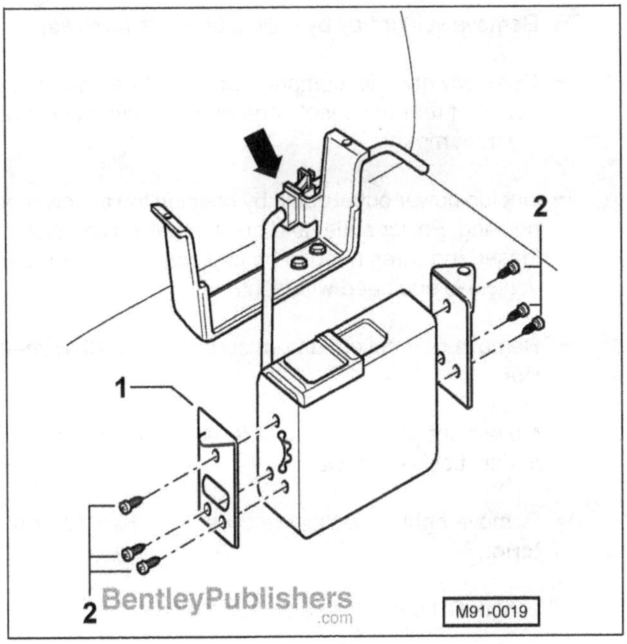

◀ Attach support brackets (**1**) to CD changer with screws (**2**).

Tightening torque
- Support brackets to CD changer . . . . 3 Nm (27 in-lb)

– Route harness connector from CD changer and connect to vehicle harness connector. Clip connector into tab (**arrow**) on CD changer mounting bracket.

– Slide CD changer into mounting bracket from top making sure that cable connector is still secured in tab.

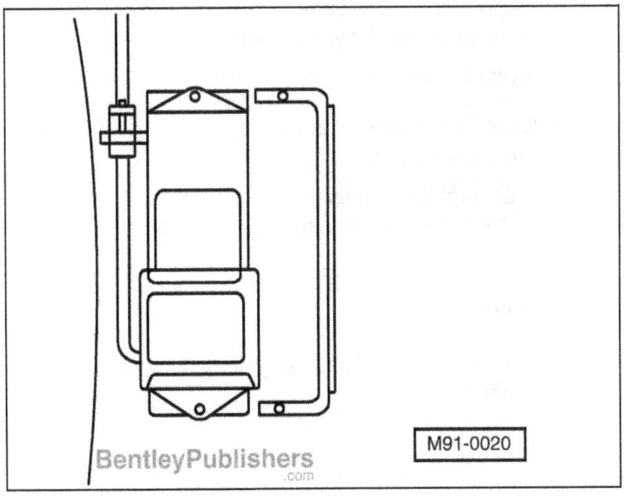

◀ Install protective cover around CD changer and secure bracket and cover with screws from top.
- Align tabs at bottom of cover with slots in mounting bracket.

NOTE —
Early CD changer kits supplied by Volkswagen did not include protective cover.

Tightening torque
- CD changer to mounting bracket. . . . . 3 Nm (27 in-lb)

– Code radio head unit for addition of CD changer using diagnostic scan tool.

– Check operation of CD changer and radio.

CD changer, installing (center console)

NOTE —

- *Always confirm CD changer and radio compatibility with your Volkswagen parts department or aftermarket parts specialist.*
- *Carefully review all of the following instructions before proceeding with installation.*
- *Never install CD changer in a location, position, or manner other than specified by these instructions.*

> ***CAUTION —***
> *Observe all cautions and warnings in* **9 Electrical System** *before starting repairs involving the electrical system.*

– Lift center console armrest to fully open position.

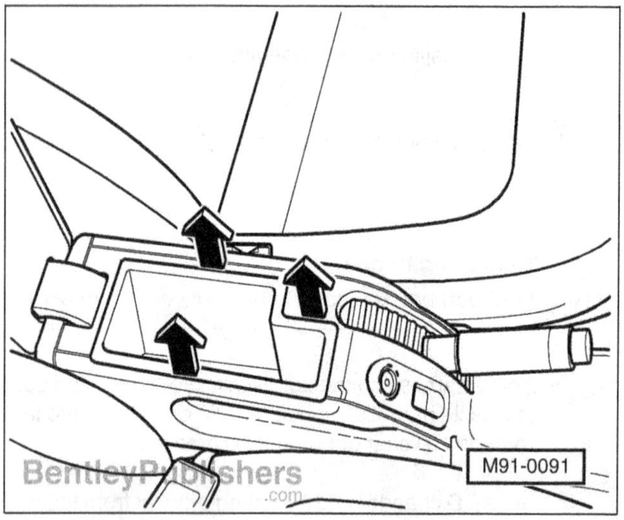

◄ Remove velour tray by pulling upwards (**arrows**).

– Open power outlet compartment or ashtray (where applicable) at rear of console and remove power outlet housing from mount.

– Unplug power outlet cable by opening top of power outlet housing. Power outlet and housing will be completely released (no wires need to be unplugged separately when vehicle is equipped with ashtray).

– Remove rear cover lid from console. See **70 Trim–Interior**.

– Move front passenger seat to its most forward position and tilt backrest forward.

– Remove right console side cover trim. See **70 Trim–Interior**.

– Locate CD cable inside center console.

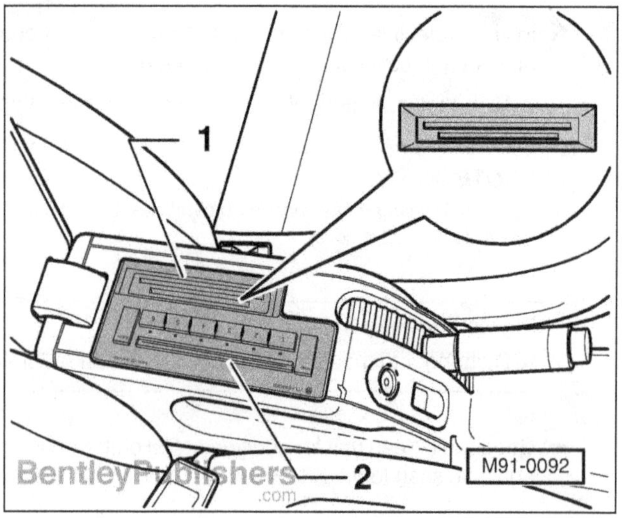

◄ Insert coin holder/storage box (**1**) into small area previously occupied by velour tray.

- Flat border must face inwards.

– Slide CD changer (**2**) into opening until mounting clips are felt to engage.

- CD slot is on passenger's seat side.
- Operation buttons face driver's seat.

– Remove protective plug from end of CD changer cable, if applicable.

– Use adapter cable to lengthen CD changer harness if necessary.

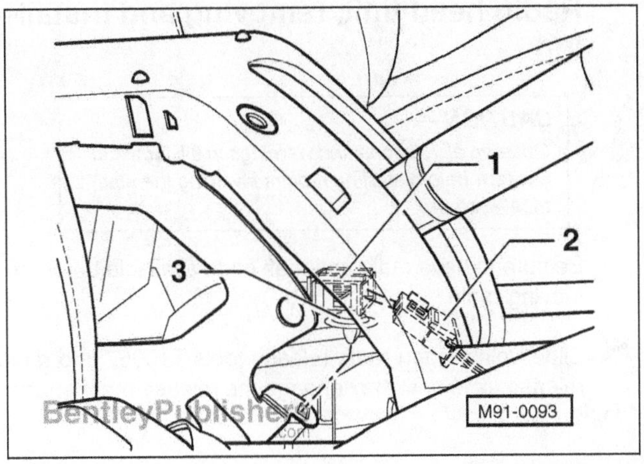

Connect CD changer cable (**2**) to changer harness connector (**1**).

- Check connections for secure fit.

– Attach harness connector (**1**) to retaining clip (**3**).

- Properly securing harness in retaining clip will help avoid noises and rattles.

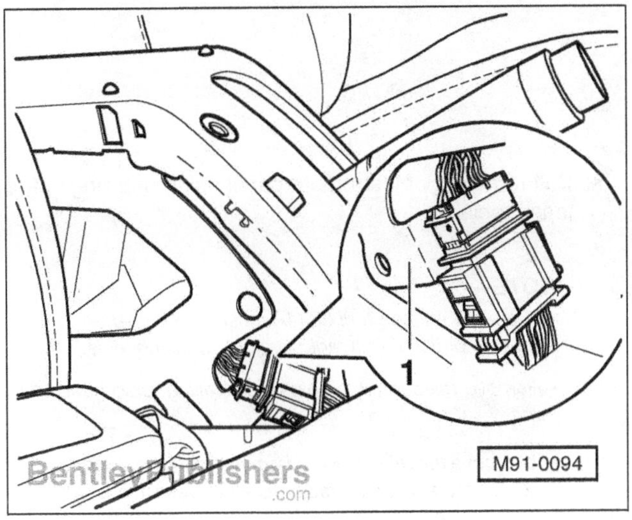

Clip connector/clip assembly to inner console frame (**1**).

– Reassemble console-mounted components and trim in reverse order of removal

– Code radio head unit for addition of CD changer using diagnostic scan tool.

– Refer to sound system and CD changer owner's guides and check proper operation of CD changer.

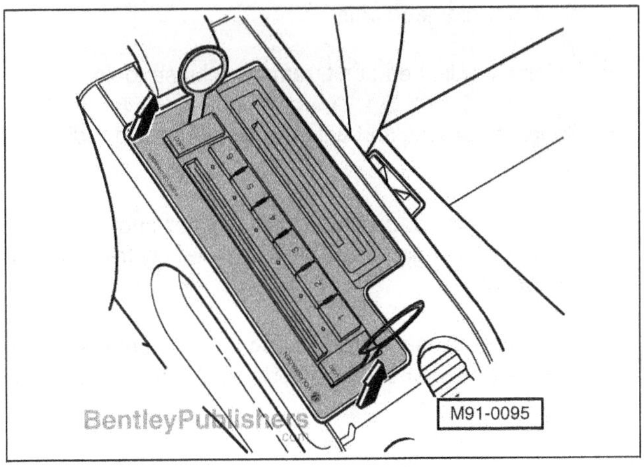

To remove CD changer, insert removal tools (special tool 3316) into CD changer as illustrated (**arrow**) until they are felt to engage.

– Lift CD changer from console using handles of special tool 3316.

– Remove right console side cover trim to disconnect electrical harness connectors. See **70 Trim–Interior**.

Radio head unit, removing and installing

> ***CAUTION—***
> *Observe all cautions and warnings in* **9 Electrical System** *before starting repairs involving the electrical system.*

– Be sure to have radio anti-theft code available before removing radio.

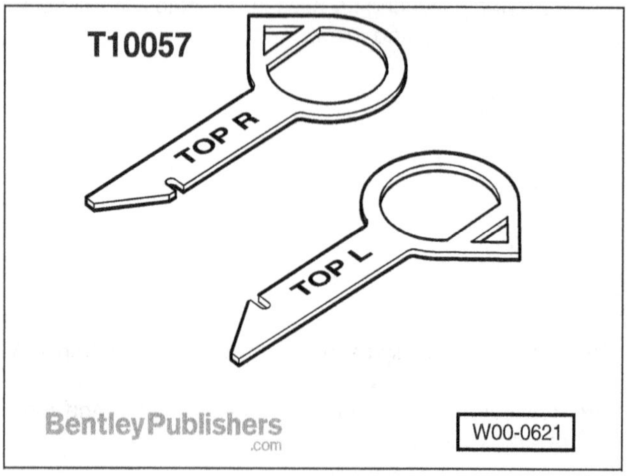

◀ Slide Volkswagen radio release tools T10057 into radio release slots until they engage the release mechanism.

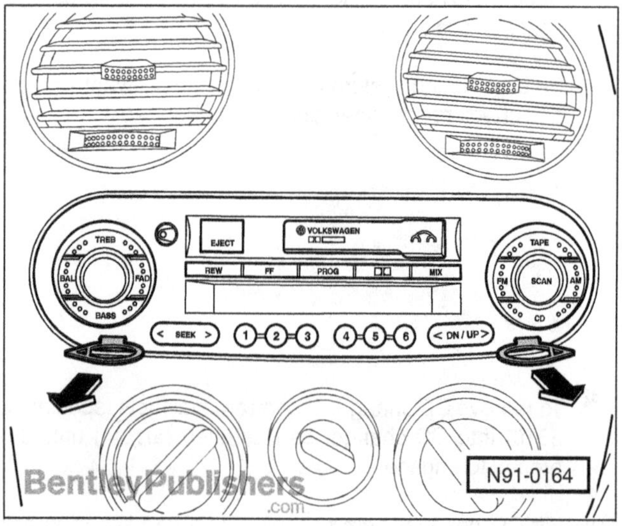

◀ Pull radio out of instrument panel using grip rings of release tools.

NOTE—
- *The radio release tool 3344A must not be pushed to the side or tilted. Pull radio straight out using tools.*
- *Remove release tools before reinstalling radio head unit.*
- *To remove the release tools from the radio the locating lugs on the side of the radio must be pressed inward.*

– Remove antenna connector and harness connectors on back of radio head unit.

– To install radio head unit, remove release tools.

– Connect harness and antenna connectors to radio head unit.

– Carefully slide radio head unit straight into instrument panel until it engages properly in assembly frame in center console.

– If new (or exchange) radio head unit has been installed, be sure to store new radio code card in a safe location.

Radio anti-theft system, description

All factory installed and supplied radios are equipped with an electronic anti-theft system. The anti-theft system is activated and electronically de-activates (locks) the radio as soon as any of the following occur:

- Voltage supply (terminal 30) drops below a predetermined voltage value.
- The radio unit is disconnected from voltage supply (terminal 30) such as when removing radio head unit or if the fuse blows.
- Vehicle battery has been disconnected.

A radio which is locked by the electronic anti-theft system has the word "SAFE" in the display window when the radio is switched on. After 3 seconds, the word "SAFE" will be replaced by the number "1000" indicating readiness to unlock. See **03 Maintenance** for more information on entering radio security code.

In addition to electronic anti-theft locking, the radio is also linked to the alarm system. The alarm horn will sound if the alarm has been set and the radio is removed.

Radio head unit terminal assignments

Table a lists the terminal assignments for the multi-pin connector on the back of the radio.

Table a. Radio head unit terminal assignments

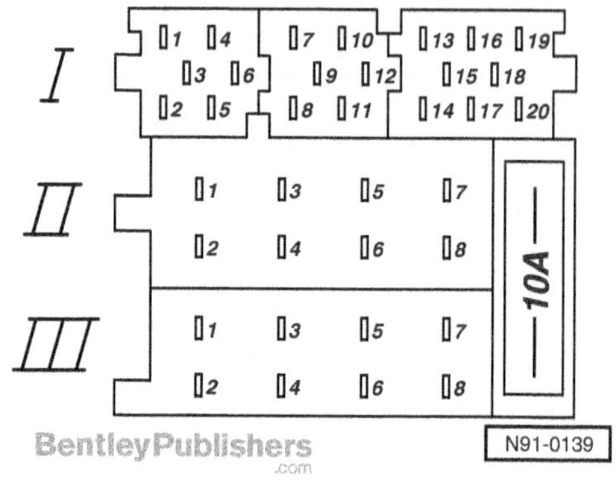

Terminal	Circuit
	Harness connector I, T20, part 1, yellow
1	Line out, L/R
2	Line out, R/R
3	Ground (GND)
4	Line out, L/F
5	Line out, R/F
6	Switched positive (B+ out) for amplifier
	Harness connector I, T20, part 2, green
7	Open
8	Clock display
9	Data display
10	Enable remote display
11	Remote control (serial input) or SWC digital bus (satellite)
12	Open
	Harness connector I, T20, part 3, blue
13	CD changer - DATA IN
14	CD changer - DATA OUT
15	CD changer - CLOCK IN
16	CD changer - Positive (B+ in), terminal 30
17	CD changer - Control signal, turn-on
18	CD changer - Ground (GND) and shield
19	CD changer - Line out, left
20	CD changer - Line out, right

Table a. Radio head unit terminal assignments

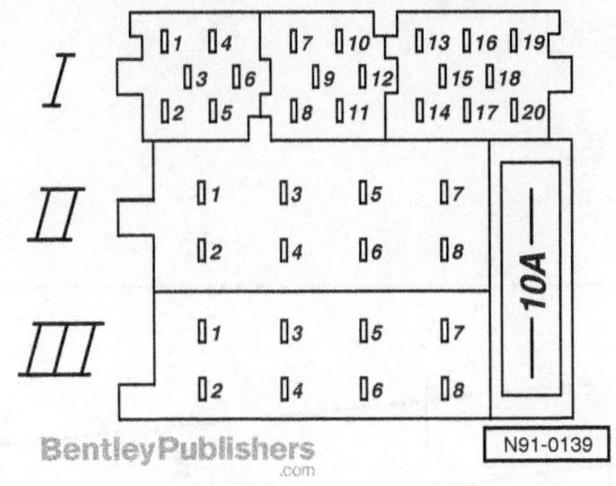

Terminal	Circuit
	Harness connector II, T8a, brown
1	R/R speaker +
2	R/R speaker -
3	R/F speaker +
4	R/F speaker -
5	L/F speaker +
6	L/F speaker -
7	L/R speaker +
8	L/R speaker -
	Harness connector III, T8, black
1	Signal for anti-theft alarm
2	Open
3	Data Link Connector (DLC), K wire
4	Ignition switch, S contact
5	Battery positive (B+), terminal 30 (control input for "SAFE" display/anti-theft)
6	Illumination, terminal 58b
7	Battery positive (B+), terminal 30
8	Battery ground, (GND), terminal 31

NOTE —

Harness connector I, T20, part 1 (yellow) and part 2 (green) are not used at this time.

Satellite receiver, removing and installing

The Satellite Digital Audio Radio Service (SDRAS) was available as an option in the New Beetle sedan from 04/2004 and the New Beetle convertible from 07/2004.

A separate satellite radio receiver is mounted under the front passenger's footwell carpet. The satellite radio uses a special antenna that receives dedicated satellite radio broadcasts. The sedan antenna is located on the roof. The convertible antenna is located in the instrument panel, under the center top cover.

Satellite radio service is provided to customers on a subscription basis. Always read the radio system owner's manual as well as any documentation provided by the satellite radio service provider.

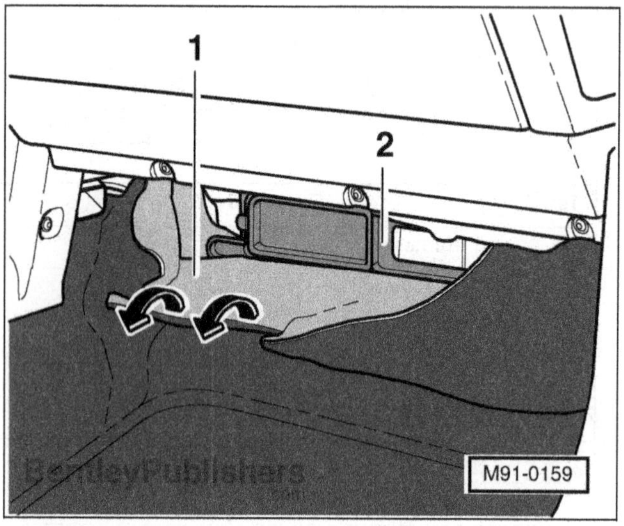

◀ Working in passenger front footwell, fold carpet (**1**) toward rear (**arrows**) until receiver (**2**) is accessible.

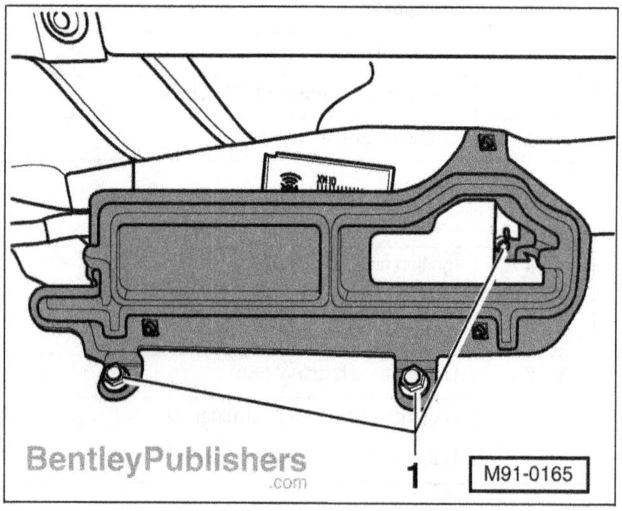

◀ Remove mounting hardware (**1**).

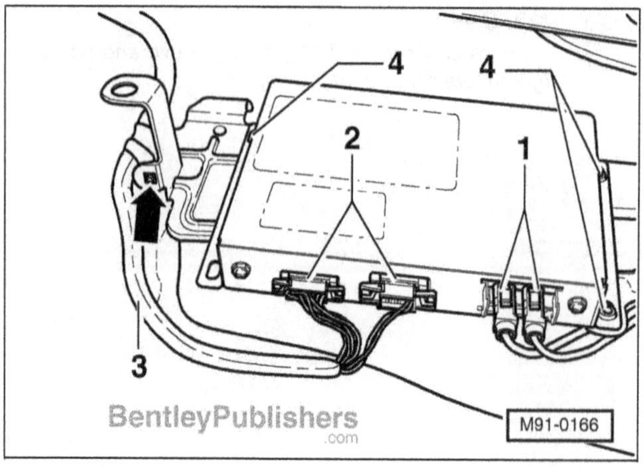

◀ Disconnect antenna connectors (**1**) and electrical connectors (**2**).

- Note satellite antenna connector (**1**) orientation: green on right and brown on left.
- Note electrical connector (**2**) orientation: two wires on right and six wires on left.

– Remove harness (**3**) from retaining bracket (**arrow**).

– Where applicable, remove screws (**4**).

– Install in reverse order of removal.

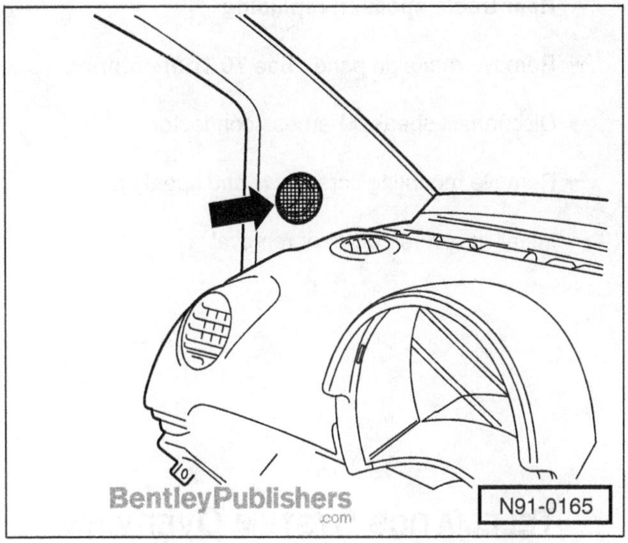

Speakers, removing and installing

A-pillar treble speaker, replacing

- The domed treble (tweeter) speakers (**arrow**) in the front A-pillars are permanently attached to the interior A-pillar trim cover.
- To replace a domed treble loudspeaker, the entire A-pillar trim cover must be replaced. See **70 Trim–Interior**.

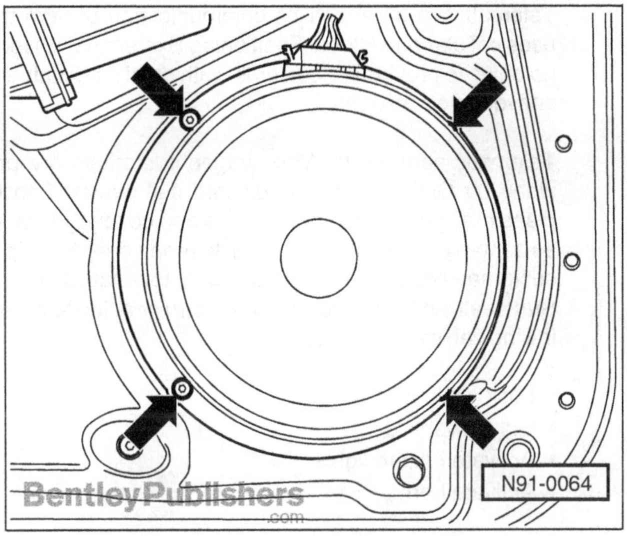

Bass speaker (woofer), front and rear, replacing

- Remove door trim panel or rear trim panel. See **70 Trim–Interior**.
- Disconnect speaker harness connector.
- Carefully drill out rivets (**arrows**) and remove speaker.
- To prevent corrosion, make sure that all metal particles from drilling are removed from inside the door.
- If the paint on door frame is damaged during drilling, touch-up immediately.
- Secure new speaker with pop rivets of proper length and diameter.
- Do not forget rubber ring between door trim and loudspeaker when reinstalling door trim panel.

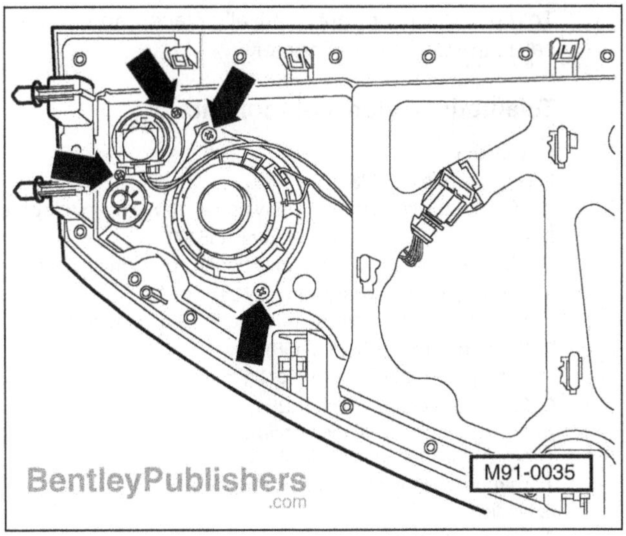

Front dash treble and midrange speakers, replacing

- Remove inner plenum trim panels from top of instrument panel (dashboard). See **70 Trim–Interior**.
  - Lift left and right inner plenum trim panels slowly and disconnect speaker harness connector.
- Carefully drill out rivets (**arrows**) or remove mounting screws and remove speaker.
- Secure new speaker with pop rivets of proper length and diameter, and/or with removed mounting screws.
- Install inner plenum trim panels.

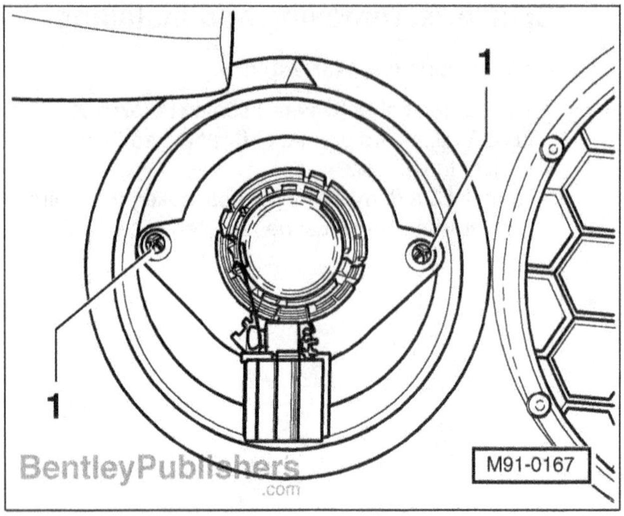

Rear treble speaker, replacing

- Remove rear side panel. See **70 Trim–Interior**.
- Disconnect speaker harness connector.
- Remove mounting screws (**1**) and speaker.
- Install in reverse order of removal.

TELEMATICS SYSTEM OVERVIEW

Telematics is an in-vehicle communication system that uses cellular and Global Positioning System (GPS) technologies to provide the customer with safety, security and convenience services.

Telematic services for Volkswagen customers are provided by OnStar® via a dedicated call center. Service plans may provide route directions and concierge services as well as emergency or vehicle repair services. Upon customer request, the Volkswagen telematics by OnStar® call center can remotely perform or activate the following vehicle functions:

- Lock vehicle
- Unlock vehicle
- Activate hazard lights
- Activate horn

During events where the airbag control module processes a crash signal, the vehicle hazard lights are activated and the doors are unlocked automatically. In addition, the Telematics system automatically places an emergency call to the OnStar® call center.

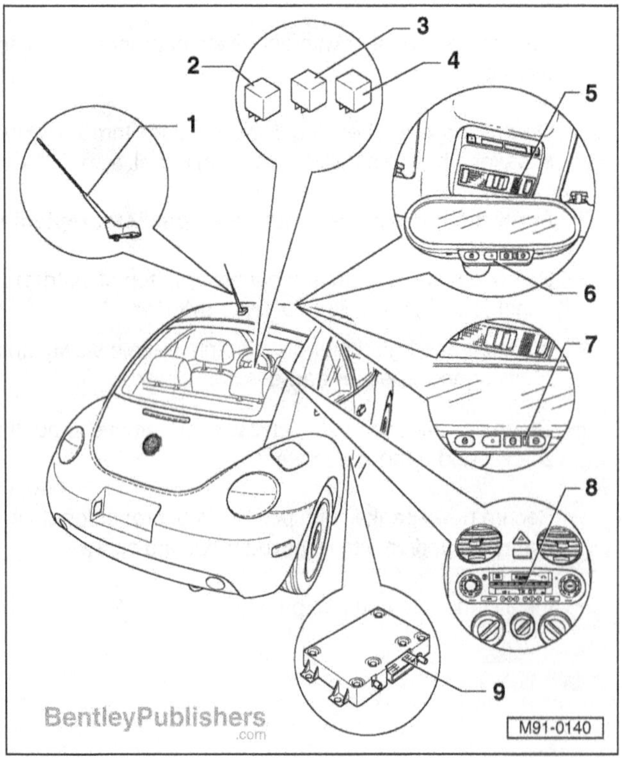

Telematic system components

1. Antenna
2. Dual horn relay (J733)
   - On 13-position auxiliary relay panel, above main relay panel
3. Auxiliary emergency flasher relay (J517)
   - On 13-position auxiliary relay panel, above main relay panel
4. Warning lamp auxiliary relay (J734)
   - On 13-position auxiliary relay panel, above main relay panel
5. Telephone microphone (R38)
   - Integrated with front interior light assembly
6. Telematics control head (E264)
   - Integrated with rear view mirror
   - Includes telematics indicator lamp
7. Telematics indicator lamp (K186)
8. Radio
9. Telematics control module (J499)

92 Wipers and Washers

GENERAL

The front windshield wipers are operated by the wiper switch on the steering column. There is a small lever on the top of the wiper switch which adjusts the interval when the wipers are in delay mode. Pulling the wiper switch back toward the passenger compartment operates the washer and the wipers automatically.

Some vehicles are equipped with heated washer nozzles which do not have a separate switch and are automatically activated (with ignition switched on) depending on outside temperature.

WIPERS AND WASHERS

CAUTION—

- *Worn or dirty wiper blades will reduce visibility, making driving hazardous. Clean blades regularly to remove road film and car wash wax build-up. Use an alcohol based cleaning solution and a lint free cloth to wipe along the rubber blade.*
- *Clean all inside and outside glass regularly. Use an alcohol based cleaning solution and wipe dry with a lint free cloth.*
- *Do not use the wiper/washer in freezing weather without first warming the windshield with the defroster. Otherwise the washer solution may freeze on the windshield and obscure your vision.*
- *To prevent scratching the glass avoid running the wiper blades over a dry windshield. A scratched windshield will reduce visibility and increase glare.*

92

Rubber wiper inserts, replacing

– Free end of wiper insert from retaining hooks by squeezing metal strip in insert with pliers.

– Slide rubber insert from retaining hooks.

– Slide rubber insert into lower retaining hooks of wiper blade.

> ***NOTE —***
> *If necessary, transfer metal strips from old insert to new insert. Notches in metal strips must face rubber and engage rubber lugs in groove.*

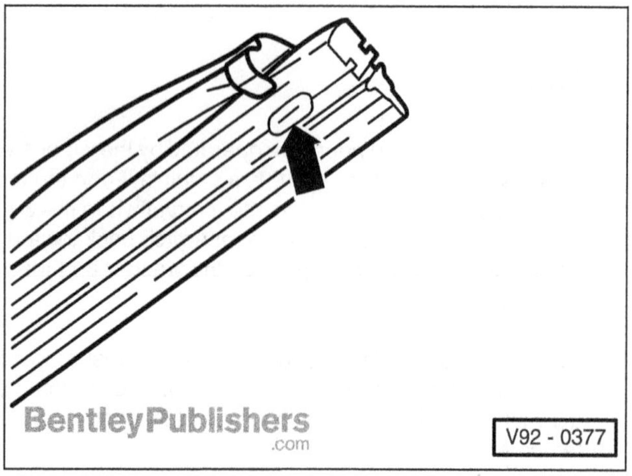

◄ Squeeze metal strips at open end of rubber insert using pliers and install insert into hooks so that hook engages rubber retaining slot (**arrow**) on insert.

Windshield wiper assembly

1. **Wiper blade**
   - Adjusting park position, see **Wiper blade park position, adjusting**
2. **Wiper blade rubber**
3. **Connecting lever**
4. **Connecting rod**
5. **Crank**
6. **Wiper motor**
7. **Hex nut M8**
   - Tighten to 20 Nm (15 ft-lb)
8. **Wiper frame**
9. **Securing bolts M6**
   - Tighten to 5 Nm (44 in-lb)
10. **Protective sleeve**
11. **Cover cap**
12. **Hex nut M8**
    - Tighten to 20 Nm (15 ft-lb)
13. **Wiper arm**
14. **Retaining clip**

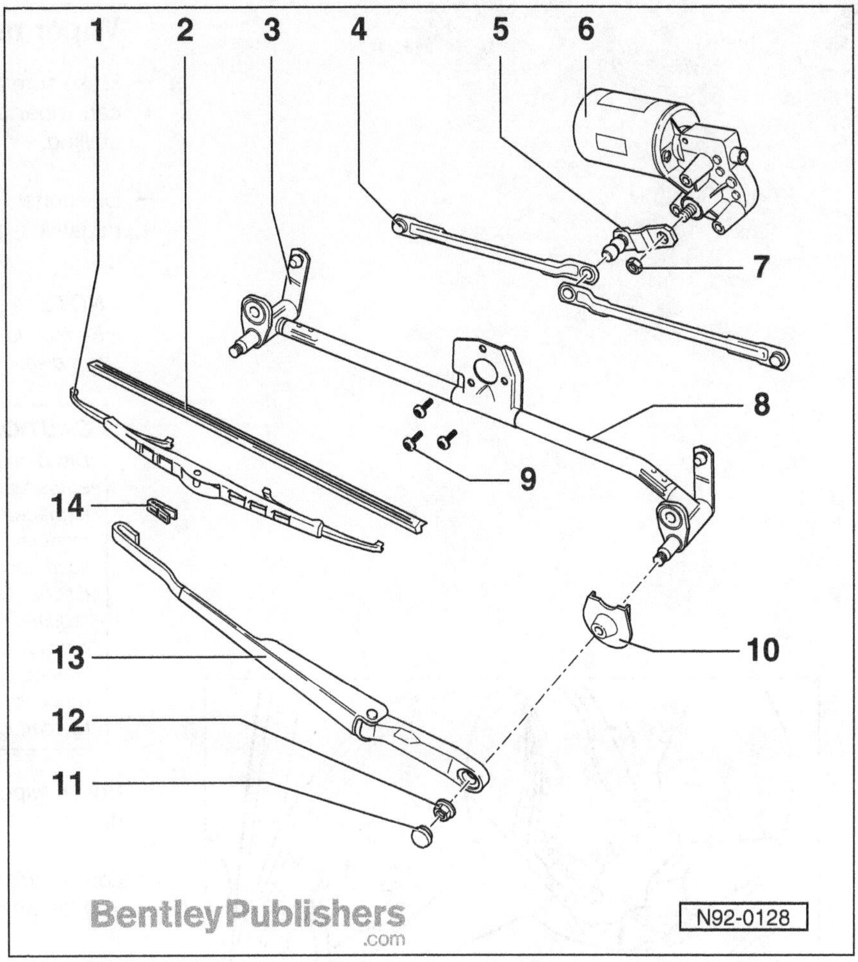

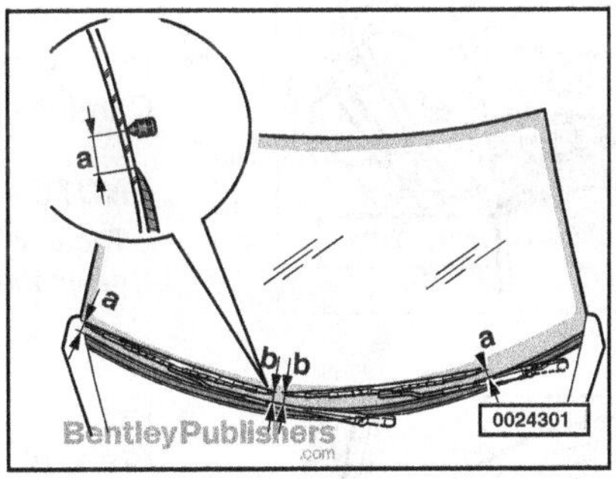

Wiper blade park position, adjusting

- Align wiper blades so they are 20 mm (¾ in.) from lower edge of windshield (**arrows a**).
- Distance between **arrows b** = 20 to 30 mm (¾ to 1-3/16 in.)

Wiper motor, removing and installing

- Make sure the wiper arms are in park position. Only then can wiper arm end position be correctly set when reinstalling.

- Disconnect battery ground (GND) strap from battery negative (–) terminal.

NOTE —
Be sure to have the anti-theft radio code on hand before disconnecting the battery.

> ***CAUTION —***
> *Disconnecting the negative (–) battery cable may erase fault codes and basic settings in the engine management and automatic transmission control modules. Some driveability problems may be noticed until the system re-adapts to operating conditions. OBD II readiness codes, which may be required for emissions testing, may also be erased. Convenience electronics (alarm system, interior light control, power locks, mirrors, and windows) may need to be re-set using a diagnostic scan tool.*

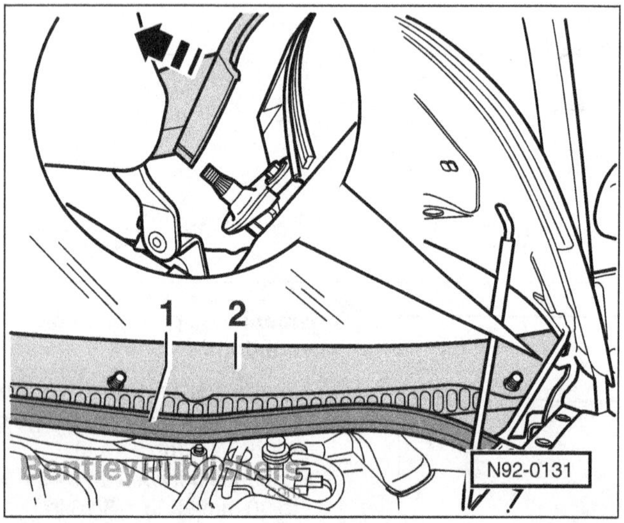

- Pry off wiper arm mounting nut cover caps using screwdriver.

- Loosen wiper arm mounting nuts and move wiper arms slightly until they release.

- Remove wiper arm mounting nuts and wiper arms.

- Pull up plenum panel rubber gasket (**1**) and remove. Carefully pry cowl panel (**2**) upward and off.

NOTE —
The cowl panel is attached to a window guide underneath the windshield.

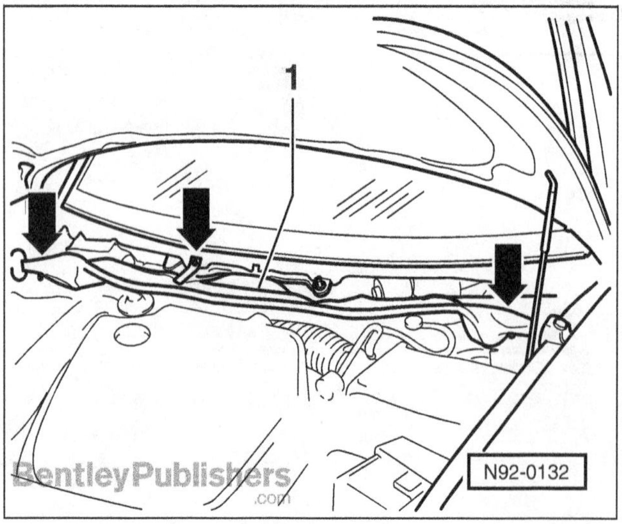

- Remove plenum close-out panel mounting screws (**arrows**) and pull close-out panel slightly forward. See also **50 Body–Front, Plenum close-out panel, removing and installing**.

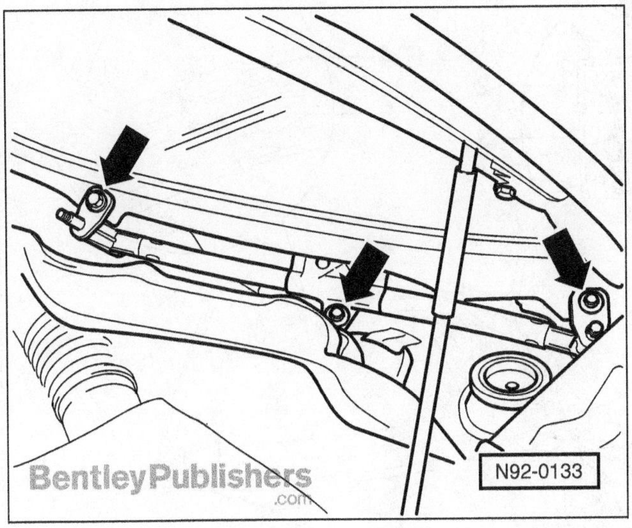

◄ Remove wiper motor assembly mounting bolts (**arrows**) and washers.

– Disconnect harness connector from wiper motor.

– Remove complete wiper assembly.

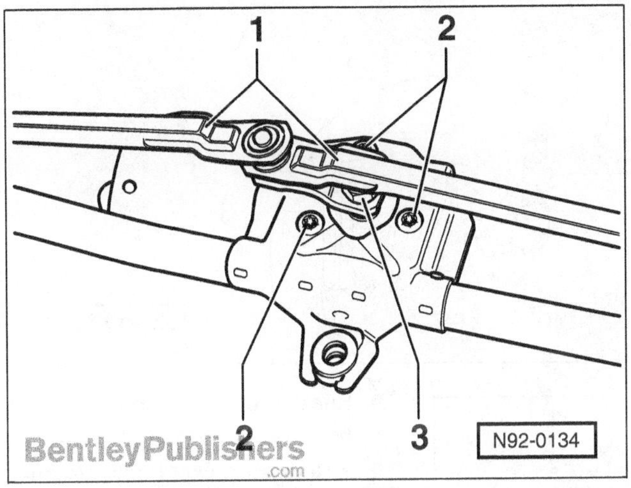

◄ Pry connecting rods (**1**) off wiper motor crank using large screwdriver.

- Unbolt wiper motor-to-frame mounting bolts (**2**).
- Remove wiper motor from wiper frame.
- Remove crank arm nut (**3**) and crank arm.

– Begin installation by connecting harness connector to wiper motor and operating wiper switch briefly to ensure that motor is in park position.

– Disconnect harness connector again and install wiper motor-to-frame mounting bolts.

Tightening torque
- Wiper motor to frame . . . . 5 Nm (44 in-lb)

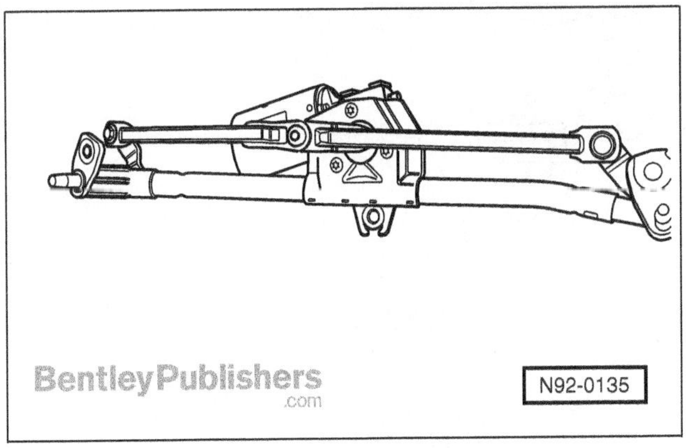

◄ Install crank and align so that connecting rods are in line.

– Tighten crank arm nut and press rods back onto crank.

Tightening torque
- Wiper motor crank arm nut. . . . . 20 Nm (15 ft-lb)

Windshield washer system overview

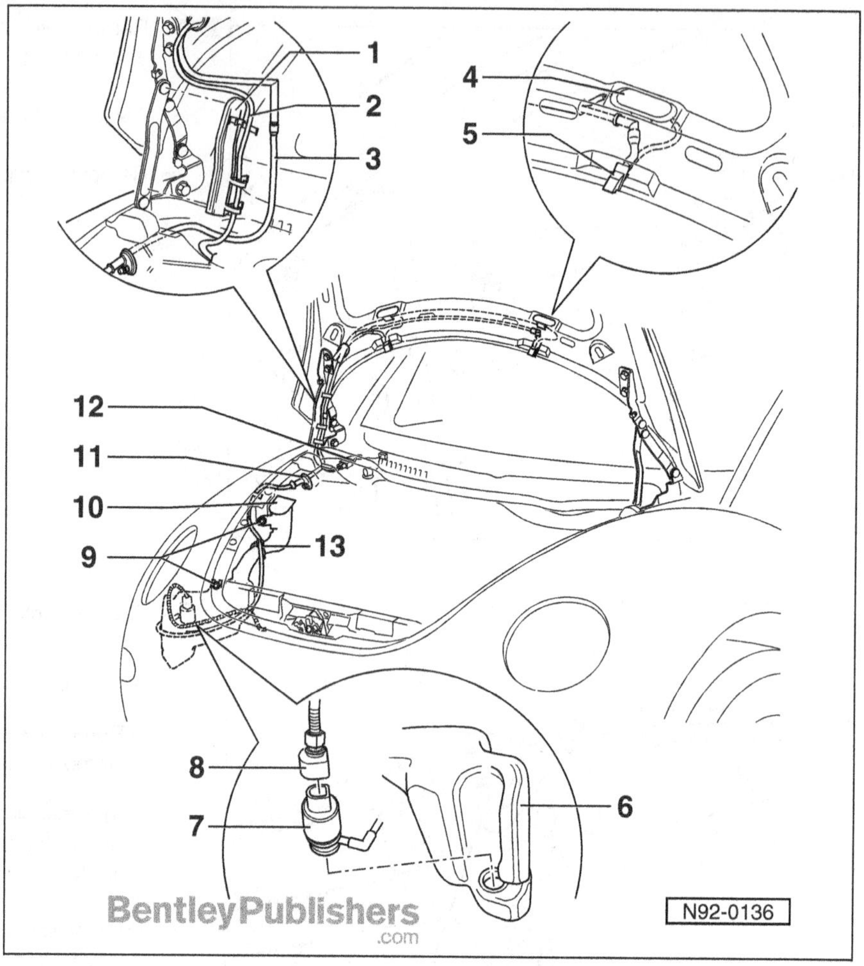

1. **Securing clip**
2. **Wiring harness for heated washer jets**
   - Optional equipment
   - Disconnecting, see
3. **Hose**
4. **Cover cap**
   - With connectors for heated washer nozzles
5. **Jets**
   - Removing, see **Windshield washer jets, removing**
   - Checking, see **Windshield washer system jets, checking**
6. **Windshield washer system reservoir**
   - Removing: remove right front wheel housing liner
7. **Windshield washer system pump**
   - Pull pump upward off rubber grommet
8. **Harness connector**
9. **Securing bolts**
   - Tighten to 5 Nm (44 in-lb)
10. **Container cap**
11. **Transfer piece**
12. **Harness connector for heated washer nozzles**
13. **Hose clip**

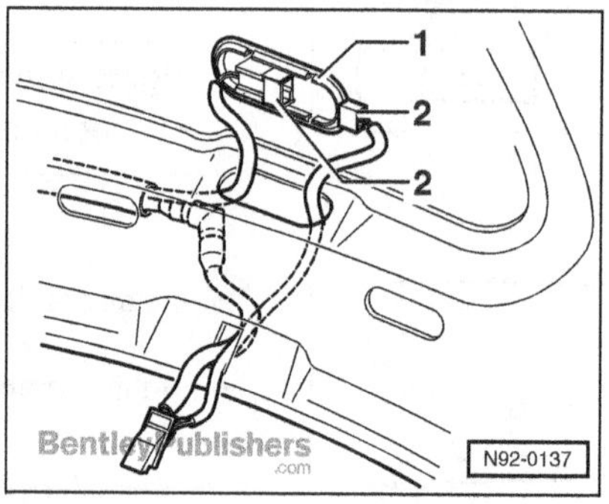

Heated washer nozzle, disconnecting harness connector

- Remove cover (**1**) and disconnect harness connector (**2**).

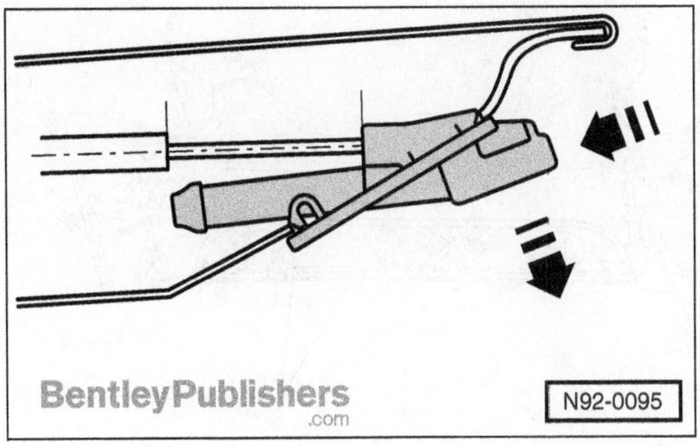

Windshield washer jets, removing

- Press jet forward and remove downward.
- Pull hose off jet and disconnect harness connector.

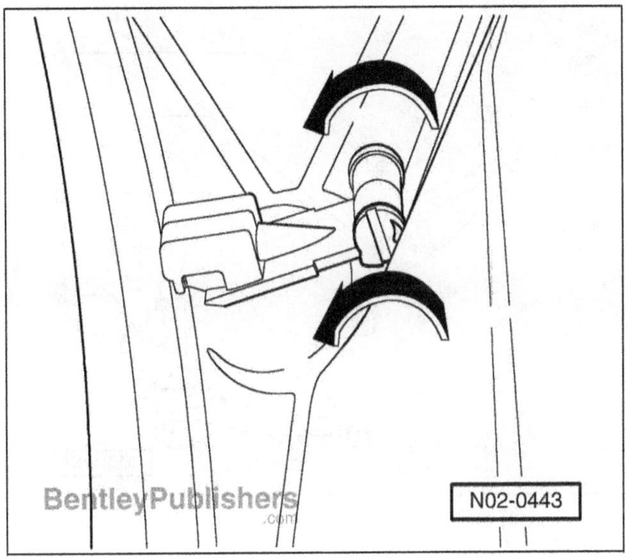

Windshield washer system jets, adjusting (through 08/1999)

- The washer jets are preset but small height adjustments can be made.
- Using screwdriver, turn eccentric on spray jet in direction of **arrow** to move spray field on windshield upward. Turn in opposite direction to move spray pattern downward.
- Do not clean spray jets by blowing through from the front.

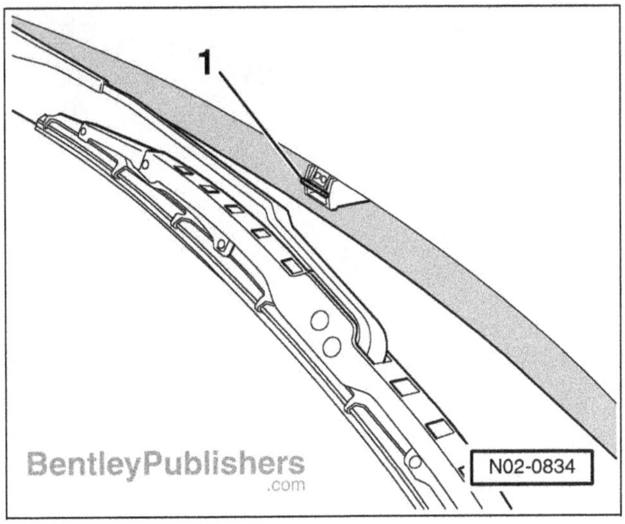

Windshield washer system jets, adjusting (from 09/1999)

- The washer jets are preset but small height adjustments can be made.
- Using a small screwdriver, raise or lower adjustment tab (**1**) to move spray field on windshield.
- Do not clean spray jets by blowing through from the front.

HEADLIGHT WASHER SYSTEM

Headlight washer system, assembly

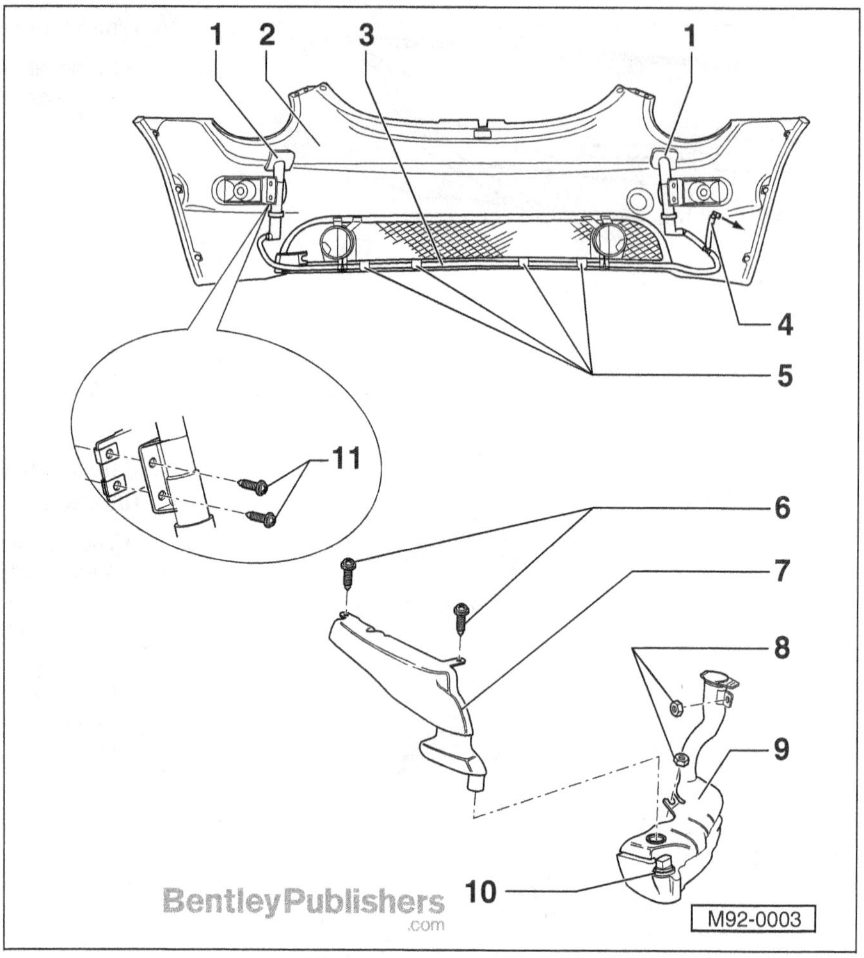

1. **Headlight washer jets**
2. **Front bumper cover**
3. **Hose**
4. **To pump for window washer and headlight washer system**
5. **Retaining springs**
6. **Screws**
   - Tighten to 5 Nm (44 in-lb)
7. **Headlight washer system reservoir**
   - Capacity: 2.5 liters (2.6 qt.)
8. **Hex Nut**
   - Tighten to 5 Nm (44 in-lb)
9. **Windshield washer system reservoir**
10. **Windshield and headlight washer system pump**
11. **Screws**
    - Tighten to 3.5 Nm (31 in-lb)

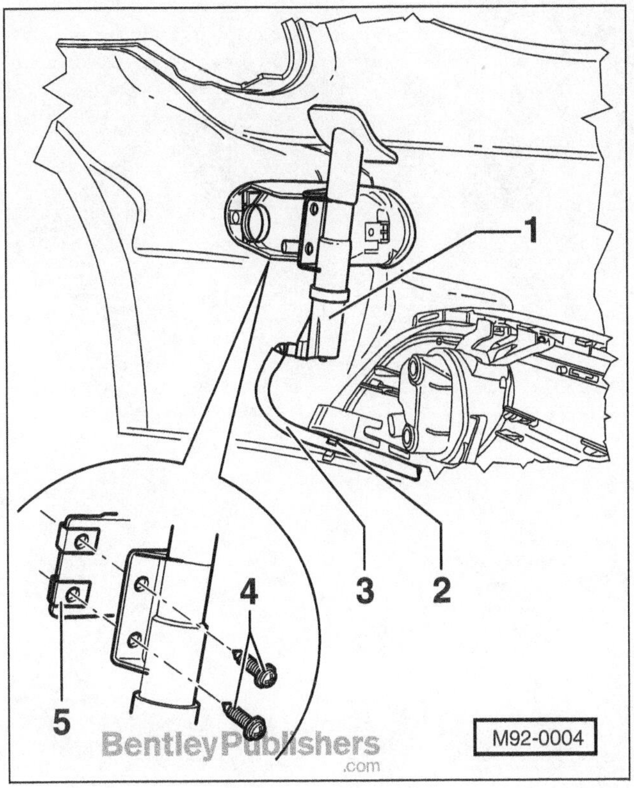

Headlight washer system jets, removing and installing

- Remove front bumper cover, see **63 Bumpers**.
- ◀ Remove screws (**4**) from speed nuts (**5**).
- Unclip headlight washer system hose (**3**) from hose clips (**2**).
- Pull hose (**3**) from spray jet carrier (**1**).
- Installation is reverse of removal.

NOTE —
Once installation is complete, bleed headlight washer system by operating system several times.

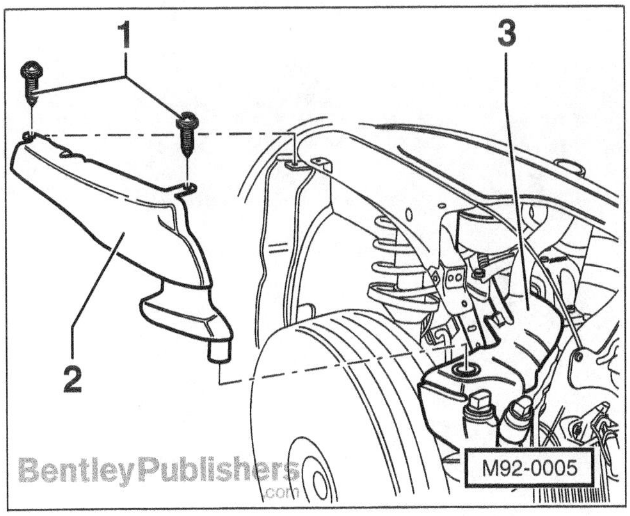

Headlight washer system reservoir, removing and installing

- Remove front bumper cover, see **63 Bumpers**.
- ◀ Remove bolts (**1**).
- Pull reservoir (**2**) out of windshield washer system reservoir (**3**).
- Installation is reverse of removal.

NOTE —
Once installation is complete, bleed headlight washer system by operating system several times.

94 Lights, Accessories–Exterior

GENERAL

This repair group covers removal, installation and replacing of the exterior lights and bulbs. A variety of different styles of exterior light assemblies were used throughout production. Be sure to have the vehicle production date on-hand to determine the correct application. High Intensity Discharge (HID) headlights, more commonly known as Xenon, are available as optional equipment from 2002.

CAUTION—

- *Before working on the electrical system disconnect battery ground (GND) strap from battery negative (–) terminal.*
- *Be sure to have the anti-theft radio code on hand before disconnecting the battery.*
- *Disconnecting the negative (–) battery cable may erase fault codes and basic settings in the engine management and automatic transmission control modules. Some driveability problems may be noticed until the system re-adapts to operating conditions. OBD II readiness codes, which may be required for emissions testing, may also be erased. Convenience electronics (alarm system, interior light control, power locks, mirrors, and windows) may need to be re-set using a diagnostic scan tool.*

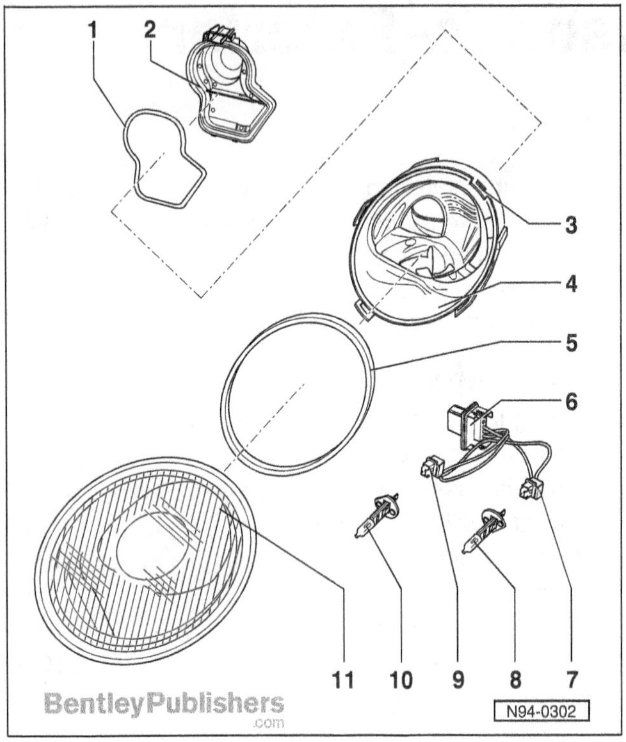

FRONT LIGHTS

Headlight assembly (through 06/2005)

1. Gasket
   - Replace if damaged
2. Cover
3. Retaining pockets
   - Qty. 4
4. Headlight housing
5. Gasket
   - Replace if damaged
6. Harness connector
   - Clipped into headlight housing
7. Harness connector for high beam bulb
8. High beam bulb
   - H1 12V, 55 watt
9. Harness connector for low beam bulb
10. Low beam bulb
    - H1 12V, 55 watt
11. Headlight glass
    - Replace if gasket is faulty
    - Unclip retaining pocket tabs to remove

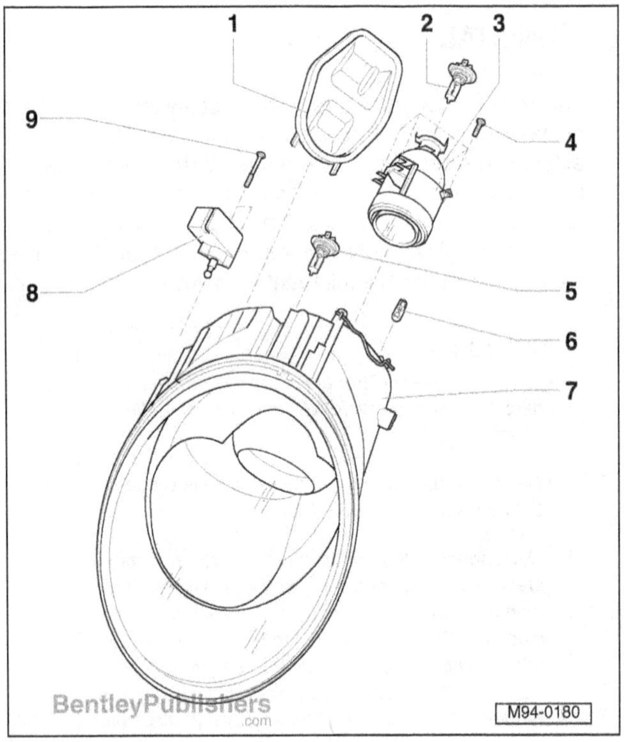

Headlight assembly (from 07/2005)

1. Cover
2. Low beam bulb
   - H7 12V, 55 watt
3. Projector/reflector
4. Screw
5. High beam bulb
   - H7 12V, 55 watt
6. Parking lamp bulb
   - 12V, 5 watt
7. Headlight housing
8. Headlight range control adjustment motor
   - Not applicable for US and Canada
9. Mounting screw

Headlight assembly (Xenon)

High Intensity Discharge (HID) headlights, commonly known as Xenon headlights, were available as optional equipment from 2002. Pay particular attention to the warnings below before working on HID/Xenon headlight components.

> ***WARNING —***
> - *Always disconnect negative (–) battery terminal prior to servicing any parts of parts of the HID lamp system marked with the yellow "High Voltage!" symbol.*
> - *After disconnecting battery, briefly turn headlight on and off in order to discharge any residual voltage.*
> - *HID bulbs are very sensitive. Internal pressure of HID glass bulb may exceed 100 bar (when operating). Bulb temperature may exceed 700°C.*
> - *Do not operate HID headlights if bulbs are not installed in housing.*
> - *Do not operate HID bulbs if they are not properly installed in housing.*
> - *Always wear safety glasses and gloves when removing and installing HID bulbs.*
> - *Never look directly at an operating HID bulb. The UV emissions of an HID bulb is approximately 2.5 times that of a comparable halogen bulb.*

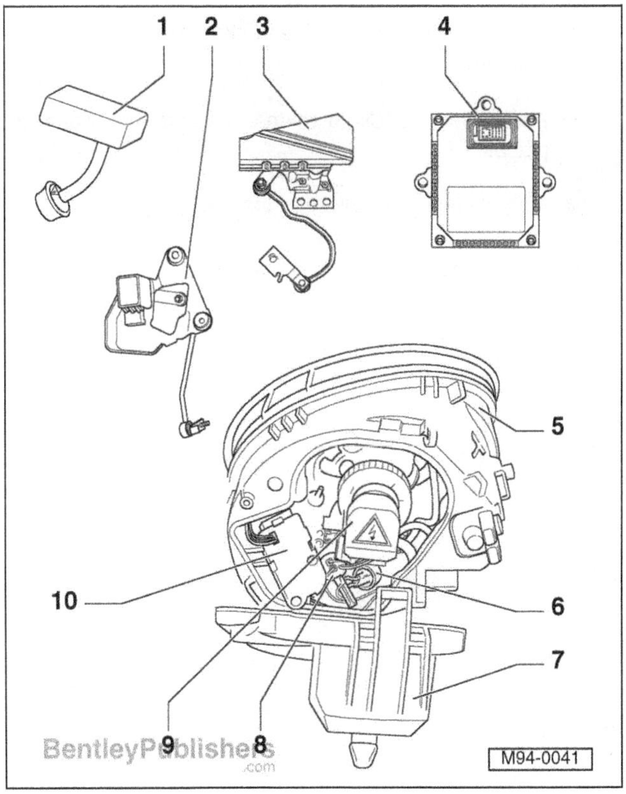

NOTE —
Volkswagen identifies electrical components by a letter and/or a number in the electrical schematics. See **EWD Electrical Wiring Diagrams**. *These electrical identifiers are listed in parenthesis as an aid to electrical troubleshooting.*

1. Left/right control module (J343/J344)
   - Integrated with headlight assembly, cannot be replaced separately
2. Left front level control system sensor (G78)
3. Left rear level control system sensor (G76)
4. Left/right control module with range control (J567/J568)
   - (J567) left control module is "master", located in left front wheel housing
   - (J568) right control module is "slave", located on right front/inner long member
5. Headlight housing
6. Parking lamp bulb
   - 12V, 6 watt
7. Sealing cap
8. High beam bulb (halogen)
   - H7 12V, 55 watt
9. Low beam bulb (Xenon)
   - Type DS2, 35 watt
   - Operating voltage: 80 - 117V
10. Left/right headlight beam adjusting motor (V48/V49)

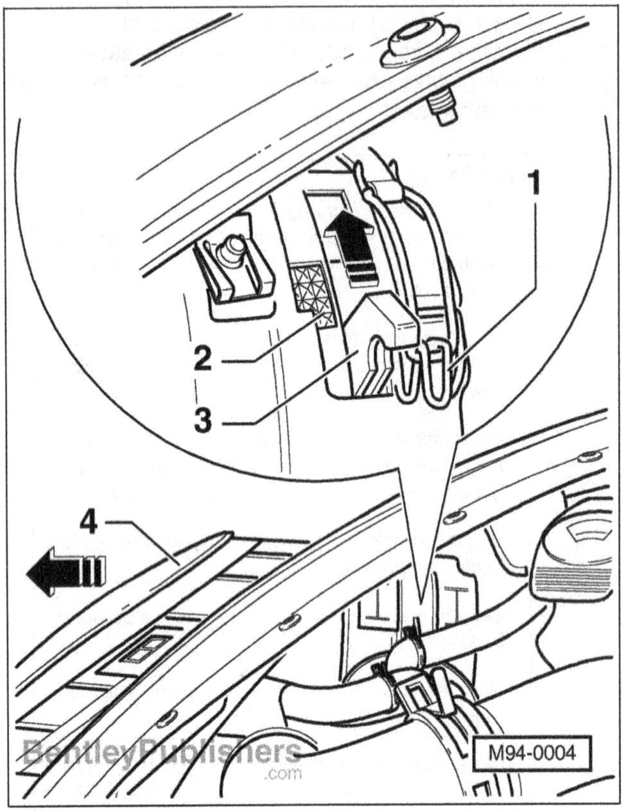

Headlight assembly, removing and installing (through 06/2005)

- Disconnect battery ground (GND) strap from battery negative (–) terminal. See the **Cautions** at the beginning of this repair group regarding battery disconnection.

◄ If applicable (through 08/1998), open retaining clamp (**1**).

- Press release button (**2**) and push up headlight release lever (**3**).

- Pull headlight housing slightly forward enough to disconnect harness connectors from headlight.

- Remove headlight housing.

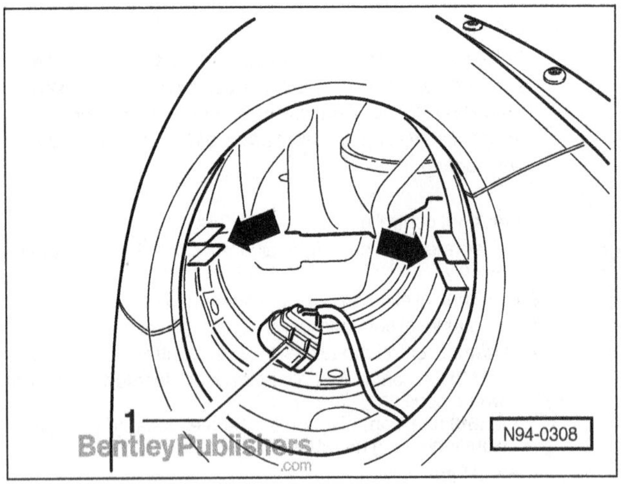

- Begin installation by connecting harness connector to headlight.

◄ Insert headlight into guide tracks (**arrows**) in fender and push inward.

- Push down headlight release lever to lock headlight in place.

- If applicable (through 08/1998), lock retaining clamp.

- Check headlight aim and function.

Headlight assembly, removing and installing (from 07/2005)

– Disconnect battery ground (GND) strap from battery negative (–) terminal. See the **Cautions** at the beginning of this repair group regarding battery disconnection.

< Insert hex wrench or socket into headlight lock (**arrow**).

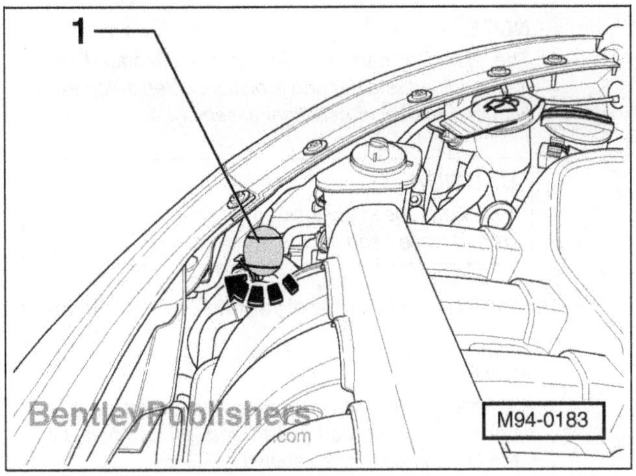

< Turn hex wrench or socket (**1**) approximately 180° in direction of **arrow** to release headlight assembly from retainer.

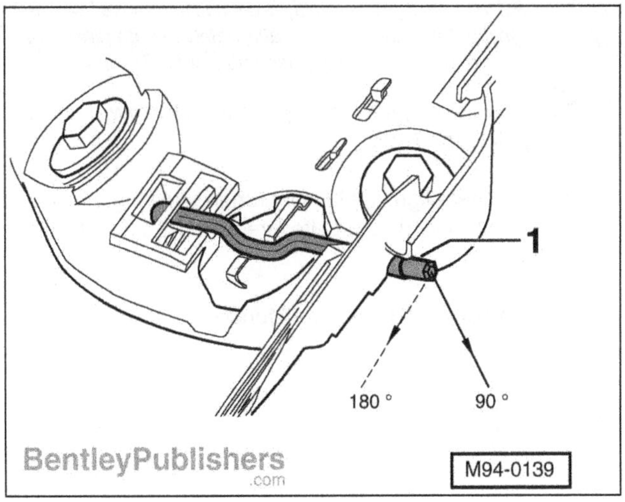

< Turn headlight lock (**1**) back 90° as shown.

NOTE —
In order for the headlight assembly to slide in an out freely, the "U" section of the headlight lock must be in the lowest position, as shown.

– If the headlight assembly is difficult to remove from vehicle, turn headlight lock (**1**) alternately in both directions while pulling headlight assembly out of retainer.

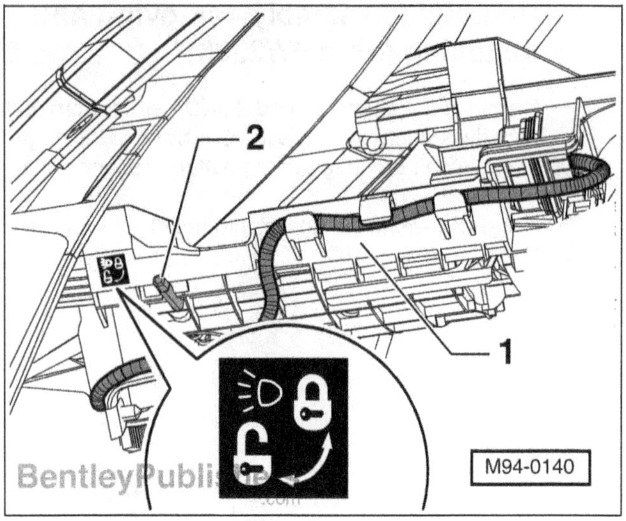

Headlight retainer (**1**) contains headlight lock position indicator (**2**).

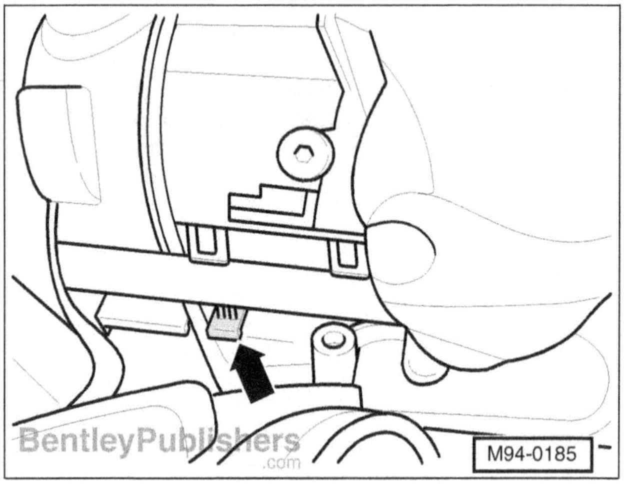

Depress lock tab (**arrow**) while pulling headlight assembly forward out of retainer.

NOTE —
The electrical harness connector is integrated within the headlight retainer and is disconnected automatically upon removal of headlight assembly.

CAUTION —
Headlight assembly lock on retainer must not be turned while headlamp is removed.

– Install in reverse order of removal, noting the following:

NOTE —
- *In order for the headlight assembly to slide in an out freely, the "U" section of the headlight lock must be in the lowest position, as shown.*
- *If the headlight assembly is difficult to remove from vehicle, turn headlight lock alternately in both directions while pulling headlight assembly out of retainer.*

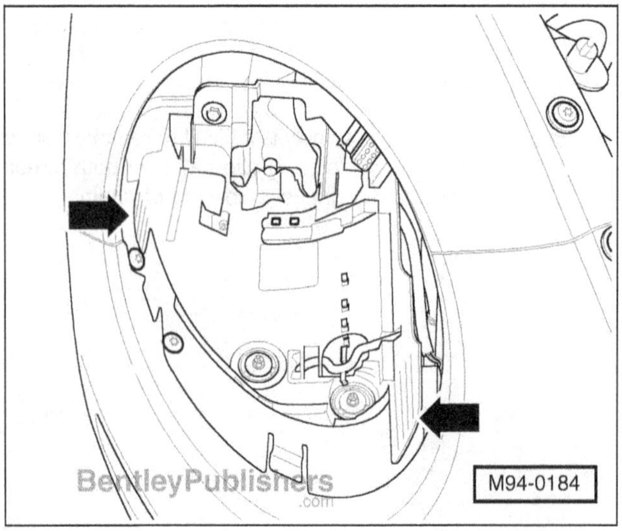

Align headlight assembly in guides (**arrows**) and carefully slide in to stop.

– Lock headlight in place by turning headlight lock approximately 180° in opposite direction used to unlock assembly.

– Check headlight aim and function.

Headlight aim, adjusting

Headlight adjustment is best accomplished with a headlight adjusting unit.

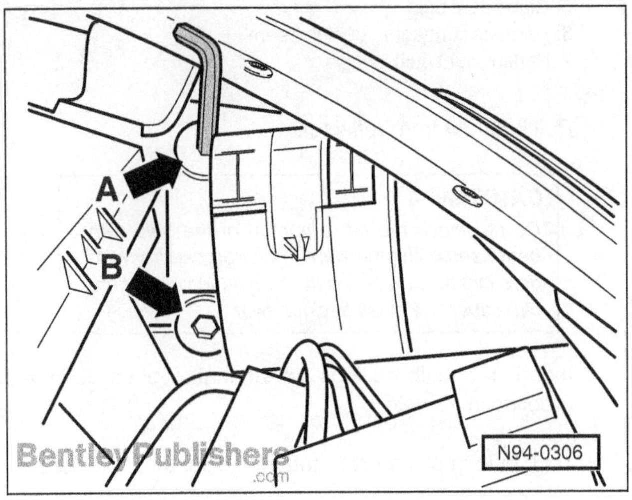

Driver's side headlight.

- Vertical adjustment (**A**)
- Lateral adjustment (**B**)

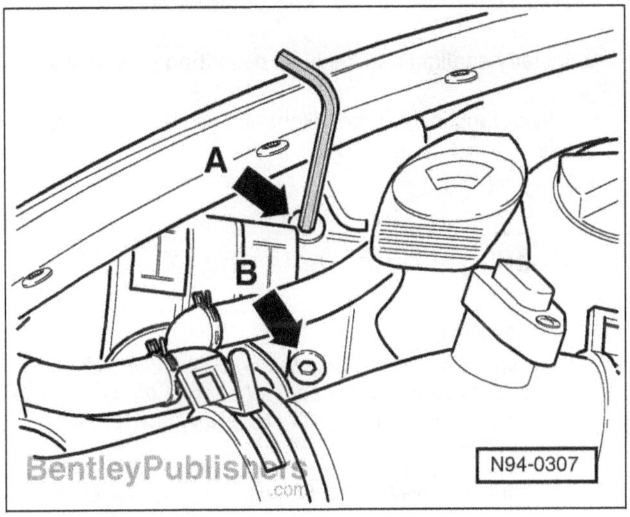

Passenger's side headlight.

- Vertical adjustment (**A**)
- Lateral adjustment (**B**)

Headlight bulb, replacing (Halogen)

– Remove headlight assembly as described previously.

– Remove cover from back of headlight assembly.

– Disconnect harness connector from bulb, as applicable.

– Push spring retaining clip over locating lugs and move clip to side.

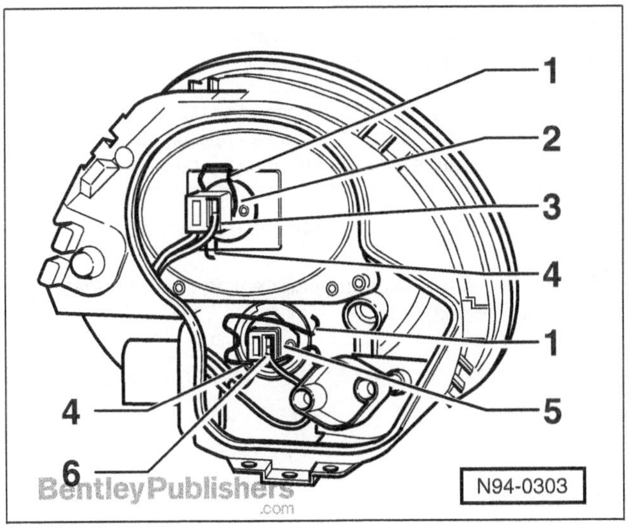

Bulb designations (through 06/2005)

1. Retaining clip
2. High beam bulb
3. Harness connector for high beam bulb
4. Locating lugs
5. Low beam bulb
6. Harness connector for low beam bulb

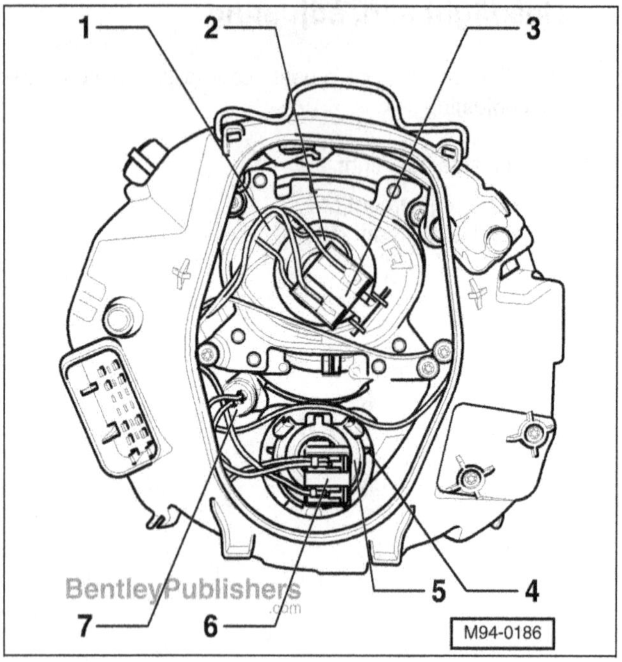

Bulb designations (from 07/2005)

1. Retaining clip
2. Low beam bulb
3. Harness connector for low beam
4. Securing ring
5. High beam bulb
6. Harness connector for high beam
7. Parking light bulb

- Pull bulb out from reflector.

> ***CAUTION —***
> *Do not touch the glass portion of the bulb with bare hands. The moisture and/or grease from fingers that evaporates on the bulb during operation, can cause the glass to cloud over.*

- Insert new bulb so lugs on laminated plate align with grooves in reflector.
- Lock bulb in place using retaining clip, as applicable.
- Install cover on back of headlight assembly.
- Install headlight assembly as described previously.
- Check function and headlight aim as described previouisly.

Headlight bulb, replacing (Xenon)

> ***WARNING —***
> - *Always disconnect negative (–) battery terminal prior to servicing any parts of parts of the HID lamp system marked with the yellow "High Voltage!" symbol.*
> - *After disconnecting battery, briefly turn headlight on and off in order to discharge any residual voltage.*
> - *Always wear safety glasses and gloves when removing and installing HID bulbs.*
> - *Never look directly at an operating HID bulb. The UV emissions of an HID bulb is approximately 2.5 times that of a comparable halogen bulb.*

- Remove headlight assembly as described previously.

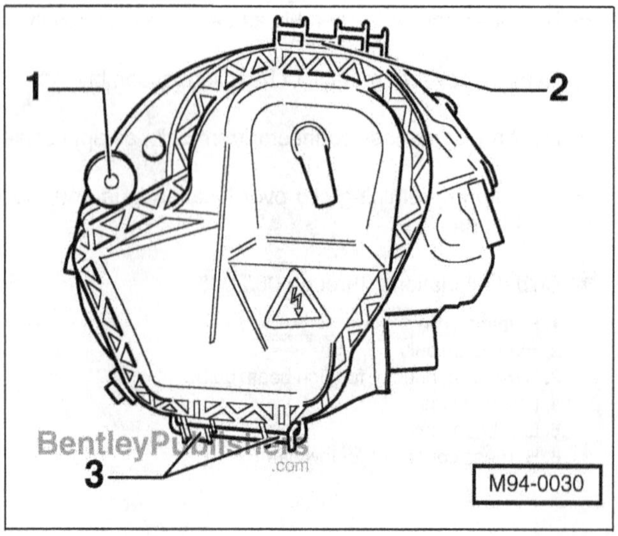

Remove screw (**1**) and open sealing cap by releasing retaining clip (**2**).

- Disengage sealing cap from lower mount (**3**) and remove.

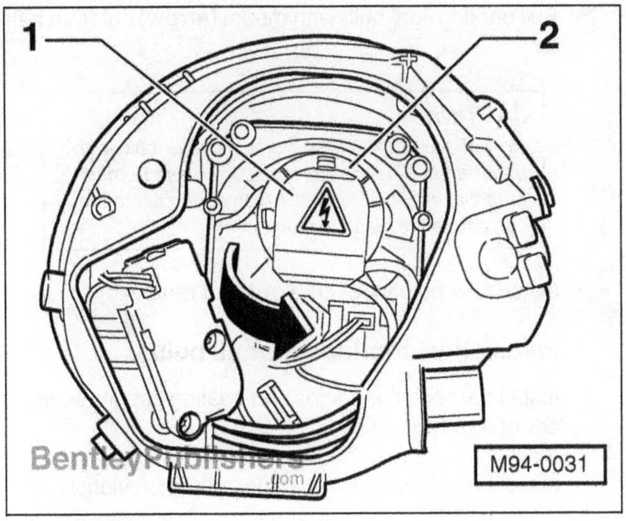

Low beam bulb (Xenon), removing

◀ Turn harness connector (**1**) 90° to left (**arrow**) and remove.

– Turn bulb retaining ring (**2**) to left and remove bulb.

> ***CAUTION—***
> *Do not touch the glass portion of the bulb with bare hands. The moisture and/or grease from fingers that evaporates on the bulb during operation, can cause the glass to cloud over.*

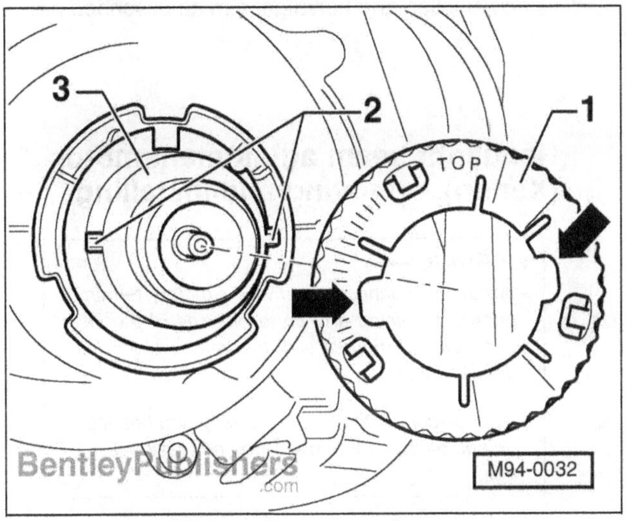

Low beam bulb (Xenon), installing

– Install new bulb into headlight assembly.

> ***CAUTION—***
> *Do not touch the glass portion of the bulb with bare hands. The moisture and/or grease from fingers that evaporates on the bulb during operation, can cause the glass to cloud over.*

◀ Position retaining ring (**1**) recesses (**arrows**) over locking tabs (**2**) of bulb (**3**) and secure by turning clockwise.

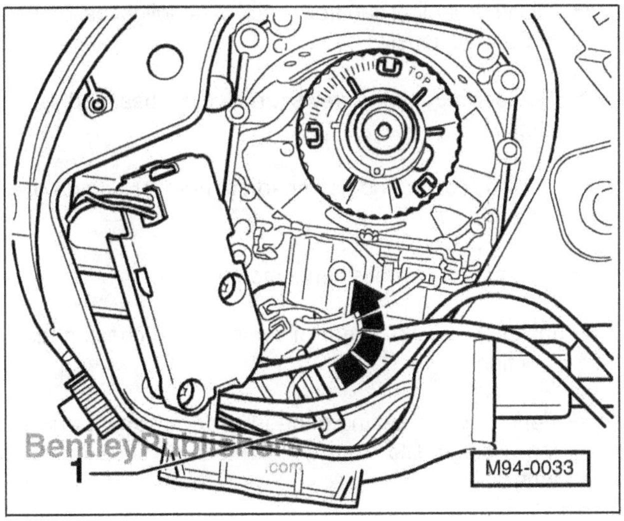

High beam bulb (Halogen), replacing

◀ Turn lever (**1**) on bulb mount to left (**arrow**) to release.

– Pull bulb out from reflector.

> ***CAUTION—***
> *Do not touch the glass portion of the bulb with bare hands. The moisture and/or grease from fingers that evaporates on the bulb during operation, can cause the glass to cloud over.*

– Install new bulb and turn lever on bulb mount to right to secure.

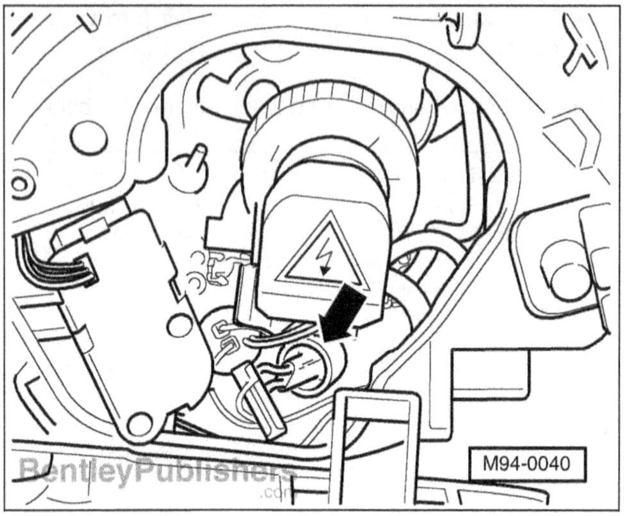

Parking light bulb, replacing

◀ Pull parking light bulb with mount (**arrow**) out from reflector.

> ***CAUTION—***
> *Do not touch the glass portion of the bulb with bare hands. The moisture and/or grease from fingers that evaporates on the bulb during operation, can cause the glass to cloud over.*

– Install new bulb in reverse order of removal.

Installation continued for all bulbs

– Install harness connector and sealing cap in reverse order of removal.

– Install headlight housing as described previously.

– Check function and headlight aim as described previouisly.

Headlight beam adjustment motor (Xenon), removing and installing

> ***WARNING—***
> - *Always disconnect negative (–) battery terminal prior to servicing any parts of parts of the HID lamp system marked with the yellow "High Voltage!" symbol.*
> - *After disconnecting battery, briefly turn headlight on and off in order to discharge any residual voltage.*

– Remove headlight assembly as described previously.

– Remove sealing cap from back of headlight assembly as shown previously.

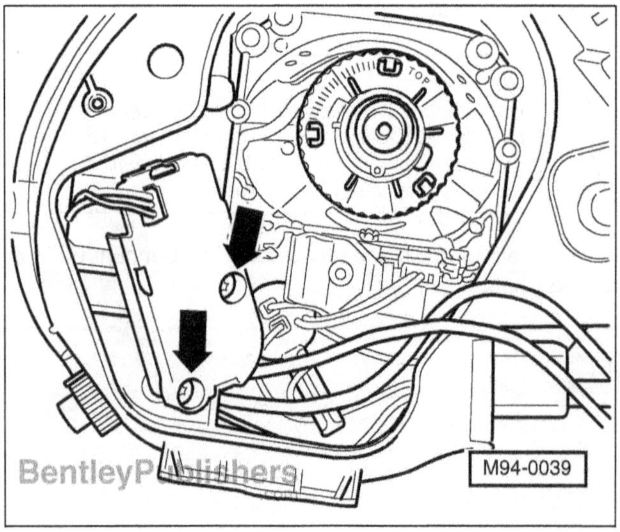

◀ Remove mounting screws (**arrows**) for beam adjustment motor.

– Pull positioning rod ball head out from locking device on reflector.

– Disconnect harness connector.

– Install in reverse order of removal and check headlight function.

– Perform basic setting of headlight adjustment motor using a diagnostic scan tool (VAS 5051/5052 or equivalent).

Fog light bulb, replacing

Through 06/2005

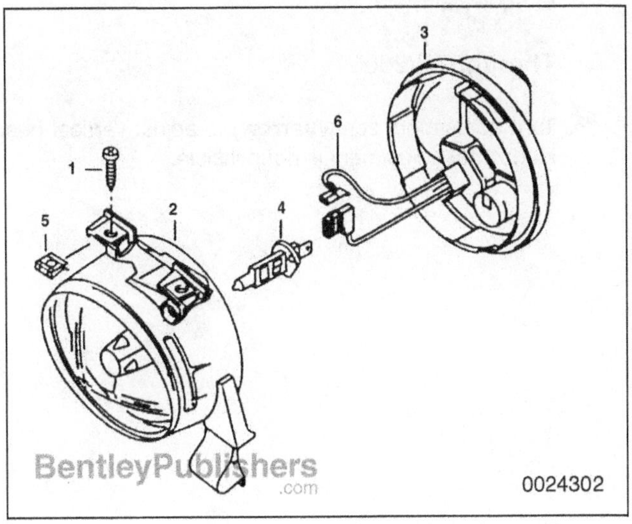

◀ Disconnect harness connector from fog light rear cap (**3**).

1. Tapping screw
2. Reflector housing
3. Rear cap
4. Halogen bulb (12V/55W)
5. Speed nut
6. Ground wire

– Turn rear cap (**3**) counterclockwise and pull down.

– Disconnect ground wire (**6**) from fog light bulb (**4**).

– Squeeze top of bulb retaining clip and rotate downward to release bulb from reflector housing.

– Remove rear cap with bulb and replace bulb.

> ***CAUTION—***
> *Do not touch the glass portion of the bulb with bare hands. The moisture and/or grease from fingers that evaporates on the bulb during operation, can cause the glass to cloud over.*

– Install in reverse order of removal.

From 07/2005

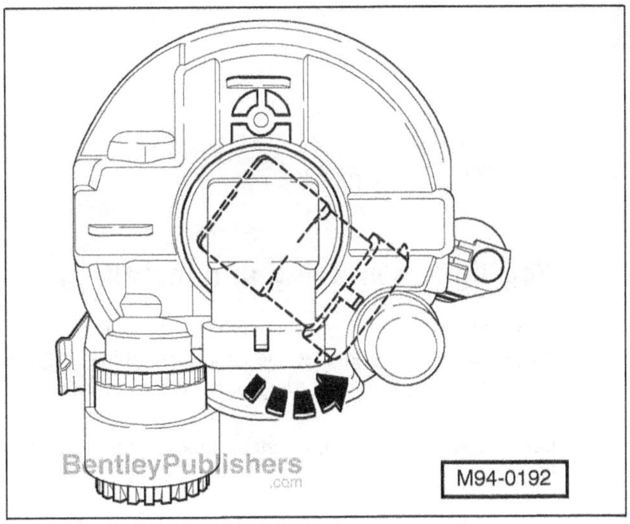

– Remove fog light assembly as shown later for best access to rear of light.

– Disconnect harness connector from rear of fog light.

◀ Turn bulb holder counter-clockwise (**arrow**) to stop and remove fog lamp.

– Replace bulb.

> ***CAUTION—***
> *Do not touch the glass portion of the bulb with bare hands. The moisture and/or grease from fingers that evaporates on the bulb during operation, can cause the glass to cloud over.*

– Install in reverse order of removal.

Fog light, adjusting

– Adjust fog lights using optical headlight aimer (VAS 5107 or equivalent).

Through 06/2005

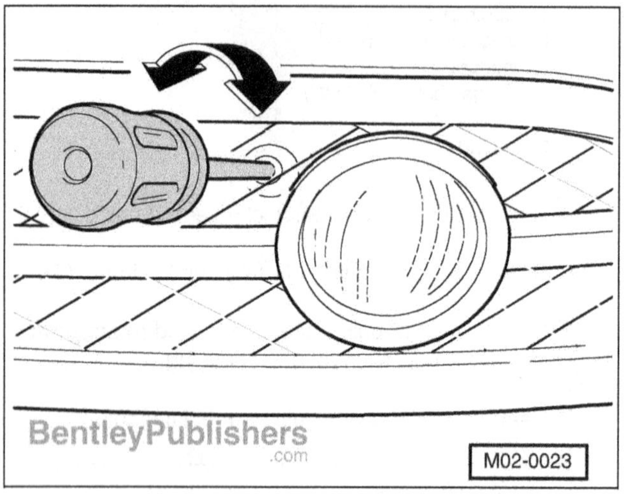

< Turn adjustment screw (**arrow**) to adjust vertical beam. Horizontal adjustment is not possible.

From 07/2005

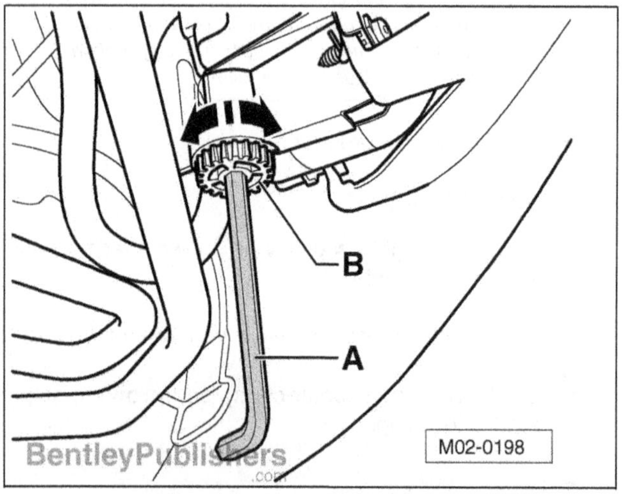

< Turn adjustment screw (**B**) using 6mm hex wrench (**A**) to adjust vertical beam. Horizontal adjustment is not possible.

Fog light, removing and installing

– Disconnect battery ground (GND) strap from battery negative (–) terminal. See the **Cautions** at the beginning of this repair group regarding battery disconnection.

– Disconnect electrical harness connector from back of fog light.

Through 06/2000

– Remove front bumper cover. See **63 Bumpers**.

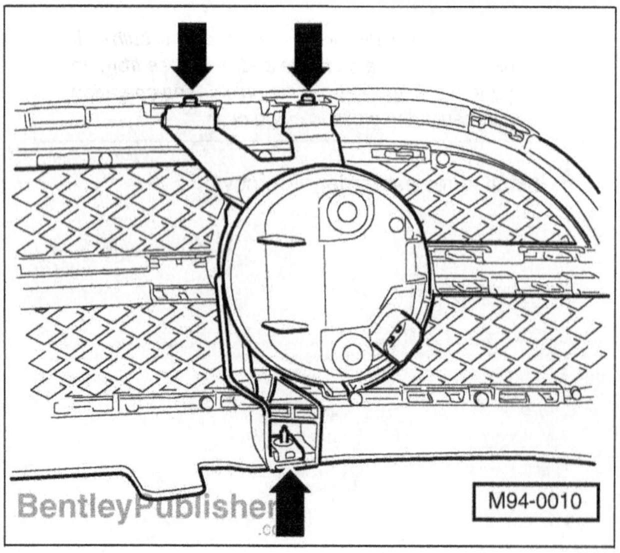

< Remove fog light mounting screws (**arrows**) from inside front bumper cover. Remove fog light assembly.

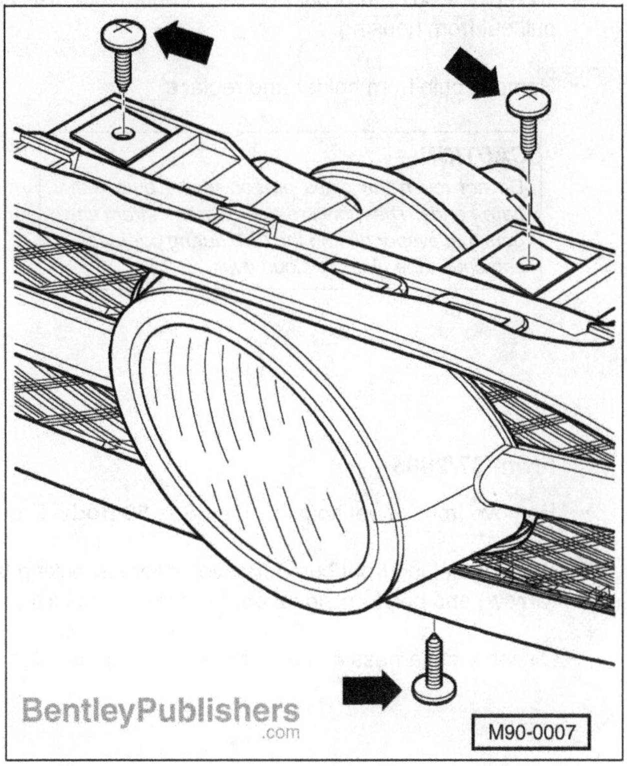

From 07/2000 through 06/2005

- Remove lower front grille from bumper cover. See **63 Bumpers**. Grille is clipped into bumper cover.
- ◄ Remove fog light mounting screws (**arrows**) from grille. Remove fog light.
- Disconnect harness connector and remove fog light.

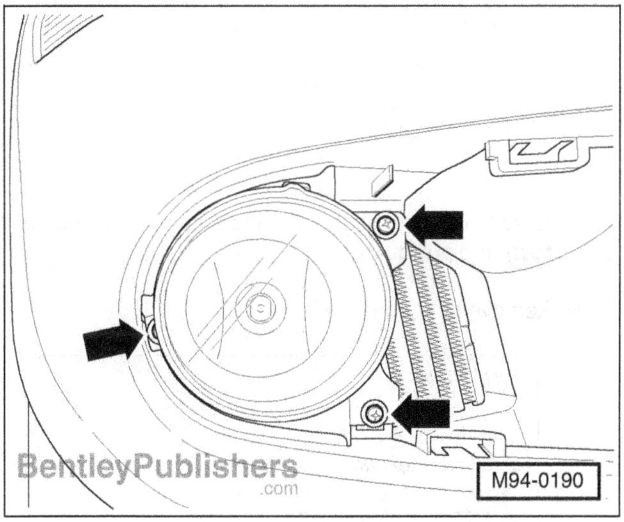

From 07/2000 through 06/2005

- Remove lower front grille from bumper cover. See **63 Bumpers**. Grille is clipped into bumper cover.
- ◄ Remove fog light mounting screws (**arrows**) from bumper cover.
- Disconnect harness connector and remove fog light.

Continued for all vehicles

- Install fog light in reverse order of removal. Check function and adjustment.

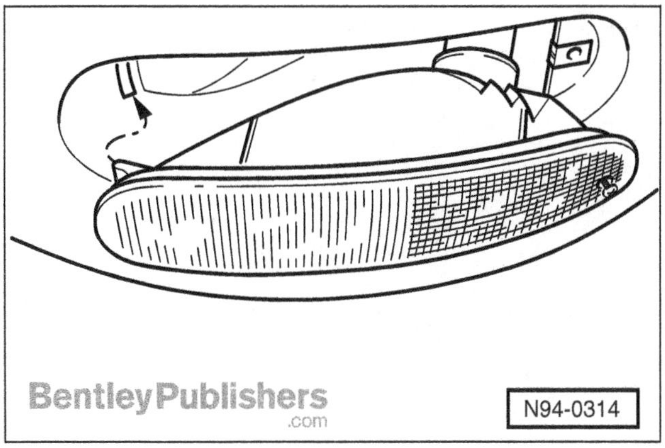

Front turn signal bulb, replacing

Through 06/2005

- Remove screw cover and securing screw for lens.
- Pull turn signal out forward and disconnect harness connector.
- ◄ When installing, slide turn signal retaining tab into mounting slot (**arrow**).

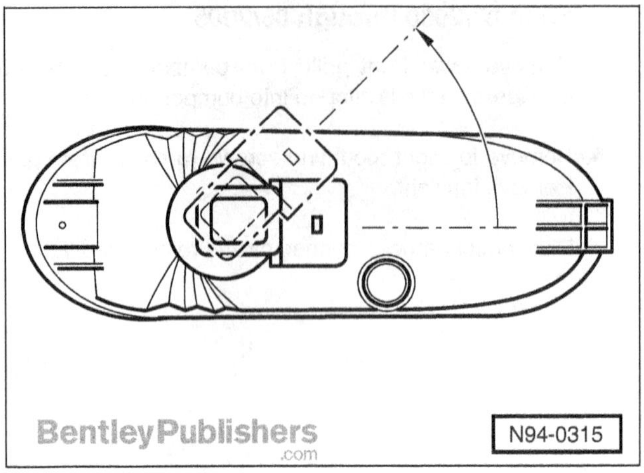

◀ To replace bulb, turn bulb holder counterclockwise and pull out from housing.

– Remove bulb from holder and replace.

> ***CAUTION —***
> *Do not touch the glass portion of the bulb with bare hands. The moisture and/or grease from fingers that evaporates on the bulb during operation, can cause the glass to cloud over.*

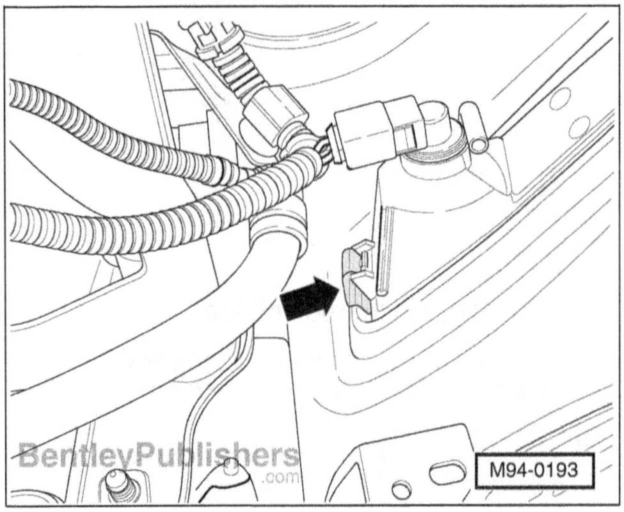

From 07/2005

– Remove front wheel housing liner. See **50 Body–Front**.

◀ Working behind front bumper cover, depress locking tab (**arrow**) and pull turn signal out from front of vehicle.

– Disconnect harness connector from back of turn signal.

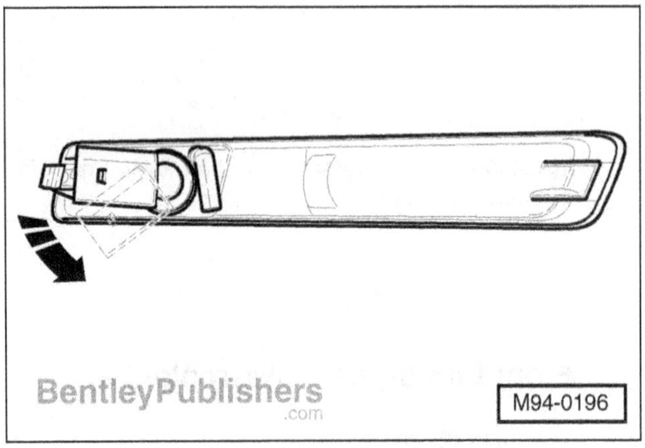

◀ To replace bulb, turn bulb holder counterclockwise (**arrow**) and pull out from housing.

– Remove bulb from holder and replace.

> ***CAUTION —***
> *Do not touch the glass portion of the bulb with bare hands. The moisture and/or grease from fingers that evaporates on the bulb during operation, can cause the glass to cloud over.*

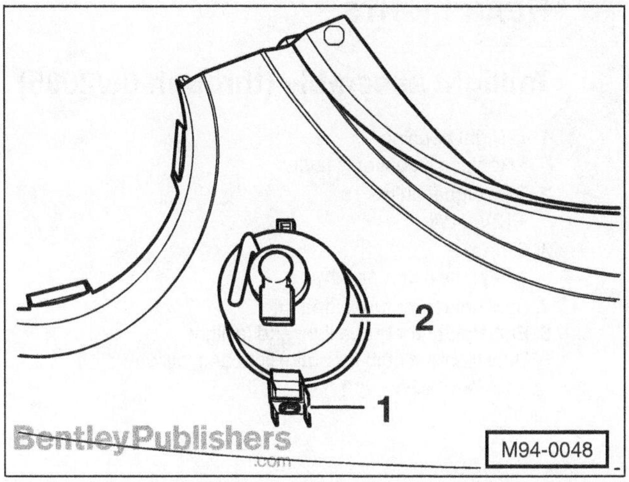

Turbo S models with round front turn signals

- Remove front bumper cover. See **63 Bumpers**.
- Disconnect harness connector from back of turn signal.
- ◄ Remove screw (**1**) and swing turn signal housing (**2**) out of mount.
- To replace bulb, turn bulb holder and pull out from housing.
  - Left light: turn bulb holder counterclockwise.
  - Right light: turn bulb holder clockwise.
- Installation is reverse of removal.

Side-mounted turn signals, removing and installing

1999 through 2003

NOTE—
- *Side-mounted turn signals that are mounted in front quarter panels must be removed from vehicle interior.*
- *Side mounted turn signals have light emitting diodes (LEDs) with a long service life.*
- *The complete turn signal assembly must be replaced if the bulbs do not operate.*

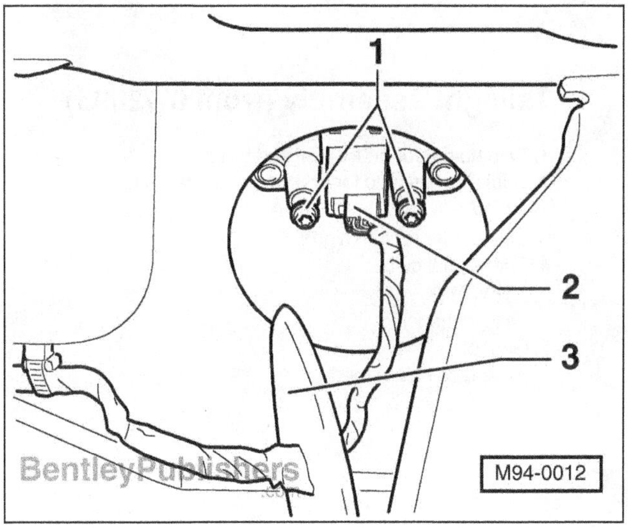

- Remove inner plenum covers from top of instrument panel. See **70 Trim–Interior**.
- Remove dust and pollen filter. See **03 Maintenance**.
- ◄ Pull rubber grommet (**3**) off and disconnect harness connector (**2**).
- Loosen captive mounting screws and remove turn signal from outside.
- Install in reverse order of removal.

From 2003

New Beetle convertibles from 2003-on and sedans from 2004-on have side mounted turn signal integrated into side view mirrors. See **66 Body–Exterior Equipment, Side-view mirror turn signal light, removing and installing**.

REAR LIGHTS

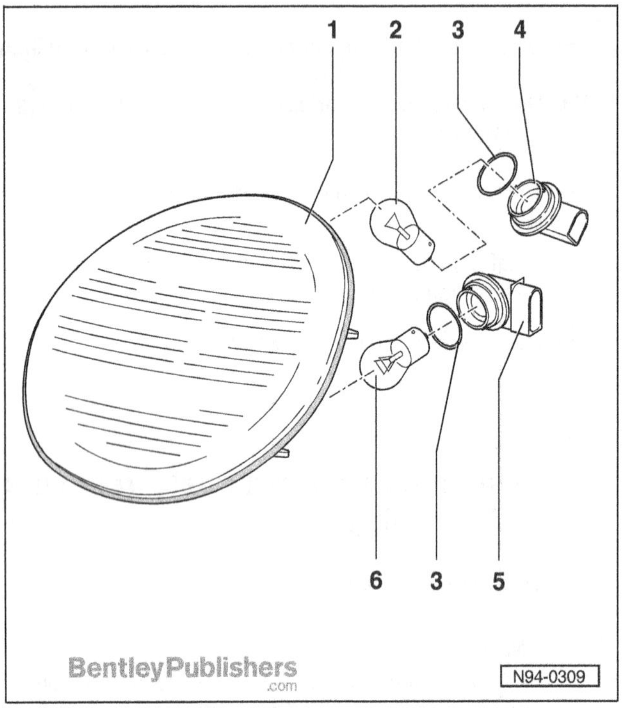

Taillight assembly (through 06/2005)

1. Taillight housing
   - Replace if gasket is faulty
2. Turn signal bulb
   - 12V/21W
3. O-ring
   - Replace if damaged
4. Bulb holder for turn signal
5. Bulb holder for brake light and taillight
6. Dual filament bulb for brake light and taillight
   - 12V/21W/5W

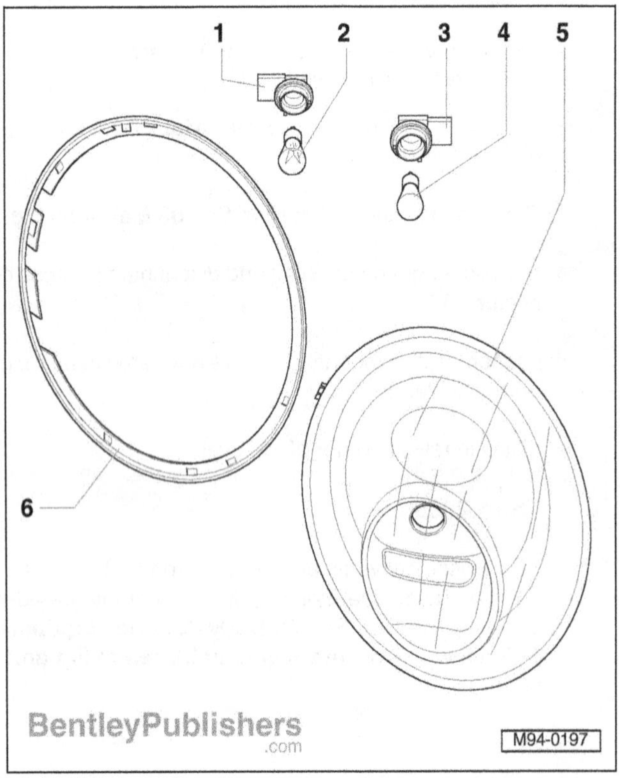

Taillight assembly (from 07/2005)

1. Bulb holder for brake light and taillight
2. Dual filament bulb for brake light and taillight
   - 12V/21W/5W
3. Bulb holder for turn signal
4. Turn signal bulb
   - 12V/21W
5. Taillight housing
6. Gasket
   - Replace if damaged

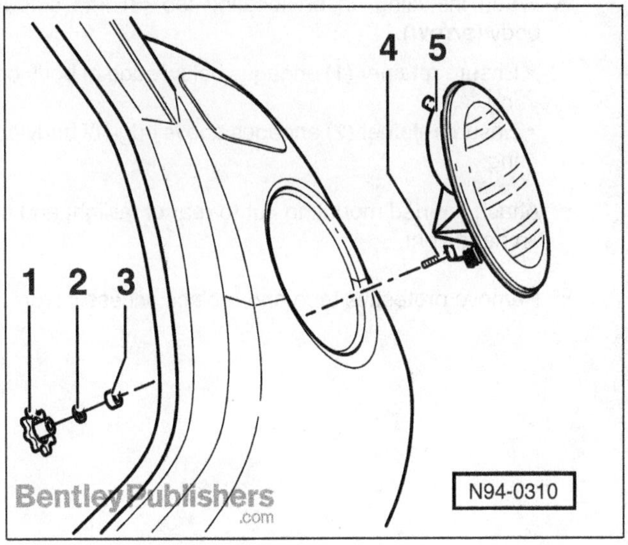

Taillight assembly, removing and installing

Through 06/2005

– Remove rear lock carrier trim, see **70 Trim–Interior**.

◄ Reach behind quarter panel trim and unscrew knurled taillight mounting nut (**1**). Make sure washer (**2**) and spacer (**3**) do not fall down behind trim panel.

> ***NOTE —***
> *It may be necessary to remove the rear compartment trim panel(s) to access taillight knurled nut, see* **70 Trim–Interior**.

– Remove taillight and disconnect harness connectors.

– When installing, line up tab on outer edge of taillight to cut-out in fender.

From 07/2005

– Use screwdriver to carefully pry out taillight access cover from rear compartment trim panel.

◄ Remove knurled mounting nut (**arrow**) from back of taillight.

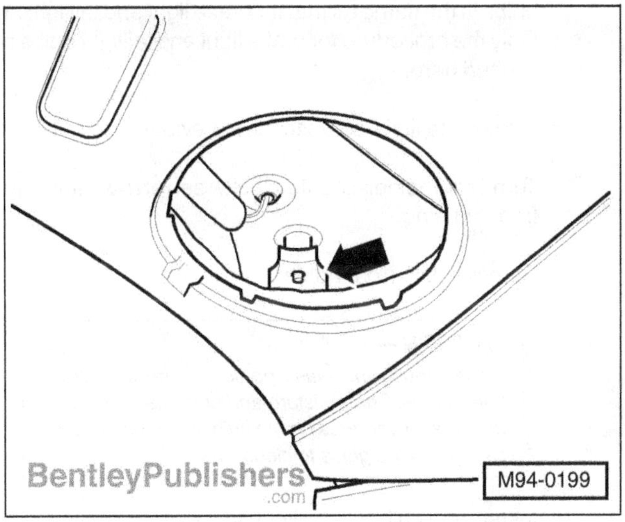

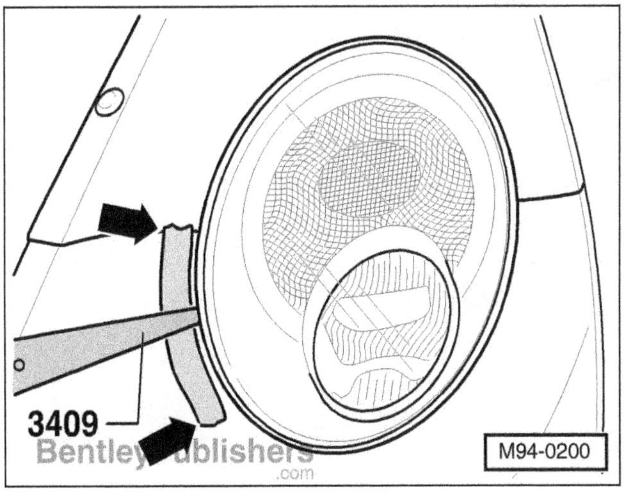

◄ Apply protective tape near edge of taillight (**arrows**).

– Carefully use plastic trim removal wedge (Volkswagen special tool **3409** shown) to pry out taillight housing.

• Position removal wedge under taillight gasket.

– Pull taillight housing out from fender and disconnect harness connectors.

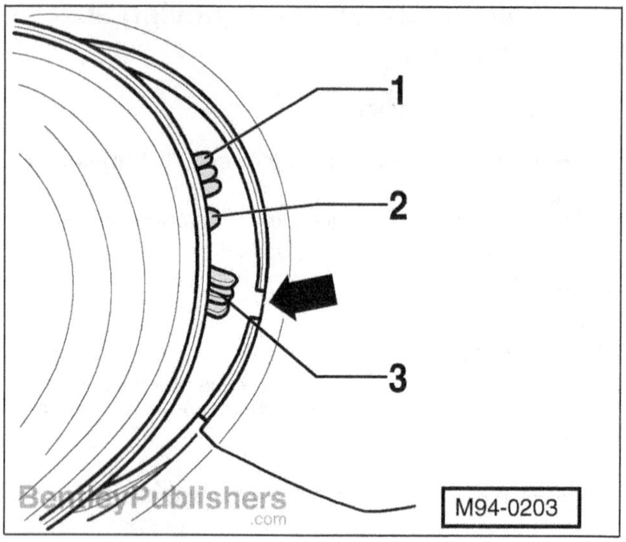

◄ When installing, insert locating tab (**3**) into cutout on body (**arrow**).

- Ensure retainer (**1**) engages below edge of body opening.
- Ensure retainer (**2**) engages above edge of body opening.

– Attach knurled mounting nut to rear of taillight and tighten hand tight.

– Remove protective tape and replace access cover.

Taillight bulbs, replacing

The procedure for removing and installing turn signal light bulbs is the same as that for brake light and taillight bulbs. Only the procedure for brake light and taillight bulbs is described here.

– Remove taillight as described previously.

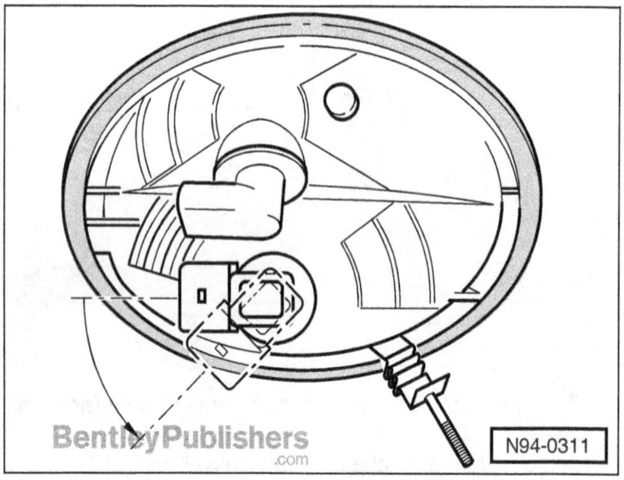

◄ Turn bulb holder counterclockwise (**arrow**) and remove from housing.

– Remove bulb from holder.

> ***CAUTION—***
> *Do not touch the glass portion of the bulb with bare hands. The moisture and/or grease from fingers that evaporates on the bulb during operation, can cause the glass to cloud over.*

– Install new bulb and taillight assembly in reverse order of removal.

Back-up lights

Through 06/2005

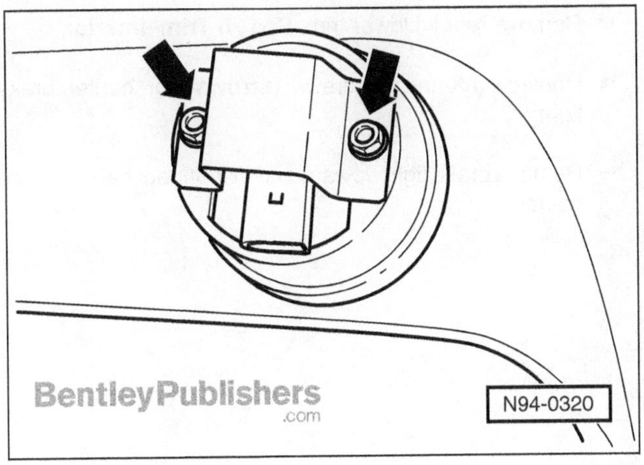
N94-0320

◄ Remove retaining nuts (**arrow**) from inside rear bumper.

– Remove back-up lights toward rear and disconnect harness connector from back-up lights.

– To remove bulb, turn bulb holder counterclockwise and pull out from housing.

> ***CAUTION—***
> *Do not touch the glass portion of the bulb with bare hands. The moisture and/or grease from fingers that evaporates on the bulb during operation, can cause the glass to cloud over.*

Through 06/2005

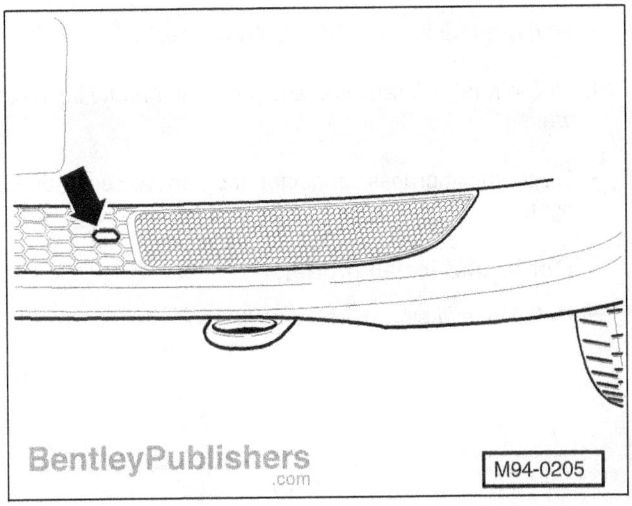
M94-0205

◄ Remove mounting screw through opening (**arrow**) in rear bumper cover.

– Disconnect harness connector and remove back-up light.

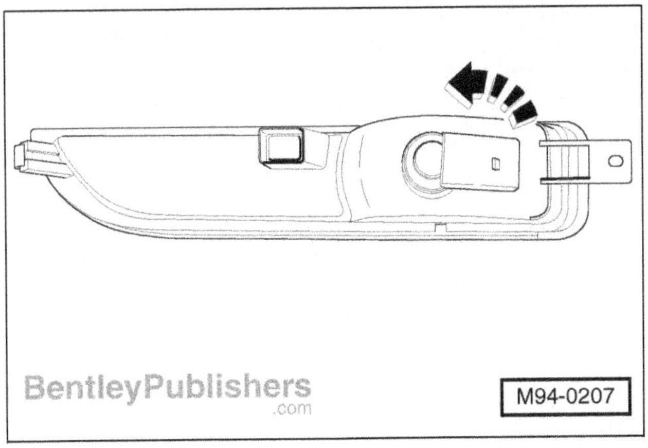
M94-0207

◄ To remove bulb, turn bulb holder counterclockwise (**arrow**) and pull out from housing.

> ***CAUTION—***
> *Do not touch the glass portion of the bulb with bare hands. The moisture and/or grease from fingers that evaporates on the bulb during operation, can cause the glass to cloud over.*

– Install in reverse order of removal.

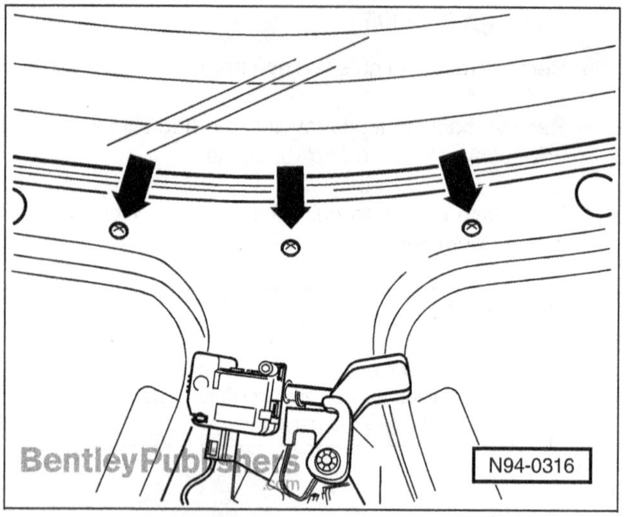

Center brake light

Sedan models

– Remove rear lid lower trim. See **70 Trim–Interior**.

◄ Unscrew mounting screws (**arrows**) for center brake light.

– Remove brake light upward and disconnect harness connector.

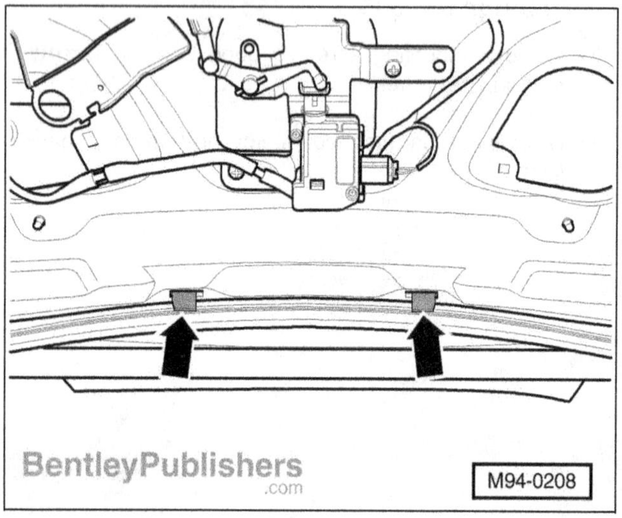

Convertible models

– Remove rear lid trim. See **70 Trim–Interior**.

◄ Unclip retainers (**arrows**) and pull center brake light outward.

– Disconnect harness connector and remove center brake light.

Continued for all models

– Install in reverse order of removal.

License plate light

Through 06/2005

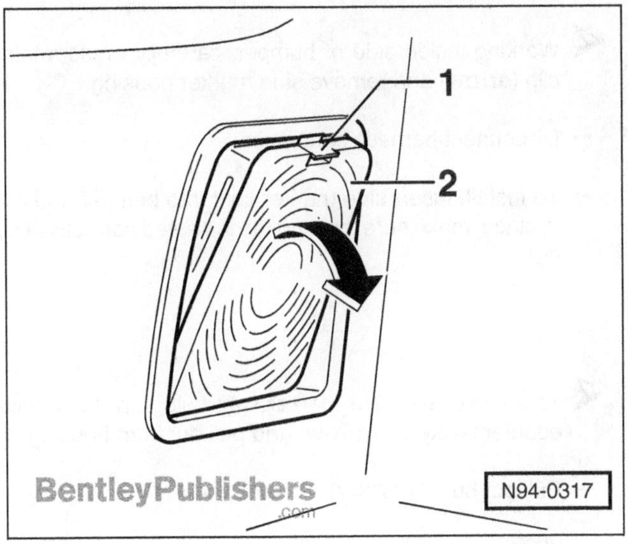

- Carefully pry out license plate light (**2**) with suitable tool, until retaining tab (**1**) releases.
- Disconnect harness connector and remove license plate light.
- To remove bulb, turn bulb holder counterclockwise and pull out from housing.

> ***CAUTION—***
> *Do not touch the glass portion of the bulb with bare hands. The moisture and/or grease from fingers that evaporates on the bulb during operation, can cause the glass to cloud over.*

From 07/2005

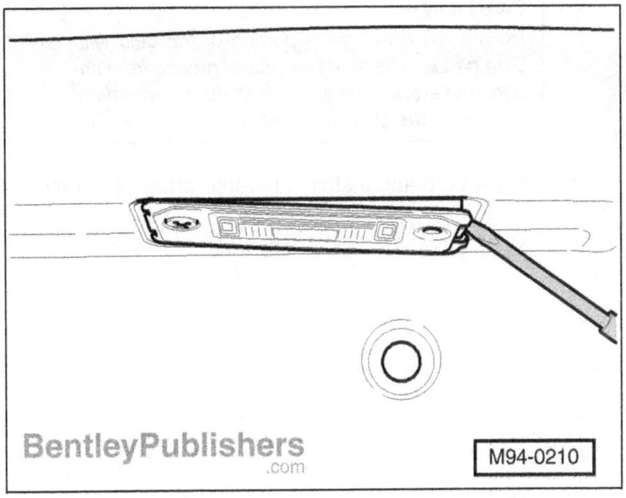

- Remove right side mounting screw from back-up light housing.
- Carefully pry out license plate light (**2**) with screwdriver.
- Disconnect harness connector and remove license plate light.

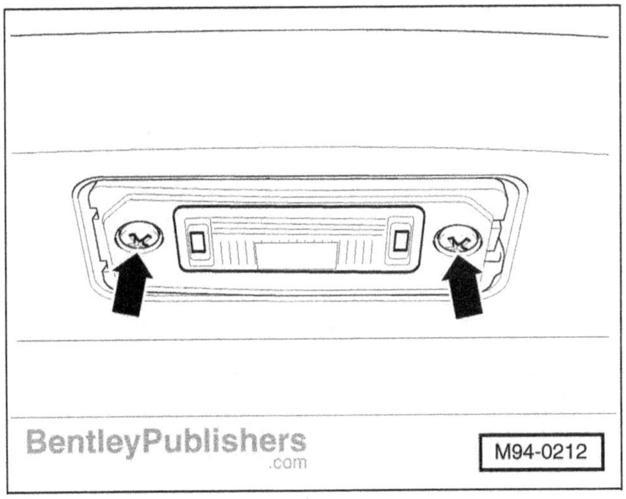

- To replace bulb, remove license plate lens mounting screws (**arrows**).
- Remove lens and pull out bulb (do not turn).

> ***CAUTION—***
> *Do not touch the glass portion of the bulb with bare hands. The moisture and/or grease from fingers that evaporates on the bulb during operation, can cause the glass to cloud over.*

- Install in reverse order of removal.

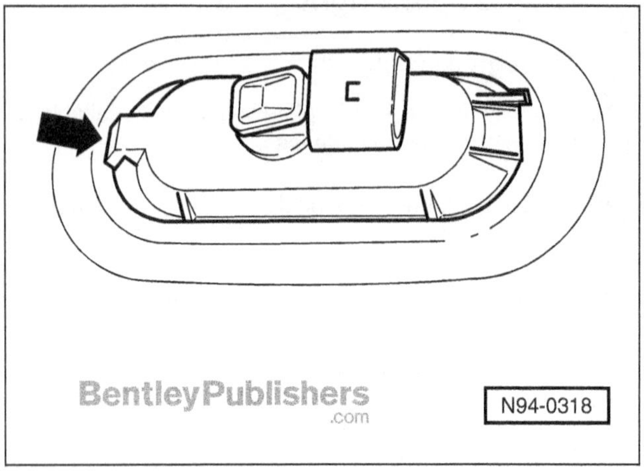

Rear side marker lights

– Remove rear wheel housing liner, see **66 Body Exterior Equipment**.

◄ Working inside side of bumper, carefully press retainer clip (**arrow**) and remove side marker housing.

– Disconnect harness connector.

– To install, insert side marker light into bumper and click in place, make sure locating tab is seated correctly in cut-out.

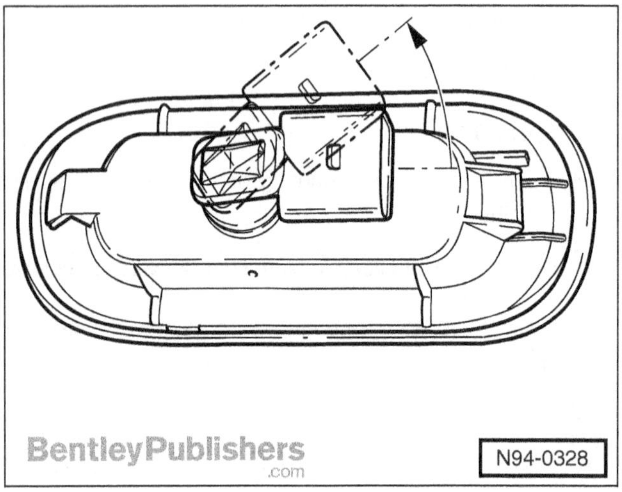

◄ To remove rear side marker light bulb, turn bulb holder counterclockwise (**arrow**) and pull out from housing.

– Pull out bulb (12V/5W), do not turn.

> ***CAUTION —***
> *Do not touch the glass portion of the bulb with bare hands. The moisture and/or grease from fingers that evaporates on the bulb during operation, can cause the glass to cloud over.*

– Replace bulb and install in reverse order of removal.

96 Lights, Accessories–Interior

GENERAL

This section covers electrical light switches and accessories located in the passenger compartment. Items such as radio, heating and A/C that require detailed coverage have their own repair groups. See **EWD Electrical Wiring Diagrams** for additional information.

> ***CAUTION—***
> - *Before working on electrical system disconnect negative (–) battery cable.*
> - *Before disconnecting battery be sure to obtain radio anti-theft code.*

INTERIOR LIGHTS AND SWITCHES

NOTE—
- *The luggage compartment light switch is integrated into the trunk lock and cannot be replaced separately. If the luggage compartment light switch is faulty the complete trunk lock must be replaced, see* **70 Trim–Interior**.
- *The door contact switch is integrated into the door lock and cannot be replaced individually. The complete door lock must be replaced if the door contact switch is faulty, see* **70 Trim–Interior**.
- *For steering column switches, see* **48 Steering**.
- *For brake light switch, see* **45 Anti-Lock Brakes (ABS).**

Light switch, removing and installing

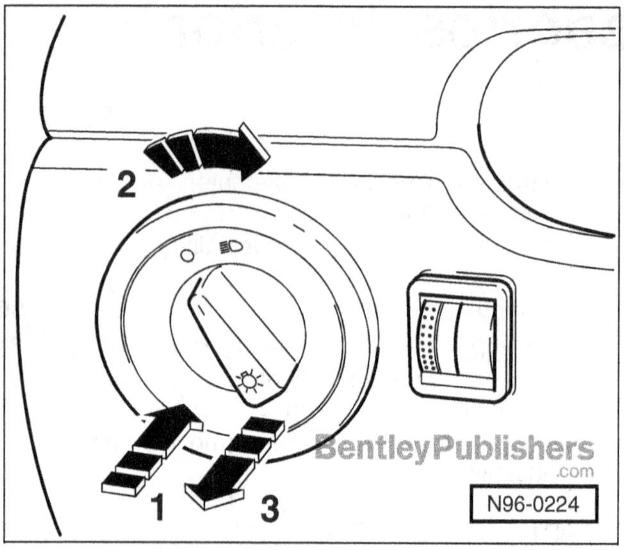

- Turn rotary knob of light switch to "0" position.
- Press rotary knob inward (**arrow 1**) and turn slightly to right (**arrow 2**). Hold rotary knob in this position, pull on rotary knob (**arrow 3**) and remove light switch from instrument panel.
- Disconnect harness connector.
- To install, first connect harness connector.

Installing

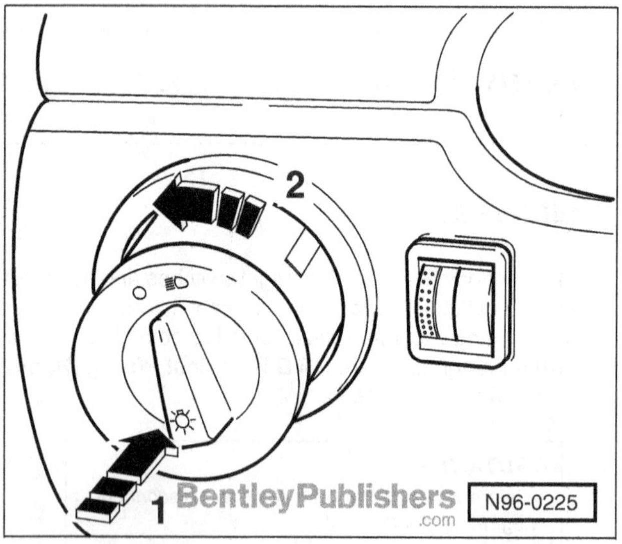

- Hold light switch and press rotary knob inward (**arrow 1**) and turn slightly to left (**arrow 2**).
- Hold rotary knob in this position and slide light switch into instrument panel.
- Turn rotary knob to "0" position, release and engage switch.

Instrument panel light dimmer switch, removing and installing

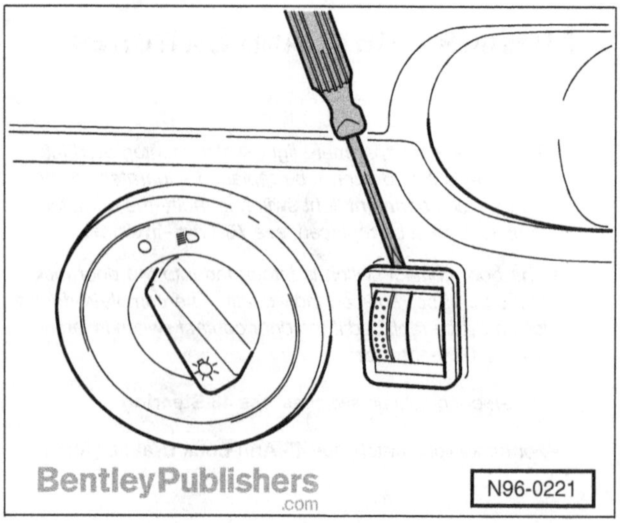

- Place screwdriver behind switch housing and pry switch carefully out of locking device.
- Disconnect harness connector.
- To install, first connect harness connector.
- Insert switch into locking device and engage.

Mirror adjustment switch, removing and installing

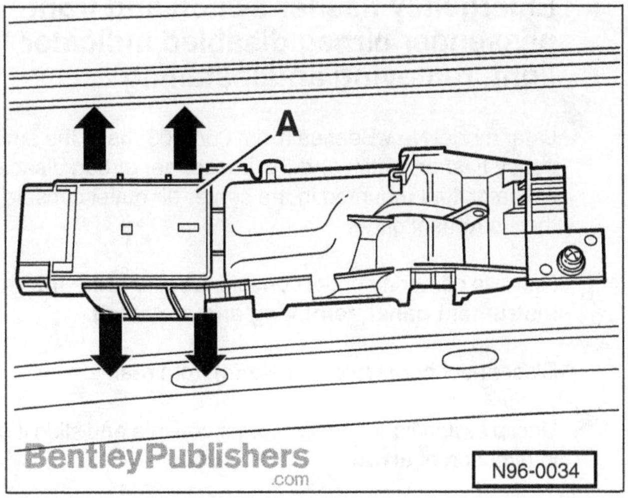

- Remove interior door panel. See **70 Trim–Interior**.
- Disconnect harness connector from switch.
- ◀ Pry mounting frame outward (**arrows**) on long side of switch and remove switches from mounting frame.
- To install, insert switch in mounting frame and engage.
- Connect harness connector.
- Install interior door trim.

Interior lock switch–passenger side, removing and installing

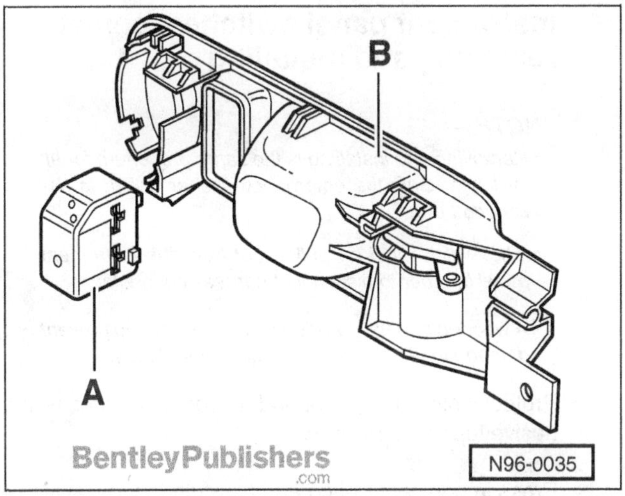

- Remove interior door panel. See **70 Trim–Interior**.
- Disconnect harness connector from switch.
- ◀ Release switch housing (**A**) by pressing sideways out from locking device (**B**) in mounting frame and remove.
- To install, first connect harness connector.
- Insert switch in recess of mounting frame and engage.
- Install door panel.

Remote fuel tank door/rear lid unlock switch, removing and installing

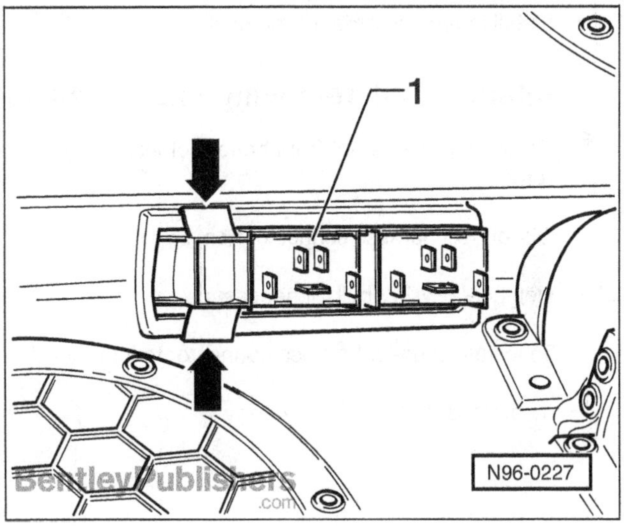

NOTE —
The switches for remote/fuel tank door and remote unlock rear trunk lid cannot be removed separately. The complete switch unit must be removed and replaced, if necessary.

- Remove interior door trim, see **70 Trim–Interior**.
- Disconnect harness connector from switch.
- ◀ Release switch housing (**1**) by pressing sideways on locking tabs (**arrows**) and remove from door trim.
- To install, insert switch into door trim and engage.
- Connect harness connector.
- Install door panel.

Emergency flasher switch and front passenger airbag disabled indicator light, removing and installing

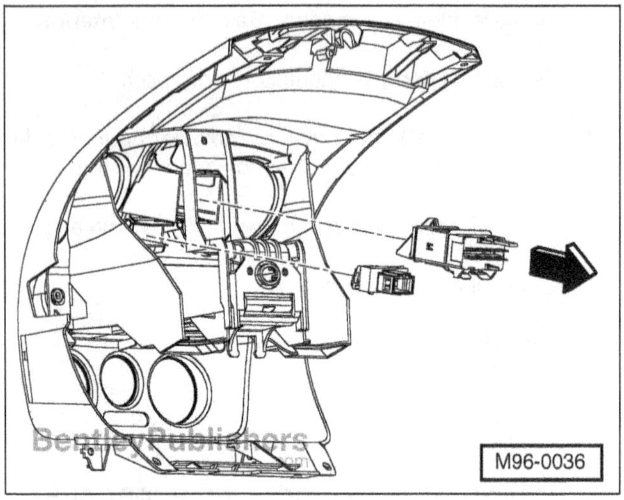
M96-0036

Later model New Beetles (from 06/2003) have the emergency flasher switch and the passenger airbag disabled indicator light mounted in the center air outlet console of the instrument panel.

– Remove center air outlet console. See **70 Trim–Interior, Instrument panel, removing and installing**.

– Disconnect harness connector(s) from rear.

– Unclip switch/light from center air console and slide it out in direction of **arrow**.

– Install in reverse order of removal. Be sure to fully engage switch/light into center air console.

Instrument panel switches–center, removing and installing

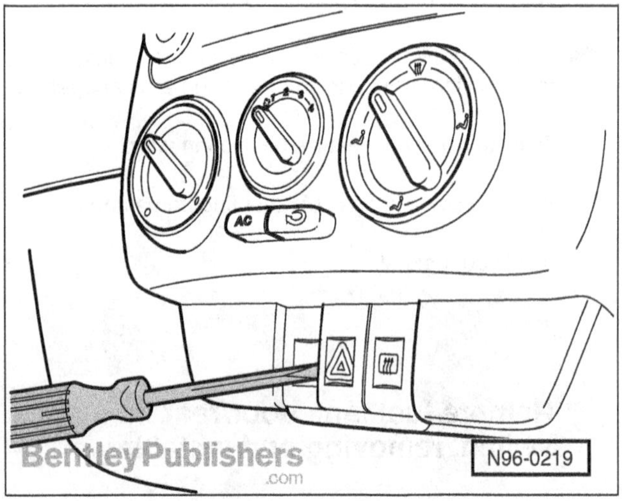
N96-0219

NOTE —

- *Removing and installing is the same procedure for all switches. Only the emergency flasher switch is described here.*
- *The switch must be pulled out from the instrument panel in order to disconnect harness connector.*
- *A wooden or plastic wedge should be used to prevent damage to plastic trim and adjacent switches.*

– Protect instrument panel and adjacent switches with adhesive tape if necessary.

– Press in adjacent switch and carefully pry out switch using flat plastic wedge or screwdriver.

– Release connections and disconnect harness connector from applicable switch.

– Install in reverse order of removal.

Interior light, removing and installing

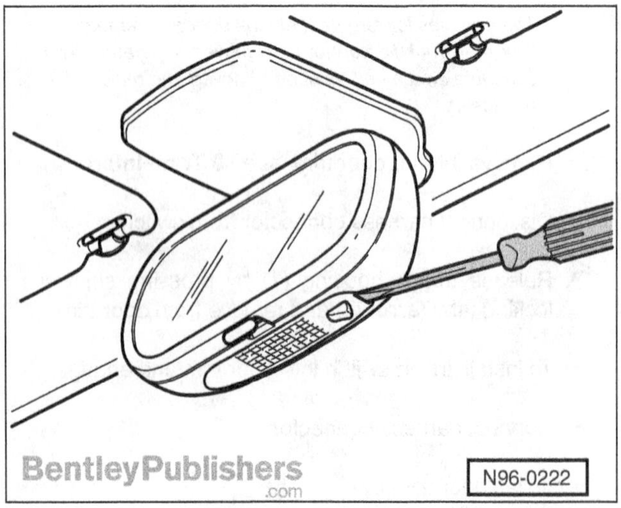
N96-0222

– Carefully pry lens out from bottom of interior mirror and remove.

– Disconnect harness connector from light.

– Replace 12V/3W bulb, if necessary.

– To install, connect harness connector to interior light.

– Insert lens and engage properly.

Glove compartment light, removing and installing

◀ Insert screwdriver behind lens and carefully pry lens out.

– Remove lens with bulb holder.

– Disconnect harness connector from light.

– Replace 12V/3W bulb, if necessary.

– To install, first connect harness connector.

– Insert lens with bulb holder into glove compartment and engage properly.

> ***NOTE—***
> *After working on the glove compartment light, ensure that the light does not stay on when the glove compartment door is closed.*

Make-up mirror light, removing and installing

> ***NOTE—***
> *The procedures for removing left make-up-mirror light and right make-up mirror light are identical.*

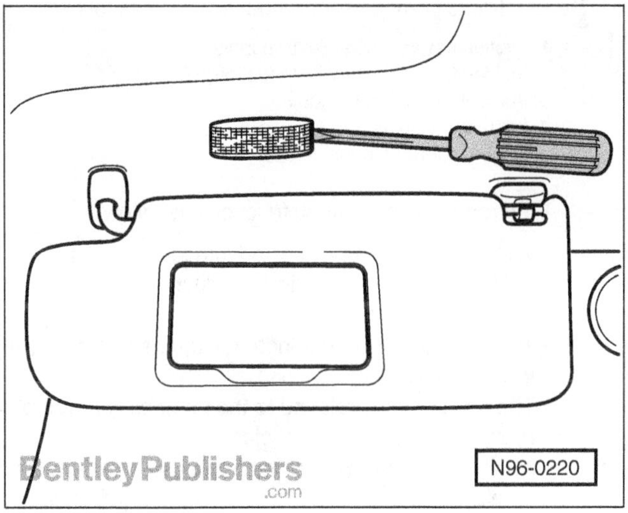

◀ Insert screwdriver behind lens and carefully pry lens out.

– Remove lens with bulb holder.

– Disconnect harness connector.

– Replace 12V/3W bulb, if necessary.

– To install, first connect harness connector.

– Insert lens with bulb holder into headliner and engage properly.

Luggage compartment light, removing and installing

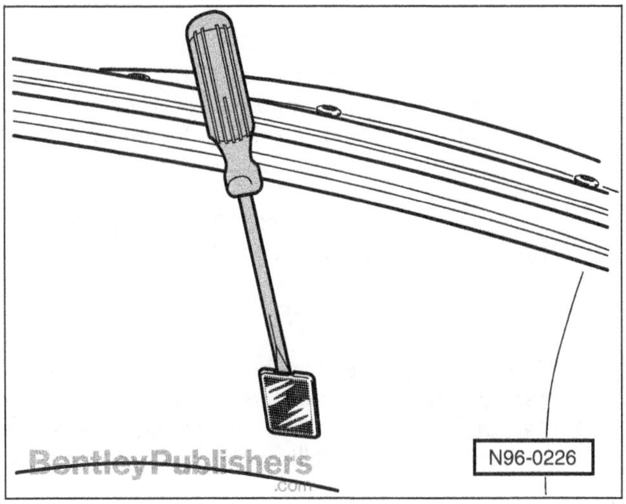

◀ Insert screwdriver behind lens and carefully pry lens out.

– Remove lens with bulb holder.

– Disconnect harness connector from light.

– Replace 12V/3W bulb, if necessary.

– To install, connect harness connector.

– Insert lens with bulb holder into mounting and engage properly.

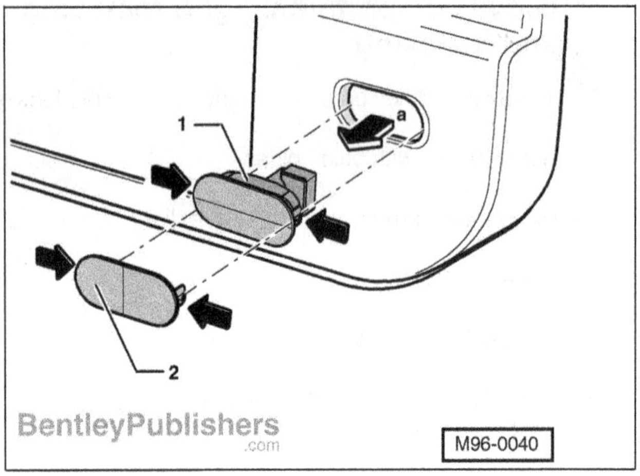

Door warning light/reflector, removing and installing

- Remove door panel. See **70 Trim–Interior**.
- Disconnect electrical connection, if applicable.
- ◄ Remove warning light (**1**) or reflector (**2**) by depressing both retainers (**arrows**). Remove in direction of **arrow a**.
- Install in reverse order of removal.

ANTI-THEFT SYSTEM

New Beetle models from model year 2000 are equipped with an anti-theft immobilizer system.

Electronic immobilizer

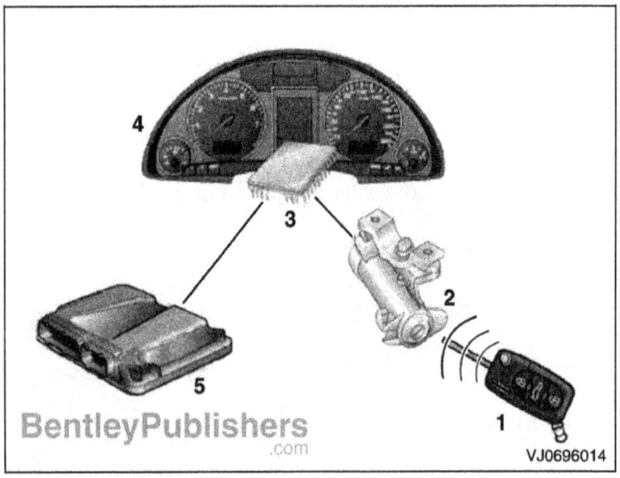

◄ Electronic immobilizer components are as follows:

1. Ignition key with wireless transponder
2. Ignition lock cylinder with induction coil
3. Immobilizer control module
4. Instrument cluster
5. Engine control module (ECM)

Electronic immobilization functions as follows:

- With the vehicle key inserted into ignition switch, a signal from the ignition cylinder induction coil energizes the key transponder.
- The key transponder sends a unique signal to the induction coil.
- The key signal is relayed to the immobilizer control module in the instrument cluster.
- If the control unit recognizes the key as valid, the signal is relayed to the engine control module (ECM) and the vehicle can be started.
- If the key is not recognized as valid, the starter and the engine management system are blocked and the vehicle cannot be started.

The immobilizer control module (J362) is integrated with the instrument cluster control module. If the anti-theft immobilizer control module is faulty, the entire instrument cluster must be replaced. See **90 Instruments**.

The immobilizer induction coil (D2) is integrated with the ignition key cylinder and cannot be replaced separately. In case of induction coil malfunction, replace the ignition cylinder. See **48 Steering**.

The key transponder is integrated with the vehicle key. In case of transponder malfunction, replace the complete key. Once a key is replaced, adapt all the vehicle keys to the vehicle using the Volkswagen scan tool or equivalent.

If only one key is adapted, the remaining keys will not start the vehicle until they are also adapted. Battery replacement for the key transponder (remote) is covered in **03 Maintenance**.

To troubleshoot the immobilizer and anti-theft alarm systems, use the Volkswagen diagnostic scan tool or equivalent.

Alarm system

The anti-theft alarm system sets off an alarm horn and activates the emergency flashers whenever unauthorized interference with the vehicle is detected.

The alarm system activates automatically when the vehicle is locked using either the remote control key or the key in driver's door. The doors, engine hood and rear hatch are monitored for opening as long as they are completely closed.

The system deactivates when the vehicle is unlocked using the remote control key or when the ignition is switched ON. More information on the anti-theft alarm system is available in the glovebox owner's manual supplied with the vehicle.

QUALITY REVIEW

When you have finished working under the hood and around other areas of the vehicle, it is advisable to take a moment to quality check or review your work. This helps to insure that the operation or repair has been completed properly with all affected systems functioning within normal parameters. These may include the following:

- Ensure that all cable ties and clamps that were removed as part of the repair are replaced.
- Ensure that all fasteners and hardware were replaced and torqued as specified.
- Make sure that all other components involved in the repair are positioned correctly, properly torqued and function properly.
- Make sure that all tools, shop cloths, fender covers and protective tapes are removed before closing the hood.
- Clean grease and fingerprints from painted surfaces, steering wheel, upholstery and shifter.

97 Fuses, Relays and Component Locations

GENERAL

This section contains fuse listings, relay positions and ground locations. It also includes an electrical component location table.

> ***WARNING —***
> *Special airbag precautions apply to any electrical system testing or repair. The airbag unit is an explosive device and must be handled with extreme care. Before starting any work on the vehicle, refer to the warnings and cautions in* **69 Seatbelts, Airbags**.

> ***CAUTION —***
> - *Before working on the electrical system always switch the ignition off and disconnect battery ground strap from battery negative (–) terminal.*
> - *Be sure to have the anti-theft radio code on hand before disconnecting the battery.*
> - *Disconnecting the negative (–) battery cable may erase fault codes and basic settings in the engine management and automatic transmission control modules. Some driveability problems may be noticed until the system re-adapts to operating conditions. OBD II readiness codes, which may be required for emissions testing, may also be erased.*
> - *Always switch test meter to the appropriate function and range before making test connections.*

For electrical wiring schematics, see **EWD Electrical Wiring Diagrams**.

> ***NOTE —***
> *The Society of Automotive Engineers (SAE) has implemented standardized terms for automotive components, including alternator, which should always be referred to as a generator.*

FUSES AND RELAYS

Fuse panels

The central electric panel (fuse panel) is located behind an access panel on left end of the instrument panel. It contains both standard size fuses and mini-fuses. The use of mini-fuses allows protection of individual circuits.

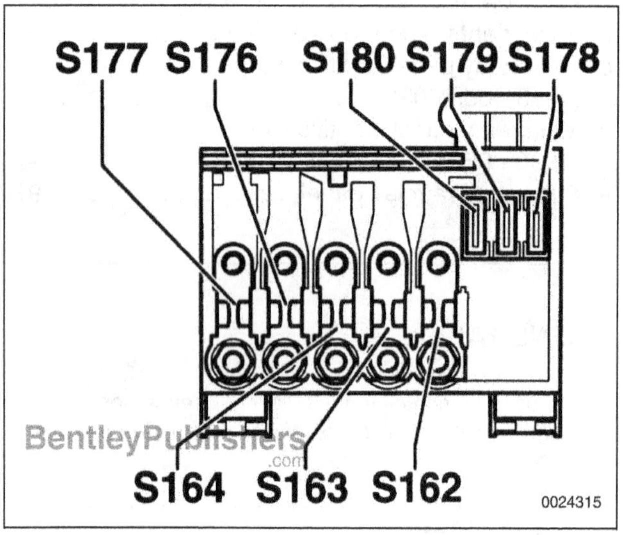

There is also a main fuse box located on top of the battery in the engine compartment. It contains special fuses for high current applications and protects the main wiring harness in the event of a short circuit. To access these fuses, squeeze tabs of plastic fuse cover and pull upward.

> ***NOTE —***
> *Volkswagen identifies electrical components by a letter and/or a number in the electrical schematics. These electrical identifiers are listed in parenthesis as an aid to electrical troubleshooting.*

Main fuse box on top of battery

- Standard fuse applications ....(fuse #) amps
  Coolant fan, single speed ....(S180) 30A
  ABS ....(S179) 30A
  ABS hydraulic pump ....(S178) 30A
- Metal fuse applications ....(fuse #) amps
  Generator (alternator)
  90 amp ....(S177) 110A
  120 amp ....(S177) 150A
  Vehicle electrical system relay panel . . .(S176) 110A
  Coolant fan, two-speed ....(S164) 40A
  Fuel pump relay or glow plug relay .....(S163) 50A
  Secondary air injection relay
  or coolant pre-heating relays ....(S162) 50A

> ***WARNING —***
> *Metal fuses in the main fuse box should be changed by an authorized Volkswagen service professional.*

> ***NOTE —***
> - *Fuse application and amperage may vary depending on equipment level.*
> - *Fuses are identified in the wiring diagrams with the prefix of S, i.e. fuse #4 is S4.*
> - *"RES" in the central electric panel refers to reserve, or places for spare fuses.*

The 1998 through May, 2003 central electric panel fuses are identified in **Table a**.

Table a. Fuse identification (1998 - 05/2003)

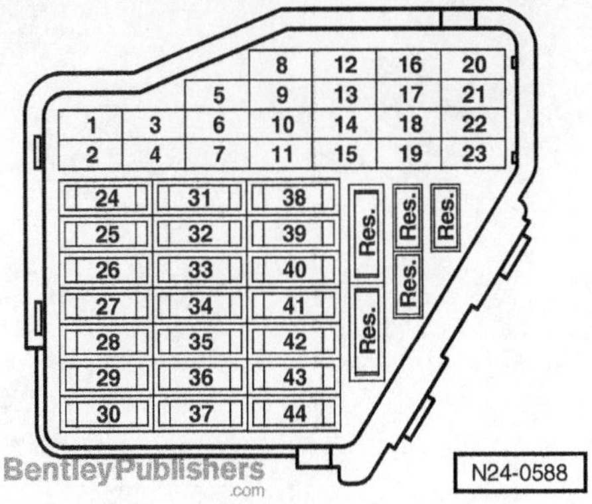

Position	Description	Amperage
1	Heated washer nozzles, mirrors	10
2	Turn signal system	10
3	Illumination	5
4	License plate light	5
5	Comfort system	7.5
6	Central locking system	5
7	Back-up lights	10
8	Telephone system (future option)	5
9	ABS	5
10	Engine control, gasoline	10
11	Instrument cluster, shift lock	5
12	B+ (battery positive voltage) for DLC	7.5
13	Brake lights	10
14	Interior lights, with central locking Interior lights, only	10 5
15	Instrument cluster, automatic trans.	5
16	A/C clutch, electric water pump	10
17	Door lock heater	7.5
18	High beam headlight, right	10
19	High beam headlight, left	10
20	Low beam headlight, right	10
21	Low beam headlight, left	10
22	Tail and side marker lights, right	5
23	Tail and side marker lights, left	5
24	Wiper system, washer pump	20

Table a. Fuse identification (1998 - 05/2003)

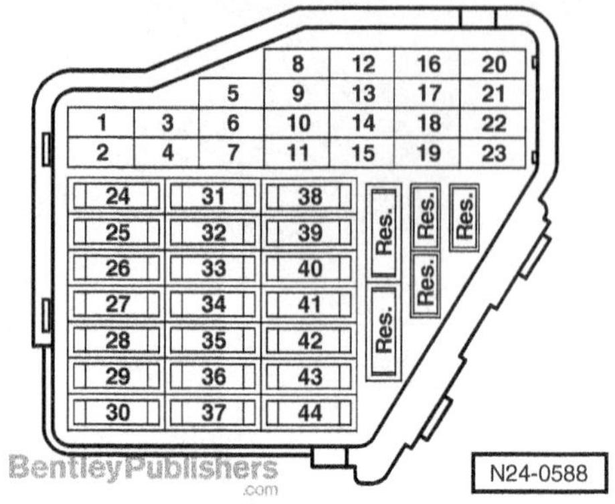

Position	Description	Amperage
25	Fresh air blower, air conditioning	25
26	Rear window defogger	25
27	Daytime running lights	10
28	Fuel pump, gasoline	15
29	Engine control, gasoline Engine control, diesel	15 10
30	Sunroof	20
31	Automatic transmission	20
32	Engine control, injectors, gasoline Engine control, injectors, diesel	10 15
33	Headlight washer system	20
34	Engine control	10
35	Trailer socket	30
36	Fog light indicator	15
37	Radio system	10
38	Luggage compartment light, central locking system	15
39	Emergency flasher system	15
40	Dual horn	20
41	Cigarette lighter, 12V power outlets	15
42	Radio system	25
43	Engine control	10
44	Heated seats	15

NOTE —

- *Fuses number 23 through 44 are identified in the wiring diagrams with an additional prefix of 2, i.e. fuse #40 is S240.*
- *Some fuse or relay positions may be empty if equipment is not installed in vehicle.*

Fuse colors and amperage

- Green 30A
- White 25A
- Yellow 20A
- Blue 15A
- Red. 10A
- Brown 7.5A
- Beige 5A
- Violet 3A

The central electric panel fuses from June, 2003 are identified in **Table b**.

Table b. Fuse identification (06/2003 - on)

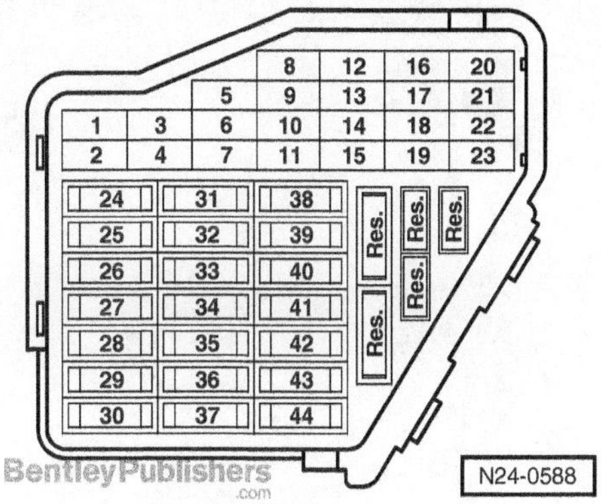

Position	Description	Amperage
1	Heated washer nozzles, mirrors, glove box light, cigarette lighter	10
2	Turn signal system	10
3	Illumination, fog light relay	5
4	License plate light	5
5	Comfort system, cruise control	7.5
6	Comfort system control module	5
7	Back-up lights	10
8	open	-
9	ABS, ESP	5
10	Instrument cluster, digital clock	10
11	Instrument cluster, shift lock	5
12	B+ (battery positive voltage) for DLC	7.5
13	Brake lights	10
14	Interior lights, with central locking Interior lights, only Comfort system control module (vehicles with power windows only)	10
15	Outside air temperature sensor	5
16	A/C clutch, electric water pump	10

Table b. Fuse identification (06/2003 - on)

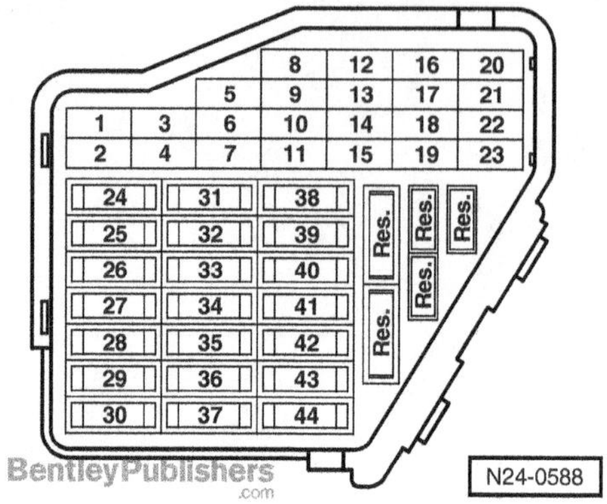

Position	Description	Amperage
17	open	-
18	High beam headlight, right	10
19	High beam headlight, left	10
20	Low beam headlight, right	10
21	Low beam headlight, left	10
22	Tail and side marker lights, right	5
23	Tail and side marker lights, left	5
24	Wiper system, washer pump	20
25	Fresh air blower, air conditioning	25
26	Rear window defogger	20
27	open	-
28	Fuel pump, gasoline	15
29	Engine control, gasoline Engine control, diesel	15 15
30	Sunroof	20
31	Automatic transmission 6-speed automatic transmission	20 20
32	Engine control, injectors, gasoline Engine control, injectors, diesel	10 15
33	open	20
34	Engine control	10
35	Trailer socket	30
36	Fog light indicator	15
37	Engine control, gasoline Engine control, diesel	20 5
38	Central locking system	15
39	Emergency flasher system	15
40	Dual horn, cigarette lighter	20
41	Cigarette lighter, 12V power outlets	5
42	Radio system	25
43	Engine control, gasoline/diesel	10
44	Heated seats	15

Relay panels

The main relay panel is located under the left side of the instrument panel. The 1998 through May, 2003 relays and additional fuse locations are identified in **Table c**.

Table c. Main relay panel identification (1998 - 05/2003)

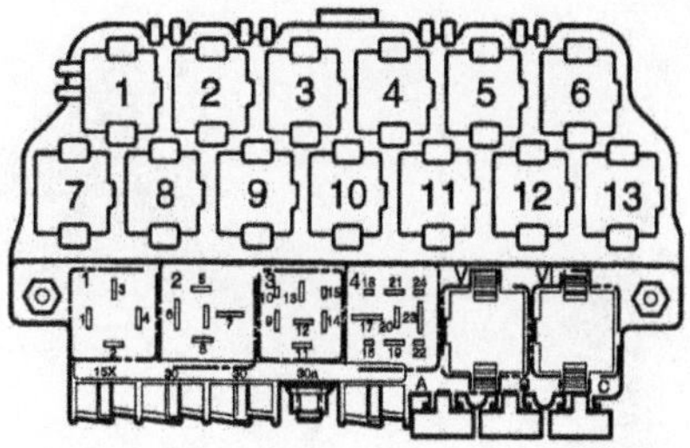

97-14163

Position	Description	Number on relay
	Lower relay panel	
1	Dual horn relay	53
2	Load reduction relay	213,100,18
3	open	
4	Fuel pump relay; TDI: always open	409
V	Wiper/Washer intermittent relay	377
VI	Wiper/Washer intermittent relay	377
	Fuses on lower relay panel	
A	Fuse for power windows (red connector)	-
B	open	-
C	Fuse for heated seats (black connector)	-
	Relays on upper (auxiliary) 13-position relay panel	
1	Fog light relay	53
2	Rear lid remote unlock motor relay	79
3	Anti-theft starter lock relay	185
4	open	-
5	open	-
6	Radio speaker/telephone change-over relay	-
7	open	
8	Daytime running lights change-over relay	173, 53
9	open	-
10	Glow plug relay	180
11	Park/Neutral position relay	175
12	Power supply (terminal 30b, B+) relay	109, 126
13	Coolant fan control relay	53

The main relay panel is located under the left side of the instrument panel. The relays and additional fuse locations are identified from June, 2003 in **Table d**.

Table d. Main relay panel identification (06/2003 - on)

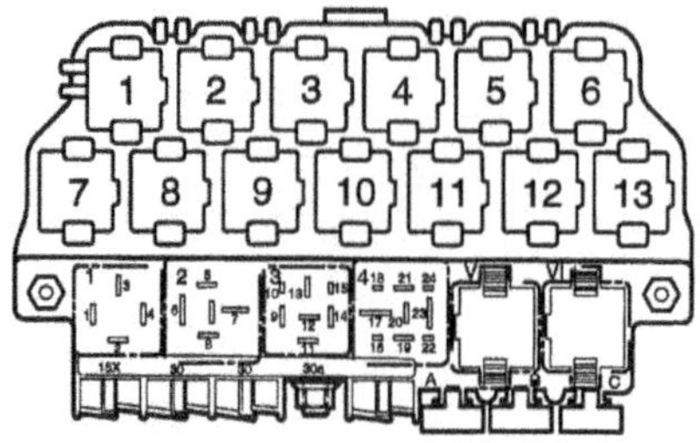

97-14163

Position	Description	Number on relay
Lower relay panel		
1	Dual horn relay	53
2	Load reduction relay	213,100,18
3	open	
4	Fuel pump relay; TDI: open	409
V	Wiper/Washer intermittent relay	377
VI	Wipe/Wash intermittent relay Wipe/Wash intermittent relay/rain sensor	377 192 389
Fuses on lower relay panel		
A	Fuse for power windows (red connector)	-
B	Brake vacuum pump	-
C	Fuse for rear spoiler motor (06/2003 - on)	-
Relays on upper (auxiliary) 13-position relay panel		
1	Starter lock/clutch pedal switch relay (BPS, manual transmission)	449
2	open	
3	Back-up light relay	53
4	Fog light relay, fog light relay (HID) Starter interlock relay (BPR, manual transmission)	53 465
5	Fuel pump relay (TDI, 2006)	53
6	open	-
7	Daytime running lights change-over relay	173
8	Heated exterior mirror relay	53
9	Alarm horn, telematic	53

Table d. Main relay panel identification (06/2003 - on)

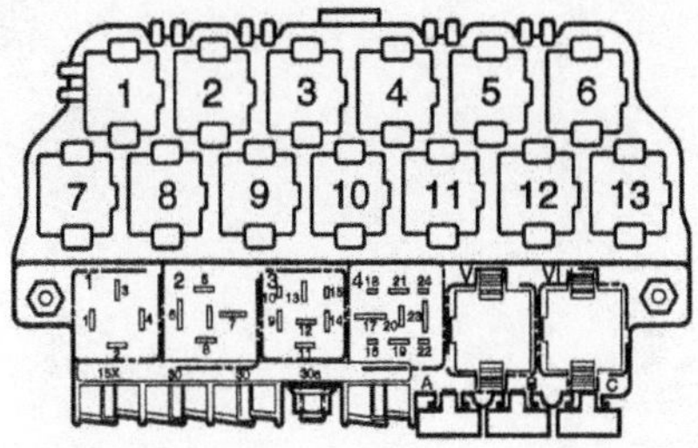

97-14163

Position	Description	Number on relay
10	Alarm system light flasher relay, telematic Electronic fuel pump relay Third speed coolant fan relay Secondary air injection (AIR) pump relay, (2.5L) Preheating coolant, low heat output relay (TDI, 06/2003 to 12/2006)	- 53 100 100 100
11	Park/Neutral position relay (4-speed auto) Park/Neutral position relay (6-speed auto) Starting interlock relay (from m.y. 2005, 6-speed auto)	126 176
12	Power supply (terminal 30b, B+) relay	109
13	Preheating coolant, high heat output relay (TDI, 06/2003 to 12/2006)	53

There are also additional fuses above the 13-position auxiliary relay panel which are identified in **Table e**.

Table e. Fuses above relay panel

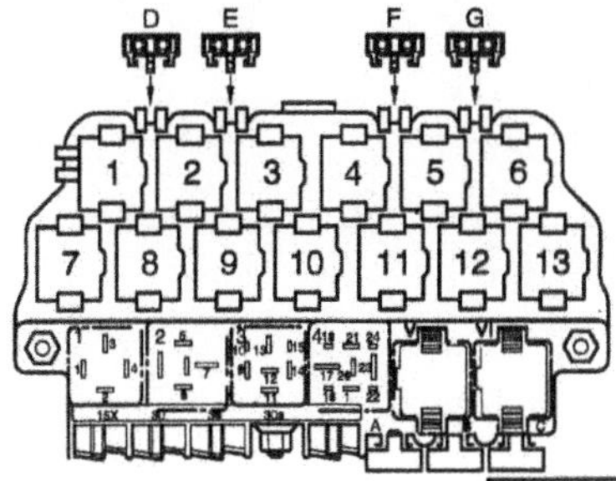

N97-0105

Position	Description (fuse #)	Amperage
D	Vacant	-
E	Power windows (S43)	30
F	Anti-theft alarm turn signal lights (S111)	15

Table e. Fuses above relay panel

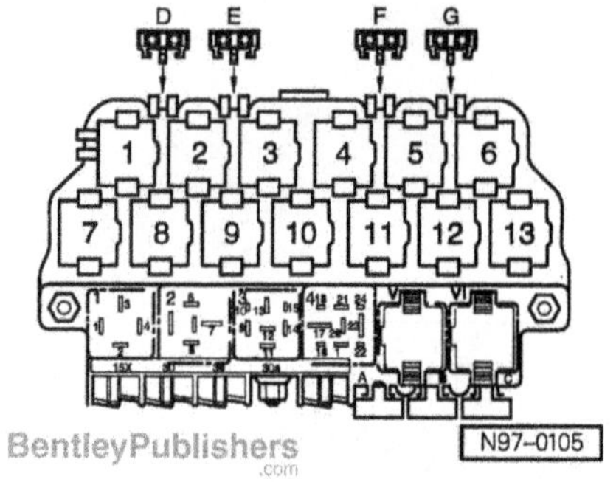

Position	Description (fuse #)	Amperage
G	Anti-theft alarm horn (S144) Mechatronic DSG fuse (S337) Brake vacuum pump fuse (S283)	15 20 20

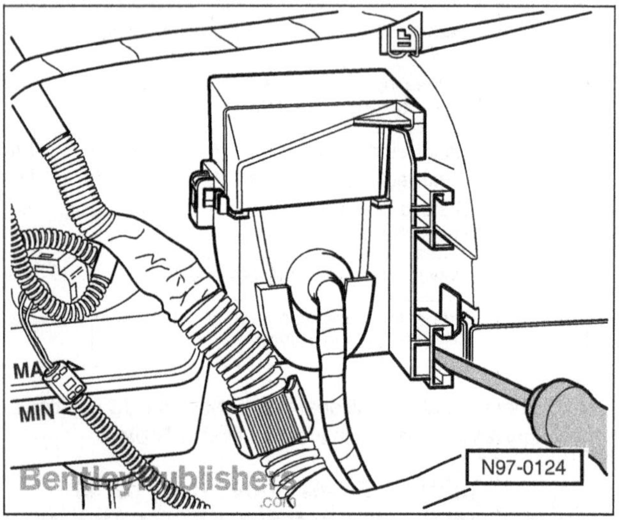

Relay panel in engine compartment

There is an additional relay panel in the rear of the engine compartment, to the left of the brake fluid reservoir. The Secondary Air Injection relay (J299) and the Engine Control Module (ECM) relay (J271) are located there. Other additional relays may also be located there. See **Component Locations** later in this repair group for a list of electrical components, relays and their locations.

< To remove the engine compartment relay panel cover, insert screwdriver behind retaining tabs, as shown, and carefully pry off.

– Swing cover forward and remove.

Fuse and relay panels, removing and installing

CAUTION—

- *Before working on the electrical system always switch the ignition off and disconnect battery ground strap from battery negative (–) terminal.*
- *Be sure to have the anti-theft radio code on hand before disconnecting the battery.*
- *Disconnecting the negative (–) battery cable may erase fault codes and basic settings in the engine management and automatic transmission control modules. Some driveability problems may be noticed until the system re-adapts to operating conditions. OBD II readiness codes, which may be required for emissions testing, may also be erased.*

– Open driver's door and remove trim panel on far left side of dash board.

– Carefully pry fuse panel cover off.

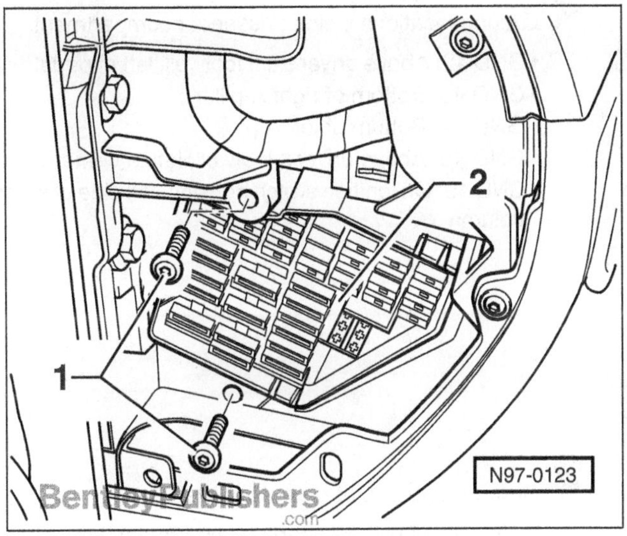

– Remove fuse panel securing screws (**1**) and fuse panel (**2**) inward and access from underneath.

– Install fuse panel in reverse order of removal.

Relay panel, removing and installing

– Remove lower instrument panel trim on driver's side, see **70 Trim-Interior**.

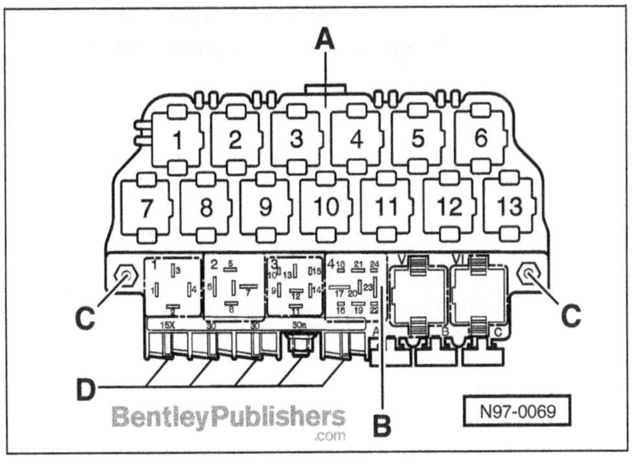

– Remove relay panel mounting screws (**C**) and loosen all bolted wire connections (**D**).

– Remove relay panels (**A** and **B**) downward.

– Install in reverse order of removal.

GROUND LOCATIONS

Various ground locations are identified below. Be sure to inspect these for clean tight connections when trouble-shooting electrical problems.

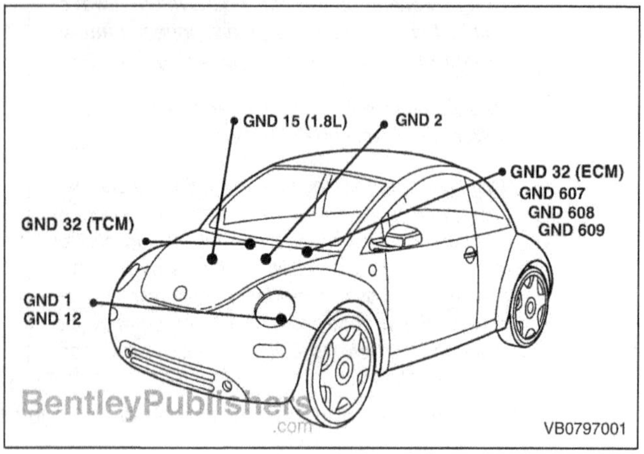

◀ Ground locations in engine compartment.

- GND 1: In engine compartment, left side, below battery tray (battery to body ground)
- GND 2: On top of transmission, near engine block
- GND 12: In engine compartment, left side, below head-light
- GND 15 (1.8L): On top of cylinder head, near ignition coil (1.8L only)
- GND 32 (TCM): In center plenum, below defroster duct
- GND 32 (ECM): In center plenum, below ECM
- GND 607: In left side plenum
- GND 608: In center plenum
- GND 609: In right side plenum

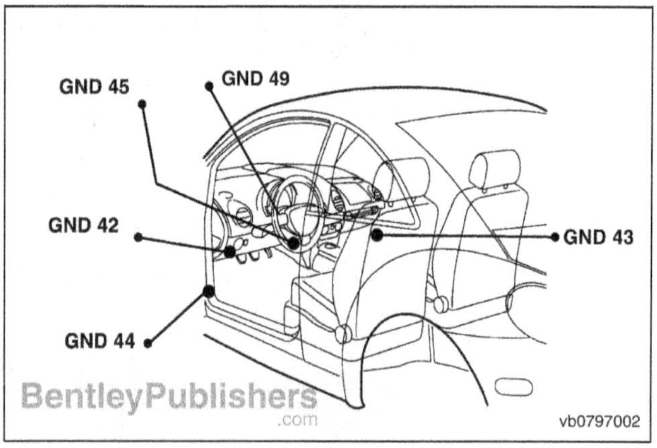

◀ Ground locations inside passenger compartment.

- GND 42: Above driver's left footwell, left of pedal cluster
- GND 43: Bottom of right A-pillar
- GND 44: Bottom of left A-pillar
- GND 45: Above driver's footwell, left of center console
- GND 49: On ignition switch housing, left side of steering column

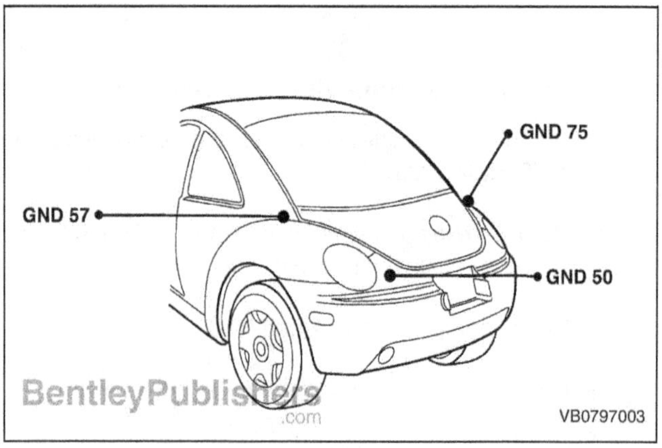

◀ Ground locations at rear of vehicle.

- GND 50: Rear of luggage compartment, left-side
- GND 57: Left C-pillar, above left rear wheelhousing
- GND 75: Right C-pillar, above right wheelhousing

COMPONENT LOCATIONS

Component	Code	OBD	Application	Location and additional information
A/C cut-out thermal switch	F163		1.9L	in "T" in coolant line, between radiator and cylinder head
A/C cut-out thermal switch	F163		2.0L	in coolant flange on cylinder head, left side, forward of ECT sensor G2/G62
A/C pressure switch	F129		all A/C	in engine compartment, right side, below plenum (some systems will have G65 instead)
ABS control module	J104		as applicable	component of ABS hydraulic control unit, in engine compartment, to left of vacuum brake booster
ABS hydraulic pump	V64		as applicable	component of ABS hydraulic control unit, in engine compartment, to left of vacuum brake booster
Airbag control module	J234		all	on central tunnel, front, beneath heater core housing
Airbag, front crash sensor, left	G283		as applicable	at front of longitudinal body frame member, left
Airbag, front crash sensor, right	G284		as applicable	at front of longitudinal body frame member, right
Airbag spiral spring/return spring with slip ring contact	F138		all	on steering column, with wiring for horn contact
Alarm horn	H8		all	in plenum, right side, to rear of water deflector (accessible through engine compartment)
Ambient temperature switch	F38		all A/C	in plenum, right side, below water deflector
Angle sensor 1 for throttle drive (electronic power control)	G187		as applicable	on throttle valve housing; component of throttle valve control module J338
Angle sensor 2 for throttle drive (electronic power control)	G188		2.0L, 2.5L	on throttle valve housing; component of throttle valve control module J338
Back-up light switch	F4		all m/t	top of manual transmission, on front of selector mechanism housing
Brake light switch	F		all	above brake pedal, integral with switch F47
Brake vacuum vent valve switch for cruise control	F47		1.9L	above brake pedal, integral with brake light switch F
Camshaft Position sensor	G40	CMP	1.8L, 2.0L	on cylinder head, right side, behind camshaft sprocket
Camshaft Position sensor 2	G163	CMP	1.8L	on cylinder head, right side, beneath toothed belt cover
CAN-BUS on board diagnostic interface	J533		as applicable	in instrument cluster
Central control module for comfort system	J393		as applicable	above pedal cluster, to left of steering column
Change-over valve for intake manifold flap	N239		1.9L	on back of intake manifold
Charge air pressure sensor	G31		1.8L	in intercooler, inside right front fender, below headlight
Circuit breaker for power windows	S43		as applicable	on thirteen position auxiliary relay panel, above relay panel, position E
Closed Throttle Position (CTP) switch	F60	CTP	1.9L	at bulkhead, above accelerator pedal, component of TP sensor G79
Closed Throttle Position (CTP) switch	F60	CTP	2.0L (early)	mounted on throttle body, component of throttle valve control module J338
Clutch Pedal Position switch	F194	CPP	all m/t	above clutch pedal, above clutch vacuum vent valve F36
Clutch vacuum vent valve switch	F36		as applicable	above clutch pedal, below CPP switch F194
Cold start injector	N108		1.9L	on side of Diesel Direct Fuel Injection (DFI) pump
Convertible top closed position switch	F202		convertible	on top right of windshield frame, beneath convertible top latch
Convertible top control module	J256		convertible	right rear corner of luggage compartment, behind trim panel
Convertible top hydraulic pump	V118		convertible	left rear corner of luggage compartment, behind trim panel
Convertible top open position switch	F171		convertible	above right rear window regulator, attached to hydraulic cylinder
Convertible top operation switch	E137		convertible	center console, above emergency brake
Convertible top tonneau cover switch	F328		convertible	under rear seatback upper trim panel, below tonneau cover latches
Coolant Fan Control control module	J293	FC	all	in engine compartment, bottom, to left of radiator
Coolant Fan Control thermal switch	F18	FC	all	on rear of radiator, bottom left

Component	Code	OBD	Application	Location and additional information
Crash sensor for side airbag, driver's side	G179		all	under driver's seat and carpet
Crash sensor for side airbag, passenger side	G180		all	under front passenger seat and carpet
Decklid central locking system motor	V53		all	in rear lid, attached to lock cylinder carrier
Diesel Direct Fuel Injection Engine Control Module	J248	DFI ECM	1.9L	in plenum, left side, beneath instrument panel cover
Differential lock valve 1	N125		all with ABS/EDL	component of ABS hydraulic control unit, in engine compartment, to left of vacuum brake booster
Differential lock valve 2	N126		all with ABS/EDL	component of ABS hydraulic control unit, in engine compartment, to left of vacuum brake booster
Door control module, driver side	J386		all	in driver's door, on component carrier, integral with window motor V147
Door control module, passenger side	J387		all	in passenger door, on component carrier, integral with window motor V148
EGR vacuum regulator solenoid valve	N18		1.9L	in engine compartment, left side, to rear of air cleaner housing, to right of wastegate bypass valve N75
Engine Coolant Level sensor	G32	ECL	all	in engine compartment, right side, component of engine coolant reservoir (expansion tank)
Engine Coolant Temperature gauge sensor	G2	ECT	1.8L, 1.9L, 2.0L	in coolant flange on cylinder head, left side; integral with ECT sensor G62
Engine Coolant Temperature sensor	G62	ECT	1.8L, 1.9L, 2.0L	in coolant flange on cylinder head, left side; integral with ECT sensor G2
Engine Coolant Temperature sensor	G62	ECT	2.5L	in coolant flange on cylinder head, left side
Engine Speed sensor	G28	RPM	1.8L	on front of engine cylinder block, lower left, near flange for transmission
Engine Speed sensor	G28	RPM	1.9L	on front of engine cylinder block, lower left, near flange for transmission
Engine Speed sensor	G28	RPM	2.0L	on front of engine cylinder block, lower left, near flange for transmission
Evaporative Emission canister purge regulator valve	N80	EVAP	1.8L	in engine compartment, right side, next to engine coolant reservoir
Evaporative Emission canister purge regulator valve	N80	EVAP	2.0L	in engine compartment, right side, next to engine coolant reservoir
Fresh air blower	V2		all	forward of glove box
Fresh air blower series resistance with fuse	N24		all	forward of glove box, beneath fresh air blower V2
Fuel cut-off valve	N109		1.9L	in top left of Diesel Direct Fuel Injection (DFI) pump
Fuel Pump	G6	FP	all except 1.9L	in fuel tank, beneath rear seat, right side, component of fuel pump module, together with sender for fuel gauge (fuel level sensor) G
Fuel temperature sensor	G81		1.9L	on top of Diesel (DFI) pump under cover
Fuse -1- (30) in fuse bracket/battery	S162		all	on battery cover, inside protective housing
Fuse -2- (30) in fuse bracket/battery	S163		all	on battery cover, inside protective housing
Fuse -3- in fuse bracket/battery	S164		all	on battery cover, inside protective housing
Fuse -4- (30) in fuse bracket/battery	S176		all	on battery cover, inside protective housing
Fuse -5- (30) in fuse bracket/battery	S177		all	on battery cover, inside protective housing
Fuse -6- (30) in fuse bracket/battery	S178		all	on battery cover, inside protective housing
Fuse -7- (30) in fuse bracket/battery	S179		all	on battery cover, inside protective housing
Fuse -8- (30) in fuse bracket/battery	S180		all	on battery cover, inside protective housing
Fuse for anti-theft alarm horn	S144		all	on thirteen position auxiliary relay panel, above relay panel, position G
Fuse for anti-theft alarm turn signal lights	S111		all	on thirteen position auxiliary relay panel, above relay panel, position F

Component	Code	OBD	Application	Location and additional information
Generator	C	GEN	all	in front of engine, right side
Glow plug control module	J370		1.9L	in plenum, beneath plastic cover
Glow plugs (engine)	Q6		1.9L	on cylinder head, next to each diesel fuel injector (4 total)
Heated mirror relay	J99		all	on thirteen position auxiliary relay panel, above relay panel, position 8
Heated Oxygen Sensor	G39	HO2S	1.8L, 2.0L, 2.5L	ahead of Three Way Catalytic Converter (TWC)
Heated Oxygen Sensor 2	G108	HO2S	1.8L, 2.0L, 2.5L	at rear of Three Way Catalytic Converter (TWC)
Heated Oxygen Sensor, center	G465	HO2S	2.5L (BPR)	at center of Three Way Catalytic Converter (TWC)
High pressure sensor	G65		all with A/C	in engine compartment, right rear, below plenum (some systems will have F129 instead)
Hood alarm switch	F120		all	on hood latch mechanism at front lock carrier
Igniter for side airbag, driver's side	N199		all	on left side of front driver's seat backrest, beneath seat cover
Igniter for side airbag, passenger side	N200		all	on right side of front passenger's seat backrest, beneath seat cover
Ignition coil	N152		2.0L	on front of engine block, left side, beneath Secondary Air Injection (AIR) pump motor
Ignition coil w/ power output stage 1	N70		1.8L, 2.5L	on cylinder head, above no. 1 cylinder
Ignition coil w/ power output stage 2	N127		1.8L, 2.5L	on cylinder head, above no. 2 cylinder
Ignition coil w/ power output stage 3	N291		1.8L, 2.5L	on cylinder head, above no. 3 cylinder
Ignition coil w/ power output stage 4	N292		1.8L, 2.5L	on cylinder head, above no. 4 cylinder
Ignition coil w/ power output stage 5	N323		2.5L	on cylinder head, above no. 5 cylinder
Induction coil for anti-theft immobilizer	D2		as applicable	on lock cylinder housing, on right side of steering column
Intake Air Temperature sensor	G42	IAT	1.8L (APH, AWP, AWV)	in intake manifold, near throttle valve
Intake Air Temperature sensor	G42	IAT	1.8L (BKF, BNU), 2.0L, 2.5L	in intake air duct, integral with MAF sensor G70
Intake Air Temperature sensor	G72	IAT	1.9L	in intake air duct between charge air cooler and intake manifold; integral with MAP sensor G71
Intake Air Temperature sensor	G72	IAT	2.0L	in intake air duct, to right of air cleaner housing. integral with MAF sensor G70
Intake Manifold change-over valve	N156		2.0L (BDC, BGD)	in engine compartment right side, on intake manifold
Intake Manifold tuning valve	N261	IMT	2.0L (BDC, BGD)	in engine compartment front, mounted on intake manifold
Kick down switch	F8		1.9L	at bulkhead, above accelerator pedal, component of TP sensor G79
Knock Sensor 1	G61	KS1	1.8L	on front of engine cylinder block, right side, near engine coolant flange
Knock Sensor 1	G61	KS1	2.0L	on front of engine cylinder block, right side
Knock Sensor 2	G66	KS2	1.8L	on front of engine cylinder block, left side, below Secondary Air Injection (AIR) pump motor V101
Knock Sensor 2	G66	KS2	2.0L	on front of engine cylinder block, behind ignition coil N152
Leak Detection Pump	V144	LDP	1.8L, 2.0L	above EVAP canister, inside right-rear wheelhousing
Leak Detection Pump	V144	LDP	2.5L	above EVAP canister, inside left-rear wheelhousing
Left convertible top closed lock switch	F324		convertible	behind left rear side panel, upper of two switches
Left convertible top contact switch	F155 G442		convertible	behind rear seat backrest, upper left
Left convertible top open lock switch	F323		convertible	behind left rear side panel, lower of two switches
Left convertible top locking motor	V291		convertible	behind left rear side panel, to left of rear woofer speaker

Component	Code	OBD	Application	Location and additional information
Left front ABS inlet valve	N101		all with ABS	component of ABS hydraulic control unit, in engine compartment, to left of vacuum brake booster
Left front ABS outlet valve	N102		all with ABS	component of ABS hydraulic control unit, in engine compartment, to left of vacuum brake booster
Left front ABS wheel speed sensor	G47		as applicable	in left front wheel bearing housing
Left front seat belt micro switch	F140		all	integral with driver's seat belt buckle (latch)
Left rear ABS inlet valve	N134		all with ABS	component of ABS hydraulic control unit, in engine compartment, to left of vacuum brake booster
Left rear ABS outlet valve	N136		all with ABS	component of ABS hydraulic control unit, in engine compartment, to left of vacuum brake booster
Left rear ABS wheel speed sensor	G46		as applicable	at left rear stub axle
Load reduction relay	J59		all	on thirteen position auxiliary relay panel, above relay panel, position 2
Lock unit for central locking, driver side	F220		all	component of door lock assembly; includes motor and switches
Lock unit for central locking, passenger side	F221		all	component of door lock assembly; includes motor and switches
Luggage compartment light switch	F5		all	in rear lid latch
Manifold Absolute Pressure sensor	G71	MAP	1.9L	in intake air duct between charge air cooler and intake manifold
Mass Air Flow sensor	G70	MAF	all	in intake air duct, to right of air cleaner housing
Modulating piston displacement sensor	G149		1.9L	on front of Diesel DFI pump integral with quantity adjuster N146
Motor for fuel tank lid unlock	V155		all	in luggage compartment, right side, above right rear wheelhousing
Motronic Engine Control Module	J220	EDM	as applicable	in plenum, right side, beneath instrument panel cover
Motronic Engine Control Module power supply relay	J271	ECM	as applicable	in engine compartment, left side, to left of brake fluid reservoir, inside housing
Multi-function Transmission Range switch	F125	TR	all 4-speed a/t	on rear of transmission housing, top
Needle lift sensor	G80		1.9L	on front of cylinder head, component of cylinder 3 fuel injector
Oil pressure switch	F1		1.8L	in front of engine, left side, on left side of engine oil cooler flange
Oil pressure switch	F1		1.9L	in front of engine, on oil filter flange, right side
Oil pressure switch	F1		2.0L	in front of engine, on oil filter flange, left side
Outside air temperature sensor	G17		as applicable	behind front bumper, left side, accessible from underneath vehicle
Park/Neutral Position Relay	J226	PNP	all 4-speed a/t	on thirteen position auxiliary relay panel, above relay panel, position 11
Parking brake warning light switch	F9		all	on central tunnel, below parking brake lever
Positive Crankcase Ventilation heating element	N79	PCV	1.9L, 2.0L	in PCV hose between cylinder head cover and intake air duct
Power sunroof control module	J245		as applicable	above headliner, forward of sunroof, center, integral with power sunroof motor V1
Power sunroof motor	V1		as applicable	above headliner, forward of sunroof, center, integral with power sunroof control module J245
Quantity adjuster	N146		1.9L	on front of Diesel DFI pump, integral with modulating piston displacement sensor G149
Rain sensor	G213		as applicable	in mounting plate for interior rear view mirror
Rear spoiler control module	J223		as applicable	in rear lid, right side, integral with rear spoiler motor V52
Rear spoiler motor	V52		as applicable	in rear lid, right side, integral with rear spoiler control module J223
Recirculating valve for turbocharger	N249		1.8L	on top of cylinder head valve cover, center
Relay for preheating coolant, high heat output	J360		1.9L	in engine compartment, left side, to left of brake fluid reservoir, inside housing

Component	Code	OBD	Application	Location and additional information
Relay for preheating coolant, low heat output	J359		1.9L	in engine compartment, left side, to left of brake fluid reservoir, inside housing
Right convertible top closed lock switch	F326		convertible	behind right rear side panel, upper of two switches
Right convertible top contact switch	F156 G443		convertible	behind rear seat backrest, right upper
Right convertible top open lock switch	F325		convertible	behind right rear side panel, lower of two switches
Right convertible top locking motor	V292		convertible	behind right rear side panel, to right of rear woofer speaker
Right front ABS inlet valve	N99		all with ABS	component of ABS hydraulic control unit, in engine compartment, to left of vacuum brake booster
Right front ABS outlet valve	N100		all with ABS	component of ABS hydraulic control unit, in engine compartment, to left of vacuum brake booster
Right front ABS wheel speed sensor	G45		as applicable	in right front wheel bearing housing
Right front seat belt micro switch	F141		all	integral with front passenger's seat belt buckle (latch)
Right rear ABS inlet valve	N133		all with ABS	component of ABS hydraulic control unit, in engine compartment, to left of vacuum brake booster
Right rear ABS outlet valve	N135		all with ABS	component of ABS hydraulic control unit, in engine compartment, to left of vacuum brake booster
Right rear ABS wheel speed sensor	G44		as applicable	at right rear stub axle
Secondary Air Injection (AIR) pump relay	J299	AIR	as applicable	in engine compartment, left side, to left of brake fluid reservoir, inside, housing
Secondary Air Injection pump motor	V101	AIR	1.8L	on front of engine, left side
Secondary Air Injection pump motor	V101	AIR	2.0L	in engine compartment, front, in front of cylinder head
Secondary Air Injection solenoid valve	N112	AIR	1.8L	on top of cylinder head valve cover, right side
Secondary Air Injection solenoid valve	N112	AIR	2.0L	in engine compartment, left side, to rear of air cleaner housing
Sender -2- for accelerator pedal position	G185		as applicable	on bulkhead, above accelerator pedal, integral with TP sensor G79
Sender for fuel gauge	G		all	in fuel tank, beneath rear seat, right side, part of fuel pump module, together with fuel pump G6 (except 1.9L)
Sender for transmission output RPM	G195		as applicable	in transmission housing, above valve body
Sensor for transmission RPM	G182			in transmission housing, above valve body
Servo motor for fresh/recirculating air door	V154		all	to right of fresh air blower housing, forward of glove box
Shift lock solenoid	N110		all 4-speed a/t	in selector mechanism housing, on central tunnel
Speedometer Vehicle Speed Sensor	G22	VSS	all transmis-sion types	on top of differential housing
Starter lock relay	J433		all	Relay position #3 on 13-position auxiliary relay panel
Throttle drive (power accelerator actuation)	G186		as applicable	on throttle valve housing; component of throttle valve control module J338
Throttle drive angle sensor 2 (power accelerator actuation)	G188		2.5L	on throttle valve housing; component of throttle valve control module J338
Throttle Position (TP) actuator	V60		as applicable	mounted on throttle body, component of throttle valve control module J338
Throttle Position sensor	G79	TPS	1.8L	on bulkhead, above accelerator pedal, integral with sender G185
Throttle Position sensor	G79	TPS	1.9L	at bulkhead, above accelerator pedal, integral with kick down switch F8 and CTP switch F60
Throttle Position sensor	G69	TPS	2.0L, 2.5L	mounted on throttle body, component of throttle valve control module J338
Throttle Position sensor	G88	TPS	2.0L	mounted on throttle body, component of throttle valve control module J338
Throttle valve control module	J338		1.8L	on throttle valve housing, integral with G186, G187, G188
Throttle valve control module	J338		2.0L	mounted on throttle body integral with CTP switch F60, TP sensors G69 and G88, and TP actuator V60
Throttle valve control module	J338		2.5L	on throttle valve housing, integral with J188, G69, G186

Component	Code	OBD	Application	Location and additional information
Transmission Control Module	J217	TCM	as applicable	in plenum, right side, beneath instrument panel cover and defroster duct
Transmission Vehicle Speed Sensor	G38	VSS	all 4-speed a/t	on top of transmission housing, to left of transmission cooler
Trunk lock alarm / central locking switch	F124		all	in rear lid, on lock cylinder carrier
Vehicle Speed Sensor	G68	VSS	all 4-speed a/t	on top of transmission housing, below transmission mount
Voltage Regulator	C1	VR	all	on Generator (GEN), in front of engine, right side
Wastegate bypass regulator valve	N75		1.8L	in intake air duct, forward of vacuum brake booster
Wastegate bypass regulator valve	N75		1.9L	in engine compartment, left side, to rear of air cleaner housing, to left of EGR solenoid valve N18
10-pin connector	T10		all	in connector station in plenum, left side, beneath instrument panel cover
10-pin connector	T10b		1.9L	in engine compartment, left side, to rear of air cleaner housing, in wiring harness guide
10-pin connector	T10b		2.0L	in engine compartment, left side, to rear of air cleaner housing, in wiring harness guide
10-pin connector	T10d		all	in connector station in plenum, left side, beneath instrument panel cover
10-pin connector	T10e		all	in connector station in plenum, left side, beneath instrument panel cover
10-pin connector	T10f		1.9L	on front of engine cylinder block, to left of Diesel Direct Fuel Injection (DFI) pump
10-pin connector	T10g		all	in connector station in plenum, left side, beneath instrument panel cover
10-pin connector	T10h		all	in connector station in plenum, left side, beneath instrument panel cover
10-pin connector	T10k		all	at bottom of left A-pillar
10-pin connector	T10l		all	at bottom of right A-pillar
10-pin connector, blue	T10c		all	in connector station in plenum, left side, beneath instrument panel cover
10-pin connector, orange	T10a		all	in connector station in plenum, left side, beneath instrument panel cover
12-pin connector	T12		all 4-speed a/t	on top of transmission final drive housing, for wiring to solenoid valves and ATF temperature sensor G93 in valve body
14-pin connector	T14		all	in plenum, center, beneath instrument panel cover
2-pin connector	T2a		1.9L	for engine glow plugs, in engine compartment, left side, behind air cleaner, in wiring harness guide
2-pin connector (for ECM/TCM)	T2b		all 4-speed a/t	in plenum, left side, beneath instrument panel cover
2-pin connector (for needle lift sensor G80)	T2c		1.9L	on front of engine cylinder block, to left of Diesel Direct Fuel Injection (DFI) pump
2-pin connector (for side airbag igniter N199)	T2b		all	on bottom of driver's seat
2-pin connector (for side airbag igniter N200)	T2c		all	on bottom of front passenger's seat
3-pin connector	T3		1.9L	on battery cover
3-pin connector	T3		2.0L	on battery cover
3-pin connector	T3f		1.8L	on front of engine cylinder block, center, above engine oil cooler
3-pin connector	T3h		all	to rear of radiator, bottom left side, for left coolant fan V7, to rear of connector T3i
3-pin connector	T3i		all	to rear of radiator, bottom left side, for right coolant fan V35, forward of connector T3h
3-pin connector (for RPM sensor G28)	T3		1.9L	on front of engine cylinder block, to left of Diesel Direct Fuel Injection (DFI) pump

Component	Code	OBD	Application	Location and additional information
3-pin connector (for RPM sensor G28)	T3b		2.0L	in front of engine cylinder block, above oil filter flange
3-pin connector (for vehicle speed sensor G68)	T3		all 4-speed a/t	on top of differential housing, near selector lever cable
4-pin connector (at starter)	T4		all	above starter, for A/C clutch and Generator (GEN)
4-pin connector (for HO2S 2 G108)	T4		1.8L	on bottom of floor panel, right side, inside protective housing
4-pin connector (for HO2S 2 G108)	T4		2.0L	brown; on bottom of right floor panel, inside protective housing
4-pin connector (for HO2S G39)	T4b		1.8L, 2.0L, 2.5L	black; on bottom of right floor panel, inside protective housing
5-pin connector	T5		all	black; on left rear C-pillar, above left rear wheelhousing
5-pin connector	T5g		all	black; in rear lid, left side, near lock cylinder carrier
5-pin connector	T5h		all	brown; in rear lid, left side, near lock cylinder carrier
6-pin connector (for HO2S G39)	T6		1.8L, 2.0L, 2.5L	on bottom of floor panel, right side, inside protective housing
6-pin connector (in connector station)	T6		all	in connector station, in plenum, left side, beneath instrument panel cover

EWD Electrical Wiring Diagrams

(continued on following page)

GENERAL

This section contains selected electrical wiring diagrams for vehicles covered by this manual. These diagrams are provided by Volkswagen of America, Inc. The publisher cannot vouch for the accuracy of these diagrams. For a list of fuses and relays, see **97 Fuses, Relays and Component Locations**.

Warnings and Cautions

WARNING —

- *Airbags and pyrotechnic seat belt tensioners utilize explosive devices. Handle with extreme care. Refer to the warnings and cautions in* **69 Seat Belts, Airbags**.
- *The ignition system of the car operates at lethal voltages. If you have a weak heart or wear a pacemaker, do not expose yourself to the ignition system electric currents. Take extra precautions when working on the ignition system or when servicing the engine while it is running or the key is ON. See* **28 Ignition System** *for additional ignition system warnings and cautions.*
- *Keep hands, clothing and other objects clear of the electric radiator cooling fan when working on a warm engine. The fan may start at any time, even when the ignition is switched OFF.*

CAUTION—

- *Do not disconnect the battery with the engine running.*
- *Switch the ignition OFF and remove the negative (-) battery cable before removing any electrical components. Connect and disconnect electrical connectors and ignition test equipment leads only while the ignition is switched OFF.*
- *Relay and fuse positions are subject to change and may vary from car to car. If questions arise, an authorized Audi dealer is the best source for the most accurate and up-to-date information.*
- *Use a digital multimeter for electrical tests. Switch multimeter to appropriate function and range before making test connections.*
- *Many control modules are static sensitive. Static discharge damages them permanently. Handle the modules using proper static prevention equipment and techniques.*
- *To avoid damaging harness connectors or relay panel sockets, use jumper wires with flat-blade connectors that are the same size as the connector or relay terminals.*
- *Do not try to start the engine of a car which has been heated above 176°F (80°C) (for example, in a paint drying booth). Allow it to cool to normal temperature.*
- *Disconnect battery before doing any electric welding on the car.*
- *Do not wash engine while it is running, or any time ignition is ON.*
- *Choose test equipment carefully. Use a digital multimeter with at least 10 MΩ input impedance, or an LED test light. An analog meter (swing-needle) or a test light with a normal incandescent bulb may draw enough current to damage sensitive electronic components.*
- *Do not use an ohmmeter to measure resistance on solid state components such as control modules.*
- *Disconnect the battery before making resistance (ohm) measurements on a circuit.*

How to read wiring diagrams

The diagram below shows how a typical wiring diagram page is laid out.

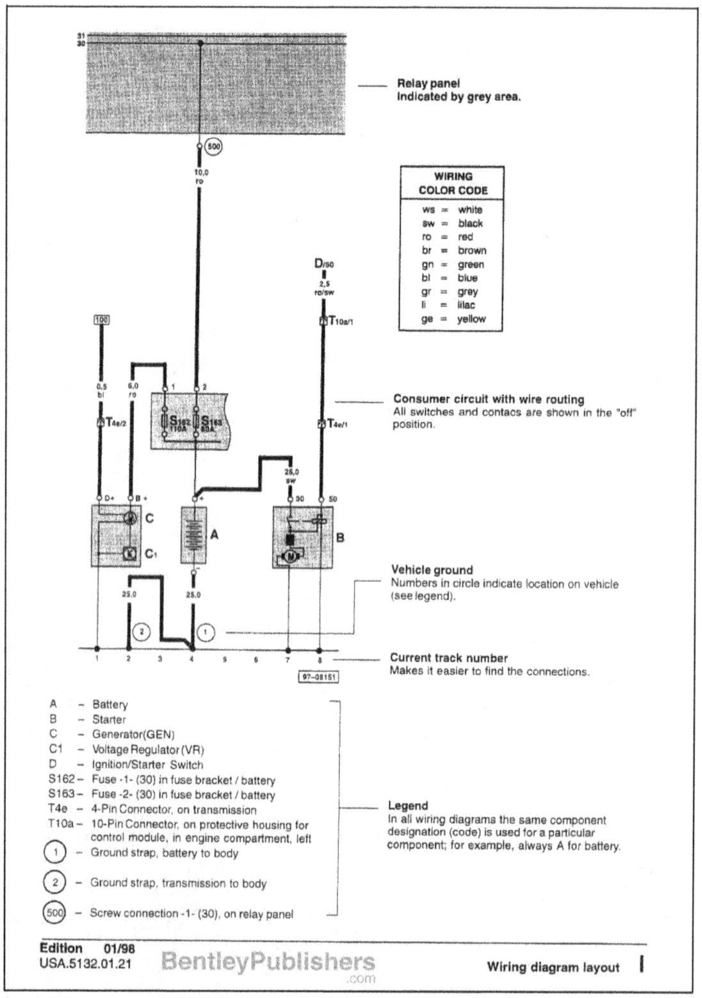

Terminal (circuit) identification

Several circuits in the vehicle's electrical system are identified with a number or letter designation. These circuits are universally identified in many wiring diagrams and may also be used to identify switch connector terminals.

Terminal	Identification
1	Ignition coil/distributor low voltage (i.e. RPM signal)
15	Switched battery positive (B+) from ignition switch
30	Battery positive (B+), hot at all times
31	Ground (–)
31b	Switched ground (–)
50	Starter control; switched B+ from ignition switch
58	Switched parking light, taillight, B+ from light switch
S (SU)	Key in ignition, switched B+ from ignition switch
X	Switched B+ from load reduction relay

Wiring diagram details

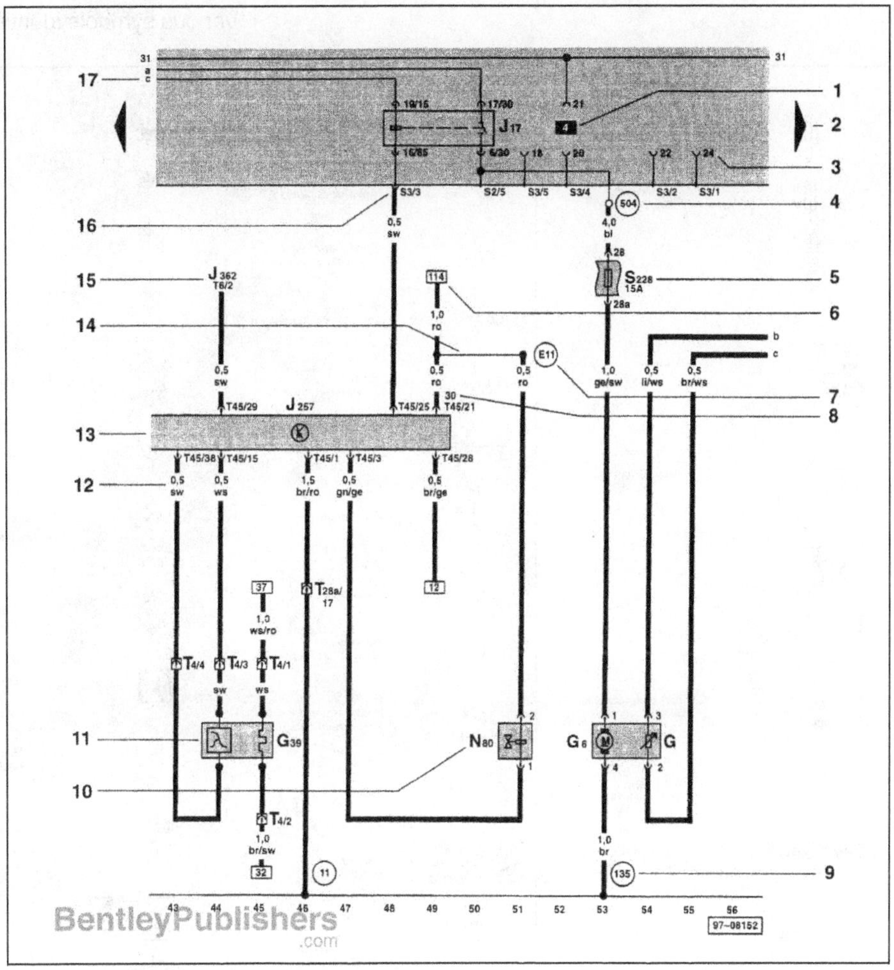

1. **Relay location number**
   - Indicates location on relay panel
2. **Arrow**
   - Indicates wiring circuit is continued on the previous and/or next page
3. **Connection designation - relay control module on relay panel**
   - Shows individual terminals in a multi-point connector
   - For example: terminal 24 of relay **4** indicated
4. **Diagram of threaded pin on relay panel**
   - White circle shows detachable connection
5. **Fuse designation**
   - For example: S228 = fuse number 228, 15 amp, in fuse holder
6. **Reference of wire continuation (current track number)**
   - Number in frame indicates current track where wire is continued on another wiring diagram page
7. **Wire connection designation in wiring harness**
   - Location of wire connections are indicated in legend
8. **Terminal designation**
   - Designation which appears on actual component and/or terminal number of a multi-point connector
9. **Ground connection designation in wire harness**
   - Locations of ground connections are indicated in legend
10. **Component designation**
    - Use legend to identify component code
11. **Component symbols**
    - See **Symbols used in wiring diagrams** in this repair group
12. **Wire cross-section size (in mm^2) and wire colors**
    - Abbreviations are explained in color chart beside wiring diagram
13. **Component symbol with open drawing side**
    - Indicated component is continued on another wiring diagram page
14. **Internal connections (thin lines)**
    - These connections are not wires
    - Internal connections are current carrying and are shown to allow tracing of current flow inside components or wiring harnesses
15. **Reference of continuation of wire to component**
    - For example: Anti-theft immobilizer (J362) control module on 6-pin connector, terminal 2
16. **Relay panel connectors**
    - Shows wiring of multi-point or single connectors on relay panel
    - For example: S3/3 = multi-point connector S3, terminal 3
17. **Reference of internal connection continuation**
    - Letters indicate where connection continues on previous and/or next page

Electrical Wiring Diagrams

Symbols used in wiring diagrams

Various symbols identified below are used in the wiring diagrams.

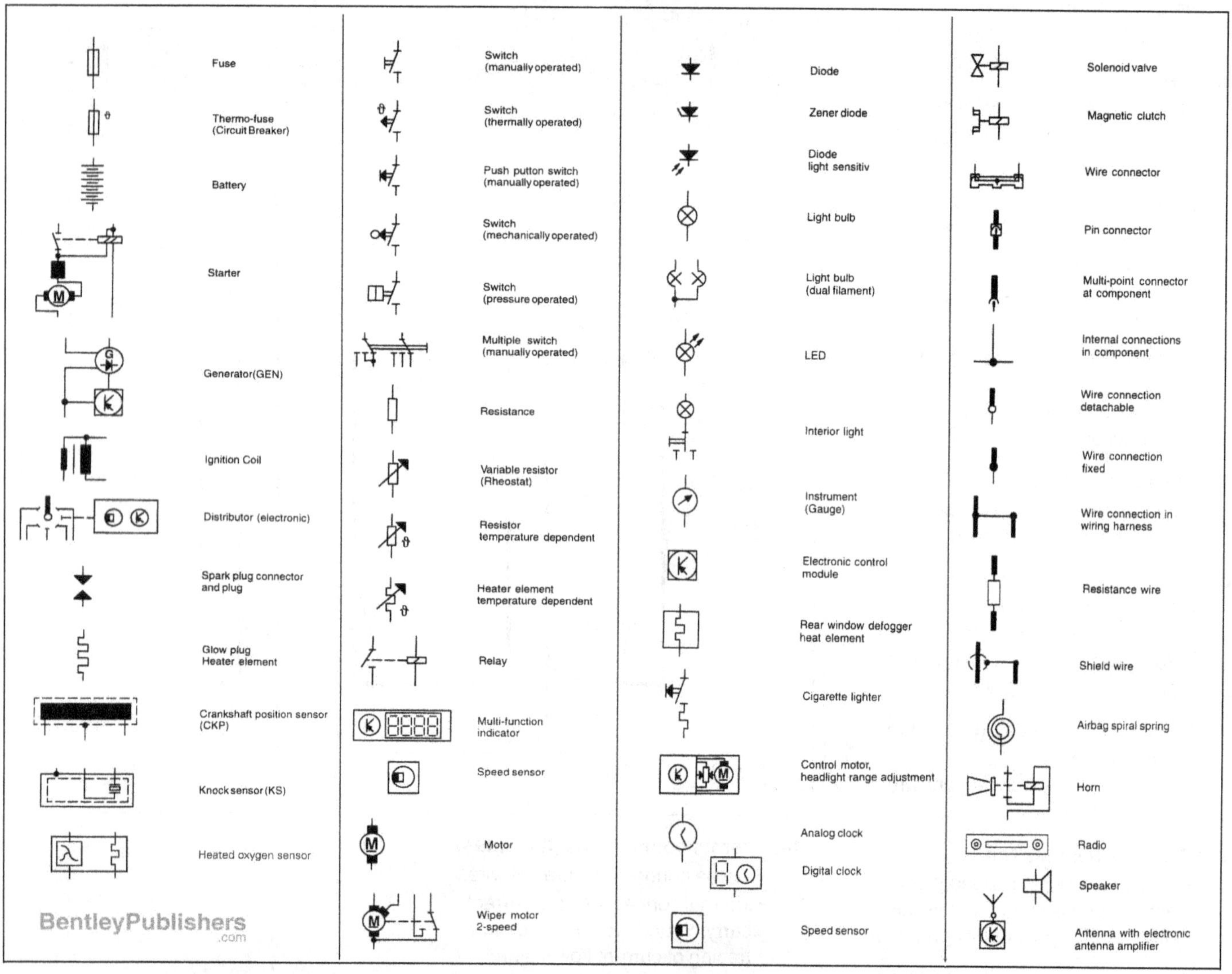

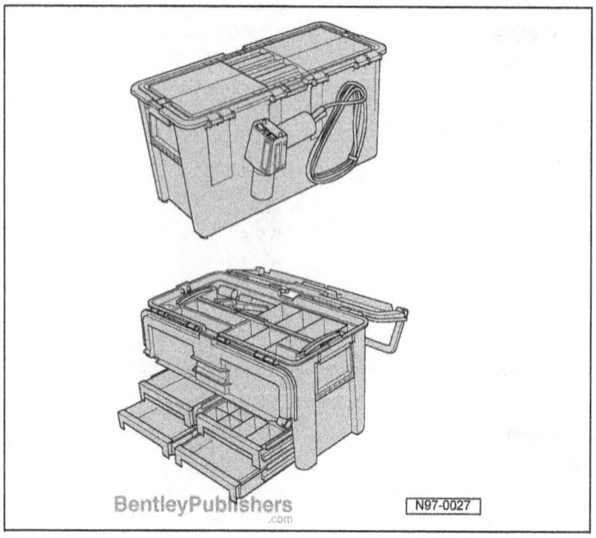

Wiring repairs

Repairs to wiring should only be done using VAS 1978A wiring harness repair kit. Various electrical malfunctions can occur using improper repair techniques (i.e. soldering, wire nuts, taping, etc.).

Vehicles covered by this manual utilize a “platform wire harness” concept, where the majority of the wiring in contained in a single wire harness. This allows for a reduction in the number of terminations, and thus in the total electrical resistance within the wiring harness. Extremely small terminals (0.64mm wide) and wires (0.34mm^2) cross-section) are used to carry very low level signals which could be damaged by improper repairs.

ELECTRICAL WIRING DIAGRAMS

Wiring diagrams have been selected from original source material provided by the vehicle manufacturer. The diagrams vary in the way they are oriented (vertical vs. horizontal) and are presented with two images per page. Running headers are located at the top of each page to indicate the title of the various diagrams.

New Beetle Wiring diagram **No. 35/1**

1,8L-Engine - 5V Turbo, Motronic Multiport Fuel Injection (MFI) 110 kW, code APH, (with cruise control)

from June 1999

Deviate relay location and fuseplacements as well as the locations of multiple connectors see section "component locations".

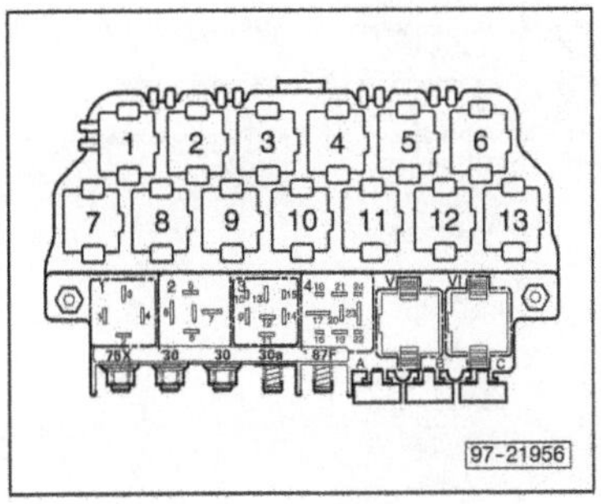

Relay location on the thirteen position auxiliary relay panel, above relay panel:

Relay panel:

- **2** Load Reduction Relay (213)
- **4** Fuel Pump Relay (409)

Fuse colors

30 A - green
25 A - white
20 A - yellow
15 A - blue
10 A - red
7,5 A - brown
5 A - beige
3 A - violet

Note: Number in parentheses indicates production control number stamped on relay housing.

New Beetle Wiring diagram **No. 35/2**

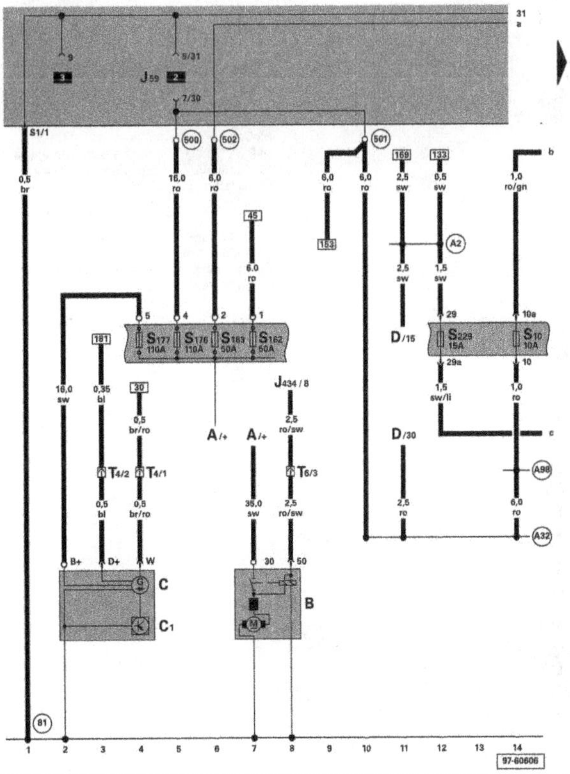

Battery, starter, generator (GEN), voltage regulator (VR), ignition/starter switch, load reduction relay, fuses in fuse bracket/battery

A - Battery
B - Starter
C - Generator (GEN)
C1 - Voltage Regulator (VR)
D - Ignition/Starter Switch
J434 - Locking relay for starter (clutch pedal switch)
J59 - Load Reduction Relay
S10 - Fuse 10 in fuse/relay panel
S162 - Fuse -1- (50), in fuse bracket/battery
S163 - Fuse -2- (50), in fuse bracket/battery
S176 - Fuse -4- (30), in fuse bracket/battery
S177 - Fuse -5- (30), in fuse bracket/battery
S229 - Fuse 29 in fuse/relay panel
T4 - 4-Pin Connector, in engine compartment left
T6 - 6-Pin Connector, brown, behind instrument panel, left

(81) - Ground connection -1-, in instrument panel wiring harness
(500) - Threaded connection -1- (30) on the relay plate
(501) - Threaded connection -2- (30) on the relay plate
(502) - Threaded connection -1- (30a) on the relay plate
(A2) - plus connection (15), in instrument panel wiring harness
(A32) - plus connection (30), in instrument panel wiring harness
(A98) - Plus connector -4- (30), in instrument panel wiring harness

ws = white
sw = black
ro = red
br = brown
gn = green
bl = blue
gr = grey
li = violet
ge = yellow
or = orange

1.8L engine management (APH) from June 1999

New Beetle — Wiring diagram — No. 35/3

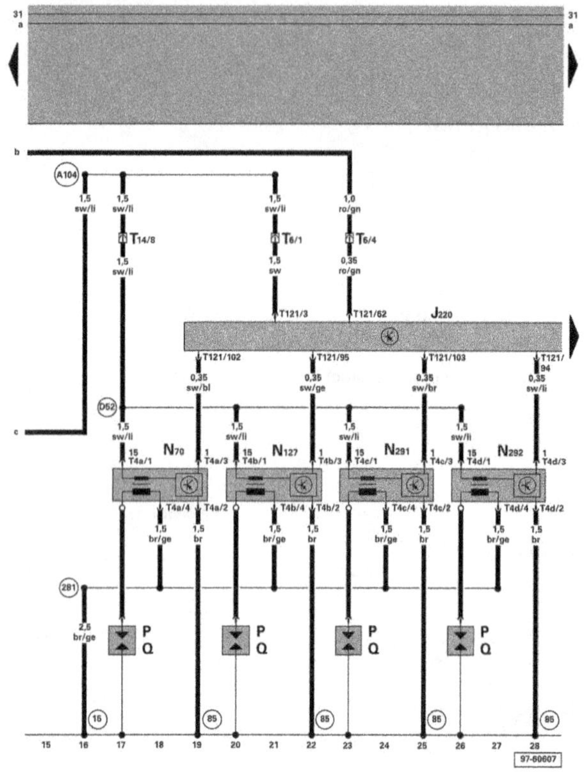

ws = white
sw = black
ro = red
br = brown
gn = green
bl = blue
gr = grey
li = violet
ge = yellow
or = orange

Motronic engine control module (ECM), ignition coil with power output stages 1, 2, 3 & 4, spark plugs

J220 - Motronic Engine Control Module (ECM)
N70 - Ignition Coil With Power Output Stage 1
N127 - Ignition Coil with Power Output Stage 2
N291 - Ignition Coil with Power Output Stage 3
N292 - Ignition Coil with Power Output Stage 4
P - Spark Plug Connectors
Q - Spark Plugs
T4a - 4-Pin Connector
T4b - 4-Pin Connector
T4c - 4-Pin Connector
T4d - 4-Pin Connector
T6 - 6-Pin Connector, brown, behind instrument panel, left
T14 - 14-Pin Connector, in engine compartment, left
T121 - 121-Pin Connector, at -J220-

(15) - Ground connection, on cylinder head
(85) - Ground connection -1-, in engine compartment wiring harness
(281) - Ground connector -1-, in wiring harness engine pre-wiring
(A104) - Plus connector -2- (15), in instrument panel wiring harness
(D52) - plus connection (15a), in engine compartment wiring harness

Edition 04.04

New Beetle — Wiring diagram — No. 35/4

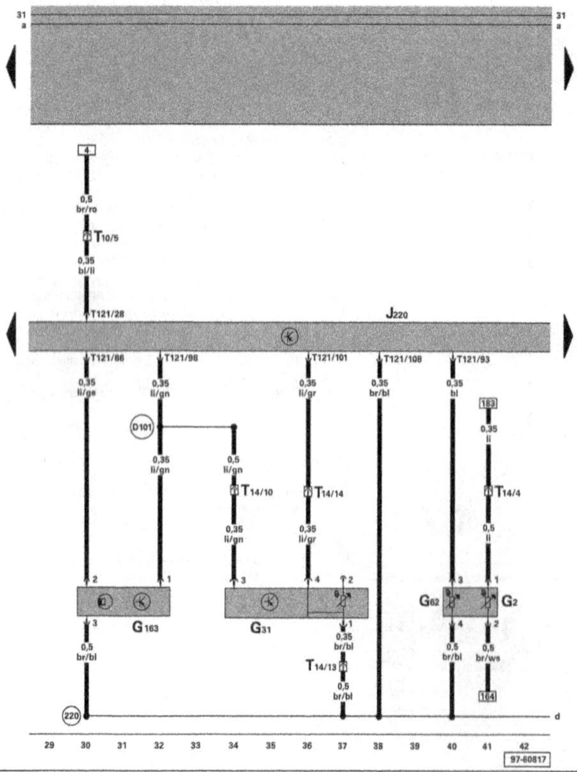

ws = white
sw = black
ro = red
br = brown
gn = green
bl = blue
gr = grey
li = violet
ge = yellow
or = orange

Motronic engine control module (ECM), engine coolant temperature (ECT) sensor, charge air pressure sensor, camshaft position (CMP) sensor 2

G2 - Engine Coolant Temperature (ECT) Sensor
G31 - Charge Air Pressure Sensor
G62 - Engine Coolant Temperature (ECT) Sensor
G163 - Camshaft Position (CMP) Sensor 2
J220 - Motronic Engine Control Module (ECM)
T14 - 14-Pin Connector, in engine compartment, left
T10 - 10-Pin Connector, white, behind instrument panel, left
T121 - 121-Pin Connector, at -J220-

(220) - Ground connection (sensor ground), in engine compartment wiring harness
(D101) - Wire connection -1-, in engine compartment wiring harness

Edition 04.04

New Beetle — Wiring diagram — No. 35/5

Motronic engine control module (ECM) heated oxygen sensor (HO2S), secondary air injection (AIR) pump relay/motor

G39 - Heated Oxygen Sensor (HO2S), black
G108 - Heated Oxygen Sensor (HO2S), brown
J220 - Motronic Engine Control Module (ECM)
J299 - Secondary Air Injection (AIR) Pump Relay
V101 - Secondary Air Injection (AIR) Pump Motor
T4 - 4-Pin Connector, brown
T6 - 6-Pin Connector, black
T121 - 121-Pin Connector, at -J220-

(609) - Ground Connection (in right plenum chamber)

(E39) - Connection-2-(87a), in wiring harness engine

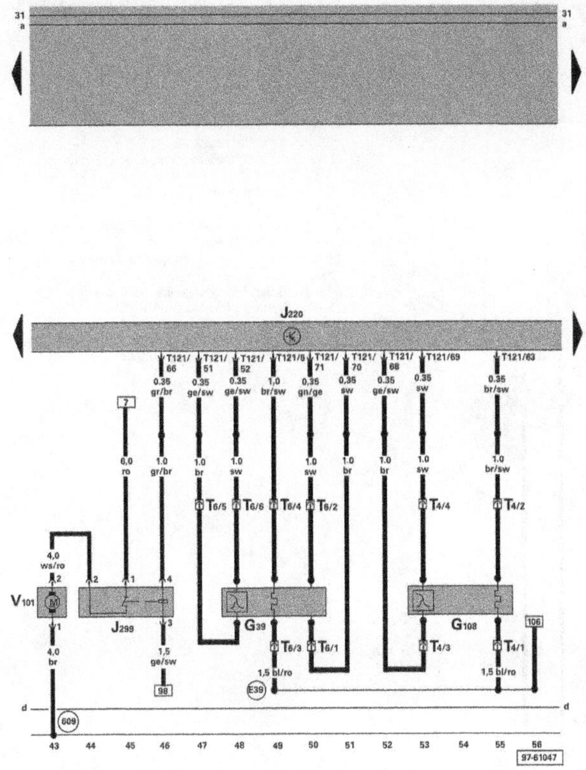

ws = white
sw = black
ro = red
br = brown
gn = green
bl = blue
gr = grey
li = violet
ge = yellow
or = orange

New Beetle — Wiring diagram — No. 35/6

Motronic engine control module (ECM), intake air temperature (IAT) sensor, knock sensor (KS) 1, throttle drive

G42 - Intake Air Temperature (IAT) Sensor
G61 - Knock Sensor (KS) 1
G186 - Throttle drive (power accelerator actuation)
G187 - Angle sensor -1- for throttle drive (power accelerator actuation)
G188 - Angle sensor -2- for throttle drive (power accelerator actuation)
J220 - Motronic Engine Control Module (ECM)
J338 - Throttle Valve Control Module
T2a - 2-Pin Connector
T6a - 6-Pin Connector
T10a - 10-Pin Connector, orange, behind instrument panel, left
T121 - 121-Pin Connector, at -J220-

(139) - Ground connection (sensor Ground), in Motronic Multiport Fuel Injection (MFI) wiring harness

* - Fuel consumtion indicator signal from Engine Control Module, only for MFA.

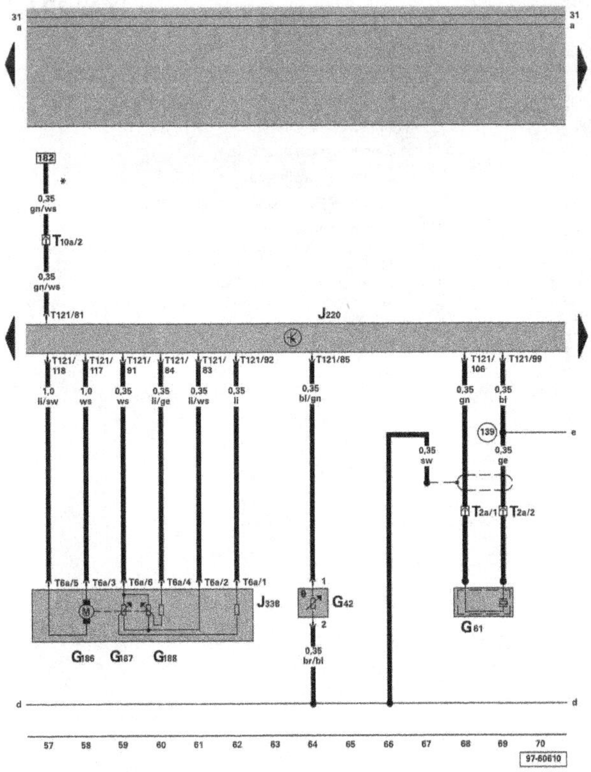

ws = white
sw = black
ro = red
br = brown
gn = green
bl = blue
gr = grey
li = violet
ge = yellow
or = orange

1.8L engine management (APH) from June 1999

New Beetle **Wiring diagram** **No. 35/7**

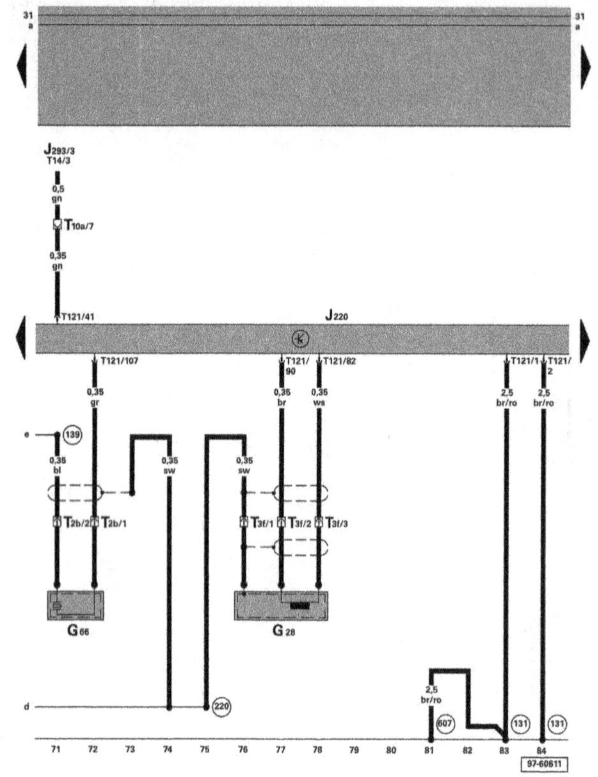

Motronic engine control module (ECM), engine speed (RPM) sensor, knock sensor (KS) 2,

G28 - Engine Speed (RPM) Sensor
G66 - Knock Sensor (KS) 2
J220 - Motronic Engine Control Module (ECM)
J293 - Coolant FC (Fan Control) Control Module
T2b - 2-Pin Connector
T3f - 3-Pin Connector
T10a - 10-Pin Connector, orange, behind instrument panel, left
T121 - 121-Pin Connector, at -J220-

(131) - Ground connection -2-, in engine compartment wiring harness
(607) - Ground Connection (in left plenum chamber)
(139) - Ground connection (sensor Ground), in Motronic Multiport Fuel Injection (MFI) wiring harness
(220) - Ground connection (sensor ground), in engine compartment wiring harness

ws = white
sw = black
ro = red
br = brown
gn = green
bl = blue
gr = grey
li = violet
ge = yellow
or = orange

Edition 04.04

New Beetle Wiring Diagram **No. 35 / 8**

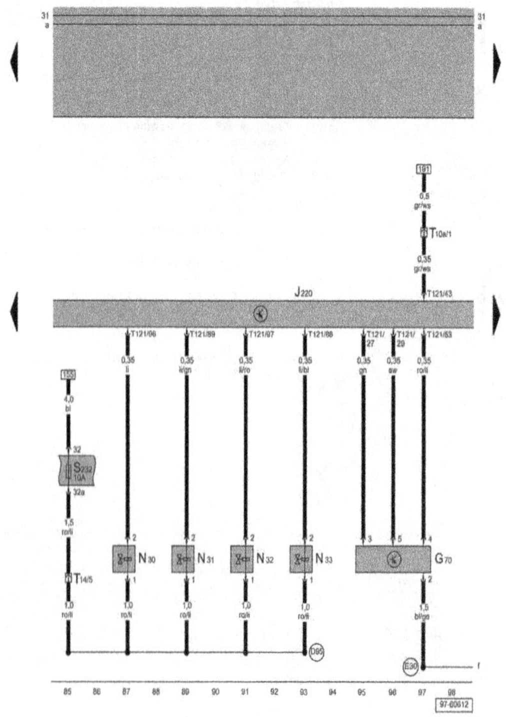

Motronic engine control module (ECM), mass air flow (MAF) sensor, fuel injectors

G70 - Mass Air Flow (MAF) Sensor
J220 - Motronic Engine Control Module (ECM)
N30 - Cylinder 1 Fuel Injector
N31 - Cylinder 2 Fuel Injector
N32 - Cylinder 3 Fuel Injector
N33 - Cylinder 4 Fuel Injector
S232 - Fuse 32 in fuse/relay panel
T10a - 10-Pin Connector, orange, behind instrument panel, left
T14 - 14-Pin Connector
T121 - 121-Pin Connector, at J220-
D95 - wire connection (injectors), in engine compartment wiring harness
E30 - Connector (87a), in wiring harness engine

ws = wt (white)
sw = bk (black)
ro = rd (red)
br = br (brown)
gn = lg (light green), dg (dark green)
bl = bl (blue), lb (light blue), db (dark blue)
gr = gy (gray)
li = lilac
ge = yl (yellow)
or = or (orange), tn (tan)
rs = vt (violet)

New Beetle — Wiring diagram — No. 35/9

Motronic engine control module (ECM), throttle position (TP) sensor, sender 2 for acelerator pedal position, evaporative emission (EVAP), leak detection pump (LDP)

G79 - Throttle Position (TP) Sensor
G185 - Sender -2- for accelerator pedal position
J220 - Motronic Engine Control Module (ECM)
N80 - Evaporative Emission (EVAP) Canister Purge Regulator Valve
V144 Leak Detection Pump (LDP)
T3d - 3-Pin Connector
T6b - 6-Pin Connector
T10c - 10-Pin Connector, blue, behind instrument panel, left
T121 - 121-Pin Connector, at -J220-

(E30) - Connector (87a), in wiring harness engine
(A101) Connector -3- (87), in instrument panel wiring harness
(E39) - Connection-2-(87a), in wiring harness engine

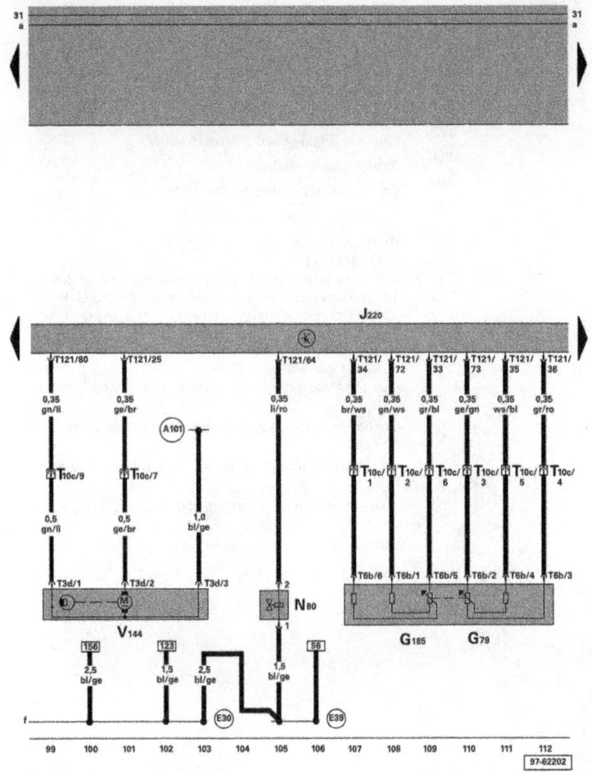

ws = white
sw = black
ro = red
br = brown
gn = green
bl = blue
gr = grey
li = violet
ge = yellow
or = orange

New Beetle — Wiring diagram — No. 35/10

Motronic engine control module (ECM), ABS control module (w/EDL), coolant FC (Fan Control) control module, secondary air injection (AIR) solenoid valve

J104 - ABS Control Module (w/EDL)
J217 - Transmission Control Module (TCM)
J220 - Motronic Engine Control Module (ECM)
J234 - Airbag Control Module
N112 - Secondary Air Injection (AIR) Solenoid Valve
S18 - Fuse 18 in fuse/relay panel
T10 - 10-Pin Connector, white, behind instrument panel, left
T25 - 25-Pin Connector, at J104
T121 - 121-Pin Connector, at -J220-

(A121) - Connection (high bus) in instrument panel wiring harness
(A122) - Connection (low bus) in instrument panel wiring harness
(A51) - wire connection (56), in instrument panel wiring harness
--- - Automatic transmission only

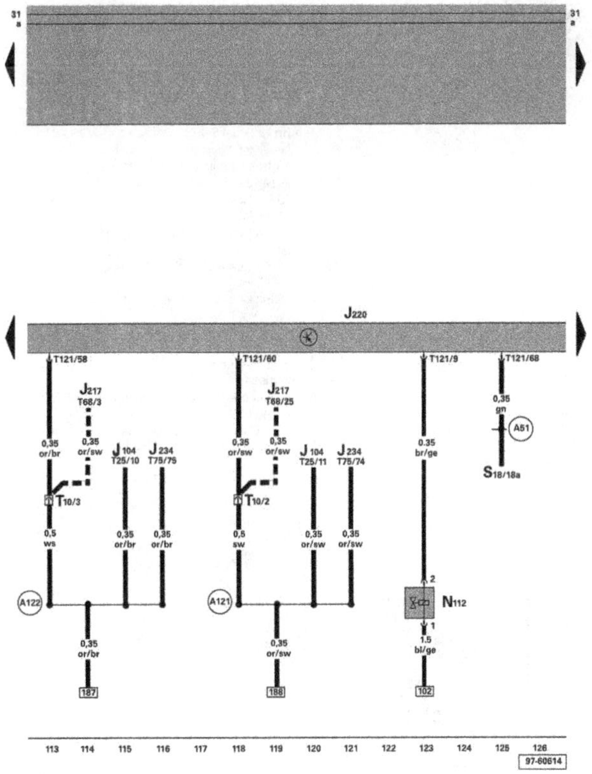

ws = white
sw = black
ro = red
br = brown
gn = green
bl = blue
gr = grey
li = violet
ge = yellow
or = orange

New Beetle

Wiring Diagram

No. 35 / 11

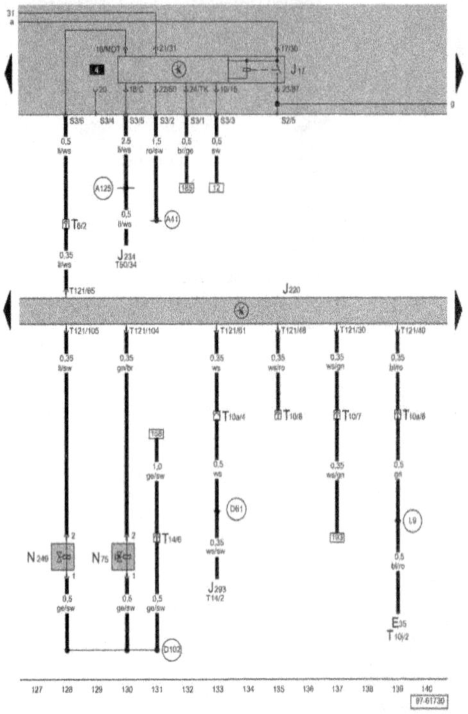

Motronic ECM, fuel pump (FP) relay, airbag control module, coolant FC (Fan control), wastegate bypass regulator valve, recirculating valve for turbocharger

D -Ignition/Starter Switch
E35 -A/C Switch
J17 -Fuel Pump (FP) Relay
J220 -Motronic Engine Control Module (ECM)
J234 -Airbag Control Module
J293 -Coolant FC (Fan Control) Control Module
N75 -Wastegate Bypass Regulator Valve
N249 -Recirculating valve for turbocharger
T3 -3-Pin Connector
T6 -6-Pin Connector, brown, behind instrument panel, left
T10 -10-Pin Connector, white, behind instrument panel, left
T10a -10-Pin Connector, orange, behind instrument panel, left
T10j -10-Pin Connector
T14 -14-Pin Connector
T121 -121-Pin Connector, at -J220-
A41 -plus connection (50), in instrument panel wiring harness
A125 -Connection (crash signal) in instrument panel wiring harness
D61 -wire connection (A/C), in engine compartment wiring harness
D102 -Wire connection -2-, in engine compartment wiring harness
L9 -Wire connection -1-, in A/C wiring harness

ws = wt (white)
sw = bk (black)
ro = rd (red)
br = br (brown)
lg (light green),
gn = dg (dark green)
bl (blue), lb (light
bl = blue), db (dark blue)
gr = gy (gray)
li = lilac
ge = yl (yellow)
or = (orange), tn (tan)
rs = vt (violet)

New Beetle

Wiring diagram

No. 35/12

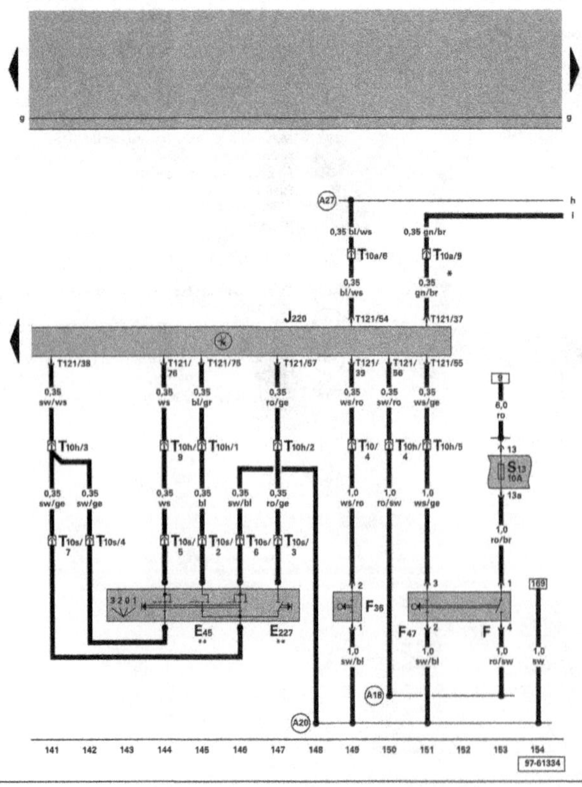

Motronic engine control module (ECM), cruise control switch, button for cruise control (set), brake light switch, clutch vacuum vent valve switch

E45 - Cruise Control Switch
E227 - Button for cruise control (set)
F - Brake Light Switch
F36 - Clutch Vacuum Vent Valve Switch
F47 - Brake Pedal Switch (Cruise control/Diesel Direct Fuel Injection)
J220 - Motronic Engine Control Module (ECM)
S13 - Fuse 13 in fuse/relay panel
T10a - 10-Pin Connector, orange, behind instrument panel, left
T10 - 10-Pin Connector, white, behind instrument panel, left
T10h - 10-Pin Connector, black, behind instrument panel, left
T10s - 10-Pin Connector, near steering column
T121 - 121-Pin Connector, at -J220-
A18 - wire connection (54), in instrument panel wiring harness
A20 - wire connection (15a), in instrument panel wiring harness
A27 - Wire Connection (vehicle speed signal), in instrument panel wiring harness
* - Vehicles without CAN-Bus only

ws = white
sw = black
ro = red
br = brown
gn = green
bl = blue
gr = grey
li = violet
ge = yellow
or = orange

1.8L engine management (APH) from June 1999

New Beetle — Wiring diagram — No. 35/13

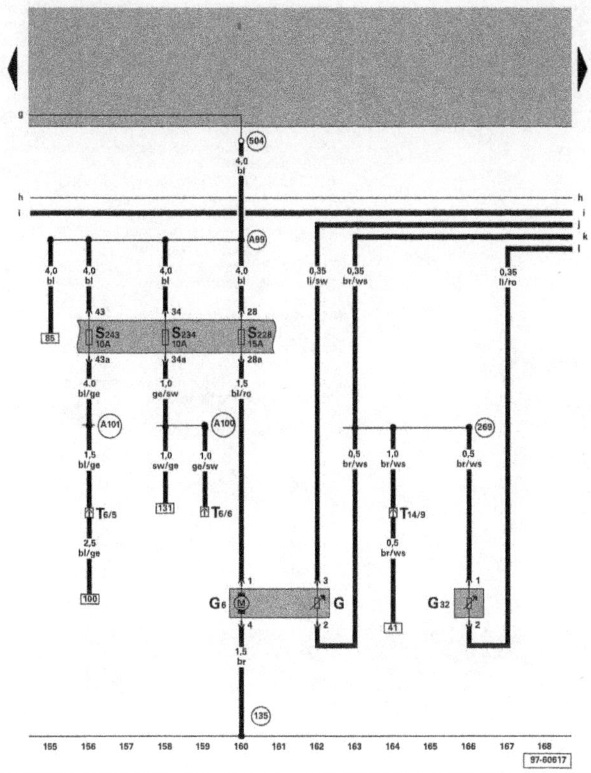

Sender for fuel gauge, fuel pump (FP), engine coolant level (ECL) sensor

G - Sender for fuel gauge
G6 - Fuel Pump (FP)
G32 - Engine Coolant Level (ECL) Sensor
S228 - Fuse 28 in fuse/relay panel
S234 - Fuse 34 in fuse/relay panel
S243 - Fuse 43 in fuse/relay panel
T6 - 6-Pin Connector, brown, behind instrument panel, left
T14 - 14-Pin Connector

(135) - Ground connection -2-, in instrument panel wiring harness
(269) - Ground connector (sensor ground) -1-, in instrument panel wiring harness
(504) - Threaded connection -1- (87) on the relay plate
(A99) - Connector -1- (87), in instrument panel wiring harness
(A100) - Connector -2- (87), in instrument panel wiring harness
(A101) - Connector -3- (87), in instrument panel wiring harness

ws = white
sw = black
ro = red
br = brown
gn = green
bl = blue
gr = grey
li = violet
ge = yellow
or = orange

New Beetle — Wiring diagram — No. 35/14

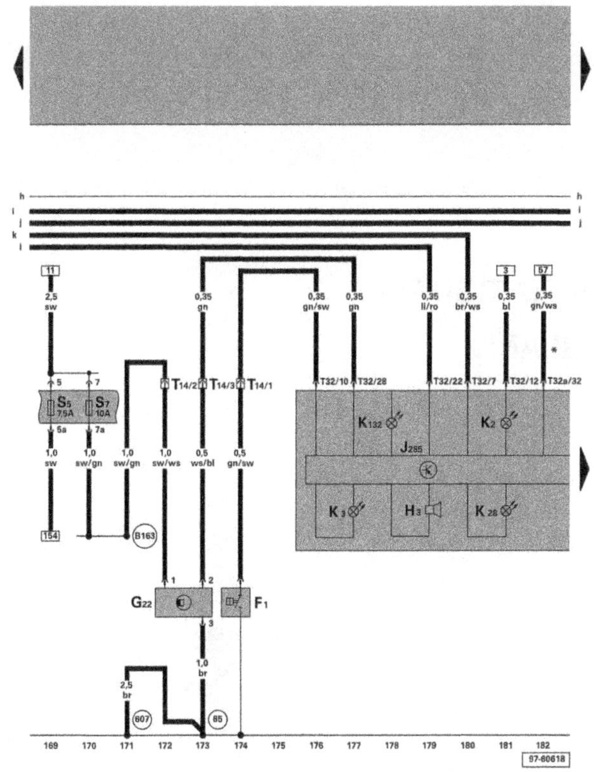

Oil pressure switch, speedometer vehicle speed sensor (VSS), instrument cluster, warning lights/ buzzer

F1 - Oil Pressure Switch
G22 - Speedometer Vehicle Speed Sensor (VSS)
H3 - Warning Buzzer
J285 - Control module with indicator unit in instrument panel insert
K2 - Generator (GEN) Warning Light
K3 - Oil Pressure Warning Light
K28 - Engine Coolant Level/Temperature (ECL/ECT) Warning Light
K132 - Electronic Power Control (EPC) Lamp
S5 - Fuse 5 in fuse/relay panel
S7 - Fuse 7 in fuse/relay panel
T14 - 14-Pin Connector
T32 - 32-Pin Connector, blue
T32a - 32-Pin Connector, green

(607) - Ground Connection (in left plenum chamber)
(85) - Ground connection -1-, in engine compartment wiring harness
(B163) - Plus connector -1- (15) in wiring harness interior

* - Fuel consumtion indicator signal from Engine Control Module, only for MFA.

ws = white
sw = black
ro = red
br = brown
gn = green
bl = blue
gr = grey
li = violet
ge = yellow
or = orange

1.8L engine management (APH) from June 1999

New Beetle Wiring diagram **No. 35/15**

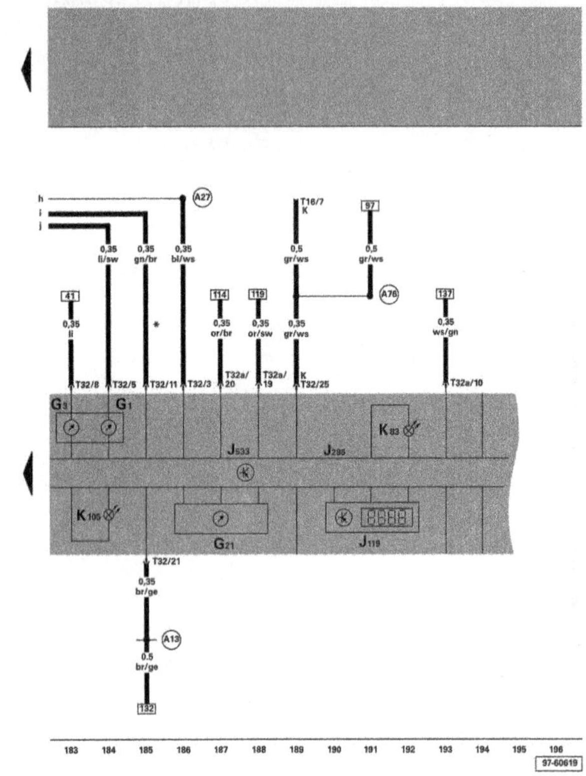

Instrument cluster, fuel gauge, engine coolant temperature (ECT) gauge, speedomter, multi-function indicator (MFA)

- G1 - Fuel Gauge
- G3 - Engine Coolant Temperature (ECT) Gauge
- G21 - Speedometer
- J119 - Multi-function Indicator (MFA)
- J285 - Control module with indicator unit in instrument panel Insert
- J533 - Data Bus On Board Diagnostic Interface
- K83 - Malfunction Indicator lamp (MIL)
- K105 - Low Fuel Level Warning Light
- T16 - 16-Pin Connector, Data link Connector (DLC), below steering column, left
- T32 - 32-Pin Connector, blue
- T32a - 32-Pin Connector, green
- A13 - wire connection (door contact switch) in instrument panel wiring harness
- A27 - Wire Connection (vehicle speed signal), in instrument panel wiring harness
- A76 - Connector (K-diagnosis wire) in instrument panel wiring harness
- * - Vehicles without CAN-Bus only

ws = white
sw = black
ro = red
br = brown
gn = green
bl = blue
gr = grey
li = violet
ge = yellow
or = orange

THIS SPACE INTENTIONALLY

LEFT BLANK

New Beetle | **Wiring diagram** | **No. 5/1**

1,8L-Engine - 5V Turbo, Motronic Multiport Fuel Injection (MFI)

- **110 kW, code AWV (with cruise control)**

from June 2000

- **132 kW, code AWP (with cruise control)**

from October 2001

Deviate relay location and fuseplacements as well as the locations of multiple connectors see section "component locations".

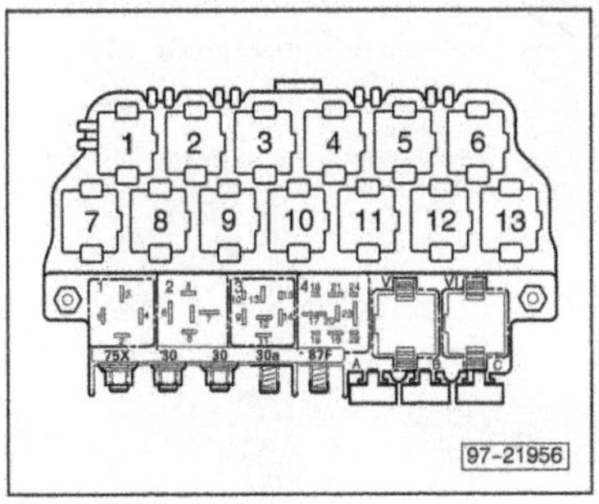

Relay location on the thirteen position auxiliary relay panel, above relay panel:

Relay panel:

- **2** Load Reduction Relay (213)
- **4** Fuel Pump (FP) Relay (409)

Fuse colors

30 A - green
25 A - white
20 A - yellow
15 A - blue
10 A - red
7,5 A - brown
5 A - beige
3 A - violet

Note: Number in parentheses indicates production control number stamped on relay housing.

New Beetle | **Wiring diagram** | **No. 5/2**

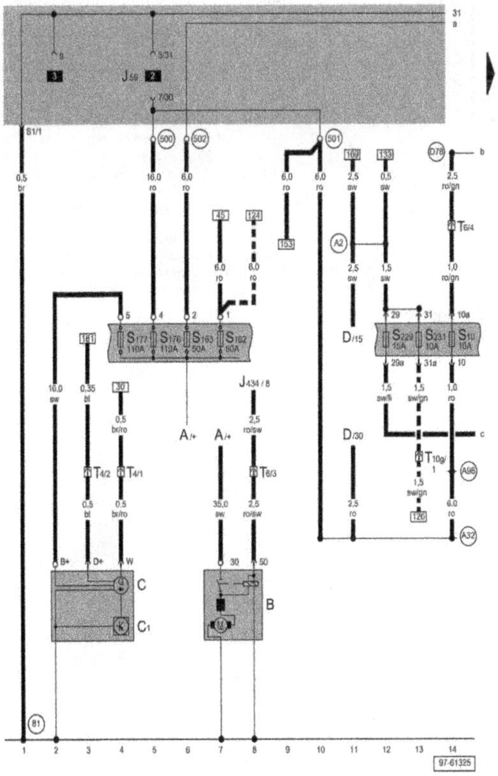

Battery, starter, generator (GEN), voltage regulator (VR), ignition/starter switch, load reduction relay, fuses in fuse bracket/battery

A - Battery
B - Starter
C - Generator (GEN)
C1 - Voltage Regulator (VR)
D - Ignition/Starter Switch
J434 - Locking relay for starter (clutch pedal switch)
J59 - Load Reduction Relay
S10 - Fuse 10 in fuse/relay panel
S162 - Fuse -1- (50), in fuse bracket/battery
S163 - Fuse -2- (50), in fuse bracket/battery
S176 - Fuse -4- (30), in fuse bracket/battery
S177 - Fuse -5- (30), in fuse bracket/battery
S229 - Fuse 29 in fuse/relay panel
S231 - Fuse 31 in fuse holder
T4 - 4-Pin Connector, in engine compartment left
T6 - 6-Pin Connector, brown, behind inst. panel, left
T10g - 10-Pin Connector, grey, in protective housing for connectors, in plenum chamber, left
81 - Ground connection -1-, in instrument panel wiring harness
500 - Threaded connection -1- (30) on the relay plate
501 - Threaded connection -2- (30) on the relay plate
502 - Threaded connection -1- (30a) on the relay plate
A2 - plus connection (15), in instrument panel wiring harness
A32 - plus connection (30), in instrument panel wiring harness
A98 - Plus connector -4- (30), in instrument panel wiring harness
D78 - Plus connector -1- (30a), in engine compartment wiring harness
--- -Automatic transmission only

ws = white
sw = black
ro = red
br = brown
gn = green
bl = blue
gr = grey
li = violet
ge = yellow
or = orange

1.8L engine management (AWV from June 2000) (AWP from Oct. 2001)

New Beetle — Wiring diagram — No. 5/3

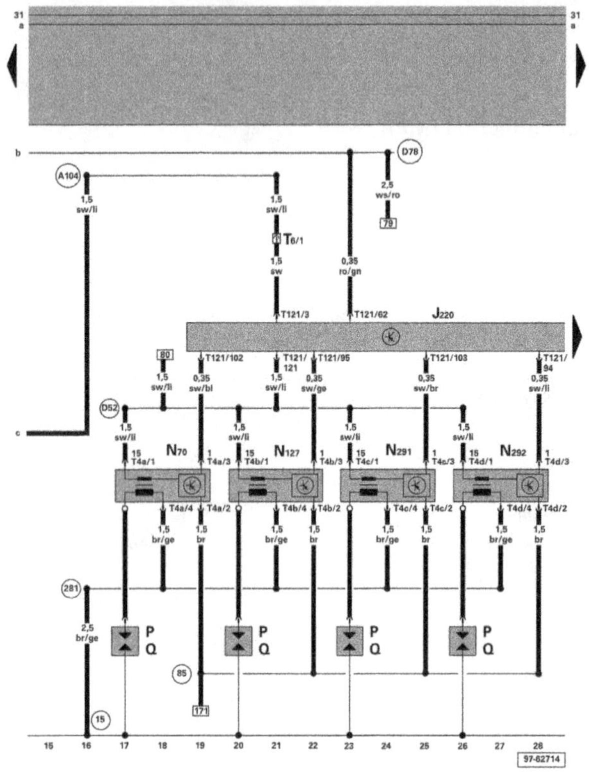

Motronic engine control module (ECM), ignition coil with power output stages 1, 2, 3 & 4, spark plugs

- J220 - Motronic Engine Control Module (ECM)
- N70 - Ignition Coil With Power Output Stage 1
- N127 - Ignition Coil with Power Output Stage 2
- N291 - Ignition Coil with Power Output Stage 3
- N292 - Ignition Coil with Power Output Stage 4
- P - Spark Plug Connectors
- Q - Spark Plugs
- T4a - 4-Pin Connector
- T4b - 4-Pin Connector
- T4c - 4-Pin Connector
- T4d - 4-Pin Connector
- T6 - 6-Pin Connector, brown, behind instrument panel, left
- T121 - 121-Pin Connector, at -J220-
- (15) - Ground connection, on cylinder head
- (85) - Ground connection -1-, in engine compartment wiring harness
- (281) - Ground connector -1-, in wiring harness engine pre-wiring
- (A104) - Plus connector -2- (15), in instrument panel wiring harness
- (D52) - plus connection (15a), in engine compartment wiring harness
- (D78) - Plus connector -1- (30a), in engine compartment wiring harness

ws = white
sw = black
ro = red
br = brown
gn = green
bl = blue
gr = grey
li = violet
ge = yellow
or = orange

97-62714

Edition 06.03

New Beetle — Wiring diagram — No. 5/4

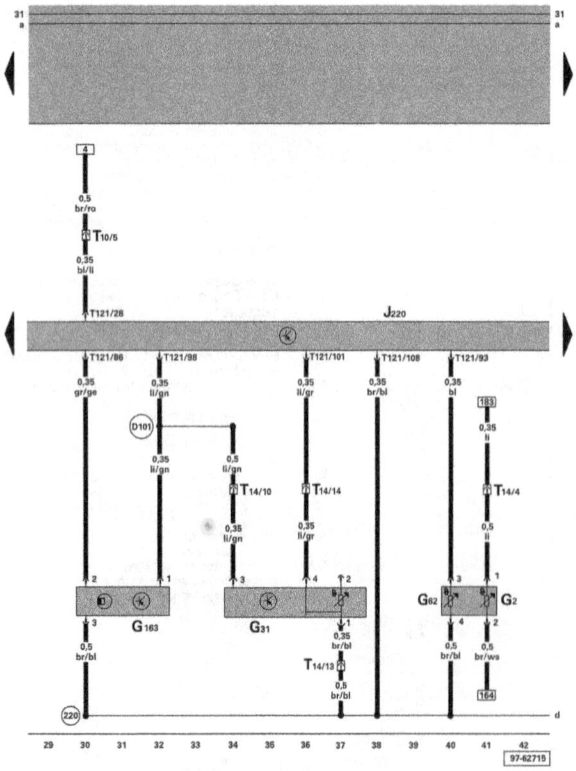

Motronic engine control module (ECM), engine coolant temperature (ECT) sensor, charge air pressure sensor, camshaft position (CMP) sensor 2

- G2 - Engine Coolant Temperature (ECT) Sensor
- G31 - Charge Air Pressure Sensor
- G62 - Engine Coolant Temperature (ECT) Sensor
- G163 - Camshaft Position (CMP) Sensor 2
- J220 - Motronic Engine Control Module (ECM)
- T14 - 14-Pin Connector, in engine compartment, left
- T10 - 10-Pin Connector, white, behind instrument panel, left
- T121 - 121-Pin Connector, at -J220-
- (220) - Ground connection (sensor ground), in engine compartment wiring harness
- (D101) - Wire connection -1-, in engine compartment wiring harness

ws = white
sw = black
ro = red
br = brown
gn = green
bl = blue
gr = grey
li = violet
ge = yellow
or = orange

97-62715

Edition 06.03

New Beetle — Wiring diagram — No. 5/5

Motronic engine control module (ECM), heated oxygen sensor (HO2S), secondary air injection (AIR) pump relay/motor

- G39 - Heated Oxygen Sensor (HO2S), black
- G108 - Heated Oxygen Sensor (HO2S), black
- J220 - Motronic Engine Control Module (ECM)
- J299 - Secondary Air Injection (AIR) Pump Relay
- T4 - 4-Pin Connector, brown
- T6 - 6-Pin Connector, black
- T121 - 121-Pin Connector, at -J220-
- V101 - Secondary Air Injection (AIR) Pump Motor
- (609) - Ground Connection (in right plenum chamber)
- (E39) - Connection-2-(87a), in wiring harness engine

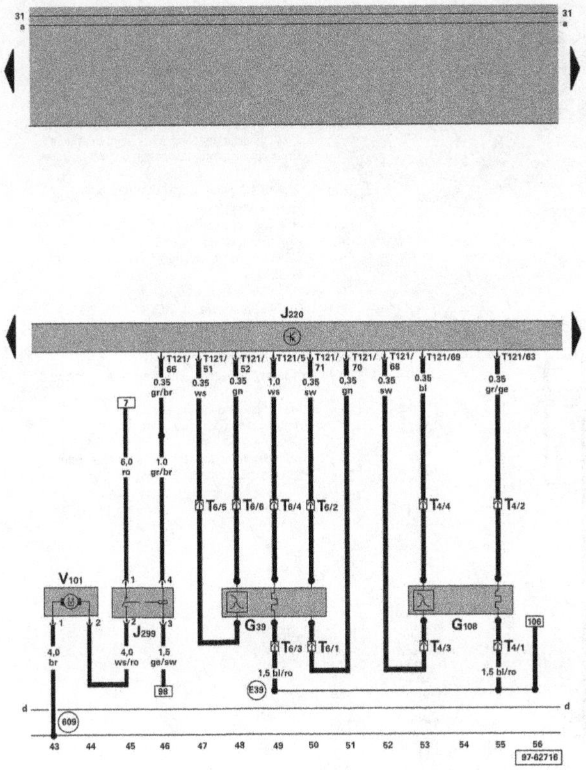

ws = white
sw = black
ro = red
br = brown
gn = green
bl = blue
gr = grey
li = violet
ge = yellow
or = orange

New Beetle — Wiring diagram — No. 5/6

Motronic engine control module (ECM), intake air temperature (IAT) sensor, knock sensor (KS) 1, throttle valve control module

- G42 - Intake Air Temperature (IAT) Sensor
- G61 - Knock Sensor (KS) 1
- G186 - Throttle drive (power accelerator actuation)
- G187 - Angle sensor -1- for throttle drive (power accelerator actuation)
- G188 - Angle sensor -2- for throttle drive (power accelerator actuation)
- J220 - Motronic Engine Control Module (ECM)
- J338 - Throttle Valve Control Module
- T2a - 2-Pin Connector
- T6a - 6-Pin Connector
- T121 - 121-Pin Connector, at -J220-
- (139) - Ground connection (sensor Ground), in Motronic Multiport Fuel Injection (MFI) wiring harness

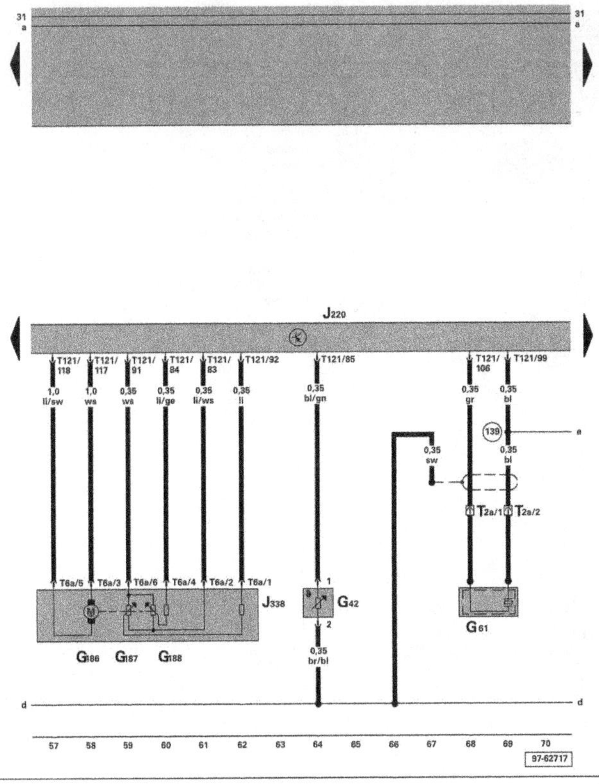

ws = white
sw = black
ro = red
br = brown
gn = green
bl = blue
gr = grey
li = violet
ge = yellow
or = orange

1.8L engine management (AWV from June 2000) (AWP from Oct. 2001)

New Beetle

Wiring diagram **No. 5/7**

Motronic engine control module (ECM) / power supply relay, engine speed (RPM) sensor, knock sensor (KS) 2

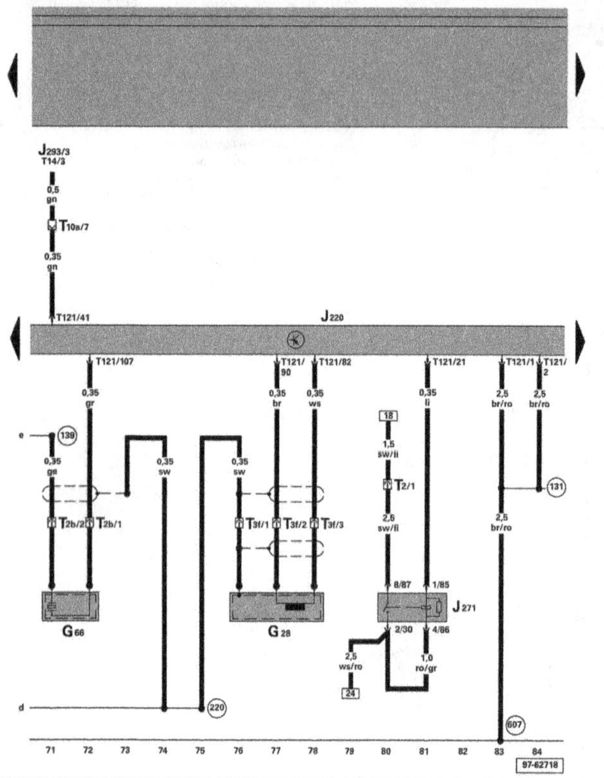

G28 - Engine Speed (RPM) Sensor
G66 - Knock Sensor (KS) 2
J220 - Motronic Engine Control Module (ECM)
J271 - Motronic Engine Control Module (ECM) power supply relay, in protective housing, in engine compartment, left, production control number (428)
J293 - Coolant FC (Fan Control) Control Module
T2 - 2-Pin Connector
T2b - 2-Pin Connector
T3f - 3-Pin Connector
T10a - 10-Pin Connector, orange, behind instrument panel, left
T14 - 14-Pin Connector, at -J293-
T121 - 121-Pin Connector, at -J220-

(131) - Ground connection -2-, in engine compartment wiring harness
(607) - Ground Connection (in left plenum chamber)

(139) - Ground connection (sensor Ground), in Motronic Multiport Fuel Injection (MFI) wiring harness
(220) - Ground connection (sensor ground), in engine compartment wiring harness

ws = white
sw = black
ro = red
br = brown
gn = green
bl = blue
gr = grey
li = violet
ge = yellow
or = orange

New Beetle

Wiring diagram **No. 5/8**

Motronic engine control module (ECM), mass air flow (MAF) sensor, fuel injectors

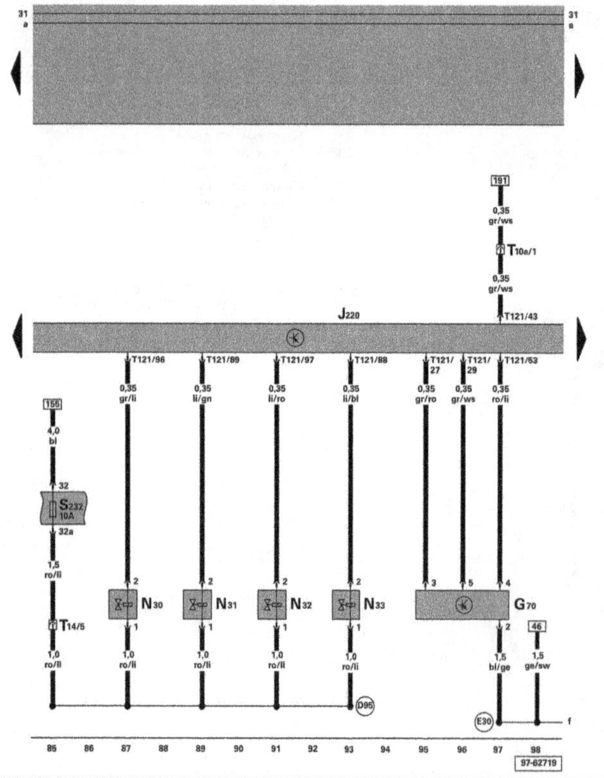

G70 - Mass Air Flow (MAF) Sensor
J220 - Motronic Engine Control Module (ECM)
N30 - Cylinder 1 Fuel Injector
N31 - Cylinder 2 Fuel Injector
N32 - Cylinder 3 Fuel Injector
N33 - Cylinder 4 Fuel Injector
S232 - Fuse 32 in fuse/relay panel
T10a - 10-Pin Connector, orange, behind instrument panel, left
T14 - 14-Pin Connector
T121 - 121-Pin Connector, at J220-

(D95) - wire connection (injectors), in engine compartment wiring harness
(E30) - Connector (87a), in wiring harness engine

ws = white
sw = black
ro = red
br = brown
gn = green
bl = blue
gr = grey
li = violet
ge = yellow
or = orange

1.8L engine management (AWV from June 2000) (AWP from Oct. 2001)

New Beetle **Wiring diagram** **No. 5/9**

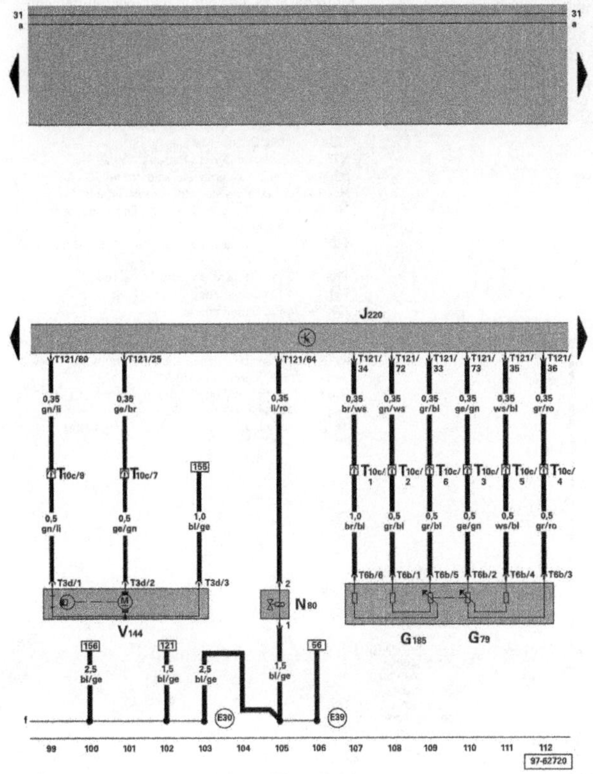

Motronic engine control module (ECM), throttle position (TP) sensor, sender 2 for acelerator pedal position, evaporative emission (EVAP), leak detection pump (LDP)

- G79 - Throttle Position (TP) Sensor
- G185 - Sender -2- for accelerator pedal position
- J220 - Motronic Engine Control Module (ECM)
- N80 - Evaporative Emission (EVAP) Canister Purge Regulator Valve
- V144 - Leak Detection Pump (LDP)
- T3d - 3-Pin Connector
- T6b - 6-Pin Connector
- T10c - 10-Pin Connector, blue, behind instrument panel, left
- T121 - 121-Pin Connector, at -J220-
- (E30) - Connector (87a), in wiring harness engine
- (E39) - Connection-2-(87a), in wiring harness engine

ws = white
sw = black
ro = red
br = brown
gn = green
bl = blue
gr = grey
li = violet
ge = yellow
or = orange

New Beetle **Wiring diagram** **No. 5/10**

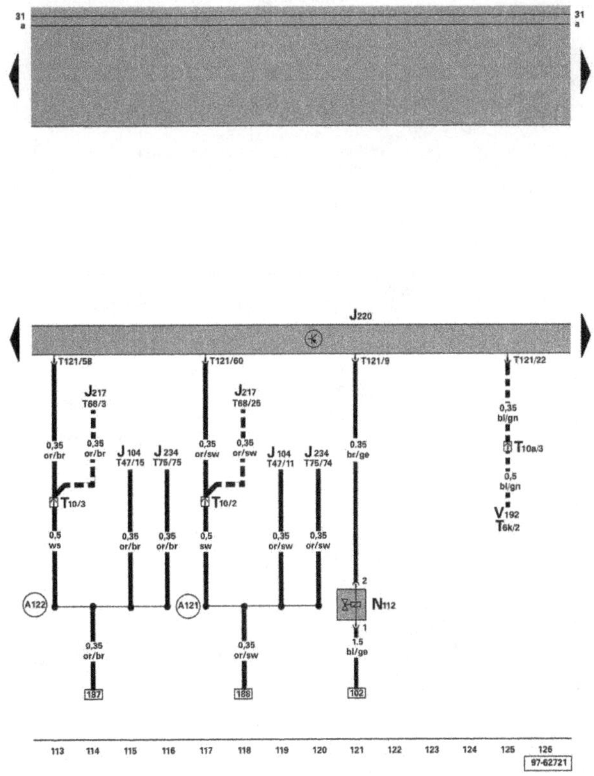

Motronic engine control module, ABS control module, transmission control module, airbag control module, secondary air injection (AIR) solenoid valve

- J104 - ABS Control Module (w/EDL)
- J217 - Transmission Control Module (TCM)
- J220 - Motronic Engine Control Module (ECM)
- J234 - Airbag Control Module
- N112 - Secondary Air Injection (AIR) Solenoid Valve
- T6K - 6-Pin Connector
- T10 - 10-Pin Connector, white, behind instrument panel, left
- T10a - 10-Pin Connector, orange, behind instrument panel, left
- T47 - 47-Pin Connector, at J104
- T68 - 68-Pin Connector, at J217
- T75 - 75-Pin Connector, at J234
- T121 - 121-Pin Connector, at J220
- V192 - Brake System Vacuum Pump
- (A121) - Connection (high bus) in instrument panel wiring harness
- (A122) - Connection (low bus) in instrument panel wiring harness
- - - - - Automatic transmission only

ws = white
sw = black
ro = red
br = brown
gn = green
bl = blue
gr = grey
li = violet
ge = yellow
or = orange

1.8L engine management (AWV from June 2000) (AWP from Oct. 2001)

New Beetle — **Wiring diagram** — **No. 5/11**

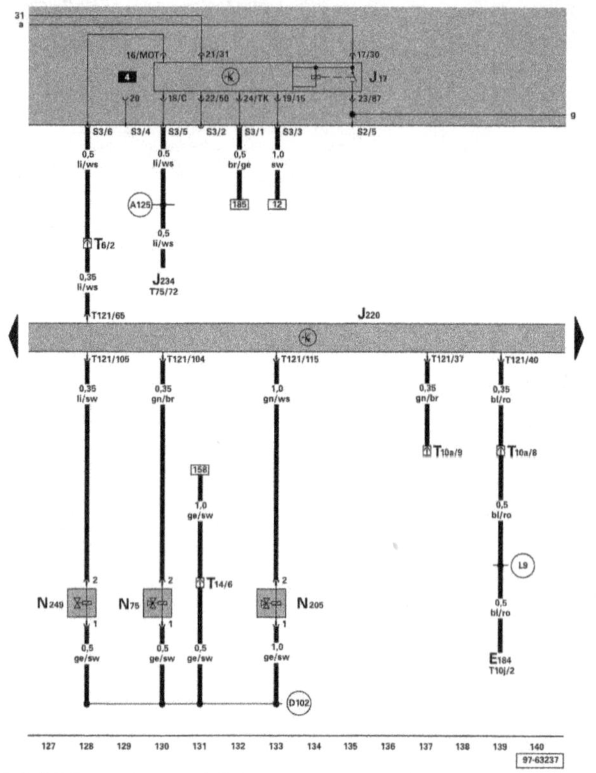

Motronic ECM, fuel pump (FP) relay, airbag control module, valve 1 for camfshaft adjustment, wastegate bypass regulator valve, recirculating valve for turbocharger

- E184 - Switch for fresh air blower and recirculated air
- J17 - Fuel Pump (FP) Relay
- J220 - Motronic Engine Control Module (ECM)
- J234 - Airbag Control Module
- N75 - Wastegate Bypass Regulator Valve
- N205 - Valve-1- for camshaft adjustment
- N249 - Recirculating valve for turbocharger
- T6 - 6-Pin Connector, brown, behind instrument panel, left
- T10a - 10-Pin Connector,orange, behind instrument panel, left
- T10j - 10-Pin Coupling conector, on -E184-
- T14 - 14-Pin Connector
- T75 - 75-Pin Connector, at -J234-
- T121 - 121-Pin Connector, at -J220-
- (A125) - Connection (crash signal) in instrument panel wiring harness
- (D102) - Wire connection -2-, in engine compartment wiring harness
- (L9) - Wire connection -1-, in A/C wiring harness

ws = white
sw = black
ro = red
br = brown
gn = green
bl = blue
gr = grey
li = violet
ge = yellow
or = orange

New Beetle — **Wiring diagram** — **No. 5/12**

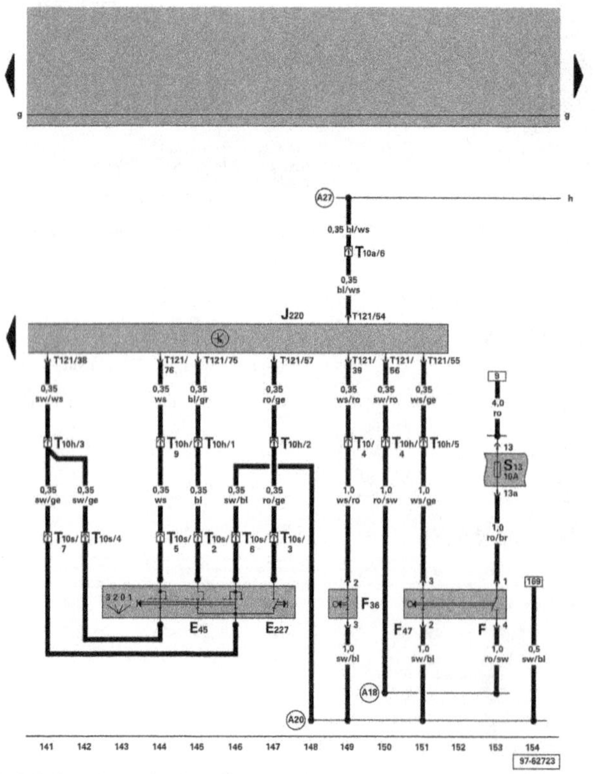

Motronic engine control module (ECM), cruise control switch, button for cruise control (set), brake light switch, clutch vacuum vent valve switch

- E45 - Cruise Control Switch
- E227 - Button for cruise control (set)
- F - Brake Light Switch
- F36 - Clutch Vacuum Vent Valve Switch
- F47 - Brake Vacuum Vent Valve Switch for cruise control
- J220 - Motronic Engine Control Module (ECM)
- S13 - Fuse 13 in fuse/relay panel
- T10 - 10-Pin Connector, white, behind instrument panel, left
- T10a - 10-Pin Connector, orange, behind instrument panel, left
- T10h - 10-Pin Connector, black, behind instrument panel, left
- T10s - 10-Pin Connector, near steering column
- T121 - 121-Pin Connector, at -J220-
- (A18) - wire connection (54), in instrument panel wiring harness
- (A20) - wire connection (15a), in instrument panel wiring harness
- (A27) - Wire Connection (vehicle speed signal), in instrument panel wiring harness

ws = white
sw = black
ro = red
br = brown
gn = green
bl = blue
gr = grey
li = violet
ge = yellow
or = orange

1.8L engine management (AWV from June 2000) (AWP from Oct. 2001)

New Beetle — **Wiring diagram** — **No. 5/13**

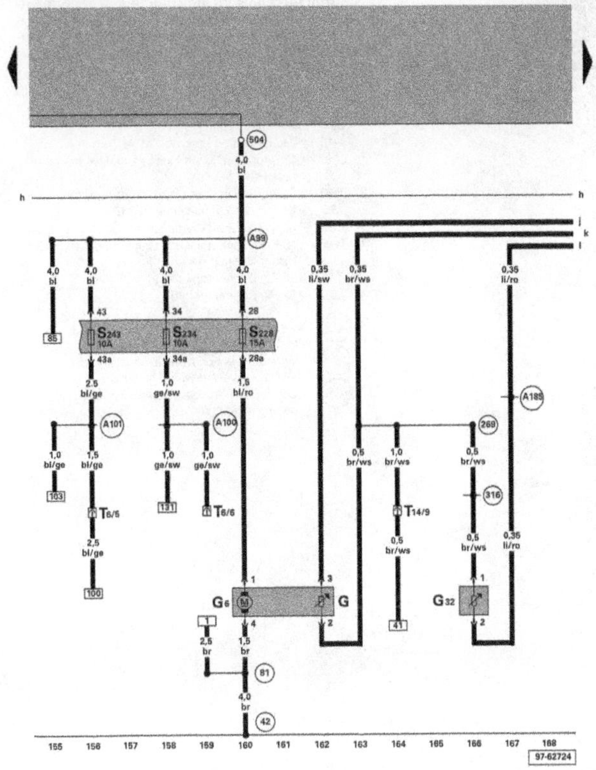

Sender for fuel gauge, fuel pump (FP), engine coolant level (ECL) sensor

- G - Sender for fuel gauge
- G6 - Fuel Pump (FP)
- G32 - Engine Coolant Level (ECL) Sensor
- S228 - Fuse 28 in fuse/relay panel
- S234 - Fuse 34 in fuse/relay panel
- S243 - Fuse 43 in fuse/relay panel
- T6 - 6-Pin Connector, brown, behind instrument panel, left
- T14 - 14-Pin Connector
- (81) - Ground connection -1-, in instrument panel wiring harness
- (42) - Ground connection, beside steering column
- (269) - Ground connector (sensor ground) -1-, in instrument panel wiring harness
- (316) - Sensor Ground Connection 2 (in engine wiring harness)
- (504) - Threaded connection -1- (87) on the relay plate
- (A99) - Connector -1- (87), in instrument panel wiring harness
- (A100) - Connector -2- (87), in instrument panel wiring harness
- (A101) - Connector -3- (87), in instrument panel wiring harness
- (A185) - Coolant Warning Connection (in instrument panel wiring harness)

ws = white
sw = black
ro = red
br = brown
gn = green
bl = blue
gr = grey
li = violet
ge = yellow
or = orange

New Beetle — **Wiring diagram** — **No. 5/14**

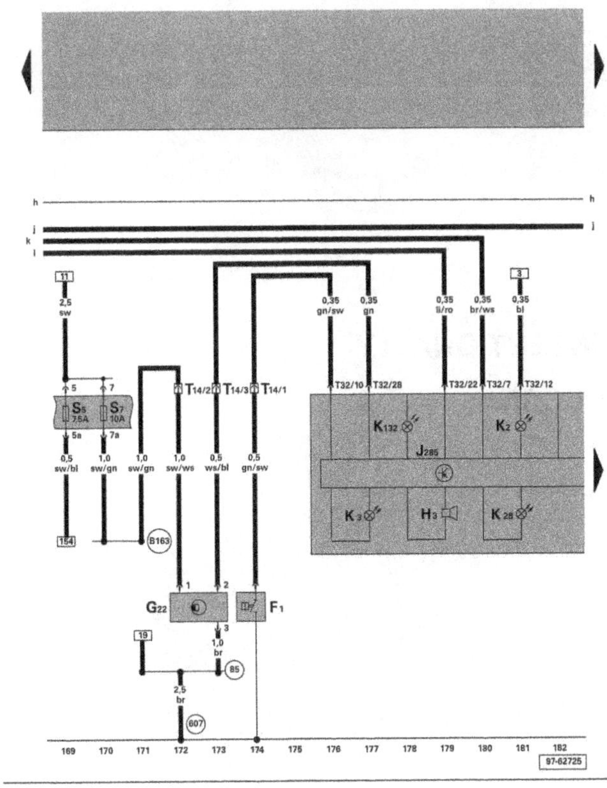

Oil pressure switch / warning light, speedometer vehicle speed sensor (VSS), warning buzzer, control module with indicator unit in instrument panel insert, generator (GEN) warning light

- F1 - Oil Pressure Switch
- G22 - Speedometer Vehicle Speed Sensor (VSS)
- H3 - Warning Buzzer
- J285 - Control module with indicator unit in instrument panel insert
- K2 - Generator (GEN) Warning Light
- K3 - Oil Pressure Warning Light
- K28 - Engine Coolant Level/Temperature (ECL/ECT) Warning Light
- K132 - Electronic Power Control (EPC) Lamp
- S5 - Fuse 5 in fuse/relay panel
- S7 - Fuse 7 in fuse/relay panel
- T14 - 14-Pin Connector
- T32 - 32-Pin Connector, blue
- (607) - Ground Connection (in left plenum chamber)
- (85) - Ground connection -1-, in engine compartment wiring harness
- (B163) - Plus connector -1- (15) in wiring harness interior

ws = white
sw = black
ro = red
br = brown
gn = green
bl = blue
gr = grey
li = violet
ge = yellow
or = orange

1.8L engine management (AWV from June 2000) (AWP from Oct. 2001)

New Beetle Wiring diagram **No. 5/15**

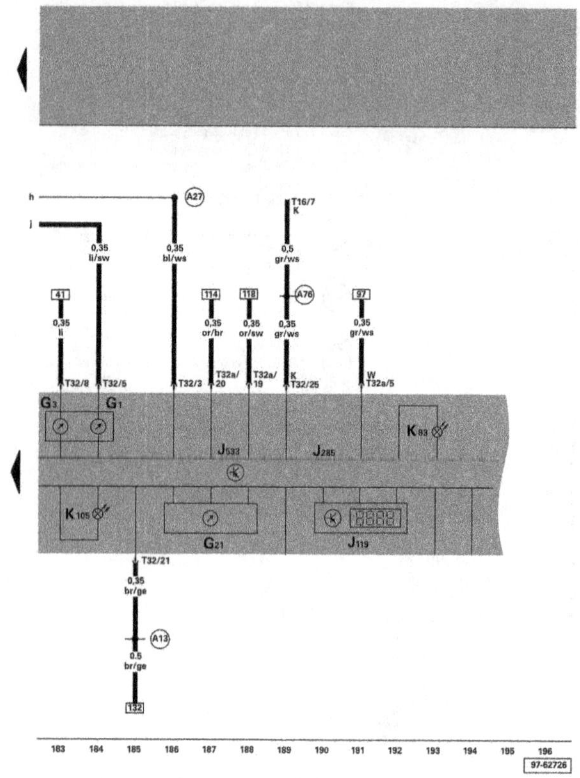

Control module with indicator unit in instrument panel insert, fuel gauge, engine coolant temperature (ECT) gauge, speedomter, multifunction indicator (MFA), low fuel level warning light

- G1 - Fuel Gauge
- G3 - Engine Coolant Temperature (ECT) Gauge
- G21 - Speedometer
- J119 - Multi-function Indicator (MFA)
- J285 - Control module with indicator unit in instrument panel insert
- J533 - Data Bus On Board Diagnostic Interface
- K83 - Malfunction Indicator lamp (MIL)
- K105 - Low Fuel Level Warning Light
- T16 - 16-Pin Connector, Data Link Connector (DLC), below steering column, left
- T32 - 32-Pin Connector, blue
- T32a - 32-Pin Connector, green
- (A13) - wire connection (door contact switch) in instrument panel wiring harness
- (A27) - Wire Connection (vehicle speed signal), in instrument panel wiring harness
- (A76) - Connector (K-diagnosis wire) in instrument panel wiring harness

ws = white
sw = black
ro = red
br = brown
gn = green
bl = blue
gr = grey
li = violet
ge = yellow
or = orange

THIS SPACE INTENTIONALLY LEFT BLANK

New Beetle Convertible Wiring diagram **No. 19/1**

1,8L-Engine - 5V Turbo, Motronic Multiport Fuel Injection (MFI)/110 kW, code AWV (with cruise control)

from January 2003

Deviate relay location and fuseplacements as well as the locations of multiple connectors see section "component locations".

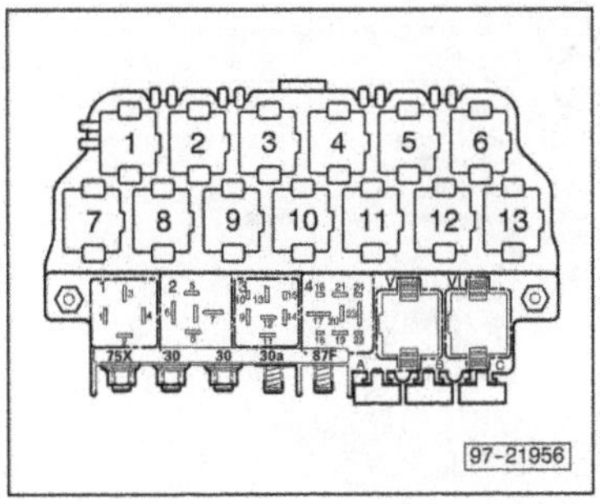

Relay location on the thirteen position auxiliary relay panel, above relay panel:

Relay panel:

- **2** Load Reduction Relay (213)
- **4** Fuel Pump (FP) Relay (409)

Fuse colors

30 A - green
25 A - white
20 A - yellow
15 A - blue
10 A - red
7,5 A - brown
5 A - beige
3 A - violet

Note: Number in parentheses indicates production control number stamped on relay housing.

New Beetle Convertible Wiring diagram **No. 19/2**

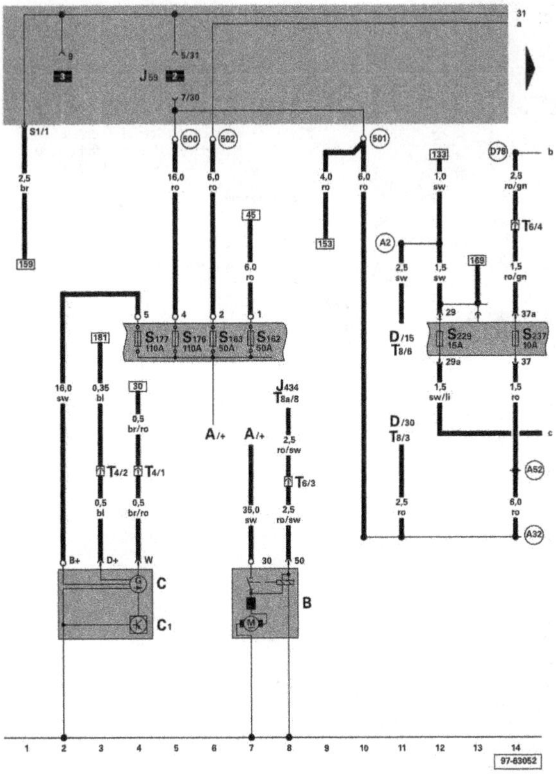

ws = white
sw = black
ro = red
br = brown
gn = green
bl = blue
gr = grey
li = violet
ge = yellow
or = orange

Battery, starter, generator (GEN), voltage regulator (VR), ignition/starter switch, load reduction relay, fuses in fuse bracket/battery

- A - Battery
- B - Starter
- C - Generator (GEN)
- C1 - Voltage Regulator (VR)
- D - Ignition/Starter Switch
- J434 - Locking relay for starter (clutch pedal switch)
- J59 - Load Reduction Relay
- S162 - Fuse -1- (50), in fuse bracket/battery
- S163 - Fuse -2- (50), in fuse bracket/battery
- S176 - Fuse -4- (30), in fuse bracket/battery
- S177 - Fuse -5- (30), in fuse bracket/battery
- S229 - Fuse 29 in fuse/relay panel
- S237 - Fuse 37 in fuse/relay panel
- T4 - 4-Pin Connector, in engine compartment left
- T6 - 6-Pin Connector, brown, behind inst. panel, left
- T8 - 8-Pin Connector, at -D-
- T8a - 8-Pin Connector, at -J434-
- (500) - Threaded connection -1- (30) on the relay plate
- (501) - Threaded connection -2- (30) on the relay plate
- (502) - Threaded connection -1- (30a) on the relay plate
- (A2) - plus connection (15), in instrument panel wiring harness
- (A32) - plus connection (30), in instrument panel wiring harness
- (A52) - Plus Connection 2 (30) (in instrument panel wiring harness)
- (D78) - Plus connector -1- (30a), in engine compartment wiring harness

1.8L engine management (AWV) from Jan. 2003, convertible

New Beetle Convertible — Wiring diagram — No. 19/3

Motronic engine control module (ECM), ignition coil with power output stages 1, 2, 3 & 4, spark plugs

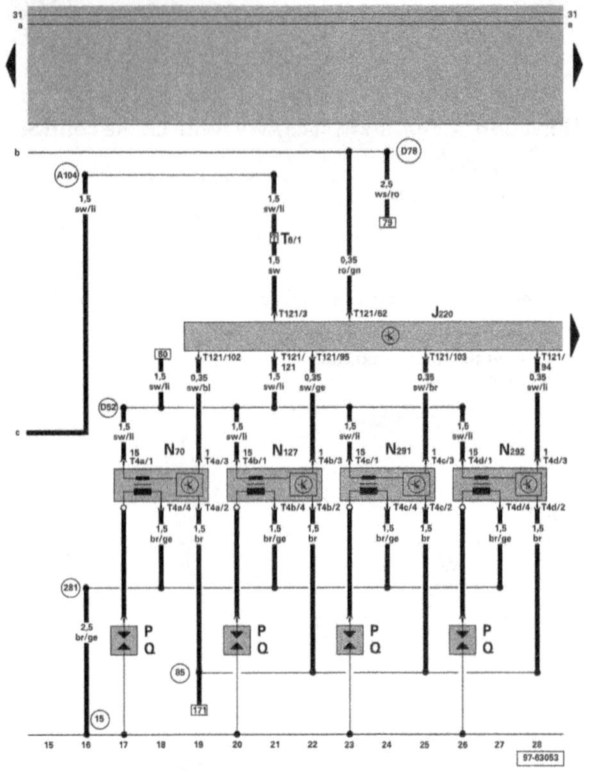

ws = white
sw = black
ro = red
br = brown
gn = green
bl = blue
gr = grey
li = violet
ge = yellow
or = orange

- J220 - Motronic Engine Control Module (ECM)
- N70 - Ignition Coil With Power Output Stage 1
- N127 - Ignition Coil with Power Output Stage 2
- N291 - Ignition Coil with Power Output Stage 3
- N292 - Ignition Coil with Power Output Stage 4
- P - Spark Plug Connectors
- Q - Spark Plugs
- T4a - 4-Pin Connector
- T4b - 4-Pin Connector
- T4c - 4-Pin Connector
- T4d - 4-Pin Connector
- T6 - 6-Pin Connector, brown, behind instrument panel, left
- T121 - 121-Pin Connector, at -J220-
- (15) - Ground connection, on cylinder head
- (85) - Ground connection -1-, in engine compartment wiring harness
- (281) - Ground connector -1-, in wiring harness engine pre-wiring
- (A104) - Plus connector -2- (15), in instrument panel wiring harness
- (D52) - plus connection (15a), in engine compartment wiring harness
- (D78) - Plus connector -1- (30a), in engine compartment wiring harness

New Beetle Convertible — Wiring diagram — No. 19/4

Motronic engine control module (ECM), engine coolant temperature (ECT) sensor, charge air pressure sensor, camshaft position (CMP) sensor 2

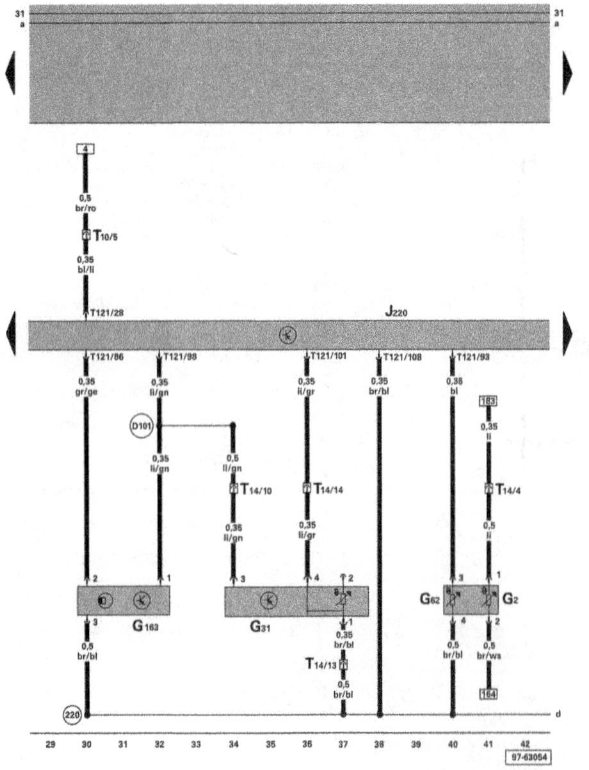

ws = white
sw = black
ro = red
br = brown
gn = green
bl = blue
gr = grey
li = violet
ge = yellow
or = orange

- G2 - Engine Coolant Temperature (ECT) Sensor
- G31 - Charge Air Pressure Sensor
- G62 - Engine Coolant Temperature (ECT) Sensor
- G163 - Camshaft Position (CMP) Sensor 2
- J220 - Motronic Engine Control Module (ECM)
- T14 - 14-Pin Connector, in engine compartment, left
- T10 - 10-Pin Connector, white, behind instrument panel, left
- T121 - 121-Pin Connector, at -J220-
- (220) - Ground connection (sensor ground), in engine compartment wiring harness
- (D101) - Wire connection -1-, in engine compartment wiring harness

New Beetle Convertible

Wiring diagram **No. 19/5**

Motronic engine control module (ECM), heated oxygen sensor (HO2S), secondary air injection (AIR) pump relay/motor

G39 - Heated Oxygen Sensor (HO2S), black
G108 - Heated Oxygen Sensor (HO2S), black
J220 - Motronic Engine Control Module (ECM)
J299 - Secondary Air Injection (AIR) Pump Relay
T4 - 4-Pin Connector, brown
T6 - 6-Pin Connector, black
T121 - 121-Pin Connector, at -J220-
V101 - Secondary Air Injection (AIR) Pump Motor

(609) - Ground Connection (in right plenum chamber)

(E39) - Connection-2-(87a), in wiring harness engine

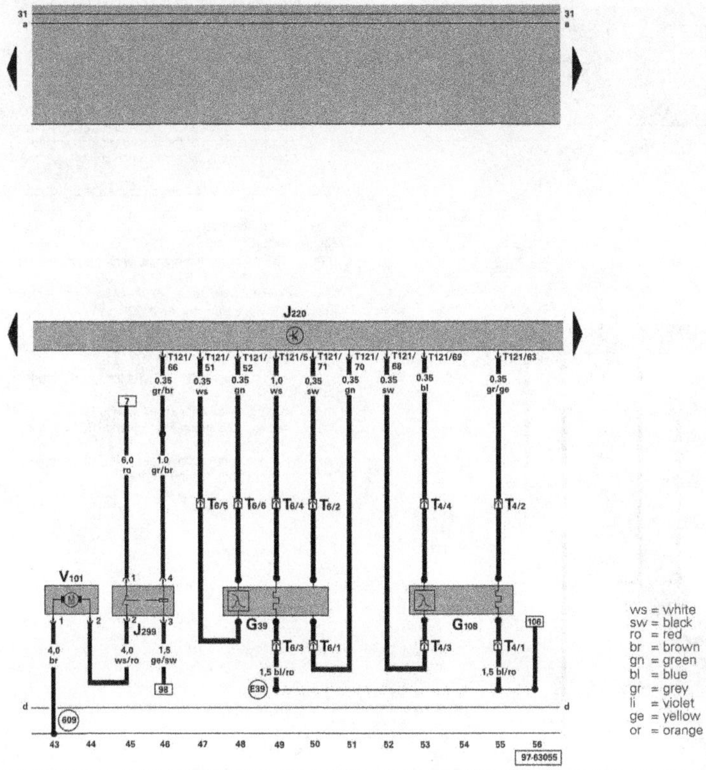

New Beetle Convertible

Wiring diagram **No. 19/6**

Motronic engine control module (ECM), intake air temperature (IAT) sensor, knock sensor (KS) 1, throttle valve control module

G42 - Intake Air Temperature (IAT) Sensor
G61 - Knock Sensor (KS) 1
G186 - Throttle drive (power accelerator actuation)
G187 - Angle sensor -1- for throttle drive (power accelerator actuation)
G188 - Angle sensor -2- for throttle drive (power accelerator actuation)
J220 - Motronic Engine Control Module (ECM)
J338 - Throttle Valve Control Module
T2a - 2-Pin Connector
T6a - 6-Pin Connector
T121 - 121-Pin Connector, at -J220-

(139) - Ground connection (sensor Ground), in Motronic Multiport Fuel Injection (MFI) wiring harness

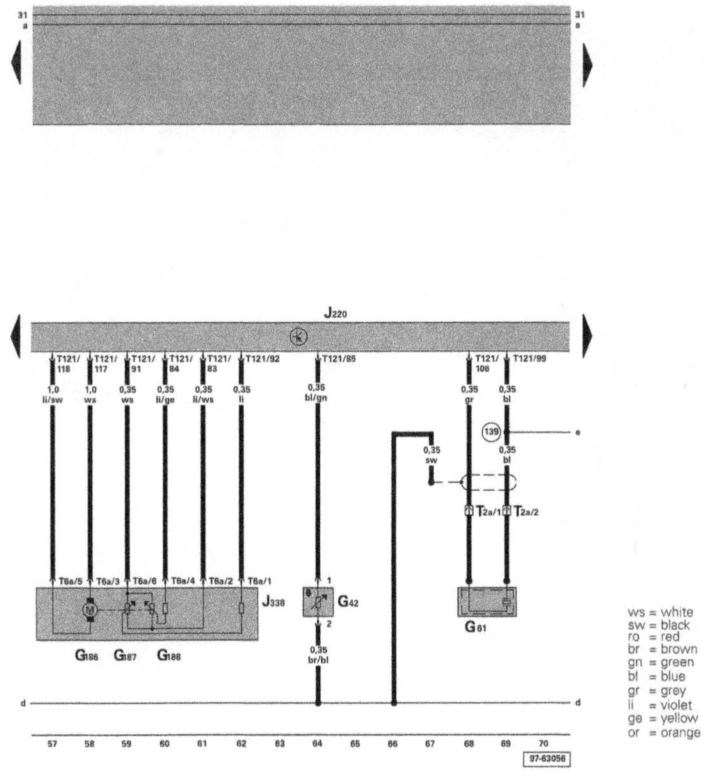

1.8L engine management (AWV) from Jan. 2003, convertible

New Beetle Convertible

Wiring diagram

No. 19/7

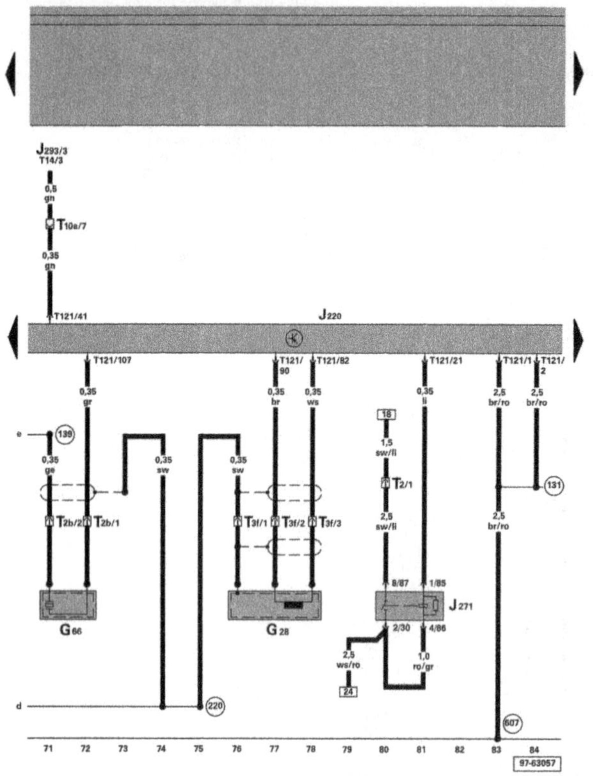

Motronic engine control module (ECM) / power supply relay, engine speed (RPM) sensor, knock sensor (KS) 2

- G28 - Engine Speed (RPM) Sensor
- G66 - Knock Sensor (KS) 2
- J220 - Motronic Engine Control Module (ECM)
- J271 - Motronic Engine Control Module (ECM) power supply relay, in protective housing, in engine compartment, left, production control number (428)
- J293 - Coolant FC (Fan Control) Control Module
- T2 - 2-Pin Connector
- T2b - 2-Pin Connector
- T3f - 3-Pin Connector
- T10a - 10-Pin Connector, orange, behind instrument panel, left
- T14 - 14-Pin Connector, at -J293-
- T121 - 121-Pin Connector, at -J220-
- (131) - Ground connection -2-, in engine compartment wiring harness
- (607) - Ground Connection (in left plenum chamber)
- (139) - Ground connection (sensor Ground), in Motronic Multiport Fuel Injection (MFI) wiring harness
- (220) - Ground connection (sensor ground), in engine compartment wiring harness

ws = white
sw = black
ro = red
br = brown
gn = green
bl = blue
gr = grey
li = violet
ge = yellow
or = orange

New Beetle Convertible

Wiring diagram

No. 19/8

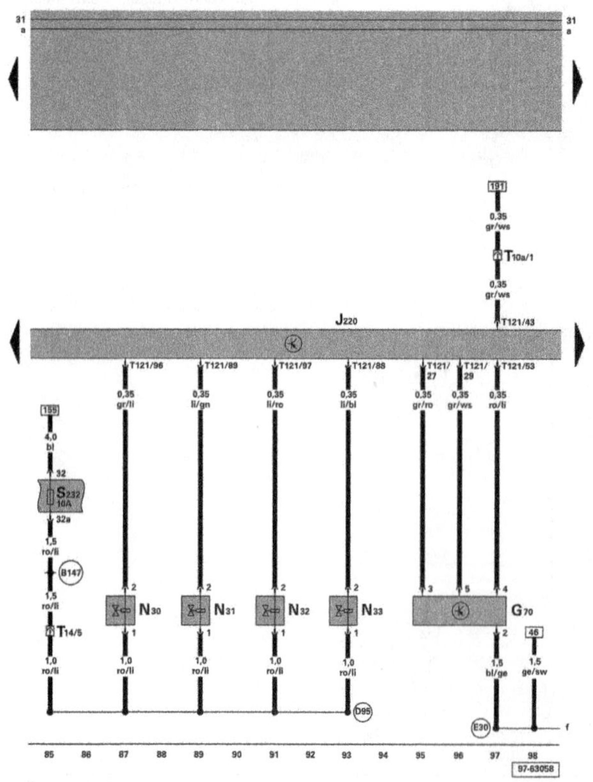

Motronic engine control module (ECM), mass air flow (MAF) sensor, fuel injectors

- G70 - Mass Air Flow (MAF) Sensor
- J220 - Motronic Engine Control Module (ECM)
- N30 - Cylinder 1 Fuel Injector
- N31 - Cylinder 2 Fuel Injector
- N32 - Cylinder 3 Fuel Injector
- N33 - Cylinder 4 Fuel Injector
- S232 - Fuse 32 in fuse/relay panel
- T10a - 10-Pin Connector, orange, behind instrument panel, left
- T14 - 14-Pin Connector
- T121 - 121-Pin Connector, at J220-
- (B147) - Plus connector -2- (87), in wiring harness interior wire connection (injectors), in engine compartment wiring harness
- (D95) - wire connection (injectors), in engine compartment wiring harness
- (E30) - Connector (87a), in wiring harness engine

ws = white
sw = black
ro = red
br = brown
gn = green
bl = blue
gr = grey
li = violet
ge = yellow
or = orange

New Beetle Convertible — Wiring diagram — No. 19/9

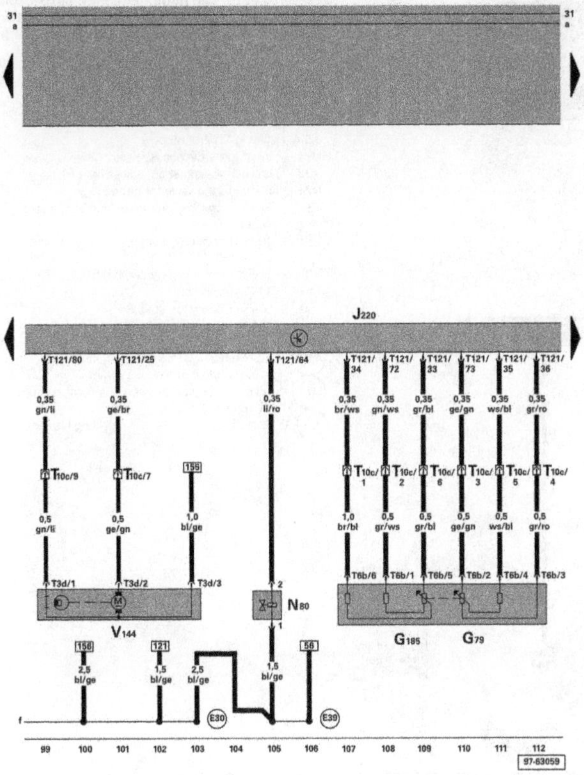

Motronic engine control module (ECM), throttle position (TP) sensor, sender 2 for acelerator pedal position, evaporative emission (EVAP), leak detection pump (LDP)

- G79 - Throttle Position (TP) Sensor
- G185 - Sender -2- for accelerator pedal position
- J220 - Motronic Engine Control Module (ECM)
- N80 - Evaporative Emission (EVAP) Canister Purge Regulator Valve
- V144 - Leak Detection Pump (LDP)
- T3d - 3-Pin Connector
- T6b - 6-Pin Connector
- T10c - 10-Pin Connector, blue, behind instrument panel, left
- T121 - 121-Pin Connector, at -J220-
- (E30) - Connector (87a), in wiring harness engine
- (E39) - Connection-2-(87a), in wiring harness engine

ws = white
sw = black
ro = red
br = brown
gn = green
bl = blue
gr = grey
li = violet
ge = yellow
or = orange

New Beetle Convertible — Wiring diagram — No. 19/10

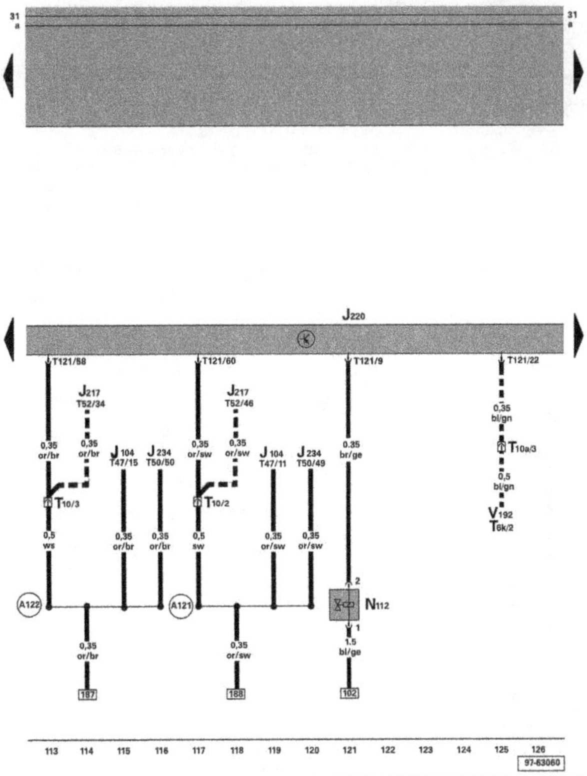

Motronic engine control module, ABS control module, transmission control module, airbag control module, secondary air injection (AIR) solenoid valve

- J104 - ABS Control Module (w/EDL)
- J217 - Transmission Control Module (TCM)
- J220 - Motronic Engine Control Module (ECM)
- J234 - Airbag Control Module
- N112 - Secondary Air Injection (AIR) Solenoid Valve
- T6K - 6-Pin Connector
- T10 - 10-Pin Connector, white, behind instrument panel, left
- T10a - 10-Pin Connector, orange, behind instrument panel, left
- T47 - 47-Pin Connector, at J104
- T50 - 50-Pin Connector, at J234
- T52 - 52-Pin Connector, at J217
- T121 - 121-Pin Connector, at J220
- V192 - Brake System Vacuum Pump
- (A121) - Connection (high bus) in instrument panel wiring harness
- (A122) - Connection (low bus) in instrument panel wiring harness
- --- - Automatic transmission only

ws = white
sw = black
ro = red
br = brown
gn = green
bl = blue
gr = grey
li = violet
ge = yellow
or = orange

1.8L engine management (AWV) from Jan. 2003, convertible

New Beetle Convertible

Wiring diagram

No. 19/11

Motronic ECM, fuel pump (FP) relay, airbag control module, valve 1 for camfshaft adjustment, wastegate bypass regulator valve, recirculating valve for turbocharger

- E184 - Switch for fresh air blower and recirculated air
- J17 - Fuel Pump (FP) Relay
- J220 - Motronic Engine Control Module (ECM)
- J234 - Airbag Control Module
- N75 - Wastegate Bypass Regulator Valve
- N205 - Valve-1- for camshaft adjustment
- N249 - Recirculating valve for turbocharger
- T6 - 6-Pin Connector, brown, behind instrument panel, left
- T10a - 10-Pin Connector,orange, behind instrument panel, left
- T10j - 10-Pin Coupling conector, on -E184-
- T14 - 14-Pin Connector
- T50 - 50-Pin Connector, at -J234-
- T121 - 121-Pin Connector, at -J220-
- (A125) - Connection (crash signal) in instrument panel wiring harness
- (D102) - Wire connection -2-, in engine compartment wiring harness
- (L9) - Wire connection -1-, in A/C wiring harness

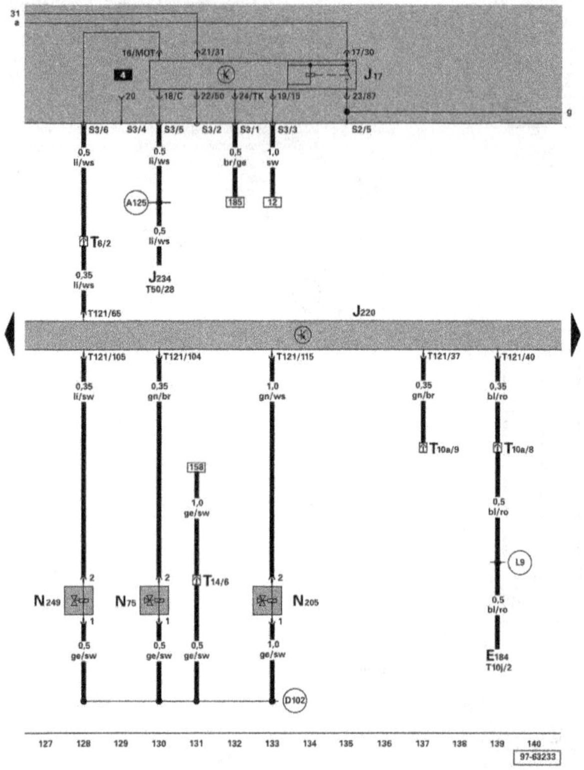

ws = white
sw = black
ro = red
br = brown
gn = green
bl = blue
gr = grey
li = violet
ge = yellow
or = orange

Edition 06.04

New Beetle Convertible

Wiring diagram

No. 19/12

Motronic engine control module (ECM), cruise control switch, button for cruise control (set), brake light switch, clutch vacuum vent valve switch

- E45 - Cruise Control Switch
- E227 - Button for cruise control (set)
- F - Brake Light Switch
- F36 - Clutch Vacuum Vent Valve Switch
- F47 - Brake Vacuum Vent Valve Switch for cruise control
- J220 - Motronic Engine Control Module (ECM)
- S13 - Fuse 13 in fuse/relay panel
- T10 - 10-Pin Connector, white, behind instrument panel, left
- T10a - 10-Pin Connector, orange, behind instrument panel, left
- T10h - 10-Pin Connector, black, behind instrument panel, left
- T10s - 10-Pin Connector, near steering column
- T121 - 121-Pin Connector, at -J220-
- (A18) - wire connection (54), in instrument panel wiring harness
- (A20) - wire connection (15a), in instrument panel wiring harness
- (A27) - Wire Connection (vehicle speed signal), in instrument panel wiring harness
- * - Vehicles with Manual transmission only

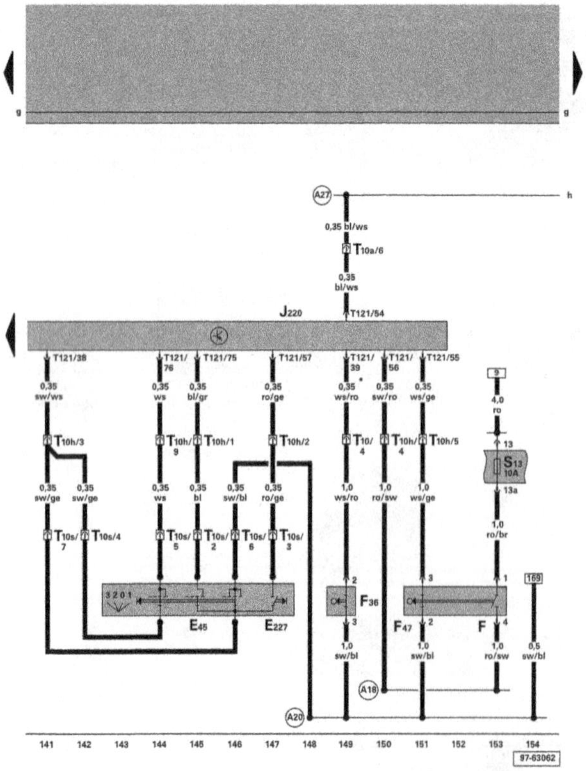

ws = white
sw = black
ro = red
br = brown
gn = green
bl = blue
gr = grey
li = violet
ge = yellow
or = orange

Edition 09.03

1.8L engine management (AWV) from Jan. 2003, convertible

New Beetle Convertible

Wiring diagram **No. 19/13**

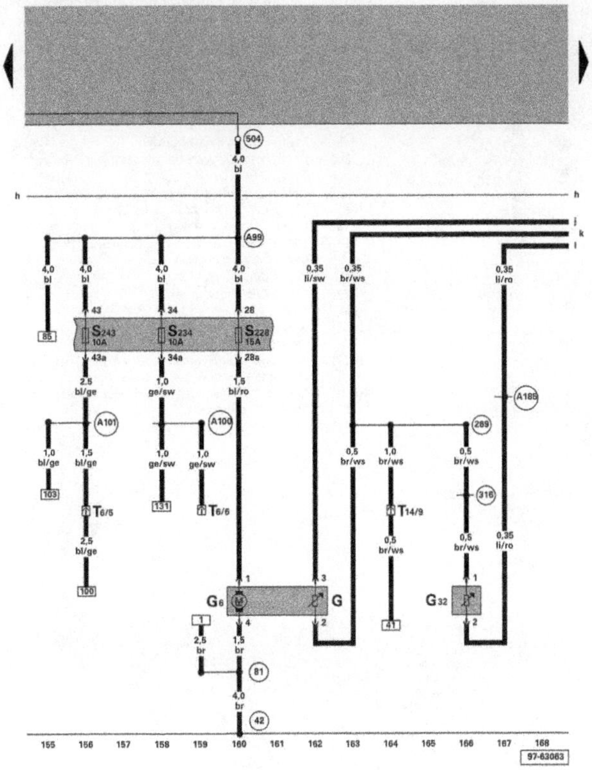

ws = white
sw = black
ro = red
br = brown
gn = green
bl = blue
gr = grey
li = violet
ge = yellow
or = orange

Sender for fuel gauge, fuel pump (FP), engine coolant level (ECL) sensor

- G - Sender for fuel gauge
- G6 - Fuel Pump (FP)
- G32 - Engine Coolant Level (ECL) Sensor
- S228 - Fuse 28 in fuse/relay panel
- S234 - Fuse 34 in fuse/relay panel
- S243 - Fuse 43 in fuse/relay panel
- T6 - 6-Pin Connector, brown, behind instrument panel, left
- T14 - 14-Pin Connector
- (81) - Ground connection -1-, in instrument panel wiring harness
- (42) - Ground connection, beside steering column
- (269) - Ground connector (sensor ground) -1-, in instrument panel wiring harness
- (316) - Sensor Ground Connection 2 (in engine wiring harness)
- (504) - Threaded connection -1- (87) on the relay plate
- (A99) - Connector -1- (87), in instrument panel wiring harness
- (A100) - Connector -2- (87), in instrument panel wiring harness
- (A101) - Connector -3- (87), in instrument panel wiring harness
- (A185) - Coolant Warning Connection (in instrument panel wiring harness)

New Beetle Convertible

Wiring diagram **No. 19/14**

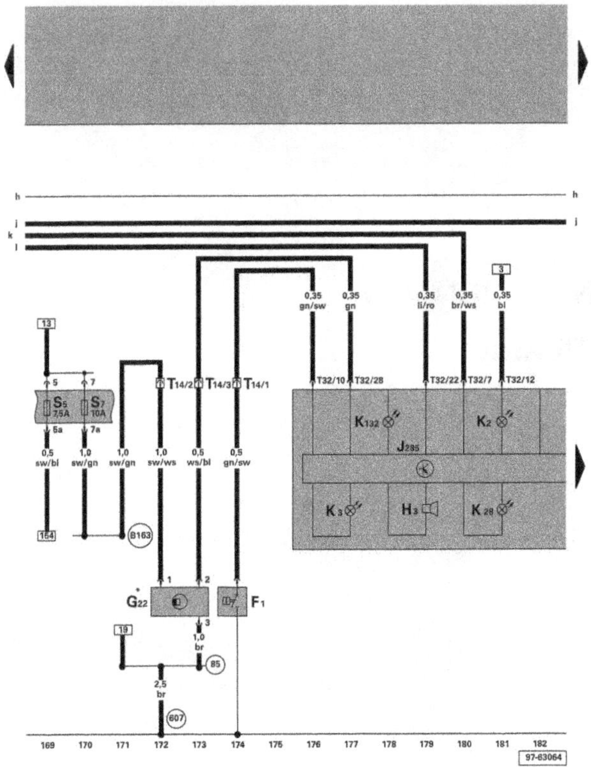

ws = white
sw = black
ro = red
br = brown
gn = green
bl = blue
gr = grey
li = violet
ge = yellow
or = orange

Oil pressure switch / warning light, speedometer vehicle speed sensor (VSS), warning buzzer, control module with indicator unit in instrument panel insert, generator (GEN) warning light

- F1 - Oil Pressure Switch
- G22 - Speedometer Vehicle Speed Sensor (VSS)
- H3 - Warning Buzzer
- J285 - Control module with indicator unit in instrument panel insert
- K2 - Generator (GEN) Warning Light
- K3 - Oil Pressure Warning Light
- K28 - Engine Coolant Level/Temperature (ECL/ECT) Warning Light
- K132 - Electronic Power Control (EPC) Lamp
- S5 - Fuse 5 in fuse/relay panel
- S7 - Fuse 7 in fuse/relay panel
- T14 - 14-Pin Connector
- T32 - 32-Pin Connector, blue
- (607) - Ground Connection (in left plenum chamber)
- (85) - Ground connection -1-, in engine compartment wiring harness
- (B163) - Plus connector -1- (15) in wiring harness interior
- * - Manual transmission only

1.8L engine management (AWV) from Jan. 2003, convertible

New Beetle Convertible Wiring diagram **No. 19/15**

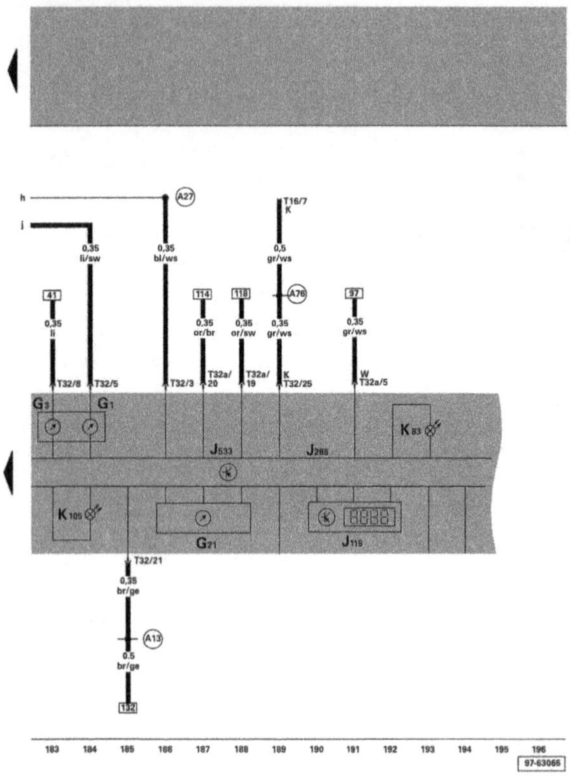

Control module with indicator unit in instrument panel insert, fuel gauge, engine coolant temperature (ECT) gauge, speedomter, multifunction indicator (MFA), low fuel level warning light

G1 - Fuel Gauge
G3 - Engine Coolant Temperature (ECT) Gauge
G21 - Speedometer
J119 - Multi-function Indicator (MFA)
J285 - Control module with indicator unit in instrument panel insert
J533 - Data Bus On Board Diagnostic Interface
K83 - Malfunction Indicator lamp (MIL)
K105 - Low Fuel Level Warning Light
T16 - 16-Pin Connector, Data Link Connector (DLC), below steering column, left
T32 - 32-Pin Connector, blue
T32a - 32-Pin Connector, green

(A13) - wire connection (door contact switch) in instrument panel wiring harness
(A27) - Wire Connection (vehicle speed signal), in instrument panel wiring harness
(A76) - Connector (K-diagnosis wire) in instrument panel wiring harness

ws = white
sw = black
ro = red
br = brown
gn = green
bl = blue
gr = grey
li = violet
ge = yellow
or = orange

THIS SPACE INTENTIONALLY

LEFT BLANK

New Beetle

Wiring Diagram

No. 35 / 1

10.2009

1,8L-Engine - 5V Turbo, Motronic Multiport Fuel Injection (MFI)

- **110 kW, code BKF (with cruise control)**
- **132 kW, code BNU (with cruise control)**

From June 2004

Additional Fuse, Relay, Control Module and Connector locations ⊐ Component Locations

Troubleshooting Procedures ⊐ Guided Fault Finding using VAS 5051/5052

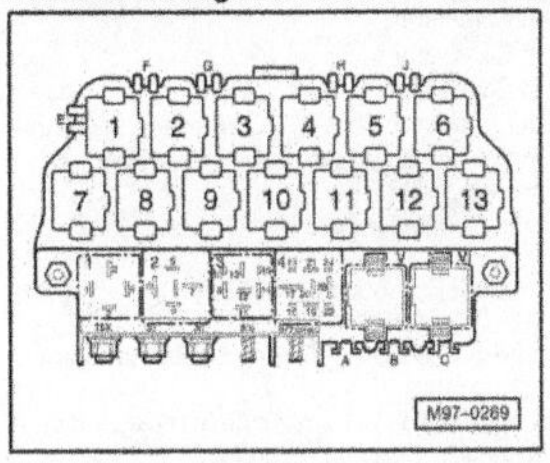

Relay locations on thirteen position auxiliary relay panel, above relay panel:
- 3 - J434 - Starter Lock/Clutch Pedal Switch Relay (53)

Relay panel:
- 1 - J4 - Horn Relay (53)
- 2 - J59 - Load Reduction Relay (100)
- 4 - J17 Fuel Pump (FP) Relay (409)

Notes:

- *J271 Motronic Engine Control Module (ECM) Power Supply Relay (428), ¹)*
- *J299 Secondary Air Injection (AIR) Pump Relay (100) ¹)*

¹) in engine compartment, left side, to the left of brake fluid reservoir, inside housing

Note:

Number in parentheses indicates production control number stamped on relay housing.

New Beetle

Wiring diagram

No. 35/2

Battery, starter, generator (GEN), voltage regulator (VR), ignition/starter switch, load reduction relay, fuses on batter fuse panel

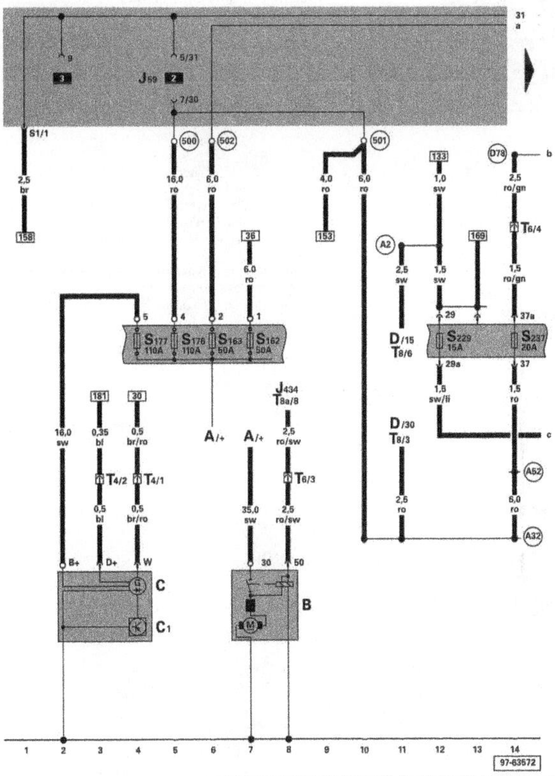

- A - Battery
- B - Starter
- C - Generator (GEN)
- C1 - Voltage Regulator (VR)
- D - Ignition/Starter Switch
- J59 - Load Reduction Relay
- J434 - Starter Lock/Clutch Pedal Switch Relay
- S237 - Fuse (on fuse panel)
- S162 - Fuse 1 - terminal 30 (on battery fuse panel)
- S163 - Fuse 1 - terminal 30 (on battery fuse panel)
- S176 - Fuse 1 - terminal 30 (on battery fuse panel)
- S177 - Fuse 1 - terminal 30 (on battery fuse panel)
- S229 - Fuse 29 in fuse/relay panel
- T4 - 4-Pin Connector, in engine compartment left
- T6 - 6-Pin Connector, brown, behind inst. panel, left
- T8 - 8-Pin Connector, at -D-
- T8a - 8-Pin Connector, at -J434-
- (500) - Threaded connection -1- (30) on the relay plate
- (501) - Threaded connection -2- (30) on the relay plate
- (502) - Threaded connection -1- (30a) on the relay plate
- (A2) - plus connection (15), in instrument panel wiring harness
- (A32) - plus connection (30), in instrument panel wiring harness
- (A52) - Plus Connection 2 (30) (in instrument panel wiring harness)
- (D78) - Plus connector -1- (30a), in engine compartment wiring harness

ws = white
sw = black
ro = red
br = brown
gn = green
bl = blue
gr = grey
li = violet
ge = yellow
or = orange

1.8L engine management (BKF, BNU) from June 2004

New Beetle

Wiring diagram

No. 35/3

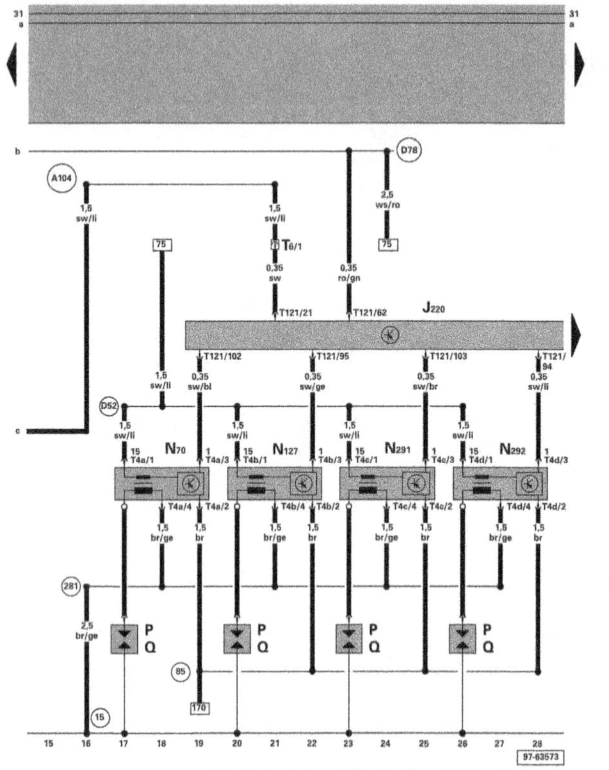

Motronic engine control module (ECM), ignition coil with power output stages 1, 2, 3 & 4, spark plugs

- J220 - Motronic Engine Control Module (ECM)
- N70 - Ignition Coil With Power Output Stage 1
- N127 - Ignition Coil with Power Output Stage 2
- N291 - Ignition Coil with Power Output Stage 3
- N292 - Ignition Coil with Power Output Stage 4
- P - Spark Plug Connectors
- Q - Spark Plugs
- T4a - 4-Pin Connector
- T4b - 4-Pin Connector
- T4c - 4-Pin Connector
- T4d - 4-Pin Connector
- T6 - 6-Pin Connector, brown, behind instrument panel, left
- T121 - 121-Pin Connector, at -J220-

- (15) - Ground connection, on cylinder head
- (85) - Ground connection -1-, in engine compartment wiring harness
- (281) - Ground connector -1-, in wiring harness engine pre-wiring
- (A104) - Plus connector -2- (15), in instrument panel wiring harness
- (D52) - plus connection (15a), in engine compartment wiring harness
- (D78) - Plus connector -1- (30a), in engine compartment wiring harness

ws = white
sw = black
ro = red
br = brown
gn = green
bl = blue
gr = grey
li = violet
ge = yellow
or = orange

Edition 11.04

New Beetle

Wiring diagram

No. 35/4

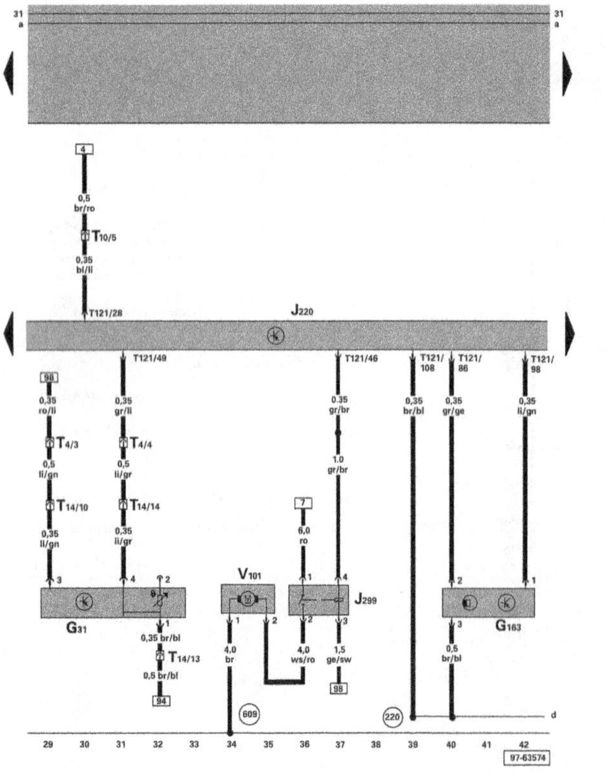

Motronic engine control module (ECM), charge air pressure sensor, camshaft position (CMP) sensor 2, secondary air injection (AIR) pump relay, secondary air injection (AIR) pump motor

- G31 - Charge Air Pressure Sensor
- G163 - Camshaft Position (CMP) Sensor 2
- J220 - Motronic Engine Control Module (ECM)
- J299 - Secondary Air Injection (AIR) Pump Relay
- T4 - 4-Pin Connector
- T10 - 10-Pin Connector, white, behind instrument panel, left
- T14 - 14-Pin Connector, in engine compartment, left
- T121 - 121-Pin Connector, at -J220-
- V101 - Secondary Air Injection (AIR) Pump Motor

- (220) - Ground connection (sensor ground), in engine compartment wiring harness
- (609) - Ground Connection (in right plenum chamber)

ws = white
sw = black
ro = red
br = brown
gn = green
bl = blue
gr = grey
li = violet
ge = yellow
or = orange

Edition 11.04

New Beetle — Wiring diagram — No. 35/5

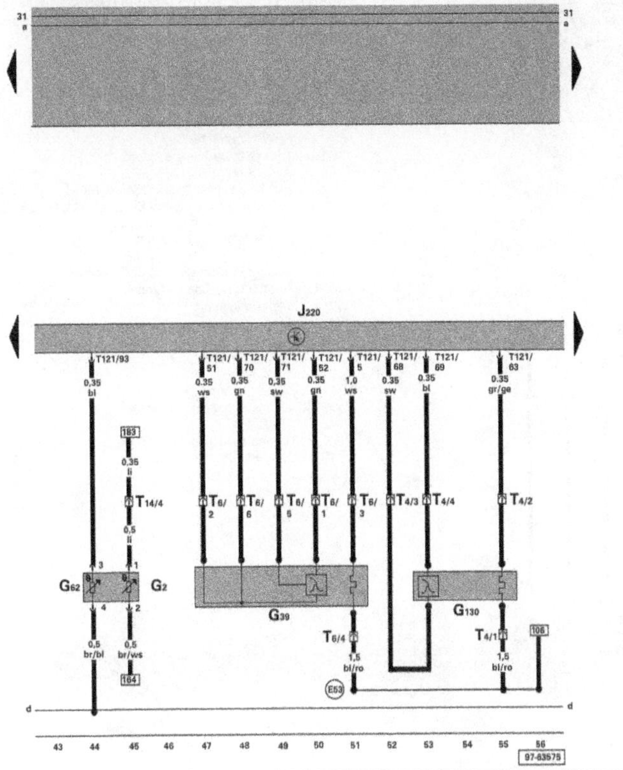

Motronic engine control module (ECM), heated oxygen sensors (HO2S), engine coolant temperature (ECT) sensor, engine coolant temperature (ECT) sensor

- G2 - Engine Coolant Temperature (ECT) Sensor
- G62 - Engine Coolant Temperature (ECT) Sensor
- G39 - Heated Oxygen Sensor (HO2S), black
- G130 - Oxygen Sensor (O2S) Behind Three Way Catalytic Converter (TWC), black
- J220 - Motronic Engine Control Module (ECM)
- T4 - 4-Pin Connector, brown
- T6 - 6-Pin Connector, black
- T14 - 14-Pin Connector, in engine compartment, left
- T121 - 121-Pin Connector, at -J220-
- (E53) - Connection 2 (87a)(in Motronic wiring harness

ws = white
sw = black
ro = red
br = brown
gn = green
bl = blue
gr = grey
li = violet
ge = yellow
or = orange

New Beetle — Wiring diagram — No. 35/6

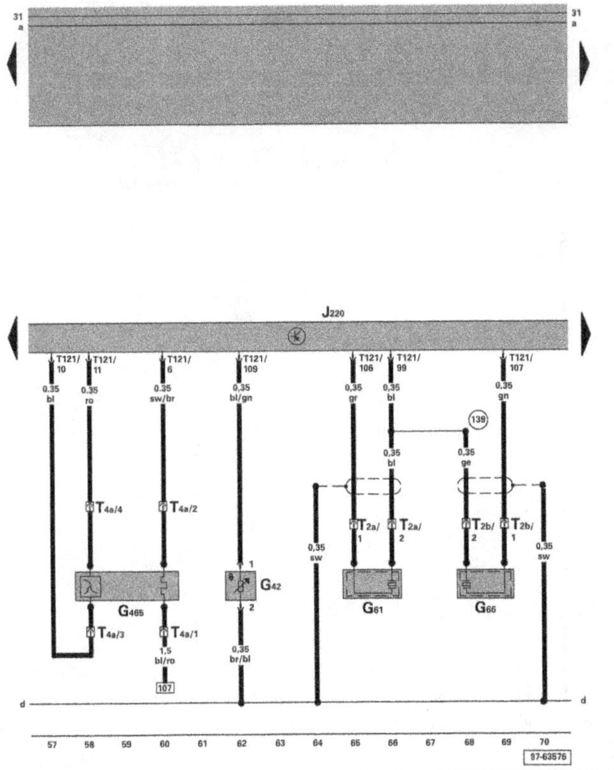

Motronic engine control module (ECM), intake air temperature (IAT) sensor, knock sensors, Oxygen sensor (O2S) in center three way catalytic converter (TWC)

- G42 - Intake Air Temperature (IAT) Sensor
- G61 - Knock Sensor (KS) 1
- G66 - Knock Sensor (KS) 2
- G465 - Oxygen Sensor (O2S) in Center Three Way Catalytic Converter (TWC), Bank 1, brown
- J220 - Motronic Engine Control Module (ECM)
- T2a - 2-Pin Connector
- T2b - 2-Pin Connector
- T4a - 4-Pin Connector, at G465
- T121 - 121-Pin Connector, at -J220-
- (139) - Ground connection (sensor Ground), in Motronic Multiport Fuel Injection (MFI) wiring harness

ws = white
sw = black
ro = red
br = brown
gn = green
bl = blue
gr = grey
li = violet
ge = yellow
or = orange

New Beetle — Wiring diagram — No. 35/7

Motronic engine control module (ECM), power supply relay, throttle valve control module, engine speed (RPM) sensor, angle sensors

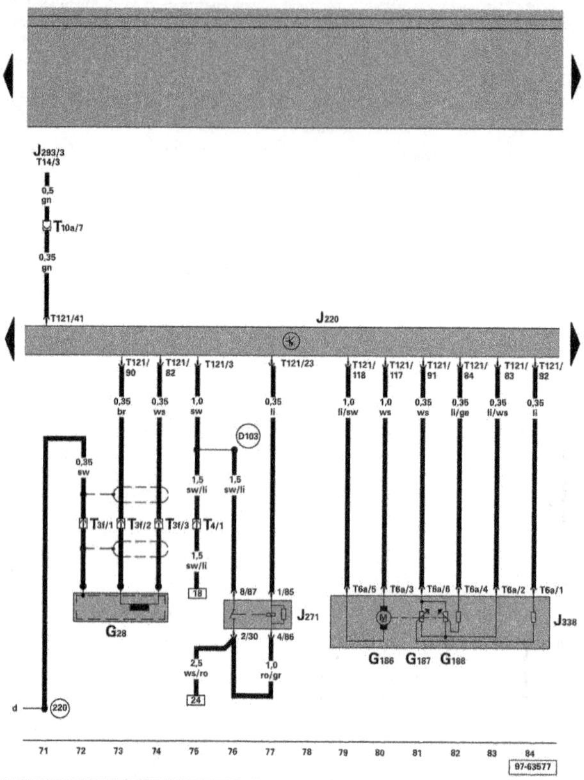

- G28 - Engine Speed (RPM) Sensor
- G186 - Throttle Drive (for Electronic Power Control (EPC))
- G187 - Throttle Drive Angle Sensor 1 (for Electronic Power Control (EPC))
- G188 - Throttle Drive Angle Sensor 2 (for Electronic Power Control (EPC))
- J220 - Motronic Engine Control Module (ECM)
- J271 - Motronic Engine Control Module (ECM) Power Supply Relay, in protective housing, in engine compartment, left, production control number (429)
- J293 - Coolant FC (Fan Control) Control Module
- J338 - Throttle Valve Control Module
- T3f - 3-Pin Connector
- T4 - 4-Pin Connector
- T6a - 6-Pin Connector
- T10a - 10-Pin Connector, orange, behind instrument panel, left
- T14 - 14-Pin Connector, at -J293-
- T121 - 121-Pin Connector, at -J220-
- (220) - Ground connection (sensor ground), in engine compartment wiring harness
- (D103) - Connection 3, in engine compartment wiring harness

ws = white
sw = black
ro = red
br = brown
gn = green
bl = blue
gr = grey
li = violet
ge = yellow
or = orange

New Beetle — Wiring diagram — No. 35/8

Motronic engine control module (ECM), mass air flow (MAF) sensor, fuel injectors

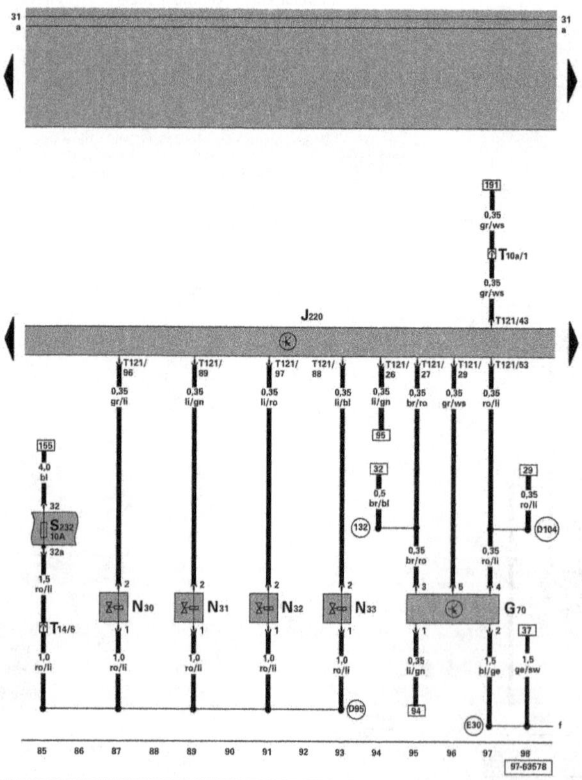

- G70 - Mass Air Flow (MAF) Sensor
- J220 - Motronic Engine Control Module (ECM)
- N30 - Cylinder 1 Fuel Injector
- N31 - Cylinder 2 Fuel Injector
- N32 - Cylinder 3 Fuel Injector
- N33 - Cylinder 4 Fuel Injector
- S232 - Fuse 32 on fuse/panel
- T10a - 10-Pin Connector, orange, behind instrument panel, left
- T14 - 14-Pin Connector
- T121 - 121-Pin Connector, at J220-
- (132) - Ground Connection 3, in engine compartment wiring harness
- (D95) - wire connection (injectors), in engine compartment wiring harness
- (D104) - Plus Connection 2 (30a), in engine compartment wiring harness
- (E30) - Connector (87a), in wiring harness engine

ws = white
sw = black
ro = red
br = brown
gn = green
bl = blue
gr = grey
li = violet
ge = yellow
or = orange

New Beetle — Wiring diagram — No. 35/9

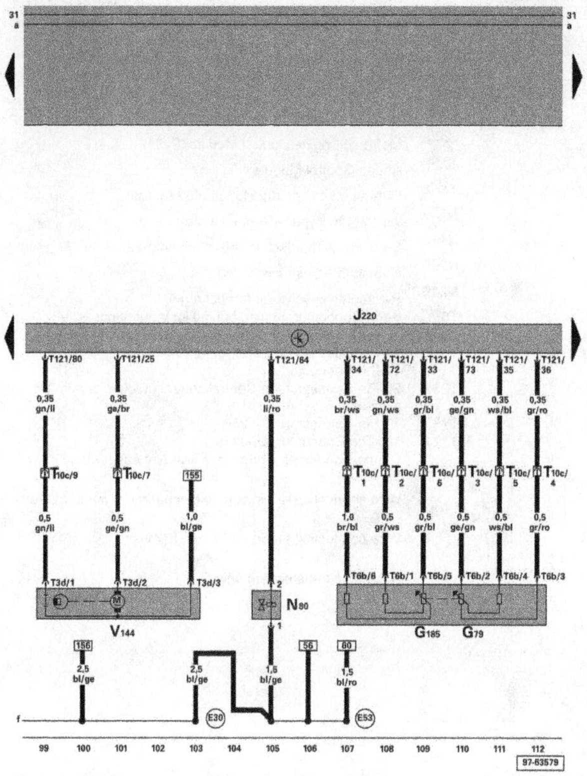

Motronic engine control module (ECM), throttle position (TP) sensor, accelerator pedal position sensor, evaporative emission (EVAP), leak detection pump (LDP)

- G79 - Throttle Position (TP) Sensor
- G185 - Accelerator Pedal Position Sensor 2
- J220 - Motronic Engine Control Module (ECM)
- N80 - Evaporative Emission (EVAP) Canister Purge Regulator Valve
- T3d - 3-Pin Connector
- T6b - 6-Pin Connector
- T10c - 10-Pin Connector, blue, behind instrument panel, left
- T121 - 121-Pin Connector, at -J220-
- V144 - Leak Detection Pump (LDP)
- (E30) - Connector (87a), in wiring harness engine
- (E53) - Connection 2 (87a)(in Motronic wiring harness

ws = white
sw = black
ro = red
br = brown
gn = green
bl = blue
gr = grey
li = violet
ge = yellow
or = orange

New Beetle — Wiring diagram — No. 35/10

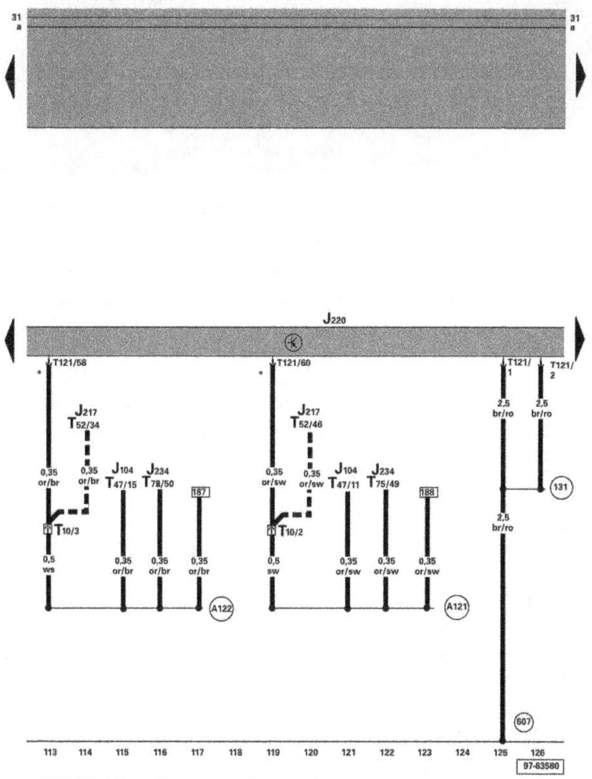

Motronic engine control module, ABS control module, transmission control module, airbag control module,

- J104 - ABS Control Module (w/EDL)
- J217 - Transmission Control Module (TCM)
- J220 - Motronic Engine Control Module (ECM)
- J234 - Airbag Control Module
- T47 - 47-Pin Connector, at J104
- T52 - 52-Pin Connector, at J217
- T75 - 75-Pin Connector, at J234
- T121 - 121-Pin Connector, at J220
- (131) - Ground connection -2-, in engine compartment wiring harness
- (607) - Ground Connection (in left plenum chamber)
- (A121) - Connection (high bus) in instrument panel wiring harness
- (A122) - Connection (low bus) in instrument panel wiring harness
- * - Manual transmission only
- --- - Automatic transmission only

ws = white
sw = black
ro = red
br = brown
gn = green
bl = blue
gr = grey
li = violet
ge = yellow
or = orange

1.8L engine management (BKF, BNU) from June 2004

New Beetle — Wiring Diagram — No. 35 / 11

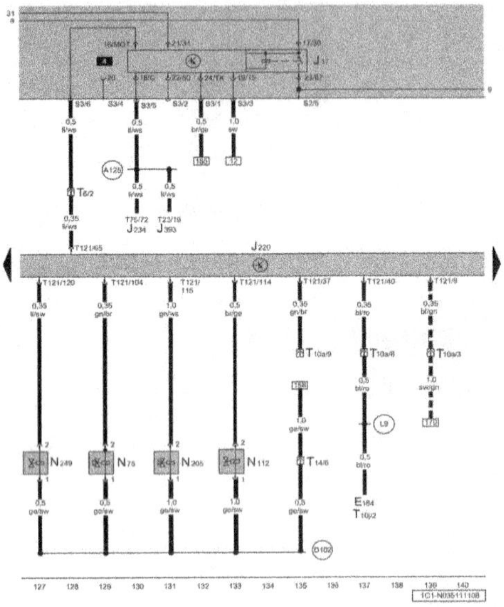

ws = wt (white)
sw = bk (black)
ro = rd (red)
br = br (brown)
gn = lg (light green), dg (dark green)
bl = bl (blue), lb (light blue), db (dark blue)
gr = gy (gray)
li = lilac
ge = yl (yellow)
or = or (orange), tn (tan)
rs = vt (violet)

Motronic ECM, fuel pump (FP) relay, fuel pump (FP) relay, airbag control module, wastegate bypass regulator valve, secondary air injection (AIR) solenoid valve, camshaft adjustment valve 1, recirculating valve for turbocharger

E184 - Fresh Air Blower/Recirculated Air Switch
J17 - Fuel Pump (FP) Relay
J220 - Motronic Engine Control Module (ECM)
J234 - Airbag Control Module
J393 - Central control module for comfort system
N75 - Wastegate Bypass Regulator Valve
N112 - Secondary Air Injection (AIR) Solenoid Valve
N205 - Camshaft Adjustment Valve 1
N249 - Recirculating valve for turbocharger
T6 - 6-Pin Connector, brown, behind instrument panel, left
T10a - 10-Pin Connector, orange, behind instrument panel, left
T10j - 10-Pin Coupling conector, on -E184-
T14 - 14-Pin Connector
T23 - 23-Pin Connector, on Central control module for comfort system
T75 - 75-Pin Connector, at -J234-
T121 - 121-Pin Connector, at -J220-
A125 - Connection (crash signal) in instrument panel wiring harness
D102 - Wire connection -2-, in engine compartment wiring harness
L9 - Wire connection -1-, in A/C wiring harness
--- - Automatic transmission only

New Beetle — Wiring diagram — No. 35/12

Motronic engine control module (ECM), cruise control switch, cruise control -set- button, brake light switch, clutch pedal switch

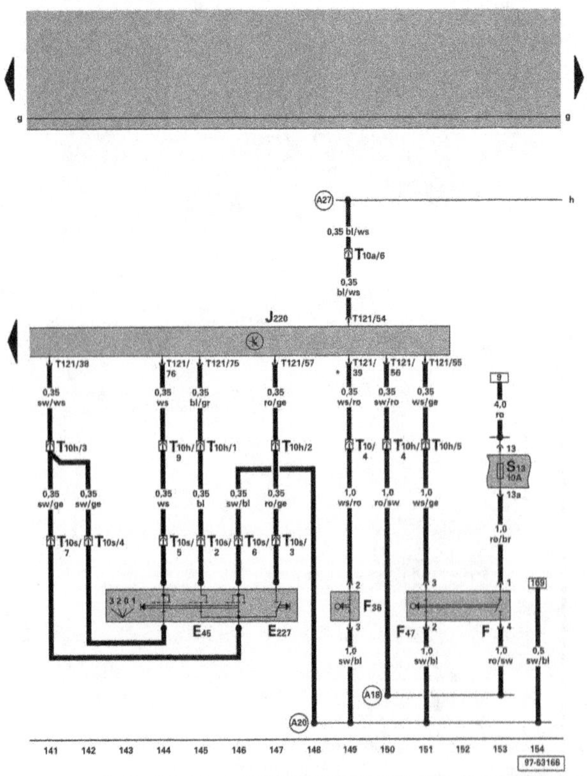

ws = white
sw = black
ro = red
br = brown
gn = green
bl = blue
gr = grey
li = violet
ge = yellow
or = orange

E45 - Cruise Control Switch
E227 - Cruise Control -SET- Button
F - Brake Light Switch
F36 - Clutch Pedal Switch
F47 - Brake Pedal Switch
J220 - Motronic Engine Control Module (ECM)
S13 - Fuse 13 (on fuse panel)
T10 - 10-Pin Connector, white, behind instrument panel, left
T10a - 10-Pin Connector, orange, behind instrument panel, left
T10h - 10-Pin Connector, black, behind instrument panel, left
T10s - 10-Pin Connector, near steering column
T121 - 121-Pin Connector, at -J220-
A18 - wire connection (54), in instrument panel wiring harness
A20 - wire connection (15a), in instrument panel wiring harness
A27 - Wire Connection (vehicle speed signal), in instrument panel wiring harness

* - Manual transmission only

New Beetle — Wiring diagram — No. 35/13

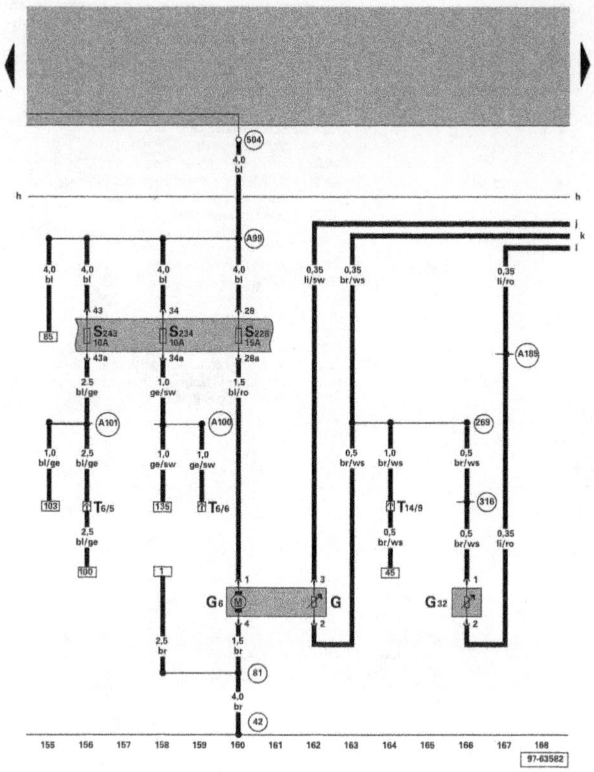

Sender for fuel gauge, fuel pump (FP), engine coolant level (ECL) sensor

- G - Fuel Level Sensor
- G6 - Fuel Pump (FP)
- G32 - Engine Coolant Level (ECL) Sensor
- S228 - Fuse 28 (on fuse panel)
- S234 - Fuse 34 (on fuse panel)
- S243 - Fuse 43 (on fuse panel)
- T6 - 6-Pin Connector, brown, behind instrument panel, left
- T14 - 14-Pin Connector
- (42) - Ground connection, beside steering column
- (81) - Ground connection -1-, in instrument panel wiring harness
- (269) - Ground connector (sensor ground) -1-, in instrument panel wiring harness
- (316) - Sensor Ground Connection 2 (in engine wiring harness)
- (504) - Threaded connection -1- (87) on the relay plate
- (A99) - Connector -1- (87), in instrument panel wiring harness
- (A100) - Connector -2- (87), in instrument panel wiring harness
- (A101) - Connector -3- (87), in instrument panel wiring harness
- (A185) - Coolant Warning Connection (in instrument panel wiring harness)

ws = white
sw = black
ro = red
br = brown
gn = green
bl = blue
gr = grey
li = violet
ge = yellow
or = orange

New Beetle — Wiring diagram — No. 35/14

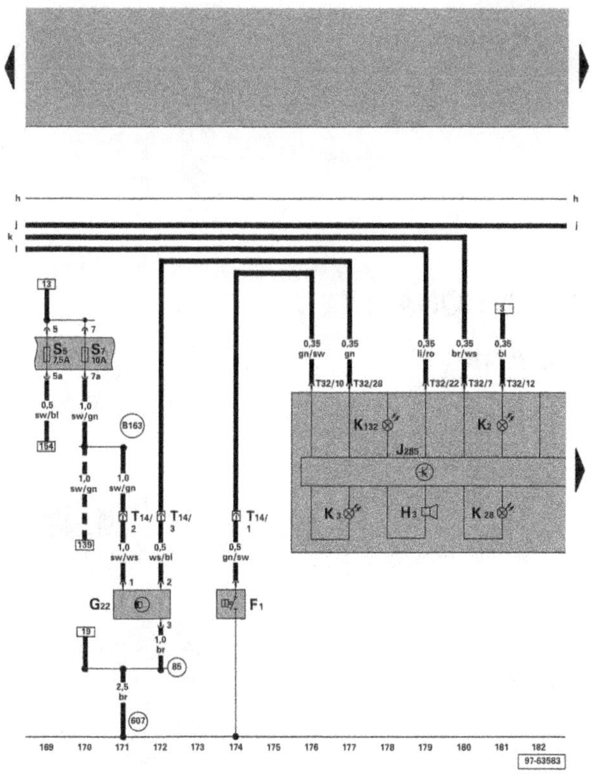

Oil pressure switch / warning light, speedometer vehicle speed sensor (VSS), warning buzzer, instrument cluster control module, generator (GEN) warning light

- F1 - Oil Pressure Switch
- G22 - Speedometer Vehicle Speed Sensor (VSS)
- H3 - Warning Buzzer
- J285 - Instrument Cluster Control Module
- K2 - Generator (GEN) Warning Light
- K3 - Oil Pressure Warning Light
- K28 - Engine Coolant Level/Temperature (ECL/ECT) Warning Light
- K132 - Electronic Power Control (EPC) Lamp
- S5 - Fuse 5 (on fuse panel)
- S7 - Fuse 7 (on fuse panel)
- T14 - 14-Pin Connector
- T32 - 32-Pin Connector, blue
- (85) - Ground connection -1-, in engine compartment wiring harness
- (607) - Ground Connection (in left plenum chamber)
- (B163) - Plus connector -1- (15) in wiring harness interior
- --- - Automatic transmission only

ws = white
sw = black
ro = red
br = brown
gn = green
bl = blue
gr = grey
li = violet
ge = yellow
or = orange

1.8L engine management (BKF, BNU) from June 2004

New Beetle Wiring diagram **No. 35/15**

Instrument cluster control module, fuel gauge, engine coolant temperature (ECT) gauge, speedometer, multifunction indicator (MFA), low fuel level warning light

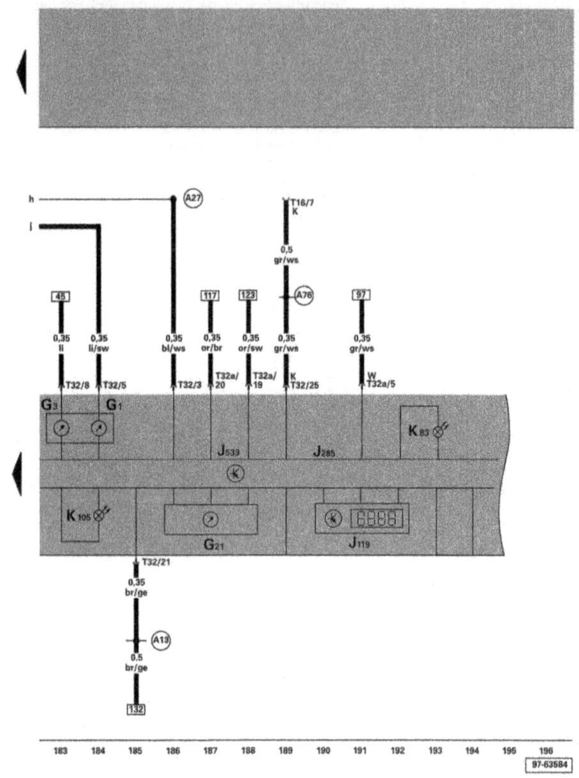

- G1 - Fuel Gauge
- G3 - Engine Coolant Temperature (ECT) Gauge
- G21 - Speedometer
- J119 - Multi-function Indicator (MFA)
- J285 - Instrument Cluster Control Module
- J533 - Data Bus On Board Diagnostic Interface
- K83 - Malfunction Indicator lamp (MIL)
- K105 - Low Fuel Level Warning Light
- T16 - 16-Pin Connector, Data Link Connector (DLC), below steering column, left
- T32 - 32-Pin Connector, blue
- T32a - 32-Pin Connector, green

- (A13) - wire connection (door contact switch) in instrument panel wiring harness
- (A27) - Wire Connection (vehicle speed signal), in instrument panel wiring harness
- (A76) - Connector (K-diagnosis wire) in instrument panel wiring harness

ws = white
sw = black
ro = red
br = brown
gn = green
bl = blue
gr = grey
li = violet
ge = yellow
or = orange

THIS SPACE INTENTIONALLY

LEFT BLANK

1,9L-Engine - Turbo Diesel Fuel Injection (DFI)/66 kW, code ALH,

from January 1998

Deviate relay location and fuseplacements as well as the locations of multiple connectors see section "component locations".

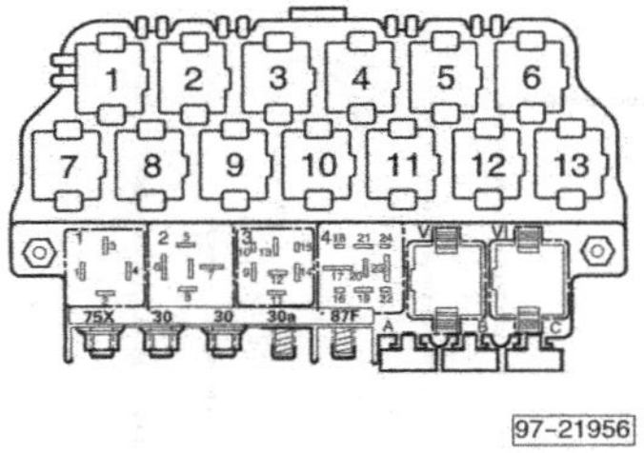

Relay location on the thirteen position auxiliary relay panel, above relay panel:

10 Glow Plug Relay (180)

12 Power Suppy (Terminal 30, B+) Relay (109))

Relay panel:

2 Load Reduction Relay (213)

Note: Number in parentheses indicates production control number stamped on relay housing.

Fuse colors
30 A - green
25 A - white
20 A - yellow
15 A - blue
10 A - red
7,5 A - brown
5 A - beige

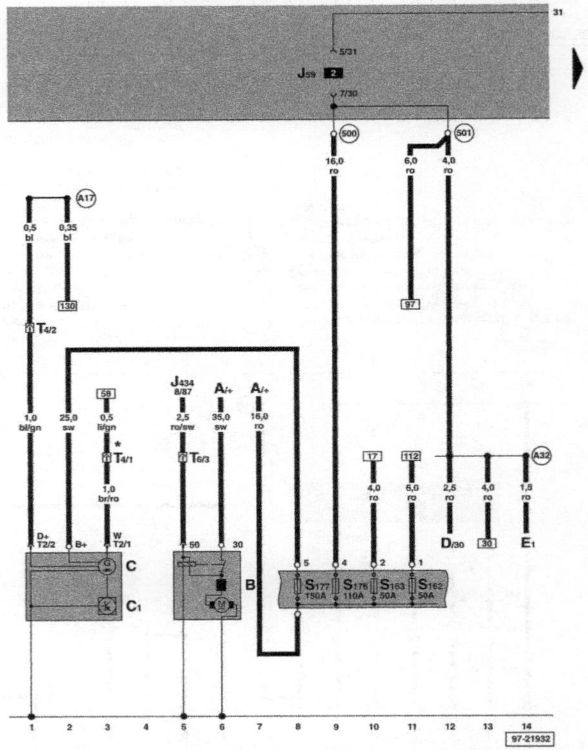

ws = white
sw = black
ro = red
br = brown
gn = green
bl = blue
gr = grey
li = violet
ge = yellow

A - Battery
B - Starter
C - Generator (GEN)
C1 - Voltage Regulator (VR)
D - Ignition/Starter Switch
E1 - Light switch
J59 - Load Reduction Relay
J434 - Locking relay for starter (clutch pedal switch)
S162 - Fuse -1- in fuse bracket/battery
S163 - Fuse -2- in fuse bracket/battery
S176 - Fuse -4- (30), in fuse bracket/battery
S177 - Fuse -5- (30), in fuse bracket/battery
T2 - Double Connector
T4 - 4-Pin Connector, in engine compartment, left
T6 - 6-Pin Connector, red, behind instrument panel, left

(500) - Threaded connection -1- (30) on the relay plate
(501) - Threaded connection -2- (30) on the relay plate
(A17) - wire connection (61), in instrument panel wiring harness
(A32) - plus connection (30), in instrument panel wiring harness
* - manual transmission only

Starter, generator (GEN)

New Beetle — Wiring diagram — No. 3/3

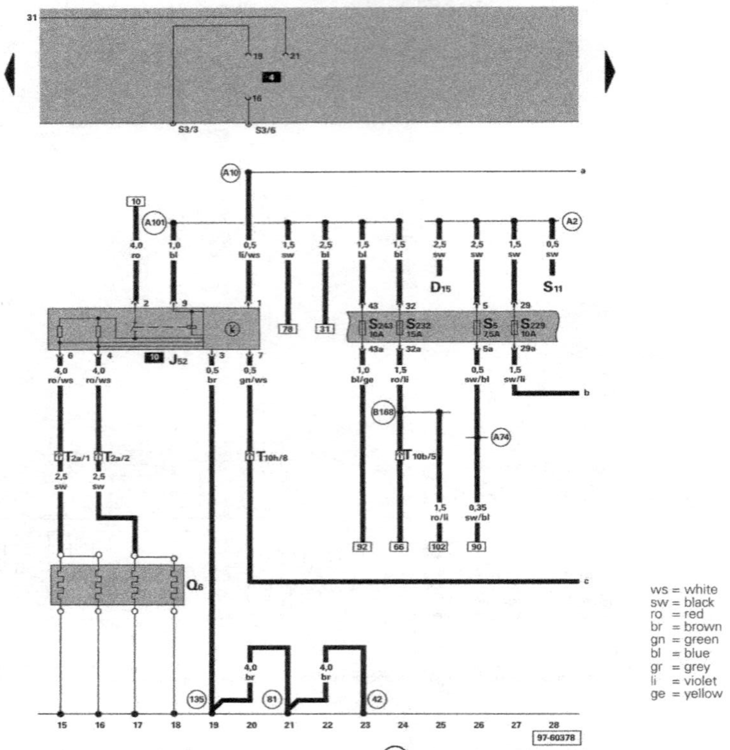

ws = white
sw = black
ro = red
br = brown
gn = green
bl = blue
gr = grey
li = violet
ge = yellow

D - Ignition/Starter Switch
J52 - Glow Plug Relay (180), on the thirteen position auxiliary relay panel, above relay panel
Q6 - Glow plugs (engine)
S5 - Fuse 5 in fuse holder
S11 - Fuse 11 in fuse holder
S229 - Fuse 29 in fuse holder
S232 - Fuse 32 in fuse holder
S243 - Fuse 43 in fuse holder
T2a - Double Connector, in engine compartment, left
T10b - 10-Pin Connector, in engine compartment, left
T10h - 10-Pin Connector, black, behind instrument panel, left
(42) - Ground connection, beside steering column
(81) - Ground connection -1-, in instrument panel wiring harness
(135) - Ground connection -2-, in instrument panel wiring harness
(A2) - plus connection (15), in instrument panel wiring harness
(A10) - wire connection (glow plug control) in instrument panel wiring harness
(A74) - Connector (15a - fuse 5), in instrument panel wiring harness
(A101) - Connector -3- (87), in instrument panel wiring harness
(B168) - Connection (86) in passenger compartment wiring harness

No. 3/4 — Wiring diagram — New Beetle

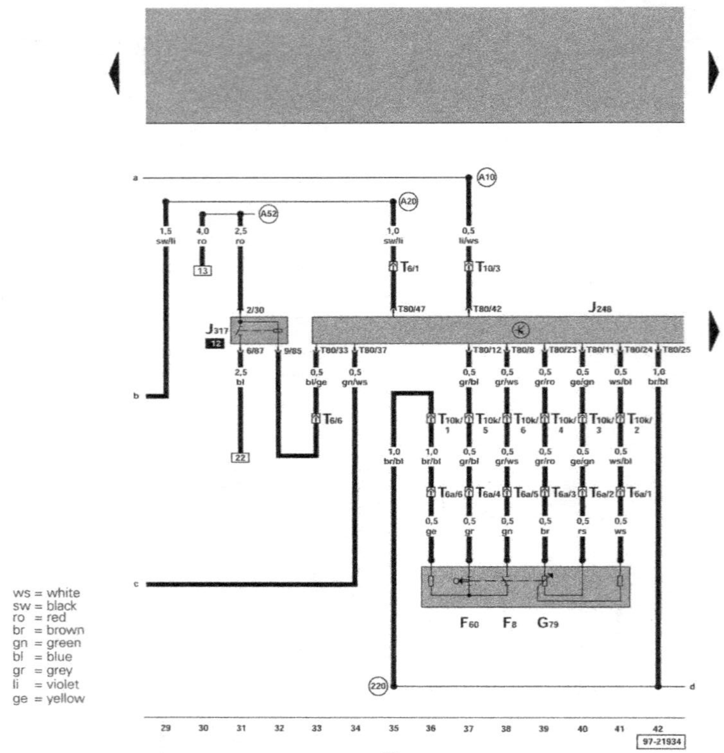

ws = white
sw = black
ro = red
br = brown
gn = green
bl = blue
gr = grey
li = violet
ge = yellow

F8 - Kick Down Switch
F60 - Closed Throttle Position (CTP) Switch
G79 - Throttle Position (TP) Sensor
J248 - Diesel Direct Fuel Injection (DFI) Engine Control Module (ECM), behind instrument panel, left
J317 - Power Supply (Terminal 30, B+) Relay
T6 - 6-Pin Connector, red, behind instrument panel, left
T6a - 6-Pin Connector, behind instrument panel, left
T10 - 10-Pin Connector, white, behind instrument panel, left
T10k - 10-Pin Connector, blue, behind instrument panel, left
T80 - Connector, 80 point, at Diesel Direct Fuel Injection (DFI) Engine Control Module (ECM)
(220) - Ground connection (sensor ground), in engine compartment wiring harness
(A10) - wire connection (glow plug control) in instrument panel wiring harness
(A20) - wire connection (15a), in instrument panel wiring harness
(A52) - plus connection (30), in instrument panel wiring harness

New Beetle — Wiring diagram — No. 3/5

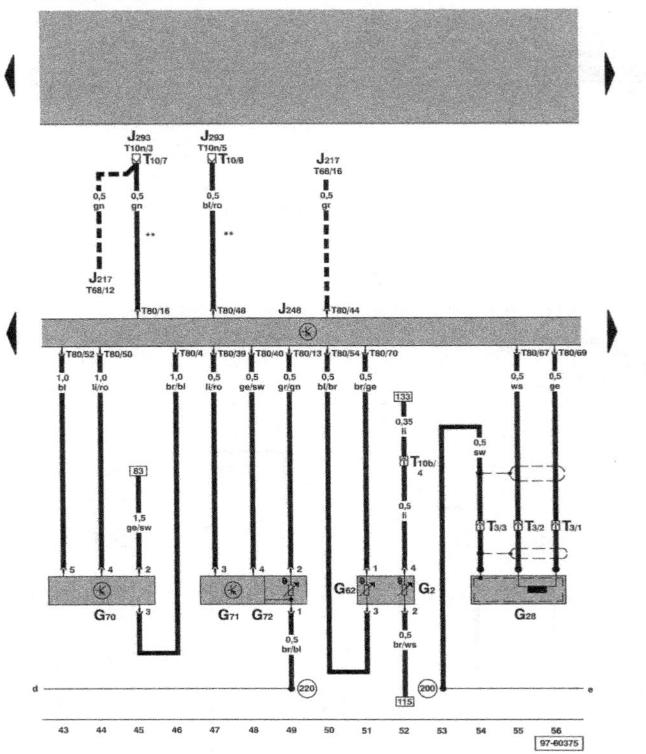

ws = white
sw = black
ro = red
br = brown
gn = green
bl = blue
gr = grey
li = violet
ge = yellow

- G2 - Engine Coolant Temperature (ECT) Sensor
- G28 - Engine Speed (RPM) Sensor
- G62 - Engine Coolant Temperature (ECT) Sensor
- G70 - Mass Air Flow (MAF) Sensor
- G71 - Manifold Absolute Pressure (MAP) Sensor
- G72 - Intake Air Temperature (IAT) Sensor
- J217 - Transmission Control Module (TCM), behind instrument panel, center
- J248 - Diesel Direct Fuel Injection (DFI) Engine Control Module (ECM), behind instrument panel, left
- J293 - Coolant FC (Fan Control) Control Module
- T3 - 3-Pin Connector, on engine, front
- T10 - 10-Pin Connector, white, behind instrument panel, left
- T10b - 10-Pin Connector, in engine compartment, left
- T10n - 10-Pin Connector, on J293
- T68 - 68-Pin Connector, on Transmission Control Module (TCM)
- T80 - Connector, 80 point
- (200) - Ground connection (shielding), in engine compartment wiring harness
- (220) - Ground connection (sensor ground), in engine compartment wiring harness
- ** - A/C Connection
- - - - Automatic transmission only

Edition 02/99
USA.5132.06.21

Diesel direct fuel injection (DFI) engine control module (ECM), mass air flow (MAF) sensor, engine speed (RPM) sensor

No. 3/6 — Wiring diagram — New Beetle

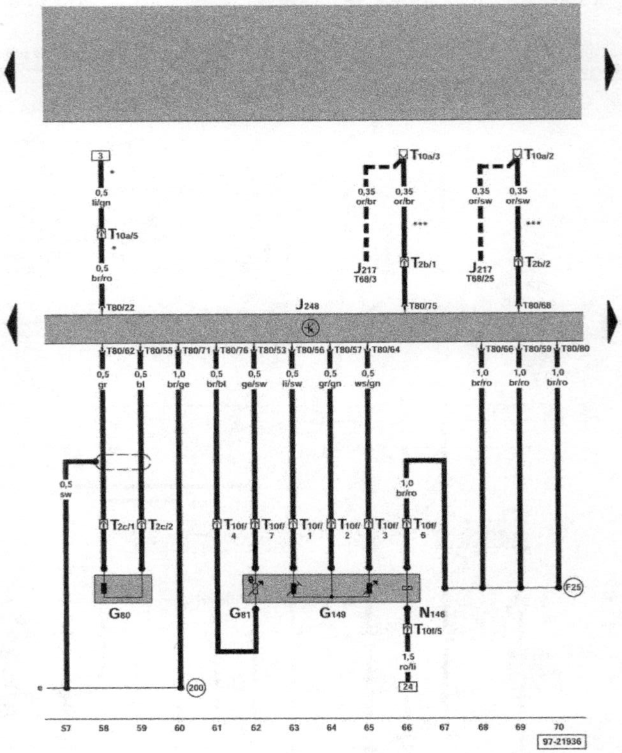

ws = white
sw = black
ro = red
br = brown
gn = green
bl = blue
gr = grey
li = violet
ge = yellow

- G80 - Needle Lift Sensor
- G81 - Fuel Temperature Sensor
- G149 - Modulating Piston Displacement Sensor
- J217 - Transmission Control Module (TCM), behind instrument panel, center
- J248 - Diesel Direct Fuel Injection (DFI) Engine Control Module (ECM), behind instrument panel, left
- N146 - Quantity Adjuster
- T2b - Double Connector, behind instrument panel, left
- T2c - Double Connector, on engine, front
- T10a - 10-Pin Connector, orange, behind instrument panel, left
- T10f - 10-Pin Connector, on engine, front
- T68 - 68-Pin Connector, on Transmission Control Module (TCM)
- T80 - Connector, 80 point
- (200) - Ground connection (shielding), in engine compartment wiring harness
- (F25) - Wire connection -1-, in Diesel Direct Fuel Injection (DFI) system wiring harness
- * - manual transmission only
- *** - connection DATA-Bus
- - - - automatic transmission only

Diesel direct fuel injection (DFI) engine control module (ECM), needle lift sensor, fuel temperature sensor, modulating piston displacement sensor

Edition 01/98
USA.5132.01.21

1.9L engine management (ALH) from Jan. 1998

New Beetle — Wiring diagram — **No. 3/7**

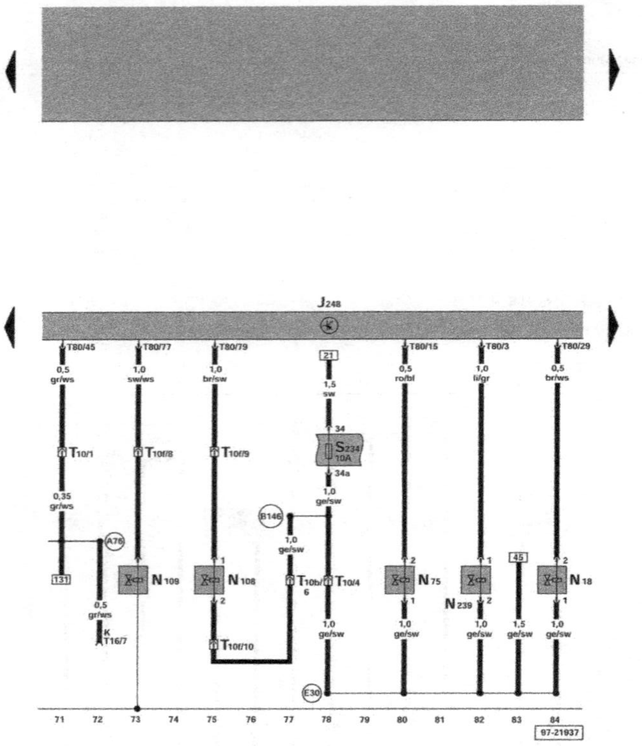

ws = white
sw = black
ro = red
br = brown
gn = green
bl = blue
gr = grey
li = violet
ge = yellow

J248 - Diesel Direct Fuel Injection (DFI) Engine Control Module (ECM), in plenum chamber, left
N18 - EGR Vacuum Regulator Solenoid Valve
N75 - Wastegate Bypass Regulator Valve
N108 - Cold Start Injector
N109 - Fuel Cut-off Valve
N239 - Change-over valve for intake manifold flap
S234 - Fuse 34 in fuse holder
T10 - 10-Pin Connector, white, behind instrument panel, left
T10b - 10-Pin Connector, in engine compartment, left
T10f - 10-Pin Connector, on engine, front
T16 - 16-Pin Connector, Data Link Connector (DLC), below steering column
T80 - Connector, 80 point

(A76) - Connector (K-diagnosis wire), in instrument panel wiring harness
(B146) - Plus connector -1- (87), in wiring harness interior
(E30) - Connector (87a), in wiring harness engine

Diesel direct fuel injection (DFI) engine control module (ECM), EGR vacuum regulator solenoid valve, cold start injector, fuel cut-off valve

No. 3/8 — Wiring diagram — **New Beetle**

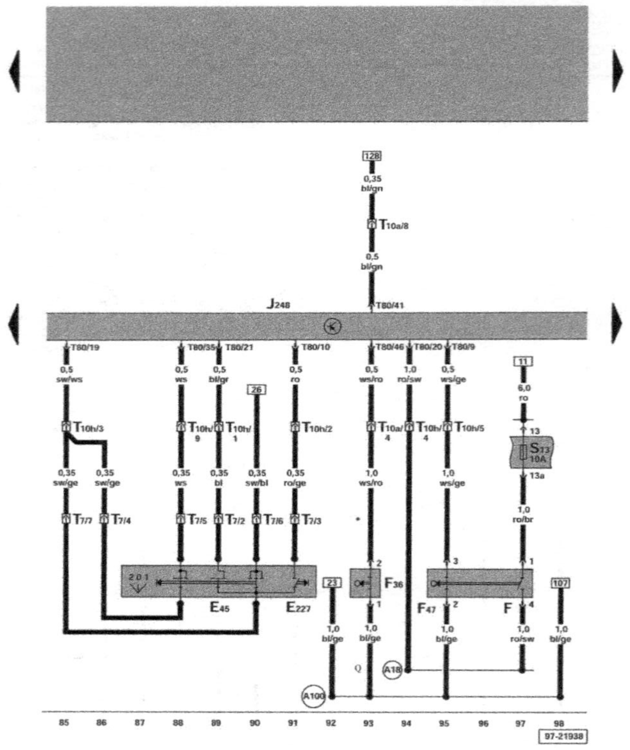

ws = white
sw = black
ro = red
br = brown
gn = green
bl = blue
gr = grey
li = violet
ge = yellow

E45 - Cruise Control Switch (if equipped)
E227 - Button for cruise control (set), (if equipped)
F - Brake Light Switch
F36 - Clutch Vacuum vent Valve Switch
F47 - Brake Vacuum Vent Valve Switch for cruise control/diesel fuel injection (DFI)
J248 - Diesel Direct Fuel Injection (DFI) Engine Control Module (ECM), behind instrument panel, left
S13 - Fuse 13 in fuse holder
T7 - 7-Pin Connector, near steering column
T10a - 10-Pin Connector, orange, behind instrument panel, left
T10h - 10-Pin Connector, black, behind instrument panel, left
T80 - Connector, 80 point

(A18) - wire connection (54), in instrument panel wiring harness
(A100) - Connector -2- (87), in instrument panel wiring harness
* - manual transmission only

Diesel direct fuel injection (DFI) engine control module (ECM), cruise control switch, brake light switch, clutch vacuum vent valve switch

New Beetle Wiring diagram **No. 3/9**

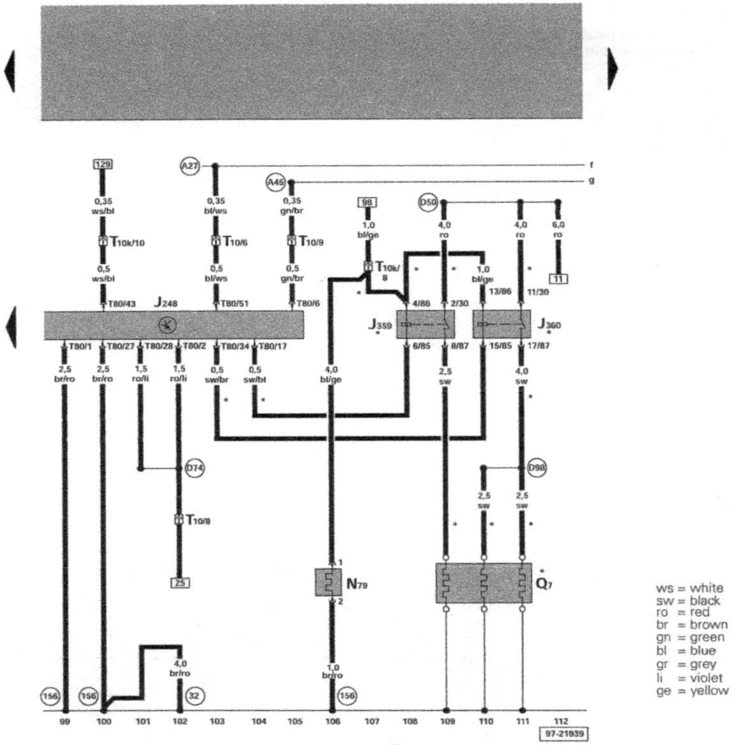

ws = white
sw = black
ro = red
br = brown
gn = green
bl = blue
gr = grey
li = violet
ge = yellow

J248 - Diesel Direct Fuel Injection (DFI) Engine Control Module (ECM), behind instrument panel, left
J359 - Relay for preheating coolant, low heat output, in engine compartment
J360 - Relay for preheating coolant, high heat output, in engine compartment
N79 - Positive Crankcase Ventilation (PCV) Heating Element
Q7 - glow plugs (coolant)
T10 - 10-Pin Connector, white, behind instrument panel, left
T10k - 10-Pin Connector, blue, behind instrument panel, left
T80 - Connector, 80 point

(32) - Ground connection, behind instrument panel, left

(156) - Ground connection, in Diesel Direct Fuel Injection (DFI) wiring harness
(A27) - wire Connection (vehicle speed signal), in instrument panel wiring harness
(A45) - wire connection (RPM-signal), in instrument panel wiring harness
(D50) - plus connection (30), in engine compartment wiring harness
(D74) - wire connection (86), in engine compartment wiring harness
(D98) - Wire connection (glow plugs), in engine compartment wiring harness

* - manual transmission only

Edition 01/98
USA.5132.01.21

Diesel direct fuel injection (DFI) engine control module (ECM), relay for preheating coolant, glow plugs (coolant)

No. 3/10 Wiring diagram **New Beetle**

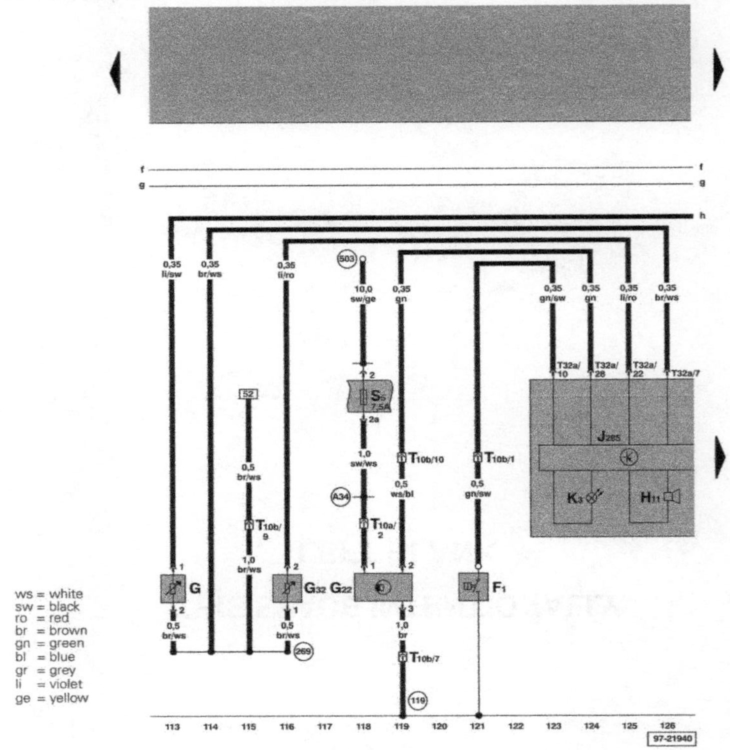

ws = white
sw = black
ro = red
br = brown
gn = green
bl = blue
gr = grey
li = violet
ge = yellow

F1 - Oil Pressure Switch
G - Sender for fuel gauge
G22 - Speedometer Vehicle Speed Sensor (VSS)
G32 - Engine Coolant Level (ECL) Sensor
H11 - Oil Pressure Warning Buzzer
J285 - Control module with indicator unit in instrument panel insert
K3 - Oil Pressure Warning Light
S2 - Fuse 2 in fuse holder
T10a - 10-Pin Connector, orange, behind instrument panel, left
T10b - 10-Pin Connector, in engine compartment, left
T32a - 32-Pin Connector, blue, on instrument cluster

(119) - Ground connection -1-, in headlight wiring harness

(269) - Ground connector (sensor ground) -1-, in instrument panel wiring harness
(503) - Threaded connection -1- (75x) on the relay plate
(A34) - wire connection (75x), in instrument panel wiring harness

Sender for fuel gauge, engine coolant level (ECL) sensor, speedometer vehicle speed sensor (VSS), visual and acoustical oil pressure control

Edition 07/98
USA.5132.03.21

New Beetle Wiring diagram **No. 3/11**

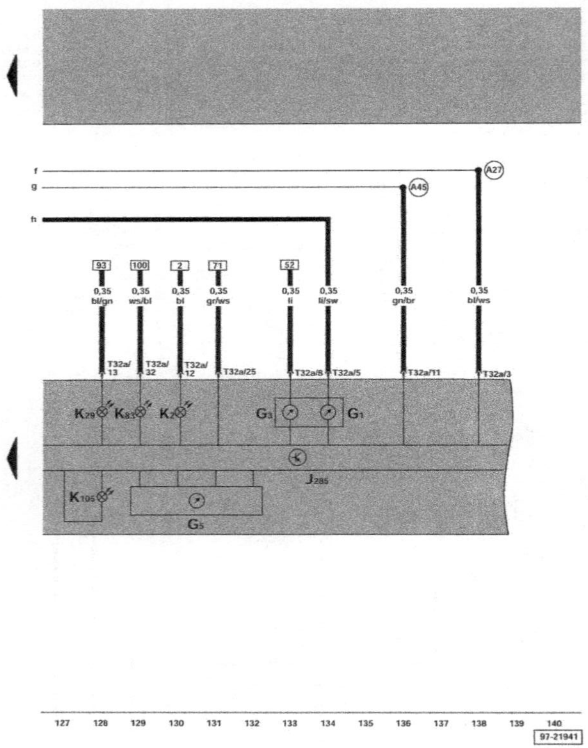

ws = white
sw = black
ro = red
br = brown
gn = green
bl = blue
gr = grey
li = violet
ge = yellow

- G1 - Fuel gauge
- G3 - Engine Coolant Temperature (ECT) Gauge
- G5 - Tachometer
- J285 - Control module with indicator unit in instrument panel insert
- K2 - Generator (GEN) Warning Light
- K83 - Exhaust Warning Light
- K29 - Glow Plug Indicator Light
- K105 - Low Fuel Level Warning Light
- T32a - 32-Pin Connector, blue, on instrument cluster

- (A27) - wire Connection (vehicle speed signal), in instrument panel wiring harness
- (A45) - wire connection (RPM-signal), in instrument panel wiring harness

Edition 01/98
USA.5132.01.21

Instrument cluster, fuel gauge, engine coolant temperature gauge, generator (GEN) warning light, low fuel level warning light

THIS SPACE INTENTIONALLY LEFT BLANK

1,9L-Engine - Turbo Diesel Fuel Injection (DFI)/66 kW, code ALH,

from June 1999

Deviate relay location and fuseplacements as well as the locations of multiple connectors see section "component locations".

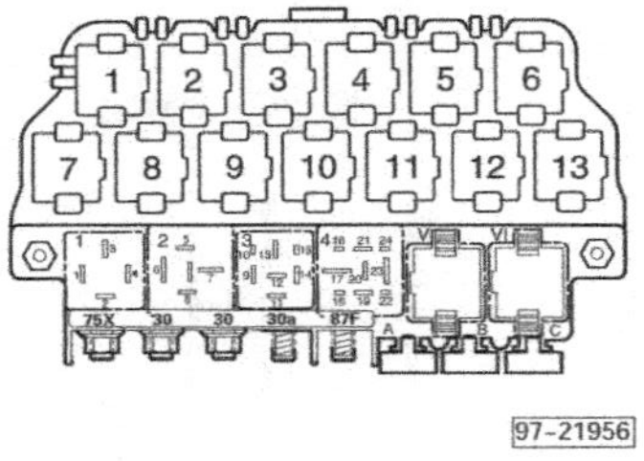

97-21956

Relay location on the thirteen position auxiliary relay panel, above relay panel:

- **10** Glow Plug Relay (180)
- **12** Power Suppy (Terminal 30, B+) Relay (109))

Relay panel:

- **2** Load Reduction Relay (100)

Note: Number in parentheses indicates production control number stamped on relay housing.

Fuse colors

30 A - green
25 A - white
20 A - yellow
15 A - blue
10 A - red
7,5 A - brown
5 A - beige

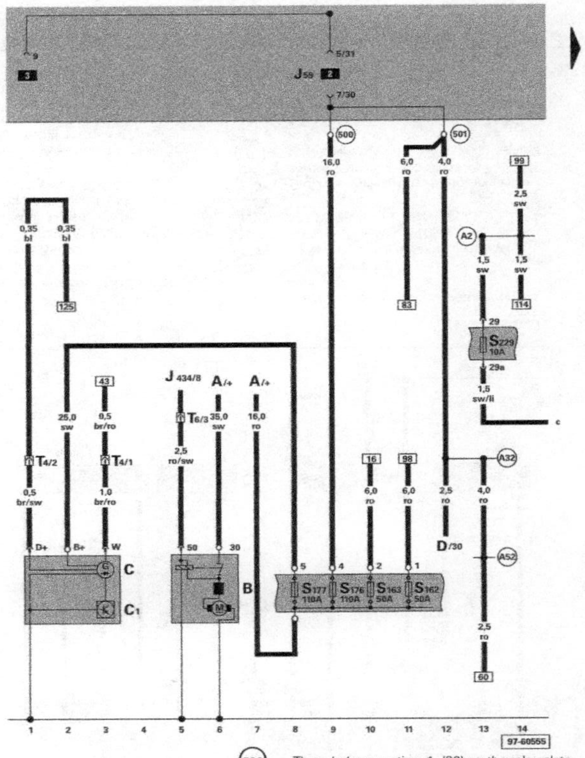

ws = white
sw = black
ro = red
br = brown
gn = green
bl = blue
gr = grey
li = violet
ge = yellow

- A - Battery
- B - Starter
- C - Generator (GEN)
- C1 - Voltage Regulator (VR)
- D - Ignition/Starter Switch
- J434 - Locking relay for starter (clutch pedal switch)
- S162 - Fuse -1- in fuse bracket/battery
- S163 - Fuse -2- in fuse bracket/battery
- S176 - Fuse -4- (30), in fuse bracket/battery
- S177 - Fuse -5- (30), in fuse bracket/battery
- S229 - Fuse 29 in fuse holder
- T4 - 4-Pin connector, in engine compartment, left
- T6 - 6-Pin Connector, brown, behind instrument panel, left

- (500) - Threaded connection -1- (30) on the relay plate
- (501) - Threaded connection -2- (30) on the relay plate
- (A2) - plus connection (15), in instrument panel wiring harness
- (A32) - plus connection (30), in instrument panel wiring harness
- (E30) - Connector (87a), in wiring harness engine
- (A52) - Plus connection (30), in instrument panel wiring harness

Battery, starter, Generator (GEN), Voltage Regulator (VR)

1.9L engine management (ALH) from June 1999

New Beetle **Wiring diagram** **No. 29/3**

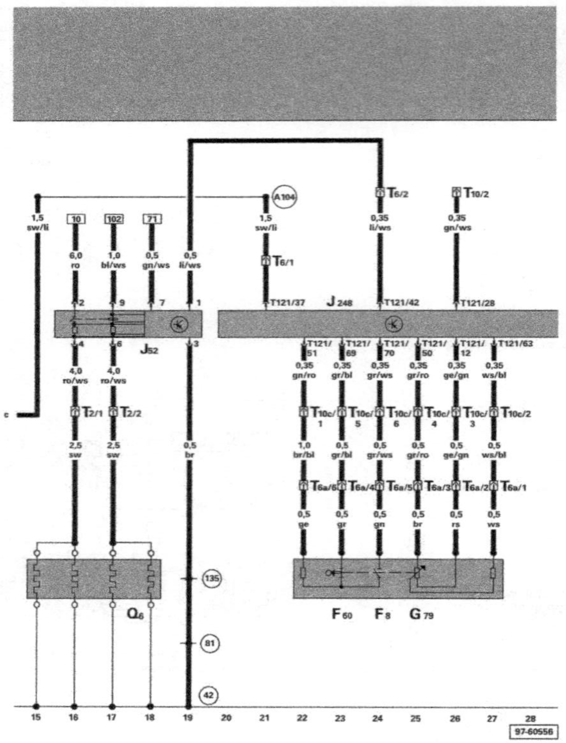

ws = white
sw = black
ro = red
br = brown
gn = green
bl = blue
gr = grey
li = violet
ge = yellow

F8 - Kick Down Switch
F60 - Closed Throttle Position (CTP) Switch
G79 - Throttle Position (TP) Sensor
J52 - Glow Plug Relay
J248 - Diesel Direct Fuel Injection (DFI) Engine Control Module (ECM)
Q6 - Glow plugs (engine)
T2 - Double Connector, in engine compartment, left
T6 - 6-Pin Connector, brown, behind instrument panel, left
T6a - 6-Pin Connector
T10 - 10-Pin Connector, orange, behind instrument panel, left
T10c - 10-Pin Connector, blue, behind instrument panel, left
T121 - 121-Pin Connector, on DFI ECM

(42) - Ground connection, beside steering column
(81) - Ground connection -1-, in instrument panel wiring harness
(135) - Ground connection -2-, in instrument panel wiring harness
(A104) - Plus connector -2- (15), in instrument panel wiring harness

 DFI Engine Control Module (ECM), kick down switch, Closed Throttle Position (CTP) Sensor, Closed Thottle Position (CTP) Switch, glow plug relay, glow plugs (engine)

No. 29/4 **Wiring diagram** **New Beetle**

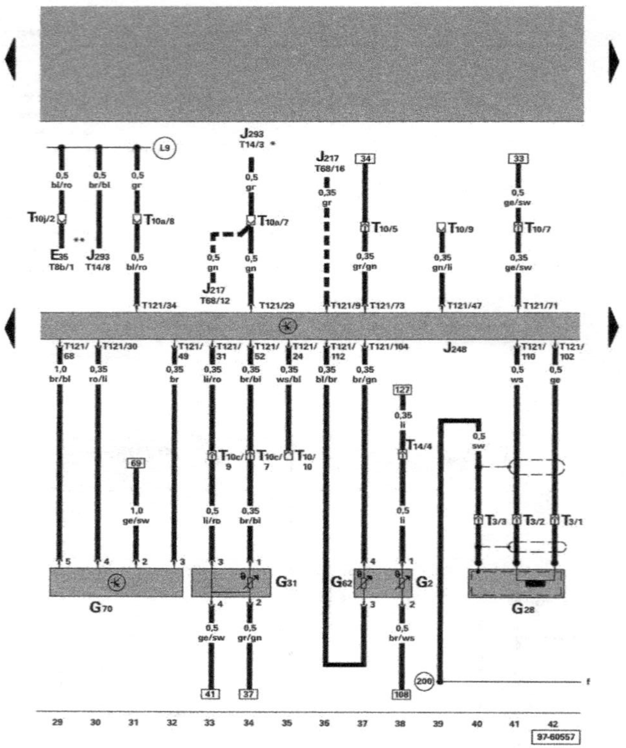

ws = white
sw = black
ro = red
br = brown
gn = green
bl = blue
gr = grey
li = violet
ge = yellow

E35 - A/C Switch
G2 - Engine Coolant Temperature (ECT) Sensor
G28 - Engine Speed (RPM) Sensor
G31 - Charge Air Pressure Sensor
G62 - Engine Coolant Temperature (ECT) Sensor
G70 - Mass Air Flow (MAF) Sensor
J217 - Transmission Control Module (TCM)
J248 - Diesel Direct Fuel Injection (DFI) Engine Control Module (ECM)
J293 - Coolant FC (Fan Control) Control Module
T3 - 3-Pin connector, on engine, front
T10 - 10-Pin connector, white, behind instrument panel, left
T10a - 10-Pin connector, orange, behind instrument panel, left
T10c - 10-Pin connector, blue, behind instrument panel, left
T10j - 10-Pin connector
T14 - 14-Pin connector, in plenum chamber
T68 - 68-Pin connector, on TCM
T121 - 121-Pin connector, on DFI ECM

(200) - Ground connection (shielding), in engine compartment wiring harness
(L9) - Wire connection -1-, in A/C wiring harness

* - A/C clutch cut-out signal from ECM/TCM
** - A/C "ON" signal from -E35-
--- - Automatic transmission only

DFI Engine Control Module (ECM), Engine Coolant Temperature (ECT) Sensor, Engine Speed (RPM) Sensor, charge air pressure sensor, Mass Air Flow (MAF) Sensor

New Beetle — Wiring diagram — No. 29/5

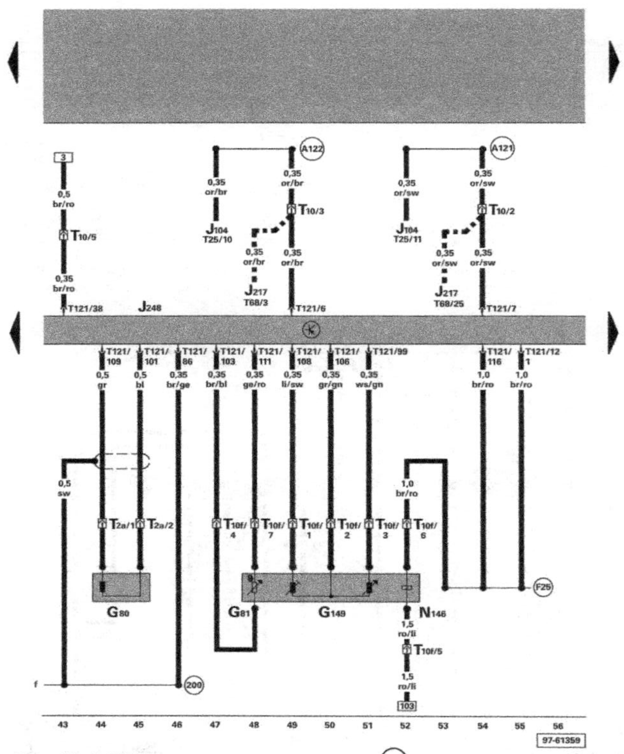

ws = white
sw = black
ro = red
br = brown
gn = green
bl = blue
gr = grey
li = violet
ge = yellow

G 80 - Needle Lift Sensor
G 81 - Fuel Temperature Sensor
G149 - Modulating Piston Displacement Sensor
J104 - ABS Control Module (w/EDL)
J217 - Transmission Control Module (TCM)
J248 - Diesel Direct Fuel Injection (DFI) Engine Control Module (ECM)
N146 - Quantity Adjuster
T2a - 2-Pin connector, on engine, front
T10 - 10-Pin connector, white, behind instrument panel, left
T10f - 10-Pin connector, on engine, front
T25 - 25-Pin connector
T68 - 68-Pin connector
T121 - 121-Pin connector, on DFI ECM

(200) - Ground connection (shielding), in engine compartment wiring harness
(A121) - Connection (high bus) in instrument panel wiring harness
(A122) - Connection (low bus) in instrument panel wiring harness
(F25) - Wire connection -1-, in Diesel Direct Fuel Injection (DFI) system wiring harness

No. 29/6 — Wiring diagram — New Beetle

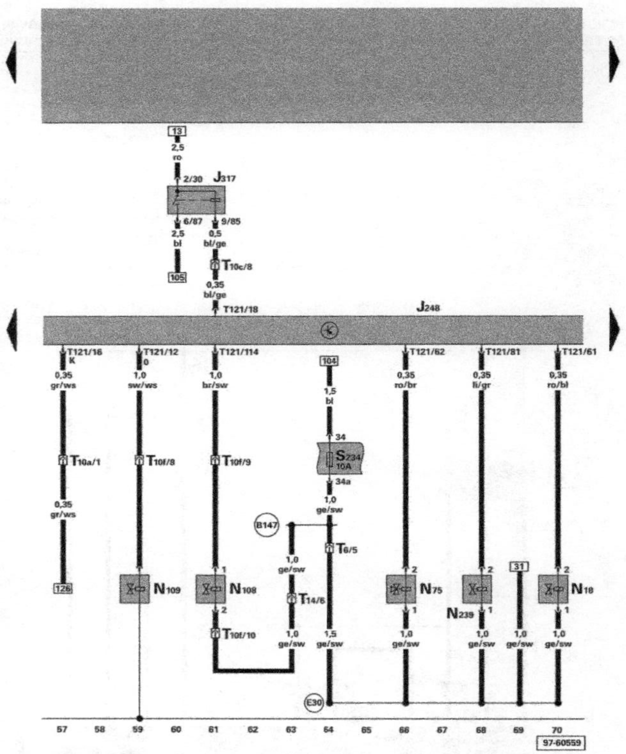

ws = white
sw = black
ro = red
br = brown
gn = green
bl = blue
gr = grey
li = violet
ge = yellow

J248 - Diesel Direct Fuel Injection (DFI) Engine Control Module (ECM)
J317 - Power Supply (Terminal 30, B+) Relay
N 18 - EGR Vacuum Regulator Solenoid Valve
N 75 - Wastegate Bypass Regulator Valve
N108 - Cold Start Injector
N109 - Fuel Cut-off Valve
N239 - Change-over valve for intake manifold flap
S234 - Fuse 34 in fuse holder
T6 - 6-Pin connector, brown, behind instrument panel, left
T10a - 10-Pin connector, orange, behind instrument panel, left
T10c - 10-Pin connector, blue, behind instrument panel, left
T10f - 10-Pin connector, black, on engine, front
T14 - 14-Pin connector, in plenum chamber
T121 - 121-Pin connector, on DFI ECM

(B147) - Plus connector -2- (87), in wiring harness
(E30) - Connector (87a), in wiring harness engine

1.9L engine management (ALH) from June 1999

New Beetle — Wiring diagram — No. 29/7

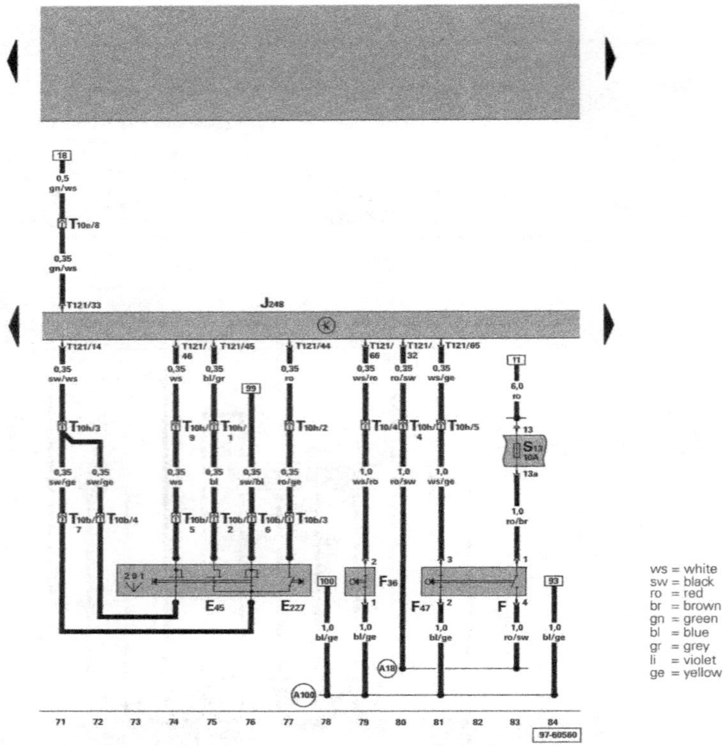

ws = white
sw = black
ro = red
br = brown
gn = green
bl = blue
gr = grey
li = violet
ge = yellow

E 45 - Cruise Control Switch
E227 - Button for cruise control (set)
F - Brake Light Switch
F36 - Clutch Vacuum Vent Valve Switch
F47 - Brake Vacuum Vent Valve Switch for cruise contro
J248 - Diesel Direct Fuel Injection (DFI) Engine Control Module (ECM)
S13 - Fuse 13 in fuse holder
T10 - 10-Pin connector, white, behind instrument panel, left
T10h - 10-Pin connector, black, behind instrument panel, left
T10b - 10-Pin connector, black, near steering column
T121 - 121-Pin connector, on DFI ECM

(A18) - wire connection (54), in instrument panel wiring harness

(A100) - Connector -2- (87), in instrument panel wiring harness

* - Manual transmission only

DFI Engine Control Module (ECM), button for cruise control (set), cruise control switch, brake light switch, clutch & brake vacuum vent valves

No. 29/8 — Wiring diagram — New Beetle

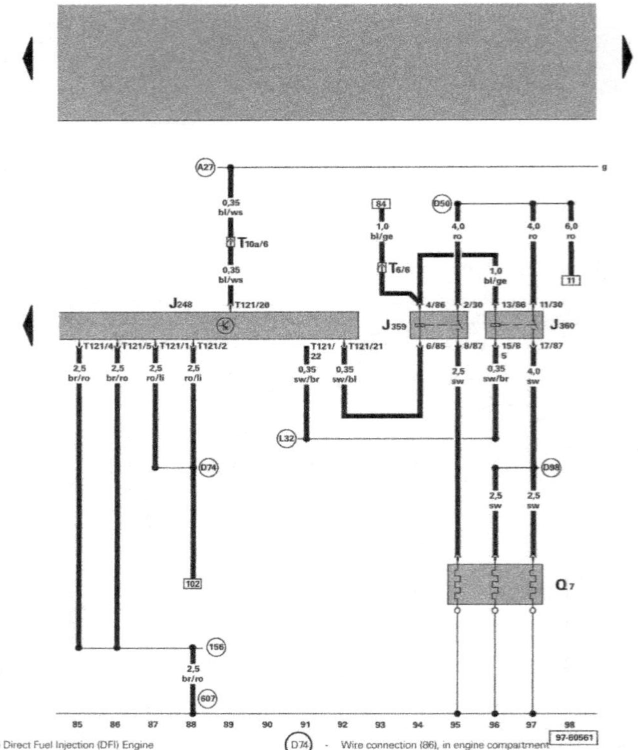

ws = white
sw = black
ro = red
br = brown
gn = green
bl = blue
gr = grey
li = violet
ge = yellow

J248 - Diesel Direct Fuel Injection (DFI) Engine Control Module (ECM
J359 - Realy for preheating coolant, low heat output
J360 - Relay for prehaeating coolant, high heat output
Q7 - Spark plugs (coolant)
T6 - 6-Pin connector, brown, behind instrument panel, left
T10a - 10-Pin connector, orange, behind instrument panel, left
T121 - 121-Pin connector, on DFI ECM

(607) - Ground Connection (in left plenum chamber)

(156) - Ground connection, in Diesel Direct Fuel Injection (DFI) wiring harness

(A27) - Wire Connection (vehicle speed signal), in instrument panel wiring harness

(D50) - Plus connection (30), in engine compartment wiring harness

(D74) - Wire connection (86), in engine compartment wiring harness

(D98) - Wire connection (glow plugs), in engine compartment wiring harness

(L32) - Wire connection -1-, in heater wiring harness

DFI Engine Control Module (ECM), relay for preheating coolant: low heat output, high heat output, glow plugs (coolant)

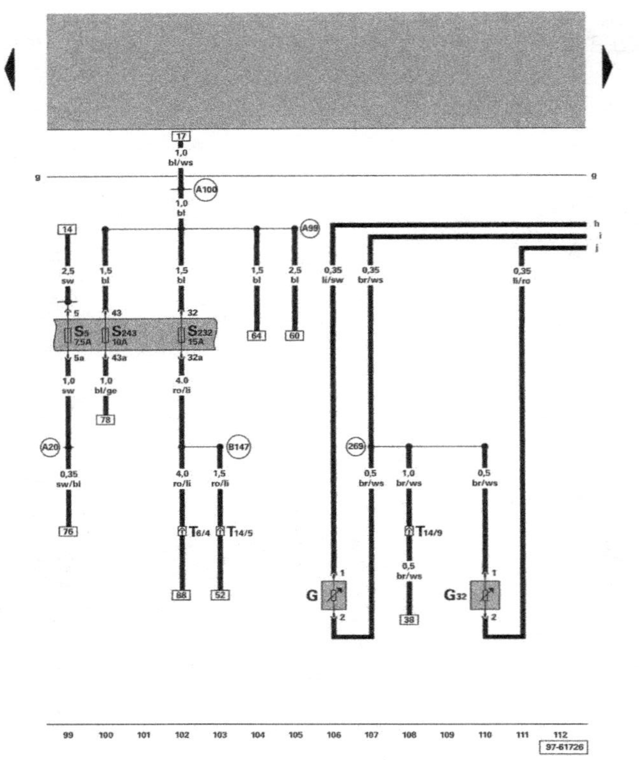

ws = white
sw = black
ro = red
br = brown
gn = green
bl = blue
gr = grey
li = violet
ge = yellow

- G - Sender for fuel gauge
- G32 - Engine Coolant Level (ECL) Sensor
- S5 - Fuse 5 in fuse holder
- S232 - Fuse 32 in fuse holder
- S243 - Fuse 43 in fuse holder
- T6 - 6-Pin connector, brown, behind instrument panel, left
- T14 - 14-Pin connector, in plenum chamber
- (269) - Ground connector (sensor ground) -1-, in instrument panel wiring harness
- (A20) - Wire connection (15a), in instrument panel wiring harness
- (A99) - Connector -1- (87), in instrument panel wiring harness
- (A100) - Connector -2- (87), in instrument panel wiring harness
- (B147) - Plus connector -2- (87), in wiring harness

Sender for fuel gauge, Engine Coolant Level (ECL) Sensor

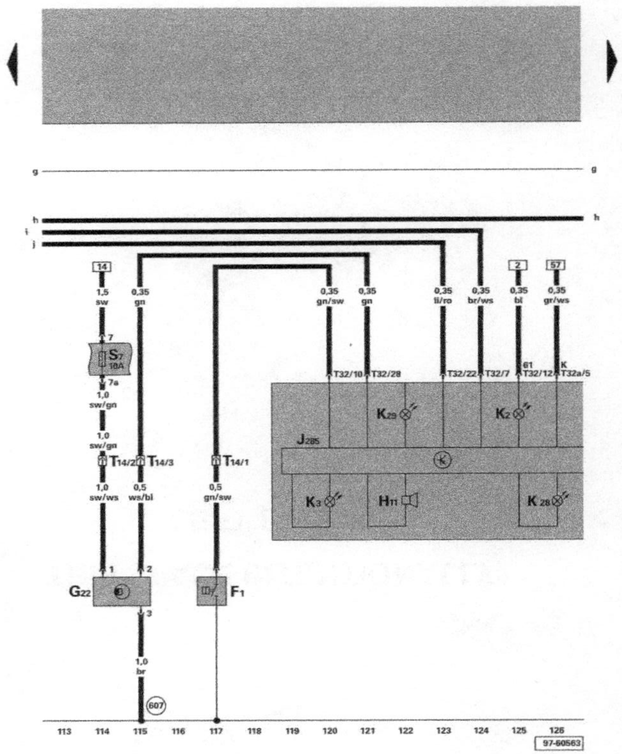

ws = white
sw = black
ro = red
br = brown
gn = green
bl = blue
gr = grey
li = violet
ge = yellow

- F1 - Oil Pressure Switch
- G22 - Speedometer Vehicle Speed Sensor (VSS)
- H11 - Oil Pressure Warning Buzzer
- J285 - Control module with indicator unit in instrument panel insert
- K2 - Generator (GEN) Warning Light
- K3 - Oil Pressure Warning Light
- K28 - Engine Coolant Level/Temperature (ECL/ECT) Warning Light
- K29 - Glow Plug Indicator Light
- S7 - Fuse 7 in fuse holder
- T14 - 14-Pin connector, in plenum chamber
- T32 - 32-Pin connector, blue, on instrument cluster
- T32a - 32-Pin connector, green, on instrument cluster
- (607) - Ground Connection (in left plenum chamber)

Oil pressure switch, Speedometer Vehicle Speed Sensor (VSS), instrument cluster, oil pressure warning buzzer, warning and indicator lights

New Beetle Wiring diagram **No. 29/11**

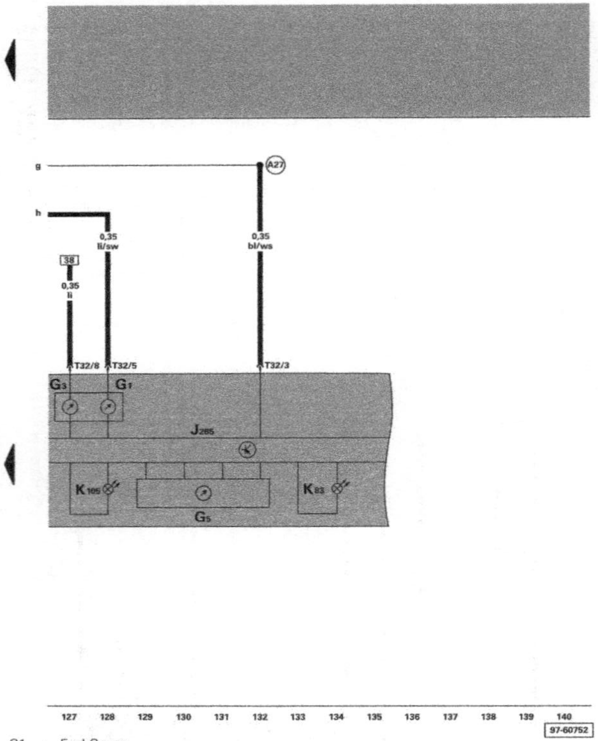

ws = white
sw = black
ro = red
br = brown
gn = green
bl = blue
gr = grey
li = violet
ge = yellow

G1 - Fuel Gauge
G3 - Engine Coolant Temperature (ECT) Gauge
G5 - Tachometer
J285 - Control module with indicator unit instrument panel insert
K83 - Malfunction Indicator Light (MIL)
K105 - Low Fuel Level Warning Light
T32 - 32-Pin connector, blue, on instrument cluster

(A27) - Wire Connection (vehicle speed signal), in instrument panel wiring harness

THIS SPACE INTENTIONALLY LEFT BLANK

New Beetle | **Wiring diagram** | **No. 4/1**

1,9L-Engine - Turbo Diesel Fuel Injection (DFI)/66 kW, code ALH,

from June 2001

Deviate relay location and fuseplacements as well as the locations of multiple connectors see section "component locations".

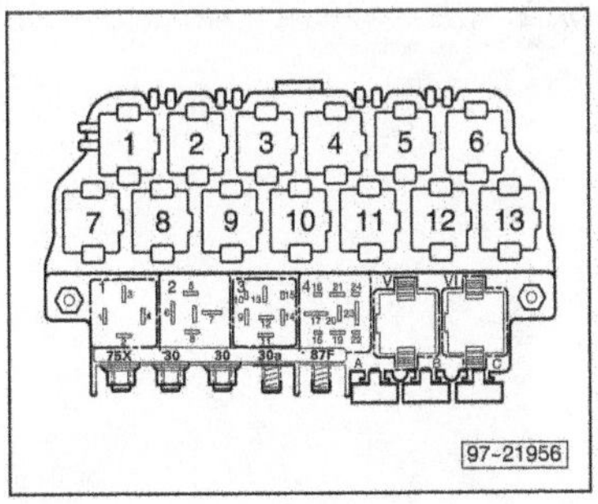

Relay location on the thirteen position auxiliary relay panel, above relay panel:

- **10** Glow Plug Relay (180)
- **12** Power Suppy (Terminal 30, B+) Relay (109))

Relay panel:

- **2** Load Reduction Relay (100)

Fuse colors

30 A - green
25 A - white
20 A - yellow
15 A - blue
10 A - red
7,5 A - brown
5 A - beige
3 A - violet

Note: Number in parentheses indicates production control number stamped on relay housing.

New Beetle | **Wiring diagram** | **No. 4/2**

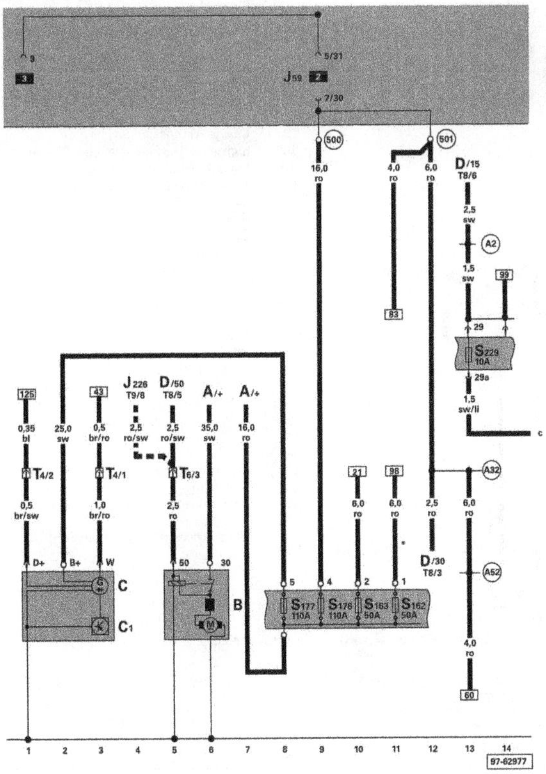

Battery, starter, Generator (GEN), Voltage Regulator (VR)

- A - Battery
- B - Starter
- C - Generator (GEN)
- C1 - Voltage Regulator (VR)
- D - Ignition/Starter Switch
- J59 - Load Reduction Relay
- J226 - Park/Neutral Position (PNP) Relay
- S162 - Fuse -1- in fuse bracket/battery
- S163 - Fuse -2- in fuse bracket/battery
- S176 - Fuse -4- (30), in fuse bracket/battery
- S177 - Fuse -5- (30), in fuse bracket/battery
- S229 - Fuse 29 in fuse holder
- T4 - 4-Pin connector, in engine compartment, left
- T6 - 6-Pin Connector, brown, behind instrument panel, left
- T8 - 8-Pin Connector, at -D-
- T9 - 9-Pin Connector
- (500) - Threaded connection -1- (30) on the relay plate
- (501) - Threaded connection -2- (30) on the relay plate
- (A2) - plus connection (15), in instrument panel wiring harness
- (A32) - plus connection (30), in instrument panel wiring harness
- (A52) - Plus connection (30), in instrument panel wiring harness
- * - Manual transmission only
- --- Automatic transmission only

ws = white
sw = black
ro = red
br = brown
gn = green
bl = blue
gr = grey
li = violet
ge = yellow
or = orange

New Beetle

Wiring diagram **No. 4/3**

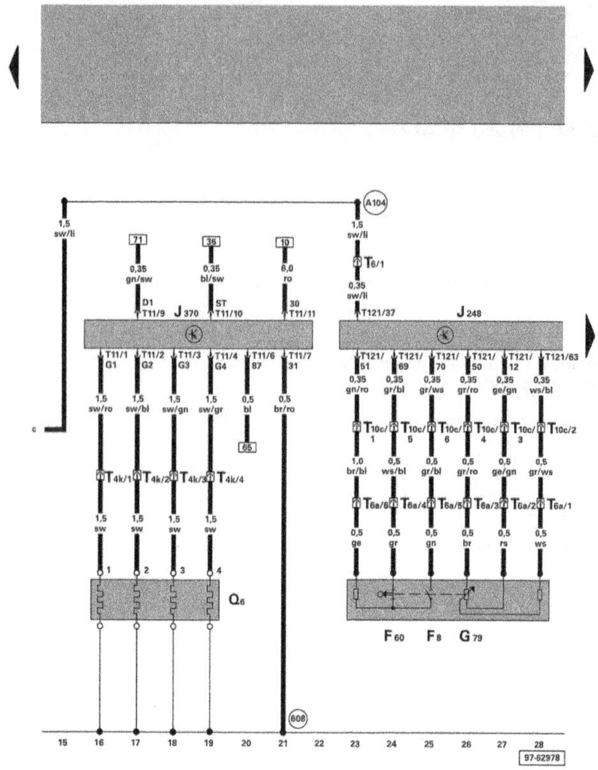

DFI Engine control module (ECM),kick down switch,throttle position (TP) sensor, closed throttle position (CTP) switch,control module for glow plug triggerin

- F8 - Kick Down Switch
- F60 - Closed Throttle Position (CTP) Switch
- G79 - Throttle Position (TP) Sensor
- J248 - Diesel Direct Fuel Injection (DFI) Engine Control Module (ECM)
- J370 - Control module for glow plug triggering
- Q6 - Glow plugs (engine)
- T4k - 4 Pin Connector
- T6 - 6-Pin Connector, brown, behind instrument panel, left
- T6a - 6-Pin Connector
- T10c - 10-Pin Connector, blue, behind instrument panel, left
- T11 - 11-Pin Connector
- T121 - 121-Pin Connector, on DFI ECM
- (608) - Ground Connection (in center plenum chamber)
- (A104) - Plus connector -2- (15), in instrument panel wiring harness

ws = white
sw = black
ro = red
br = brown
gn = green
bl = blue
gr = grey
li = violet
ge = yellow
or = orange

97-62978

New Beetle

Wiring diagram **No. 4/4**

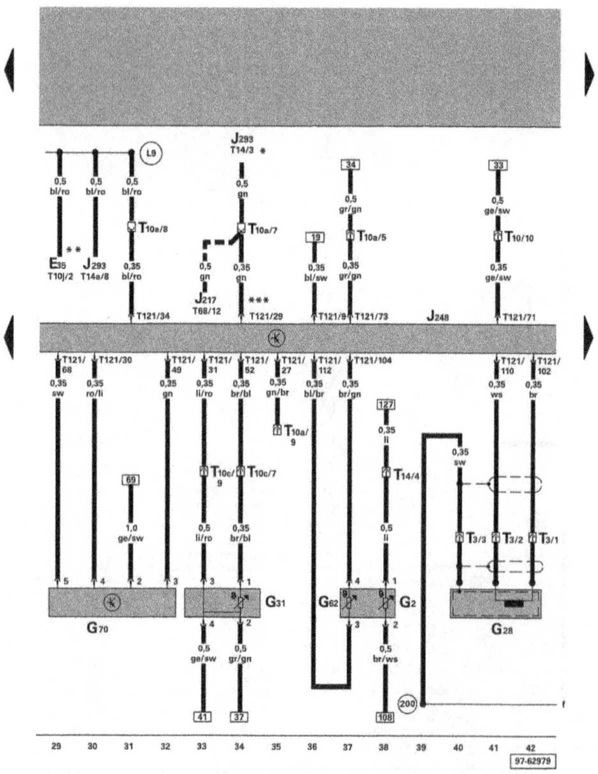

DFI Engine control module (ECM), engine coolant temperature (ECT) sensor, engine speed (RPM) sensor, charge air pressure sensor, mass air flow (MAF) sensor

- E 35 - A/C Switch
- G2 - Engine Coolant Temperature (ECT) Sensor
- G28 - Engine Speed (RPM) Sensor
- G31 - Charge Air Pressure Sensor
- G62 - Engine Coolant Temperature (ECT) Sensor
- G70 - Mass Air Flow (MAF) Sensor
- J217 - Transmission Control Module (TCM)
- J248 - Diesel Direct Fuel Injection (DFI) Engine Control Module (ECM)
- J293 - Coolant FC (Fan Control) Control Module
- T3 - 3-Pin connector, on engine, front
- T10 - 10-Pin connector, white, behind instrument panel, left
- T10a - 10-Pin connector, orange, behind instrument panel, left
- T10c - 10-Pin connector, blue, behind instrument panel, left
- T10j - 10-Pin connector
- T14 - 14-Pin connector, in plenum chambe
- T14a - 14-Pin connector, at -J293-
- T68 - 68-Pin connector, on TCM
- T121 - 121-Pin connector, on DFI ECM
- (200) - Ground connection (shielding), in engine compartment wiring harness
- (L9) - Wire connection -1-, in A/C wiring harness
- * - A/C clutch cut-out signal from ECM/TCM
- ** - A/C "ON" signal from -E35-
- *** - Manual Transmission only
- - - - - Automatic transmission only

ws = white
sw = black
ro = red
br = brown
gn = green
bl = blue
gr = grey
li = violet
ge = yellow
or = orange

97-62979

New Beetle Wiring diagram **No. 4/5**

DFI Engine control module (ECM), needle lift sensor, fuel temperature sensor, modulating piston displacement sensor, quantity adjuster

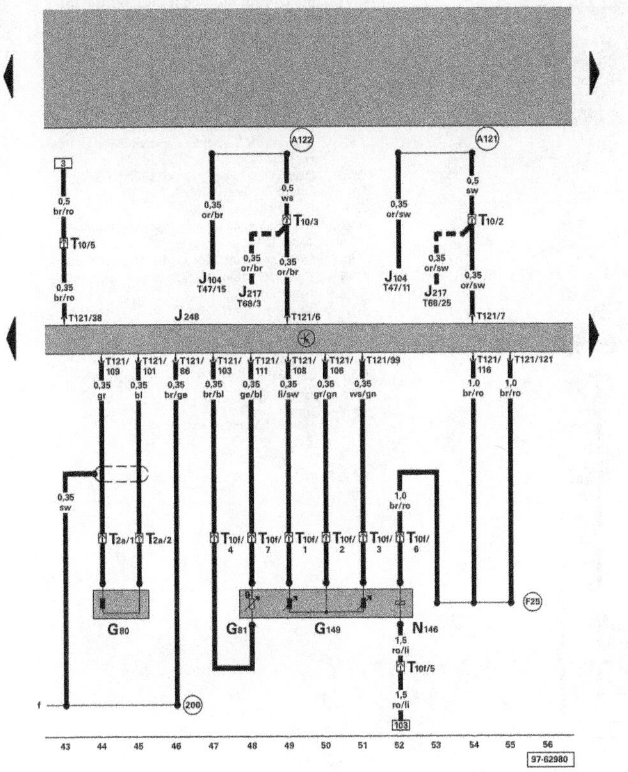

G 80 - Needle Lift Sensor
G 81 - Fuel Temperature Sensor
G149 - Modulating Piston Displacement Sensor
J104 - ABS Control Module (w/EDL)
J217 - Transmission Control Module (TCM)
J248 - Diesel Direct Fuel Injection (DFI) Engine Control Module (ECM)
N146 - Quantity Adjuster
T2a - 2-Pin connector, on engine, front
T10 - 10-Pin connector, white, behind instrument panel, left
T10f - 10-Pin connector, on engine, front
T47 - 47-Pin connector
T68 - 68-Pin connector
T121 - 121-Pin connector, on DFI ECM

(200) - Ground connection (shielding), in engine compartment wiring harness
(A121) - Connection (high bus) in instrument panel wiring harness
(A122) - Connection (low bus) in instrument panel wiring harness
(F25) - Wire connection -1-, in Diesel Direct Fuel Injection (DFI) system wiring harness

- - - - Automatic Transmission only

ws = white
sw = black
ro = red
br = brown
gn = green
bl = blue
gr = grey
li = violet
ge = yellow
or = orange

New Beetle Wiring diagram **No. 4/6**

DFI ECM, power supply relay, EGR vacuum regulator solenoid valve, wastegate bypass regulator valve, fuel cut-off valve, change-over valve for int. man. flap.

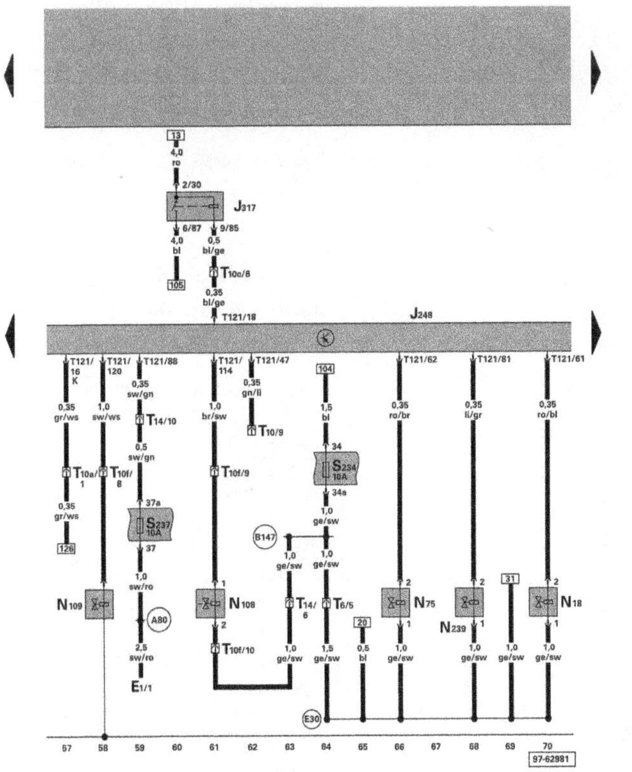

E 1 - Light switch
J248 - Diesel Direct Fuel Injection (DFI) Engine Control Module (ECM)
J317 - Power Supply (Terminal 30, B+) Relay
N 18 - EGR Vacuum Regulator Solenoid Valve
N 75 - Wastegate Bypass Regulator Valve
N108 - Cold Start Injector
N109 - Fuel Cut-off Valve
N239 - Change-over valve for intake manifold flap
S234 - Fuse 34 in fuse holder
S237 - Fuse 37 in fuse holder
T6 - 6-Pin connector, brown, behind instrument panel, left
T10 - 10-Pin connector, white, behind instrument panel, left
T10a - 10-Pin connector, orange, behind instrument panel, left
T10c - 10-Pin connector, blue, behind instrument panel, left
T10f - 10-Pin connector, black, on engine, front
T14 - 14-Pin connector, in plenum chamber
T121 - 121-Pin connector, on DFI ECM

(A80) - Connector -1- (X) in instrument panel wiring harness
(B147) - Plus connector -2- (87), in wiring harness
(E30) - Connector (87a), in wiring harness engine

ws = white
sw = black
ro = red
br = brown
gn = green
bl = blue
gr = grey
li = violet
ge = yellow
or = orange

New Beetle **Wiring diagram** **No. 4/7**

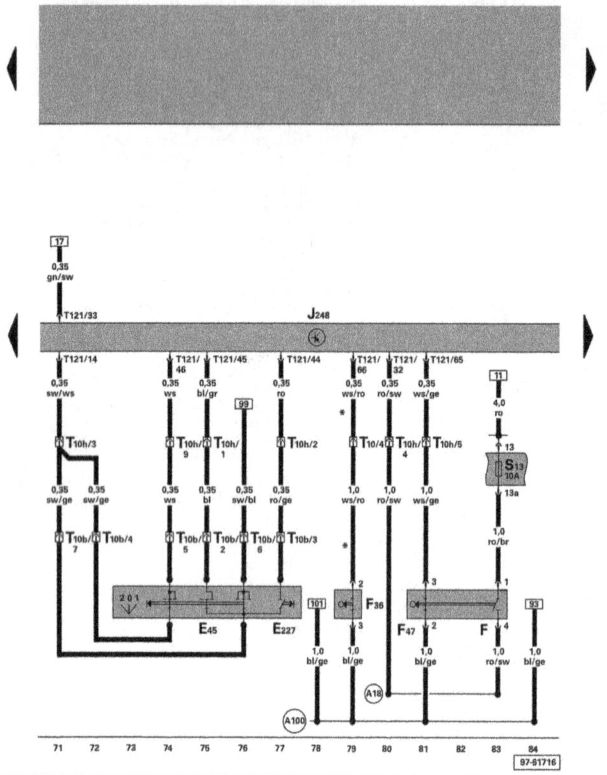

DFI Engine control module (ECM), button for cruise control (set), cruise control switch, brake light switch, clutch & brake switches

- E 45 - Cruise Control Switch
- E227 - Button for cruise control (set)
- F - Brake Light Switch
- F36 - Clutch Vacuum Vent Valve Switch
- F47 - Brake Pedal Switch (cruise control/Diesel Direct Fuel Injection)
- J248 - Diesel Direct Fuel Injection (DFI) Engine Control Module (ECM)
- S13 - Fuse 13 in fuse holder
- T10 - 10-Pin connector, white, behind instrument panel, left
- T10b - 10-Pin connector, black, near steering column
- T10h - 10-Pin connector, black, behind instrument panel, left
- T121 - 121-Pin connector, on DFI ECM
- (A18) - wire connection (54), in instrument panel wiring harness
- (A100) - Connector -2- (87), in instrument panel wiring harness
- * - Manual transmission only

ws = white
sw = black
ro = red
br = brown
gn = green
bl = blue
gr = grey
li = violet
ge = yellow
or = orange

New Beetle **Wiring diagram** **No. 4/8**

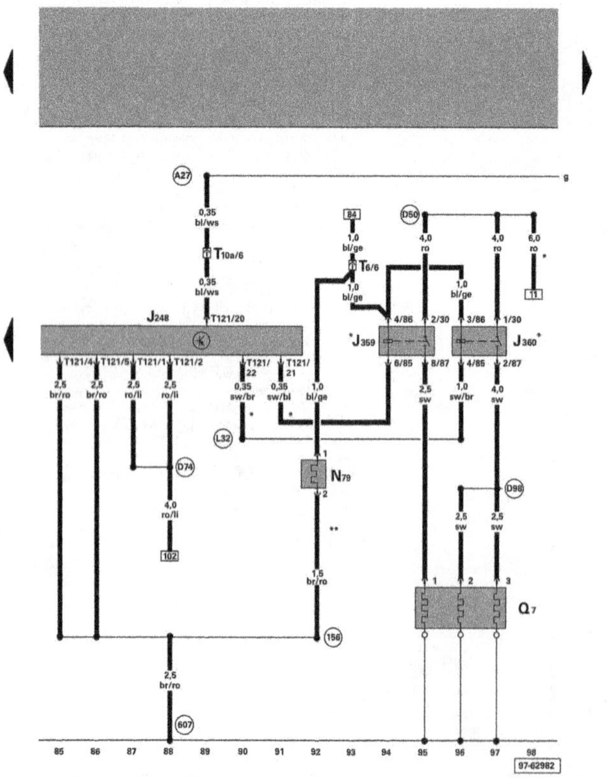

DFI Engine control module (ECM), relay for preheating coolant: low heat output, high heat output, spark plugs (coolant)

- J248 - Diesel Direct Fuel Injection (DFI) Engine Control Module (ECM
- J359 - Realy for preheating coolant, low heat output *
- J360 - Relay for prehaeating coolant, high heat output*
- N 79 - Positive Crankcase Ventilation (PCV) Heating Element
- Q7 - Spark plugs (coolant)
- T6 - 6-Pin connector, brown, behind instrument panel, left
- T10a - 10-Pin connector, orange, behind instrument panel, left
- T121 - 121-Pin connector, on DFI ECM
- (156) - Ground connection, in Diesel Direct Fuel Injection (DFI) wiring harness
- (607) - Ground Connection (in left plenum chamber)
- (A27) - Wire Connection (vehicle speed signal), in instrument panel wiring harness
- (D50) - Plus connection (30), in engine compartment wiring harness
- (D74) - Wire connection (86), in engine compartment wiring harness
- (D98) - Wire connection (glow plugs), in engine compartment wiring harness
- (L32) - Wire connection -1-, in heater wiring harness
- * - Manual Transmission only
- ** - Vehicles with Positive Crankcase Ventilation (PCV) Heating Element and Automatic Transmission

ws = white
sw = black
ro = red
br = brown
gn = green
bl = blue
gr = grey
li = violet
ge = yellow
or = orange

BentleyPublishers.com

New Beetle

Wiring diagram

No. 4/9

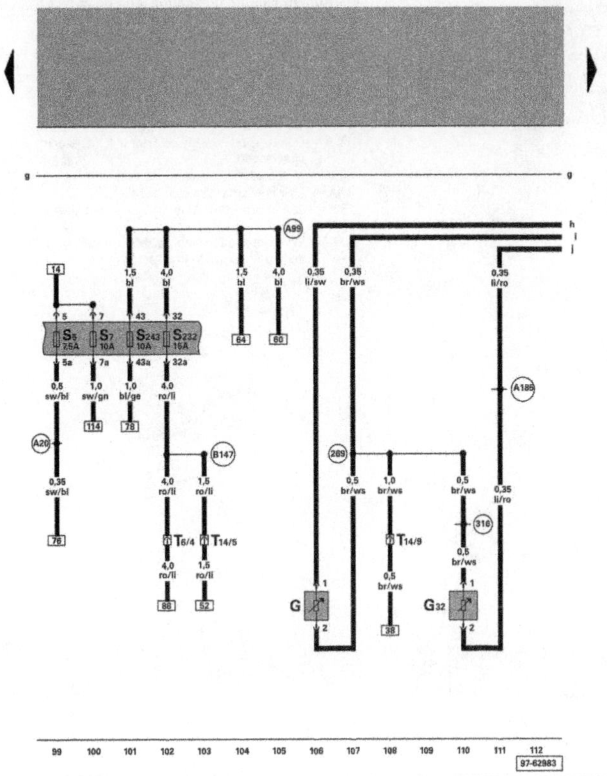

Sender for fuel gauge, engine coolant level (ECL) sensor

- G - Sender for fuel gauge
- G32 - Engine Coolant Level (ECL) Sensor
- S5 - Fuse 5 in fuse holder
- S7 - Fuse 7 in fuse holder
- S232 - Fuse 32 in fuse holder
- S243 - Fuse 43 in fuse holder
- T6 - 6-Pin connector, brown, behind instrument panel, left
- T14 - 14-Pin connector, in plenum chamber

- (269) - Ground connector(sensor ground) - 1 -,in instrument panel wiring harness
- (316) - Sensor Ground Connection 2 (in engine wiring harness)
- (A20) - Wire connection (15a), in instrument panel wiring harness
- (A99) - Connector -1- (87), in instrument panel wiring harness
- (A186) - Coolant Warning Connection (in instrument panel wiring harness)
- (B147) - Plus connector -2- (87), in wiring harness

ws = white
sw = black
ro = red
br = brown
gn = green
bl = blue
gr = grey
li = violet
ge = yellow
or = orange

New Beetle

Wiring diagram

No. 4/10

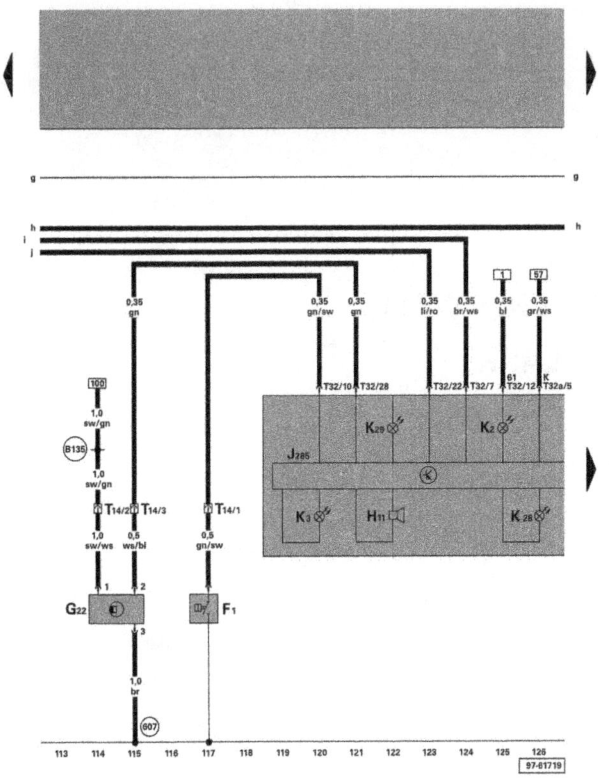

Oil pressure switch, Speedometer Vehicle Speed Sensor (VSS), instrument cluster, oil pressure warning buzzer, warning and indicator lights

- F1 - Oil Pressure Switch
- G22 - Speedometer Vehicle Speed Sensor (VSS)
- H11 - Oil Pressure Warning Buzzer
- J285 - Control module with indicator unit in instrument panel insert
- K2 - Generator (GEN) Warning Light
- K3 - Oil Pressure Warning Light
- K28 - Engine Coolant Level/Temperature (ECL/ECT) Warning Light
- K29 - Glow Plug Indicator Light
- T14 - 14-Pin connector, in plenum chamber
- T32 - 32-Pin connector, blue, on instrument cluster
- T32a - 32-Pin connector, green, on instrument cluster

- (607) - Ground Connection (in left plenum chamber)
- (B135) - Plus connector -1- (15a) in wiring harness interior

ws = white
sw = black
ro = red
br = brown
gn = green
bl = blue
gr = grey
li = violet
ge = yellow
or = orange

1.9L engine management (ALH) from June 2001

New Beetle Wiring diagram **No. 4/11**

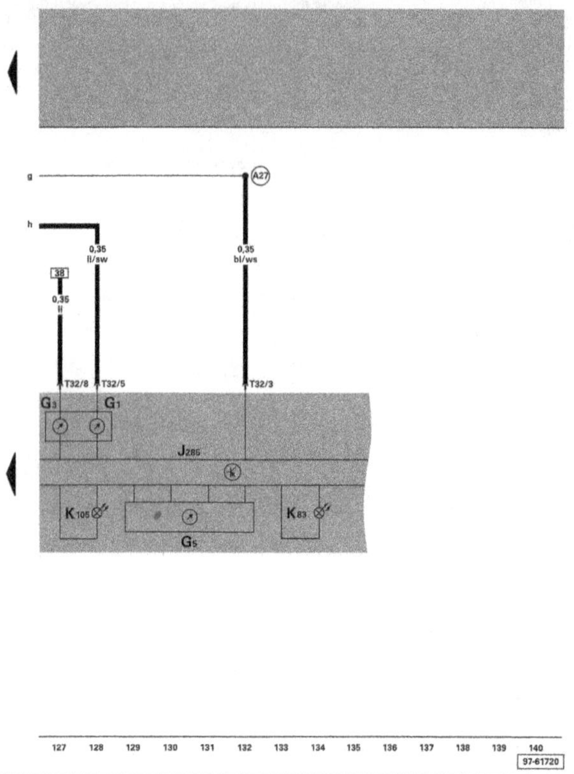

Instrument cluster, fuel gauge, engine coolant temperature (ECT) gauge, tachometer, Malfunction Indicator Lamp (MIL) low fuel level warning light

- G1 - Fuel Gauge
- G3 - Engine Coolant Temperature (ECT) Gauge
- G5 - Tachometer
- J285 - Control module with indicator unit instrument panel insert
- K83 - Malfunction Indicator Lamp (MIL)
- K105 - Low Fuel Level Warning Light
- T32 - 32-Pin connector, blue, on instrument cluster
- (A27) - Wire Connection (vehicle speed signal), in instrument panel wiring harness

ws = white
sw = black
ro = red
br = brown
gn = green
bl = blue
gr = grey
li = violet
ge = yellow
or = orange

THIS SPACE INTENTIONALLY LEFT BLANK

New Beetle Wiring Diagram **No. 32 / 1**

10.2009

1.9L - Engine - Turbo Diesel PD Fuel Injection (DFI)/74 kW, code BEW
From July 2003,

Additional Fuse, Relay, Control Module and Connector locations ⮕ Component Locations

Troubleshooting Procedures ⮕ Guided Fault Finding using VAS 5051/5052

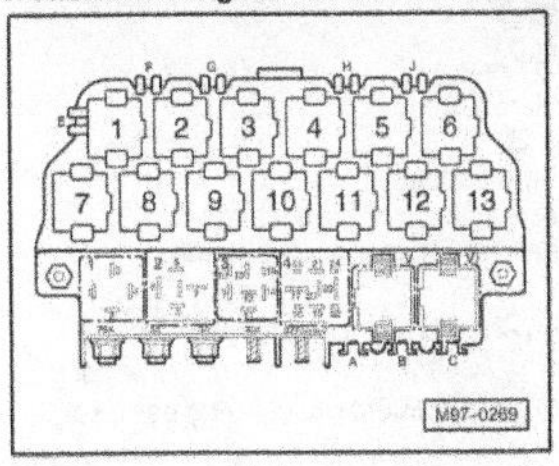

Relay locations on thirteen position auxiliary relay panel, above relay panel:
- 3 - J434 - Starter Lock/Clutch Pedal Switch Relay (53)
- 5 - J17 - Fuel Pump (FP) Relay (53)
- 12 - J317 - Power Supply (terminal 30, B+) Relay (109)

Relay panel:
- 1 - J4 - Horn Relay (53)
- 2 - J59 - Load Reduction Relay (100)

Notes:

- *J179 - Automatic Glow Time Control Module, in plenum chamber, center*
- **J360 - Preheating Coolant, High Heat Output Relay (100)*
- **J359 - Preheating Coolant, Low Heat Output Relay (53), (429),*
- **in engine compartment, left side, to the left of brake fluid reser voir, inside housing*
- *Number in parentheses indicates production control number stamped on relay housing.*

New Beetle **Wiring diagram** **No. 32/2**

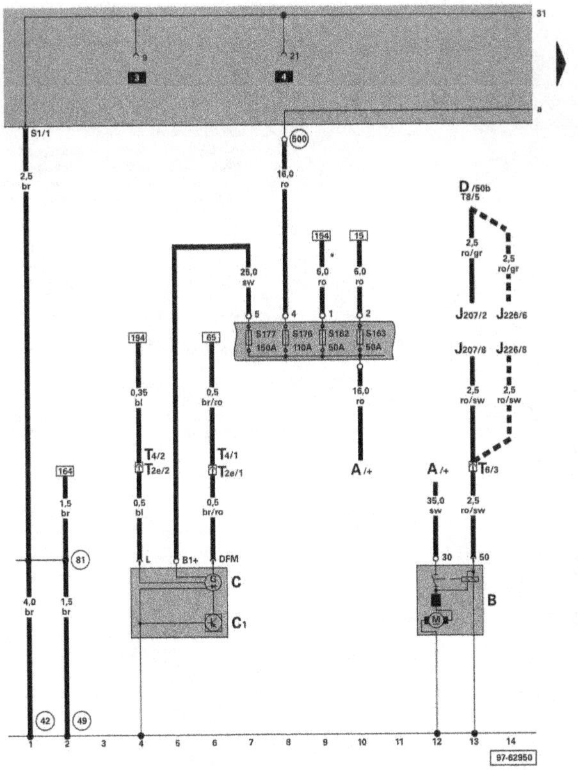

Generator (GEN), starter

- A - Battery
- B - Starter
- C - Generator (GEN)
- C1 - Voltage Regulator (VR)
- D - Ignition/Starter Switch
- J207 - Starting Interlock Relay
- J226 - Park/Neutral Position (PNP) Relay
- S162 - Fuse -1- (30) in fuse bracket/battery
- S163 - Fuse -2- (30) in fuse bracket/battery
- S176 - Fuse -4- (30) in fuse bracket/battery
- S177 - Fuse -5- (30) in fuse bracket/battery
- T2e - Double Connector, near starter (vehicles without air conditioning)
- T4 - 4-Pin Connector, near starter (vehicles with air conditioning only)
- T8 - 8-Pin Connector, at -D-
- T6 - 6-Pin Connector, brown, in protective housing for connectors, in plenum chamber, left
- (42) - Ground connection, beside steering column
- (49) - Ground connection, on steering column
- (81) - Ground connection -1-, in instrument panel wiring harness
- (500) - Threaded connection -1- (30) on the relay plate
- * - Manual Transmission only
- --- - Automatic transmission only

ws = white
sw = black
ro = red
br = brown
gn = green
bl = blue
gr = grey
li = violet
ge = yellow
or = orange

New Beetle

Wiring Diagram

No. 32 / 3

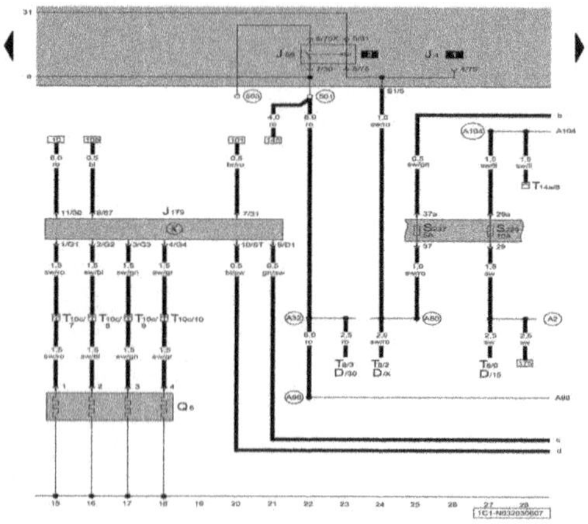

Control module for glow plug triggering, glow plugs (engine)

- D -Ignition/Starter Switch
- J4 -Dual Horn Relay
- J59 -Load Reduction Relay
- J179 -Automatic Glow Time Control Module, in plenum chamber, center
- Q6 -Glow plugs (engine)
- S229 -Fuse 29 in fuse holder
- S237 -Fuse 37 in fuse holder
- T8 -8-Pin Connector, at -D-
- T14a -14-Pin Connector, near battery
- T10c -10-Pin Connector, in engine compartment, left
- 501 -Threaded connection -2- (30) on the relay plate
- 503 -Threaded connection -2- (75x) on the relay plate
- A2 -Plus connection (15), in instrument panel wiring harness
- A32 -Plus connection (30), in instrument panel wiring harness
- A80 -Connection -1- (X) in instrument panel wiring harness
- A98 -Plus connection -4- (30), in instrument panel wiring harness
- A104 -Plus connection -2- (15), in instrument panel wiring harness

ws = wt (white)
sw = bk (black)
ro = rd (red)
br = br (brown)
gn = lg (light green), dg (dark green)
bl = bl (blue), lb (light blue), db (dark blue)
gr = gy (gray)
li = lilac
ge = yl (yellow)
or = or (orange), tn (tan)
rs = vt (violet)

New Beetle

Wiring diagram

No. 32/4

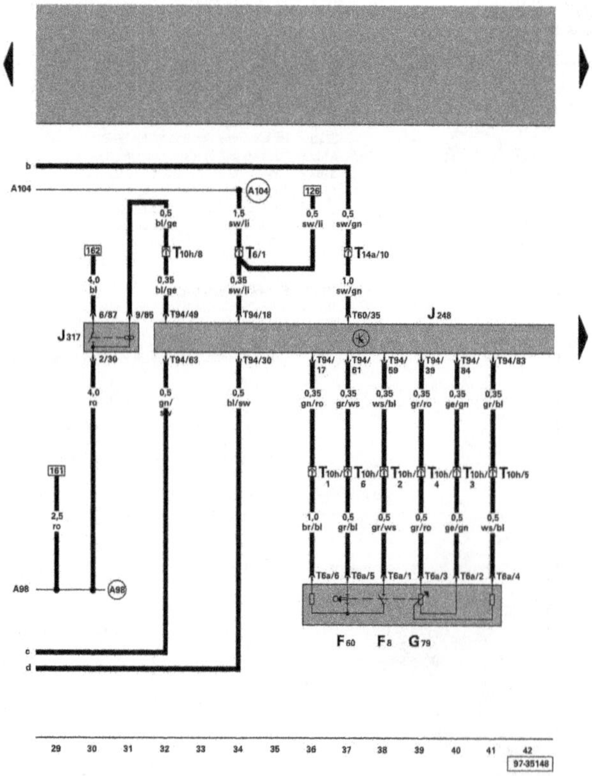

Diesel direct fuel injection (DFI) engine control module (ECM), power supply (terminal 30, B+) relay, kick down switch, (CTP) switch, throttle position (TP) sensor

- F8 - Kick Down Switch
- F60 - Closed Throttle Position (CTP) Switch
- G79 - Throttle Position (TP) Sensor
- J248 - Diesel Direct Fuel Injection (DFI) Engine Control Module (ECM), in plenum chamber, center
- J317 - Power Supply (Terminal 30, B+) Relay, on the thirteenfold auxiliary relay panel, above relay panel
- T6 - 6-Pin Connector, brown, in protective housing for connectors, in plenum chamber, left
- T6a - 6-Pin Connector
- T10h - 10-Pin Connector, blue, in protective housing for connectors, in plenum chamber, left
- T14a - 14-Pin Connector, near battery
- T60 - 60-Pin Connector
- T94 - 94-Pin Connector
- (A98) - Plus connection -4- (30), in instrument panel wiring harness
- (A104) - Plus connection -2- (15), in instrument panel wiring harness

ws = white
sw = black
ro = red
br = brown
gn = green
bl = blue
gr = grey
li = violet
ge = yellow
or = orange

New Beetle — Wiring diagram — No. 32/5

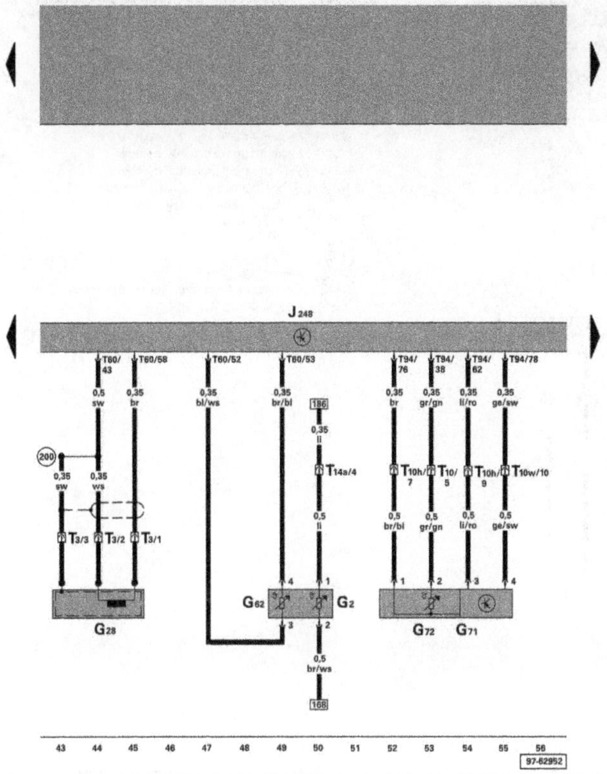

Diesel direct fuel injection (DFI) engine control module (ECM), engine speed (RPM) sensor, engine coolant temperature (ECT) sensor, intake air temperature (IAT) sensor, manifold absolute pressure (MAP) sensor

- G2 - Engine Coolant Temperature (ECT) Sensor
- G28 - Engine Speed (RPM) Sensor
- G62 - Engine Coolant Temperature (ECT) Sensor
- G71 - Manifold Absolute Pressure (MAP) Sensor
- G72 - Intake Air Temperature (IAT) Sensor
- J248 - Diesel Direct Fuel Injection (DFI) Engine Control Module (ECM), in plenum chamber, center
- T3 - 3-Pin Connector, in engine compartment, front
- T10 - 10-Pin Connector, orange, in protective housing for connectors, in plenum chamber, left
- T10h - 10-Pin Connector, blue, in protective housing for connectors, in plenum chamber, left
- T10w - 10-Pin Connector, white, in protective housing for connectors, in plenum chamber, left
- T14a - 14-Pin Connector, near battery
- T60 - 60-Pin Connector
- T94 - 94-Pin Connector
- (200) - Ground connection (shielding), in engine compartment wiring harness

ws = white
sw = black
ro = red
br = brown
gn = green
bl = blue
gr = grey
li = violet
ge = yellow
or = orange

New Beetle — Wiring diagram — No. 32/6

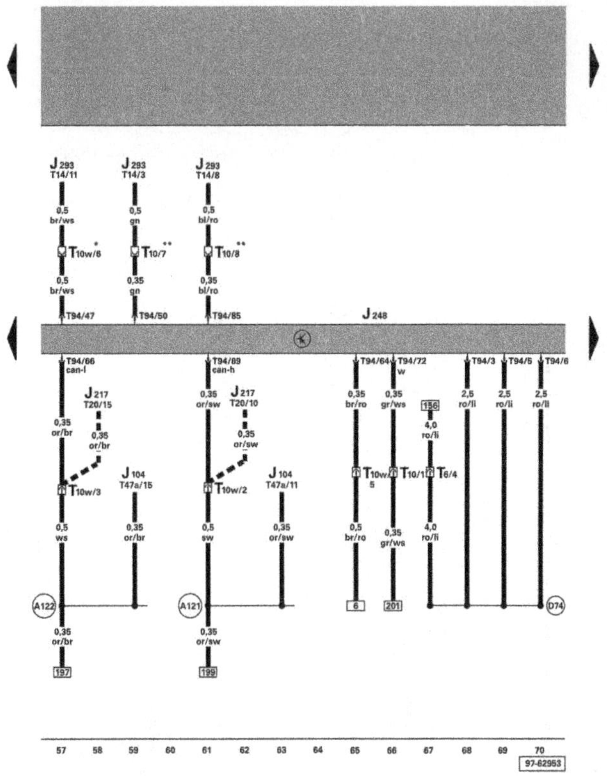

Diesel direct fuel injection (DFI) engine control module (ECM)

- J104 - ABS Control Module (w/EDL), in engine compartment, left
- J217 - Transmission Control Module (TCM)
- J248 - Diesel Direct Fuel Injection (DFI) Engine Control Module (ECM), in plenum chamber, center
- J293 - Coolant Fan Control (FC)Control Module
- T6 - 6-Pin Connector, brown, in protective housing for connectors, in plenum chamber, left
- T10 - 10-Pin Connector, orange, in protective housing for connectors, in plenum chamber, left
- T10w - 10-Pin Connector, white, in protective housing for connectors, in plenum chamber, left
- T14 - 14-Pin Connector, at -J293-
- T20 - 20-Pin Connector, at -J217-
- T47a - 47-Pin Connector, on ABS Control Module
- T94 - 94-Pin Connector
- (A121) - Connection (high bus), in instrument panel wiring harness
- (A122) - Connection (low bus), in instrument panel wiring harness
- (D74) - Wire connection (86), in engine compartment wiring harness
- * - Coolant Fan Control (FC) Control Module connection
- ** - A/C connection
- - - - - Automatic transmission only

ws = white
sw = black
ro = red
br = brown
gn = green
bl = blue
gr = grey
li = violet
ge = yellow
or = orange

1.9L engine management (BEW) from July 2003

New Beetle — Wiring diagram — No. 32/7

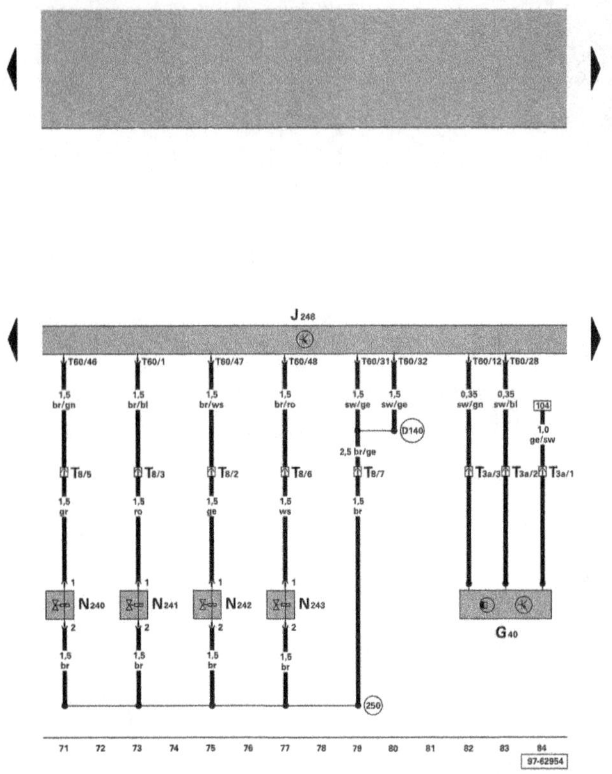

ws = white
sw = black
ro = red
br = brown
gn = green
bl = blue
gr = grey
li = violet
ge = yellow
or = orange

Diesel direct fuel injection (DFI) engine control module (ECM), valves for pump/injectors, camshaft position (CMP) sensor

- G40 - Camshaft Position (CMP) Sensor
- J248 - Diesel Direct Fuel Injection (DFI) Engine Control Module (ECM), in plenum chamber, center
- N240 - Valve for pump/injector, cylinder 1
- N241 - Valve for pump/injector, cylinder 2
- N242 - Valve for pump/injector, cylinder 3
- N243 - Valve for pump/injector, cylinder 4
- T3a - 3-Pin Connector, in engine compartment, front
- T8 - 8-Pin Connector, on left cylinder head
- T60 - 60-Pin Connector

- (250) - Ground connection in wiring harness engine pre-wiring - Diesel
- (D140) - Connection (injectors), in wiring harness, engine pre-wiring

Edition 09.03

New Beetle — Wiring diagram — No. 32/8

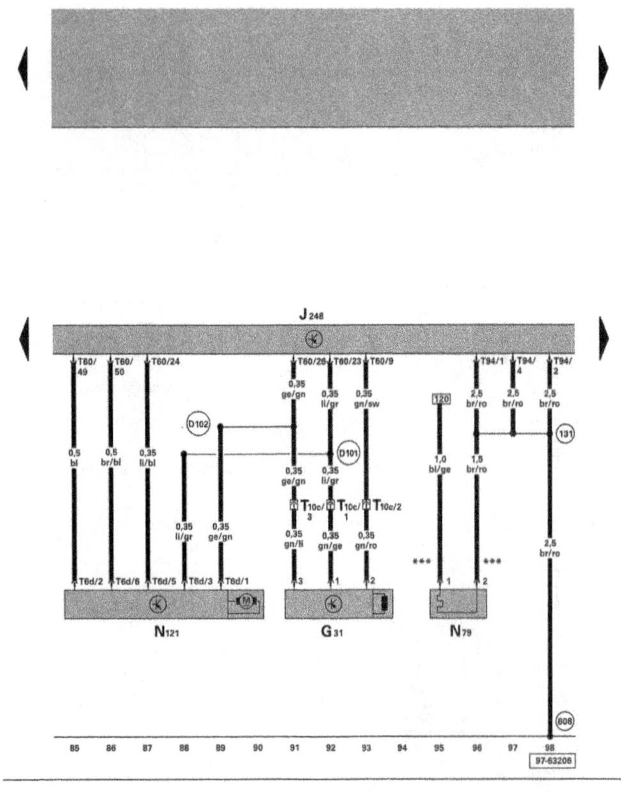

ws = white
sw = black
ro = red
br = brown
gn = green
bl = blue
gr = grey
li = violet
ge = yellow
or = orange

Diesel direct fuel injection (DFI) engine control module (ECM), charge air pressure sensor, EGR vacuum regulator solenoid valve, positive crankcase ventilation (PCV) heating element

- G31 - Charge Air Pressure Sensor
- J248 - Diesel Direct Fuel Injection (DFI) Engine Control Module (ECM), in plenum chamber, center
- N121 - EGR Vacuum Regulator Valve
- N79 - Positive Crankcase Ventilation (PCV) Heating Element***
- T6d - 6-Pin Connector
- T10c - 10-Pin Connector, in engine compartment, left
- T60 - 60-Pin Connector
- T94 - 94-Pin Connector

- (131) - Ground connection -2-, in engine compartment wiring harness
- (608) - Ground connection (in center plenum chamber)
- (D101) - Wire connection -1-, in engine compartment wiring harness
- (D102) - Wire connection -2-, in engine compartment wiring harness

- *** - Vehicles with positive crankcase ventilation (PCV) heating element

Edition 04.04

New Beetle

Wiring diagram

No. 32/9

Diesel direct fuel injection (DFI) engine control module (ECM), motor for intake flap, change-over valve for intake manifold flap, EGR vacuum regulator solenoid valve, wastegate bypass regulator valve

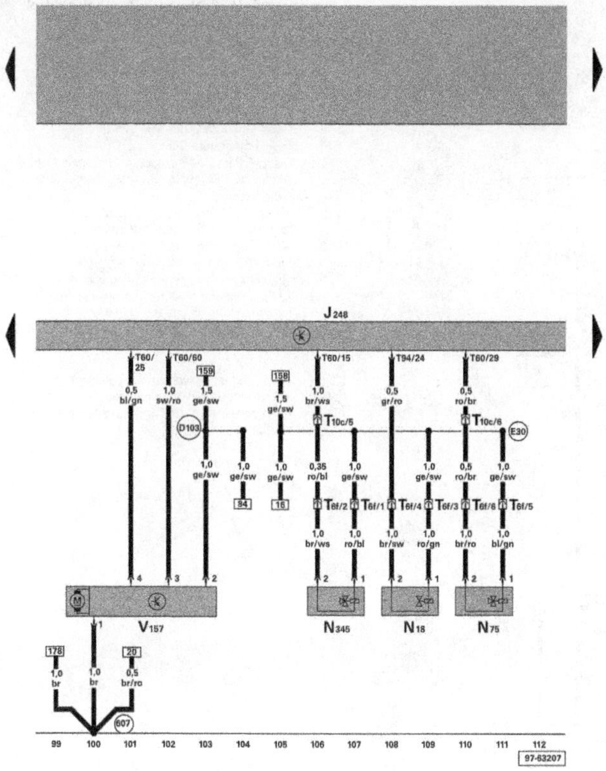

- J248 - Diesel Direct Fuel Injection (DFI) Engine Control Module (ECM), in plenum chamber, center
- N18 - EGR Vacuum Regulator Solenoid Valve
- N75 - Wastegate Bypass Regulator Valve
- N345 - Exhaust Gas Recirculation (EGR) Cooler Switch-over Valve
- T6f - 6-Pin Connector, in engine compartment, front
- T10c - 10-Pin Connector, in engine compartment, left
- T60 - 60-Pin Connector
- T94 - 94-Pin Connector
- V157 - Motor for intake flap
- (607) - Ground Connection (in left plenum chamber)
- (D103) - wire connection -3-, in engine compartment wiring harness
- (E30) - Connector (87a), in wiring harness engine

ws = white
sw = black
ro = red
br = brown
gn = green
bl = blue
gr = grey
li = violet
ge = yellow
or = orange

New Beetle

Wiring diagram

No. 32/10

Diesel direct fuel injection (DFI) engine control module (ECM), heated oxygen sensor (HO2S), mass air flow (MAF) sensor

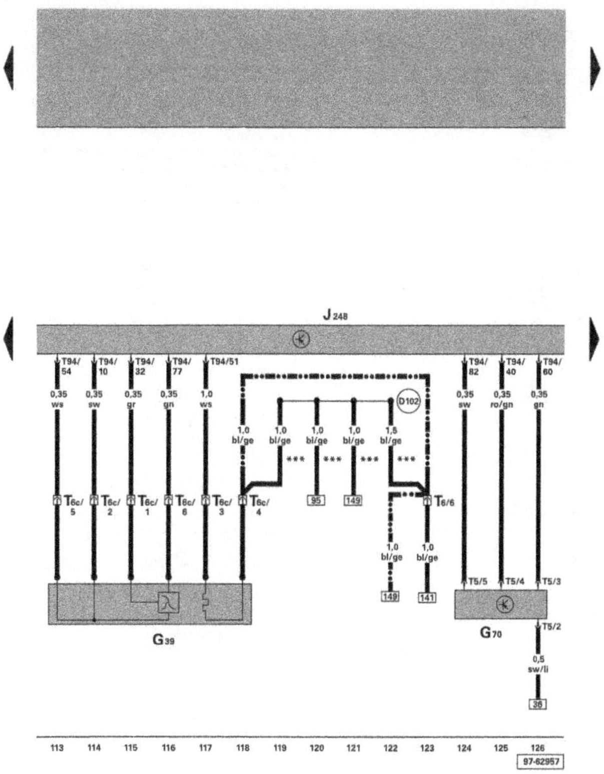

- G39 - Heated Oxygen Sensor (HO2S)
- G70 - Mass Air Flow (MAF) Sensor
- J248 - Diesel Direct Fuel Injection (DFI) Engine Control Module (ECM), in plenum chamber, center
- T5 - 5-Pin Connector
- T6 - 6-Pin Connector, brown, in protective housing for connectors, in plenum chamber, left
- T6c - 6-Pin Connector, in protective housing for connectors under right floor
- T94 - 94-Pin Connector
- (D102) - Wire connection -2-, in engine compartment wiring harness
- *** - Vehicles with positive crankcase ventilation (PCV) heating element
- -**- - Vehicles without positive crankcase ventilation (PCV) heating element

ws = white
sw = black
ro = red
br = brown
gn = green
bl = blue
gr = grey
li = violet
ge = yellow
or = orange

1.9L engine management (BEW) from July 2003

New Beetle — Wiring diagram — No. 32/11

Diesel direct fuel injection (DFI) engine control module (ECM), cruise control switch, clutch vacuum vent valve switch, fuel temperature sensor

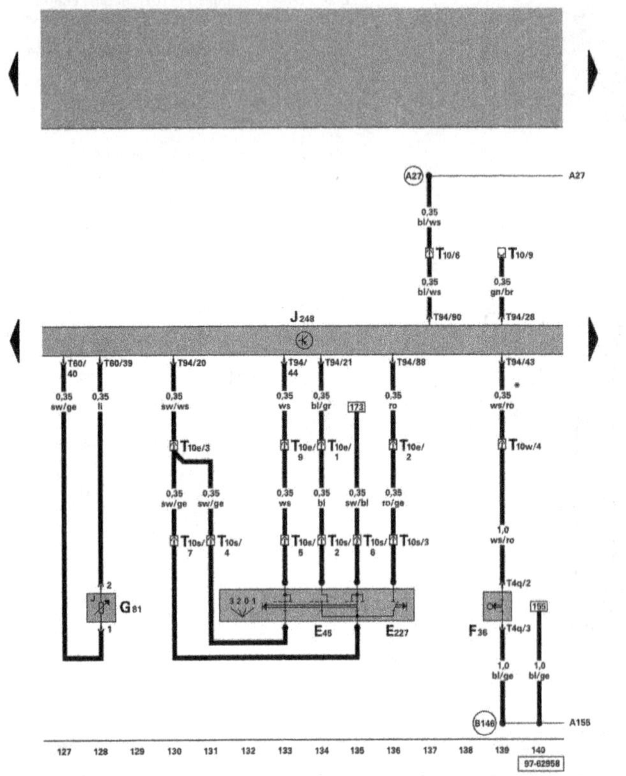

ws = white
sw = black
ro = red
br = brown
gn = green
bl = blue
gr = grey
li = violet
ge = yellow
or = orange

- E45 - Cruise Control Switch
- E227 - Cruise Control Push Button (SET)
- F36 - Clutch Vacuum Vent Valve Switch
- G81 - Fuel Temperature Sensor
- J248 - Diesel Direct Fuel Injection (DFI) Engine Control Module (ECM), in plenum chamber, center
- T4q - 4-Pin Connector
- T10 - 10-Pin Connector, orange, in protective housing for connectors, in plenum chamber, left
- T10e - 10-Pin Connector, black, in protective housing for connectors, in plenum chamber, left
- T10s - 10-Pin Connector, near steering column
- T10w - 10-Pin Connector, white, in protective housing for connectors, in plenum chamber, left
- T60 - 60-Pin Connector
- T94 - 94-Pin Connector
- (A27) - Wire connection (vehicle speed signal), in instrument panel wiring harness
- (B146) - Plus connector -1- (87), in wiring harness interior
- * Manual transmission only

Edition 09.03

New Beetle — Wiring diagram — No. 32/12

Diesel direct fuel injection (DFI) engine control module (ECM), brake light switch, brake vacuum vent valve switch for cruise control/diesel, Relay for preheating coolant, glow plugs (coolant)

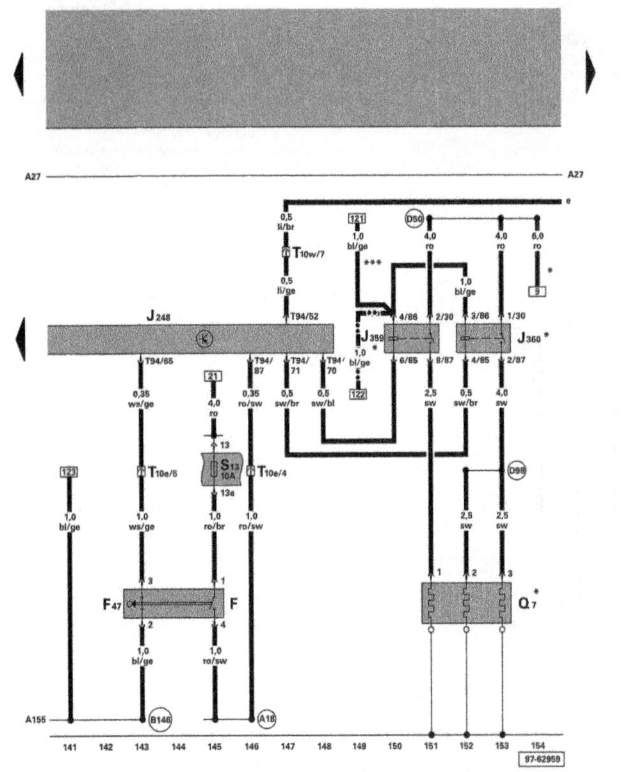

ws = white
sw = black
ro = red
br = brown
gn = green
bl = blue
gr = grey
li = violet
ge = yellow
or = orange

- F - Brake Light Switch
- F47 - Brake Vacuum Vent Valve Switch for cruise control/diesel
- J248 - Diesel Direct Fuel Injection (DFI) Engine Control Module (ECM), in plenum chamber, center
- J359 - Relay for preheating coolant, low heat output, in protective housing, in engine compartment, left, production control number (429)
- J360 - Relay for preheating coolant, high heat output, in protective housing, in engine compartment, left, production control number (100)
- Q7 - Glow plugs (coolant)
- S13 - Fuse 13 in fuse holder
- T10e - 10-Pin Connector, black, in protective housing for connectors, in plenum chamber, left
- T10w - 10-Pin Connector, white, in protective housing for connectors, in plenum chamber, left
- T94 - 94-Pin Connector
- (A18) - Wire connection (54), in instrument panel wiring harness
- (B146) - Plus connector -1- (87), in wiring harness interior
- (D50) - Plus connection (30), in engine compartment wiring harness
- (D98) - Wire connection (glow plugs), in engine compartment wiring harness
- -•••- - Vehicles without positive crankcase ventilation (PCV) heating element
- * - Manual transmission only
- *** - Vehicles with positive crankcase ventilation (PCV) heating element

Edition 09.03

New Beetle — Wiring diagram — No. 32/13

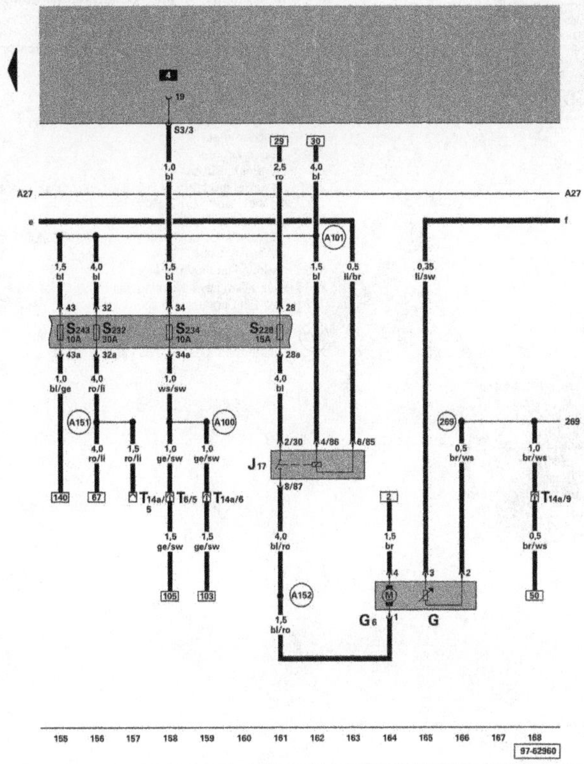

Fuel pump (FP) relay, fuel pump (FP), Fuel level Sensor

- G - Fuel Level Sensor
- G6 - Fuel Pump (FP)
- J17 - Fuel Pump (FP) Relay
- S228 - Fuse 28 in fuse holder
- S232 - Fuse 32 in fuse holder
- S234 - Fuse 34 in fuse holder
- S243 - Fuse 43 in fuse holder
- T6 - 6-Pin Connector, brown, in protective housing for connectors, in plenum chamber, left
- T14a - 14-Pin Connector, near battery
- (269) - Ground connection (sensor ground) -1-, in instrument panel wiring harness
- (A100) - Connection -2- (87), in instrument panel wiring harness
- (A101) - Connector -3- (87), in instrument panel wiring harness
- (A151) - Connection 4 (87) (in instrument panel wiring harness)
- (A152) - Connection 5 (87) (in instrument panel wiring harness)

ws = white
sw = black
ro = red
br = brown
gn = green
bl = blue
gr = grey
li = violet
ge = yellow
or = orange

New Beetle — Wiring diagram — No. 32/14

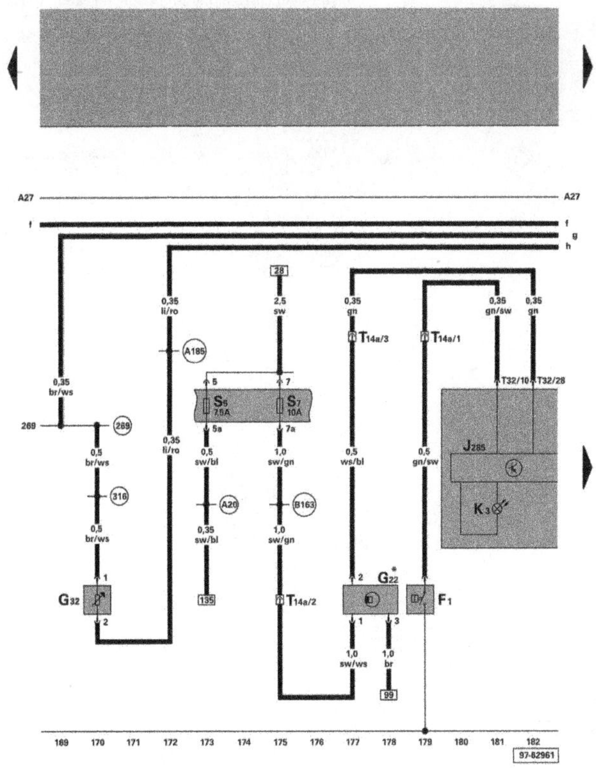

Instrument cluster, oil pressure switch, speedo meter vehicle speed sensor (VSS), oil pressure warning light, engine coolant level (ECL) sensor

- F1 - Oil Pressure Switch
- G22 - Speedometer Vehicle Speed Sensor (VSS)
- G32 - Engine Coolant Level (ECL) Sensor
- J285 - Control module with indicator unit in instrument panel insert
- K3 - Oil Pressure Warning Light
- S5 - Fuse 5 in fuse holder
- S7 - Fuse 7 in fuse holder
- T14a - 14-Pin Connector, near battery
- T32 - 32-Pin Connector, blue
- (269) - Ground connection (sensor ground) -1-, in instrument panel wiring harness
- (316) - Sensor Ground Connection 2 (in engine wiring harness)
- (A20) - Wire connection (15a), in instrument panel wiring harness
- (A185) - Coolant Warning Connection (in instrument panel wiring harness)
- (B163) - Plus connection -1- (15) in wiring harness interior

* Manual transmission only

ws = white
sw = black
ro = red
br = brown
gn = green
bl = blue
gr = grey
li = violet
ge = yellow
or = orange

New Beetle — Wiring diagram — No. 32/15

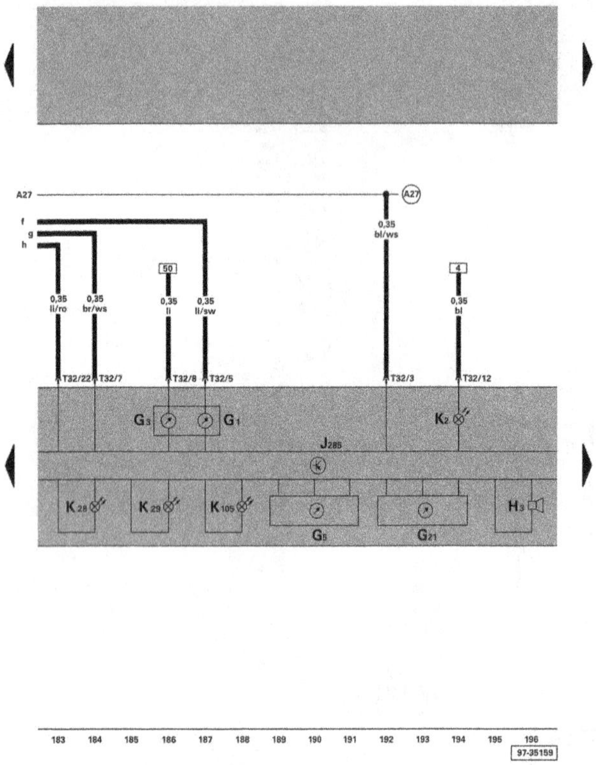

Instrument cluster, engine coolant temperature (ECL/ECT) gauge, fuel gauge, tachometer, speedometer, generator (GEN) warning light, low fuel level warning light, glow plug indicator light

- G1 - Fuel gauge
- G3 - Engine Coolant Temperature (ECT) Gauge
- G5 - Tachometer
- G21 - Speedometer
- H3 - Warning Buzzer
- J285 - Control module with indicator unit in instrument panel insert
- K2 - Generator (GEN) Warning Light
- K28 - Engine Coolant Level/Temperature (ECL/ECT) Warning Light
- K29 - Glow Plug Indicator Light
- K105 - Low Fuel Level Warning Light
- T32 - 32-Pin Connector, blue
- (A27) - Wire connection (vehicle speed signal), in instrument panel wiring harness

ws = white
sw = black
ro = red
br = brown
gn = green
bl = blue
gr = grey
li = violet
ge = yellow
or = orange

New Beetle — Wiring diagram — No. 32/16

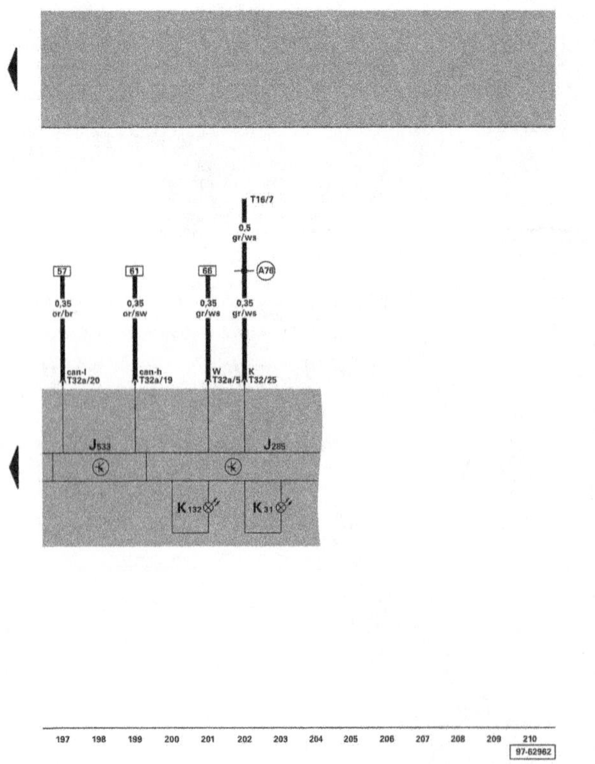

Instrument cluster, multi-function indicator (MFI), outside air temperature sensor, electronic power control (EPC) warning lamp, cruise control indicator light, malfunction indicator lamp

- J285 - Control module with indicator unit in instrument panel insert
- J533 - Data Bus On Board Diagnostic Interface
- K31 - Cruise Control Indicator Light
- K132 - Electronic Power Control (EPC) Warning Lamp
- T16 - Data Link Connector (DLC), below instrument panel, left
- T32 - 32-Pin Connector, blue
- T32a - 32-Pin Connector, green
- (A76) - Connection (K-diagnosis wire), in instrument panel wiring harness

ws = white
sw = black
ro = red
br = brown
gn = green
bl = blue
gr = grey
li = violet
ge = yellow
or = orange

2,0L-Engine - Motronic Multiport Fuel Injection (MFI)/85 kW, code AEG, (without cruise control)

from January 1998

Deviate relay location and fuseplacements as well as the locations of multiple connectors see section "component locations".

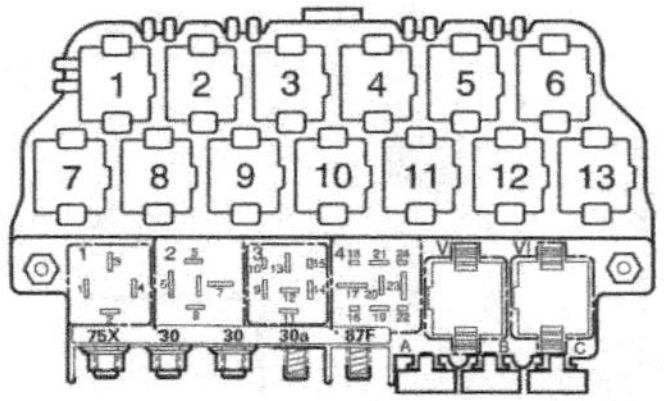

97-21956

Relay location on the thirteen position auxiliary relay panel, above relay panel:

Relay panel:

2 Load Reduction Relay (213), (18)

4 Fuel Pump Relay (409)

Note: Number in parentheses indicates production control number stamped on relay housing.

Fuse colors
30 A - green
25 A - white
20 A - yellow
15 A - blue
10 A - red
7,5 A - brown
5 A - beige

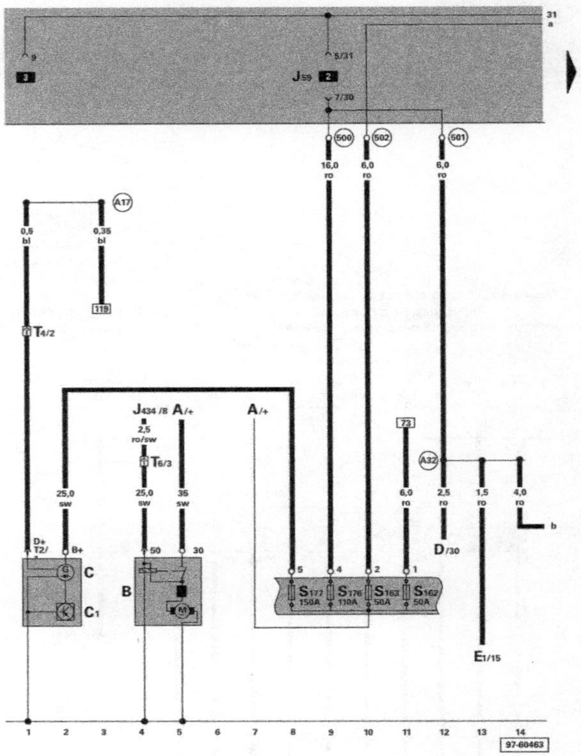

ws = white
sw = black
ro = red
br = brown
gn = green
bl = blue
gr = grey
li = violet
ge = yellow

- A - Battery
- B - Starter
- C - Generator (GEN)
- C1 - Voltage Regulator (VR)
- D - Ignition/Starter Switch
- E1 - Light Switch
- J59 - Load Reduction Relay
- J434 - Locking relay for starter (clutch pedal switch)
- S163 - Fuse -2- (50)in fuse bracket/battery
- S176 - Fuse -4- (30), in fuse bracket/battery
- S177 - Fuse -5- (30), in fuse bracket/battery
- S162 - Fuse -1- in fuse bracket/battery
- T2 - Double Connector
- T4 - 4-Pin Connector, in engine compartment left
- T6 - 6-Pin Connector, brown, behind instrument panel, left

- (500) - Threaded connection -1- (30) on the relay plate
- (501) - Threaded connection -2- (30) on the relay plate
- (502) - Threaded connection -1- (30a) on the relay plate
- (A17) - wire connection (61), in instrument panel wiring harness
- (A32) - plus connection (30), in instrument panel wiring harness

Starter, Generator (GEN)

2.0L engine management (AEG) from Jan. 1998

New Beetle Wiring diagram **No. 2/3**

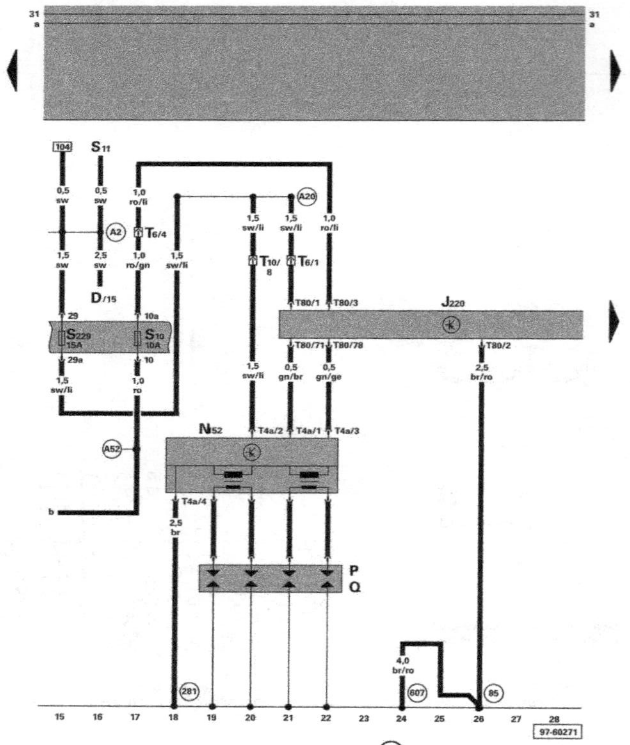

ws = white
sw = black
ro = red
br = brown
gn = green
bl = blue
gr = grey
li = violet
ge = yellow

- D - Ignition/Starter Switch
- J220 - Motronic Engine Control Module (ECM), behind istrument panel, left
- N152 - Ignition Coil
- P - Spark Plug Connectors
- Q - Spark Plugs
- S10 - Fuse 10 in fuse holder
- S11 - Fuse 11 in fuse holder
- S229 - Fuse 29 in fuse holder
- T4a - 4-Pin Connector, black
- T6 - 6-Pin Connector, brown, behind instrument panel, left
- T10 - 10-Pin Connector, in engine compartment, left
- T80 - 80-Pin Connector, at Engine Control Module (ECM)

- (85) - Ground connection -1-, in engine compartment wiring harness
- (281) - Ground connector -1-, in wiring harness engine pre-wiring
- (A2) - plus connection (15), in instrument panel wiring harness
- (A20) - wite connection (15a), in instrument panel wiring harness
- (A52) - plus connection (30), in instrument panel wiring harness
- (607) - Ground Connection (in left plenum chamber)

Edition 02/99 USA.5132.06.21 **Motronic engine control module (ECM), ignition system**

No. 2/4 Wiring diagram **New Beetle**

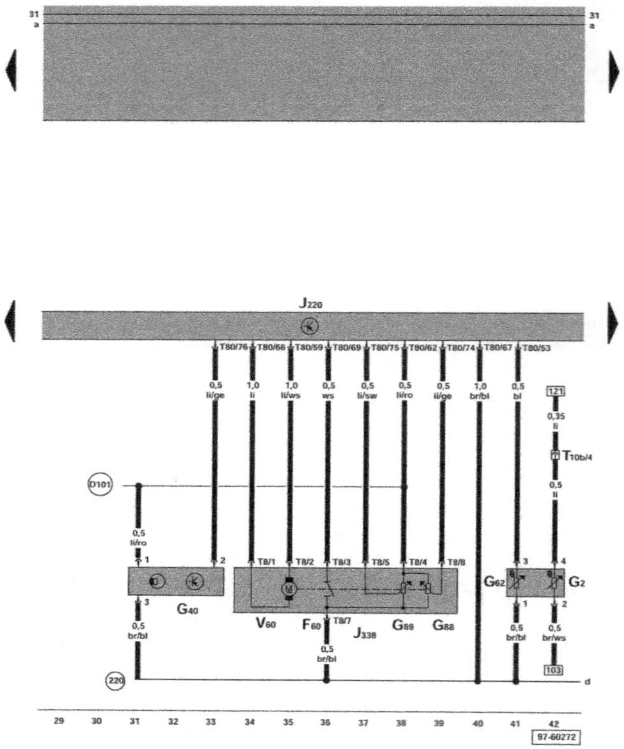

ws = white
sw = black
ro = red
br = brown
gn = green
bl = blue
gr = grey
li = violet
ge = yellow

- F60 - Closed Throttle Position (CTP) Switch
- G2 - Engine Coolant Temperature (ECT) Sensor
- G40 - Camshaft Position (CMP) Sensor
- G62 - Engine Coolant Temperature (ECT) Sensor
- G69 - Throttle Position (TP) Sensor
- G88 - Throttle Position (TP) Sensor
- J220 - Motronic Engine Control Module (ECM)
- J338 - Throttle Valve Control Module
- V60 - Throttle Position (TP) Actuator
- T8 - 8-Pin Connector
- T10b - 10-Pin Connector, behind instrument panel, left
- T80 - Connector, 80-point, at Engine Control Module (ECM)

- (220) - Ground connection (sensor ground), in engine compartment wiring harness
- (D101) - Connection 1 (in engine compartment wiring harness)

Motronic Engine Control Module (ECM), Throttle Valve Control Module, Camshaft Position (CMP) Sensor, Engine Coolant Temperature (ECT) Sensor **Edition 09/98** USA.5132.05.21

New Beetle Wiring diagram **No. 2/5**

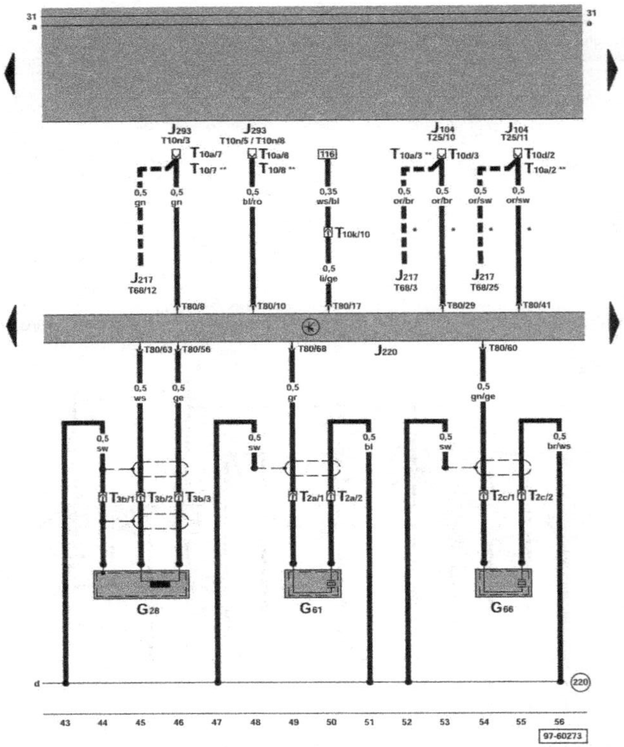

or = orange
ws = white
sw = black
ro = red
br = brown
gn = green
bl = blue
gr = grey
li = violet
ge = yellow

G28 - Engine Speed (RPM) Sensor
G61 - Knock Sensor (KS) 1
G66 - Knock Sensor (KS) 2
J104 - ABS Control Module (w/EDL)
J217 - Transmission Control Module (TCM), behind instrument panel
J220 - Motronic Engine Control Module (ECM), behind instrument panel, left
J255 - Climatronic Control Module
J293 - Coolant FC (Fan Control) Control Module
T2a - 3-Pin Connector
T3b - 3-Pin Connector
T2c - 3-Pin Connector
T6 - 6-Pin Connector, red, behind instrument panel
T10 - 10-Pin Connector, wihite, behind instrument panel, left
T10a - 10-Pin Connector, orange, behind instrument panel, left
T10d - 10-Pin Connector, green, behind instrument panel, left
T10k - 10-Pin Connector, blue, behind instrument panel, left
T25 - 25-Pin Connector, at J104
T68 - Connector, 68-point, on Transmisson Control Module (TCM)
T80 - Connector, 80-point, at Engine Control Module (ECM)
(220) - Ground connection (sensor ground), in engine compartment wiring harness
* - connection for DATA-bus
--- - automatic transmission only
** - early vehicles only

Edition 09/98
USA.5132.05.21

Motronic engine control module (ECM), engine speed (RPM) sensor, knock sensor (KS) 1, knock sensor (KS) 2

No. 2/6 Wiring diagram **New Beetle**

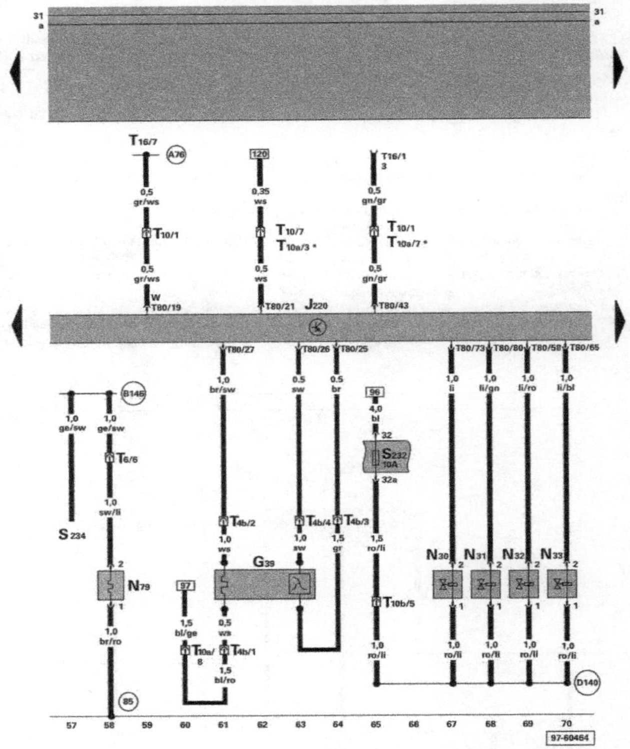

ws = white
sw = black
ro = red
br = brown
gn = green
bl = blue
gr = grey
li = violet
ge = yellow

G39 - Heated Oxygen Sensor (HO2S)
J220 - Motronic Engine Control Module (ECM), behind instrument panel, left
N30 - Cylinder 1 Fuel Injector
N31 - Cylinder 2 Fuel Injector
N32 - Cylinder 3 Fuel Injector
N33 - Cylinder 4 Fuel Injector
N79 - Positive Crankcase Ventilation (PCV) Heating Element
S232 - Fuse 32 in fuse holder
S234 - Fuse 34 in fuse holder
T4b - 4-Pin Connector
T6 - 6-Pin Connector, red, behind instrument panel, left
T10 - 10-Pin Connector, white, behind instrument panel, left
T10a - 10-Pin Connector, orange, behind instrument panel, left
T10b - 10-Pin Connector, in engine compartment, left
T16 - 16-Pin Connector, Data Link Connector (DLC), below steering column
T80 - 80-Pin Connector, at Engine Control Module (ECM)
(85) - Ground connection -1-, in engine compartment wiring harness
(A76) - Connector (K-diagnosis wire), in instrument panel wiring harness
(D140) - Connector (injectors), in wiring harness, engine pre-wiring
(B146) - Plus connector -1- (87), in wiring harness interior
* - early vehicles only

Motronic engine control module (ECM), fuel injectors, heated oxygen sensor (HO2S) 2, positive crankcase ventilation (PCV) heating element

Edition 06/99
USA.5132.07.21

2.0L engine management (AEG) from Jan. 1998

New Beetle — Wiring diagram — No. 2/7

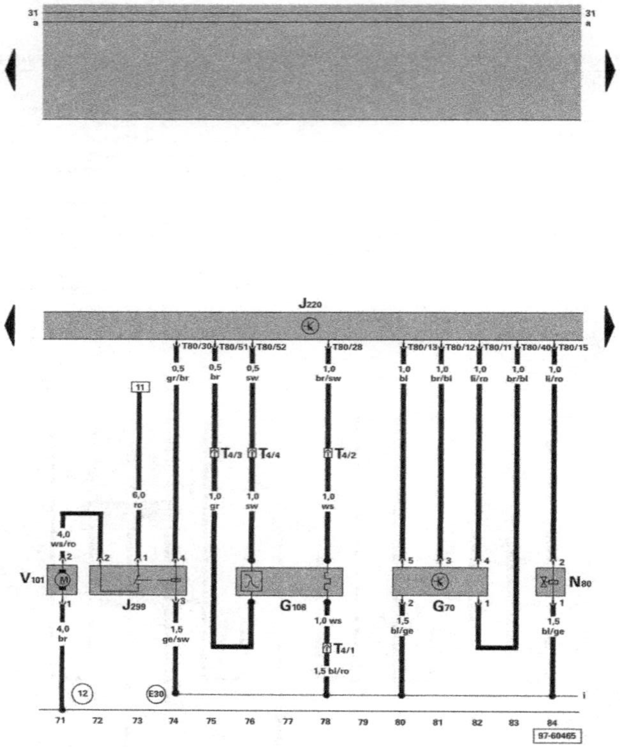

ws = white
sw = black
ro = red
br = brown
gn = green
bl = blue
gr = grey
li = violet
ge = yellow

G108 - Heated Oxygen Sensor (HO2S) 2
G70 - Mass Air Flow (MAF) Sensor
J220 - Motronic Engine Control Module (ECM), behind instrument panel, left
J299 - Secondary Air Injection (AIR) Pump Relay, in engine compartment, left
N80 - Evaporative Emission (EVAP) Canister Purge Regulator Valve
V101 - Secondary Air Injection (AIR) Pump Motor
T4 - 4-Pin Connector
T80 - 80-Pin Connector

(12) - Ground connection, in engine compartment, left
(E30) - Connector (87a), in wiring harness engine

Edition 06/99
USA.5132.07.21

Aoxygen sensor (HO2S), mass air flow (MAF) sensor, evaporative emission (EVAP) canister purge regulator valve, secondary air injection (AIR) system

No. 2/8 — Wiring diagram — New Beetle

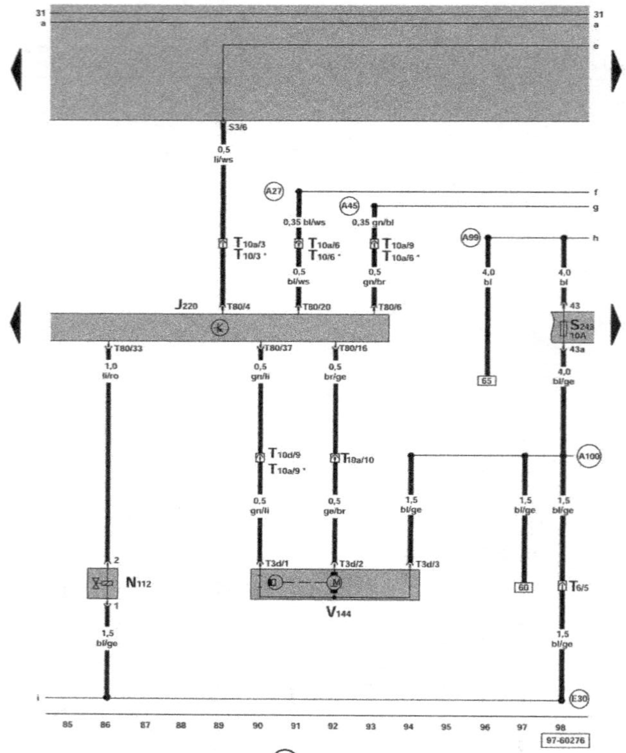

ws = white
sw = black
ro = red
br = brown
gn = green
bl = blue
gr = grey
li = violet
ge = yellow

J220 - Motronic Engine Control Module (ECM), behind instrument panel, left
N112 - Secondary Air Injection (AIR) Solenoid Valve
S243 - Fuse 43 in fuse holder
V144 - Leak Detection Pump (LDP)
T3d - 3-Pin Connector
T6 - 6-Pin Connector, red, behind instrument panel, left
T10 - 10-Pin Connector,white, behind instrument panel, left
T10a - 10-Pin Connector,orange, behind instrument panel, left
T10d - 10-Pin Connector, green, behind instrument panel, left
T80 - Connector, 80-point, at Engine Control Module (ECM)

(A27) - wire Connection (vehicle speed signal), in instrument panel wiring harness
(A45) - wire connection (RPM-signal), in instrument panel wiring harness
(A99) - Connector -1- (87), in instrument panel wiring harness
(A100) - Connector -2- (87), in instrument panel wiring harness
(E30) - Connector (87a), in wiring harness engine
* - early vehicles only

Motronic engine control module (ECM), diagnosis pump for fuel system, secondary air injection solenoid valve

Edition 09/98
USA.5132.05.21

New Beetle — Wiring diagram — No. 2/9

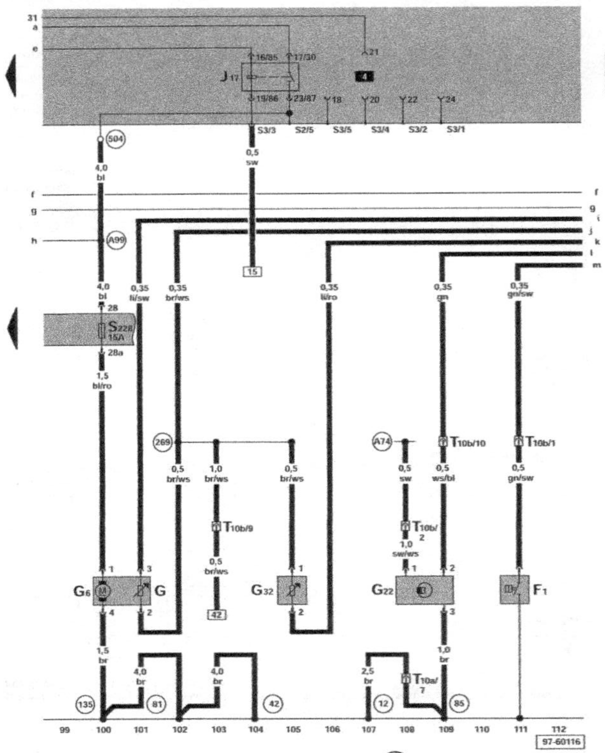

ws = white
sw = black
ro = red
br = brown
gn = green
bl = blue
gr = grey
li = violet
ge = yellow

F1 - Oil Pressure Switch
G - Sender for fuel gauge
G6 - Fuel Pump (FP)
G22 - Speedometer Vehicle Speed Sensor (VSS)
G32 - Engine Coolant Level (ECL) Sensor
J17 - Fuel Pump (FP) Relay
S228 - Fuse 28 in fuse holder
T10a - 10-Pin Connector, orange, behind instrument panel, left
T10b - 10-Pin Connector, in engine compartment, left

(42) - Ground connection, beside steering column

(81) - Ground connection -1-, in instrument panel wiring harness

(85) - Ground connection -1-, in engine compartment wiring harness

(12) - Ground connection, in engine compartment, left

(135) - Ground connection -2-, in instrument panel wiring harness

(269) - Ground connector (sensor ground) -1-, in instrument panel wiring harness

(503) - Threaded connection -1- (75x) on the relay plate

(504) - Threaded connection -1- (87) on the relay plate

(A74) - wire connection (75x), in instrument panel wiring harness

(A99) - Connector -1- (87), in instrument panel wiring harness

Oil pressure switch, fuel pump (FP), fuel pump (FP) relay, engine coolant level sensor, speedomter vehicle speed sensor (VSS)

No. 2/10 — Wiring diagram — New Beetle

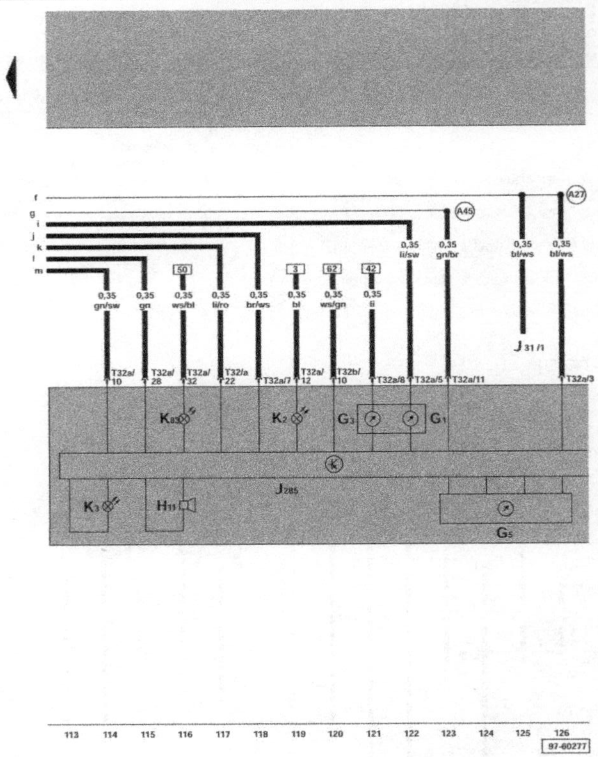

ws = white
sw = black
ro = red
br = brown
gn = green
bl = blue
gr = grey
li = violet
ge = yellow

G1 - Fuel gauge
G3 - Engine Coolant Temperature (ECT) Gauge
G5 - Tachometer
H11 - Oil Pressure Warning Buzzer
J31 - Wiper/Washer Intermittent Relay
J285 - Control module with indicator unit in instrument panel insert
K2 - Generator (GEN) Warning Light
K3 - Oil Pressure Warning Light
K83 - Malfunction Indicator Lamp (MIL)
T32a - 32-Pin Connector, blue, on instrument cluster
T32b - 32-Pin Connector, green, on instrument cluster

(A27) - wire Connection (vehicle speed signal), in instrument panel wiring harness

(A45) - wire connection (RPM-signal), in instrument panel wiring harness

Instrument cluster, fuel gauge, engine coolant temperature (ECT) gauge, tachometer, generator (GEN) warning light, oil pressure warning light

2,0L-Engine - Motronic Multiport Fuel Injection (MFI)/85 kW, code AEG, (without cruise control)

from August 1998

Deviate relay location and fuseplacements as well as the locations of multiple connectors see section "component locations".

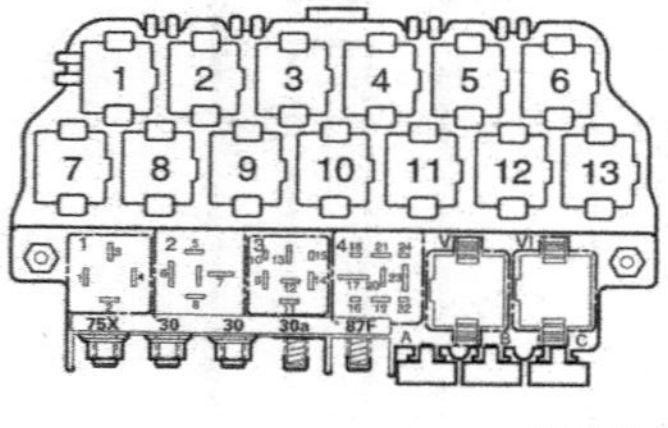

97-21956

Relay location on the thirteen position auxiliary relay panel, above relay panel:

Relay panel:

- **2** Load Reduction Relay (213), (18)
- **4** Fuel Pump Relay (409)

Note: Number in parentheses indicates production control number stamped on relay housing.

Fuse colors
30 A - green
25 A - white
20 A - yellow
15 A - blue
10 A - red
7,5 A - brown
5 A - beige

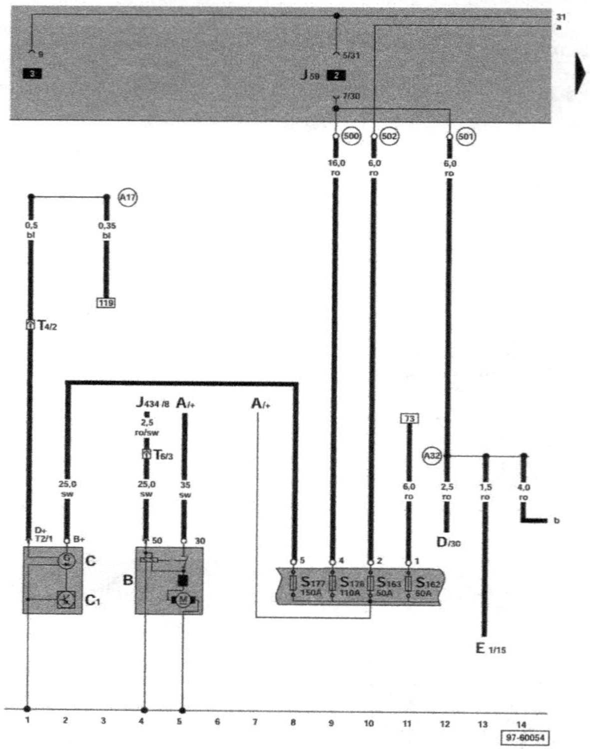

ws = white
sw = black
ro = red
br = brown
gn = green
bl = blue
gr = grey
li = violet
ge = yellow

A - Battery
B - Starter
C - Generator (GEN)
C1 - Voltage Regulator (VR)
D - Ignition/Starter Switch
E1 - Light Switch
J59 - Load Reduction Relay
J434 - Locking relay for starter (clutch pedal switch)
S162 - Fuse -1- in fuse bracket/battery
S163 - Fuse -2- (50)in fuse bracket/battery
S176 - Fuse -4- (30), in fuse bracket/battery
S177 - Fuse -5- (30), in fuse bracket/battery
T2 - Double Connector
T4 - 4-Pin Connector, in engine compartment left
T6 - 6-Pin Connector, brown, behind instrument panel, left

(500) - Threaded connection -1- (30) on the relay plate
(501) - Threaded connection -2- (30) on the relay plate
(502) - Threaded connection -1- (30a) on the relay plate
(A17) - wire connection (61), in instrument panel wiring harness
(A32) - plus connection (30), in instrument panel wiring harness

Starter, generator (GEN)

New Beetle — Wiring diagram — **No. 19/3**

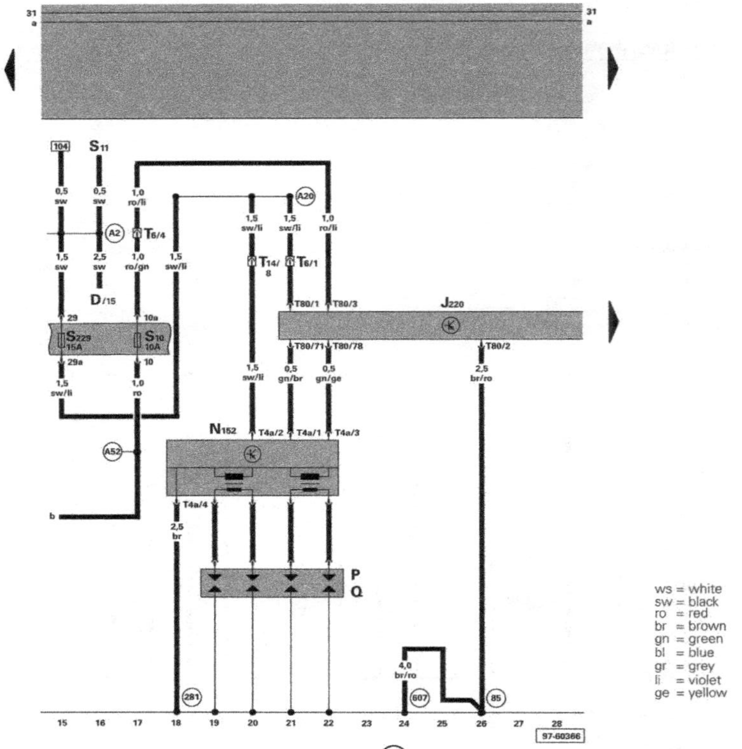

ws = white
sw = black
ro = red
br = brown
gn = green
bl = blue
gr = grey
li = violet
ge = yellow

D - Ignition/Starter Switch
J220 - Motronic Engine Control Module (ECM), behind istrument panel, left
N152 - Ignition Coil
P - Spark Plug Connectors
Q - Spark Plugs
S10 - Fuse 10 in fuse holder
S11 - Fuse 11 in fuse holder
S229 - Fuse 29 in fuse holder
T4a - 4-Pin Connector, black
T6 - 6-Pin Connector, brown, behind instrument panel, left
T14 - 14-Pin Connector, in engine compartment, left
T80 - 80-Pin Connector, at Motronic Engine Control Module (ECM)
(85) - Ground connection -1-, in engine compartment wiring harness
(281) - Ground connector -1-, in wiring harness engine pre-wiring
(607) - Ground Connection (in left plenum chamber)
(A2) - plus connection (15), in instrument panel wiring harness
(A20) - wire connection (15a), in instrument panel wiring harness
(A52) - plus connection (30), in instrument panel wiring harness

No. 19/4 — Wiring diagram — **New Beetle**

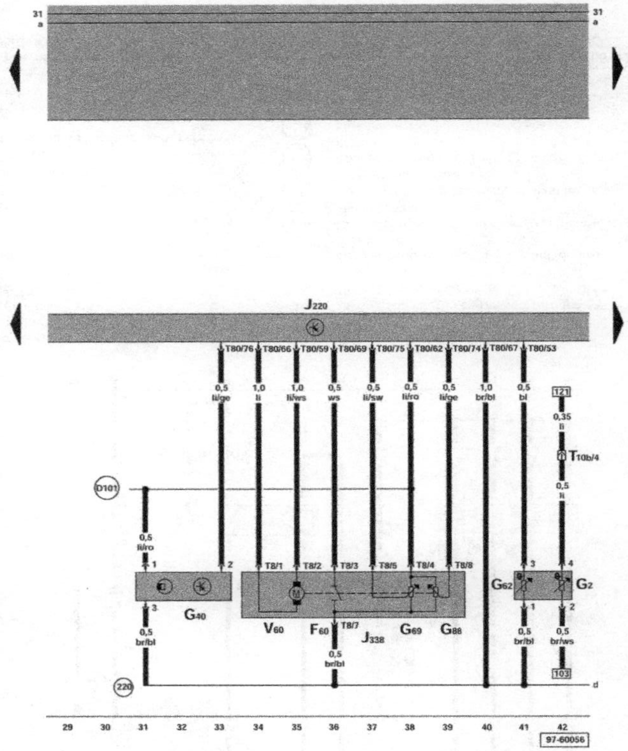

ws = white
sw = black
ro = red
br = brown
gn = green
bl = blue
gr = grey
li = violet
ge = yellow

F60 - Closed Throttle Position (CTP) Switch
G2 - Engine Coolant Temperature (ECT) Sensor
G40 - Camshaft Position (CMP) Sensor
G62 - Engine Coolant Temperature (ECT) Sensor
G69 - Throttle Position (TP) Sensor
G88 - Throttle Position (TP) Sensor
J220 - Motronic Engine Control Module (ECM)
J338 - Throttle Valve Control Module
V60 - Throttle Position (TP) Actuator
T8 - 8-Pin Connector
T10b - 10-Pin Connector, behind instrument panel, left
T80 - Connector, 80-point, at Engine Control Module (ECM)
(220) - Ground connection (sensor ground), in engine compartment wiring harness
(D101) - Connection 1 (in engine compartment wiring harness)

2.0L engine management (AEG) from Aug. 1998

New Beetle — Wiring diagram — No. 19/5

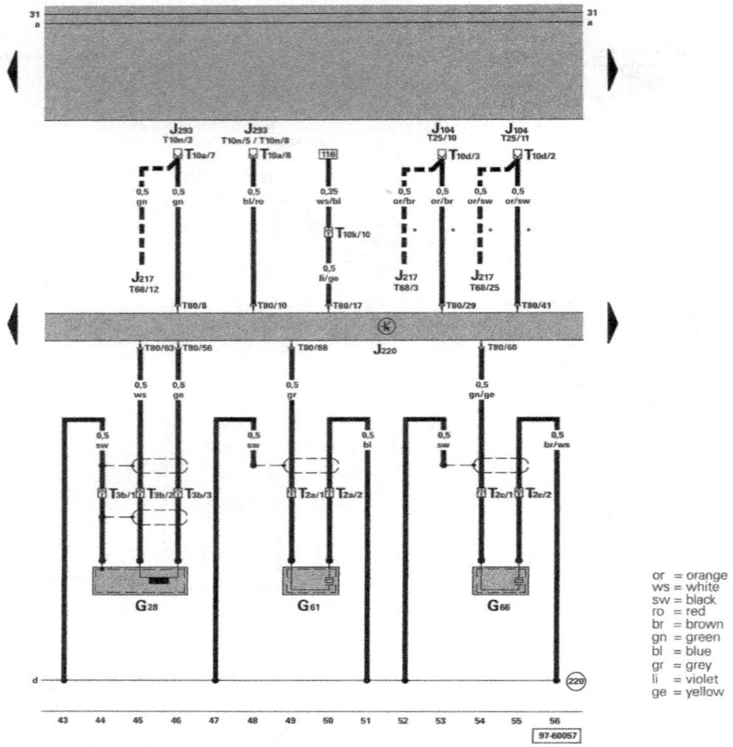

or = orange
ws = white
sw = black
ro = red
br = brown
gn = green
bl = blue
gr = grey
li = violet
ge = yellow

- G28 - Engine Speed (RPM) Sensor
- G61 - Knock Sensor (KS) 1
- G66 - Knock Sensor (KS) 2
- J104 - ABS Control Module (w/EDL)
- J217 - Transmission Control Module (TCM), behind instrument panel
- J220 - Motronic Engine Control Module (ECM), behind instrument panel, left
- J293 - Coolant FC (Fan Control) Control Module
- T2a - 2-Pin Connector
- T3b - 3-Pin Connector
- T2c - 2-Pin Connector
- T6 - 6-Pin Connector, red, behind instrument panel
- T10 - 10-Pin Connector, white, behind instrument panel, left
- T10a - 10-Pin Connector, orange, behind instrument panel, left
- T10d - 10-Pin Connector, green, behind instrument panel, left
- T10k - 10-Pin Connector, blue, behind instrument panel, left
- T25 - 25-Pin Connector, at J104
- T68 - Connector, 68-point, on Transmisson Control Module (TCM)
- T80 - Connector, 80-point, at Engine Control Module (ECM)
- (220) - Ground connection (sensor ground), in engine compartment wiring harness
- * - Connection for DATA-bus
- --- - Automatic transmission only

Edition 02/99
USA.5132.06.21

Motronic engine control module (ECM), engine speed (RPM) sensor, knock sensor (KS) 1, knock sensor (KS) 2

No. 19/6 — Wiring diagram — New Beetle

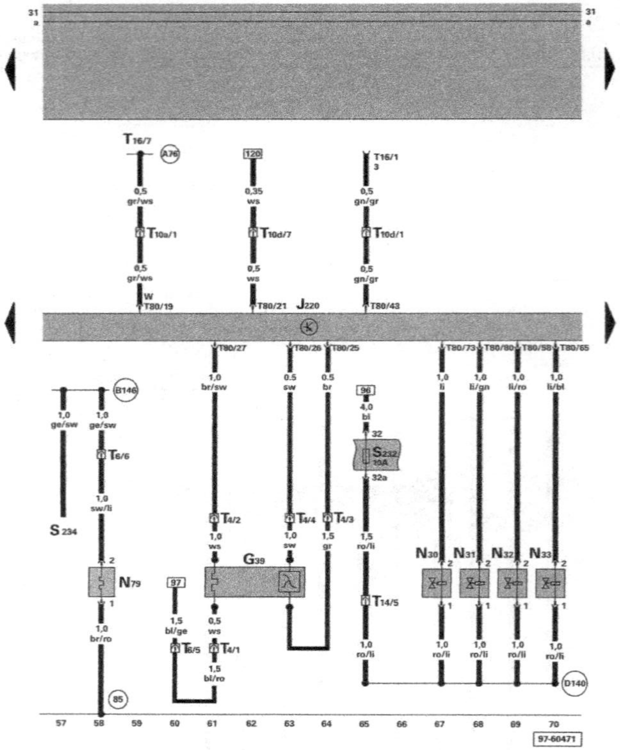

ws = white
sw = black
ro = red
br = brown
gn = green
bl = blue
gr = grey
li = violet
ge = yellow

- G39 - Heated Oxygen Sensor (HO2S)
- J220 - Motronic Engine Control Module (ECM), behind instrument panel, left
- N30 - Cylinder 1 Fuel Injector
- N31 - Cylinder 2 Fuel Injector
- N32 - Cylinder 3 Fuel Injector
- N33 - Cylinder 4 Fuel Injector
- N79 - Positive Crankcase Ventilation (PCV) Heating Element
- S232 - Fuse 32 in fuse holder
- S234 - Fuse 34 in fuse holder
- T4 - 4-Pin Connector
- T6 - 6-Pin Connector, red, behind instrument panel, left
- T10a - 10-Pin Connector, orange, behind instrument panel, left
- T10d - 10-Pin Connector, green, behind instrument panel, left
- T14 - 14-Pin Connector, in engine compartment, left
- T16 - 16-Pin Connector, Data Link Connector (DLC), below steering column
- T80 - 80-Pin Connector, at Engine Control Module (ECM)
- (85) - Ground connection -1-, in engine compartment wiring harness
- (A76) - Connector (K-diagnosis wire), in instrument panel wiring harness
- (D140) - Connector (injectors), in wiring harness, engine pre-wiring
- (B146) - Plus connector -1- (87), in wiring harness interior

Motronic engine control module (ECM), fuel injectors, heated oxygen sensor (HO2S) 2, positive crankcase ventilation (PCV) heating element

Edition 06/99
USA.5132.07.21

New Beetle — Wiring diagram — No. 19/7

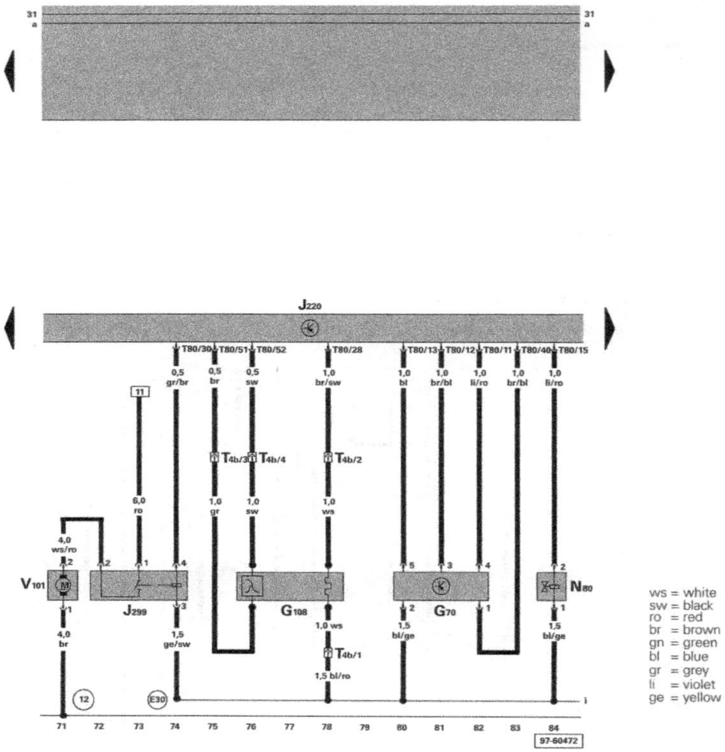

ws = white
sw = black
ro = red
br = brown
gn = green
bl = blue
gr = grey
li = violet
ge = yellow

- G108 - Heated Oxygen Sensor (HO2S) 2
- G70 - Mass Air Flow (MAF) Sensor
- J220 - Motronic Engine Control Module (ECM), behind instrument panel, left
- J299 - Secondary Air Injection (AIR) Pump Relay, in engine compartment, left
- N80 - Evaporative Emission (EVAP) Canister Purge Regulator Valve
- V101 - Secondary Air Injection (AIR) Pump Motor
- T4b - 4-Pin Connector
- T80 - 80-Pin Connector
- (12) - Ground connection, in engine compartment, left
- (E30) - Connector (87a), in wiring harness engine

Edition 06/99 USA.5132.07.21

Heated oxygen sensor (HO2S), mass air flow (MAF) sensor, evaporative emission (EVAP) canister purge regulator valve, secondary air injection (AIR) system

No. 19/8 — Wiring diagram — New Beetle

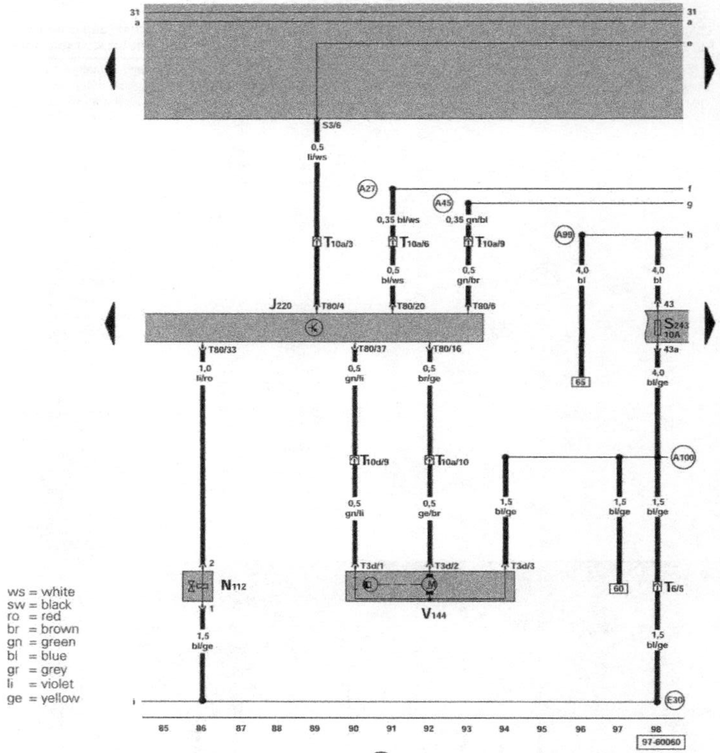

ws = white
sw = black
ro = red
br = brown
gn = green
bl = blue
gr = grey
li = violet
ge = yellow

- J220 - Motronic Engine Control Module (ECM), behind instrument panel, left
- N112 - Secondary Air Injection (AIR) Solenoid Valve
- S243 - Fuse 43 in fuse holder
- V144 - Leak Detection Pump (LDP)
- T3d - 3-Pin Connector
- T6 - 6-Pin Connector, red, behind instrument panel, left
- T10a - 10-Pin Connector,orange, behind instrument panel, left
- T10d - 10-Pin Connector, green, behind instrument panel, left
- T80 - Connector, 80-point, at Engine Control Module (ECM)
- (A27) - wire Connection (vehicle speed signal), in instrument panel wiring harness
- (A45) - wire connection (RPM-signal), in instrument panel wiring harness
- (A99) - Connector -1- (87), in instrument panel wiring harness
- (A100) - Connector -2- (87), in instrument panel wiring harness
- (E30) - Connector (87a), in wiring harness engine

Motronic engine control module (ECM), leak detection pump (LDP) secondary air injection solenoid valve

Edition 02/99 USA.5132.06.21

New Beetle — Wiring diagram — No. 19/9

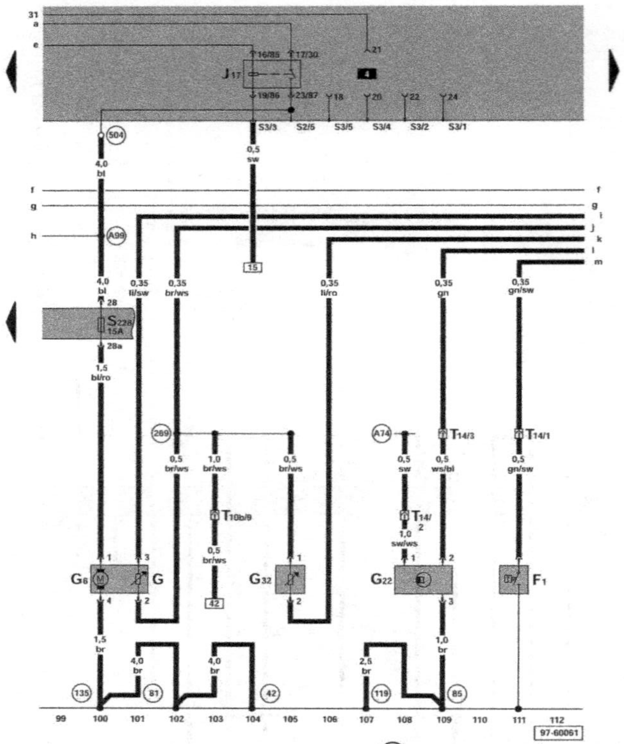

ws = white
sw = black
ro = red
br = brown
gn = green
bl = blue
gr = grey
li = violet
ge = yellow

- F1 - Oil Pressure Switch
- G - Sender for fuel gauge
- G6 - Fuel Pump (FP)
- G22 - Speedometer Vehicle Speed Sensor (VSS)
- G32 - Engine Coolant Level (ECL) Sensor
- J17 - Fuel Pump (FP) Relay
- S228 - Fuse 28 in fuse holder
- T14 - 14-Pin Connector, in engine compartment, left
- (42) - Ground connection, beside steering column
- (81) - Ground connection -1-, in instrument panel wiring harness
- (85) - Ground connection -1-, in engine compartment wiring harness
- (119) - Ground connection -1-, in headlight wiring harness
- (135) - Ground connection -2-, in instrument panel wiring harness
- (269) - Ground connector (sensor ground) -1-, in instrument panel wiring harness
- (504) - Threaded connection -1- (87) on the relay plate
- (A74) - wire connection (75x), in instrument panel wiring harness
- (A99) - Connector -1- (87), in instrument panel wiring harness

Oil pressure switch, fuel pump (FP), fuel pump (FP) relay, engine coolant level sensor, speedomter vehicle speed sensor (VSS)

No. 19/10 — Wiring diagram — New Beetle

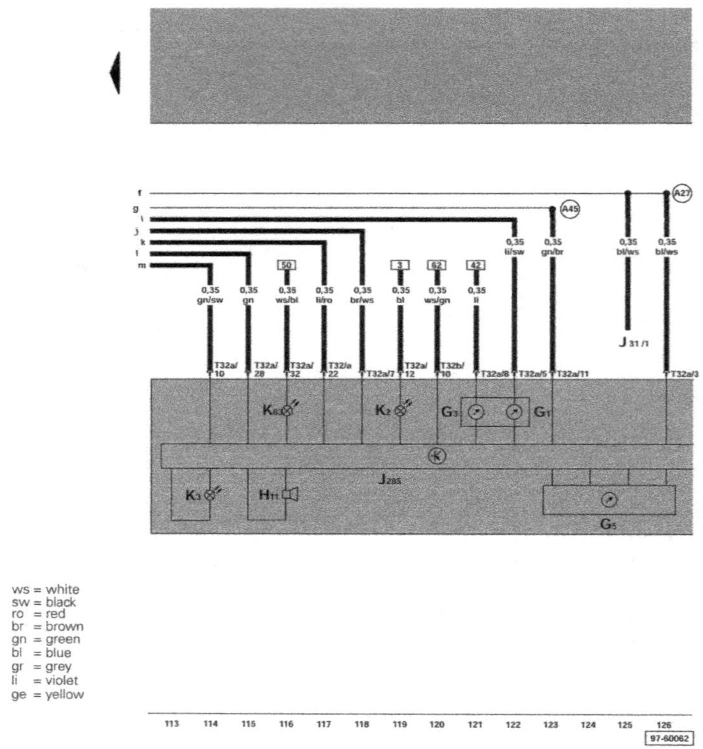

ws = white
sw = black
ro = red
br = brown
gn = green
bl = blue
gr = grey
li = violet
ge = yellow

- G1 - Fuel gauge
- G3 - Engine Coolant Temperature (ECT) Gauge
- G5 - Tachometer
- H11 - Oil Pressure Warning Buzzer
- J31 - Wiper/Washer Intermittent Relay
- J285 - Control module with indicator unit in instrument panel insert
- K2 - Generator (GEN) Warning Light
- K3 - Oil Pressure Warning Light
- K83 - Malfunction Indicator Lamp (MIL)
- T32a - 32-Pin Connector, blue, on instrument cluster
- T32b - 32-Pin Connector, green, on instrument cluster
- (A27) - wire Connection (vehicle speed signal), in instrument panel wiring harness
- (A45) - wire connection (RPM-signal), in instrument panel wiring harness

Instrument cluster, fuel gauge, engine coolant temperature (ECT) gauge, tachometer, generator (GEN) warning light, oil pressure warning light

New Beetle **Wiring diagram** **No. 3/1**

2,0L-Engine - Motronic Multiport Fuel Injection (MFI)/85 kW, code AVH/AZG, (with cruise control)

from August 2000

Deviate relay location and fuseplacements as well as the locations of multiple connectors see section "component locations".

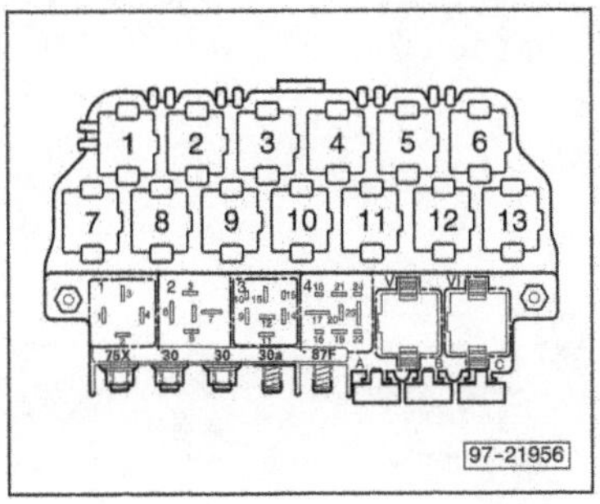

Relay location on the thirteen position auxiliary relay panel, above relay panel:

Relay panel:

- **2** Load Reduction Relay (213)
- **4** Fuel Pump Relay (409)

Fuse colors

30 A - green
25 A - white
20 A - yellow
15 A - blue
10 A - red
7,5 A - brown
5 A - beige
3 A - violet

Note: Number in parentheses indicates production control number stamped on relay housing.

New Beetle **Wiring diagram** **No. 3/2**

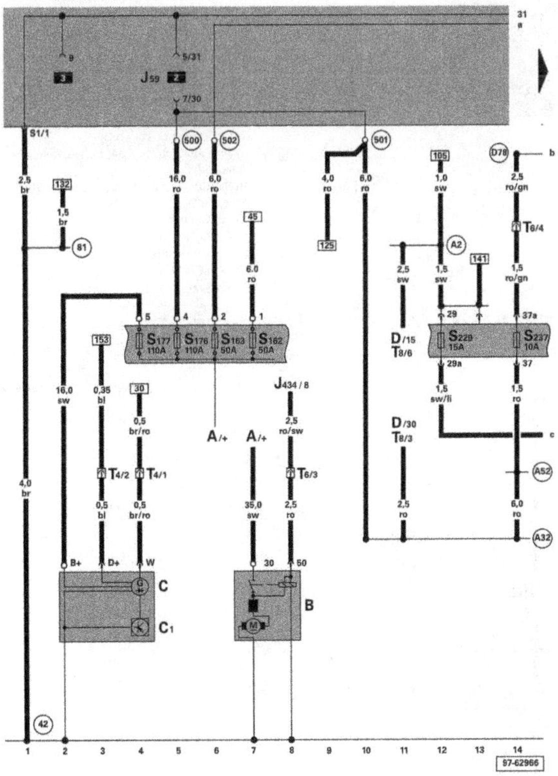

Battery, starter, generator (GEN), voltage regulator (VR), ignition/starter switch, load reduction relay, fuses in fuse bracket/battery

A - Battery
B - Starter
C - Generator (GEN)
C1 - Voltage Regulator (VR)
D - Ignition/Starter Switch
J434 - Locking relay for starter (clutch pedal switch)
J59 - Load Reduction Relay
S162 - Fuse -1- (50), in fuse bracket/battery
S163 - Fuse -2- (50), in fuse bracket/battery
S176 - Fuse -4- (30), in fuse bracket/battery
S177 - Fuse -5- (30), in fuse bracket/battery
S229 - Fuse 29 in fuse/relay panel
S237 - Fuse 37 in fuse/relay panel
T4 - 4-Pin Connector, in engine compartment left
T6 - 6-Pin Connector, brown, behind instrument panel, left
T8 - 8-Pin Connector, at -D-

(42) - Ground connection, beside steering column
(81) - Ground connection -1-, in instrument panel wiring harness
(500) - Threaded connection -1- (30) on the relay plate
(501) - Threaded connection -2- (30) on the relay plate
(502) - Threaded connection -1- (30a) on the relay plate
(A2) - plus connection (15), in instrument panel wiring harness
(A32) - plus connection (30), in instrument panel wiring harness
(A52) - Plus Connection 2 (30) (in instrument panel wiring harness)
(D78) - Plus connector -1- (30a), in engine compartment wiring harness

ws = white
sw = black
ro = red
br = brown
gn = green
bl = blue
gr = grey
li = violet
ge = yellow
or = orange

New Beetle — Wiring diagram — **No. 3/3**

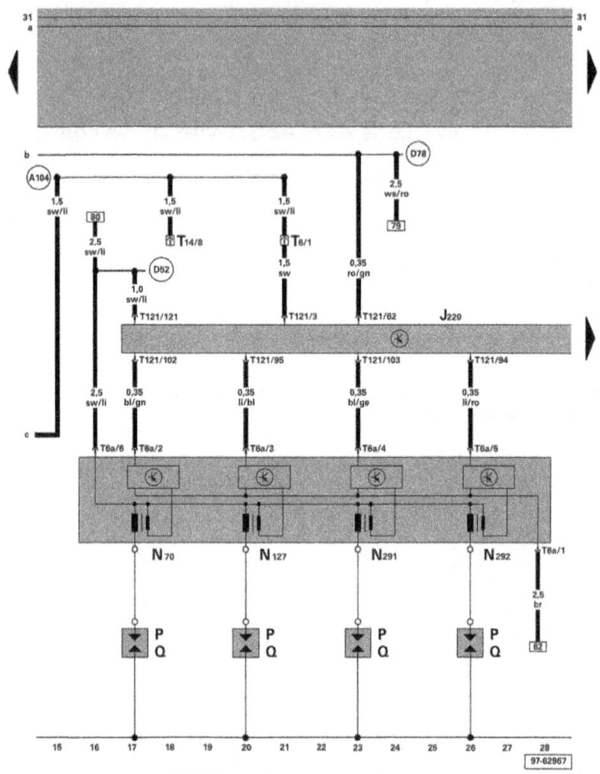

Motronic en gine control module (ECM), ignition coil, spark plugs

- J220 - Motronic Engine Control Module (ECM)
- N70 - Ignition Coil 1 with Power Output Stage
- N127 - Ignition Coil 2 with Power Output Stage
- N291 - Ignition Coil 3 with Power Output Stage
- N292 - Ignition Coil 4 with Power Output Stage
- P - Spark Plug Connectors
- Q - Spark Plugs
- T6 - 6-Pin Connector, brown, behind instrument panel, left
- T6a - 6-Pin Connector
- T14 - 14-Pin Connector, in left plenum chamber
- T121 - 121-Pin Connector, at -J220-
- (A104) - Plus connector -2- (15), in instrument panel wiring harness
- (D52) - plus connection (15a), in engine compartment wiring harness
- (D78) - Plus connector -1- (30a), in engine compartment wiring harness

ws = white
sw = black
ro = red
br = brown
gn = green
bl = blue
gr = grey
li = violet
ge = yellow
or = orange

New Beetle — Wiring diagram — **No. 3/4**

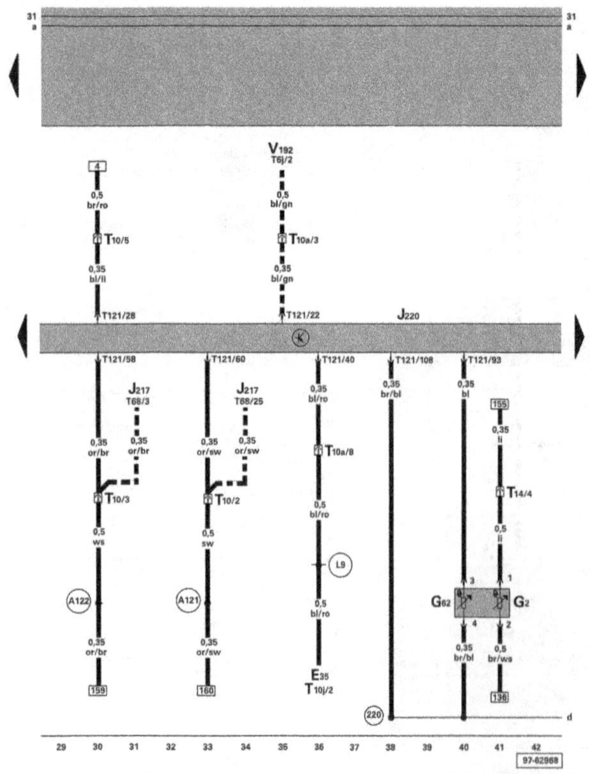

Motronic engine control module (ECM), engine coolant temperature (ECT) sensor

- E35 - A/C Switch
- G2 - Engine Coolant Temperature (ECT) Sensor
- G62 - Engine Coolant Temperature (ECT) Sensor
- J217 - Transmission Control Module (TCM)
- J220 - Motronic Engine Control Module (ECM)
- T6j - 6-Pin Connector at -V192-
- T10 - 10-Pin Connector, white, behind instrument panel, left
- T10a - 10-Pin Connector, orange, behind instrument panel, left
- T10j - 10-Pin Connector
- T14 - 14-Pin Connector, in left plenum chamber
- T68 - 68-Pin Connector
- T121 - 121-Pin Connector, at -J220-
- V192 - Brake System Vacuum Pump
- (220) - Ground connection (sensor ground), in engine compartment wiring harness
- (A121) - Connection (high bus) in instrument panel wiring harness
- (A122) - Connection (low bus) in instrument panel wiring harness
- (L9) - Wire connection -1-, in A/C wiring harness
- - - - - Only vehicles code AVH with Automatic transmission

ws = white
sw = black
ro = red
br = brown
gn = green
bl = blue
gr = grey
li = violet
ge = yellow
or = orange

New Beetle — Wiring diagram — No. 3/5

Motronic engine control module (ECM), heated oxygen sensor (HO2S), secondary air injection (AIR) pump relay/motor

- G39 - Heated Oxygen Sensor (HO2S), black
- G130 - Oxygen Sensor (O2S) behind Three Way Catalytic Converter (TWC), brown
- J220 - Motronic Engine Control Module (ECM)
- J299 - Secondary Air Injection (AIR) Pump Relay
- V101 - Secondary Air Injection (AIR) Pump Motor
- T4 - 4-Pin Connector, brown
- T6 - 6-Pin Connector, black
- T121 - 121-Pin Connector, at -J220-
- (85) - Ground connection -1-, in engine compartment wiring harness
- (609) - Ground Connection (in right plenum chamber)
- (E30) - Connection (87a), in wiring harness engine

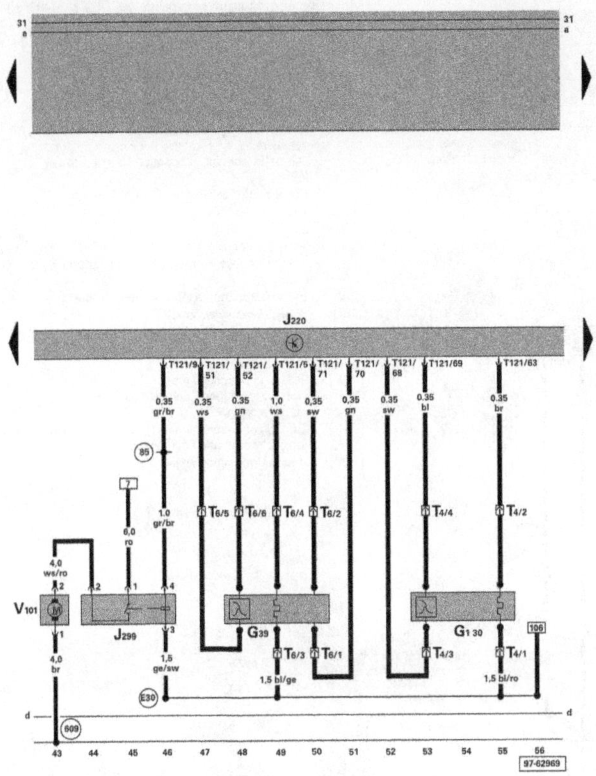

ws = white
sw = black
ro = red
br = brown
gn = green
bl = blue
gr = grey
li = violet
ge = yellow
or = orange

New Beetle — Wiring diagram — No. 3/6

Camshaft Position (CMP) Sensor, knock sensor (KS) 1, throttle drive (power accelerator actuation), angle sensor 1/2 for throttle drive, motronic engine control module

- G40 - Camshaft Position (CMP) Sensor
- G61 - Knock Sensor (KS) 1
- G186 - Throttle drive (power accelerator actuation)
- G187 - Angle sensor -1- for throttle drive (power accelerator actuation)
- G188 - Angle sensor -2- for throttle drive (power accelerator actuation)
- J220 - Motronic Engine Control Module (ECM)
- J338 - Throttle Valve Control Module
- T2a - 2-Pin Connector
- T6a - 6-Pin Connector
- T121 - 121-Pin Connector, at -J220-
- (139) - Ground connection (sensor Ground), in Motronic Multiport Fuel Injection (MFI) wiring harness

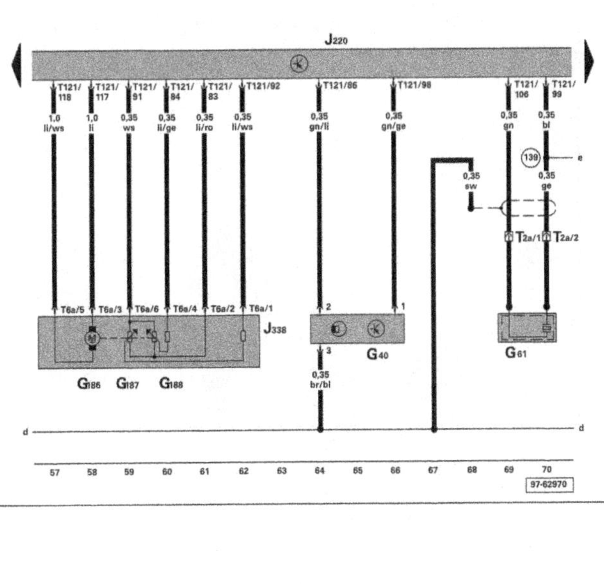

ws = white
sw = black
ro = red
br = brown
gn = green
bl = blue
gr = grey
li = violet
ge = yellow
or = orange

2.0L engine management (AVH, AZG) from Aug. 2000

New Beetle — **Wiring diagram** — **No. 3/7**

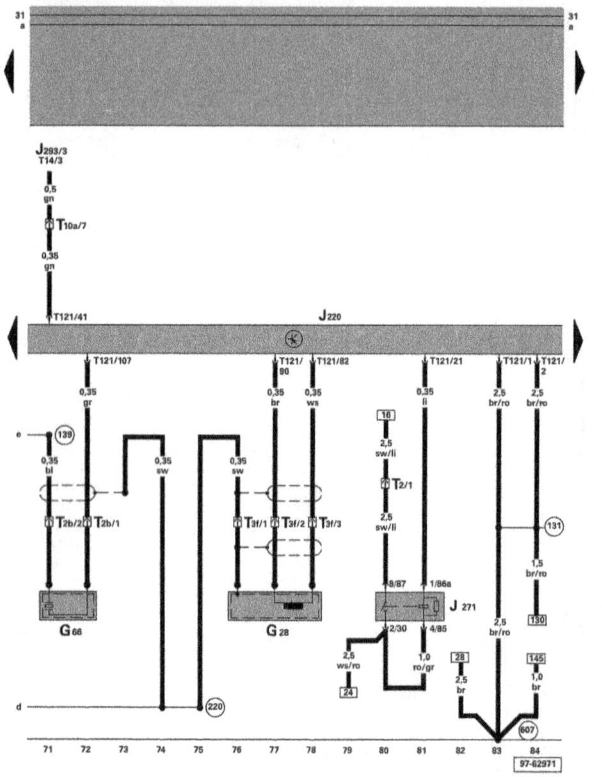

Motronic engine control module (ECM) / power supply relay, engine speed (RPM) sensor, knock sensor (KS) 2

G28 - Engine Speed (RPM) Sensor
G66 - Knock Sensor (KS) 2
J220 - Motronic Engine Control Module (ECM)
J271 - Motronic Engine Control Module (ECM) power supply relay, in protective housing, in engine compartment, left, production control number (428)
J293 - Coolant FC (Fan Control) Control Module
T2 - 2-Pin Connector
T2b - 2-Pin Connector
T3f - 3-Pin Connector
T10a - 10-Pin Connector, orange, behind instrument panel, left
T14 - 14-Pin Connector in left plenum chamber
T121 - 121-Pin Connector, at J220-

(131) - Ground connection -2-, in engine compartment wiring harness
(139) - Ground connection (sensor Ground), in Motronic Multiport Fuel Injection (MFI) wiring harness
(220) - Ground connection (sensor ground), in engine compartment wiring harness
(607) - Ground Connection (in left plenum chamber)

ws = white
sw = black
ro = red
br = brown
gn = green
bl = blue
gr = grey
li = violet
ge = yellow
or = orange

97-62971

New Beetle — **Wiring diagram** — **No. 3/8**

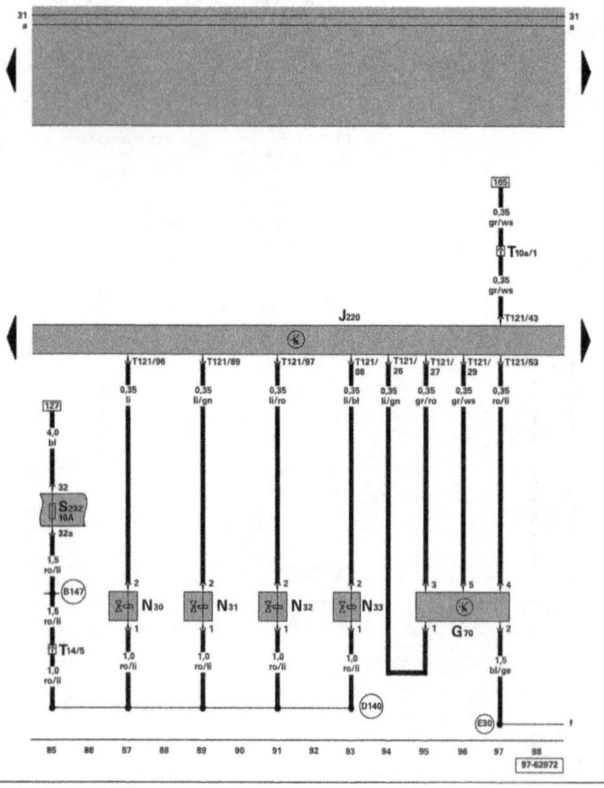

Motronic engine control module (ECM), mass air flow (MAF) sensor, fuel injectors

G70 - Mass Air Flow (MAF) Sensor
J220 - Motronic Engine Control Module (ECM)
N30 - Cylinder 1 Fuel Injector
N31 - Cylinder 2 Fuel Injector
N32 - Cylinder 3 Fuel Injector
N33 - Cylinder 4 Fuel Injector
S232 - Fuse 32 in fuse/relay panel
T10a - 10-Pin Connector, orange, behind instrument panel, left
T14 - 14-Pin Connector
T121 - 121-Pin Connector, at J220-

(B147) - Plus connector -2- (87), in wiring harness interior
(D140) - Connector (injectors), in wiring harness, engine pre-wiring
(E30) - Connector (87a), in wiring harness engine

ws = white
sw = black
ro = red
br = brown
gn = green
bl = blue
gr = grey
li = violet
ge = yellow
or = orange

97-62972

New Beetle — Wiring diagram — No. 3/9

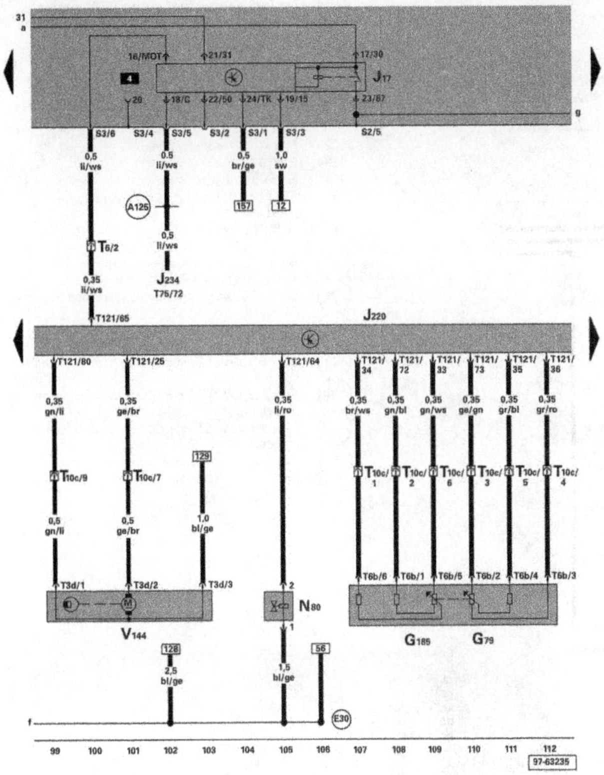

Motronic engine control module (ECM), throttle position (TP) sensor, sender 2 for acelerator pe dal position, evaporative emission (EVAP), leak detection pump (LDP)

- G79 - Throttle Position (TP) Sensor
- G185 - Sender -2- for acelerator pedal position
- J17 - Fuel Pump (FP) Relay
- J220 - Motronic Engine Control Module (ECM)
- J234 - Airbag Control Module
- N80 - Evaporative Emission (EVAP) Canister Purge Regulator Valve
- V144 - Leak Detection Pump (LDP)
- T3d - 3-Pin Connector
- T6 - 6-Pin Connector, brown, behind instrument panel, left
- T6b - 6-Pin Connector
- T10c - 10-Pin Connector, blue, behind instrument panel, left
- T75 - 75-Pin Connector
- T121 - 121-Pin Connector, at -J220-
- (A125) - Connection (crash signal) in instrument panel wiring harness
- (E30) - Connector (87a), in wiring harness engine

ws = white
sw = black
ro = red
br = brown
gn = green
bl = blue
gr = grey
li = violet
ge = yellow
or = orange

Edition 06.04

New Beetle — Wiring diagram — No. 3/10

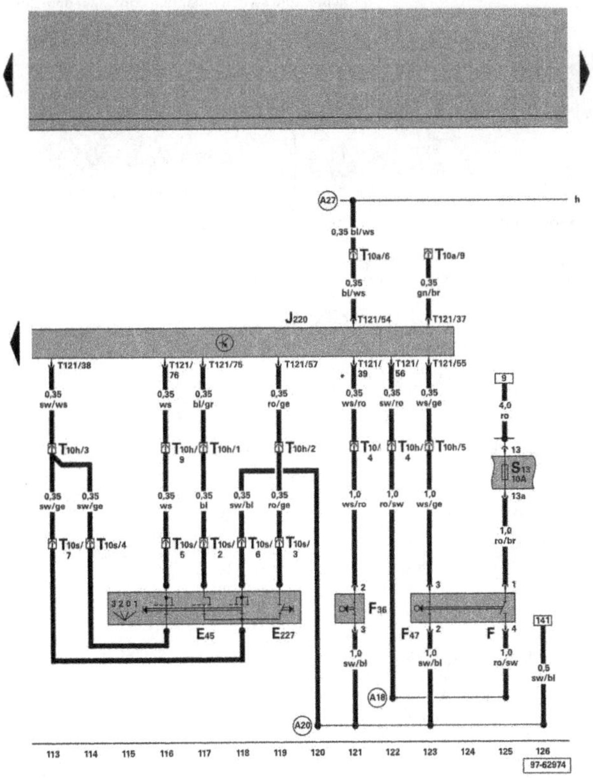

Motronic engine control module (ECM), cruise control switch, button for cruise control (set), brake light switch, clutch vacuum vent valve switch

- E45 - Cruise Control Switch
- E227 - Button for cruise control (set)
- F - Brake Light Switch
- F36 - Clutch Vacuum Vent Valve Switch
- F47 - Brake Pedal Switch (Cruise control/Diesel Direct Injection)
- J220 - Motronic Engine Control Module (ECM)
- S13 - Fuse 13 in fuse/relay panel
- T10 - 10-Pin Connector, white, behind instrument panel, left
- T10a - 10-Pin Connector, orange, behind instrument panel, left
- T10h - 10-Pin Connector, black, behind instrument panel, left
- T10s - 10-Pin Connector, near steering column
- T121 - 121-Pin Connector, at -J220-
- (A18) - wire connection (54), in instrument panel wiring harness
- (A20) - wire connection (15a), in instrument panel wiring harness
- (A27) - Wire Connection (vehicle speed signal), in instrument panel wiring harness
- * - Manual transmission only

ws = white
sw = black
ro = red
br = brown
gn = green
bl = blue
gr = grey
li = violet
ge = yellow
or = orange

Edition 09.03

2.0L engine management (AVH, AZG) from Aug. 2000

New Beetle — Wiring diagram — No. 3/11

Sender for fuel gauge, fuel pump (FP), engine coolant level (ECL) sensor

G - Sender for fuel gauge
G6 - Fuel Pump (FP)
G32 - Engine Coolant Level (ECL) Sensor
N79 - Positive crankcase ventilation (PVC) heating element
S228 - Fuse 28 in fuse/relay panel
S234 - Fuse 34 in fuse/relay panel
S243 - Fuse 43 in fuse/relay panel
T6 - 6-Pin Connector, brown, behind instrument panel, left
T14 - 14-Pin Connector

(269) - Ground connector (sensor ground) -1-, in instrument panel wiring harness
(316) - Sensor Ground Connection 2 (in engine wiring harness)
(504) - Threaded connection -1- (87) on the relay plate
(A99) - Connector -1- (87), in instrument panel wiring harness
(A100) - Connector -2- (87), in instrument panel wiring harness
(A101) - Connector -3- (87), in instrument panel wiring harness
(A188) - Coolant Warning Connection (in instrument panel wiring harness)

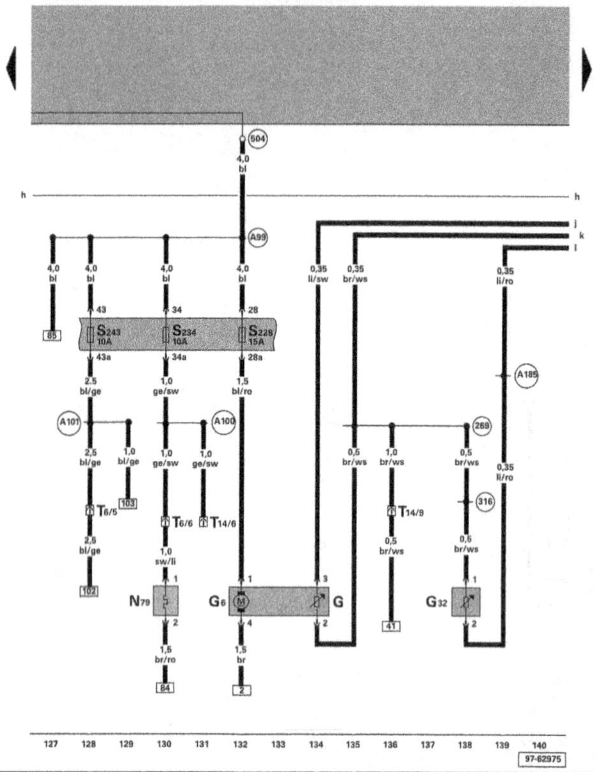

ws = white
sw = black
ro = red
br = brown
gn = green
bl = blue
gr = grey
li = violet
ge = yellow
or = orange

New Beetle — Wiring diagram — No. 3/12

Oil pressure switch / warning light, speedometer vehicle speed sensor (VSS), warning buzzer, instrument cluster, generator (GEN) warning light

F1 - Oil Pressure Switch
G22 - Speedometer Vehicle Speed Sensor (VSS)
H3 - Warning Buzzer
J285 - Control module with indicator unit in instrument panel insert
K2 - Generator (GEN) Warning Light
K3 - Oil Pressure Warning Light
K28 - Engine Coolant Level/Temperature (ECL/ECT) Warning Light
K132 - Electronic Power Control (EPC) Lamp
S5 - Fuse 5 in fuse/relay panel
S7 - Fuse 7 in fuse/relay panel
T14 - 14-Pin Connector
T32 - 32-Pin Connector, blue

(B163) - Plus connector -1- (15) in wiring harness interior

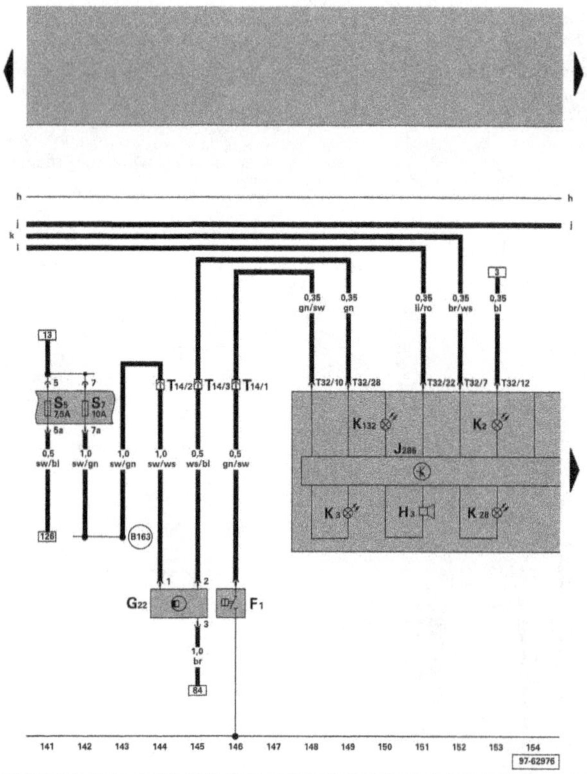

ws = white
sw = black
ro = red
br = brown
gn = green
bl = blue
gr = grey
li = violet
ge = yellow
or = orange

New Beetle **Wiring diagram** **No. 3/13**

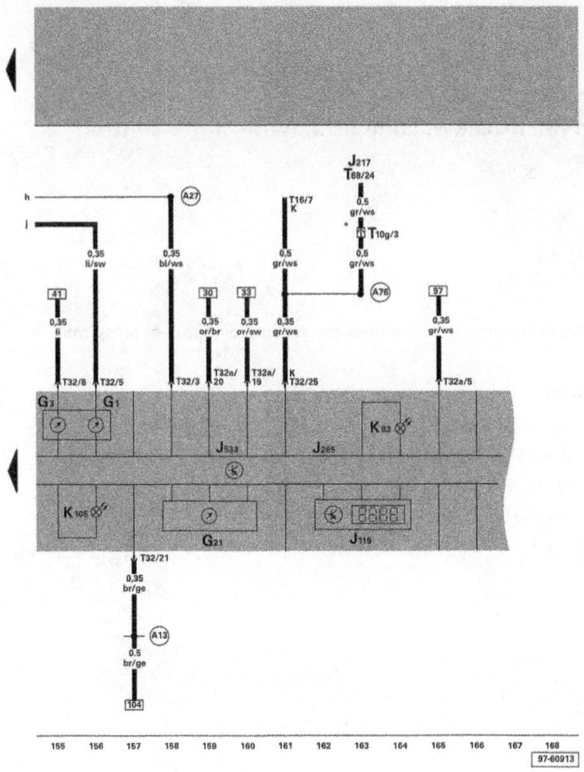

Instrument cluster, fuel gauge, engine coolant temperature (ECT) gauge, speedomter, multifunc tion indicator (MFA), low fuel level warning light

G1 - Fuel Gauge
G3 - Engine Coolant Temperature (ECT) Gauge
G21 - Speedometer
J119 - Multi-function Indicator (MFA)
J217 - Transmission Control Module (TCM), behind instrument panel, center
J285 - Control module with indicator unit in instrument panel insert
J533 - Data Bus On Board Diagnostic Interface
K83 - Malfunction Indicator lamp (MIL)
K105 - Low Fuel Level Warning Light
T10g - 10-Pin Connector, grey, behind instrument panel, left
T16 - 16-Pin Connector, Data link Connector (DLC), below steering column, left
T32 - 32-Pin Connector, blue
T32a - 32-Pin Connector, green
T68 - 68-Pin Connector

A13 - wire connection (door contact switch) in instrument panel wiring harness
A27 - Wire Connection (vehicle speed signal), in instrument panel wiring harness
A76 - Connector (K-diagnosis wire) in instrument panel wiring harness

* - Automatic transmission only

ws = white
sw = black
ro = red
br = brown
gn = green
bl = blue
gr = grey
li = violet
ge = yellow
or = orange

THIS SPACE INTENTIONALLY LEFT BLANK

New Beetle Convertible — Wiring diagram — No. 2/1

2,0L-Engine - Motronic Multiport Fuel Injection (MFI)/85 kW, code BDC, (with cruise control)

from August 2002

Deviate relay location and fuseplacements as well as the locations of multiple connectors see section "component locations".

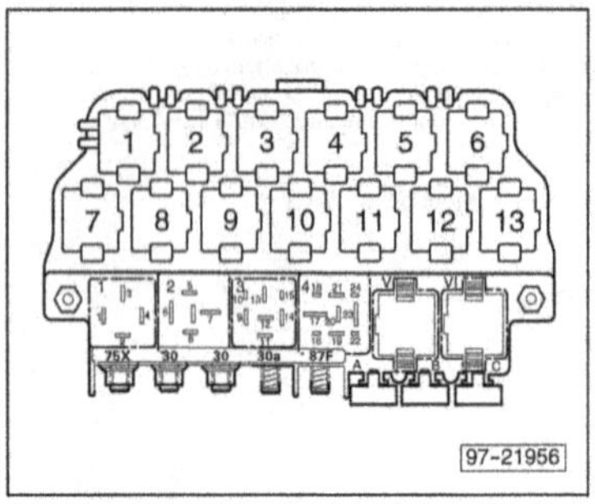

Relay panel:

- **1** Dual Horn Relay (53)
- **2** Load Reduction Relay (100)
- **4** Fuel Pump (FP) Relay (409)
- **V** Wiper/Washer Intermittent Relay (377)
- **VI** Wiper/Washer Intermittent Relay (377)

Note: Number in parentheses indicates production control number stamped on relay housing.

Fuse colors

30 A - green
25 A - white
20 A - yellow
15 A - blue
10 A - red
7,5 A - brown
5 A - beige
3 A - violet

New Beetle Convertible — Wiring diagram — No. 2/2

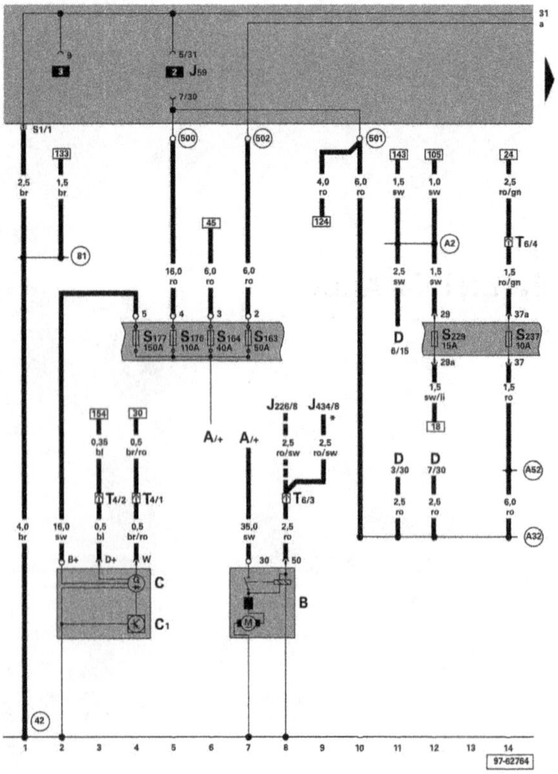

Battery, starter, generator (GEN), voltage regulator (VR), ignition/starter switch, load reduction relay, fuses in fuse bracket/battery

- A - Battery
- B - Starter
- C - Generator (GEN)
- C1 - Voltage Regulator (VR)
- D - Ignition/Starter Switch
- J59 - Load Reduction Relay
- J226 - Park/Neutral Position (PNP) Relay (126)
- J434 - Locking relay for starter (clutch pedal switch)
- S163 - Fuse -2- , in fuse bracket/battery
- S164 - Fuse -3- , in fuse bracket/battery
- S176 - Fuse -4- (30), in fuse bracket/battery
- S177 - Fuse -5- (30), in fuse bracket/battery
- S229 - Fuse 29 in fuse holder
- S237 - Fuse 37 in in fuse holder
- T4 - 4-Pin Connector, in engine compartment left
- T6 - 6-Pin Connector, brown, behind instrument panel, left
- (42) - Ground connection, beside steering column
- (81) - Ground connection -1-, in instrument panel wiring harness
- (500) - Threaded connection -1- (30) on the relay plate
- (501) - Threaded connection -2- (30) on the relay plate
- (502) - Threaded connection -1- (30a) on the relay plate
- (A2) - plus connection (15), in instrument panel wiring harness
- (A32) - plus connection (30), in instrument panel wiring harness
- (A52) - Plus connector -2- (30), in instrument panel wiring harness
- * - Manual transmission only
- --- - Automatic transmission only

ws = white
sw = black
ro = red
br = brown
gn = green
bl = blue
gr = grey
li = violet
ge = yellow

New Beetle Convertible — Wiring diagram — No. 2/3

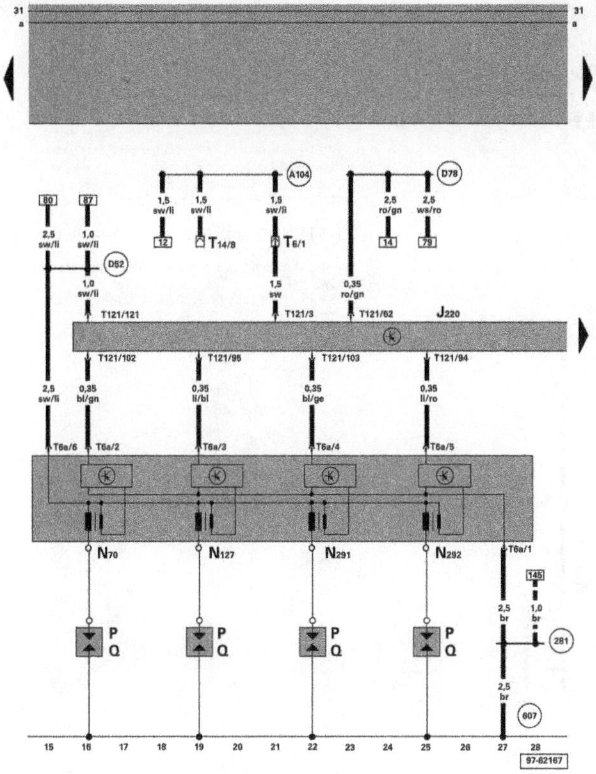

Motronic engine control module (ECM), ignition coil, spark plugs

- J220 - Motronic Engine Control Module (ECM)
- N70 - Ignition Coil 1 with Power Output Stage
- N127 - Ignition Coil 2 with Power Output Stage
- N291 - Ignition Coil 3 with Power Output Stage
- N292 - Ignition Coil 4 with Power Output Stage
- P - Spark Plug Connectors
- Q - Spark Plugs
- T6 - 6-Pin Connector, brown, behind instrument panel, left
- T6a - 6-Pin Connector
- T14 - 14-Pin Connector, in plenum chamber, center
- T121 - 121-Pin Connector, at -J220-
- (281) - Ground connector -1-, in wiring harness engine pre-wiring
- (607) - Ground Connection (in left plenum chamber)
- (A104) - Plus connector -2- (15), in instrument panel wiring harness
- (D52) - plus connection (15a), in engine compartment wiring harness
- (D78) - Plus connector -1- (30a), in engine compartment wiring harness
- --- - Manual transmission only

ws = white
sw = black
ro = red
br = brown
gn = green
bl = blue
gr = grey
li = violet
ge = yellow

New Beetle Convertible — Wiring diagram — No. 2/4

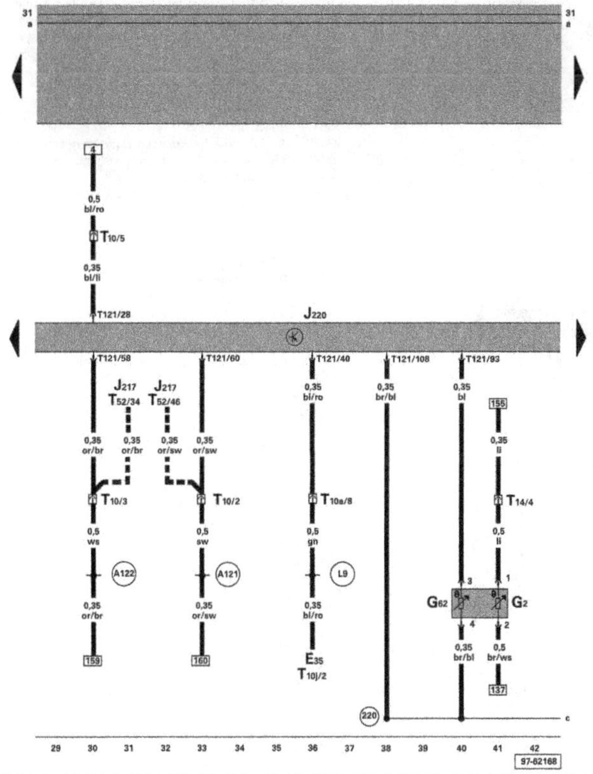

Motronic engine control module (ECM), engine coolant temperature (ECT) sensor

- E35 - A/C Switch
- G2 - Engine Coolant Temperature (ECT) Sensor
- G62 - Engine Coolant Temperature (ECT) Sensor
- J217 - Transmission Control Module (TCM) (tiptronic)
- J220 - Motronic Engine Control Module (ECM) (Tiptronic)
- T10 - 10-Pin Connector, white, behind instrument panel, left
- T10a - 10-Pin Connector, orange, behind instrument panel, left
- T10j - 10-Pin Connector
- T14 - 14-Pin Connector, in plenum chamber, center
- T52 - 52-Pin Connector
- T121 - 121-Pin Connector, at -J220-
- (220) - Ground connection (sensor ground), in engine compartment wiring harness
- (A121) - Connection (high bus) in instrument panel wiring harness
- (A122) - Connection (low bus) in instrument panel wiring harness
- (L9) - Wire connection -1-, in A/C wiring harness
- --- - Automatic transmission only

ws = white
sw = black
ro = red
br = brown
gn = green
bl = blue
gr = grey
li = violet
ge = yellow

2.0L engine management (BDC) from Aug. 2002, convertible

New Beetle Convertible Wiring diagram **No. 2/5**

Motronic engine control module (ECM), heated oxygen sensors (HO2S), secondary air injection (AIR) pump relay/motor

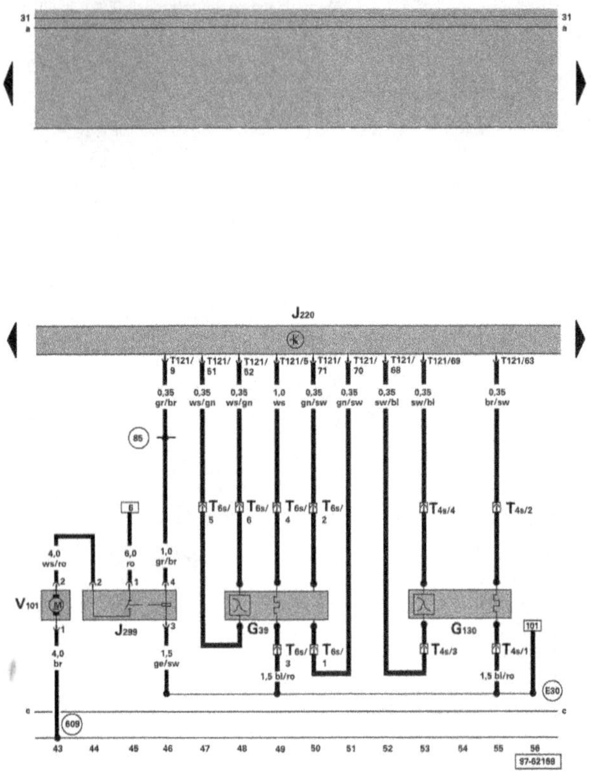

- G39 - Heated Oxygen Sensor (HO2S), black
- G130 - Oxygen Sensor (02S) behind Three Way Catalytic Converter (TWC), brown
- J220 - Motronic Engine Control Module (ECM)
- J299 - Secondary Air Injection (AIR) Pump Relay
- V101 - Secondary Air Injection (AIR) Pump Motor
- T4s - 4-Pin Connector, brown
- T6s - 6-Pin Connector, black
- T121 - 121-Pin Connector, at -J220-

- (85) - Ground connection -1-, in engine compartment wiring harness
- (609) - Ground Connection (in right plenum chamber)
- (E30) - Connector (87a), in wiring harness engine

ws = white
sw = black
ro = red
br = brown
gn = green
bl = blue
gr = grey
li = violet
ge = yellow

New Beetle Convertible Wiring diagram **No. 2/6**

Motronic engine control module (ECM), camshaft position (CMP) sensor, knock sensor (KS) 1, throttle drive (power accelerator actuation), angle sensors for throttle drive, throttle valve control module, motronic engine control module camshaft position (CMP) sensor

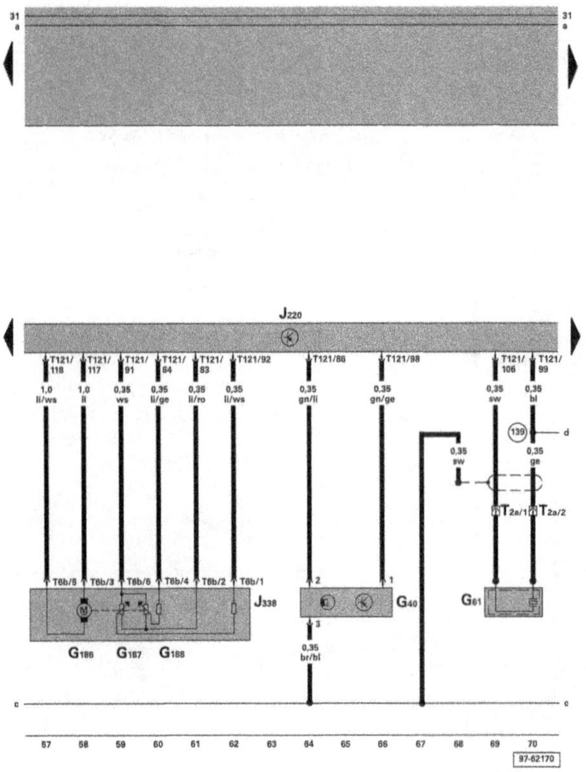

- G40 - Camshaft Position (CMP) Sensor
- G61 - Knock Sensor (KS) 1
- G186 - Throttle drive (power accelerator actuation)
- G187 - Angle sensor -1- for throttle drive (power accelerator actuation)
- G188 - Angle sensor -2- for throttle drive (power accelerator actuation)
- J220 - Motronic Engine Control Module (ECM)
- J338 - Throttle Valve Control Module
- T2a - 2-Pin Connector
- T6b - 6-Pin Connector
- T121 - 121-Pin Connector, at -J220-

- (139) - Ground connection (sensor Ground), in Motronic Multiport Fuel Injection (MFI) wiring harness

ws = white
sw = black
ro = red
br = brown
gn = green
bl = blue
gr = grey
li = violet
ge = yellow

New Beetle Convertible

Wiring diagram **No. 2/7**

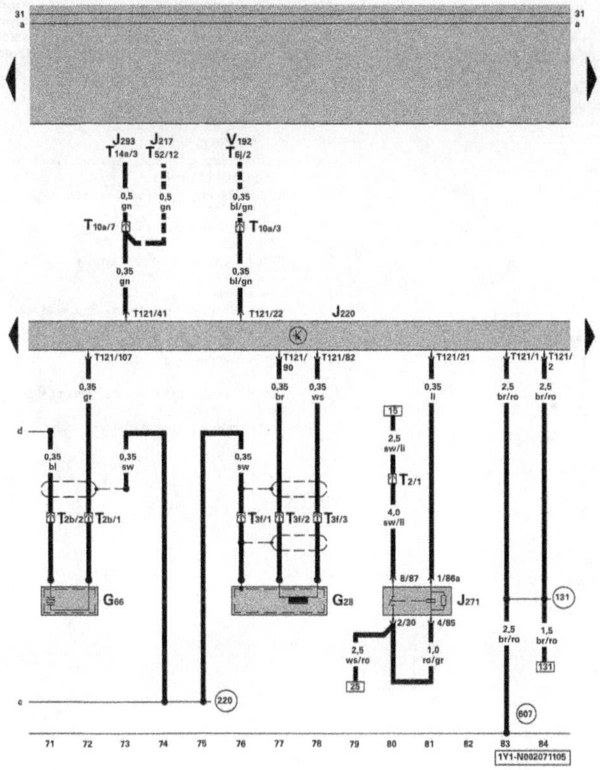

Motronic engine control module (ECM)/power supply relay, engine speed (RPM) sensor, knock sensor (KS) 2

- G28 - Engine Speed (RPM) Sensor
- G66 - Knock Sensor (KS) 2
- J217 - Transmission Control Module (TCM) (tiptronic)
- J220 - Motronic Engine Control Module (ECM)
- J271 - Motronic Engine Control Module (ECM) power supply relay, in protective housing, in engine compartment, left, production control number (429)
- J293 - Coolant FC (Fan Control) Control Module
- T2 - 2-Pin Connector
- T2b - 2-Pin Connector
- T3f - 3-Pin Connector
- T6j - 6-Pin Connector
- T10a - 10-Pin Connector, orange, behind instrument panel, left
- T14a - 14-Pin Connector
- T52 - 52-Pin Connector
- T121 - 121-Pin Connector, at -J220-
- V192 - Brake System Vacuum Pump

- (131) - Ground connection -2-, in engine compartment wiring harness
- (220) - Ground connection (sensor ground), in engine compartment wiring harness
- (607) - Ground Connection (in left plenum chamber)

- --- - Automatic transmission only

ws = white
sw = black
ro = red
br = brown
gn = green
bl = blue
gr = grey
li = violet
ge = yellow

New Beetle Convertible

Wiring diagram **No. 2/8**

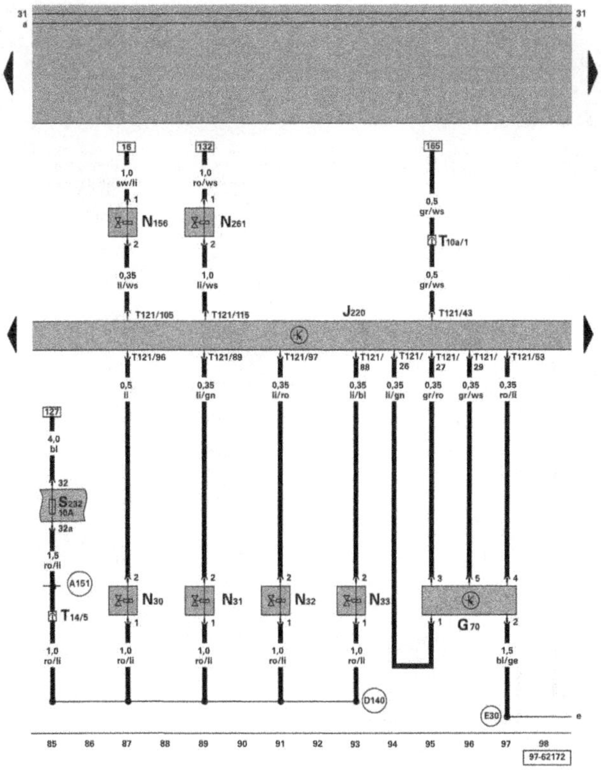

Motronic engine control module (ECM), mass air flow (MAF) sensor, fuel injectors, intake manifold change-over valve, intake manifold tuning (IMT) valve -2-

- G70 - Mass Air Flow (MAF) Sensor
- J220 - Motronic Engine Control Module (ECM)
- N30 - Cylinder 1 Fuel Injector
- N31 - Cylinder 2 Fuel Injector
- N32 - Cylinder 3 Fuel Injector
- N33 - Cylinder 4 Fuel Injector
- N156 - Intake Manifold Change-Over Valve
- N261 - Intake Manifold Tuning (IMT) Valve -2-
- S232 - Fuse 32 in fuse holder
- T10a - 10-Pin Connector, orange, behind instrument panel, left
- T14 - 14-Pin Connector, in plenum chamber, center
- T121 - 121-Pin Connector, at -J220-

- (A151) - Connection 4 (87) (in instrument panel wiring harness)
- (D140) - Connector (injectors), in wiring harness, engine pre-wiring
- (E30) - Connector (87a), in wiring harness engine

ws = white
sw = black
ro = red
br = brown
gn = green
bl = blue
gr = grey
li = violet
ge = yellow

2.0L engine management (BDC) from Aug. 2002, convertible

New Beetle Convertible — Wiring diagram — No. 2/9

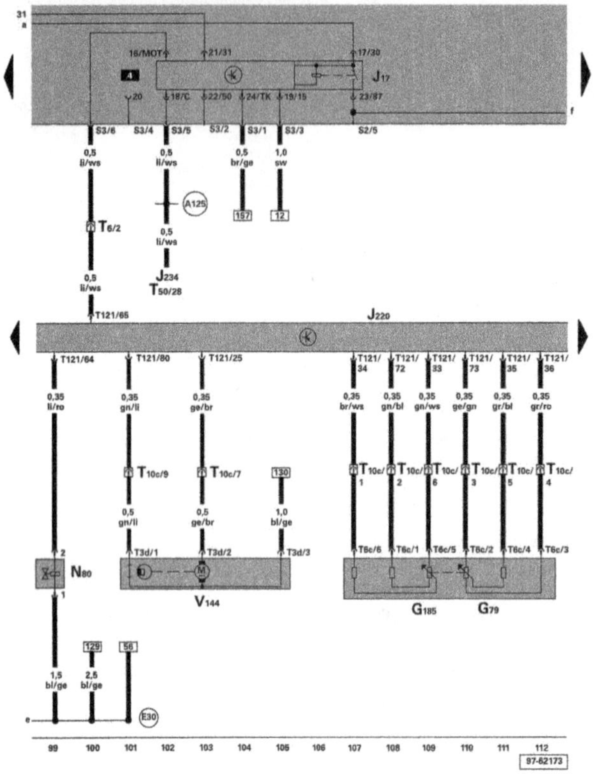

Motronic engine control module (ECM), throttle position (TP) sensor, sender -2- for acelerator pedal position, fuel pump (FP) relay, evaporative emission (EVAP) canister purge regulator valve, leak detection pump (LDP)

G79 - Throttle Position (TP) Sensor
G185 - Sender -2- for accelerator pedal position
J17 - Fuel Pump (FP) Relay
J220 - Motronic Engine Control Module (ECM)
J234 - Airbag Control Module
N80 - Evaporative Emission (EVAP) Canister Purge Regulator Valve
V144 - Leak Detection Pump (LDP)
T3d - 3-Pin Connector
T6 - 6-Pin Connector, brown, behind instrument panel, left
T6c - 6-Pin Connector
T10c - 10-Pin Connector, blue, behind instrument panel, left
T50 - 50-Pin Connector, at -J234-
T121 - 121-Pin Connector, at -J220-

(A125) - Connection (crash signal) in instrument panel wiring harness
(E30) - Connector (87a), in wiring harness engine

ws = white
sw = black
ro = red
br = brown
gn = green
bl = blue
gr = grey
li = violet
ge = yellow

Edition 11.02

New Beetle Convertible — Wiring diagram — No. 2/10

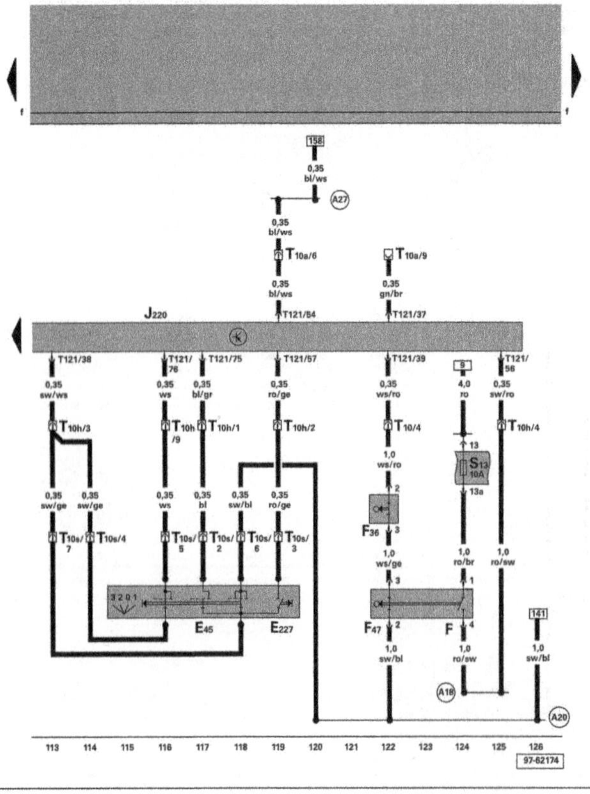

Motronic engine control module (ECM), cruise control switch, button for cruise control (set), brake light switch, clutch vacuum vent valve switch, brake pedal switch

E45 - Cruise Control Switch
E227 - Button for cruise control (set)
F - Brake Light Switch
F36 - Clutch Vacuum Vent Valve Switch
F47 - Brake Pedal Switch (Cruise control/Diesel Direct Injection)
J220 - Motronic Engine Control Module (ECM)
S13 - Fuse 13 in fuse holder
T10 - 10-Pin Connector, white, behind instrument panel, left
T10a - 10-Pin Connector, orange, behind instrument panel, left
T10h - 10-Pin Connector, black, behind instrument panel, left
T10s - 10-Pin Connector, near steering column

T121 - 121-Pin Connector, at -J220-

(A18) - wire connection (54), in instrument panel wiring harness
(A20) - wire connection (15a), in instrument panel wiring harness
(A27) - Wire Connection (vehicle speed signal), in instrument panel wiring harness

ws = white
sw = black
ro = red
br = brown
gn = green
bl = blue
gr = grey
li = violet
ge = yellow

Edition 11.02

New Beetle Convertible — Wiring diagram — No. 2/11

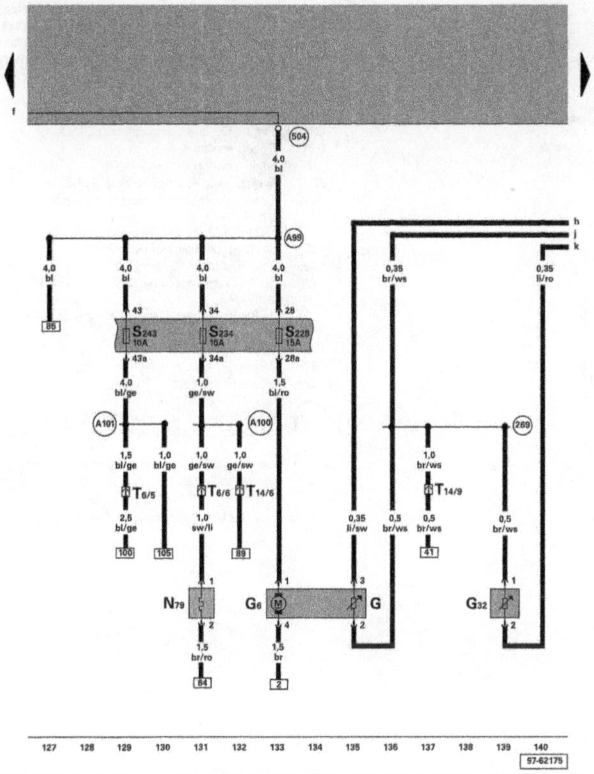

Sender for fuel gauge, fuel pump (FP), engine coolant level (ECL) sensor, positive crankcase ventilation (PVC) heating element

- G - Sender for fuel gauge
- G6 - Fuel Pump (FP)
- G32 - Engine Coolant Level (ECL) Sensor
- N79 - Positive crankcase ventilation (PVC) heating element
- S228 - Fuse 28 in fuse holder
- S234 - Fuse 34 in fuse holder
- S243 - Fuse 43 in fuse holder
- T6 - 6-Pin Connector, brown, behind instrument panel, left
- T14 - 14-Pin Connector, in plenum chamber, center
- (269) - Ground connector (sensor ground) -1-, in instrument panel wiring harness
- (504) - Threaded connection -1- (87) on the relay plate
- (A99) - Connector -1- (87), in instrument panel wiring harness
- (A100) - Connector -2- (87), in instrument panel wiring harness
- (A101) - Connector -3- (87), in instrument panel wiring harness

ws = white
sw = black
ro = red
br = brown
gn = green
bl = blue
gr = grey
li = violet
ge = yellow

New Beetle Convertible — Wiring diagram — No. 2/12

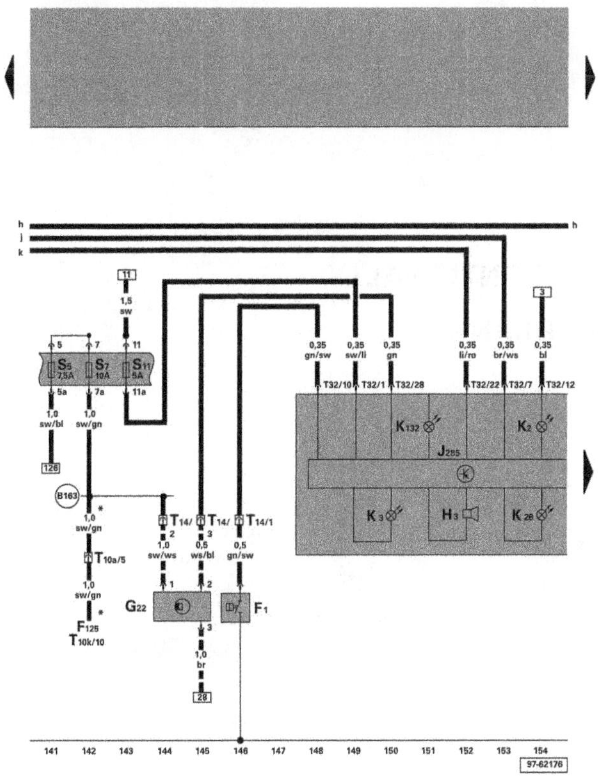

Oil pressure switch / warning light, speedometer vehicle speed sensor (VSS), warning buzzer, control module with indicator unit in instrument panel insert, generator (GEN) warning light, engine coolant level/temperature (ECL/ECT) warning light, electronic power control (EPC) warning lamp

- F1 - Oil Pressure Switch
- F125 - Multi-Function Transmission Range (TR) Switch
- G22 - Speedometer Vehicle Speed Sensor (VSS)
- H3 - Warning Buzzer
- J285 - Control module with indicator unit in instrument panel insert
- K2 - Generator (GEN) Warning Light
- K3 - Oil Pressure Warning Light
- K28 - Engine Coolant Level/Temperature (ECL/ECT) Warning Light
- K132 - Electronic Power Control (EPC) Warning Lamp
- S5 - Fuse 5 in fuse holder
- S7 - Fuse 7 in fuse holder
- S11 - Fuse 11 in fuse holder
- T10a - 10-Pin Connector
- T10k - 10-Pin Connector
- T14 - 14-Pin Connector, in left plenum chamber
- T32 - 32-Pin Connector, blue
- (B163) - Plus connector -1- (15) in wiring harness interior

- * - Automatic transmission only
- --- - Manual transmission only

ws = white
sw = black
ro = red
br = brown
gn = green
bl = blue
gr = grey
li = violet
ge = yellow

2.0L engine management (BDC) from Aug. 2002, convertible

New Beetle Convertible Wiring diagram **No. 2/13**

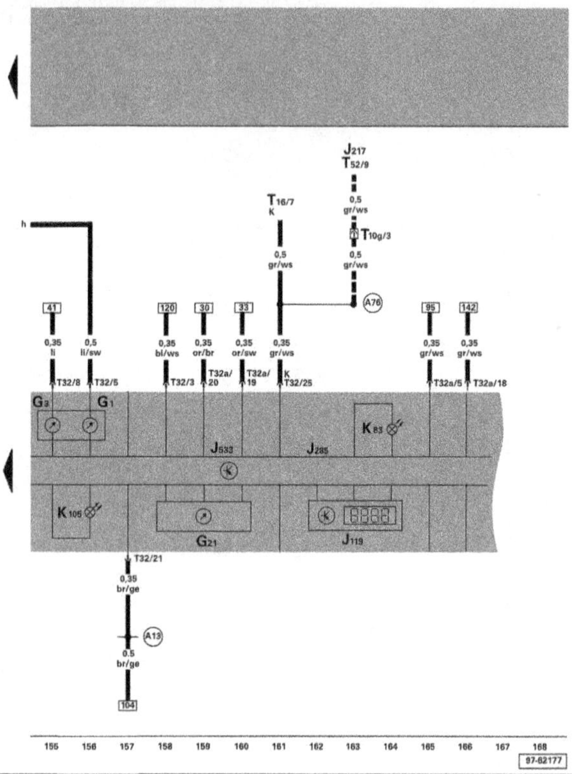

Fuel gauge, engine coolant temperature (ECT) gauge, speedometer, multi-function indicator (MFA), control module with indicator unit in instrument panel insert, data bus on board diagnostic interface, malfunction indicator lamp (MIL), low fuel level warning light

- G1 - Fuel Gauge
- G3 - Engine Coolant Temperature (ECT) Gauge
- G21 - Speedometer
- J119 - Multi-function Indicator (MFA)
- J217 - Transmission Control Module (TCM), Tiptronic
- J285 - Control module with indicator unit in instrument panel insert
- J533 - Data Bus On Board Diagnostic Interface
- K83 - Malfunction Indicator lamp (MIL)
- K105 - Low Fuel Level Warning Light
- T10g - 10-Pin Connector, grey, behind instrument panel, left
- T16 - 16-Pin Connector, Data link Connector (DLC), below steering column, left
- T32 - 32-Pin Connector, blue
- T32a - 32-Pin Connector, green
- T52 - 52-Pin Connector
- (A13) - wire connection (door contact switch) in instrument panel wiring harness
- (A76) - Connector (K-diagnosis wire) in instrument panel wiring harness
- --- - Only for 6-speed Automatic transmission

ws = white
sw = black
ro = red
br = brown
gn = green
bl = blue
gr = grey
li = violet
ge = yellow

THIS SPACE INTENTIONALLY LEFT BLANK

New Beetle — Wiring Diagram — No. 631 / 1

10.2009

2,0L-Engine - Motronic Multiport Fuel Injection (MFI)/85 kW, code BEV, (with cruise control)
From June 2003

Additional Fuse, Relay, Control Module and Connector locations ⇨ Component Locations

Troubleshooting Procedures ⇨ Guided Fault Finding using VAS 5051/5052

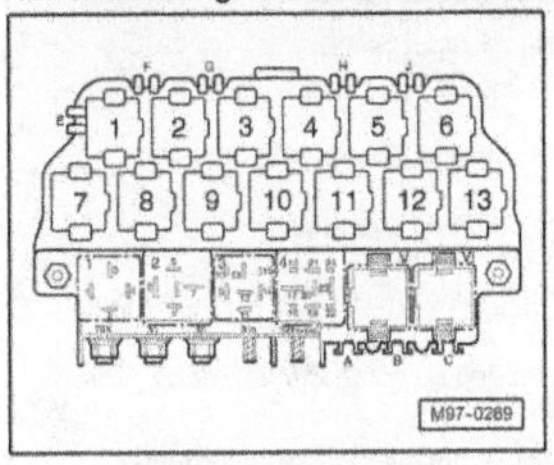

Relay locations on thirteen position auxiliary relay panel, above relay panel:
- 3 - J434 - Starter Lock/Clutch Pedal Switch Relay (53)

Relay panel:
- 1 - J4 - Horn Relay (53)
- 2 - J59 - Load Reduction Relay (100)
- 4 - J17 Fuel Pump (FP) Relay (409)

Notes:

- *J271 Motronic Engine Control Module (ECM) Power Supply Relay (428), ¹)*
- *J299 Secondary Air Injection (AIR) Pump Relay (100) ¹)*

¹) in engine compartment, left side, to the left of brake fluid reservoir, inside housing

Note:

Number in parentheses indicates production control number stamped on relay housing.

New Beetle — Wiring Diagram — No. 631 / 2

Battery, starter, generator (GEN), voltage regulator (VR), ignition/starter switch, load reduction relay, fuses in fuse bracket/battery

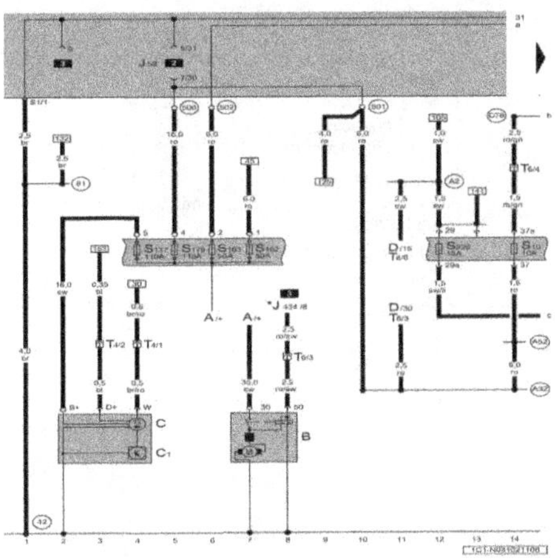

A -Battery
B -Starter
C -Generator (GEN)
C1 -Voltage Regulator (VR)
D -Ignition/Starter Switch
J434 -Locking relay for starter (clutch pedal switch)
J59 -Load Reduction Relay
S10 -Fuse 10 in fuse/relay panel
S162 -Fuse -1- (50), in fuse bracket/battery
S163 -Fuse -2- (50), in fuse bracket/battery
S176 -Fuse -4- (30), in fuse bracket/battery
S177 -Fuse -5- (30), in fuse bracket/battery
S229 -Fuse 29 in fuse/relay panel
T4 -4-Pin Connector, in engine compartment left
T6 -6-Pin Connector, brown, behind instrument panel, left
T8 -8-Pin Connector, at -D-
42 -Ground connection, beside steering column
81 -Ground connection -1-, in instrument panel wiring harness
500 -Threaded connection -1- (30) on the relay plate
501 -Threaded connection -2- (30) on the relay plate
502 -Threaded connection -1- (30a) on the relay plate
A2 -plus connection (15), in instrument panel wiring harness
A32 -plus connection (30), in instrument panel wiring harness
A52 -Plus Connection 2 (30) (in instrument panel wiring harness)
D78 -Plus connector -1- (30a), in engine compartment wiring harness
* -Manual transmission only

ws = wt (white)
sw = bk (black)
ro = rd (red)
br = br (brown)
gn = lg (light green), dg (dark green)
bl = bl (blue), lb (light blue), db (dark blue)
gr = gy (gray)
li = lilac
ge = yl (yellow)
or = or (orange), tn (tan)
rs = vt (violet)

New Beetle

Wiring Diagram

No. 631 / 3

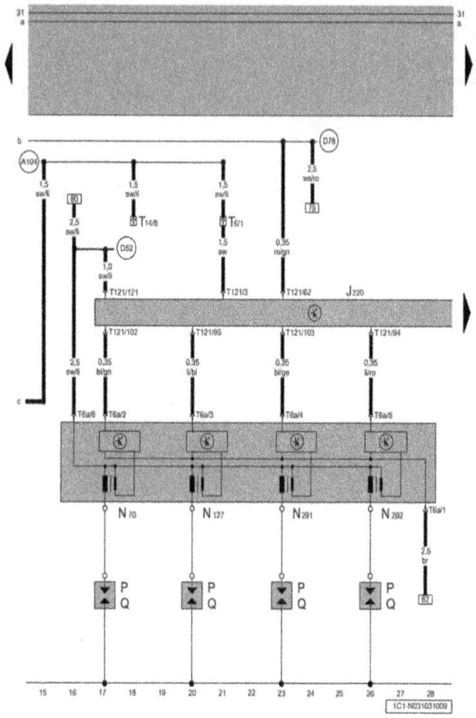

ws = wt (white)
sw = bk (black)
ro = rd (red)
br = br (brown)
gn = lg (light green), dg (dark green)
bl = bl (blue), lb (light blue) ,db (dark blue)
gr = gy (gray)
li = lilac
ge = yl (yellow)
or = or (orange), tn (tan)
rs = vt (violet)

Motronic en gine control module (ECM), ignition coil, spark plugs

J220 -Motronic Engine Control Module (ECM)
N70 -Ignition Coil 1 with Power Output Stage
N127 -Ignition Coil 2 with Power Output Stage
N291 -Ignition Coil 3 with Power Output Stage
N292 -Ignition Coil 4 with Power Output Stage
P -Spark Plug Connectors
Q -Spark Plugs
T6 -6-Pin Connector, brown, behind instrument panel, left
T6a -6-Pin Connector
T14 -14-Pin Connector, in left plenum chamber
T121 -121-Pin Connector, at -J220-
A104 -Plus connector -2- (15), in instrument panel wiring harness
D52 -plus connection (15a), in engine compartment wiring harness
D78 -Plus connector -1- (30a), in engine compartment wiring harness

New Beetle

Wiring Diagram

No. 631 / 4

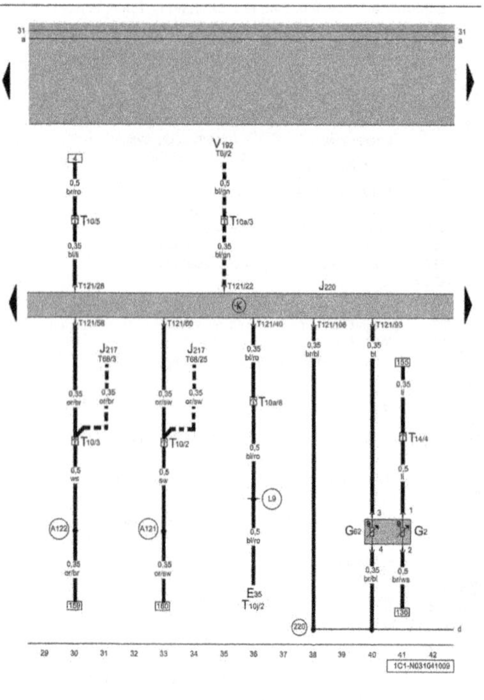

ws = wt (white)
sw = bk (black)
ro = rd (red)
br = br (brown)
gn = lg (light green), dg (dark green)
bl = bl (blue), lb (light blue) ,db (dark blue)
gr = gy (gray)
li = lilac
ge = yl (yellow)
or = or (orange), tn (tan)
rs = vt (violet)

Motronic engine control module (ECM), engine coolant temperature (ECT) sensor

E35 -A/C Switch
G2 -Engine Coolant Temperature (ECT) Sensor
G62 -Engine Coolant Temperature (ECT) Sensor
J217 -Transmission Control Module (TCM)
J220 -Motronic Engine Control Module (ECM)
T6j -6-Pin Connector at -V192-
T10 -10-Pin Connector, white, behind instrument panel, left
T10a -10-Pin Connector, orange, behind instrument panel, left
T10j -10-Pin Connector
T14 -14-Pin Connector, in left plenum chamber
T68 -68-Pin Connector
T121 -121-Pin Connector, at -J220-
V192 -Brake System Vacuum Pump
220 -Ground connection (sensor ground), in engine compartment wiring harness
A121 -Connection (high bus) in instrument panel wiring harness
A122 -Connection (low bus) in instrument panel wiring harness
L9 -Wire connection -1-, in A/C wiring harness
--- -Automatic transmission only

New Beetle

Wiring Diagram

No. 631 / 5

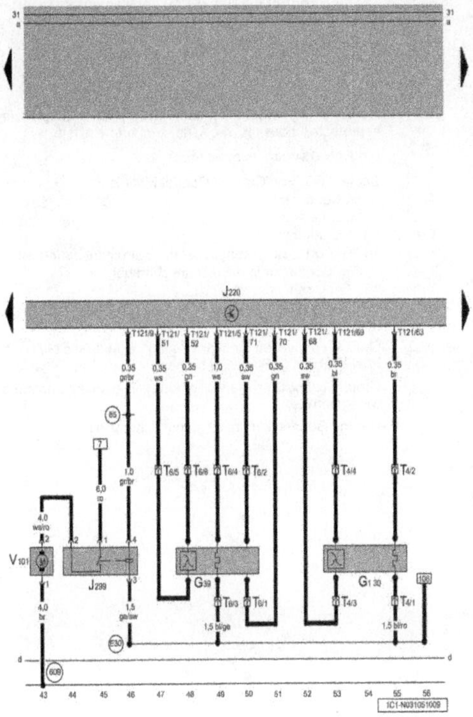

ws	= wt (white)
sw	= bk (black)
ro	= rd (red)
br	= br (brown)
gn	= lg (light green), dg (dark green)
bl	= bl (blue), lb (light blue), db (dark blue)
gr	= gy (gray)
li	= lilac
ge	= yl (yellow)
or	= or (orange), tn (tan)
rs	= vt (violet)

Motronic engine control module (ECM), heated oxygen sensor (HO2S), secondary air injection (AIR) pump relay/motor

- G39 -Heated Oxygen Sensor (HO2S), black
- G130 -Oxygen Sensor (02S) behind Three Way Catalytic Converter (TWC), brown
- J220 -Motronic Engine Control Module (ECM)
- J299 -Secondary Air Injection (AIR) Pump Relay
- V101 -Secondary Air Injection (AIR) Pump Motor
- T4 -4-Pin Connector, brown
- T6 -6-Pin Connector, black
- T121 -121-Pin Connector, at -J220-
- 85 -Ground connection -1-, in engine compartment wiring harness
- 609 -Ground Connection (in right plenum chamber)
- E30 -Connection (87a), in wiring harness engine

New Beetle

Wiring Diagram

No. 631 / 6

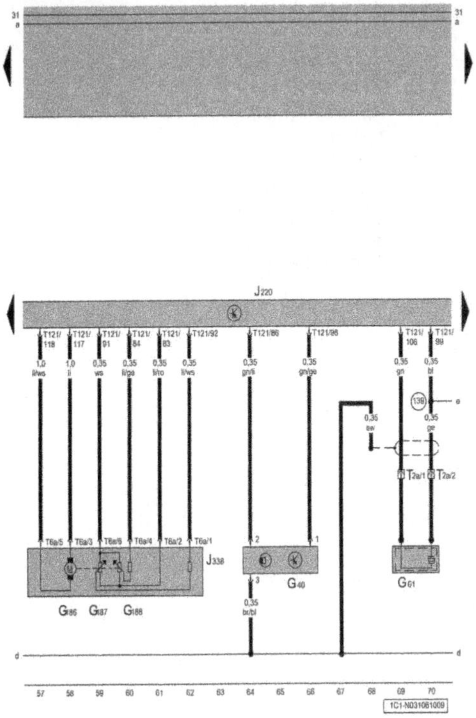

ws	= wt (white)
sw	= bk (black)
ro	= rd (red)
br	= br (brown)
gn	= lg (light green), dg (dark green)
bl	= bl (blue), lb (light blue), db (dark blue)
gr	= gy (gray)
li	= lilac
ge	= yl (yellow)
or	= or (orange), tn (tan)
rs	= vt (violet)

Camshaft Position (CMP) Sensor, knock sensor (KS) 1, throttle drive (power accelerator actuation), angle sensor 1/2 for throttle drive, motronic engine control module

- G40 -Camshaft Position (CMP) Sensor
- G61 -Knock Sensor (KS) 1
- G186 -Throttle drive (power accelerator actuation)
- G187 -Angle sensor -1- for throttle drive (power accelerator actuation)
- G188 -Angle sensor -2- for throttle drive (power accelerator actuation)
- J220 -Motronic Engine Control Module (ECM)
- J338 -Throttle Valve Control Module
- T2a -2-Pin Connector
- T6a -6-Pin Connector
- T121 -121-Pin Connector, at -J220-
- 139 -Ground connection (sensor Ground), in Motronic Multiport Fuel Injection (MFI) wiring harness

New Beetle

Wiring Diagram

No. 631 / 7

Motronic engine control module (ECM) / power supply relay, engine speed (RPM) sensor, knock sensor (KS) 2

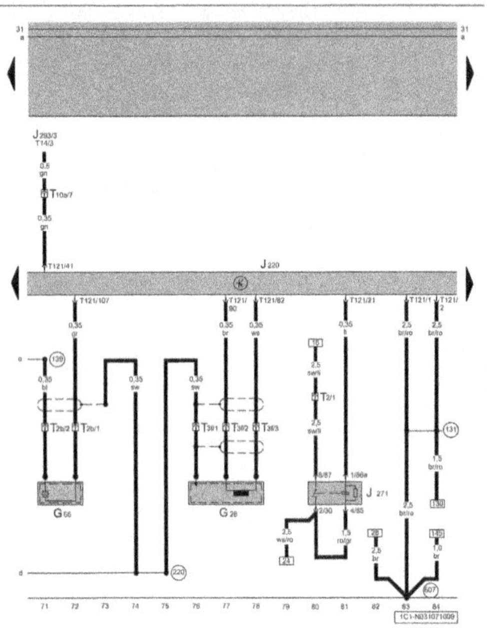

G28 -Engine Speed (RPM) Sensor
G66 -Knock Sensor (KS) 2
J220 -Motronic Engine Control Module (ECM)
J271 -Motronic Engine Control Module (ECM) power supply relay, in protective housing, in engine compartment, left, production control number (428)
J293 -Coolant FC (Fan Control) Control Module
T2 -2-Pin Connector
T2b -2-Pin Connector
T3f -3-Pin Connector
T10a -10-Pin Connector, orange, behind instrument panel, left
T14 -14-Pin Connector in left plenum chamber
T121 -121-Pin Connector, at -J220-
131 -Ground connection -2-, in engine compartment wiring harness
139 -Ground connection (sensor Ground), in Motronic Multiport Fuel Injection (MFI) wiring harness
220 -Ground connection (sensor ground), in engine compartment wiring harness
607 -Ground Connection (in left plenum chamber)

ws = wt (white)
sw = bk (black)
ro = rd (red)
br = br (brown)
gn = lg (light green), dg (dark green)
bl = bl (blue), lb (light blue), db (dark blue)
gr = gy (gray)
li = lilac
ge = yl (yellow)
or = or (orange), tn (tan)
rs = vt (violet)

New Beetle

Wiring Diagram

No. 631 / 8

Motronic engine control module (ECM), mass air flow (MAF) sensor, fuel injectors

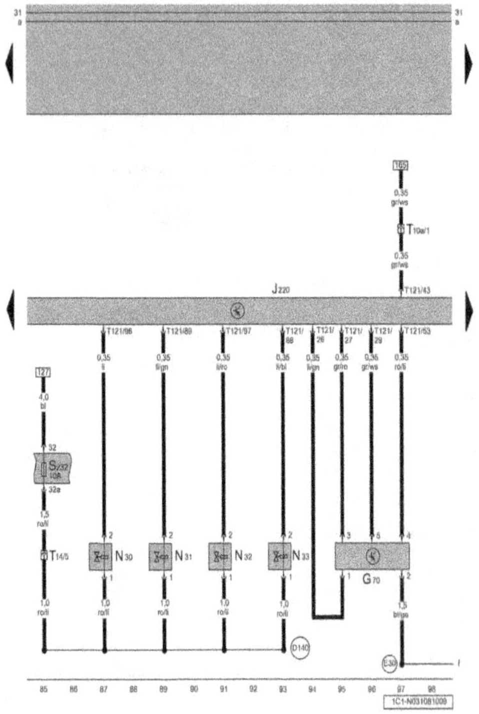

G70 -Mass Air Flow (MAF) Sensor
J220 -Motronic Engine Control Module (ECM)
N30 -Cylinder 1 Fuel Injector
N31 -Cylinder 2 Fuel Injector
N32 -Cylinder 3 Fuel Injector
N33 -Cylinder 4 Fuel Injector
S232 -Fuse 32 in fuse/relay panel
T10a -10-Pin Connector, orange, behind instrument panel, left
T14 -14-Pin Connector
T121 -121-Pin Connector, at J220-
D140 -Connector (injectors), in wiring harness, engine pre-wiring
E30 -Connector (87a), in wiring harness engine

ws = wt (white)
sw = bk (black)
ro = rd (red)
br = br (brown)
gn = lg (light green), dg (dark green)
bl = bl (blue), lb (light blue), db (dark blue)
gr = gy (gray)
li = lilac
ge = yl (yellow)
or = or (orange), tn (tan)
rs = vt (violet)

New Beetle

Wiring Diagram

No. 631 / 9

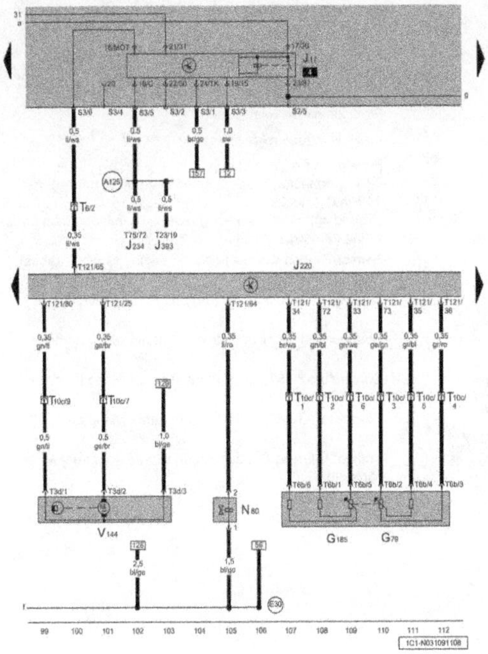

ws = wt (white)
sw = bk (black)
ro = rd (red)
br = br (brown)
gn = lg (light green), dg (dark green)
bl = bl (blue), lb (light blue), db (dark blue)
gr = gy (gray)
li = lilac
ge = yl (yellow)
or = or (orange), tn (tan)
rs = vt (violet)

Motronic engine control module (ECM), throttle position (TP) sensor, sender 2 for acelerator pe dal position, evaporative emission (EVAP), leak detection pump (LDP)

G79 -Throttle Position (TP) Sensor
G185 -Sender -2- for accelerator pedal position
J17 -Fuel Pump (FP) Relay
J220 -Motronic Engine Control Module (ECM)
J234 -Airbag Control Module
J393 -Central control module for comfort system
N80 -Evaporative Emission (EVAP) Canister Purge Regulator Valve
V144 -Leak Detection Pump (LDP)
T3d -3-Pin Connector
T6 -6-Pin Connector, brown, behind instrument panel, left
T6b -6-Pin Connector
T10c -10-Pin Connector, blue, behind instrument panel, left
T23 -23-Pin Connector, on Central control module for comfort system
T75 -75-Pin Connector
T121 -121-Pin Connector, at -J220-
A125 -Connection (crash signal) in instrument panel wiring harness
E30 -Connector (87a), in wiring harness engine

New Beetle

Wiring Diagram

No. 631 / 10

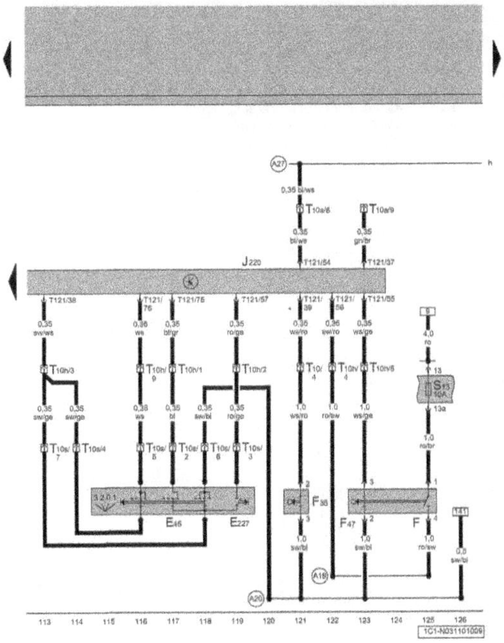

ws = wt (white)
sw = bk (black)
ro = rd (red)
br = br (brown)
gn = lg (light green), dg (dark green)
bl = bl (blue), lb (light blue), db (dark blue)
gr = gy (gray)
li = lilac
ge = yl (yellow)
or = or (orange), tn (tan)
rs = vt (violet)

Motronic engine control module (ECM), cruise control switch, button for cruise control (set), brake light switch, clutch vacuum vent valve switch

E45 -Cruise Control Switch
E227 -Button for cruise control (set)
F -Brake Light Switch
F36 -Clutch Vacuum Vent Valve Switch
F47 -Brake Pedal Switch (Cruise control/Diesel Direct Injection)
J220 -Motronic Engine Control Module (ECM)
S13 -Fuse 13 in fuse/relay panel
T10 -10-Pin Connector, white, behind instrument panel, left
T10a -10-Pin Connector, orange, behind instrument panel, left
T10h -10-Pin Connector, black, behind instrument panel, left
T10s -10-Pin Connector, near steering column
T121 -121-Pin Connector, at -J220-
A18 -wire connection (54), in instrument panel wiring harness
A20 -wire connection (15a), in instrument panel wiring harness
A27 -Wire Connection (vehicle speed signal), in instrument panel wiring harness
* -Manual transmission only

2.0L engine management (BEV) from June 2003

New Beetle

Wiring Diagram

No. 631 / 11

Sender for fuel gauge, fuel pump (FP), engine coolant level (ECL) sensor

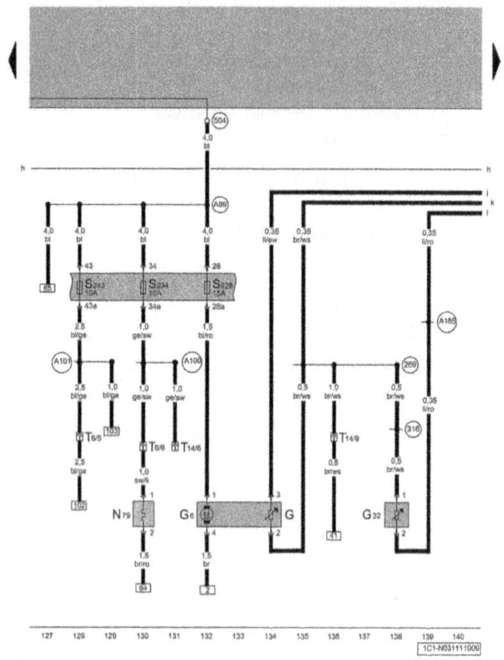

- G -Sender for fuel gauge
- G6 -Fuel Pump (FP)
- G32 -Engine Coolant Level (ECL) Sensor
- N79 -Positive crankcase ventilation (PVC) heating element
- S228 -Fuse 28 in fuse/relay panel
- S234 -Fuse 34 in fuse/relay panel
- S243 -Fuse 43 in fuse/relay panel
- T6 -6-Pin Connector, brown, behind instrument panel, left
- T14 -14-Pin Connector
- 269 -Ground connector (sensor ground) -1-, in instrument panel wiring harness
- 316 -Sensor Ground Connection 2 (in engine wiring harness)
- 504 -Threaded connection -1- (87) on the relay plate
- A99 -Connector -1- (87), in instrument panel wiring harness
- A100 -Connector -2- (87), in instrument panel wiring harness
- A101 -Connector -3- (87), in instrument panel wiring harness
- A185 -Coolant Warning Connection (in instrument panel wiring harness)

ws = wt (white)
sw = bk (black)
ro = rd (red)
br = br (brown)
gn = lg (light green), dg (dark green)
bl = bl (blue), lb (light blue), db (dark blue)
gr = gy (gray)
li = lilac
ge = yl (yellow)
or = or (orange), tn (tan)
rs = vt (violet)

New Beetle

Wiring Diagram

No. 631 / 12

Oil pressure switch / warning light, speedometer vehicle speed sensor (VSS), warning buzzer, instrument cluster, generator (GEN) warning light

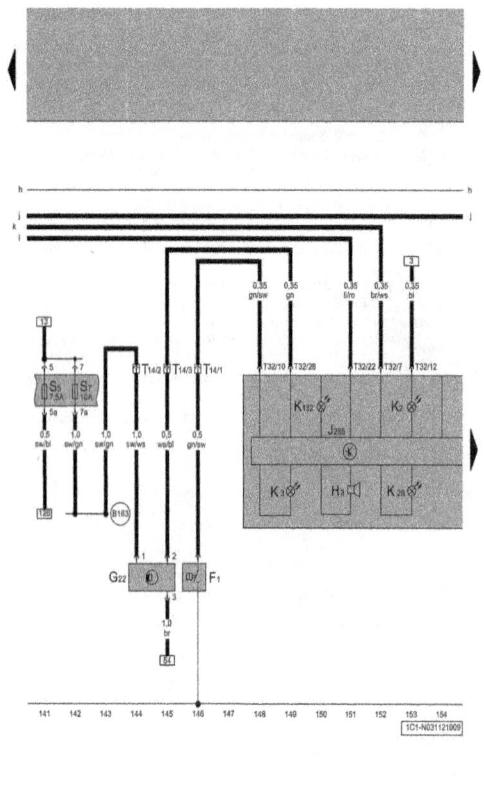

- F1 -Oil Pressure Switch
- G22 -Speedometer Vehicle Speed Sensor (VSS)
- H3 -Warning Buzzer
- J285 -Control module with indicator unit in instrument panel insert
- K2 -Generator (GEN) Warning Light
- K3 -Oil Pressure Warning Light
- K28 -Engine Coolant Level/Temperature (ECL/ECT) Warning Light
- K132 -Electronic Power Control (EPC) Lamp
- S5 -Fuse 5 in fuse/relay panel
- S7 -Fuse 7 in fuse/relay panel
- T14 -14-Pin Connector
- T32 -32-Pin Connector, blue
- B163 -Plus connector -1- (15) in wiring harness interior

ws = wt (white)
sw = bk (black)
ro = rd (red)
br = br (brown)
gn = lg (light green), dg (dark green)
bl = bl (blue), lb (light blue), db (dark blue)
gr = gy (gray)
li = lilac
ge = yl (yellow)
or = or (orange), tn (tan)
rs = vt (violet)

New Beetle | Wiring Diagram | No. 631 / 13

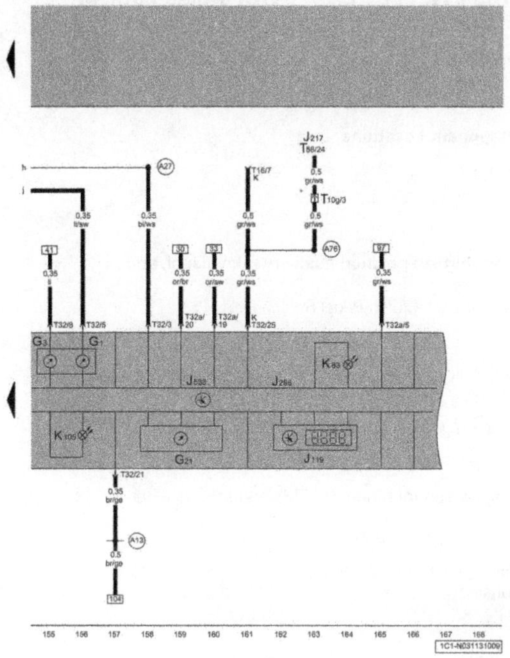

ws	=	wt (white)
sw	=	bk (black)
ro	=	rd (red)
br	=	br (brown)
gn	=	lg (light green), dg (dark green)
bl	=	bl (blue), lb (light blue), db (dark blue)
gr	=	gy (gray)
li	=	lilac
ge	=	yl (yellow)
or	=	or (orange), tn (tan)
rs	=	vt (violet)

Instrument cluster, fuel gauge, engine coolant temperature (ECT) gauge, speedomter, multifunc tion indicator (MFA), low fuel level warning light

G1	-Fuel Gauge
G3	-Engine Coolant Temperature (ECT) Gauge
G21	-Speedometer
J119	-Multi-function Indicator (MFA)
J217	-Transmission Control Module (TCM), behind instrument panel, center
J285	-Control module with indicator unit in instrument panel insert
J533	-Data Bus On Board Diagnostic Interface
K83	-Malfunction Indicator lamp (MIL)
K105	-Low Fuel Level Warning Light
T10g	-10-Pin Connector, grey, behind instrument panel, left
T16	-16-Pin Connector, Data link Connector (DLC), below steering column, left
T32	-32-Pin Connector, blue
T32a	-32-Pin Connector, green
T68	-68-Pin Connector
A13	-wire connection (door contact switch) in instrument panel wiring harness
A27	-Wire Connection (vehicle speed signal), in instrument panel wiring harness
A76	-Connector (K-diagnosis wire) in instrument panel wiring harness
*	-Automatic transmission only

THIS SPACE INTENTIONALLY

LEFT BLANK

New Beetle | **Wiring diagram** | **No. 38/1**

2,0L-Engine - Motronic Multiport Fuel Injection (MFI)/85 kW, code BEV, (with cruise control),

from June 2004,

Additional Fuse, Relay, Control Module and Connector locations ⇒ Component Locations

Troubleshooting Procedures ⇒ Guided Fault Finding using VAS 5051/5052

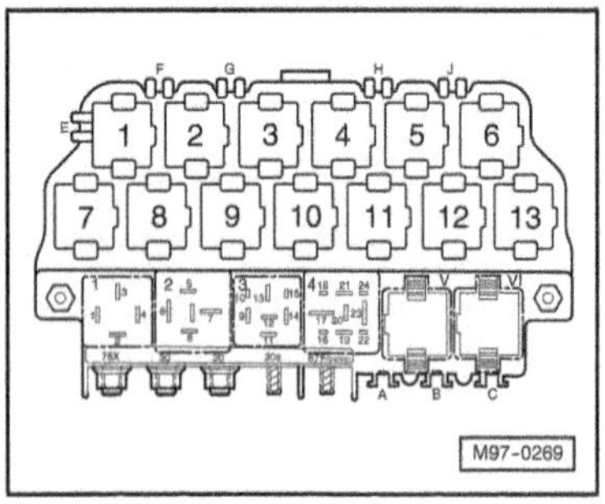

Relay locations on thirteen position auxiliary relay panel, above relay panel:

3 J434 - Starter Lock/Clutch Pedal Switch Relay (53)

Relay panel:

1 J4 - Horn Relay (53)

2 J59 - Load Reduction Relay (100)

4 J17 Fuel Pump (FP) Relay (409)

Notes:

J271 Motronic Engine Control Module (ECM) Power Supply Relay (428), [1)]

J299 Secondary Air Injection (AIR) Pump Relay (100) [1)]

[1)] in engine compartment, left side, to the left of brake fluid reservoir, inside housing

Number in parentheses indicates production control number stamped on relay housing.

New Beetle | **Wiring diagram** | **No. 38/2**

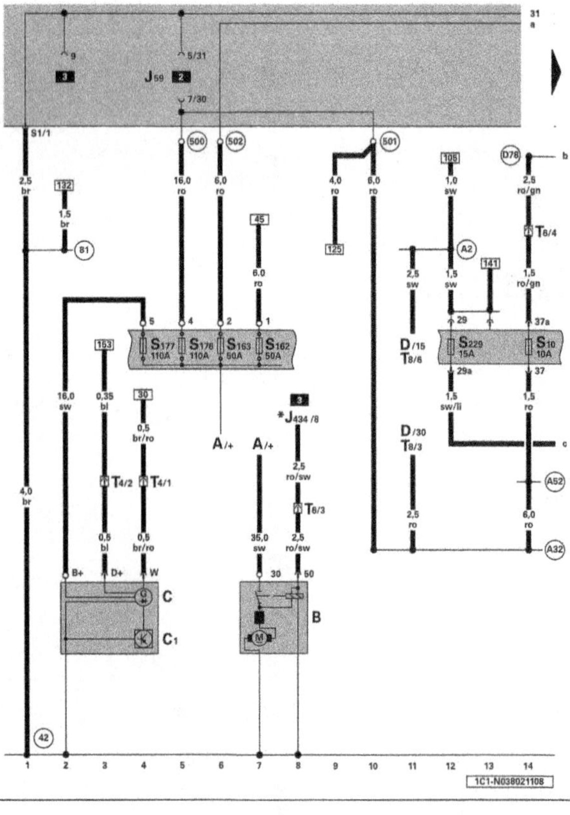

Battery, starter, generator (GEN), voltage regulator (VR), ignition/starter switch, load reduction relay, fuses in fuse bracket/battery

- A - Battery
- B - Starter
- C - Generator (GEN)
- C1 - Voltage Regulator (VR)
- D - Ignition/Starter Switch
- J59 - Load Reduction Relay
- J434 - Locking relay for starter (clutch pedal switch)
- S10 - Fuse 10 in fuse/relay panel
- S162 - Fuse -1- (50), in fuse bracket/battery
- S163 - Fuse -2- (50), in fuse bracket/battery
- S176 - Fuse -4- (30), in fuse bracket/battery
- S177 - Fuse -5- (30), in fuse bracket/battery
- S229 - Fuse 29 in fuse/relay panel
- T4 - 4-Pin Connector, in engine compartment left
- T6 - 6-Pin Connector, brown, behind instrument panel, left
- T8 - 8-Pin Connector, at -D-
- (42) - Ground connection, beside steering column
- (81) - Ground connection -1-, in instrument panel wiring harness
- (500) - Threaded connection -1- (30) on the relay plate
- (501) - Threaded connection -2- (30) on the relay plate
- (502) - Threaded connection -1- (30a) on the relay plate
- (A2) - plus connection (15), in instrument panel wiring harness
- (A32) - plus connection (30), in instrument panel wiring harness
- (A52) - Plus Connection 2 (30) (in instrument panel wiring harness)
- (D78) - Plus connector -1- (30a), in engine compartment wiring harness
- * - Manual transmission only

ws = white
sw = black
ro = red
br = brown
gn = green
bl = blue
gr = grey
li = violet
ge = yellow
or = orange

New Beetle — Wiring diagram — No. 38/3

Motronic en gine control module (ECM), ignition coil, spark plugs

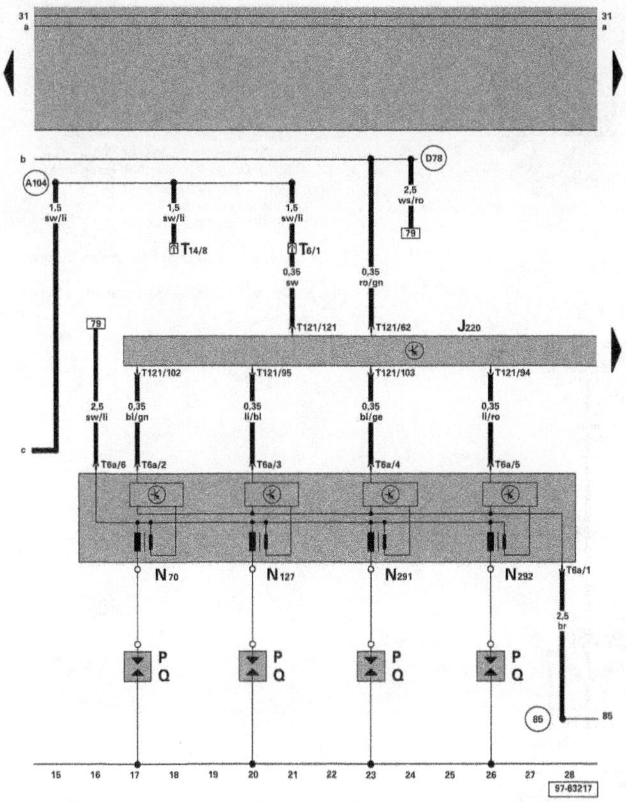

- J220 - Motronic Engine Control Module (ECM)
- N70 - Ignition Coil 1 with Power Output Stage
- N127 - Ignition Coil 2 with Power Output Stage
- N291 - Ignition Coil 3 with Power Output Stage
- N292 - Ignition Coil 4 with Power Output Stage
- P - Spark Plug Connectors
- Q - Spark Plugs
- T6 - 6-Pin Connector, brown, behind instrument panel, left
- T6a - 6-Pin Connector
- T14 - 14-Pin Connector, in left plenum chamber
- T121 - 121-Pin Connector, at -J220-

- (85) - Ground Connection 1, in engine compartment wiring harness
- (A104) - Plus connector -2- (15), in instrument panel wiring harness
- (D78) - Plus connector -1- (30a), in engine compartment wiring harness

ws = white
sw = black
ro = red
br = brown
gn = green
bl = blue
gr = grey
li = violet
ge = yellow
or = orange

New Beetle — Wiring diagram — No. 38/4

Motronic engine control module (ECM), engine coolant temperature (ECT) sensor

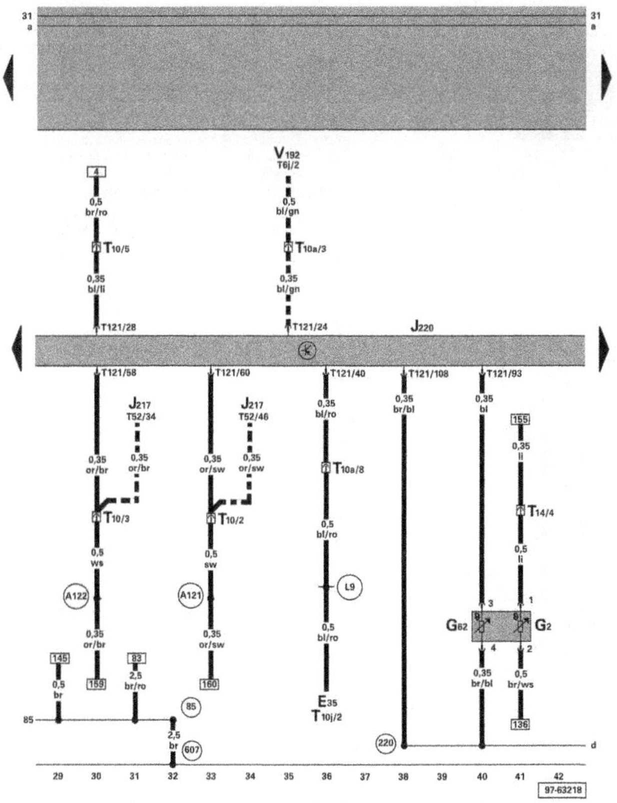

- E35 - A/C Switch
- G2 - Engine Coolant Temperature (ECT) Sensor
- G62 - Engine Coolant Temperature (ECT) Sensor
- J217 - Transmission Control Module (TCM)
- J220 - Motronic Engine Control Module (ECM)
- T6j - 6-Pin Connector at -V192-
- T10 - 10-Pin Connector, white, behind instrument panel, left
- T10a - 10-Pin Connector, orange, behind instrument panel, left
- T10j - 10-Pin Connector
- T14 - 14-Pin Connector, in left plenum chamber
- T52 - 52-Pin Connector, at -J217-
- T121 - 121-Pin Connector, at -J220-
- V192 - Brake System Vacuum Pump

- (85) - Ground Connection 1, in engine compartment wiring harness
- (220) - Ground connection (sensor ground), in engine compartment wiring harness
- (607) - Ground Connection (in left plenum chamber)
- (A121) - Connection (high bus) in instrument panel wiring harness
- (A122) - Connection (low bus) in instrument panel wiring harness
- (L9) - Wire connection -1-, in A/C wiring harness

- - - - Automatic transmission only

ws = white
sw = black
ro = red
br = brown
gn = green
bl = blue
gr = grey
li = violet
ge = yellow
or = orange

2.0L engine management (BEV) from June 2004

New Beetle — Wiring diagram — **No. 38/5**

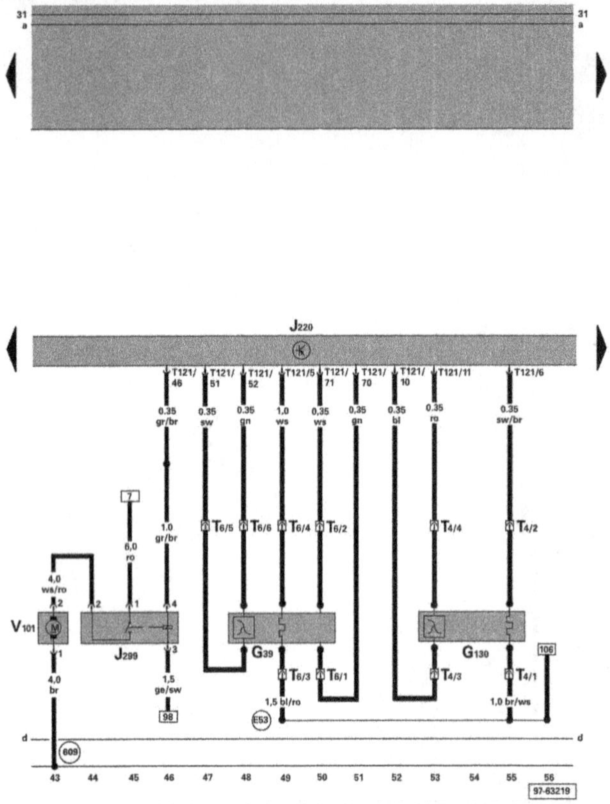

Motronic engine control module (ECM), heated oxygen sensor (HO2S), secondary air injection (AIR) pump relay/motor

- G39 - Heated Oxygen Sensor (HO2S), black
- G130 - Oxygen Sensor (02S) behind Three Way Catalytic Converter (TWC), brown
- J220 - Motronic Engine Control Module (ECM)
- J299 - Secondary Air Injection (AIR) Pump Relay
- V101 - Secondary Air Injection (AIR) Pump Motor
- T4 - 4-Pin Connector, brown
- T6 - 6-Pin Connector, black
- T121 - 121-Pin Connector, at -J220-

(609) - Ground Connection (in right plenum chamber)

(E53) - Connection 2 (87a)(in Motronic wiring harness

ws = white
sw = black
ro = red
br = brown
gn = green
bl = blue
gr = grey
li = violet
ge = yellow
or = orange

New Beetle — Wiring diagram — **No. 38/6**

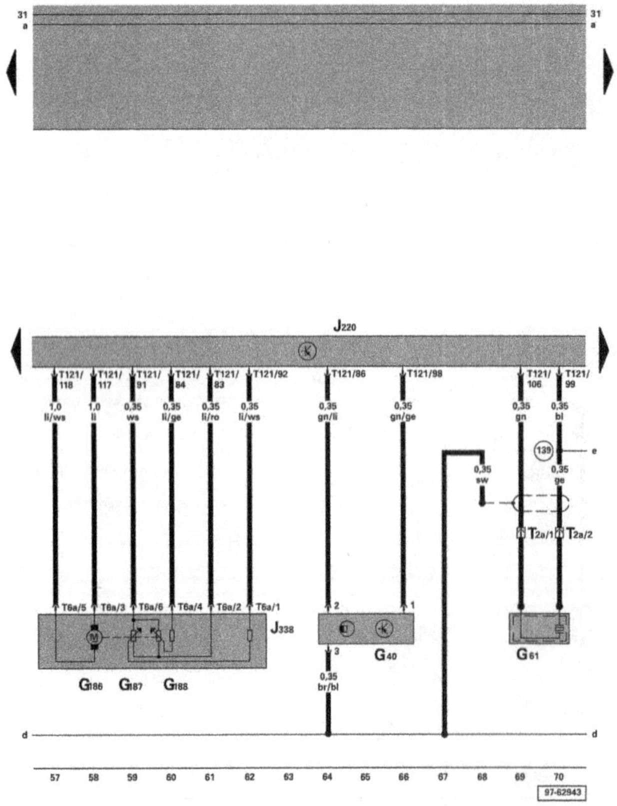

Camshaft Position (CMP) Sensor, knock sensor (KS) 1, throttle drive (power accelerator actuation), angle sensor 1/2 for throttle drive, motronic engine control module

- G40 - Camshaft Position (CMP) Sensor
- G61 - Knock Sensor (KS) 1
- G186 - Throttle drive (power accelerator actuation)
- G187 - Angle sensor -1- for throttle drive (power accelerator actuation)
- G188 - Angle sensor -2- for throttle drive (power accelerator actuation)
- J220 - Motronic Engine Control Module (ECM)
- J338 - Throttle Valve Control Module
- T2a - 2-Pin Connector
- T6a - 6-Pin Connector
- T121 - 121-Pin Connector, at -J220-

(139) - Ground connection (sensor Ground), in Motronic Multiport Fuel Injection (MFI) wiring harness

ws = white
sw = black
ro = red
br = brown
gn = green
bl = blue
gr = grey
li = violet
ge = yellow
or = orange

New Beetle — Wiring diagram — No. 38/7

Motronic engine control module (ECM) / power supply relay, engine speed (RPM) sensor, knock sensor (KS) 2

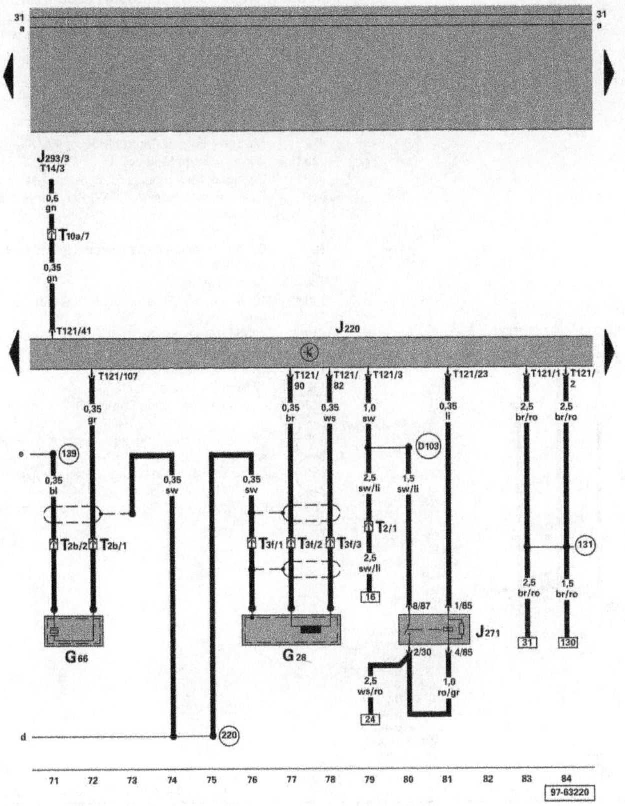

- G28 - Engine Speed (RPM) Sensor
- G66 - Knock Sensor (KS) 2
- J220 - Motronic Engine Control Module (ECM)
- J271 - Motronic Engine Control Module (ECM) power supply relay, in protective housing, in engine compartment, left, production control number (428)
- J293 - Coolant FC (Fan Control) Control Module
- T2 - 2-Pin Connector
- T2b - 2-Pin Connector
- T3f - 3-Pin Connector
- T10a - 10-Pin Connector, orange, behind instrument panel, left
- T14 - 14-Pin Connector in left plenum chamber
- T121 - 121-Pin Connector, at -J220-
- (131) - Ground connection -2-, in engine compartment wiring harness
- (139) - Ground connection (sensor Ground), in Motronic Multiport Fuel Injection (MFI) wiring harness
- (220) - Ground connection (sensor ground), in engine compartment wiring harness
- (D103) - Connection 3, in engine compartment wiring harness

ws = white
sw = black
ro = red
br = brown
gn = green
bl = blue
gr = grey
li = violet
ge = yellow
or = orange

New Beetle — Wiring diagram — No. 38/8

Motronic engine control module (ECM), mass air flow (MAF) sensor, fuel injectors

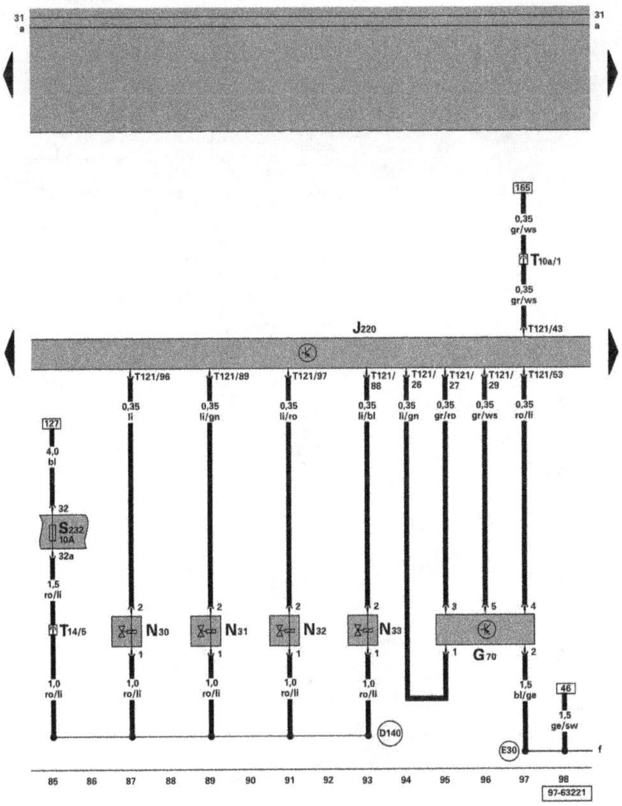

- G70 - Mass Air Flow (MAF) Sensor
- J220 - Motronic Engine Control Module (ECM)
- N30 - Cylinder 1 Fuel Injector
- N31 - Cylinder 2 Fuel Injector
- N32 - Cylinder 3 Fuel Injector
- N33 - Cylinder 4 Fuel Injector
- S232 - Fuse 32 in fuse/relay panel
- T10a - 10-Pin Connector, orange, behind instrument panel, left
- T14 - 14-Pin Connector
- T121 - 121-Pin Connector, at J220-
- (D140) - Connector (injectors), in wiring harness, engine pre-wiring
- (E30) - Connector (87a), in wiring harness engine

ws = white
sw = black
ro = red
br = brown
gn = green
bl = blue
gr = grey
li = violet
ge = yellow
or = orange

New Beetle — Wiring diagram — No. 38/9

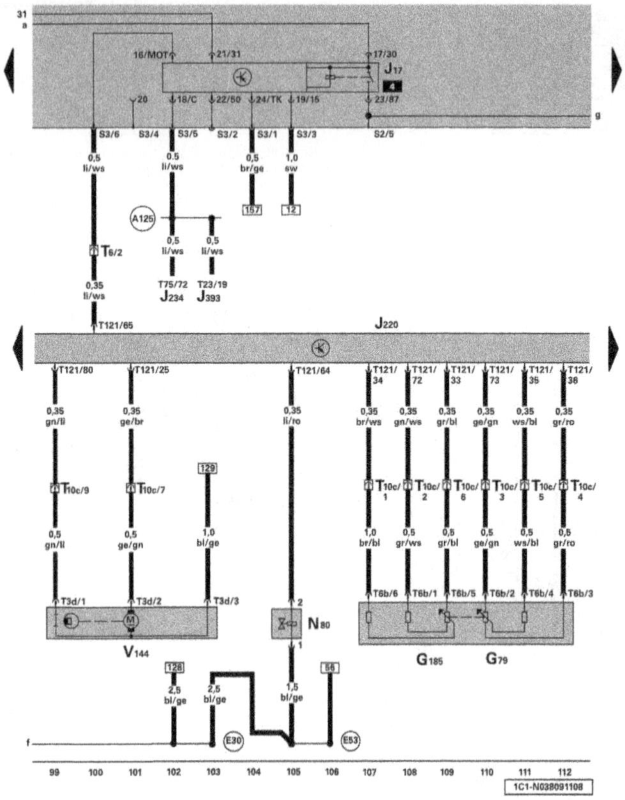

ws = white
sw = black
ro = red
br = brown
gn = green
bl = blue
gr = grey
li = violet
ge = yellow
or = orange

Motronic engine control module (ECM), throttle position (TP) sensor, sender 2 for acelerator pe dal position, evaporative emission (EVAP), leak detection pump (LDP)

G79 - Throttle Position (TP) Sensor
G185 - Sender -2- for accelerator pedal position
J17 - Fuel Pump (FP) Relay
J220 - Motronic Engine Control Module (ECM)
J234 - Airbag Control Module
J393 - Central control module for comfort system
N80 - Evaporative Emission (EVAP) Canister Purge Regulator Valve
T3d - 3-Pin Connector
T6 - 6-Pin Connector, brown, behind instrument panel left
T6b - 6-Pin Connector
T10c - 10-Pin Connector, blue, behind instrument panel, left
T23 - 23-Pin Connector, on Central control module for comfort system
T75 - 75-Pin Connector
T121 - 121-Pin Connector, at -J220-
V144 - Leak Detection Pump (LDP)

A125 - Connection (crash signal) in instrument panel wiring harness
E30 - Connector (87a), in wiring harness engine
E53 - Connection 2 (87a)(in Motronic wiring harness

New Beetle — Wiring diagram — No. 38/10

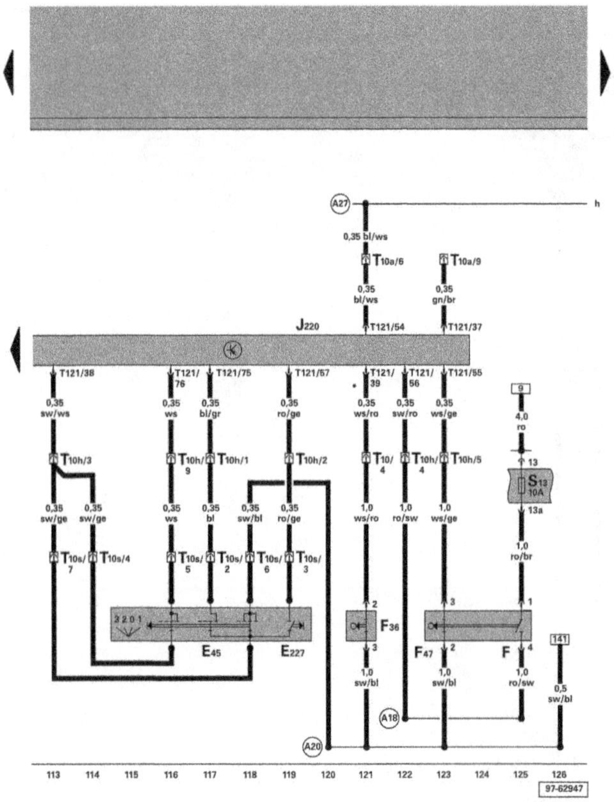

ws = white
sw = black
ro = red
br = brown
gn = green
bl = blue
gr = grey
li = violet
ge = yellow
or = orange

Motronic engine control module (ECM), cruise control switch, button for cruise control (set), brake light switch, clutch vacuum vent valve switch

E45 - Cruise Control Switch
E227 - Button for cruise control (set)
F - Brake Light Switch
F36 - Clutch Vacuum Vent Valve Switch
F47 - Brake Pedal Switch (Cruise control/Diesel Direct Injection)
J220 - Motronic Engine Control Module (ECM)
S13 - Fuse 13 in fuse/relay panel
T10 - 10-Pin Connector, white, behind instrument panel, left
T10a - 10-Pin Connector, orange, behind instrument panel, left
T10h - 10-Pin Connector, black, behind instrument panel, left
T10s - 10-Pin Connector, near steering column
T121 - 121-Pin Connector, at -J220-

A18 - wire connection (54), in instrument panel wiring harness
A20 - wire connection (15a), in instrument panel wiring harness
A27 - Wire Connection (vehicle speed signal), in instrument panel wiring harness

* - Manual transmission only

New Beetle Wiring diagram **No. 38/11**

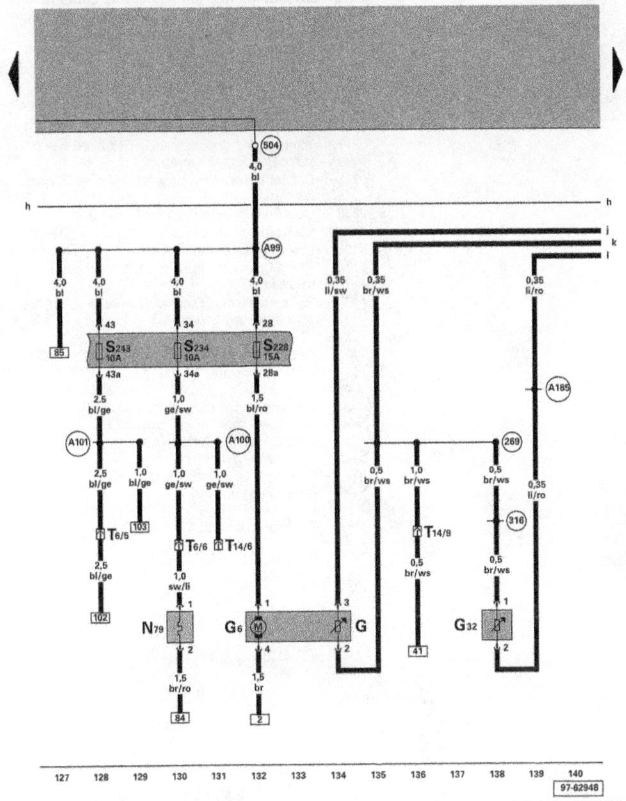

Sender for fuel gauge, fuel pump (FP), engine coolant level (ECL) sensor

- G - Sender for fuel gauge
- G6 - Fuel Pump (FP)
- G32 - Engine Coolant Level (ECL) Sensor
- N79 - Positive crankcase ventilation (PVC) heating element
- S228 - Fuse 28 in fuse/relay panel
- S234 - Fuse 34 in fuse/relay panel
- S243 - Fuse 43 in fuse/relay panel
- T6 - 6-Pin Connector, brown, behind instrument panel, left
- T14 - 14-Pin Connector
- 269 - Ground connector (sensor ground) -1-, in instrument panel wiring harness
- 316 - Sensor Ground Connection 2 (in engine wiring harness)
- 504 - Threaded connection -1- (87) on the relay plate
- A99 - Connector -1- (87), in instrument panel wiring harness
- A100 - Connector -2- (87), in instrument panel wiring harness
- A101 - Connector -3- (87), in instrument panel wiring harness
- A185 - Coolant Warning Connection (in instrument panel wiring harness)

ws = white
sw = black
ro = red
br = brown
gn = green
bl = blue
gr = grey
li = violet
ge = yellow
or = orange

New Beetle Wiring diagram **No. 38/12**

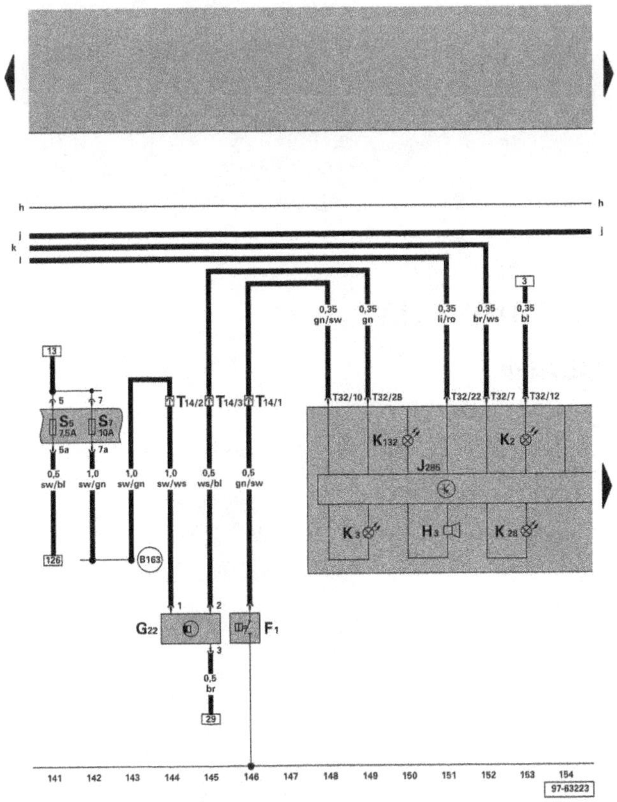

Oil pressure switch / warning light, speedometer vehicle speed sensor (VSS), warning buzzer, instrument cluster, generator (GEN) warning light

- F1 - Oil Pressure Switch
- G22 - Speedometer Vehicle Speed Sensor (VSS)
- H3 - Warning Buzzer
- J285 - Control module with indicator unit in instrument panel insert
- K2 - Generator (GEN) Warning Light
- K3 - Oil Pressure Warning Light
- K28 - Engine Coolant Level/Temperature (ECL/ECT) Warning Light
- K132 - Electronic Power Control (EPC) Lamp
- S5 - Fuse 5 in fuse/relay panel
- S7 - Fuse 7 in fuse/relay panel
- T14 - 14-Pin Connector
- T32 - 32-Pin Connector, blue
- B163 - Plus connector -1- (15) in wiring harness interior

ws = white
sw = black
ro = red
br = brown
gn = green
bl = blue
gr = grey
li = violet
ge = yellow
or = orange

New Beetle Wiring diagram **No. 38/13**

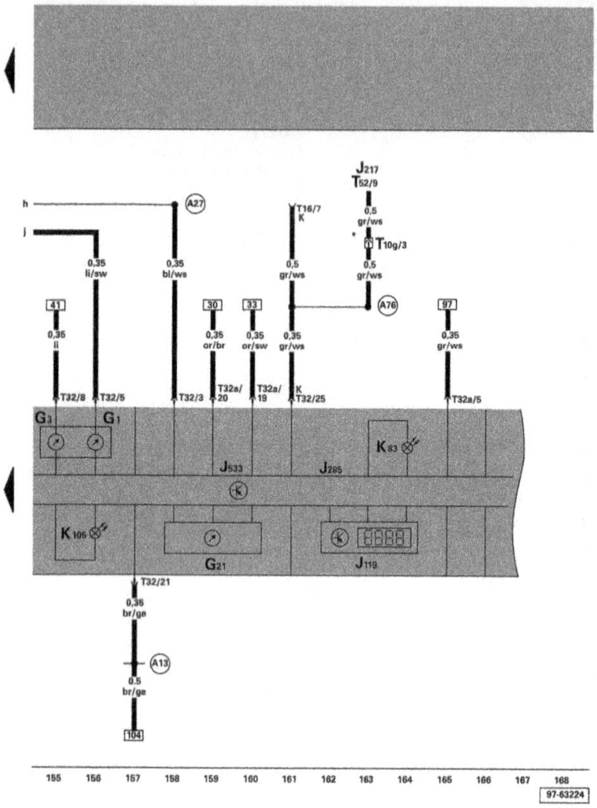

ws = white
sw = black
ro = red
br = brown
gn = green
bl = blue
gr = grey
li = violet
ge = yellow
or = orange

Instrument cluster, fuel gauge, engine coolant temperature (ECT) gauge, speedomter, multifunc tion indicator (MFA), low fuel level warning light

G1 - Fuel Gauge
G3 - Engine Coolant Temperature (ECT) Gauge
G21 - Speedometer
J119 - Multi-function Indicator (MFA)
J217 - Transmission Control Module (TCM), behind instrument panel, center
J285 - Control module with indicator unit in instrument panel insert
J533 - Data Bus On Board Diagnostic Interface
K83 - Malfunction Indicator lamp (MIL)
K105 - Low Fuel Level Warning Light
T10g - 10-Pin Connector, grey, behind instrument panel, left
T16 - 16-Pin Connector, Data link Connector (DLC), below steering column, left
T32 - 32-Pin Connector, blue
T32a - 32-Pin Connector, green
T52 - 52-Pin Connector, at -J217-

(A13) - wire connection (door contact switch) in instrument panel wiring harness
(A27) - Wire Connection (vehicle speed signal), in instrument panel wiring harness
(A76) - Connector (K-diagnosis wire) in instrument panel wiring harness

* - Automatic transmission only

THIS SPACE INTENTIONALLY

LEFT BLANK

New Beetle Convertible — Wiring diagram — No. 18/1

2,0L-Engine - Motronic Multiport Fuel Injection (MFI)/85 kW, code BGD, (with cruise control)

from June 2003

Deviate relay location and fuseplacements as well as the locations of multiple connectors see section "component locations".

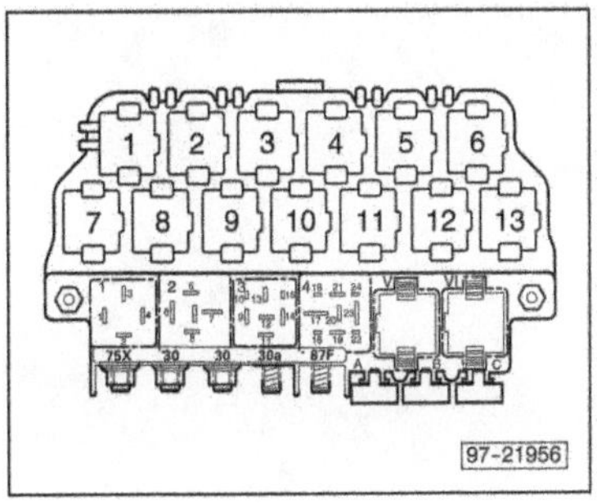

Relay panel:

- **1** Dual Horn Relay (53)
- **2** Load Reduction Relay (100)
- **4** Fuel Pump (FP) Relay (409)
- **V** Wiper/Washer Intermittent Relay (377)
- **VI** Wiper/Washer Intermittent Relay (377)

Fuse colors

30 A - green
25 A - white
20 A - yellow
15 A - blue
10 A - red
7,5 A - brown
5 A - beige
3 A - violet

Note: Number in parentheses indicates production control number stamped on relay housing.

New Beetle Convertible — Wiring diagram — No. 18/2

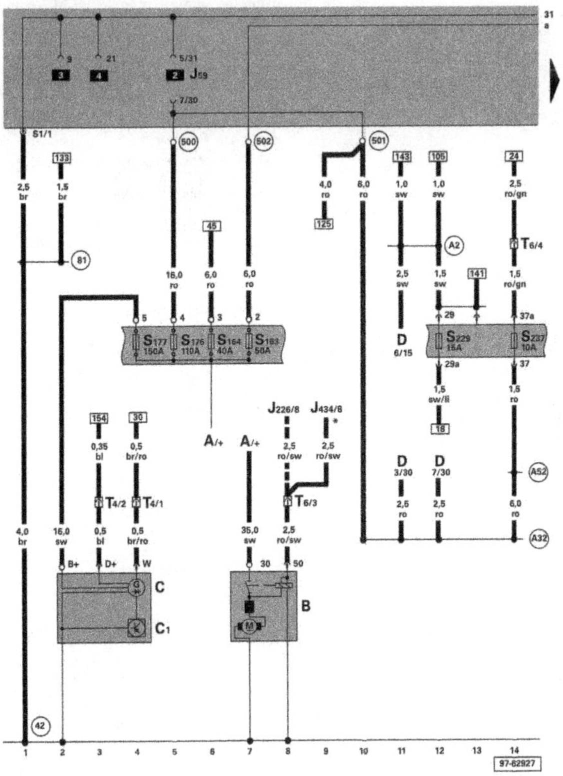

Battery, starter, generator (GEN), voltage regulator (VR), ignition/starter switch, load reduction relay, fuses in fuse bracket/battery

A - Battery
B - Starter
C - Generator (GEN)
C1 - Voltage Regulator (VR)
D - Ignition/Starter Switch
J59 - Load Reduction Relay
J226 - Park/Neutral Position (PNP) Relay (126)
J434 - Locking relay for starter (clutch pedal switch)
S163 - Fuse -2- , in fuse bracket/battery
S164 - Fuse -3- , in fuse bracket/battery
S176 - Fuse -4- (30), in fuse bracket/battery
S177 - Fuse -5- (30), in fuse bracket/battery
S229 - Fuse 29 in fuse holder
S237 - Fuse 37 in in fuse holder
T4 - 4-Pin Connector, in engine compartment left
T6 - 6-Pin Connector, brown, behind instrument panel, left

(42) - Ground connection, beside steering column
(81) - Ground connection -1-, in instrument panel wiring harness
(500) - Threaded connection -1- (30) on the relay plate
(501) - Threaded connection -2- (30) on the relay plate
(502) - Threaded connection -1- (30a) on the relay plate
(A2) - plus connection (15), in instrument panel wiring harness
(A32) - plus connection (30), in instrument panel wiring harness
(A52) - Plus connector -2- (30), in instrument panel wiring harness

* - Manual transmission only
--- - Automatic transmission only

ws = white
sw = black
ro = red
br = brown
gn = green
bl = blue
gr = grey
li = violet
ge = yellow

2.0L engine management (BGD) from June 2003, convertible

New Beetle Convertible — Wiring diagram — **No. 18/3**

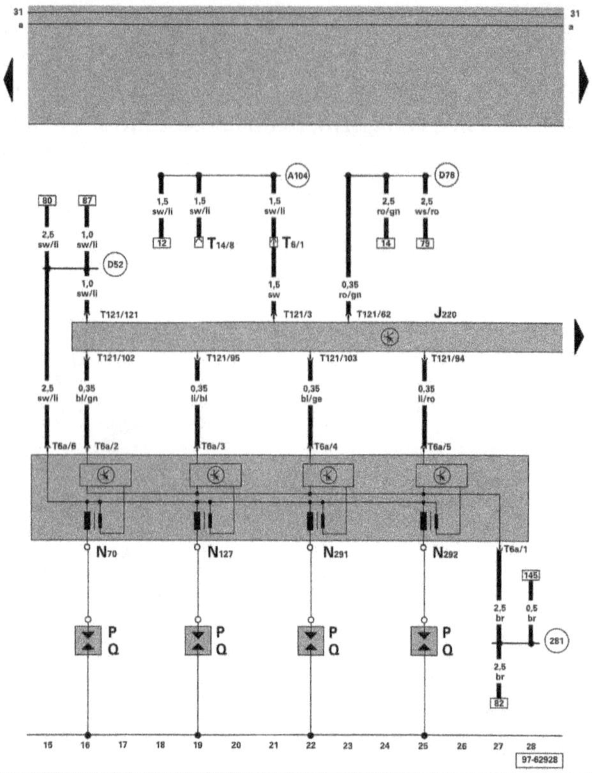

Motronic engine control module (ECM), ignition coil, spark plugs

- J220 - Motronic Engine Control Module (ECM)
- N70 - Ignition Coil 1 with Power Output Stage
- N127 - Ignition Coil 2 with Power Output Stage
- N291 - Ignition Coil 3 with Power Output Stage
- N292 - Ignition Coil 4 with Power Output Stage
- P - Spark Plug Connectors
- Q - Spark Plugs
- T6 - 6-Pin Connector, brown, behind instrument panel, left
- T6a - 6-Pin Connector
- T14 - 14-Pin Connector, in plenum chamber, center
- T121 - 121-Pin Connector, at -J220-

- (281) - Ground connector -1-, in wiring harness engine pre-wiring
- (A104) - Plus connector -2- (15), in instrument panel wiring harness
- (D52) - plus connection (15a), in engine compartment wiring harness
- (D78) - Plus connector -1- (30a), in engine compartment wiring harness

ws = white
sw = black
ro = red
br = brown
gn = green
bl = blue
gr = grey
li = violet
ge = yellow

New Beetle Convertible — Wiring diagram — **No. 18/4**

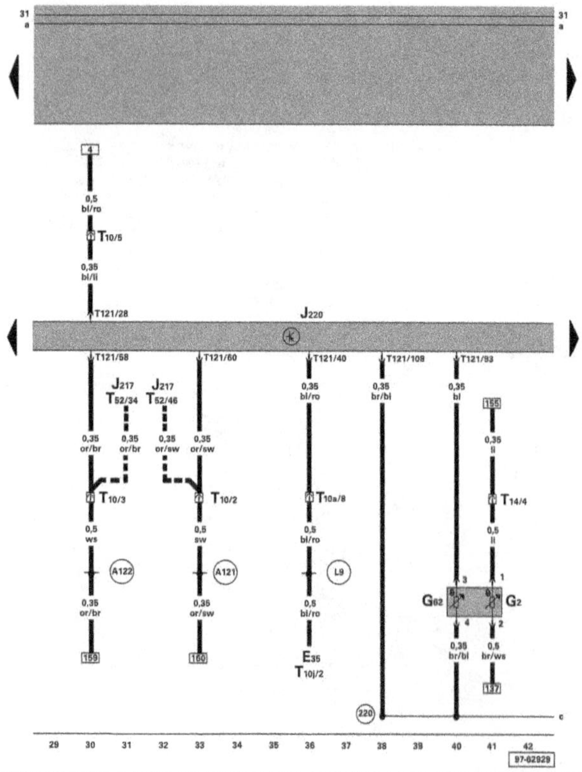

Motronic engine control module (ECM), engine coolant temperature (ECT) sensor

- E35 - A/C Switch
- G2 - Engine Coolant Temperature (ECT) Sensor
- G62 - Engine Coolant Temperature (ECT) Sensor
- J217 - Transmission Control Module (TCM) (tiptronic)
- J220 - Motronic Engine Control Module (ECM) (Tiptronic)
- T10 - 10-Pin Connector, white, behind instrument panel, left
- T10a - 10-Pin Connector, orange, behind instrument panel, left
- T10j - 10-Pin Connector
- T14 - 14-Pin Connector, in plenum chamber, center
- T52 - 52-Pin Connector
- T121 - 121-Pin Connector, at -J220-

- (220) - Ground connection (sensor ground), in engine compartment wiring harness
- (A121) - Connection (high bus) in instrument panel wiring harness
- (A122) - Connection (low bus) in instrument panel wiring harness
- (L9) - Wire connection -1-, in A/C wiring harness

- - - - Automatic transmission only

ws = white
sw = black
ro = red
br = brown
gn = green
bl = blue
gr = grey
li = violet
ge = yellow

New Beetle Convertible

Wiring diagram

No. 18/5

Motronic engine control module (ECM), heated oxygen sensors (HO2S), secondary air injection (AIR) pump relay/motor

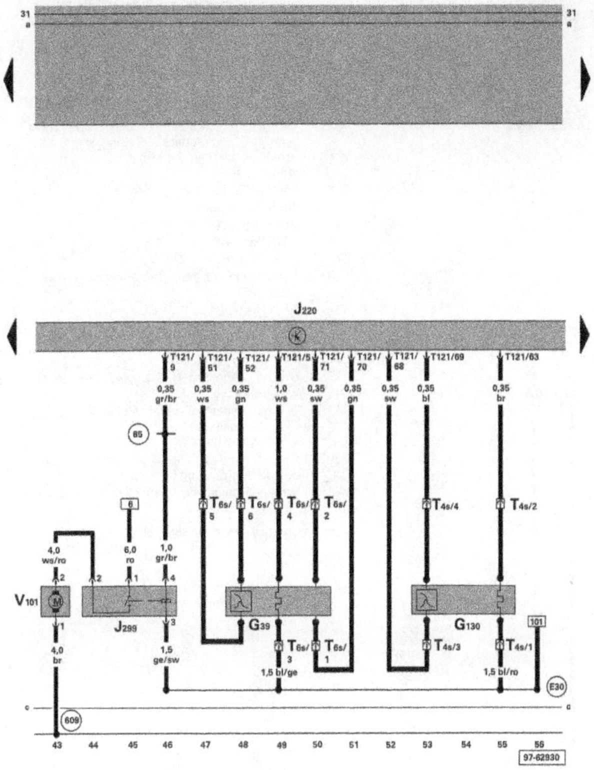

- G39 - Heated Oxygen Sensor (HO2S), black
- G130 - Oxygen Sensor (02S) behind Three Way Catalytic Converter (TWC), brown
- J220 - Motronic Engine Control Module (ECM)
- J299 - Secondary Air Injection (AIR) Pump Relay
- V101 - Secondary Air Injection (AIR) Pump Motor
- T4s - 4-Pin Connector, brown
- T6s - 6-Pin Connector, black
- T121 - 121-Pin Connector, at -J220-
- (85) - Ground connection -1-, in engine compartment wiring harness
- (609) - Ground Connection (in right plenum chamber)
- (E30) - Connector (87a), in wiring harness engine

ws = white
sw = black
ro = red
br = brown
gn = green
bl = blue
gr = grey
li = violet
ge = yellow

Edition 09.03

New Beetle Convertible

Wiring diagram

No. 18/6

Motronic engine control module (ECM), camshaft position (CMP) sensor, knock sensor (KS) 1, throttle drive (power accelerator actuation), angle sensors for throttle drive, throttle valve control module

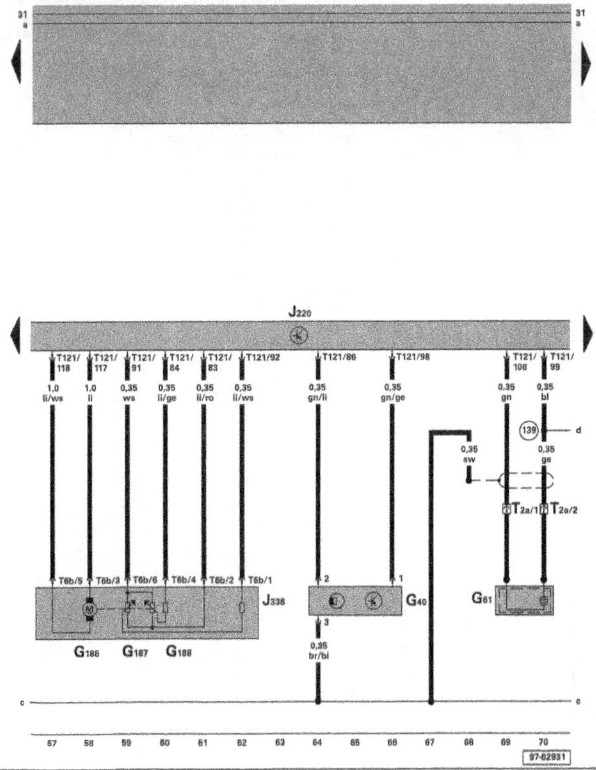

- G40 - Camshaft Position (CMP) Sensor
- G61 - Knock Sensor (KS) 1
- G186 - Throttle drive (power accelerator actuation)
- G187 - Angle sensor -1- for throttle drive (power accelerator actuation)
- G188 - Angle sensor -2- for throttle drive (power accelerator actuation)
- J220 - Motronic Engine Control Module (ECM)
- J338 - Throttle Valve Control Module
- T2a - 2-Pin Connector
- T6b - 6-Pin Connector
- T121 - 121-Pin Connector, at -J220-
- (139) - Ground connection (sensor Ground), in Motronic Multiport Fuel Injection (MFI) wiring harness

ws = white
sw = black
ro = red
br = brown
gn = green
bl = blue
gr = grey
li = violet
ge = yellow

Edition 09.03

2.0L engine management (BGD) from June 2003, convertible

New Beetle Convertible — Wiring diagram — No. 18/7

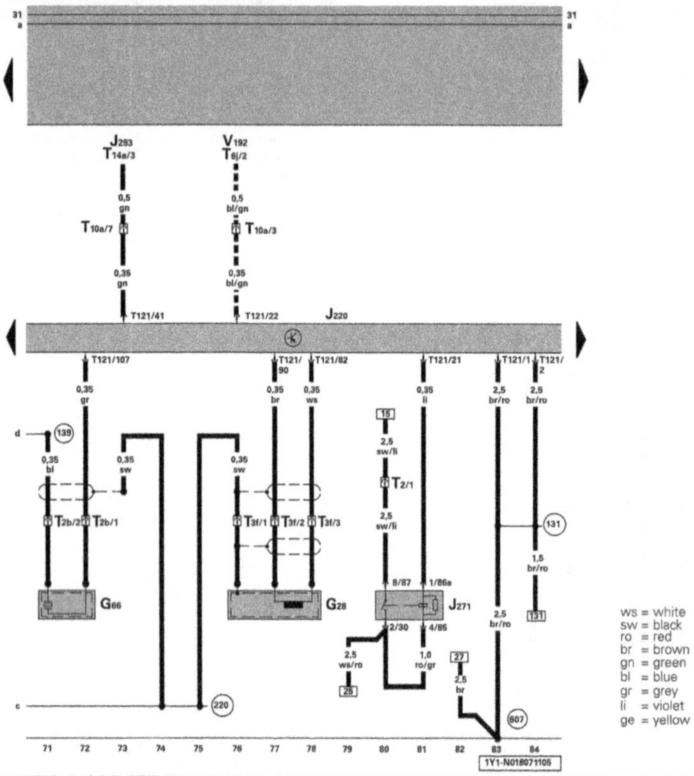

Motronic engine control module (ECM)/power supply relay, engine speed (RPM) sensor, knock sensor (KS) 2

- G28 - Engine Speed (RPM) Sensor
- G66 - Knock Sensor (KS) 2
- J220 - Motronic Engine Control Module (ECM)
- J271 - Motronic Engine Control Module (ECM) power supply relay, in protective housing, in engine compartment, left, production control number (429)
- J293 - Coolant FC (Fan Control) Control Module
- T2 - 2-Pin Connector
- T2b - 2-Pin Connector
- T3f - 3-Pin Connector
- T6j - 6-Pin Connector
- T10a - 10-Pin Connector, orange, behind instrument panel, left
- T14a - 14-Pin Connector
- T121 - 121-Pin Connector, at -J220-
- V192 - Brake System Vacuum Pump

- (131) - Ground connection -2-, in engine compartment wiring harness
- (139) - Ground connection (sensor Ground), in Motronic Multiport Fuel Injection (MFI) wiring harness
- (220) - Ground connection (sensor ground), in engine compartment wiring harness
- (607) - Ground Connection (in left plenum chamber)

--- - Automatic transmission only

ws = white
sw = black
ro = red
br = brown
gn = green
bl = blue
gr = grey
li = violet
ge = yellow

Edition 11.05

New Beetle Convertible — Wiring diagram — No. 18/8

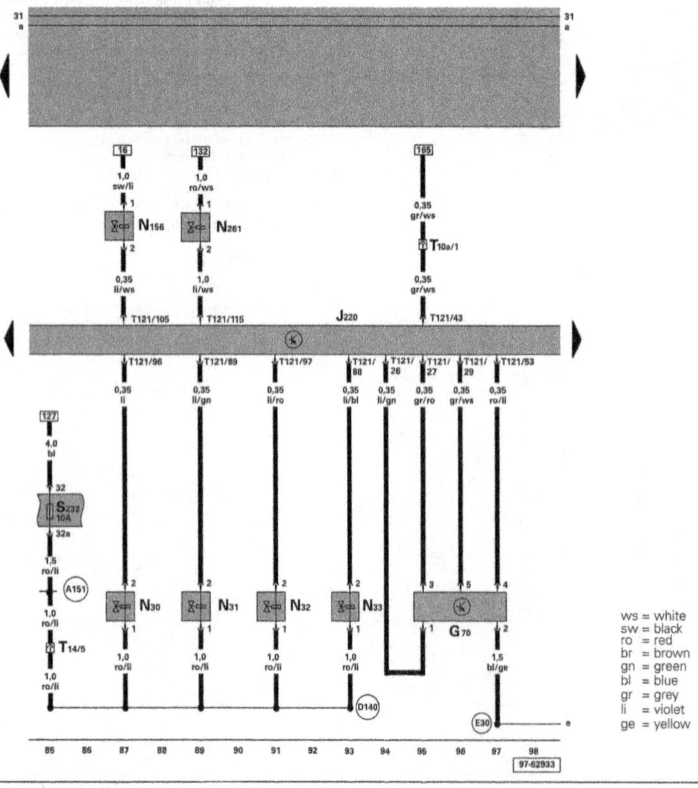

Motronic engine control module (ECM), mass air flow (MAF) sensor, fuel injectors, intake manifold change-over valve, intake manifold tuning (IMT) valve -2-

- G70 - Mass Air Flow (MAF) Sensor
- J220 - Motronic Engine Control Module (ECM)
- N30 - Cylinder 1 Fuel Injector
- N31 - Cylinder 2 Fuel Injector
- N32 - Cylinder 3 Fuel Injector
- N33 - Cylinder 4 Fuel Injector
- N156 - Intake Manifold Change-Over Valve
- N261 - Intake Manifold Tuning (IMT) Valve -2-
- S232 - Fuse 32 in fuse holder
- T10a - 10-Pin Connector, orange, behind instrument panel, left
- T14 - 14-Pin Connector, in plenum chamber, center
- T121 - 121-Pin Connector, at -J220-

- (A151) - Connection 4 (87) (in instrument panel wiring harness)
- (D140) - Connector (injectors), in wiring harness, engine pre-wiring
- (E30) - Connector (87a), in wiring harness engine

ws = white
sw = black
ro = red
br = brown
gn = green
bl = blue
gr = grey
li = violet
ge = yellow

Edition 09.03

New Beetle Convertible — Wiring diagram — No. 18/9

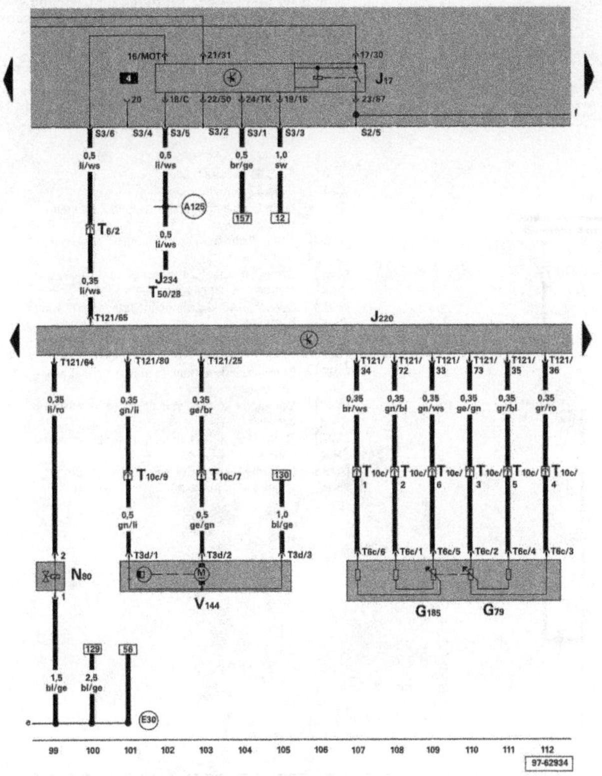

Motronic engine control module (ECM), throttle position (TP) sensor, sender -2- for acelerator pedal position, fuel pump (FP) relay, evaporative emission (EVAP) canister purge regulator valve, leak detection pump (LDP)

G79 - Throttle Position (TP) Sensor
G185 - Sender -2- for accelerator pedal position
J17 - Fuel Pump (FP) Relay
J220 - Motronic Engine Control Module (ECM)
J234 - Airbag Control Module
N80 - Evaporative Emission (EVAP) Canister Purge Regulator Valve
V144 - Leak Detection Pump (LDP)
T3d - 3-Pin Connector
T6 - 6-Pin Connector, brown, behind instrument panel, left
T6c - 6-Pin Connector
T10c - 10-Pin Connector, blue, behind instrument panel, left
T50 - 50-Pin Connector, at -J234-
T121 - 121-Pin Connector, at -J220-

(A125) - Connection (crash signal) in instrument panel wiring harness
(E30) - Connector (87a), in wiring harness engine

ws = white
sw = black
ro = red
br = brown
gn = green
bl = blue
gr = grey
li = violet
ge = yellow

New Beetle Convertible — Wiring diagram — No. 18/10

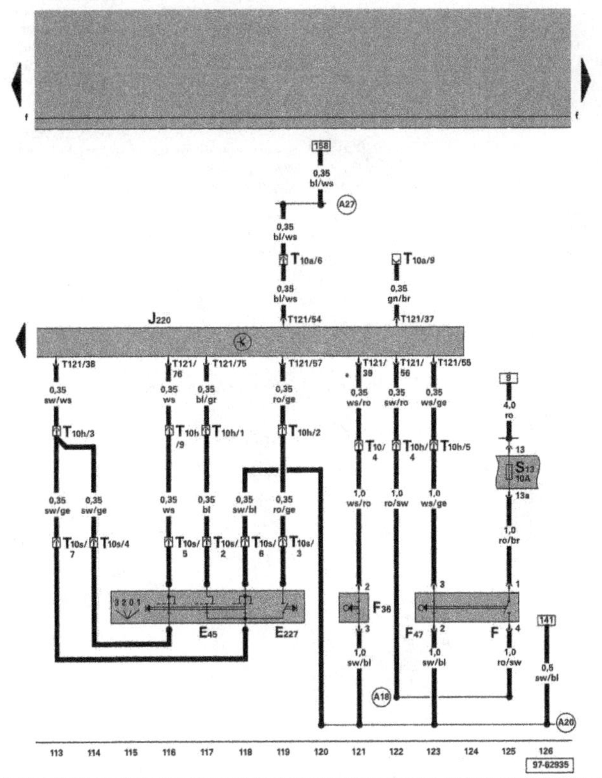

Motronic engine control module (ECM), cruise control switch, button for cruise control (set), brake light switch, clutch vacuum vent valve switch, brake pedal switch

E45 - Cruise Control Switch
E227 - Button for cruise control (set)
F - Brake Light Switch
F36 - Clutch Vacuum Vent Valve Switch
F47 - Brake Pedal Switch (Cruise control/Diesel Direct Injection)
J220 - Motronic Engine Control Module (ECM)
S13 - Fuse 13 in fuse holder
T10 - 10-Pin Connector, white, behind instrument panel, left
T10a - 10-Pin Connector, orange, behind instrument panel, left
T10h - 10-Pin Connector, black, behind instrument panel, left
T10s - 10-Pin Connector, near steering column

T121 - 121-Pin Connector, at -J220-

(A18) - wire connection (54), in instrument panel wiring harness
(A20) - wire connection (15a), in instrument panel wiring harness
(A27) - Wire Connection (vehicle speed signal), in instrument panel wiring harness

* - Manual transmission only

ws = white
sw = black
ro = red
br = brown
gn = green
bl = blue
gr = grey
li = violet
ge = yellow

2.0L engine management (BGD) from June 2003, convertible

New Beetle Convertible — Wiring diagram — No. 18/11

Sender for fuel gauge, fuel pump (FP), engine coolant level (ECL) sensor, positive crankcase ventilation (PCV) heating element

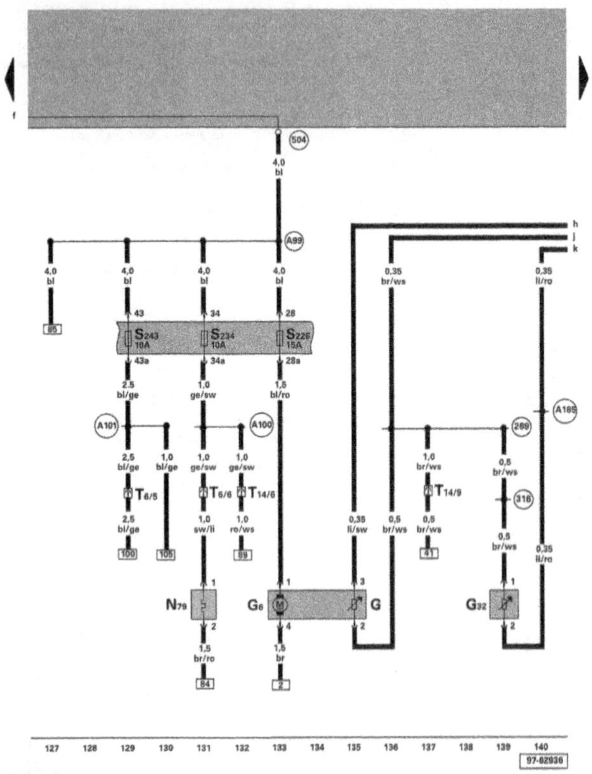

- G - Sender for fuel gauge
- G6 - Fuel Pump (FP)
- G32 - Engine Coolant Level (ECL) Sensor
- N79 - Positive Crankcase Ventilation (PCV) Heating Element
- S228 - Fuse 28 in fuse holder
- S234 - Fuse 34 in fuse holder
- S243 - Fuse 43 in fuse holder
- T6 - 6-Pin Connector, brown, behind instrument panel, left
- T14 - 14-Pin Connector, in plenum chamber, center
- (269) - Ground connector (sensor ground) -1-, in instrument panel wiring harness
- (316) - Sensor Ground Connection 2 (in engine wiring harness)
- (504) - Threaded connection -1- (87) on the relay plate
- (A99) - Connector -1- (87), in instrument panel wiring harness
- (A100) - Connector -2- (87), in instrument panel wiring harness
- (A101) - Connector -3- (87), in instrument panel wiring harness
- (A185) - Coolant Warning Connection (in instrument panel wiring harness)

ws = white
sw = black
ro = red
br = brown
gn = green
bl = blue
gr = grey
li = violet
ge = yellow

New Beetle Convertible — Wiring diagram — No. 18/12

Oil pressure switch / warning light, speedometer vehicle speed sensor (VSS), warning buzzer, control module with indicator unit in instrument panel insert, generator (GEN) warning light, engine coolant level/temperature (ECL/ECT) warning light, electronic power control (EPC) warning lamp

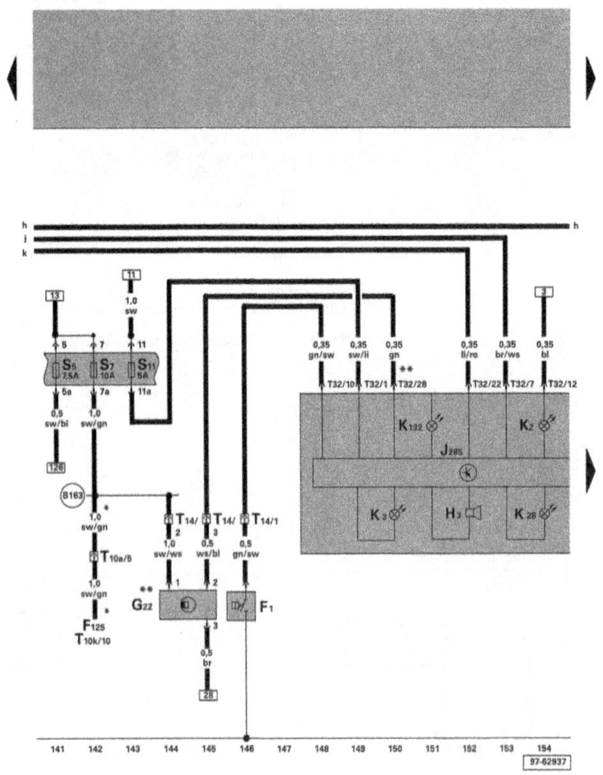

- F1 - Oil Pressure Switch
- F125 - Multi-Function Transmission Range (TR) Switch
- G22 - Speedometer Vehicle Speed Sensor (VSS)
- H3 - Warning Buzzer
- J285 - Control module with indicator unit in instrument panel insert
- K2 - Generator (GEN) Warning Light
- K3 - Oil Pressure Warning Light
- K28 - Engine Coolant Level/Temperature (ECL/ECT) Warning Light
- K132 - Electronic Power Control (EPC) Warning Lamp
- S5 - Fuse 5 in fuse holder
- S7 - Fuse 7 in fuse holder
- S11 - Fuse 11 in fuse holder
- T10a - 10-Pin Connector
- T10k - 10-Pin Connector
- T14 - 14-Pin Connector, in left plenum chamber
- T32 - 32-Pin Connector, blue
- (B163) - Plus connector -1- (15) in wiring harness interior
- * - Automatic transmission only
- ** - Manual transmission only

ws = white
sw = black
ro = red
br = brown
gn = green
bl = blue
gr = grey
li = violet
ge = yellow

New Beetle Convertible | **Wiring diagram** | **No. 18/13**

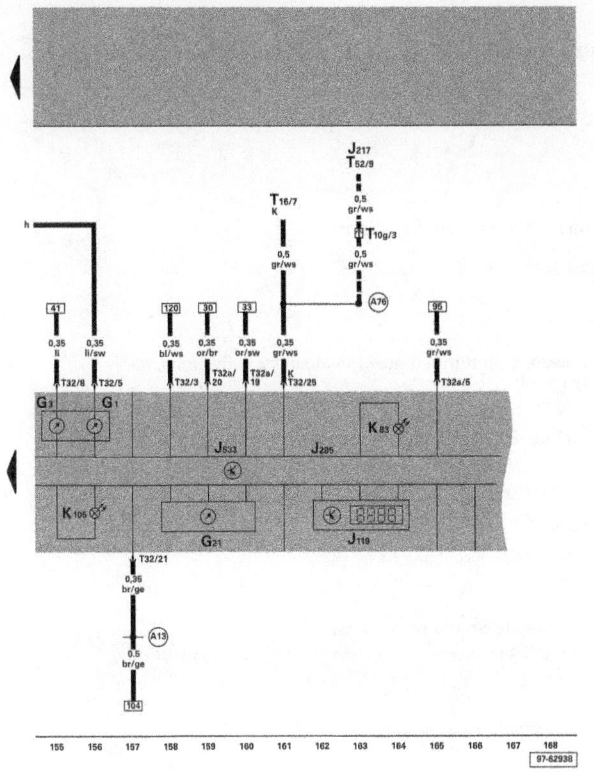

Fuel gauge, engine coolant temperature (ECT) gauge, speedometer, multi-function indicator (MFA), control module with indicator unit in instrument panel insert, data bus on board diagnostic interface, malfunction indicator lamp (MIL), low fuel level warning light

- G1 - Fuel Gauge
- G3 - Engine Coolant Temperature (ECT) Gauge
- G21 - Speedometer
- J119 - Multi-function Indicator (MFA)
- J217 - Transmission Control Module (TCM), Tiptronic
- J285 - Control module with indicator unit in instrument panel insert
- J533 - Data Bus On Board Diagnostic Interface
- K83 - Malfunction Indicator lamp (MIL)
- K105 - Low Fuel Level Warning Light
- T10g - 10-Pin Connector, grey, behind instrument panel, left
- T16 - 16-Pin Connector, Data link Connector (DLC), below steering column, left
- T32 - 32-Pin Connector, blue
- T32a - 32-Pin Connector, green
- T52 - 52-Pin Connector

- (A13) - wire connection (door contact switch) in instrument panel wiring harness
- (A76) - Connector (K-diagnosis wire) in instrument panel wiring harness

- - - - 6-speed Automatic transmission only

ws = white
sw = black
ro = red
br = brown
gn = green
bl = blue
gr = grey
li = violet
ge = yellow

THIS SPACE INTENTIONALLY

LEFT BLANK

New Beetle — Wiring diagram — **No. 44/1**

2.5L - Engine - Motronic Multiport Fuel Injection (MFI)/ 110 kw, code BPR (with cruise control),

from July 2005,

Additional Fuse, Relay, Control Module and Connector locations ⇒ Component Locations

Troubleshooting Procedures ⇒ Guided Fault Finding using VAS 5051/5052

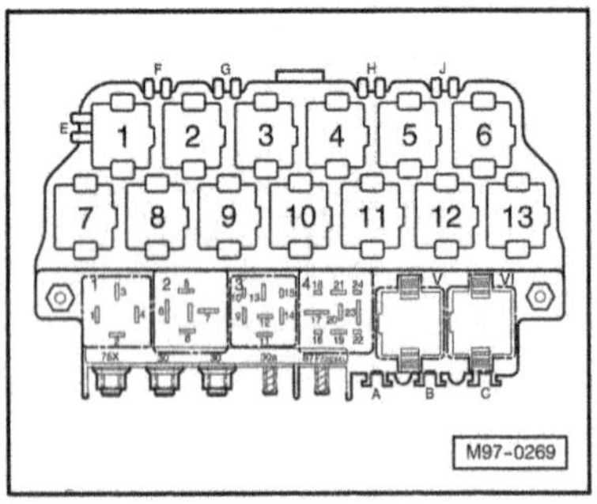

Relay locations on thirteen position auxiliary relay panel, above relay panel:

4 J207 - Starting Interlock Relay (465) (only to BPR engines)

10 J299 - Secondary Air Injection (AIR) Pump Relay (100)

13 J271 - Motronic Engine Control Module (ECM) Power Supply Relay (428)

Relay panel:

2 J59 - Load Reduction Relay (100)

4 J17 - Fuel Pump (FP) Relay (409)

Fuse placements on the relay panel:

C S130 - Secondary Air Injection (AIR) Pump Fuse (10A)

Note:

Number in parentheses indicates production control number stamped on relay housing.

New Beetle — Wiring diagram — **No. 44/2**

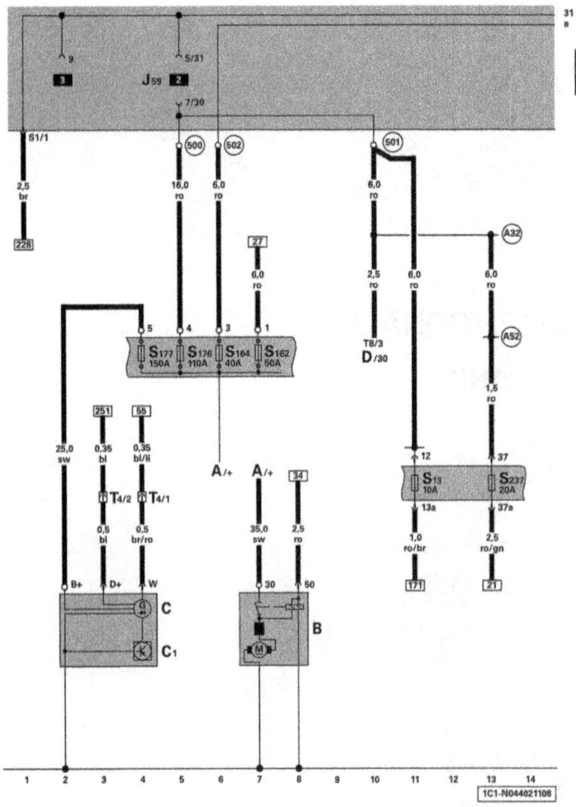

Starter, generator (GEN), Voltage Regulator (VR), Load Reduction Relay, Fuses on battery fuse panel

- A - Battery
- B - Starter
- C - Generator (GEN)
- C1 - Voltage Regulator (VR)
- D - Ignition/Starter Switch
- J59 - Load Reduction Relay
- S13 - Fuse (on fuse/relay panel)
- S162 - Fuse 1 - terminal 30 (on battery fuse panel)
- S164 - Fuse 3 - terminal 30 (on battery fuse panel)
- S176 - Fuse 4 - terminal 30 (on battery fuse holder)
- S177 - Fuse 5 - terminal 30 (on battery fuse panel)
- S237 - Fuse (on fuse panel)
- T4 - 4-Pin Connector, in engine compartment left
- T8 - 8-Pin Connector, at -D-
- (500) - Threaded Connection 1 - terminal 30 (on relay panel)
- (501) - Threaded Connection 2 - terminal 30 (on relay panel)
- (502) - Threaded Connection 1 - terminal 30a (on relay panel)
- (A32) - Plus Connection (30) (in instrument panel wiring harness)
- (A52) - Plus Connection 2 (30) (in instrument panel wiring harness)

ws = white
sw = black
ro = red
br = brown
gn = green
bl = blue
gr = grey
li = violet
ge = yellow
or = orange

New Beetle — Wiring diagram — No. 44/3

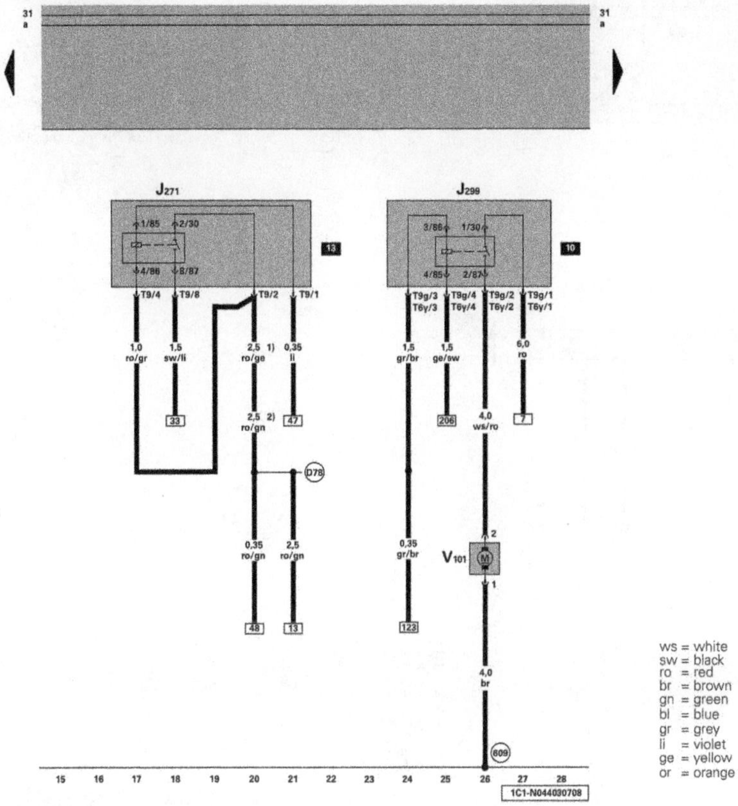

Motronic Engine Control Module (ECM) Pow er Supply Relay, Secondary Air Injection (AIR) Pump Relay, Secondary Air Injection (AIR) Pu mp Motor

- J271 - Motronic Engine Control Module (ECM) Power Supply Relay
- J299 - Secondary Air Injection (AIR) Pump Relay
- T6y - 6-Pin Connector 2)
- T9 - 9-Pin Connector
- T9g - 9-Pin Connector 1)
- V101 - Secondary Air Injection (AIR) Pump Motor
- (609) - Ground Connection (in right plenum chamber)
- (D78) - Plus Connection 1 (30a) (in engine compartment wiring harness)
- 1) - Through June 2007
- 2) - From July 2007

New Beetle — Wiring diagram — No. 44/4

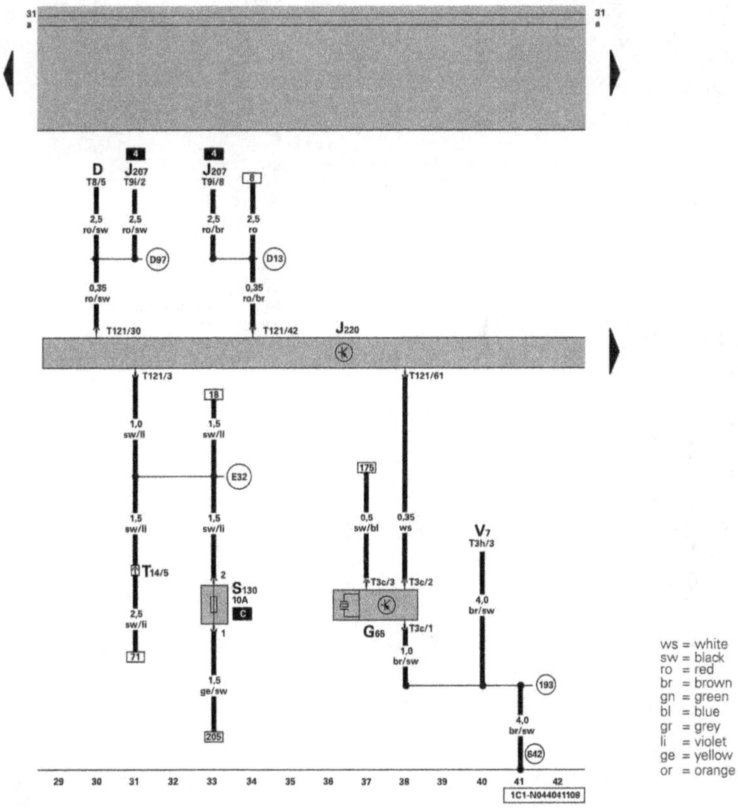

Motronic Engine Contr ol Module (ECM), High Pressure Sensor, Starter Lock/Clutch Pedal Switch Relay, Coolant Fan

- D - Ignition/Starter Switch
- G65 - High Pressure Sensor
- J207 - Starting Interlock Relay (465)
- J220 - Motronic Engine Control Module (ECM)
- S130 - Secondary Air Injection (AIR) Pump Fuse
- T3c - 3-Pin connector
- T3h - 3-Pin Connector
- T8 - 8-Pin Connector
- T9i - 9-Pin Connector
- T14 - 14-Pin Connector
- T121 - 121-Pin Connector
- V7 - Coolant Fan
- (193) - Ground Connection 1 (in coolant fan wiring harness)
- (642) - Ground Connection (on electronic fan control)
- (D13) - Connection (50) (in right front wiring harness)
- (D97) - Connection (50) (in right engine compartment wiring harness)
- (E32) - Connection (87) (in Motronic wiring harness)

New Beetle — Wiring diagram — No. 44/5

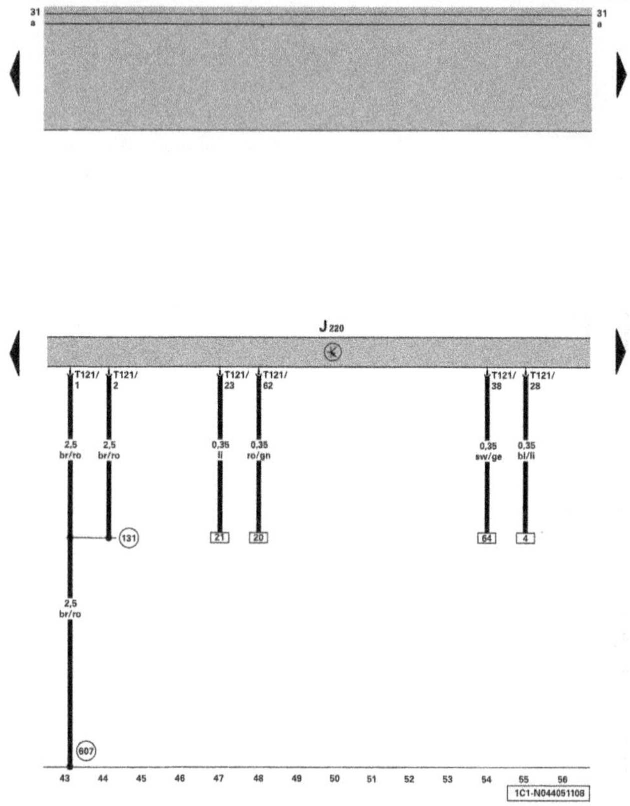

Motronic Engine Control Module (ECM)

J220 - Motronic Engine Control Module (ECM)
T121 - 121-Pin Connector

(131) - Ground Connection 2 (in engine compartment wiring harness)
(607) - Ground Connection (in left plenum chamber)

ws = white
sw = black
ro = red
br = brown
gn = green
bl = blue
gr = grey
li = violet
ge = yellow
or = orange

New Beetle — Wiring diagram — No. 44/6

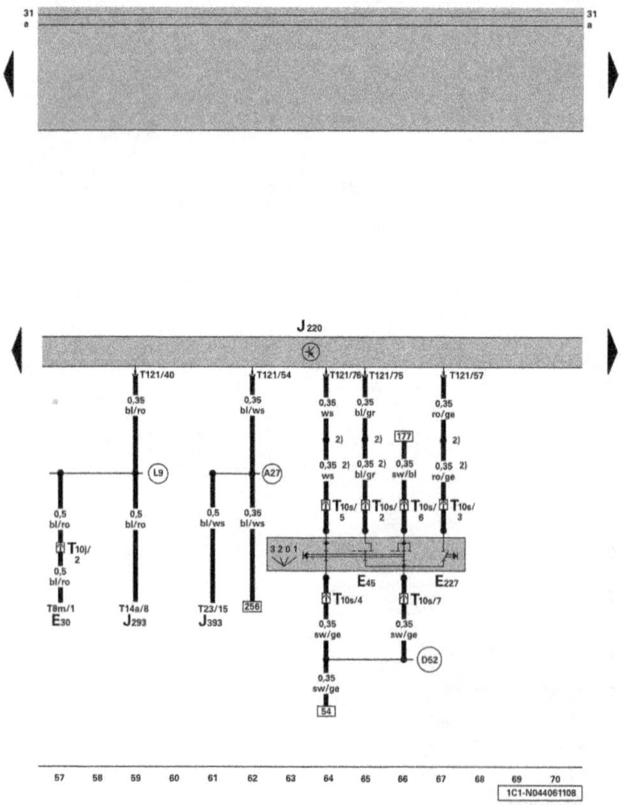

Cruise Control Switch, Cruise Control -SET- Button, Motronic Engine Control Module (ECM)

E30 - A/C Switch
E45 - Cruise Control Switch
E227 - Cruise Control -SET- Button
J220 - Motronic Engine Control Module (ECM)
J293 - Coolant Fan Control (FC) Control Module
J393 - Comfort System Central Control Module
T8m - 8-Pin Connector
T10j - 10-Pin Connector
T10s - 10-Pin Connector
T14a - 14-Pin Connector
T23 - 23-Pin Connector
T121 - 121-Pin Connector

(A27) - Vehicle Speed Signal (VSS) Connection (in instrument panel wiring harness)
(D52) - Plus Connection (15a) (in engine compartment wiring harness)
(L9) - Connection 1 (in A/C wiring harness)

2) - From July 2007

ws = white
sw = black
ro = red
br = brown
gn = green
bl = blue
gr = grey
li = violet
ge = yellow
or = orange

New Beetle **Wiring diagram** **No. 44/7**

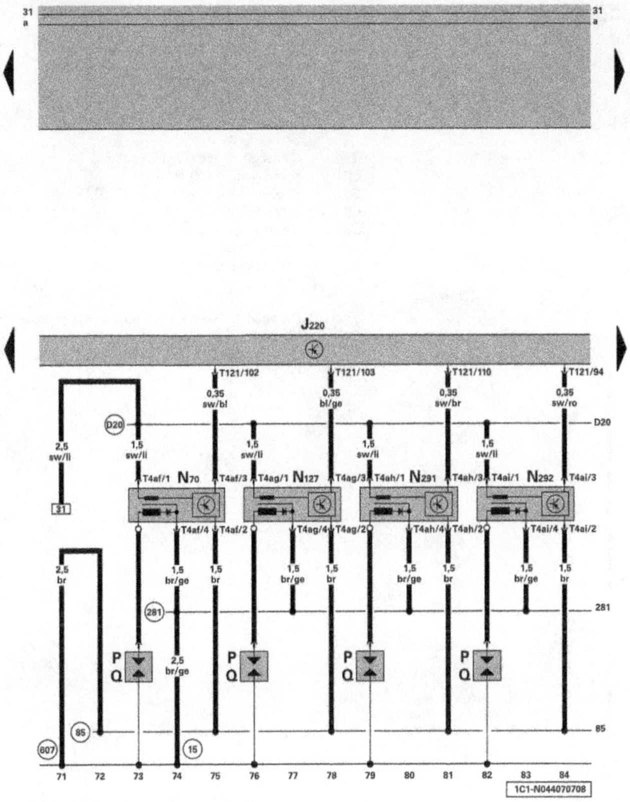

Motronic Engine Control Module (ECM), Ignition Coils, Spark Plug Connectors, Spark Plugs

- J220 - Motronic Engine Control Module (ECM)
- N70 - Ignition Coil 1 with Power Output Stage
- N127 - Ignition Coil 2 with Power Output Stage
- N291 - Ignition Coil 3 with Power Output Stage
- N292 - Ignition Coil 4 with Power Output Stage
- P - Spark Plug Connectors
- Q - Spark Plugs
- T4af - 4-Pin Connector
- T4ag - 4-Pin Connector
- T4ah - 4-Pin Connector
- T4ai - 4-Pin Connector
- T121 - 121-Pin Connector

- (15) - Ground Connection (on cylinder head)
- (85) - Ground Connection 1 (in engine compartment wiring harness)
- (281) - Ground Connection 1 (in engine wiring harness)
- (607) - Ground Connection (in left plenum chamber)
- (D20) - Connection (87a) (in right front wiring harness)

ws = white
sw = black
ro = red
br = brown
gn = green
bl = blue
gr = grey
li = violet
ge = yellow
or = orange

New Beetle **Wiring diagram** **No. 44/8**

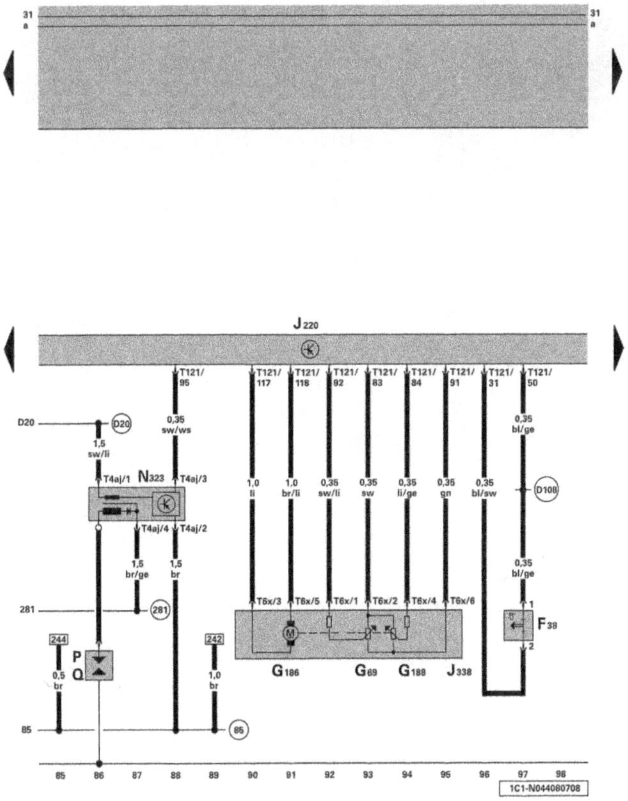

Ambient Temperature Switch, Motronic Engine Control Module (ECM), Throttle Valve Control Module, Ignition Coil 5 with Power Output Stage, Spark Plugs

- F38 - Ambient Temperature Switch
- G69 - Throttle Position (TP) Sensor
- G186 - Throttle Drive (for Electronic Power Control (EPC))
- G188 - Throttle Drive Angle Sensor 2 (for Electronic Power Control (EPC))
- J220 - Motronic Engine Control Module (ECM)
- J338 - Throttle Valve Control Module
- N323 - Ignition Coil 5 with Power Output Stage
- P - Spark Plug Connectors
- Q - Spark Plugs
- T4aj - 4-Pin Connector
- T6x - 6-Pin Connector
- T121 - 121-Pin Connector

- (85) - Ground Connection 1 (in engine compartment wiring harness)
- (281) - Ground Connection 1 (in engine wiring harness)
- (D20) - Connection (87a) (in right front wiring harness)
- (D108) - Connection 6 (in engine compartment wiring harness)

ws = white
sw = black
ro = red
br = brown
gn = green
bl = blue
gr = grey
li = violet
ge = yellow
or = orange

2.5L engine management (BPR) from July 2005

New Beetle **Wiring diagram** **No. 44/9**

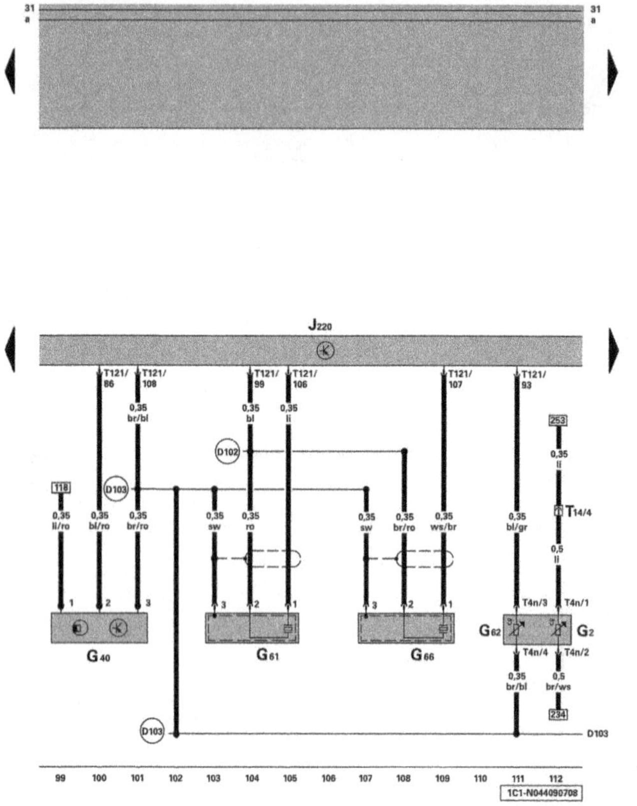

Engine Coolant Temperature (ECT) Gauge Sensor, Camshaft Position (CMP) Sensor, Knock Sensor (KS) 1, Engine Coolant Temperature (ECT) Sensor, Knock Sensor (KS) 2, Motronic Engine Control Module (ECM)

- G2 - Engine Coolant Temperature (ECT) Gauge Sensor
- G40 - Camshaft Position (CMP) Sensor
- G61 - Knock Sensor (KS) 1
- G62 - Engine Coolant Temperature (ECT) Sensor
- G66 - Knock Sensor (KS) 2
- J220 - Motronic Engine Control Module (ECM)
- T4n - 4-Pin Connector
- T14 - 14-Pin Connector
- T121 - 121-Pin Connector

- D102 - Connection 2 (in engine compartment wiring harness)
- D103 - Connection 3 (in engine compartment wiring harness)

ws = white
sw = black
ro = red
br = brown
gn = green
bl = blue
gr = grey
li = violet
ge = yellow
or = orange

New Beetle **Wiring diagram** **No. 44/10**

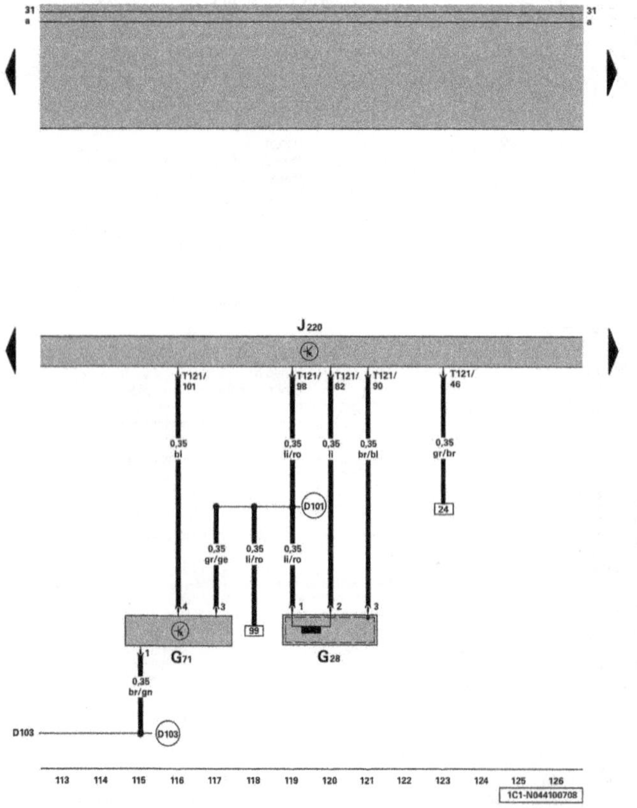

Engine Speed (RPM) Sensor, Manifold Absolute Pressure (MAP) Sensor, Motronic Engine Control Module (ECM)

- G28 - Engine Speed (RPM) Sensor
- G71 - Manifold Absolute Pressure (MAP) Sensor
- J220 - Motronic Engine Control Module (ECM)
- T121 - 121-Pin Connector

- D101 - Connection 1 (in engine compartment wiring harness)
- D103 - Connection 3 (in engine compartment wiring harness)

ws = white
sw = black
ro = red
br = brown
gn = green
bl = blue
gr = grey
li = violet
ge = yellow
or = orange

New Beetle

Wiring diagram **No. 44/11**

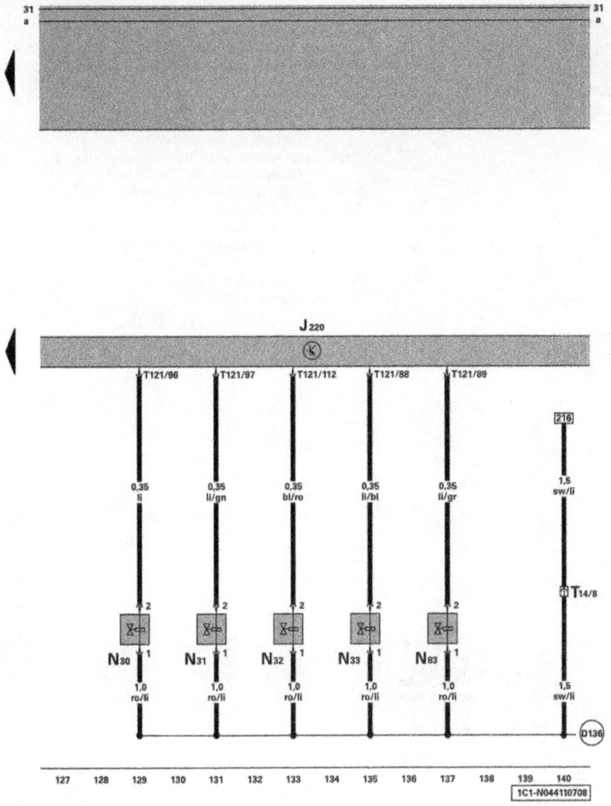

Motronic Engine Control Module (ECM), Fuel Injectors

J220 - Motronic Engine Control Module (ECM)
N30 - Cylinder 1 Fuel Injector
N31 - Cylinder 2 Fuel Injector
N32 - Cylinder 3 Fuel Injector
N33 - Cylinder 4 Fuel Injector
N83 - Cylinder 5 Fuel Injector
T14 - 14-Pin Connector
T121 - 121-Pin Connector

(D136) - Plus Connection 2 (15a (in engine compartment wiring harness)

ws = white
sw = black
ro = red
br = brown
gn = green
bl = blue
gr = grey
li = violet
ge = yellow
or = orange

New Beetle

Wiring diagram **No. 44/12**

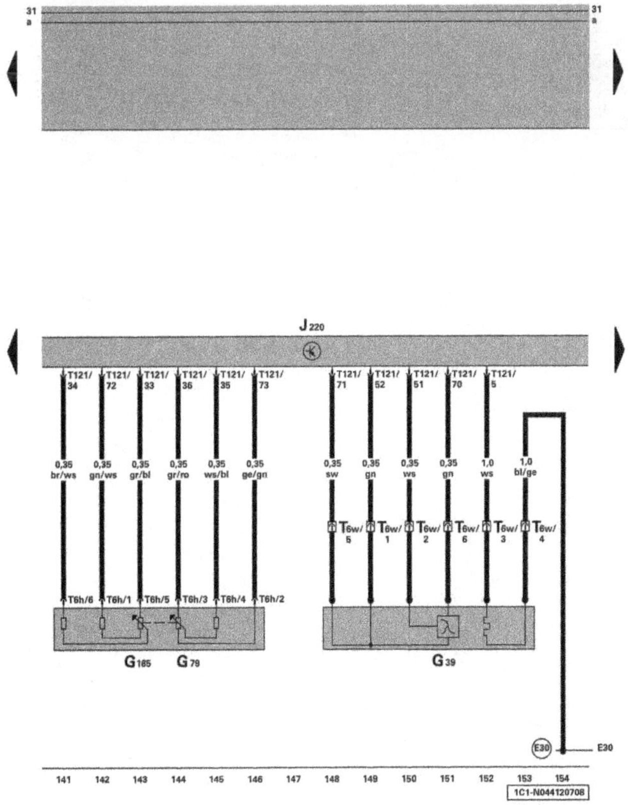

Heated Oxygen Sensor (HO2S), Throttle Position (TP) Sensor, Accelerator Pedal Position Sensor 2, Motronic Engine Control Module (ECM)

G39 - Heated Oxygen Sensor (HO2S)
G79 - Throttle Position (TP) Sensor
G185 - Accelerator Pedal Position Sensor 2
J220 - Motronic Engine Control Module (ECM)
T6h - 6-Pin Connector
T6w - 6-Pin Connector
T121 - 121-Pin Connector

(E30) - Connection (87a) (in engine wiring harness)

ws = white
sw = black
ro = red
br = brown
gn = green
bl = blue
gr = grey
li = violet
ge = yellow
or = orange

New Beetle

Wiring diagram

No. 44/13

Oxygen Sensor (O2S) Behind Three Way Catalytic Converter (TWC), Oxygen Sensor (O2S) in Center Three Way Catalytic Converter (TWC), Bank 1, Motronic Engine Control Module (ECM)

G130 - Oxygen Sensor (O2S) Behind Three Way Catalytic Converter (TWC)
G465 - Oxygen Sensor (O2S) in Center Three Way Catalytic Converter (TWC), Bank 1
J220 - Motronic Engine Control Module (ECM)
T4ao - 4-Pin Connector
T4v - 4-Pin Connector
T121 - 121-Pin Connector

1) - Through June 2007
2) - From July 2007

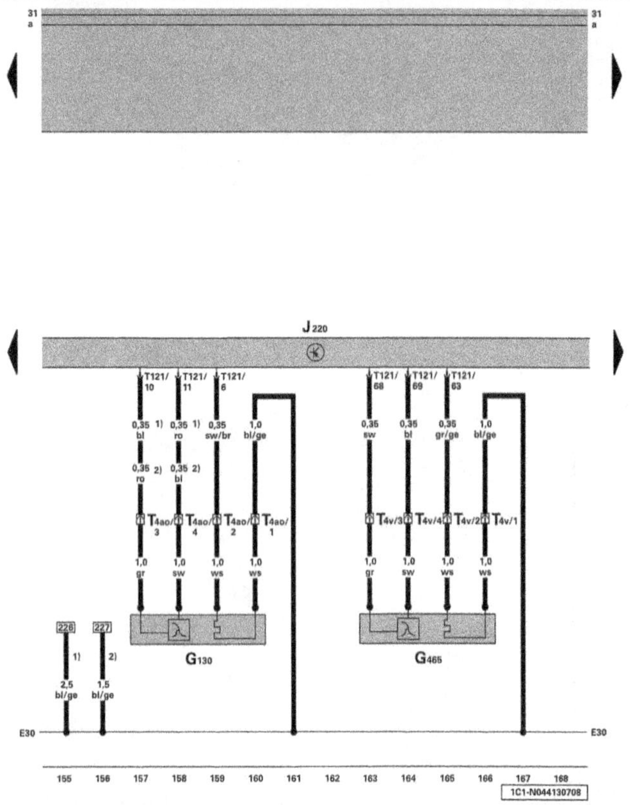

ws = white
sw = black
ro = red
br = brown
gn = green
bl = blue
gr = grey
li = violet
ge = yellow
or = orange

Edition 07.08

New Beetle

Wiring diagram

No. 44/14

Brake Light Switch, Clutch Pedal Switch, Brake Pedal Switch, Motronic Engine Control Module (ECM)

F - Brake Light Switch
F36 - Clutch Pedal Switch
F47 - Brake Pedal Switch
J220 - Motronic Engine Control Module (ECM)
T4b - 4-Pin Connector
T4m - 4-Pin Connector
T4y - 4-Pin Connector
T121 - 121-Pin Connector
V35 - Right Coolant Fan

(A20) - Plus Connection (15a) (in instrument panel wiring harness)
(B131) - Connection (54) (in interior wiring harness)

*** - Coolant fan connection
2) - From July 2007

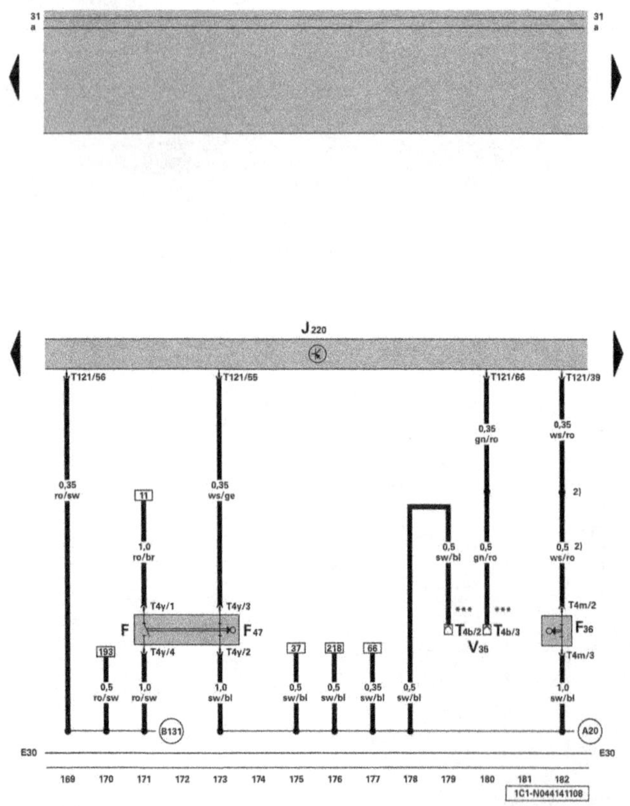

ws = white
sw = black
ro = red
br = brown
gn = green
bl = blue
gr = grey
li = violet
ge = yellow
or = orange

Edition 11.08

New Beetle — Wiring diagram — No. 44/15

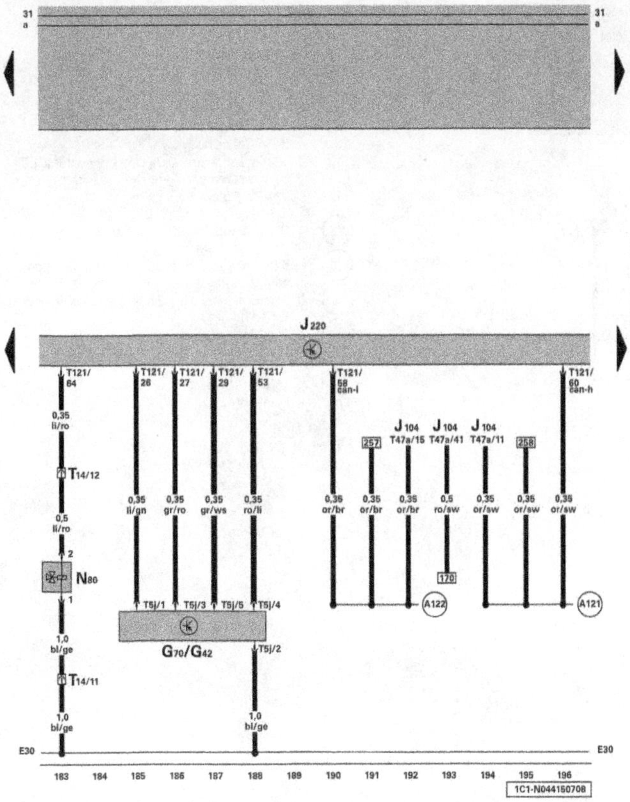

Intake Air Temperature (IAT) Sensor, Mass Air Flow (MAF) Sensor, Motronic Engine Control Module (ECM), Evaporative Emission (EVAP) Canister Purge Regulator Valve

- G42 - Intake Air Temperature (IAT) Sensor
- G70 - Mass Air Flow (MAF) Sensor
- J104 - ABS Control Module
- J220 - Motronic Engine Control Module (ECM)
- N80 - Evaporative Emission (EVAP) Canister Purge Regulator Valve
- T5j - 5-Pin Connector
- T14 - 14-Pin Connector
- T47a - 47-Pin Connector
- T121 - 121-Pin Connector
- (A121) - High-Bus Connection (in instrument panel wiring harness)
- (A122) - Low-Bus Connection (in instrument panel wiring harness)

ws = white
sw = black
ro = red
br = brown
gn = green
bl = blue
gr = grey
li = violet
ge = yellow
or = orange

New Beetle — Wiring diagram — No. 44/16

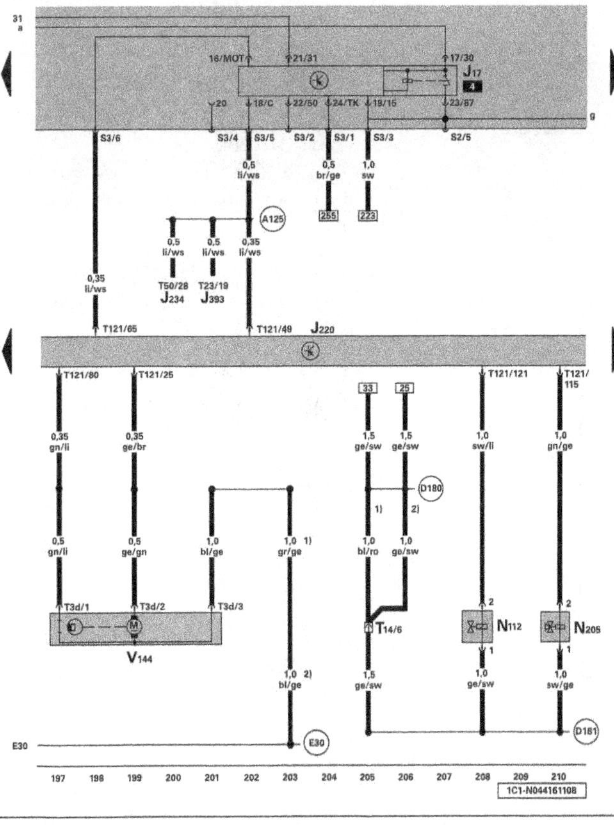

Motronic Engine Control Module (ECM), Fuel Pump (FP) Relay, Secondary Air Injection (AIR) Solenoid Valve, Camshaft Adjustment Valve 1, Leak Detection Pump (LDP)

- J17 - Fuel Pump (FP) Relay
- J220 - Motronic Engine Control Module (ECM)
- J234 - Airbag Control Module
- J393 - Comfort System Central Control Module
- N112 - Secondary Air Injection (AIR) Solenoid Valve
- N205 - Camshaft Adjustment Valve 1
- T3d - 3-Pin Connector
- T14 - 14-Pin Connector, near battery
- T23 - 23-Pin Connector, on Central control module for comfort system
- T50 - 50-Pin Connector
- T121 - 121-Pin Connector
- V144 - Leak Detection Pump (LDP)
- (A125) - Crash Signal Connection (in instrument panel wiring harness)
- (D180) - Connection (87a) (in engine wiring harness)
- (D181) - Connection 2 (87a) (in engine wiring harness)
- (E30) - Connection (87a) (in engine wiring harness)

1) - Through June 2007
2) - From July 2007

ws = white
sw = black
ro = red
br = brown
gn = green
bl = blue
gr = grey
li = violet
ge = yellow
or = orange

2.5L engine management (BPR) from July 2005

New Beetle — Wiring diagram — No. 44/17

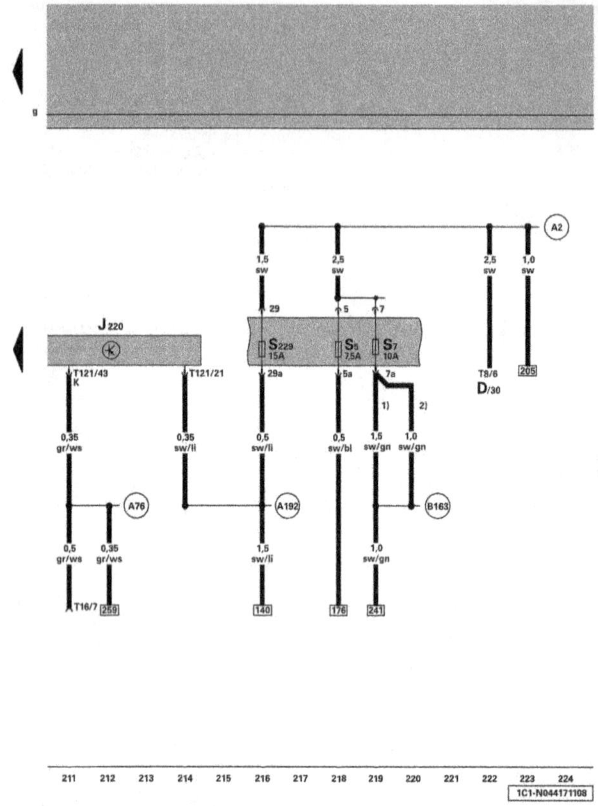

Motronic Engine Control Module (ECM)

- D - Ignition/Starter Switch
- J220 - Motronic Engine Control Module (ECM)
- S5 - Fuse (on fuse/relay panel)
- S7 - Fuse (on fuse/relay panel)
- S229 - Fuse (on fuse panel)
- T8 - 8-Pin Connector
- T16 - 16-Pin Connector, Data Link Connector (DLC), bellow steering column, left
- T121 - 121-Pin Connector

- (A2) - Plus Connection (15) (in instrument panel wiring harness)
- (A76) - K-Diagnostic Connection (in instrument panel wiring harness)
- (A192) - Plus Connection 3 (15a) (in instrument panel wiring harness)
- (B163) - Plus Connection 1 (15) (in interior wiring harness)

- 1) - Through June 2007
- 2) - From July 2007

ws = white
sw = black
ro = red
br = brown
gn = green
bl = blue
gr = grey
li = violet
ge = yellow
or = orange

New Beetle — Wiring diagram — No. 44/18

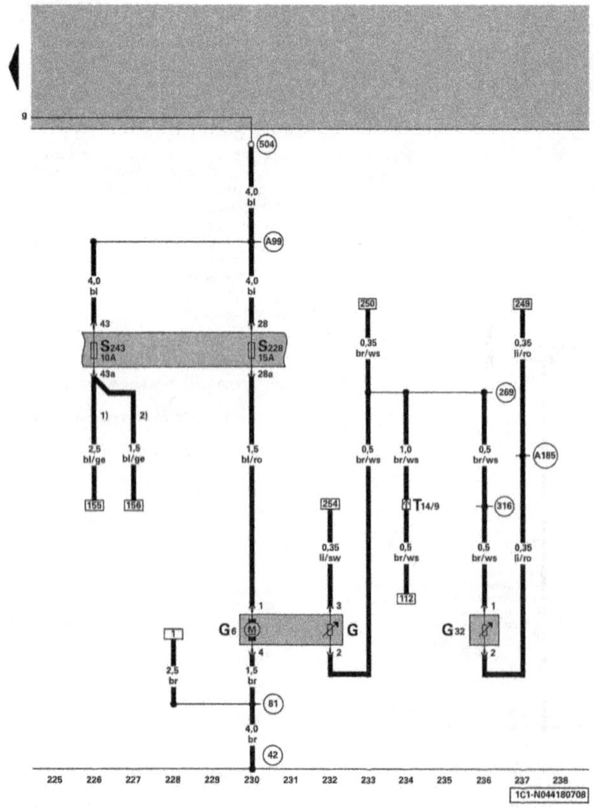

Fuel Level Sensor, Transfer Fuel Pump (FP), Engine Coolant Level (ECL) Sensor

- G - Fuel Level Sensor
- G6 - Transfer Fuel Pump (FP)
- G32 - Engine Coolant Level (ECL) Sensor
- S228 - Fuse (on fuse panel)
- S243 - Fuse (on fuse panel)
- T14 - 14-Pin Connector

- (42) - Ground Connection (beside steering column)
- (81) - Ground Connection 1 (in instrument panel wiring harness)
- (269) - Sensor Ground Connection 1 (in instrument panel wiring harness)
- (316) - Sensor Ground Connection 2 (in engine wiring harness)
- (504) - Threaded Connection 1 - terminal 87 (on relay panel)
- (A99) - Connection 1 (87) (in instrument panel wiring harness)
- (A185) - Coolant Warning Connection (in instrument panel wiring harness)

- 1) - Through June 2007
- 2) - From July 2007

ws = white
sw = black
ro = red
br = brown
gn = green
bl = blue
gr = grey
li = violet
ge = yellow
or = orange

New Beetle

Wiring diagram

No. 44/19

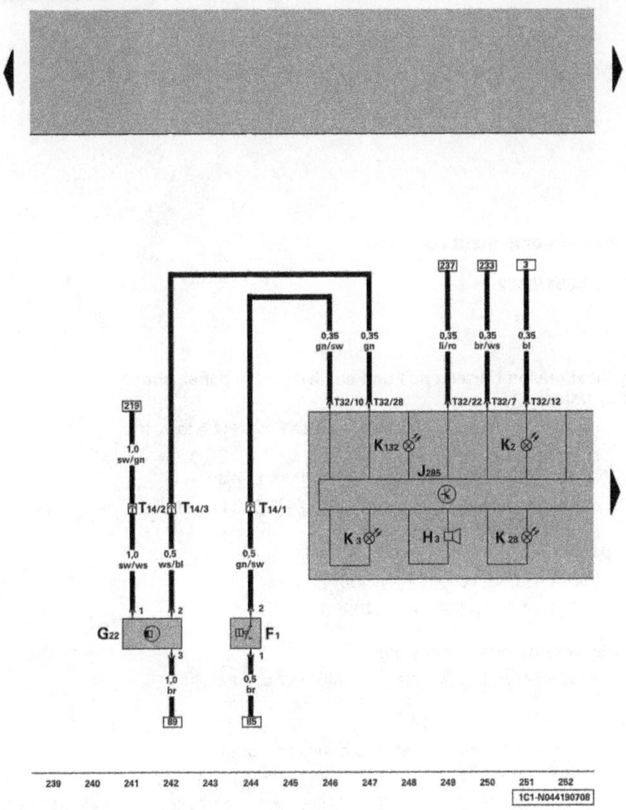

Oil Pressure Switch, Speedometer Vehicle Speed Sensor (VSS), Warning Buzzer and Tone, Instrument Cluster Control Module, Generator (GEN) Warning Lamp, Oil Pressure Warning Lamp, Engine Coolant Level/Tempe rature (ECL/ECT) Warning Lamp, Electronic Power Control (EPC) Warning Lamp

- F1 - Oil Pressure Switch
- G22 - Speedometer Vehicle Speed Sensor (VSS)
- H3 - Warning Buzzer and Tone
- J285 - Instrument Cluster Control Module
- K2 - Generator (GEN) Warning Lamp
- K3 - Oil Pressure Warning Lamp
- K28 - Engine Coolant Level/Temperature (ECL/ECT) Warning Lamp
- K132 - Electronic Power Control (EPC) Warning Lamp
- T14 - 14-Pin Connector
- T32 - 32-Pin Connector, blue

ws = white
sw = black
ro = red
br = brown
gn = green
bl = blue
gr = grey
li = violet
ge = yellow
or = orange

New Beetle

Wiring diagram

No. 44/20

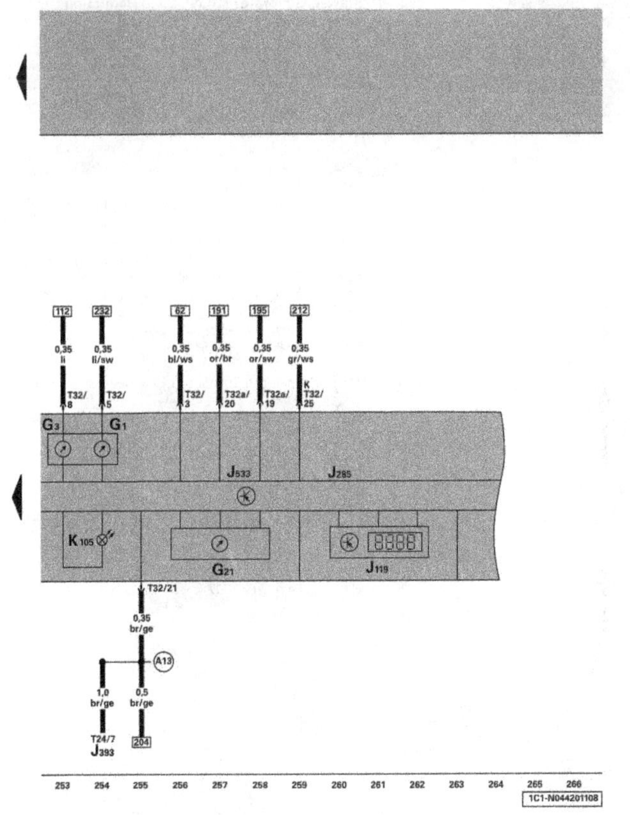

Fuel Gauge, Engine Coolant Temperature (ECT) Gauge, Speedometer, Multi-function Indicator (MFA), Instrument Cluster Control Module, Data Bus On Board Diagnostic Inter face

- G1 - Fuel Gauge
- G3 - Engine Coolant Temperature (ECT) Gauge
- G21 - Speedometer
- J119 - Multi-function Indicator (MFA)
- J285 - Instrument Cluster Control Module
- J393 - Comfort System Central Control Module
- J533 - Data Bus On Board Diagnostic Interface
- K105 - Low Fuel Level Warning Lamp
- T24 - 24-Pin Connector
- T32 - 32-Pin Connector, blue
- T32a - 32-Pin Connector, green
- (A13) - Door Contact Switch Connection (in instrument panel wiring harness)

ws = white
sw = black
ro = red
br = brown
gn = green
bl = blue
gr = grey
li = violet
ge = yellow
or = orange

New Beetle Wiring diagram **No. 43/1**

2.5L - Engine - Motronic Multiport Fuel Injection (MFI)/110 kw, code BPS (with cruise control),

from July 2005,

Additional Fuse, Relay, Control Module and Connector locations ⇒ Component Locations

Troubleshooting Procedures ⇒ Guided Fault Finding using VAS 5051/5052

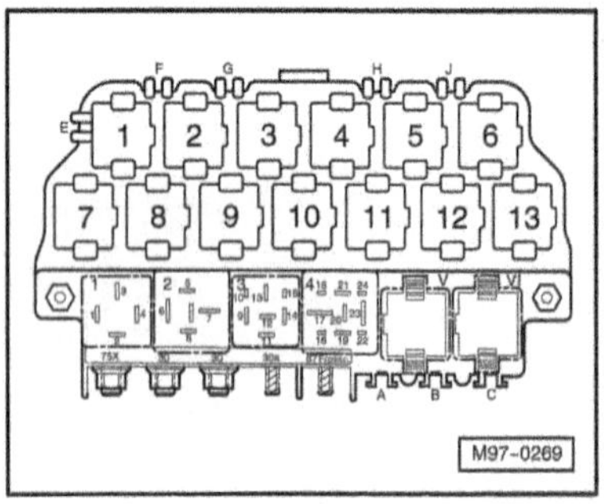

Relay locations on thirteen position auxiliary relay panel, above relay panel:

1 J434 - Starter Lock/Clutch Pedal Switch Relay (449) (only to BPS engines)

10 J299 - Secondary Air Injection (AIR) Pump Relay (100)

13 J271 - Motronic Engine Control Module (ECM) Power Supply Relay (428)

Relay panel:

2 J59 - Load Reduction Relay (100)

4 J17 - Fuel Pump (FP) Relay (409)

Fuse placements on the relay panel:

C S130 - Secondary Air Injection (AIR) Pump Fuse (10A)

Note:

Number in parentheses indicates production control number stamped on relay housing.

New Beetle Wiring diagram **No. 43/2**

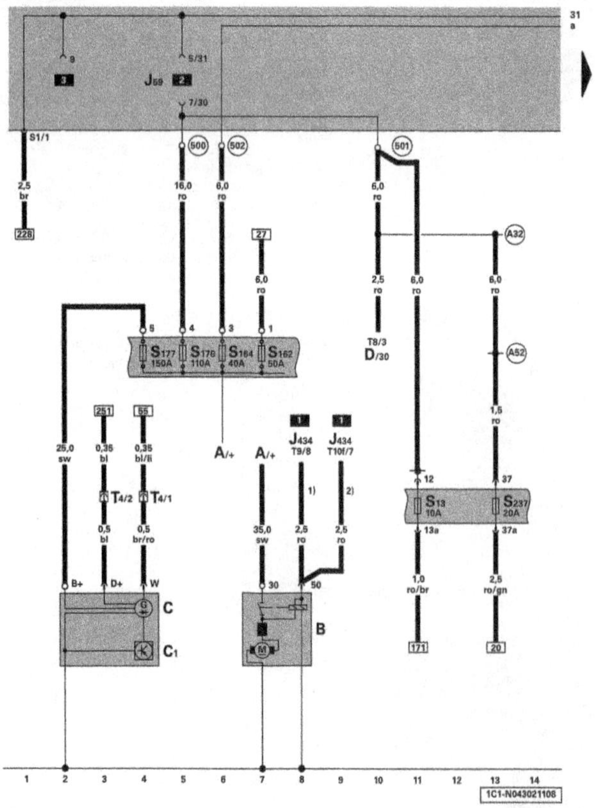

Starter, Generator (GEN), Load Reduction Relay, Fuses on battery fuse panel

- A - Battery
- B - Starter
- C - Generator (GEN)
- C1 - Voltage Regulator (VR)
- D - Ignition/Starter Switch
- J59 - Load Reduction Relay
- J434 - Starter Lock/Clutch Pedal Switch Relay
- S13 - Fuse (on fuse/relay panel)
- S162 - Fuse 1 - terminal 30 (on battery fuse panel)
- S164 - Fuse 3 - terminal 30 (on battery fuse panel)
- S176 - Fuse 4 - terminal 30 (on battery fuse holder)
- S177 - Fuse 5 - terminal 30 (on battery fuse panel)
- S237 - Fuse (on fuse panel)
- T4 - 4-Pin Connector, in engine compartment left
- T8 - 8-Pin Connector, at -D-
- T9 - 9-Pin Connector, at -J434-
- T10f - 10-Pin Connector
- (500) - Threaded Connection 1 - terminal 30 (on relay panel)
- (501) - Threaded Connection 2 - terminal 30 (on relay panel)
- (502) - Threaded Connection 1 - terminal 30a (on relay panel)
- (A32) - Plus Connection (30) (in instrument panel wiring harness)
- (A52) - Plus Connection 2 (30) (in instrument panel wiring harness)
- 1) - Through June 2007
- 2) - From July 2007

ws = white
sw = black
ro = red
br = brown
gn = green
bl = blue
gr = grey
li = violet
ge = yellow
or = orange

New Beetle

Wiring diagram **No. 43/3**

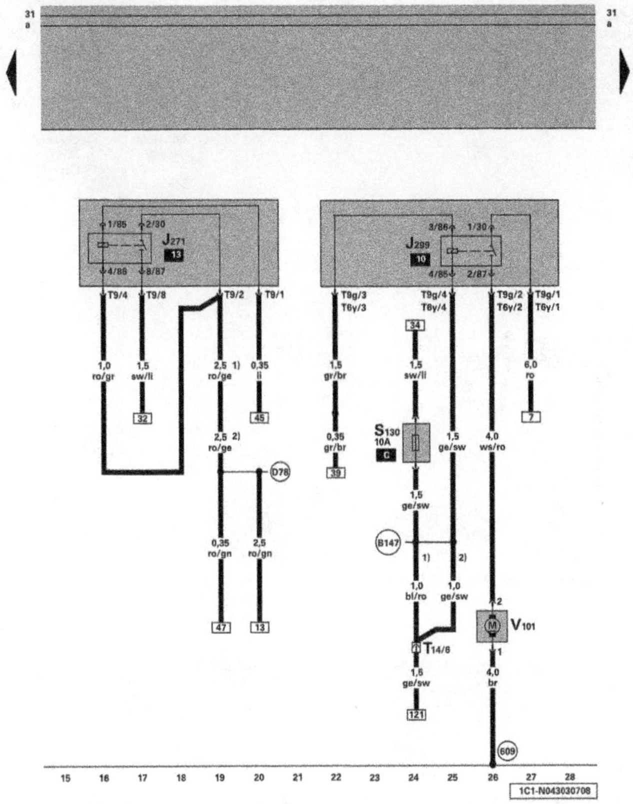

Motronic Engine Control Module (ECM) Power Supply Relay, Secondary Air Injection (AIR) Pump Fuse, Secondary Air Injection (AIR) Pump Motor

- J271 - Motronic Engine Control Module (ECM) Power Supply Relay
- J299 - Secondary Air Injection (AIR) Pump Relay
- T6y - 6-Pin Connector 2)
- T9 - 9-Pin Connector
- T9g - 9-Pin Connector 1)
- T14 - 14-Pin Connector
- S130 - Secondary Air Injection (AIR) Pump Fuse
- V101 - Secondary Air Injection (AIR) Pump Motor

- (609) - Ground Connection (in right plenum chamber)
- (D78) - Plus Connection 1 (30a) (in engine compartment wiring harness)
- (B147) - Plus Connection 2 (87) (in interior wiring harness)

- 1) - Through June 2007
- 2) - From July 2007

ws = white
sw = black
ro = red
br = brown
gn = green
bl = blue
gr = grey
li = violet
ge = yellow
or = orange

New Beetle

Wiring diagram **No. 43/4**

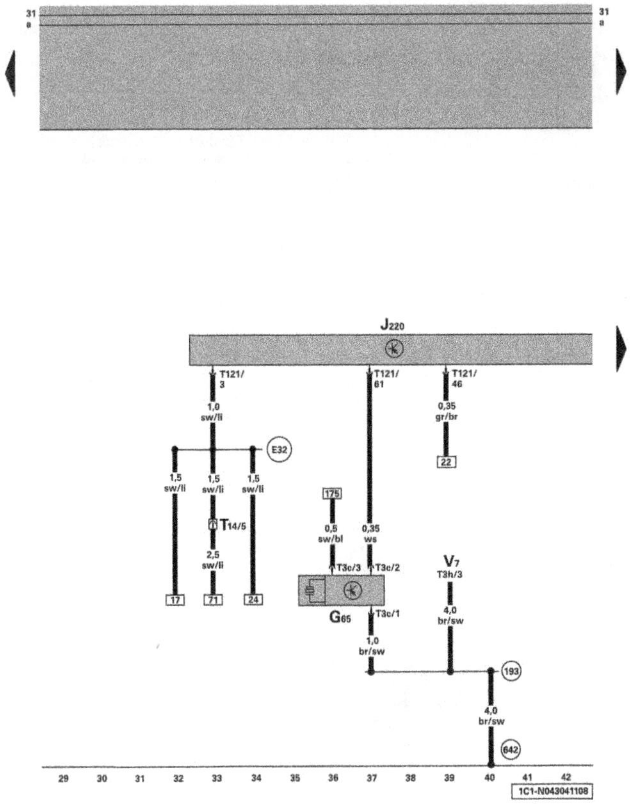

High Pressure Sensor, Motronic Engine Control Module (ECM)

- G65 - High Pressure Sensor
- J220 - Motronic Engine Control Module (ECM)
- T3c - 3-Pin connector
- T3h - 3-Pin Connector
- T14 - 14-Pin Connector
- T121 - 121-Pin Connector
- V7 - Kühlerlüfter

- (193) - Ground Connection 1 (in coolant fan wiring harness)
- (642) - Ground Connection (on electronic fan control)
- (E32) - Connection (87) (in Motronic wiring harness)

ws = white
sw = black
ro = red
br = brown
gn = green
bl = blue
gr = grey
li = violet
ge = yellow
or = orange

2.5L engine management (BPS) from July 2005

New Beetle — Wiring diagram — No. 43/5

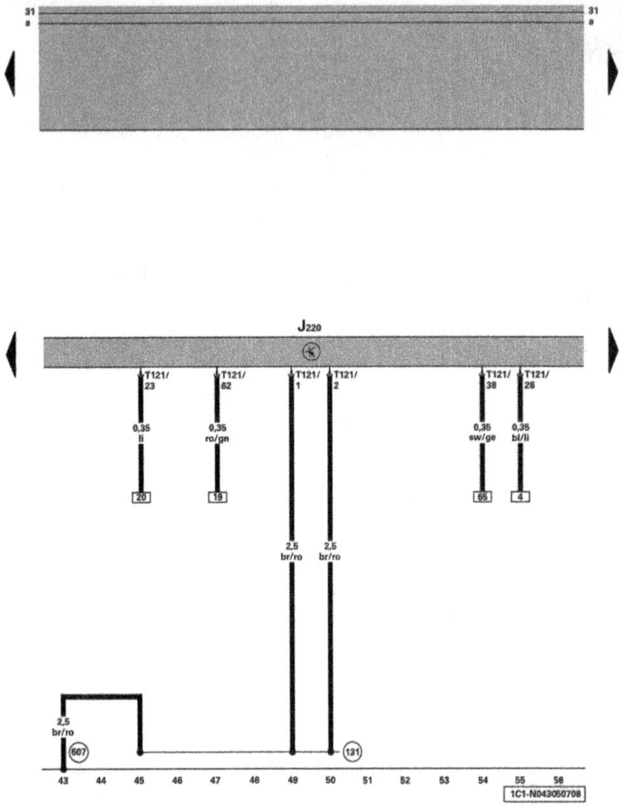

Motronic Engine Control Module (ECM)

J220 - Motronic Engine Control Module (ECM)
T121 - 121-Pin Connector

(131) - Ground Connection 2 (in engine compartment wiring harness)
(607) - Ground Connection (in left plenum chamber)

ws = white
sw = black
ro = red
br = brown
gn = green
bl = blue
gr = grey
li = violet
ge = yellow
or = orange

New Beetle — Wiring diagram — No. 43/6

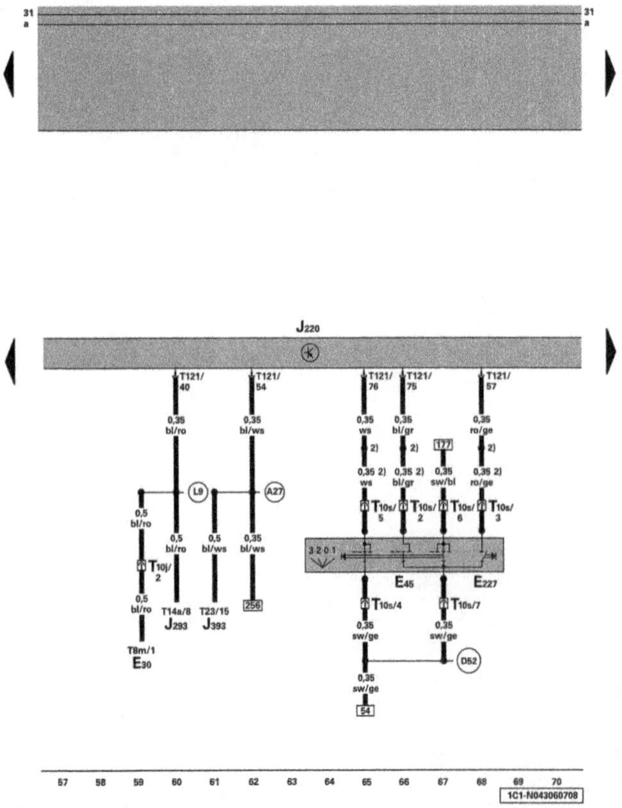

Cruise Control Switch, Cruise Control -SET- Button, Motronic Engine Control Module (ECM)

E30 - A/C Switch
E45 - Cruise Control Switch
E227 - Cruise Control -SET- Button
J220 - Motronic Engine Control Module (ECM)
J293 - Coolant Fan Control (FC) Control Module
J393 - Comfort System Central Control Module
T8m - 8-Pin Connector
T10j - 10-Pin Connector
T10s - 10-Pin Connector
T14a - 14-Pin Connector
T23 - 23-Pin Connector
T121 - 121-Pin Connector

(A27) - Vehicle Speed Signal (VSS) Connection (in instrument panel wiring harness)
(D52) - Plus Connection (15a) (in engine compartment wiring harness)
(L9) - Connection 1 (in A/C wiring harness)

2) - From July 2007

ws = white
sw = black
ro = red
br = brown
gn = green
bl = blue
gr = grey
li = violet
ge = yellow
or = orange

New Beetle

Wiring diagram

No. 43/7

Motronic Engine Control Module (ECM), Ignition Coils, Spark Plug Connectors, Spark Plugs

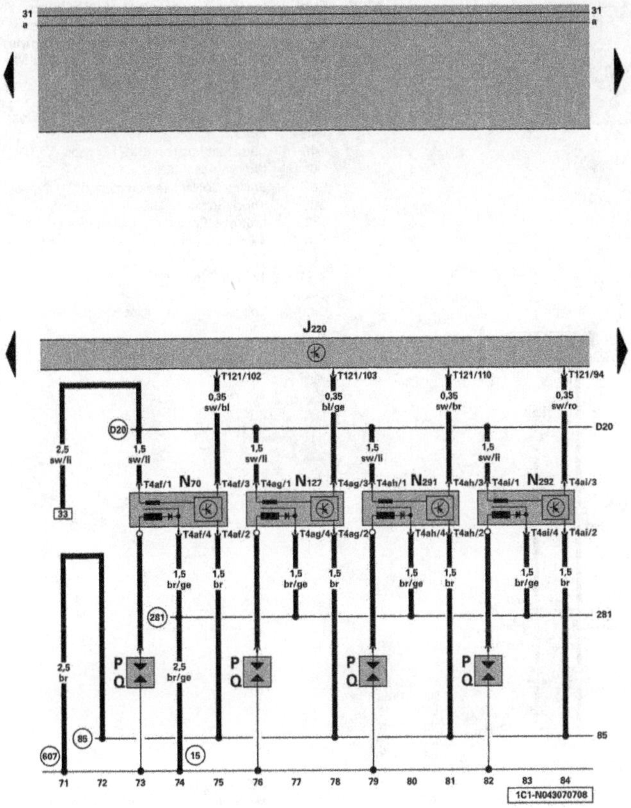

ws = white
sw = black
ro = red
br = brown
gn = green
bl = blue
gr = grey
li = violet
ge = yellow
or = orange

- J220 - Motronic Engine Control Module (ECM)
- N70 - Ignition Coil 1 with Power Output Stage
- N127 - Ignition Coil 2 with Power Output Stage
- N291 - Ignition Coil 3 with Power Output Stage
- N292 - Ignition Coil 4 with Power Output Stage
- P - Spark Plug Connectors
- Q - Spark Plugs
- T4af - 4-Pin Connector
- T4ag - 4-Pin Connector
- T4ah - 4-Pin Connector
- T4ai - 4-Pin Connector
- T121 - 121-Pin Connector

- (15) - Ground Connection (on cylinder head)
- (85) - Ground Connection 1 (in engine compartment wiring harness)
- (281) - Ground Connection 1 (in engine wiring harness)
- (607) - Ground Connection (in left plenum chamber)
- (D20) - Connection (87a) (in right front wiring harness)

New Beetle

Wiring diagram

No. 43/8

Ambient Temperature Switch, Motronic Engine Control Module (ECM), Throttle Valve Control Module, Ignition Coil 5 with Power Output Stage, Spark Plugs

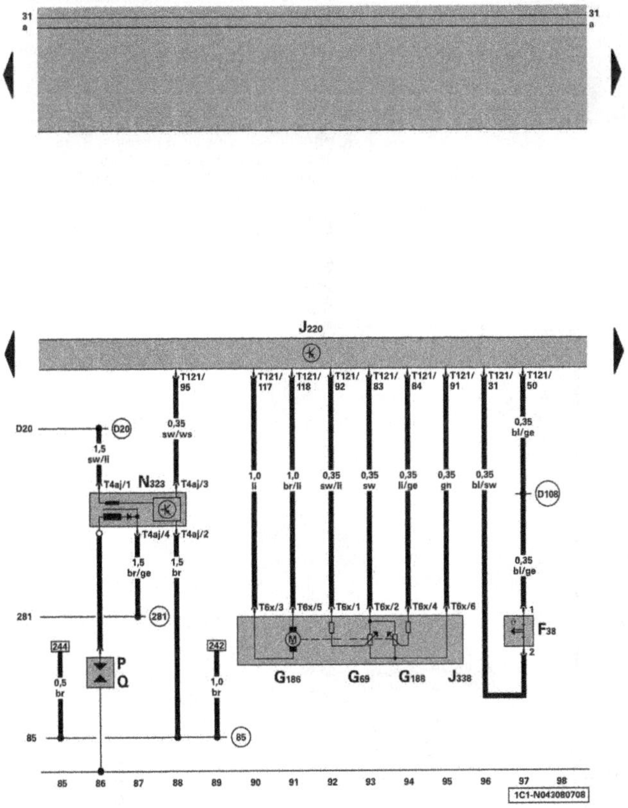

ws = white
sw = black
ro = red
br = brown
gn = green
bl = blue
gr = grey
li = violet
ge = yellow
or = orange

- F38 - Ambient Temperature Switch
- G69 - Throttle Position (TP) Sensor
- G186 - Throttle Drive (for Electronic Power Control (EPC))
- G188 - Throttle Drive Angle Sensor 2 (for Electronic Power Control (EPC))
- J220 - Motronic Engine Control Module (ECM)
- J338 - Throttle Valve Control Module
- N323 - Ignition Coil 5 with Power Output Stage
- P - Spark Plug Connectors
- Q - Spark Plugs
- T4aj - 4-Pin Connector
- T6x - 6-Pin Connector
- T121 - 121-Pin Connector

- (85) - Ground Connection 1 (in engine compartment wiring harness)
- (281) - Ground Connection 1 (in engine wiring harness)
- (D20) - Connection (87a) (in right front wiring harness)
- (D108) - Connection 6 (in engine compartment wiring harness)

New Beetle — Wiring diagram — No. 43/9

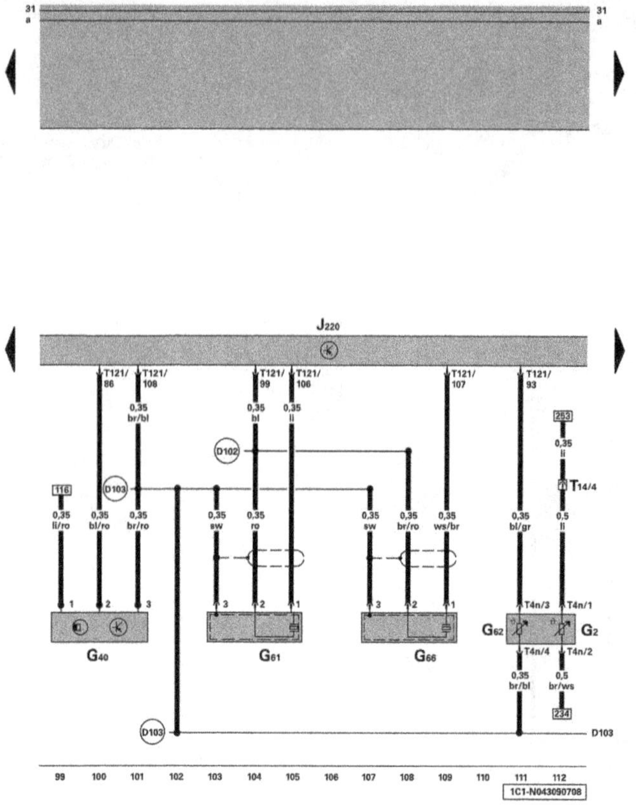

Engine Coolant Temperature (ECT) Gauge Sensor, Camshaft Position (CMP) Sensor, Knock Sensor (KS) 1, Engine Coolant Temperature (ECT) Sensor, Knock Sensor (KS) 2, Motronic Engine Control Module (ECM)

- G2 - Engine Coolant Temperature (ECT) Gauge Sensor
- G40 - Camshaft Position (CMP) Sensor
- G61 - Knock Sensor (KS) 1
- G62 - Engine Coolant Temperature (ECT) Sensor
- G66 - Knock Sensor (KS) 2
- J220 - Motronic Engine Control Module (ECM)
- T4n - 4-Pin Connector
- T14 - 14-Pin Connector
- T121 - 121-Pin Connector

- (D102) - Connection 2 (in engine compartment wiring harness)
- (D103) - Connection 3 (in engine compartment wiring harness)

ws = white
sw = black
ro = red
br = brown
gn = green
bl = blue
gr = grey
li = violet
ge = yellow
or = orange

New Beetle — Wiring diagram — No. 43/10

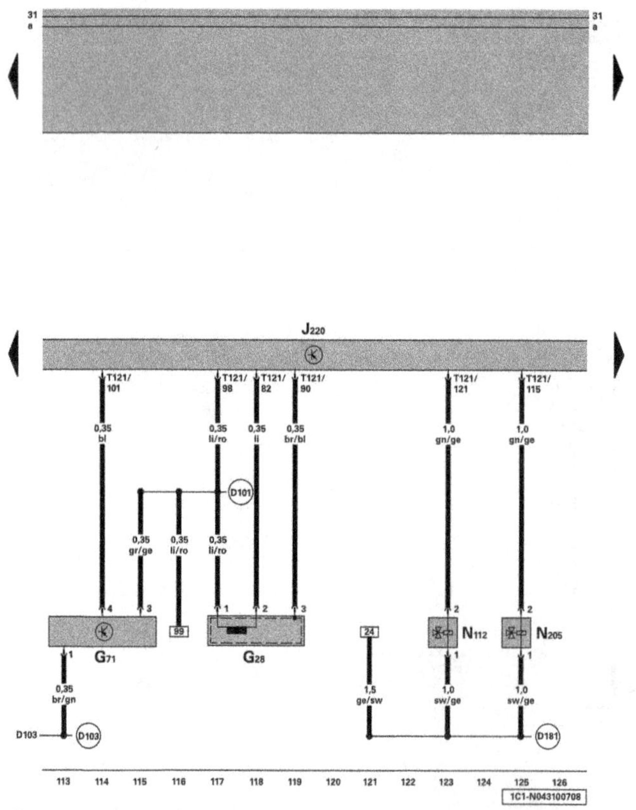

Engine Speed (RPM) Sensor, Manifold Absolute Pressure (MAP) Sensor, Secondary Air Injection (AIR) Solenoid Valve, Camshaft Adjustment Valve 1, Motronic Engine Control Module (ECM)

- G28 - Engine Speed (RPM) Sensor
- G71 - Manifold Absolute Pressure (MAP) Sensor
- N112 - Secondary Air Injection (AIR) Solenoid Valve
- N205 - Camshaft Adjustment Valve 1
- J220 - Motronic Engine Control Module (ECM)
- T121 - 121-Pin Connector

- (D101) - Connection 1 (in engine compartment wiring harness)
- (D103) - Connection 3 (in engine compartment wiring harness)
- (D181) - Connection 2 (87a) (in engine wiring harness)

ws = white
sw = black
ro = red
br = brown
gn = green
bl = blue
gr = grey
li = violet
ge = yellow
or = orange

New Beetle

Wiring diagram

No. 43/11

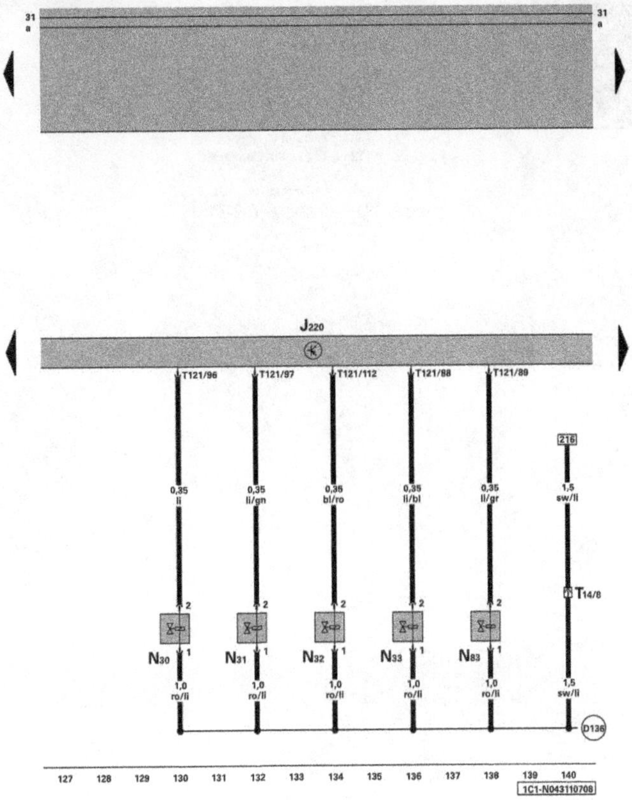

Motronic Engine Control Module, Fuel Injectors

J220 - Motronic Engine Control Module (ECM)
N30 - Cylinder 1 Fuel Injector
N31 - Cylinder 2 Fuel Injector
N32 - Cylinder 3 Fuel Injector
N33 - Cylinder 4 Fuel Injector
N83 - Cylinder 5 Fuel Injector
T14 - 14-Pin Connector
T121 - 121-Pin Connector

(D136) - Plus Connection 2 (15a (in engine compartment wiring harness)

ws = white
sw = black
ro = red
br = brown
gn = green
bl = blue
gr = grey
li = violet
ge = yellow
or = orange

New Beetle

Wiring diagram

No. 43/12

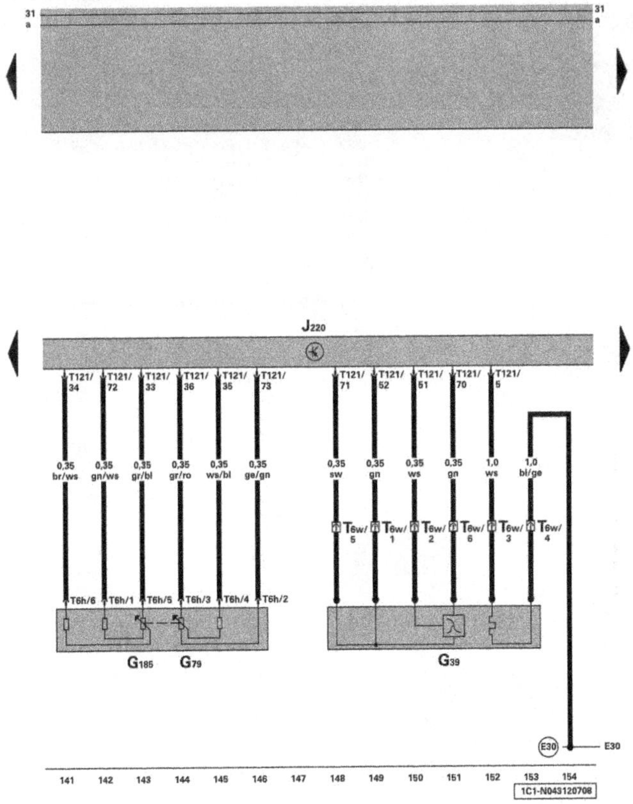

Heated Oxygen Sensor (HO2S), Throttle Position (TP) Sensor, Accelerator Pedal Position Sensor 2, Motronic Engine Control Module (ECM)

G39 - Heated Oxygen Sensor (HO2S)
G79 - Throttle Position (TP) Sensor
G185 - Accelerator Pedal Position Sensor 2
J220 - Motronic Engine Control Module (ECM)
T6h - 6-Pin Connector
T6w - 6-Pin Connector
T121 - 121-Pin Connector

(E30) - Connection (87a) (in engine wiring harness)

ws = white
sw = black
ro = red
br = brown
gn = green
bl = blue
gr = grey
li = violet
ge = yellow
or = orange

2.5L engine management (BPS) from July 2005

New Beetle

Wiring diagram **No. 43/13**

Oxygen Sensor (O2S) Behind Three Way Catalytic Converter (TWC), Motronic Engine Control Module (ECM)

G130 - Oxygen Sensor (O2S) Behind Three Way Catalytic Converter (TWC)
J220 - Motronic Engine Control Module (ECM)
T4ao - 4-Pin Connector
T121 - 121-Pin Connector

1) - Through June 2007
2) - From July 2007

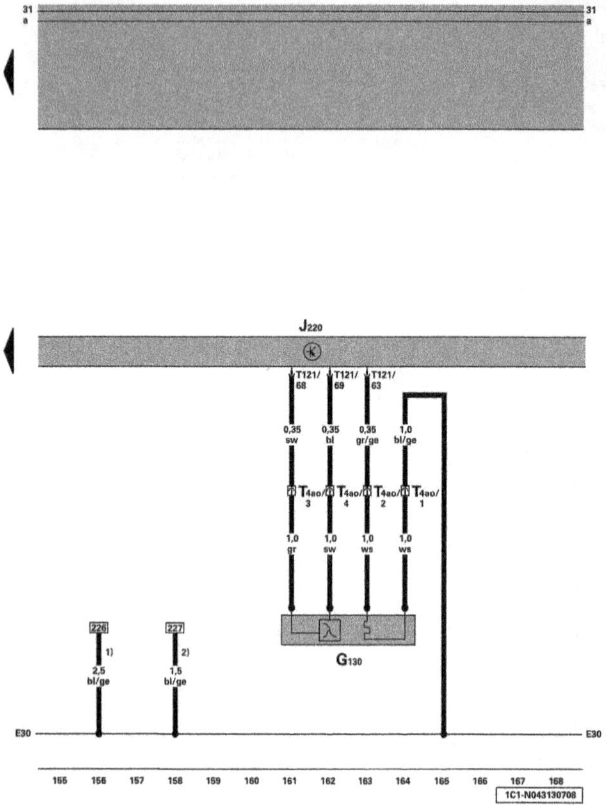

ws = white
sw = black
ro = red
br = brown
gn = green
bl = blue
gr = grey
li = violet
ge = yellow
or = orange

New Beetle

Wiring diagram **No. 43/14**

Brake Light Switch, Clutch Pedal Switch, Brake Pedal Switch, Motronic Engine Control Module (ECM)

F - Brake Light Switch
F36 - Clutch Pedal Switch
F47 - Brake Pedal Switch
J220 - Motronic Engine Control Module (ECM)
T4b - 4-Pin Connector
T4m - 4-Pin Connector
T4y - 4-Pin Connector
T121 - 121-Pin Connector
V35 - Right Coolant Fan

(A20) - Plus Connection (15a) (in instrument panel wiring harness)
(B131) - Connection (54) (in interior wiring harness)

*** - Coolant fan connection
2) - From July 2007

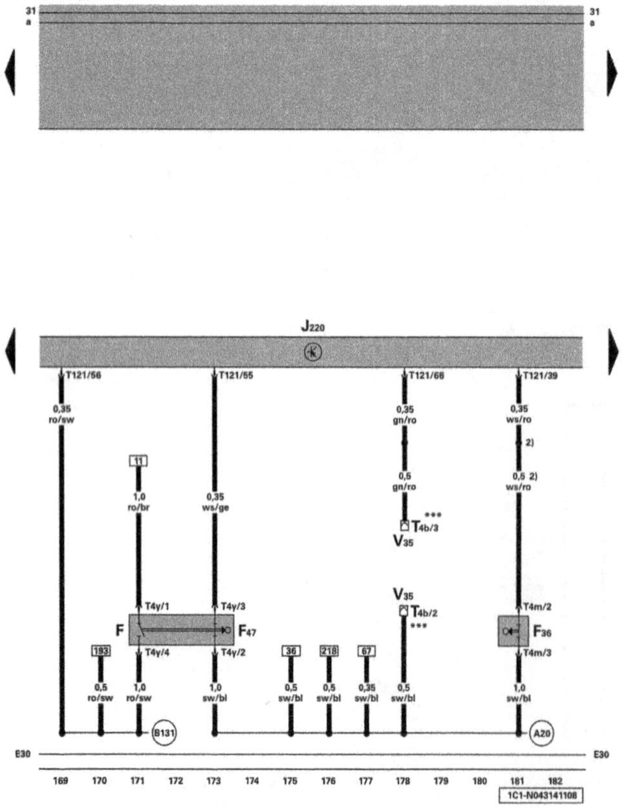

ws = white
sw = black
ro = red
br = brown
gn = green
bl = blue
gr = grey
li = violet
ge = yellow
or = orange

New Beetle — Wiring diagram — No. 43/15

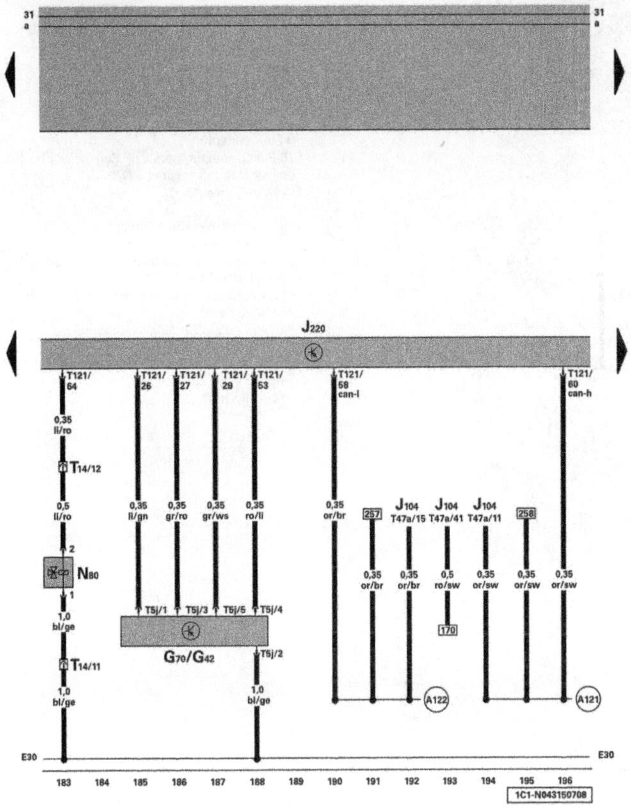

Intake Air Temperature (IAT) Sensor, Mass Air Flow (MAF) Sensor, Motronic Engine Control Module, Evaporative Emission (EVAP) Canister Purge Regulator Valve

- G42 - Intake Air Temperature (IAT) Sensor
- G70 - Mass Air Flow (MAF) Sensor
- J104 - ABS Control Module
- J220 - Motronic Engine Control Module (ECM)
- N80 - Evaporative Emission (EVAP) Canister Purge Regulator Valve
- T5j - 5-Pin Connector
- T14 - 14-Pin Connector
- T47a - 47-Pin Connector
- T121 - 121-Pin Connector
- (A121) - High-Bus Connection (in instrument panel wiring harness)
- (A122) - Low-Bus Connection (in instrument panel wiring harness)

ws = white
sw = black
ro = red
br = brown
gn = green
bl = blue
gr = grey
li = violet
ge = yellow
or = orange

New Beetle — Wiring diagram — No. 43/16

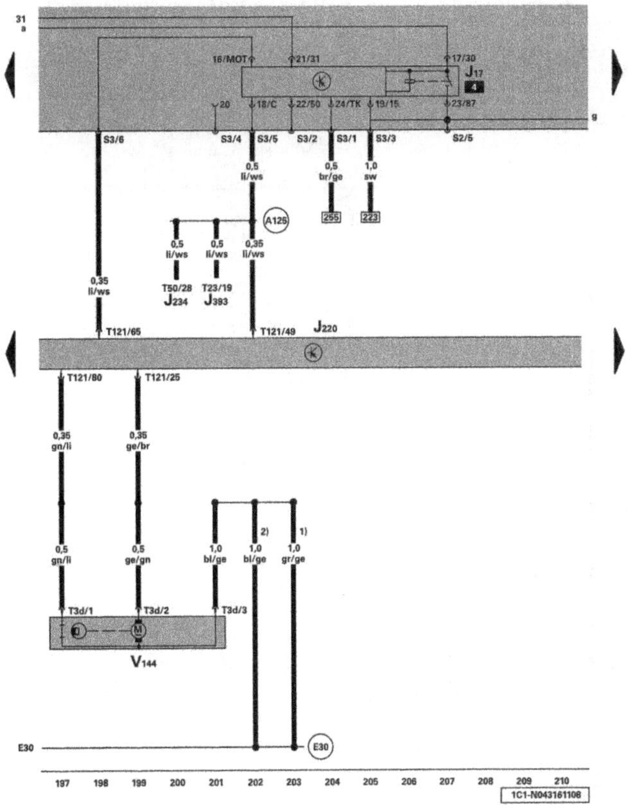

Motronic Engine Control Module (ECM), Fuel Pump (FP) Relay, Leak Detection Pump (LDP)

- J17 - Fuel Pump (FP) Relay
- J220 - Motronic Engine Control Module (ECM)
- J234 - Airbag Control Module
- J393 - Comfort System Central Control Module
- T3d - 3-Pin Connector
- T23 - 23-Pin Connector, on Central control module for comfort system
- T50 - 50-Pin Connector
- T121 - 121-Pin Connector
- V144 - Leak Detection Pump (LDP)
- (A125) - Crash Signal Connection (in instrument panel wiring harness)
- (E30) - Connection (87a) (in engine wiring harness)

1) - Through June 2007
2) - From July 2007

ws = white
sw = black
ro = red
br = brown
gn = green
bl = blue
gr = grey
li = violet
ge = yellow
or = orange

New Beetle — Wiring diagram — No. 43/17

Motronic Engine Control Module (ECM)

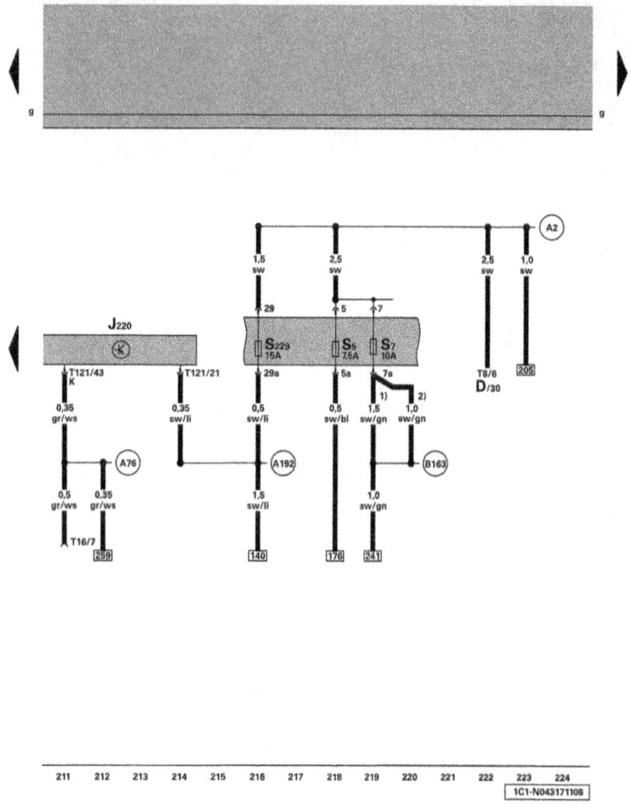

- D - Ignition/Starter Switch
- J220 - Motronic Engine Control Module (ECM)
- S5 - Fuse (on fuse/relay panel)
- S7 - Fuse (on fuse/relay panel)
- S229 - Fuse (on fuse panel)
- T8 - 8-Pin Connector
- T16 - 16-Pin Connector, Data Link Connector (DLC), bellow steering column, left
- T121 - 121-Pin Connector
- (A2) - Plus Connection (15) (in instrument panel wiring harness)
- (A76) - K-Diagnostic Connection (in instrument panel wiring harness)
- (A192) - Plus Connection 3 (15a) (in instrument panel wiring harness)
- (B163) - Plus Connection 1 (15) (in interior wiring harness)
- 1) - Through June 2007
- 2) - From July 2007

ws = white
sw = black
ro = red
br = brown
gn = green
bl = blue
gr = grey
li = violet
ge = yellow
or = orange

New Beetle — Wiring diagram — No. 43/18

Fuel Level Sensor, Transfer Fuel Pump (FP), Engine Coolant Level (ECL) Sensor

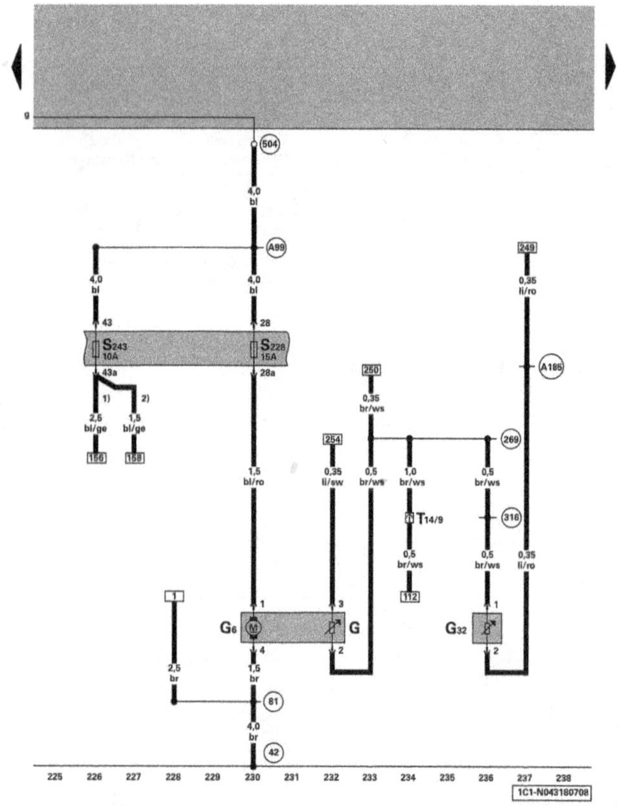

- G - Fuel Level Sensor
- G6 - Transfer Fuel Pump (FP)
- G32 - Engine Coolant Level (ECL) Sensor
- S228 - Fuse (on fuse panel)
- S243 - Fuse (on fuse panel)
- T14 - 14-Pin Connector
- (42) - Ground Connection (beside steering column)
- (81) - Ground Connection 1 (in instrument panel wiring harness)
- (269) - Sensor Ground Connection 1 (in instrument panel wiring harness)
- (316) - Sensor Ground Connection 2 (in engine wiring harness)
- (504) - Threaded Connection 1 - terminal 87 (on relay panel)
- (A99) - Connection 1 (87) (in instrument panel wiring harness)
- (A185) - Coolant Warning Connection (in instrument panel wiring harness)
- 1) - Through June 2007
- 2) - From July 2007

ws = white
sw = black
ro = red
br = brown
gn = green
bl = blue
gr = grey
li = violet
ge = yellow
or = orange

New Beetle

Wiring diagram

No. 43/19

Oil Pressure Switch, Speedometer Vehicle Speed Sensor (VSS), Warning Buzzer and Tone, Instrument Cluster Control Module, Generator (GEN) Warning Lamp, Oil Pressure Warning Lamp, Engine Coolant Level/Temperature (ECL /ECT) Warning Lamp, Electronic Power Control (EPC) Warning Lamp

F1 - Oil Pressure Switch
G22 - Speedometer Vehicle Speed Sensor (VSS)
H3 - Warning Buzzer and Tone
J285 - Instrument Cluster Control Module
K2 - Generator (GEN) Warning Lamp
K3 - Oil Pressure Warning Lamp
K28 - Engine Coolant Level/Temperature (ECL/ECT) Warning Lamp
K132 - Electronic Power Control (EPC) Warning Lamp
T14 - 14-Pin Connector
T32 - 32-Pin Connector

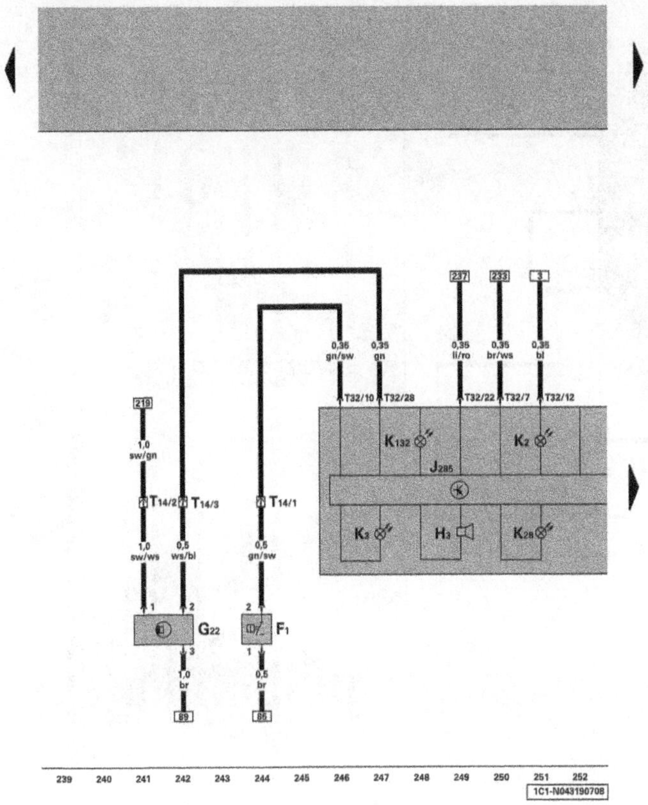

ws = white
sw = black
ro = red
br = brown
gn = green
bl = blue
gr = grey
li = violet
ge = yellow
or = orange

New Beetle

Wiring diagram

No. 43/20

Fuel Gauge, Engine Coolant Temperature (ECT) Gauge, Speedometer, Multi-function Indicator (MFA), Instrument Cluster Control Module, Data Bus On Board Diagnostic Interface

G1 - Fuel Gauge
G3 - Engine Coolant Temperature (ECT) Gauge
G21 - Speedometer
J119 - Multi-function Indicator (MFA)
J285 - Instrument Cluster Control Module
J393 - Comfort System Central Control Module
J533 - Data Bus On Board Diagnostic Interface
K105 - Low Fuel Level Warning Lamp
T24 - 24-Pin Connector
T32 - 32-Pin Connector
T32a - 32-Pin Connector

(A13) - Door Contact Switch Connection (in instrument panel wiring harness)

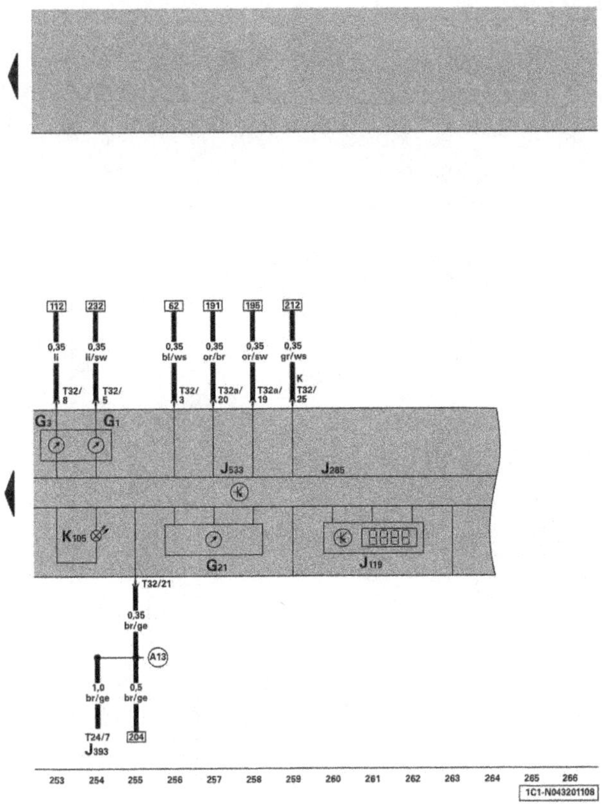

ws = white
sw = black
ro = red
br = brown
gn = green
bl = blue
gr = grey
li = violet
ge = yellow
or = orange

Anti-lock brake system (ABS) with electronic differential lock (EDL),

from January 1998

Deviate relay location and fuseplacements as well as the locations of multiple connectors see section "component locations".

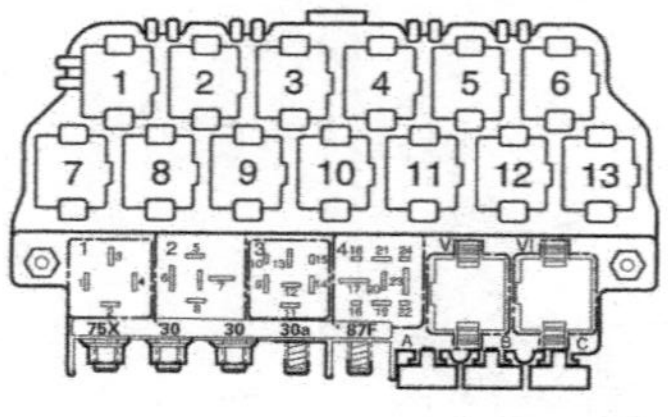

97-21956

Fuse colors
30 A - green
25 A - white
20 A - yellow
15 A - blue
10 A - red
7,5 A - brown
5 A - beige

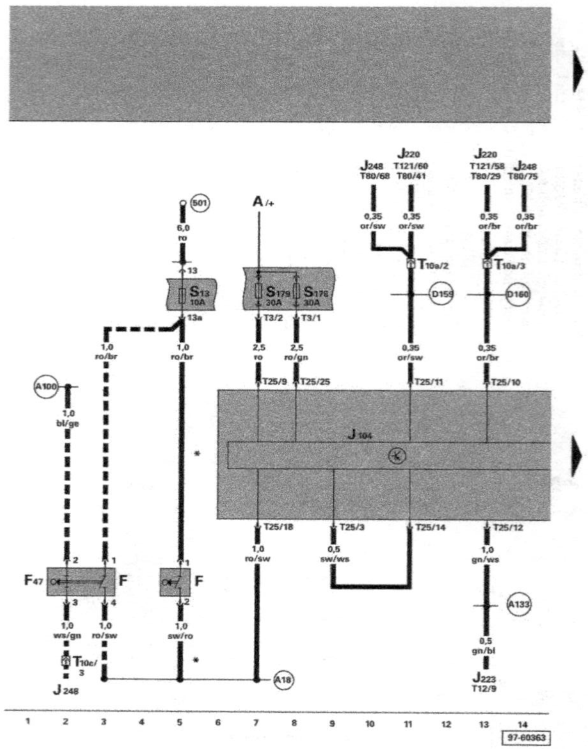

ws = white
sw = black
ro = red
br = brown
gn = green
bl = blue
gr = grey
li = violet
ge = yellow

A - Battery
F - Brake Light Switch
F47 - Brake Vacuum Vent Valve Switch for cruise control/diesel fuel injection (DFI)
J104 - ABS Control Module (w/EDL)
J220 - Motronic Engine Control Module (TCM)
J223 - Rear Spoiler Control Module
J248 - Diesel Direct Fuel Injection (DFI) Engine Control Module (ECM)
S13 - Fuse 13 in fuse holder
S178 - Fuse -6- (30), in fuse bracket/battery
S179 - Fuse -7- (30), in fuse bracket/battery
T3 - 3-Pin Connector,
T10a - 10-Pin Connector, orange, behind instrument panel,left
T10c - 10-Pin Connector, blue, behind instrument panel, left
T12 - 12-pin Connector
T25 - 25-Pin Connector
T80 - 80-Pin Connector, on Control Module
T121 - 121-Pin Connector, on Control Module
(A18) - wire connection (54) , in instrument panel wiring harness
(A100) - Connector -2- (87) in instrument panel wiring harness
(A133) - Connection (ABS-signal left) in instrument panel wiring harness
(501) - Threaded connection -2- (30) on the relay plate
(D159) - CAN - bus High, in engine wiring harness
(D160) - CAN - bus Low, in engine wiring harness
- - - Diesel only
* - Gasoline only

ABS control module, brake light switch light switch

New Beetle — Wiring diagram — **No. 5/3**

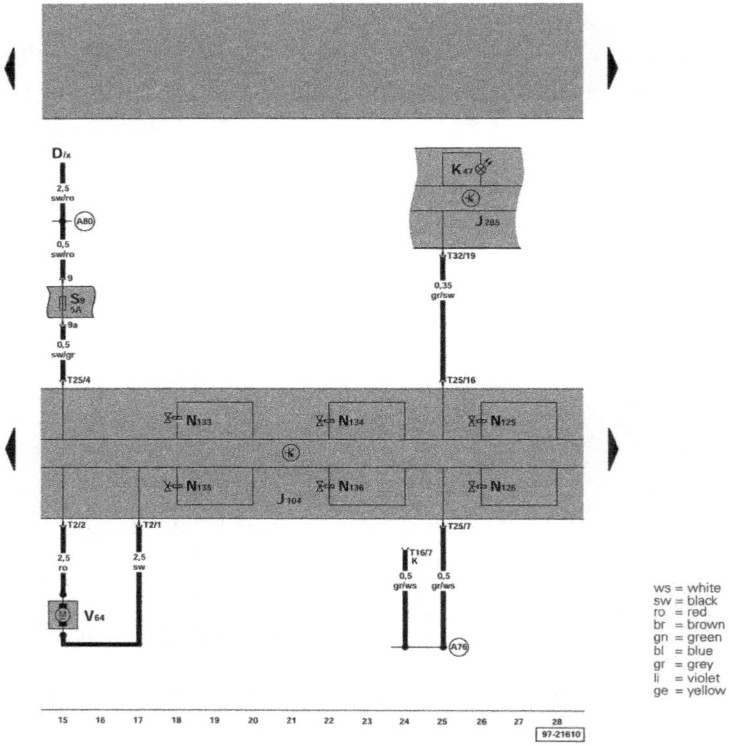

ws = white
sw = black
ro = red
br = brown
gn = green
bl = blue
gr = grey
li = violet
ge = yellow

- D - Ignition/Starter Switch
- J104 - ABS Control Module (w/EDL)
- J285 - Control module with indicator unit in instrument panel insert
- K47 - ABS Warning Light
- N125 - Differential Lock Valve 1
- N126 - Differential Lock Valve 2
- N133 - Right Rear ABS Inlet Valve
- N134 - Left Rear ABS Inlet Valve
- N135 - Right Rear ABS Outlet Valve
- N136 - Left Rear ABS Outlet Valve
- S9 - Fuse 9 in fuse holder
- T2 - Double Connector
- T16 - 16-Pin Connector, Data Link Connector (DLC) below steering column
- T25 - 25-Pin Connector
- T32 - 32-Pin Connector, blue, on instrument cluster
- V64 - ABS Hydraulic Pump
- (A76) - Connector (K-diagnosis wire), in instrument panel wiring harness
- (A80) - Connector -1- (X), in instrument panel wiring harness

No. 5/4 — Wiring diagram — **New Beetle**

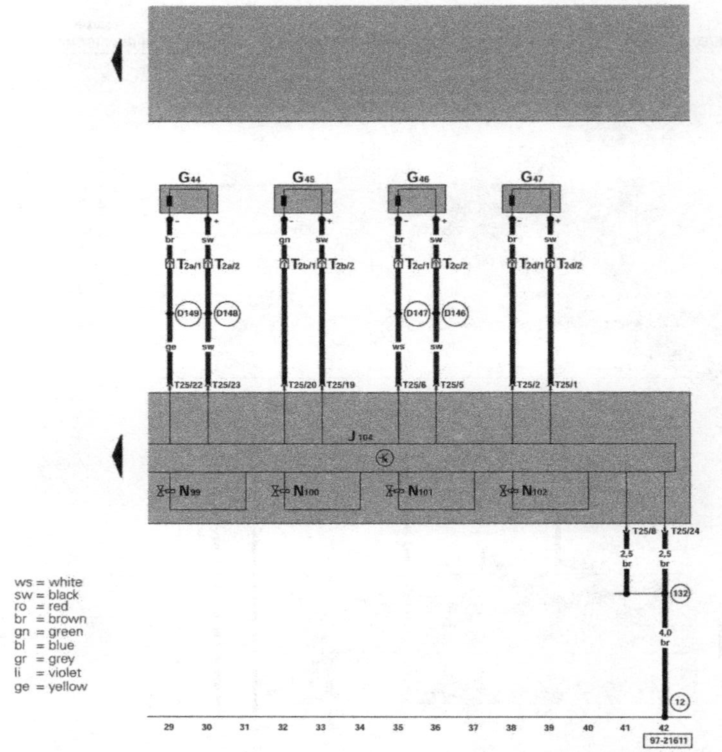

ws = white
sw = black
ro = red
br = brown
gn = green
bl = blue
gr = grey
li = violet
ge = yellow

- G44 - Right Rear ABS Wheel Speed Sensor
- G45 - Right Front ABS Wheel Speed Sensor
- G46 - Left Rear ABS Wheel Speed Sensor
- G47 - Left Front ABS Wheel Speed Sensor
- J104 - ABS Control Module (w/EDL)
- N99 - Right Front ABS Inlet Valve
- N100 - Right Front ABS Outlet Valve
- N101 - Left Front ABS Inlet Valve
- N101 - Right Front ABS Inlet Valve
- T2a - Double Connector, on Rear Right ABS Wheel Sensor
- T2b - Double Connector, on Front Right ABS Wheel Sensor
- T2c - Double Connector, on Rear Left ABS Wheel Sensor
- T2d - Double Connector, on Front Left ABS Wheel Sensor
- T25 - 25-Pin Connector, at ABS Control Module
- (12) - Ground connection, in engine compartment, left
- (132) - Ground connection -3-, in engine compartment wiring harness
- (D146) - Connector(speed sensor rear, left +), in wiring harness engine compartment
- (D147) - Connector(speed sensor rear, left -), in wiring harness engine compartment
- (D148) - Connector(speed sensor rear, right +), in wiring harness engine compartment
- (D149) - Connector(speed sensor rear, right -), in wiring harness engine compartment

BentleyPublishers.com

Anti-lock brake system (ABS), Anti-lock brake system (ABS) with electronic differential lock (EDL) and anti-slip regulation (ASR)

from June 1999

Deviate relay location and fuseplacements as well as the locations of multiple connectors see section "component locations".

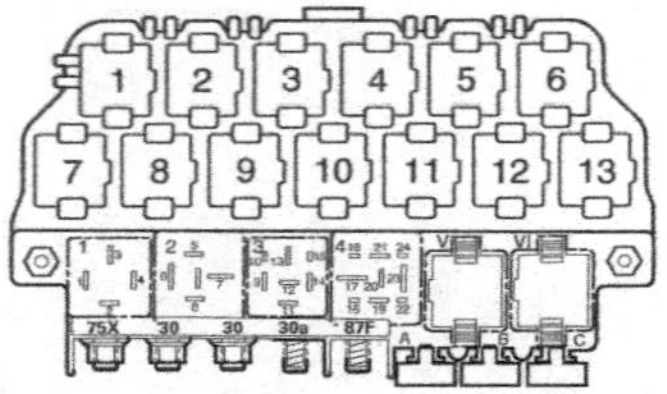

97-21956

Fuse colors
30 A - green
25 A - white
20 A - yellow
15 A - blue
10 A - red
7,5 A - brown
5 A - beige

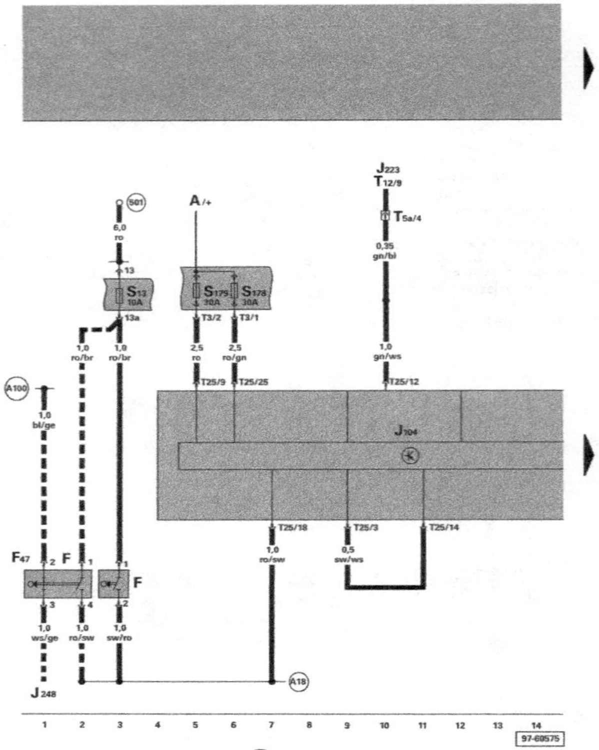

ws = white
sw = black
ro = red
br = brown
gn = green
bl = blue
gr = grey
li = violet
ge = yellow

A - Battery
F - Brake Light Switch
F47 - Brake Vacuum Vent Valve Switch for cruise control/diesel fuel injection (DFI)
J104 - ABS Control Module (w/EDL)
J223 - Rear Spoiler Control Module
J248 - Diesel Direct Fuel Injection (DFI) Engine Control Module (ECM)
S13 - Fuse 13 in fuse holder
S178 - Fuse -6- (30), in fuse bracket/battery
S179 - Fuse -7- (30), in fuse bracket/battery
T3 - 3-Pin Connector,
T5a - 5-Pin Connector
T12 - 12-Pin Connector
T25 - 25-Pin Connector

(A18) - wire connection (54) , in instrument panel wiring harness

(A100) - Connector -2- (87) in instrument panel wiring harness
(501) - Threaded connection -2- (30) on the relay plate
- - - Diesel only / cruise control

ABS control module, brake light switch, brake vacuum vent valve switch for cruise control

New Beetle Wiring diagram **No.31/3**

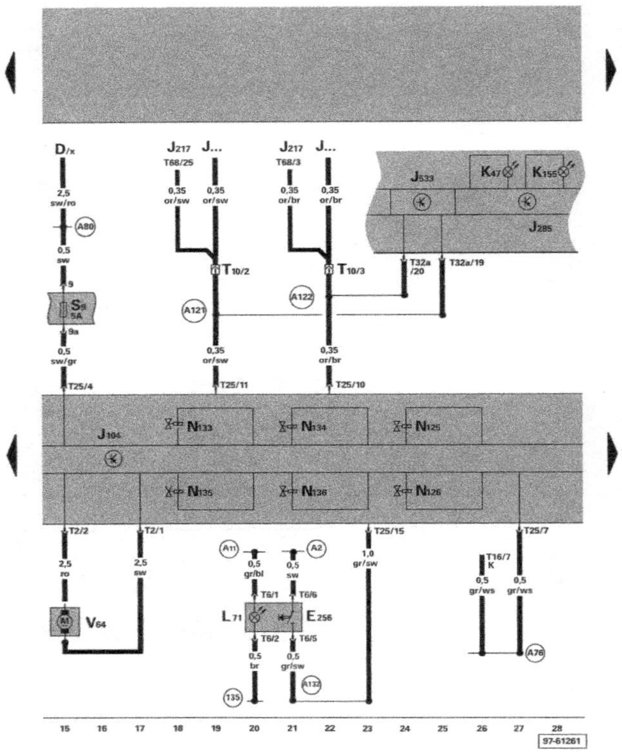

ws = white
sw = black
ro = red
br = brown
gn = green
bl = blue
gr = grey
li = violet
ge = yellow

- D - Ignition/Starter Switch
- E256 - ASR/ESP Button
- J... - Engine Control Module (ECM)
- J104 - ABS Control Module (w/EDL)
- J217 - Transmission Control Module (TCM),
- J285 - Control module with indicator unit in instrument panel insert
- J533 - Data Bus On Board Diagnostic Interface
- K47 - ABS Warning Light
- K155 - ASR Control Lamp
- L71 - Anti-slip Control Switch Illumination
- N125 - Differential Lock Valve 1
- N126 - Differential Lock Valve 2
- N133 - Right Rear ABS Inlet Valve
- N134 - Left Rear ABS Inlet Valve
- N135 - Right Rear ABS Outlet Valve
- N136 - Left Rear ABS Outlet Valve
- S9 - Fuse 9 in fuse holder
- T2 - 2-Pin Connector
- T10 - 10-Pin Connector, white, behind instrument panel,left
- T16 - 16-Pin Connector, Data Link Connector (DLC)
- T25 - 25-Pin Connector
- T32a - 32-Pin Connector, green, on instrument cluster
- V64 - ABS Hydraulic Pump
- T68 - 68-Pin Connector
- (A2) - Plus connection (15) in instrument panel wiring harness
- (A11) - Plus connection (58b) in instrument panel wiring harness
- (A76) - Connector (K-diagnosis wire), in instrument panel wiring
- (A80) - Connector -1- (X), in instrument panel wiring harness
- (135) - Ground connection -2-, in instrument panel wiring harness
- (A121) - Connection (high bus) in instrument panel wiring harness
- (A122) - Connection (low bus) in instrument panel wiring harness
- (A132) - ASR/ESP Connection (in instrument panel wiring harness

No. 31/4 Wiring diagram **New Beetle**

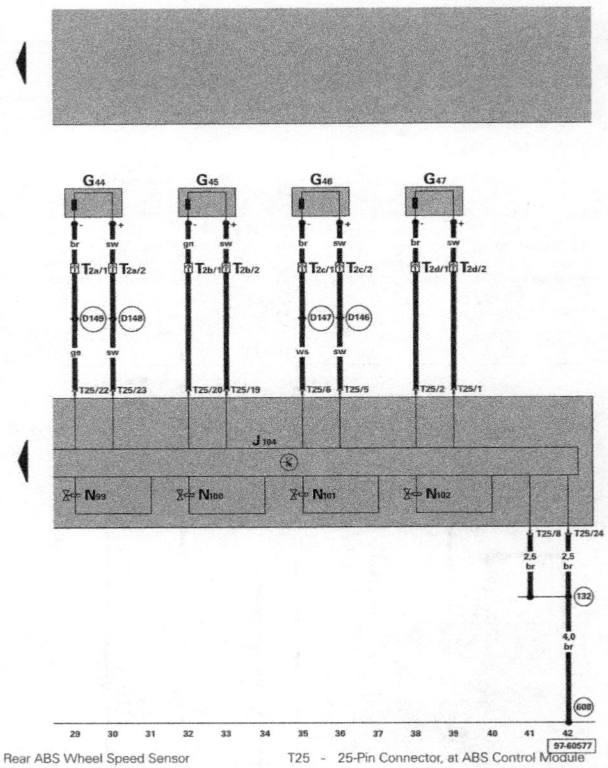

ws = white
sw = black
ro = red
br = brown
gn = green
bl = blue
gr = grey
li = violet
ge = yellow

- G44 - Right Rear ABS Wheel Speed Sensor
- G45 - Right Front ABS Wheel Speed Sensor
- G46 - Left Rear ABS Wheel Speed Sensor
- G47 - Left Front ABS Wheel Speed Sensor
- J104 - ABS Control Module (w/EDL)
- N99 - Right Front ABS Inlet Valve
- N100 - Right Front ABS Outlet Valve
- N101 - Left Front ABS Inlet Valve
- N102 - Right Front ABS Inlet Valve
- T2a - Double Connector, on Rear Right ABS Wheel Sensor
- T2b - Double Connector, on Front Right ABS Wheel Sensor
- T2c - Double Connector, on Rear Left ABS Wheel Sensor
- T2d - Double Connector, on Front Left ABS Wheel Sensor
- T25 - 25-Pin Connector, at ABS Control Module
- (132) - Ground connection -3-, in engine compartment wiring harness
- (608) - Ground Connection (in center plenum chamber)
- (D146) - Connector(speed sensor rear, left +), in wiring harness engine compartment
- (D147) - Connector(speed sensor rear, left -), in wiring harness engine compartment
- (D148) - Connector(speed sensor rear, right +), in wiring harness engine compartment
- (D149) - Connector(speed sensor rear, right -), in wiring harness engine compartment

Anti-lock brake system (ABS), Anti-lock brake system (ABS) with electronic differential lock (EDL) and anti-slip control (ASC),

from October 2000

Deviate relay location and fuseplacements as well as the locations of multiple connectors see section "component locations".

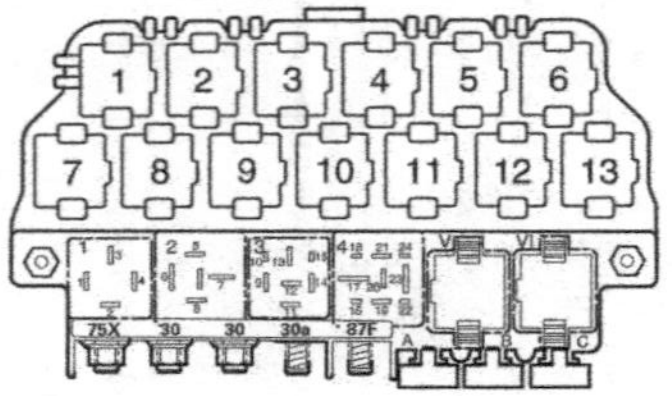

97-21956

Fuse colors
30 A - green
25 A - white
20 A - yellow
15 A - blue
10 A - red
7.5 A - brown
5 A - beige

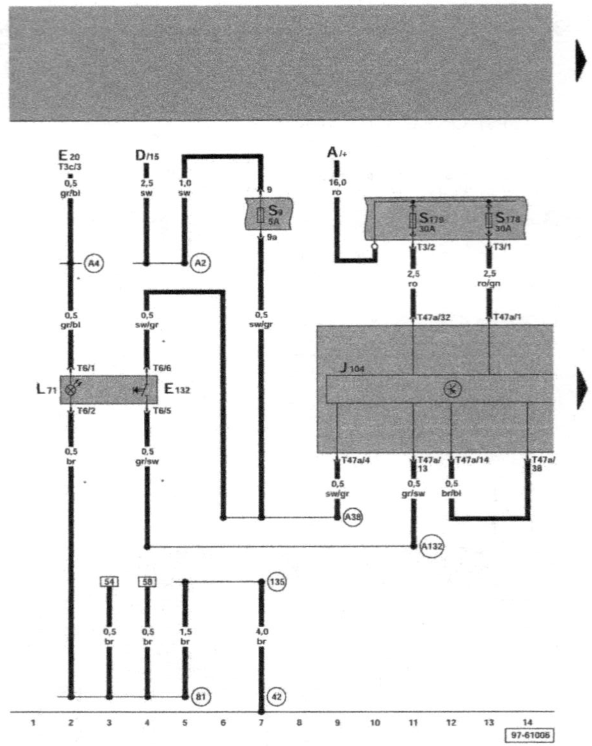

ws = white
sw = black
ro = red
br = brown
gn = green
bl = blue
gr = grey
li = lilac
ge = yellow

- A - Battery
- D - Ignition/Starter Switch
- E20 - Instrument Panel Light Dimmer Switch
- E132 - Anti-Slip Control Switch*
- J104 - ABS Control Module (w/EDL), in engine compartment, left
- L71 - Anti-Slip Control Switch Illumination*
- S9 - Fuse 9 in fuse holder
- S178 - Fuse -6- (30) in fuse bracket/battery
- S179 - Fuse -7- (30) in fuse bracket/battery
- T3 - 3-Pin Connector
- T3c - 3-Pin Connector
- T6 - 6-Pin Connector*
- T47a - 47-Pin Connector

- (42) - Ground connection, beside steering column
- (81) - Ground connection -1-, in instrument panel wiring harness
- (135) - Ground connection -2-, in instrument panel wiring harness
- (A2) - Plus connection (15), in instrument panel wiring harness
- (A4) - Plus connection (58b), in instrument panel wiring harness
- (A38) - Plus connection -2- (15a), in instrument panel wiring harness
- (A132) - Connection (ASC), in instrument panel wiring harness

* - Vehicles with Anti-Slip Control (ASC) only

ABS control module (w/EDL), anti-slip control switch

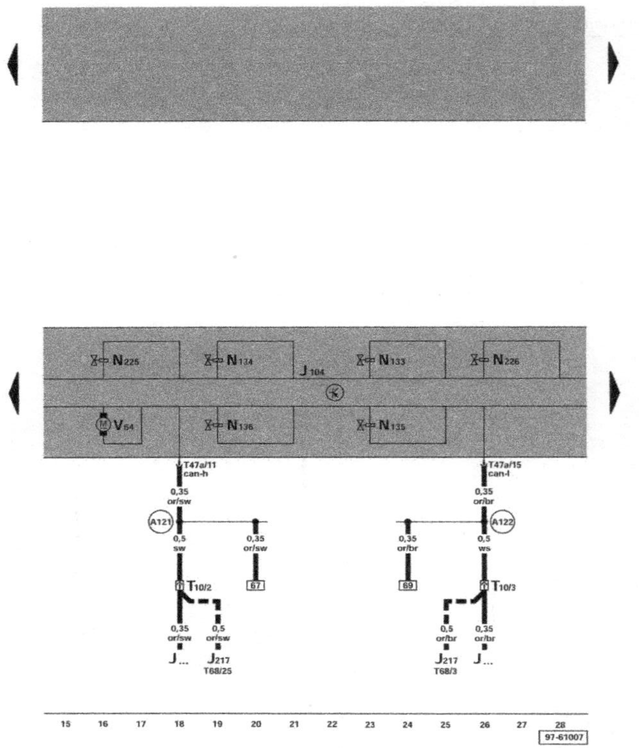

ws = white
sw = black
ro = red
br = brown
gn = green
bl = blue
gr = grey
li = lilac
ge = yellow
or = orange

J104 - ABS Control Module (w/EDL), in engine compartment, left
J217 - Transmission Control Module (TCM), in plenum chamber, center
J... - Engine Control Module (ECM)
N133 - Right Rear ABS Inlet Valve
N134 - Left Rear ABS Inlet Valve
N135 - Right Rear ABS Outlet Valve
N136 - Left Rear ABS Outlet Valve
N225 - Pilot valve -1- traction control
N226 - Pilot valve -2- traction control
T10 - 10-Pin Connector, white, in protective housing for connectors, in plenum chamber, left
T47a - 47-Pin Connector
T68 - 68-Pin Connector, on Transmission Control Module (TCM)
V64 - ABS Hydraulic Pump

(A121) - Connection (high bus), in instrument panel wiring harness
(A122) - Connection (low bus), in instrument panel wiring harness

--- Automatic transmission only

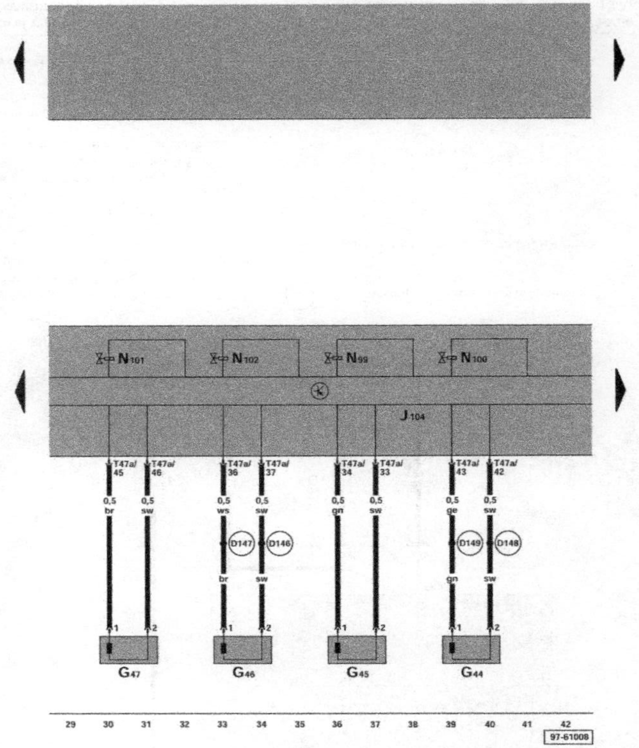

ws = white
sw = black
ro = red
br = brown
gn = green
bl = blue
gr = grey
li = lilac
ge = yellow

G44 - Right Rear ABS Wheel Speed Sensor
G45 - Right Front ABS Wheel Speed Sensor
G46 - Left Rear ABS Wheel Speed Sensor
G47 - Left Front ABS Wheel Speed Sensor
J104 - ABS Control Module (w/EDL), in engine compartment, left
N99 - Right Front ABS Inlet Valve
N100 - Right Front ABS Outlet Valve
N101 - Left Front ABS Inlet Valve
N102 - Left Front ABS Outlet Valve
T47a - 47-Pin Connector

(D146) - Connector (speed sensor rear, left +), in wiring harness engine compartment
(D147) - Connector (speed sensor rear, left -), in wiring harness engine compartment
(D148) - Connector (speed sensor rear, right +), in wiring harness engine compartment
(D149) - Connector (speed sensor rear, right -), in wiring harness engine compartment

New Beetle Wiring diagram **No. 44/5**

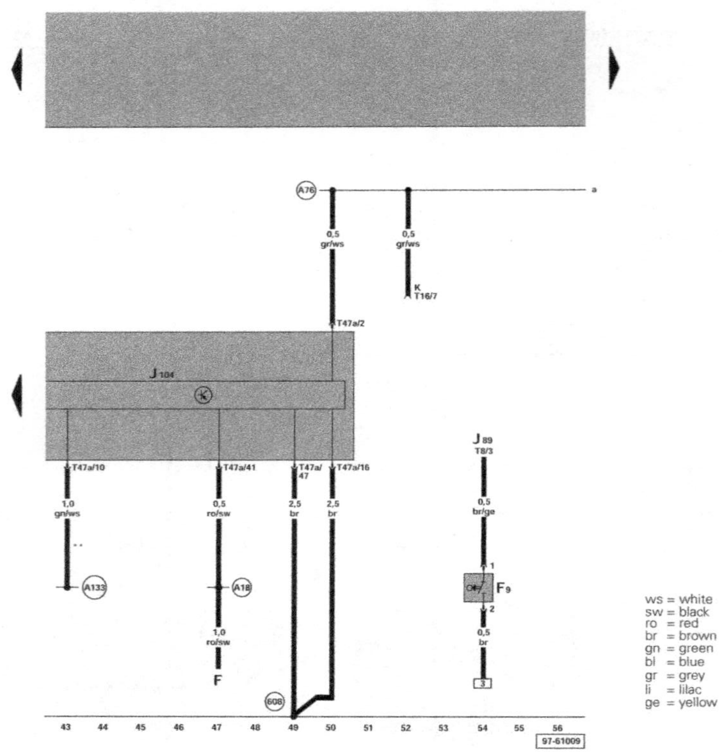

ws = white
sw = black
ro = red
br = brown
gn = green
bl = blue
gr = grey
li = lilac
ge = yellow

- F - Brake Light Switch
- F9 - Parking Brake Warning Light Switch
- J 89 - Daytime Running Lights Change-over Relay
- J104 - ABS Control Module (w/EDL), in engine compartment, left
- T8 - 8-Pin Connector
- T16 - Data Link Connector (DLC), below instrument panel, left
- T47a - 47-Pin Connector
- (608) - Ground Connection (in center plenum chamber)
- (A18) - Wire connection (54), in instrument panel wiring harness
- (A76) - Connector (K-diagnosis wire) in instrument panel wiring harness
- (A133) - Connection (ABS-signal left) in instrument panel wiring harness
- ** - Vehicles with ESP and Spoiler only

Edition 10/00 USA.5132.13.21 **ABS control module (w/EDL), parking brake warning light switch**

No. 44/6 Wiring diagram **New Beetle**

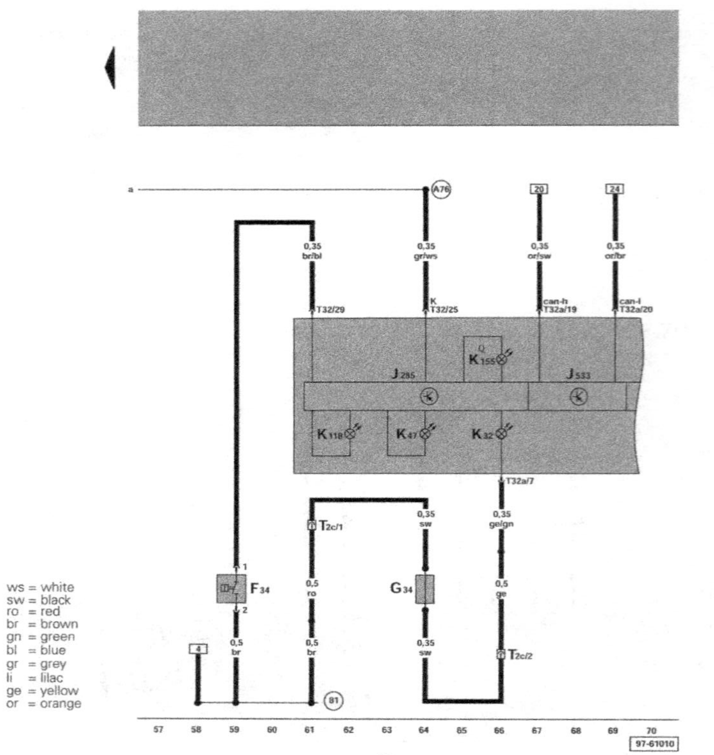

ws = white
sw = black
ro = red
br = brown
gn = green
bl = blue
gr = grey
li = lilac
ge = yellow
or = orange

- F34 - Brake Fluid Level Warning Switch
- G34 - Left Front Brake Pad Wear Sensor
- J285 - Control module with indicator unit in instrument panel insert
- J533 - Data Bus On Board Diagnostic Interface
- K32 - Brake Pad Wear Indicator Light
- K47 - ABS Warning Light
- K118 - Warning light for brake system
- K155 - Warning light for Anti-Slip Control (ASC)*
- T2c - Double Connector, near left front brake pad wear sensor
- T32 - 32-Pin Connector, blue
- T32a - 32-Pin Connector, green
- (81) - Ground connection -1-, in instrument panel wiring harness
- (A76) - Connector (K-diagnosis wire) in instrument panel wiring harness
- * - Vehicles with Anti-Slip Control (ASC) only

Brake fluid level warning switch, left front brake pad wear sensor, instrument cluster, ABS warning light, warning light for: brake system/anti-slip control ASC), **Edition 10/00** USA.5132.13.21

Anti-lock brake system (ABS), Anti-lock brake system (ABS) with electronic differential lock (EDL), anti-slip regulation (ASR) and electronic stabilization program (ESP)

from December 2001

Deviate relay location and fuseplacements as well as the locations of multiple connectors see section "component locations".

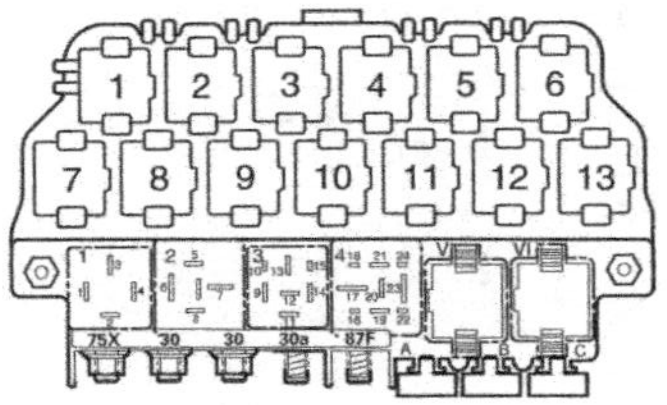

97-21956

Fuse colors
30 A - green
25 A - white
20 A - yellow
15 A - blue
10 A - red
7,5 A - brown
5 A - beige

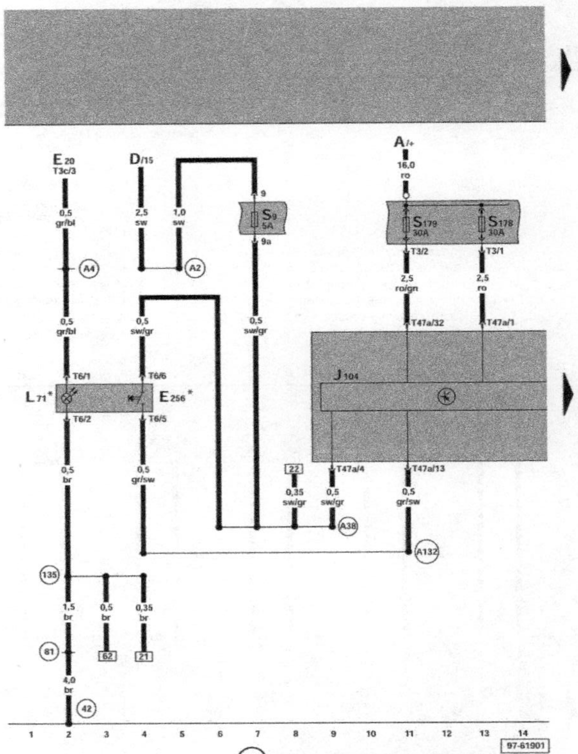

97-61901

ws = white
sw = black
ro = red
br = brown
gn = green
bl = blue
gr = grey
li = violet
ge = yellow
or = orange

A - Battery
D - Ignition/Starter Switch
E20 - Instrument Panel Light Dimmer Switch
E256 - ASR/ESP Button *
J104 - ABS Control Module (w/EDL), in engine compartment, left
L71 - Anti-Slip Control Switch Illumination*
S9 - Fuse 9 in fuse holder
S178 - Fuse -6- (30) in fuse bracket/battery
S179 - Fuse -7- (30) in fuse bracket/battery
T3 - 3-Pin Connector in fuse holder/battery
T3c - 3-Pin Connector
T6 - 6-Pin Connector*
T47a - 47-Pin Connector, on ABS Control Module

(42) - Ground connection, beside steering column

(81) - Ground connection -1-, in instrument panel wiring harness
(135) - Ground connection -2-, in instrument panel wiring harness
(A2) - Plus connection (15), in instrument panel wiring harness
(A4) - Plus connection (58b), in instrument panel wiring harness
(A38) - Plus connection -2- (15a), in instrument panel wiring harness
(A132) - Connection (ASC), in instrument panel wiring harness

* - Only vehicles with ASR/ESP

ABS control module (w/EDL), anti-slip control switch, ASR/ESP button

New Beetle Wiring diagram **No. 52/3**

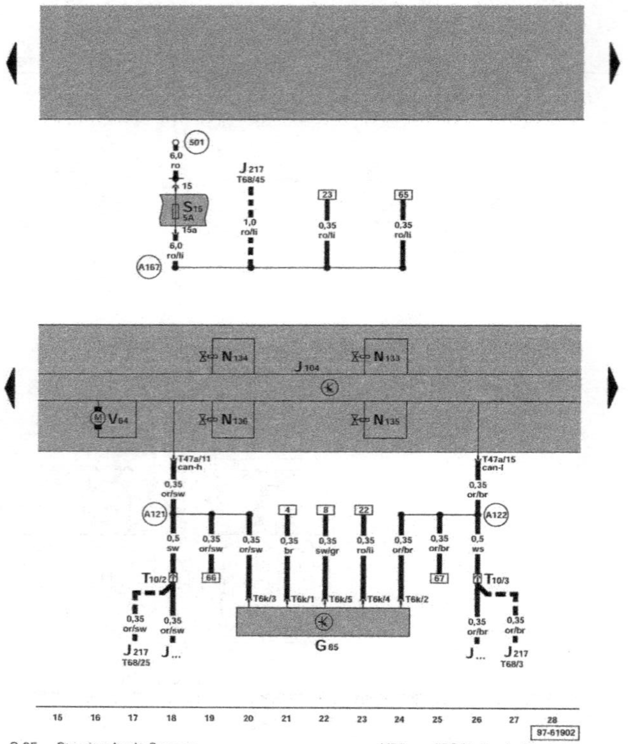

ws = white
sw = black
ro = red
br = brown
gn = green
bl = blue
gr = grey
li = violet
ge = yellow
or = orange

G 85 - Steering Angle Sensor
J104 - ABS Control Module (w/EDL), in engine compartment, left
J217 - Transmission Control Module (TCM), in plenum chamber, center
J... - Engine Control Module (ECM)
N133 - Right Rear ABS Inlet Valve
N134 - Left Rear ABS Inlet Valve
N135 - Right Rear ABS Outlet Valve
N136 - Left Rear ABS Outlet Valve
S 15 - Fuse 15 in fuse holder
T6k - 6-Pin Connector,
T10 - 10-Pin Connector, white, in protective housing for connectors, in plenum chamber, left
T47a - 47-Pin Connector, on ABS Control Module
T68 - 68-Pin Connector, on Transmission Control Module (TCM)

V64 - ABS Hydraulic Pump

(501) - Threaded connection -2- (30) on the relay plate

(A121) - Connection (high bus), in instrument panel wiring harness

(A122) - Connection (low bus), in instrument panel wiring harness

(A167) - Plus Connection 3 (30a) (in instrument panel wiring harness)

--- Automatic transmission only

Edition 08/02 **ABS control module (w/EDL), ABS hydraulic pump, steering angle sensor**
USA.5132.17.21

No. 52/4 Wiring diagram **New Beetle**

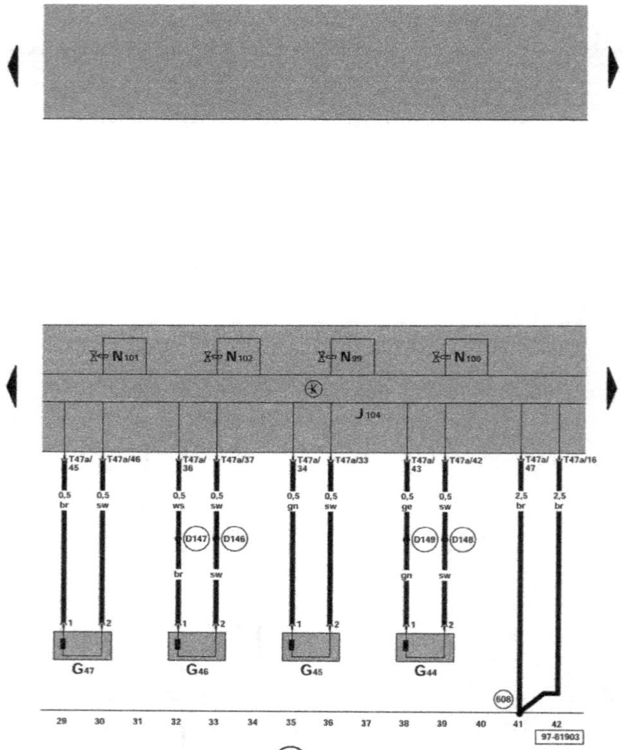

ws = white
sw = black
ro = red
br = brown
gn = green
bl = blue
gr = grey
li = violet
ge = yellow
or = orange

G44 - Right Rear ABS Wheel Speed Sensor
G45 - Right Front ABS Wheel Speed Sensor
G46 - Left Rear ABS Wheel Speed Sensor
G47 - Left Front ABS Wheel Speed Sensor
J104 - ABS Control Module (w/EDL), in engine compartment, left
N99 - Right Front ABS Inlet Valve
N100 - Right Front ABS Outlet Valve
N101 - Left Front ABS Inlet Valve
N102 - Left Front ABS Outlet Valve
T47a - 47-Pin Connector, on ABS Control Module

(608) - Ground Connection (in center plenum chamber)

(D146) - Connector (speed sensor rear, left +), in wiring harness engine compartment

(D147) - Connector (speed sensor rear, left -), in wiring harness engine compartment

(D148) - Connector (speed sensor rear, right +), in wiring harness engine compartment

(D149) - Connector (speed sensor rear, right -), in wiring harness engine compartment

ABS control module (w/EDL), front and rear ABS wheel speed sensors **Edition 08/02**
USA.5132.17.21

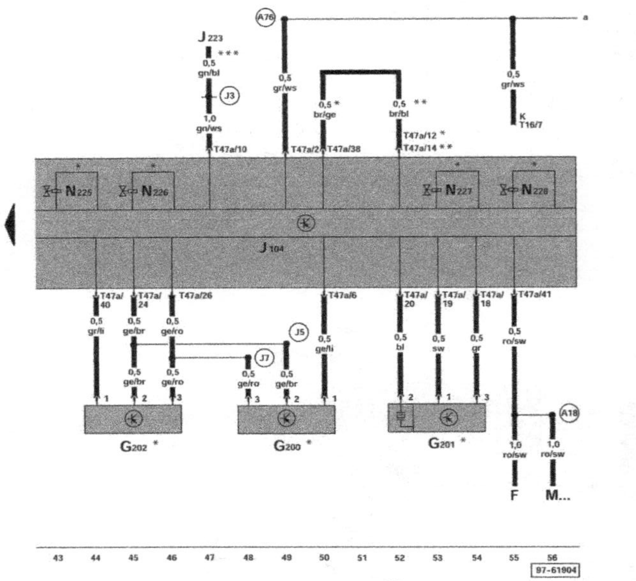

ws = white
sw = black
ro = red
br = brown
gn = green
bl = blue
gr = grey
li = violet
ge = yellow
or = orange

F - Brake Light Switch
G200 - Sensor for transverse acceleration
G201 - Sender 1 for brake booster
G202 - Sender for Rotation Rate
J104 - ABS Control Module (w/EDL), in engine compartment, left
J223 - Rear Spoiler Control Module
M - Brake lamps
N225 - Pilot valve -1- traction control
N226 - Pilot valve -2- traction control
N227 - High pressure switch valve -1- traction control
N228 - High pressure switch valve -2- traction control
T16 - Data Link Connector (DLC), below instrument panel, left
T47a - 47-Pin Connector, on ABS Control Module

(A18) - Wire connection (54), in instrument panel wiring harness
(A76) - Connector (K-diagnosis wire) in instrument panel wiring harness
(J3) - wire connection -1-, in ABS wiring harness
(J5) - wire connection -2-, in ABS wiring harness
(J7) - wire connection -3-, in ABS wiring harness

* - Only vehicles with ESP
** - Only vehicles without ESP
*** - Only vehicles with rear spoiler

ABS control module (w/EDL), sensor for transverse acceleration, sender 1 for brake booster, sender for rotation rate

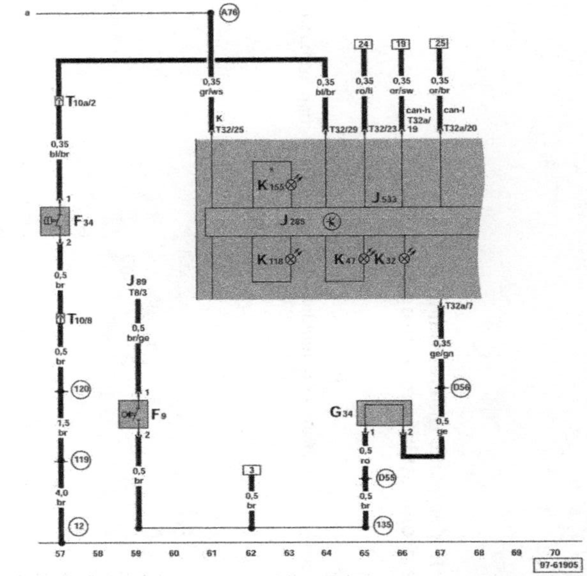

ws = white
sw = black
ro = red
br = brown
gn = green
bl = blue
gr = grey
li = violet
ge = yellow
or = orange

F9 - Parking Brake Warning Light Switch
F34 - Brake Fluid Level Warning Switch
G34 - Left Front Brake Pad Wear Sensor
J 89 - Daytime Running Lights Change-over Relay
J285 - Control module with indicator unit in instrument panel insert
J533 - Data Bus On Board Diagnostic Interface
K32 - Brake Pad Wear Indicator Light
K47 - ABS Warning Light
K118 - Warning light for brake system
K155 - Warning light for Anti-Slip Control (ASC)*
T8 - 8-Pin Connector
T10 - 10-Pin Connector, white behind intrument panel , left
T10a - 10-Pin Connector, orange behind instrument panel, left
T32 - 32-Pin Connector, blue, on instrument cluster
T32a - 32 Pin Connector, green, on instrument cluster

(12) - Ground connection, in engine compartment, left
(119) - Ground connection -1-, in headlight wiring harness
(120) - Ground connection -2-, in headlight wiring harness
(135) - Ground connection -2-, in instrument panel wiring harness
(A76) - Connector (K-diagnosis wire) in instrument panel wiring harness
(D55) - wire connection (left brake wear indicator), in engine compartment wiring harness
(D56) - wire connection (left brake wear indicator, in engine compartment wiring harness
* - Vehicles with Anti-Slip Regulation (ASR) only

Parking brake warning light switch, brake fluid level warning switch, left front brake pad wear sensor, control module with indicator unit in instrument panel insert

New Beetle Wiring diagram **No. 9/1**

Anti-lock brake system (ABS), Anti-lock brake system (ABS) with electronic differential lock (EDL), anti-slip regulation (ASR) and electronic stabilization program (ESP),

from June 2002

Deviate relay location and fuseplacements as well as the locations of multiple connectors see section "component locations".

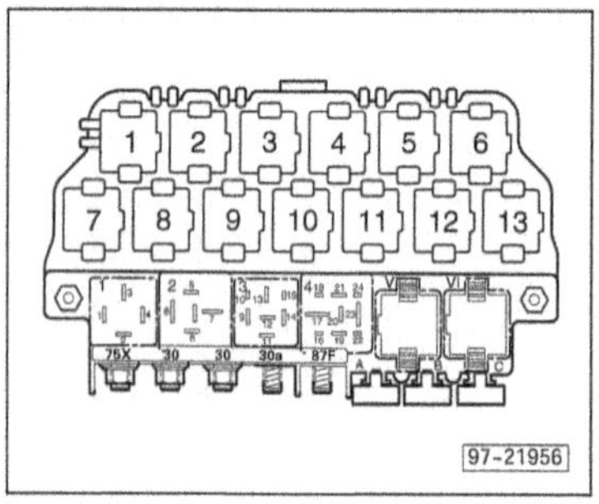

Fuse colors

30 A - green
25 A - white
20 A - yellow
15 A - blue
10 A - red
7,5 A - brown
5 A - beige
3 A - violet

New Beetle Wiring diagram **No. 9/2**

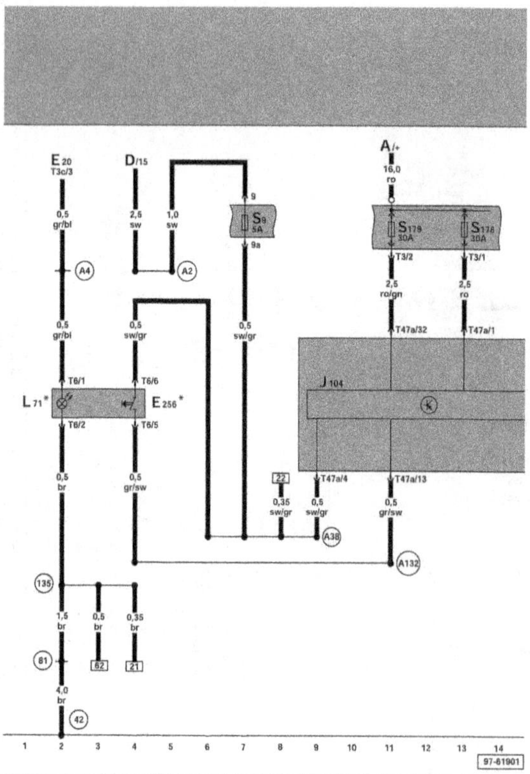

ABS control module (w/EDL), anti-slip control switch, ASR/ESP button

A - Battery
D - Ignition/Starter Switch
E20 - Instrument Panel Light Dimmer Switch
E256 - ASR/ESP Button *
J104 - ABS Control Module (w/EDL), in engine compartment, left
L71 - Anti-Slip Control Switch Illumination*
S9 - Fuse 9 in fuse holder
S178 - Fuse -6- (30) in fuse bracket/battery
S179 - Fuse -7- (30) in fuse bracket/battery
T3 - 3-Pin Connector in fuse holder/battery
T3c - 3-Pin Connector
T6 - 6-Pin Connector*
T47a - 47-Pin Connector, on ABS Control Module

(42) - Ground connection, beside steering column
(81) - Ground connection -1-, in instrument panel wiring harness
(135) - Ground connection -2-, in instrument panel wiring harness
(A2) - Plus connection (15), in instrument panel wiring harness
(A4) - Plus connection (58b), in instrument panel wiring harness
(A38) - Plus connection -2- (15a), in instrument panel wiring harness
(A132) - Connection (ASC), in instrument panel wiring harness

* - Only vehicles with ASR/ESP

ws = white
sw = black
ro = red
br = brown
gn = green
bl = blue
gr = grey
li = violet
ge = yellow
or = orange

New Beetle — Wiring diagram — No. 9/3

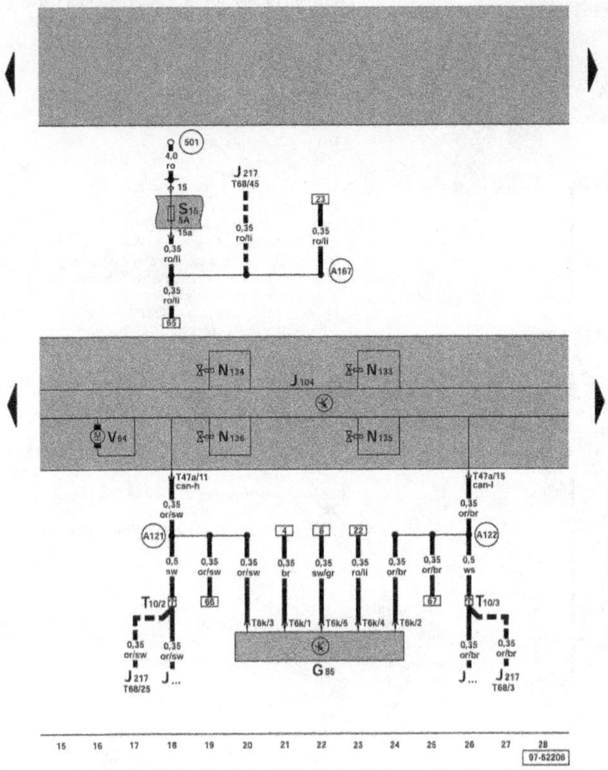

ABS control module (w/EDL), ABS hydraulic pump, steering angle sensor

- G 85 - Steering Angle Sensor
- J104 - ABS Control Module (w/EDL), in engine compartment, left
- J217 - Transmission Control Module (TCM), in plenum chamber, center
- J... - Engine Control Module (ECM)
- N133 - Right Rear ABS Inlet Valve
- N134 - Left Rear ABS Inlet Valve
- N135 - Right Rear ABS Outlet Valve
- N136 - Left Rear ABS Outlet Valve
- S 15 - Fuse 15 in fuse holder
- T6k - 6-Pin Connector,
- T10 - 10-Pin Connector, white, in protective housing for connectors, in plenum chamber, left
- T47a - 47-Pin Connector, on ABS Control Module
- T68 - 68-Pin Connector, on Transmission Control Module (TCM)
- V64 - ABS Hydraulic Pump

- (501) - Threaded connection -2- (30) on the relay plate
- (A121) - Connection (high bus), in instrument panel wiring harness
- (A122) - Connection (low bus), in instrument panel wiring harness
- (A167) - Plus Connection 3 (30a) (in instrument panel wiring harness)
- --- Automatic transmission only

ws = white
sw = black
ro = red
br = brown
gn = green
bl = blue
gr = grey
li = violet
ge = yellow
or = orange

New Beetle — Wiring diagram — No. 9/4

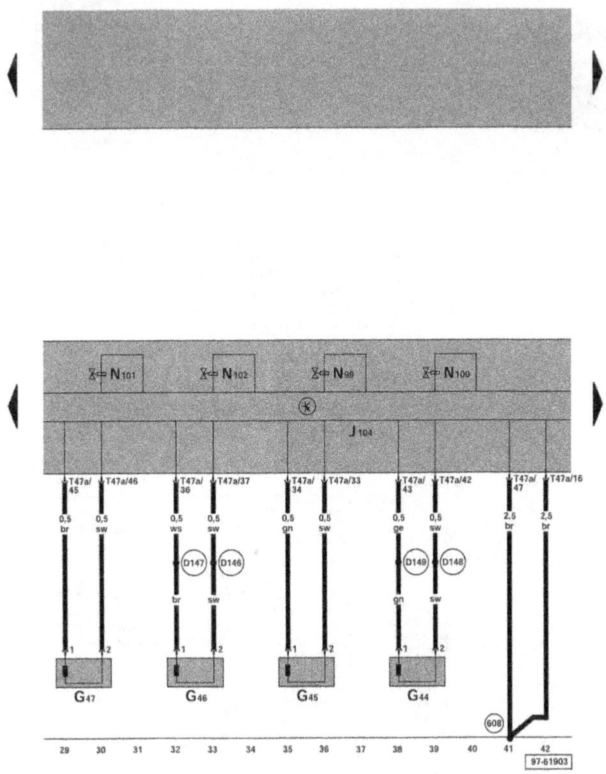

ABS control module (w/EDL), front and rear ABS wheel speed sensors

- G44 - Right Rear ABS Wheel Speed Sensor
- G45 - Right Front ABS Wheel Speed Sensor
- G46 - Left Rear ABS Wheel Speed Sensor
- G47 - Left Front ABS Wheel Speed Sensor
- J104 - ABS Control Module (w/EDL), in engine compartment, left
- N99 - Right Front ABS Inlet Valve
- N100 - Right Front ABS Outlet Valve
- N101 - Left Front ABS Inlet Valve
- N102 - Left Front ABS Outlet Valve
- T47a - 47-Pin Connector, on ABS Control Module

- (608) - Ground Connection (in center plenum chamber)
- (D146) - Connector (speed sensor rear, left +), in wiring harness engine compartment
- (D147) - Connector (speed sensor rear, left -), in wiring harness engine compartment
- (D148) - Connector (speed sensor rear, right +), in wiring harness engine compartment
- (D149) - Connector (speed sensor rear, right -), in wiring harness engine compartment

ws = white
sw = black
ro = red
br = brown
gn = green
bl = blue
gr = grey
li = violet
ge = yellow
or = orange

New Beetle — Wiring diagram — **No. 9/5**

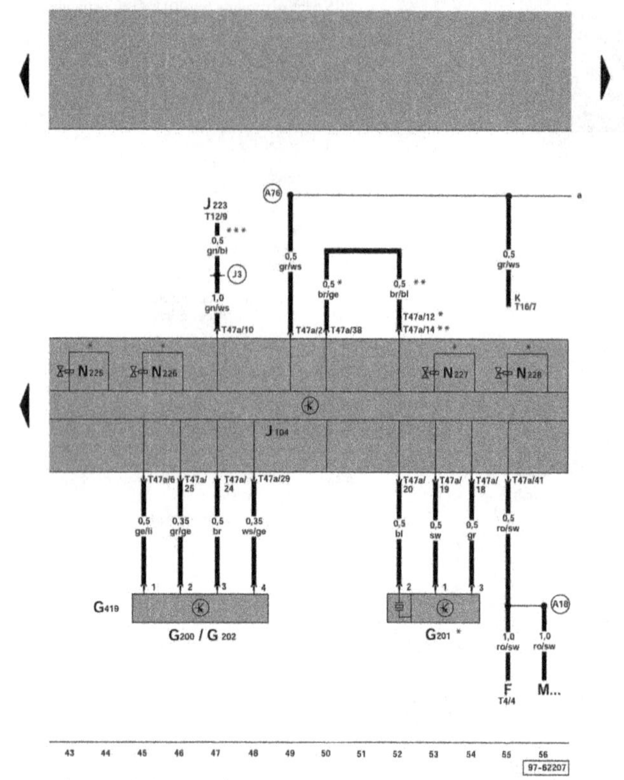

ABS control module (w/EDL), sensor for transverse acceleration, sender 1 for brake booster, sender for rotation rate

- F - Brake Light Switch
- G200 - Sensor for transverse acceleration
- G201 - Sender 1 for brake booster
- G202 - Sender for Rotation Rate
- G419 - ESP-Sensor Unit
- J104 - ABS Control Module (w/EDL), in engine compartment, left
- J223 - Rear Spoiler Control Module
- M - Brake lamps
- N225 - Pilot valve -1- traction control
- N226 - Pilot valve -2- traction control
- N227 - High pressure switch valve -1- traction control
- N228 - High pressure switch valve -2- traction control
- T4 - 4-Pin Connector, at F
- T12 - 12-Pin Connector, at J223
- T16 - Data Link Connector (DLC), below instrument panel, left
- T47a - 47-Pin Connector, on ABS Control Module

- (A18) - Wire connection (54), in instrument panel wiring harness
- (A76) - Connector (K-diagnosis wire) in instrument panel wiring harness
- (J3) - wire connection -1-, in ABS wiring harness
- (J5) - wire connection -2-, in ABS wiring harness
- (J7) - wire connection -3-, in ABS wiring harness

- * - Only vehicles with ESP
- ** - Only vehicles without ESP
- *** - Only vehicles rear spoiler

ws = white
sw = black
ro = red
br = brown
gn = green
bl = blue
gr = grey
li = violet
ge = yellow
or = orange

New Beetle — Wiring diagram — **No. 9/6**

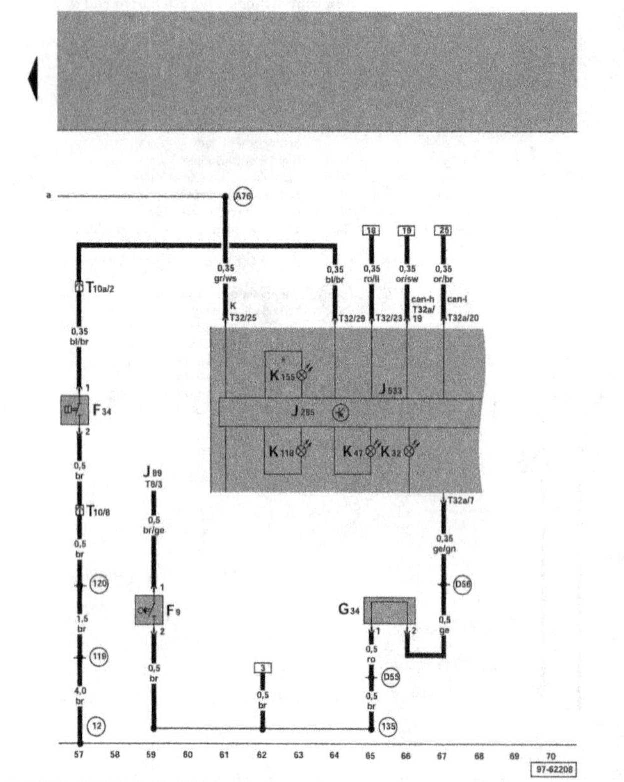

Parking brake warning light switch, brake fluid level warning switch, left front brake pad wear sensor, control module with indicator unit in instrument panel insert

- F9 - Parking Brake Warning Light Switch
- F34 - Brake Fluid Level Warning Switch
- G34 - Left Front Brake Pad Wear Sensor
- J 89 - Daytime Running Lights Change-over Relay
- J285 - Control module with indicator unit in instrument panel insert
- J533 - Data Bus On Board Diagnostic Interface
- K32 - Brake Pad Wear Indicator Light
- K47 - ABS Warning Light
- K118 - Warning light for brake system
- K155 - Warning light for Anti-Slip Control (ASC)*
- T8 - 8-Pin Connector
- T10 - 10-Pin Connector, white behind intrument panel , left
- T10a - 10-Pin Connector, orange behind instrument panel, left
- T32 - 32-Pin Connector, blue, on instrument cluster
- T32a - 32 Pin Connector, green, on instrument cluster

- (12) - Ground connection, in engine compartment, left
- (119) - Ground connection -1-, in headlight wiring harness
- (120) - Ground connection -2-, in headlight wiring harness
- (135) - Ground connection -2-, in instrument panel wiring harness
- (A76) - Connector (K-diagnosis wire) in instrument panel wiring harness
- (D55) - wire connection (left brake wear indicator), in engine compartment wiring harness
- (D56) - wire connection (left brake wear indicator, in engine compartment wiring harness

- * - Vehicles with Anti-Slip Regulation (ASR - ASC) only

ws = white
sw = black
ro = red
br = brown
gn = green
bl = blue
gr = grey
li = violet
ge = yellow
or = orange

New Beetle Wiring diagram **No. 52/1**

Anti-lock brake system (ABS) with electronic differential lock (EDL), anti-slip regulation (ASR) and electronic stabilization program (ESP),

from July 2005,

Additional Fuse, Relay, Control Module and Connector locations ⇒ Component Locations

Electrical troubleshooting ⇒ Guided Fault Finding using VAS 5051/5052

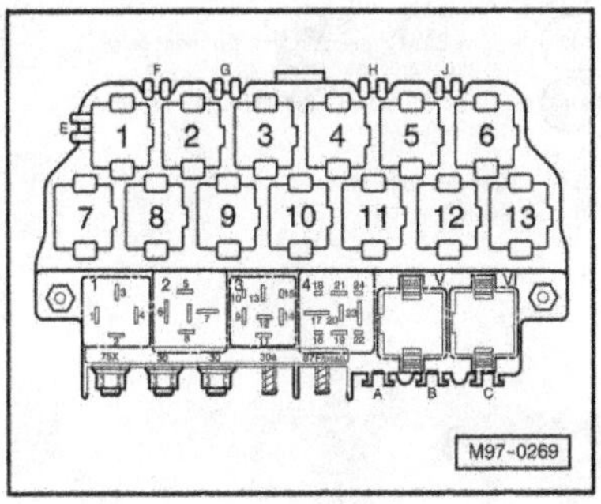

Note:

Number in parentheses indicates production control number stamped on relay housing.

New Beetle Wiring diagram **No. 52/2**

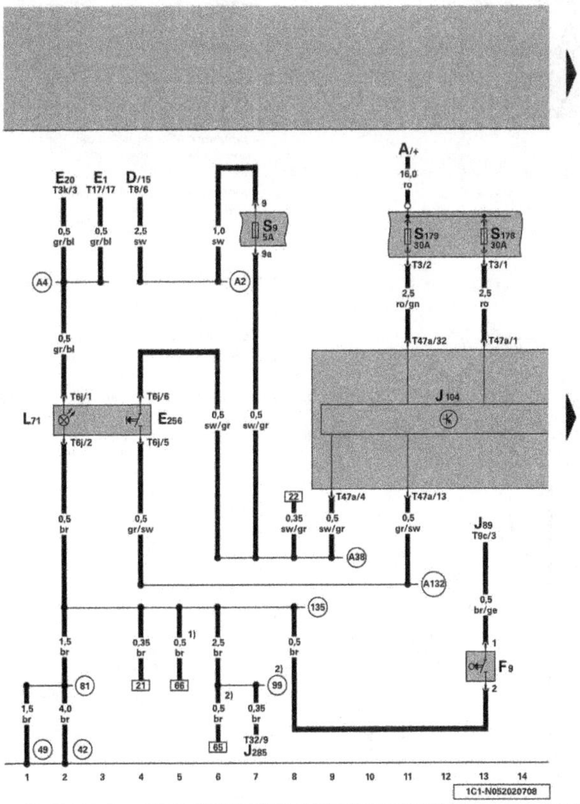

ASR/ESP Button, ABS Control Module

A	-	Battery
D	-	Ignition/Starter Switch
E1	-	Light Switch
E20	-	Instrument Panel Illumination Dimmer Switch
E256	-	ASR/ESP Button
F9	-	Parking Brake Warning Light Switch
J89	-	Daytime Running Lamps Change-over Relay
J104	-	ABS Control Module
J285	-	Instrument Cluster Control Module
L71	-	Anti-Slip Regulation (ASR) Switch Illumination
S9	-	Fuse (on fuse/relay panel)
S178	-	Fuse 6 - terminal 30 (on battery fuse panel)
S179	-	Fuse 7 - terminal 30 (on battery fuse panel)
T3	-	3-Pin Connector in fuse holder/battery
T3k	-	3-Pin Connector
T6j	-	6-Pin Connector
T8	-	8-Pin Connector
T9c	-	9-Pin Connector
T17	-	17-Pin Connector
T32	-	32-Pin Connector
T17	-	17-Pin Connector
T47a	-	47-Pin Connector, on ABS Control Module
(42)	-	Ground Connection (beside steering column)
(49)	-	Ground Connection (near steering column)
(81)	-	Ground Connection 1 (in instrument panel wiring harness)
(99)	-	Ground Connection (in console wiring harness)
(135)	-	Ground Connection 2 (in instrument panel wiring harness)
(A2)	-	Plus Connection (15) (in instrument panel wiring harness)
(A4)	-	Plus Connection (58b) (in instrument panel wiring harness)
(A38)	-	Plus Connection 2 (15a) (in instrument panel wiring harness)
(A132)	-	Plus Connection 2 (15a) (in instrument panel wiring harness)
1)	-	Through June 2007
2)	-	From July 2007

ws = white
sw = black
ro = red
br = brown
gn = green
bl = blue
gr = grey
li = violet
ge = yellow
or = orange

New Beetle — Wiring diagram — No. 52/3

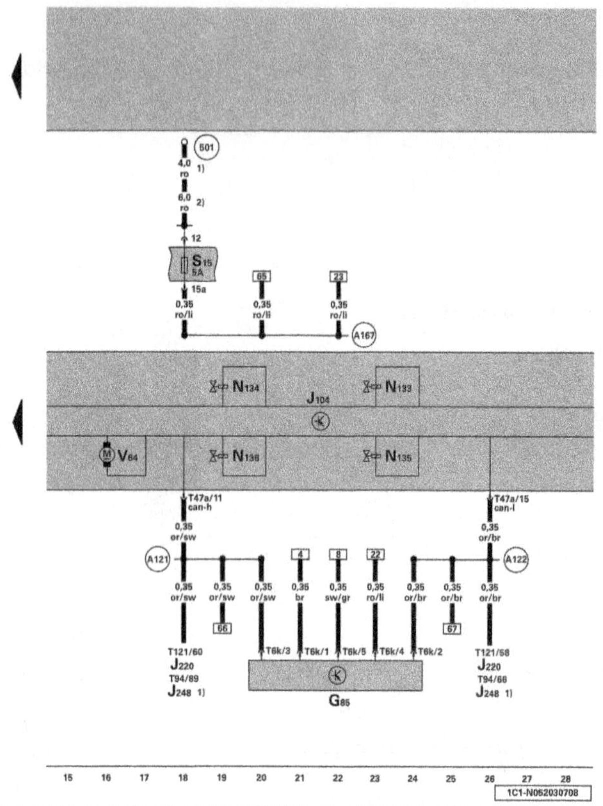

Steering Angle Sensor, ABS Control Module, ABS Hydraulic Pump

- G85 - Steering Angle Sensor
- J104 - ABS Control Module
- J220 - Motronic Engine Control Module (ECM)
- J248 - Diesel Direct Fuel Injection (DFI) Engine Control Module (ECM)
- N133 - Right Rear ABS Inlet Valve
- N134 - Left Rear ABS Inlet Valve
- N135 - Right Rear ABS Outlet Valve
- N136 - Left Rear ABS Outlet Valve
- S15 - Fuse (on fuse/relay panel)
- T6k - 6-Pin Connector,
- T47a - 47-Pin Connector, on ABS Control Module
- T94 - 94-Pin Connector
- T121 - 121-Pin Connector
- V64 - ABS Hydraulic Pump
- (501) - Threaded Connection 2 - terminal 30 (on relay panel)
- (A121) - High-Bus Connection (in instrument panel wiring harness)
- (A122) - Low-Bus Connection (in instrument panel wiring harness)
- (A167) - Plus Connection 3 (30a) (in instrument panel wiring harness)

1) - Through June 2007
2) - From July 2007

ws = white
sw = black
ro = red
br = brown
gn = green
bl = blue
gr = grey
li = violet
ge = yellow
or = orange

New Beetle — Wiring diagram — No. 52/4

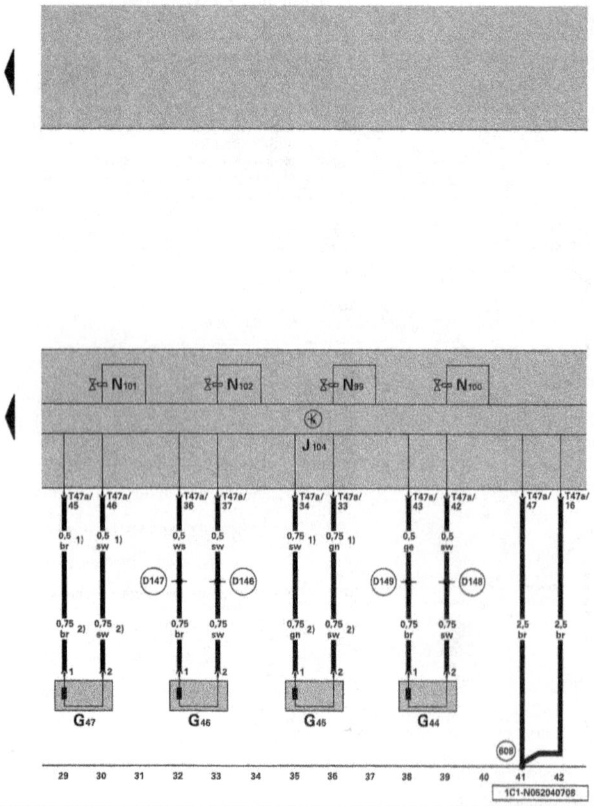

Right Rear ABS Wheel Speed Sensor, Right Front ABS Wheel Speed Sensor, Left Rear ABS Wheel Speed Sensor, Left Front ABS Wheel Speed Sensor, ABS Control Module

- G44 - Right Rear ABS Wheel Speed Sensor
- G45 - Right Front ABS Wheel Speed Sensor
- G46 - Left Rear ABS Wheel Speed Sensor
- G47 - Left Front ABS Wheel Speed Sensor
- J104 - ABS Control Module
- N99 - Right Front ABS Inlet Valve
- N100 - Right Front ABS Outlet Valve
- N101 - Left Front ABS Inlet Valve
- N102 - Left Front ABS Outlet Valve
- T47a - 47-Pin Connector, on ABS Control Module
- (608) - Ground Connection (in center plenum chamber)
- (D146) - Connection (speed sensor rear, left +) (in engine compartment wiring harness)
- (D147) - Connection (speed sensor rear, left -) (in engine compartment wiring harness)
- (D148) - Connection (speed sensor rear, right +) (in engine compartment wiring harness)
- (D149) - Connection (speed sensor rear, right -) (in engine compartment wiring harness)

1) - Through June 2007
2) - From July 2007

ws = white
sw = black
ro = red
br = brown
gn = green
bl = blue
gr = grey
li = violet
ge = yellow
or = orange

New Beetle Wiring diagram **No. 52/5**

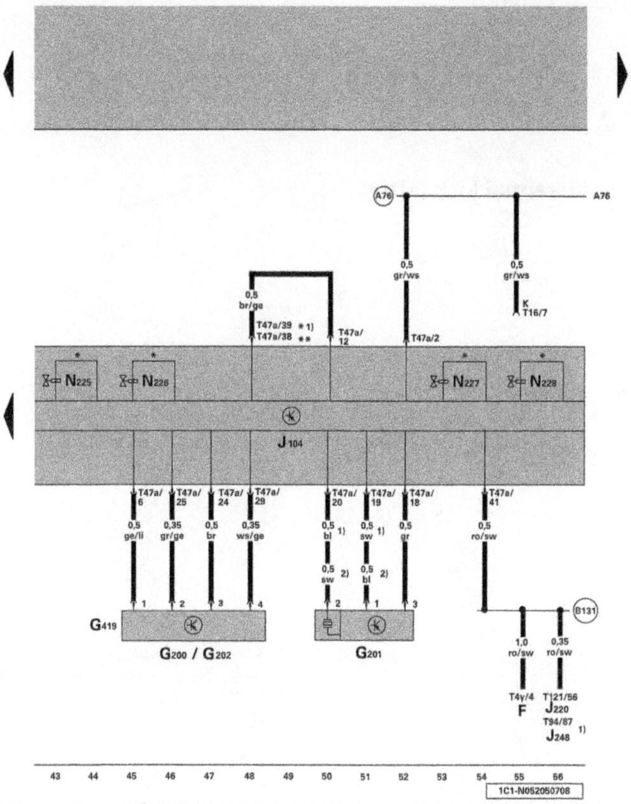

ws = white
sw = black
ro = red
br = brown
gn = green
bl = blue
gr = grey
li = violet
ge = yellow
or = orange

Brake Pressure Sensor 1, ESP Sensor Unit, AB S Control Module

- F - Brake Light Switch
- G200 - Transverse Acceleration Sensor
- G201 - Brake Pressure Sensor 1
- G202 - Rotation Rate Sensor
- G419 - ESP Sensor Unit
- J104 - ABS Control Module
- J220 - Motronic Engine Control Module (ECM)
- J248 - Diesel Direct Fuel Injection (DFI) Engine Control Module (ECM)
- N225 - Driving Dynamics Regulation Switch Valve 1
- N226 - Driving Dynamics Regulation Switch Valve 2
- N227 - Driving Dynamics Regulation High Pressure Switch Valve 1
- N228 - Driving Dynamics Regulation High Pressure Switch Valve 2
- T4y - 4-Pin Connector, at F
- T16 - 16-pin connector, Data Link Connector (DLC), below steering column, left
- T47a - 47-Pin Connector, on ABS Control Module
- T94 - 94-Pin Connector
- T121 - 121-Pin Connector
- (A76) - K-Diagnostic Connection (in instrument panel wiring harness)
- (B131) - Connection (54) (in interior wiring harness)

- * - Only Engine code BPR and BPS,
- ** - Only Engine code BEW, 1)

1) - Through June 2007
2) - From July 2007

New Beetle Wiring diagram **No. 52/6**

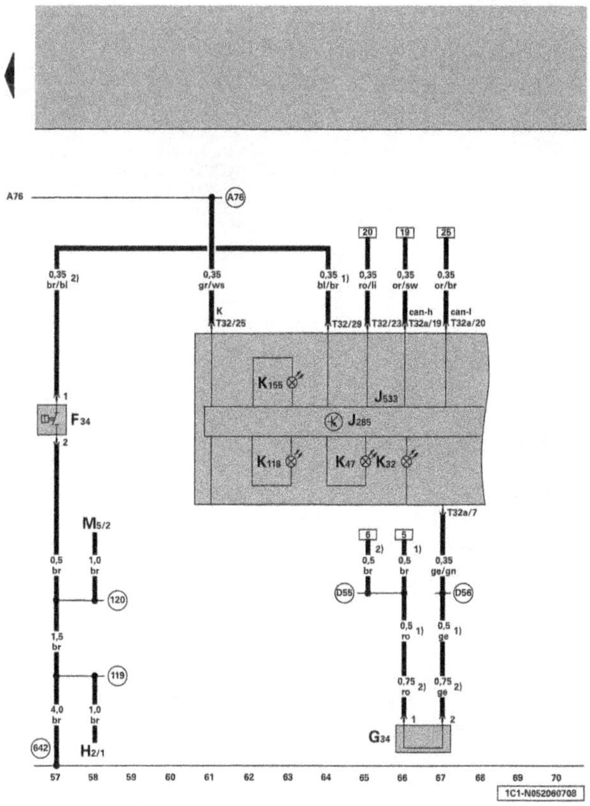

ws = white
sw = black
ro = red
br = brown
gn = green
bl = blue
gr = grey
li = violet
ge = yellow
or = orange

Brake Fluid Level Warning Switch, Left Front Bra ke Pad Wear Sensor, Instrument Cluster Control Module

- F34 - Brake Fluid Level Warning Switch
- G34 - Left Front Brake Pad Wear Sensor
- H2 - High Tone Horn
- J285 - Instrument Cluster Control Module
- J533 - Data Bus On Board Diagnostic Interface
- K32 - Brake Pad Wear Indicator Lamp
- K47 - ABS Warning Lamp
- K118 - Brake System Warning Lamp
- K155 - ASR/ESP Control Lamp
- M5 - Left Front Turn Signal Lamp
- T32 - 32-Pin Connector, blue, on instrument cluster
- T32a - 32 Pin Connector, green, on instrument cluster
- (119) - Ground Connection 1 (in headlamp wiring harness)
- (120) - Ground Connection 2 (in headlamp wiring harness)
- (642) - Ground Connection (on electronic fan control)
- (A76) - K-Diagnostic Connection (in instrument panel wiring harness)
- (D55) - Left Brake Wear Indicator Connection (in engine compartment wiring harness)
- (D56) - Left Brake Wear Indicator Connection (in engine compartment wiring harness)

1) - Through June 2007
2) - From July 2007

New Beetle Convertible Wiring diagram **No. 34/1**

Anti-lock brake system (ABS) with electronic differential lock (EDL),

from July 2005

Additional Fuse, Relay, Control Module and Connector locations ⇒ Component Locations

Electrical troubleshooting ⇒ Guided Fault Finding using VAS 5051/5052

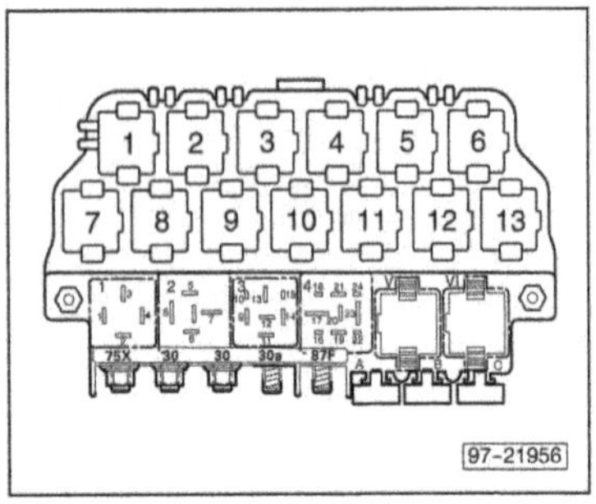

New Beetle Convertible Wiring diagram **No. 34/2**

ABS control module (w/EDL)

A	-	Battery
D	-	Ignition/Starter Switch
J104	-	ABS Control Module (w/EDL), in engine compartment, left
S9	-	Fuse 9 in fuse holder
S178	-	Fuse -6- (30) in fuse bracket/battery
S179	-	Fuse -7- (30) in fuse bracket/battery
T3	-	3-Pin Connector in fuse holder/battery
T8	-	8-Pin Connector
T47a	-	47-Pin Connector, on ABS Control Module
(42)	-	Ground connection, beside steering column
(81)	-	Ground connection -1-, in instrument panel wiring harness
(135)	-	Ground connection -2-, in instrument panel wiring harness
(A2)	-	Plus connection (15), in instrument panel wiring harness
(A38)	-	Plus connection -2- (15a), in instrument panel wiring harness

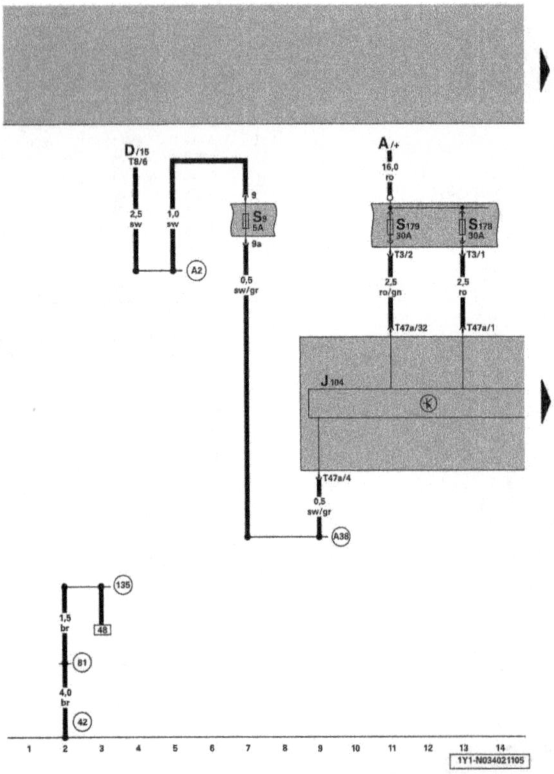

ws = white
sw = black
ro = red
br = brown
gn = green
bl = blue
gr = grey
li = violet
ge = yellow
or = orange

New Beetle Convertible — Wiring diagram — No. 34/3

Right Rear ABS Wheel Speed Sensor, Right Front ABS Wheel Speed Sensor, Left Rear ABS Wheel Speed Sensor, Left Front ABS Wheel Speed Sensor, ABS Control Module (w/EDL), in engine compartment, left

- G44 - Right Rear ABS Wheel Speed Sensor
- G45 - Right Front ABS Wheel Speed Sensor
- G46 - Left Rear ABS Wheel Speed Sensor
- G47 - Left Front ABS Wheel Speed Sensor
- J104 - ABS Control Module (w/EDL), in engine compartment, left
- N99 - Right Front ABS Inlet Valve
- N100 - Right Front ABS Outlet Valve
- N101 - Left Front ABS Inlet Valve
- N102 - Left Front ABS Outlet Valve
- T47a - 47-Pin Connector, on ABS Control Module

- (608) - Ground Connection (in center plenum chamber)
- (D146) - Connector (speed sensor rear, left +), in wiring harness engine compartment
- (D147) - Connector (speed sensor rear, left -), in wiring harness engine compartment
- (D148) - Connector (speed sensor rear, right +), in wiring harness engine compartment
- (D149) - Connector (speed sensor rear, right -), in wiring harness engine compartment

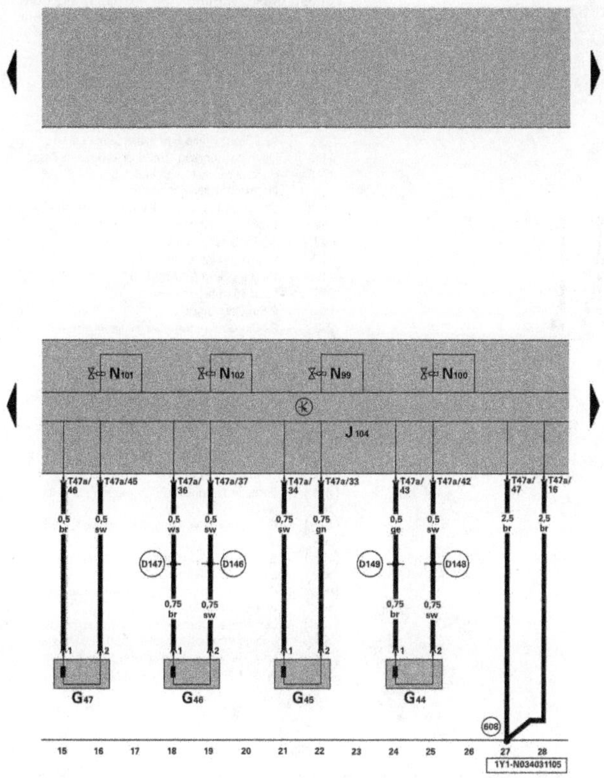

ws = white
sw = black
ro = red
br = brown
gn = green
bl = blue
gr = grey
li = violet
ge = yellow
or = orange

Edition 11.05

New Beetle Convertible — Wiring diagram — No. 34/4

ABS control module (w/EDL)

- F - Brake Light Switch
- J104 - ABS Control Module (w/EDL), in engine compartment, left
- J220 - Motronic Engine Control Module (ECM)
- J256 - Convertible Top Control Module
- N133 - Right Rear ABS Inlet Valve
- N134 - Left Rear ABS Inlet Valve
- N135 - Right Rear ABS Outlet Valve
- N136 - Left Rear ABS Outlet Valve
- T4y - 4-Pin Connector, at F
- T16 - Data Link Connector (DLC), below instrument panel, left
- T26 - 26-Pin connector, connector B, black, on J256
- T47a - 47-Pin Connector, on ABS Control Module
- T121 - 121-Pin Connector
- V64 - ABS Hydraulic Pump

- (A76) - Connector (K-diagnosis wire) in instrument panel wiring harness
- (A121) - Connection (high bus), in instrument panel wiring harness
- (A122) - Connection (low bus), in instrument panel wiring harness
- (B131) - Connection (54) (in interior wiring harness)
- (J3) - Connection 1 (in ABS wiring harness)

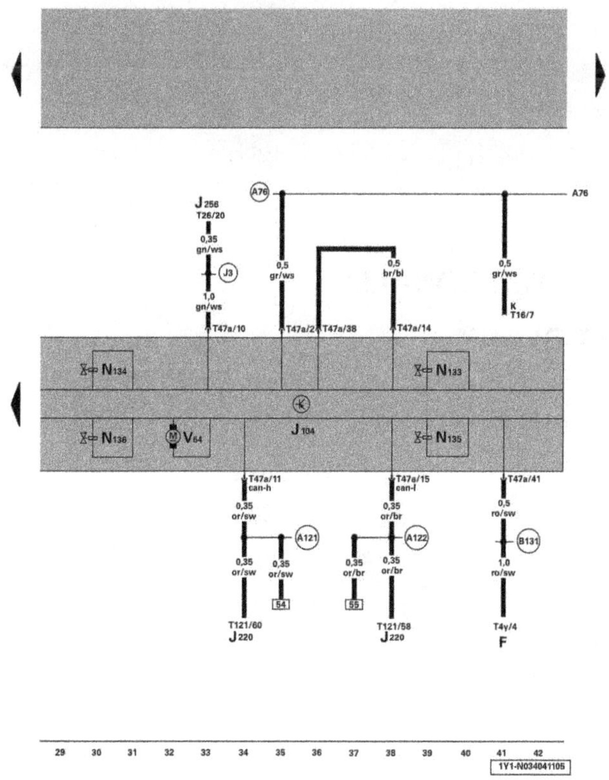

ws = white
sw = black
ro = red
br = brown
gn = green
bl = blue
gr = grey
li = violet
ge = yellow
or = orange

Edition 11.05

New Beetle Convertible Wiring diagram **No. 34/5**

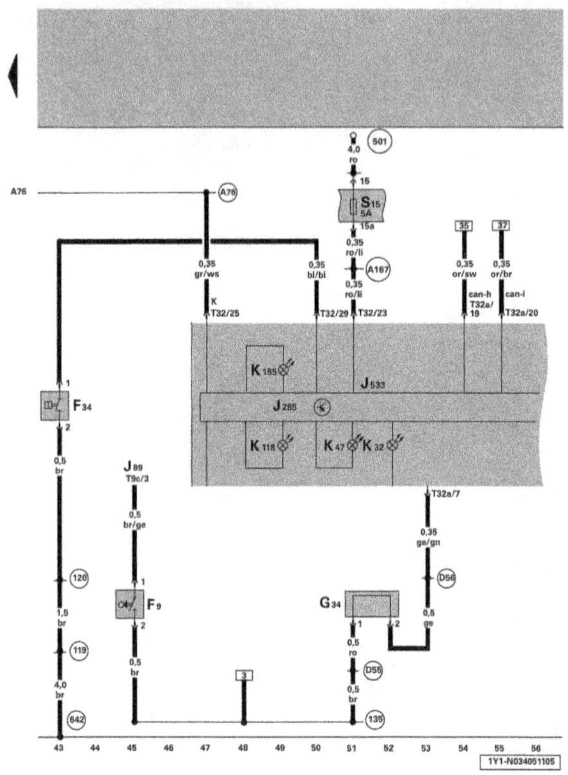

Parking brake warning light switch, brake fluid level warning switch, left front brake pad wear sensor, control module with indicator unit in instrument panel insert, Data Bus On Board Diagnostic Interface

- F9 - Parking Brake Warning Light Switch
- F34 - Brake Fluid Level Warning Switch
- G34 - Left Front Brake Pad Wear Sensor
- J 89 - Daytime Running Lights Change-over Relay
- J285 - Control module with indicator unit in instrument panel insert
- J533 - Data Bus On Board Diagnostic Interface
- K32 - Brake Pad Wear Indicator Light
- K47 - ABS Warning Light
- K118 - Warning light for brake system
- K155 - Warning light for Anti-Slip Control (ASC)
- S15 - Fuse 15 in fuse holder
- T9c - 9-Pin Connector
- T32 - 32-Pin Connector, blue, on instrument cluster
- T32a - 32 Pin Connector, green, on instrument cluster

- (119) - Ground connection -1-, in headlight wiring harness
- (120) - Ground connection -2-, in headlight wiring harness
- (135) - Ground connection -2-, in instrument panel wiring harness
- (501) - Threaded connection -2- (30) on the relay plate
- (642) - Ground Connection (on electronic fan control)

- (A76) - Connector (K-diagnosis wire) in instrument panel wiring harness
- (A167) - Plus Connection 3 (30a) (in instrument panel wiring harness)
- (D55) - Left Brake Wear Indicator Connection (in engine compartment wiring harness)
- (D56) - Left Brake Wear Indicator Connection (in engine compartment wiring harness)

ws = white
sw = black
ro = red
br = brown
gn = green
bl = blue
gr = grey
li = violet
ge = yellow
or = orange

Edition 11.05

THIS SPACE INTENTIONALLY

LEFT BLANK

Air conditioning,

from January 1998

Deviate relay location and fuseplacements as well as the locations of multiple connectors see section "component locations".

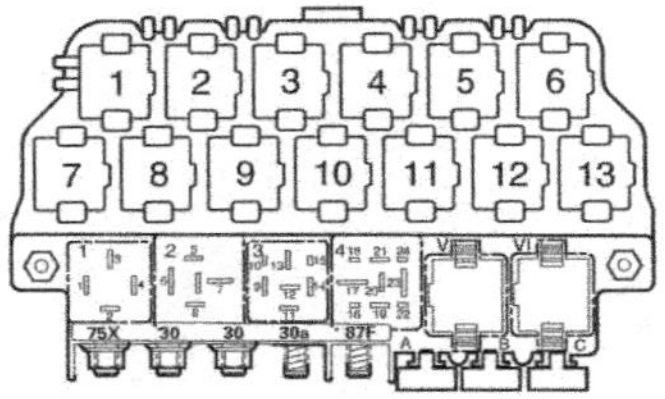

97-21956

Fuse colors
30 A - green
25 A - white
20 A - yellow
15 A - blue
10 A - red
7,5 A - brown
5 A - beige

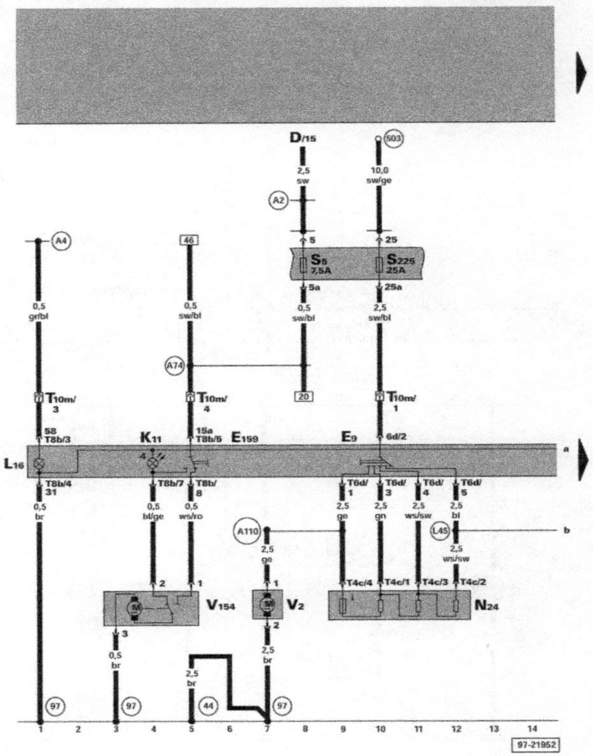

ws = white
sw = black
ro = red
br = brown
gn = green
bl = blue
gr = grey
li = violet
ge = yellow

D - Ignition/Starter Switch
E9 - Fresh Air Blower Switch
E159 - Fresh Air/Recirculating Flap Switch
K114 - Fresh Air and Recirculating Air Mode Indicator Light
L16 - Fresh Air Control Lever Light
N24 - Fresh Air Blower Series Resistance With Fuse
S5 - Fuse 5 in fuse panel
S225 - Fuse 25 in fuse panel
T4c - 4-Pin Connector
T6d - 6-Pin Connector
T8b - 8-Pin Connector
T10m- 10-Pin Connector
V2 - Fresh Air Blower
V154 - Servo motor for fresh-/recirculating air door

(44) - Ground connection, on left A-pillar, lower part
(97) - Ground connection -1-, in A/C wiring harness
(503) - Screw Connection (75x), on relay panel
(A2) - plus connection (15), in instrument panel wiring harness
(A4) - plus connection (58b), in instrument panel wiring harness
(A74) - Connector (15a - fuse 5), in instrument panel wiring harness
(A110) - Connection (fresh air blower), in instrument panel wiring harness
(L45) - Wire connection, in A/C wiring harness

Fresh air blower, fresh air blower switch, fresh air/recirculating flap switch, fresh air blower series resistance with fuse

New Beetle Wiring diagram **No. 8/3**

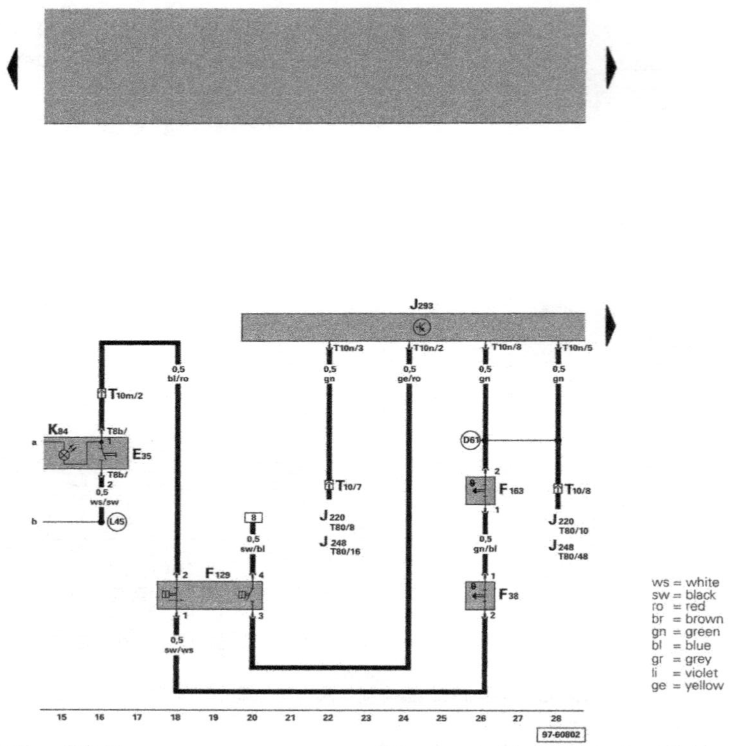

ws = white
sw = black
ro = red
br = brown
gn = green
bl = blue
gr = grey
li = violet
ge = yellow

E35 - A/C Switch
F38 - Ambient Temperature Switch
F129 - A/C Pressure Switch
F163 - A/C Cut-Out Thermal Switch
J220 - Motronic Engine Control Module (ECM), behind instrument panel , left
J248 - Diesel Direct Fuel Injection (DFI) Engine Control Module (ECM), behind instrument panel, left
J293 - Coolant FC (Fan Control) Control Module, in engine compartment
K84 - A/C Indicator Light
T6 - 6-Pin Connector, red, behind instrument panel, left
T10 - 10-pin Connector
T10m- 10-Pin Connector
T10n - 10-Pin Connector
T80 - Connector, 80 point, on Engine Control Module (ECM)
(D61) - wire connection (A/C), in engine compartment wiring harness
(L45) - Wire connection, in A/C system wiring harness

Coolant FC (fan control) control module, A/C cut-out thermal switch, ambient temperature sensor, A/C indicator light

No. 8/4 Wiring diagram **New Beetle**

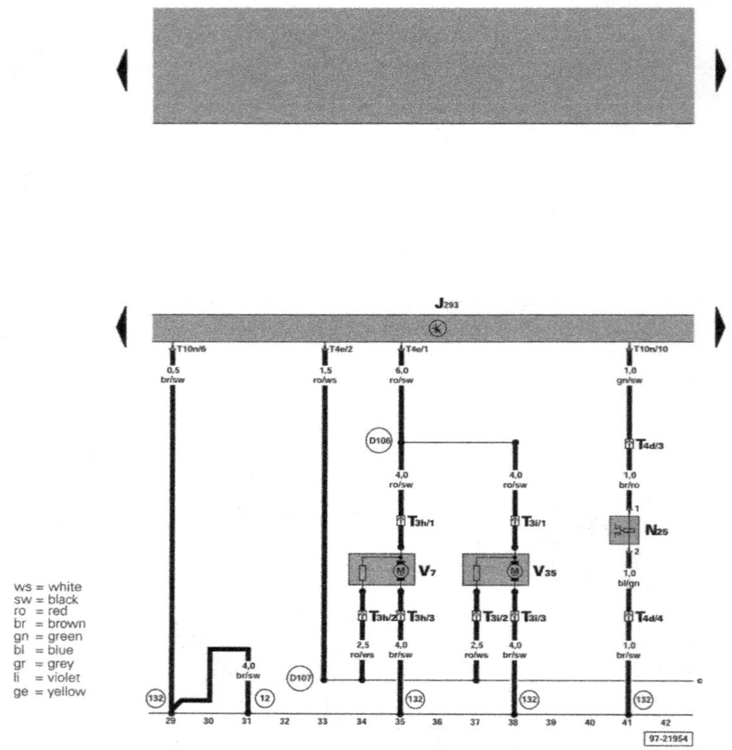

ws = white
sw = black
ro = red
br = brown
gn = green
bl = blue
gr = grey
li = violet
ge = yellow

J293 - Coolant FC (Fan Control) Control Module, in engine compartment
N25 - A/C Clutch
T3h - 3-Pin Connector, in engine compartment
T3i - 3-Pin Connector, in engine compartment
T4d - 4-Pin Connector, near A/C Clutch
T4e - 4-Pin Connector, on Coolant FC (Fan Control) Control Module
T10n - 10-Pin Connector
V7 - Coolant Fan
V35 - Right Coolant Fan
(12) - Ground connection, in engine compartment, left
(132) - Ground connection -3-, in engine compartment wiring harness
(D106) - wire connection -4-, in engine compartment wiring harness
(D107) - wire connection -5-, in engine compartment wiring harness

Coolant FC (fan control) control module, A/C clutch, coolant fan, right coolant fan

New Beetle Wiring diagram **No. 8/5**

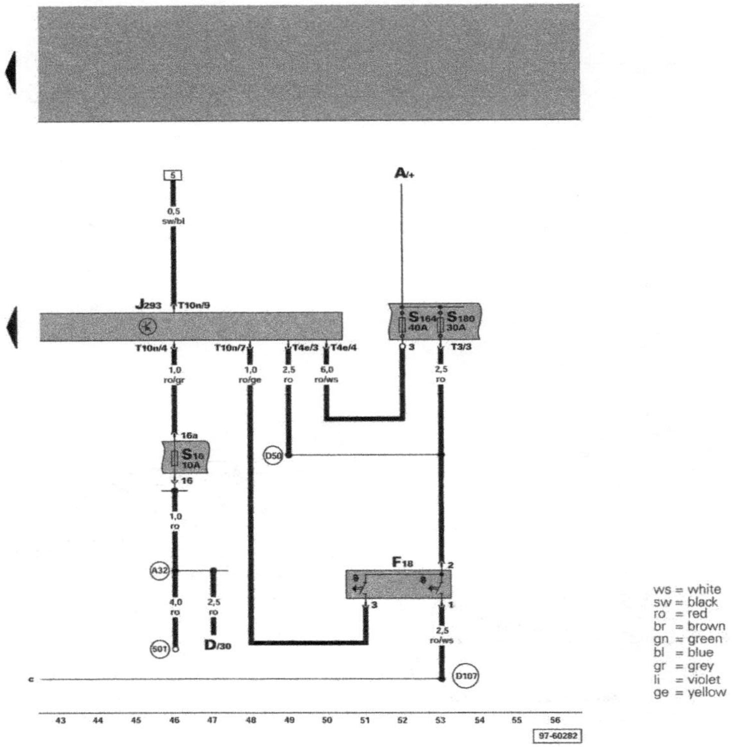

ws = white
sw = black
ro = red
br = brown
gn = green
bl = blue
gr = grey
li = violet
ge = yellow

A - Battery
D - Ignition/Starter Switch
F18 - Coolant Fan Control (FC) Thermal Switch
J293 - Coolant FC (Fan Control) Control Module
S16 - Fuse 16 in fuse holder
S164 - Fuse -3- in fuse bracket/battery
S180 - Fuse -8- (30), in fuse bracket/battery
T4e - 4-Pin Connector, on Coolant FC (Fan Control) Control Module
T10n - 10-Pin Connector, on Coolant FC (Fan Control) Control Module
(501) - Threaded connection -2- (30), on the relay plate
(A32) - plus connection (30), in instrument panel wiring harness
(D50) - plus connection (30), in engine compartment wiring harness
(D107) - wire connection -5-, in engine compartment wiring harness

Edition 09/98 **Coolant fan control (FC) module, coolant fan control (FC) thermal switch**
USA.5132.05.21

THIS SPACE INTENTIONALLY
LEFT BLANK

Air conditioning,

S 1,8L-Engine - Motronic Multiport Fuel Injection (MFI) / 110 kW, code APH,

from June 1999

Deviate relay location and fuseplacements as well as the locations of multiple connectors see section "component locations".

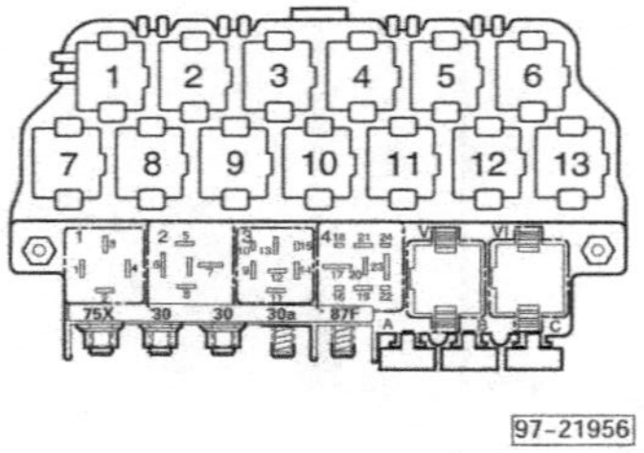

Relay location on the thirteen position auxiliary relay panel, above relay panel:

10 Third Speed Coolant Fan Control (FC) Relay (100)

Fuse colors
30 A - green
25 A - white
20 A - yellow
15 A - blue
10 A - red
7,5 A - brown
5 A - beige

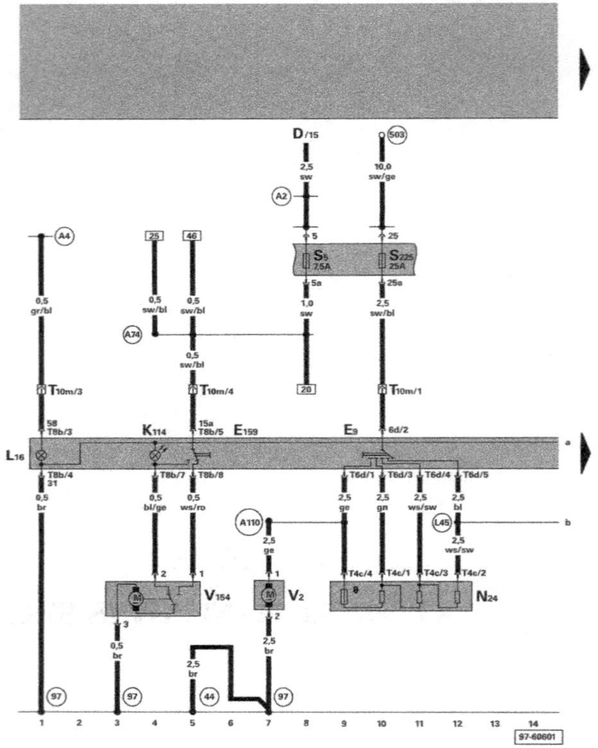

ws = white
sw = black
ro = red
br = brown
gn = green
bl = blue
gr = grey
li = violet
ge = yellow

D - Ignition/Starter Switch
E9 - Fresh Air Blower Switch
E159 - Fresh Air/Recirculating Flap Switch
K114 - Fresh Air and Recirculating Air Mode Indicator Light
L16 - Fresh Air Control Lever Light
N24 - Fresh Air Blower Series Resistance With Fuse
S5 - Fuse 5 in fuse holder
S225 - Fuse 25 in fuse holder
T4c - 4-Pin Connector
T6d - 6-Pin Connector
T8b - 8-Pin Connector
T10m - 10-Pin Connector
V2 - Fresh Air Blower
V154 - Servo motor for fresh-/recirculating air door

(44) - Ground connection, on left A-pillar, lower part
(97) - Ground connection -1-, in A/C wiring harness
(503) - Screw Connection (75x), on relay panel
(A2) - plus connection (15), in instrument panel wiring harness
(A4) - plus connection (58b), in instrument panel wiring harness
(A74) - Connector (15a - fuse 5), in instrument panel wiring harness
(A110) - Connection (fresh air blower), in instrument panel wiring harness
(L45) - Wire connection, in A/C wiring harness

Fresh air blower, fresh air blower switch, fresh air/recirculating flap switch, fresh air blower series resistance with fuse

Air conditioning (1.8L APH from June 1999)

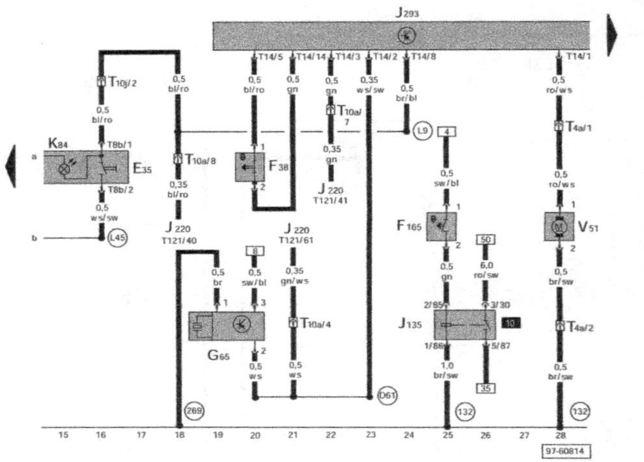

ws = white
sw = black
ro = red
br = brown
gn = green
bl = blue
gr = grey
li = violet
ge = yellow

E 35 - A/C Switch
G 65 - High Pressure Sensor
F38 - Ambient Temperature Switch
F165 - Third Speed Coolant Fan Control (FC) Thermal Switch
J135 - Third Speed Coolant Fan Control (FC) Relay
J220 - Motronic Engine Control Module (ECM)
J293 - Coolant FC (Fan Control) Control Module, in engine compartment
K84 - A/C Indicator Light
T4a - 4-pin Connector
T8b - 8-pin Connector
T10a - 10-pin Connector, orange, behind instrument panel left
T10j - 10-Pin Connector
T14 - 14-Pin Connector, on Coolant FC (Fan Control) Control Module
T121 - Connector, 121point, on Engine Control Module (ECM)
V51 - After-Run Coolant Pump
(L9) - Wire connection -1-, in A/C wiring harness
(L45) - Wire connection, in A/C system wiring harness
(132) - Ground connection -3-, in engine compartment wiring harness
(269) - Ground connector (sensor ground) -1-, in instrument panel wiring harness
(D61) - wire connection (A/C), in engine compart ment wiring harness

Coolant FC (fan control) control module, A/C cut-out thermal switch, ambient temperature sensor, A/C indicator light, After-run coolant pump

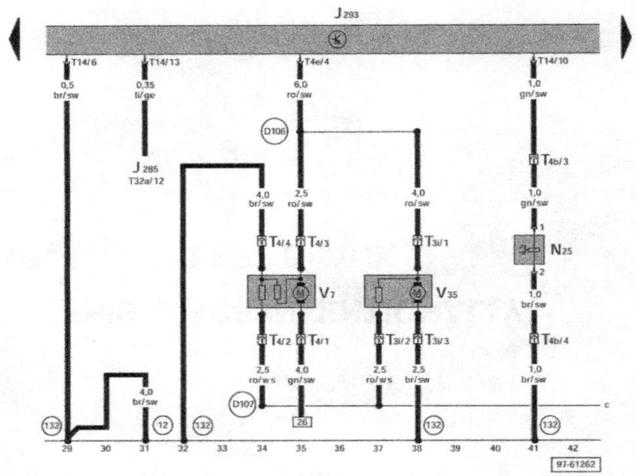

ws = white
sw = black
ro = red
br = brown
gn = green
bl = blue
gr = grey
li = violet
ge = yellow

J285 - Control module with indicator unit in instrument panel insert
J293 - Coolant FC (Fan Control) Control Module, in engine compartment
N25 - A/C Clutch
T3h - 3-Pin Connector, in engine compartment
T3i - 3-Pin Connector, in engine compartment
T4 4-Pin Connector
T4b - 4-Pin Connector, near A/C Clutch
T4e - 4-Pin Connector, on Coolant FC (Fan Control) Control Module
T14 - 14-Pin Connector, on Coolant FC (Fan Control) Control Module
T32a 32-Pin Connector, green
V7 - Coolant Fan
V35 - Right Coolant Fan
(12) - Ground connection, in engine compartment, left
(132) - Ground connection -3-, in engine compartment wiring harness
(D106) - wire connection -4-, in engine compartment wiring harness
(D107) - wire connection -5-, in engine compartment wiring harness

Coolant FC (fan control) control module, A/C clutch, coolant fan, right coolant fan

Air conditioning (1.8L APH from June 1999)

New Beetle Wiring diagram **No. 34/5**

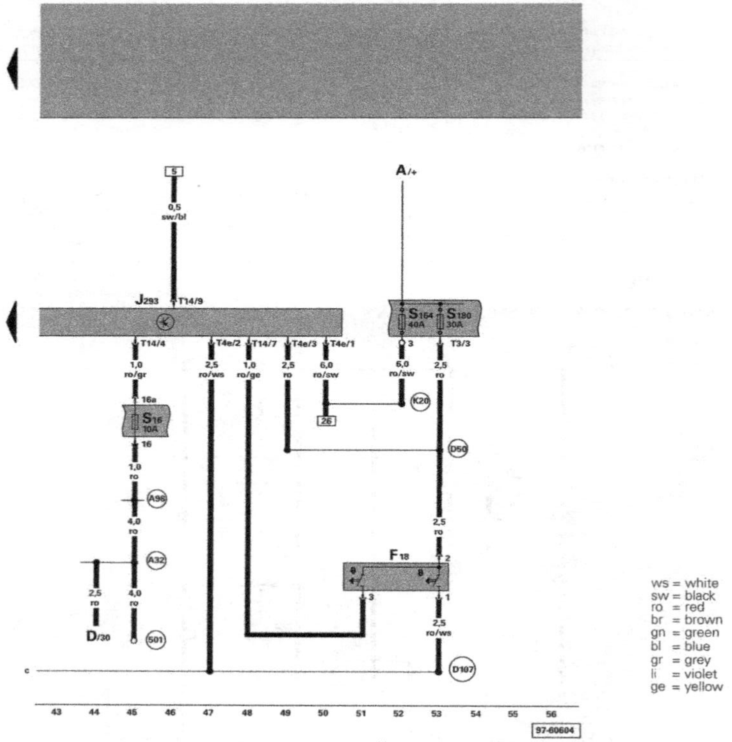

ws = white
sw = black
ro = red
br = brown
gn = green
bl = blue
gr = grey
li = violet
ge = yellow

- A - Battery
- D - Ignition/Starter Switch
- F18 - Coolant Fan Control (FC) Thermal Switch
- J293 - Coolant FC (Fan Control) Control Module
- S16 - Fuse 16 in fuse holder
- S164 - Fuse -3- in fuse bracket/battery
- S180 - Fuse -8- (30), in fuse bracket/battery
- T4e - 4-Pin Connector, on Coolant FC (Fan Control) Control Module
- T14 - 14-Pin Connector, on Coolant FC (Fan Control) Control Module
- (501) - Threaded connection -2- (30), on the relay plate
- (A32) - plus connection (30), in instrument panel wiring harness
- (A96) - Plus connector -4- (30), in instrument panel wiring harness
- (D50) - plus connection (30), in engine compartment wiring harness
- (D107) - wire connection -5-, in engine compartment wiring harness
- (K20) - plus connection -1-(30), in coolant fan wiring harness

Edition 10/99 **Coolant fan control (FC) module, coolant fan control (FC) thermal switch**
USA.5132.09.21

THIS SPACE INTENTIONALLY LEFT BLANK

Air conditioning,

- **1,8L-Engine - Motronic Multiport Fuel Injection (MFI) / 110 kW, code AWV,**

from June 2000

- **1,8L-Engine - Motronic Multiport Fuel Injection (MFI) / 132 kW, code AWP,**

from October 2001

Deviate relay location and fuseplacements as well as the locations of multiple connectors see section "component locations".

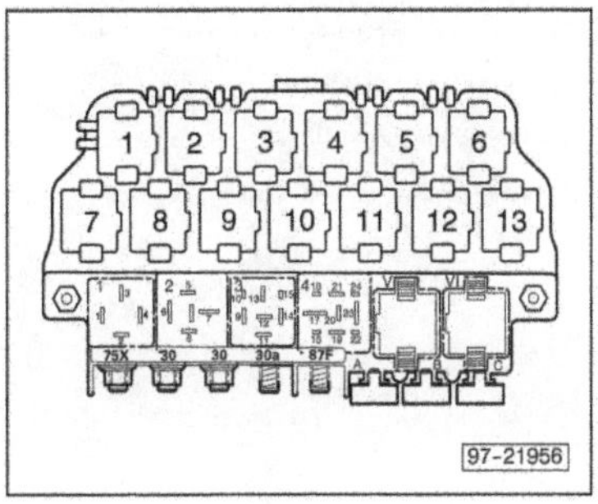

Relay location on the thirteen position auxiliary relay panel, above relay panel:

10 - Third Speed Coolant Fan Control (FC) Relay (100)

Fuse colors

30 A - green
25 A - white
20 A - yellow
15 A - blue
10 A - red
7,5 A - brown
5 A - beige
3 A - violet

Note: Number in parentheses indicates production control number stamped on relay housing.

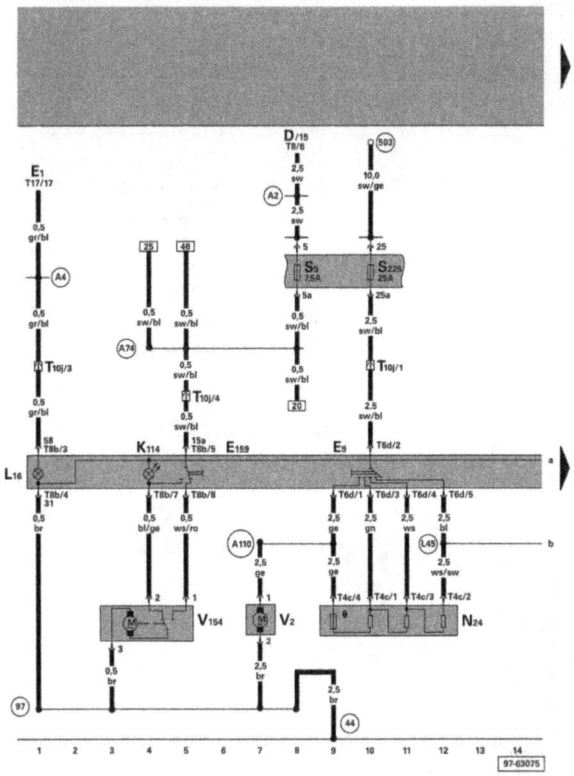

ws = white
sw = black
ro = red
br = brown
gn = green
bl = blue
gr = grey
li = violet
ge = yellow
or = orange

Fresh air blower, fresh air blower switch, fresh air/recirculating flap switch, fresh air blower series resistance with fuse

D - Ignition/Starter Switch
E1 - Light switch
E9 - Fresh Air Blower Switch
E159 - Fresh Air/Recirculating Flap Switch
K114 - Fresh Air and Recirculating Air Mode Indicator Light
L16 - Fresh Air Control Lever Light
N24 - Fresh Air Blower Series Resistance With Fuse
S5 - Fuse 5 in fuse holder
S225 - Fuse 25 in fuse holder
T4c - 4-Pin Connector
T6d - 6-Pin Connector
T8 - 8-Pin Connector, at -D-
T8b - 8-Pin Connector
T10j - 10-Pin Connector
T17 - 17-Pin Connector, at -E1-
V2 - Fresh Air Blower
V154 - Servo motor for fresh-/recirculating air door

(44) - Ground connection, on left A-pillar, lower part
(97) - Ground connection -1-, in A/C wiring harness
(503) - Screw Connection (75x), on relay panel
(A2) - plus connection (15), in instrument panel wiring harness
(A4) - plus connection (58b), in instrument panel wiring harness
(A74) - Connector (15a - fuse 5), in instrument panel wiring harness
(A110) - Connection (fresh air blower), in instrument panel wiring harness
(L45) - Wire connection, in A/C wiring harness

Air conditioning (1.8L AWV from June 1999) (1.8L AWP from Oct. 2001)

New Beetle **Wiring diagram** **No. 8/3**

Coolant FC (fan control) control module, A/C cut-out thermal switch, ambient temperature switch, A/C indicator light, After-run coolant pump

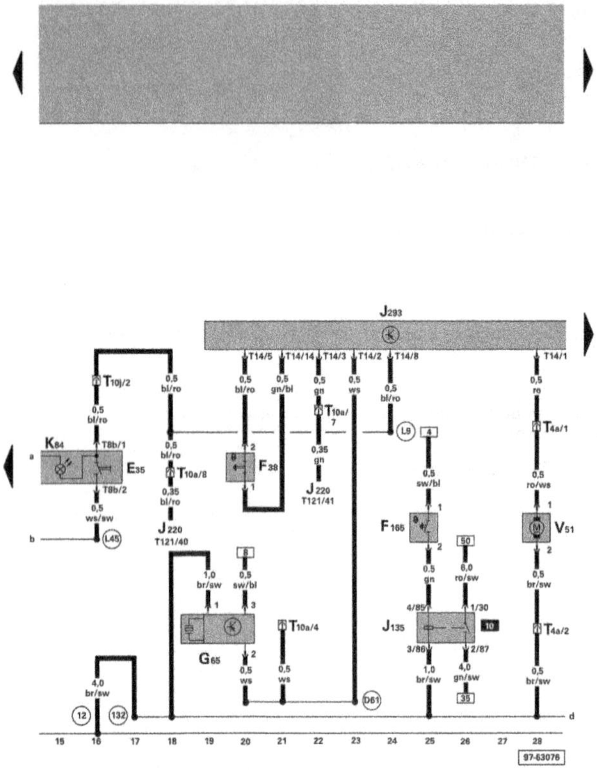

E 35 - A/C Switch
G 65 - High Pressure Sensor
F38 - Ambient Temperature Switch
F165 - Third Speed Coolant Fan Control (FC) Thermal Switch
J135 - Third Speed Coolant Fan Control (FC) Relay
J220 - Motronic Engine Control Module (ECM)
J293 - Coolant FC (Fan Control) Control Module, in engine compartment
K84 - A/C Indicator Light
T4a - 4-pin Connector
T8b - 8-pin Connector
T10a - 10-pin Connector, orange, behind instrument panel left
T10j - 10-Pin Connector
T14 - 14-Pin Connector, on Coolant FC (Fan Control) Control Module
T121 - Connector, 121point, on Engine Control Module (ECM)
V51 - After-Run Coolant Pump

(12) - Ground connection, in engine compartment, left
(132) - Ground connection -3-, in engine compartment wiring harness
(D61) - wire connection (A/C), in engine compart ment wiring harness
(L9) - Wire connection -1-, in A/C wiring harness
(L45) - Wire connection, in A/C system wiring harness

ws = white
sw = black
ro = red
br = brown
gn = green
bl = blue
gr = grey
li = violet
ge = yellow
or = orange

Edition 11.03

New Beetle **Wiring diagram** **No. 8/4**

Coolant FC (fan control) control module, A/C clutch, coolant fan, right coolant fan

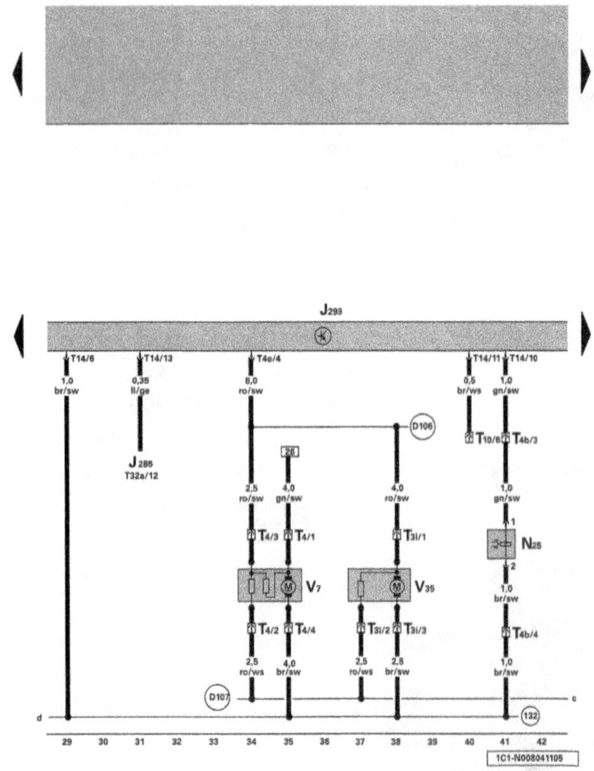

J285 - Control module with indicator unit in instrument panel insert
J293 - Coolant FC (Fan Control) Control Module, in engine compartment
N25 - A/C Clutch
T3i - 3-Pin Connector, in engine compartment
T4 4-Pin Connector
T4b - 4-Pin Connector, near A/C Clutch
T4e - 4-Pin Connector, on Coolant FC (Fan Control) Control Module
T10 - 10-Pin Connector, white, behind instrument panel, left
T14 - 14-Pin Connector, on Coolant FC (Fan Control) Control Module
T32a 32-Pin Connector, green
V7 - Coolant Fan
V35 - Right Coolant Fan

(132) - Ground connection -3-, in engine compartment wiring harness
(D106) - wire connection -4-, in engine compartment wiring harness
(D107) - wire connection -5-, in engine compartment wiring harness

ws = white
sw = black
ro = red
br = brown
gn = green
bl = blue
gr = grey
li = violet
ge = yellow
or = orange

Edition 11.05

New Beetle **Wiring diagram** **No. 8/5**

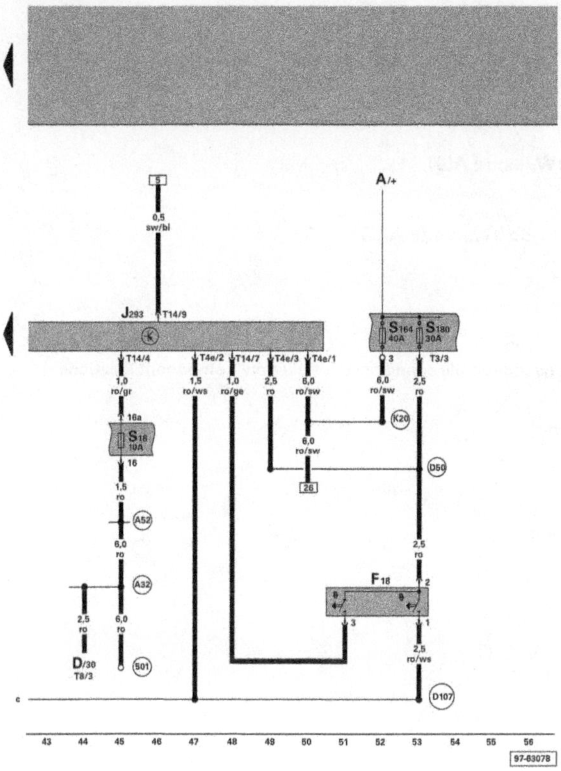

Coolant fan control (FC) module, coolant fan control (FC) thermal switch

A - Battery
D - Ignition/Starter Switch
F18 - Coolant Fan Control (FC) Thermal Switch
J293 - Coolant FC (Fan Control) Control Module
S16 - Fuse 16 in fuse holder
S164 - Fuse -3- in fuse bracket/battery
S180 - Fuse -8- (30), in fuse bracket/battery
T3 - 3-Pin Connector
T4e - 4-Pin Connector, on Coolant FC (Fan Control) Control Module
T8 - 8-Pin Connector, at -D-
T14 - 14-Pin Connector, on Coolant FC (Fan Control) Control Module
(501) - Threaded connection -2- (30), on the relay plate
(A32) - plus connection (30), in instrument panel wiring harness
(A52) - Plus Connection 2 (30) (in instrument panel wiring harness)
(D50) - plus connection (30), in engine compartment wiring harness
(D107) - wire connection -5-, in engine compartment wiring harness
(K20) - plus connection -1-(30), in coolant fan wiring harness

ws = white
sw = black
ro = red
br = brown
gn = green
bl = blue
gr = grey
li = violet
ge = yellow
or = orange

THIS SPACE INTENTIONALLY LEFT BLANK

Air conditioning (1.9L ALH from June 1999) (2.0L AZG from Aug. 2000)

New Beetle Wiring diagram **No. 7/1**

Air conditioning,

- **1,9L-Engine - Turbo Diesel Fuel Injection (DFII) / 66 kW, code ALH,**

from June 1999

- **2,0L-Engine - Motronic Multiport Fuel Injection (MFI) / 85 kW, code AZG**

from August 2000

Deviate relay location and fuseplacements as well as the locations of multiple connectors see section "component locations".

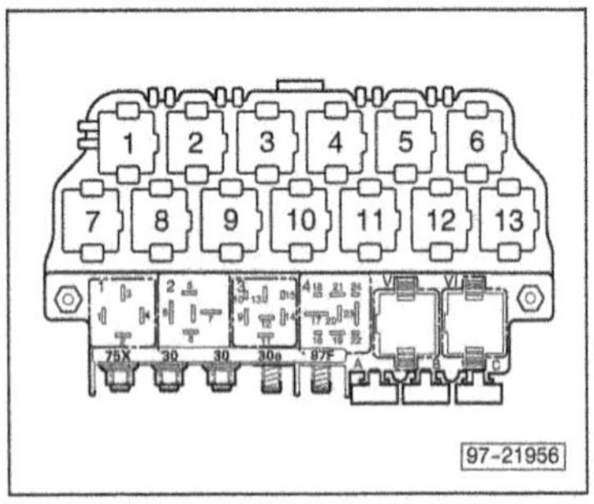

Fuse colors

30 A - green
25 A - white
20 A - yellow
15 A - blue
10 A - red
7,5 A - brown
5 A - beige
3 A - violet

New Beetle Wiring diagram **No. 7/2**

Fresh air blower switch, fresh air/recirculating flap, fresh air control level light, servo motor for fresh/recirculating air door, fresh air blower series resistance with fuse

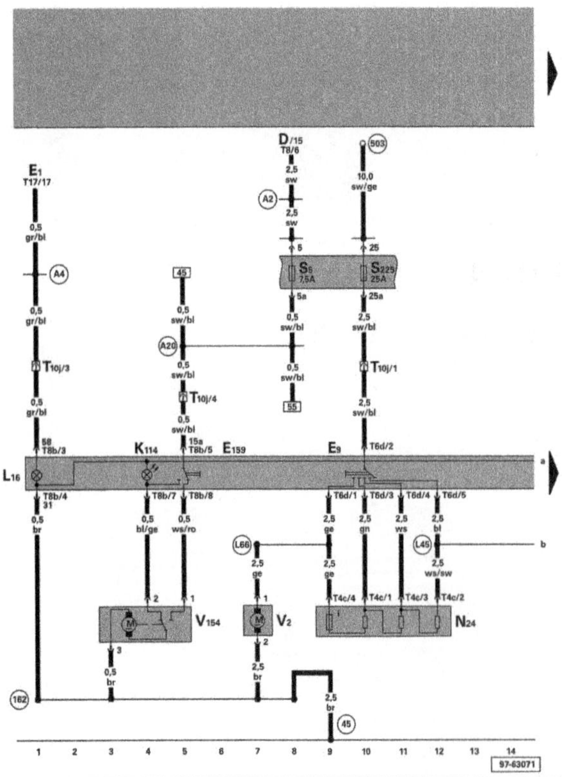

- D - Ignition/Starter Switch
- E1 - Light switch
- E9 - Fresh Air Blower Switch
- E159 - Fresh Air/Recirculating Flap Switch
- K114 - Fresh/Recirculating Air Mode Indicator Light
- L16 - Fresh Air Control Lever Light
- N24 - Fresh Air Blower Series Resistance With Fuse
- S5 - Fuse 5 in fuse holder
- S225 - Fuse in fuse holder
- T4c - 4-Pin Connector
- T6d - 6-Pin Connector
- T8 - 8-Pin Connector, at -D-
- T8b - 8-Pin Connector
- T10j - 10-Pin Connector
- T17 - 17-Pin Connector, at -E1-
- V2 - Fresh Air Blower
- V154 - Servo motor for fresh/recirculating air door

- (45) - Ground connection, behind instrument panel, center
- (162) - Ground connection, in blower motor wiring harness
- (503) - Threaded connection -2- (75x) on the relay plate
- (A2) - plus connection (15), in instrument panel wiring harness
- (A4) - plus connection (58b), in instrument panel wiring harness
- (A20) - wire connection (15a), in instrument panel wiring harness
- (L45) - Wire connection, in A/C system wiring harness
- (L66) - Connector, in wiring harness heater blower

ws = white
sw = black
ro = red
br = brown
gn = green
bl = blue
gr = grey
li = violet
ge = yellow
or = orange

New Beetle

Wiring diagram

No. 7/3

A/C Switch, ambient temperature switch, Coolant FC (Fan Control) Control Module, A/C indicator light.

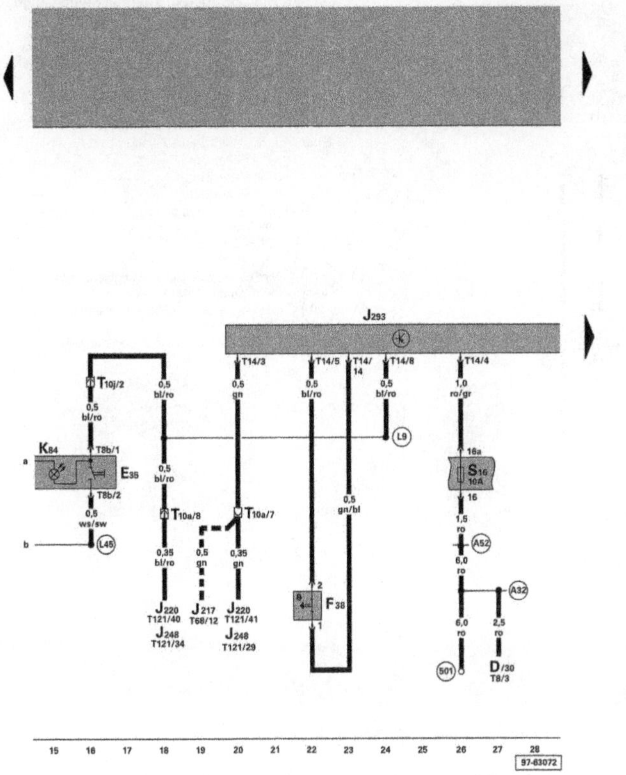

- D - Ignition/Starter Switch
- E35 - A/C Switch
- F38 - Ambient Temperature Switch
- J217 - Automatic Transmission Control Module (TCM)
- J220 - Motronic Engine Control Module (ECM)
- J248 - Diesel Direct Fuel Injection (DFI) Engine Control Module (ECM)
- J293 - Coolant FC (Fan Control) Control Module
- K84 - A/C Indicator Light
- S16 - wire connection (passenger door), in central
- T8 - 8-Pin connector, at -D-
- T8b - 8-Pin connector
- T10a - 10-Pin connector, orange, behind instrument panel, left.
- T10j - 10-Pin connector
- T14 - 14-Pin connector
- T68 - 68-Pin connector, at TCM
- T121 - 121-Pin connector, at ECM
- (501) - Threaded connection -2- (30) on the relay plate
- (A32) - Plus connection (30), in instrument panel wiring harness
- (A52) - Plus Connection 2 (30) (in instrument panel wiring harness)
- (L9) - Wire connection -1-, in A/C wiring harness
- (L45) - Wire connection, in A/C system wiring harness
- --- - Automatic transmission only

ws = white
sw = black
ro = red
br = brown
gn = green
bl = blue
gr = grey
li = violet
ge = yellow
or = orange

New Beetle

Wiring diagram

No. 7/4

Coolant Fan Control (FC) Module, A/C clutch, right coolant fan

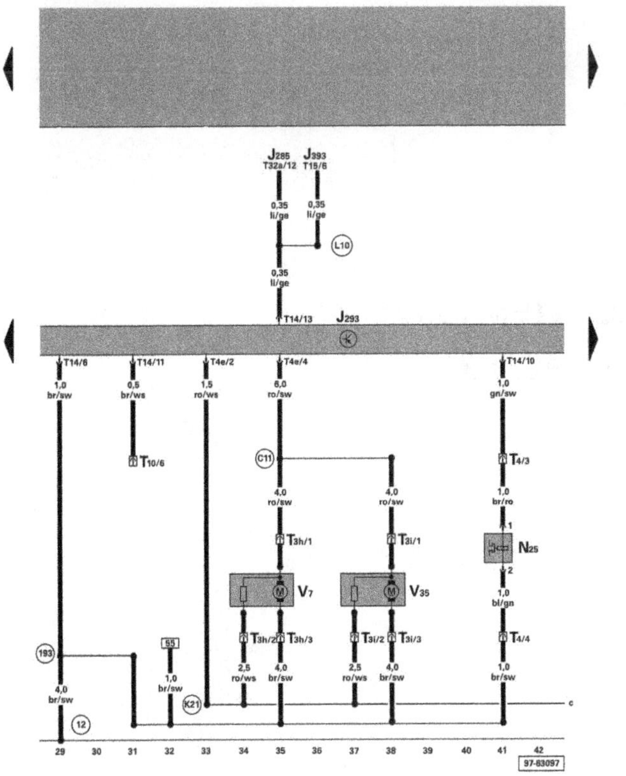

- J285 - Control module with indicator unit in instrument panel insert
- J293 - Coolant FC (Fan Control) Control Module
- J393 - Central control module for comfort system
- N25 - A/C Clutch
- T3h - 3-Pin Connector, in engine compartment
- T3i - 3-Pin Connector, in engine compartment
- T4 - 4-Pin Connector, near A/C Clutch
- T4e - 4-Pin Connector, on coolant FC (Fan Control) Control Module
- T10 - 10-Pin Connector, white, behind instrument panel, left
- T14 - 14-Pin Connector
- T15 - 15-Pin Connector, on -J393-
- T32a - 32-Pin Connector, green, on instrument cluster
- V7 - Coolant Fan
- V35 - Right Coolant Fan
- (12) - Ground connection, in engine compartment, left
- (193) - Ground connection -1-, in coolant fan wiring harness
- (C11) - Wire connection, in coolant fan wiring harness
- (K21) - wire connection -1-, in coolant fan wiring harness
- (L10) - wire connection -2-, in A/C wiring harness

ws = white
sw = black
ro = red
br = brown
gn = green
bl = blue
gr = grey
li = violet
ge = yellow
or = orange

Air conditioning (1.9L ALH from June 1999) (2.0L AZG from Aug. 2000)

New Beetle Wiring diagram **No. 7/5**

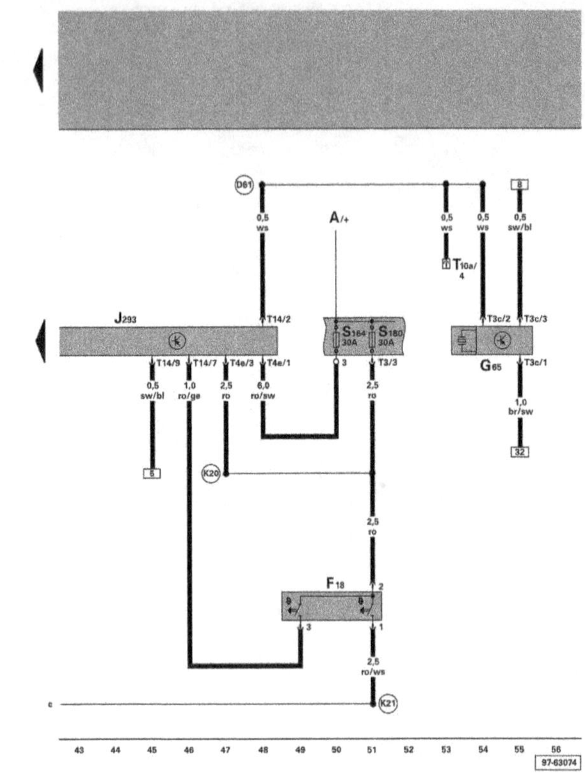

Coolant Fan Control (FC) Thermal Switch, high pressure sensor

- A - Battery
- F18 - Coolant Fan Control (FC) Thermal Switch
- G65 - High Pressure Sensor
- J293 - Coolant FC (Fan Control) Control Module
- S164 - Fuse -3- in fuse bracket/battery
- S180 - Fuse -8- (30), in fuse bracket/battery
- T3 - 3-Pin Connector
- T3c - 3-Pin Connector
- T4e - 4-Pin Connector
- T10a - 10-Pin connector, orange, behind instrument panel, left.
- T14 - 14-Pin Connector
- (D61) - wire connection (A/C), in engine compartment wiring harness
- (K20) - plus connection -1-(30), in coolant fan wiring harness
- (K21) - wire connection -1-, in coolant fan wiring harness

ws = white
sw = black
ro = red
br = brown
gn = green
bl = blue
gr = grey
li = violet
ge = yellow
or = orange

THIS SPACE INTENTIONALLY

LEFT BLANK

New Beetle **Wiring diagram** **No. 39/1**

Air conditioning,

- **1,9L-Engine - Turbo Diesel Fuel Injection (DFII) / 74 kW, code BEW,**

from June 2003

Additional Fuse, Relay, Control Module and Connector locations ⇒ Compontent Locations

Troubleshooting Procedures ⇒ Guided Fault Finding using VAS 5051/5052

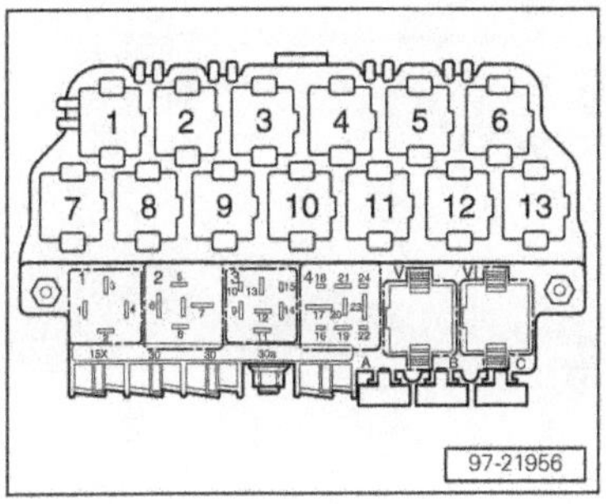

Relay location on the thirteenfold auxiliary relay panel, above relay panel:

Relay panel:

New Beetle **Wiring diagram** **No. 39/2**

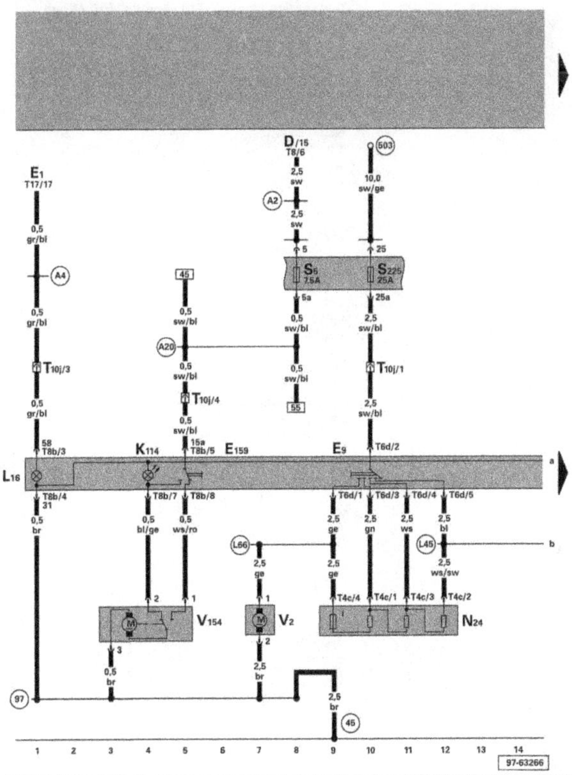

ws = white
sw = black
ro = red
br = brown
gn = green
bl = blue
gr = grey
li = violet
ge = yellow
or = orange

Fresh air blower switch, fresh air/recirculating flap swith, fresh air control level light, fresh air blower series resistance with fuse, servo motor for fresh/recirculating air door

- D - Ignition/Starter Switch
- E1 - Light Switch
- E9 - Fresh Air Blower Switch
- E159 - Fresh Air/Recirculating Flap Switch
- K114 - Fresh/Recirculating Air Mode Indicator Light
- L16 - Fresh Air Control Lever Light
- N24 - Fresh Air Blower Series Resistance With Fuse
- S5 - Fuse 5 in fuse holder
- S225 - Fuse in fuse holder
- T4c - 4-Pin Connector
- T6d - 6-Pin Connector
- T8 - 8-Pin Connector, at -D-
- T8b - 8-Pin Connector
- T10j - 10-Pin Connector
- T17 - 17-Pin Connector, at -E1-
- V2 - Fresh Air Blower
- V154 - Servo motor for fresh/recirculating air door

- (45) - Ground connection, behind instrument panel, center
- (97) - Ground Connection 1, in A/C wiring harness
- (503) - Threaded connection -2- (75x) on the relay plate
- (A2) - plus connection (15), in instrument panel wiring harness
- (A4) - plus connection (58b), in instrument panel wiring harness
- (A20) - wire connection (15a), in instrument panel wiring harness
- (L45) - Wire connection, in A/C system wiring harness
- (L66) - Connector, in wiring harness heater blower

Air conditioning (1.9L BEW from June 2003)

New Beetle — Wiring diagram — No. 39/3

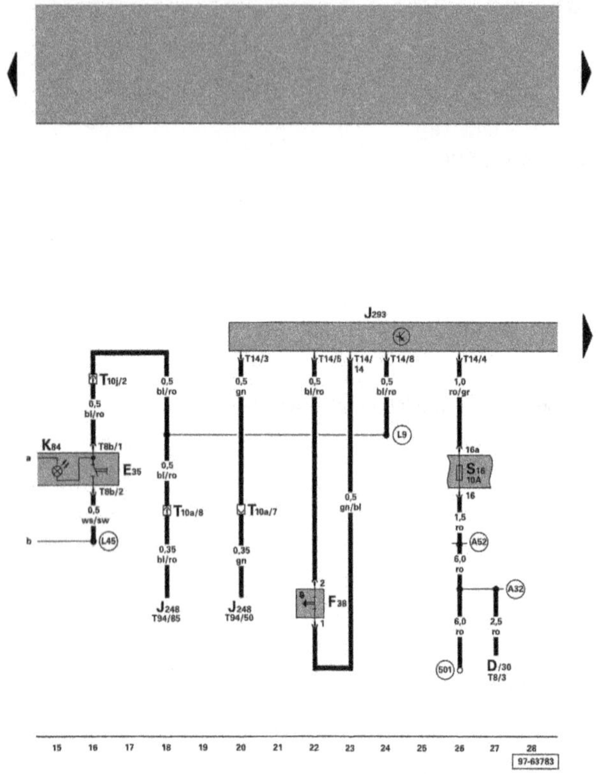

ws = white
sw = black
ro = red
br = brown
gn = green
bl = blue
gr = grey
li = violet
ge = yellow
or = orange

A/C Switch, ambient temperature switch, Coolant FC (Fan Control) Control Module, A/C indicator light.

- D - Ignition/Starter Switch
- E35 - A/C Switch
- F38 - Ambient Temperature Switch
- J248 - Diesel Direct Fuel Injection (DFI) Engine Control Module (ECM)
- J293 - Coolant FC (Fan Control) Control Module
- K84 - A/C Indicator Light
- S16 - wire connection (passenger door), in central
- T8 - 8-Pin connector, at -D-
- T8b - 8-Pin connector
- T10a - 10-Pin connector, orange, behind instrument panel, left.
- T10j - 10-Pin connector
- T14 - 14-Pin connector
- T94 - 94-Pin connector, at J248
- (501) - Threaded connection -2- (30) on the relay plate
- (A32) - Plus connection (30), in instrument panel wiring harness
- (A52) - Plus Connection 2 (30) (in instrument panel wiring harness)
- (L9) - Wire connection -1-, in A/C wiring harness
- (L45) - Wire connection, in A/C system wiring harness

New Beetle — Wiring diagram — No. 39/4

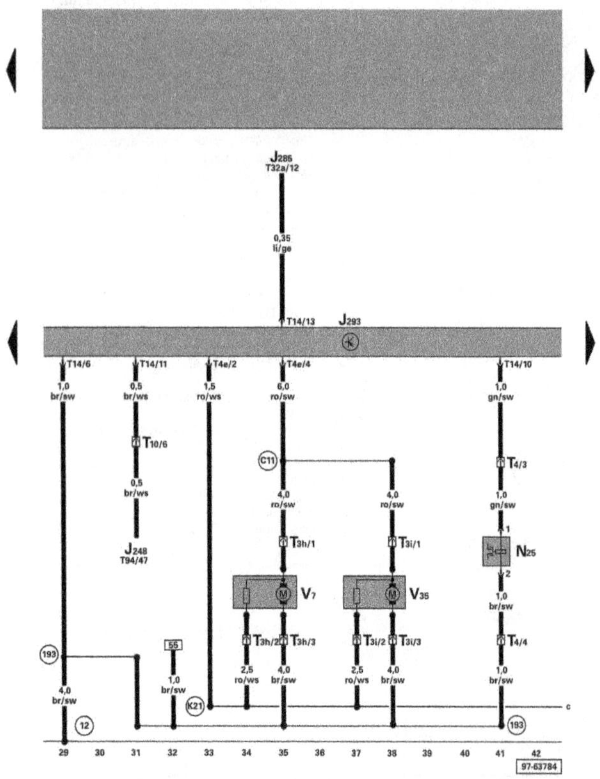

ws = white
sw = black
ro = red
br = brown
gn = green
bl = blue
gr = grey
li = violet
ge = yellow
or = orange

Coolant Fan Control (FC) Module, A/C clutch, Coolant Fan, right coolant fan

- J248 - Diesel Direct Fuel Injection (DFI) Engine Control Module (ECM)
- J285 - Instrument Cluster Control Module
- J293 - Coolant FC (Fan Control) Control Module
- N25 - A/C Clutch
- T3h - 3-Pin Connector, in engine compartment
- T3i - 3-Pin Connector, in engine compartment
- T4 - 4-Pin Connector, near A/C Clutch
- T4e - 4-Pin Connector, on coolant FC (Fan Control) Control Module
- T10 - 10-Pin Connector, white, behind instrument panel, left
- T14 - 14-Pin Connector
- T32a - 32-Pin Connector, green, on instrument cluster
- T94 - 94-Pin Connector, at J248
- V7 - Coolant Fan
- V35 - Right Coolant Fan
- (12) - Ground connection, in engine compartment, left
- (193) - Ground connection -1-, in coolant fan wiring harness
- (C11) - Wire connection, in coolant fan wiring harness
- (K21) - wire connection -1-, in coolant fan wiring harness

New Beetle **Wiring diagram** **No. 39/5**

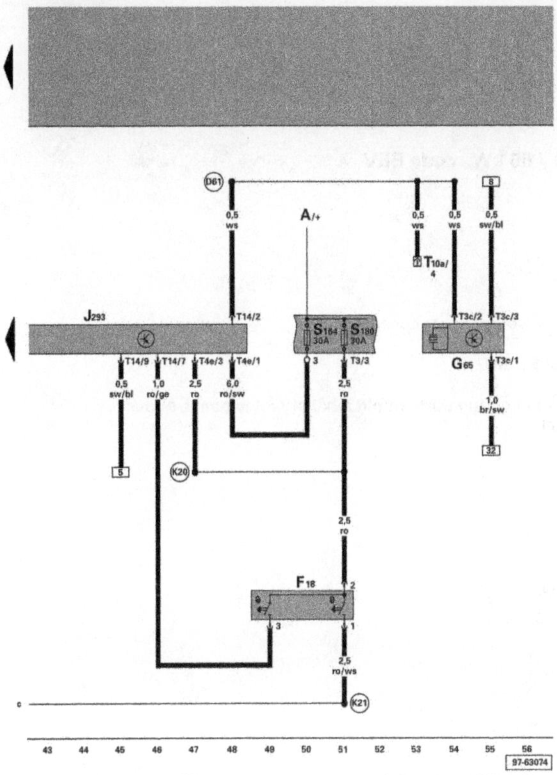

Coolant Fan Control (FC) Thermal Switch, high pressure sensor, Coolant FC (Fan Control) Control Module

- A - Battery
- F18 - Coolant Fan Control (FC) Thermal Switch
- G65 - High Pressure Sensor
- J293 - Coolant FC (Fan Control) Control Module
- S164 - Fuse -3- in fuse bracket/battery
- S180 - Fuse -8- (30), in fuse bracket/battery
- T3 - 3-Pin Connector
- T3c - 3-Pin Connector
- T4e - 4-Pin Connector
- T10a - 10-Pin connector, orange, behind instrument panel, left.
- T14 - 14-Pin Connector
- (D61) - wire connection (A/C), in engine compartment wiring harness
- (K20) - plus connection -1-(30), in coolant fan wiring harness
- (K21) - wire connection -1-, in coolant fan wiring harness

ws = white
sw = black
ro = red
br = brown
gn = green
bl = blue
gr = grey
li = violet
ge = yellow
or = orange

THIS SPACE INTENTIONALLY LEFT BLANK

New Beetle Wiring diagram **No. 42/1**

Air conditioning,

- **2,0L-Engine - Motronic Multiport Fuel Injection (MFI) / 85 kW, code BEV**

from June 2003

Additional Fuse, Relay, Control Module and Connector locations ⇒ Compontent Locations

Troubleshooting Procedures ⇒ Guided Fault Finding using VAS 5051/5052

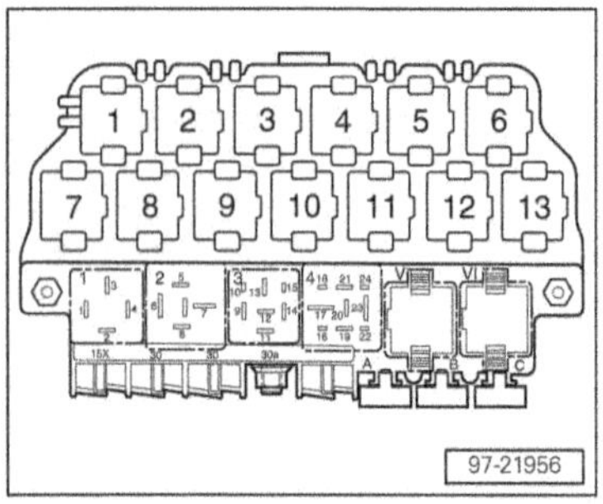

Relay location on the thirteenfold auxiliary relay panel, above relay panel:

Relay panel:

New Beetle Wiring diagram **No. 42/2**

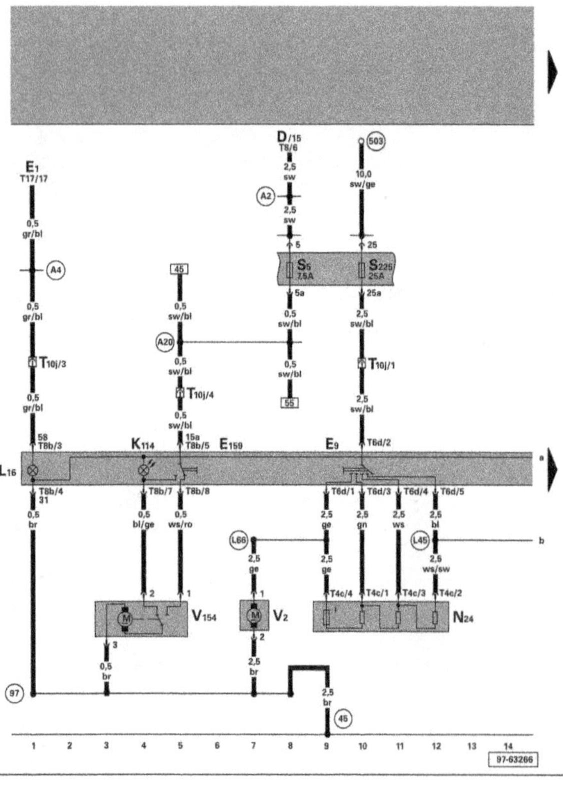

Fresh air blower switch, fresh air/recirculating flap switch, fresh air control level light, Fresh Air Blower Series Resistance With Fuse, Fresh Air Blower, Servo motor for fresh/recirculating air door

- D - Ignition/Starter Switch
- E1 - Light switch
- E9 - Fresh Air Blower Switch
- E159 - Fresh Air/Recirculating Flap Switch
- K114 - Fresh/Recirculating Air Mode Indicator Light
- L16 - Fresh Air Control Lever Light
- N24 - Fresh Air Blower Series Resistance With Fuse
- S5 - Fuse 5 in fuse holder
- S225 - Fuse in fuse holder
- T4c - 4-Pin Connector
- T6d - 6-Pin Connector
- T8 - 8-Pin Connector, at -D-
- T8b - 8-Pin Connector
- T10j - 10-Pin Connector
- T17 - 17-Pin Connector, at -E1-
- V2 - Fresh Air Blower
- V154 - Servo motor for fresh/recirculating air door

- (45) - Ground connection, behind instrument panel, center
- (97) - Ground Connection 1, in A/C wiring harness
- (503) - Threaded connection -2- (75x) on the relay plate
- (A2) - plus connection (15), in instrument panel wiring harness
- (A4) - plus connection (58b), in instrument panel wiring harness
- (A20) - wire connection (15a), in instrument panel wiring harness
- (L45) - Wire connection, in A/C system wiring harness
- (L66) - Connector, in wiring harness heater blower

ws = white
sw = black
ro = red
br = brown
gn = green
bl = blue
gr = grey
li = violet
ge = yellow
or = orange

New Beetle — Wiring diagram — No. 42/3

A/C Switch, ambient temperature switch, Coolant FC (Fan Control) Control Module, A/C indicator light.

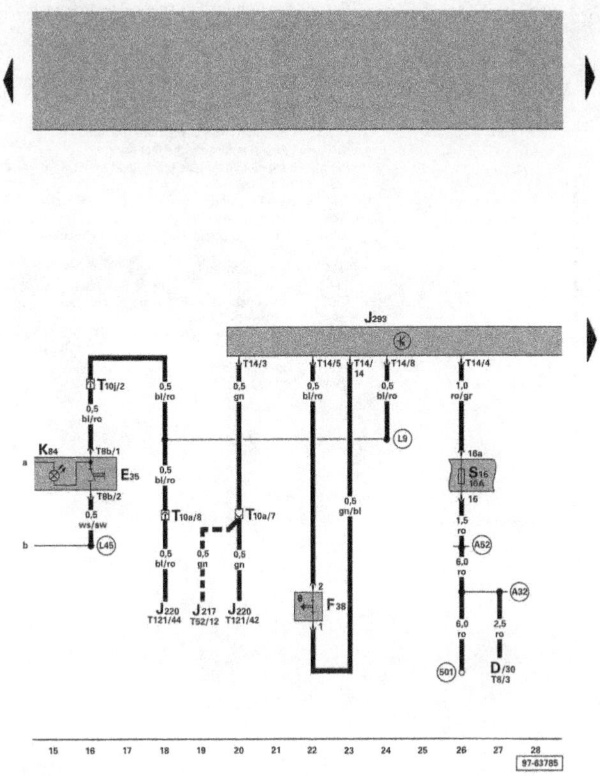

- D - Ignition/Starter Switch
- E35 - A/C Switch
- F38 - Ambient Temperature Switch
- J217 - Automatic Transmission Control Module (TCM)
- J220 - Motronic Engine Control Module (ECM)
- J293 - Coolant FC (Fan Control) Control Module
- K84 - A/C Indicator Light
- S16 - wire connection (passenger door), in central
- T8 - 8-Pin connector, at -D-
- T8b - 8-Pin connector
- T10a - 10-Pin connector, orange, behind instrument panel, left.
- T10j - 10-Pin connector
- T14 - 14-Pin connector
- T52 - 52-Pin connector, at J217
- T121 - 121-Pin connector, at J220
- (501) - Threaded connection -2- (30) on the relay plate
- (A32) - Plus connection (30), in instrument panel wiring harness
- (A52) - Plus Connection 2 (30) (in instrument panel wiring harness)
- (L9) - Wire connection -1-, in A/C wiring harness
- (L45) - Wire connection, in A/C system wiring harness
- --- - Automatic transmission, only

ws = white
sw = black
ro = red
br = brown
gn = green
bl = blue
gr = grey
li = violet
ge = yellow
or = orange

Edition 11.04

New Beetle — Wiring diagram — No. 42/4

Coolant Fan Control (FC) Module, A/C clutch, Coolant Fan, right coolant fan

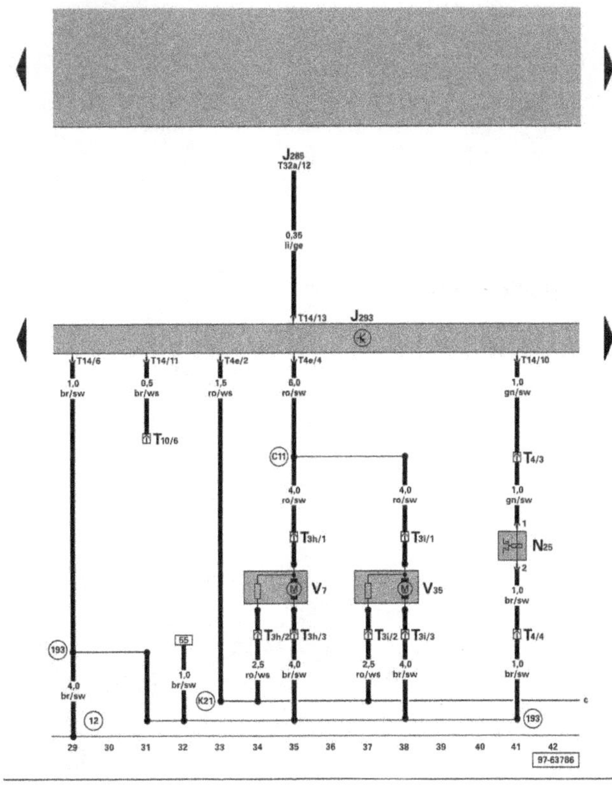

- J285 - Instrument Cluster Control Module
- J293 - Coolant FC (Fan Control) Control Module
- N25 - A/C Clutch
- T3h - 3-Pin Connector, in engine compartment
- T3i - 3-Pin Connector, in engine compartment
- T4 - 4-Pin Connector, near A/C Clutch
- T4e - 4-Pin Connector, on coolant FC (Fan Control) Control Module
- T10 - 10-Pin Connector, white, behind instrument panel, left
- T14 - 14-Pin Connector
- T32a - 32-Pin Connector, green, on instrument cluster
- V7 - Coolant Fan
- V35 - Right Coolant Fan
- (12) - Ground connection, in engine compartment, left
- (193) - Ground connection -1-, in coolant fan wiring harness
- (C11) - Wire connection, in coolant fan wiring harness
- (K21) - wire connection -1-, in coolant fan wiring harness

ws = white
sw = black
ro = red
br = brown
gn = green
bl = blue
gr = grey
li = violet
ge = yellow
or = orange

Edition 11.04

Air conditioning (2.0L BEV from June 2003)

New Beetle Wiring diagram **No. 42/5**

Coolant Fan Control (FC) Thermal Switch, high pressure sensor, Coolant FC (Fan Control) Control Module

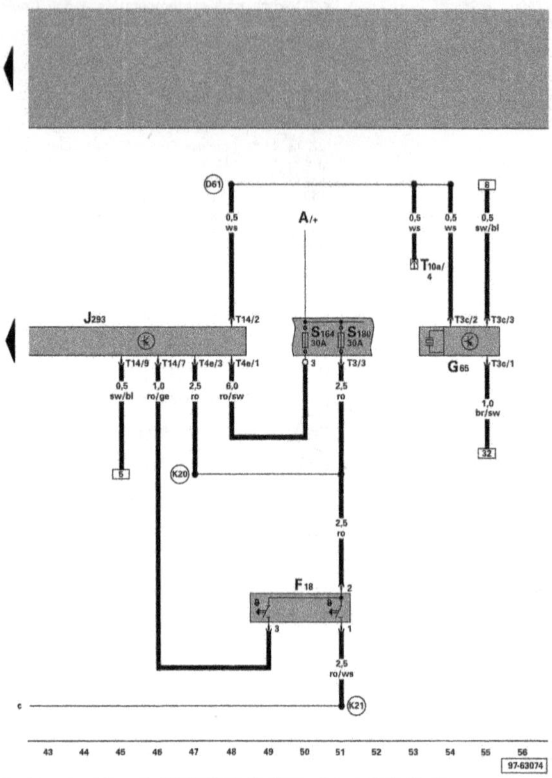

A - Battery
F18 - Coolant Fan Control (FC) Thermal Switch
G65 - High Pressure Sensor
J293 - Coolant FC (Fan Control) Control Module
S164 - Fuse -3- in fuse bracket/battery
S180 - Fuse -8- (30), in fuse bracket/battery
T3 - 3-Pin Connector
T3c - 3-Pin Connector
T4e - 4-Pin Connector
T10a - 10-Pin connector, orange, behind instrument panel, left.
T14 - 14-Pin Connector

D61 - wire connection (A/C), in engine compartment wiring harness
K20 - plus connection -1-(30), in coolant fan wiring harness
K21 - wire connection -1-, in coolant fan wiring harness

ws = white
sw = black
ro = red
br = brown
gn = green
bl = blue
gr = grey
li = violet
ge = yellow
or = orange

THIS SPACE INTENTIONALLY

LEFT BLANK

New Beetle **Wiring diagram** **No. 40/1**

Air conditioning,

- **1,8L-Engine - Motronic Multiport Fuel Injection (MFI) / 110 kW, code BKF,**
- **1,8L-Engine - Motronic Multiport Fuel Injection (MFI) / 132 kW, code BNU,**

from June 2004

Additional Fuse, Relay, Control Module and Connector locations ⇒ Compontent Locations

Troubleshooting Procedures ⇒ Guided Fault Finding using VAS 5051/5052

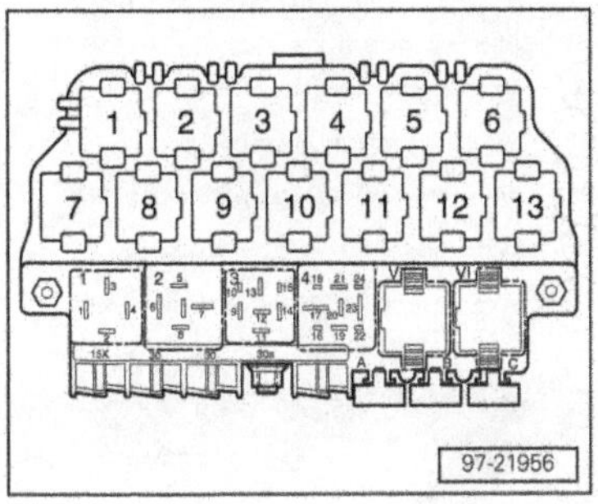

Relay location on the thirteenfold auxiliary relay panel, above relay panel:

10 J135 - Third Speed Coolant Fan Control (FC) Relay (100)

Relay panel:

New Beetle **Wiring diagram** **No. 40/2**

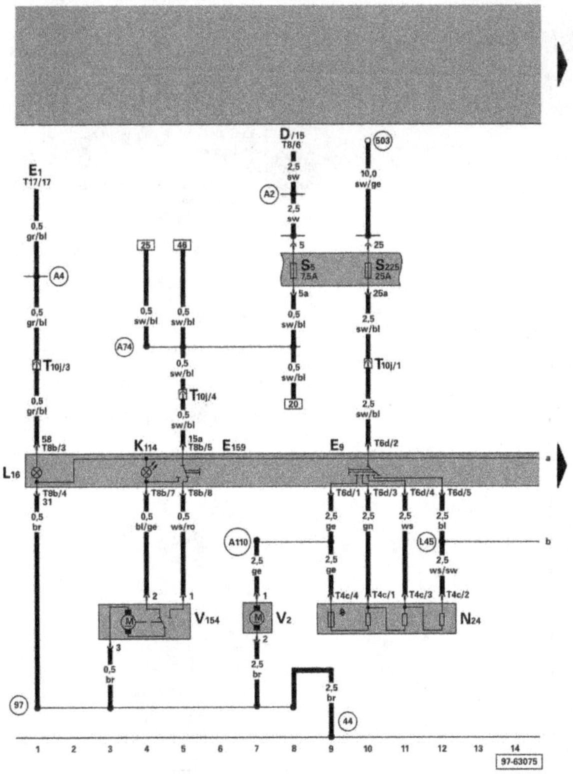

ws = white
sw = black
ro = red
br = brown
gn = green
bl = blue
gr = grey
li = violet
ge = yellow
or = orange

Fresh air blower, fresh air blower switch, fresh air/recirculating flap switch, fresh air blower series resistance with fuse

D - Ignition/Starter Switch
E1 - Light switch
E9 - Fresh Air Blower Switch
E159 - Fresh Air/Recirculating Flap Switch
K114 - Fresh Air and Recirculating Air Mode Indicator Light
L16 - Fresh Air Control Lever Light
N24 - Fresh Air Blower Series Resistance With Fuse
S5 - Fuse 5 in fuse holder
S225 - Fuse 25 in fuse holder
T4c - 4-Pin Connector
T6d - 6-Pin Connector
T8 - 8-Pin Connector, at -D-
T8b - 8-Pin Connector
T10j - 10-Pin Connector
T17 - 17-Pin Connector, at -E1-
V2 - Fresh Air Blower
V154 - Servo motor for fresh-/recirculating air door

(44) - Ground connection, on left A-pillar, lower part
(97) - Ground connection -1-, in A/C wiring harness
(503) - Screw Connection (75x), on relay panel
(A2) - plus connection (15), in instrument panel wiring harness
(A4) - plus connection (58b), in instrument panel wiring harness
(A74) - Connector (15a - fuse 5), in instrument panel wiring harness
(A110) - Connection (fresh air blower), in instrument panel wiring harness
(L45) - Wire connection, in A/C wiring harness

Air conditioning (1.8L BKF, BNU from June 2004)

New Beetle

Wiring diagram

No. 40/3

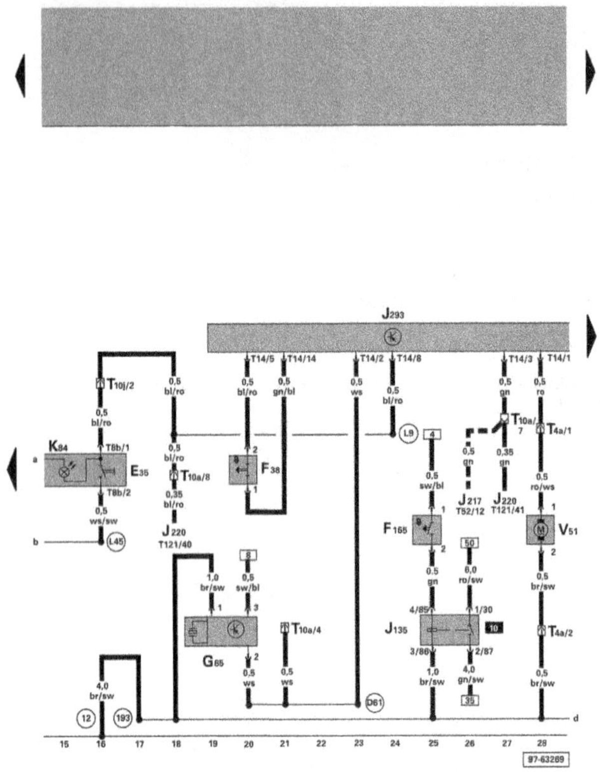

Coolant FC (fan control) control module, A/C switch, ambient temperature switch, A/C indicator light, After-run coolant pump

- E 35 - A/C Switch
- G 65 - High Pressure Sensor
- F38 - Ambient Temperature Switch
- F165 - Third Speed Coolant Fan Control (FC) Thermal Switch
- J135 - Third Speed Coolant Fan Control (FC) Relay
- J217 - Transmission Control Module (TCM)
- J220 - Motronic Engine Control Module (ECM)
- J293 - Coolant FC (Fan Control) Control Module, in engine compartment
- K84 - A/C Indicator Light
- T4a - 4-pin Connector
- T8b - 8-pin Connector
- T10a - 10-pin Connector, orange, behind instrument panel left
- T10j - 10-Pin Connector
- T14 - 14-Pin Connector, on Coolant FC (Fan Control) Control Module
- T52 - 52-Pin Connector, at J217
- T121 - Connector, 121point, on Engine Control Module (ECM)
- V51 - After-Run Coolant Pump
- (12) - Ground connection, in engine compartment, left
- (193) - Ground Connection 1, in coolant fan wiring harness
- (D61) - wire connection (A/C), in engine compart ment wiring harness
- (L9) - Wire connection -1-, in A/C wiring harness
- (L45) - Wire connection, in A/C system wiring harness
- --- - Automatic transmission only

ws = white
sw = black
ro = red
br = brown
gn = green
bl = blue
gr = grey
li = violet
ge = yellow
or = orange

Edition 06.04

New Beetle

Wiring diagram

No. 40/4

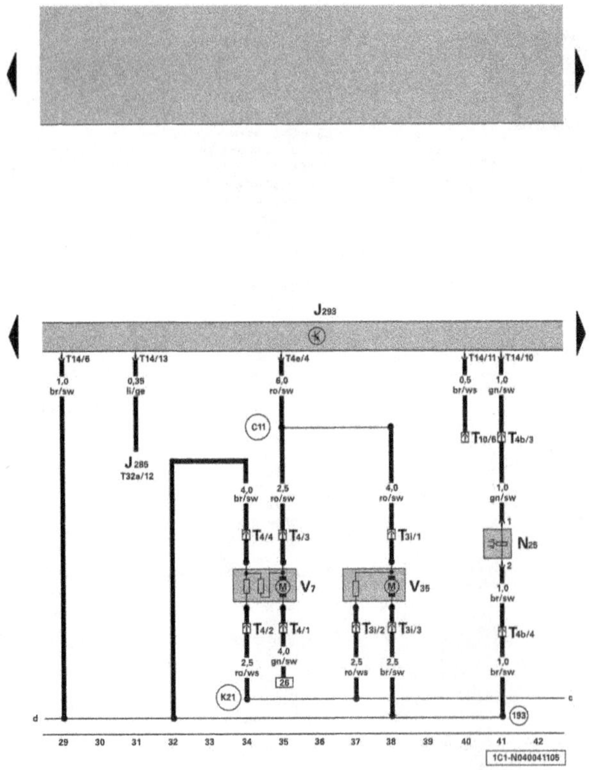

Coolant FC (fan control) control module, A/C clutch, coolant fan, right coolant fan

- J285 - Control module with indicator unit in instrument panel insert
- J293 - Coolant FC (Fan Control) Control Module, in engine compartment
- N25 - A/C Clutch
- T3i - 3-Pin Connector, in engine compartment
- T4 4-Pin Connector
- T4b - 4-Pin Connector, near A/C Clutch
- T4e - 4-Pin Connector, on Coolant FC (Fan Control) Control Module
- T10 - 10-Pin Connector, white, behind instrument panel, left
- T14 - 14-Pin Connector, on Coolant FC (Fan Control) Control Module
- T32a 32-Pin Connector, green
- V7 - Coolant Fan
- V35 - Right Coolant Fan
- (193) - Ground Connection 1, in coolant fan wiring harness
- (C11) - Connection, in coolant fan wiring harness
- (K21) - Connection 1, in coolant fan wiring harness

ws = white
sw = black
ro = red
br = brown
gn = green
bl = blue
gr = grey
li = violet
ge = yellow
or = orange

Edition 11.05

New Beetle **Wiring diagram** **No. 40/5**

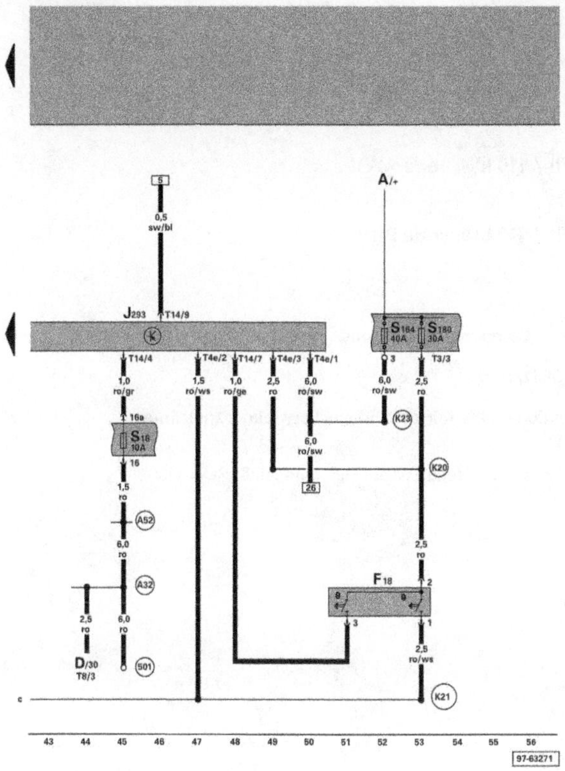

Coolant fan control (FC) module, coolant fan control (FC) thermal switch

A - Battery
D - Ignition/Starter Switch
F18 - Coolant Fan Control (FC) Thermal Switch
J293 - Coolant FC (Fan Control) Control Module
S16 - Fuse 16 in fuse holder
S164 - Fuse -3- in fuse bracket/battery
S180 - Fuse -8- (30), in fuse bracket/battery
T3 - 3-Pin Connector
T4e - 4-Pin Connector, on Coolant FC (Fan Control) Control Module
T8 - 8-Pin Connector, at -D-
T14 - 14-Pin Connector, on Coolant FC (Fan Control) Control Module
(501) - Threaded connection -2- (30), on the relay plate
(A32) - plus connection (30), in instrument panel wiring harness
(A52) - Plus Connection 2 (30) (in instrument panel wiring harness)
(K20) - Plus Connection 1(30), in coolant fan wiring harness
(K21) - Connection 1, in coolant fan wiring harness
(K23) - Plus Connection 2(30), in coolant fan wiring harness

ws = white
sw = black
ro = red
br = brown
gn = green
bl = blue
gr = grey
li = violet
ge = yellow
or = orange

THIS SPACE INTENTIONALLY

LEFT BLANK

New Beetle Convertible Wiring diagram **No. 26/1**

Air conditioning,

- **1,8L-Engine - Motronic Multiport Fuel Injection (MFI) / 110 kW, code AWV,**

from January 2003

- **1,8L-Engine - Motronic Multiport Fuel Injection (MFI) / 110 kW, code BKF,**

from June 2004

Additional Fuse, Relay, Control Module and Connector locations ⇒ Compontent Locations

Troubleshooting Procedures ⇒ Guided Fault Finding using VAS 5051/5052

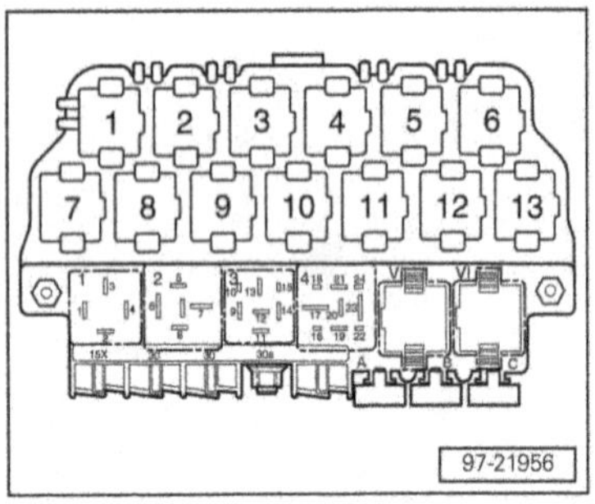

Relay location on the thirteenfold auxiliary relay panel, above relay panel:

10 J135 - Third Speed Coolant Fan Control (FC) Relay (100)

Relay panel:

New Beetle Convertible Wiring diagram **No. 26/2**

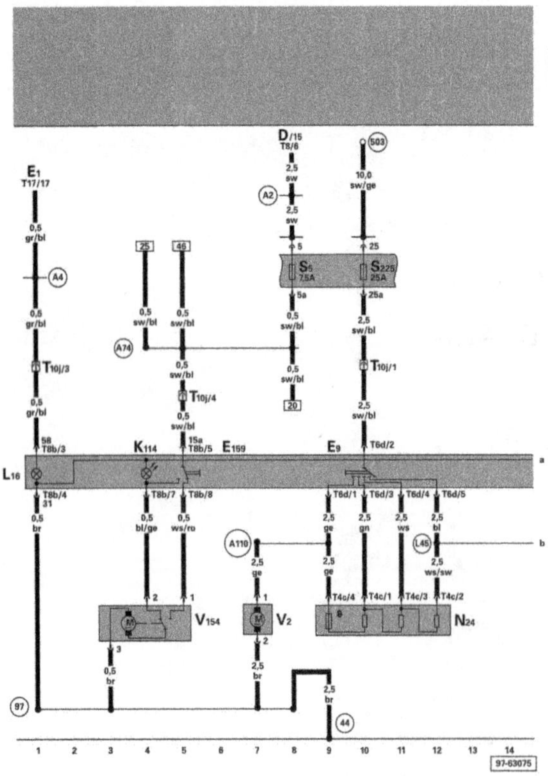

ws = white
sw = black
ro = red
br = brown
gn = green
bl = blue
gr = grey
li = violet
ge = yellow
or = orange

Fresh air blower, fresh air blower switch, fresh air/recirculating flap switch, fresh air blower series resistance with fuse

- D - Ignition/Starter Switch
- E1 - Light switch
- E9 - Fresh Air Blower Switch
- E159 - Fresh Air/Recirculating Flap Switch
- K114 - Fresh Air and Recirculating Air Mode Indicator Light
- L16 - Fresh Air Control Lever Light
- N24 - Fresh Air Blower Series Resistance With Fuse
- S5 - Fuse 5 in fuse holder
- S225 - Fuse 25 in fuse holder
- T4c - 4-Pin Connector
- T6d - 6-Pin Connector
- T8 - 8-Pin Connector, at -D-
- T8b - 8-Pin Connector
- T10j - 10-Pin Connector
- T17 - 17-Pin Connector, at -E1-
- V2 - Fresh Air Blower
- V154 - Servo motor for fresh-/recirculating air door
- (44) - Ground connection, on left A-pillar, lower part
- (97) - Ground connection -1-, in A/C wiring harness
- (503) - Screw Connection (75x), on relay panel
- (A2) - plus connection (15), in instrument panel wiring harness
- (A4) - plus connection (58b), in instrument panel wiring harness
- (A74) - Connector (15a - fuse 5), in instrument panel wiring harness
- (A110) - Connection (fresh air blower), in instrument panel wiring harness
- (L45) - Wire connection, in A/C wiring harness

New Beetle Convertible

Wiring diagram **No. 26/3**

Coolant FC (fan control) control module, A/C switch, ambient temperature switch, A/C indicator light, After-run coolant pump

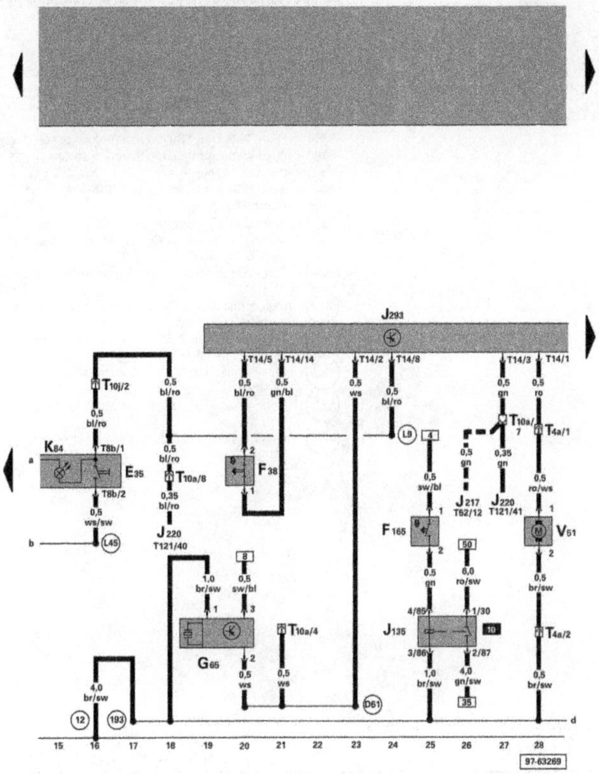

E 35 - A/C Switch
G 65 - High Pressure Sensor
F38 - Ambient Temperature Switch
F165 - Third Speed Coolant Fan Control (FC) Thermal Switch
J135 - Third Speed Coolant Fan Control (FC) Relay
J217 - Transmission Control Module (TCM)
J220 - Motronic Engine Control Module (ECM)
J293 - Coolant FC (Fan Control) Control Module, in engine compartment
K84 - A/C Indicator Light
T4a - 4-pin Connector
T8b - 8-pin Connector
T10a - 10-pin Connector, orange, behind instrument panel left
T10j - 10-Pin Connector
T14 - 14-Pin Connector, on Coolant FC (Fan Control) Control Module
T52 - 52-Pin Connector, at J217
T121 - Connector, 121point, on Engine Control Module (ECM)
V51 - After-Run Coolant Pump

(12) - Ground connection, in engine compartment, left
(193) - Ground Connection 1, in coolant fan wiring harness
(D61) - wire connection (A/C), in engine compart ment wiring harness
(L9) - Wire connection -1-, in A/C wiring harness
(L45) - Wire connection, in A/C system wiring harness

--- - Automatic transmission only

ws = white
sw = black
ro = red
br = brown
gn = green
bl = blue
gr = grey
li = violet
ge = yellow
or = orange

New Beetle Convertible

Wiring diagram **No. 26/4**

Coolant FC (fan control) control module, A/C clutch, coolant fan, right coolant fan

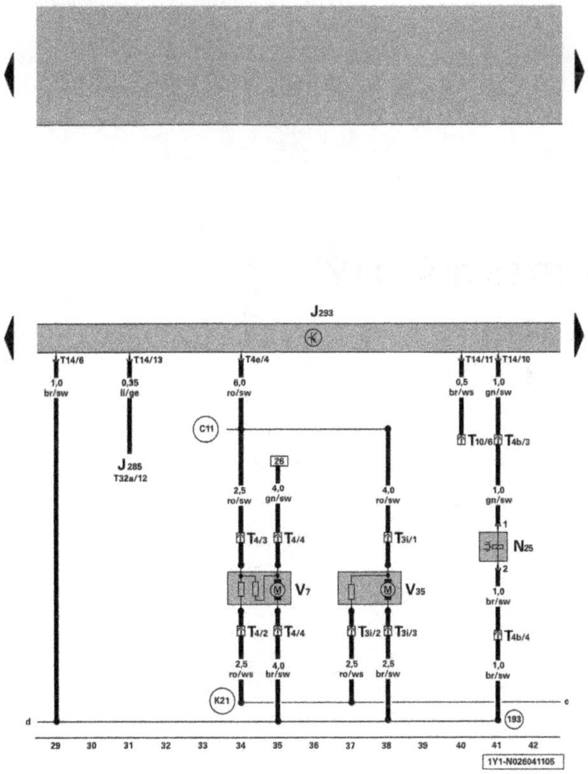

J285 - Instrument Cluster Control Module
J293 - Coolant FC (Fan Control) Control Module, in engine compartment
N25 - A/C Clutch
T3i - 3-Pin Connector, in engine compartment
T4 - 4-Pin Connector
T4b - 4-Pin Connector, near A/C Clutch
T4e - 4-Pin Connector, on Coolant FC (Fan Control) Control Module
T10 - 10-Pin Connector, white, behind instrument panel, left
T14 - 14-Pin Connector, on Coolant FC (Fan Control) Control Module
T32a 32-Pin Connector, green
V7 - Coolant Fan
V35 - Right Coolant Fan

(193) - Ground Connection 1, in coolant fan wiring harness
(C11) - Connection, in coolant fan wiring harness
(K21) - Connection 1, in coolant fan wiring harness

ws = white
sw = black
ro = red
br = brown
gn = green
bl = blue
gr = grey
li = violet
ge = yellow
or = orange

Air conditioning (1.8L AWV from Jan. 2003) (1.8L BKF from June 2004)

New Beetle Convertible Wiring diagram **No. 26/5**

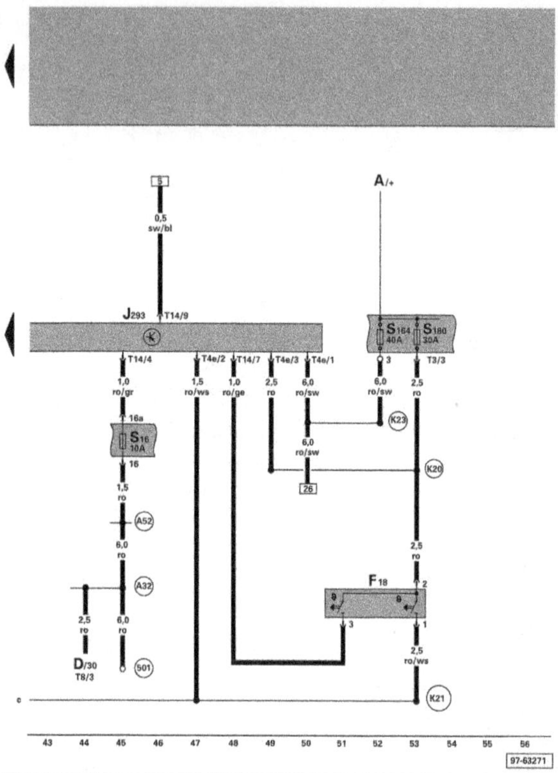

Coolant fan control (FC) module, coolant fan control (FC) thermal switch

- A - Battery
- D - Ignition/Starter Switch
- F18 - Coolant Fan Control (FC) Thermal Switch
- J293 - Coolant FC (Fan Control) Control Module
- S16 - Fuse 16 (on fuse panel)
- S164 - Fuse 3 - termnial 30 (on battery fuse panel)
- S180 - Fuse 4 - terminal 30 (on battery fuse panel)
- T3 - 3-Pin Connector
- T4e - 4-Pin Connector, on Coolant FC (Fan Control) Control Module
- T8 - 8-Pin Connector, at -D-
- T14 - 14-Pin Connector, on Coolant FC (Fan Control) Control Module
- (501) - Threaded connection -2- (30), on the relay plate
- (A32) - plus connection (30), in instrument panel wiring harness
- (A52) - Plus Connection 2 (30) (in instrument panel wiring harness)
- (K20) - Plus Connection 1(30), in coolant fan wiring harness
- (K21) - Connection 1, in coolant fan wiring harness
- (K23) - Plus Connection 2(30), in coolant fan wiring harness

ws = white
sw = black
ro = red
br = brown
gn = green
bl = blue
gr = grey
li = violet
ge = yellow
or = orange

THIS SPACE INTENTIONALLY LEFT BLANK

New Beetle Convertible

Wiring diagram **No. 25/1**

Air conditioning,

- **2,0L-Engine - Motronic Multiport Fuel Injection (MFI) / 85 kW, code BGD,**

from June 2003

- **2,0L-Engine - Motronic Multiport Fuel Injection (MFI) / 85 kW, code BEV,**

from June 2004

Additional Fuse, Relay, Control Module and Connector locations ⇒ Compontent Locations

Troubleshooting Procedures ⇒ Guided Fault Finding using VAS 5051/5052

Relay location on the thirteenfold auxiliary relay panel, above relay panel:

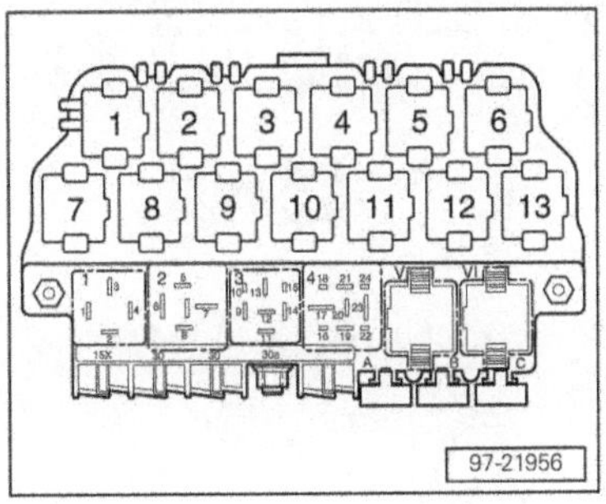

Relay panel:

New Beetle Convertible

Wiring diagram **No. 25/2**

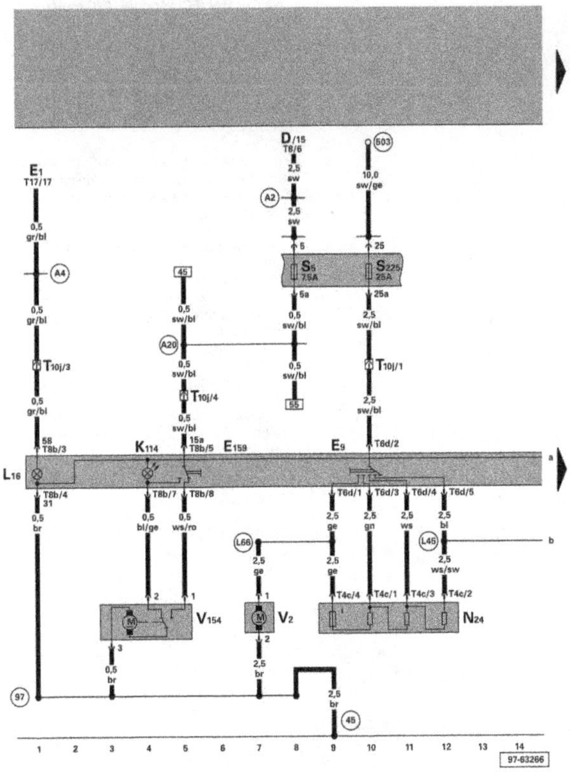

Fresh air blower switch, fresh air/recirculating flap, fresh air control level light, servo motor for fresh/recirculating air door, fresh air blower series resistance with fuse

- D - Ignition/Starter Switch
- E1 - Light switch
- E9 - Fresh Air Blower Switch
- E159 - Fresh Air/Recirculating Flap Switch
- K114 - Fresh/Recirculating Air Mode Indicator Light
- L16 - Fresh Air Control Lever Light
- N24 - Fresh Air Blower Series Resistance With Fuse
- S5 - Fuse 5 in fuse holder
- S225 - Fuse in fuse holder
- T4c - 4-Pin Connector
- T6d - 6-Pin Connector
- T8 - 8-Pin Connector, at -D-
- T8b - 8-Pin Connector
- T10j - 10-Pin Connector
- T17 - 17-Pin Connector, at -E1-
- V2 - Fresh Air Blower
- V154 - Servo motor for fresh/recirculating air door

- (45) - Ground connection, behind instrument panel, center
- (97) - Ground Connection 1, in A/C wiring harness
- (503) - Threaded connection -2- (75x) on the relay plate
- (A2) - plus connection (15), in instrument panel wiring harness
- (A4) - plus connection (58b), in instrument panel wiring harness
- (A20) - wire connection (15a), in instrument panel wiring harness
- (L45) - Wire connection, in A/C system wiring harness
- (L66) - Connector, in wiring harness heater blower

ws = white
sw = black
ro = red
br = brown
gn = green
bl = blue
gr = grey
li = violet
ge = yellow
or = orange

Air conditioning (2.0L BGD from June 2003) (2.0L BEV from June 2004)

New Beetle Convertible Wiring diagram **No. 25/3**

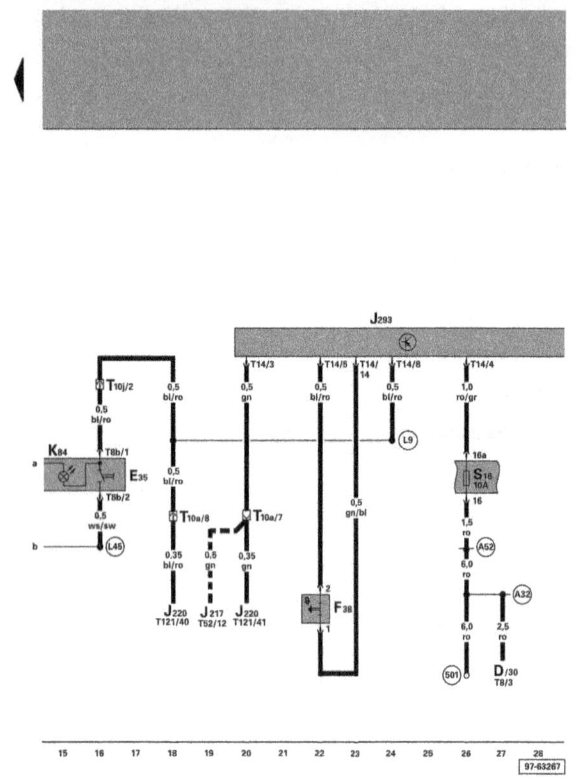

ws = white
sw = black
ro = red
br = brown
gn = green
bl = blue
gr = grey
li = violet
ge = yellow
or = orange

A/C Switch, ambient temperature switch, Coolant FC (Fan Control) Control Module, A/C indicator light.

- D - Ignition/Starter Switch
- E35 - A/C Switch
- F38 - Ambient Temperature Switch
- J217 - Automatic Transmission Control Module (TCM)
- J220 - Motronic Engine Control Module (ECM)
- J293 - Coolant FC (Fan Control) Control Module
- K84 - A/C Indicator Light
- S16 - wire connection (passenger door), in central
- T8 - 8-Pin connector, at -D-
- T8b - 8-Pin connector
- T10a - 10-Pin connector, orange, behind instrument panel, left.
- T10j - 10-Pin connector
- T14 - 14-Pin connector
- T52 - 52-Pin connector, at J217
- T121 - 121-Pin connector, at J220
- (501) - Threaded connection -2- (30) on the relay plate
- (A32) - Plus connection (30), in instrument panel wiring harness
- (A52) - Plus Connection 2 (30) (in instrument panel wiring harness)
- (L9) - Wire connection -1-, in A/C wiring harness
- (L45) - Wire connection, in A/C system wiring harness
- --- - Automatic transmission only

Edition 06.04

New Beetle Convertible Wiring diagram **No. 25/4**

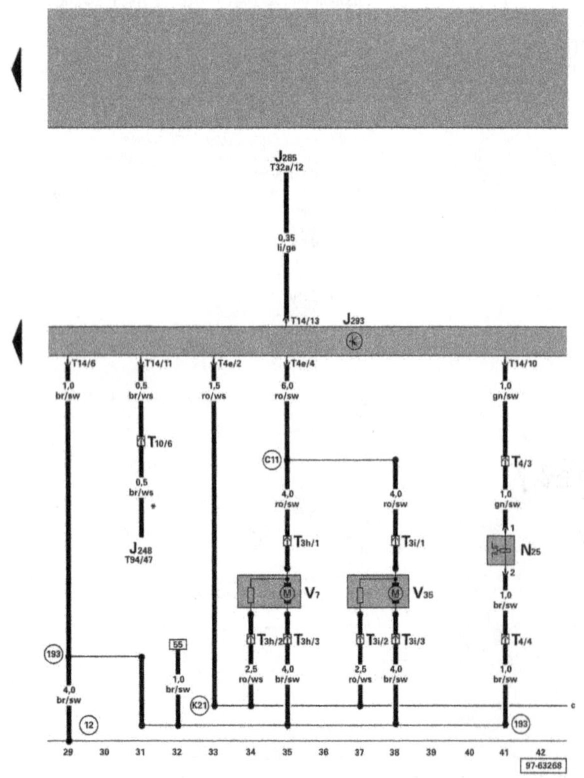

ws = white
sw = black
ro = red
br = brown
gn = green
bl = blue
gr = grey
li = violet
ge = yellow
or = orange

Coolant Fan Control (FC) Module, A/C clutch, right coolant fan

- J248 - Diesel Direct Fuel Injection (DFI) Engine Control Module (ECM)
- J285 - Control module with indicator unit in instrument panel insert
- J293 - Coolant FC (Fan Control) Control Module
- N25 - A/C Clutch
- T3h - 3-Pin Connector, in engine compartment
- T3i - 3-Pin Connector, in engine compartment
- T4 - 4-Pin Connector, near A/C Clutch
- T4e - 4-Pin Connector, on coolant FC (Fan Control) Control Module
- T10 - 10-Pin Connector, white, behind instrument panel, left
- T14 - 14-Pin Connector
- T32a - 32-Pin Connector, green, on instrument cluster
- T94 - 94-Pin Connector, at J248
- V7 - Coolant Fan
- V35 - Right Coolant Fan
- (12) - Ground connection, in engine compartment, left
- (193) - Ground connection -1-, in coolant fan wiring harness
- (C11) - Wire connection, in coolant fan wiring harness
- (K21) - wire connection -1-, in coolant fan wiring harness
- * - not applicable to USA/CDN

Edition 06.04

Air conditioning (2.0L BGD from June 2003) (2.0L BEV from June 2004)

New Beetle Convertible **Wiring diagram** **No. 25/5**

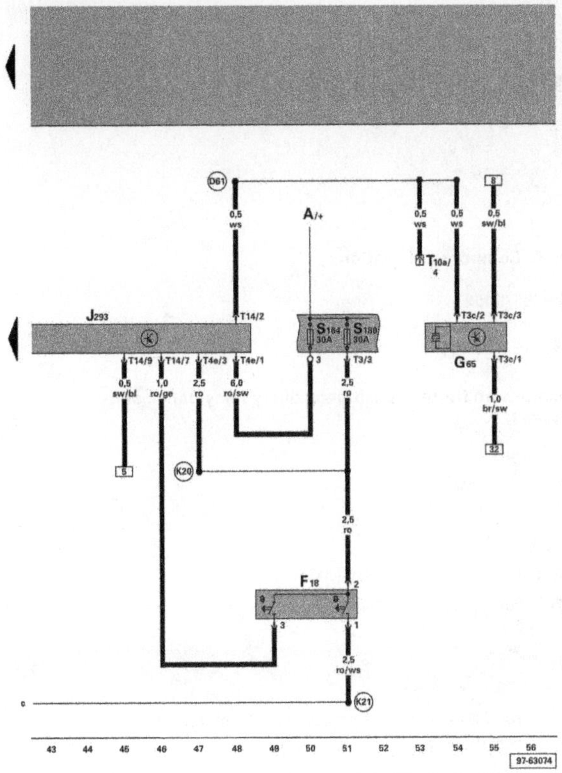

Coolant Fan Control (FC) Thermal Switch, high pressure sensor

A - Battery
F18 - Coolant Fan Control (FC) Thermal Switch
G65 - High Pressure Sensor
J293 - Coolant FC (Fan Control) Control Module
S164 - Fuse -3- in fuse bracket/battery
S180 - Fuse -8- (30), in fuse bracket/battery
T3 - 3-Pin Connector
T3c - 3-Pin Connector
T4e - 4-Pin Connector
T10a - 10-Pin connector, orange, behind instrument panel, left.
T14 - 14-Pin Connector
(D61) - wire connection (A/C), in engine compartment wiring harness
(K20) - plus connection -1-(30), in coolant fan wiring harness
(K21) - wire connection -1-, in coolant fan wiring harness

ws = white
sw = black
ro = red
br = brown
gn = green
bl = blue
gr = grey
li = violet
ge = yellow
or = orange

THIS SPACE INTENTIONALLY LEFT BLANK

New Beetle **Wiring diagram** **No. 47/1**

Air conditioning,

I **2.5L -Engine - Motronic Multiport Fuel Injection (MFI)/110 kw, code BPR/BPS,**

from July 2005,

Additional Fuse, Relay, Control Module and Connector locations ⇒ Component Locations

Troubleshooting Procedures ⇒ Guided Fault Finding using VAS 5051/5052

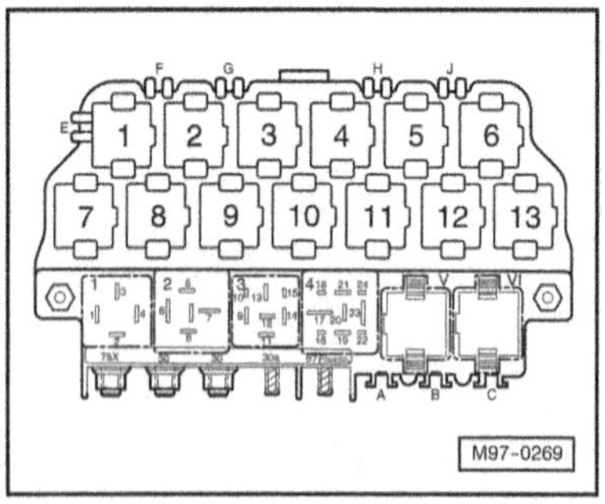

Relay locations on thirteen positionauxiliary relay panel, above relay panel:

Relay panel:

2 J59 - Load Reduction Relay (100)

4 J17 - Fuel Pump (FP) Relay (409)

Note:

Number in parentheses indicates production control number stamped on relay housing.

New Beetle **Wiring diagram** **No. 47/2**

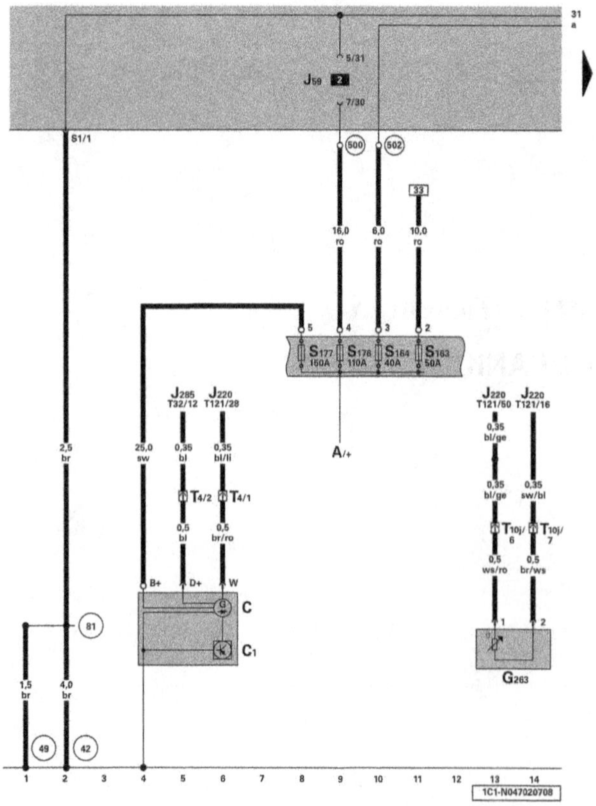

Generator (GEN), Voltage regulator (VR), Load Reduction Relay, Fuses on battery fuse panel

- A - Battery
- C - Generator (GEN)
- C1 - Voltage Regulator (VR)
- G263 - Evaporator Vent Temperature Sensor
- J59 - Load Reduction Relay
- J220 - Motronic Engine Control Module (ECM)
- J285 - Instrument Cluster Control Module
- S163 - Fuse 2 - terminal 30 (on battery fuse panel)
- S164 - Fuse 3 - terminal 30 (on battery fuse panel)
- S176 - Fuse 4 - terminal 30 (on battery fuse holder)
- S177 - Fuse 5 - terminal 30 (on battery fuse panel)
- T4 - 4-Pin Connector, in engine compartment left
- T10j - 10-Pin Connector
- T32 - 32-Pin Connector, at J285
- T121 - 121-Pin Connector, at J220
- (42) - Ground Connection (beside steering column)
- (49) - Ground Connection (near steering column)
- (81) - Ground Connection 1 (in instrument panel wiring harness)
- (500) - Threaded Connection 1 - terminal 30 (on relay panel)
- (502) - Threaded Connection 1 - terminal 30a (on relay panel)

ws = white
sw = black
ro = red
br = brown
gn = green
bl = blue
gr = grey
li = violet
ge = yellow
or = orange

New Beetle — Wiring diagram — No. 47/3

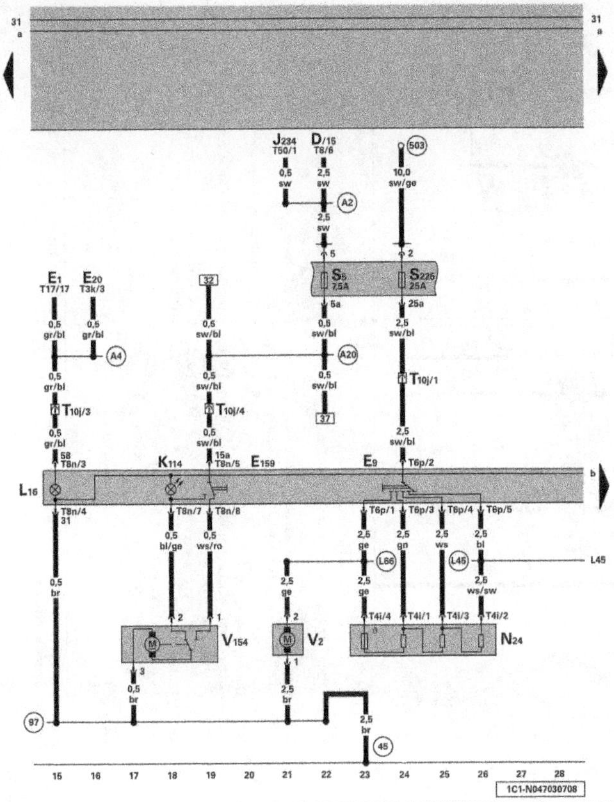

Fresh Air Blower Switch, Fresh Air/Recirculating Flap Switch, Fresh Air and Recirculating Air Mode Indicator Lamp, Fresh Air Control Illumination, Fresh Air Blower Series Resistor With Fuse, Fresh Air Blower, Fresh/Recirculating Air Door Motor

- D - Ignition/Starter Switch
- E1 - Light Switch
- E9 - Fresh Air Blower Switch
- E20 - Instrument Panel Illumination Dimmer Switch
- E159 - Fresh Air/Recirculating Flap Switch
- J234 - Airbag Control Module
- K114 - Fresh Air and Recirculating Air Mode Indicator Lamp
- L16 - Fresh Air Control Illumination
- N24 - Fresh Air Blower Series Resistor With Fuse
- S5 - Fuse (on fuse/relay panel)
- S225 - Fuse (on fuse panel)
- T3k - 3-Pin Connector
- T4i - 4-Pin Connector
- T6p - 6-Pin Connector
- T8 - 8-Pin Connector, at -D-
- T8n - 8-Pin Connector
- T10j - 10-Pin Connector
- T17 - 17-Pin Connector, at -E1-
- T50 - 50-Pin Connector
- V2 - Fresh Air Blower
- V154 - Fresh/Recirculating Air Door Motor
- (45) - Ground Connection (behind instrument panel, center)
- (97) - Ground Connection 1 (in A/C wiring harness)
- (503) - Treaded Connection - terminal 75x (on relay panel)
- (A2) - Plus Connection (15) (in instrument panel wiring harness)
- (A4) - Plus Connection (58b) (in instrument panel wiring harness)
- (A20) - Plus Connection (15a) (in instrument panel wiring harness)
- (L45) - Connection (in A/C system wiring harness)
- (L66) - Connection (in wiring harness) heater blower

ws = white
sw = black
ro = red
br = brown
gn = green
bl = blue
gr = grey
li = violet
ge = yellow
or = orange

New Beetle — Wiring diagram — No. 47/4

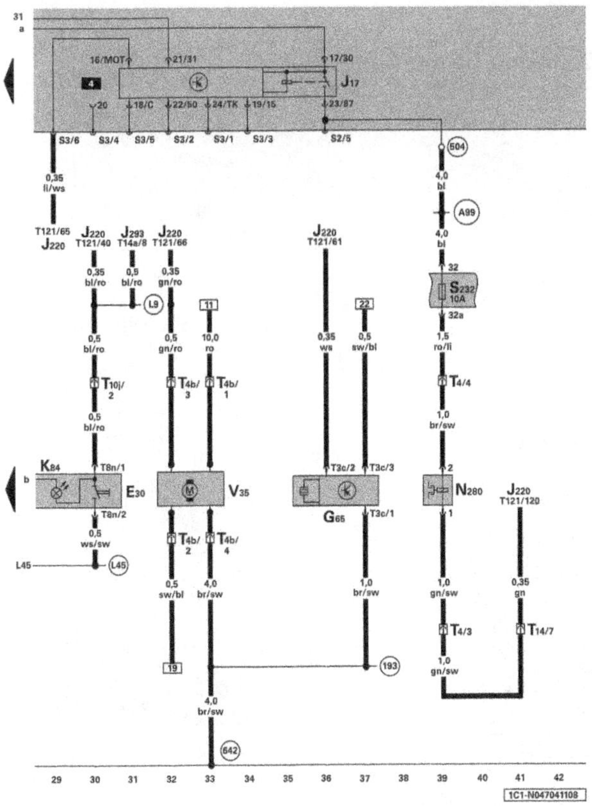

A/C Switch, High Pressure Sensor, Fuel Pump (FP) Relay, A/C Indicator Lamp, A/C Compressor Regulator Valve, Right Coolant Fan

- E30 - A/C Switch
- G65 - High Pressure Sensor
- J17 - Fuel Pump (FP) Relay, (409)
- J220 - Motronic Engine Control Module (ECM)
- J293 - Coolant Fan Control (FC) Control Module
- K84 - A/C Indicator Lamp
- N280 - A/C Compressor Regulator Valve
- S232 - Fuse (on fuse panel)
- T3c - 3-Pin connector
- T4 - 4-Pin connector
- T4b - 4-Pin connector
- T8n - 8-Pin connector
- T10j - 10-Pin connector
- T14 - 14-Pin connector
- T14a - 14-Pin connector
- T121 - 121-Pin connector, at J220
- V35 - Right Coolant Fan
- (193) - Ground Connection 1 (in coolant fan wiring harness)
- (504) - Threaded Connection 1 - terminal 87 (on relay panel)
- (642) - Ground Connection (on electronic fan control)
- (A99) - Connection 1 (87) (in instrument panel wiring harness)
- (L9) - Connection 1 (in A/C wiring harness)
- (L45) - Connection (in A/C system wiring harness)

ws = white
sw = black
ro = red
br = brown
gn = green
bl = blue
gr = grey
li = violet
ge = yellow
or = orange

Airbag systems,

from January 1998

Deviate relay location and fuseplacements as well as the locations of multiple connectors see section "component locations".

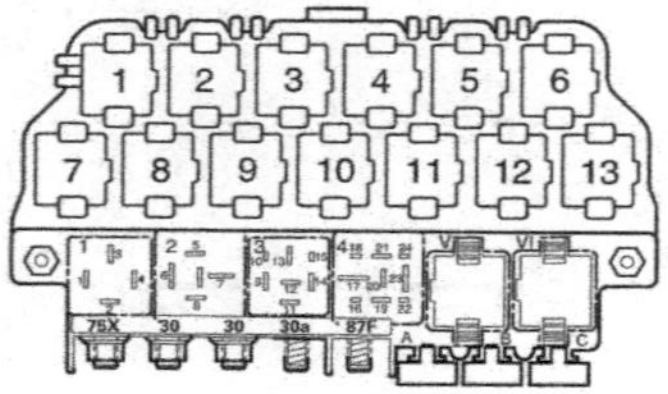

97-21956

Fuse colors
30 A - green
25 A - white
20 A - yellow
15 A - blue
10 A - red
7.5 A - brown
5 A - beige

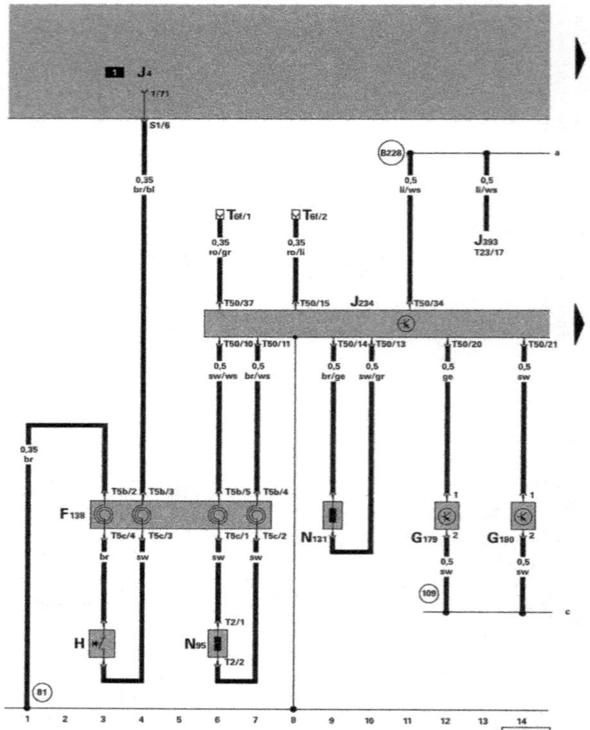

ws = white
sw = black
ro = red
br = brown
gn = green
bl = blue
gr = grey
li = violet
ge = yellow

F138 - Airbag Spiral Spring/Return Spring With Slip Ring
G179 - Crash sensor for side airbag, driver's side
G180 - Crash sensor for side airbag, passenger side
H - Horn Button
J4 - Dual Horn Relay
J234 - Airbag Control Module, behind console
J393 - Central control module for comfort system
N95 - Driver's Side Airbag Igniter
N131 - Passenger's Side Airbag Igniter 1
T2 - Double Connector, yellow, on Airbag Igniter
T5b - 5-Pin Connector, yellow, near steering column on Airbag Spiral Spring/Return Spring
T5c - 5-Pin Connector, black, near steering column on Airbag Spiral Spring/Return Spring
T6f - 6-Pin Connector, below passenger's seat, red
T23 - 23-Pin Connector, on Central control module for comfort system
T50 - 50-Pin Connector
(109) - Ground connection, in airbag wiring harness
(81) - Ground connection -1-, in instrument panel wiring harness
(B228) - wire connection (crash), in instrument panel wiring harness

Airbag control module, airbag spiral spring, igniters for front airbags, crash sensors for side airbags

New Beetle **Wiring diagram** **No. 6/3**

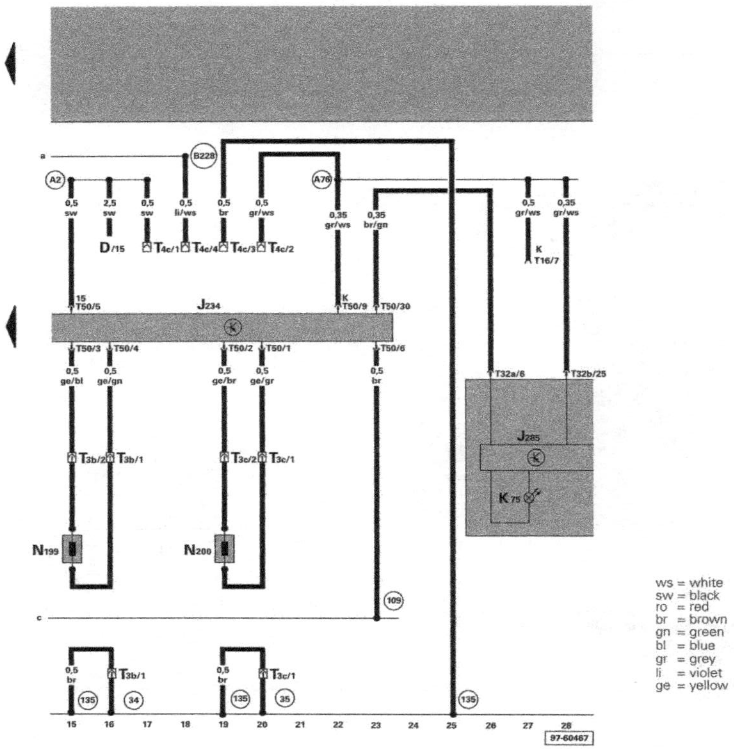

ws = white
sw = black
ro = red
br = brown
gn = green
bl = blue
gr = grey
li = violet
ge = yellow

- D - Ignition/Starter Switch
- J234 - Airbag Control Module, behind console
- J285 - Control module with indicator unit in instrument panel insert
- K75 - Airbag Malfunction Indicator Lamp (MIL)
- N199 - Igniter for side airbag, driver's side
- N200 - Igniter for side airbag, passenger side
- T3b - 3 -Pin Connector
- T3c - 3 -Pin Connector
- T4c - 4-Pin Connector
- T16 - 16-Pin Connector, Data Link Connector (DLC) below steering column
- T32a - 32-Pin Connector, blue, on instrument cluster
- T32b - 32-Pin Connector, green, on instrument cluster
- T50 - 50-Pin Connector

- (34) - Ground connection, below driver's seat
- (35) - Ground connection, below passenger'seat
- (109) - Ground connection, in airbag wiring harness
- (135) - Ground connection -2-, in instrument panel wiring harness
- (A2) - plus connection (15), in instrument panel wiring harness
- (A76) - Connector (K-diagnosis wire), in instrument panel wiring harness
- (B228) - wire connection (crash), in instrument panel wiring harness

Edition 06/99
USA.5132.07.21

Airbag control module, igniter for side airbag, airbag malfunction indicator lamp (MIL)

THIS SPACE INTENTIONALLY

LEFT BLANK

Airbag systems,

from June 1999

Deviate relay location and fuseplacements as well as the locations of multiple connectors see section "component locations".

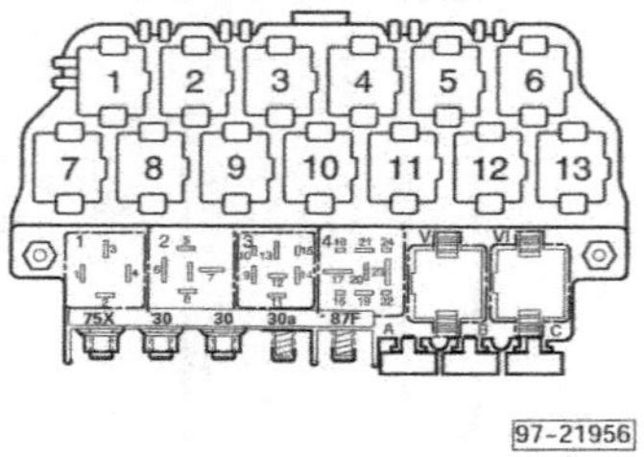

Relay location on the thirteen position auxiliary relay panel, above relay panel:

Relay panel:

1 Dual Horn Relay (53)

Note: Number in parentheses indicates production control number stamped on relay housing.

Fuse colors
30 A - green
25 A - white
20 A - yellow
15 A - blue
10 A - red
7,5 A - brown
5 A - beige

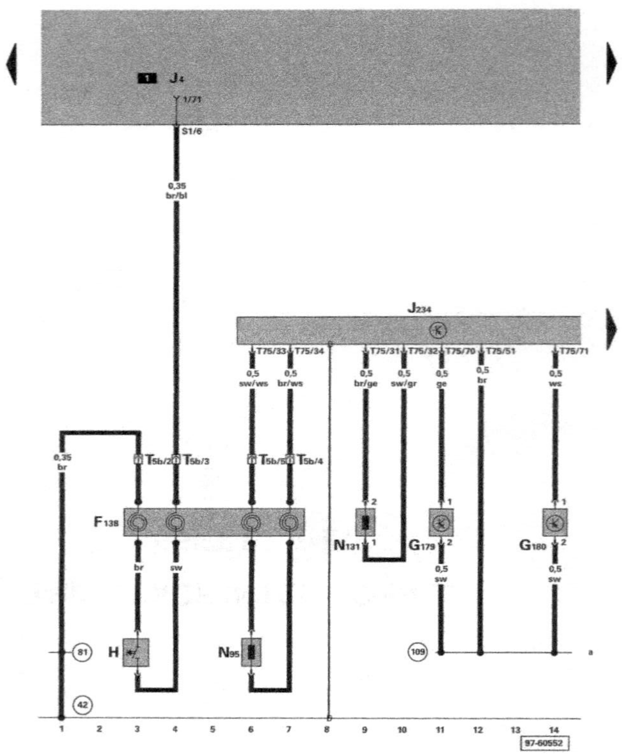

ws = white
sw = black
ro = red
br = brown
gn = green
bl = blue
gr = grey
li = violet
ge = yellow

F138 - Airbag Spiral Spring/Return Spring With Slip Ring
G179 - Crash sensor for side airbag, driver's side
G180 - Crash sensor for side airbag, passenger side
H - Signal horn activation
J4 - Dual Horn Relay
J234 - Airbag Control Module
N 95 - Driver's Side Airbag Igniter
N131 - Passenger's Side Airbag Igniter 1
T5b - 5-Pin Connector, near steering column
T75 - 75-Pin Connector

(42) - Ground connection, beside steering column
(81) - Ground connection -1-, in instrument panel wiring harness
(109) - Ground connection, in airbag wiring harness

Airbag control module, airbag spiral spring, igniters for front airbags, crash sensors for side airbags

New Beetle — Wiring diagram — No. 28/3

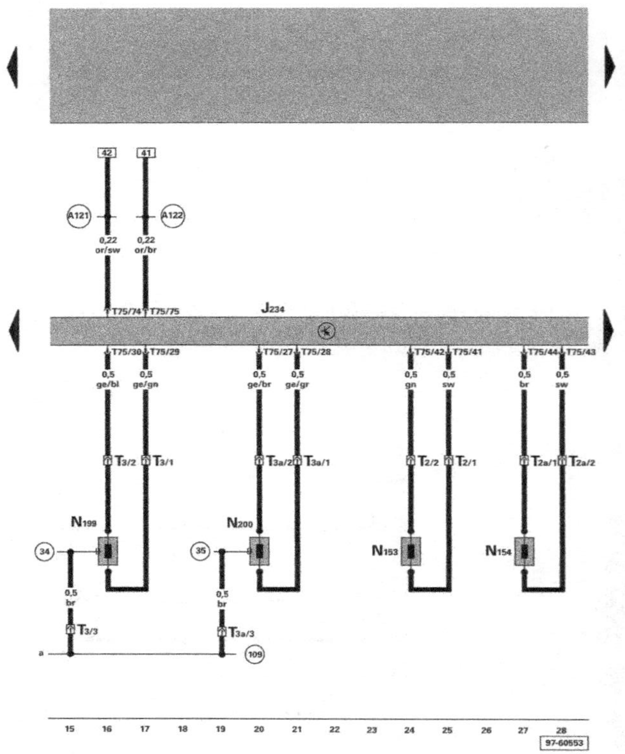

ws = white
sw = black
ro = red
br = brown
gn = green
bl = blue
gr = grey
li = violet
ge = yellow

N199 - Igniter for side airbag, driver's side
N200 - Igniter for side airbag, passenger side
J234 - Airbag Control Module
N153 - Left Seat Belt Tensioner Igniter
N154 - Right Seat Belt Tensioner Igniter
T2 - 2-Pin Connector
T2a - 2-Pin Connector
T3 - 3-Pin Connector
T3a - 3-Pin Connector
T75 - 75-Pin Connector

(34) - Ground connection, below driver's seat
(35) - Ground connection, below passenger's seat
(109) - Ground connection, in airbag wiring harness
(A121) - Connection (high bus) in instrument panel wiring harness
(A122) - Connection (low bus) in instrument panel wiring harness

Airbag control module, igniters for side airbag, left/right seat belt tensioner igniter,

No. 28/4 — Wiring diagram — New Beetle

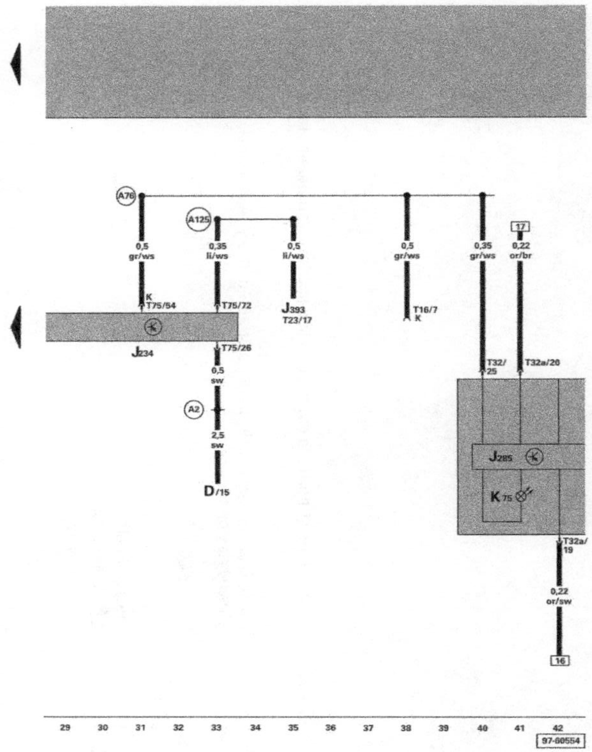

ws = white
sw = black
ro = red
br = brown
gn = green
bl = blue
gr = grey
li = violet
ge = yellow

D - Ignition/Starter Switch
J234 - Airbag Control Module
J285 - Control module with indicator unit in instrument panel insert
J393 - Central control module for comfort system
K75 - Airbag Malfunction Indicator Lamp (MIL)

T16 - 16- Pin Connector, Data Link Connector (DLC), below steering column, left
T23 - 23-Pin Connector
T32 - 32-Pin Connector, blue, on instrument cluster
T32a - 32-Pin Connector, green, on instrument cluster

(A2) - plus connection (15), in instrument panel wiring harness
(A76) - Connection (K-diagnosis wire) in instrument
(A125) - Connection (crash signal) in instrument panel wiring harness

Airbag control module, Airbag Malfunction Indicator Lamp (MIL), Data Link Connector (DLC)

New Beetle — Wiring diagram — No. 6/1

Airbag systems,

from October 2000

Deviate relay location and fuseplacements as well as the locations of multiple connectors see section "component locations".

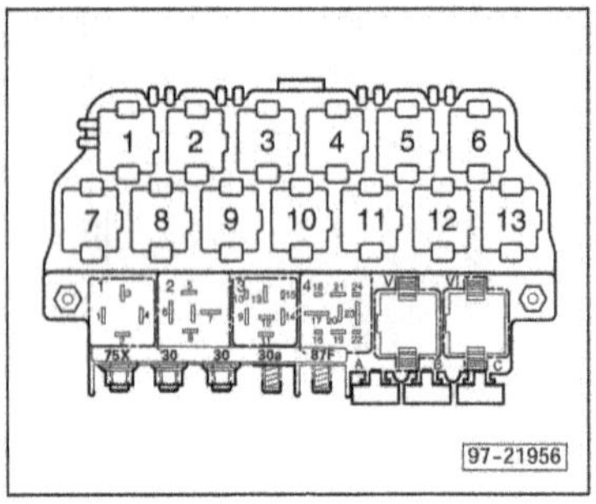

Relay location on the thirteen position auxiliary relay panel, above relay panel:

Relay panel:

1 Dual Horn Relay (53)

Fuse colors

30 A - green
25 A - white
20 A - yellow
15 A - blue
10 A - red
7,5 A - brown
5 A - beige
3 A - violet

Note: Number in parentheses indicates production control number stamped on relay housing.

New Beetle — Wiring diagram — No. 6/2

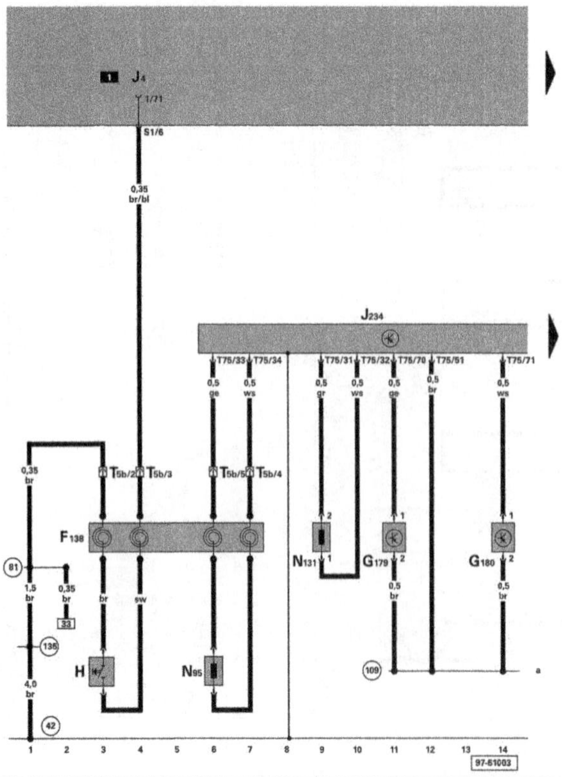

Airbag control module, airbag spiral spring, igniters for front airbags, crash sensors for side airbags

- F138 - Airbag Spiral Spring/Return Spring With Slip Ring
- G179 - Crash sensor for side airbag, driver's side
- G180 - Crash sensor for side airbag, passenger side
- H - Signal horn activation
- J4 - Dual Horn Relay
- J234 - Airbag Control Module
- N 95 - Driver's Side Airbag Igniter
- N131 - Passenger's Side Airbag Igniter 1
- T5b - 5-Pin Connector, near steering column
- T75 - 75-Pin Connector
- (42) - Ground connection, beside steering column
- (81) - Ground connection -1-, in instrument panel wiring harness
- (109) - Ground connection, in airbag wiring harness
- (135) - Ground connection -2-, in instrument panel wiring harness

ws = white
sw = black
ro = red
br = brown
gn = green
bl = blue
gr = grey
li = violet
ge = yellow
or = orange

New Beetle — Wiring diagram — No. 6/3

Airbag control module, igniters for side airbag, left/right seat belt tensioner igniter,

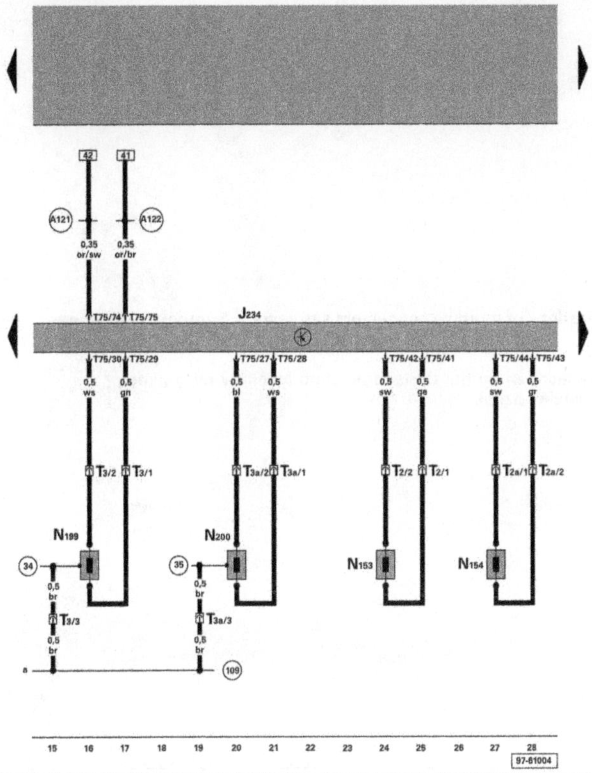

- N199 - Igniter for side airbag, driver's side
- N200 - Igniter for side airbag, passenger side
- J234 - Airbag Control Module
- N153 - Left Seat Belt Tensioner Igniter
- N154 - Right Seat Belt Tensioner Igniter
- T2 - 2-Pin Connector
- T2a - 2-Pin Connector
- T3 - 3-Pin Connector
- T3a - 3-Pin Connector
- T75 - 75-Pin Connector
- (34) - Ground connection, below driver's seat
- (35) - Ground connection, below passenger's seat
- (109) - Ground connection, in airbag wiring harness
- (A121) Connection (high bus) in instrument panel wiring harness
- (A122) Connection (low bus) in instrument panel wiring harness

ws = white
sw = black
ro = red
br = brown
gn = green
bl = blue
gr = grey
li = violet
ge = yellow
or = orange

New Beetle — Wiring diagram — No. 6/4

Airbag control module, Aarbag malfunction indicator lamp (MIL), data link connector (DLC)

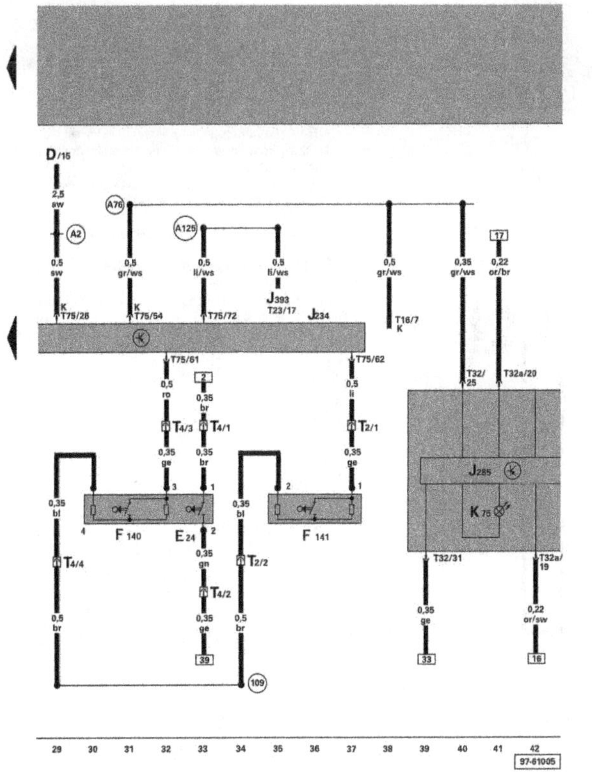

- D - Ignition/Starter Switch
- E24 - Left Seat Belt Switch
- F140 - Left front seat belt micro switch
- F141 - Right front seat belt micro switch
- J234 - Airbag Control Module
- J285 - Control module with indicator unit in instrument panel insert
- J393 - Central control module for comfort system
- K75 - Airbag Malfunction Indicator Lamp (MIL)
- T2 - 2-Pin Connector
- T4 - 4-Pin Connector
- T16 - 16- Pin Connector, Data Link Connector (DLC), below steering column, left
- T23 - 23-Pin Connector
- T32 - 32-Pin Connector, blue, on instrument cluster
- T32a - 32-Pin Connector, green, on instrument cluster
- T75 - 75-Pin Connector
- (A2) - plus connection (15), in instrument panel wiring harness
- (A76) Connection (K-diagnosis wire) in instrument
- (A125) Connection (crash signal) in instrument panel wiring harness
- (109) Ground connection, in airbag wiring harness

ws = white
sw = black
ro = red
br = brown
gn = green
bl = blue
gr = grey
li = violet
ge = yellow
or = orange

New Beetle — Wiring diagram — No. 15/1

Airbag systems,

from September 2003

Deviate relay location and fuseplacements as well as the locations of multiple connectors see section "component locations".

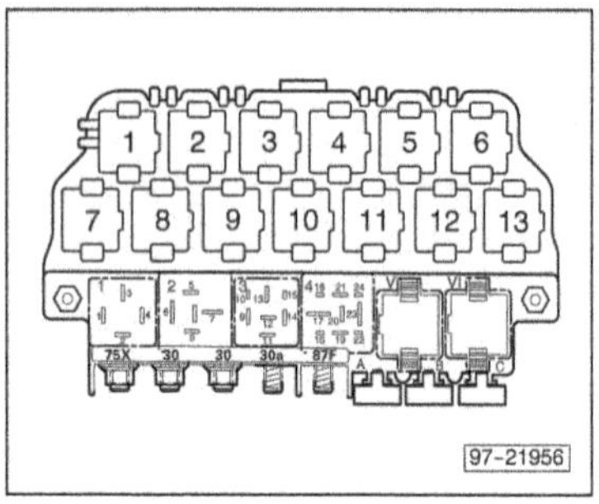

Relay location on the thirteen position auxiliary relay panel, above relay panel:

Relay panel:

1 Dual Horn Relay (53)

Fuse colors

30 A - green
25 A - white
20 A - yellow
15 A - blue
10 A - red
7,5 A - brown
5 A - beige
3 A - violet

Note: Number in parentheses indicates production control number stamped on relay housing.

New Beetle — Wiring diagram — No. 15/2

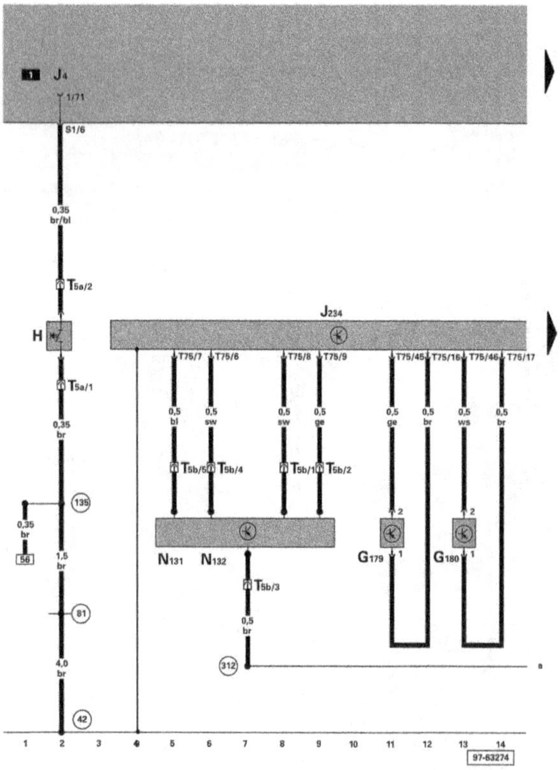

Airbag control module, crash sensors for side airbags

G179 - Crash sensor for side airbag, driver's side
G180 - Crash sensor for side airbag, passenger side
H - Signal horn activation
J4 - Dual Horn Relay
J234 - Airbag Control Module
N131 - Front Passenger's Side Airbag Igniter 1
N132 - Front Passenger's Side Airbag Igniter 2
T5a - 5-Pin Connector
T5b - 5-Pin Connector, yellow, near steering column
T75 - 75-Pin Connector, at -J234-

(42) - Ground connection, beside steering column

(81) - Ground connection -1-, in instrument panel wiring harness

(135) - Ground connection -2-, in instrument panel wiring harness

(312) Ground Connection 2 (in Airbag wiring harness)

ws = white
sw = black
ro = red
br = brown
gn = green
bl = blue
gr = grey
li = violet
ge = yellow
or = orange

Airbag system from Oct. 2000

New Beetle — Wiring diagram — No. 6/3

Airbag control module, igniters for side airbag, left/right seat belt tensioner igniter,

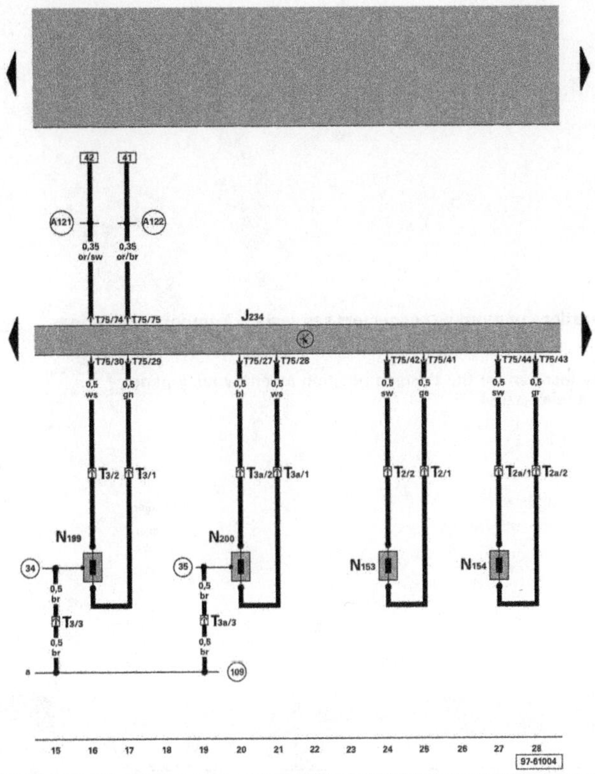

- N199 - Igniter for side airbag, driver's side
- N200 - Igniter for side airbag, passenger side
- J234 - Airbag Control Module
- N153 - Left Seat Belt Tensioner Igniter
- N154 - Right Seat Belt Tensioner Igniter
- T2 - 2-Pin Connector
- T2a - 2-Pin Connector
- T3 - 3-Pin Connector
- T3a - 3-Pin Connector
- T75 - 75-Pin Connector
- (34) - Ground connection, below driver's seat
- (35) - Ground connection, below passenger's seat
- (109) - Ground connection, in airbag wiring harness
- (A121) Connection (high bus) in instrument panel wiring harness
- (A122) Connection (low bus) in instrument panel wiring harness

ws = white
sw = black
ro = red
br = brown
gn = green
bl = blue
gr = grey
li = violet
ge = yellow
or = orange

New Beetle — Wiring diagram — No. 6/4

Airbag control module, Aarbag malfunction indicator lamp (MIL), data link connector (DLC)

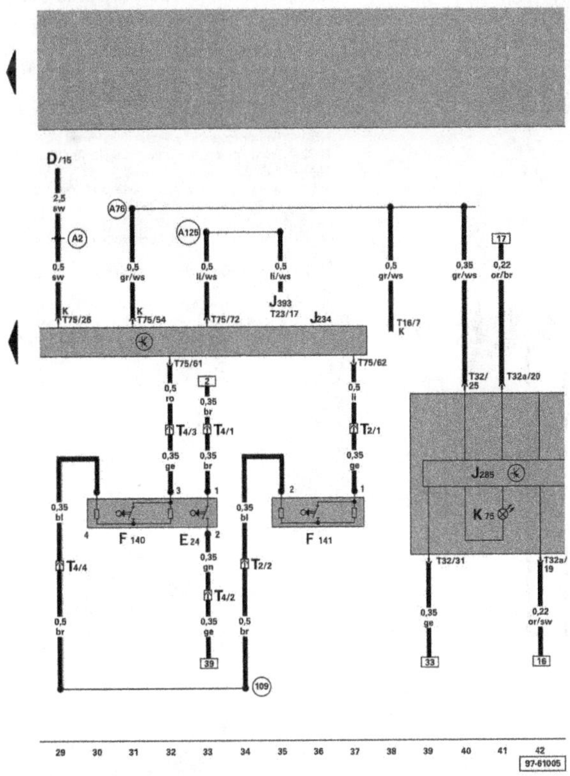

- D - Ignition/Starter Switch
- E24 - Left Seat Belt Switch
- F140 - Left front seat belt micro switch
- F141 - Right front seat belt micro switch
- J234 - Airbag Control Module
- J285 - Control module with indicator unit in instrument panel insert
- J393 - Central control module for comfort system
- K75 - Airbag Malfunction Indicator Lamp (MIL)
- T2 - 2-Pin Connector
- T4 - 4-Pin Connector
- T16 - 16- Pin Connector, Data Link Connector (DLC), below steering column, left
- T23 - 23-Pin Connector
- T32 - 32-Pin Connector, blue, on instrument cluster
- T32a - 32-Pin Connector, green, on instrument cluster
- T75 - 75-Pin Connector
- (A2) - plus connection (15), in instrument panel wiring harness
- (A76) Connection (K-diagnosis wire) in instrument
- (A125) Connection (crash signal) in instrument panel wiring harness
- (109) Ground connection, in airbag wiring harness

ws = white
sw = black
ro = red
br = brown
gn = green
bl = blue
gr = grey
li = violet
ge = yellow
or = orange

New Beetle — Wiring diagram — No. 15/1

Airbag systems,

from September 2003

Deviate relay location and fuseplacements as well as the locations of multiple connectors see section "component locations".

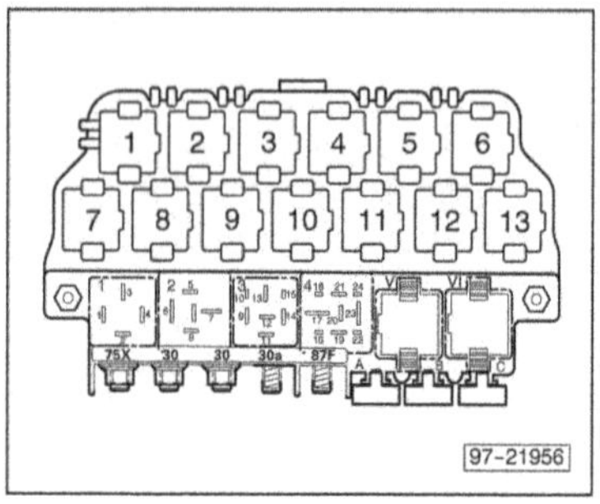

Relay location on the thirteen position auxiliary relay panel, above relay panel:

Relay panel:

1 Dual Horn Relay (53)

Fuse colors

30 A - green
25 A - white
20 A - yellow
15 A - blue
10 A - red
7,5 A - brown
5 A - beige
3 A - violet

Note: Number in parentheses indicates production control number stamped on relay housing.

New Beetle — Wiring diagram — No. 15/2

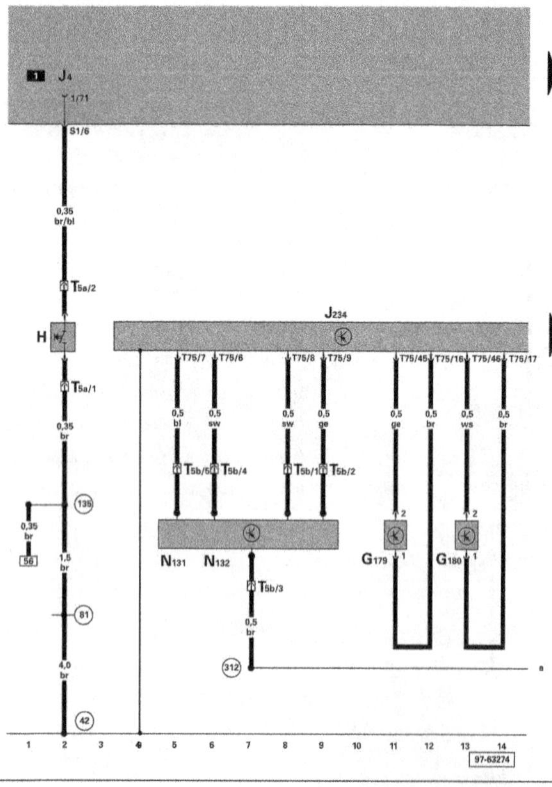

Airbag control module, crash sensors for side airbags

G179 - Crash sensor for side airbag, driver's side
G180 - Crash sensor for side airbag, passenger side
H - Signal horn activation
J4 - Dual Horn Relay
J234 - Airbag Control Module
N131 - Front Passenger's Side Airbag Igniter 1
N132 - Front Passenger's Side Airbag Igniter 2
T5a - 5-Pin Connector
T5b - 5-Pin Connector, yellow, near steering column
T75 - 75-Pin Connector, at -J234-

(42) - Ground connection, beside steering column

(81) - Ground connection -1-, in instrument panel wiring harness

(135) - Ground connection -2-, in instrument panel wiring harness

(312) - Ground Connection 2 (in Airbag wiring harness)

ws = white
sw = black
ro = red
br = brown
gn = green
bl = blue
gr = grey
li = violet
ge = yellow
or = orange

New Beetle — Wiring diagram — No. 15/3

Airbag control module, igniters for side airbag, left/right seat belt tensioner igniter,

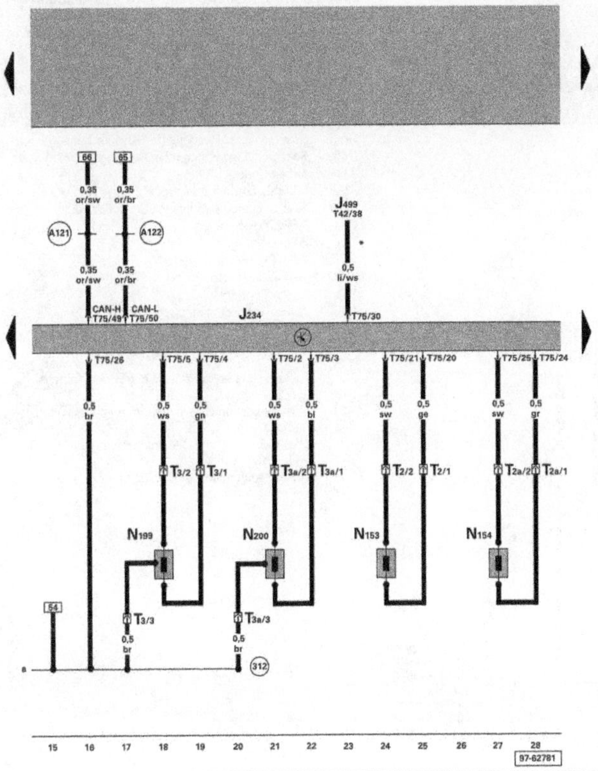

- J234 - Airbag Control Module
- J499 - Telematic Control Module *
- N153 - Left Seat Belt Tensioner Igniter
- N154 - Right Seat Belt Tensioner Igniter
- N199 - Igniter for side airbag, driver's side
- N200 - Igniter for side airbag, passenger side
- T2 - 2-Pin Connector
- T2a - 2-Pin Connector
- T3 - 3-Pin Connector
- T3a - 3-Pin Connector
- T42 - 42-Pin Connector, at -J499-
- T75 - 75-Pin Connector, at -J234-

- (312) Ground Connection 2 (in Airbag wiring harness)
- (A121) Connection (high bus) in instrument panel wiring harness
- (A122) Connection (low bus) in instrument panel wiring harness

- * - Only vehicles with Telematics

ws = white
sw = black
ro = red
br = brown
gn = green
bl = blue
gr = grey
li = violet
ge = yellow
or = orange

New Beetle — Wiring diagram — No. 15/4

Airbag control module, Key switch for switching off passenger side airbag, Airbag Spiral Spring/ Return Spring With Slip Ring, Drivers/Passengers Front Airbag Crash Sensor

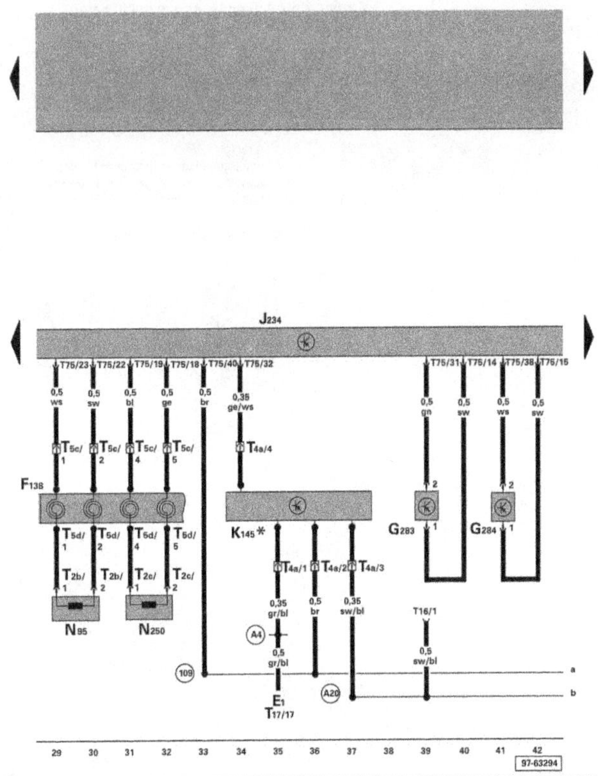

- E 1 - Light switch
- F138 - Airbag Spiral Spring/Return Spring With Slip Ring
- G283 - Drivers Front Airbag Crash Sensor
- G284 - Passengers Front Airbag Crash Sensor
- J234 - Airbag Control Module
- K145 - Warning light for airbag off, passenger side*
- N95 - Driver's Side Airbag Igniter
- N250 - Igniter 2 for airbag, driver side
- T2b - 2-Pin Connector, green, behind steering wheel airbag
- T2c - 2-Pin Connector, orange, behind steering wheel airbag
- T4a - 4-Pin Connector
- T5c - 5-Pin Connector, yellow, behind steering column switch cover
- T5d - 5-Pin Connector, black, behind steering wheel airbag
- T16 - 16-Pin Connector, Data Link Connector (DLC), bellow steering column, left
- T17 - 17-Pin Connector
- T75 - 75-Pin Connector, at -J234-

- (109) Ground connection, in airbag wiring harness
- (A4) - plus connection (58b), in instrument panel wiring harness
- (A20) - Plus Connection (15a) (in instrument panel wiring harness)

- * - Not applicable to USA/CDN

ws = white
sw = black
ro = red
br = brown
gn = green
bl = blue
gr = grey
li = violet
ge = yellow
or = orange

Airbag system from Sept. 2003

New Beetle — Wiring diagram — No. 15/5

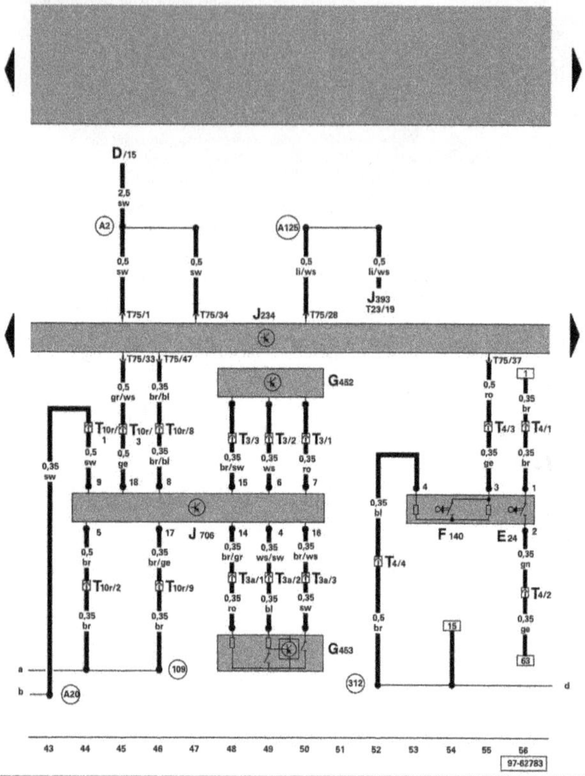

Airbag Control Module, Seat Occupied Recognition Control Module, Seat Occupied Recognition Pressure Sensor, Seat Occupied Recognition Belt Force Sensor, Left front seat belt micro switch

- D - Ignition/Starter Switch
- E24 - Left Seat Belt Switch
- F140 - Left front seat belt micro switch
- G452 - Seat Occupied Recognition Pressure Sensor
- G453 - Seat Occupied Recognition Belt Force Sensor
- J234 - Airbag Control Module
- J393 - Central control module for comfort system
- J706 - Seat Occupied Recognition Control Module
- T3 - 3-Pin Connector
- T3a - 3-Pin Connector
- T4 - 4-Pin Connector
- T10r - 10- Pin Connector
- T23 - 23-Pin Connector
- T75 - 75-Pin Connector, at -J234-
- (109) Ground connection, in airbag wiring harness
- (312) Ground Connection 2 (in Airbag wiring harness)
- (A2) - plus connection (15), in instrument panel wiring harness
- (A20) Plus Connection (15a) (in instrument panel wiring harness)
- (A125) Connection (crash signal) in instrument panel wiring harness

ws = white
sw = black
ro = red
br = brown
gn = green
bl = blue
gr = grey
li = violet
ge = yellow
or = orange

New Beetle — Wiring diagram — No. 15/6

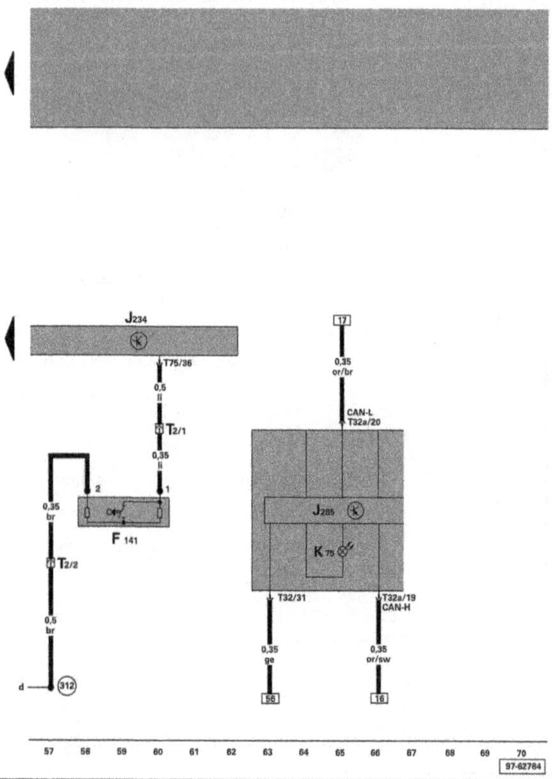

Airbag control module, Airbag malfunction indicator lamp (MIL)

- F141 - Right front seat belt micro switch
- J234 - Airbag Control Module
- J285 - Control module with indicator unit in instrument panel insert
- K75 - Airbag Malfunction Indicator Lamp (MIL)
- T2 - 2-Pin Connector
- T32 - 32-Pin Connector, blue, on instrument cluster
- T32a - 32-Pin Connector, green, on instrument cluster
- T75 - 75-Pin Connector, at -J234-
- (312) Ground Connection 2 (in Airbag wiring harness)

ws = white
sw = black
ro = red
br = brown
gn = green
bl = blue
gr = grey
li = violet
ge = yellow
or = orange

New Beetle **Wiring diagram** **No. 53/1**

Airbag systems,

from July 2005,

Additional Fuse, Relay, Control Module and Connector locations ⇒ Component Locations

Electrical troubleshooting ⇒ Guided Fault Finding using VAS 5051/5052

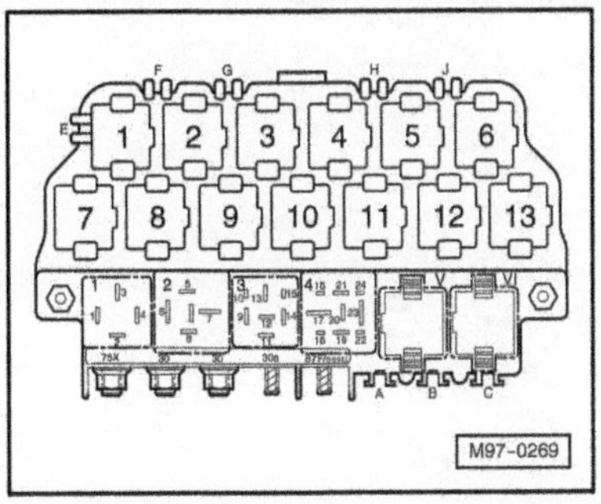

Relay location on the thirteen position auxiliary relay panel, above relay panel:

Relay panel::

1 J4 - Dual Horn Relay (53)

4 J17 - Fuel Pump (FP) Relay (409)

Note:

Number in parentheses indicates production control number stamped on relay housing.

New Beetle **Wiring diagram** **No. 53/2**

Crash sensor for side airbags, Airbag Control Module

G179 - Driver Side Airbag Crash Sensor
G180 - Crash sensors for side airbags
H - Signal horn activation
J4 - Horn Relay
J234 - Airbag Control Module
N131 - Front Passenger Airbag Igniter 1
N132 - Front Passenger Airbag Igniter 2
T2k - Double Connector
T5o - 5-Pin Connector, yellow, near steering column
T50 - 50-Pin Connector, at -J234-

(42) - Ground Connection (beside steering column)
(49) - Ground Connection (near steering column)
(81) - Ground Connection 1 (in instrument panel wiring harness)
(135) - Ground Connection 2 (in instrument panel wiring harness)
(312) - Ground Connection 2 (in airbag wiring harness)

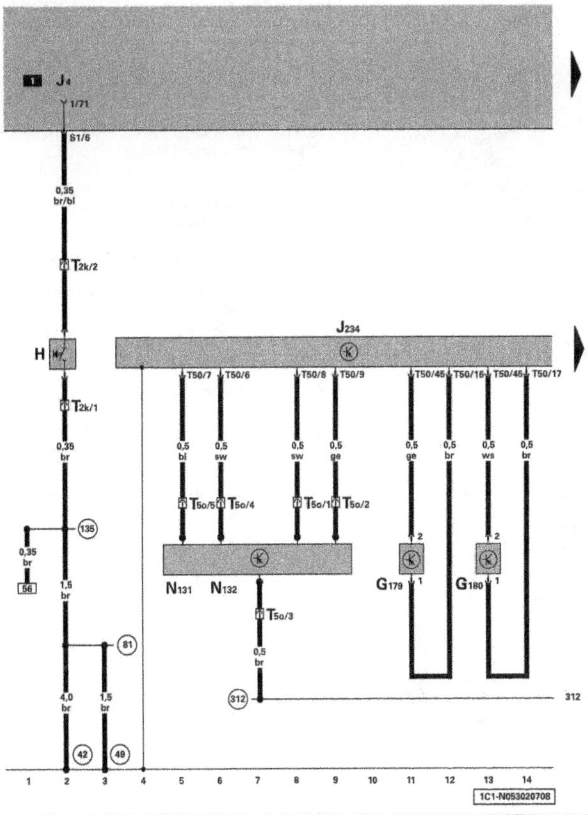

ws = white
sw = black
ro = red
br = brown
gn = green
bl = blue
gr = grey
li = violet
ge = yellow
or = orange

New Beetle

Wiring diagram

No. 53/3

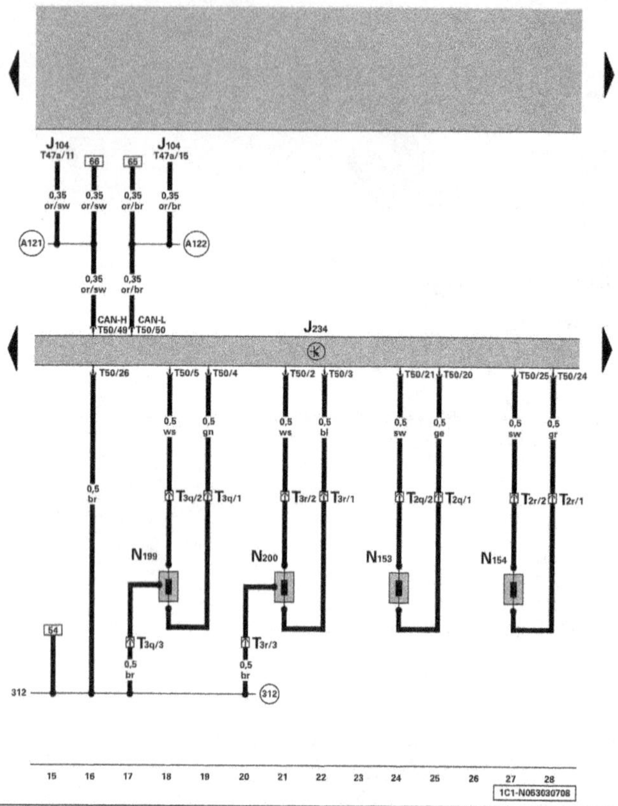

Airbag Control Module, Left/right seat belt tensioner igniter, Igniters for side airbag,

- J104 - ABS Control Module
- J234 - Airbag Control Module
- N153 - Driver Seat Belt Tensioner Igniter 1
- N154 - Front Passenger Seat Belt Tensioner Igniter 1
- N199 - Driver Side Airbag Igniter
- N200 - Front Passenger Side Airbag Igniter
- T2q - Double Connector
- T2r - Double Connector
- T3q - 3-Pin Connector
- T3r - 3-Pin Connector
- T47a - 47-Pin Connector
- T50 - 50-Pin Connector, at -J234-
- (312) - Ground Connection 2 (in airbag wiring harness)
- (A121) - High-Bus Connection (in instrument panel wiring harness)
- (A122) - Low-Bus Connection (in instrument panel wiring harness)

ws = white
sw = black
ro = red
br = brown
gn = green
bl = blue
gr = grey
li = violet
ge = yellow
or = orange

New Beetle

Wiring diagram

No. 53/4

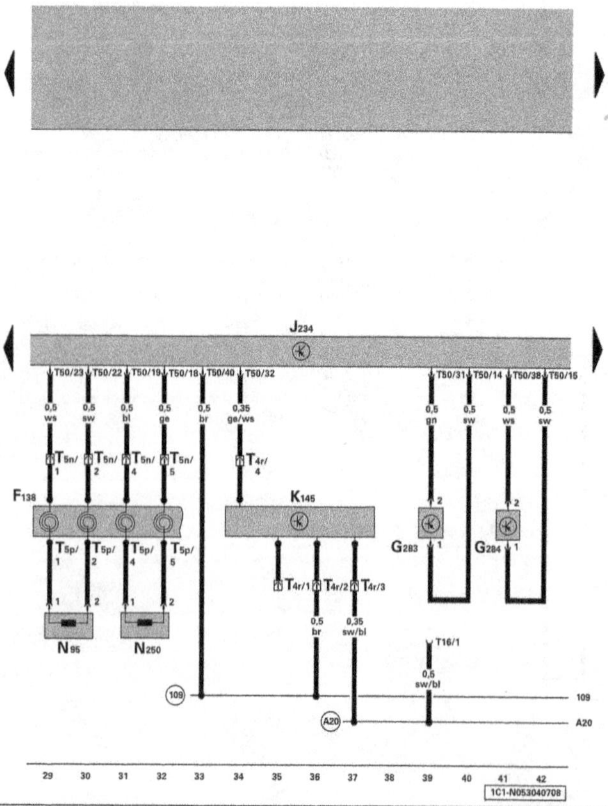

Airbag Spiral Spring/Return Spring With Slip Ring, Driver Front Airbag Crash Sensor, Airbag Control Module

- F138 - Airbag Spiral Spring/Return Spring With Slip Ring
- G283 - Driver Front Airbag Crash Sensor
- G284 - Front Passenger Front Airbag Crash Sensor
- J234 - Airbag Control Module
- K145 - Front Passenger Airbag Disabled Indicator Lamp
- N95 - Driver Airbag Igniter
- N250 - Driver Airbag Igniter 2
- T4r - 4-Pin Connector
- T5n - 5-Pin Connector, yellow, behind steering column switch cover
- T5p - 5-Pin Connector, black, behind steering wheel airbag
- T16 - 16-Pin Connector, Data Link Connector (DLC), bellow steering column, left
- T50 - 50-Pin Connector, at -J234-
- (109) - Ground Connection (in airbag wiring harness)
- (A20) - Plus Connection (15a) (in instrument panel wiring harness)

ws = white
sw = black
ro = red
br = brown
gn = green
bl = blue
gr = grey
li = violet
ge = yellow
or = orange

New Beetle

Wiring diagram

No. 53/5

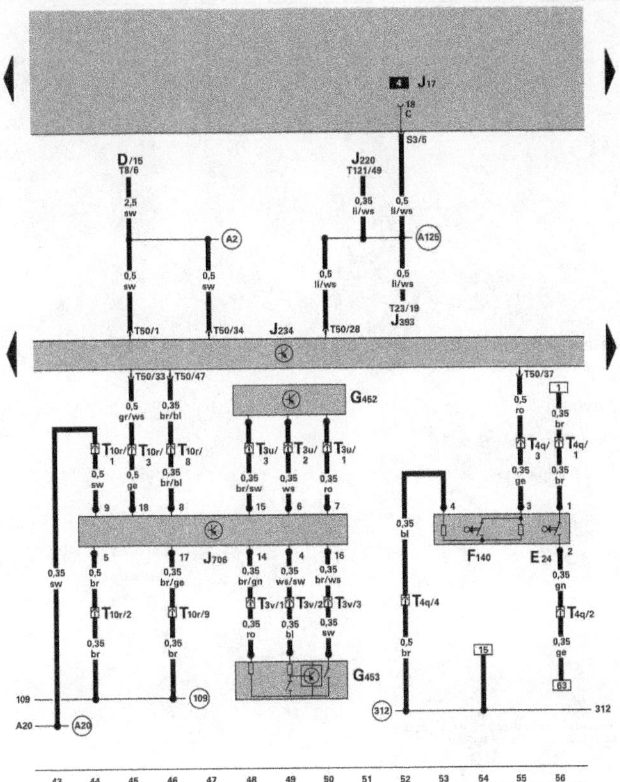

Driver Seat Belt Switch, Seat Occupied Recog nition Pressure Sensor, Seat Occupied Recog nition Seat Belt Force Sensor, Airbag Control Module

- D - Ignition/Starter Switch
- E24 - Driver Seat Belt Switch
- F140 - Left front seat belt micro switch
- G452 - Seat Occupied Recognition Pressure Sensor
- G453 - Seat Occupied Recognition Seat Belt Force Sensor
- J17 - Fuel Pump (FP) Relay
- J220 - Motronic Engine Control Module (ECM)
- J234 - Airbag Control Module
- J393 - Central control module for comfort system
- J706 - Seat Occupied Recognition Control Module
- T3u - 3-Pin Connector
- T3v - 3-Pin Connector
- T4q - 4-Pin Connector
- T8 - 8-Pin Connector
- T10r - 10- Pin Connector
- T23 - 23-Pin Connector
- T50 - 50-Pin Connector, at -J234-
- T121 - 121-Pin Connector

- (109) - Ground Connection (in airbag wiring harness)
- (312) - Ground Connection 2 (in airbag wiring harness)
- (A2) - Plus Connection (15) (in instrument panel wiring harness)
- (A20) - Plus Connection (15a) (in instrument panel wiring harness)
- (A125) - Crash Signal Connection (in instrument panel wiring harness)

ws = white
sw = black
ro = red
br = brown
gn = green
bl = blue
gr = grey
li = violet
ge = yellow
or = orange

43 44 45 46 47 48 49 50 51 52 53 54 55 56

1C1-N053050708

New Beetle

Wiring diagram

No. 53/6

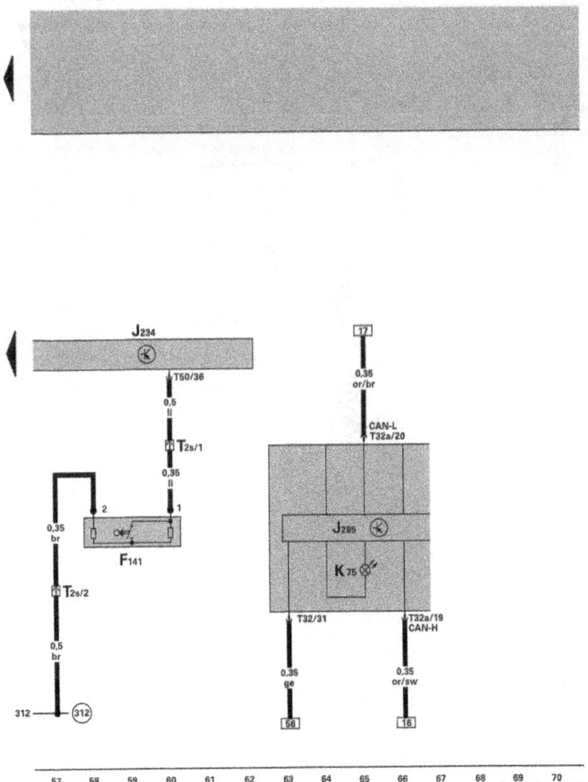

Airbag Control Module, Airbag Malfunction Indicator Lamp (MIL)

- F141 - Right front seat belt micro switch
- J234 - Airbag Control Module
- J285 - Instrument Cluster Control Module
- K75 - Airbag Malfunction Indicator Lamp (MIL)
- T2s - Double Connector
- T32 - 32-Pin Connector, blue, on instrument cluster
- T32a - 32-Pin Connector, green, on instrument cluster
- T50 - 50-Pin Connector, at -J234-

- (312) - Ground Connection 2 (in airbag wiring harness)

ws = white
sw = black
ro = red
br = brown
gn = green
bl = blue
gr = grey
li = violet
ge = yellow
or = orange

57 58 59 60 61 62 63 64 65 66 67 68 69 70

1C1-N053060708

New Beetle Convertible Wiring diagram **No. 6/1**

Airbag systems, Automatic Rollover Supports™

from August 2002

Deviate relay location and fuseplacements as well as the locations of multiple connectors see section "component locations".

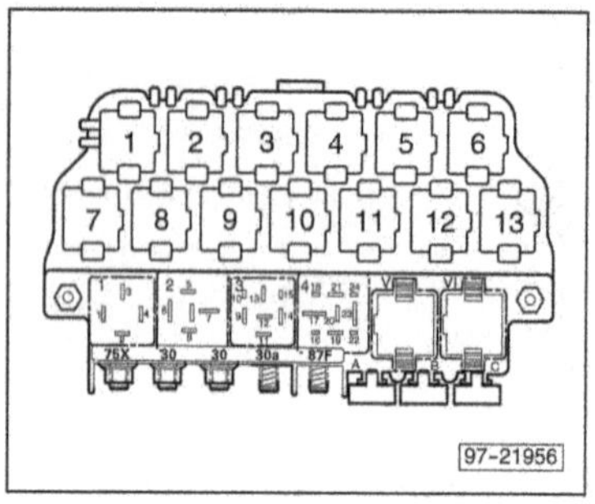

Relay panel:

1 Dual Horn Relay (53)

4 Fuel Pump (FP) Relay (409)

Fuse colors

30 A - green
25 A - white
20 A - yellow
15 A - blue
10 A - red
7,5 A - brown
5 A - beige
3 A - violet

Note: Number in parentheses indicates production control number stamped on relay housing.

New Beetle Convertible Wiring diagram **No. 6/2**

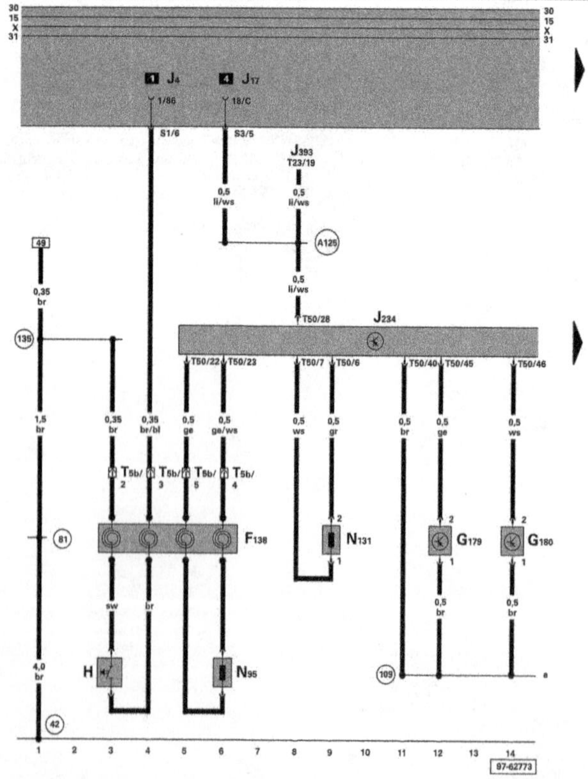

Airbag spiral spring/return spring with slip ring, crash sensor for side airbag drivers side and passenger side, horn relay, signal horn activation

F138 - Airbag Spiral Spring/Return Spring With Slip Ring
G179 - Crash sensor for side airbag, driver's side
G180 - Crash sensor for side airbag, passenger side
H - Signal horn activation
J4 - Horn Relay
J17 - Fuel Pump (FP) Relay
J234 - Airbag Control Module
J393 - Central control module for comfort system
N95 - Driver's Side Airbag Igniter
N131 - Passenger's Side Airbag Igniter 1
T5b - 5-Pin Connector, beside steering column
T23 - 23-Pin Connector, A connector, central module
T50 - 50-Pin Connector, at J234

(42) - Ground connection, beside steering column

(81) - Ground connection -1-, in instrument panel wiring harness

(109) - Ground connection, in airbag wiring harness

(135) - Ground connection -2-, in instrument panel wiring harness

(A125) - Connection (crash signal) in instrument panel wiring harness

ws = white
sw = black
ro = red
br = brown
gn = green
bl = blue
gr = grey
li = violet
ge = yellow
or = orange

New Beetle Convertible — Wiring diagram — No. 6/3

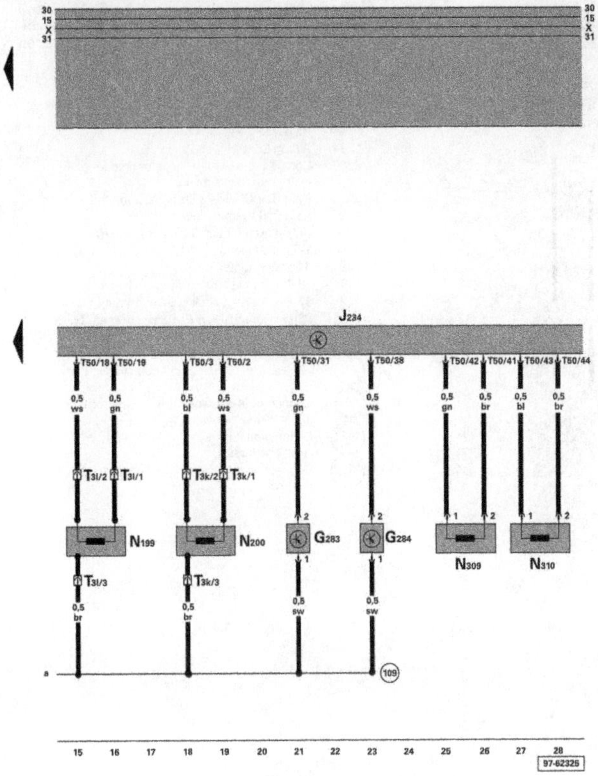

Drivers front airbag crash sensor, passengers front airbag crash sensor, airbag control module, igniter for side airbag drivers and passengers side, left and right rollover solenoid

G283 - Drivers Front Airbag Crash Sensor
G284 - Passengers Front Airbag Crash Sensor
J234 - Airbag Control Module
N199 - Igniter for side airbag, driver's side
N200 - Igniter for side airbag, passenger side
N309 - Left Rollover Solenoid
N310 - Right Rollover Solenoid
T3k - 3-Pin Connector, under passenger's seat
T3l - 3-Pin Connector, under driver's seat
T50 - 50-Pin Connector, in airbag control module

(109) - Ground connection, in airbag wiring harness

ws = white
sw = black
ro = red
br = brown
gn = green
bl = blue
gr = grey
li = violet
ge = yellow
or = orange

New Beetle Convertible — Wiring diagram — No. 6/4

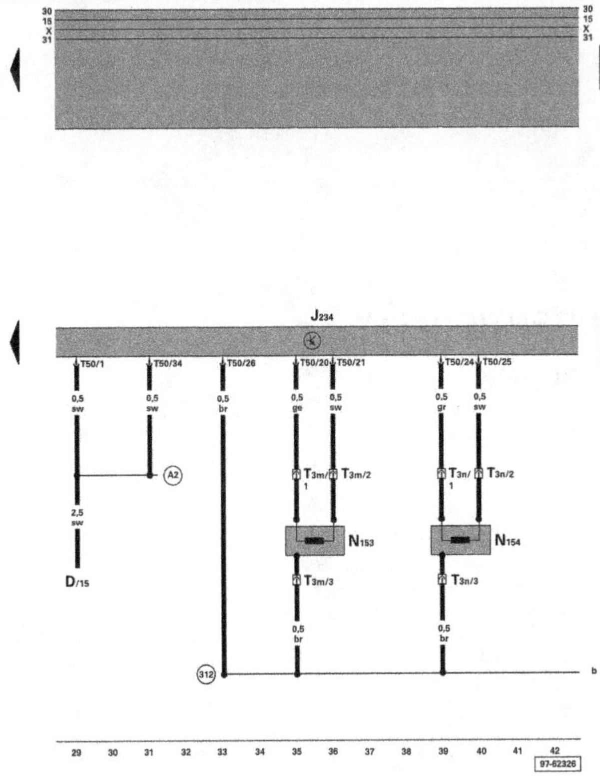

Airbag control module, drivers and passengers seat belts tensioner igniter 1

D - Ignition/Starter Switch
J234 - Airbag Control Module
N153 - Driver's Seat Belt Tensioner Igniter 1
N154 - Passenger's Seat Belt Tensioner Igniter 1
T3m - 3-Pin Connector, under driver's seat
T3n - 3-Pin Connector, under passenger's seat
T50 - 50-Pin Connector, in airbag control module

(312) - Ground Connection 2 (in Airbag wiring harness)

(A2) - plus connection (15), in instrument panel wiring harness

ws = white
sw = black
ro = red
br = brown
gn = green
bl = blue
gr = grey
li = violet
ge = yellow
or = orange

Airbag system and rollover supports from Aug. 2002, convertible

New Beetle Convertible Wiring diagram **No. 6/5**

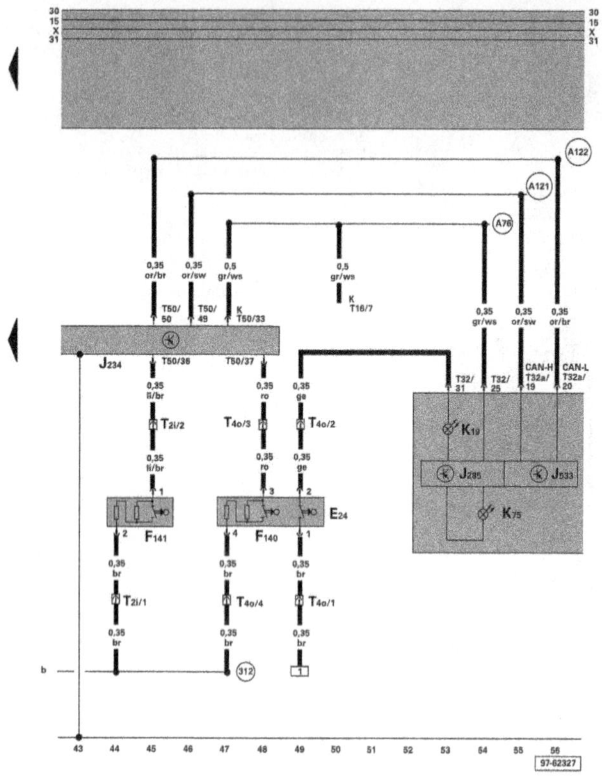

Left seat belt switch, airbag control module, control module with indicator unit in instrument panel insert, airbag malfunction indicator, data bus on board diagnostic interface

- E24 - Left Seat Belt Switch
- F140 - Left Front Seatbelt Microswitch
- F141 - Right Front Seatbelt Microswitch
- J234 - Airbag Control Module
- J285 - Control module with indicator unit in instrument panel insert
- J533 - Data Bus On Board Diagnostic Interface
- K19 - Seat Belt Warning Light
- K75 - Airbag Malfunction Indicator Lamp (MIL)
- T2i - 2-Pin Connector
- T4o - 4-Pin Connector
- T16 - 16-Pin Connector
- T32 - 32-Pin Connector, blue, in instrument panel
- T32a - 32-Pin Connector, green, in instrument panel
- T50 - 50-Pin Connector, in airbag control unit
- (312) - Ground Connection 2 (in Airbag wiring harness)
- (A76) - Connector (K-diagnosis wire) in instrument panel wiring harness
- (A121) - Connection (high bus) in instrument panel wiring harness
- (A122) - Connection (low bus) in instrument panel wiring harness

ws = white
sw = black
ro = red
br = brown
gn = green
bl = blue
gr = grey
li = violet
ge = yellow
or = orange

THIS SPACE INTENTIONALLY LEFT BLANK

New Beetle Convertible **Wiring diagram** **No. 17/1**

Airbag systems, Automatic Rollover Supports(TM),

from September 2003

Additional Fuse, Relay, Control Module and Connector locations ⇒ Component Locations

Electrical troubleshooting ⇒ Guided Fault Finding using VAS 5051/5052

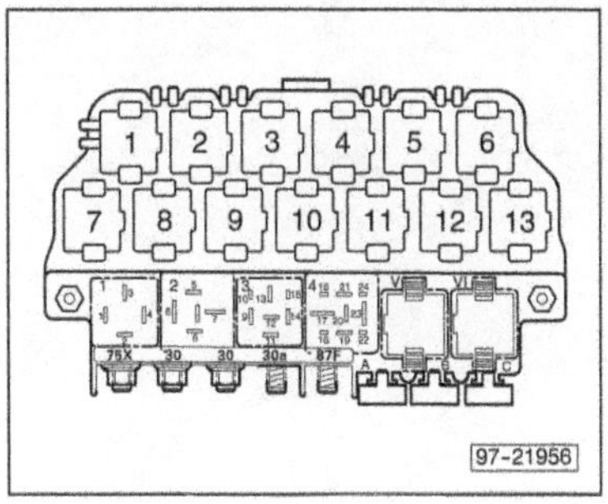

Relay location on the thirteen position auxiliary relay panel, above relay panel:

Relay panel::

1 - Dual Horn Relay (53)

Note:

Number in parentheses indicates production control number stamped on relay housing.

New Beetle Convertible **Wiring diagram** **No. 17/2**

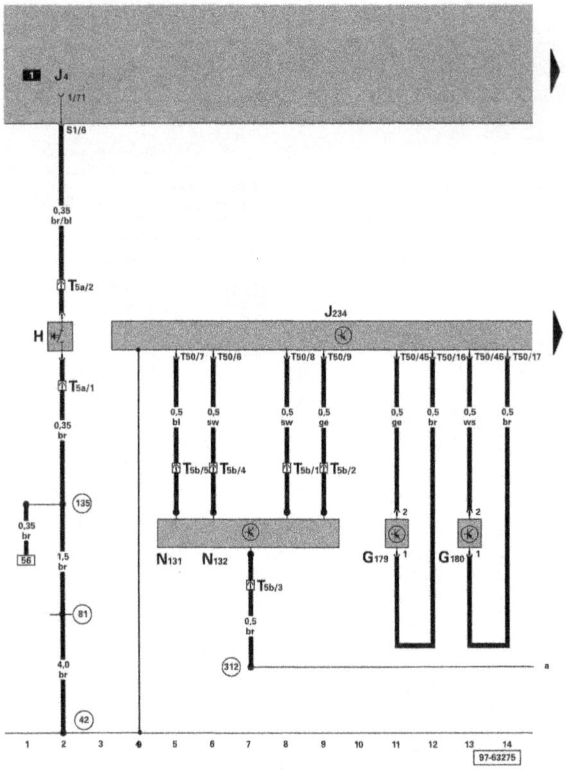

Airbag control module, crash sensors for side airbags

G179 - Crash sensor for side airbag, driver's side
G180 - Crash sensor for side airbag, passenger side
H - Signal horn activation
J4 - Dual Horn Relay
J234 - Airbag Control Module
N131 - Front Passenger's Side Airbag Igniter 1
N132 - Front Passenger's Side Airbag Igniter 2
T5a - 5-Pin Connector
T5b - 5-Pin Connector, yellow
T50 - 50-Pin Connector, at -J234-

(42) - Ground connection, beside steering column

(81) - Ground connection -1-, in instrument panel wiring harness

(135) - Ground connection -2-, in instrument panel wiring harness

(312) - Ground Connection 2 (in Airbag wiring harness)

ws = white
sw = black
ro = red
br = brown
gn = green
bl = blue
gr = grey
li = violet
ge = yellow
or = orange

Airbag system and rollover supports from Sept. 2003, convertible

New Beetle Convertible — Wiring diagram — No. 17/3

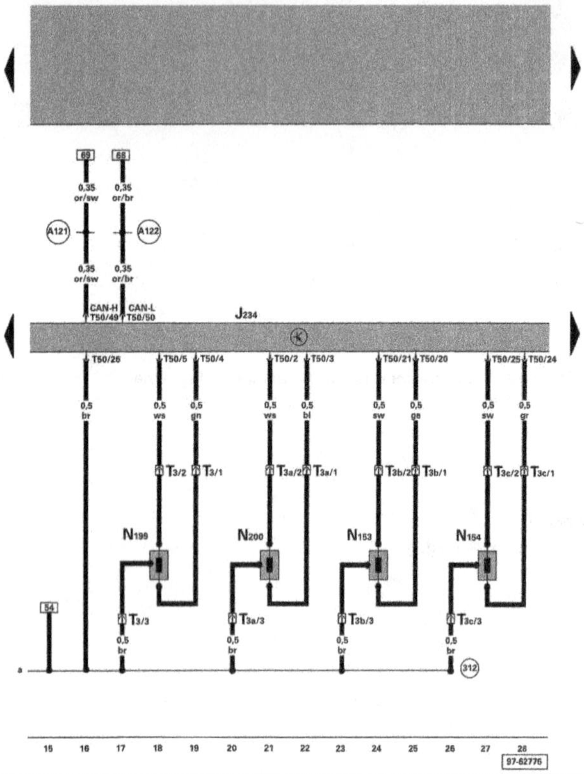

Airbag control module, igniters for side airbag, left/right seat belt tensioner igniter,

J234 - Airbag Control Module
N153 - Left Seat Belt Tensioner Igniter
N154 - Right Seat Belt Tensioner Igniter
N199 - Igniter for side airbag, driver's side
N200 - Igniter for side airbag, passenger side
T3 - 3-Pin Connector
T3a - 3-Pin Connector
T3b - 3-Pin Connector
T3c - 3-Pin Connector
T50 - 50-Pin Connector, at -J234-

(312) Ground Connection 2 (in Airbag wiring harness)

(A121) Connection (high bus) in instrument panel wiring harness

(A122) Connection (low bus) in instrument panel wiring harness

ws = white
sw = black
ro = red
br = brown
gn = green
bl = blue
gr = grey
li = violet
ge = yellow
or = orange

Edition 06.03

New Beetle Convertible — Wiring diagram — No. 17/4

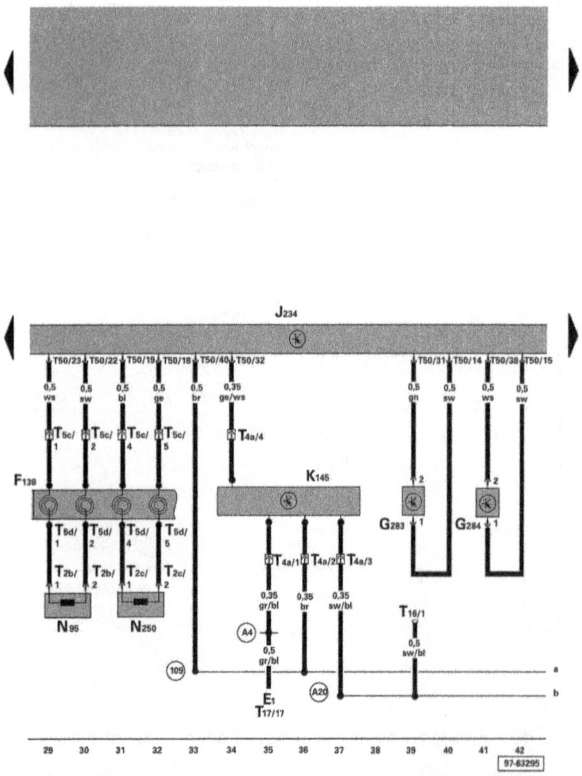

Airbag control module, Key switch for switching off passenger side airbag, Airbag Spiral Spring/Return Spring With Slip Ring, Drivers/Passengers Front Airbag Crash Sensor

E 1 - Light Switch
F138 - Airbag Spiral Spring/Return Spring With Slip Ring
G283 - Drivers Front Airbag Crash Sensor
G284 - Passengers Front Airbag Crash Sensor
J234 - Airbag Control Module
K145 - Warning light for airbag off, passenger side
N95 - Driver's Side Airbag Igniter
N250 - Igniter 2 for airbag, driver side
T2b - 2-Pin Connector, green, behind steering wheel airbag
T2c - 2-Pin Connector, orange, behind steering wheel airbag
T4a - 4-Pin Connector
T5c - 5-Pin Connector, yellow, behind steering column switch cover
T5d - 5-Pin Connector, black, behind steering wheel airbag
T16 - 16-Pin Connector, Data Link Connector (DLC), bellow steering column, left
T17 - 17-Pin Connector
T50 - 50-Pin Connector, at -J234-

(109) Ground connection, in airbag wiring harness

(A4) - plus connection (58b), in instrument panel wiring harness

(A20) - Plus Connection (15a) (in instrument panel wiring harness)

ws = white
sw = black
ro = red
br = brown
gn = green
bl = blue
gr = grey
li = violet
ge = yellow
or = orange

Edition 06.04

New Beetle Convertible — Wiring diagram — No. 17/5

Airbag Control Module, Seat Occupied Recognition Control Module, Seat Occupied Recognition Pressure Sensor, Seat Occupied Recognition Belt Force Sensor, Left front seat belt micro switch

- D - Ignition/Starter Switch
- E24 - Left Seat Belt Switch
- F140 - Left front seat belt micro switch
- G452 - Seat Occupied Recognition Pressure Sensor
- G453 - Seat Occupied Recognition Belt Force Sensor
- J234 - Airbag Control Module
- J393 - Central control module for comfort system
- J706 - Seat Occupied Recognition Control Module
- T3 - 3-Pin Connector
- T3a - 3-Pin Connector
- T4 - 4-Pin Connector
- T10r - 10- Pin Connector
- T23 - 23-Pin Connector
- T50 - 50-Pin Connector, at -J234-
- 109 - Ground connection, in airbag wiring harness
- 312 - Ground Connection 2 (in Airbag wiring harness)
- A2 - plus connection (15), in instrument panel wiring harness
- A20 - Plus Connection (15a) (in instrument panel wiring harness)
- A125 - Connection (crash signal) in instrument panel wiring harness

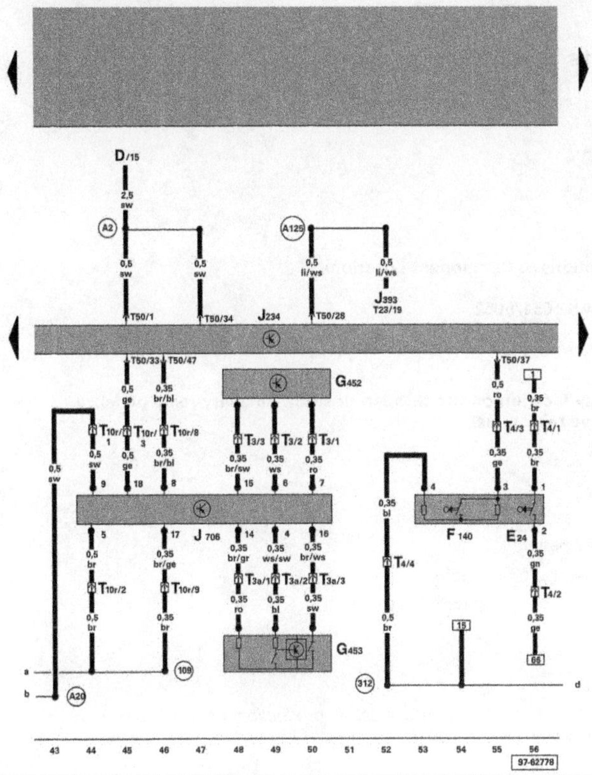

ws = white
sw = black
ro = red
br = brown
gn = green
bl = blue
gr = grey
li = violet
ge = yellow
or = orange

New Beetle Convertible — Wiring diagram — No. 17/6

Airbag control module, Airbag malfunction indicator lamp (MIL)

- F141 - Right front seat belt micro switch
- J234 - Airbag Control Module
- J285 - Control module with indicator unit in instrument panel insert
- K75 - Airbag Malfunction Indicator Lamp (MIL)
- N309 - Left Rollover Solenoid
- N310 - Right Rollover Solenoid
- T2 - 2-Pin Connector
- T32 - 32-Pin Connector, blue, on instrument cluster
- T32a - 32-Pin Connector, green, on instrument cluster
- T50 - 50-Pin Connector, at -J234-
- 312 - Ground Connection 2 (in Airbag wiring harness)

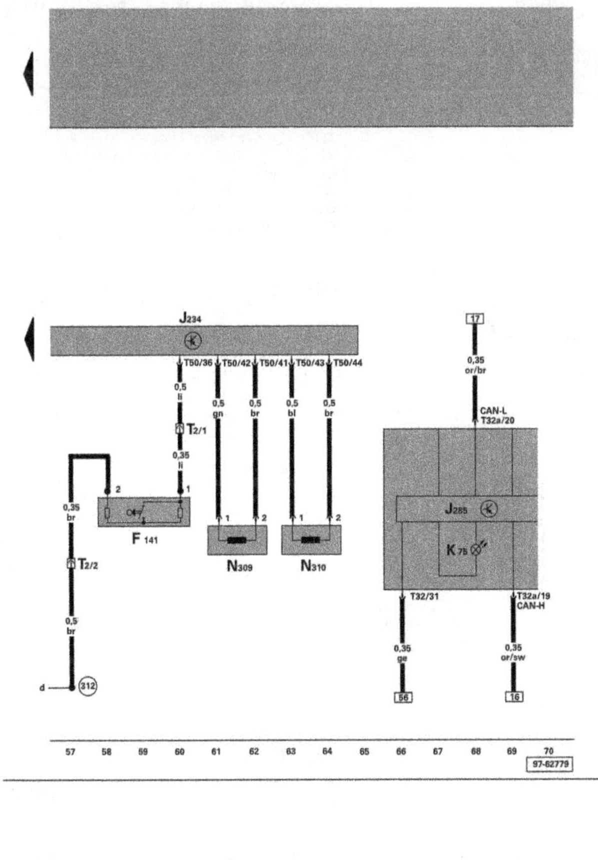

ws = white
sw = black
ro = red
br = brown
gn = green
bl = blue
gr = grey
li = violet
ge = yellow
or = orange

New Beetle Convertible Wiring diagram **No. 36/1**

Airbag systems, Automatic Rollover Supports™,

from July 2005,

Additional Fuse, Relay, Control Module and Connector locations ⇒ Component Locations

Electrical troubleshooting ⇒ Guided Fault Finding using VAS 5051/5052

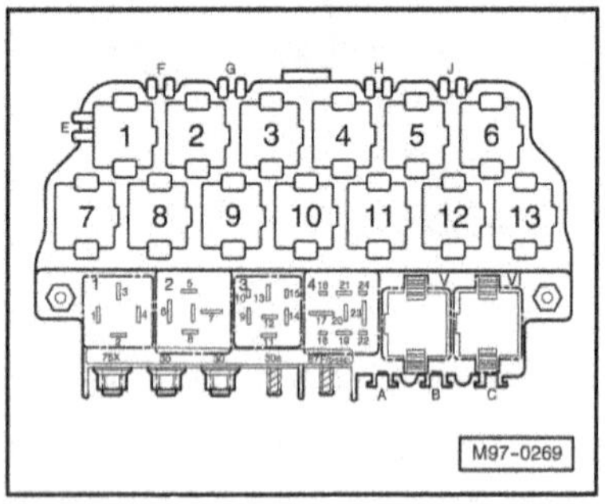

Relay location on the thirteen position auxiliary relay panel, above relay panel:

Relay panel:

1 J4 - Horn Relay (53)

4 J17 - Fuel Pump (FP) Relay (409)

Note:

Number in parentheses indicates production control number stamped on relay housing.

New Beetle Convertible Wiring diagram **No. 36/2**

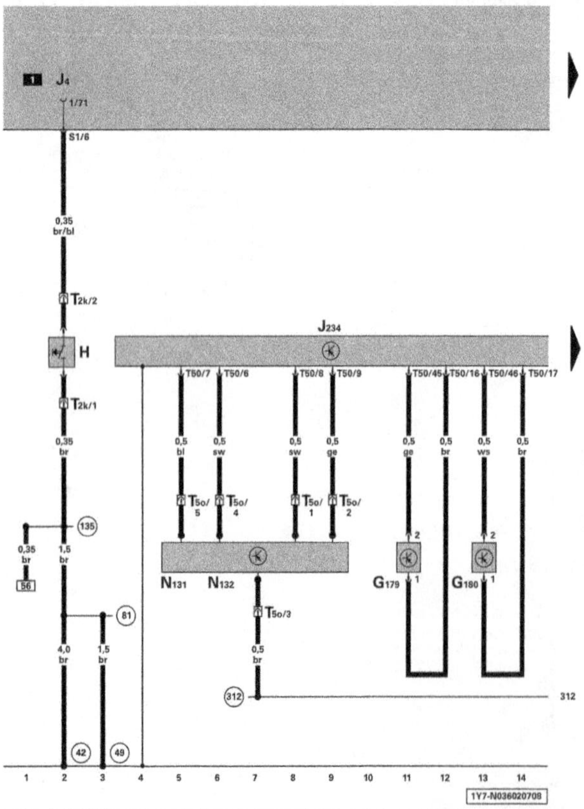

Driver Side Airbag Crash Sensor, Front Passenger Side Airbag Crash Sensor, Signal Horn Activation, Airbag Control Module, Front Passenger Airbag Igniter 1, Front Passenger Airbag Igniter 2

- G179 - Driver Side Airbag Crash Sensor
- G180 - Front Passenger Side Airbag Crash Sensor
- H - Signal Horn Activation
- J4 - Horn Relay
- J234 - Airbag Control Module
- N131 - Front Passenger Airbag Igniter 1
- N132 - Front Passenger Airbag Igniter 2
- T2k - Double Connector
- T5o - 5-Pin Connector, yellow
- T50 - 50-Pin Connector, at -J234-
- (42) - Ground Connection (beside steering column)
- (49) - Ground Connection (near steering column)
- (81) - Ground Connection 1 (in instrument panel wiring harness)
- (135) - Ground Connection 2 (in instrument panel wiring harness)
- (312) - Ground Connection 2 (in Airbag wiring harness)

ws = white
sw = black
ro = red
br = brown
gn = green
bl = blue
gr = grey
li = violet
ge = yellow
or = orange

New Beetle Convertible — Wiring diagram — No. 36/3

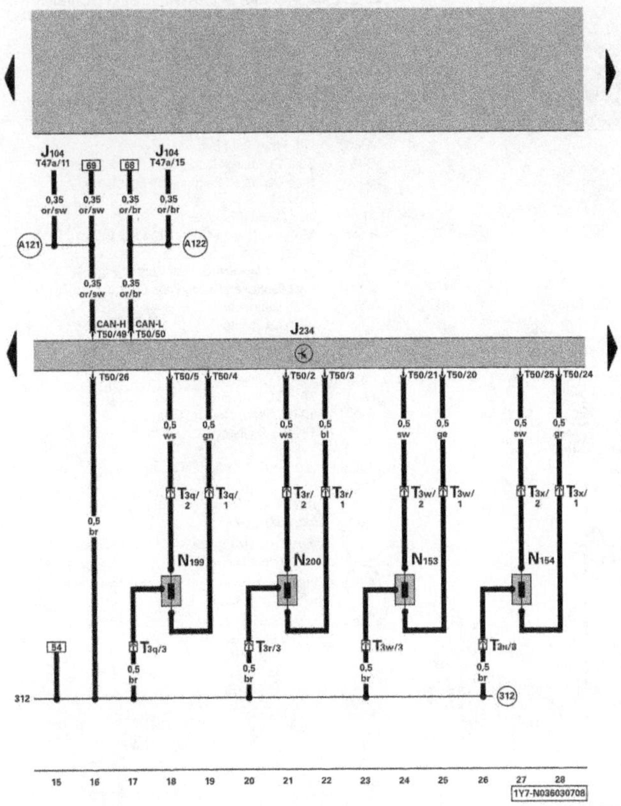

Airbag Control Module, Driver/Front Passenger Seat Belt Tensioner Igniter 1, Driver/Front Passenger Side Airbag Igniter

- J104 - ABS Control Module
- J234 - Airbag Control Module
- N153 - Driver Seat Belt Tensioner Igniter 1
- N154 - Front Passenger Seat Belt Tensioner Igniter 1
- N199 - Driver Side Airbag Igniter
- N200 - Front Passenger Side Airbag Igniter
- T3q - 3-Pin Connector
- T3r - 3-Pin Connector
- T3w - 3-Pin Connector
- T3x - 3-Pin Connector
- T47a - 47-Pin Connector
- T50 - 50-Pin Connector, at -J234-
- (312) - Ground Connection 2 (in Airbag wiring harness)
- (A121) - High-Bus Connection (in instrument panel wiring harness)
- (A122) - Low-Bus Connection (in instrument panel wiring harness)

ws = white
sw = black
ro = red
br = brown
gn = green
bl = blue
gr = grey
li = violet
ge = yellow
or = orange

New Beetle Convertible — Wiring diagram — No. 36/4

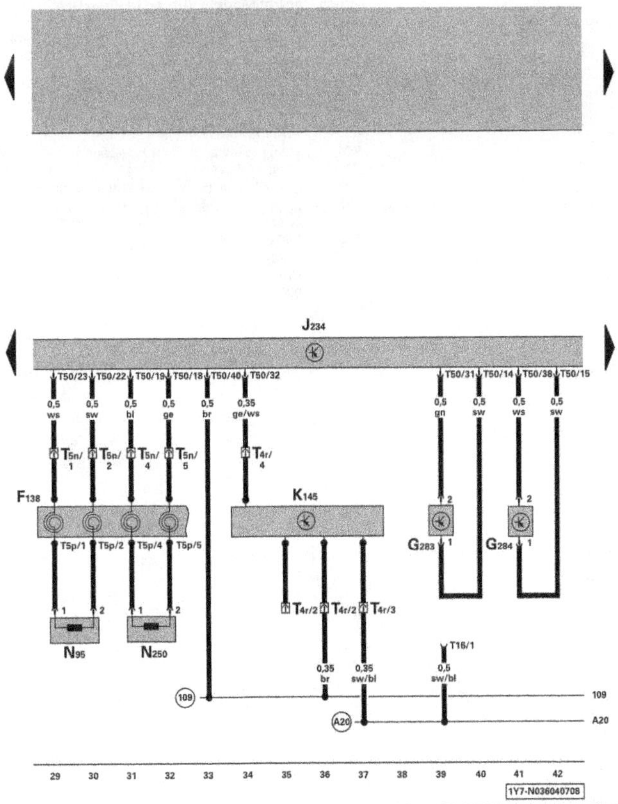

Airbag Spiral Spring/Return Spring With Slip Ring, Driver/Front Passenger Front Airbag Crash Sensor, Airbag Control Module

- F138 - Airbag Spiral Spring/Return Spring With Slip Ring
- G283 - Driver Front Airbag Crash Sensor
- G284 - Front Passenger Front Airbag Crash Sensor
- J234 - Airbag Control Module
- K145 - Front Passenger Airbag Disabled Indicator Lamp
- N95 - Driver Airbag Igniter
- N250 - Driver Airbag Igniter 2
- T4r - 4-Pin Connector
- T5n - 5-Pin Connector, yellow, behind steering column switch cover
- T5p - 5-Pin Connector, black, behind steering wheel airbag
- T16 - 16-Pin Connector, Data Link Connector (DLC), bellow steering column, left
- T50 - 50-Pin Connector, at -J234-
- (109) - Ground Connection (in airbag wiring harness)
- (A20) - Plus Connection (15a) (in instrument panel wiring harness)

ws = white
sw = black
ro = red
br = brown
gn = green
bl = blue
gr = grey
li = violet
ge = yellow
or = orange

Airbag system and rollover supports from July 2005, convertible

New Beetle Convertible — Wiring diagram — No. 36/5

Left Front Seat Belt Microswitch, Seat Occupied Recognition Pressure Sensor, Seat Occupied Recognition Seat Belt Force Sensor, Air Bag Control Module, Seat Occupied Recognition Control Module

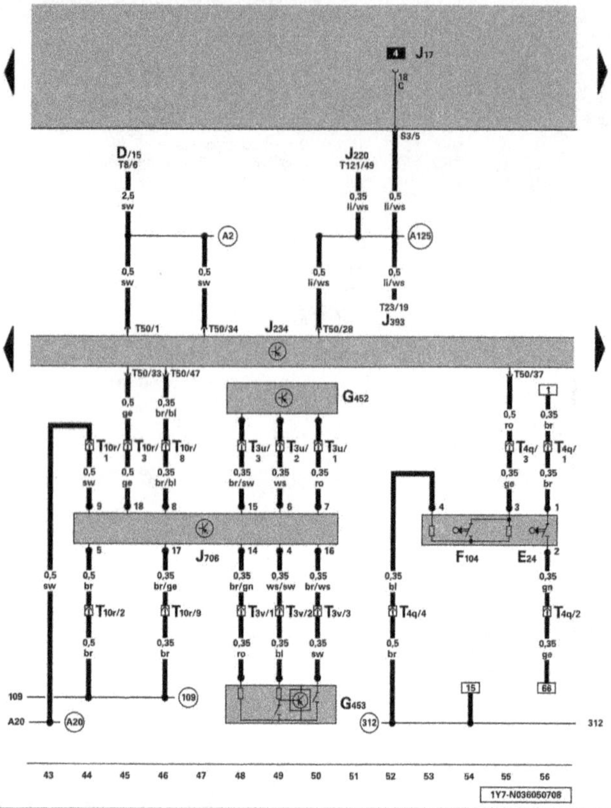

ws = white
sw = black
ro = red
br = brown
gn = green
bl = blue
gr = grey
li = violet
ge = yellow
or = orange

- D - Ignition/Starter Switch
- E24 - Driver Seat Belt Switch
- F140 - Left Front Seat Belt Microswitch
- G452 - Seat Occupied Recognition Pressure Sensor
- G453 - Seat Occupied Recognition Seat Belt Force Sensor
- J17 - Fuel Pump (FP) Relay
- J220 - Motronic Engine Control Module (ECM)
- J234 - Airbag Control Module
- J393 - Comfort System Central Control Module
- J706 - Seat Occupied Recognition Control Module
- T3u - 3-Pin Connector
- T3v - 3-Pin Connector
- T4q - 4-Pin Connector
- T8 - 8-Pin Connector
- T10r - 10- Pin Connector
- T23 - 23-Pin Connector
- T50 - 50-Pin Connector, at -J234-
- T121 - 121-Pin Connector
- (109) - Ground Connection (in airbag wiring harness)
- (312) - Ground Connection 2 (in Airbag wiring harness)
- (A2) - Plus Connection (15) (in instrument panel wiring harness)
- (A20) - Plus Connection (15a) (in instrument panel wiring harness)
- (A125) - Crash Signal Connection (in instrument panel wiring harness)

New Beetle Convertible — Wiring diagram — No. 36/6

Airbag Control Module, Airbag Malfunction Indicator Lamp (MIL)

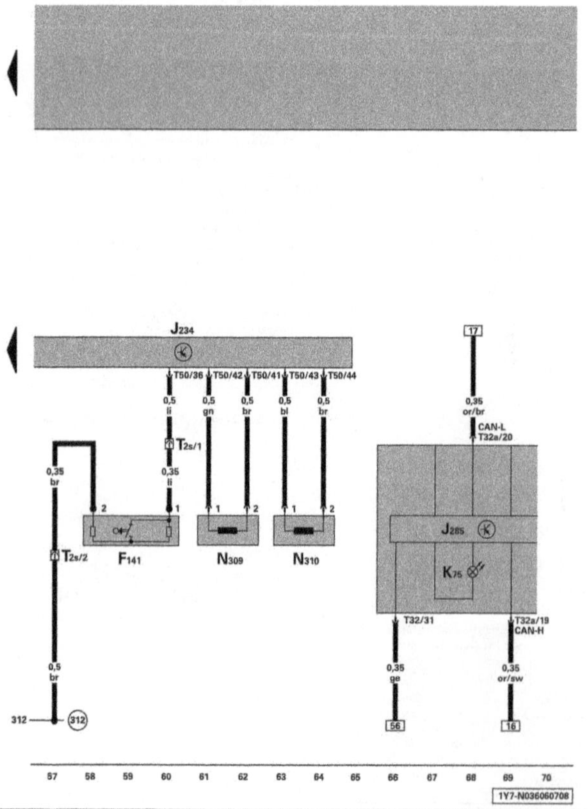

ws = white
sw = black
ro = red
br = brown
gn = green
bl = blue
gr = grey
li = violet
ge = yellow
or = orange

- F141 - Right front seat belt micro switch
- J234 - Airbag Control Module
- J285 - Instrument Cluster Control Module
- K75 - Airbag Malfunction Indicator Lamp (MIL)
- N309 - Driver Rollover Protection Solenoid
- N310 - Front Passenger Rollover Protection Solenoid
- T2s - Double Connector
- T32 - 32-Pin Connector, blue, on instrument cluster
- T32a - 32-Pin Connector, green, on instrument cluster
- T50 - 50-Pin Connector, at -J234-
- (312) - Ground Connection 2 (in Airbag wiring harness)

Automatic transmission,

from January 1998

Deviate relay location and fuseplacements as well as the locations of multiple connectors see section "component locations".

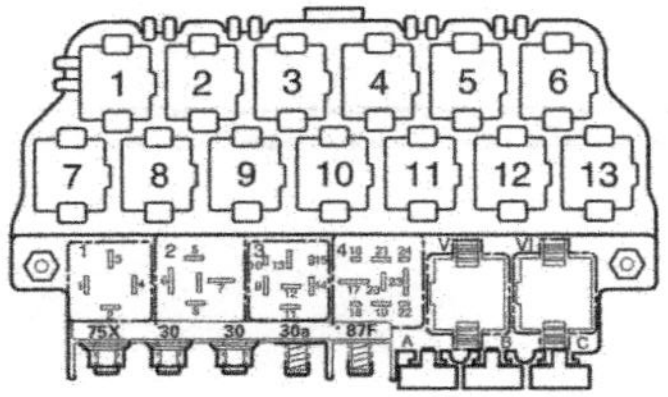

97-21956

Relay location on the thirteen position auxiliary relay panel, above relay panel:

11 Park/Neutral Position (PNP) Relay (175)

Relay panel:

Note: Number in parentheses indicates production control number stamped on relay housing.

Fuse colors
30 A - green
25 A - white
20 A - yellow
15 A - blue
10 A - red
7,5 A - brown
5 A - beige

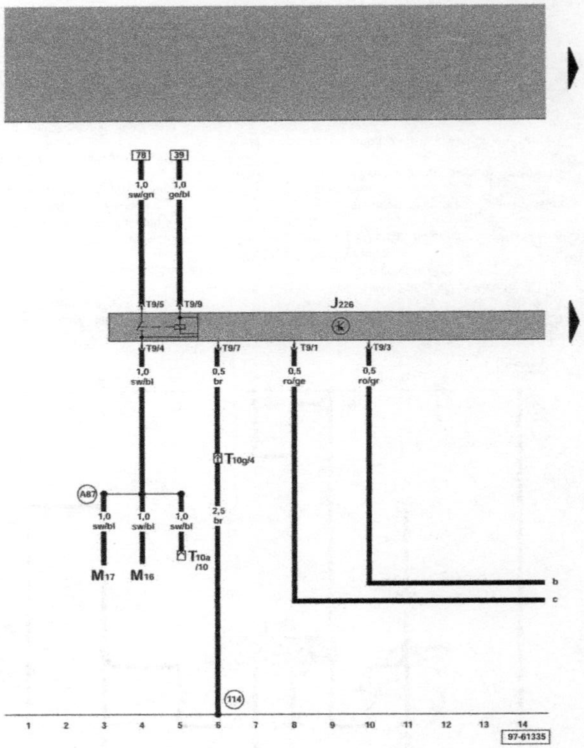

ws = white
sw = black
ro = red
br = brown
gn = green
bl = blue
gr = grey
li = violet
ge = yellow

J226 - Park/Neutral Position (PNP) Relay (175), on the thirteen position auxiliary relay panel
M16 - Left Back-Up Light
M17 - Right Back-Up Light
T9 - 9-Pin Connector
T10a - 10-Pin Connector, orange, in protective housing for connectors, in plenum chamber, left
T10g - 10-Pin Connector, grey, in protective housing for connectors, in plenum chamber, left
(114) - Ground connection, in automatic transmission wiring harness
(A87) - Connector (reverse lamp), in instrument panel wiring harness

Park/Neutral position (PNP) relay

New Beetle — Wiring diagram — No. 4/3

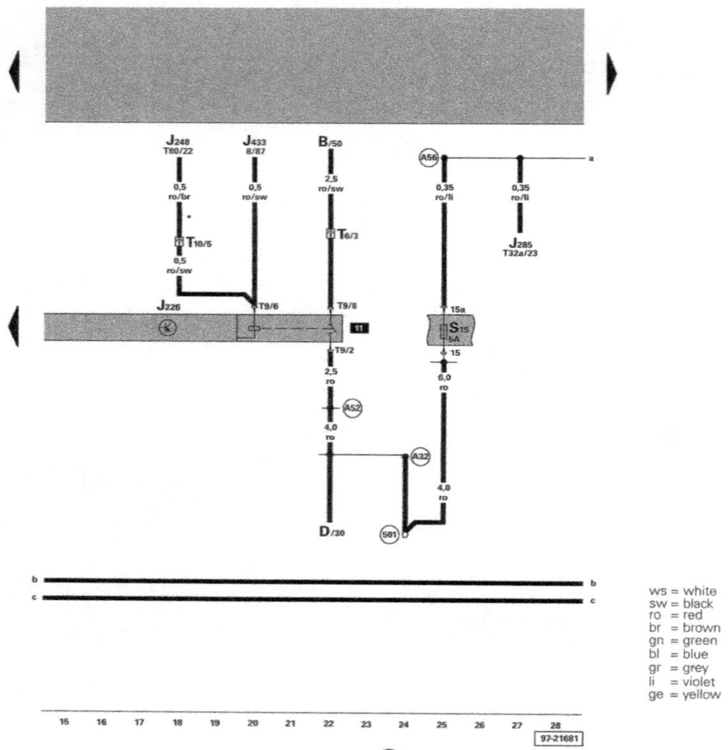

ws = white
sw = black
ro = red
br = brown
gn = green
bl = blue
gr = grey
li = violet
ge = yellow

B - Starter
D - Ignition/Starter Switch
J248 - Diesel Direct Injection Control Module
J226 - Park/Neutral Position (PNP) Relay (175), on the thirteen position auxiliary relay panel
J285 - Control module with indicator unit in instrument panel insert
J433 - Lock relay for starter (theft warning system)
S15 - Fuse 15 in fuse holder
T6 - 6-Pin Connector, red, behind instrument panel
T9 - 9-Pin Connector
T10d - 10-Pin Connector, green, behind instrument panel, left
T32a - 32-Pin Connector, blue, on instrument cluster
T80 - 80-Pin Connector, on engine Control Module

(501) - Threaded connection -2- (30) on the relay plate
(A32) - plus connection (30), in instrument panel wiring harness
(A52) - plus connection (30), in instrument panel wiring harness
(A56) - plus connection -2- (30), in instrument panel wiring harness

* - Diesel only

Edition 02/98
USA.5132.02.21

Park/Neutral position (PNP) relay

No. 4/4 — Wiring diagram — New Beetle

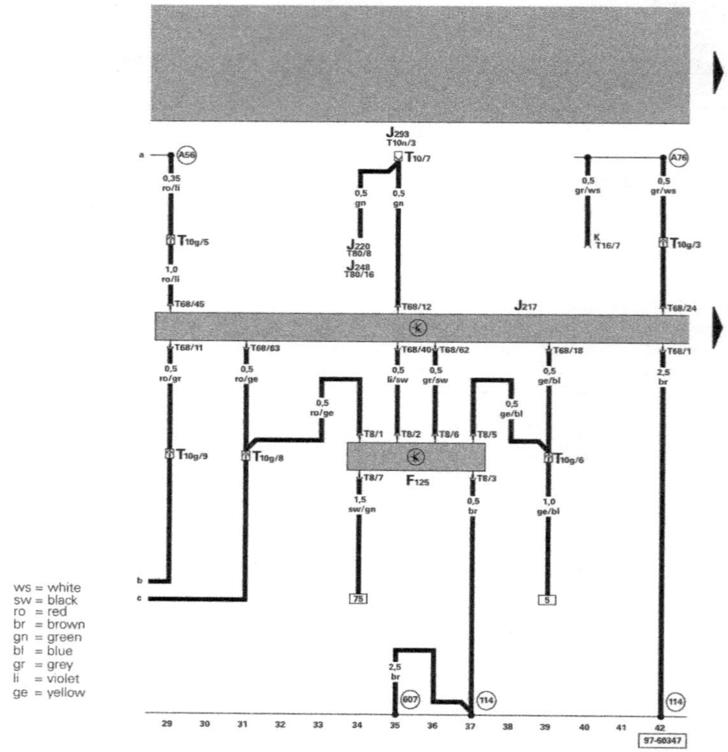

ws = white
sw = black
ro = red
br = brown
gn = green
bl = blue
gr = grey
li = violet
ge = yellow

F125 - Multi-Function Transmission Range (TR) Switch
J217 - Transmission Control Module (TCM), behind instrument panel, center
J220 - Motronic Engine Control Module (ECM), behind instrument panel, left
J248 - Diesel Direct Fuel Injection (DFI) Engine Control Module (ECM), behind instrument panel, left
J293 - Coolant FC (Fan Control) Control Module
T6 - 6-Pin Connector, red, behind instrument panel, left
T8 - 8-Pin Connector, on Mult-Function Transmission Range (TR) Switch
T10 - 10-Pin Connector, white, behind instrument panel, left
T10g - 10-Pin Connector, grey, behind instrument panel, left
T10n - 10 -Pin Connector, at Coolant Fan Control (FC) Module
T16 - 16-Pin Connector, Data Link Connector (DLC), below steering column
T68 - 68-Pin Connector
T80 - Connector, 80 point, on Engine Control Module (ECM)
(114) - Ground connection, in automatic transmission wiring harness
(607) - Ground Connection (in left plenum chamber)
(A56) - plus connection -2- (30), in instrument panel wiring harness
(A76) - Connector (K-diagnosis wire), in instrument panel wiring harness

Transmission control module (TCM), kick down switch, multi-function transmission range (TR) switch

Edition 02/99
USA.5132.06.21

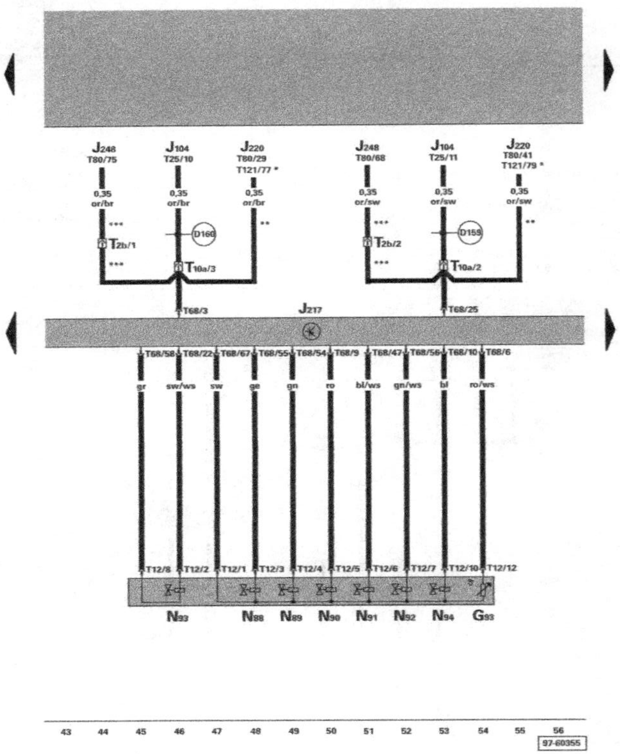

ws = white
sw = black
ro = red
br = brown
gn = green
bl = blue
gr = grey
li = violet
ge = yellow

G93 - Transmission Fluid Temperature Sensor
J104 - ABS Control Module (w/EDL)
J217 - Transmission Control Module (TCM), behind instrument panel, center
J220 - Motronic Engine Control Module (ECM), behind instrument panel, left
J248 - Diesel Direct Fuel Injection (DFI) Engine Control Module (ECM), behind instrument panel, left
N88 - Solenoid Valve 1
N89 - Solenoid Valve 2
N90 - Solenoid Valve 3
N91 - Solenoid Valve 4
N92 - Solenoid Valve 5
N93 - Solenoid Valve 6
N94 - Solenoid Valve 7
T2b - Double Connector (Diesel only)
T10a - 10-Pin Connector, orange, behind instrument panel, left
T12 - 12-Pin Connector
T25 - 25-Pin Connector, on ABS Control Module (w/EDL)
T68 - 68-Pin Connector
T80 - Connector, 80 point, on Engine Control Module (ECM), 2.0l
T121 - Connector, 121 point, on Engine Control Module (ECM), 1.8l
(D159) - CAN - bus High, in engine wiring harness
(D160) - CAN - bus Low, in engine wiring harness
* - Gasoline 1.8l only
** - Gasoline 2.0l only
*** - Diesel only

Transmission control module, solenoid valves, transmission fluid temperature sensor

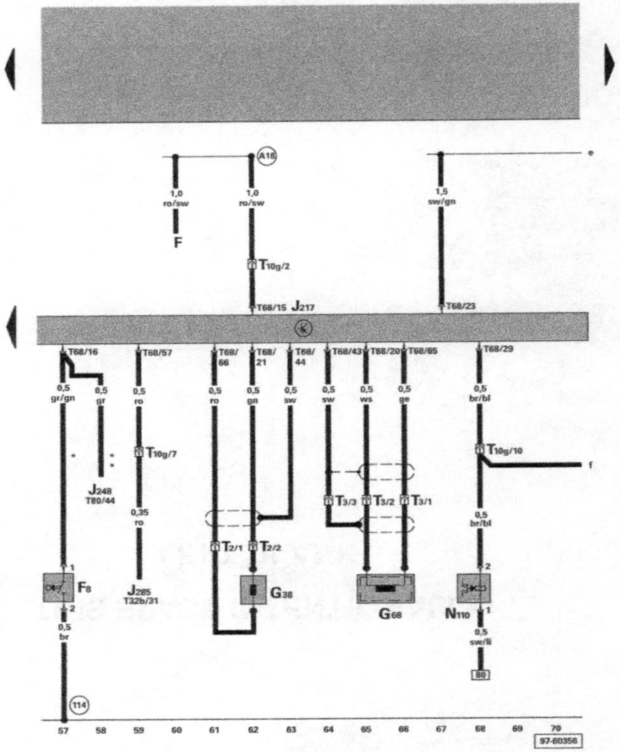

ws = white
sw = black
ro = red
br = brown
gn = green
bl = blue
gr = grey
li = violet
ge = yellow

F - Brake Light Switch
F8 - Kick-Down Switch
G38 - Transmission Vehicle Speed Sensor
G68 - Vehicle Speed Sensor (VSS)
J217 - Transmission Control Module (TCM), behind instrument panel, center
J248 - Diesel Direct Fuel Injection (DFI) Engine Control Module (ECM)
J285 - Control module with indicator unit in instrument panel insert
N110 - Shift Lock Solenoid
T2 - Double-Connector
T3 - 3-Pin Connector, on transmission
T10g - 10-Pin Connector, grey, behind instrument panel, left
T32b - 32-Pin Connector, blue, on instrument cluster
T68 - 68-Pin Connector
T80 - 80-Pin Connector
(114) - Ground connection in automatic transmission wiring harness
(A18) - wire connection (54), in instrument panel wiring harness
* - Gasoline 2.0l only
** - Diesel only

Transmission control module (TCM), vehicle speed sensor, warning light for selector lever position P/N

Automatic transmission from Jan. 1998

New Beetle Wiring diagram **No. 4/7**

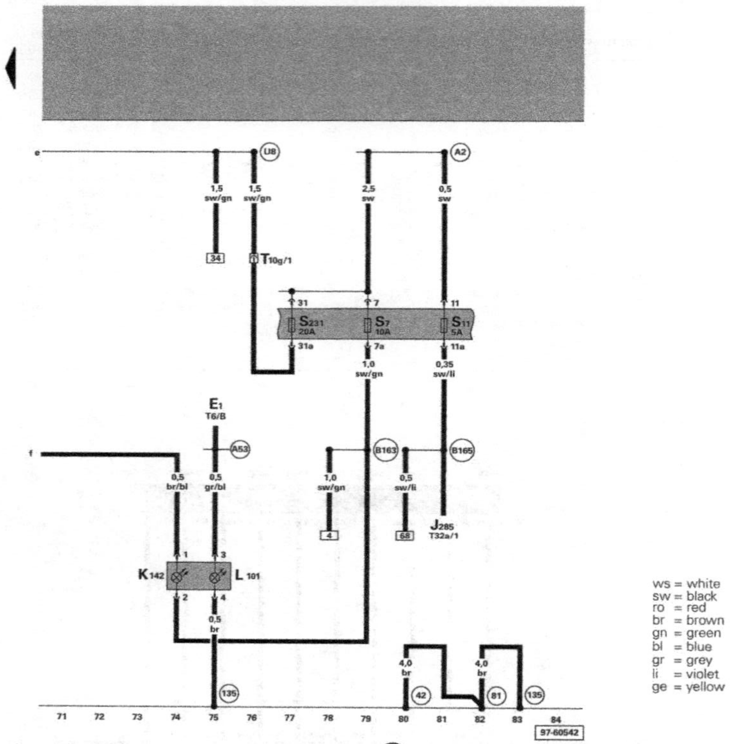

ws = white
sw = black
ro = red
br = brown
gn = green
bl = blue
gr = grey
li = violet
ge = yellow

E1 - Light switch
J285 - Control module with indicator unit in instrument panel insert
K142 - Warning light for selector lever position P/N
L101 - Illumination for selector lever scale
S7 - Fuse 7 in fuse holder
S11 - Fuse 11 in fuse holder
S231 - Fuse 31 in fuse holder
T6 - 6-Pin Connector
T10g - 10-Pin Connector, grey, behind instrument panel, left
T32a - 32-Pin Connector, blue, on instrument cluster

(42) - Ground connection, beside steering column

(81) - Ground connection -1-, in instrument panel wiring harness

(135) - Ground connection -2-, in instrument panel wiring harness
(A2) - plus connection (15), in instrument panel wiring harness
(A53) - plus connection (58b), in instrument panel wiring harness
(B163) - Plus connector -1- (15), in wiring harness interior
(B165) - Plus connector -2- (15), in wiring harness interior
(U8) - Connector (15a), in automatic transmission wiring harness

Edition 06/99
USA.5132.07.21

Warning light for selector lever position P/N, illumination for selector lever scale

THIS SPACE INTENTIONALLY LEFT BLANK

Automatic transmission,

from June 1999

Deviate relay location and fuseplacements as well as the locations of multiple connectors see section "component locations".

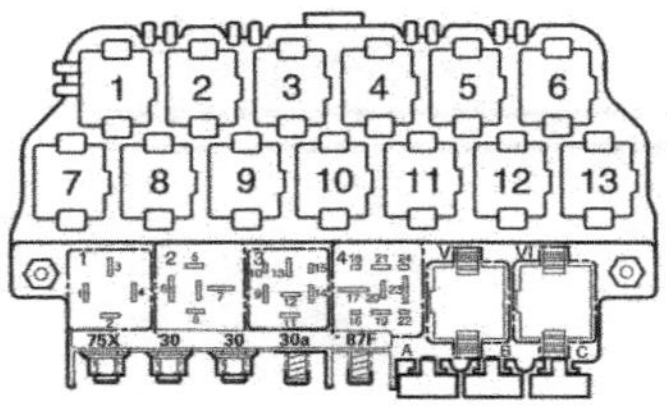

97-21956

Relay location on the thirteen position auxiliary relay panel, above relay panel:

11 Park/Neutral Position (PNP) Relay (175)

Relay panel:

Note: Number in parentheses indicates production control number stamped on relay housing.

Fuse colors
30 A - green
25 A - white
20 A - yellow
15 A - blue
10 A - red
7,5 A - brown
5 A - beige

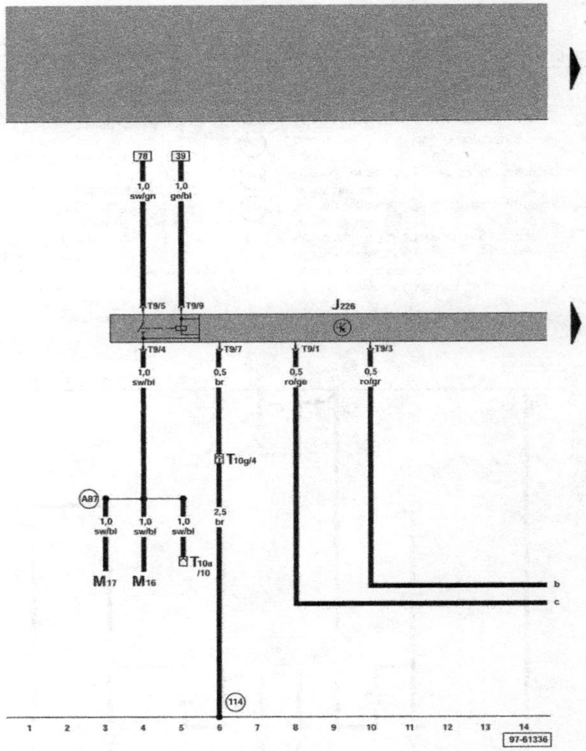

ws = white
sw = black
ro = red
br = brown
gn = green
bl = blue
gr = grey
li = violet
ge = yellow

- J226 - Park/Neutral Position (PNP) Relay (175), on the thirteen position auxiliary relay panel
- M16 - Left Back-Up Light
- M17 - Right Back-Up Light
- T9 - 9-Pin Connector
- T10a - 10-Pin Connector, orange, in protective housing for connectors, in plenum chamber, left
- T10g - 10-Pin Connector, grey, in protective housing for connectors, in plenum chamber, left
- (114) - Ground connection, in automatic transmission wiring harness
- (A87) - Connector (reverse lamp), in instrument panel wiring harness

New Beetle Wiring diagram **No. 33/3**

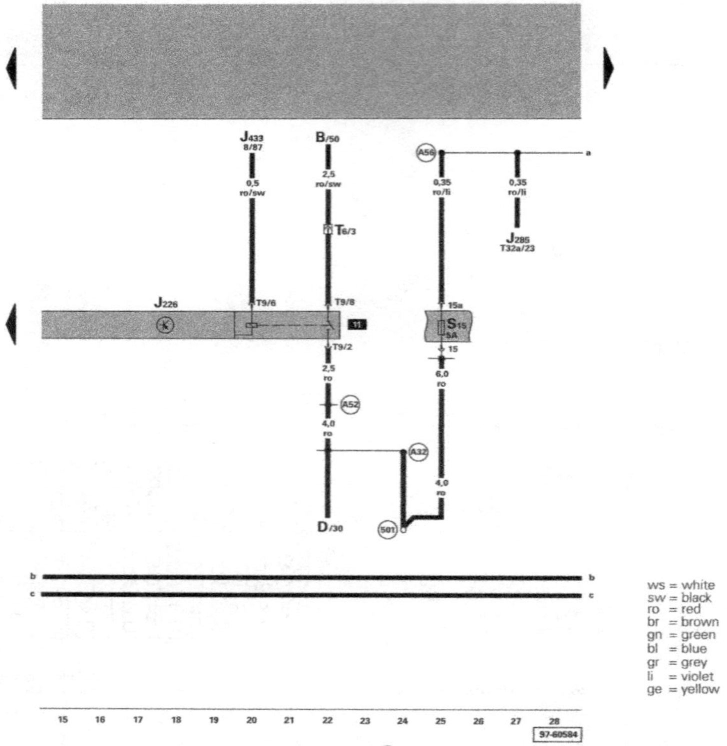

ws = white
sw = black
ro = red
br = brown
gn = green
bl = blue
gr = grey
li = violet
ge = yellow

- B - Starter
- D - Ignition/Starter Switch
- J248 - Diesel Direct Injection Control Module
- J226 - Park/Neutral Position (PNP) Relay (175), on the thirteen position auxiliary relay panel
- J285 - Control module with indicator unit in instrument panel insert
- J433 - Locking relay for starter (anti-theft immobilizer)
- S15 - Fuse 15 in fuse holder
- T6 - 6-Pin Connector, red, behind instrument panel
- T9 - 9-Pin Connector
- T32a - 32-Pin Connector, blue, on instrument cluster
- T80 - 80-Pin Connector, at ECM
- (A32) - plus connection (30), in instrument panel wiring harness
- (A52) - plus connection (30), in instrument panel wiring harness
- (A56) - plus connection -2- (30), in instrument panel wiring harness
- (501) - Threaded connection -2- (30) on the relay plate

Edition 10/00
USA.5132.13.21

Park/Neutral Position (PNP) Relay

No. 33/4 Wiring diagram **New Beetle**

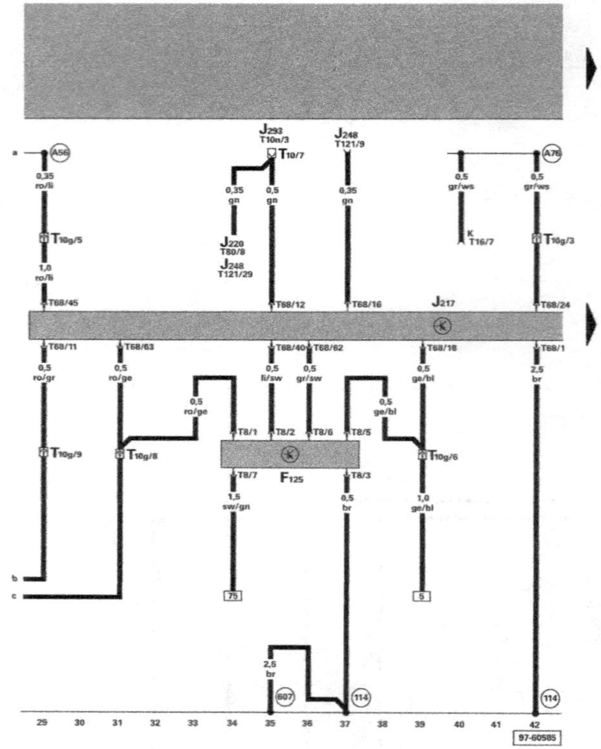

ws = white
sw = black
ro = red
br = brown
gn = green
bl = blue
gr = grey
li = violet
ge = yellow

- F125 - Multi-Function Transmission Range (TR) Switch
- J217 - Transmission Control Module (TCM), behind instrument panel, center
- J220 - Motronic Engine Control Module (ECM), behind instrument panel, left
- J248 - Diesel Direct Fuel Injection (DFI) Engine Control Module (ECM), behind instrument panel, left
- J293 - Coolant FC (Fan Control) Control Module
- T6 - 6-Pin Connector, red, behind instrument panel, left
- T8 - 8-Pin Connector, on Mult-Function Transmission Range (TR) Switch
- T10 - 10-Pin Connector, white, behind instrument panel, left
- T10g - 10-Pin Connector, grey, behind instrument panel, left
- T10n - 10 -Pin Connector, at Coolant Fan Control (FC) Module
- T16 - 16-Pin Connector, Data Link Connector (DLC), below steering column
- T68 - 68-Pin Connector
- T121 - Connector, 121 point, on Engine Control Module (ECM)
- (114) - Ground connection, in automatic transmission wiring harness
- (607) - Ground Connection (in left plenum chamber)
- (A56) - plus connection -2- (30), in instrument panel wiring harness
- (A76) - Connector (K-diagnosis wire), in instrument panel wiring harness

Transmission Control Module (TCM), kick down switch, Multi-function Transmission Range (TR) Switch

Edition 10/00
USA.5132.13.21

Automatic transmission from June 1999

New Beetle — Wiring diagram — No. 33/5

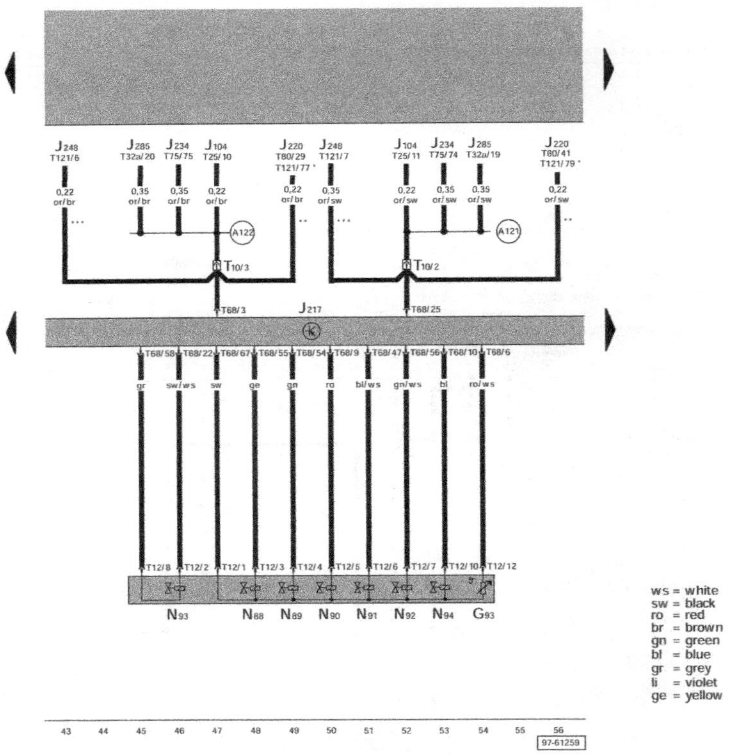

ws = white
sw = black
ro = red
br = brown
gn = green
bl = blue
gr = grey
li = violet
ge = yellow

G93 - Transmission Fluid Temperature Sensor
J104 - ABS Control Module (w/EDL)
J217 - Transmission Control Module (TCM), behind instrument panel, center
J220 - Motronic Engine Control Module (ECM), behind instrument panel, left
J234 - Airbag Control Module, behind console
J248 - Diesel Direct Fuel Injection (DFI) Engine Control Module (ECM), behind instrument panel, left
J285 - Control module with indicator unit in instrument panel insert
N88 - Solenoid Valve 1
N89 - Solenoid Valve 2
N90 - Solenoid Valve 3
N91 - Solenoid Valve 4
N92 - Solenoid Valve 5
N93 - Solenoid Valve 6
N94 - Solenoid Valve 7
T10 - 10-Pin Connector, white, behind inst. panel, left
T12 - 12-Pin Connector
T25 - 25-Pin Connector, on ABS Control Module
T32a - 32-Pin Connector, green on instrument cluster
T68 - 68-Pin Connector on TCM
T80 - 80-Pin Connector on ECM
T121 - 121-Pin Connector on ECM
(A121) - Connection (high bus) in instrument panel wiring harness
(A122) - Connection (low bus) in instrument panel wiring harness
* - Gasoline 1.8L only
** - Gasoline 2.0L only
*** - Diesel 1.9L only

Edition 03/01
USA.5132.14.21

Transmission Control Module (TCM), solenoid valves, transmission fluid temperature sensor

No. 33/6 — Wiring diagram — New Beetle

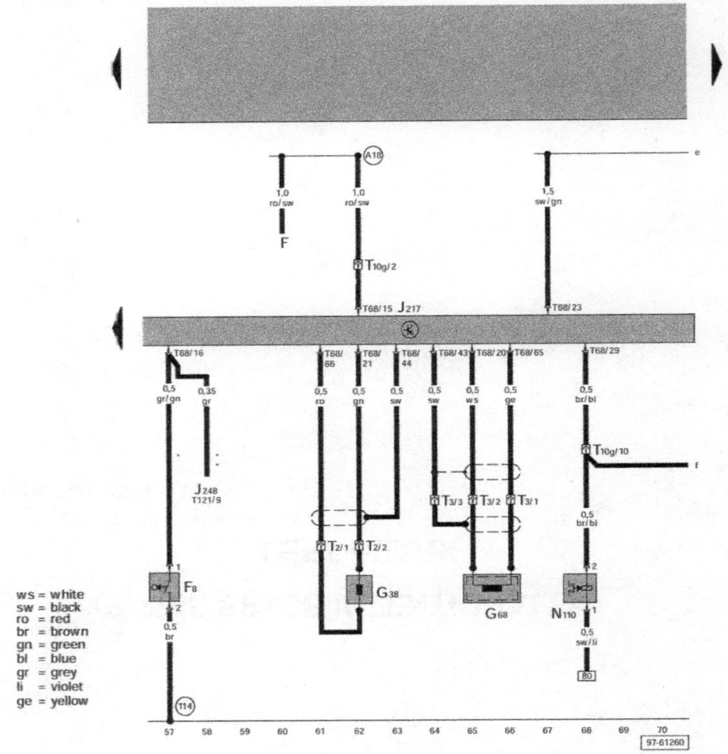

ws = white
sw = black
ro = red
br = brown
gn = green
bl = blue
gr = grey
li = violet
ge = yellow

F - Brake Light Switch
F8 - Kick-Down Switch
G38 - Transmission Vehicle Speed Sensor
G68 - Vehicle Speed Sensor (VSS)
J217 - Transmission Control Module (TCM), behind instrument panel, center
J248 - Diesel Direct Fuel Injection (DFI) Engine Control Module (ECM)
N110 - Shift Lock Solenoid
T2 - 2-Pin Connector
T3 - 3-Pin Connector, on transmission
T10g - 10-Pin Connector, grey, behind instrument panel, left
T68 - 68-Pin Connector on TCM
T121 - 121-Pin Connector on ECM
(114) - Ground connection in automatic transmission wiring harness
(A18) - wire connection (54), in instrument panel wiring harness
* - Gasoline 2.0L only
** - Diesel 1.9L only

Transmission Control Module (TCM), Vehicle Speed Sensor (VSS), warning light for selector lever position P/N

Edition 03/01
USA.5132.14.21

New Beetle Wiring diagram **No. 33/7**

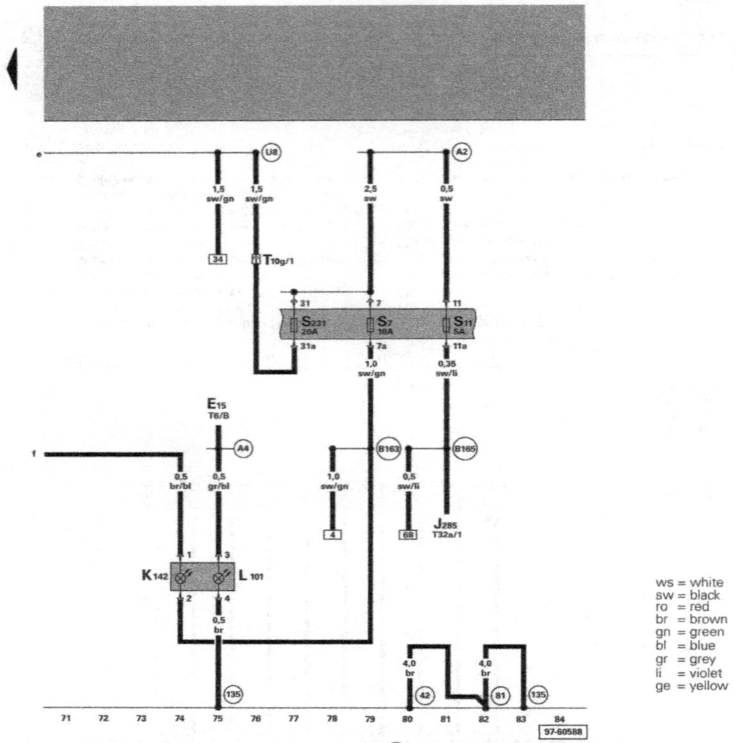

ws = white
sw = black
ro = red
br = brown
gn = green
bl = blue
gr = grey
li = violet
ge = yellow

- E15 - Rear window defogger switch
- J285 - Control module with indicator unit in instrument panel insert
- K142 - Warning light for selector lever position P/N
- L101 - Illumination for selector lever scale
- S7 - Fuse 7 in fuse holder
- S11 - Fuse 11 in fuse holder
- S231 - Fuse 31 in fuse holder
- T6 - 6-Pin Connector
- T10g - 10-Pin Connector, grey, behind instrument panel, left
- T32a - 32-Pin Connector, blue, on instrument cluster
- (42) - Ground connection, beside steering column
- (81) - Ground connection -1-, in instrument panel wiring harness
- (135) - Ground connection -2-, in instrument panel wiring harness
- (A2) - plus connection (15), in instrument panel wiring harness
- (A4) - plus connection (58b), in instrument panel wiring harness
- (B163) - Plus connector -1- (15), in wiring harness interior
- (B165) - Plus connector -2- (15), in wiring harness interior
- (U8) - Connector (15a), in automatic transmission wiring harness

Edition 08/99
USA.5132.08.21

Warning light for selector lever position P/N, illumination for selector lever scale

THIS SPACE INTENTIONALLY LEFT BLANK

New Beetle **Wiring diagram** **No. 11/1**

Automatic transmission,

from February 2001

Deviate relay location and fuseplacements as well as the locations of multiple connectors see section "component locations".

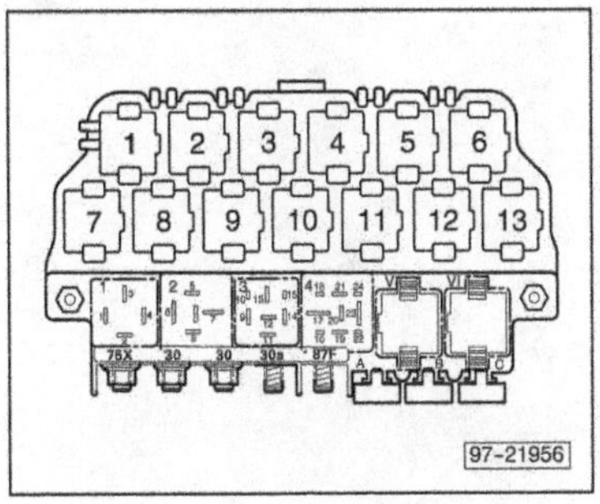

Relay location on the thirteen position auxiliary relay panel, above relay panel:

11 Park/Neutral Position (PNP) Relay (175)

Fuse colors

30 A - green
25 A - white
20 A - yellow
15 A - blue
10 A - red
7,5 A - brown
5 A - beige
3 A - violet

Note: Number in parentheses indicates production control number stamped on relay housing.

New Beetle **Wiring diagram** **No. 11/2**

Park/Neutral Position (PNP) Relay

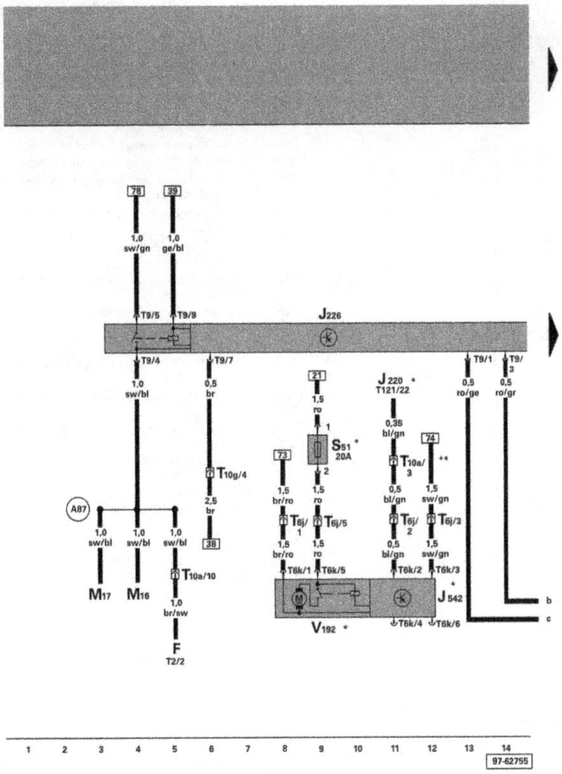

- F - Brake Light Switch
- J220 - Motronic Engine Control Module (ECM) *
- J226 - Park/Neutral Position (PNP) Relay (175), on the thirteen position auxiliary relay panel
- J542 - Brake Booster Control Module *
- M16 - Left Back-Up Light
- M17 - Right Back-Up Light
- S51 - Fuse, in protective housing, in engine compartment, left *
- T2 - 2-Pin Connector at -F-
- T6j - 6-Pin Connector
- T6K - 6-Pin Connector
- T9 - 9-Pin Connector
- T10a - 10-Pin Connector, orange, in protective housing for connectors, in plenum chamber, left
- T10g - 10-Pin Connector, grey, in protective housing for connectors, in plenum chamber, left
- T121 - 121-Pin Connector, at J220
- V192 - Brake System Vacuum Pump *
- (A87) - Connector (reverse lamp), in instrument panel wiring harness

- * - Only vehicles 1,8 L
- ** - Through January 2003 only

ws = white
sw = black
ro = red
br = brown
gn = green
bl = blue
gr = grey
li = violet
ge = yellow
or = orange

New Beetle

Wiring diagram

No. 11/3

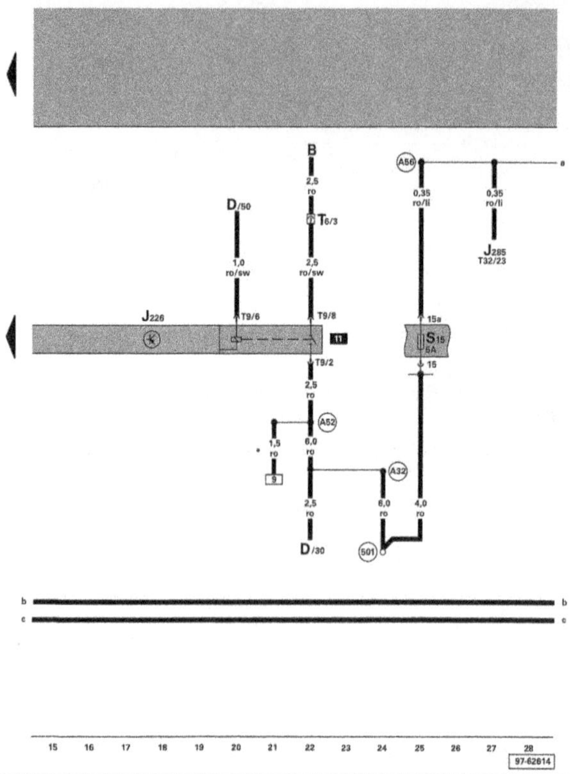

Park/Neutral Position (PNP) Relay

- B - Starter
- D - Ignition/Starter Switch
- J226 - Park/Neutral Position (PNP) Relay (175), on the thirteen position auxiliary relay panel
- J285 - Control module with indicator unit in instrument panel insert
- S15 - Fuse 15 in fuse holder
- T6 - 6-Pin Connector,brown, behind instrument panel
- T9 - 9-Pin Connector
- T32 - 32-Pin Connector, blue, on instrument cluster
- (501) - Threaded connection -2- (30) on the relay plate
- (A32) - plus connection (30), in instrument panel wiring harness
- (A52) - plus connection (30), in instrument panel wiring harness
- (A56) - plus connection -2- (30), in instrument panel wiring harness
- * - Only vehicles 1,8 L

ws = white
sw = black
ro = red
br = brown
gn = green
bl = blue
gr = grey
li = violet
ge = yellow
or = orange

New Beetle

Wiring diagram

No. 11/4

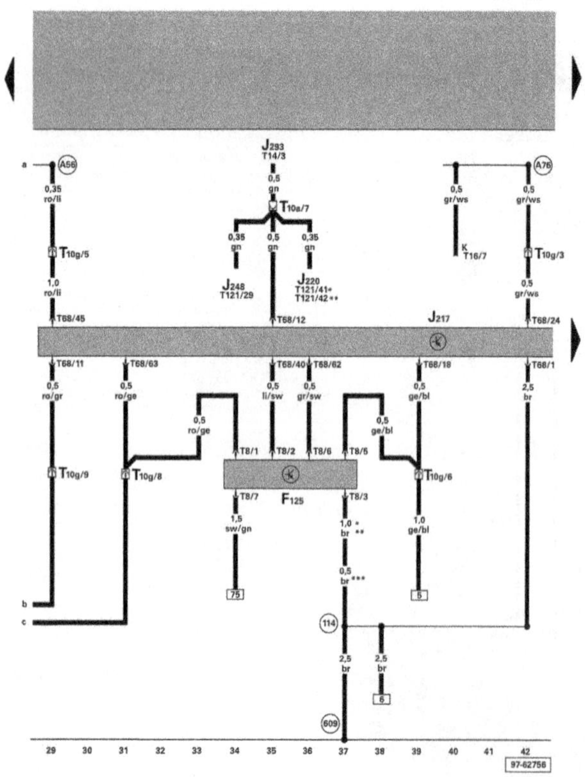

Transmission Control Module (TCM), kick down switch, Multi-function Transmission Range (TR) Switch

- F125 - Multi-Function Transmission Range (TR) Switch
- J217 - Transmission Control Module (TCM), behind instrument panel, center
- J220 - Motronic Engine Control Module (ECM), behind instrument panel, left
- J248 - Diesel Direct Injection Control Module
- J293 - Coolant FC (Fan Control) Control Module
- T8 - 8-Pin Connector, on Mult-Function Transmission Range (TR) Switch
- T10a - 10-Pin Connector, orange, behind instrument panel, left
- T10g - 10-Pin Connector, grey, behind instrument panel, left
- T14 - 14 -Pin Connector, at Coolant Fan Control (FC) Module
- T16 - 16-Pin Connector, Data Link Connector (DLC), below steering column
- T68 - 68-Pin Connector
- T121 - Connector, 121 point, on Engine Control Module (ECM)
- (114) - Ground connection, in automatic transmission wiring harness
- (609) - Ground Connection (in right plenum chamber)
- (A56) - plus connection -2- (30), in instrument panel wiring harness
- (A76) - Connector (K-diagnosis wire), in instrument panel wiring harness
- * - Gasoline engines - 1,8 L
- ** - Gasoline engines - 2,0 L
- *** - Diesel engines - 1,9 L

ws = white
sw = black
ro = red
br = brown
gn = green
bl = blue
gr = grey
li = violet
ge = yellow
or = orange

New Beetle

Wiring diagram

No. 11/5

Transmission Control Module (TCM), solenoid valves, transmission fluid temperature sensor

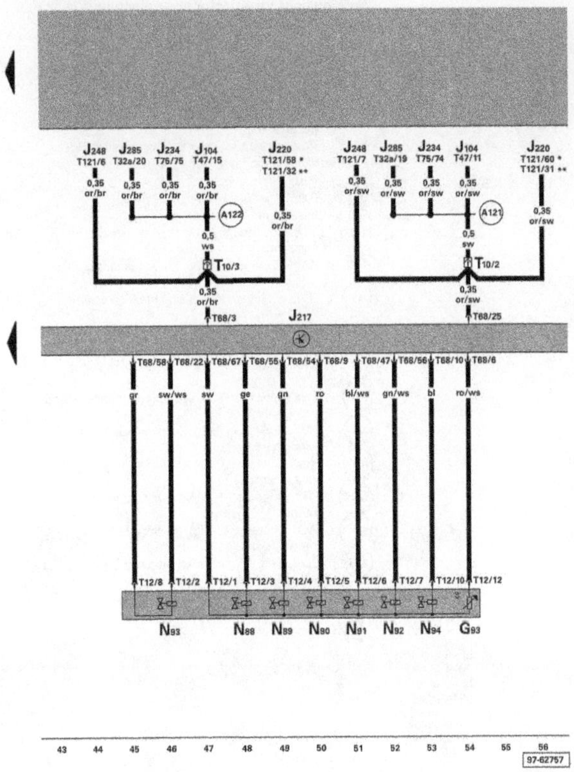

ws = white
sw = black
ro = red
br = brown
gn = green
bl = blue
gr = grey
li = violet
ge = yellow
or = orange

G93 - Transmission Fluid Temperature Sensor
J104 - ABS Control Module (w/EDL)
J217 - Transmission Control Module (TCM), behind instrument panel, center
J220 - Motronic Engine Control Module (ECM), behind instrument panel, left
J234 - Airbag Control Module, behind console
J248 - Diesel Direct Fuel Injection (DFI) Engine Control Module (ECM), behind instrument panel, left
J285 - Control module with indicator unit in instrument panel insert
N88 - Solenoid Valve 1
N89 - Solenoid Valve 2
N90 - Solenoid Valve 3
N91 - Solenoid Valve 4
N92 - Solenoid Valve 5
N93 - Solenoid Valve 6
N94 - Solenoid Valve 7
T10 - 10-Pin Connector, white, behind instrument panel, left
T12 - 12-Pin Connector
T47 - 47-Pin Connector, on ABS Control Module
T32a - 32-Pin Connector, green on instrument cluster
T68 - 68-Pin Connector on TCM
T75 - 75-Pin Connector on Airbag Control Module
T121 - 121-Pin Connector on ECM

(A121) - Connection (high bus) in instrument panel wiring harness
(A122) - Connection (low bus) in instrument panel wiring harness

* - Gasoline engines - 1,8L
** - Gasoline engines - 2,0L

New Beetle

Wiring diagram

No. 11/6

Transmission Control Module (TCM), Vehicle Speed Sensor (VSS)

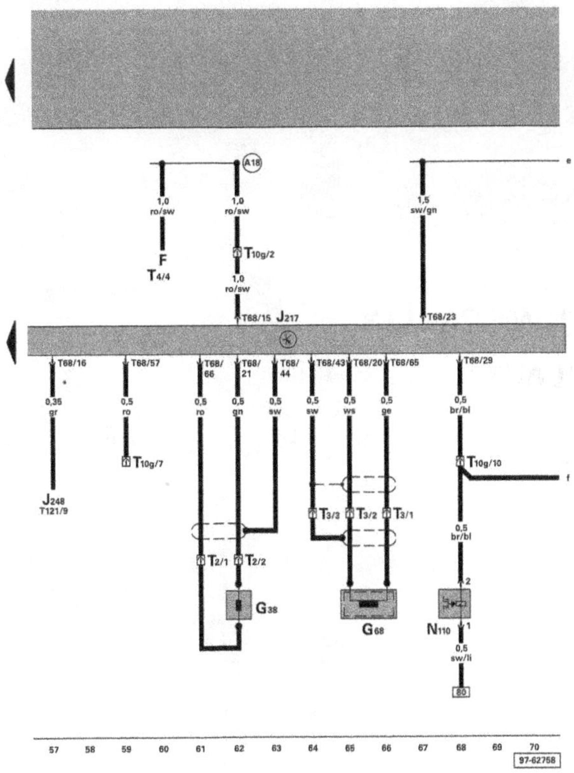

ws = white
sw = black
ro = red
br = brown
gn = green
bl = blue
gr = grey
li = violet
ge = yellow
or = orange

F - Brake Light Switch
G38 - Transmission Vehicle Speed Sensor
G68 - Vehicle Speed Sensor (VSS)
J217 - Transmission Control Module (TCM), behind instrument panel, center
J248 - Diesel Direct Fuel Injection (DFI) Engine Control Module (ECM)
N110 - Shift Lock Solenoid
T2 - 2-Pin Connector
T4 - 4-Pin Connector at -F-
T3 - 3-Pin Connector, on transmission
T10g - 10-Pin Connector, grey, behind instrument panel, left
T68 - 68-Pin Connector on TCM
T121 - 121-Pin Connector on ECM

(A18) - wire connection (54), in instrument panel wiring harness

* - Diesel 1.9L only

New Beetle — Wiring diagram — **No. 11/7**

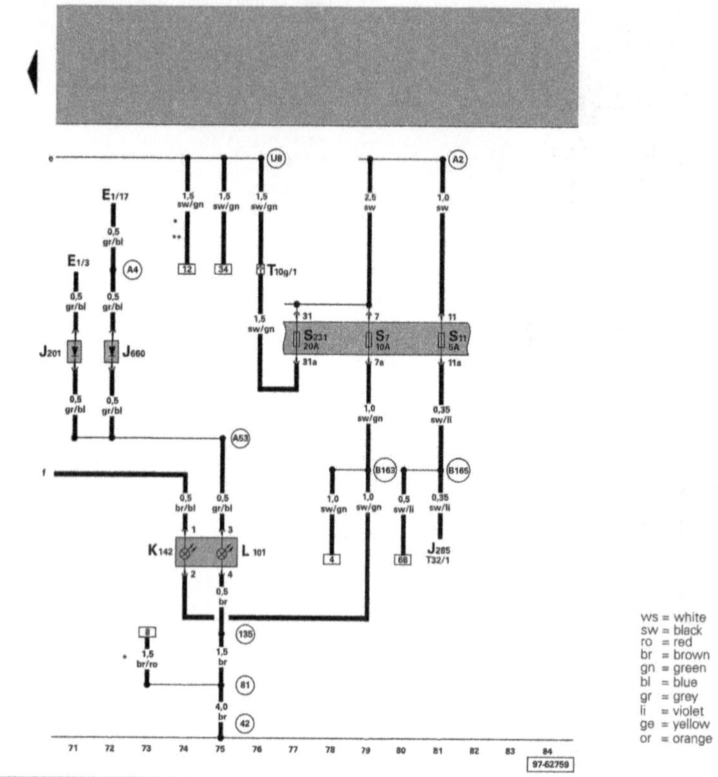

ws = white
sw = black
ro = red
br = brown
gn = green
bl = blue
gr = grey
li = violet
ge = yellow
or = orange

Warning light for selector lever position P/N, illumination for selector lever scale

- E1 - Light switch
- J201 - Protection Diode
- J660 - Protection Diode 2
- J285 - Control module with indicator unit in instrument panel insert
- K142 - Warning light for selector lever position P/N
- L101 - Illumination for selector lever scale
- S7 - Fuse 7 in fuse holder
- S11 - Fuse 11 in fuse holder
- S231 - Fuse 31 in fuse holder
- T10g - 10-Pin Connector, grey, behind instrument panel, left
- T32 - 32-Pin Connector, blue, on instrument cluster
- (42) - Ground connection, beside steering column
- (81) - Ground connection -1-, in instrument panel wiring harness
- (135) - Ground connection -2-, in instrument panel wiring harness
- (A2) - plus connection (15), in instrument panel wiring harness
- (A4) - plus connection (58b), in instrument panel wiring harness
- (A53) - plus connection (58b), in instrument panel wiring harness
- (B163) - Plus connector -1- (15), in wiring harness interior
- (B165) - Plus connector -2- (15), in wiring harness interior
- (U8) - Connector (15a), in automatic transmission wiring harness

* - Only vehicles 1,8 L
** - Through January 2003 only

Edition 06.03

THIS SPACE INTENTIONALLY LEFT BLANK

New Beetle Convertible Wiring diagram **No. 3/1**

Automatic transmission, 6-speed automatic with Tiptronic

from August 2002

Deviate relay location and fuseplacements as well as the locations of multiple connectors see section "component locations".

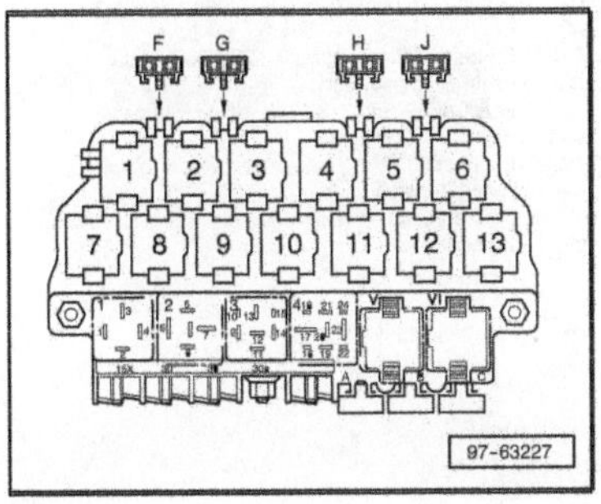

Relay location on the thirteen position auxiliary relay panel, above relay panel:

11 Starting Interlock Relay (126)

Relay panel:

2 Load Reduction Relay (100)

4 Fuel Pump (FP) Relay (409)

Fuse placements on the relay panel:

B Brake Vacuum Pump Fuse (S283)

J Brake Vacuum Pump Fuse (S283)*

* M.Y.2004➤

Note: Number in parentheses indicates production control number stamped on relay housing.

Fuse colors

30 A - green
25 A - white
20 A - yellow
15 A - blue
10 A - red
7,5 A - brown
5 A - beige
3 A - violet

New Beetle Convertible Wiring diagram **No. 3/2**

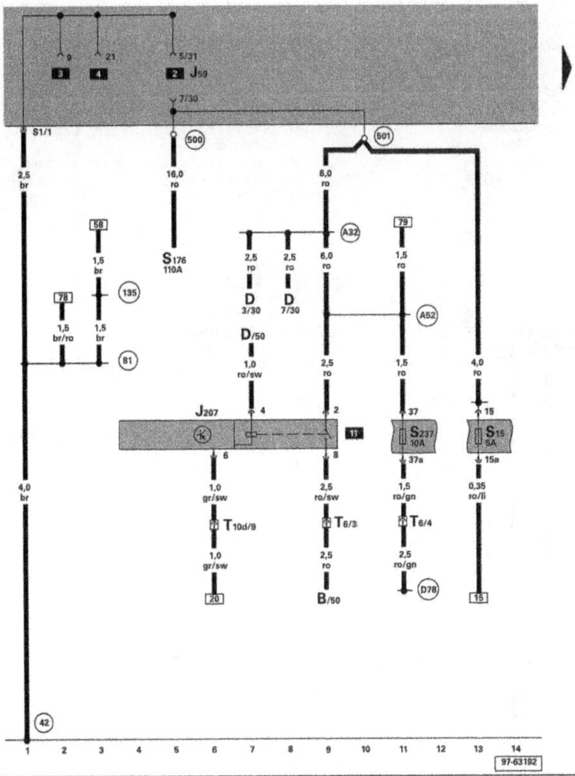

Park/Neutral position (PNP) relay

B - Starter
D - Ignition/Starter Switch
J59 - Load Reduction Relay
J207 - Starting Interlock Relay (126)
S15 - Fuse 15 in fuse holder
S176 - Fuse -4- (30), in fuse bracket/battery
S237 - Fuse 37 in in fuse holder
T6 - 6-Pin Connector, brown, in plenum chamber, left
T10d - 10- pin connector, grey, in plenum chamber, left

(42) - Ground connection, beside steering column
(81) - Ground connection -1-, in instrument panel wiring harness
(135) - Ground connection -2-, in instrument panel wiring harness
(500) - Threaded connection -1- (30) on the relay plate
(501) - Threaded connection -2- (30) on the relay plate
(A32) - Plus connection (30), in instrument panel wiring harness
(A52) - Plus Connection 2 (30) (in instrument panel wiring harness)
(D78) - Plus connector -1- (30a), in engine compartment wiring harness

ws = white
sw = black
ro = red
br = brown
gn = green
bl = blue
gr = grey
li = violet
ge = yellow

Automatic transmission, 6-speed Tiptronic from Aug. 2002, convertible

New Beetle Convertible

Wiring diagram

No. 3/3

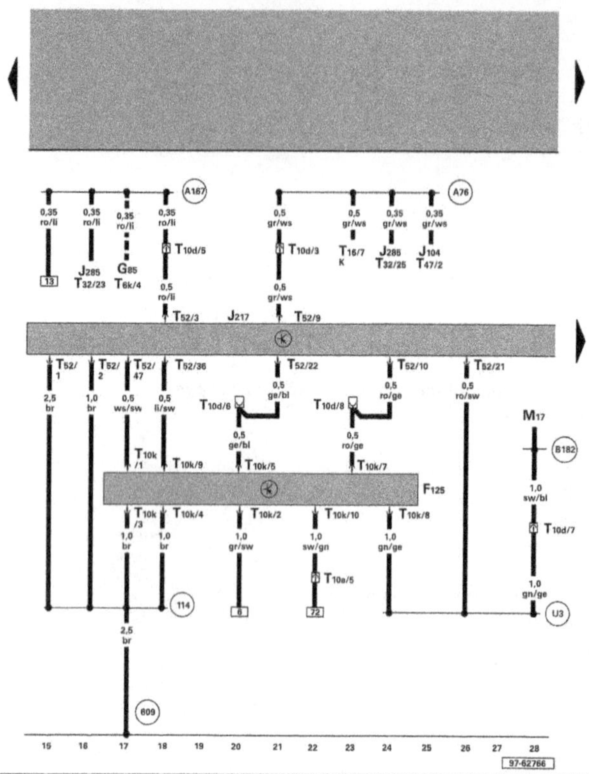

Transmission Control Module (TCM), multi-function transmission range (TR) switch , ABS Control module (w/EDL)

F125 - Multi-Function Transmission Range (TR) Switch
G85 - Steering Angle Sensor
J104 - ABS Control Module (w/EDL)
J217 - Transmission Control Module (TCM)
J285 - Control module with indicator unit in instrument panel insert
M17 - Right Back-Up Light
T6k - 6-pin connector
T10a - 10-pin connector, orange, in plenum chamber, left
T10d - 10-pin connector, grey, in plenum chamber, left
T10k - 10-pin connector
T16 - 16-pin connector, Data Link Connector (DCL), bellow steering column, left
T32 - 32-Pin Connector, blue, on instrument cluster
T47 - 47-pin connector
T52 - 52-pin connector

(114) - Ground connection, in automatic transmission wiring harness
(609) - Ground Connection (in right plenum chamber)
(A76) - Connector (K-diagnosis wire) in instrument panel wiring harness
(A167) - Plus Connection 3 (30a) (in instrument panel wiring harness)
(B182) - Connection (RF) in passenger compartment wiring harness
(U3) - Wire connection -1-, in automatic transmission wiring harness

--- - Vehicles with ESP only

ws = white
sw = black
ro = red
br = brown
gn = green
bl = blue
gr = grey
li = violet
ge = yellow

New Beetle Convertible

Wiring diagram

No. 3/4

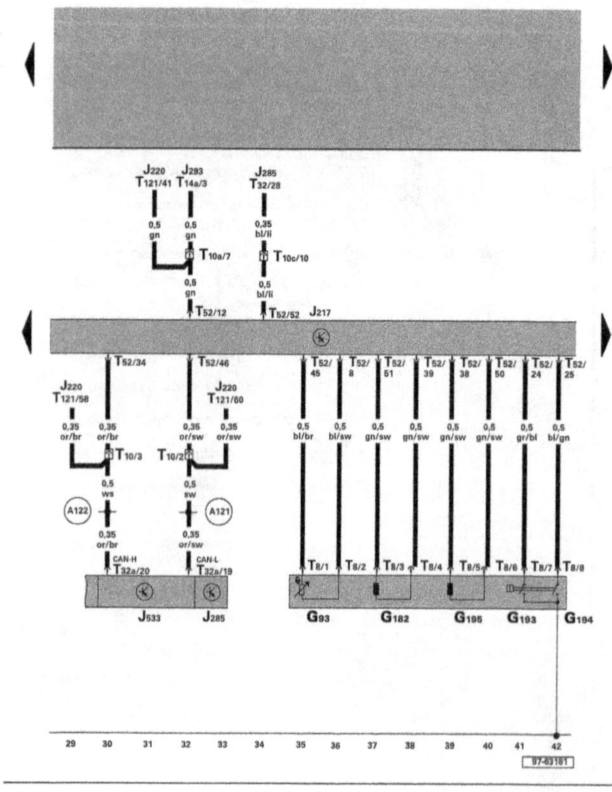

Transmission control module (TCM), transmission fluid temperature sensor, sensor for transmission RPM, senders 1 and 2 for hydraulic pressure, automatic transmission, sender for transmission output RPM

G93 - Transmission Fluid Temperature Sensor
G182 - Sensor for transmission RPM
G193 - Sender 1 for hydraulic pressure, automatic transmission
G194 - Sender 2 for hydraulic pressure, automatic transmission
G195 - Sender for transmission output RPM
J217 - Transmission Control Module (TCM)
J220 - Motronic Engine Control Module (ECM)
J285 - Control module with indicator unit in instrument panel insert
J293 - Coolant FC (Fan Control) Control Module
J533 - Data Bus On Board Diagnostic Interface
T8 - 8-pin connector
T10 - 10-pin connector
T10a - 10-pin connector
T10c - 10-pin connector, blue, behind instrument panel, left.
T14a - 14-pin connector
T32 - 32-pin connector, blue, on instrument cluster
T32a - 32-pin connector, green, on instrument cluster
T52 - 52-pin connector
T121 - 121- pin connector

(A121) - Connection (high bus) in instrument panel wiring harness
(A122) - Connection (low bus) in instrument panel wiring harness

ws = white
sw = black
ro = red
br = brown
gn = green
bl = blue
gr = grey
li = violet
ge = yellow

New Beetle Convertible

Wiring diagram **No. 3/5**

Transmission control module (TCM), solenoid valves

J217 - Transmission Control Module (TCM)
T14 - 14-pin connector
T52 - 52- pin connector
N88 - Solenoid Valve 1
N89 - Solenoid Valve 2
N90 - Solenoid Valve 3
N91 - Solenoid Valve 4
N92 - Solenoid Valve 5
N93 - Solenoid Valve 6
N282 - Solenoid Valve 9
N283 - Solenoid Valve 10

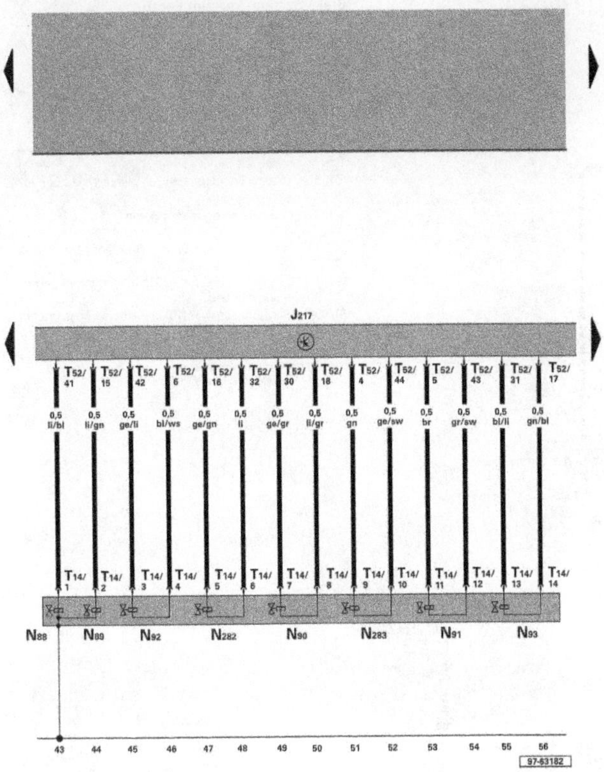

ws = white
sw = black
ro = red
br = brown
gn = green
bl = blue
gr = grey
li = violet
ge = yellow

Edition 04.04

New Beetle Convertible

Wiring diagram **No. 3/6**

Tiptronic switch, warning light for selector lever position P/N, illumination for selector lever scale, shift lock solenoid

E1 - Light switch
F189 - Tiptronic Switch
J29 - Blocking Diode
J217 - Transmission Control Module (TCM)
K142 - Warning light for selector lever position P/N
L101 - Illumination for selector lever scale
N110 - Shift Lock Solenoid
T8 - 8-pin connector
T10d - 10- pin connector, grey, in plenum chamber, left
T10h - 10- pin connector, black, in plenum chamber, left
T52 - 52-pin connector

(A4) - Plus connection (58b), in instrument panel wiring harness panel wiring harness
(A53) - plus connection (58b), in instrument panel wiring harness
(U8) - Plus Connection (15a) (automatic transmission wiring harness)

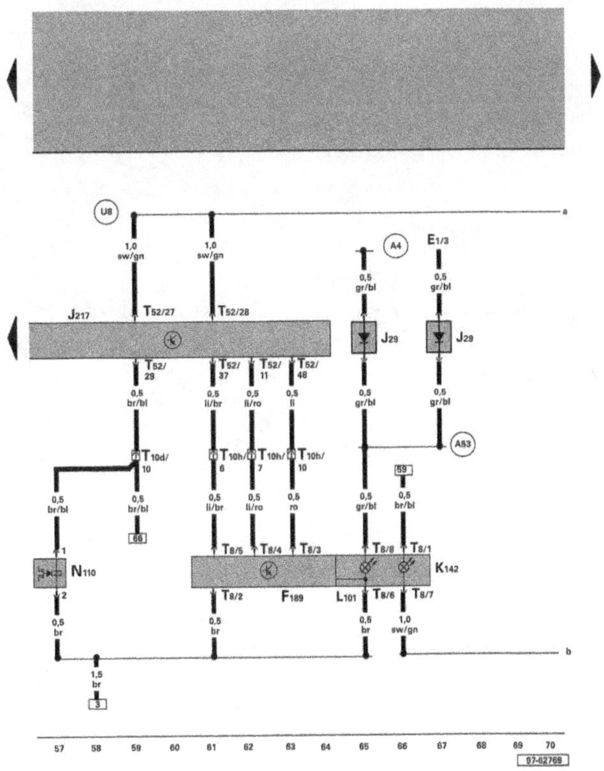

ws = white
sw = black
ro = red
br = brown
gn = green
bl = blue
gr = grey
li = violet
ge = yellow

Edition 06.03

Automatic transmission, 6-speed Tiptronic from Aug. 2002, convertible

New Beetle Convertible Wiring diagram **No. 3/7**

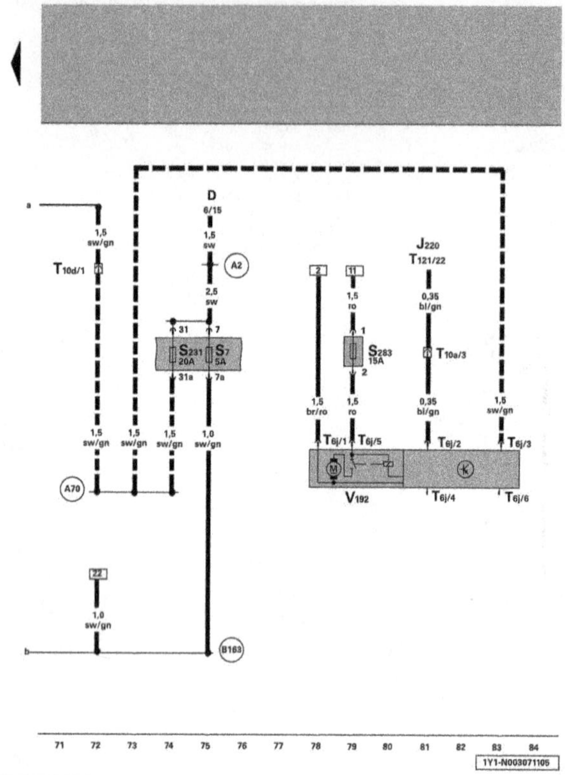

Brake System Vacuum Pump

D - Ignition/Starter Switch
J220 - Motronic Engine Control Module (ECM)
S7 - Fuse 7 in fuse holder
S231 - Fuse 31 in fuse holder
S283 - Brake Vacuum Pump Fuse, on the relay panel
T6j - 6-Pin Connector, black
T10a - 10-Pin Connector, orange, behind instrument panel, left
T10d - 10- pin connector, grey, in plenum chamber, left
T121 - 121- pin connector
V192 - Brake System Vacuum Pump

(A2) - Plus connection (15), in instrument panel wiring harness
(A70) - Connector (15a, fuse 231), in instrument panel wiring harness
(B163) - Plus connector -1- (15) in wiring harness interior

_ _ - Through February 2003 only

ws = white
sw = black
ro = red
br = brown
gn = green
bl = blue
gr = grey
li = violet
ge = yellow

THIS SPACE INTENTIONALLY

LEFT BLANK

New Beetle Wiring diagram **No. 36/1**

Automatic transmission, 6-speed automatic with Tiptronic, 09G,

from June 2004,

Additional Fuse, Relay, Control Module and Connector locations ⇒ Component Locations

Troubleshooting Procedures ⇒ Guided Fault Finding using VAS 5051/5052

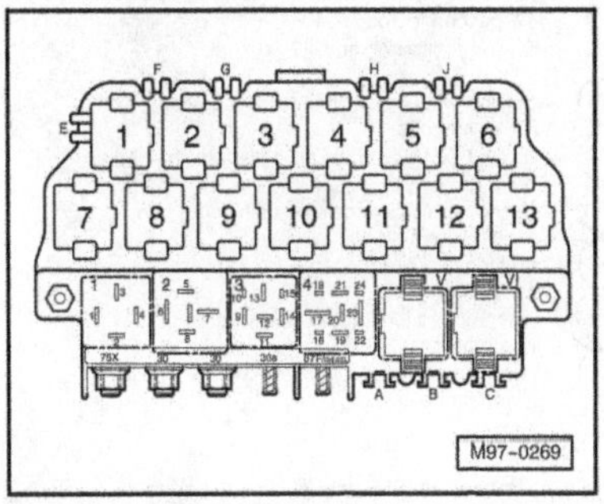

Relay locations on thirteen position auxiliary relay panel, above relay panel:

11 - J226 - Park/Neutral Position (PNP) Relay (175)

Fuse locations on thirteen position auxiliary relay panel, above relay panel:

J S283 - Brake Vacuum Pump Fuse (20A)

Relay panel:

2 J59 - Load Reduction Relay (100)

Notes:

Number in parentheses indicates production control number stamped on relay housing.

New Beetle Wiring diagram **No. 36/2**

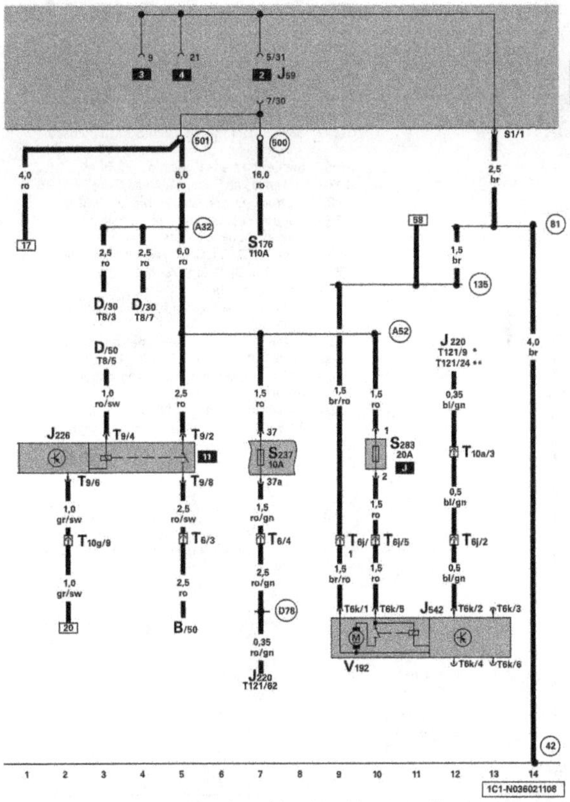

ws = white
sw = black
ro = red
br = brown
gn = green
bl = blue
gr = grey
li = violet
ge = yellow

Park/Neutral Position (PNP) Relay, Brake System Vacuum Pump

- B - Starter
- D - Ignition/Starter Switch
- J59 - Load Reduction Relay
- J226 - Park/Neutral Position (PNP) Relay (175)
- J220 - Motronic Engine Control Module (ECM)
- J542 - Brake Booster Control Module
- S176 - Fuse -4- (30), in fuse bracket/battery
- S237 - Fuse 37 in in fuse holder
- S283 - Brake Vacuum Pump Fuse
- T6 - 6-Pin Connector, brown, behind instrument panel, left
- T6j - 6-pin connector
- T6k - 8-pin connector
- T8 - 8- pin connector, at D
- T9 - 9- pin connector, at J226
- T10a - 10-pin connector
- T10g - 10-Pin Connector, grey, behind instrument panel, left
- T121 - 121-Pin Connector, at J220
- V192 - Brake System Vacuum Pump
- (42) - Ground connection, beside steering column
- (81) - Ground connection -1-, in instrument panel wiring harness
- (135) - Ground connection -2-, in instrument panel wiring harness
- (500) - Threaded connection -1- (30) on the relay plate
- (501) - Threaded connection -2- (30) on the relay plate
- (A32) - Plus connection (30), in instrument panel wiring harness
- (A52) - Plus Connection 2 (30) (in instrument panel wiring harness)
- (D78) - Plus connector -1- (30a), in engine compartment wiring harness
- * - 1.8 L engine
- ** - 2.0 L engine

New Beetle — Wiring diagram — No. 36/3

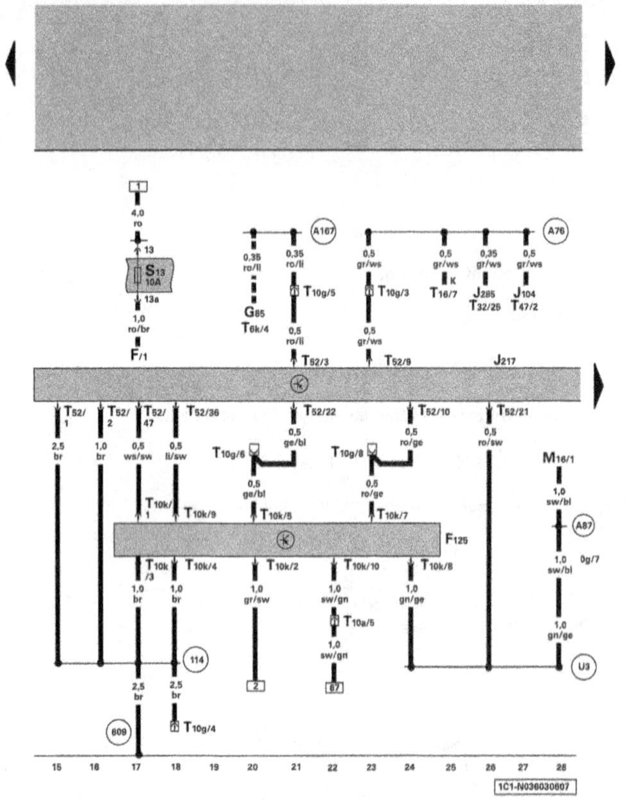

ws = white
sw = black
ro = red
br = brown
gn = green
bl = blue
gr = grey
li = violet
ge = yellow

Transmission Control Module (TCM), multi-function transmission range (TR) switch

- F - Brake Light Switch
- F125 - Multi-Function Transmission Range (TR) Switch
- G85 - Steering Angle Sensor
- J104 - ABS Control Module (w/EDL)
- J217 - Transmission Control Module (TCM)
- J285 - Control module with indicator unit in instrument panel insert
- M16 - Left Back-Up Light
- S13 - Fuse 13 in fuse holder
- T6k - 6-pin connector, at G85
- T10a - 10-Pin Connector, orange, behind instrument panel, left
- T10g - 10-Pin Connector, grey, behind instrument panel, left
- T10k - 10- pin connector, at F125
- T16 - 16- pin connector, Data Link Connector (DCL), bellow steering column, left
- T32 - 32-Pin Connector, blue, on instrument cluster
- T47 - 47- pin connector, at J104
- T52 - 52- pin connector, at J217

- (114) - Ground connection, in automatic transmission wiring harness
- (609) - Ground Connection (in right plenum chamber)
- (A76) - Connector (K-diagnosis wire) in instrument panel wiring harness
- (A87) - Connection (reverse lamp), in instrument panel wiring harness
- (A167) - Plus Connection 3 (30a) (in instrument panel wiring harness)
- (U3) - Wire connection -1-, in automatic transmission wiring harness

--- - Vehicles with ESP only

New Beetle — Wiring diagram — No. 36/4

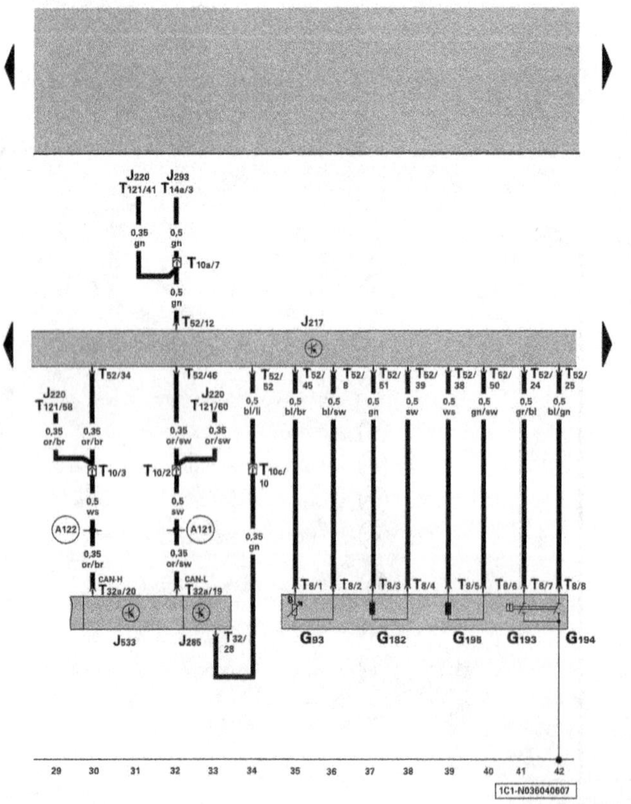

ws = white
sw = black
ro = red
br = brown
gn = green
bl = blue
gr = grey
li = violet
ge = yellow

Transmission vehicle speed sensor, vehicle speed sensor, Transmission Fluid Temperature Sensor, Transmission Control Module (TCM)

- G93 - Transmission Fluid Temperature Sensor
- G182 - Sensor for transmission RPM
- G193 - Sender 1 for hydraulic pressure, automatic transmission
- G194 - Sender 2 for hydraulic pressure, automatic transmission
- G195 - Sender for transmission output RPM
- J217 - Transmission Control Module (TCM)
- J220 - Motronic Engine Control Module (ECM)
- J285 - Control module with indicator unit in instrument panel insert
- J293 - Coolant FC (Fan Control) Control Module
- J533 - Data Bus On Board Diagnostic Interface
- T8 - 8-pin connector
- T10 - 10-Pin Connector, white, behind instrument panel, left
- T10a - 10-Pin Connector, orange, behind instrument panel, left
- T10c - 10-Pin Connector, blue, behind instrument panel, left
- T14a - 14-pin connector, at J293
- T32 - 32-pin connector, blue, on instrument cluster
- T32a - 32-pin connector, green, on instrument cluster
- T52 - 52-pin connector, at J217
- T121 - 121- pin connector, at J220

- (A121) - Connection (high bus) in instrument panel wiring harness
- (A122) - Connection (low bus) in instrument panel wiring harness

Automatic transmission, 6-speed Tiptronic (2.0L BEV, 1.8L BKF) from June 2004

New Beetle

Wiring diagram

No. 36/5

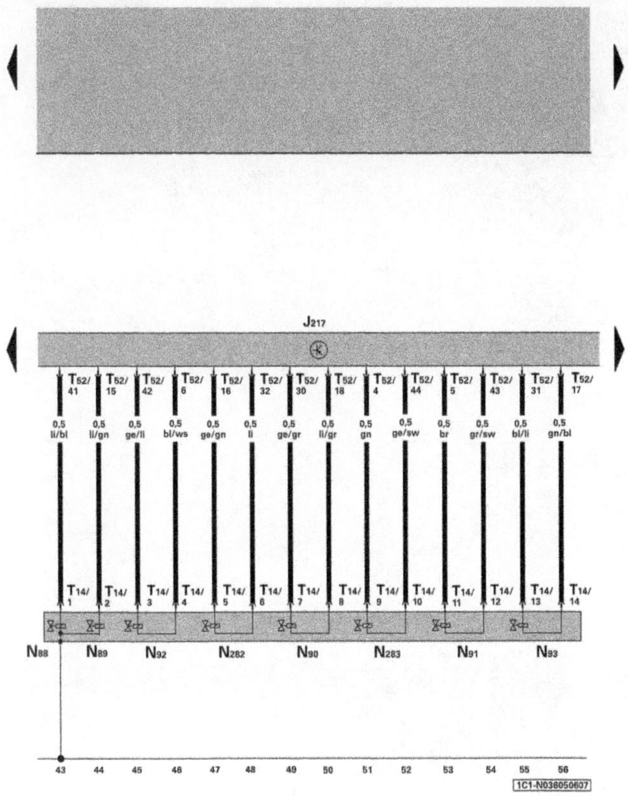

Transmission control module (TCM), solenoid valves

J217	-	Transmission Control Module (TCM)
N88	-	Solenoid Valve 1
N89	-	Solenoid Valve 2
N90	-	Solenoid Valve 3
N91	-	Solenoid Valve 4
N92	-	Solenoid Valve 5
N93	-	Solenoid Valve 6
N282	-	Solenoid Valve 9
N283	-	Solenoid Valve 10
T14	-	14-pin connector
T52	-	52- pin connector, at J217

ws = white
sw = black
ro = red
br = brown
gn = green
bl = blue
gr = grey
li = violet
ge = yellow

New Beetle

Wiring diagram

No. 36/6

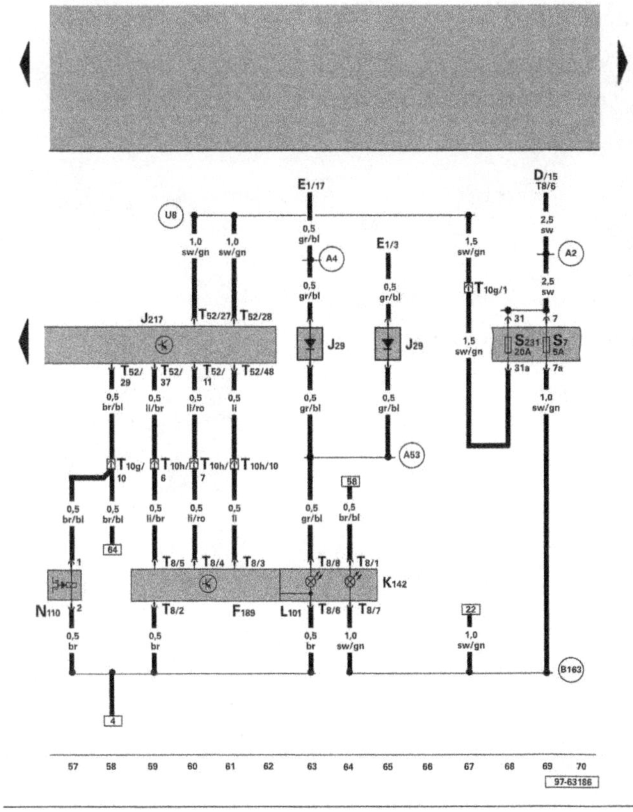

Tiptronic switch, warning light for selector lever position P/N, illumination for selector lever scale, shift lock solenoid

D	-	Ignition/Starter Switch
E1	-	Light switch
F189	-	Tiptronic Switch
J29	-	Blocking Diode
J217	-	Transmission Control Module (TCM)
K142	-	Warning light for selector lever position P/N
L101	-	Illumination for selector lever scale
N110	-	Shift Lock Solenoid
S7	-	Fuse 7 in fuse holder
S231	-	Fuse 31 in fuse holder
T8	-	8-pin connector, at F189
T10g	-	10-Pin Connector, grey, behind instrument panel, left
T10h	-	10-Pin Connector, black, behind instrument panel, left
T52	-	52-pin connector, at J217
(A2)	-	Plus connection (15), in instrument panel wiring harness
(A4)	-	Plus connection (58b), in instrument panel wiring harness panel wiring harness
(A53)	-	plus connection (58b), in instrument panel wiring harness
(B163)	-	Plus connector -1- (15) in wiring harness interior
(U8)	-	Plus Connection (15a) (automatic transmission wiring harness)

ws = white
sw = black
ro = red
br = brown
gn = green
bl = blue
gr = grey
li = violet
ge = yellow

Automatic transmission, 6-speed Tiptronic (2.0L BEV, 1.8L BKF) from June 2004

New Beetle **Wiring diagram** **No. 36/7**

Brake Booster Control Module, Brake System Vacuum Pump

J220 - Motronic Engine Control Module (ECM)
J542 - Brake Booster Control Module
S51 -
T6J - 6-pin connector
T6K - 8-pin connector
T10a - 10-pin connector
T121 - 121- pin connector, at J220
V192 - Brake System Vacuum Pump

* - 1.8 L engine
** - 2.0 L engine

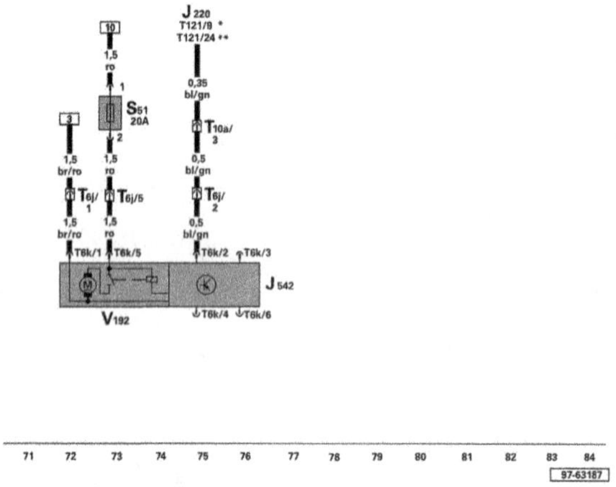

ws = white
sw = black
ro = red
br = brown
gn = green
bl = blue
gr = grey
li = violet
ge = yellow

THIS SPACE INTENTIONALLY LEFT BLANK

New Beetle — Wiring diagram — **No. 49/1**

Automatic Transmission, 6-speed with Tiptronic, 09G,

- **2.5L - Engine - Motronic Multiport Fuel Injection (MFI)/110 kW, code BPR,**
- **2.5L - Engine - Motronic Multiport Fuel Injection (MFI)/110 kW, code BPS,**

from July 2005,

Additional Fuse, Relay, Control Module and Connector locations ⇒ Component Locations

Electrical troubleshooting ⇒ Guided Fault Finding using VAS 5051/5052

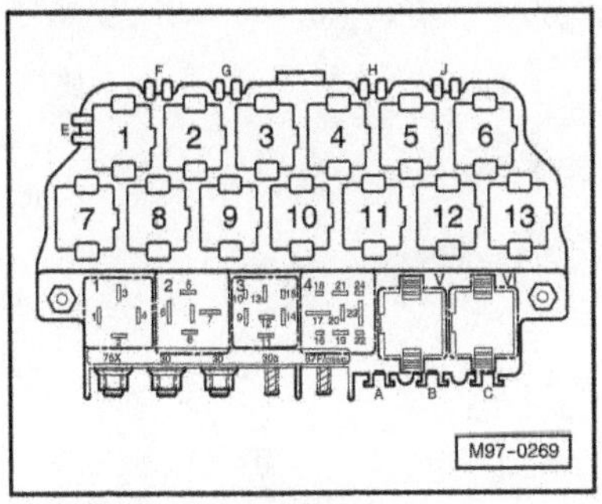

Relay location on the thirteen position auxiliary relay panel, above relay panel:

11 - J226 - Park/Neutral Position (PNP) Relay (175)

Relay panel:

2 - J59 - Load reduction Relay (100)

Note:

Number in parentheses indicates production control number stamped on relay housing.

New Beetle — Wiring diagram — **No. 49/2**

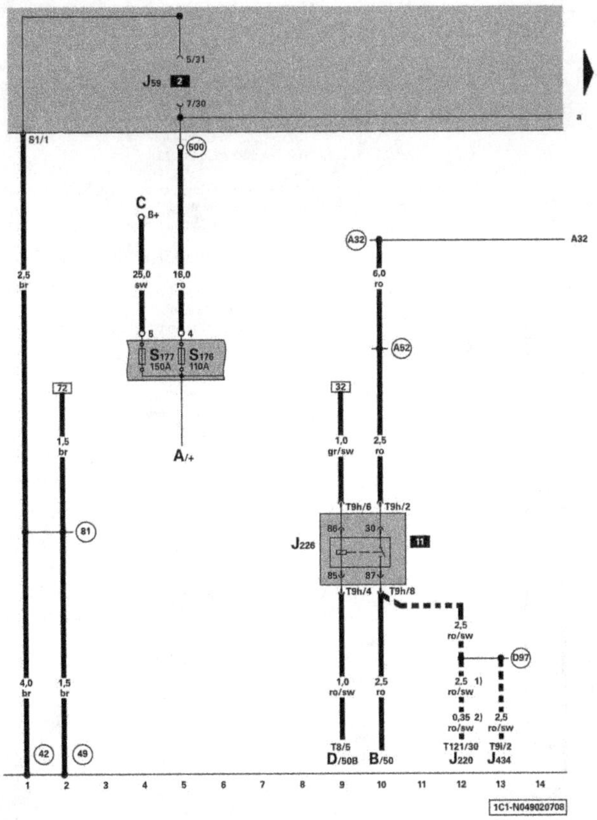

Load Reduction Relay, Park/Neutral Position (PNP) Relay

A	-	Battery
B	-	Starter
C	-	Generator (GEN)
D	-	Ignition/Starter Switch
J59	-	Load Reduction Relay
J220	-	Motronic Engine Control Module (ECM)
J226	-	Park/Neutral Position (PNP) Relay
J434	-	Starter Lock/Clutch Pedal Switch Relay
S176	-	Fuse 4 - terminal 30 (on battery fuse holder)
S177	-	Fuse 5 - terminal 30 (on battery fuse panel)
T8	-	8-Pin Connector, at -D-
T9h	-	9-Pin Connector, at -J226-
T9i	-	9-Pin Connector, at -J434-
T121	-	121-Pin Connector, at -J220-
(42)	-	Ground Connection (beside steering column)
(49)	-	Ground Connection (near steering column)
(81)	-	Ground Connection 1 (in instrument panel wiring harness)
(500)	-	Threaded Connection 1 - terminal 30 (on relay panel)
(A32)	-	Plus Connection (30) (in instrument panel wiring harness)
(A52)	-	Plus Connection 2 (30) (in instrument panel wiring harness)
(D97)	-	Connection (50) (in right engine compartment wiring harness)
- - -	-	Only Engine code BPR
1)	-	Through June 2007
2)	-	From July 2007

ws = white
sw = black
ro = red
br = brown
gn = green
bl = blue
gr = grey
li = violet
ge = yellow
or = orange

Automatic transmission, 6-speed Tiptronic (2.5L BPR, BPS) from July 2005

New Beetle — Wiring diagram — No. 49/3

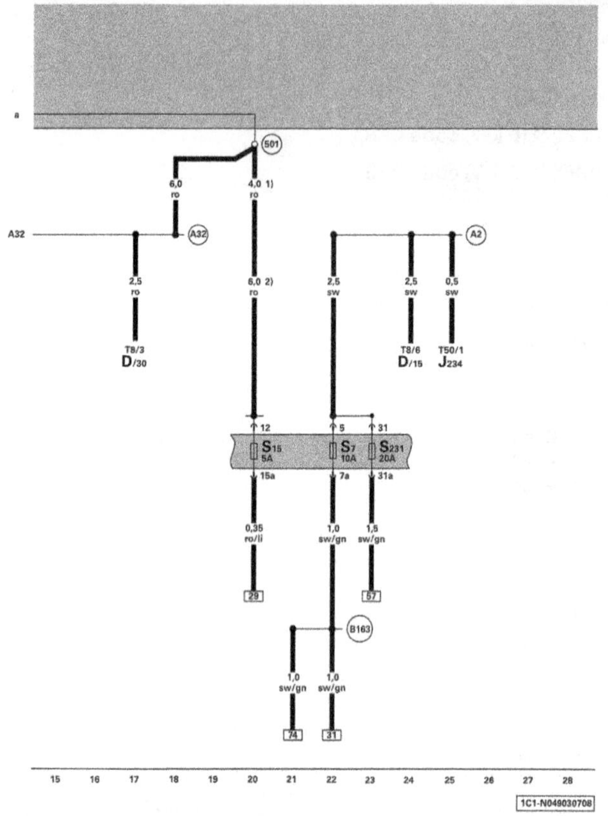

Fuses

D - Ignition/Starter Switch
J234 - Airbag Control Module
S7 - Fuse (on fuse/relay panel)
S15 - Fuse (on fuse/relay panel)
S231 - Fuse (on fuse panel)
T8 - 8-Pin Connector, at -D-
T50 - 50-Pin Connector

(501) - Threaded Connection 2 - terminal 30 (on relay panel)
(A2) - Plus Connection (15) (in instrument panel wiring harness)
(A32) - Plus Connection (30) (in instrument panel wiring harness)
(B163) - Plus Connection 1 (15) (in interior wiring harness)

1) - Through June 2007
2) - From July 2007

ws = white
sw = black
ro = red
br = brown
gn = green
bl = blue
gr = grey
li = violet
ge = yellow
or = orange

New Beetle — Wiring diagram — No. 49/4

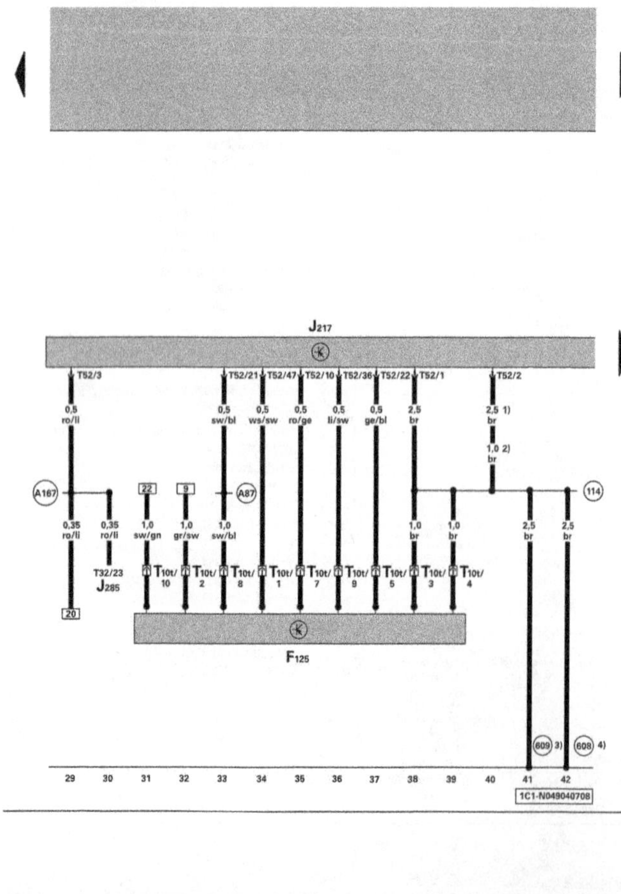

Multi-Function Transmission Range (TR) Switch, Transmission Control Module (TCM)

F125 - Multi-Function Transmission Range (TR) Switch
J217 - Transmission Control Module (TCM)
J285 - Instrument Cluster Control Module
T10t - 10-Pin Connector
T32 - 32-Pin Connector
T52 - 52-Pin Connector

(114) - Ground Connection (in automatic transmission wiring harness)
(608) - Ground Connection (in center plenum chamber)
(609) - Ground Connection (in right plenum chamber)
(A87) - Back-up Light Connection (in instrument panel wiring harness)
(A167) - Plus Connection 3 (30a) (in instrument panel wiring harness)

1) - Through June 2007
2) - From July 2007
3) - Through June 2008
4) - From July 2008

ws = white
sw = black
ro = red
br = brown
gn = green
bl = blue
gr = grey
li = violet
ge = yellow
or = orange

Automatic transmission, 6-speed Tiptronic (2.5L BPR, BPS) from July 2005

New Beetle

Wiring diagram

No. 49/5

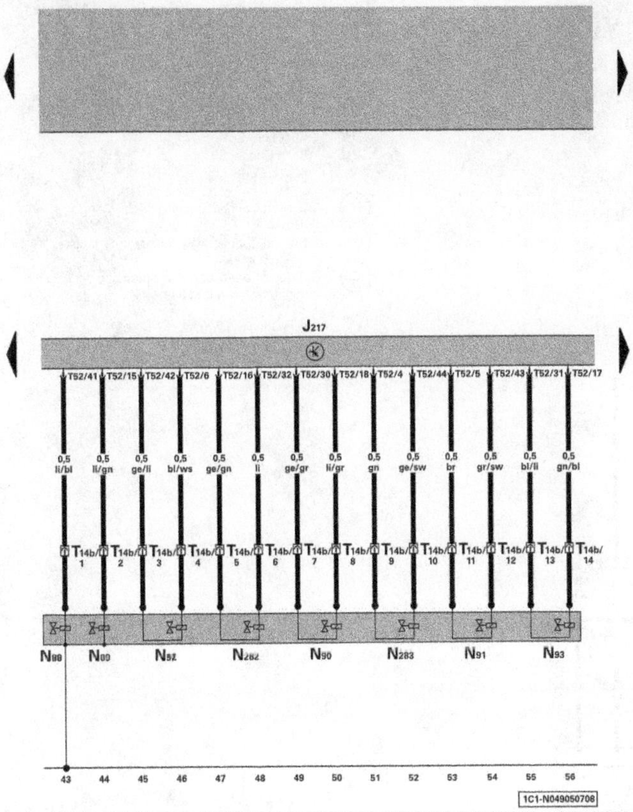

Transmission Control Module (TCM), Solenoid Valves

- J217 - Transmission Control Module (TCM)
- N88 - Solenoid Valve 1
- N89 - Solenoid Valve 2
- N90 - Solenoid Valve 3
- N91 - Solenoid Valve 4
- N92 - Solenoid Valve 5
- N93 - Solenoid Valve 6
- N282 - Solenoid Valve 9
- N283 - Solenoid Valve 10
- T14b - 14-Pin Connector
- T52 - 52-Pin Connector

ws = white
sw = black
ro = red
br = brown
gn = green
bl = blue
gr = grey
li = violet
ge = yellow
or = orange

New Beetle

Wiring diagram

No. 49/6

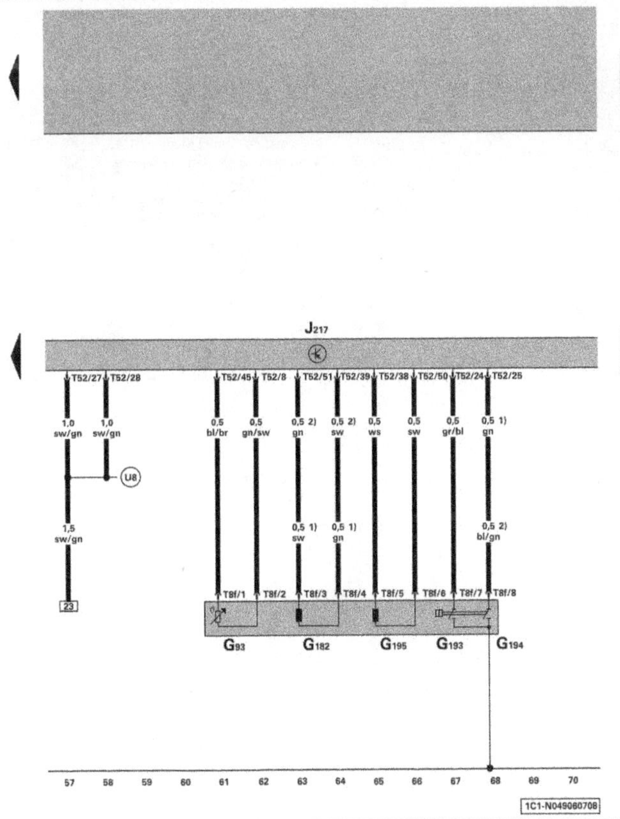

Transmission Fluid Temperature Sensor, Transmission Input Speed (RPM) Sensor, Automatic Transmission Hydraulic Pressure Sensor 1, Automatic Transmission Hydraulic Pressure Sensor 2, Transmission Output Speed (RPM) Sensor, Transmission Control Module (TCM)

- G93 - Transmission Fluid Temperature Sensor
- G182 - Transmission Input Speed (RPM) Sensor
- G193 - Automatic Transmission Hydraulic Pressure Sensor 1
- G194 - Automatic Transmission Hydraulic Pressure Sensor 2
- G195 - Transmission Output Speed (RPM) Sensor
- J217 - Transmission Control Module (TCM)
- T8f - 8-Pin Connector
- T52 - 52-Pin Connector
- (U8) - Plus Connection (15a) (automatic transmission wiring harness)

1) - Through June 2007
2) - From July 2007

ws = white
sw = black
ro = red
br = brown
gn = green
bl = blue
gr = grey
li = violet
ge = yellow
or = orange

Automatic transmission, 6-speed Tiptronic (2.5L BPR, BPS) from July 2005

New Beetle

Wiring diagram

No. 49/7

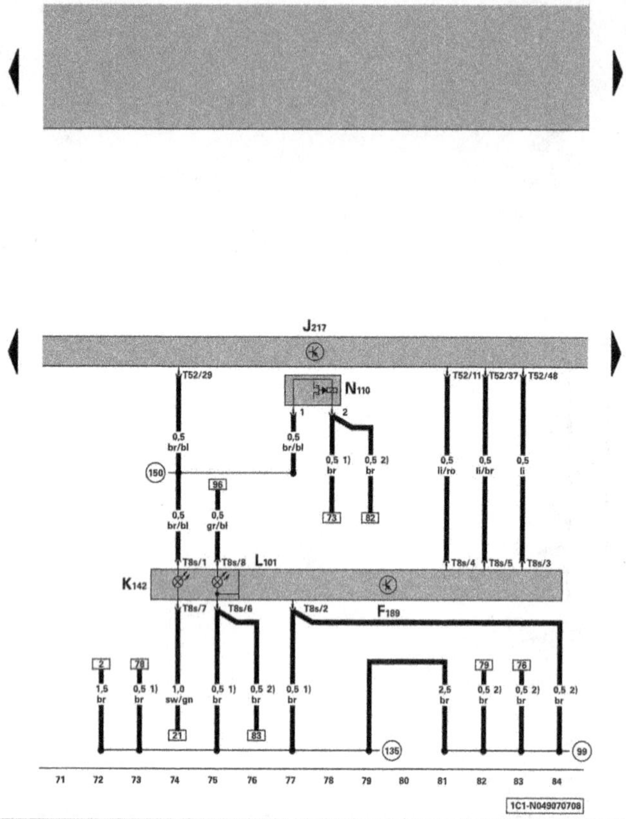

Tiptronic Switch, Transmission Control Module (TCM), Shift Lock Solenoid

- F189 - Tiptronic Switch
- J217 - Transmission Control Module (TCM)
- K142 - Selector Lever Position P/N Indicator Lamp
- L101 - Selector Lever Scale Illumination
- N110 - Shift Lock Solenoid
- T8s - 8-Pin Connector
- T52 - 52-Pin Connector

- (99) - Ground Connection (in console wiring harness)
- (135) - Ground Connection 2 (in instrument panel wiring harness)
- (150) - Ground Connection 2 (in automatic transmission wiring harness)

1) - Through June 2007
2) - From July 2007

ws = white
sw = black
ro = red
br = brown
gn = green
bl = blue
gr = grey
li = violet
ge = yellow
or = orange

New Beetle

Wiring diagram

No. 49/8

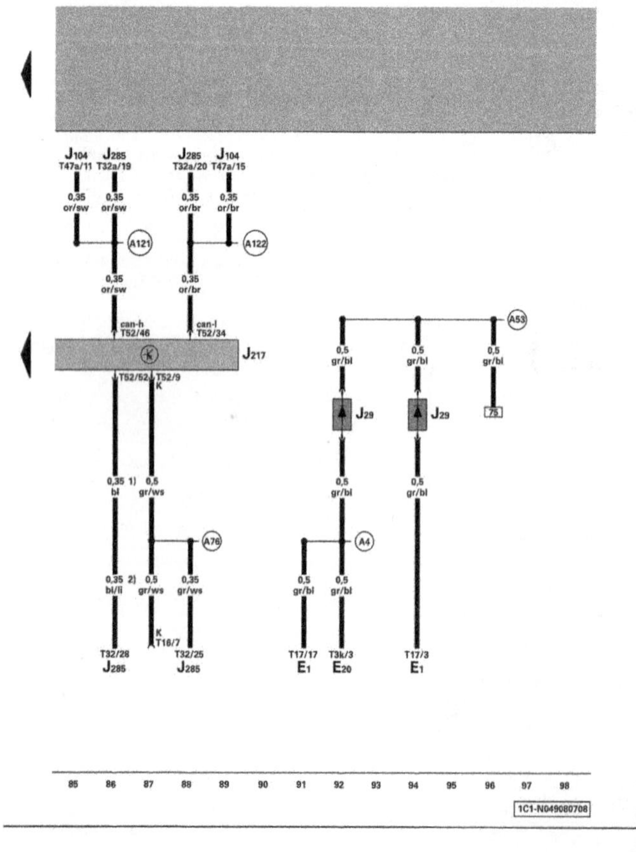

Blocking Diode, Transmission Control Module (TCM)

- E1 - Light Switch
- E20 - Instrument Panel Illumination Dimmer Switch
- J29 - Blocking Diode
- J104 - ABS Control Module
- J217 - Transmission Control Module (TCM)
- J285 - Instrument Cluster Control Module
- T3k - 3-Pin Connector
- T16 - 16-Pin Connector, Data Link Connector (DLC), bellow steering column, left
- T17 - 17-Pin Connector
- T32 - 32-Pin Connector
- T32a - 32-Pin Connector
- T47a - 47-Pin Connector
- T52 - 52-Pin Connector

- (A4) - Plus Connection (58b) (in instrument panel wiring harness)
- (A53) - Plus Connection (58b) (in instrument panel wiring harness)
- (A76) - K-Diagnostic Connection (in instrument panel wiring harness)
- (A121) - High-Bus Connection (in instrument panel wiring harness)
- (A122) - Low-Bus Connection (in instrument panel wiring harness)

1) - Through June 2007
2) - From July 2007

ws = white
sw = black
ro = red
br = brown
gn = green
bl = blue
gr = grey
li = violet
ge = yellow
or = orange

New Beetle | **Wiring diagram** | **No. 39/1**

CAN-Bus Powertrain

from June 1999

Deviate relay location and fuseplacements as well as the locations of multiple connectors see section "component locations".

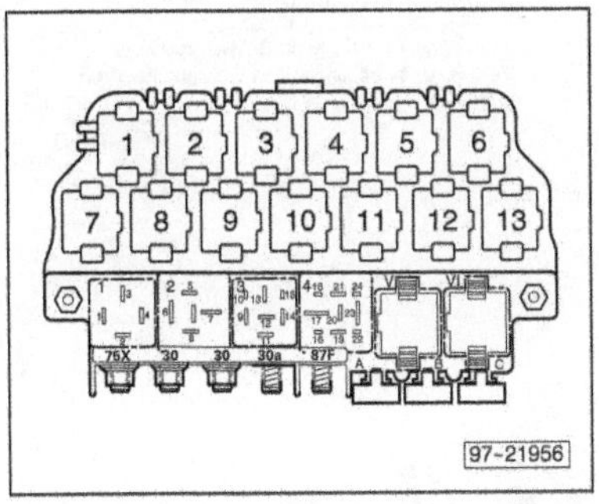

Fuse colors

30 A - green
25 A - white
20 A - yellow
15 A - blue
10 A - red
7,5 A - brown
5 A - beige
3 A - violet

Note: Number in parentheses indicates production control number stamped on relay housing.

New Beetle | **Wiring diagram** | **No. 39/2**

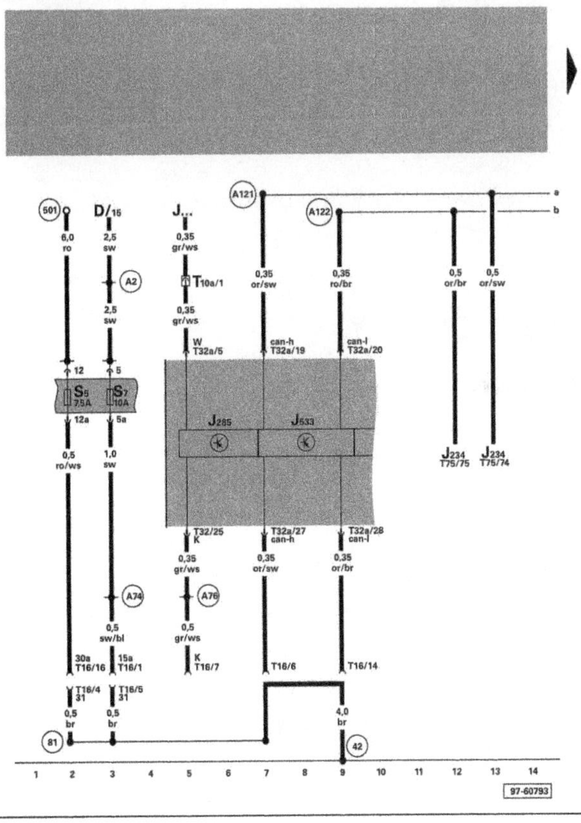

Data link connector (DLC), Airbag Control Module, control module with indicator unit in instrument panel insert, data bus on board diagnostic interface.

- D - Ignition/Starter Switch
- J234 - Airbag Control Module
- J285 - Control module with indicator unit in instrument panel insert
- J533 - Data bus on board diagnostic interface
- J... - Engine Control Module (ECM)
- S5 - Fuse 5 in fuse holder
- S12 - Fuse 12 in fuse holder
- T10a - 10-Pin Connector,orange, behind instrument panel, left
- T16 - 16-Pin Connector, Data Link Connector (DLC), below steering column, left
- T32 - 32-Pin Connector, blue
- T32a - 32-Pin Connector, green
- T75 - 75-Pin Connector
- (42) - Ground connection, beside steering column
- (81) - Ground connection -1-, in instrument panel wiring harness
- (501) - Threaded connection -2- (30) on the relay plate
- (A2) - Plus connection (15), in instrument panel wiring harness
- (A74) - Connector (15a - fuse 5) in instrument panel wiring harness
- (A76) - Connector (K-diagnosis wire) in instrument panel wiring harness
- (A121) - Connection (high bus) in instrument panel wiring harness
- (A122) - Connection (low bus) in instrument panel wiring harness

ws = white
sw = black
ro = red
br = brown
gn = green
bl = blue
gr = grey
li = violet
ge = yellow
or = orange

New Beetle Wiring diagram **No. 39/3**

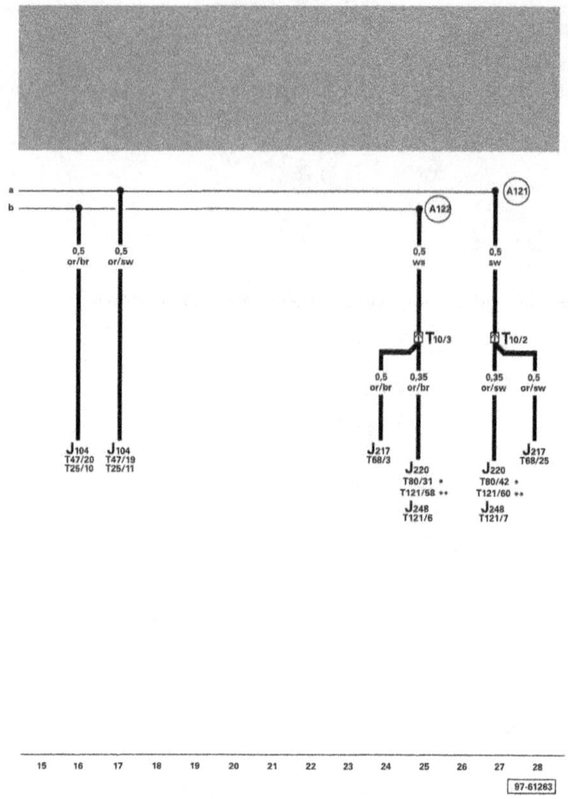

ABS Control Module (w/EDL), Transmission Control Module (TCM), Motronic Engine Control Module (ECM)

J104 - ABS Control Module (w/EDL)
J217 - Transmission Control Module (TCM)
J220 - Motronic Engine Control Module (ECM)
J248 - Diesel Direct Fuel Injection (DFI) Engine Control Module (ECM)
T10 - 10-Pin Connector, white, behind instrument panel, left
T25 - 25-Pin Connector
T47 - 47-Pin Connector
T68 - 68-Pin Connector
T80 - 80-Pin Connector
T121 - 121-Pin Connector

(A121) - Connection (high bus) in instrument panel wiring harness
(A122) - Connection (low bus) in instrument panel wiring harness

* - 2.0L engine through model year 2000
** - All 1.8L engines, as well as 2.0L engines from model year 2001

ws = white
sw = black
ro = red
br = brown
gn = green
bl = blue
gr = grey
li = violet
ge = yellow
or = orange

THIS SPACE INTENTIONALLY LEFT BLANK

New Beetle Wiring diagram **No. 54/1**

CAN-Bus - Powertrain and Data Link Connector (DLC),

from July 2005,

Additional Fuse, Relay, Control Module and Connector locations ⇒ Component Locations

Electrical troubleshooting ⇒ Guided Fault Finding using VAS 5051/5052

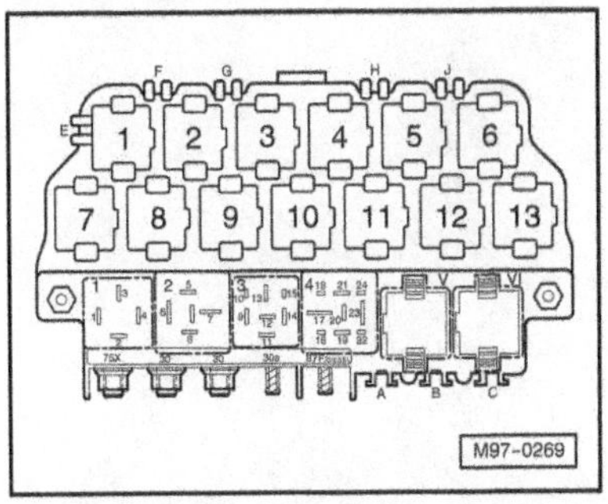

Note:

Number in parentheses indicates production control number stamped on relay housing.

New Beetle Wiring diagram **No. 54/2**

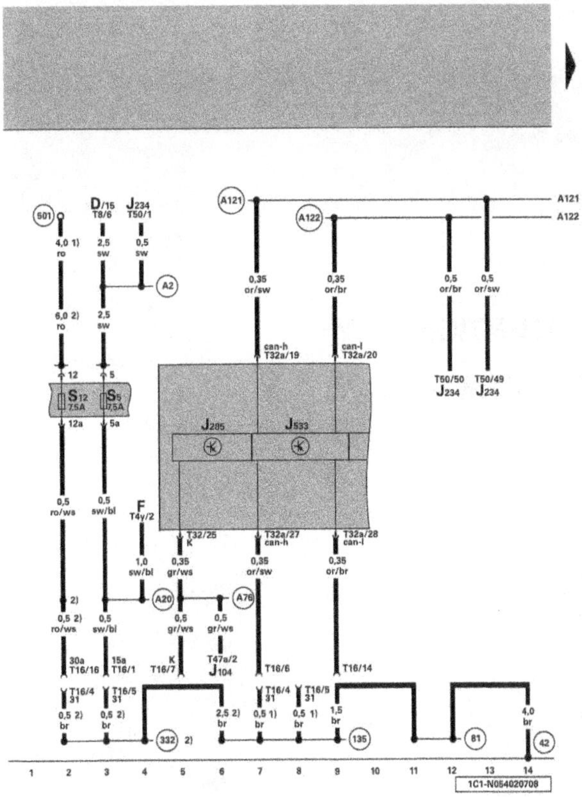

ws = white
sw = black
ro = red
br = brown
gn = green
bl = blue
gr = grey
li = violet
ge = yellow
or = orange

Instrument Cluster Control Module, Data Bus On Board Diagnostic Interface, Data Link Connector (DLC)

- D - Ignition/Starter Switch
- F - Brake Light Switch
- J104 - ABS Control Module
- J234 - Airbag Control Module
- J285 - Instrument Cluster Control Module
- J533 - Data Bus On Board Diagnostic Interface
- S5 - Fuse (on fuse/relay panel)
- S12 - Fuse 12 in fuse holder
- T4y - 4-Pin Connector
- T8 - 8-Pin Connector
- T16 - 16-Pin Connector, Data Link Connector (DLC), below steering column, left
- T32 - 32-Pin Connector, blue
- T32a - 32-Pin Connector, green
- T47a - 47-Pin Connector
- T50 - 50-Pin Connector
- (42) - Ground Connection (beside steering column)
- (81) - Ground Connection 1 (in instrument panel wiring harness)
- (135) - Ground Connection 2 (in instrument panel wiring harness)
- (332) - Ground Connection 4 (in instrument panel wiring harness)
- (501) - Threaded Connection 2 - terminal 30 (on relay panel)
- (A2) - Plus Connection (15) (in instrument panel wiring harness)
- (A20) - Plus Connection (15a) (in instrument panel wiring harness)
- (A76) - K-Diagnostic Connection (in instrument panel wiring harness)
- (A121) - High-Bus Connection (in instrument panel wiring harness)
- (A122) - Low-Bus Connection (in instrument panel wiring harness)

1) - Through June 2008
2) - From July 2008

New Beetle Wiring diagram **No. 54/3**

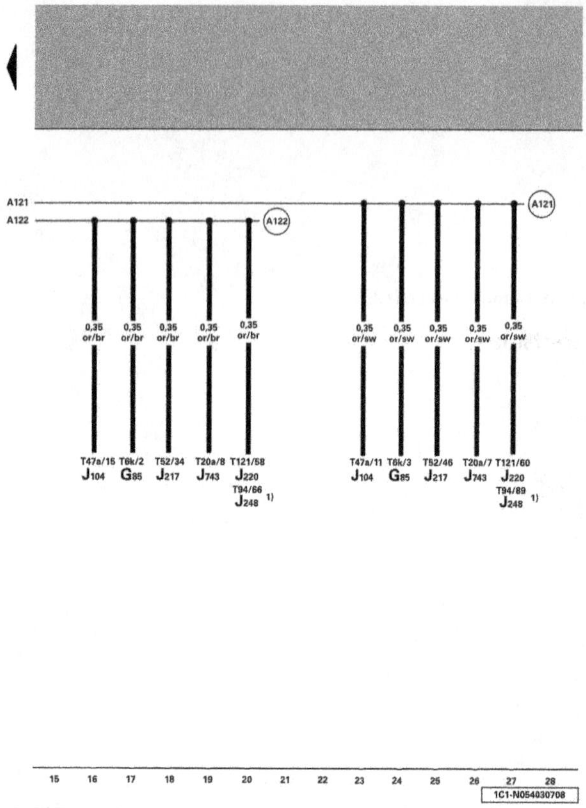

High-Bus and Low-Bus Connections

G85 - Steering Angle Sensor
J104 - ABS Control Module
J217 - Transmission Control Module (TCM)
J220 - Motronic Engine Control Module (ECM)
J248 - Diesel Direct Fuel Injection (DFI) Engine Control Module (ECM)
J743 - Direct Shift Gearbox (DSG) Mechatronic
T6k - 6-Pin Connector
T20a - 20-Pin Connector, at TCM
T47a - 47-Pin Connector, at ABS Control Module
T52 - 52-Pin Connector, at TCM
T121 - 121-Pin Connector, at ECM

(A121) - High-Bus Connection (in instrument panel wiring harness)
(A122) - Low-Bus Connection (in instrument panel wiring harness)

1) - Through June 2008

ws = white
sw = black
ro = red
br = brown
gn = green
bl = blue
gr = grey
li = violet
ge = yellow
or = orange

THIS SPACE INTENTIONALLY LEFT BLANK

Comfort system, with anti-theft alarm system (vehicles with power windows)

from January 1998

Deviate relay location and fuseplacements as well as the locations of multiple connectors see section "component locations".

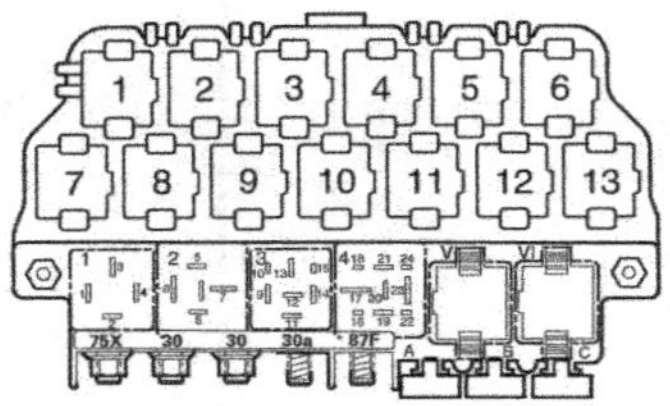

97-21956

Fuse colors
30 A - green
25 A - white
20 A - yellow
15 A - blue
10 A - red
7,5 A - brown
5 A - beige

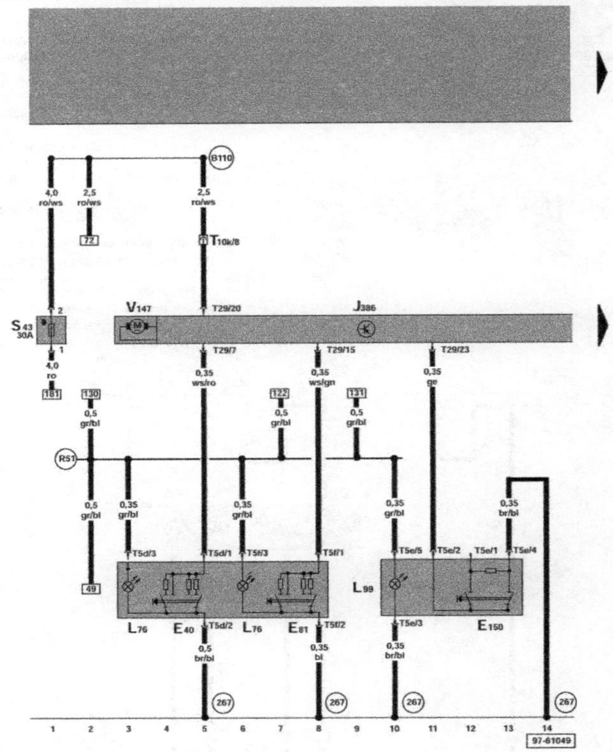

ws = white
sw = black
ro = red
br = brown
gn = green
bl = blue
gr = grey
li = violet
ge = yellow

E40 - Left Front Window Switch
E81 - Switch for window regulator front right, driver
E150 - Switch for interior lock, driver side
J386 - Door control module, driver side
L76 - Push Button Light
L99 - Lighting for switch interior lock
S43 - Power Window Circuit Breaker
T5d - 5-Pin Connector, brown
T5f - 5-Pin Connector, black
T5e - 5-Pin Connector, black
T10k - 10-Pin Connector, black, on A-Pillar left
T29 - 29-Pin Connector, on Door control module
V147 - Motor for window regulator, driver side

(267) - Ground connector -2-, in wiring harness door cable

(B110) - Connection (30, window regulator) in wiring harness interior
(R51) - Connection (58b), in door wiring harness, driver's side

Door control module (driver side), left front window switch, switch for window regulator front right (driver), switch for interior lock (driver side)

Comfort system and alarm from Jan. 1998

New Beetle — Wiring diagram — No. 9/3

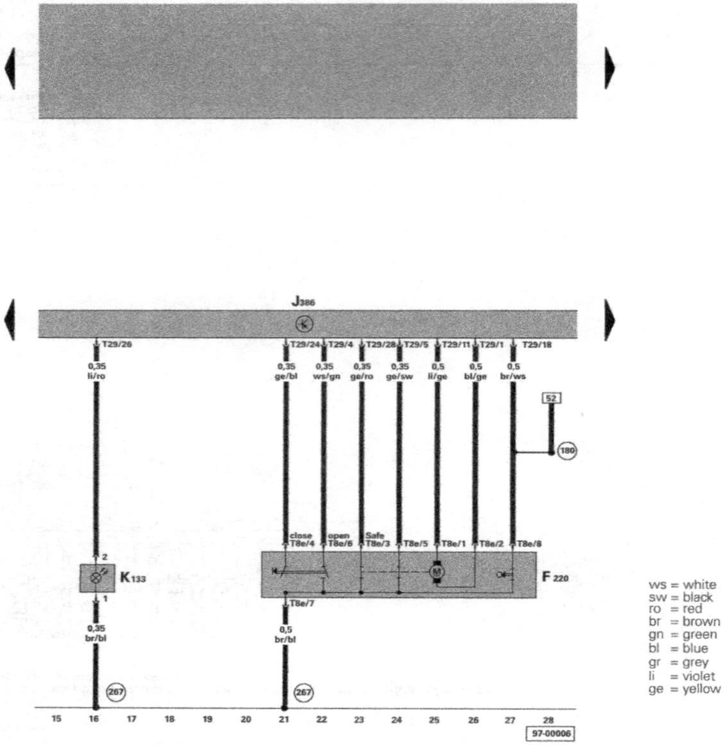

ws = white
sw = black
ro = red
br = brown
gn = green
bl = blue
gr = grey
li = violet
ge = yellow

F220 - Lock unit for central locking, driver side
J386 - Door control module, driver side
K133 - Warning light for central locking -SAFE-
T8e - 8-Pin Connector, on Lock unit for central locking, driver side
T29 - 29-Pin Connector, on Door control module

(180) - Ground connection -2-, in central locking system wiring harness
(267) - Ground connector -2-, in wiring harness door cable

Edition 08/98
USA.5132.04.21

Door control module (driver side), lock unit for central locking (driver side)

No. 9/4 — Wiring diagram — New Beetle

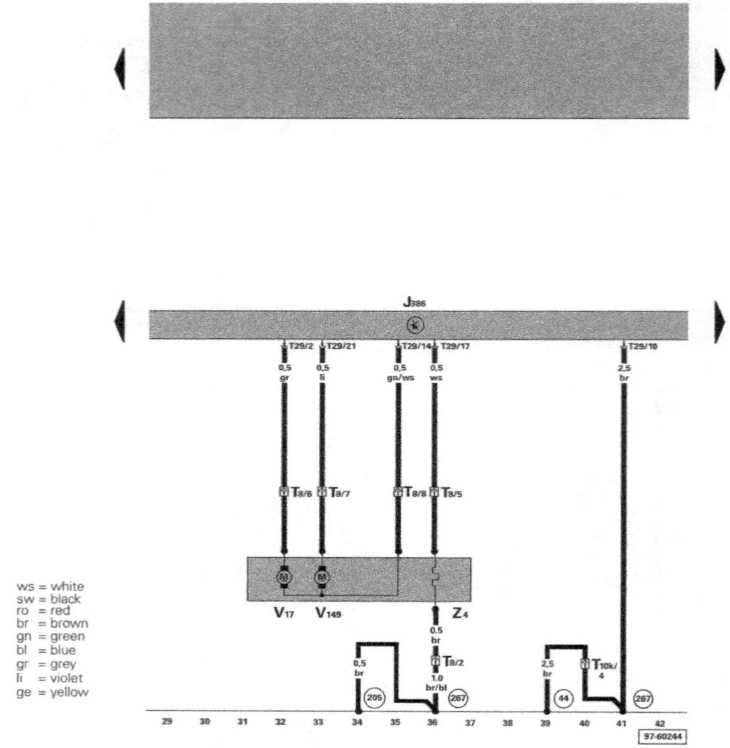

ws = white
sw = black
ro = red
br = brown
gn = green
bl = blue
gr = grey
li = violet
ge = yellow

J386 - Door control module, driver side
T8 - 8-Pin Connector, in driver's door wiring
T10k - 10-Pin Connector, on A-Pillar left, black
T29 - 29-Pin Connector, on Door control module - driver side
V17 - Driver's Side Mirror Adjustment Motor
V149 - Motor for mirror adjustment, driver side

Z4 - Heated outside mirror, driver side

(44) - Ground connection, on left A-pillar, lower part

(205) - Ground connection, in driver's door wiring harness

(267) - Ground connector -2-, in wiring harness door cable

Door control module (driver side), driver's side mirror adjustment motor, motor for mirror adjustment (driver side), heated outside mirror (driver side)

Edition 09/98
USA.5132.05.21

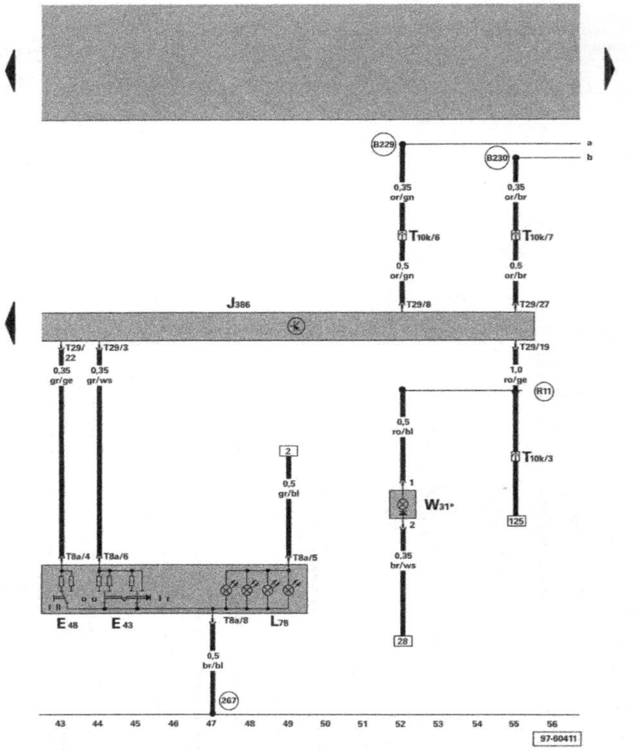

ws = white
sw = black
ro = red
br = brown
gn = green
bl = blue
gr = grey
li = violet
ge = yellow

E43 - Mirror Adjustment Switch
E48 - Mirror Selector Switch
J386 - Door control module, driver side
L78 - Mirror Adjusting Switch Light
T8 - 8-Pin Connector
T10k - 10-Pin Connector, on A-Pillar left, black
T29 - 29-Pin Connector, on Door control module

(W3) - Entry light front, left

(267) - Ground connector -2-,in wiring harness door cable
(B229) - connection (High-Bus), in wiring harness, interior
(B230) - connection (Low-Bus), in wiring harness, interior

(R11) - plus connection -2- (30), in driver's door wiring harness

* - Where applicable

Door control module (driver side) mirror adjustment switch, mirror selector switch, door warning lights

ws = white
sw = black
ro = red
br = brown
gn = green
bl = blue
gr = grey
li = violet
ge = yellow

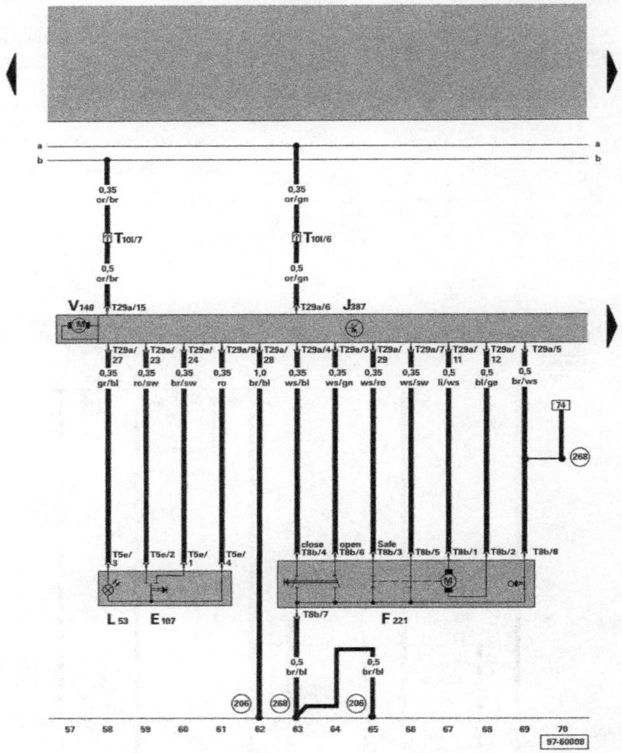

E107 - Switch for window regulator, in passenger door
F221 - Lock unit for central locking, passenger side
J387 - Door control module, passenger side
T5e - 5-Pin Connector
T8b - 8-Pin Connector
T10l - 10-Pin Connector, black, on A-Pillar right
T29a - 29-Pin Connector, on Door control module
L53 - Power Window Switch Light
V148 - Motor for window regulator, passenger side

(206) - Ground connection, in passenger's door wiring harness
(268) - Ground connector -2-, in wiring harness door cable - passenger side

Door control module (passenger side), lock unit for central locking (passenger side), switch for window regulator (passenger side)

Comfort system and alarm from Jan. 1998

New Beetle — Wiring diagram — No. 9/7

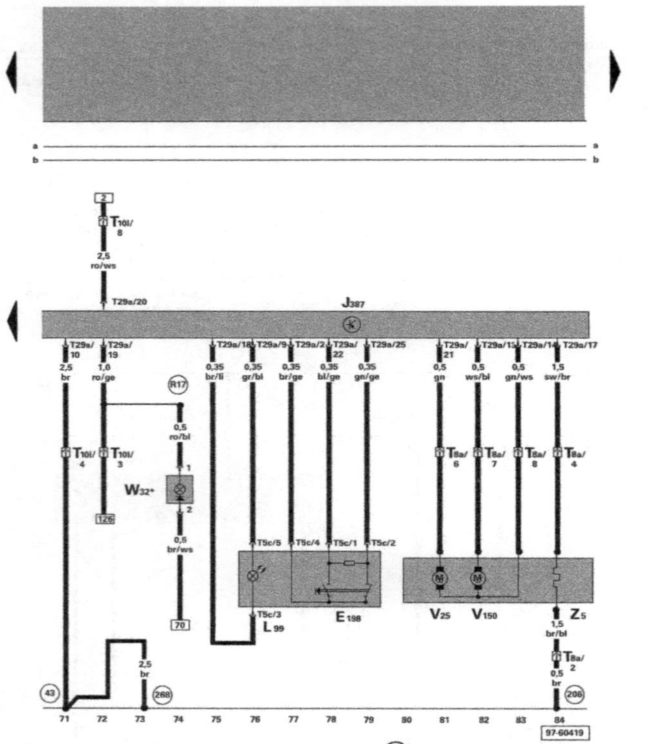

ws = white
sw = black
ro = red
br = brown
gn = green
bl = blue
gr = grey
li = violet
ge = yellow

E198 - Switch for interior lock, passenger side
J387 - Door control module, passenger side
L99 - Lighting for switch interior lock
T5c - 5-Pin Connector, black
T10l - 10-Pin Connector, black, on A-Pillar right
T8a - 8-Pin Connector
T29a - 29-Pin Connector, on Door control module
V25 - Passenger's Side Mirror Adjustment Motor
V150 - Motor for mirror adjustment, passenger side
W32 - Entry light front, right

Z5 - Heated outside mirror, passenger side

(R17) - Plus connection -2- (30), in passenger's door wiring harness
(43) - Ground connection, on right A-pillar, lower part
(206) - Ground connection, in passenger's door wiring harness
(268) - Ground connector -2-, in wiring harness door cable - passenger side
* - Where applicable

Door control module (passenger side), switch for interior lock (passenger side), passenger's side mirror adjustment motor, heated outside mirror

No. 9/8 — Wiring diagram — New Beetle

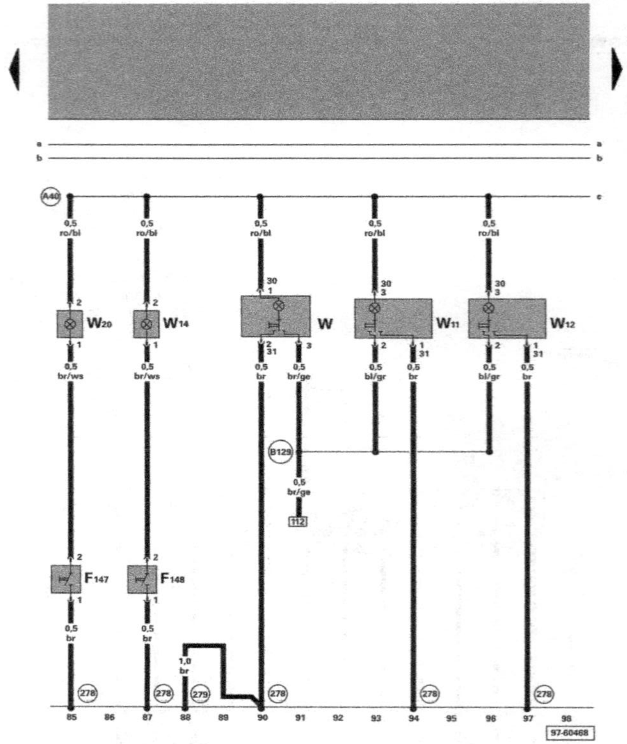

ws = white
sw = black
ro = red
br = brown
gn = green
bl = blue
gr = grey
li = violet
ge = yellow

F147 - Left Make-Up Mirror Light Switch
F148 - Right Make-Up Mirror Light Switch
W - Front Interior Light
W11 - Left Rear Reading Light
W12 - Right Rear Reading Light
W14 - Right Make-up Mirror Light
W20 - Left Make-up Mirror Light

(278) - Ground connector -4-, in wiring harness interior
(279) - Ground connector -5, in wiring harness interior
(A40) - plus connection -1- (30), in instrument cluster wiring harness
(B129) - Connector (interior light 3/L), in wiring harness interior

Front interior light, left rear reading light, right rear reading light, right make-up mirror light, left make-up mirror light

Comfort system and alarm from Jan. 1998

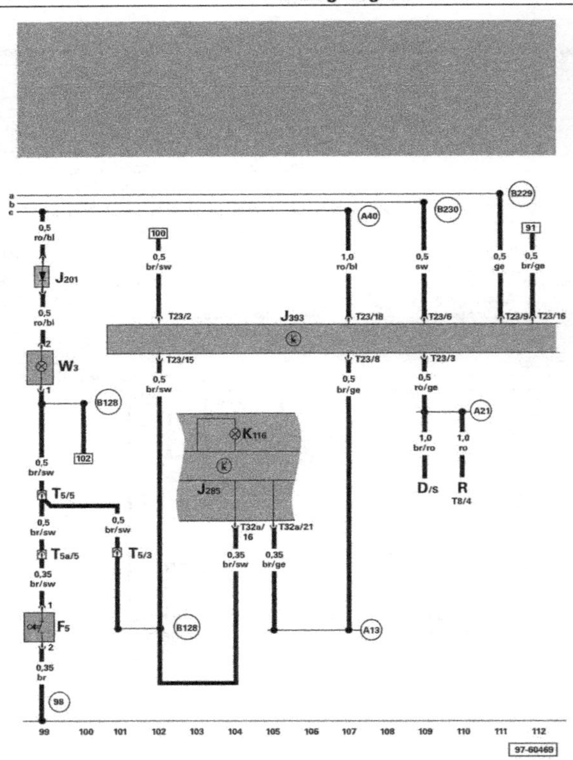

ws = white
sw = black
ro = red
br = brown
gn = green
bl = blue
gr = grey
li = violet
ge = yellow

D - Ignition/Starter Switch
F5 - Luggage Compartment Light Switch
J201 - Protection Diode
J285 - Control module with indicator unit in instrument panel insert
J393 - Central control module for comfort system
K116 - Warning light for rear lid unlocked
R - Radio
T5 - 5-Pin Connector, black, on C-pillar, left
T5a - 5-Pin Connector, black, in rear lid
T8 - 8-Pin Connector, on radio
T23 - 23-Pin Connector, on Central control module for comfort system
T32a - 32-Pin Connector, blue, on instrument cluster
W3 - Luggage compartment Light

(98) - Ground connection, in rear lid wiring harness
(A13) - wire connection (door contact switch) in instrument panel wiring harness
(A21) - wire connection (86s), in instrument panel wiring harness
(A40) - plus connection -1- (30), in instrument panel wiring harness
(B128) - Connector (Luggage compartment lamp) in wiring harness interior
(B229) - High-bus Connection (in interior wiring harness)
(B230) - Low-bus Connection (in interior wiring harness)

Central control module for comfort system, luggage compartment light, luggage compartment light switch

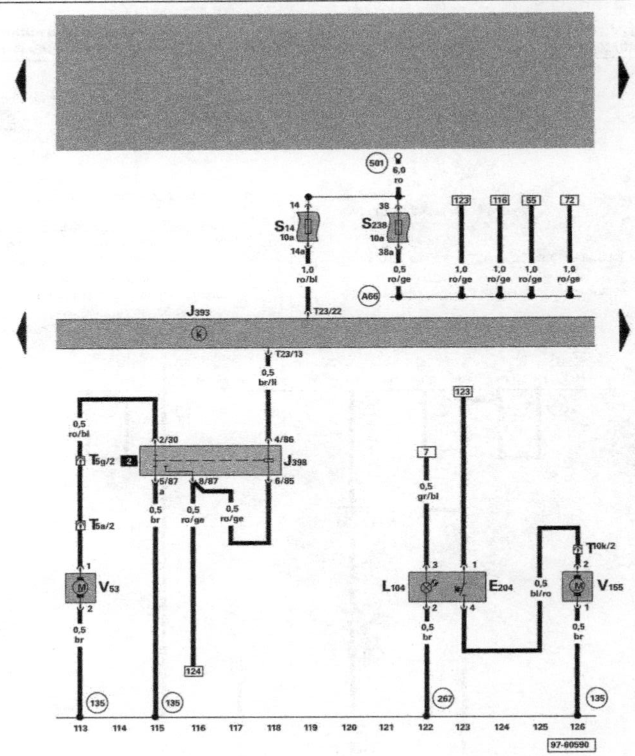

ws = white
sw = black
ro = red
br = brown
gn = green
bl = blue
gr = grey
li = violet
ge = yellow

E204 - Switch for remote/fuel tank door
J393 - Central control module for comfort system
J398 - Relay for motor remote unlock rear lid
L104 - Illumination for switch remote unlock
S14 - Fuse in fuse holder
S238 - Fuse in fuse holder
T5a - 5-Pin Connector, black, in rear lid
T5g - 5-Pin Connector, brown, in rear lid
T10k - 10-Pin Connector, on A-Pillar left, black
T23 - 23-Pin Connector, on Central control module for comfort system
V53 - Decklid Central Locking System Motor
V155 - Motor for fuel tank lid unlock

(135) - Ground connection -2-, in instrument panel wiring harness

(267) - Ground connector -2-, in wiring harness door cable - driver side
(501) - Screw connection -2- (30), on relay panel
(A66) - Connector (30a, central locking/anti-theft warning system/IR) in instrument panel wiring harness

Central control module for comfort system, relay for motor remote unlock rear lid, switch for remote unlock, rear lid

Comfort system and alarm from Jan. 1998

New Beetle Wiring diagram **No. 9/11**

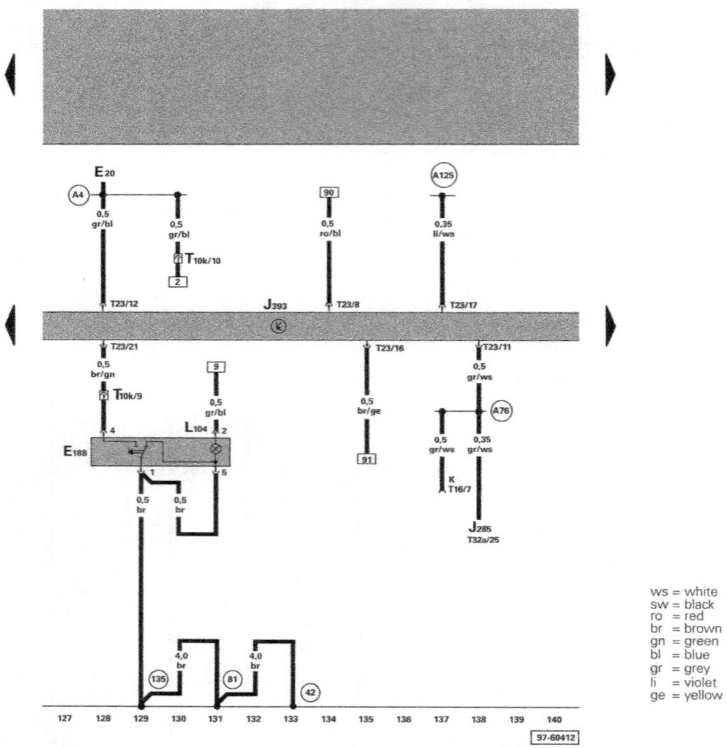

ws = white
sw = black
ro = red
br = brown
gn = green
bl = blue
gr = grey
li = violet
ge = yellow

E20 - Instrument Panel Light Dimmer Switch
E188 - Switch for remote unlock, rear lid
J285 - Control module with indicator unit in instrument panel insert
J393 - Central control module for comfort system
L104 - Illumination for switch remote unlock
T10k - 10-Pin Connector, on A-Pillar left, black
T23 23-Pin Connector, on Central control module for comfort system
T32a - 32-Pin Connector, blue, on instrument cluster

(42) - Ground connection, beside steering column

(81) - Ground connection -1-, in instrument panel wiring harness

(135) - Ground connection -2-, in instrument panel wiring harness

(A4) - plus connection (58b) in instrument panel wiring harness

(A76) - Connector (K-diagnosis wire) in instrument panel wiring harness

(A125) - wire connection (crash), in instrument panel wiring harness

Edition 07/00
USA.5132.11.21

Central control module for comfort system, switch for remote/fuel tank door, motor for fuel tank lid unlock

No. 9/12 Wiring diagram **New Beetle**

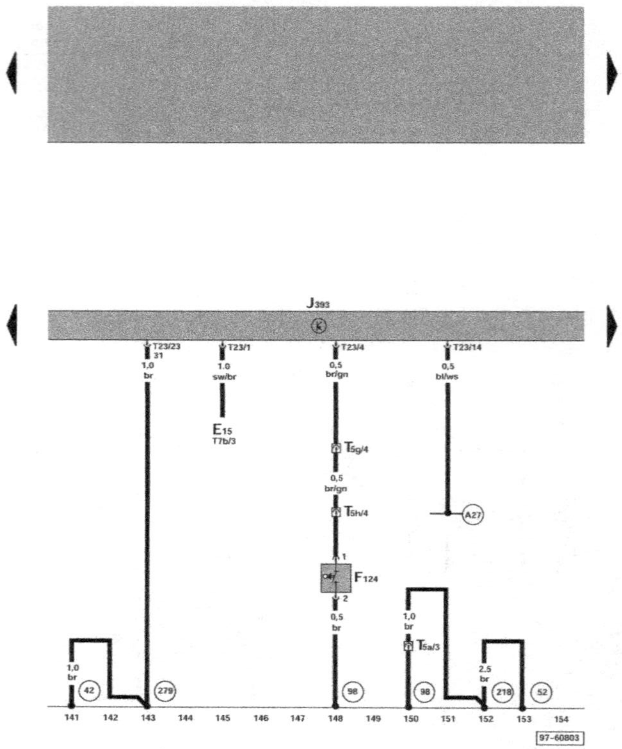

ws = white
sw = black
ro = red
br = brown
gn = green
bl = blue
gr = grey
li = violet
ge = yellow

E15 - Rear window defogger switch
F124 - Trunk Lock Alarm/Central Locking Switch
J393 - Central control module for comfort system
T5a 5-Pin Connector, black, in rear lid
T5g 5-Pin Connector, brown, in rear lid
T5h 5-Pin Connector, brown, in rear lid
T7b 7-Pin Connector, on rear window defogger switch
T23 23-Pin Connector, on Central control module for comfort system

(42) - Ground connection, beside steering column

(52) - Ground connection, in rear lid, left

(98) - Ground connection, in rear lid wiring harness

(218) - Ground connection -1-, in rear lid wiring harness

(279) - Ground connector -5-, in wiring harness interior

(A27) - wire Connection (vehicle speed signal), in instrument panel wiring harness

Central control module for comfort system, decklid central locking servo motor, switch for central locking (trunk lid)

Edition 07/00
USA.5132.11.21

Comfort system and alarm from Jan. 1998

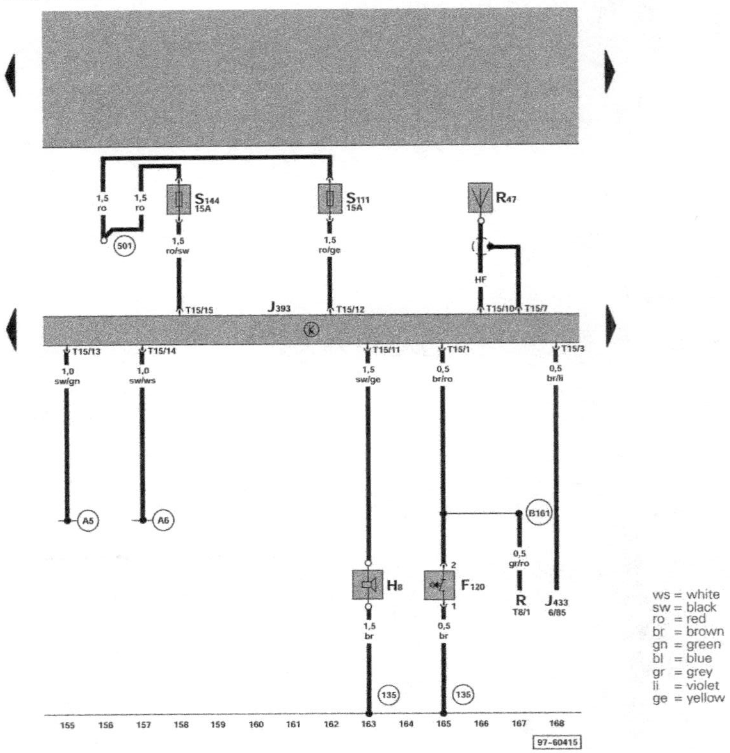

ws = white
sw = black
ro = red
br = brown
gn = green
bl = blue
gr = grey
li = violet
ge = yellow

F120 - Hood Alarm Switch
H8 - Alarm Horn
J393 - Central control module for comfort system
J433 - Locking relay for starter (clutch pedal switch)
R - Radio
R47 - Antenna wire for central locking and antitheft warning system
S111 - Alarm System and Anti-Theft Engine Disable Fuse on thirteen position relay panel
S144 - Fuse for central locking/anti-theft warning system on thirteen position relay panel
T8 - 8-Pin Connector, on radio
T15 - 15-Pin Connector, on Central control module for comfort system

(135) - Ground connection -2-, in instrument panel wiring harness
(501) - Threaded connection -1- (30) on the relay plate
(A5) - plus connection (right turn signal), in instrument panel wiring harness
(A6) - plus connection (left turn signal), in instrument panel wiring harness
(B161) - connection (Alarm system), in wiring harness, interior

Central control module for comfort system, hood alarm switch, alarm horn

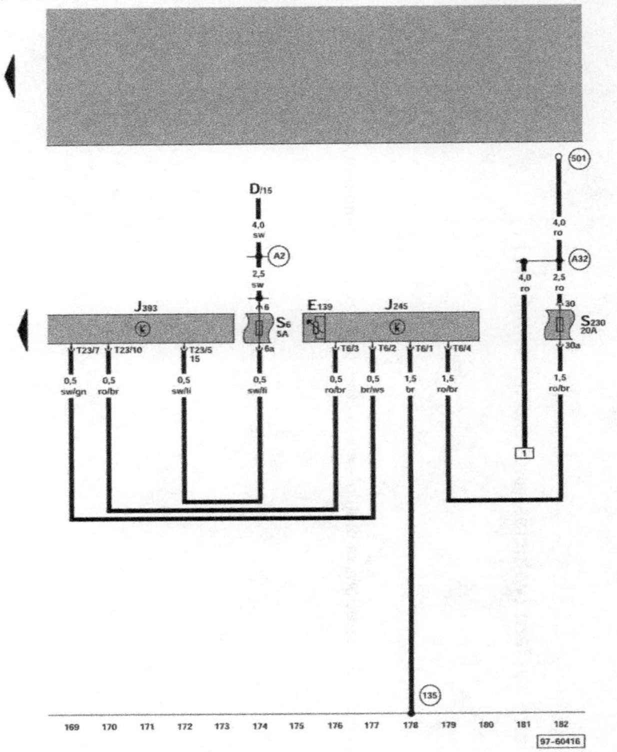

ws = white
sw = black
ro = red
br = brown
gn = green
bl = blue
gr = grey
li = violet
ge = yellow

D - Ignition/Starter Switch
E139 - Sunroof Regulator
J245 - Power Sunroof Control Module
J393 - Central control module for comfort system
S6 - Fuse 6 in fuse holder
S230 - Fuse 30 in fuse holder
T6 - 6-Pin Connector
T23 - 23-Pin Connector

(135) - Ground connection -2-, in instrument panel wiring harness
(501) - Threaded connection -2- (30), on relay panel

(A2) - plus connection (15), in instrument panel wiring harness
(A32) - plus connection (30), in instrument panel wiring harness

Central control module for comfort system, power sunroof

New Beetle Wiring diagram **No. 48/1**

Comfort system, with anti-theft alarm system (vehicles with power windows)

from June 2001

Deviate relay location and fuseplacements as well as the locations of multiple connectors see section "component locations".

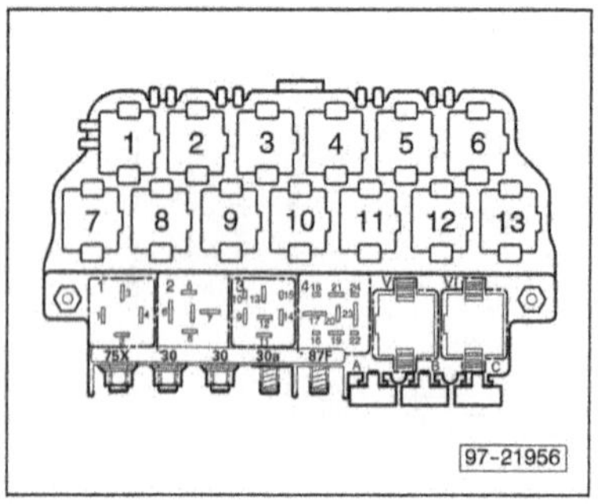

Fuse colors

30 A - green
25 A - white
20 A - yellow
15 A - blue
10 A - red
7,5 A - brown
5 A - beige
3 A - violet

Note: Number in parentheses indicates production control number stamped on relay housing.

New Beetle Wiring diagram **No. 48/2**

Door control module (driver side), left front window switch, switch for window regulator front right (driver), switch for interior lock (driver side)

E40 - Left Front Window Switch
E81 - Switch for window regulator front right, driver
E150 - Switch for interior lock, driver side
J386 - Door control module, driver side
L76 - Push Button Light
L99 - Lighting for switch interior lock
S43 - Power Window Circuit Breaker
T5d - 5-Pin Connector, brown
T5f - 5-Pin Connector, black
T5e - 5-Pin Connector, black
T10k - 10-Pin Connector, black, on A-Pillar left
T29 - 29-Pin Connector, on Door control module
V147 - Motor for window regulator, driver side

(267) - Ground connection -2-, in wiring harness door cable
(A32) - plus connection (30), in instrument panel wiring harness
(A52) - Plus Connection 2 (30) (in instrument panel wiring harness)
(B110) - Connection (30, window regulator) in wiring harness interior
(R51) - Connection (58b), in door wiring harness, driver's side

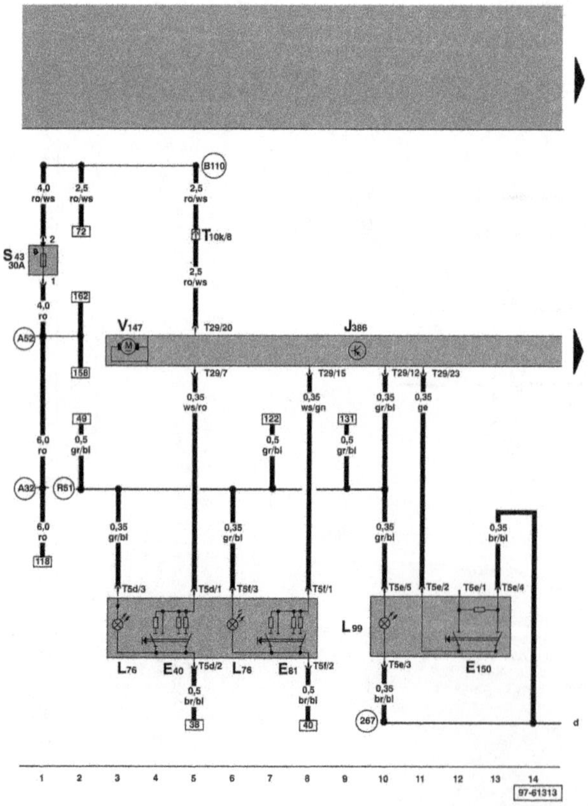

ws = white
sw = black
ro = red
br = brown
gn = green
bl = blue
gr = grey
li = violet
ge = yellow
or = orange

New Beetle — Wiring diagram — No. 48/3

Door control module (driver side), lock unit for central locking (driver side), warning light for central locking

- F220 - Lock unit for central locking, driver side
- J386 - Door control module, driver side
- K133 - Warning light for central locking -SAFE-
- T8e - 8-Pin Connector, on Lock unit for central locking, driver side
- T29 - 29-Pin Connector, on Door control module
- (304) - Ground Connection -3- (driver's door wiring harness)

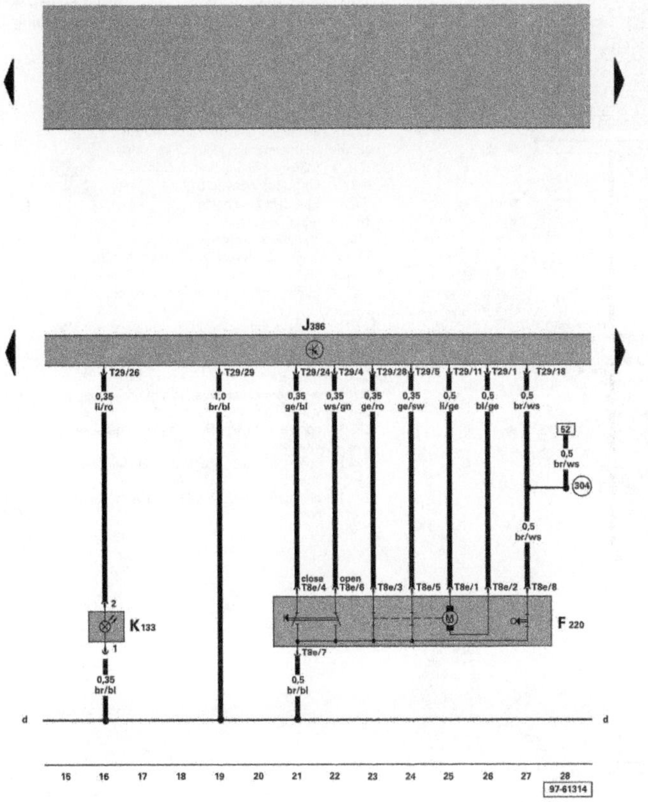

ws = white
sw = black
ro = red
br = brown
gn = green
bl = blue
gr = grey
li = violet
ge = yellow
or = orange

New Beetle — Wiring diagram — No. 48/4

Door control module (driver side), driver's side mirror adjustment motor, motor for mirror adjustment (driver side), heated outside mirror (driver side)

- J386 - Door control module, driver side
- T8 - 8-Pin Connector, in driver's door wiring
- T10k - 10-Pin Connector, on A-Pillar left, black
- T29 - 29-Pin Connector, on Door control module - driver side
- V17 - Driver's Side Mirror Adjustment Motor
- V149 - Motor for mirror adjustment, driver side
- Z4 - Heated outside mirror, driver side
- (44) - Ground connection, on left A-pillar, lower part
- (205) - Ground connection, in driver's door wiring harness

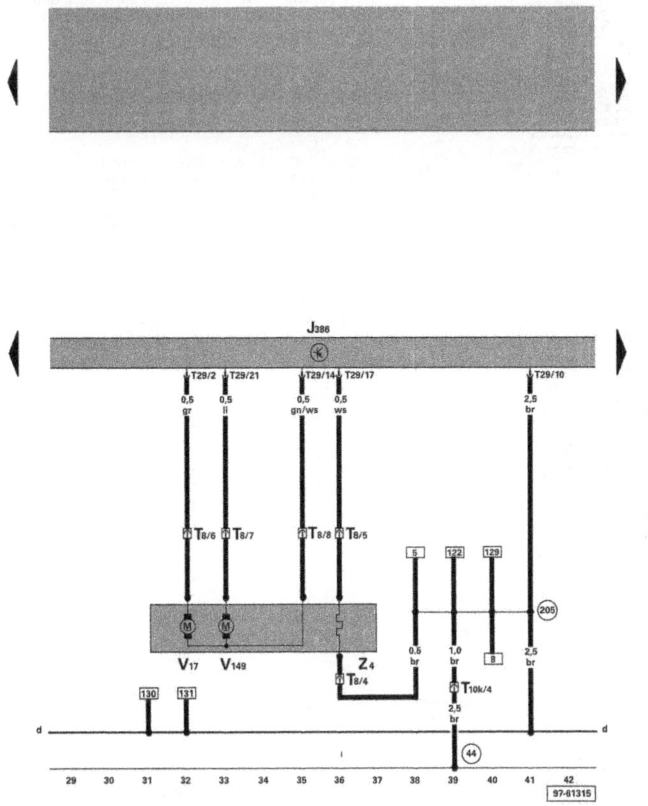

ws = white
sw = black
ro = red
br = brown
gn = green
bl = blue
gr = grey
li = violet
ge = yellow
or = orange

Comfort system and alarm from June 2001

New Beetle

Wiring diagram

No. 48/5

Door control module (driver side), mirror adjustment switch, mirror selector switch, door warning lights

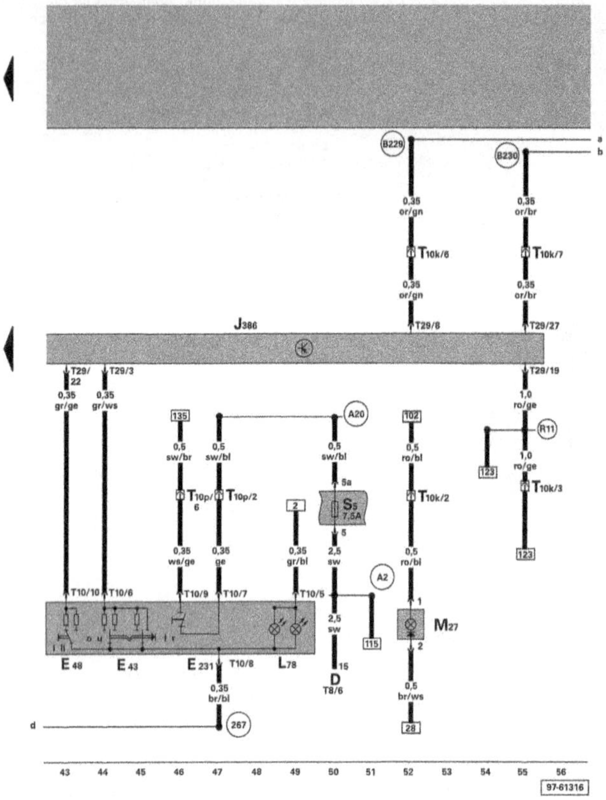

D - Ignition/Starter Switch
E43 - Mirror Adjustment Switch
E48 - Mirror Selector Switch
E231 - Switch for outside mirror heating
J386 - Door control module, driver side
L78 - Mirror Adjusting Switch Light
M27 - Left Door Warning Light
S5 - Fuse 5 in fuse holder
T8 - 8-Pin Connector
T10 - 10-Pin Connector
T10k - 10-Pin Connector, on A-Pillar left, black
T10p - 10-Pin Connector, on A-Pillar left, blue
T29 - 29-Pin Connector, on Door control module

(267) - Ground connector -2-, in wiring harness door cable - driver side
(A2) - plus connection (15), in instrument panel wiring harness
(A20) - wire connection (15a), in instrument panel wiring harness
(B229) - connection (High-Bus), in wiring harness, interior
(B230) - connection (Low-Bus), in wiring harness, interior
(R11) - plus connection -2- (30), in driver's door wiring harness

ws = white
sw = black
ro = red
br = brown
gn = green
bl = blue
gr = grey
li = violet
ge = yellow
or = orange

New Beetle

Wiring diagram

No. 48/6

Door control module (passenger side), lock unit for central locking (passenger side), switch for window regulator (passenger side)

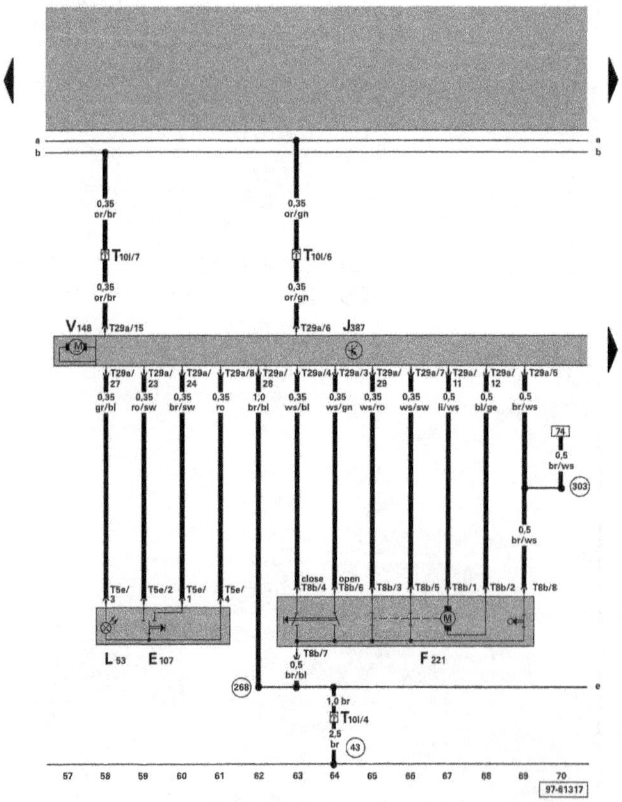

E107 - Switch for window regulator, in passenger door
F221 - Lock unit for central locking, passenger side
J387 - Door control module, passenger side
T5e - 5-Pin Connector
T8b - 8-Pin Connector
T10l - 10-Pin Connector, black, on A-Pillar right
T29a - 29-Pin Connector, on Door control module
L53 - Power Window Switch Light
V148 - Motor for window regulator, passenger side

(43) - Ground connection, on right A-pillar, lower part
(268) - Ground connector -2-, in wiring harness door cable passenger side
(303) - Ground Connection -3- (passenger's door wiring harness)

ws = white
sw = black
ro = red
br = brown
gn = green
bl = blue
gr = grey
li = violet
ge = yellow
or = orange

New Beetle

Wiring diagram

No. 48/7

Door control module (passenger side), switch for interior lock (passenger side), passenger's side mirror adjustment motor, heated outside mirror

E198 - Switch for interior lock, passenger side
J387 - Door control module, passenger side
L99 - Lighting for switch interior lock
M28 - Right Door Warning Light
T5c - 5-Pin Connector
T8a - 8-Pin Connector
T10l - 10-Pin Connector, black, on A-Pillar right
T29a - 29-Pin Connector, on Door control module
V25 - Passenger's Side Mirror Adjustment Motor
V150 - Motor for mirror adjustment, passenger side
Z5 - Heated outside mirror, passenger side

(268) - Ground connector -2-, in wiring harness door cable - passenger side

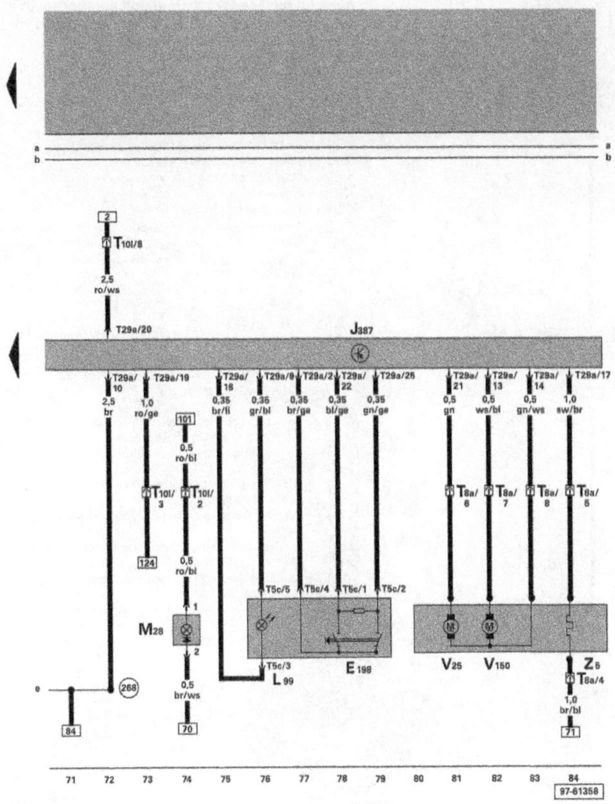

ws = white
sw = black
ro = red
br = brown
gn = green
bl = blue
gr = grey
li = violet
ge = yellow
or = orange

New Beetle

Wiring diagram

No. 48/8

Front interior light, left rear reading light, right rear reading light, right make-up mirror light, left make-up mirror light

F147 - Left Make-Up Mirror Light Switch
F148 - Right Make-Up Mirror Light Switch
W - Front Interior Light
W11 - Left Rear Reading Light
W12 - Right Rear Reading Light
W14 - Right Make-up Mirror Light
W20 - Left Make-up Mirror Light
T12 - 12-Pin Connector
T12a - 12-Pin Connector

(128) - Ground connection -1-, in interior light wiring harness
(163) - Ground connection -2-, in interior light wiring harness
(B129) - Connector (interior light 3/L), in wiring harness interior
(R6) - plus connection -1-, in interior light wiring harness
(R7) - plus connection -2-, in interior light wiring harness

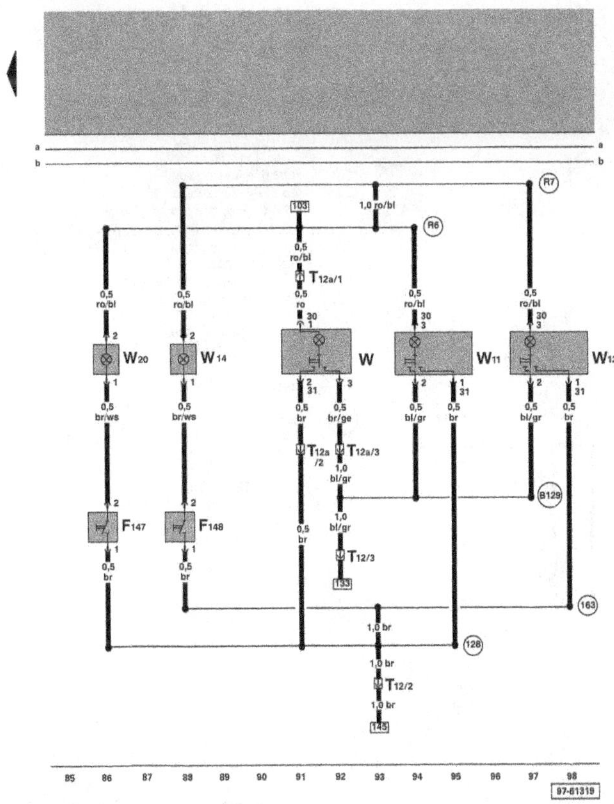

ws = white
sw = black
ro = red
br = brown
gn = green
bl = blue
gr = grey
li = violet
ge = yellow
or = orange

New Beetle — Wiring diagram — No. 48/9

Trunk lid release switch, luggage compartment light switch, central control module for comfort system, luggage compartment light

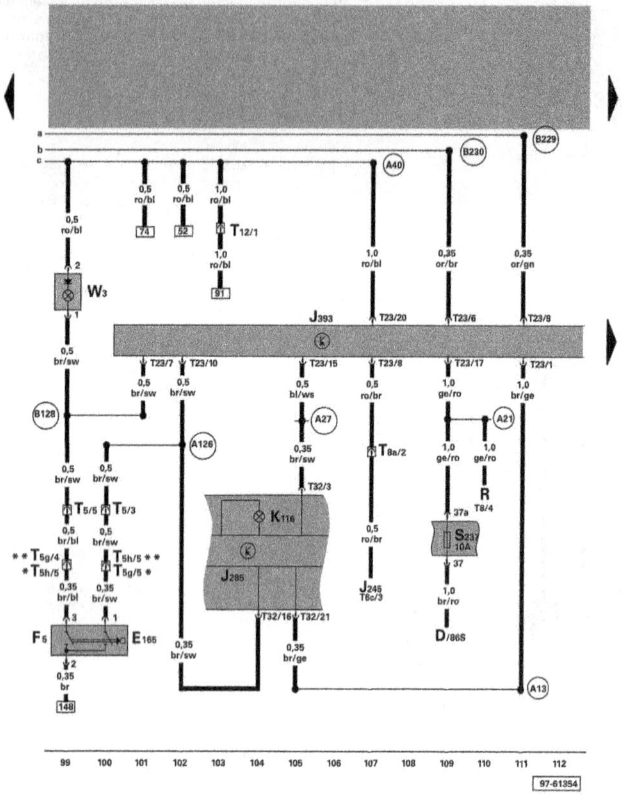

- D - Ignition/Starter Switch
- E165 - Trunk Lid Release Switch
- F5 - Luggage Compartment Light Switch
- J285 - Control module with indicator unit in instr. insert
- J393 - Central control module for comfort system
- J245 - Power Sunroof Control Module
- K116 - Warning light for rear lid unlocked
- R - Radio
- S237 - Fuse 37 in fuse holder
- T5 - 5-Pin Connector, black, on C-pillar, left
- T5g - 5-Pin Connector, black, rear lid - left, near lock cylinder
- T5h - 5-Pin Connector, brown, rear lid - left, near lock cylinder
- T6c - 6-Pin Connector
- T8 - 8-Pin Connector, on radio
- T8a - 8-Pin Connector
- T12 - 12Pin Connector, on A pillar left, central part
- T23 - 23-Pin Connector, on J393
- T32 - 32-Pin Connector, blue, on instrument cluster
- W3 - Luggage compartment Light
- (A13) - wire connection (door contact switch) in instrument panel wiring harness
- (A21) - wire connection (86s), in instrument panel wiring harness
- (A27) - Wire Connection (vehicle speed signal), in instrument panel wiring harness
- (A40) - plus connection -1- (30), in instrument panel wiring harness
- (A126) - Connection (contact switch in rear lid), in instrument panel wiring harness
- (B128) - Connector (Luggage compartment lamp) in wiring harness interior
- (B229) - High-bus Connection (in interior wiring harness)
- (B230) - Low-bus Connection (in interior wiring harness)
- * - to 09.01 (with rear lid lock cylinder)
- ** - from 09.01 (without rear lid lock cylinder)

ws = white
sw = black
ro = red
br = brown
gn = green
bl = blue
gr = grey
li = violet
ge = yellow
or = orange

New Beetle — Wiring diagram — No. 48/10

Central control module for comfort system, switch for remote/fuel tank door, illumination for switch remote unlock, motor for fuel tank lid unlock

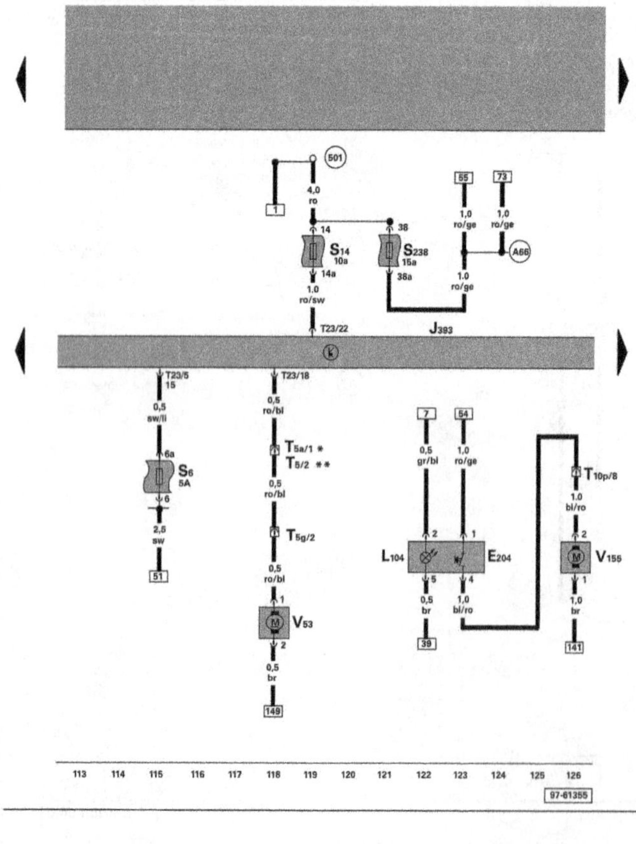

- E204 - Switch for remote/fuel tank door
- J393 - Central control module for comfort system
- L104 - Illumination for switch remote unlock
- S14 - Fuse 14 in fuse holder
- S6 - Fuse 6 in fuse holder
- S238 - Fuse 38 in fuse holder
- T5 - 5-Pin Connector, black, on C pillar, left
- T5a - 5-Pin Connector, brown, on C pillar, left
- T5g - 5-Pin Connector, black, in rear lid, left side, near lock cylinder carrier
- T10p - 10-Pin Connector, on A-Pillar left, blue
- T23 - 23-Pin Connector, on Central control module for comfort system
- V 53 - Rear Lid Central Locking System Motor
- V155 - Motor for fuel tank lid unlock
- (501) - Screw connection -2- (30), on relay panel
- (A66) - Connector (30a, central locking/anti-theft warning system/IR) in instrument panel wiring harness
- * - to 09.01 (with rear lid lock cylinder)
- ** - from 09.01 (without rear lid lock cylinder)

ws = white
sw = black
ro = red
br = brown
gn = green
bl = blue
gr = grey
li = violet
ge = yellow
or = orange

New Beetle — Wiring diagram — No. 48/11

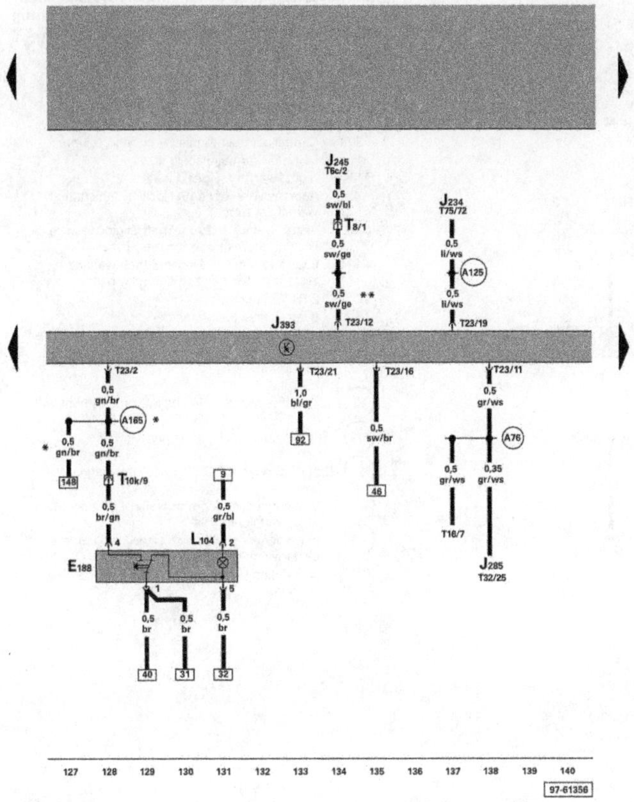

Central control module for comfort system, switch for remote unlock, rear lid, illumination for switch remote unlock

- E188 - Switch for remote unlock, rear lid
- J234 - Airbag Control Module
- J245 - Power Sunroof Control Module
- J285 - Control module with indicator unit in instrument panel insert
- J393 - Central control module for comfort system
- L104 - Illumination for switch remote unlock
- T6c - 6-Pin Connector
- T8 - 8-Pin Connector
- T10k - 10-Pin Connector, on A-Pillar left, black
- T16 - 16-Pin Connector, Data Link Connector (DLC), below steering coulm, left
- T23 23-Pin Connector, on Central control module for comfort system
- T32 - 32-Pin Connector, blue, on instrument cluster
- T75 - 75-Pin Connector
- A76 - Connector (K-diagnosis wire) in instrument panel wiring harness
- A125 - wire connection (crash), in instrument panel wiring harness
- A165 - Rear Lid Contact Switch Connection (in instrument panel wiring harness)

* - to 09.01 (with rear lid lock cylinder)
** - only vehicles with power sunroof

ws = white
sw = black
ro = red
br = brown
gn = green
bl = blue
gr = grey
li = violet
ge = yellow
or = orange

New Beetle — Wiring diagram — No. 48/12

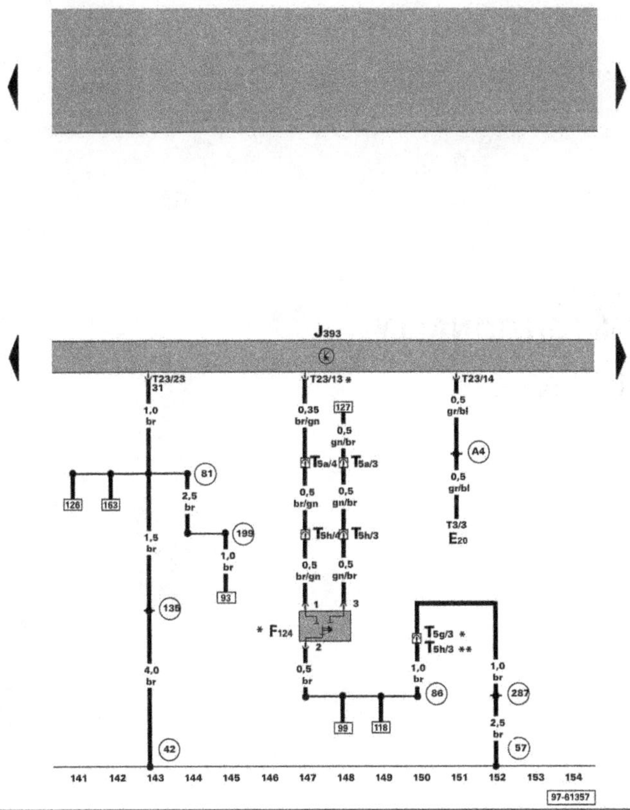

Central control module for comfort system, trunk lock alarm/central locking switch

- E20 - Instrument Panel Light Dimmer Switch
- F124 - Trunk Lock Alarm/Central Locking Switch *
- J393 - Central control module for comfort system
- T3 - 3-Pin Connector
- T5a - 5-Pin Connector, brown, on C pillar, left
- T5g - 5-Pin Connector, black, in rear lid, left side, near lock cylinder carrier
- T5h - 5-Pin Connector, brown, in rear lid, left side, near lock cylinder carrier
- T23 - 23-Pin Connector, on Central control module for comfort system
- 42 - Ground connection, beside steering column
- 57 - Ground connection, on left rear pillar
- 81 - Ground connection -1-, in instrument panel wiring harness
- 86 - Ground connection -1-, in rear wiring harness
- 135 - Ground connection -2-, in instrument panel wiring harness
- 199 - Ground connection -3-, in instrument panel wiring harness
- 287 - Ground connection, in wiring harness rear hatch hydraulics
- A4 - plus connection (58b), in instrument panel wiring harness

* - to 09.01 (with rear lid lock cylinder)
** - from 09.01 (without rear lid lock cylinder)

ws = white
sw = black
ro = red
br = brown
gn = green
bl = blue
gr = grey
li = violet
ge = yellow
or = orange

New Beetle Wiring diagram **No. 48/13**

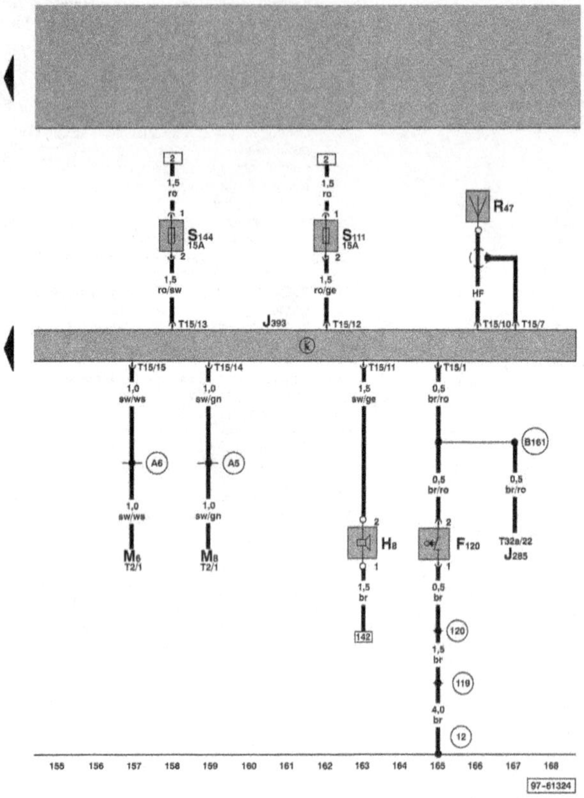

Central control module for comfort system, hood alarm switch, alarm horn, antenna wire for central locking and antitheft warning system

F120 - Hood Alarm Switch
H8 - Alarm Horn
J285 - Control module with indicator unit in instrument panel insert
J393 - Central control module for comfort system
M6 - Left Rear Turn Signal Light
M8- - Right Rear Turn Signal Light
R47 - Antenna wire for central locking and antitheft warning system
S111 - Alarm System and Anti-Theft Engine Disable Fuse on thirteen position relay panel
S144 - Fuse for central locking/anti-theft warning system on thirteen position relay panel
T2 - 2-Pin Connector
T8 - 8-Pin Connector, on radio
T32a - 32-Pin Connector, green , on instrument closter
T15 - 15-Pin Connector, on Central control module for comfort system

(12) - Ground connection, in engine compartment, left
(119) - Ground connection -1-, in headlight wiring harness
(120) - Ground connection -2-, in headlight wiring harness
(A5) - plus connection (right turn signal), in instrument panel wiring harness
(A6) - plus connection (left turn signal), in instrument panel wiring harness
(B161) - connection (Alarm system), in wiring harness, interior

ws = white
sw = black
ro = red
br = brown
gn = green
bl = blue
gr = grey
li = violet
ge = yellow
or = orange

THIS SPACE INTENTIONALLY LEFT BLANK

New Beetle Convertible — Wiring diagram — No. 7/1

Comfort system, with anti-theft alarm system (vehicles with power windows)

from August 2002

Deviate relay location and fuseplacements as well as the locations of multiple connectors see section "component locations".

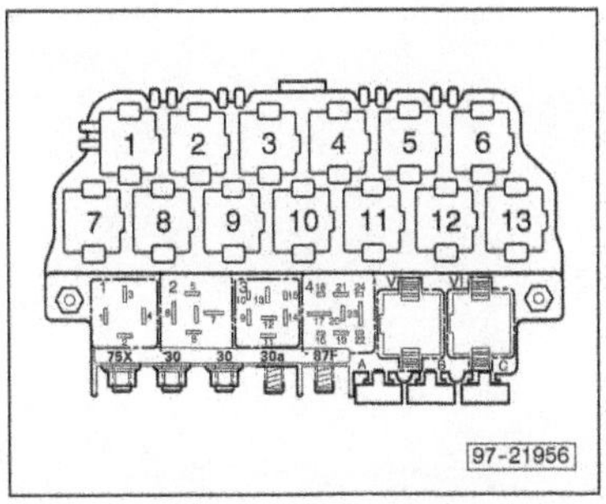

Fuse colors

30 A - green
25 A - white
20 A - yellow
15 A - blue
10 A - red
7,5 A - brown
5 A - beige
3 A - violet

New Beetle Convertible — Wiring diagram — No. 7/2

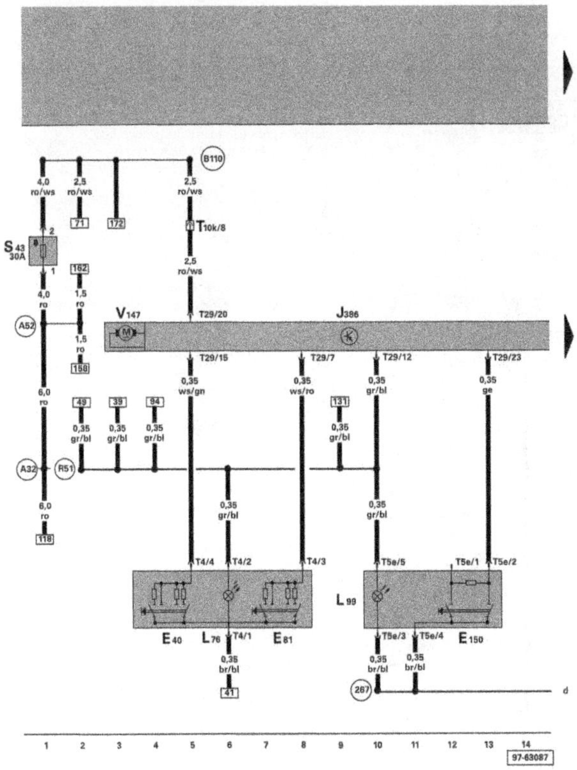

Door control module (driver side), left front window switch, switch for window regulator front right (driver), switch for interior lock (driver side)

- E40 - Left Front Window Switch
- E81 - Switch for window regulator front right, driver
- E150 - Switch for interior lock, driver side
- J386 - Door control module, driver side
- L76 - Push Button Light
- L99 - Lighting for switch interior lock
- S43 - Power Window Circuit Breaker
- T4 - 4-Pin Connector, at -E40-
- T5e - 5-Pin Connector, black
- T10k - 10-Pin Connector, black, on A-Pillar left
- T29 - 29-Pin Connector, on Door control module
- V147 - Motor for window regulator, driver side

- (267) - Ground connection -2-, in wiring harness door cable
- (A32) - plus connection (30), in instrument panel wiring harness
- (A52) - Plus Connection 2 (30) (in instrument panel wiring harness)
- (B110) - Connection (30, window regulator) in wiring harness interior
- (R51) - Connection (58b), in door wiring harness, driver's side

ws = white
sw = black
ro = red
br = brown
gn = green
bl = blue
gr = grey
li = violet
ge = yellow
or = orange

Comfort system and alarm from Aug. 2002, convertible

New Beetle Convertible — Wiring diagram — No. 7/3

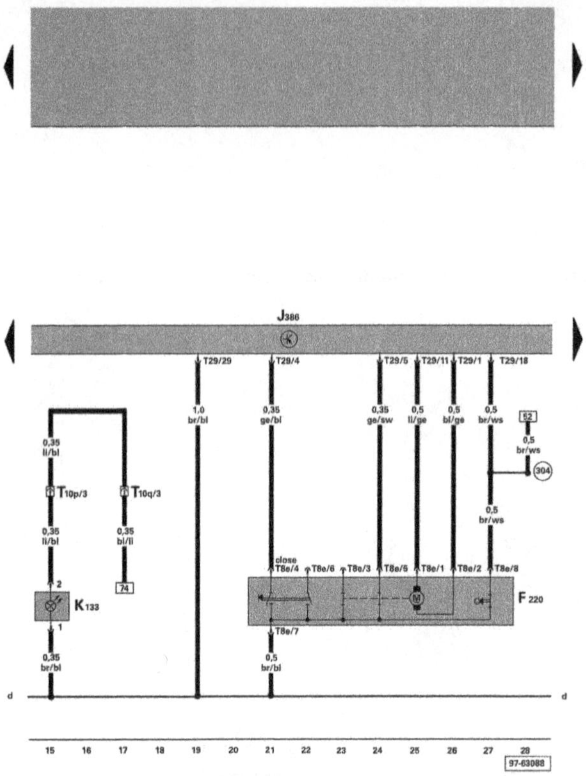

Door control module (driver side), lock unit for central locking (driver side), warning light for central locking

- F220 - Lock unit for central locking, driver side
- J386 - Door control module, driver side
- K133 - Warning light for central locking -SAFE-
- T8e - 8-Pin Connector, on Lock unit for central locking, driver side
- T10p - 10-Pin Connector, on A-Pillar left, blue
- T10q - 10-Pin Connector, on A-Pillar right, blue
- T29 - 29-Pin Connector, on Door control module
- (304) - Ground Connection -3- (driver's door wiring harness)

ws = white
sw = black
ro = red
br = brown
gn = green
bl = blue
gr = grey
li = violet
ge = yellow
or = orange

New Beetle Convertible — Wiring diagram — No. 7/4

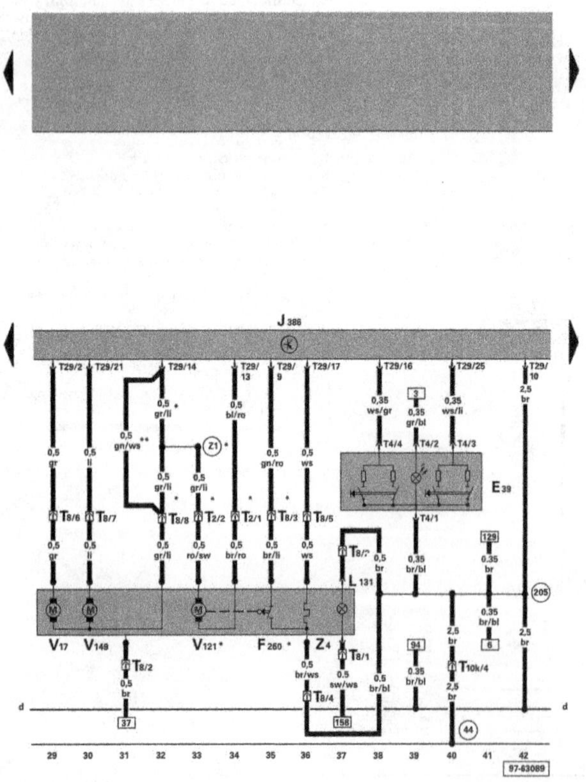

Safety switch for window regulator rear, door control module (driver side), driver's side mirror adjustment motor, motor for mirror adjustment (driver side), heated outside mirror (driver side)

- E39 - Safety switch for window regulator rear
- F260 - Driver's Folding Mirror Contact Switch
- J386 - Door control module, driver side
- L131 - Left Mirror Turn Signal Lamp
- T2 - 2-Pin Coupling Connector for -V121-
- T4 - 4-Pin Connector, at -E39-
- T8 - 8-Pin Connector, in driver's door wiring
- T10k - 10-Pin Connector, on A-Pillar left, black
- T29 - 29-Pin Connector, on Door control module - driver side
- V17 - Driver's Side Mirror Adjustment Motor
- V121 - Motor for fold-in mirror, driver side
- V149 - Motor for mirror adjustment, driver side
- Z4 - Heated outside mirror, driver side
- (44) - Ground connection, on left A-pillar, lower part
- (205) - Ground connection, in driver's door wiring harness
- (Z1) - Wire connection -1-, in power heated mirror wiring harness
- * - from m.y. 2004
- ** - Only vehicles without motor for fold in mirror

ws = white
sw = black
ro = red
br = brown
gn = green
bl = blue
gr = grey
li = violet
ge = yellow
or = orange

New Beetle Convertible — Wiring diagram — No. 7/5

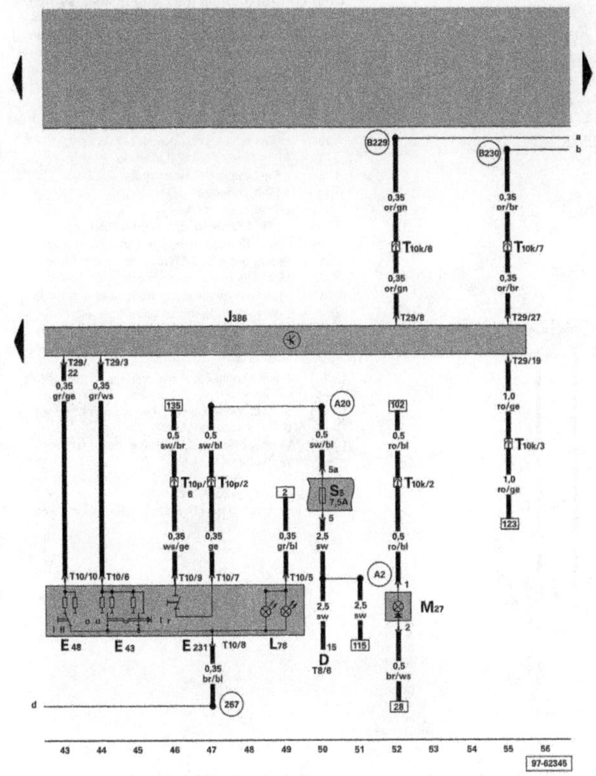

Door control module (driver side), mirror adjustment switch, mirror selector switch, door war ning lights, switch for outside mirror heating

- D - Ignition/Starter Switch
- E43 - Mirror Adjustment Switch
- E48 - Mirror Selector Switch
- E231 - Switch for outside mirror heating
- J386 - Door control module, driver side
- L78 - Mirror Adjusting Switch Light
- M27 - Left Door Warning Light
- S5 - Fuse 5 in fuse holder
- T8 - 8-Pin Connector
- T10 - 10-Pin Connector
- T10k - 10-Pin Connector, on A-Pillar left, black
- T10p - 10-Pin Connector, on A-Pillar left, blue
- T29 - 29-Pin Connector, on Door control module

- (267) - Ground connector -2-, in wiring harness door cable - driver side
- (A2) - plus connection (15), in instrument panel wiring harness
- (A20) - wire connection (15a), in instrument panel wiring harness
- (B229) - connection (High-Bus), in wiring harness, interior
- (B230) - connection (Low-Bus), in wiring harness, interior

ws = white
sw = black
ro = red
br = brown
gn = green
bl = blue
gr = grey
li = violet
ge = yellow
or = orange

Edition 11.03

New Beetle Convertible — Wiring diagram — No. 7/6

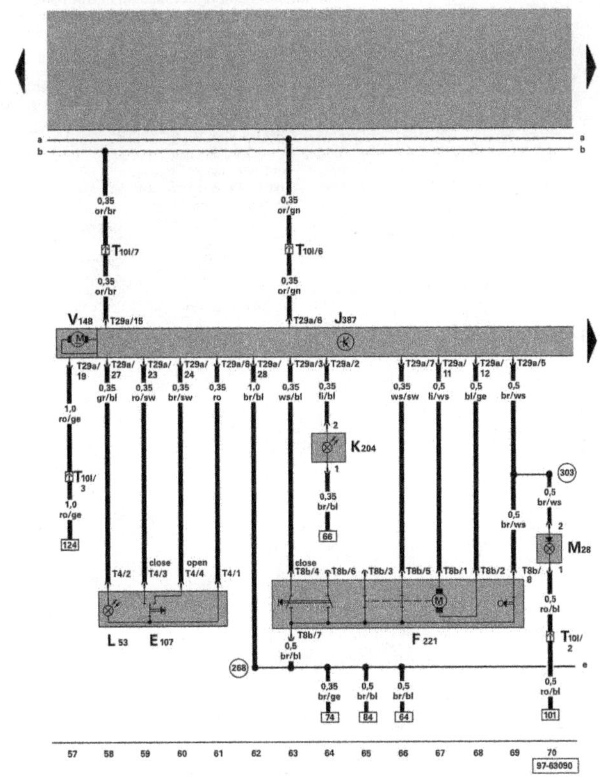

Door control module (passenger side), lock unit for central locking (passenger side), switch for window regulator (passenger side), power window switch light, motor for window regulator, passenger side, front Passenger's Central Locking -SAFE- Indicator Lamp

- E107 - Switch for window regulator, in passenger door
- F221 - Lock unit for central locking, passenger side
- J387 - Door control module, passenger side
- K204 - Front Passenger's Central Locking -SAFE- Indicator Lamp
- L53 - Power Window Switch Light
- M28 - Right Door Warning Light
- T4 - 4-Pin Connector, at -E107-
- T8b - 8-Pin Connector
- T10l - 10-Pin Connector, black, on A-Pillar right
- T29a - 29-Pin Connector, on Door control module
- V148 - Motor for window regulator, passenger side

- (268) - Ground connector -2-, in wiring harness door cable passenger side
- (303) - Ground Connection -3- (passenger's door wiring harness)

ws = white
sw = black
ro = red
br = brown
gn = green
bl = blue
gr = grey
li = violet
ge = yellow
or = orange

Edition 11.03

Comfort system and alarm from Aug. 2002, convertible

New Beetle Convertible — Wiring diagram — No. 7/7

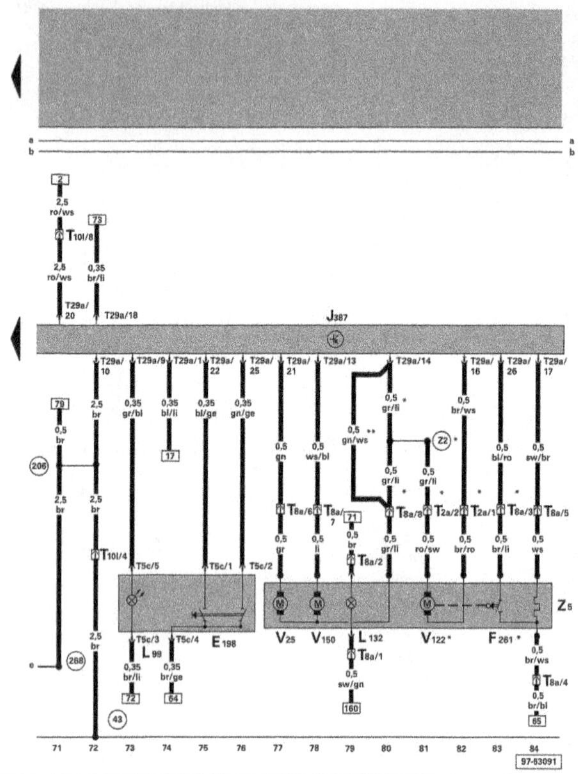

Door control module (passenger side), switch for interior lock (passenger side), passenger's side mirror adjustment motor, heated outside mirror, motor for mirror adjustment, passenger side

- E198 - Switch for interior lock, passenger side
- F261 - Passenger's Folding Mirror Contact Switch
- J387 - Door control module, passenger side
- L99 - Lighting for switch interior lock
- L132 - Right Mirror Turn Signal Lamp
- T2a - 2-Pin Coupling Connector for -V122-
- T5c - 5-Pin Connector
- T8a - 8-Pin Connector
- T10l - 10-Pin Connector, black, on A-Pillar right
- T29a - 29-Pin Connector, on Door control module
- V25 - Passenger's Side Mirror Adjustment Motor
- V122 - Motor for fold-in mirror, passenger side
- V150 - Motor for mirror adjustment, passenger side
- Z5 - Heated outside mirror, passenger side

- (43) - Ground connection, on right A-pillar, lower part
- (206) - Ground connection, in passenger's door wiring harness
- (268) - Ground connector -2-, in wiring harness door cable - passenger side
- (Z2) - Wire connection -2-, in power heated mirror wiring harness

- * - From m.y. 2004
- ** - Only vehicles without motor for fold in mirror

ws = white
sw = black
ro = red
br = brown
gn = green
bl = blue
gr = grey
li = violet
ge = yellow
or = orange

Edition 11.03

New Beetle Convertible — Wiring diagram — No. 7/8

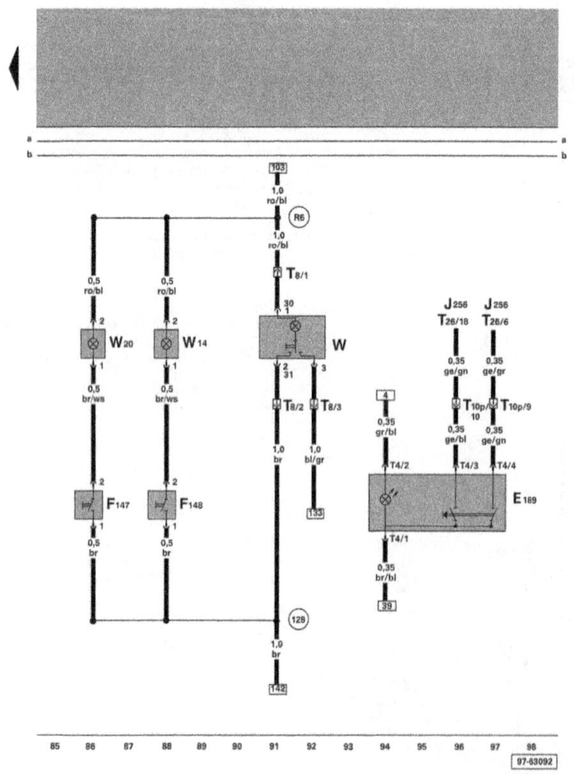

Front interior light, right make-up mirror light, left make-up mirror light, central switch for window regulator, in driver's door, right make-up mirror light switch

- E189 - Central switch for window regulator, in driver's door
- F147 - Left Make-Up Mirror Light Switch
- F148 - Right Make-Up Mirror Light Switch
- J256 - Convertible Top Control Module
- W - Front Interior Light
- W14 - Right Make-up Mirror Light
- W20 - Left Make-up Mirror Light
- T4 - 4-Pin Connector, at -E189-
- T8 - 8-Pin Connector
- T10p - 10-Pin Connector, on A-Pillar left, blue
- T26 - 26-Pin Connector, at -J256-

- (128) - Ground connection -1-, in interior light wiring harness
- (R6) - plus connection -1-, in interior light wiring harness

ws = white
sw = black
ro = red
br = brown
gn = green
bl = blue
gr = grey
li = violet
ge = yellow
or = orange

Edition 11.03

New Beetle Convertible

Wiring diagram **No. 7/9**

Trunk lid release switch, luggage compartment light switch, central control module for comfort system, luggage compartment light

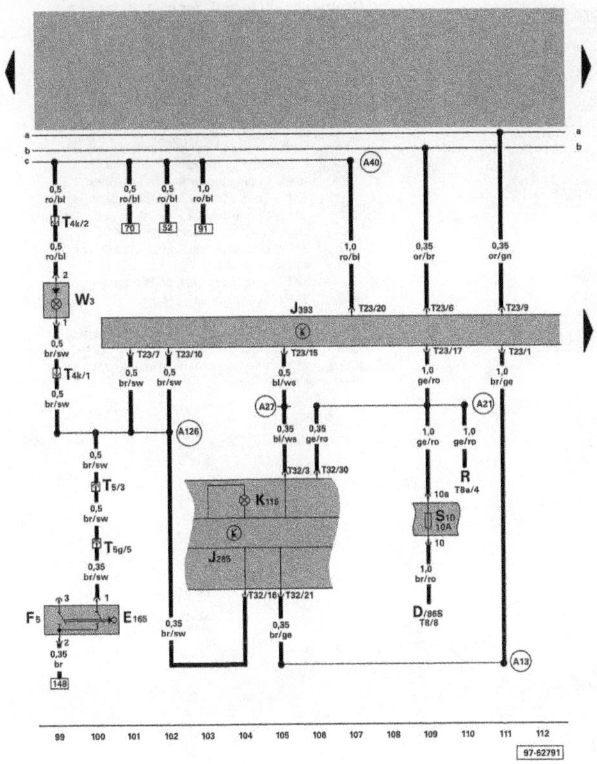

- D - Ignition/Starter Switch
- E165 - Trunk Lid Release Switch
- F5 - Luggage Compartment Light Switch
- J285 - Control module with indicator unit in instr. insert
- J393 - Central control module for comfort system
- K116 - Warning light for rear lid unlocked
- R - Radio
- S10 - Fuse 10 in fuse holder
- T4k - 4-Pin connector, in luggage compartment, right-side
- T5 - 5-Pin Connector, black, on C-pillar, left
- T5g - 5-Pin Connector, black, rear lid - left, near lock cylinder
- T8 - 8-Pin Connector, at - D -
- T8a - 8-Pin Connector, on radio
- T23 - 23-Pin Connector, on J393
- T32 - 32-Pin Connector, blue, on instrument cluster
- W3 - Luggage compartment Light

- (A13) - wire connection (door contact switch) in instrument panel wiring harness
- (A21) - wire connection (86s), in instrument panel wiring harness
- (A27) - Wire Connection (vehicle speed signal), in instrument panel wiring harness
- (A40) - plus connection -1- (30), in instrument panel wiring harness
- (A126) - Connection (contact switch in rear lid), in instrument panel wiring harness

ws = white
sw = black
ro = red
br = brown
gn = green
bl = blue
gr = grey
li = violet
ge = yellow
or = orange

New Beetle Convertible

Wiring diagram **No. 7/10**

Comfort System Central Control Module, Fuel Tank Lid Unlock Motor, Fuel Filler Lid Unlock Relay, Rear Lid Central Locking System Motor

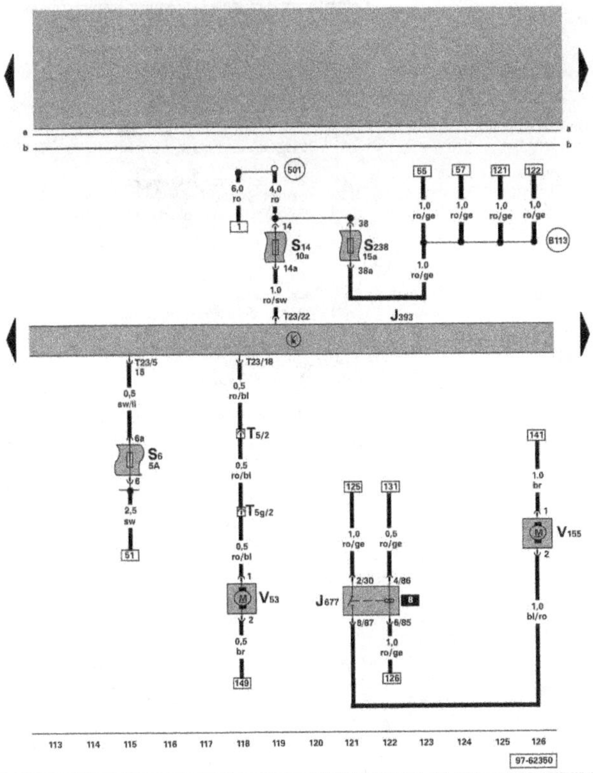

- J393 - Comfort System Central Control Module
- J677 - Fuel Filler Lid Unlock Relay (53)
- S14 - Fuse 14 in fuse holder
- S6 - Fuse 6 in fuse holder
- S238 - Fuse 38 in fuse holder
- T5 - 5-Pin Connector, black, on C pillar, left
- T5g - 5-Pin Connector, black, in rear lid, left side, near lock cylinder carrier
- T23 - 23-Pin Connector, on Central control module for comfort system
- V 53 - Rear Lid Central Locking System Motor
- V155 - Fuel Tank Lid Unlock Motor

- (501) - Screw connection -2- (30), on relay panel
- (B113) - Plus connector (30a), in wiring harness interior, left

ws = white
sw = black
ro = red
br = brown
gn = green
bl = blue
gr = grey
li = violet
ge = yellow
or = orange

Comfort system and alarm from Aug. 2002, convertible

New Beetle Convertible

Wiring diagram **No. 7/11**

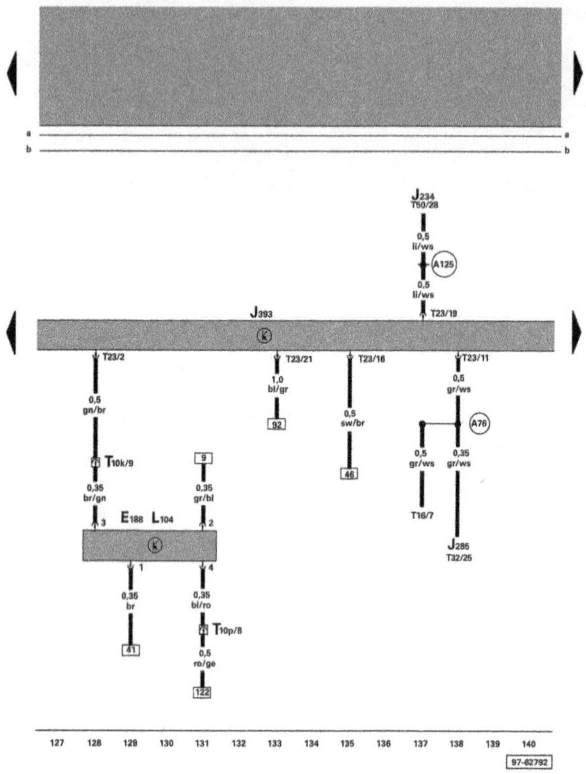

Central control module for comfort system, switch for remote unlock, rear lid, illumination for switch remote unlock

- E188 - Switch for remote unlock, rear lid
- J234 - Airbag Control Module
- J285 - Control module with indicator unit in instrument panel insert
- J393 - Central control module for comfort system
- L104 - Illumination for switch remote unlock
- T10k - 10-Pin Connector, on A-Pillar left, black
- T10p - 10-Pin Connector, on A-Pillar left, blue
- T16 - 16-Pin Connector, Data Link Connector (DLC), black, under instrument panel, left
- T23 - 23-Pin Connector, on Central control module for comfort system
- T32 - 32-Pin Connector, blue, on instrument cluster
- T50 - 50-Pin Connector, at J234
- (A76) - Connector (K-diagnosis wire) in instrument panel wiring harness
- (A125) - wire connection (crash), in instrument panel wiring harness

ws = white
sw = black
ro = red
br = brown
gn = green
bl = blue
gr = grey
li = violet
ge = yellow
or = orange

New Beetle Convertible

Wiring diagram **No. 7/12**

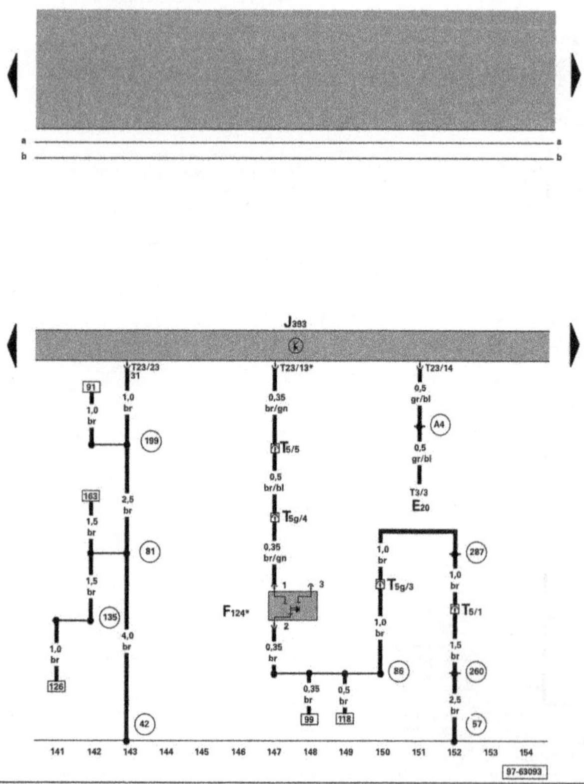

Central control module for comfort system, trunk lock alarm/central locking switch

- E20 - Instrument Panel Light Dimmer Switch
- F124 - Trunk Lock Alarm/Central Locking Switch*
- J393 - Central control module for comfort system
- T3 - 3-Pin Connector
- T5 - 5-Pin Connector, black, on C pillar, left
- T5g - 5-Pin Connector, black, in rear lid, left side, near lock cylinder carrier
- T23 - 23-Pin Connector, on Central control module for comfort system
- (42) - Ground connection, beside steering column
- (57) - Ground connection, on left rear pillar
- (81) - Ground connection -1-, in instrument panel wiring harness
- (86) - Ground connection -1-, in rear wiring harness
- (135) - Ground connection -2-, in instrument panel wiring harness
- (199) - Ground connection -3-, in instrument panel wiring harness
- (260) - Ground Connection, in wiring harness taillights
- (287) - Ground connection, in wiring harness rear hatch hydraulics
- (A4) - plus connection (58b), in instrument panel wiring harness
- * - Where applicable through December 2003 production

ws = white
sw = black
ro = red
br = brown
gn = green
bl = blue
gr = grey
li = violet
ge = yellow
or = orange

New Beetle Convertible

Wiring diagram **No. 7/13**

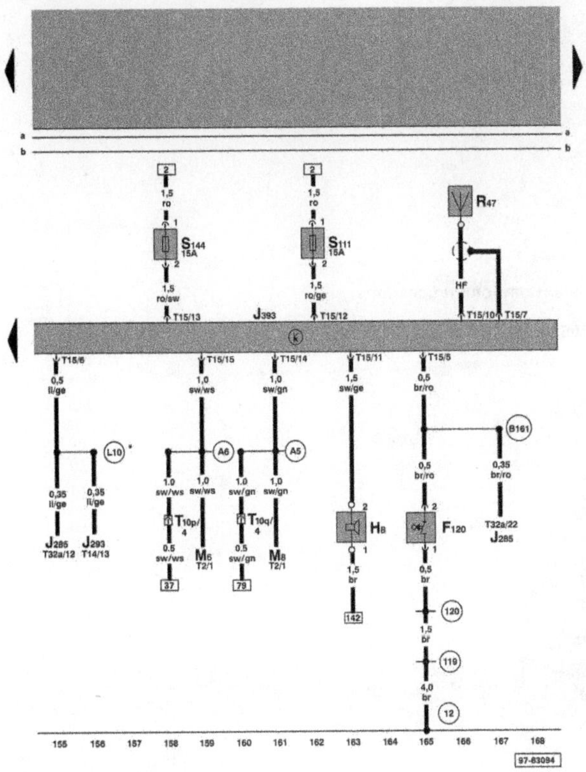

Central control module for comfort system, hood alarm switch, alarm horn, antenna wire for cen tral locking and antitheft warning system

F120 - Hood Alarm Switch
H8 - Alarm Horn
J285 - Control module with indicator unit in instrument panel insert
J293 - Coolant Fan Control (FC)Control Module
J393 - Central control module for comfort system
M6 - Left Rear Turn Signal Light
M8- - Right Rear Turn Signal Light
R47 - Antenna wire for central locking and antitheft warning system
S111 - Alarm System and Anti-Theft Engine Disable Fuse on thirteen position relay panel
S144 - Fuse for central locking/anti-theft warning system on thirteen position relay panel
T2 - 2-Pin Connector
T10p - 10-Pin Connector, on A-Pillar left, blue
T10q - 10-Pin Connector, on A-Pillar right, blue
T14 - 14-Pin Connector, on -J293-
T15 - 15-Pin Connector, on Central control module for comfort system
T32a - 32-Pin Connector, green , on instrument closter

(12) - Ground connection, in engine compartment, left
(119) - Ground connection -1-, in headlight wiring harness
(120) - Ground connection -2-, in headlight wiring harness
(A5) - plus connection (right turn signal), in instrument panel wiring harness
(A6) - plus connection (left turn signal), in instrument panel wiring harness
(B161) - connection (Alarm system), in wiring harness, interior
(L10) - wire connection -2-, in A/C wiring harness *

ws = white
sw = black
ro = red
br = brown
gn = green
bl = blue
gr = grey
li = violet
ge = yellow
or = orange

* - Only vehicles with A/C

New Beetle Convertible

Wiring diagram **No. 7/14**

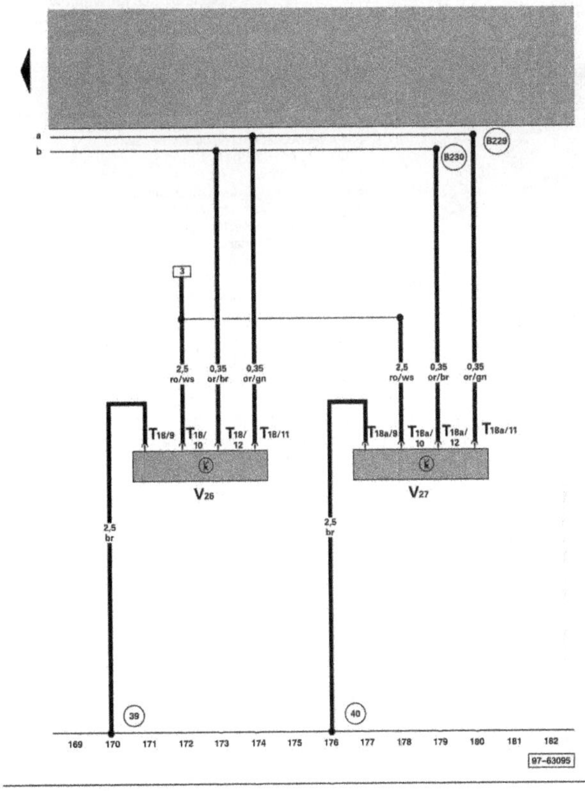

Motor for window regulator, rear, left and right

T18 - 18-Pin Connector, at -V26-
T18a - 18-Pin Connector, at - V27-
V26 - Motor for window regulator, rear, left
V27 - Motor for window regulator, rear, right

(39) - Ground connection, below rear seat bench, left
(40) - Ground connection, below rear seat bench, right
(B229) - Connection (High-Bus) (interior wiring harness)
(B230) - Connection (Low-Bus) (interior wiring harness)

ws = white
sw = black
ro = red
br = brown
gn = green
bl = blue
gr = grey
li = violet
ge = yellow
or = orange

New Beetle Wiring diagram **No. 51/1**

Comfort system, with anti-theft alarm system,

from July 2005,

Additional Fuse, Relay, Control Module and Connector locations ⇒ Component Locations

Electrical troubleshooting ⇒ Guided Fault Finding using VAS 5051/5052

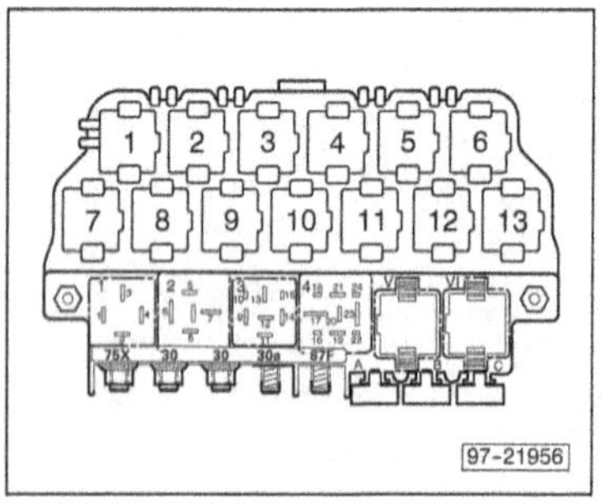

New Beetle Wiring diagram **No. 51/2**

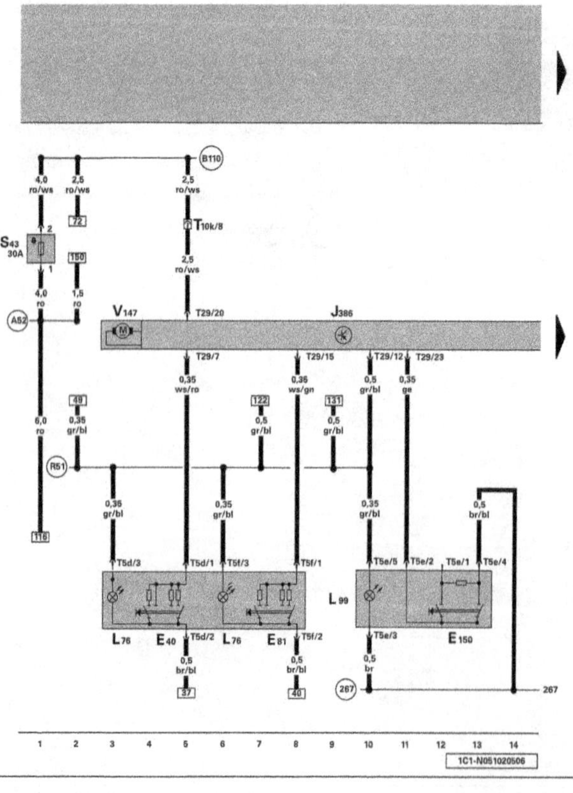

Left Front Window Switch, Switch for window regulator front right, driver, Switch for interior lock, driver side, Door control module, driver side, Motor for window regulator, driver side

- E40 - Left Front Window Switch
- E81 - Switch for window regulator front right, driver
- E150 - Switch for interior lock, driver side
- J386 - Door control module, driver side
- L76 - Push Button Light
- L99 - Lighting for switch interior lock
- S43 - Power Window Circuit Breaker
- T5d - 5-Pin Connector, brown
- T5e - 5-Pin Connector, black
- T5f - 5-Pin Connector, black
- T10k - 10-Pin Connector, black, on A-Pillar left
- T29 - 29-Pin Connector, on Door control module
- V147 - Motor for window regulator, driver side

- (267) - Ground connection -2-, in wiring harness door cable
- (A52) - Plus Connection 2 (30) (in instrument panel wiring harness)
- (B110) - Connection (30, window regulator) in wiring harness interior
- (R51) - Connection (58b), in door wiring harness, driver's side

ws = white
sw = black
ro = red
br = brown
gn = green
bl = blue
gr = grey
li = violet
ge = yellow
or = orange

New Beetle

Wiring diagram

No. 51/3

Lock unit for central locking, driver side, Door control module, driver side, Warning light for central locking -SAFE-

- F220 - Lock unit for central locking, driver side
- J386 - Door control module, driver side
- K133 - Warning light for central locking -SAFE-
- T8p - 8-Pin Connector, on Lock unit for central locking, driver side
- T29 - 29-Pin Connector, on Door control module
- (304) - Ground Connection -3- (driver's door wiring harness)

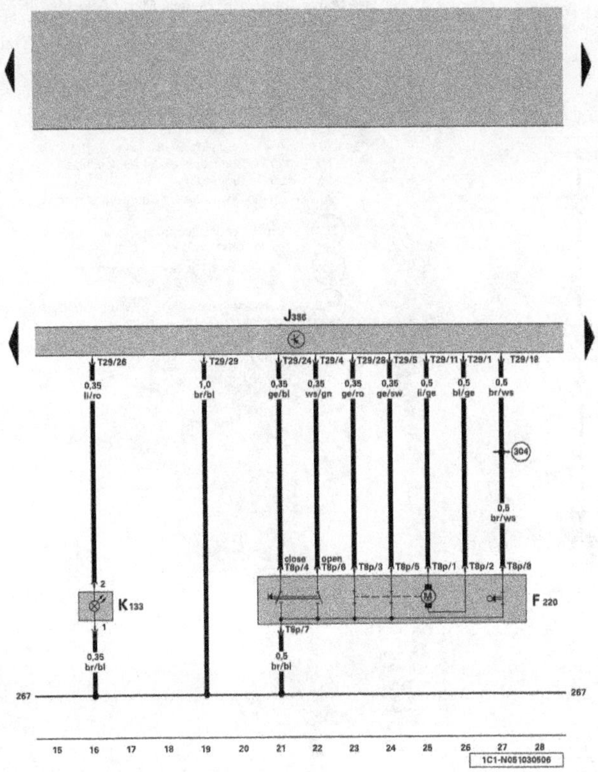

ws = white
sw = black
ro = red
br = brown
gn = green
bl = blue
gr = grey
li = violet
ge = yellow
or = orange

Edition 05.06

New Beetle

Wiring diagram

No. 51/4

Door control module (driver side), driver's side mirror adjustment motor, motor for mirror adjustment (driver side), heated outside mirror (driver side)

- J386 - Door control module, driver side
- T8l - 8-Pin Connector, in driver's door wiring
- T10k - 10-Pin Connector, on A-Pillar left, black
- T10p - 10-Pin Connector, on A-Pillar left, blue
- T29 - 29-Pin Connector, on Door control module - driver side
- V17 - Driver's Side Mirror Adjustment Motor
- V149 - Motor for mirror adjustment, driver side
- Z4 - Heated outside mirror, driver side
- (44) - Ground connection, on left A-pillar, lower part
- (205) - Ground connection, in driver's door wiring harness

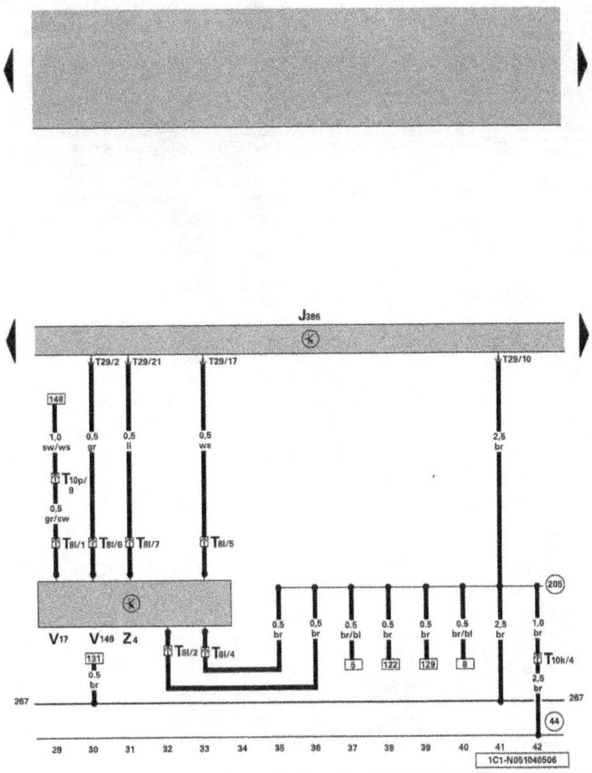

ws = white
sw = black
ro = red
br = brown
gn = green
bl = blue
gr = grey
li = violet
ge = yellow
or = orange

Edition 05.06

New Beetle — Wiring diagram — No. 51/5

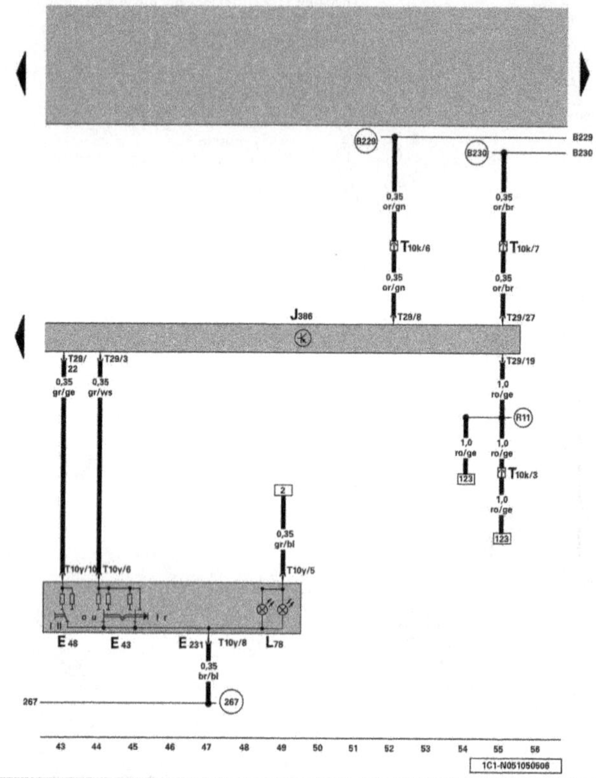

Mirror Adjustment Switch, Mirror Selector Switch, Switch for outside mirror heating, Door control module, driver side

- E43 - Mirror Adjustment Switch
- E48 - Mirror Selector Switch
- E231 - Switch for outside mirror heating
- J386 - Door control module, driver side
- L78 - Mirror Adjusting Switch Light
- T10k - 10-Pin Connector, on A-Pillar left, black
- T10y - 10-Pin Connector
- T29 - 29-Pin Connector, on Door control module
- (267) - Ground connector -2-, in wiring harness door cable - driver side
- (B229) - Connection (High-Bus), in wiring harness, interior
- (B230) - Connection (Low-Bus), in wiring harness, interior
- (R11) - Plus connection -2- (30), in driver's door wiring harness

ws = white
sw = black
ro = red
br = brown
gn = green
bl = blue
gr = grey
li = violet
ge = yellow
or = orange

New Beetle — Wiring diagram — No. 51/6

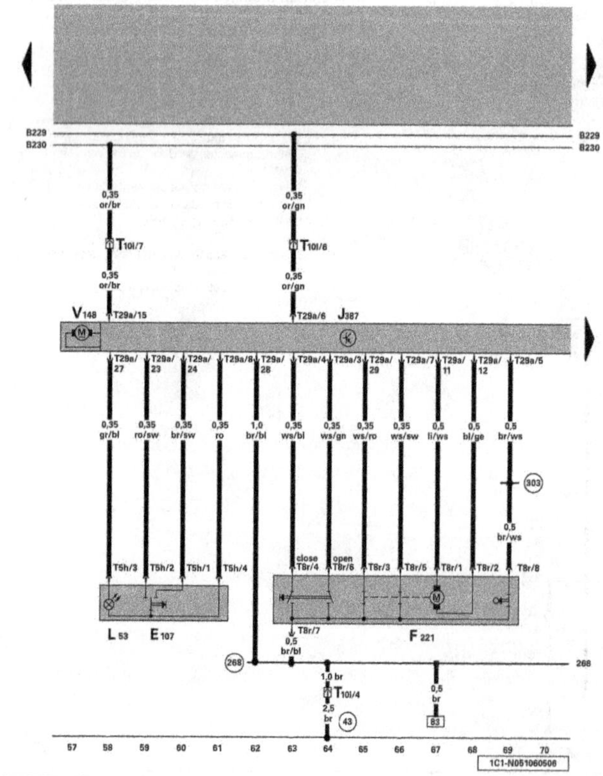

Switch for window regulator, in passenger door, Lock unit for central locking, passenger side, Door control module, passenger side, Motor for window regulator, passenger side

- E107 - Switch for window regulator, in passenger door
- F221 - Lock unit for central locking, passenger side
- J387 - Door control module, passenger side
- T5h - 5-Pin Connector
- T8r - 8-Pin Connector
- T10l - 10-Pin Connector, black, on A-Pillar right
- T29a - 29-Pin Connector, on Door control module
- L53 - Power Window Switch Light
- V148 - Motor for window regulator, passenger side
- (43) - Ground connection, on right A-pillar, lower part
- (268) - Ground connector -2-, in wiring harness door cable passenger side
- (303) - Ground Connection -3- (passenger's door wiring harness)

ws = white
sw = black
ro = red
br = brown
gn = green
bl = blue
gr = grey
li = violet
ge = yellow
or = orange

New Beetle — Wiring diagram — No. 51/7

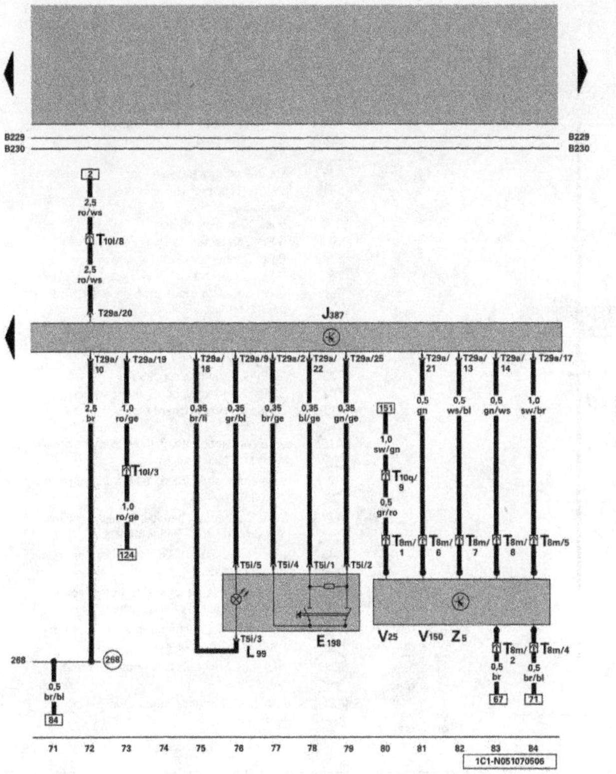

Switch for interior lock, passenger side, Door control module, passenger side, Passenger's Side Mirror Adjustment Motor, Motor for mirror adjustment, passenger side, Heated outside mirror, passenger side

- E198 - Switch for interior lock, passenger side
- J387 - Door control module, passenger side
- L99 - Lighting for switch interior lock
- T5i - 5-Pin Connector
- T8m - 8-Pin Connector
- T10l - 10-Pin Connector, black, on A-Pillar right
- T10q - 10-Pin Connector, blue, on A-Pillar right
- T29a - 29-Pin Connector, on Door control module
- V25 - Passenger's Side Mirror Adjustment Motor
- V150 - Motor for mirror adjustment, passenger side
- Z5 - Heated outside mirror, passenger side
- (268) - Ground connector -2-, in wiring harness door cable - passenger side

ws = white
sw = black
ro = red
br = brown
gn = green
bl = blue
gr = grey
li = violet
ge = yellow
or = orange

New Beetle — Wiring diagram — No. 51/8

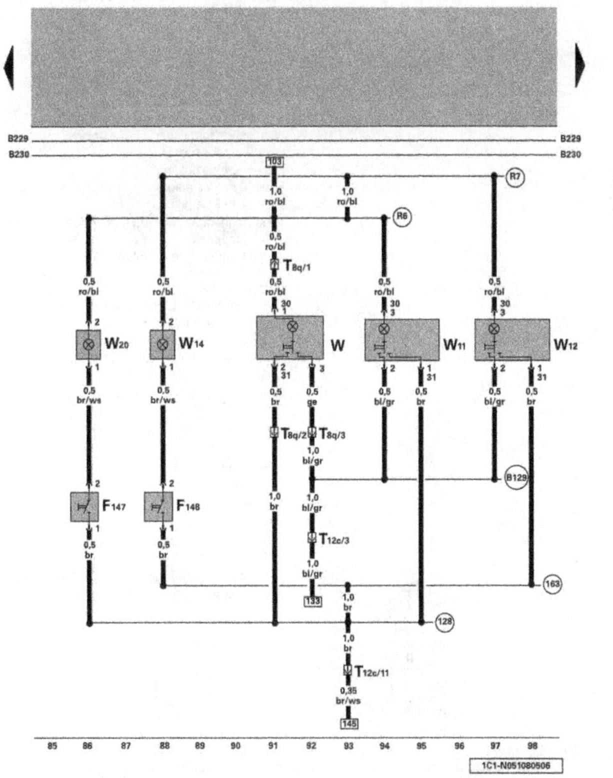

Left Make-Up Mirror Light Switch, Right Make-Up Mirror Light Switch, Front Interior Light, Left Rear Reading Light, Right Rear Reading Light, Right Make-up Mirror Light, Left Make-up Mirror Light

- F147 - Left Make-Up Mirror Light Switch
- F148 - Right Make-Up Mirror Light Switch
- T8q - 8-Pin Connector at W
- T12c - 12-Pin Coupling Connector, on A pillar left, central part
- W - Front Interior Light
- W11 - Left Rear Reading Light
- W12 - Right Rear Reading Light
- W14 - Right Make-up Mirror Light
- W20 - Left Make-up Mirror Light
- (128) - Ground connection -1-, in interior light wiring harness
- (163) - Ground connection -2-, in interior light wiring harness
- (B129) - Connector (interior light 3/L), in wiring harness interior
- (R6) - Plus connection -1-, in interior light wiring harness
- (R7) - Plus connection -2-, in interior light wiring harness 3

ws = white
sw = black
ro = red
br = brown
gn = green
bl = blue
gr = grey
li = violet
ge = yellow
or = orange

New Beetle

Wiring diagram

No.51/9

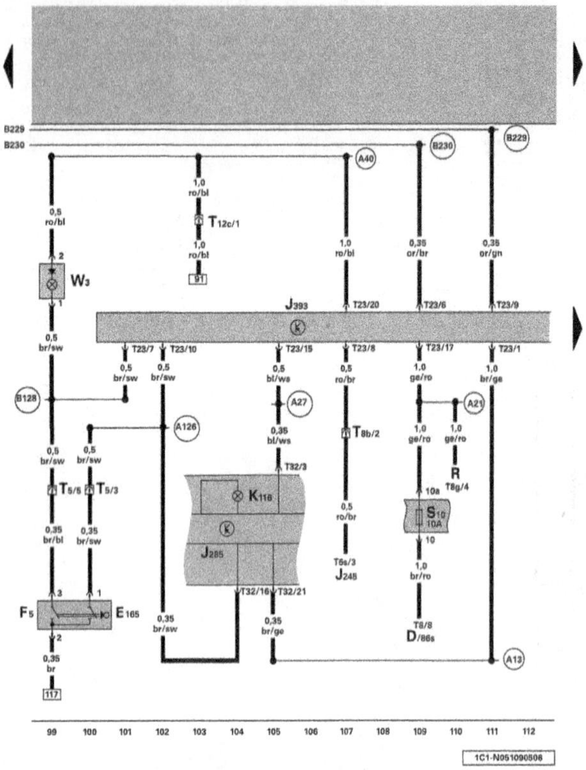

Trunk lid release switch, luggage compartment light switch, central control module for comfort system, luggage compartment light

- D - Ignition/Starter Switch
- E165 - Trunk Lid Release Switch
- F5 - Luggage Compartment Light Switch
- J245 - Power Sunroof Control Module
- J285 - Control module with indicator unit in instr. insert
- J393 - Central control module for comfort system
- K116 - Warning light for rear lid unlocked
- R - Radio
- S10 - Fuse 10 in fuse holder
- T5 - 5-Pin Connector, black, on C-pillar, left
- T6s - 6-Pin Connector
- T8 - 8-Pin Connector
- T8b - 8-Pin Coupling Connector, on A pillar left, central part
- T8g - 8-Pin Connector on radio
- T12c - 12-Pin Coupling Connector, on A pillar left, central part
- T23 - 23-Pin Connector, on J393
- T32 - 32-Pin Connector, blue, on instrument cluster
- W3 - Luggage compartment Light

- (A13) - wire connection (86s), in instrument panel wiring harness
- (A21) - wire connection (86s), in instrument panel wiring harness
- (A27) - Wire Connection (vehicle speed signal), in instrument panel wiring harness
- (A40) - Plus connection -1- (30), in instrument panel wiring harness
- (A126) - Connection (contact switch in rear lid), in instrument panel wiring harness
- (B128) - Connector (Luggage compartment lamp) in wiring harness interior
- (B229) - High-bus Connection (in interior wiring harness)
- (B230) - Low-bus Connection (in interior wiring harness)

ws = white
sw = black
ro = red
br = brown
gn = green
bl = blue
gr = grey
li = violet
ge = yellow
or = orange

New Beetle

Wiring diagram

No.51/10

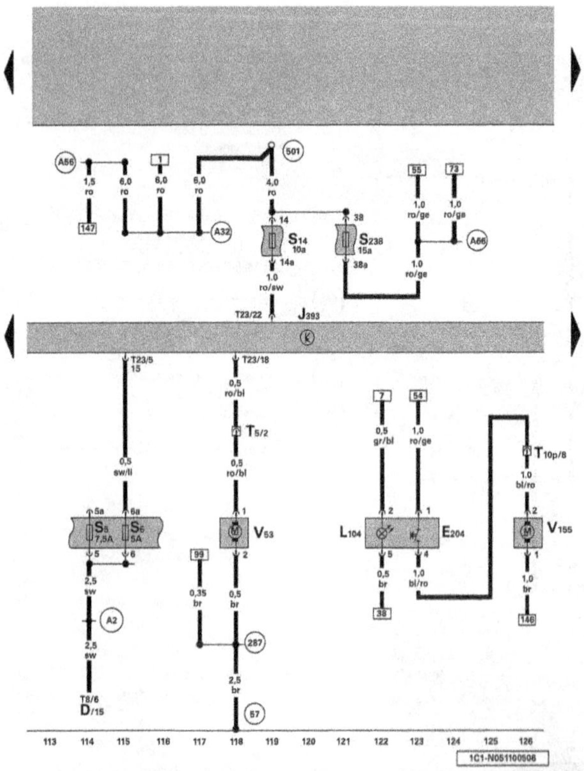

Switch for remote/fuel tank door, Central control module for comfort system, Rear Lid Central Locking System Motor, Motor for fuel tank lid unlock

- D - Ignition/Starter Switch
- E204 - Switch for remote/fuel tank door
- J393 - Central control module for comfort system
- L104 - Illumination for switch remote unlock
- S5 - Fuse 5 in fuse holder
- S6 - Fuse 6 in fuse holder
- S14 - Fuse 14 in fuse holder
- S238 - Fuse 38 in fuse holder
- T5 - 5-Pin Connector, black, on C pillar, left
- T8 - 8-Pin Connector
- T10p - 10-Pin Connector, on A-Pillar left, blue
- T23 - 23-Pin Connector, on Central control module for comfort system
- V 53 - Rear Lid Central Locking System Motor
- V155 - Motor for fuel tank lid unlock

- (57) - Ground connection, on left rear pillar
- (287) - Ground connection, in wiring harness rear hatch hydraulics
- (501) - Screw connection -2- (30), on relay panel
- (A2) - Plus connection (15), in instrument panel wiring harness
- (A32) - Plus connection (30), in instrument panel wiring harness
- (A56) - Plus Connection 2 (30) (in instrument cluster wiring harness)
- (A66) - Connector (30a, central locking/anti-theft warning system/IR) in instrument panel wiring harness

ws = white
sw = black
ro = red
br = brown
gn = green
bl = blue
gr = grey
li = violet
ge = yellow
or = orange

New Beetle — Wiring diagram — No.51/11

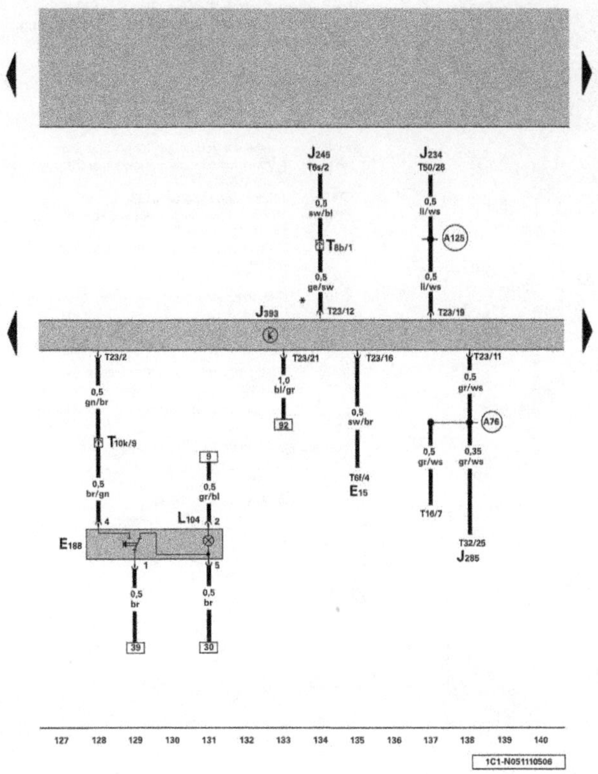

Switch for remote unlock, rear lid, Central control module for comfort system

- E15 - Rear Window Defogger Switch
- E188 - Switch for remote unlock, rear lid
- J234 - Airbag Control Module
- J245 - Power Sunroof Control Module
- J285 - Control module with indicator unit in instrument panel insert
- J393 - Central control module for comfort system
- L104 - Illumination for switch remote unlock
- T6f - 6-Pin Connector
- T6s - 6-Pin Connector
- T8b - 8-Pin Connector
- T10k - 10-Pin Connector, on A-Pillar left, black
- T16 - 16-Pin Connector, Data Link Connector (DLC), below steering coulm, left
- T23 - 23-Pin Connector, on Central control module for comfort system
- T32 - 32-Pin Connector, blue, on instrument cluster
- T50 - 50-Pin Connector
- (A76) - Connector (K-diagnosis wire) in instrument panel wiring harness
- (A125) - wire connection (crash), in instrument panel wiring harness
- * - Only vehicles with power sunroof

ws = white
sw = black
ro = red
br = brown
gn = green
bl = blue
gr = grey
li = violet
ge = yellow
or = orange

New Beetle — Wiring diagram — No. 51/12

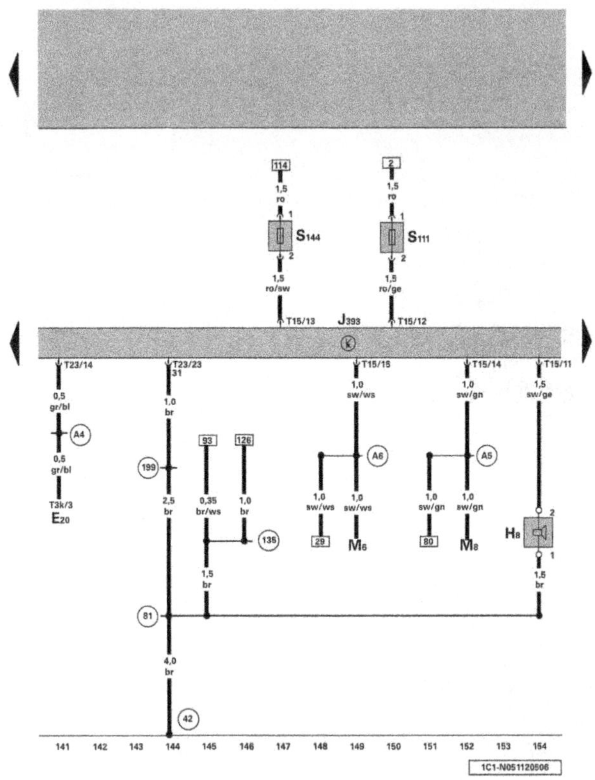

Alarm Horn, Central control module for comfort system

- E20 - Instrument Panel Light Dimmer Switch
- H8 - Alarm Horn
- J393 - Central control module for comfort system
- M6 - Left Rear Turn Signal Light
- M8- - Right Rear Turn Signal Light
- S111 - Alarm System and Anti-Theft Engine Disable Fuse on thirteen position relay panel
- S144 - Fuse for central locking/anti-theft warning system on thirteen position relay panel
- T3k - 3-Pin Connector at E20
- T15 - 15-Pin Connector, on Central control module for comfort system
- T23 - 23-Pin Connector, on Central control module for comfort system
- (42) - Ground connection, beside steering column
- (81) - Ground connection -1-, in instrument panel wiring harness
- (135) - Ground connection -2-, in instrument panel wiring harness
- (199) - Ground connection -3-, in instrument panel wiring harness
- (A4) - Plus connection (58b), in instrument panel wiring harness
- (A5) - Plus connection (right turn signal), in instrument panel wiring harness
- (A6) - Plus connection (left turn signal), in instrument panel wiring harness

ws = white
sw = black
ro = red
br = brown
gn = green
bl = blue
gr = grey
li = violet
ge = yellow
or = orange

New Beetle Wiring diagram **No. 51/13**

Hood Alarm Switch, Coolant Fan Control (FC) Control Module, Antenna wire for central locking and antitheft warning system

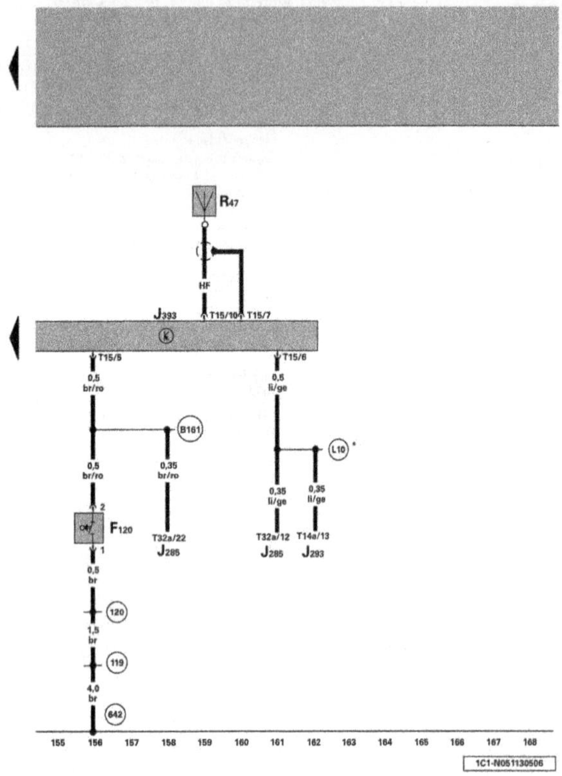

ws = white
sw = black
ro = red
br = brown
gn = green
bl = blue
gr = grey
li = violet
ge = yellow
or = orange

- F120 - Hood Alarm Switch
- J285 - Control module with indicator unit in instrument panel insert
- J293 - Coolant Fan Control (FC)Control Module
- J393 - Central control module for comfort system
- R47 - Antenna wire for central locking and antitheft warning system
- T14a - 14-Pin Connector, on -J293-
- T15 - 15-Pin Connector, on Central control module for comfort system
- T32a - 32-Pin Connector, green, on instrument closter

- (119) - Ground connection -1-, in headlight wiring harness
- (120) - Ground connection -2-, in headlight wiring harness
- (642) - Punto de masa, ventilador de control electrónico
- (B161) - Connection (Alarm system), in wiring harness, interior
- (L10) - Connection 2 (in A/C wiring harness)*

* - Only vehicles with A/C

THIS SPACE INTENTIONALLY

LEFT BLANK

New Beetle Convertible **Wiring diagram** **No. 20/1**

Manual Convertible Top

from January 2004

Additional Fuse, Relay, Control Module and Connector locations ⇒ Component Locations

Electrical troubleshooting ⇒ Guided Fault Finding using VAS 5051/5052

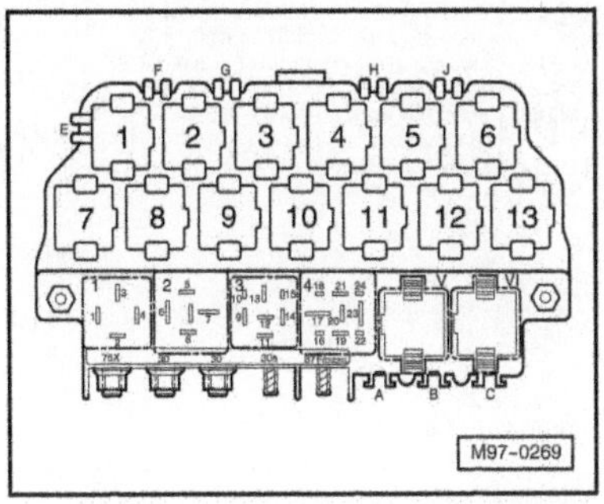

Relay location on the thirteen position auxiliary relay panel, above relay panel:

5/6 J256 - Convertible Top Control Module (477) [2]

Note:

[2] Only vehicles with manual convertible top

Note:

Number in parentheses indicates production control number stamped on relay housing.

New Beetle Convertible **Wiring diagram** **No. 20/2**

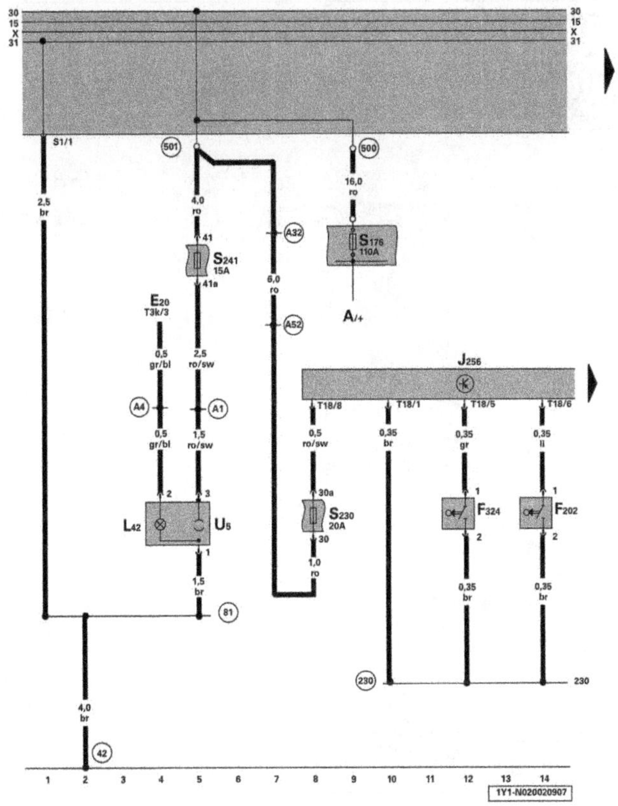

ws = white
sw = black
ro = red
br = brown
gn = green
bl = blue
gr = grey
li = violet
ge = yellow
or = orange

Convertible Top Closed Position Switch, Left Convertible Top Closed Lock Switch, Convertible Top Control Module, 12 V Socket

- A - Battery
- E20 - Instrument Panel Illumination Dimmer Switch
- F202 - Convertible Top Closed Position Switch
- F324 - Left Convertible Top Closed Lock Switch
- J256 - Convertible Top Control Module
- L42 - Socket Light Bulb
- S176 - Fuse -4- (30), in fuse bracket/battery
- S230 - Fuse 30 in fuse holder
- S241 - Fuse 41 in fuse holder
- U 5 - 12 V Socket
- T3k - 3-Pin connector
- T18 - 18-Pin connector, on J256
- (42) - Ground connection, beside steering column
- (81) - Ground connection -1-, in instrument panel wiring harness
- (230) - Ground connection -1-, in power window/central locking system and convertible top wiring harness
- (500) - Threaded connection -1- (30) on the relay plate
- (501) - Threaded connection -2- (30) on the relay plate
- (A1) - Plus connection (30a), in instrument panel wiring harness
- (A4) - Plus connection (58b), in instrument panel wiring harness
- (A32) - Plus connection (30), in instrument panel wiring harness
- (A52) - Plus Connection 2 (30) (in instrument panel wiring harness)

New Beetle Convertible

Wiring diagram

No. 20/3

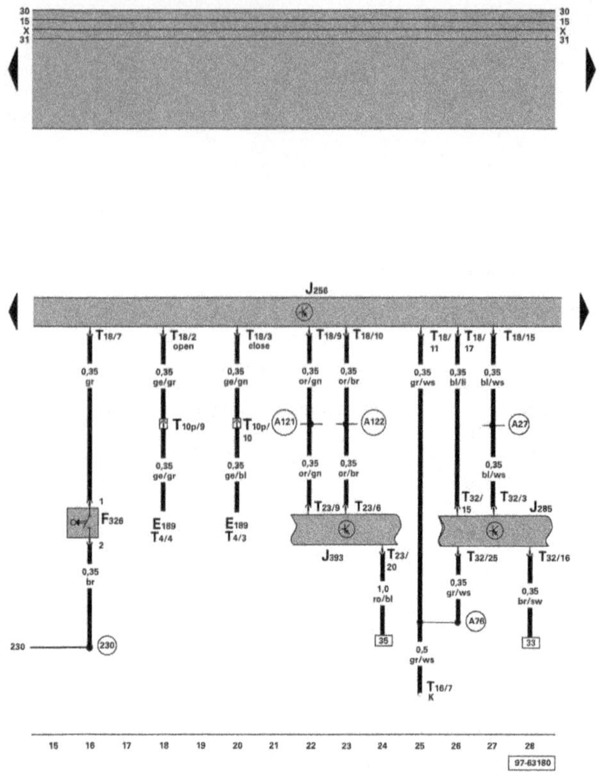

Right Convertible Top Open Lock Switch, Convertible Top Control Module

- E189 - Central switch for window regulator, in driver's door
- F326 - Right Convertible Top Closed Lock Switch
- J256 - Convertible Top Control Module
- J285 - Control module with indicator unit in instrument panel insert
- J393 - Central control module for comfort system
- T4 - 4-Pin connector, on E189
- T10p - 10 pin connector, blue, at bottom of left A-pillar
- T16 - 16-Pin connector, Data Link Connector (DLC), bellow steering column
- T18 - 18-Pin connector, on J256
- T23 - 23-Pin connector, on J393
- T32 - 32-Pin connector, blue

- (230) - Ground connection -1-, in power window/central locking system and convertible top wiring harness
- (A27) - Wire Connection (vehicle speed signal), in instrument panel wiring harness
- (A76) - Connector (K-diagnosis wire) in instrument panel wiring harness
- (A121) - Connection (high bus) in instrument panel wiring harness
- (A122) - Connection (low bus) in instrument panel wiring harness

ws = white
sw = black
ro = red
br = brown
gn = green
bl = blue
gr = grey
li = violet
ge = yellow
or = orange

Edition 04.04

New Beetle Convertible

Wiring diagram

No. 20/4

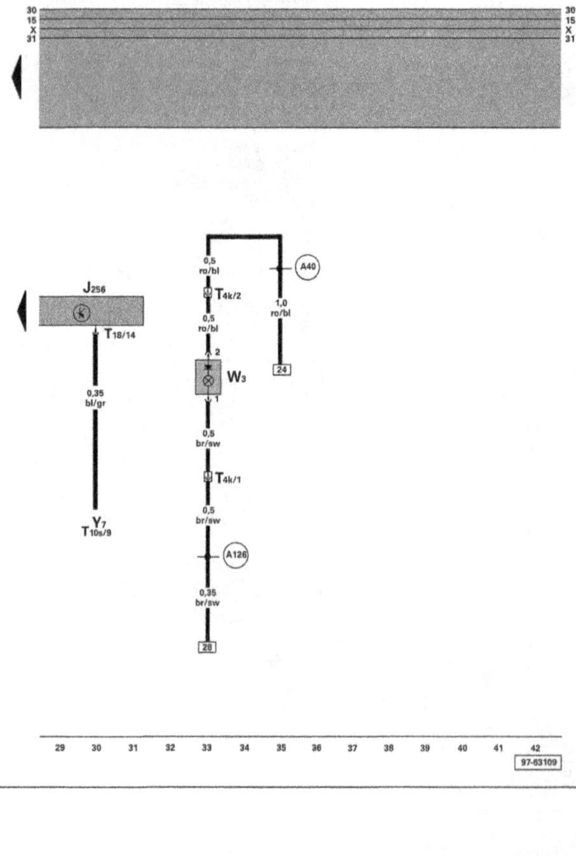

Convertible top control module, Luggage compartment Light

- J256 - Convertible Top Control Module
- T4k - 4-Pin connector, in luggage compartment, right-side
- T10s - 10-Pin connector, black, at interior mirror
- T18 - 18-Pin connector, on J256
- W3 - Luggage compartment Light
- Y7 - Automatic Day/Night Interior Mirror (LED display console at mirror base)

- (A40) - plus connection -1- (30), in instrument panel wiring harness
- (A126) - Connection (contact switch in rear lid), in instrument panel wiring harness

ws = white
sw = black
ro = red
br = brown
gn = green
bl = blue
gr = grey
li = violet
ge = yellow
or = orange

Edition 02.04

New Beetle Convertible

Wiring diagram **No. 4/1**

Power Convertible Top

from August 2002

Deviate relay location and fuseplacements as well as the locations of multiple connectors see section "component locations".

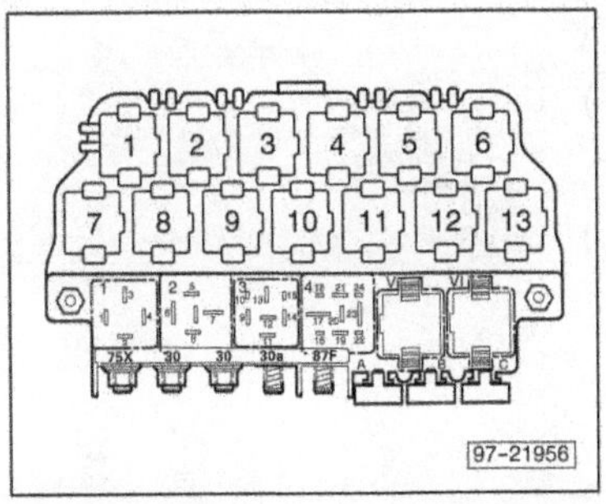

Fuse colors

- 30 A - green
- 25 A - white
- 20 A - yellow
- 15 A - blue
- 10 A - red
- 7,5 A - brown
- 5 A - beige
- 3 A - violet

Note: Number in parentheses indicates production control number stamped on relay housing.

New Beetle Convertible

Wiring diagram **No. 4/2**

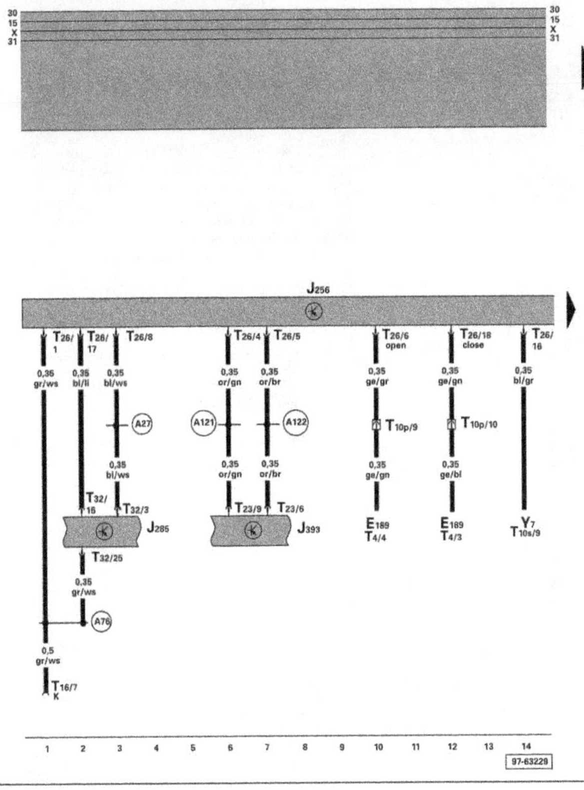

Convertible top control module, Control module with indicator unit in instrument panel insert, Central control module for comfort system

- E189 - Central switch for window regulator, in driver's door
- J256 - Convertible Top Control Module
- J285 - Control module with indicator unit in instrument panel insert
- J393 - Central control module for comfort system
- T4 - 10-Pin connector, black, in driver's door
- T10p - 10-Pin connector, blue, at bottom of left A-pillar
- T10s - 10-Pin connector, black, at interior mirror
- T16 - 16-Pin Connector, Data Link Connector (DLC), bellow steering column
- T23 - 23-Pin connector
- T26 - 26-Pin connector, connector B, black, on J256
- T32 - 32-Pin connector, blue
- Y7 - Automatic Day/Night Interior Mirror

- (A27) - Wire Connection (vehicle speed signal), in instrument panel wiring harness
- (A76) - Connector (K-diagnosis wire) in instrument panel wiring harness
- (A121) - Connection (high bus) in instrument panel wiring harness
- (A122) - Connection (low bus) in instrument panel wiring harness

ws = white
sw = black
ro = red
br = brown
gn = green
bl = blue
gr = grey
li = violet
ge = yellow
or = orange

New Beetle Convertible

Wiring diagram

No. 4/3

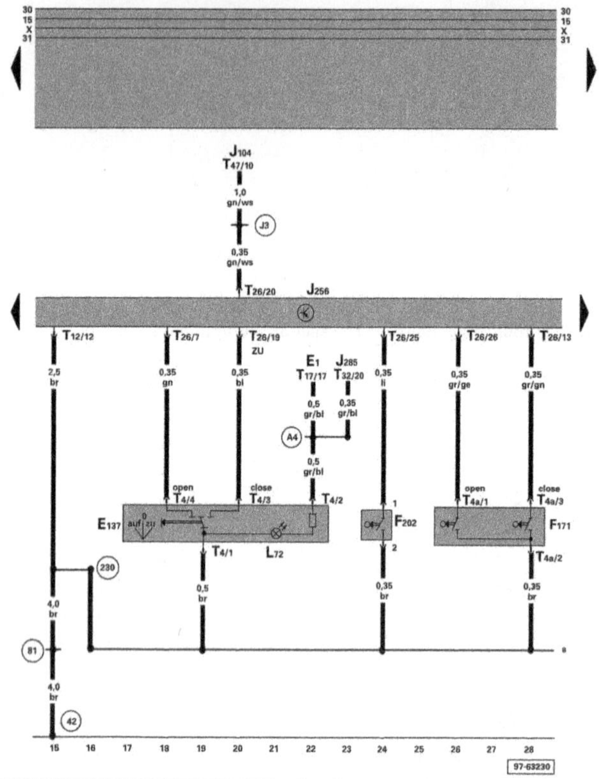

Convertible top control module, convertible top operation switch, convertible top open position switch, convertible top closed position switch, convertible top switch light

- E1 - Light switch
- E137 - Convertible Top Operation Switch
- F171 - Convertible Top Open Position Switch
- F202 - Convertible Top Closed Position Switch
- J104 - ABS Control Module (w/EDL)
- J256 - Convertible Top Control Module
- J285 - Control module with indicator unit in instrument panel insert
- T4 - 4-Pin connector
- T4a - 4-Pin connector
- T17 - 17-Pin connector, at E1
- T12 - 12-Pin connector, connector A, black, on J256
- T26 - 26-Pin connector, connector B, black, on J256
- T32 - 32-Pin Connector, blue, on instrument cluster
- T47 - 47-Pin connector
- L72 - Convertible Top Switch Light
- (42) - Ground Connection, beside steering column
- (81) - Ground Connection 1, in instrument panel wiring harness
- (230) - Ground connection -1-, in power window/central locking system and convertible top wiring harness
- (A4) - Plus connection (58b), in instrument panel wiring harness
- (J3) - Wire connection -1-, in ABS wiring harness

ws = white
sw = black
ro = red
br = brown
gn = green
bl = blue
gr = grey
li = violet
ge = yellow
or = orange

Edition 06.04

New Beetle Convertible

Wiring diagram

No. 4/4

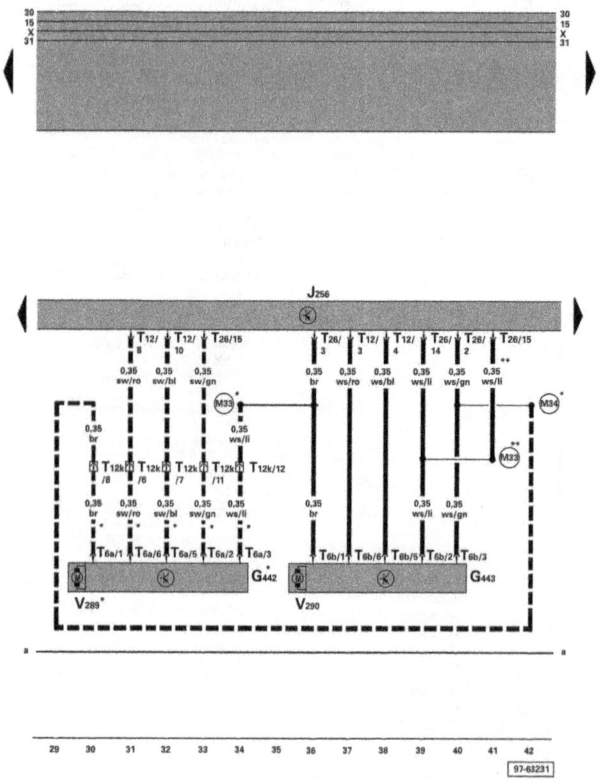

Convertible top control module, left / right linkage flap position sensor, left/right linkage flap motor

- G442 - Left Linkage Flap Position Sensor
- G443 - Right Linkage Flap Position Sensor
- J256 - Convertible Top Control Module
- T6a - 6-Pin connector
- T6b - 6-Pin connector
- T12 - 12-Pin connector, connector A, black, on J256
- T12k - 12-Pin connector, in luggage compartment, right-side
- T26 - 26-Pin connector, connector B, black, on J256
- V289 - Left Linkage Flap motor
- V290 - Right Linkage Flap Motor
- (M33) - Wire connection -2-, in convertible top wiring harness*
- (M33) - Wire connection -2-, in convertible top wiring harness**
- (M34) - Wire connection -3-, in convertible top wiring harness

* - Through May 2004

** - From June 2004

ws = white
sw = black
ro = red
br = brown
gn = green
bl = blue
gr = grey
li = violet
ge = yellow
or = orange

Edition 11.04

New Beetle Convertible — Wiring diagram — No. 4/5

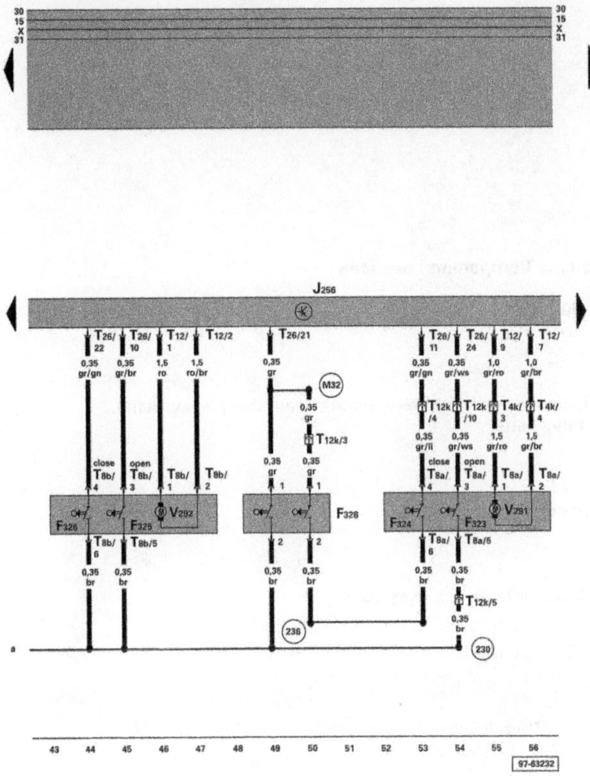

Convertible top control module, left/right convertible top open lock switch, left/right convertible top closed lock switch, convertible top tonneau cover switch, left/right top locking motor

- F323 - Left Convertible Top Open Lock Switch
- F324 - Left Convertible Top Closed Lock Switch
- F325 - Right Convertible Top Open Lock Switch
- F326 - Right Convertible Top Closed Lock Switch
- F328 - Convertible Top Tonneau Cover Switch
- J256 - Convertible Top Control Module
- T4k - 4-Pin connector, in luggage compartment, right-side
- T8a - 8-Pin connector
- T8b - 8-Pin connector
- T12 - 12-Pin connector, connector A, black, on J256
- T12k - 12-Pin connector, in luggage compartment, right-side
- T26 - 26-Pin connector, connector B, black, on J256
- V291 - Left Top Locking Motor
- V292 - Right Top Locking Motor
- (230) - Ground connection -1-, in power window/central locking system and convertible top wiring harness
- (236) - Ground connector -2- in wiring harness, window regulator/central locking and convertible top
- (M32) - Wire connection -1-, in convertible top wiring harness

ws = white
sw = black
ro = red
br = brown
gn = green
bl = blue
gr = grey
li = violet
ge = yellow
or = orange

New Beetle Convertible — Wiring diagram — No. 4/6

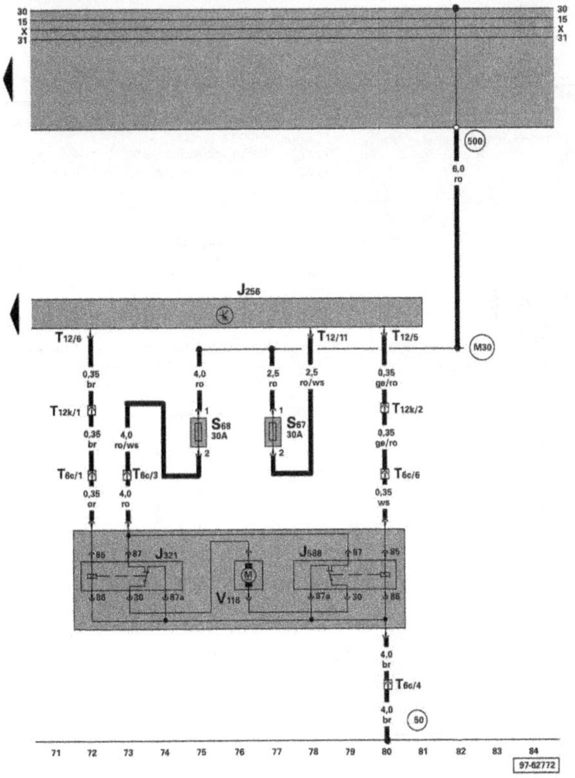

Convertible top control module, convertible top fuse, convertible top circuit breaker

- J256 - Convertible Top Control Module
- J321 - Convertible Top Hydraulic Pump Relay
- J588 - Convertible Top Hydraulic Pump Relay 2
- S67 - Convertible Top Operation Fuse
- S68 - Convertible Top Motor Circuit Breaker
- T6c - 6-Pin connector, black, on top of hydraulic Pump , in luggage compartment, left-side
- T12 - 12-Pin connector, connector A, black, on J256
- T12k - 12-Pin connector, in luggage compartment, right-side
- V118 - Convertible Top Hydraulic Pump
- (50) - Ground connection, in luggage compartment, left
- (500) - Threaded connection -1- (30) on the relay plate
- (M30) - plus connection (30), in power convertible top wiring harnes

ws = white
sw = black
ro = red
br = brown
gn = green
bl = blue
gr = grey
li = violet
ge = yellow
or = orange

New Beetle Convertible — Wiring diagram — No. 41/1

Power Convertible Top,

from July 2005,

Additional Fuse, Relay, Control Module and Connector locations ⇒ Component Locations

Electrical troubleshooting ⇒ Guided Fault Finding using VAS 5051/5052

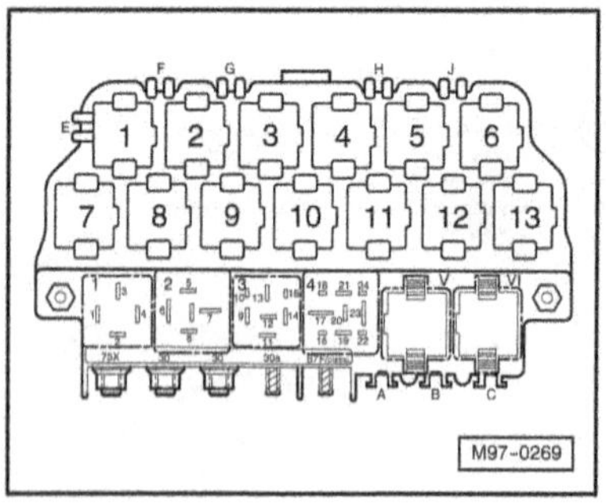

Relay location on the thirteen position auxiliary relay panel, above relay panel:

Relay panel:

Fuse placements on the relay panel:

C S68 - Convertible Top Motor Circuit Breaker (30A)

H S67 - Convertible Top Operation Fuse (30A)

Note:

Number in parentheses indicates production control number stamped on relay housing.

New Beetle Convertible — Wiring diagram — No. 41/2

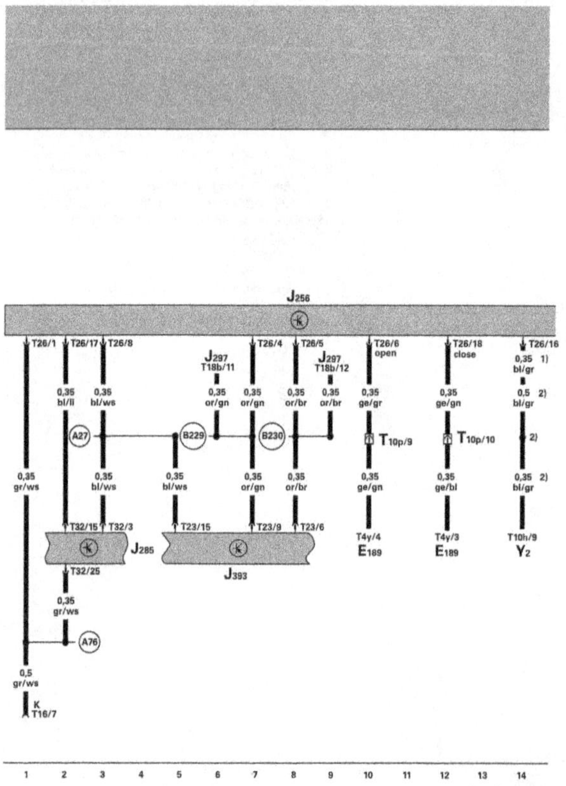

Convertible Top Control Module

E189 - Central Window Regulator Switch (in Driver door)
J256 - Convertible Top Control Module
J285 - Alarm System Control Module
J297 - Left Rear Power Window Control Module
J393 - Comfort System Central Control Module
T4y - 4-Pin Connector
T10h - 10-Pin connector, black
T10p - 10-Pin connector, blue, at bottom of left A-pillar
T16 - 16-Pin Connector, Data Link Connector (DLC), bellow steering column
T18b - 18-Pin Connector
T23 - 23-Pin Connector
T26 - 26-Pin Connector, connector B, black, on J256
T32 - 32-Pin Connector, blue
Y2 - Digital Clock

(A27) - Vehicle Speed Signal (VSS) Connection (in instrument panel wiring harness)
(A76) - K-Diagnostic Connection (in instrument panel wiring harness)
(B229) - High-Bus Connection (in interior wiring harness)
(B230) - Low-Bus Connection (in interior wiring harness)

1) - Through June 2008
2) - From July 2008

ws = white
sw = black
ro = red
br = brown
gn = green
bl = blue
gr = grey
li = violet
ge = yellow
or = orange

New Beetle Convertible — Wiring diagram — No. 41/3

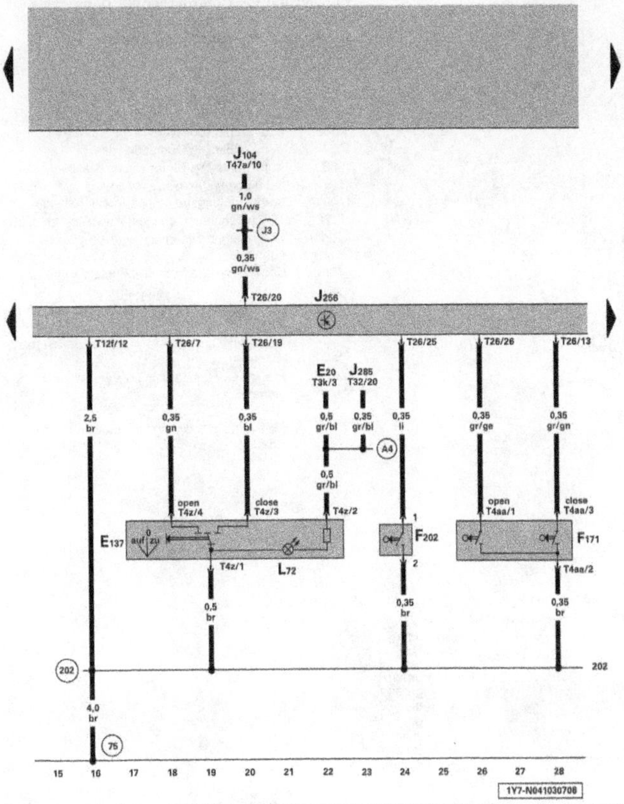

Convertible Top Operation Switch, Convertible Top Open Position Switch, Convertible Top Closed Position Switch, Convertible Top Con trol Module

- E20 - Instrument Panel Illumination Dimmer Switch
- E137 - Convertible Top Operation Switch
- F171 - Convertible Top Open Position Switch
- F202 - Convertible Top Closed Position Switch
- J104 - ABS Control Module
- J256 - Convertible Top Control Module
- J285 - Instrument Cluster Control Module
- T3k - 3-Pin connector
- T4z - 4-Pin connector
- T4aa - 4-Pin connector
- T12f - 12-Pin connector, connector A, black, on J256
- T26 - 26-Pin connector, connector B, black, on J256
- T32 - 32-Pin Connector, blue, on instrument cluster
- T47a - 47-Pin connector
- L72 - Convertible Top Switch Illumination
- (75) - Ground Connection (on right rear pillar)
- (202) - Ground Connection (in convertible top wiring harness)
- (A4) - Plus Connection (58b) (in instrument panel wiring harness)
- (J3) - Connection 1 (in ABS wiring harness)

ws = white
sw = black
ro = red
br = brown
gn = green
bl = blue
gr = grey
li = violet
ge = yellow
or = orange

New Beetle Convertible — Wiring diagram — No. 41/4

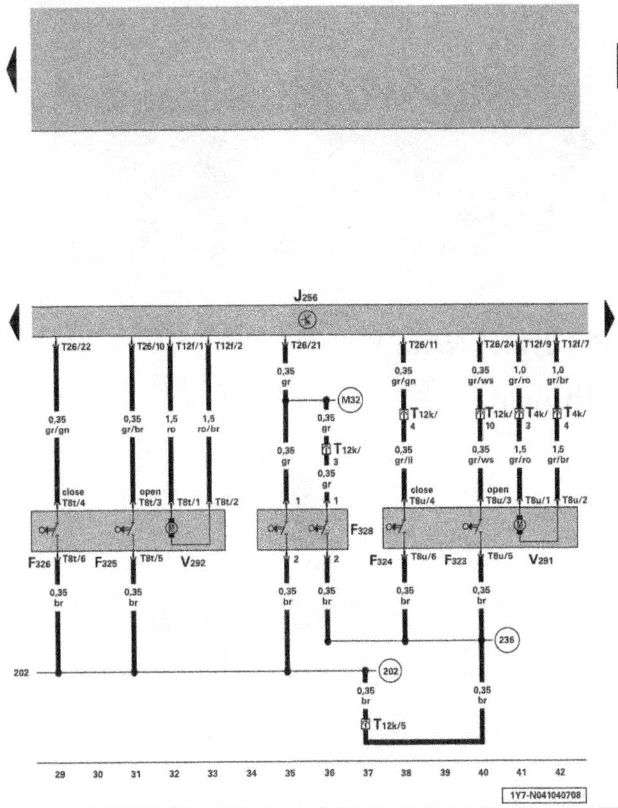

Left Convertible Top Open/Closed Lock Switch, Right Convertible Top Open/Closed Lock Switch, Convertible Top Tonneau Cover Switch, Convert ible Top Control Module, Left Top Locking Motor, Right Top Locking Motor

- F323 - Left Convertible Top Open Lock Switch
- F324 - Left Convertible Top Closed Lock Switch
- F325 - Right Convertible Top Open Lock Switch
- F326 - Right Convertible Top Closed Lock Switch
- F328 - Convertible Top Tonneau Cover Switch
- J256 - Convertible Top Control Module
- T4k - 4-Pin connector, in luggage compartment, right-side
- T8t - 8-Pin connector
- T8u - 8-Pin connector
- T12f - 12-Pin connector, connector A, black, on J256
- T12k - 12-Pin connector, in luggage compartment, right-side
- T26 - 26-Pin connector, connector B, black, on J256
- V291 - Left Top Locking Motor
- V292 - Right Top Locking Motor
- (202) - Ground Connection (in convertible top wiring harness)
- (236) - Ground Connection 2 (in window regulator/central locking and convertible top wiring harness)
- (M32) - Connection 1 (in convertible top wiring harness)

ws = white
sw = black
ro = red
br = brown
gn = green
bl = blue
gr = grey
li = violet
ge = yellow
or = orange

Convertible top (power) from July 2005

New Beetle Convertible

Wiring diagram

No. 41/5

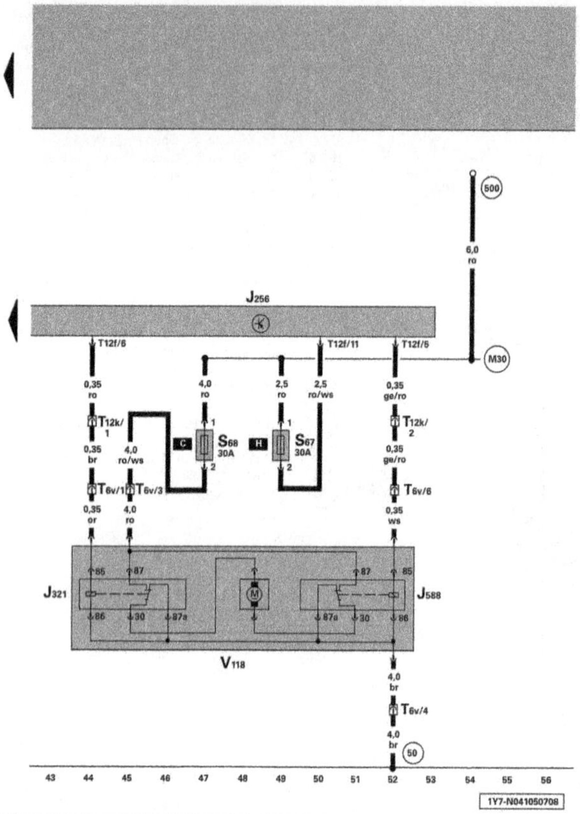

Convertible Top Control Module, Convertible Top Hydraulic Pump Relay, Convertible Top Hydraulic Pump Relay 2, Convertible Top Hydraulic Pump

J256	-	Convertible Top Control Module
J321	-	Convertible Top Hydraulic Pump Relay*
J588	-	Convertible Top Hydraulic Pump Relay 2*
S67	-	Convertible Top Operation Fuse
S68	-	Convertible Top Motor Circuit Breaker
T6v	-	6-Pin connector, black, on top of hydraulic Pump , in luggage compartment, left-side
T12f	-	12-Pin connector, connector A, black, on J256
T12k	-	12-Pin connector, in luggage compartment, right-side
V118	-	Convertible Top Hydraulic Pump*
(50)	-	Ground Connection (in luggage compartment, left)
(500)	-	Threaded Connection 1 - terminal 30 (on relay panel)
(M30)	-	Plus connection (30) (in convertible top wiring harness)
*	-	On hydraulic Pump, in luggage compartment, left-side

ws = white
sw = black
ro = red
br = brown
gn = green
bl = blue
gr = grey
li = violet
ge = yellow
or = orange

THIS SPACE INTENTIONALLY LEFT BLANK

Cruise control, 2.0L-Engine - code AEG only

from January 1998

Deviate relay location and fuseplacements as well as the locations of multiple connectors see section "component locations".

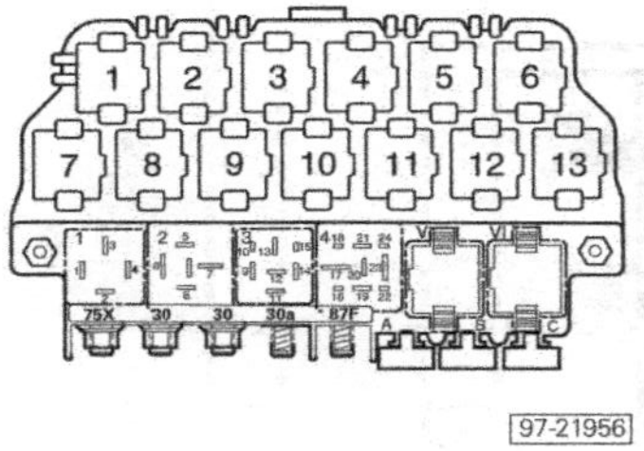

Relay panel:

- **2** Load Reduction Relay (100)
- **4** Fuel Pump (FP) Relay (409)

Note: Number in parentheses indicates production control number stamped on relay housing.

Fuse colors
30 A - green
25 A - white
20 A - yellow
15 A - blue
10 A - red
7,5 A - brown
5 A - beige

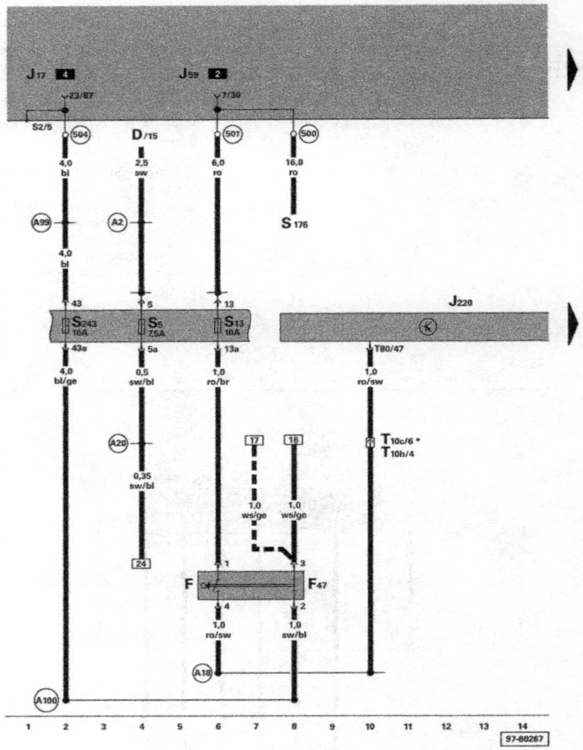

ws = white
sw = black
ro = red
br = brown
gn = green
bl = blue
gr = grey
li = violet
ge = yellow

- D - Ignition/Starter Switch
- F - Brake Light Switch
- F47 - Brake Pedal Switch
- J17 - Fuel Pump (FP) Relay
- J59 - Load Reduction Relay
- J220 - Motronic Engine Control Module (ECM)
- S5 - Fuse 5 in fuse holder
- S13 - Fuse 13 in fuse holder
- S176 - Fuse -4- (30), in fuse bracket/battery
- S243 - Fuse 43 in fuse holder
- T10c - 10-Pin Connector, blue, behind instrument panel, left
- T10h - 10-Pin Connector, black, behind instrument panel, left
- T80 - Connector, 80 point
- (500) - Threaded connection -1- (30) on the relay plate
- (501) - Threaded connection -2- (30) on the relay plate
- (504) - Threaded connection -1- (87) on the relay plate
- (A2) - plus connection (15), in instrument panel wiring harness
- (A18) - plus connection (54), in rear wiring harness
- (A20) - wire connection (15a), in instrument panel wiring harness
- (A99) - Connector -1- (87), in instrument panel
- (A100) - Connector -2- (87), in instrument panel wiring harness
- * - early vehicles only
- --- Automatic transmission only

Brake light switch, brake pedal switch, fuel pump relay, Motronic engine control module

Cruise control (2.0L AEG) from Jan. 1998

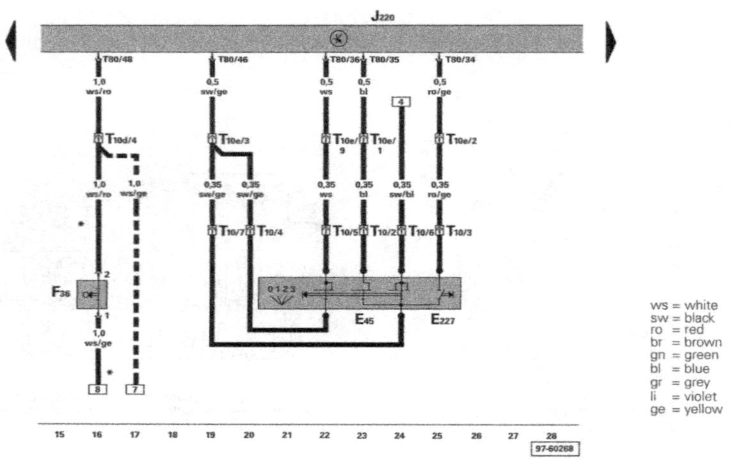

ws = white
sw = black
ro = red
br = brown
gn = green
bl = blue
gr = grey
li = violet
ge = yellow

E45 - Cruise Control Switch
E227 - Button for cruise control (set)
F36 - Clutch Vacuum Vent Valve Switch
J220 - Motronic Engine Control Module (ECM)
T10 - 10-pin connector, to right of steering column
T10d - 10-pin Connector, green, behind instrument panel, left
T10e - 10-pin Connector, black, behind instrument panel, left
T80 - 80-pin Connector, in Motronic Engine Control Module

(A100) - Connection 2 (87) (in instrument panel wiring harness)

* - Standard transmission only

--- Automatic transmission only

Cruise control switch, button for cruise control, clutch vacuum vent valve switch, Motronic engine control module.

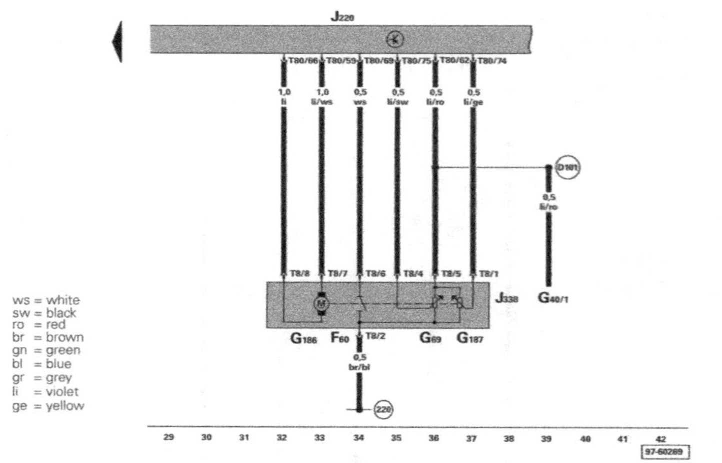

ws = white
sw = black
ro = red
br = brown
gn = green
bl = blue
gr = grey
li = violet
ge = yellow

F60 - Closed Throttle Position (CTP) Switch
G40 - Camshaft Position (CMP) Sensor
G69 - Throttle Position (TP) Sensor
G186 - Throttle Drive (electronic throttle)
G187 - Throttle Drive Angle Sensor 1 (electronic throttle)
J220 - Motronic Engine Control Module
J338 - Throttle Valve Control Module
T8 - 8-pin Connector
T80 - 80-pin Connector, in Motronic Engine Control Module

(220) - Ground Connection, Sensor Ground (in engine compartment wiring harness)

(D101) - Connection 1 (in engine compartment wiring harness)

Closed throttle position, Camshaft position sensor, Throttle position sensor, Throttle drive (electronic), Throttle drive angle sensor 1.

Cruise control, 2.0L-Engine - code AEG only

from August 1998

Deviate relay location and fuseplacements as well as the locations of multiple connectors see section "component locations".

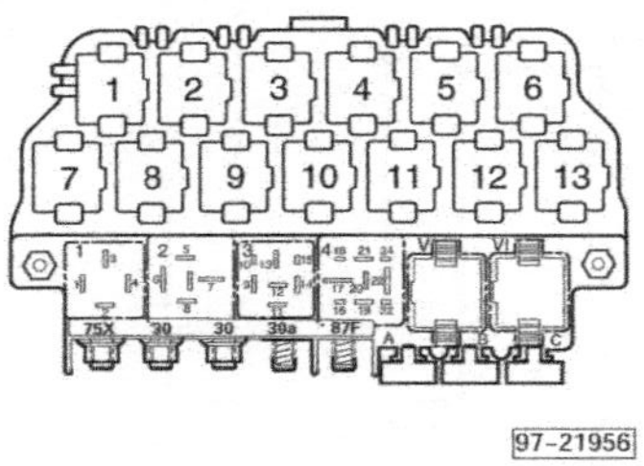

97-21956

Relay panel:

- **2** Load Reduction Relay (100)
- **4** Fuel Pump (FP) Relay (409)

Note: Number in parentheses indicates production control number stamped on relay housing.

Fuse colors
30 A - green
25 A - white
20 A - yellow
15 A - blue
10 A - red
7,5 A - brown
5 A - beige

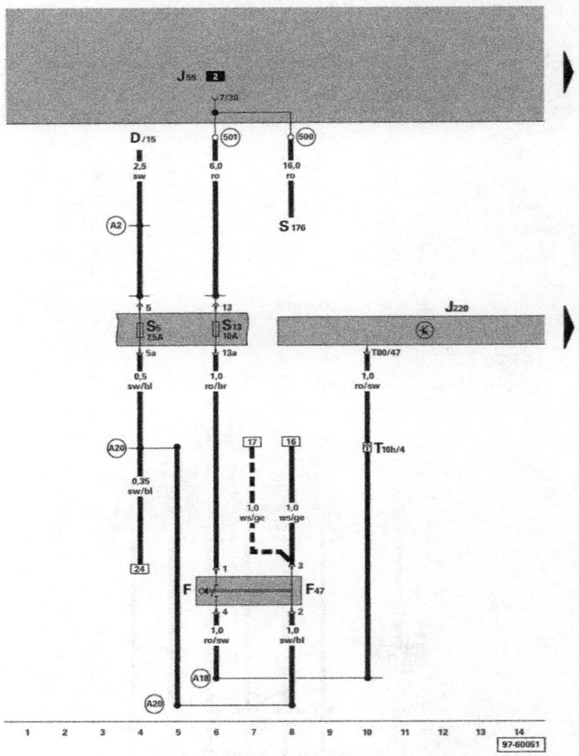

ws = white
sw = black
ro = red
br = brown
gn = green
bl = blue
gr = grey
li = violet
ge = yellow

D - Ignition/Starter Switch
F - Brake Light Switch
F47 - Brake Pedal Switch
J59 - Load Reduction Relay
J220 - Motronic Engine Control Module (ECM)
S5 - Fuse 5 in fuse holder
S13 - Fuse 13 in fuse holder
S176 - Fuse -4- (30), in fuse bracket/battery
T10h - 10-Pin Connector, black, behind instrument panel, left
T80 - Connector, 80 point
(500) - Threaded connection -1- (30) on the relay plate
(501) - Threaded connection -2- (30) on the relay plate
(504) - Threaded connection -1- (87) on the relay plate
(A2) - plus connection (15), in instrument panel wiring harness
(A18) - plus connection (54), in rear wiring harness
(A20) - wire connection (15a), in instrument panel wiring harness
(A99) - Connector -1- (87), in instrument panel
(A100) - Connector -2- (87), in instrument panel wiring harness
--- Automatic transmission only

Cruise control (2.0L AEG) from Aug. 1998

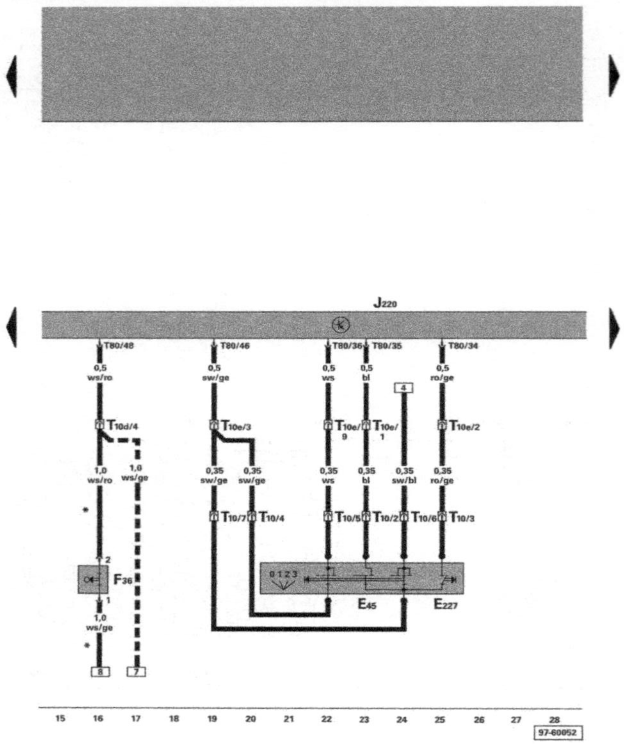

ws = white
sw = black
ro = red
br = brown
gn = green
bl = blue
gr = grey
li = violet
ge = yellow

E45 - Cruise Control Switch
E227 - Button for cruise control (set)
F36 - Clutch Vacuum Vent Valve Switch
J220 - Motronic Engine Control Module (ECM)
T10 - 10-pin connector, to right of steering column
T10d - 10-pin Connector, green, behind instrument panel, left
T10e - 10-pin Connector, black, behind instrument panel, left
T80 - 80-pin Connector, in Motronic Engine Control Module

(A100) - Connection 2 (87) (in instrument panel wiring harness)

* - Standard transmission only

--- Automatic transmission only

Cruise control switch, button for cruise control, clutch vacuum vent valve switch, Motronic engine control module.

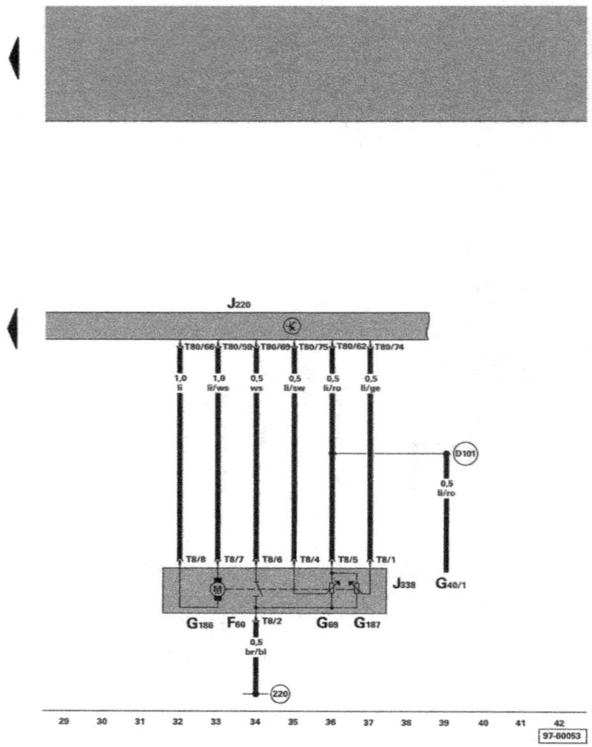

ws = white
sw = black
ro = red
br = brown
gn = green
bl = blue
gr = grey
li = violet
ge = yellow

F60 - Closed Throttle Position (CTP) Switch
G40 - Camshaft Position (CMP) Sensor
G69 - Throttle Position (TP) Sensor
G186 - Throttle Drive (electronic throttle)
G187 - Throttle Drive Angle Sensor 1 (electronic throttle)
J220 - Motronic Engine Control Module
J338 - Throttle Valve Control Module
T8 - 8-pin Connector
T80 - 80-pin Connector, in Motronic Engine Control Module

(220) - Ground Connection, Sensor Ground (in engine compartment wiring harness)

(D101) - Connection 1 (in engine compartment wiring harness)

Closed throttle position, Camshaft position sensor, Throttle position sensor, Throttle drive (electronic), Throttle drive angle sensor 1.

New Beetle **Wiring diagram** **No. 16/1**

Digital clock with ambient temperature display and front interior light, rear reading lights. (vehicles with power windows)

from August 1998

Deviate relay location and fuseplacements as well as the locations of multiple connectors see section "component locations".

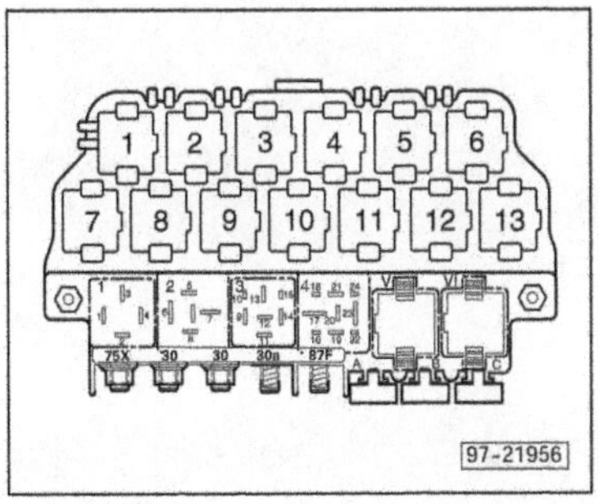

Fuse colors

30 A - green
25 A - white
20 A - yellow
15 A - blue
10 A - red
7,5 A - brown
5 A - beige
3 A - violet

New Beetle **Wiring diagram** **No. 16/2**

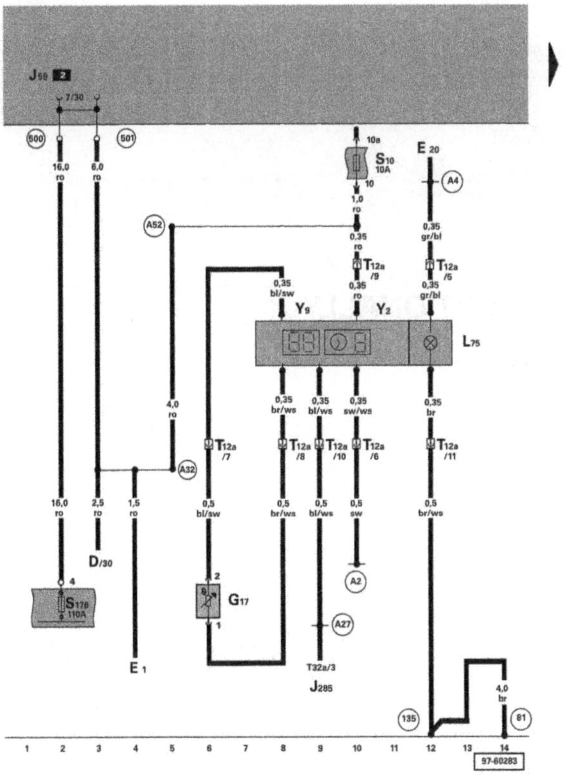

Digital clock, digital temperature display, light switch

- D - Ignition/Starter Switch
- E1 - Light Switch
- E20 - Switches and instruments - lighting control
- G17 - Outside Air Temperature Sensor
- J59 - Load Reduction Relay
- J285 - Control module w. indicator unit in instrument panel insert
- L75 - Light for digital display
- S10 - Fuse in fuse box
- S176 - Fuse -4- (30), in fuse bracket/battery
- T12a - 12-Pin Conenctor
- T32a - 32-Pin Connector
- Y2 - Digital Clock
- Y9 - Digital temperature display

- (A2) - Positive (+) connection (15), in instrument panel wiring harness
- (A4) - Positive (+) connection (58b), in instrument panel wiring harness
- (A27) - Wire Connection (vehicle speed signal), in instrument panel wiring harness
- (A32) - plus connection (30), in instrument panel wiring harness
- (A52) - Positive connection (30), in instrument panel wiring harness
- (81) - Ground connection -1-, in instrument panel wiring harness
- (135) - Ground connection -2-, in instrument panel wiring harness
- (500) - Threaded connection -1- (30), on the relay plate
- (501) - Threaded connection -2- (30), on the relay plate

ws = white
sw = black
ro = red
br = brown
gn = green
bl = blue
gr = grey
li = violet
ge = yellow
or = orange

Digital clock and interior lights from Aug. 1998

New Beetle — Wiring diagram — **No. 16/3**

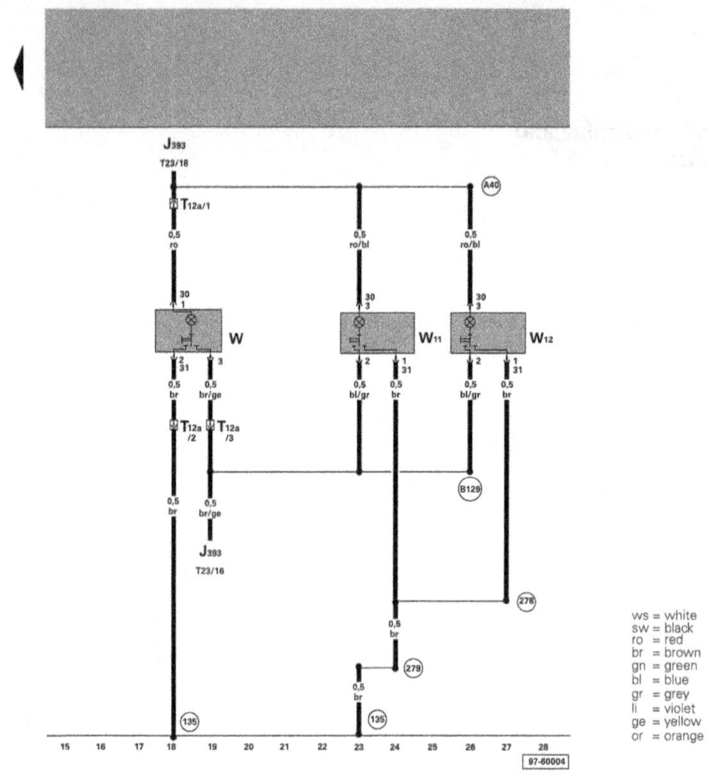

Front interior light, left rear reading light, right rear reading light

- J393 - Central control module for comfort system
- W - Front Interior Light
- W11 - Left Rear Reading Light
- W12 - Right Rear Reading Light
- T23 - 23-Pin Connector, on Central control module for comfort system
- (135) - Ground connection -2-, in instrument panel wiring harness
- (278) - Ground connector -4-, in wiring harness interior
- (279) - Ground connector -5, in wiring harness interior
- (A40) - plus connection -1- (30), in instrument cluster wiring harness
- (B129) - Connector (interior light 3/L), in wiring harness interior

THIS SPACE INTENTIONALLY LEFT BLANK

New Beetle Convertible Wiring diagram **No. 24/1**

Digital clock with ambient temperature display and front interior light,

from June 2003

Deviate relay location and fuseplacements as well as the locations of multiple connectors see section "component locations".

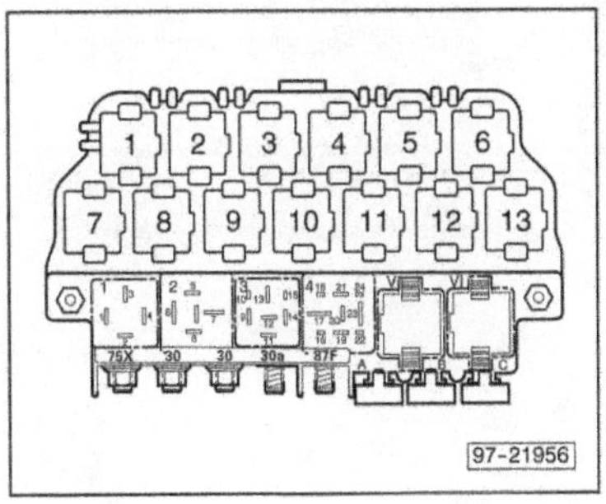

Fuse colors

30 A - green
25 A - white
20 A - yellow
15 A - blue
10 A - red
7,5 A - brown
5 A - beige
3 A - violet

New Beetle Convertible Wiring diagram **No. 24/2**

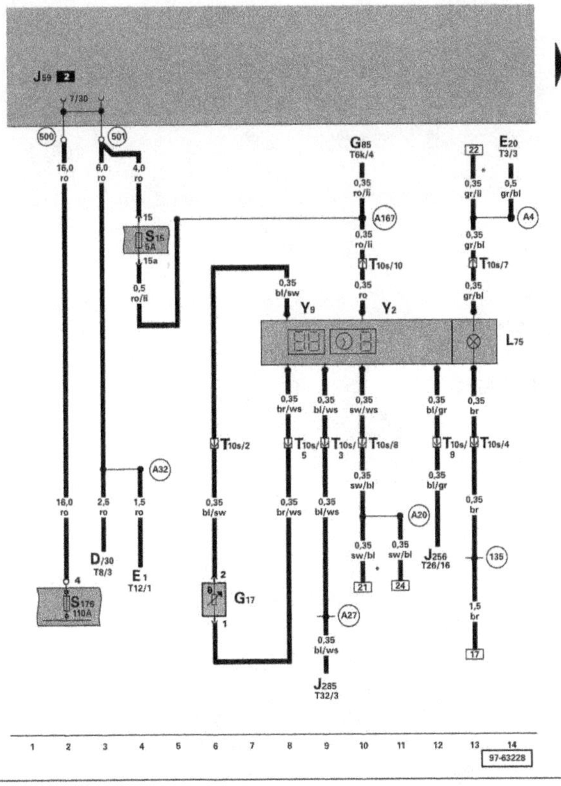

Digital clock, digital temperature display, Light for digital display

D - Ignition/Starter Switch
E1 - Light Switch
E20 - Switches and instruments - lighting control
G17 - Outside Air Temperature Sensor
G85 - Steering Angle Sensor
J59 - Load Reduction Relay
J256 - Convertible Top Control Module
J285 - Control module w. indicator unit in instrument panel insert
L75 - Light for digital display
S15 - Fuse 15 in fuse holder
S176 - Fuse -4- (30), in fuse bracket/battery
T3 - 3-Pin Conenctor, at E20
T6k - 6-Pin Conenctor, at G85
T8 - 8-Pin Conenctor, at D
T10s - 10-Pin Conenctor
T12 - 12-Pin Conenctor, at E1
T26 - 26-Pin Conenctor, at J256
T32 - 32-Pin Connector
Y2 - Digital Clock
Y9 - Digital temperature display

(A4) - Positive (+) connection (58b), in instrument panel wiring harness
(A20) - Plus Connection (15a) (in instrument panel wiring harness)
(A27) - Wire Connection (vehicle speed signal), in instrument panel wiring harness
(A32) - plus connection (30), in instrument panel wiring harness
(A167) - Plus Connection 3 (30a) (in instrument panel wiring harness)
(135) - Ground connection -2-, in instrument panel wiring harness
(500) - Threaded connection -1- (30), on the relay plate
(501) - Threaded connection -2- (30), on the relay plate

* - Only vehicles with Automatic day/night interior mirror

ws = white
sw = black
ro = red
br = brown
gn = green
bl = blue
gr = grey
li = violet
ge = yellow
or = orange

Digital clock and interior lights from June 2003, convertible

New Beetle Convertible Wiring diagram **No. 24/3**

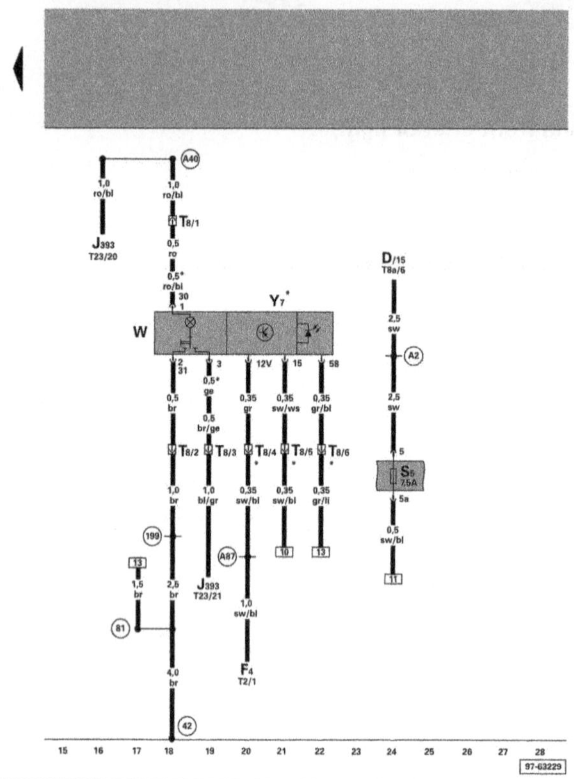

Front interior light, Automatic Day/Night Interior Mirror

- D - Ignition/Starter Switch
- F4 - Back-Up Light Switch
- J393 - Central control module for comfort system
- S5 - Fuse 5 in fuse holder
- T2 - Double Connector
- T8 - 8-Pin Connector
- T8a - 8-Pin Connector, at D
- T23 - 23-Pin Connector, on Central control module for comfort system
- W - Front Interior Light
- Y7 - Automatic Day/Night Interior Mirror *
- (42) - Ground Connection, beside steering column
- (81) - Ground Connection 1, in instrument panel wiring harness
- (199) - Ground Connection 3, in instrument panel wiring harness
- (A2) - Plus Connection (15), in instrument panel wiring harness
- (A40) - plus connection -1- (30), in instrument cluster wiring harness
- (A87) - Connection (reverse lamp), in instrument panel wiring harness
- * - Only vehicles with Automatic day/night interior mirror

ws = white
sw = black
ro = red
br = brown
gn = green
bl = blue
gr = grey
li = violet
ge = yellow
or = orange

THIS SPACE INTENTIONALLY LEFT BLANK

New Beetle **Wiring diagram** **No. 61/1**

Digital clock with ambient temperature display, and front interior light, rear reading lights,

from July 2005

Additional Fuse, Relay, Control Module and Connector locations ⇒ Component Locations

Electrical troubleshooting ⇒ Guided Fault Finding using VAS 5051/5052

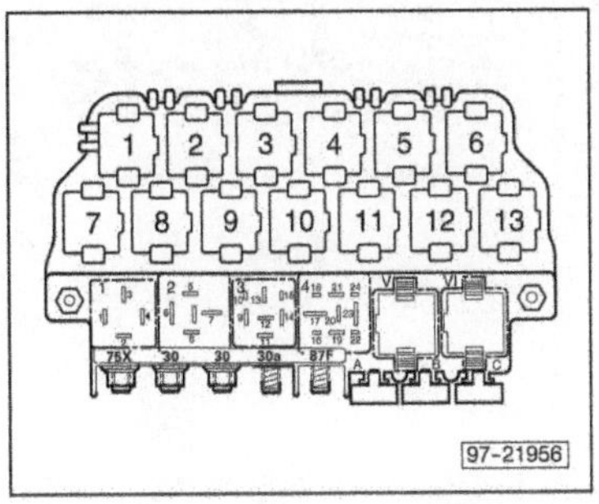

New Beetle **Wiring diagram** **No. 61/2**

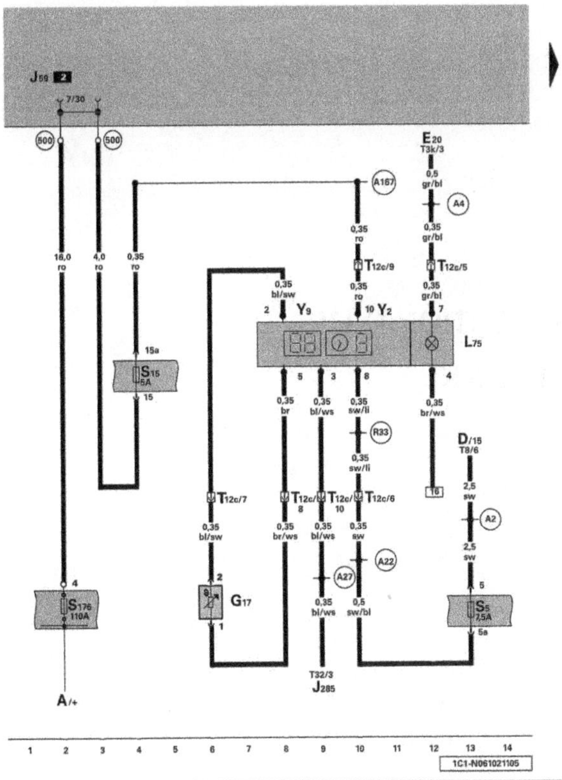

ws = white
sw = black
ro = red
br = brown
gn = green
bl = blue
gr = grey
li = violet
ge = yellow
or = orange

Digital clock, digital temperature display

- D - Ignition/Starter Switch
- E20 - Switches and instruments - lighting control
- G17 - Outside Air Temperature Sensor
- J59 - Load Reduction Relay
- J285 - Control module w. indicator unit in instrument panel insert
- L75 - Light for digital display
- S5 - Fuse (on fuse/relay panel)
- S15 - Fuse (on fuse/relay panel)
- S176 - Fuse -4- (30), in fuse bracket/battery
- T3k - 3-Pin Conenctor
- T8 - 8-Pin Conenctor
- T12c - 12-Pin Conenctor
- T32 - 32-Pin Connector
- Y2 - Digital Clock
- Y9 - Digital temperature display

- (500) - Threaded connection -1- (30), on the relay plate
- (501) - Threaded connection -2- (30), on the relay plate
- (A2) - Positive (+) connection (15), in instrument panel wiring harness
- (A4) - Positive (+) connection (58b), in instrument panel wiring harness
- (A22) - Connection (75s) (in instrument panel wiring harness)
- (A27) - Wire Connection (vehicle speed signal), in instrument panel wiring harness
- (A167) - Plus Connection 3 (30a) (in instrument panel wiring harness)
- (R33) - Plus Connection 3 (in interior light wiring harnss)

New Beetle **Wiring diagram** **No. 61/3**

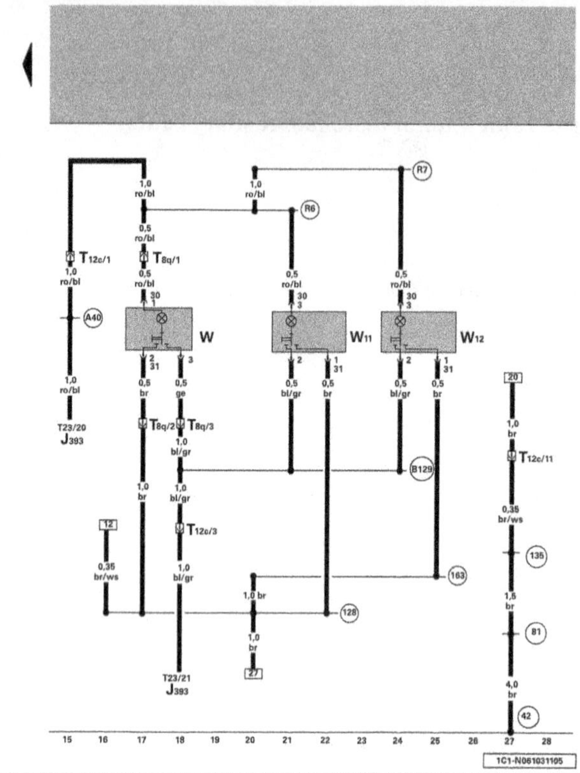

Front Interior Light, Left Rear Reading Light, Right Rear Reading Light

- J393 - Comfort System Central Control Module
- T8q - 8-Pin Connector at W
- T12c - 12-Pin Coupling Connector, on A pillar left, central part
- T23 - 23-Pin Connector, at J393
- W - Front Interior Light
- W11 - Left Rear Reading Light
- W12 - Right Rear Reading Light
- 42 - Ground Connection (beside steering column)
- 81 - Ground Connection (beside steering column)
- 128 - Ground connection -1-, in interior light wiring harness
- 135 - Ground connection -2-, in instrument panel wiring harness
- 163 - Ground connection -2-, in interior light wiring harness
- A40 - Plus Connection 1 (30) (in instrument cluster wiring harness)
- B129 - Connector (interior light 3/L), in wiring harness interior
- R6 - Plus connection -1-, in interior light wiring harness
- R7 - Plus connection -2-, in interior light wiring harness

ws = white
sw = black
ro = red
br = brown
gn = green
bl = blue
gr = grey
li = violet
ge = yellow
or = orange

THIS SPACE INTENTIONALLY LEFT BLANK

New Beetle **Wiring diagram** **No. 33/1**

Direct Shift Automatic Transmission (DSG) 02E

from August 2003

Deviate relay location and fuseplacements as well as the locations of multiple connectors see section "component locations".

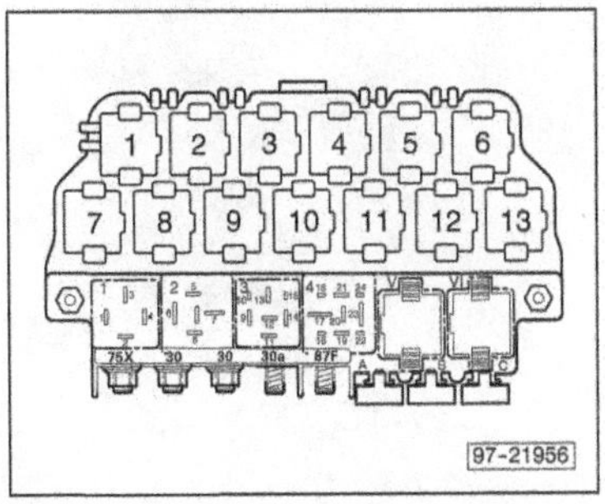

Relay location on the thirteen position auxiliary relay panel, above relay panel:

- **11** Park/Neutral Position (PNP) Relay (126)
- **3** Back -Up Light Relay (53)

Relay panel:

- **2** Load Reduction Relay (100)
- **4** Fuel Pump (FP) Relay (409)

Fuse colors

30 A - green
25 A - white
20 A - yellow
15 A - blue
10 A - red
7,5 A - brown
5 A - beige
3 A - violet

Note: Number in parentheses indicates production control number stamped on relay housing.

New Beetle **Wiring diagram** **No. 33/2**

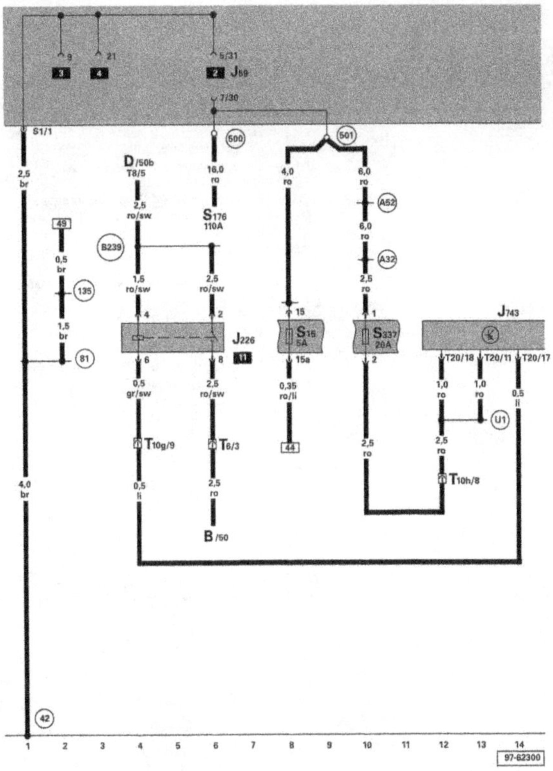

Transmission Control Module (TCM), Park/Neutral Position (PNP) Relay

B - Starter
D - Ignition/Starter Switch
J59 - Load Reduction Relay
J743 - Direct Shift Gearbox (DSG) Mechatronic
J226 - Park/Neutral Position (PNP) Relay (126)
S15 - Fuse 15 in fuse holder
S176 - Fuse -4- (30), in fuse bracket/battery
S340 - Fuse for direct shift gearbox mechatronic
T6 - 6-Pin Connector, brown, in plenum chamber, left
T8 - 8-Pin Connector, at -D-
T10g - 10-Pin Connector, grey, behind instrument panel, left
T10h - 10-Pin Connector, black, behind instrument panel, left
T20 - 20-Pin Connector, at -J217-
(42) - Ground connection, beside steering column
(81) - Ground connection -1-, in instrument panel wiring harness
(135) - Ground connection -2-, in instrument panel wiring harness
(500) - Threaded connection -1- (30) on the relay plate
(501) - Threaded connection -2- (30) on the relay plate
(A32) - Plus connection (30), in instrument panel wiring harness
(A52) - Plus Connection 2 (30) (in instrument panel wiring harness)
(B239) - Plus Connection 1 (50) (in interior wiring harness)
(U1) - wire connection (50), in automatic transmission wiring harness

ws = white
sw = black
ro = red
br = brown
gn = green
bl = blue
gr = grey
li = violet
ge = yellow
or = orange

New Beetle

Wiring diagram

No. 33/3

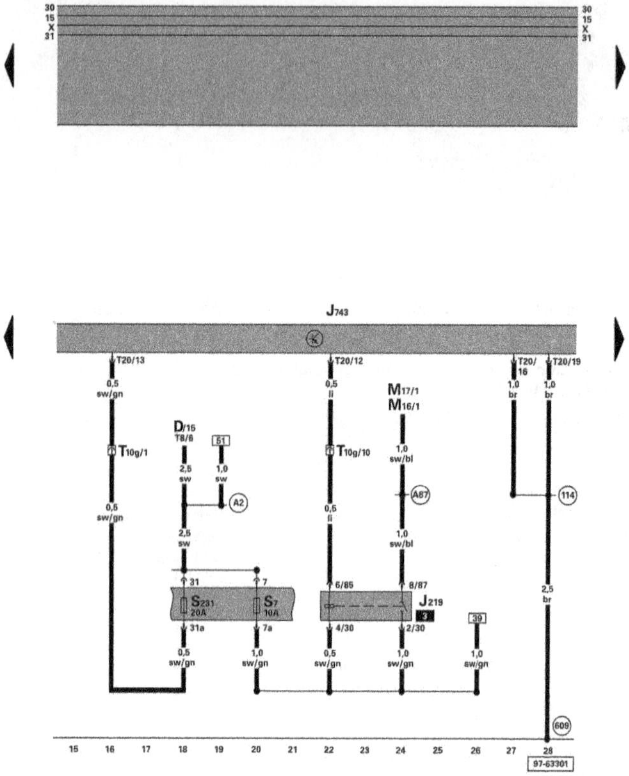

Transmission Control Module (TCM), Back-Up Light Relay

- D - Ignition/Starter Switch
- J743 - Direct Shift Gearbox (DSG) Mechatronic
- J219 - Back-Up Light Relay
- M16 - Left Back-Up Light
- M17 - Right Back-Up Light
- S7 - Fuse 7 in fuse/relay panel
- S231 - Fuse 31 in fuse/relay panel
- T8 - 8-Pin Connector, at -D-
- T10g - 10-Pin Connector, grey, behind instrument panel, left
- T20 - 20-Pin Connector, at -J217-

- (114) - Ground connection, in automatic transmission wiring harness
- (609) - Ground Connection (in right plenum chamber)
- (A2) - plus connection (15), in instrument panel wiring harness
- (A87) - Connector (reverse lamp), in instrument panel wiring harness

ws = white
sw = black
ro = red
br = brown
gn = green
bl = blue
gr = grey
li = violet
ge = yellow
or = orange

New Beetle

Wiring diagram

No. 33/4

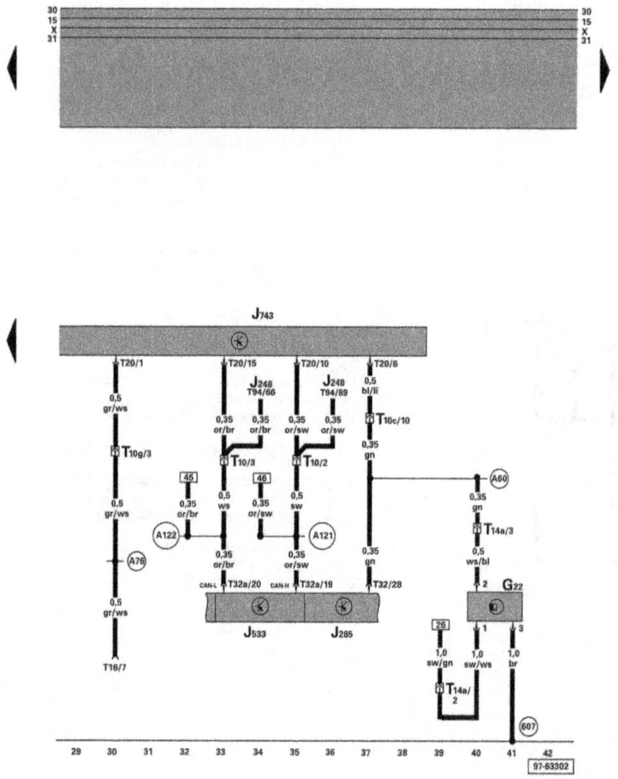

Speedometer Vehicle Speed Sensor, Transmission Control Module (TCM), Control module with indicator unit in instrument panel insert, Data Bus On Board Diagnostic Interface

- G22 - Speedometer Vehicle Speed Sensor
- J743 - Direct Shift Gearbox (DSG) Mechatronic
- J248 - Diesel Direct Fuel Injection (DFI) Engine Control Module (ECM)
- J285 - Control module with indicator unit in instrument panel insert
- J533 - Data Bus On Board Diagnostic Interface
- T10 - 10-Pin Connector, white, behind instrument panel, left
- T10c - 10-Pin Connector, blue, behind instrument panel, left
- T10g - 10-Pin Connector, grey, behind instrument panel, left
- T14a - 14-Pin Connector, near battery
- T16 - 16-Pin Connector, Data link Connector (DLC), below steering column, left
- T20 - 20-Pin Connector, at -J217-
- T32 - 32-Pin Connector, blue
- T32a - 32-Pin Connector, green
- T94 - 94-Pin Connector at -J248-

- (607) - Ground Connection (in left plenum chamber)
- (A60) - Wire connection (Vehicle speed signal), in instrument cluster wiring harness
- (A76) - Connector (K-diagnosis wire) in instrument panel wiring harness
- (A121) - Connection (high bus) in instrument panel wiring harness
- (A122) - Connection (low bus) in instrument panel wiring harness

ws = white
sw = black
ro = red
br = brown
gn = green
bl = blue
gr = grey
li = violet
ge = yellow
or = orange

New Beetle **Wiring diagram** **No. 33/5**

Tiptronic switch, warning light for selector lever position P/N, illumination for selector lever scale

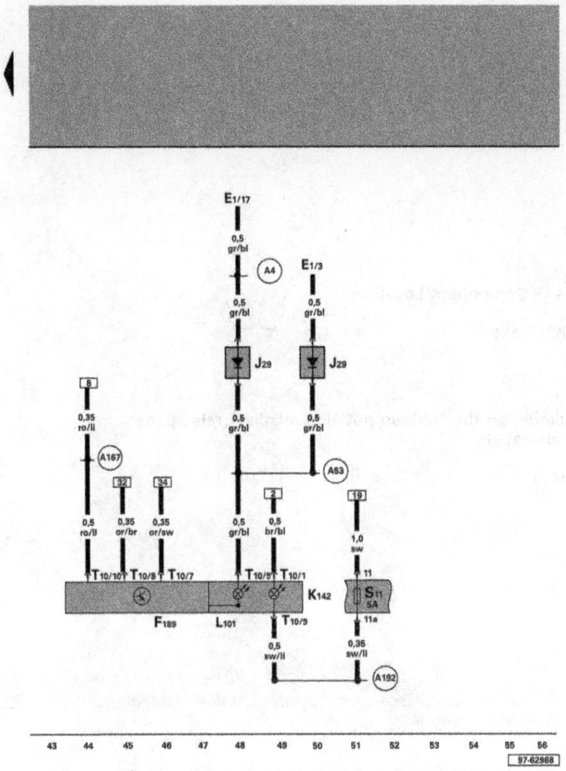

E1 - Light switch
F189 - Tiptronic Switch
J29 - Blocking Diode
K142 - Warning light for selector lever position P/N
L101 - Illumination for selector lever scale
T10 - 10-Pin Connector, white, behind instrument panel, left

(A4) - Plus connection (58b), in instrument panel wiring harness panel wiring harness
(A53) - plus connection (58b), in instrument panel wiring harness
(A167) - Plus Connection 3 (30a) (in instrument panel wiring harness)
(A192) - Plus Connection 3 (15a) (in instrument panel wiring harness)

ws = white
sw = black
ro = red
br = brown
gn = green
bl = blue
gr = grey
li = violet
ge = yellow
or = orange

THIS SPACE INTENTIONALLY

LEFT BLANK

New Beetle Wiring diagram **No. 50/1**

Direct Shift Automatic Transmission (DSG) 02E,

from July 2005

Additional Fuse, Relay, Control Module and Connector locations ⇒ Component Locations

Electrical troubleshooting ⇒ Guided Fault Finding using VAS 5051/5052

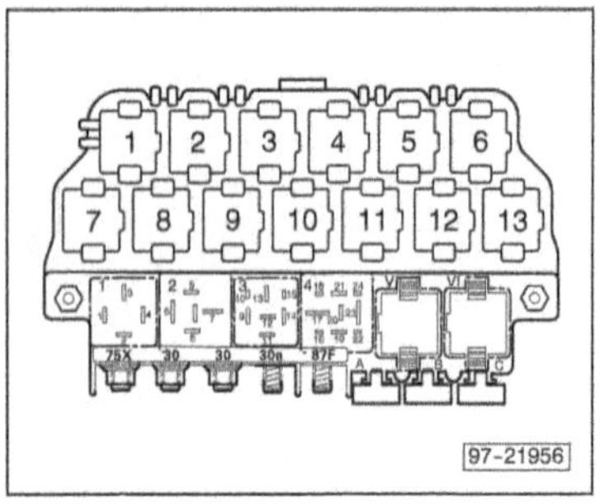

Relay location on the thirteen position auxiliary relay panel, above relay panel:

3 - J219 - Back-up Light Relay (53)

Relay panel:

2 - J59 - Load reduction Relay (100)

Note:

Number in parentheses indicates production control number stamped on relay housing.

New Beetle Wiring diagram **No. 50/2**

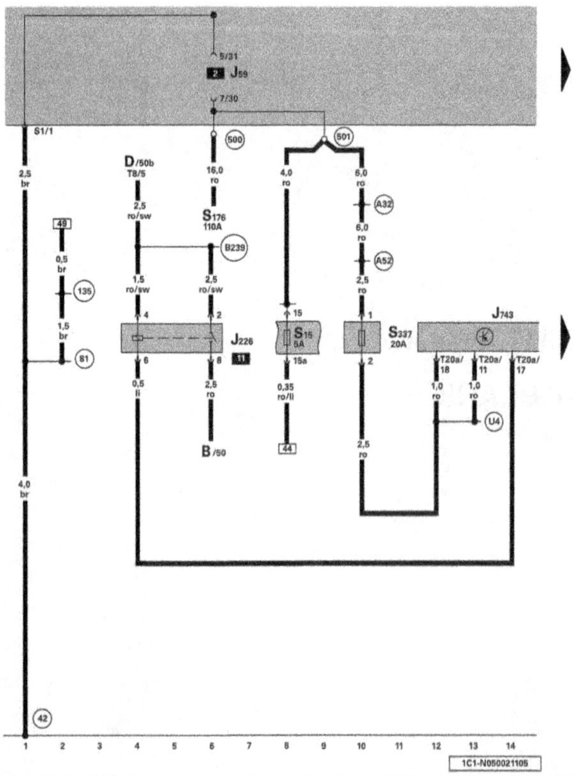

ws = white
sw = black
ro = red
br = brown
gn = green
bl = blue
gr = grey
li = violet
ge = yellow
or = orange

Transmission Control Module (TCM), Park/Neutral Position (PNP) Relay

- B - Starter
- D - Ignition/Starter Switch
- J59 - Load Reduction Relay
- J743 - Direct Shift Gearbox (DSG) Mechatronic
- J226 - Park/Neutral Position (PNP) Relay (126)
- S15 - Fuse 15 in fuse holder
- S176 - Fuse -4- (30), in fuse bracket/battery
- S337 - Mechatronic Direct Shift Gearbox Fuse
- T8 - 8-Pin Connector, at -D-
- T20a - 20-Pin Connector, at -J743-
- (42) - Ground connection, beside steering column
- (81) - Ground connection -1-, in instrument panel wiring harness
- (135) - Ground connection -2-, in instrument panel wiring harness
- (500) - Threaded connection -1- (30) on the relay plate
- (501) - Threaded connection -2- (30) on the relay plate
- (A32) - Plus connection (30), in instrument panel wiring harness
- (A52) - Plus Connection 2 (30) (in instrument panel wiring harness)
- (B239) - Plus Connection 1 (50) (in interior wiring harness)
- (U4) - Connection (30) (in automatic transmission wiring harness)

New Beetle

Wiring diagram **No. 50/3**

Transmission Control Module (TCM), Back-Up Light Relay

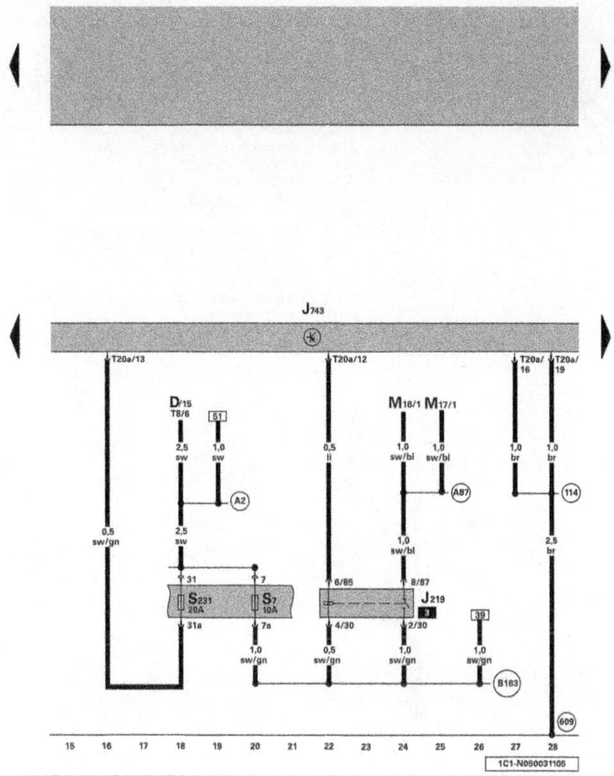

- D - Ignition/Starter Switch
- J743 - Direct Shift Gearbox (DSG) Mechatronic
- J219 - Back-Up Light Relay
- M16 - Left Back-Up Light
- M17 - Right Back-Up Light
- S7 - Fuse 7 in fuse/relay panel
- S231 - Fuse 31 in fuse/relay panel
- T8 - 8-Pin Connector, at -D-
- T20a - 20-Pin Connector, at -J743-

- (114) - Ground connection, in automatic transmission wiring harness
- (609) - Ground Connection (in right plenum chamber)
- (A2) - Plus connection (15), in instrument panel wiring harness
- (A87) - Connector (reverse lamp), in instrument panel wiring harness
- (B163) - Plus Connection 1 (15) (in interior wiring harness)

ws = white
sw = black
ro = red
br = brown
gn = green
bl = blue
gr = grey
li = violet
ge = yellow
or = orange

New Beetle

Wiring diagram **No. 50/4**

Speedometer Vehicle Speed Sensor, Transmission Control Module (TCM), Instrument Cluster Control Module, Data Bus On Board Diagnostic Interface

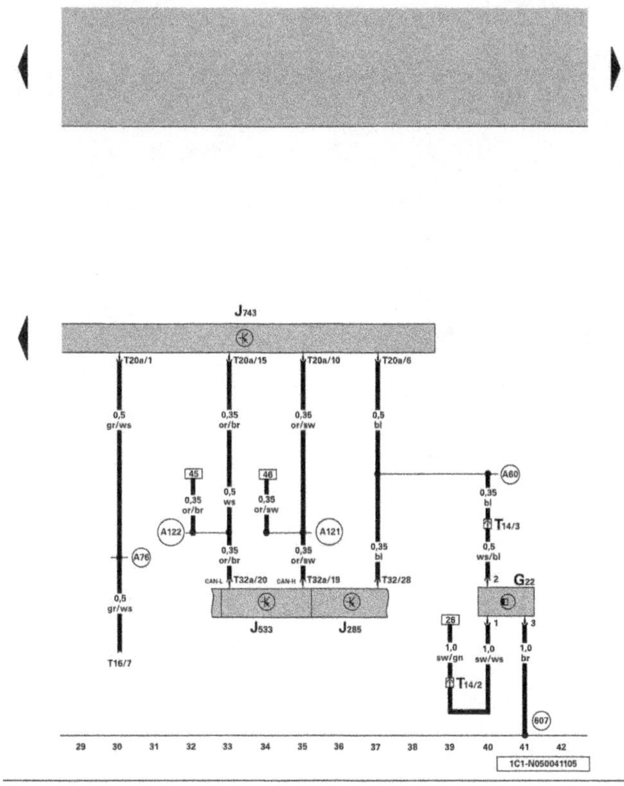

- G22 - Speedometer Vehicle Speed Sensor (VSS)
- J743 - Direct Shift Gearbox (DSG) Mechatronic
- J285 - Instrument Cluster Control Module
- J533 - Data Bus On Board Diagnostic Interface
- T14 - 14-Pin Connector, near battery
- T16 - 16-Pin Connector, Data link Connector (DLC), below steering column, left
- T20a - 20-Pin Connector, at -J743-
- T32 - 32-Pin Connector, blue
- T32a - 32-Pin Connector, green

- (607) - Ground Connection (in left plenum chamber)
- (A60) - Wire connection (Vehicle speed signal), in instrument cluster wiring harness
- (A76) - Connector (K-diagnosis wire) in instrument panel wiring harness
- (A121) - Connection (high bus) in instrument panel wiring harness
- (A122) - Connection (low bus) in instrument panel wiring harness

ws = white
sw = black
ro = red
br = brown
gn = green
bl = blue
gr = grey
li = violet
ge = yellow
or = orange

New Beetle Wiring diagram **No. 50/5**

Tiptronic switch, warning light for selector lever position P/N, illumination for selector lever scale

- E1 - Light switch
- E20 - Instrument Panel Illumination Dimmer Switch
- F189 - Tiptronic Switch
- J29 - Blocking Diode
- K142 - Warning light for selector lever position P/N
- L101 - Illumination for selector lever scale
- T3k - 3-Pin Connector
- T10u - 10-Pin Connector
- T17 - 17-Pin Connector

- (A4) - Plus connection (58b), in instrument panel wiring harness panel wiring harness
- (A53) - Plus Connection (58b) (in instrument panel wiring harness)
- (A167) - Plus Connection 3 (30a) (in instrument panel wiring harness)
- (B165) - Conexión positiva -2- (15), en mazo de cables habitáculo

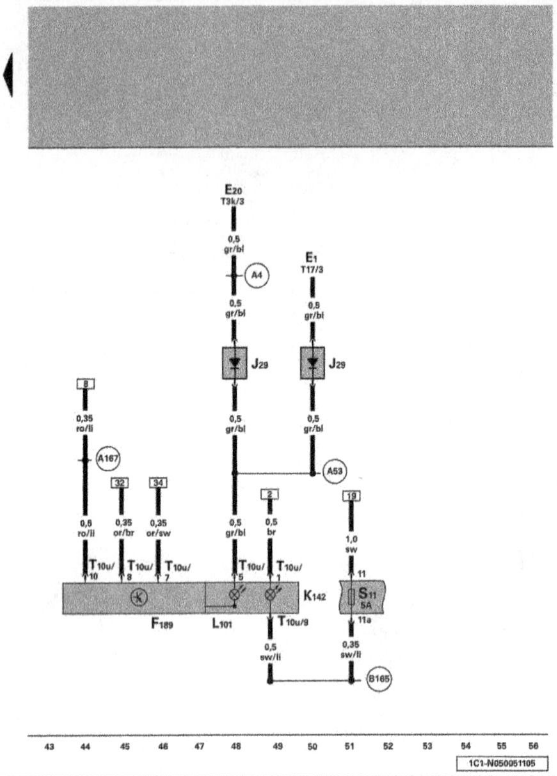

- ws = white
- sw = black
- ro = red
- br = brown
- gn = green
- bl = blue
- gr = grey
- li = violet
- ge = yellow
- or = orange

THIS SPACE INTENTIONALLY LEFT BLANK

New Beetle

Wiring diagram

No. 18/1

Front foglights

from January 1998

Deviate relay location and fuseplacements as well as the locations of multiple connectors see section "component locations".

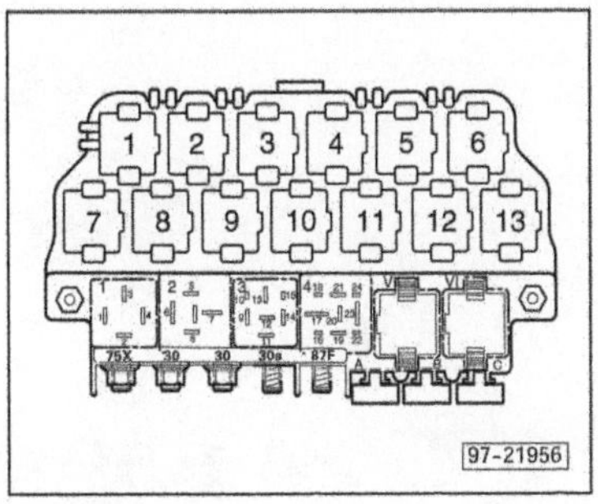

Fuse colors

30 A - green
25 A - white
20 A - yellow
15 A - blue
10 A - red
7,5 A - brown
5 A - beige
3 A - violet

New Beetle

Wiring diagram

No. 18/2

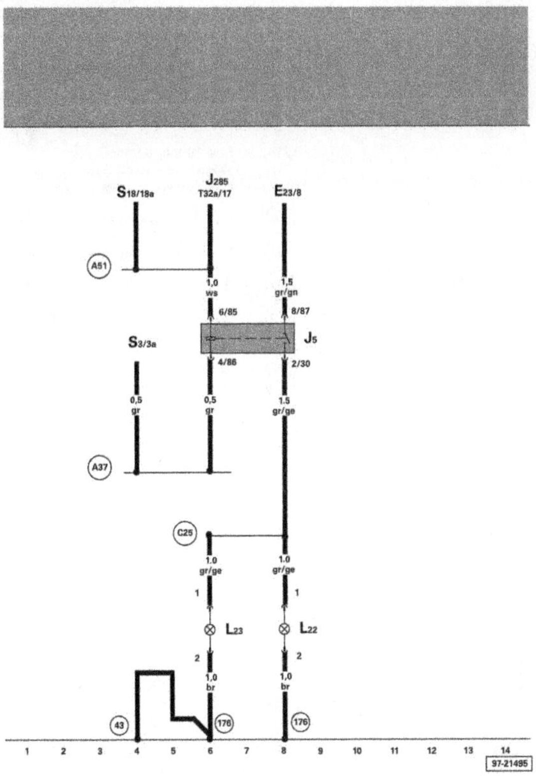

Front foglights

E23 - Fog light switch
J5 - Fog light relay
J285 - Control module with indicator unit in instrument panel insert
L22 - Left front foglight
L23 - Right front foglight
S3 - Fuse 3 in fuse holder
S18 - Fuse 18 in fuse holder
T32a - 32-Pin Connector, blue, on J285

(43) - Ground connection, on right A pillar, lower part

(176) - Ground connection, in right headlight wiring harness

(A37) - wire connection (58a), in instrument panel wiring harness

(A51) - wire connection (56), in instrument panel wiring harness

(C25) - wire connection, fog lights, in front light wiring harness

ws = white
sw = black
ro = red
br = brown
gn = green
bl = blue
gr = grey
li = violet
ge = yellow
or = orange

New Beetle — Wiring diagram — No. 55/1

Fog Lamps,

from July 2005

Additional Fuse, Relay, Control Module and Connector locations ⇒ Component Locations

Electrical troubleshooting ⇒ Guided Fault Finding using VAS 5051/5052

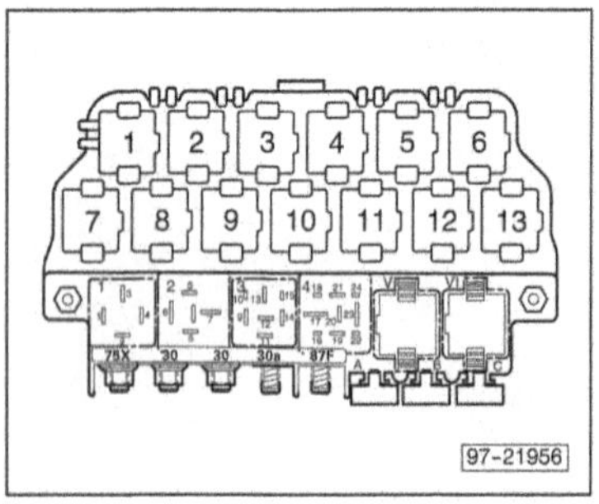

New Beetle — Wiring diagram — No. 55/2

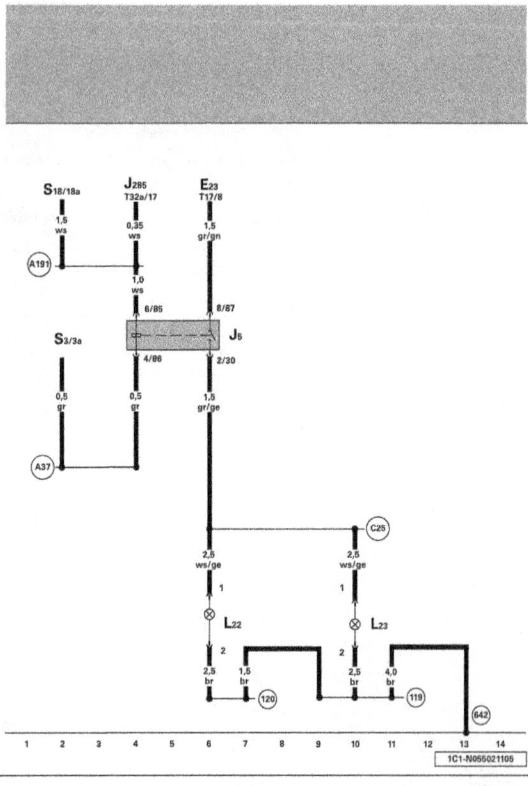

Fog light relay, Front foglights

- E23 - Fog light switch
- J5 - Fog light relay
- J285 - Control module with indicator unit in instrument panel insert
- L22 - Left front foglight
- L23 - Right front foglight
- S3 - Fuse 3 in fuse holder
- S18 - Fuse 18 in fuse holder
- T17 - 17-Pin Connector, on E23
- T32a - 32-Pin Connector, blue, on J285

- (119) - Ground Connection 1 (in headlamp wiring harness)
- (120) - Ground Connection 2 (in headlamp wiring harness)
- (642) - Ground Connection (on electronic fan control)

- (A37) - Wire connection (58a), in instrument panel wiring harness
- (A191) - Connection (56aR) (in instrument panel wiring harness)
- (C25) - wire connection, fog lights, in front light wiring harness

ws = white
sw = black
ro = red
br = brown
gn = green
bl = blue
gr = grey
li = violet
ge = yellow
or = orange

New Beetle Wiring diagram **No. 64/1**

Garage door opener - HomeLink®,

from July 2005,

Additional Fuse, Relay, Control Module and Connector locations ⇒ Compontent Locations

Troubleshooting Procedures ⇒ Guided Fault Finding using VAS 5051/5052

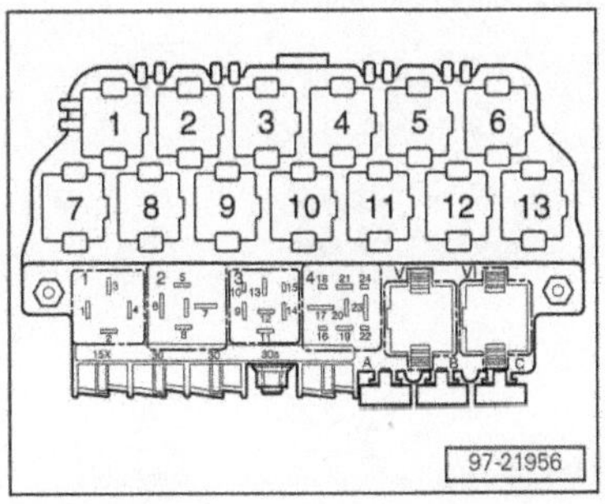

Relay location on the thirteenfold auxiliary relay panel, above relay panel:

Relay panel:

New Beetle Wiring diagram **No. 64/2**

Driver's Make-Up Mirror Light Contact Switch, Garage Door Opener Control Module

F147 - Driver's Make-Up Mirror Light Contact Switch
J393 - Comfort System Central Control Module
J530 - Garage Door Opener Control Module
T3g - 3-Pin Connector
T12c - 12-Pin Connector
T23 - 23-Pin Connector
W20 - Driver's Illuminated Make-up Mirror Light

(42) - Ground Connection (beside steering column)
(81) - Ground Connection 1 (in instrument panel wiring harness)
(128) - Ground Connection 1 (in interior light wiring harness)
(135) - Ground Connection 2 (in instrument panel wiring harness)
(347) - Ground Connection (in roof wiring harness)
(A40) - Plus Connection 1 (30) (in instrument cluster wiring harness)
(R6) - Plus Connection 1 (in interior light wiring harness)

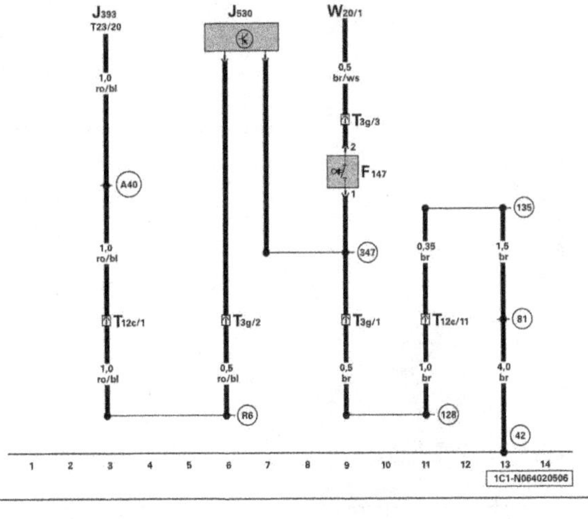

ws = white
sw = black
ro = red
br = brown
gn = green
bl = blue
gr = grey
li = violet
ge = yellow
or = orange

New Beetle **Wiring diagram** **No. 19**

Headlight washer,

from November 2000

Deviate relay location and fuseplacements as well as the locations of multiple connectors see section "component locations".

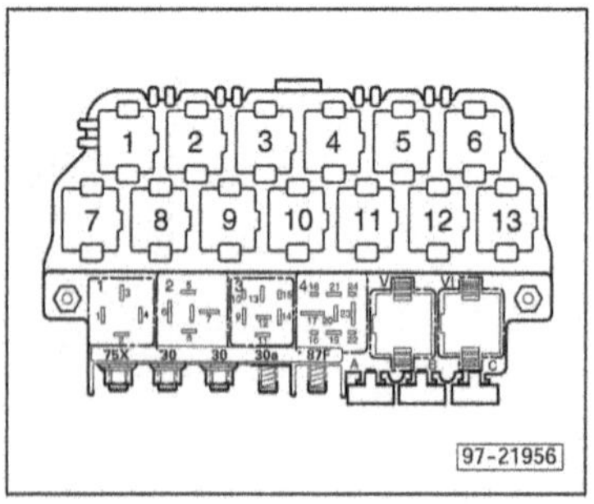

Relay location on the thirteen position auxiliary relay panel, above relay panel:

Relay panel:

- V
- VI — Wiper/Washer Intermittent Relay (389)

Fuse colors

30 A - green
25 A - white
20 A - yellow
15 A - blue
10 A - red
7,5 A - brown
5 A - beige
3 A - violet

Note: Number in parentheses indicates production control number stamped on relay housing.

New Beetle **Wiring diagram** **No. 19/2**

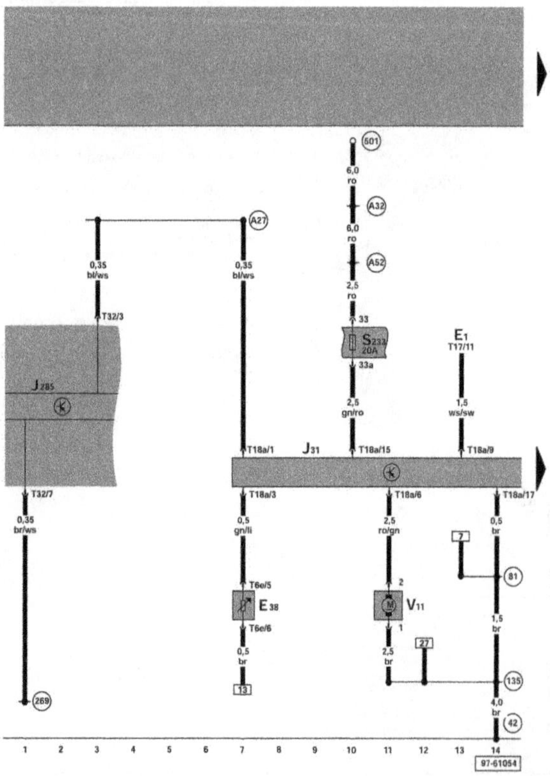

ws = white
sw = black
ro = red
br = brown
gn = green
bl = blue
gr = grey
li = violet
ge = yellow
or = orange

Wiper/washer intermittent relay, windshield wiper intermittent regulator, , headlight washer pump

- E1 - Light switch
- E38 - Windshield Wiper Intermittent Regulator
- J31 - Wiper/Washer Intermittent Relay, on relais panel
- J285 - Control module with indicator unit in instrument panel insert
- S233 - Fuse 33 in fuse holder
- T6e - 6-Pin Connector
- T17 - 17-Pin Connector
- T18a - 18-Pin Connector
- T32 - 32-Pin Connector, blue
- V11 - Headlight Washer Pump

- (42) - Ground connection, beside steering column
- (81) - Ground connection -1-, in instrument panel wiring harness
- (135) - Ground connection -2-, in instrument panel wiring harness
- (269) - Ground connector (sensor ground) -1-, in instrument panel wiring harness
- (501) - Threaded connection -2- (30) on the relay plate
- (A27) - Wire connection (vehicle speed signal), in instrument panel wiring harness
- (A32) - Plus connection (30), in instrument panel wiring harness
- (A52) - Plus Connection 2 (30) (in instrument panel wiring harness)

New Beetle

Wiring diagram **No. 19/3**

Wiper/washer intermittent relay, windshield wiper/washer switch, windshield wiper motor

E22 - Windshield Wiper/Washer Switch
J31 - Wiper/Washer Intermittent Relay, on relay panel
S224 - Fuse 24 in fuse holder
S227 - Fuse 27 in fuse holder
T5c - 5-Pin Connector, on windshield wiper motor
T8c - 8-Pin Connector
T18a - 18-Pin Connector
V - Windshield Wiper Motor

(179) - Ground connection, in left headlight wiring harness
(503) - Threaded connection -2- (75x) on the relay plate

(A96) - Connector (53a), in instrument panel wiring harness
(A97) - Connector (53), in instrument panel wiring harness
(A102) - Connector (windshield wiper), in instrument panel wiring harness

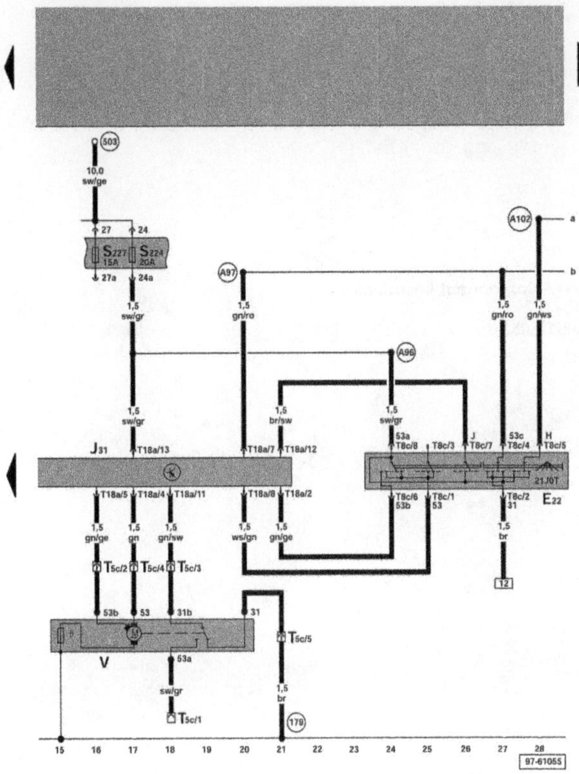

ws = white
sw = black
ro = red
br = brown
gn = green
bl = blue
gr = grey
li = violet
ge = yellow
or = orange

Edition 08.02

New Beetle

Wiring diagram **No. 19/4**

Windshield and rear window washer pump

V59 - Windshield and Rear Window Washer Pump

(A97) - Connector (53), in instrument panel wiring harness
(A102) - Connector (windshield wiper), in instrument panel wiring harness

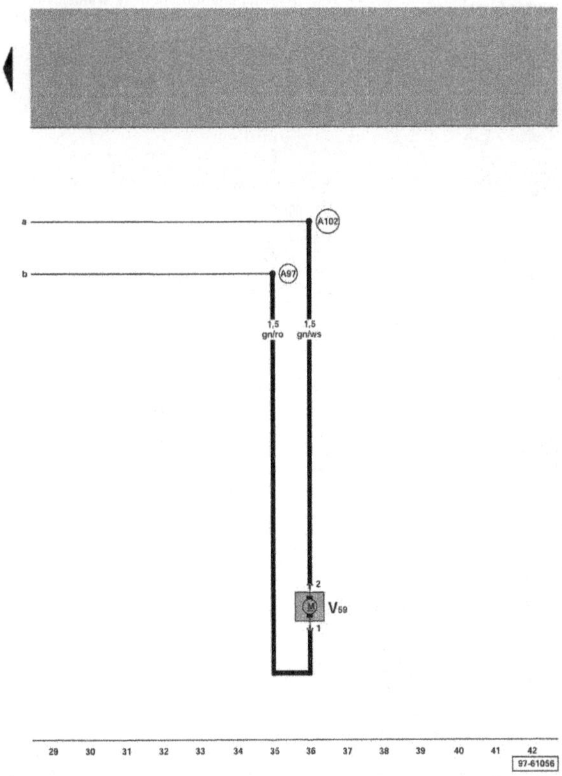

ws = white
sw = black
ro = red
br = brown
gn = green
bl = blue
gr = grey
li = violet
ge = yellow
or = orange

Edition 08.02

New Beetle Convertible Wiring diagram **No. 48/1**

Headlight washer,

from July 2005

Additional Fuse, Relay, Control Module and Connector locations ⇒ Component Locations

Electrical troubleshooting ⇒ Guided Fault Finding using VAS 5051/5052

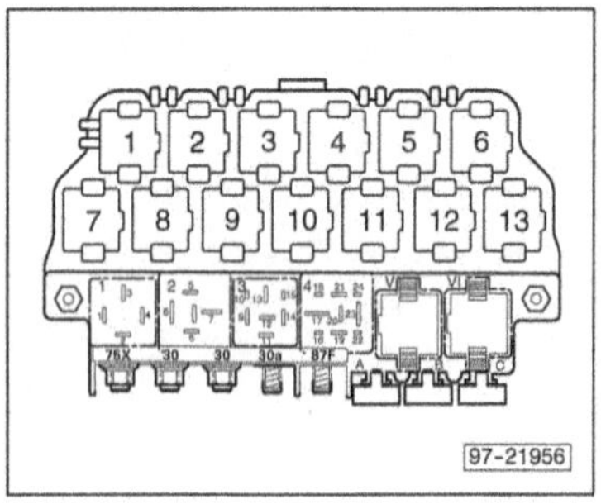

New Beetle Convertible Wiring diagram **No. 48/2**

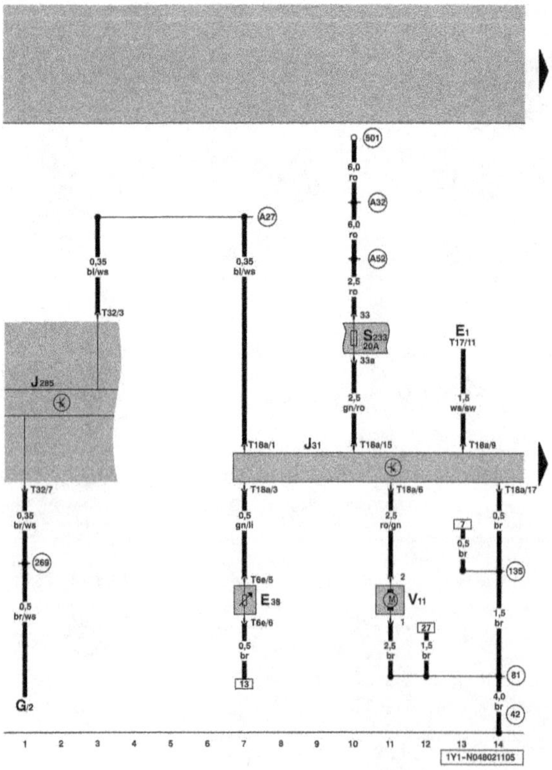

ws = white
sw = black
ro = red
br = brown
gn = green
bl = blue
gr = grey
li = violet
ge = yellow
or = orange

Wiper/washer intermittent relay, windshield wiper intermittent regulator, headlight washer pump

- E1 - Light switch
- E38 - Windshield Wiper Intermittent Regulator
- G - Fuel Level Sensor
- J31 - Wiper/Washer Intermittent Relay, on relais panel
- J285 - Control module with indicator unit in instrument panel insert
- S233 - Fuse 33 in fuse holder
- T6e - 6-Pin Connector
- T17 - 17-Pin Connector
- T18a - 18-Pin Connector
- T32 - 32-Pin Connector, blue
- V11 - Headlight Washer Pump
- (42) - Ground connection, beside steering column
- (81) - Ground connection -1-, in instrument panel wiring harness
- (135) - Ground connection -2-, in instrument panel wiring harness
- (269) - Ground connector (sensor ground) -1-, in instrument panel wiring harness
- (501) - Threaded connection -2- (30) on the relay plate
- (A27) - Vehicle Speed Signal (VSS) Connection (in instrument panel wiring harness
- (A32) - Plus connection (30), in instrument panel wiring harness
- (A52) - Plus Connection 2 (30) (in instrument panel wiring harness)

New Beetle Convertible

Wiring diagram **No. 48/3**

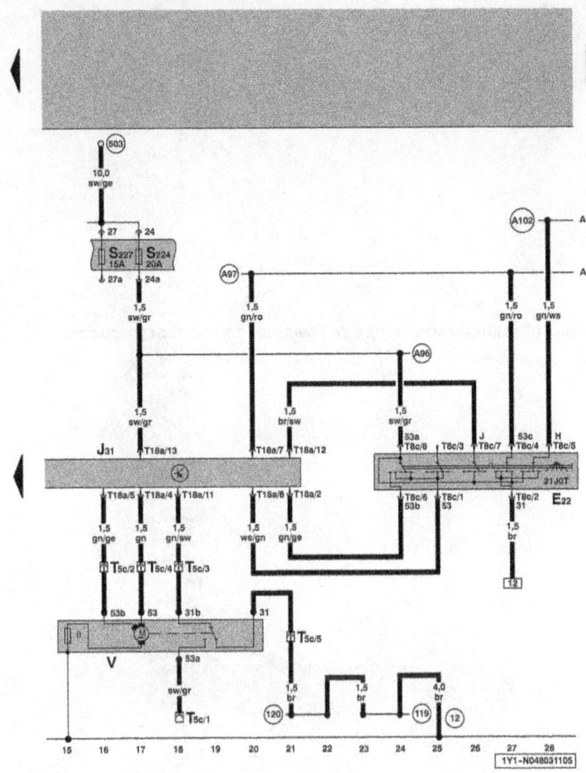

Wiper/washer intermittent relay, windshield wiper/washer switch, windshield wiper motor

- E22 - Windshield Wiper/Washer Switch
- J31 - Wiper/Washer Intermittent Relay, on relay panel
- S224 - Fuse 24 in fuse holder
- S227 - Fuse 27 in fuse holder
- T5c - 5-Pin Connector, on windshield wiper motor
- T8c - 8-Pin Connector
- T18a - 18-Pin Connector
- V - Windshield Wiper Motor
- (12) - Ground Connection (in engine compartment, left)
- (119) - Ground Connection 1 (in headlamp wiring harness)
- (120) - Ground Connection 2 (in headlamp wiring harness)
- (503) - Threaded connection -2- (75x) on the relay plate
- (A96) - Connector (53a), in instrument panel wiring harness
- (A97) - Connector (53), in instrument panel wiring harness
- (A102) - Connector (windshield wiper), in instrument panel wiring harness

ws = white
sw = black
ro = red
br = brown
gn = green
bl = blue
gr = grey
li = violet
ge = yellow
or = orange

New Beetle Convertible

Wiring diagram **No. 48/4**

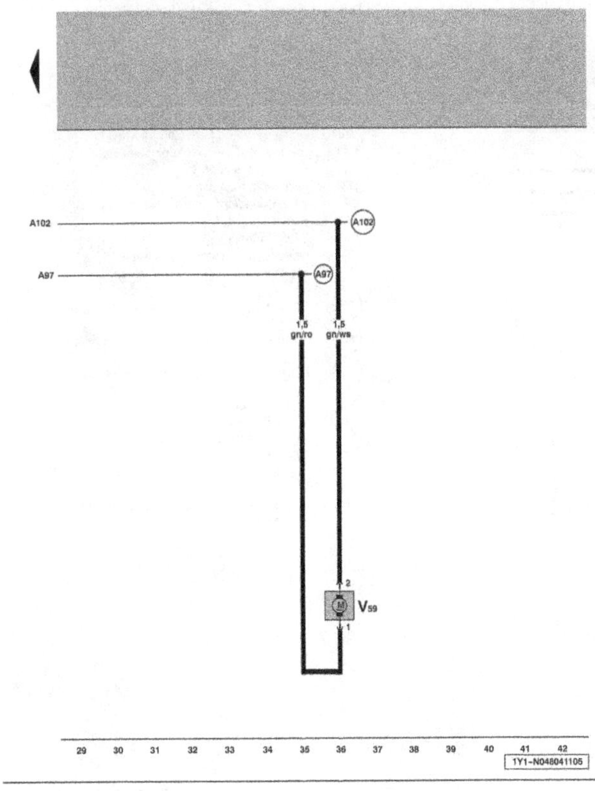

Windshield and rear window washer pump

- V59 - Windshield and Rear Window Washer Pump
- (A97) - Connector (53), in instrument panel wiring harness
- (A102) - Connector (windshield wiper), in instrument panel wiring harness

ws = white
sw = black
ro = red
br = brown
gn = green
bl = blue
gr = grey
li = violet
ge = yellow
or = orange

New Beetle **Wiring diagram** **No. 20/1**

Heated seats

from January 1998

Deviate relay location and fuseplacements as well as the locations of multiple connectors see section "component locations".

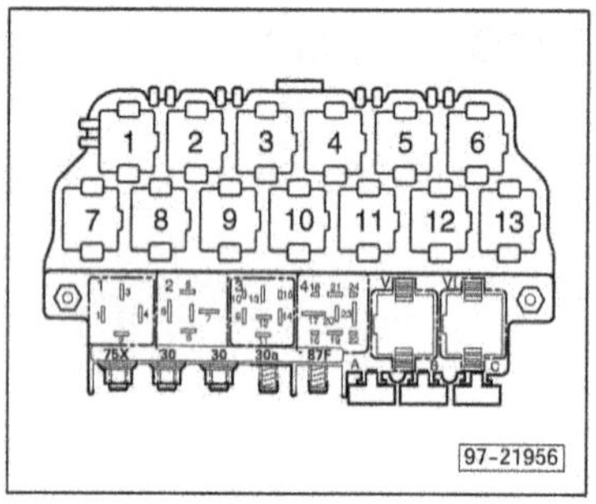

Fuse colors

30 A - green
25 A - white
20 A - yellow
15 A - blue
10 A - red
7,5 A - brown
5 A - beige
3 A - violet

New Beetle **Wiring diagram** **No. 20/2**

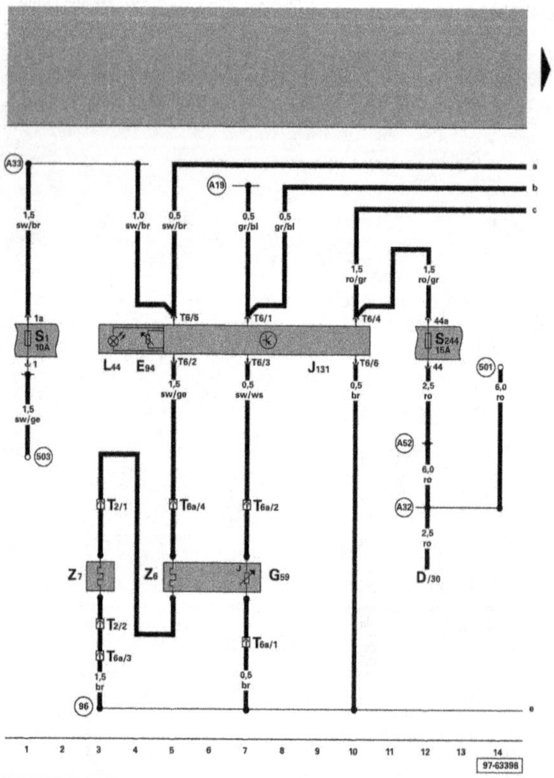

Driver's heated seat control module, driver's heated seat temperature sensor, driver's seat heating element

- D - Ignition/Starter Switch
- E94 - Adjuster for heated driver's seat
- G59 - Driver's Heated Seat Temperature Sensor
- J131 - Driver's Heated Seat Control Module
- L44 - Heated Seat Switch Illumination Light
- S1 - Fuse 1 in fuse panel
- S244 - Fuse 44 in fuse panel
- T2 - 2-Pin Connector, under driver's seat
- T6 - 6-Pin Connector
- T6a - 6-Pin Connector, green, below driver's seat
- Z6 - Driver's Seat Heating Element
- Z7 - Driver's Backrest Heating Element
- (96) - Ground connection -1-, in heated seats wiring harness
- (501) - Threaded connection -2- (30) on the relay plate
- (503) - Bolted Connection (75x) (at relay panel)
- (A19) - wire connection (58d), in instrument panel wiring harness
- (A32) - Plus connection (30), in instrument panel wiring harness
- (A33) - Connection (75), in instrument panel wiring harness
- (A52) - Plus connector -4- (30), in instrument panel wiring harness

ws = white
sw = black
ro = red
br = brown
gn = green
bl = blue
gr = grey
li = violet
ge = yellow
or = orange

New Beetle **Wiring diagram** **No. 20/3**

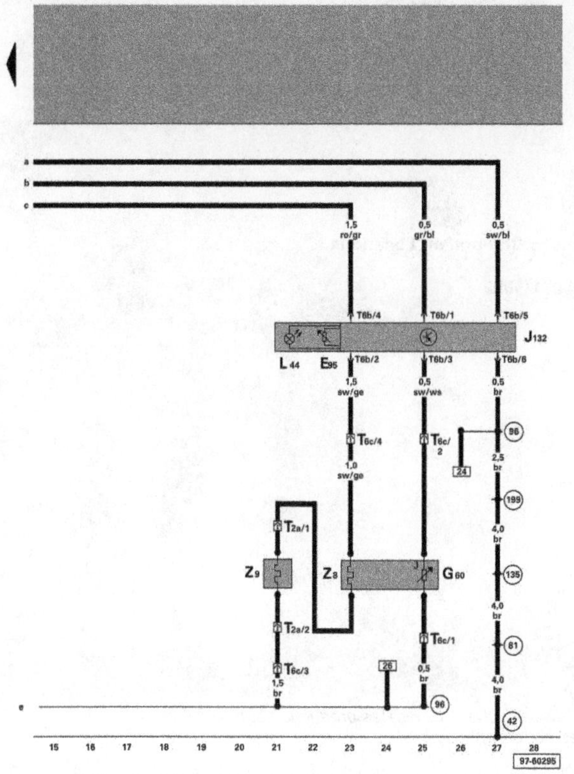

ws = white
sw = black
ro = red
br = brown
gn = green
bl = blue
gr = grey
li = violet
ge = yellow
or = orange

Passenger's heated seat control module, passenger's heated seat temperature sensor, passenger's seat heating element, passenger's backrest heating element

- E95 - Adjuster for heated passenger seat
- G60 - Passenger's Heated Seat Temperature Sensor
- J132 - Passenger's Heated Seat Control Module
- L44 - Heated Seat Switch Illumination Light
- T2a - 2-Pin Connector, under passenger's seat
- T6b - 6-Pin Connector
- T6c - 6-Pin Connector, green, below passenger's seat
- Z8 - Passenger's Seat Heating Element
- Z9 - Passenger's Backrest Heating Element

- (42) - Ground connection, beside steering column
- (81) - Ground connection -1-, in instrument panel wiring harness
- (96) - Ground connection -1-, in heated seats wiring harness
- (135) - Ground connection -2-, in instrument panel wiring harness
- (199) - Ground connection -3-, in instrument panel wiring harness

THIS SPACE INTENTIONALLY

LEFT BLANK

New Beetle

Wiring diagram **No. 60/1**

Heated seats,

from July 2005,

Additional Fuse, Relay, Control Module and Connector locations ⇒ Component Locations

Electrical troubleshooting ⇒ Guided Fault Finding using VAS 5051/5052

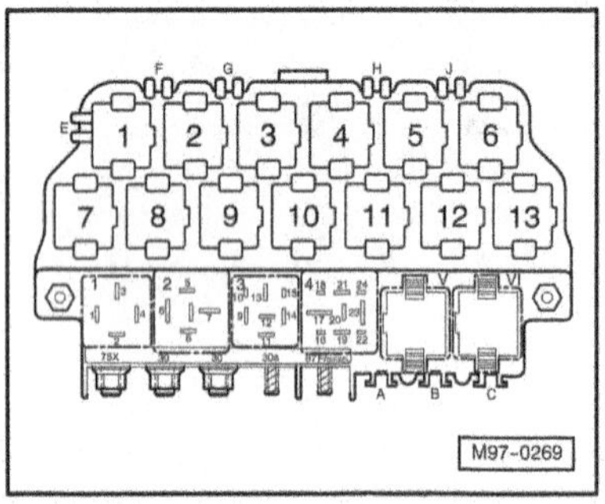

Note:

Number in parentheses indicates production control number stamped on relay housing.

New Beetle

Wiring diagram **No. 60/2**

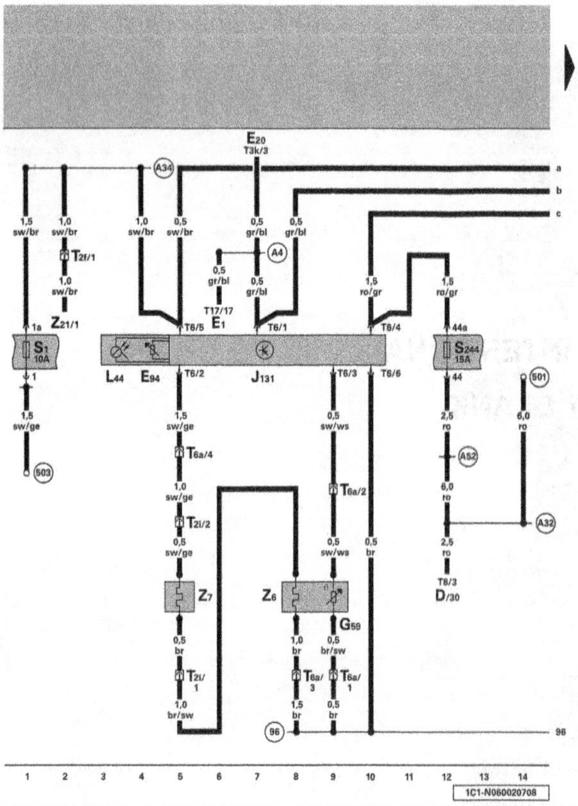

ws = white
sw = black
ro = red
br = brown
gn = green
bl = blue
gr = grey
li = violet
ge = yellow
or = orange

Heated Driver Seat Adjuster, Heated Driver Seat Temperature Sensor, Heated Driver Seat Control Module, Driver Seat Heating Element, Driver Backrest Heating Element

D	-	Ignition/Starter Switch
E1	-	Light Switch
E20	-	Instrument Panel Illumination Dimmer Switch
E94	-	Heated Driver Seat Adjuster
G59	-	Heated Driver Seat Temperature Sensor
J131	-	Heated Driver Seat Control Module
L44	-	Heated Seat Switch Illumination
S1	-	Fuse (on fuse/relay panel)
S244	-	Fuse (on fuse panel)
T2f	-	Double Connector
T2i	-	Double Connector, under driver's seat
T3k	-	3-Pin Connector
T6	-	6-Pin Connector
T6a	-	6-Pin Connector, green, below driver's seat
T8	-	8-Pin Connector
T17	-	17-Pin Connector
Z6	-	Driver Seat Heating Element
Z7	-	Driver Backrest Heating Element
Z21	-	Right Washer Nozzle Heater
(96)	-	Ground Connection 1 (in heated seats wiring harness)
(501)	-	Threaded Connection 2 - terminal 30 (on relay panel)
(503)	-	Treaded Connection - terminal 75x (on relay panel)
(A4)	-	Plus Connection (58b) (in instrument panel wiring harness)
(A32)	-	Plus Connection (30) (in instrument panel wiring harness)
(A34)	-	Connection (75x) (in instrument panel wiring harness)
(A52)	-	Plus Connection 2 (30) (in instrument panel wiring harness)

New Beetle **Wiring diagram** **No. 60/3**

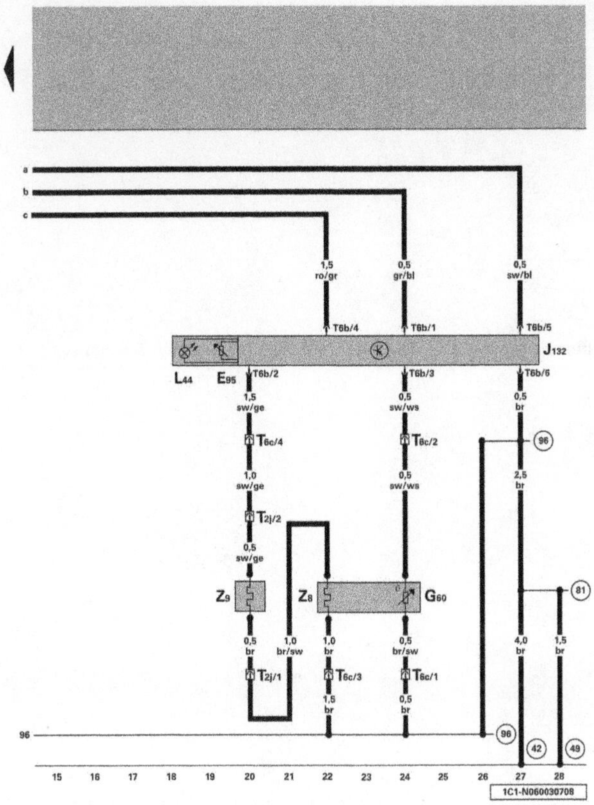

ws = white
sw = black
ro = red
br = brown
gn = green
bl = blue
gr = grey
li = violet
ge = yellow
or = orange

Heated Front Passenger Seat Adjuster, Front Passenger Heated Seat Temperature Sensor, Heated Front Passenger Seat Control Module, Front Passenger Seat Heating Element, Front Passenger Backrest Heating Element

E95	-	Heated Front Passenger Seat Adjuster
G60	-	Front Passenger Heated Seat Temperature Sensor
J132	-	Heated Front Passenger Seat Control Module
L44	-	Heated Seat Switch Illumination
T2j	-	Double Connector, under passenger's seat
T6b	-	6-Pin Connector
T6c	-	6-Pin Connector, green, below passenger's seat
Z8	-	Front Passenger Seat Heating Element
Z9	-	Front Passenger Backrest Heating Element
(42)	-	Ground Connection (beside steering column)
(49)	-	Ground Connection (near steering column)
(81)	-	Ground Connection 1 (in instrument panel wiring harness)
(96)	-	Ground Connection 1 (in heated seats wiring harness)

THIS SPACE INTENTIONALLY
LEFT BLANK

New Beetle Wiring diagram **No. 21/1**

Heated washer nozzles

from January 1998

Deviate relay location and fuseplacements as well as the locations of multiple connectors see section "component locations".

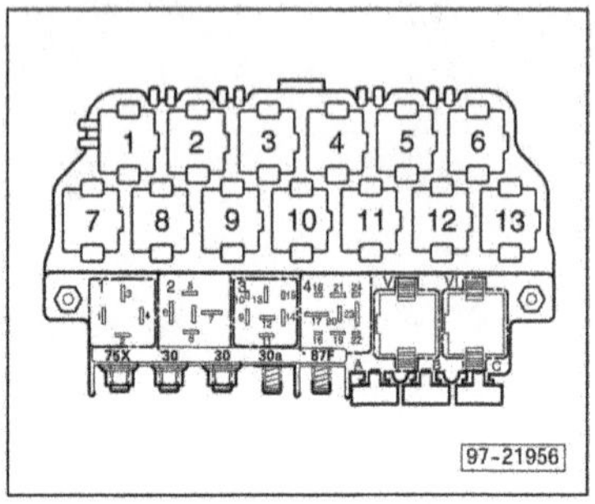

Fuse colors

30 A - green
25 A - white
20 A - yellow
15 A - blue
10 A - red
7,5 A - brown
5 A - beige
3 A - violet

New Beetle Wiring diagram **No. 21/2**

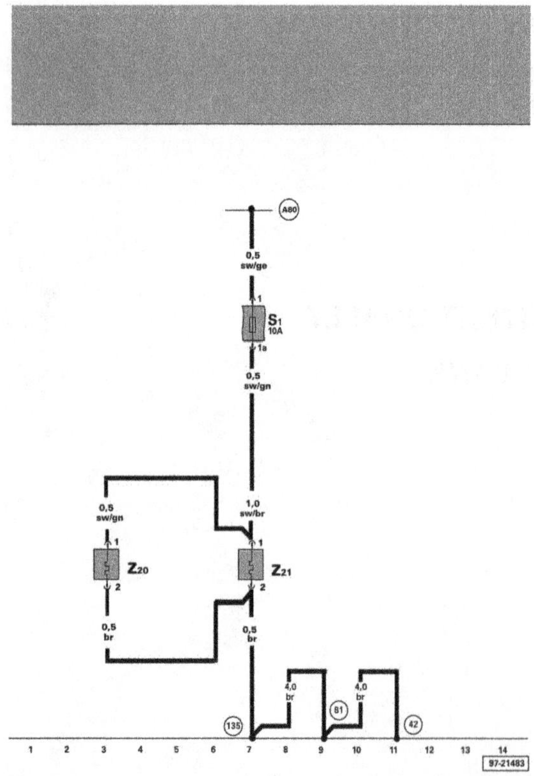

Heated washer nozzles

S1 - Fuse 1 in fuse holder
Z20 - Left washer nozzle heater
Z21 - Right washer nozzle heater

(42) - Ground connection, beside steering column

(81) - Ground connection -1-, in instrument panel wiring harness

(135) - Ground connection -2-, in instrument panel wiring harness

(A80) - Connector -1- (X) in instrument panel wiring harness

ws = white
sw = black
ro = red
br = brown
gn = green
bl = blue
gr = grey
li = violet
ge = yellow
or = orange

New Beetle **Wiring diagram** **No. 63/1**

Heated washer nozzles,

from July 2005

Additional Fuse, Relay, Control Module and Connector locations ⇒ Component Locations

Electrical troubleshooting ⇒ Guided Fault Finding using VAS 5051/5052

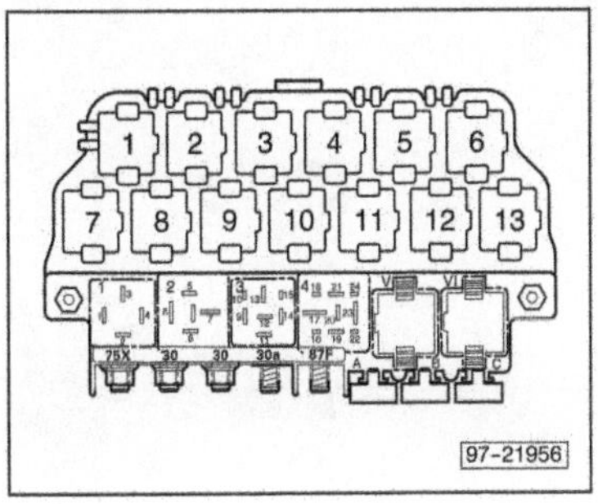

New Beetle **Wiring diagram** **No. 63/2**

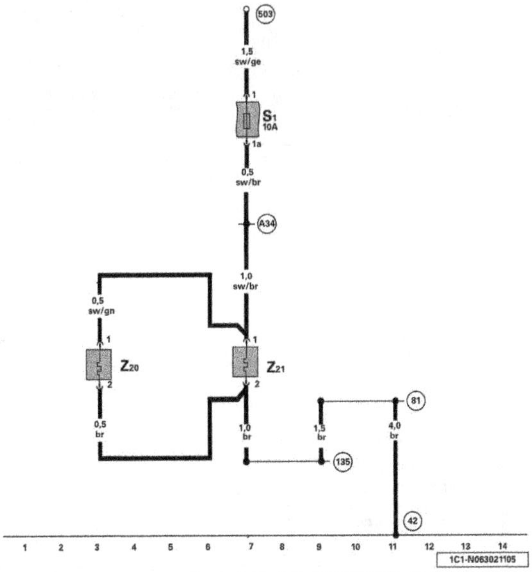

Heated washer nozzles

S1 - Fuse 1 in fuse holder
Z20 - Left washer nozzle heater
Z21 - Right washer nozzle heater
(42) - Ground connection, beside steering column
(81) - Ground connection -1-, in instrument panel wiring harness
(135) - Ground connection -2-, in instrument panel wiring harness
(503) - Treaded Connection - terminal 75x (on relay panel)
(A34) - Connection (75x) (in instrument panel wiring harness)

ws = white
sw = black
ro = red
br = brown
gn = green
bl = blue
gr = grey
li = violet
ge = yellow
or = orange

New Beetle Wiring diagram **No. 2/1**

High Intensity Gas Discharge (HID) Headlights,

from June 2003

Deviate relay location and fuseplacements as well as the locations of multiple connectors see section "component locations".

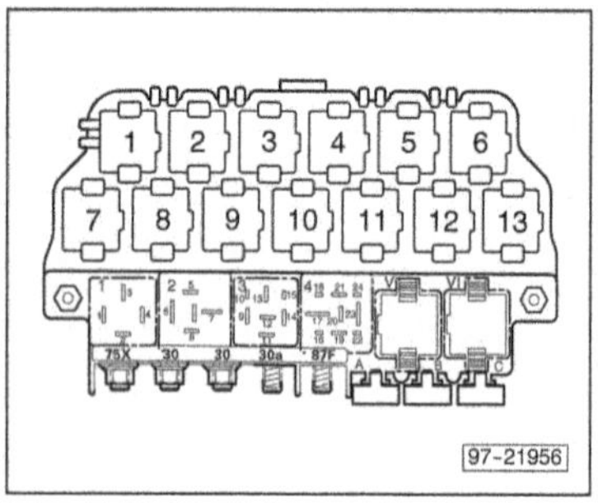

Fuse colors

30 A - green
25 A - white
20 A - yellow
15 A - blue
10 A - red
7,5 A - brown
5 A - beige
3 A - violet

New Beetle Wiring diagram **No. 2/2**

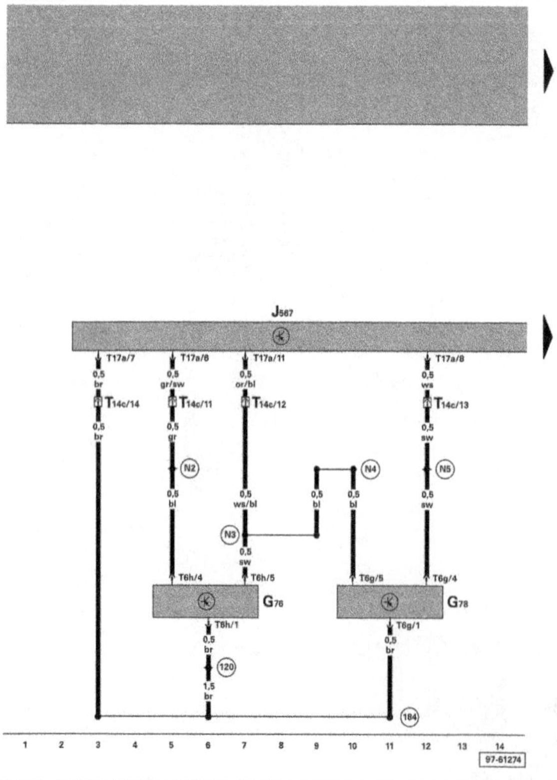

Left rear level control system sensor, left front level control system sensor, ballast unit for gas discharge headlight left

G 76 - Left Rear Level Control System Sensor
G 78 - Left Front Level Control System Sensor
J567 - Left Headlight Range Control Module
T6g - 6-Pin Connector, on vehicle level sensor, front left
T6h - 6-Pin Connector, on vehicle level sensor, rear left
T14c - 14-Pin Connector, on Control module for gas discharge lamp, left
T17a - 17-Pin Connector, on Control module for gas discharge lamp, left

(120) - Ground connection -2-, in headlight wiring harness
(184) - Ground connection (headlight beam control), in engine compartment wiring harness
(N2) - Connection 1 (in headlight range control wiring harness)
(N3) - Connection 2 (in headlight range control wiring harness)
(N4) - Connection 3 (in headlight range control wiring harness)
(N5) - Connection 4 (in headlight range control wiring harness)

ws = white
sw = black
ro = red
br = brown
gn = green
bl = blue
gr = grey
li = violet
ge = yellow
or = orange

New Beetle

Wiring diagram

No. 2/3

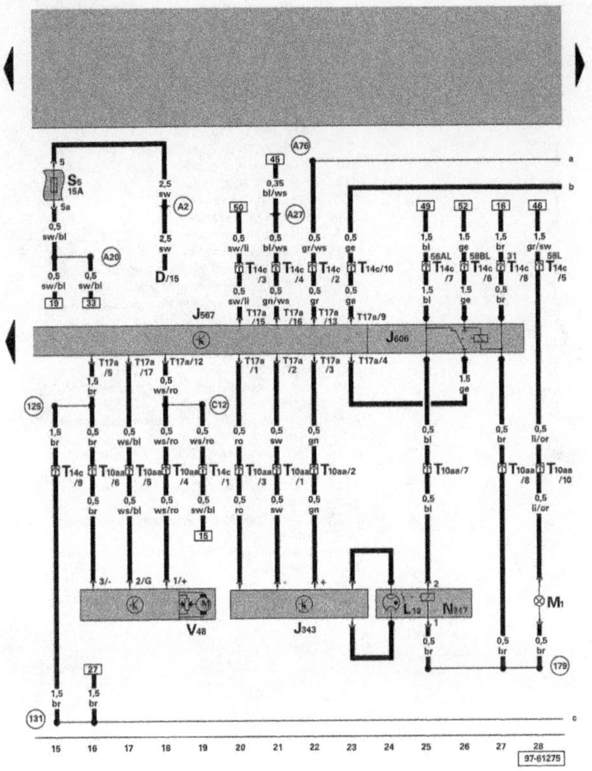

ws = white
sw = black
ro = red
br = brown
gn = green
bl = blue
gr = grey
li = violet
ge = yellow
or = orange

Control mod.high-intensity gas disch.lamp, ballast unit gas disch.headlight left, left high/low beam switch-over relay, high intensity gas disch.lamp, left parking light

- D - Ignition/Starter Switch
- J343 - Control Module for High-intensity gas discharge lamp
- J567 - Left Headlight Range Control Module
- J606 - Left High/Low Beam Switch-over Relay
- L 13 - High intensity gas discharge lamp
- M 1 - Left Parking Light
- N347 - Left High/Low Beam Switch-over Solenoid
- S5 - Fuse 5 in fuse holder
- T10aa - 10-Pin Connector, on headlight with gas dicharge lamp, left
- T14c - 14-Pin Connector, on Control module for gas discharge lamp, left
- T17a - 17-Pin Connector, on Control module for gas discharge lamp, left
- V 48 - Left Headlight Beam Adjusting Motor
- (125) - Ground connection -3-, in headlight wiring harness
- (131) - Ground connection -2-, in engine compartment wiring harness
- (179) - Ground connection, in left headlight wiring harness
- (A2) - plus connection (15), in instrument panel wiring harness
- (A20) - wire connection (15a), in instrument panel wiring harness
- (A27) - Wire Connection (vehicle speed signal), in instrument panel wiring harness
- (A76) - Connector (K-diagnosis wire) in instrument panel wiring harness
- (C12) - Plus connection -1-, in headlight wiring harness

New Beetle

Wiring diagram

No. 2/4

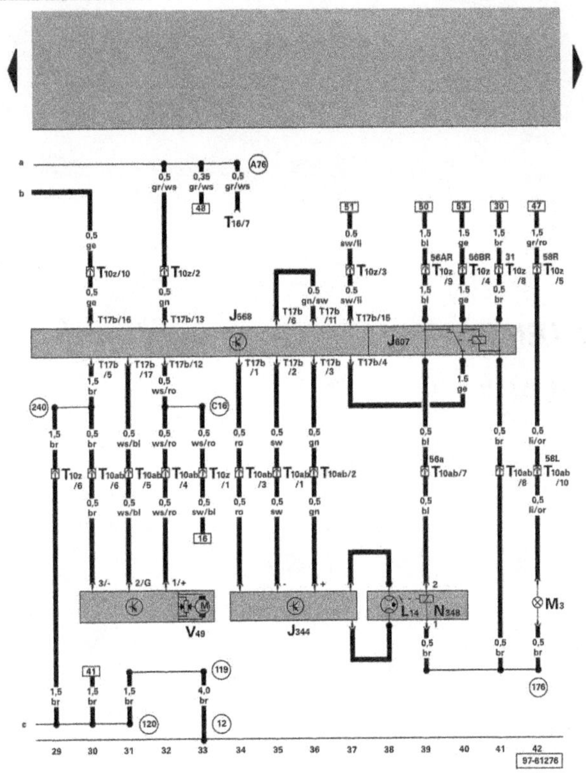

ws = white
sw = black
ro = red
br = brown
gn = green
bl = blue
gr = grey
li = violet
ge = yellow
or = orange

Control module high intensity gas disch.lamp, ballast unit gas disch.headlight right, right high/low beam switch-over relay, high intensity disch.lamp, right parking light

- J344 - Control Module for High intensity gas dis charge lamp
- J568 - Right Headlight Range Control Module
- J607 - Right High/Low Beam Switch-over Relay
- L 14 - High intensity discharge lamp
- M 3 - Right Parking Light
- N348 - Right High/Low Beam Switch-over Solenoid
- T10z - 10-Pin Connector, on Control module for gas discharge lamp, right
- T10ab - 10-Pin Connector, on headlight with gas discharge lamp, right
- T16 - 16-Pin Connector, Data Link Connector (DLC), below steering column, left
- T17b - 17-Pin Connector, on Control module for gas discharge lamp, right
- V 49 - Right Headlight Beam Adjusting Motor
- (12) - Ground connection, in engine compartment, left
- (119) - Ground connection -1-, in headlight wiring harness
- (120) - Ground connection -2-, in headlight wiring harness
- (176) - Ground connection, in right headlight wiring harness
- (240) - Ground connection -4- in wiring harness headlamp
- (A76) - Connector (K-diagnosis wire) in instrument panel wiring harness
- (C16) - Wire connection -2-, in headlight wiring harness

HID (Xenon) headlights from June 2003

New Beetle Wiring diagram **No. 2/5**

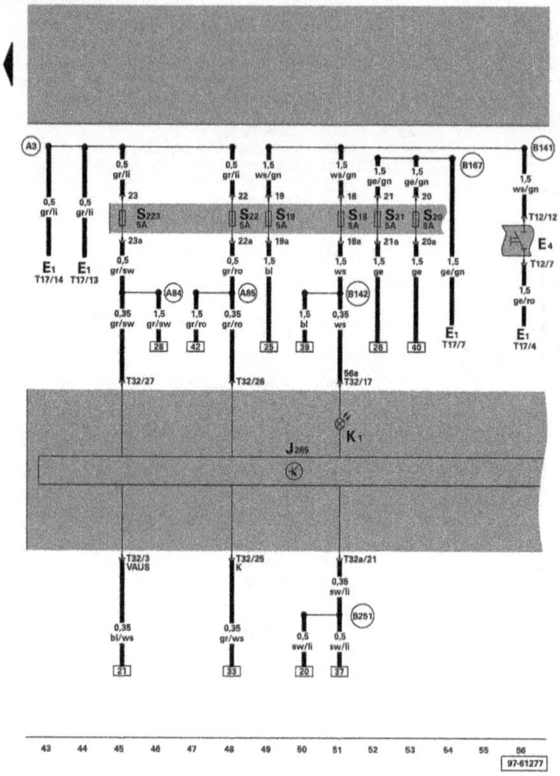

Headlight dimmer/flasher switch, control module with indicator unit in instrument panel insert, headlight high beam indicator light

- E 1 - Light switch
- E 4 - Headlight Dimmer/Flasher Switch
- J285 - Control module with indicator unit in instrument panel insert
- K 1 - Headlight High Beam Indicator Light
- S18 - Fuse 18 in fuse holder
- S19 - Fuse 19 in fuse holder
- S20 - Fuse 20 in fuse holder
- S21 - Fuse 21 in fuse holder
- S22 - Fuse 22 in fuse holder
- S223 - Fuse 23 in fuse holder
- T12 - 12-Pin Connector, near steering column
- T17 - 17-Pin Connector, on light switch
- T32 - 32-Pin Connector, blue, on instrument cluster
- T32a - 32-Pin Connector, green, on instrument cluster

- (A3) - plus connection (58), in instrument panel wiring harness
- (A84) - Connector (58L) in Instrument panel wiring harness
- (A85) - Connector (58R) in Instrument panel wiring harness
- (B141) - Plus connector -1- (56a), in wiring harness interior
- (B142) - Plus connector -2- (56a), in wiring harness interior
- (B167) - Connection (56b) in passenger compartment wiring harness
- (B251) - Error Message Connection (in interior wiring harness)

ws = white
sw = black
ro = red
br = brown
gn = green
bl = blue
gr = grey
li = violet
ge = yellow
or = orange

THIS SPACE INTENTIONALLY

LEFT BLANK

New Beetle Wiring diagram **No. 22/1**

Power outlets

from January 1998

Deviate relay location and fuseplacements as well as the locations of multiple connectors see section "component locations".

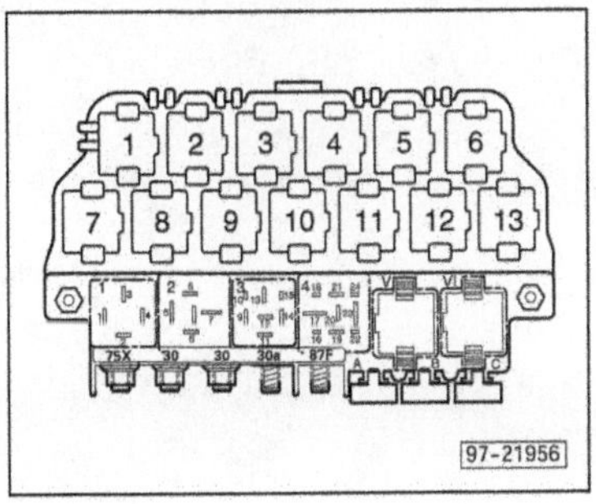

Fuse colors

30 A - green
25 A - white
20 A - yellow
15 A - blue
10 A - red
7.5 A - brown
5 A - beige
3 A - violet

New Beetle Wiring diagram **No. 22/2**

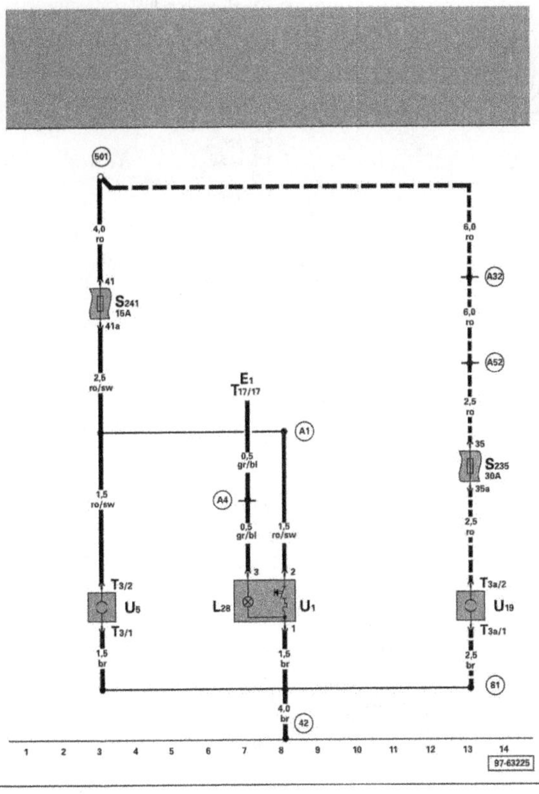

12 V socket, cigarette lighter

E1 - Light switch
L28 - Cigarette lighter light
S235 - Fuse 35 in fuse holder
S241 - Fuse 41 in fuse holder
T3 - 3-Pin Connector
T3a - 3-Pin Connector
T17 - 17-Pin Connector, at -E1-
U1 - Cigarette Lighter
U5 - 12 V socket
U19 - 12V outlet -3- (in luggage compartment)

(42) - Ground connection, beside steering column
(81) - Ground connection -1-, in instrument panel wiring harness
(501) - Threaded connection -2- (30) on the relay plate
(A1) - plus connection (30a), in instrument panel wiring harness
(A4) - plus connection (58b), in instrument panel wiring harness
(A32) - plus connection (30), in instrument panel wiring harness
(A52) - Plus Connection 2 (30) (in instrument panel wiring harness)

- - - - through 06.04 only

ws = white
sw = black
ro = red
br = brown
gn = green
bl = blue
gr = grey
li = violet
ge = yellow
or = orange

New Beetle Convertible — Wiring diagram — No. 15/1

Power outlets

from August 2002

Deviate relay location and fuseplacements as well as the locations of multiple connectors see section "component locations".

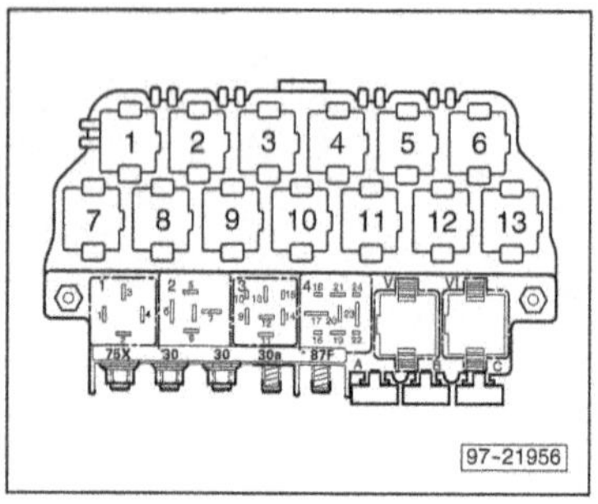

Fuse colors

30 A - green
25 A - white
20 A - yellow
15 A - blue
10 A - red
7,5 A - brown
5 A - beige
3 A - violet

New Beetle Convertible — Wiring diagram — No. 15/2

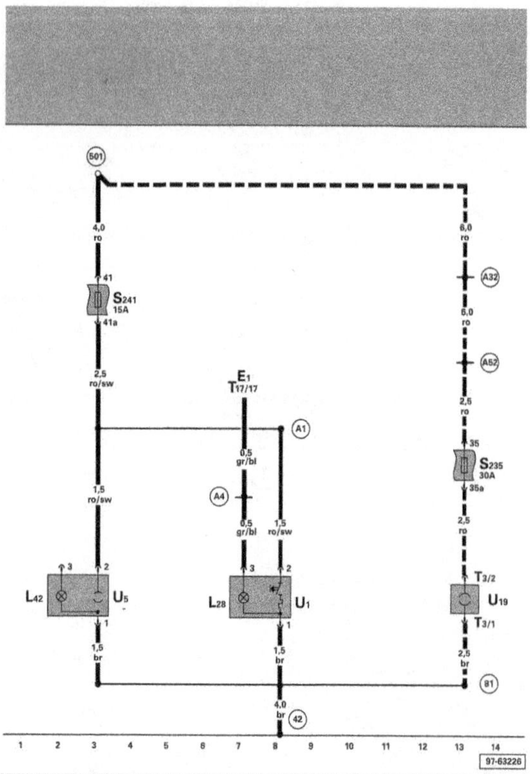

12 V socket, cigarette lighter

- E1 - Light Switch
- L28 - Cigarette lighter light
- L42 - Socket light bulb
- S235 - Fuse 35 in fuse holder
- S241 - Fuse 41 in fuse holder
- T3 - 3-Pin Connector
- U1 - Cigarette Lighter
- U5 - 12 V socket
- U19 - 12V outlet -3- (in luggage compartment)
- T17 - 17-Pin Connector, at -E1-
- (42) - Ground connection, beside steering column
- (81) - Ground connection -1-, in instrument panel wiring harness
- (501) - Threaded connection -2- (30) on the relay plate
- (A1) - plus connection (30a), in instrument panel wiring harness
- (A4) - plus connection (58b), in instrument panel wiring harness
- (A32) - plus connection (30), in instrument panel wiring harness
- (A52) - Plus Connection 2 (30) (in instrument panel wiring harness)

- - - - Through June 2004 only

ws = white
sw = black
ro = red
br = brown
gn = green
bl = blue
gr = grey
li = violet
ge = yellow
or = orange

New Beetle Wiring diagram **No. 62/1**

Power outlets,

from July 2005 to June 2007

Additional Fuse, Relay, Control Module and Connector locations ⇒ Component Locations

Electrical troubleshooting ⇒ Guided Fault Finding using VAS 5051/5052

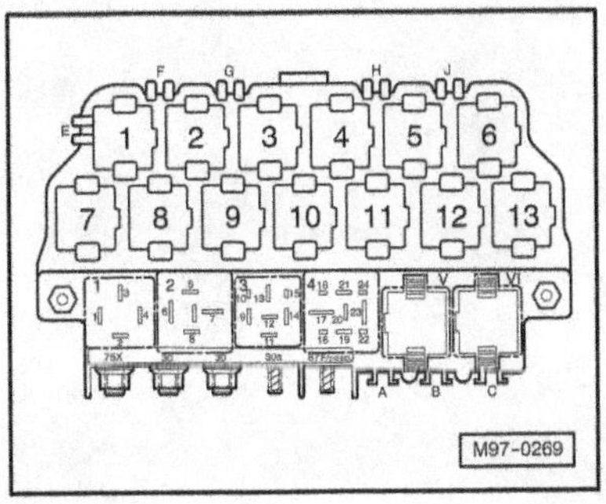

Note:

Number in parentheses indicates production control number stamped on relay housing.

New Beetle Wiring diagram **No. 62/2**

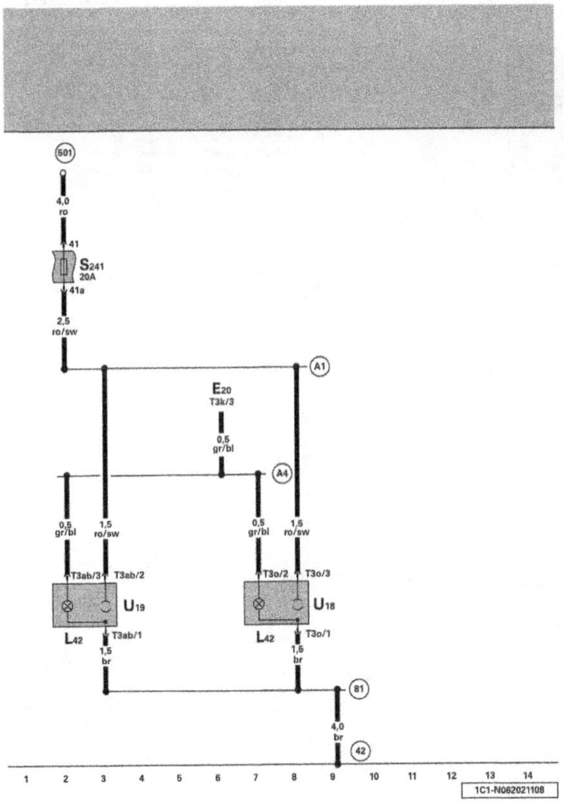

12 V sockets

E20 - Instrument Panel Illumination Dimmer Switch
L42 - Socket Illumination
S27 - Rear Fog Lamp Fuse
S241 - Fuse (on fuse panel)
T3k - 3-Pin Connector
T3o - 3-Pin Connector
T3ab - 3-Pin Connector
U18 - 12V Socket 2
U19 - 12V Socket 3

(42) - Ground Connection (beside steering column)
(81) - Ground Connection 1 (in instrument panel wiring harness)
(135) - Ground Connection 2 (in instrument panel wiring harness)
(501) - Threaded Connection 2 - terminal 30 (on relay panel)
(503) - Treaded Connection - terminal 75x (on relay panel)
(A1) - Plus Connection (30a) (in instrument panel wiring harness)
(A4) - Plus Connection (58b) (in instrument panel wiring harness)
(A36) - Connection (75a) (in instrument panel wiring harness)

ws = white
sw = black
ro = red
br = brown
gn = green
bl = blue
gr = grey
li = violet
ge = yellow
or = orange

New Beetle Wiring diagram **No. 24/1**

Radio system,

from January 1998

Deviate relay location and fuseplacements as well as the locations of multiple connectors see section "component locations".

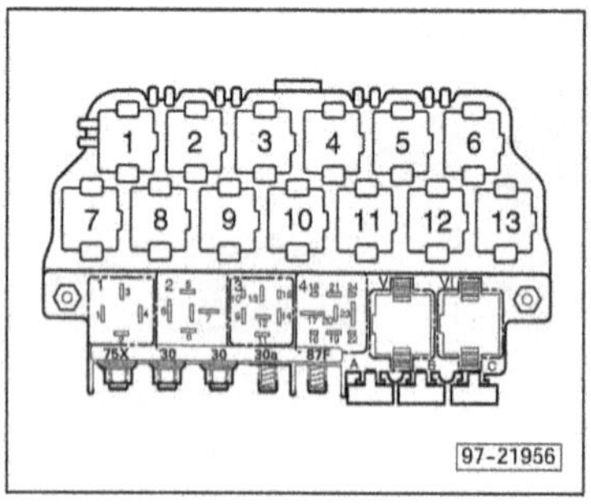

Fuse colors

30 A - green
25 A - white
20 A - yellow
15 A - blue
10 A - red
7,5 A - brown
5 A - beige
3 A - violet

New Beetle Wiring diagram **No. 24/2**

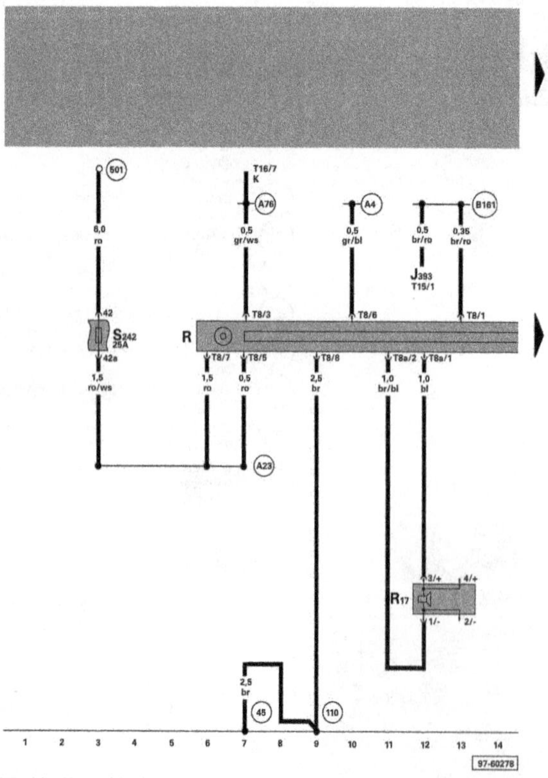

Radio, right rear speaker

- J393 - Central control system for comfort system
- R - Radio
- R17 - Right Rear Woofer, in right rear door
- S242 - Fuse 42 in fuse holder
- T8 - 8-Pin Connector, black, on radio
- T8a - 8-Pin Connector, brown, on radio
- T15 - 15-Pin Connector, on Central control module for comfort system
- T16 - 16-Pin Connector, Data Link Connector (DLC), below steering column

- (45) - Ground connection, behind instrument panel, center
- (110) - Ground connection -2-, in instrument panel wiring harness
- (501) - Screw Connection -2- (30), on relay panel
- (A4) - plus connection (58b), in instrument panel wiring harness
- (A23) - wire connection (30al), in instrument panel wiring harness
- (A76) - Connector (K-diagnosis wire), in instrument panel wiring harness
- (B161) - connection (Alarm system), in wiring harness, interior

ws = white
sw = black
ro = red
br = brown
gn = green
bl = blue
gr = grey
li = violet
ge = yellow
or = orange

New Beetle

Wiring diagram **No. 24/3**

Radio, front speakers, left rear speaker

R - Radio
R15 - Left Rear Woofer, in left rear door
R20 - Left Front Tweeter, in driver's door
R21 - Left Front Woofer, in driver's door
R22 - Right Front Tweeter, in passenger's door
R23 - Right Front Woofer, in passenger's door
T8a - 8-Pin Connector, brown, on radio
T10k - 10Pin Connector, black, connector station A-pillar, left
T10l - 10Pin Connector, black, connector station A-pillar, right

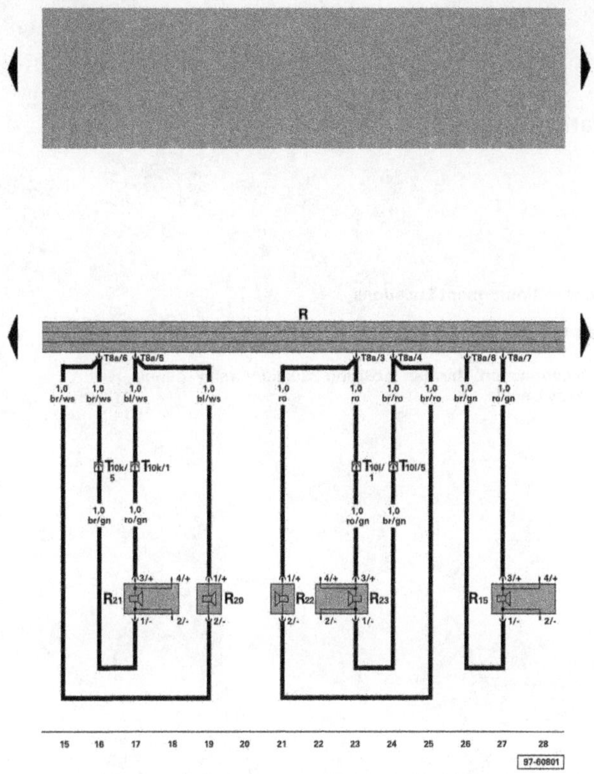

ws = white
sw = black
ro = red
br = brown
gn = green
bl = blue
gr = grey
li = violet
ge = yellow
or = orange

New Beetle

Wiring diagram **No. 24/4**

Radio, Antenna, Antenna Amplifier, CD Changer Unit

D - Ignition/Starter Switch
J285 - Control module with indicator unit in instrument panel insert
R - Radio
R11 - Antenna
R24 - Antenna Amplifier
R41 - CD Changer Unit
S237 - Fuse 37 in fuse holder
T1 - 1-Pin Connector, black, below console
T8 - 8-Pin Connector, black, on radio
T10a - 10-Pin Connector, on CD Changer Unit
T20 - 20-Pin Connector, on radio
T32a - 32-Pin Connector, blue, on instrument cluster

(110) - Ground connection -2-, in instrument panel wiring harnesses
(A21) - wire connection (86s), in instrument panel wiring harnes

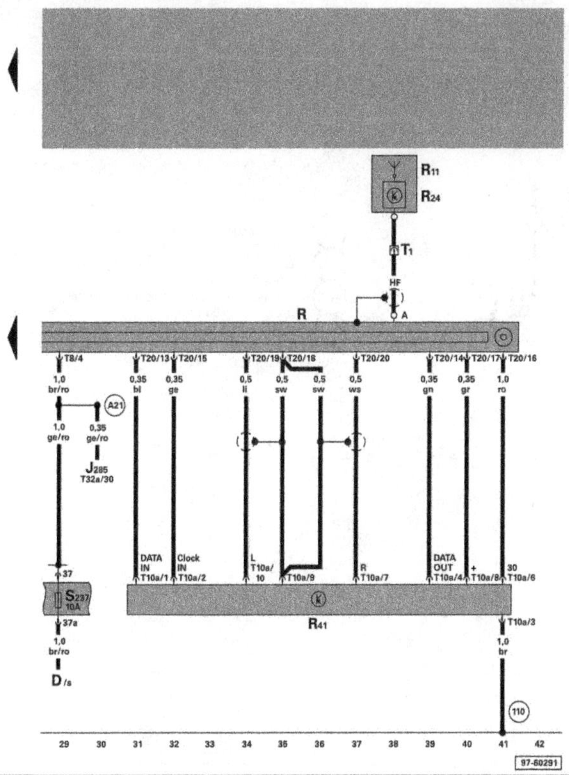

ws = white
sw = black
ro = red
br = brown
gn = green
bl = blue
gr = grey
li = violet
ge = yellow
or = orange

New Beetle Convertible Wiring diagram **No. 28/1**

Radio system (with CD changer wiring preparation)

from November 2004

Additional Fuse, Relay, Control Module and Connector locations ⇒ Component Locations

Troubleshooting Procedures ⇒ Guided Fault Finding using VAS 5051/5052

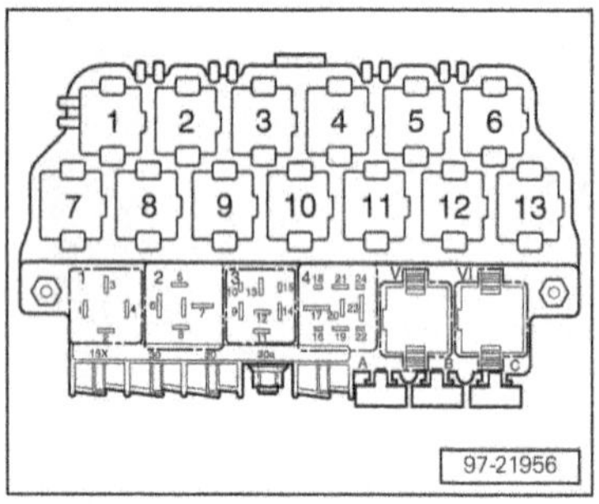

Relay locations on thirteen position auxiliary relay panel, above relay panel:

Relay panel:

New Beetle Convertible Wiring diagram **No. 28/2**

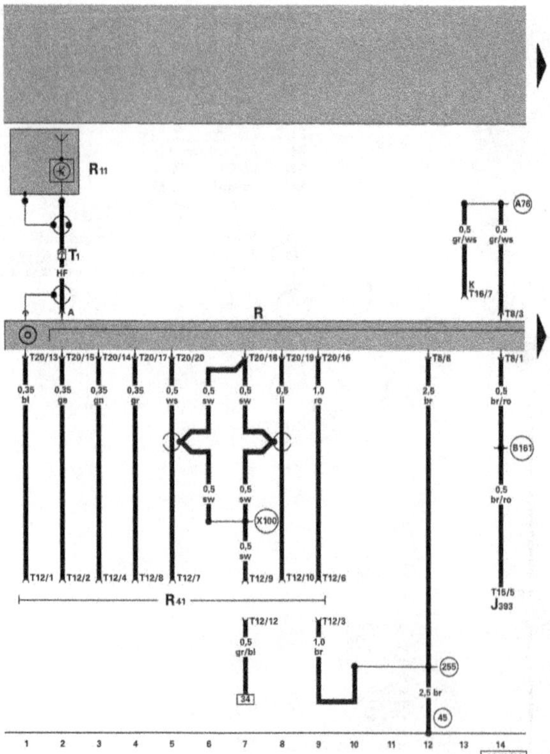

Radio, Antenna

- J393 - Comfort System Central Control Module
- R - Radio
- R11 - Antenna
- R41 - CD Changer
- T1 - Single Connector
- T8 - 8-Pin Connector
- T12 - 12-Pin Connector
- T15 - 15-Pin Connector
- T16 - 16-Pin Connector, black, Data Link Connector (DLC), below instrument panel, left
- T20 - 20-Pin Connector, on Radio
- (45) - Ground connection, behind instrument panel, center
- (255) - Ground connection, -1-, in wiring harness radio
- (A76) - Connector (K-diagnosis wire) in instrument panel wiring harness
- (B161) - Connector (anti-theft warning system), in wiring harness interior
- (X100) - Shielding Connection (in optional equipment wiring harness)

ws = white
sw = black
ro = red
br = brown
gn = green
bl = blue
gr = grey
li = violet
ge = yellow
or = orange

New Beetle Convertible — Wiring diagram — No. 28/3

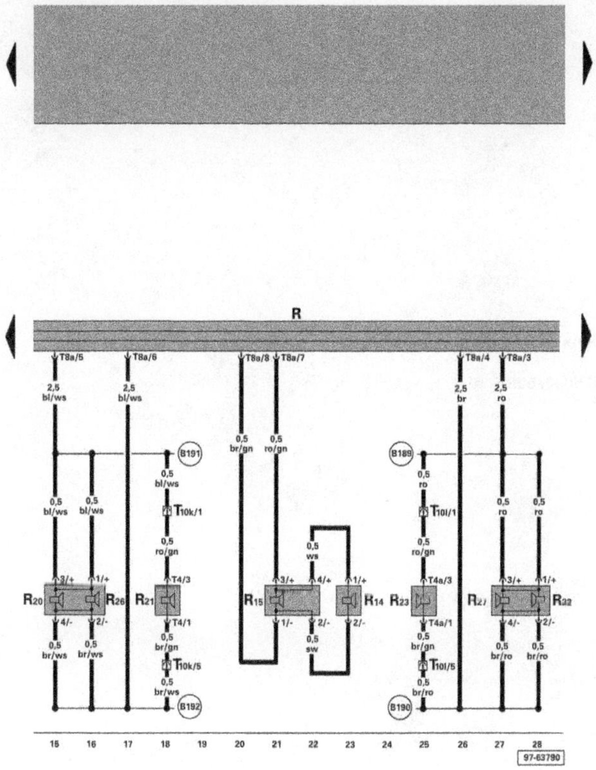

Radio, Treble speakers, Bass speakers, Midrange speakers

- R - Radio
- R14 - Left Rear Treble Speaker
- R15 - Left Rear Bass Speaker
- R20 - Left Front Treble Speaker
- R21 - Left Front Bass Speaker
- R22 - Right Front Treble Speaker
- R23 - Right Front Bass Speaker
- R26 - Left Front Midrange Speaker
- R27 - Right Front Midrange Speaker
- T4 - 4-Pin Connector
- T4a - 4-Pin Connector
- T8a - 8-Pin Connector
- T10k - 10-Pin Connector, black, connector station A-pillar, left
- T10l - 10-Pin Connector, black, connector station A-pillar, right
- (B189) - Connection (positive, RF speaker) in passenger compartment wiring harness
- (B190) - Connection (negative, RF speaker) in passenger compartment wiring harness
- (B191) - Connection (positive, LF speaker) in passenger compartment wiring harness
- (B192) - Connection (negative, LF speaker) in passenger compartment wiring harness

ws = white
sw = black
ro = red
br = brown
gn = green
bl = blue
gr = grey
li = violet
ge = yellow
or = orange

New Beetle Convertible — Wiring diagram — No. 28/4

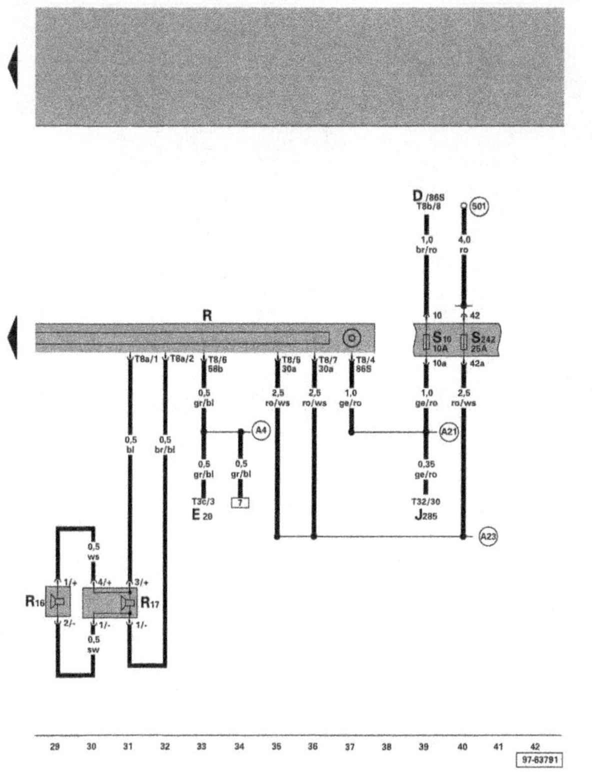

Radio, right rear treble speaker, right rear bass speaker

- D - Ignition/Starter Switch
- E20 - Instrument Panel Light Dimmer Switch
- J285 - Instrument Cluster Control Module
- R - Radio
- R16 - Right Rear Treble Speaker
- R17 - Right Rear Bass Speaker
- S10 - Fuse 10 (on fuse panel)
- S242 - Fuse 42 (on fuse panel)
- T3c - 3-Pin Connector
- T8 - 8-Pin Connector
- T8a - 8-Pin Connector
- T8b - 8-Pin Connector
- T32 - 32-Pin Connector, blue, on instrument cluster
- (501) - Threaded connection -2- (30) on the relay plate
- (A4) - Plus connection (58b), in instrument panel wiring harness
- (A21) - Wire connection (86s), in instrument panel wiring harness
- (A23) - Wire connection (30a), in instrument panel wiring harness

ws = white
sw = black
ro = red
br = brown
gn = green
bl = blue
gr = grey
li = violet
ge = yellow
or = orange

New Beetle Wiring diagram **No. 58/1**

Radio system,

from July 2005

Additional Fuse, Relay, Control Module and Connector locations ⇒ Component Locations

Electrical troubleshooting ⇒ Guided Fault Finding using VAS 5051/5052

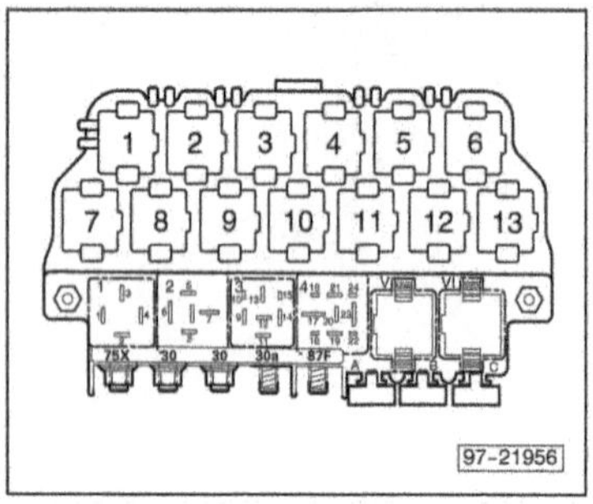

New Beetle Wiring diagram **No. 58/2**

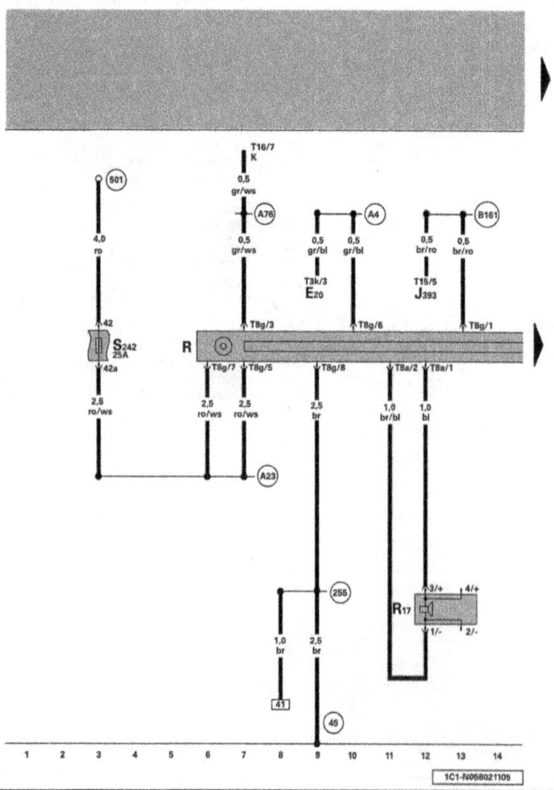

Radio, right rear speaker

E20 - Central control system for comfort system
J393 - Central control system for comfort system
R - Radio
R17 - Right Rear Woofer, in right rear door
S242 - Fuse 42 in fuse holder
T3k - 3-Pin Connector
T8a - 8-Pin Connector, brown, on radio
T8g - 8-Pin Connector, black, on radio
T15 - 15-Pin Connector, on Central control module for comfort system
T16 - 16-Pin Connector, Data Link Connector (DLC), below steering column

(45) - Ground connection, behind instrument panel, center
(255) - Ground connection, behind instrument panel, center
(501) - Screw Connection -2- (30), on relay panel
(A4) - Plus connection (58b), in instrument panel wiring harness
(A23) - wire connection (30al), in instrument panel wiring harness
(A76) - Connector (K-diagnosis wire), in instrument panel wiring harness
(B161) - connection (Alarm system), in wiring harness, interior

ws = white
sw = black
ro = red
br = brown
gn = green
bl = blue
gr = grey
li = violet
ge = yellow
or = orange

New Beetle — Wiring diagram — No. 58/3

Radio, front speakers, left rear speaker

- R - Radio
- R15 - Left Rear Woofer, in left rear door
- R20 - Left Front Tweeter, in driver's door
- R21 - Left Front Woofer, in driver's door
- R22 - Right Front Tweeter, in passenger's door
- R23 - Right Front Woofer, in passenger's door
- T8a - 8-Pin Connector, brown, on radio
- T10k - 10Pin Connector, black, connector station A-pillar, left
- T10l - 10Pin Connector, black, connector station A-pillar, right

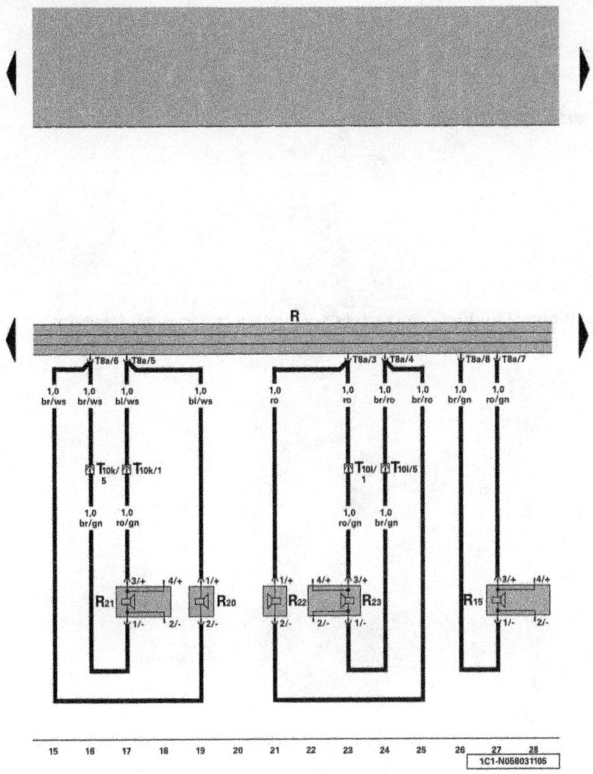

ws = white
sw = black
ro = red
br = brown
gn = green
bl = blue
gr = grey
li = violet
ge = yellow
or = orange

New Beetle — Wiring diagram — No. 58/4

Radio, Antenna, Antenna Amplifier, CD Changer Unit

- D - Ignition/Starter Switch
- J285 - Control module with indicator unit in instrument panel insert
- R - Radio
- R11 - Antenna
- R24 - Antenna Amplifier
- R41 - CD Changer Unit
- S10 - Fuse (on fuse/relay panel)
- T1 - 1-Pin Connector, black, below console
- T8 - 8-Pin Connector, black, on radio
- T8g - 8-Pin Connector
- T12d - 12-Pin Connector
- T20 - 20-Pin Connector, on radio
- T32 - 32-Pin Connector, blue, on instrument cluster
- (A21) - Connection (86s) (in instrument panel wiring harness)
- (V14) - Connection (shielding) (in CD-changer wiring harness)

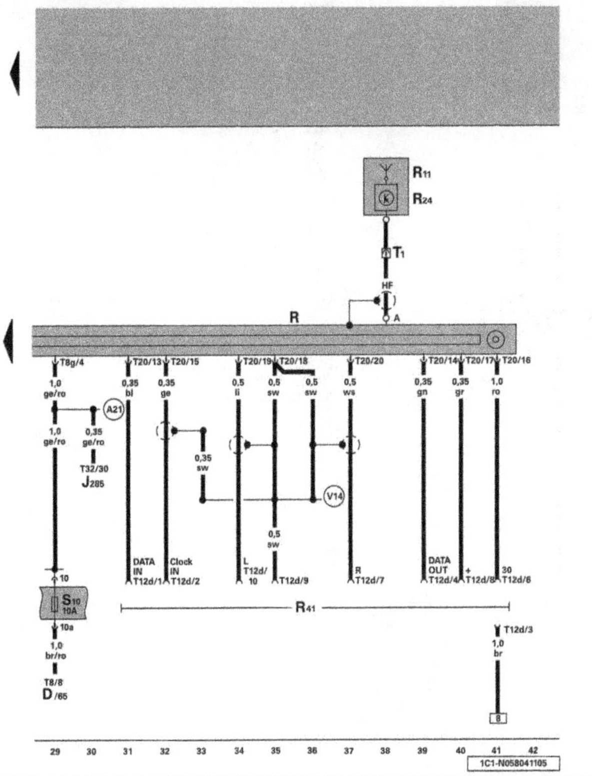

ws = white
sw = black
ro = red
br = brown
gn = green
bl = blue
gr = grey
li = violet
ge = yellow
or = orange

New Beetle — Wiring diagram — No. 25/1

Radio system "Monsoon" (with CD changer wiring preparation)

from October 2000

Deviate relay location and fuseplacements as well as the locations of multiple connectors see section "component locations".

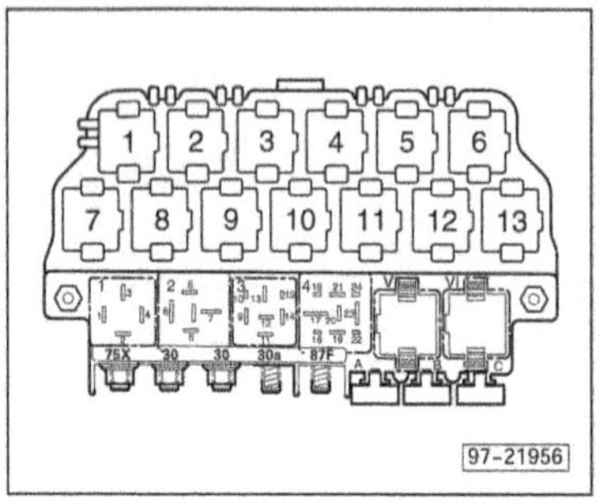

Fuse colors

30 A - green
25 A - white
20 A - yellow
15 A - blue
10 A - red
7,5 A - brown
5 A - beige
3 A - violet

New Beetle — Wiring diagram — No. 25/2

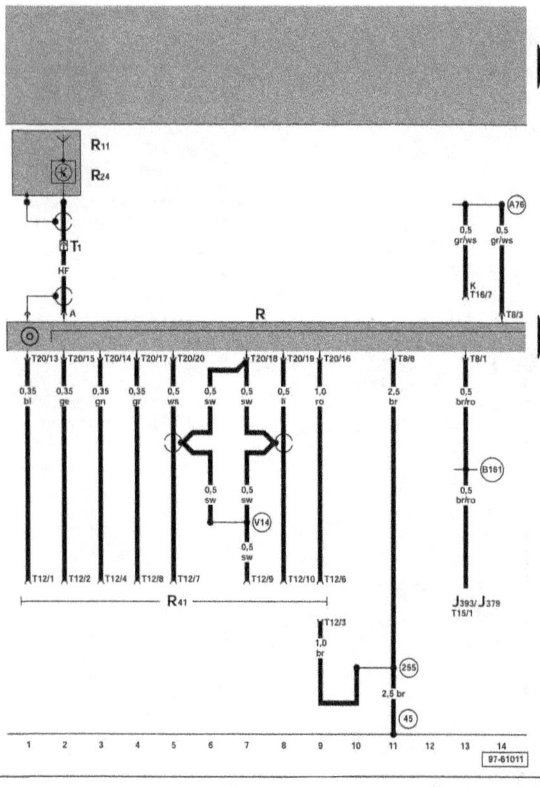

Radio, wiring harness from radio to CD changer unit

J379 - Control module for central locking and anti-theft-system
J393 - Central control module for comfort system
R - Radio
R11 - Antenna
R24 - Antenna Amplifier
R41 - CD Changer Unit
T1 - Single Connector
T8 - 8-Pin Connector
T12 - 12-Pin Connector
T15 - 15-Pin Connector
T16 - Data Link Connector (DLC), below instrument panel, left
T20 - 20-Pin Connector, on Radio

(45) - Ground connection, behind instrument panel, center
(255) - Ground connection, -1-, in wiring harness radio
(A76) - Connector (K-diagnosis wire) in instrument panel wiring harness
(B161) - Connector (anti-theft warning system), in wiring harness interior
(V14) - Wire connection (shielding), in CD-changer wiring harness

ws = white
sw = black
ro = red
br = brown
gn = green
bl = blue
gr = grey
li = violet
ge = yellow
or = orange

New Beetle — Wiring diagram — No. 25/3

Radio

- D - Ignition/Starter Switch
- E20 - Instrument Panel Light Dimmer Switch
- J285 - Control module with indicator unit in instrument panel insert
- R - Radio
- S237 - Fuse 37 in fuse holder
- S242 - Fuse 42 in fuse holder
- T3c - 3-Pin Connector
- T8 - 8-Pin Connector
- T8a - 8-Pin Connector
- T32 - 32-Pin Connector, blue

- (501) - Threaded connection -2- (30) on the relay plate
- (A4) - Plus connection (58b), in instrument panel wiring harness
- (A21) - Wire connection (86s), in instrument panel wiring harness
- (A23) - Wire connection (30a), in instrument panel wiring harness

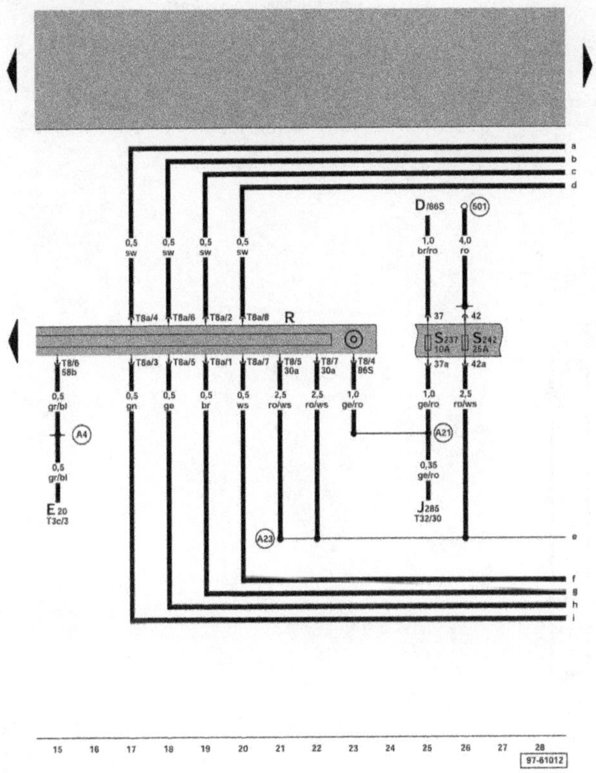

ws = white
sw = black
ro = red
br = brown
gn = green
bl = blue
gr = grey
li = violet
ge = yellow
or = orange

New Beetle — Wiring diagram — No. 25/4

Amplifier

- T23 - 23-Pin Connector
- T24 - 24-Pin Connector
- R12 - Amplifier, in luggage compartment, left

- (45) - Ground connection, behind instrument panel, center
- (256) - Ground connection, -2-, in wiring harness radio
- (A23) - Wire connection (30a), in instrument panel wiring harness
- (V21) - Plus connector (30), in wiring harness Radio-Amplifier

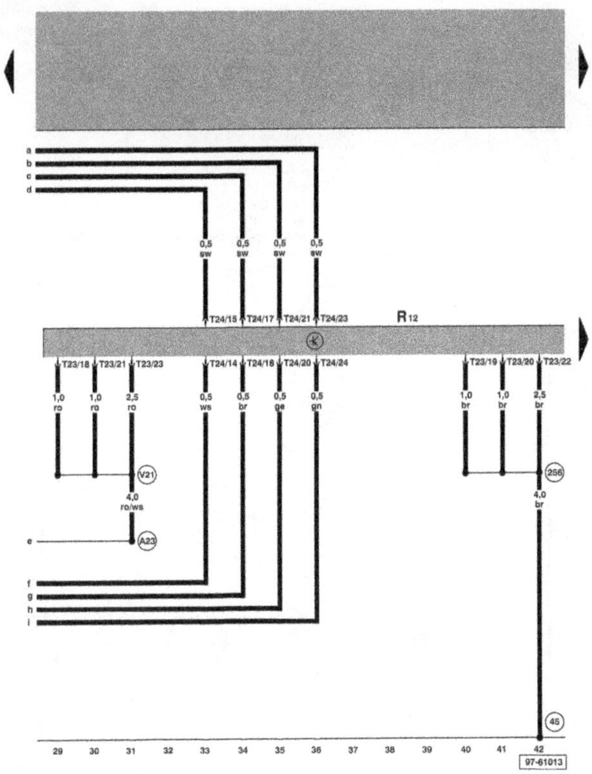

ws = white
sw = black
ro = red
br = brown
gn = green
bl = blue
gr = grey
li = violet
ge = yellow
or = orange

New Beetle

Wiring diagram

No. 25/5

Amplifier, loud speaker front

- R 12 - Amplifier, in luggage compartment , left
- R20 - Treble speaker, left front
- R22 - Treble speaker, right front
- R 26 - Left Front Midrange Speaker
- R 27 - Right Front Midrange Speaker
- T4 - 4-Pin Connector
- T4a - 4-Pin Connector
- T23 - 23-Pin Connector

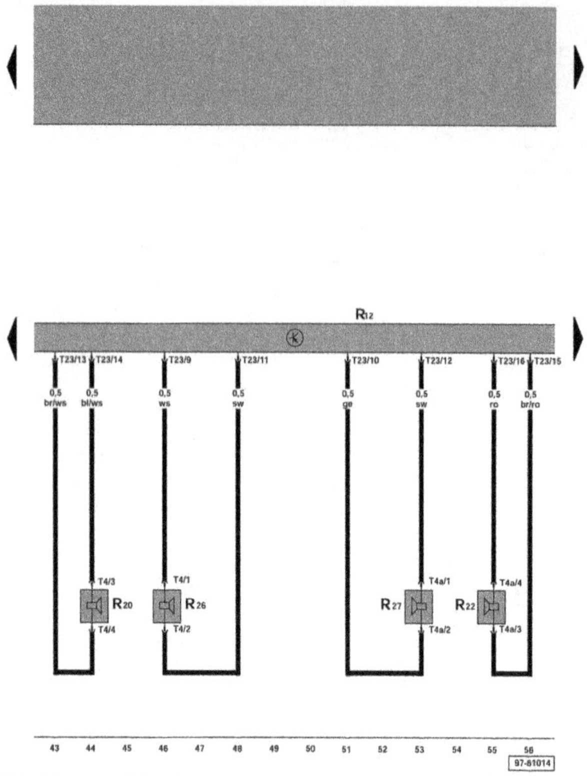

ws = white
sw = black
ro = red
br = brown
gn = green
bl = blue
gr = grey
li = violet
ge = yellow
or = orange

Edition 08.02

New Beetle

Wiring diagram

No. 25/6

Amplifier, Bass speakers

- R12 - Amplifier, in luggage compartment, left
- R15 - Left Rear Bass Speaker
- R17 - Right Rear Bass Speaker
- R 21 - Left Front Bass Speaker
- R 23 - Right Front Bass Speaker
- T4b - 4-Pin Connector
- T4c - 4-Pin Connector
- T4d - 4-Pin Connector in driver's door wiring
- T4e - 4-Pin Connector in passenger's door wiring
- T10k - 10-Pin Connector, black, connector station A-pillar, left
- T10l - 10-Pin Connector, black, connector station A-pillar, right
- T23 - 23-Pin Connector

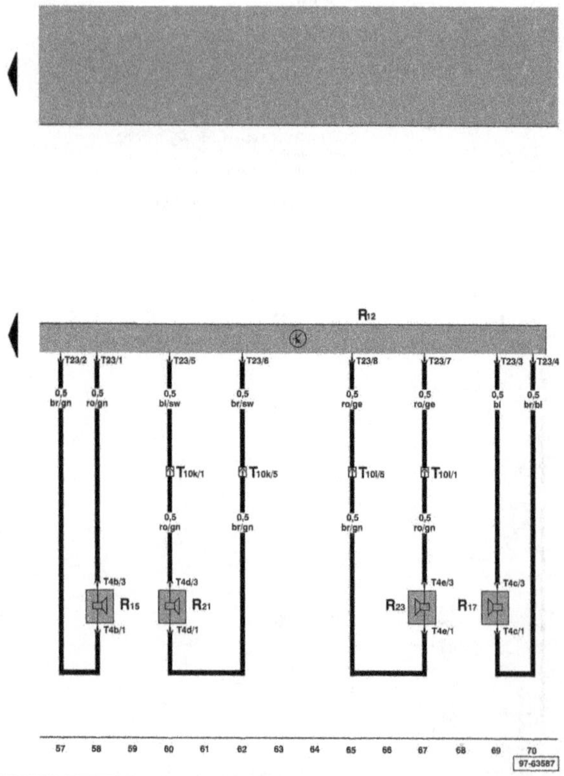

ws = white
sw = black
ro = red
br = brown
gn = green
bl = blue
gr = grey
li = violet
ge = yellow
or = orange

Edition 11.04

New Beetle Convertible Wiring diagram **No. 5/1**

Radio system "Monsoon" (with CD changer wiring preparation)

from August 2002

Deviate relay location and fuseplacements as well as the locations of multiple connectors see section "component locations".

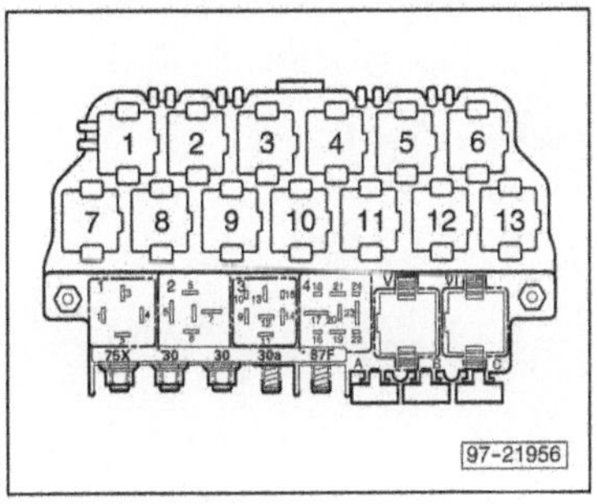

Fuse colors

30 A - green
25 A - white
20 A - yellow
15 A - blue
10 A - red
7,5 A - brown
5 A - beige
3 A - violet

New Beetle Convertible Wiring diagram **No. 5/2**

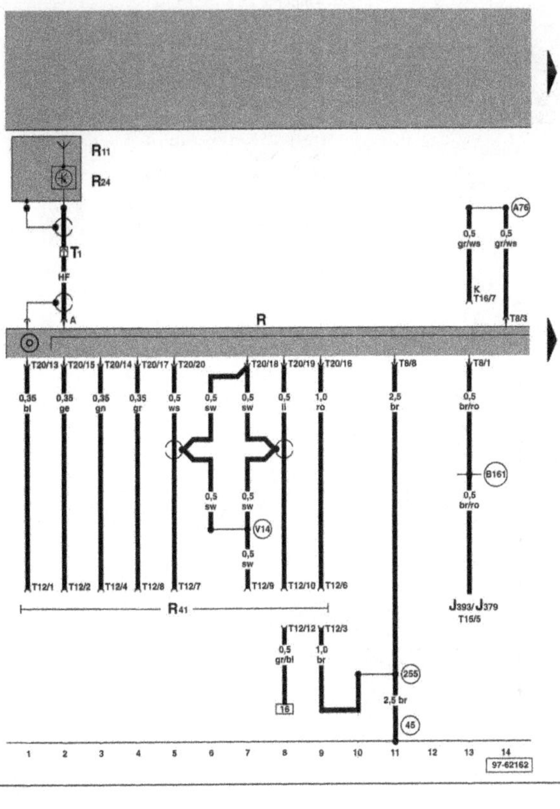

Radio, wiring harness from radio to CD changer unit, antenna, antenna amplifier

J379 - Control module for central locking and anti-theft-system
J393 - Central control module for comfort system
R - Radio
R11 - Antenna
R24 - Antenna Amplifier
R41 - CD Changer Unit
T1 - Single Connector
T8 - 8-Pin Connector
T12 - 12-Pin Connector
T15 - 15-Pin Connector
T16 - Data Link Connector (DLC), below instrument panel, left
T20 - 20-Pin Connector, on Radio

(45) - Ground connection, behind instrument panel, center
(255) - Ground connection, -1-, in wiring harness radio
(A76) - Connector (K-diagnosis wire) in instrument panel wiring harness
(B161) - Connector (anti-theft warning system), in wiring harness interior
(V14) - Wire connection (shielding), in CD-changer wiring harness

ws = white
sw = black
ro = red
br = brown
gn = green
bl = blue
gr = grey
li = violet
ge = yellow
or = orange

Radio system Monsoon from Aug. 2002, convertible

New Beetle Convertible

Wiring diagram **No. 5/3**

Radio

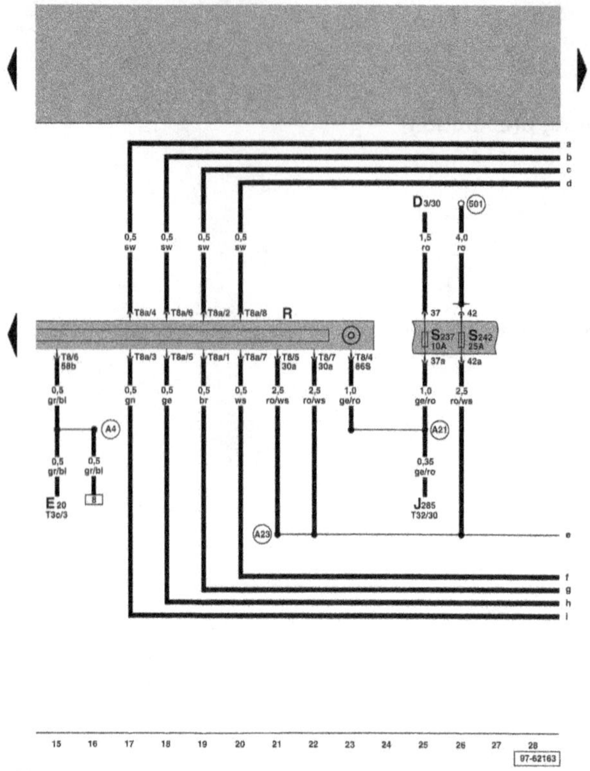

- D - Ignition/Starter Switch
- E20 - Instrument Panel Light Dimmer Switch
- J285 - Control module with indicator unit in instrument panel insert
- R - Radio
- S237 - Fuse 37 in fuse holder
- S242 - Fuse 42 in fuse holder
- T3c - 3-Pin Connector
- T8 - 8-Pin Connector
- T8a - 8-Pin Connector
- T32 - 32-Pin Connector, blue
- (501) - Threaded connection -2- (30) on the relay plate
- (A4) - Plus connection (58b), in instrument panel wiring harness
- (A21) - Wire connection (86s), in instrument panel wiring harness
- (A23) - Wire connection (30a), in instrument panel wiring harness

ws = white
sw = black
ro = red
br = brown
gn = green
bl = blue
gr = grey
li = violet
ge = yellow
or = orange

New Beetle Convertible

Wiring diagram **No. 5/4**

Amplifier

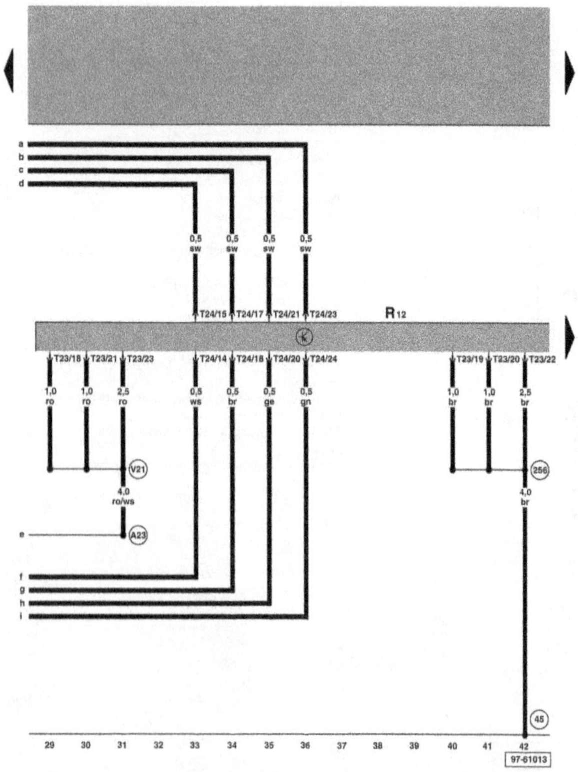

- R12 - Amplifier, in luggage compartment, left
- T23 - 23-Pin Connector
- T24 - 24-Pin Connector
- (45) - Ground connection, behind instrument panel, center
- (256) - Ground connection, -2-, in wiring harness radio
- (A23) - Wire connection (30a), in instrument panel wiring harness
- (V21) - Plus connector (30), in wiring harness Radio-Amplifier

ws = white
sw = black
ro = red
br = brown
gn = green
bl = blue
gr = grey
li = violet
ge = yellow
or = orange

New Beetle Convertible

Wiring diagram

No. 5/5

Amplifier, loud speaker front

R 12 - Amplifier, in luggage compartment , left
R 20 - Treble speaker, left front
R 21 - Bass speaker, left front
R 22 - Treble speaker, right front
R 23 - Bass speaker, right front
T4 - 4-Pin Connector
T4a - 4-Pin Connector
T23 - 23-Pin Connector

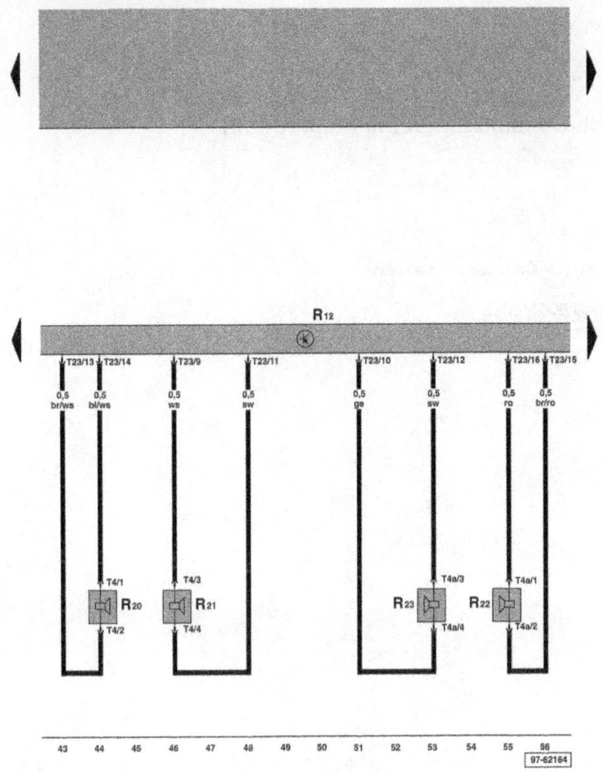

ws = white
sw = black
ro = red
br = brown
gn = green
bl = blue
gr = grey
li = violet
ge = yellow
or = orange

Edition 11.02

New Beetle Convertible

Wiring diagram

No. 5/6

Amplifier, front loudspeakers, rear loudspeakers

R12 - Amplifier, in luggage compartment, left
R14 - Left Rear Treble Speaker
R15 - Left Rear Bass Speaker
R16 - Right Rear Treble Speaker
R17 - Right Rear Bass Speaker
R21 - Left Front Bass Speaker
R23 - Right Front Bass Speaker
T4b - 4-Pin Connector, left rear
T4c - 4-Pin Connector, right rear
T4d - 4-Pin Connector in driver's door wiring
T4e - 4-Pin Connector in passenger's door wiring
T4f - 4-Pin Connector, left rear
T4g - 4-Pin Connector, right rear
T10k - 10-Pin Connector, black, connector station A-pillar, left
T10l - 10-Pin Connector, black, connector station A-pillar, right
T23 - 23-Pin Connector

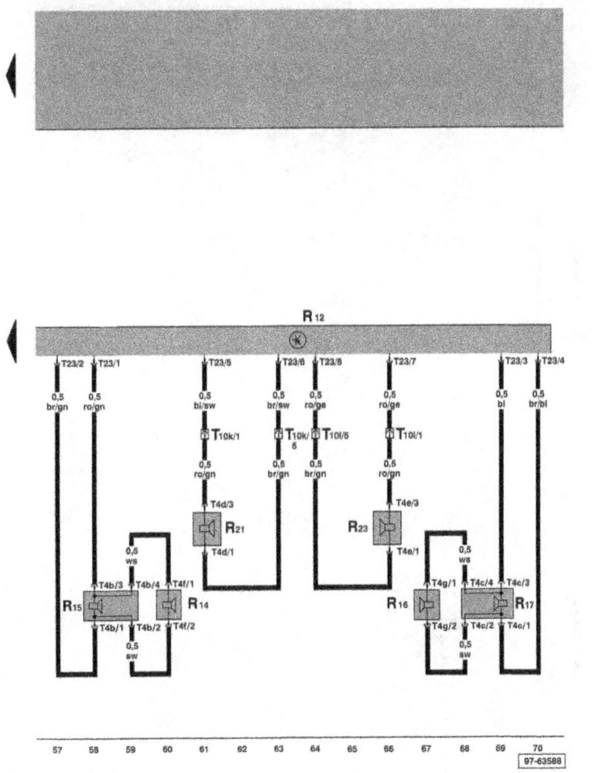

ws = white
sw = black
ro = red
br = brown
gn = green
bl = blue
gr = grey
li = violet
ge = yellow
or = orange

Edition 11.04

New Beetle Wiring diagram **No. 34/1**

Radio system "Monsoon", Satellite Radio (with CD changer wiring preparation)

from March 2004

Additional Fuse, Relay, Control Module and Connector locations ⇒ Component Locations

Troubleshooting Procedures ⇒ Guided Fault Finding using VAS 5051/5052

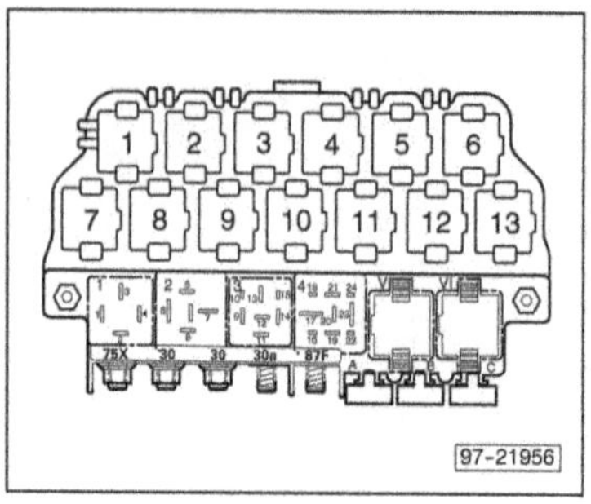

Fuse colors

30 A - green
25 A - white
20 A - yellow
15 A - blue
10 A - red
7,5 A - brown
5 A - beige
3 A - violet

New Beetle Wiring diagram **No. 34/2**

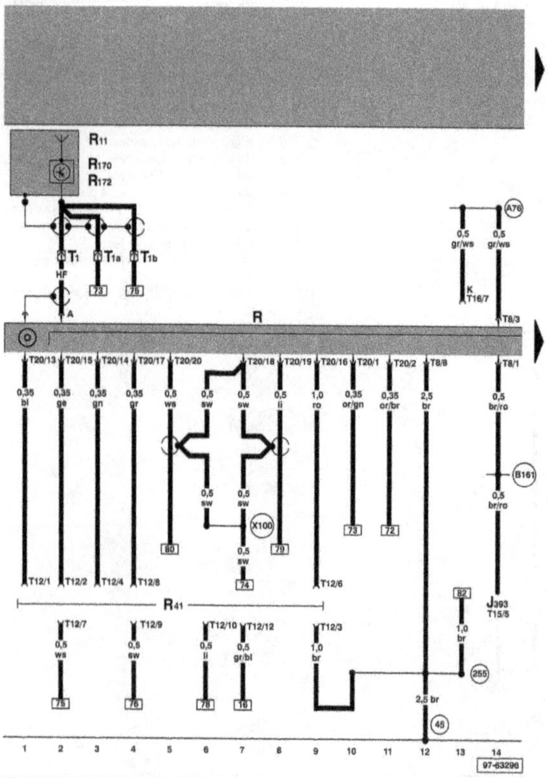

Radio, wiring harness from radio to CD changer unit

J393 - Central control module for comfort system
R - Radio
R11 - Antenna
R41 - CD Changer
R170 - Satellite Antenna
R172 - Satellite Tuner Antenna
T1 - Single Connector
T1a - Single Connector
T1b - Single Connector
T8 - 8-Pin Connector
T12 - 12-Pin Connector
T15 - 15-Pin Connector
T16 - Data Link Connector (DLC), below instrument panel, left
T20 - 20-Pin Connector, on Radio

(45) - Ground connection, behind instrument panel, center
(255) - Ground connection, -1-, in wiring harness radio
(A76) - Connector (K-diagnosis wire) in instrument panel wiring harness
(B161) - Connector (anti-theft warning system), in wiring harness interior
(X100) - Shielding Connection (in optional equipment wiring harness)

ws = white
sw = black
ro = red
br = brown
gn = green
bl = blue
gr = grey
li = violet
ge = yellow
or = orange

Radio system Monsoon satellite from March 2004

New Beetle — Wiring diagram — No. 34/3

Radio

D - Ignition/Starter Switch
E20 - Instrument Panel Light Dimmer Switch
J285 - Control module with indicator unit in instrument panel insert
R - Radio
S10 - Fuse 10 in fuse holder
S242 - Fuse 42 in fuse holder
T3c - 3-Pin Connector
T8 - 8-Pin Connector
T8a - 8-Pin Connector
T32 - 32-Pin Connector, blue

(501) - Threaded connection -2- (30) on the relay plate
(A4) - Plus connection (58b), in instrument panel wiring harness
(A21) - Wire connection (86s), in instrument panel wiring harness
(A23) - Wire connection (30a), in instrument panel wiring harness

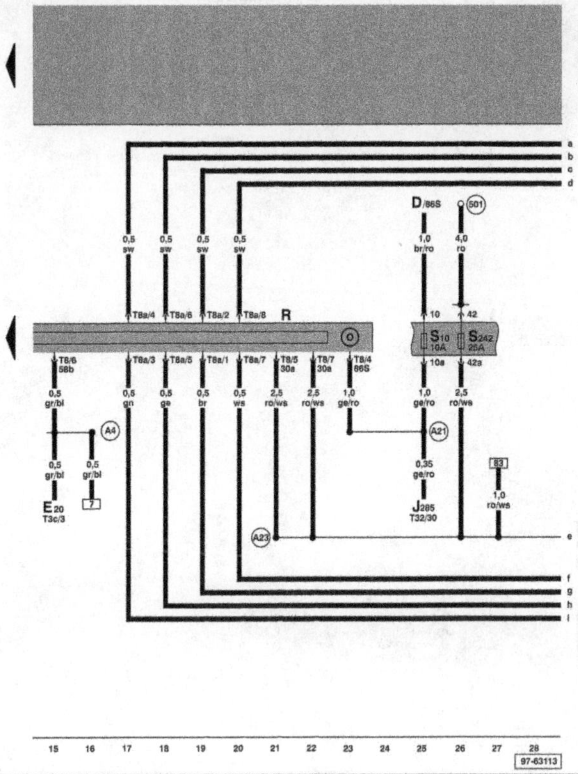

ws = white
sw = black
ro = red
br = brown
gn = green
bl = blue
gr = grey
li = violet
ge = yellow
or = orange

New Beetle — Wiring diagram — No. 34/4

Amplifier

T23 - 23-Pin Connector
T24 - 24-Pin Connector
R12 - Amplifier, in luggage compartment, left

(45) - Ground connection, behind instrument panel, center
(256) - Ground connection, -2-, in wiring harness radio
(A23) - Wire connection (30a), in instrument panel wiring harness
(V21) - Plus connector (30), in wiring harness Radio-Amplifier

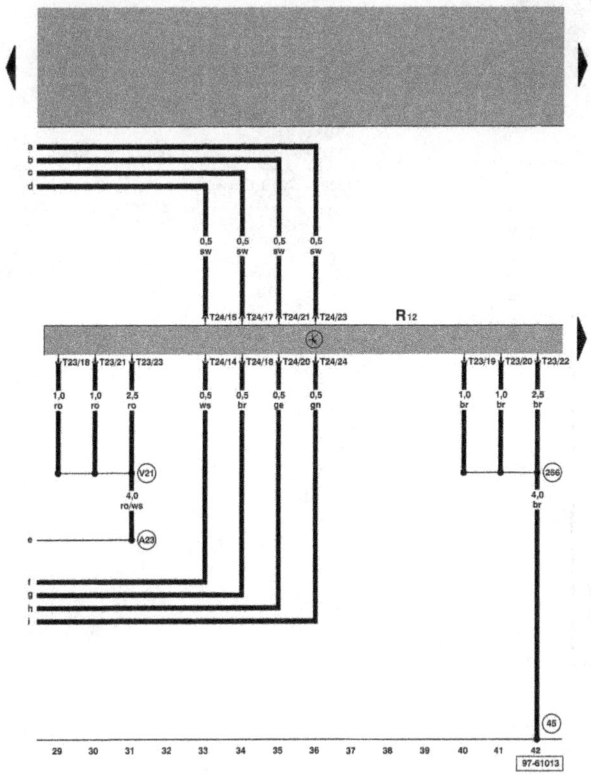

ws = white
sw = black
ro = red
br = brown
gn = green
bl = blue
gr = grey
li = violet
ge = yellow
or = orange

Radio system Monsoon satellite from March 2004

New Beetle — Wiring diagram — No. 34/5

Amplifier, loudspeakers, left

- R12 - Amplifier, in luggage compartment , left
- R14 - Left Rear Treble Speaker
- R15 - Left Rear Bass Speaker
- R20 - Left Front Treble Speaker
- R21 - Left Front Bass Speaker
- R26 - Left Front Midrange Speaker
- T4 - 4-Pin Connector
- T10k - 10-Pin Connector, black, connector station A-pillar, left
- T23 - 23-Pin Connector, at R12

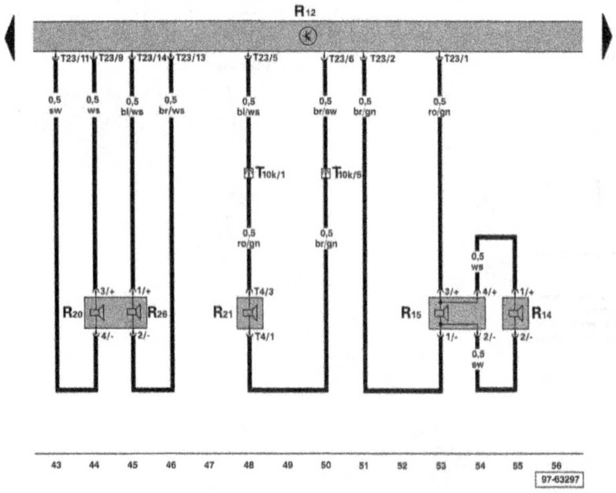

ws = white
sw = black
ro = red
br = brown
gn = green
bl = blue
gr = grey
li = violet
ge = yellow
or = orange

New Beetle — Wiring diagram — No. 34/6

Amplifier, loudspeakers, right

- R12 - Amplifier, in luggage compartment , left
- R16 - Right Rear Treble Speaker
- R17 - Right Rear Bass Speaker
- R22 - Right Front Treble Speaker
- R23 - Right Front Bass Speaker
- R27 - Right Front Midrange Speaker
- T4a - 4-Pin Connector
- T10l - 10-Pin Connector, black, connector station A-pillar, right
- T23 - 23-Pin Connector, at R12

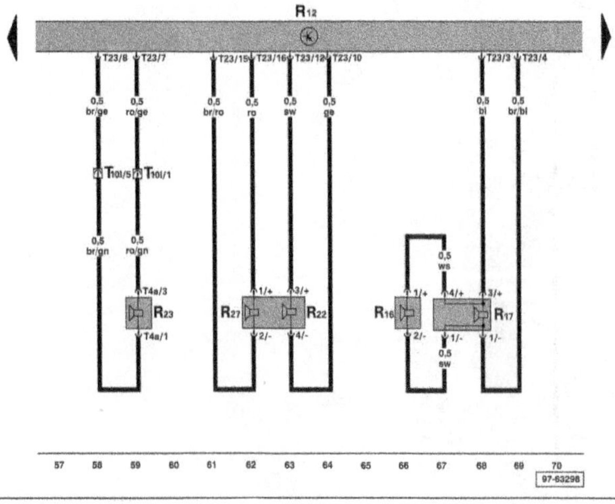

ws = white
sw = black
ro = red
br = brown
gn = green
bl = blue
gr = grey
li = violet
ge = yellow
or = orange

New Beetle **Wiring diagram** **No. 34/7**

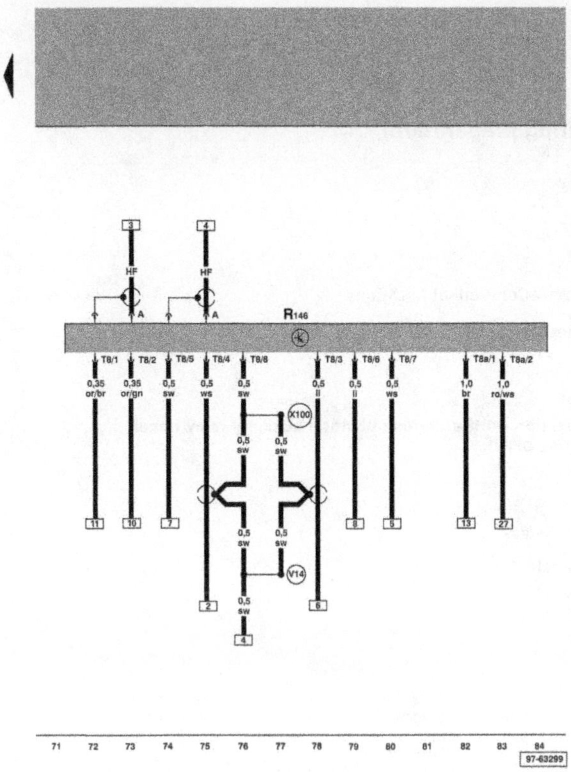

Satellite Radio

R146 - Satellite Radio
T8 - 8-Pin Connector on R146
T8a - 8-Pin Connector on R146

(V14) - wire connection (shielding), in CD-changer wiring harness
(X100) - Shielding Connection (in optional equipment wiring harness)

ws = white
sw = black
ro = red
br = brown
gn = green
bl = blue
gr = grey
li = violet
ge = yellow
or = orange

THIS SPACE INTENTIONALLY

LEFT BLANK

New Beetle Wiring diagram **No. 65/1**

Radio system "Monsoon" (with CD changer wiring preparation),

from July 2005,

Additional Fuse, Relay, Control Module and Connector locations ⇒ Component Locations

Electrical troubleshooting ⇒ Guided Fault Finding using VAS 5051/5052

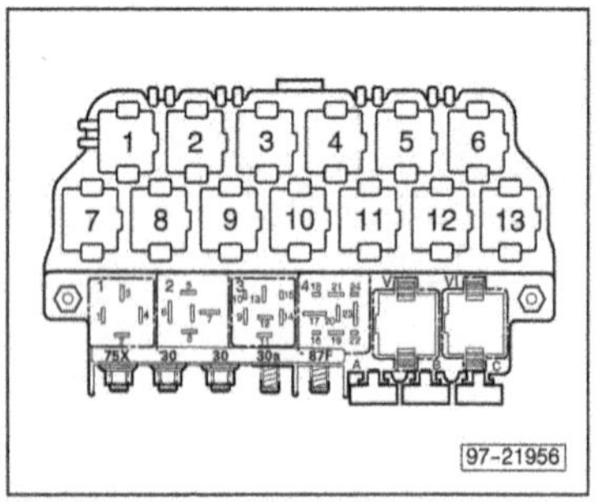

Relay location on the thirteen position auxiliary relay panel, above relay panel:

Relay panel:

Note:

Number in parentheses indicates production control number stamped on relay housing.

New Beetle Wiring diagram **No. 65/2**

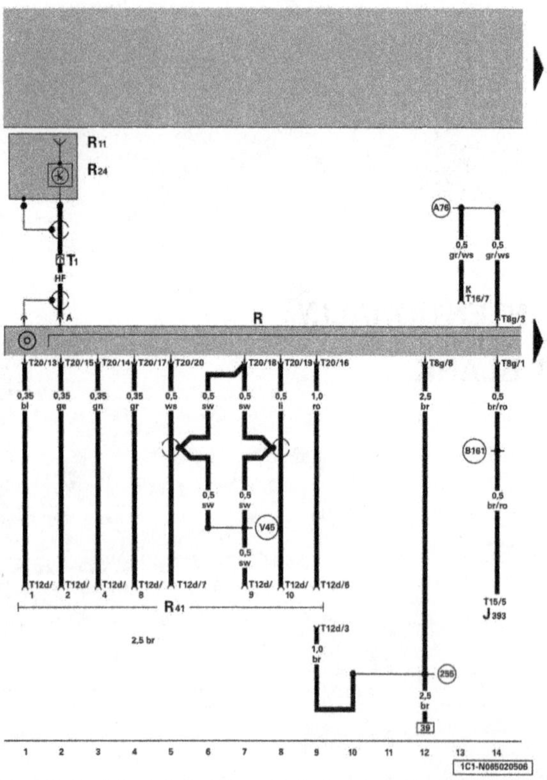

Radio, wiring harness from radio to CD changer unit

- J393 - Comfort System Central Control Module
- R - Radio
- R11 - Antenna
- R24 - Antenna Amplifier
- R41 - CD Changer
- T1 - Single Connector
- T8g - 8-Pin Connector
- T12d - 12-Pin Connector
- T15 - 15-Pin Connector
- T16 - Data Link Connector (DLC), below instrument panel, left
- T20 - 20-Pin Connector, on Radio
- (255) - Ground Connection 1 (in radio wiring harness)
- (A76) - K-Diagnosis Wire Connection (in instrument panel wiring harness)
- (B161) - Anti-Theft Alarm System Connection (in interior wiring harness)
- (V45) - Connection 2 (shielding)(in CD changer wiring harness)

ws = white
sw = black
ro = red
br = brown
gn = green
bl = blue
gr = grey
li = violet
ge = yellow
or = orange

New Beetle — Wiring diagram — No. 65/3

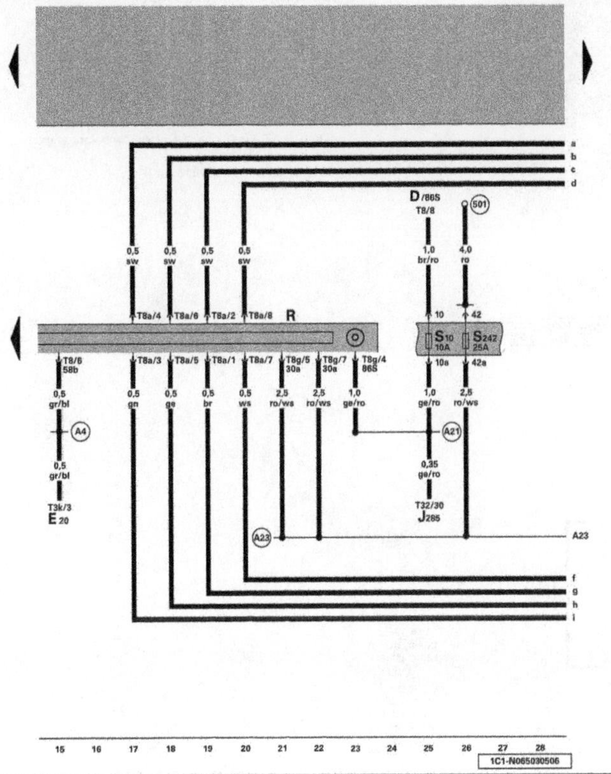

ws = white
sw = black
ro = red
br = brown
gn = green
bl = blue
gr = grey
li = violet
ge = yellow
or = orange

Radio

D - Ignition/Starter Switch
E20 - Instrument Panel Illumination Dimmer Switch
J285 - Instrument Cluster Control Module
R - Radio
S10 - Fuse (on fuse/relay panel)
S242 - Fuse (on fuse panel)
T3k - 3-Pin Connector
T8 - 8-Pin Connector, at D
T8a - 8-Pin Connector
T8g - 8-Pin Connector
T32 - 32-Pin Connector, blue

Note: - Fuse 23 in fuse panel is identified with additional prefix 2, Example: fuse 23 on fuse panel is listed as 223 in wiring diagrams

(501) - Threaded Connection 2 - terminal 30 (on relay panel)
(A4) - Plus Connection (58b) (in instrument panel wiring harness)
(A21) - Connection (86s) (in instrument panel wiring harness)
(A23) - Connection (30al) (in instrument panel wiring harness)

New Beetle — Wiring diagram — No. 65/4

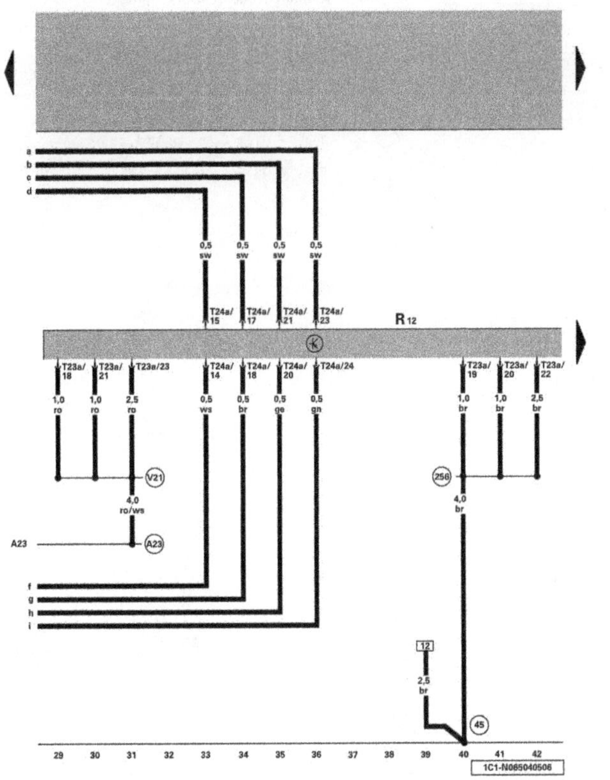

ws = white
sw = black
ro = red
br = brown
gn = green
bl = blue
gr = grey
li = violet
ge = yellow
or = orange

Amplifier

R12 - Amplifier
T23a - 23-Pin Connector
T24a - 24-Pin Connector

(45) - Ground Connection (behind instrument panel, center)
(256) - Ground Connection 2 (in radio wiring harness)
(A23) - Connection (30al) (in instrument panel wiring harness)
(V21) - Plus Connection (30) (in wiring Radio - Amplifier wiring harness)

Radio system Monsoon from July 2005

New Beetle — Wiring diagram — No. 65/5

Amplifier, loudspeakers, left

- R12 - Amplifier
- R14 - Left Rear Treble Speaker
- R15 - Left Rear Bass Speaker
- R20 - Left Front Treble Speaker
- R21 - Left Front Bass Speaker
- R26 - Left Front Midrange Speaker
- T10k - 10-Pin Connector, black, connector station A-pillar, left
- T23a - 23-Pin Connector, at R12

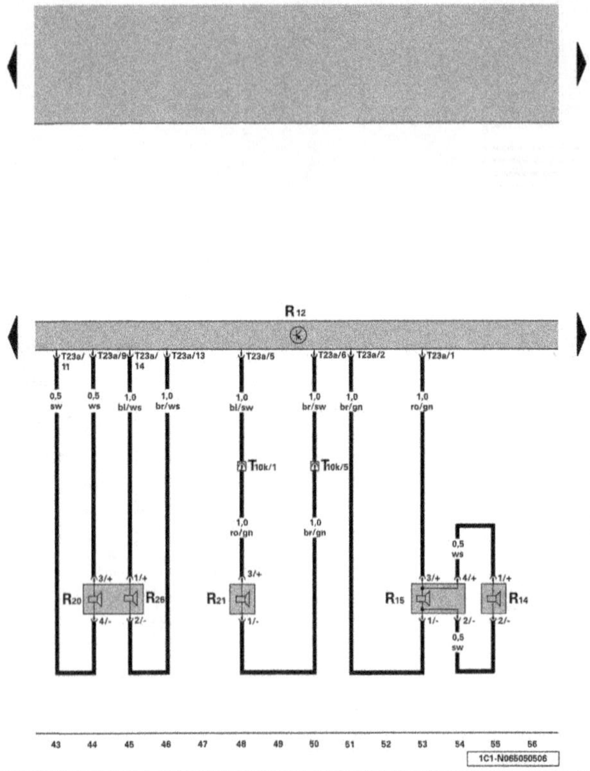

ws = white
sw = black
ro = red
br = brown
gn = green
bl = blue
gr = grey
li = violet
ge = yellow
or = orange

New Beetle — Wiring diagram — No. 65/6

Amplifier, loudspeakers, right

- R12 - Amplifier
- R16 - Right Rear Treble Speaker
- R17 - Right Rear Bass Speaker
- R22 - Right Front Treble Speaker
- R23 - Right Front Bass Speaker
- R27 - Right Front Midrange Speaker
- T10l - 10-Pin Connector, black, connector station A-pillar, right
- T23a - 23-Pin Connector, at R12

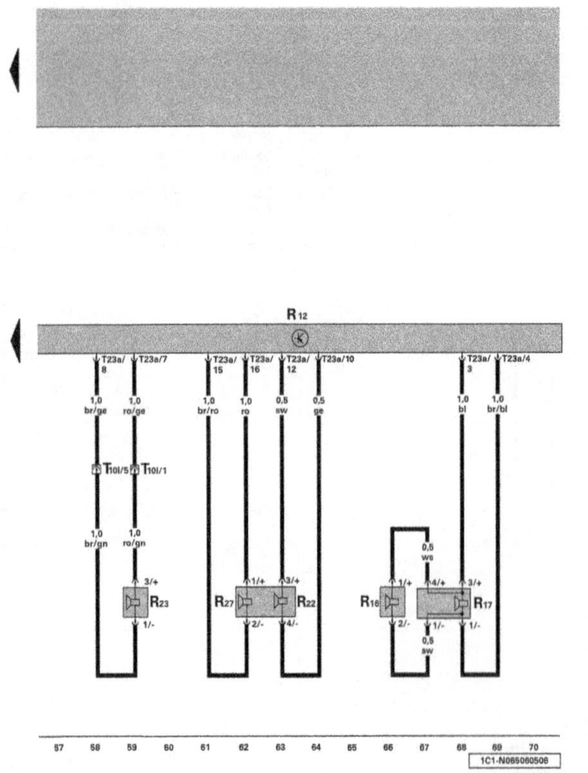

ws = white
sw = black
ro = red
br = brown
gn = green
bl = blue
gr = grey
li = violet
ge = yellow
or = orange

New Beetle Wiring diagram **No. 37/1**

Radio system "Monsoon", Satellite Radio (with CD changer wiring preparation)

from June 2004

Additional Fuse, Relay, Control Module and Connector locations ⇒ Component Locations

Troubleshooting Procedures ⇒ Guided Fault Finding using VAS 5051/5052

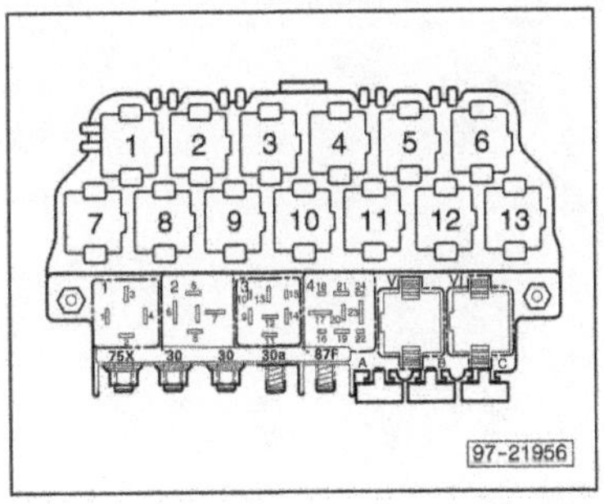

Fuse colors

30 A - green
25 A - white
20 A - yellow
15 A - blue
10 A - red
7,5 A - brown
5 A - beige
3 A - violet

New Beetle Wiring diagram **No. 37/2**

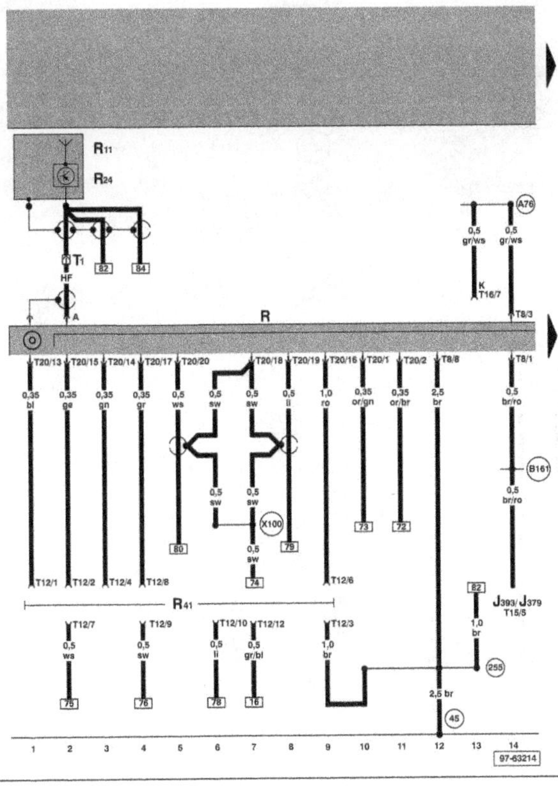

Radio, wiring harness from radio to CD changer unit

- J379 - Control module for central locking and anti-theft-system
- J393 - Central control module for comfort system
- R - Radio
- R11 - Antenna
- R24 - Antenna Amplifier
- R41 - CD Changer Unit
- T1 - Single Connector
- T8 - 8-Pin Connector
- T12 - 12-Pin Connector
- T15 - 15-Pin Connector
- T16 - Data Link Connector (DLC), below instrument panel, left
- T20 - 20-Pin Connector, on Radio
- (45) - Ground connection, behind instrument panel, center
- (255) - Ground connection, -1-, in wiring harness radio
- (A76) - Connector (K-diagnosis wire) in instrument panel wiring harness
- (B161) - Connector (anti-theft warning system), in wiring harness interior
- (X100) - Shielding Connection (in optional equipment wiring harness)

ws = white
sw = black
ro = red
br = brown
gn = green
bl = blue
gr = grey
li = violet
ge = yellow
or = orange

Radio system Monsoon satellite from June 2004

New Beetle

Wiring diagram

No. 37/3

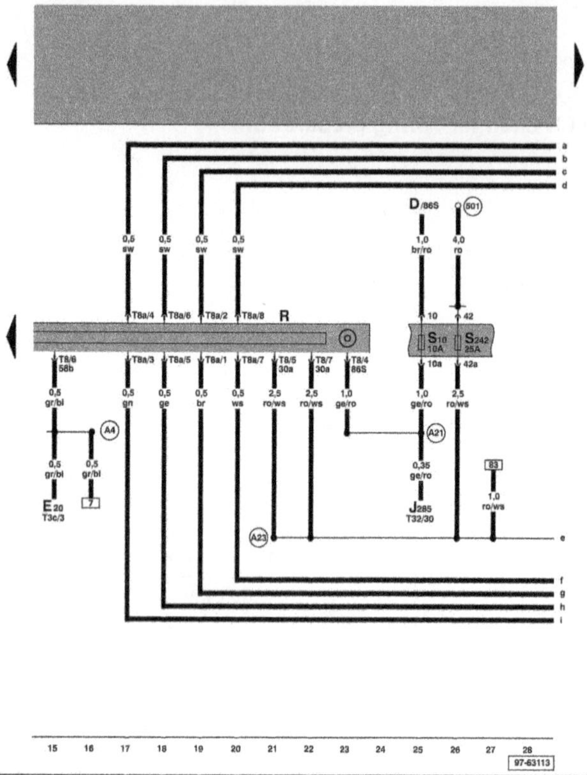

Radio

- D - Ignition/Starter Switch
- E20 - Instrument Panel Light Dimmer Switch
- J285 - Control module with indicator unit in instrument panel insert
- R - Radio
- S10 - Fuse 10 in fuse holder
- S242 - Fuse 42 in fuse holder
- T3c - 3-Pin Connector
- T8 - 8-Pin Connector
- T8a - 8-Pin Connector
- T32 - 32-Pin Connector, blue

- (501) - Threaded connection -2- (30) on the relay plate
- (A4) - Plus connection (58b), in instrument panel wiring harness
- (A21) - Wire connection (86s), in instrument panel wiring harness
- (A23) - Wire connection (30a), in instrument panel wiring harness

ws = white
sw = black
ro = red
br = brown
gn = green
bl = blue
gr = grey
li = violet
ge = yellow
or = orange

New Beetle

Wiring diagram

No. 37/4

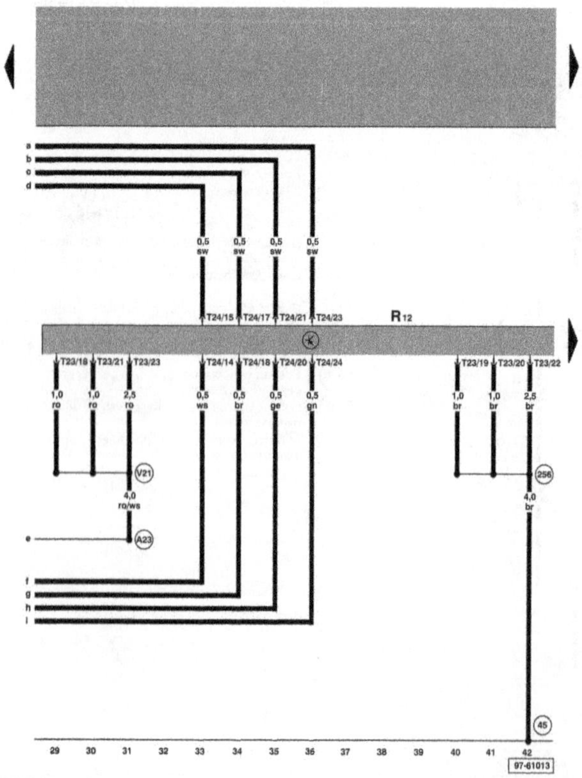

Amplifier

- T23 - 23-Pin Connector
- T24 - 24-Pin Connector
- R12 - Amplifier, in luggage compartment, left

- (45) - Ground connection, behind instrument panel, center
- (256) - Ground connection, -2-, in wiring harness radio
- (A23) - Wire connection (30a), in instrument panel wiring harness
- (V21) - Plus connector (30), in wiring harness Radio-Amplifier

ws = white
sw = black
ro = red
br = brown
gn = green
bl = blue
gr = grey
li = violet
ge = yellow
or = orange

New Beetle — Wiring diagram — No. 37/5

Amplifier, loud speaker front

R 12 - Amplifier, in luggage compartment , left
R20 - Treble speaker, left front
R22 - Treble speaker, right front
R 26 - Left Front Midrange Speaker
R 27 - Right Front Midrange Speaker
T4 - 4-Pin Connector
T4a - 4-Pin Connector
T23 - 23-Pin Connector

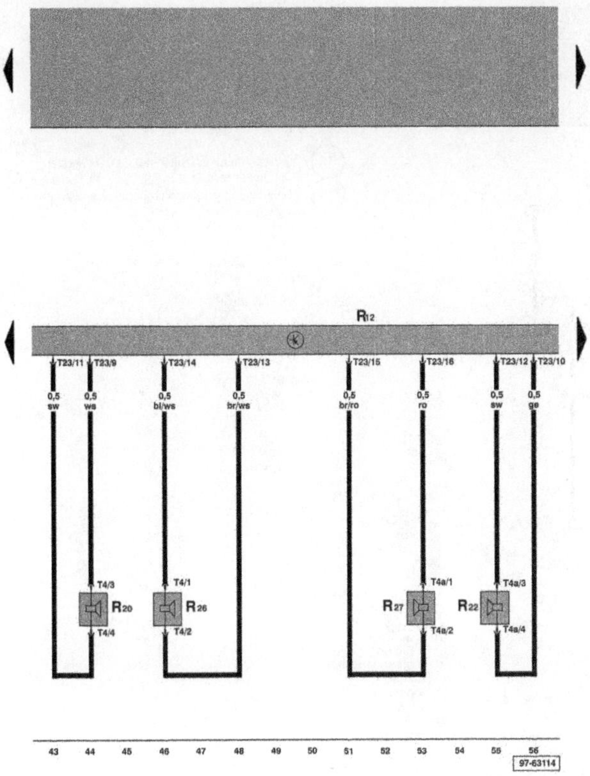

ws = white
sw = black
ro = red
br = brown
gn = green
bl = blue
gr = grey
li = violet
ge = yellow
or = orange

New Beetle — Wiring diagram — No. 37/6

Amplifier, loud speaker rear

R 4 - Left Rear Speaker
R 5 - Right Rear Speaker
R12 - Amplifier, in luggage compartment, left
R 21 - Bass speaker, left front in driver's door
R 23 - Bass speaker, right front in passenger's door
T4b - 4-Pin Connector
T4c - 4-Pin Connector
T4d - 4-Pin Connector in driver's door wiring
T4e - 4-Pin Connector in passenger's door wiring
T10k - 10-Pin Connector, black, connector station A-pillar, left
T10l - 10-Pin Connector, black, connector station A-pillar, right
T23 - 23-Pin Connector

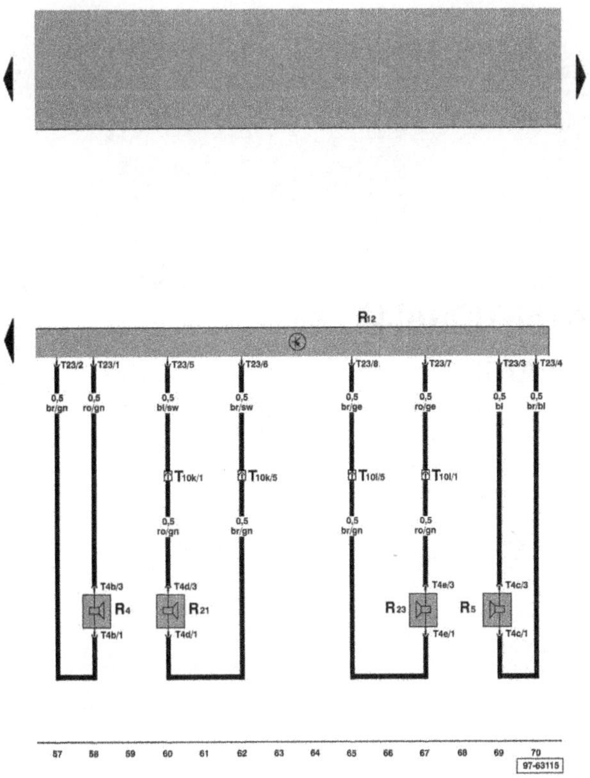

ws = white
sw = black
ro = red
br = brown
gn = green
bl = blue
gr = grey
li = violet
ge = yellow
or = orange

New Beetle Wiring diagram **No. 37/7**

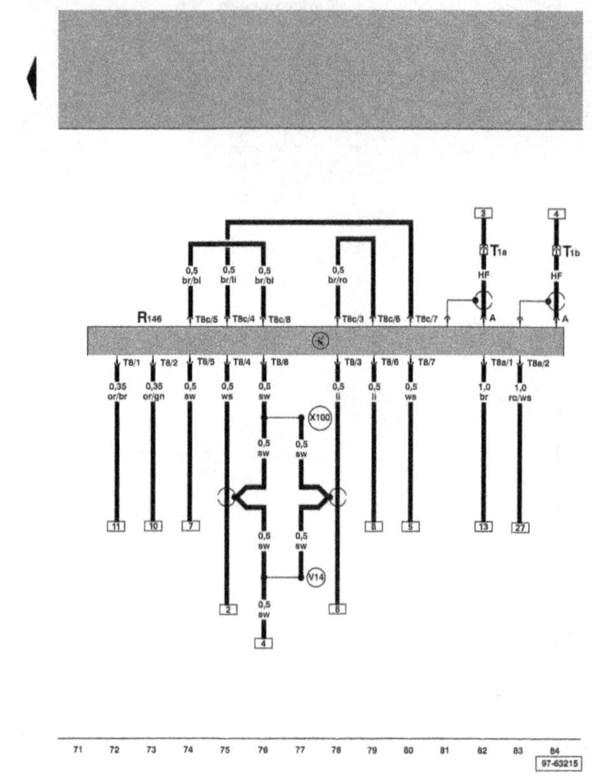

Satellite Radio

R146 - Satellite Radio
T1a - Single Connector
T1b - Single Connector
T8 - 8-Pin Connector on R146
T8a - 8-Pin Connector on R146
T8c - 8-Pin Connector on R146

(V14) - wire connection (shielding), in CD-changer wiring harness
(X100) - Shielding Connection (in optional equipment wiring harness)

ws = white
sw = black
ro = red
br = brown
gn = green
bl = blue
gr = grey
li = violet
ge = yellow
or = orange

THIS SPACE INTENTIONALLY LEFT BLANK

New Beetle Wiring diagram **No. 59/1**

Radio system "Monsoon", Satellite Radio (with CD changer wiring preparation),

from July 2005

Additional Fuse, Relay, Control Module and Connector locations ⇒ Component Locations

Electrical troubleshooting ⇒ Guided Fault Finding using VAS 5051/5052

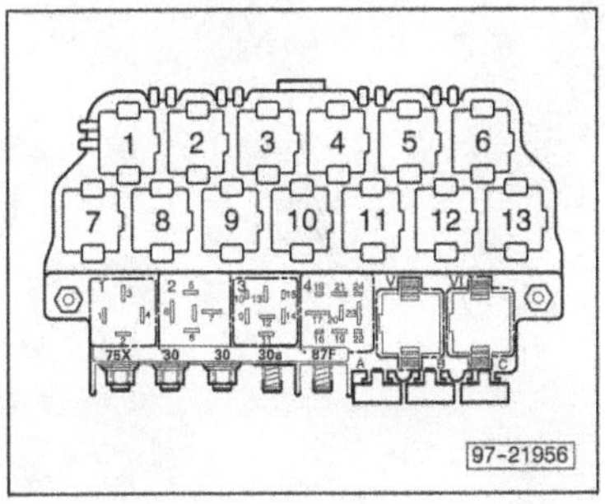

Edition 11.05

New Beetle Wiring diagram **No. 59/2**

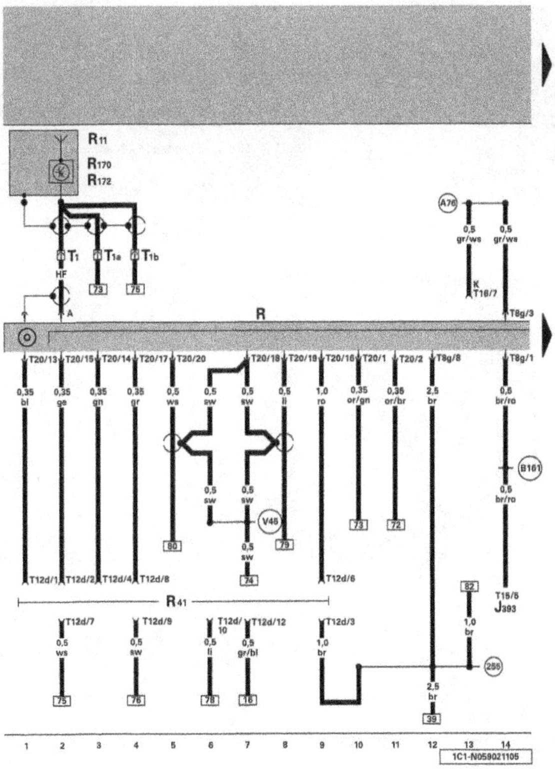

Radio, wiring harness from radio to CD changer unit

- J393 - Central control module for comfort system
- R - Radio
- R11 - Antenna
- R41 - CD Changer
- R170 - Satellite Antenna
- R172 - Satellite Tuner Antenna
- T1 - Single Connector
- T1a - Single Connector
- T1b - Single Connector
- T8g - 8-Pin Connector
- T12d - 12-Pin Connector
- T15 - 15-Pin Connector
- T16 - Data Link Connector (DLC), below instrument panel, left
- T20 - 20-Pin Connector, on Radio
- (255) - Ground connection, -1-, in wiring harness radio
- (A76) - Connector (K-diagnosis wire) in instrument panel wiring harness
- (B161) - Connector (anti-theft warning system), in wiring harness interior
- (V45) - Connection 2 (shielding)(in CD changer wiring harness)

ws = white
sw = black
ro = red
br = brown
gn = green
bl = blue
gr = grey
li = violet
ge = yellow
or = orange

Edition 11.05

New Beetle

Wiring diagram

No. 59/3

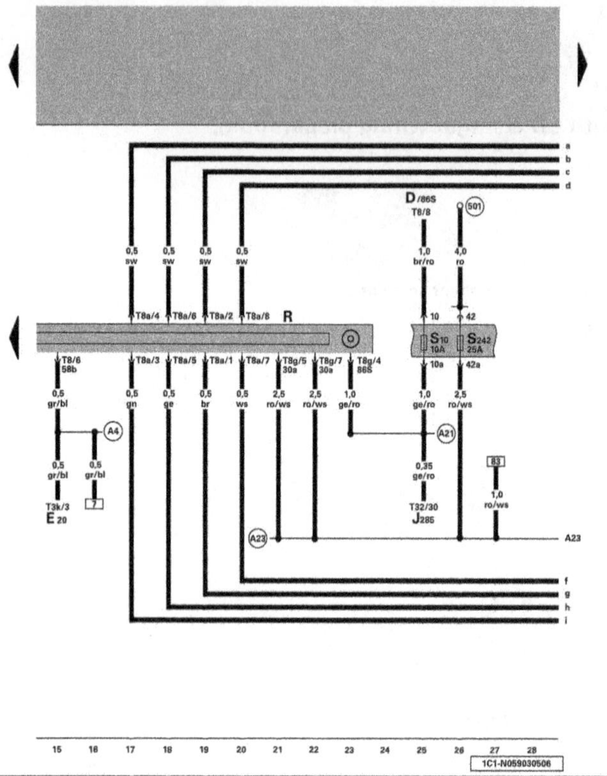

Radio

- D - Ignition/Starter Switch
- E20 - Instrument Panel Illumination Dimmer Switch
- J285 - Instrument Cluster Control Module
- R - Radio
- S10 - Fuse (on fuse/relay panel)
- S242 - Fuse (on fuse panel)
- T3k - 3-Pin Connector
- T8 - 8-Pin Connector
- T8a - 8-Pin Connector
- T8g - 8-Pin Connector
- T32 - 32-Pin Connector, blue
- (501) - Threaded Connection 2 - terminal 30 (on relay panel)
- (A4) - Plus Connection (58b) (in instrument panel wiring harness)
- (A21) - Connection (86s) (in instrument panel wiring harness)
- (A23) - Connection (30al) (in instrument panel wiring harness)

ws = white
sw = black
ro = red
br = brown
gn = green
bl = blue
gr = grey
li = violet
ge = yellow
or = orange

Edition 05.06

New Beetle

Wiring diagram

No. 59/4

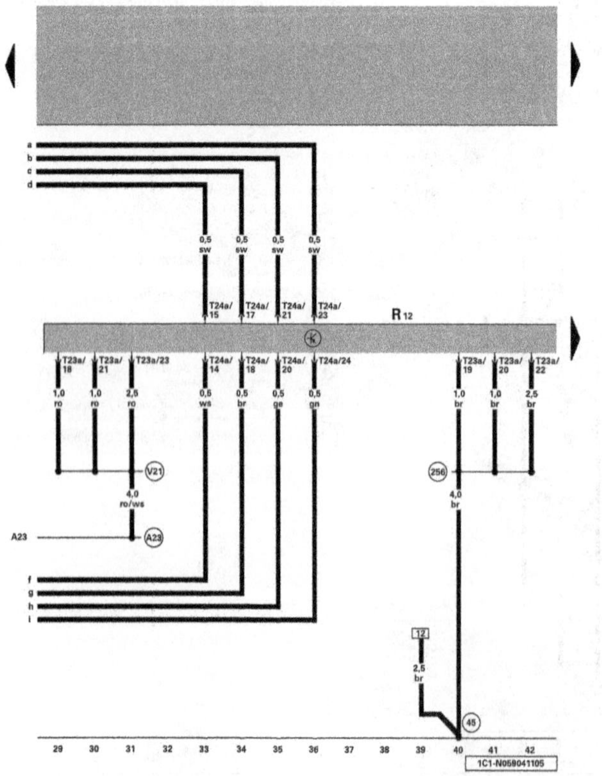

Amplifier

- T23a - 23-Pin Connector
- T24a - 24-Pin Connector
- R12 - Amplifier, in luggage compartment, left
- (45) - Ground connection, behind instrument panel, center
- (256) - Ground connection, -2-, in wiring harness radio
- (A23) - Connection (30al) (in instrument panel wiring harness)
- (V21) - Plus connector (30), in wiring harness Radio-Amplifier

ws = white
sw = black
ro = red
br = brown
gn = green
bl = blue
gr = grey
li = violet
ge = yellow
or = orange

Edition 11.05

New Beetle Wiring diagram **No. 59/5**

Amplifier, loudspeakers, left

- R12 - Amplifier, in luggage compartment , left
- R14 - Left Rear Treble Speaker
- R15 - Left Rear Bass Speaker
- R20 - Left Front Treble Speaker
- R21 - Left Front Bass Speaker
- R26 - Left Front Midrange Speaker
- T10k - 10-Pin Connector, black, connector station A-pillar, left
- T23a - 23-Pin Connector, at R12

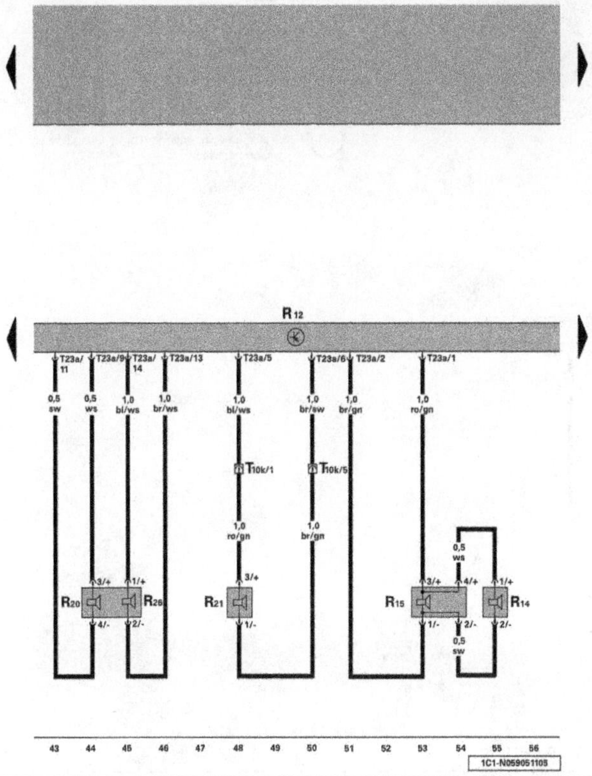

ws = white
sw = black
ro = red
br = brown
gn = green
bl = blue
gr = grey
li = violet
ge = yellow
or = orange

New Beetle Wiring diagram **No. 59/6**

Amplifier, loudspeakers, right

- R12 - Amplifier, in luggage compartment , left
- R16 - Right Rear Treble Speaker
- R17 - Right Rear Bass Speaker
- R22 - Right Front Treble Speaker
- R23 - Right Front Bass Speaker
- R27 - Right Front Midrange Speaker
- T10l - 10-Pin Connector, black, connector station A-pillar, right
- T23a - 23-Pin Connector, at R12

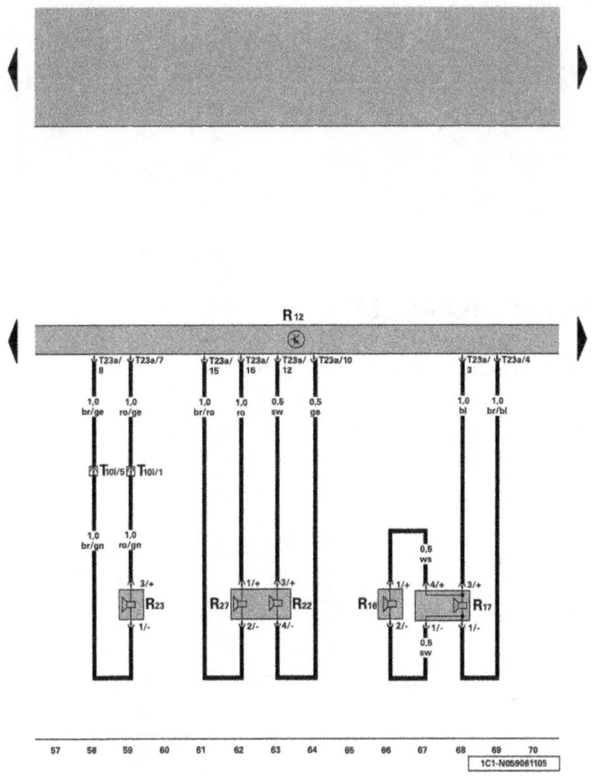

ws = white
sw = black
ro = red
br = brown
gn = green
bl = blue
gr = grey
li = violet
ge = yellow
or = orange

Radio sytem Monsoon satellite from July 2005

New Beetle

Wiring diagram

No. 59/7

Satellite Radio

R146 - Satellite Radio
T8h - 8-Pin Connector on R146
T8i - 8-Pin Connector on R146

(V14) - wire connection (shielding), in CD-changer wiring harness
(X100) - Shielding Connection (in optional equipment wiring harness)

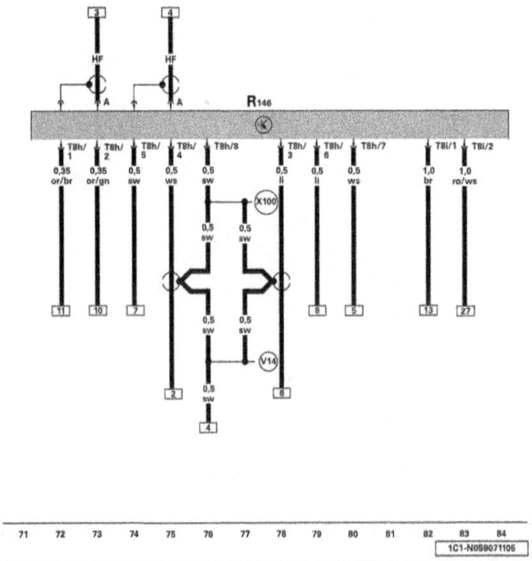

ws = white
sw = black
ro = red
br = brown
gn = green
bl = blue
gr = grey
li = violet
ge = yellow
or = orange

THIS SPACE INTENTIONALLY

LEFT BLANK

New Beetle Wiring diagram **No. 41/1**

Radio system, Satellite Radio (with CD changer wiring preparation)

from June 2004

Additional Fuse, Relay, Control Module and Connector locations ⇒ Compontent Locations

Troubleshooting Procedures ⇒ Guided Fault Finding using VAS 5051/5052

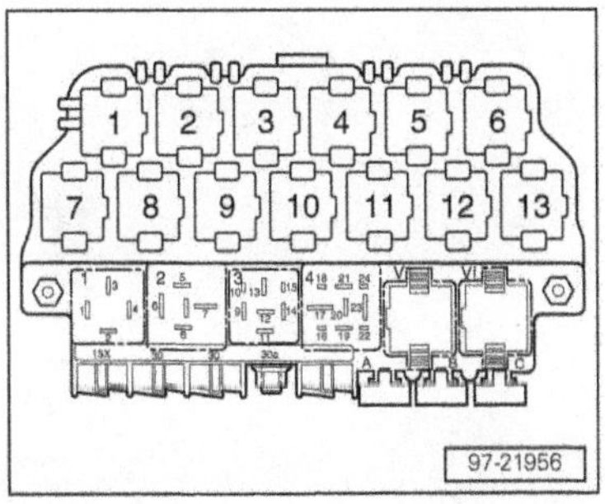

Relay location on the thirteenfold auxiliary relay panel, above relay panel:

Relay panel:

New Beetle Wiring diagram **No. 41/2**

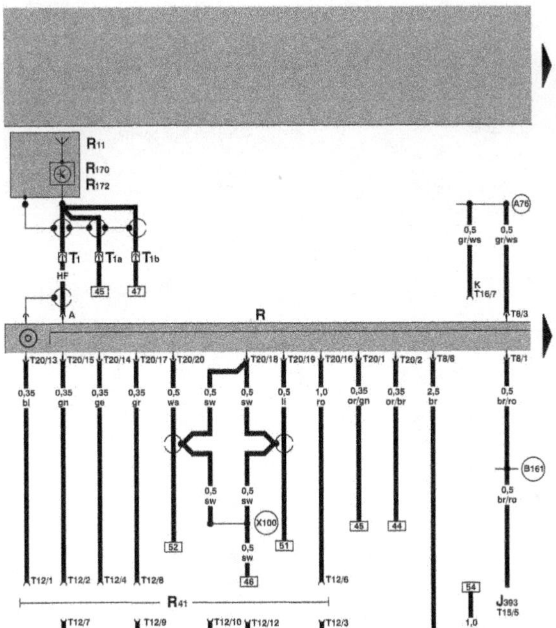

Radio, wiring harness from radio to CD changer unit

- J393 - Central control module for comfort system
- R - Radio
- R11 - Antenna
- R41 - CD Changer
- R170 - Satellite Antenna
- R172 - Satellite Tuner Antenna
- T1 - Single Connector
- T1a - Single Connector
- T1b - Single Connector
- T8 - 8-Pin Connector
- T12 - 12-Pin Connector
- T15 - 15-Pin Connector
- T16 - Data Link Connector (DLC), below instrument panel, left
- T20 - 20-Pin Connector, on Radio
- (45) - Ground connection, behind instrument panel, center
- (255) - Ground connection, -1-, in wiring harness radio
- (A76) - Connector (K-diagnosis wire) in instrument panel wiring harness
- (B161) - Connector (anti-theft warning system), in wiring harness interior
- (X100) - Shielding Connection (in optional equipment wiring harness)

ws = white
sw = black
ro = red
br = brown
gn = green
bl = blue
gr = grey
li = violet
ge = yellow
or = orange

New Beetle

Wiring diagram

No. 41/3

Radio, loudspeakers

- R - Radio
- R15 - Left Rear Bass Speaker
- R20 - Left Front Treble Speaker
- R21 - Left Front Bass Speaker
- R22 - Right Front Treble Speaker
- R23 - Right Front Bass speaker
- T8a - 8-Pin Connector
- T10k - 10-Pin Connector, black, connector station A-pillar, left
- T10l - 10-Pin Connector, black, connector station A-pillar, right

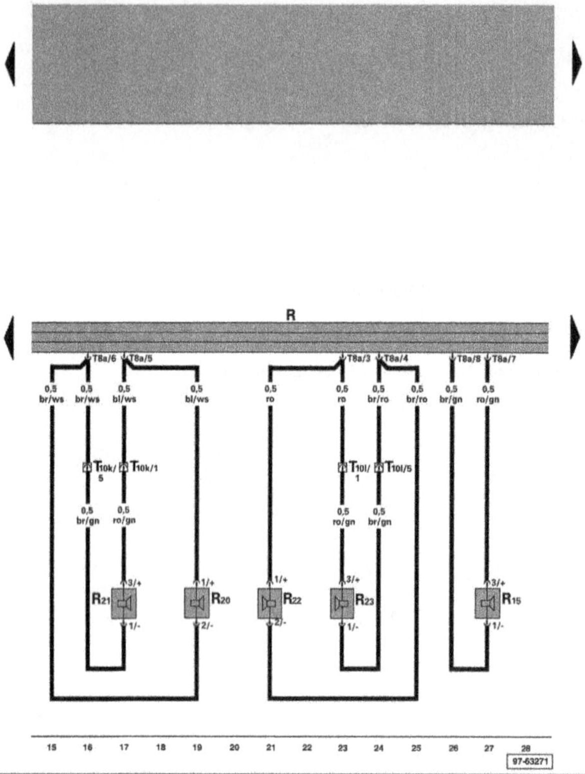

ws = white
sw = black
ro = red
br = brown
gn = green
bl = blue
gr = grey
li = violet
ge = yellow
or = orange

New Beetle

Wiring diagram

No. 41/4

Radio, loudspeaker

- D - Ignition/Starter Switch
- E20 - Instrument Panel Light Dimmer Switch
- J285 - Control module with indicator unit in instrument panel insert
- R - Radio
- R17 - Right Rear Bass Speaker
- S10 - Fuse 10 in fuse holder
- S242 - Fuse 42 in fuse holder
- T3c - 3-Pin Connector
- T8 - 8-Pin Connector
- T8a - 8-Pin Connector
- T8b - 8-Pin Connector
- T32 - 32-Pin Connector, blue
- (501) - Threaded connection -2- (30) on the relay plate
- (A4) - Plus connection (58b), in instrument panel wiring harness
- (A21) - Wire connection (86s), in instrument panel wiring harness
- (A23) - Wire connection (30a), in instrument panel wiring harness

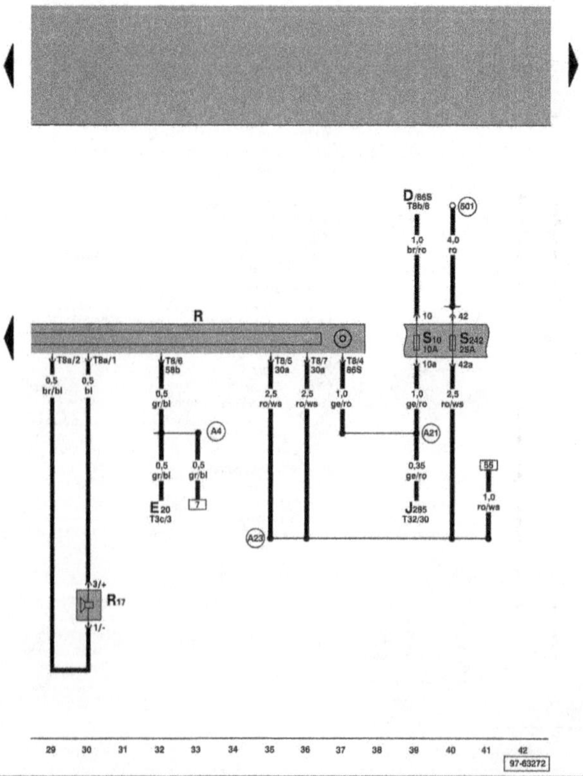

ws = white
sw = black
ro = red
br = brown
gn = green
bl = blue
gr = grey
li = violet
ge = yellow
or = orange

New Beetle **Wiring diagram** **No. 41/5**

Satellite Radio

R146 - Satellite Radio
T8 - 8-Pin Connector on R146
T8a - 8-Pin Connector on R146

(V14) - wire connection (shielding), in CD-changer wiring harness
(X100) - Shielding Connection (in optional equipment wiring harness)

* - Wire taped to wiring harness (blind, without continuity or connection)

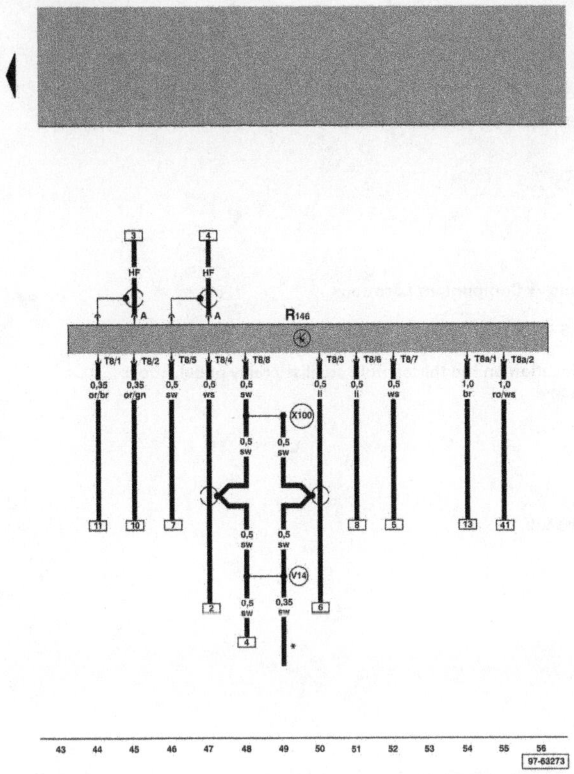

ws = white
sw = black
ro = red
br = brown
gn = green
bl = blue
gr = grey
li = violet
ge = yellow
or = orange

THIS SPACE INTENTIONALLY LEFT BLANK

New Beetle Convertible — Wiring diagram — No. 27/1

Radio system, Satellite Radio (with CD changer wiring preparation)

from June 2004

Additional Fuse, Relay, Control Module and Connector locations ⇒ Compontent Locations

Troubleshooting Procedures ⇒ Guided Fault Finding using VAS 5051/5052

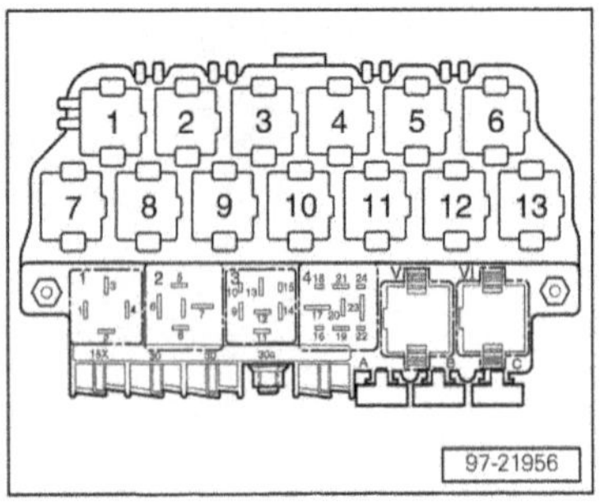

Relay location on the thirteenfold auxiliary relay panel, above relay panel:

Relay panel:

New Beetle Convertible — Wiring diagram — No. 27/2

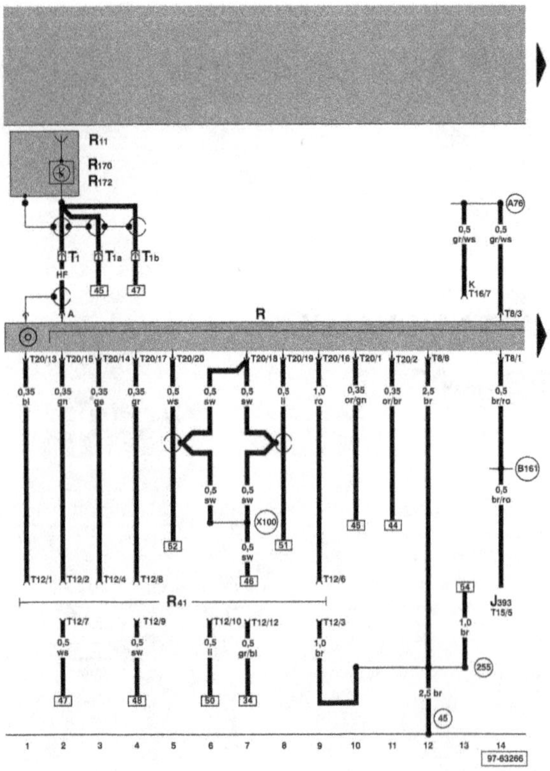

Radio, wiring harness from radio to CD changer unit

- J393 - Central control module for comfort system
- R - Radio
- R11 - Antenna
- R41 - CD Changer
- R170 - Satellite Antenna
- R172 - Satellite Tuner Antenna
- T1 - Single Connector
- T1a - Single Connector
- T1b - Single Connector
- T8 - 8-Pin Connector
- T12 - 12-Pin Connector
- T15 - 15-Pin Connector
- T16 - Data Link Connector (DLC), below instrument panel, left
- T20 - 20-Pin Connector, on Radio

- (45) - Ground connection, behind instrument panel, center
- (255) - Ground connection, -1-, in wiring harness radio
- (A76) - Connector (K-diagnosis wire) in instrument panel wiring harness
- (B161) - Connector (anti-theft warning system), in wiring harness interior
- (X100) - Shielding Connection (in optional equipment wiring harness)

ws = white
sw = black
ro = red
br = brown
gn = green
bl = blue
gr = grey
li = violet
ge = yellow
or = orange

New Beetle Convertible

Wiring diagram **No. 27/3**

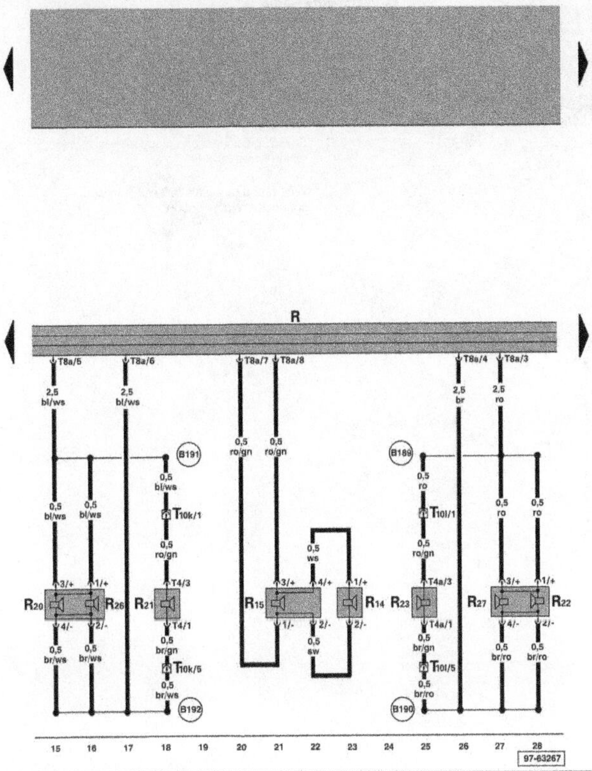

Radio, loudspeakers

- R - Radio
- R14 - Left Rear Treble Speaker
- R15 - Left Rear Bass Speaker
- R20 - Left Front Treble Speaker
- R21 - Left Front Bass speaker
- R22 - Right Front Treble Speaker
- R23 - Right Front Bass Speaker
- R26 - Left Front Midrange Speaker
- R27 - Right Front Midrange Speaker
- T4 - 4-Pin Connector
- T4a - 4-Pin Connector
- T8a - 8-Pin Connector
- T10k - 10-Pin Connector, black, connector station A-pillar, left
- T10l - 10-Pin Connector, black, connector station A-pillar, right
- (B189) - Connection (positive, RF speaker) in passenger compartment wiring harness
- (B190) - Connection (negative, RF speaker) in passenger compartment wiring harness
- (B191) - Connection (positive, LF speaker) in passenger compartment wiring harness
- (B192) - Connection (negative, LF speaker) in passenger compartment wiring harness

ws = white
sw = black
ro = red
br = brown
gn = green
bl = blue
gr = grey
li = violet
ge = yellow
or = orange

New Beetle Convertible

Wiring diagram **No. 27/4**

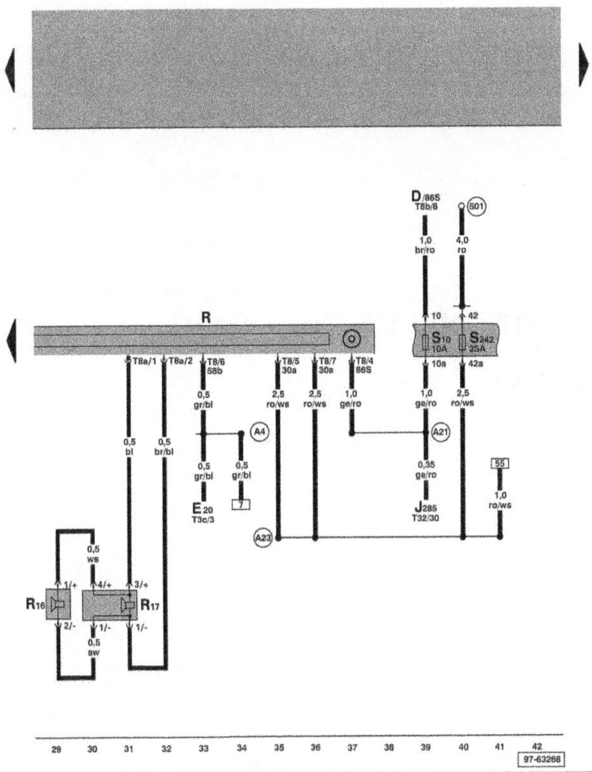

Radio

- D - Ignition/Starter Switch
- E20 - Instrument Panel Light Dimmer Switch
- J285 - Control module with indicator unit in instrument panel insert
- R - Radio
- R16 - Right Rear Treble Speaker
- R17 - Right Rear Bass Speaker
- S10 - Fuse 10 in fuse holder
- S242 - Fuse 42 in fuse holder
- T3c - 3-Pin Connector
- T8 - 8-Pin Connector
- T8a - 8-Pin Connector
- T8b - 8-Pin Connector
- T32 - 32-Pin Connector, blue
- (501) - Threaded connection -2- (30) on the relay plate
- (A4) - Plus connection (58b), in instrument panel wiring harness
- (A21) - Wire connection (86s), in instrument panel wiring harness
- (A23) - Wire connection (30a), in instrument panel wiring harness

ws = white
sw = black
ro = red
br = brown
gn = green
bl = blue
gr = grey
li = violet
ge = yellow
or = orange

New Beetle Convertible Wiring diagram **No. 27/5**

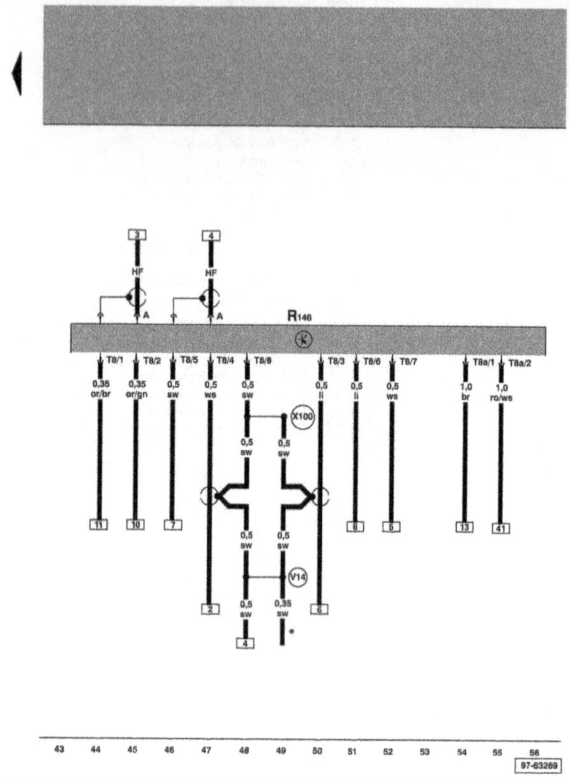

Satellite Radio

R146 - Satellite Radio
T8 - 8-Pin Connector on R146
T8a - 8-Pin Connector on R146

(V14) - wire connection (shielding), in CD-changer wiring harness
(X100) - Shielding Connection (in optional equipment wiring harness)

* - Wire taped to wiring harness (without continuity or connection)

ws = white
sw = black
ro = red
br = brown
gn = green
bl = blue
gr = grey
li = violet
ge = yellow
or = orange

THIS SPACE INTENTIONALLY LEFT BLANK

New Beetle **Wiring diagram** **No. 66/1**

Radio system, Satellite Radio (with CD changer wiring preparation),

from July 2005,

Additional Fuse, Relay, Control Module and Connector locations ⇒ Component Locations

Electrical troubleshooting ⇒ Guided Fault Finding using VAS 5051/5052

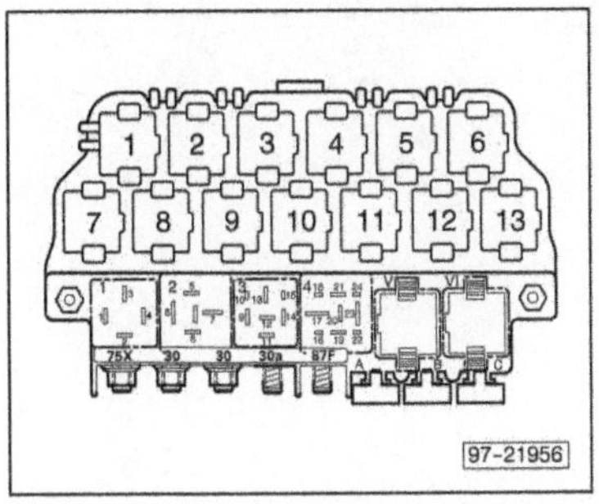

Relay location on the thirteen position auxiliary relay panel, above relay panel:

Relay panel:

Note:

Number in parentheses indicates production control number stamped on relay housing.

New Beetle **Wiring diagram** **No. 66/2**

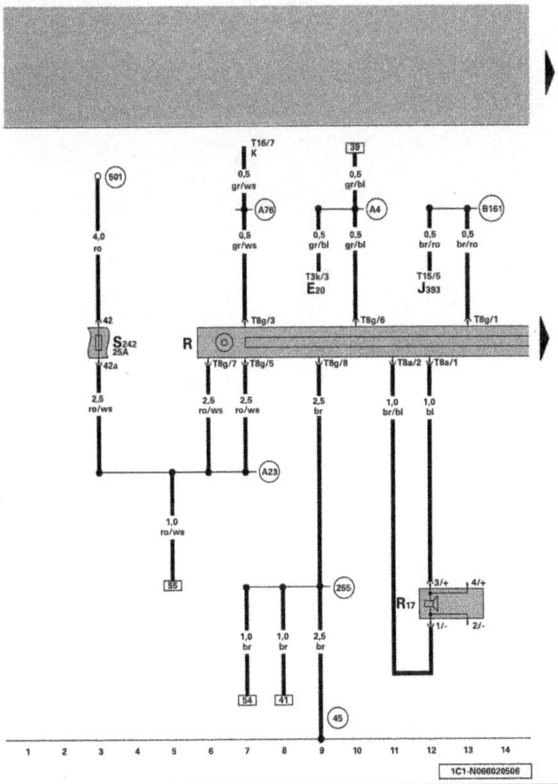

ws = white
sw = black
ro = red
br = brown
gn = green
bl = blue
gr = grey
li = violet
ge = yellow
or = orange

Radio, right rear speaker

- E20 - Instrument Panel Illumination Dimmer Switch
- J393 - Comfort System Central Control Module
- R - Radio
- R17 - Right Rear Bass Speaker
- S242 - Fuse (on fuse panel)
- T3k - 3-Pin Connector
- T8a - 8-Pin Connector, brown, on radio
- T8g - 8-Pin Connector, black, on radio
- T15 - 15-Pin Connector, on Central control module for comfort system
- T16 - 16-Pin Connector, Data Link Connector (DLC), below steering column

Note: - Fuse 23 in fuse panel is identified with additional prefix 2, Example: fuse 23 on fuse panel is listed as 223 in wiring diagrams

- (45) - Ground Connection (behind instrument panel, center)
- (255) - Ground Connection 1 (in radio wiring harness)
- (501) - Threaded Connection 2 - terminal 30 (on relay panel)
- (A4) - Plus Connection (58b) (in instrument panel wiring harness)
- (A23) - Connection (30al) (in instrument panel wiring harness)
- (A76) - K-Diagnosis Wire Connection (in instrument panel wiring harness)
- (B161) - Anti-Theft Alarm System Connection (in interior wiring harness)

New Beetle

Wiring diagram **No. 66/3**

Radio, front speakers, left rear speaker

- R - Radio
- R15 - Left Rear Bass Speaker
- R20 - Left Front Treble Speaker
- R21 - Left Front Bass Speaker
- R22 - Right Front Treble Speaker
- R23 - Right Front Bass Speaker
- T8a - 8-Pin Connector, brown, on radio
- T10k - 10-Pin Connector, black, connector station A-pillar, left
- T10l - 10-Pin Connector, black, connector station A-pillar, right

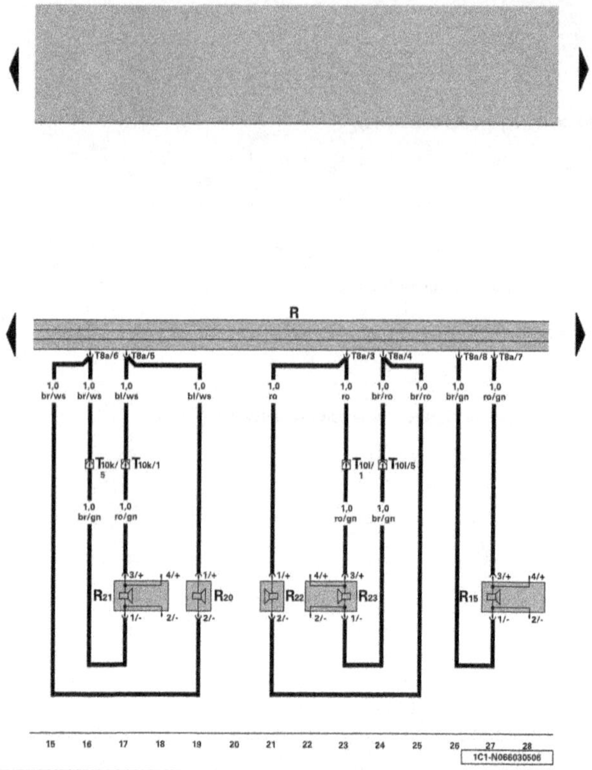

ws = white
sw = black
ro = red
br = brown
gn = green
bl = blue
gr = grey
li = violet
ge = yellow
or = orange

New Beetle

Wiring diagram **No. 66/4**

Radio, Antenna, Antenna Amplifier, CD Changer Unit, Satellite Antenna, Satellite Tuner Antenna

- D - Ignition/Starter Switch
- J285 - Instrument Cluster Control Module
- R - Radio
- R11 - Antenna
- R24 - Antenna Amplifier
- R41 - CD Changer
- R170 - Satellite Antenna
- R172 - Satellite Tuner Antenna
- S10 - Fuse (on fuse/relay panel)
- T1 - Single Connector, black, below console
- T1a - Single Connector
- T1b - Single Connector
- T8 - 8-Pin Connector, black, on radio
- T8g - 8-Pin Connector
- T12d - 12-Pin Connector
- T20 - 20-Pin Connector, on radio
- T32 - 32-Pin Connector, blue, on instrument cluster

Note: - Fuse 23 in fuse panel is identified with additional prefix 2. Example: fuse 23 on fuse panel is listed as 223 in wiring diagrams

- (A21) - Connection (86s) (in instrument panel wiring harness)
- (V45) - Connection 2 (shielding)(in CD changer wiring harness)

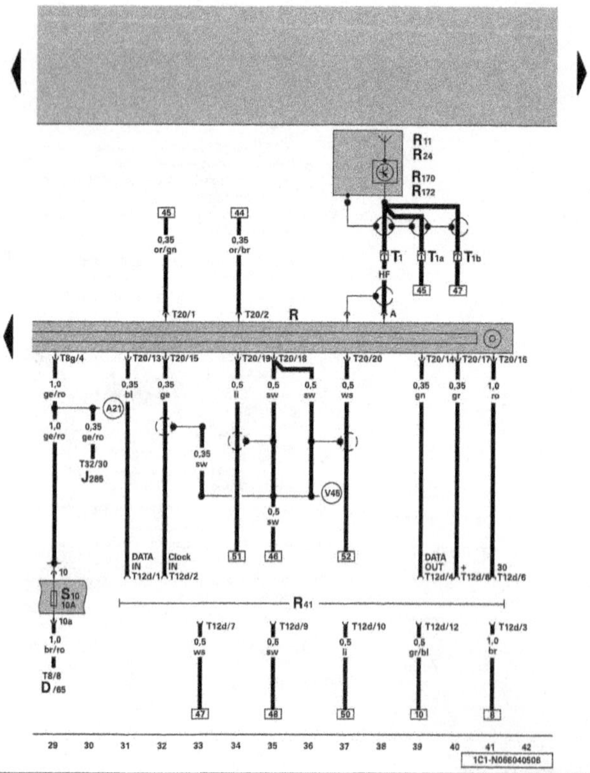

ws = white
sw = black
ro = red
br = brown
gn = green
bl = blue
gr = grey
li = violet
ge = yellow
or = orange

New Beetle Wiring diagram **No. 66/5**

Satellite Radio

R146 - Satellite Radio
T8h - 8-Pin Connector on R146
T8i - 8-Pin Connector on R146

V14 - Connection (shielding) (in CD-changer wiring harness)
X100 - Shielding Connection (in optional equipment wiring harness)

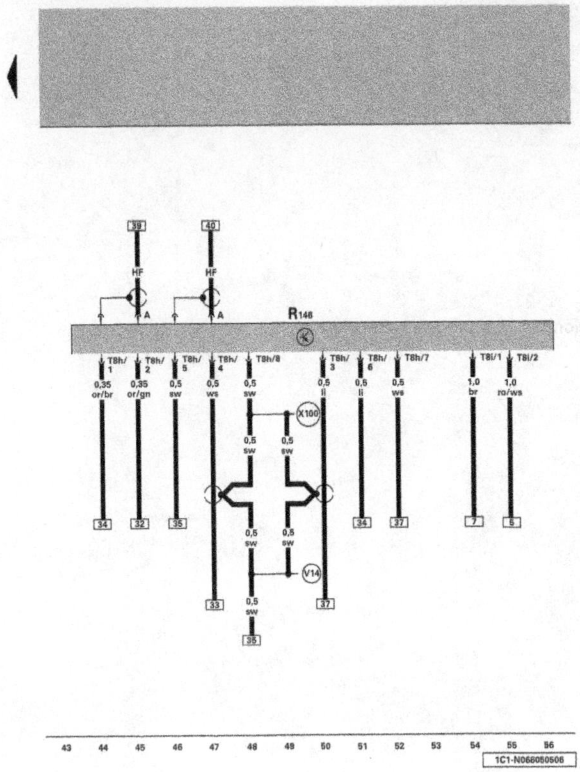

ws = white
sw = black
ro = red
br = brown
gn = green
bl = blue
gr = grey
li = violet
ge = yellow
or = orange

THIS SPACE INTENTIONALLY

LEFT BLANK

New Beetle Wiring diagram **No. 26/1**

Rear spoiler

from March 1999

Deviate relay location and fuseplacements as well as the locations of multiple connectors see section "component locations".

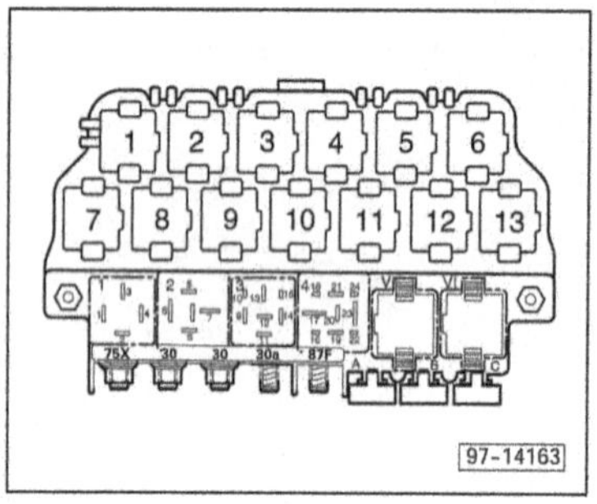

Fuse colors

30 A - green
25 A - white
20 A - yellow
15 A - blue
10 A - red
7.5 A - brown
5 A - beige
3 A - violet

New Beetle Wiring diagram **No. 26/2**

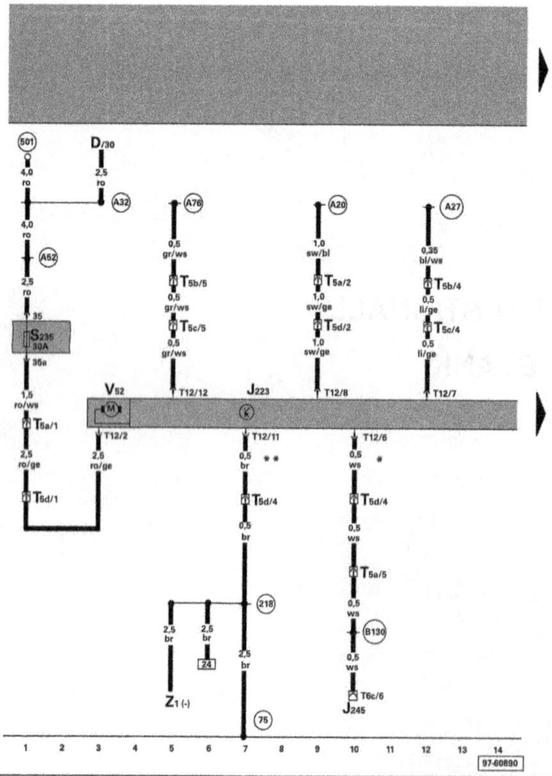

ws = white
sw = black
ro = red
br = brown
gn = green
bl = blue
gr = grey
li = violet
ge = yellow
or = orange

Rear spoiler control module, rear spoiler motor

D - Ignition/Starter Switch
J223 - Rear Spoiler Control Module
J245 - Power Sunroof Control Module
S235 - Fuse 35 in fuse holder
V52 - Rear Spoiler Motor
Z1 - Heated rear window
T5a - 5-Pin Connector
T5b - 5-Pin Connector
T5c - 5-Pin Connector, black
T5d - 5-Pin Connector, brown
T6c - 6-Pin Connector
T12 - 12-Pin Connector

(75) - Ground connection, on right rear pillar

(218) - Ground connection -1-, in rear lid wiring harness

(501) - Threaded connection -2- (30) on the relay plate

(A20) - Wire connection (15a), in instrument panel wiring harness

(A27) - Wire Connection (vehicle speed signal), in inst. panel wiring harness

(A32) - Plus connection (30), in instrument panel wiring harness

(A52) - Plus connection (30), in instrument panel wiring harness

(A76) - Connector (K-diagnosis wire) in instrument panel wiring harness

(B130) - Connector (sliding sunroof) in wiring harness interior

* - Only vehicles with sunroof
** - Only vehicles without sunroof

Rear spoiler from March 1999

New Beetle **Wiring diagram** **No. 26/3**

Rear spoiler control module, rear spoiler switch

E127 - Rear Spoiler Switch
J104 - ABS Control Module (w/EDL)
J223 - Rear Spoiler Control Module
J245 - Power Sunroof Control Module
J285 - Cont. module with indi. unit in inst. panel insert
T5 - 5-Pin Connector
T5a - 5-Pin Connector
T5b - 5-Pin Connector
T5c - 5-Pin Connector, black
T5d - 5-Pin Connector, brown
T6c - 6-Pin Connector
T12 - 12-Pin Connector
T25 - 25-Pin Connector
T32 - 32-Pin Connector, blue, on instrument cluster
(238) - Ground connection -1-, in wiring harness interior
(A133) - Connection (ABS-signal left) in instrument panel wiring harness

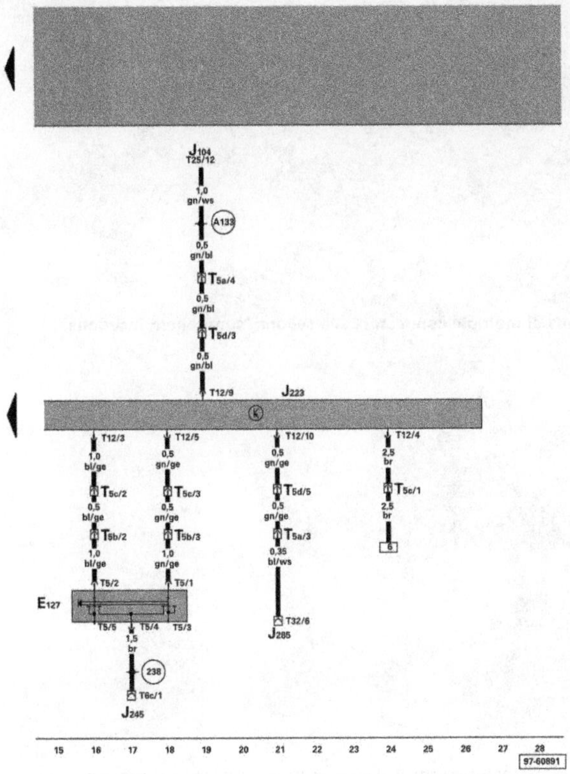

ws = white
sw = black
ro = red
br = brown
gn = green
bl = blue
gr = grey
li = violet
ge = yellow
or = orange

THIS SPACE INTENTIONALLY

LEFT BLANK

New Beetle — Wiring diagram — **No. 29/1**

Rear spoiler

from December 2002

Deviate relay location and fuseplacements as well as the locations of multiple connectors see section "component locations".

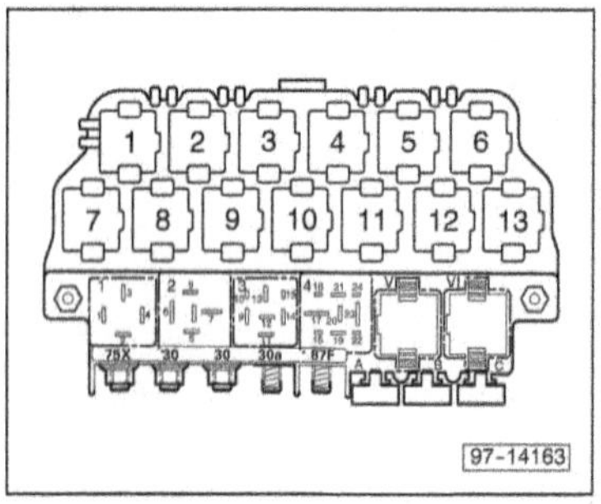

Fuse colors

30 A - green
25 A - white
20 A - yellow
15 A - blue
10 A - red
7,5 A - brown
5 A - beige
3 A - violet

New Beetle — Wiring diagram — **No. 29/2**

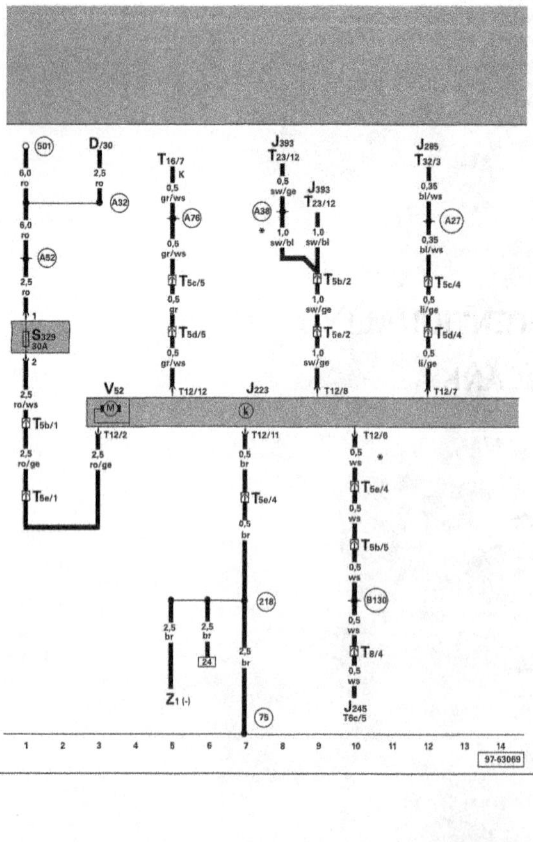

Rear spoiler control module, rear spoiler motor

- D - Ignition/Starter Switch
- J223 - Rear Spoiler Control Module
- J245 - Power Sunroof Control Module
- J285 - Control module with indicator unit in instrument panel insert
- J393 - Central control module for comfort system
- S329 - Rear Spoiler motor Fuse
- T5b - 5-Pin Connector, black, on C-Pillar right
- T5c - 5-Pin Connector, brown, on C-Pillar right
- T5d - 5-Pin Connector, black, on rear lid right
- T5e - 5-Pin Connector, brown, on rear lid right
- T6c - 6-Pin Connector
- T8 - 8-Pin Coupling Connector near front interior light
- T12 - 12-Pin Connector at J223
- T16 - 16-Pin Connector, Data Link Connector (DLC), bellow steering column, left
- T23 - 23-Pin Connector at J393
- T32 - 32-Pin Connector, blue, on instrument cluster
- V52 - Rear Spoiler Motor
- Z1 - Heated rear window
- (75) - Ground connection, on right rear pillar
- (218) - Ground connection -1-, in rear lid wiring harness
- (501) - Threaded connection -2- (30) on the relay plate
- (A27) - Wire Connection (vehicle speed signal), in inst. panel wiring harness
- (A32) - Plus connection (30), in instrument panel wiring harness
- (A38) - plus connection -2- (15a), in instrument panel wiring harness
- (A52) - Plus connection (30), in instrument panel wiring harness
- (A76) - Connector (K-diagnosis wire) in instrument panel wiring harness
- (B130) - Connector (sliding sunroof) in wiring harness interior
- * - Only vehicles with sunroof

ws = white
sw = black
ro = red
br = brown
gn = green
bl = blue
gr = grey
li = violet
ge = yellow
or = orange

New Beetle **Wiring diagram** **No. 29/3**

Rear spoiler control module, rear spoiler switch

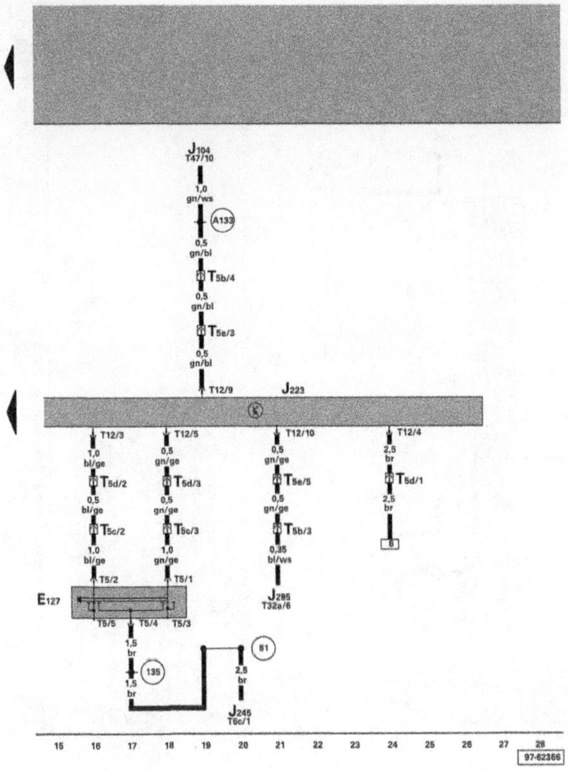

E127 - Rear Spoiler Switch
J104 - ABS Control Module (w/EDL)
J223 - Rear Spoiler Control Module
J245 - Power Sunroof Control Module
J285 - Cont. module with indi. unit in inst. panel insert
T5 - 5-Pin Connector at E127
T5b - 5-Pin Connector, black, on C-Pillar right
T5c - 5-Pin Connector, brown, on C-Pillar right
T5d - 5-Pin Connector, black, on rear lid right
T5e - 5-Pin Connector, brown, on rear lid right
T6c - 6-Pin Connector at J245
T12 - 12-Pin Connector at J223
T32a - 32-Pin Connector, green, on instrument cluster
T47 - 47-Pin Connector at J104

(81) - Ground connection -1-, in instrument panel wiring harness
(135) - Ground connection -2-, in instrument panel wiring harness
(A133) - Connection (ABS-signal left) in instrument panel wiring harness

ws = white
sw = black
ro = red
br = brown
gn = green
bl = blue
gr = grey
li = violet
ge = yellow
or = orange

THIS SPACE INTENTIONALLY

LEFT BLANK

New Beetle - Standard Equipment,

from January 1998

Deviate relay location and fuseplacements as well as the locations of multiple connectors see section "component locations".

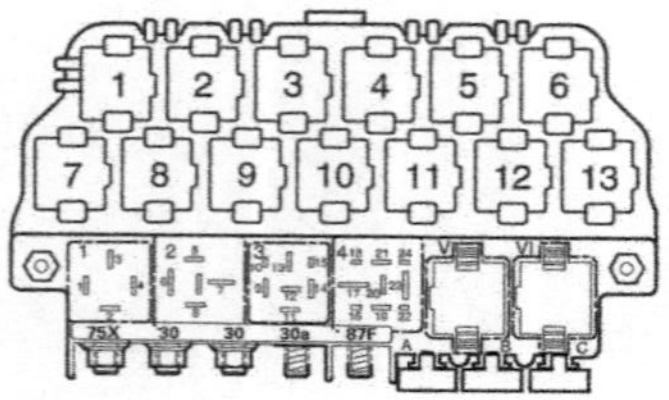

97-21956

Relay location on the thirteen position auxiliary relay panel, above relay panel:

- **3** Lock relay for starter (theft warning system) (185)
- **4** Lock relay for starter (clutch pedal switch) (53)
- **8** Daytime Running Lights Change-over Relay (173)

Relay panel:

- **1** Dual Horn Relay (53)
- **2** Load Reduction Relay (100)
- **4** Fuel Pump (FP) Relay (409)
- **V** Wiper/Washer Intermittent Relay (377)
- **VI** Wiper/Washer Intermittent Relay (377)

Note: Number in parentheses indicates production control number stamped on relay housing.

Fuse colors
30 A - green
25 A - white
20 A - yellow
15 A - blue
10 A - red
7.5 A - brown
5 A - beige

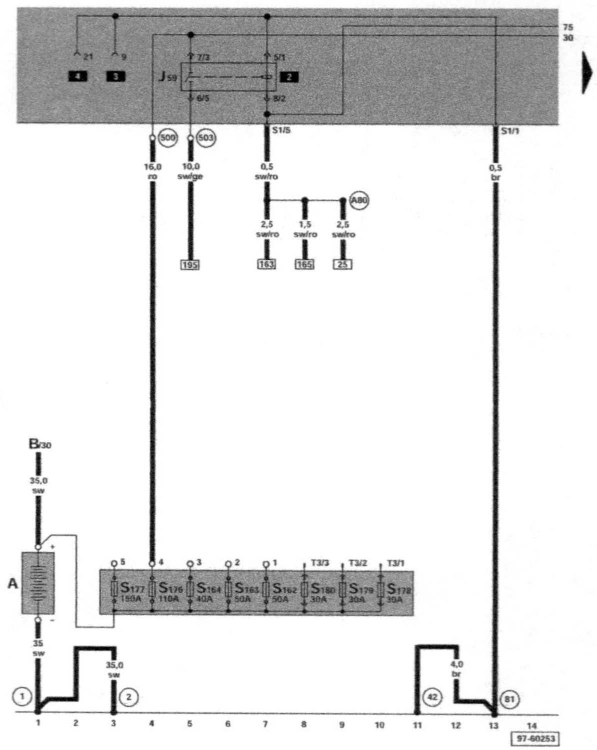

ws = white
sw = black
ro = red
br = brown
gn = green
bl = blue
gr = grey
li = violet
ge = yellow

A - Battery
B - Starter
J59 - Load Reduction Relay
S162 - Fuse -1- in fuse bracket/battery
S163 - Fuse -2- in fuse bracket/battery
S164 - Fuse -3- in fuse bracket/battery
S176 - Fuse -4- (30), in fuse bracket/battery
S177 - Fuse -5- (30), in fuse bracket/battery
S178 - Fuse -6- (30), in fuse bracket/battery
S179 - Fuse -7- (30), in fuse bracket/battery
S180 - Fuse -8- (30), in fuse bracket/battery

(1) - Ground strap, battery to body
(2) - Ground strap, transmission to body

(42) - Ground connection, beside steering column
(81) - Ground connection -1-, in instrument panel wiring harness
(500) - Threaded connection -1- (30) on the relay plate
(503) - Threaded connection -1- (75x) on the relay plate
(A80) - Connector -1- (X), in instrument panel wiring harness

Battery, load reduction relay, fuses in fuse bracket/battery

Standard equipment from Jan. 1998

New Beetle Wiring diagram **No. 1/3**

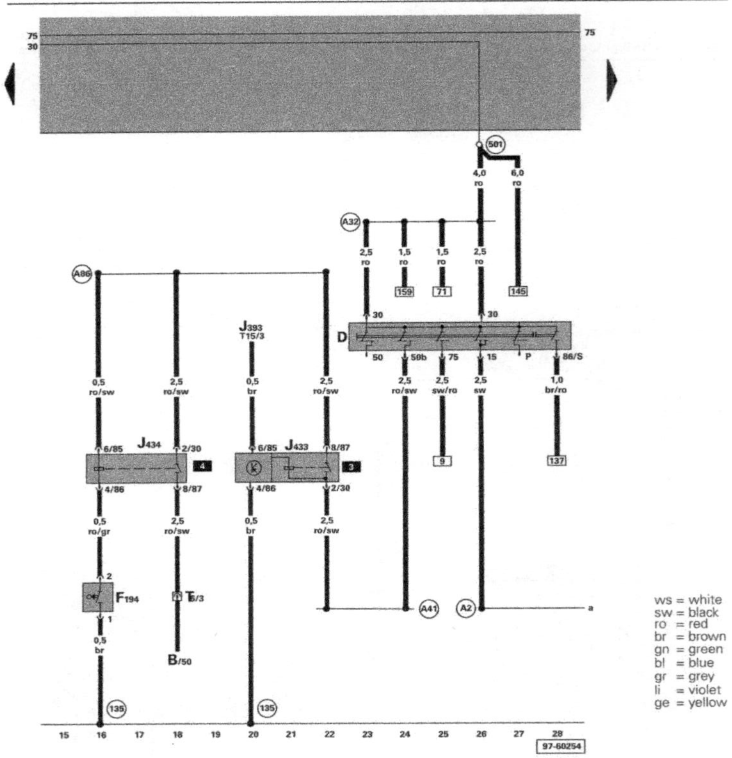

ws = white
sw = black
ro = red
br = brown
gn = green
bl = blue
gr = grey
li = violet
ge = yellow

B - Starter
D - Ignition/Starter Switch
F194 - Clutch Pedal Position (CPP) Switch
J393 - Central control module for comfort system
J433 - Lock relay for starter (theft warning system) (185), on the thirteen position relay panel
J434 - Locking relay for starter (clutch pedal switch)
T6 - 6-Pin Connector, red, behind instrument panel, left
T15 - 15-Pin Connector, on Central control module for comfort system

(135) - Ground connection -2-, in instrument panel wiring harness
(501) - Threaded connection -2- (30) on the relay plate

(A2) - plus connection (15), in instrument panel wiring harness
(A32) - plus connection (30), in instrument panel wiring harness
(A41) - plus connection (50), in instrument panel wiring harness
(A86) - Conenction (50a), in instrument panel wiring harness

Ignition/starter switch, lock relay for starter (theft warning system), locking relay for starter (clutch pedal switch)

No. 1/4 Wiring diagram **New Beetle**

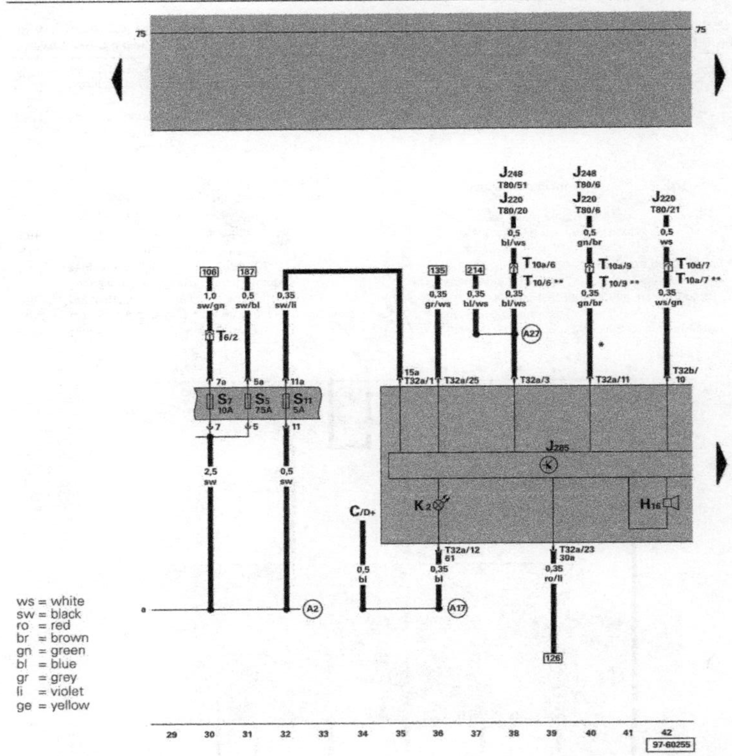

ws = white
sw = black
ro = red
br = brown
gn = green
bl = blue
gr = grey
li = violet
ge = yellow

C - Generator (GEN)
H16 - Warning buzzer for light switched on
J220 - Motronic Engine Control Module (ECM), behind instrument panel, left
J248 - Diesel Direct Fuel Injection (DFI) Engine Control Module (ECM), behind instrument panel, left
J285 - Control module with indicator unit in instrument panel insert
K2 - Generator (GEN) Warning Light
S5 - Fuse 5 in fuse holder
S7 - Fuse 7 in fuse holder
S11 - Fuse 11 in fuse holder
T6 - 6-Pin Connector, brown, behind instrument panel, left
T10 - 10-Pin Connector, white, behind instrument panel, left
T10a - 10-Pin Connector, orange, behind instrument panel, left
T10d - 10-Pin Connector, green, behind instrument panel, left
T32a - 32-Pin Connector, blue, on instrument cluster
T32b - 32-Pin Connector, green, on instrument cluster
T80 - Connector, 80 point, on Engine Control Module (ECM)

(A2) - plus connection (15), in instrument panel wiring harness
(A17) - wire connection (61), in instrument panel wiring harness
(A27) - wire Connection (vehicle speed signal), in instrument panel wiring harness
* - speed signal from engine control module
** - early vehicles only

Instrument cluster, warning buzzer for light swiched on, generator (GEN)

Standard equipment from Jan. 1998

New Beetle — Wiring diagram — No. 1/5

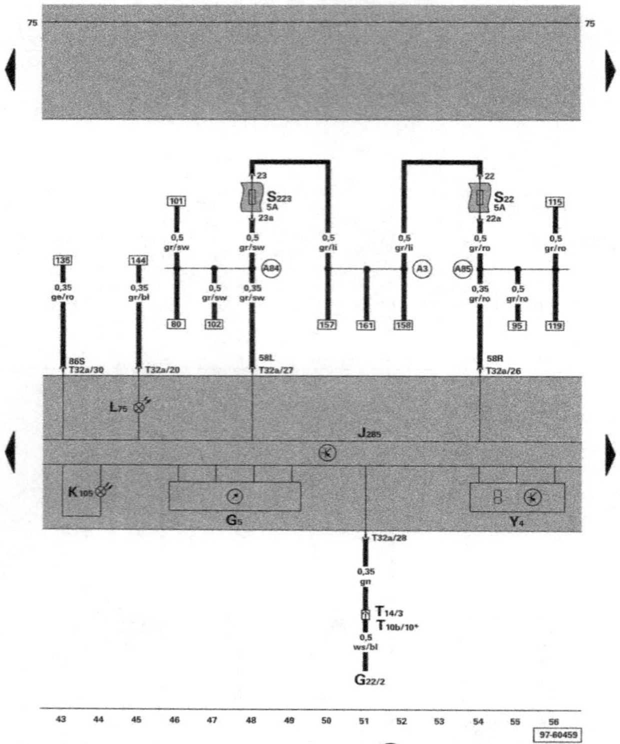

ws = white
sw = black
ro = red
br = brown
gn = green
bl = blue
gr = grey
li = violet
ge = yellow

G5 - Tachometer
G22 - Speedometer Vehicle Speed Sensor (VSS)
J285 - Control module with indicator unit in instrument panel insert
K105 - Low Fuel Level Warning Light
L75 - Digital Display Light
S22 - Fuse 22 in fuse holder
S223 - Fuse 23 in fuse holder
T10b - 10-Pin Connector, in engine compartment, left
T14 - 14-Pin Connector, in engine compartment, left
T32a - 32-Pin Connector, blue, on instrument cluster
Y4 - Odometer Display

(A3) - plus connection (58), in instrument panel wiring harness
(A84) - Connector (58L), in instrument panel wiring harness
(A85) - Connector (58R), in instrument panel wiring harness

* - early vehicles only

Instrument cluster, tachometer, low fuel level warning light, odometer display

No. 1/6 — Wiring diagram — New Beetle

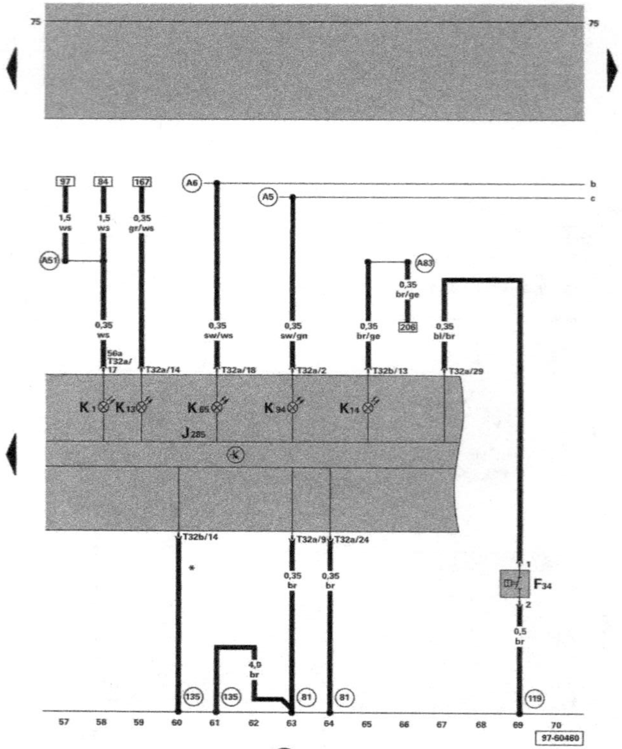

ws = white
sw = black
ro = red
br = brown
gn = green
bl = blue
gr = grey
li = violet
ge = yellow

F34 - Brake Fluid Level Warning Switch
J285 - Control module with indicator unit in instrument panel insert
K1 - Headlight High Beam Indicator Light
K13 - Rear Fog Light Indicator Light
K14 - Parking Brake Indicator Light
K65 - Left Turn Signal Indicator Light
K94 - Right Turn Signal Indicator Light
T32a - 32-Pin Connector, blue, on instrument cluster
T32b - 32-Pin Connector, green, on instrument cluster

(81) - Ground connection -1-, in instrument panel wiring harness
(119) - Ground connection -1-, in headlight wiring harness
(135) - Ground connection -2-, in instrument panel wiring harness

(A5) - plus connection (right turn signal), in instrument panel wiring harness
(A6) - plus connection (left turn signal), in instrument panel wiring harness
(A51) - wire connection (56), in instrument panel wiring harness
(A83) - Connector (daytime running lights), in instrument panel wiring harness

* - Only vehicles without ABS

Instrument cluster, brake fluid level warning switch, turn signal indicator lights, headlight high beam indicator light, rear fog light indicator light

Standard equipment from Jan. 1998

New Beetle — Wiring diagram — No. 1/7

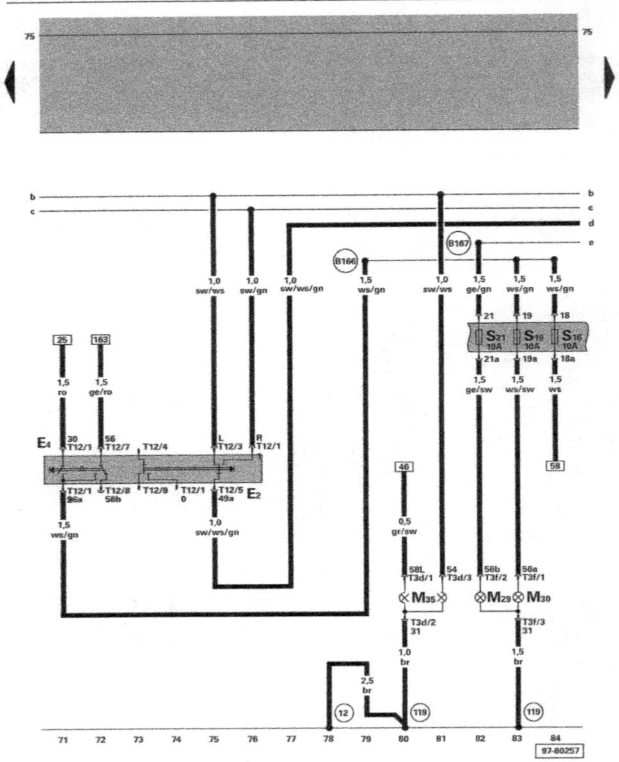

ws = white
sw = black
ro = red
br = brown
gn = green
bl = blue
gr = grey
li = violet
ge = yellow

- E2 - Turn Signal Switch
- E4 - Headlight Dimmer/Flasher Switch
- M29 - Left Low Beam Headlight
- M30 - Left High Beam Headlight
- M35 - Light for turn signal and side marker, front, left
- S18 - Fuse 18 in fuse holder
- S19 - Fuse 19 in fuse holder
- S21 - Fuse 21 in fuse holder
- T3d - 3-Pin Connector, on light for turn signal and side marker, front, left
- T3f - 3-Pin Connector, on headlight left
- T12 - 12-Pin Connector, near steering column
- (12) - Ground connection, in engine compartment, left
- (119) - Ground connection -1-, in headlight wiring harness
- (B166) - Connection (56a) in passenger compartment wiring harness
- (B167) - Connection (56b) in passenger compartment wiring harness

Turn signal switch, headlight dimmer/flasher switch

No. 1/8 — Wiring diagram — New Beetle

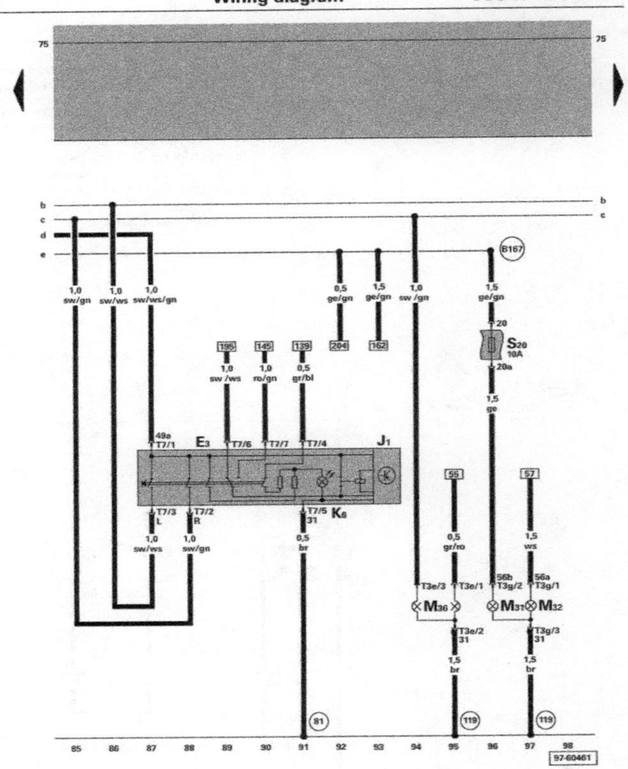

ws = white
sw = black
ro = red
br = brown
gn = green
bl = blue
gr = grey
li = violet
ge = yellow

- E3 - Emergency Flasher Switch
- J1 - Turn Signal Relay
- K6 - Emergency Flasher Warning Light
- M31 - Right Low Beam Headlight
- M32 - Right High Beam Headlight
- M36 - Light for turn signal and side marker, front, right
- S20 - Fuse 20 in fuse holder
- T3e - 3-Pin Connector
- T3g - 3-Pin Connector
- T7 - 7-Pin Connector
- (81) - Ground connection -1-, in instrument panel wiring harness
- (119) - Ground connection -1-, in headlight wiring harness
- (B167) - Connection (56b) in passenger compartment wiring harness

Emergency flasher switch with turn signal relay, right low and high beam headlights, light for turn signal and side marker, front, right

New Beetle — Wiring diagram — No. 1/9

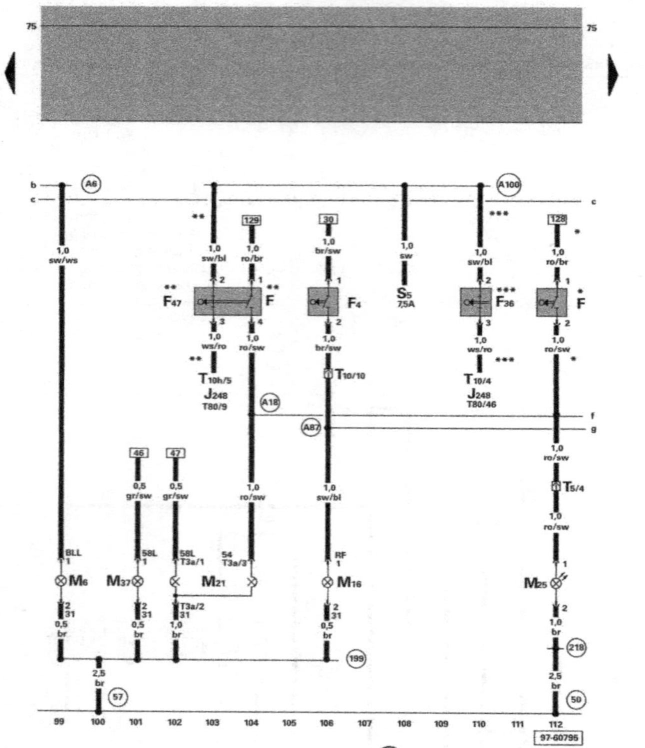

ws = white
sw = black
ro = red
br = brown
gn = green
bl = blue
gr = grey
li = violet
ge = yellow

F - Brake Light Switch
F4 - Back-Up light Switch
F36 - Clutch Vacuum Vent Valve Switch
F47 - Brake Vacuum Vent Valve Switch for cruise control/ (DFI)
J248 - Diesel Direct Fuel Injection (DFI) Engine Control Module (ECM)
M6 - Left Rear Turn Signal Light
M16 - Left Back-Up Light
M21 - Left Brake/Tail Light
M25 - High-mount Brake Light
M37 - Side marker light, left rear
S5 - Fuse 5 in fuse/relay panel
T3a - 3-Pin Connector
T5 - 5-Pin Connector, black, on C-pillar, left
T10 - 10-Pin Connector, white, behind instrument panel, left
T80 - 80-Pin Connector, on Diesel (DFI) Control Module

(50) - Ground connection, in luggage compartment, left
(57) - Ground connection, on left rear pillar
(199) - Ground Connection -3-, in instrument panel wiring harness
(218) - Ground connection -1- in rear lid wiring harness
(A6) - Plus connection (left turn signal), in instrument panel wiring harness
(A18) - Wire connection (54), in instrument panel wiring harness
(A20) - Wire connection (15a) in instrument panel wiring harness
(A87) - Connector (reverse lamp) in instrument panel wiring harness
* - Vehicles without cruise control (2.0L w/cruise control ⇒ diagram No.13)
** - Diesel 1.9L with cruise control
*** - Diesel 1.9L with manual transmission only

Edition 03.00 **Brake light switch, brake/clutch vacuum vent valve/switch, back-up light**
USA.5132.10.21 **switch, left back-up light, left brake/tail light, hight-mount brake light**

No. 1/10 — Wiring diagram — New Beetle

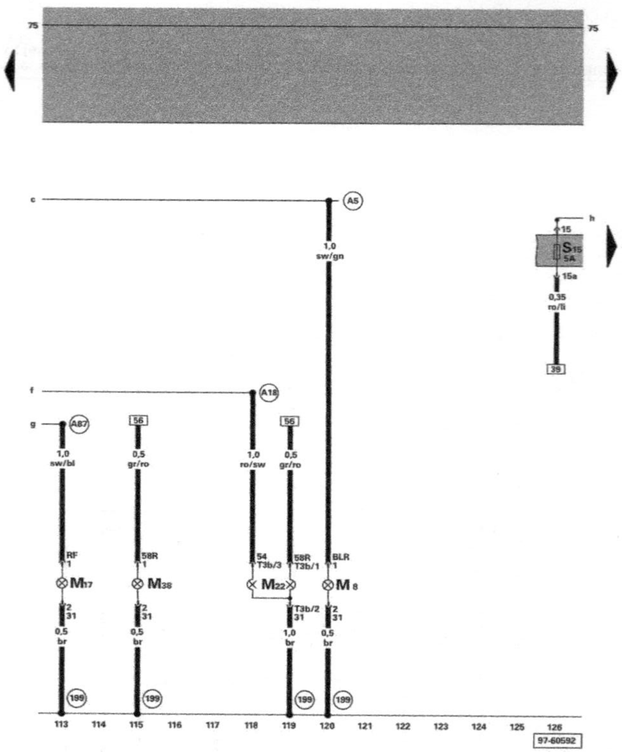

ws = white
sw = black
ro = red
br = brown
gn = green
bl = blue
gr = grey
li = violet
ge = yellow

M8 - Right Rear Turn Signal Light
M17 - Right Back-Up Light
M22 - Right Brake/Tail Light
M38 - Side marker light, right rear
S15 - Fuse 15 in fuse holder
T3b - 3-Pin Connector
T23 - 23-Pin Connector, on Central control module for comfort system
(199) - Ground connection -3-, in instrument panel wiring harness
(A5) - plus connection (right turn signal), in instrument panel wiring harness
(A18) - wire connection (54), in instrument panel wiring harness
(A87) - Connector (reverse lamp), in instrument panel wiring harness

Luggage compartment light, right rear turnsignal light, right back-up light, right brake/tail light, side marker light, right, rear **Edition 03/00**
USA.5132.10.21

New Beetle — Wiring diagram — No. 1/11

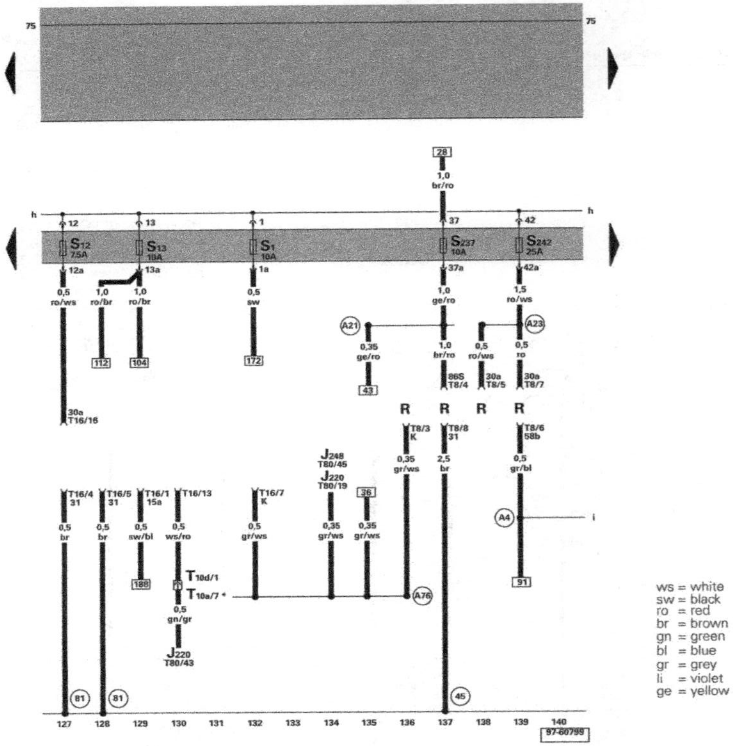

ws = white
sw = black
ro = red
br = brown
gn = green
bl = blue
gr = grey
li = violet
ge = yellow

- J220 - Motronic Engine Control Module (ECM), behind instrument panel, left
- J248 - Diesel Direct Fuel Injection (DFI) Engine Control Module (ECM), behind instrument panel, left
- R - Radio
- S12 - Fuse 12 in fuse holder
- S13 - Fuse 13 in fuse holder
- S237 - Fuse 37 in fuse holder
- S242 - Fuse 42 in fuse holder
- T8 - 8-Pin Connector, on radio
- T10a - 10-Pin Connector, orange, behind instrument panel, left * (Does apply on early vehicles)
- T10d - 10-Pin Connector, green, behind instrument panel, left
- T16 - 16-Pin Connector, Data Link Connector (DLC), below steering column
- (T80) - Connector, 80 point, on Engine Control Module (ECM)
- (45) - Ground connection, behind instrument panel, center
- (81) - Ground connection -1-, in instrument panel wiring harness
- (A4) - plus connection (58b) in instrument panel wiring harness
- (A21) - wire connection (86s), in instrument panel wiring harness
- (A23) - wire connection (30al), in instrument panel wiring harness
- (A76) - Connector (K-diagnosis wire), in instrument panel wiring harness

No. 1/12 — Wiring diagram — New Beetle

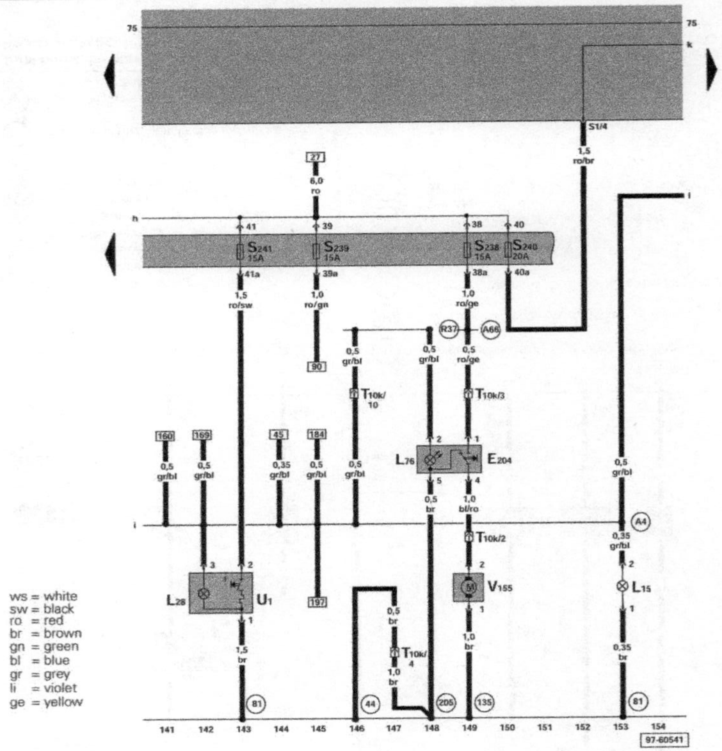

ws = white
sw = black
ro = red
br = brown
gn = green
bl = blue
gr = grey
li = violet
ge = yellow

- E204 - Switch for remote/fuel tank door
- L15 - Ashtray Light
- L28 - Cigarette Lighter Light
- L76 - Push Button Light
- S238 - Fuse 38 in fuse holder
- S239 - Fuse 39 in fuse holder
- S240 - Fuse 40 in fuse holder
- S241 - Fuse 41 in fuse holder
- T10k - 10-pin connector, black, on conector station A-pillar, left
- U1 - Cigarette Lighter
- V155 - Motor for fuel tank lid unlock
- (44) - Ground connection, on left A-pillar, lower part
- (81) - Ground connection -1-, in instrument panel wiring harness
- (135) - Ground connection -2-, in instrument panel wiring harness
- (205) - Ground connection, in driver's door wiring harness
- (A4) - plus connection (58b) in instrument panel wiring harness
- (A66) - Connector (30a, central locking/anti-theft warning system/IR), in instrument panel wiring harness
- (R37) - Connector (58b), in driver's door wiring harness

Standard equipment from Jan. 1998

New Beetle — Wiring diagram — No. 1/13

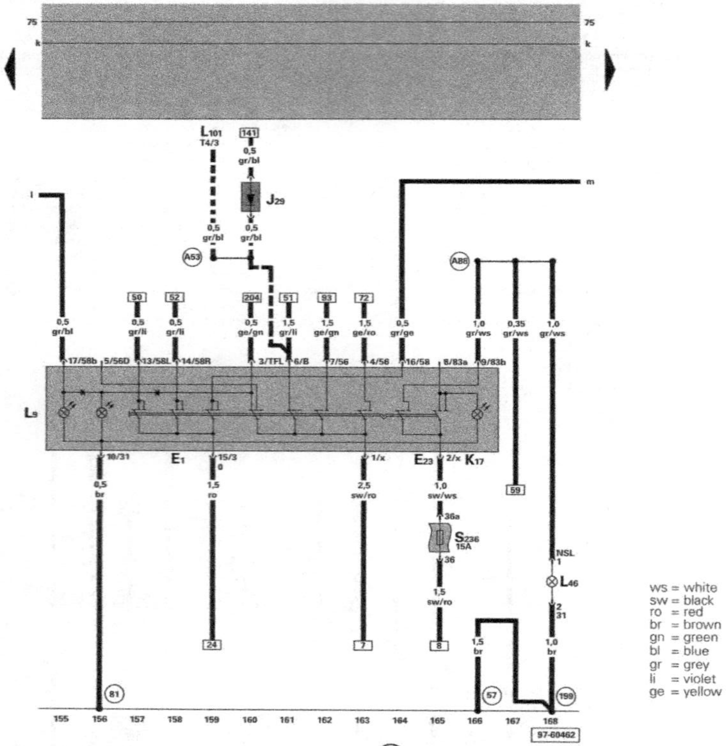

ws = white
sw = black
ro = red
br = brown
gn = green
bl = blue
gr = grey
li = violet
ge = yellow

E1 - Light switch
E23 - Fog Light Switch
J29 - Blocking diode
K17 - Fog Light Indicator Light
L9 - Headlight Switch Licht
L101 - Illumination for selector lever scale
L46 - Left Rear Fog Light
S236 - Fuse 36 in fuse holder

(57) - Ground connection, on left rear pillar

(81) - Ground connection -1-, in instrument panel wiring harness

(199) - Ground connection -3-, in instrument panel wiring harness

(A53) - plus connection (58b), in instrument panel wiring harness

(A88) - Connector fog light, in instrument panel wiring harness

(B167) - Connection (56b) in passenger compartment wiring harness

- - - - - automatic transmission only

Edition 06/99
USA.5132.07.21

Light switch, left rear fog light

No. 1/14 — Wiring diagram — New Beetle

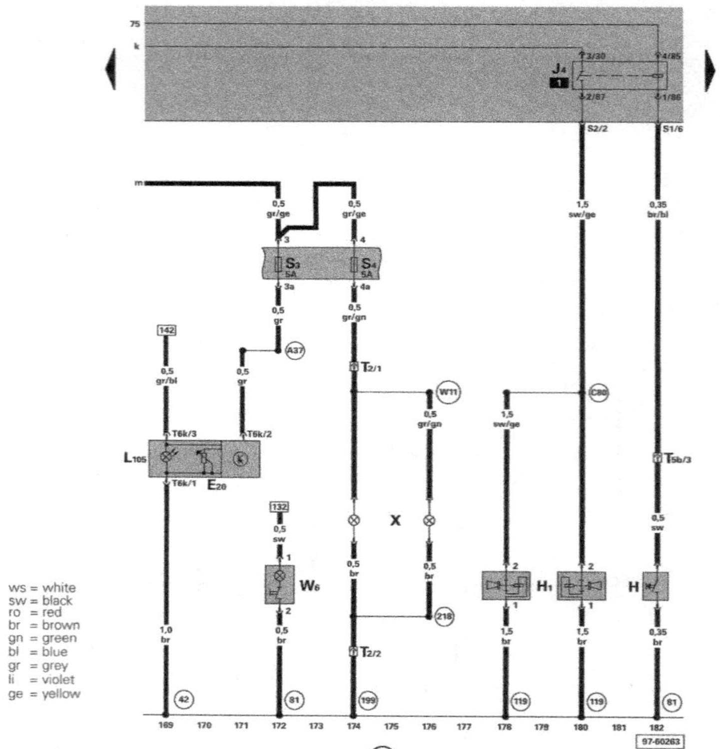

ws = white
sw = black
ro = red
br = brown
gn = green
bl = blue
gr = grey
li = violet
ge = yellow

E20 - Instrument Panel Light Dimmer Switch
H - Horn Button
H1 - Dual Horns
J4 - Dual Horn Relay
L105 - Illumination for lighting controller
S3 - Fuse 3 in fuse holder
S4 - Fuse 4 in fuse holder
T2 - Double Connector, in rear lid
T5b - 5-Pin Connector, near steering column
T6K - 6-Pin Connector
W6 - Glove Compartment Light
X - License Plate Light

(42) - Ground connection, beside steering column

(81) - Ground connection -1-, in instrument panel wiring harness

(119) - Ground connection -1-, in headlight wiring harness

(199) - Ground connection -3-, in instrument panel wiring harness

(218) - Ground connection -1-, in rear lid wiring harness

(A37) - wire connection (58a), in instrument panel wiring harness

(C80) - wire connection (58b), in wiring harness headlamp

(W11) - wire connection (58), in rear lid wiring harness

Dual horns, instrument light dimmer switch, glove compartment light, license plate light

Edition 09/98
USA.5132.05.21

New Beetle — Wiring diagram — No. 1/15

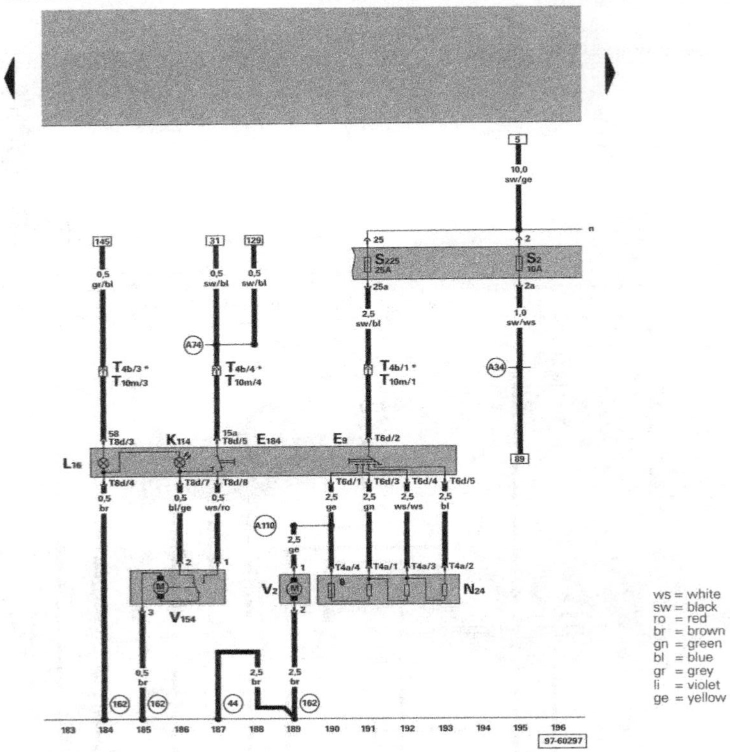

ws = white
sw = black
ro = red
br = brown
gn = green
bl = blue
gr = grey
li = violet
ge = yellow

E9 - Fresh Air Blower Switch
E184 - Switch for fresh air blower and recirculated air
K114 - Fresh Air and Recirculating Air Mode Indicator Light
L16 - Fresh Air Control Lever Light
N24 - Fresh Air Blower Series Resistance With Fuse
S2 - Fuse 2 in fuse holder
S225 - Fuse 25 in fuse holder
T4a - 4-Pin Connector
T4b - 4-Pin Connector, behind console
T6d - 6-Pin Connector
T8d - 8-Pin Connector
T10m - 10-Pin Connector
V2 - Fresh Air Blower
V154 - Servo motor for fresh-/recirculating air door

(44) - Ground connection, on left A-pillar, lower part
(162) - Ground connection, in blower motor wiring harness
(A34) - wire connection (75x), in instrument panel wiring harness
(A74) - Connector (15a - fuse 5), in instrument panel wiring harness
(A110) - Connector (fresh air blower), in instrument panel wiring harness

* Applies only to early models

Fresch air blower, fresh air blower switch

No. 1/16 — Wiring diagram — New Beetle

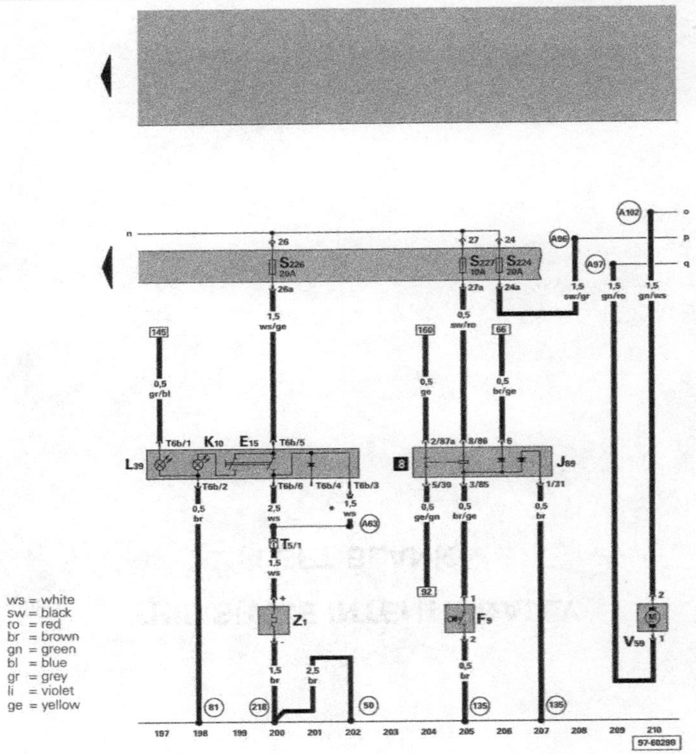

ws = white
sw = black
ro = red
br = brown
gn = green
bl = blue
gr = grey
li = violet
ge = yellow

E15 - Rear window defogger switch
F9 - Parking Brake Warning Light Switch
J89 - Daytime Running Lights Change-over Relay (173), on the thirteen position relay panel
K10 - Rear Window Defogger Indicator Light
L39 - Rear Window Defogger Switch Light
S224 - Fuse 24 in fuse holder
S226 - Fuse 26 in fuse holder
S227 - Fuse 27 in fuse holder
T5 - 5-Pin Connector, black, on C-pillar, left
T6b - 6-Pin Connector
V59 - Windshield and Rear Window Washer Pump
Z1 - Heated rear window

(50) - Ground connection, in left rear luggage compartment
(81) - Ground connection -1-, in instrument panel wiring harness
(135) - Ground connection -2-, in instrument panel wiring harness
(218) - Ground connection -1-, in rear wiring harness
(A63) - Connector (mirror adjustment/-heated) in instrument panel wiring harness
(A96) - Connector (53a), in instrument panel wiring harness
(A97) - Connector (53), in instrument panel wiring harness
(A102) - Connector (windshield wiper), in instrument panel wiring harness

* Applies only to later models with manual windows

Heated rear window, rear window defogger switch, parking brake warning light switch, daytime running lights change-over relay

Standard equipment from Jan. 1998

New Beetle Wiring diagram **No. 1/17**

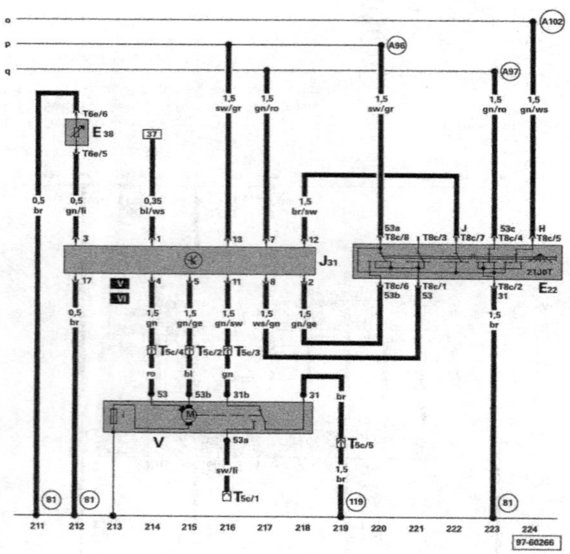

ws = white
sw = black
ro = red
br = brown
gn = green
bl = blue
gr = grey
li = violet
ge = yellow

E22 - Winshield Wiper/Washer Switch
E38 - Windshield Wiper Intermittent Regulator
J31 - Wiper/Washer Intermittent Relay, on the thirteen position relay panel, production control number (377)
T5c - 5-Pin Connector, on Windshield Wiper Motor
T6e - 6-Pin Connector, near steering column
T8c - 8-Pin Connector
V - Windshield Wiper Motor

(81) - Ground connection -1-, in instrument panel wiring harness
(119) - Ground connection -1-, in headlight wiring harness
(A96) - Connector (53a), in instrument panel wiring harness
(A97) - Connector (53), in instrument panel wiring harness
(A102) - Connector (windshield wiper), in instrument panel wiring harness

THIS SPACE INTENTIONALLY LEFT BLANK

New Beetle - Standard Equipment,

from August 1998

Deviate relay location and fuseplacements as well as the locations of multiple connectors see section "component locations".

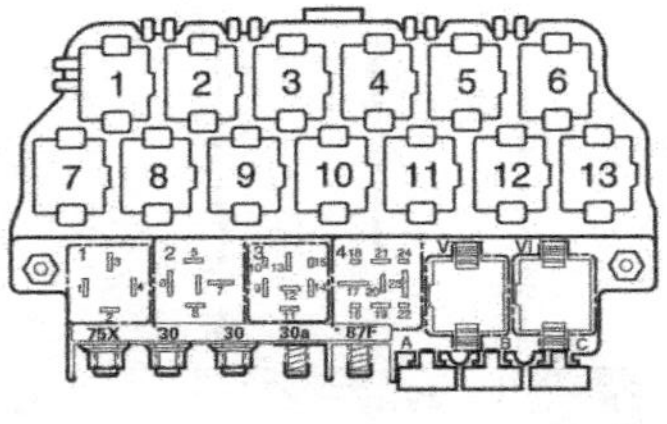

97-21956

Relay location on the thirteen position auxiliary relay panel, above relay panel:

- **3** Lock relay for starter (theft warning system) (185)
- **4** Lock relay for starter (clutch pedal switch) (53)
- **8** Daytime Running Lights Change-over Relay (173)

Relay panel:

- **1** Dual Horn Relay (53)
- **2** Load Reduction Relay (100)
- **4** Fuel Pump (FP) Relay (409)
- **V** Wiper/Washer Intermittent Relay (377)
- **VI** Wiper/Washer Intermittent Relay (377)

Note: Number in parentheses indicates production control number stamped on relay housing.

Fuse colors
30 A - green
25 A - white
20 A - yellow
15 A - blue
10 A - red
7,5 A - brown
5 A - beige

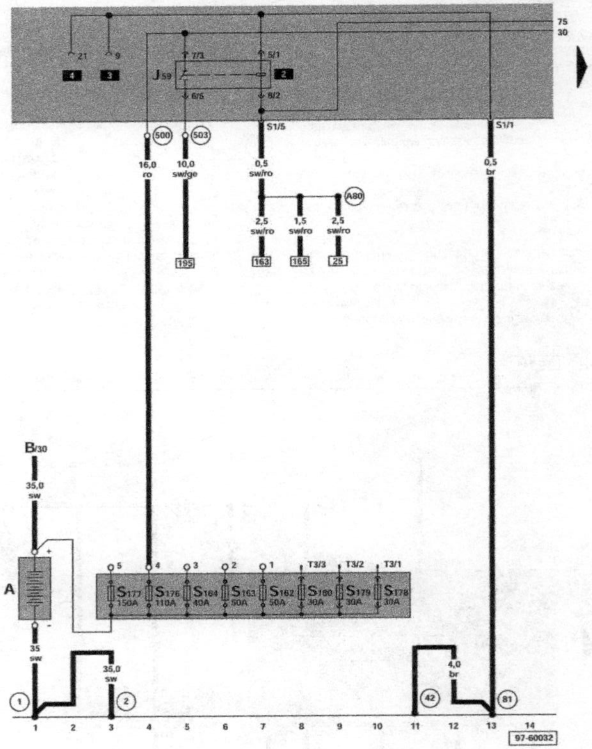

ws = white
sw = black
ro = red
br = brown
gn = green
bl = blue
gr = grey
li = violet
ge = yellow

A - Battery
B - Starter
J59 - Load Reduction Relay
S162 - Fuse -1- in fuse bracket/battery
S163 - Fuse -2- in fuse bracket/battery
S164 - Fuse -3- in fuse bracket/battery
S176 - Fuse -4- (30), in fuse bracket/battery
S177 - Fuse -5- (30), in fuse bracket/battery
S178 - Fuse -6- (30), in fuse bracket/battery
S179 - Fuse -7- (30), in fuse bracket/battery
S180 - Fuse -8- (30), in fuse bracket/battery

(1) - Ground strap, battery to body

(2) - Ground strap, transmission to body

(42) - Ground connection, beside steering column

(81) - Ground connection -1-, in instrument panel wiring harness

(500) - Threaded connection -1- (30) on the relay plate

(503) - Threaded connection -1- (75x) on the relay plate

(A80) - Connector -1- (X), in instrument panel wiring harness

Battery, load reduction relay, fuses in fuse bracket/battery

Standard equipment from Aug. 1998

New Beetle Wiring diagram **No. 21/3**

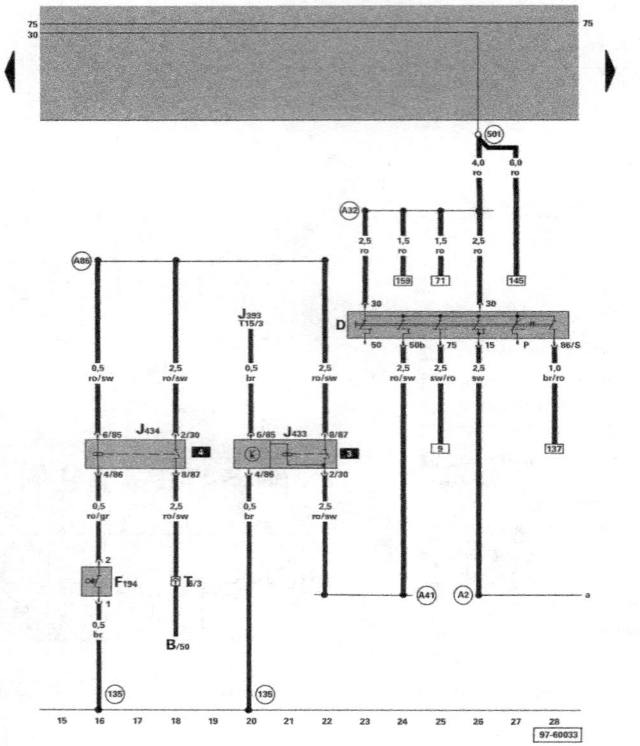

ws = white
sw = black
ro = red
br = brown
gn = green
bl = blue
gr = grey
li = violet
ge = yellow

B - Starter
D - Ignition/Starter Switch
F194 - Clutch Pedal Position (CPP) Switch
J393 - Central control module for comfort system
J433 - Lock relay for starter (theft warning system) (185), on the thirteen position relay panel
J434 - Locking relay for starter (clutch pedal switch)
T6 - 6-Pin Connector, red, behind instrument panel, left
T15 - 15-Pin Connector, on Central control module for comfort system

(135) - Ground connection -2-, in instrument panel wiring harness
(501) - Threaded connection -2- (30) on the relay plate

(A2) - plus connection (15), in instrument panel wiring harness
(A32) - plus connection (30), in instrument panel wiring harness
(A41) - plus connection (50), in instrument panel wiring harness
(A86) - Conenction (50a), in instrument panel wiring harness

Edition 03/00
USA.5132.10.21

Ignition/starter switch, lock relay for starter (theft warning system), locking relay for starter (clutch pedal switch)

No. 21/4 Wiring diagram **New Beetle**

ws = white
sw = black
ro = red
br = brown
gn = green
bl = blue
gr = grey
li = violet
ge = yellow

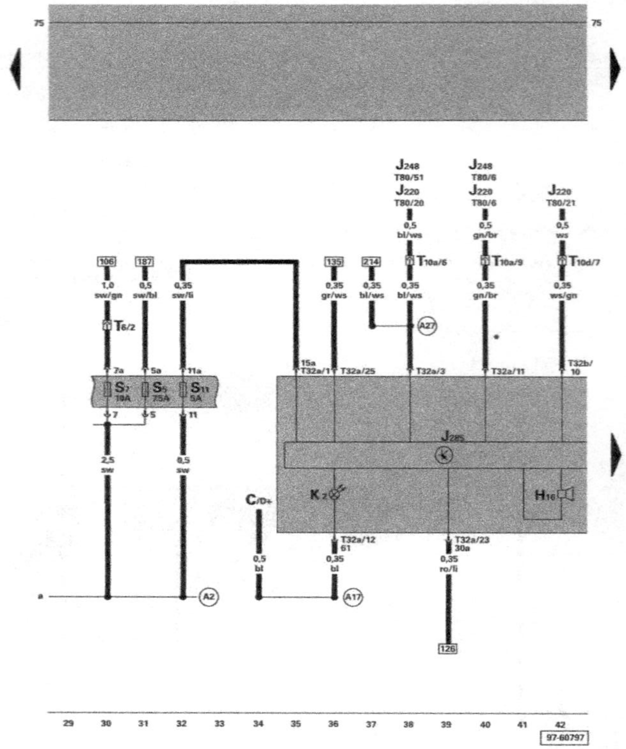

C - Generator (GEN)
H16 - Warning buzzer for light switched on
J220 - Motronic Engine Control Module (ECM), behind instrument panel, left
J248 - Diesel Direct Fuel Injection (DFI) Engine Control Module (ECM), behind instrument panel, left
J285 - Control module with indicator unit in instrument panel insert
K2 - Generator (GEN) Warning Light
S5 - Fuse 5 in fuse holder
S7 - Fuse 7 in fuse holder
S11 - Fuse 11 in fuse holder
T6 - 6-Pin Connector, brown, behind instrument panel, left
T10d - 10-Pin Connector, green, behind instrument panel, left
T10a - 10-Pin Connector, orange, behind instrument panel, left
T32a - 32-Pin Connector, blue, on instrument cluster
T32b - 32-Pin Connector, green, on instrument cluster
T80 - Connector, 80 point, on Engine Control Module (ECM)

(A2) - plus connection (15), in instrument panel wiring harness
(A17) - wire connection (61), in instrument panel wiring harness
(A27) - wire Connection (vehicle speed signal), in instrument panel wiring harness
* - speed signal from engine control module

Instrument cluster, warning buzzer for light swiched on, generator (GEN)

Edition 03/00
USA.5132.10.21

New Beetle Wiring diagram **No. 21/5**

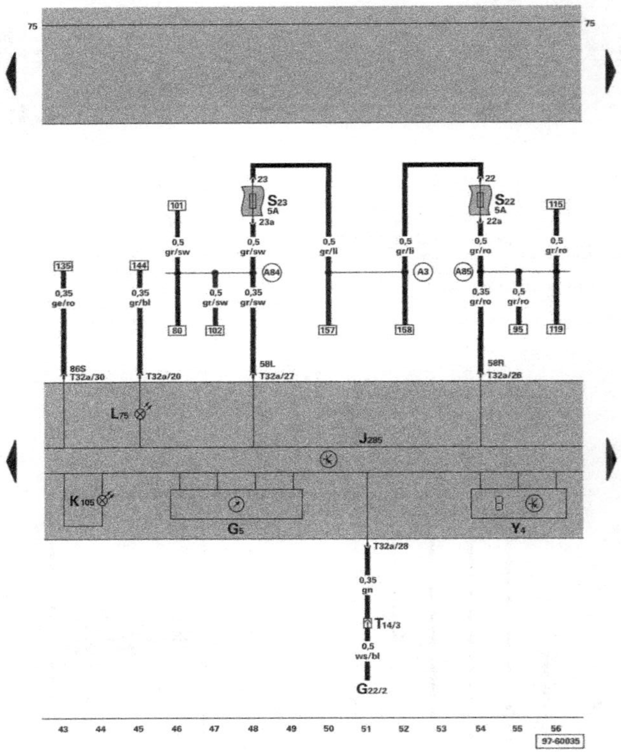

ws = white
sw = black
ro = red
br = brown
gn = green
bl = blue
gr = grey
li = violet
ge = yellow

G5 - Tachometer
G22 - Speedometer Vehicle Speed Sensor (VSS)
J285 - Control module with indicator unit in instrument panel insert
K105 - Low Fuel Level Warning Light
L75 - Digital Display Light
S22 - Fuse 22 in fuse holder
S23 - Fuse 23 in fuse holder
T14 - 14-Pin Connector, in engine compartment, left
T32a - 32-Pin Connector, blue, on instrument cluster
Y4 - Odometer Display

(A3) - plus connection (58), in instrument panel wiring harness
(A84) - Connector (58L), in instrument panel wiring harness
(A85) - Connector (58R), in instrument panel wiring harness

Instrument cluster, tachometer, low fuel level warning light, odometer display

No. 21/6 Wiring diagram **New Beetle**

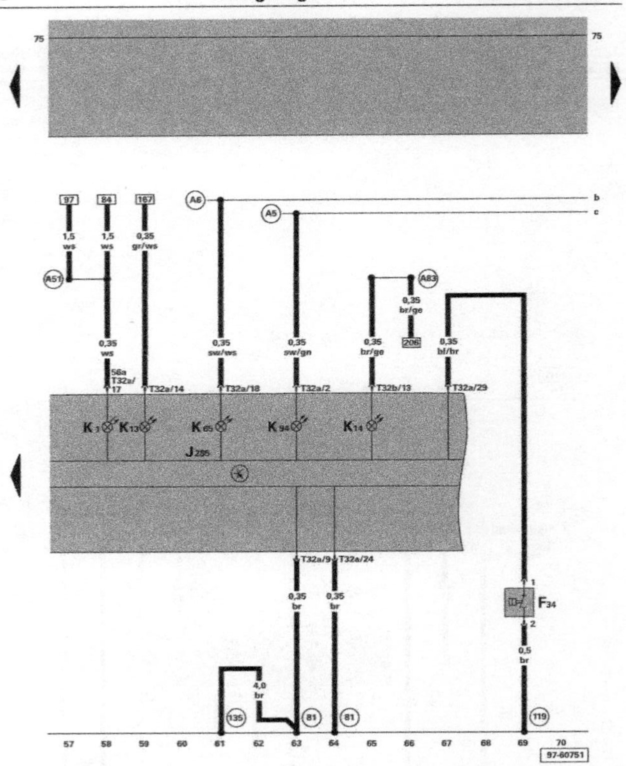

ws = white
sw = black
ro = red
br = brown
gn = green
bl = blue
gr = grey
li = violet
ge = yellow

F34 - Brake Fluid Level Warning Switch
J285 - Control module with indicator unit in instrument panel insert
K1 - Headlight High Beam Indicator Light
K13 - Rear Fog Light Indicator Light
K14 - Parking Brake Indicator Light
K65 - Left Turn Signal Indicator Light
K94 - Right Turn Signal Indicator Light
T32a - 32-Pin Connector, blue, on instrument cluster
T32b - 32-Pin Connector, green, on instrument cluster

(81) - Ground connection -1-, in instrument panel wiring harness
(119) - Ground connection -1-, in headlight wiring harness
(135) - Ground connection -2-, in instrument panel wiring harness

(A5) - plus connection (right turn signal), in instrument panel wiring harness
(A6) - plus connection (left turn signal), in instrument panel wiring harness
(A51) - wire connection (56), in instrument panel wiring harness
(A83) - Connector (daytime running lights), in instrument panel wiring harness

Instrument cluster, brake fluid level warning switch, turn signal indicator lights, headlight high beam indicator light, rear fog light indicator light

Standard equipment from Aug. 1998

New Beetle Wiring diagram **No. 21/7**

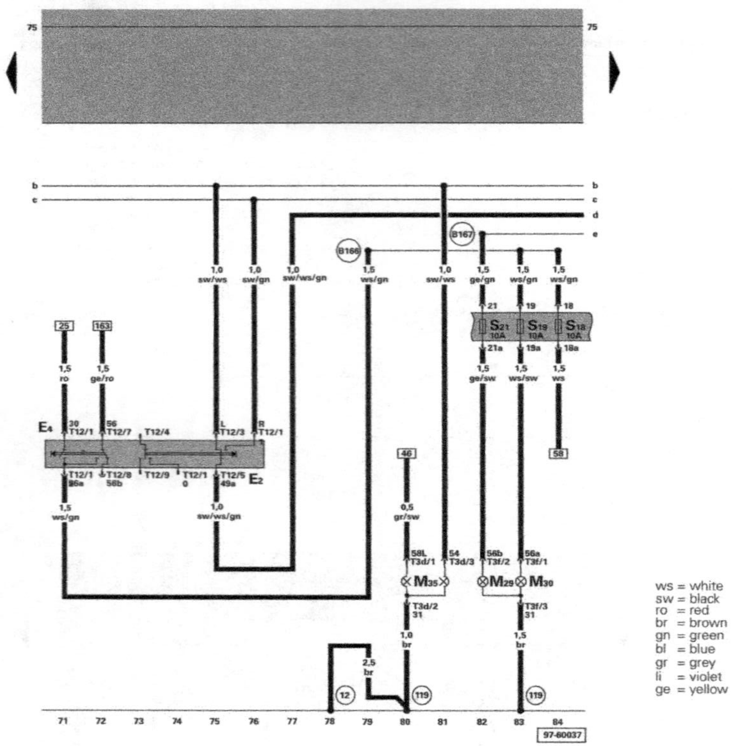

ws = white
sw = black
ro = red
br = brown
gn = green
bl = blue
gr = grey
li = violet
ge = yellow

- E2 - Turn Signal Switch
- E4 - Headlight Dimmer/Flasher Switch
- M29 - Left Low Beam Headlight
- M30 - Left High Beam Headlight
- M35 - Light for turn signal and side marker, front, left
- S18 - Fuse 18 in fuse holder
- S19 - Fuse 19 in fuse holder
- S21 - Fuse 21 in fuse holder
- T3d - 3-Pin Connector, on light for turn signal and side marker, front, left
- T3f - 3-Pin Connector, on headlight left
- T12 - 12-Pin Connector, near steering column
- (12) - Ground connection, in engine compartment, left
- (119) - Ground connection -1-, in headlight wiring harness
- (B166) - Connection (56a) in passenger compartment wiring harness
- (B167) - Connection (56b) in passenger compartment wiring harness

Edition 09/98
USA.5132.05.21

Turn signal switch, headlight dimmer/flasher switch

No. 21/8 Wiring diagram **New Beetle**

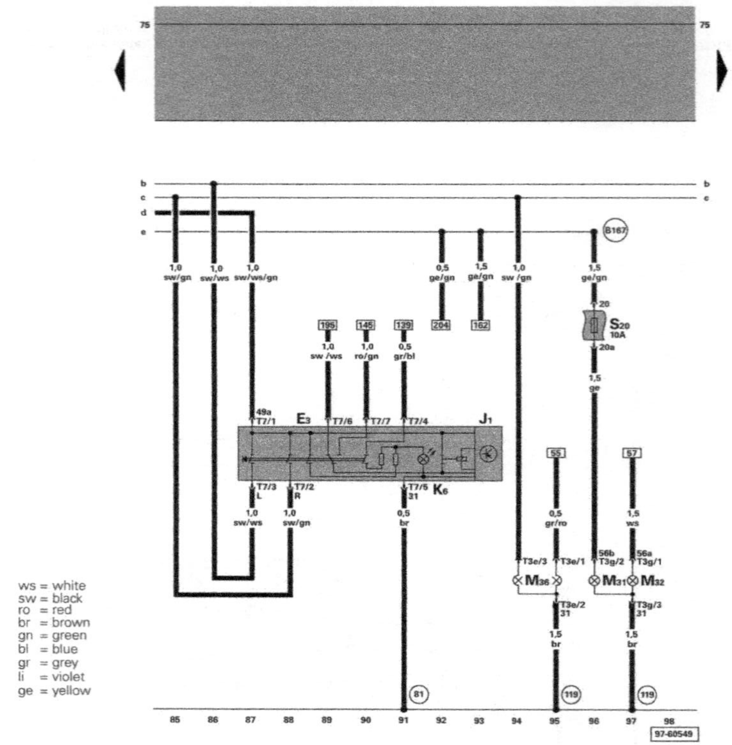

ws = white
sw = black
ro = red
br = brown
gn = green
bl = blue
gr = grey
li = violet
ge = yellow

- E3 - Emergency Flasher Switch
- J1 - Turn Signal Relay
- K6 - Emergency Flasher Warning Light
- M31 - Right Low Beam Headlight
- M32 - Right High Beam Headlight
- M36 - Light for turn signal and side marker, front, right
- S20 - Fuse 20 in fuse holder
- T3e - 3-Pin Connector
- T3g - 3-Pin Connector
- T7 - 7-Pin Connector
- (81) - Ground connection -1-, in instrument panel wiring harness
- (119) - Ground connection -1-, in headlight wiring harness
- (B167) - Connection (56b) in passenger compartment wiring harness

Emergency flasher switch with turn signal relay, right low and high beam headlights, light for turn signal and side marker, front, right

Edition 06/99
USA.5132.07.21

Standard equipment from Aug. 1998

New Beetle — Wiring diagram — No. 21/9

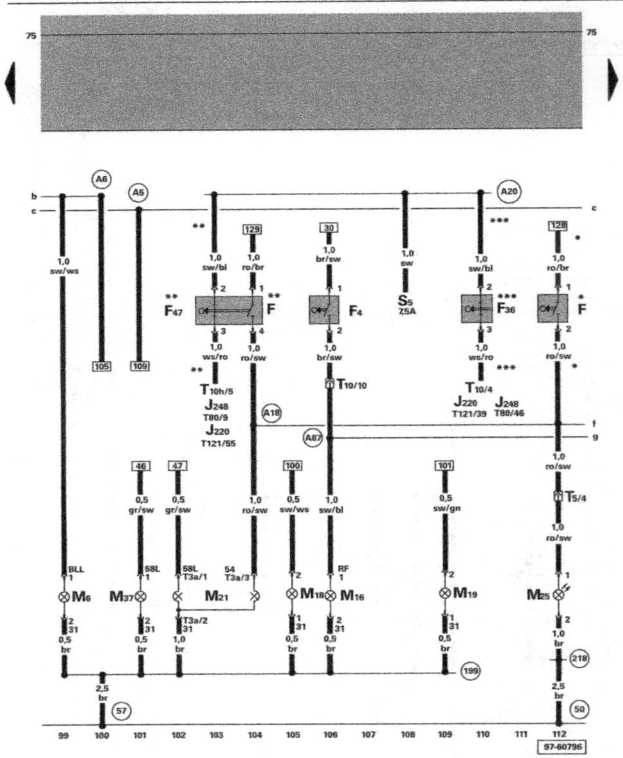

ws = white
sw = black
ro = red
br = brown
gn = green
bl = blue
gr = grey
li = violet
ge = yellow

- F - Brake Light Switch
- F4 - Back-Up light Switch
- F36 - Clutch Vacuum Vent Valve Switch
- F47 - Brake Vacuum Vent Valve Switch for cruise control/ (DFI)
- J220 - Motronic Engine Control Module (ECM)
- J248 - Diesel Direct Fuel Injection (DFI) Engine Control Module (ECM)
- M6 - Left Rear Turn Signal Light
- M16 - Left Back-Up Light
- M18 - Left, Side Turn Signal Light
- M19 - Right, Side Turn Signal Light
- M21 - Left Brake/Tail Light
- M25 - High-mount Brake Light
- M37 - Side marker light, left rear
- S5 - Fuse 5 in fuse/relay panel
- T3a - 3-Pin Connector
- T5 - 5-Pin Connector, black, on C-pillar, left
- T10 - 10-Pin Connector, white, behind instrument panel, left
- T80 - 80-Pin Connector, on Diesel (DFI) Control Module
- T121 - 121-Pin Connector, on Motronic Control Module (ECM)
- (50) - Ground connection, in luggage compartment, left
- (57) - Ground connection, on left rear pillar
- (199) - Ground Connection -3-, in instrument panel wiring harness
- (218) - Ground connection -1- in rear lid wiring harness
- (A5) - Plus connection (right turn signal), in instrument panel wiring harness
- (A6) - Plus connection (left turn signal), in instrument panel wiring harness
- (A18) - Wire connection (54), in instrument panel wiring harness
- (A20) - Wire connection (15a) in instrument panel wiring harness
- (A87) - Connector (reverse lamp) in instrument panel wiring harness
- * - Vehicles without cruise control (2.0L w/cruise control ⇒ diagram No.22)
- ** - Gasoline 1.8L and Diesel 1.9L with cruise control
- *** - Gasoline 1.8L and Diesel 1.9L with manual transmission only

Edition 03.00 **Brake light switch, brake/clutch vacuum vent valve/switch, back-up light switch, left back-up light, left brake/tail light, hight-mount brake light**
USA.5132.10.21

No. 21/10 — Wiring diagram — New Beetle

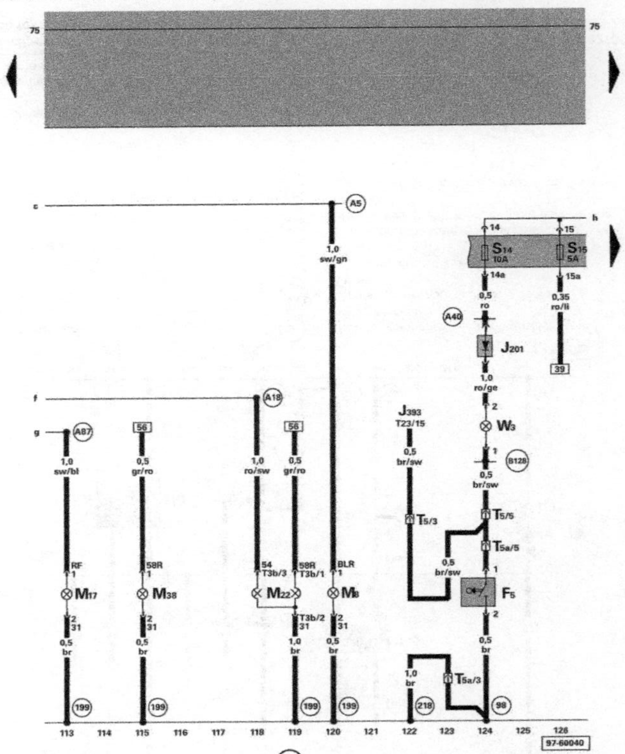

ws = white
sw = black
ro = red
br = brown
gn = green
bl = blue
gr = grey
li = violet
ge = yellow

- F5 - Luggage Compartment Light Switch
- J201 - Protection Diode
- J393 - Central control module for comfort system
- M8 - Right Rear Turn Signal Light
- M17 - Right Back-Up Light
- M22 - Right Brake/Tail Light
- M38 - Side marker light, right rear
- S14 - Fuse 14 in fuse holder
- S15 - Fuse 15 in fuse holder
- T3b - 3-Pin Connector
- T5 - 5-Pin Connector, black, on C-pillar, left
- T5a - 5-Pin Connector, black, in rear lid
- T23 - 23-Pin Connector, on Central control module for comfort system
- W3 - Luggage compartment Light
- (98) - Ground connection, in rear lid wiring harness
- (199) - Ground connection -3-, in instrument panel wiring harness
- (218) - Ground connection -1-, in rear lid wiring harness
- (A5) - plus connection (right turn signal), in instrument panel wiring harness
- (A18) - wire connection (54), in instrument panel wiring harness
- (A40) - plus connection -1- (30), in instrument panel wiring harness
- (A87) - Connector (reverse lamp), in instrument panel wiring harness
- (B128) - Connector (Luggage compartment lamp) in wiring harness interior

Luggage compartment light, right rear turnsignal light, right back-up light, right brake/tail light, side marker light, right, rear

Edition 03/00
USA.5132.10.21

New Beetle Wiring diagram **No. 21/11**

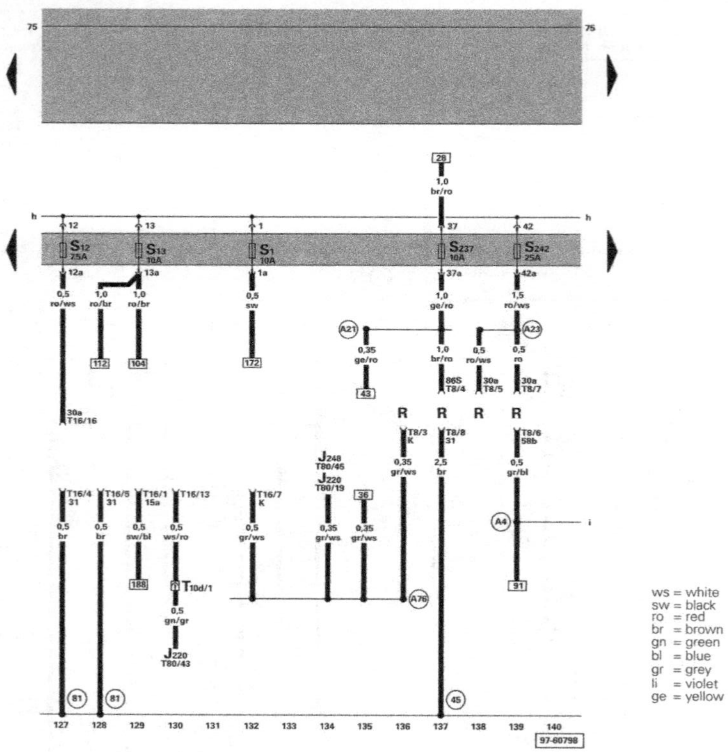

ws = white
sw = black
ro = red
br = brown
gn = green
bl = blue
gr = grey
li = violet
ge = yellow

J220 - Motronic Engine Control Module (ECM), behind instrument panel, left
J248 - Diesel Direct Fuel Injection (DFI) Engine Control Module (ECM), behind instrument panel, left
R - Radio
S12 - Fuse 12 in fuse holder
S13 - Fuse 13 in fuse holder
S237 - Fuse 37 in fuse holder
S242 - Fuse 42 in fuse holder
T8 - 8-Pin Connector, on radio
T10d - 10-Pin Connector, green, behind instrument panel, left
T16 - 16-Pin Connector, Data Link Connector (DLC), below steering column
T80 - Connector, 80 point, on Engine Control Module (ECM)

(45) - Ground connection, behind instrument panel, center
(81) - Ground connection -1-, in instrument panel wiring harness
(A4) - plus connection (58b) in instrument panel wiring harness
(A21) - wire connection (86s), in instrument panel wiring harness
(A23) - wire connection (30al), in instrument panel wiring harness
(A76) - Connector (K-diagnosis wire), in instrument panel wiring harness

Edition 03/00
USA.5132.10.21

Data Link Connector (DLC), radio connection

No. 21/12 Wiring diagram **New Beetle**

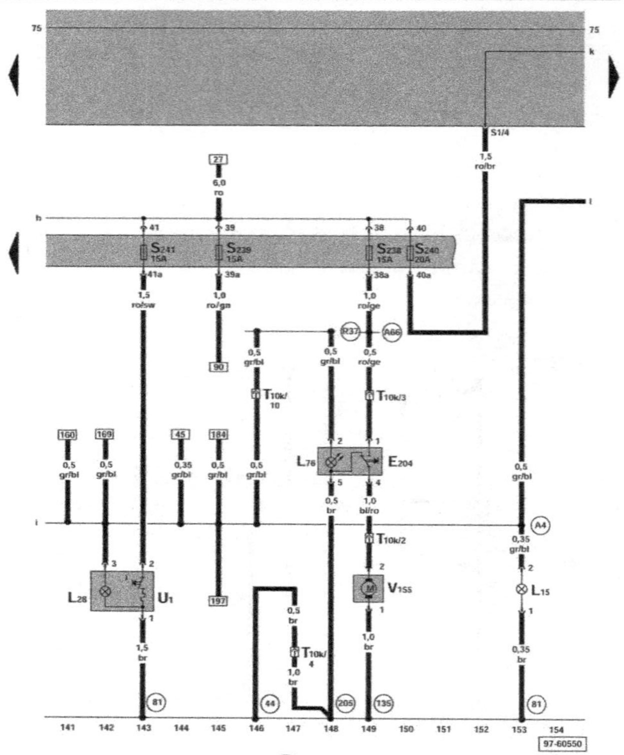

ws = white
sw = black
ro = red
br = brown
gn = green
bl = blue
gr = grey
li = violet
ge = yellow

E204 - Switch for remote/fuel tank door
L15 - Ashtray Light
L28 - Cigarette Lighter Light
L76 - Push Button Light
S238 - Fuse 38 in fuse holder
S239 - Fuse 39 in fuse holder
S240 - Fuse 40 in fuse holder
S241 - Fuse 41 in fuse holder
T10k - 10-pin connector, black, on conector station A-pillar, left
U1 - Cigarette Lighter
V155 - Motor for fuel tank lid unlock

(44) - Ground connection, on left A-pillar, lower part

(81) - Ground connection -1-, in instrument panel wiring harness

(135) - Ground connection -2-, in instrument panel wiring harness
(205) - Ground connection, in driver's door wiring harness
(A4) - plus connection (58b) in instrument panel wiring harness
(A66) - Connector (30a, central locking/anti-theft warning system/IR), in instrument panel wiring harness
(R37) - Connector (58b), in driver's door wiring harness

Ashtray light, cigarette lighter, cigarette lighter light, motor for fuel tank lid unlock, switch for remote/fuel tank door

Edition 03/00
USA.5132.10.21

New Beetle — Wiring diagram — No. 21/13

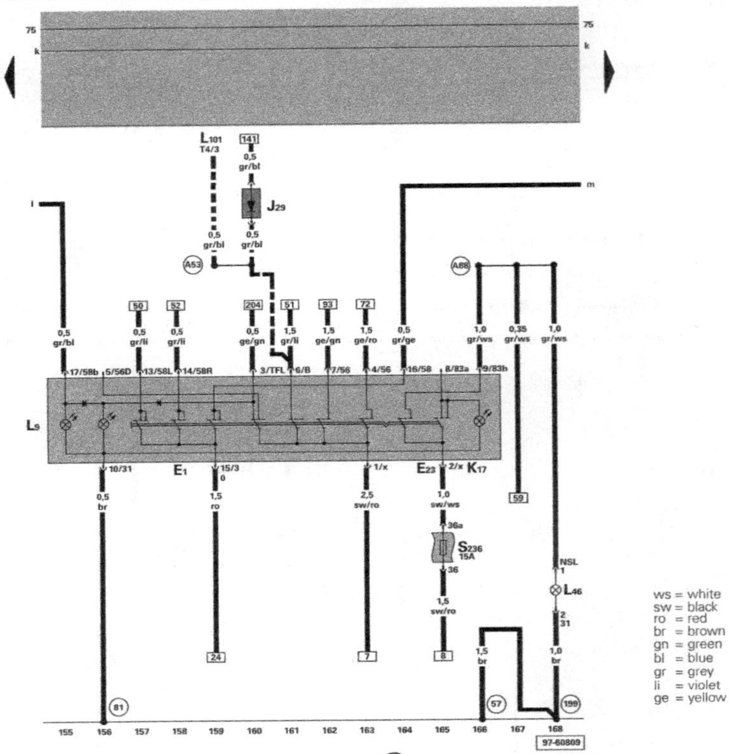

ws = white
sw = black
ro = red
br = brown
gn = green
bl = blue
gr = grey
li = violet
ge = yellow

- E1 - Light switch
- E23 - Fog Light Switch
- J29 - Blocking diode in wiring harness under driver door sill
- K17 - Fog Light Indicator Light
- L9 - Headlight Switch Licht
- L46 - Left Rear Fog Light
- L101 - Illumination for selector lever scale
- S236 - Fuse 36 in fuse holder

- (A53) - plus connection (58b), in instrument panel wiring harness
- (A88) - Connector fog light, in instrument panel wiring harness
- (B167) - Connection (56b) in passenger compartment wiring harness

------ Only automatic transmission

- (57) - Ground connection, on left rear pillar
- (81) - Ground connection -1-, in instrument panel wiring harness
- (199) - Ground connection -3-, in instrument panel wiring harness

Light switch, left rear fog light

No. 21/14 — Wiring diagram — New Beetle

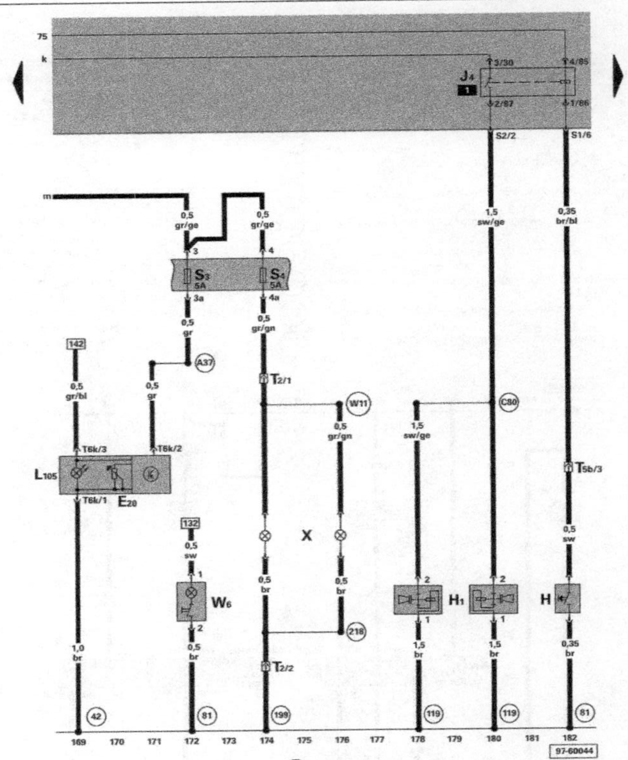

ws = white
sw = black
ro = red
br = brown
gn = green
bl = blue
gr = grey
li = violet
ge = yellow

- E20 - Instrument Panel Light Dimmer Switch
- H - Horn Button
- H1 - Dual Horns
- J4 - Dual Horn Relay
- L105 - Illumination for lighting controller
- S3 - Fuse 3 in fuse holder
- S4 - Fuse 4 in fuse holder
- T2 - Double Connector, in rear lid
- T5b - 5-Pin Connector, near steering column
- T6K - 6-Pin Connector
- W6 - Glove Compartment Light
- X - License Plate Light

- (42) - Ground connection, beside steering column
- (81) - Ground connection -1-, in instrument panel wiring harness

- (119) - Ground connection -1-, in headlight wiring harness
- (199) - Ground connection -3-, in instrument panel wiring harness
- (218) - Ground connection -1-, in rear lid wiring harness
- (A37) - wire connection (58a), in instrument panel wiring harness
- (C80) - wire connection (58b), in wiring harness headlamp
- (W11) - wire connection (58), in rear lid wiring harness

Dual horns, instrument light dimmer switch, glove compartment light, license plate light

New Beetle Wiring diagram **No. 21/15**

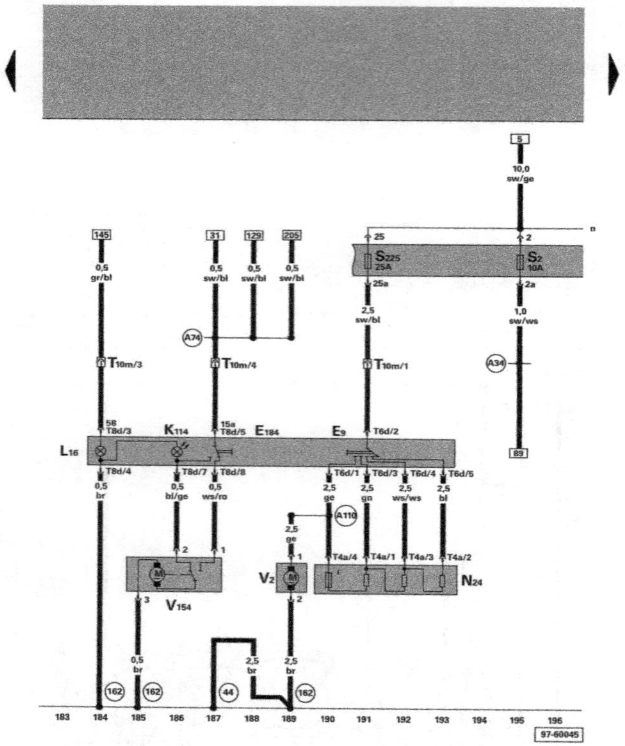

ws = white
sw = black
ro = red
br = brown
gn = green
bl = blue
gr = grey
li = violet
ge = yellow

E9 - Fresh Air Blower Switch
E184 - Switch for fresh air blower and recirculated air
K114 - Fresh Air and Recirculating Air Mode Indicator Light
L16 - Fresh Air Control Lever Light
N24 - Fresh Air Blower Series Resistance With Fuse
S2 - Fuse 2 in fuse holder
S225 - Fuse 25 in fuse holder
T4a - 4-Pin Connector
T6d - 6-Pin Connector
T8d - 8-Pin Connector
T10m - 10-Pin Connector
V2 - Fresh Air Blower
V154 - Servo motor for fresh-/recirculating air door

(44) - Ground connection, on left A-pillar, lower part
(162) - Ground connection, in blower motor wiring harness
(A34) - wire connection (75x), in instrument panel wiring harness
(A74) - Connector (15a - fuse 5), in instrument panel wiring harness
(A110) - Connector (fresh air blower), in instrument panel wiring harness

No. 21/16 Wiring diagram **New Beetle**

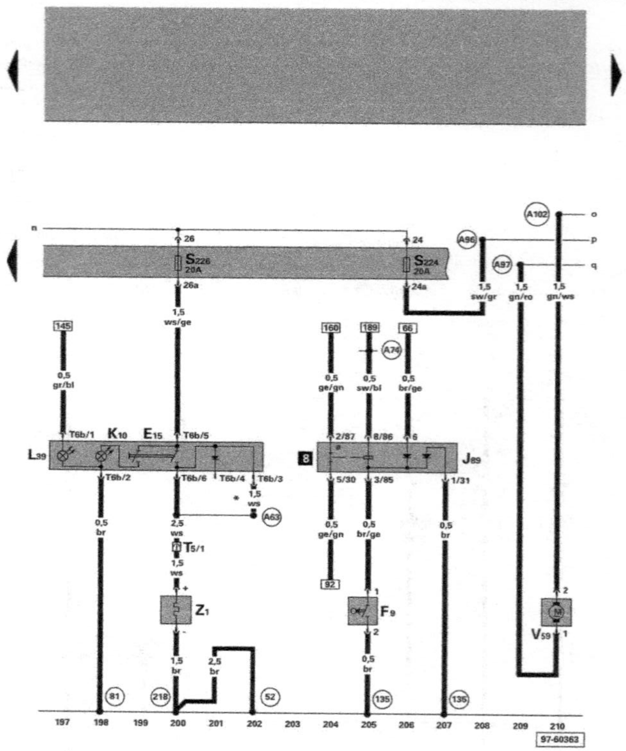

ws = white
sw = black
ro = red
br = brown
gn = green
bl = blue
gr = grey
li = violet
ge = yellow

E15 - Rear window defogger switch
F9 - Parking Brake Warning Light Switch
J89 - Daytime Running Lights Change-over Relay (173), on the thirteen position relay panel
K10 - Rear Window Defogger Indicator Light
L39 - Rear Window Defogger Switch Light
S224 - Fuse 24 in fuse holder
S226 - Fuse 26 in fuse holder
T5 - 5-Pin Connector, black, on C-pillar, left
T6b - 6-Pin Connector
V59 - Windshield and Rear Window Washer Pump
Z1 - Heated rear window
(52) - Ground connection, in rear lid, left
(81) - Ground connection -1-, in instrument panel wiring harness

(135) - Ground connection -2-, in instrument panel wiring harness
(218) - Ground connection -1-, in rear wiring harness
(A63) - Connector (mirror adjustment/ -heated) in instrument panel wiring harness
(A74) - Connector (15a -fuse 5), in instrument panel
(A96) - Connector (53a), in instrument panel wiring harness
(A97) - Connector (53), in instrument panel wiring harness
(A102) - Connector (windshield wiper), in instrument panel wiring harness
* Manual windows, only

New Beetle Wiring diagram **No. 21/17**

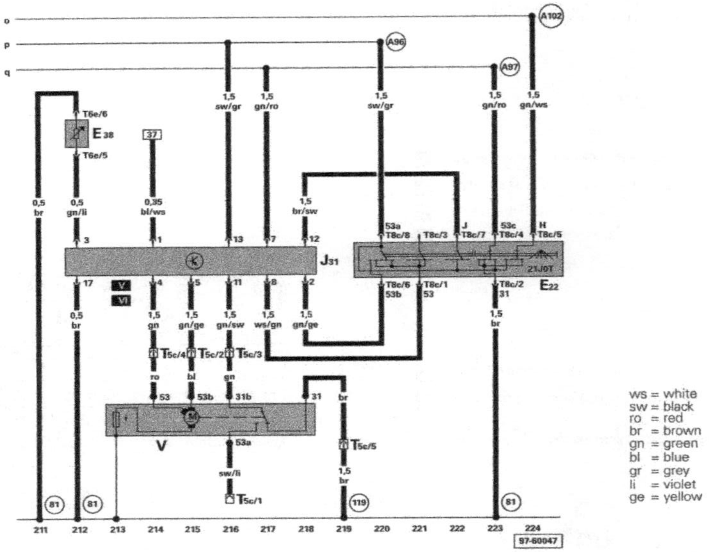

E22 - Winshield Wiper/Washer Switch
E38 - Windshield Wiper Intermittent Regulator
J31 - Wiper/Washer Intermittent Relay, on the thirteen position relay panel, production control number (377)
T5c - 5-Pin Connector, on Windshield Wiper Motor
T6e - 6-Pin Connector, near steering column
T8c - 8-Pin Connector
V - Windshield Wiper Motor

(81) - Ground connection -1-, in instrument panel wiring harness
(119) - Ground connection -1-, in headlight wiring harness
(A96) - Connector (53a), in instrument panel wiring harness
(A97) - Connector (53), in instrument panel wiring harness
(A102) - Connector (windshield wiper), in instrument panel wiring harness

Edition 09/98
USA.5132.05.21

Windshield wiper/washer switch, windshield wiper motor, windshield wiper/washer intermittent relay

THIS SPACE INTENTIONALLY LEFT BLANK

New Beetle - Standard Equipment,

from June 1999

Deviate relay location and fuseplacements as well as the locations of multiple connectors see section "component locations".

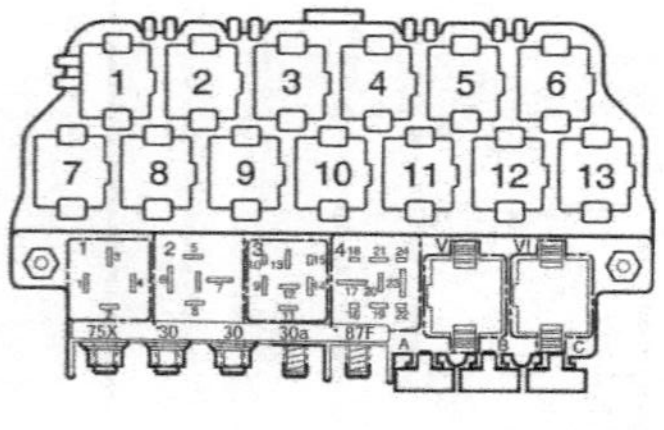

97-21956

Relay location on the thirteen position auxiliary relay panel, above relay panel:

- **4** Lock relay for starter (clutch pedal switch) (53)
- **7** Daytime Running Lights Change-over Relay (173)

Relay panel:

- **1** Dual Horn Relay (53)
- **2** Load Reduction Relay (100)
- **4** Fuel Pump (FP) Relay (409)
- **V** Wiper/Washer Intermittent Relay (377)
- **VI** Wiper/Washer Intermittent Relay (377)

Note: Number in parentheses indicates production control number stamped on relay housing.

Fuse colors
30 A - green
25 A - white
20 A - yellow
15 A - blue
10 A - red
7,5 A - brown
5 A - beige

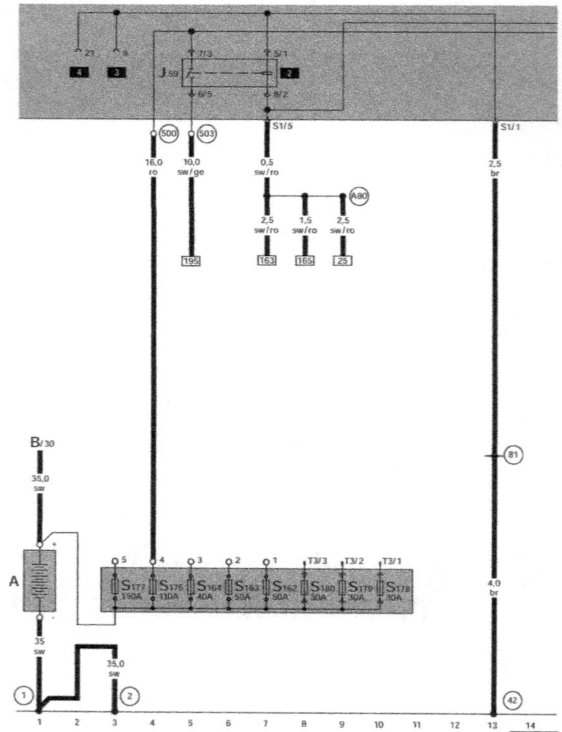

ws = white
sw = black
ro = red
br = brown
gn = green
bl = blue
gr = grey
li = violet
ge = yellow

A - Battery
B - Starter
J59 - Load Reduction Relay
S162 - Fuse -1- in fuse bracket/battery
S163 - Fuse -2- in fuse bracket/battery
S164 - Fuse -3- in fuse bracket/battery
S176 - Fuse -4- (30), in fuse bracket/battery
S177 - Fuse -5- (30), in fuse bracket/battery
S178 - Fuse -6- (30), in fuse bracket/battery
S179 - Fuse -7- (30), in fuse bracket/battery
S180 - Fuse -8- (30), in fuse bracket/battery

(1) - Ground strap, battery to body
(2) - Ground strap, transmission to body

(42) - Ground connection, beside steering column
(81) - Ground connection -1-, in instrument panel wiring harness
(500) - Threaded connection -1- (30) on the relay plate
(503) - Threaded connection -1- (75x) on the relay plate
(A80) - Connector -1- (X), in instrument panel wiring harness

Battery, load reduction relay, fuses in fuse bracket/battery

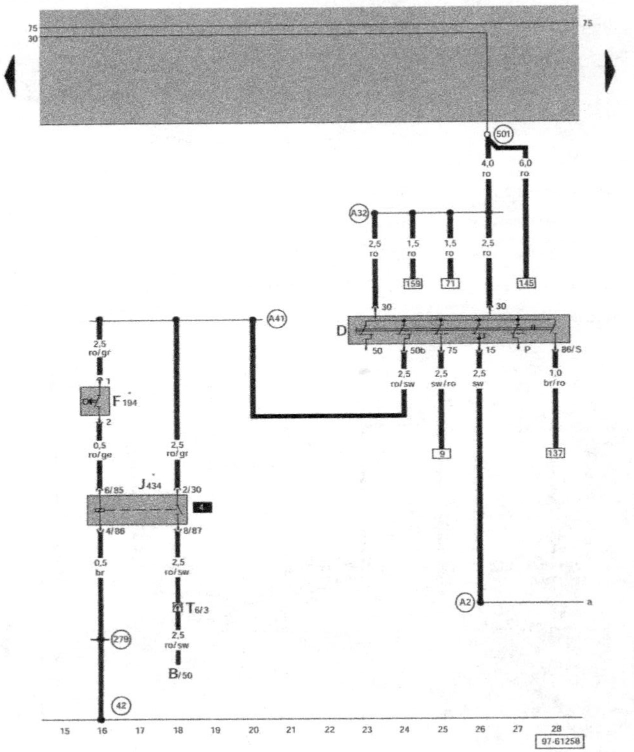

ws = white
sw = black
ro = red
br = brown
gn = green
bl = blue
gr = grey
li = violet
ge = yellow

- B - Starter
- D - Ignition/Starter Switch
- F194 - Clutch Pedal Position (CPP) Switch
- J434 - Locking relay for starter (clutch pedal switch)
- T6 - 6-Pin Connector, brown, behind instrument panel, left
- (42) - Ground connection, beside steering column
- (279) - Ground connector -5-, in wiring harness interior
- (501) - Threaded connection -2- (30) on the relay plate
- (A2) - plus connection (15), in instrument panel wiring harness
- (A32) - plus connection (30), in instrument panel wiring harness
- (A41) - plus connection (50), in instrument panel wiring harness
- * - Manual transmission only

Ignition/starter switch, lock relay for starter (theft warning system), locking relay for starter (clutch pedal switch)

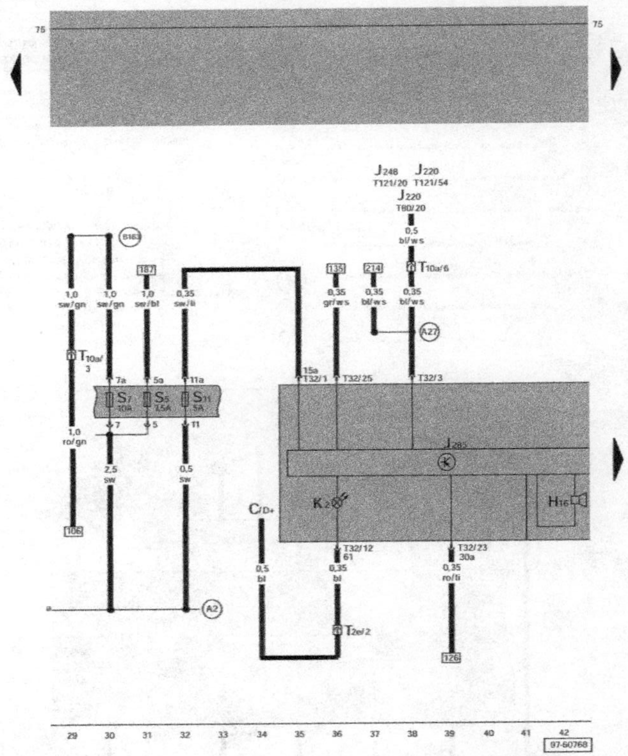

ws = white
sw = black
ro = red
br = brown
gn = green
bl = blue
gr = grey
li = violet
ge = yellow

- C - Generator (GEN)
- H16 - Warning buzzer for light switched on
- J220 - Motronic Engine Control Module (ECM)
- J248 - Diesel Direct Fuel Injection (DFI) Engine Control Module (ECM)
- J285 - Control module with indicator unit in instrument panel insert
- K2 - Generator (GEN) Warning Light
- S5 - Fuse 5 in fuse holder
- S7 - Fuse 7 in fuse holder
- S11 - Fuse 11 in fuse holder
- T2e - Double Connector, near starter (only vehicles with A/C)
- T6 - 6-Pin Connector, brown, behind instrument panel, left
- T10d - 10-Pin Connector, green, behind instrument panel, left
- T10a - 10-Pin Connector, orange, behind instrument panel, left
- T32 - 32-Pin Connector, blue, on instrument cluster
- T80 - 80-Pin Connector, on Motronic Engine Control Module (ECM)
- T121 - 121-Pin Connector, on Motronic/Diesel DFI Control Module
- (A2) - Plus connection (15), in instrument panel wiring harness
- (A27) - Connection (vehicle speed signal), in instrument panel wiring harness
- (B163) - Plus connector -1- (15) in wiring harness interior

Instrument cluster, warning buzzer for light swiched on, generator (GEN) warning lamp

Standard equipment from June 1999

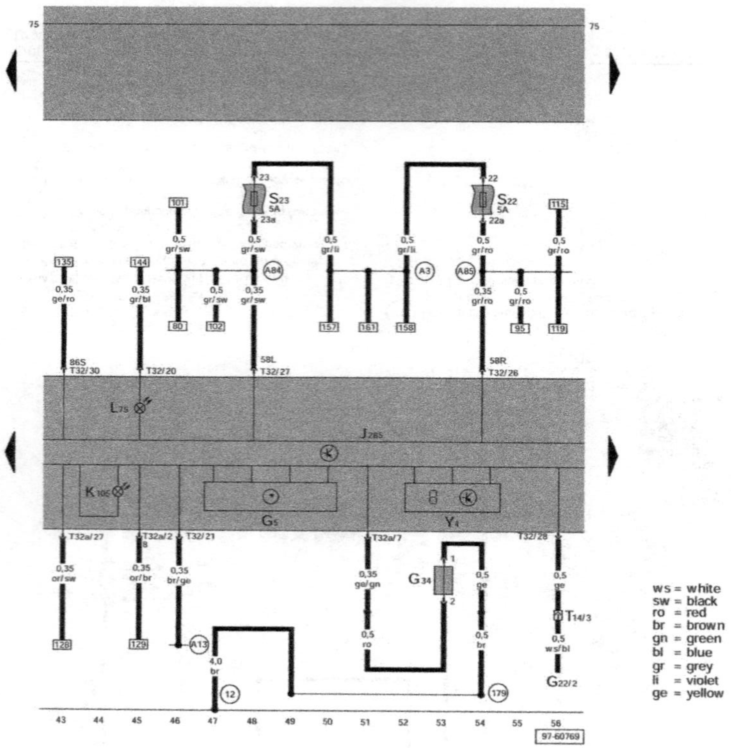

ws = white
sw = black
ro = red
br = brown
gn = green
bl = blue
gr = grey
li = violet
ge = yellow

- G5 - Tachometer
- G34 - Left Front Brake Pad Wear Sensor
- G22 - Speedometer Vehicle Speed Sensor (VSS)
- J285 - Control module with indicator unit in instrument panel insert
- K105 - Low Fuel Level Warning Light
- L75 - Digital Display Light
- S22 - Fuse 22 in fuse holder
- S23 - Fuse 23 in fuse holder
- T14 - 14-Pin Connector, in plenum chamber
- T32 - 32-Pin Connector, blue, on instrument cluster
- T32a - 32-Pin Connector, green, on instrument cluster
- Y4 - Odometer Display
- (12) - Ground connection, in engine compartment, left
- (179) - Ground connection, in left headlight wiring harness
- (A3) - plus connection (58), in instrument panel wiring harness
- (A13) - Wire connection (door contact switch) in instrument panel wiring harness
- (A84) - Connector (58L), in instrument panel wiring harness
- (A85) - Connector (58R), in instrument panel wiring harness

Instrument cluster, tachometer, low fuel level warning light, odometer display

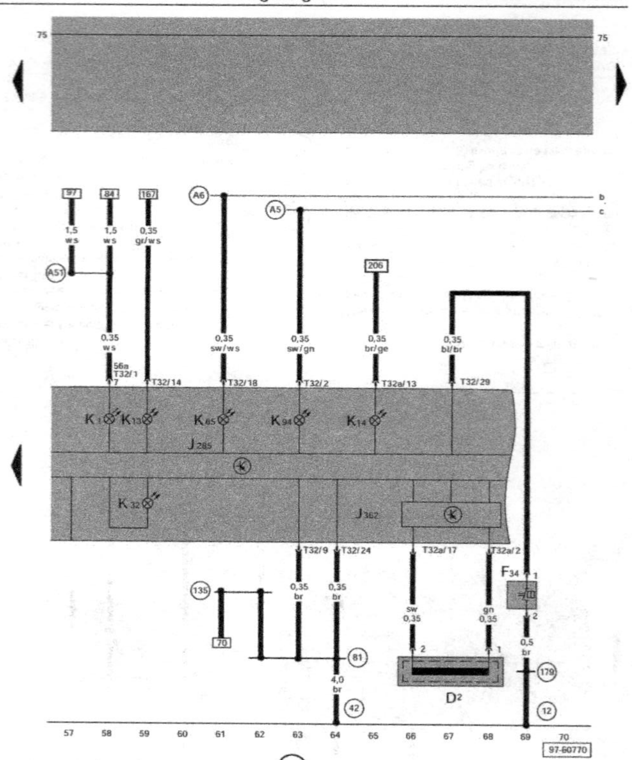

ws = white
sw = black
ro = red
br = brown
gn = green
bl = blue
gr = grey
li = violet
ge = yellow

- D2 - Induction coil of antitheft immobilizer
- F34 - Brake Fluid Level Warning Switch
- G34 - Left Front Brake Pad Wear Sensor
- J285 - Control module with indicator unit in inst. panel insert
- J362 - Control module for anti-theft immobilizer
- K1 - Headlight High Beam Indicator Light
- K13 - Rear Fog Light Indicator Light
- K14 - Parking Brake Indicator Light
- K32 - Brake Pad Wear Indicator Light
- K65 - Left Turn Signal Indicator Light
- K94 - Right Turn Signal Indicator Light
- T9 - 9-Pin Connector
- T32 - 32-Pin Connector, blue, on instrument cluster
- T32a - 32-Pin Connector, green, on instrument cluster
- (12) - Ground connection, in engine compartment, left
- (42) - Ground connection, beside steering column
- (81) - Ground connection -1-, in instrument panel wiring harness
- (135) - Ground connection -2-, instrument panel wiring harness
- (179) - Ground connection, in left headlight wiring harness
- (A5) - plus connection (right turn signal), in instrument panel wiring harness
- (A6) - plus connection (left turn signal), in instrument panel wiring harness
- (A51) - wire connection (56), in instrument panel wiring harness
- (A83) - Connector (daytime running lights), in instrument panel wiring harness

Instrument cluster, indicator lights, induction coil of anti-theft immobilizer, brake fluid level warning switch, left front brake pad wear sensor

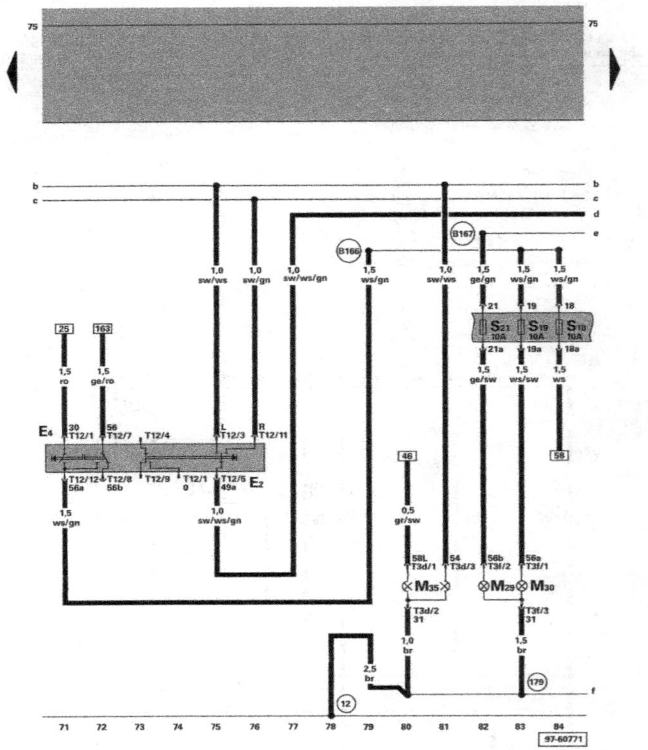

ws = white
sw = black
ro = red
br = brown
gn = green
bl = blue
gr = grey
li = violet
ge = yellow

- E2 - Turn Signal Switch
- E4 - Headlight Dimmer/Flasher Switch
- M29 - Left Low Beam Headlight
- M30 - Left High Beam Headlight
- M35 - Light for turn signal and side marker, front, left
- S18 - Fuse 18 in fuse holder
- S19 - Fuse 19 in fuse holder
- S21 - Fuse 21 in fuse holder
- T3d - 3-Pin Connector, on light for turn signal and side marker, front, left
- T3f - 3-Pin Connector, on headlight left
- T12 - 12-Pin Connector, near steering column

- (12) - Ground connection, in engine compartment, left
- (179) - Ground connection, in left headlight wiring harness
- (B166) - Connection (56a) in passenger compartment wiring harness
- (B167) - Connection (56b) in passenger compartment wiring harness

Turn signal switch, headlight dimmer/flasher switch

ws = white
sw = black
ro = red
br = brown
gn = green
bl = blue
gr = grey
li = violet
ge = yellow

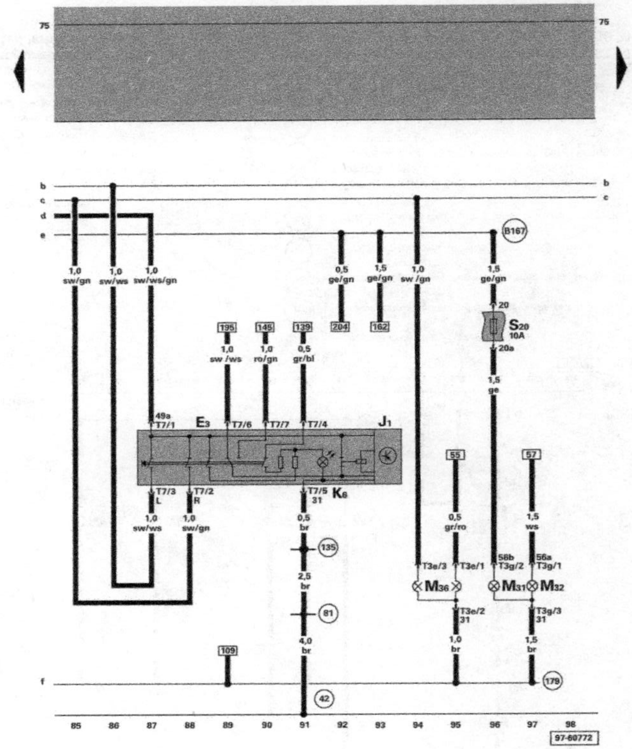

- E3 - Emergency Flasher Switch
- J1 - Turn Signal Relay
- K6 - Emergency Flasher Warning Light
- M31 - Right Low Beam Headlight
- M32 - Right High Beam Headlight
- M36 - Light for turn signal and side marker, front, right
- S20 - Fuse 20 in fuse holder
- T3e - 3-Pin Connector
- T3g - 3-Pin Connector
- T7 - 7-Pin Connector

- (42) - Ground connection, beside steering column
- (81) - Ground connection -1-, in instrument panel wiring harness
- (135) - Ground connection -2-, in instrument panel wiring harness
- (179) - Ground connection , in left headlight wiring harness
- (B167) - Connection (56b) in passenger compartment wiring harness

Emergency flasher switch with turn signal relay, right low and high beam headlights, light for turn signal and side marker, front, right

New Beetle Wiring diagram **No. 37/9**

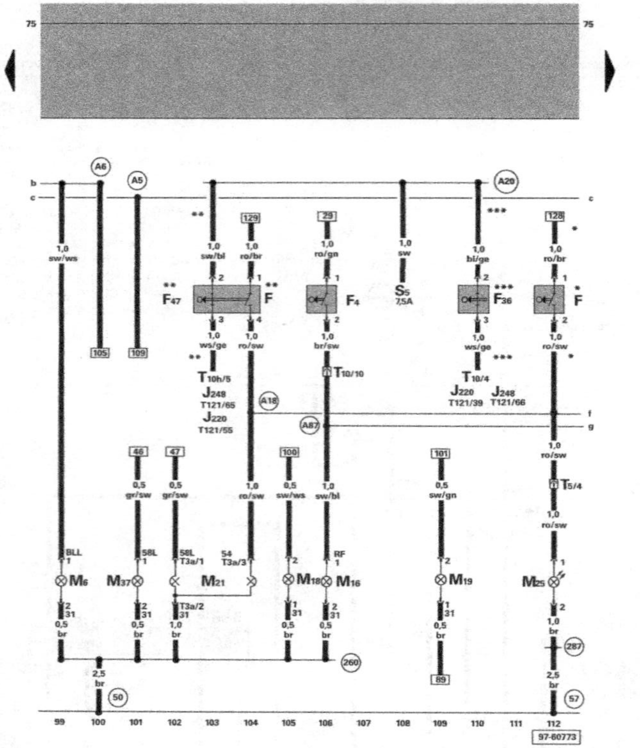

ws = white
sw = black
ro = red
br = brown
gn = green
bl = blue
gr = grey
li = violet
ge = yellow

- F - Brake Light Switch
- F4 - Back-Up light Switch
- F36 - Clutch Vacuum Vent Valve Switch
- F47 - Brake Vacuum Vent Valve Switch for cruise control/ (DFI)
- J220 - Motronic Engine Control Module (ECM)
- J248 - Diesel Direct Fuel Injection (DFI) Engine Control Module (ECM)
- M6 - Left Rear Turn Signal Light
- M16 - Left Back-Up Light
- M18 - Left, Side Turn Signal Light
- M19 - Right, Side Turn Signal Light
- M21 - Left Brake/Tail Light
- M25 - High-mount Brake Light
- M37 - Side marker light, left rear
- S5 - Fuse 5 in fuse/relay panel
- T3a - 3-Pin Connector
- T5 - 5-Pin Connector, black, on C-pillar, left
- T10 - 10-Pin Connector, white, behind instrument panel, left
- T121 - 121-Pin Connector, on Motronic/Diesel (DFI) ECM

- (50) - Ground connection, in luggage compartment, left
- (57) - Ground connection, on left rear pillar
- (260) - Ground Connection, in wiring harness taillights
- (287) - Ground connection, in wiring harness rear hatch hydraulics
- (A5) - Plus connection (right turn signal), in instrument panel wiring harness
- (A6) - Plus connection (left turn signal), in instrument panel wiring harness
- (A18) - Wire connection (54), in instrument panel wiring harness
- (A20) - Wire connection (15a) in instrument panel wiring harness
- (A87) - Connector (reverse lamp) in instrument panel wiring harness
- (A99) - Connector -1- (87), in instrument panel wiring harness
- * - 2.0L w/o cruise control only (2.0L w/cruise control ⇒ diagram No. 22)
- ** - Gasoline 1.8L and Diesel 1.9L
- *** - Gasoline 1.8L and Diesel 1.9L with manual transmission only

Edition 03.00 **Brake light switch, brake/clutch vacuum vent valve/switch, back-up light switch, left back-up light, left brake/tail light, hight-mount brake light**

USA.5132.10.21

No. 37/10 Wiring diagram **New Beetle**

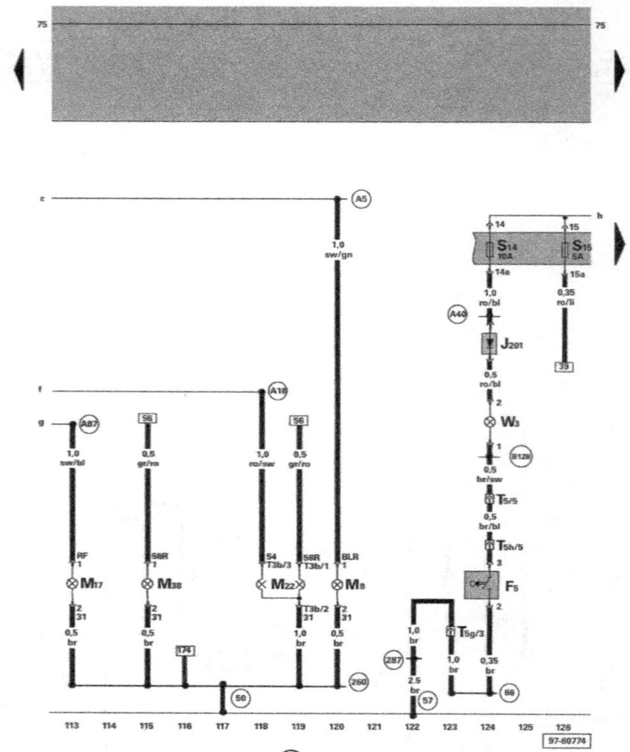

ws = white
sw = black
ro = red
br = brown
gn = green
bl = blue
gr = grey
li = violet
ge = yellow

- F5 - Luggage Compartment Light Switch
- J201 - Protection Diode
- J393 - Central control module for comfort system
- M8 - Right Rear Turn Signal Light
- M17 - Right Back-Up Light
- M22 - Right Brake/Tail Light
- M38 - Side marker light, right rear
- S14 - Fuse 14 in fuse holder
- S15 - Fuse 15 in fuse holder
- T3b - 3-Pin Connector
- T5 - 5-Pin Connector, black, on C-pillar, left
- T5g - 5-Pin Connector, black, in rear lid
- T5h - 5 Pin Connector, brown, in rear lid
- T23 - 23-Pin Connector, on Central control module for comfort sys.
- W3 - Luggage compartment Light
- (50) - Ground connection, in luggage compartment, left

- (57) - Ground connection, on left rear pillar
- (86) - Ground connection -1-, in rear wiring harness
- (260) - Ground Connection, in wiring harness taillights
- (287) - Ground connection, in wiring harness rear hatch hydraulics
- (A5) - Plus connection (right turn signal), in instrument panel wiring harness
- (A18) - Wire connection (54), in instrument panel wiring harness
- (A40) - Plus connection -1- (30), in instrument panel wiring harness
- (A87) - Connector (reverse lamp), in instrument panel wiring harness
- (B128) - Connector (Luggage compartment lamp) in wiring harness interior

Luggage compartment light, right rear turnsignal light, right back-up light, right brake/tail light, side marker light, right, rear **Edition 03.00**

USA.5132.10.21

New Beetle — Wiring diagram — No. 37/11

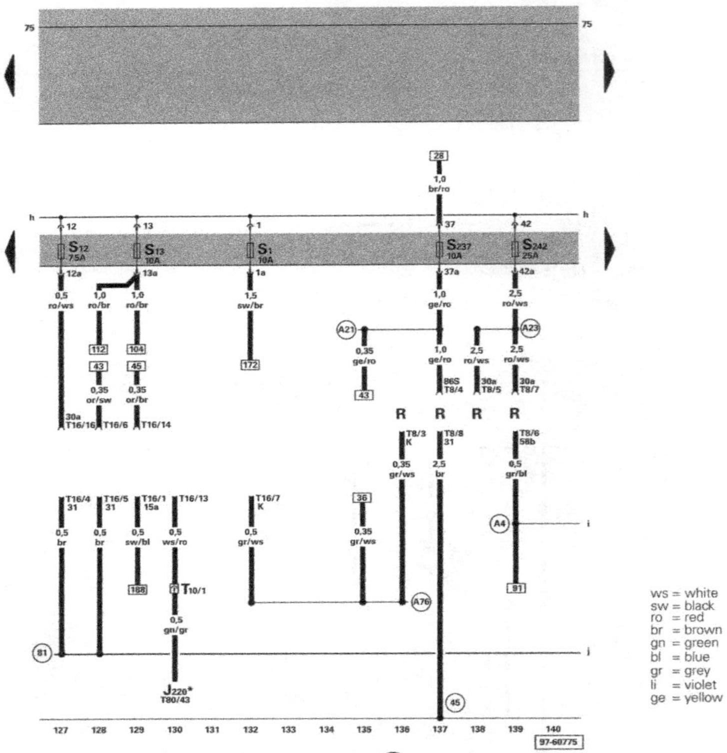

ws = white
sw = black
ro = red
br = brown
gn = green
bl = blue
gr = grey
li = violet
ge = yellow

J220 - Motronic Engine Control Module (ECM)
R - Radio
S12 - Fuse 12 in fuse holder
S13 - Fuse 13 in fuse holder
S237 - Fuse 37 in fuse holder
S242 - Fuse 42 in fuse holder
T8 - 8-Pin Connector, on radio
T10 - 10-Pin Connector, white, behind instrument panel, left
T16 - 16-Pin Connector, Data Link Connector (DLC), below steering column, left
T80 - 80-Pin Connector, on Motronic ECM

(45) - Ground connection, behind instrument panel, center
(81) - Ground connection -1-, in instrument panel wiring harness

(A4) - plus connection (58b) in instrument panel wiring harness
(A21) - wire connection (86s), in instrument panel wiring harness
(A23) - wire connection (30al), in instrument panel wiring harness
(A76) - Connector (K-diagnosis wire), in instrument panel wiring harness

* - 2.0L Gasoline engine only

Edition 06/01
USA.5132.15.21

Data Link Connector (DLC), radio connection

No. 37/12 — Wiring diagram — New Beetle

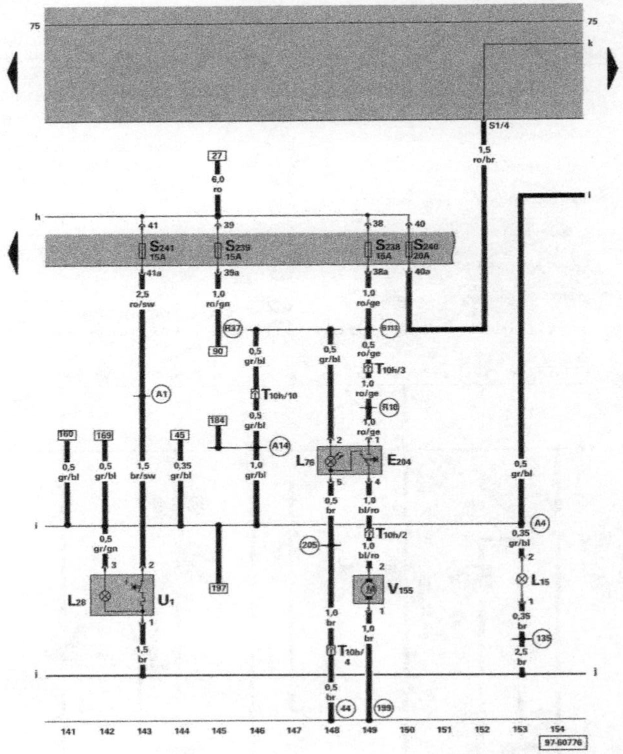

ws = white
sw = black
ro = red
br = brown
gn = green
bl = blue
gr = grey
li = violet
ge = yellow

E204 - Switch for remote/fuel tank door
L15 - Ashtray Light
L28 - Cigarette Lighter Light
L76 - Push Button Light
S238 - Fuse 38 in fuse holder
S239 - Fuse 39 in fuse holder
S240 - Fuse 40 in fuse holder
S241 - Fuse 41 in fuse holder
T10h - 10-pin connector, black, on conec. station A-pillar, left
U1 - Cigarette Lighter
V155 - Motor for fuel tank lid unlock

(44) - Ground connection, on left A-pillar, lower part
(135) - Ground connection -2-, in instrument panel wiring harness

(199) - Ground connection -3-, in instrument panel wiring harness
(205) - Ground connection, in driver's door wiring harness
(A1) - plus connection (30a), in instrument panel wiring harness
(A4) - plus connection (58b) in instrument panel wiring harness
(A14) - plus connection (58b) in cigarette lighter wiring harness
(B113) - Plus connector (30a), in wiring harness interior, left
(R10) - Plus connection -1- (30), in driver's door
(R37) - Connector (58b), in driver's door wiring harness

Ashtray light, cigarette lighter, cigarette lighter light, motor for fuel tank lid unlock, switch for remote/fuel tank door

Edition 06/01
USA.5132.15.21

Standard equipment from June 1999

New Beetle Wiring diagram **No. 37/13**

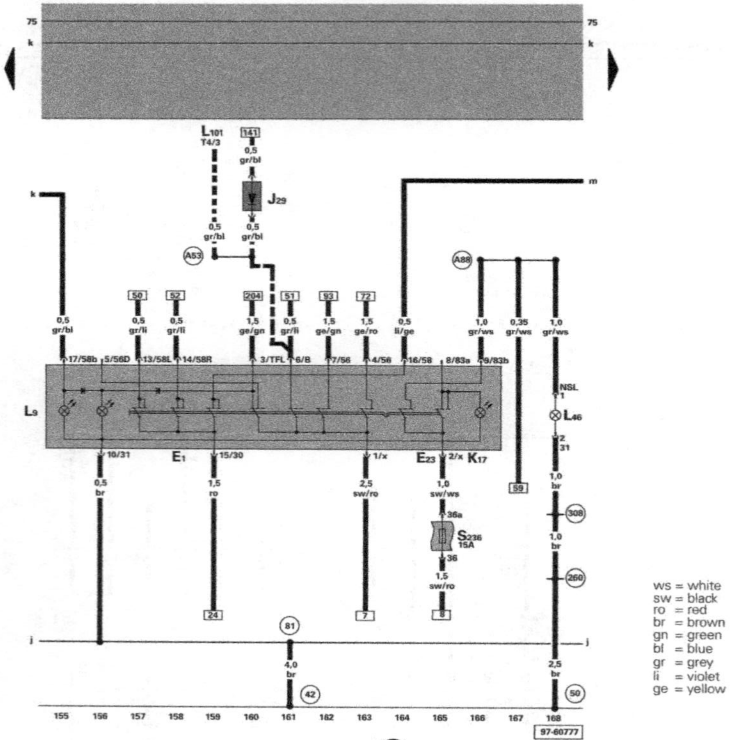

ws = white
sw = black
ro = red
br = brown
gn = green
bl = blue
gr = grey
li = violet
ge = yellow

E1 - Light switch
E23 - Fog Light Switch
J29 - Blocking diode
K17 - Fog Light Indicator Light
L9 - Headlight Switch Licht
L46 - Left Rear Fog Light
L101 - Illumination for selector lever scale
S236 - Fuse 36 in fuse holder

(42) - Ground connection, beside steering column

(50) - Ground connection, in luggage compartment, left

(81) - Ground connection -1-, in instrument panel wiring harness

(260) - Ground Connection, in wiring harness taillights

(308) - Ground Connection -2- (tail light wiring harness)

(A53) - plus connection (58b), in instrument panel wiring harness

(A88) - Connector fog light, in instrument panel wiring harness

------ - Only automatic transmistion

Edition 03.00
USA.5132.10.21

Light switch, left rear fog light

No. 37/14 Wiring diagram **New Beetle**

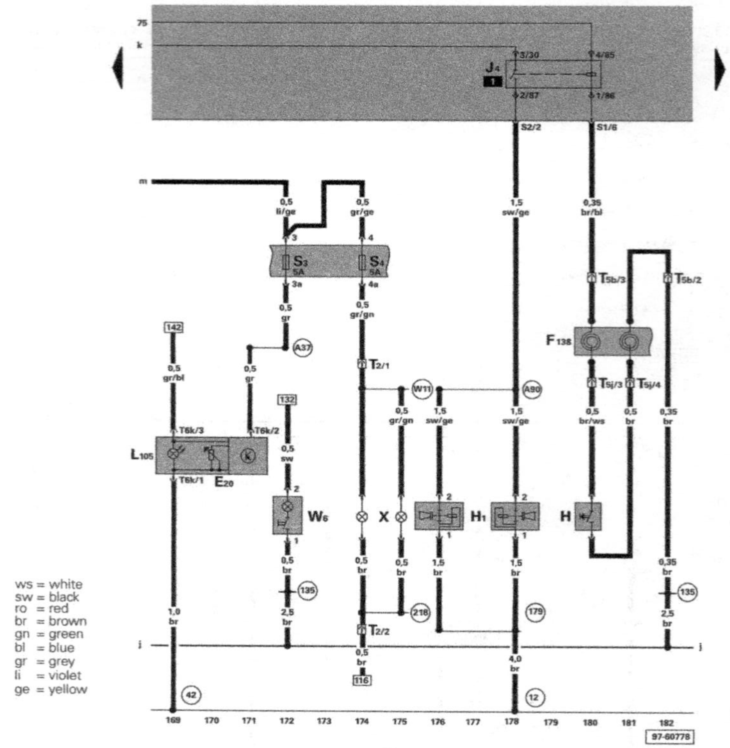

ws = white
sw = black
ro = red
br = brown
gn = green
bl = blue
gr = grey
li = violet
ge = yellow

E20 - Instrument Panel Light Dimmer Switch
F138 - Airbag Spiral Spring/Return Spring With Slip Ring
H - Horn Button
H1 - Dual Horns
J4 - Dual Horn Relay
L105 - Illumination for lighting controller
S3 - Fuse 3 in fuse holder
S4 - Fuse 4 in fuse holder
T2 - Double Connector, in rear lid
T5b - 5-Pin Connector, on airbag spiral spring
T5J - 5-Pin Connector, behind driver's airbag
T6k - 6-Pin Connector
W6 - Glove Compartment Light
X - License Plate Light

(12) - Ground connection, in engine compartment, left

(42) - Ground connection, beside steering column

(135) - Ground connection -2-, in instrument panel wiring harness

(179) - Ground connection -1-, in headlight wiring harness

(218) - Ground connection -1-, in rear lid wiring harness

(260) - Ground Connection, in wiring harness taillights

(A37) - Connection (58a), in instrument panel wiring harness

(A90) - Connection, (dual tone horn), in instrument panel wiring harness

(W11) - Connection (58), in rear lid wiring harness

Dual horns, instrument light dimmer switch, glove compartment light, license plate light

Edition 03.00
USA.5132.10.21

Standard equipment from June 1999

New Beetle — Wiring diagram — No. 37/15

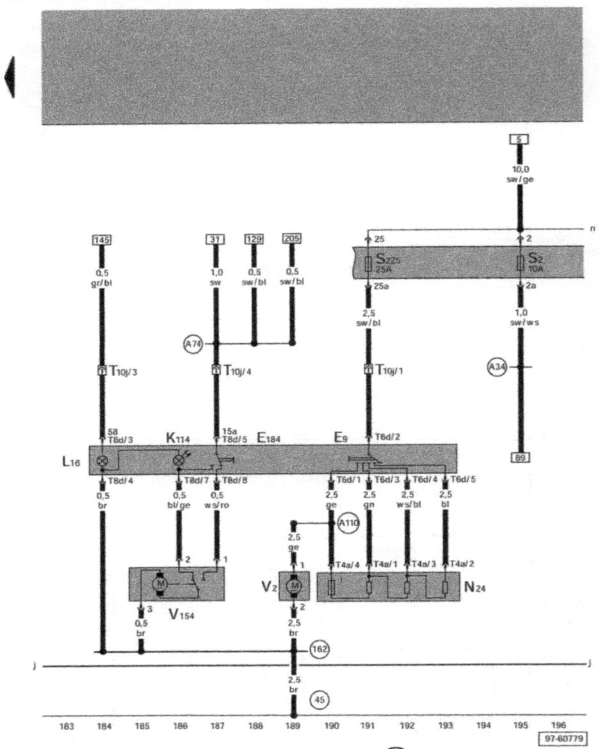

ws = white
sw = black
ro = red
br = brown
gn = green
bl = blue
gr = grey
li = violet
ge = yellow

- E9 - Fresh Air Blower Switch
- E184 - Switch for fresh air blower and recirculated air
- K114 - Fresh Air and Recirculating Air Mode Indicator Light
- L16 - Fresh Air Control Lever Light
- N24 - Fresh Air Blower Series Resistance With Fuse
- S2 - Fuse 2 in fuse holder
- S225 - Fuse 25 in fuse holder
- T4a - 4-Pin Connector
- T6d - 6-Pin Connector
- T8d - 8-Pin Connector
- T10j - 10-Pin Connector
- V2 - Fresh Air Blower
- V154 - Servo motor for fresh-/recirculating air door
- (45) - Ground connection, behind instrument panel, center
- (162) - Ground connection, in blower motor wiring harness
- (A34) - wire connection (75x), in instrument panel wiring harness
- (A74) - Connector (15a - fuse 5), in instrument panel wiring harness
- (A110) - Connector (fresh air blower), in instrument panel wiring harness

No. 37/16 — Wiring diagram — New Beetle

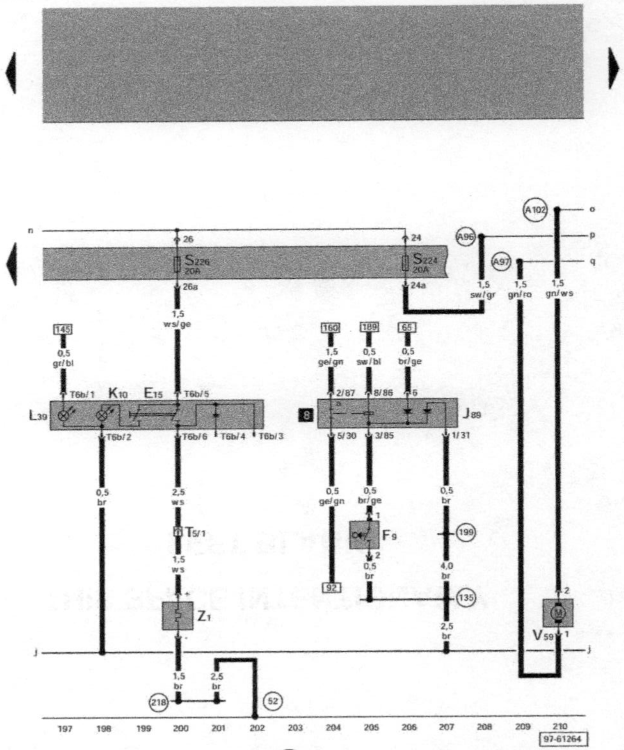

ws = white
sw = black
ro = red
br = brown
gn = green
bl = blue
gr = grey
li = violet
ge = yellow

- E15 - Rear window defogger switch
- F9 - Parking Brake Warning Light Switch
- J89 - Daytime Running Lights Change-over Relay (173), on the thirteen position relay panel
- K10 - Rear Window Defogger Indicator Light
- L39 - Rear Window Defogger Switch Light
- S224 - Fuse 24 in fuse holder
- S226 - Fuse 26 in fuse holder
- T5 - 5-Pin Connector, black, on C-pillar, left
- T6b - 6-Pin Connector
- V59 - Windshield and Rear Window Washer Pump
- Z1 - Heated rear window
- (52) - Ground connection, in rear lid, left
- (135) - Ground connection -2-, in instrument panel wiring harness
- (199) - Ground connection -3-, in instrument panel wiring harness
- (218) - Ground connection -1-, in rear wiring harness
- (A74) - Connector (15a -fuse 5), in instrument panel
- (A96) - Connector (53a), in instrument panel wiring harness
- (A97) - Connector (53), in instrument panel wiring harness
- (A102) - Connector (windshield wiper), in instrument panel wiring harness

Standard equipment from June 1999

New Beetle Wiring diagram **No. 37/17**

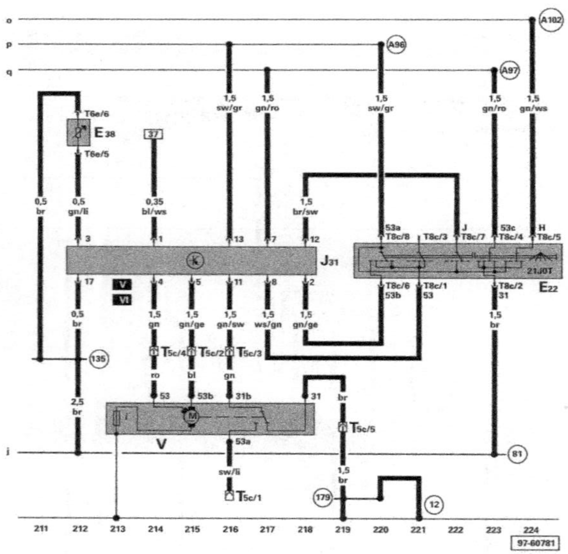

ws = white
sw = black
ro = red
br = brown
gn = green
bl = blue
gr = grey
li = violet
ge = yellow

E22 - Winshield Wiper/Washer Switch
E38 - Windshield Wiper Intermittent Regulator
J31 - Wiper/Washer Intermittent Relay, on the thirteen position relay panel, production control number (377)
T5c - 5-Pin Connector, on Windshield Wiper Motor
T6e - 6-Pin Connector, near steering column
T8c - 8-Pin Connector
V - Windshield Wiper Motor

(81) - Ground connection -1-, in instrument panel wiring harness
(135) - Ground connection -2-, in instrument panel wiring harness
(179) - Ground connection, in left headlight wiring harness

(A96) - Connector (53a), in instrument panel wiring harness
(A97) - Connector (53), in instrument panel wiring harness
(A102) - Connector (windshield wiper), in instrument panel wiring harness

Edition 03.00 USA.5132.10.21

Windshield wiper/washer switch, windshield wiper motor, windshield wiper/washer intermittent relay

THIS SPACE INTENTIONALLY LEFT BLANK

New Beetle **Wiring diagram** **No. 1/1**

Standard Equipment,

from August 2002,

Additional Fuse, Relay, Control Module and Connector locations ⇒ Component Locations

Troubleshooting Procedures ⇒ Guided Fault Finding using VAS 5051/5052

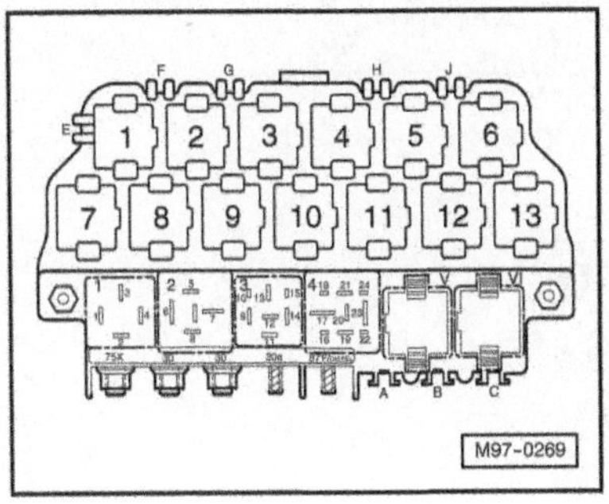

Relay locations on thirteen position auxiliary relay panel, above relay panel:

3 J434 Starter Lock/Clutch Pedal Switch Relay (53)

7 J89 Daytime Running Lights Change-over Relay (173)

Relay panel:

1 J4 Dual horn Relay (53)

2 J59 Load Reduction Relay (100)

V/V1 J31 Wiper/Washer Intermitten Relay (377)

Note:

Number in parentheses indicates production control number stamped on relay housing.

New Beetle **Wiring diagram** **No. 1/2**

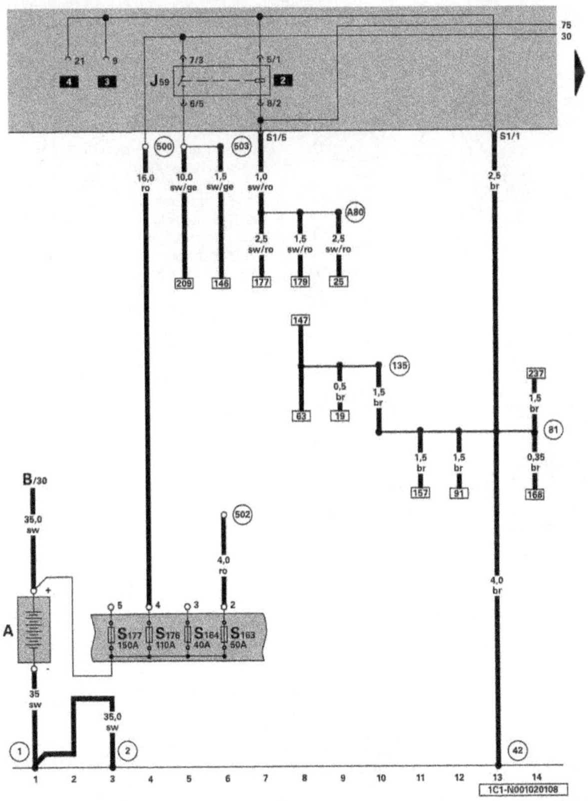

Battery, load reduction relay, fuses in fuse bracket/battery

A - Battery
B - Starter
J59 - Load Reduction Relay
S163 - Fuse -2- in fuse bracket/battery
S164 - Fuse -3- in fuse bracket/battery
S176 - Fuse -4- (30), in fuse bracket/battery
S177 - Fuse -5- (30), in fuse bracket/battery

(1) - Ground strap, battery to body
(2) - Ground strap, transmission to body
(42) - Ground connection, beside steering column
(81) - Ground connection -1-, in instrument panel wiring harness
(135) - Ground connection -2-, in instrument panel wiring harness
(500) - Threaded connection -1- (30) on the relay plate
(502) - Threaded connection -1- (30a) on the relay plate
(503) - Threaded connection -1- (75x) on the relay plate
(A80) - Connector -1- (X), in instrument panel wiring harness

ws = white
sw = black
ro = red
br = brown
gn = green
bl = blue
gr = grey
li = violet
ge = yellow
or = orange

Standard equipment from Aug. 2002

New Beetle

Wiring diagram

No. 1/3

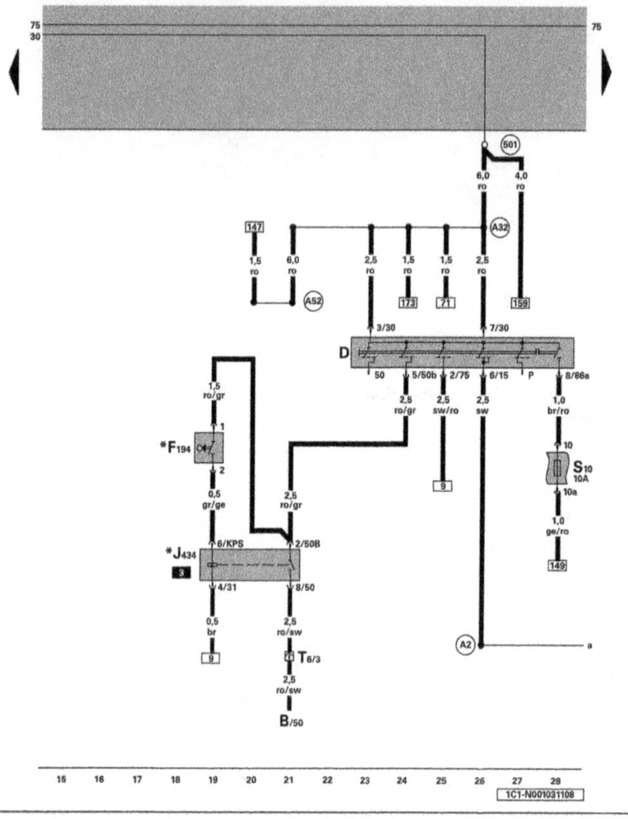

Ignition/starter switch, clutch pedal position (CPP) switch, locking relay for starter (clutch pedal switch)

B - Starter
D - Ignition/Starter Switch
F194 - Clutch Pedal Position (CPP) Switch
J434 - Locking relay for starter (clutch pedal switch)
S10 - Fuse 10 in fuse holder
T6 - 6-Pin Connector, brown, behind instrument panel, left

(501) - Threaded connection -2- (30) on the relay plate

(A2) - plus connection (15), in instrument panel wiring harness
(A32) - plus connection (30), in instrument panel wiring harness
(A52) - Plus Connection 2 (30) (in instrument panel wiring harness)

* - Manual transmission only

ws = white
sw = black
ro = red
br = brown
gn = green
bl = blue
gr = grey
li = violet
ge = yellow
or = orange

New Beetle

Wiring diagram

No. 1/4

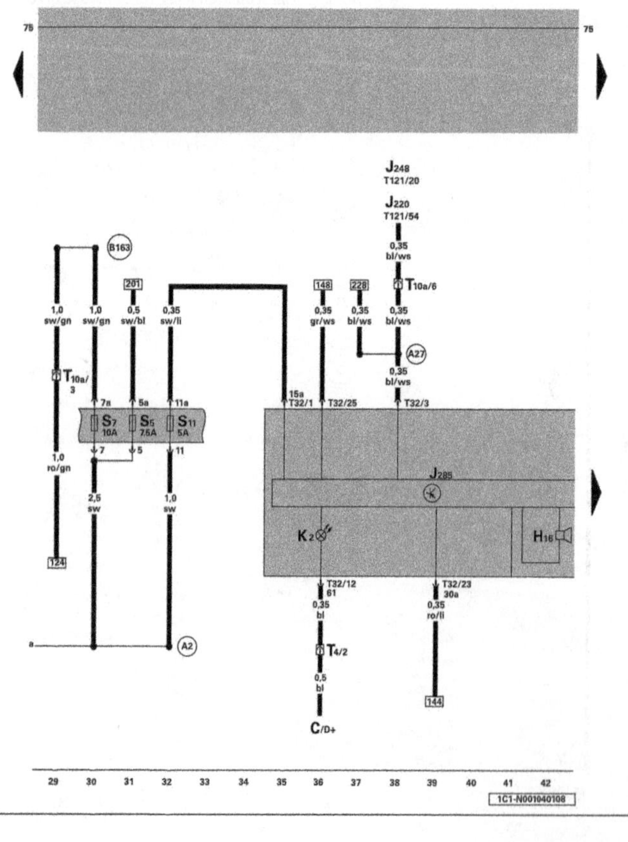

Instrument cluster, warning buzzer for light swiched on, generator (GEN) warning light

C - Generator (GEN)
H16 - Warning buzzer for light switched on
J220 - Motronic Engine Control Module (ECM)
J248 - Diesel Direct Fuel Injection (DFI) Engine Control Module (ECM)
J285 - Control module with indicator unit in instrument panel insert
K2 - Generator (GEN) Warning Light
S5 - Fuse 5 in fuse holder
S7 - Fuse 7 in fuse holder
S11 - Fuse 11 in fuse holder
T4 - 4-Pin Connector, in engine compartment left
T6 - 6-Pin Connector, brown, behind instrument panel, left
T10d - 10-Pin Connector, green, behind instrument panel, left
T10a - 10-Pin Connector, orange, behind instrument panel, left
T32 - 32-Pin Connector, blue, on instrument cluster
T121 - 121-Pin Connector, on Motronic/Diesel DFI Control Module

(A2) - Plus connection (15), in instrument panel wiring harness
(A27) - Connection (vehicle speed signal), in instrument panel wiring harness
(B163) - Plus connector -1- (15) in wiring harness interior

ws = white
sw = black
ro = red
br = brown
gn = green
bl = blue
gr = grey
li = violet
ge = yellow
or = orange

New Beetle — Wiring diagram — No. 1/5

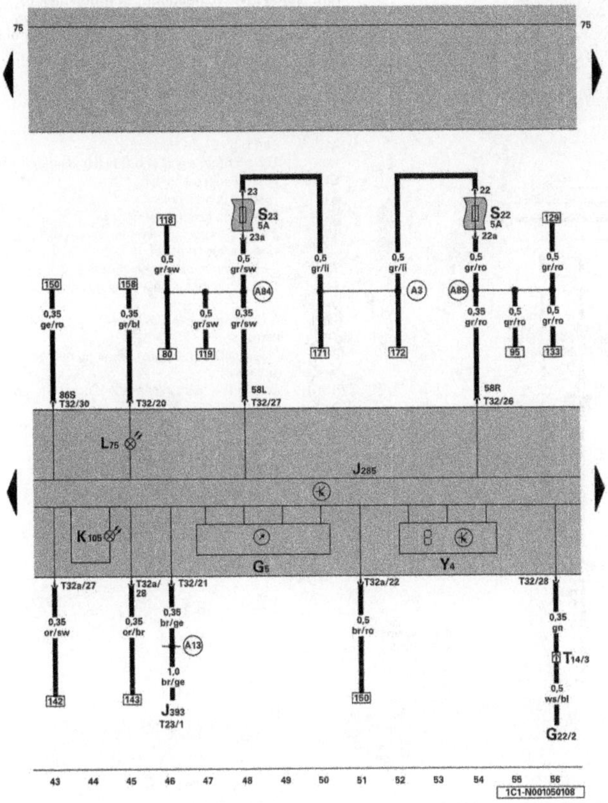

Instrument cluster, tachometer, low fuel level warning light, odometer display

- G5 - Tachometer
- G22 - Speedometer Vehicle Speed Sensor (VSS)
- J285 - Control module with indicator unit in instrument panel insert
- J393 - Central control module for comfort system
- K105 - Low Fuel Level Warning Light
- L75 - Digital Display Light
- S22 - Fuse 22 in fuse holder
- S23 - Fuse 23 in fuse holder
- T14 - 14-Pin Connector, in plenum chamber
- T23 - 23-Pin Connector, at J393
- T32 - 32-Pin Connector, blue, on instrument cluster
- T32a - 32-Pin Connector, green, on instrument cluster
- Y4 - Odometer Display

- (12) - Ground connection, in engine compartment, left
- (179) - Ground connection, in left headlight wiring harness
- (A3) - plus connection (58), in instrument panel wiring harness
- (A13) - Wire connection (door contact switch) in instrument panel wiring harness
- (A84) - Connector (58L), in instrument panel wiring harness
- (A85) - Connector (58R), in instrument panel wiring harness

ws = white
sw = black
ro = red
br = brown
gn = green
bl = blue
gr = grey
li = violet
ge = yellow
or = orange

New Beetle — Wiring diagram — No. 1/6

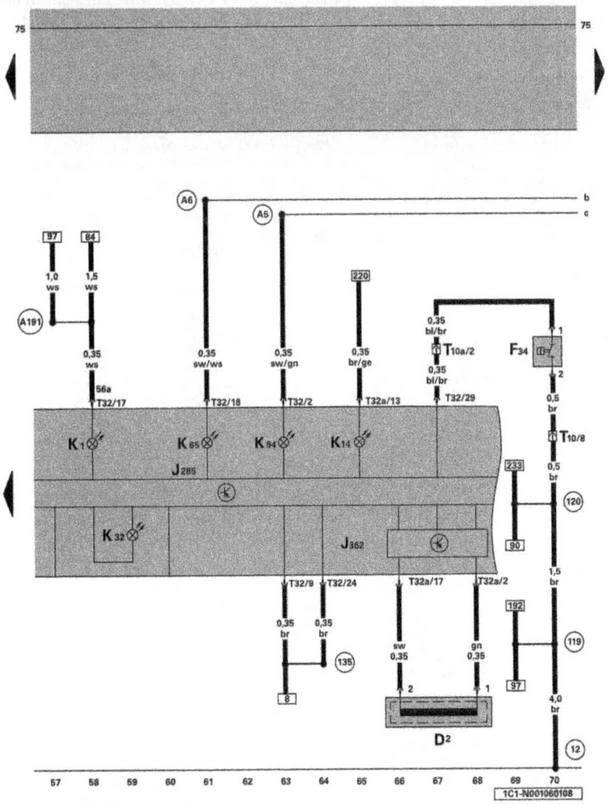

Instrument cluster, indicator lights, induction coil of anti-theft immobilizer, brake fluid level warning switch

- D2 - Induction coil of antitheft immobilizer
- F34 - Brake Fluid Level Warning Switch
- J285 - Control module with indicator unit in inst. panel insert
- J362 - Control module for anti-theft immobilizer
- K1 - Headlight High Beam Indicator Light
- K14 - Parking Brake Indicator Light
- K32 - Brake Pad Wear Indicator Light
- K65 - Left Turn Signal Indicator Light
- K94 - Right Turn Signal Indicator Light
- T10 - 10-Pin Connector, white, behind instrument panel, left
- T10a - 10-Pin Connector, orange, behind instrument panel, left
- T32 - 32-Pin Connector, blue, on instrument cluster
- T32a - 32-Pin Connector, green, on instrument cluster

- (12) - Ground connection, in engine compartment, left
- (119) - Ground connection -1-, in headlight wiring harness
- (120) - Ground connection -2-, in headlight wiring harness
- (135) - Ground connection -2-, instrument panel wiring harness
- (A5) - plus connection (right turn signal). in instrument panel wiring harness
- (A6) - plus connection (left turn signal), in instrument panel wiring harness
- (A191) - Connection (56aR) (in instrument panel wiring harness

ws = white
sw = black
ro = red
br = brown
gn = green
bl = blue
gr = grey
li = violet
ge = yellow
or = orange

New Beetle — Wiring diagram — No. 1/7

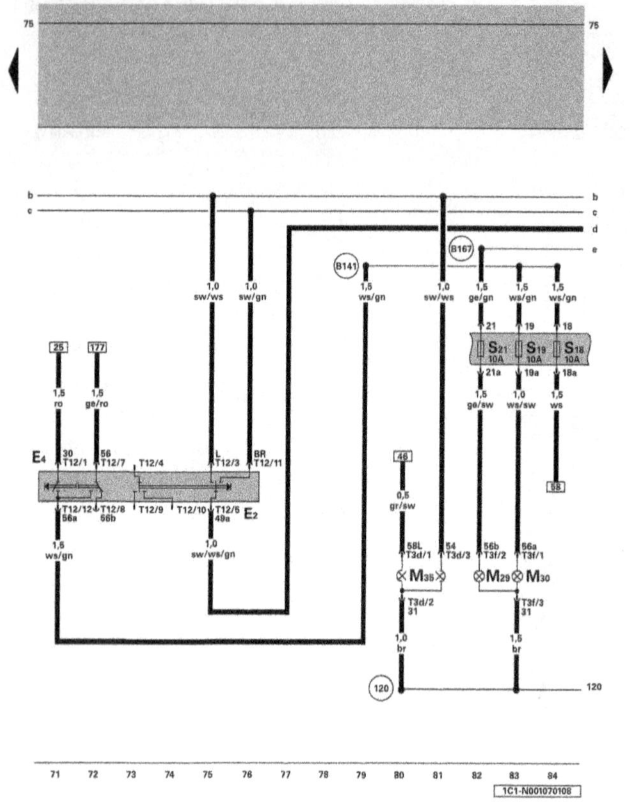

Turn signal switch, headlight dimmer/flasher switch, Left Low Beam Headlight, Left High Beam Headlight, Light for turn signal and side marker, front, left

- E2 - Turn Signal Switch
- E4 - Headlight Dimmer/Flasher Switch
- M29 - Left Low Beam Headlight
- M30 - Left High Beam Headlight
- M35 - Light for turn signal and side marker, front, left
- S18 - Fuse 18 in fuse holder
- S19 - Fuse 19 in fuse holder
- S21 - Fuse 21 in fuse holder
- T3d - 3-Pin Connector, on light for turn signal and side marker, front, left
- T3f - 3-Pin Connector, on headlight left
- T12 - 12-Pin Connector, near steering column
- (120) - Ground connection -2-, in headlight wiring harness
- (B141) - Plus connector -1- (56a), in wiring harness interior
- (B167) - Connection (56b) in passenger compartment wiring harness

ws = white
sw = black
ro = red
br = brown
gn = green
bl = blue
gr = grey
li = violet
ge = yellow
or = orange

New Beetle — Wiring diagram — No. 1/8

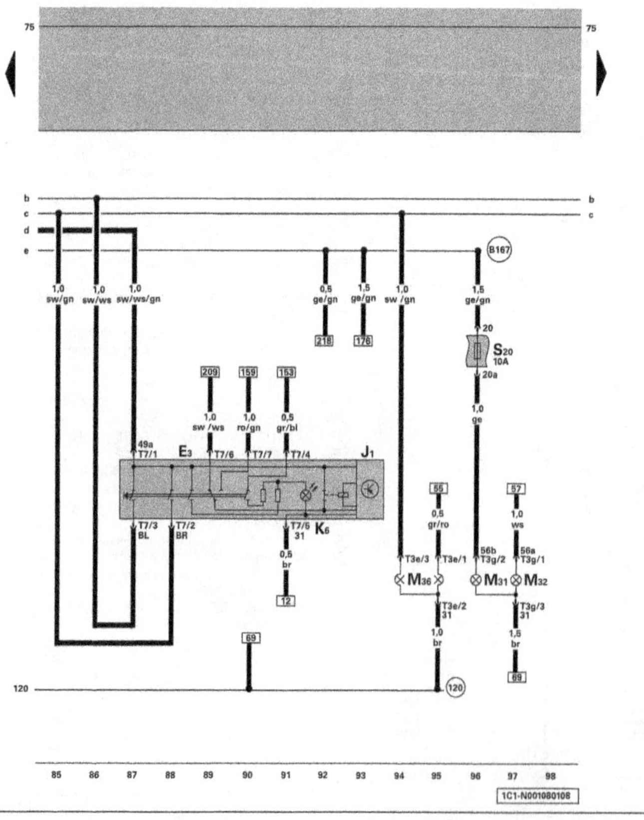

Emergency flasher switch with turn signal relay, right low and high beam headlights, light for turn signal and side marker, front, right

- E3 - Emergency Flasher Switch
- J1 - Turn Signal Relay
- K6 - Emergency Flasher Warning Light
- M31 - Right Low Beam Headlight
- M32 - Right High Beam Headlight
- M36 - Light for turn signal and side marker, front, right
- S20 - Fuse 20 in fuse holder
- T3e - 3-Pin Connector
- T3g - 3-Pin Connector
- T7 - 7-Pin Connector
- (120) - Ground connection -2-, in headlight wiring harness
- (B167) - Connection (56b) in passenger compartment wiring harness

ws = white
sw = black
ro = red
br = brown
gn = green
bl = blue
gr = grey
li = violet
ge = yellow
or = orange

New Beetle — Wiring diagram — No. 1/9

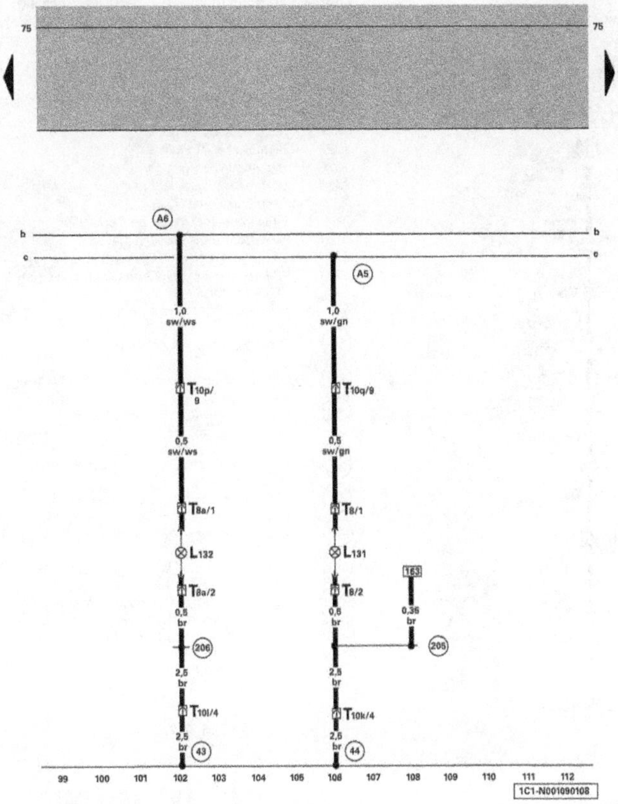

Mirror Turn Signal Lamps

- L131 - Left Mirror Turn Signal Lamp
- L132 - Right Mirror Turn Signal Lamp
- T8 - 8-Pin Connector, at L131
- T8a - 8-Pin Connector, at L132
- T10k - 10-Pin Connector, black, on a Pillar left
- T10 l - 10-Pin Connector, black, on a Pillar right
- T10p - 10-Pin Connector, blue, on a Pillar left
- T10q - 10-Pin Connector, blue, on a Pillar right
- (43) - Ground connection, on right A-pillar, lower part
- (44) - Ground connection, on left A-pillar, lower part
- (205) - Ground connection, in driver's door wiring harness
- (206) - Ground connection, in passenger's door wiring harness
- (A5) - Plus connection (right turn signal), in instrument panel wiring harness
- (A6) - Plus connection (left turn signal), in instrument panel wiring harness

ws = white
sw = black
ro = red
br = brown
gn = green
bl = blue
gr = grey
li = violet
ge = yellow
or = orange

New Beetle — Wiring diagram — No. 1/10

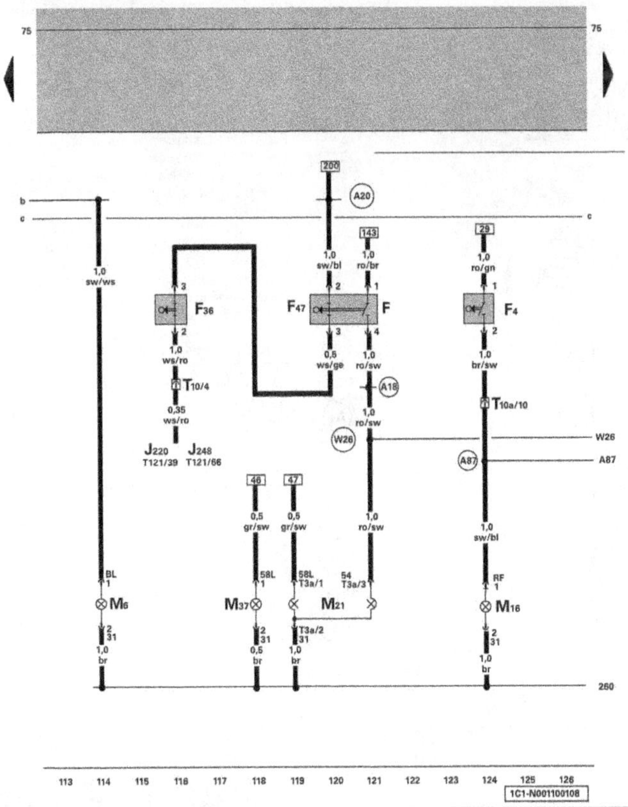

Brake Light Switch, Back-Up light Switch, Brake Pedal Switch (cruise control/Diesel Direct Fuel Injection, Clutch Vacuum Vent Valve Switch

- F - Brake Light Switch
- F4 - Back-Up light Switch
- F36 - Clutch Vacuum Vent Valve Switch
- F 47 - Brake Pedal Switch (cruise control/Diesel Direct Fuel Injection
- J220 - Motronic Engine Control Module (ECM)
- J248 - Diesel Direct Fuel Injection (DFI) Engine Control Module (ECM)
- M6 - Left Rear Turn Signal Light
- M16 - Left Back-Up Light
- M21 - Left Brake/Tail Light
- M37 - Side marker light, left rear
- T10a - 10-Pin Connector, orange, behind instrument panel, left
- T121 - 121-Pin Connector, on Motronic/Diesel (DFI) ECM
- (A18) - Wire connection (54), in instrument panel wiring harness
- (A20) - Wire connection (15a) in instrument panel wiring harness
- (A87) - Connector (reverse lamp) in instrument panel wiring harness
- (W26) - Instrument Panel Reading Light

ws = white
sw = black
ro = red
br = brown
gn = green
bl = blue
gr = grey
li = violet
ge = yellow
or = orange

New Beetle — Wiring diagram — No. 1/11

Luggage compartment light, right rear turn signal light, right back-up light, right brake/tail light, side marker light, right, rear

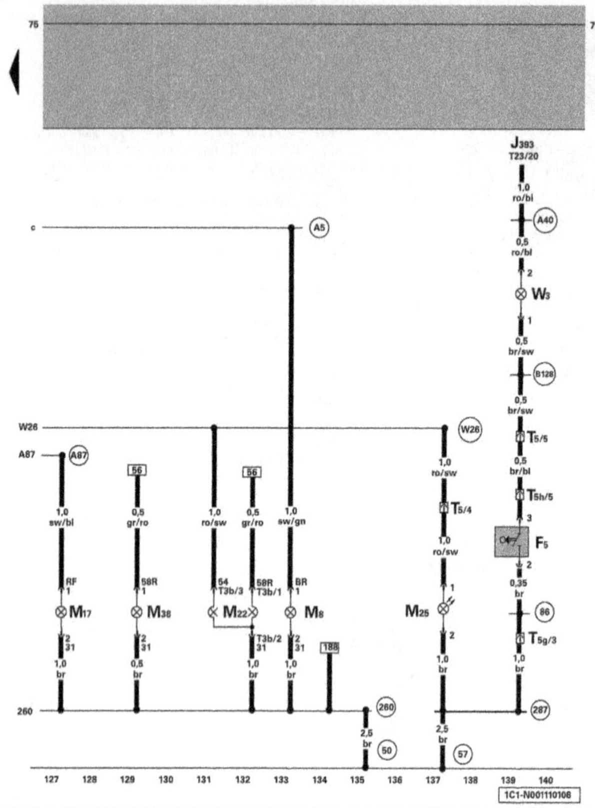

- F5 - Luggage Compartment Light Switch
- J393 - Central control module for comfort system
- M8 - Right Rear Turn Signal Light
- M17 - Right Back-Up Light
- M22 - Right Brake/Tail Light
- M25 - High-mount Brake Light
- M38 - Side marker light, right rear
- T3b - 3-Pin Connector
- T5 - 5-Pin Connector, black, on C-pillar, left
- T5g - 5-Pin Connector, black, in rear lid
- T5h - 5 Pin Connector, brown, in rear lid
- T23 - 23-Pin Connector, on Central control module for comfort system.
- W3 - Luggage compartment Light
- (57) - Ground connection, on left rear pillar
- (86) - Ground connection -1-, in rear wiring harness
- (260) - Ground Connection, in wiring harness taillights
- (287) - Ground connection, in wiring harness rear hatch hydraulics
- (A5) - Plus connection (right turn signal), in instrument panel wiring harness
- (A40) - Plus connection -1- (30), in instrument panel wiring harness
- (A87) - Connector (reverse lamp), in instrument panel wiring harness
- (B128) - Connector (Luggage compartment lamp) in wiring harness interior
- (W26) - Instrument Panel Reading Light

ws = white
sw = black
ro = red
br = brown
gn = green
bl = blue
gr = grey
li = violet
ge = yellow
or = orange

New Beetle — Wiring diagram — No. 1/12

Data Link Connector (DLC), radio connection

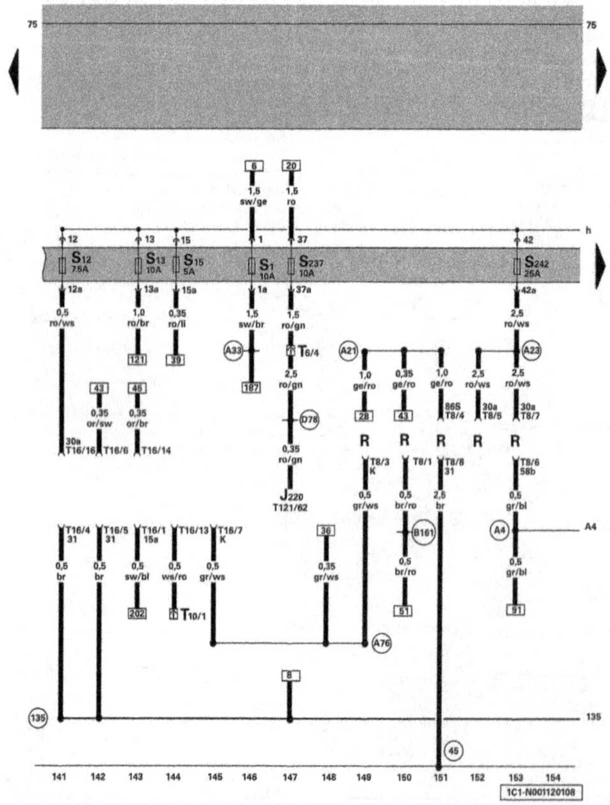

- J220 - Motronic Engine Control Module (ECM)
- R - Radio
- S1 - Fuse 1 in fuse holder
- S12 - Fuse 12 in fuse holder
- S13 - Fuse 13 in fuse holder
- S15 - Fuse 15 in fuse holder
- S237 - Fuse 37 in fuse holder
- S242 - Fuse 42 in fuse holder
- T6 - 6-Pin Connector, brown, behind instrument panel, left
- T8 - 8-Pin Connector, on radio
- T10 - 10-Pin Connector, white, behind instrument panel, left
- T16 - 16-Pin Connector, Data Link Connector (DLC), bellow steering column, left
- T121 - 121-Pin Connector, at J220
- (45) - Ground connection, behind instrument panel, center
- (135) - Ground connection -2-, in instrument panel wiring harness
- (A4) - plus connection (58b) in instrument panel wiring harness
- (A21) - wire connection (86s), in instrument panel wiring harness
- (A23) - wire connection (30al), in instrument panel wiring harness
- (A33) - wire connection (75), in instrument panel wiring harness
- (A76) - Connector (K-diagnosis wire), in instrument panel wiring harness
- (B161) - Connector (anti-theft warning system), in wiring harness interior
- (D78) - plus connection -1- (30a), in engine compartment wiring harness

ws = white
sw = black
ro = red
br = brown
gn = green
bl = blue
gr = grey
li = violet
ge = yellow
or = orange

New Beetle — Wiring diagram — No. 1/13

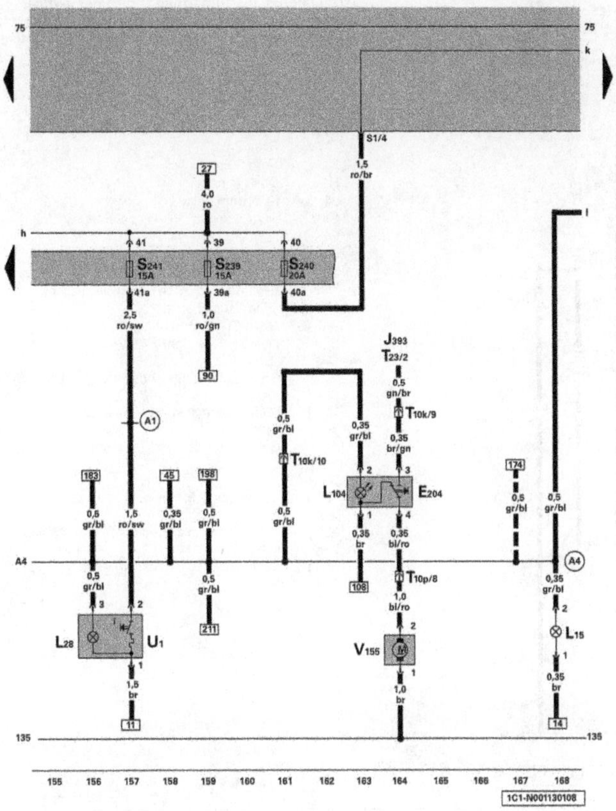

Ashtray light, cigarette lighter, cigarette lighter light, motor for fuel tank lid unlock, switch for remote/fuel tank door

- E204 - Switch for remote/fuel tank door
- J393 - Central control module for comfort system
- L15 - Ashtray Light
- L28 - Cigarette Lighter Light
- L104 - Remote Unlock Switch Illumination
- S239 - Fuse 39 in fuse holder
- S240 - Fuse 40 in fuse holder
- S241 - Fuse 41 in fuse holder
- T10k - 10-pin connector, black, on A-pillar, left
- T10p - 10-pin connector, blue, on A-pillar, left
- T23 - 23-pin connector, at J393
- U1 - Cigarette Lighter
- V155 - Motor for fuel tank lid unlock
- (A1) - plus connection (30a), in instrument panel wiring harness
- (A4) - plus connection (58b) in instrument panel wiring harness
- --- - Automatic transmission only

ws = white
sw = black
ro = red
br = brown
gn = green
bl = blue
gr = grey
li = violet
ge = yellow
or = orange

New Beetle — Wiring diagram — No. 1/14

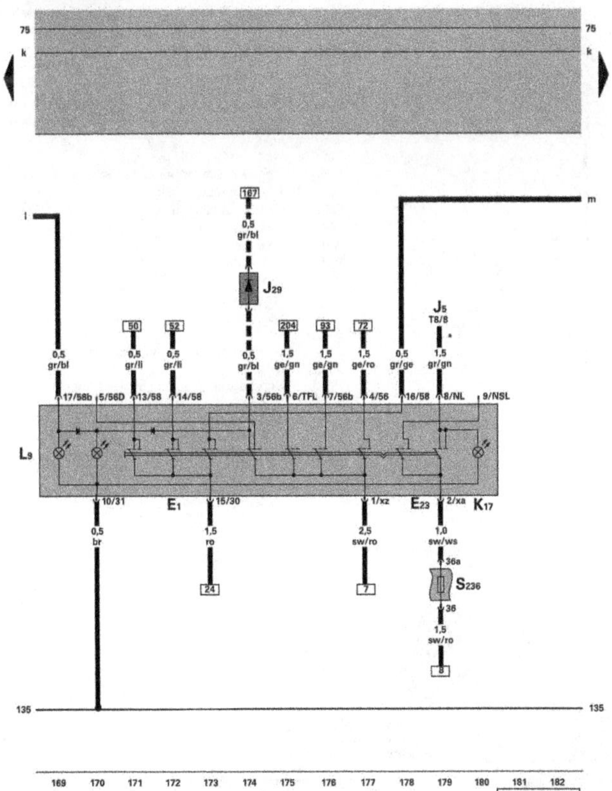

Light switch, left rear fog light

- E1 - Light switch
- E23 - Fog Light Switch
- J5 - Fog Light Relay
- J29 - Blocking diode
- K17 - Fog Light Indicator Light
- L9 - Headlight Switch Licht
- S236 - Fuse 36 in fuse holder
- T8 - 8-Pin Connector at J5
- * - Only vehicles with fog lights
- --- - Automatic transmission only

ws = white
sw = black
ro = red
br = brown
gn = green
bl = blue
gr = grey
li = violet
ge = yellow
or = orange

New Beetle — Wiring diagram — No. 1/15

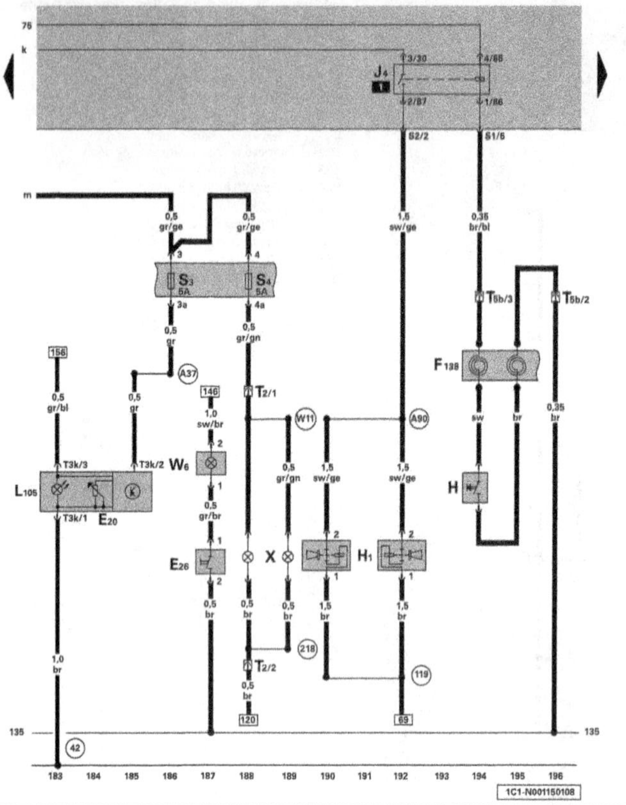

Dual horns, instrument light dimmer switch, glove compartment light, license plate light, Glove compartment Switch, Illumination for lighting controller

- E20 - Instrument Panel Light Dimmer Switch
- E 26 - Glove compartment Switch
- F138 - Airbag Spiral Spring/Return Spring With Slip Ring
- H - Horn Button
- H1 - Dual Horns
- J4 - Dual Horn Relay
- L105 - Illumination for lighting controller
- S3 - Fuse 3 in fuse holder
- S4 - Fuse 4 in fuse holder
- T2 - Double Connector, behind rear bumper cover
- T5b - 5-Pin Connector, on airbag spiral spring
- T3k - 3-Pin Connector, at E20
- W6 - Glove Compartment Light
- X - License Plate Light
- (42) - Ground connection, beside steering column
- (119) - Ground connection -1-, in headlight wiring harness
- (218) - Ground connection -1-, in rear lid wiring harness
- (A37) - Connection (58a), in instrument panel wiring harness
- (A90) - Connection, (dual tone horn), in instrument panel wiring harness
- (W11) - Connection (58), in rear lid wiring harness

ws = white
sw = black
ro = red
br = brown
gn = green
bl = blue
gr = grey
li = violet
ge = yellow
or = orange

New Beetle — Wiring diagram — No. 1/16

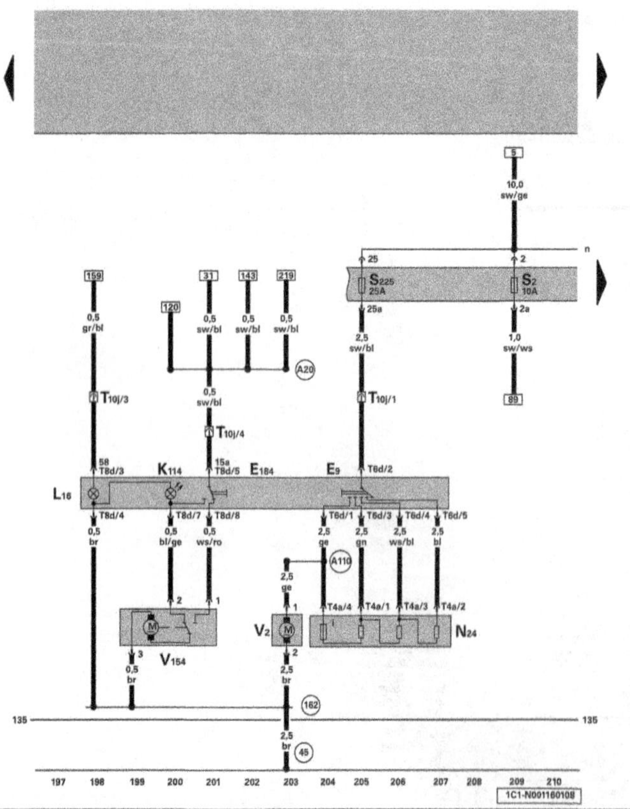

Switch for fresh air blower and recirculated air, Fresh Air Blower Series Resistance With Fus, Fresh Air Blower, Servo motor for fresh-/recirculating air door

- E9 - Fresh Air Blower Switch
- E184 - Switch for fresh air blower and recirculated air
- K114 - Fresh Air and Recirculating Air Mode Indicator Light
- L16 - Fresh Air Control Lever Light
- N24 - Fresh Air Blower Series Resistance With Fuse
- S2 - Fuse 2 in fuse holder
- S225 - Fuse 25 in fuse holder
- T4a - 4-Pin Connector
- T6d - 6-Pin Connector
- T8d - 8-Pin Connector
- T10j - 10-Pin Connector
- V2 - Fresh Air Blower
- V154 - Servo motor for fresh-/recirculating air door
- (45) - Ground connection, behind instrument panel, center
- (162) - Ground connection, in blower motor wiring harness
- (A20) - Plus Connection (15a) (in instrument panel wiring harness)
- (A110) - Connector (fresh air blower), in instrument panel wiring harness

ws = white
sw = black
ro = red
br = brown
gn = green
bl = blue
gr = grey
li = violet
ge = yellow
or = orange

New Beetle — Wiring diagram — No. 1/17

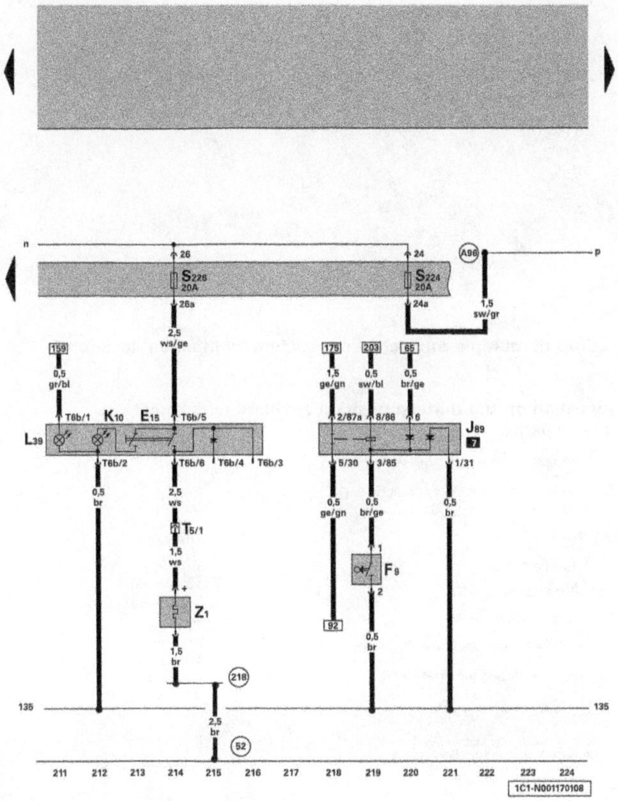

Heated rear window, rear window defogger switch, parking brake warning light switch, Ddytime running lights change-over relay,

- E15 - Rear window defogger switch
- F9 - Parking Brake Warning Light Switch
- J89 - Daytime Running Lights Change-over Relay (173), on the thirteen position relay panel
- K10 - Rear Window Defogger Indicator Light
- L39 - Rear Window Defogger Switch Light
- S224 - Fuse 24 in fuse holder
- S226 - Fuse 26 in fuse holder
- T5 - 5-Pin Connector, black, on C-pillar, left
- T6b - 6-Pin Connector
- Z1 - Heated rear window
- (52) - Ground connection, in rear lid, left
- (218) - Ground connection -1-, in rear lid wiring harness
- (A96) - Connector (53a), in instrument panel wiring harness

ws = white
sw = black
ro = red
br = brown
gn = green
bl = blue
gr = grey
li = violet
ge = yellow
or = orange

New Beetle — Wiring diagram — No. 1/18

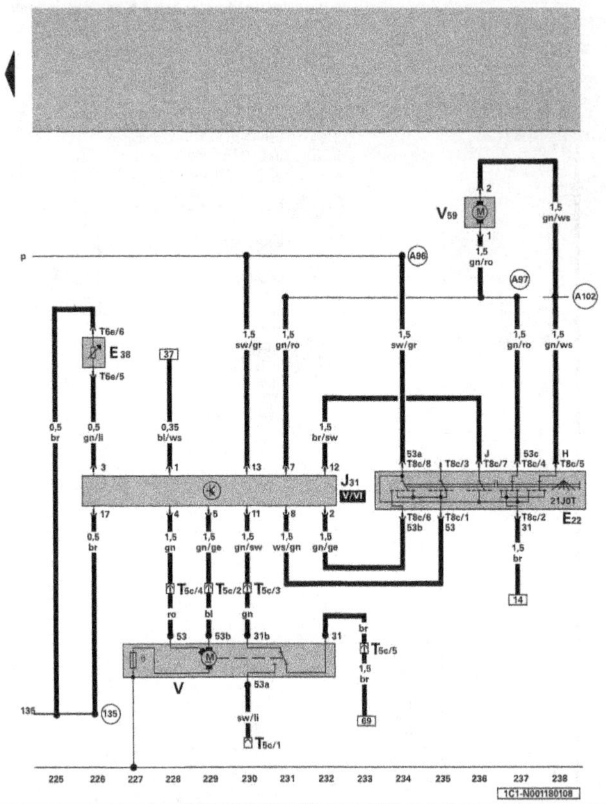

Windshield wiper/washer switch, windshield wiper motor, windshield wiper/washer intermittent relay, Windshield and Rear Window Washer Pump

- E22 - Winshield Wiper/Washer Switch
- E38 - Windshield Wiper Intermittent Regulator
- J31 - Wiper/Washer Intermittent Relay, on the thirteen position relay panel, production control number (377)
- T5c - 5-Pin Connector, on Windshield Wiper Motor
- T6e - 6-Pin Connector, near steering column
- T8c - 8-Pin Connector
- V - Windshield Wiper Motor
- V59 - Windshield and Rear Window Washer Pump
- (135) - Ground connection -2-, in instrument panel wiring harness
- (A96) - Connector (53a), in instrument panel wiring harness
- (A97) - Connector (53), in instrument panel wiring harness
- (A102) Connector (windshield wiper), in instrument panel wiring harness

ws = white
sw = black
ro = red
br = brown
gn = green
bl = blue
gr = grey
li = violet
ge = yellow
or = orange

New Beetle Convertible — Wiring diagram — No. 1/1

Standard Equipment,

from August 2002

Deviate relay location and fuseplacements as well as the locations of multiple connectors see section "component locations".

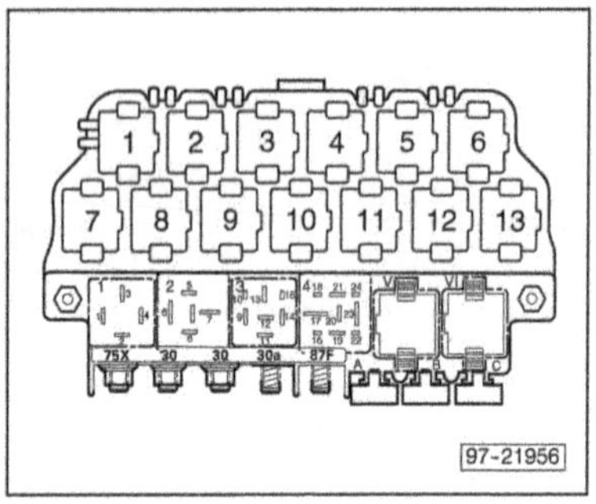

Relay location on the thirteen position auxiliary relay panel, above relay panel:

- **3** Lock relay for starter (clutch pedal switch) (53)
- **7** Daytime Running Lights Change-over Relay (173)

Relay panel:

- **1** Dual Horn Relay (53)
- **2** Load Reduction Relay (100)
- **4** Fuel Pump (FP) Relay (409)
- **V** Wiper/Washer Intermittent Relay (377)
- **VI** Wiper/Washer Intermittent Relay (377)

Fuse colors

30 A - green
25 A - white
20 A - yellow
15 A - blue
10 A - red
7.5 A - brown
5 A - beige
3 A - violet

Note: Number in parentheses indicates production control number stamped on relay housing.

New Beetle Convertible — Wiring diagram — No. 1/2

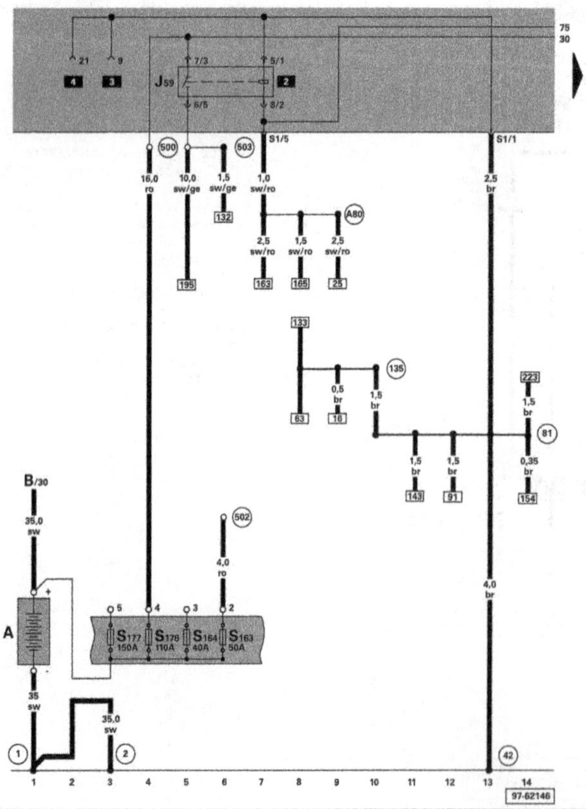

Battery, load reduction relay, fuses in fuse bracket/battery

A - Battery
B - Starter
J59 - Load Reduction Relay
S163 - Fuse -2- in fuse bracket/battery
S164 - Fuse -3- in fuse bracket/battery
S176 - Fuse -4- (30), in fuse bracket/battery
S177 - Fuse -5- (30), in fuse bracket/battery

(1) - Ground strap, battery to body
(2) - Ground strap, transmission to body
(42) - Ground connection, beside steering column
(81) - Ground connection -1-, in instrument panel wiring harness
(135) - Ground connection -2-, in instrument panel wiring harness
(500) - Threaded connection -1- (30) on the relay plate
(502) - Threaded connection -1- (30a) on the relay plate
(503) - Threaded connection -1- (75x) on the relay plate
(A80) - Connector -1- (X), in instrument panel wiring harness

ws = white
sw = black
ro = red
br = brown
gn = green
bl = blue
gr = grey
li = violet
ge = yellow
or = orange

New Beetle Convertible

Wiring diagram **No. 1/3**

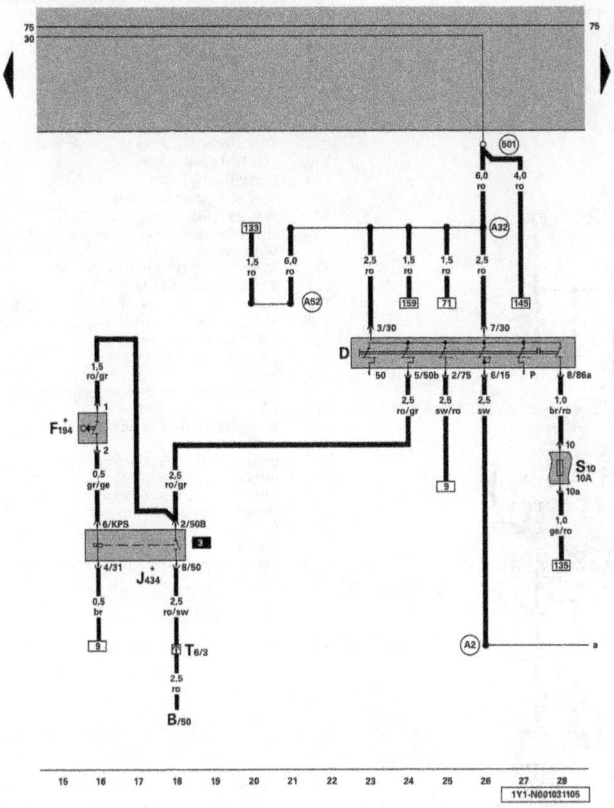

Ignition/starter switch, clutch pedal position (CPP) switch, locking relay for starter (clutch pedal switch)

- B - Starter
- D - Ignition/Starter Switch
- F194 - Clutch Pedal Position (CPP) Switch
- J434 - Locking relay for starter (clutch pedal switch)
- S10 - Fuse 10 in fuse holder
- T6 - 6-Pin Connector, brown, behind instrument panel, left

- (501) - Threaded connection -2- (30) on the relay plate
- (A2) - plus connection (15), in instrument panel wiring harness
- (A32) - plus connection (30), in instrument panel wiring harness
- (A52) - Plus Connection 2 (30) (in instrument panel wiring harness)

- * - Manual transmission only

ws = white
sw = black
ro = red
br = brown
gn = green
bl = blue
gr = grey
li = violet
ge = yellow
or = orange

New Beetle Convertible

Wiring diagram **No. 1/4**

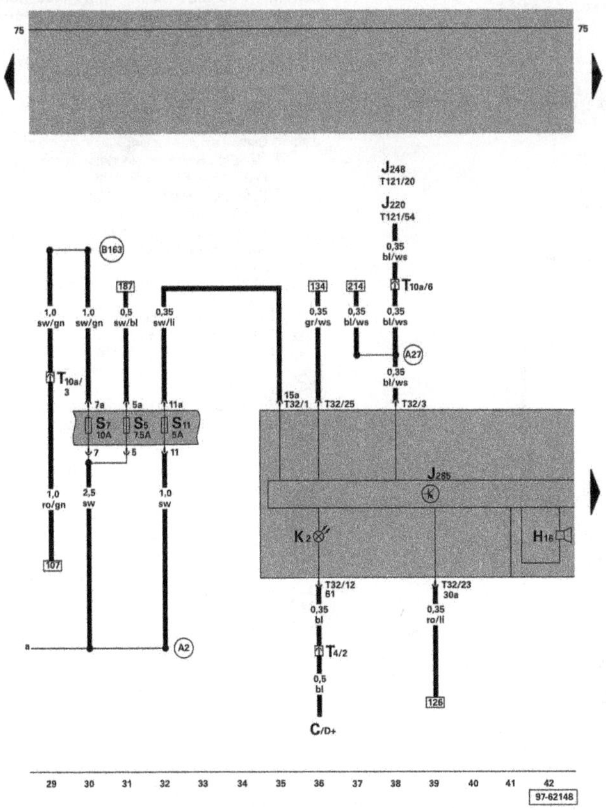

Instrument cluster, warning buzzer for light swiched on, generator (GEN) warning light

- C - Generator (GEN)
- H16 - Warning buzzer for light switched on
- J220 - Motronic Engine Control Module (ECM)
- J248 - Diesel Direct Fuel Injection (DFI) Engine Control Module (ECM)
- J285 - Control module with indicator unit in instrument panel insert
- K2 - Generator (GEN) Warning Light
- S5 - Fuse 5 in fuse holder
- S7 - Fuse 7 in fuse holder
- S11 - Fuse 11 in fuse holder
- T4 - 4-Pin Connector, in engine compartment left
- T6 - 6-Pin Connector, brown, behind instrument panel, left
- T10d - 10-Pin Connector, green, behind instrument panel, left
- T10a - 10-Pin Connector, orange, behind instrument panel, left
- T32 - 32-Pin Connector, blue, on instrument cluster
- T121 - 121-Pin Connector, on Motronic/Diesel DFI Control Module

- (A2) - Plus connection (15), in instrument panel wiring harness
- (A27) - Connection (vehicle speed signal), in instrument panel wiring harness
- (B163) - Plus connector -1- (15) in wiring harness interior

ws = white
sw = black
ro = red
br = brown
gn = green
bl = blue
gr = grey
li = violet
ge = yellow
or = orange

Standard equipment from Aug. 2002, convertible

New Beetle Convertible — Wiring diagram — No. 1/5

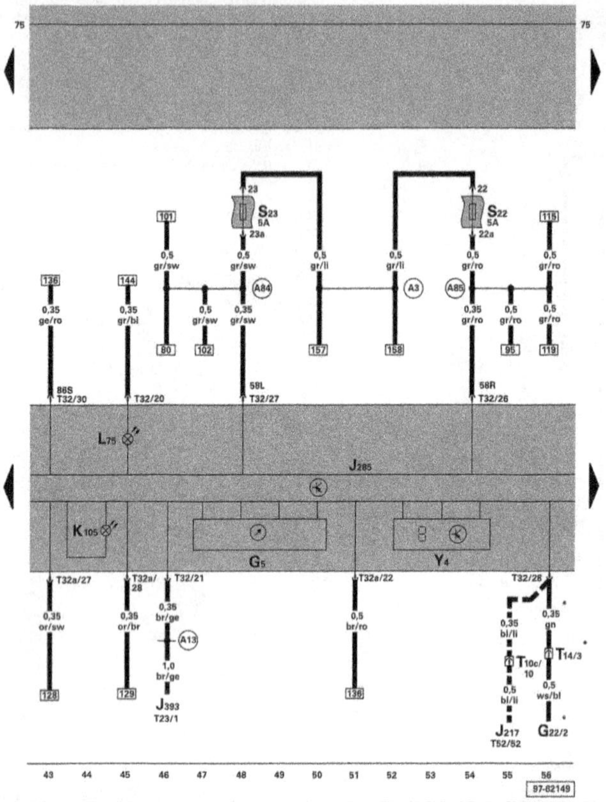

Instrument cluster, tachometer, low fuel level warning light, odometer display

- G5 - Tachometer
- G22 - Speedometer Vehicle Speed Sensor (VSS)
- J217 - Transmission Control Module (TCM)
- J285 - Control module with indicator unit in instrument panel insert
- J393 - Central control module for comfort system
- K105 - Low Fuel Level Warning Light
- L75 - Digital Display Light
- S22 - Fuse 22 in fuse holder
- S23 - Fuse 23 in fuse holder
- T10c - 10-pin Connector, blue, behind instrument panel, left.
- T14 - 14-Pin Connector, in plenum chamber
- T23 - 23-Pin Connector, at J393
- T32 - 32-Pin Connector, blue, on instrument cluster
- T32a - 32-Pin Connector, green, on instrument cluster
- T52 - 52-Pin Connector
- Y4 - Odometer Display
- (12) - Ground connection, in engine compartment, left
- (179) - Ground connection, in left headlight wiring harness
- (A3) - plus connection (58), in instrument panel wiring harness
- (A13) - Wire connection (door contact switch) in instrument panel wiring harness
- (A84) - Connector (58L), in instrument panel wiring harness
- (A85) - Connector (58R), in instrument panel wiring harness
- • - Manual transmission only
- - - - - Automatic transmission only

ws = white
sw = black
ro = red
br = brown
gn = green
bl = blue
gr = grey
li = violet
ge = yellow
or = orange

Edition 11.03

New Beetle Convertible — Wiring diagram — No. 1/6

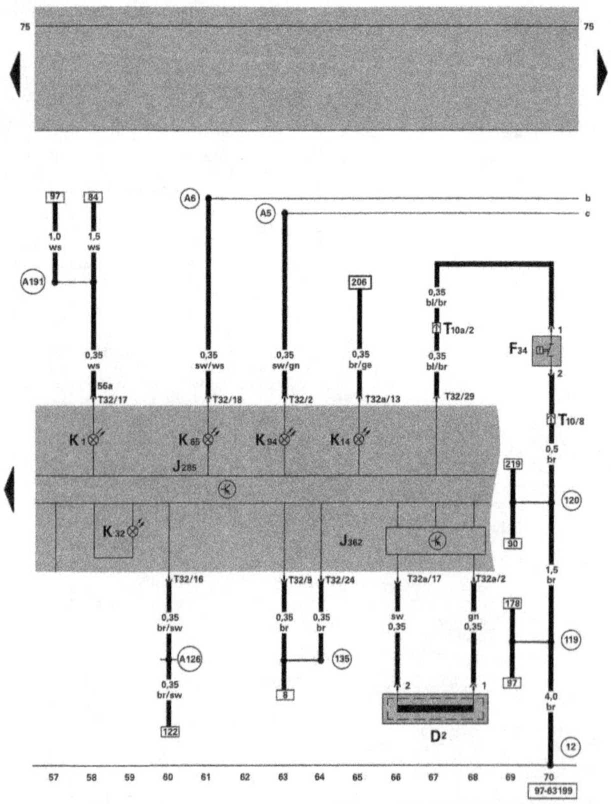

Instrument cluster, indicator lights, induction coil of anti-theft immobilizer, brake fluid level warning switch

- D2 - Induction coil of antitheft immobilizer
- F34 - Brake Fluid Level Warning Switch
- J285 - Control module with indicator unit in inst. panel insert
- J362 - Control module for anti-theft immobilizer
- K1 - Headlight High Beam Indicator Light
- K14 - Parking Brake Indicator Light
- K32 - Brake Pad Wear Indicator Light
- K65 - Left Turn Signal Indicator Light
- K94 - Right Turn Signal Indicator Light
- T10 - 10-Pin Connector, white, behind instrument panel, left
- T10a - 10-Pin Connector, orange, behind instrument panel, left
- T32 - 32-Pin Connector, blue, on instrument cluster
- T32a - 32-Pin Connector, green, on instrument cluster
- (12) - Ground connection, in engine compartment, left
- (119) - Ground connection -1-, in headlight wiring harness
- (120) - Ground connection -2-, in headlight wiring harness
- (135) - Ground connection -2-, instrument panel wiring harness
- (A5) - plus connection (right turn signal), in instrument panel wiring harness
- (A6) - plus connection (left turn signal), in instrument panel wiring harness
- (A126) - Connection (contact switch in rear lid) in instrument panel wiring harness
- (A191) - Connection (56aR) (in instrument panel wiring harness

ws = white
sw = black
ro = red
br = brown
gn = green
bl = blue
gr = grey
li = violet
ge = yellow
or = orange

Edition 04.04

New Beetle Convertible

Wiring diagram **No. 1/7**

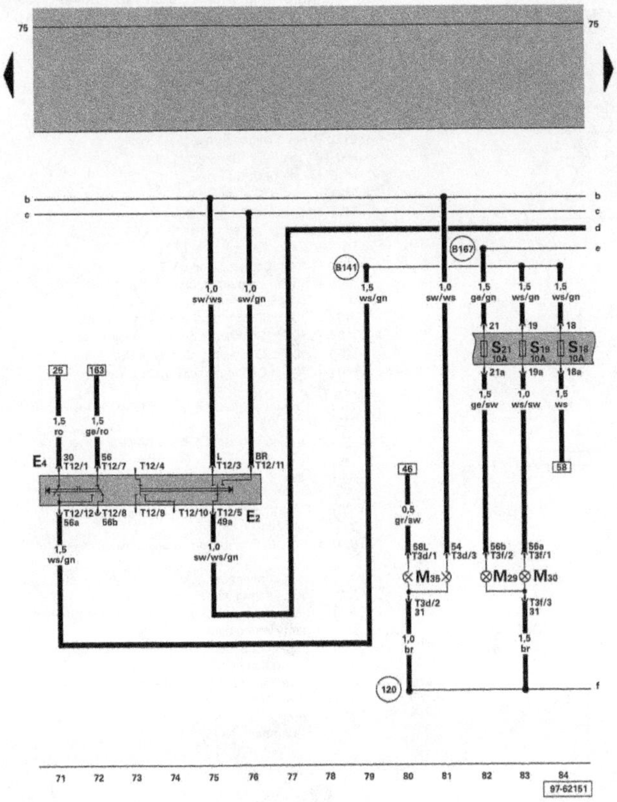

Turn signal switch, headlight dimmer/flasher switch, left low beam headlight, left high beam headlight, light for turn signal and side marker, front, left

- E2 - Turn Signal Switch
- E4 - Headlight Dimmer/Flasher Switch
- M29 - Left Low Beam Headlight
- M30 - Left High Beam Headlight
- M35 - Light for turn signal and side marker, front, left
- S18 - Fuse 18 in fuse holder
- S19 - Fuse 19 in fuse holder
- S21 - Fuse 21 in fuse holder
- T3d - 3-Pin Connector, on light for turn signal and side marker, front, left
- T3f - 3-Pin Connector, on headlight left
- T12 - 12-Pin Connector, near steering column

- (120) - Ground connection -2-, in headlight wiring harness
- (B141) - Plus connector -1- (56a), in wiring harness interior
- (B167) - Connection (56b) in passenger compartment wiring harness

ws = white
sw = black
ro = red
br = brown
gn = green
bl = blue
gr = grey
li = violet
ge = yellow
or = orange

New Beetle Convertible

Wiring diagram **No. 1/8**

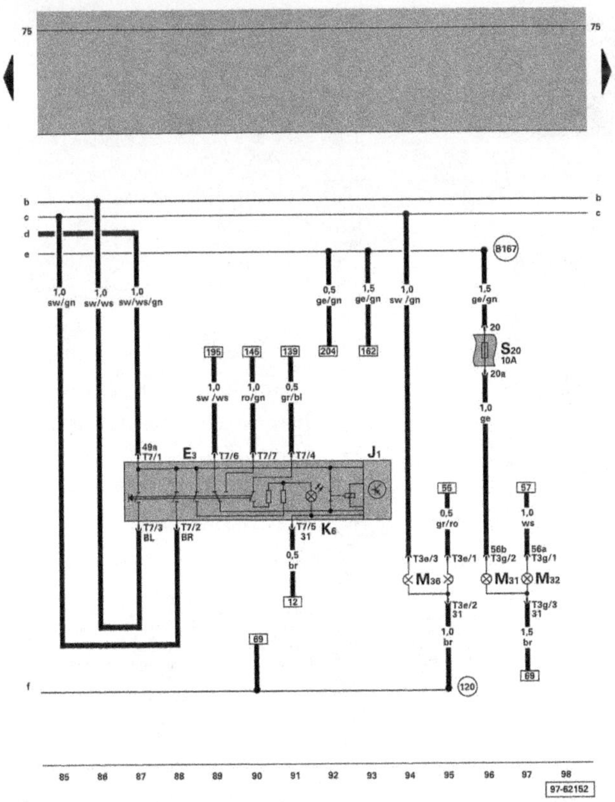

Emergency flasher switch with turn signal relay, right low and high beam headlights, light for turn signal and side marker, front, right

- E3 - Emergency Flasher Switch
- J1 - Turn Signal Relay
- K6 - Emergency Flasher Warning Light
- M31 - Right Low Beam Headlight
- M32 - Right High Beam Headlight
- M36 - Light for turn signal and side marker, front, right
- S20 - Fuse 20 in fuse holder
- T3e - 3-Pin Connector
- T3g - 3-Pin Connector
- T7 - 7-Pin Connector

- (120) - Ground connection -2-, in headlight wiring harness
- (B167) - Connection (56b) in passenger compartment wiring harness

ws = white
sw = black
ro = red
br = brown
gn = green
bl = blue
gr = grey
li = violet
ge = yellow
or = orange

New Beetle Convertible — Wiring diagram — No. 1/9

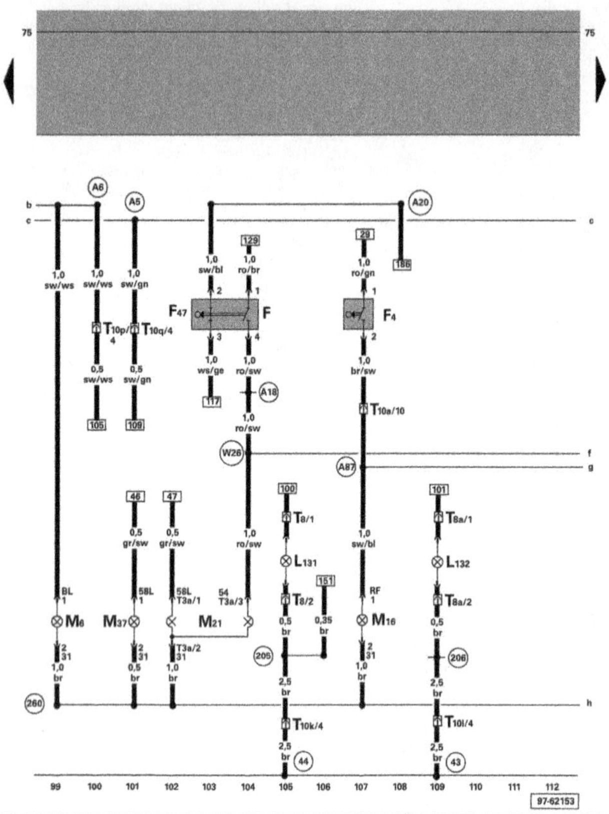

Brake light switch, brake/clutch vacuum vent valve/switch, back-up light switch, left back-up light, left brake/tail light, hight-mount brake light

- F - Brake Light Switch
- [F4] - Back-Up light Switch
- F 47 - Brake Pedal Switch (cruise control/Diesel Direct Fuel Injection
- L131 - Left Mirror Turn Signal Lamp
- L132 - Right Mirror Turn Signal Lamp
- M6 - Left Rear Turn Signal Light
- M16 - Left Back-Up Light
- M21 - Left Brake/Tail Light
- M37 - Side marker light, left rear
- T3a - 3-Pin Connector
- T8 - 8-Pin Connector, at L131
- T8a - 8-Pin Connector, at L132
- T10a - 10-Pin Connector, orange, behind instrument panel, left
- T10k - 10-Pin Connector, black, on a Pillar left
- T10 l - 10-Pin Connector, black, on a Pillar right
- T10p - 10-Pin Connector, blue, on a Pillar left
- T10q - 10-Pin Connector, blue, on a Pillar right
- (43) - Ground connection, on right A-pillar, lower part
- (44) - Ground connection, on left A-pillar, lower part
- (205) - Ground connection, in driver's door wiring harness
- (206) - Ground connection, in passenger's door wiring harness
- (260) - Ground Connection, in wiring harness taillights
- (A5) - Plus connection (right turn signal), in instrument panel wiring harness
- (A6) - Plus connection (left turn signal), in instrument panel wiring harness
- (A18) - Wire connection (54), in instrument panel wiring harness
- (A20) - Wire connection (15a) in instrument panel wiring harness
- (A87) - Connector (reverse lamp) in instrument panel wiring harness

ws = white
sw = black
ro = red
br = brown
gn = green
bl = blue
gr = grey
li = violet
ge = yellow
or = orange

New Beetle Convertible — Wiring diagram — No. 1/10

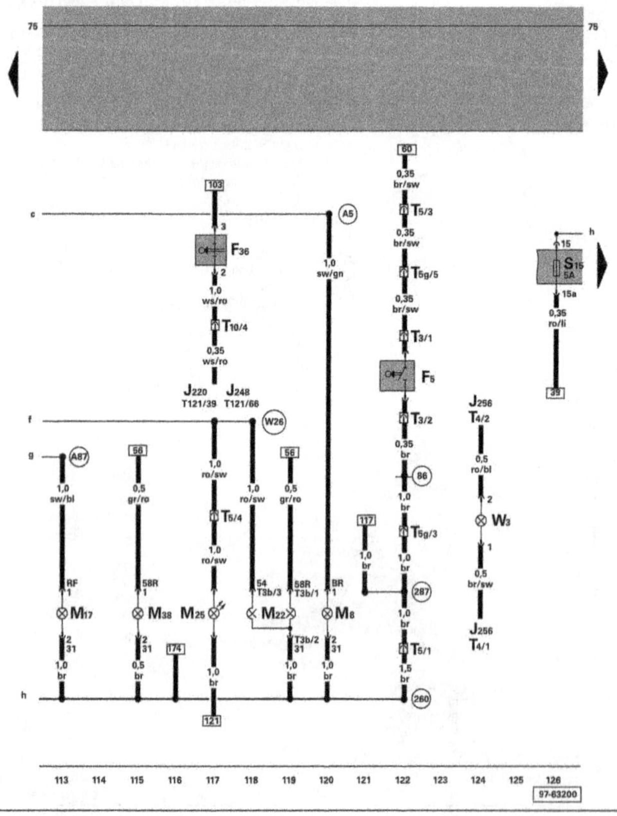

Luggage compartment light, right rear turn signal light, right back-up light, right brake/tail light, side marker light, right, rear

- [F5] - Luggage Compartment Light Switch
- [F36] - Clutch Vacuum Vent Valve Switch
- J220 - Motronic Engine Control Module (ECM)
- J248 - Diesel Direct Fuel Injection (DFI) Engine Control Module (ECM)
- [J256] - Convertible Top Control Module
- M8 - Right Rear Turn Signal Light
- M17 - Right Back-Up Light
- M22 - Right Brake/Tail Light
- M25 - High-mount Brake Light
- M38 - Side marker light, right rear
- S15 - Fuse 15 in fuse holder
- T3 - 3-Pin Connector, at F5
- T3b - 3-Pin Connector
- T4 - 4-Pin Connector, at J256
- T5 - 5-Pin Connector, black, on C-pillar, left
- T5g - 5-Pin Connector, black, in rear lid
- T10 - 10-Pin Connector, white, behind instrument panel, left
- T121 - 121-Pin Connector, on Motronic/Diesel (DFI) ECM
- W3 - Luggage compartment Light
- (86) - Ground connection -1-, in rear wiring harness
- (260) - Ground Connection, in wiring harness taillights
- (287) - Ground connection, in wiring harness rear hatch hydraulics
- (A5) - Plus connection (right turn signal), in instrument panel wiring harness
- (A87) - Connector (reverse lamp), in instrument panel wiring harness
- (B128) - Connector (Luggage compartment lamp) in wiring harness interior
- (W26) - Instrument Panel Reading Light

ws = white
sw = black
ro = red
br = brown
gn = green
bl = blue
gr = grey
li = violet
ge = yellow
or = orange

New Beetle Convertible — Wiring diagram — No. 1/11

Data Link Connector (DLC), radio connection

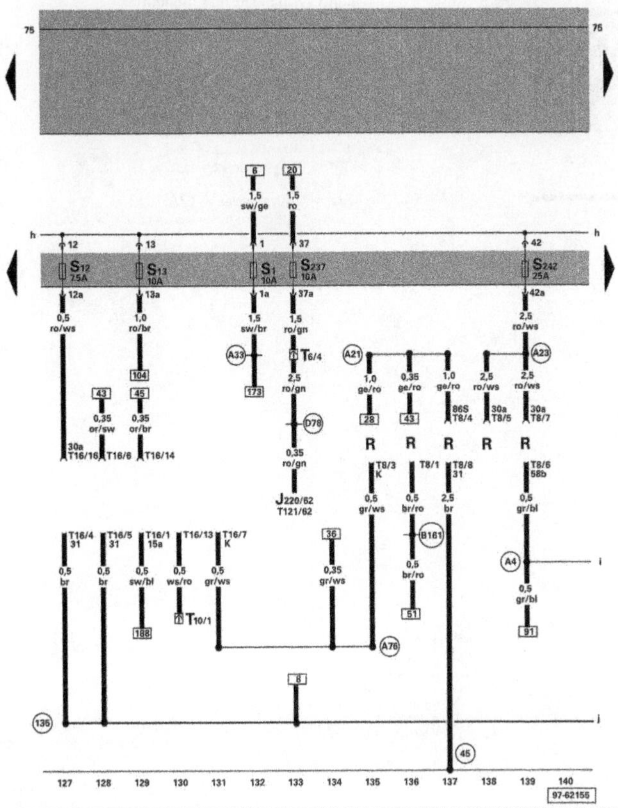

- J220 - Motronic Engine Control Module (ECM)
- R - Radio
- S1 - Fuse 1 in fuse holder
- S12 - Fuse 12 in fuse holder
- S13 - Fuse 13 in fuse holder
- S237 - Fuse 37 in fuse holder
- S242 - Fuse 42 in fuse holder
- T6 - 6-Pin Connector, brown, behind instrument panel, left
- T8 - 8-Pin Connector, on radio
- T10 - 10-Pin Connector, white, behind instrument panel, left
- T16 - 16-Pin Connector, Data Link Connector (DLC), bellow steering column, left
- T121 - 121-Pin Connector, at J220
- (45) - Ground connection, behind instrument panel, center
- (135) - Ground connection -2-, in instrument panel wiring harness
- (A4) - plus connection (58b) in instrument panel wiring harness
- (A21) - wire connection (86s), in instrument panel wiring harness
- (A23) - wire connection (30al), in instrument panel wiring harness
- (A33) - wire connection (75), in instrument panel wiring harness
- (A76) - Connector (K-diagnosis wire), in instrument panel wiring harness
- (B161) - Connector (anti-theft warning system), in wiring harness interior
- (D78) - plus connection -1- (30a), in engine compartment wiring harness

ws = white
sw = black
ro = red
br = brown
gn = green
bl = blue
gr = grey
li = violet
ge = yellow
or = orange

New Beetle Convertible — Wiring diagram — No. 1/12

Ashtray light, cigarette lighter, cigarette lighter light, motor for fuel tank lid unlock, switch for remote/fuel tank door

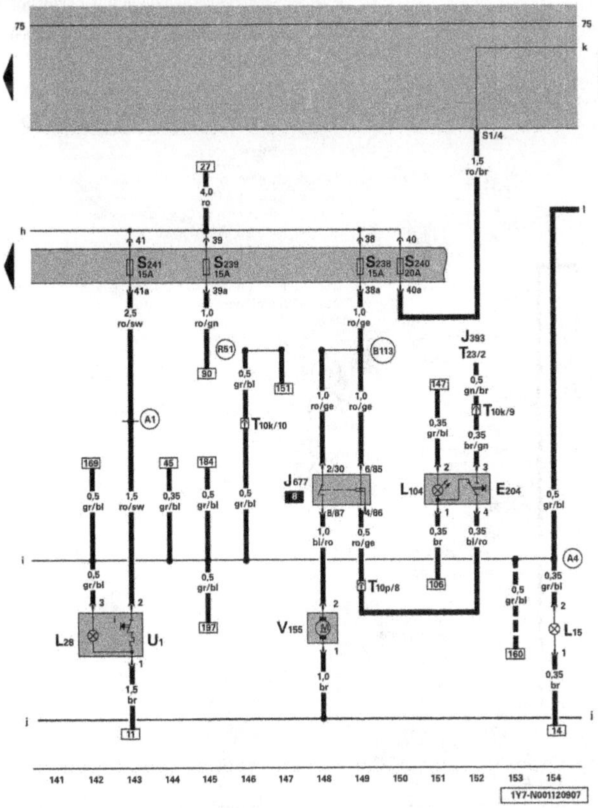

- E204 - Switch for remote/fuel tank door
- J393 - Central control module for comfort system
- J677 - Fuel Filler Lid Unlock Relay
- L15 - Ashtray Light
- L28 - Cigarette Lighter Light
- L104 - Remote Unlock Switch Illumination
- S238 - Fuse 38 in fuse holder
- S239 - Fuse 39 in fuse holder
- S240 - Fuse 40 in fuse holder
- S241 - Fuse 41 in fuse holder
- T10k - 10-pin connector, black, on A-pillar, left
- T10p - 10-pin connector, blue, on A-pillar, left
- T23 - 23-pin connector, at J393
- U1 - Cigarette Lighter
- V155 - Motor for fuel tank lid unlock
- (A1) - plus connection (30a), in instrument panel wiring harness
- (A4) - plus connection (58b) in instrument panel wiring harness
- (B113) - Plus connector (30a), in wiring harness interior, left
- (R10) - Plus connection -1- (30), in driver's door
- (R51) - Connection (58b) (driver's side door wiring harness)
- - - - Automatic transmission only

ws = white
sw = black
ro = red
br = brown
gn = green
bl = blue
gr = grey
li = violet
ge = yellow
or = orange

New Beetle Convertible

Wiring diagram **No. 1/13**

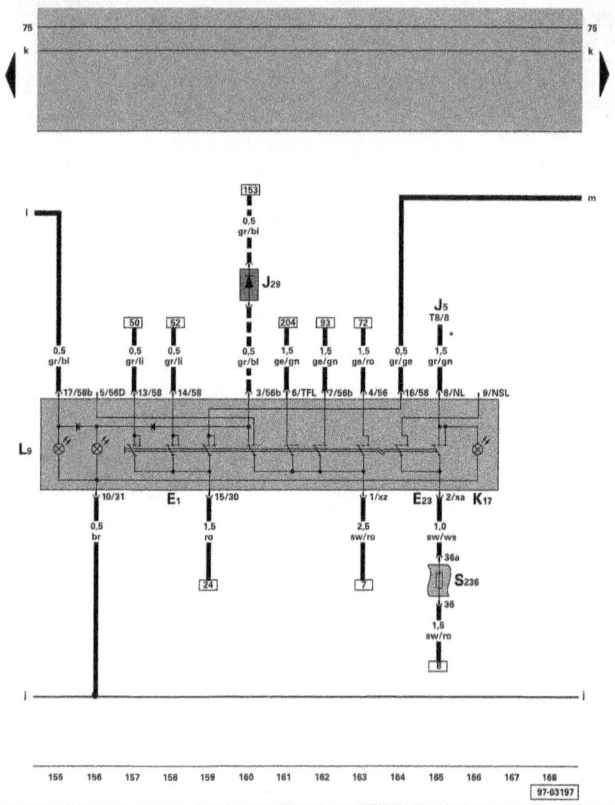

Light switch, left rear fog light

E1 - Light switch
E23 - Fog Light Switch
J5 - Fog Light Relay
J29 - Blocking diode
K17 - Fog Light Indicator Light
L9 - Headlight Switch Licht
S236 - Fuse 36 in fuse holder
T8 - 8-Pin Connector at J5

* - Only vehicles with fog lights
--- - Automatic transmission only

ws = white
sw = black
ro = red
br = brown
gn = green
bl = blue
gr = grey
li = violet
ge = yellow
or = orange

Edition 04.04

New Beetle Convertible

Wiring diagram **No. 1/14**

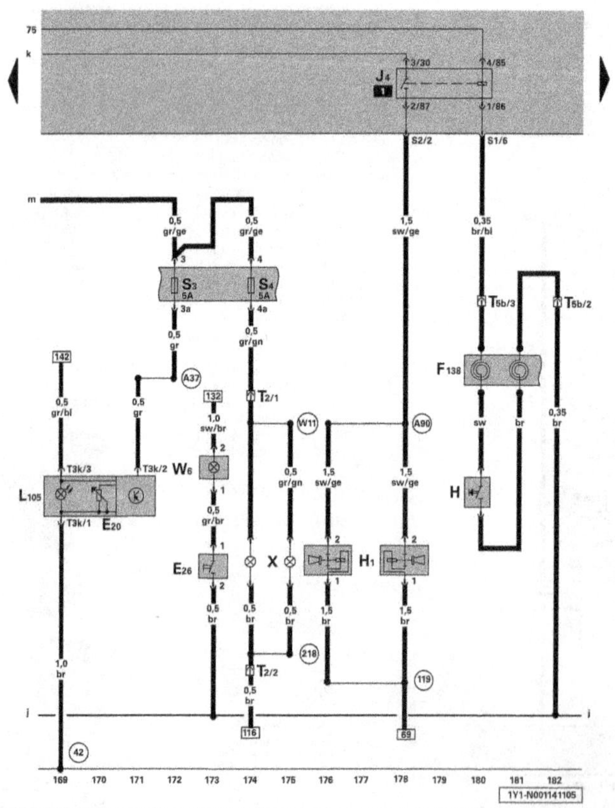

Dual horns, instrument light dimmer switch, glove compartment light, license plate light, Glove compartment Switch, Illumination for lighting controller

E20 - Instrument Panel Light Dimmer Switch
E 26 - Glove compartment Switch
F138 - Airbag Spiral Spring/Return Spring With Slip Ring
H - Horn Button
H1 - Dual Horns
J4 - Dual Horn Relay
L105 - Illumination for lighting controller
S3 - Fuse 3 in fuse holder
S4 - Fuse 4 in fuse holder
T2 - Double Connector, behind rear bumper cover
T5b - 5-Pin Connector, on airbag spiral spring
T3k - 3-Pin Connector, at E20
W6 - Glove Compartment Light
X - License Plate Light

(42) - Ground connection, beside steering column
(119) - Ground connection -1-, in headlight wiring harness
(218) - Ground connection -1-, in rear lid wiring harness
(A37) - Connection (58a), in instrument panel wiring harness
(A90) - Connection, (dual tone horn), in instrument panel wiring harness
(W11) - Connection (58), in rear lid wiring harness

ws = white
sw = black
ro = red
br = brown
gn = green
bl = blue
gr = grey
li = violet
ge = yellow
or = orange

Edition 11.05

New Beetle Convertible — Wiring diagram — No. 1/15

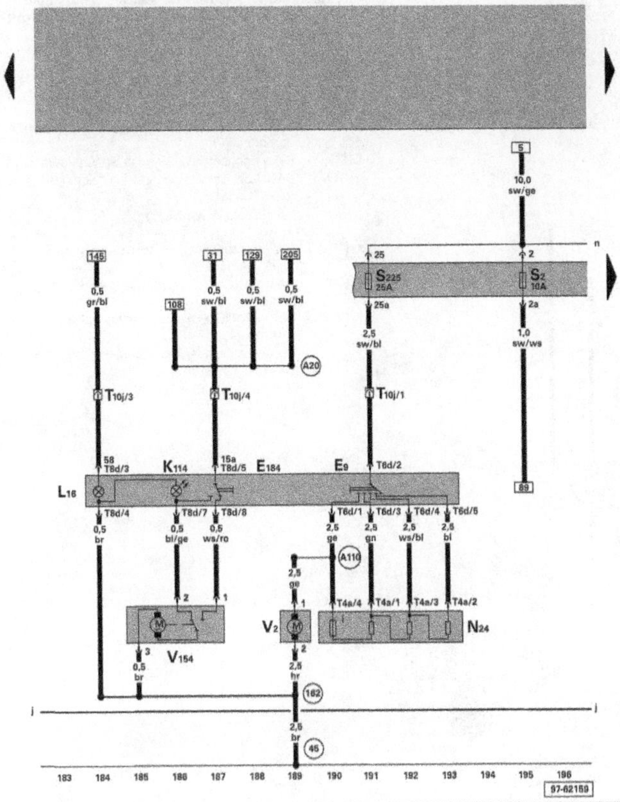

Switch for fresh air blower and recirculated air, fresh air blower series resistance with fuse, fresh air blower, servo motor for fresh-/recirculating air door

- E9 - Fresh Air Blower Switch
- E184 - Switch for fresh air blower and recirculated air
- K114 - Fresh Air and Recirculating Air Mode Indicator Light
- L16 - Fresh Air Control Lever Light
- N24 - Fresh Air Blower Series Resistance With Fuse
- S2 - Fuse 2 in fuse holder
- S225 - Fuse 25 in fuse holder
- T4a - 4-Pin Connector
- T6d - 6-Pin Connector
- T8d - 8-Pin Connector
- T10j - 10-Pin Connector
- V2 - Fresh Air Blower
- V154 - Servo motor for fresh-/recirculating air door

- (45) - Ground connection, behind instrument panel, center
- (162) - Ground connection, in blower motor wiring harness
- (A20) - Plus Connection (15a) (in instrument panel wiring harness)
- (A110) - Connector (fresh air blower), in instrument panel wiring harness

ws = white
sw = black
ro = red
br = brown
gn = green
bl = blue
gr = grey
li = violet
ge = yellow
or = orange

New Beetle Convertible — Wiring diagram — No. 1/16

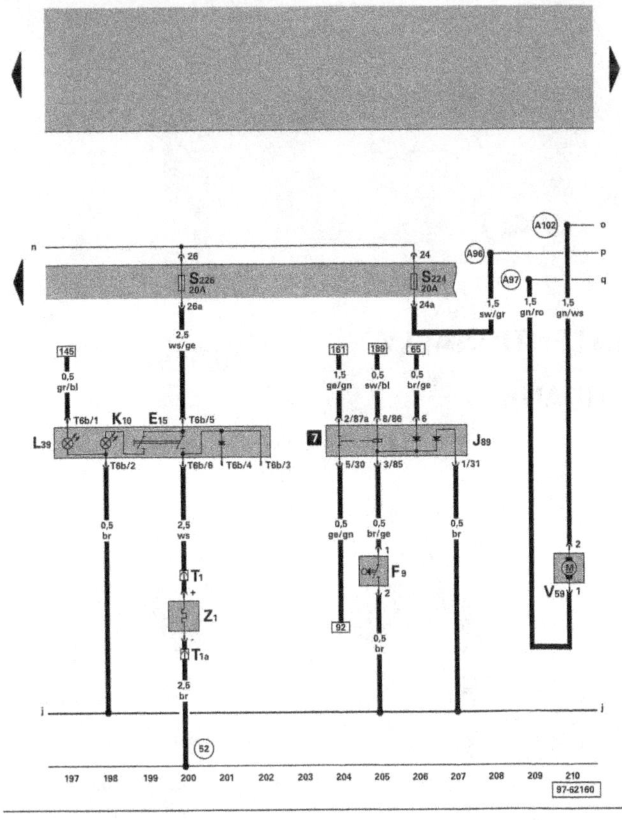

Heated rear window, rear window defogger switch, parking brake warning light switch, daytime running lights change-over relay, windshield and rear window washer pump

- E15 - Rear window defogger switch
- F9 - Parking Brake Warning Light Switch
- J89 - Daytime Running Lights Change-over Relay (173), on the thirteen position relay panel
- K10 - Rear Window Defogger Indicator Light
- L39 - Rear Window Defogger Switch Light
- S224 - Fuse 24 in fuse holder
- S226 - Fuse 26 in fuse holder
- T1 - 1-Pin Connector, at Z1+
- T1a - 1-Pin Connector, at Z1-
- T6b - 6-Pin Connector
- V59 - Windshield and Rear Window Washer Pump
- Z1 - Heated rear window

- (52) - Ground connection, in rear lid, left
- (A96) - Connector (53a), in instrument panel wiring harness
- (A97) - Connector (53), in instrument panel wiring harness
- (A102) - Connector (windshield wiper), in instrument panel wiring harness

ws = white
sw = black
ro = red
br = brown
gn = green
bl = blue
gr = grey
li = violet
ge = yellow
or = orange

New Beetle Convertible **Wiring diagram** **No. 1/17**

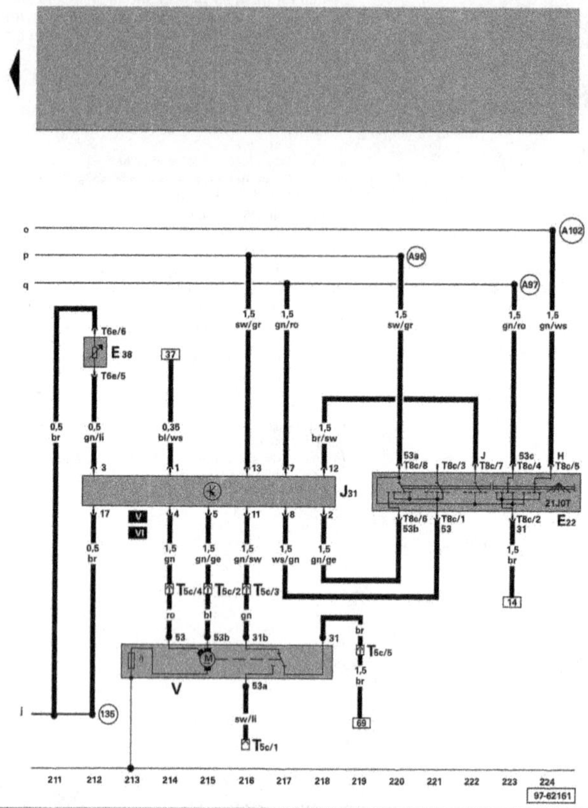

Windshield wiper/washer switch, windshield wiper motor, windshield wiper/washer intermittent relay

E22 - Winshield Wiper/Washer Switch
E38 - Windshield Wiper Intermittent Regulator
J31 - Wiper/Washer Intermittent Relay, on the thirteen position relay panel, production control number (377)
T5c - 5-Pin Connector, on Windshield Wiper Motor
T6e - 6-Pin Connector, near steering column
T8c - 8-Pin Connector
V - Windshield Wiper Motor

(135) - Ground connection -2-, in instrument panel wiring harness
(A96) - Connector (53a), in instrument panel wiring harness
(A97) - Connector (53), in instrument panel wiring harness
(A102) Connector (windshield wiper), in instrument panel wiring harness

ws = white
sw = black
ro = red
br = brown
gn = green
bl = blue
gr = grey
li = violet
ge = yellow
or = orange

THIS SPACE INTENTIONALLY LEFT BLANK

New Beetle **Wiring diagram** **No. 46/1**

Standard Equipment,

from July 2005,

Additional Fuse, Relay, Control Module and Connector locations ⇒ Component Locations

Troubleshooting Procedures ⇒ Guided Fault Finding using VAS 5051/5052

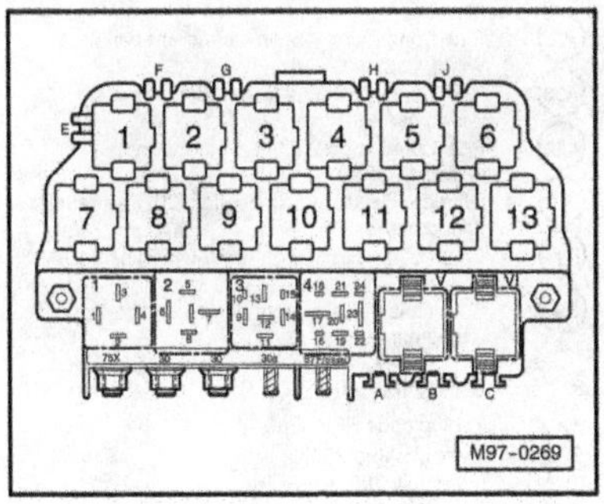

Relay locations on thirteen position auxiliary relay panel, above relay panel:

1 J434 Starter Lock/Clutch Pedal Switch Relay (449)

4 J434 Starter Lock/Clutch Pedal Switch Relay (465) [1)]

7 J89 Daytime Running Lamps Change-over Relay (173)

Relay panel:

1 J4 Horn Relay (53)

2 J59 Load Reduction Relay (100)

4 J17 Fuel Pump (FP) Relay (409)

V/VI J31 Wiper/Washer Intermitten Relay (377)

A S43 Power Window Circuit Breaker (30A)

[1)] Only to BPR engine

Note:

Number in parentheses indicates production control number stamped on relay housing.

New Beetle **Wiring diagram** **No. 46/2**

Battery, Load Reduction Relay, Fuses on battery

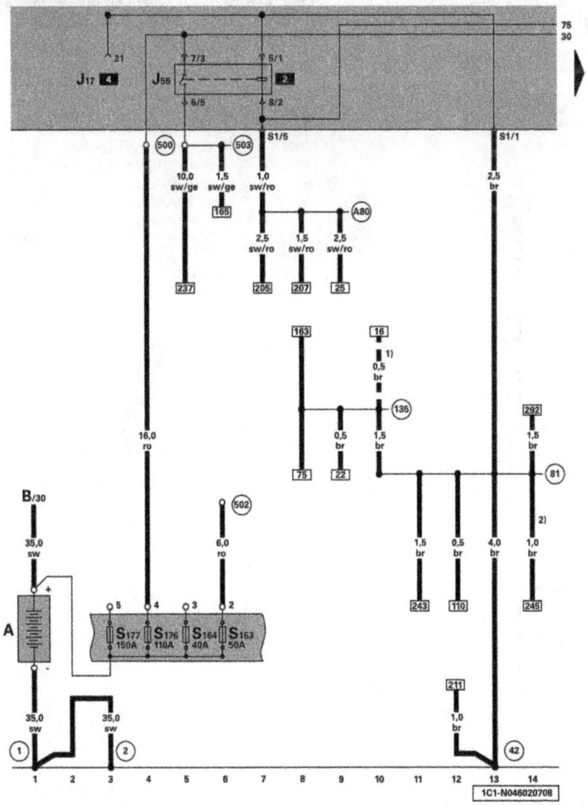

A - Battery
B - Starter
J17 - Fuel Pump (FP) Relay
J59 - Load Reduction Relay
S163 - Fuse 2 -terminal 30 (on battery fuse panel)
S164 - Fuse 3 -terminal 30 (on battery fuse panel)
S176 - Fuse 4 - terminal 30 (on battery fuse holder)
S177 - Fuse 5 -terminal 30 (on battery fuse panel)
(1) - Ground strap, battery to body
(2) - Ground strap, transmission to body
(42) - Ground Connection (beside steering column)
(81) - Ground Connection 1 (in instrument panel wiring harness)
(135) - Ground Connection 2 (in instrument panel wiring harness)
(500) - Threaded Connection 1 - terminal 30 (on relay panel)
(502) - Threaded Connection 1 - terminal 30a (on relay panel)
(503) - Treaded Connection - terminal 75x (on relay panel)
(A80) - Connection 1 (X) (in instrument panel wiring harness)

- - - - Engine code BPR, only
1) - Through June 2007
2) - From July 2007

ws = white
sw = black
ro = red
br = brown
gn = green
bl = blue
gr = grey
li = violet
ge = yellow
or = orange

New Beetle — Wiring diagram — No. 46/3

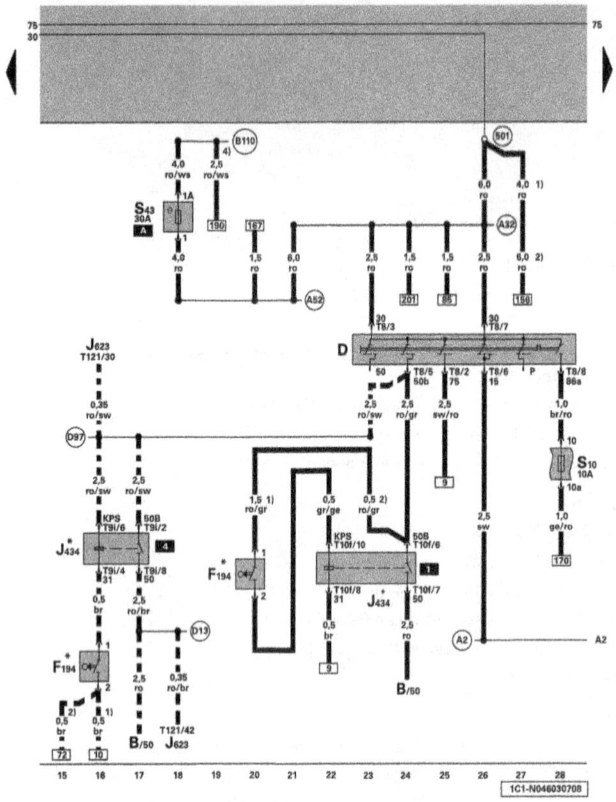

ws = white
sw = black
ro = red
br = brown
gn = green
bl = blue
gr = grey
li = violet
ge = yellow
or = orange

Ignition/Starter Switch, Clutch Pedal Starter Interlock Switch, Starter Lock/Clutch Pedal Switch Relay

- B - Starter
- D - Ignition/Starter Switch
- F194 - Clutch Pedal Starter Interlock Switch
- J434 - Starter Lock/Clutch Pedal Switch Relay
- J623 - Engine Control Module (ECM)
- S10 - Fuse (on fuse/relay panel)
- S43 - Power Window Circuit Breaker
- T8 - 8-Pin Connector
- T9i - 9-Pin Connector
- T10f - 10-Pin Connector
- T121 - 121-Pin Connector

- (501) - Threaded Connection 2 - terminal 30 (on relay panel)
- (A2) - Plus Connection (15) (in instrument panel wiring harness)
- (A32) - Plus Connection (30) (in instrument panel wiring harness)
- (A52) - Plus Connection 2 (30) (in instrument panel wiring harness)
- (B110) - Window Regulator Connection (30) (in interior wiring harness)
- (D13) - Connection (50) (in right front wiring harness)
- (D97) - Connection (50) (in right engine compartment wiring harness)

- * - Manual transmission only
- - - - - Engine code BPR, Only
- 1) - Through June 2007
- 2) - From July 2007
- 4) - From July 2008

New Beetle — Wiring diagram — No. 46/4

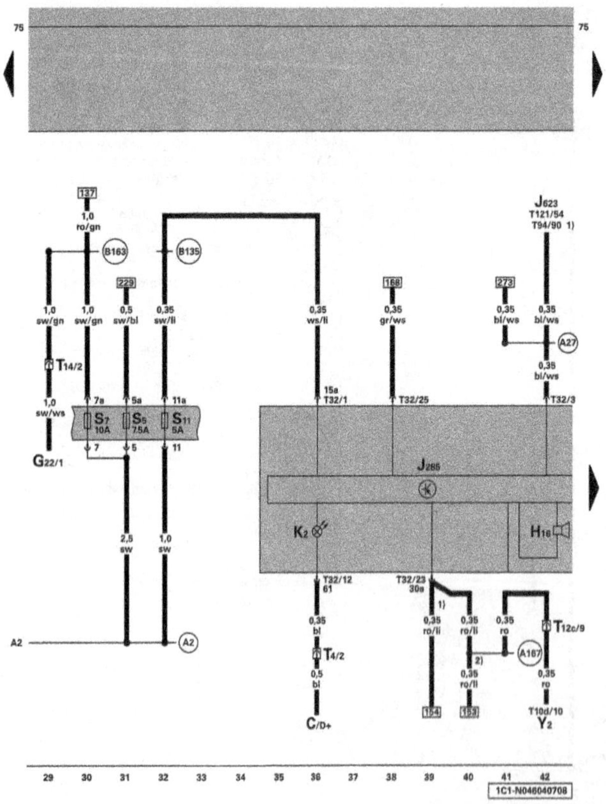

ws = white
sw = black
ro = red
br = brown
gn = green
bl = blue
gr = grey
li = violet
ge = yellow
or = orange

Light Switched -On- Warning Buzzer, Instrument Cluster Control Module, Generator (GEN) Warning Lamp

- C - Generator (GEN)
- G22 - Speedometer Vehicle Speed Sensor (VSS)
- H16 - Light Switched -On- Warning Buzzer
- J285 - Instrument Cluster Control Module
- J623 - Engine Control Module (ECM)
- K2 - Generator (GEN) Warning Lamp
- S5 - Fuse (on fuse/relay panel)
- S7 - Fuse (on fuse/relay panel)
- S11 - Fuse (on fuse/relay panel)
- T4 - 4-Pin Connector, in engine compartment left
- T10d - 10-Pin Connector
- T12c - 12-Pin Connector
- T14 - 14-Pin Connector
- T32 - 32-Pin Connector, blue, on instrument cluster
- T94 - 94-Pin Connector
- T121 - 121-Pin Connector, on Motronic/Diesel DFI Control Module
- Y2 - Digital Clock

- (A2) - Plus Connection (15) (in instrument panel wiring harness)
- (A27) - Vehicle Speed Signal (VSS) Connection (in instrument panel wiring harness)
- (A167) - Plus Connection 3 (30a) (in instrument panel wiring harness)
- (B135) - Plus Connection 1 (15a) (in interior wiring harness)
- (B163) - Plus Connection 1 (15) (in interior wiring harness)

- 1) - Through June 2007
- 2) - From July 2007

New Beetle

Wiring diagram

No. 46/5

Instrument Cluster Control Module, Low Fuel Level Warning Lamp, Odometer Display

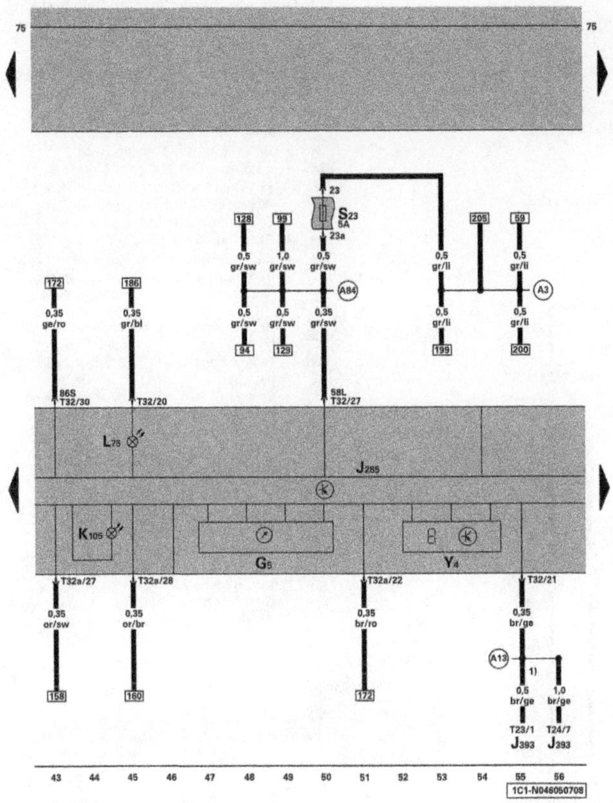

G5 - Tachometer
J285 - Instrument Cluster Control Module
J393 - Comfort System Central Control Module
K105 - Low Fuel Level Warning Lamp
L75 - Digital Display Illumination
S23 - Heater Fuse
T23 - 23-Pin Connector, at J393
T24 - 24-Pin Connector, at J393
T32 - 32-Pin Connector, blue, on instrument cluster
T32a - 32-Pin Connector, green, on instrument cluster
Y4 - Odometer Display

(A3) - Plus Connection (58) (in instrument panel wiring harness)
(A13) - Wire connection (door contact switch) in instrument panel wiring harness
(A84) - Door Contact Switch Connection (in instrument panel wiring harness)

1) - Through June 2007

ws = white
sw = black
ro = red
br = brown
gn = green
bl = blue
gr = grey
li = violet
ge = yellow
or = orange

New Beetle

Wiring diagram

No. 46/6

Instrument Cluster Control Module, High Beam Indicator Lamp, Brake Pad Wear Indicator Lamp

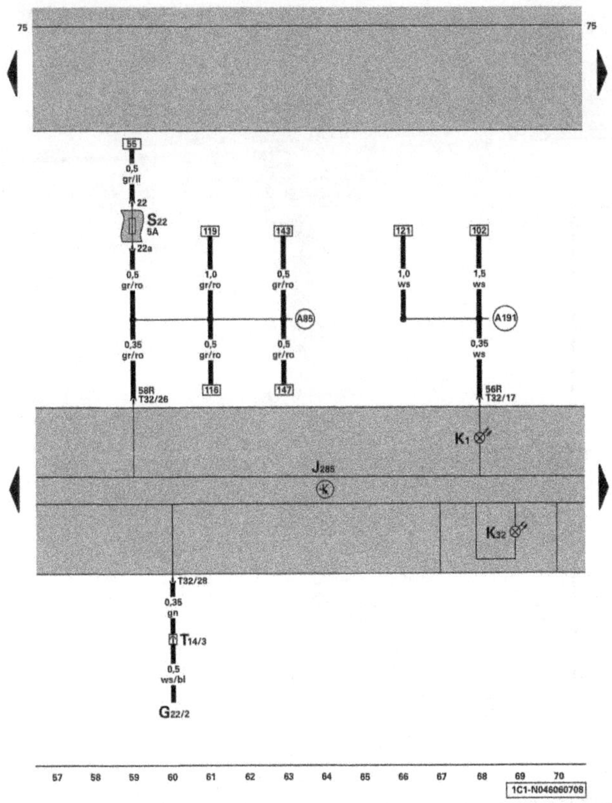

G22 - Speedometer Vehicle Speed Sensor (VSS)
J285 - Instrument Cluster Control Module
K1 - High Beam Indicator Lamp
K32 - Brake Pad Wear Indicator Lamp
S22 - Fuse (on fuse/relay panel)
T14 - 14-Pin Connector, in plenum chamber
T32 - 32-Pin Connector, blue, on instrument cluster

(A85) - Connection (58R) (in Instrument panel wiring harness)
(A191) - Connection (56aR) (in instrument panel wiring harness)

ws = white
sw = black
ro = red
br = brown
gn = green
bl = blue
gr = grey
li = violet
ge = yellow
or = orange

Standard equipment from July 2005

New Beetle

Wiring diagram

No. 46/7

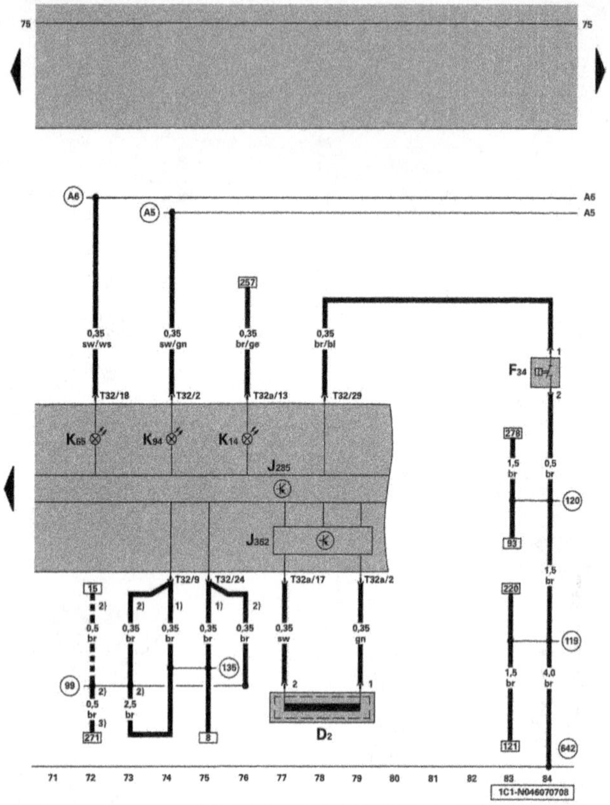

Anti-Theft Immobilizer Reading Coil, Brake Fluid Level Warning Switch, Instrument Cluster Control Module, Anti-Theft Immobilizer Control Module, Parking Brake Indicator Lamp, Left Turn Signal Indicator Lamp, Right Turn Signal Indicator Lamp

- D2 - Anti-Theft Immobilizer Reading Coil
- F34 - Brake Fluid Level Warning Switch
- J285 - Instrument Cluster Control Module
- J362 - Anti-Theft Immobilizer Control Module
- K14 - Parking Brake Indicator Lamp
- K65 - Left Turn Signal Indicator Lamp
- K94 - Right Turn Signal Indicator Lamp
- T32 - 32-Pin Connector, blue, on instrument cluster
- T32a - 32-Pin Connector, green, on instrument cluster

- (99) - Ground Connection (in console wiring harness)
- (119) - Ground Connection 1 (in headlamp wiring harness)
- (120) - Ground Connection 2 (in headlamp wiring harness)
- (135) - Ground Connection 2 (in instrument panel wiring harness)
- (642) - Ground Connection (on electronic fan control)

- (A5) - Right Turn Signal Plus Connection (in instrument panel wiring harness)
- (A6) - Left Turn Signal Plus Connection (in instrument panel wiring harness)

- 1) - Through June 2007
- 2) - From July 2007
- 3) - Through June 2008

ws = white
sw = black
ro = red
br = brown
gn = green
bl = blue
gr = grey
li = violet
ge = yellow
or = orange

Edition 07.08

New Beetle

Wiring diagram

No. 46/8

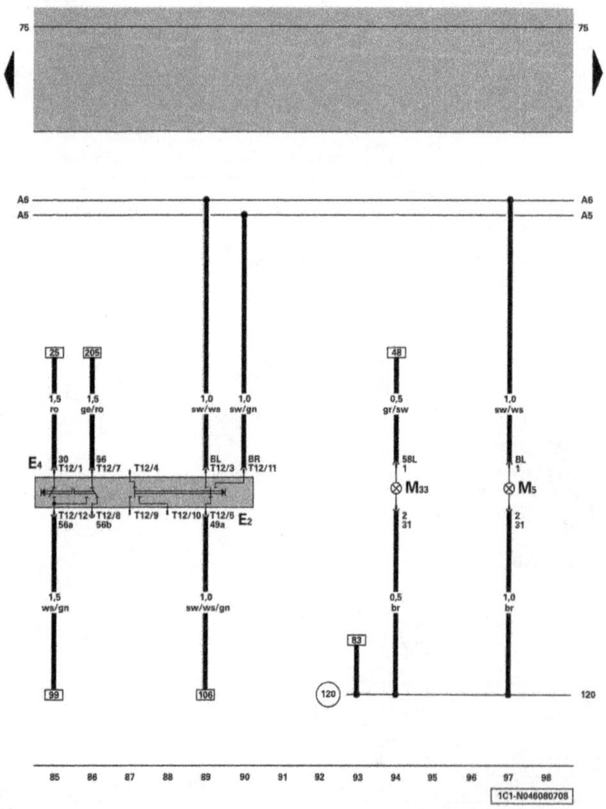

Turn Signal Switch, Headlamp Dimmer/Flasher Switch, Left Front Turn Signal Lamp, Left Front Side Marker Lamp

- E2 - Turn Signal Switch
- E4 - Headlamp Dimmer/Flasher Switch
- M5 - Left Front Turn Signal Lamp
- M33 - Left Front Side Marker Lamp
- T12 - 12-Pin Connector, near steering column

- (120) - Ground Connection 2 (in headlamp wiring harness)

ws = white
sw = black
ro = red
br = brown
gn = green
bl = blue
gr = grey
li = violet
ge = yellow
or = orange

Edition 07.08

New Beetle — Wiring diagram — No. 46/9

Emergency Flasher Switch, Turn Signal Relay, Left Parking Lamp, Left Low Beam Headlamp, Left High Beam Headlamp

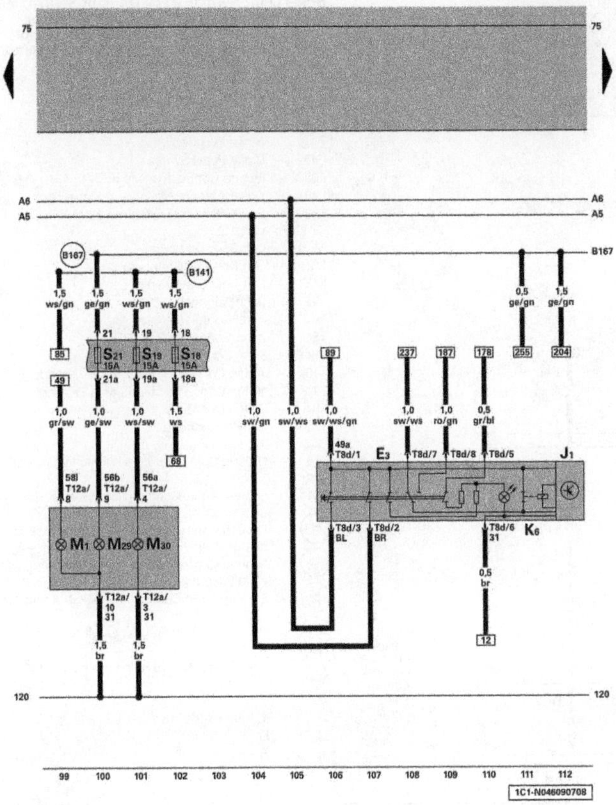

E3 - Emergency Flasher Switch
J1 - Turn Signal Relay
K6 - Emergency Flasher Indicator Lamp
M1 - Left Parking Lamp
M29 - Left Low Beam Headlamp
M30 - Left High Beam Headlamp
S18 - Fuse (on fuse/relay panel)
S19 - Fuse (on fuse/relay panel)
S21 - Fuse (on fuse/relay panel)
T8d - 8-Pin Connector
T12a - 12-Pin Connector

(B141) - Plus Connection 1 (56a) (in interior wiring harness)
(B167) - Connection (56b) (in passenger compartment wiring harness)

ws = white
sw = black
ro = red
br = brown
gn = green
bl = blue
gr = grey
li = violet
ge = yellow
or = orange

New Beetle — Wiring diagram — No. 46/10

Right Parking Lamp, Left Rear Turn Signal Lamp, Right Front Turn Signal Lamp, Right Low Beam Headlamp, Right High Beam Headlamp, Right Front Side Marke Lamp

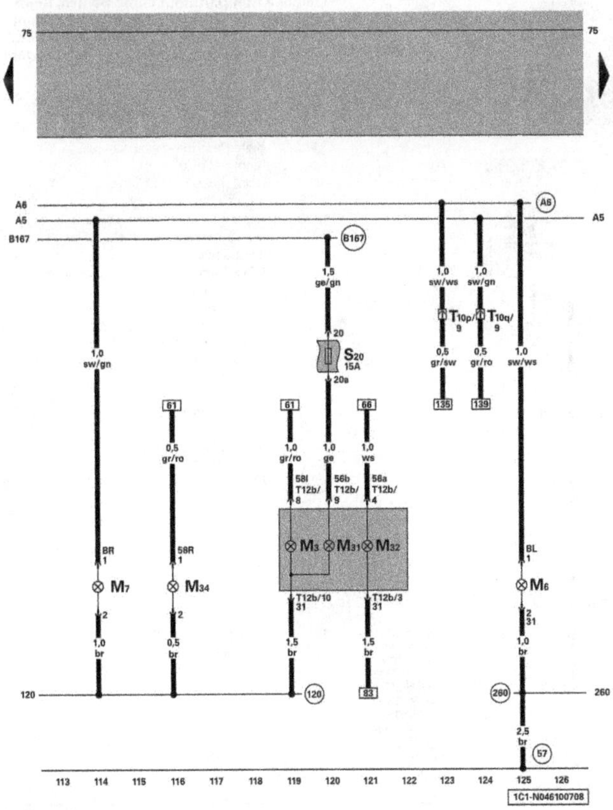

M3 - Right Parking Lamp
M6 - Left Rear Turn Signal Lamp
M7 - Right Front Turn Signal Lamp
M31 - Right Low Beam Headlamp
M32 - Right High Beam Headlamp
M34 - Right Front Side Marker Lamp
S20 - Fuse (on fuse/relay panel)
T10p - 10-Pin Connector
T10q - 10-Pin Connector
T12b - 12-Pin Connector

(57) - Ground Connection (on left rear pillar)
(120) - Ground Connection 2 (in headlamp wiring harness)
(260) - Ground Connection (in taillamps wiring harness)
(A6) - Left Turn Signal Plus Connection (in instrument panel wiring harness)
(B167) - Connection (56b) (in passenger compartment wiring harness)

ws = white
sw = black
ro = red
br = brown
gn = green
bl = blue
gr = grey
li = violet
ge = yellow
or = orange

Standard equipment from July 2005

New Beetle — Wiring diagram — No. 46/11

Brake Light Switch, Back-Up light Switch, Brake Pedal Switch, Driver Exterior Mirror Turn Signal Lamp, Front Passenger Exterior Mirror Turn Signal Lamp, Left Back-Up Lamp, Left Brake/Taillight Lamp, Left Rear Side Marker Lamp

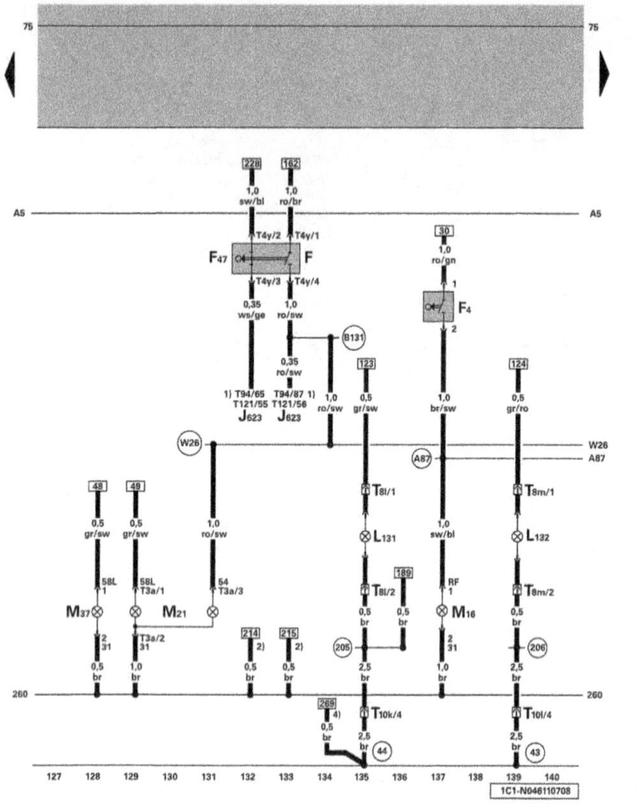

- F - Brake Light Switch
- F4 - Back-Up light Switch
- F47 - Brake Pedal Switch
- J623 - Engine Control Module (ECM)
- L131 - Driver Exterior Mirror Turn Signal Lamp
- L132 - Front Passenger Exterior Mirror Turn Signal Lamp
- M16 - Left Back-Up Lamp
- M21 - Left Brake/Taillight Lamp
- M37 - Left Rear Side Marker Lamp
- T3a - 3-Pin Connector
- T4y - 4-Pin Connector
- T8l - 8-Pin Connector, at L131
- T8m - 8-Pin Connector, at L132
- T10k - 10-Pin Connector, black, on a Pillar left
- T10l - 10-Pin Connector, black, on a Pillar right
- T94 - 94-Pin Connector
- T121 - 121-Pin Connector
- (43) - Ground Connection (on right lower A-pillar)
- (44) - Ground Connection (on left lower A-pillar)
- (205) - Ground Connection (in driver's door wiring harness)
- (206) - Ground Connection (in Front Passenger's door wiring harness)
- (A87) - Back-up Light Connection (in instrument panel wiring harness)
- (B131) - Connection (54) (in interior wiring harness)
- (W26) - Plus Connection (54) (in taillamp wiring harness)

1) - Through June 2007
2) - From July 2007
4) - From July 2008

ws = white
sw = black
ro = red
br = brown
gn = green
bl = blue
gr = grey
li = violet
ge = yellow
or = orange

New Beetle — Wiring diagram — No. 46/12

Luggage Compartment Lamp Switch, Right Rear Turn Signal Lamp, Right Back-Up Lamp, Right Brake/Taillight Lamp, High-mount Brake Light, Right Rear Side Marker Lamp, Luggage Compartment Light

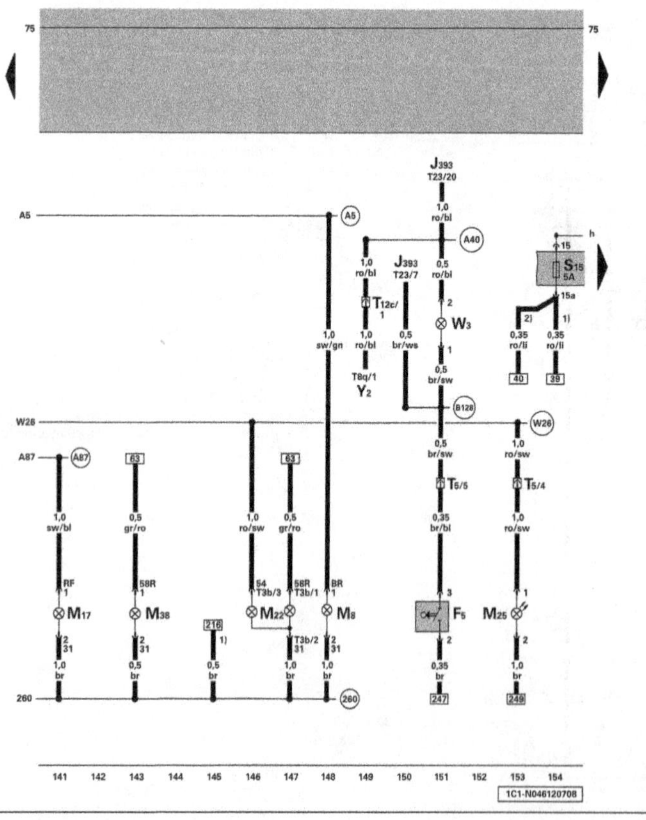

- F5 - Luggage Compartment Lamp Switch
- J393 - Comfort System Central Control Module
- M8 - Right Rear Turn Signal Lamp
- M17 - Right Back-Up Lamp
- M22 - Right Brake/Taillight Lamp
- M25 - High-mount Brake Light
- M38 - Right Rear Side Marker Lamp
- S15 - Fuse (on fuse/relay panel)
- T3b - 3-Pin Connector
- T5 - 5-Pin Connector, black, on C-pillar, left
- T8q - 8-Pin Connector
- T12c - 12-Pin Connector
- T23 - 23-Pin Connector, on Central control module for comfort system.
- W3 - Luggage Compartment Light
- Y2 - Digital Clock
- (260) - Ground Connection (in taillamps wiring harness)
- (A5) - Right Turn Signal Plus Connection (in instrument panel wiring harness)
- (A40) - Plus Connection 1 (30) (in instrument cluster wiring harness)
- (A87) - Back-up Light Connection (in instrument panel wiring harness)
- (B128) - Luggage Compartment Lamp Connection (in interior wiring harness)
- (W26) - Plus Connection (54) (in taillamp wiring harness)

1) - Through June 2007
2) - From July 2007

ws = white
sw = black
ro = red
br = brown
gn = green
bl = blue
gr = grey
li = violet
ge = yellow
or = orange

New Beetle — Wiring diagram — No. 46/13

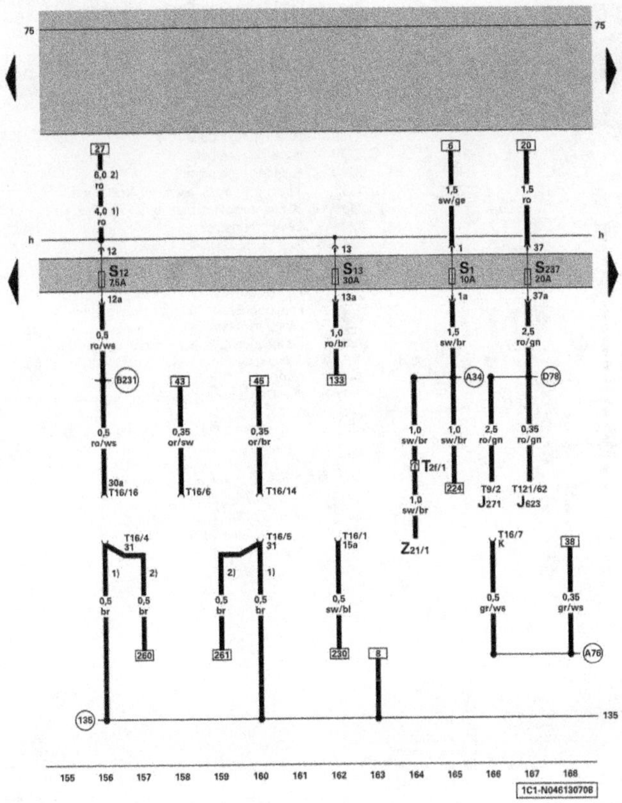

Data Link Connector (DLC)

- J271 - Motronic Engine Control Module (ECM) Power Supply Relay
- J623 - Engine Control Module (ECM)
- S1 - Fuse (on fuse/relay panel)
- S12 - Fuse (on fuse/relay panel)
- S13 - Fuse (on fuse/relay panel)
- S237 - Fuse (on fuse panel)
- T2f - 2-Pin Connector
- T9 - 9-Pin Connector
- T16 - 16-Pin Connector, black, Data Link Connector (DLC), below steering column, left
- T121 - 121-Pin Connector, at J220
- Z21 - Right Washer Nozzle Heater

- (135) - Ground Connection 2 (in instrument panel wiring harness)
- (A34) - Connection (75x) (in instrument panel wiring harness)
- (A76) - K-Diagnostic Connection (in instrument panel wiring harness)
- (B231) - Plus Connection 5 (30a) (in interior wiring harness)
- (D78) - Plus Connection 1 (30a) (in engine compartment wiring harness)

1) - Through June 2007
2) - From July 2007

ws = white
sw = black
ro = red
br = brown
gn = green
bl = blue
gr = grey
li = violet
ge = yellow
or = orange

New Beetle — Wiring diagram — No. 46/14

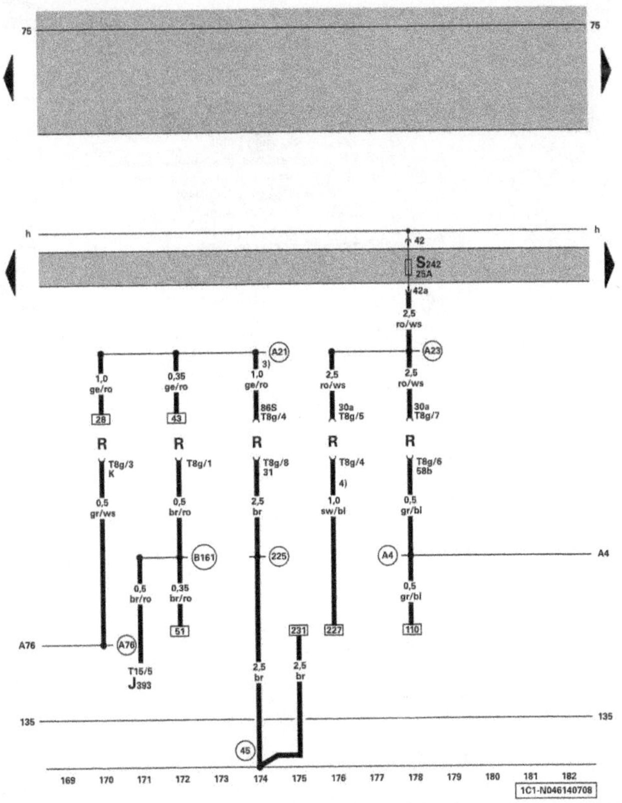

Radio Connection

- J393 - Comfort System Central Control Module
- R - Radio
- S242 - Fuse (on fuse panel)
- T8g - 8-Pin Connector, on radio
- T15 - 15-Pin Connector, on radio

- (45) - Ground Connection (behind instrument panel, center)
- (255) - Ground Connection 1 (in radio wiring harness)
- (A4) - Plus Connection (58b) (in instrument panel wiring harness)
- (A21) - Connection (86s) (in instrument panel wiring harness)
- (A23) - Connection (30al) (in instrument panel wiring harness)
- (A76) - K-Diagnostic Connection (in instrument panel wiring harness)
- (B161) - Anti-Theft Alarm System Connection (in interior wiring harness)

3) - Through June 2008
4) - From July 2008

ws = white
sw = black
ro = red
br = brown
gn = green
bl = blue
gr = grey
li = violet
ge = yellow
or = orange

New Beetle — Wiring diagram — No. 46/15

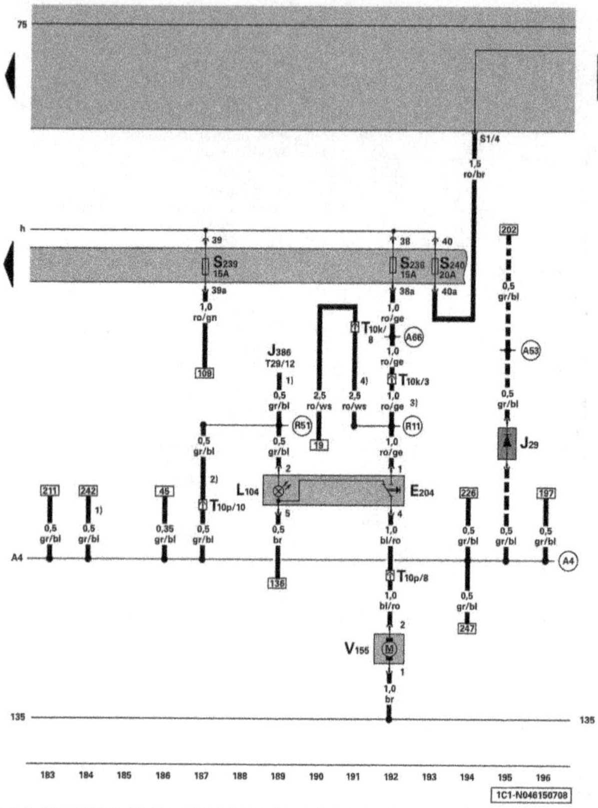

Remote Fuel Tank Flap Switch, Blocking Diode, Fuel Tank Lid Unlock Motor

- E204 - Remote Fuel Tank Flap Switch
- J29 - Blocking Diode
- J386 - Driver Door Control Module
- L104 - Remote Unlock Switch Illumination
- S238 - Fuse (on fuse panel)
- S239 - Fuse (on fuse panel)
- S240 - Fuse (on fuse panel)
- T10k - 10-pin connector, black, on A-pillar, left
- T10p - 10-pin connector, blue, on A-pillar, left
- T29 - 29-pin connector
- V155 - Fuel Tank Lid Unlock Motor

- (A4) - Plus connection (58b) in instrument panel wiring harness
- (A53) - Plus Connection (58b) (in instrument panel wiring harness)
- (A66) - Connection (30a, central locking/Anti-Theft warning system/IR (in instrument panel wiring harness)
- (R11) - Plus Connection 2 (30) (in driver's door wiring harness)
- (R51) - Connection (58b) (in driver's door wiring harness)

- --- - Automatic transmission only
- 1) - Through June 2007
- 2) - From July 2007
- 3) - Through June 2008
- 4) - From July 2008

ws = white
sw = black
ro = red
br = brown
gn = green
bl = blue
gr = grey
li = violet
ge = yellow
or = orange

New Beetle — Wiring diagram — No. 46/16

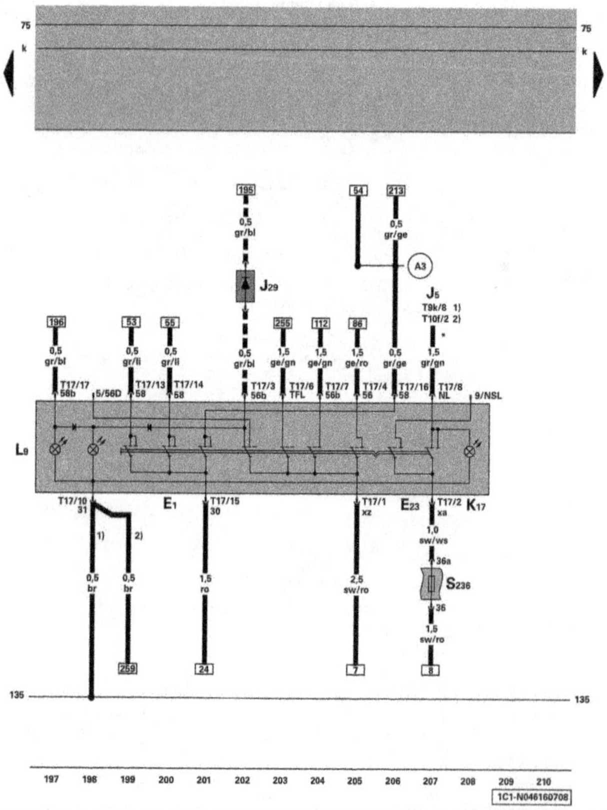

Light Switch, Fog Lamp Switch, Blocking Diode, Fog Lamp Indicator Lamp, Headlamp Sw itch Illumination

- E1 - Light Switch
- E23 - Fog Lamp Switch
- J5 - Fog Lamp Relay
- J29 - Blocking Diode
- K17 - Fog Lamp Indicator Lamp
- L9 - Headlamp Switch Illumination
- S236 - Fuse (on fuse panel)
- T9k - 9-Pin Connector
- T10f - 10-Pin Connector
- T17 - 17-Pin Connector

- (A3) - Plus Connection (58) (in instrument panel wiring harness)

- * - Vehicles with fog lamps, only
- --- - Automatic transmission, only
- 1) - Through June 2007
- 2) - From July 2007

ws = white
sw = black
ro = red
br = brown
gn = green
bl = blue
gr = grey
li = violet
ge = yellow
or = orange

New Beetle

Wiring diagram

No. 46/17

Instrument Panel Illumination Dimmer Switch, Signal Horn Activation, High Tone Horn, Low Tone Horn, Horn Relay, Glove Compartment Light, License Plate Light

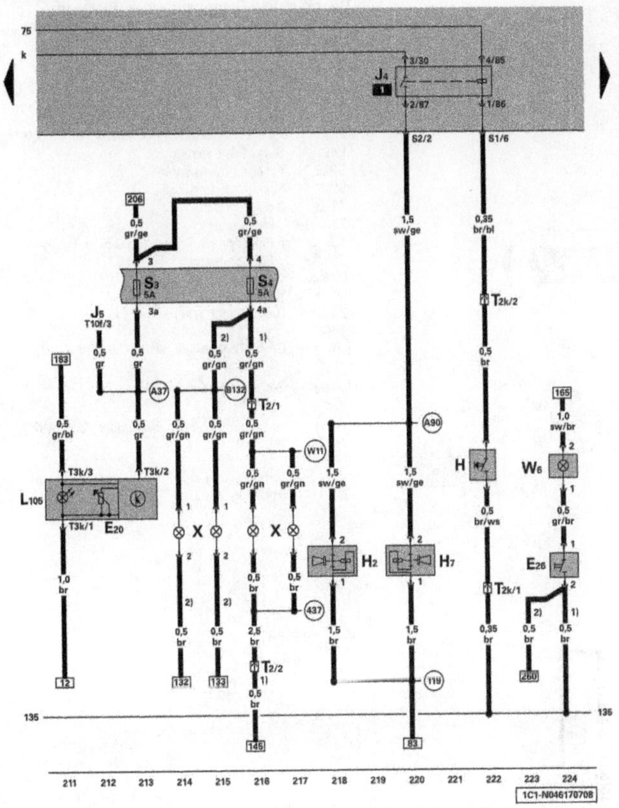

E20 - Instrument Panel Illumination Dimmer Switch
E26 - Glove Compartment Lamp Switch
H - Signal Horn Activation
H2 - High Tone Horn
H7 - Low Tone Horn
J4 - Horn Relay
J5 - Fog Lamp Relay
L105 - Lighting Control Illumination
S3 - Fuse (on fuse/relay panel)
S4 - Fuse (on fuse/relay panel)
T2 - Double Connector, behind rear bumpercover
T2k - Double Connector
T3k - 3-Pin Connector, at E20
T10f - 10-Pin Connector
W6 - Glove Compartment Light
X - License Plate Light

(119) - Ground Connection 1 (in headlamp wiring harness)
(437) - Ground Connection (in license plate light wiring harness)
(A37) - Connection (58a) (in instrument panel wiring harness)
(A90) - Dual-tone Horn Connection (in instrument panel wiring harness)
(B132) - License Plate Light Connection (in interior wiring harness)
(W11) - Connection (58) (in rear lid wiring harness)

1) - Through June 2007
2) - From July 2007

ws = white
sw = black
ro = rod
br = brown
gn = green
bl = blue
gr = grey
li = violet
ge = yellow
or = orange

New Beetle

Wiring diagram

No. 46/18

Fresh Air Blower/Recirculated Air Switch, Fresh Air Blower Series Resistor With Fuse, Fresh Air Blower, Fresh/Recirculating Air Door Motor

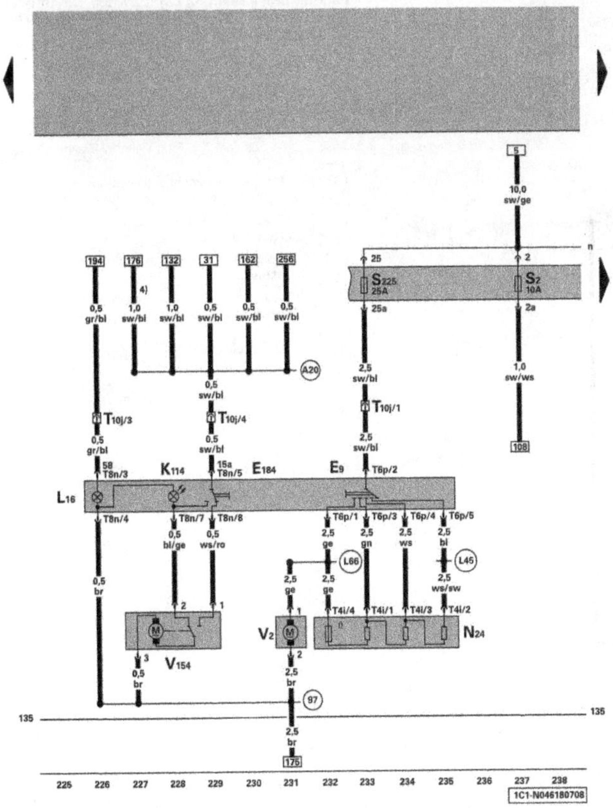

E9 - Fresh Air Blower Switch
E184 - Fresh Air Blower/Recirculated Air Switch
K114 - Fresh Air and Recirculating Air Mode Indicator Lamp
L16 - Fresh Air Control Illumination
N24 - Fresh Air Blower Series Resistor With Fuse
S2 - Fuse (on fuse/relay panel)
S225 - Fuse (on fuse panel)
T4i - 4-Pin Connector
T6p - 6-Pin Connector
T8n - 8-Pin Connector
T10j - 10-Pin Connector
V2 - Fresh Air Blower
V154 - Fresh/Recirculating Air Door Motor

(97) - Ground Connection 1 (in A/C wiring harness)
(A20) - Plus Connection (15a) (in instrument panel wiring harness)
(L45) - Connection (in A/C system wiring harness)
(L66) - Connection (in wiring harness) heater blower

4) - From July 2008

ws = white
sw = black
ro = red
br = brown
gn = green
bl = blue
gr = grey
li = violet
ge = yellow
or = orange

Standard equipment from July 2005

New Beetle — Wiring diagram — No. 46/19

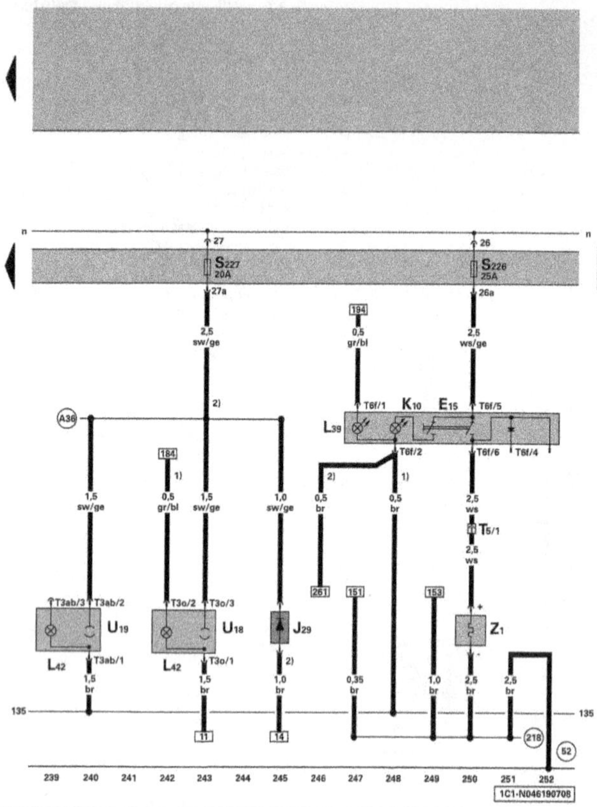

Rear Window Defogger Switch, Blocking Diode, Rear Window Defogger Indicator Lamp, 12V Socket, Heated Rear Window

- E15 - Rear Window Defogger Switch
- J29 - Blocking Diode
- K10 - Rear Window Defogger Indicator Lamp
- L39 - Rear Window Defogger Switch Illumination
- L42 - Socket Illumination
- S226 - Fuse (on fuse panel)
- S227 - Fuse (on fuse panel)
- T3ab - 3-pin connector
- T3o - 3-pin connector
- T5 - 5-Pin Connector, black, on C-pillar, left
- T6f - 6-Pin Connector
- U18 - 12V Socket 2
- U19 - 12V Socket 3
- Z1 - Heated Rear Window
- (52) - Ground connection (in rear lid, left)
- (218) - Ground connection 1 (in rear lid wiring harness)
- (A36) - Connection (75a) (in instrument panel wiring harness)

1) - Through June 2007
2) - From July 2007

ws = white
sw = black
ro = red
br = brown
gn = green
bl = blue
gr = grey
li = violet
ge = yellow
or = orange

New Beetle — Wiring diagram — No. 46/20

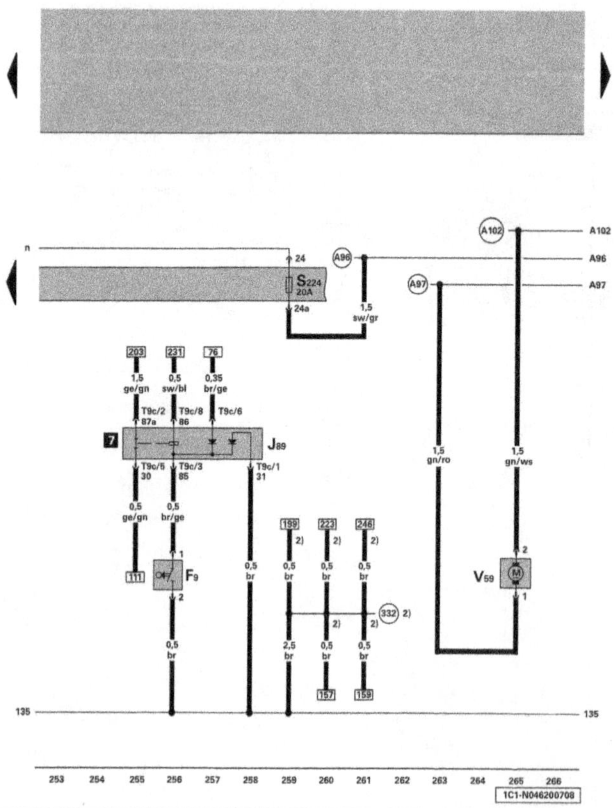

Parking Brake Warning Light Switch, Daytime Running Lamps Change-over Relay, Windshield And Rear Window Washer Pump

- F9 - Parking Brake Warning Light Switch
- J89 - Daytime Running Lamps Change-over Relay
- S224 - Fuse (on fuse panel)
- T9c - 9-Pin Connector
- V59 - Windshield And Rear Window Washer Pump
- (332) - Ground Connection 4 (in instrument panel wiring harness)
- (A96) - Connection (53a) (in instrument panel wiring harness)
- (A97) - Connection (53) (in instrument panel wiring harness)
- (A102) - Windsield Wiper Connection (in instrument panel wiring harness)

2) - From July 2007

ws = white
sw = black
ro = red
br = brown
gn = green
bl = blue
gr = grey
li = violet
ge = yellow
or = orange

New Beetle

Wiring diagram

No. 46/21

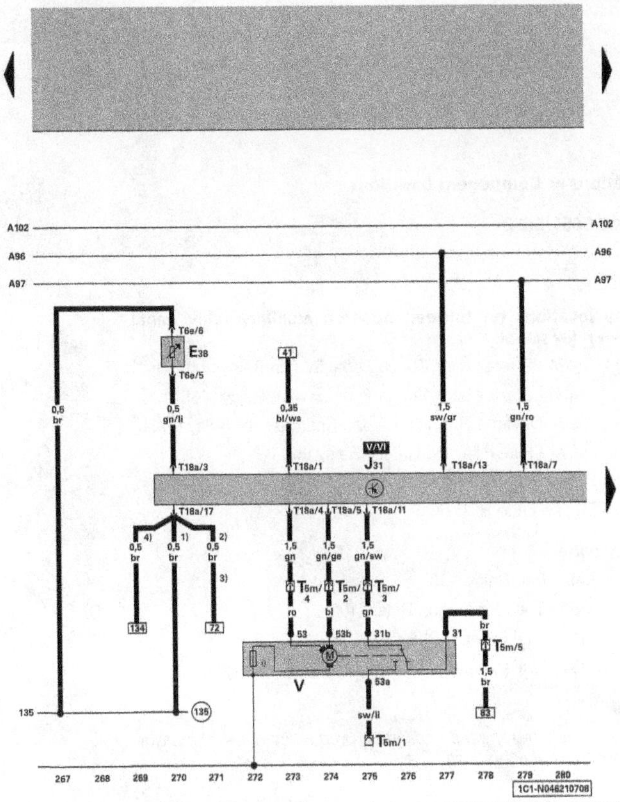

Windshield Wiper Intermittent Regulator, Wiper/Washer Intermittent Relay, Windshield Wiper Motor

E38 - Windshield Wiper Intermittent Regulator
J31 - Wiper/Washer Intermittent Relay
T5m - 5-Pin Connector, on Windshield Wiper Motor
T6e - 6-Pin Connector, near steering column
T18a - 18-Pin Connector
V - Windshield Wiper Motor

(135) - Ground Connection 2 (in instrument panel wiring harness)

1) - Through June 2007
2) - From July 2007
3) - Through June 2008
4) - From July 2008

ws = white
sw = black
ro = red
br = brown
gn = green
bl = blue
gr = grey
li = violet
ge = yellow
or = orange

New Beetle

Wiring diagram

No. 46/22

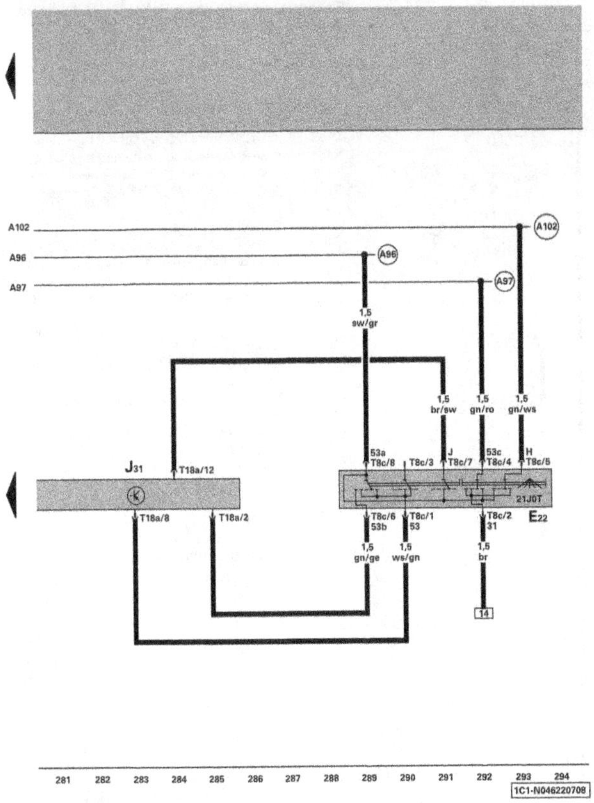

Winshield Wiper/Washer Switch, Wiper/Washer Intermittent Relay

E22 - Winshield Wiper/Washer Switch
J31 - Wiper/Washer Intermittent Relay
T8c - 8-Pin Connector
T18a - 18-Pin Connector

(A96) - Connection (53a) (in instrument panel wiring harness)
(A97) - Connection (53) (in instrument panel wiring harness)
(A102) - Windsield Wiper Connection (in instrument panel wiring harness)

ws = white
sw = black
ro = red
br = brown
gn = green
bl = blue
gr = grey
li = violet
ge = yellow
or = orange

New Beetle Convertible **Wiring diagram** **No. 30/1**

Standard Equipment,

from July 2005,

Additional Fuse, Relay, Control Module and Connector locations ⇒ Component Locations

Electrical troubleshooting ⇒ Guided Fault Finding using VAS 5051/5052

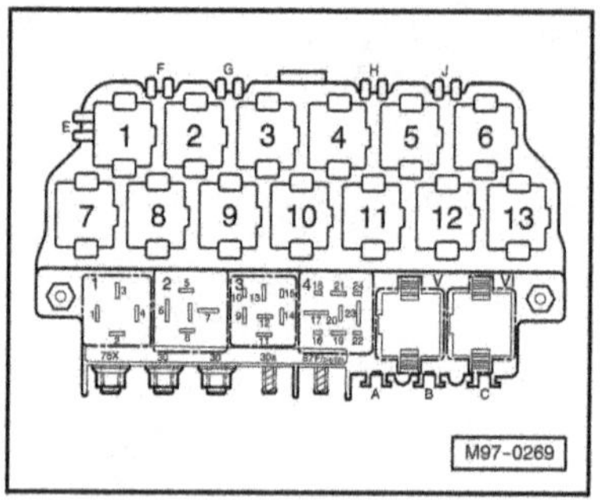

Relay locations on thirteen position auxiliary relay panel, above relay panel:

1 J434 - Starter Lock/Clutch Pedal Switch Relay (449) 5)

4 J434 - Starter Lock/Clutch Pedal Switch Relay (465) 6)

7 J89 - Daytime Running Lamps Change-over Relay (173)

8 J677 - Fuel Filler Lid Unlock Relay (53)

5) - Engine code BPS only

6) - Engine code BPR only

Relay panel:

1 J4 - Horn Relay (53)

2 J59 - Load Reduction Relay (100)

4 J17 - Fuel Pump (FP) Relay (409)

V/VI J31 - Wiper/Washer Intermittent Relay (192)

Note:

Number in parentheses indicates production control number stamped on relay housing.

New Beetle Convertible **Wiring diagram** **No. 30/2**

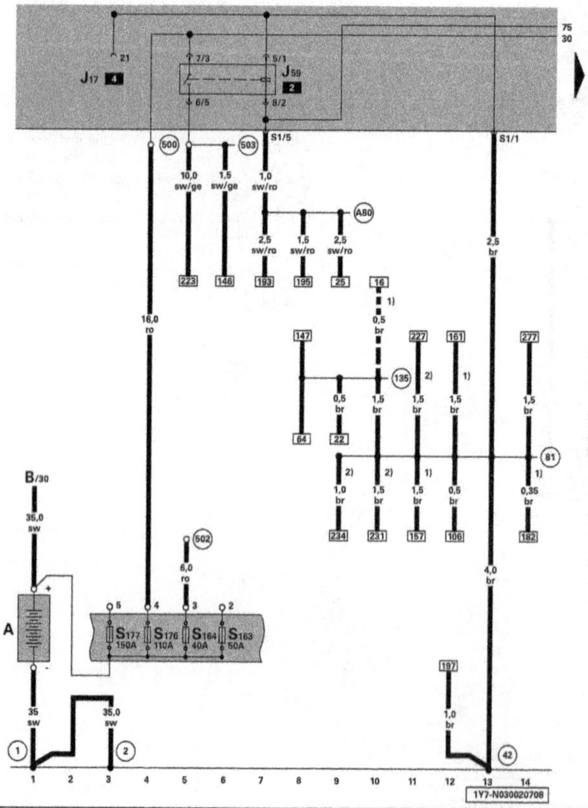

Battery, Load Reduction Relay, Fuses on battery fuse panel

A - Battery
B - Starter
J17 - Fuel Pump (FP) Relay
J59 - Load Reduction Relay
S163 - Fuse 2 -terminal 30 (on battery fuse panel)
S164 - Fuse 3 -terminal 30 (on battery fuse panel)
S176 - Fuse 4 - terminal 30 (on battery fuse holder)
S177 - Fuse 5 - terminal 30 (on battery fuse panel)

(1) - Ground strap, battery to body
(2) - Ground strap, transmission to body
(42) - Ground Connection (beside steering column)
(81) - Ground Connection 1 (in instrument panel wiring harness)
(135) - Ground Connection 2 (in instrument panel wiring harness)
(500) - Threaded Connection 1 - terminal 30 (on relay panel)
(502) - Threaded Connection 1 - terminal 30a (on relay panel)
(503) - Treaded Connection - terminal 75x (on relay panel)
(A80) - Connection 1 (X) (in instrument panel wiring harness)

- - - - Engine code BPR only
1) - Through June 2007
2) - From July 2007

ws = white
sw = black
ro = red
br = brown
gn = green
bl = blue
gr = grey
li = violet
ge = yellow
or = orange

New Beetle Convertible — Wiring diagram — No. 30/3

Ignition/Starter Switch, Clutch Pedal Starter Interlock Switch, Starter Lock/Clutch Pedal Switch Relay

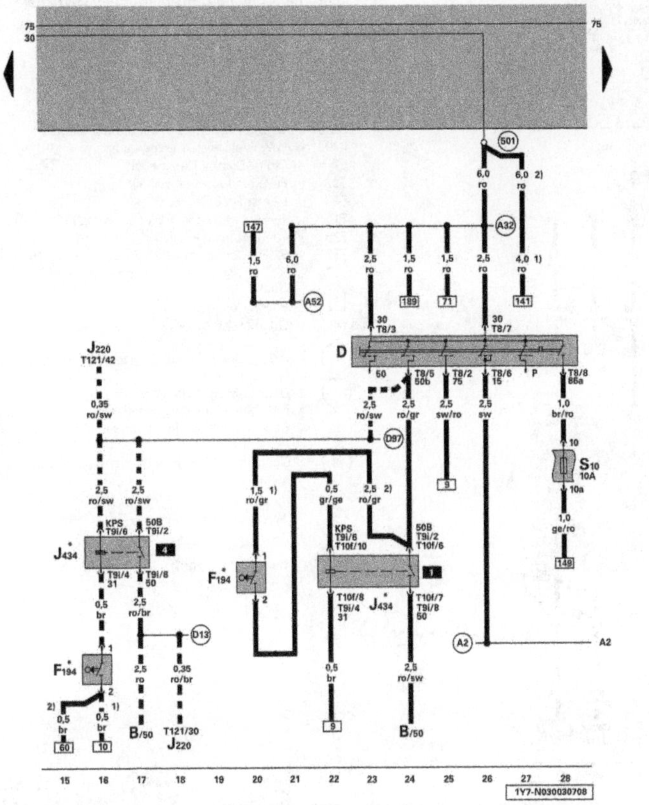

- B - Starter
- D - Ignition/Starter Switch
- F194 - Clutch Pedal Starter Interlock Switch
- J220 - Motronic Engine Control Module (ECM)
- J434 - Starter Lock/Clutch Pedal Switch Relay
- S10 - Fuse (on fuse/relay panel)
- T8 - 8-Pin Connector
- T9i - 9-Pin Connector 1)
- T10f - 10-Pin Connector 2)
- T121 - 121-Pin Connector

- 501 - Threaded Connection 2 - terminal 30 (on relay panel)
- A2 - Plus Connection (15) (in instrument panel wiring harness)
- A32 - Plus Connection (30) (in instrument panel wiring harness)
- A52 - Plus Connection 2 (30) (in instrument panel wiring harness)
- D13 - Connection (50) (in right front wiring harness)
- D97 - Connection (50) (in right engine compartment wiring harness)

- * - Manual transmission only
- --- - Engine code BPR only
- 1) - Through June 2007
- 2) - From July 2007

ws = white
sw = black
ro = red
br = brown
gn = green
bl = blue
gr = grey
li = violet
ge = yellow
or = orange

New Beetle Convertible — Wiring diagram — No. 30/4

Light Switched -On- Warning Buzzer, Instrument Cluster Control Module, Generator (GEN) Warning Lamp

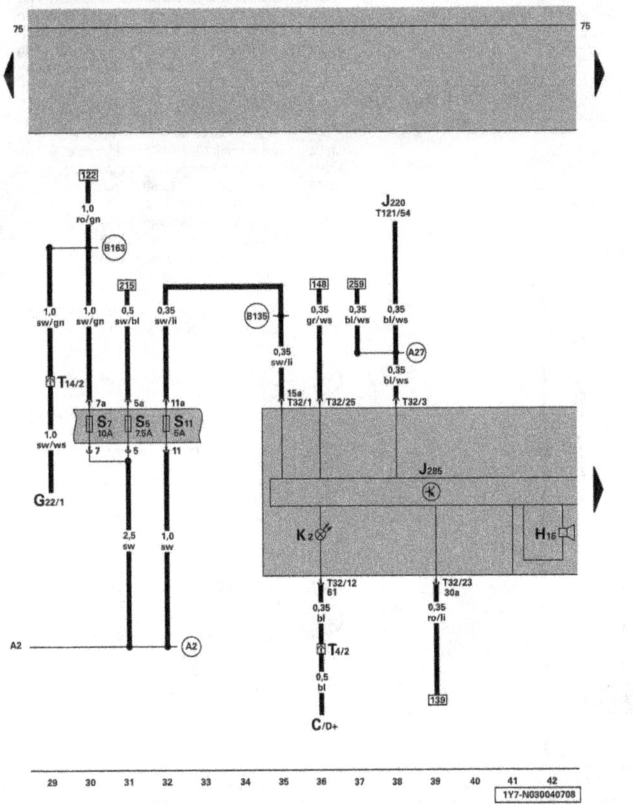

- C - Generator (GEN)
- G22 - Speedometer Vehicle Speed Sensor (VSS)
- H16 - Light Switched -On- Warning Buzzer
- J220 - Motronic Engine Control Module (ECM)
- J285 - Instrument Cluster Control Module
- K2 - Generator (GEN) Warning Lamp
- S5 - Fuse (on fuse/relay panel)
- S7 - Fuse (on fuse/relay panel)
- S11 - Fuse (on fuse/relay panel)
- T4 - 4-Pin Connector, in engine compartment left
- T14 - 14-Pin Connector
- T32 - 32-Pin Connector, blue, on instrument cluster
- T121 - 121-Pin Connector, on Motronic/Diesel DFI Control Module

- A2 - Plus Connection (15) (in instrument panel wiring harness)
- A27 - Vehicle Speed Signal (VSS) Connection (in instrument panel wiring harness)
- B135 - Plus Connection 1 (15a) (in interior wiring harness)
- B163 - Plus Connection 1 (15) (in interior wiring harness)

ws = white
sw = black
ro = red
br = brown
gn = green
bl = blue
gr = grey
li = violet
ge = yellow
or = orange

New Beetle Convertible

Wiring diagram

No. 30/5

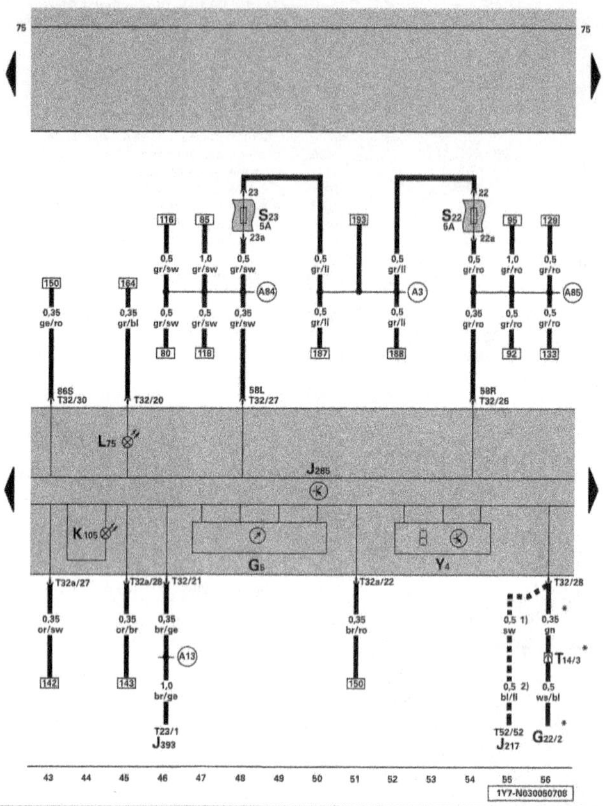

Instrument Cluster Control Module, Low Fuel Level Warning Lamp, Odometer Display

- G5 - Tachometer
- G22 - Speedometer Vehicle Speed Sensor (VSS)
- J217 - Transmission Control Module (TCM)
- J285 - Instrument Cluster Control Module
- J393 - Comfort System Central Control Module
- K105 - Low Fuel Level Warning Lamp
- L75 - Digital Display Illumination
- S22 - Fuse (on fuse/relay panel)
- S23 - Heater Fuse
- T14 - 14-Pin Connector, in plenum chamber
- T23 - 23-Pin Connector, at J393
- T32 - 32-Pin Connector, blue, on instrument cluster
- T32a - 32-Pin Connector, green, on instrument cluster
- T52 - 52-Pin Connector
- Y4 - Odometer Display

- (A3) - Plus Connection (58) (in instrument panel wiring harness)
- (A13) - Door Contact Switch Connection (in instrument panel wiring harness)
- (A84) - Connection (58L) (in Instrument panel wiring harness)
- (A85) - Connection (58R) (in instrument panel wiring harness)

- * - Manual transmission only
- --- - Automatic transmission only
- 1) - Through June 2007
- 2) - From July 2007

ws = white
sw = black
ro = red
br = brown
gn = green
bl = blue
gr = grey
li = violet
ge = yellow
or = orange

New Beetle Convertible

Wiring diagram

No. 30/6

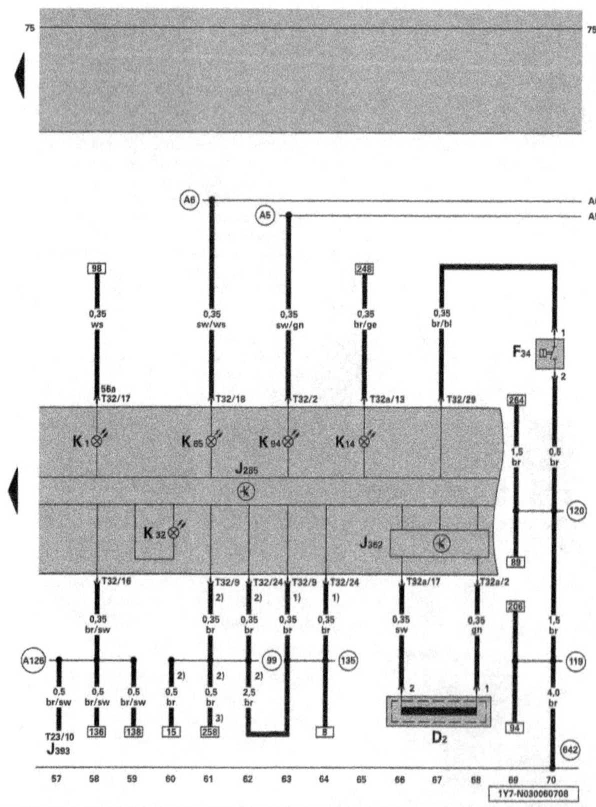

Instrument Cluster Control Module, Anti-Theft Immobilizer Control Module, High Beam Indicator Lamp, Parking Brake Indicator Lamp, Brake Pad Wear Indicator Lamp, Left Turn Signal Indicator Lamp, Right Turn Signal Indicator Lamp

- D2 - Anti-Theft Immobilizer Reading Coil
- F34 - Brake Fluid Level Warning Switch
- J285 - Instrument Cluster Control Module
- J362 - Anti-Theft Immobilizer Control Module
- J393 - Comfort System Central Control Module
- K1 - High Beam Indicator Lamp
- K14 - Parking Brake Indicator Lamp
- K32 - Brake Pad Wear Indicator Lamp
- K65 - Left Turn Signal Indicator Lamp
- K94 - Right Turn Signal Indicator Lamp
- T23 - 23-Pin Connector, at J393
- T32 - 32-Pin Connector, blue, on instrument cluster
- T32a - 32-Pin Connector, green, on instrument cluster

- (99) - Ground Connection (in console wiring harness)
- (119) - Ground Connection 1 (in headlamp wiring harness)
- (120) - Ground Connection 2 (in headlamp wiring harness)
- (135) - Ground Connection 2 (in instrument panel wiring harness)
- (642) - Ground Connection (on electronic fan control)

- (A5) - Right Turn Signal Plus Connection (in instrument panel wiring harness)
- (A6) - Left Turn Signal Plus Connection (in instrument panel wiring harness)
- (A126) - Rear Lid Contact Switch Connection (in instrument panel wiring harness)

- 1) - Through June 2007
- 2) - From July 2007
- 3) - Through June 2008

ws = white
sw = black
ro = red
br = brown
gn = green
bl = blue
gr = grey
li = violet
ge = yellow
or = orange

New Beetle Convertible

Wiring diagram

No. 30/7

Turn Signal Switch, Headlamp Dimmer/Flasher Switch, Left Front Turn Signal Lamp, Left Front Side Marker Lamp

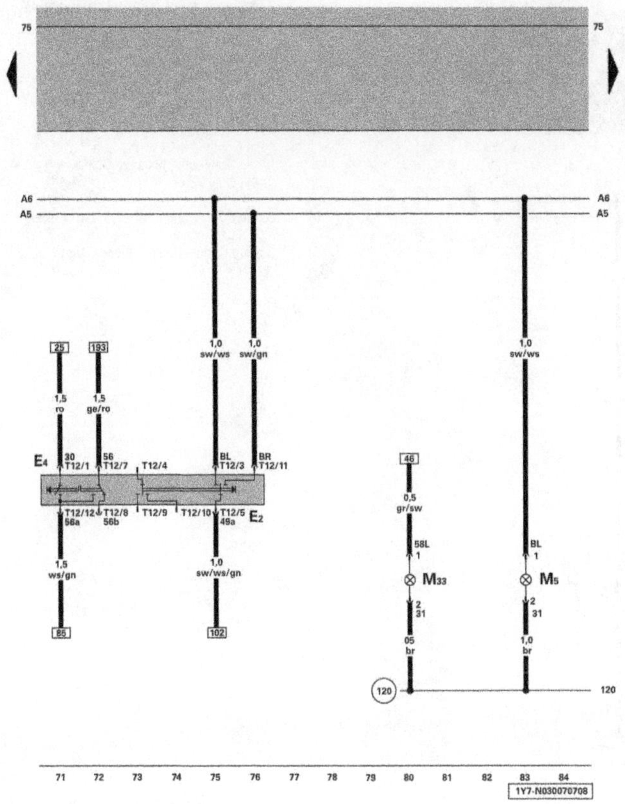

E2 - Turn Signal Switch
E4 - Headlamp Dimmer/Flasher Switch
M5 - Left Front Turn Signal Lamp
M33 - Left Front Side Marker Lamp
T12 - 12-Pin Connector, near steering column

(120) - Ground Connection 2 (in headlamp wiring harness)

ws = white
sw = black
ro = red
br = brown
gn = green
bl = blue
gr = grey
li = violet
ge = yellow
or = orange

New Beetle Convertible

Wiring diagram

No. 30/8

Left Parking Lamp, Right Parking Lamp, Right Front Turn Signal Lamp, Left Low Beam Head Lamp, Left High Beam Headlamp, Right Low Beam Headlamp, Right High Beam Head lamp, Right Front Side Marker Lamp

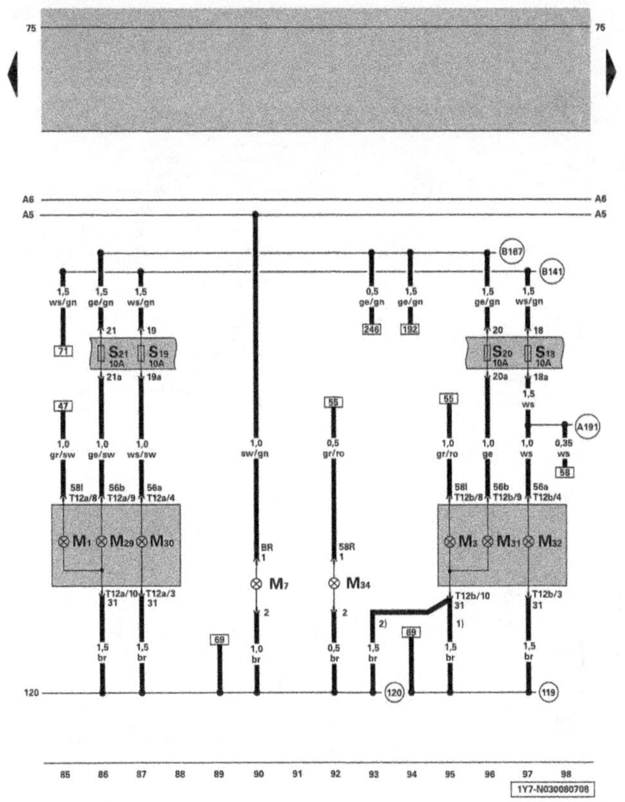

M1 - Left Parking Lamp
M3 - Right Parking Lamp
M7 - Right Front Turn Signal Lamp
M29 - Left Low Beam Headlamp
M30 - Left High Beam Headlamp
M31 - Right Low Beam Headlamp
M32 - Right High Beam Headlamp
M34 - Right Front Side Marker Lamp
S18 - Fuse (on fuse/relay panel)
S19 - Fuse (on fuse/relay panel)
S20 - Fuse (on fuse/relay panel)
S21 - Fuse (on fuse/relay panel)
T12a - 12-Pin Connector
T12b - 12-Pin Connector

(119) - Ground Connection 1 (in headlamp wiring harness)
(120) - Ground Connection 2 (in headlamp wiring harness)
(A191) - Connection (56aR) (in instrument panel wiring harness)
(B141) - Plus Connection 1 (56a) (in interior wiring harness)
(B167) - Connection (56b) (in passenger compartment wiring harness)

1) - Through June 2007
2) - From July 2007

ws = white
sw = black
ro = red
br = brown
gn = green
bl = blue
gr = grey
li = violet
ge = yellow
or = orange

BentleyPublishers.com
Edition 07.08

New Beetle Convertible — Wiring diagram — No. 30/9

Emergency Flasher Switch, Turn Signal Relay, Driver Exterior Mirror Turn Signal Lamp

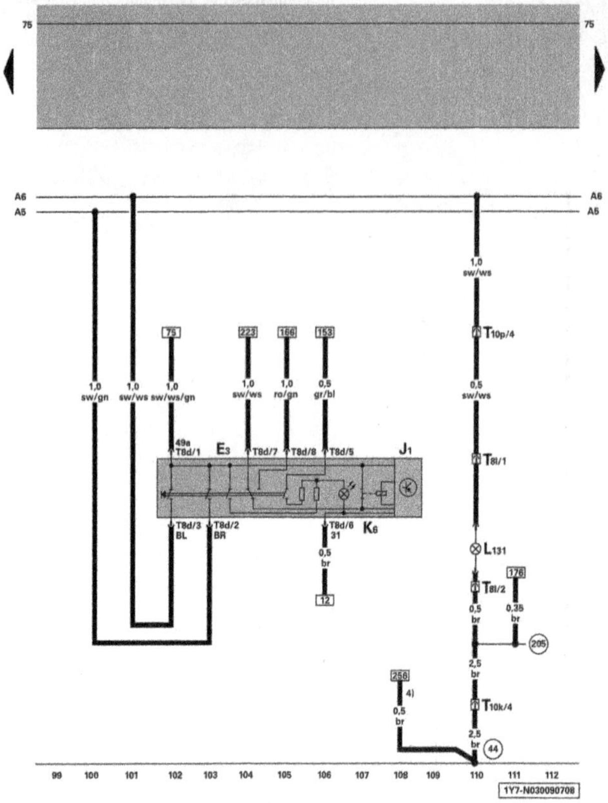

- E3 - Emergency Flasher Switch
- J1 - Turn Signal Relay
- K6 - Emergency Flasher Indicator Lamp
- L131 - Driver Exterior Mirror Turn Signal Lamp
- T8d - 8-Pin Connector
- T8l - 8-Pin Connector, at L131
- T10k - 10-Pin Connector, black, on A Pillar left
- T10p - 10-Pin Connector
- (44) - Ground Connection (on left lower A-pillar)
- (205) - Ground Connection (in driver's door wiring harness)
- 4) - From July 2008

ws = white
sw = black
ro = red
br = brown
gn = green
bl = blue
gr = grey
li = violet
ge = yellow
or = orange

New Beetle Convertible — Wiring diagram — No. 30/10

Brake Light Switch, Back-Up Light Switch, Brake Pedal Switch, Front Passenger Exterior Mirror Turn Signal Lamp, Left Rear Turn Signal Lamp, Left Back-Up Lamp, Left Brake/Taillight Lamp, Left Rear Side Marker Lamp

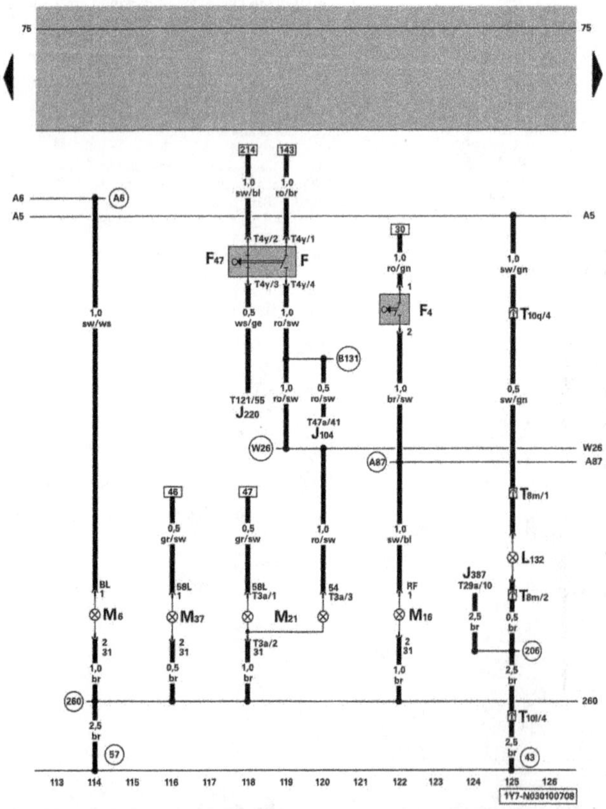

- F - Brake Light Switch
- F4 - Back-Up Light Switch
- F47 - Brake Pedal Switch
- J104 - ABS Control Module
- J220 - Motronic Engine Control Module (ECM)
- J387 - Front Passenger Door Control Module
- L132 - Front Passenger Exterior Mirror Turn Signal Lamp
- M6 - Left Rear Turn Signal Lamp
- M16 - Left Back-Up Lamp
- M21 - Left Brake/Taillight Lamp
- M37 - Left Rear Side Marker Lamp
- T3a - 3-Pin Connector
- T4y - 4-Pin Connector
- T8m - 8-Pin Connector, at L1320
- T10l - 10-Pin Connector, black, on A Pillar right
- T10q - 10-Pin Connector
- T29a - 29-Pin Connector
- T47a - 47-Pin Connector
- T121 - 121-Pin Connector
- (43) - Ground Connection (on right lower A-pillar)
- (57) - Ground Connection (on left rear pillar)
- (206) - Ground Connection (in Front Passenger's door wiring harness)
- (260) - Ground Connection (in taillamps wiring harness)
- (A6) - Left Turn Signal Plus Connection (in instrument panel wiring harness)
- (A87) - Back-up Light Connection (in instrument panel wiring harness)
- (B131) - Connection (54) (in interior wiring harness)
- (W26) - Plus Connection (54) (in taillamp wiring harness)

ws = white
sw = black
ro = red
br = brown
gn = green
bl = blue
gr = grey
li = violet
ge = yellow
or = orange

New Beetle Convertible

Wiring diagram **No. 30/11**

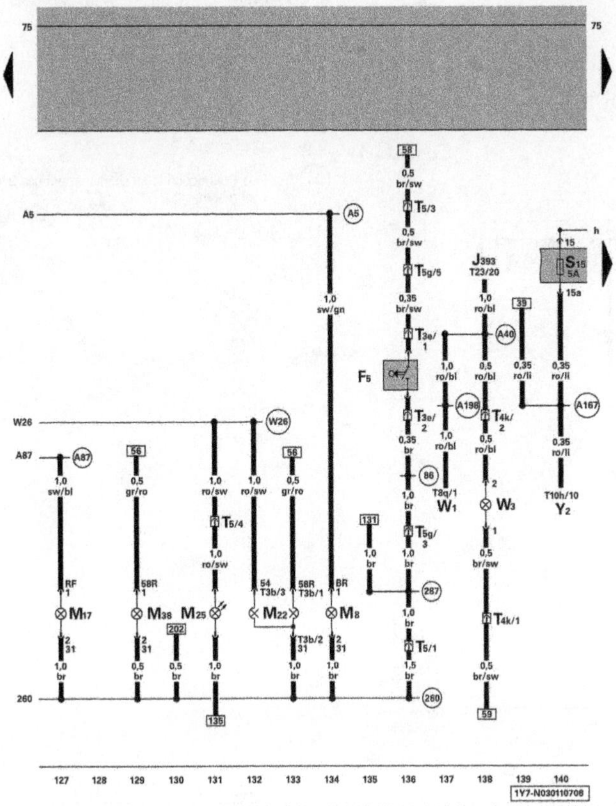

Luggage Compartment Lamp Switch, Right Rear Turn Signal Lamp, Right Back-Up Lamp, Right Brake/Taillight Lamp, High-mount Brake Light, Right Rear Side Marker Lamp

F5 - Luggage Compartment Lamp Switch
J393 - Comfort System Central Control Module
M8 - Right Rear Turn Signal Lamp
M17 - Right Back-Up Lamp
M22 - Right Brake/Taillight Lamp
M25 - High-mount Brake Light
M38 - Right Rear Side Marker Lamp
S15 - Fuse (on fuse/relay panel)
T3b - 3-Pin Connector
T3e - 3-Pin Connector, at F5
T4k - 4-Pin Connector, at J256
T5 - 5-Pin Connector, black, on C-pillar, left
T5g - 5-Pin Connector, black, in rear lid
T8q - 8-Pin Connector
T10h - 10-Pin Connector, at Y2
T23 - 23-Pin Connector
W1 - Front Interior Light
W3 - Luggage Compartment Light
Y2 - Digital Clock

(86) - Ground Connection 1 (in rear wiring harness)
(260) - Ground Connection (in taillamps wiring harness)
(287) - Ground Connection (in rear lid hydraulics wiring harness)
(A5) - Right Turn Signal Plus Connection (in instrument panel wiring harness)
(A40) - Plus Connection 1 (30) (in instrument cluster wiring harness)
(A87) - Back-up Light Connection (in instrument panel wiring harness)
(A167) - Plus Connection 3 (30a) (in instrument panel wiring harness)
(A198) - Connection 6 (30a) (in instrument panel wiring harness)
(W26) - Plus Connection (54) (in taillamp wiring harness)

ws = white
sw = black
ro = red
br = brown
gn = green
bl = blue
gr = grey
li = violet
ge = yellow
or = orange

New Beetle Convertible

Wiring diagram **No. 30/12**

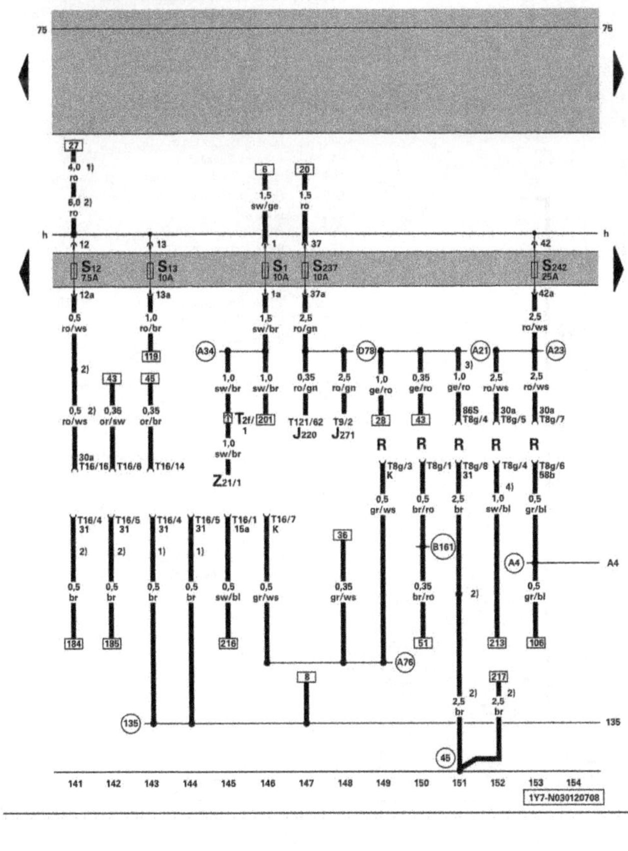

Data Link Connector (DLC), Radio Connection

J220 - Motronic Engine Control Module (ECM)
J271 - Motronic Engine Control Module (ECM) Power Supply Relay
R - Radio
S1 - Fuse (on fuse/relay panel)
S12 - Fuse (on fuse/relay panel)
S13 - Fuse (on fuse/relay panel)
S237 - Fuse (on fuse panel)
S242 - Fuse (on fuse panel)
T2f - 2-Pin Connector
T8g - 8-Pin Connector, on radio
T9 - 9-Pin Connector
T16 - 16-Pin Connector, black, Data Link Connector (DLC), bellow steering column, left
T121 - 121-Pin Connector, at J220
Z21 - Heated Rear Window

(45) - Ground Connection (behind instrument panel, center)
(135) - Ground Connection 2 (in instrument panel wiring harness)
(A4) - Plus Connection (58b) (in instrument panel wiring harness)
(A21) - Connection (86s) (in instrument panel wiring harness)
(A23) - Connection (30al) (in instrument panel wiring harness)
(A34) - Connection (75x) (in instrument panel wiring harness)
(A76) - K-Diagnostic Connection (in instrument panel wiring harness)
(B161) - Anti-Theft Alarm System Connection (in interior wiring harness)
(D78) - Plus Connection 1 (30a) (in engine compartment wiring harness)

1) - Through June 2007
2) - From July 2007
3) - Through June 2008
4) - From July 2008

ws = white
sw = black
ro = red
br = brown
gn = green
bl = blue
gr = grey
li = violet
ge = yellow
or = orange

New Beetle Convertible

Wiring diagram

No. 30/13

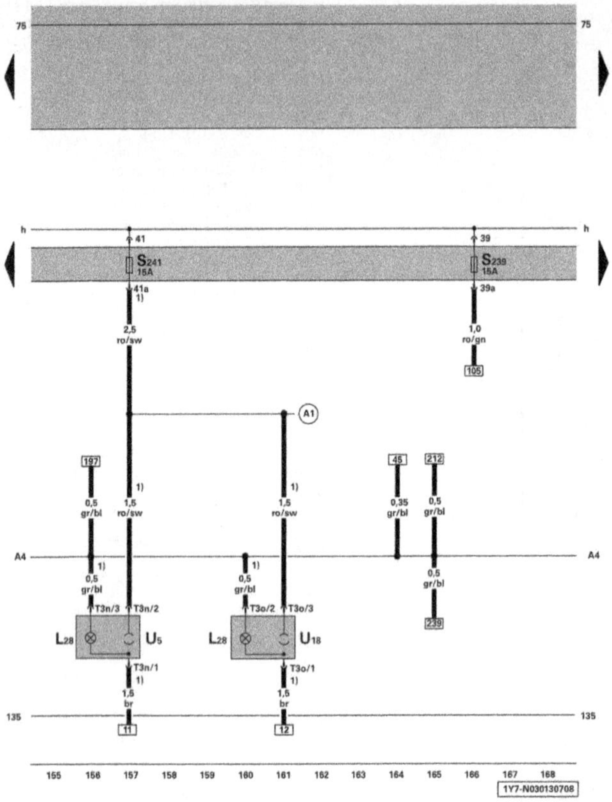

12 V Sockets

L28 - Cigarette Lighter Illumination
S239 - Fuse (on fuse panel)
S241 - Fuse (on fuse panel)
T3n - 3-pin connector, at U5
T3o - 3-pin connector, at U18
U5 - 12 V Socket
U18 - 12V Socket 2

(A1) - Plus Connection (30a) (in instrument panel wiring harness)

1) - Through June 2007

ws = white
sw = black
ro = red
br = brown
gn = green
bl = blue
gr = grey
li = violet
ge = yellow
or = orange

New Beetle Convertible

Wiring diagram

No. 30/14

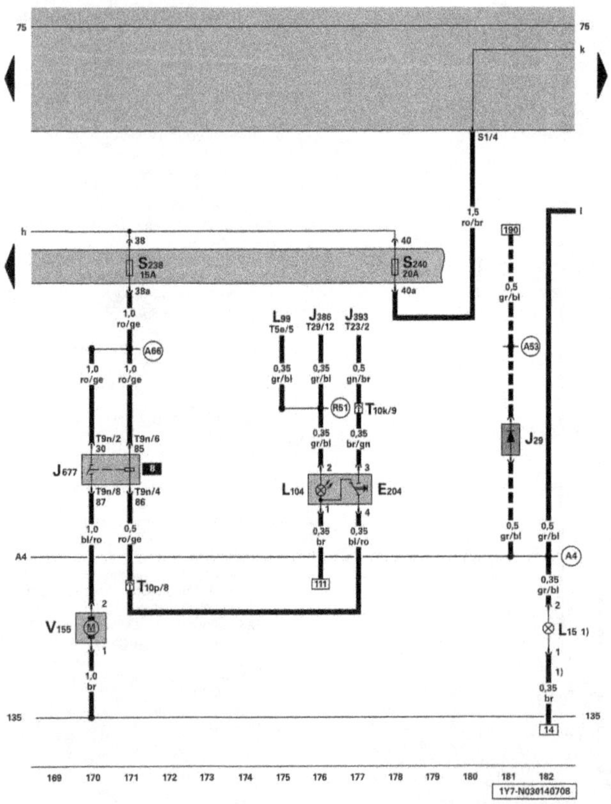

Remote Fuel Tank Flap Switch, Blocking Dio de, Ashtray Illumination, Remote Unlock Swi tch Illumination, Fuel Tank Lid Unlock Motor

E204 - Remote Fuel Tank Flap Switch
J29 - Blocking Diode
J386 - Driver Door Control Module
J393 - Comfort System Central Control Module
J677 - Fuel Filler Lid Unlock Relay
L15 - Ashtray Illumination
L99 - Interior Lock Switch Illumination
L104 - Remote Unlock Switch Illumination
S238 - Fuse (on fuse panel)
S240 - Fuse (on fuse panel)
T5e - 5-Pin Connector
T9n - 9-pin Connector, at J677
T10k - 10-pin Connector, black, on A-pillar, left
T10p - 10-pin Connector, blue, on A-pillar, left
T23 - 23-pin Connector, at J393
T29 - 29-pin Connector
V155 - Fuel Tank Lid Unlock Motor

(A4) - Plus Connection (58b) (in instrument panel wiring harness)
(A53) - Plus Connection (58b) (in instrument panel wiring harness)
(A66) - Connection (30a, central locking/Anti-Theft warning system/IR (in instrument panel wiring harness)
(R51) - -Connection (58b) (in driver's door wiring harness)

--- - Automatic transmission only
1) - Through June 2007

ws = white
sw = black
ro = red
br = brown
gn = green
bl = blue
gr = grey
li = violet
ge = yellow
or = orange

New Beetle Convertible — Wiring diagram — No. 30/15

Light Switch, Blocking Diode, Headlamp Switch Illumination

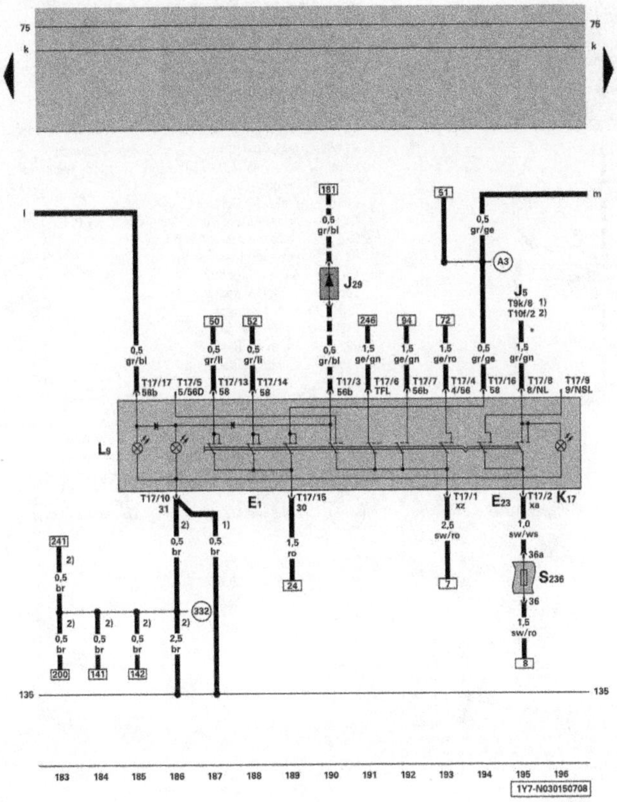

- E1 - Light Switch
- E23 - Fog Lamp Switch
- J5 - Fog Lamp Relay
- J29 - Blocking Diode
- K17 - Fog Lamp Indicator Lamp
- L9 - Headlamp Switch Illumination
- S236 - Fuse (on fuse panel)
- T9k - 9-Pin Connector, at J5
- T10f - 10-Pin Connector, at J5
- T17 - 17-Pin Connector, at E1

- (332) - Ground Connection 4 (in instrument panel wiring harness)
- (A3) - Plus Connection (58) (in instrument panel wiring harness)

- * - Only vehicles with fog lights
- - - - - Automatic transmission only
- 1) - Through June 2007
- 2) - From July 2007

ws = white
sw = black
ro = red
br = brown
gn = green
bl = blue
gr = grey
li = violet
ge = yellow
or = orange

New Beetle Convertible — Wiring diagram — No.30/16

Instrument Panel Illumination Dimmer Switch, Glove Compartment Lamp Switch, Signal Horn Activation, Signal Horn/Dual Tone Horn, Horn Relay, Glove Compartment Light, License Plate Light

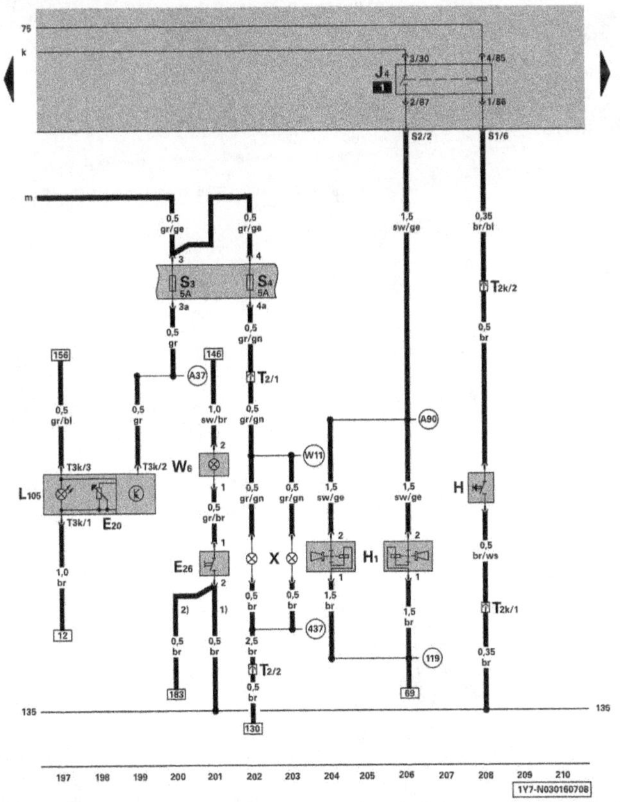

- E20 - Instrument Panel Illumination Dimmer Switch
- E26 - Glove Compartment Lamp Switch
- H - Signal Horn Activation
- H1 - Signal Horn/Dual Tone Horn
- J4 - Horn Relay
- L105 - Lighting Control Illumination
- S3 - Fuse (on fuse/relay panel)
- S4 - Fuse (on fuse/relay panel)
- T2 - Double Connector, behind rear bumper cover
- T2k - Double Connector
- T3k - 3-Pin Connector, at E20
- W6 - Glove Compartment Light
- X - License Plate Light

- (119) - Ground Connection 1 (in headlamp wiring harness)
- (437) - Ground Connection (in license plate light wiring harness)
- (A37) - Connection (58a) (in instrument panel wiring harness)
- (A90) - Dual-tone Horn Connection (in instrument panel wiring harness)
- (W11) - Connection (58) (in rear lid wiring harness)

ws = white
sw = black
ro = red
br = brown
gn = green
bl = blue
gr = grey
li = violet
ge = yellow
or = orange

Standard equipment from July 2005, convertible

New Beetle Convertible — Wiring diagram — No. 30/17

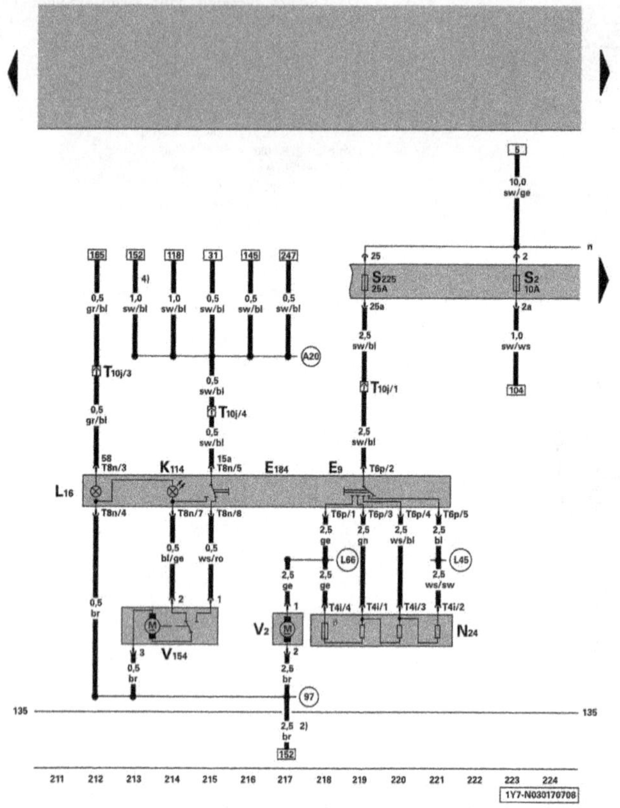

Fresh Air Blower Switch, Fresh Air Blower/Recirculated Air Switch, Fresh Air and Recirculating Air Mode Indicator Lamp, Fresh Air Control Illumination, Fresh Air Blower Series Resistor With Fuse, Fresh Air Blower, Fresh/Recirculating Air Door Motor

- E9 - Fresh Air Blower Switch
- E184 - Fresh Air Blower/Recirculated Air Switch
- K114 - Fresh Air and Recirculating Air Mode Indicator Lamp
- L16 - Fresh Air Control Illumination
- N24 - Fresh Air Blower Series Resistor With Fuse
- S2 - Fuse (on fuse/relay panel)
- S225 - Fuse (on fuse panel)
- T4i - 4-Pin Connector
- T6p - 6-Pin Connector
- T8n - 8-Pin Connector
- T10j - 10-Pin Connector
- V2 - Fresh Air Blower
- V154 - Fresh/Recirculating Air Door Motor

- (97) - Ground Connection 1 (in A/C wiring harness)
- (A20) - Plus Connection (15a) (in instrument panel wiring harness)
- (L45) - Connection (in A/C system wiring harness)
- (L66) - Connection (in wiring harness) heater blower

- 2) - From July 2007
- 4) - From July 2008

ws = white
sw = black
ro = red
br = brown
gn = green
bl = blue
gr = grey
li = violet
ge = yellow
or = orange

New Beetle Convertible — Wiring diagram — No. 30/18

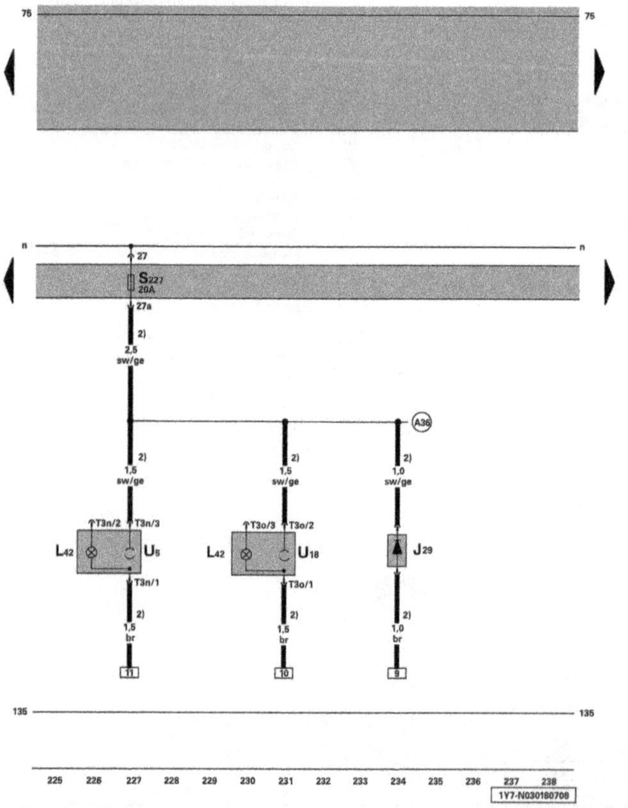

Blocking Diode, 12 V Sockets

- J29 - Blocking Diode
- L42 - Socket Illumination
- S227 - Fuse (on fuse panel)
- T3n - 3-Pin Connector
- T3o - 3-Pin Connector
- U5 - 12 V Socket
- U18 - 12V Socket 2

- (A36) - Connection (75a) (in instrument panel wiring harness)

- 2) - From July 2007

ws = white
sw = black
ro = red
br = brown
gn = green
bl = blue
gr = grey
li = violet
ge = yellow
or = orange

New Beetle Convertible

Wiring diagram

No. 30/19

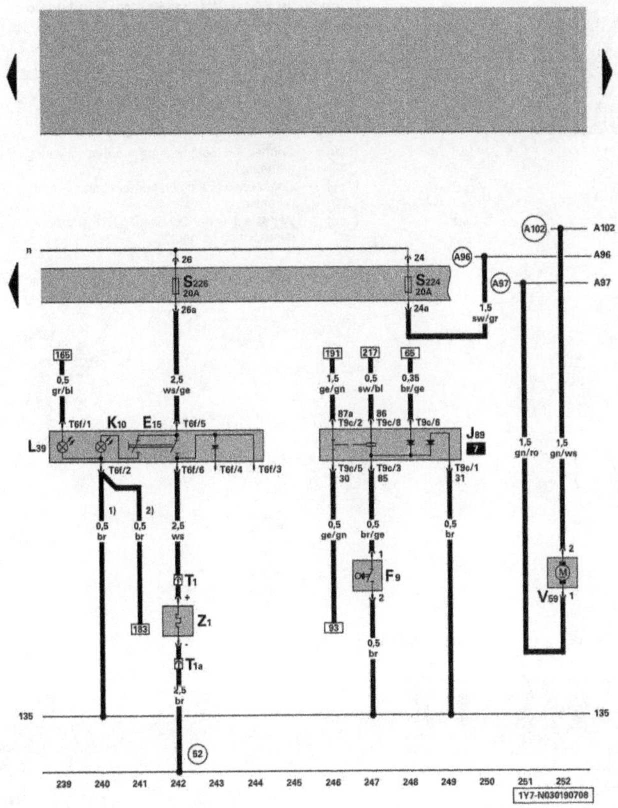

Rear Window Defogger Switch, Parking Brake Warning Light Switch, Daytime Running Lamps Change-over Relay, Windshield and Rear Window Washer Pump, Heated Rear Window

E15 - Rear Window Defogger Switch
F9 - Parking Brake Warning Light Switch
J89 - Daytime Running Lamps Change-over Relay
K10 - Rear Window Defogger Indicator Lamp
L39 - Rear Window Defogger Switch Illumination
S224 - Fuse (on fuse panel)
S226 - Fuse (on fuse panel)
T1 - Single Connector, at Z1+
T1a - Single Connector, at Z1-
T6f - 6-Pin Connector, at E15
T9c - 9-Pin Connector, at J89
V59 - Windshield and Rear Window Washer Pump
Z1 - Heated Rear Window

(52) - Ground Connection (in rear lid, left)

(A96) - Connection (53a) (in instrument panel wiring harness)

(A97) - Connection (53) (in instrument panel wiring harness)

(A102) - Windsield Wiper Connection (in instrument panel wiring harness)

1) - Through June 2007
2) - From July 2007

ws = white
sw = black
ro = rod
br = brown
gn = green
bl = blue
gr = grey
li = violet
ge = yellow
or = orange

New Beetle Convertible

Wiring diagram

No. 30/20

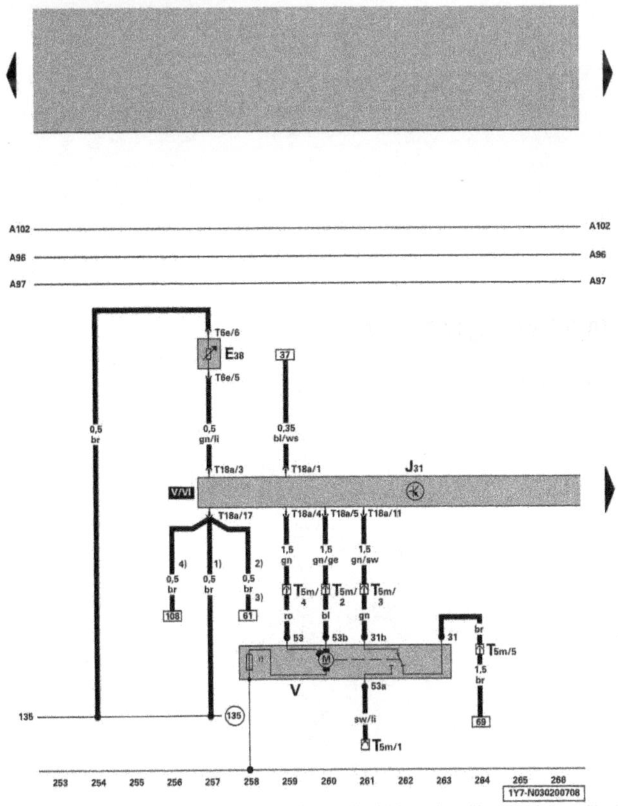

Windshield Wiper Intermittent Regulator, Windshield Wiper/Washer Intermittent Relay, Windshield Wiper Motor

E38 - Windshield Wiper Intermittent Regulator
J31 - Wiper/Washer Intermittent Relay
T5m - 5-Pin Connector, on Windshield Wiper Motor
T6e - 6-Pin Connector, near steering column
T18a - 18-Pin Connector, at J31
V - Windshield Wiper Motor

(135) - Ground Connection 2 (in instrument panel wiring harness)

1) - Through June 2007
2) - From July 2007
3) - Through June 2008
4) - From July 2008

ws = white
sw = black
ro = red
br = brown
gn = green
bl = blue
gr = grey
li = violet
ge = yellow
or = orange

Standard equipment from July 2005, convertible

New Beetle Convertible Wiring diagram **No. 30/21**

Windshield Wiper/Washer Switch, Windshield Wiper/Washer Intermittent Relay

E22 - Winshield Wiper/Washer Switch
J31 - Wiper/Washer Intermittent Relay
T8c - 8-Pin Connector, at E22
T18a - 18-Pin Connector, at J31

(A96) - Connection (53a) (in instrument panel wiring harness)
(A97) - Connection (53) (in instrument panel wiring harness)
(A102) - Windsield Wiper Connection (in instrument panel wiring harness)

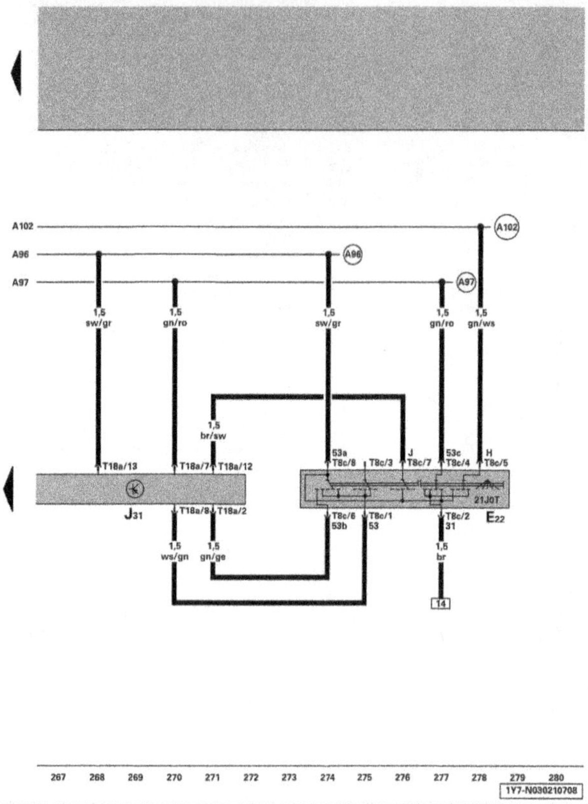

ws = white
sw = black
ro = red
br = brown
gn = green
bl = blue
gr = grey
li = violet
ge = yellow
or = orange

THIS SPACE INTENTIONALLY

LEFT BLANK

Power sunroof

from August 1998

Deviate relay location and fuseplacements as well as the locations of multiple connectors see section "component locations".

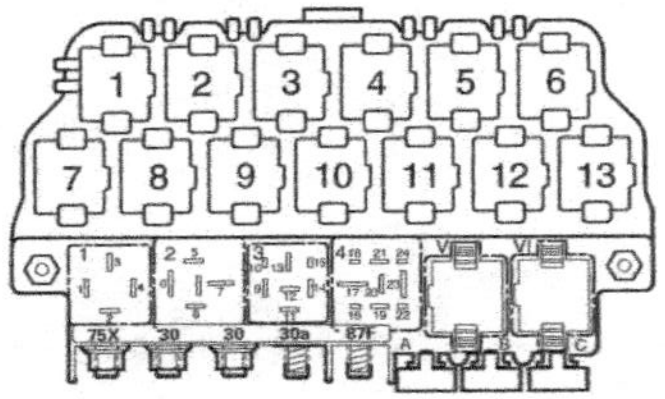

97-21956

Fuse colors
30 A - green
25 A - white
20 A - yellow
15 A - blue
10 A - red
7,5 A - brown
5 A - beige

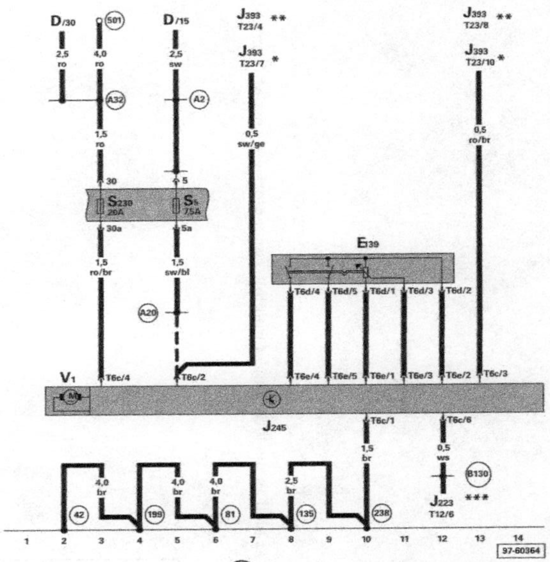

97-60364

ws = white
sw = black
ro = red
br = brown
gn = green
bl = blue
gr = grey
li = violet
ge = yellow

D - Ignition/Starter Switch
E139 - Sunroof Regulator
J223 - Rear Spoiler Control Module ***
J393 - Central control module for comfort system
J245 - Power Sunroof Control Module
S5 - Fuse 5 in fuse holder
S230 - Fuse 30 in fuse holder
T6c - 6-Pin Connector
T6d - 6-Pin Connector
T6e - 6-Pin Connector
T12 - 12-Pin Connector
V1 - Sunroof Motor

(42) - Ground connection, beside steering column
(81) - Ground connection -1-, in instrument panel wiring harness
(135) - Ground connection -2-, in instrument panel wiring harness
(199) - Ground connection -3-, in instrument panel wiring harness
(238) - Ground connection -1-, in wiring harness interior
(501) - Threaded connection -2- (30) on the relay plate
(A2) - plus connection (15), in instrument panel wiring harness
(A20) - wire connection (15a), in instrument panel wiring harness
(A32) - plus connection (30), in instrument panel wiring harness
(B130) - Connector (sliding sunroof) in wiring harness interior
- - - without central locking (where applicable)
* - with power windows
** - with manual windows
*** - Where applicable

Sunroof regulator, power sunroof control module, sunroof motor

New Beetle Wiring diagram **No. 23/1**

Power sunroof

from June 2001

Deviate relay location and fuseplacements as well as the locations of multiple connectors see section "component locations".

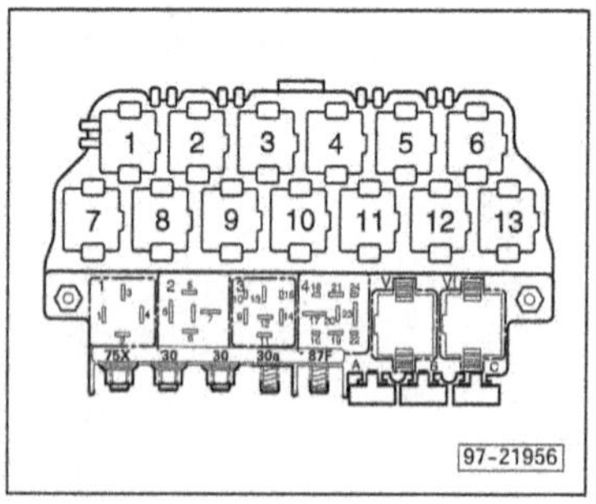

Fuse colors

30 A - green
25 A - white
20 A - yellow
15 A - blue
10 A - red
7,5 A - brown
5 A - beige
3 A - violet

New Beetle Wiring diagram **No. 23/2**

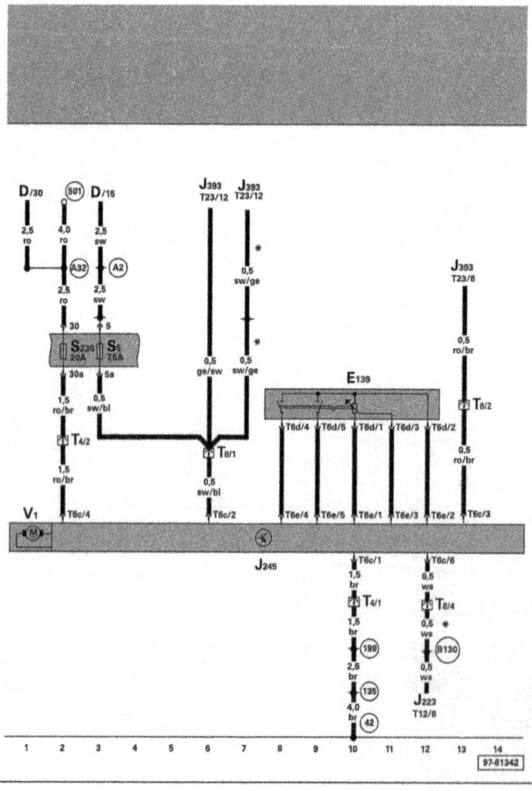

Sunroof regulator, power sunroof control module, sunroof motor

- D - Ignition/Starter Switch
- E139 - Sunroof Regulator
- J223 - Rear Spoiler Control Module
- J393 - Central control module for comfort system
- J245 - Power Sunroof Control Module
- S5 - Fuse 5 in fuse holder
- S230 - Fuse 30 in fuse holder
- T4 - 4-Pin Connector
- T6c - 6-Pin Connector
- T6d - 6-Pin Connector
- T6e - 6-Pin Connector
- T8 - 8-Pin Connector, near front interior light
- T12 - 12-Pin Connector
- T23 - 23-Pin Connector
- V1 - Sunroof Motor

- (42) - Ground connection, beside steering column
- (135) - Ground connection -2-, in instrument panel wiring harness
- (199) - Ground connection -3-, in instrument panel wiring harness
- (501) - Threaded connection -2- (30) on the relay plate
- (A2) - plus connection (15), in instrument panel wiring harness
- (A32) - plus connection (30), in instrument panel wiring harness
- (B130) - Connector (sliding sunroof) in wiring harness interior
- * - Only vehicles with rear spoiler

ws = white
sw = black
ro = red
br = brown
gn = green
bl = blue
gr = grey
li = violet
ge = yellow
or = orange

New Beetle Wiring diagram **No. 28/1**

Power sunroof

from December 2002

Deviate relay location and fuseplacements as well as the locations of multiple connectors see section "component locations".

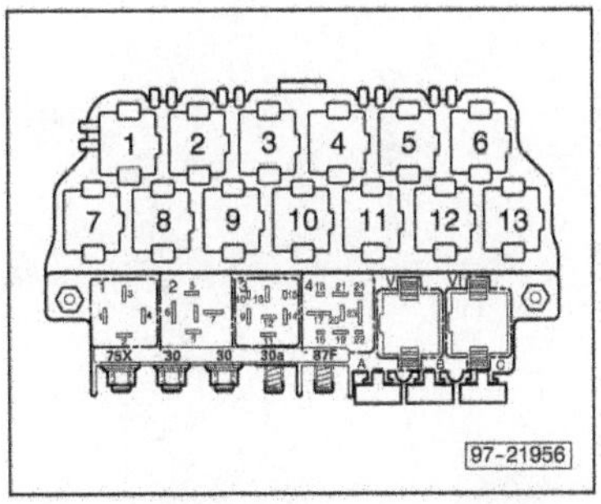

Fuse colors

30 A - green
25 A - white
20 A - yellow
15 A - blue
10 A - red
7,5 A - brown
5 A - beige
3 A - violet

New Beetle Wiring diagram **No. 28/2**

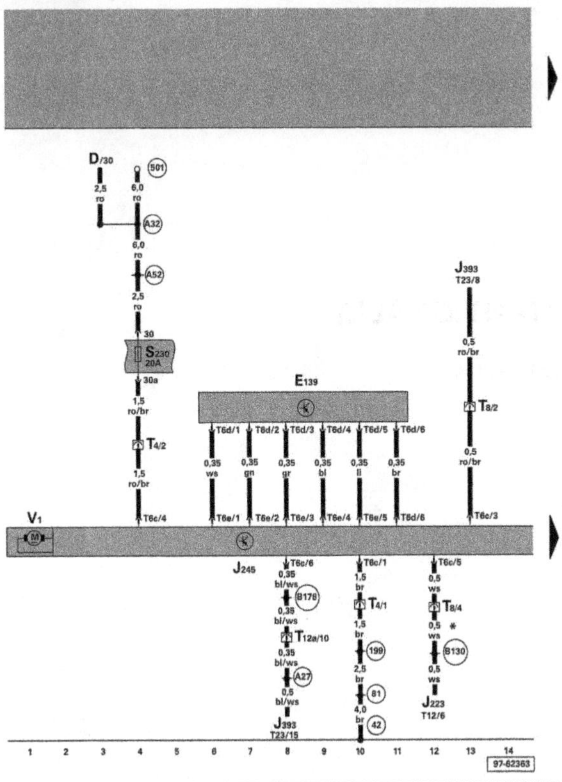

Sunroof regulator, power sunroof control module, sunroof motor

- D - Ignition/Starter Switch
- E139 - Sunroof Regulator
- J223 - Rear Spoiler Control Module
- J245 - Power Sunroof Control Module
- J393 - Central control module for comfort system
- S230 - Fuse 30 in fuse holder
- T4 - 4-Pin Coupling Connector, near front interior light
- T6c - 6-Pin Connector at J245
- T6d - 6-Pin Connector at E139
- T6e - 6-Pin Connector at J245
- T8 - 8-Pin Coupling Connector, near front interior light
- T12 - 12-Pin Connector at J223
- T12a - 12-Pin Coupling Connector, near front interior light
- T23 - 23-Pin Connector at J393
- V1 - Sunroof Motor
- (42) - Ground connection, beside steering column
- (81) - Ground connection -1-, in instrument panel wiring harness
- (199) - Ground connection -3-, in instrument panel wiring harness
- (501) - Threaded connection -2- (30) on the relay plate
- (A27) - Wire Connection (vehicle speed signal), in instrument panel wiring harness
- (A32) - plus connection (30), in instrument panel wiring harness
- (A52) - Plus Connection 2 (30) (in instrument panel wiring harness)
- (B130) - Connector (sliding sunroof) in wiring harness interior
- (B178) - Connection (speed signal) in passenger compartment wiring harness
- * - Only vehicles with rear spoiler

ws = white
sw = black
ro = red
br = brown
gn = green
bl = blue
gr = grey
li = violet
ge = yellow
or = orange

Sunroof from Dec. 2002

New Beetle Wiring diagram **No. 28/3**

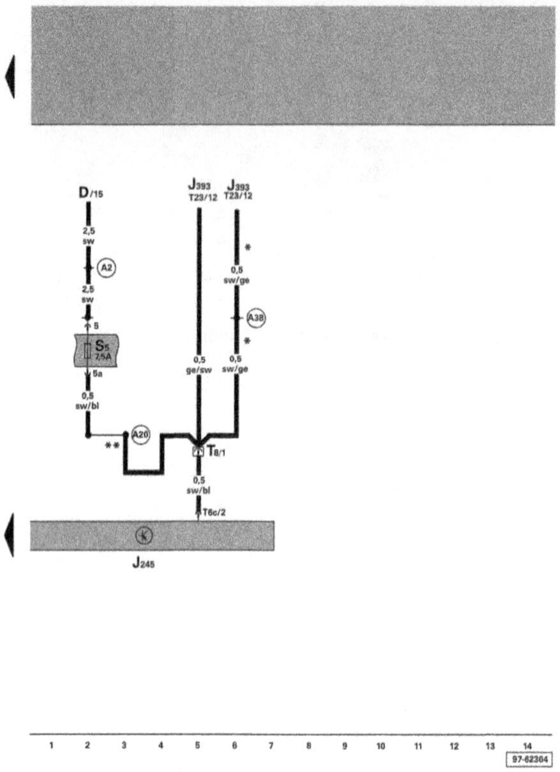

Power sunroof control module

D - Ignition/Starter Switch
J393 - Central control module for comfort system
J245 - Power Sunroof Control Module
S5 - Fuse 5 in fuse holder
T6c - 6-Pin Connector
T8 - 8-Pin Connector, near front interior light
T23 - 23-Pin Connector

(A2) - plus connection (15), in instrument panel wiring harness
(A20) - Plus Connection (15a) (in instrument panel wiring harness)
(A38) - plus connection -2- (15a), in instrument panel wiring harness

* - Only vehicles with rear spoiler
** - Only vehicles without power windows and without central locking

ws = white
sw = black
ro = red
br = brown
gn = green
bl = blue
gr = grey
li = violet
ge = yellow
or = orange

THIS SPACE INTENTIONALLY

LEFT BLANK

New Beetle **Wiring diagram** **No. 57/1**

Power sunroof,

from July 2005,

Additional Fuse, Relay, Control Module and Connector locations ⇒ Component Locations

Electrical troubleshooting ⇒ Guided Fault Finding using VAS 5051/5052

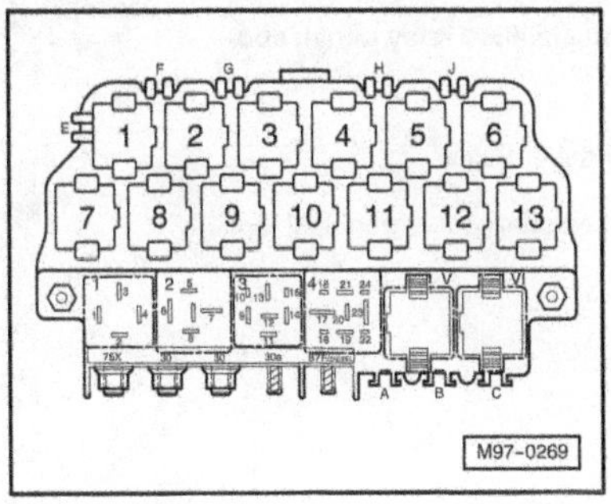

Note:

Number in parentheses indicates production control number stamped on relay housing.

New Beetle **Wiring diagram** **No. 57/2**

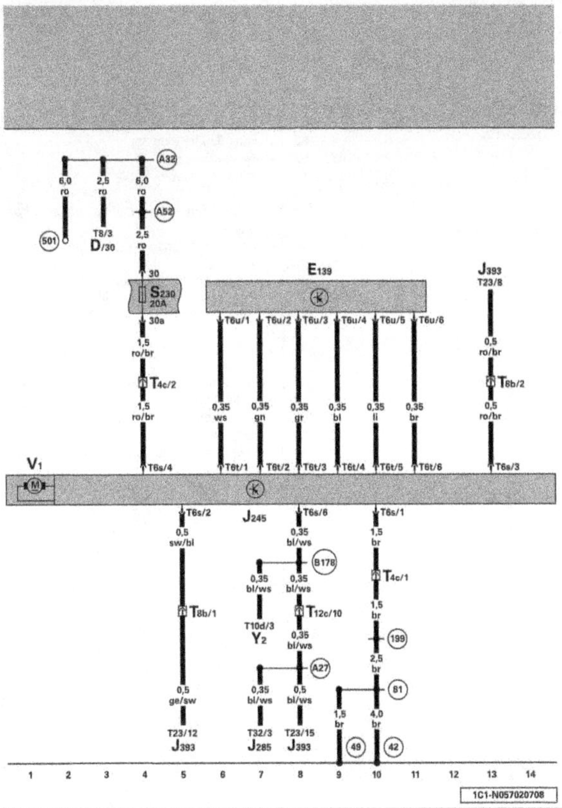

Sunroof Regulator, Power Sunroof Control Module, Sunroof Motor

- D - Ignition/Starter Switch
- E139 - Sunroof Regulator
- J245 - Power Sunroof Control Module
- J285 - Instrument Cluster Control Module
- J393 - Comfort System Central Control Module
- S230 - Fuse (on fuse panel)
- T4c - 4-Pin Connector, near front interior light
- T6s - 6-Pin Connector at J245
- T6t - 6-Pin Connector at J245
- T6u - 6-Pin Connector at E139
- T8 - 8-Pin Connector
- T8b - 8-Pin Connector, near front interior light
- T10d - 10-Pin Connector
- T12c - 12-Pin Connector, near front interior light
- T23 - 23-Pin Connector at J393
- T32 - 32-Pin Connector
- V1 - Sunroof Motor
- Y2 - Digital Clock
- (42) - Ground Connection (beside steering column)
- (49) - Ground Connection (near steering column)
- (81) - Ground Connection 1 (in instrument panel wiring harness)
- (199) - Ground Connection 3 (in instrument panel wiring harness)
- (501) - Threaded Connection 2 - terminal 30 (on relay panel)
- (A27) - Vehicle Speed Signal (VSS) Connection (in instrument panel wiring harness)
- (A32) - Plus Connection (30) (in instrument panel wiring harness)
- (A52) - Plus Connection 2 (30) (in instrument panel wiring harness)
- (B178) - Speed Signal Connection (in passenger compartment wiring harness)

ws = white
sw = black
ro = red
br = brown
gn = green
bl = blue
gr = grey
li = violet
ge = yellow
or = orange

New Beetle — Wiring diagram — No. 30/1

Telematics,

from October 2003,

Additional Fuse, Relay, Control Module and Connector locations ⇒ Component Locations

Troubleshooting Procedures ⇒ Guided Fault Finding using VAS 5051/5052

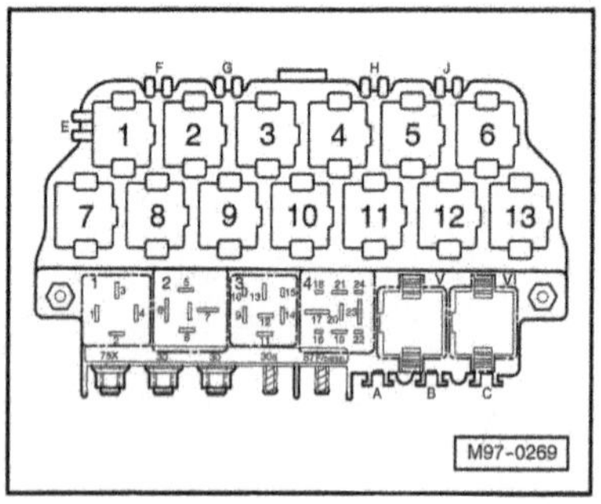

Relay locations on thirteen position auxiliary relay panel, above relay panel:

9 J641 - Alarm Horn Relay

10 J237 - Alarm System Light Flasher Relay

Fuse locations on thirteen position auxiliary relay panel, above relay panel:

H S338 - Telematics Fuse, (25 A)

Relay panel:

1 J4 - Horn Relay (53)

Notes:

Number in parentheses indicates production control number stamped on relay housing.

New Beetle — Wiring diagram — No. 30/2

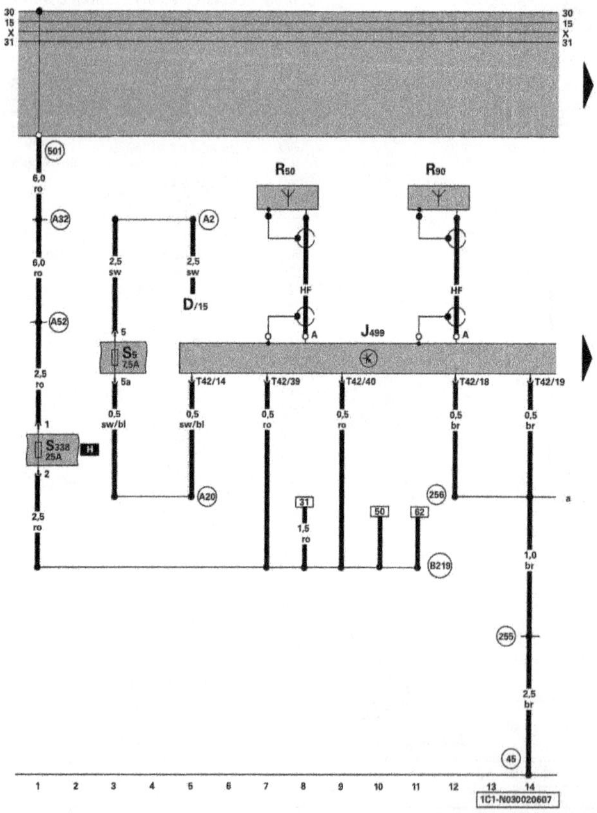

Telematic Control Module, Antenna for Navigation System (GPS), Auxiliary Antenna (Telematic)

D - Ignition/Starter Switch
J499 - Telematic Control Module
R50 - Antenna for Navigation System (GPS)
R90 - Auxiliary Antenna (Telematic)
S5 - Fuse 5 in fuse holder
S338 - Telematic fuse
T42 - 42-Pin Connector at J499

(45) - Ground connection, behind instrument panel, center
(255) - Ground connection, -1-, in wiring harness Radio
(256) - Ground connection, -2-, in wiring harness Radio
(501) - Threaded connection -2- (30) on the relay plate
(A2) - plus connection (15), in instrument panel wiring harness
(A20) - Plus Connection (15a) (in instrument panel wiring harness)
(A32) - plus connection (30), in instrument panel wiring harness
(A52) - Plus Connection 2 (30) (in instrument panel wiring harness)
(B219) - Plus Connection -4- (30a) (interior wiring harness)

ws = white
sw = black
ro = red
br = brown
gn = green
bl = blue
gr = grey
li = violet
ge = yellow
or = orange

New Beetle

Wiring diagram

No. 30/3

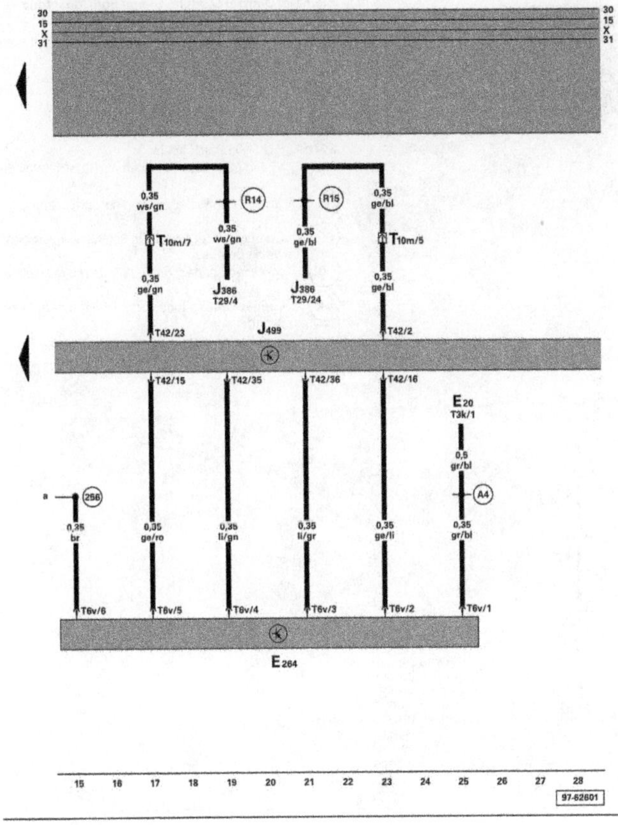

Telematic Control Module, Telematic Control Head

E20 - Instrument Panel Light Dimmer Switch
E264 - Telematic Control Head
J386 - Door control module, driver side
J499 - Telematic Control Module
T3k - 3-Pin Connector at E20
T6v - 3-Pin Connector for Telematic Control Head
T10m - 10-Pin Connector, brown, on A-Pillar left
T29 - 29-Pin Connector, on Door control module
T42 - 3-Pin Connector for Telematic Control Module

(256) - Ground connection, -2-, in wiring harness Radio

(A4) - plus connection (58b), in instrument panel wiring harness

(R14) - wire connection -1- (open), in driver's door wiring harness

(R15) - plus connection -2- (closed), in driver's door wiring harness

ws = white
sw = black
ro = red
br = brown
gn = green
bl = blue
gr = grey
li = violet
ge = yellow
or = orange

New Beetle

Wiring diagram

No. 30/4

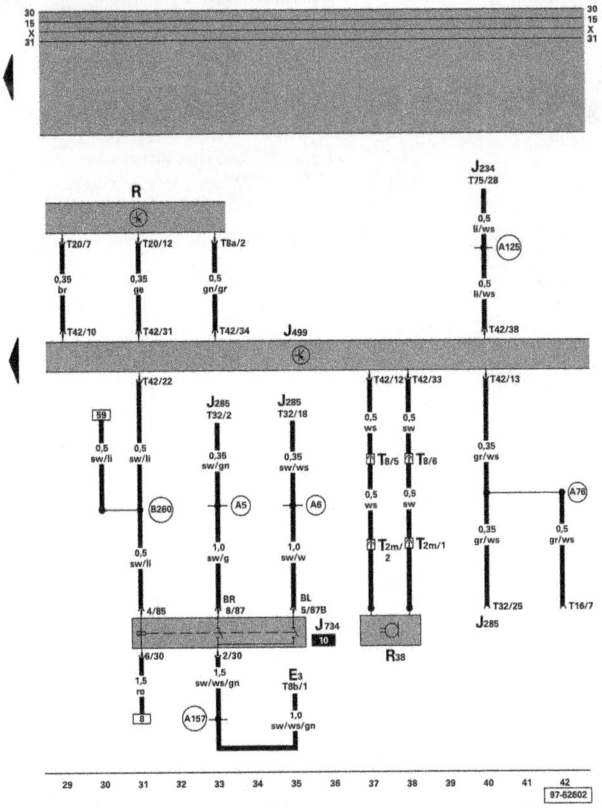

Telematic Control Module, Alarm System Light Flasher Relay, Radio, Telephone Microphone

E3 - Emergency Flasher Switch
J234 - Airbag Control Module
J285 - Control module with indicator unit in instrument panel insert
J499 - Telematic Control Module
J734 - Warning Lamp Auxiliary Relay
R - Radio
R38 - Telephone Microphone
T2m - 2-Pin Connector for Telephone Microphone
T8 - 8-Pin Coupling Connector, near front interior light
T8a - 8-Pin Connector at Radio
T8b - 8-Pin Connector at E3
T16 - 16-Pin Connector, Data Link Connector(DLC), below steering column, left
T20 - 20-Pin Connector at Radio
T32 - 32-Pin Connector, blue, on instrument cluster
T42 - 42-Pin Connector for Telematic Control Module
T75 - 75-Pin Connector for Airbag Control Module

(A5) - plus connection (right turn signal), in instrument panel wiring harness

(A6) - plus connection (left turn signal), in instrument panel wiring harness

(A76) - Connector (K-diagnosis wire) in instrument panel wiring harness

(A125) - Connection (crash signal) in instrument panel wiring harness

(A157) - Connection (49a) (in instrument panel wiring harness)

(B260) - Warning Lamp Connection (in interior wiring harness)

ws = white
sw = black
ro = red
br = brown
gn = green
bl = blue
gr = grey
li = violet
ge = yellow
or = orange

New Beetle — Wiring diagram — No. 30/5

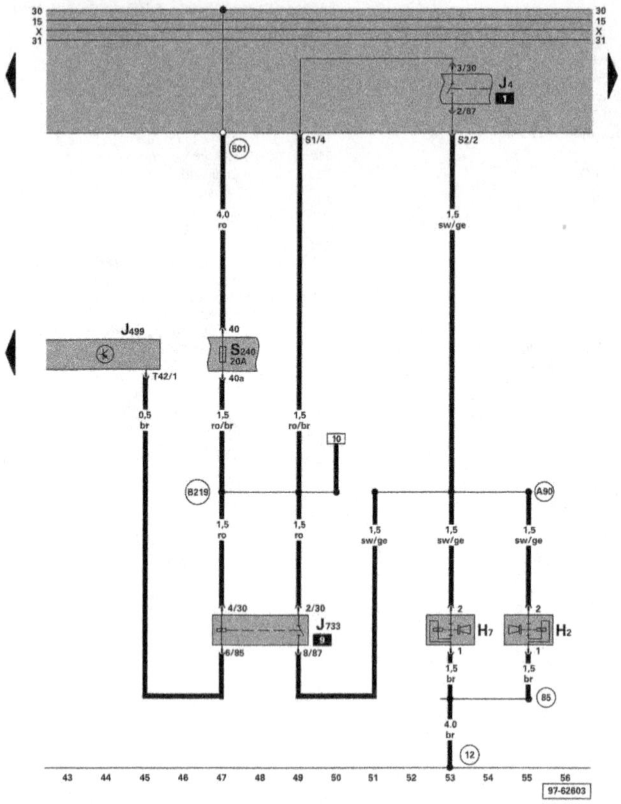

Telematic Control Module, Signal horn/dual tone horn, Low Tone Horn, Alarm Horn Relay

- H2 - Signal horn/dual tone horn
- H7 - Low Tone Horn
- J4 - Horn Relay
- J499 - Telematic Control Module
- J733 - Dual Horn Auxiliary Relay
- S240 - Fuse 40 in fuse holder
- T42 - 42-Pin Connector for Telematic Control Module

- (12) - Ground connection, in engine compartment, left
- (85) - Ground connection -1-, in engine compartment wiring harness
- (501) - Threaded connection -2- (30) on the relay plate
- (A90) - Connector (two-tone horn), in instrument panel wiring harness
- (B219) - Plus Connection -4- (30a) (interior wiring harness)

ws = white
sw = black
ro = red
br = brown
gn = green
bl = blue
gr = grey
li = violet
ge = yellow
or = orange

New Beetle — Wiring diagram — No. 30/6

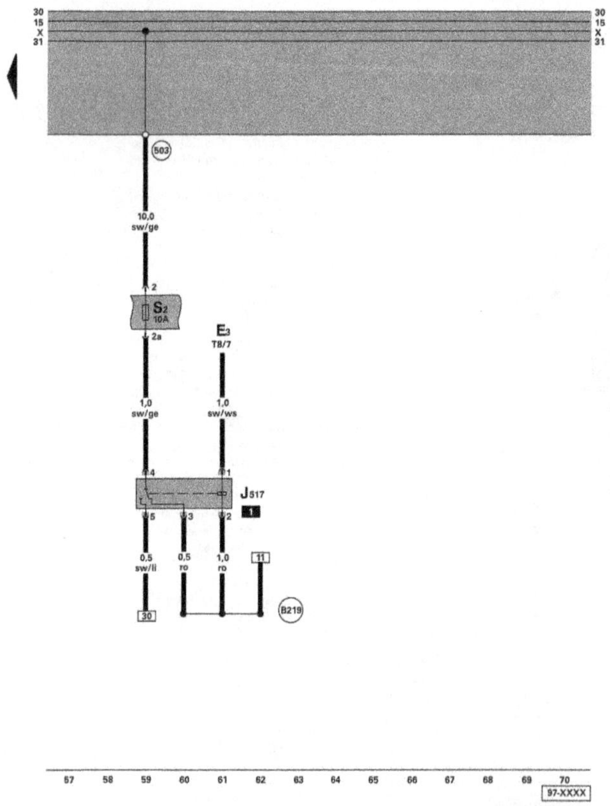

Alarm System Light Flasher Relay

- E3 - Emergency Flasher Switch
- J517 - Auxiliary Emergency Flasher Relay
- S2 - Fuse 2 in fuse holder
- T8 - 8-Pin Connector at E3

- (503) - Bolted Connection (75x) (at relay panel)
- (B219) - Plus Connection -4- (30a) (interior wiring harness)

ws = white
sw = black
ro = red
br = brown
gn = green
bl = blue
gr = grey
li = violet
ge = yellow
or = orange

New Beetle Wiring diagram **No. 67/1**

Tire pressure monitoring,

from July 2007,

Aditional Fuse, Relay, Control Module and Connector locations ⇒ Component Locations

Troubleshooting Procedures ⇒ Guide Fault Finding using VAS 5051/5052

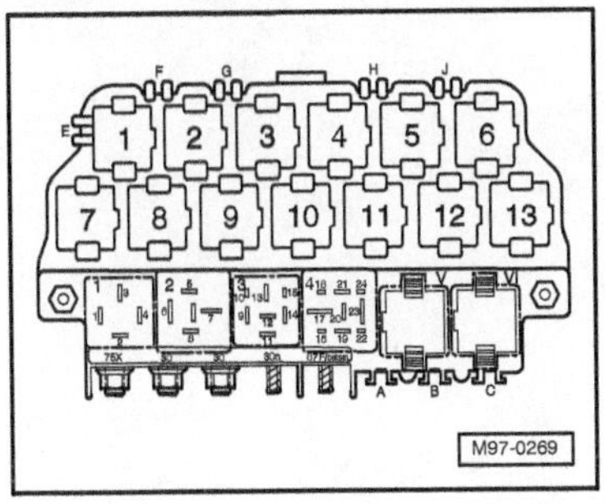

Relay location on the thirteen position auxiliary relay panel, above relay panel:

Relay panel:

2 J59 - Load Reduction Relay (100)

Note:

Number in parentheses indicates production control number stamped on relay housing.

New Beetle Wiring diagram **No. 67/2**

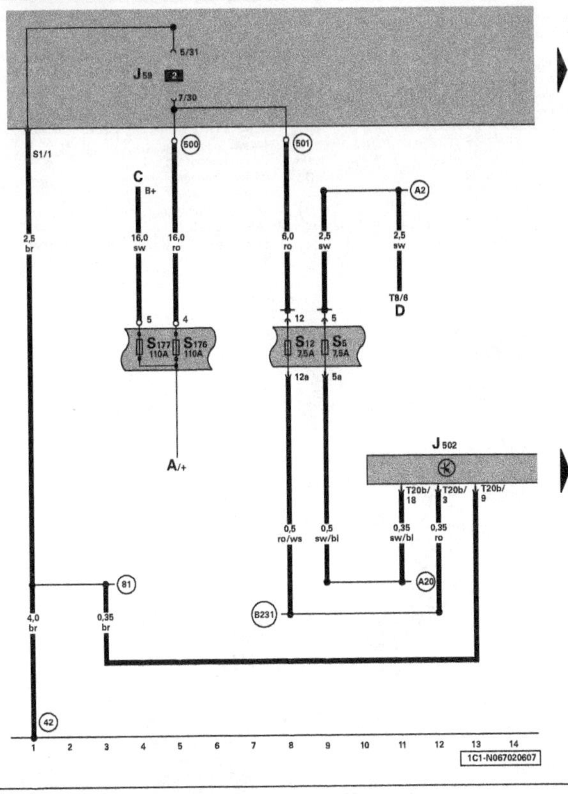

Tire Pressure Monitoring Control Module

- A - Batterie
- C - Generator (GEN)
- D - Ignition/Starter Switch
- J59 - Load Reduction Relay
- J502 - Tire Pressure Monitoring Control Module
- S5 - Fuse (on fuse/relay panel)
- S12 - Fuse (on fuse/relay panel)
- S176 - Fuse 4 - terminal 30 (on battery fuse holder)
- S177 - Fuse 5 - terminal 30 (on battery fuse panel)
- T8 - 8-Pin Connector
- T20b - 20-Pin Connector
- (42) - Ground Connection (beside steering column)
- (81) - Ground Connection 1 (in instrument panel wiring harness)
- (500) - Threaded Connection 1 - terminal 30 (on relay panel)
- (501) - Threaded Connection 2 - terminal 30 (on relay panel)
- (A2) - Plus Connection (15) (in instrument panel wiring harness)
- (A20) - Plus Connection (15a) (in instrument panel wiring harness)
- (B231) - Plus Connection 5 (30a) (in interior wiring harness)

ws = white
sw = black
ro = red
br = brown
gn = green
bl = blue
gr = grey
li = violet
ge = yellow
or = orange

Tire pressure monitoring from July 2007

New Beetle — Wiring diagram — No. 67/3

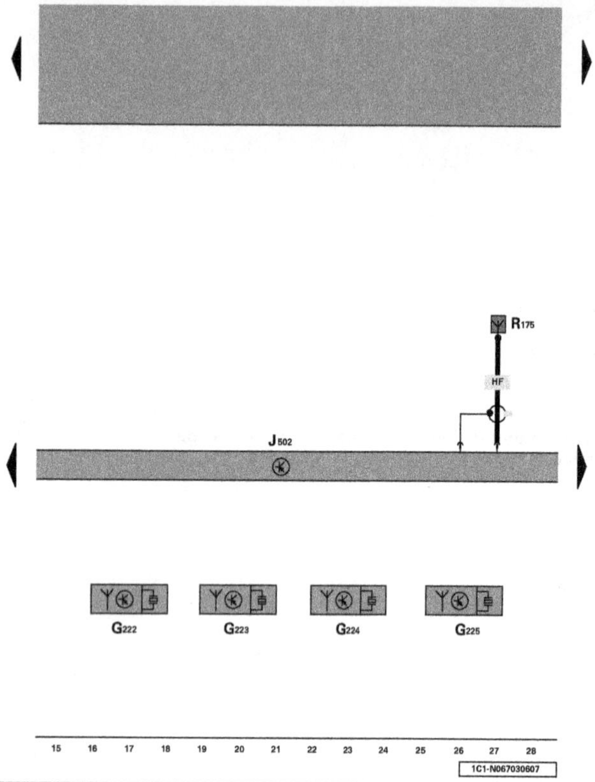

Tire Pressure Monitoring Control Module, Tire Pressure Monitoring Antenna

G222 - Left Front Tire Pressure Monitoring Sensor
G223 - Right Front Tire Pressure Monitoring Sensor
G224 - Left Rear Tire Pressure Monitoring Sensor
G225 - Right Rear Tire Pressure Monitoring Sensor
J502 - Tire Pressure Monitoring Control Module
R175 - Tire Pressure Monitoring Antenna

ws = white
sw = black
ro = red
br = brown
gn = green
bl = blue
gr = grey
li = violet
ge = yellow
or = orange

New Beetle — Wiring diagram — No. 67/4

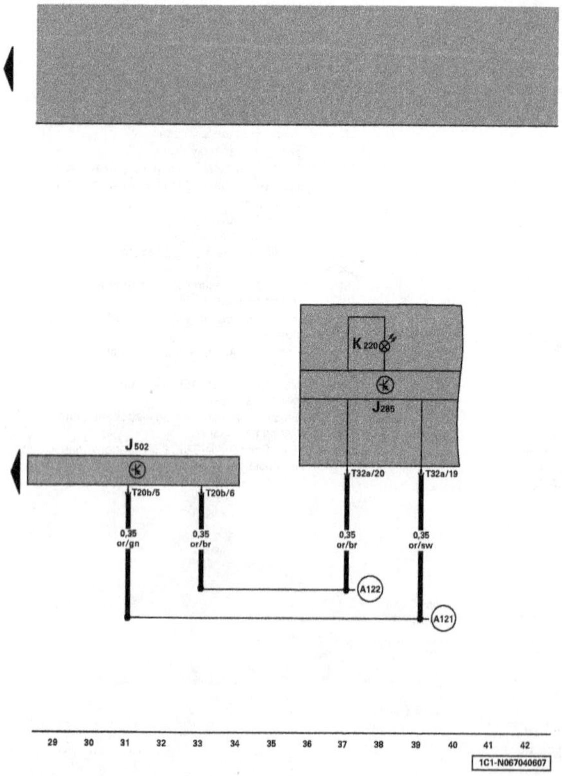

Instrument Cluster Control Module, Tire Pressure Monitoring Control Module, Tire Pressure Monitoring Display Indicator Lamp

J285 - Instrument Cluster Control Module
J502 - Tire Pressure Monitoring Control Module
K220 - Tire Pressure Monitoring Display Indicator Lamp
T20b - 20-Pin Connector
T32a - 32-Pin Connector

(A121) - High-Bus Connection (in instrument panel wiring harness)
(A122) - Low-Bus Connection (in instrument panel wiring harness)

ws = white
sw = black
ro = red
br = brown
gn = green
bl = blue
gr = grey
li = violet
ge = yellow
or = orange

New Beetle Wiring diagram **No. 68/1**

Comfort system, with anti-theft alarm system,

from July 2007,

Additional Fuse, Relay, Control Module and Connector locations ⇒ Component Locations

Troubleshooting Procedures ⇒ Guided Fault Finding using VAS 5051/5052

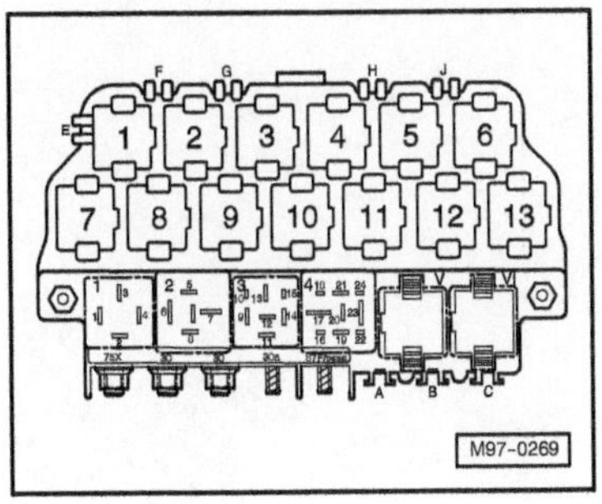

Relay locations on thirteen position auxiliary relay panel, above relay panel:

Fuse locations on thirteen position auxiliary relay panel, above relay panel:

J S144 - Central Locking/Anti-Theft Warning System Fuse (15A)

H S111 - Alarm System and Anti-Theft Immobilizer Fuse (15A)

Relay panel:

Fuse placements on the relay panel:

A S43 - Power Window Circuit Breaker (30A)

Notes:

Number in parentheses indicates production control number stamped on relay housing.

New Beetle Wiring diagram **No. 68/2**

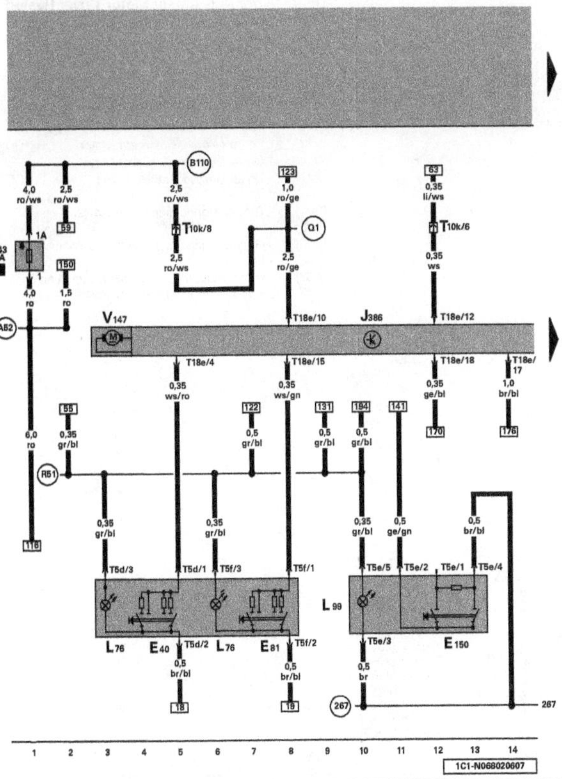

Left Front Window Regulator Switch, Driver Right Front Window Regulator Switch, Driver Interior Locking Switch, Driver Door Control Module , Driver Window Regulator Motor

E40	-	Left Front Window Regulator Switch
E81	-	Driver Right Front Window Regulator Switch
E150	-	Driver Interior Locking Switch
J386	-	Driver Door Control Module
L76	-	Push Button Illumination
L99	-	Interior Lock Switch Illumination
S43	-	Power Window Circuit Breaker
T5d	-	5-Pin Connector, brown
T5e	-	5-Pin Connector, black
T5f	-	5-Pin Connector, black
T10k	-	10-Pin Connector, black, on A-Pillar left
V147	-	Driver Window Regulator Motor
(267)	-	Ground Connection 2 (in driver's door wiring harness)
(A52)	-	Plus Connection 2 (30) (in instrument panel wiring harness)
(B110)	-	Window Regulator Connection (30) (in interior wiring harness)
(Q1)	-	Plus Connection (30) (in power window wiring harness)
(R51)	-	Connection (58b) (in driver's door wiring harness)

ws = white
sw = black
ro = red
br = brown
gn = green
bl = blue
gr = grey
li = violet
ge = yellow
or = orange

New Beetle

Wiring diagram

No. 68/3

Driver Door Central Locking Lock Unit , Driver Door Control Module , Central Locking -SAFE- Indicator Lamp

F220 - Driver Door Central Locking Lock Unit
J386 - Driver Door Control Module
K133 - Central Locking -SAFE- Indicator Lamp
T18e - 18-Pin Connector
T8p - 8-Pin Connector, on Lock unit for central locking, driver side

(304) - Ground Connection 3 (in driver's door wiring harness)

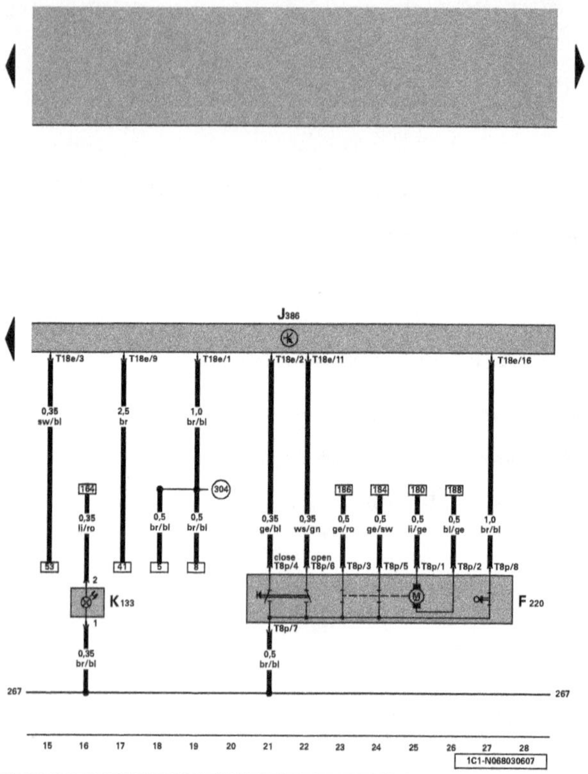

ws = white
sw = black
ro = red
br = brown
gn = green
bl = blue
gr = grey
li = violet
ge = yellow
or = orange

New Beetle

Wiring diagram

No. 68/4

Driver Mirror Adjustment Motor, Driver Heated Outside Mirror

E15 - Rear Window Defogger Switch
T6f - 6-Pin Connector
T8l - 8-Pin Connector, in driver's door wiring
T10k - 10-Pin Connector, on A-Pillar left, black
T10p - 10-Pin Connector, on A-Pillar left, blue
V17 - Driver Mirror Adjustment Motor
V149 - Driver Mirror Adjustment Motor
Z4 - Driver Heated Outside Mirror

(44) - Ground Connection (on left lower A-pillar)

(205) - Ground Connection (in driver's door wiring harness)

(A63) - Mirror Adjustment/Heating Connection (in instrument panel wiring harness)

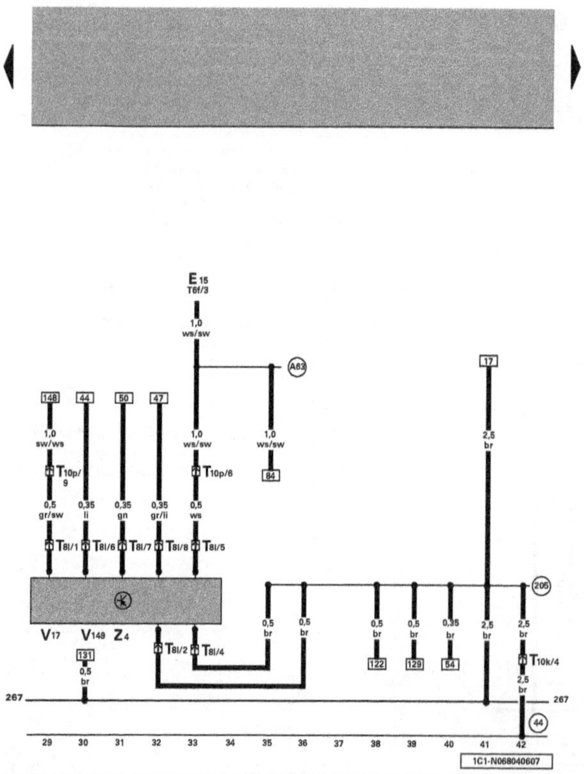

ws = white
sw = black
ro = red
br = brown
gn = green
bl = blue
gr = grey
li = violet
ge = yellow
or = orange

New Beetle — Wiring diagram — No. 68/5

Mirror Adjustment Switch, Mirror Selector Switch, Outside Mirror Heating Switch

- E43 - Mirror Adjustment Switch
- E48 - Mirror Selector Switch
- E231 - Outside Mirror Heating Switch
- L78 - Mirror Adjustment Switch Illumination
- T10p - 10-Pin Connector
- T10y - 10-Pin Connector
- (A20) - Plus Connection (15a) (in instrument panel wiring harness)
- (Q19) - Connection (heated mirror) (in power window wiring harness)
- (R27) - Power Mirror Connection (in driver's door wiring harness)

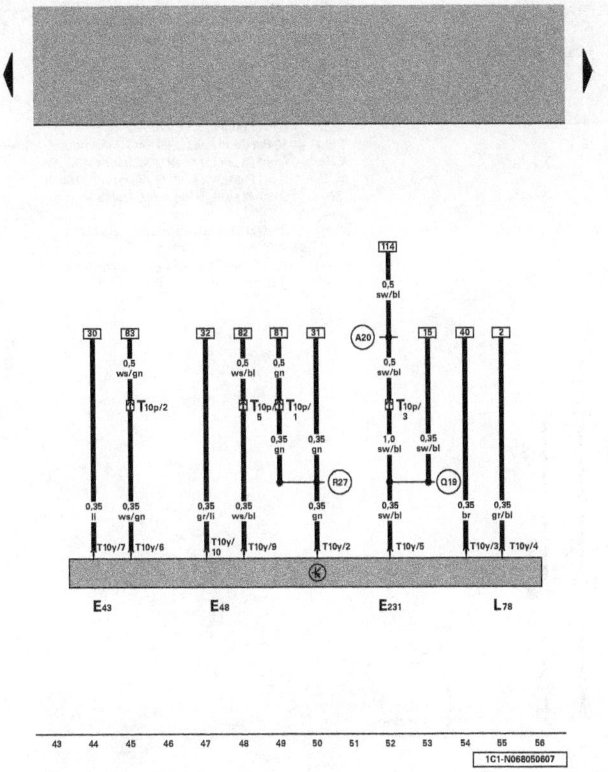

ws = white
sw = black
ro = red
br = brown
gn = green
bl = blue
gr = grey
li = violet
ge = yellow
or = orange

New Beetle — Wiring diagram — No. 68/6

Passenger Door Window Regulator Switch, Front Passenger Door Central Locking Lock Unit, Front Passenger Door Control Module, Front Passenger Window Regulator Motor

- E107 - Passenger Door Window Regulator Switch
- F221 - Front Passenger Door Central Locking Lock Unit
- J387 - Front Passenger Door Control Module
- T5h - 5-Pin Connector
- T8r - 8-Pin Connector
- T10l - 10-Pin Connector, black, on A-Pillar right
- T18f - 18-Pin Connector
- L53 - Power Window Switch Illumination
- V148 - Front Passenger Window Regulator Motor
- (43) - Ground Connection (on right lower A-pillar)
- (268) - Ground Connection 2 (in front passenger's door wiring harness)

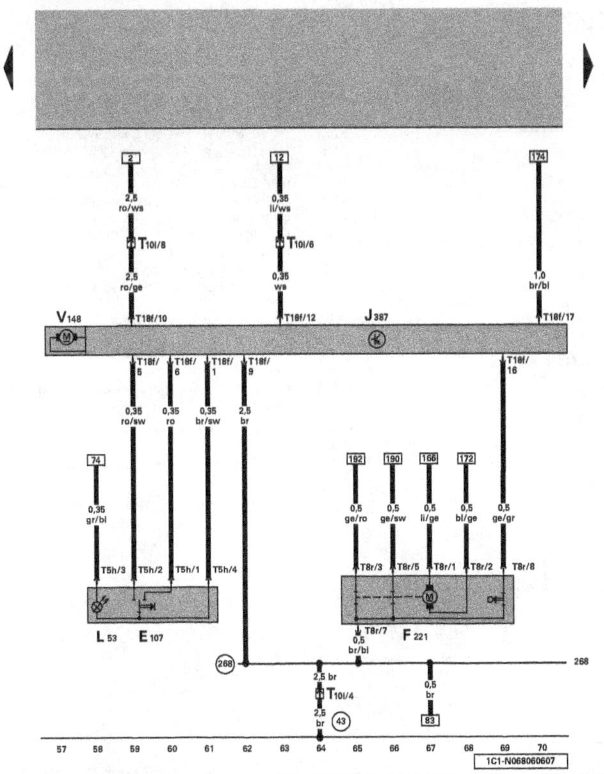

ws = white
sw = black
ro = red
br = brown
gn = green
bl = blue
gr = grey
li = violet
ge = yellow
or = orange

New Beetle — Wiring diagram — No. 68/7

Front Passenger Interior Lock Switch, Front Passenger Mirror Adjustment Motor, Front Passenger Heated Outside Mirror

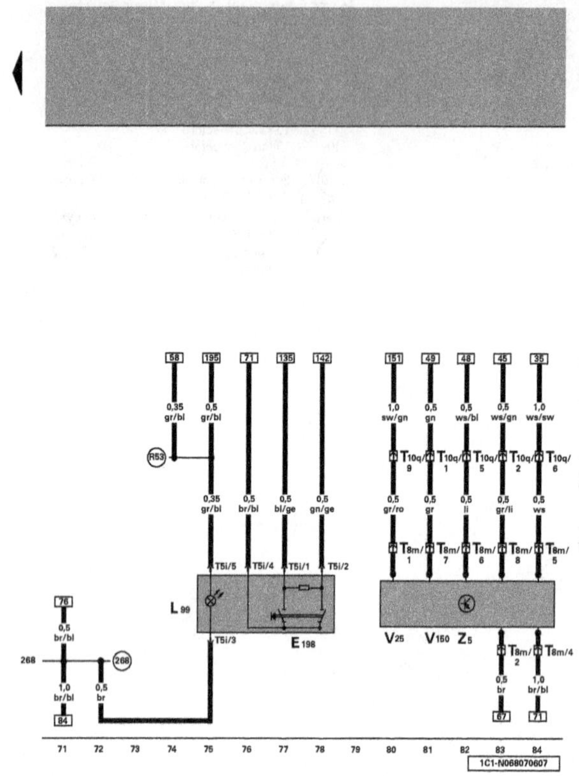

- E198 - Front Passenger Interior Lock Switch
- L99 - Interior Lock Switch Illumination
- T5i - 5-Pin Connector
- T8m - 8-Pin Connector
- T10q - 10-Pin Connector, blue, on A-Pillar right
- V25 - Front Passenger Mirror Adjustment Motor
- V150 - Front Passenger Mirror Adjustment Motor
- Z5 - Front Passenger Heated Outside Mirror
- (268) - Ground Connection 2 (in front passenger's door wiring harness)
- (R53) - Connection (58s) (in front passenger's door wiring harness)

ws = white
sw = black
ro = red
br = brown
gn = green
bl = blue
gr = grey
li = violet
ge = yellow
or = orange

New Beetle — Wiring diagram — No. 68/8

Driver Vanity Mirror Light Contact Switch, Front Passenger Vanity Mirror Light Contact Switch, Front Interior Light, Left Rear Reading Light, Right Rear Reading Light, Front Passenger Illuminated Vanity Mirror Light, Driver Illuminated Vanity Mirror Light

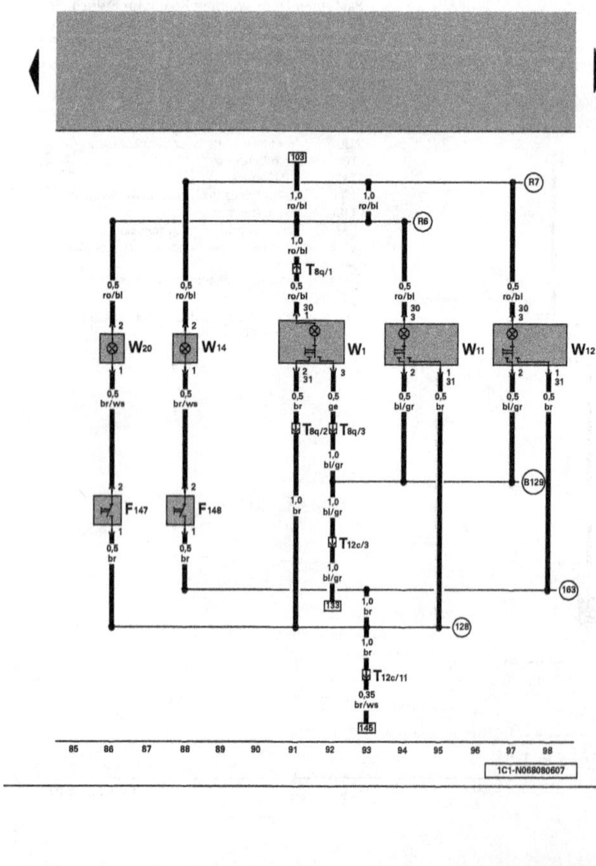

- F147 - Driver Vanity Mirror Light Contact Switch
- F148 - Front Passenger Vanity Mirror Light Contact Switch
- T8q - 8-Pin Connector at W
- T12c - 12-Pin Coupling Connector, on A pillar left, central part
- W1 - Front Interior Light
- W11 - Left Rear Reading Light
- W12 - Right Rear Reading Light
- W14 - Front Passenger Illuminated Vanity Mirror Light
- W20 - Driver Illuminated Vanity Mirror Light
- (128) - Ground Connection 1 (in interior light wiring harness)
- (163) - Ground Connection 2 (in interior light wiring harness)
- (B129) - Interior Light Connection (3/L) (in interior wiring harness)
- (R6) - Plus Connection 1 (in interior light wiring harness)
- (R7) - Plus Connection 2 (in interior light wiring harness)

ws = white
sw = black
ro = red
br = brown
gn = green
bl = blue
gr = grey
li = violet
ge = yellow
or = orange

New Beetle — Wiring diagram — No.68/9

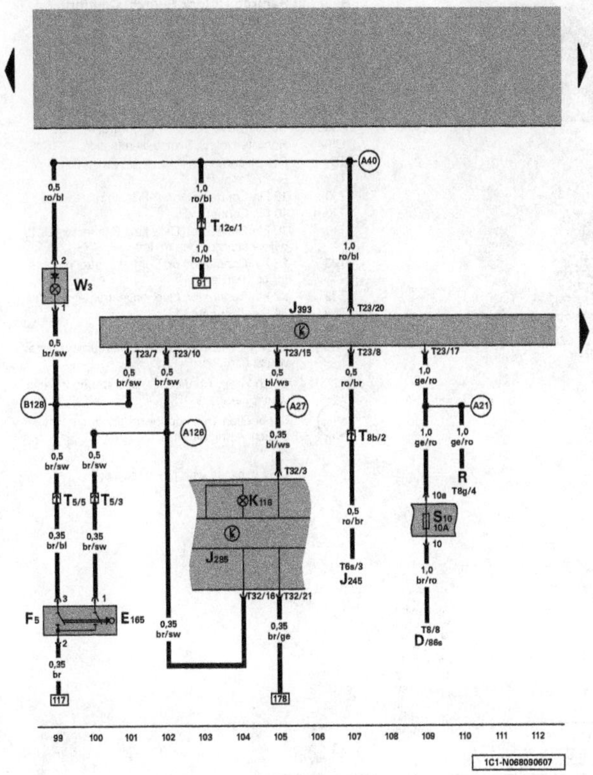

ws = white
sw = black
ro = red
br = brown
gn = green
bl = blue
gr = grey
li = violet
ge = yellow
or = orange

Rear Lid Unlock Switch, Luggage Compart ment Light Switch, Comfort System Central Control Module , Luggage compartment Light

- D - Ignition/Starter Switch
- E165 - Rear Lid Unlock Switch
- F5 - Luggage Compartment Light Switch
- J245 - Power Sunroof Control Module
- J285 - Instrument Cluster Control Module
- J393 - Comfort System Central Control Module
- K116 - Rear Lid Unlocked Warning Lamp
- R - Radio
- S10 - Fuse (on fuse/relay panel)
- T5 - 5-Pin Connector, black, on C-pillar, left
- T6s - 6-Pin Connector
- T8 - 8-Pin Connector
- T8b - 8-Pin Coupling Connector. on A pillar left, central part
- T8g - 8-Pin Connector on radio
- T12c - 12-Pin Coupling Connector, on A pillar left, central part
- T23 23-Pin Connector, on J393
- T32 - 32-Pin Connector, blue, on instrument cluster
- W3 - Luggage compartment Light
- (A21) - Connection (86s) (in instrument panel wiring harness)
- (A27) - Vehicle Speed Signal (VSS) Connection (in instrument panel wiring harness)
- (A40) - Plus Connection 1 (30) (in instrument cluster wiring harness)
- (A126) - Rear Lid Contact Switch Connection (in instrument panel wiring harness)
- (B128) - Right Rear Speed Sensor Connection (in instrument panel wiring harness)

New Beetle — Wiring diagram — No.68/10

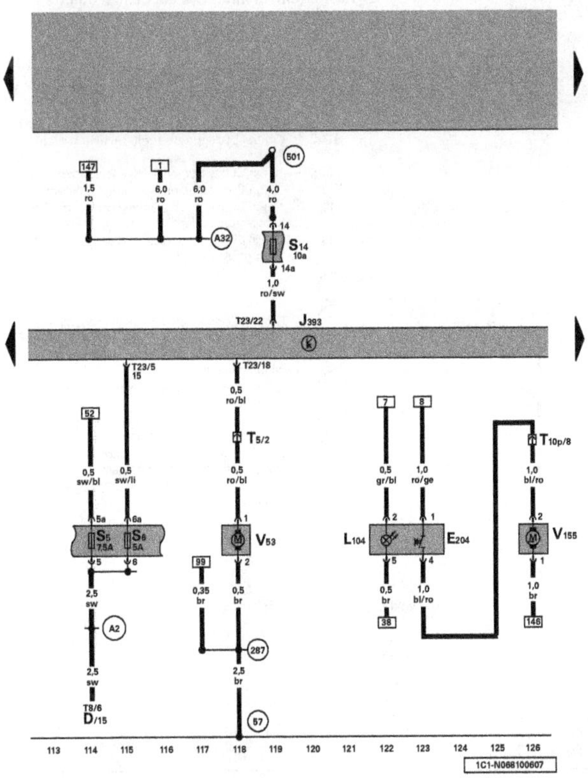

ws = white
sw = black
ro = red
br = brown
gn = green
bl = blue
gr = grey
li = violet
ge = yellow
or = orange

Remote Fuel Tank Flap Switch, Comfort System Central Control Module, Rear Lid Cen tral Locking System Motor, Fuel Tank Lid Unlock Motor

- D - Ignition/Starter Switch
- E204 - Remote Fuel Tank Flap Switch
- J393 - Comfort System Central Control Module
- L104 - Remote Unlock Switch Illumination
- S5 - Fuse (on fuse/relay panel)
- S6 - Fuse (on fuse/relay panel)
- S14 - Fuse (on fuse/relay panel)
- T5 - 5-Pin Connector, black, on C pillar, left
- T8 - 8-Pin Connector
- T10p - 10-Pin Connector, on A-Pillar left, blue
- T23 23-Pin Connector, on Central control module for comfort system
- V 53 - Rear Lid Central Locking System Motor
- V155 - Fuel Tank Lid Unlock Motor
- (57) - Ground Connection (on left rear pillar)
- (287) - Ground Connection (in rear lid hydraulics wiring harness)
- (501) - Threaded Connection 2 - terminal 30 (on relay panel)
- (A2) - Plus Connection (15) (in instrument panel wiring harness)
- (A32) - Plus Connection (30) (in instrument panel wiring harness)

New Beetle — Wiring diagram — No.68/11

Rear Lid Remote Unlock Switch, Comfort System Central Control Module

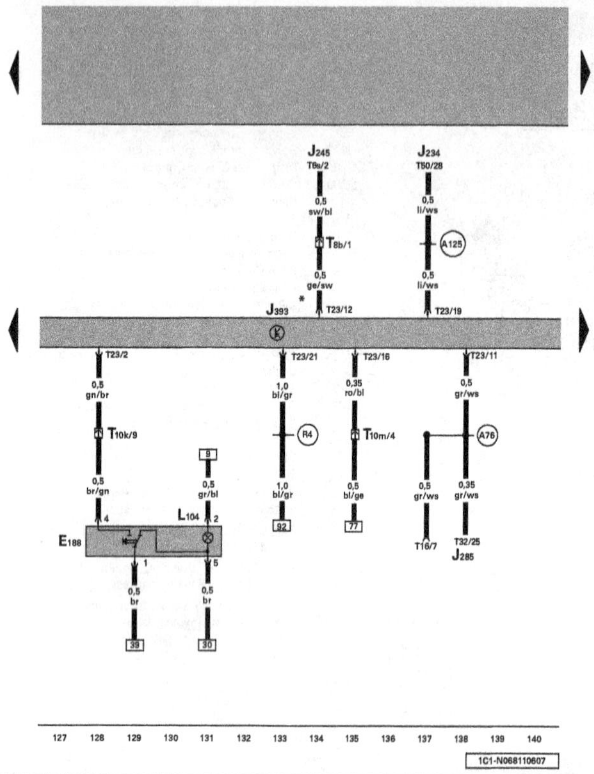

- E188 - Rear Lid Remote Unlock Switch
- J234 - Airbag Control Module
- J245 - Power Sunroof Control Module
- J285 - Instrument Cluster Control Module
- J393 - Comfort System Central Control Module
- L104 - Remote Unlock Switch Illumination
- T6s - 6-Pin Connector
- T8b - 8-Pin Connector
- T10k - 10-Pin Connector, on A-Pillar left, black
- T10m - 10-Pin Connector
- T16 - 16-Pin Connector, Data Link Connector (DLC), below steering coulm, left
- T23 - 23-Pin Connector, on Central control module for comfort system
- T32 - 32-Pin Connector, blue, on instrument cluster
- T50 - 50-Pin Connector
- (A76) - K-Diagnostic Connection (in instrument panel wiring harness)
- (A125) - Crash Signal Connection (in instrument panel wiring harness)
- (R4) - Connection 1 (in interior light/door contact switch wiring harness)
- * - Only vehicles with power sunroof

ws = white
sw = black
ro = red
br = brown
gn = green
bl = blue
gr = grey
li = violet
ge = yellow
or = orange

New Beetle — Wiring diagram — No. 68/12

Alarm Horn, Comfort System Central Control Module

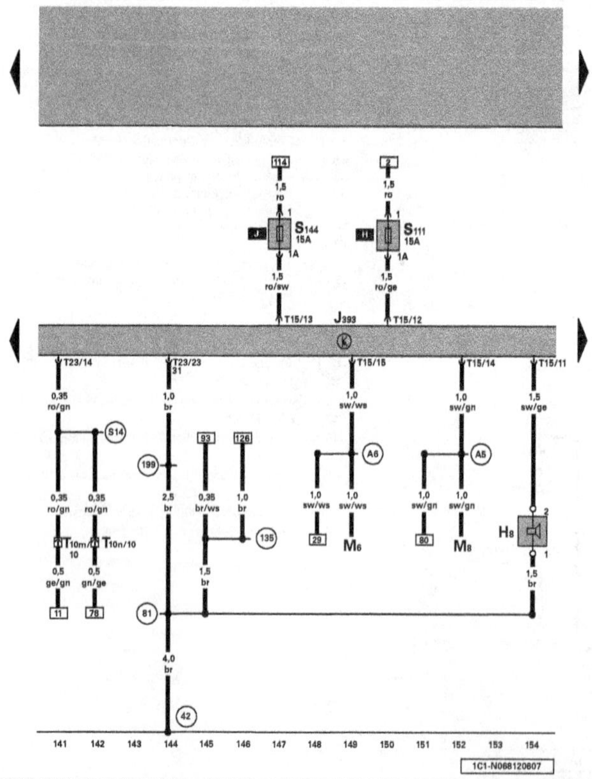

- H8 - Alarm Horn
- J393 - Comfort System Central Control Module
- M6 - Left Rear Turn Signal Lamp
- M8 - Right Rear Turn Signal Lamp
- S111 - Alarm System and Anti-Theft Immobilizer Fuse
- S144 - Central Locking/Anti-Theft Warning System Fuse
- T10m - 10-Pin Connector, on A-Pillar left, brown
- T10n - 10-Pin Connector, on A-Pillar right, brown
- T15 - 15-Pin Connector, on Central control module for comfort system
- T23 - 23-Pin Connector, on Central control module for comfort system
- (42) - Ground Connection (beside steering column)
- (81) - Ground Connection 1 (in instrument panel wiring harness)
- (135) - Ground Connection 2 (in instrument panel wiring harness)
- (199) - Ground Connection 3 (in instrument panel wiring harness)
- (A5) - Right Turn Signal Plus Connection (in instrument panel wiring harness)
- (A6) - Left Turn Signal Plus Connection (in instrument panel wiring harness)
- (S14) - -Open- Connection (in central locking system wiring harness)

ws = white
sw = black
ro = red
br = brown
gn = green
bl = blue
gr = grey
li = violet
ge = yellow
or = orange

New Beetle

Wiring diagram

No. 68/13

Hood Alarm Switch, Coolant Fan Control (FC) Control Module, Central Locking and Anti-Theft Alarm System Antenna

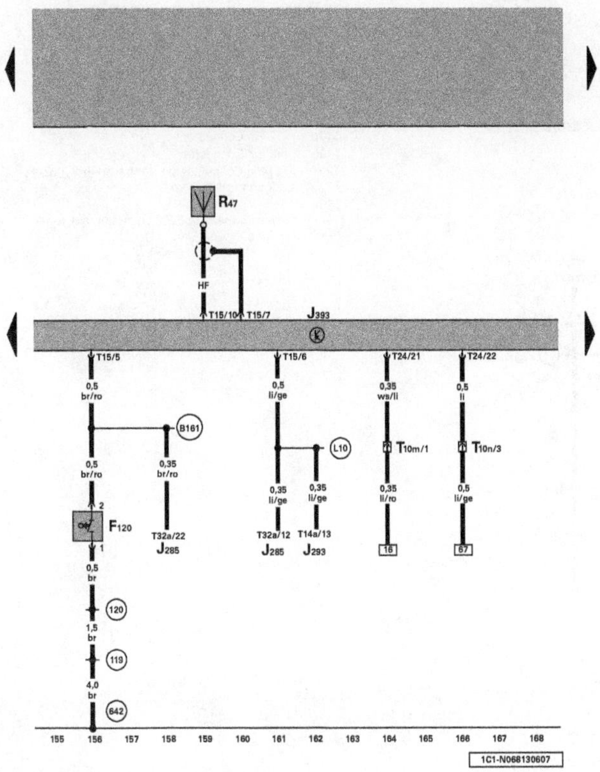

F120 - Hood Alarm Switch
J285 - Instrument Cluster Control Module
J293 - Coolant Fan Control (FC)Control Module
J393 - Comfort System Central Control Module
R47 - Central Locking and Anti-Theft Alarm System Antenna
T10m - 10-Pin Connector, on A-Pillar left, brown
T10n - 10-Pin Connector, on A-Pillar right, brown
T14a - 14-Pin Connector, on -J293-
T15 - 15-Pin Connector, on Central control module for comfort system
T24 - 24-Pin Connector, on Central control module for comfort system
T32a - 32-Pin Connector, green, on instrument closter

(119) - Ground Connection 1 (in headlamp wiring harness)
(120) - Ground Connection 2 (in headlamp wiring harness)
(642) - Ground Connection (on electronic fan control)
(B161) - Anti-Theft Alarm System Connection (in interior wiring harness)
(L10) - Connection 2 (in A/C wiring harness)*

* - Only vehicles with A/C

ws = white
sw = black
ro = red
br = brown
gn = green
bl = blue
gr = grey
li = violet
ge = yellow
or = orange

New Beetle

Wiring diagram

No. 68/14

Comfort System Central Control Module

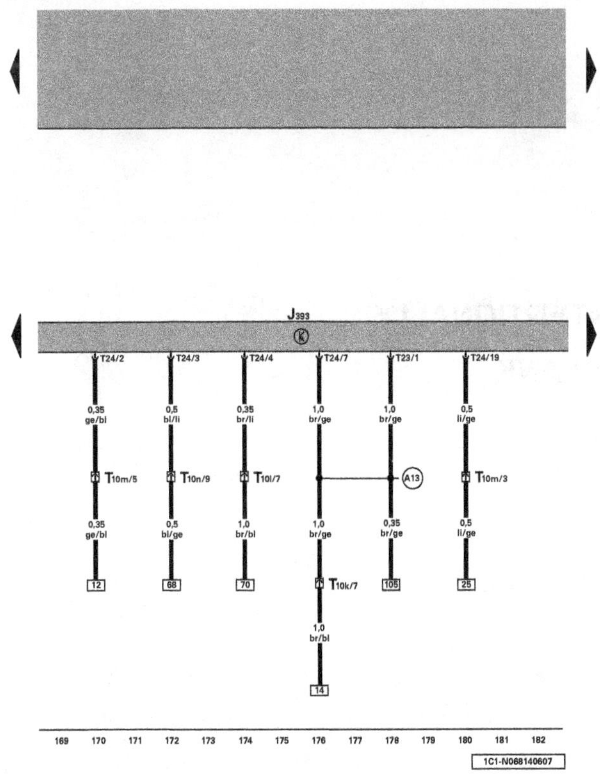

J393 - Comfort System Central Control Module
T10k - 10-Pin Connector, on A-Pillar left, black
T10l - 10-Pin Connector, on A-Pillar right, black
T10m - 10-Pin Connector, on A-Pillar left, brown
T10n - 10-Pin Connector, on A-Pillar right, brown
T23 - 23-Pin Connector, on Central control module for comfort system
T24 - 24-Pin Connector, on Central control module for comfort system

(A13) - Door Contact Switch Connection (in instrument panel wiring harness)

ws = white
sw = black
ro = red
br = brown
gn = green
bl = blue
gr = grey
li = violet
ge = yellow
or = orange

New Beetle Wiring diagram **No. 68/15**

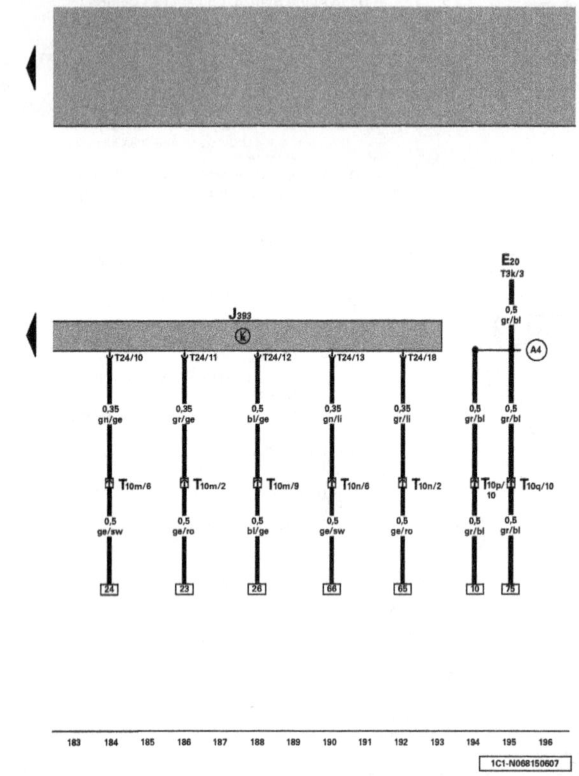

Comfort System Central Control Module

- E20 - Instrument Panel Illumination Dimmer Switch
- J393 - Comfort System Central Control Module
- T3k - 3-Pin Connector
- T10m - 10-Pin Connector, on A-Pillar left, brown
- T10n - 10-Pin Connector, on A-Pillar right, brown
- T10p - 10-Pin Connector
- T10q - 10-Pin Connector
- T24 - 24-Pin Connector, on Central control module for comfort system
- (A4) - Plus Connection (58b) (in instrument panel wiring harness)

ws = white
sw = black
ro = red
br = brown
gn = green
bl = blue
gr = grey
li = violet
ge = yellow
or = orange

Edition 06.07

THIS SPACE INTENTIONALLY LEFT BLANK

New Beetle **Wiring diagram** **No. 72/1**

Comfort system, with anti-theft alarm system,

from July 2008,

Additional Fuse, Relay, Control Module and Connector locations ⇒ Component Locations

Troubleshooting Procedures ⇒ Guided Fault Finding using VAS 5051/5052

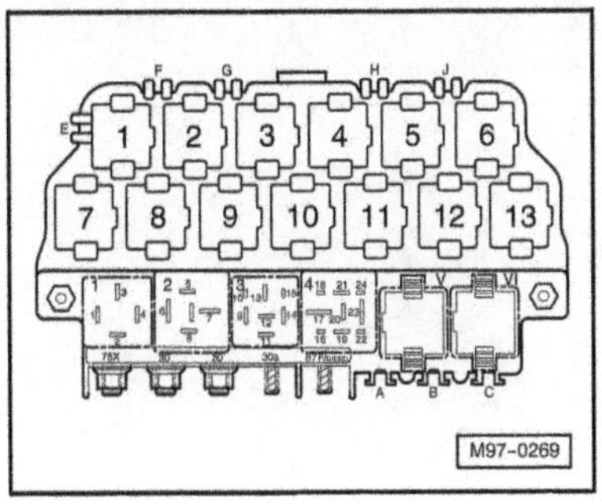

Relay locations on thirteen position auxiliary relay panel, above relay panel:

Relay panel:

4 J17 - Fuel Pump (FP) Relay (409)

Fuse placements on the relay panel:

A S43 - Power Window Circuit Breaker (30A)

H S111 - Alarm System and Anti-Theft Immobilizer Fuse (15A)

J S144 - Central Locking/Anti-Theft Warning System Fuse (15A)

Number in parentheses indicates production control number stamped on relay housing.

New Beetle **Wiring diagram** **No. 72/2**

Left Front Window Regulator Switch, Driver Interior Lock Switch, Driver Door Control Module, Driver Window Regulator Motor

E40 - Left Front Window Regulator Switch
E150 - Driver Interior Lock Switch
J386 - Driver Door Control Module
L76 - Push Button Illumination
L99 - Interior Lock Switch Illumination
S43 - Power Window Circuit Breaker
T4ac - 4-Pin Connector
T5e - 5-Pin Connector, black
T10k - 10-Pin Connector, black, on A-Pillar left
T18e - 18-Pin Connector
V147 - Driver Window Regulator Motor

(267) - Ground Connection 2 (in driver's door wiring harness)
(A52) - Plus Connection 2 (30) (in instrument panel wiring harness)
(B110) - Window Regulator Connection (30) (in interior wiring harness)
(Q1) - Plus Connection (30) (in power window wiring harness)
(R51) - Connection (58b) (in driver's door wiring harness)

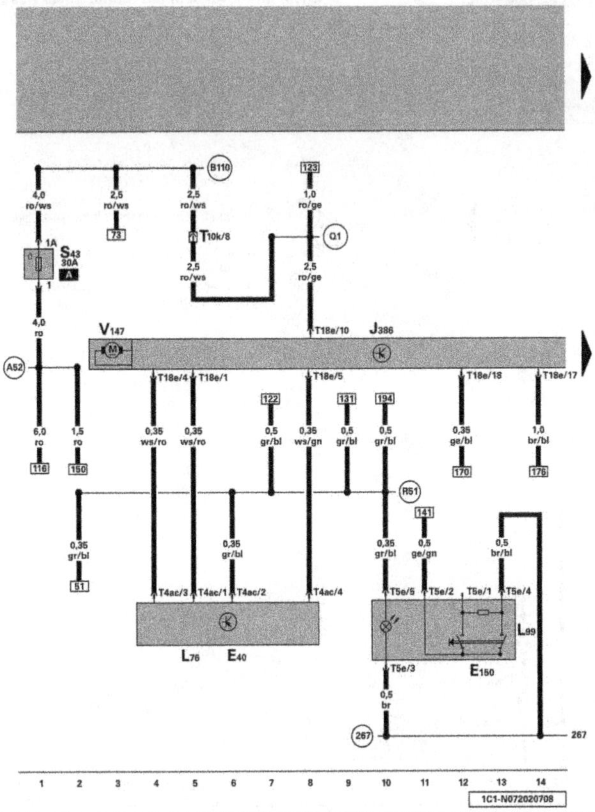

ws = white
sw = black
ro = red
br = brown
gn = green
bl = blue
gr = grey
li = violet
ge = yellow
or = orange

New Beetle — Wiring diagram — No. 72/3

Driver Door Central Locking Lock Unit, Driver Door Control Module, Central Locking -SAFE- Indicator Lamp

- F220 - Driver Door Central Locking Lock Unit
- J386 - Driver Door Control Module
- K133 - Central Locking -SAFE- Indicator Lamp
- T8p - 8-Pin Connector, on Lock unit for central locking, driver side
- T10k - 10-Pin Connector
- T18e - 18-Pin Connector

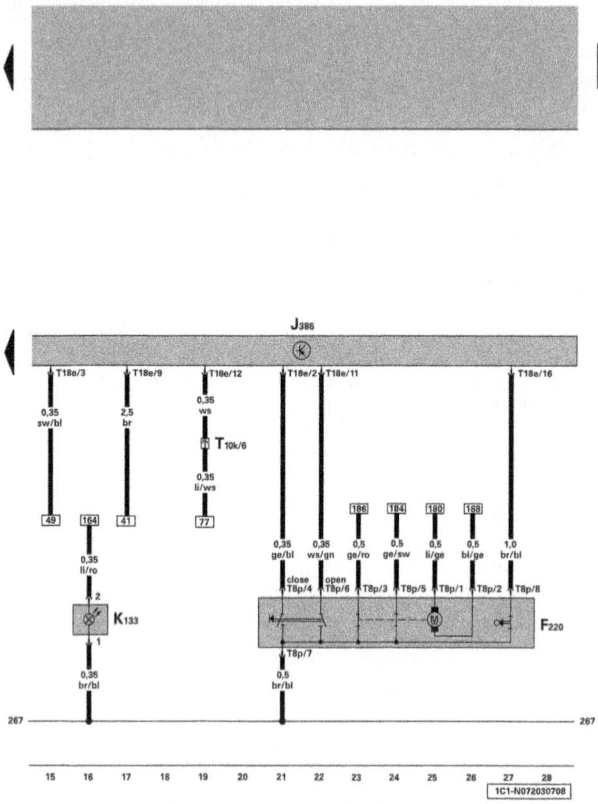

ws = white
sw = black
ro = red
br = brown
gn = green
bl = blue
gr = grey
li = violet
ge = yellow
or = orange

New Beetle — Wiring diagram — No. 72/4

Driver Mirror Adjustment Motor, Driver Heated Outside Mirror

- E15 - Rear Window Defogger Switch
- T6f - 6-Pin Connector
- T8l - 8-Pin Connector, in driver's door wiring
- T10k - 10-Pin Connector, on A-Pillar left, black
- T10p - 10-Pin Connector, on A-Pillar left, blue
- V17 - Driver Mirror Adjustment Motor
- V149 - Driver Mirror Adjustment Motor
- Z4 - Driver Heated Outside Mirror
- (44) - Ground Connection (on left lower A-pillar)
- (205) - Ground Connection (in driver's door wiring harness)
- (A63) - Mirror Adjustment/Heating Connection (in instrument panel wiring harness)

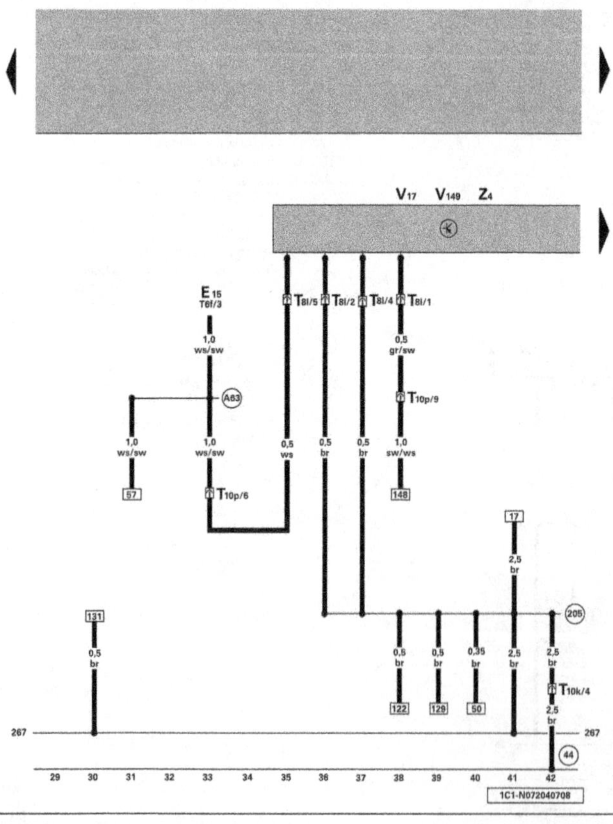

ws = white
sw = black
ro = red
br = brown
gn = green
bl = blue
gr = grey
li = violet
ge = yellow
or = orange

New Beetle

Wiring diagram

No. 72/5

Mirror Adjustment Switch, Mirror Selector Switch, Outside Mirror Heating Switch, Driver Mirror Adjustment Motor, Front Passenger Mirror Adjustment Motor, Driver Mirror Adjustment Motor, Front Passenger Mirror Adjustment Motor, Driver Heated Outside Mirror, Front Passenger Heated Outside Mirror

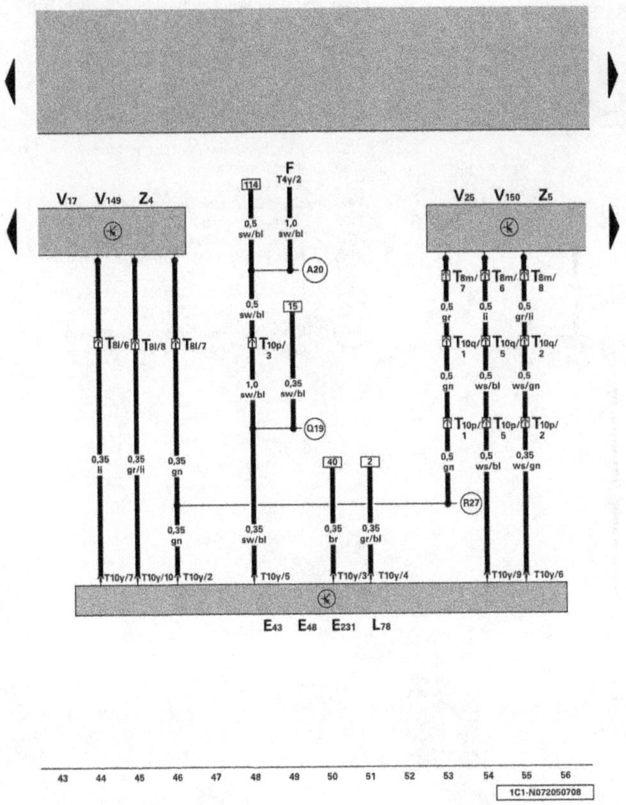

E43 - Mirror Adjustment Switch
E48 - Mirror Selector Switch
E231 - Outside Mirror Heating Switch
F - Brake Light Switch
L78 - Mirror Adjustment Switch Illumination
T4y - 4-Pin Connector
T10p - 10-Pin Connector
T10q - 10-Pin Connector, blue, on A-Pillar right
T10y - 10-Pin Connector
V17 - Driver Mirror Adjustment Motor
V25 - Front Passenger Mirror Adjustment Motor
V149 - Driver Mirror Adjustment Motor
V150 - Front Passenger Mirror Adjustment Motor
Z4 - Driver Heated Outside Mirror
Z5 - Front Passenger Heated Outside Mirror

(A20) - Plus Connection (15a) (in instrument panel wiring harness)
(Q19) - Connection (heated mirror) (in power window wiring harness)
(R27) - Power Mirror Connection (in driver's door wiring harness)

ws = white
sw = black
ro = red
br = brown
gn = green
bl = blue
gr = grey
li = violet
ge = yellow
or = orange

New Beetle

Wiring diagram

No. 72/6

Front Passenger Interior Lock Switch, Front Passenger Mirror Adjustment Motor, Front Passenger Heated Outside Mirror

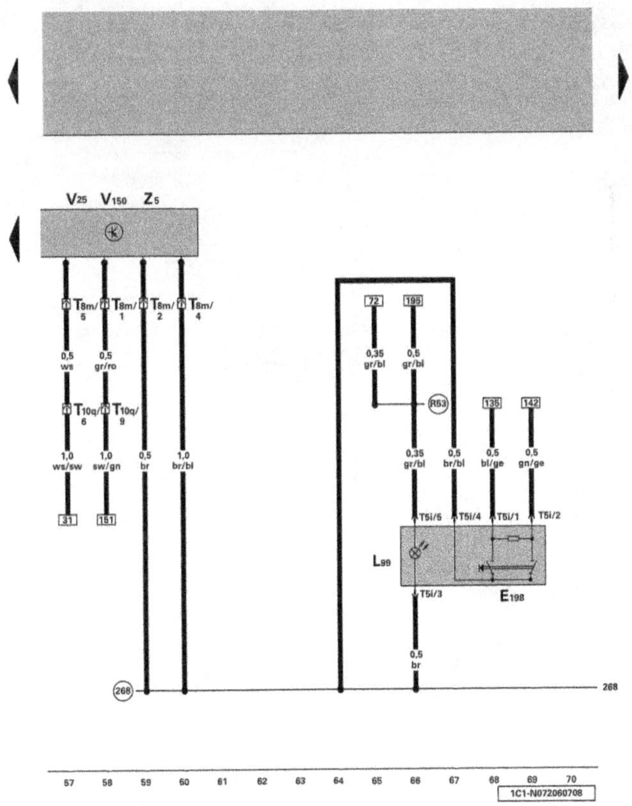

E198 - Front Passenger Interior Lock Switch
L99 - Interior Lock Switch Illumination
T5i - 5-Pin Connector
T8m - 8-Pin Connector
T10q - 10-Pin Connector, blue, on A-Pillar right
V25 - Front Passenger Mirror Adjustment Motor
V150 - Front Passenger Mirror Adjustment Motor
Z5 - Front Passenger Heated Outside Mirror

(268) - Ground Connection 2 (in front passenger's door wiring harness)
(R53) - Connection (58s) (in front passenger's door wiring harness)

ws = white
sw = black
ro = red
br = brown
gn = green
bl = blue
gr = grey
li = violet
ge = yellow
or = orange

New Beetle

Wiring diagram

No. 72/7

Passenger Door Window Regulator Switch, Fro nt Passenger Door Central Locking Lock Unit, Front Passenger Door Control Module, Front Pas senger Window Regulator Motor

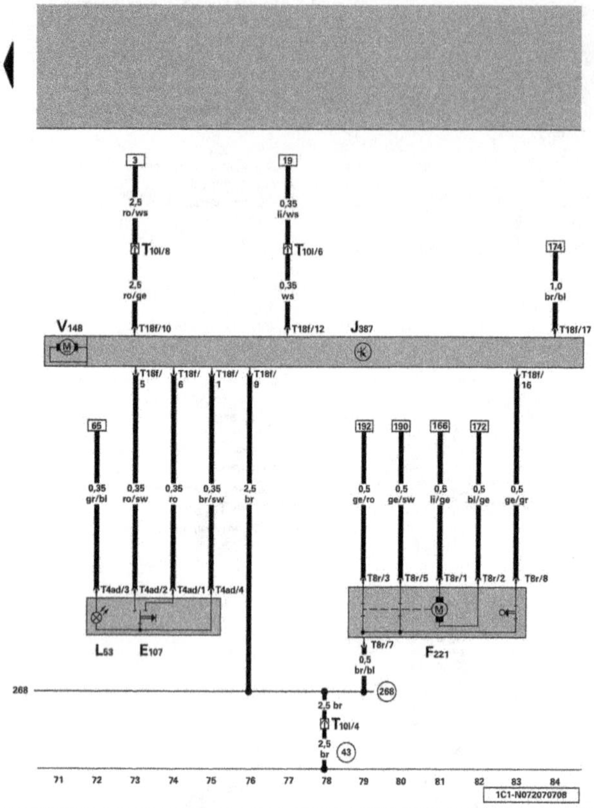

- E107 - Passenger Door Window Regulator Switch
- F221 - Front Passenger Door Central Locking Lock Unit
- J387 - Front Passenger Door Control Module
- T4ad - 4-Pin Connector
- T8r - 8-Pin Connector
- T10l - 10-Pin Connector, black, on A-Pillar right
- T18f - 18-Pin Connector
- L53 - Power Window Switch Illumination
- V148 - Front Passenger Window Regulator Motor
- (43) - Ground Connection (on right lower A-pillar)
- (268) - Ground Connection 2 (in front passenger's door wiring harness)

ws = white
sw = black
ro = red
br = brown
gn = green
bl = blue
gr = grey
li = violet
ge = yellow
or = orange

New Beetle

Wiring diagram

No. 72/8

Driver Vanity Mirror Lamp Contact Switch, Fro nt Passenger Vanity Mirror Lamp Contact Swit ch, Front Interior Light, Left Rear Reading Lam p, Right Rear Reading Lamp, Front Passenger Illuminated Vanity Mirror Lamp, Driver Illumin ated Vanity Mirror Lamp

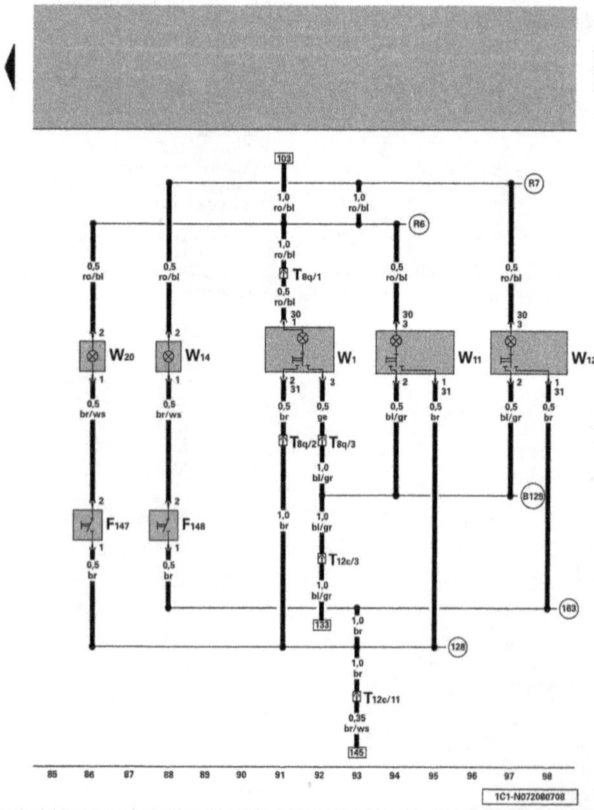

- F147 - Driver Vanity Mirror Lamp Contact Switch
- F148 - Front Passenger Vanity Mirror Lamp Contact Switch
- T8q - 8-Pin Connector at W
- T12c - 12-Pin Coupling Connector, on A pillar left, central part
- W1 - Front Interior Light
- W11 - Left Rear Reading Lamp
- W12 - Right Rear Reading Lamp
- W14 - Front Passenger Illuminated Vanity Mirror Lamp
- W20 - Driver Illuminated Vanity Mirror Lamp
- (128) - Ground Connection 1 (in interior light wiring harness)
- (163) - Ground Connection 2 (in interior light wiring harness)
- (B129) - Interior Light Connection (3/L) (in interior wiring harness)
- (R6) - Plus Connection 1 (in interior light wiring harness)
- (R7) - Plus Connection 2 (in interior light wiring harness)

ws = white
sw = black
ro = red
br = brown
gn = green
bl = blue
gr = grey
li = violet
ge = yellow
or = orange

New Beetle — Wiring diagram — No.72/9

Rear Lid Unlock Switch, Luggage Compartme nt Lamp Switch, Comfort System Central Con trol Module, Rear Lid Unlocked Warning Lam p,Luggage Compartment Light

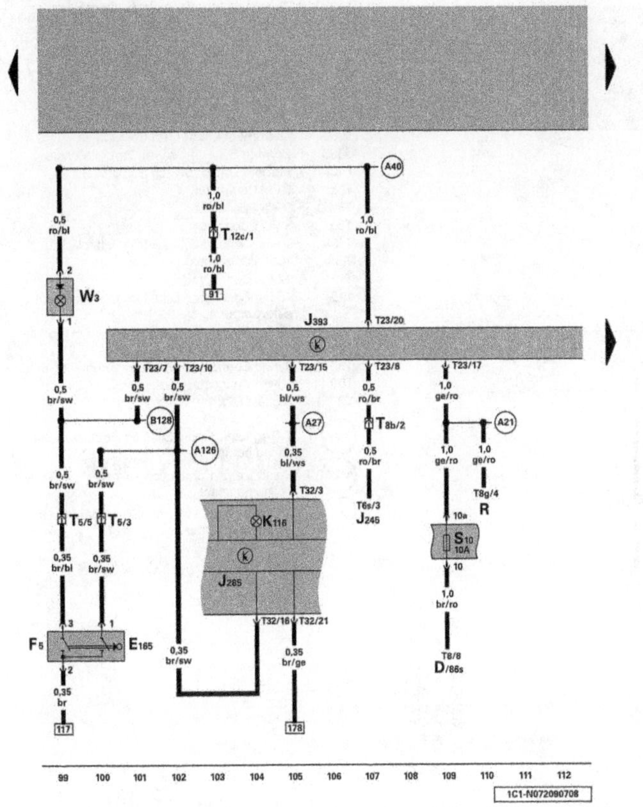

D - Ignition/Starter Switch
E165 - Rear Lid Unlock Switch
F5 - Luggage Compartment Lamp Switch
J245 - Power Sunroof Control Module
J285 - Instrument Cluster Control Module
J393 - Comfort System Central Control Module
K116 - Rear Lid Unlocked Warning Lamp
R - Radio
S10 - Fuse (on fuse/relay panel)
T5 - 5-Pin Connector, black, on C-pillar, left
T6s - 6-Pin Connector
T8 - 8-Pin Connector
T8b - 8-Pin Coupling Connector, on A pillar left, central part
T8g - 8-Pin Connector on radio
T12c - 12-Pin Coupling Connector, on A pillar left, central part
T23 - 23-Pin Connector, on J393
T32 - 32-Pin Connector, blue, on instrument cluster
W3 - Luggage Compartment Light

(A21) - Connection (86s) (in instrument panel wiring harness)
(A27) - Vehicle Speed Signal (VSS) Connection (in instrument panel wiring harness)
(A40) - Plus Connection 1 (30) (in instrument cluster wiring harness)
(A126) - Rear Lid Contact Switch Connection (in instrument panel wiring harness)
(B128) - Luggage Compartment Lamp Connection (in interior wiring harness)

ws = white
sw = black
ro = red
br = brown
gn = green
bl = blue
gr = grey
li = violet
ge = yellow
or = orange

New Beetle — Wiring diagram — No.72/10

Remote Fuel Tank Flap Switch, Comfort Syste m Central Control Module, Rear Lid Central Lo cking System Motor, Fuel Tank Lid Unlock Mot or

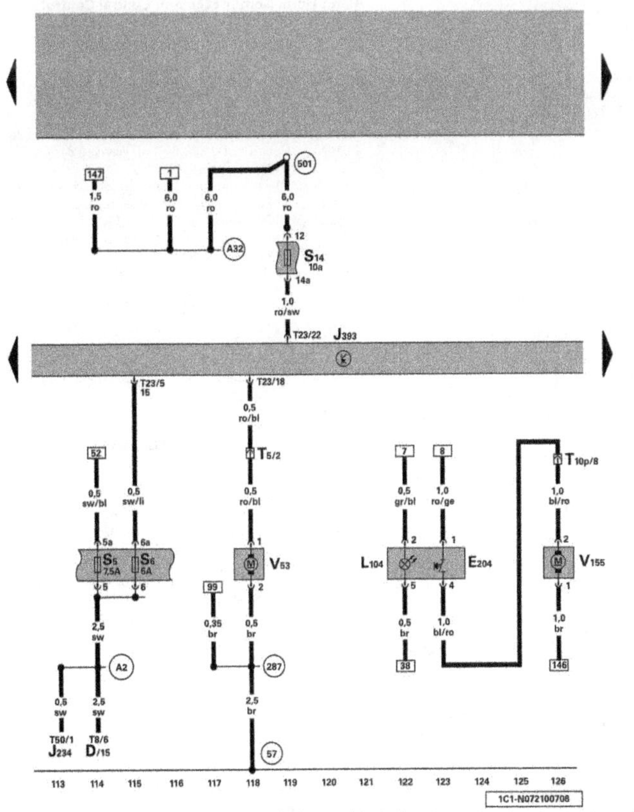

D - Ignition/Starter Switch
E204 - Remote Fuel Tank Flap Switch
J234 - Airbag Control Module
J393 - Comfort System Central Control Module
L104 - Remote Unlock Switch Illumination
S5 - Fuse (on fuse/relay panel)
S6 - Fuse (on fuse/relay panel)
S14 - Fuse (on fuse/relay panel)
T5 - 5-Pin Connector, black, on C pillar, left
T8 - 8-Pin Connector
T10p - 10-Pin Connector, on A-Pillar left, blue
T23 - 23-Pin Connector, on Central control module for comfort system
T50 - 50-Pin Connector
V 53 - Rear Lid Central Locking System Motor
V155 - Fuel Tank Lid Unlock Motor

(57) - Ground Connection (on left rear pillar)
(287) - Ground Connection (in rear lid hydraulics wiring harness)
(501) - Threaded Connection 2 - terminal 30 (on relay panel)
(A2) - Plus Connection (15) (in instrument panel wiring harness)
(A32) - Plus Connection (30) (in instrument panel wiring harness)

ws = white
sw = black
ro = red
br = brown
gn = green
bl = blue
gr = grey
li = violet
ge = yellow
or = orange

New Beetle

Wiring diagram

No.72/11

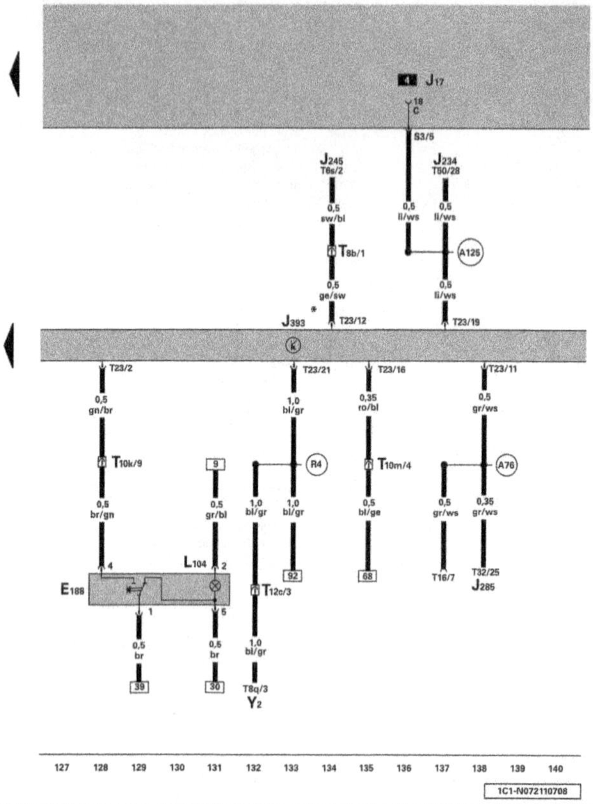

Rear Lid Remote Unlock Switch, Comfort System Central Control Module

- E188 - Rear Lid Remote Unlock Switch
- J17 - Fuel Pump (FP) Relay
- J234 - Airbag Control Module
- J245 - Power Sunroof Control Module
- J285 - Instrument Cluster Control Module
- J393 - Comfort System Central Control Module
- L104 - Remote Unlock Switch Illumination
- T6s - 6-Pin Connector
- T8b - 8-Pin Connector
- T8q - 8-Pin Connector
- T10k - 10-Pin Connector, on A-Pillar left, black
- T10m - 10-Pin Connector
- T12c - 12-Pin Connector
- T16 - 16-Pin Connector, Data Link Connector (DLC), below steering coulm, left
- T23 - 23-Pin Connector, on Central control module for comfort system
- T32 - 32-Pin Connector, blue, on instrument cluster
- T50 - 50-Pin Connector
- Y2 - Digital Clock
- (A76) - K-Diagnostic Connection (in instrument panel wiring harness)
- (A125) - Crash Signal Connection (in instrument panel wiring harness)
- (R4) - Connection 1 (in interior light/door contact switch wiring harness)
- * - Only vehicles with power sunroof

ws = white
sw = black
ro = red
br = brown
gn = green
bl = blue
gr = grey
li = violet
ge = yellow
or = orange

New Beetle

Wiring diagram

No. 72/12

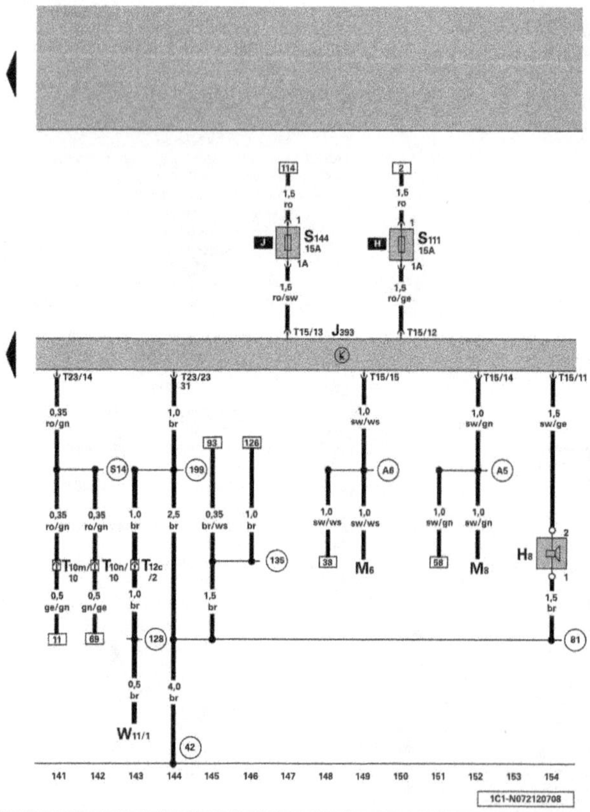

Alarm Horn, Comfort System Central Control Module

- H8 - Alarm Horn
- J393 - Comfort System Central Control Module
- M6 - Left Rear Turn Signal Lamp
- M8 - Right Rear Turn Signal Lamp
- S111 - Alarm System and Anti-Theft Immobilizer Fuse
- S144 - Central Locking/Anti-Theft Warning System Fuse
- T10m - 10-Pin Connector, on A-Pillar left, brown
- T10n - 10-Pin Connector, on A-Pillar right, brown
- T12c - 12-Pin Connector
- T15 - 15-Pin Connector, on Central control module for comfort system
- T23 - 23-Pin Connector, on Central control module for comfort system
- W11 - Left Rear Reading Lamp
- (42) - Ground Connection (beside steering column)
- (81) - Ground Connection 1 (in instrument panel wiring harness)
- (128) - Ground Connection 1 (in interior light wiring harness)
- (135) - Ground Connection 2 (in instrument panel wiring harness)
- (199) - Ground Connection 3 (in instrument panel wiring harness)
- (A5) - Right Turn Signal Plus Connection (in instrument panel wiring harness)
- (A6) - Left Turn Signal Plus Connection (in instrument panel wiring harness)
- (S14) - -Open- Connection (in central locking system wiring harness)

ws = white
sw = black
ro = red
br = brown
gn = green
bl = blue
gr = grey
li = violet
ge = yellow
or = orange

New Beetle — Wiring diagram — No. 72/13

Hood Alarm Switch, Coolant Fan Control (FC) Control Module, Central Locking and Anti-Theft Alarm System Antenna

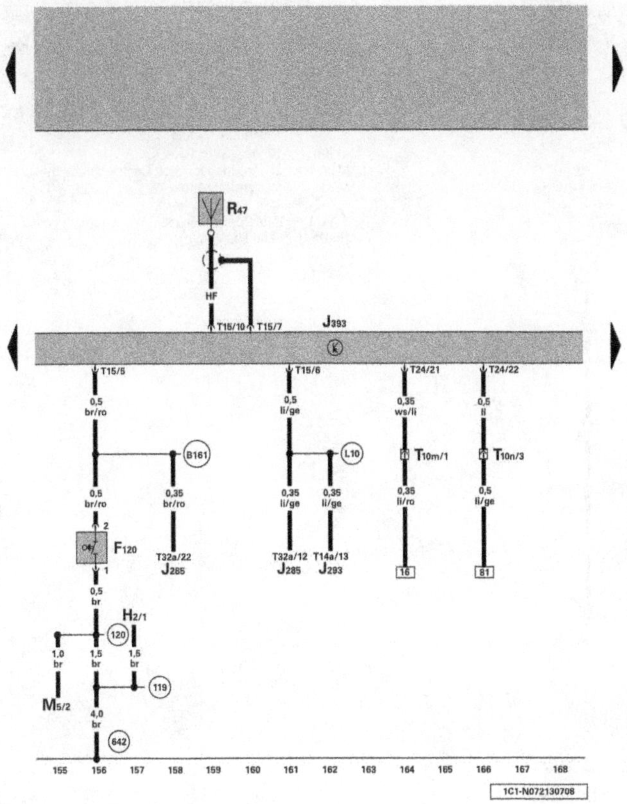

F120	-	Hood Alarm Switch
H2	-	High Tone Horn
J285	-	Instrument Cluster Control Module
J293	-	Coolant Fan Control (FC)Control Module
J393	-	Comfort System Central Control Module
M5	-	Left Front Turn Signal Lamp
R47	-	Central Locking and Anti-Theft Alarm System Antenna
T10m	-	10-Pin Connector, on A-Pillar left, brown
T10n	-	10-Pin Connector, on A-Pillar right, brown
T14a	-	14-Pin Connector, on -J293-
T15	-	15-Pin Connector, on Central control module for comfort system
T24	-	24-Pin Connector, on Central control module for comfort system
T32a	-	32-Pin Connector, green, on instrument closter
(119)	-	Ground Connection 1 (in headlamp wiring harness)
(120)	-	Ground Connection 2 (in headlamp wiring harness)
(642)	-	Ground Connection (on electronic fan control)
(B161)	-	Anti-Theft Alarm System Connection (in interior wiring harness)
(L10)	-	Connection 2 (in A/C wiring harness)*
*	-	Only vehicles with A/C

ws = white
sw = black
ro = red
br = brown
gn = green
bl = blue
gr = grey
li = violet
ge = yellow
or = orange

New Beetle — Wiring diagram — No. 72/14

Comfort System Central Control Module

J393	-	Comfort System Central Control Module
T10k	-	10-Pin Connector, on A-Pillar left, black
T10l	-	10-Pin Connector, on A-Pillar right, black
T10m	-	10-Pin Connector, on A-Pillar left, brown
T10n	-	10-Pin Connector, on A-Pillar right, brown
T24	-	24-Pin Connector, on Central control module for comfort system
(A13)	-	Door Contact Switch Connection (in instrument panel wiring harness)

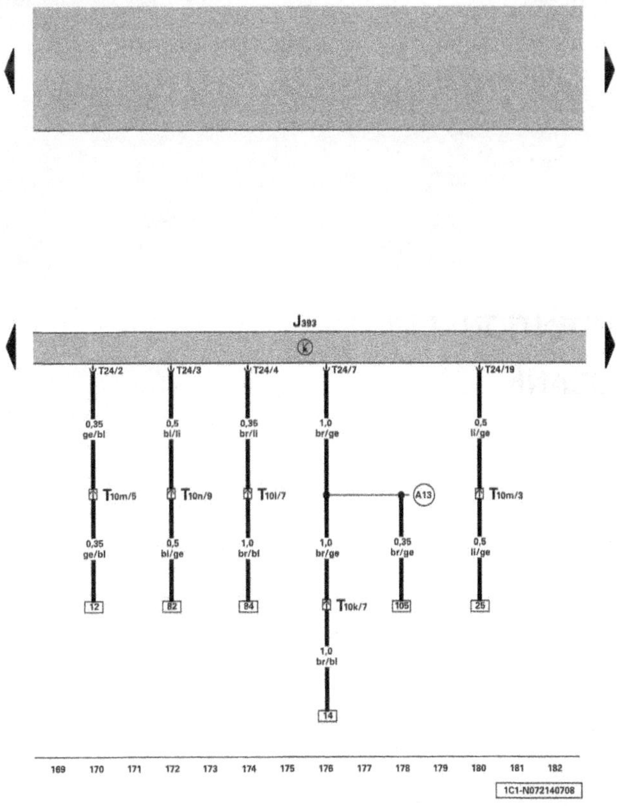

ws = white
sw = black
ro = red
br = brown
gn = green
bl = blue
gr = grey
li = violet
ge = yellow
or = orange

New Beetle Wiring diagram **No. 72/15**

Comfort System Central Control Module

- E20 - Instrument Panel Illumination Dimmer Switch
- J393 - Comfort System Central Control Module
- T3k - 3-Pin Connector
- T10m - 10-Pin Connector, on A-Pillar left, brown
- T10n - 10-Pin Connector, on A-Pillar right, brown
- T10p - 10-Pin Connector
- T10q - 10-Pin Connector
- T24 - 24-Pin Connector, on Central control module for comfort system
- (A4) - Plus Connection (58b) (in instrument panel wiring harness)

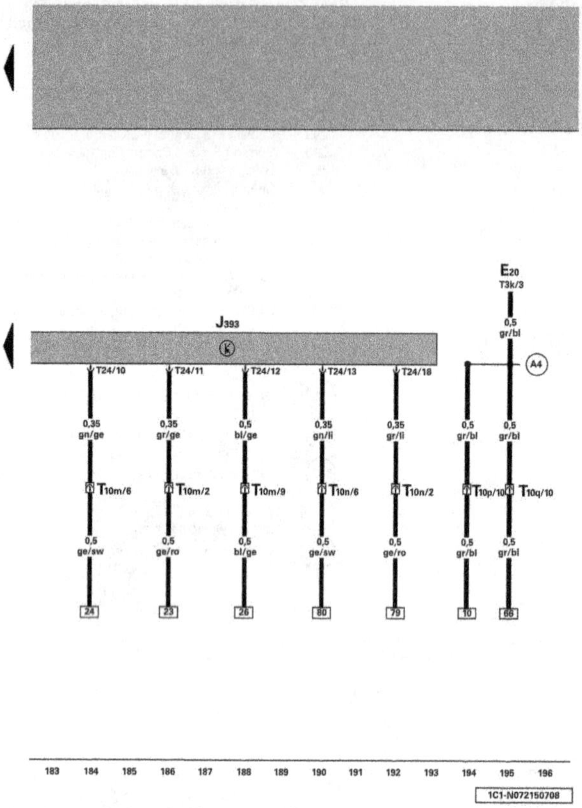

ws = white
sw = black
ro = red
br = brown
gn = green
bl = blue
gr = grey
li = violet
ge = yellow
or = orange

THIS SPACE INTENTIONALLY LEFT BLANK

New Beetle Convertible — Wiring diagram — No. 52/1

Tire pressure monitoring,

from July 2007,

Additional Fuse, Relay, Control Module and Connector locations ⇒ Component Locations

Troubleshooting Procedures ⇒ Guided Fault Finding using VAS 5051/5052

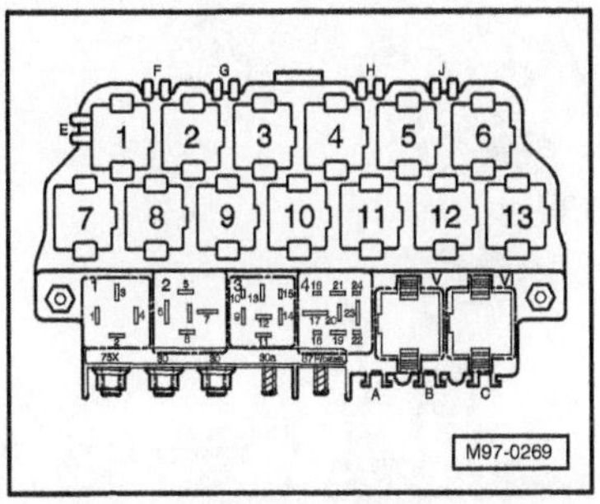

Relay location on the thirteen position auxiliary relay panel, above relay panel:

Relay panel:

2 J59 - Load Reduction Relay (100)

Note:

Number in parentheses indicates production control number stamped on relay housing.

Edition 06.07

New Beetle Convertible — Wiring diagram — No. 52/2

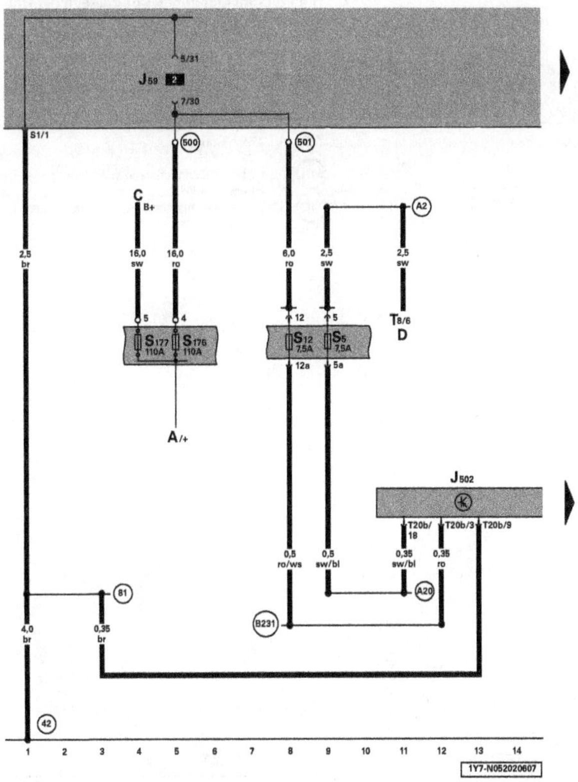

Tire Pressure Monitoring Control Module

- A - Battery
- C - Generator (GEN)
- D - Conmutador de encendido y arranque
- J59 - Load Reduction Relay
- J502 - Tire Pressure Monitoring Control Module
- S5 - Fuse 5 (on fuse/relay panel)
- S12 - Fuse 12 (on fuse/relay panel)
- S176 - Fuse 4 - terminal 30 (on battery fuse holder)
- S177 - Fuse 5 - terminal 30 (on battery fuse panel)
- T8 - 8-Pin Connector
- T20b - 20-Pin Connector
- (42) - Ground Connection (beside steering column)
- (81) - Ground Connection 1 (in instrument panel wiring harness)
- (500) - Threaded Connection 1 - terminal 30 (on relay panel)
- (501) - Threaded Connection 2 - terminal 30 (on relay panel)
- (A2) - Plus Connection (15) (in instrument panel wiring harness)
- (A20) - Plus Connection (15a) (in instrument panel wiring harness)
- (B231) - Plus Connection 5 (30a) (in interior wiring harness)

ws = white
sw = black
ro = red
br = brown
gn = green
bl = blue
gr = grey
li = violet
ge = yellow
or = orange

Edition 06.07

Tire pressure monitoring, convertible from July 2007

New Beetle Convertible — Wiring diagram — No. 52/3

Tire Pressure Monitoring Sensors, Tire Pressure Monitoring Control Module,

G222 - Left Front Tire Pressure Monitoring Sensor
G223 - Right Front Tire Pressure Monitoring Sensor
G224 - Left Rear Tire Pressure Monitoring Sensor
G225 - Right Rear Tire Pressure Monitoring Sensor
J502 - Tire Pressure Monitoring Control Module
R175 - Tire Pressure Monitoring Antenna

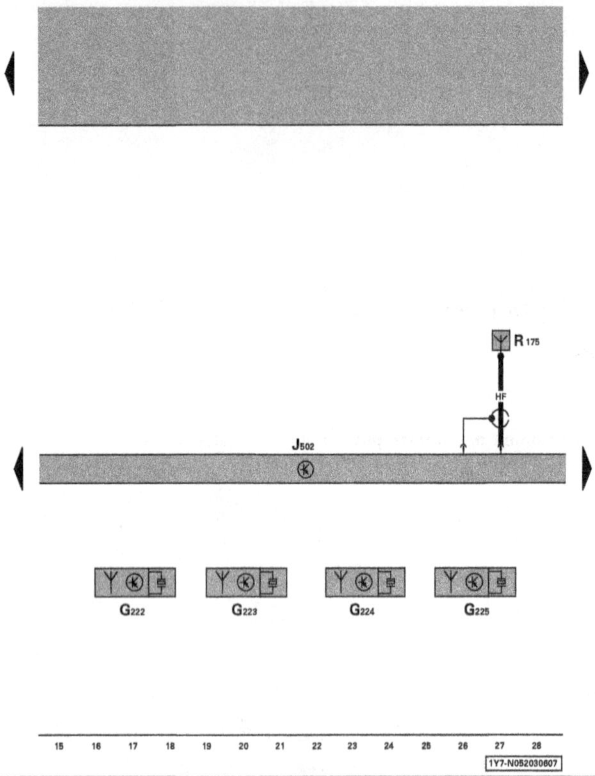

ws = white
sw = black
ro = red
br = brown
gn = green
bl = blue
gr = grey
li = violet
ge = yellow
or = orange

New Beetle Convertible — Wiring diagram — No. 52/4

Tire Pressure Monitoring Control Module, Tire Pressure Monitoring Display Indicator Lamp

J285 - Instrument Cluster Control Module
J502 - Tire Pressure Monitoring Control Module
K220 - Tire Pressure Monitoring Display Indicator Lamp
T20b - 20-Pin Connector
T32a - 32-Pin Connector

(A121) - High-Bus Connection (in instrument panel wiring harness)
(A122) - Low-Bus Connection (in instrument panel wiring harness)

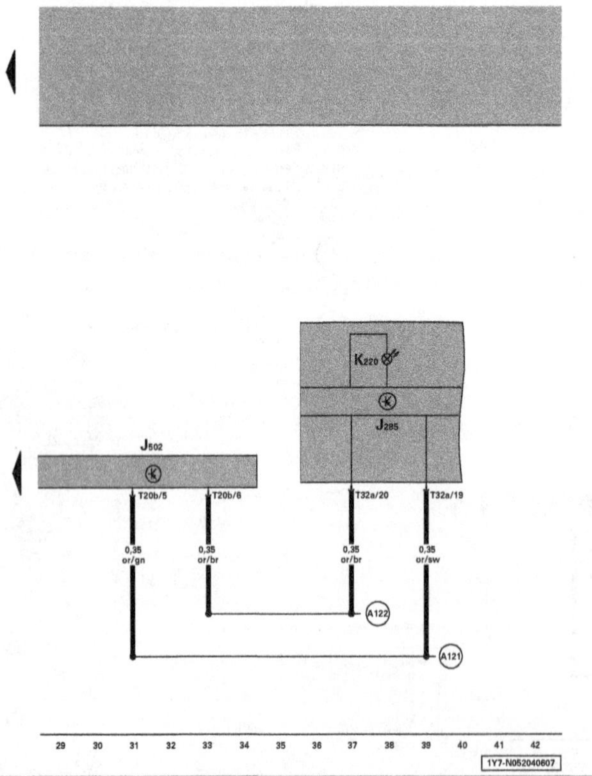

ws = white
sw = black
ro = red
br = brown
gn = green
bl = blue
gr = grey
li = violet
ge = yellow
or = orange

OBD On Board Diagnostics

GENERAL

This section gives a brief overview of On Board Diagnostics and a list of Diagnostic Trouble Codes (DTCs) applicable to the powertrain (P-codes) and related network communications (U-codes).

While most DTCs can be accessed with a Generic Scan Tool (GST), full and complete vehicle system information can only be obtained using a Volkswagen supplied diagnostic scan tool computer such as the VAS 5051/5052 and derivatives, or an equivalent aftermarket scan tool or computer based scan tool program. The function for advanced diagnosis with these tools is called guided fault finding. A listing of scan tool and special tool suppliers is at the end of this section.

ON BOARD DIAGNOSTICS

All vehicles covered by this manual are equipped with On Board Diagnostics (OBD). All system access is by way of the Data Link Connector (DLC) located under the left lower side of the dashboard. The DLC uses the industry standard SAE 16-pin connector.

On Board Diagnostics

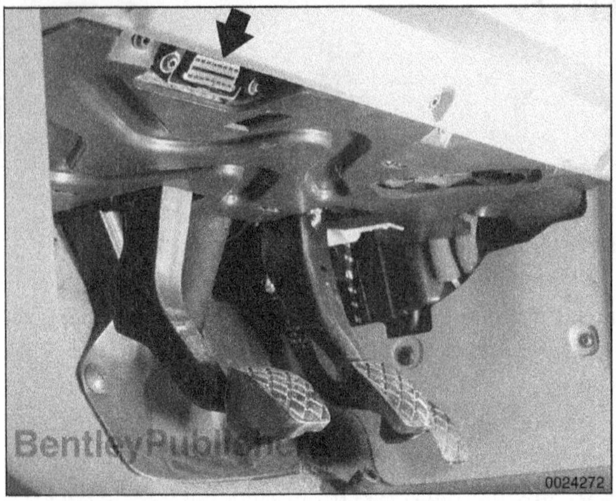

Data link connector (**arrow**) is located under left lower side of dashboard. Connector is recessed into panel as shown and does not use a cover panel.

On Board Diagnostics include the United States government mandated second generation vehicle emissions monitoring system (OBD II) and Volkswagen factory system diagnostics. OBD II codes are identified as P-codes and U-codes. Volkswagen factory codes are identified as VAG codes. Most OBD II codes also have a corresponding VAG code. Both codes are listed in the DTC tables in this section as an aid to identification.

The vehicles covered by this manual are equipped with diagnostics that are far more extensive than the federal requirements. Communication is by way of twisted pair wiring called a "bus". Controller Area Network (CAN) communication over the twisted pair wiring is called CAN-bus. There are different CAN-bus lines covering most of the electronically controlled systems in the vehicle.

Terminal assignment for 16-pin DLC (T16). Terminals not listed are not currently in use.

1 - Terminal 15 (power with ignition on and in crank position)
4 - Terminal 31 (ground)
5 - Terminal 31 (ground)
7 - Diagnosis K-wire
16 - Terminal 30 (battery positive power)

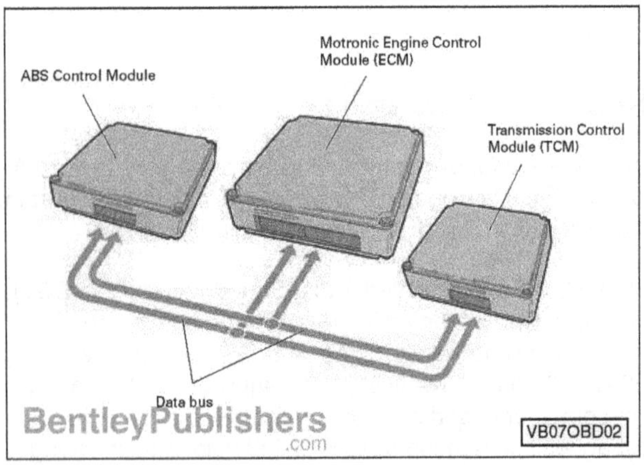

CAN-bus interface links ABS, engine, and transmission control modules. A similar CAN-bus connects control units of the convenience system.

Access to systems other than OBD II may require an advanced scan tool (VAS 5051/5052) and is outside the scope of this manual.

OBD II

On Board Diagnostics II monitors operation and function of engine management, and automatic transmission. Other emissions-related systems or parts of them, are monitored to insure compliance with specified emission levels.

Vehicle emission levels are continually monitored by the OBD II system and problems are recognized and recorded. A Malfunction Indicator Light (MIL) in the instrument cluster alerts the driver to the problem and the need to have the system checked. When a DTC is stored, or a potential problem exists, any of several indicator lights in the instrument cluster could be illuminated to notify the driver.

It is important to note that a diagnostic trouble code does not necessarily mean that an electronic component has failed. Even running low on gasoline can set a DTC as the system tries to cope with the ever-lessening fuel volume and pressure.

Diagnostic trouble code structure

When a system operates outside assigned operating parameters, a code is stored denoting that failure mode. This code is known as a diagnostic trouble code or DTC.

Each DTC is assigned two codes. The first code is a numerical code assigned by Volkswagen (VAG code). The second code is referred to as a P-code or U-code and follows a structure required by law and defined by the Society of Automotive Engineers (SAE). Diagnostic Trouble Codes (DTCs) follow a standard format. This standard uses a letter to designate the system and four numbers or letters to further identify and detail the malfunction as listed below.

First digit structure is as follows:
- Pxxxx for powertrain
- Uxxxx for vehicle network communications
- Bxxxx for body
- Cxxxx for chassis

Second digit structure is:
- P0xxx Government required codes
- P1xxx Manufacturer codes for additional emission system function; not required but reported to government
- P2xxx Fuel and air mixture. Additional emissions regulations
- P3xxx Fuel and air mixture. Additional emissions regulations

Third digit structure is:
- Px1xx measurement of air and fuel
- Px2xx measurement of air and fuel
- Px3xx ignition system
- Px4xx additional emission control
- Px5xx speed and idle regulation
- Px6xx computer and output signals
- Px7xx transmission
- Px8xx transmission
- Px9xx control modules, input and output signals

The fourth and fifth digits denote the individual components and systems. In some cases, the fifth digit can be a letter. Currently A through D are used.

The digit structure shown above has “P” as the first digit, however, letters “B”, “C”, and “U” could be used where applicable. Only “P” and “U” DTCs are currently used in vehicles covered by this manual.

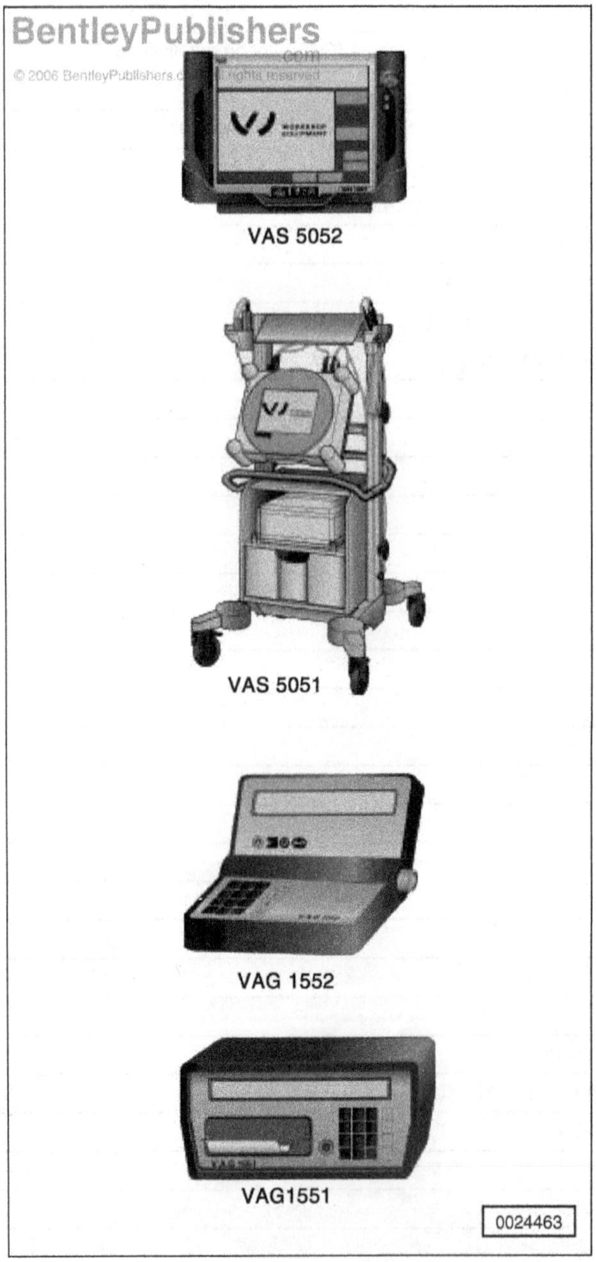

Sampling of Volkswagen supplied scan tools. VAS 5051 has been supplemented by a “B” version. Also note that VAG 1551 and VAG 1552 may not have the ability to access all vehicle systems and may not display the latest DTC text.

Diagnostic Trouble Codes

NOTE—
- *The following table contains a list of available scan tool codes for Volkswagen and Audi vehicles available at the time of publication. Some of the codes do not apply to vehicles covered by this manual.*

Diagnostic Trouble Codes (Powertrain)

SAE P-code	V.A.G. code	Description
P0010	16394	Intake "A" Camshaft Position Actuator Circuit Bank 1 Malfunction
P0011	16395	Intake "A" Camshaft Position, Timing Over Advanced or System Performance Bank 1
P0012	16396	Intake "A" Camshaft Position, Timing Over Retarded Bank 1
P0013	16397	Intake "B" Camshaft Position Actuator Circuit Bank 1 Malfunction
P0014	16398	Intake "B" Camshaft Position, Timing Over Advanced or System Performance Bank 1
P0015	16399	Intake "B" Camshaft Position, Timing Over Retarded Bank 1
P0016	16100	Camshaft Position, Camshaft Position Correlation B1 Sensor "A"
P0017	16401	Bank 1 Camshaft Position Sensor 3 (G300) or Engine Speed Sensor (G28) - Incorrect allocation
P0018	16402	Camshaft Position, Camshaft Position Correlation B1 Sensor "B"
P0019	16403	Bank 2 Camshaft Position Sensor 4 (G301) or Engine Speed Sensor (G28) - Incorrect allocation
P0020	16404	-A- Camshaft Pos. Actuator Circ. Bank 2 Malfunction
P0021	16405	-A- Camshaft Position Bank 2 - Timing Over-Advanced or System Performance
P0022	16406	-A- Camshaft Position Bank 2 - Timing Over-Retarded
P0023	16407	-B- Camshaft Position Actuator Circuit Bank 2 - Malfunction
P0024	16408	-B- Camshaft Position Bank 2 - Timing Over Advanced or System Performance
P0025	16409	-B- Camshaft Position Bank 2 - Timing Over-Retarded
P0030	16414	HO2S Heater Control Circuit, B1S1
P0031	16415	HO2S Heater Control Circuit Low, B1S1
P0032	16416	HO2S Heater Control Circuit High, B1S1
P0036	16420	HO2S Heater Control Circuit B1S2
P0037	16421	HO2S Heater Control Circuit Low B1S2
P0038	16422	HO2S Heater Control Circuit High B1S2
P0040	16424	O2 Sensor Signals Swapped B1S1/B2S1
P0041	16425	O2 Sensor Signals Swapped B2S1/B2S2
P0042	16426	HO2S Heater Control Circuit B1S3
P0043	16427	HO2S Heater Control Circuit Low B1S3
P0044	16428	HO2S Heater Control Circuit High B1S3
P0050	16434	HO2S Heater Control Circuit B2S1
P0051	16435	HO2S Heater Control Circuit Low B2S1
P0052	16436	HO2S Heater Control Circuit High B2S1
P0056	16440	HO2S Heater Control Circuit B2S2
P0057	16441	HO2S Heater Control Circuit Low B2S2
P0058	16442	HO2S Heater Control Circuit High B2S2
P0065	16449	Air Assisted Injector Control Range/Performance
P0066	16450	Air Assisted Injector Control Low Input/Short to ground
P0067	16451	Air Assisted Injector Control Input/Short to B+
P0068	16452	MAP/MAF Throttle Position Correlation
P0069	16453	MAP/Barometric Pressure Correlation

Diagnostic Trouble Codes (Powertrain)

SAE P-code	V.A.G. code	Description
P0087		Fuel Rail/System Pressure - Too Low
P0088		Fuel Rail/System Pressure - Too High
P0089	16473	Fuel Pressure Regulator 1 - Performance
P0097	16481	Intake Air Temperature Sensor 2 Circuit - Low
P0098	16482	Intake Air Temperature Sensor 2 Circuit - High
P0100	16484	Mass or Volume Air Flow Circuit
P0101	16485	Mass or Volume Air Flow Circ Range/Performance
P0102	16486	Mass or Volume Air Flow Circ Low Input
P0103	16487	Mass or Volume Air Flow Circ High Input
P0105	16489	Manifold Abs.Pressure or Bar.Pressure Voltage supply
P0106	16490	Manifold Abs.Pressure or Bar.Pressure Range/Performance
P0107	16491	Manifold Abs.Pressure or Bar.Pressure Low Input
P0108	16492	Manifold Abs.Pressure or Bar.Pressure High Input
P0111	16495	Intake Air Temperature Sensor Circuit Range/Performance
P0112	16496	Intake Air Temp.Circ Low Input
P0113	16497	Intake Air Temp.Circ High Input
P0116	16500	Engine Coolant Temp.Circ Range/Performance
P0117	16501	Engine Coolant Temp.Circ Low Input
P0118	16502	Engine Coolant Temp.Circ High Input
P0120	16504	Throttle/Pedal Pos.Sensor A Circ Malfunction
P0121	16505	Throttle/Pedal Pos.Sensor A Circ Range/Performance
P0122	16506	Throttle/Pedal Pos.Sensor A Circ Low Input
P0123	16507	Throttle/Pedal Pos.Sensor A Circ High Input
P0125	16509	Insufficient Coolant Temp.for Closed Loop Fuel Control
P0128	16512	Coolant Thermostat/Valve Temperature below control range
P0130	16514	O2 Sensor Circ., B1S1 Malfunction
P0131	16515	O2 Sensor Circ., B1S1 Low Voltage
P0132	16516	O2 Sensor Circ., B1S1 High Voltage
P0133	16517	O2 Sensor Circ., B1S1 Slow Response
P0134	16518	O2 Sensor Circ., B1S1 No Activity Detected
P0135	16519	O2 Sensor Heater Circ.,Bank 1-Sensor 1 Malfunction
P0136	16520	O2 Sensor Circ., B1S2 Malfunction
P0137	16521	O2 Sensor Circ., B1S2 Low Voltage
P0138	16522	O2 Sensor Circ., B1S2 High Voltage
P0139	16523	O2 Sensor Circ., B1S2 Slow Response
P0140	16524	O2 Sensor Circ., B1S2 No Activity Detected
P0141	16525	O2 Sensor Heater Circ., B1S2 Malfunction
P0142	16526	O2 Sensor Circ., B1S3
P0143	16427	O2 Sensor Circ., Low Voltage B1S3
P0144	16428	O2 Sensor Circ., High Voltage B1S3
P0145	16429	O2 Sensor Circ., Slow Response B1S3
P0146	16530	O2 Sensor Circ., No Activity Detected B1S3
P0147	16531	O2 Sensor Heater Circ. B1S3
P0150	16534	O2 Sensor Circ.,Bank2-Sensor1 Malfunction

Diagnostic Trouble Codes (Powertrain)

SAE P-code	V.A.G. code	Description
P0151	16535	O2 Sensor Circ.,Bank2-Sensor1 Low Voltage
P0152	16536	O2 Sensor Circ.,Bank2-Sensor1 High Voltage
P0153	16537	O2 Sensor Circ.,Bank2-Sensor1 Slow Response
P0154	16538	O2 Sensor Circ.,Bank2-Sensor1 No Activity Detected
P0155	16539	O2 Sensor Heater Circ.,Bank2-Sensor1 Malfunction
P0156	16540	O2 Sensor Circ.,Bank2-Sensor2 Malfunction
P0157	16541	O2 Sensor Circ.,Bank2-Sensor2 Low Voltage
P0158	16542	O2 Sensor Circ.,Bank2-Sensor2 High Voltage
P0159	16543	O2 Sensor Circ.,Bank2-Sensor2 Slow Response
P0160	16544	O2 Sensor Circ.,Bank2-Sensor2 No Activity Detected
P0161	16545	O2 Sensor Heater Circ.,Bank2-Sensor2 Malfunction
P0169		Incorrect Fuel Composition
P0170	16554	Fuel Trim,Bank1 Malfunction
P0171	16555	Fuel Trim,Bank1 System too Lean
P0172	16556	Fuel Trim,Bank1 System too Rich
P0173	16557	Fuel Trim,Bank2 Malfunction
P0174	16558	Fuel Trim,Bank2 System too Lean
P0175	16559	Fuel Trim,Bank2 System too Rich
P0181	16565	Fuel temperature sender A Circuit Range/Performance
P0182	16566	Fuel temperature sender A Circuit Low
P0183	16567	Fuel temperature sender A Circuit High
P0188	16572	Fuel Temperature Sensor 2 - Short circuit to B+
P0190	16574	Fuel Pressure Sensor (G247) - Electrical Malfunction
P0192	16576	Fuel Rail Pressure Sensor Circuit Low
P0197	16581	Engine Oil Temperature Circuit Low Input
P0198	16582	Engine Oil Temperature Circuit High Input
P0200	16584	Injector Circuit Open
P0201	16585	Cyl.1, Injector Circuit Fault in electrical circuit
P0202	16586	Cyl.2, Injector Circuit Fault in electrical circuit
P0203	16587	Cyl.3, Injector Circuit Fault in electrical circuit
P0204	16588	Cyl.4, Injector Circuit Fault in electrical circuit
P0205	16589	Cyl.5 Injector Circuit Fault in electrical circuit
P0206	16590	Cyl.6 Injector Circuit Fault in electrical circuit
P0207	16591	Cyl.7 Injector Circuit Fault in electrical circuit
P0208	16592	Cyl.8 Injector Circuit Fault in electrical circuit
P0215	16599	Engine Shut-Off Solenoid Malfunction
P0216	16600	Injector/Injection Timing Control Malfunction
P0219	16603	Engine Overspeed Condition
P0221	16605	Throttle Pos. Sensor -B- Circuit Range/Performance
P0222	16606	Throttle Pos. Sensor -B- Circuit Low Input
P0223	16607	Throttle Pos. Sensor -B- Circuit High Input
P0225	16609	Throttle Pos. Sensor -C- Circuit Voltage supply
P0226	16610	Throttle Pos. Sensor -C- Circuit Range/Performance
P0227	16611	Throttle Pos. Sensor -C- Circuit Low Input

Diagnostic Trouble Codes (Powertrain)

SAE P-code	V.A.G. code	Description
P0228	16612	Throttle Pos. Sensor -C- Circuit High Input
P0230	16614	Fuel Pump Primary Circuit Fault in electrical circuit
P0231	16615	Fuel Pump Secondary Circuit Low
P0232	16616	Fuel Pump Secondary Circuit High
P0234	16618	Turbocharger Overboost Condition Control limit exceeded
P0235	16619	Turbocharger Boost Sensor (A) Circ Control limit not reached
P0236	16620	Turbocharger Boost Sensor (A) Circ Range/Performance
P0237	16621	Turbocharger Boost Sensor (A) Circ Low Input
P0238	16622	Turbocharger Boost Sensor (A) Circ High Input
P0240	16624	Turbo/Super Charger Boost Sensor B Circuit - Range/Performance
P0241	16625	Turbo/Super Charger Boost Sensor B Circuit - Low
P0242	16626	Turbo/Super Charger Boost Sensor B Circuit - High
P0243	16627	Turbocharger Wastegate Solenoid (A) Open/Short Circuit to Ground
P0245	16629	Turbocharger Wastegate Solenoid (A) Low Input/Short to ground
P0246	16630	Turbocharger Wastegate Solenoid (A) High Input/Short to B+
P0247		Turbocharger Wastegate Solenoid (B)
P0249	16633	Boost Pressure Control Valve 2 - Open/Short to Ground
P0250	16634	Wastegate Bypass Regulator Valve 2 - Short circuit to B+
P0251	16635	Injection Pump Fuel Metering Control A (Cam/Rotor/Injector)
P0252	16636	Injection Pump Fuel Metering Control A - Range/Performance
P0261	16645	Cyl.1 Injector Circuit Low Input/Short to ground
P0262	16646	Cyl.1 Injector Circuit High Input/Short to B+
P0263	16647	Cyl. 1 Contribution/Balance
P0264	16648	Cyl.2 Injector Circuit Low Input/Short to ground
P0265	16649	Cyl.2 Injector Circuit High Input/Short to B+
P0266	16650	Cyl. 2 Contribution/Balance
P0267	16651	Cyl.3 Injector Circuit Low Input/Short to ground
P0268	16652	Cyl.3 Injector Circuit High Input/Short to B+
P0269	16653	Cyl. 3 Contribution/Balance
P0270	16654	Cyl.4 Injector Circuit Low Input/Short to ground
P0271	16655	Cyl.4 Injector Circuit High Input/Short to B+
P0272	16656	Cyl.4 Contribution/Balance
P0273	16657	Cyl.5 Injector Circuit Low Input/Short to ground
P0274	16658	Cyl.5 Injector Circuit High Input/Short to B+
P0275		Cyl.5 Contribution/Balance
P0276	16660	Cyl.6 Injector Circuit Low Input/Short to ground
P0277	16661	Cyl.6 Injector Circuit High Input/Short to B+
P0299	16683	Turbo/Supercharger underboost
P0300	16684	Random/Multiple Cylinder Misfire Detected
P0301	16685	Cyl.1 Misfire Detected
P0302	16686	Cyl.2 Misfire Detected
P0303	16687	Cyl.3 Misfire Detected
P0304	16688	Cyl.4 Misfire Detected
P0305	16689	Cyl.5 Misfire Detected

Diagnostic Trouble Codes (Powertrain)

SAE P-code	V.A.G. code	Description
P0306	16690	Cyl.6 Misfire Detected
P0313	16697	Misfire Detected Low Fuel Level
P0314	16698	Single Cylinder Misfire
P0318	16702	Rough Road Sensor "A" Signal Circuit
P0321	16705	Ign./Distributor Eng.Speed Inp.Circ Range/Performance
P0322	16706	Ign./Distributor Eng.Speed Inp.Circ No Signal
P0324	16708	Knock Control System Error
P0325	16709	Knock Sensor 1 Circuit Electrical Fault in Circuit
P0326	16710	Knock Sensor 1 Circuit Range/Performance
P0327	16711	Knock Sensor 1 Circ Low Input
P0328	16712	Knock Sensor 1 Circ High Input
P0332	16716	Knock Sensor 2 Circ Low Input
P0333	16717	Knock Sensor 2 Circ High Input
P0335	16719	Crankshaft Pos. Sensor (A) Circ Malfunction
P0336	16720	Crankshaft Pos. Sensor (A) Circ Range/Performance/Missing tooth
P0337	16721	Crankshaft Pos.Sensor (A) Circ Low Input
P0340	16724	Camshaft Pos. Sensor (A) Circ Incorrect allocation
P0341	16725	Camshaft Pos.Sensor Circ Range/Performance
P0342	16726	Camshaft Pos.Sensor Circ Low Input
P0343	16727	Camshaft Pos.Sensor Circ High Input
P0345	16729	Camshaft Position Sensor "A" Circuit Bank 2
P0346	16730	Camshaft Position Sensor "A" Circuit Range/Performance Bank 2
P0347	16731	Camshaft Position Sensor "A" Circuit Low Bank 2
P0348	16732	Camshaft Position Sensor "A" Circuit High Bank 2
P0351	16735	Ignition Coil (A) Cyl.1 Prim./Sec. Circ Malfunction
P0352	16736	Ignition Coil (B) Cyl.2 Prim./Sec. Circ Malfunction
P0353	16737	Ignition Coil (C) Cyl.3 Prim./Sec. Circ Malfunction
P0354	16738	Ignition Coil (D) Cyl.4 Prim./Sec. Circ Malfunction
P0355	16739	Ignition Coil (E) Cyl.5 Prim./Sec. Circ Malfunction
P0356	16740	Ignition Coil (F) Cyl.6 Prim./Sec. Circ Malfunction
P0365	16749	Camshaft Position Sensor (B) Circuit (Bank1) - Malfunction
P0366	16750	Camshaft Position Sensor (B) Circuit (Bank1) - Range/Performance
P0367	16751	Camshaft Position Sensor (B) Circuit (Bank1) - Low
P0368	16752	Camshaft Position Sensor (B) Circuit (Bank1) - High
P0380	16764	Glow Plug/Heater Circuit (A) Electrical Fault in Circuit
P0381	16765	Glow Plug/Heater indicator Circuit
P0391	16775	Camshaft Position Sensor (B) Circuit (Bank2) - Range/Performance
P0392	16776	Camshaft Position Sensor (B) Circuit (Bank2) - Low
P0393	16777	Camshaft Position Sensor (B) Circuit (Bank2) - High
P0400	16784	Exhaust Gas Recirc.Flow Malfunction
P0401	16785	Exhaust Gas Recirc.Flow Insufficient Detected
P0402	16786	Exhaust Gas Recirc.Flow Excessive Detected
P0403	16787	Exhaust Gas Recirc. Contr. Circ Malfunction
P0404	16788	Exhaust Gas Recirc. Contr. Circ Range/Performance

Diagnostic Trouble Codes (Powertrain)

SAE P-code	V.A.G. code	Description
P0405	16789	Exhaust Gas Recirc. Sensor (A) Circ Low Input
P0406	16790	Exhaust Gas Recirc. Sensor (A) Circ High Input
P0407	16791	Exhaust Gas Recirc. Sensor (B) Circ Low Input
P0408	16792	Exhaust Gas Recirc. Sensor (B) Circ High Input
P0410	16794	Sec.Air Inj.Sys Malfunction
P0411	16795	Sec.Air Inj.Sys. Incorrect Flow Detected
P0412	16796	Sec.Air Inj.Sys.Switching Valve A Circ Malfunction
P0413	16797	Secondary Air Injection System Switching Valve "A" Circuit Open
P0414	16798	Secondary Air Injection System Switching Valve "A" Circuit Shorted
P0416	16800	Secondary Air Injection System Switching Valve (B) Circuit - Open
P0417	16801	Secondary Air Injection System Switching Valve (B) Circuit - Shorted
P0418	16802	Sec. Air Inj. Sys. Relay (A) Contr. Circ Malfunction
P0420	16804	Catalyst System,Bank1 Efficiency Below Threshold
P0422	16806	Main Catalyst,Bank1 Below Threshold
P0423		Heated Catalyst Efficiency Below Threshold, Bank 1
P0424		Heated Catalyst Temperature Below Threshold, Bank 1
P0425		Catalyst Temperature Sensor Circuit, Bank 1 Sensor 1
P0426		Catalyst Temperature Sensor Circuit Range/Performance, Bank 1 Sensor 1
P0427	16811	Catalyst Temperature Sensor, Bank 1 Low Input/Short to ground
P0428	16812	Catalyst Temperature Sensor, Bank 1 High Input/Open/Short Circuit to B+
P0429		Catalyst Heater Control Circuit, Bank 1
P0430	16814	Catalyst System Efficiency Below Threshold Bank 2
P0431	16815	Warm Up Catalyst Bank 2 - Efficiency Below Threshold
P0432	16816	Main Catalyst,Bank2 Efficiency Below Threshold
P0436	16820	Catalyst Temperature Sensor, Bank 2 Range/Performance
P0437	16821	Catalyst Temperature Sensor, Bank 2 Low Input/Short to ground
P0438	16822	Catalyst Temperature Sensor, Bank 2 High Input/Open/Short Circuit to B+
P0440	16824	EVAP Emission Contr.Sys. Malfunction
P0441	16825	EVAP Emission Contr.Sys.Incorrect Purge Flow
P0442	16826	EVAP Emission Contr.Sys.(Small Leak) Leak Detected
P0443	16827	EVAP Emiss. Contr. Sys. Purge Valve Circ Electrical Fault in Circuit
P0444	16828	Evaporative Emission System Purge Control Valve Circuit Open
P0445	16829	Evaporative Emission System Purge Control Valve Circuit Shorted
P0449	16833	Evaporative Emission System Vent Valve/Solenoid Circuit
P0452	16836	EVAP Emission Contr.Sys.Press.Sensor Low Input
P0453	16837	EVAP Emission Contr.Sys.Press.Sensor High Input
P0455	16839	EVAP Emission Contr.Sys. Gross Leak Detected
P0456	16840	EVAP Emission Contr.Sys.Very Small Leak Detected
P0458	16842	Evaporative Emission System Purge Regulator Valve (N80) Short to Ground
P0459	16843	Evaporative Emission System Purge Regulator Valve (N80) Short to Positive
P0461	16845	Fuel Level Sensor Circ Range/Performance
P0462	16846	Fuel Level Sensor Circuit Low Input
P0463	16847	Fuel Level Sensor Circuit High Input
P0480	16864	Fan 1 Control Circuit

Diagnostic Trouble Codes (Powertrain)

SAE P-code	V.A.G. code	Description
P0481	16865	Fan 2Control Circuit
P0489	16873	Exhaust Gas Recirculation Control Circuit Low
P0490	16874	Exhaust Gas Recirculation Control Circuit High
P0491	16875	Secondary Air Injection System Insufficient Flow Bank 1
P0492	16876	Secondary Air Injection System Insufficient Flow Bank 2
P0498	16882	Evaporative Emission System Vent Valve Control Circuit Low
P0499	16883	Evaporative Emission System Vent Valve Control Circuit High
P0501	16885	Vehicle Speed Sensor Range/Performance
P0503	16887	Vehicle Speed Sensor Intermittent/Erratic/High Input
P0505	16889	Idle Control System Malfunction
P0506	16890	Idle Control System RPM Lower than Expected
P0507	16891	Idle Control System Higher than Expected
P0510	16894	Closed Throttle Pos.Switch Malfunction
P0513	16897	Incorrect Immobilizer Key
P0531	16915	A/C Refrigerant Pressure Sensor Circuit Range/Performance
P0532	16916	A/C Refrigerant Pressure Sensor Circuit Low Input
P0533	16917	A/C Refrigerant Pressure Sensor Circuit High Input
P0544	16928	Exhaust Gas Temperature Sensor Circuit B1S1
P0545	16929	Exhaust Gas Temperature Sensor Circuit Low B1S1
P0546	16930	Exhaust Gas Temperature Sensor Circuit High B1S1
P0551	16935	Power Steering Pressure Sensor Circuit Range/Performance
P0560	16944	System Voltage Malfunction
P0562	16946	System Voltage Low Voltage
P0563	16947	System Voltage High Voltage
P0564		Cruise Control Multi-Function Input “A” Circuit
P0568	16952	Cruise Control Set Signal Incorrect Signal
P0571	16955	Cruise/Brake Switch (A) Circ Malfunction
P0597	16981	Map Controlled Engine Cooling Thermostat - Open
P0598	16982	Map Controlled Engine Cooling Thermostat - Short to ground
P0599	16983	Map Controlled Engine Cooling Thermostat - Short to B+
P0600	16984	Serial Comm. Link (Data Bus) Message Missing
P0601	16985	Internal Contr.Module Memory Check Sum Error
P0602	16986	Control Module Programming Error/Malfunction
P0603	16987	Internal Contr.Module (KAM) Error
P0604	16988	Internal Contr.Module Random Access Memory (RAM) Error
P0605	16989	Internal Contr.Module ROM Test Error
P0606	16990	ECM/PCM Processor
P0607	16991	Control Module performance
P0608	16992	Control Module Voltage Supply “A” Sensors/Actuators
P0609	16993	Control Module Voltage Supply “B” Sensors/Actuators
P0610	16994	Control Module Vehicle Options Error
P0613	16997	TCM Processor Malfunction
P0614	16998	ECM/TCM Incompatible (mismatch)
P0627	17011	Fuel Pump (FP) Relay Electrical Malfunction/Open Circuit

Diagnostic Trouble Codes (Powertrain)

SAE P-code	V.A.G. code	Description
P0628	17012	Fuel Pump (FP) Relay Short Circuit to Ground
P0629	17013	Fuel Pump Relay - Short circuit to B+
P0634	17084	PCM/ECM/TCM Internal Temperature Too High
P0638	17022	Throttle Actuator Control Range/Performance B1
P0639	17023	Throttle Actuator Control Range/Performance B2
P0641	17025	Sensor Reference Voltage “A” Circuit Open
P0642	17026	Sensor Reference Voltage “A” Circuit Low
P0643	17027	Sensor Reference Voltage “A” Circuit High
P0645	17029	A/C Clutch Relay Control Circuit
P0650	17034	MIL Control Circuit Electrical Fault in Circuit
P0651	17035	Sensor Reference Voltage “B” Circuit Open
P0652	17036	Sensor Reference Voltage “B” Circuit Low
P0653	17037	Sensor Reference Voltage “B” Circuit High
P0654	17038	Engine RPM Output Circuit Electrical Fault in Circuit
P0656	17040	Fuel Level Output Circuit Electrical Fault in Circuit
P0657	17041	Supply for engine components - Open circuit
P0658	17042	Supply for engine components - Too low
P0659	17043	Supply voltage for engine components - Too large
P0660	17044	Intake Manifold Tuning Valve Control Circuit Open Bank 1
P0661	17045	Intake Manifold Tuning Valve Control Circuit Low Bank 1
P0662	17046	Intake Manifold Tuning Valve Control Circuit High Bank 1
P0668	17052	PCM/ECM/TCM Internal Temperature Sensor Circuit Voltage Too Low
P0669	17053	PCM/ECM/TCM Internal Temperature Sensor Circuit Voltage Too High
P0670	17054	Glow Plug Module Control Circuit
P0671	17055	Cylinder 1 Glow Plug Circuit
P0672	17056	Cylinder 2 Glow Plug Circuit
P0673	17057	Cylinder 3 Glow Plug Circuit
P0674	17058	Cylinder 4 Glow Plug Circuit
P0684	17068	Glow Plug Module to PCM Communication Circuit Range/Performance
P0685	17069	ECM/PCM Power Relay Control Circuit Open
P0686	17070	ECM/PCM Power Relay Control Low
P0687	17071	ECM/PCM Power Relay Control High
P0688	17072	ECM/PCM Power Relay Sense Circuit Open
P0691	17075	Fan Control Circuit Low
P0692	17076	Fan Control Circuit High
P0693	17077	Fan 2 Control Circuit - Low
P0694	17078	Fan 2 Control Circuit - High
P0697	17081	Sensor Reference Voltage “C” Circuit Open
P0698	17082	Sensor Reference Voltage “C” Circuit Low
P0699	17083	Sensor Reference Voltage “C” Circuit High
P0700	17084	Transm.Contr.System Malfunction
P0701	17085	Transmission Control System Range/Performance
P0702	17086	Transm.Contr.System Electrical
P0703	17087	Torque Converter/Brake Switch B Circ Malfunction

Diagnostic Trouble Codes (Powertrain)

SAE P-code	V.A.G. code	Description
P0704	17088	Clutch Switch Input Circuit Malfunction
P0705	17089	Transm.Range Sensor Circ.(PRNDL Inop.) Malfunction
P0706	17090	Transm.Range Sensor Circ Range/Performance
P0707	17091	Transm.Range Sensor Circ Low Input
P0708	17092	Transm.Range Sensor Circ High Input
P0710	17094	Transm.Fluid Temp.Sensor Circ. Malfunction
P0711	17095	Transm.Fluid Temp.Sensor Circ. Range/Performance
P0712	17096	Transm.Fluid Temp.Sensor Circ. Low Input
P0713	17097	Transm.Fluid Temp.Sensor Circ. High Input
P0714	17098	Transm.Fluid Temp.Sensor Circ. Intermittent
P0715	17099	Input Turbine/Speed Sensor Circ. Malfunction
P0716	17100	Input Turbine/Speed Sensor Circ. Range/Performance
P0717	17101	Input Turbine/Speed Sensor Circ. No Signal
P0720	17104	Output Speed Sensor Circ Malfunction
P0721	17105	Output Speed Sensor Circ Range/Performance
P0722	17106	Output Speed Sensor Circ No Signal
P0725	17109	Engine Speed Inp.Circ. Malfunction
P0726	17110	Engine Speed Inp.Circ. Range/Performance
P0727	17111	Engine Speed Inp.Circ. No Signal
P0729	17113	Gear 6 Incorrect Ratio
P0730	17114	Gear Incorrect Ratio
P0731	17115	Gear 1 Incorrect Ratio
P0732	17116	Gear 2 Incorrect Ratio
P0733	17117	Gear 3 Incorrect Ratio
P0734	17118	Gear 4 Incorrect Ratio
P0735	17119	Gear 5 Incorrect Ratio
P0736	17120	Reverse gear incorrect gear ratio
P0740	17124	Torque Converter Clutch Circuit Malfunction
P0741	17125	Torque Converter Clutch Circuit Performance or Stuck Off
P0743	17127	Torque Converter Clutch Circuit Electrical
P0746	17130	Pressure Control Solenoid “A” Performance or Stuck Off
P0747	17131	Pressure Control Solenoid “A” Stuck On
P0748	17132	Pressure Control Solenoid “A” Electrical
P0749	17133	Pressure Control Solenoid “A” Intermittent
P0750	17134	Shift Solenoid A malfunction
P0751	17135	Shift Solenoid A Performance or Stuck Off
P0752	17136	Shift Solenoid A Stuck On
P0753	17137	Shift Solenoid A Electrical
P0756	17140	Shift Solenoid B Performance or Stuck Off
P0757	17141	Shift Solenoid B Stuck On
P0758	17142	Shift Solenoid B Electrical
P0761	17145	Shift Solenoid C Performance or Stuck Off
P0762	17146	Shift Solenoid C Stuck On
P0763	17147	Shift Solenoid C Electrical

Diagnostic Trouble Codes (Powertrain)

SAE P-code	V.A.G. code	Description
P0766	17150	Shift Solenoid D Performance or Stuck Off
P0768	17152	Shift Solenoid D Electrical
P0771	17155	Shift Solenoid E Performance or Stuck Off
P0773	17157	Shift Solenoid E Electrical
P0776	17160	Pressure Control Solenoid B Performance or Stuck Off
P0777	17161	Pressure Control Solenoid B Stuck On
P0778	17162	Pressure Control Solenoid B Electrical
P0780	17164	Shift monitoring
P0781	17165	1-2 Shift
P0782	17166	2-3 Shift
P0783	17167	3-4 Shift
P0784	17168	4-5 Shift
P0785	17169	Shift/Timing Solenoid "A" Electrical
P0790	17174	Normal/Performance Switch Circ Malfunction
P0791	17175	Intermediate Shaft Speed Sensor "A" Circuit
P0796	17180	Pressure Control Solenoid "C" Performance or Stuck Off
P0797	17181	Pressure Control Solenoid "C" Stuck On
P0798	17182	Pressure Control Solenoid "C" Electrical
P0799	17183	Pressure Control Solenoid "C" Intermittent
P0811	17195	Excessive Clutch Slippage
P0820	17204	Gear Lever X-Y Position Sensor Circuit
P0821		Gear Lever X Position Circuit
P0822		Gear Lever Y Position Circuit
P0829	17213	5-6 Shift
P0838		Four Wheel Drive (4WD) Switch Circuit Low
P0840	17224	Transmission Fluid Pressure Sensor/Switch "A" Circuit Electrical Malfunction
P0841	17225	Transmission Fluid Pressure Sensor/Switch "A" Circuit Range/Performance
P0842	17226	Transmission Fluid Pressure Sensor/Switch "A" Circuit Low
P0843	17227	Transmission Fluid Pressure Sensor/Switch "A" Circuit High
P0844	17228	Transmission Fluid Pressure Sensor/Switch "A" Circuit Intermittent
P0845	17229	Transmission Fluid Pressure Sensor/Switch "B" Circuit Electrical Malfunction
P0846	17230	Transmission Fluid Pressure Sensor/Switch "B" Circuit Range/Performance
P0863	17247	TCM Communication Circuit
P0864	17248	TCM Communication Circuit Range/Performance
P0865	17249	TCM Communication Circuit Low
P0884		TCM Poer Input Signal Intermittent
P0886		TCM Communication Circuit Low
P0887		TCM Communication Circuit High
P0889	17273	TCM Power Relay Sense Circuit Range/Performance
P0890	17274	TCM Power Relay Sense Circuit Low
P0891	17275	TCM Power Relay Sense Circuit High
P0892	17276	TCM Power Relay Sense Circuit Intermittent
P0914	17298	Gear Shift Position Circuit Electrical Malfunction
P0915		Gear Shift Position Circuit Range/Performance

Diagnostic Trouble Codes (Powertrain)

SAE P-code	V.A.G. code	Description
P0918		Gear Shift Position Circuit Intermittent
P0919	17303	Gear Shift Position Control Error
P0929	17313	Gear Shift Lock Solenoid Control Circuit Range/Performance
P1009	17417	Air mass meter 1/2 - implausible signal from load detection
P1025	17433	Fuel Pressure Regulator Valve (N276) Mechanical malfunction
P1026	17434	Activation intake manifold flap for airstream regulation, Short Circuit to B+
P1027	17435	Activation intake manifold flap for airstream regulation, Short Circuit to Ground
P1028	17436	Activation intake manifold flap for airstream regulation, Open Circuit
P1073	17481	Mass Air Flow (MAF) Sensor 2 (G246) - Signal too small
P1074	17482	Mass Air Flow (MAF) Sensor 2 (G246) - Signal too large
P1093	17501	Bank 1, fuel measuring system 2 Malfunction
P1098	17506	Intake Manifold Tuning (IMT) Valve Position Sensor (G336) Open circuit/short circuit to B+
P1101	17509	O2 Sensor Circ.,Bank1-Sensor1Voltage too Low/Air Leak
P1102	17510	O2 Sensor Heating Circ.,Bank1-Sensor1 Short to B+
P1103	17511	O2 Sensor Heating Circ.,Bank1-Sensor1 Output too Low
P1104	17512	Bank1-Sensor2 Voltage too Low/Air Leak
P1105	17513	O2 Sensor Heating Circ.,Bank1-Sensor 2 High
P1106	17514	O2 Sensor Circ.,Bank2-Sensor1 Voltage too Low/Air Leak
P1107	17515	O2 Sensor Heating Circ.,Bank2-Sensor1 Short to B+
P1108	17516	O2 Sensor Heating Circ.,Bank2-Sensor1 Output too Low
P1109	17517	O2 Sensor Circ.,Bank2-Sensor2 Voltage too Low/Air Leak
P1110	17518	O2 Sensor Heating Circ.,Bank2-Sensor2 Short to B+
P1111	17519	O2 Control (Bank 1) System too lean
P1112	17520	O2 Control (Bank 1) System too rich
P1113	17521	Bank1-Sensor1 Internal Resistance too High
P1114	17522	Bank1-Sensor2 Internal Resistant too High
P1115	17523	O2 Sensor Heater Circ.,Bank1-Sensor1 Short to Ground
P1116	17524	O2 Sensor Heater Circ.,Bank1-Sensor1 Open
P1117	17525	O2 Sensor Heater Circ.,Bank1-Sensor2 Short to Ground
P1118	17526	O2 Sensor Heater Circ.,Bank1-Sensor2 Open
P1119	17527	O2 Sensor Heater Circ.,Bank2-Sensor1 Short to Ground
P1120	17528	O2 Sensor Heater Circ.,Bank2-Sensor1 Open
P1121	17529	O2 Sensor Heater Circ.,Bank2-Sensor2 Short to Ground
P1122	17530	O2 Sensor Heater Circ.,Bank2-Sensor2 Open
P1123	17531	Long Term Fuel Trim Add.Air.,Bank1 System too Rich
P1124	17532	Long Term Fuel Trim Add.Air.,Bank1 System too Lean
P1125	17533	Long Term Fuel Trim Add.Air.,Bank2 System too Rich
P1126	17534	Long Term Fuel Trim Add.Air.,Bank2 System too Lean
P1127	17535	Long Term Fuel Trim mult.,Bank1 System too Rich
P1128	17536	Long Term Fuel Trim mult.,Bank1 System too Lean
P1129	17537	Long Term Fuel Trim mult.,Bank2 System too Rich
P1130	17538	Long Term Fuel Trim mult.,Bank2 System too Lean
P1131	17539	Bank2-Sensor1 Internal Resistance too High
P1132	17540	O2 Sensor Heating Circ.,Bank1+2-Sensor1 Short to B+

Diagnostic Trouble Codes (Powertrain)

SAE P-code	V.A.G. code	Description
P1133	17541	O2 Sensor Heating Circ.,Bank1+2-Sensor1 Electrical Malfunction
P1134	17542	O2 Sensor Heating Circ.,Bank1+2-Sensor2 Short to B+
P1135	17543	O2 Sensor Heating Circ.,Bank1+2-Sensor2 Electrical Malfunction
P1136	17544	Long Term Fuel Trim Add.Fuel,Bank1 System too Lean
P1137	17545	Long Term Fuel Trim Add.Fuel,Bank1 System too Rich
P1138	17546	Long Term Fuel Trim Add.Fuel,Bank2 System too Lean
P1139	17547	Long Term Fuel Trim Add.Fuel,Bank2 System too Rich
P1140	17548	Bank2-Sensor2 Internal Resistance too High
P1141	17549	Load Calculation Cross Check Range/Performance
P1142	17550	Load Calculation Cross Check Lower Limit Exceeded
P1143	17551	Load Calculation Cross Check Upper Limit Exceeded
P1144	17552	Mass or Volume Air Flow Circ Open/Short to Ground
P1145	17553	Mass or Volume Air Flow Circ Short to B+
P1146	17554	Mass or Volume Air Flow Circ Supply Malfunction
P1147	17555	O2 Control (Bank 2) System too lean
P1148	17556	O2 Control (Bank 2) System too rich
P1149	17557	O2 Control (Bank 1) Out of range
P1150	17558	O2 Control (Bank 2) Out of range
P1151	17559	Bank1, Long Term Fuel Trim, Range 1 Leanness Lower Limit Exceeded
P1152	17560	Bank1, Long Term Fuel Trim, Range 2 Leanness Lower Limit Exceeded
P1154	17562	Manifold Switch Over Malfunction
P1155	17563	Manifold Abs.Pressure Sensor Circ. Short to B+
P1156	17564	Manifold Abs.Pressure Sensor Circ. Open/Short to Ground
P1157	17565	Manifold Abs.Pressure Sensor Circ. Power Supply Malfunction
P1158	17566	Manifold Abs.Pressure Sensor Circ. Range/Performance
P1160	17568	Manifold Temp.Sensor Circ. Short to Ground
P1161	17569	Manifold Temp.Sensor Circ. Open/Short to B+
P1162	17570	Fuel Temp.Sensor Circ. Short to Ground
P1163	17571	Fuel Temp.Sensor Circ. Open/Short to B+
P1164	17572	Fuel Temperature Sensor Range/Performance/Incorrect Signal
P1165	17573	Bank1, Long Term Fuel Trim, Range 1 Rich Limit Exceeded
P1166	17574	Bank1, Long Term Fuel Trim, Range 2 Rich Limit Exceeded
P1167	17575	Mass Air Flow (MAF) Sensor 2 (G246) - Implausible signal
P1171	17579	Throttle Actuation Potentiometer Sign.2 Range/Performance
P1172	17580	Throttle Actuation Potentiometer Sign.2 Signal too Low
P1173	17581	Throttle Actuation Potentiometer Sign.2 Signal too High
P1174	17582	Fuel Trim, Bank 1 Different injection times
P1175	17583	Pre-injection switched off
P1176	17584	O2 Correction Behind Catalyst,B1 Limit Attained ; Note - 1999m.y. (2.0L) may also display DTC B0423
P1177	17585	O2 Correction Behind Catalyst,B2 Limit Attained
P1178	17586	Linear 02 Sensor / Pump Current Open Circuit
P1179	17587	Linear 02 Sensor / Pump Current Short to ground
P1180	17588	Linear 02 Sensor / Pump Current Short to B+
P1181	17589	Linear 02 Sensor / Reference Voltage Open Circuit

Diagnostic Trouble Codes (Powertrain)

SAE P-code	V.A.G. code	Description
P1182	17590	Linear 02 Sensor / Reference Voltage Short to ground
P1183	17591	Linear 02 Sensor / Reference Voltage Short to B+
P1184	17592	Linear 02 Sensor / Common Ground Wire Open Circuit
P1185	17593	Linear 02 Sensor / Common Ground Wire Short to ground
P1186	17594	Linear 02 Sensor / Common Ground Wire Short to B+
P1187	17595	Linear 02 Sensor / Compens. Resistor Open Circuit
P1188	17596	Linear 02 Sensor / Compens. Resistor Short to ground
P1189	17597	Linear 02 Sensor / Compens. Resistor Short to B+
P1190	17598	Linear 02 Sensor / Reference Voltage Incorrect Signal
P1196	17604	O2 Sensor Heater Circ.,Bank1-Sensor1 Electrical Malfunction
P1197	17605	O2 Sensor Heater Circ.,Bank2-Sensor1 Electrical Malfunction
P1198	17606	O2 Sensor Heater Circ.,Bank1-Sensor2 Electrical Malfunction
P1199	17607	O2 Sensor Heater Circ.,Bank2-Sensor2 Electrical Malfunction
P1201	17609	Cyl.1-Fuel Inj.Circ. Electrical Malfunction
P1202	17610	Cyl.2-Fuel Inj.Circ. Electrical Malfunction
P1203	17611	Cyl.3-Fuel Inj.Circ. Electrical Malfunction
P1204	17612	Cyl.4-Fuel Inj.Circ. Electrical Malfunction
P1205	17613	Cyl.5-Fuel Inj.Circ. Electrical Malfunction
P1206	17614	Cyl.6-Fuel Inj.Circ. Electrical Malfunction
P1209	17617	Intake valves for cylinder shut-off Short circuit to ground
P1210	17618	Intake valves for cylinder shut-off Short to B+
P1211	17619	Intake valves for cylinder shut-off Open circuit
P1213	17621	Cyl.1-Fuel Inj.Circ. Short to B+
P1214	17622	Cyl.2-Fuel Inj.Circ. Short to B+
P1215	17623	Cyl.3-Fuel Inj.Circ. Short to B+
P1216	17624	Cyl.4-Fuel Inj.Circ. Short to B+
P1217	17625	Cyl.5-Fuel Inj.Circ. Short to B+
P1218	17626	Cyl.6-Fuel Inj.Circ. Short to B+
P1221	17629	Cylinder shut-off exhaust valves Short circuit to ground
P1222	17630	Cylinder shut-off exhaust valves Short to B+
P1223	17631	Cylinder shut-off exhaust valves Open circuit
P1225	17633	Cyl.1-Fuel Inj.Circ. Short to Ground
P1226	17634	Cyl.2-Fuel Inj.Circ. Short to Ground
P1227	17635	Cyl.3-Fuel Inj.Circ. Short to Ground
P1228	17636	Cyl.4-Fuel Inj.Circ. Short to Ground
P1229	17637	Cyl.5-Fuel Inj.Circ. Short to Ground
P1230	17638	Cyl.6-Fuel Inj.Circ. Short to Ground
P1237	17645	Cyl.1-Fuel Inj.Circ. Open Circ.
P1238	17646	Cyl.2-Fuel Inj.Circ. Open Circ.
P1239	17647	Cyl.3-Fuel Inj.Circ. Open Circ.
P1240	17648	Cyl.4-Fuel Inj.Circ. Open Circ.
P1241	17649	Cyl.5-Fuel Inj.Circ. Open Circ.
P1242	17650	Cyl.6-Fuel Inj.Circ. Open Circ.
P1245	17653	Needle Lift Sensor (G80) - Short circuit to Ground (GND)

Diagnostic Trouble Codes (Powertrain)

SAE P-code	V.A.G. code	Description
P1246	17654	Needle Lift Sender (G80) - implausible signal
P1247	17655	Needle Lift Sensor (G80) - Open circuit/short circuit to B+
P1248	17656	Injection Start Control Deviation
P1249	17657	Fuel consumption signal Electrical Fault in Circuit
P1250	17658	Fuel Level Too Low
P1251	17659	Start of Injection Solenoid Circ Short to B+
P1252	17660	Start of Injection Solenoid Circ Open/Short to Ground
P1253	17661	Fuel consumption signal Short to ground
P1254	17662	Fuel consumption signal Short to B+
P1255	17663	Engine Coolant Temp.Circ Short to Ground
P1256	17664	Engine Coolant Temp.Circ Open/Short to B+
P1257	17665	Engine Coolant System Valve Open
P1258	17666	Engine Coolant System Valve Short to B+
P1259	17667	Engine Coolant System Valve Short to Ground
P1280	17688	Fuel Inj.Air Contr.Valve Circ. Flow too Low
P1283	17691	Fuel Inj.Air Contr.Valve Circ. Electrical Malfunction
P1284	17692	Fuel Inj.Air Contr.Valve Circ. Open
P1285	17693	Fuel Inj.Air Contr.Valve Circ. Short to Ground
P1286	17694	Fuel Inj.Air Contr.Valve Circ. Short to B+
P1287	17695	Turbocharger bypass valve open
P1288	17696	Turbocharger bypass valve short to B+
P1289	17697	Turbocharger bypass valve short to ground
P1295	17703	Bypass Turbocharger Throughput Faulty
P1296	17704	Cooling system malfunction
P1297	17705	Connection turbocharger - throttle valve pressure hose
P1300	17708	Misfire detected Reason: Fuel level too low
P1319	17721	Knock Sensor 1 Circ. Short to Ground
P1320	17728	Knock Sensor 2 Circ. Short to Ground
P1321	17729	Knock Sensor 3 Circ. Low Input
P1322	17730	Knock Sensor 3 Circ. High Input
P1323	17731	Knock Sensor 4 Circ. Low Input
P1324	17732	Knock Sensor 4 Circ. High Input
P1325	17733	Cyl.1-Knock Contr. Limit Attained
P1326	17734	Cyl.2-Knock Contr. Limit Attained
P1327	17735	Cyl.3-Knock Contr. Limit Attained
P1328	17736	Cyl.4-Knock Contr. Limit Attained
P1329	17737	Cyl.5-Knock Contr. Limit Attained
P1330	17738	Cyl.6-Knock Contr. Limit Attained
P1335	17743	Engine Torque Monitoring 2 Control Limit Exceeded
P1336	17744	Engine Torque Monitoring Adaptation at limit
P1337	17745	Camshaft Pos.Sensor,Bank1 Short to Ground
P1338	17746	Camshaft Pos.Sensor,Bank1 Open Circ./Short to B+
P1339	17747	Crankshaft Pos./Engine Speed Sensor Cross Connected
P1340	17748	Crankshaft-/Camshaft Pos.Sens.Signals Out of Sequence

Diagnostic Trouble Codes (Powertrain)

SAE P-code	V.A.G. code	Description
P1341	17749	Ignition Coil Power Output Stage 1 Short to Ground
P1342	17750	Ignition Coil Power Output Stage 1 Short to B+
P1343	17751	Ignition Coil Power Output Stage 2 Short to Ground
P1344	17752	Ignition Coil Power Output Stage 2 Short to B+
P1345	17753	Ignition Coil Power Output Stage 3 Short to Ground
P1346	17754	Ignition Coil Power Output Stage 3 Short to B+
P1347	17755	Bank2,Crankshaft-/Camshaft os.Sens.Sign. Out of Sequence
P1348	17756	Ignition Coil Power Output Stage 1 Open Circuit
P1349	17757	Ignition Coil Power Output Stage 2 Open Circuit
P1350	17758	Ignition Coil Power Output Stage 3 Open Circuit
P1354	17762	Modulation Piston Displ.Sensor Circ. Malfunction
P1355	17763	Cyl. 1, ignition circuit Open Circuit
P1356	17764	Cyl. 1, ignition circuit Short to B+
P1357	17765	Cyl. 1, ignition circuit Short to ground
P1358	17766	Cyl. 2, ignition circuit Open Circuit
P1359	17767	Cyl. 2, ignition circuit Short Circuit to B+
P1360	17768	Cyl. 2, ignition circuit Short Circuit to Ground
P1361	17769	Cyl. 3, ignition circuit Open Circuit
P1362	17770	Cyl. 3, ignition circuit Short Circuit to B+
P1363	17771	Cyl. 3, ignition circuit Short Circuit to ground
P1364	17772	Cyl. 4 ignition circuit Open Circuit
P1365	17773	Cyl. 4 ignition circuit Short circuit to B+
P1366	17774	Cyl. 4 ignition circuit Short circuit to ground
P1367	17775	Cyl. 5, ignition circuit Open Circuit
P1368	17776	Cyl. 5, ignition circuit Short Circuit to B+
P1369	17777	Cyl. 5, ignition circuit short to ground
P1370	17778	Cyl. 6, ignition circuit Open Circuit
P1371	17779	Cyl. 6, ignition circuit Short Circuit to B+
P1372	17780	Cyl. 6, ignition circuit short to ground
P1385	17793	Internal Control Module Memory Check Sum Error
P1386	17794	Internal Control Module Knock Control Circ.Error
P1387	17795	Internal Contr. Module altitude sensor error
P1388	17796	Internal Contr. Module drive by wire error
P1391	17799	Camshaft Pos.Sensor,Bank2 Short to Ground
P1392	17800	Camshaft Pos.Sensor,Bank2 Open Circ./Short to B+
P1393	17801	Ignition Coil Power Output Stage 1 Electrical Malfunction
P1394	17802	Ignition Coil Power Output Stage 2 Electrical Malfunction
P1395	17803	Ignition Coil Power Output Stage 3 Electrical Malfunction
P1396	17804	Engine Speed Sensor Missing Tooth
P1397	17805	Engine speed wheel Adaptation limit reached
P1398	17806	Engine RPM signal, TD Short to ground
P1399	17807	Engine RPM signal, TD Short Circuit to B+
P1400	17808	EGR Valve Circ Electrical Malfunction
P1401	17809	EGR Valve Circ Short to Ground

Diagnostic Trouble Codes (Powertrain)

SAE P-code	V.A.G. code	Description
P1402	17810	EGR Valve Circ Short to B+
P1403	17811	EGR Flow Deviation
P1404	17812	EGR Flow Basic Setting not carried out
P1406	17814	EGR Temp.Sensor Range/Performance
P1407	17815	EGR Temp.Sensor Signal too Low
P1408	17816	EGR Temp.Sensor Signal too High
P1409	17817	Tank Ventilation Valve Circ. Electrical Malfunction
P1410	17818	Tank Ventilation Valve Circ. Short to B+
P1411	17819	Sec.Air Inj.Sys.,Bank2 Flow too Flow
P1412	17820	EGR Different.Pressure Sensor Signal too Low
P1413	17821	EGR Different.Pressure Sensor Signal too High
P1414	17822	Sec.Air Inj.Sys.,Bank2 Leak Detected
P1417	17825	Fuel Level Sensor Circ Signal too Low
P1418	17826	Fuel Level Sensor Circ Signal too High
P1420	17828	Sec.Air Inj.Valve Circ Electrical Malfunction
P1421	17829	Sec.Air Inj.Valve Circ Short to Ground
P1422	17830	Sec.Air Inj.Sys.Contr.Valve Circ Short to B+
P1423	17831	Sec.Air Inj.Sys.,Bank1 Flow too Low
P1424	17832	Sec.Air Inj.Sys.,Bank1 Leak Detected
P1425	17833	Evaporative Emission (EVAP) Canister Purge Regulator Valve 1 Short To Ground
P1426	17834	Evaporative Emission (EVAP) Canister Purge Regulator Valve 1 Open Circuit
P1432	17840	Sec.Air Inj.Valve Open
P1433	17841	Sec.Air Inj.Sys.Pump Relay Circ. open
P1434	17842	Sec.Air Inj.Sys.Pump Relay Circ. Short to B+
P1435	17843	Sec.Air Inj.Sys.Pump Relay Circ. Short to ground
P1436	17844	Sec.Air Inj.Sys.Pump Relay Circ. Electrical Malfunction
P1439	17847	EGR Potentiometer Error in Basic Setting
P1440	17848	EGR Valve Power Stage Open
P1441	17849	EGR Valve Circ Open/Short to Ground
P1442	17850	EGR Valve Position Sensor Signal too high
P1443	17851	EGR Valve Position Sensor Signal too low
P1444	17852	EGR Valve Position Sensor range/performance
P1445	17853	Catalyst Temp.Sensor 2 Circ. Range/Performance
P1446	17854	Catalyst Temp.Circ Short to Ground
P1447	17855	Catalyst Temp.Circ Open/Short to B+
P1448	17856	Catalyst Temp.Sensor 2 Circ. Short to Ground
P1449	17857	Catalyst Temp.Sensor 2 Circ. Open/Short to B+
P1450	17858	Sec.Air Inj.Sys.Circ Short to B+
P1451	17859	Sec.Air Inj.Sys.Circ Short to Ground
P1452	17860	Sec.Air Inj.Sys. Open Circ.
P1453	17861	Exhaust gas temperature sensor 1 open/short to B+
P1454	17862	Exhaust gas temperature sensor short 1 to ground
P1455	17863	Exhaust gas temperature sensor 1 range/performance
P1456	17864	Exhaust gas temperature control bank 1 limit attained

Diagnostic Trouble Codes (Powertrain)

SAE P-code	V.A.G. code	Description
P1457	17865	Exhaust gas temperature sensor 2 open/short to B+
P1458	17866	Exhaust gas temperature sensor 2 short to ground
P1459	17867	Exhaust gas temperature sensor 2 range/performance
P1460	17868	Exhaust gas temperature control bank 2 limit attained
P1461	17869	Exhaust gas temperature control bank 1 Range/Performance
P1462	17870	Exhaust gas temperature control bank 2 Range/Performance
P1465	17873	Additive Pump Short Circuit to B+
P1466	17874	Additive Pump Open/Short to Ground
P1467	17875	EVAP Canister Purge Solenoid Valve Short Circuit to B+
P1468	17876	EVAP Canister Purge Solenoid Valve Short Circuit to Ground
P1469	17877	EVAP Canister Purge Solenoid Valve Open Circuit
P1470	17878	EVAP Emission Contr.LDP Circ Electrical Malfunction
P1471	17879	EVAP Emission Contr.LDP Circ Short to B+
P1472	17880	EVAP Emission Contr.LDP Circ Short to Ground
P1473	17881	EVAP Emission Contr.LDP Circ Open Circ.
P1474	17882	EVAP Canister Purge Solenoid Valve electrical malfunction
P1475	17883	EVAP Emission Contr.LDP Circ Malfunction/Signal Circ.Open
P1476	17884	EVAP Emission Contr.LDP Circ Malfunction/Insufficient Vacuum
P1477	17885	EVAP Emission Contr.LDP Circ Malfunction
P1478	17886	EVAP Emission Contr.LDP Circ Clamped Tube Detected
P1490	17898	Evaporative Emission (EVAP) Canister Purge Regulator Valve 2 Short To Ground
P1491	17899	Evaporative Emission (EVAP) Canister Purge Regulator Valve 2 Open Circuit
P1495	17903	Radiator Exhaust Gas Recirculation Switch-over Valve (N345), Open circuit/short circuit to ground
P1496	17904	Radiator Exhaust Gas Recirculation Switch-over Valve (N345), Short to B+
P1500	17908	Fuel Pump Relay Circ. Electrical Malfunction
P1501	17909	Fuel Pump Relay Circ. Short to Ground
P1502	17910	Fuel Pump Relay Circ. Short to B+
P1503	17911	Load signal from Alternator Term. DF Range/performance/Incorrect Signal
P1504	17912	Intake Air Sys.Bypass Leak Detected
P1505	17913	Closed Throttle Pos. Does Not Close/Open Circ
P1506	17914	Closed Throttle Pos.Switch Does Not Open/Short to Ground
P1507	17915	Idle Sys.Learned Value Lower Limit Attained
P1508	17916	Idle Sys.Learned Value Upper Limit Attained
P1509	17917	Idle Air Control Circ. Electrical Malfunction
P1510	17918	Idle Air Control Circ. Short to B+
P1511	17919	Intake Manifold Changeover Valve circuit electrical malfunction
P1512	17920	Intake Manifold Changeover Valve circuit Short to B+
P1513	17921	Intake Manifold Changeover Valve2 circuit Short to B+
P1514	17922	Intake Manifold Changeover Valve2 circuit Short to ground
P1515	17923	Intake Manifold Changeover Valve circuit Short to Ground
P1516	17924	Intake Manifold Changeover Valve circuit Open
P1517	17925	Main Relay Circ. Electrical Malfunction
P1518	17926	Main Relay Circ. Short to B+
P1519	17927	Intake Camshaft Contr.,Bank1 Malfunction

Diagnostic Trouble Codes (Powertrain)

SAE P-code	V.A.G. code	Description
P1520	17928	Intake Manifold Changeover Valve2 circuit Open
P1521	17929	Intake Manifold Changeover Valve2 circuit electrical malfunction
P1522	17930	Intake Camshaft Contr.,Bank2 Malfunction
P1523	17931	Crash Signal from Airbag Control Unit range/performance
P1525	17933	Intake Camshaft Contr.Circ.,Bank1 Electrical Malfunction
P1526	17934	Intake Camshaft Contr.Circ.,Bank1 Short to B+
P1527	17935	Intake Camshaft Contr.Circ.,Bank1 Short to Ground
P1528	17936	Intake Camshaft Contr.Circ.,Bank1 Open
P1529	17937	Camshaft Control Circuit Short to B+
P1530	17938	Camshaft Control Circuit Short to ground
P1531	17939	Camshaft Control Circuit open
P1533	17941	Intake Camshaft Contr.Circ.,Bank2 Electrical Malfunction
P1534	17942	Intake Camshaft Contr.Circ.,Bank2 Short to B+
P1535	17943	Intake Camshaft Contr.Circ.,Bank2 Short to Ground
P1536	17944	Intake Camshaft Contr.Circ.,Bank2 Open
P1537	17945	Engine Shutoff Solenoid Malfunction
P1538	17946	Engine Shutoff Solenoid Open/Short to Ground
P1539	17947	Clutch Vacuum Vent Valve Switch Incorrect signal
P1540	17948	Vehicle Speed Sensor High Input
P1541	17949	Fuel Pump Relay Circ Open
P1542	17950	Throttle Actuation Potentiometer Range/Performance
P1543	17951	Throttle Actuation Potentiometer Signal too Low
P1544	17952	Throttle Actuation Potentiometer Signal too High
P1545	17953	Throttle Pos.Contr Malfunction
P1546	17954	Boost Pressure Contr.Valve Short to B+
P1547	17955	Boost Pressure Contr.Valve Short to Ground
P1548	17956	Boost Pressure Contr.Valve Open
P1549	17957	Boost Pressure Contr.Valve Short to Ground
P1550	17958	Charge Pressure Deviation
P1551	17959	Barometric Pressure Sensor Circ. Short to B+
P1552	17960	Barometric Pressure Sensor Circ. Open/Short to Ground
P1553	17961	Barometric/manifold pressure signal ratio out of range
P1554	17962	Idle Speed Contr.Throttle Pos. Basic Setting Conditions not met
P1555	17963	Charge Pressure Upper Limit exceeded
P1556	17964	Charge Pressure Contr. Negative Deviation
P1557	17965	Charge Pressure Contr. Positive Deviation
P1558	17966	Throttle Actuator Electrical Malfunction
P1559	17967	Idle Speed Contr.Throttle Pos. Adaptation Malfunction
P1560	17968	Maximum Engine Speed Exceeded
P1561	17969	Quantity Adjuster Deviation
P1562	17970	Quantity Adjuster Upper Limit Attained
P1563	17971	Quantity Adjuster Lower Limit Attained
P1564	17972	Idle Speed Contr.Throttle Pos. Low Voltage During Adaptation
P1565	17973	Idle Speed Control Throttle Position lower limit not attained

Diagnostic Trouble Codes (Powertrain)

SAE P-code	V.A.G. code	Description
P1566	17974	Load signal from A/C compressor range/performance
P1567	17975	Load signal from A/C compressor no signal
P1568	17976	Idle Speed Contr.Throttle Pos. mechanical Malfunction
P1569	17977	Cruise control switch Incorrect signal
P1570	17978	Contr.Module Locked
P1571	17979	Left Eng. Mount Solenoid Valve Short to B+
P1572	17980	Left Eng. Mount Solenoid Valve Short to ground
P1573	17981	Left Eng. Mount Solenoid Valve Open circuit
P1574	17982	Left Eng. Mount Solenoid Valve Electrical fault in circuit
P1575	17983	Right Eng. Mount Solenoid Valve Short to B+
P1576	17984	Right Eng. Mount Solenoid Valve Short to ground
P1577	17985	Right Eng. Mount Solenoid Valve Open circuit
P1578	17986	Right Eng. Mount Solenoid Valve Electrical fault in circuit
P1579	17987	Idle Speed Contr.Throttle Pos. Adaptation not started
P1580	17988	Throttle Actuator B1 Malfunction
P1581	17989	Idle Speed Contr.Throttle Pos. Basic Setting Not Carried Out
P1582	17990	Idle Adaptation at Limit
P1583	17991	Transmission mount valves Short to B+
P1584	17992	Transmission mount valves Short to ground
P1585	17993	Transmission mount valves Open circuit
P1586	17994	Engine mount solenoid valves Short to B+
P1587	17995	Engine mount solenoid valves Short to ground
P1588	17996	Engine mount solenoid valves Open circuit
P1592	18000	Height sensor/charge pressure sensor, Unplausible ratio
P1600	18008	Power Supply (B+) Terminal 15 Low Voltage
P1601	18009	Power Supply Relay (J317), Unplausible signal
P1602	18010	Power Supply (B+) Terminal 30 Low Voltage
P1603	18011	Internal Control Module Malfunction
P1604	18012	Internal Control Module Driver Error
P1605	18013	Rough Road/Acceleration Sensor Electrical Malfunction
P1606	18014	Rough Road Spec Engine Torque ABS-ECU Electrical Malfunction
P1607	18015	Vehicle speed signal Error message from instrument cluster
P1608	18016	Steering angle signal Error message from steering angle sensor
P1609	18017	Crash shut-down activated
P1610	18018	Internal Control Module Memory Check Sum Error
P1611	18019	MIL Call-up Circ./Transm.Contr.Module Short to Ground
P1612	18020	Electronic Control Module Incorrect Coding
P1613	18021	MIL Call-up Circ Open/Short to B+
P1614	18022	MIL Call-up Circ./Transm.Contr.Module Range/Performance
P1615	18023	Engine Oil Temperature Sensor Circuit range/performance
P1618	18026	Glow Plug Relay (J52) - Short circuit to B+
P1619	18027	Glow Plug Relay (J52) - Open circuit/short circuit to Ground (GND)
P1620	18028	Engine coolant temperature signal open/short to B+
P1621	18029	Engine coolant temperature signal short to ground

Diagnostic Trouble Codes (Powertrain)

SAE P-code	V.A.G. code	Description
P1622	18030	Engine coolant temperature signal range/performance
P1623	18031	Data Bus Powertrain No Communication
P1624	18032	MIL Request Sign.active
P1625	18033	Data-Bus Powertrain Implausible Message from Transm.Contr.
P1626	18034	Data-Bus Powertrain Missing Message from Transm.Contr.
P1627	18035	Data-Bus Powertrain missing message from fuel injection pump
P1628	18036	Data-Bus Powertrain missing message from steering sensor
P1629	18037	Data-Bus Powertrain missing message from distance control
P1630	18038	Accelera.Pedal Pos.Sensor 1 Signal too Low
P1631	18039	Accelera.Pedal Pos.Sensor 1 Signal too High
P1632	18040	Accelera.Pedal Pos.Sensor 1 Power Supply Malfunction
P1633	18041	Accelera.Pedal Pos.Sensor 2 Signal too Low
P1634	18042	Accelera.Pedal Pos.Sensor 2 Signal too High
P1635	18043	Data Bus Powertrain missing message f.air condition control
P1636	18044	Data Bus Powertrain missing message from Airbag control
P1637	18045	Data Bus Powertrain missing message f.central electr.control
P1638	18046	Data Bus Powertrain missing message from clutch control
P1639	18047	Accelera.Pedal Pos.Sensor 1+2 Range/Performance
P1640	18048	Internal Contr.Module (EEPROM) Error
P1641	18049	Please check DTC Memory of Air Condition ECU
P1642	18050	Please check DTC Memory of Airbag ECU
P1643	18051	Please check DTC Memory of central electric ECU
P1644	18052	Please check DTC Memory of clutch ECU
P1645	18053	Data Bus Powertrain missing message f.all wheel drive contr.
P1646	18054	Check DTC Memory of all wheel drive ECU
P1647	18055	Check coding of ECUs in Data Bus Powertrain
P1648	18056	Data Bus Powertrain Malfunction
P1649	18057	Data Bus Powertrain Missing message from ABS Control Module
P1650	18058	Data Bus Powertrain Missing message fr.instrument panel ECU
P1651	18059	Data Bus Powertrain missing messages
P1652	18060	Please check DTC Memory of transmission ECU
P1653	18061	Please check DTC Memory of ABS Control Module
P1654	18062	Please check DTC Memory of control panel ECU
P1655	18063	Please check DTC Memory of ADR Control Module
P1656	18064	A/C clutch relay circuit short to ground
P1657	18065	A/C clutch relay circuit short to B+
P1658	18066	Data Bus Powertrain Incorrect signal from ADR Control Module
P1673	18081	Speed signal malfunction message from ABS control module
P1674	18082	Data-bus-drivetrain implausible message from instrument panel
P1676	18084	Drive by Wire-MIL Circ. Electrical Malfunction
P1677	18085	Drive by Wire-MIL Circ. Short to B+
P1678	18086	Drive by Wire-MIL Circ. Short to Ground
P1679	18087	Drive by Wire-MIL Circ. Open
P1681	18089	Contr.Unit Programming, Programming not Finished

Diagnostic Trouble Codes (Powertrain)

SAE P-code	V.A.G. code	Description
P1682	18090	Powertrain databus Unplausible message from ABS control module
P1683	18091	Powertrain databus Unplausible message from Airbag Control Modules
P1684	18092	Contr.Unit Programming Communication Error
P1685	180093	Data-bus-drivetrain, implausible message from steering angle sensor
P1686	18094	Contr.Unit Error Programming Error
P1687	18095	Data-bus-driving gear, Implausible message from climate control module
P1689	18097	Data-bus-drivetrain, implausible message from electronic central electric
P1690	18098	Malfunction Indication Light Malfunction
P1691	18099	Malfunction Indication Light Open
P1692	18100	Malfunction Indication Light Short to Ground
P1693	18101	Malfunction Indication Light Short to B+
P1694	18102	Malfunction Indication Light Open/Short to Ground
P1696	18104	Powertrain databus, implausible signal from Steering Column Electronic
P1698	18106	Read fault memory of Steering Column Electronic
P1702	18110	Malfunction cannot activate control module replacement function, since another malfunction with equal priority is present
P1704	18112	Kick Down Switch (F8) Malfunction
P1705	18113	Gear/Ratio Monitoring Adaptation limit reached
P1706	18114	Transmission faulty
P1707	18115	Interference in Mechatronic unit
P1711	18119	Wheel Speed Signal 1 Range/Performance
P1716	18124	Wheel Speed Signal 2 Range/Performance
P1721	18129	Wheel Speed Signal 3 Range/Performance
P1723	18131	Starter Interlock Circ. Open
P1724	18132	Starter Interlock Circ. Short to Ground
P1726	18134	Wheel Speed Signal 4 Range/Performance
P1728	18136	Different Wheel Speed Signals Range/Performance
P1729	18137	Starter Interlock Circ. Short to B+
P1733	18141	Tiptronic Switch Down Circ. Short to Ground
P1739	18147	Tiptronic Switch up Circ. Short to Ground
P1740	18148	Clutch temperature control
P1741	18149	Clutch pressure adaptation at limit
P1742	18150	Clutch torque adaptation at limit
P1743	18151	Clutch slip control signal too high
P1744	18152	Tiptronic Switch Recognition Circ. Short to Ground
P1745	18153	Transm.Contr.Unit Relay Short to B+
P1746	18154	Transm.Contr.Unit Relay Malfunction
P1747	18155	Transm.Contr.Unit Relay Open/Short to Ground
P1748	18156	Transm.Contr.Unit Self-Check
P1749	18157	Transm.Contr.Unit Incorrect Coded
P1750	18158	Power Supply Voltage Low Voltage
P1751	18159	Power Supply Voltage High Voltage
P1752	18160	Power Supply Malfunction
P1753	18161	Tiptronic Switch (F189), Unplausible signal

Diagnostic Trouble Codes (Powertrain)

SAE P-code	V.A.G. code	Description
P1754	18162	Tiptronic Switch (F189), up function, Open circuit/short circuit to B+
P1755	18163	Tiptronic Switch (F189), down function, Open circuit/short circuit to B+
P1756	18164	Tiptronic Switch (F189), recognition function, Open circuit/short circuit to B+
P1759	18167	Transmission fluid temperature, Exceeded multiple times!
P1760	18168	Shift Lock Solenoid (N110) Malfunction
P1761	18169	Shift Lock Solenoid (N110) Short to Ground
P1762	18170	Shift Lock Solenoid (N110) Short to B+
P1763	18171	Shift Lock Solenoid (N110) Open
P1764	18172	Transmission temperature control
P1765	18173	Hydraulic Pressure Sensor 2 adaptation at limit
P1766	18174	Throttle Angle Signal Stuck Off
P1767	18175	Throttle Angle Signal Stuck On
P1768	18176	Hydraulic Pressure Sensor 2 Too High
P1769	18177	Hydraulic Pressure Sensor 2 Too Low
P1770	18178	Load Signal Range/Performance
P1771	18179	Load Signal Stuck Off
P1772	18180	Load Signal Stuck On
P1773	18181	Hydraulic Pressure Sensor 1 Too High
P1774	18182	Hydraulic Pressure Sensor 1 Too Low
P1775	18183	Hydraulic Pressure Sensor 1 adaptation at limit
P1776	18184	Hydraulic Pressure Sensor 1 range/performance
P1777	18185	Hydraulic Pressure Sensor 2 range/performance
P1778	18186	Solenoid EV7 Electrical Malfunction
P1780	18188	Torque retraction malfunctioning
P1781	18189	Engine Torque Reduction Open/Short to Ground
P1782	18190	Engine Torque Reduction Short to B+
P1784	18192	Shift up/down Wire Open/Short to Ground
P1785	18193	Shift up/down Wire Short to B+
P1786	18194	Reversing Light Circ. Open
P1787	18195	Reversing Light Circ. Short to Ground
P1788	18196	Reversing Light Circ. Short to B+
P1789	18197	Idle Speed Intervention Circ. Error Message from Engine Contr.
P1790	18198	Transmission Range Display Circ. Open
P1791	18199	Transmission Range Display Circ. Short to Ground
P1792	18200	Transmission Range Display Circ. Short to B+
P1793	18201	Output Speed Sensor 2 Circ. No Signal
P1795	18203	Vehicle Speed Signal Circ. Open
P1796	18204	Vehicle Speed Signal Circ. Short to Ground
P1797	18205	Vehicle Speed Signal Circ. Short to B+
P1798	18206	Output Speed Sensor 2 Circ. Range/Performance
P1799	18207	Output Speed Sensor 2 Circ. Rpm too High
P1813	18221	Pressure Contr.Solenoid 1 Electrical
P1814	18222	Pressure Contr.Solenoid 1 Open/Short to Ground
P1815	18223	Pressure Contr.Solenoid 1 Short to B+

Diagnostic Trouble Codes (Powertrain)

SAE P-code	V.A.G. code	Description
P1818	18226	Pressure Contr.Solenoid 2 Electrical
P1819	18227	Pressure Contr.Solenoid 2 Open/Short to Ground
P1820	18228	Pressure Contr.Solenoid 2 Short to B+
P1823	18231	Pressure Contr.Solenoid 3 Electrical
P1824	18232	Pressure Contr.Solenoid 3 Open/Short to Ground
P1825	18233	Pressure Contr.Solenoid 3 Short to B+
P1828	18236	Pressure Contr.Solenoid 4 Electrical
P1829	18237	Pressure Contr.Solenoid 4 Open/Short to Ground
P1830	18238	Pressure Contr.Solenoid 4 Short to B+
P1833	18241	Pressure regulating valve 5 for automatic transmission, electrical malfunction in circuit
P1834	18242	Pressure Contr.Solenoid 5 Open/Short to Ground
P1835	18243	Pressure Contr.Solenoid 5 Short to B+
P1838	18246	Pressure regulating valve 6 (N371), electrical malfunction in circuit
P1839	18247	Pressure regulating valve 6 (N371), Open circuit/short circuit to Ground (GND)
P1840	18248	Pressure Regulating Valve 6 (N371), Short circuit to B+
P1841	18249	Engine/Transmission Control Modules Versions do not match
P1842	18250	Please check DTC Memory of instrument panel ECU
P1843	18251	Please check DTC Memory of ADR Control Module
P1844	18252	Please check DTC Memory of central electric control ECU
P1845	18253	Read DTC memory of Steering Column Electronic Systems Control Module (J527)
P1847	18255	Please check DTC Memory of brake system ECU
P1848	18256	Please check DTC Memory of engine ECU
P1849	18257	Please check DTC Memory of transmission ECU
P1850	18258	Data-Bus Powertrain Missing Message from Engine Contr.
P1851	18259	Data-Bus Powertrain Missing Message from Brake Contr.
P1852	18260	Data-Bus Powertrain Implausible Message from Engine Contr.
P1853	18261	Data-Bus Powertrain Implausible Message from Brake Contr.
P1854	18262	Data-Bus Powertrain Hardware Defective
P1855	18263	Data-Bus Powertrain Software version Contr.
P1856	18264	Throttle/Pedal Pos.Sensor A Circ. Error Message from Engine Contr.
P1857	18265	Load Signal Error Message from Engine Contr.
P1858	18266	Engine Speed Input Circ. Error Message from Engine Contr.
P1859	18267	Brake Switch Circ. Error Message from Engine Contr.
P1860	18268	Kick Down Switch Error Message from Engine Contr.
P1861	18269	Throttle Position (TP) sensor Error Message from ECM
P1862	18270	Data Bus Powertrain Missing message from instr. panel ECU
P1863	18271	Data Bus Powertrain Missing Message from St. Angle Sensor
P1864	18272	Data Bus Powertrain Missing message from ADR control module
P1865	18273	Data Bus Powertrain Missing message from central electronics
P1866	18274	Data Bus Powertrain Missing messages
P1912	18320	Brake Booster pressure Sensor Short To Positive
P1913	18321	Brake Booster pressure Sensor Short To Ground
P1914	18322	Brake Booster pressure Sensor Implausible Signal
P1924	18332	Control Module pin coding, Unplausible signal

Diagnostic Trouble Codes (Powertrain)

SAE P-code	V.A.G. code	Description
P2004	18436	Intake Manifold Runner Control bank 1 - Stuck open
P2008	18440	Intake Manifold Runner Control Circuit Open Circuit Bank 1
P2014	18446	Intake Manifold Runner Position Sensor - Electrical Malfunction
P2015	18447	Intake Manifold Runner Position Sensor - Range/Performance
P2016	18448	Intake Manifold Runner Position Sensor - Short to Ground (GND)
P2088	18520	Intake “A” Camshaft Position Actuator Control Circuit Low Bank 1
P2089	18521	Intake “A” Camshaft Position Actuator Control Circuit High Bank 1
P2090	18522	“B” Camshaft Position Actuator Control Circuit Low Bank 1
P2091	18523	“B” Camshaft Position Actuator Control Circuit High Bank 1
P2092	18524	'A' Camshaft Position Actuator Circuit Bank 2 - Short circuit to Ground (GND)
P2093	18525	'A' Camshaft Position Actuator Circuit Bank 2 - Short circuit to B+
P2094	18526	'B' Camshaft Position Actuator Circuit Bank 2 - Short circuit to Ground (GND)
P2095	18527	'B' Camshaft Position Actuator Circuit Bank 2 - Short circuit to B+
P2096	18528	Post Catalyst Fuel Trim System Too Lean Bank 1
P2097	18529	Post Catalyst Fuel Trim System Too Rich Bank 1
P2098	18530	Bank 2 oxygen sensor correction behind catalyst - Lean control limit exceeded
P2099	18531	Bank 2 oxygen sensor correction behind catalyst - Rich control limit exceeded
P2100	18532	Throttle Actuator Control Motor Circuit Open (Motor for Intake Flap V157 implausible signal)
P2101	18533	Throttle Actuator Control Motor Circuit Range/Performance
P2102	18534	Throttle Actuator Control Motor Circuit Low (Motor for Intake Flap V157 signal Too Low)
P2103	18535	Throttle Actuator Control Motor Circuit High (Motor for Intake Flap V157 signal Too High)
P2106	18538	Throttle Actuator Control Motor Circuit Forced Limited Power
P2108	18540	Throttle Actuator Control Module Performance (Motor for Intake Flap V157 Malfunction)
P2110		Throttle Actuator Control System - Forced Limited RPM
P2112	18544	Throttle Actuator Control System, Stuck Closed
P2122	18554	Throttle/Pedal position Sensor/Switch “D” Circuit Low Input
P2123	18555	Throttle/Pedal position Sensor/Switch “D” Circuit High Input
P2127	18559	Throttle/Pedal position Sensor/Switch “E” Circuit Low Input
P2128	18560	Throttle/Pedal position Sensor/Switch “E” Circuit High Input
P2133	18565	Throttle/Pedal position Sensor/Switch “F” Circuit High Input
P2138	18570	Throttle/Pedal position Sensor/Switch “D”/“E” Circuit Voltage Correlation
P2146	18578	Fuel Injector Group A Supply Voltage Circuit/Open
P2149	18581	Fuel Injector Group B Supply Voltage Circuit/Open
P2177	18609	System Too Lean Off Idle, Bank 1
P2178	18610	System Too Rich Off Idle, Bank 1
P2179	18611	System Too Lean Off Idle, Bank 2
P2180	18612	System Too Rich Off Idle, Bank 2
P2181	18613	Cooling System Performance
P2184	18616	Engine Coolant Temperature (ECT) Sensor (on radiator) Circuit 2, Signal too low
P2185	18617	Engine Coolant Temperature (ECT) Sensor (on radiator) Circuit 2, Signal too high
P2187	18619	System Too Lean At Idle, Bank 1
P2188	18620	System Too Rich At Idle, Bank 1
P2189	18621	System Too Lean At Idle, Bank 2
P2190	18622	System Too Rich At Idle, Bank 2

Diagnostic Trouble Codes (Powertrain)

SAE P-code	V.A.G. code	Description
P2191	18623	System Too Lean At Higher Load, Bank 1
P2192	18624	System Too Rich At Higher Load, Bank 1
P2193	18625	System Too Lean At Higher Load, Bank 2
P2194	18626	System Too Rich At Higher Load, Bank 2
P2195	18627	O2 Sensor Signal Stuck Lean B1S1
P2196	18628	O2 Sensor Signal Stuck Rich B1S1
P2197	18629	O2 Sensor Signal Stuck Lean Bank 2 Sensor 1
P2198	18630	O2 Sensor Signal Stuck Rich Bank 2 Sensor 1
P2199		Intake Air Temperature Sensor 1/2 Correlation
P2231	18663	O2 Sensor Signal Circuit Shorted to Heater Circuit B1S1
P2234	18666	O2 Sensor Signal Circuit Shorted to Heater Circuit B2S1
P2237	18669	O2 Sensor Positive Current Control Circuit Open B1S1
P2238	18670	O2 Sensor Positive Current Control Circuit Low B1S1
P2239	18671	O2 Sensor Positive Current Control Circuit High B1S1
P2240	18672	O2 Sensor Positive Current Control Circuit Open B2S1
P2241	18673	O2 Sensor Positive Current Control Circuit Bank 2 Sensor 1 - Low
P2242	18674	O2 Sensor Positive Current Control Circuit Bank 2 Sensor 1 - High
P2243	18675	O2 Sensor Reference Voltage Circuit Open B1S1
P2245	18677	O2 Sensor Reference Voltage Circuit Low B1S1
P2246	18678	O2 Sensor Reference Voltage Circuit High B1S1
P2247	18679	O2 Sensor Reference Voltage Circuit Open B2S1
P2249	18681	O2 Sensor Reference Voltage Circuit Bank 2 Sensor 1 - Low
P2250	18682	O2 Sensor Reference Voltage Circuit Bank 2 Sensor 1 - High
P2251	18683	O2 Sensor Negative Current Control Circuit Open B1S1
P2252	18684	O2 Sensor Negative Current Control Circuit Low B1S1
P2253	18685	O2 Sensor Negative Current Control Circuit High B1S1
P2254	18686	O2 Sensor Negative Current Control Circuit Open B2S1
P2255	18687	O2 Sensor Negative Current Control Circuit Bank 2 Sensor 1 - Low
P2256	18688	O2 Sensor Negative Current Control Circuit Bank 2 Sensor 1 - High
P2257	18689	Secondary Air Injection System Control "A" Circuit Low
P2258	18690	Secondary Air Injection System Control "A" Circuit High
P2261	18693	Turbo Boost Pressure Not Detected, Mechanical
P2263	18695	Turbo/Super Charger Boost System Performance
P2270	18702	O2 Sensor Signal Stuck Lean B1S2
P2271	18703	O2 Sensor Signal Stuck Rich B1S2
P2272	18704	O2 Sensor Signal Stuck Lean B2S2
P2273	18705	O2 Sensor Signal Stuck Rich B2S2
P2274	18706	O2 Sensor Signal Stuck Lean B1S3
P2275	18707	O2 Sensor Signal Stuck Rich B1S3
P2279	18711	Intake Air System Leak
P2293	18725	Fuel Pressure Regulator (N276) Mechanical Malfunction
P2294	18726	Fuel Pressure Regulator (N276) Open circuit
P2295	18727	Fuel Pressure Regulator (N276) Short Circuit to Ground
P2296	18728	Fuel Pressure Regulator (N276) Short Circuit to B+

Diagnostic Trouble Codes (Powertrain)

SAE P-code	V.A.G. code	Description
P2300		Ignition coil “A” Primary Control Circuit Low
P2301		Ignition coil “A” Primary Control Circuit High
P2303		Ignition coil “B” Primary Control Circuit Low
P2304		Ignition coil “B” Primary Control Circuit High
P2306		Ignition coil “C” Primary Control Circuit Low
P2307		Ignition coil “C” Primary Control Circuit High
P2309		Ignition coil “D” Primary Control Circuit Low
P2310		Ignition coil “D” Primary Control Circuit High
P2312		Ignition coil “E” Primary Control Circuit Low
P2313		Ignition coil “E” Primary Control Circuit High
P2336	18768	Cylinder #1 Above Knock Threshold
P2337	18769	Cylinder #2 Above Knock Threshold
P2338	18770	Cylinder #3 Above Knock Threshold
P2339	18771	Cylinder #4 Above Knock Threshold
P2340	18772	Cylinder #5 Above Knock Threshold
P2341	18773	Cylinder #6 Above Knock Threshold
P2400	18832	Evaporative Emission System Leak Detection Pump Control Circuit Open
P2401	18833	Evaporative Emission System Leak Detection Pump Control Circuit Low
P2402	18834	Evaporative Emission System Leak Detection Pump Control Circuit High
P2403	18835	Evaporative Emission System Leak Detection Pump Sense Circuit Open
P2404	18836	Evaporative Emission System Leak Detection Pump Sense Circuit Range/Performance
P2413	18845	Exhaust Gas Recirculation System Performance
P2414	18846	O2 Sensor Exhaust Sample Error B1S1
P2415	18847	Bank 2 sensor 1 - voltage too low/leakage air
P2416	18848	Bank 1 oxygen sensors before and behind main cat swapped
P2422	18854	Evaporative Emission System Vent Valve Stuck Closed
P2425	18857	Exhaust Gas Recirculation Cooling Valve Control Circuit Open
P2426	18858	Exhaust Gas Recirculation Cooling Valve Control Circuit Low
P2427	18859	Exhaust Gas Recirculation Cooling Valve Control Circuit High
P2539	18971	Low Fuel Pressure Sensor (G410) - Open circuit
P2540	18972	Low Fuel Pressure Sensor (G410) - implausible signal
P2541	18973	Low Fuel Pressure Sensor (G410) Open Circuit
P2564	18996	Turbocharger Boost Control Position Sensor Circuit Low
P2565	18997	Turbocharger Boost Control Position Sensor Circuit High
P2600	19032	Coolant pump - Open
P2602	19034	Coolant pump - Short to ground
P2603	19035	Coolant pump - Short to B+
P2626	19058	O2 Sensor Pumping Current Trim Circuit Open B1S1
P2629	19061	O2 Sensor Pumping Current Trim Circuit Open B2S1
P2632	19064	Fuel Pump 2 Relay (J49) - Open circuit
P2634	19066	Fuel Pump 2 Relay (J49) - Short circuit to B+
P2637	19069	Torque Management Feedback Signal “A”
P2708	19140	Shift Solenoid "F" Stuck On
P2711	19143	Unexpected Mechanical Gear Disengagement

Diagnostic Trouble Codes (Powertrain)

SAE P-code	V.A.G. code	Description
P2714	19146	Pressure Control Solenoid "D" Performance or Stuck Off
P2715	19147	Pressure Control Solenoid "D" Stuck On
P2716	19141	Pressure Control Solenoid “D” Electrical
P2717	19142	Pressure Control Solenoid "D" Intermittent
P2723	19155	Pressure Control Solenoid ”E” Performance or Stuck Off
P2724	19156	Pressure Control Solenoid ”E” Stuck On
P2725	19157	Pressure Control Solenoid ”E” Electrical
P2726	19158	Pressure Control Solenoid ”E” Intermittent
P2732	19164	Pressure Control Solenoid ”F” Performance or Stuck On
P2733	19165	Pressure Control Solenoid ”F” Stuck On
P2734	19166	Pressure Control Solenoid ”F” Electrical
P2735	19167	Pressure Control Solenoid ”F” Intermittent
P3000	19456	Glow Plug Indictor Light (K29) - Malfunction message from instrument cluster
P3011	19467	Fuel Pump 2 Relay (J49) - Short circuit to B+
P3012	19468	Fuel Pump 2 Relay (J49) - Open circuit/short circuit to Ground (GND)
P3017	19473	Electronic Differential Lock Intake Valve for Left Front and Right Rear Short to Positive
P3018	19474	Exhaust Flap Valve 1 Short to Ground
P3019	19475	Exhaust Flap Valve 1 Open Circuit
P3025	19481	Angle sensor 1 (G297) - Implausible signal
P3026	19482	Angle sensor 1 (G297) - Signal too small
P3027	19483	Angle sensor 1 (G297) - Signal too large
P3028	19484	Angle sensor 2 (on throttle drive 2 power accelerator actuation) (G298), Unplausible Signal
P3029	19485	Angle sensor 2 (G298) - Signal too small
P3030	19486	Angle sensor 2 (G298) - Signal too large
P3031	19487	Throttle Drive 2 (power accelerator actuation) (G296), Electrical malfunction in circuit
P3032	19488	Throttle Valve Control Module 2 (J544), Malfunction in basic setting
P3033	19489	Intake Air Temperature Sensor 2 (G299) - Signal too small
P3034	19490	Intake Air Temperature Sensor 2 (G299) - Signal too large
P3035	19491	Throttle Valve Control Module 2 (J544), Mechanical malfunction
P3038	19494	Throttle Valve Control Module 2 (J544) - Lower impact is not being reached
P3045	19501	Fuel Pump Electronic Faulty
P3047	19503	Activation Starter Relay 2, Short to Positive
P3048	19504	Activation Starter Relay 2, Short to Negative
P3049	19505	Activation Starter Relay 2, Open Circuit
P3050	19506	Starter Relay 2 Electrical Malfunction in Circuit (Stuck/Inoperative Relay)
P3057	19513	Exhaust Pressure Sensor 2 High
P3058	19514	Exhaust Gas Recirculation Flow - Excessive flow detected
P3059	19515	Exhaust Gas Recirculation Flow - Excessive flow detected
P3060	19516	Cooling Bypass Valve Exhaust Gas Recirculation Flow - Electrical
P3061	19517	Cooling Bypass Valve 2 Exhaust Gas Recirculation Flow - Electrical
P3062	19518	Turbo/Super Charger - Overboost condition
P3063	19519	Turbo/Super Charger - Underboost condition
P3078	19534	Thrfottle Valve Control Module Airflow at Idle Too Low
P3081	19537	Engine Temperature Too Low

Diagnostic Trouble Codes (Powertrain)

SAE P-code	V.A.G. code	Description
P3092	19548	Internal Control Module Memory Checksum Error
P3093	19549	Internal Control Module Memory Checksum Error
P3096	19552	Internal Control Module Memory Checksum Error
P3097	19553	Internal Control Module Memory Checksum Error
P3130	19586	Exhaust Gas Recirculation system - Control limit exceeded
P3131	19587	Exhaust Gas Recirculation system - Lowest control limit exceeded
P3137	19593	Intake Manifold Runner Control (IMRC) Basic setting not carried out
P3138	19594	Intake Manifold Runner Control (IMRC) Deviation
P3191	19647	Intake Manifold Runner Control (IMRC) Bank 1 Control, Basic setting aborted with mechanical impact open
P3192	19648	Intake Manifold Runner Control (IMRC) Bank 1 Control, Basic setting aborted with mechanical impact closed
P3193	19649	Intake Manifold Runner Control (IMRC) Bank 1 Control Circuit outside valid range
P3211	19667	Exhaust B1S1 Heater return Connection
P3212	19668	Exhaust B2S1 Heater return Connection
P3255	19711	Bank 1, oxygen sensor before catalytic converter, heating circuit Regulation at upper impact
P3256	19712	Bank 1, oxygen sensor before catalytic converter, heating circuit Regulation at lower impact
P3262	19718	Exhaust bank 1/2 oxygen sensors behind catalytic converter - swapped
P3266	19722	Bank 1, sensor 1 Internal resistance too large
P3328	19784	Intake Flap Motor 2 - Short to B+
P3329	19785	Intake Flap motor 2 - Open/short to ground
P3330	19786	Intake Flap Motor 2 - no signal
P3331	19787	Intake Flap Motor 2 - faulty
P3339	19795	Automatic Glow Time Control Relay 2 - electrical malfunction in circuit

Diagnostic Trouble Codes (Vehicle Network Communications)

V.A.G. code	SAE U-code	Description
28673	U0001	High Speed CAN Communication Bus
28674	U0002	High Speed CAN Communication Bus Performance
	U0073	Control Module Communication Bus Off
28772	U0100	Lost Communication With ECM/PCM "A"
28773	U0101	Lost Communication with Transmission Control Module (TCM)
28775	U0103	Lost Communication With Gear Shift Control Module
	U0115	Lost Communication With ECM/PCM "B"
28793	U0121	Lost Communication with Anti-Lock Brake System (ABS) Control Module
28798	U0126	Lost Communication with Steering Angle Sensor (G85)
28818	U0146	Lost Communication with Diagnostic Interface
28823	U0151	Lost Communication with Airbag Control Module
28827	U0155	Lost Communication with Instrument Panel Cluster (IPC) Control Module
2836	U0164	Lost Communication with A/C Control Module
28974	U0302	Software Incompatibility with Transmission Control Module (TCM)
28987	U0315	Software Incompatibility with Anti-Lock Brake System (ABS) Control Module
29074	U0402	Invalid Data Received From Transmission Control (TCM) Module
29076	U0404	Invalid Data Received From Gear Shift Control Module

Diagnostic Trouble Codes (Vehicle Network Communications)

V.A.G. code	SAE U-code	Description
29087	U0415	Invalid Data Received From Anti-Lock Brake System (ABS) Control Module
29705	U1009	Gateway, Range/Performance
65535	n/a	Internal control module memory check sum error

Diagnostic Trouble Codes (P-Hex Codes)

SAE P-code	Description
P000A	"A" Camshaft Position Slow Response, Bank 1
P000B	"B" Camshaft Position Slow Response, Bank 1
P000C	"A" Camshaft Position Slow Response, Bank 2
P000D	"B" Camshaft Position Slow Response, Bank 2
P010B	Mass or Volume Air Flow "B" Circuit Range/Performance
P010C	Mass or Volume Air Flow "B" Circuit Low
P010D	Mass or Volume Air Flow "B" Circuit High
P016A	ISO/SAE reserved
P025A	Fuel Pump Module Control Circuit/Open
P025C	Fuel Pump Module Control Circuit Low
P025D	Fuel Pump Module Control Circuit High
P050A	Cold Start Idle Air Control System Performance
P062B	Internal Control Module Fuel Injector Control Performance
P117A	Bank 1, Lambda Correction Catalyst Center Control Limit Reached
P12A1	Fuel Rail Pressure Sensor Inappropriately Low
P12A2	Fuel Rail Pressure Sensor Inappropriately High
P12A4	Fuel Rail Pump Control Valve Stuck Closed

Diagnostic Scan Tool and Special Tool Suppliers

The following suppliers offer diagnostic scan tools or computerized scan tool programs. Some of them also offer special tools and equipment for specialized repair work.

Assenmacher Specialty Tools, Inc.
6440 Odell Place
Boulder, CO 80301
(800) 525-2943
www.asttool.com

Baum Tools Unlimited, Inc.
PO Box 5867
Sarasota, FL 34277
(800) 848-6657
www.baumtools.com

Equipment Solutions
P.O. Box 1450
Kenosha, WI 53141-1450
(800) 892-9650

Metalnerd
113 Crestview Drive, Suite B
Greensburg, PA, 15601 USA
(412) 601-4270
www.metalnerd.com

Ross-Tech
1881 Suumneytown Pike
Lansdale, PA 19446
267-638-2300
www.ross-tech.com

Samstag Sales
115 Main St. N., Suite 216
Carthage, TN 37030
(615) 735-3388
www.samstagsales.com

Shade Tree Software
4186 Culebra Ct.
Boulder, CO 80301
(800) 419-6896
www.shadetreesoftware.com

ZDMAK Tools
P.O. Box 5100
Sarasota, FL 34277
(877) 938-6657
www.zdmak.com

Zelenda Automotive, Inc.
27-12 37th Ave.
Long Island City, NY 11101
(888) 892-8348
www.zelenda.com

Most of the above suppliers offer aftermarket diagnostic scan tools. Equipment Solutions and Zelenda Automotive have been official suppliers for Volkswagen and Audi, including VAG 1551/1552 and VAS 5051/5052 factory diagnostic scan tools.

The older VAG 1551/1552 scan tools will not access all functions on vehicles covered by this manual. However the 1551/1552 tools can still be helpful in troubleshooting numerous diagnostic trouble codes. The flow chart on the last page of this repair group is useful when using the older VAG 1551/1552 tools.

Readiness Code

The readiness code will be set after complete diagnostic monitoring of all components and systems has been carried out. Most of the time, the readiness code will set itself after the appropriate conditions have been met. These may include a cold or a hot start, operation at a certain load or speed for a certain period of time, or operation at different temperatures. After repairs, it is usually advantageous to verify the proper operation of a component or system. Sometimes this can be accomplished by driving the vehicle and observing the Malfunction Indicator Light (MIL), but a more efficient way to confirm a repair is by setting the readiness code.

Setting the readiness code requires the use of Volkswagen scan tool VAG 1551/1552, Volkswagen diagnostic computer VAS 5051/5052, or an equivalent aftermarket scan tool or scan tool computer program.

A thorough understanding of your equipment operation and it's capabilities is also required. The readiness code is then generated according to instructions provided by the factory and the scan tool manufacturer.

Readiness code values[2]								
Digit position[1]								Diagnostic Function
1	2	3	4	5	6	7	8	
							0	Three Way Catalyst Efficiency
						0		Catalyst Heating (currently always 0)
					0			Evaporative Emissions System (Fuel tank venting system)
				0				Secondary Air Injection System
			0					Air Conditioning
		0						Oxygen Sensors
	0							Oxygen Sensor Heaters
0								Exhaust Gas Recirculation (EGR)

Notes:
1. Value can only be a 1 or a 0.
   A value of 1 indicates that test has not yet run or has run and failed.
   A value of 0 indicates that test has run and passed or function is not applicable.
2. Readiness code is properly set when all 8 positions are zero, i.e. "00000000".

The readiness code can be displayed in any of several areas within the ECM's diagnostic memory depending on the individual scan tool. It will always have a format consisting of 8 digits. If the 8 digit code contains characters other than the number 0, such as 10100110, than the readiness code is NOT set. When the code displays all zeros, i.e. 00000000, the readiness code IS set and no further action is required. While the format is required to be the same for all engines, certain equipment may not be used on each engine. For example, some New Beetle engines do not use EGR and as such, digit position number 8 will always show 0. Other Volkswagen engines that use EGR will show either a 1 or a 0 as appropriate.

The tables on the following pages outline the procedures for setting the readiness code for various engine types. The code can be read from function 08 in the appropriate display group. Most ECM versions will also display the readiness code separately through function 15.

Quick Reference: Readiness code setting (1.8L engine code: APH)

				Procedure VAG 1551/5051 Settings after successful Diagnosis			
Step	Diagnostic Status	Engine Condition	Function/ Display-Group	Channel 1	Channel 2	Channel 3	Channel 4
1	Check DTC Memory	Ignition ON Engine NOT on	02				
2	Clear Memory	Ignition ON Engine NOT on	05				
3	Throttle Body Adaption	Ignition ON Engine NOT on	04 060			Counter counts to 9 during adaption	ADP. OK
4	Oxygen Sensor Aging*	Engine ON at idle	04 034	820-860 RPM			B1 S1 OK
Install Special Tool VAG 1788/10 and set engine speed to 2280-2760 RPM							
5	Catalytic Converter*	Engine ON	04 046	2280-2760 RPM	Min. 400 °C	Man Trans max 0.33 Auto Trans max 0.40	CAT B1 OK
Remove Special Tool VAG 1788/10 and let engine idle (820-860 RPM)							
6	Secondary Air	Engine ON at idle	04 077	820-860 RPM			SYST OK
7	EVAP System*	Engine ON at idle	04 070				EVAP OK
8	Leak Test EVAP System*	Engine ON at idle	04 071			Check end	SYST OK
9	Check Readiness Code	Engine ON at idle	15	00000000			

Automatic transmission selector lever in "P" or "N" position. All electrical consumers shut off

Requirement: Coolant temp must be 5-95°C, intake temp less than 60°C. (See display group 04, display field 3 and 4)

*** These tests may take several minutes**

Quick Reference: Readiness code setting (1.8L engine code: AWV, AWP)

			Procedure VAG 1551 Settings after successful Diagnosis				
Step	**Diagnostic Status**	**Engine Condition**	**Function/ Display-Group**	**Channel 1**	**Channel 2**	**Channel 3**	**Channel 4**
1	Check DTC Memory	Engine OFF	02	-	-	-	-
2	Clear DTC Memory	Ignition ON	05				
3	Conditions fulfilled before generating RC	Engine ON 1800....2500/min**	08/004	1800 - 2500 RPM**	more than 8V	more than 60°C	Less than 80 °C
4	Conditions fulfilled before generating RC	Engine ON 1800....2500/min**	08/030	- - - - - - 1 1	-	-	-
5	Fuel Supply System	Engine ON Idle	04/107	640 – 840 RPM	-23-+23%	-	Syst. OK
6	EVAP Canister Purge Regulator Valve Check	Engine ON Idle	04/070	0% - 10%	-	More than 30%	EVAP OK
7	Operational Readiness of Oxygen Sen. behind Cat. Conv.	Engine ON Idle	04/036	smaller 0.4V or larger 0.5V	B1-S2 OK	-	-
8	Oxygen Sensor aging behind Cat. Conv.	Engine ON 1800.....2200/min**	04/043	1800 - 2200 RPM**	more than 300°C	0.1 – 0.9V	B1-S2 OK
9	Diagnosis Camshaft Timing Check	Engine ON 1800.....2200/min**	04/094	1800 - 2200 RPM**	CS-ctr.1 ON/OFF	Syst. OK	-
10	Diagnosis Knock Sensors	Engine ON 1800.....2200/min**	04/028	1800 - 2200 RPM**	12–25%	more than 40°C	Syst. OK
11	Response Time Front Heated Oxygen Sensor	Engine ON 1800.....2200/min**	04/034	1800 - 2200 RPM**	more than 350°C	more than 0.5	B1-S1 OK
12	Oxygen Sensor Control System	Engine ON Idle	04/037	12% - 30%	0.1V–0.9 V	-1200ms – +1200ms	Syst. OK
13	Three Way Catalytic Converter	Engine ON 1800.....2200/min**	04/046	1800 - 2200 RPM**	more than 400°C	0.0 - 0.32	CATB1 OK
14	Fuel System Leakage Check	Engine ON Idle	04/071	-	-	Check END	Syst. OK
15	Secondary Air System	Engine ON Idle	04/077	640 – 820 RPM	5 – 12 g/s	more than - 45%	Syst. OK
16	Misfire Activity Check	Engine ON Idle	08/014	640 – 820 RPM	12 – 30%	0	activated
17	Check Readiness Code Setting	Engine ON Idle	08/086	00000000	00000000	00000000	00000000

Requirement: Coolant Temperature more than 80°C (08/001/2)
*** to warm up engine if necessary ** only during Diagnostic Performance**

Quick Reference: Readiness code setting (1.8L engine code: BKF)

Auto Transmission in " P " & Engine Hood Open

Step	Diagnosis	Condition	Function / MVB	Channel 1	Channel 2	Channel 3	Channel 4	Comments
1	Check DTC memory Clear DTC memory	Engine off Ignition. On	01/02....01/05	x	x	x	x	Check fault memory 01 - 02 Delete fault memory 01 - 05
2	Throttle adaptation	Engine off Ignition. On	04/060	x	x	x	Adapt I.O.	Do not touch accelerator pedal
3	Kick-down adaptation (Auto trans.)	Engine off Ignition. On	04/063	x	x	Kick-down	Adapt I.O.	Depress accelerator fully until I.O.
4	LDP	Idle	04/202	x	x	Text	xxxxx100	Fuel Cap Loosen then re-fasten prior to test.
5	Engine warm-up	3000 RPM	08/04	RPM	Batt. Volts	ECT > 80 c.	IATS	Run until Channel 3 > 80 deg. c.
6	Oxygen sensor before cat. Dynamic check	Brake / Gas	04/206	x	x	x	xxxxx100	Diagnosis could take several minutes
7	OXYS wires	RPM briefly to 6,000 return to idle	08/208	N/A	x	x	00111111	Repeat short RPM increase to 6,000 RPM and returning to idle. Repeat until Ch. 4, 3rd value from the right = 1
8	OXYS Heater	Idle	04/205	x	x	x	xxxx1100	Diagnosis takes approx 70 seconds
9	Filling registration	Brake / Gas	04/209	IATS	x	x	xxxxx100	IATS <75 deg. Cels.
10	MAF registration	Brake / Gas	04/210	x	x	x	xxxxx100	Diagnosis can take approx 5 minutes
11	OXYS 1 Readiness	Brake / Gas	04/213	x	x	x	xxxxxx100	Diagnosis can take approx 70 seconds
12	OXYS 1 Aging	Brake / Gas	04/215	x	x	x	xxxxxx100	Diagnosis can take approx 70 seconds
13	OXYS 1 Heater	Idle	04/211	x	x	x	xxxxxx100	Diagnosis could take several minutes
14	OXYS plausability	Brake / Gas	04/207	x	x	x	xxxxxx100	Diagnosis can take approx. 70 sec. if HOXY is ready. Check MVB 30 Ch.1 = xxx11
15	OXYS Readiness	Brake / Gas	04/216	x	x	x	xxxxxx100	Diagnosis can take approx 70 seconds
16	OXYS 2 Readiness	Brake / Gas	04/214	x	x	x	xxxxxx100	Diagnosis can take approx 70 seconds
17	Sensor dynamic	Brake / Gas	04/220	x	x	x	xxxxx100	Diagnosis could take several minutes
18	OXYS 2 Heater	Idle	04/212	x	x	x	xxxxxx100	Diagnosis can take approx 70 seconds
19	Front CAT	Brake / Gas	04/218	x	x	x	xxxxxx100	Diagnosis could take aprox. 90 seconds
20	EVAP purge valve	Idle	04/221	x	x	x	xxxxxx100	Diagnosis can take approx 70 seconds
21	Secondary air	Idle	04/222	x	x	x	xxxxx100	Diagnosis can take approx 70 seconds
22	Readiness status	Idle	08/086	00000000	00000000	00000000	x	Readiness set if all Zeros

If Readiness Code not set check DTC's 2/28/2005

Quick Reference: Readiness code setting (1.9L engine code: ALH)

Work Step	Instructions
Preliminary	Raise engine speed with vehicle stationary. Apply hand brake. Manual transmission in **Neutral.** Automatic Transmission range selector lever in "**N**". No DTCs stored in DTC memory. Engine Coolant Temperature 50°F (10°C) minimum.
Work Step 1	Start engine and let idle for 35 seconds minimum.
Work Step 2	Test drive vehicle in a safe location and maintain 2000 RPM in 3rd gear (manual or automatic transmission) for 5 seconds, minimum.
Work Step 3	Accelerate from an engine speed of 2200 RPM (tachometer) to Wide Open Throttle (WOT) in 2nd or 3rd gear for 8 seconds.
Work Step 4	Stop vehicle, switch OFF the engine, and wait a minimum of 10 seconds.
Work Step 5	Repeat Work Steps 1, 2 and 3.
Work Step 6	Read out Readiness Code, Scan Tool Function 15.
Work Step 7	If Readiness code not OK, check for malfunctions in DTC memory and eliminate as necessary.
Work Step 8	Check DTC memory again and then repeat work steps as needed.
Notes: Do NOT switch engine OFF during test sequence Work Steps, except as noted. If test sequence is interrupted or not followed, Readiness Code will not be set.	

Quick Reference: Readiness code setting (1.9L engine code: BEW)

Turbo Diesel PD MY2004; Engine Codes BHW , BEW and BKW*

			Procedure VAG 5051/52		Settings after successful Diagnosis		
Step	Diagnostic Status	Engine Condition	Function Display-Group	Channel 1	Channel 2	Channel 3	Channel 4
1	Check DTC memory	Ignition ON/Engine not running	02				
2	Erase DTC memory	Ignition ON/Engine not running	05				
3		Run engine to operating temperature	08/01				80 °C
4		Turn ignition OFF for 10 seconds min.	N/A				
5		Turn ignition ON for 10 seconds min.	N/A				
6	Check glow plug system	1) Turn ignition OFF. 2) Turn ignition ON. 3) Activate Glow plug.	03 08/86			System ok x1xxxxxx	
7	Check EGR system	Engine running at 1600-1700 RPM for 10 seconds min.	86	XX1XXXXX			
		Check is complete	86	XXX1XXXX			

Check oxygen sensor/oxygen sensor heater and knock sensor - Continued at address word 01 engine

Step	Diagnostic Status	Engine Condition	Function Display-Group	Channel 1	Channel 2	Channel 3	Channel 4
8	Check oxygen sensor is active	Engine running at 3500 RPM	08/31 08/40		>10%	 xxxxxxx1	
9		Idle for 60 seconds min.	08/86 08/87			 xxxx1111	1xxxxxxx
10	Check oxygen sensor and knock sensor	**Road test:** Drive vehicle in 2nd Gear and decelerate vehicle from 25 MPH for at least 2 seconds min.	08/89 with newer software				xx0xxxxx
11	Continued	Turn Ignition OFF and ON					

Step	Diagnostic Status	Engine Condition	Function Display-Group	Channel 1	Channel 2	Channel 3	Channel 4
12	Check readiness code	Ignition ON/Engine not running	01/17	00000000	00000111	11100000	00000000
13	Repeat this procedure 2 times						

Note: * The V10 TDI has a possibility for the oxygen sensor. After idling for a longer time the oxygen sensor will be inactive.
MB 031 Channel 2 will show a value <10% than repeat Step 8.

Quick Reference: Readiness code setting (2.0L engine code: AEG, AZG)

				Procedure VAG 1551/5051 Settings after successful Diagnosis			
Step	Diagnostic Status	Engine Condition	Function/ Display-Group	Channel 1	Channel 2	Channel 3	Channel 4
1	Check DCT Memory	Ignition ON Engine NOT on	02				
2	Clear Memory	Ignition ON Engine NOT on	05				
3	Throttle Body Adaption	Ignition ON Engine NOT on	04 098			Idle	ADP. OK
4	Secondary Air	Engine ON at idle	04 077	Engine Temp 6-96 °C			SYST. OK
5	EVAP System*	Engine ON at idle	04 070				EVAP OK
6	Leak Test EVAP System*	Engine ON at idle	04 071		Abort	Check end	SYST OK
Install Special Tool VAG 1788/10 and set engine speed to 2200-2800 RPM							
7	Oxygen Sensor Aging*	Engine ON	04 034	2200-2800 RPM	>352.0 °C	max. 2.20 s	B1 S1 OK
Increase Engine RPM to 2800-3200 RPM							
8	Catalytic Converter*	Engine ON	04 046	2800-3200 RPM	>352.0 °C	max. 0.55	CAT B1 OK
Remove Special Tool VAG 1788/10 and let engine idle							
9	Check Readiness Code	Engine ON at idle	15	00000000			

Automatic transmission selector lever in "P" or "N" position. All electrical consumers shut off

Requirement: Coolant temp must be 5-95°C, intake temp less than 60°C. (See display group 04, display field 3 and 4)

* These tests may take several minutes

Quick Reference: Readiness code setting (2.0L engine code: AVH)

				Procedure VAG 1551/5051 Settings after successful Diagnosis			
Step	Diagnostic Status	Engine Condition	Function/ Display-Group	Channel 1	Channel 2	Channel 3	Channel 4
1	Check DTC Memory	Ignition ON Engine NOT on	02				
2	Clear Memory	Ignition ON Engine NOT on	05				
3	Throttle Body Adaption	Ignition ON Engine NOT on	04, 60			8	ADAP. OK
4	Leak Detection*	Engine ON, at Idle	04, 202				xx0xx1x0
5	Knock Sensor Diag.	Road test with sec. Person Engine RPM >3000 RPM	08, 203				xxxxx100
6	Vehicle Speed Diag.	Road test with sec. Person Engine RPM >3000 RPM	08, 204				xxxxx100
7	Diag. O2 Sensor Heating Before Cat	Depress Brake Pedal and Hold Press Accelerator to Full Throttle	04, 205				xxxxx100
8	Diag. O2 Sensor Aging Before Cat	Depress Brake Pedal and Hold Press Accelerator to Full Throttle	04, 212				xxxxx100
9	Fuel Mixture Diag. Partial Load	Depress Brake Pedal and Hold Press Accelerator to Full Throttle	04, 216				xxxxx100
10	Air Mass Filling Detection Diag.	Depress Brake Pedal and Hold Press Accelerator to Full Throttle	04, 218				00000x00
11	Diag. O2 Sensor Heating After Cat	Depress Brake Pedal and Hold Press Accelerator to Full Throttle	04, 219			Between 200°C and 600°C	xxxxx100
12	Diag. O2 Sensor Voltage After Cat	Depress Brake Pedal and Hold Press Accelerator to Full Throttle	04, 221			Between 200°C and 600°C	000xx100
13	Diag. O2 Sensor After Cat	Depress Brake Pedal and Hold Press Accelerator to Full Throttle	04, 224				xxxxx100
14	Catalytic Converter	Depress Brake Pedal and Hold Press Accelerator to Full Throttle	04, 226			Between 450°C and 740°C	xxxxx100
15	Fuel Mixture at Idle	Engine ON, at Idle	04, 216				xxxx1100
16	EVAP System	Engine ON, at Idle	04, 228				xxxxx100
17	Secondary Air	Engine ON, at Idle	04, 229				xxxxx100
18	Fuel Mixture at Idle	Engine ON, at Idle	04, 231				xxxxx100
19	Check Readiness Code	Engine ON, at Idle	15	00000000			

Automatic transmission selector lever in "P" or "N" position. All electrical consumers shut off
Requirement: Coolant temp must be at least 80°C, intake temp less than 60°C. (See display group 04, display field 3 and 4)
The Catalyst Temperature must be at least 200°C and below 740°C (See display group 34 display field 2)
<u>If Readiness Code cannot be set Check DTCs.</u>
*** These tests may take several minutes**

Quick Reference: Readiness code setting (2.0L engine code: BEV, BGD) (2004 m.y.)

			Procedure VAG 5051		Settings after successful Diagnosis		
Step	Diagnostic Status	Engine Condition	Function Display-Group	Channel 1	Channel 2	Channel 3	Channel 4
1 2	Check DTC Memory Clear DTC Memory	Engine OFF Ignition ON		-	-	-	-
3	Adaptation Throttle Valve Control Module	Engine OFF Ignition ON	04/060	xx%	xx%	0 / 8	ADP.OK
4	Adaptation Kick Down (only with Automatic Transmission)	Engine OFF Ignition ON	04/063	xx%	xx%	Kick Down	ADP.OK
5	Check Fuel System Leakage *	Engine ON at Idle	04/202			Text	xxxxx100
6	O2 Sensor b. C. Heating LSU	Engine ON at Idle	04/205				xxxxx100
7	Diagnostic of Catalytic Converter *	Depress Brake Pedal and Hold Press Accelerator to Full Throttle	04/226			>1	xxxxx100
8	To activate O2 Sensor b.C. LSU <u>**only BGD engine**</u>	Press Accelerator to 6000 RPM in a steady rate ; hold 5 sec. *If Step 10 after idling not xxxxx100 repeat Step 8 or Test Drive vehicle in 1st Gear to 5000 RPM and decelerate for 2 seconds min.*	08/005	1/min			Idle Enrich Part Throttle Full Throttle Overrun
9	O2 Sensor b. C. Aging LSU	Depress Brake Pedal and Hold Press Accelerator to Full Throttle	04/212	0....30 **			xxxxxx00
10	O2 Sensor b. C. Aging LSU	Engine ON at Idle	04/212	30	<700°C	>4%	xxxxx100
11	Check Fuel Delivery System	Engine ON at Idle	04/216			-7.5 - 7.5%	xxxxxx00
12	Check Fuel Delivery System Partly Throttle	Depress Brake Pedal and Hold Press Accelerator to Full Throttle	04/216		-20.0 - 20.0% both 216 test for 30 seconds		xxxxxx00
13	Air Mass Filling Detection Diag.	Depress Brake Pedal and Hold Press Accelerator to Full Throttle	04/218	-0.20 - 0.20	0.84 - 1.15	0.8 - 1.20	xxxxx100
14	O2 Sensor a. C. Heating LSF	Depress Brake Pedal and Hold Press Accelerator to Full Throttle	04/219			200 - 600°C	xxxxx100 xxxxx100
15	O2 Sensor a. C. LSF	Depress Brake Pedal and Hold Press Accelerator to Full Throttle	04/221			300 - 800°C	xxxxx100
16	O2 Sensor a. C. Aging LSF	Depress Brake Pedal and Hold Press Accelerator to Full Throttle	04/224				xxxxx100
17	EVAP System	Engine ON at Idle	04/228				xxxxx100
18	Secondary Air Bank	Engine ON at Idle	04/229				xxxxx100
19	Check Readybits/Zyklusbits	Engine ON at Idle	08/086	00000000	00000000	00000000	

Automatic transmission selector lever in "P" or "N" position. All electrical consumers shut off.

Requirement: Coolant temp must be at least 80°C,intake temp less than 60°C. (See display group 04 field 3 and 4)

<u>If Readiness Code cannot be set Check DTCs. For more Details Refer to Repair Manual "Readiness Code Generating".</u>

*These tests may take several minutes

** After 04/212 Channel 1 reach 30 let engine run at idle until test is finished (xxxxx100).

Quick Reference: Readiness code setting (2.0L engine code: BEV) (2005 m.y.)

2005 2.0L ULEV Engine Code BEV

Not Applicable to 2004 BEV 2.0L

Step	Diagnosis	Condition	Function Display group	Channel 1	Channel 2	Channel 3	Channel 4	Coments
1	Check DTC memory Clear DTC memory	Engine off Ignition. On	01/02 01/05	x	x	x	x	Check fault memory 01 - 02 Delete fault memory 01 - 05
2	Throttle adaptation	Engine off Ignition. On	04/060	x	x	x	Adapt I.O.	Do not touch accelerator pedal
3	Kick-down adaptation (Auto trans. Only)	Engine off Ignition. On	04/063	x	x	Kick-down	Adapt I.O.	Depress accelerator fully until I.O.
4	LDP	Idle	04/202	x	x	x	xxxxx100	Loosen then re-fasten prior to test.
5	Engine warm - up	3000 rpm	08/04	RPM	Volts	CTS	IATS	Run until Channel 3 > 80 deg. C
6	Oxygen sensor before cat. Heating	Idle	04/204	x	x	x	xxxxx100	
7	Diag. Fuel filling	Brake / Gas	04/209	x	x	x	xxxxx100	
8	Oxygen sensor before cat. Dynamic check	Brake / Gas	04/206	x	x	x	xxxxx100	
9	Plausibilty oxygen sensor before cat.	Brake / Gas	04/207	x	x	x	xxxxx100	
10	Catalyst Diagnosis Quick Pass Check	RPM briefly to 6,000 rpm return to idle	08/217	x	x	x	xxxxx100	Repeat until channel 4 = xxxxx100
11	Filling registration	Brake / Gas	04/210	x	x	x	xxxxx100	
12	Oxygen sensor after cat. Heater	Brake / Gas	04/212	x	x	x	xxxxx100	
13	Oxygen sensor after cat. Diag.	Brake / Gas	04/214	x	x	x	xxxxx100	
14	Oxygen sensor after cat. Diag.	Brake / Gas	04/216	x	x	x	xxxxx100	
15	EVAP purge valve	Idle	04/220	x	x	x	xxxxx100	
16	Oxygen sensor before cat. Wiring check	2nd gear selected, accelerate to 3000 rpm. Decelerate for > 2 sec	08/208	x	x	x	xx111111	Driving required Repeat until channel 4 = xx111111
17	Secondary air	Idle	04/221	x	x	x	xxxxx100	
18	Readiness status	Idle	08/086	00000000	00000000	00000000	x	Readiness code set if all Zeros

3/9/2005
1 of 1

Some test could take several minutes

Quick Reference: Readiness code setting (2.5L engine code: BPR)

Jetta & New Beetle 2.5L SULEV : Engine Code BGQ & BPR

Test Step	Diagnose	Conditions	Fuction/ MVB	Channel 1	Channel 2	Channel 3	Channel 4	Remarks
1	Fault memory	Key on, engine off	01-02 01-05	x	x	x	x	Check fault memory 01 - 02 Delete fault memory 01 - 05
2	Throttle Adaptation	Key on, engine off	04/060	x	X	X	ADP. O.K.	Do not touch accelerator pedal
3	Kick-Down adaptation	Key on, engine off	04/063	X	X	Kick-down	ADP. O.K.	Press accelerator pedal to floor
4	EVAP Leakage Check	Idle	04/202	x	x	x	xxxxx100	If Diagnosis does not start loosen / re-fasten gas cap
5	Engine Warm-Up	2,500 RPM	08/004	RPM	x	ECTS	IATS	Run until Channel 3 = 80 deg. c.
6	OXYS Heater B1-S1	Idle	04/203	x	x	x	xxxxx100	Diagnosis runs after 120 seconds
7	OXYS Dynamic B1-S1	Brake / Gas	04/208	x	x	x	xxxxxx100	Intake air temp. < 60 deg.C. See note
8	OXYS plausibility B1-S1	Brake / Gas	04/209	x	x	x	xxxxxx100	Diagnosis runs if HOXY is ready. Check MVB 30 Ch.1 = xxx11
9	OXYS Heater B1-S2	Brake / Gas	04/210	x	x	x	xxxxxx100	Diagnosis could take several minutes
10	OXYS Heater B1-S3	Brake / Gas	04/211	x	x	x	xxxxxx100	Diagnosis could take several minutes
11	OXYS Readiness B1-S2	Brake / Gas	04/212	x	x	x	xxxxxx100	Diagnosis runs after 20 seconds
12	OXYS Readiness B1-S3	Brake / Gas	04/213	x	x	x	xxxxxx100	Diagnosis runs after 20 seconds
13	OXYS Aging B1-S2	Brake / Gas	04/214	x	x	x	xxxxxx100	Diagnosis runs after 20 seconds
14	OXYS Aging B1-S3	Brake / Gas	04/215	x	x	x	xxxxxx100	Diagnosis runs after 20 seconds
15	Front CAT	Brake / Gas	04/217	x	x	x	xxxxxx100	Diagnosis could take several minutes
16	Intake Leak Diagnosis	Brake / Gas	04/221	x	x	x	xxxxxx100	Diagnosis runs after 20 seconds
17	OXYS Decel Check B1-S1	RPM briefly to 6,000 / return to idle	04/222	x	x	x	00111111	Repeat short RPM increase to 6,000 RPM and returning to idle. Repeat until Ch. 4, 3rd value from the right = 1
18	Secondary air system	Idle	04/223	x	x	x	xxxxxx100	Requirements must be met. OXYS must be ok. 5 < IATS < 100 deg. cels. 5 < ECT < 100 deg. cels. CAT temp. < 800 deg. Cels.
19	EVAP purge valve	Idle	04/224	x	x	x	xxxxxx100	Diagnosis runs after 30 seconds ECT>50c
20	Readiness status	Ignition on	08/086	00000000	Readiness Code Set if all zeros			

Automatic transmission in "P" position
Test step 7: In some cases, an open hood and use of a shop fan may be needed to keep intake air temp. < 60 deg. C.
If RC not set, check for DTC's

5/30/2006
Page 1 of 1
Version 2

Quick Reference: Readiness code setting (2.5L engine code: BPS)

Jetta & New Beetle 2.5L ULEV: Engine Code BGP & BPS

Test Step	Diagnose	Conditions	Fuction/ MVB	Channel 1	Channel 2	Channel 3	Channel 4	Remarks
1	Fault memory	Key on, engine off	01-02 01-05	x	x	x	x	Check fault memory 01 - 02 Delete fault memory 01 - 05
2	Throttle Adaptation	Key on, engine off	04/060	x	X	X	ADP. O.K.	Do not touch accelerator pedal
3	Kick-Down adaptation	Key on, engine off	04/063	X	X	Kick-down	ADP. O.K.	Press accelerator pedal to floor
4	EVAP Leakage Check	Idle	04/202	x	x	x	xxxxx100	If Diagnosis does not start loosen / re-fasten gas cap
5	Engine Warm-Up	2,500 RPM	08/004	RPM	x	ECTS	IATS	Run until Channel 3 = 80 deg. c.
6	OXYS Heater B1-S1	Idle	04/203	x	x	x	xxxxx100	Diagnosis runs after 120 seconds
7	OXYS Dynamic B1-S1	Brake / Gas	04/208	x	x	x	xxxxxx100	Intake air temp. < 60 deg. C. See note
8	OXYS plausibility B1-S1	Brake / Gas	04/209	x	x	x	xxxxxx100	Diagnosis runs if HOXY is ready. Check MVB 30 Ch.1 = xxx11
9	OXYS Heater B1-S2	Brake / Gas	04/211	x	x	x	xxxxxx100	Diagnosis could take several minutes
10	OXYS Readiness B1-S2	Brake / Gas	04/213	x	x	x	xxxxxx100	Diagnosis runs after 20 seconds
11	OXYS Aging B1-S2	Brake / Gas	04/215	x	x	x	xxxxxx100	Diagnosis runs after 20 seconds
12	Front CAT	Brake / Gas	04/217	x	x	x	xxxxxx100	Diagnosis could take several minutes
13	OXYS Decel Check B1-S1	RPM briefly to 6,000 / return to idle	04/222	x	x	x	00111111	Repeat short RPM increase to 6,000 RPM and returning to idle. Repeat until Ch. 4, 3rd value from the right = 1
14	Secondary air system	Idle	04/223	x	x	x	xxxxxx100	Requirements must be met. OXYS must be ok. 5 < IATS < 100 deg. cels. 5 < ECT < 100 deg. cels. CAT temp. < 800 deg. Cels.
15	EVAP purge valve	Idle	04/224	x	x	x	xxxxxx100	Diagnosis runs after 30 seconds ECT>50c
16	Readiness status	Ignition on	08/086	00000000	Readiness Code Set if all zeros			

Automatic Transmission in "P" position
Step 7: In some cases, an open hood and use of a shop fan may be needed to keep Intake air < 60 deg. C
If RC not set, check for DTC's

5/30/2006
Page 1 of 1
version 2

V.A.G. 1551/1552 OPERATING OVERVIEW FLOW CHART

MODE MENU

1 - Rapid Data =>
2 - Blink Code
3 - Self Test
4 - Dealership Code
5 - LT - Diagnosis

ADDRESS WORDS MENU (V9)

01 - Engine Electronics =>
11 - Engine Electronics II
41 - Diesel Pump Electronics
51 - Electric Drive
61 - Battery Control
71 - Battery Charger
02 - Transmission Electronics
12 - Clutch Electronics
22 - 4 Wheel Drive Electronics
03 - Brake Electronics
13 - Distance Control
33 - OBD II
14 - Suspension Electronics
24 - Anti - Slip Control
34 - Level Control
44 - Steering Assist
54 - Rear Spoiler
15 - Airbag
25 - Anti-Theft Immobilizer Sensor
35 - Central Locking
45 - Interior Monitor
55 - Headlamp Vert. Aim Control
65 - Tire Pressure Monitoring
75 - Emergency Call Module
16 - Steering Wheel Electronics
26 - Automatic Roof Control
36 - Seat Adjustment Driver's Side
46 - Comfort System (AC CM)
56 - Radio
66 - Seat & Mirror Adjustment
76 - Parking Aid
17 - Instrument Cluster
37 - Navigation System
47 - Sound System
57 - Airbag (Cabriolet only)
67 - Language Control
08 - A/C & Heating Electronics
18 - Auxiliary Heater
09 - Electronic Control Module
19 - Diagnostic Interface for Data Bus
29 - Left Light Control
39 - Right Light Control
49 - Auto Light Switch
00 - Automatic Test Sequence

FUNCTIONS MENU

01 - Check Control Module Versions
02 - Check DTC Memory
03 - Output Diagnostic Test Mode
04 - Basic Setting
Display Group xxx =>
05 - Erase DTC Memory
06 - End Output
07 - Code Control Module
08 - Read Measuring Value Block
Display Group xxx =>
09 - Read Individual Measuring Values
Channel xx
10 - Adaptation
Channel xx
11 - Log-In Procedure
15 - Readiness Code

DISPLAY GROUP FORMAT

Display Group 000 (only)
xx xx xx xx xx xx xx xx xx xx

Display Group 001 (and higher)
xxxxx xxxxx xxxxx xxxxx

Service Training

Self-Study Program

for the

1998 New Beetle

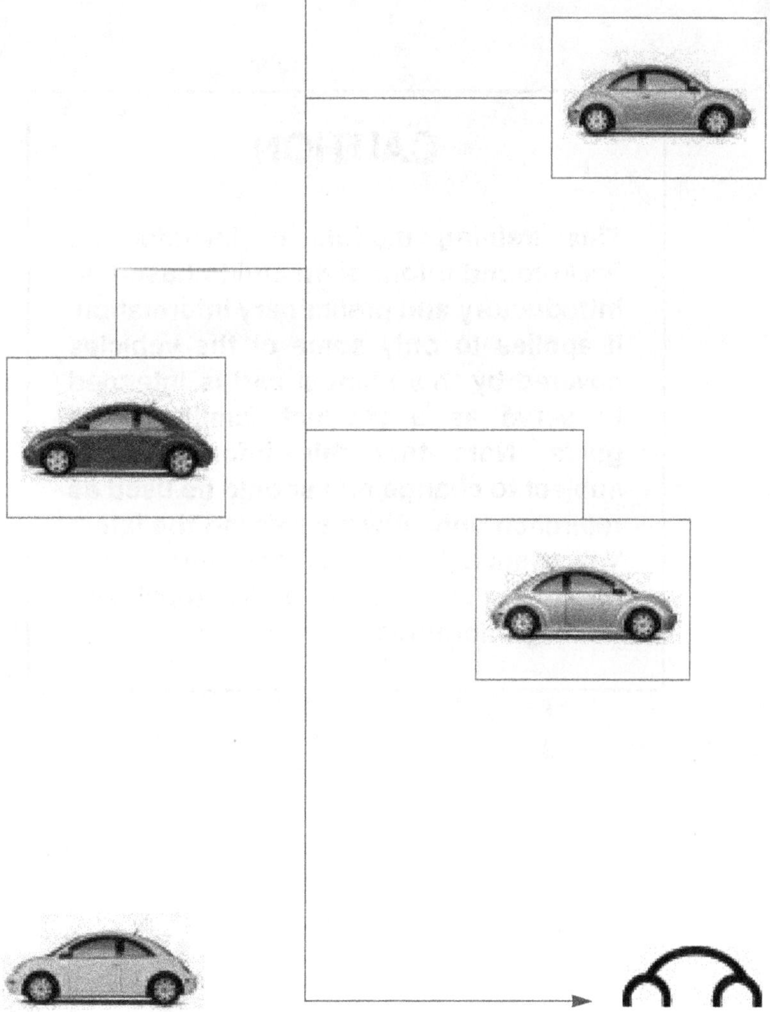

Course Number
891803

CAUTION

This training module is included as background information and is based on introductory and preliminary information. It applies to only some of the vehicles covered by this manual and is intended to serve as a product familiarization guide. Note that this information is subject to change and should be used as reference only. Always refer to the latest Volkswagen factory service information for the most up to date service, repair and training information.

Table of Contents

Introduction

Concept 1, the hit of the 1994 North American International Auto Show in Detroit, has become an automotive reality in the form of the *New Beetle*. Few show cars in history have generated as much interest and speculation as Concept 1. The decision to produce the *New Beetle*, in a form so close to the original design, has been followed with intense interest by the Volkswagen faithful and the automotive press alike. "Spy" photos and industry rumors have appeared in virtually every important automotive publication for the last four years. Why the intense interest?

Has there ever been another automobile as distinctive as the original "Beetle?" People the world over recognize the shape, and it is one that evokes memories of good times and fun. The *New Beetle* brings back those memories. It makes people smile. But it is more than just an exercise in nostalgia. The *New Beetle* is engineered with the same practicality and dependability as the original, in a thoroughly modern package.

Market research shows the *New Beetle* will occupy a unique position among its rivals in the marketplace. It has all the practical appeal of popular cars in its "Small Car" market segment such as the Honda Civic, Dodge Neon and Ford Escort. At the same time, it shares the emotional appeal and visual excitement of sporty cars like the Toyota RAV4, Mazda Miata and BMW Z3.

The power train, chassis and accessories of the *New Beetle* will be familiar to technicians servicing current model Volkswagens but, as this Self-Study Program describes, there are changes and refinements to almost every system in the vehicle. The *New Beetle* showcases this evolution as the first new-generation vehicle produced for North America.

There are some other significant firsts for the *New Beetle* – it's introduction to the public at the North American International Auto Show in Detroit in January, 1998, exactly four years after Concept 1 stole the same show, makes it the first all-new Volkswagen model ever to be introduced in the United States.

The Volkswagen factory in Puebla, Mexico is the first to produce the *New Beetle* – the first time that the Mexican factory is manufacturing an all-new model, not already in production elsewhere in the world.

SSP 418/251

New Beetle Overview

The New Beetle is a thoroughly modern design that captures the essential styling of the original.

From the slope of the hood and the curve of the fenders to the charming interior appointments, this new shape recalls the best-loved qualities of the old.

Available engines include 2.0 liter (gas) and 1.9 liter TDI Diesel.

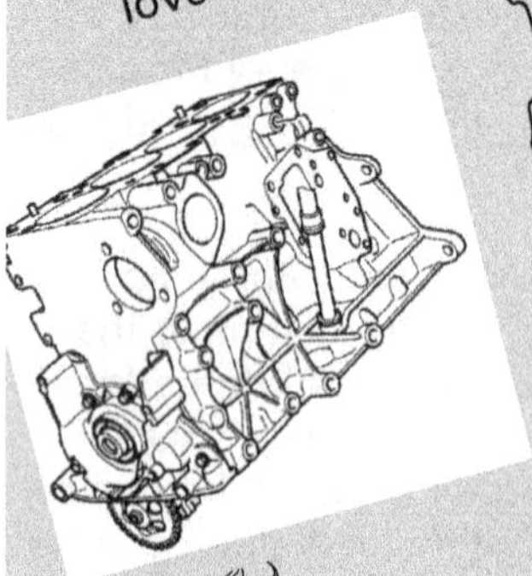

They have much in common with other current Volkswagen engines, but the New Beetle engines introduce significant new developments in engine efficiency and in overall performance.

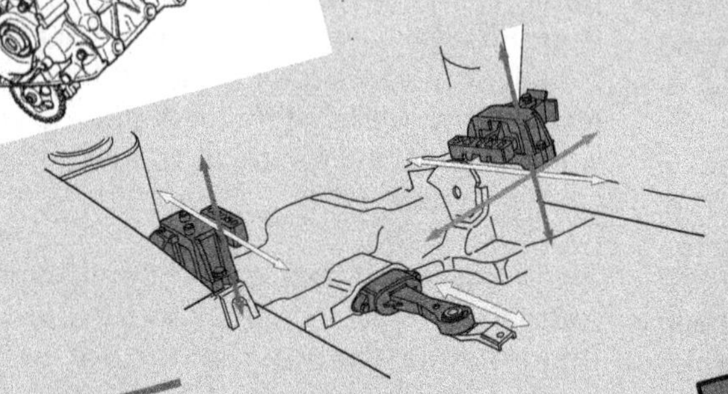

All-new "pendulum" type engine and transmission mounts deliver the best control of FWD power train noise and vibration ever in a Volkswagen.

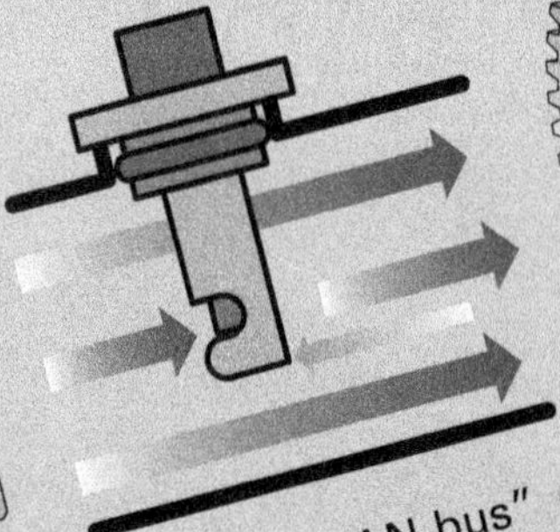

State-of-the-art Motronic and Diesel Direct Fuel Injection (DFI) systems have been refined for better performance and lower exhaust emissions.

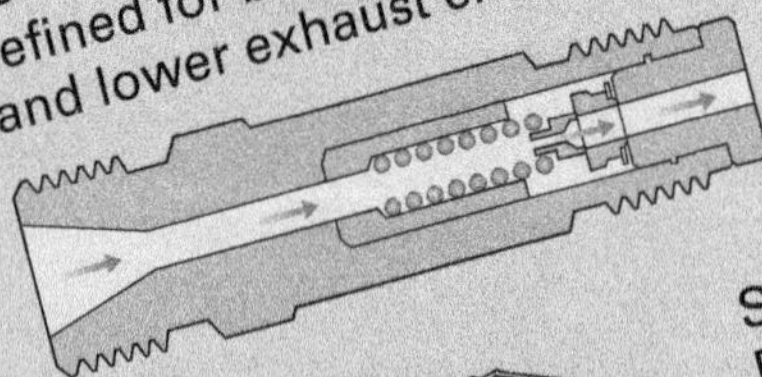

Sophisiticated "CAN-bus" electronics let the Engine Control Module (ECM), the automatic Transmission Control Module (TCM) and ABS Control Modules share important information.

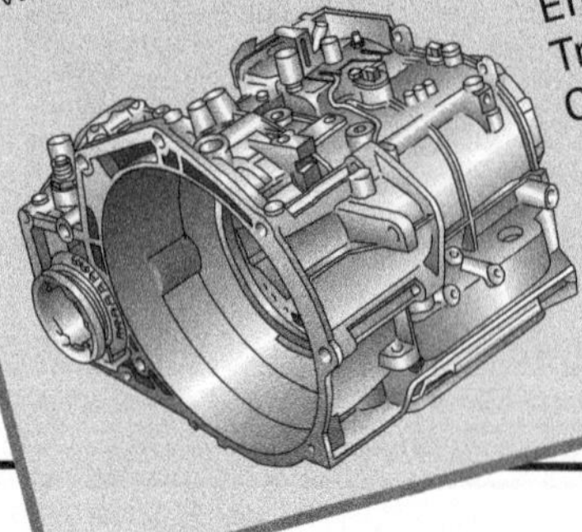

New Beetle Overview

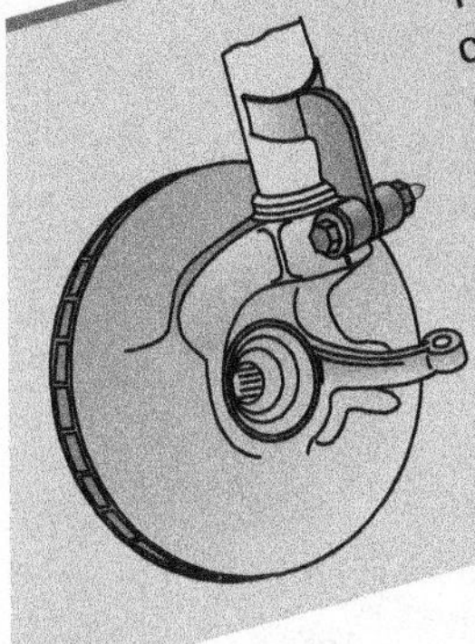

The well-proven strut and lower control arm front suspension has evolved from the Golf/Jetta.

A new strut and wheel bearing housing design, along with other enhancements, helps optimize stability, steering effort, comfort, handling and service.

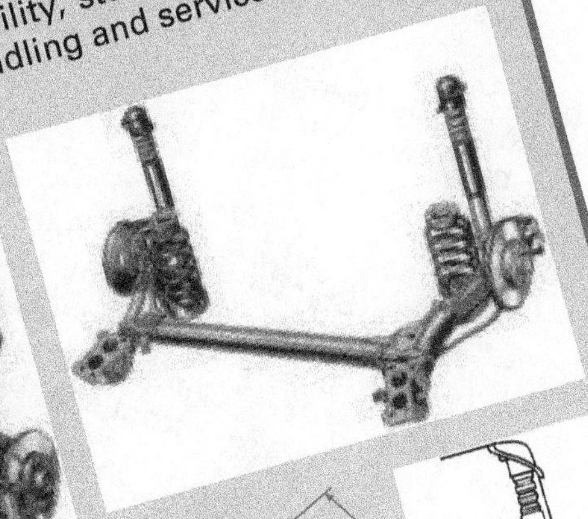

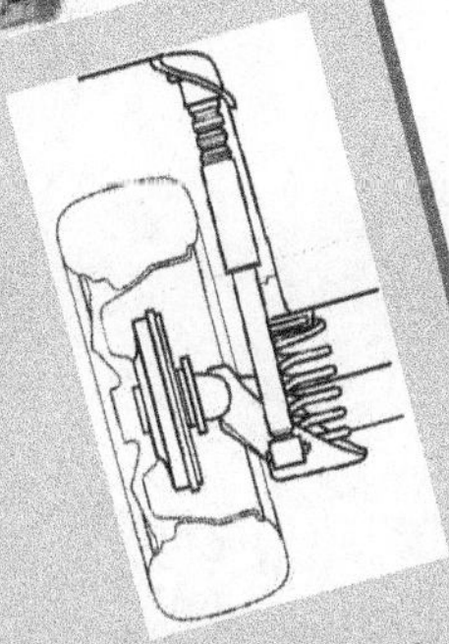

Advanced beam axle/trailing arm rear suspension with special mounts and geometry provides unparalleled control of track and toe change during cornering.

Its a clever design that keeps the springs and shock absorbers separated, increases trunk space, and reduces road noise inside.

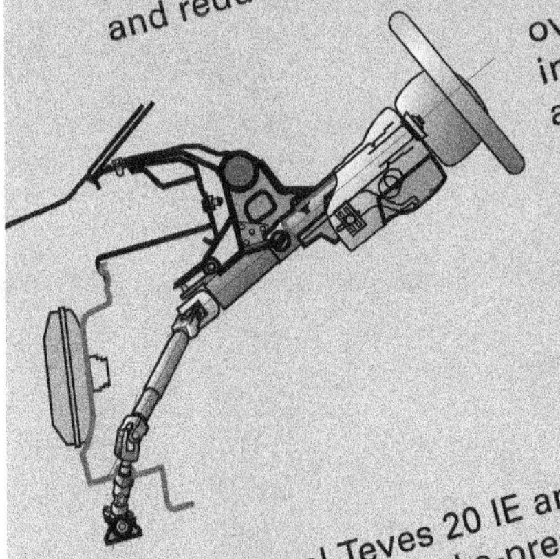

The new steering column is part of the overall engineering for collision protection, integrated with the surrounding structure and designed for controlled deformation on impact.

Tilt and telescopic adjustments, using a single lever, are standard on every New Beetle.

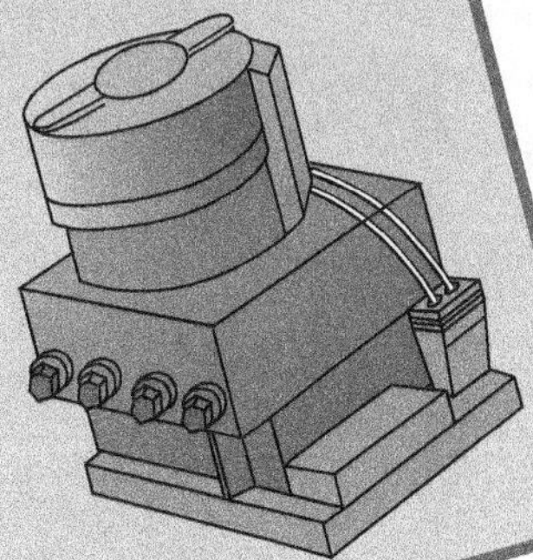

Optional 3-channel Teves 20 IE anti-lock brakes (ABS) includes an Electronic Brake-pressure Distribution (EBD) system, which eliminates the mechanical brake pressure regulator at the rear wheels.

New Beetle Overview

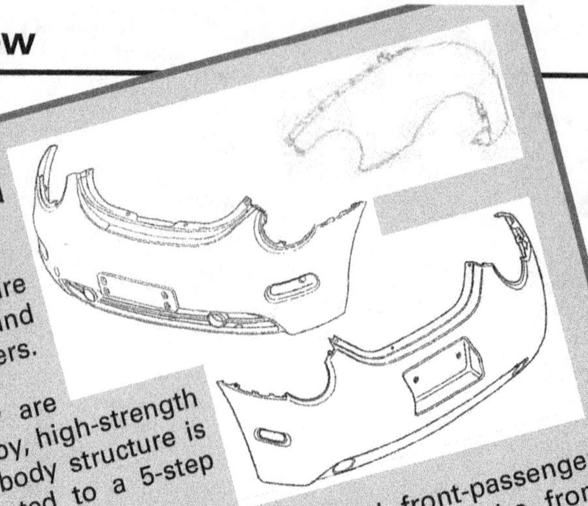

The complete body has been designed for durability and corrosion protection.

Molded plastic panels are used for all four fenders, and front and rear bumper covers.

Other parts of the body are constructed of special alloy, high-strength steel panels. The steel body structure is fully-galvanized and treated to a 5-step finishing/painting process.

Standard driver and front-passenger safety equipment includes the front airbags, pyrotechnic seat belt tensioners, and side airbags in the front seat backrests.

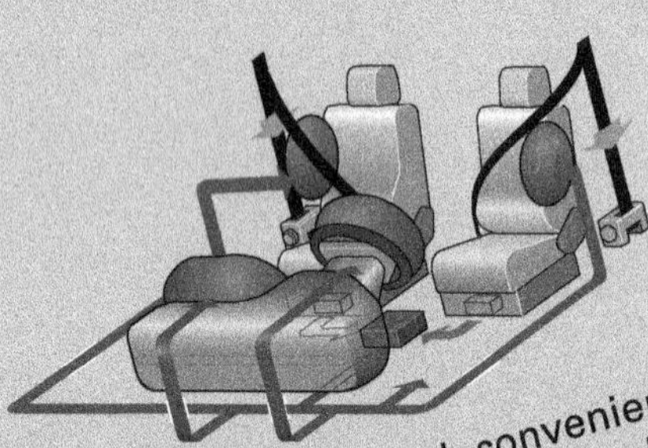

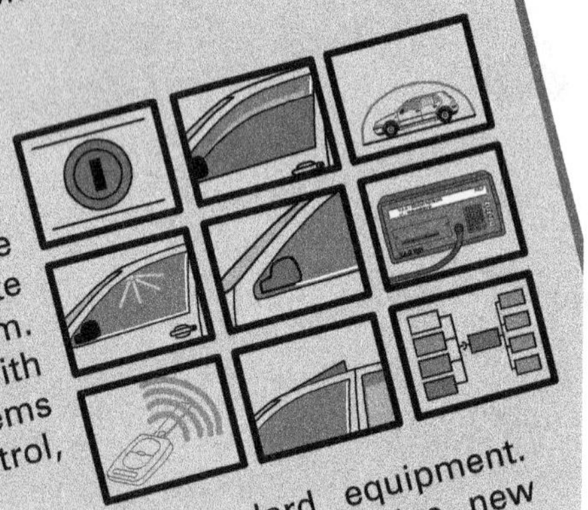

Several comfort and convenience features are standard, including central locking with remote control, interior light control, and anti-theft alarm. Optional power windows are also integrated with central locking and interior light control systems using "CAN-bus" technology for system control, communications, and On Board Diagnostic.

Air conditioning is standard equipment. The system includes an innovative new plate-type evaporator, and air recirculation flap control.

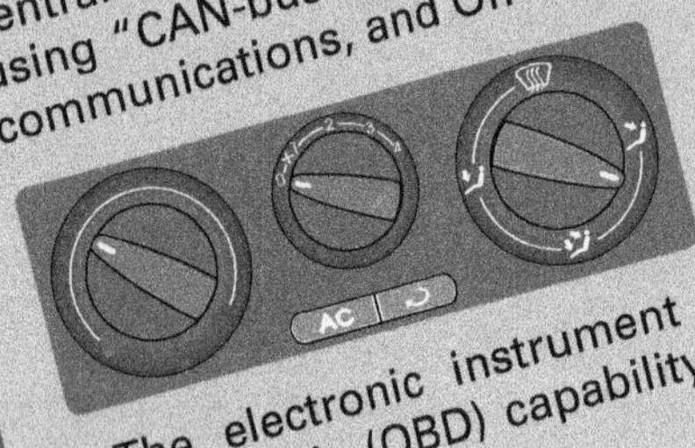

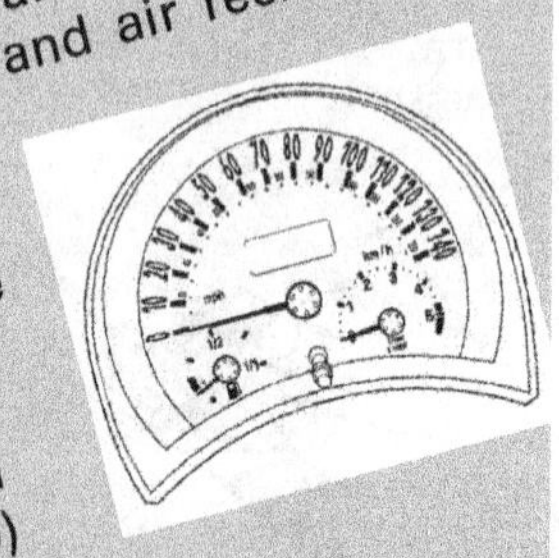

The electronic instrument cluster has On Board Diagnostic (OBD) capability, and new red and blue illumination.

A new sound system has new control features, and On Board Diagnostic (OBD) capability.

State-of-the-art construction, sophisticated power train and running gear, a vast array of comfort/convenience features, and more than a passing resemblance to one of the world's all-time great cars, make the New Beetle a great value for its time ... just like the original!

New Beetle Overview

Dimensions

Overall height A	59.5 in. (1511 mm)
Track B: front	59.6 in. (1514 mm
rear	58.7 in. (1491 mm)
Overall width C	67.9 in. (1724 mm)
Wheelbase D	98.8 in. (2512 mm)
Overall length E	161.1 in. (4089 mm)
Ground clearance	4.2 in. (102 mm)
Turning circle	32.8 (10.0 m)
Curb weight 5-spd. manual	2712 lb. (1230 kg)
4-spd. automatic	2778 lb. (1260 kg)
Drag coefficient (C_d)	0.38

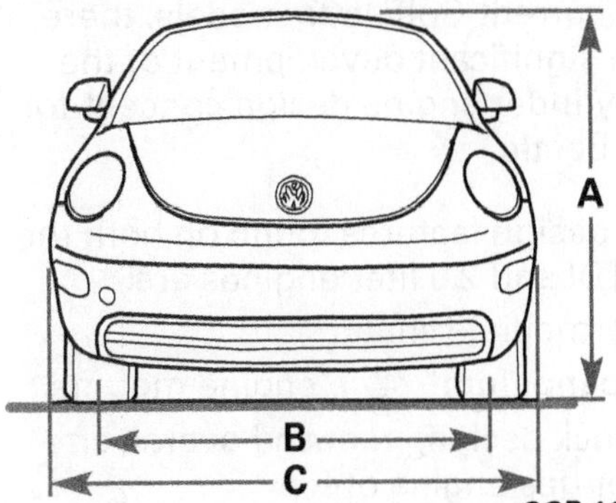

SSP 418/165

Passenger volume	84 ft^3 (2.4 m^3)
Trunk volume	12 ft^3 (0.9 m^3)
Seating capacity	Four

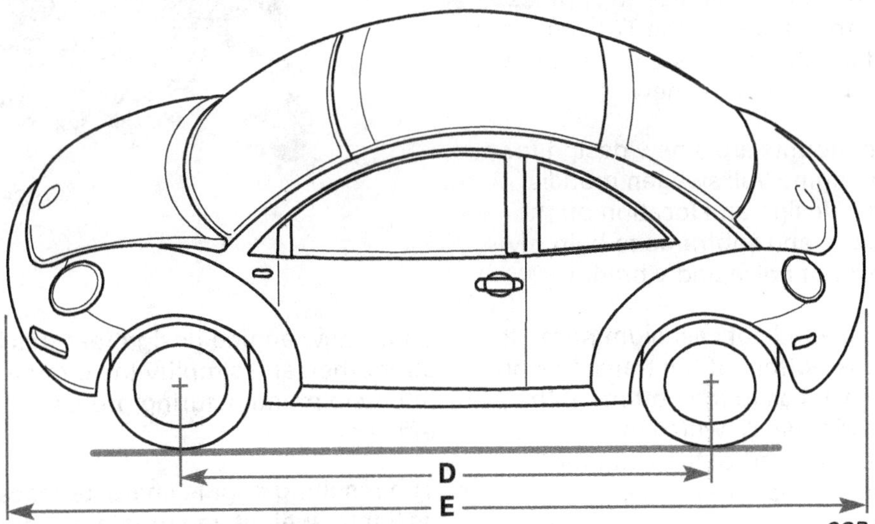

SSP 418/238

SSP 418/250a

Engines — New Design Features

The New Beetle is being introduced with a choice of two engines — the 2.0 liter (gas) engine and the 1.9 liter TDI Diesel.

Although both engines might at first look familiar, and appear to be carried over from the current Golf/Jetta models, there has been significant development of the basic 4-cylinder engine design concept for the New Beetle.

The new design features found on both the 1.9 liter TDI and 2.0 liter engines are:

- No intermediate shaft.
- New "pendulum" style engine mounts.
- Low-block design; reduced overall size (2.0 liter gas engine only).
- Cast aluminum oil pan.
- Lightweight valvetrain.
- Chain-driven oil pump.
- New coolant pump housing.
- New thermostat housing.

Both engine cylinder blocks have been designed to eliminate the intermediate shaft used on current Golf/Jetta engines. As a result, the oil pump, the TDI vacuum pump, and the 2.0 liter gasoline engine's spark-ignition system are new.

The engine mounts are a new design for transverse-engine Volkswagen models, employing a design and location on the body structure and subframe to help give better control of noise and vibration.

The oil pan is made of cast aluminum. It fastens at the bottom of the transmission housing, as well as to the bottom of the engine cylinder block, which makes the pan a more rigid part of the combined engine-transmission assembly.

The coolant pump and thermostat are both mounted in integrated housings that are cast directly into the engine cylinder block.

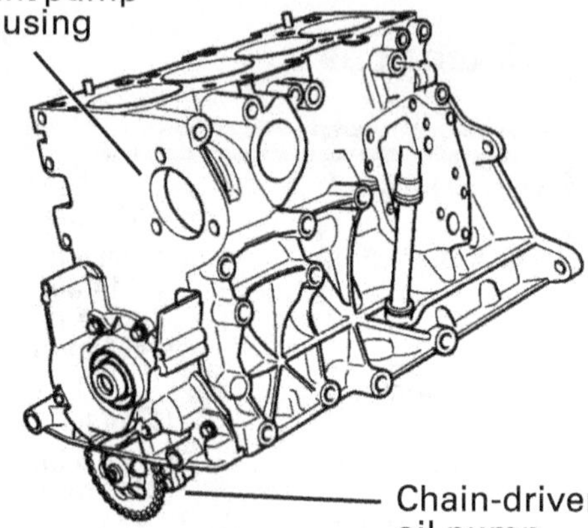

SSP 418/172

SSP 418/003

These new engine design features help strengthen and simplify the overall design, and help manufacturing processes more efficient.

The resulting engine characteristics, such as lighter weight and greater rigidity, boost overall performance and help reduce the transmission of noise and vibration into the vehicle interior.

Engines — New Design Features

Engine and transmission mounts

The new engine and transmission mount design suspends the power train assembly on its rotational axis. It works on the same principle as a pendulum.

The Transmission Mount on the left side and the Pendulum Support are bonded rubber mounts. The Engine Mount on the right side is a combination bonded rubber mount and hydraulic dampener.

CAUTION

This training module applies to only some of the vehicles covered by this manual. This information is subject to change and should be used as reference only. Always refer to the latest Volkswagen factory service information for the most up to date service, repair and training information.

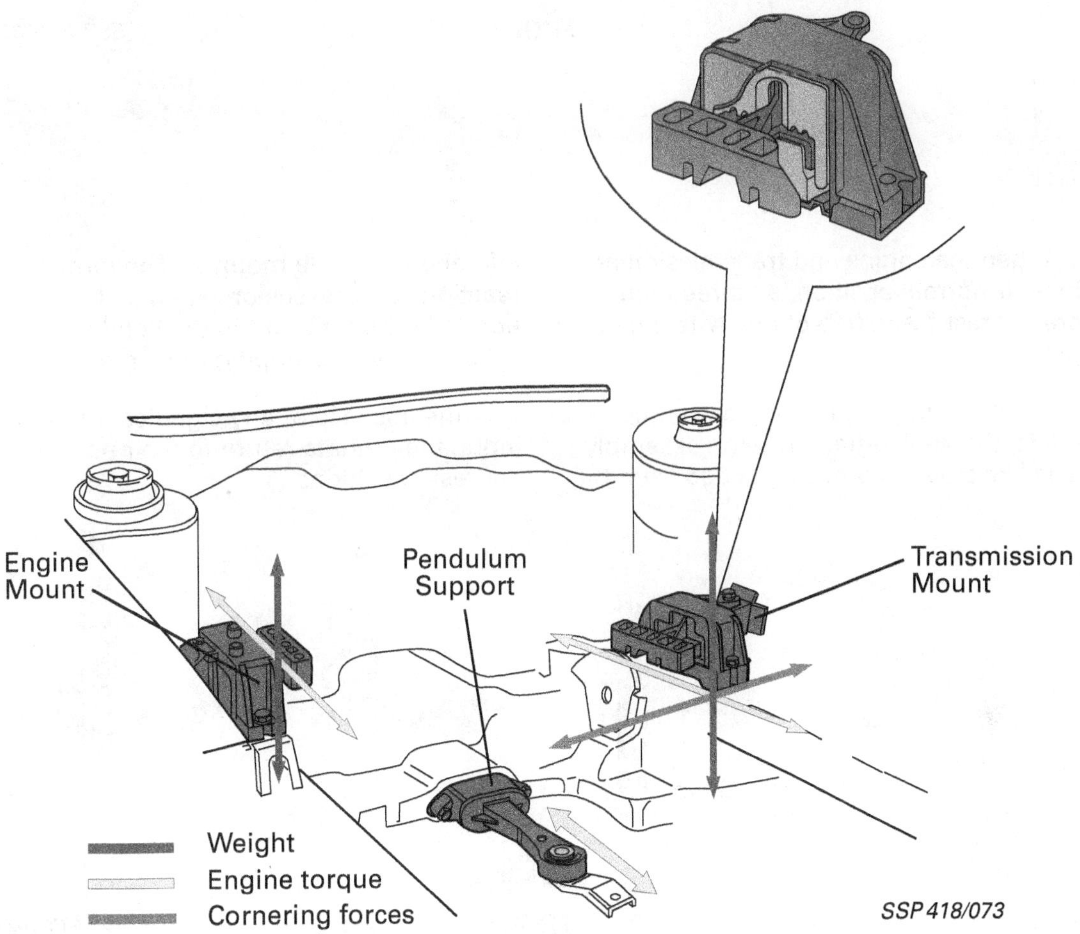

SSP 418/073

Engines — New Design Features

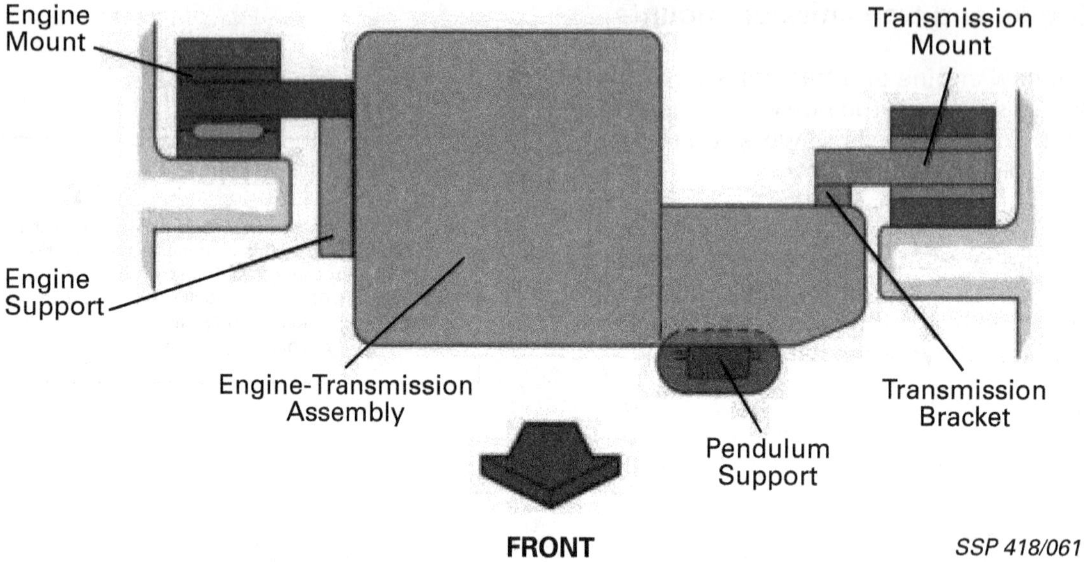

SSP 418/061

1 When the engine and transmission are in their normal position, all three mounts are "at rest," and free of any twisting or pre-load.

2, **3** When subjected to engine torque loads, the engine-transmission assembly acts like a pendulum, suspended from the left- and right-side mounts. The torque reaction compresses or expands the bonded rubber mount in the Pendulum Support below, damping movement.

4 In the absence of any significant torque input, the mounts return to their normal "at rest" positions.

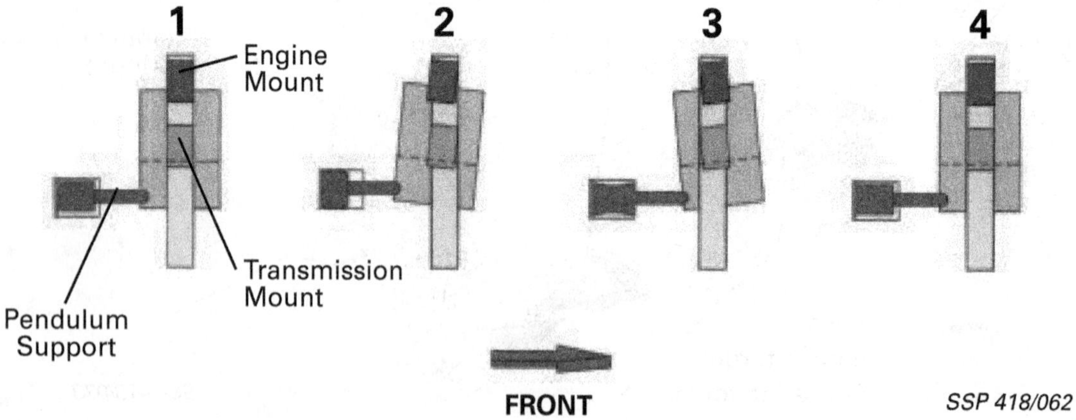

SSP 418/062

Engines – New Design Features

"Low-block" engine design
(2.0 liter gas engine only)

Overall engine height, when compared to the 2.0 liter engine used in the current Golf and Jetta, has been reduced by 18 mm. To accomodate the shorter block, connecting rod length has been reduced accordingly. This reduction in height allows for a lower hood line.

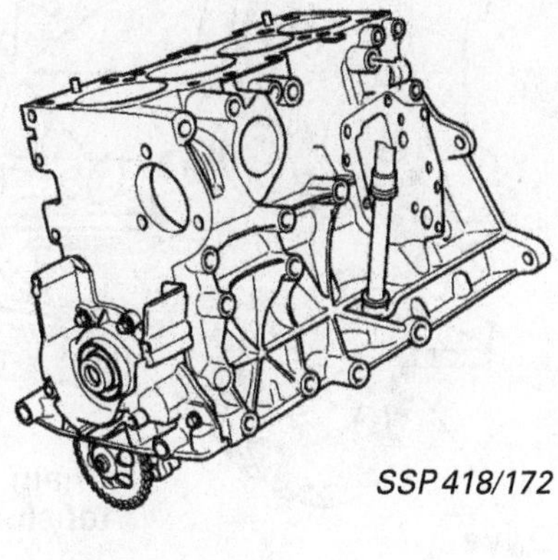

SSP 418/172

Cast aluminum oil pan

The oil pan is now made of cast aluminum, with the following features:
- Lighter weight.
- More rigid power train assembly.
- Silicone sealant replaces "conventional" gasket.

SSP 418/153

Lightweight valve train

Single-coil valve springs are used, which reduce the overall weight of the valve train.

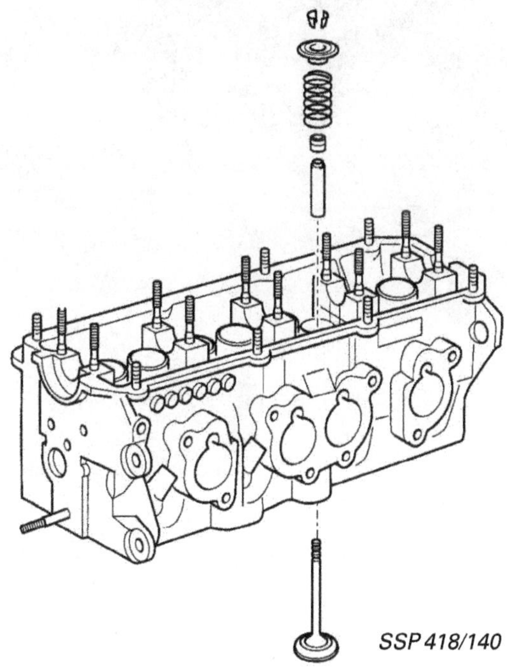

SSP 418/140

Engines – New Design Features

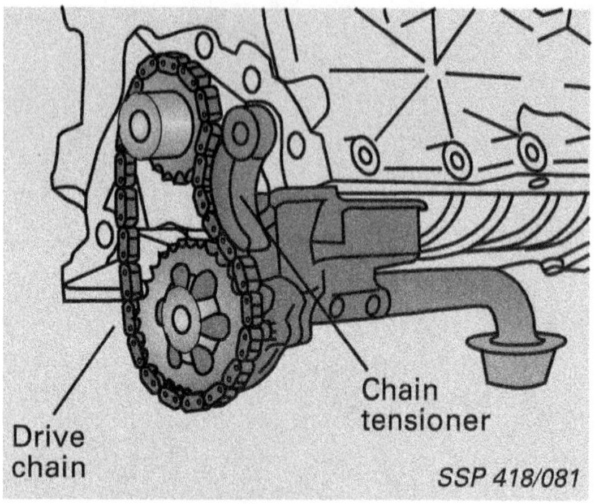

SSP 418/081

Lubrication system

One of the most significant change to the engine cylinder block is the chain-driven Oil Pump.

A spring-loaded chain tensioner takes up any slack in the chain.

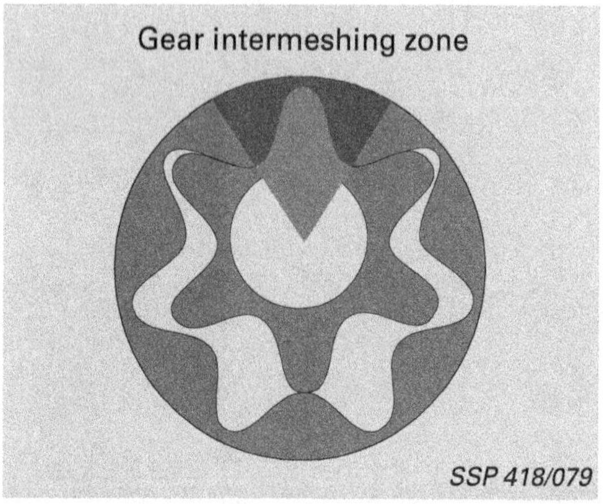

SSP 418/079

Oil Pump

The Oil Pump is an internal gear pump. The advantages are:
- Small gear intermeshing zone to minimize contact area for less friction.
- Large operating volume and greater efficiency.
- Fewer moving parts.

Oil flow circuit

The complete oil flow circuit is shown on the opposite page. The significant features are:
- Oil Pressure Control Valve is located downstream from the Oil Filter.
- Engine oil pressure is monitored by a single Oil Pressure Switch.
- Oil Return Cut-off Valve is integrated with the Oil Filter connection.

Engines – New Design Features

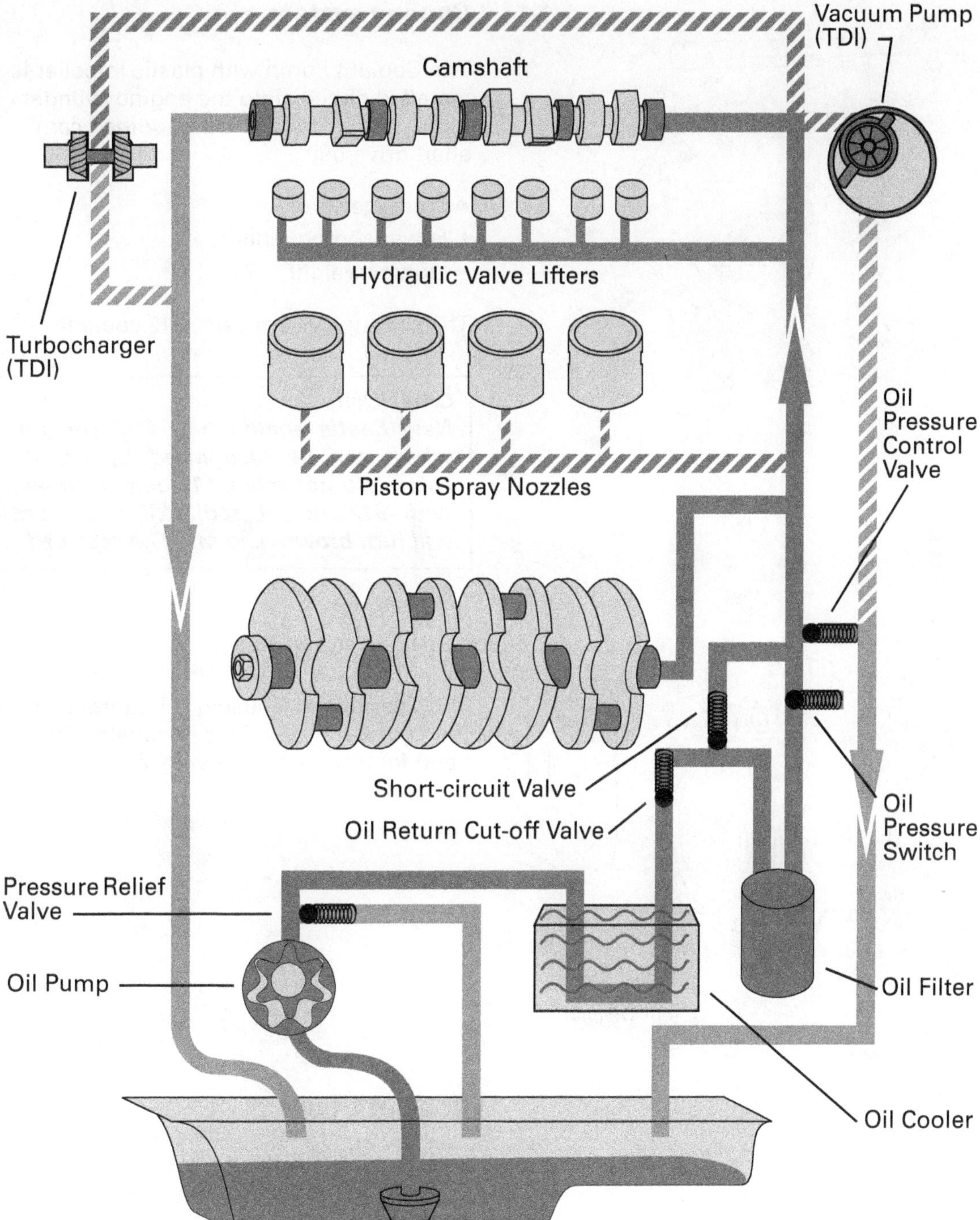

SSP 418/080

Engines — New Design Features

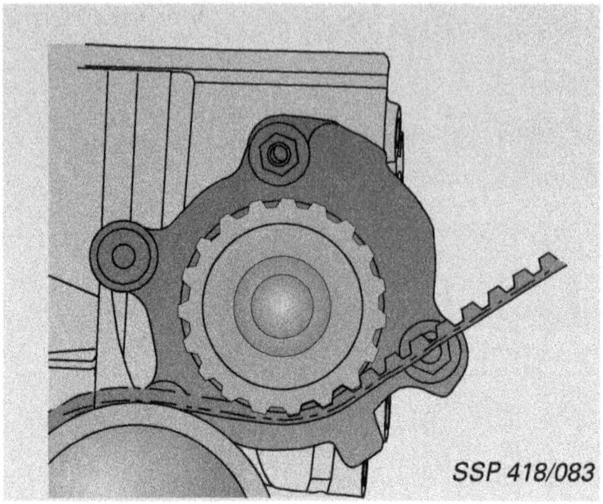

SSP 418/083

Cooling system

The Coolant Pump with plastic impeller is installed directly into the engine cylinder block, and is driven by the toothed cam-shaft drive belt.

Advantages are:

- Fewer components.
- Lighter weight.

The cooling system uses G12 coolant.

CAUTION!
New Beetle engines use G12 coolant, which can be recognized by its red color. Do not mix G11 coolant (green) with G12 coolant (red). Mixed coolant will turn brown, and must be replaced.

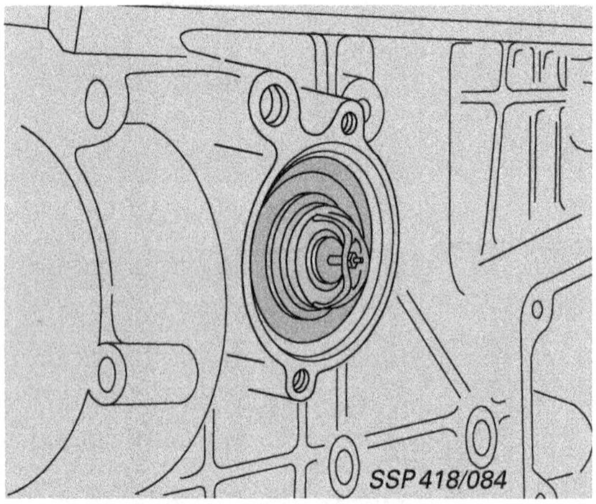

SSP 418/084

Coolant Thermostat

The Thermostat Housing is integral with the Cylinder Block. This eliminates the need for a separate housing, and saves weight.

Coolant circuit

The coolant circuit is shown in schematic form on the opposite page. It reflects the coolant flow in the new engine cylinder block.

Engines – New Design Features

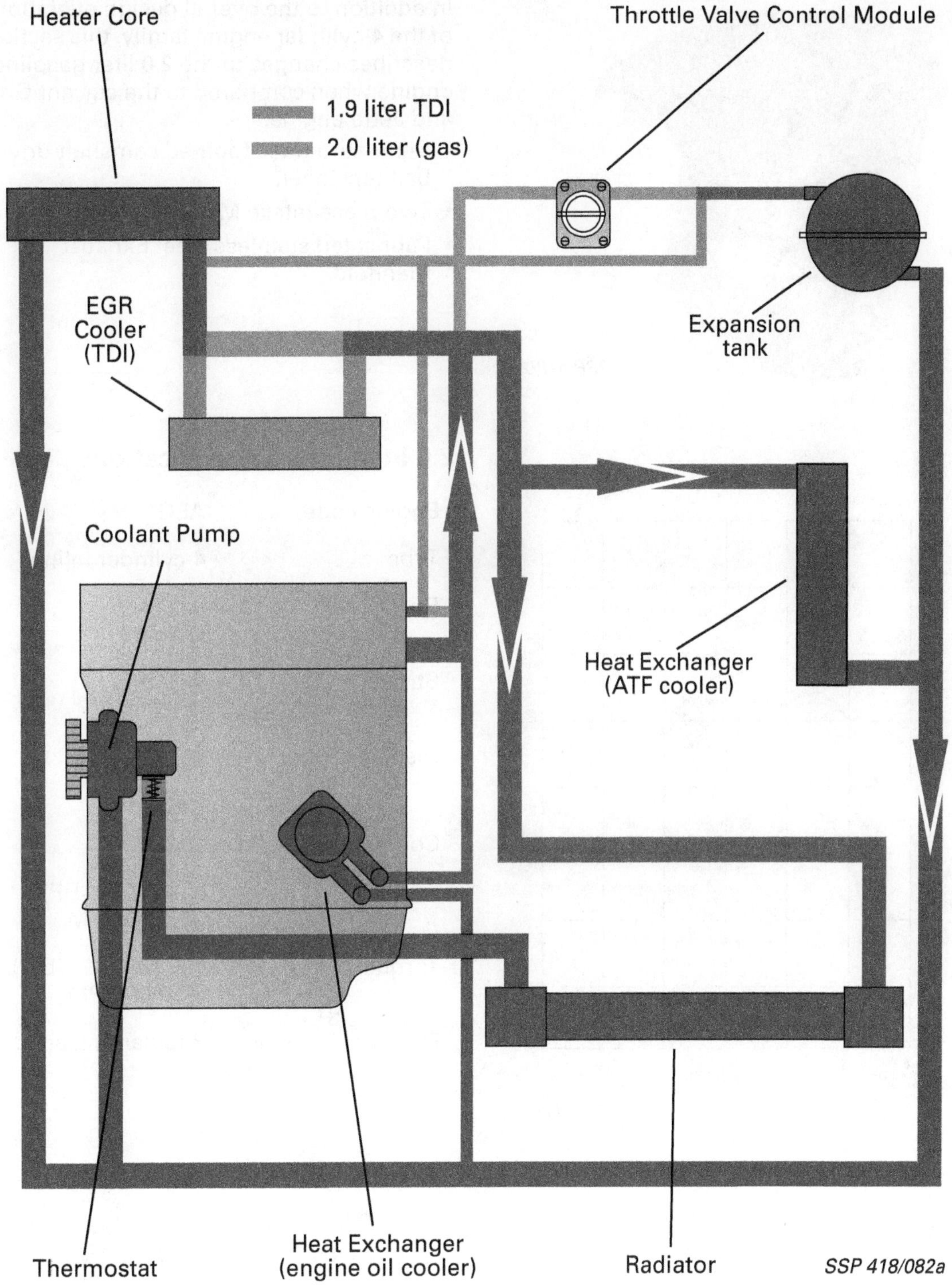

SSP 418/082a

2.0 Liter Engine – AEG

SSP 418/219

In addition to the overall design evolution of the 4-cylinder engine family, this section describes changes to the 2.0 liter gasoline engine when compared to the current Golf and Jetta engine:

- Semi-automatic toothed camshaft drive belt tensioner.
- Two-piece Intake Manifold.
- Fabricated stainless steel Exhaust Manifold.

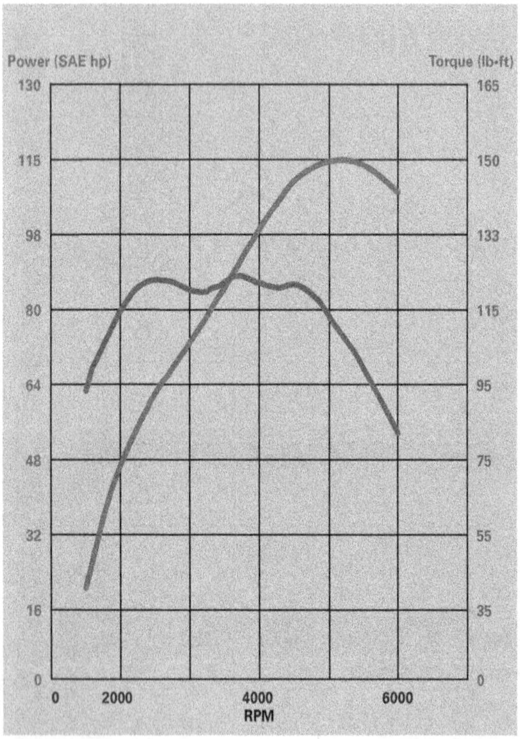

SSP 418/002

2.0 liter engine specifications

Engine code	AEG
Type	4-cylinder inline
Bore	82.5 mm (3.25 in)
Stroke	92.8 mm (3.65 in)
Displacement	1984 cm^3 (121.1 in^3)
Compression ratio	10.0 : 1
Power	85 kW (115 HP) @ 5200 RPM
Torque	165 Nm (122 lb ft) @ 2600 RPM
Fuel requirement	Regular unleaded

2.0 Liter Engine – AEG

Semi-automatic toothed camshaft drive belt tensioner

The 2.0 liter engine is equipped with a semi-automatic belt tensioner. A spring in the tensioner pre-loads the belt. Tension is adjusted according to indicator marks on the tensioner.

> ***CAUTION!***
> ***Belt tension should only be adjusted on a cold or slightly warm engine.***

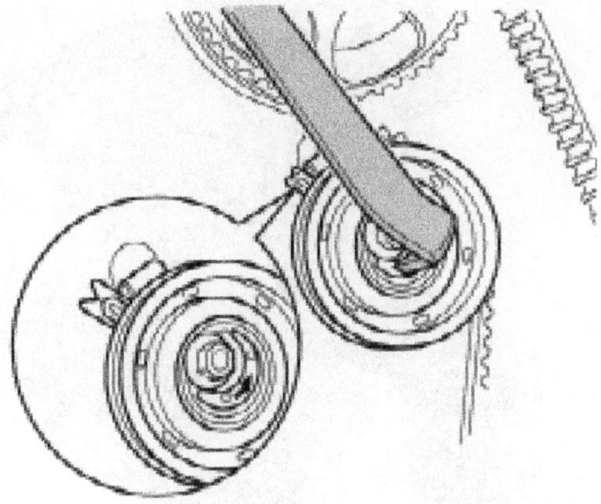

SSP 418/174

Intake Manifold

The 2-piece aluminium Intake Manifold was designed to allow a lower hood line. It weighs less than the manifold used on the current Golf and Jetta 2.0 liter engine.

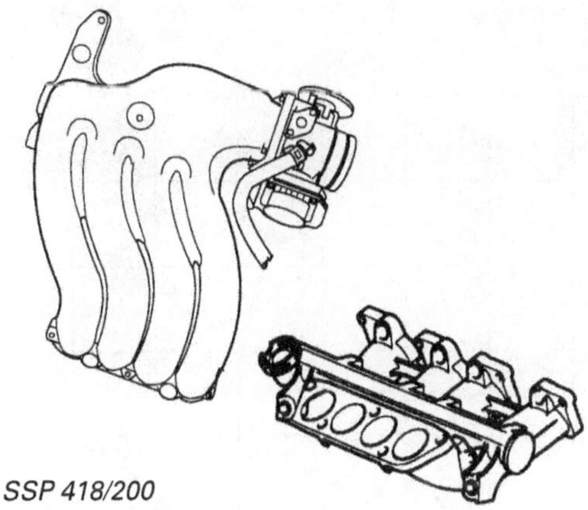

SSP 418/200

Exhaust Manifold

The Exhaust Manifold is fabricated from stainless steel tubing.

A cast aluminum Heat Shield, mounted above the Exhaust Manifold, includes a mounting for the Secondary Air Injection (AIR) Valve.

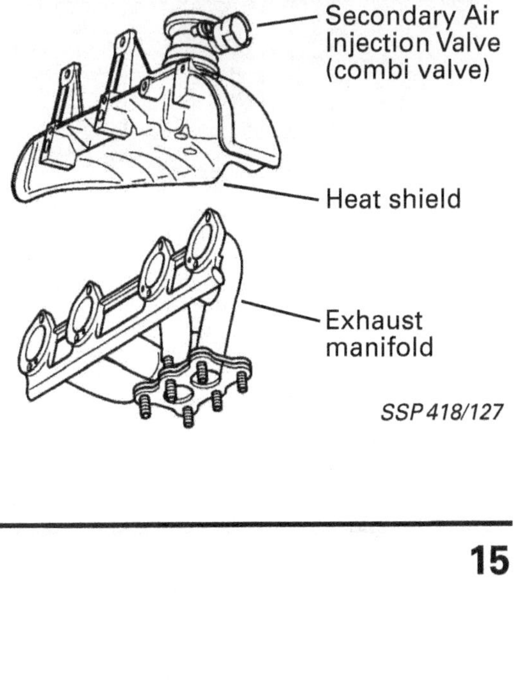

SSP 418/127

1.9 Liter TDI Diesel Engine — ALH

SSP 418/003

Overview

In addition to the overall design evolution of the 4-cylinder engine family, this latest 1.9 liter TDI Diesel engine includes the following updates and new features:

- Pistons with revised top ring position.
- Magnesium Cylinder Head Cover with new gasket and new mounting points around outer edge.
- Camshaft-driven Vacuum Pump.
- Cartridge-type Oil Filter.

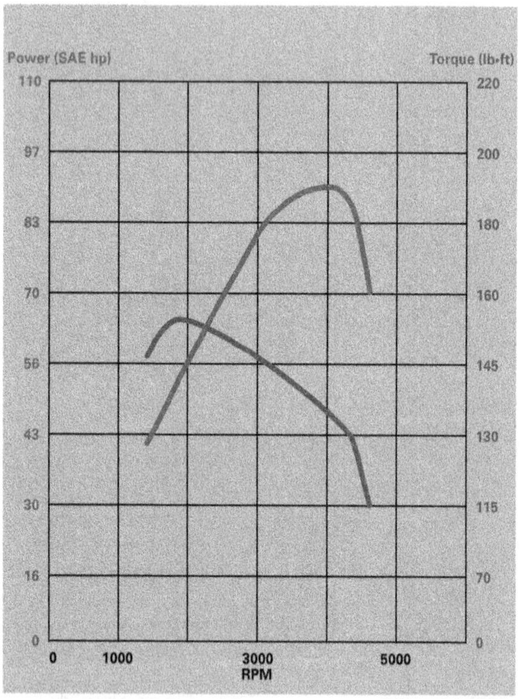

SSP 418/004

1.9 liter TDI engine specifications

Engine code	ALH
Type	4-cylinder inline
Bore	79.9 mm (3.13 in)
Stroke	95.5 mm (3.76 in)
Displacement	1896 cm^3 (115.7 in^3)
Compression ratio	19.5 : 1
Power	66 kW (90 HP) @ 4000 RPM
Torque	202 Nm (149 lb ft) @ 1900 RPM
Fuel requirement	Diesel, 45 cetane

1.9 Liter TDI Diesel Engine — ALH

Pistons

Pistons for the New Beetle TDI engine are a new design compared to those for the TDI engine in current Volkswagen models. The distance from top of the piston to the first ring land has been reduced, from 9 mm to 6 mm.

This change helps reduce particulate and hydrocarbon exhaust emissions.

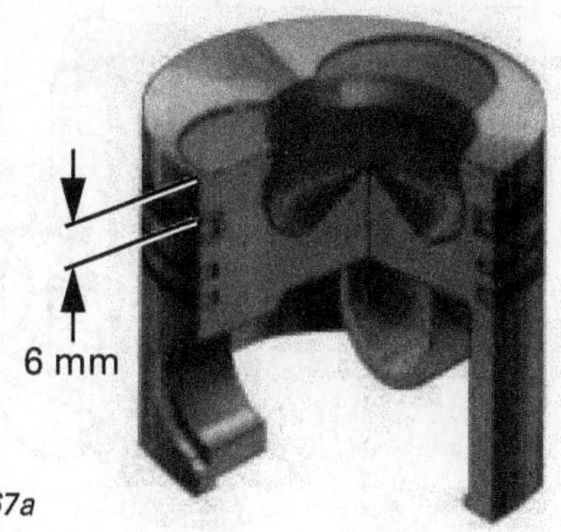

SSP 418/067a

Cylinder head cover

The cast magnesium cylinder head cover is mounted with a series of fasteners around the outer perimeter, instead of the 3 center bolts on the previous TDI engine.

Instead of a conventional separate gasket (inset, left), the gasket is bonded to the cover (inset, right). The rigid cover and the positively located gasket give a reliable seal between the cover and cylinder head.

A wire mesh oil separator inside the cover helps provide better oil separation for crankcase ventilation.

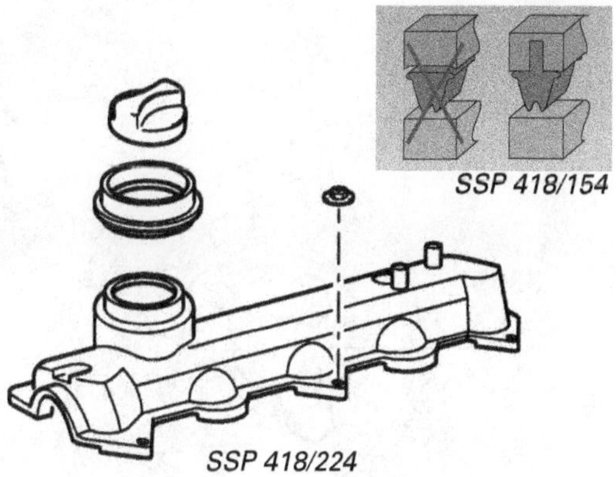

SSP 418/154

SSP 418/224

Toothed camshaft drive belt

The toothed camshaft drive belt has two guide pulleys to improve belt wrap around the sprockets. More belt wrap reduces belt slippage and helps increase belt life

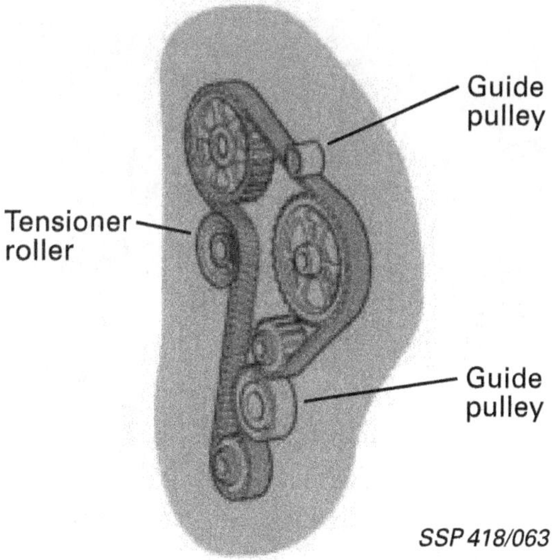

SSP 418/063

1.9 Liter TDI Diesel Engine – ALH

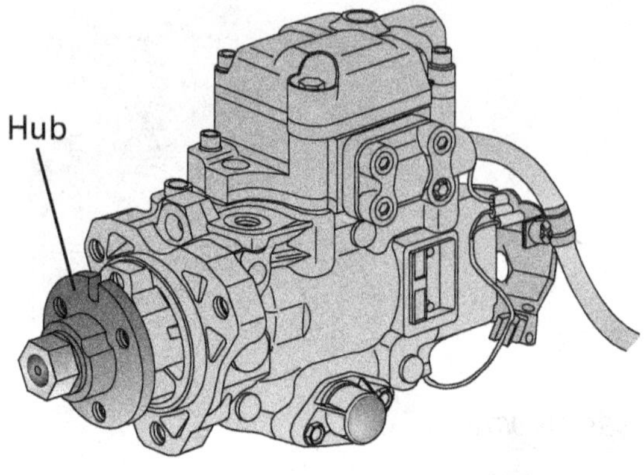

SSP 418/088

TDI Diesel Direct Fuel Injection (DFI) pump

The TDI Diesel DFI pump sprocket bolts to a separate hub, which is pressed onto the pump drive shaft.

A new special tool locks the hub in place for replacing the toothed belt.

The three sprocket bolts can be loosened so the sprocket can move slightly on the hub when re-installing the belt.

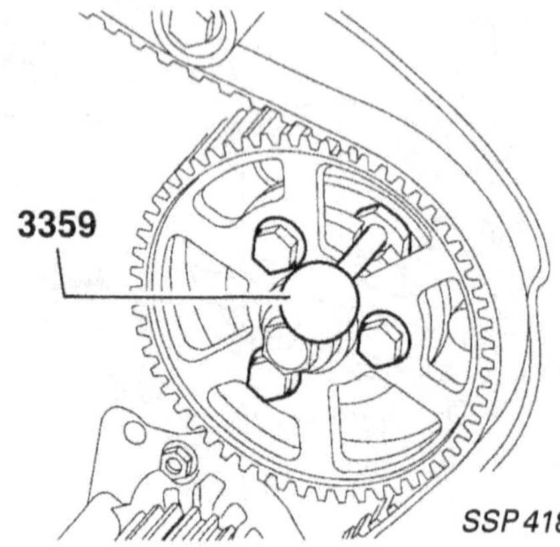

SSP 418/198

CAUTION!
Do not loosen or remove the hub from the Diesel DFI pump. It is pressed on only and NOT keyed to the pump drive shaft! If removed, it will no longer be correctly indexed to the pump.

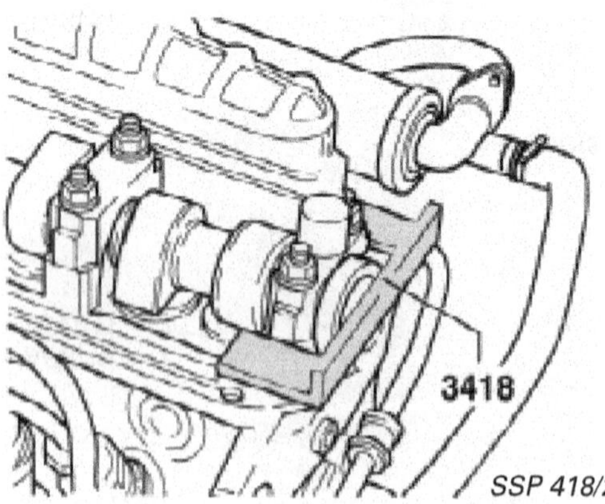

SSP 418/234

3418 Setting Bar

The engine cylinder head and end-most camshaft bearing cap have been designed to provide a mount for the vacuum pump.

This requires a new setting bar for use when replacing and adjusting the toothed camshaft drive belt. For this new TDI engine version, setting bar 3418 is used to precisely establish camshaft position.

1.9 Liter TDI Diesel Engine — ALH

Cartridge-type Oil Filter

In the interest of environmental friendliness, a replaceable Oil Filter Element is used on the New Beetle TDI engine. The element is removed from the top of the upright Filter Housing.

Replacing the Oil Filter Element is easier and neater with this design. When the cap is unscrewed from the top of the Filter Housing, it also opens a drain that allows oil in the housing to drain back into the oil pan before removing the element.

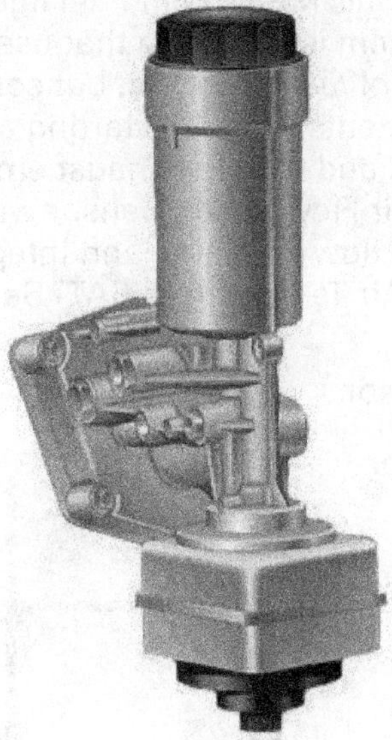

SSP 418/011

Vacuum Pump

The Vacuum Pump is driven directly by the camshaft. It is a vane-type pump, with a higher output than the current Golf/Jetta TDI pump.

A Vacuum Pump is required on TDI models because the Turbo-Diesel engine does not produce the manifold vacuum necessary to operate auxiliary components.

A consistent vacuum source is required to to operate:

- Vacuum Brake Booster.
- Changeover Valve for Intake Manifold Flap -N239-.
- EGR Valve -N18-.

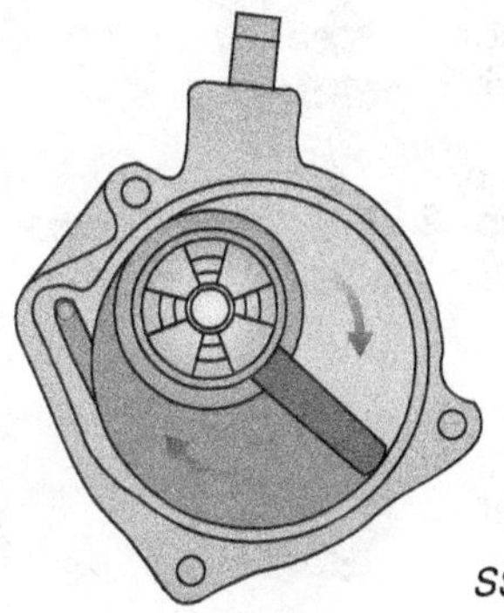

SSP 418/092

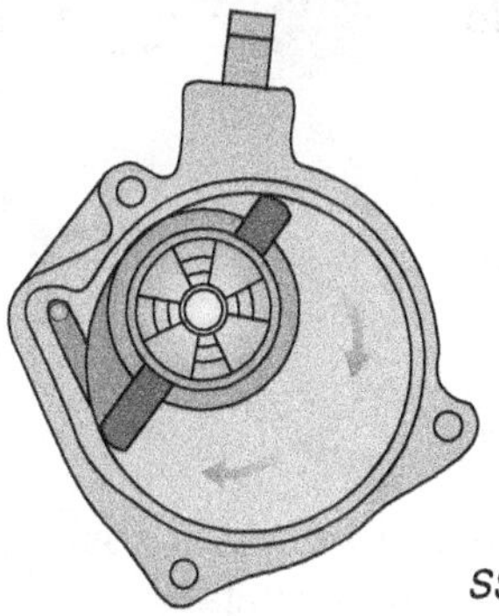

SSP 418/093

Motronic 5.9.2 Engine Management System

The Motronic Multi-point Fuel Injection (MFI) system is similar to that used on the 2.0 liter Golf/Jetta models, but some enhancements improve starting and fuel economy, and reduce exhaust emissions:

- Mass Air Flow (MAF) Sensor with reverse flow detection and integral Intake Air Temperature (IAT) Sensor.
- Camshaft Position (CMP) Sensor.
- Distributorless Ignition.
- Throttle Valve Control Module supports ECM operation of Cruise Control.
- Enhanced Evaporative Emission (EVAP) control system.

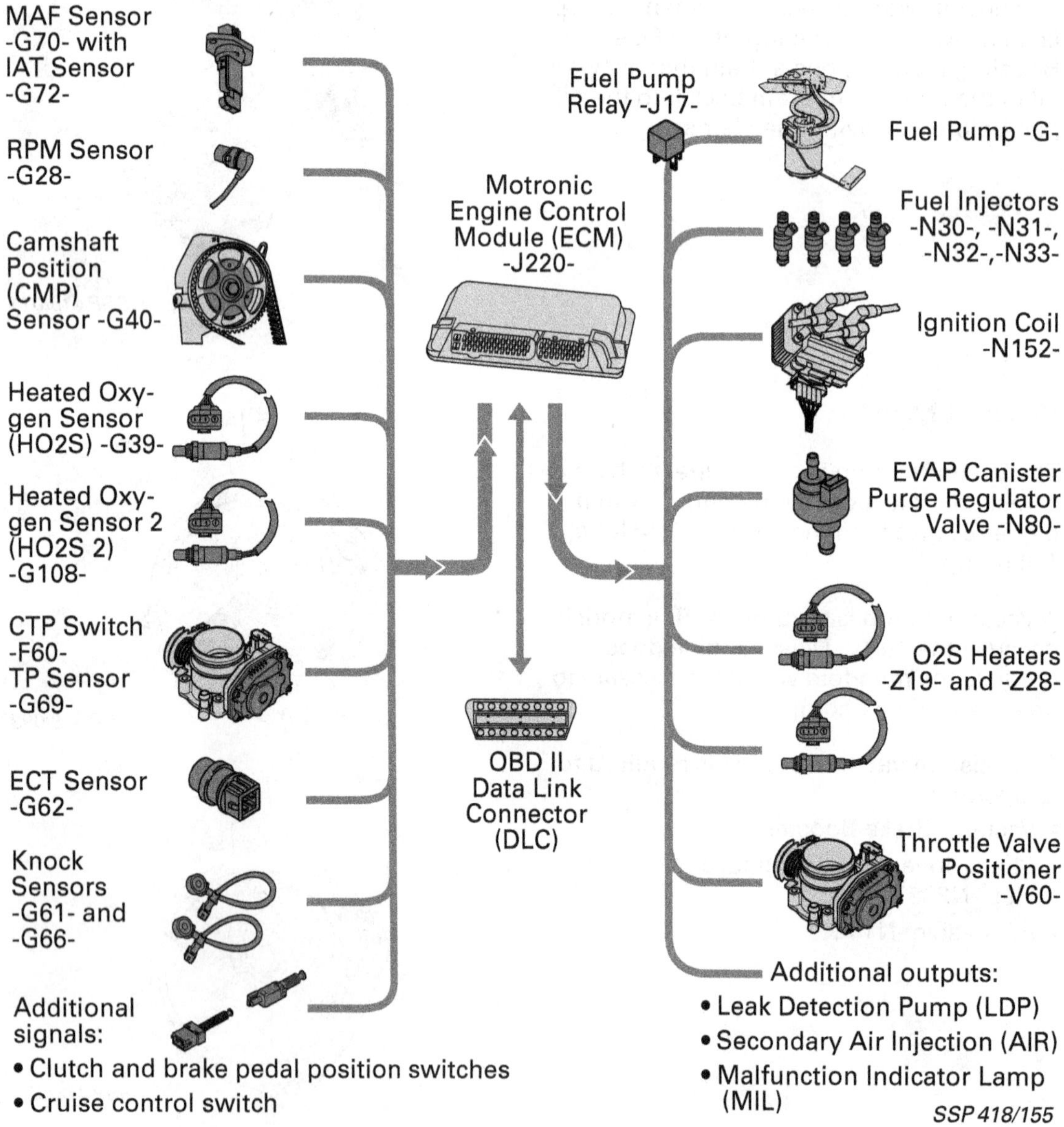

SSP 418/155

Motronic 5.9.2 Engine Management System

Mass Air Flow (MAF) Sensor

The MAF Sensor is a new hot-film sensor that more accurately measures intake air flow. It allows more precise control of the air/fuel mixture, reducing harmful exhaust emissions and improving fuel economy.

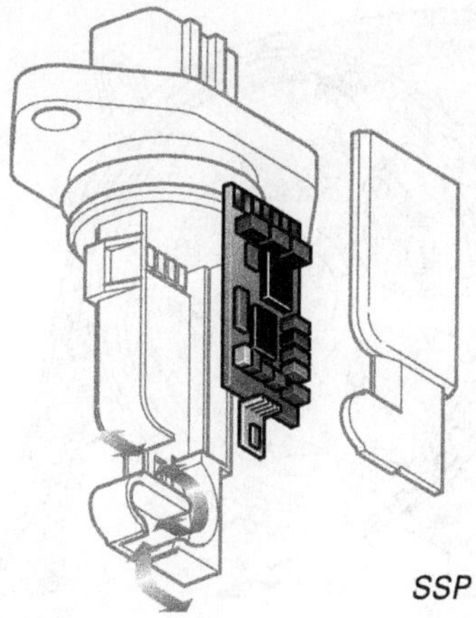

SSP 418/006

The new sensor element is encased in a glass membrane for better protection from contaminants in the intake air flow—an improvement over other MAF Sensors, which are vulnerable to contamination from oil vapor and moisture.

The new MAF Sensor actually uses two sensors, T_1 and T_2, to measure upstream and downstream of a heating element. With the engine running, normal intake air flow results in T_1 upstream of the heating element being cooler than T_2 downstream. The ECM evaluates the difference in the two signals to calculate mass air flow.

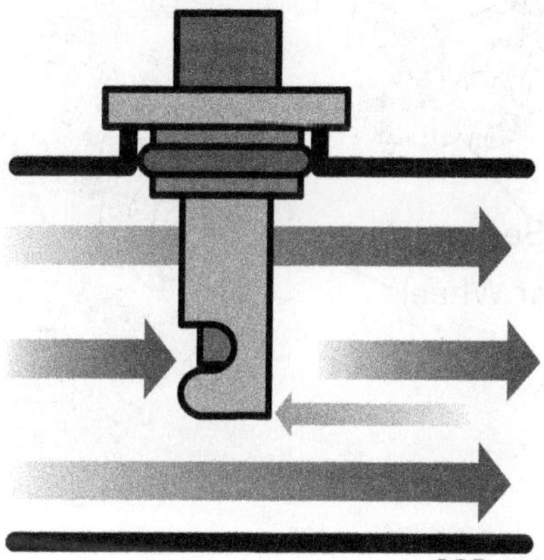

SSP 418/005

A significant new feature is Reverse Flow Detection. Intake air turbulence and complex flow dynamics, caused in part by the opening and closing of the intake valves, mean there is some reverse air flow in the Intake Air Ducting. Other MAF sensors will measure intake air flow regardless of direction, with a result that is slightly different than actual mass air flow.

In the case of reverse air flow, T_2 is now "upstream" of T_1. This has the opposite effect on the overall measurement, and allows the ECM to factor out reverse flow data from its mass air flow calculations.

Intake Air Temperature (IAT) Sensor

The IAT Sensor is now integrated with the Mass Air Flow (MAF) Sensor. It is mounted on the sensor housing that projects into the Intake Air Duct.

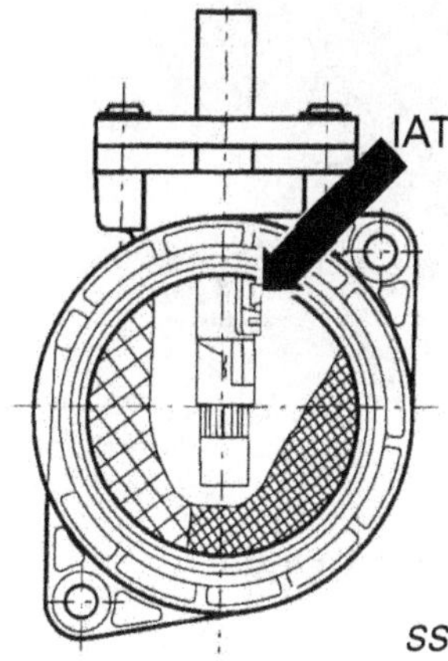

SSP 418/197

Motronic 5.9.2 Engine Management System

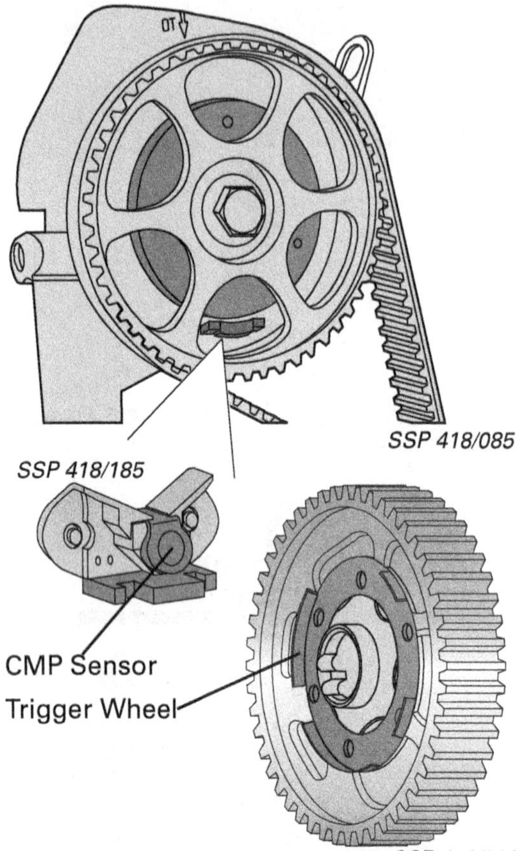

SSP 418/085

SSP 418/185

SSP 418/210

Camshaft Position (CMP) Sensor

The Intermediate Shaft and the Distributor have both been eliminated on the AEG engine. This required the CMP Sensor to be moved to a new position. It is now mounted behind the camshaft drive belt sprocket, with the Trigger Wheel attached to the back of the sprocket.

The four Trigger Wheel windows—two wide and two narrow—pass the sensor in a distinctive sequence that in turn produces a distinct pulse width pattern for each 90° of camshaft rotation.

The ECM identifies Camshaft Position and determines the Fuel Injection and Ignition sequence before the engine has completed 1/2-revolution.

This results in improved engine starting and reduced cold-start emissions.

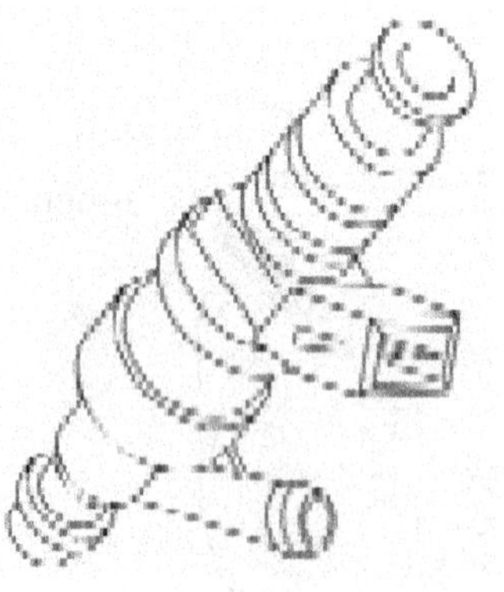
SSP 418/196

Air-shrouded Fuel Injectors

Intake Manifold Vacuum draws air in and directs it past each Fuel Injector tip to help atomize the fuel more thoroughly, for more efficient combustion and reduced exhaust emissions. This is especially effective at low RPM, so idle quality is improved. The system uses measured Intake Air, drawn in downstream of the MAF Sensor.

The air is supplied by a line from the Intake Air Duct, through a manifold near the Fuel Rail, to passages at the base of each Fuel Injector.

Motronic 5.9.2 Engine Management System

ECM-operated Cruise Control

The Cruise Control system, available as an option on the 2.0 liter New Beetle, is a new system that no longer uses a vacuum pump and vacuum actuator to control throttle position.

The Motronic ECM operates the Throttle Valve Control Module. The Throttle Valve Positioner used for Idle Air Control has an operating range that allows it maintain the vehicle speed that has been selected by the driver.

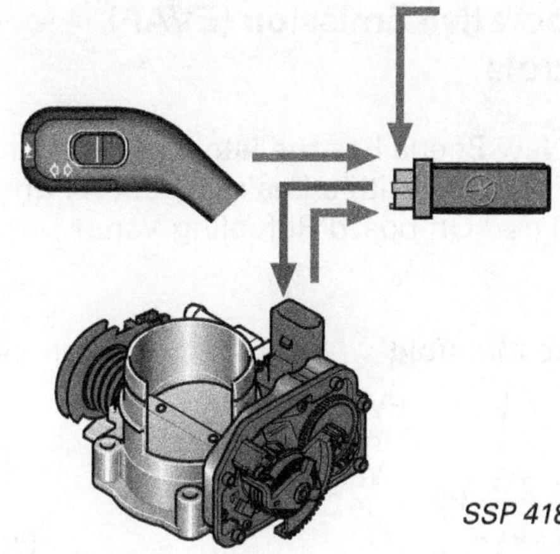

SSP 418/032

Clutch Pedal Position (CPP) Switches

Two Clutch Pedal Position (CPP) Switches are used on vehicles with cruise control. CPP Switch -F194- operates Locking Relay for Starter -J434- (starter interlock).

A second switch provides a Clutch Pedal Position input to the Motronic ECM to disengage the Cruise Control. It also cancels the Throttle Closing Dampener function.

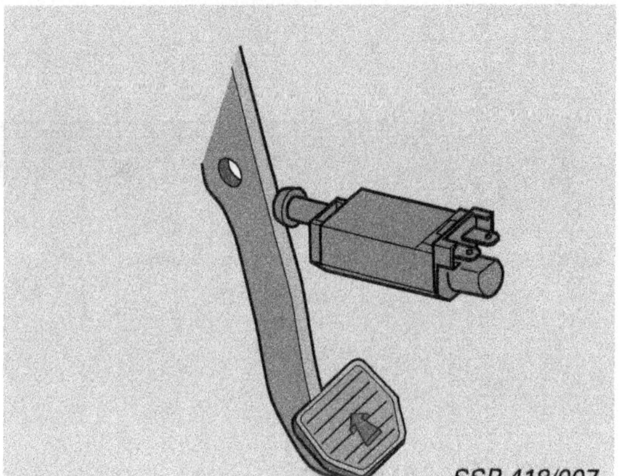

SSP 418/007

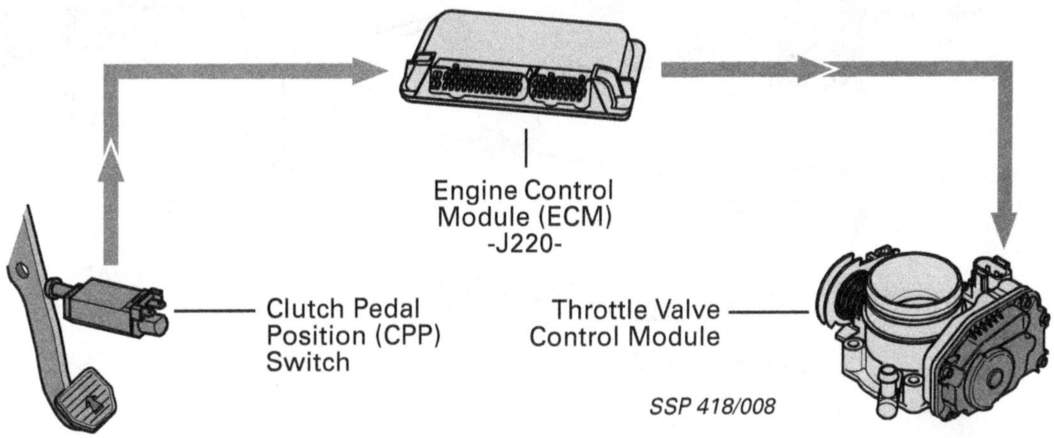

SSP 418/008

Motronic 5.9.2 Engine Management System

Evaporative Emission (EVAP) controls

The New Beetle has the latest generation EVAP system, with a Leak Detection Pump (LDP) and On-board Refueling Vapor Recovery (ORVR) system mounted in the right-rear wheelhousing. A system overview and normal venting are shown below.

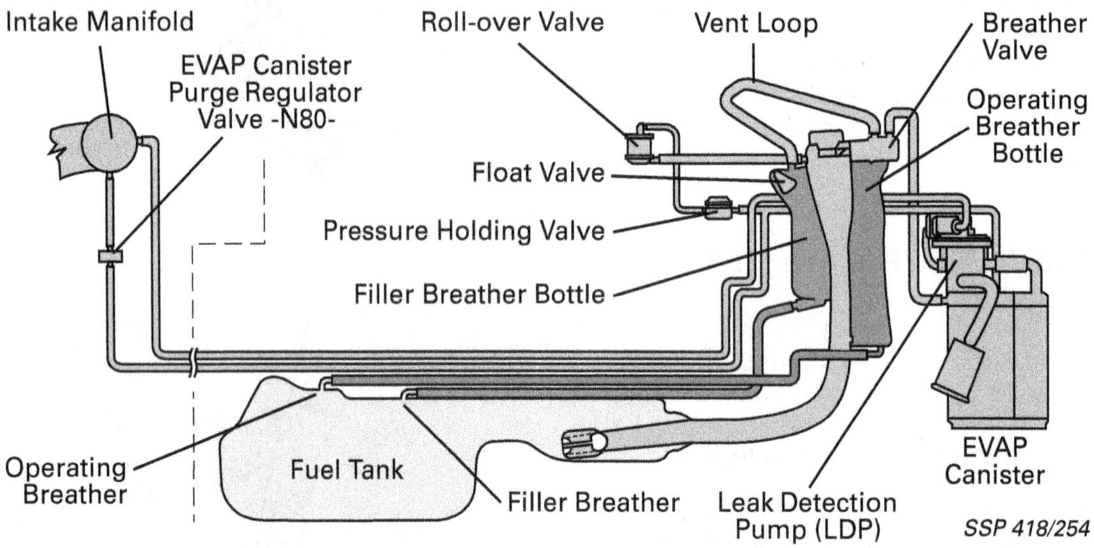

SSP 418/254

Normal operation

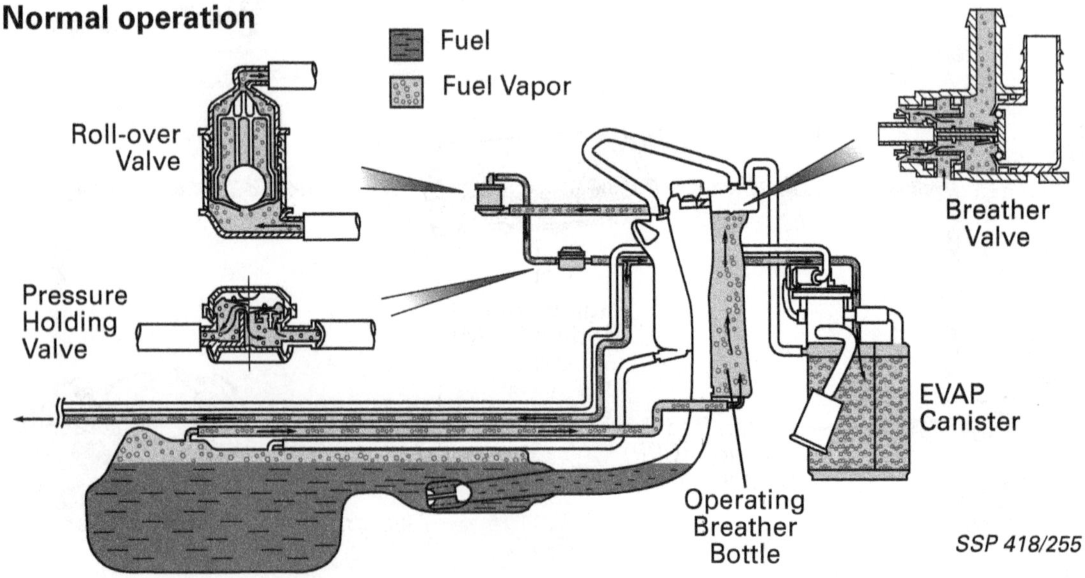

SSP 418/255

Motronic 5.9.2 Engine Management System

ORVR system operation

The On-board Refueling Vapor Recovery (ORVR) system is a type of Evaporative Emission (EVAP) control system required for new cars, starting in model year 1998. When refueling, vapor in the Fuel Tank is displaced by incoming fuel, and this fuel vapor, which used to be vented to atmosphere, must be recovered. The ORVR system requires a larger EVAP Canister, larger Vent Lines, and a new Breather Valve.

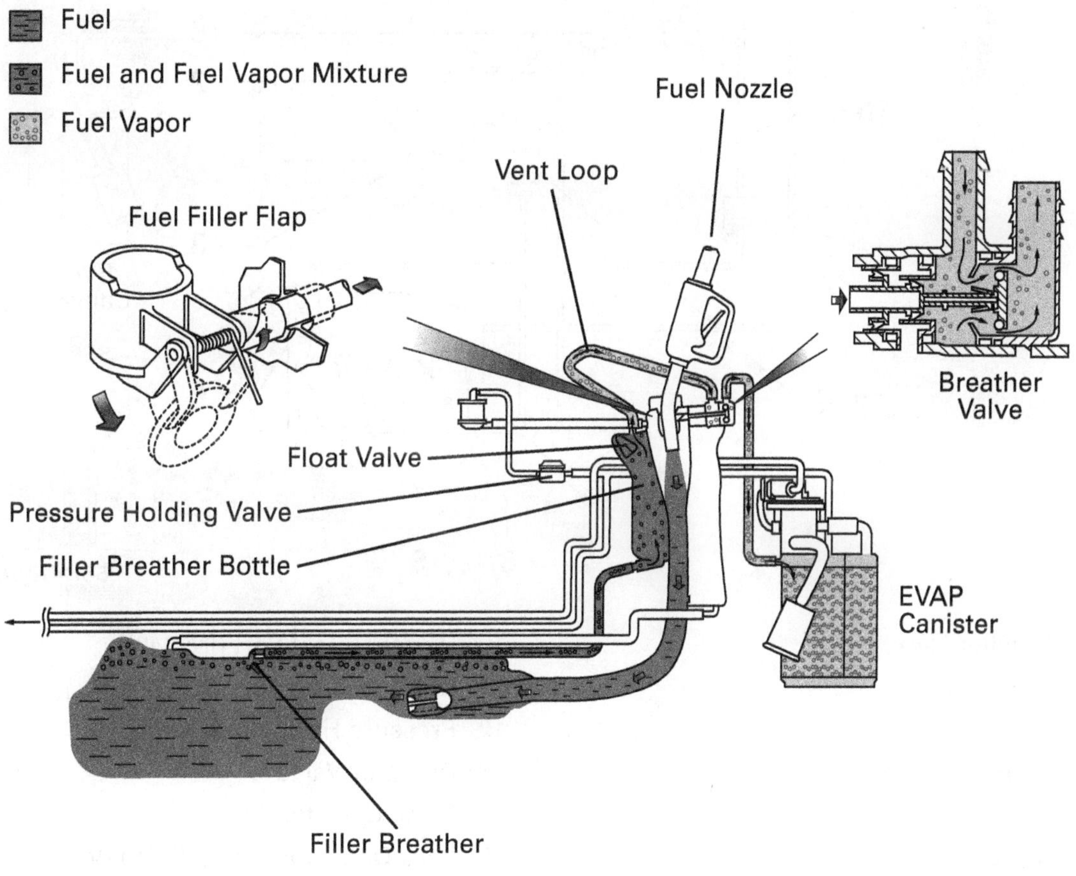

SSP 418/256

Motronic 5.9.2 Engine Management System

Evaporative Emission (EVAP) controls — schematic diagram

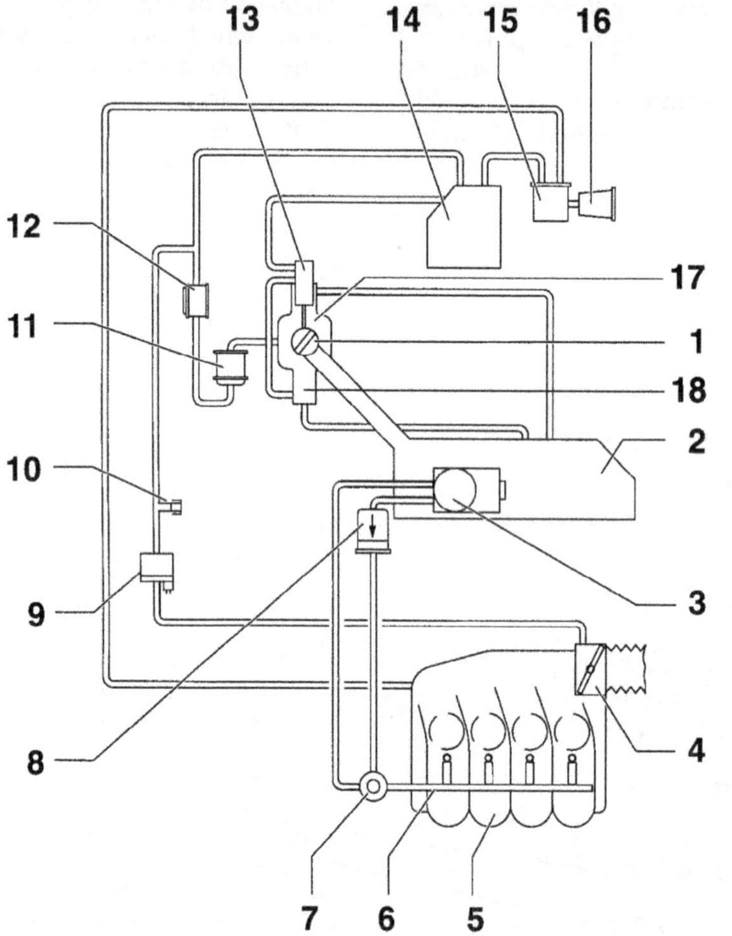

SSP 418/168

1 Filler Flap
2 Fuel Tank
3 Fuel Pump
4 Throttle Valve
5 Intake Manifold
6 Fuel Rail with Fuel Injectors
7 Fuel Pressure Regulator
8 Fuel Filter
9 EVAP Canister Purge Regulator Valve -N80-
10 Pressure Test Port (IM 240)
11 Roll-over Valve
12 Pressure Holding Valve
13 Breather Valve
14 EVAP Canister
15 Leak Detection Pump (LDP) -V144-
16 LDP Filter
17 Operating Breather Bottle
18 Filler Breather Bottle

Motronic 5.9.2 Engine Management System

Breather Valve

The Breather Valve (-13-) allows venting from both the Operating Breather Bottle (-17-) and the Filler Breather Bottle (-18-).

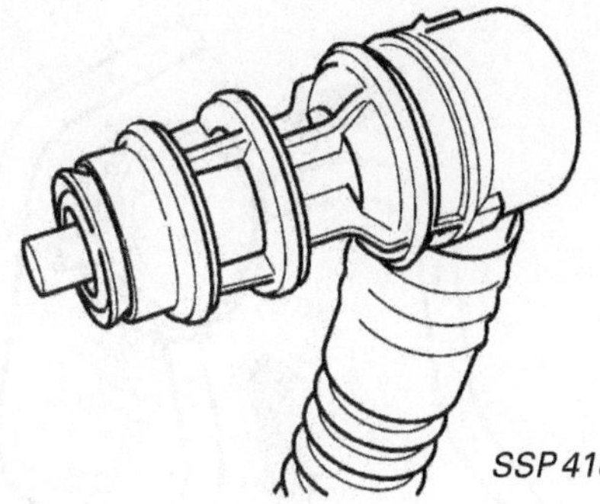

SSP 418/184

During refueling, the Filler Flap (-1-) operates the Breather Valve, closing off the vent in the Operating Breather Bottle, and venting only from the Filler Breather Bottle.

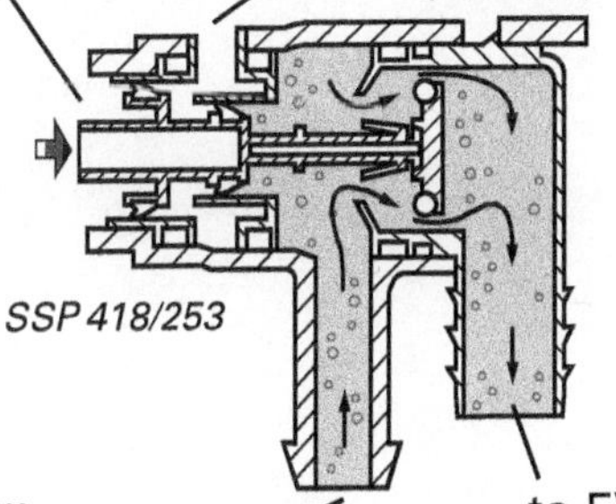

SSP 418/253

During normal operation, the Breather Valve (-13-) is in position to vent fuel vapor from the Operating Breather Bottle (-17-) into the Filler Neck, and out the Vent Line to the Roll-over Valve (-11-).

In this position the Vent Loop allows venting of the Filler Breather Bottle through the Breather Valve into the Filler Neck and out the Vent Line to the Roll-over Valve.

All fuel vapor passes through the Roll-over Valve before going on to either the EVAP Canister (-14-) or the Intake Manifold (-5-).

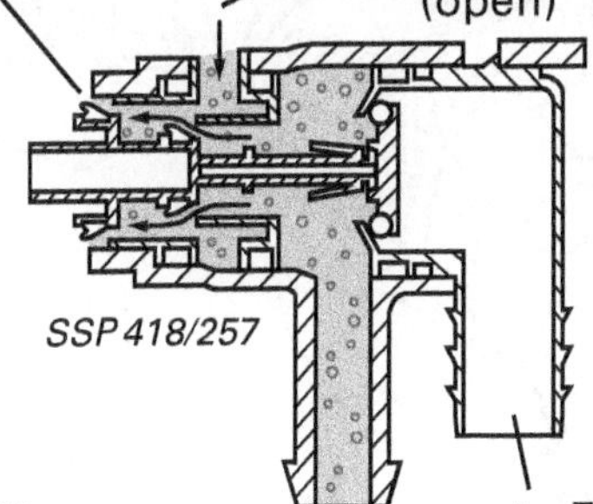

SSP 418/257

Motronic 5.9.2 Engine Management System

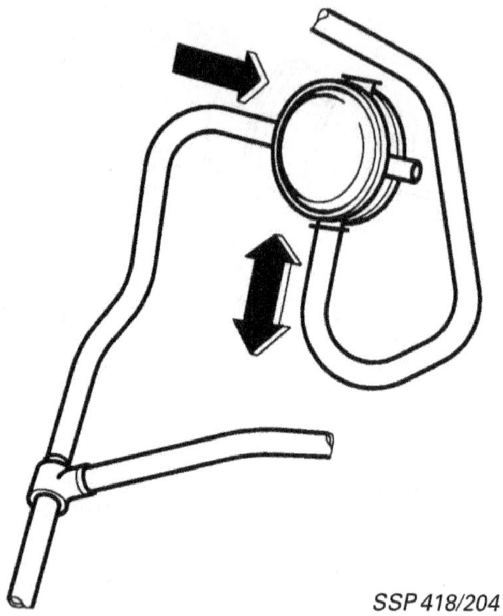

SSP 418/204

Pressure Holding Valve

Vapor is vented from the Roll-over Valve to the Pressure Holding Valve (-12-). It closes, causing slight pressure build-up in the Fuel Tank (-2-) to minimize fuel vaporization.

Once pressure reaches a pre-determined limit (approx. 0.25 mbar), fuel vapor is vented through a tee connection to the EVAP Canister and EVAP Canister Purge Regulator Valve -N80- (-9-).

The Pressure Holding Valve also opens to allow air back into the tank as fuel is used. Air is drawn in through the LDP Filter (-16-) and EVAP Canister.

When the EVAP Canister Purge Regulator Valve -N80- opens, engine vacuum creates low pressure in the vent lines to draw fuel vapor out of the EVAP Canister. The Pressure Holding Valve closes to prevent fuel vapor from being drawn out of the tank.

During purging of the EVAP canister, fresh air is drawn in through the Leak Detection Pump (-15-) by way of the LDP Filter.

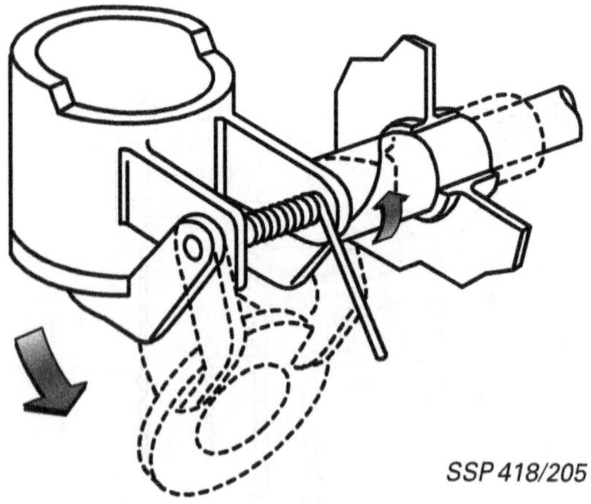

SSP 418/205

Fuel Filler Flap

A spring-loaded flap normally closes off the top of the Fuel Filler Neck, and is opened when a Fuel Filler Nozzle is inserted into the filler neck.

The narrow Fuel Filler Neck helps create a "liquid seal" as fuel flows into the tank, preventing vapor from venting out through the Fuel Filler. The shape of the fuel filler insert also creates a venturi effect in the filler neck to keep vapor from escaping.

When the flap in the Filler Neck is opened, a Rotary Slide attached to the flap operates the Breather Valve.

Motronic 5.9.2 Engine Management System

The Breather Valve opens to ventilate the Filler Breather Bottle through the Vent Loop. At the same time, the Breather Valve closes the vent from the Operating Breather Bottle (like the older vent valve).

As the Fuel Tank fills, the Breather Valve directs displaced Fuel Vapor to the EVAP Canister via a large diameter Vent Line.

The large quantity of displaced air and fuel vapor passes through the new, larger EVAP Canister where Fuel Vapor is collected and thus removed from the air. The air is then exhausted through the Leak Detection Pump (-15-) and the LDP Filter (-16-)

As the Fuel Tank fills, the fuel level rises in the Filler Breather Bottle. A Float Valve in the bottle closes, preventing liquid fuel from entering the Vent Lines and the EVAP Canister.

The fuel rises in the Filler Neck until it shuts off the filler nozzle. When the filler nozzle is removed, the flap switches the Breather Valve back to normal Fuel Tank venting by way of both the Operating Breather Bottle and Filler Breather Bottle, and all Fuel Vapor is vented through the Roll-over Valve.

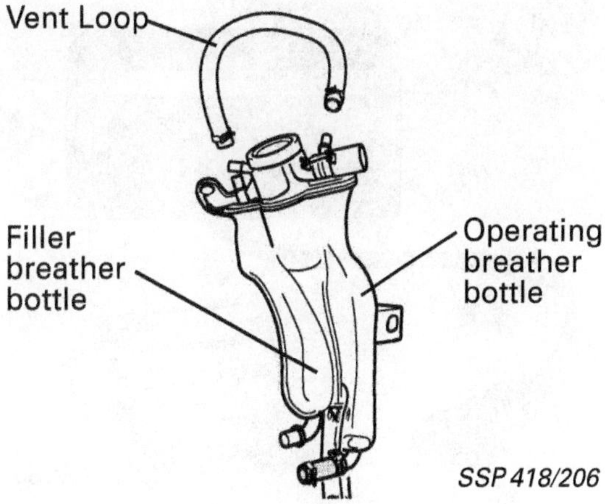

SSP 418/206

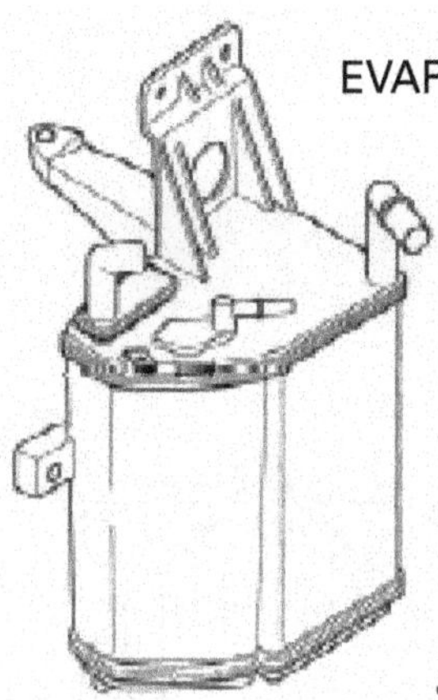

SSP 418/207

Another new EVAP system component is a tee fitting with a One-way Valve (arrow) installed in the line just before the EVAP Canister Purge Regulator Valve -N80-.

A protective cap helps identify the fitting, which is located in the engine compartment just in front of the right strut housing.

States that have enacted IM240 emissions testing procedures require a pressure test of the EVAP system. This valve is used for that purpose, and for system testing during production.

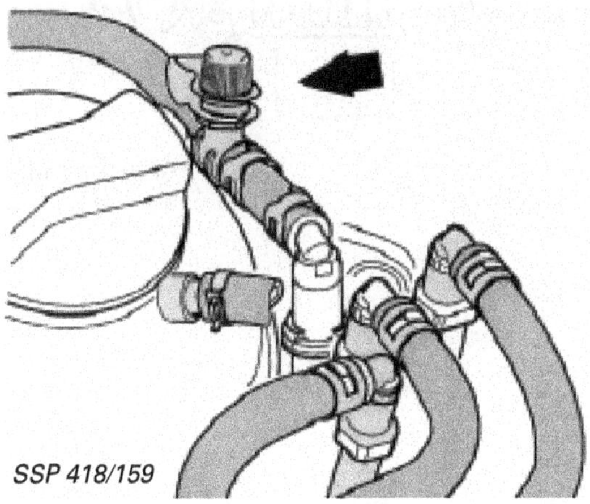
SSP 418/159

Motronic 5.9.2 Engine Management System

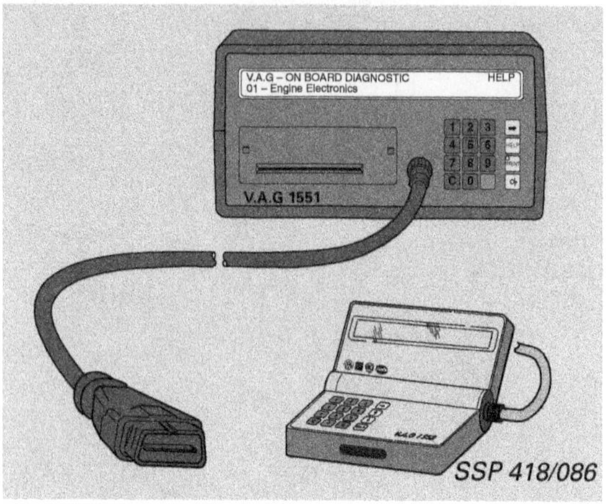

SSP 418/086

EVAP On Board Diagnostic (OBD)

The integrity of the EVAP System is monitored by the OBD II function of the Motronic ECM.

It is not possible to check the operation of the ORVR function using the VAG1551 or VAG1552 scan tool, or the OBD II function.

CAUTION

This training module applies to only some of the vehicles covered by this manual. This information is subject to change and should be used as reference only. Always refer to the latest Volkswagen factory service information for the most up to date service, repair and training information.

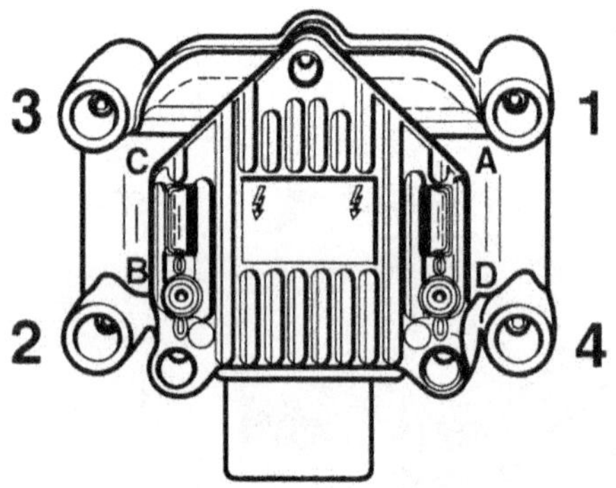

SSP 418/216

Distributorless ignition

Ignition Coil -N152- with two double-ended coils and Power Output Stage replaces the Distributor and single ignition coil used on the 2.0 liter engine with engine code ABA. The coil is mounted on the engine cylinder block, above the oil filter housing.

Each double-ended Ignition Coil supplies Ignition Secondary Voltage to two Spark Plugs simultaneously, one for Cylinders 1 and 4, and the other for Cylinders 2 and 3.

Motronic 5.9.2 Engine Management System

The distributorless ignition system operates on the "waste spark" principle:

As pistons in one pair of cylinders get near Top Dead Center (TDC), the ECM signals the Power Output Stage to discharge the Ignition Secondary Voltage from the appropriate double-ended Ignition Coil.

Only one of the two pistons will be on its compression stroke; the other will be on the exhaust stroke.

Both Spark Plugs fire simultaneously; one ignites the mixture during that cylinder's compression stroke; the other spark is "wasted" with no effect in the cylinder that is on its exhaust stroke.

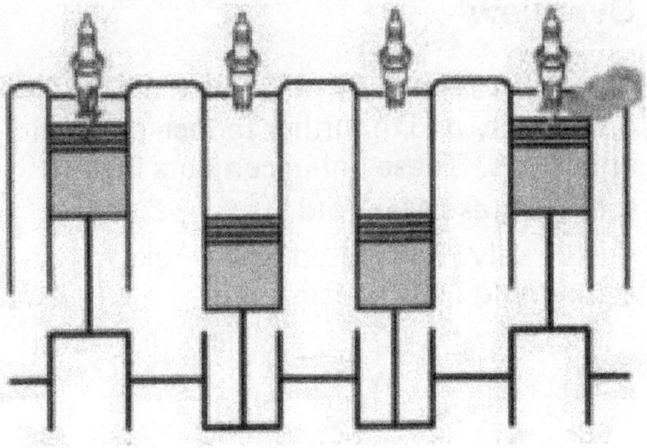

SSP 418/115

After the next 360° of crankshaft rotation the ignition event is repeated for the same two pistons, now with the other of the two on its compression stroke. That cylinder's Spark Plug now ignites the air/fuel mixture; the other fires without effect.

This ignition cycle works for the other two cylinders in the same way.

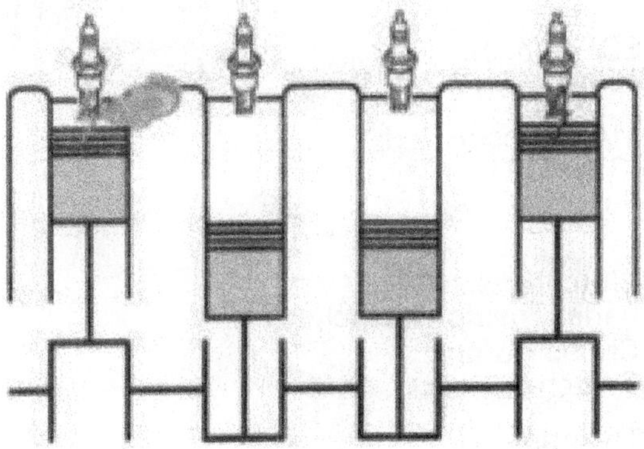

SSP 418/116

TDI Diesel Direct Fuel Injection

Overview

The new TDI has been enhanced to improve driveability and to further reduce exhaust emissions. These enhancements include:

- New Intake Manifold Flap.
- EGR Valve integrated with Intake Manifold Flap housing.
- Higher Fuel Injector opening pressure.
- Smaller diameter Fuel Injector lines and new Non-return Valves.
- Intake Air Silencer.
- MAF Sensor with reverse flow detection.

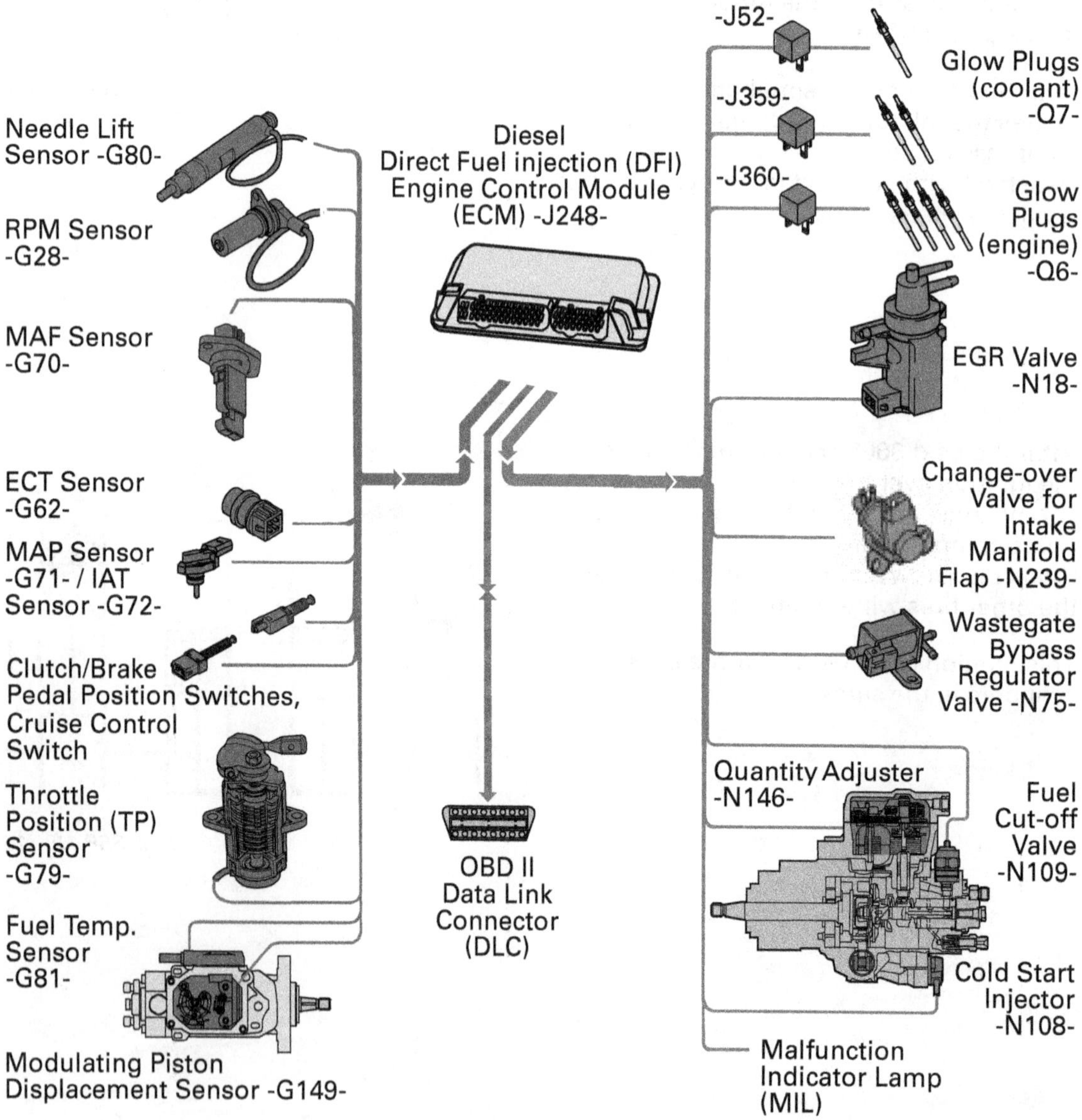

SSP 418/220

TDI Diesel Direct Fuel Injection

Intake Manifold Flap

Diesel engines will tend to stop running abruptly due to their high compression ratio. Resistance from high compression pressure stops the pistons, and this reaction can cause the engine to "shudder."

An Intake Manifold Flap has been added to help reduce compression pressure during engine shut-down. The flap stops intake air flow into the engine.

The Intake Manifold Flap is operated by a vacuum diaphragm, which is actuated by Change-over Valve for Intake Manifold Flap -N239-. When the engine stops, vacuum is applied and the flap closes.

The TDI's Diesel Direct Fuel Injection (DFI) Engine Control Module (ECM) controls operation of Change-over Valve for Intake Manifold Flap -N239-.

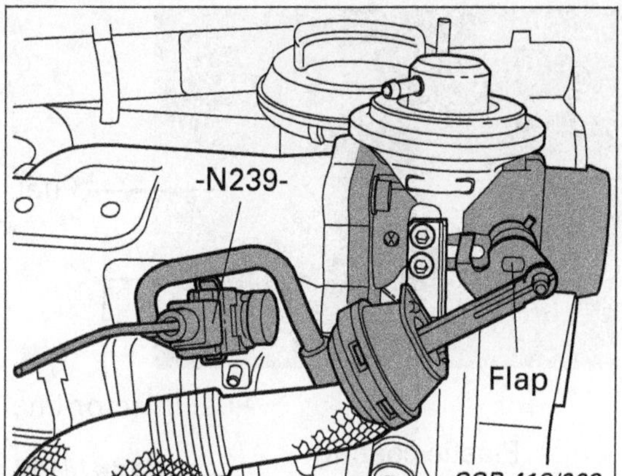

SSP 418/009

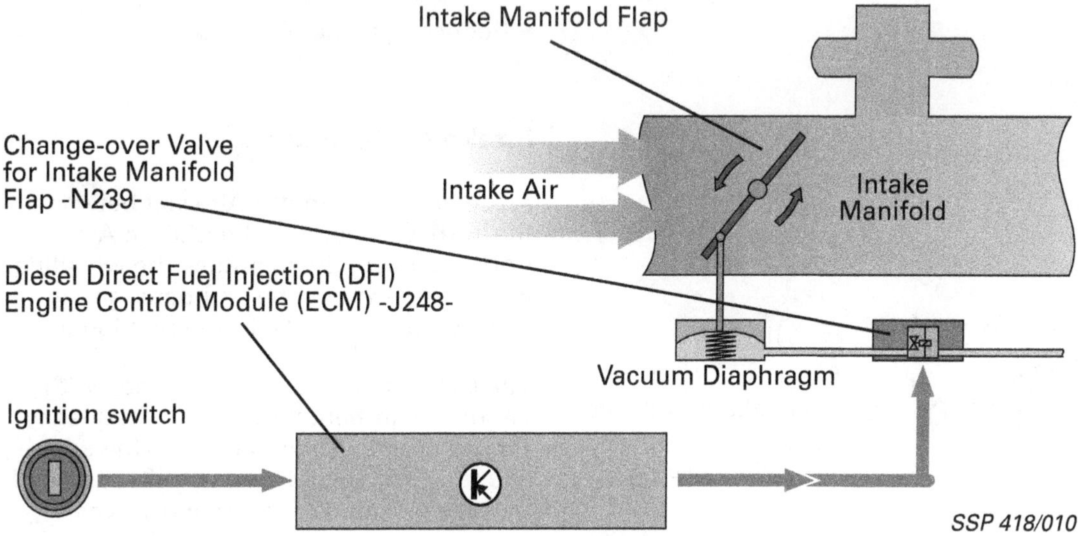

SSP 418/010

TDI Diesel Direct Fuel Injection

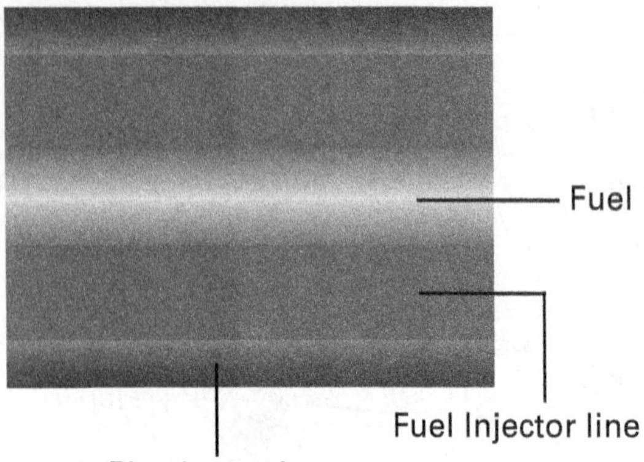

SSP 418/091

Fuel Injector lines

The outside of the Fuel Injector lines are plastic coated for corrosion protection.

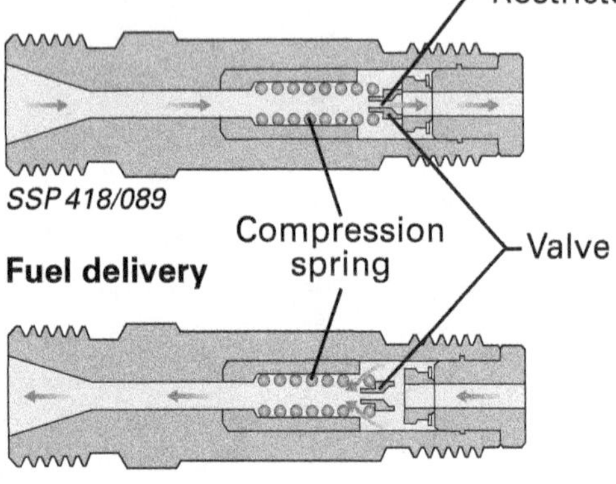

SSP 418/089

SSP 418/090

Non-return valves

Non-return valves replace the check valves formerly used in the Fuel Distributor of the Diesel Direct Fuel Injection (DFI) pump. A restrictor allows residual fuel pressure to bleed off after the Fuel Injector closes.

Relieving fuel pressure between injection pulses yields some significant performance gains:

- Lower fuel consumption.
- Reduced "black smoke" particulate emissions.
- Reduced engine noise.

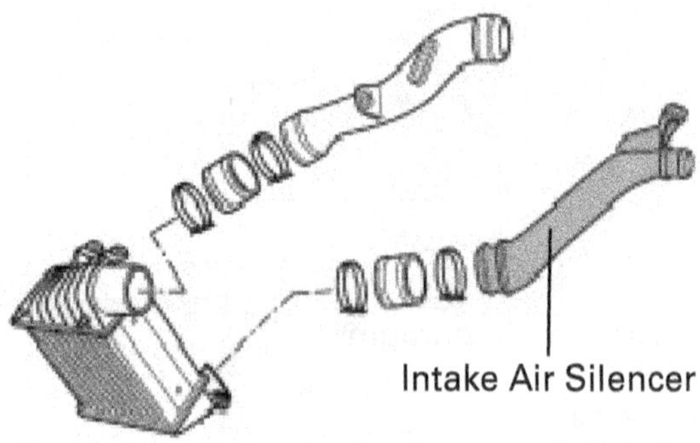

SSP 418/241

Intake Air Silencer

A silencer in the Intake Air Duct between the Turbocharger and the Charge Air Cooler is designed to reduce the whistling noise produced by the Turbocharger at Engine Speed of 1700 RPM and higher.

The silencer is "sausage" shaped, with various chambers tuned to the frequency ranges of the whistling noise. Unwanted sound energy is reduced by reflection and by superposition of the sound waves.

TDI Diesel Direct Fuel Injection

Manifold Absolute Pressure (MAP) Sensor -G71- and Intake Air Temperature (IAT) Sensor -G72-

The MAP Sensor is no longer located in the DFI ECM housing. For the TDI application in the New Beetle it has been combined with the IAT Sensor, and both are located in the Intake Air Duct between the Charge Air Cooler and Intake Manifold.

The signals that the MAP Sensor and IAT Sensor provide to the DFI ECM are used to calculate intake air density.

The two sensors are located together to ensure uniform signals that correspond to each other.

Locating the MAP Sensor in the Intake Air Duct also eliminates the need for a hose connection to the DFI ECM.

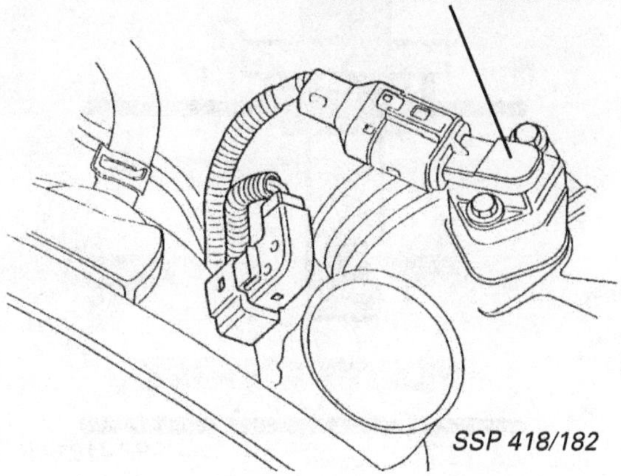

SSP 418/182

Coolant Glow Plugs

Coolant Glow Plugs for pre-heating the engine coolant are mounted in a housing at the left end of the cylinder head.

Only manual transmission TDI vehicles are equipped with coolant Glow Plugs.

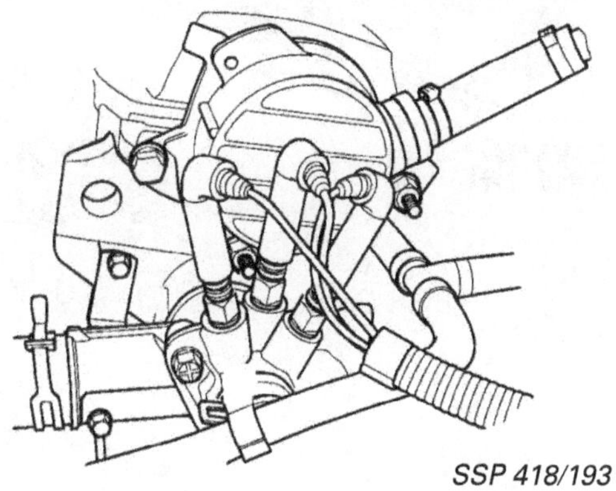
SSP 418/193

TDI Diesel Direct Fuel Injection

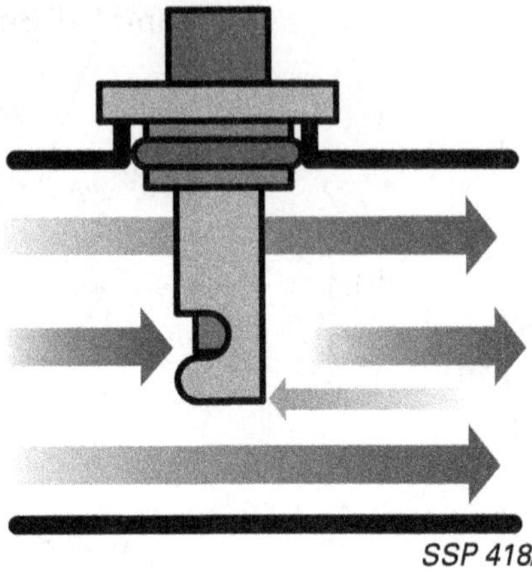

SSP 418/005

Mass Air Flow (MAF) Sensor -G70- with Reverse Flow Detection

This is similar to the MAF Sensor on the 2.0 liter Motronic engine, but does not use an integrated Intake Air Temperature (IAT) Sensor.

The new MAF Sensor with reverse flow detection uses two sensors to measure intake air flow upstream and downstream of a Heating Element. The ECM evaluates the difference in the two signals to calculate mass air flow. For a more detailed explanation, see page 21.

Reverse air flow changes the relationship between the two signals. This allows the ECM to factor out reverse flow data from its mass air flow calculations.

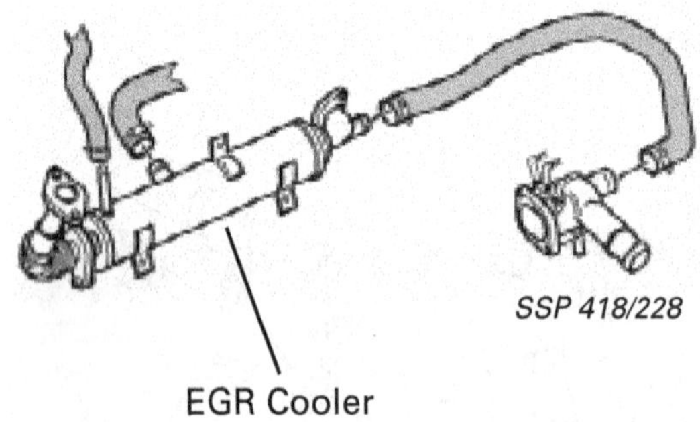

SSP 418/228

EGR Cooler

Water-cooled Exhaust Gas Recirculation (EGR) has been a feature of TDI engines since 1997. An Engine Coolant Heat Exchanger reduces the temperature of the recirculated exhaust gas.

Recirculated exhaust gas temperature is reduced by as much as 50°C (122°F). This reduces NO_x emission levels.

CAUTION

This training module applies to only some of the vehicles covered by this manual. This information is subject to change and should be used as reference only. Always refer to the latest Volkswagen factory service information for the most up to date service, repair and training information.

TDI Diesel Direct Fuel Injection

Diesel Direct Fuel Injection (DFI) pump pressure

DFI pump outlet pressure is higher for automatic transmission vehicles only. Pressure has been increased from 850 bar to 950 bar (from 12,325 psi to 13,775 psi).

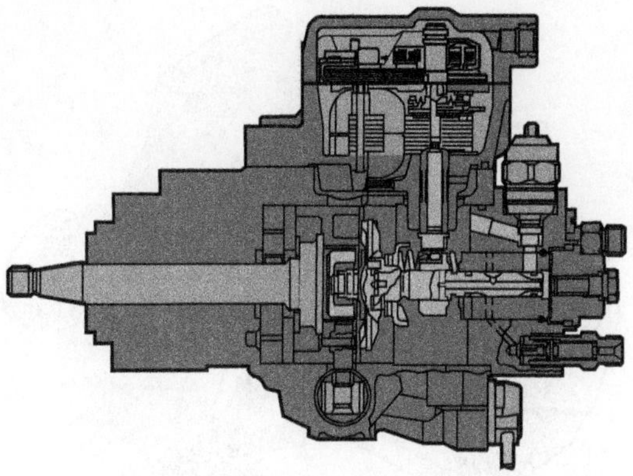

SSP 418/225

Fuel Cooler

Due to the high pressure created at the fuel distributor on the TDI Diesel DFI pump, the fuel is heated. Unused, heated fuel is returned to the fuel tank. For this reason, a Fuel Cooler is added to the supply line on automatic transmission TDI vehicles. Fuel passes though a finned aluminum cooler on its way to the pump.

Injection quantity calculations are modified for fuel and coolant temperatures. Lower fuel temperature helps the DFI ECM more accurately meter the correct quantity of fuel, which helps reduce emissions.

Fuel Injectors

The Fuel Injectors are also different for TDI engines mated to automatic transmissions. Stiffer springs increase the fuel injector opening pressure from 190 to 220 bar (from 2,755 psi to 3,190 psi).

SSP 418/066

5-spd. Manual Transmission 02J

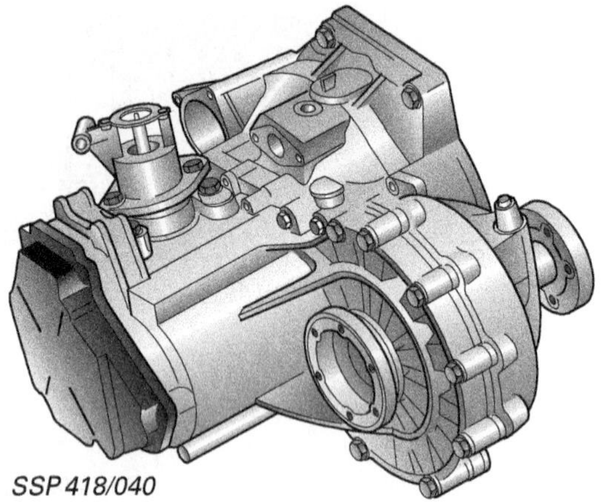

SSP 418/040

The 5-speed manual transmission 02J is the latest version from the 02A family of transmissions.

Shift mechanism

The 5-speed manual transmission 02J uses a cable gear shift mechanism.

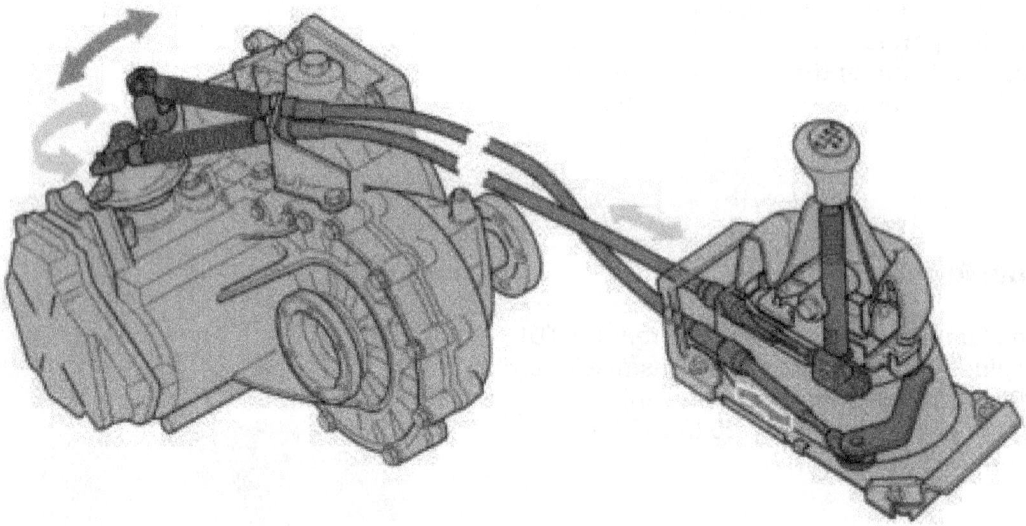

SSP 418/114

5-spd. Manual Transmission 02J

Mounting points

The transmission housing itself includes a mounting point for the pendulum-style left-side engine/transmission mount, and has the necessary provisions for attaching the cast aluminum oil pan.

Drive flange seal protection

The transmission housing design has also been modified to help prevent dirt from entering the drive flange oil seals.

A protective sleeve extends from the seal bore to surround the drive flange shaft. There is only a 1 mm gap between the protective sleeve and the drive flange shaft.

The protective sleeve on the left side is part of the transmission housing. A separate plastic sleeve is used on the right side.

CV joints

New Beetles with manual transmissions are equipped with conventional ball-type CV joints.

CAUTION

This training module applies to only some of the vehicles covered by this manual. This information is subject to change and should be used as reference only. Always refer to the latest Volkswagen factory service information for the most up to date service, repair and training information.

5-spd. Manual Transmission 02J

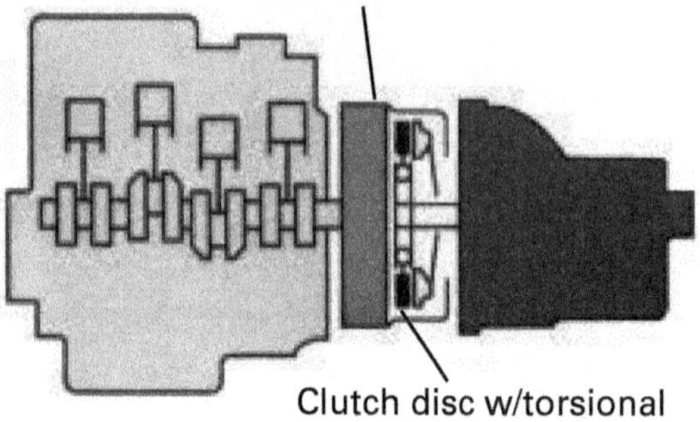

SSP 418/053

Dual-mass flywheel (TDI)

New Beetles with TDI Diesel engine and manual transmission are equipped with a sophisticated dual-mass flywheel. It is designed to isolate the transmission from vibrations created by the engine, and help smooth out crankshaft oscillations better than a conventional flywheel.

The advantages are:

- Less engine vibration transmitted to the passenger compartment.
- Elimination of "gear rattling" noise.
- Smoother clutch engagement.
- Smoother shifting.

A conventional flywheel uses simple mass to counteract the effects of engine firing pulses and smooth the natural oscillations of a running engine. A spring-loaded clutch disc hub helps dampen torsional vibration for clutch engagement, but has little effect on crankshaft oscillations.

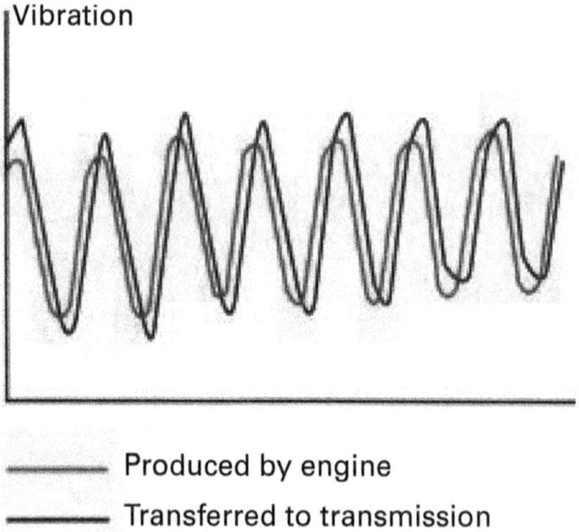

SSP 418/054

This graph illustrates the most common vibration frequencies for a vehicle with a conventional flywheel and clutch.

The direct mechanical connection between crankshaft, flywheel, clutch, and transmission input shaft means that most engine crankshaft oscillations are transferred directly through to the transmission.

The result is that excessive noise, vibration and harshness is transferred to the rest of the vehicle, and to the occupants.

5-spd. Manual Transmission 02J

With a dual-mass flywheel, the primary mass is mounted directly to the engine crankshaft, and it helps smooth crankshaft oscillations in the same manner as a more conventional flywheel.

A secondary mass is attached to the primary through a spring and damper mechanism, which helps to isolate it from the remaining engine oscillations.

The clutch is mounted to the secondary mass of the dual-mass flywheel.

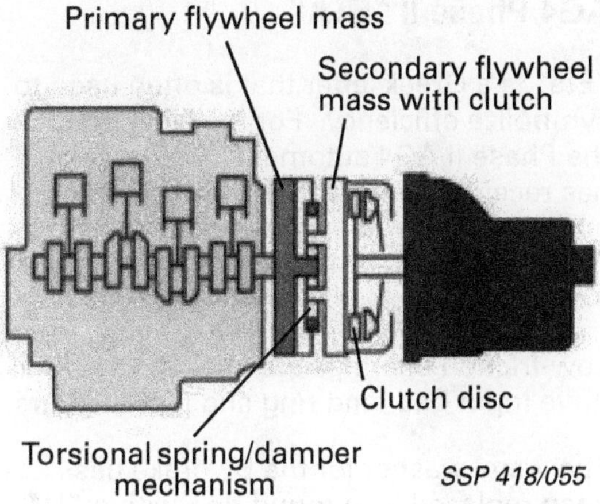

SSP 418/055

This graph illustrates the change in normal vibration frequencies for a vehicle with a dual-mass flywheel.

Even when the clutch is fully engaged, the clutch and transmission input shaft are in part isolated from most of the engine's crankshaft oscillations by the flywheel's internal spring and damper mechanism. Thus, much less of the force and vibration of crankshaft oscillation is transferred through to the transmission.

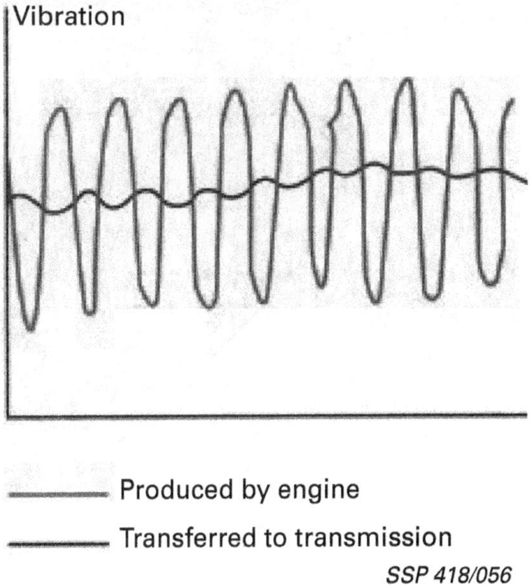

SSP 418/056

4-spd. AG4 Automatic Transmission 01M

AG4 Phase II "ETA"

"Eta" is a Greek letter that is often used to symbolize efficiency. For the New Beetle, the Phase II AG4 automatic transmission has received internal hardware changes and control module software upgrades that help to reduce internal friction and yield more efficient overall operation.

Low-friction bearings are used for the final drive input gear and ring and pinion gears.

The wave washer for the B2 brake has been replaced by a round ring with a "U" shaped cross-section. Inside are 2 metal rings that are in contact on their outside edges and function as a spring. All B2 plates have a slightly wavy shape.

Compared to earlier versions of the Phase II transmission, the B2 brake and K2 clutch use fewer inner and outer plates in each clutch pack for less drag.

	Phase II	Phase II ETA
B2 brake	6 plates	4 plates
K2 clutch	5 plates	4 plates

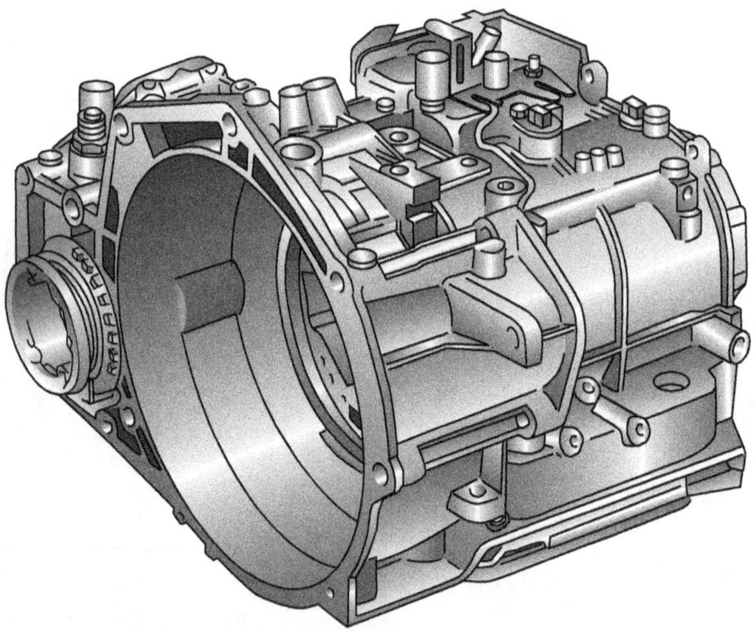

SSP 418/041

4-spd. AG4 Automatic Transmission 01M

Transmission Control Module (TCM) software

TCM software changes compensate for the reduction in the number of clutch plates. Operating pressure has been increased, but the shifts are actually smoother in the ETA transmission.

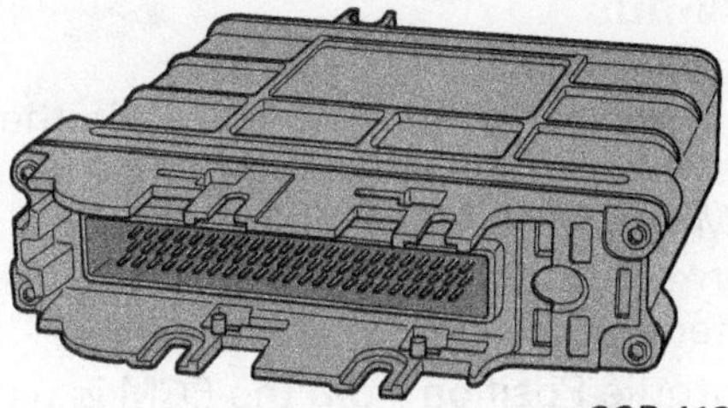

SSP 418/098

> ***CAUTION!***
> ***Transmission Control Modules (TCM) for the Phase II and Phase II ETA transmissions are not interchangeable. The software differences affect mechanical functions, and substituting one for the other will damage the transmission!***

Valve Body

The function of Solenoid Valve EV7 in the Valve Body has been changed.

In the earlier Phase II transmissions, the solenoid valve was used to change the position of the Diverter Valve for the B2 brake and K3 clutch.

On Phase II ETA transmissions, the solenoid valve changes the position of the Diverter Valve for the K1 clutch, relieving oil pressure rapidly during the 3rd Gear to 4th Gear shift to improve shift quality.

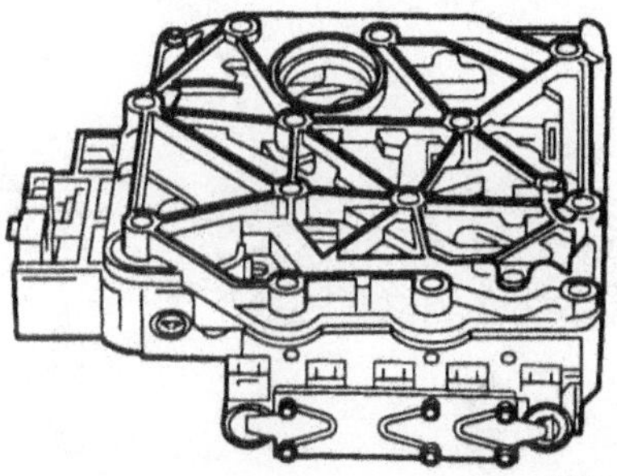

SSP 418/217

4-spd. AG4 Automatic Transmission 01M

CAN-bus

The Engine Control Module (ECM), the Automatic Transmission Control Module (TCM), and the ABS Control Module share information. Examples include, but are not limited to:

- Throttle Position from the ECM is used by the TCM to adjust Shift Points.
- Vehicle Speed from the TCM is used by the ECM for Closed Throttle Position deceleration.
- Transmission shift points are used by the ECM to reduce engine torque.

In current Golf/Jetta models, individual wires are needed to carry the information for each signal. In the New Beetle, signals are now carried over a two-wire Controller Area Network bus, or "CAN-bus."

The capability for serial communications between Control Modules on the CAN-bus reduces the number of wires running between these modules.

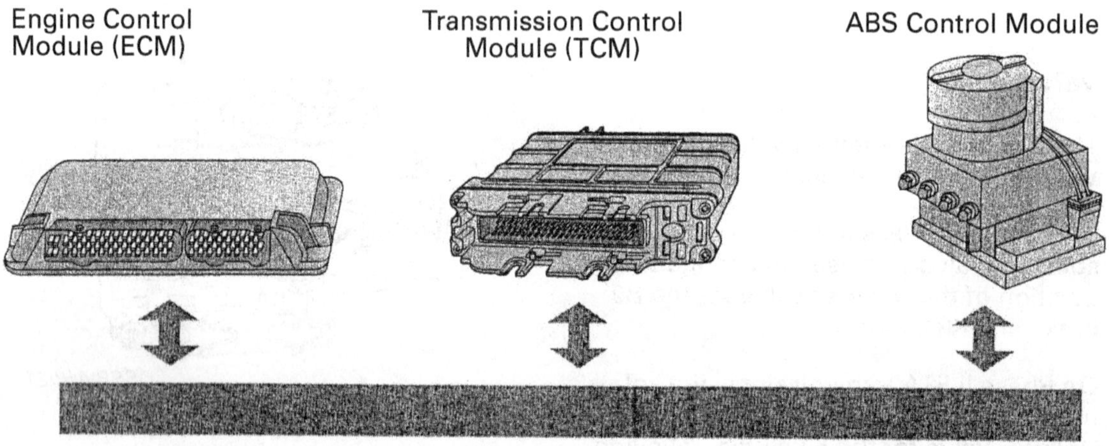

SSP 418/033

4-spd. AG4 Automatic Transmission 01M

Diagnosis of the CAN-bus

The CAN-bus is an internal function of the control modules, and the wires are the only external components.

A malfunction in the CAN-bus will be identified and stored as a Diagnostic Trouble Code (DTC) in DTC memory for one or more of the control modules. Any stored malfunctions can be identified using the VAG1551 Scan Tool.

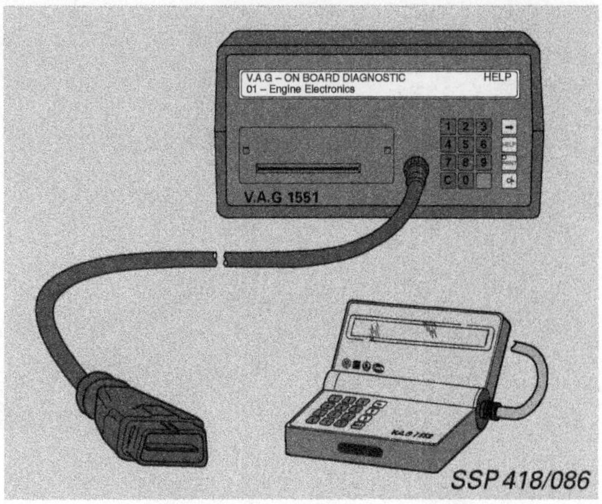

SSP 418/086

Tripod CV joints

Tripod CV joints are used to isolate power train vibrations from the body.

All New Beetles with automatic transmissions are equipped with tripod inner CV joints. These joints are better than the more conventional ball-type joints at dampening typical engine vibrations.

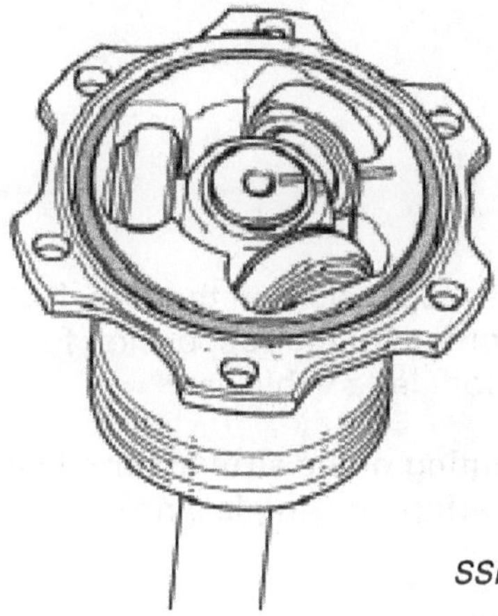

SSP 418/144

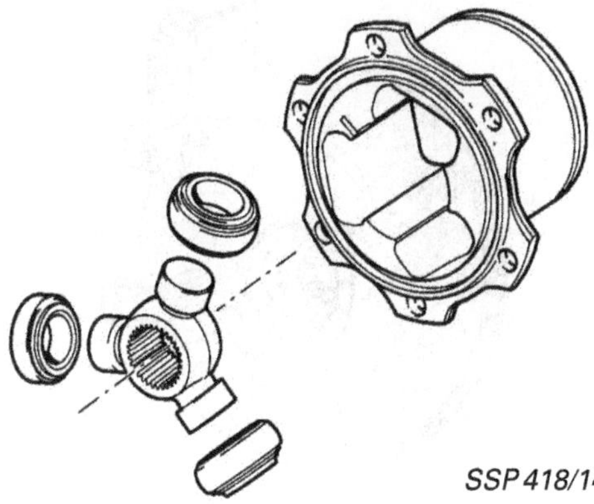

SSP 418/143

Front Suspension

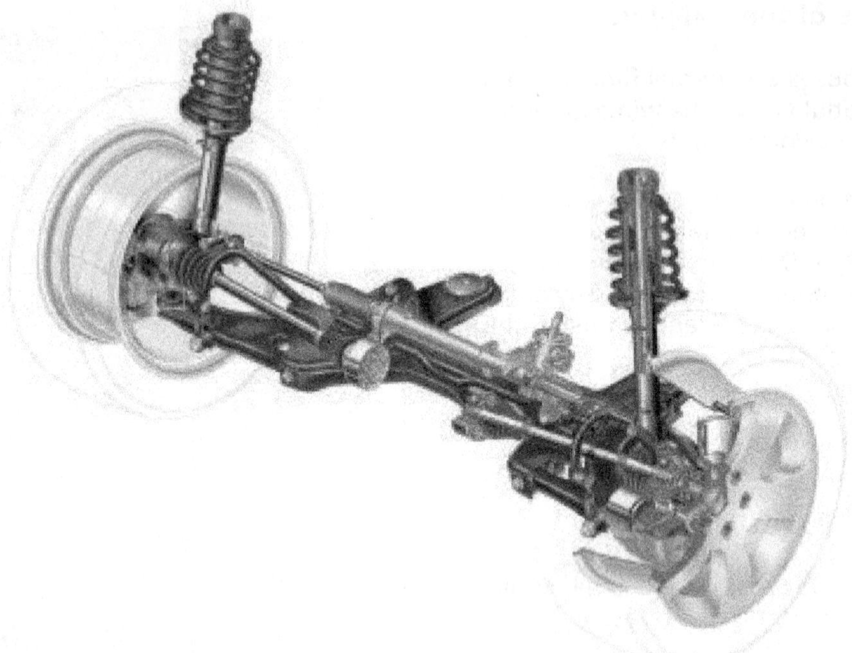

SSP 418/070

The front suspension on the New Beetle is a development of the suspension found on current Golf/Jetta vehicles.

The mounting of the strut assembly to the wheel bearing housing is a new design.

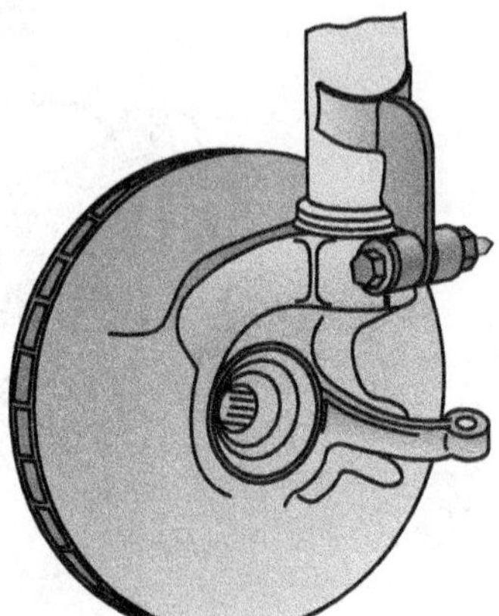

SSP 418/013a

The stabilizer bar is also a new design, with new links at the control arms.

5-bolt, 16-inch wheels are the standard equipment, and sporty 16-inch alloy wheels are optional.

Wheel bearing housing

The wheel bearing housings are a special cast iron with high graphite content. One benefit is that no additional lubricant is necessary when installing a new wheel bearing in the housing.

Front Suspension

Front wheel alignment

The front strut attachment to the wheel bearing housing no longer allows camber adjustment. Camber can be equalized, however, by moving the subframe from side to side.

Caster offset of 40 mm helps ensure good directional stability. Power-assisted steering, which is standard equipment on the New Beetle, compensates for what would otherwise be higher steering effort.

Toe can be adjusted by loosening and turning either tie rod.

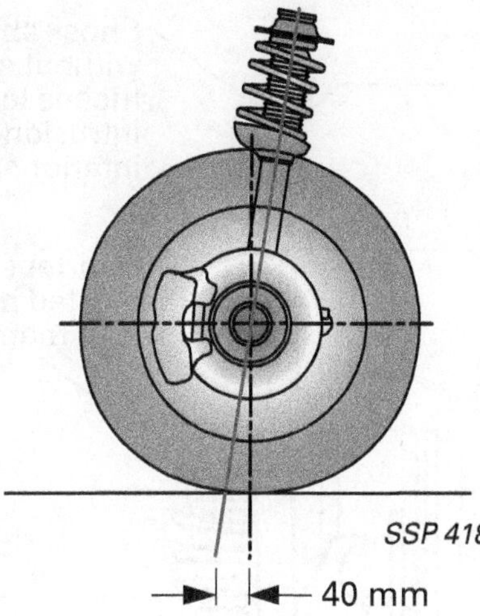

SSP 418/013b

Front strut

The two-piece strut assembly consists of a separate shock absorber housing and wheel bearing housing.

The shock absorber portion of the strut is clamped in the wheel bearing housing with a single pinch-bolt.

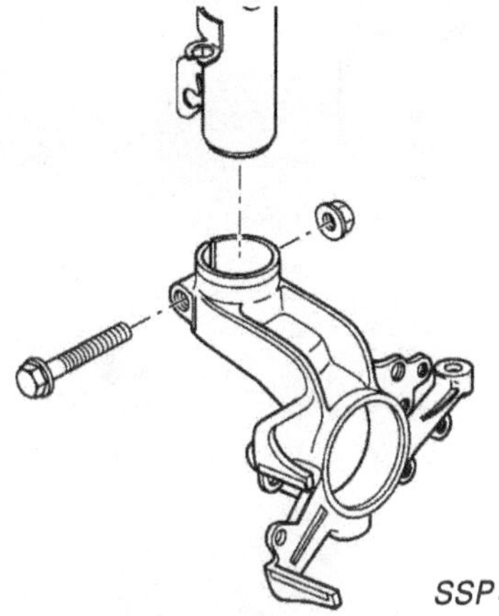

SSP 418/223

A new special tool, spreader socket 3424, is used to spread the housing to remove and install the strut.

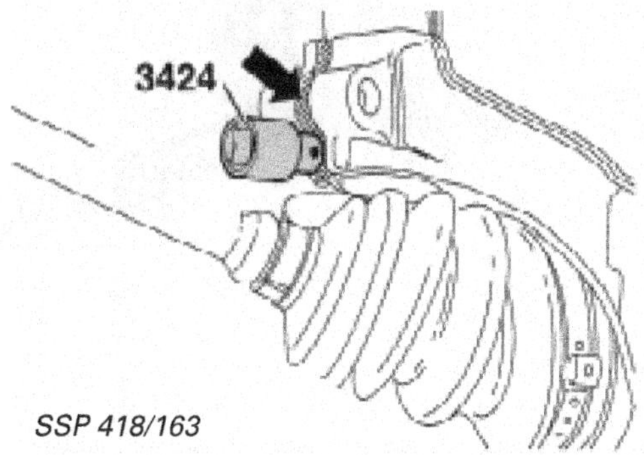

SSP 418/163

Rear Suspension

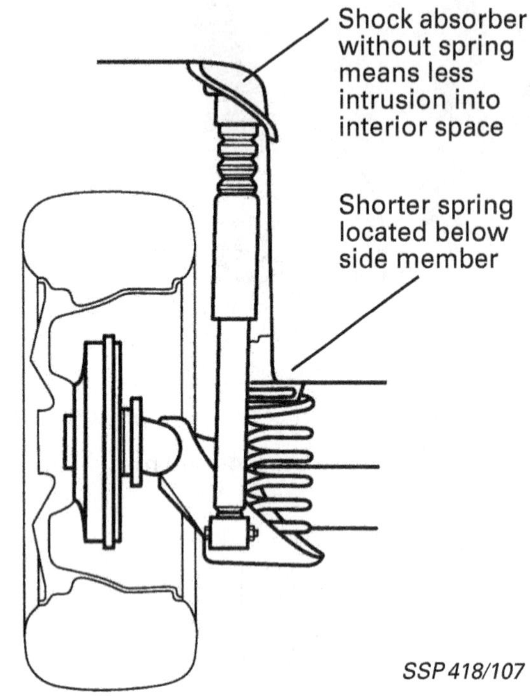

SSP 418/107

Rear axle beam

The rear suspension is a torsion beam and trailing arm design with built-in stabilizer bar, similar to the one used on the current Golf/Jetta vehicles. The coil springs and gas-filled shock absorbers, however, are now mounted separately.

The springs mount between the trailing arms and the frame side members. The shock absorbers are mounted behind the springs.

There are two main advantages to this arrangement. The springs do not intrude into the interior space. This allows for a wider luggage compartment. The tops of shock absorbers mount outside the interior compartment in the wheelhousing. This means less road noise is transmitted into the vehicle interior.

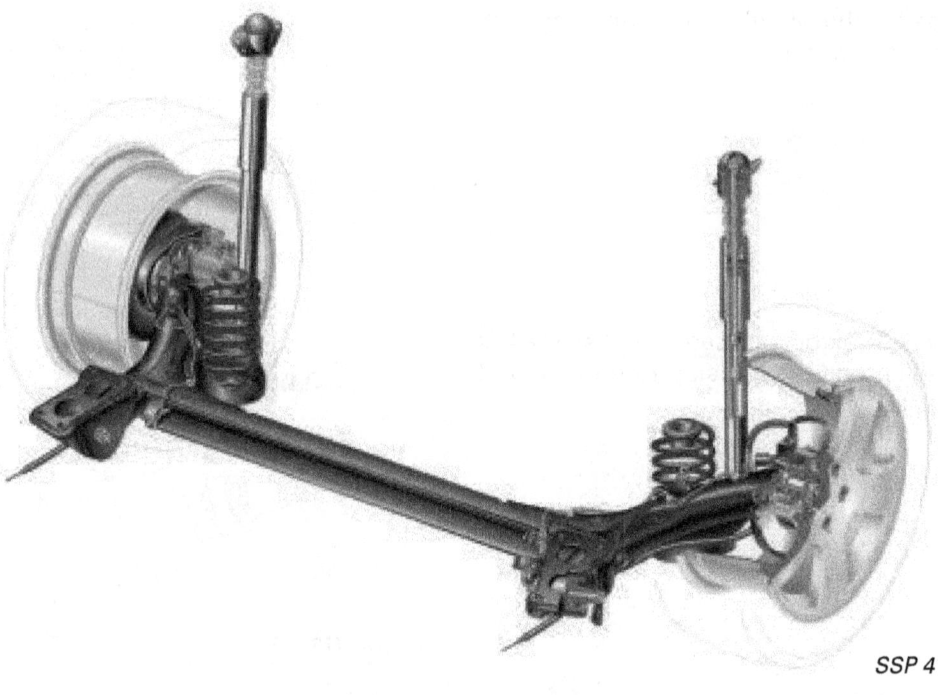

SSP 418/071

Rear Suspension

Rear axle mounts

The rear axle beam is attached to the body with new mounts aligned at a 25° angle to the transverse axis of the vehicle.

When cornering, lateral forces acting on the suspension alter the track and cause a self-steering effect. Angled mounts reduce the tendency to self-steer and increase vehicle stability as a turn is completed.

The trailing arm bushings and mounting brackets have markings that have to be aligned during bushing installation.

> ***CAUTION!***
> ***The self-steering correction will not be effective if the bushings have not been installed correctly.***

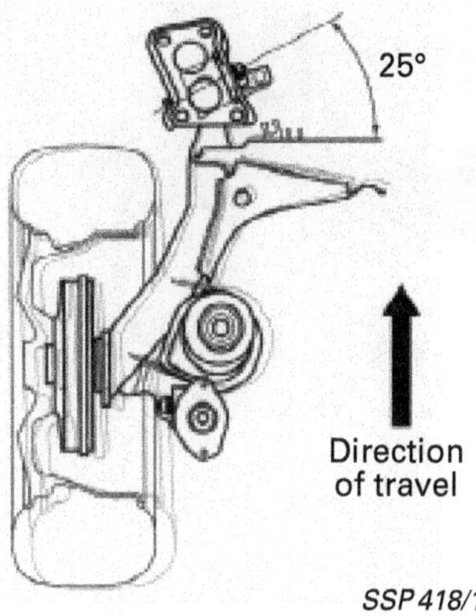

SSP 418/112

Wheel bearings

Each new double-row rear wheel bearing consists of two inner races, two sets of ball bearings, and a hub, which doubles as the bearing outer race. This assembly is pressed onto the stub axle, and the brake disc mounts to the hub.

Special tool 3420 is used to press the bearing onto the stub axle, where it is secured with a 12-point nut and thrust washer.

The bearings cannot be disassembled, and no adjustment is necessary. The thrust adjustment is designed into the bearing and is set when the axle nut is torqued.

> ***CAUTION!***
> ***Proper axle nut tightening torque is critical to set the bearing adjustment. The bearing is destroyed and must be replaced when it is removed from the stub axle.***

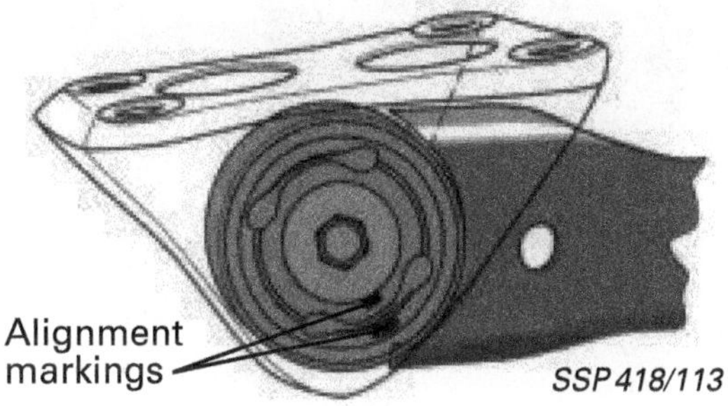

SSP 418/113

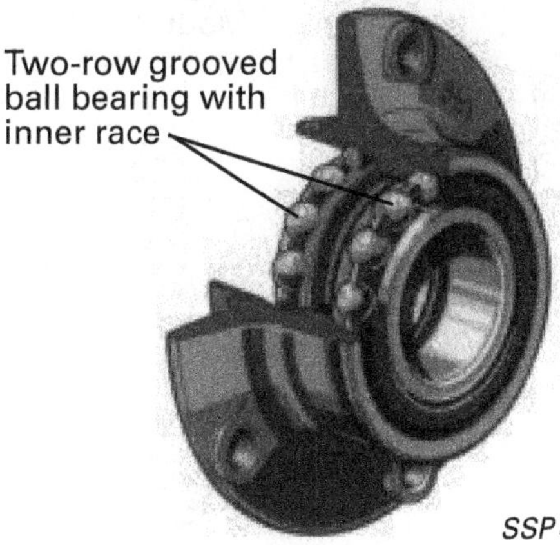

SSP 418/100

Brakes

SSP 418/050

The New Beetle is equipped with the customary dual-diagonal brake hydraulic system, with disc brakes front and rear as standard equipment. Rear brake calipers are aluminum to reduce unsprung weight. Vehicles without ABS use a mechanical brake pressure regulator.

SSP 418/051

Anti-lock Brake System (ABS)

ABS is optional equipment on the New Beetle. The Teves 20 IE 3-channel system features:

- New Inlet Valves for better pressure regulation and improved pedal feel.
- Electronic Brake-pressure Distribution (EBD).

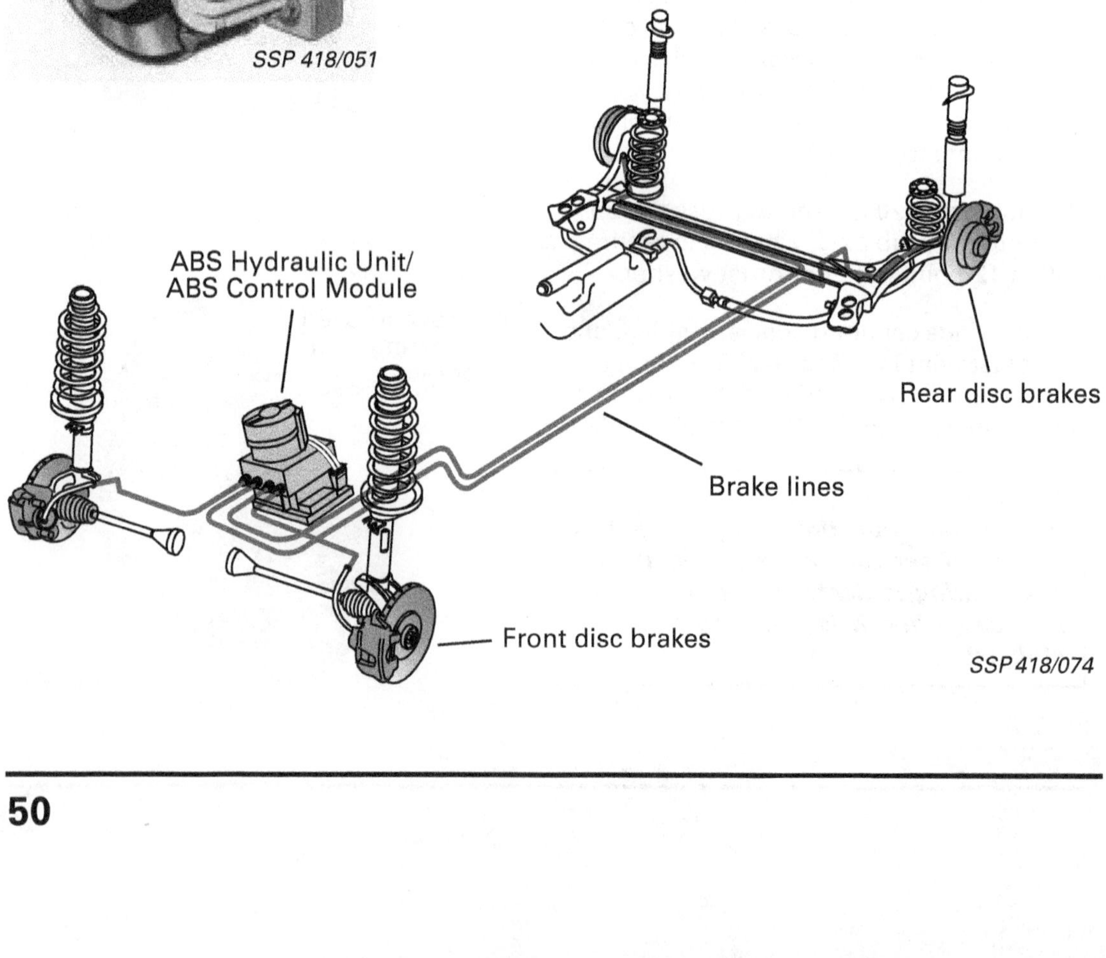

SSP 418/074

Brakes

Electronic Brake-pressure Distribution (EBD)

EBD eliminates the mechanical rear Brake Pressure Regulator. ABS controls brake pressure to the rear wheels using the ABS Solenoid Valves. The operating range of EBD ends at the ABS cut-in point.

If a malfunction causes ABS to shut down, the basic hydraulic braking still functions normally.

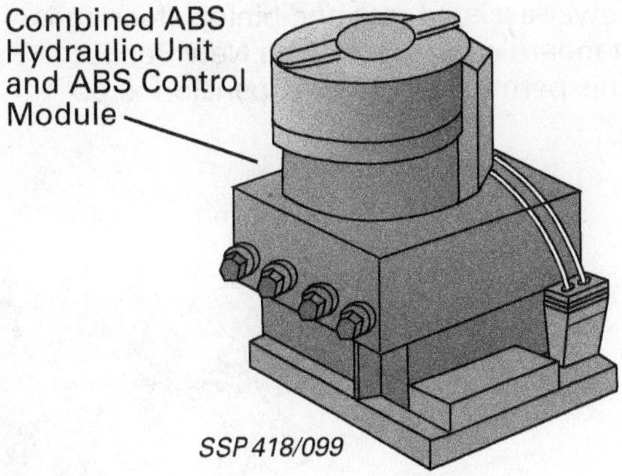

SSP 418/099

> **CAUTION**
>
> **This training module applies to only some of the vehicles covered by this manual. This information is subject to change and should be used as reference only. Always refer to the latest Volkswagen factory service information for the most up to date service, repair and training information.**

Steering

Power-assisted rack and pinion steering is standard equipment on all New Beetles. This permits the front suspension to be designed with a higher caster angle, which contributes to greater directional stability.

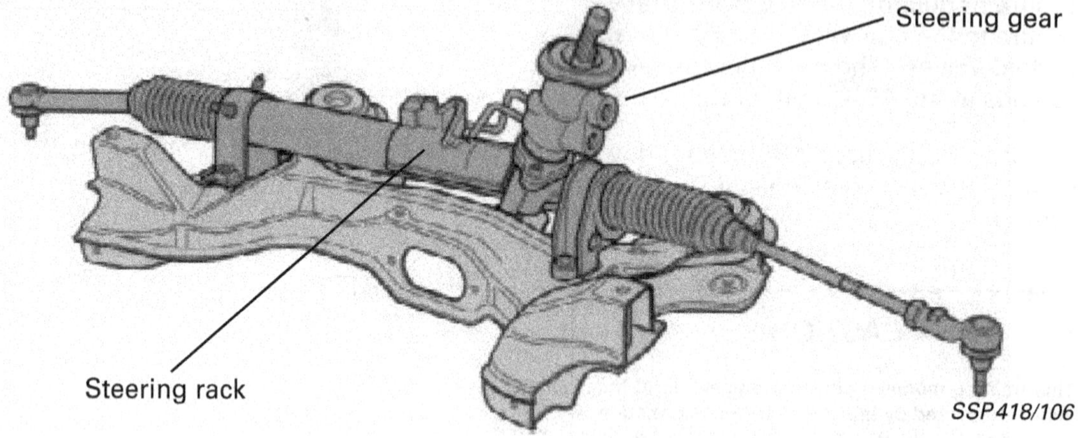

SSP418/106

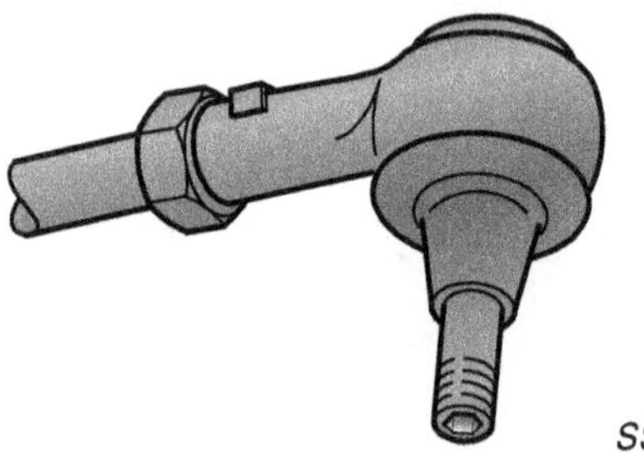

SSP 418/013c

The taper on the tie rod ends is no longer designed for a self-locking fit with the socket in the steering arm.

A hex socket incorporated into the stud can be used to counter-hold the tie rod end when loosening and tightening the nut.

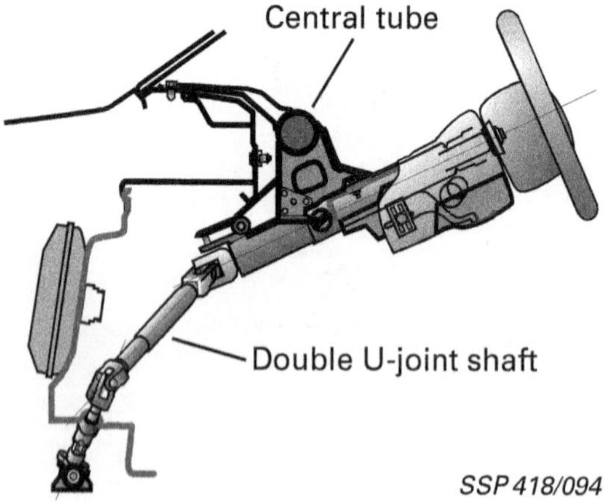

SSP418/094

The steering column is mounted to the central tube that runs between the A-pillars and supports the instrument panel.

It attaches to the steering pinion through a double universal joint (U-joint) shaft. The double U-joint shaft telescopes, which allows for easy removal of the lower joint from the steering gear.

Steering

The New Beetle has a manually adjustable tilting and telescoping steering column. The tilt function allows a 44 mm range of height adjustment, and the telescoping function allows 45 mm of fore-and-aft adjustment.

A single lever located under the steering column loosens to allow both adjustments simultaneously.

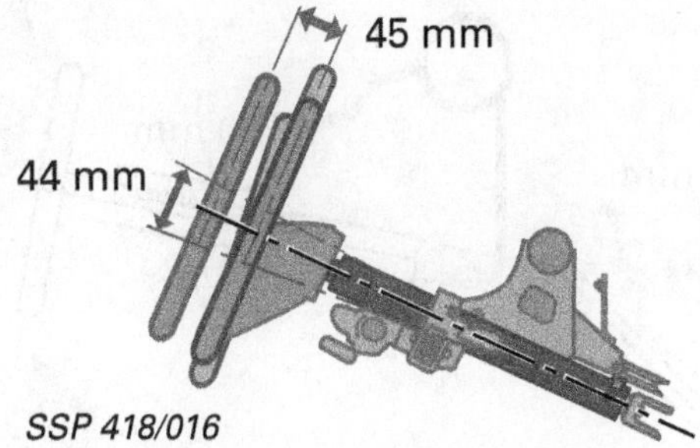

SSP 418/016

Anti-theft steering column design

The steering wheel lock is a new design. A slotted locking ring is pressed onto the steering column with a corrugated tube in between.

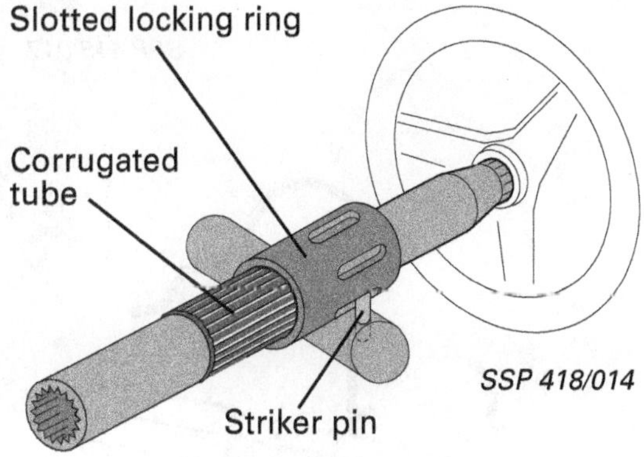

SSP 418/014

When the ignition key is removed from the lock cylinder, the striker pin engages in one of the slots in the ring. When the striker pin engages the locking ring, the steering wheel can still be turned. However, because of the friction created between the corrugated tube and locking ring, turning the wheel requires a force greater than 150 Nm (111 lb-ft).

Not only is the vehicle difficult to steer, the ignition lock will not be damaged during an attempted theft.

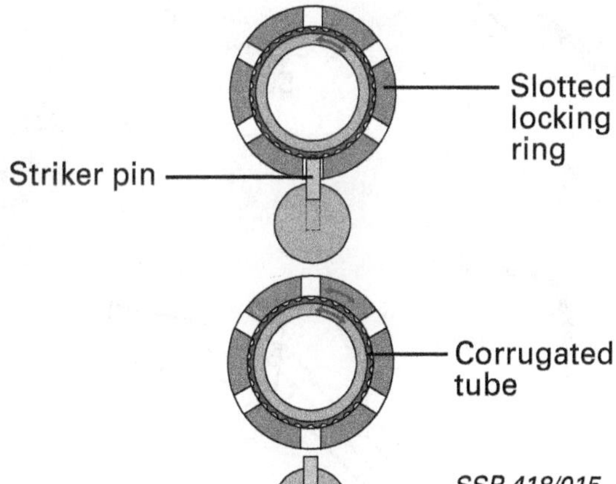

SSP 418/015

CAUTION!
Do not use the steering lock to counter-hold the column when removing the steering wheel bolt. If necessary, have a second technician hold the steering wheel while loosening the bolt.

Steering

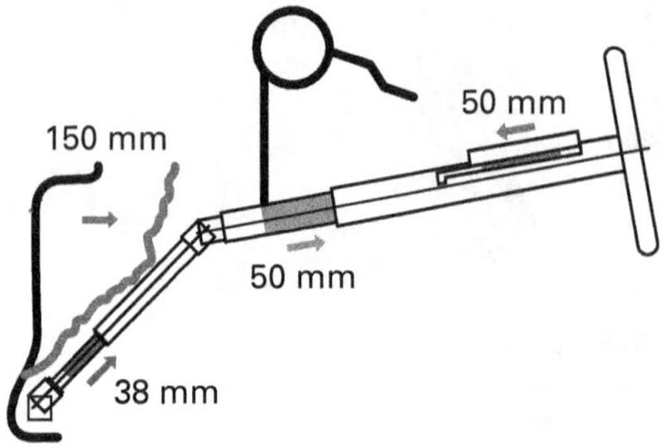

SSP 418/017

Collision protection

In the event of a collision, the steering column is designed to collapse in three ways to prevent it from being forced into the passenger compartment.

The column's universal joints allow the bulkhead to be displaced rearward by as much as 150 mm without damaging the column.

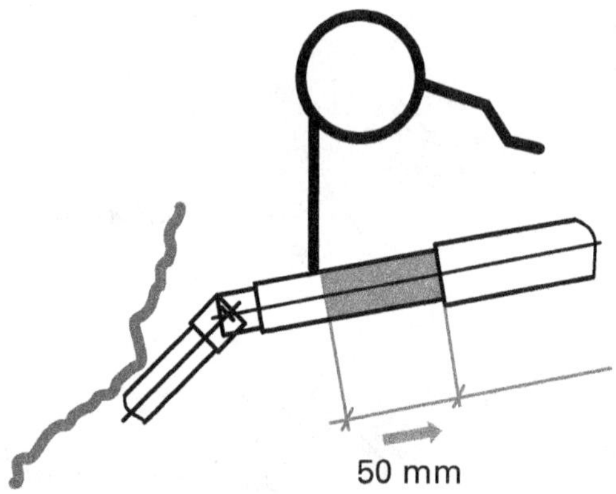

SSP 418/211

If the bulkhead is displaced far enough to reach the steering column, the column can telescope up to 50 mm.

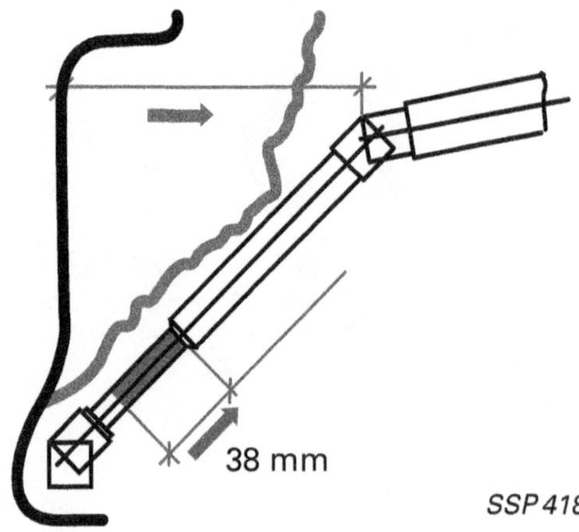

SSP 418/213

If the impact causes a force to come from below the steering rack, the double universal joint shaft can collapse up to 38 mm.

Steering

The mounting bolts hold the column to the central tube (crossmember) through tapered slots in the mounting pedestal.

If the driver's body makes contact with the inflated airbag, the steering column mount is displaced forward by 50 mm.

The force of the driver's impact is dissipated when the tapered slots of the mounting pedestal are forced against the mounting bolts.

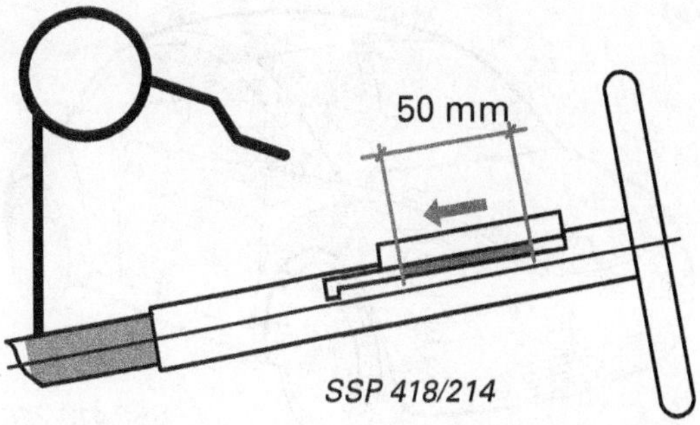

SSP 418/214

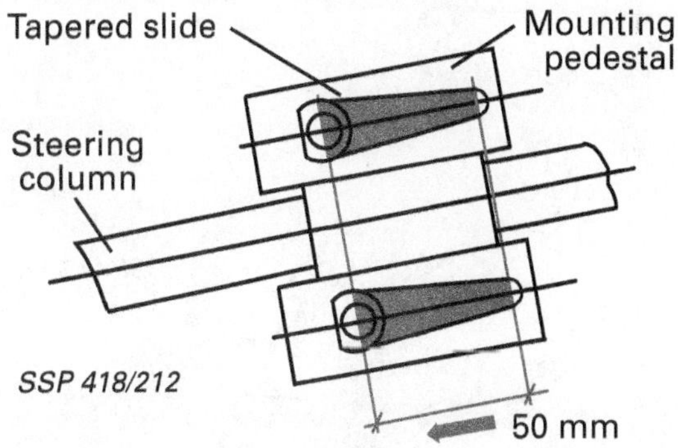

SSP 418/212

Servicing the steering column

To provide for the telescope function, the steering column is two piece. An inner shaft is splined to the outer column. The two shafts should not be separated when removing or installing the steering column.

New steering columns come with a transportation pin installed in the column to prevent separation. If damaged, the steering column can not be repaired and must be replaced as a unit.

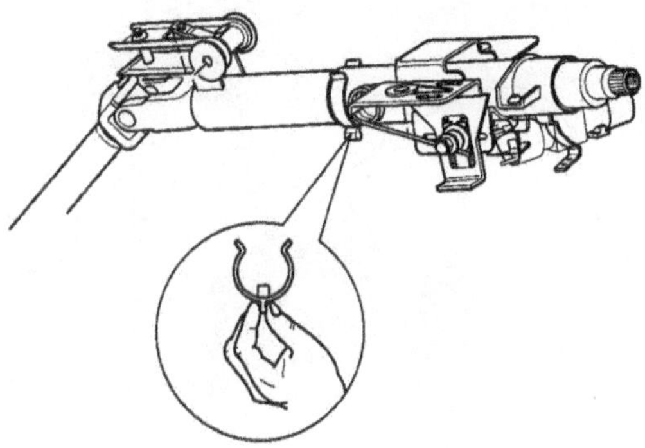

SSP 418/139

> ***CAUTION!***
> ***Do not separate the steering column shafts. Column shafts that are separated and reassembled may produce a knocking noise if the splines are not matched as originally assembled.***

Body—Exterior

SSP 418/226

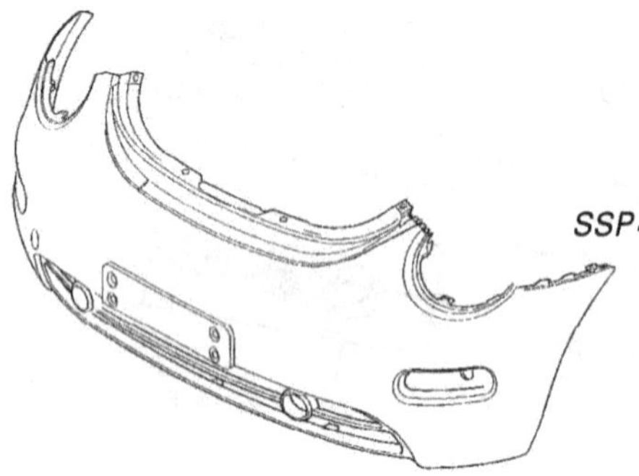

SSP 418/176

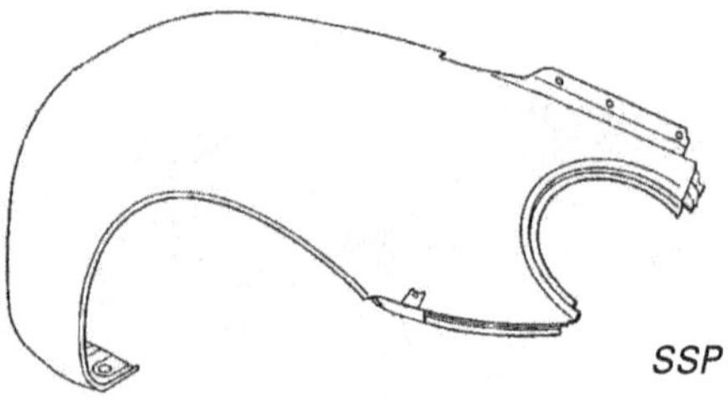

SSP 418/179

SSP 418/023

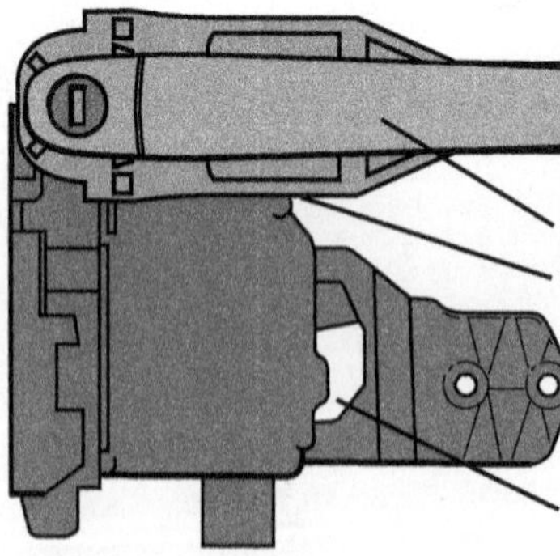

The body of the New Beetle benefits from construction techniques that strengthen it and help protect it from corrosion. The torsional rigidity is higher than that of current Golf/Jetta vehicles. New materials and manufacturing processes increase both durability and safety.

High-strength steel panels

Special alloy steel is used for some of the body panels which allows them to be thinner than normal. This steel is relatively soft when formed and welded but achieves its high strength when the body is baked at approximately 170°C (338°F) during the paint curing process. This leaves the thin panels as strong as the thicker ones and saves a great deal of weight in the body structure.

TPO body parts

The New Beetle uses TPO (thermoplastic olefin) fenders and front and rear bumper covers. The obvious advantage of plastic is its superior resistance to minor dents and corrosion.

> ***CAUTION!***
> ***Plastic parts are susceptible to damage from some auto body refinishing materials. When repainting plastic body parts, only use approved refinishing materials and paints.***

Door lock reinforcement

The door locks on the New Beetle are encased in a metal reinforcement plate. This reinforcement makes it more difficult for thieves to pierce a hole in the door to unlock the vehicle.

Body—Exterior

Fully galvanized body

The New Beetle has a fully galvanized body. Panels that are not externally visible are hot-dip galvanized. Because the hot-dip galvanizing process leaves a textured surface, electrolytically galvanized panels, which have a smoother surface, are used for the exterior body panels.

The use of fully galvanized sheet metal construction allows Volkswagen to offer a 12-year rust-through warranty.

Five-step paint process

Electrolytic galvanizing provides a good base for the five-step paint process.

A zinc phosphate coating is followed by an electro-chemical primer dip coating.

A filler is applied, and then the top color coat. The metallic finishes also receive a clear coat.

The filler and top coats have elastic properties that make them chip resistant.

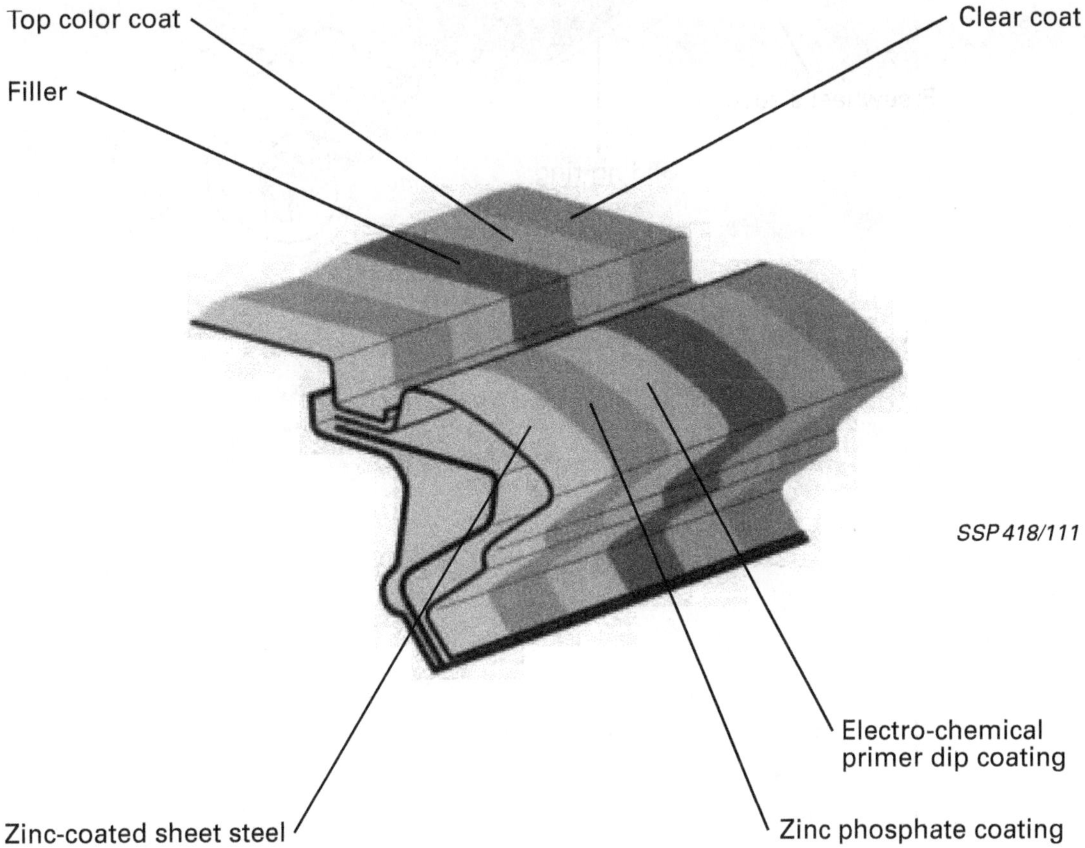

SSP 418/111

Body—Exterior

Door locks

The lock cylinders for the doors, rear lid, and ignition are "freewheeling." This feature prevents the locks from being forced and limits damage caused by attempted break-ins. The lock cylinder unit consists of the following components:

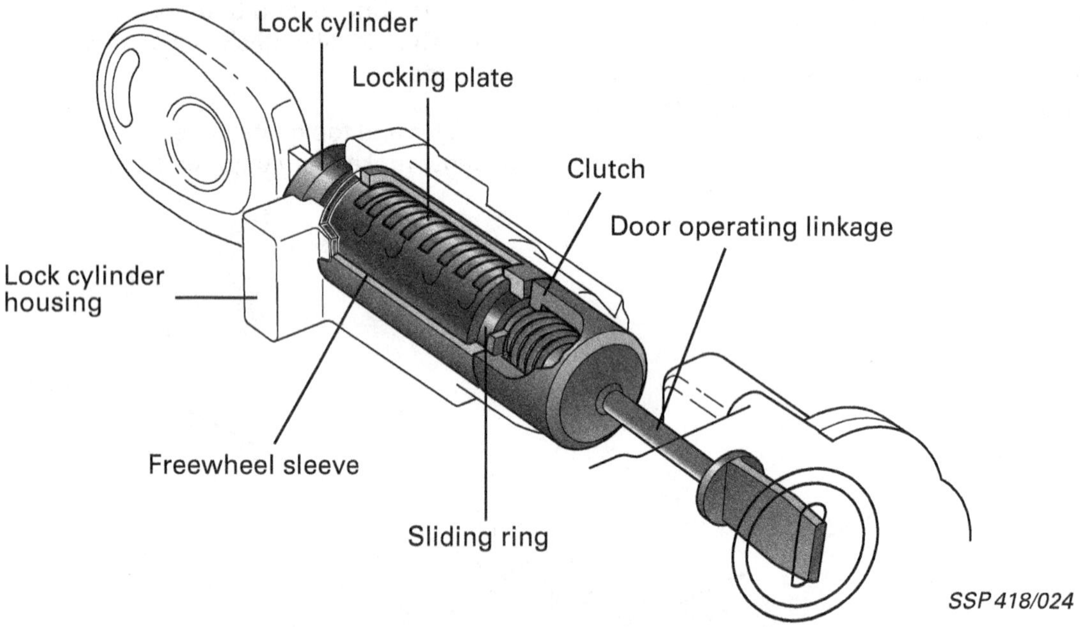

SSP 418/024

Body—Exterior

Freewheeling lock cylinder function

When the key is not inserted into the lock cylinder the locking plates engage in the freewheel sleeve.

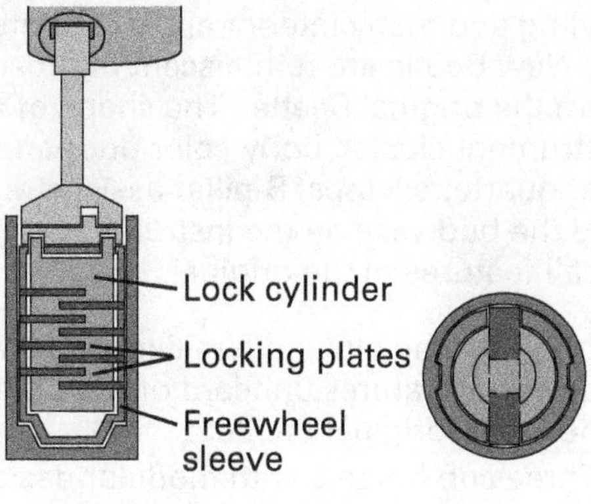

SSP 418/025

Inserting the correct key in the lock draws the locking plates into the cylinder.

When the key is turned, the lock cylinder rotates inside the freewheel sleeve, turning the sliding ring and operating linkage.

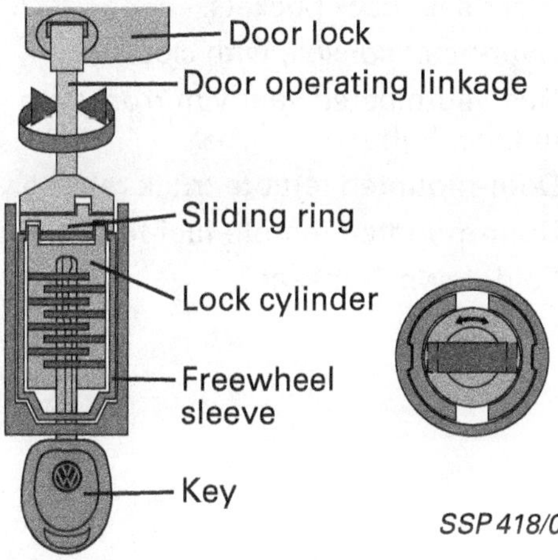

SSP 418/026

If the wrong key or other object is inserted into the lock the locking plates stay engaged in the freewheel sleeve.

When the key is turned, the freewheel sleeve is also turned. Two cams in the lock cylinder housing push the freewheel sleeve into the lock cylinder housing.

The freewheel sleeve disengages the sliding ring from the lock cylinder, preventing the ring from turning the operating linkage. The door, lid or ignition stays locked.

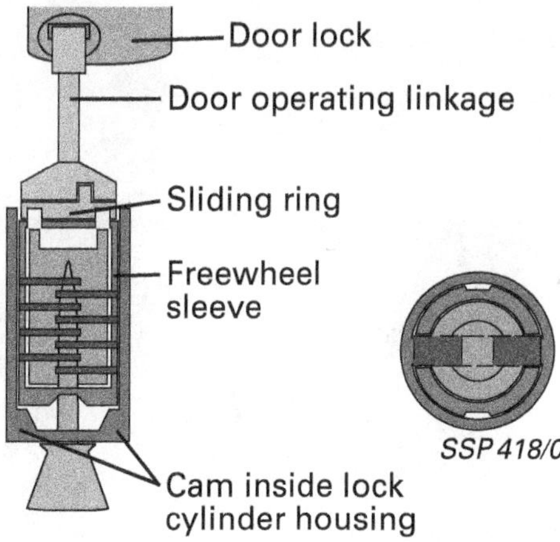

SSP 418/027

Body—Interior

Styling and many interior appointments of the New Beetle are reminiscent of those from the original Beetle. The shape of the instrument cluster, body-color door and rear quarter sill tops, B-pillar assist straps and the bud vase on the instrument panel recall features of the original.

Along with the old and familiar style are many new features unheard of during the time of the original Beetle:

- Three cup holders with modular design.
- Smoker's option replaces one cup holder.
- Door storage nets.
- Front seat back pockets.
- Over-head console with clock.
- Dual illuminated vanity mirrors with indirect lighting.
- Door-mounted remote trunk release.
- Door-mounted remote fuel-filler release.
- Fold-down rear seat.

Front seats

The front seats have a mechanical ratcheting seat height adjustment feature similar to that found on the new Passat. A lever on the outboard side of each seat can be pulled up to raise seat height or pushed down to lower the height.

The easy-entry system built into the front seats works automatically to help improve access to the rear seat. When the front seat back is tilted forward, the seat slides forward on the adjustment track to provide easier entry to the rear seat area.

When the seat back is pushed back again, the seat automatically returns to its previous adjustment position.

SSP 418/124

Body—Interior

Door panels

The door component carrier is made of sheet metal and is a structural part of the door. They are designed to help prevent injury to the occupants in the event of a side impact.

In addition to a side-impact beam in the door, there are also blocks of high-density foam padding to protect the occupants' ribs and pelvis.

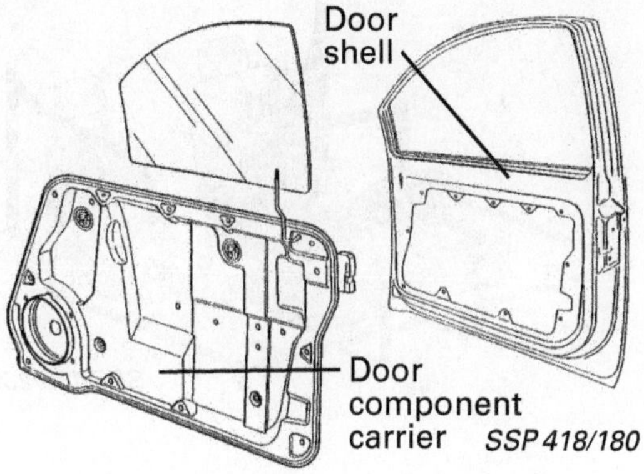

SSP 418/180

> **CAUTION**
>
> **This training module applies to only some of the vehicles covered by this manual. This information is subject to change and should be used as reference only. Always refer to the latest Volkswagen factory service information for the most up to date service, repair and training information.**

Airbags and Seat Belts

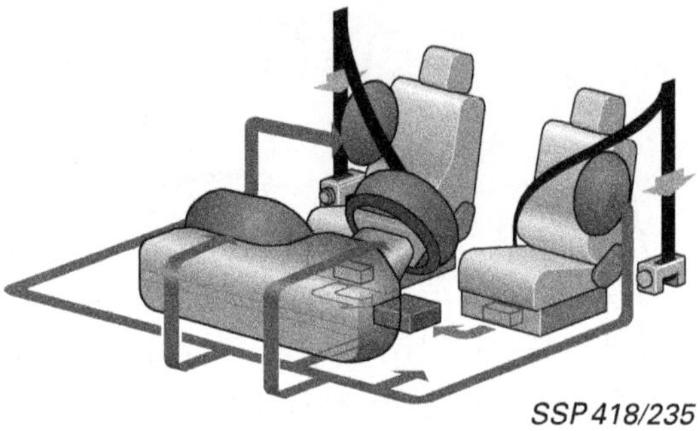

SSP 418/235

Airbags

Driver and passenger front and side airbags are standard equipment. The Airbag Control Module is mounted on the center tunnel in the front of the passenger compartment. It contains the Crash Sensor for the front airbags. The Side Airbag Crash Sensors are mounted on the floor under the front seats.

The system is designed to only deploy the airbags facing the direction of the collision. The front airbags only deploy in a frontal collision if there is sufficient force and the angle of impact falls within the 60° span of the forward direction of travel.

A Side Airbag will deploy in response to a signal from its Crash Sensor. An impact on one side of the vehicle will not deploy the Side Airbag on the opposite side.

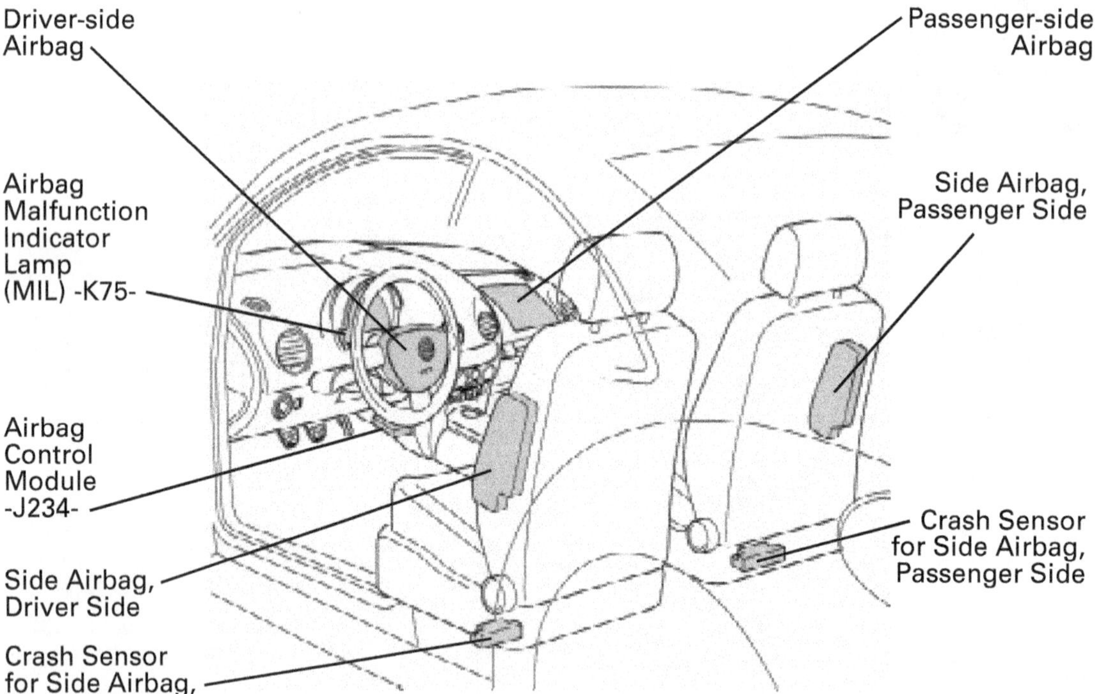

SSP 418/236

Airbags and Seat Belts

In addition to activating Airbag Igniters, monitoring Crash Sensors and checking the circuitry for malfunctions, the Airbag Control Module sends a signal to the Central Control Module to unlock the doors, switch on the interior light, and switch on the emergency flashers if airbags are deployed.

- J234 Airbag Control Module.
- J393 Central Control Module for Comfort System.
- F220 Lock Unit for Central Locking, Driver Side.
- F221 Lock Unit for Central Locking, Passenger Side.

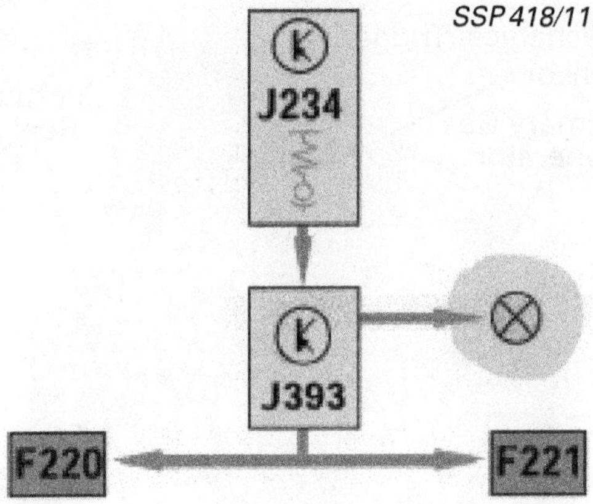

SSP 418/117

> ***WARNING!***
> ***When removing the front seats, steps must be taken to prevent static electricity from deploying the Side Airbags.***
>
> ***First , touch the door striker pin to discharge any build up of static electricity before disconnecting the connectors.***
>
> ***Then, connect Adapter for Side Airbag VAS5061 to the side airbag harness connector to establish a Ground circuit for the system.***

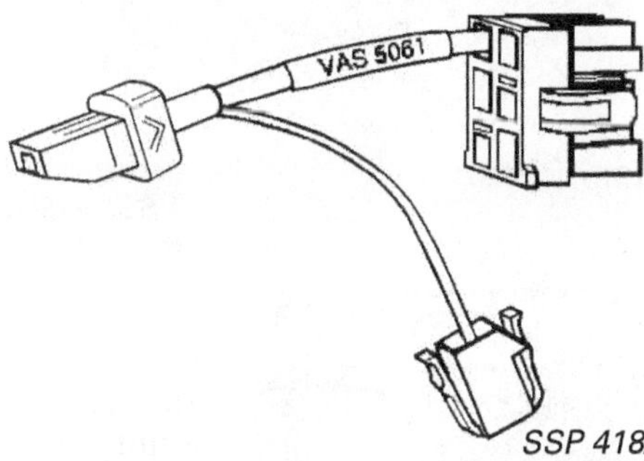

SSP 418/052

Airbag On Board Diagnostic (OBD)

On Board Diagnostic capability now includes an output Diagnostic Test Mode (DTM) for the airbag system.

After beginning the output DTM, a signal from the Airbag Control Module will unlock the doors, switch on the emergency flashers, and switch on the interior lights (if the switch is set to the door contact position).

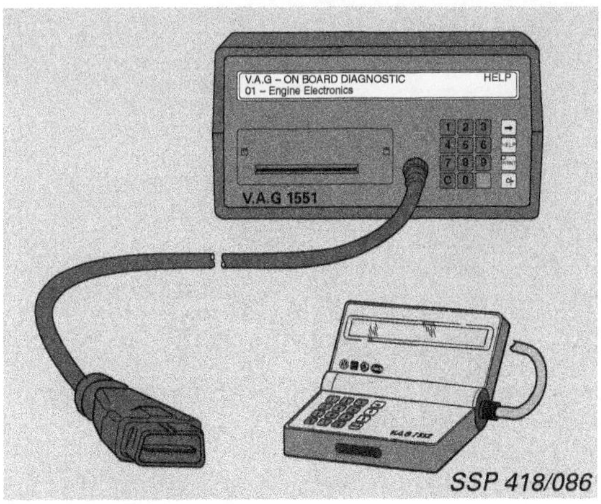

SSP 418/086

Airbags and Seat Belts

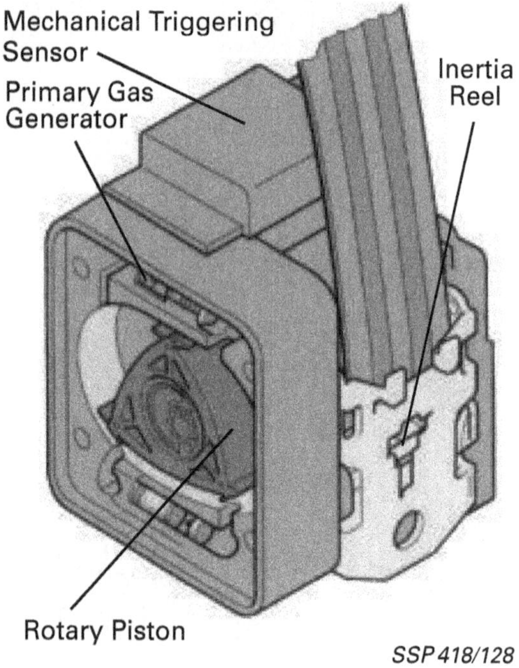

SSP418/128

Pyrotechnic Seat Belt Tensioners

Pyrotechnic seat belt tensioners are used on the front seats only. They are a Rotary Piston type, and are mechanically activated by the force of deceleration in a collision.

When the tensioners are deployed, the seat belts are retracted 200 mm, taking up normal slack in the belt webbing, and pulling the occupants back into the seats.

There are three working chambers in the Seat Belt Tensioner.

The following sequence describes the operation of the Pyrotechnic System:

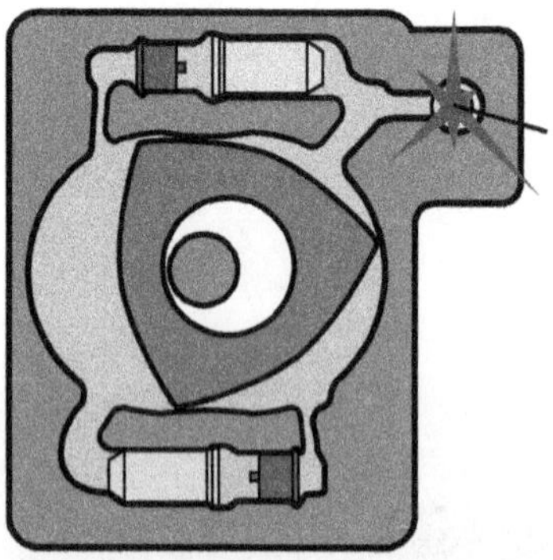

SSP418/129

The Mechanical Triggering Sensor ignites the Primary Gas Generator with an impact bolt and igniter.

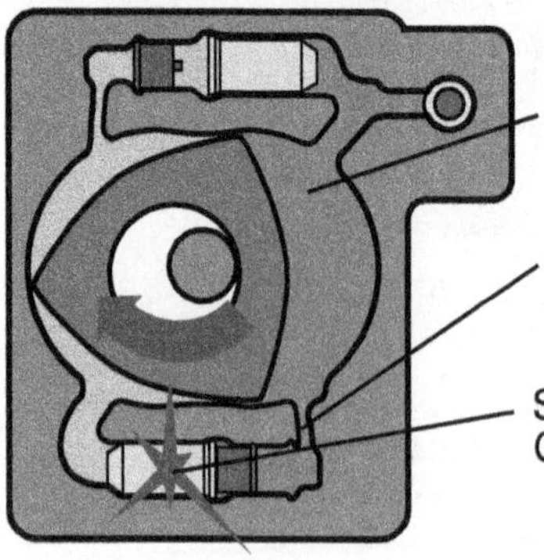

SSP418/130

The expanding gas in the first working chamber starts to turn the Rotary Piston.

Once the charge reaches the 1st Overflow Port, it ignites the Secondary Gas Generator by means of an impact pin.

Airbags and Seat Belts

The piston is rotated further as a result the secondary gas propellant charge.

This charge, in turn, reaches the 2nd Overflow Port and ignites the (3rd) Third Gas Generator.

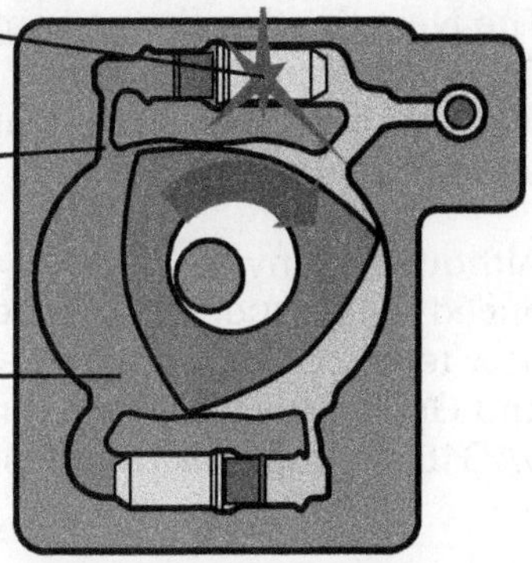

SSP 418/131

The expanding gas from the third charge drives the Rotary Piston through the final part of its rotation.

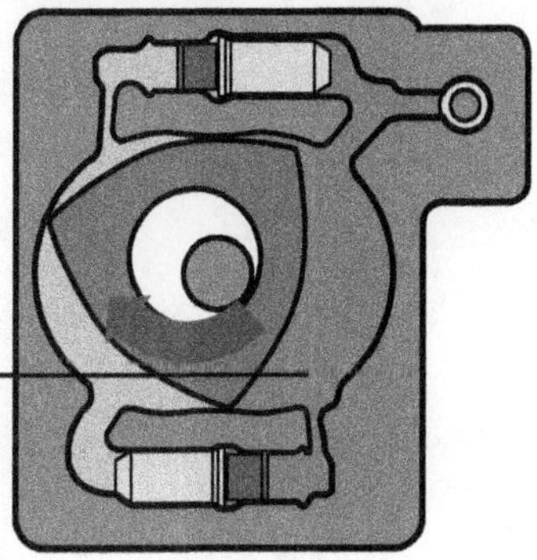

SSP 418/132

Belt force limiter

A belt force limiter is built into the inertia reel to prevent bruising or internal injuries to the vehicle occupants from the force of impact and the retraction of the seat belts.

A torsion shaft in the belt reel mechanism acts like a spring to limit the tensile force.

SSP 418/170

Child seat feature

All seat belts other than the driver's have a child safety feature that is activated by fully extending the belt. When activated, a ratchet allows the belt to retract only, enabling children or child safety seats to be restrained more securely.

Comfort and Convenience Features

The New Beetle offers a comprehensive list of standard comfort and convenience features that have often been optional equipment on other models.

Although many technicians are experienced in the operation of these systems, new technology controls their operation, and On Board Diagnostic through the VAG1551 simplifies troubleshooting.

New Beetle standard equipment:
- Central Locking with Radio-frequency Remote Control.
- Heated Power Mirrors.
- Anti-theft Alarm System.
- Interior Light Control.

Optional power accessories:
- Power Windows with one-touch down/up and Pinch Protection.
- Convenience-open for Power Windows.

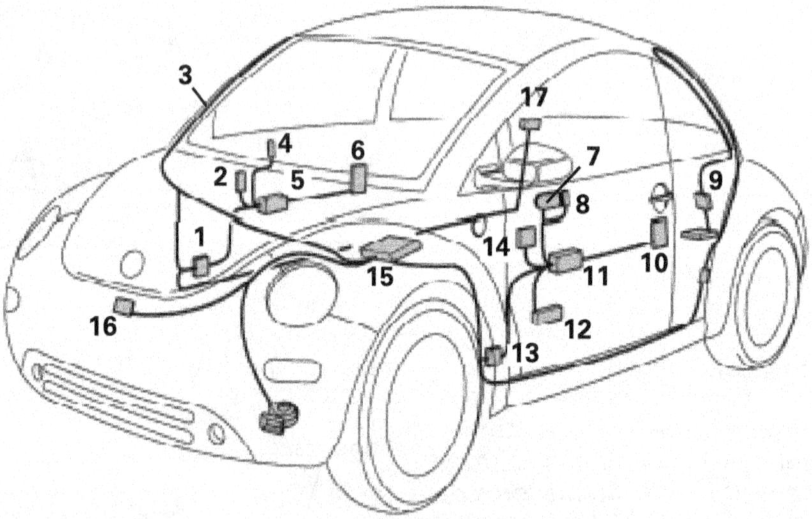

SSP418/237

1 A-pillar Connector Station
2 Switch for Window Regulator, in Passenger Door -E107-
3 Antenna Wire for Central Locking and Anti-theft Warning System -R47-
4 Switch for Interior Lock, Passenger Side -E198-
5 Door Control Module, Passenger Side -J387-
6 Lock Unit for Central Locking, Passenger Side -F221-
7 Mirror Adjustment Switch -E43-
Mirror Selector Switch -E48-
8 Switch for Interior Lock, Driver Side -E150-
9 Motor for Remote Unlock Rear Lid -V151-
10 Lock Unit for Central Locking, Driver Side -F220-
11 Door Control Module, Driver Side -J386-
12 Switch for Remote Unlock, Rear Lid -E188-
Switch for Remote/Fuel Tank Door -E204-
13 A-pillar Connector Station
14 Left Front Window Switch -E40-;
Right Front Window Switch, Driver Side -E81-
15 Central Control Module for Comfort System -J393-
16 Hood Alarm Switch -F120-
17 Motor for Fuel Tank Lid Unlock -V155-

Comfort and Convenience Features

Comfort and convenience systems

Although these features can be found on other Volkswagen models, what is new on the New Beetle is the way that the comfort and convenience systems are linked together.

The system includes the Central Control Module mounted under the instrument panel near the steering column, and two Door Control Modules mounted with the power window motors.

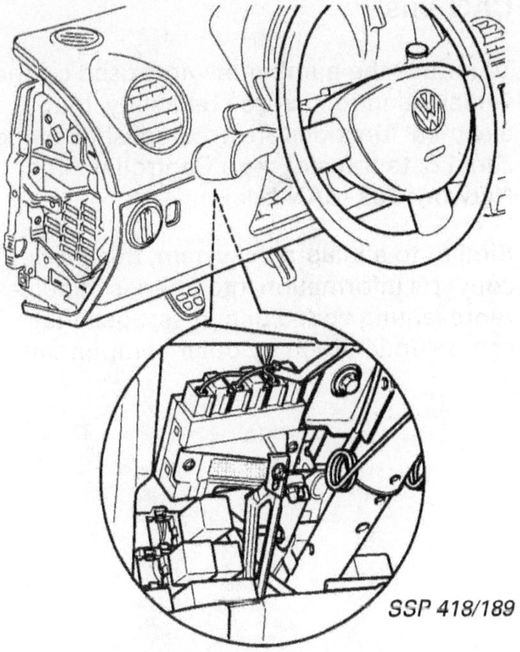

SSP 418/189

The Central Control Module and the Door Control Modules communicate with each other using a Controller Area Network bus (CAN-bus). Vehicles not equipped with power windows use a different Central Control Module.

The Central Control Module controls the following:

- Interior Light Control.
- Anti-theft Warning System.
- Radio-frequency Remote Control for Central Locking.
- Electrical system interface.

The Door Control Modules control the following:

- Central Locking.
- Power Windows.
- Power Mirror Adjustments.

Central Control Module (with Power Windows)

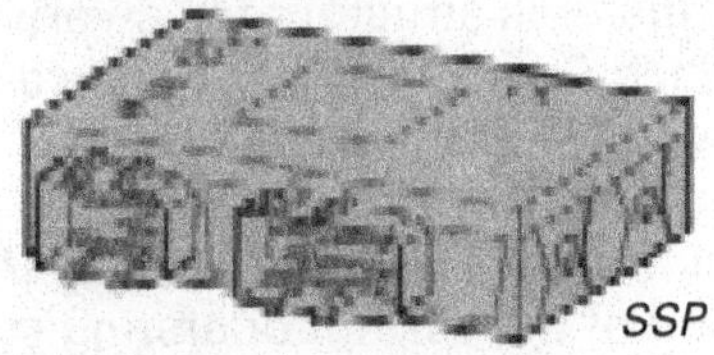

SSP 418/104

Central Control Module (without Power Windows)

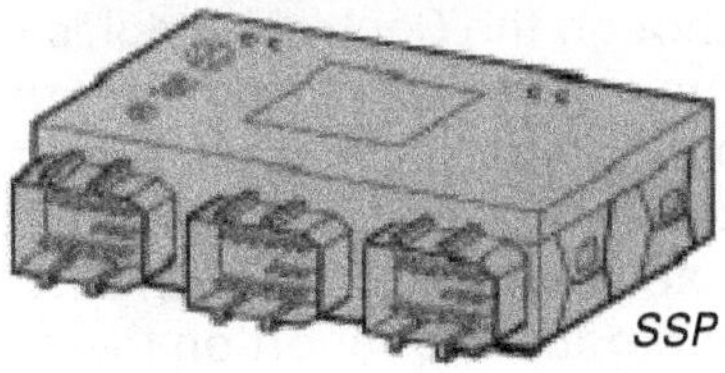

SSP 418/103

Comfort and Convenience Features

CAN-bus

To reduce the number of wires used on the vehicle, and to increase reliability, many electrical functions are accomplished using serial data transfer via a Controller Area Network bus (CAN-bus).

Similar to a telegraph system, this process converts information into a series of pulses representing coded data messages that can be understood by other components.

There are several advantages to using a CAN-bus to control the operation of the comfort and convenience features.

Wiring is greatly simplified. Only short runs of wire are needed between the Door Control Modules and the components in the doors. Only 10 wires pass between the door and the A-pillar, where up to 30 wires were used to control these components on current Golf/Jetta models.

Data transfer

Data transfer is structured in a very simple way. Each Control Module sends a message every 20 ms. This message contains the status data and commands. All of the Control Modules on the CAN-bus are informed of the current operating status of the system at these intervals.

These signals are binary coded messages and although the Control Modules receive the same signals, only those programmed to recognise specific messages respond. Because the system is bi-directional, messages can travel back and forth between all Control Modules to report on their status.

Although just a single wire is sufficient to transmit the data, two wires are used for greater resistance to interference. Digital messages are transmitted along both cables simultaneously, but inverted to one another. When the signals are analyzed by the Control Modules, the voltage of the inverted signal is subtracted from the other signal.

This makes it possible to distinguish the normal signals from interference, since an interference spike would travel along both wires with the same polarity.

The messages, divided into seven fields, include a status field to identify their importance. If two messages require transmitting at the same time, the message with the highest priority is transmitted first.

Comfort and Convenience Features

Power windows

Power windows are available as optional equipment, and are controlled by the Door Control Modules. These are part of the comfort system, and built into the Window Motors. The motor can be removed without removing the carrier assembly.

The power windows are operated by a five-position switch. This system also has one-touch up and one-touch down functions.

The power windows can be raised or lowered with or without the one-touch feature.

To raise or lower the window, move the switch to the first detent and hold until the window reaches the desired position.

To fully raise or lower the window using the one-touch feature, move the switch to the second detent and release.

Pinch Protection only operates during the one-touch up function. If an obstruction prevents a window from closing, the window motor reverses itself. This feature is only operational with the interior window switches.

If a window binds while being closed using the one-touch up function, the Pinch Protection feature may prevent it from fully closing (a build-up of ice or dirt on the window or door frame may cause this to happen). Closing the window a second time, the window will stop at the obstruction but will not retract. Remove any obstruction, then close the window.

Each motor also includes a position sensor. The Door Control Module uses this signal to determine window-closed and window-open positions. The information from the sensors is stored in memory for use with the one-touch function.

If the power supply to the Door Control Module is lost, the one-touch function will be deactivated. To reset the one-touch function, run the window all the way down, then up completely. The Control Module will lock the sensor input position into memory, and the one-touch function will be restored.

The power windows remain functional for 10 minutes after the the ignition is switched off, provided that a door is not opened. The one-touch up function is deactivated when the ignition is switched off.

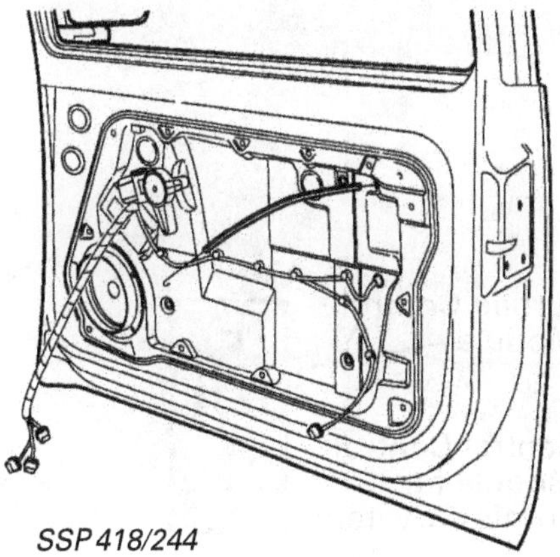

SSP 418/244

Comfort and Convenience Features

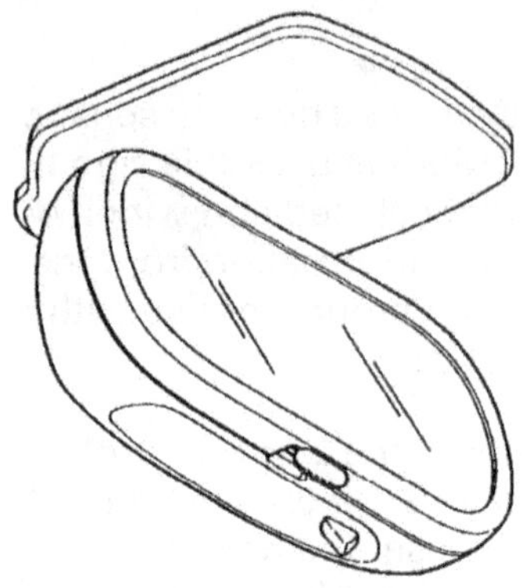

SSP 418/181

CAUTION

This training module applies to only some of the vehicles covered by this manual. This information is subject to change and should be used as reference only. Always refer to the latest Volkswagen factory service information for the most up to date service, repair and training information.

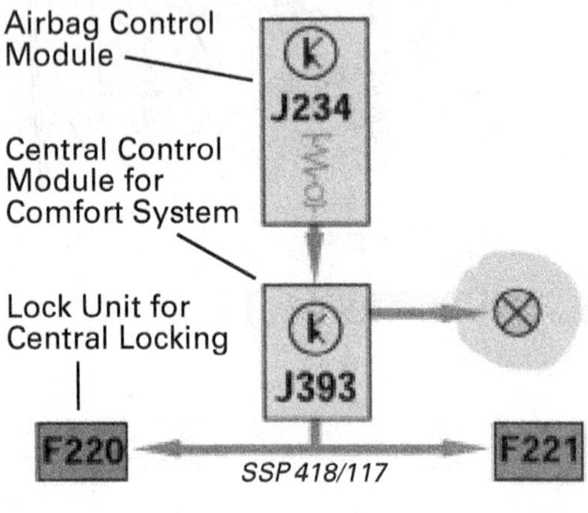

SSP 418/117

Interior Light Control

The Central Control Module for Comfort System -J393- controls the switching and dimming of the interior lights. It also limits the length of time the lights remain on if the doors are left open with the ignition switched off.

The Central Control Module for Comfort System -J393- receives signals from the Door Contact Switches and the Rear Lid Contact Switch. They are microswitches built into the lock assemblies.

If the interior light has been switched on, the control module will switch it off again after one hour. If the light is switched off, then on again, the sequence repeats.

Interior lights will be switched on for a maximum of 30 seconds, then dimmed to off, if the control module receives any of the following inputs:

- Car unlocked with key or remote control.
- One door opened from inside.
- Ignition key removed from ignition with doors closed.

When the doors are locked, or when a door is closed after the doors have been locked, the interior lights are switched off.

If an accident deploys an airbag, the Airbag Control Module -J234- sends a signal to the Central Control Module for Comfort System -J393-.

- The interior lights are switched on for 10 minutes (if the light switches are set to door contact position).
- The doors are unlocked.
- The emergency flashers are activated.

Comfort and Convenience Features

Rear lid and fuel tank lid locks

The Rear Lid is always mechanically locked. It can be opened electrically using the switch in the driver's door, or the remote transmitter. In the event of an electrical system malfunction, it can be opened using the key.

The trunk release switch, located in the driver's door panel, signals the Central Control Module for Comfort System -J393-. The relay supplies current to the motor to unlock the rear lid.

The rear lid lock cylinder and lifting handle are located under the VW emblem on the rear lid. The emblem swivels to the side to reveal the lock and handle. As the emblem moves the handle pops out of its recess. When the key is turned, a microswitch in the lock is activated by a cam on the lock cylinder.

The microswitch sends a signal to Central Control Module for Comfort System -J393-, which activates the Relay -J398-, switching current to the unlocking motor.

CAUTION!
The rear lid handle does not unlatch the rear lid. Only the switch in the driver's door, the key in the rear lid lock, or the remote control will open the lid.

The motor for the fuel tank lid lock is powered directly through the switch in the drivers door. If an electrical malfunction prevents the fuel tank lid from opening it can be opened mechanically. Removing a round access panel inside the luggage compartment (right side) reveals the lock cable. Pulling on the cable releases the lock.

Sleep Mode

To reduce current draw from the comfort/ convenience electronics, a Sleep Mode is programmed into the system. The system draws 150 mA in an active state. When the ignition is switched off, the system goes into a "wait" state.

When all door lock swiches are recognized as closed and power window function timers have run out, the Door Control Module sends a "set sleep bit" via the CAN-bus.

Once all Control Modules on the bus recognize the set sleep bits of the other Control Modules, they cease data transfer and go into Sleep Mode.

In Sleep Mode, only the Anti-theft Alarm, Radio-frequency Remote Control, and the Alarm LED are operational. Current draw drops to 7 mA. Any input that requires the system to react, such as a door-open signal from the remote control, results in a wake-up command and the system becomes active within 100 ms.

Comfort and Convenience Features

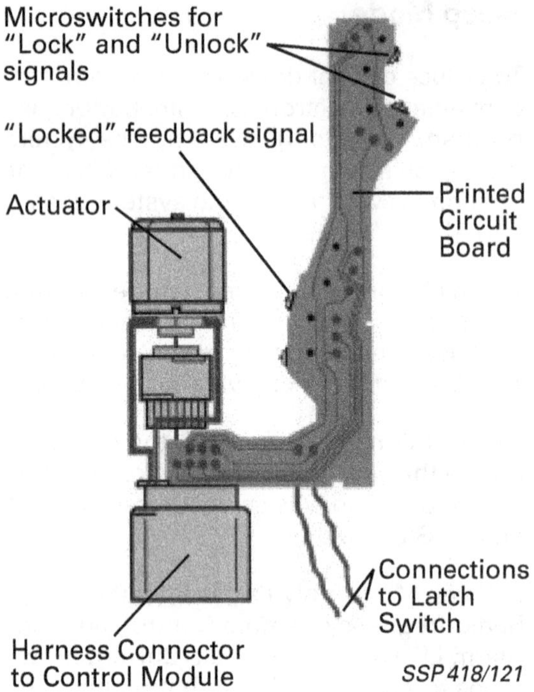

SSP 418/121

Door lock electronics

The door lock assemblies contain microswitches used to operate the Anti-theft Alarm, Central Locking and Interior Light Control. The switches provide input to the Door Control Modules where their signals are processed and sent out over the CAN-bus. The switches are not replaceable.

The microswitches provide the following signals:

- DOOR OPEN signal.
- LOCK signal.
- UNLOCK signal.
- LOCKED feedback signal.

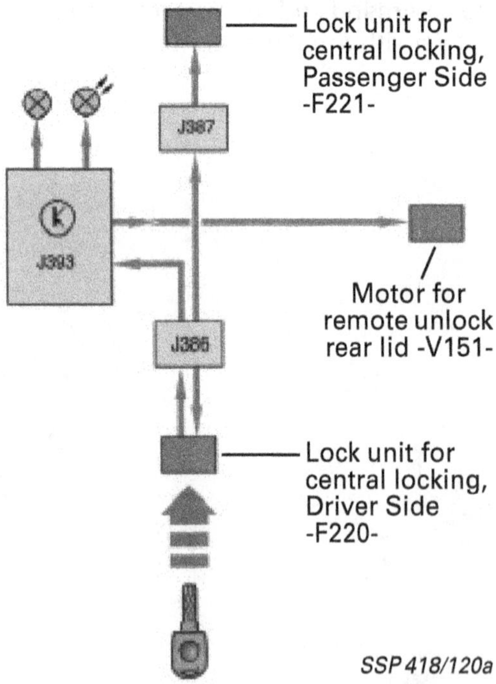

SSP 418/120a

Central Locking System function

The vacuum-controlled Central Locking System from current Golf/Jetta models has been replaced by electrically operated door locks. Each Door Control Module signals the Central Control Module and the other Door Control Module by way of the CAN-bus when a door is locked or unlocked.

The block diagram shows an example of using the driver's door lock to lock the vehicle, close the windows and sunroof and set the anti-theft alarm.

The Door Control Module, Driver Side -J386- is activated by the Lock Module for Central Locking, Driver Side -F220- when the key is turned in the lock cylinder.

- -J386- checks status of -F220- to determine whether door is closed.
- If closed, -F220- locks the driver's door.

Comfort and Convenience Features

The lock signal is sent over the CAN-bus to the Central Control Module for Comfort System -J393- and to the Door Control Module, Passenger Door -J387-.

- -J387- independently initiates the lock function in the passenger door.
- -J393- switches off the Interior Light, and activates the Anti-theft Alarm.

This sequence can be initiated by using the Radio-frequency Remote Control.

The Central Locking System can also be activated using the Switch for Interior Lock, Driver Side -E150- or the Switch for Interior Lock, Passenger Side -E198-. If a door lock button is pressed, only the lock for that door is activated.

The Central Locking System can also be unlocked by the Airbag Control Module. In the event of a collision with enough force to deploy the airbags, the Airbag Control Module -J234- also signals the Central Control Module for Comfort System -J393- to unlock the doors and switch on the interior light.

Central Locking Radio-frequency Remote Control

All New Beetles come with remote control for the central locks. It is built into the Control Module for Comfort System, or the Central Locking System Control Module for vehicles without power windows. The remote control has the following functions:

- Selective door unlocking.
- Locking of both doors.
- Setting of the Anti-theft Alarm.
- Rear lid opening.
- "Panic" function.

Selective unlocking unlocks the driver's door with one press of the unlock button. A second press of the unlock button will unlock the passenger door. If a door is not opened within 30 seconds of an unlock command being received by the Central Locking Control Module or Control Module for Comfort System, the doors will lock again.

The Anti-theft Alarm is set when the doors are locked. Confirmation is by a short sounding of the Anti-theft Warning Horn and a brief flash of the emergency flashers.

Pressing the rear lid button will unlatch the lid electrically.

The "Panic" function will set off the anti-theft warning horn and switch on the emergency flashers.

Comfort and Convenience Features

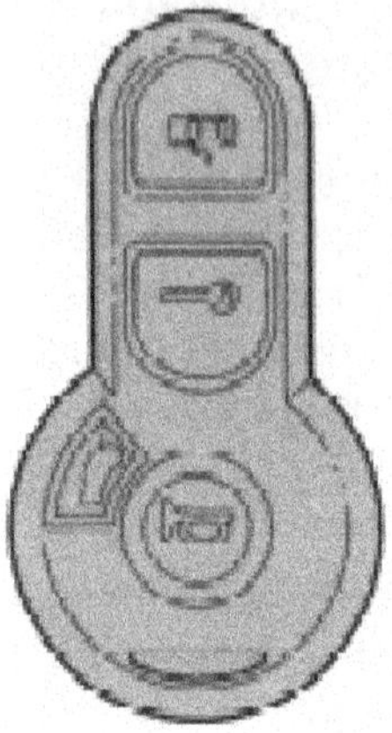

SSP 418/201

Remote Control coding

Up to four Remote Control units can be coded for use on one vehicle. There are two procedures available for coding. Codes can be programmed into individual remote controls using the procedures in the Owner's Manual. These procedures require the use of a spare key.

The VAG1551 Scan Tool can be used to program up to four keys at once. Refer to information on "Adaptation" Function 10 under Address Word 46 for the Comfort System or Address Word 35 for Central Locking, in the Repair Manual or the Volkswagen Electronic Service Information System (VESIS).

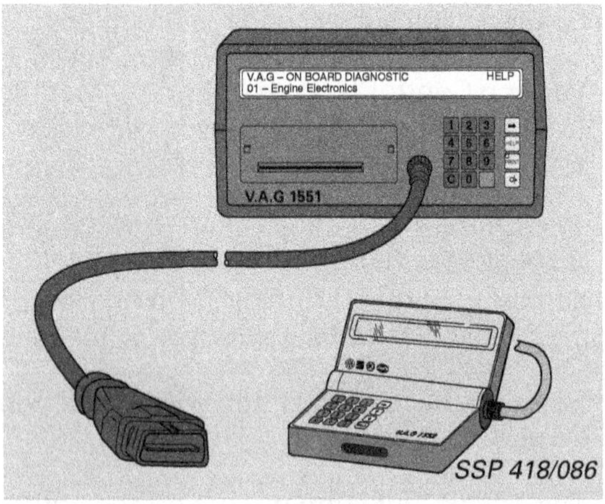

SSP 418/086

On Board Diagnostic (OBD)

The comfort and convenience system for the New Beetle has On Board Diagnostic (OBD) capability using the VAG1551 scan tool. Address Word 46 will provide the following functions:

01 - Check Control Module Version

02 - Check DTC Memory

03 - Output Diagnostic Test Mode

05 - Erase DTC Memory

06 - End Output

07 - Code Control Module

08 - Read Measuring Value Block

10 - Adaptation

Comfort and Convenience Features

Because there are three Control Modules in this system, Address Word 46 gives a slightly different display sequence. First the Central Control Module for Comfort System information is displayed. It is similar to other Control Module displays.

When the arrow button is pressed, however, "Select function" is not displayed. Instead, the Arrow button advances to the left Door Control Module display, and pressing the Arrow button again goes to the right Door Control Module display. "Select function" is displayed the next time the Arrow button is pressed.

"Automatic Test Sequence" Address Word 00 will print out all three Control Module displays before printing the status of DTC memory.

"Check DTC Memory" Function 02 helps pinpoint malfunctions and "Output Diagnostic Test Mode" Function 03 checks the Outputs for proper function.

"Read Measuring Value Block" Function 08 can be used to check all switch functions and the CAN-bus for the Comfort System. It will also display system status:

- Locked.
- Unlocked.
- Door OPEN.

Although the Central Locking System used on vehicles without power windows does not have a CAN-bus, these systems still have On Board Diagnostic capability. "Central Locking" Address Word 35 is used for access to the Control Module.

Refer to the Repair Manual or the Volkswagen Electronic Service Information System for diagnostic procedures using the VAG1551 Scan Tool.

With Power Windows

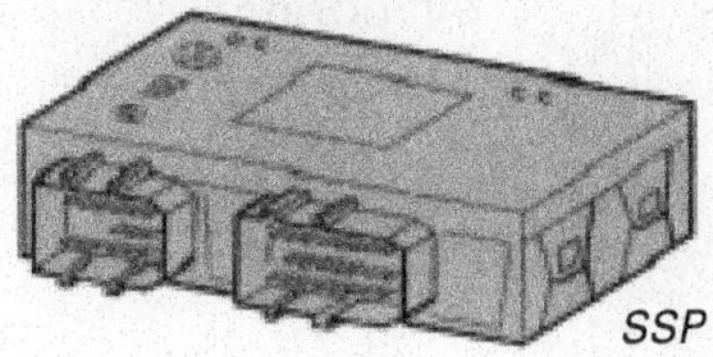

SSP 418/104

Without Power Windows

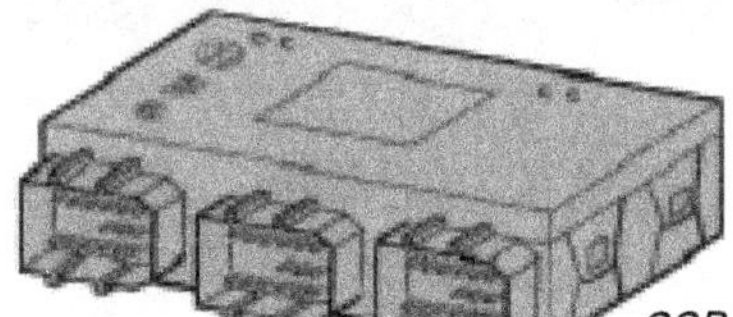

SSP 418/103

Heating and Air Conditioning

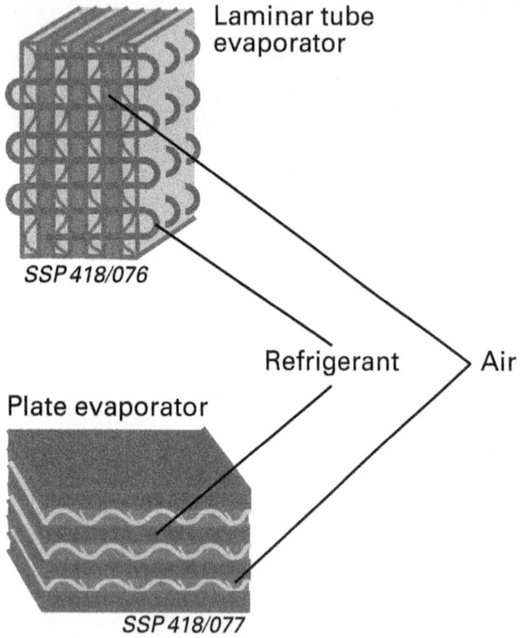

Air conditioning is standard equipment on all New Beetles. The system uses a blend door for temperature regulation. Air flow controls feature a new air recirculation function.

Refrigerant system

The New Beetle comes with the familiar expansion valve air conditioning refrigerant system using a variable displacement compressor to move the R134a refrigerant through the system.

What is new in the system is a "plate" evaporator that replaces the tube and fin version currently used on other Volkswagen vehicles. The new evaporator has the same capacity and cooling surface area as the old one while taking up 40% less space.

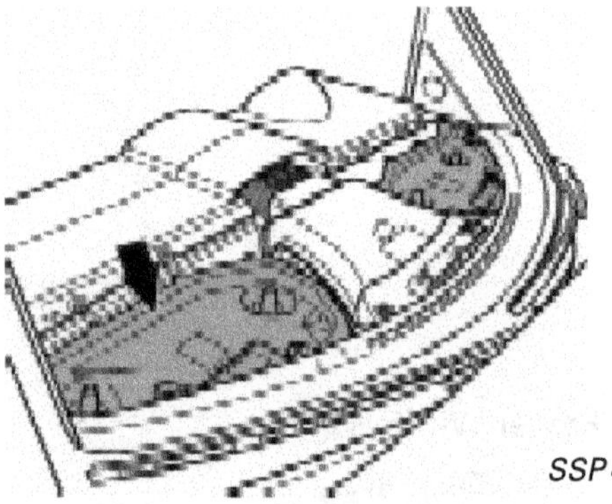

SSP 418/247

Dust and pollen filter

A new dust and pollen filter is used in the New Beetle.

The pollen filter is located in the plenum chamber. Access to the filter is from the inside of the vehicle. The top cover in the instrument panel covers the plenum and is removable.

Heating and Air Conditioning

Air Recirculation Mode

A servo motor is used to control the Fresh Air/Recirculation Flap.

The switch is on the heating and A/C control panel. If the system is in Recirculation Mode when the air flow controls are set to defrost, the Fresh Air/Recirculation Flap is switched to the fresh air function.

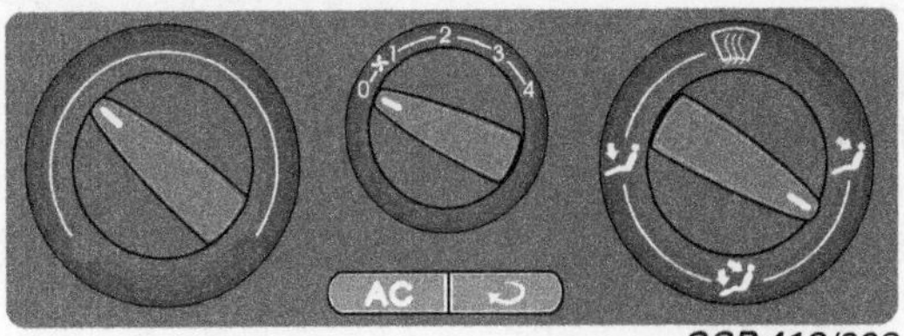

SSP 418/028

Defrost Mode

When the air flow controls are set to Defrost Mode, the Central Flap closes off air flow to the center instrument panel vent, but a small amount of air is directed to the side air outlets to help defrost the side windows.

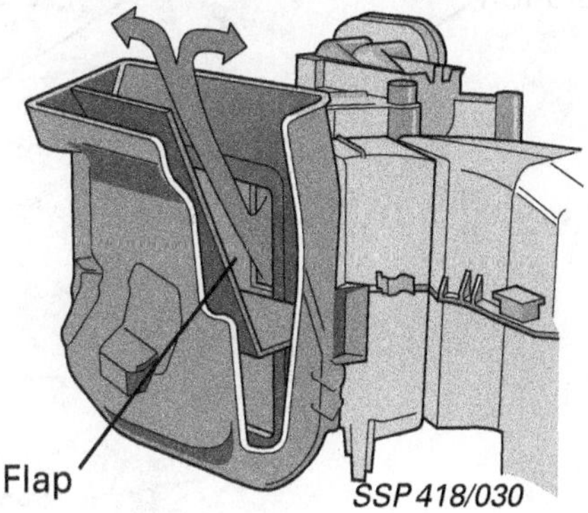

SSP 418/030

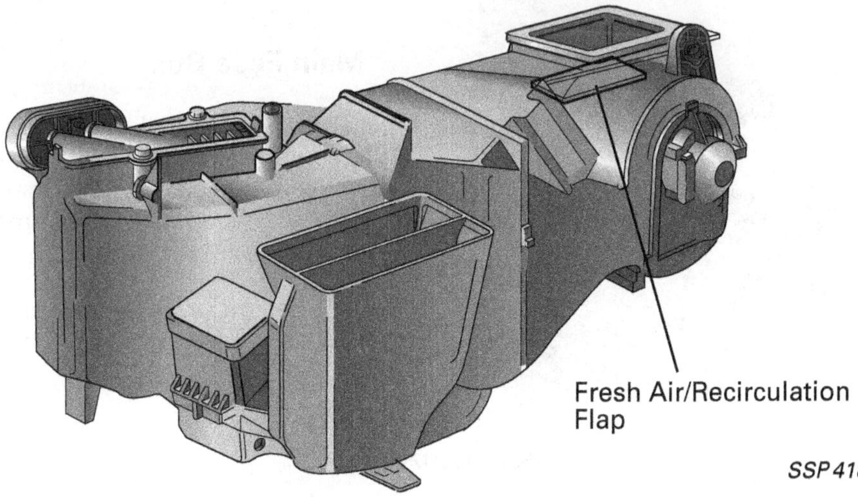

SSP 418/029

Electrical System

The New Beetle uses a decentralized electrical system. It's primary feature is that the electrical system is subdivided into separate connector stations, relay carriers and fuse boxes.

This system provides the technician with easier access to test points when troubleshooting electrical problems.

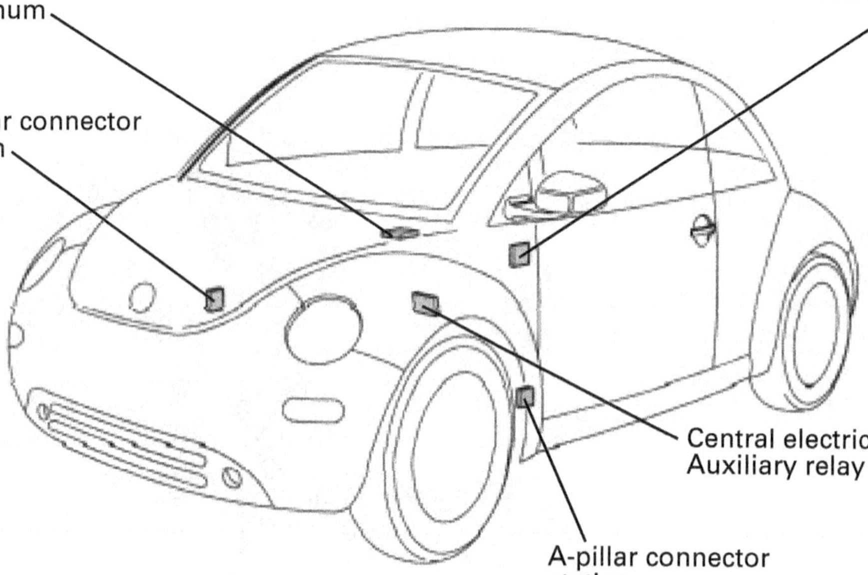

SSP 418/252

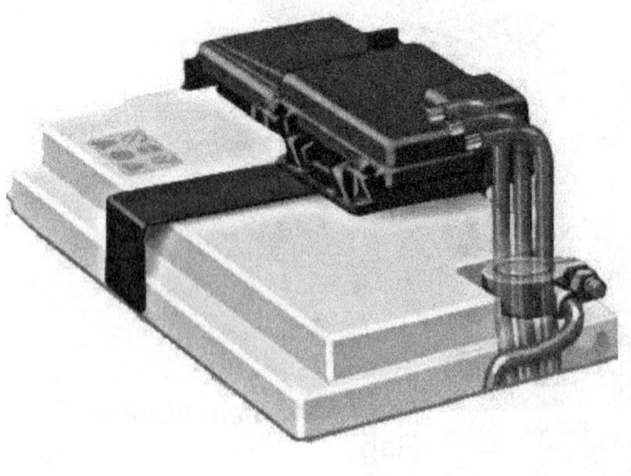

SSP 418/047

Main Fuse Box

A Main Fuse Box is located at the battery. It contains special fuses for high-current applications such as the Charging System, Engine Management Systems, and the radiator Coolant Fans.

This design protects the main wiring harnesses in the event of a short circuit.

Electrical System

Central Electric Panel with Auxiliary Relay Carrier

A Central Electric Panel is located where the fuse/relay panel is on current Golf and Jetta models. It has sockets for six relays. Below the relay sockets are several screw terminals for battery (terminal 30), ignition (terminal 15), and load reduction (terminal 75 or X) power.

An Auxiliary Relay Panel above the Central Electric Panel contains the relays for power accessories and optional equipment.

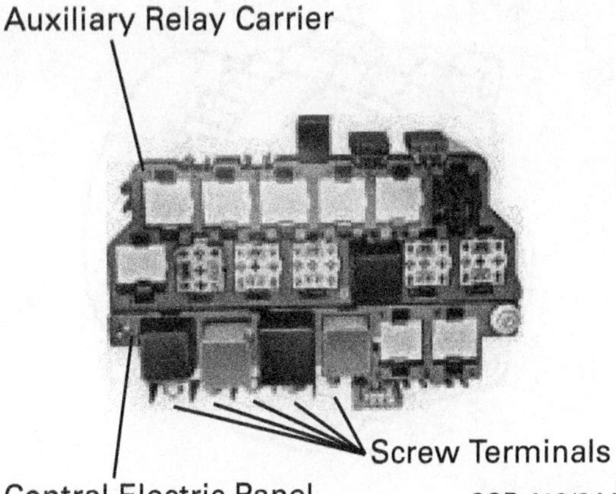

SSP 418/044

Fuse Box

The Fuse Box is located on the left end of the Instrument Panel behind a cover. It contains both standard size fuses and mini-fuses. The use of mini-fuses allows protection of individual circuits.

SSP 418/045

Connector Stations

The decentralized electrical system requires less wiring and reduces vehicle weight. The Connector Stations are located in the plenum and at the A-pillars.

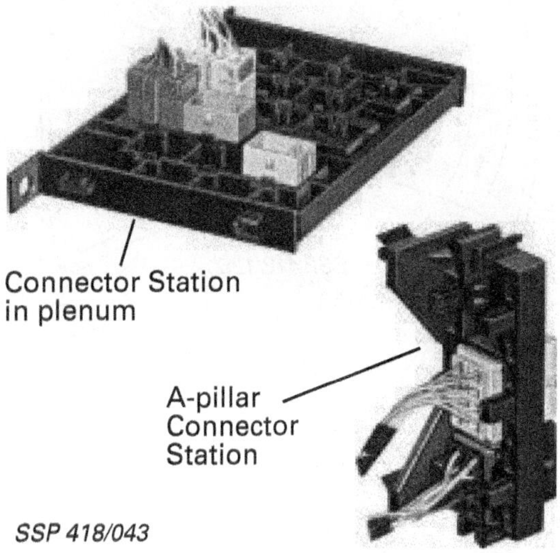

SSP 418/043

Electrical System

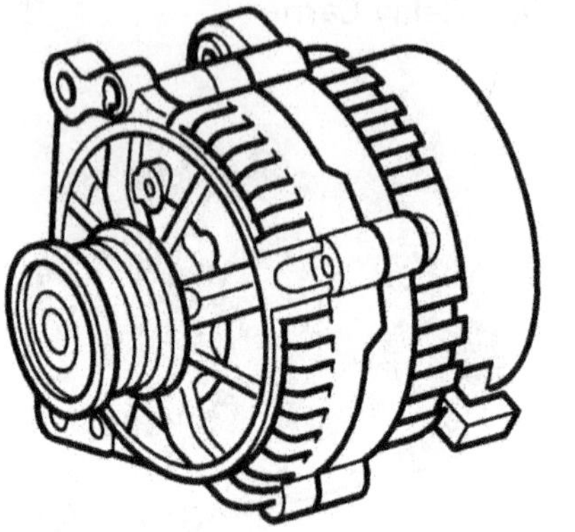

SSP 418/240

TDI Generator

Bosch Generators used on TDI engines with manual transmission are equipped with a Freewheel Disc on the ribbed belt pulley. This feature applies only during deceleration when the Glow Plugs are operating and there is a heavy load on the charging system. It temporarily cuts out the charging system and allows the Generator rotor to freewheel.

The Freewheel Disc can be retro-fitted to Bosch generators not originally equipped with them. Some TDI engines may be equipped with Valeo generators, which do not have the freewheel feature, and cannot be retro-fitted.

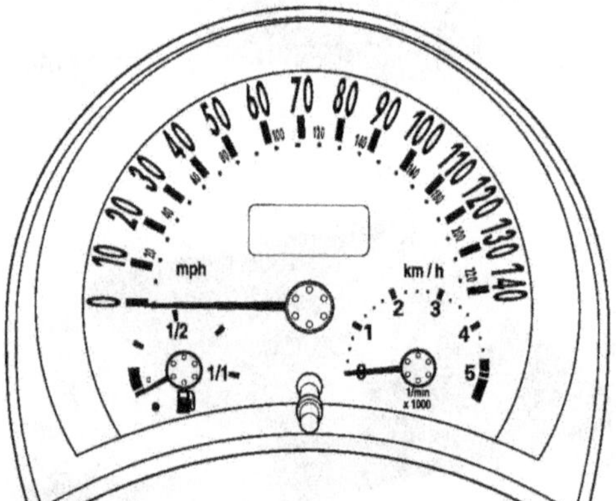

SSP 418/125

Instruments

The New Beetle Instrument Cluster is styled to resemble the one in the original Beetle. A large round Speedometer dial surrounds the smaller Fuel Gauge and Tachometer. Warning Lights are used for Coolant Level, CoolantTemperature and Oil Pressure.

Instrument Cluster Lighting is new. The displays are in blue and the indicators are red. The various warning lights are located on the outer perimeter of the cluster. Light bulbs are not replaceable. The cluster can be removed with the steering wheel still installed.

Electrical System

Instrument Cluster On Board Diagnostic (OBD)

The Instrument Cluster may be supplied by either VDO or Magneti Marelli, and have OBD capability. Using the VAG1551 Scan Tool and address word 17, the following functions are accessible:

01 - Check Control Module Versions

02 - Check DTC Memory

03 - Output Diagnostic Test Mode

05 - Erase DTC Memory

06 - End Output

07 - Code Control Module

08 - Read Measuring Value Block

10 - Adaptation

Diagnostic Trouble Code (DTC) memory can detect malfunctions in the inputs to the Instrument Cluster Control Module. The output DTM can range the analog indicators, activate the mileage display, warning lights and buzzer.

The instrument cluster can be coded for country, engine and instrument cluster manufacturer. Adaptation is used to enter the vehicle mileage into a new instrument cluster, and to calibrate the fuel gauge. Refer to *Electrical Equipment* Repair Group 01 in the Repair Manual or the Volkswagen Electronic Service Information System (VESIS) for diagnosis of the Instrument Cluster Control Module.

Clock

The Digital Clock is located in the roof above the rear view mirror.

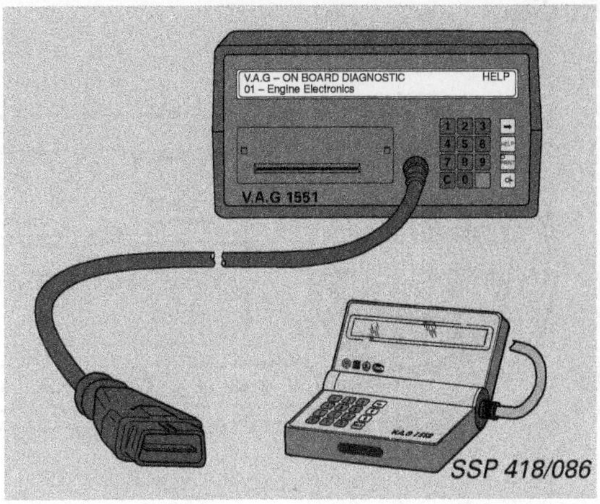

SSP 418/086

CAUTION

This training module applies to only some of the vehicles covered by this manual. This information is subject to change and should be used as reference only. Always refer to the latest Volkswagen factory service information for the most up to date service, repair and training information.

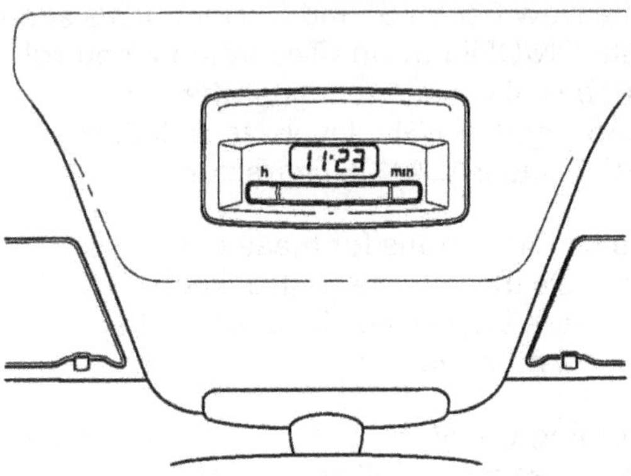

SSP 418/242

Electrical System

Radio

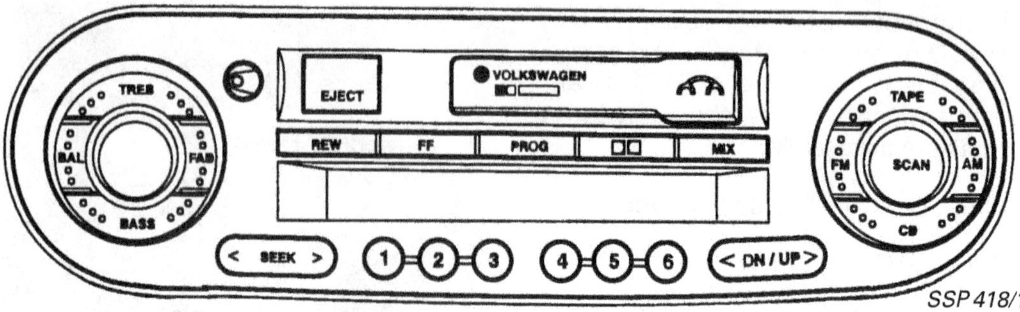

SSP 418/167

Volkswagen Sound System

The New Beetle has a 6-speaker sound system. The unique shape of the radio face plate matches the interior styling and makes the radio less desirable to steal. Switches and controls are illuminated in red, with the display field in blue. The colors match the instrument cluster lighting.

A 6-disc CD changer is available. It mounts in the luggage compartment. Every New Beetle is pre-wired at the factory for CD changer installation at the dealer.

Operation

The New Beetle Sound System has a separate ON/OFF button. The volume control knob is also used to adjust the radio's BASS and TREBLE levels as well as the FADE and BALANCE characteristics.

Selection buttons for these characteristics encircle the volume control knob. Pressing a button displays its function on the radio with a numeric value.

Holding the SCAN button down for more than 2 seconds will activate the custom switch-on volume level feature. This feature reverts to a chosen volume setting each time the radio is switched on, regardless of the volume setting when it was switched off.

In CD mode, holding the SCAN button down for more than 2 seconds allows the CD player to be set for random play on a single disc or all six discs. Selection is made by pressing the MIX button.

Anti-theft coding

If the radio is disconnected from its power source it must have its anti-theft code re-entered. When the radio is first turned on the word "SAFE" appears. After 3 seconds the display changes to the number 1000. It is not necessary to press a combination of switches to get this display.

The code is entered by pressing the right side ">" button for the SEEK function and entering the code using the four preset buttons. After the proper code has been entered, an audible "beep" will signal that function is resumed.

Electrical System

Radio On Board Diagnostic (OBD)

The following functions are possible using the VAG1551 Scan Tool using "Radio" Address Word 56:

01 - Check Control Module Versions

02 - Check DTC Memory

03 - Output Diagnostic Test Mode

05 - Erase DTC Memory

06 - End Output

07 - Code Control Module

08 - Read Measuring Value Block

The VAG1551 Scan Tool will indicate malfunctions in speakers, amplifiers and antenna circuits. The Output DTM will test the speakers. The radio can be coded for CD and instrument cluster radio display in the vehicle in which it is installed.

Refer to Repair Group 01 in the *Electrical Equipment* Repair Manual or Volkswagen Electronic Service Information System to diagnose the Volkswagen Sound System.

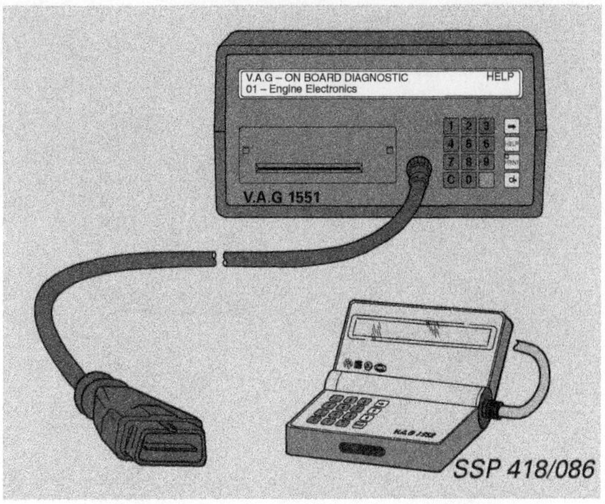

SSP 418/086

Radio removal

New radio removal tools are required to remove the radio from the instrument panel. Special tools 3344A fit into two horizontal slots in the lower edge of the radio.

3344 A

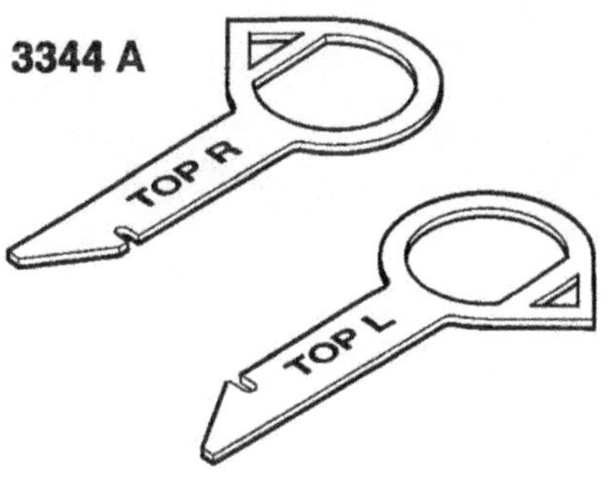

SSP 418/137

Electrical System

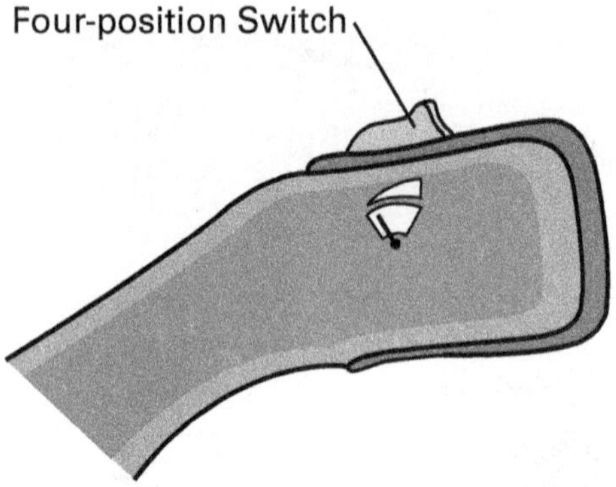

SSP 418/034

Windshield wipers and washers

Intermittent wipers

A four-position switch is mounted in the top of the wiper switch on the steering column stalk. The pause intervals are vehicle speed dependent. For each setting, the faster the vehicle speed, the shorter the pause interval.

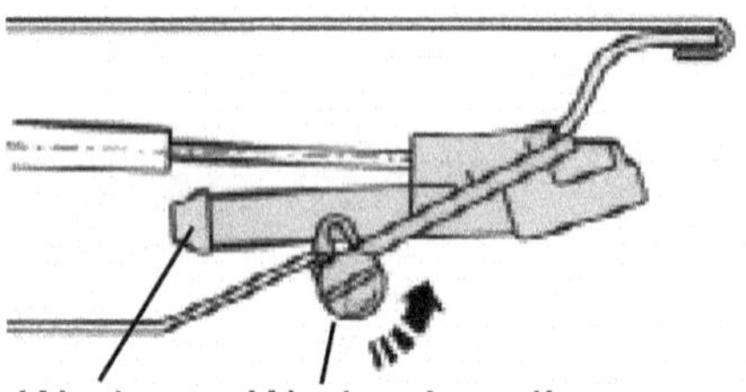

SSP 418/157

Windshield washers

The windshield washer jets are mounted in the hood. Heated washer jets are available in the cold weather option package.

Washer jet aim is preset, but slight adjustment is possible by turning an eccentric on the jet using a small screw driver.

Lighting

Headlights

New Beetles are equipped with projector beam headlights. A clamp and locking lever system allows for easy removal and installation.

Adjustment is carried out with two socket-head screws. The lens stays flush with the fender while the reflector is adjusted inside the light housing.

Electrical System

Taillights

The taillight housings are also easy to remove to replace light bulbs.

Remove the nut and the light housing slides out toward the rear.

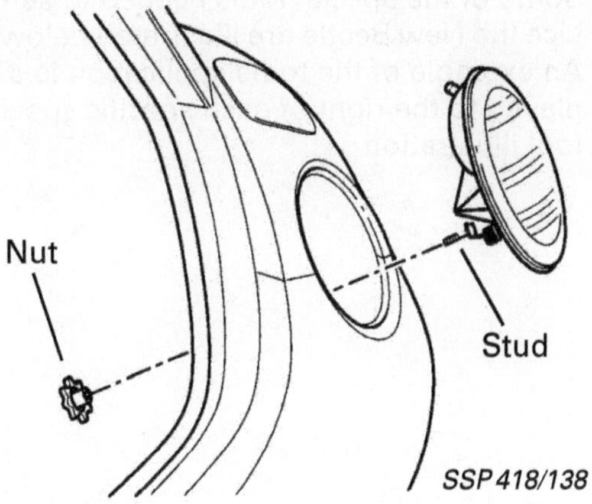

SSP 418/138

Interior lighting

A combination interior light and reading light is mounted in the bottom of the rear view mirror housing. Vanity mirrors located in the sun visors have indirect lighting by mounting the lights in the headliner. The glove box and luggage compartments have lights as well.

Third brake light

The third brake light is mounted in the rear lid. It is a sealed unit containing a row of LEDs and circuit protection resistors. If an LED goes out the entire unit is replaced.

CAUTION!
A thin rubber gasket seals the third brake light to an opening in the rear lid. Care must be taken when installing the light to ensure a proper seal. Improper installation could cause a water leak in the luggage compartment.

Special Tools

Some of the special tools needed to service the New Beetle are illustrated below. An example of the tools application is displayed to the right of each specific special tool illustration.

This is not a complete list of all the tools that may be in the New Beetle special tool package. Some of these tools may already be in service at dealerships.

10-222A/8 Engine Support Adapters

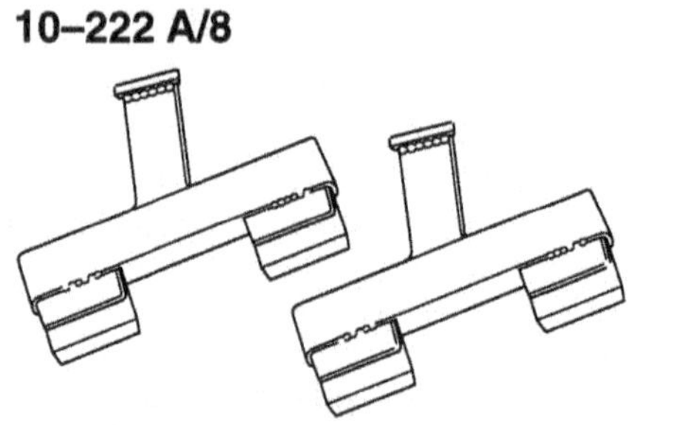

SSP 418/149

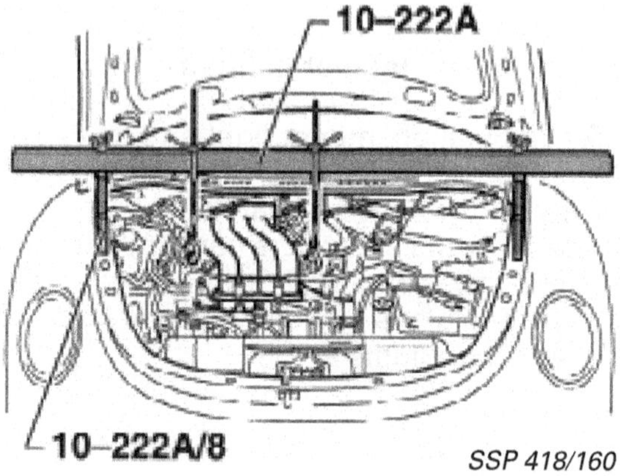

SSP 418/160

3344A Radio Removal Tool

SSP 418/137

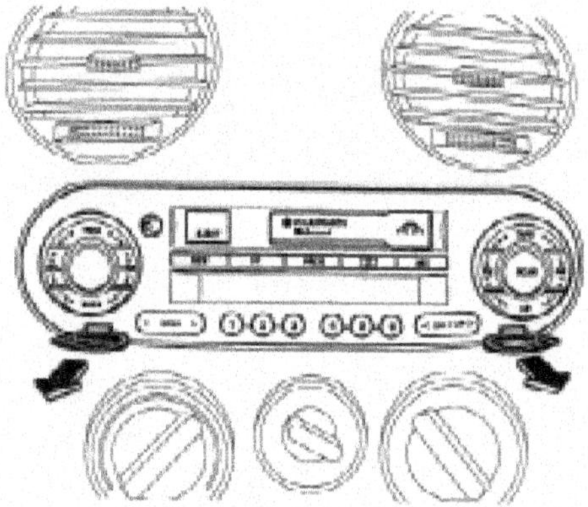

SSP418/246

Special Tools

3415 Retainer

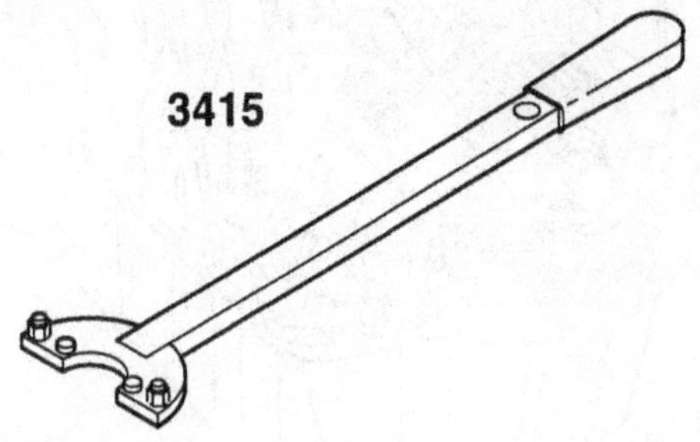

SSP 418/150

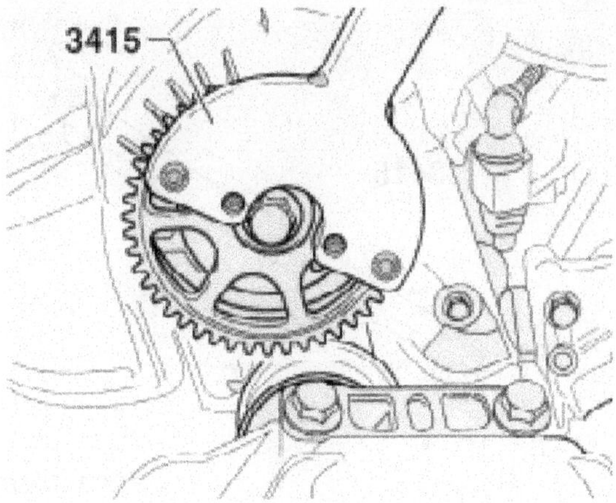

3417 Oil Filter Wrench (TDI cartridge filter)

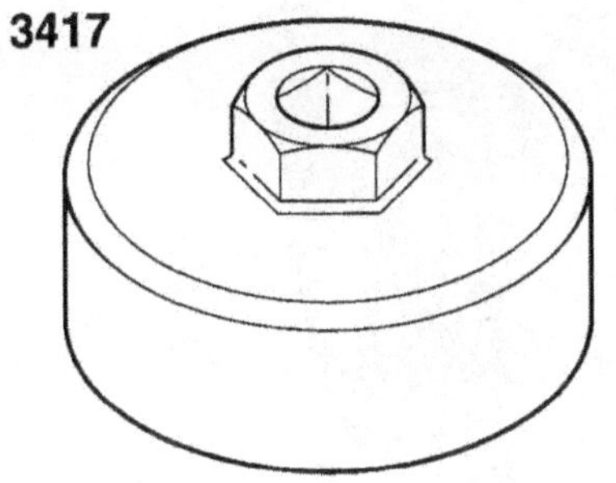

SSP 418/230

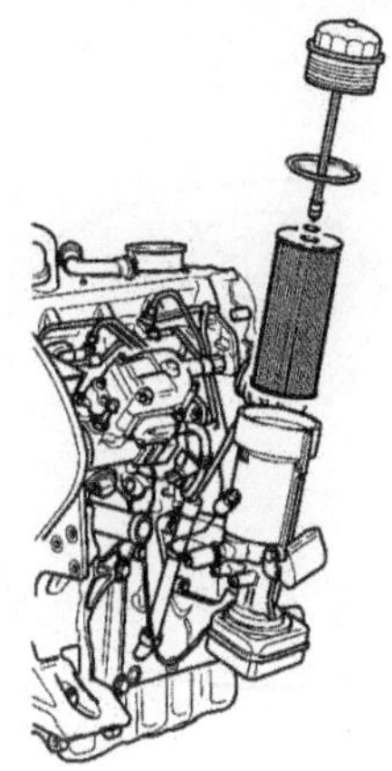

SSP 418/231

3418 Setting Bar (TDI camshaft)

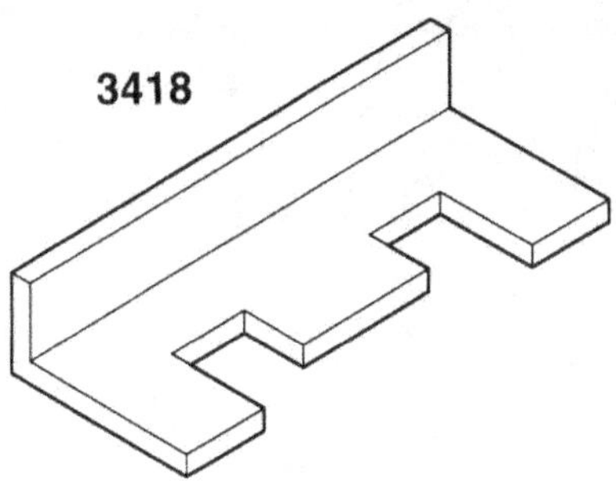

SSP 418/191

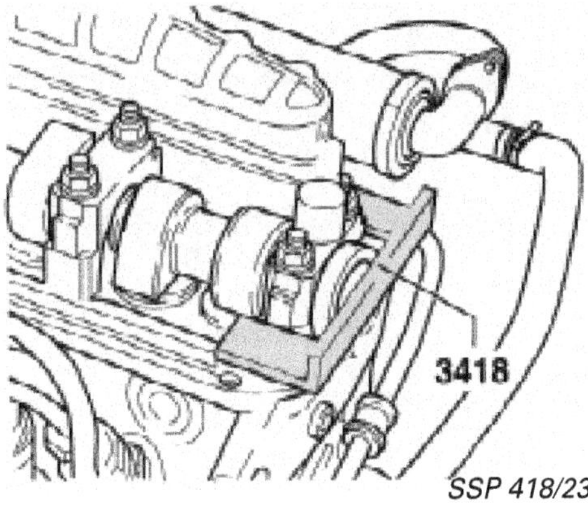

SSP 418/234

Special Tools

3415 Retainer

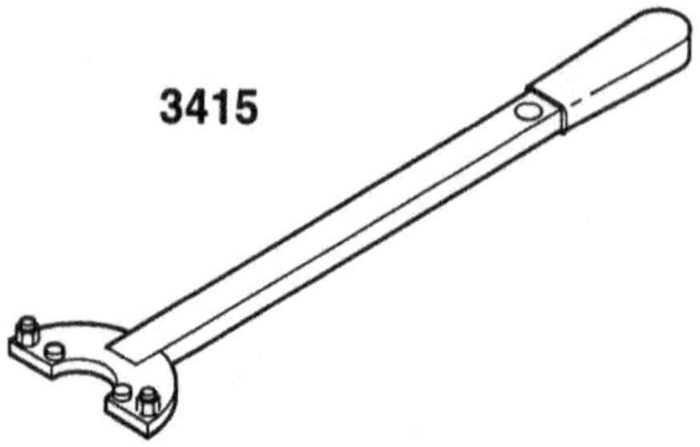

SSP 418/150

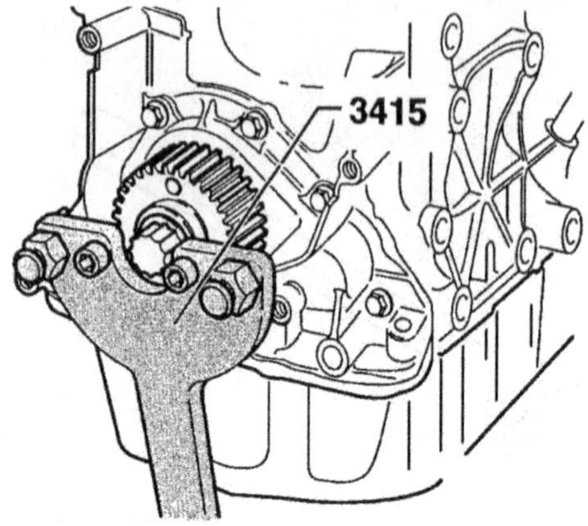

SSP 418/156

3417 Oil Filter Wrench (TDI cartridge filter)

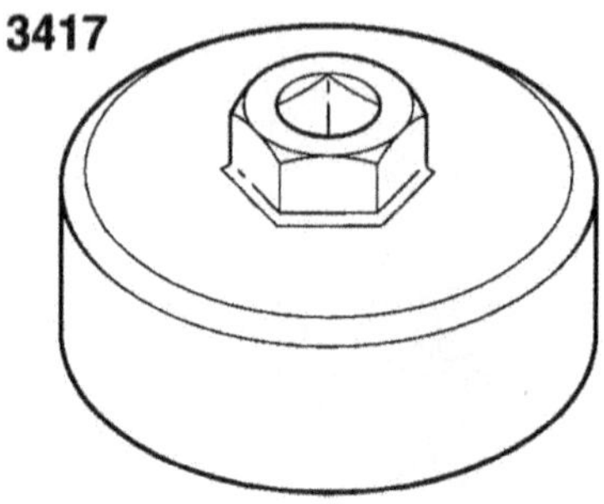

SSP 418/230

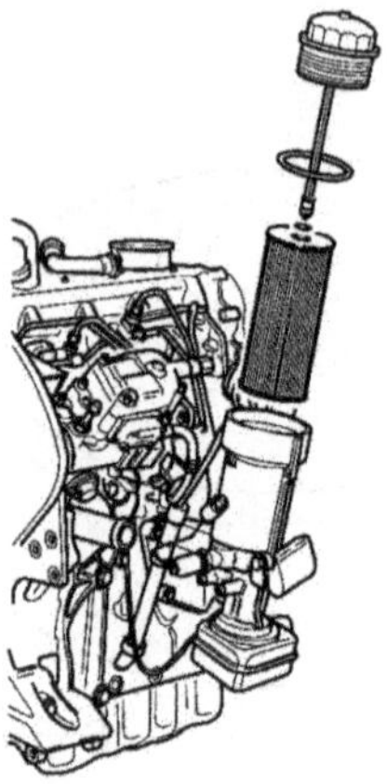

SSP 418/231

3418 Setting Bar (TDI camshaft)

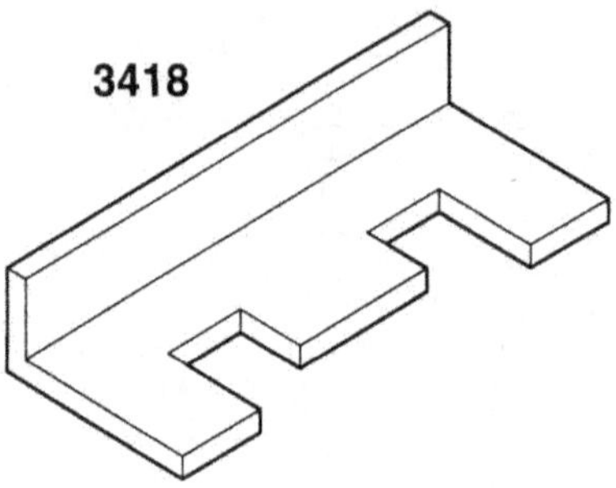

SSP 418/191

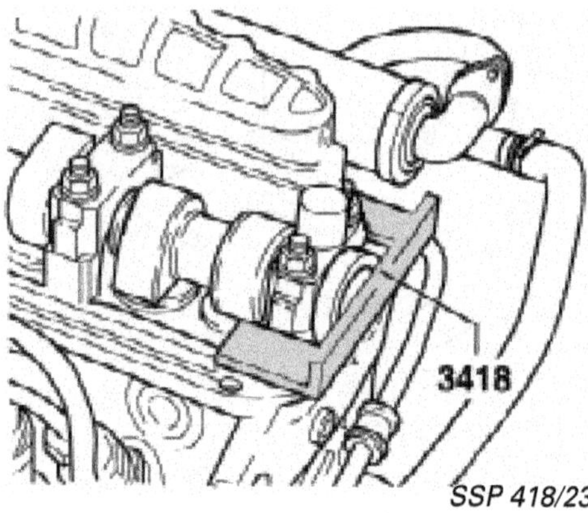

SSP 418/234

Special Tools

3420 Wheel Bearing Tool (rear)

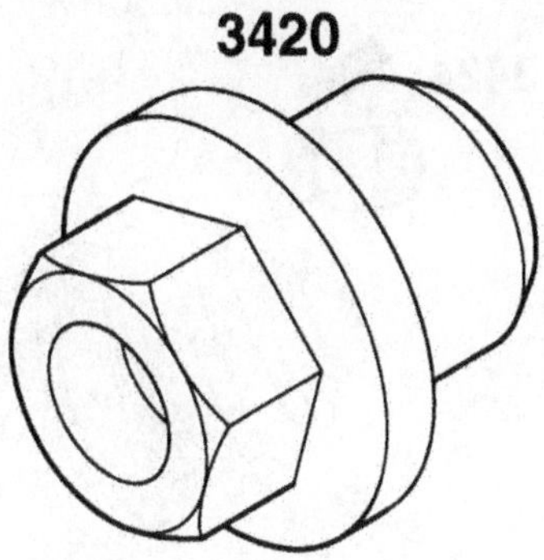

SSP 418/141

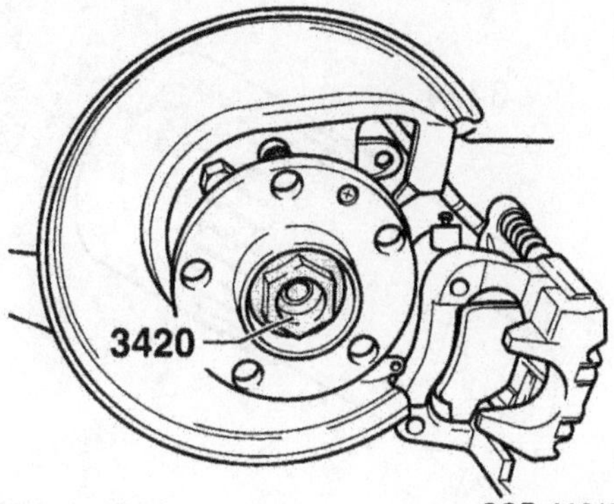

SSP 418/146

3422 Shift Linkage Gauge (02J trans.)

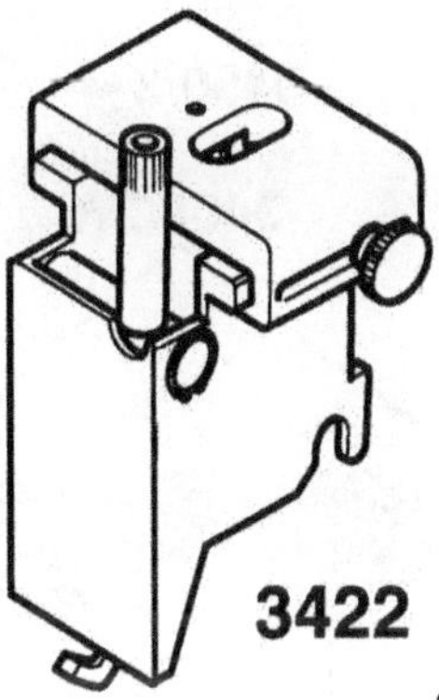

SSP 418/229

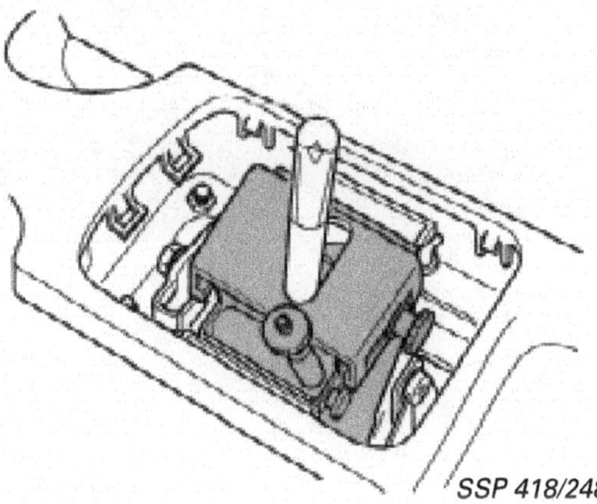

SSP 418/248

423 Puller (front wheel bearing race)

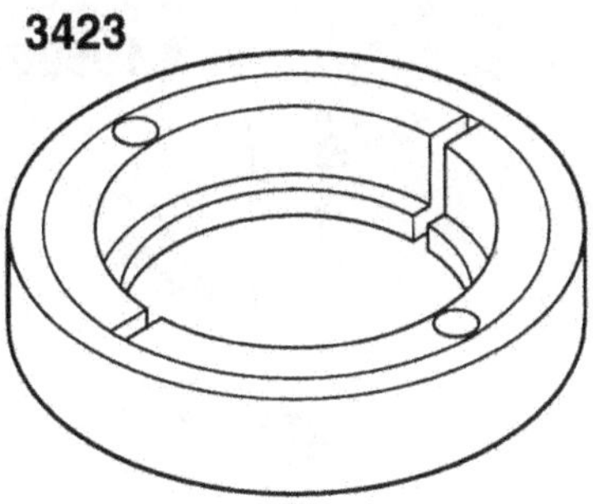

SSP 418/147

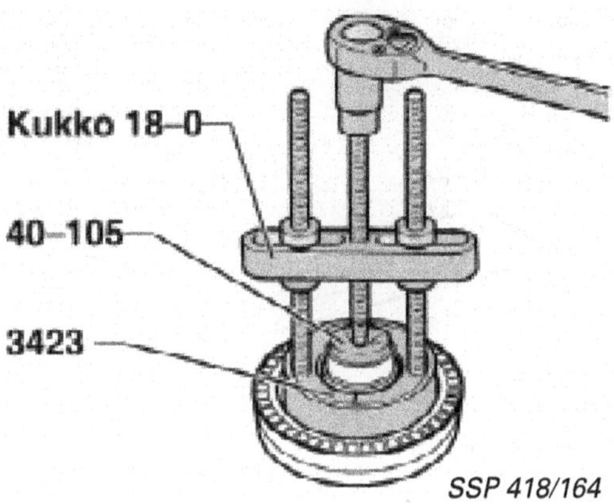

SSP 418/164

Special Tools

3424 Spreader Tool (strut mount)

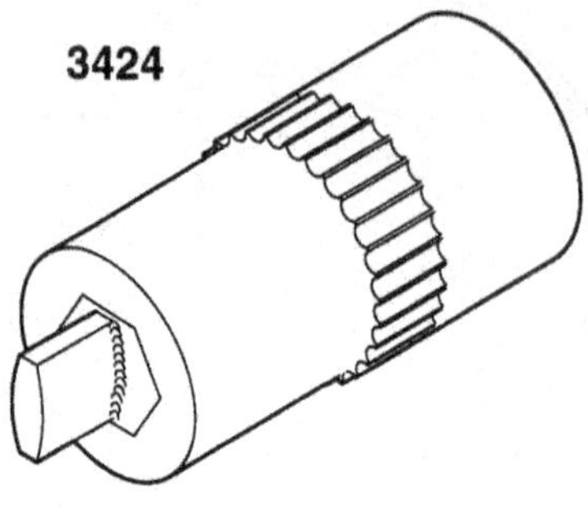

SSP 418/142

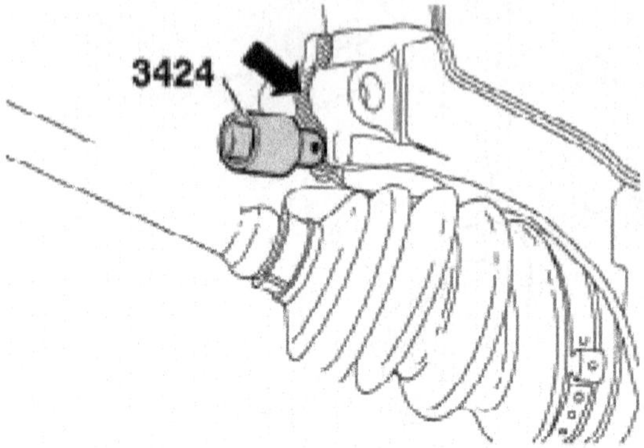

SSP 418/163

3450/2A, 3450/3 Cylinder Head Guide Pins

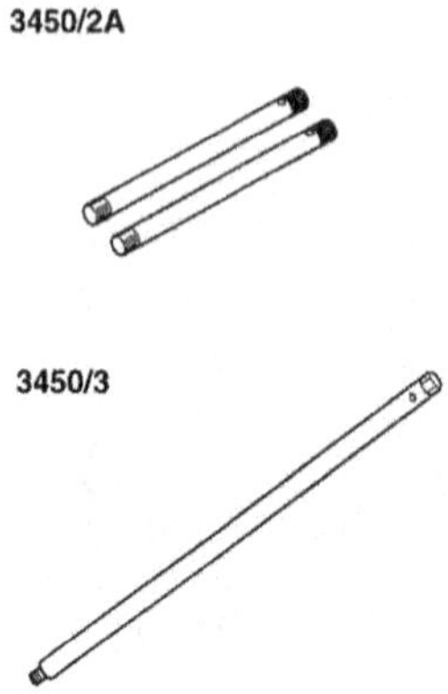

SSP 418/148

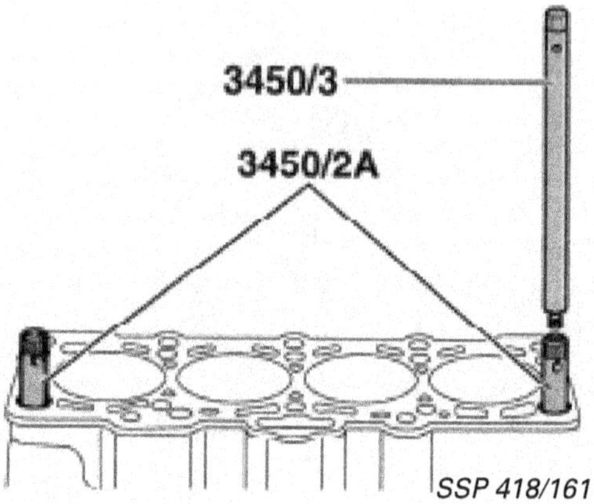

SSP 418/161

T10006 Brake Push Rod Tool

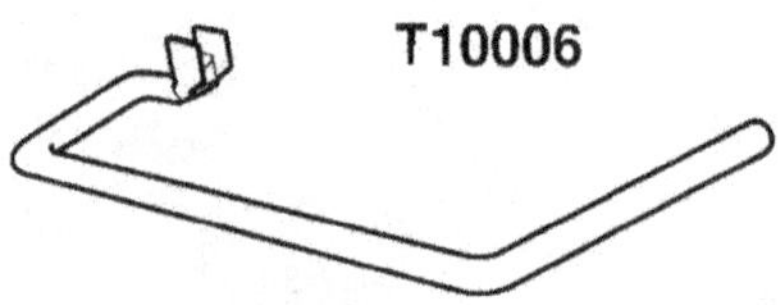

SSP 418/195

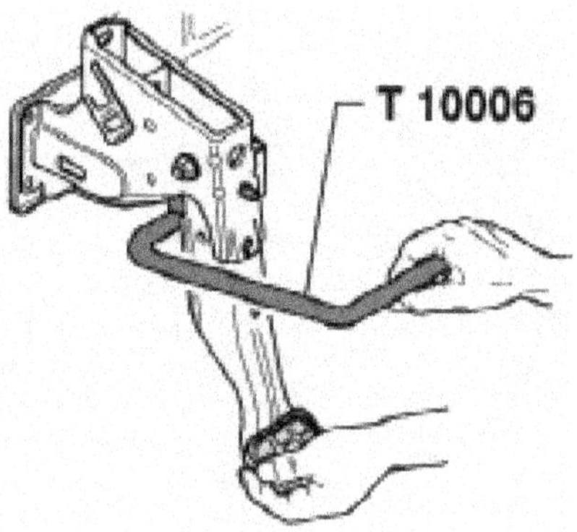

SSP 418/202

Special Tools

T00011 12-Point Socket Head Screwdriver

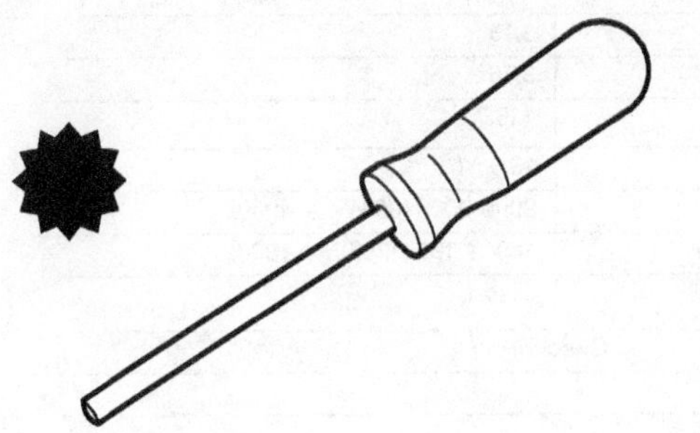

SSP 418/038

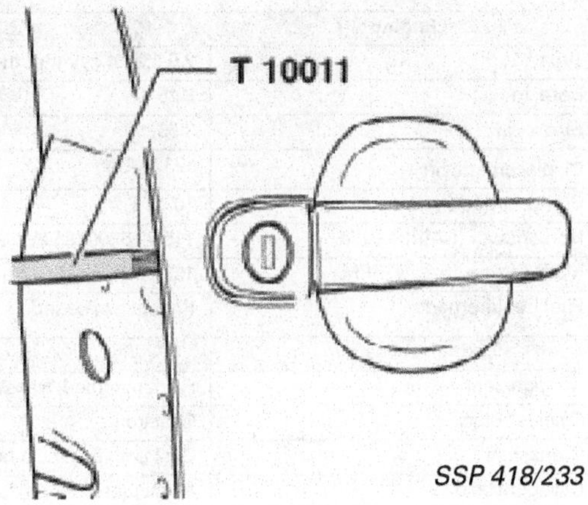

SSP 418/233

T10020 Pin Wrench (belt tensioner)

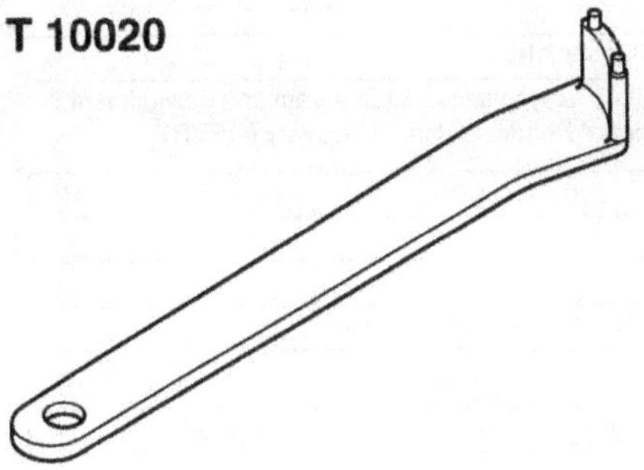

SSP 418/149a

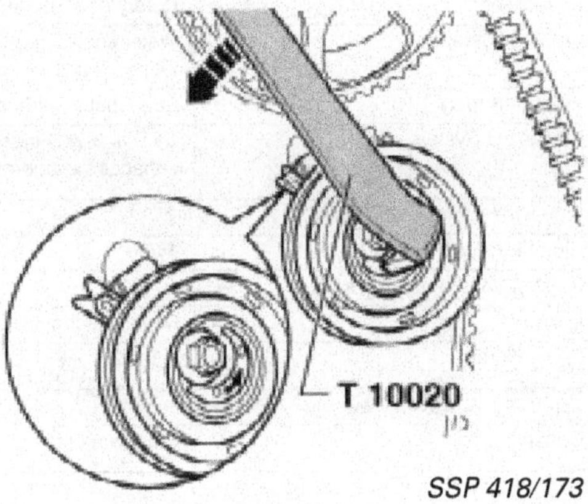

SSP 418/173

VAS5061 Side Airbag Ground Harness

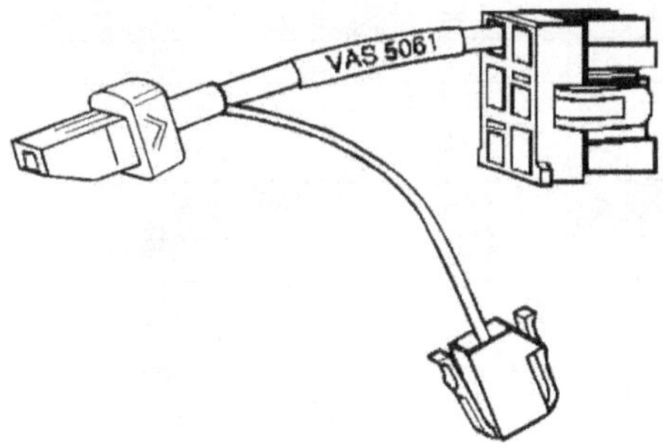

SSP 418/052

1998 New Beetle Technical Specifications

1998 New Beetle Technical Specifications — USA

Engine	Gas	Diesel
Type	2.0 liter, 4-cylinder in-line	1.9 liter, 4-cylinder in-line
Bore, in	3.25	3.13
Stroke, in	3.65	3.76
Displacement, in^3	121.1	115.7
Compression Ratio	10.0 : 1	19.5 : 1
Horsepower (SAE) @ RPM	115 @ 5200 (85 kW @ 5200)	90 @ 4000 (66 kW @ 4000)
Max. Torque, lb-ft @ RPM	122 @ 2600 (165 Nm @ 2600)	149 @ 1900 (202 Nm 1900)
Fuel Requirement	Regular unleaded	Diesel

Engine Design	Gasoline
Arrangement	Front mounted, transverse
Cylinder Block	Cast iron
Crankshaft	Cast iron, five main bearings
Cylinder Head	Aluminum alloy, cross flow
Valvetrain	Single overhead camshaft, spur belt driven, 2 valves per cylinder, maintenance free hydraulic lifters
Lubrication	Rotary gear pump, chain driven, oil cooler
Cooling System	Water cooled, water pump, cross flow radiator, thermostatically controlled electric 2-speed radiator coolant fan
Fuel/Air Supply	Sequential multi-port fuel injection (Motronic)
Emissions System	OBD II, 3-way catalytic converter with 2 oxygen sensors (upstream and downstream), enhanced evaporation system, Onboard Refueling Vapor Recovery (ORVR)

Electrical System	
Alternator, Volts/Amps	14/90
Battery, Volts (Ampere-Hours)	12 (60)
Ignition	Digital electronic, Distributorless coil block, with knock sensor
Firing Order	1-3-4-2

Drive Train	
Configuration	Front wheel drive

Transmission Gear Ratios	5-Speed Manual		4-Speed Automatic	
	Gasoline	Diesel	Gasoline	Diesel
1st	3.78 : 1	3.50 : 1	2.74 : 1	2.71 :1
2nd	2.12 : 1	1.94 : 1	1.55 : 1	1.44 : 1
3rd	1.36 : 1	1.23 : 1	1.00 : 1	1.00 : 1
4th	1.03 : 1	0.84 : 1	0.68 : 1	0.74 : 1
5th	0.84 : 1	0.68 : 1		
Reverse	3.60 : 1	3.60 : 1	2.11 : 1	2.88 : 1
Final drive	4.24 : 1	3.89 : 1	4.88 : 1	3.63 : 1

Capacities	
Engine Oil (with filter), qt	4.8
Fuel Tank, gal	14.5
Cooling system, qt	6.7

Steering	
Type	Rack and pinion, power assisted
Turns (lock to lock)	3.2
Turning circle (curb to curb), ft.	32.8
Ratio	17.8 : 1

New Beetle Technical Specifications

Interior Volume - SAE		
EPA class	Subcompact	EPA
Passenger Volume, ft^3	84.3	84
Trunk Volume, ft^3	12	12
Seating Capacity	Four	
	Front	Rear
Volume, ft^3	49.7	34.6
Head Room, in.	41.3	36.7
Shoulder Room, in.	52.8	49.4
Leg Room, in.	39.4	33.0

Body, Chassis and Suspension	
Type	Unitized construction, bolt-on front fenders
Front Suspension	Independent MacPherson struts, coil springs, telescopic shock absorbers, stabilizer bar
Rear Suspension	Independent torsion beam axle, coil springs, telescopic shock absorbers, stabilizer bar
Service Brakes	Power assisted, dual diagonal hydraulic circuits, 280 mm vented front discs and 239 mm solid rear discs
Anti-lock Braking System	Optional, all 4 wheels
Parking Brake	Mechanical, effective on rear wheels
Wheels	6.5 J x 16 steel, with full wheel covers, 5 bolts
Tires	205/55 R 16 H all season
Drag Coefficient	0.38
Frontal Area, ft^3	TBD

Dimensions		
Wheelbase, in	98.9	
Track: Front, in	59.6	
Rear, in	58.7	
Suspension Height: Front, in	n/a	
Rear, in	n/a	
Overall Length, in	161.1	
Overall Width, in	67.9	
Overall Height, in	59.5	
Ground Clearance, in	4.2	
	5-speed manual	4-speed automatic
Curb Weight, lb	2,712	2,778
Payload, lb	992	936

Fuel Consumption (EPA Estimates)				
	5-speed manual		4-speed automatic	
	Gasoline	Diesel	Gasoline	Diesel
City, mpg	23	41	22	34
Highway, mpg	29	48	27	44

New Beetle Technical Specifications

1998 New Beetle Technical Specifications – Canada

Engine	Gasoline	Diesel
Type	2.0 liter, 4-cylinder in-line	1.9 liter, 4-cylinder in-line
Bore, mm	82.5	79.5
Stroke, mm	92.7	95.5
Displacement, cm^3	1984	1896
Compression Ratio	10.0 : 1	19.5 : 1
Power (kW) @ RPM	85 @ 5200 (115 HP @ 5200)	66 @ 4000 (90 HP @ 4000)
Max. Torque, Nm @ RPM	165 @ 2600 (122 lb-ft @ 2600)	202 @ 1900 (149 lb-ft @ 1900)
Fuel Requirement	Regular unleaded	Diesel

Engine Design	Gasoline
Arrangement	Front mounted, transverse
Cylinder Block	Cast iron
Crankshaft	Cast iron, five main bearings
Cylinder Head	Aluminum alloy, cross flow
Valvetrain	Single overhead camshaft, spur belt driven, 2 valves per cylinder, maintenance free hydraulic lifters
Lubrication	Rotary gear pump, chain driven, oil cooler
Cooling System	Water cooled, water pump, cross flow radiator, thermostatically controlled electric 2-speed radiator fan
Fuel/Air Supply	Sequential multi port fuel injection (Motronic)
Emissions System	OBD II, 3-way catalytic converter with 2 oxygen sensors (upstream and downstream), enhanced evaporation system, Onboard Refueling Vapor Recovery (ORVR)

Electrical System	
Alternator, Volts/Amps	14/90
Battery, Volts (Ampere-Hours)	12 (60)
Ignition	Digital electronic, Distributorless coil block, with knock sensor
Firing Order	1-3-4-2

Drive Train	
Configuration	Front wheel drive

Transmission Gear Ratios	5-Speed Manual		4-Speed Automatic	
	Gasoline	Diesel	Gasoline	Diesel
1st	3.78 : 1	3.50 : 1	2.74 : 1	2.71 :1
2nd	2.12 : 1	1.94 : 1	1.55 : 1	1.44 : 1
3rd	1.36 : 1	1.23 : 1	1.00 : 1	1.00 : 1
4th	1.03 : 1	0.84 : 1	0.68 : 1	0.74 : 1
5th	0.84 : 1	0.68 : 1		
Reverse	3.60 : 1	3.60 : 1	2.11 : 1	2.88 : 1
Final drive	4.24 : 1	3.89 : 1	4.88 : 1	3.63 : 1

Capacities	
Engine Oil (with filter), liter	4.5
Fuel Tank, liter	55.0
Cooling system, liter	6.3
Steering	
Type	Rack and pinion, power assisted
Turns (lock to lock)	3.2
Turning Circle (curb to curb), m	10.0
Ratio	17.8 : 1

New Beetle Technical Specifications

Interior Volume - SAE		
EPA class	Subcompact	EPA
Passenger Volume, ft^3	84.3	84
Trunk Volume, ft^3	12	12
Seating Capacity	Four	
	Front	Rear
Volume, m^3	1.41	.98
Head Room, mm	1498	932
Shoulder Room, mm	1341	1255
Leg Room, mm	1001	831

Body, Chassis and Suspension	
Type	Unitized construction, bolt-on front fenders
Front Suspension	Independent MacPherson struts, coil springs, telescopic shock absorbers, stabilizer bar
Rear Suspension	Independent torsion beam axle, coil springs, telescopic shock absorbers, stabilizer bar
Service Brakes	Power assisted, dual diagonal hydraulic circuits, 280 mm vented front discs and 239 mm solid rear discs
Anti-lock Braking System	Optional, all 4 wheels
Parking Brake	Mechanical, effective on rear wheels
Wheels	6.5 J x 16 steel, with full wheel covers, 5 bolts
Tires	205/55 R 16 H all season
Drag Coefficient	0.38
Frontal Area, m^3	TBD

Dimensions		
Wheelbase, mm	2512	
Track: Front, mm	1514	
Rear, mm	1491	
Suspension Height: Front, mm	n/a	
Rear, mm	n/a	
Overall Length, mm	4089	
Overall Width, mm	1724	
Overall Height, mm	1511	
Ground Clearance, mm	102.0	
	5-Speed Manual	4-Speed Automatic
Curb Weight, kg	1230	1260
Payload, kg	450	424

Fuel Consumption				
	5-Speed Manual		4-Speed Automatic	
	Gasoline	Diesel	Gasoline	Diesel
City, L / 100 km	10.1	5.8	10.9	6.9
Highway, L / 100 km	7.3	4.5	8.1	5.0
City, miles/imperial gal.	28	49	26	41
Highway, miles/imperial gal.	39	63	35	56

8.25 in

New Beetle Convertible

Top and Window Repair

Self Study Program
Course Number 861303

CAUTION

This training module is included as background information and is based on introductory and preliminary information. It applies to only some of the vehicles covered by this manual and is intended to serve as a product familiarization guide. Note that this information is subject to change and should be used as reference only. Always refer to the latest Volkswagen factory service information for the most up to date service, repair and training information.

Table of Contents

The Self-Study Program provides you with information regarding repair and adjustment procedures.

The Self-Study Program is not a Repair Manual.

For maintenance and repair work, always refer to current service repair information.

Introduction

Welcome to this New Beetle Convertible Self-Study Program, the print companion piece to the New Beetle Convertible Top and Window Repair video.

The purpose of this Self-Study Program is to address some of the features that are unique to the New Beetle Convertible, features that you will be asked to repair or adjust in your dealerships.

These features were chosen because they are new. This training follows the step-by-step procedures found in service repair information, while also offering additional insights.

The objectives of this training are to help you become adept at:

- Door window, regulator and motor removal, installtion and adjustment
- Rear side window, regulator and motor removal, installation and adjustment
- Convertible top outer shell removal, installation and adjustment

None of these features are particularly difficult to repair, replace or adjust. Hopefully, this Self-Study Program will take some of the mystery out of the new repair procedures.

In particular, you will see why it may make some sense to keep the convertible top repair business in your dealership instead of perhaps automatically sending it out to a trim shop.

We would hope that when used in conjunction with the New Beetle Convertible Top and Window Repair video, you will feel both comfortable and knowledgeable when it comes to performing these new procedures.

This Self-Study Program is a supplement to current service repair information, not a replacement for it.

Door Window, Regulator and Motor

One of the repairs you might see will involve one of the new electric window assemblies.

The procedures that we will use to repair these windows are pretty straightforward, and are in keeping with service repair information.

Unlike the rear side window, we will need to take a lot of the door apart, including the assembly carrier, in order to take out the door window.

Door Window Removal

- First, tape off the chrome trim to protect it from damage.
- Leave the battery connected so you can move the window once you have the inner door panel removed.
- Put the windows DOWN.

Taking off the inner door panel is the same as taking one off the New Beetle Sedan and is a fairly standard procedure. You do not have to remove the carrier plate to remove the door window.

- Remove the door handle trim using a nylon bone so you do not damage the soft plastic parts.
- Remove the two Phillips head screws under the door handle trim.
- Remove the three torx screws at the bottom of the door.

Door Window, Regulator and Motor

- Pry the door panel out, lifting it up and pulling it out of the upper seal, but not removing it from the door until you have disconnected the door release cable and the electrical connectors.
- Disconnect the power window switch from the door panel and reconnect it so you can operate the window later.
- Remove the inner door seal so you can take the window out of the door. Tape it to the top of the door so you do not have to remove the rear end of the seal that is secured to the door.

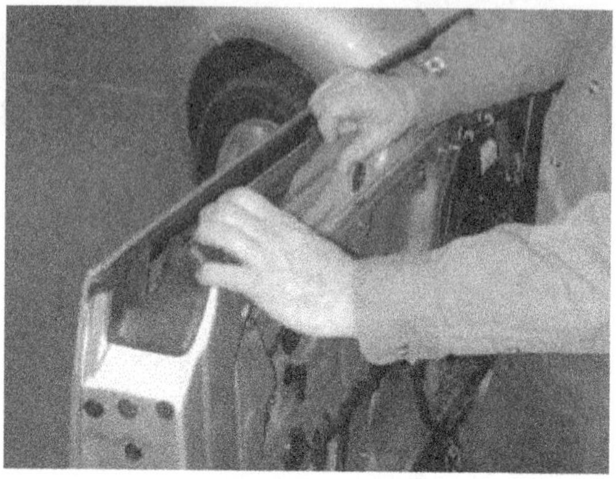

- Using a bone, lever out the two cover caps for the rear and front upper assembly openings in the carrier plate.

Door Window, Regulator and Motor

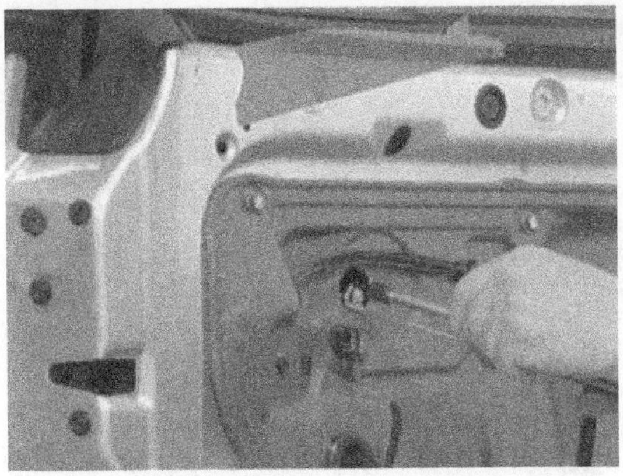

- Move the window up until you can see the bolts through the two assembly openings. These bolts retain the window.
- Unlike the New Beetle Sedan, you need to remove the rear bolt. Do not apply much pressure, because there is a metal retaining plate attached to the window. This plate can fall off into the door. Use a magnet to ensure that you also do not drop the bolt into the door, which would require you to take off the carrier plate.
- You only need to loosen the front bolt, just like on other Volkswagen models.

- If you cannot move the window with the power window motor, just remove the three bolts that attach it to the door. Then operate the window manually to get the window into position.

When putting in a new power window motor, line up the spline driver by moving the glass mounting brackets up and down to line up while fitting the motor into position.

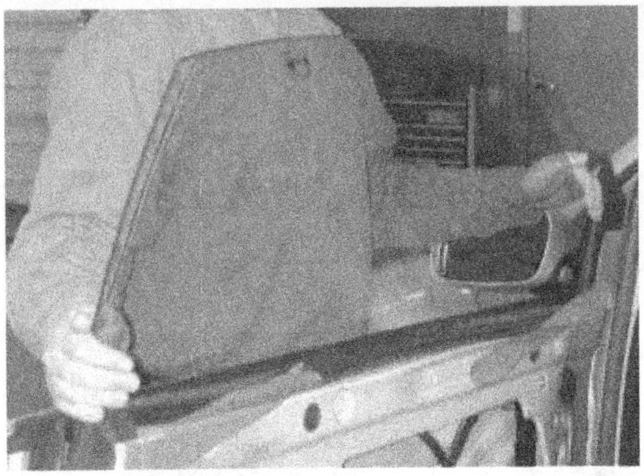

- Now you can pull the window out of the door, taking care not to damage the seal. Set the window aside.
- If you were only replacing the window glass, you would just reverse the removal procedures.
- If the window is broken, you will still need to remove the carrier plate to vacuum the glass from the bottom of the door.

Door Window, Regulator and Motor

Carrier Plate and Regulator Removal

- Remove the door lock cylinder, then remove the door handle cable.
- Remove the two bolts that hold in the door lock assembly on the end of the door, located below the door lock cylinder screw. Otherwise, you will break the door lock assembly plastic guide.

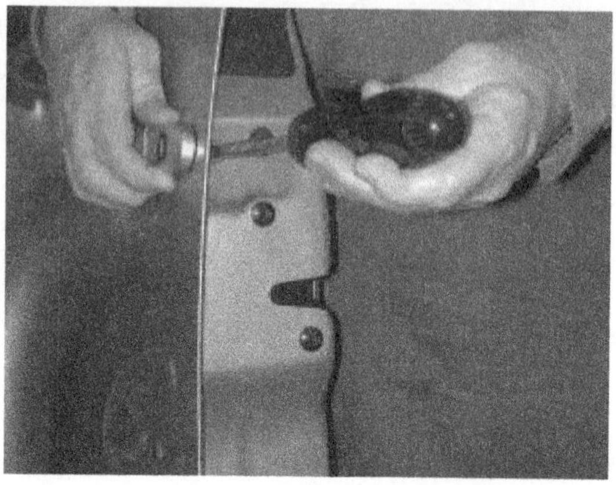

- To avoid breaking clips in the process of disconnecting the wiring harness from the carrier plate, disconnect the wiring harness from the A-pillar station.

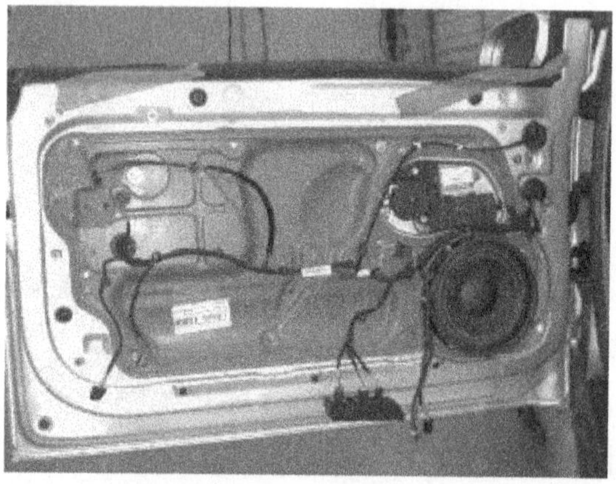

- Just as with other Volkswagen models, you can accomplish this by removing the underside of the instrument panel, removing the brake release handle and sill plate, and threading the harness through the A-pillar.

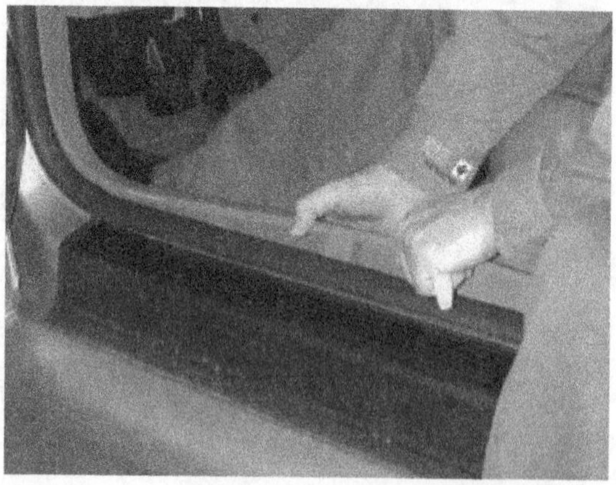

Door Window, Regulator and Motor

- Unbolt the strain relief for the wiring harness at the front of the door.

- Disconnect the power window motor from the wiring harness.
- Disconnect the power mirror harness.
- Remove all the screws on the carrier plate.

- Remove the two white clips at the top of the carrier plate by pushing them in from the backside of the carrier plate. It is not possible to pull them out from the front.

Door Window, Regulator and Motor

- Pull the carrier plate carefully away from the door frame.

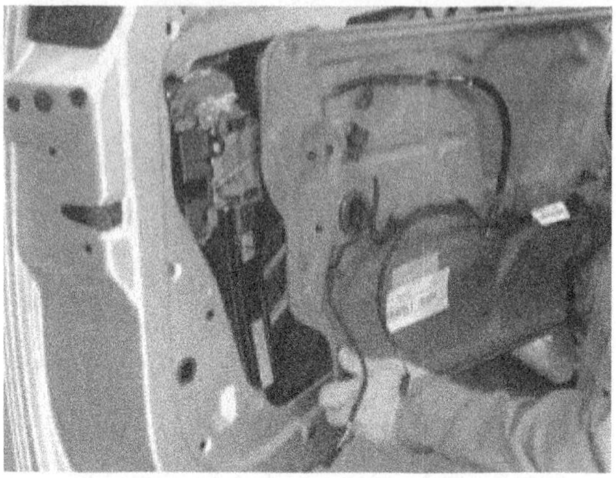

- Remove the four bolts that hold in the regulator.
- Mark the slot positions of the two bottom bolts with a felt tip pen. These bolts adjust the angle of the window. Your marks will prove helpful during reinstallation and adjustment.

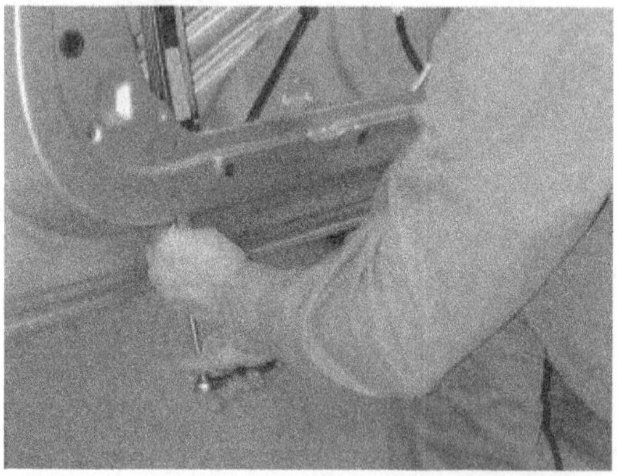

- Remove the regulator.

Door Window, Regulator and Motor

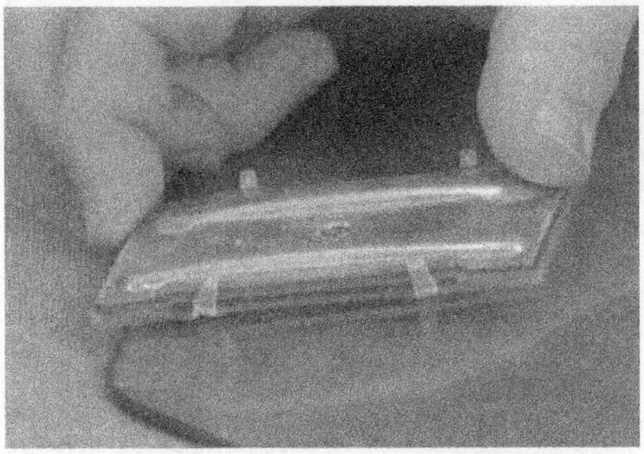

Re-InstallationTips

- Make sure the metal retaining plate on the window glass is positioned squarely. Otherwise, you could break the window when tightening the bolt during re-installation.

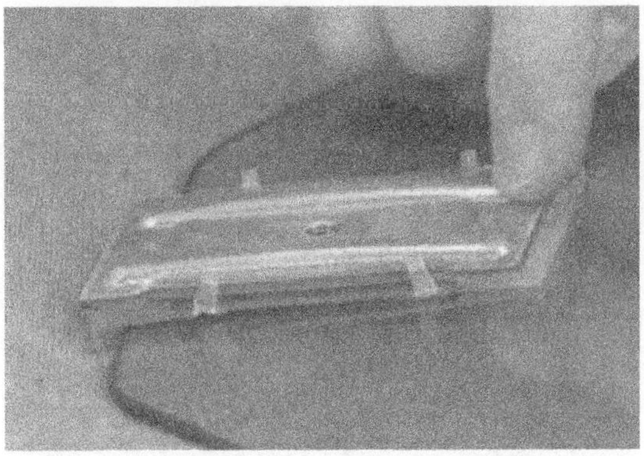

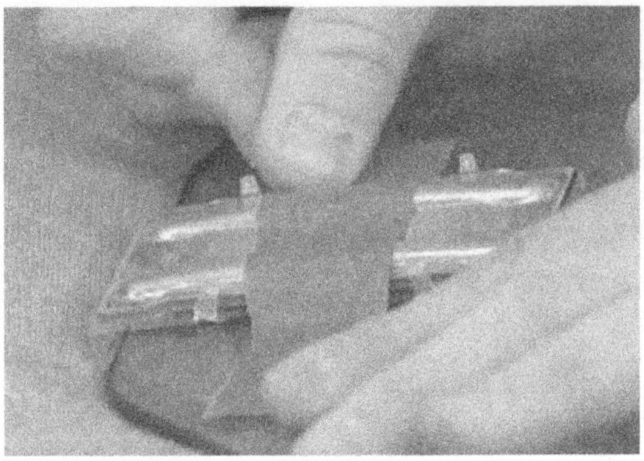

- Tape the plate down to hold it in the proper position. The tape will not be seen after the window is installed.
- Carefully re-install the window, making sure not to catch the retaining plate on the door seal.

Door Window, Regulator and Motor

- The regulator has a forward support clamp, which is similar to the New Beetle Sedan, Golf and Jetta regulators. It has an adjustment screw on the bottom of the clamp.

- The rear support does not have a clamp. Instead, the window sits on two aluminum feet, with an adjustment screw underneath.
- These two regulator support adjustment screws adjust the height of the window from the front to the back.
- Adjustments can be made after the regulator is installed, and is lowered to the bottom of the door.

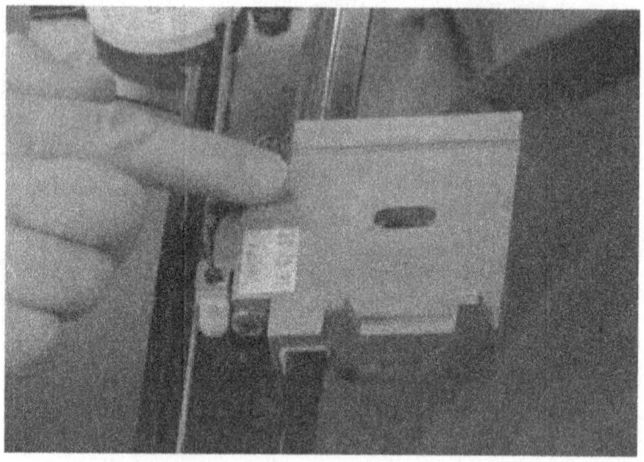

CAUTION

This training module applies to only some of the vehicles covered by this manual. This information is subject to change and should be used as reference only. Always refer to the latest Volkswagen factory service information for the most up to date service, repair and training information.

Door Window, Regulator and Motor

- Adjust the window by sliding it back and then forward so that it fits comfortably in the front seal. You do not want to push too hard, which could create drag, activating the Pinch Protection later.
- When doing final window adjustments, put three pieces of tape across the top of the window. Draw a line on them where the seal meets the window.

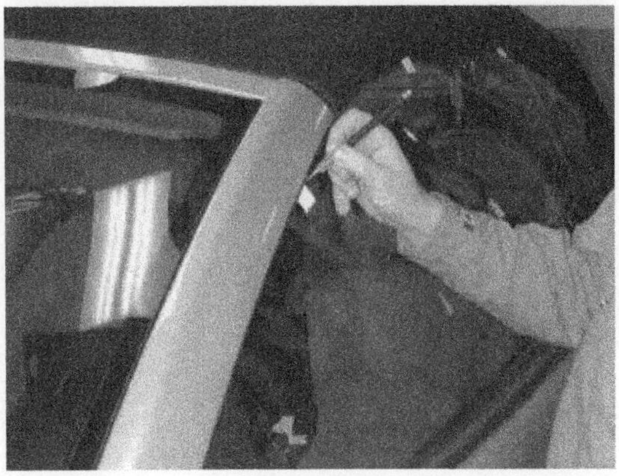

- When you open the door, you should see 8 mm to 10 mm between the top of the window and the line you drew.
- If your lines are not even, adjust the two height adjustment screws on the regulator supports. Have the regulator all the way down to the bottom of the door.
- If the window is a little too far out or too far in, check the adjustment by closing the door and seeing if the window goes squarely up into the seal.
- If the window is hitting on the outside or the inside of the seal, you need to loosen the two bottom mounting points (the ones in slots) for the regulator, moving them in or out.

Rear Side Window, Regulator and Motor

Removing, installing and adjusting the rear side window, regulator, and motor are similar to the procedures for the door windows. But there are more differences than similarities.

Some of these procedures involve removing the rear interior side panel, which also needs to be removed when replacing the convertible top outer shell.

Rear Side Window Removal

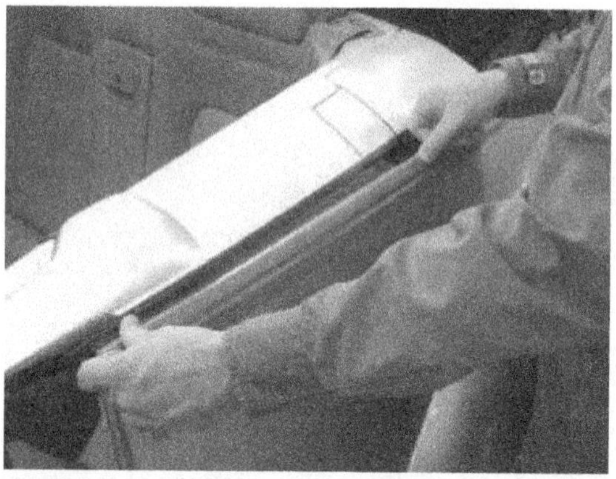

- Lower the convertible top.
- Lower the door and rear side windows.
- Tape off the chrome trim to protect it from damage.
- Leave the battery connected. You will need to operate the window during these procedures.

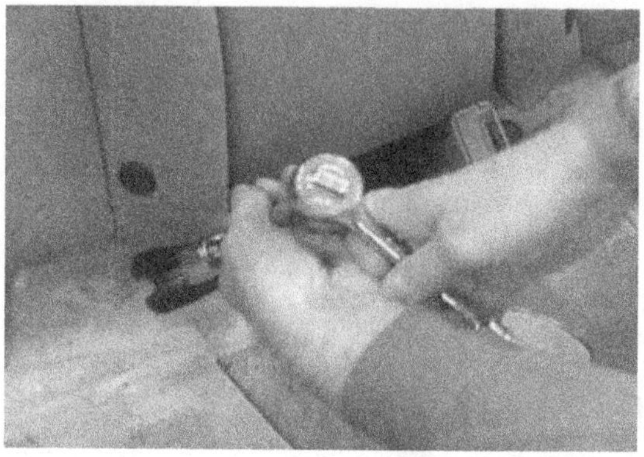

- Remove the seat bottom.
- Remove the seatbelt latches so you do not scratch them when removing the seat back.
- Pull the headrests off.
- Take the bolts off the seatback.
- Open the pass through door for the ski sack, then pull it off toward the front.
- Pull out the bottom of the seatback, lifting straight up as you remove it.

Rear Side Window, Regulator and Motor

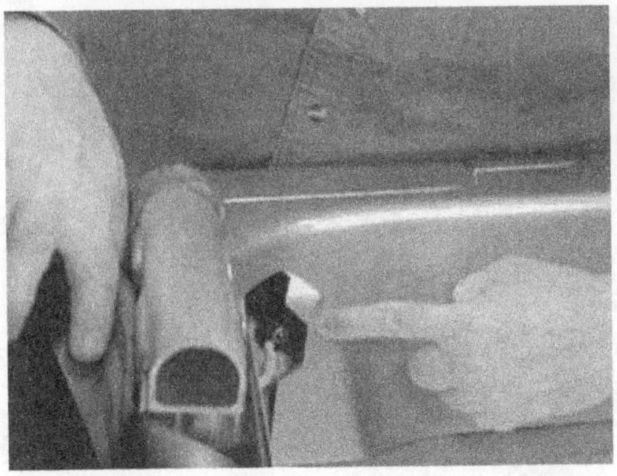

- Tape up the end of the rollover protection so that you will not scratch the side panel when you remove it.
- Remove the upper screw at the rear of the side panel.

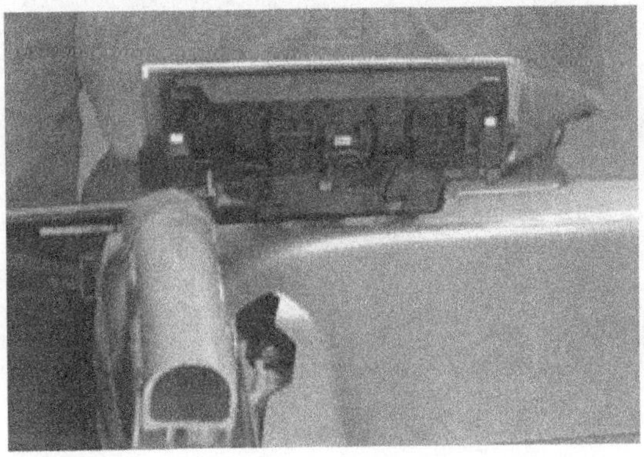

- Operate the side flaps until you hear the motor for the locks run. This positions the arm for the motor facing up so that it will clear the body.

- Use a bone to pop the arm free from the rod, then close the flap.

Rear Side Window, Regulator and Motor

- Pull the rear lower edge of the side panel inward.
- Use a bone to release the side panel from the sill plate. Pry and pull up at the same time. If the clip comes off the backside of the panel, just pop it back in.
- Twist the bone as you pry the side panel from the body.
- Pull the side panel up out of the window sill.

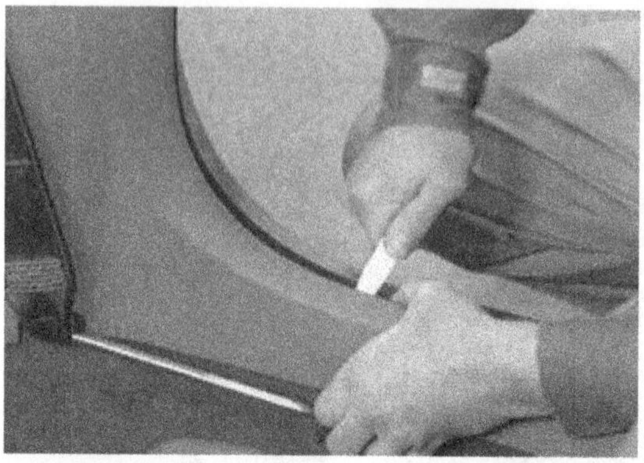

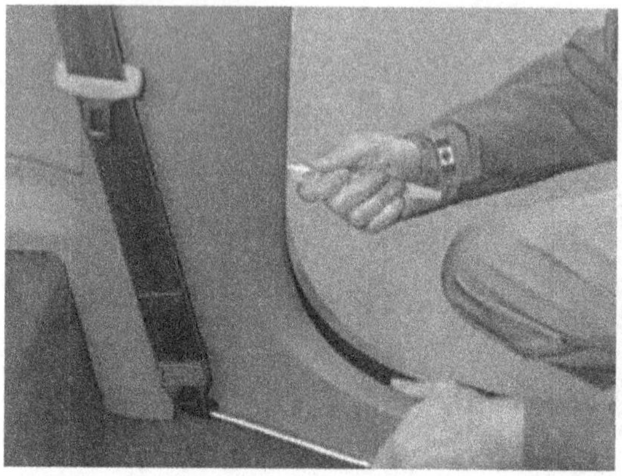

CAUTION

This training module applies to only some of the vehicles covered by this manual. This information is subject to change and should be used as reference only. Always refer to the latest Volkswagen factory service information for the most up to date service, repair and training information.

Rear Side Window, Regulator and Motor

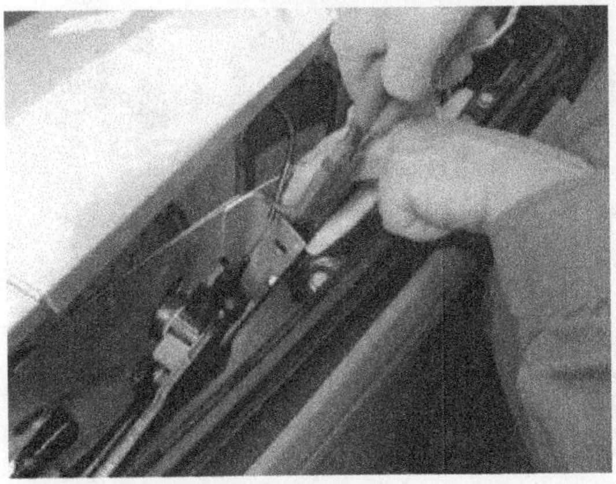

- Remove the clip from the side flap cable inside the body, popping the socket off the ball.
- Remove the cable from the bracket.

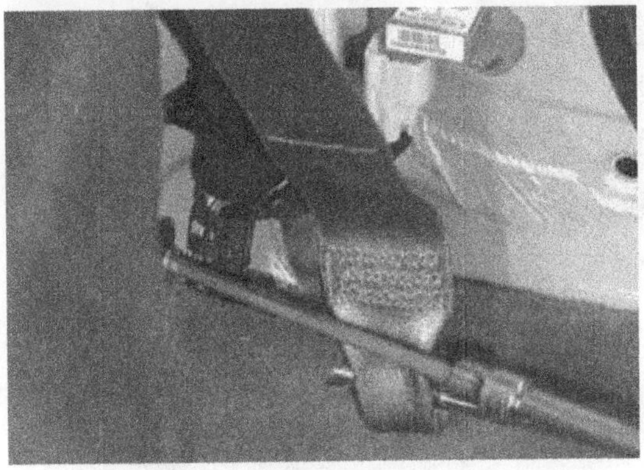

- Disconnect the electrical connectors.
- Remove the rear and front seatbelt trim pieces.
- Unbolt the rear seatbelt, then pull it through the side panel.
- Remove the bolt for the front seatbelt retaining bracket.
- Slide the seatbelt off the bracket, then pull it through the side panel.
- Remove the side panel.

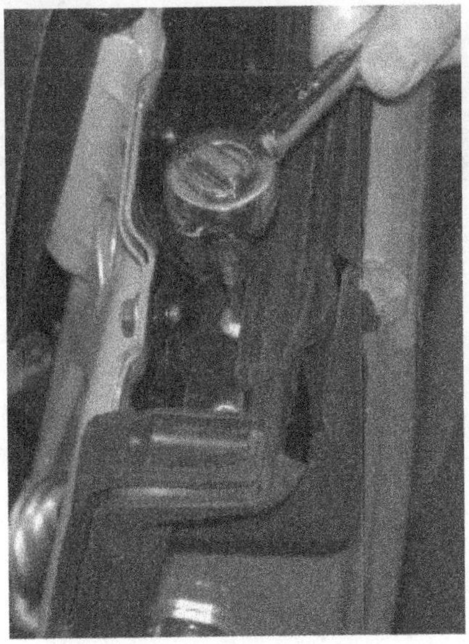

- Pull off the inner window slot seal.
- Remove the two 6 mm bolts in the upper retaining plate, then remove the retaining plate.

Rear Side Window, Regulator and Motor

- Remove the retainer for the loudspeaker.
- Push the grommet for the loudspeaker connector wire through the carrier plate from the front so it will be slack. Do not disconnect the speaker. *Pix loudspeaker*
- Carefully lower the speaker down and onto a towel or soft cloth so the speaker cone does not get damaged.
- Raise the window all the way up.

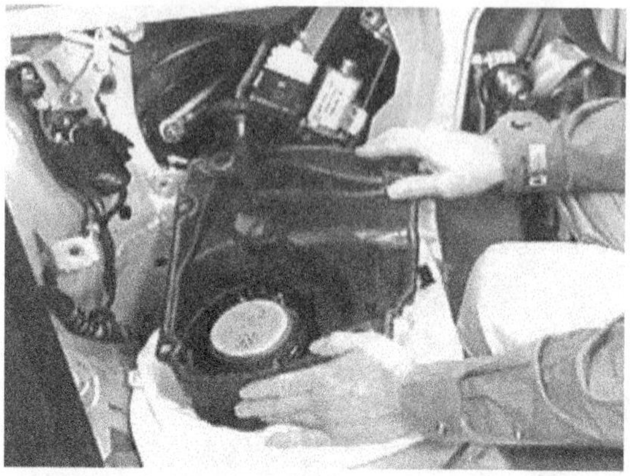

- Remove the two bolts that hold the window to the window regulator. They are located in the two upper front assembly openings. Use a magnet so you do not drop the bolts into the body.

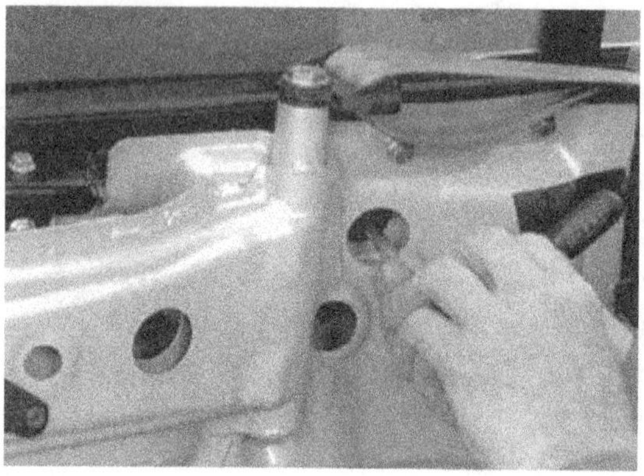

- Lower the window a bit until you can see the lower bolt through the hole in the regulator, then remove it.
- Pull the window out of the body, being careful not to scrape the outer seal.

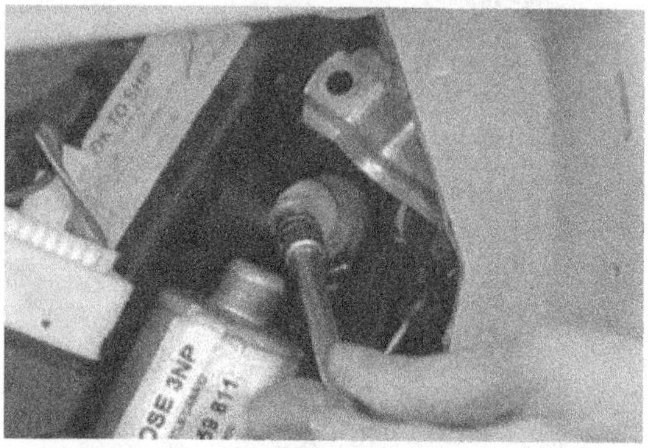

Rear Side Window, Regulator and Motor

Regulator Removal

- Move the regulator to its top position so you will not damage the window seal when you remove the regulator from the body.
- Disconnect the power window motor.

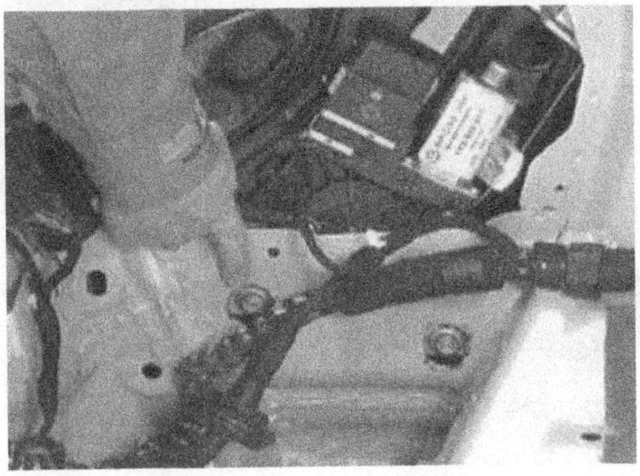

- There are three mounting/adjustment combi-nuts on the regulator.
- Remove the two lower assembly opening caps, loosening the combi-nuts inside both openings with special tool VAG 1739.

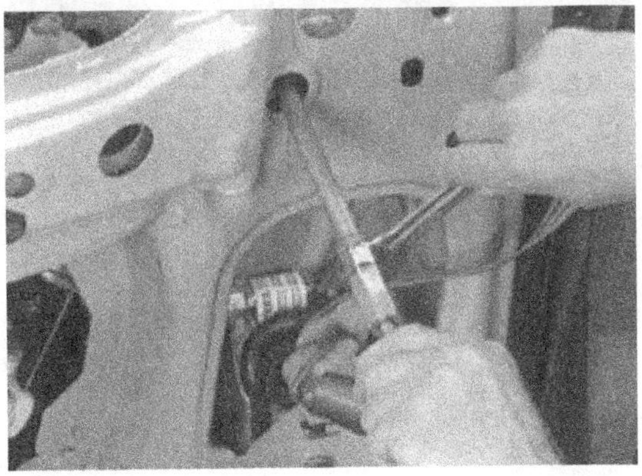

- Loosen the top combi-nut in the assembly opening located at the top front of the body, again using the VAG 1739.

Rear Side Window, Regulator and Motor

- Carefully pull the regulator out of the body, making sure not to scrape the outer seal.

- Each of the three regulator mounting/ adjustment combi-nuts is comprised of a stud used to make the adjustment, and a locking nut.
- Special tool VAG 1739 is actually two tools. On the inside is a 4 mm hex, a pretty standard tool. This hex lives inside a 13 mm deep socket.

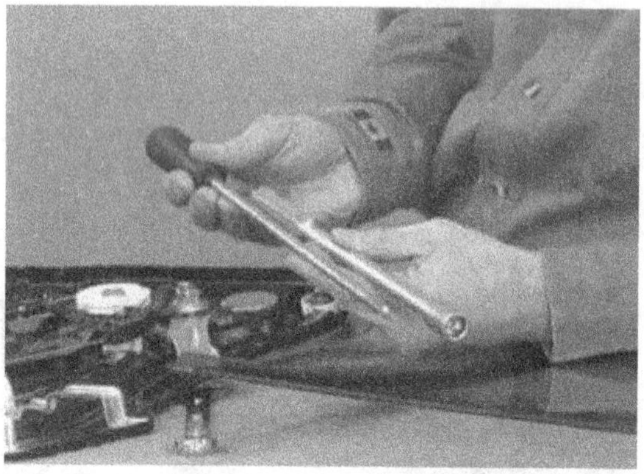

- You use the 4 mm hex to make the adjustment on the stud. The 13 mm socket is used to tighten down the nut and lock it into position.

Rear Side Window, Regulator and Motor

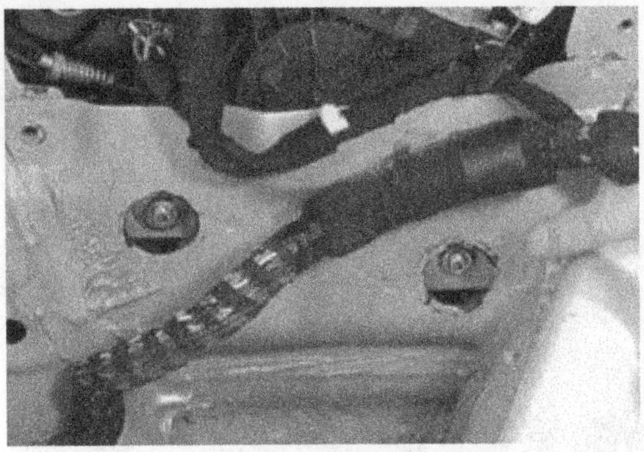

- The two mounting points at the bottom of the regulator should not be moved. They should be tight. If they are loose, put a drop of Loctite on them.
- The top adjustment bolt adjusts where the window sits in the body, either inside or outside in relationship to the body.

- At the bottom of the window is an adjustment bolt that you can move either in or out to adjust where the window sits in the roof seal.

- There are two studs above the adjustment bolt. These serve as mounting points, accessed through two slots in the regulator. The slots that the studs fit into allow for a lot of movement. You can tilt the window back, forward, up, or down.
- Adjusting the window is not an exact science. You do have to make the window adjustments with both the window and regulator installed in the car.

Rear Side Window, Regulator and Motor

Window Installation

- Lower the window regulator, without the window attached, into the body with the regulator run all the way to the top.
- Snug up all three regulator mounting bolts, but not too tight.
- Reconnect the power window motor.
- Lower the regulator so that you will be able to see the attachment point for the bottom of the window when you put it it.
- Put the window in.
- Hand-tighten the lower bolt on the bottom fo the window.

- Raise the window carefully so you can see the two window adjustment points on the bottom of the window through the assembly openings at the top of the body.
- Put the nuts on these two adjustment points, just hand-tightening them so you can move the window around later.
- Hand-tighten the nut on the lower portion of the window.

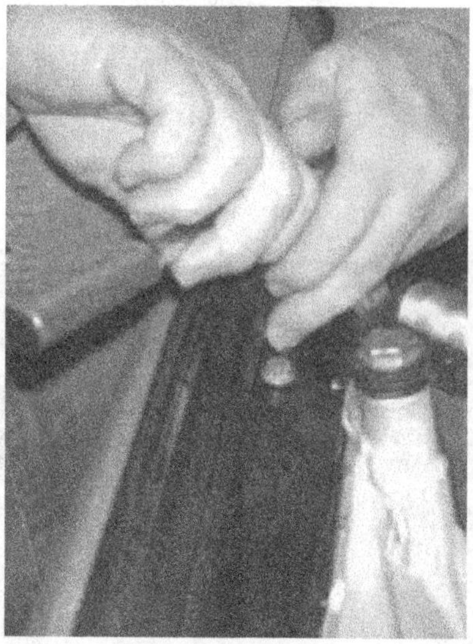

- Install the upper retaining plate. First, thread the two 6 mm bolts that go through the regulator so they will align the placement of the retaining plate. Then push the retaining plate down and snug up the other bolts, finger tight at most.
- With the retaining plate now holding the regulator in its proper position, tighten the lower regulator mounting points. Leave the top regulator adjustment bolt loose beause it is an adjustment point that you will lock in later.

Rear Side Window, Regulator and Motor

- Install the window seal. The back edge of the seal can be a little difficult to install because it must go on two brackets then into another bracket. Install the inner part of the seal first, then work to the outside.
- Next, make sure the door is adjusted. There is no use trying to adjust the windows if the door is out of adjustment.
- Open up the door a little and raise the rear window. Remember, this window is still loose.

- Raise the door window and move the rear window back a bit to make sure the two windows will not hit each other. Carefully close the door.
- With the door closed, move the rear window to the approximate position it needs to be.

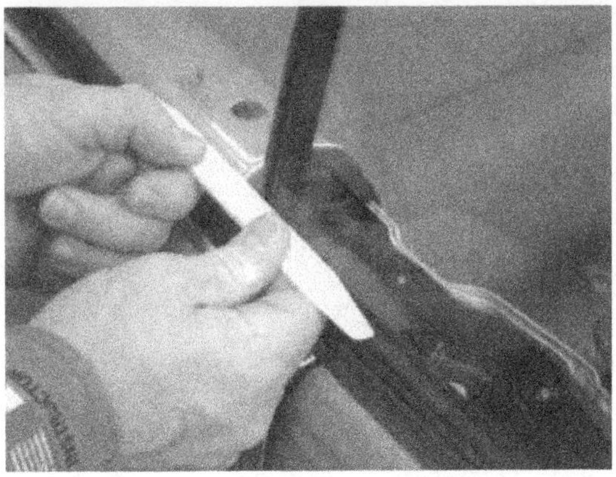

- You will need to adjust the bottom of the window first, making sure it is not too far out or in. Press against the outside bottom of the rear window with a bone to bring it in line with the door window.

The rear window should be in approximately .5 mm compared to the front window.

Rear Side Window, Regulator and Motor

- Use the VAG 1739 to adjust the upper regulator bolt. When the bottom of the windows line up, tighten this bolt.
- Tighten up the two 6 mm bolts that were left loose in the retaining plate.

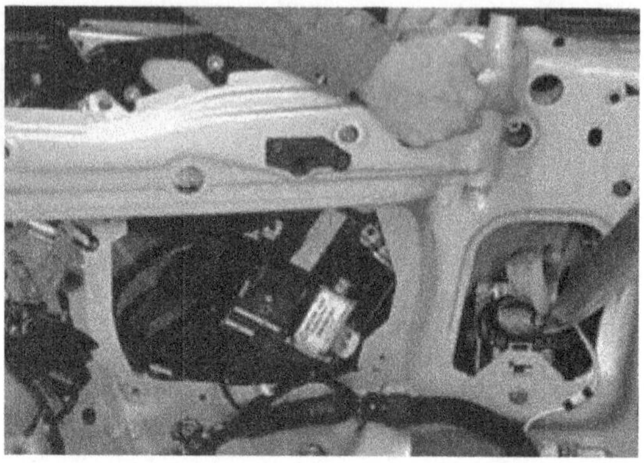

- Now the angle and height of the rear window needs to be adjusted. The specification for the gap between the two windows is 9 mm even, from top to bottom.
- The top edge of the door and rear windows should be smooth and even.
- Manipulate the still loose rear window to achieve this, then snug up the window adjustment bolts through the upper body assembly openings. Do not over-tighten them or you will break the window.

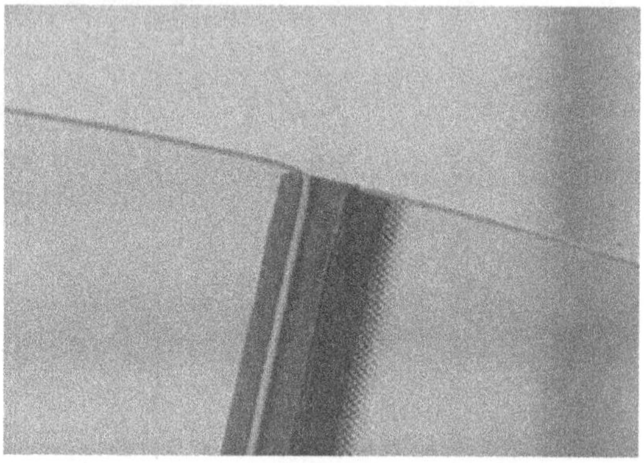

CAUTION

This training module applies to only some of the vehicles covered by this manual. This information is subject to change and should be used as reference only. Always refer to the latest Volkswagen factory service information for the most up to date service, repair and training information.

Rear Side Window, Regulator and Motor

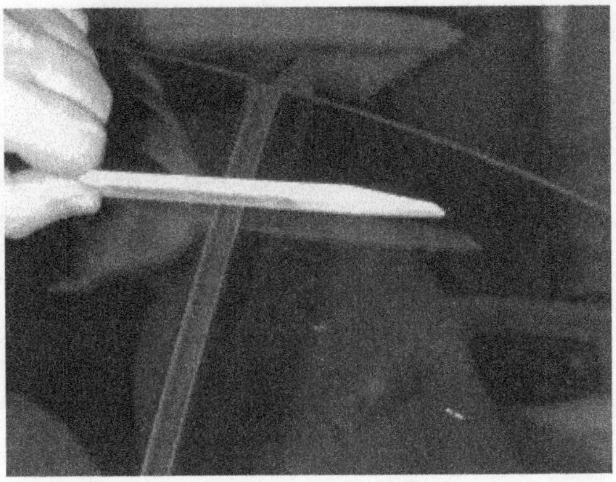

- Now place the bone across the top of the door window and rear window, where they meet. This will show you how in or out of alignment the rear window is to the body at the top of the window.
- This adjustment does not have to be absolutely perfect because as the window comes up into the seal, it will go into the right position.

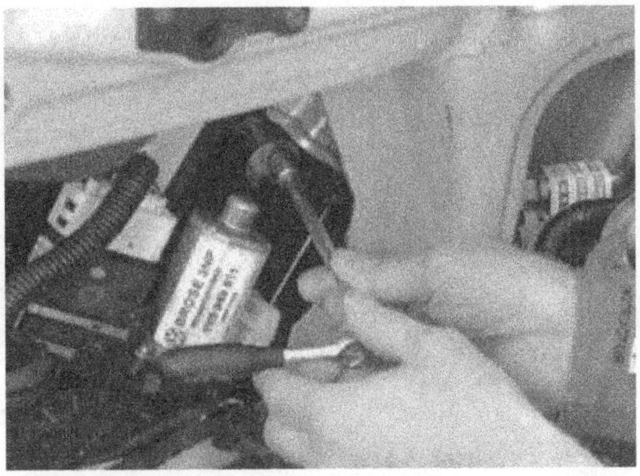

- If the rear window is too far in or out, you can adjust the lower window bolt in the regulator hole. To access this bolt, lower the window until you can see the bolt, then make your adjustment.

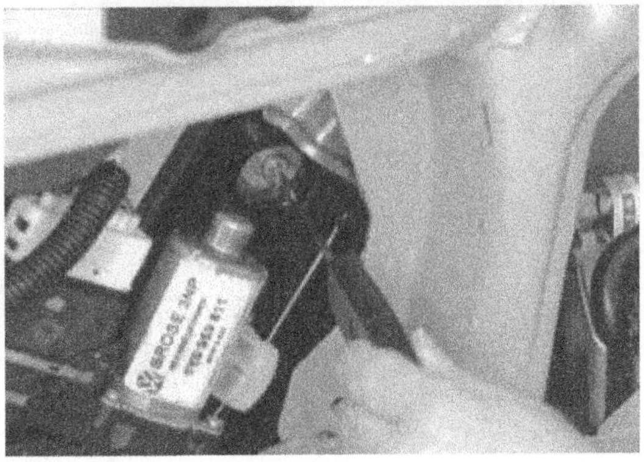

- If the window is in, you need to screw out the adjuster. Make sure the lock nut is loose, then use a 3 mm Allen to adjust the bolt.

Rear Side Window, Regulator and Motor

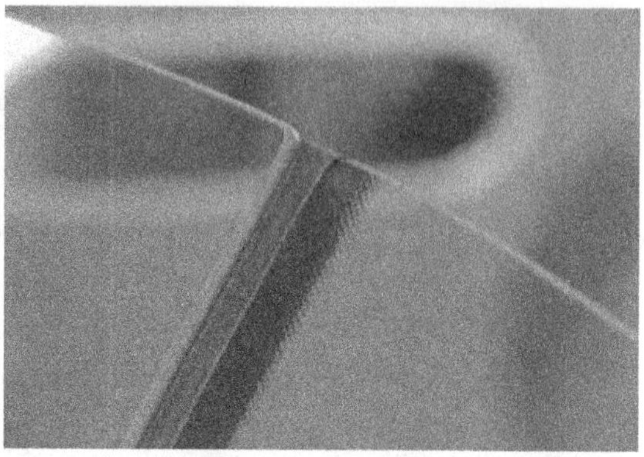

- Raise the window again to check your adjustment, repeating this procedure until the window is properly adjusted.
- Snug the window adjustment bolt.
- Raise the window and check once more.
- Then, just as with the door window, place and mark three pieces of tape on top of the rear window to check the depth of the window going into the roof seal.

Check Side Flap Operation

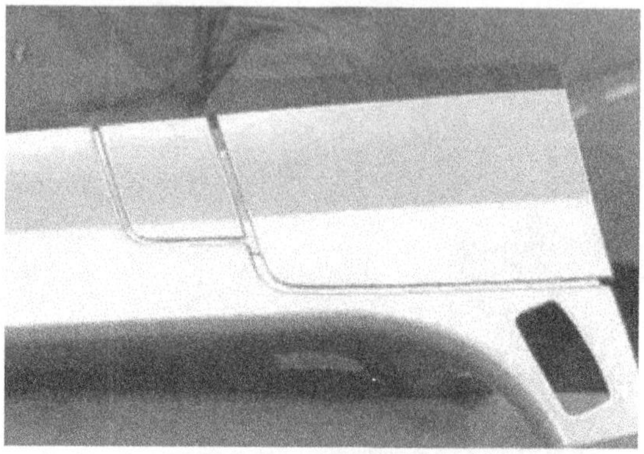

Once the side panel is reinstalled, you cannot adjust it. So check side flap operation before putting the panel back in the car.

- The side flap, which is actually a pair of flaps, front and rear, moves two ways.
- The rear flap lifts up, controlled by a motor.
- The front flap pulls forward, controlled by a cable that is attached to the roof. When the roof moves, it pulls the front flap forward and into the side panel.

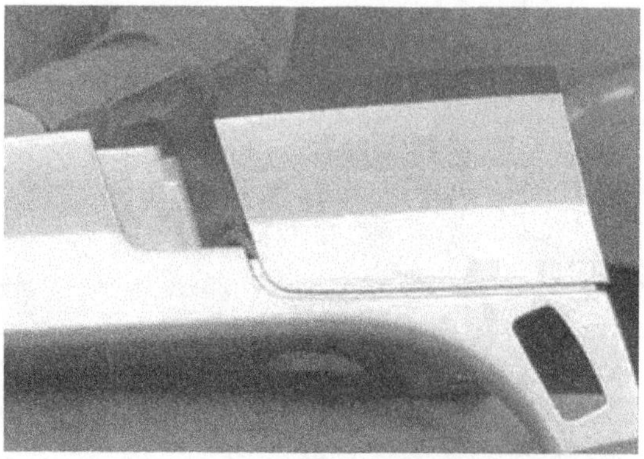

- Critical to proper side flap operation is making sure all gaps around the flaps are even, and the height of the flaps match the height of the side panel.
- It is important that when the front portion slides forward it is low enough to clear the side panel, so it is not scratched.

Rear Side Window, Regulator and Motor

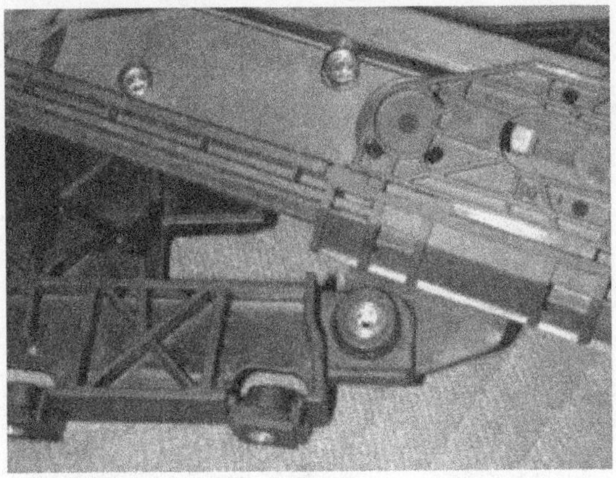

Side Flap Adjustment Tips

Here are some adjustment tips for the side flaps. Check service repair information for updates.

- There are three adjustments for the forward flap, located under the flap assembly. To adjust these, loosen the screws underneath, then use an Allen key to turn the inserts. This will raise or lower the front flap.
- While these screws are loose, you can move the assembly back and forth to adjust the gap between the flaps.

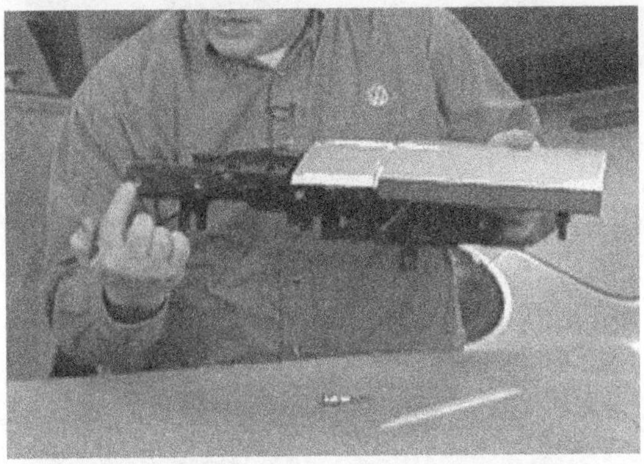

- To adjust the height of the rear flap, there are three slotted screws that hold the flap assembly to the side panel. By loosening these, you can raise or lower the entire assembly, including tilt.

- Again, make sure the flaps are level and even with the side panel, and the gaps are even all around.
- Do not adjust the length of the rod that operates the rear flap.

Convertible Top Outer Shell

Removing the convertible top outer shell will take about two hours, but it is definitely something you can do in your dealership. If you follow these procedures, the top will come off and go on fairly easily.

These procedures will also put you in a good position to remove the roof frame.

- First tape off the area around the perimeter of the roof to protect the chrome strip and paint.
- Leave the battery connected. You are going to need to operate the roof during these procedures.
- Open the roof halfway, with the windows down.
- Remove the rear seat and rear interior side panels.
- Remove the servo motors from the side panels and then reconnect them to the wiring harness. You will need them to operate the roof.

- Pull off the convertible top leading edge trim with a bone. Some clips may come off but they can be reused in re-assembly.
- Turn the convertible top handle 90° to pull the edge trim off the roof.

Convertible Top Outer Shell

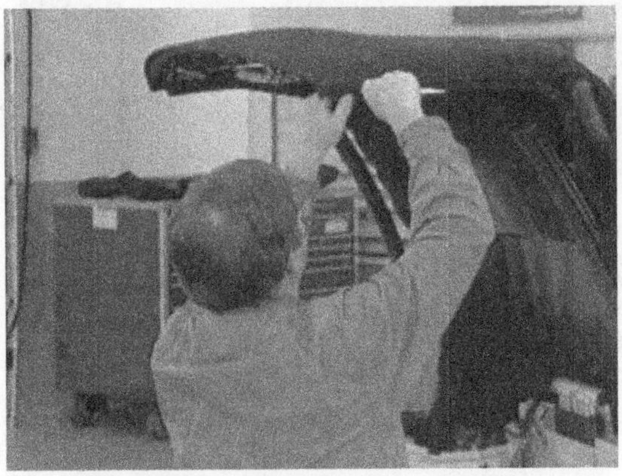

- Remove the side seals.
- Remove the bolts that secure the front corners to the roof and the front beading that attaches to the roof.

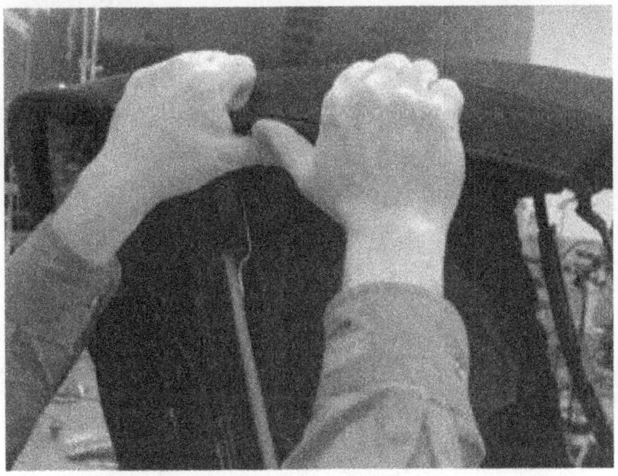

- Pull out the side beading.

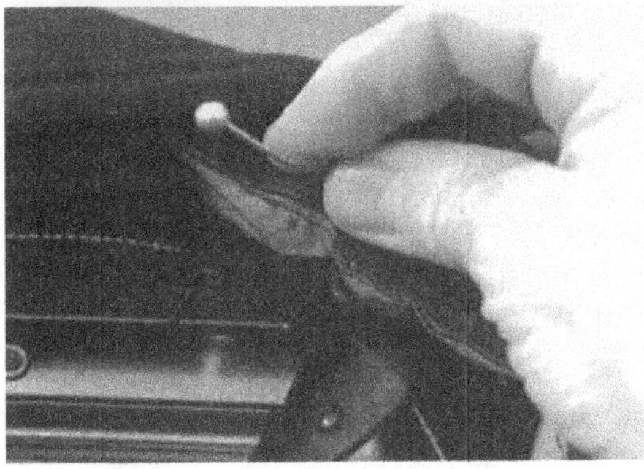

- Remove the small cable from the corner of the channel.

Convertible Top Outer Shell

- Clamp the headliner out of the way on either side of the inside of the C-pillar so that you can more easily drill out the two rivets on either side of the roof. Be extremely careful not to damage the headliner in the process, so use rubber-tipped clamps and do not apply much force when drilling.

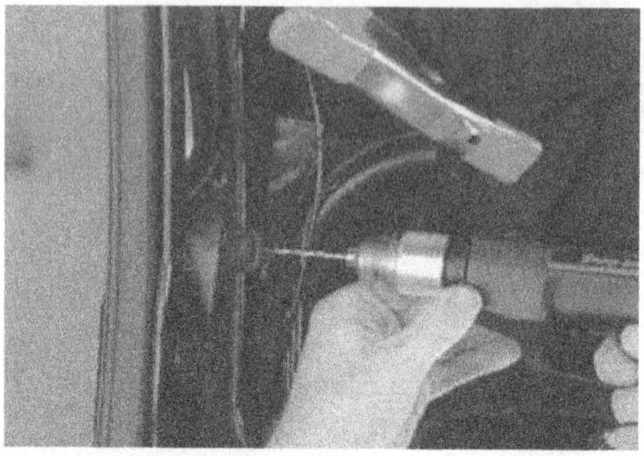

- Remove the window seal from the C-pillar. Also pull the roof beading out of the C-pillar channel. this can be difficult in the area of the metal clips. It helps to pull up as much slack of the roof fabric as you can.
- Close the roof.

Convertible Top Outer Shell

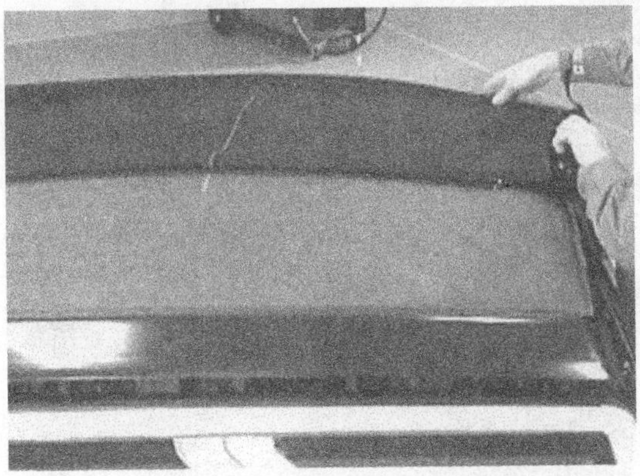

- The convertible top outer shell is fastened to the inner lining using three zipper fasteners.

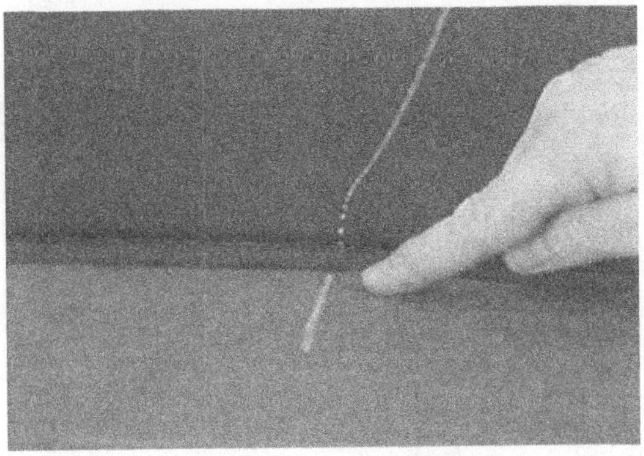

- There are alignment marks on the outer shell and inner lining to help in realigning the outer shell during re-assembly.

CAUTION

This training module applies to only some of the vehicles covered by this manual. This information is subject to change and should be used as reference only. Always refer to the latest Volkswagen factory service information for the most up to date service, repair and training information.

Convertible Top Outer Shell

- The zippers are attached with staples at the end of each zipper. There are alignment marks for these staples that will aid you during reassembly.
- Cut the staples with a pair of pliers, being careful not to cut through the zipper.

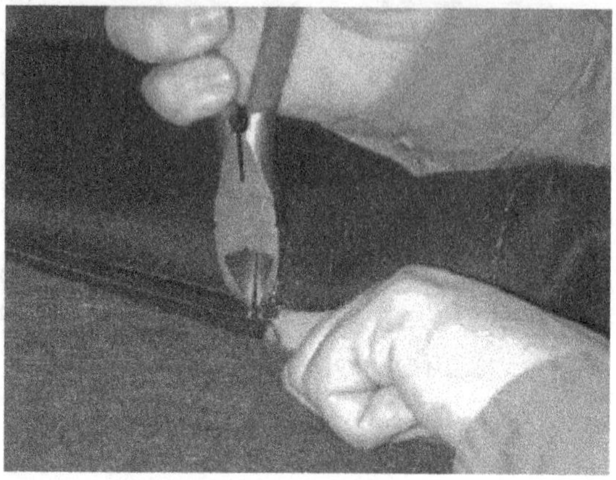

- Unzip the first zipper.

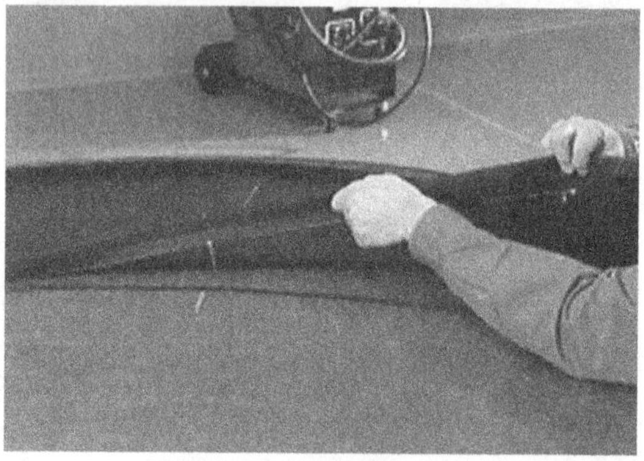

Convertible Top Outer Shell

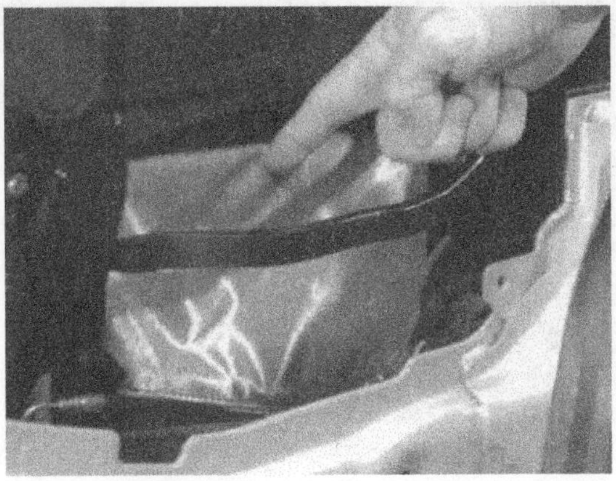

- Remove the water guides inside the body panels on both sides of the vehicle. Reach behind the water guide and pull up at the bottom of the water deflector. Be careful not to cut yourself on the two metal clips that hold it down.

- Pull the beading on the lower edge of the headliner out of its groove.
- Using a bone, pull the beading from around the rear window out of its groove.

Be careful not to break the rear window beading.

- Pull the headliner away from the rear window.

Convertible Top Outer Shell

- Clamp the headliner out of your way so that you can carefully disconnect the ribbon from the connector for the rear window defroster, which is located to the left of the rear window.
- Drill out the rivet that holds in the retaining clip. Be careful not to drill through the roof fabric.

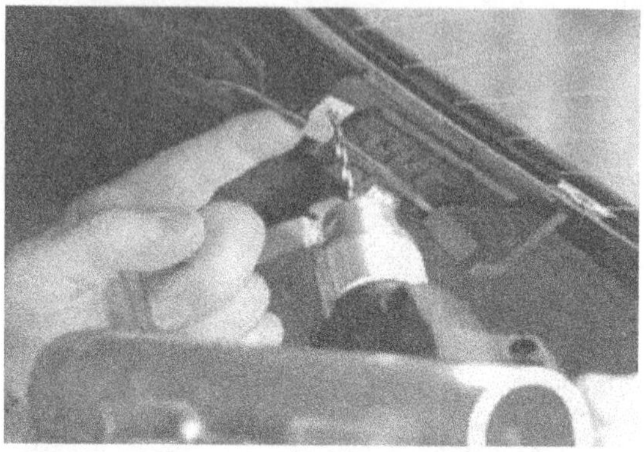

- Remove the securing clip from the ball socket, then separate the hinge. Do this on both sides of the rear window.

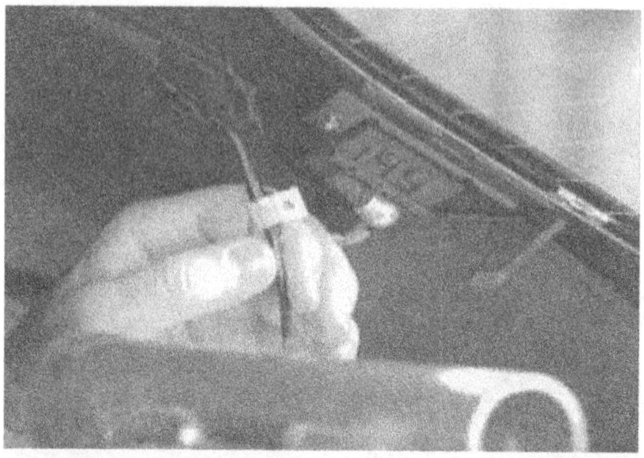

- Under the rear window, remove all of the bolts that secure the outer shell fabric to the body.
- Remove the retaining bracket.

Convertible Top Outer Shell

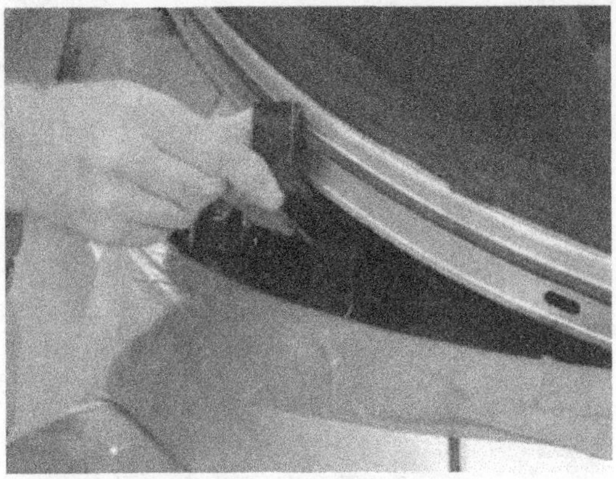

- Pull the retaining trim away from the body and lift it up, being careful not to scratch the body paint or chrome trim. Secure it away from the body, using brackets VAG 1887. Having a second technician assist can be helpful in preventing any damage.

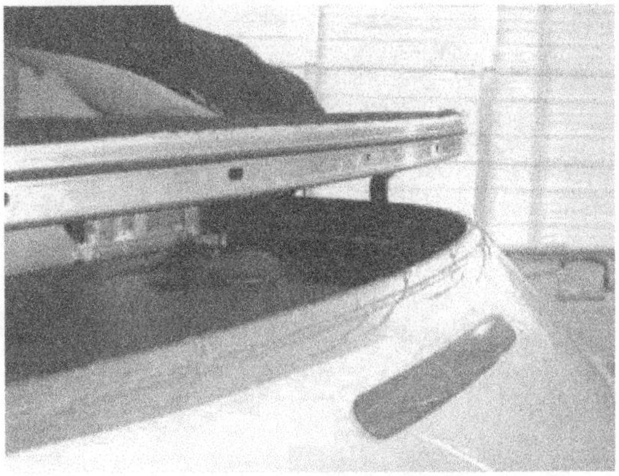

- Release the center and rear zip fasteners.

Convertible Top Outer Shell

- From inside the C-pillar area, remove the screw for the convertible top tensioning cable. It's the upper of the two screws.

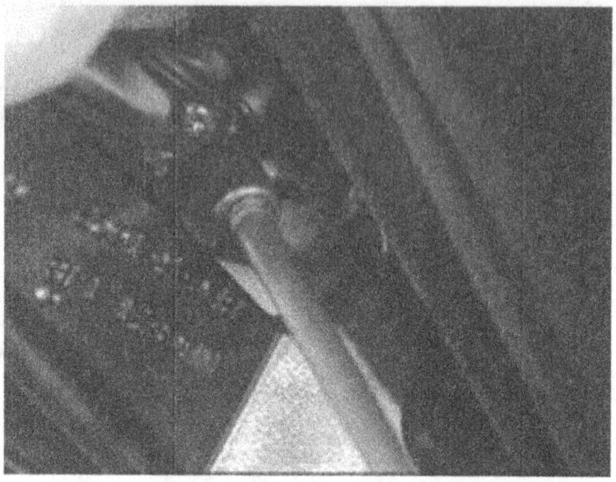

- Slide the outer shell out of the retaining frame, moving it toward the right side of the car.

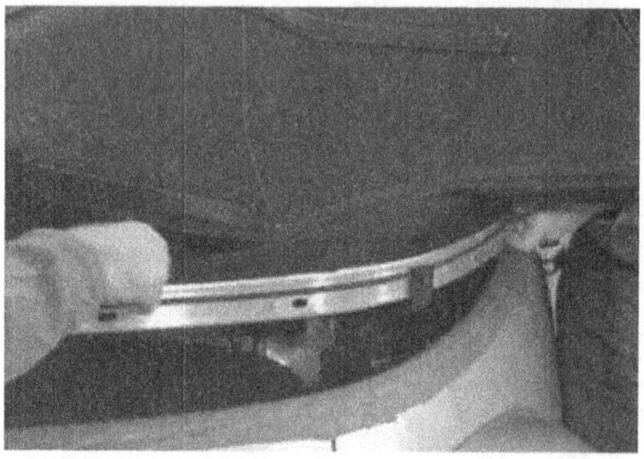

CAUTION

This training module applies to only some of the vehicles covered by this manual. This information is subject to change and should be used as reference only. Always refer to the latest Volkswagen factory service information for the most up to date service, repair and training information.

Convertible Top Outer Shell

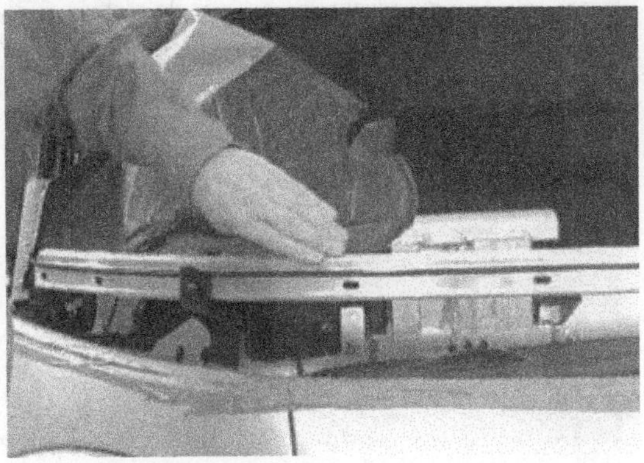

Re-InstallationTips

- During re-assembly, check the rubber sealing cord in the roof retainer. Replace it if it's dried out or cracked.

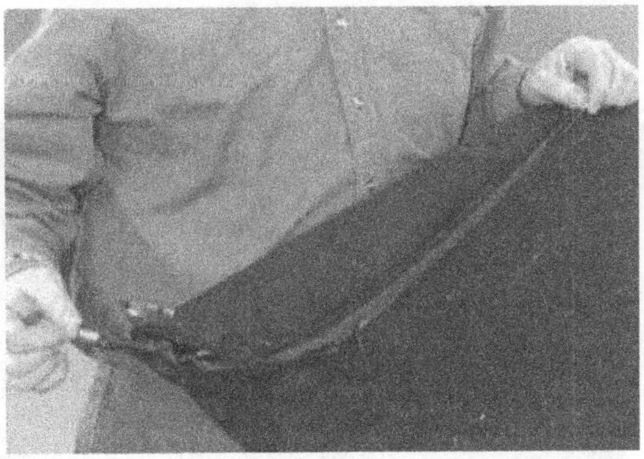

- If you are replacing the outer shell, you will need to remove the tensioning cable from the old shell and thread it into the new one.

- With a second technician, slide the outer shell into the retaining frame. If it does not slide easily, spray some silicone onto the track.

Convertible Top Outer Shell

- When re-attaching the zippers, the first step is to realign the marks on the outer shell and the inner liner.

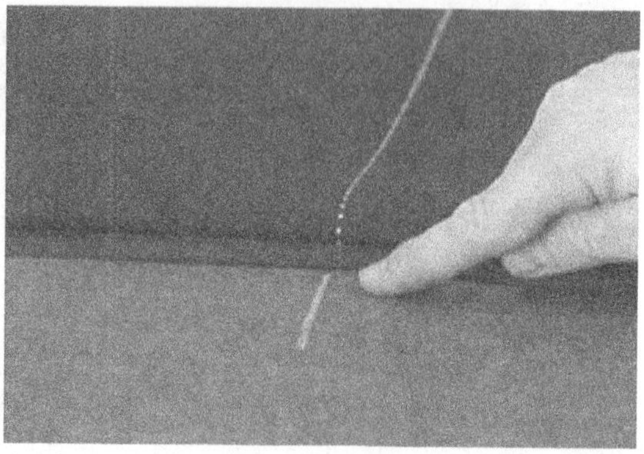

- Then start the zipper pliers (VAS 6138 Closing Pliers) in the center of the zipper. Push it back and forth so that you can feel the click when you have engaged the zipper properly.
- Zip the fastener all the way to one end, then come around to the other side of the car. Again, engage the zipper pliers, pushing them toward the already fastened zipper area before pulling the zipper closed.

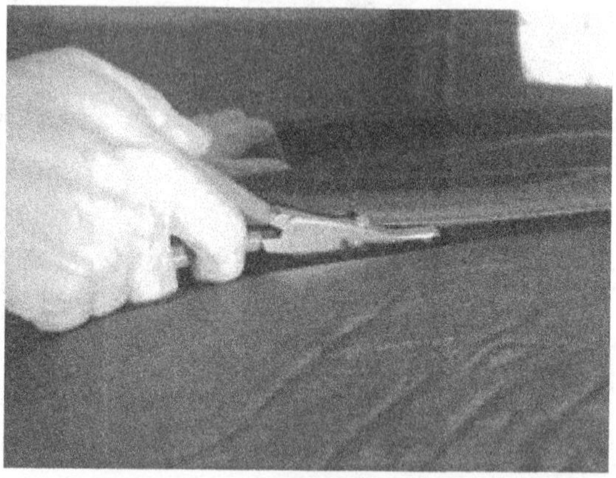

Convertible Top Outer Shell

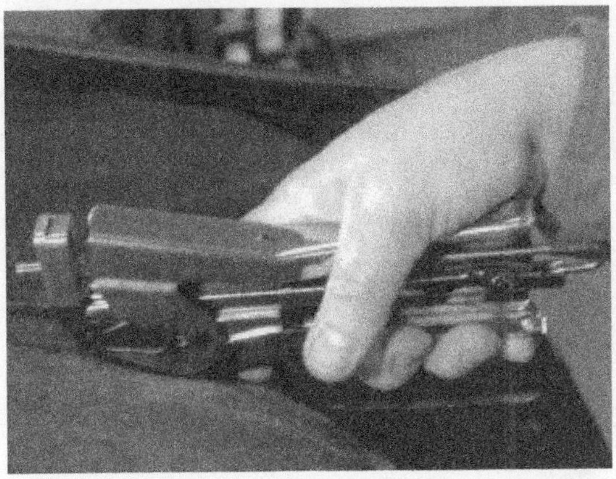

- Fasten the ends of the zippers with VAS 6148 Staple Pliers and VAS 6148/1 Staples. Make sure that you get a good bite on an area of the zipper where the fabric is not damaged from previous staples.

Convertible Top Outer Shell

There are also two areas on the C-pillar that require special attention. The first deals with putting the outer shell back on the frame.

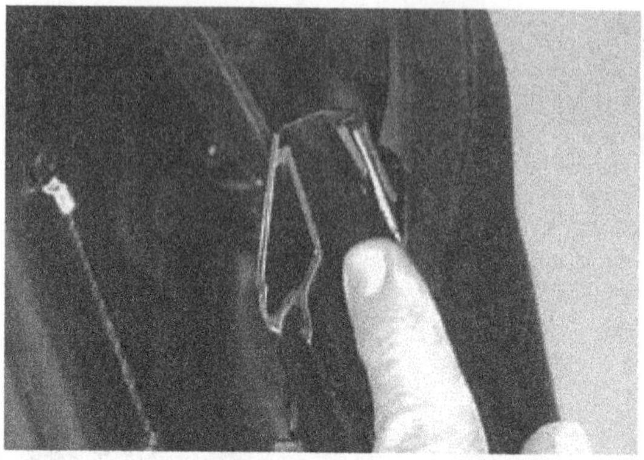

- It can be difficult inserting the C-pillar beading. You need to position the beading at the top of the C-pillar rail. The best way to get the beading started is to pull it down until it is even. Then press it into place. Having the beading too loose or too tight can cause bunching of the roof or too much strain on the roof.

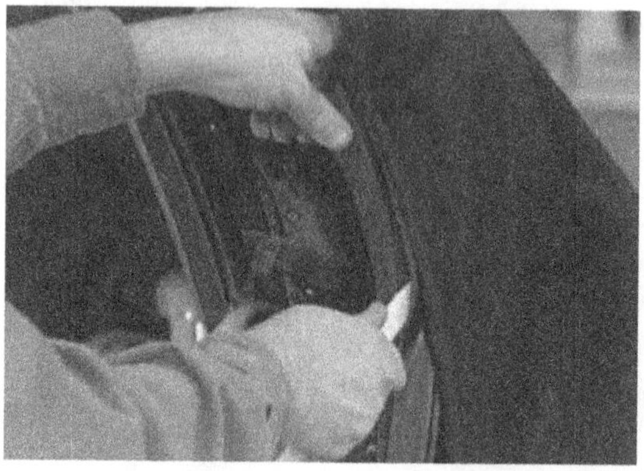

- The window seal that goes into the roof frame at the C-pillar can also be difficult to reinstall. This is because it has two areas that it needs to engage in the rail, on the inside and on the outside. It is suggested that you spray a little silicone in the seal grooves, then use a bone to press it into the window slot. Engage the inside portion of the seal first, then the outside.

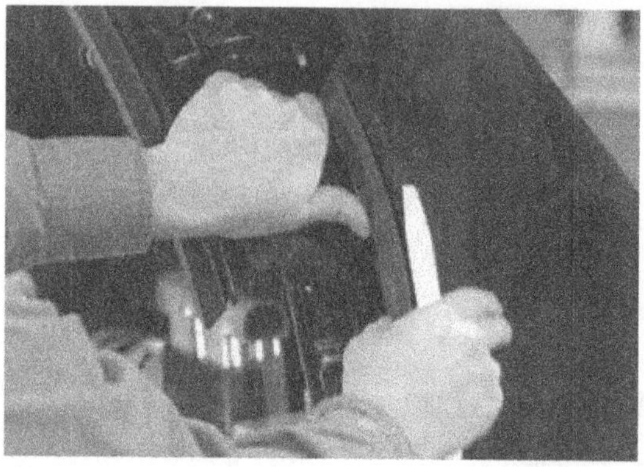

Convertible Top Outer Shell

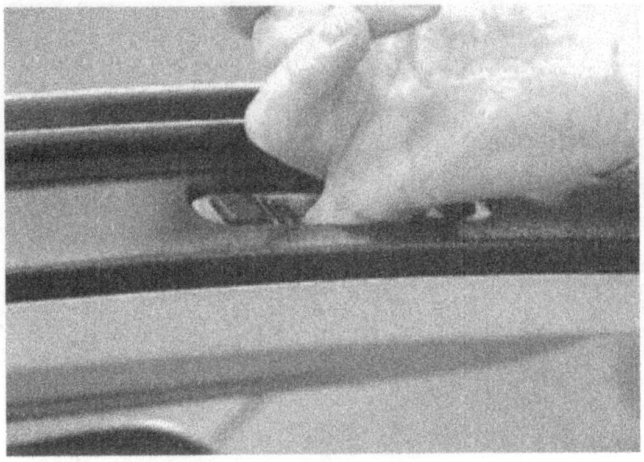

One of the adjustments you may be asked to make in your dealership is to the roof latch assembly. This is because the latch on the righthand side activates a microswitch on top of the windshield header frame. If this microswitch is not activated, the roof control module will think the roof is always open. These procedures are pretty easy.

- First check to see if the microswitch is free by placing a finger into the right latch opening on top of the windshield header frame.
- If the microswitch appears to be working properly and there is still an activation problem, then remove the roof header trim panel.
- Also check the two rods that operate the latch. Make sure they are both straight.

CAUTION

This training module applies to only some of the vehicles covered by this manual. This information is subject to change and should be used as reference only. Always refer to the latest Volkswagen factory service information for the most up to date service, repair and training information.

Convertible Top Outer Shell

If the two rods are straight but the righthand latch is not going far enough to trip the microswitch, you can lengthen the right rod.

- First remove the clip, then pop the socket off the ball.

- Use two wrenches to loosen up the ball, then lengthen the rod.
- Once you have made sure that the latch is now operating the microswitch properly, tighten everything up, reinstall the clip, and reinstall the roof trim panel.

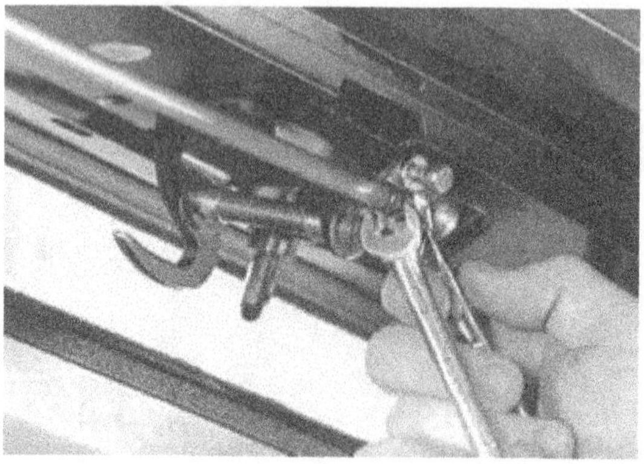

Remember:

This Self-Study Program was designed to complement the New Beetle Convertible Top Window Repair video. Use these two learning aids to supplement current service repair information.

This Self-Study Program does not replace service repair information.

Always refer to service repair information for the most current repair and adjustment procedures.

New Beetle Convertible

Technician's Reference Guide
Course Number 892202

CAUTION

This training module is included as background information and is based on introductory and preliminary information. It applies to only some of the vehicles covered by this manual and is intended to serve as a product familiarization guide. Note that this information is subject to change and should be used as reference only. Always refer to the latest Volkswagen factory service information for the most up to date service, repair and training information.

Table of Contents

The Technician's Reference Guide provides you with information regarding designs and functions.

The Technician's Reference Guide is not a Repair Manual.

For maintenance and repair work, always refer to the current technical literature.

Course Goal and Objectives

Course Goal

The goal of this course is to provide technicians with the necessary skills to diagnose and repair the New Beetle Convertible.

Course Objectives

After completing this course, technicians will be able to:

- Identify the structural and functional differences between the New Beetle Sedan and the New Beetle Convertible
- Perform repair and maintenance procedures unique to the New Beetle Convertible, including:
  - Top cover removal and installation
  - Removal, installation, and adjustment of front and rear windows
  - Removal and installation of windshield trim
  - Trunk lid adjustment
  - Rollover protection extension and retraction
- Perform diagnostic procedures unique to the New Beetle Convertible, including:
  - Read convertible top related measuring value blocks
  - Read rollover protection related measuring value blocks
  - Convertible top diagnosis
  - Rollover protection diagnosis
- Perform the PDI checklist items unique to the New Beetle Convertible

Introduction

Introduction

The wait is finally over.

Ever since the introduction of the New Beetle Sedan in 1998, New Beetle fans have wanted a New Beetle Convertible. Some wanted it so badly that they parted with their Volkswagen factory warranty and a few thousand dollars in exchange for an aftermarket conversion. But when a converted New Beetle started showing up in television and movies, the numerous inquiries at Volkswagen dealerships made it obvious that the public had waited long enough.

Today, about one percent of the vehicles sold in North America are convertibles. Although small in number, this market has demonstrated stable growth, even in years when total vehicle sales were lower.

Over the next few years, the small convertible segment is expected to grow more than the midsize and premium convertible segments. For the long term, the New Beetle Convertible will be well-positioned to take a significant piece of this market.

The short term prospects are favorable too. Some of Volkswagen's competitors have opted to cancel their lower-priced convertibles in order to offer vehicles in the midsize and premium range. (They probably knew they couldn't compete with the New Beetle Convertible!) This leaves practically the entire small convertible market to Volkswagen.

Convertibles appeal to drivers who are more out-going and social. Their #1 reason for purchasing a convertible is that it's fun to drive. They are non-conformists, and they are proud of it. Their bold style is reflected in the cars that they drive. They like the feeling of open air motoring, and they enjoy being seen and attracting attention.

Fun, bold, attention-getting style made the New Beetle Sedan a great success. The introduction of the New Beetle Convertible is designed to enhance the New Beetle brand, building on the momentum begun by the Sedan.

Introduction

Convertible Heritage

The original Type 15 Volkswagen Beetle Convertible is an automotive design classic. In terms of both sales and popularity, it also was a tremendous success in the marketplace during the years it was produced, 1949 to 1980. More than 330,000 of them were built, creating a niche market for Volkswagen that would carry over to the company's Karmann Ghia and Golf convertibles.

In partnership with venerable German coachbuilder Wilhelm Karmann GmbH since the earliest days of the Beetle convertible, Volkswagen has produced over one million convertibles. The New Beetle Convertible continues this rich heritage with proven design and engineering principles that will excite casual and enthusiast drivers alike.

Volkswagen Convertible Timeline

- 1949-1980 Volkswagen Beetle Convertible, 331,847 units built
- 1957-1974 Volkswagen Karmann Ghia Convertible, 80,877 units built
- 1978-1993 Volkswagen Golf A1 Convertible, 388,522 units built
- 1993-1997 Volkswagen Golf A3 Convertible, 139,578 units built
- 1998- Volkswagen Golf A4 Convertible, 57,786 units built (as of 31 Dec 00)

Convertible Top

Starting at the Top

The most obvious new feature of the New Beetle Convertible is the top. Like the original Beetle convertible, the New Beetle Convertible's top folds back and remains outside, rather than concealing itself inside the body.

Despite similar design cues, the New Beetle Sedan and New Beetle Convertible have remarkably different technology, construction, and safety features.

Because a rigid top contributes so much to the structural integrity of a vehicle, making it quiet, smooth, comfortable, and safe, it's difficult for an automaker to meet modern expectations without the top. The New Beetle Convertible's forward thinking design and engineering overcomes these difficulties.

CAUTION

This training module applies to only some of the vehicles covered by this manual. This information is subject to change and should be used as reference only. Always refer to the latest Volkswagen factory service information for the most up to date service, repair and training information.

Convertible Top

Top Fully Closed

Composition of the Convertible Top

The New Beetle Convertible's top is constructed of three main layers, including:

- Exterior textile top shell layer
- Center .59 inch (15 mm) insulating layer
- Interior lining

The exterior textile shell itself is also a three-layer design. The exterior shell's outer layer is poly-acrylic fiber yarn. Its center layer is a rubber coating that bonds the inner and outer fabrics and ensures leak-free operation. The lower layer of the exterior shell is a polyester fiber yarn that, when pulled taut, helps ensure that the convertible top is secure and correctly aligned.

The convertible top frame is constructed of hinged bows, with the locking assembly attached to the front of the convertible top frame.

Top Fully Open, with Boot Secured

Convertible Top

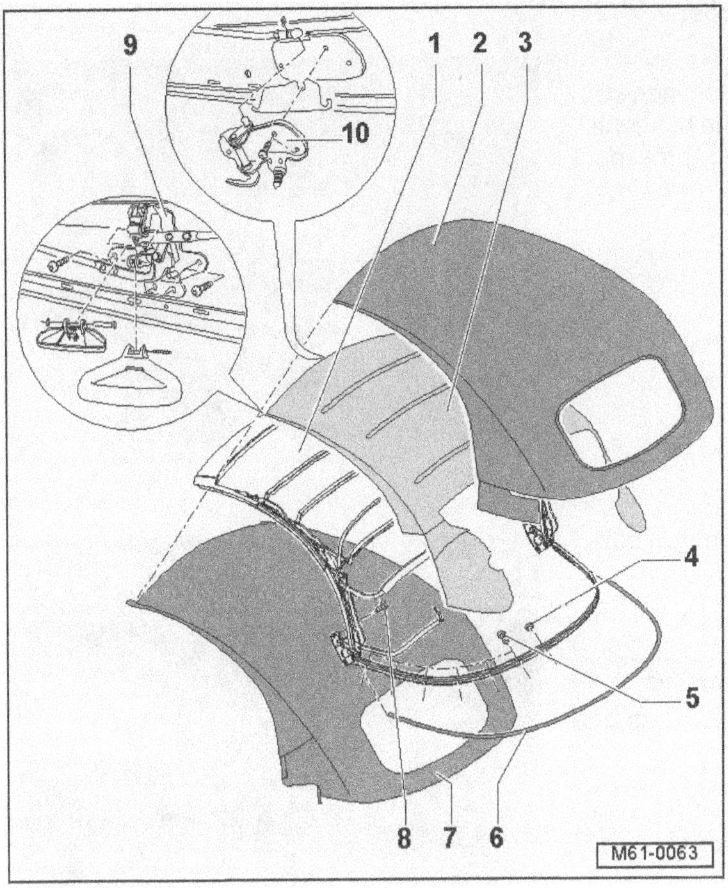

Convertible Top Components

1 Canopy frame

2 Canopy fabric

3 Padding matting

4 Nut

5 Bolt/screw

6 Sealing band

7 Canopy lining

8 Bolt/screw

9 Locking assembly

A heated rear glass window is integrated with the rear cover. It is bonded to the cover and cannot be replaced.

Caution: When removing the top frame, do not remove these bolts, or factory frame adjustments will be lost.

Convertible Top

Closing the Top If Power is Lost

If for any reason it is not possible to close the soft top by pressing the button in the center console (for instance if the battery is drained) the following emergency procedure must be followed.

- Switch ignition OFF
- Engage parking brake
- Pull plastic D-ring below side trim panel lid in direction of arrow
  - Lid mechanism releases
- Lift side trim panel lid
- Pull red rope upwards to unlatch soft top locking hook
- Open trunk
- Remove plastic cover protecting hydraulic pump on left side of luggage compartment trim panel
- Turn screw on hydraulic pump counterclockwise until it stops
- Close trunk
- Swing top forward by grabbing with both hands, lifting and pushing it forward
- Lock roof in place
- Turn screw on hydraulic pump clockwise until it stops

Rollover Protection and Body Reinforcements

Design and Engineering Features

Convertibles by their very nature pose numerous challenges to designers and engineers, primarily with regard to rollover protection and vehicle stability.

The New Beetle Convertible features an ultra rigid windshield crossmember, reinforced A-pillars, and twin rollover protection bars that create a secure space, minimizing the possibility of injury in a rollover accident.

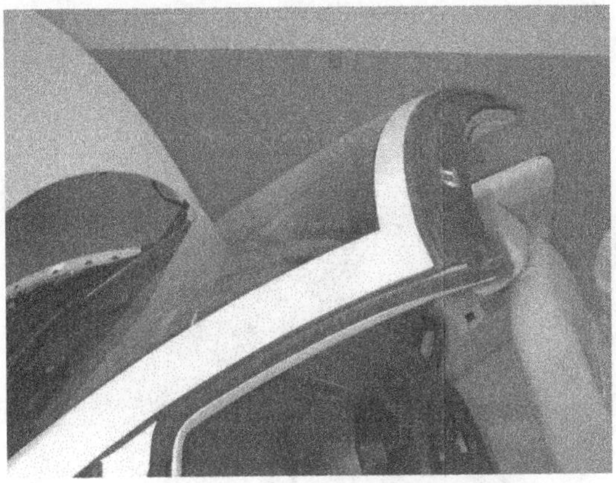

Reinforced A-Pillars and Windshield Crossmember

- The stiffness of the A-pillars, particularly important for the safety of the driver and passengers in the event of a rollover, is reinforced in the New Beetle Convertible with high-tensile strength tubes and an optimized design.

Twin Rollover Protection

- Rollover protection is spring activated by the airbag control module in either a side, rear-end, or rollover collision.

Rollover Protection and Body Reinforcements

New Shoulder Belt Retractor Position

- This rollover protection makes it possible to eliminate the B-pillars, which improves visibility. As a result, the shoulder belt retractors are securely mounted under the rear windowsills.

Underbody Supports

- To regain the body stiffness and integrity that is present in a sedan, additional structural support has been added to the New Beetle Convertible.
- New diagonal braces bolted to the undercarriage have been added to the convertible to absorb flexing about the logitudinal and lateral axes.

CAUTION

This training module applies to only some of the vehicles covered by this manual. This information is subject to change and should be used as reference only. Always refer to the latest Volkswagen factory service information for the most up to date service, repair and training information.

Rollover Protection and Body Reinforcements

Additional Structural Enhancements

- Augmented chassis sheetmetal
- Additional layer of sheetmetal in front footwells
- Enhanced door supports
- Strengthened rear body
- Taller, stronger door sills

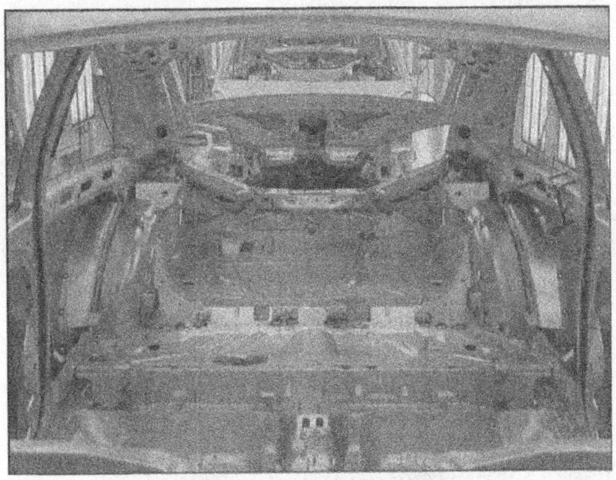

New Beetle Sedan chassis

New Beetle Convertible chassis

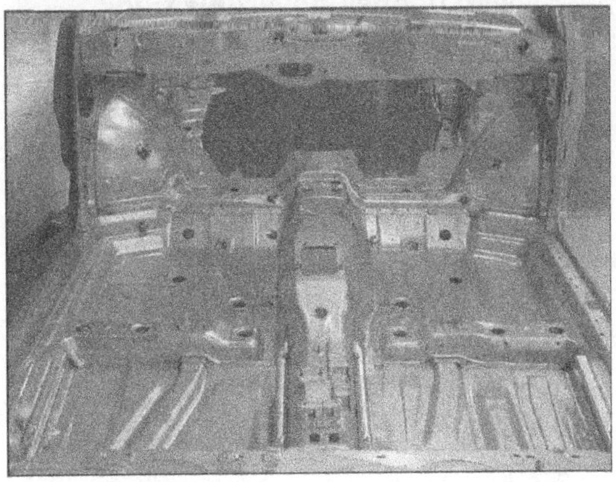

New Beetle Sedan front footwells

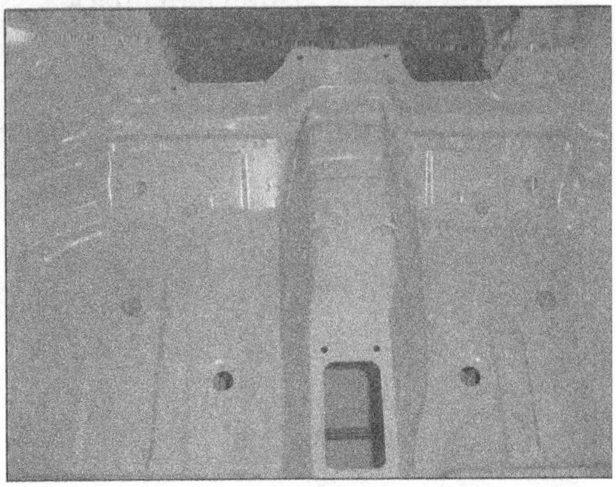

New Beetle Convertible front footwells

Rollover Protection and Body Reinforcements

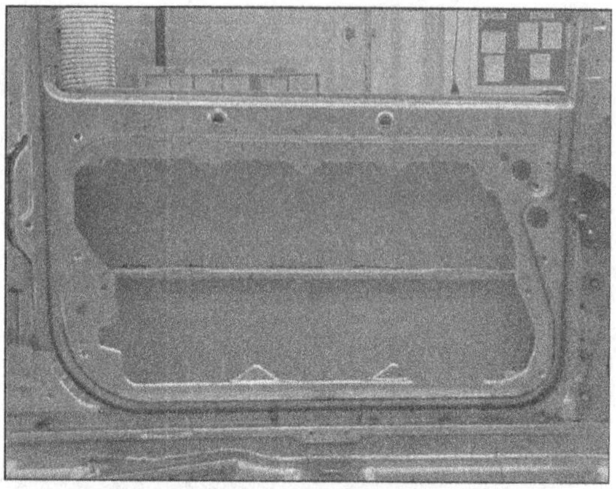

New Beetle Sedan door beam

New Beetle Convertible door beam

New Beetle Sedan hatchback

New Beetle Convertible trunk

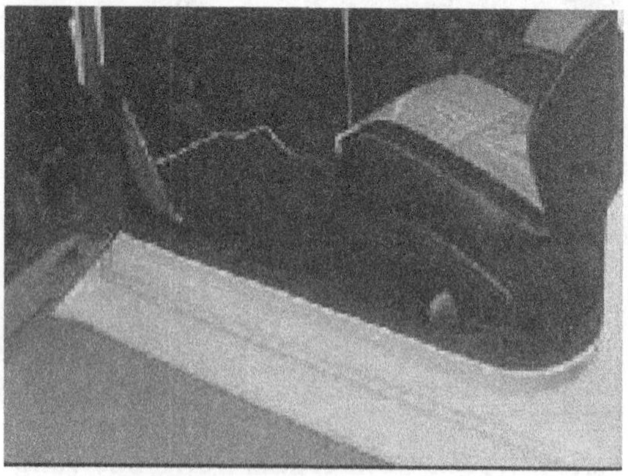

New Beetle Sedan door sill

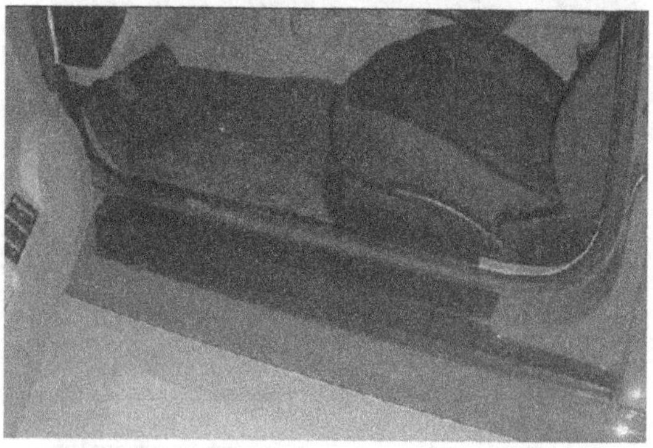

New Beetle Convertible door sill

New Features

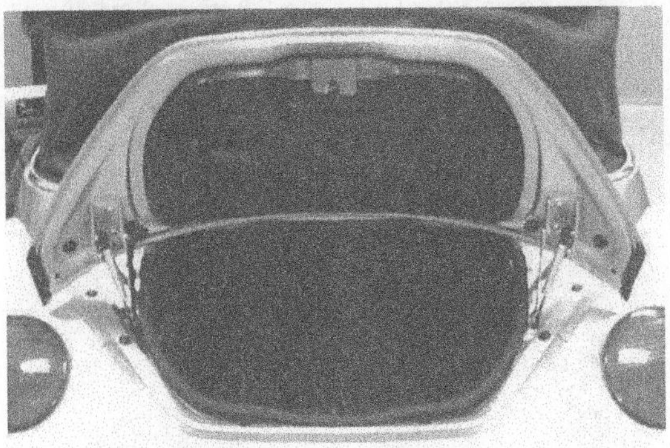

The New Beetle Convertible has a trunk lid in place of the hatchback on the sedan. The trunk lid lifter struts, seal, trunk interior trim, and center brake light are also unique to the New Beetle Convertible. Also new is an anti-entrapment lever, a safety feature located on the inside of the trunk lid, which enables the trunk lid to be opened from inside the trunk.

There are several changes to the doors, as well as new locations and switches for the power windows, trunk release, and fuel door. A lock cylinder now secures the trunk and fuel door.

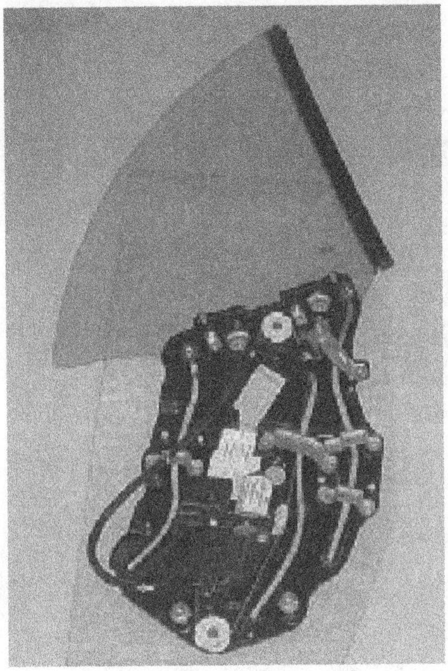

Unlike the Sedan, the New Beetle Convertible is equipped with four power windows with window regulators. Its power windows also feature Pinch Protection, which reverses window travel if an obstruction is detected while the window is closing.

New Features

The door lock rod has been replaced with an LED status indicator located on top of the door sills. There are three status states. When the LED lights are OFF, the doors are unlocked. When the LED lights are ON, the doors are locked. When the car alarm is ON, the LED lights flash.

Door locks can be operated with the key, the remote, or the interior switch. A redesigned window assembly carrier accommodates the numerous trim changes.

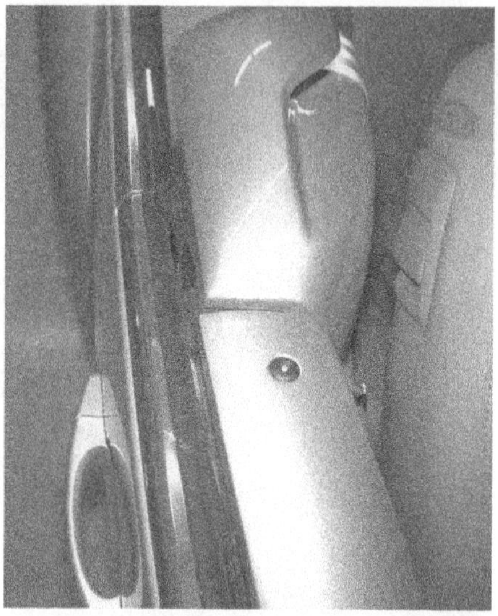

Though the rollover protection system eliminates the possibility of a fold-down rear seat, a pass-through "ski bag" is provided for long items.

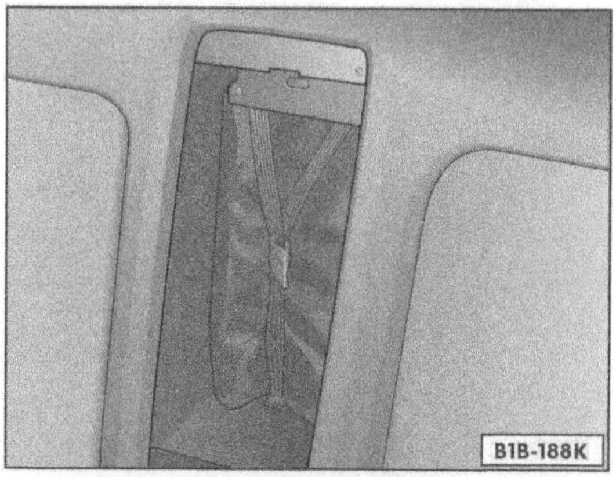

B1B-188K

Other exterior and interior changes include:

- Front seat backrest release
- Windshield frame
- Exterior chrome windowsill trim
- Rear side panels
- Door sill trim
- Side view mirrors equipped with turn signal indicators

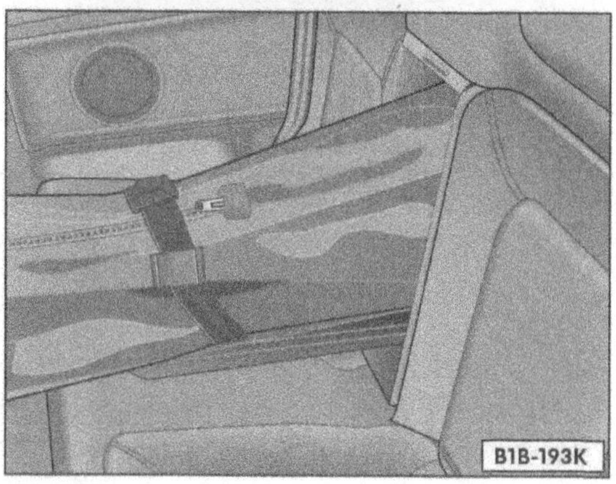

B1B-193K

New Features

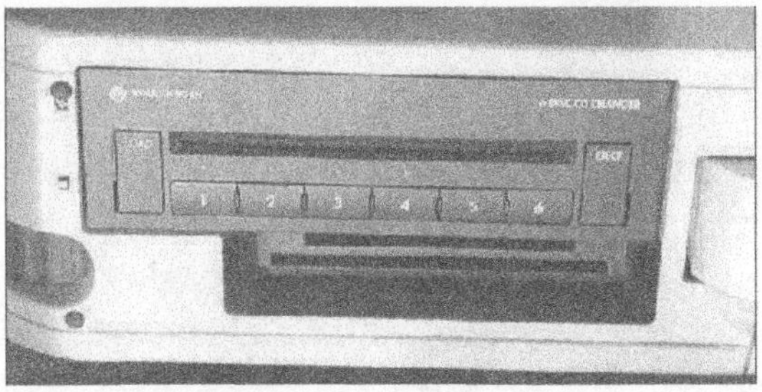

The Jumbo Box center console is available in the New Beetle Convertible in GL trim. A power point is located at the back of the center console. An optional 6-disc CD changer may be vertically mounted in the Jumbo Box.

The 10-speaker Monsoon® audio system available in the New Beetle Convertible will easily compete with the additional noise of open-air motoring. Two subwoofers, two midrange, and two tweeters are in the front seat area, and two woofers and two tweeters are located in the rear seat area.

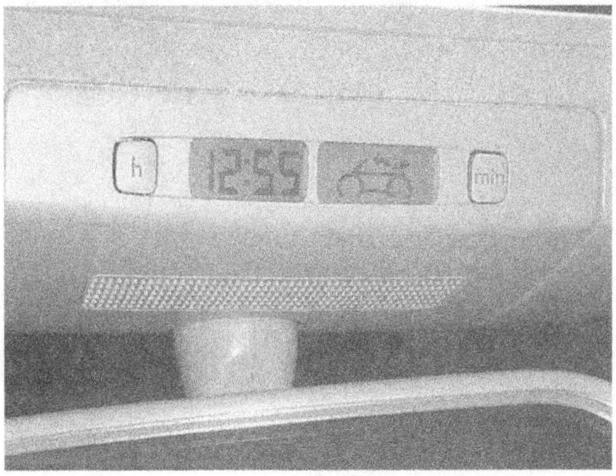

The temperature display above the rearview mirror doubles as a convertible top status indicator.

Other electronic improvements include convertible top diagnostics and changes to the controller area network (CAN) to allow diagnostics and communication between the convertible top module, the airbag module, and the central control module for the comfort system J393.

CAUTION

This training module applies to only some of the vehicles covered by this manual. This information is subject to change and should be used as reference only. Always refer to the latest Volkswagen factory service information for the most up to date service, repair and training information.

Powertrain

New 2.0L Engine (BDC)

A new 2.0L engine is being introduced with the New Beetle Convertible. The new engine code is BDC and represents an updated version of the familiar 2.0L gasoline engine. It will eventually replace the base engines for the New Beetle Sedan, Golf, and Jetta models.

This engine has lighter pistons and connecting rods, resulting in smoother operation and increased torque. In addition, the one piece aluminum cylinder head cover results in improved sealing qualities, as well as a 50% improvement in liquid-vapor separation. The engine no longer requires a crankcase breather.

The 150 hp 1.8L turbocharged engine, engine code AWP (or similar) will be available some time in 2003.

Powertrain

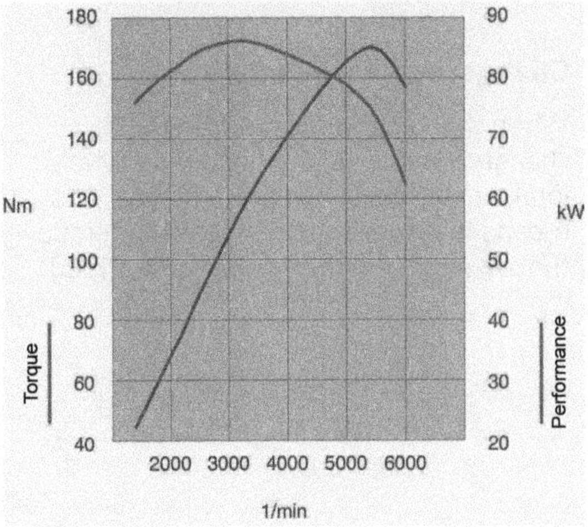

2.0 Liter Two-Valve Engine (BDC)

- Balancer shaft gear assembly
- Twin path intake manifold
- EPC
- Trapezoidal rods
- Lightweight moving parts
- Aluminum cylinder head cover
- New crankshaft bearing caps
- 4 in 1 exhaust pipe for improved torque output (was 4 in 2)

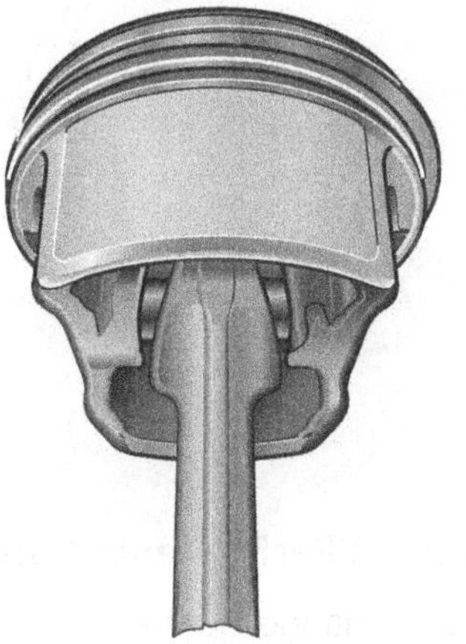

Lightweight Operation

The following design characteristics provide a significant reduction in engine noise and friction:

- Industrially cracked design (optimized weight)
- Piston pins in diameter from 20 mm to 19 mm
- Milling of small end (trapezoidal shape)
- Pistons adapted to narrower small end

Powertrain

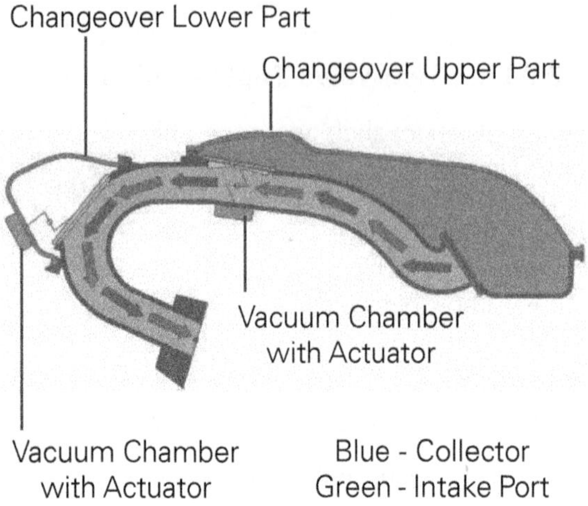

New Multi-Path Intake Manifold

Changeover: Torque Path

When the variable intake manifold changeover valves are closed the whole intake manifold (green) is available. This means that there is an increase in torque in the lower torque range from 780 – 4000 rpm.

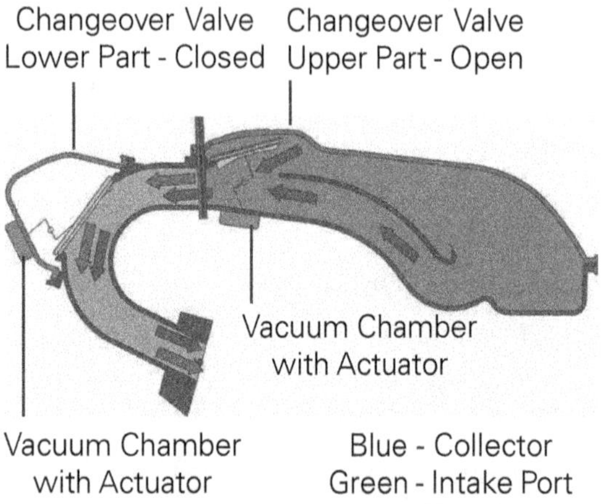

Changeover: Performance Path — Long

The first changeover valve opens at 4000 rpm in the full throttle range.

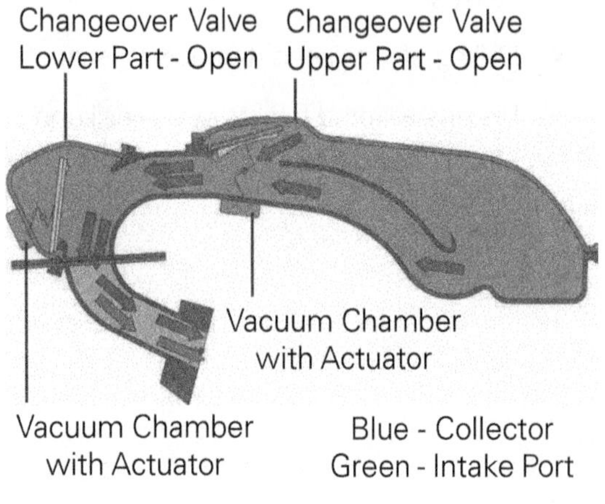

Changeover: Performance Path — Short

The second changeover valve causes further shortening of the intake port. The changeover valve in the lower part opens at 4800 rpm in the full throttle range. This results in an increase in performance.

Powertrain

Balancer Shaft Gear Assembly Operation

By installing a balancer shaft gear assembly, inertia forces and engine vibration are balanced and there is also a reduction in internal noise.

- Rumbling and roaring noises at high revs from about 5000 rpm are reduced considerably.
- Engine vibration is reduced.
- The drive for the balancer shaft and the oil pump comes from an endless chain driven by the crankshaft.
- The balancer shaft is driven directly by an input shaft.
- There is a 2:1 ratio on the two balancer gear pairs. The balance shafts spin at twice the crankshaft speed.
- The balancer gears feature plastic covers on the "light" side to prevent foaming of the oil.
- The oil pump is integrated in the balancer shafts. Parts of the oil pump can be renewed in case of repair.
- The balancer shaft with drive and balancer weights cannot be dismantled. Renewal is only possible as a complete unit with oil pump.
- The chain is tensioned via a hydraulic chain tensioner.
- Please read the repair instructions in the workshop manual.
- For installation and/or removal, the engine and the balancer shaft must be set to cylinder 1. The balancer shaft is locked in that position using special tool T10060.

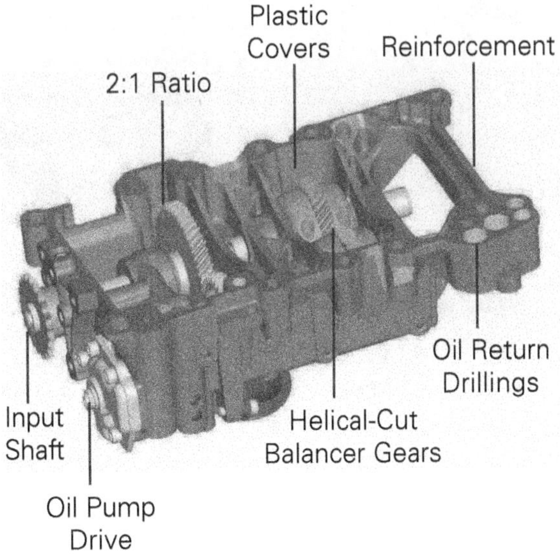

Powertrain

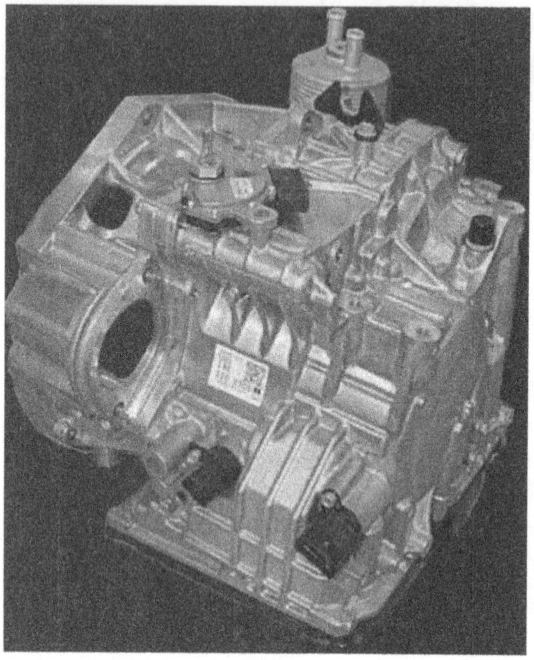

New 6-Speed Automatic Transmission

The New Beetle Convertible features a 6-speed automatic transmission with Tiptronic. This transmission uses a new Sport shifting concept that extends the useful shift range between economy and sports performance driving.

Power transferred through the torque converter is passed on to the planetary gearsets. The transmission has one Simpson (front) and one Ravigneaux (rear) planetary gearsets.

Simpson Front Planetary Gearset

Ravigneaux Rear Planetary Gearset

Powertrain

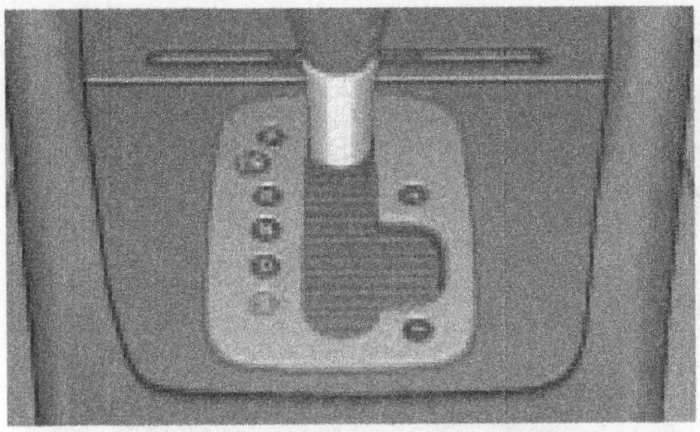

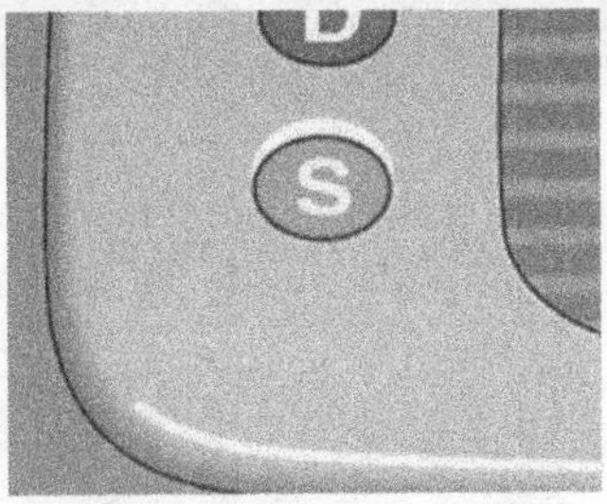

New Sport Shifting Concept

One of the most striking features of the New Beetle Convertible's 6-speed automatic transmission is its new shift lever gate, and new shifting concept.

Selector lever positions 4, 3, and 2 have been discontinued, here replaced by the "S" (Sport) position. The DSP (dynamic shift program) and Tiptronic function make separate 4, 3, and 2 positions unnecessary.

With the selector lever in the "S" position, a Sports shift program provides relevant driving dynamics. In position "S," the DSP also provides adaptation of the set values to the driving conditions. Shifts will occur at higher RPMs, and shift timing is shortened, for firmer shifts.

CAUTION

This training module applies to only some of the vehicles covered by this manual. This information is subject to change and should be used as reference only. Always refer to the latest Volkswagen factory service information for the most up to date service, repair and training information.

Convertible Top Shell Removal

Removing the Convertible Top

- Position the convertible top so that it is half open.
- Remove the roof leading trim edge.
- Pull the roof frame front seal off the roof frame.
- Remove the convertible top fabric securing bolt/screw.

- Pull the convertible top fabric clamping beading out to the side from the roof leading edge clamping channel.
- Pull the hooking beading out of the front of the roof leading edge.
- Unhook the cable with ball.

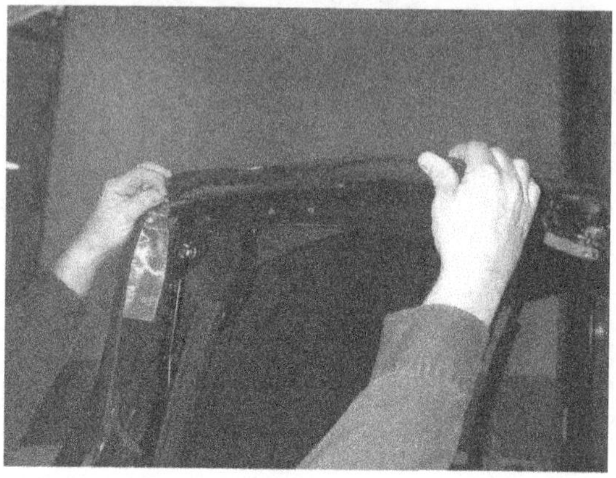

Convertible Top Shell Removal

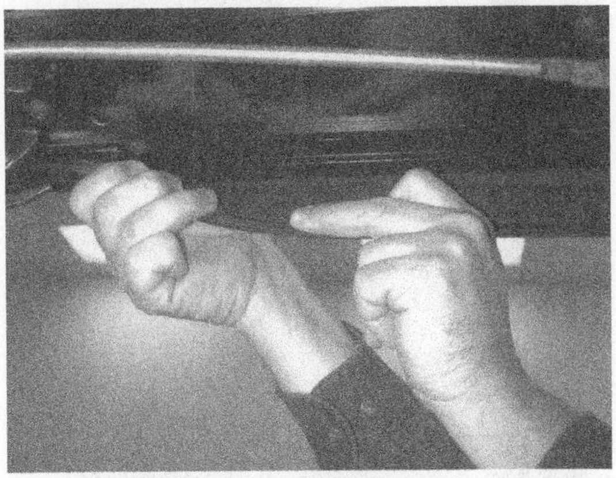

- Pull the convertible top fabric clamping beading out to the side from the roof leading edge clamping channel.
- Pull the hooking beading out of the front of the roof leading edge.
- Unhook the cable with ball.

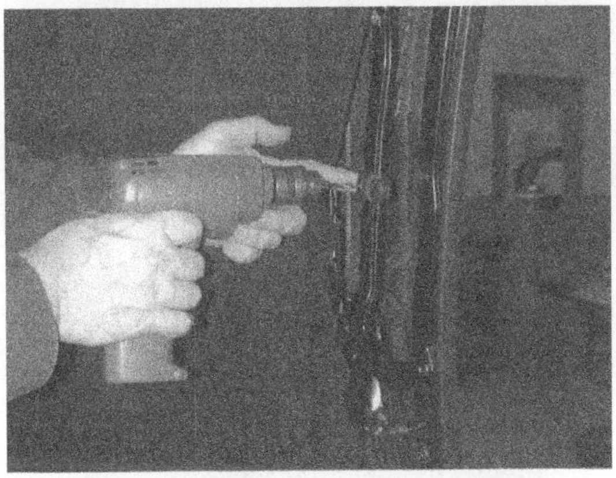

- Drill out the pop rivet from the interior for side tensioning belt.
- Pull off the C-pillar roof frame seal, pulling down and inward.

- Pull the beading out of the C-pillar channel, pulling down and outward.
- Fold the convertible top forward.

Convertible Top Shell Removal

- Cut through the staple using side cutting pliers.
- Open the front of the zip fastener.

- Pull the insert beading off the water guide.
- Pull the insert beading off the convertible lining on the convertible top retaining frame, pulling down.
- Pull the beading out of the rear window insert channel.
- Disconnect the connector.

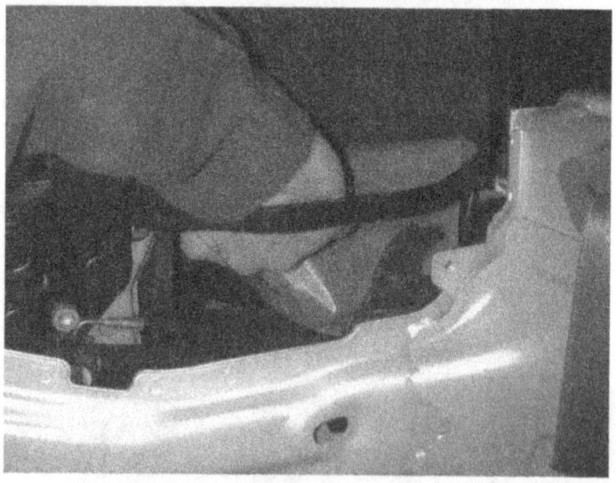

- Remove the bolts and nuts that fasten the convertible top frame to the body, removing them from the inside.
- Press the convertible fabric retaining frame off the body.

Convertible Top Shell Removal

- Secure the convertible top fabric retaining frame with the two assembly brackets (VAG 1887) to the body.

  When removing the convertible top, use cloth-backed adhesive tape to protect the chrome trim and body paint.

- Cut through the staple of the second zipper.
- Release the padding material bonding on the C-pillar.
- Remove the bolt/screw for the convertible top cover tensioning cable.
- Pull the tensioning cable out.
- Pull the convertible top fabric out of the convertible top retaining rail, pulling right (a second technician is needed to assist).
- Remove convertible top from its frame.

CAUTION

This training module applies to only some of the vehicles covered by this manual. This information is subject to change and should be used as reference only. Always refer to the latest Volkswagen factory service information for the most up to date service, repair and training information.

Component Locations

Convertible Top Control Module J256

- This control module is located in the trunk area, right rear.

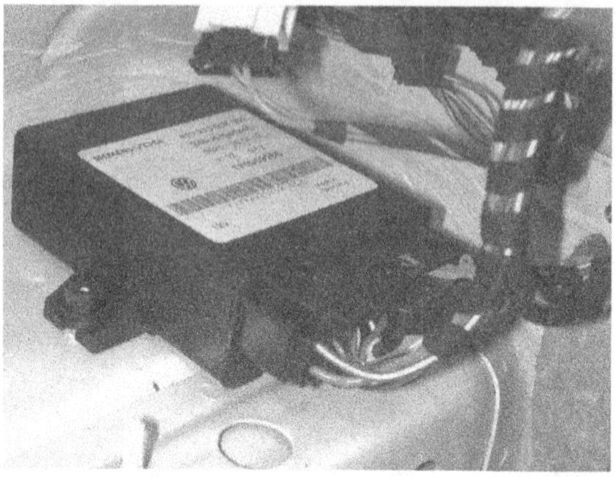

Convertible Top Hydraulic Pump

- This hydraulic pump is mounted in the trunk area, left rear.
- The lower photo shows the relays on the back of the hydraulic pump.

Component Locations

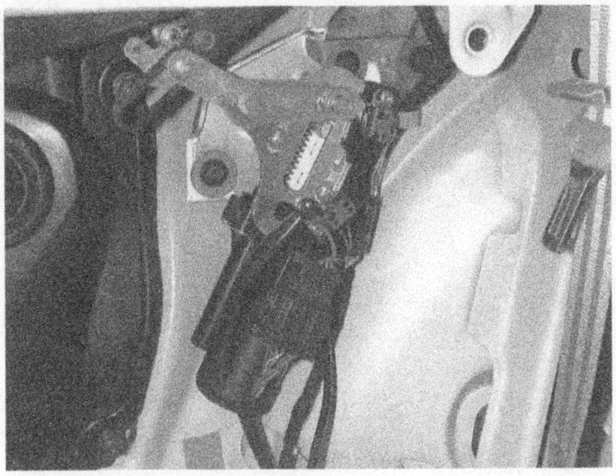

ConvertibleTop Locking Motor

- There are two convertible top locking electric motors: one located under the inside left rear interior panel, and the other located under the inside right rear interior panel.

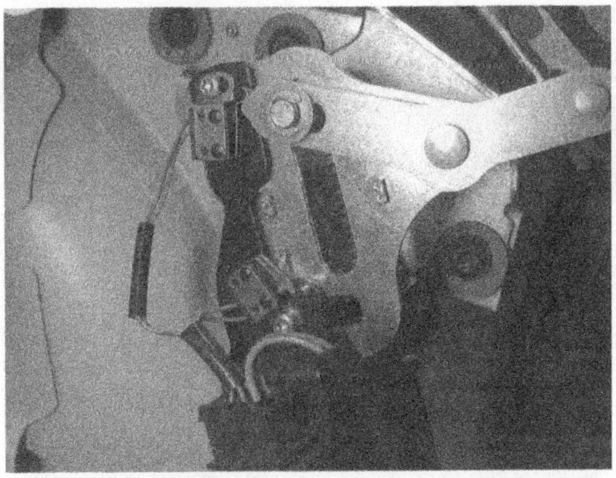

Electric Motor Locked Microswitches (two each, left and right)

- These two switches are located under the inside left and right rear interior panels.

Combined Rod Flap Electric Motor and Position Sensor

- There are two combined rod flap electric motors/position sensors: one mounted in the left side flap assembly, and one mounted in the right side flap assembly.
- These motors/sensors are not interchangeable: the left side motor/ sensor does not fit on the right side, and vice versa.

Component Locations

Two Rollover Solenoids

- These two solenoids are located behind the left rear seat, in the rear bulkhead.

Convertible Top Operation Microswitch E137

- This microswitch is located in the center console, and is used to open and close the convertible top.

Right Hydraulic Cylinder Sliding Microswitch for Convertible Top Position

- This microswitch is located next to the hydraulic cylinder under the right upper door panel cap.

Component Locations

ConvertibleTop Latched Microswitch

- This microswitch is located on the right side top of the windshield crossmember.

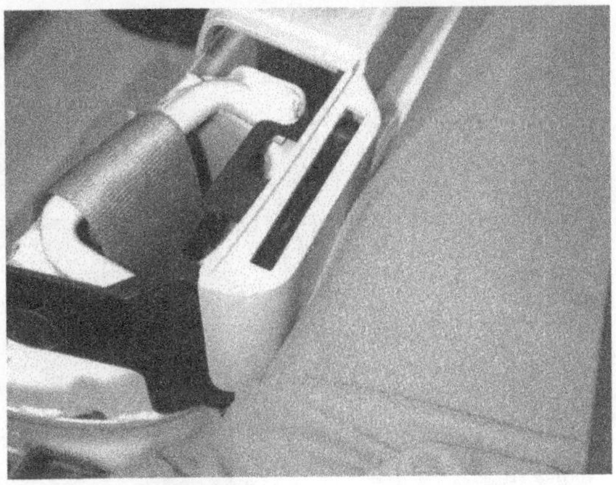

Boot Latch Microswitches (Boot Cover Installed) F155 and F156

- These microswitches are located in the left and right boot latches: F155 in the left, and F156 in the right.

Front Crash Sensor

- There are two front crash sensors, with both located under the front clip, just inside the headlamps.

Component Locations

Side Impact Crash Sensor

- There are two side impact crash sensors, one at the inside base of each B-pllar.

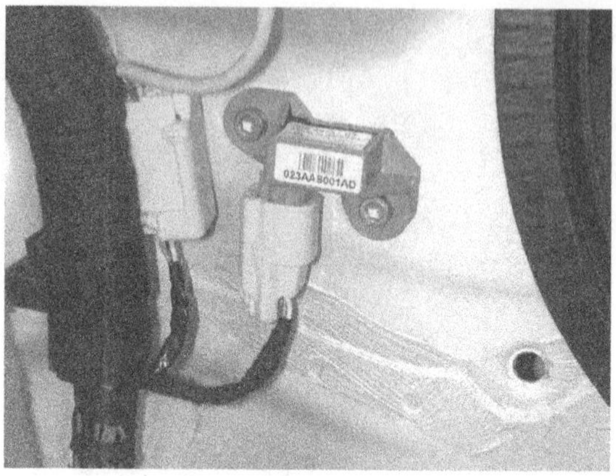

Antenna

- The antenna is located in the front windshield.
- The lower photo shows the antenna amplifier, which is located in the windshield crossmember.

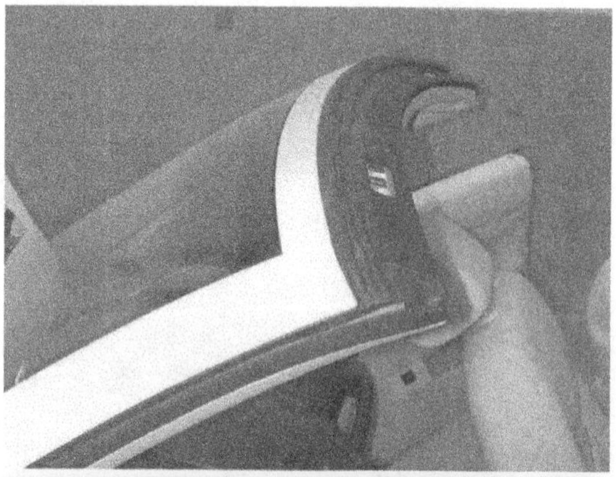

Campaign Circulars — Appendix D

GENERAL

This section contains a list of campaign circulars issued by Volkswagen of America. Also known as recall campaigns, they contain important information for owners of vehicles covered by this manual. In most cases this information will have been sent to owners of New Beetles purchased from authorized Volkswagen dealers. Please see an authorized Volkswagen dealer for more information.

Table a. Campaign Circulars

Code	Date	Model	Action	Description
M6/M600	Feb. 2006	2004 - 2005 with TDI engine	Replace EGR cooler	The EGR cooler on affected vehicles has the potential to develop an internal leak. If this happens, the MIL light may come on, and the vehicle may exceed Federal and/or California emissions standards.
M8/M800	Feb. 2005	2006 with 2.5L and 17 inch wheels	Install new tire information label	Some affected vehicles were produced with a tire information label that lists an incorrect front tire size. The correct front tire size should be 225/45 R17.
N3/26D8	April 2007	2001 with 1.8L engine	Inspect catalytic converter, replace if necessary	A single-brick metallic catalytic converter (part # 1J0 254 502 GX) with TLEV certification was called out and used as a replacement part on a small population of 2001 m.y. 1.8T engine vehicles instead of a double-brick metallic catalytic converter with LEV certification. Vehicles with the incorrect catalytic converter may exceed Federal and/or California emissions standards.
N4/N400	Aug. 2007	1998 - 2007 New Beetle 2003 - 2007 New Beetle Convertible	Replace brake light switch	Affected vehicles have a brake light switch that may malfunction. If this happens, the brake lights could become inoperative or remain on. Failure to provide the proper signal when braking could lead to a crash without warning.
Q9/66D1	Feb. 2008	2008 New Beetle special edition triple white sedan	Replace roof spoiler	Some affected vehicles used an incorrect adhesive to attach the roof spoiler to rear glass. The spoiler could become loose or fall from vehicle without warning, even though initially the spoiler may feel secure.
UA/UA00	May 1998	1998 New Beetle	Install new battery tray and secure engine compartment wiring harness	The engine compartment wiring harness may be too close to the battery tray and could become chafed and cause an electrical short.
UW/UW00	April 2003	2000 - 2001 Volkswagens	Identify rear seat child restraint anchorage points	Affected vehicles do not fully comply with Federal Motor Vehicle Safety Standard F.M.V.S.S. No. 225 (Isofix). The standard requires that the Child Restraint Anchorage location (CR), which was inadvertently omitted during production, must be identified.

Table a. Campaign Circulars

Code	Date	Model	Action	Description
VB/ZH/ VB00	May 2004	2004 New Beetle and New Beetle Convertible	Replace tire information label	Some affected vehicles were produced or supplied with an incorrect tire information label that identifies the vehicle as a five-passenger vehicle instead of a four-passenger vehicle. Exceeding the correct seating capacity in a vehicle puts occupants at risk for injury in a crash, and can result in vehicle overloading, which can affect vehicle handling and put you at risk for a crash.
VC/VC00	May 2005	2004 - 2005 New Beetle	Install mercury label on vehicles registered in VT, ME or CT	Affected vehicles were sold without a mercury content information label that is required by some state laws. This label is intended, in part, to remind salvage personnel to remove and properly dispose of any mercury-containing components before the vehicle is scrapped. While tiny amounts of mercury may be incorporated in some vehicle components such as xenon headlights, backlit displays and navigation systems, they do not pose a health risk to anyone using the vehicle.
VK	Aug. 2004	2004 New Beetle with Monsoon sound system	Check/replace amplifier	Amplifier on affected vehicles may not completely shut off when ignition is turned off and internal vehicle temperature exceeds 98°F. If the vehicle is parked for 12 hours or more, the steady current draw from the amplifier may drain the vehicle's battery.
VY/VY00	Aug. 2005	2004 - 2005 New Beetle with 1.9L	Check/install diesel pump kit	A number of high pressure diesel fuel pumps from one supplier were produced with an improper fastener during a certain production period. In rare cases, one of these improper fasteners may yield due to repeated cyclic loading and allow diesel fuel to escape from the high pressure pump. Diesel fuel in the presence of an ignition source may lead to a fire. Additionally, due to the quantity of fuel that may be released, there is a crash risk posed to vehicles traveling behind the subject vehicle.
WA/WA00	May 2007	1998 -2001 New Beetle with 2.0L	Reprogram ECM and replace oxygen sensor	The oxygen sensor in the upper part of the exhaust system could fail under certain conditions, potentially causing vehicle to exceed Federal and/or California emissions standards, or to experience increased fuel consumption. "Thermal shock" within the sensor caused by moisture accumulation could cause oxygen sensor failure. Check the emission control system for proper function, reset electronically the engine control module and replace oxygen sensor, if necessary.
WG/WG00	June 2004	2001 - 2003 Volkswagens	Replace ignition coils	Vehicles identified in campaign should have ignition coils replaced. A failed ignition coil could result in rough idling or reduced driving power, in addition to the illumination of the MIL light.
WJ/WJ00	March 2002	2001 - 2002	Check and replace ABS control unit	An electronic chip in the ABS control unit could experience an electrical short causing the connector wire insulation to become scorched.
WU/WU00	Jan. 2005	2004 New Beetle and New Beetle Convertible	Replace PODS assembly	Passive Occupant Detection System (PODS) may become disabled in affected vehicles due to a problem with the software in the PODS control unit.
ZH/ZH00	May 2004	2004 New Beetle and New Beetle Convertible	Replace tire information label	Some affected vehicles were produced or supplied with an incorrect tire information label that identifies the vehicle as a five-passenger vehicle instead of a four-passenger vehicle. Exceeding the correct seating capacity in a vehicle puts occupants at risk for injury in a crash, and can result in vehicle overloading, which can affect vehicle handling and put you at risk for a crash.

> ***WARNING—***
> - *Your common sense, good judgement, and general alertness are crucial to safe and successful service work. Before attempting any work on your Volkswagen, read* **00 Warnings and Cautions** *and the copyright page at the front of the manual. Review these warnings and cautions each time you prepare to work on your Volkswagen. Please also read any warnings and cautions that accompany the procedures in the manual.*

A

B

C

> **WARNING—**
> - *Your common sense, good judgement, and general alertness are crucial to safe and successful service work. Before attempting any work on your Volkswagen, read* **00 Warnings and Cautions** *and the copyright page at the front of the manual. Review these warnings and cautions each time you prepare to work on your Volkswagen. Please also read any warnings and cautions that accompany the procedures in the manual.*

WARNING—

- *Your common sense, good judgement, and general alertness are crucial to safe and successful service work. Before attempting any work on your Volkswagen, read* **00 Warnings and Cautions** *and the copyright page at the front of the manual. Review these warnings and cautions each time you prepare to work on your Volkswagen. Please also read any warnings and cautions that accompany the procedures in the manual.*

W

Selected Books and Repair Information From Bentley Publishers

Engineering

The Hack Mechanic Guide to European Automotive Electrical Systems *Rob Siegel*

ISBN 978-0-8376-1751-0

Mechanical Ignition Handbook: The Hack Mechanic Guide to Vintage Ignition Systems *Rob Siegel*
ISBN 978-0-8376-1767-1

Physics for Gearheads
Randy Beikmann
ISBN 978-0-8376-1615-5

Bosch Automotive Handbook
Robert Bosch GmbH
ISBN 978-0-8376-1732-9

Bosch Fuel Injection and Engine Management *Charles O. Probst, SAE*
ISBN 978-0-8376-0300-1

Maximum Boost: Designing, Testing, and Installing Turbocharger Systems
Corky Bell ISBN 978-0-8376-0160-1

Motorsports

Alex Zanardi: My Sweetest Victory
Alex Zanardi and Gianluca Gasparini
ISBN 978-0-8376-1249-2

The Unfair Advantage *Mark Donohue and Paul van Valkenburgh*
ISBN 978-0-8376-0069-7

Equations of Motion - Adventure, Risk and Innovation
William F. Milliken
ISBN 978-0-8376-1570-7

Audi

Audi A4 Service Manual: 2002-2008, 1.8L Turbo, 2.0L Turbo, 3.0L, 3.2L
Bentley Publishers
ISBN 978-0-8376-1574-5

Audi TT Service Manual: 2000-2006, 1.8L turbo, 3.2 L *Bentley Publishers*
ISBN 978-0-8376-1625-4

BMW

Memoirs of a Hack Mechanic
Rob Siegel ISBN 978-0-8376-1720-6

BMW 4 Series (F32, F33, F36) Service Manual: 2014-2016 *Bentley Publishers*
ISBN 978-0-8376-1765-7

BMW X3 (E83) Service Manual: 2004-2010 *Bentley Publishers*
ISBN 978-0-8376-1731-2

BMW X5 (53) Service Manual: 2000-2006 *Bentley Publishers*
ISBN 978-0-8376-1643-8

BMW 3 Series (F30, F31, F34) Service Manual: 2012-2015 *Bentley Publishers*
ISBN 978-0-8376-1752-7

BMW 3 Series (E90, E91, E92, E93) Service Manual: 2006-2011
Bentley Publishers
ISBN 978-0-8376-1723-7

BMW 3 Series (E46) Service Manual: 1999-2005 *Bentley Publishers*
ISBN 978-0-8376-1657-5

BMW 3 Series (E36) Service Manual: 1992-1998 *Bentley Publishers*
ISBN 978-0-8376-1709-1

BMW 3 Series (E30) Service Manual: 1984-1990 *Bentley Publishers*
ISBN 978-0-8376-1647-6

BMW Z3 (E36/7) Service Manual: 1996-2002 *Bentley Publishers*
ISBN 978-0-8376-1617-9

BMW 5 Series (E60, E61) Service Manual: 2004-2010 *Bentley Publishers*
ISBN 978-0-8376-1689-6

Corvette

Corvette: America's Star-Spangled Sports Car *Karl Ludvigsen*
ISBN 978-0-8376-1659-9

Corvette by the Numbers: The Essential Corvette Parts Reference 1955-1982 *Alan Covin*
ISBN 978-0-8376-0228-2

Zora Arkus-Duntov: The Legend Behind Corvette *Jerry Burton*
ISBN 978-08376-0858-7

Porsche

Porsche 911 (996) Service Manual: 1999-2005 *Bentley Publishers*
ISBN 978-0-8376-1710-7

Porsche 911 (993) Service Manual: 1995-1998 *Bentley Publishers*
ISBN 978-0-8376-1719-0

Porsche Boxster Service Manual: 1997-2004 *Bentley Publishers*
ISBN 978-0-8376-1645-2

Porsche 911 Carrera Service Manual: 1984-1989 *Bentley Publishers*
ISBN 978-0-8376-1696-4

Porsche 911 SC Service Manual: 1978-1983 *Bentley Publishers*
ISBN 978-0-8376-1705-3

Porsche: Excellence Was Expected
Karl Ludvigsen ISBN 978-0-8376-0235-6

Porsche — Origin of the Species
Karl Ludvigsen ISBN 978-0-8376-1331-4

Volkswagen

Volkswagen Rabbit, GTI Service Manual: 2006-2009 *Bentley Publishers*
ISBN 978-0-8376-1664-3

Volkswagen Jetta Service Manual: 2005-2010 *Bentley Publishers*
ISBN 978-0-8376-1616-2

Volkswagen Jetta, Golf, GTI Service Manual: 1999-2005 *Bentley Publishers*
ISBN 978-0-8376-1678-0

Volkswagen Jetta, Golf, GTI: 1993-1999, Cabrio: 1995-2002 Service Manual *Bentley Publishers*
ISBN 978-0-8376-1660-5

Volkswagen GTI, Golf, Jetta Service Manual: 1985-1992 *Bentley Publishers*
ISBN 978-0-8376-1637-7

Volkswagen Corrado Repair Manual: 1990-1994 *Bentley Publishers*
ISBN 978-0-8376-1699-5

Volkswagen Passat, Passat Wagon Service Manual: 1998-2005
Bentley Publishers
ISBN 978-0-8376-1669-8

MINI Repair Manuals

MINI Cooper Service Manual: 2007-2013 *Bentley Publishers*
ISBN 978-0-8376-1730-5

MINI Cooper Service Manual: 2002-2006 *Bentley Publishers*
ISBN 978-0-8376-1639-1

BentleyPublishers®
Automotive Reference

Bentley Publishers has published service manuals and automobile books since 1950. For more information, please contact Bentley Publishers at 1734 Massachusetts Avenue, Cambridge, MA 02138 USA, or visit our web site at
`BentleyPublishers.com`

Notes

Notes

Notes

Notes

Notes

Notes

Notes